The Periodic Table of Elements

1 **H** 1.008																	2 **He** 4.003
3 **Li** 6.94	4 **Be** 9.01											5 **B** 10.81	6 **C** 12.011	7 **N** 14.01	8 **O** 16.00	9 **F** 19.00	10 **Ne** 20.18
11 **Na** 22.99	12 **Mg** 24.31											13 **Al** 26.98	14 **Si** 28.09	15 **P** 30.97	16 **S** 32.06	17 **Cl** 35.45	18 **Ar** 39.95
19 **K** 39.10	20 **Ca** 40.08	21 **Sc** 44.96	22 **Ti** 47.90	23 **V** 50.94	24 **Cr** 52.00	25 **Mn** 54.94	26 **Fe** 55.85	27 **Co** 58.93	28 **Ni** 58.71	29 **Cu** 63.55	30 **Zn** 65.37	31 **Ga** 69.72	32 **Ge** 72.59	33 **As** 74.92	34 **Se** 78.96	35 **Br** 79.90	36 **Kr** 83.80
37 **Rb** 85.47	38 **Sr** 87.62	39 **Y** 88.91	40 **Zr** 91.22	41 **Nb** 92.91	42 **Mo** 95.94	43 **Tc** 98.91	44 **Ru** 101.07	45 **Rh** 102.91	46 **Pd** 106.4	47 **Ag** 107.87	48 **Cd** 112.40	49 **In** 114.82	50 **Sn** 118.69	51 **Sb** 121.75	52 **Te** 127.60	53 **I** 126.90	54 **Xe** 131.30
55 **Cs** 132.91	56 **Ba** 137.34		72 **Hf** 178.49	73 **Ta** 180.95	74 **W** 183.85	75 **Re** 186.2	76 **Os** 190.2	77 **Ir** 192.2	78 **Pt** 195.09	79 **Au** 196.97	80 **Hg** 200.59	81 **Tl** 204.37	82 **Pb** 207.19	83 **Bi** 208.98	84 **Po** (209)	85 **At** (210)	86 **Rn** (222)
87 **Fr** (229)	88 **Ra** 226.03																

Lanthanides

Actinides

Principles of Biochemistry

SECOND EDITION

Principles of Biochemistry

Albert L. Lehninger

Late University Professor of Medical Sciences
The Johns Hopkins University

David L. Nelson

Professor of Biochemistry
University of Wisconsin—Madison

Michael M. Cox

Professor of Biochemistry
University of Wisconsin—Madison

WORTH PUBLISHERS

Principles of Biochemistry

Second Edition

Albert L. Lehninger, David L. Nelson, and Michael M. Cox

Copyright © 1993, 1982 by Worth Publishers, Inc.

All rights reserved

Printed in the United States of America

Library of Congress Catalog Card No. 91-67492

ISBN: 0-87901-500-4

Printing: 5 4 3 2 1   Year: 97 96 95 94 93

Development Editor: Valerie Neal

Design: Malcolm Grear Designers

Art Director: George Touloumes

Project Editor: Elizabeth Geller

Production Supervisor: Sarah Segal

Layout: Patricia Lawson

Picture Editor: Stuart Kenter

Illustration Design: Susan Tilberry

Illustrators: Susan Tilberry, Alan Landau, and Joan Waites

Computer Art: Laura Pardi Duprey and York Graphic Services

Composition: York Graphic Services

Printing and binding: R.R. Donnelley and Sons

Cover: The active site of the proteolytic enzyme chymotrypsin, showing the substrate (blue and purple) and the amino acid residues (red and orange) critical to catalysis. Determination of the detailed reaction mechanism of this enzyme (described on pp. 223–226) helped to establish the general principles of enzyme action.

Frontispiece: A view of tobacco ribulose-1,5-bisphosphate carboxylase (rubisco). This enzyme is central to photosynthetic carbon dioxide fixation; it is the most abundant enzyme in the biosphere. Different subunits are shown in blues and grays. Important active site residues are shown in red. Sulfates bound at the active site (an artifact of the crystallization procedure) are shown in yellow.

Cover, frontispiece, and part opening images produced by Alisa Zapp (see *Molecular Modeling* credits, p. IC-4) and enhanced by Academy Arts.

Worth Publishers

33 Irving Place

New York, NY 10003

Preface

This revision of Lehninger's *Principles of Biochemistry* was conceived late in 1987, as we (Nelson and Cox) were reluctantly concluding that the first edition of *Principles* would soon be out of date for the introductory biochemistry course we teach at the University of Wisconsin–Madison. Our long and productive experience with this text and its predecessors (the first and second editions of *Biochemistry*), the positive appraisals of our students, and our belief that none of the other available texts complemented our course and teaching philosophy quite so well, provided the inspiration for the book now in your hands.

Principles of Biochemistry, Second Edition, is an introduction to our favorite subject—and our attempt to make it yours. It has been designed for one- and two-semester courses for undergraduates majoring in biochemistry and related disciplines, as well as for graduate students who require a broad introduction to biochemistry. It is also suitable for courses at medical, dental, veterinary, pharmacy, and other professional schools. The book will be used most successfully by students who have completed two years of college-level chemistry, including organic chemistry, and have received at least an introduction to biology. While some background in physics and physical chemistry would be useful, all relevant principles are introduced in a manner that should make them accessible to most students.

Although more than a decade has passed since Albert Lehninger wrote the first edition of *Principles of Biochemistry,* our goals in revising it have remained true to the original. A look at the Contents will confirm that Lehninger's ground-breaking organization, in which a discussion of biomolecules is followed by metabolism and then information pathways, has been retained; we believe it is still pedagogically sound. At every other level, however, this second edition is a re-creation, rather than a revision, of the original text. Every chapter has been comprehensively overhauled, not just by adding and deleting information, but by completely reorganizing its presentation and content, and in some cases its relationship to other chapters. More than half of the chapters were written wholly or nearly from scratch.

The need for fundamental changes arose from the advance of biochemical knowledge and the continued emergence of basic principles. Recent advances have profoundly influenced the manner in which biochemistry can and should be taught. A decade ago, for example, only a few well-studied proteins were available to illustrate the principles of protein structure and function; now hundreds vie for attention.

We have resisted the temptation to allow the book to become encyclopedic, focusing instead on subjects appropriate to its ambitious title. The decision to include or delete a topic was generally based on the degree to which it illustrated or clarified a broader theme. Our major objective has been to highlight principles that rationalize and systematize large bodies of information.

Although these principles are often surprisingly simple, the body of information and experimental advances that must be understood to appreciate them can be daunting. Biochemistry first and foremost seeks to explain biology in chemical terms.

We wrote with the following goals in mind:

- to introduce the language of biochemistry, with careful explanations of the meaning, origin, and significance of terms;

- to provide a balanced understanding of the physical, chemical, and biological context in which each biomolecule, reaction, or pathway operates (here, for instance, you will not only read that ATP is a cell's energy currency, but you will also find out what chemical properties make ATP—rather than pyrophosphate, say—suitable for this function);

- to project a clear and repeated emphasis on major themes, especially those relating to evolution, thermodynamics, regulation, and the relationship between structure and function;

- to explain and to place in context the most important techniques that have brought us to our current understanding of biochemistry; and

- to sustain the student's interest by developing topics in a logical and stepwise manner; taking every opportunity to point out connections between processes; identifying gaps in our knowledge that promise to challenge future generations of scientists; supplying the historical context of selected major discoveries, when such context is useful; and highlighting the implications of biochemical advances for society.

The information in *Principles of Biochemistry,* Second Edition, is still presented in four parts, but the organization, content, and function of each have undergone substantial changes from the previous edition. Themes that unify groups of chapters are introduced in the opening text for each part and reinforced in individual chapters. The organization within and across chapters will help students to maintain the focus on major themes and essential information and so will enhance their understanding.

Part I represents an acknowledgment of the broad range of backgrounds that students bring to their study of biochemistry. Many years of teaching have convinced us of the value of introducing certain fundamental concepts, even for those students who have already been exposed to them. Students will find these chapters useful for filling gaps in their understanding of biology and chemistry, and for consolidating information that they have never fully assimilated. Part I introduces the central concepts of bioenergetics and thermodynamics—in the first stages of an incremental build-up in the student's understanding and appreciation of the cellular transformations of energy and matter. The structure and evolution of cells and organelles are described; in later chapters frequent reference is made to this material. A simplified overview of nucleic acids and information pathways is also provided in Part I so that key concepts can be used to support the discussion of protein structure and function.

In **Part II** we describe each major class of biomolecule, with the focus in each chapter on a detailed description of molecular properties. A special effort has been made to gather information of practical significance into easily accessible reference tables. This material is not simply a list of molecules to memorize; biomolecules are beautiful and the principles underlying their structures and functions are fundamental to everything else in biochemistry. Thermodynamics and the critical role of water and noncovalent interactions in the structure and function of macromolecules provide unifying themes that are developed gradually throughout Part II. Computer-generated images of biomolecules, with clear didactic rather than aesthetic functions, play a prominent role in these chapters.

Although derived from Chapters 5 to 12 of the previous edition and still presented in eight chapters, Part II has undergone major changes. The chapters on fibrous proteins and vitamins are gone. Fibrous proteins take their rightful place in the coverage of three-dimensional protein structure in Chapter 7. Vitamins and cofactors are individually integrated wherever their biochemical functions are first introduced. Two new chapters have been added, one dealing with membrane structure and function including transport (Chapter 10) and the other with nucleotides and nucleic acids (Chapter 12). By presenting membranes and transport in Part II, we are able to integrate the role of transmembrane ion gradients, membrane-bound enzymes, and cellular compartmentation into the discussion of metabolism in Part III. The placement of the chapter on nucleotides and nucleic acids at the end of Part II permits a thorough treatment of the subject, drawing on principles of macromolecular structure and carbohydrate chemistry introduced in Chapters 7 and 11. The discussion of nucleotide chemistry in Chapter 12 also provides a critical foundation for several aspects of thermodynamics and metabolism that are elaborated upon in Part III. In addition to these new chapters, extensive changes have been made to the presentation of protein structure and enzymatic catalysis in Chapters 7 and 8.

Part III covers bioenergetics and intermediary metabolism. Thermodynamics takes center stage as Chapter 13 builds on the development of this theme woven through Parts I and II. This chapter also introduces oxidation and reduction reactions, permitting quantitative descriptions of redox changes in metabolic pathways.

Catabolic processes are grouped together in Chapters 14 to 18, and the biosynthetic pathways are grouped in Chapters 19 to 21. Although this organization necessarily separates the coverage of some related reactions, it facilitates the parallel presentation of many more related processes (such as the similar types of reactions and intertwined pathways that link amino acid biosynthesis to nucleotide biosynthesis). Most important, it maintains a focus on the larger metabolic patterns and themes. This fundamental organization was Lehninger's own, a natural outgrowth of his perspective on bioenergetics and the integration of metabolism.

The light-dependent and light-independent reactions of photosynthesis, treated together in a single chapter in the first edition, are now divided between two adjacent chapters. Photophosphorylation is presented along with oxidative phosphorylation in Chapter 18, which has been extensively revised to emphasize the fundamental mechanistic similarity between the two processes. Photosynthetic carbon fixation is presented in Chapter 19, where it finds a logical home within a larger discussion of carbohydrate biosynthesis. The first-edition chapter on human nutrition is gone; relevant material and updated examples have been integrated into the remaining chapters. Part III culminates in Chapter 22 on the integration of metabolism and an overview of regulation, including the biochemical basis of hormone action. This largely new chapter includes some material from the chapters entitled "Hormones" and "Digestion, Transport, and the Integration of Metabolism" in the first edition; much new information has been added on the mechanism of hormone action, on the role of protein phosphorylation in regulation, and on oncogenes as defective regulatory proteins.

Major enzymatic cofactors and classes of biochemical reactions are described where they are first encountered. Selected examples of reaction mechanisms are provided to help convey the chemical rationale of each pathway. The division of labor between organs and cellular organelles is carefully traced so that these elegant schemes and their importance to life can be readily appreciated. Differences in pathways utilized by different classes of organisms are regularly examined; examples relevant to vertebrates (and particularly to human medicine) are balanced by the addition of more complete descriptions of processes in plants and microorganisms. These features are complemented by the careful application of color in pathway diagrams; colors used for certain classes of processes are standardized throughout Part III.

Part IV consists of six chapters devoted to information pathways. Here, very little has been retained from the first edition. Part IV is not designed as an encyclopedic survey of molecular biology, but as a thorough introduction to the *biochemistry* underlying the conversion of information contained in DNA to cellular macromolecules. The theme of thermodynamics recurs throughout Part IV, with emphasis on the special requirements encountered in the synthesis of a macromolecule containing information.

Chapter 23 covers higher-order structure in DNA, building on material in Chapter 12. The next three chapters cover all important aspects of the synthesis, modification, and degradation of DNA, RNA, and proteins. Regulation of gene expression is presented in Chapter 27, emphasizing principles and patterns that have a firm biochemical foundation. Each of these chapters covers the key advances to date in both prokaryotic and eukaryotic biochemistry. Combining the two is somewhat nontraditional, but advances in biochemistry have consistently demonstrated that fundamental principles are universal. The few significant biochemical distinctions between prokaryotes and eukaryotes are more readily recognized and appreciated within a parallel discussion.

The book concludes with a discussion of recombinant DNA technology (Chapter 28). The biochemical foundation of the technology is emphasized, as well as its impact on biochemical advances and on society as a whole. This chapter is designed as a portal through which the student can move to more advanced courses in many areas of biochemistry, molecular biology, and cell biology.

An understanding of modern biochemistry is impossible without an introduction to the experimental methods that made possible each major advance. Descriptions of many of these important techniques are woven into the presentations of the concepts and principles they have revealed. A complete listing of the experimental methods described in the book can be found in the index, under "Techniques."

Within each chapter, many new tables have been added. Boxed essays enhance and extend the text material by providing historical information, descriptions of methods, relevant stories and examples, in-depth discussions of related topics, and some material that is simply interesting. Each chapter concludes with a summary of important points, an up-to-date list of suggested readings, and a set of problems. The over 300 problems include many that require the student to work with and interpret original data.

The end-papers now feature a number of useful tables summarizing information that must be referred to often. The improved appendices include an extensive list of abbreviations found in the biochemical literature and solutions to the end-of-chapter problems, many of which include detailed explanations. The glossary has been revised and expanded to provide definitions of nearly 750 important terms.

The text is complemented by a new art program, providing over 900 figures in full color. These illustrations elucidate and extend the text, substituting meticulously crafted images for the proverbial thousand words. Over 100 computer-generated images illustrate biomolecules ranging from simple metabolites to large protein complexes. In all of these illustrations, whether drawn by hand or by computer, great care has been taken to devise and consistently apply color and styling codes that enhance clarity. Photographs of some of the pioneers of biochemistry have been included to give flesh and blood to the chemistry.

The relevance of biochemistry to society has never been greater; the impact of biotechnology on advances in medicine, agriculture, environmental sciences, forensic science, and a host of other fields has profound implications for our future. As scientists increasingly advance the capability to alter life itself, we emphatically reiterate Albert Lehninger's argument that a knowledge of biochemistry is useful for all well-informed citizens, whatever their callings—quite apart from the very real intellectual excitement it offers.

The two new authors are longtime colleagues at the University of Wisconsin–Madison who have collaborated in the teaching of undergraduate biochemistry for over a decade. Lehninger's textbooks have played a formative role not only in our teaching, but also in our own educations. Our interests and areas of expertise (Nelson: in intermediary metabolism and its regulation, comparative biochemistry, cell biology, and evolution; Cox: in enzyme mechanisms, macromolecular structure and topology, and molecular biology at the enzymatic level) complement each other, as well as the expertise in metabolism and bioenergetics that Lehninger brought to the first edition. While each of us bore primary responsibility for the revision of half the chapters, each chapter bears the marks of the close consultation that has characterized the planning and execution of this book.

In presenting this new edition of *Principles of Biochemistry,* we welcome your criticisms, suggestions, and comments of all kinds.

Acknowledgments

Our writing has been supported by many people in a project that evolved into an international effort. Their advice and encouragement were critical throughout these five years, and we are indebted to all of them for their contributions to this book and for the education they have given us in textbook development.

The overall organization of the project and management of its development were overseen from the earliest stages by Valerie Neal (first in New York, then Cambridge, England). Valerie also read and commented on the entire manuscript. Additional editorial and organizational input involving every phase of the project was provided by Brook Soltvedt (first in Madison, then Hartford, Connecticut). Major improvements in content and organization of the first-draft manuscript were suggested by James Funston. The final manuscript benefited from the superb copyediting of Linda Hansford. Our organization in Madison was supported and enhanced by Alisa Zapp.

The outstanding production staff at Worth Publishers gave new meaning to the phrase "grace under pressure." The critical and demanding role of project editor was very capably executed by Elizabeth Geller. In supporting roles, Alison Brower facilitated communication in this far-flung enterprise, and Maja Lorkovic augmented the proofreading of galleys and page proofs. Anne Vinnicombe's active support and participation also played a vital role in the project's success. The overall book design was conceived by Malcolm Grear Designers, the beautiful page layouts were the work of George Touloumes and Pat Lawson, and the entire production process was supervised by Sarah Segal.

The extraordinary art program was conceived and overseen by Susan Tilberry and her associates in Washington, DC. Sue personally supervised and collaborated with every artist who contributed to the book, ensuring pedagogic unity in the art program, as well as aesthetic and stylistic continuity. Some of Sue's concepts were rendered by Laura Pardi Duprey in New York using computer graphics. Contributions to this enterprise were also made by Alan Landau in New York. We have been impressed and gratified to witness the process by which our often indecipherable scribbles have been transformed into the elegant figures that grace this book.

Virtually all of the computer-generated images of biomolecules were custom designed by the authors and executed under their supervision by Alisa Zapp in Madison. Alisa's mastery of modern graphics workstations made the generation of these images both a pleasure and an educational experience for the authors. The molecular graphics effort was greatly aided by several colleagues at Madison; the Silicon Graphics system was generously loaned to us by Ron Raines, and Andrzej Krezel and Ivan Rayment provided valuable advice and assistance. The molecular images were generated using the MidasPlus software package from the University of California, San Francisco, where Alisa received training and guidance from Julie Newdall. Laura Vanderploeg photographed the images. In generating these images, we are indebted not only to the many researchers who deposited coordinates in the Brookhaven and Cambridge databases, but also to numerous scientists who freely shared coordinates for macromolecular structures very recently generated, and in some cases before their publication elsewhere.

Sean Carroll contributed valuable material and ideas on the biochemistry of development for Chapter 27.

A number of the end-of-chapter problems were derived in whole or in part from the excellent *Guide to Lehninger's Principles of Biochemistry with Solutions to Problems,* written by Paul van Eikeren. Two of the problems in Chapter 19 were contributed by Roger Persell of Hunter College.

Sarah Green entered the entire first edition into a word processor to provide a starting point. The manuscript was subsequently retyped many times by Karen Davis. Kerri Phillips typed the appendices. Scarlett Presley checked and corrected the suggested readings and many other details. The photographs used in this edition were tracked down by Stuart Kenter.

The book has benefited enormously from the combined wisdom and experience of over 100 reviewers. Every chapter was critiqued at one stage or another by a dozen or more biochemists with interests and expertise in both the chapter topic and biochemical education. For their indispensible advice and criticism, we thank:

Roger A. Acey
California State University,
Long Beach

Hugh Akers
Lamar University, Beaumont

Richard L. Anderson
Michigan State University

Laurens Anderson
University of Wisconsin–Madison

Dean R. Appling
University of Texas, Austin

John N. Aronson
University of Arizona

John L. Aull
Auburn University

Stephen J. Benkovic
Pennsylvania State University

Loran L. Bieber
Michigan State University

R. D. Blake
University of Maine

Robert E. Blankenship
Arizona State University

Richard P. Boyce
University of Florida

Michael D. Brown
Emory University School of Medicine

Barbara K. Burgess
University of California, Irvine

Robert H. Burris
University of Wisconsin–Madison

Joseph M. Calvo
Cornell University

Rita A. Calvo
Cornell University

James W. Campbell
Rice University

Sean B. Carroll
University of Wisconsin–Madison

Thomas R. Cech
University of Colorado, Boulder

James J. Champoux
University of Washington

Halvor N. Christensen
University of California, San Diego
School of Medicine

Randolph A. Coleman
College of William and Mary

Thomas Conway
University of Iowa

James B. Courtright
Marquette University

Nicholas R. Cozzarelli
University of California, Berkeley

T. E. Creighton
European Molecular Biology
Laboratory, Heidelberg, Germany

Michael A. Cusanovich
University of Arizona

Glyn Dawson
University of Chicago

Richard A. Dilley
Purdue University

David M. Dooley
Amherst College

William Dowhan
University of Texas Health Science
Center, Medical School

Ruth L. Dusenbery
Wayne State University

P. Leslie Dutton
University of Pennsylvania
School of Medicine

Elliot L. Elson
Washington University
School of Medicine

Martha F. Ferger
Cornell University

Michael E. Friedman
Auburn University

Herbert C. Friedmann
University of Chicago

James A. Fuchs
University of Minnesota

Edward A. Funkhouser
Texas A & M University

Wilbert Gamble
Oregon State University

Michael Glaser
University of Illinois,
Champaign-Urbana

David Goodman
Princeton University

Lawrence Grossman
The Johns Hopkins University

Hans M. Gunderson
Northern Arizona University

James H. Hagemann
New Mexico State University

Richard W. Hanson
Case Western Reserve University

Franklin M. Harold
Colorado State University

Alfred E. Harper
University of Wisconsin–Madison

Pamela Camp Hay
Davidson College

Mark Hermodson
Purdue University

Richard H. Himes
University of Kansas

Peter C. Hinkle
Cornell University

Lowell E. Hokin
University of Wisconsin–Madison
Medical School

H. David Husic
Lafayette College

Larry L. Jackson
Montana State University

André T. Jagendorf
Cornell University

William P. Jencks
Brandeis University

Richard G. Jensen
University of Arizona

Eric R. Johnson
Ball State University

Dean P. Jones
Emory University School of Medicine

Eugene P. Kennedy
Harvard Medical School

John W. Kimball

Hans L. Kornberg
Cambridge University, England

Kenneth N. Kreuzer
Duke University Medical Center

M. Daniel Lane
*The Johns Hopkins University
School of Medicine*

Henry Lardy
University of Wisconsin–Madison

I. Robert Lehman
*Stanford University
School of Medicine*

John J. Lemasters
*University of North Carolina,
Chapel Hill*

Stuart M. Linn
University of California, Berkeley

George Lorimer
*Dupont Central Research &
Development, Wilmington, Delaware*

Salvador Luria
*Massachusetts Institute of
Technology*

Dawn S. Luthe
Mississippi State University

Jane M. Magill
Texas A & M University

Rusty J. Mans
*University of Florida
College of Medicine*

Thomas F. J. Martin
University of Wisconsin–Madison

C. R. Matthews
Pennsylvania State University

Mark McNamee
University of California, Davis

Alan H. Mehler
*Howard University
College of Medicine*

Dale L. Oxender
*Parke-Davis Research
Ann Arbor, Michigan*

C. N. Pace
Texas A & M University

Peter L. Pedersen
*The Johns Hopkins University
School of Medicine*

Roger Persell
Hunter College

Christian R. H. Raetz
*Merck Research Laboratories
Rahway, New Jersey*

Ronald T. Raines
University of Wisconsin–Madison

Ivan Rayment
University of Wisconsin–Madison

M. Thomas Record
University of Wisconsin–Madison

Hans Ris
University of Wisconsin–Madison

Robert W. Roxby
University of Maine

Gerald Schatten
University of Wisconsin–Madison

Paul R. Schimmel
*Massachusetts Institute of
Technology*

Leslie O. Schulz
University of Wisconsin–Milwaukee

Bruce R. Selman
University of Wisconsin–Madison

Thomas D. Sharkey
University of Wisconsin–Madison

Jessup M. Shively
Clemson University

Ram P. Singhal
Wichita State University

Thomas W. Sneider
Colorado State University

Ronald Somerville
Purdue University

James Spudich
*Stanford University
School of Medicine*

Jane Starling
University of Missouri, St. Louis

Theodore L. Steck
University of Chicago

Charles J. Stewart
San Diego State University

JoAnne Stubbe
*Massachusetts Institute of
Technology*

Susan S. Taylor
University of California, San Diego

Jeremy Thorner
University of California, Berkeley

Carl L. Tipton
Iowa State University

Bert M. Tolbert
University of Colorado, Boulder

Ronald Vale
*University of California,
San Francisco*

Paul van Eikeren
*Sepracor
Marlboro, Massachusetts*

Emile van Schaftingen
*Université Catholique de Louvain
Brussels, Belgium*

Dennis E. Vance
University of Alberta

Janna P. Wehrle
*The Johns Hopkins University
School of Medicine*

Eric S. Weinberg
University of Pennsylvania

James Wells
*Genentech
South San Francisco, California*

We want particularly to acknowledge the crucial assistance of Roger Persell of Hunter College, who insightfully reviewed multiple drafts and page proofs of every chapter. We lack the space here to acknowledge all of the other individuals whose special efforts went into this book. We offer instead our sincere thanks, and the finished book that they helped guide to completion. We, of course, assume full responsibility for errors of fact or emphasis.

We are grateful to our students at the University of Wisconsin–Madison and at Spelman College in Atlanta who provided inspiration; to the students and staff of our research groups, who helped us balance the competing demands of research, teaching, administration, and textbook writing; and to our colleagues in the Department of Biochemistry at Madison, who patiently answered our questions, corrected our misconceptions, and in many cases reviewed chapters. Dave Nelson is indebted to his colleagues in the Departments of Chemistry and Biology at Spelman College for their hospitality throughout his sabbatical, during which some of the final chapter drafts were produced. We also wish to thank the entire staff of Worth Publishers, who allowed us the freedom we needed and exhibited extraordinary patience as we added late-breaking advances to "finalized" chapters, and whose dedication to producing a fine-quality book inspired our best efforts and made this a rewarding experience.

Finally, we express our deepest appreciation to our wives, Toya and Beth, who endured the long evenings and weekends we devoted to book writing with extraordinary grace and provided constant encouragement.

Madison, Wisconsin   David L. Nelson
August 1992   Michael M. Cox

To our teachers:

Paul R. Burton

Albert Finholt

William P. Jencks

Eugene P. Kennedy

Homer Knoss

Arthur Kornberg

I. Robert Lehman

David E. Sheppard

Harold B. White III

About the Authors

David L. Nelson, born in Fairmont, Minnesota, received his BS in Chemistry and Biology from St. Olaf College in 1964. He earned his PhD in Biochemistry at Stanford Medical School under Arthur Kornberg, and was a postdoctoral fellow at the Harvard Medical School with Eugene P. Kennedy, who was one of Lehninger's first graduate students. Nelson went to the University of Wisconsin–Madison in 1971 and became a full professor of biochemistry in 1982.

Nelson's thesis research at Stanford was on the intermediary metabolism of sporulating and germinating bacteria. At Harvard he studied the energetics, genetics, and biochemistry of ion transport in *E. coli*. At Wisconsin his research has focused on the signal transductions that regulate ciliary motion and exocytosis in the protozoan *Paramecium*. The enzymes of signal transductions, including a variety of protein kinases, are primary targets of study. His research group uses enzyme purification, immunological techniques, electron microscopy, genetics, molecular biology, and electrophysiology to study these processes.

Dr. Nelson has a distinguished record as a lecturer and research supervisor. For 20 years he has taught an intensive survey of biochemistry for advanced biochemistry undergraduates and graduate students in the life sciences (using Lehninger's *Biochemistry* and *Principles of Biochemistry* for much of that time). He has also taught a survey of biochemistry for nursing students, a graduate course on membrane structure and function, and a graduate seminar on membranes and sensory transductions. He has sponsored numerous PhD, MS, and undergraduate honors theses, and has received awards for his outstanding teaching, including the Dreyfus Teacher–Scholar Award and the Atwood Distinguished Professorship. In 1991–1992 he was a visiting professor of chemistry and biology at Spelman College.

Michael M. Cox was born in Wilmington, Delaware. In his first biochemistry course, Lehninger's *Biochemistry* was a major influence in refocusing his fascination with biology and inspiring him to pursue a career in biochemistry. After graduating from the University of Delaware in 1974, Cox went to Brandeis University to do his doctoral work with William Jencks, and then to Stanford in 1979 for postdoctoral study with I. Robert Lehman, moving to the University of Wisconsin–Madison in 1983. He became a full professor of biochemistry in 1992.

His doctoral research was on general acid and base catalysis as a model for enzyme-catalyzed reactions. At Stanford, Cox began work on the enzymes involved in genetic recombination, designing still-used purification and assay methods, illuminating the process of DNA branch migration, and cloning the gene for a site-specific recombinase from yeast. Exploration of the enzymes of genetic recombination has remained the central theme of his research.

Dr. Cox has coordinated a large and active research team at Wisconsin, investigating the enzymology, topology, and energetics of genetic recombination. A primary focus has been the mechanism of DNA strand exchange and the role of ATP in the RecA system. The research team has also concentrated on the FLP recombinase of yeast and the process it controls, and is developing chromosomal targeting systems based on the FLP recombinase. For the past 8 years he has taught (with Dave Nelson) the survey of biochemistry and has lectured in graduate courses on DNA structure and topology, protein–DNA interactions, and the biochemistry of recombination. He has received awards for both his teaching and his research, including the Dreyfus Teacher–Scholar Award and the 1989 Eli Lilly Award in Biological Chemistry.

Contents in Brief

Contents

Foundations of Biochemistry

Fifteen to twenty billion years ago the universe arose with a cataclysmic explosion that hurled hot, energy-rich subatomic particles into all space. Within seconds, the simplest elements (hydrogen and helium) were formed. As the universe expanded and cooled, galaxies condensed under the influence of gravity. Within these galaxies, enormous stars formed and later exploded as supernovae, releasing the energy needed to fuse simpler atomic nuclei into the more complex elements. Thus were produced, over billions of years, the chemical elements found on earth today. Biochemistry asks how the thousands of different biomolecules formed from these elements interact with each other to confer the remarkable properties of living organisms.

In Part I we will summarize the biological and chemical background to biochemistry. Living organisms operate within the same physical laws that apply to all natural processes, and we begin by discussing those laws and several axioms that flow from them (Chapter 1). These axioms make up the molecular logic of life. They define the means by which cells transform energy to accomplish work, catalyze the chemical transformations that typify them, assemble molecules of great complexity from simpler subunits, form supramolecular complexes that are the machinery of life, and store and pass on the instructions for the assembly of all future generations of organisms from simple, nonliving precursors.

Cells, the units of all living organisms, share certain features; but the cells of different organisms, and the various cell types within a single organism, are remarkably diverse in structure and function. Chapter 2 is a brief description of the common features and the diverse specializations of cells, and of the evolutionary processes that lead to such diversity.

Nearly all of the organic compounds from which living organisms are constructed are products of biological activity. These **biomolecules** were selected during the course of biological evolution for their fitness in performing specific biochemical and cellular functions. The biomolecules can be characterized and understood in the same terms that apply to the molecules of inanimate matter: the types of bonds between atoms, the factors that contribute to bond formation and bond strength, the three-dimensional structure of molecules, and chemical reactivities. Three-dimensional structure is especially important in biochemistry; the specificity of biological interactions, such as those between enzyme and substrate, antibody and antigen, hormone and receptor, is achieved by close steric complementarity between molecules. Prominent among the forces that stabilize three-dimensional

Facing page: Supernova SN 1987a (the bright "star" at the lower right) resulted from the explosion of a blue supergiant star in the Large Magellanic Cloud, a galaxy near the Milky Way. Energy released by nuclear explosions in such supernovae brought about the fusion of simple atomic nuclei, forming the more complex elements of which the earth, its atmosphere, and all living things are composed.

structure are noncovalent interactions, individually weak but with significant cumulative effects on the structure of biological macromolecules. Chapter 3 provides the chemical basis for later discussions of the structure, catalysis, and metabolic interconversions of individual classes of biomolecules.

Water is the medium in which the first cells arose, and the solvent in which most biochemical transformations occur. The properties of water have shaped the course of evolution and exert a decisive influence on the structure of biomolecules in aqueous solution. Many of the weak interactions within and between biomolecules are strongly affected by the solvent properties of water. Even water-insoluble components of cells, such as membrane lipids, interact with each other in ways dictated by the polar properties of water. In Chapter 4 we consider the properties of water, the weak noncovalent interactions that occur in aqueous solutions of biomolecules, and the ionization of water and of solutes in aqueous solution.

These initial chapters are intended to provide a chemical backdrop for the later discussions of biochemical structures and reactions, so that whatever your background in chemistry or biology, you can immediately begin to follow, and to enjoy, the action.

The Molecular Logic of Life

The Chemical Unity of Diverse Living Organisms

Living organisms are composed of lifeless molecules. When these molecules are isolated and examined individually, they conform to all the physical and chemical laws that describe the behavior of inanimate matter. Yet living organisms possess extraordinary attributes not shown by any random collection of molecules. In this chapter, we first consider the properties of living organisms that distinguish them from other collections of matter. After arriving at a broad definition of life, we can describe a set of principles that characterize all living organisms. These principles underlie the organization of organisms and the cells that make them up, and they provide the framework for this book. They will help you to keep the larger picture in mind while exploring the illustrative examples presented in the text.

Living Matter Has Several Characteristics

What distinguishes all living organisms from all inanimate objects? First, they are structurally complicated and highly organized. They possess intricate internal structures (Fig. 1–1a) and contain many kinds of complex molecules. By contrast, the inanimate matter in our environment—clay, sand, rocks, seawater—usually consists of mixtures of relatively simple chemical compounds.

Second, living organisms extract, transform, and use energy from their environment (Fig. 1–1b), usually in the form of either chemical nutrients or the radiant energy of sunlight. This energy enables living organisms to build and maintain their own intricate structures and to do mechanical, chemical, osmotic, and other types of work. By contrast, inanimate matter does not use energy in a systematic way to maintain structure or to do work. Inanimate matter tends to decay toward a more disordered state, to come to equilibrium with its surroundings.

The third and most characteristic attribute of living organisms is the capacity for precise self-replication and self-assembly (Fig. 1–1c), a property that can be regarded as the quintessence of the living state. A single bacterial cell placed in a sterile nutrient medium can give rise to a billion identical "daughter" cells in 24 hours. Each of the cells contains thousands of different molecules, some extremely complex; yet each bacterium is a faithful copy of the original, constructed entirely from information contained within the genetic material of the original cell. By contrast, mixtures of inanimate matter show no capacity to grow and reproduce in forms identical in mass, shape, and internal structure, generation after generation.

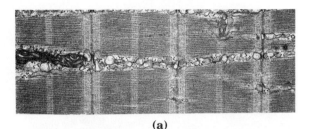

(a)

(b)

(c)

Figure 1–1 Some characteristics of living matter. **(a)** Microscopic complexity and organization are apparent in this thin section of vertebrate muscle tissue, viewed with the electron microscope. **(b)** The lion uses organic compounds obtained by eating other animals to fuel intense bursts of muscular activity. The zebra derives energy from compounds in the plants it consumes; the plants derive their energy from sunlight. **(c)** Biological reproduction occurs with near-perfect fidelity.

Erwin Schrödinger
1887–1961

Figure 1–2 Diverse living organisms share common chemical features. The eagle, the oak tree, the soil bacterium, and the human share the same basic structural units (cells), the same kinds of macromolecules (DNA, RNA, proteins) made up of the same kinds of monomeric subunits (nucleotides, amino acids), the same pathways for synthesis of cellular components, and the same genetic code and evolutionary ancestors.

The ability to self-replicate has no true analog in the nonliving world, but there is an instructive analogy in the growth of crystals in saturated solutions. Crystallization produces more material identical in lattice structure with the original "seed" crystal. Crystals are much less complex than the simplest living organisms, and their structure is static, not dynamic as are living cells. Nonetheless, the ability of crystals to "reproduce" themselves led the physicist Erwin Schrödinger to propose in his famous essay "What Is Life?" that the genetic material of cells must have some of the properties of a crystal. Schrödinger's 1944 notion (years before the modern understanding of gene structure was achieved) describes rather accurately some of the properties of deoxyribonucleic acid, the material of genes.

Each component of a living organism has a specific function. This is true not only of macroscopic structures such as leaves and stems or hearts and lungs, but also of microscopic intracellular structures such as the nucleus or chloroplast. Even individual chemical compounds in cells have specific functions. The interplay among the chemical components of a living organism is dynamic; changes in one component cause coordinating or compensating changes in another, with the result that the whole ensemble displays a character beyond that of the individual constituents. The collection of molecules carries out a program, the end result of which is the reproduction of the program and the self-perpetuation of that collection of molecules.

Biochemistry Seeks to Explain Life in Chemical Terms

The molecules of which living organisms are composed conform to all the familiar laws of chemistry, but they also interact with each other in accordance with another set of principles, which we shall refer to collectively as *the molecular logic of life*. These principles do not involve new or as yet undiscovered physical laws or forces. Instead, they are a set of relationships characterizing the nature, function, and interactions of biomolecules.

If living organisms are composed of molecules that are intrinsically inanimate, how do these molecules confer the remarkable combination of characteristics we call life? How is it that a living organism appears to be more than the sum of its inanimate parts? Philosophers once answered that living organisms are endowed with a mysterious and divine life force, but this doctrine (vitalism) has been firmly rejected by modern science. The basic goal of the science of biochemistry is to determine how the collections of inanimate molecules that constitute living organisms interact with each other to maintain and perpetuate life. Although biochemistry yields important insights and practical applications in medicine, agriculture, nutrition, and industry, it is ultimately concerned with the wonder of life itself.

Chemical Unity Underlies Biological Diversity

A massive oak tree, an eagle that soars above it, and a soil bacterium that grows among its roots appear superficially to have very little in common. However, a hundred years of biochemical research has revealed that living organisms are remarkably alike at the microscopic and chemical levels (Fig. 1–2). Biochemistry seeks to describe in molecular terms those structures, mechanisms, and chemical processes shared by all organisms and to discover the organizing principles that underlie life in all of its diverse forms.

Although there is a fundamental unity to life, it is important to recognize at the outset that very few generalizations about living organisms are absolutely correct for every organism under every condition. The range of habitats in which organisms live, from hot springs to Arctic tundra, from animal intestines to college dormitories, is matched by a correspondingly wide range of specific biochemical adaptations. These adaptations are integrated within the fundamental chemical framework shared by all organisms. Although generalizations are not perfect, they remain useful. In fact, exceptions often illuminate scientific generalizations.

All Macromolecules Are Constructed from a Few Simple Compounds

Most of the molecular constituents of living systems are composed of carbon atoms covalently joined with other carbon atoms and with hydrogen, oxygen, or nitrogen. The special bonding properties of carbon permit the formation of a great variety of molecules. Organic compounds of molecular weight (M_r) less than about 500, such as amino acids, nucleotides, and monosaccharides, serve as **monomeric subunits** of proteins, nucleic acids, and polysaccharides, respectively. A single protein molecule may have 1,000 or more amino acids, and deoxyribonucleic acid has millions of nucleotides.

Each cell of the bacterium *Escherichia coli* (*E. coli*) contains more than 6,000 different kinds of organic compounds, including about 3,000 different proteins and a similar number of different nucleic acid molecules. In humans there may be tens of thousands of different kinds of proteins, as well as many types of polysaccharides (chains of simple sugars), a variety of lipids, and many other compounds of lower molecular weight.

To purify and to characterize thoroughly all of these molecules would be an insuperable task were it not for the fact that each class of **macromolecules** (proteins, nucleic acids, polysaccharides) is composed of a small, common set of monomeric subunits. These monomeric subunits can be covalently linked in a virtually limitless variety of sequences (Fig. 1–3), just as the 26 letters of the English alphabet can be arranged into a limitless number of words, sentences, or books.

Deoxyribonucleic acids (DNA) are constructed from only four different kinds of simple monomeric subunits, the deoxyribonucleotides, and **ribonucleic acids (RNA)** are composed of just four types of ribonucleotides. **Proteins** are composed of 20 different kinds of amino acids. The eight kinds of nucleotides from which all nucleic acids are built and the 20 different kinds of amino acids from which all proteins are built are identical in all living organisms.

Most of the monomeric subunits from which all macromolecules are constructed serve more than one function in living cells. The nucleotides serve not only as subunits of nucleic acids, but also as energy-carrying molecules. The amino acids are subunits of protein molecules, and also precursors of hormones, neurotransmitters, pigments, and many other kinds of biomolecules.

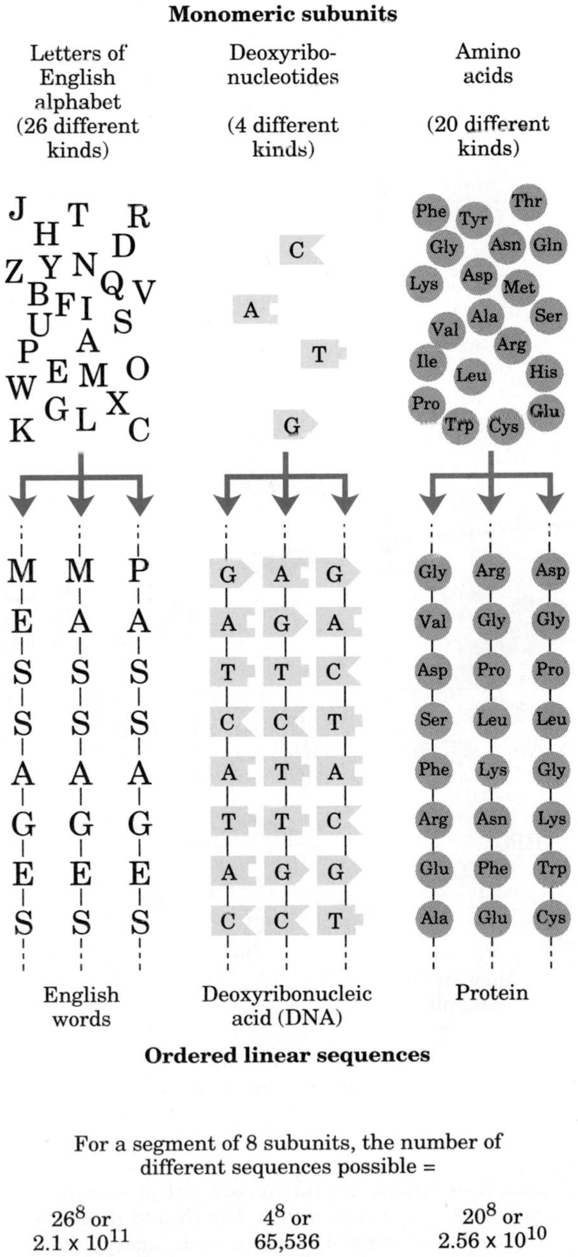

Figure 1–3 Monomeric subunits in linear sequences can spell infinitely complex messages. The number of different sequences possible (S) depends on the number of different kinds of subunits (N) and the length of the linear sequence (L): $S = N^L$. For polymers the size of proteins ($L \approx 1,000$), S is very large, and for nucleic acids, for which L may be many millions, S is astronomical.

From these considerations we can now set out some of the principles in the molecular logic of life:

All living organisms have the same kinds of monomeric subunits.

There are underlying patterns in the structure of biological macromolecules.

The identity of each organism is preserved by its possession of distinctive sets of nucleic acids and of proteins.

Energy Production and Consumption in Metabolism

Energy is a central theme in biochemistry: cells and organisms depend upon a constant supply of energy to oppose the inexorable tendency in nature for decay to the lowest energy state. The synthetic reactions that occur within cells, like the synthetic processes in any factory, require the input of energy. Energy is consumed in the motion of a bacterium or an Olympic sprinter, in the flashing of a firefly or the electrical discharge of an eel. The storage and expression of information cost energy, without which structures rich in information inevitably become disordered and meaningless. Cells have evolved highly efficient mechanisms for capturing the energy of sunlight, or extracting the energy of oxidizable fuels, and coupling the energy thus obtained to the many energy-consuming processes they carry out.

Organisms Are Never at Equilibrium with Their Surroundings

In the course of biological evolution, one of the first developments must have been an oily membrane that enclosed the water-soluble molecules of the primitive cell, segregating them and allowing them to accumulate to relatively high concentrations. The molecules and ions contained within a living organism differ in kind and in concentration from those in the organism's surroundings. The cells of a freshwater fish contain certain inorganic ions at concentrations far different from those in the surrounding water (Fig. 1–4). Proteins, nucleic acids, sugars, and fats are present in the fish but essentially absent from the surrounding water, which instead contains carbon, hydrogen, and oxygen atoms only in simpler molecules such as carbon dioxide and water. When the fish dies, its contents eventually come to equilibrium with those of its surroundings.

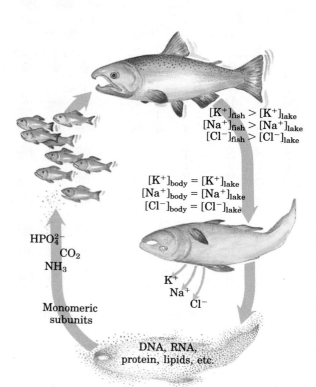

$[K^+]_{fish} > [K^+]_{lake}$
$[Na^+]_{fish} > [Na^+]_{lake}$
$[Cl^-]_{fish} > [Cl^-]_{lake}$

$[K^+]_{body} = [K^+]_{lake}$
$[Na^+]_{body} = [Na^+]_{lake}$
$[Cl^-]_{body} = [Cl^-]_{lake}$

HPO_4^{2-}
CO_2
NH_3

Monomeric subunits

K^+
Na^+
Cl^-

DNA, RNA, protein, lipids, etc.

Figure 1–4 Living organisms are not at equilibrium with their surroundings. Death and decay restore the equilibrium. During growth, energy from food is used to build complex molecules and to concentrate ions from the surroundings. When the organism dies, it loses its ability to derive energy from food. Without energy, the dead body cannot maintain concentration gradients; ions leak out. Inexorably, macromolecular components decay to simpler compounds. These simple compounds serve as nutritional sources for phytoplankton, which are then eaten by larger organisms. (By convention, square brackets denote concentration—in this case, of ionic species.)

Molecular Composition Reflects a Dynamic Steady State

Although the chemical composition of an organism may be almost constant through time, the population of molecules within a cell or organism is far from static. Molecules are synthesized and then broken down by continuous chemical reactions, involving a constant flux of mass and energy through the system. The hemoglobin molecules carrying oxygen from your lungs to your brain at this moment were synthesized within the past month; by next month they will have been degraded and replaced with new molecules. The glucose you ingested with your most recent meal is now circulating in your bloodstream; before the day is over these particular glucose molecules will have been converted into something else, such as carbon dioxide or fat, and will have been replaced with a fresh supply of glucose. The amount of hemoglobin and glucose in the blood remains nearly constant because the rate of synthesis or intake of each just balances the rate of its breakdown, consumption, or conversion into some other product (Fig. 1–5). The constancy of concentration does not, therefore, reflect chemical inertness of the components, but is rather the result of a **dynamic steady state.**

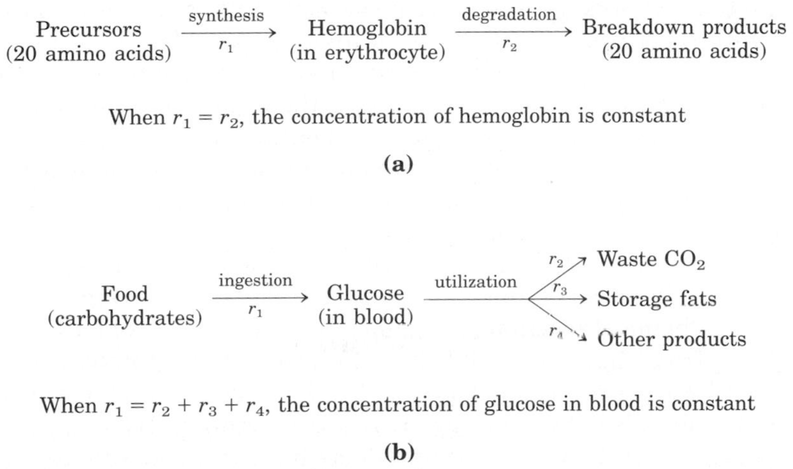

$$\text{Precursors (20 amino acids)} \xrightarrow[r_1]{\text{synthesis}} \text{Hemoglobin (in erythrocyte)} \xrightarrow[r_2]{\text{degradation}} \text{Breakdown products (20 amino acids)}$$

When $r_1 = r_2$, the concentration of hemoglobin is constant

(a)

$$\text{Food (carbohydrates)} \xrightarrow[r_1]{\text{ingestion}} \text{Glucose (in blood)} \xrightarrow[]{\text{utilization}} \begin{matrix} \xrightarrow{r_2} \text{Waste } CO_2 \\ \xrightarrow{r_3} \text{Storage fats} \\ \xrightarrow{r_4} \text{Other products} \end{matrix}$$

When $r_1 = r_2 + r_3 + r_4$, the concentration of glucose in blood is constant

(b)

Figure 1–5 A dynamic steady state results when the rate of appearance of a cellular component is exactly matched by the rate of its disappearance. In **(a),** a protein (hemoglobin) is synthesized, then degraded. In **(b),** glucose derived from food (or from carbohydrate stores) enters the bloodstream in some tissues (intestine, liver), then leaves the blood to be consumed by metabolic processes in other tissues (heart, brain, skeletal muscle). In this scheme, r_1, r_2, etc., represent the rates of the various processes. The dynamic steady-state concentrations of hemoglobin and glucose are maintained by complex mechanisms regulating the relative rates of the processes shown here.

Organisms Exchange Energy and Matter with Their Surroundings

Living cells and organisms must perform work to stay alive and to reproduce themselves. The continual synthesis of cellular components requires chemical work; the accumulation and retention of salts and various organic compounds against a concentration gradient involves osmotic work; and the contraction of a muscle or the motion of a bacterial flagellum represents mechanical work. Biochemistry examines the processes by which energy is extracted, channeled, and consumed, so it is essential to develop an understanding of the fundamental principles of bioenergetics.

Consider the simple mechanical example shown in Figure 1–6. An object at the top of an inclined plane has a certain amount of potential energy as a result of its elevation. It tends spontaneously to slide down the plane, losing its potential energy of position as it approaches the ground. When an appropriate string-and-pulley device is attached to the object, the spontaneous downward motion can accomplish a certain

Mechanical example

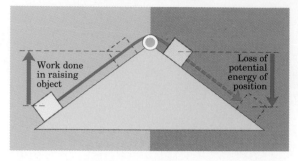

Chemical example

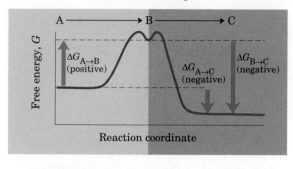

■ **Endergonic**   ■ **Exergonic**

Figure 1–6 (Top) The downward motion of an object releases potential energy that can do work. The potential energy made available by spontaneous downward motion (an exergonic process, represented by the pink box) can be coupled to the upward movement of another object (an endergonic process, represented by the blue box). (Bottom) A spontaneous (exergonic) chemical reaction (B→C) releases free energy, which can pull or drive an endergonic reaction (A→B) when the two reactions share a common intermediate, B. The exergonic reaction B→C has a large, negative free-energy change ($\Delta G_{B\rightarrow C}$), and the endergonic reaction A→B has a smaller, positive free-energy change ($\Delta G_{A\rightarrow B}$). The free-energy change for the overall reaction A→C is the arithmetic sum of these two values ($\Delta G_{A\rightarrow C}$). Because the value of $\Delta G_{A\rightarrow C}$ is negative, the overall reaction is exergonic and proceeds spontaneously.

amount of work, an amount never greater than the change in potential energy of position. The amount of energy actually available to do work (called the **free energy**) will always be somewhat less than the total change in energy, because some energy is dissipated as the heat of friction. The greater the elevation of the object relative to its final position, the greater the change in energy as it slides downward, and the greater the amount of work that can be accomplished.

In the chemical analog of this mechanical example (Fig. 1–6, bottom), a reactant, B, is converted into a product, C. The compounds B and C each contain a certain amount of potential energy, related to the kind and number of bonds in each type of molecule. This energy is analogous to the potential energy in an elevated object. Some of the energy is available to do work when B is converted into C by a chemical reaction that involves no change in temperature or pressure. This portion of the energy, the free energy, is designated G (for J. Willard Gibbs, who developed much of the theory of chemical energetics), and the change in free energy during the conversion of B to C is ΔG.

We can define a **system** as all of the reactants and products, the solvent, and the immediate atmosphere—in short, everything within a defined region of space. The system and its surroundings together constitute the **universe.** If the system exchanges neither matter nor energy with its surroundings, it is said to be **closed.** The magnitude of the free-energy change for a process proceeding toward equilibrium depends upon how far from equilibrium the system was in its initial state. In the mechanical example, no spontaneous sliding will occur once the object has reached the ground; the object is then at equilibrium with its surroundings, and the free-energy change for sliding along the horizontal surface is zero.

In chemical reactions in closed systems, the process also proceeds spontaneously until equilibrium is reached. The **free-energy change** (ΔG) for a chemical reaction is a quantitative expression of how far the system is from chemical equilibrium. Reactions that proceed with the release of free energy are **exergonic,** and because the products of such reactions have less free energy than the reactants, ΔG is negative. Chemical reactions in which the products have more free energy than the reactants are **endergonic,** and for these reactions ΔG is positive. When all of the chemical species in the system are at equilibrium, the free-energy change for the reaction is zero, and no further net conversion of reactants into products will occur without the input of energy or matter from outside the system.

As in the mechanical example, some of the energy released in a spontaneous process can accomplish work—chemical work in this case. In living systems, as in mechanical processes, part of the total energy change in the chemical reaction is unavailable to accomplish work. Some is dissipated as heat, and some is lost as **entropy,** a measure of energy due to randomness, which we will define more rigorously later.

How is free energy from a chemical reaction channeled into energy-requiring processes in living organisms? In the mechanical example in Figure 1–6, it is clear that if one sliding object is coupled to another object on another inclined plane, the energy released by the spontaneous downward sliding of one may be harnessed to produce upward motion of the other, a motion that cannot occur spontaneously. This is a direct analogy to a biochemical process in which the energy released in an exergonic chemical reaction can be used to drive another reaction that is endergonic and would not proceed spontaneously. The reactions in this system are coupled because the product of one (compound B) is a reactant in the other. This **coupling** of an exergonic reaction with an endergonic one is absolutely central to the free-energy exchanges that occur in all living systems. In biological energy coupling, the simultaneous occurrence of two reactions is not enough. The two reactions must be coupled in the sense of Figure 1–6 (bottom); the two reactions share an intermediate, B.

A living organism is an open system; it exchanges both matter and energy with its surroundings. Living organisms use either of two strategies to derive free energy from their surroundings: (1) they take up chemical components from the environment (fuels), extract free energy by means of exergonic reactions involving these fuels, and couple these reactions to endergonic reactions; or (2) they use energy absorbed from sunlight to bring about exergonic photochemical reactions, to which they couple endergonic reactions.

Living organisms create and maintain their complex, orderly structures at the expense of free energy from their environment.

Exergonic chemical or photochemical reactions are coupled to endergonic processes through shared chemical intermediates, channeling the free energy to do work.

Cells and Organisms Interconvert Different Forms of Energy

The first law of thermodynamics, developed from physics and chemistry but fully valid for biological systems as well, describes the energy conservation principle:

In any physical or chemical change, the total amount of energy in the universe remains constant, although the form of the energy may change.

Not until the nineteenth century did physicists discover that energy can be **transduced** (converted from one form to another), yet living cells have been using that principle for eons. Cells are consummate transducers of energy, capable of interconverting chemical, electromagnetic, mechanical, and osmotic energy with great efficiency (Fig. 1–7). Biological energy transducers differ from many familiar machines that depend on temperature or pressure differences. The steam engine, for example, converts the chemical energy of fuel into heat, raising the temperature of water to its boiling point to produce steam pressure that drives a mechanical device. The internal combustion engine, similarly, depends upon changes in temperature and pressure. By contrast, all parts of a living organism must operate at about the same temperature and pressure, and heat flow is therefore not a useful source of energy. Cells are **isothermal,** or constant-temperature, systems.

Living cells are chemical engines that function at constant temperature.

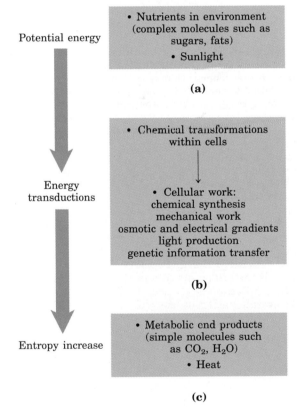

Figure 1–7 During metabolic transductions, entropy increases as the potential energy of complex nutrient molecules decreases. Living organisms **(a)** extract energy from their environment, **(b)** convert some of it into useful forms of energy to produce work, and **(c)** return some energy to the environment as heat, together with end-product molecules that are less well organized than the starting fuel, increasing the entropy of the universe.

The Flow of Electrons Provides Energy for Organisms

Virtually all of the energy transductions in cells can be traced to a flow of electrons from one molecule to another, in the oxidation of fuel or in the trapping of light energy during photosynthesis. This electron flow is "downhill," from higher to lower electrochemical potential; as such, it is formally analogous to the flow of electrons in an electric circuit driven by an electrical battery. Nearly all living organisms derive their energy, directly or indirectly, from the radiant energy of sunlight, which arises from the thermonuclear fusion reactions that form helium in the sun (Fig. 1–8). Photosynthetic cells absorb the sun's radiant energy and use it to drive electrons from water to carbon dioxide, forming energy-rich products such as starch and sucrose. In doing so, most photosynthetic organisms release molecular oxygen into the atmosphere. Ultimately, nonphotosynthetic organisms obtain energy for their needs by oxidizing the energy-rich products of photosynthesis, passing electrons to atmospheric oxygen to form water, carbon dioxide, and other end products, which are recycled in the environment. All of these reactions involving electron flow are **oxidation–reduction reactions.** Thus, other principles of the living state emerge:

The energy needs of virtually all organisms are provided, directly or indirectly, by solar energy.

The flow of electrons in oxidation–reduction reactions underlies energy transduction and energy conservation in living cells.

All living organisms are dependent on each other through exchanges of energy and matter via the environment.

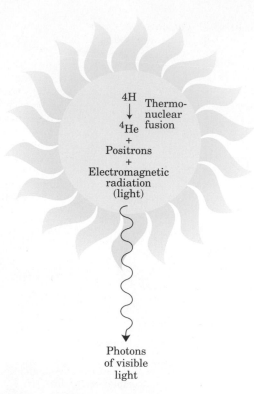

$$4H$$
$$\downarrow$$
$$^4He$$
$$+$$
Positrons
$$+$$
Electromagnetic
radiation
(light)

Thermo-
nuclear
fusion

Photons
of visible
light

Figure 1–8 Sunlight is the ultimate source of all biological energy. Thermonuclear reactions in the sun produce energy that is transmitted to the earth as light and converted into chemical energy by plants and certain microorganisms.

Enzymes Promote Sequences of Chemical Reactions

The fact that a reaction is exergonic does not mean that it will necessarily proceed rapidly. The reaction coordinate diagram in Figure 1–6 (bottom) is actually an oversimplification. The path from reactant to product almost invariably involves an energy barrier, called the activation barrier (Fig. 1–9), that must be surmounted for any reaction to occur. The breaking and joining of bonds generally requires the prior bending or stretching of existing bonds, creating a **transition state** of higher free energy than either reactant or product. The highest point in the reaction coordinate diagram represents the transition state.

Activation barriers are crucial to the stability of biomolecules in living systems. Although, when isolated from other cellular components, most biomolecules are stable for days or even years, inside cells they often undergo chemical transformations within milliseconds. Without activation barriers, biomolecules within cells would rapidly break down to simple, low-energy forms. The lifetime of complex molecules would be very short, and the extraordinary continuity and organization of life would be impossible.

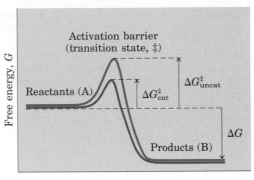

Reaction coordinate (A → B)

Figure 1–9 The energetic course of a chemical reaction. A high activation barrier, representing the transition state, must be overcome in the conversion of reactants (A) into products (B), even though the products are more stable than the reactants—as indicated by a large, negative free-energy change (ΔG). The energy required to overcome the activation barrier is the activation energy ($\Delta G^{\ddagger}$). Enzymes catalyze reactions by lowering the activation barrier. They bind the transition-state intermediates tightly, and the binding energy of this interaction effectively reduces the activation energy from $\Delta G^{\ddagger}_{uncat}$ to $\Delta G^{\ddagger}_{cat}$. (Note that the activation energy is unrelated to the free-energy change of the reaction, ΔG.)

Virtually every cellular chemical reaction occurs because of **enzymes**—catalysts that are capable of greatly enhancing the rate of specific chemical reactions without being consumed in the process (Fig. 1–10). Enzymes, as catalysts, act by lowering this energy barrier between reactant and product. The **activation energy** ($\Delta G^{\ddagger}$; Fig. 1–9) required to overcome this energy barrier could in principle be supplied by heating the reaction mixture, but this option is not available in living cells. Instead, during a reaction, enzymes bind reactant molecules in the transition state, thereby lowering the activation energy and enormously accelerating the rate of the reaction. The relationship between the activation energy and reaction rate is exponential; a small decrease in $\Delta G^{\ddagger}$ results in a very large increase in reaction rate. Enzyme-catalyzed reactions commonly proceed at rates up to 10^{10}- to 10^{14}-fold greater than the uncatalyzed rates.

Enzymes are, with a few exceptions we will consider later, proteins. Each enzyme protein is specific for the catalysis of a specific reaction, and each reaction in a cell is catalyzed by a different enzyme. Thousands of different types of enzymes are therefore required by each cell. The multiplicity of enzymes, their high specificity for reactants, and their susceptibility to regulation give cells the capacity to lower activation barriers selectively. This selectivity is crucial in the effective regulation of cellular processes.

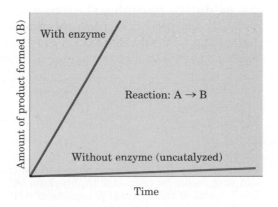

Time

Figure 1–10 An enzyme increases the rate of a specific chemical reaction. In the presence of an enzyme specific for the conversion of reactant A into product B, the rate of the reaction may increase a millionfold or more over that of the uncatalyzed reaction. The enzyme is not consumed in the process; one enzyme molecule can act repeatedly to convert many molecules of A to B.

OH
|
$CH_3—CH—CH—COO^-$ Threonine (A)
|
$^+NH_3$

↓ enzyme 1

$CH_3—CH_2—C—COO^-$ α-Ketobutyrate (B)
‖
O

↓ enzyme 2

OH
|
$CH_3—CH_2—C—COO^-$ α-Aceto-α-hydroxy-
| butyrate (C)
C=O
|
CH_3

↓ enzyme 3

CH_3 H
| |
$CH_3—CH_2—C———C—COO^-$ α,β-Dihydroxy-β-
| | methylvalerate (D)
OH   OH

↓ enzyme 4

CH_3
|
$CH_3—CH_2—C—C—COO^-$ α-Keto-β-methyl-
| ‖   valerate (E)
H   O

↓ enzyme 5

CH_3 H
| |
$CH_3—CH_2—CH—C—COO^-$ Isoleucine (F)
|
$^+NH_3$

Figure 1–11 An example of a typical synthetic (anabolic) pathway. In the bacterium *E. coli*, threonine is converted to isoleucine in five steps, each catalyzed by a separate enzyme. (Only the main reactants and products are shown here.) Threonine, in turn, was synthesized from a simpler precursor. Both threonine and isoleucine are precursors of much larger and more complex molecules: the proteins. (The letters A to F correspond to those in Fig. 1–14.)

The thousands of enzyme-catalyzed chemical reactions in cells are functionally organized into many different sequences of consecutive reactions called **pathways,** in which the product of one reaction becomes the reactant in the next (Fig. 1–11). Some of these sequences of enzyme-catalyzed reactions degrade organic nutrients into simple end products, in order to extract chemical energy and convert it into a form useful to the cell. Together these degradative, free-energy-yielding reactions are designated **catabolism.** Other enzyme-catalyzed pathways start from small precursor molecules and convert them to progressively larger and more complex molecules, including proteins and nucleic acids; such synthetic pathways invariably require the input of energy, and taken together represent **anabolism.** The network of enzyme-catalyzed pathways constitutes cellular **metabolism.**

ATP Is the Universal Carrier of Metabolic Energy, Linking Catabolism and Anabolism

Cells capture, store, and transport free energy in a chemical form. **Adenosine triphosphate (ATP)** (Fig. 1–12) functions as the major carrier of chemical energy in all cells. ATP carries energy between metabolic pathways by serving as the shared intermediate that couples endergonic reactions to exergonic ones. The terminal phosphate group of ATP is transferred to a variety of acceptor molecules, which are thereby activated for further chemical transformation. The adenosine diphosphate (ADP) that remains after the phosphate transfer is recycled to become ATP, at the expense of either chemical energy (during **oxidative phosphorylation**) or solar energy in photosynthetic cells (by the process of **photophosphorylation**). ATP is the major connecting link (the shared intermediate) between the catabolic and anabolic networks of enzyme-catalyzed reactions in the cell (Fig. 1–13).

(a)

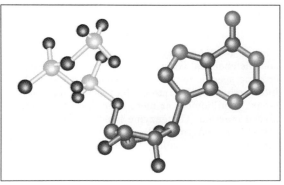

(b)

Figure 1–12 (a) Structural formula and **(b)** ball-and-stick model for adenosine triphosphate (ATP). The removal of the terminal phosphate of ATP is highly exergonic, and this reaction is coupled to many endergonic reactions in the cell.

Figure 1–13 ATP is the chemical intermediate linking energy-releasing to energy-requiring cell processes. Its role in the cell is analogous to that of money in an economy: it is "earned/produced" in exergonic reactions and "spent/consumed" in endergonic ones.

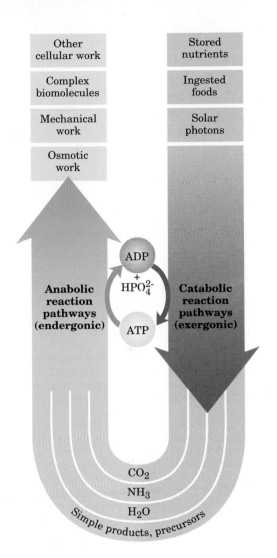

These linked networks of enzyme-catalyzed reactions are virtually identical in all living organisms.

Metabolism Is Regulated to Achieve Balance and Economy

Not only can living cells simultaneously synthesize thousands of different kinds of carbohydrate, fat, protein, and nucleic acid molecules and their simpler subunits, they can also do so in the precise proportions required by the cell. For example, when rapid cell growth occurs, the precursors of proteins and nucleic acids must be made in large quantities, whereas in nongrowing cells the requirement for these precursors is much reduced. Key enzymes in each metabolic pathway are regulated so that each type of precursor molecule is produced in a quantity appropriate to the current requirements of the cell. Consider the pathway shown in Figure 1–14 (see also Fig. 1–11), which leads to the synthesis of isoleucine (one of the amino acids, the monomeric subunits of proteins). If a cell begins to produce more isoleucine than is needed for protein synthesis, the unused isoleucine accumulates. High concentrations of isoleucine inhibit the catalytic activity of the first enzyme in the pathway, immediately slowing the production of the amino acid. Such **negative feedback** keeps the production and utilization of each metabolic intermediate in balance.

Living cells also regulate the synthesis of their own catalysts, the enzymes. Thus a cell can switch off the synthesis of an enzyme required to make a given product whenever that product is available ready-made in the environment. These self-adjusting and self-regulating properties allow cells to maintain themselves in a dynamic steady state, despite fluctuations in the external environment.

Living cells are self-regulating chemical engines, adjusted for maximum economy.

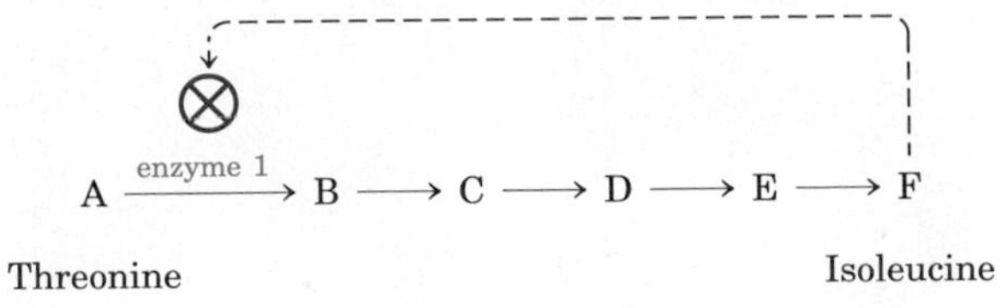

Figure 1–14 Regulation of a biosynthetic pathway by feedback inhibition. In the pathway by which isoleucine is formed in five steps from threonine (Fig. 1–11), the accumulation of the product isoleucine (F) causes inhibition of the first reaction in the pathway by binding to the enzyme catalyzing this reaction and reducing its activity. (The letters A to F represent the corresponding compounds shown in Fig. 1–11.)

Biological Information Transfer

The continued existence of a biological species requires that its genetic information be maintained in a stable form and, at the same time, expressed with very few errors. Effective storage and accurate expression of the genetic message defines individual species, distinguishes them from one another, and assures their continuity over successive generations.

Among the seminal discoveries of twentieth-century biology are the chemical nature and the three-dimensional structure of the genetic material, DNA. The sequence of deoxyribonucleotides in this linear polymer encodes the instructions for forming all other cellular components and provides a template for the production of identical DNA molecules to be distributed to progeny when a cell divides.

Genetic Continuity Is Vested in DNA Molecules

Perhaps the most remarkable of all the properties of living cells and organisms is their ability to reproduce themselves with nearly perfect fidelity for countless generations. This continuity of inherited traits implies constancy, over thousands or millions of years, in the structure of the molecules that contain the genetic information. Very few historical records of civilization, even those etched in copper or carved in stone, have survived for a thousand years (Fig. 1–15). But there is good evidence that the genetic instructions in living organisms have remained nearly unchanged over very much longer periods; many bacteria have nearly the same size, shape, and internal structure and contain the same kinds of precursor molecules and enzymes as those that lived a billion years ago.

Figure 1–15 Two ancient scripts. **(a)** The Prism of Sennacherib, inscribed in about 700 B.C., describes in characters of the Assyrian language some historical events during the reign of King Sennacherib. The Prism contains about 20,000 characters, weighs about 50 kg, and has survived almost intact for about 2,700 years. **(b)** The single DNA molecule of the bacterium *E. coli,* seen leaking out of a disrupted cell, is hundreds of times longer than the cell itself and contains all of the encoded information necessary to specify the cell's structure and functions. The bacterial DNA contains about 10 million characters (nucleotides), weighs less than 10^{-10} g, and has undergone only relatively minor changes during the past several million years. The black spots and white specks are artifacts of the preparation.

(a)

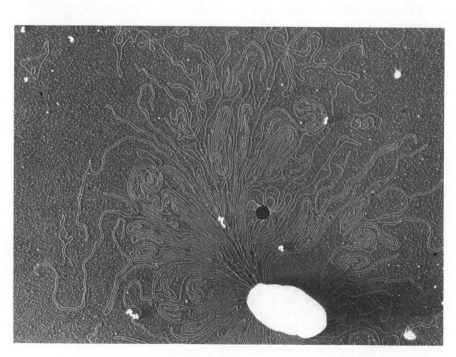

(b)

Hereditary information is preserved in DNA, a long, thin organic polymer so fragile that it will fragment from the shear forces arising in a solution that is stirred or pipetted. A human sperm or egg, carrying the accumulated hereditary information of millions of years of evolution, transmits these instructions in the form of DNA molecules, in which the linear sequence of covalently linked nucleotide subunits encodes the genetic message.

The Structure of DNA Allows for Its Repair and Replication with Near-Perfect Fidelity

The capacity of living cells to preserve their genetic material and to duplicate it for the next generation results from the structural complementarity between the two halves of the DNA molecule (Fig. 1–16). The basic unit of DNA is a linear polymer of four different monomeric subunits, **deoxyribonucleotides** (see Fig. 1–3), arranged in a precise linear sequence. It is this linear sequence that encodes the genetic information. Two of these polymeric strands are twisted about each other to form the DNA double helix, in which each monomeric subunit in one strand pairs specifically with the complementary subunit in the opposite strand. In the enzymatic replication or repair of DNA, one of the two strands serves as a template for the assembly of another, structurally complementary DNA strand. Before a cell divides, the two DNA strands separate and each serves as a template for the synthesis of a complementary strand, generating two identical double-helical molecules, one for each daughter cell. If one strand is damaged, continuity of information is assured by the information present on the other strand.

Genetic information is encoded in the linear sequence of four kinds of subunits of DNA.

The double-helical DNA molecule contains an internal template for its own replication and repair.

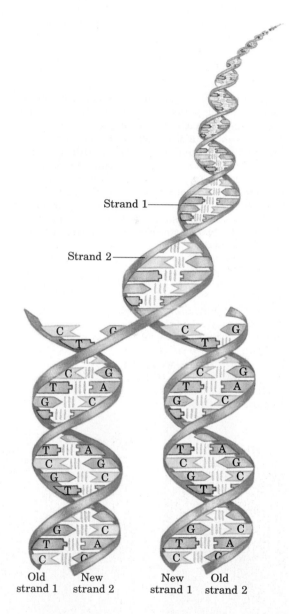

Strand 1

Strand 2

Old strand 1 · New strand 2 · New strand 1 · Old strand 2

Figure 1–16 The complementary structure of double-stranded DNA accounts for its accurate replication. DNA is a linear polymer of four subunits, the deoxyribonucleotides deoxyadenylate (A), deoxyguanylate (G), deoxycytidylate (C), and deoxythymidylate (T), joined covalently. Each nucleotide has the intrinsic ability, due to its precise three-dimensional structure, to associate very specifically but noncovalently with one other nucleotide: A always associates with its complement T, and G with its complement C. In the double-stranded DNA molecule, the sequence of nucleotides in one strand is **complementary** to the sequence in the other; wherever G occurs in strand 1, C occurs in strand 2; wherever A occurs in strand 1, T occurs in strand 2. The two strands of the DNA, held together by a large number of hydrogen bonds (represented here by vertical blue lines) between the pairs of complementary nucleotides, twist about each other to form the DNA double helix. In DNA replication, prior to cell division, the two strands of the original DNA separate and two new strands are synthesized, each with a sequence complementary to one of the original strands. The result is two double-helical DNA molecules, each identical to the original DNA.

Changes in the Hereditary Instructions Allow Evolution

Despite the near-perfect fidelity of genetic replication, infrequent, unrepaired mistakes in the replication process produce changes in the nucleotide sequence of DNA, representing a genetic **mutation** (Fig. 1–17). Incorrectly repaired damage to one of the DNA strands has the same effect. Mutations can change the instructions for producing cellular components. Many mutations are deleterious or even lethal to the organism; they may, for example, cause the synthesis of a defective enzyme that is not able to catalyze an essential metabolic reaction.

Figure 1–17 The gradual accumulation of mutations over long periods of time results in new biological species, each with a unique DNA sequence. At top is shown a short segment of a gene in a hypothetical progenitor organism. With the passage of time, changes in nucleotide sequence (mutations, indicated here by colored boxes) occur, *one at a time,* resulting in progeny with different DNA sequences. These mutant progeny themselves undergo occasional mutations, yielding their own progeny differing by two or more nucleotides from the original sequence.

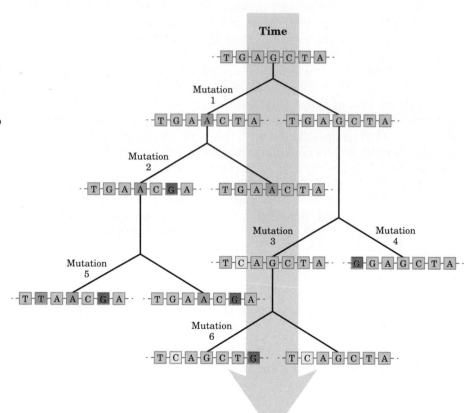

Occasionally the mutation better equips an organism or cell to survive in its environment. The mutant enzyme might, for example, have acquired a slightly different specificity, so that it is now able to use as a reactant some compound that the cell was previously unable to metabolize. If a population of cells were to find itself in an environment where that compound was the only available source of fuel, the mutant cell would have an advantage over the other, unmutated (**wild-type**) cells in the population. The mutant cell and its progeny would survive in the new environment, whereas wild-type cells would starve and be eliminated.

Chance genetic variations in individuals in a population, combined with natural selection (survival of the fittest individuals in a challenging or changing environment), have resulted in the evolution of an enormous variety of organisms, each adapted to life in a particular ecological niche.

Molecular Anatomy Reveals Evolutionary Relationships

Biochemistry has confirmed and greatly extended evolutionary theory. Carolus Linnaeus recognized the anatomic similarities and differences among living organisms and provided a framework for assessing the relatedness of different species. Charles Darwin gave us a unifying hypothesis to explain the phylogeny of modern organisms—the origin of different species from a common ancestor. Biochemistry has begun to reveal the molecular anatomy of cells of different species—the sequences of subunits in nucleic acids and proteins and the three-dimensional structures of individual molecules of nucleic acid and protein. There is a reasonable prospect that when the twenty-first century dawns, we will know the entire nucleotide sequence of all of the genes that make up the biological heritage of a human.

At the molecular level, evolution is the emergence over time of different sequences of nucleotides within genes. With new genetic sequences being experimentally determined almost daily, biochemists have an enormously rich treasury of evidence with which to analyze evolutionary relationships and to refine evolutionary theory. The molecular phylogeny derived from gene sequences is consistent with, but in many cases more precise than, the classical phylogeny based on macroscopic structures.

Molecular structures and mechanisms have been conserved in evolution even though organisms have continuously diverged at the level of gross anatomy. At the molecular level, the basic unity of life is readily apparent; crucial molecular structures and mechanisms are remarkably similar from the simplest to the most complex organisms. Biochemistry makes it possible to discover the unifying features common to all life. This book examines many of these features: the mechanisms for energy conservation, biosynthesis, gene replication, and gene expression.

Carolus Linnaeus
1707–1778

Charles Darwin
1809–1882

The Linear Sequence in DNA Encodes Proteins with Three-Dimensional Structures

The information in DNA is encoded as a linear (one-dimensional) sequence of the nucleotide units of DNA, but the expression of this information results in a three-dimensional cell. This change from one to three dimensions occurs in two phases. A linear sequence of deoxyribonucleotides in DNA codes (through the intermediary, RNA) for the production of a protein with a corresponding linear sequence of amino acids (Fig. 1–18). The protein folds itself into a particular three-dimensional shape, dictated by its amino acid sequence. The precise three-dimensional structure (**native conformation**) is crucial to the protein's function as either catalyst or structural element. This principle emerges:

> The linear sequence of amino acids in a protein leads to the acquisition of a unique three-dimensional structure by a self-assembly process.

Once a protein has folded into its native conformation, it may associate noncovalently with other proteins, or with nucleic acids or lipids,

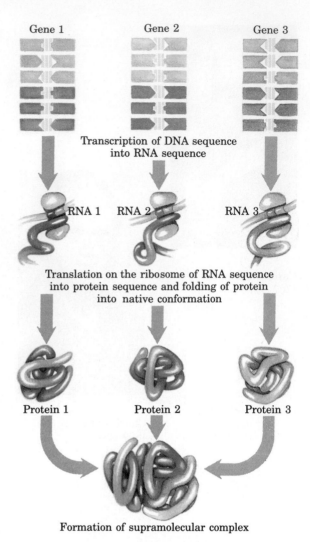

Gene 1   Gene 2   Gene 3

Transcription of DNA sequence
into RNA sequence

RNA 1   RNA 2   RNA 3

Translation on the ribosome of RNA sequence
into protein sequence and folding of protein
into native conformation

Protein 1   Protein 2   Protein 3

Formation of supramolecular complex

Figure 1–18 Linear sequences of deoxyribonucleotides in DNA, arranged into units known as genes, are transcribed into ribonucleic acid (RNA) molecules with complementary ribonucleotide sequences. The RNA sequences are then translated into linear protein chains, which fold spontaneously into their native three-dimensional shapes. Individual proteins sometimes associate with other proteins to form supramolecular complexes, stabilized by numerous weak interactions.

to form supramolecular complexes such as chromosomes, ribosomes, and membranes (Fig. 1–18). These complexes are in many cases self-assembling. The individual molecules of these complexes have specific, high-affinity binding sites for each other, and within the cell they spontaneously form functional complexes.

Individual macromolecules with specific affinity for other macromolecules self-assemble into supramolecular complexes.

Noncovalent Interactions Stabilize Three-Dimensional Structures

The forces that provide stability and specificity to the three-dimensional structures of macromolecules and supramolecular complexes are mostly noncovalent interactions. These interactions, individually weak but collectively strong, include hydrogen bonds, ionic interactions among charged groups, van der Waals interactions, and hydrophobic interactions among nonpolar groups. These weak interactions are transient; individually they form and break in small fractions of a second. The transient nature of noncovalent interactions confers a flexibility on macromolecules that is critical to their function. Furthermore, the large number of noncovalent interactions in a single macromolecule makes it unlikely that at any given moment all the interactions will be broken; thus macromolecular structures are stable over time.

Three-dimensional biological structures combine the properties of flexibility and stability.

The flexibility and stability of the double-helical structure of DNA are due to the complementarity of its two strands and the many weak interactions between them. The flexibility of these interactions allows strand separation during DNA replication (see Fig. 1–16); the complementarity of the double helix is essential to genetic continuity.

Noncovalent interactions are also central to the specificity and catalytic efficiency of enzymes. Enzymes bind transition-state intermediates through numerous weak but precisely oriented interactions. Because the weak interactions are flexible, the complex survives the structural distortions as the reactant is converted into product.

The formation of noncovalent interactions provides the energy for self-assembly of macromolecules by stabilizing native conformations relative to unfolded, random forms. The native conformation of a protein is that in which the energetic advantages of forming weak interactions counterbalance the tendency of the protein chain to assume random forms. Given a specific linear sequence of amino acids and a specific set of conditions (temperature, ionic conditions, pH), a protein will assume its native conformation spontaneously, without a template or scaffold to direct the folding.

The Physical Roots of the Biochemical World

We can now summarize the various principles of the molecular logic of life:

A living cell is a self-contained, self-assembling, self-adjusting, self-perpetuating isothermal system of molecules that extracts free energy and raw materials from its environment.

The cell carries out many consecutive reactions promoted by specific catalysts, called enzymes, which it produces itself.

The cell maintains itself in a dynamic steady state, far from equilibrium with its surroundings. There is great economy of parts and processes, achieved by regulation of the catalytic activity of key enzymes.

Self-replication through many generations is ensured by the self-repairing, linear information-coding system. Genetic information encoded as sequences of nucleotide subunits in DNA and RNA specifies the sequence of amino acids in each distinct protein, which ultimately determines the three-dimensional structure and function of each protein.

Many weak (noncovalent) interactions, acting cooperatively, stabilize the three-dimensional structures of biomolecules and supramolecular complexes.

At no point in our examination of the molecular logic of living cells have we encountered any violation of known physical laws; nor have we needed to define new physical laws. The organic machinery of living cells functions within the same set of laws that governs the operation of inanimate machines, but the chemical reactions and regulatory processes of cells have been highly refined during evolution.

This set of principles has been most thoroughly validated in studies of unicellular organisms (such as the bacterium *E. coli*), which are exceptionally amenable to biochemical and genetic study. Although multicellular organisms must solve certain problems not encountered by unicellular organisms, such as the differentiation of the fertilized egg into specialized cell types, the same principles have been found to apply. Can such simple and mechanical statements apply to humans as well, with their extraordinary capacity for thought, language, and creativity? The pace of recent biochemical progress toward understanding such processes as gene regulation, cellular differentiation, communication among cells, and neural function has been extraordinarily fast, and is accelerating. The success of biochemical methods in solving and redefining these problems justifies the hope that the most complex functions of the most highly developed organism will eventually be explicable in molecular terms.

The relevant facts of biochemistry are many; the student approaching this subject for the first time may occasionally feel overwhelmed. Perhaps the most encouraging development in twentieth-century biology is the realization that, for all of the enormous diversity in the biological world, there is a fundamental unity and simplicity to life. The organizing principles, the biochemical unity, and the evolutionary perspective of diversity, provided at the molecular level, will serve as helpful frames of reference for the study of biochemistry.

Further Reading

Asimov, I. (1962) *Life and Energy: An Exploration of the Physical and Chemical Basis of Modern Biology,* Doubleday & Co., Inc., New York.

An engaging account of the role of energy transformations in biology, written for the intelligent layman.

Blum, H.F. (1968) *Time's Arrow and Evolution,* 3rd edn, Princeton University Press, Princeton, NJ.

An excellent discussion of the way the second law of thermodynamics has influenced biological evolution.

Dulbecco, R. (1987) *The Design of Life,* Yale University Press, New Haven, CT.

An unusual and excellent introduction to biology.

Fruton, J.S. (1972) *Molecules and Life. Historical Essays on the Interplay of Chemistry and Biology,* Wiley Interscience, New York.

This series of essays describes the development of biochemistry from Pasteur's studies of fermentation to the present studies of metabolism and information transfer. You may want to refer to these essays through this textbook.

Hawking, S. (1988) *A Brief History of Time,* Bantam Books, Inc., New York.

Jacob, F. (1974) *The Logic of Life: A History of Heredity,* Pantheon Books, Inc., New York. Originally published (1970) as *La logique du vivant: une histoire de l'hérédité,* Editions Gallimard, Paris.

A fascinating historical and philosophical account of the route by which we came to the present molecular understanding of life.

Kornberg, A. (1987) The two cultures: chemistry and biology. *Biochem.* **26,** 6888–6891.

The importance of applying chemical tools to biological problems, described by an eminent practitioner.

Monod, J. (1971) *Chance and Necessity,* Alfred A. Knopf, Inc., New York. [Paperback version (1972) Vintage Books, New York.] Originally published (1970) as *Le hasard et la nécessité,* Editions du Seuil, Paris.

An exploration of the philosophical implications of biological knowledge.

Schrödinger, E. (1944) *What is Life?* Cambridge University Press, New York. [Reprinted (1956) in *What is Life? and Other Scientific Essays,* Doubleday Anchor Books, Garden City, NY.]

A thought-provoking look at life, written by a prominent physical chemist.

Cells

Cells are the structural and functional units of all living organisms. The smallest organisms consist of single cells and are microscopic, whereas larger organisms are multicellular. The human body, for example, contains at least 10^{14} cells. Unicellular organisms are found in great variety throughout virtually every environment from Antarctica to hot springs to the inner recesses of larger organisms. Multicellular organisms contain many different types of cells, which vary in size, shape, and specialized function. Yet no matter how large and complex the organism, each of its cells retains some individuality and independence.

Despite their many differences, cells of all kinds share certain structural features (Fig. 2–1). The **plasma membrane** defines the periphery of the cell, separating its contents from the surroundings. It is composed of enormous numbers of lipids and protein molecules, held together primarily by noncovalent hydrophobic interactions (p. 18), forming a thin, tough, pliable, hydrophobic bilayer around the cell. The membrane is a barrier to the free passage of inorganic ions and most other charged or polar compounds; transport proteins in the plasma membrane allow the passage of certain ions and molecules. Other membrane proteins are receptors that transmit signals from the outside to the inside of the cell, or are enzymes that participate in membrane-associated reaction pathways.

Because the individual lipid and protein subunits of the plasma membrane are not covalently linked, the entire structure is remarkably flexible, allowing changes in the shape and size of the cell. As a cell grows, newly made lipid and protein molecules are inserted into its plasma membrane; cell division produces two cells, each with its own membrane. Growth and fission occur without loss of membrane integrity. In a reversal of the fission process, two separate membrane surfaces can fuse, also without loss of integrity. Membrane fusion and fission are central to mechanisms of transport known as endocytosis and exocytosis.

The internal volume bounded by the plasma membrane, the **cytoplasm,** is composed of an aqueous solution, the **cytosol,** and a variety of insoluble, suspended particles (Fig. 2–1). The cytosol is not simply a dilute aqueous solution; it has a complex composition and gel-like consistency. Dissolved in the cytosol are many enzymes and the RNA molecules that encode them; the monomeric subunits (amino acids and nucleotides) from which these macromolecules are assembled; hundreds of small organic molecules called **metabolites,** intermediates in biosynthetic and degradative pathways; **coenzymes,** compounds of

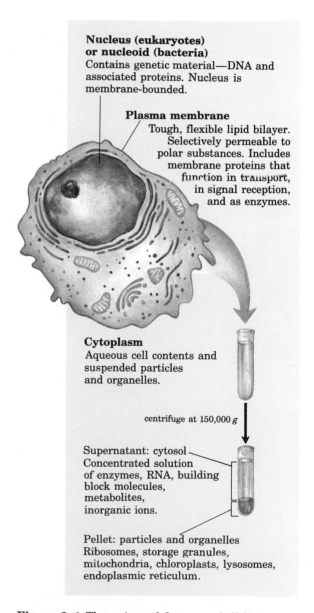

Nucleus (eukaryotes) or nucleoid (bacteria)
Contains genetic material—DNA and associated proteins. Nucleus is membrane-bounded.

Plasma membrane
Tough, flexible lipid bilayer. Selectively permeable to polar substances. Includes membrane proteins that function in transport, in signal reception, and as enzymes.

Cytoplasm
Aqueous cell contents and suspended particles and organelles.

centrifuge at 150,000 *g*

Supernatant: cytosol
Concentrated solution of enzymes, RNA, building block molecules, metabolites, inorganic ions.

Pellet: particles and organelles
Ribosomes, storage granules, mitochondria, chloroplasts, lysosomes, endoplasmic reticulum.

Figure 2–1 The universal features of all living cells: a nucleus or nucleoid, a plasma membrane, and cytoplasm. The cytosol is that portion of the cytoplasm that remains in the supernatant after centrifugation of a cell extract of 150,000 *g* for 1 h.

intermediate molecular weight (M_r 200 to 1,000) that are essential participants in many enzyme-catalyzed reactions; and inorganic ions.

Among the particles suspended in the cytosol are supramolecular complexes and, in higher organisms but not in bacteria, a variety of membrane-bounded organelles in which specialized metabolic machinery is localized. **Ribosomes,** complexes of over 50 different protein and RNA molecules, are small particles, 18 to 22 nm in diameter. Ribosomes are the enzymatic machines on which protein synthesis occurs; they often occur in clusters called **polysomes** (polyribosomes) held together by a strand of messenger RNA. Also present in the cytoplasm of many cells are granules containing stored nutrients such as starch and fat. Nearly all living cells have either a **nucleus** or a **nucleoid,** in which the **genome** (the complete set of genes, composed of DNA) is stored and replicated. The DNA molecules are always very much longer than the cells themselves, and are tightly folded and packed within the nucleus or nucleoid as supramolecular complexes of DNA with specific proteins. The bacterial nucleoid is not separated from the cytoplasm by a membrane, but in higher organisms, the nuclear material is enclosed within a double membrane, the nuclear envelope. Cells with nuclear envelopes are called **eukaryotes** (Greek *eu,* "true," and *karyon,* "nucleus"); those without nuclear envelopes—bacterial cells—are **prokaryotes** (Greek *pro,* "before"). Unlike bacteria, eukaryotes have a variety of other membrane-bounded organelles in their cytoplasm, including mitochondria, lysosomes, endoplasmic reticulum, Golgi complexes, and, in photosynthetic cells, chloroplasts.

In this chapter we review briefly the evolutionary relationships among some commonly studied cells and organisms, and the structural features that distinguish cells of various types. Our main focus is on eukaryotic cells. Also discussed in brief are the cellular parasites known as viruses.

Cellular Dimensions

Most cells are of microscopic size. Animal and plant cells are typically 10 to 30 μm in diameter, and many bacteria are only 1 to 2 μm long.

What limits the dimensions of a cell? The lower limit is probably set by the minimum number of each of the different biomolecules required by the cell. The smallest complete cells, certain bacteria known collectively as mycoplasma, are 300 nm in diameter and have a volume of about 10^{-14} mL. A single ribosome is about 20 nm in its longest dimension, so a few ribosomes take up a substantial fraction of the cell's volume. In a cell of this size, a 1 μM solution of a compound represents only 6,000 molecules.

The upper limit of cell size is set by the rate of diffusion of solute molecules in aqueous systems. The availability of fuels and essential nutrients from the surrounding medium is sometimes limited by the rate of their diffusion to all regions of the cell. A bacterial cell that depends upon oxygen-consuming reactions for energy production (an **aerobic** cell) must obtain molecular oxygen (O_2) from the surrounding medium by diffusion through its plasma membrane. The cell is so small, and the ratio of its surface area to its volume is so large, that every part of its cytoplasm is easily reached by O_2 diffusing into the cell. As the size of a cell increases, its surface-to-volume ratio decreases (Fig. 2–2), until metabolism consumes O_2 faster than diffusion can

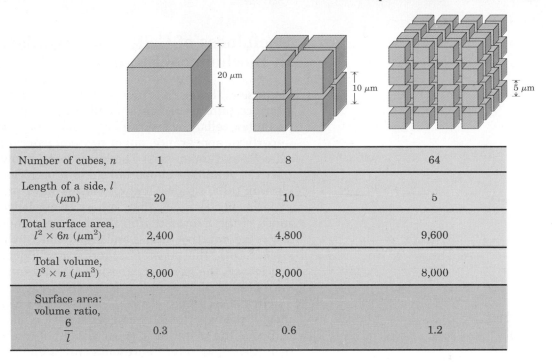

Number of cubes, n	1	8	64
Length of a side, l (μm)	20	10	5
Total surface area, $l^2 \times 6n$ (μm^2)	2,400	4,800	9,600
Total volume, $l^3 \times n$ (μm^3)	8,000	8,000	8,000
Surface area: volume ratio, $\dfrac{6}{l}$	0.3	0.6	1.2

supply it. Aerobic metabolism thus becomes impossible as cell size increases beyond a certain point, placing a theoretical upper limit on the size of the aerobic cell.

There are interesting exceptions to this generalization that cells must be small. The giant alga *Nitella* has cells several centimeters long. To assure the delivery of nutrients, metabolites, and genetic information (RNA) to all of its parts, each cell is vigorously "stirred" by active cytoplasmic streaming (p. 43). The shape of a cell can also help to compensate for its large size. A smooth sphere has the smallest surface-to-volume ratio possible for a given volume. Many large cells, although roughly spherical, have highly convoluted surfaces (Fig. 2–3a), creating larger surface areas for the same volume and thus facilitating the uptake of fuels and nutrients and release of waste products to the surrounding medium. Other large cells (neurons, for example) have large surface-to-volume ratios because they are long and thin, star-shaped, or highly branched (Fig. 2–3b), rather than spherical.

Figure 2–2 Smaller cells have larger ratios of surface area to volume, and their interiors are therefore more accessible to substances diffusing into the cell through the surface. When the large cube (representing a large cell) is subdivided into many smaller cubes (cells), the total surface area increases greatly without a change in the total volume, and the surface-to-volume ratio increases accordingly.

Figure 2–3 Convolutions of the plasma membrane, or long, thin extensions of the cytoplasm, increase the surface-to-volume ratio of cells. **(a)** Cells of the intestinal mucosa (the inner lining of the small intestine) are covered with microvilli, increasing the area for absorption of nutrients from the intestine. **(b)** Neurons of the hippocampus of the human brain are several millimeters long, but the long extensions (axons) are only about 10 nm wide.

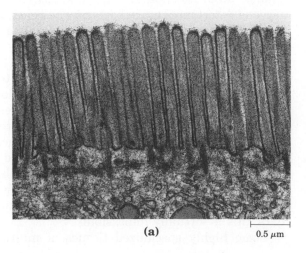

(a)   0.5 μm

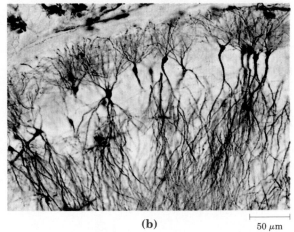

(b)   50 μm

Usefulness of Cells and Organisms in Biochemical Studies

Because all living cells have evolved from the same progenitors, they share certain fundamental similarities. Careful biochemical study of just a few cells, however different in biochemical details and varied in superficial appearance, ought to yield general principles applicable to all cells and organisms. The burgeoning knowledge in biology in the past 150 years has supported these propositions over and over again. Certain cells, tissues, and organisms have proved more amenable to experimental studies than others. Knowledge in biochemistry, and much of the information in this book, continues to be derived from a few representative tissues and organisms, such as the bacterium *Escherichia coli,* the yeast *Saccharomyces,* photosynthetic algae, spinach leaves, the rat liver, and the skeletal muscle of several different vertebrates.

In the isolation of enzymes and other cellular components, it is ideal if the experimenter can begin with a plentiful and homogeneous source of the material. The component of interest (such as an enzyme or nucleic acid) often represents only a miniscule fraction of the total material, and many grams of starting material are needed to obtain a few micrograms of the purified component. Certain types of physical and chemical studies of biomolecules are precluded if only microgram quantities of the pure substance are available. A homogeneous source of an enzyme or nucleic acid, in which all of the cells are genetically and biochemically identical, leaves no doubt about which cell type yielded the purified component, and makes it safer to extrapolate the results of in vitro studies to the situation in vivo. A large culture of bacterial or protistan cells (*E. coli, Saccharomyces,* or *Chlamydomonas,* for example), all derived by division from the same parent and therefore genetically identical, meets the requirement for a plentiful and homogeneous source. Individual tissues from laboratory animals (rat liver, pig brain, rabbit muscle) are plentiful sources of similar, though not identical, cells. Some animal and plant cells proliferate in cell culture, producing populations of identical (cloned) cells in quantities suitable for biochemical analysis.

Genetic mutants, in which a defect in a single gene produces a specific functional defect in the cell or organism, are extremely useful in establishing that a certain cellular component is essential to a particular cellular function. Because it is technically much simpler to produce and detect mutants in bacteria and yeast, these organisms (*E. coli* and *Saccharomyces cerevisiae,* for example) have been favorite experimental targets for biochemical geneticists.

An organism that is easy to culture in the laboratory, with a short generation time, offers significant advantages to the research biochemist. An organism that requires only a few simple precursor molecules in its growth medium can be cultured in the presence of a radioisotopically labeled precursor, and the metabolic fate of that precursor can then be conveniently traced by following the incorporation of the radioactive atoms into its metabolic products. The short generation time (minutes or hours) of microorganisms allows the investigator to follow a labeled precursor or a genetic defect through many generations in a few days. In higher organisms with generation times of months or years, this is virtually impossible.

Some highly specialized tissues of multicellular organisms are

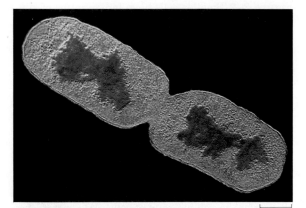

A dividing *Escherichia coli* cell.   0.25 μm

Saccharomyces cerevisiae (baker's yeast).   2 μm

remarkably enriched in some particular component related to their specialized function. Vertebrate skeletal muscle is a rich source of actin and myosin; pancreatic secretory cells contain high concentrations of rough endoplasmic reticulum; sperm cells are rich in DNA and in flagellar proteins; liver (the major biosynthetic organ of vertebrates) contains high concentrations of many enzymes of biosynthetic pathways; spinach leaves contain large numbers of chloroplasts; and so on. For studies on such specific components or processes, biochemists commonly choose a specialized tissue for their experimental systems.

Sometimes simplicity of structure or function makes a particular cell or organism attractive as an experimental system. For studies of plasma membrane structure and function, the mature erythrocyte (red blood cell) has been a favorite; it has no internal membranes to complicate purification of the plasma membrane. Some bacterial viruses (bacteriophages) have few genes. Their DNA molecules are therefore smaller and much simpler than those of humans or corn plants, and it has proved easier to study replication in these viruses than in human or corn chromosomes.

The biochemical description of living cells in this book is a composite, based on studies of many types of cells. The biochemist must always exercise caution in generalizing from results obtained in studies of selected cells, tissues, and organisms, and in relating what is observed in vitro to what happens within the living cell.

Evolution and Structure of Prokaryotic Cells

All of the organisms alive today are believed to have evolved from ancient, unicellular progenitors. Two large groups of extant prokaryotes evolved from these early forms: **archaebacteria** (Greek, *archē,* "origin") and **eubacteria.** Eubacteria inhabit the soil, surface waters, and the tissues of other living or decaying organisms. Most common and well-studied bacteria, including *E. coli* and the **cyanobacteria** (formerly called blue–green algae), are eubacteria. The archaebacteria are more recently discovered and less well studied. They inhabit more extreme environments—salt brines, hot acid springs, bogs, and the deep regions of the ocean.

Within each of these two large groups of bacteria are subgroups distinguished by the habitats to which they are best adapted. In some habitats there is a plentiful supply of oxygen, and the resident organisms live by aerobic metabolism; their catabolic processes ultimately result in the transfer of electrons from fuel molecules to oxygen. Other environments are virtually devoid of oxygen, forcing resident organisms to conduct their catabolic business without it. Many of the organisms that have evolved in these **anaerobic** environments are obligate anaerobes; they die when exposed to oxygen.

All organisms, including bacteria, can be classified as either **chemotrophs** (those obtaining their energy from a chemical fuel) or **phototrophs** (those using sunlight as their primary energy source). Certain organisms can synthesize some or all of their monomeric subunits, metabolic intermediates, and macromolecules from very simple starting materials such as CO_2 and NH_3; these are the **autotrophs.** Others must acquire some of their nutrients from the environment preformed (by autotrophic organisms, for example); these are **heterotrophs.** There are therefore four general modes of obtaining fuel and energy, and four general groups of organisms distinguished by these

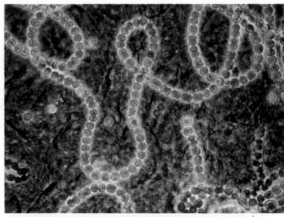

Nostoc sp., a photosynthetic cyanobacterium. ⊢———⊣ 0.25 μm

Figure 2–4 Organisms can be classified according to their source of energy (shaded red) and the form in which they obtain carbon atoms (shaded blue) for the synthesis of cellular material. Organic compounds are both energy source and carbon source for chemoheterotrophs such as ourselves. Some, but not all, chemoheterotrophs consume O_2 and produce CO_2, and some photoautotrophs produce O_2 (shaded green).

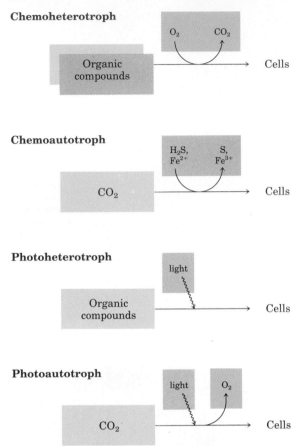

Chemoheterotroph

Chemoautotroph

Photoheterotroph

Photoautotroph

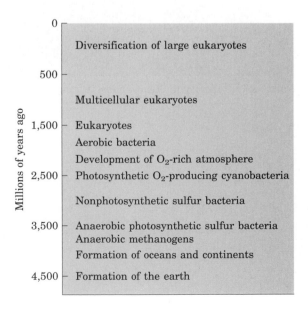

Figure 2–5 Landmarks in the evolution of life on earth.

modes: chemoheterotrophs, chemoautotrophs, photoheterotrophs, and photoautotrophs (Fig. 2–4).

As shown in Figure 2–5, the earliest cells probably arose about 3.5 billion (3.5×10^9) years ago in the rich mixture of organic compounds, the "primordial soup," of prebiotic times; they were almost certainly chemoheterotrophs. The organic compounds were originally synthesized from such components of the early earth's atmosphere as CO, CO_2, N_2, and CH_4 by the nonbiological actions of volcanic heat and lightning (Chapter 3). Primitive heterotrophs gradually acquired the capability to derive energy from certain compounds in their environment and to use that energy to synthesize more and more of their own precursor molecules, thereby becoming less dependent on outside sources of these molecules—less extremely heterotrophic. A very significant evolutionary event was the development of pigments capable of capturing visible light from the sun and using the energy to reduce or "fix" CO_2 into more complex organic compounds. The original electron (hydrogen) donor for these **photosynthetic** organisms was probably H_2S, yielding elemental sulfur as the byproduct, but at some point cells developed the enzymatic capacity to use H_2O as the electron donor in photosynthetic reactions, producing O_2. The cyanobacteria are the modern descendants of these early photosynthetic O_2 producers.

The atmosphere of the earth in the earliest stages of biological evolution was nearly devoid of O_2, and the earliest cells were therefore anaerobic. With the rise of O_2-producing photosynthetic cells, the earth's atmosphere became progressively richer in O_2, allowing the evolution of aerobic organisms, which obtained energy by passing electrons from fuel molecules to O_2 (that is, by oxidizing organic compounds). Because electron transfers involving O_2 yield energy (they are very exergonic; see Chapter 1), aerobic organisms enjoyed an energetic advantage over their anaerobic counterparts when both competed in an environment containing O_2. This advantage translated into the predominance of aerobic organisms in O_2-rich environments.

Modern bacteria inhabit almost every ecological niche in the biosphere, and there are bacterial species capable of using virtually every type of organic compound as a source of carbon and energy. Perhaps three-fourths of all the living matter on the earth consists of microscopic organisms, most of them bacteria.

Bacteria play an important role in the biological exchanges of matter and energy. Photosynthetic bacteria in both fresh and marine waters trap solar energy and use it to generate carbohydrates and other cell materials, which are in turn used as food by other forms of life. Some bacteria can capture molecular nitrogen (N_2) from the atmosphere and use it to form biologically useful nitrogenous compounds, a process known as **nitrogen fixation.** Because animals and most plants cannot do this, bacteria form the starting point of many food chains in the biosphere. They also participate as ultimate consumers, degrading the organic structures of dead plants and animals and recycling the end products to the environment.

Escherichia coli Is the Best-Studied Prokaryotic Cell

Bacterial cells share certain common structural features, but also show group-specific specializations (Fig. 2–6). *E. coli* is a usually harmless inhabitant of the intestinal tract of human beings and many other mammals. The *E. coli* cell is about 2 μm long and a little less than 1 μm in diameter. It has a protective outer membrane and an inner plasma membrane that encloses the cytoplasm and the nucleoid. Between the inner and outer membranes is a thin but strong layer of peptidoglycans (sugar polymers cross-linked by amino acids), which gives the cell its shape and rigidity. The plasma membrane and the layers outside it constitute the **cell envelope.** Differences in the cell envelope account for the different affinities for the dye Gentian violet, which is the basis for Gram's stain; gram-positive bacteria retain the dye, and gram-negative bacteria do not. The outer membrane of *E. coli*, like that of other gram-negative eubacteria, is similar to the plasma membrane in structure but is different in composition. In gram-positive bacteria (*Bacillus subtilis* and *Staphylococcus aureus,* for example) there is no outer membrane, and the peptidoglycan layer surrounding the plasma membrane is much thicker than that in gram-negative bacteria. The plasma membranes of eubacteria consist of a thin bilayer of lipid molecules penetrated by proteins. Archaebacterial membranes have a similar architecture, although their lipids differ from those of the eubacteria.

The plasma membrane contains proteins capable of transporting certain ions and compounds into the cell and carrying products and waste out. Also in the plasma membrane of most eubacteria are electron-carrying proteins (cytochromes) essential in the formation of ATP from ADP (Chapter 1). In the photosynthetic bacteria, internal membranes derived from the plasma membrane contain chlorophyll and other light-trapping pigments.

From the outer membrane of *E. coli* cells and some other eubacteria protrude short, hairlike structures called **pili,** by which cells adhere to the surfaces of other cells. Strains of *E. coli* and other motile bacteria have one or more long **flagella,** which can propel the bacterium through its aqueous surroundings. Bacterial flagella are thin, rigid, helical rods, 10 to 20 nm thick. They are attached to a protein structure that spins in the plane of the cell surface, rotating the flagellum.

The cytoplasm of *E. coli* contains about 15,000 ribosomes, thousands of copies of each of several thousand different enzymes, numerous metabolites and cofactors, and a variety of inorganic ions. Under some conditions, granules of polysaccharides or droplets of lipid accumulate. The nucleoid contains a single, circular molecule of DNA. Although the DNA molecule of an *E. coli* cell is almost 1,000 times longer

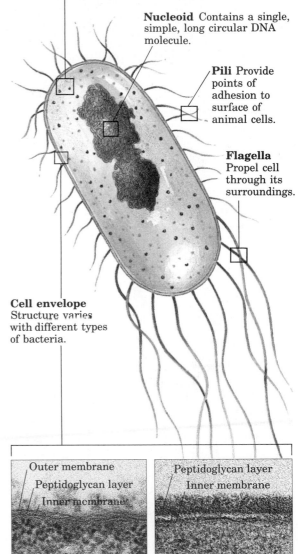

Ribosomes Bacterial ribosomes are smaller than eukaryotic ribosomes, but serve the same function— protein synthesis from an RNA message.

Nucleoid Contains a single, simple, long circular DNA molecule.

Pili Provide points of adhesion to surface of animal cells.

Flagella Propel cell through its surroundings.

Cell envelope Structure varies with different types of bacteria.

Gram-negative bacteria Outer membrane and peptidoglycan layer

(Outer membrane / Peptidoglycan layer / Inner membrane)

Gram-positive bacteria Thicker peptidoglycan layer; outer membrane absent

(Peptidoglycan layer / Inner membrane)

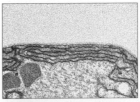

Cyanobacteria Type of gram-negative bacteria with tougher peptidoglycan layer and extensive internal membrane system containing photosynthetic pigments

Archaebacteria No peptidoglycan layer

Figure 2–6 Common structural features of bacterial cells. Because of differences in cell envelope structure, some eubacteria (gram-positive bacteria) retain Gram's stain, and others (gram-negative bacteria) do not. *E. coli* is gram-negative. Cyanobacteria are also eubacteria, but are distinguished by their extensive internal membrane system, in which photosynthetic pigments are localized.

than the cell itself, it is packaged with proteins and tightly folded into the nucleoid, which is less than 1 μm in its longest dimension. As in all bacteria, no membrane surrounds the genetic material. In addition to the DNA in the nucleoid, the cytoplasm of most bacteria contains many smaller, circular segments of DNA called **plasmids.** These nonessential segments of DNA are especially amenable to experimental manipulation and are extremely useful to the molecular geneticist. In nature, some plasmids confer resistance to toxins and antibiotics in the environment.

There is a primitive division of labor within the bacterial cell. The cell envelope regulates the flow of materials into and out of the cell, and protects the cell from noxious environmental agents. The plasma membrane and the cytoplasm contain a variety of enzymes essential to energy metabolism and the synthesis of precursor molecules; the ribosomes manufacture proteins; and the nucleoid stores and transmits genetic information. Most bacteria lead existences that are nearly independent of other cells, but some bacterial species tend to associate in clusters or filaments, and a few (the myxobacteria, for example) demonstrate primitive social behavior. Only eukaryotic cells, however, form true multicellular organisms with a division of labor among cell types.

Evolution of Eukaryotic Cells

Fossils older than 1.5 billion years are limited to those from small and relatively simple organisms, similar in size and shape to modern prokaryotes. Starting about 1.5 billion years ago, the fossil record begins to show evidence of larger and more complex organisms, probably the earliest eukaryotic cells (see Fig. 2–5). Details of the evolutionary path from prokaryotes to eukaryotes cannot be deduced from the fossil record alone, but morphological and biochemical comparison of modern organisms has suggested a reasonable sequence of events consistent with the fossil evidence.

Eukaryotic Cells Evolved from Prokaryotes in Several Stages

Three major changes must have occurred as prokaryotes gave rise to eukaryotes (Fig. 2–7). First, as cells acquired more DNA (Table 2–1), mechanisms evolved to fold it compactly into discrete complexes with specific proteins and to divide it equally between daughter cells at cell division. These DNA–protein complexes, **chromosomes,** (Greek *chroma,* "color" and *soma,* "body"), become especially compact at the time of cell division, when they can be visualized with the light microscope as threads of **chromatin.** Second, as cells became larger, a system of intracellular membranes developed, including a double membrane surrounding the DNA. This membrane segregated the nuclear process of RNA synthesis using a DNA template from the cytoplasmic process of protein synthesis on ribosomes. Finally, primitive eukaryotic cells, which were incapable of photosynthesis or of aerobic metabolism, pooled their assets with those of aerobic bacteria or photosynthetic bacteria to form **symbiotic associations** that became permanent. Some aerobic bacteria evolved into the mitochondria of modern eukaryotes, and some photosynthetic cyanobacteria became the chloroplasts of modern plant cells. Prokaryotic and eukaryotic cells are compared in Table 2–2.

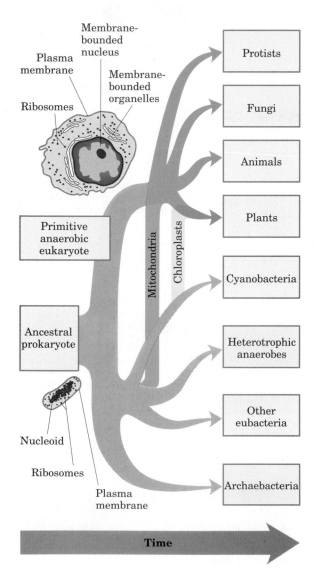

Figure 2–7 One view of how modern plants, animals, fungi, protists, and bacteria share a common evolutionary precursor.

Table 2–1 DNA content and genome complexity

	Genome size (nucleotide pairs)	Relative genome size (*E. coli* = 1)	Length of DNA (mm)
Viruses			
SV40	5×10^3	0.00125	0.0017
T7	4×10^4	0.01	0.014
T2	2×10^5	0.05	0.68
Prokaryotes			
Mycoplasma	3×10^5	0.075	0.10
Bacillus	3×10^6	0.75	1.02
E. coli	4×10^6	1.00	1.36
Fungi			
Yeast	2×10^7	5	68
Animals			
Fruit fly	2×10^8	50	6.8
Chicken	2×10^9	500	680
Human	5×10^9	1,250	1,700
Plants			
Peas	9×10^9	2,250	3,100
Trillium	1×10^{11}	30,000	34,000

Source: From Becker, W.M. & Deamer, D.W. (1991) *The World of the Cell,* 2nd edn, p. 363, The Benjamin/Cummings Publishing Company, Menlo Park, CA.

Table 2–2 Comparison of prokaryotic and eukaryotic cells

Characteristic	Prokaryotic cell	Eukaryotic cell
Size	Generally small (1–10 μm)	Generally large (10–100 μm)
Genome	DNA with nonhistone protein; genome in nucleoid, not surrounded by membrane	DNA complexed with histone and nonhistone proteins in chromosomes; chromosomes in nucleus with membranous envelope
Cell division	Fission or budding; no mitosis	Mitosis including mitotic spindle; centrioles in many
Membrane-bounded organelles	Absent	Mitochondria, chloroplasts (in plants), endoplasmic reticulum, Golgi complexes, lysosomes, etc.
Nutrition	Absorption; some photosynthesis	Absorption, ingestion; photosynthesis by some
Energy metabolism	No mitochondria; oxidative enzymes bound to plasma membrane; great variation in metabolic pattern	Oxidative enzymes packaged in mitochondria; more unified pattern of oxidative metabolism
Cytoskeleton	None	Complex, with microtubules, intermediate filaments, actin filaments
Intracellular movement	None	Cytoplasmic streaming, endocytosis, phagocytosis, mitosis, axonal transport

Source: Modified from Hickman, C.P., Roberts, L.S., & Hickman, F.M. (1990) *Biology of Animals,* 5th edn, p. 30, Mosby–Yearbook, Inc. St. Louis, MO.

Early Eukaryotic Cells Gave Rise to Diverse Protists

With the rise of primitive eukaryotic cells, further evolution led to a tremendous diversity of unicellular eukaryotic organisms **(protists).** Some of these (those with chloroplasts) resembled modern photosynthetic protists such as *Euglena* and *Chlamydomonas;* other, nonphotosynthetic protists were more like *Paramecium* or *Dictyostelium.* Unicellular eukaryotes are abundant, and the cells of all multicellular animals, plants, and fungi are eukaryotic; there are only a few thousand prokaryotic species, but millions of species of eukaryotic organisms.

Major Structural Features of Eukaryotic Cells

Typical eukaryotic cells (Fig. 2–8) are much larger than prokaryotic cells—commonly 10 to 30 μm in diameter, with cell volumes 1,000 to 10,000 times larger than those of bacteria. The distinguishing characteristic of eukaryotes is the nucleus with a complex internal structure, surrounded by a double membrane. The other striking difference between eukaryotes and prokaryotes is that eukaryotes contain a number of other membrane-bounded organelles. The following sections describe the structures and roles of the components of eukaryotic cells in more detail.

Figure 2–8 Schematic illustration of the two types of eukaryotic cell: a representative animal cell **(a)** and a representative plant cell **(b).**

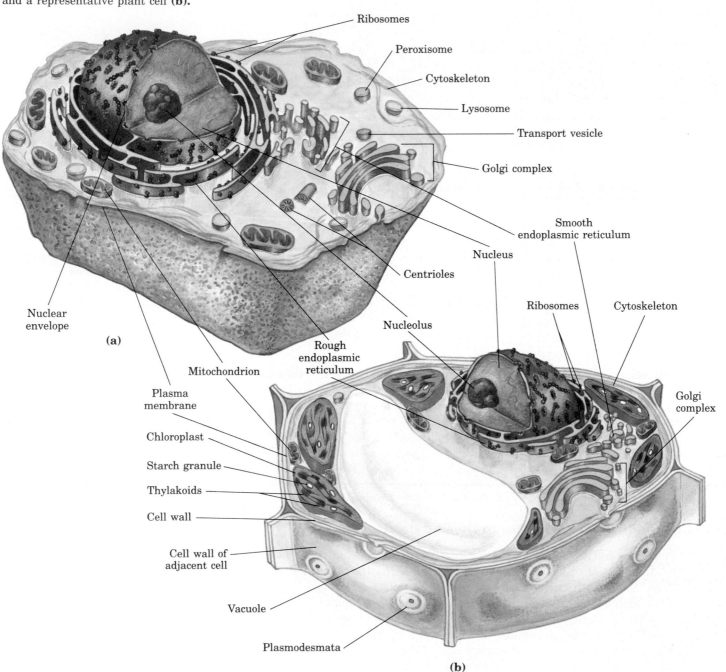

Ribosomes
Peroxisome
Cytoskeleton
Lysosome
Transport vesicle
Golgi complex
Smooth endoplasmic reticulum
Nucleus
Centrioles
Nucleolus
Nuclear envelope
(a)
Mitochondrion
Plasma membrane
Chloroplast
Starch granule
Thylakoids
Cell wall
Cell wall of adjacent cell
Vacuole
Plasmodesmata
Rough endoplasmic reticulum
Ribosomes
Cytoskeleton
Golgi complex
(b)

The Plasma Membrane Contains Transporters and Receptors

The external surface of a cell is in contact with other cells, the extracellular fluid, and the solutes, nutrient molecules, hormones, neurotransmitters, and antigens in that fluid. The plasma membranes of all cells contain a variety of **transporters,** proteins that span the width of the membrane and carry nutrients into and waste products out of the cell. Cells also have surface membrane proteins **(signal receptors)** that present highly specific binding sites for extracellular signaling molecules (receptor ligands). When an external ligand binds to its specific receptor, the receptor protein transduces the signal carried by that ligand into an intracellular message (Fig. 2–9). For example, some surface receptors are associated with **ion channels** that open when the receptor is occupied; others span the membrane and activate or inhibit cellular enzymes on the inner membrane surface. Whatever the mode of **signal transduction,** surface receptors characteristically act as signal amplifiers—a single ligand molecule bound to a single receptor may cause the flux of thousands of ions through an opened channel, or the synthesis of thousands of molecules of an intracellular messenger molecule by an activated enzyme.

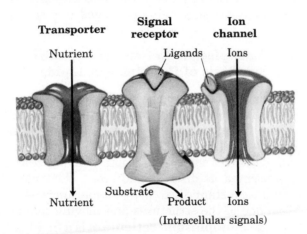

Transporter   **Signal receptor**   **Ion channel**

Nutrient   Ligands   Ions

Nutrient   Substrate   Product   Ions

(Intracellular signals)

Figure 2–9 Proteins in the plasma membrane serve as transporters, signal receptors, and ion channels. Extracellular signals are amplified by receptors, because binding of a single ligand molecule to the surface receptor causes many molecules of an intracellular signal molecule to be formed, or many ions to flow through the opened channel. Transporters carry substances into and out of the cell, but do not act as signal amplifiers.

Some surface receptors recognize ligands of low molecular weight, and others recognize macromolecules. For example, binding of acetylcholine (M_r 146) to its receptor begins a cascade of cellular events that underlie the transmission of signals for muscle contraction. Blood proteins ($M_r > 20,000$) that carry lipids (lipoproteins) are recognized by specific cell surface receptors and then transported into the cells. Antigens (proteins, viruses, or bacteria, recognized by the immune system as foreign) bind to specific receptors and trigger the production of antibodies. During the development of multicellular organisms, neighboring cells influence each other's developmental paths, as signal molecules from one cell type react with receptors of other cells. Thus the surface membrane of a cell is a complex mosaic of different kinds of highly specific "molecular antennae" through which cells receive, amplify, and react to external signals.

Most cells of higher plants have a **cell wall** outside the plasma membrane (Fig. 2–8b), which serves as a rigid, protective shell. The cell wall, composed of cellulose and other carbohydrate polymers, is thick but porous. It allows water and small molecules to pass readily, but swelling of the cell due to the accumulation of water is resisted by the rigidity of the wall.

Endocytosis and Exocytosis Carry Traffic across the Plasma Membrane

Endocytosis is a mechanism for transporting components of the surrounding medium deep into the cytoplasm. In this process (Fig. 2–10), a region of the plasma membrane invaginates, enclosing a small volume of extracellular fluid within a bud that pinches off inside the cell by membrane fission. The resulting small vesicle (**endosome**) can move into the interior of the cell, delivering its contents to another organelle bounded by a single membrane (a lysosome, for example; see p. 34) by fusion of the two membranes. The endosome thus serves as an intracellular extension of the plasma membrane, effectively allowing intimate contact between components of the extracellular medium and regions deep within the cytoplasm, which could not be reached by diffusion alone. **Phagocytosis** is a special case of endocytosis, in which the material carried into the cell (within a phagosome) is particulate, such as a cell fragment or even another, smaller cell. The inverse of endocytosis is **exocytosis** (Fig. 2–10), in which a vesicle in the cytoplasm moves to the inside surface of the plasma membrane and fuses with it, releasing the vesicular contents outside the membrane. Many proteins destined for secretion into the extracellular space are released by exocytosis after being packaged into secretory vesicles.

The Endoplasmic Reticulum Organizes the Synthesis of Proteins and Lipids

The small transport vesicles moving to and from the plasma membrane in exocytosis and endocytosis are parts of a dynamic system of intracellular membranes (Fig. 2–10), which includes the endoplasmic reticulum, the Golgi complexes, the nuclear envelope, and a variety of small vesicles such as lysosomes and peroxisomes. Although generally represented as discrete and static elements, these structures are in fact in constant flux, with membrane vesicles continually budding from one of the structures and moving to and merging with another.

The **endoplasmic reticulum** is a highly convoluted, three-dimensional network of membrane-enclosed spaces extending throughout the cytoplasm and enclosing a subcellular compartment (the lumen of the endoplasmic reticulum) separate from the cytoplasm. The many flattened branches (cisternae) of this compartment are continuous with each other and with the nuclear envelope. In cells specialized for the secretion of proteins into the extracellular space, such as the pancreatic cells that secrete the hormone insulin, the endoplasmic reticulum is particularly prominent. The ribosomes that synthesize proteins destined for export attach to the outer (cytoplasmic) surface of the endoplasmic reticulum, and the secretory proteins are passed through the membrane into the lumen as they are synthesized. Proteins destined for sequestration within lysosomes, or for insertion into the nuclear or plasma membranes, are also synthesized on ribosomes attached to the endoplasmic reticulum. By contrast, proteins that will remain and function within the cytosol are synthesized on cytoplasmic ribosomes unassociated with the endoplasmic reticulum.

The attachment of thousands of ribosomes (usually in regions of large cisternae) gives the **rough endoplasmic reticulum** its granular appearance (Fig. 2–10) and thus its name. In other regions of the cell, the endoplasmic reticulum is free of ribosomes. This **smooth endoplasmic reticulum,** which is physically continuous with the rough

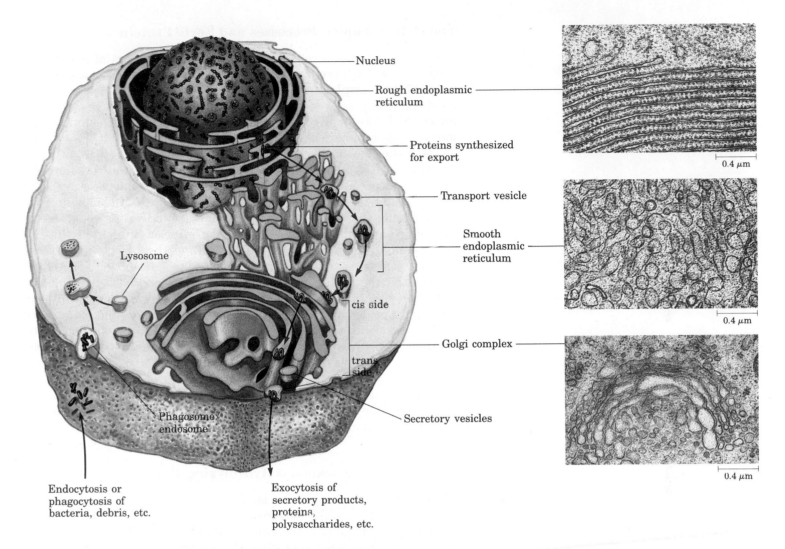

Nucleus

Rough endoplasmic reticulum

Proteins synthesized for export

Transport vesicle

Smooth endoplasmic reticulum

Lysosome

cis side

Golgi complex

trans side

Phagosome/ endosome

Secretory vesicles

Endocytosis or phagocytosis of bacteria, debris, etc.

Exocytosis of secretory products, proteins, polysaccharides, etc.

0.4 μm

0.4 μm

0.4 μm

Figure 2–10 The endomembrane system includes the nuclear envelope, endoplasmic reticulum, Golgi complex, and several types of small vesicles. This system encloses a compartment (the lumen) distinct from the cytosol. Contents of the lumen move from one region of the endomembrane system to another as small transport vesicles bud from one component and fuse with another. High-magnification electron micrographs of a sectioned cell show rough endoplasmic reticulum, studded with ribosomes, smooth endoplasmic reticulum, and the Golgi complex.

The endomembrane system is dynamic; newly synthesized proteins move into the lumen of the rough endoplasmic reticulum and thus to the smooth endoplasmic reticulum, then to the Golgi complex via transport vesicles. In the Golgi complex, molecular "addresses" are added to specific proteins to direct them to the cell surface, lysosomes, or secretory vesicles. The contents of secretory vesicles are released from the cell by exocytosis. Endocytosis and phagocytosis bring extracellular materials into the cell. Fusion of endosomes (or phagosomes) with lysosomes, which are full of digestive enzymes, results in the degradation of the extracellular materials.

endoplasmic reticulum, is the site of lipid biosynthesis and of a variety of other important processes, including the metabolism of certain drugs and toxic compounds. Smooth endoplasmic reticulum is generally tubular, in contrast to the long, flattened cisternae typical of rough endoplasmic reticulum. In some tissues (skeletal muscle, for example) the endoplasmic reticulum is specialized for the storage and rapid release of calcium ions. Ca^{2+} release is the trigger for many cellular events, including muscle contraction.

The Golgi Complex Processes and Sorts Protein

Nearly all eukaryotic cells have characteristic clusters of membrane vesicles called **dictyosomes.** Several connected dictyosomes constitute a **Golgi complex.** A Golgi complex (also called Golgi apparatus) is most commonly seen as a stack of flattened membrane vesicles (cisternae) (Fig. 2–10). Near the ends of these cisternae are numerous, much smaller, spherical vesicles (transport vesicles) that bud off the edges of the cisternae.

The Golgi complex is asymmetric, structurally and functionally. The cis side faces the rough endoplasmic reticulum, and the trans side, the plasma membrane; between these are the medial elements. Proteins, during their synthesis on ribosomes bound to the rough endoplasmic reticulum, are inserted into the interior (lumen) of the cisternae. Small membrane vesicles containing the newly synthesized proteins bud from the endoplasmic reticulum and move to the Golgi complex, fusing with the cis side. As the proteins pass through the Golgi complex to the trans side, enzymes in the complex modify the protein molecules by adding sulfate, carbohydrate, or lipid moieties to side chains of certain amino acids. One of the functions of this modification of a newly synthesized protein is to "address" it to its proper destination as it leaves the Golgi complex in a transport vesicle budding from the trans side. Certain proteins are enclosed in secretory vesicles, eventually to be released from the cell by exocytosis. Others are targeted for intracellular organelles such as lysosomes, or for incorporation into the plasma membrane during cell growth.

Lysosomes Are Packets of Hydrolyzing Enzymes

Lysosomes, found in the cytoplasm of animal cells, are spherical vesicles bounded by a single membrane. They are usually about 1 μm in diameter, about the size of a small bacterium (Fig. 2–10). Lysosomes contain enzymes capable of digesting proteins, polysaccharides, nucleic acids, and lipids. They function as cellular recycling centers for complex molecules brought into the cell by endocytosis, fragments of foreign cells brought in by phagocytosis, or worn-out organelles from the cell's own cytoplasm. These materials selectively enter the lysosomes by fusion of the lysosomal membrane with endosomes, phagosomes, or defective organelles, and are then degraded to their simple components (amino acids, monosaccharides, fatty acids, etc.), which are released into the cytosol to be recycled into new cellular components or further catabolized.

The degradative enzymes within lysosomes would be harmful if not confined by the lysosomal membrane; they would be free to act on all cellular components. The lysosomal compartment is more acidic (pH $\leq$ 5) than the cytoplasm (pH $\approx$ 7); the acidity is due to the action of an ATP-fueled proton pump in the lysosomal membrane. Lysosomal enzymes are much less active at pH 7 than at pH $\leq$ 5, which provides a second line of defense against destruction of cytosolic macromolecules, should these enzymes escape into the cytosol.

Vacuoles of Plant Cells Play Several Important Roles

Plant cells do not have organelles identical to lysosomes, but their **vacuoles** carry out similar degradative reactions as well as other func-

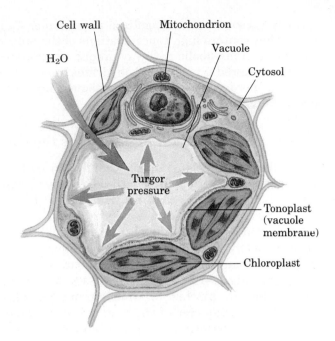

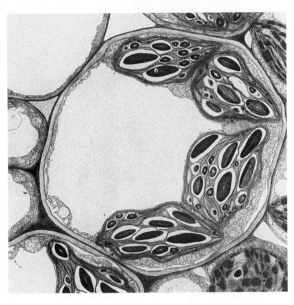

Figure 2–11 The vacuole of a plant cell contains high concentrations of a variety of stored compounds and waste products. Water enters the vacuole by osmosis and increases the vacuolar volume. The resulting turgor pressure forces the cytoplasm out against the cell wall. The rigidity of the cell wall prevents expansion and rupture of the plasma membrane.

tions not found in animal cells. Growing plant cells contain several small vacuoles, vesicles bounded by a single membrane, which fuse and become one large vacuole in the center of the mature cell (Fig. 2–11; see also Fig. 2–8b). The surrounding membrane, the **tonoplast,** regulates the entry into the vacuole of ions, metabolites, and cellular structures destined for degradation. In the mature cell, the vacuole may represent as much as 90% of the total cell volume, pressing the cytoplasm into a thin layer between the tonoplast and the plasma membrane. The liquid within the vacuole, the cell sap, contains digestive enzymes that degrade and recycle macromolecular components no longer useful to the cell. In some plant cells, the vacuole contains high concentrations of pigments (anthocyanins) that give the deep purple and red colors to the flowers of roses and geraniums and the fruits of grapes and plums. Like the contents of lysosomes, the cell sap is generally more acidic than the surrounding cytosol. In addition to its role in storage and degradation of cellular components, the vacuole also provides physical support to the plant cell. Water passes into the vacuole by osmosis because of the high solute concentration of the cell sap, creating outward pressure on the cytosol and the cell wall. This turgor pressure within cells stiffens the plant tissue (Fig. 2–11).

Peroxisomes Destroy Hydrogen Peroxide, and Glyoxysomes Convert Fats to Carbohydrates

Some of the oxidative reactions in the breakdown of amino acids and fats produce free radicals and hydrogen peroxide (H_2O_2), very reactive chemical species that could damage cellular machinery. To protect the cell from these destructive byproducts, such reactions are segregated within small membrane-bounded vesicles called **peroxisomes.** The hydrogen peroxide is degraded by catalase, an enzyme present in large quantities in peroxisomes and glyoxysomes; it catalyzes the reaction $2H_2O_2 \longrightarrow 2H_2O + O_2$.

Glyoxysomes are specialized peroxisomes found in certain plant cells. They contain high concentrations of the enzymes of the **glyoxylate cycle,** a metabolic pathway unique to plants that allows the conversion of stored fats into carbohydrates during seed germination. Lysosomes, peroxisomes, and glyoxysomes are sometimes referred to collectively as **microbodies.**

The Nucleus of Eukaryotes Contains the Genome

The eukaryotic nucleus is very complex in both its structure and its biological activity, compared with the relatively simple nucleoid of prokaryotes. The nucleus contains nearly all of the cell's DNA, typically 1,000 times more than is present in a bacterial cell; a small amount of DNA is also present in mitochondria and chloroplasts. The nucleus is surrounded by a **nuclear envelope,** composed of two membranes separated by a narrow space and continuous with the rough endoplasmic reticulum (Fig. 2–12; see also Fig. 2–10). At intervals the two nuclear membranes are pinched together around openings **(nuclear pores),** which have a diameter of about 90 nm. Associated with the pores are protein structures (nuclear pore complexes), specific macromolecule transporters that allow only certain molecules to pass between the cytoplasm and the aqueous phase of the nucleus (the **nucleoplasm**), such as enzymes synthesized in the cytoplasm and required in the nucleoplasm for DNA replication, transcription, or repair. Messenger RNA precursors and associated proteins also pass out of the nucleus through the nuclear pore complexes, to be translated on ribosomes in the cytoplasm; the nucleoplasm contains no ribosomes.

Figure 2–12 The nucleus and nuclear envelope. **(a)** Scanning electron micrograph of the surface of the nuclear envelope, showing numerous nuclear pores. **(b)** Electron micrograph of the nucleus of the alga *Chlamydomonas*. The dark body in the center of the nucleus is the nucleolus, and the granular material that fills the rest of the nucleus is chromatin. The nuclear envelope has paired membranes with nuclear pores; two are shown by arrows.

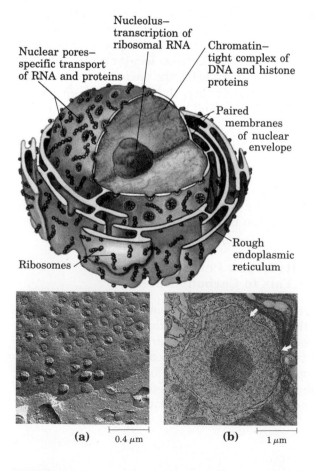

Nucleolus–transcription of ribosomal RNA

Nuclear pores–specific transport of RNA and proteins

Chromatin–tight complex of DNA and histone proteins

Paired membranes of nuclear envelope

Rough endoplasmic reticulum

Ribosomes

(a)   0.4 μm   **(b)**   1 μm

Inside the nucleus is the **nucleolus,** which appears dense in electron micrographs (Fig. 2–12b) because of its high content of RNA. The nucleolus is a specific region of the nucleus, in which the DNA contains many copies of the genes encoding ribosomal RNA. To produce the large number of ribosomes needed by the cell, these genes are continually copied into RNA (transcribed). The nucleolus is the visible evidence of the transcriptional machinery and the RNA product. Ribosomal RNA produced in the nucleolus passes into the cytoplasm through the nuclear pores. The rest of the nucleus contains chromatin, so called because early microscopists found that it stained brightly with certain dyes. Chromatin consists of DNA and proteins bound tightly to the DNA, and represents the chromosomes, which are decondensed in the interphase (nondividing) nucleus and not individually visible.

Before division of the cell **(cytokinesis),** nuclear division **(mitosis)** occurs. The chromatin condenses into discrete bodies, the chromosomes (Fig. 2–13). Cells of each species have a characteristic number of chromosomes with specific sizes and shapes. The protist *Tetrahymena* has 4; cabbage has 20, humans have 46, and the plant *Ophioglossum,* about 1,250! Usually each cell has two copies of each chromosome; such cells are called **diploid.** Gametes (egg and sperm, for example) produced by meiosis (Chapter 24) have only one copy of each chromosome and are called **haploid.** During sexual reproduction, two haploid gametes combine to regenerate a diploid cell in which each chromosome pair consists of a maternal and a paternal chromosome.

Chromosomes and chromatin are composed of DNA and a family of positively charged proteins, **histones,** which associate strongly with DNA by ionic interactions with its many negatively charged phosphate groups. About half of the mass of chromatin is DNA and half is histones. When DNA replicates prior to cell division, large quantities of histones are also synthesized to maintain this 1:1 ratio. The histones and DNA associate in complexes called **nucleosomes,** in which the DNA strand winds around a core of histone molecules (Fig. 2–13). The DNA of a single human chromosome forms about a million nucleosomes; nucleosomes associate to form very regular and compact supramolecular complexes. The resulting chromatin fibers, about 30 nm in diameter, condense further by forming a series of looped regions, which cluster with adjacent looped regions to form the chromosomes visible during cell division. This tight packing of DNA into nucleosomes achieves a remarkable condensation of the DNA molecules. The DNA in the chromosomes of a single diploid human cell would have a combined length of about 2 m if fully stretched as a DNA double helix, but the combined length of all 46 chromosomes is only about 200 nm.

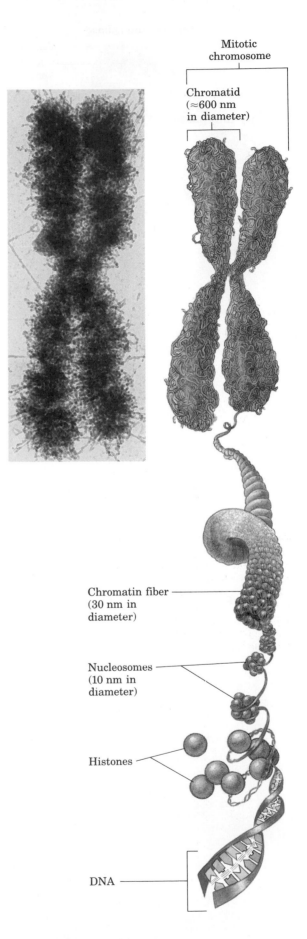

Figure 2–13 Chromosomes are visible in the electron microscope during mitosis. Shown here is one of the 46 human chromosomes. Every chromosome is composed of two chromatids, each consisting of tightly folded chromatin fibers. Each chromatin fiber is in turn formed by the packaging of a DNA molecule wrapped about histone proteins to form a series of nucleosomes. (Adapted from Becker, W.M. & Deamer, D.W. (1991) *The World of the Cell,* 2nd edn, Fig. 13–20, The Benjamin/Cummings Publishing Company, Menlo Park, CA.)

Mitotic
chromosome

Chromatid
(≈600 nm
in diameter)

Chromatin fiber
(30 nm in
diameter)

Nucleosomes
(10 nm in
diameter)

Histones

DNA

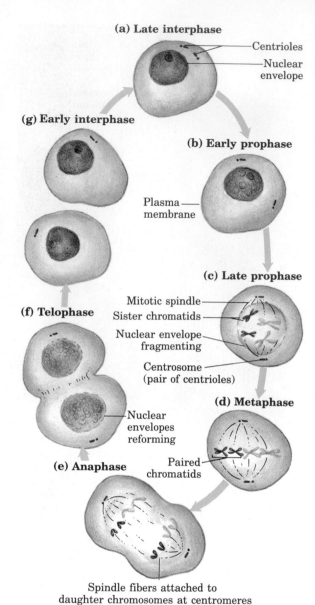

(a) Late interphase

Centrioles

Nuclear envelope

(g) Early interphase

(b) Early prophase

Plasma membrane

(c) Late prophase

Mitotic spindle

Sister chromatids

Nuclear envelope fragmenting

Centrosome (pair of centrioles)

(f) Telophase

(d) Metaphase

Nuclear envelopes reforming

(e) Anaphase

Paired chromatids

Spindle fibers attached to daughter chromosomes at centromeres

Figure 2–14 Mitosis and cell division in animal cells. In the interphase (nondividing) nucleus **(a),** the chromosomes are in the form of dispersed chromatin. As mitosis begins **(b),** chromatin condenses into chromosomes and the mitotic spindle begins to form; centrosomes, which typically contain centriole pairs, dictate the orientation of the spindle. The nuclear envelope disintegrates and the nucleolus disappears **(c),** and the chromosomes align at the center of the cell **(d).** The chromatids of each chromosome move to opposite poles of the cell, pulled by spindle fibers attached to their centromeres **(e),** and a nuclear envelope forms around each new set of chromosomes **(f).** Finally, two daughter cells form by cell division (cytokinesis) **(g).** Although the same basic process occurs in all eukaryotes, there are differences in details of mitosis in plants, fungi, and protists.

Before the beginning of mitosis, each chromosome is duplicated to form paired, identical **chromatids,** each of which is a double helix of DNA. During mitosis (Fig. 2–14), the two chromatids move to opposite ends (poles) of the cell, each becoming a new chromosome. Small cylindrical particles called centrioles, composed of the protein tubulin, provide the spatial organization for the migration of chromatids to opposite ends of the dividing cell. To allow the separation of chromatids, the nuclear envelope breaks down, dispersing into membrane vesicles. When the separation of the two sets of chromosomes is complete, a nuclear envelope derived from the endoplasmic reticulum re-forms around each set. Finally, the two halves of the cell are separated by cytokinesis, and each daughter cell has a complete diploid complement of chromosomes. After mitosis is complete the chromosomes decondense to form dispersed chromatin, and the nucleoli, which disappeared early in mitosis, reappear.

Mitochondria Are the Power Plants of Aerobic Eukaryotic Cells

Mitochondria (singular, **mitochondrion**) are very conspicuous in the cytoplasm of most eukaryotic cells (Fig. 2–15). These membrane-bounded organelles vary in size, but typically have a diameter of about $1\ \mu m$, similar to that of bacterial cells. Mitochondria also vary widely in shape, number, and location, depending on the cell type or tissue function. Most plant and animal cells contain several hundred to a thousand mitochondria. Generally, cells in more metabolically active tissues devote a larger proportion of their volume to mitochondria.

Each mitochondrion has two membranes. The outer membrane is unwrinkled and completely surrounds the organelle. The inner membrane has infoldings called **cristae,** which give it a large surface area. The inner compartment of mitochondria, the **matrix,** is a very concentrated aqueous solution of many enzymes and chemical intermediates involved in energy-yielding metabolism. Mitochondria contain many enzymes that together catalyze the oxidation of organic nutrients by molecular oxygen (O_2); some of these enzymes are in the matrix and some are embedded in the inner membrane. The chemical energy released in mitochondrial oxidations is used to generate ATP, the major energy-carrying molecule of cells. In aerobic cells, mitochondria are the principal producers of ATP, which diffuses to all parts of the cell and provides the energy for cellular work.

Unlike other membranous structures such as lysosomes, Golgi complexes, and the nuclear envelope, mitochondria are produced only by division of previously existing mitochondria; each mitochondrion contains its own DNA, RNA, and ribosomes. Mitochondrial DNA codes for certain proteins specific to the mitochondrial inner membrane, but other mitochondrial proteins are encoded in nuclear DNA. This and other evidence supports the theory that mitochondria are the descendants of aerobic bacteria that lived symbiotically with early eukaryotic cells.

Chloroplasts Convert Solar Energy into Chemical Energy

Plastids are specialized organelles in the cytoplasm of plants; they have two surrounding membranes. Most conspicuous of the plastids and characteristically present in all green plant cells and eukaryotic algae are the **chloroplasts** (Fig. 2–16). Like mitochondria, the chloroplasts may be considered power plants, with the important difference that chloroplasts use solar energy, whereas mitochondria use the chemical energy of oxidizable molecules. Pigment molecules in chloroplasts absorb the energy of light and use it to make ATP and, ultimately, to reduce carbon dioxide to form carbohydrates such as starch and sucrose. The photosynthetic process in eukaryotes and in cyanobacteria produces O_2 as a byproduct of the light-capturing reactions. Photosynthetic plant cells contain both chloroplasts and mitochondria. Chloroplasts transduce energy only in the light, but mitochondria function independently of light, oxidizing carbohydrates generated by photosynthesis during daylight hours.

Chloroplasts are generally larger (diameter 5 μm) than mitochondria and occur in many different shapes. Because chloroplasts contain a high concentration of the pigment **chlorophyll,** photosynthetic cells are usually green, but their color depends on the relative amounts of other pigments present. These pigment molecules, which together can absorb light energy over much of the visible spectrum, are localized in the internal membranes of the chloroplast, which form stacks of closed cisternae known as **thylakoids** (Fig. 2–16). Like mitochondria, chloroplasts contain DNA, RNA, and ribosomes. Chloroplasts appear to have had their evolutionary origin in symbiotic ancestors of the cyanobacteria.

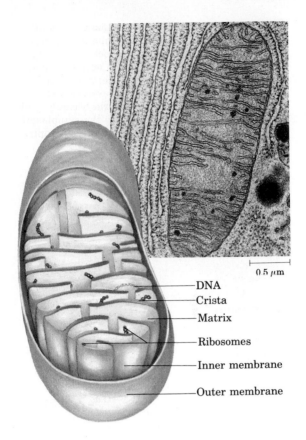

0.5 μm

DNA
Crista
Matrix
Ribosomes
Inner membrane
Outer membrane

Figure 2–15 Structure of a mitochondrion. This electron micrograph of a mitochondrion shows the smooth outer membrane and the numerous infoldings of the inner membrane, called cristae. (Note the extensive rough endoplasmic reticulum surrounding the mitochondrion.)

Outer membrane
Inner membrane
DNA Ribosomes Thylakoids

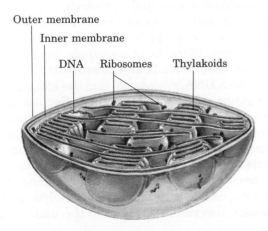

Figure 2–16 A chloroplast in a photosynthetic cell. The thylakoids are flattened membranous sacs that contain chlorophyll, the light-harvesting pigment.

10 μm

Figure 2–17 A plausible theory for the evolutionary origin of mitochondria and chloroplasts. It is based on a number of striking biochemical and genetic similarities between certain aerobic bacteria and mitochondria, and between certain cyanobacteria and chloroplasts. During the evolution of eukaryotic cells, the invading bacteria became symbiotic with the host cell. Ultimately the cytoplasmic bacteria became the mitochondria and chloroplasts of modern cells.

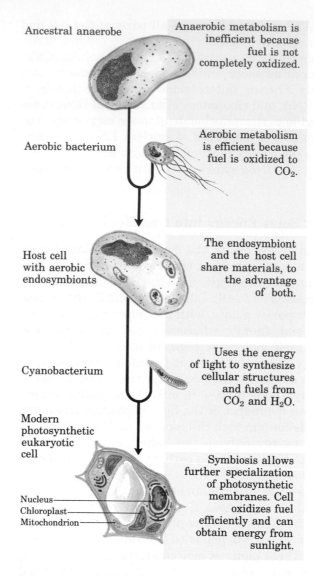

Ancestral anaerobe — Anaerobic metabolism is inefficient because fuel is not completely oxidized.

Aerobic bacterium — Aerobic metabolism is efficient because fuel is oxidized to CO_2.

Host cell with aerobic endosymbionts — The endosymbiont and the host cell share materials, to the advantage of both.

Cyanobacterium — Uses the energy of light to synthesize cellular structures and fuels from CO_2 and H_2O.

Modern photosynthetic eukaryotic cell — Symbiosis allows further specialization of photosynthetic membranes. Cell oxidizes fuel efficiently and can obtain energy from sunlight.

Nucleus
Chloroplast
Mitochondrion

Mitochondria and Chloroplasts Probably Evolved from Endosymbiotic Bacteria

Several independent lines of evidence suggest that the mitochondria and chloroplasts of modern eukaryotes were derived during evolution from aerobic bacteria and cyanobacteria that took up endosymbiotic residence in early eukaryotic cells (Fig. 2–17; see also Fig. 2–7). Mitochondria are always derived from preexisting mitochondria, and chloroplasts from chloroplasts, by simple fission, just as bacteria multiply by fission. Mitochondria and chloroplasts are in fact semiautonomous; they contain DNA, ribosomes, and the enzymatic machinery to synthesize proteins encoded in their DNA. Sequences in mitochondrial DNA are strikingly similar to sequences in certain aerobic bacteria, and chloroplast DNA shows strong sequence homology with the DNA of certain cyanobacteria. The ribosomes found in mitochondria and chloroplasts are more similar in size, overall structure, and ribosomal RNA sequences to those of bacteria than to those in the cytoplasm of the eukaryotic cell. The enzymes that catalyze protein synthesis in these organelles also resemble those of the bacteria more closely.

If mitochondria and chloroplasts are the descendants of early bacterial endosymbionts, some of the genes present in the original free-living bacteria must have been transferred into the nuclear DNA of the host eukaryote over the course of evolution. Neither mitochondria nor chloroplasts contain all of the genes necessary to specify all of their proteins. Most of the proteins of both organelles are encoded in nuclear genes, translated on cytoplasmic ribosomes, and subsequently imported into the organelles.

The Cytoskeleton Stabilizes Cell Shape, Organizes the Cytoplasm, and Produces Motion

Several types of protein filaments visible with the electron microscope crisscross the eukaryotic cell, forming an interlocking three-dimensional meshwork throughout the cytoplasm, the **cytoskeleton.** There are three general types of cytoplasmic filaments: actin filaments, microtubules, and intermediate filaments (Fig. 2–18). They differ in width (from about 6 to 22 nm), composition, and specific function, but all apparently provide structure and organization to the cytoplasm and shape to the cell. Actin filaments and microtubules also help to produce the motion of organelles or of the whole cell.

Each of the cytoskeletal components is composed of simple protein subunits that polymerize to form filaments of uniform thickness. These filaments are not permanent structures; they undergo constant disassembly into their monomeric subunits and reassembly into filaments. Their locations in cells are not rigidly fixed, but may change dramatically with mitosis, cytokinesis, or changes in cell shape. All types of filaments associate with other proteins that cross-link filaments to themselves or to other filaments, influence assembly or disassembly, or move cytoplasmic organelles along the filaments.

Figure 2–18 The three types of cytoplasmic filaments. The upper panels show epithelial cells photographed after treatment with antibodies that bind to and specifically stain **(a)** actin filaments bundled together to form "stress fibers," **(b)** microtubules radiating from the cell center, and **(c)** intermediate filaments, extending throughout the cytoplasm. For these experiments, antibodies that specifically recognize actin, tubulin, or intermediate filament proteins are covalently attached to a fluorescent compound. When the cell is viewed with a fluorescence microscope, only the stained structures are visible. The lower panels show each type of filament as visualized by electron microscopy.

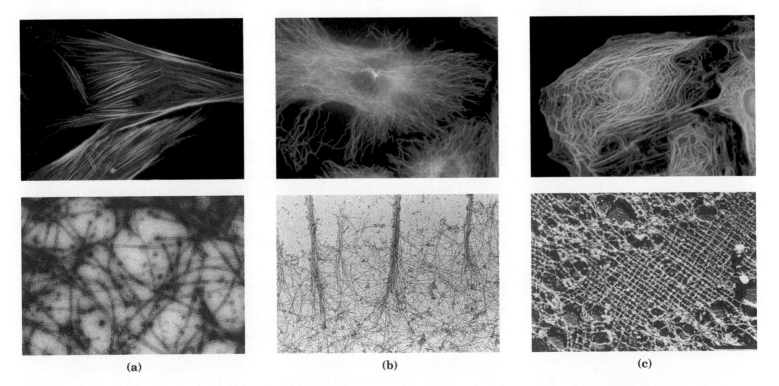

(a)   (b)   (c)

Actin Filaments Are Ubiquitous in Eukaryotic Cells

Actin is a protein present in virtually all eukaryotes, from the protists to the vertebrates. In the presence of ATP, the monomeric protein spontaneously associates into linear, helical polymers, 6 to 7 nm in diameter, called actin filaments or microfilaments (Fig. 2–19).

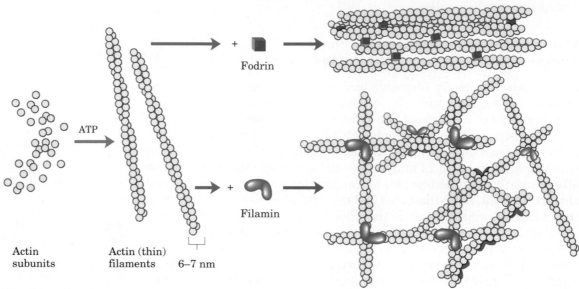

Actin   Actin (thin)
subunits   filaments   6–7 nm

Figure 2–19 Individual subunits of actin polymerize to form actin filaments. The protein filamin holds two filaments together where they cross at right angles. The filaments are cross-linked by another protein, fodrin, to form side-by-side aggregates or bundles.

The importance of actin polymerization and depolymerization is clear from the effects of cytochalasins, compounds that bind to actin and block polymerization. Cells treated with a cytochalasin lose actin filaments and their ability to carry out cytokinesis, phagocytosis, and amoeboid movement. However, chromatid separation at mitosis is not affected, ruling out an essential role for actin in this process. Compounds such as cytochalasins, which are naturally occurring poisons or specific toxins, are often very helpful in experimental studies in pinpointing the important participants in a biological process.

Cells contain proteins that bind to actin monomers or filaments and influence the state of actin aggregation (Fig. 2–19). Filamin and fodrin cross-link actin filaments to each other, stabilizing the meshwork and greatly increasing the viscosity of the medium in which the filaments are suspended; a concentrated solution of actin in the presence of filamin is a gel too viscous to pour. Large numbers of actin filaments bound to specific plasma membrane proteins lie just beneath and more or less parallel to the plasma membrane, conferring shape and rigidity on the cell surface.

Myosins Move along Actin Filaments Using the Energy of ATP

Actin filaments bind to a family of proteins called myosins, enzymes that use the energy of ATP breakdown to move themselves along the actin filament in one direction. The simplest members of this family, such as myosin I, have a globular head and a short tail (Fig. 2–20). The head binds to and moves along an actin filament, driven by the breakdown of ATP. The tail region binds to the membrane of a cytoplasmic organelle, dragging the organelle behind as the myosin head moves along the actin filament. It appears likely that myosins of this type bind to various organelles, providing specific transport systems to move each type of organelle through the cytoplasm. This motion is readily seen in living cells such as the giant green alga *Nitella;* endoplasmic reticulum, as well as mitochondria, nucleus, and other membrane-bound organelles and vesicles move uniformly around the cell at 50 to 75 μm/s in a process called cytoplasmic streaming (Fig. 2–20). This motion has the effect of mixing the cytoplasmic contents of the enormous algal cell much more efficiently than would occur by diffusion alone.

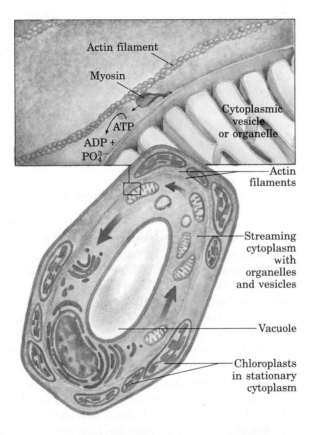

Figure 2–20 Myosin molecules move along actin filaments using energy from ATP. Cytoplasmic streaming is produced in the giant green alga *Nitella* as myosin pulls organelles around a track of actin filaments. The chloroplasts of *Nitella* are located in the layer of stationary cytoplasm that lies between the actin filaments and the cell membrane.

A larger form of myosin is found in muscle cells, and also in the cytoplasm of many nonmuscle cells. This type of myosin also has a globular head that binds to and moves along actin filaments in an ATP-driven reaction, but it has a longer tail, which permits myosin molecules to associate side by side to form thick filaments (see Fig. 7–31). Contractile systems composed of actin and myosin occur in a wide variety of organisms, from slime molds to humans. Actin–myosin complexes form the contractile ring that squeezes the cytoplasm in two during cytokinesis in all eukaryotes. In multicellular animals, muscle cells are filled with highly organized arrays of actin (thin) filaments and myosin (thick) filaments, which produce a coordinated contractile force by ATP-driven sliding of actin filaments past stationary myosin filaments.

Figure 2–21 Microtubules are formed from dimers of the proteins α- and β-tubulin. Colchicine blocks the assembly of microtubules, and can be used to arrest mitosis in cells.

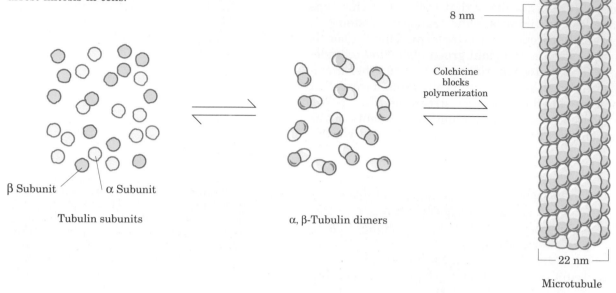

β Subunit   α Subunit

Tubulin subunits

α, β-Tubulin dimers

Colchicine blocks polymerization

Tubulin { α β

8 nm

22 nm

Microtubule

Microtubules Are Rigid, Hollow Rods Composed of Tubulin Subunits

Like actin filaments, **microtubules** form spontaneously from their monomeric subunits, but the polymeric structure of microtubules is slightly more complex. Dimers of α- and β-tubulin form linear polymers (protofilaments), 13 of which associate side by side to form the hollow microtubule, about 22 nm in diameter (Fig. 2–21). Most microtubules undergo continual polymerization and depolymerization in cells by addition of tubulin subunits primarily at one end and dissociation at the other. Microtubules are present throughout the cytoplasm, but are concentrated in specific regions at certain times. For example, when sister chromatids move to opposite poles of a dividing cell during mitosis, a highly organized array of microtubules (the mitotic spindle; Fig. 2–14) provides the framework and probably the motive force for the separation of chromatids. Colchicine, a poisonous alkaloid from meadow saffron, prevents tubulin polymerization. Colchicine treatment reversibly blocks the movement of chromatids during mitosis, demonstrating that microtubules are required for this process.

Microtubules, like actin filaments, associate with a variety of proteins that move along them, form cross-bridges, or influence their state of polymerization. Kinesin and cytoplasmic dynein, proteins found in the cytoplasm of many cells, bind to microtubules and move along them using the energy of ATP to drive their motion (Fig. 2–22). Each protein is capable of associating with specific organelles and pulling them along the microtubule over long distances at rates of about 1 μm/s. The beating motion of cilia and eukaryotic flagella also involves dynein and microtubules.

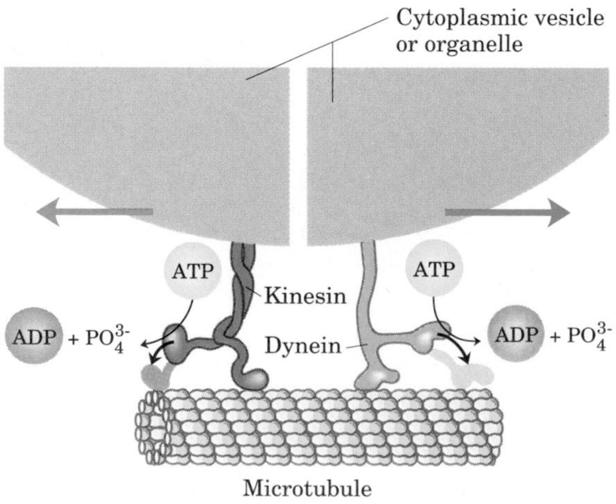

Cytoplasmic vesicle or organelle

ATP   ATP

ADP + PO_4^{3-} Kinesin   ADP + PO_4^{3-}

Dynein

Microtubule

Figure 2–22 Kinesin and dynein are ATP-driven molecular engines that move along microtubular "rails."

The Motion of Cilia and Flagella Results from Movement of Dynein along Microtubules

Cilia and **flagella,** motile structures extending from the surface of many protists and certain cells of animals and plants, are all constructed on the same microtubule-based architectural plan (Fig. 2–23). (Although they bear the same name, the flagella of bacteria (p. 28) are completely different in structure and in action from the flagella of eukaryotes.) Eukaryotic cilia and flagella, which are sheathed in an extension of the plasma membrane, contain nine fused pairs of microtubules arranged around two central microtubules (the 9 + 2 arrangement; Fig. 2–23). Ciliary and flagellar motion results from the coordinated sliding of outer doublet microtubules relative to their neighbors, driven by ATP. The motions of cilia and flagella propel protists through their surrounding medium, in search of food, or light, or some condition essential to their survival. Sperm are also propelled by flagellar beating. Ciliated cells in tissues such as the trachea and oviduct move extracellular fluids past the surface of the ciliated tissue.

The contraction of skeletal muscle, the propelling action of cilia and flagella, and the intracellular transport of organelles all rely on the same fundamental mechanism: the splitting of ATP by proteins such as kinesin, myosin, and dynein drives sliding motion along microfilaments or microtubules.

Intermediate Filaments Provide Structure in the Cytoplasm

The third type of cytoplasmic filament is a family of structures with dimensions (diameter 8 to 10 nm) intermediate between actin filaments and microtubules. Several different types of monomeric protein subunits form **intermediate filaments.** Some cells contain large amounts of one type; some types of intermediate filament are absent from certain cells; and some cell types apparently lack intermediate filaments altogether. As is the case for actin filaments and microtubules, intermediate filament formation is reversible, and the cytoplasmic distribution of these structures is subject to regulated changes.

The function of intermediate filaments is probably to provide internal mechanical support for the cell and to position its organelles. Vimentin (M_r 57,000) is the monomeric subunit of the intermediate filaments found in the endothelial cells that line blood vessels, and in adipocytes (fat cells). Vimentin fibers appear to anchor the nucleus and fat droplets in specific cellular locations. Intermediate filaments composed of desmin (M_r 55,000) hold the Z disks of striated muscle tissue in place. Neurofilaments are constructed of three different protein subunits (M_r 70,000, 150,000, and 210,000), and provide rigidity to the long extensions (axons) of neurons. In the glial cells that surround neurons, intermediate filaments are constructed from glial fibrillary acidic protein (M_r 50,000).

The intermediate filaments composed of keratins, a family of structural proteins, are particularly prominent in certain epidermal cells of vertebrates, and form covalently cross-linked meshworks that persist even after the cell dies. Hair, fingernails, and feathers are among the structures composed primarily of keratins.

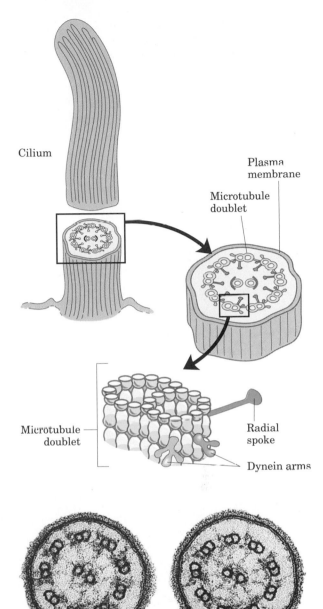

Cilium

Plasma membrane

Microtubule doublet

Microtubule doublet

Radial spoke

Dynein arms

0.1 μm

Figure 2–23 Cilia and eukaryotic flagella have the same architecture: nine microtubule doublets surround a central pair of microtubules. Cross section of cilia shows the 9 + 2 arrangement of microtubules.

The Cytoplasm Is Crowded, Highly Ordered, and Dynamic

The picture that emerges from this brief survey is of a eukaryotic cell with a cytoplasm crisscrossed by a meshwork of structural fibers, throughout which extends a complex system of membrane-bounded compartments (see Fig. 2–8). Both the filaments and the organelles are dynamic: the filaments disassemble and reassemble elsewhere; membranous vesicles bud from one organelle, move to and join another. Transport vesicles, mitochondria, chloroplasts, and other organelles move through the cytoplasm along protein filaments, drawn by kinesin, cytoplasmic dynein, myosin, and perhaps other similar proteins. Exocytosis and endocytosis provide paths between the cell interior and the surrounding medium, allowing for the secretion of proteins and other components produced within the cell and the uptake of extracellular components. The intracellular membrane systems segregate specific metabolic processes, and provide surfaces on which certain enzyme-catalyzed reactions occur.

Although complex, this organization of the cytoplasm is far from random. The motion and positioning of organelles and cytoskeletal elements are under tight regulation, and at certain stages in a eukaryotic cell's life, dramatic, finely orchestrated reorganizations occur, such as spindle formation, chromatid migration to the poles, and nuclear envelope disintegration and re-formation during mitosis. The interactions between the cytoskeleton and organelles are noncovalent, reversible, and subject to regulation in response to various intracellular and extracellular signals. Cytoskeletal rearrangements are modulated by Ca^{2+} and by a variety of proteins.

Organelles Can Be Isolated by Centrifugation

A major advance in the biochemical study of cells was the development of methods for separating organelles from the cytosol and from each other. In a typical cellular fractionation, cells or tissues are disrupted by gentle homogenization in a medium containing sucrose (about 0.2 M). This treatment ruptures the plasma membrane but leaves most of the organelles intact. (The sucrose creates a medium with an osmotic pressure similar to that within organelles; this prevents diffusion of water into the organelles, which would cause them to swell, burst, and spill their contents.)

Organelles such as nuclei, mitochondria, and lysosomes differ in size and therefore sediment at different rates during centrifugation. They also differ in specific gravity, and they "float" at different levels in a density gradient (Fig. 2–24). Differential centrifugation results in a rough fractionation of the cytoplasmic contents, which may be further purified by isopycnic centrifugation. In this procedure, organelles of different buoyant densities (the result of different ratios of lipid and protein in each type of organelle) are separated on a density gradient. By carefully removing material from each region of the gradient and observing it with a microscope, the biochemist can establish the position of each organelle and obtain purified organelles for further study. In this way it was established, for example, that lysosomes contain degradative enzymes, mitochondria contain oxidative enzymes, and chloroplasts contain photosynthetic pigments. The isolation of an organelle enriched in a certain enzyme is often the first step in the purification of that enzyme.

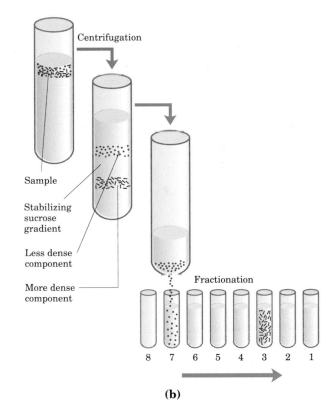

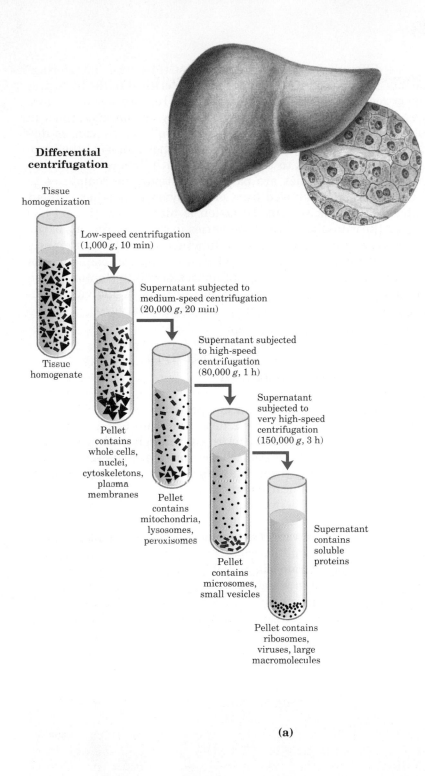

Differential centrifugation

Tissue homogenization

Low-speed centrifugation (1,000 *g*, 10 min)

Tissue homogenate

Pellet contains whole cells, nuclei, cytoskeletons, plasma membranes

Supernatant subjected to medium-speed centrifugation (20,000 *g*, 20 min)

Pellet contains mitochondria, lysosomes, peroxisomes

Supernatant subjected to high-speed centrifugation (80,000 *g*, 1 h)

Pellet contains microsomes, small vesicles

Supernatant subjected to very high-speed centrifugation (150,000 *g*, 3 h)

Supernatant contains soluble proteins

Pellet contains ribosomes, viruses, large macromolecules

(a)

Isopycnic (sucrose-density) centrifugation

Centrifugation

Sample

Stabilizing sucrose gradient

Less dense component

More dense component

Fractionation

8 7 6 5 4 3 2 1

(b)

Figure 2–24 A tissue such as liver is mechanically homogenized to break cells and disperse their contents in an aqueous buffer. The large and small particles in this suspension can be separated by centrifugation at different speeds **(a),** or particles of different density can be separated by isopycnic centrifugation **(b).** In isopycnic centrifugation, a centrifuge tube is filled with a solution, the density of which increases from top to bottom; some solute such as sucrose is dissolved at different concentrations to produce this density gradient. When a mixture of organelles is layered on top of the density gradient and the tube is centrifuged at high speed, individual organelles sediment until their buoyant density exactly matches that in the gradient. Each layer can be collected separately.

In Vitro Studies May Overlook Important Interactions among Molecules

One of the most effective approaches to understanding a biological process is to study purified individual molecules such as enzymes, nucleic acids, or structural proteins. The purified components are amenable to detailed characterization in vitro; their physical properties and catalytic activities can be studied without "interference" from other molecules present in the intact cell. Although this approach has been remarkably revealing, it must always be remembered that the inside of a cell is quite different from the inside of a test tube. The "interfering" components eliminated by purification may be critical to the biological function or regulation of the molecule purified. In vitro studies of pure enzymes are commonly done at very low enzyme concentrations in thoroughly stirred aqueous solutions. In the cell, an enzyme is dissolved or suspended in a gel-like cytosol with thousands of other proteins, some of which bind to that enzyme and influence its activity. Within cells, some enzymes are parts of multienzyme complexes in which reactants are channeled from one enzyme to another without ever entering the bulk solvent. Diffusion is hindered in the gel-like cytosol, and the cytosolic composition varies in different regions of the cell. In short, a given molecule may function somewhat differently within the cell than it does in vitro. One of the central challenges of biochemistry is to understand the influences of cellular organization and macromolecular associations on the function of individual enzymes—to understand function in vivo as well as in vitro.

Evolution of Multicellular Organisms and Cellular Differentiation

All modern unicellular eukaryotes—the protists—contain the organelles and mechanisms that we have described, indicating that these organelles and mechanisms must have evolved relatively early. The protists are extraordinarily versatile. The ciliated protist *Paramecium*, for example, moves rapidly through its aqueous surroundings by beating its cilia; senses mechanical, chemical, and thermal stimuli from its environment, and responds by changing its path; finds, engulfs, and digests a variety of food organisms, and excretes the indigestable fragments; eliminates excess water that leaks through its membrane; and finds and mates with sexual partners. Nonetheless, being unicellular has its disadvantages. Paramecia probably live out their lives in a very small region of the pond in which they began life, because their motility is limited by the small thrust of their microscopic cilia, and their ability to detect a better environment at a distance is limited by the short range of their sensory apparatus.

Figure 2–25 A gallery of differentiated cells. **(a)** Secretory cells of the pancreas, with an extensive endoplasmic reticulum. **(b)** Portion of a skeletal muscle cell, with organized actin and myosin filaments. **(c)** Collenchyma cells of a plant stem. **(d)** Rabbit sperm cells, with long flagella for motility. **(e)** Human erythrocyte. **(f)** Human embryo at the two-celled stage.

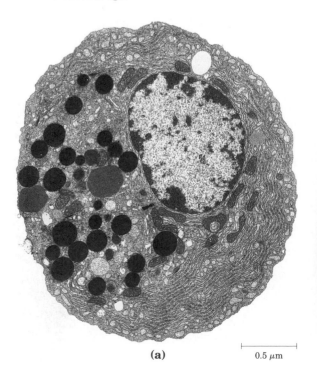

(a)   0.5 μm

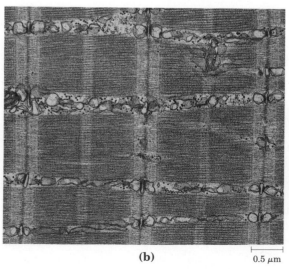

(b)   0.5 μm

(c)   0.1 μm

At some later stage of evolution, unicellular organisms found it advantageous to cluster together, thereby acquiring greater motility, efficiency, or reproductive success than their free-living single-celled competitors. Further evolution of such clustered organisms led to permanent associations among individual cells and eventually to specialization within the colony—to cellular differentiation.

The advantages of cellular specialization led to the evolution of ever more complex and highly differentiated organisms, in which some cells carried out the sensory functions, others the digestive, photosynthetic, or reproductive functions. Many modern multicellular organisms contain hundreds of different cell types, each specialized for some function that supports the entire organism. Fundamental mechanisms that evolved early have been further refined and embellished through evolution. The simple mechanism responsible for the motion of myosin along actin filaments in slime molds has been conserved and elaborated in vertebrate muscle cells, which are literally filled with actin, myosin, and associated proteins that regulate muscle contraction. The same basic structure and mechanism that underlie the beating motion of cilia in *Paramecium* and flagella in *Chlamydomonas* are employed by the highly differentiated vertebrate sperm cell. Figure 2–25 illustrates the range of cellular specializations encountered in multicellular organisms.

The individual cells of a multicellular organism remain delimited by their plasma membranes, but they have developed specialized surface structures for attachment to and communication with each other (Fig. 2–26). At **tight junctions,** the plasma membranes of adjacent cells are closely apposed, with no extracellular fluid separating them. **Desmosomes** (occurring only in plant cells) hold two cells together; the small extracellular space between them is filled with fibrous, presumably adhesive, material. **Gap junctions** provide small, reinforced openings between adjacent cells, through which electric currents, ions, and small molecules can pass. In higher plants, **plasmodesmata** form channels resembling gap junctions; they provide a path through the cell wall for the movement of small molecules between adjacent cells. Each of these junctions is reinforced by membrane proteins or cytoskeletal filaments. The type of junction(s) between neighboring cells varies from tissue to tissue.

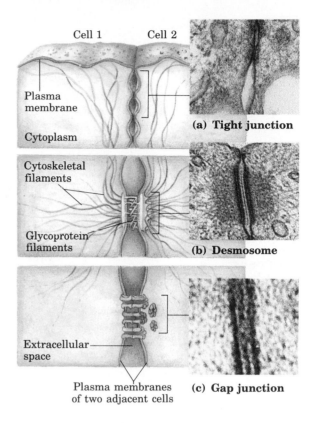

Cell 1   Cell 2

Plasma membrane

Cytoplasm

(a) Tight junction

Cytoskeletal filaments

Glycoprotein filaments

(b) Desmosome

Extracellular space

Plasma membranes of two adjacent cells

(c) Gap junction

Figure 2–26 Three types of junctions between cells. **(a)** Tight junctions produce a seal between adjacent cells. **(b)** Desmosomes, typical of plant cells, weld adjacent cells together and are reinforced by various cytoskeletal elements. **(c)** Gap junctions allow ions and electric currents to flow between adjacent cells.

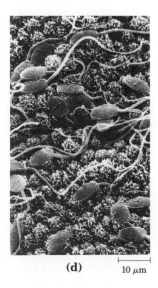

(d)   10 μm

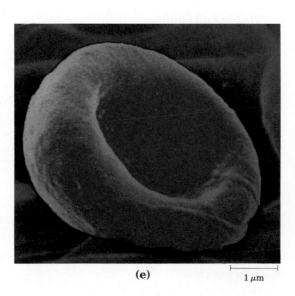

(e)   1 μm

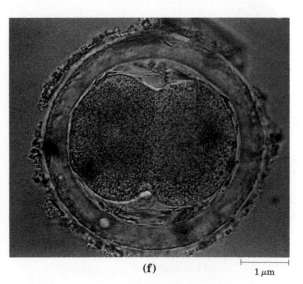

(f)   1 μm

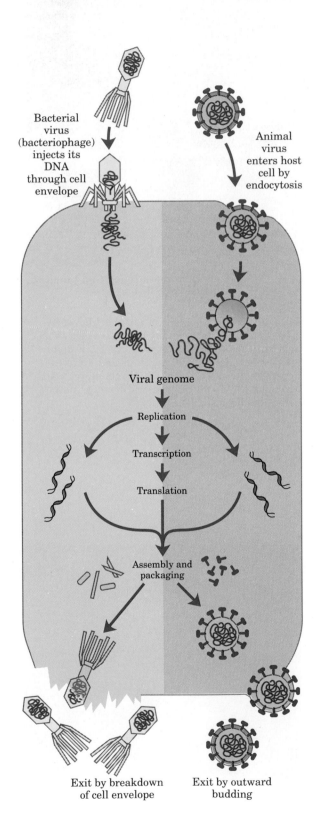

Bacterial
virus
(bacteriophage)
injects its
DNA
through cell
envelope

Animal
virus
enters host
cell by
endocytosis

Viral genome

Replication

Transcription

Translation

Assembly and
packaging

Exit by breakdown
of cell envelope

Exit by outward
budding

Viruses: Parasites of Cells

Viruses are supramolecular complexes that can replicate themselves in appropriate host cells. They consist of a nucleic acid (DNA or RNA) molecule surrounded by a protective shell, or capsid, made up of protein molecules and, in some cases, a membranous envelope. Viruses exist in two states. Outside the host cells that formed them, viruses are simply nonliving particles called **virions,** which are regular in size, shape, and composition and can be crystallized. Once a virus or its nucleic acid component gains entry into a specific host cell, it becomes an intracellular parasite. The viral nucleic acid carries the genetic message specifying the structure of the intact virion. It diverts the host cell's enzymes and ribosomes from their normal cellular roles to the manufacture of many new daughter viral particles. As a result, hundreds of progeny viruses may arise from the single virion that infected the host cell (Fig. 2–27). In some host–virus systems, the progeny virions escape through the host cell's plasma membrane. Other viruses cause cell lysis (membrane breakdown and host cell death) as they are released.

A different type of response results from some viral infections, in which viral DNA becomes integrated into the host's chromosome and is replicated with the host's own genes. Integrated viral genes may have little or no effect on the host's survival, but they often cause profound changes in the host cell's appearance and activity.

Hundreds of different viruses are known, each more or less specific for a host cell (Table 2–3), which may be an animal, plant, or bacterial cell. Viruses specific for bacteria are known as **bacteriophages,** or simply **phages** (Greek *phagein,* "to eat"). Some viruses contain only one kind of protein in their capsid—the tobacco mosaic virus, for example, a simple plant virus and the first to be crystallized. Other viruses contain dozens or hundreds of different kinds of proteins. Even some of these large and complex viruses have been crystallized, and their detailed molecular structures are known (Fig. 2–28). Viruses differ greatly in size. Bacteriophage ϕX174, one of the smallest, has a diameter of 18 nm. Vaccinia virus is one of the largest; its virions are almost as large as the smallest bacteria. Viruses also differ in shape and complexity of structure. The human immunodeficiency virus (HIV) (Fig. 2–29) is relatively simple in structure, but devastating in action; it causes AIDS.

Table 2–3 summarizes the type and size of the nucleic acid components of a number of viruses. Some viruses are highly pathogenic in humans; for example, those causing poliomyelitis, influenza, herpes, hepatitis, AIDS, the common cold, infectious mononucleosis, shingles, and certain types of cancer.

Biochemistry has profited enormously from the study of viruses, which has provided new information about the structure of the genome, the enzymatic mechanisms of nucleic acid synthesis, and the regulation of the flow of genetic information.

Figure 2–27 Infection of a bacterial cell by a bacteriophage (left), and of an animal cell by a virus (right) results in the formation of many copies of the infecting virus.

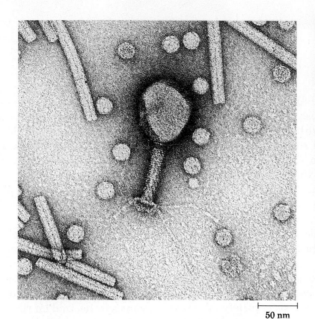

50 nm

Figure 2–28 The structures of several viruses, viewed with the electron microscope. Turnip yellow mosaic virus (small, spherical particles), tobacco mosaic virus (long cylinders), and bacteriophage T4 (shaped like a hand mirror).

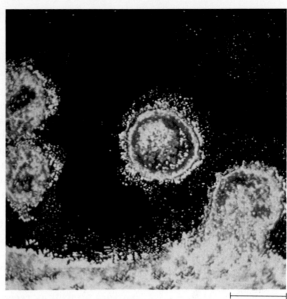

0.1 μm

Figure 2–29 Human immunodeficiency viruses (HIV), the causative agent of AIDS, leaving an infected T lymphocyte of the immune system.

Table 2–3 Some well-studied animal viruses

Virus	Known hosts	Genomic material	Genome size (kilobases)*
Adenoviruses	Vertebrates	DNA	36
SV40	Primates	DNA	5
Herpes	Vertebrates	DNA	150
Vaccinia	Vertebrates	DNA	200
Parvoviruses	Vertebrates	DNA	1–2
Retroviruses	Vertebrates and (?)	RNA/DNA	5–8
Reoviruses	Vertebrates	RNA	1.2–4.0[†]
Influenza	Mammals	RNA	1.0–3.3[†]
Vesicular stomatitis	Vertebrates	RNA	12
Sindbis	Insects and vertebrates	RNA	10
Poliomyelitis	Primates	RNA	7
Human immuno-deficiency (HIV)	Primates	RNA	9.7

Source: From Darnell, J., Lodish, H., & Baltimore D. (1990) *Molecular Cell Biology,* 2nd edn, p. 183, Scientific American Books, Inc., New York.

* Size is given in kilobases (1 kilobase = 1,000 nucleotides) °or single-stranded nucleic acids, or kilobase pairs for double-stranded molecules.

† Reoviruses have ten double-stranded RNA segments, and influenza has eight single-stranded RNA segments; the length of each segment is in the range indicated.

Summary

Cells, the structural and functional units of living organisms, are of microscopic dimensions. Their small size, combined with convolutions of their surfaces, results in high surface-to-volume ratios, facilitating the diffusion of fuels, nutrients, and waste products between the cell and its surroundings. All cells share certain features: DNA containing the genetic information, ribosomes, and a plasma membrane that surrounds the cytoplasm. In eukaryotes the genetic material is surrounded by a nuclear envelope; prokaryotes have no such membrane.

The plasma membrane is a tough, flexible permeability barrier, which contains numerous transporters as well as receptors for a variety of extracellular signals. The cytoplasm consists of the cytosol and organelles. The cytosol is a concentrated solution of proteins, RNA, metabolic intermediates and cofactors, and inorganic ions, in which are suspended various particles. Ribosomes are supramolecular complexes on which protein synthesis occurs; bacterial ribosomes are slightly smaller than those of eukaryotic cells, but are similar in structure and function.

Certain organisms, tissues, and cells offer advantages for biochemical studies. *E. coli* and yeast can be cultured in large quantities, have short generation times, and are especially amenable to genetic manipulation. The specialized functions of liver, muscle, and fat tissue, and of erythrocytes, make them attractive for the study of specific processes.

The first living cells were prokaryotic and anaerobic; they probably arose about 3.5 billion years ago, when the atmosphere was devoid of oxygen. With the passage of time, biological evolution led to cells capable of photosynthesis, with O_2 as a byproduct. As O_2 accumulated, prokaryotic cells capable of the aerobic oxidation of fuels evolved. The two major groups of bacteria, eubacteria and archaebacteria, diverged early in evolution. The cell envelope of some types of bacteria includes layers outside the plasma membrane that provide rigidity or protection. Some bacteria have flagella for propulsion. The cytoplasm of bacteria contains no membrane-bounded organelles but does contain ribosomes and granules of nutrients, as well as a nucleoid which contains the cell's DNA. Some photosynthetic bacteria have extensive intracellular membranes that contain light-capturing pigments.

About 1.5 billion years ago, eukaryotic cells emerged. They were larger than bacteria, and their genetic material was more complex. These early cells established symbiotic relationships with prokaryotes that lived in their cytoplasm; modern mitochondria and chloroplasts are derived from these early endosymbionts. Mitochondria and chloroplasts are intracellular organelles surrounded by a double membrane. They are the principal sites of ATP synthesis in eukaryotic, aerobic cells. Chloroplasts are found only in photosynthetic organisms, but mitochondria are ubiquitous among eukaryotes.

Modern eukaryotic cells have a complex system of intracellular membranes. This endomembrane system consists of the nuclear envelope, rough and smooth endoplasmic reticulum, the Golgi complex, transport vesicles, lysosomes, and endosomes. Proteins synthesized on ribosomes bound to the rough endoplasmic reticulum pass into the endomembrane system, traveling through the Golgi complex on their way to organelles or to the cell surface, where they are secreted by exocytosis. Endocytosis brings extracellular materials into the cell, where they can be digested by degradative enzymes in the lysosomes. In plants, the central vacuole is the site of degradative processes; it also serves as a storage depot for a variety of side products of metabolism and maintains cell turgor.

The genetic material in eukaryotic cells is organized into chromosomes, highly ordered complexes of DNA and histone proteins. Before cell division (cytokinesis), each chromosome is replicated, and the duplicate chromosomes are separated by the process of mitosis.

The cytoskeleton is an intracellular meshwork of actin filaments, microtubules, and intermediate filaments of several types. The cytoskeleton confers shape on the cell, and reorganization of cytoskeletal filaments results in the shape changes accompanying amoeboid movement and cell division. Intracellular organelles move along filaments of the cytoskeleton, propelled by proteins such as dynein, kinesin, and myosin, using the energy of ATP. Dynein and tubulin are central to the motion and structure of cilia and flagella, and myosin and actin are responsible for the contractile motion of skeletal muscle. The organelles can be separated by differential centrifugation and by isopycnic centrifugation.

In multicellular organisms, there is a division of labor among several types of cells. Individual cells are joined to each other by tight junctions or gap junctions, and (in plants) desmosomes or plasmodesmata. Viruses are parasites of living cells, capable of subverting the cellular machinery for their own replication.

Further Reading

General

Alberts, B., Bray, D., Lewis, J., Raff, M., Roberts, K., & Watson, J.D. (1989) *Molecular Biology of the Cell,* 2nd edn, Garland Publishing, Inc., New York.
A superb textbook on cell structure and function, covering the topics considered in this chapter, and a useful reference for many of the following chapters.

Becker, W.M. & Deamer, D.W. (1991) *The World of the Cell,* 2nd edn, The Benjamin/Cummings Publishing Company, Redwood City, CA.
An excellent introductory textbook of cell biology.

Curtis, H. & Barnes, N.S. (1989) *Biology,* 5th edn, Worth Publishers, Inc., New York.
A beautifully written and illustrated general biology textbook.

Darnell, J., Lodish, H., & Baltimore, D. (1990) *Molecular Cell Biology,* 2nd edn, Scientific American Books, Inc., New York.
Like the book by Alberts and coauthors, a superb text useful for this and later chapters.

Prescott, D.M. (1988) *Cells,* Jones and Bartlett Publishers, Boston, MA.
A short, well-illustrated introductory textbook on cell structure and function, with emphasis on structure.

Evolution of Cells

Evolution of Catalytic Function. (1987) *Cold Spring Harb. Symp. Quant. Biol.* **52.**
A collection of excellent papers on many aspects of molecular and cellular evolution.

Knoll, A.H. (1991) End of the proterozoic eon. *Sci. Am.* **265** (October), 64–73.
Discussion of the evidence that an increase in atmospheric oxygen led to the development of multicellular organisms, including large animals.

Margulis, L. (1992) *Symbiosis in Cell Evolution. Microbial Evolution in the Archean and Proterozoic Eons,* 2nd edn, W.H. Freeman and Company, New York.
Clear discussion of the hypothesis that mitochondria and chloroplasts are descendants of bacteria that became symbiotic with primitive eukaryotic cells.

Schopf, J.W. (1978) The evolution of the earliest cells. *Sci. Am.* **239** (September), 110–139.

Vidal, G. (1984) The oldest eukaryotic cell. *Sci. Am.* **250** (February), 48–57.

Structure of Cells and Organelles

Bloom, W. & Fawcett, D.W. (1986) *A Textbook of Histology,* 11th edn, W.B. Saunders Company, Philadelphia, PA.
A standard textbook, containing detailed descriptions of the structures of animal cells, tissues, and organs.

de Duve, C. (1984) *A Guided Tour of the Living Cell,* Scientific American Books, Inc., New York.
An easy-to-read, well-illustrated description of the structure and functions of the organelles of the eukaryotic cell.

Margulis, L. & Schwartz, K.V. (1987) *Five Kingdoms: An Illustrated Guide to the Phyla of Life on Earth,* 2nd edn, W.H. Freeman and Company, New York.
Description of unicellular and multicellular organisms, beautifully illustrated with electron micrographs and drawings showing the diversity of structure and function.

Rothman, J.E. (1985) The compartmental organization of the Golgi apparatus. *Sci. Am.* **253** (September), 74–89.

Cytoskeleton

Gelfand, V. & Bershadsky, A.D. (1991) Microtubule dynamics: mechanism, regulation, and function. *Annu. Rev. Cell Biol.* **7,** 93–116.

Organization of the Cytoplasm. (1981) *Cold Spring Harb. Symp. Quant. Biol.* **46.**
More than 90 excellent papers on microtubules, microfilaments, and intermediate filaments and their biological roles.

Schroer, T.A. & Sheetz, M.P. (1991) Functions of microtubule-based motors. *Annu. Rev. Physiol.* **53,** 629–652.

Steinert, P.M. & Parry, D.A.D. (1985) Intermediate filaments: conformity and diversity of expression and structure. *Annu. Rev. Cell Biol.* **1,** 41–65.

Stossel, T.P. (1989) From signal to pseudopod: how cells control cytoplasmic actin assembly. *J. Biol. Chem.* **264,** 18261–18264.

Vale, R.D. (1990) Microtubule-based motor proteins. *Curr. Opinion Cell Biol.* **2,** 15–22.

Vallee, R.B. & Shpetner, H.S. (1990) Motor proteins of cytoplasmic microtubules. *Annu. Rev. Biochem.* **59,** 909–932.

Problems

Some problems on the contents of Chapter 2 follow. They involve simple geometrical and numerical relationships concerning cell structure and activities. (For your reference in solving these problems, please see the tables printed on the inside of the back cover.) Each problem has a title for easy reference and discussion.

1. *The Size of Cells and Their Components* Given their approximate diameters, calculate the approximate number of (a) hepatocytes (diameter 20 μm), (b) mitochondria (1.5 μm), and (c) actin molecules (3.6 nm) that can be placed in a single layer on the head of a pin (diameter 0.5 mm). Assume each structure is spherical. The area of a circle is πr^2, where $\pi = 3.14$.

2. *Number of Solute Molecules in the Smallest Known Cells* Mycoplasmas are the smallest known cells. They are spherical and have a diameter of about 0.33 μm. Because of their small size they readily pass through filters designed to trap larger bacteria. One species, *Mycoplasma pneumoniae,* is the causative organism of the disease primary atypical pneumonia.

(a) D-Glucose is the major energy-yielding nutrient of mycoplasma cells. Its concentration within such cells is about 1.0 mM. Calculate the number of glucose molecules in a single mycoplasma cell. Avogadro's number, the number of molecules in 1 mol of a nonionized substance, is 6.02×10^{23}. The volume of a sphere is $\frac{4}{3}\pi r^3$.

(b) The first enzyme required for the energy-yielding metabolism of glucose is hexokinase (M_r 100,000). Given that the intracellular fluid of mycoplasma cells contains 10 g of hexokinase per liter, calculate the molar concentration of hexokinase.

3. *Components of* E. coli *E. coli* cells are rod-shaped, about 2 μm long and 0.8 μm in diameter. The volume of a cylinder is $\pi r^2 h$, where h is the height of the cylinder.

(a) If the average density of *E. coli* (mostly water) is 1.1×10^3 g/L, what is the weight of a single cell?

(b) The protective cell wall of *E. coli* is 10 nm thick. What percentage of the total volume of the bacterium does the wall occupy?

(c) *E. coli* is capable of growing and multiplying rapidly because of the inclusion of some 15,000 spherical ribosomes (diameter 18 nm) in each cell, which carry out protein synthesis. What percentage of the total cell volume do the ribosomes occupy?

4. *Genetic Information in* E. coli *DNA* The genetic information contained in DNA consists of a linear sequence of successive code words, known as codons. Each codon is a specific sequence of three nucleotides (three nucleotide pairs in double-stranded DNA), and each codon codes for a single amino acid unit in a protein. The molecular weight of an *E. coli* DNA molecule is about 2.5×10^9. The average molecular weight of a nucleotide pair is 660, and each nucleotide pair contributes 0.34 nm to the length of DNA.

(a) Calculate the length of an *E. coli* DNA molecule. Compare the length of the DNA molecule with the actual cell dimensions. How does the DNA molecule fit into the cell?

(b) Assume that the average protein in *E. coli* consists of a chain of 400 amino acids. What is the maximum number of proteins that can be coded by an *E. coli* DNA molecule?

5. *The High Rate of Bacterial Metabolism* Bacterial cells have a much higher rate of metabolism than animal cells. Under ideal conditions some bacteria will double in size and divide in 20 min, whereas most animal cells require 24 h. The high rate of bacterial metabolism requires a high ratio of surface area to cell volume.

(a) Why would the surface-to-volume ratio have an effect on the maximum rate of metabolism?

(b) Calculate the surface-to-volume ratio for the spherical bacterium *Neisseria gonorrhoeae* (diameter 0.5 μm), responsible for the disease gonorrhea. Compare it with the surface-to-volume ratio for globular amoeba, a large eukaryotic cell of diameter 150 μm. The surface area of a sphere is $4\pi r^2$.

6. *A Strategy to Increase the Surface Area of Cells* Certain cells whose function is to absorb nutrients, e.g., the cells lining the small intestine or the root hair cells of a plant, are optimally adapted to their role because their exposed surface area is increased by microvilli. Consider a spherical epithelial cell (diameter 20 μm) lining the small intestine. Since only a part of the cell surface faces the interior of the intestine, assume that a "patch" corresponding to 25% of the cell area is covered with microvilli. Furthermore, assume that the microvilli are cylinders 0.1 μm in diameter, 1.0 μm long, and spaced in a regular grid 0.2 μm on center.

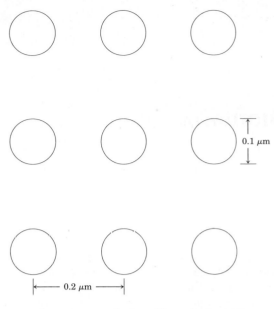

0.1 μm

0.2 μm

Arrangement of microvilli on the "patch"

(a) Calculate the number of microvilli on the patch.

(b) Calculate the surface area of the patch, assuming it has no microvilli.

(c) Calculate the surface area of the patch, assuming it does have microvilli.

(d) What percentage improvement of the absorptive capacity (reflected by the surface-to-volume ratio) does the presence of microvilli provide?

7. *Fast Axonal Transport* Some neurons have long, thin extensions (axons) as long as 2 m. Small membrane vesicles carrying materials essential to axonal function move along microtubules from the cell body to the tip of the axon by kinesin-dependent "fast axonal transport." If the average velocity of a vesicle is 1 μm/s, how long does it take a vesicle to move the 2 m from cell body to axonal tip? What are the possible advantages of this ATP-dependent process over simple diffusion to move materials to the axonal tip?

8. *Toxic Effects of Phalloidin* Phalloidin is a toxin produced by the mushroom *Amanita phalloides*. It binds specifically to actin microfilaments and blocks their disassembly. Cytochalasin B is another toxin, which blocks microfilament assembly from actin monomers (see p. 42).

(a) Predict the effect of phalloidin on cytokinesis, phagocytosis, and amoeboid movement, given the effects of cytochalasins on these processes.

(b) A specific antibody (a protein of $M_r \approx$ 150,000) binds actin tightly and is found to block microfilament assembly in vitro (in the test tube). Would you expect this antibody to mimic the effects of cytochalasin in vivo (in living cells)?

9. *Osmotic Breakage of Organelles* In the isolation of cytosolic enzymes, cells are often broken in the presence of 0.2 M sucrose to prevent osmotic swelling and bursting of the intracellular organelles. If the desired enzymes are in the cytosol, why is it necessary to be concerned about possible damage to particulate organelles?

Biomolecules

The chemical composition of living material, such as this jellyfish, differs from that of its physical environment, which for this organism is salt water.

Biochemistry aims to explain biological form and function in chemical terms. One of the most fruitful approaches to understanding biological phenomena has been to purify an individual chemical component, such as a protein, from a living organism and to characterize its chemical structure or catalytic activity. As we begin the study of biomolecules and their interactions, some basic questions deserve attention. What chemical elements are found in cells? What kinds of molecules are present in living matter? In what proportions do they occur? How did they come to be there? In what ways are the kinds of molecules found in living cells especially suited to their roles?

We review here some of the chemical principles that govern the properties of biological molecules: the covalent bonding of carbon with itself and with other elements, the functional groups that occur in common biological molecules, the three-dimensional structure and stereochemistry of carbon compounds, and the common classes of chemical reactions that occur in living organisms. Next, we discuss the monomeric units and the contribution of entropy to the free-energy changes of reactions in which these units are polymerized to form macromolecules. Finally, we consider the origin of the monomeric units from simple compounds in the earth's atmosphere during prebiological times—that is, chemical evolution.

Chemical Composition

By the beginning of the nineteenth century, it had become clear to chemists that the composition of living matter is strikingly different from that of the inanimate world. Antoine Lavoisier (1743–1794) noted the relative chemical simplicity of the "mineral world," and contrasted it with the complexity of the "plant and animal worlds"; the latter, he knew, were composed of compounds rich in the elements carbon, oxygen, nitrogen, and phosphorus. The development of organic chemistry preceded, and provided invaluable insights for, the development of biochemistry.

We will briefly review some fundamental concepts of organic chemistry: the nature of bonding between atoms of carbon and of hydrogen, oxygen, and nitrogen; the functional groups that result from these combinations; and the diversity of organic compounds that are derived from these elements.

1 **H** 1.008																	2 **He** 4.003
3 **Li** 6.94	4 **Be** 9.01											5 **B** 10.81	6 **C** 12.011	7 **N** 14.01	8 **O** 16.00	9 **F** 19.00	10 **Ne** 20.18
11 **Na** 22.99	12 **Mg** 24.31											13 **Al** 26.98	14 **Si** 28.09	15 **P** 30.97	16 **S** 32.06	17 **Cl** 35.45	18 **Ar** 39.95
19 **K** 39.10	20 **Ca** 40.08	21 **Sc** 44.96	22 **Ti** 47.90	23 **V** 50.94	24 **Cr** 52.00	25 **Mn** 54.94	26 **Fe** 55.85	27 **Co** 58.93	28 **Ni** 58.71	29 **Cu** 63.55	30 **Zn** 65.37	31 **Ga** 69.72	32 **Ge** 72.59	33 **As** 74.92	34 **Se** 78.96	35 **Br** 79.90	36 **Kr** 83.80
37 **Rb** 85.47	38 **Sr** 87.62	39 **Y** 88.91	40 **Zr** 91.22	41 **Nb** 92.91	42 **Mo** 95.94	43 **Tc** 98.91	44 **Ru** 101.07	45 **Rh** 102.91	46 **Pd** 106.4	47 **Ag** 107.87	48 **Cd** 112.40	49 **In** 114.82	50 **Sn** 118.69	51 **Sb** 121.75	52 **Te** 127.60	53 **I** 126.90	54 **Xe** 131.30
55 **Cs** 132.91	56 **Ba** 137.34	Lanthanides	72 **Hf** 178.49	73 **Ta** 180.95	74 **W** 183.85	75 **Re** 186.2	76 **Os** 190.2	77 **Ir** 192.2	78 **Pt** 195.09	79 **Au** 196.97	80 **Hg** 200.59	81 **Tl** 204.37	82 **Pb** 207.19	83 **Bi** 208.98	84 **Po** (209)	85 **At** (210)	86 **Rn** (222)
87 **Fr** (223)	88 **Ra** 226.03	Actinides															

Figure 3–1 Elements essential to animal life and health. Bulk elements (shaded orange) are structural components of cells and tissues and are required in the diet in gram quantities daily. For trace elements (shaded yellow), the requirements are much smaller: for humans, a few milligrams per day of Fe, Cu, and Zn, even less of the others. The elemental requirements for plants and microorganisms are very similar to those shown here.

Living Matter Is Composed Mostly of the Lighter Elements

Only about 30 of the more than 90 naturally occurring chemical elements are essential to living organisms. Most of the elements in living matter have relatively low atomic numbers; only five have atomic numbers above that of selenium, 34 (Fig. 3–1). The four most abundant elements in living organisms, in terms of the percentage of the total number of atoms, are hydrogen, oxygen, nitrogen, and carbon, which together make up over 99% of the mass of most cells. They are the lightest elements capable of forming one, two, three, and four bonds, respectively (Fig. 3–2). In general, the lightest elements form the strongest bonds. Six of the eight most abundant elements in the

Atom	Number of unpaired electrons (in red)	Number of electrons in complete outer shell
H·	1	2
:Ö·	2	8
:N·	3	8
·C·	4	8
:S·	2	8
:P·	3	8

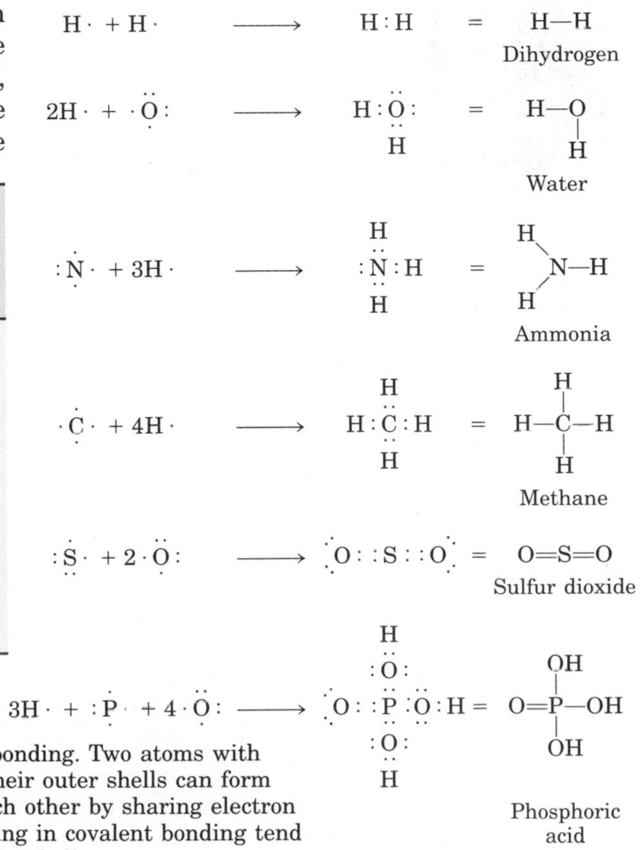

Figure 3–2 Covalent bonding. Two atoms with unpaired electrons in their outer shells can form covalent bonds with each other by sharing electron pairs. Atoms participating in covalent bonding tend to fill their outer electron shells.

Table 3–1 Elemental abundance in seawater, the human body, and the earth's crust*

Seawater (%)		Human body (%)		Earth's crust (%)	
H	66	H	63	O	47
O	33	O	25.5	Si	28
Cl	0.33	C	9.5	Al	7.9
Na	0.28	N	1.4	Fe	4.5
Mg	0.033	Ca	0.31	Ca	3.5
S	0.017	P	0.22	Na	2.5
Ca	0.0062	Cl	0.08	K	2.5
K	0.0060	K	0.06	Mg	2.2
C	0.0014				

* Values are given as percentage of total number of atoms.

Table 3–2 The biological functions of some trace elements

Element	Example of biological function
Fe	Electron carrier in oxidation–reduction reactions
Cu	Component of mitochondrial oxidase
Mn	Cofactor of the enzyme arginase and other enzymes
Zn	Cofactor of dehydrogenases
Co	Component of vitamin B_{12}
Mo	Component of N_2-fixing enzyme
Se	Component of the enzyme glutathione peroxidase
V	Cofactor of the enzyme nitrate reductase
Ni	Cofactor of the enzyme urease
I	Component of thyroid hormone
Mg	Cofactor in photosynthesis

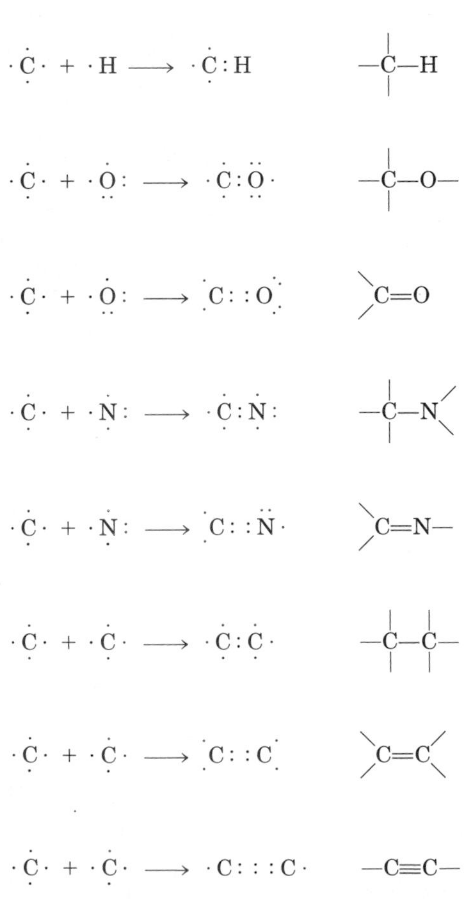

human body are also among the nine most abundant elements in seawater (Table 3–1), and several of the elements abundant in humans are components of the atmosphere and were probably present in the atmosphere before the appearance of life on earth. Primitive seawater was most likely the liquid medium in which living organisms first arose, and the primitive atmosphere was probably a source of methane, ammonia, water, and hydrogen, the starting materials for the evolution of life. The trace elements (Fig. 3–1) represent a miniscule fraction of the weight of the human body, but all are absolutely essential to life, usually because they are essential to the function of specific enzymes (Table 3–2).

Biomolecules Are Compounds of Carbon

The chemistry of living organisms is organized around the element carbon, which accounts for more than one-half the dry weight of cells. In methane (CH_4), a carbon atom shares four electron pairs with four hydrogen atoms; each of the shared electron pairs forms a single bond. Carbon can also form single and double bonds to oxygen and nitrogen atoms (Fig. 3–3). Of greatest significance in biology is the ability of carbon atoms to share electron pairs with each other to form very stable carbon–carbon single bonds. Each carbon atom can form single bonds with one, two, three, or four other carbon atoms. Two carbon atoms also can share two (or three) electron pairs, thus forming carbon–carbon double (or triple) bonds (Fig. 3–3). Covalently linked carbon atoms can form linear chains, branched chains, and cyclic and cagelike structures. To these carbon skeletons are added groups of other atoms, called **functional groups,** which confer specific chemical properties on the molecule. Molecules containing covalently bonded carbon backbones are called **organic compounds;** they occur in an almost limitless variety. Most biomolecules are organic compounds; we can therefore infer that the bonding versatility of carbon was a major factor in the selection of carbon compounds for the molecular machinery of cells during the origin and evolution of living organisms.

Figure 3–3 Versatility of carbon in forming covalent single, double, and triple bonds (in red), particularly between carbon atoms. Triple bonds occur only rarely in biomolecules.

Organic Compounds Have Specific Shapes and Dimensions

The four covalent single bonds that can be formed by a carbon atom are arranged tetrahedrally, with an angle of about 109.5° between any two bonds (Fig. 3–4) and an average length of 0.154 nm. There is free rotation around each carbon–carbon single bond unless very large or highly charged groups are attached to both carbon atoms, in which case rotation may be restricted. A carbon–carbon double bond is shorter (about 0.134 nm long) and rigid and allows little rotation about its axis. (Fig. 3–4). No other chemical element can form molecules of such widely different sizes and shapes or with such a variety of functional groups.

Functional Groups Determine Chemical Properties

Most biomolecules can be regarded as derivatives of hydrocarbons, compounds with a covalently linked carbon backbone to which only hydrogen atoms are bonded. The backbones of hydrocarbons are very stable. The hydrogen atoms may be replaced by a variety of functional groups to yield different families of organic compounds. Typical families of organic compounds are the alcohols, which have one or more hydroxyl groups; amines, which have amino groups; aldehydes and ketones, which have carbonyl groups; and carboxylic acids, which have carboxyl groups (Fig. 3–5).

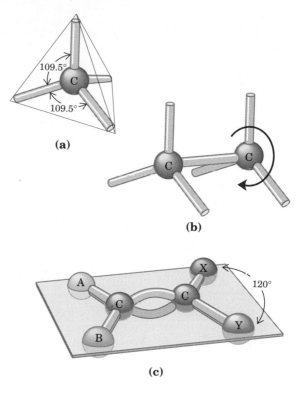

(a)

(b)

(c)

Figure 3–4 (a) Carbon atoms have a characteristic tetrahedral arrangement of their four single bonds, which are about 0.154 nm long and at an angle of 109.5° to each other. **(b)** Carbon–carbon single bonds have freedom of rotation, shown for the compound ethane (CH_3—CH_3). **(c)** Carbon–carbon double bonds are shorter and do not allow free rotation. The single bonds on each doubly bonded carbon make an angle of 120° with each other. The two doubly bonded carbons and the atoms designated A, B, X, and Y all lie in the same rigid plane.

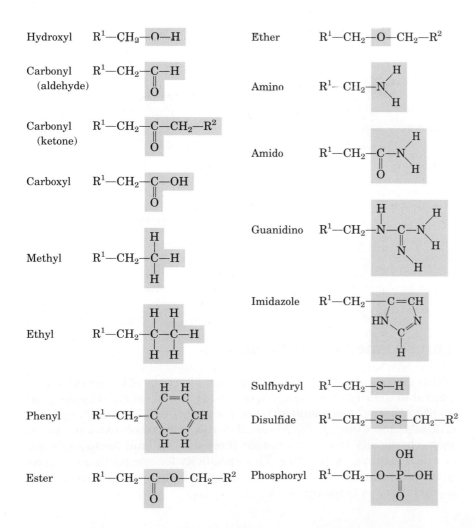

Figure 3–5 Some functional groups frequently encountered in biomolecules. All groups are shown in their uncharged (un-ionized) form.

Many biomolecules are polyfunctional, containing two or more different kinds of functional groups (Fig. 3–6), each with its own chemical characteristics and reactions. Amino acids, an important family of molecules that serve primarily as monomeric subunits of proteins, contain at least two different kinds of functional groups: an amino group and a carboxyl group, as shown for histidine in Figure 3–6. The ability of an amino acid to condense (see Fig. 3–14e) with other amino acids to form proteins is dependent on the chemical properties of these two functional groups.

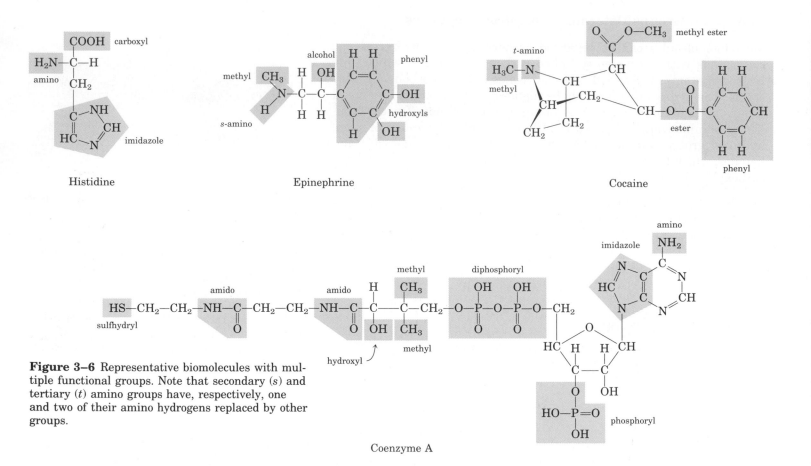

Histidine   Epinephrine   Cocaine

Coenzyme A

Figure 3–6 Representative biomolecules with multiple functional groups. Note that secondary (*s*) and tertiary (*t*) amino groups have, respectively, one and two of their amino hydrogens replaced by other groups.

Three-Dimensional Structure

Although the covalent bonds and functional groups of biomolecules are central to their function, they do not tell the whole story. The arrangement in three-dimensional space of the atoms of a biomolecule is also crucially important. Compounds of carbon can often exist in two or more chemically indistinguishable three-dimensional forms, only one of which is biologically active. This specificity for one particular molecular configuration is a universal feature of biological interactions. All biochemistry is three-dimensional.

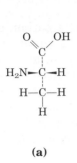
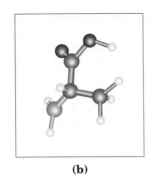
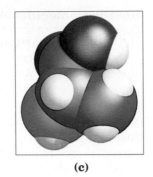

(a)   (b)   (c)

Figure 3–7 Models of the structure of the amino acid alanine. **(a)** Structural formula in perspective form. The symbol ➝ represents a bond in which the atom at the wide end projects out of the plane of the paper, toward the reader; dashes represent a bond extending behind the plane of the paper. **(b)** Ball-and-stick model, showing relative bond lengths and the bond angles. The balls indicate the approximate size of the atomic nuclei. **(c)** Space-filling model, in which each atom is shown having its correct van der Waals radius (see Table 3–3).

Each Cellular Component Has a Characteristic Three-Dimensional Structure

Biomolecules have characteristic sizes and three-dimensional structures, which derive from their backbone structures and their substituent functional groups. Figure 3–7 shows three ways to illustrate the three-dimensional structures of molecules. The perspective diagram specifies unambiguously the three-dimensional structure (stereochemistry) of a compound. Bond angles and center-to-center bond lengths are best represented with ball-and-stick models, whereas the outer contours of molecules are better represented by space-filling models. In space-filling models, the radius of each atom is proportional to its van der Waals radius (Table 3–3), and the contours of the molecule represent the outer limits of the region from which atoms of other molecules are excluded.

The three-dimensional conformation of biomolecules is of the utmost importance in their interactions; for example, in the binding of a substrate (reactant) to the catalytic site of an enzyme (Fig. 3–8), the two molecules must fit each other closely, in a complementary fashion, for biological function. Such complementarity also is required in the binding of a hormone molecule to its receptor on a cell surface, or in the recognition of an antigen by a specific antibody.

The study of the three-dimensional structure of biomolecules with precise physical methods is an important part of modern research on cell structure and biochemical function. The most informative method is x-ray crystallography. If a compound can be crystallized, the diffraction of x rays by the crystals can be used to determine with great precision the position of every atom in the molecule relative to every other atom. The structures of most small biomolecules (those with less than about 50 atoms), and of many larger molecules such as proteins, have been deduced by this means. X-ray crystallography yields a static picture of the molecule within the confines of the crystal. However, biomolecules almost never exist within cells as crystals; rather, they are dissolved in the cytosol or associated with some other component(s) of the cell. Molecules have more freedom of intramolecular motion in solution than in a crystal. In large molecules such as proteins, the small variations allowed in the three-dimensional structures of their monomeric subunits add up to extensive flexibility. Techniques such as nuclear magnetic resonance (NMR) spectroscopy complement x-ray crystallography by providing information about the three-dimensional structure of biomolecules in solution. Knowledge of the detailed three-dimensional structure of a molecule often sheds light on the mechanisms of the reactions in which the molecule participates.

Table 3–3 van der Waals radii and covalent (single-bond) radii of some elements*

Element	van der Waals radius (nm)	Covalent radius for single bond (nm)
H	0.1	0.030
O	0.14	0.074
F	0.14	0.071
N	0.15	0.073
C	0.17	0.077
S	0.18	0.103
Cl	0.18	0.099
P	0.19	0.110
Br	0.20	0.114
I	0.22	0.133

* The van der Waals radius is about twice the covalent radius for each element. The distance between nuclei in a van der Waals interaction or a covalent bond is about equal to the sum of the values for the two atoms. Thus the length of a carbon–carbon single bond is about 0.077 + 0.077 = 0.154 nm.

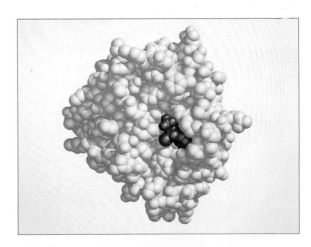

Figure 3–8 Complementary fit of a substrate molecule to the active or catalytic site on an enzyme molecule. The enzyme shown here is chymotrypsin, an enzyme that acts in the intestine to degrade dietary protein. Its substrate (shown in red) fits into a groove at the active site of the enzyme.

Figure 3–9 Molecular asymmetry: chiral and achiral molecules. **(a)** When a carbon atom has four different substituent groups (A, B, X, Y), they can be arranged in two ways that represent nonsuperimposable mirror images of each other (enantiomers). Such a carbon atom is asymmetric and is called a chiral atom or chiral center. **(b)** When there are only three dissimilar groups around the carbon atom (i.e., the same group occurs twice), only one configuration in space is possible and the molecule is symmetric, or achiral. In this case the molecule is superimposable on its mirror image: the molecule on the left can be rotated counterclockwise (when looking down its vertical bond from A to C) to create the molecule on the right.

Most Biomolecules Are Asymmetric

The tetrahedral arrangement of single bonds around a carbon atom confers on some organic compounds another property of great importance in biology. When four *different* atoms or functional groups are bonded to a carbon atom in an organic molecule, the carbon atom is said to be asymmetric; it can exist in two different isomeric forms **(stereoisomers)** that have different configurations in space. A special class of stereoisomers, called **enantiomers,** are nonsuperimposable mirror images of each other (Fig. 3–9). The two enantiomers of a compound have identical chemical properties, but differ in a characteristic physical property, the ability to rotate the plane of plane-polarized light. A solution of one enantiomer rotates the plane of such light to the right, and a solution of the other, to the left. Compounds without an asymmetric carbon atom do not rotate the plane of plane-polarized light.

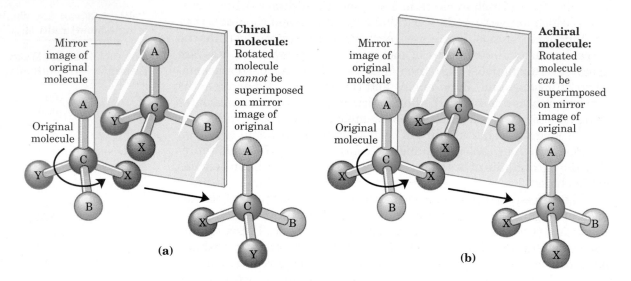

(a)   (b)

Louis Pasteur, in 1843, was the first to arrive at the correct explanation for this phenomenon of **optical activity.** Investigating the crystalline material that accumulated in wine casks ("paratartaric acid," also called racemic acid, from Latin *racemus,* "grape"), he had used a fine forceps to separate two types of crystals identical in shape, but mirror images of each other (Fig. 3–10). Both proved to have all of the chemical properties of tartaric acid, but one type rotated polarized light to the left, the other, to the right, but to the same extent. He later described the experiment and its interpretation:

> In isomeric bodies, the elements and the proportions in which they are combined are the same, only the arrangement of the atoms is different. . . . We know, on the one hand, that the molecular arrangements of the two tartaric acids are asymmetric, and, on the other hand, that these arrangements are absolutely identical, excepting that they exhibit asymmetry in opposite directions. Are the atoms of the dextro acid grouped in the form of a right-handed spiral, or are they placed at the apex of an irregular tetrahedron, or are they disposed according to this or that asymmetric arrangement? We do not know.*

Louis Pasteur
1822–1895

---

* From Pasteur's lecture to the Société Chimique de Paris in 1883, quoted in DuBos, R. (1976) *Louis Pasteur: Free Lance of Science,* p. 95, Charles Scribner's Sons, New York.

Now we do know. X-ray crystallographic studies in 1951 confirmed that the levorotatory and dextrorotatory forms of tartaric acid are mirror images of each other, and established the absolute configuration of each (Fig. 3–10). The same approach has been used to demonstrate that the amino acid alanine exists in two enantiomeric forms (Chapter 5). The central carbon atom of the alanine molecule is bonded to four different substituent groups: a methyl group, an amino group, a carboxyl group, and a hydrogen atom. The two stereoisomers of alanine are nonsuperimposable mirror images of each other, and thus are enantiomers.

Compounds with asymmetric carbon atoms can be regarded as occurring in left- and right-handed forms, and are therefore called **chiral compounds** (Greek *chiros*, "hand"). Correspondingly, the asymmetric atom or center of chiral compounds is called the **chiral atom** or **chiral center** (Fig. 3–9). All but one of the 20 amino acids have chiral centers; glycine is the exception.

More generally, variations in the three-dimensional structure of biomolecules are described in terms of configuration and conformation. These terms are not synonyms. **Configuration** denotes the spatial arrangement of an organic molecule that is conferred by the presence of either (1) double bonds, around which there is no freedom of rotation, or (2) chiral centers, around which substituent groups are arranged in a specific sequence. The identifying characteristic of configurational isomers is that they cannot be interconverted without breaking one or more covalent bonds.

Figure 3–11a shows the configurations of maleic acid, which occurs in some plants, and its isomer fumaric acid, an intermediate in sugar metabolism. These compounds are **geometric** or **cis–trans isomers;** they differ in the arrangement of their substituent groups with respect to the nonrotating double bond. Maleic acid is the cis isomer and fumaric acid the trans isomer; each is a well-defined compound that can be isolated and purified. These two compounds are stereoisomers but not enantiomers; they are not mirror images of each other.

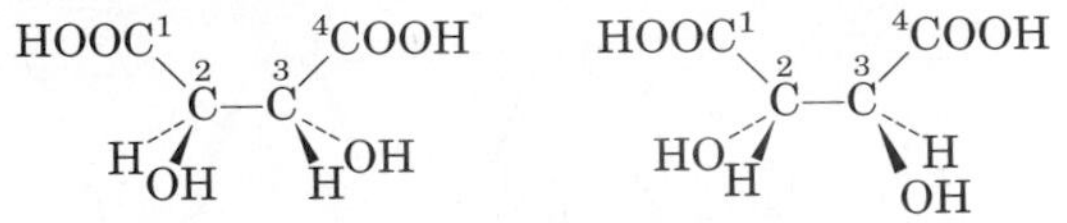

2R,3R-Tartaric acid
(dextrorotatory)

2S,3S-Tartaric acid
(levorotatory)

Figure 3–10 Pasteur separated crystals of two stereoisomers of tartaric acid and showed that solutions of the separated forms each rotated polarized light to the same extent but in opposite directions. Pasteur's dextrorotatory and levorotatory forms were later shown to be the R,R and S,S isomers shown here. For compounds with more than one chiral center, the RS system of nomenclature is often more useful than the D and L system described in Chapter 5. In the RS system, each group attached to a chiral carbon is assigned a *priority*. The priorities of some common substituents are: $-OR > -OH > -NH_2 > -COOH > -CHO > -CH_2OH > -CH_3 > -H$. The chiral carbon atom is viewed with the group of lowest priority pointing away from the viewer. If the priority of the other three groups decreases in counterclockwise order, the configuration is S; if in clockwise order, R. In this way each chiral carbon is designated as either R or S, and the inclusion of these designations in the name of the compound provides an unambiguous description of the stereochemistry at each chiral center.

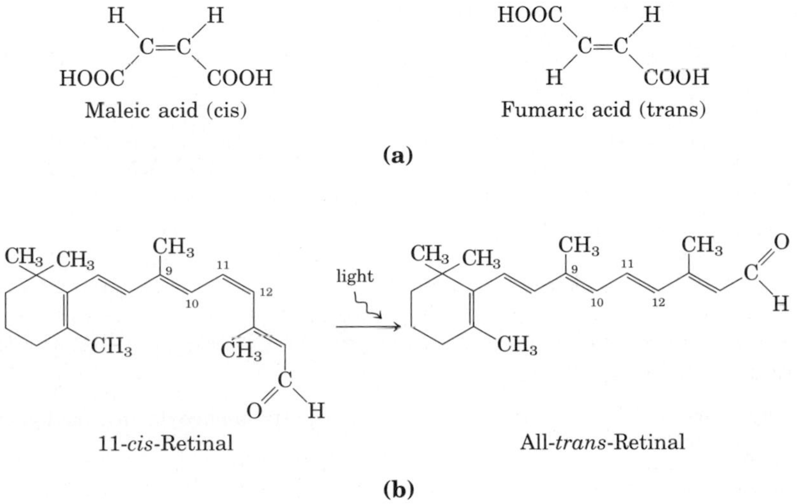

Maleic acid (cis)

Fumaric acid (trans)

(a)

11-*cis*-Retinal

All-*trans*-Retinal

(b)

Figure 3–11 Configurations of stereoisomers. **(a)** Isomers such as maleic acid and fumaric acid cannot be interconverted without breaking covalent bonds, which requires the input of much energy. **(b)** In the vertebrate retina, the initial event in light detection is the absorption of visible light by 11-*cis*-retinal. The energy of the absorbed light (about 250 kJ/mol) converts 11-*cis*-retinal to all-*trans*-retinal, triggering electrical changes in the retinal cell that lead to a nerve impulse.

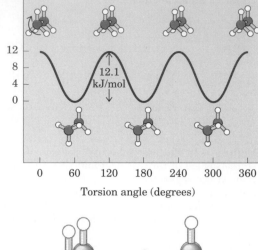

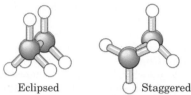

Eclipsed   Staggered

Figure 3–12 Many conformations of ethane are possible because of freedom of rotation around the carbon–carbon single bond. When the front carbon atom (as viewed by the reader) and its three attached hydrogens are rotated relative to the rear carbon atom, the potential energy of the molecule rises in the fully eclipsed conformation (torsion angle 0°, 120°, etc.), then falls in the fully staggered conformation (torsion angle 60°, 180°, etc.). The energy differences are small enough to allow rapid interconversion of the two forms (millions of times per second), thus the eclipsed and staggered forms cannot be isolated separately.

Figure 3–13 Stereoisomers that are distinguished by sensory receptors for smell and taste in humans. **(a)** Two stereoisomers of carvone, designated R and S (see Fig. 3–10, legend). R-carvone (from spearmint oil) has the characteristic fragrance of spearmint; S-carvone (from caraway seed oil) smells like caraway. **(b)** Aspartame, the artificial sweetener sold under the trade name NutraSweet, is easily distinguishable by taste from its bitter-tasting stereoisomer, although the two differ only in the configuration about one of the two chiral carbon atoms (in red).

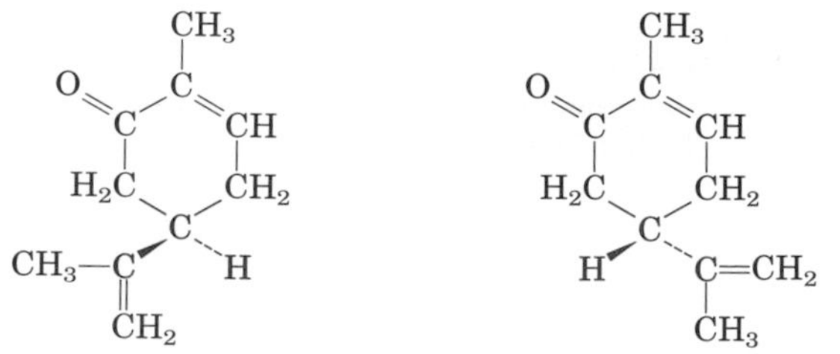

Molecular **conformation** refers to the spatial arrangement of substituent groups that are free to assume different positions in space, without breaking any bonds, because of the freedom of bond rotation. In the simple hydrocarbon ethane, for example, there is nearly complete freedom of rotation around the carbon–carbon single bond. Many different, interconvertible conformations of the ethane molecule are therefore possible, depending upon the degree of rotation (Fig. 3–12). Two conformations are of special interest: the staggered conformation, which is more stable than all others and thus predominates, and the eclipsed form, which is least stable. It is not possible to isolate either of these conformational forms, because they are freely interconvertible and in equilibrium with each other. However, when one or more of the hydrogen atoms on each carbon is replaced by a functional group that is either very large or electrically charged, freedom of rotation around the carbon–carbon single bond is hindered. This limits the number of stable conformations of the ethane derivative.

Interactions between Biomolecules Are Stereospecific

Many biomolecules besides amino acids are chiral, containing one or more asymmetric carbon atoms. The chiral molecules in living organisms are usually present in only one of their chiral forms. For example, the amino acids occur in proteins only as the L isomers. Glucose, the monomeric subunit of starch, has five asymmetric carbons, but occurs biologically in only one of its chiral forms, the D isomer. (The conventions for naming stereoisomers of the amino acids are described in Chapter 5; those for sugars, in Chapter 11). In contrast, when a compound having an asymmetric carbon atom is chemically synthesized in the laboratory, the nonbiological reactions usually produce all possible chiral forms in an equimolar mixture that does not rotate polarized light (a **racemic mixture**). The chiral forms in such a mixture can be separated only by painstaking physical methods. Chiral compounds in living cells are produced in only one chiral form because the enzymes that synthesize them are also chiral molecules.

Stereospecificity, the ability to distinguish between stereoisomers, is a common property of enzymes and other proteins and a characteristic feature of the molecular logic of living cells. If the binding site on a protein is complementary to one isomer of a chiral compound, it will not be complementary to the other isomer, for the same reason that a left glove does not fit a right hand. Two striking examples of the ability of biological systems to distinguish stereoisomers are shown in Figure 3–13.

R-Carvone
(spearmint)

S-Carvone
(caraway)

(a)

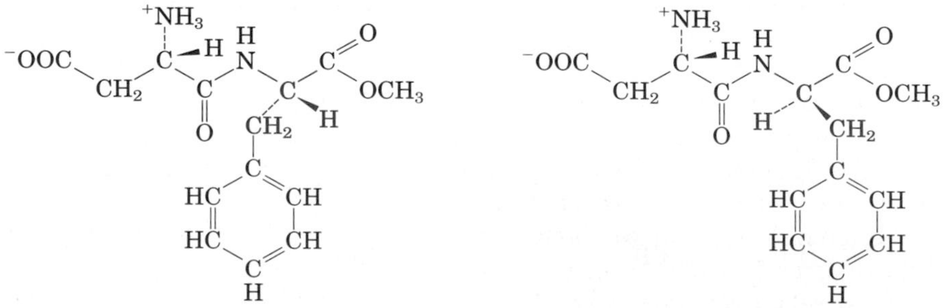

L-Aspartyl-L-phenylalanyl methyl ester
(aspartame) (sweet)

L-Aspartyl-D-phenylalanyl methyl ester
(bitter)

(b)

Chemical Reactivity

Saturated hydrocarbons—molecules with carbon–carbon single bonds and without double bonds or substituent groups—are not easily attacked by most chemical reagents; biomolecules, with their various functional groups, are much more chemically reactive. Functional groups alter the electron distribution and the geometry of neighboring atoms and thus affect the chemical reactivity of the entire molecule. The breakage and formation of chemical bonds during cellular metabolism release energy, some in the form of heat.

It is possible to analyze and predict the chemical behavior and reactions of biomolecules from the functional groups they bear. Enzymes recognize a specific pattern of functional groups in a biomolecule and catalyze characteristic chemical changes in the compound that contains these groups. Although a large number of different chemical reactions occur in a typical cell, these reactions are of only a few types, readily understandable in terms that apply to all reactions of organic compounds.

Bond Strength is Related to the Electronegativities of the Bonded Atoms

When the two atoms sharing electrons in a covalent bond have equal affinities for the electrons, as in the case of two carbon atoms, the resulting bond is nonpolar. When two elements that differ in electron affinity, or electronegativity (Table 3–4), form a covalent bond (e.g., C and O), that bond is polarized; the shared electrons are more likely to be in the region of the more electronegative atom (O) than of the less electronegative (C). In the extreme case of two atoms of very different electronegativity (Na and Cl, for example), one of the atoms actually gives up the electron(s) to the other atom, resulting in the formation of ions and ionic interactions such as those in solid NaCl.

The strength of chemical bonds (Table 3–5) depends upon the relative electronegativities of the elements involved, the distance of the bonding electrons from each nucleus, and the nuclear charge. The number of electrons shared also influences bond strength; double bonds are stronger than single bonds, and triple bonds are stronger yet. The strength of a bond is expressed as bond energy, in joules. (In biochemistry, calories have often been used as units of energy—bond energy and free energy, for example. The joule is the unit of energy in the International System of Units, and is used throughout this book. For conversions, 1 cal is equal to 4.18 J.) Bond energy can be thought of as either the amount of energy required to break a bond or the amount of energy gained by the surroundings when two atoms form the bond. One way to put energy into a system is to heat it, which gives the molecules more kinetic energy; temperature is a measurement of the average kinetic energy of a population of molecules. When molecular motion is sufficiently violent, intramolecular vibrations and intermolecular collisions sometimes break chemical bonds. Heating raises the fraction of molecules with energies high enough to react.

In chemical reactions, bonds are broken and new ones are formed. The difference between the energy from the surroundings used to break bonds and the energy gained by the surroundings in the formation of new ones is virtually identical to the **enthalpy change** for the reaction, ΔH. (The energy difference becomes exactly equal to the enthalpy change after a slight correction for any volume change in the

Table 3–4 The electronegativities of some elements

Element	Electronegativity*
F	4.0
O	3.5
Cl	3.0
N	3.0
Br	2.8
S	2.5
C	2.5
I	2.5
Se	2.4
P	2.1
H	2.1
Cu	1.9
Fe	1.8
Co	1.8
Ni	1.8
Mo	1.8
Zn	1.6
Mn	1.5
Mg	1.2
Ca	1.0
Li	1.0
Na	0.9
K	0.8

* The higher the number, the more electronegative is the element.

Table 3–5 Strengths of bonds common in biomolecules

Type of bond	Bond dissociation energy (kJ/mol)	Type of bond	Bond dissociation energy (kJ/mol)
Single bonds		*Double bonds*	
O—H	461	C=O	712
H—H	435	C=N	615
P—O	419	C=C	611
C—H	414	P=O	502
N—H	389		
C—O	352	*Triple bonds*	
C—C	348	C≡C	816
S—H	339	N≡N	930
C—N	293		
C—S	260	*Noncovalent bonds or interactions*	
N—O	222	Hydrogen bonds	
S—S	214	van der Waals interactions	4–20
		Hydrophobic interactions	
		Ionic interactions	

system at constant pressure.) If heat energy is absorbed by the system as the change occurs (that is, if the reaction is **endothermic**), then ΔH has, by definition, a positive value; when heat is produced, as in **exothermic** reactions, ΔH is negative. In short, the change in enthalpy for a covalent reaction reflects the kinds and numbers of bonds that are made and broken. As we shall see later in this chapter, the enthalpy change is one of three factors that determine the free-energy change for a reaction; the other two are the temperature and the change in entropy.

Five Types of Chemical Transformations Occur in Cells

Most cells have the capacity to carry out thousands of specific, enzyme-catalyzed reactions: transformation of simple nutrients such as glucose into amino acids, nucleotides, or lipids; extraction of energy from fuels by oxidation; or polymerization of subunits into macromolecules, for example. Fortunately for the student of biochemistry, there is a pattern in this multitude of reactions; we do not need to learn all of these reactions to comprehend the molecular logic of life.

Most of the reactions in living cells fall into one of five general categories (Fig. 3–14): functional-group transfers (a), oxidations and reductions (b), reactions that rearrange the bond structure around one or more carbons (c), reactions that form or break carbon–carbon bonds (d), and reactions in which two molecules condense, with the elimination of a molecule of water (e). Reactions within one category generally occur by similar mechanisms.

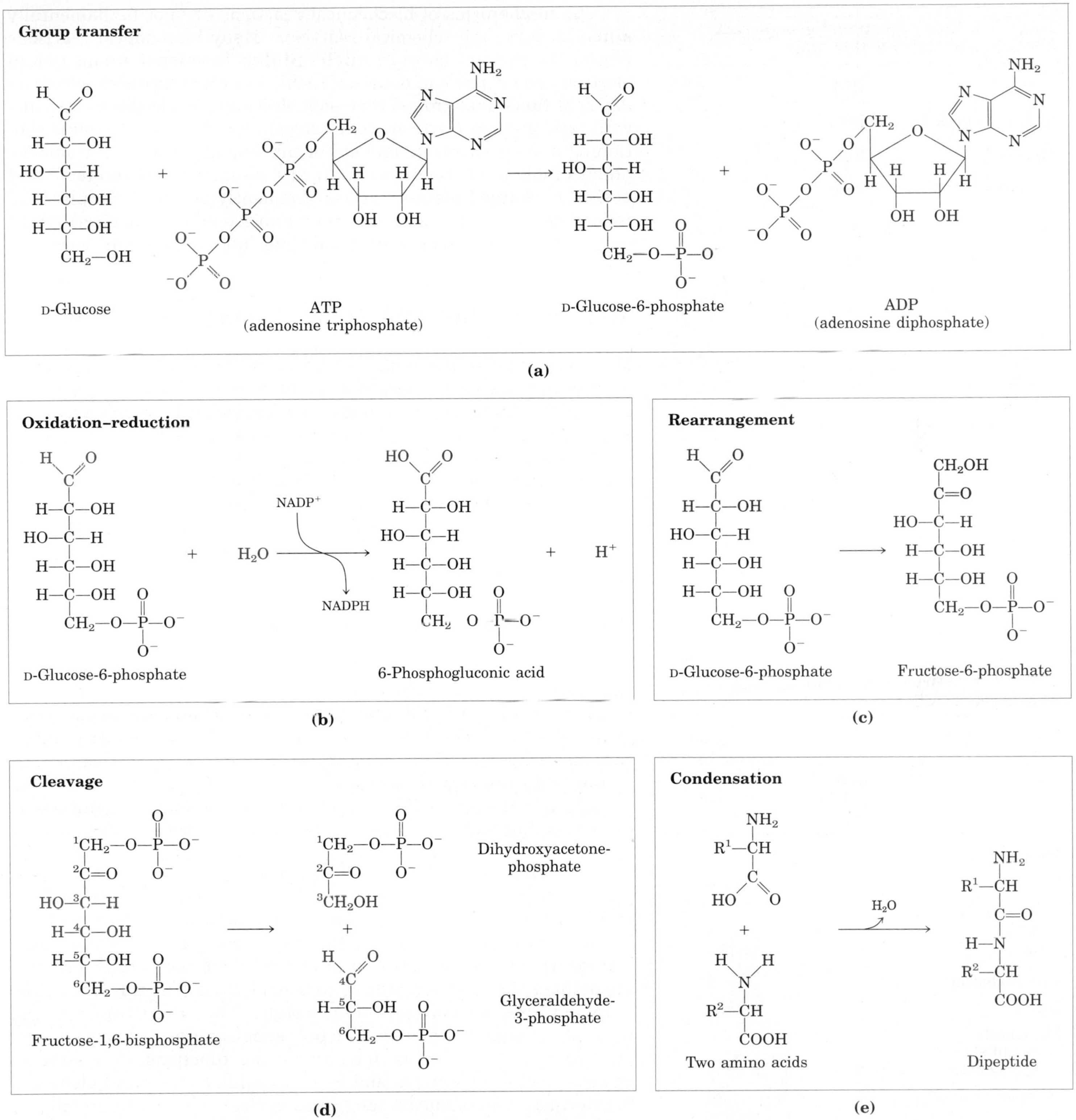

Figure 3–14 Examples of five general types of chemical transformations that occur in cells. The reactions (a) through (d) are enzyme-catalyzed reactions that take place in your tissues as you use glucose as a source of energy (Chapter 14). In (a) a phosphoryl group is transferred from ATP to glucose; (b) an aldehyde is oxidized to a carboxylic acid and an oxidized electron carrier (NADP$^+$) is reduced; (c) a rearrangement converts an aldehyde to a ketone; (d) a molecule is cleaved to form two smaller molecules. Reaction (e) represents the condensation of two amino acids with the elimination of H_2O to form a peptide bond; condensation reactions occur in many cellular processes in which larger molecules are assembled from small precursors.

Table 3–6 Some functional groups that act as nucleophiles within cells

Water	H$\overset{..}{\underset{..}{O}}$H
Hydroxyl (alcohol)	R$\overset{..}{\underset{..}{O}}$H
Alkoxyl	R$\overset{..}{\underset{..}{O}}$:$^-$
Sulfhydryl	RSH
Sulfide	R$\overset{..}{\underset{..}{S}}$$^-$
Amino	R$\overset{..}{N}$H$_2$
Carboxylate	R—C$\overset{\displaystyle O}{\underset{\displaystyle O^-}{\parallel}}$
Imidazole	(imidazole ring structure with N and NH)

The mechanisms of biochemical reactions are not fundamentally different from other chemical reactions. Many biochemical reactions involve interactions between **nucleophiles,** functional groups rich in electrons and capable of donating them, and **electrophiles,** electron-deficient functional groups that seek electrons. Nucleophiles combine with, and give up electrons to, electrophiles. Functional groups containing oxygen, nitrogen, and sulfur are important biological nucleophiles (Table 3–6). Positively charged hydrogen atoms (protons) and positively charged metals (cations) frequently act as electrophiles in cells. A carbon atom can act as either a nucleophilic or an electrophilic center, depending upon which bonds and functional groups surround it.

Macromolecules and Their Monomeric Subunits

Many of the molecules found within cells are macromolecules, polymers of high molecular weight assembled from relatively simple precursors. Polysaccharides, proteins, and nucleic acids, which may have molecular weights ranging from tens of thousands to (in the case of DNA) billions, are produced by the polymerization of relatively small subunits with molecular weights of 500 or less. The synthesis of macromolecules is a major energy-consuming activity of cells. Macromolecules themselves may be further assembled into supramolecular complexes, forming functional units such as ribosomes, membranes, and organelles.

The Major Constituents of Organisms Are Macromolecules

Table 3–7 shows the major classes of biomolecules in a representative single-celled organism, *Escherichia coli*. Water is the most abundant single compound in *E. coli* and in all other cells and organisms. Inorganic salts and mineral elements, on the other hand, constitute only a very small fraction of the total dry weight, but many of them are in approximate proportion to their distribution in seawater (see Table 3–1). Nearly all of the solid matter in all kinds of cells is organic and is present in four forms: proteins, nucleic acids, polysaccharides, and lipids.

Proteins, long polymers of amino acids, constitute the largest fraction (besides water) of cells. Some proteins have catalytic activity and function as enzymes, others serve as structural elements, and still others carry specific signals (in the case of receptors) or specific substances (in the case of transport proteins) into or out of cells. Proteins are perhaps the most versatile of all biomolecules. The **nucleic acids,** DNA and RNA, are polymers of nucleotides. They store, transmit, and translate genetic information. The **polysaccharides,** polymers of simple sugars such as glucose, have two major functions: they serve as energy-yielding fuel stores and as extracellular structural elements. Shorter polymers of sugars (oligosaccharides) attached to proteins or lipids at the cell surface serve as specific cellular signals. The **lipids,** greasy or oily hydrocarbon derivatives, serve as structural components of membranes, as a storage form of energy-rich fuel, and in other roles. These four classes of large biomolecules are all synthesized in condensation reactions (Fig. 3–14e). In macromolecules—proteins, nucleic acids, and polysaccharides—the number of monomeric subunits is

Table 3–7 Molecular components of an *E. coli* cell

	Percentage of total weight of cell	Approximate number of different molecular species
Water	70	1
Proteins	15	3,000
Nucleic acids		
DNA	1	1
RNA	6	>3,000
Polysaccharides	3	5
Lipids	2	20
Monomeric subunits and intermediates	2	500
Inorganic ions	1	20

very large. Proteins have molecular weights in the range of 5,000 to over 1 million; the nucleic acids have molecular weights ranging up to several billion; and polysaccharides, such as starch, also have molecular weights into the millions. Individual lipid molecules are much smaller (M_r 750 to 1,500), and are not classed as macromolecules. However, when large numbers of lipid molecules associate noncovalently, very large structures result. Cellular membranes are built of enormous aggregates containing millions of lipid molecules.

Macromolecules Are Constructed from Monomeric Subunits

Although living organisms contain a very large number of different proteins and different nucleic acids, a fundamental simplicity underlies their structure (Chapter 1). The simple monomeric subunits from which all proteins and all nucleic acids are constructed are few in number and identical in all living species. Proteins and nucleic acids are **informational macromolecules:** each protein and each nucleic acid has a characteristic information-rich subunit sequence (Fig. 3–15).

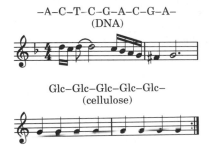

Polysaccharides built from only a single kind of unit, or from two different alternating units, are not informational molecules in the same sense as are proteins and nucleic acids (Fig. 3–15). However, complex polysaccharides made up of six or more different kinds of sugars connected in branched chains do have the structural and stereochemical variety that enables them to carry information recognizable by other macromolecules.

Monomeric Subunits Have Simple Structures

Figure 3–16 shows the structures of some monomeric units, arranged in families. We have already seen that the most abundant polysaccharides in nature, starch and cellulose, are constructed of repeating units of D-glucose. The monomeric subunits of proteins are 20 different amino acids; all have an amino group (an imino group in the case of proline) and a carboxyl group attached to the same carbon atom, called, by convention, the α carbon. These α-amino acids differ from each other only in their side chains (Fig. 3–16).

The recurring structural units of all nucleic acids are eight different nucleotides; four kinds of nucleotides are the structural units of DNA, and four others are the units of RNA. Each nucleotide is made up of three components: (1) a nitrogenous organic base, (2) a five-carbon sugar, and (3) phosphate (Fig. 3–16). The eight different nucleotides of DNA and RNA are built from five different organic bases combined with two different sugars.

Figure 3–15 Informational and structural macromolecules. A, T, C, and G represent the four deoxynucleotides of DNA, and glucose (Glc) is the repeating monomeric subunit of starch and cellulose. The number of possible permutations and combinations of four deoxynucleotides is virtually limitless, as is the number of melodies possible with a few musical notes. A polymer of one subunit type is information-poor and monotonous.

The 20 amino acids of proteins

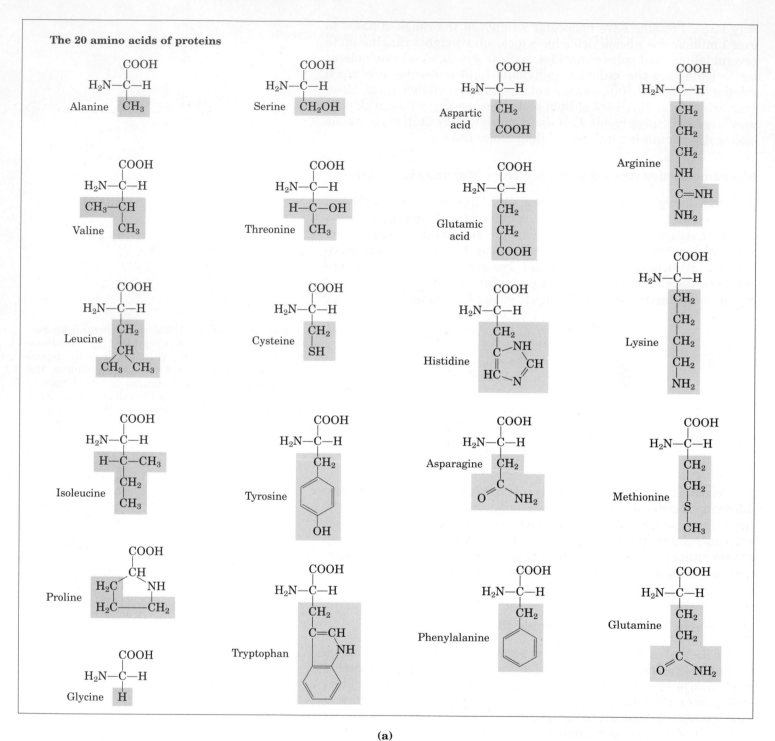

(a)

Figure 3–16 The organic compounds from which most larger structures in cells are constructed: the ABCs of biochemistry. Shown on these two pages are **(a)** the 20 amino acids from which the proteins of all organisms are built (the side chains are shaded red), **(b)** the five nitrogenous bases, two five-carbon sugars, and phosphoric acid from which all nucleic acids are built, **(c)** five components found in many membrane lipids, and **(d)** α-D-glucose, the parent sugar from which most carbohydrates are derived. Note that phosphoric acid is a subunit of both nucleic acids and membrane lipids. The five-carbon and six-carbon sugars are shown here in their ring forms rather than their straight-chain forms (Chapter 11). All components are shown in their un-ionized form.

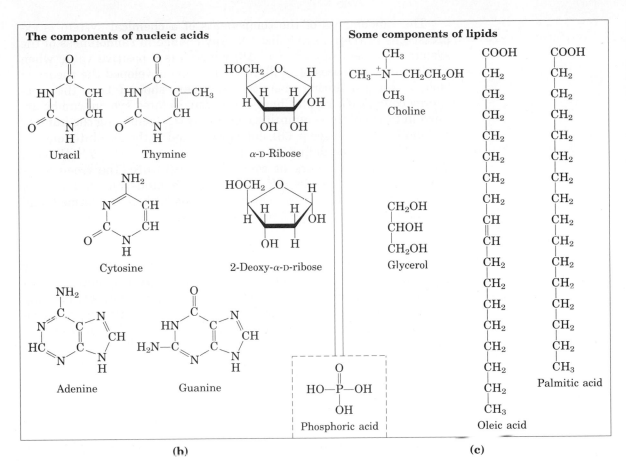

(b)   **(c)**   **(d)**

Lipids also are constructed from relatively few kinds of subunits. Most lipid molecules contain one or more long-chain fatty acids, of which palmitic acid and oleic acid are parent compounds. Many lipids also contain an alcohol, e.g., glycerol, and some contain phosphate (Fig. 3–16). Thus, only three dozen different organic compounds are the parents of most biomolecules.

Each of the compounds in Figure 3–16 has multiple functions in living organisms (Fig. 3–17). Amino acids are not only the monomeric subunits of proteins; some also act as neurotransmitters and as precursors of hormones and toxins. Adenine serves both as a subunit in the structure of nucleic acids and of ATP, and as a neurotransmitter. Fatty acids serve as components of complex membrane lipids, energy-rich fuel-storage fats, and the protective waxy coats on leaves and fruits. D-Glucose is the monomeric subunit of starch and cellulose, and also is the precursor of other sugars such as D-mannose and sucrose.

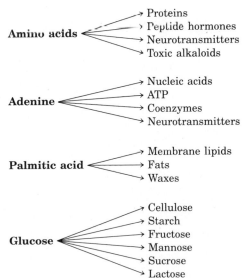

Figure 3–17 Each simple component in Fig. 3–16 is a precursor of many other kinds of biomolecules.

Subunit Condensation Creates Order and Requires Energy

It is extremely improbable that amino acids in a mixture would spontaneously condense into a protein with a unique sequence. This would represent increased order in a population of molecules; but according to the second law of thermodynamics (Chapter 13) the tendency is toward ever-greater disorder in the universe. To bring about the synthesis of macromolecules from their monomeric subunits, free energy must be supplied to the system (the cell).

J. Willard Gibbs
1839–1903

The randomness of the components of a chemical system is expressed as **entropy,** symbolized S. Any change in randomness of the system is the entropy change, ΔS, which has a positive value when randomness increases. J. Willard Gibbs, who developed the theory of energy changes during chemical reactions, showed that the free-energy content (G; recall Chapter 1) of any isolated system can be defined in terms of three quantities: enthalpy (H) (reflecting the number and kinds of bonds; see p. 66), entropy (S), and T, the absolute temperature (Kelvin). The definition of free energy is: $G = H - TS$. When a chemical reaction occurs at constant temperature, the free-energy change is determined by ΔH, reflecting the kinds and numbers of chemical bonds and noncovalent interactions broken and formed, and ΔS, the change in the system's randomness:

$$\Delta G = \Delta H - T\,\Delta S$$

Recall from Chapter 1 that a process tends to occur spontaneously only if ΔG is negative. How, then, can cells synthesize polymers such as proteins and nucleic acids, if the free-energy change for polymerizing subunits is positive? They couple these thermodynamically unfavorable (endergonic) reactions to other cellular reactions that liberate free energy (exergonic reactions), so that the *sum* of the free-energy changes is negative:

Amino acids $\longrightarrow$	proteins	ΔG_1 is positive (endergonic)
ATP $\longrightarrow$	AMP + $2PO_4^{3-}$	ΔG_2 is negative (exergonic)

Sum: Amino acids + ATP $\longrightarrow$ proteins + AMP + $2PO_4^{3-}$

The sum of ΔG_1 and ΔG_2 is negative (the overall process is exergonic).

There Is a Hierarchy in Cell Structure

The monomeric subunits in Figure 3–16 are very small compared with biological macromolecules. An amino acid molecule such as alanine is less than 0.5 nm long. Hemoglobin, the oxygen-carrying protein of erythrocytes, consists of nearly 600 amino acid units covalently linked into four long chains, which are folded into globular shapes and associated in a tetrameric structure with a diameter of 5.5 nm. Protein molecules in turn are small compared with ribosomes (about 20 nm in diameter), which contain about 70 different proteins and several different RNA molecules. Ribosomes, in their turn, are much smaller than organelles such as mitochondria, typically 1,000 nm in diameter. It is a long jump from the simple biomolecules to the larger cellular structures that can be seen with the light microscope. Figure 3–18 illustrates the structural hierarchy in cellular organization.

In proteins, nucleic acids, and polysaccharides, the individual subunits are joined by covalent bonds. By contrast, in supramolecular complexes, the different macromolecules are held together by noncovalent interactions—much weaker, individually, than covalent bonds. Among these are hydrogen bonds (between polar groups), ionic interactions (between charged groups), hydrophobic interactions (between nonpolar groups), and van der Waals interactions, all of which have energies of only a few kilojoules, compared with covalent bonds, which have bond energies of 200 to 900 kJ/mol (see Table 3–5). The nature of these noncovalent interactions will be described in the next chapter.

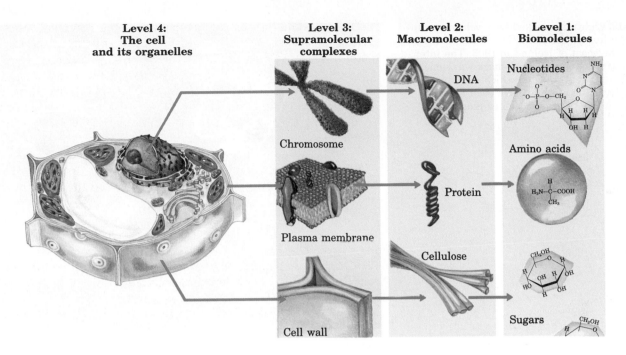

| Level 4: The cell and its organelles | Level 3: Supramolecular complexes | Level 2: Macromolecules | Level 1: Biomolecules |

Chromosome — DNA — Nucleotides

Plasma membrane — Protein — Amino acids

Cell wall — Cellulose — Sugars

Figure 3–18 The structural hierarchy in the molecular organization of cells. The nucleus of this plant cell, for example, contains several types of supramolecular complexes, including chromosomes. Chromosomes consist of macromolecules—DNA and many different proteins. Each type of macromolecule is constructed from simple subunits—DNA from the deoxyribonucleotides, for example. (Adapted from Becker, W.M. & Deamer, D.W. (1991) *The World of the Cell,* 2nd edn, Fig. 2–15, The Benjamin/Cummings Publishing Company, Menlo Park, CA.)

The large numbers of weak interactions between macromolecules in supramolecular complexes stabilize the resulting noncovalent structures.

Although the monomeric subunits of macromolecules are so much smaller than cells and organelles, they influence the shape and function of these much larger structures. In sickle-cell anemia, a hereditary human disorder, the hemoglobin molecule is defective. In the two β chains of hemoglobin from healthy individuals, a glutamic acid residue occurs at position 6. In people with sickle-cell anemia, a valine residue occurs at position 6. This single difference in the sequence of the 146 amino acids of the β chain affects only a tiny portion of the molecule, yet it causes the hemoglobin to form large aggregates within the erythrocytes, which become deformed (sickled) and function abnormally.

Prebiotic Evolution

Because all biological macromolecules are made from the same three dozen subunits, it seems likely that all living organisms descended from a single primordial cell line. These subunits are proposed to have had, singly and collectively, the most successful combination of chemical and physical properties for their function as the raw materials of biological macromolecules and for carrying out the basic energy-transforming and self-replicating features of a living cell. These primordial organic compounds may have been retained during biological evolution over billions of years because of their unique fitness.

Figure 3–19 Lightning evoked by a volcanic eruption that resulted in the formation of the island of Surtsey off the coast of Iceland in 1963. The intense fields of electrical, thermal, and shock-wave energy generated by such cataclysms, which were frequent on the primitive earth, could have been a major factor in the origin of organic compounds.

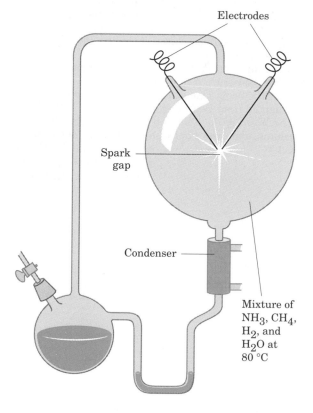

Figure 3–20 Spark-discharge apparatus of the type used by Miller and Urey in experiments demonstrating abiotic formation of organic compounds under primitive atmospheric conditions. After subjecting the gaseous contents of the system to electrical sparks, products were collected by condensation. Biomolecules such as amino acids were among the products (see Table 3–8).

Biomolecules First Arose by Chemical Evolution

We come now to a puzzle. Apart from their occurrence in living organisms, organic compounds, including the basic biomolecules, occur only in trace amounts in the earth's crust, the sea, and the atmosphere. How did the first living organisms acquire their characteristic organic building blocks? In 1922, the biochemist Aleksandr I. Oparin proposed a theory for the origin of life early in the history of the earth, postulating that the atmosphere was once very different from that of today. Rich in methane, ammonia, and water, and essentially devoid of oxygen, it was a reducing atmosphere, in contrast to the oxidizing environment of our era. In Oparin's theory, electrical energy of lightning discharges or heat energy from volcanoes (Fig. 3–19) caused ammonia, methane, water vapor, and other components of the primitive atmosphere to react, forming simple organic compounds. These compounds then dissolved in the ancient seas, which over many millenia became enriched with a large variety of simple organic compounds. In this warm solution (the "primordial soup") some organic molecules had a greater tendency than others to associate into larger complexes. Over millions of years, these in turn assembled spontaneously to form membranes and catalysts (enzymes), which came together to become precursors of the first primitive cells. For many years, Oparin's views remained speculative and appeared untestable.

Chemical Evolution Can Be Simulated in the Laboratory

A classic experiment on the abiotic (nonbiological) origin of organic biomolecules was carried out in 1953 by Stanley Miller in the laboratory of Harold Urey. Miller subjected gaseous mixtures of NH_3, CH_4, water vapor, and H_2 to electrical sparks produced across a pair of electrodes (to simulate lightning) for periods of a week or more (Fig. 3–20), then analyzed the contents of the closed reaction vessel. The gas phase of the resulting mixture contained CO and CO_2, as well as the starting materials. The water phase contained a variety of organic compounds, including some amino acids, hydroxy acids, aldehydes, and hydrogen cyanide (HCN). This experiment established the possibility of abiotic production of biomolecules in relatively short times under relatively mild conditions.

Several developments have allowed more refined studies of the type pioneered by Miller and Urey, and have yielded strong evidence that a wide variety of biomolecules, including proteins and nucleic acids, could have been produced spontaneously from simple starting materials probably present on the earth at the time life arose.

Modern extensions of the Miller experiments have employed "atmospheres" that include CO_2 and HCN, and much improved technology for identifying small quantities of products. The formation of hundreds of organic compounds has been demonstrated (Table 3–8). These compounds include more than ten of the common amino acids, a variety of mono-, di-, and tricarboxylic acids, fatty acids, adenine, and formaldehyde. Under certain conditions, formaldehyde polymerizes to form sugars containing three, four, five, and six carbons. The sources of energy that are effective in bringing about the formation of these compounds include heat, visible and ultraviolet (UV) light, x rays, gamma radiation, ultrasound and shock waves, and alpha and beta particles.

In addition to the many monomers that form in these experiments, polymers of nucleotides (nucleic acids) and of amino acids (proteins) also form. Some of the products of the self-condensation of HCN are effective promoters of such polymerization reactions (Fig. 3–21), and inorganic ions present in the earth's crust (Cu^{2+}, Ni^{2+}, and Zn^{2+}) also enhance the rate of polymerization.

Table 3–8 Some of the products shown to form under prebiotic conditions	
Amino acids	*Carboxylic acids*
Glycine	Formic acid
Alanine	Acetic acid
α-Aminobutyric acid	Propionic acid
Valine	Straight and branched
Leucine	fatty acids (C_4–C_{10})
Isoleucine	Glycolic acid
Proline	Lactic acid
Aspartic acid	Succinic acid
Glutamic acid	
Serine	*Nucleic acid bases*
Threonine	Adenine
	Guanine
Sugars	Xanthine
Straight and branched	Hypoxanthine
pentoses and hexoses	Cytosine
	Uracil

Source: From Miller, S.L. (1987) Which organic compounds could have occurred on the prebiotic earth? *Cold Spring Harb. Symp. Quant. Biol.* **52,** 17–27.

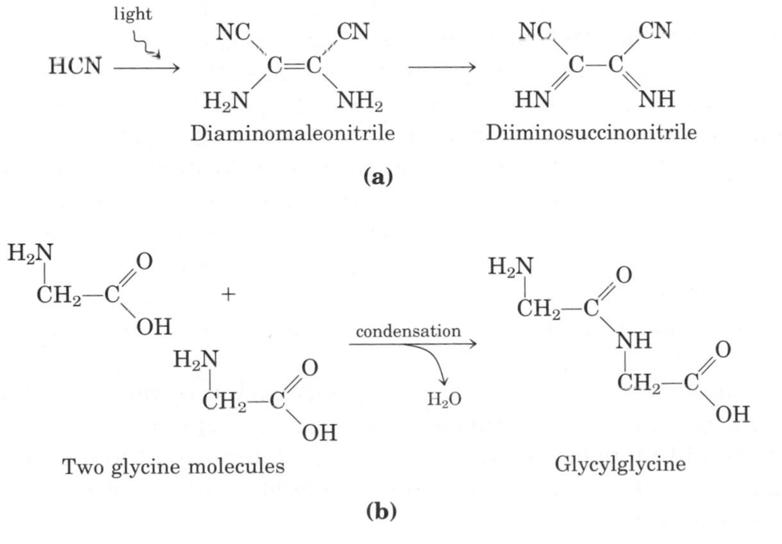

Diaminomaleonitrile   Diiminosuccinonitrile

(a)

Two glycine molecules   Glycylglycine

(b)

Figure 3–21 Among the products of electrical discharge through an atmosphere containing HCN are compounds such as those in **(a)**. These compounds promote the polymerization of monomers such as amino acids into polymers **(b)**.

In short, laboratory experiments on the spontaneous formation of biomolecules under prebiotic conditions have provided good evidence that many of the chemical components of living cells, including proteins and RNA, can form under these conditions. Short polymers of RNA can act as catalysts in biologically significant reactions (Chapter 25), and it seems likely that RNA played a crucial role in prebiotic evolution, both as catalyst and as information repository.

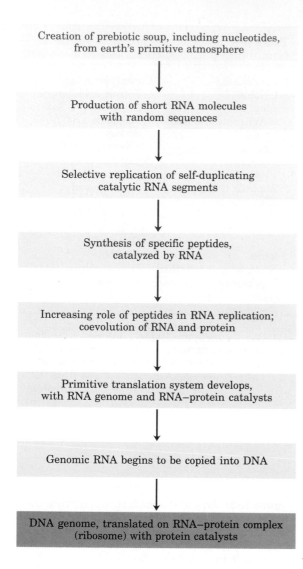

Creation of prebiotic soup, including nucleotides, from earth's primitive atmosphere

Production of short RNA molecules with random sequences

Selective replication of self-duplicating catalytic RNA segments

Synthesis of specific peptides, catalyzed by RNA

Increasing role of peptides in RNA replication; coevolution of RNA and protein

Primitive translation system develops, with RNA genome and RNA–protein catalysts

Genomic RNA begins to be copied into DNA

DNA genome, translated on RNA–protein complex (ribosome) with protein catalysts

Figure 3–22 One possible "RNA world" scenario, showing the transition from the prebiotic RNA world (shades of yellow) to the biotic DNA world (orange).

RNA Molecules May Have Been the First Genes and Catalysts

In modern organisms, nucleic acids encode the genetic information that specifies the structure of enzymes, and enzymes have the ability to catalyze the replication and repair of nucleic acids. The mutual dependence of these two classes of biomolecules poses the perplexing question: which came first, DNA or protein?

The answer may be: neither. The discovery that RNA molecules can act as catalysts in their own formation suggests that RNA may have been the first gene *and* the first catalyst. According to this scenario (Fig. 3–22), one of the earliest stages of biological evolution was the chance formation, in the primordial soup, of an RNA molecule that had the ability to catalyze the formation of other RNA molecules of the same sequence—a self-replicating, self-perpetuating RNA. The concentration of a self-replicating RNA molecule would increase exponentially, as one molecule formed two, two formed four, and so on. The fidelity of self-replication was presumably less than perfect, so the process would generate variants of the RNA, some of which might be even better able to self-replicate. In the competition for nucleotides, the most efficient of the self-replicating sequences would win, and less efficient replicators would fade from the population.

The division of function between DNA (genetic information storage) and protein (catalysis) was, according to the "RNA world" hypothesis, a later development (Fig. 3–22). New variants of self-replicating RNA molecules developed, with the additional ability to catalyze the condensation of amino acids into peptides. Occasionally, the peptide(s) thus formed would reinforce the self-replicating ability of the RNA, and the pair—RNA molecule and helping peptide—could undergo further modifications in sequence, generating even more efficient self-replicating systems. Sometime after the evolution of this primitive protein-synthesizing system, there was a further development: DNA molecules with sequences complementary to the self-replicating RNA molecules took over the function of conserving the "genetic" information, and RNA molecules evolved to play roles in protein synthesis. Proteins proved to be versatile catalysts, and over time, assumed that function. Lipidlike compounds in the primordial soup formed relatively impermeable layers surrounding self-replicating collections of molecules. The concentration of proteins and nucleic acids within these lipid enclosures favored the molecular interactions required in self-replication.

This "RNA world" hypothesis is plausible but by no means universally accepted. The hypothesis does make testable predictions, and to the extent that experimental tests are possible within finite times (less than or equal to the life span of a scientist!), the hypothesis will be tested and refined.

Biological Evolution Began More Than Three Billion Years Ago

The earth was formed about 4.5 billion years ago, and the first definitive evidence of life dates to about 3.5 billion years ago. An international group of scientists showed in 1980 that certain ancient rock formations (stromatolites; Fig. 3–23) in western Australia contained fossils of primitive microorganisms. Somewhere on earth during that first billion-year period, there arose the first simple organism, capable

(a)

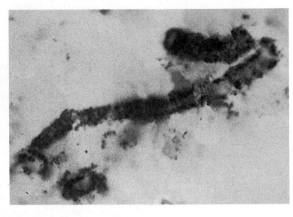

(b)

Figure 3–23 Ancient reefs in Australia contain fossil evidence of microbial life in the sea of 3.5 billion years ago. Bits of sand and limestone became trapped in the sticky extracellular coats of cyanobacteria, gradually building up these stromatolites found in Hamelin Bay, Western Australia **(a)**. Microscopic examination of thin sections of stromatolite reveals microfossils of filamentous bacteria **(b)**.

of replicating its own structure from a template (RNA?) that was the first genetic material. Because the terrestrial atmosphere at the dawn of life was nearly devoid of oxygen, and because there were few microorganisms to scavenge organic compounds formed by natural processes, these compounds were relatively stable. Given this stability and eons of time, the improbable became inevitable: the organic compounds were incorporated into evolving cells to produce more and more effective self-reproducing catalysts. The process of biological evolution had begun. Organisms developed mechanisms for harnessing the energy of sunlight through photosynthesis, to make sugars and other organic molecules from carbon dioxide, and to convert molecular nitrogen from the atmosphere into nitrogenous biomolecules such as amino acids. By developing their own capacities to synthesize biomolecules, cells became independent of the random processes by which such compounds had first appeared on earth. As evolution proceeded, organisms began to interact and to derive mutual benefits from each other's products, forming increasingly complex ecological systems.

Summary

Most of the dry weight of living organisms consists of organic compounds, molecules containing covalently bonded carbon backbones to which other carbon, hydrogen, oxygen, or nitrogen atoms may be attached. Carbon appears to have been selected in the course of biological evolution because of the ability of carbon atoms to form single and double bonds with each other, making possible formation of linear, cyclic, and branched backbone structures in great variety. To these backbones are attached different kinds of functional groups, which determine the chemical properties of the molecules. Organic biomolecules also have characteristic shapes (configurations and conformations) in three dimensions. Many biomolecules occur in asymmetric or chiral forms called enantiomers, stereoisomers that are nonsuperimposable mirror images of each other. Usually, only one of a pair of enantiomers has biological activity.

The strength of covalent chemical bonds, measured in joules, depends on the electronegativities and sizes of the atoms that share electrons. The enthalpy change (ΔH) for a chemical reaction reflects the number and kind of bonds made and broken. For endothermic reactions, ΔH is positive; for exothermic reactions, negative. The many different chemical reactions that occur within a cell fall into five general categories: group transfers, oxidation—

reduction reactions, rearrangements of the bonds around carbon atoms, breakage or formation of carbon–carbon bonds, and condensations.

Most of the organic matter in living cells consists of macromolecules: nucleic acids, proteins, and polysaccharides. Each type of macromolecule is composed of small, covalently linked monomeric subunits of relatively few kinds. Proteins are polymers of 20 different kinds of amino acids, nucleic acids are polymers of different nucleotide units (four in DNA, four in RNA), and polysaccharides are polymers of recurring sugar units. Nucleic acids and proteins are informational macromolecules; the characteristic sequences of their subunits constitute the genetic individuality of a species. Simple polysaccharides act as structural components, but some complex polysaccharides also are informational macromolecules.

There is a structural hierarchy in the molecular organization of cells. Cells contain organelles, such as nuclei, mitochondria, and chloroplasts, which in turn contain supramolecular complexes, such as membranes and ribosomes, and these consist in turn of clusters of macromolecules that are bound together by many relatively weak, noncovalent forces. The macromolecules consist of covalently linked subunits. The formation of macromolecules from simple subunits creates order (decreases entropy); this synthesis requires energy and therefore must be coupled to exergonic reactions.

The small biomolecules such as amino acids and sugars probably first arose spontaneously from atmospheric gases and water under the influence of electrical energy (lightning) during the early history of the earth. Such processes, called chemical evolution, can be simulated in the laboratory. The monomeric subunits of cellular macromolecules appear to have been selected during early biological evolution as being the most fit for their biological functions. These subunit molecules are relatively few in number, but are very versatile; evolution has combined small biomolecules to yield macromolecules capable of diverse biological functions. The first macromolecules may have been RNA molecules that were capable of catalyzing their own replication. Later in evolution, DNA took over the function of storing genetic information, proteins became the cellular catalysts, and RNA mediated between these, allowing the expression of genetic information as proteins.

Further Reading

General

Baker, J.J. & Allen, G.E. (1981) *Matter, Energy, and Life: An Introduction for Biology Students,* 4th edn, Addison-Wesley Publishing Co., Inc., Reading, MA.

Callewaert, D.M. & Genyea, J. (1980) *Basic Chemistry: General, Organic, Biological,* Worth Publishers, Inc., New York.

Dickerson, R.E. & Geis, I. (1976) *Chemistry, Matter, and the Universe,* The Benjamin/Cummings Publishing Company, Menlo Park, CA.

Frieden, E. (1972) The chemical elements of life. *Sci. Am.* **227** (July), 52–64.

The Molecules of Life. (1985) *Sci. Am.* **253** (October).
An entire issue devoted to the structure and function of biomolecules. It includes articles on DNA, RNA, and proteins, and their subunits.

Chemistry and Stereochemistry

Brewster, J.H. (1986) Stereochemistry and the origins of life. *J. Chem. Educ.* **8,** 667–670.

An interesting and lucid discussion of the ways in which evolution could have selected only one of two stereoisomers for the construction of proteins and other molecules.

Hegstrom, R.A. & Kondepudi, D.K. (1990) The handedness of the universe. *Sci. Am.* **262** (January), 108–115.
Stereochemistry and the asymmetry of biomolecules, viewed in the context of the universe.

Loudon, M. (1988) *Organic Chemistry,* 2nd edn, The Benjamin/Cummings Publishing Company, Menlo Park, CA.

This and the following two books provide details on stereochemistry and the chemical reactivity of functional groups. All excellent textbooks.

Morrison, R.T. & Boyd, R.N. (1988) *Organic Chemistry,* 5th edn, Allyn & Bacon, Inc., Boston, MA.

Streitweiser, A. Jr. & Heathcock, C.H. (1981) *Introduction to Organic Chemistry,* 2nd edn, Macmillan Publishing Co., Inc., New York.

Prebiotic Evolution

Cavalier-Smith, T. (1987) The origin of cells: a symbiosis between genes, catalysts, and membranes. *Cold Spring Harb. Symp. Quant. Biol.* **52,** 805–824.

Darnell, J.E. & Doolittle, W.F. (1986) Speculations on the early course of evolution. *Proc. Natl. Acad. Sci. USA* **83,** 1271–1275.
A clear statement of the RNA world scenario.

Evolution of Catalytic Function. (1987) *Cold Spring Harb. Symp. Quant. Biol.* **52.**
A collection of almost 100 articles on all aspects of prebiotic and early biological evolution; probably the single best source on molecular evolution.

Ferris, J.P. (1984) The chemistry of life's origin. *Chem. Eng. News* **62,** 21–35.

A short, clear description of the experimental evidence for the synthesis of biomolecules under prebiotic conditions.

Horgan, J. (1991) In the beginning . . . *Sci. Am.* **264** (February), 116–125.
A brief, clear statement of current theories regarding prebiotic evolution.

Miller, S.L. (1987) Which organic compounds could have occurred on the prebiotic earth? *Cold Spring Harb. Symp. Quant. Biol.* **52,** 17–27.

Schopf, J.W. (ed) (1983) *Earth's Earliest Biosphere,* Princeton University Press, Princeton, NJ.
A comprehensive discussion of geologic history and its relation to the development of life.

Problems

1. *Vitamin C: Is the Synthetic Vitamin as Good as the Natural One?* One claim put forth by purveyors of health foods is that vitamins obtained from natural sources are more healthful than those obtained by chemical synthesis. For example, it is claimed that pure L-ascorbic acid (vitamin C) obtained from rose hips is better for you than pure L-ascorbic acid manufactured in a chemical plant. Are the vitamins from the two sources different? Can the body distinguish a vitamin's source?

2. *Identification of Functional Groups* Figure 3–5 shows the common functional groups of biomolecules. Since the properties and biological activities of biomolecules are largely determined by their functional groups, it is important to be able to identify them. In each of the molecules at right, identify the constituent functional groups.

3. *Drug Activity and Stereochemistry* The quantitative differences in biological activity between the two enantiomers of a compound are sometimes quite large. For example, the D isomer of the drug isoproterenol, used to treat mild asthma, is 50 to 80 times more effective as a bronchodilator than the L isomer. Identify the chiral center in isoproterenol. Why would the two enantiomers have such radically different bioactivity?

Isoproterenol

Ethanolamine
(a)

Glycerol
(b)

Phosphoenolpyruvic acid, an intermediate in glucose metabolism
(c)

Threonine, an amino acid
(d)

Pantothenic acid, a vitamin
(e)

D-Glucosamine
(f)

Problem 2

4. *Drug Action and Shape of Molecules*   Some years ago two drug companies marketed a drug under the trade names Dexedrine and Benzedrine. The structure of the drug is shown below.

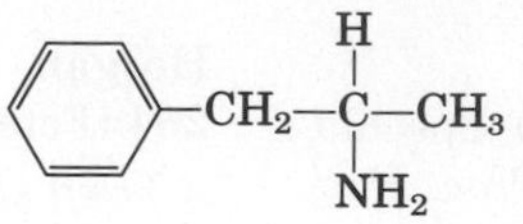

The physical properties (C, H, and N analysis, melting point, solubility, etc.) of Dexedrine and Benzedrine were identical. The recommended oral dosage of Dexedrine (which is still available) was 5 mg/d, but the recommended dosage of Benzedrine was significantly higher. Apparently it required considerably more Benzedrine than Dexedrine to yield the same physiological response. Explain this apparent contradiction.

5. *Components of Complex Biomolecules*   Figure 3–16 shows the structures of the major components of complex biomolecules. For each of the three important biomolecules below (shown in their ionized forms at physiological pH), identify the constituents.

   (a) Guanosine triphosphate (GTP), an energy-rich nucleotide that serves as precursor to RNA:

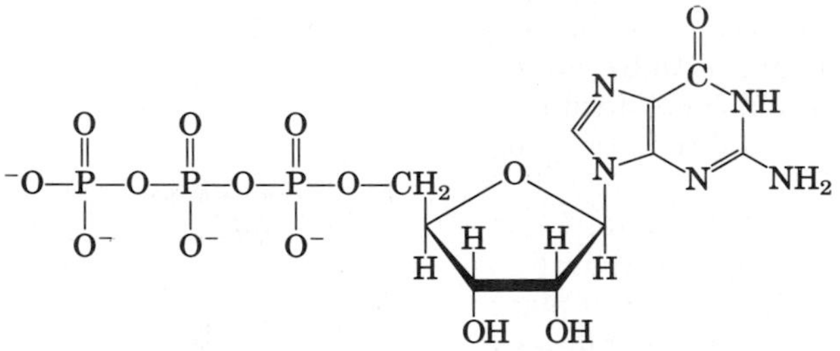

   (b) Phosphatidylcholine, a component of many membranes:

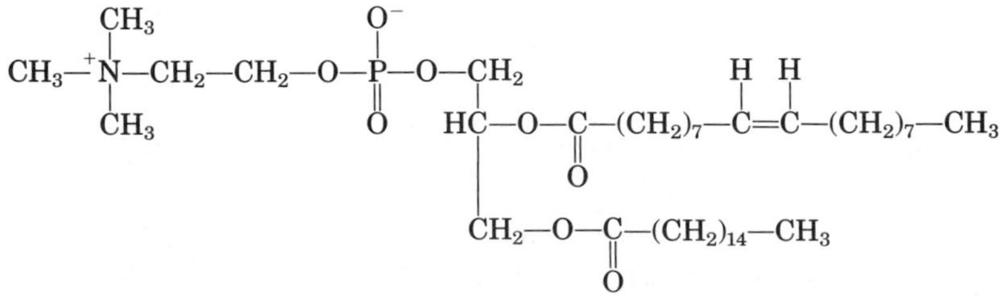

   (c) Methionine enkephalin, the brain's own opiate:

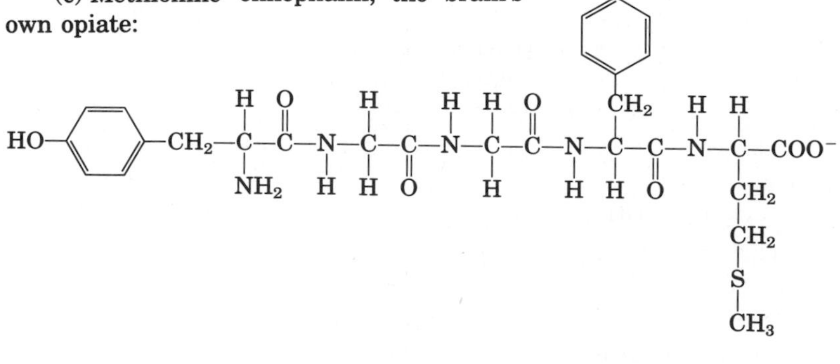

6. *Determination of the Structure of a Biomolecule* An unknown substance, X, was isolated from rabbit muscle. The structure of X was determined from the following observations and experiments. Qualitative analysis showed that X was composed entirely of C, H, and O. A weighed sample of X was completely oxidized, and the amount of H_2O and CO_2 produced was measured. From this quantitative analysis, it was concluded that X contains 40.00% C, 6.71% H, and 53.29% O by weight. The molecular mass of X was determined by a mass spectrometer and found to be 90.00. An infrared spectrum of X showed that it contained one double bond. X dissolved readily in water to give an acidic solution. A solution of X was tested in a polarimeter and demonstrated optical activity.

   (a) Determine the empirical and molecular formula of X.

   (b) Draw the possible structures of X that fit the molecular formula and contain one double bond. Consider *only* linear or branched structures and disregard cyclic structures. Note that oxygen makes very poor bonds to itself.

   (c) What is the structural significance of the observed optical activity? Which structures in (b) does this observation eliminate? Which structures are consistent with the observation?

   (d) What is the structural significance of the observation that a solution of X was acidic? Which structures in (b) are now eliminated? Which structures are consistent with the observation?

   (e) What is the structure of X? Is more than one structure consistent with all the data?

Water: Its Effect on Dissolved Biomolecules

Water is the most abundant substance in living systems, making up 70% or more of the weight of most organisms. Water pervades all portions of every cell and is the medium in which the transport of nutrients, the enzyme-catalyzed reactions of metabolism, and the transfer of chemical energy occur. The first living organisms probably arose in the primeval oceans; evolution was shaped by the properties of the medium in which it occurred. All aspects of cell structure and function are adapted to the physical and chemical properties of water. This chapter begins with descriptions of these physical and chemical properties. The strong attractive forces between water molecules result in water's solvent properties. The slight tendency of water to ionize is also of crucial importance to the structure and function of biomolecules, and we will review the topic of ionization in terms of equilibrium constants, pH, and titration curves. Finally, we will consider the way in which aqueous solutions of weak acids or bases and their salts act as buffers against pH changes in biological systems. The water molecule and its ionization products, H^+ and OH^-, profoundly influence the structure, self-assembly, and properties of all cellular components, including enzymes and other proteins, nucleic acids, and lipids. The noncovalent interactions responsible for the specificity of "recognition" among biomolecules are decisively influenced by the solvent properties of water.

Weak Interactions in Aqueous Systems

Hydrogen bonds between water molecules provide the cohesive forces that make water a liquid at room temperature and that favor the extreme ordering of molecules typical of crystalline water (ice). Polar biomolecules dissolve readily in water because they can replace energetically favorable water–water interactions with even more favorable water–solute interactions (hydrogen bonds and electrostatic interactions). In contrast, nonpolar biomolecules interfere with favorable water–water interactions and are poorly soluble in water. In aqueous solutions, these molecules tend to cluster together to minimize the energetically unfavorable effects of their presence.

    Hydrogen bonds and ionic, hydrophobic (Greek, "water-fearing"), and van der Waals interactions, although individually weak, are numerous in biological macromolecules and collectively have a very significant influence on the three-dimensional structures of proteins, nucleic acids, polysaccharides, and membrane lipids. Before we begin a

This view of the earth from space shows that most of the planet's surface is covered with water. The seas, where living organisms probably first arose, are today the habitat of countless modern organisms.

detailed discussion of these biomolecules in the following chapters, it is useful to review the properties of the solvent, water, in which they are assembled and carry out their functions.

Hydrogen Bonding Gives Water Its Unusual Properties

Water has a higher melting point, boiling point, and heat of vaporization than most other common liquids (Table 4–1). These unusual properties are a consequence of strong attractions between adjacent water molecules, which give liquid water great internal cohesion.

Table 4–1 Melting point, boiling point, and heat of vaporization of some common liquids

	Melting point (°C)	Boiling point (°C)	Heat of vaporization (J/g)*
Water	0	100	2,260
Methanol (CH_3OH)	−98	65	1,100
Ethanol (CH_3CH_2OH)	−117	78	854
Propanol ($CH_3CH_2CH_2OH$)	−127	97	687
Butanol ($CH_3(CH_2)_2CH_2OH$)	−90	117	590
Acetone (CH_3COCH_3)	−95	56	523
Hexane ($CH_3(CH_2)_4CH_3$)	−98	69	423
Benzene (C_6H_6)	6	80	394
Butane ($CH_3(CH_2)_2CH_3$)	−135	−0.5	381
Chloroform ($CHCl_3$)	−63	61	247

* The heat energy required to convert 1.0 g of a liquid at its boiling point, at atmospheric pressure, into its gaseous state at the same temperature. It is a direct measure of the energy required to overcome attractive forces between molecules in the liquid phase.

What is the cause of these strong intermolecular attractions in liquid water? Each hydrogen atom of a water molecule shares an electron pair with the oxygen atom. The geometry of the water molecule is dictated by the shapes of the outer electron orbitals of the oxygen atom, which are similar to the bonding orbitals of carbon (see Fig. 3–4a). These orbitals describe a rough tetrahedron, with a hydrogen atom at each of two corners and unshared electrons at the other two (Fig. 4–1). The H—O—H bond angle is 104.5°, slightly less than the 109.5° of a perfect tetrahedron; the nonbonding orbitals of the oxygen atom slightly compress the orbitals shared by hydrogen.

The oxygen nucleus attracts electrons more strongly than does the hydrogen nucleus (i.e., the proton); oxygen is more electronegative (see Table 3–4). The sharing of electrons between H and O is therefore unequal; the electrons are more often in the vicinity of the oxygen atom than of the hydrogen. The result of this unequal electron sharing is two electric dipoles in the water molecule, one along each of the H—O bonds; the oxygen atom bears a partial negative charge (δ^-), and each hydrogen a partial positive charge (δ^+). The resulting electrostatic attraction between the oxygen atom of one water molecule and the hydrogen of another (Fig. 4–1c) constitutes a **hydrogen bond.**

Hydrogen bonds are weaker than covalent bonds. The hydrogen bonds in liquid water have a **bond energy** (the energy required to break a bond) of only about 20 kJ/mol, compared with 460 kJ/mol for the covalent O—H bond. At room temperature, the thermal energy of an aqueous solution (the kinetic energy resulting from the motion of individual atoms and molecules) is of the same order as that required to break hydrogen bonds. When water is heated, its temperature

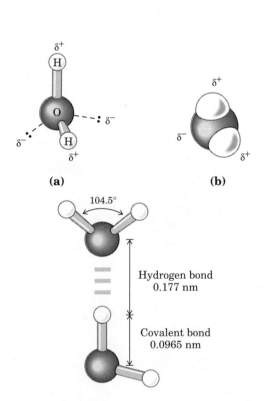

(a)   **(b)**

(c)

Figure 4–1 The dipolar nature of the H_2O molecule, shown **(a)** by ball-and-stick and **(b)** by space-filling models. The dashed lines in **(a)** represent the nonbonding orbitals. There is a nearly tetrahedral arrangement of the outer shell electron pairs around the oxygen atom; the two hydrogen atoms have localized partial positive charges and the oxygen atom has two localized partial negative charges. **(c)** Two H_2O molecules joined by a hydrogen bond (designated by three blue lines) between the oxygen atom of the upper molecule and a hydrogen atom of the lower one. Hydrogen bonds are longer and weaker than covalent O—H bonds.

increase reflects the faster motion of individual water molecules. Although at any given time most of the molecules in liquid water are hydrogen-bonded, the lifetime of each hydrogen bond is less than 1×10^{-9} s. The apt phrase "flickering clusters" has been applied to the short-lived groups of hydrogen-bonded molecules in liquid water. The very large *number* of hydrogen bonds between molecules nevertheless confers great internal cohesion on liquid water.

The nearly tetrahedral arrangement of the orbitals about the oxygen atom (Fig. 4–1a) allows each water molecule to form hydrogen bonds with as many as four neighboring water molecules. At any given instant in liquid water at room temperature, each water molecule forms hydrogen bonds with an average of 3.4 other water molecules. The water molecules are in continuous motion in the liquid state, so hydrogen bonds are constantly and rapidly being broken and formed. In ice, however, each water molecule is fixed in space and forms hydrogen bonds with four other water molecules, to yield a regular lattice structure (Fig. 4–2). To break the large numbers of hydrogen bonds in such a lattice requires much thermal energy, which accounts for the relatively high melting point of water (Table 4–1). When ice melts or water evaporates, heat is taken up by the system:

$$H_2O(s) \longrightarrow H_2O(l) \qquad \Delta H = +5.9 \text{ kJ/mol}$$

$$H_2O(l) \longrightarrow H_2O(g) \qquad \Delta H = +44.0 \text{ kJ/mol}$$

During melting or evaporation, the entropy of the aqueous system increases as more highly ordered arrays of water molecules relax into the less orderly hydrogen-bonded arrays in liquid water, or the wholly disordered water molecules in the gaseous state. At room temperature, both the melting of ice and the evaporation of water occur spontaneously; the tendency of the water molecules to associate through hydrogen bonds is outweighed by the energetic push toward randomness. Recall that the free-energy change (ΔG) must have a negative value for a process to occur spontaneously: $\Delta G = \Delta H - T\Delta S$, where ΔG represents the driving force, ΔH the energy from making and breaking bonds, and ΔS the increase in randomness. Since ΔH is positive for melting and evaporation, it is clearly the increase in entropy (ΔS) that makes ΔG negative and drives these transformations.

Water Forms Hydrogen Bonds with Solutes

Hydrogen bonds are not unique to water. They readily form between an electronegative atom (usually oxygen or nitrogen) and a hydrogen atom covalently bonded to another electronegative atom in the same or another molecule (Fig. 4–3). However, hydrogen atoms covalently bonded to carbon atoms, which are not electronegative, do not participate in hydrogen bonding. The distinction explains why butanol ($CH_3CH_2CH_2CH_2OH$) has a relatively high boiling point of 117 °C, whereas butane ($CH_3CH_2CH_2CH_3$) has a boiling point of only −0.5 °C. Butanol has a polar hydroxyl group and thus can form hydrogen bonds with other butanol molecules.

Uncharged but polar biomolecules such as sugars dissolve readily in water because of the stabilizing effect of the many hydrogen bonds that form between the hydroxyl groups or the carbonyl oxygen of the sugar and the polar water molecules. Alcohols, aldehydes, and ketones all form hydrogen bonds with water, as do compounds containing N—H bonds (Fig. 4–4), and molecules containing such groups tend to be soluble in water.

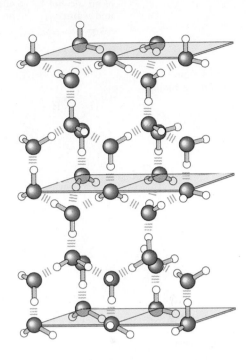

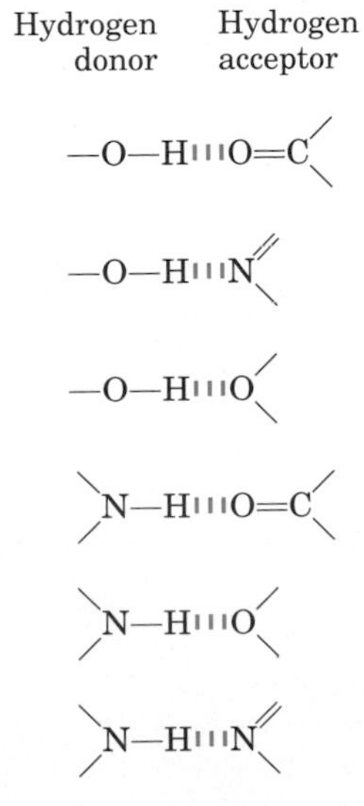

Figure 4–2 In ice, each water molecule forms the maximum of four hydrogen bonds, creating a regular crystal lattice. In liquid water at room temperature, by contrast, each water molecule forms an average of 3.4 hydrogen bonds with other water molecules. The crystal lattice of ice occupies more space than the same number of H_2O molecules occupy in liquid water; ice is less dense than liquid water, and thus floats.

Hydrogen   Hydrogen
donor   acceptor

—O—H⋯O=C

—O—H⋯N

—O—H⋯O

N—H⋯O=C

N—H⋯O

N—H⋯N

Figure 4–3 Common types of hydrogen bonds. In biological systems, the electronegative atom (the hydrogen acceptor) is usually oxygen or nitrogen. The distance between two hydrogen-bonded atoms varies from 0.26 to 0.31 nm.

84

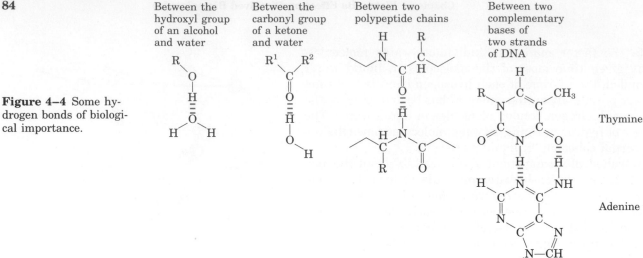

Figure 4–4 Some hydrogen bonds of biological importance.

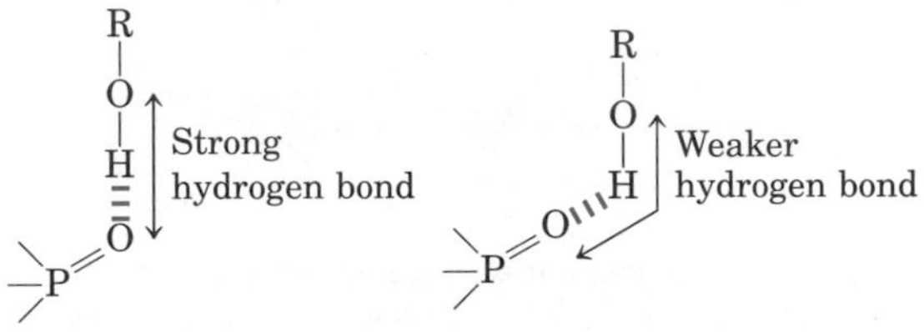

Between the hydroxyl group of an alcohol and water

Between the carbonyl group of a ketone and water

Between two polypeptide chains

Between two complementary bases of two strands of DNA

Thymine

Adenine

Figure 4–5 Directionality of the hydrogen bond. The attraction between the partial electric charges (see Fig. 4–1) is greatest when the three atoms involved (in this case O, H, and O) lie in a straight line.

Hydrogen bonds are strongest when the bonded molecules are oriented to maximize electrostatic interaction, which occurs when the hydrogen atom and the two atoms that share it are in a straight line (Fig. 4–5). Hydrogen bonds are thus highly directional and capable of holding two hydrogen-bonded molecules or groups in a specific geometric arrangement. We shall see later that this property of hydrogen bonds confers very precise three-dimensional structures upon protein and nucleic acid molecules, in which there are many intramolecular hydrogen bonds.

Water Interacts Electrostatically with Charged Solutes

Water is a polar solvent. It readily dissolves most biomolecules, which are generally charged or polar compounds (Table 4–2); compounds that dissolve easily in water are **hydrophilic** (Greek, "water-loving"). In contrast, nonpolar solvents such as chloroform and benzene are poor solvents for polar biomolecules, but easily dissolve nonpolar biomolecules such as lipids and waxes.

Water dissolves salts such as NaCl by hydrating and stabilizing the Na^+ and Cl^- ions, weakening their electrostatic interactions and thus counteracting their tendency to associate in a crystalline lattice (Fig. 4–6). The solubility of charged biomolecules in water is also a result of hydration and charge screening. Compounds with functional groups such as ionized carboxylic acids ($-COO^-$), protonated amines ($-NH_3^+$), and phosphate esters or anhydrides are generally soluble in water for the same reason.

Water is especially effective in screening the electrostatic interactions between dissolved ions. The strength, or force (F), of these **ionic interactions** depends upon the magnitude of the charges (Q), the distance between the charged groups (r), and the dielectric constant (ϵ) of the solvent through which the interactions occur:

$$F = \frac{Q_1 Q_2}{\epsilon r^2}$$

The dielectric constant is a physical property reflecting the number of dipoles in a solvent. For water at 25 °C, ϵ (which is dimensionless) is 78.5, and for the very nonpolar solvent benzene, ϵ is 4.6. Thus ionic interactions are much stronger in less polar environments. The dependence on r^2 is such that ionic attractions or repulsions operate over limited distances, in the range of 10 to 40 nm (depending on the electrolyte concentration) when the solvent is water.

Table 4–2 Some examples of polar, nonpolar, and amphipathic biomolecules

Biomolecule	Ionic form at pH 7
Polar	

Glucose

CH$_2$OH
O
H H OH
HO OH H H
H OH

Glycine $\overset{+}{N}H_3$—CH$_2$—COO$^-$

Aspartic acid
$\overset{+}{N}H_3$
$^-$OOC—CH$_2$—CH—COO$^-$

Lactic acid
CH$_3$—CH—COO$^-$
OH

Glycerol
OH
HOCH$_2$—CH—CH$_2$OH

Nonpolar

Typical wax
O
CH$_3$(CH$_2$)$_7$ CH—CH—(CH$_2$)$_6$—CH$_2$—C
O
CH$_3$—(CH$_2$)$_7$—CH=CH—(CH$_2$)$_7$—CH$_2$

Amphipathic

Phenylalanine
$\overset{+}{N}H_3$
CH$_2$—CH—COO$^-$

Phosphatidyl-
choline
O
CH$_3$(CH$_2$)$_{15}$CH$_2$—C—O—CH$_2$
CH$_3$(CH$_2$)$_{15}$CH$_2$—C—O—CH O $^+$N(CH$_3$)$_3$
O CH$_2$—P—O—CH$_2$—CH$_2$
O$^-$

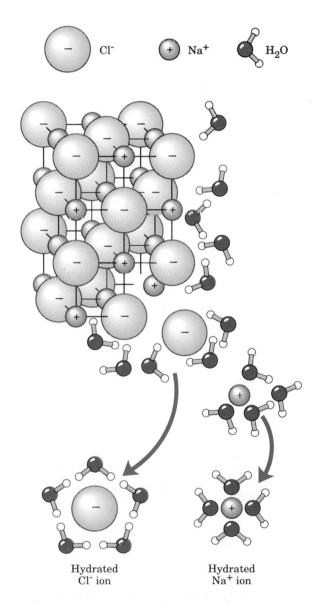

Cl$^-$ Na$^+$ H$_2$O

Hydrated
Cl$^-$ ion

Hydrated
Na$^+$ ion

Entropy Increases as Crystalline Substances Dissolve

As a salt such as NaCl dissolves, the Na$^+$ and Cl$^-$ ions leaving the crystal lattice acquire far greater freedom of motion (Fig. 4–6). The resulting increase in the entropy (randomness) of the system is largely responsible for the ease of dissolving salts such as NaCl in water. In thermodynamic terms, formation of the solution occurs with a favorable change in free energy: $\Delta G = \Delta H - T\Delta S$, where ΔH has a small positive value and $T\Delta S$ a large positive value; thus ΔG is negative.

Figure 4–6 Water dissolves many crystalline salts by hydrating their component ions. The NaCl crystal lattice is disrupted as water molecules cluster about the Cl$^-$ and Na$^+$ ions. The ionic charges are thus partially neutralized, and the electrostatic attractions necessary for lattice formation are weakened.

Nonpolar Gases Are Poorly Soluble in Water

The biologically important gases CO_2, O_2, and N_2 are nonpolar. In the diatomic molecules O_2 and N_2, electrons are shared equally by both atoms. In CO_2, each C=O bond is polar, but the two dipoles are oppositely directed and cancel each other (Table 4–3). The movement of these molecules from the disordered gas phase into aqueous solution constrains their motion and therefore represents a decrease in entropy. These gases are consequently very poorly soluble in water (Table 4–3). Some organisms have water-soluble carrier proteins (hemoglobin and myoglobin, for example) that facilitate the transport of O_2. Carbon dioxide forms carbonic acid (H_2CO_3) in aqueous solution, and is transported in that form.

Two other gases, NH_3 and H_2S, also have biological roles in some organisms; these are polar and dissolve readily in water (Table 4–3).

Table 4–3 Solubilities of some gases in water

Gas	Structure*	Polarity	Solubility in water (g/L)	Temperature (°C)
Nitrogen	N≡N	Nonpolar	0.018	40
Oxygen	O=O	Nonpolar	0.035	50
Carbon dioxide	O=C=O	Nonpolar	0.97	45
Ammonia	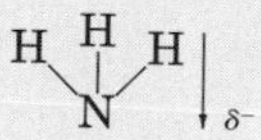 H—N(H)(H)	Polar	900	10
Hydrogen sulfide	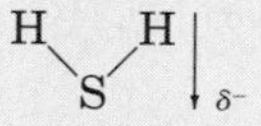 H—S—H	Polar	1,860	40

* The arrows represent electric dipoles; there is a partial negative charge (δ^-) at the head of the arrow, a partial positive charge (δ^+; not shown here) at the tail.

Nonpolar Compounds Force Energetically Unfavorable Changes in the Structure of Water

When water is mixed with a hydrocarbon such as benzene or hexane, two phases form; neither liquid is soluble in the other. Shorter hydrocarbons such as ethane have small but measurable solubility in water. Nonpolar compounds such as benzene, hexane, and ethane are **hydrophobic**—they are unable to undergo energetically favorable interactions with water molecules, and they actually interfere with the hydrogen bonding among water molecules. All solute molecules or ions dissolved in water interfere with the hydrogen bonding of some water molecules in their immediate vicinity, but polar or charged solutes (such as NaCl) partially compensate for lost hydrogen bonds by forming new solute–water interactions. The net change in enthalpy (ΔH) for dissolving these solutes is generally small. Hydrophobic solutes offer no such compensation, and their addition to water may therefore result in a small gain of enthalpy; the breaking of hydrogen bonds requires the addition of energy to the system. Furthermore, dissolving hydrophobic solutes in water results in a measurable decrease in entropy. Water molecules in the immediate vicinity of a nonpolar solute are constrained in their possible orientations, resulting in a shell of

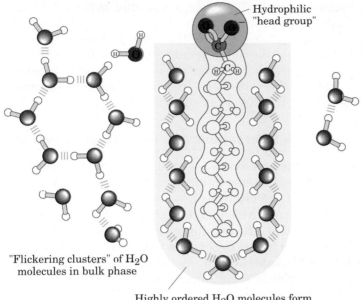

Hydrophilic "head group"

"Flickering clusters" of H₂O molecules in bulk phase

Highly ordered H₂O molecules form "cages" around the hydrophobic alkyl chains

(a)

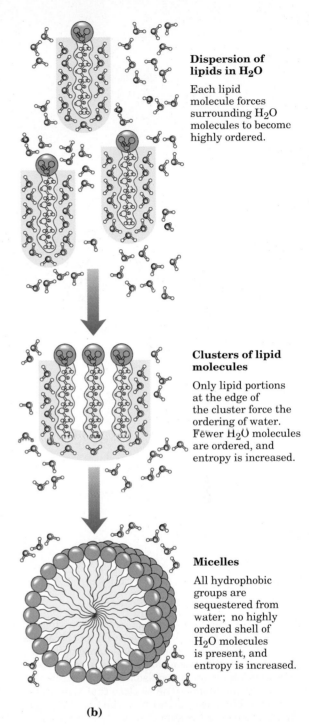

Dispersion of lipids in H₂O

Each lipid molecule forces surrounding H₂O molecules to become highly ordered.

Clusters of lipid molecules

Only lipid portions at the edge of the cluster force the ordering of water. Fewer H₂O molecules are ordered, and entropy is increased.

Micelles

All hydrophobic groups are sequestered from water; no highly ordered shell of H₂O molecules is present, and entropy is increased.

(b)

Figure 4–7 (a) The long-chain fatty acids have very hydrophobic alkyl chains, each of which is surrounded by a layer of highly ordered water molecules. **(b)** By clustering together in micelles, the fatty acid molecules expose a smaller hydrophobic surface area to the water, and fewer water molecules are found in the shell of ordered water. The energy gained by freeing immobilized water molecules stabilizes the micelle.

highly ordered water molecules around each solute molecule. The number of water molecules in the highly ordered shell is proportional to the surface area of the hydrophobic solute. The free-energy change for dissolving a nonpolar solute in water is thus unfavorable: $\Delta G = \Delta H - T\Delta S$, where ΔH has a positive value, ΔS a negative value, and thus ΔG is positive.

Amphipathic compounds contain regions that are polar (or charged) and regions that are nonpolar (Table 4–2). When amphipathic compounds are mixed with water, the two regions of the solute molecule experience conflicting tendencies; the polar or charged, hydrophilic region interacts favorably with the solvent and tends to dissolve, but the nonpolar, hydrophobic region has the opposite tendency, to avoid contact with the water (Fig. 4–7a). The nonpolar regions of the molecules cluster together to present the smallest hydrophobic area to the solvent, and the polar regions are arranged to maximize their interaction with the aqueous solvent (Fig. 4–7b). These stable structures of amphipathic compounds in water, called **micelles,** may contain hundreds or thousands of molecules. The forces that hold the nonpolar regions of the molecules together are called **hydrophobic interactions.** The strength of these interactions is not due to any intrinsic attraction between nonpolar molecules. Rather, it results from the system's achieving greatest thermodynamic stability by minimizing the entropy decrease that results from the ordering of water molecules around hydrophobic portions of the solute molecule.

Many biomolecules are amphipathic (Table 4–2); proteins, pigments, certain vitamins, and the sterols and phospholipids of membranes all have polar and nonpolar surface regions. Structures composed of these molecules are stabilized by hydrophobic interactions among the nonpolar regions. Hydrophobic interactions among lipids, and between lipids and proteins, are the most important determinants of structure in biological membranes; and hydrophobic interactions between nonpolar amino acids stabilize the three-dimensional folding patterns of proteins.

Hydrogen bonding between water and polar solutes also causes some ordering of water molecules, but the effect is less significant than with nonpolar solutes. Part of the driving force for the binding of a polar substrate to the complementary polar surface of an enzyme is the entropy increase resulting from the disordering of ordered water molecules around the substrate (reactant), as the enzyme displaces hydrogen-bonded water from the substrate.

van der Waals Interactions Are Weak Interatomic Attractions

When two uncharged atoms are brought very close together, their surrounding electron clouds influence each other. Random variations in the positions of the electrons around one nucleus may create a transient electric dipole, which induces a transient, opposite electric dipole in the nearby atom. The two dipoles are weakly attracted to each other, bringing the two nuclei closer. The force of this weak attraction is the **van der Waals interaction.** As the two nuclei draw closer together, their electron clouds begin to repel each other, and at some point the van der Waals attraction exactly balances this repulsive force (Fig. 4–8); the nuclei cannot be brought closer, and are said to be in van der Waals contact. For each atom, there is a characteristic **van der Waals radius,** a measure of how close that atom will allow another to approach (see Table 3–3).

Weak Interactions Are Crucial to Macromolecular Structure and Function

The noncovalent interactions we have described (hydrogen bonds and ionic, hydrophobic, and van der Waals interactions) are much weaker than covalent bonds (see Table 3–5). The input of about 350 kJ of energy is required to break a mole (6×10^{23}) of C—C single bonds, and of about 410 kJ to break a mole of C—H bonds, but only 4 to 8 kJ is sufficient to disrupt a mole of typical van der Waals interactions (Table 4–4). Hydrophobic interactions are similarly weak, and ionic interactions and hydrogen bonds are only a little stronger; a typical hydrogen bond in aqueous solvent can be broken by the input of about 20 kJ/mol.

Figure 4–8 The changes in energy as two atoms approach. Two opposite forces operate on the atoms, plotted here as a function of the distance between the atoms: an attraction that increases as the two approach (blue), and a repulsion that increases very sharply as the atoms come so close that their outer electron orbitals overlap (black). The net energy of the interaction is the sum of these two (red); an energy minimum occurs just before the repulsive effect dominates (at r_{me}). The closest approach that is energetically feasible, r_v, defines the van der Waals radii; it is the sum of the van der Waals radii of the two atoms.

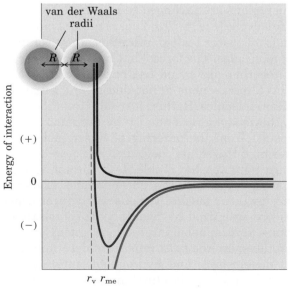

Distance between centers of atoms

Table 4–4 Four weak interactions among biomolecules in aqueous solvent

Weak interaction		Stabilization energy (kJ/mol)
Hydrogen bonds		
Between neutral groups	$\diagdown C\!=\!O\cdots H\!-\!O\!-$	8–21
Between peptide bonds	$\diagdown C\!=\!O\cdots H\!-\!N\diagup$	8–21
Ionic interactions		
Attraction	$-{}^+NH_3 \rightarrow\leftarrow {}^-O\!-\!\overset{\overset{O}{\|}}{C}\!-$	42
Repulsion	$-{}^+NH_3 \longleftrightarrow H_3N^+\!-$	$\approx\!-21$
Hydrophobic interactions	$\begin{array}{cc} CH_3\diagup\!\diagdown CH_3 & CH_3\diagup\!\diagdown CH_3 \\ CH & CH \\ CH_2 & CH_2 \end{array}$	4–8
van der Waals interactions	Any two atoms in close proximity	4

In aqueous solvent at 25 °C, the available thermal energy is of the same order as the strength of these weak interactions. Furthermore, the interaction between solute and solvent (water) molecules is nearly as favorable as solute–solute interactions. Consequently, hydrogen bonds and ionic, hydrophobic, and van der Waals interactions are continually formed and broken.

Although these four types of interactions are individually weak relative to covalent bonds, the cumulative effect of many such interactions in a protein or nucleic acid can be very significant. For example, the noncovalent binding of an enzyme to its substrate may involve several hydrogen bonds and one or more ionic interactions, as well as hydrophobic and van der Waals interactions. The formation of each of these weak bonds contributes to a net decrease in free energy; this binding free energy is released as bond formation stabilizes the system. The stability of a noncovalent interaction such as that of a small molecule hydrogen-bonded to its macromolecular partner is calculable from the binding energy. Stability, as measured by the equilibrium constant (see below) of the binding reaction, varies *exponentially* with binding energy. The unfolding of a molecule stabilized by numerous weak interactions requires many of these interactions to be disrupted at the same time; because the interactions fluctuate randomly, such simultaneous disruptions are very unlikely. The molecular stability bestowed by two or five or 20 weak interactions is therefore much greater than would be expected from a simple addition of binding energies.

Macromolecules such as proteins, DNA, and RNA contain so many sites of potential hydrogen bonding or ionic, van der Waals, or hydrophobic interactions that the cumulative effect of the many small binding forces is enormous. The most stable (native) structure of most macromolecules is that in which weak-bonding possibilities are maximized. The folding of a single polypeptide or polynucleotide chain into its three-dimensional shape is determined by this principle. The binding of an antigen to a specific antibody depends on the cumulative effects of many weak interactions. The energy released when an enzyme binds noncovalently to its substrate is the main source of catalytic power for the enzyme. The binding of a hormone or a neurotransmitter to its cellular receptor protein is the result of weak interactions. One consequence of the size of enzymes and receptors is that their large surfaces provide many opportunities for weak interactions. At the molecular level, the complementarity between interacting biomolecules reflects the complementarity and weak interactions between polar, charged, and hydrophobic groups on the surfaces of the molecules.

Ionization of Water, Weak Acids, and Weak Bases

Although many of the solvent properties of water can be explained in terms of the uncharged H_2O molecule, the small degree of ionization of water to hydrogen ions (H^+) and hydroxide ions (OH^-) must also be taken into account. Like all reversible reactions, the ionization of water can be described by an equilibrium constant. When weak acids or weak bases are dissolved in water, they can contribute H^+ by ionizing (if acids) or consume H^+ by being protonated (if bases); these processes are also governed by equilibrium constants. The total hydrogen ion concentration from all sources is experimentally measurable; it is expressed as the pH of the solution. To predict the state of ionization of solutes in water, we must take into account the relevant equilibrium constants for each ionization reaction. We therefore turn now to a brief discussion of the ionization of water and of weak acids and bases dissolved in water.

The Equilibrium Point of Reversible Reactions Is Expressed by an Equilibrium Constant

Water molecules have a slight tendency to undergo reversible ionization to yield a hydrogen ion and a hydroxide ion, giving the equilibrium

$$H_2O \rightleftharpoons H^+ + OH^- \tag{4-1}$$

This reversible ionization is crucial to the role of water in cellular function, so we must have a means of expressing the extent of ionization of water in quantitative terms. A brief review of some properties of reversible chemical reactions will show how this can be done.

The position of equilibrium of any chemical reaction is given by its **equilibrium constant.** For the generalized reaction

$$A + B \rightleftharpoons C + D \tag{4-2}$$

an equilibrium constant can be defined in terms of the concentrations of reactants (A and B) and products (C and D) present at equilibrium:

$$K_{eq} = \frac{[C][D]}{[A][B]}$$

(Strictly speaking, the concentration terms should be the *activities,* or effective concentrations in nonideal solutions, of each species. Except in very accurate work, the equilibrium constant may be approximated by measuring the *concentrations* at equilibrium.)

The equilibrium constant is fixed and characteristic for any given chemical reaction at a specified temperature. It defines the composition of the final equilibrium mixture of that reaction, regardless of the starting amounts of reactants and products. Conversely, one can calculate the equilibrium constant for a given reaction at a given temperature if the equilibrium concentrations of all its reactants and products are known. We will show in a later chapter that the standard free-energy change (ΔG°) is directly related to K_{eq}.

The Ionization of Water Is Expressed by an Equilibrium Constant

The degree of ionization of water at equilibrium (Eqn 4–1) is small; at 25 °C only about one of every 10^7 molecules in pure water is ionized at any instant. The equilibrium constant for the reversible ionization of water (Eqn 4–1) is

$$K_{eq} = \frac{[H^+][OH^-]}{[H_2O]} \tag{4–3}$$

In pure water at 25 °C, the concentration of water is 55.5 M (grams of H_2O in 1 L divided by the gram molecular weight, or $1000/18 = 55.5$ M), and is essentially constant in relation to the very low concentrations of H^+ and OH^-, namely, 1×10^{-7} M. Accordingly, we can substitute 55.5 M in the equilibrium constant expression (Eqn 4–3) to yield

$$K_{eq} = \frac{[H^+][OH^-]}{55.5 \text{ M}}$$

which, on rearranging, becomes

$$(55.5 \text{ M})(K_{eq}) = [H^+][OH^-] = K_w \tag{4–4}$$

where K_w designates the product $(55.5 \text{ M})(K_{eq})$, the **ion product of water** at 25 °C.

The value for K_{eq} has been determined by electrical-conductivity measurements of pure water (in which only the ions arising from the dissociation of H_2O can carry current) and found to be 1.8×10^{-16} M at 25 °C. Substituting this value for K_{eq} in Equation 4–4 gives

$$(55.5 \text{ M})(1.8 \times 10^{-16} \text{ M}) = [H^+][OH^-]$$

$$99.9 \times 10^{-16} \text{ M}^2 = [H^+][OH^-]$$

$$1.0 \times 10^{-14} \text{ M}^2 = [H^+][OH^-] = K_w$$

Thus the product $[H^+][OH^-]$ in aqueous solutions at 25 °C always equals 1×10^{-14} M^2. When there are exactly equal concentrations of both H^+ and OH^-, as in pure water, the solution is said to be at **neutral pH.** At this pH, the concentration of H^+ and OH^- can be calculated from the ion product of water as follows:

$$K_w = [H^+][OH^-] = [H^+]^2$$

Solving for $[H^+]$ gives

$$[H^+] = \sqrt{K_w} = \sqrt{1 \times 10^{-14} \text{ M}^2}$$

$$[H^+] = [OH^-] = 10^{-7} \text{ M}$$

BOX 4-1   The Ion Product of Water: Two Illustrative Problems

The ion product of water makes it possible to calculate the concentration of H^+, given the concentration of OH^-, and vice versa; the following problems demonstrate this.

1. What is the concentration of H^+ in a solution of 0.1 M NaOH?

$$K_w = [H^+][OH^-]$$

Solving for $[H^+]$ gives

$$[H^+] = \frac{K_w}{[OH^-]} = \frac{1 \times 10^{-14} \text{ M}^2}{0.1 \text{ M}}$$

$$= \frac{10^{-14} \text{ M}^2}{10^{-1} \text{ M}} = 10^{-13} \text{ M} \quad (answer)$$

2. What is the concentration of OH^- in a solution in which the H^+ concentration is 0.00013 M?

$$K_w = [H^+][OH^-]$$

Solving for $[OH^-]$ gives

$$[OH^-] = \frac{K_w}{[H^+]} = \frac{1 \times 10^{-14} \text{ M}^2}{0.00013 \text{ M}}$$

$$= \frac{1 \times 10^{-14} \text{ M}^2}{1.3 \times 10^{-4} \text{ M}}$$

$$= 7.7 \times 10^{-11} \text{ M} \quad (answer)$$

As the ion product of water is constant, whenever the concentration of H^+ ions is greater than 1×10^{-7} M, the concentration of OH^- must become less than 1×10^{-7} M, and vice versa. When the concentration of H^+ is very high, as in a solution of hydrochloric acid, the OH^- concentration must be very low. From the ion product of water we can calculate the H^+ concentration if we know the OH^- concentration, and vice versa (Box 4–1).

The pH Scale Designates the H⁺ and OH⁻ Concentrations

The ion product of water, K_w, is the basis for the **pH scale** (Table 4–5). It is a convenient means of designating the actual concentration of H^+ (and thus of OH^-) in any aqueous solution in the range between 1.0 M H^+ and 1.0 M OH^-. The term pH is defined by the expression

$$pH = \log \frac{1}{[H^+]} = -\log [H^+]$$

The symbol p denotes "negative logarithm of." For a precisely neutral solution at 25 °C, in which the concentration of hydrogen ions is 1.0×10^{-7} M, the pH can be calculated as follows:

$$pH = \log \frac{1}{1 \times 10^{-7}} = \log (1 \times 10^7) = \log 1.0 + \log 10^7$$

$$= 0 + 7.0$$

$$= 7.0$$

The value of 7.0 for the pH of a precisely neutral solution is not an arbitrarily chosen figure; it is derived from the absolute value of the ion product of water at 25 °C, which by convenient coincidence is a round number. Solutions having a pH greater than 7 are alkaline or basic; the concentration of OH^- is greater than that of H^+. Conversely, solutions having a pH less than 7 are acidic (Table 4–5).

Note that the pH scale is logarithmic, not arithmetic. To say that two solutions differ in pH by 1 pH unit means that one solution has ten times the H^+ concentration of the other, but it does not tell us the

Table 4-5 The pH scale

$[H^+]$ (M)	pH	$[OH^-]$ (M)	pOH*
$10^0(1)$	0	10^{-14}	14
10^{-1}	1	10^{-13}	13
10^{-2}	2	10^{-12}	12
10^{-3}	3	10^{-11}	11
10^{-4}	4	10^{-10}	10
10^{-5}	5	10^{-9}	9
10^{-6}	6	10^{-8}	8
10^{-7}	7	10^{-7}	7
10^{-8}	8	10^{-6}	6
10^{-9}	9	10^{-5}	5
10^{-10}	10	10^{-4}	4
10^{-11}	11	10^{-3}	3
10^{-12}	12	10^{-2}	2
10^{-13}	13	10^{-1}	1
10^{-14}	14	$10^{-0}(1)$	0

* The expression pOH is sometimes used to describe the basicity, or OH^- concentration, of a solution; pOH is defined by the expression $pOH = -\log [OH^-]$, which is analogous to the expression for pH. Note that for all cases, $pH + pOH = 14$.

absolute magnitude of the difference. Figure 4–9 gives the pH of some common aqueous fluids. A cola drink (pH 3.0) or red wine (pH 3.7) has an H$^+$ concentration approximately 10,000 times greater than that of blood (pH 7.4).

The pH of an aqueous solution can be approximately measured using various indicator dyes, including litmus, phenolphthalein, and phenol red, which undergo color changes as a proton dissociates from the dye molecule. Accurate determinations of pH in the chemical or clinical laboratory are made with a glass electrode that is selectively sensitive to H$^+$ concentration but insensitive to Na$^+$, K$^+$, and other cations. In a pH meter the signal from such an electrode is amplified and compared with the signal generated by a solution of accurately known pH.

Measurement of pH is one of the most important and frequently used procedures in biochemistry. The pH affects the structure and activity of biological macromolecules; for example, the catalytic activity of enzymes. Measurements of the pH of the blood and urine are commonly used in diagnosing disease. The pH of the blood plasma of severely diabetic people, for example, is often lower than the normal value of 7.4; this condition is called acidosis. In certain other disease states the pH of the blood is higher than normal, the condition of alkalosis.

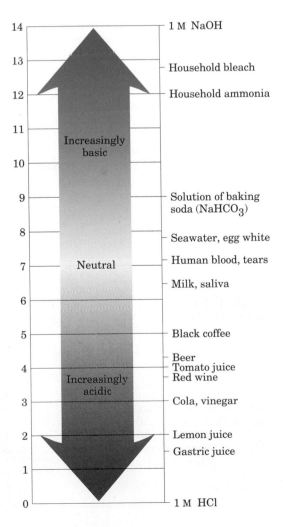

Figure 4–9 The pH of some aqueous fluids.

Weak Acids and Bases Have Characteristic Dissociation Constants

Hydrochloric, sulfuric, and nitric acids, commonly called strong acids, are completely ionized in dilute aqueous solutions; the strong bases NaOH and KOH are also completely ionized.

Biochemists are often more concerned with the behavior of weak acids and bases—those not completely ionized when dissolved in water. These are common in biological systems and play important roles in metabolism and its regulation. The behavior of aqueous solutions of weak acids and bases is best understood if we first define some terms.

Acids may be defined as proton donors and bases as proton acceptors. A proton donor and its corresponding proton acceptor make up a **conjugate acid–base pair** (Table 4–6). Acetic acid (CH_3COOH), a proton donor, and the acetate anion (CH_3COO^-), the corresponding proton acceptor, constitute a conjugate acid–base pair, related by the reversible reaction

$$CH_3COOH \rightleftharpoons H^+ + CH_3COO^-$$

Each acid has a characteristic tendency to lose its proton in an aqueous solution. The stronger the acid, the greater its tendency to lose its proton. The tendency of any acid (HA) to lose a proton and form its conjugate base (A^-) is defined by the equilibrium constant (K) for the reversible reaction

$$HA \rightleftharpoons H^+ + A^-$$

which is

$$K = \frac{[H^+][A^-]}{[HA]}$$

Equilibrium constants for ionization reactions are more usually called ionization or **dissociation constants.** The dissociation constants of some acids, often designated K_a, are given in Table 4–7. Stronger acids, such as formic and lactic acids, have higher dissociation constants; weaker acids, such as dihydrogen phosphate ($H_2PO_4^-$), have lower dissociation constants.

Also included in Table 4–7 are values of pK_a, which is analogous to pH and is defined by the equation

$$pK_a = \log \frac{1}{K_a} = -\log K_a$$

The more strongly dissociated the acid, the lower its pK_a. As we shall now see, the pK_a of any weak acid can be determined quite easily.

Table 4–6 Some conjugate acid–base pairs*

Proton donor	Proton acceptor
CH_3COOH (acetic acid)	CH_3COO^-
H_3PO_4 (phosphoric acid)	$H_2PO_4^-$
$H_2PO_4^-$ (dihydrogen phosphate)	HPO_4^{2-}
HPO_4^{2-} (hydrogen phosphate)	PO_4^{3-}
NH_4^+ (ammonium)	NH_3
H_2CO_3 (carbonic acid)	HCO_3^-
HCO_3^- (bicarbonate)	CO_3^{2-}

* Each pair consists of a proton donor and a proton acceptor. Some compounds, such as acetic acid, are monoprotic; they can give up only one proton. Others are diprotic (H_2CO_3 and glycine) or triprotic (H_3PO_4).

Table 4–7 Dissociation constant and pK_a of some common weak acids (proton donors) at 25 °C

Acid	K_a (M)	pK_a
HCOOH (formic acid)	1.78×10^{-4}	3.75
CH_3COOH (acetic acid)	1.74×10^{-5}	4.76
CH_3CH_2COOH (propionic acid)	1.35×10^{-5}	4.87
$CH_3CH(OH)COOH$ (lactic acid)	1.38×10^{-4}	3.86
H_3PO_4 (phosphoric acid)	7.25×10^{-3}	2.14
$H_2PO_4^-$ (dihydrogen phosphate)	1.38×10^{-7}	6.86
HPO_4^{2-} (monohydrogen phosphate)	3.98×10^{-13}	12.4
H_2CO_3 (carbonic acid)	1.70×10^{-4}	3.77
HCO_3^- (bicarbonate)	6.31×10^{-11}	10.2
NH_4^+ (ammonium)	5.62×10^{-10}	9.25

Titration Curves Reveal the pK_a of Weak Acids

Titration is used to determine the amount of an acid in a given solution. In this procedure, a measured volume of the acid is titrated with a solution of a strong base, usually sodium hydroxide (NaOH), of known concentration. The NaOH is added in small increments until the acid is consumed (neutralized), as determined with an indicator dye or with a pH meter. The concentration of the acid in the original solution can be calculated from the volume and concentration of NaOH added.

A plot of the pH against the amount of NaOH added (a **titration curve**) reveals the pK_a of the weak acid. Consider the titration of a 0.1 M solution of acetic acid (IIAc) with 0.1 M NaOH at 25 °C (Fig. 4–10). Two reversible equilibria are involved in the process:

$$H_2O \rightleftharpoons H^+ + OH^- \qquad (4\text{--}5)$$

$$HAc \rightleftharpoons H^+ + Ac^- \qquad (4\text{--}6)$$

The equilibria must simultaneously conform to their characteristic equilibrium constants, which are, respectively,

$$K_w = [H^+][OH^-] = 1 \times 10^{-14} \ M^2 \qquad (4\text{--}7)$$

$$K_a = \frac{[H^+][Ac^-]}{[HAc]} = 1.74 \times 10^{-5} \ M \qquad (4\text{--}8)$$

At the beginning of the titration, before any NaOH is added, the acetic acid is already slightly ionized, to an extent that can be calculated from its dissociation constant (Eqn 4–8).

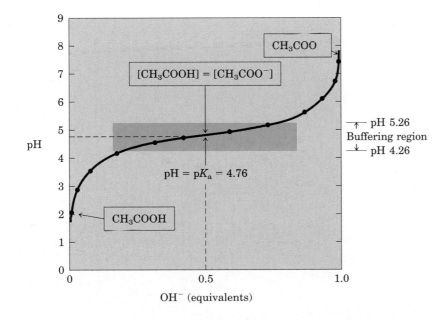

OH$^-$ (equivalents)

Figure 4–10 The titration curve of acetic acid. After the addition of each increment of NaOH to the acetic acid solution, the pH of the mixture is measured. This value is plotted against the fraction of the total amount of NaOH required to neutralize the acetic acid (i.e., to bring it to pH ≈ 7). The points so obtained yield the titration curve. Shown in the boxes are the predominant ionic forms at the points designated. At the midpoint of the titration, the concentrations of the proton donor and proton acceptor are equal. The pH at this point is numerically equal to the pK_a of acetic acid. The shaded zone is the useful region of buffering power.

As NaOH is gradually introduced, the added OH$^-$ combines with the free H$^+$ in the solution to form H$_2$O, to an extent that satisfies the equilibrium relationship in Equation 4–7. As free H$^+$ is removed, HAc dissociates further to satisfy its own equilibrium constant (Eqn 4–8). The net result as the titration proceeds is that more and more HAc ionizes, forming Ac$^-$, as the NaOH is added. At the midpoint of the titration (Fig. 4–10), at which exactly 0.5 equivalent of NaOH has been added, one-half of the original acetic acid has undergone dissociation, so that the concentration of the proton donor, [HAc], now equals that of the proton acceptor, [Ac$^-$]. At this midpoint a very important relationship holds: the pH of the equimolar solution of acetic acid and acetate is exactly equal to the pK_a of acetic acid (pK_a = 4.76; see Table 4–7 and Fig. 4–10). The basis for this relationship, which holds for all weak acids, will soon become clear.

As the titration is continued by adding further increments of NaOH, the remaining undissociated acetic acid is gradually converted into acetate. The end point of the titration occurs at about pH 7.0: all the acetic acid has lost its protons to OH$^-$, to form H_2O and acetate. Throughout the titration the two equilibria (Eqns 4–5 and 4–6) co-exist, each always conforming to its equilibrium constant.

Figure 4-11 compares the titration curves of three weak acids with very different dissociation constants: acetic acid (pK_a = 4.76); dihydrogen phosphate (pK_a = 6.86); and ammonium ion, or NH_4^+ (pK_a = 9.25). Although the titration curves of these acids have the same shape, they are displaced along the pH axis because these acids have different strengths. Acetic acid is the strongest and loses its proton most readily, since its K_a is highest (pK_a lowest) of the three. Acetic acid is already half dissociated at pH 4.76. $H_2PO_4^-$ loses a proton less readily, being half dissociated at pH 6.86. NH_4^+ is the weakest acid of the three and does not become half dissociated until pH 9.25.

The most important point about the titration curve of a weak acid is that it shows graphically that a weak acid and its anion—a conjugate acid–base pair—can act as a buffer.

Figure 4–11 Comparison of the titration curves of three weak acids, CH_3COOH, $H_2PO_4^-$, and NH_4^+. The predominant ionic forms at designated points in the titration are given in boxes. The regions of buffering capacity are indicated at the right. Conjugate acid–base pairs are effective buffers between approximately 25 and 75% neutralization of the proton-donor species.

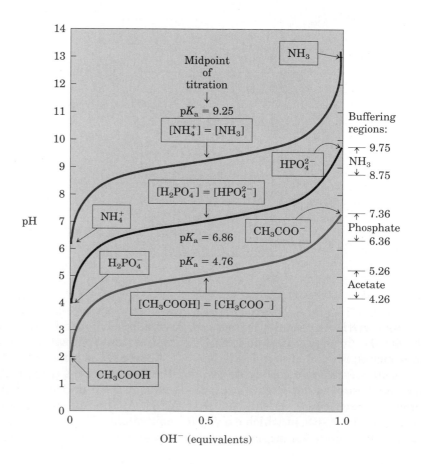

Buffering against pH Changes in Biological Systems

Almost every biological process is pH dependent; a small change in pH produces a large change in the rate of the process. This is true not only for the many reactions in which the H^+ ion is a direct participant, but also for those in which there is no apparent role for H^+ ions. The enzymes that catalyze cellular reactions, and many of the molecules on which they act, contain ionizable groups with characteristic pK_a values. The protonated amino ($-NH_3^+$) and carboxyl groups of amino acids and the phosphate groups of nucleotides, for example, function as weak acids; their ionic state depends upon the pH of the solution in which they are dissolved. As we noted above, ionic interactions are among the forces that stabilize a protein molecule and allow an enzyme to recognize and bind its substrate.

Cells and organisms maintain a specific and constant cytosolic pH, keeping biomolecules in their optimal ionic state, usually near pH 7. In multicellular organisms, the pH of the extracellular fluids (blood, for example) is also tightly regulated. Constancy of pH is achieved primarily by biological buffers: mixtures of weak acids and their conjugate bases.

We describe here the ionization equilibria that account for buffering, and show the quantitative relationship between the pH of a buffered solution and the pK_a of the buffer. Biological buffering is illustrated by the phosphate and carbonate buffering systems of humans.

Buffers Are Mixtures of Weak Acids and Their Conjugate Bases

Buffers are aqueous systems that tend to resist changes in their pH when small amounts of acid (H^+) or base (OH^-) are added. A buffer system consists of a weak acid (the proton donor) and its conjugate base (the proton acceptor). As an example, a mixture of equal concentrations of acetic acid and acetate ion, found at the midpoint of the titration curve in Figure 4–10, is a buffer system. The titration curve of acetic acid has a relatively flat zone extending about 0.5 pH units on either side of its midpoint pH of 4.76. In this zone there is only a small change in pH when increments of either H^+ or OH^- are added to the system. This relatively flat zone is the buffering region of the acetic acid–acetate buffer pair. At the midpoint of the buffering region, where the concentration of the proton donor (acetic acid) exactly equals that of the proton acceptor (acetate), the buffering power of the system is maximal; that is, its pH changes least on addition of an increment of H^+ or OH^-. The pH at this point in the titration curve of acetic acid is equal to its pK_a. The pH of the acetate buffer system does change slightly when a small amount of H^+ or OH^- is added, but this change is very small compared with the pH change that would result if the same amount of H^+ (or OH^-) were added to pure water or to a solution of the salt of a strong acid and strong base, such as NaCl, which have no buffering power.

Buffering results from two reversible reaction equilibria occurring in a solution of nearly equal concentrations of a proton donor and its conjugate proton acceptor. Figure 4–12 helps to explain how a buffer system works. Whenever H^+ or OH^- is added to a buffer, the result is a small change in the ratio of the relative concentrations of the weak

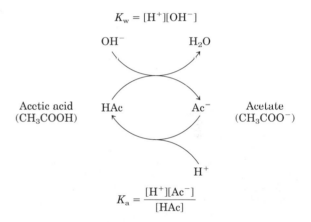

$$K_w = [H^+][OH^-]$$

$$K_a = \frac{[H^+][Ac^-]}{[HAc]}$$

Figure 4–12 Capacity of the acetic acid–acetate couple to act as a buffer system, capable of absorbing either H^+ or OH^- through the reversibility of the dissociation of acetic acid. The proton donor, in this case acetic acid (HAc), contains a reserve of bound H^+, which can be released to neutralize an addition of OH^- to the system, forming H_2O. This happens because the product $[H^+][OH^-]$ transiently exceeds K_w (1×10^{-14} M^2). The equilibrium quickly adjusts so that this product equals 1×10^{-14} M^2 (at 25 °C), thus transiently reducing the concentration of H^+. But now the quotient $[H^+][Ac^-]/[HAc]$ is less then K_a, so HAc dissociates further to restore equilibrium. Similarly, the conjugate base, Ac^-, can react with H^+ ions added to the system; again, the two ionization reactions simultaneously come to equilibrium. Thus a conjugate acid–base pair, such as acetic acid and acetate ion, tends to resist a change in pH when small amounts of acid or base are added. Buffering action is simply the consequence of two reversible reactions taking place simultaneously and reaching their points of equilibrium as governed by their equilibrium constants, K_w and K_a.

acid and its anion and thus a small change in pH. The decrease in concentration of one component of the system is balanced exactly by an increase in the other. The sum of the buffer components does not change, only their ratio.

Each conjugate acid–base pair has a characteristic pH zone in which it is an effective buffer (Fig. 4–11). The $H_2PO_4^-/HPO_4^{2-}$ pair has a pK_a of 6.86 and thus can serve as a buffer system near pH 6.86; the NH_4^+/NH_3 pair, with a pK_a of 9.25, can act as a buffer near pH 9.25.

A Simple Expression Relates pH, pK, and Buffer Concentration

The quantitative relationship among pH, the buffering action of a mixture of weak acid with its conjugate base, and the pK_a of the weak acid is given by the **Henderson–Hasselbalch equation.** The titration curves of acetic acid, $H_2PO_4^-$, and NH_4^+ (Fig. 4–11) have nearly identical shapes, suggesting that they all reflect a fundamental law or relationship. This is indeed the case. The shape of the titration curve of any weak acid is expressed by the Henderson–Hasselbalch equation, which is important for understanding buffer action and acid–base balance in the blood and tissues of the vertebrate organism. This equation is simply a useful way of restating the expression for the dissociation constant of an acid. For the dissociation of a weak acid HA into H^+ and A^-, the Henderson–Hasselbalch equation can be derived as follows:

$$K_a = \frac{[H^+][A^-]}{[HA]}$$

First solve for $[H^+]$:

$$[H^+] = K_a \frac{[HA]}{[A^-]}$$

Then take the negative logarithm of both sides:

$$-\log [H^+] = -\log K_a - \log \frac{[HA]}{[A^-]}$$

Substitute pH for $-\log [H^+]$ and pK_a for $-\log K_a$:

$$pH = pK_a - \log \frac{[HA]}{[A^-]}$$

Now invert $-\log [HA]/[A^-]$, which involves changing its sign, to obtain the Henderson–Hasselbalch equation:

$$pH = pK_a + \log \frac{[A^-]}{[HA]}$$

which is stated more generally as

$$pH = pK_a + \log \frac{[\text{proton acceptor}]}{[\text{proton donor}]}$$

This equation fits the titration curve of all weak acids and enables us to deduce a number of important quantitative relationships. For example, it shows why the pK_a of a weak acid is equal to the pH of the solution at the midpoint of its titration. At this point $[HA] = [A^-]$, and

$$pH = pK_a + \log 1.0 = pK_a + 0 = pK_a$$

BOX 4–2 Solving Problems with the Henderson–Hasselbalch Equation

1. Calculate the pK_a of lactic acid, given that when the concentration of free lactic acid is 0.010 M and the concentration of lactate is 0.087 M, the pH is 4.80.

$$pH = pK_a + \log \frac{[\text{lactate}]}{[\text{lactic acid}]}$$

$$pK_a = pH - \log \frac{[\text{lactate}]}{[\text{lactic acid}]}$$

$$= 4.80 - \log \frac{0.087}{0.010} = 4.80 - \log 8.7$$

$$= 4.80 - 0.94 = 3.86 \quad (answer)$$

2. Calculate the pH of a mixture of 0.1 M acetic acid and 0.2 M sodium acetate. The pK_a of acetic acid is 4.76.

$$pH = pK_a + \log \frac{[\text{acetate}]}{[\text{acetic acid}]}$$

$$= 4.76 + \log \frac{0.2}{0.1} = 4.76 + 0.301$$

$$= 5.06 \quad (answer)$$

3. Calculate the ratio of the concentrations of acetate and acetic acid required in a buffer system of pH 5.30.

$$pH = pK_a + \log \frac{[\text{acetate}]}{[\text{acetic acid}]}$$

$$\log \frac{[\text{acetate}]}{[\text{acetic acid}]} = pH - pK_a$$

$$= 5.30 - 4.76 = 0.54$$

$$\frac{[\text{acetate}]}{[\text{acetic acid}]} = \text{antilog } 0.54 = 3.47 \quad (answer)$$

The Henderson–Hasselbalch equation also makes it possible to calculate the pK_a of any acid from the molar ratio of proton-donor and proton-acceptor species at any given pH; to calculate the pH of a conjugate acid–base pair of a given pK_a and a given molar ratio; and to calculate the molar ratio of proton donor and proton acceptor at any pH given the pK_a of the weak acid (Box 4–2).

Weak Acids or Bases Buffer Cells and Tissues against pH Changes

The cytoplasm of most cells contains high concentrations of proteins, which contain many amino acids with functional groups that are weak acids or weak bases. The side chain of the amino acid histidine (Fig. 4–13) has a pK_a of 6.0, and proteins containing histidine residues can therefore buffer effectively near neutral pH. Nucleotides such as ATP, as well as many low molecular weight metabolites, contain ionizable groups that can contribute buffering power to the cytoplasm. Some highly specialized organelles and extracellular compartments have high concentrations of compounds that contribute buffering capacity: organic acids buffer the vacuoles of plant cells; ammonia buffers urine.

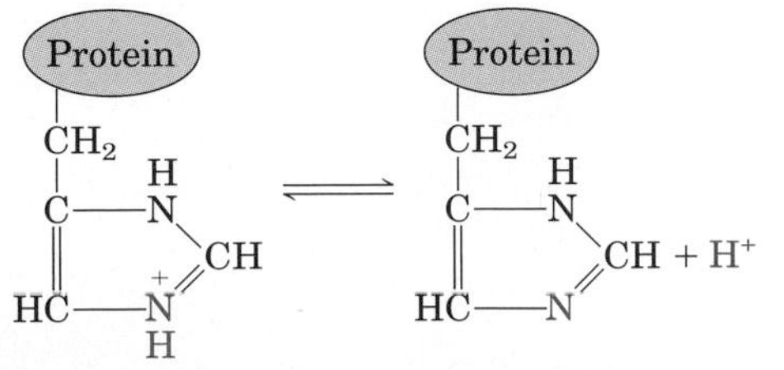

Figure 4–13 The amino acid histidine, a component of proteins, is a weak acid. The pK_a of the protonated nitrogen of the side chain is 6.0.

Phosphate and Bicarbonate Are Important Biological Buffers

The intracellular and extracellular fluids of all multicellular organisms have a characteristic and nearly constant pH, which is regulated by various biological activities. The organism's first line of defense against changes in internal pH is provided by buffer systems. Two important biological buffers are the phosphate and bicarbonate systems. The phosphate buffer system, which acts in the cytoplasm of all cells, consists of $H_2PO_4^-$ as proton donor and HPO_4^{2-} as proton acceptor:

$$H_2PO_4^- \rightleftharpoons H^+ + HPO_4^{2-}$$

The phosphate buffer system works exactly like the acetate buffer system, except for the pH range in which it functions. The phosphate buffer system is maximally effective at a pH close to its pK_a of 6.86 (see Table 4–7 and Fig. 4–11), and thus tends to resist pH changes in the range between about 6.4 and 7.4. It is therefore effective in providing buffering power in intracellular fluids; in mammals, for example, extracellular fluids and most cytoplasmic compartments have a pH in the range of 6.9 to 7.4.

Blood plasma is buffered in part by the bicarbonate system, consisting of carbonic acid (H_2CO_3) as proton donor and bicarbonate (HCO_3^-) as proton acceptor:

$$H_2CO_3 \rightleftharpoons H^+ + HCO_3^-$$

This system has an equilibrium constant

$$K_1 = \frac{[H^+][HCO_3^-]}{[H_2CO_3]}$$

and functions as a buffer in the same way as other conjugate acid–base pairs. It is unique, however, in that one of its components, carbonic acid (H_2CO_3), is formed from dissolved (d) carbon dioxide and water, according to the reversible reaction

$$CO_2(d) + H_2O \rightleftharpoons H_2CO_3$$

which has an equilibrium constant given by the expression

$$K_2 = \frac{[H_2CO_3]}{[CO_2(d)][H_2O]}$$

Carbon dioxide is a gas under normal conditions, and the concentration of dissolved CO_2 is the result of equilibration with CO_2 of the gas phase:

$$CO_2(g) \rightleftharpoons CO_2(d)$$

This process has an equilibrium constant given by

$$K_3 = \frac{[CO_2(d)]}{[CO_2(g)]}$$

The pH of a bicarbonate buffer system depends on the concentration of H_2CO_3 and HCO_3^-, the proton donor and acceptor components. The concentration of H_2CO_3 in turn depends on the concentration of dissolved CO_2, which in turn depends on the concentration or partial pressure of CO_2 in the gas phase; thus the pH of a bicarbonate buffer exposed to a gas phase is ultimately determined by the concentration of HCO_3^- in the aqueous phase and the partial pressure of CO_2 in the gas phase (Box 4–3).

BOX 4–3   Blood, Lungs, and Buffer: The Bicarbonate Buffer System

In animals with lungs, the bicarbonate buffer system is an effective physiological buffer near pH 7.4 because the H_2CO_3 of the blood plasma is in equilibrium with a large reserve capacity of $CO_2(g)$ in the air space of the lungs. This buffer system involves three reversible equilibria between gaseous CO_2 in the lungs and bicarbonate (HCO_3^-) in the blood plasma (Fig. 1). When H^+ is added to blood as it passes through the tissues, reaction 1 proceeds toward a new equilibrium, in which the concentration of H_2CO_3 is increased. This increases the concentration of $CO_2(d)$ in the blood (reaction 2), and

thus increases the pressure of $CO_2(g)$ in the air space of the lungs (reaction 3); the extra CO_2 is exhaled.

Conversely, when OH^- is added to the blood plasma, the opposite events occur: the H^+ concentration is lowered, causing more H_2CO_3 to dissociate into H^+ and HCO_3^-. This in turn causes more $CO_2(g)$ from the lungs to dissolve in the blood plasma. The rate of breathing, that is, the rate of inhaling and exhaling CO_2, can quickly adjust these equilibria to keep the blood pH nearly constant.

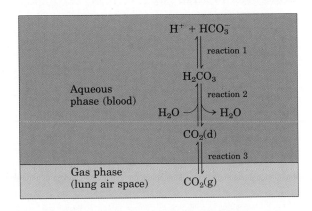

Figure 1 The CO_2 in the air space of the lungs is in equilibrium with the bicarbonate buffer in the blood plasma passing through the lung capillaries. Because the concentration of dissolved CO_2 can be adjusted rapidly through changes in the rate of breathing, the bicarbonate buffer system of the blood is in near-equilibrium with a large potential reservoir of CO_2.

Human blood plasma normally has a pH close to 7.40. Should the pH-regulating mechanisms fail or be overwhelmed, as may happen in severe uncontrolled diabetes when an overproduction of metabolic acids causes acidosis, the pH of the blood can fall to 6.8 or below, leading to irreparable cell damage and death. In other diseases the pH may rise to lethal levels. Although many aspects of cell structure and function are influenced by pH, it is the catalytic activity of enzymes that is especially sensitive. Enzymes typically show maximal catalytic activity at a characteristic pH, called the **optimum pH** (Fig. 4–14). On either side of the optimum pH their catalytic activity often declines sharply. Thus a small change in pH can make a large difference in the rate of some crucial enzyme-catalyzed reaction. Biological control of the pH of cells and body fluids is therefore of central importance in all aspects of metabolism and cellular activities.

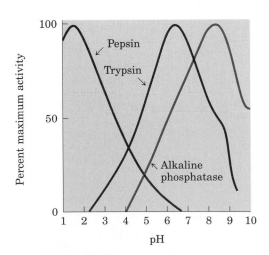

Figure 4–14 The pH optima of some enzymes: pepsin, a digestive enzyme secreted into gastric juice (black); trypsin, a digestive enzyme that acts in the small intestine (red); alkaline phosphatase of bone tissue (blue).

Water as a Reactant

Water is not just the solvent in which the chemical reactions of living cells occur; it is very often a direct participant in those reactions. The formation of ATP from ADP and inorganic phosphate is a condensation reaction (see Fig. 3–14) in which the elements of water are eliminated (Fig. 4–15a). The compound formed by this condensation is called a phosphate **anhydride. Hydrolysis reactions** are responsible for the enzymatic depolymerization of proteins, carbohydrates, and nucleic acids ingested in the diet. Hydrolytic enzymes **(hydrolases)** catalyze the addition of the elements of water to the bonds that connect monomeric subunits in these macromolecules (Fig. 4–15). Hydrolysis reactions are almost invariably exergonic, and the formation of cellular polymers from their subunits by simple reversal of hydrolysis would be endergonic and as such does not occur. We shall see that cells circumvent this thermodynamic obstacle by coupling the endergonic condensation reactions to exergonic processes, such as breakage of the anhydride bond in ATP.

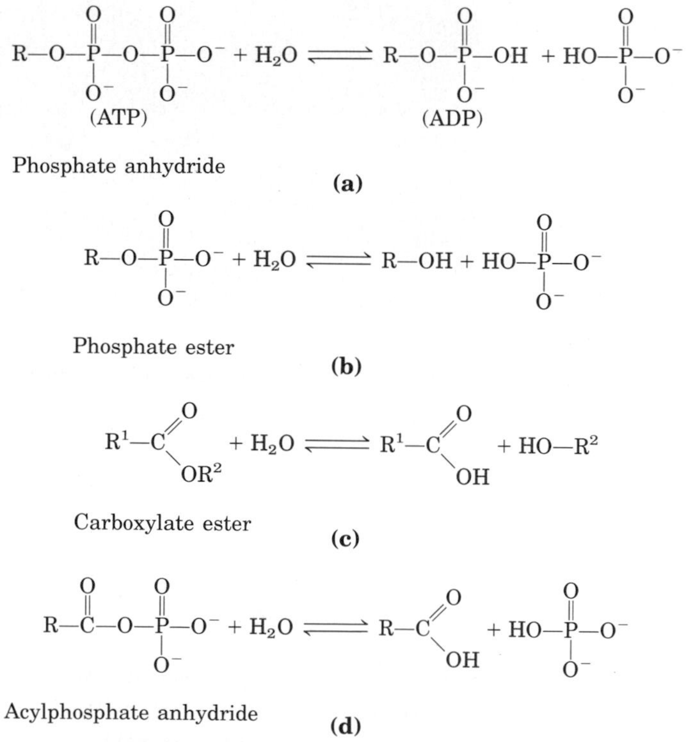

Figure 4–15 Water participates directly in a variety of reactions. **(a)** ATP is a phosphate anhydride formed by a condensation reaction (loss of the elements of water) between ADP and phosphate. R represents adenosine monophosphate (AMP). This condensation reaction requires energy. The hydrolysis (addition of the elements of water) of ATP releases an equivalent amount of energy. **(b)**, **(c)**, and **(d)** represent similar condensation and hydrolysis reactions common in biological systems.

Phosphate anhydride **(a)**

Phosphate ester **(b)**

Carboxylate ester **(c)**

Acylphosphate anhydride **(d)**

You are (we hope!) consuming oxygen as you read. Water and carbon dioxide are the end products of the oxidation of fuels such as glucose. The overall reaction of this process can be summarized by the equation:

$$C_6H_{12}O_6 + 6O_2 \longrightarrow 6CO_2 + 6H_2O$$
Glucose

The "metabolic water" thus formed from stored fuels is actually enough to allow some animals in very dry habitats (gerbils, kangaroo rats, camels) to survive without drinking water for extended periods.

Green plants and algae use the energy of sunlight (represented by $h\nu$, the energy of light of frequency ν; h is Planck's constant) to split water in the process of photosynthesis:

$$2H_2O + 2A \xrightarrow{h\nu} O_2 + 2AH_2$$

In this reaction, A is an electron-accepting species, which varies with the type of photosynthetic organism.

The Fitness of the Aqueous Environment for Living Organisms

Organisms have effectively adapted to their aqueous environment and have even evolved means of exploiting the unusual properties of water. The high specific heat of water (the heat energy required to raise the temperature of 1 g of water by 1 °C) is useful to cells and organisms because it allows water to act as a "heat buffer," permitting the temperature of an organism to remain relatively constant as the temperature of the air fluctuates and as heat is generated as a byproduct of metabolism. Furthermore, some vertebrates exploit the high heat of vaporization of water (see Table 4–1) by using (thus losing) excess body heat to evaporate sweat. The high degree of internal cohesion of liquid water, due to hydrogen bonding, is exploited by plants as a means of transporting dissolved nutrients from the roots to the leaves during the process of transpiration. Even the lower density of ice than of liquid water has important biological consequences in the life cycles of aquatic organisms. Ponds freeze from the top down, and the layer of ice at the top insulates the water below from frigid air, preventing the pond (and the organisms in it) from freezing solid. Most fundamental to all living organisms is the fact that many physical and biological properties of cell macromolecules, particularly the proteins and nucleic acids, derive from their interactions with water molecules of the surrounding medium. The influence of water on the course of biological evolution has been profound and determinative. If life forms have evolved elsewhere in the universe, it is unlikely that they resemble those of earth, unless their extraterrestrial origin is also a place in which plentiful liquid water is available as solvent.

Aqueous environments support a myriad of species. Soft corals, sponges, bryozoans, and algae compete for space on this reef substrate off the Philippine Islands.

Summary

Water is the most abundant compound in living organisms. Its relatively high freezing point, boiling point, and heat of vaporization are the result of strong intermolecular attractions in the form of hydrogen bonding between adjacent water molecules. Liquid water has considerable short-range order and consists of short-lived hydrogen-bonded clusters. The polarity and hydrogen-bonding properties of water make it a potent solvent for many ionic compounds and other polar molecules. Nonpolar compounds, including the gases CO_2, O_2, and N_2, are poorly soluble in water. Water disperses amphipathic molecules to form micelles, clusters of molecules in which the hydrophobic groups are hidden from water and the polar groups are exposed on the external surface.

Four types of weak interactions occur within and between biomolecules in an aqueous solvent: hydrogen bonds and ionic, hydrophobic, and van der Waals interactions. Although weak individually, these interactions collectively create a very strong stabilizing force for proteins, nucleic acids, and membranes. Weak (noncovalent) interactions are also at the heart of enzyme catalysis, antibody function, and receptor–ligand interactions.

Water ionizes very slightly to form H^+ and OH^- ions. In dilute aqueous solutions, the concentrations of H^+ and OH^- ions are inversely related by the expression $K_w = [H^+][OH^-] = 1 \times 10^{-14}$ M^2 (at 25 °C). The hydrogen-ion concentration of biological systems is usually expressed in terms of pH, defined as pH $= -\log [H^+]$. The pH of aqueous solutions is measured by means of glass electrodes sensitive to H^+ concentration.

Acids are defined as proton donors and bases as proton acceptors. A conjugate acid–base pair consists of a proton donor (HA) and its corresponding proton acceptor (A^-). The tendency of an acid HA to donate protons is expressed by its dissociation constant ($K_a = [H^+][A^-]/[HA]$) or by the function pK_a, defined as $-\log K_a$, which can be determined from an experimental titration curve. The pH of a solution of a weak acid is quantitatively related to its pK_a and to the ratio of the concentrations of its proton-donor and proton-acceptor species by the Henderson–Hasselbalch equation.

A conjugate acid–base pair can act as a buffer and resist changes in pH; its capacity to do so is greatest at a pH equal to its pK_a. Many types of biomolecules have functional groups that contribute buffering capacity. H_2CO_3/HCO_3^- and $H_2PO_4^-/HPO_4^{2-}$ are important biological buffer systems. The catalytic activity of enzymes is strongly influenced by pH, and it is essential that the environments in which they function be buffered against large pH changes.

Water is not only the solvent in which metabolic reactions occur; it participates directly in many of the reactions, including hydrolysis and condensation reactions.

The physical and chemical properties of water are central to biological structure and function. The evolution of life on earth was doubtless influenced greatly by both the solvent and reactant properties of water.

Further Reading

General

Dick, D.A.T. (1966) *Cell Water,* Butterworth Publishers, Inc., Stoneham, MA.

A classic description of the properties and functions of water in living organisms.

Edsall, J.T. & Wyman, J. (1958) *Biophysical Chemistry,* Vol. 1, Academic Press, Inc., New York.

An excellent discussion of water and its fitness as a biological solvent.

Eisenberg, D. & Kauzmann, W. (1969) *The Structure and Properties of Water,* Oxford University Press, New York.

An advanced treatment of the physical chemistry of water.

Franks, R. (ed.) (1975) *Water—A Comprehensive Treatise,* Vol. 4, Plenum Press, New York.

Franks, R. & Mathias, S.F. (eds.) (1982) *Biophysics of Water,* John Wiley & Sons, Inc., New York.

A large collection of papers on the structure of pure water and of the cytoplasm.

Henderson, L.J. (1927) *The Fitness of the Environment,* Beacon Press, Boston, MA. [Reprinted (1958).]
This book is a classic; it includes a discussion of the suitability of water as the solvent for life on earth.

Kuntz, I.D. & Zipp, I. (1977) Water in biological systems. *New Engl. J. Med.* **297,** 262–266.
A brief review of the physical state of cytosolic water and its interactions with dissolved biomolecules.

Solomon, A.K. (1971) The state of water in red cells. *Sci. Am.* **244** (February), 88–96.
A description of research on the structure of water within cells.

Stillinger, F.H. (1980) Water revisited. *Science* **209,** 451–457.
A short review of the physical structure of water, including the importance of hydrogen bonding and the nature of hydrophobic interactions.

Symons, M.C.R. (1981) Water structure and reactivity. *Acc. Chem. Res.* **14,** 179–187.

Wiggins, P.M. (1990) Role of water in some biological processes. *Microbiol. Rev.* **54,** 432–449.
A recent and excellent review of water in biology, including discussion of the physical structure of liquid water, its interaction with biomolecules, and the state of water in living cells.

Weak Interactions in Aqueous Systems

Fersht, A.R. (1987) The hydrogen bond in molecular recognition. *Trends Biochem. Sci.* **12,** 301–304.
A clear, brief, quantitative discussion of the contribution of hydrogen bonding to molecular recognition and enzyme catalysis.

Frieden, E. (1975) Non-covalent interactions: key to biological flexibility and specificity. *J. Chem. Educ.* **52,** 754–761.
Review of the four kinds of weak interactions that stabilize macromolecules and confer biological specificity, with clear examples.

Tanford, C. (1978) The hydrophobic effect and the organization of living matter. *Science* **200,** 1012–1018.
An excellent review of the chemical and energetic basis for hydrophobic interactions between biomolecules in aqueous solutions.

Weak Acids, Weak Bases, and Buffers

Montgomery, R. & Swenson, C.A. (1976) *Quantitative Problems in Biochemical Sciences,* 2nd edn, W.H. Freeman and Company, New York.
This and the following book are excellent compilations of solved problems, many of which concern pH, the ionization of weak acids and bases, and buffers.

Segel, I.H. (1976) *Biochemical Calculations,* 2nd edn, John Wiley & Sons, Inc., New York.

Problems

1. *Artificial Vinegar* One way to make vinegar (*not* the preferred way) is to prepare a solution of acetic acid, the sole acid component of vinegar, at the proper pH (see Fig. 4–9) and add appropriate flavoring agents. Acetic acid (M_r 60) is a liquid at 25 °C with a density of 1.049 g/mL. Calculate the amount (volume) that must be added to distilled water to make 1 L of simulated vinegar (see Table 4–7).

2. *Acidity of Gastric HCl* In a hospital laboratory, a 10.0 mL sample of gastric juice, obtained several hours after a meal, was titrated with 0.1 M NaOH to neutrality; 7.2 mL of NaOH was required. The stomach contained no ingested food or drink, thus assume that no buffers were present. What was the pH of the gastric juice?

3. *Measurement of Acetylcholine Levels by pH Changes* The concentration of acetylcholine, a neurotransmitter, can be determined from the pH changes that accompany its hydrolysis. When incubated with a catalytic amount of the enzyme acetylcholinesterase, acetylcholine is quantitatively converted into choline and acetic acid, which dissociates to yield acetate and a hydrogen ion:

In a typical analysis, 15 mL of an aqueous solution containing an unknown amount of acetylcholine had a pH of 7.65. When incubated with acetylcholinesterase, the pH of the solution decreased to a final value of 6.87. Assuming that there was no buffer in the assay mixture, determine the number of moles of acetylcholine in the 15 mL of unknown.

4. Significance of the pK_a of an Acid One common description of the pK_a of an acid is that it represents the pH at which the acid is half ionized, that is, the pH at which it exists as a 50:50 mixture of the acid and the conjugate base. Demonstrate this relationship for an acid HA, starting from the equilibrium-constant expression.

5. Properties of a Buffer The amino acid glycine is often used as the main ingredient of a buffer in biochemical experiments. The amino group of glycine, which has a pK_a of 9.6, can exist either in the protonated form ($-NH_3^+$) or as the free base ($-NH_2$) because of the reversible equilibrium

$$R-NH_3^+ \rightleftharpoons R-NH_2 + H^+$$

(a) In what pH range can glycine be used as an effective buffer due to its amino group?

(b) In a 0.1 M solution of glycine at pH 9.0, what fraction of glycine has its amino group in the $-NH_3^+$ form?

(c) How much 5 M KOH must be added to 1.0 L of 0.1 M glycine at pH 9.0 to bring its pH to exactly 10.0?

(d) In order to have 99% of the glycine in its $-NH_3^+$ form, what must the numerical relation be between the pH of the solution and the pK_a of the amino group of glycine?

6. The Effect of pH on Solubility The strongly polar hydrogen-bonding nature of water makes it an excellent solvent for ionic (charged) species. By contrast, un-ionized, nonpolar organic molecules, such as benzene, are relatively insoluble in water. In principle, the aqueous solubility of all organic acids or bases can be increased by deprotonation or protonation of the molecules, respectively, to form charged species. For example, the solubility of benzoic acid in water is low. The addition of sodium bicarbonate raises the pH of the solution and deprotonates the benzoic acid to form benzoate ion, which is quite soluble in water.

Benzoic acid
$pK_a \approx 5$

Benzoate ion

Are the molecules in (a) to (c) (below) more soluble in an aqueous solution of 0.1 M NaOH or 0.1 M HCl? (The dissociable protons are shown in red.)

Pyridine ion
$pK_a \approx 5$

(a)

β-Naphthol
$pK_a \approx 10$

(b)

N-Acetyltyrosine methyl ester
$pK_a \approx 10$

(c)

7. Treatment of Poison Ivy Rash Catechols substituted with long-chain alkyl groups are the components of poison ivy and poison oak that produce the characteristic itchy rash.

$(CH_2)_n-CH_3$
$pK_a \approx 8$

If you were exposed to poison ivy, which of the treatments below would you apply to the affected area? Justify your choice.

(a) Wash the area with cold water.

(b) Wash the area with dilute vinegar or lemon juice.

(c) Wash the area with soap and water.

(d) Wash the area with soap, water, and baking soda (sodium bicarbonate).

8. pH and Drug Absorption Aspirin is a weak acid with a pK_a of 3.5.

It is absorbed into the blood through the cells lining the stomach and the small intestine. Absorption requires passage through the cell membrane, which is determined by the polarity of the molecule: charged and highly polar molecules pass slowly, whereas neutral hydrophobic ones pass rapidly. The pH of the gastric juice in the stomach is about 1.5 and the pH of the contents of the small intestine is about 6. Is more aspirin absorbed into the bloodstream from the stomach or from the small intestine? Clearly justify your choice.

9. *Preparation of Standard Buffer for Calibration of a pH Meter* The glass electrode used in commercial pH meters gives an electrical response proportional to the hydrogen-ion concentration. To convert these responses into pH, glass electrodes must be calibrated against standard solutions of known hydrogen-ion concentration. Determine the weight in grams of sodium dihydrogen phosphate ($NaH_2PO_4 \cdot H_2O$; formula weight (FW) 138.01) and disodium hydrogen phosphate (Na_2HPO_4; FW 141.98) needed to prepare 1 L of a standard buffer at pH 7.00 with a total phosphate concentration of 0.100 M (see Table 4–7).

10. *Control of Blood pH by the Rate of Respiration*

(a) The partial pressure of CO_2 in the lungs can be varied rapidly by the rate and depth of breathing. For example, a common remedy to alleviate hiccups is to increase the concentration of CO_2 in the lungs. This can be achieved by holding one's breath, by very slow and shallow breathing (hypoventilation), or by breathing in and out of a paper bag. Under such conditions, the partial pressure of CO_2 in the air space of the lungs rises above normal. Qualitatively explain the effect of these procedures on the blood pH.

(b) A common practice of competitive short-distance runners is to breathe rapidly and deeply (hyperventilation) for about half a minute to remove CO_2 from their lungs just before running in, say, a 100 m dash. Their blood pH may rise to 7.60. Explain why the blood pH goes up.

(c) During a short-distance run the muscles produce a large amount of lactic acid from their glucose stores. In view of this fact, why might hyperventilation before a dash be useful?

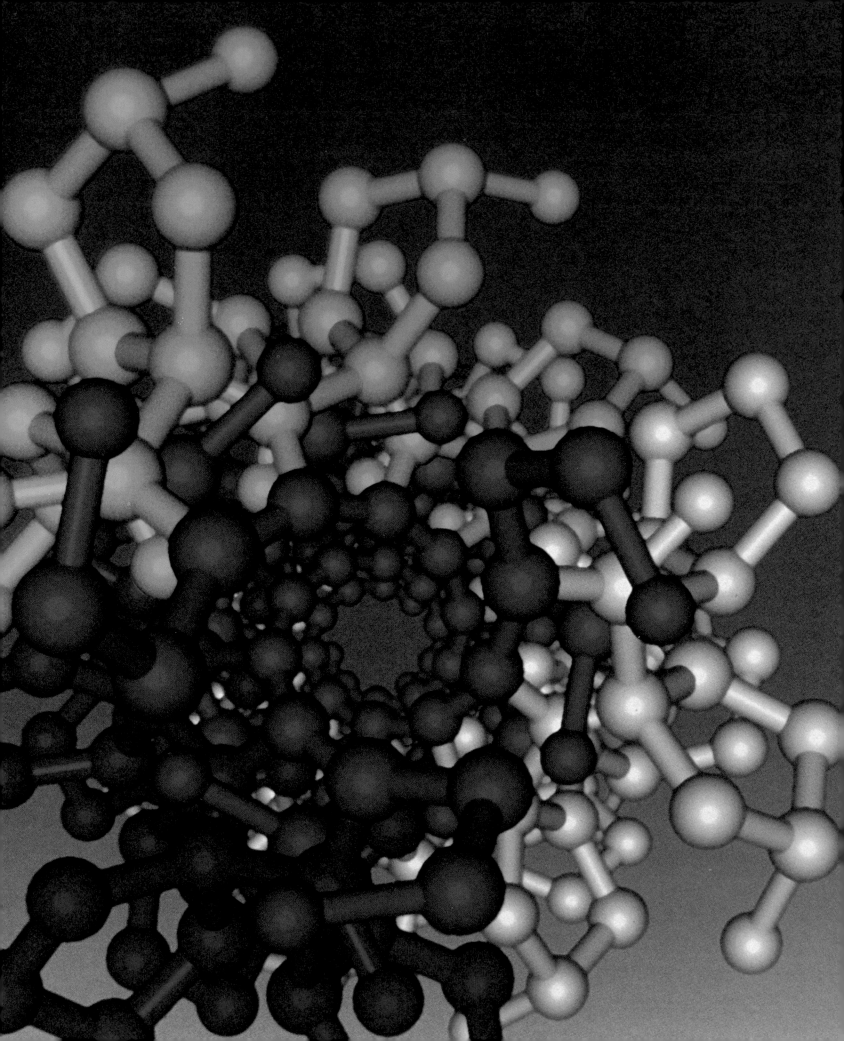

Structure and Catalysis

In Part I we contrasted the complex structure and function of living cells with the relative simplicity of the monomeric units from which the enzymes, supramolecular complexes, and organelles of the cells are constructed. Part II is devoted to the structure and function of the major classes of cellular constituents: amino acids and proteins (Chapters 5 through 8), fatty acids, lipids, and membranes (Chapters 9 and 10), sugars and polysaccharides (Chapter 11), and nucleotides and nucleic acids (Chapter 12). We begin in each case by considering the covalent structure of the simple subunits (amino acids, fatty acids, monosaccharides, and nucleotides). These subunits are a major part of the language of biochemistry; familiarity with them is a prerequisite for understanding more advanced topics covered in this book, as well as the rapidly growing and exciting literature of biochemistry.

After describing the covalent chemistry of the monomeric units, we consider the structure of the macromolecules and supramolecular complexes derived from them. An overriding theme is that the polymeric macromolecules in living systems, though large, are highly ordered chemical entities, with specific sequences of monomeric subunits giving rise to discrete structures and functions. This fundamental theme can be broken down into three interrelated principles: (1) the unique structure of each macromolecule determines its function; (2) noncovalent interactions play a critical role in the structure and function of macromolecules; and (3) the specific sequences of monomeric subunits in polymeric macromolecules contain the information upon which the ordered living state depends. Each of these principles deserves further comment.

The relationship between structure and function is especially evident in proteins, which exhibit an extraordinary diversity of functions. One particular polymeric sequence of amino acids produces a strong, fibrous structure found in hair and wool; another produces a protein that transports oxygen in the blood. Similarly, the special functions of lipids, polysaccharides, and nucleic acids can be understood as a direct manifestation of their chemical structure, with their characteristic monomeric subunits linked in precise functional groups or polymers. Lipids aggregate to form membranes; sugars linked together become energy stores and structural fibers; nucleotides in a polymer become the blueprint for an entire organism.

As we move from monomeric units to larger and larger polymers, the chemical focus shifts from covalent bonds to noncovalent interactions. The covalent nature of monomeric units, and of the bonds that connect them in polymers, places strong constraints upon the shapes

Facing page: End-on view of the triple-stranded collagen superhelix. Collagen, a component of connective tissue, provides tensile strength and resiliency. Its strength is derived in part from the three tightly wrapped identical helical strands (shown in gray, purple, and blue), much the way a length of rope is stronger than its constituent fibers. The tight wrapping is made possible by the presence of glycine, shown in red, at every third position along each strand, where the strands are in contact. Glycine's small size allows for very close contact.

assumed by large molecules. It is the numerous noncovalent interactions, however, that dictate the stable native conformation and provide the flexibility necessary for the biological function of these large molecules. We will see that noncovalent interactions are essential to the catalytic power of enzymes, the arrangement and properties of lipids in a membrane, and the critical interaction of complementary base pairs in nucleic acids.

The principle that sequences of monomeric subunits are information-rich emerges fully in the discussion of nucleic acids in Chapter 12. However, proteins and some polysaccharides are also information-rich molecules. The amino acid sequence is a form of information that directs the folding of the protein into its unique three-dimensional structure, and ultimately determines the function of the protein. Some polysaccharides also have unique sequences and three-dimensional structures that can be recognized by other macromolecules.

For each class of molecules we find a similar structural hierarchy, in which subunits of fixed structure are connected by bonds of limited flexibility, to form macromolecules with three-dimensional structures determined by noncovalent interactions. Together, the molecules described in Part II are the "stuff" of life. We begin with the amino acids.

Amino Acids and Peptides

Proteins are the most abundant macromolecules in living cells, occurring in all cells and all parts of cells. Proteins also occur in great variety; thousands of different kinds may be found in a single cell. Moreover, proteins exhibit great diversity in their biological function. Their central role is made evident by the fact that proteins are the most important final products of the information pathways discussed in Part IV of this book. In a sense, they are the molecular instruments through which genetic information is expressed. It is appropriate to begin the study of biological macromolecules with the proteins, whose name derives from the Greek *prōtos*, meaning "first" or "foremost."

Relatively simple monomeric subunits provide the key to the structure of the thousands of different proteins. All proteins, whether from the most ancient lines of bacteria or from the most complex forms of life, are constructed from the same ubiquitous set of 20 amino acids, covalently linked in characteristic linear sequences. Because each of these amino acids has a distinctive side chain that determines its chemical properties, this group of 20 precursor molecules may be regarded as the alphabet in which the language of protein structure is written.

Proteins are chains of amino acids, each joined to its neighbor by a specific type of covalent bond. What is most remarkable is that cells can produce proteins that have strikingly different properties and activities by joining the same 20 amino acids in many different combinations and sequences. From these building blocks different organisms can make such widely diverse products as enzymes, hormones, antibodies, the lens protein of the eye, feathers, spider webs, rhinoceros horns (Fig. 5–1), milk proteins, antibiotics, mushroom poisons, and a myriad of other substances having distinct biological activities.

Protein structure and function is the topic for the next four chapters. In this chapter we begin with a description of amino acids and the covalent bonds that link them together in peptides and proteins.

Figure 5–1 The protein keratin is formed by all vertebrates. It is the chief structural component of hair, scales, horn, wool, nails, and feathers. The black rhinoceros is nearing extinction in the wild because of the myths prevalent in some parts of the world that a powder derived from its horn has aphrodisiac properties. In reality, the chemical properties are no different from those of powdered bovine hooves or human fingernails.

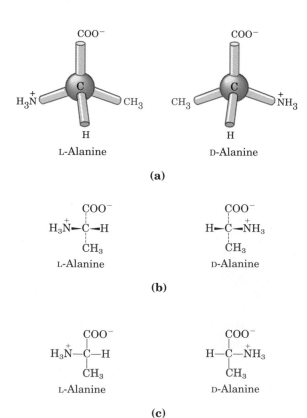

Figure 5–2 General structure of the amino acids found in proteins. With the exception of the nature of the R group, this structure is common to all the α-amino acids of proteins (except proline). The α carbon is shown in blue. R (in red) represents the R group or side chain, which is different in each amino acid. In all amino acids except glycine (shown for comparison) the α-carbon atom has four different substituent groups.

Amino Acids

Proteins can be reduced to their constituent amino acids by a variety of methods, and the earliest studies of proteins naturally focused on the free amino acids derived from them. The first amino acid to be discovered in proteins was asparagine, in 1806. The last of the 20 to be found, threonine, was not identified until 1938. All the amino acids have trivial or common names, in some cases derived from the source from which they were first isolated. Asparagine was first found in asparagus, as one might guess; glutamate was found in wheat gluten; tyrosine was first isolated from cheese (thus its name is derived from the Greek *tyros,* "cheese"); and glycine (Greek *glykos,* "sweet") was so named because of its sweet taste.

Amino Acids Have Common Structural Features

All of the 20 amino acids found in proteins have a carboxyl group and an amino group bonded to the same carbon atom (the α carbon) (Fig. 5–2). They differ from each other in their side chains, or R groups, which vary in structure, size, and electric charge, and influence the solubility of amino acids in water. When the R group contains additional carbons in a chain, they are designated β, γ, δ, ϵ, etc., proceeding out from the α carbon. The 20 amino acids of proteins are often referred to as the standard, primary, or normal amino acids, to distinguish them from amino acids within proteins that are modified after the proteins are synthesized, and from many other kinds of amino acids present in living organisms but not in proteins. The standard amino acids have been assigned three-letter abbreviations and one-letter symbols (Table 5–1), which are used as shorthand to indicate the composition and sequence of amino acids in proteins.

We note in Figure 5–2 that for all the standard amino acids except one (glycine) the α carbon is asymmetric, bonded to four different substituent groups: a carboxyl group, an amino group, an R group, and a hydrogen atom. The α-carbon atom is thus a **chiral center** (see Fig. 3–9). Because of the tetrahedral arrangement of the bonding orbitals around the α-carbon atom of amino acids, the four different substituent groups can occupy two different arrangements in space, which are nonsuperimposable mirror images of each other (Fig. 5–3). These two forms are called **enantiomers** or **stereoisomers** (see Fig. 3–9). All molecules with a chiral center are also **optically active**—i.e., they can rotate plane-polarized light, with the direction of the rotation differing for different stereoisomers.

Figure 5–3 (a) The two stereoisomers of alanine. L- and D-alanine are nonsuperimposable mirror images of each other. **(b, c)** Two different conventions for showing the configurations in space of stereoisomers. In perspective formulas **(b)** the wedge-shaped bonds project out of the plane of the paper, the dashed bonds behind it. In projection formulas **(c)** the horizontal bonds are assumed to project out of the plane of the paper, the vertical bonds behind. However, projection formulas are often used casually without reference to stereochemical configuration.

Table 5–1 Properties and conventions associated with the standard amino acids

Amino acid	Abbreviated names		M_r	pK_1 (—COOH)	pK_2 (—NH_3^+)	pK_R (R group)	pI	Hydropathy index*	Occurrence in Proteins (%)†
Nonpolar, aliphatic R groups									
Glycine	Gly	G	75	2.34	9.60		5.97	−0.4	7.5
Alanine	Ala	A	89	2.34	9.69		6.01	1.8	9.0
Valine	Val	V	117	2.32	9.62		5.97	4.2	6.9
Leucine	Leu	L	131	2.36	9.60		5.98	3.8	7.5
Isoleucine	Ile	I	131	2.36	9.68		6.02	4.5	4.6
Proline	Pro	P	115	1.99	10.96		6.48	−1.6	4.6
Aromatic R groups									
Phenylalanine	Phe	F	165	1.83	9.13		5.48	2.8	3.5
Tyrosine	Tyr	Y	181	2.20	9.11	10.07	5.66	−1.3	3.5
Tryptophan	Trp	W	204	2.38	9.39		5.89	−0.9	1.1
Polar, uncharged R groups									
Serine	Ser	S	105	2.21	9.15	13.60	5.68	−0.8	7.1
Threonine	Thr	T	119	2.11	9.62	13.60	5.87	−0.7	6.0
Cysteine	Cys	C	121	1.96	8.18	10.28	5.07	2.5	2.8
Methionine	Met	M	149	2.28	9.21		5.74	1.9	1.7
Asparagine	Asn	N	132	2.02	8.80		5.41	−3.5	4.4
Glutamine	Gln	Q	146	2.17	9.13		5.65	−3.5	3.9
Negatively charged R groups									
Aspartate	Asp	D	133	1.88	9.60	3.65	2.77	−3.5	5.5
Glutamate	Glu	E	147	2.19	9.67	4.25	3.22	−3.5	6.2
Positively charged R groups									
Lysine	Lys	K	146	2.18	8.95	10.53	9.74	−3.9	7.0
Arginine	Arg	R	174	2.17	9.04	12.48	10.76	−4.5	4.7
Histidine	His	H	155	1.82	9.17	6.00	7.59	−3.2	2.1

* A scale combining hydrophobicity and hydrophilicity; can be used to predict which amino acids will be found in an aqueous environment (− values) and which will be found in a hydrophobic environment (+ values). See Box 10-2. From Kyte, J. & Doolittle, R.F. (1982) *J. Mol. Biol.* **157,** 105–132.

† Average occurrence in over 200 proteins. From Klapper, M.H. (1977) *Biochem. Biophys. Res. Commun.* **78,** 1018–1024.

The classification and naming of stereoisomers is based on the **absolute configuration** of the four substituents of the asymmetric carbon atom. For this purpose a reference compound has been chosen, to which all other optically active compounds are compared. This reference compound is the 3-carbon sugar glyceraldehyde (Fig. 5–4), the smallest sugar to have an asymmetric carbon atom. The naming of configurations of both simple sugars and amino acids is based on the absolute configuration of glyceraldehyde, as established by x-ray diffraction analysis. The stereoisomers of all chiral compounds having a configuration related to that of L-glyceraldehyde are designated L (for levorotatory, derived from *levo,* meaning "left"), and the stereoisomers related to D-glyceraldehyde are designated D (for dextrorotatory, derived from *dextro,* meaning "right"). The symbols L and D thus refer to the absolute configuration of the four substituents around the chiral carbon.

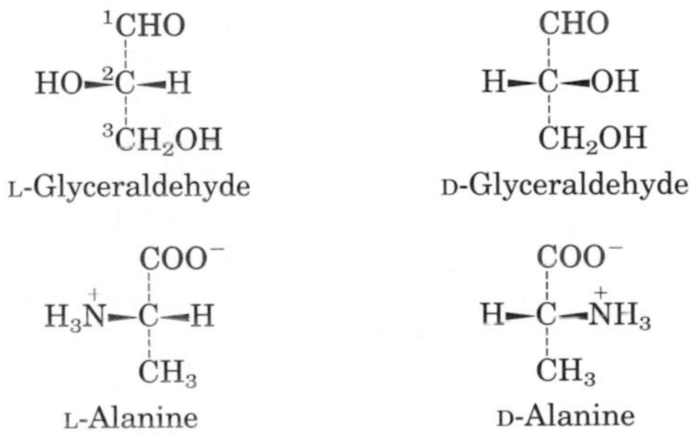

Figure 5–4 Steric relationship of the stereoisomers of alanine to the absolute configuration of L- and D-glyceraldehyde. In these perspective formulas, the carbons are lined up vertically, with the chiral atom in the center. The carbons in these molecules are numbered beginning with the aldehyde or carboxyl carbons on the end, or 1 to 3 from top to bottom as shown. When presented in this way, the R group of the amino acid (in this case the methyl group of alanine) is always below the α carbon. L-Amino acids are those with the α-amino group on the left, and D-amino acids have the α-amino group on the right.

Proteins Contain L-Amino Acids

Nearly all biological compounds with a chiral center occur naturally in only one stereoisomeric form, either D or L. The amino acids in protein molecules are the L stereoisomers. D-Amino acids have been found only in small peptides of bacterial cell walls and in some peptide antibiotics (see Fig. 5–19).

It is remarkable that the amino acids of proteins are all L stereoisomers. As we noted in Chapter 3, when chiral compounds are formed by ordinary chemical reactions, a racemic mixture of D and L isomers results. Whereas the L and D forms of chiral molecules are difficult for a chemist to distinguish and isolate, they are as different as night and day to a living system. The ability of cells to specifically synthesize the L isomer of amino acids reflects one of many extraordinary properties of enzymes (Chapter 8). The stereospecificity of the reactions catalyzed by some enzymes is made possible by the asymmetry of their active sites. The characteristic three-dimensional structures of proteins (Chapter 7), which dictate their diverse biological activities, require that *all* their constituent amino acids be of one stereochemical series.

Amino Acids Are Ionized in Aqueous Solutions

Amino acids in aqueous solution are ionized and can act as acids or bases. Knowledge of the acid–base properties of amino acids is extremely important in understanding the physical and biological properties of proteins. Moreover, the technology of separating, identifying, and quantifying the different amino acids, which are necessary steps in determining the amino acid composition and sequence of protein molecules, is based largely on their characteristic acid–base behavior.

Those α-amino acids having a single amino group and a single carboxyl group crystallize from neutral aqueous solutions as fully ionized species known as **zwitterions** (German for "hybrid ions"), each having both a positive and a negative charge (Fig. 5–5). These ions are electrically neutral and remain stationary in an electric field. The dipolar nature of amino acids was first suggested by the observation that crystalline amino acids have melting points much higher than those of other organic molecules of similar size. The crystal lattice of amino acids is held together by strong electrostatic forces between positively and negatively charged functional groups of neighboring molecules, resembling the stable ionic crystal lattice of NaCl (see Fig. 4–6).

Figure 5–5 Nonionic and zwitterionic forms of amino acids. Note the separation of the + and − charges in the zwitterion, which makes it an electric dipole. The nonionic form does not occur in significant amounts in aqueous solutions. The zwitterion predominates at neutral pH.

Amino Acids Can Be Classified by R Group

An understanding of the chemical properties of the standard amino acids is central to an understanding of much of biochemistry. The topic can be simplified by grouping the amino acids into classes based on the properties of their R groups (Table 5–1), in particular, their **polarity** or tendency to interact with water at biological pH (near pH 7.0). The polarity of the R groups varies widely, from totally nonpolar or hydrophobic (water-insoluble) to highly polar or hydrophilic (water-soluble).

The structures of the 20 standard amino acids are shown in Figure 5–6, and many of their properties are listed in Table 5–1. There are five main classes of amino acids, those whose R groups are: nonpolar and aliphatic; aromatic (generally nonpolar); polar but uncharged; negatively charged; and positively charged. Within each class there are gradations of polarity, size, and shape of the R groups.

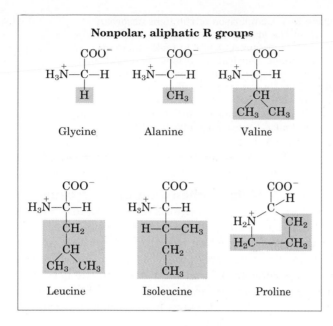

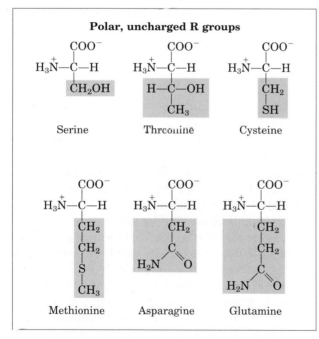

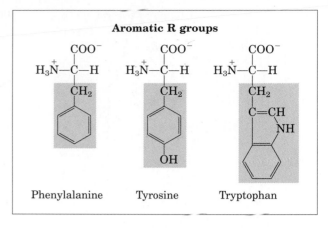

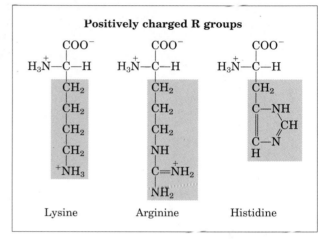

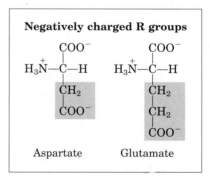

Figure 5–6 The 20 standard amino acids of proteins. They are shown with their amino and carboxyl groups ionized, as they would occur at pH 7.0. The portions in black are those common to all the amino acids; the portions shaded in red are the R groups.

Nonpolar, Aliphatic R Groups The hydrocarbon R groups in this class of amino acids are nonpolar and hydrophobic (Fig. 5–6). The bulky side chains of **alanine, valine, leucine,** and **isoleucine,** with their distinctive shapes, are important in promoting hydrophobic interactions within protein structures. **Glycine** has the simplest amino acid structure. Where it is present in a protein, the minimal steric hindrance of the glycine side chain allows much more structural flexibility than the other amino acids. **Proline** represents the opposite structural extreme. The secondary amino (imino) group is held in a rigid conformation that reduces the structural flexibility of the protein at that point.

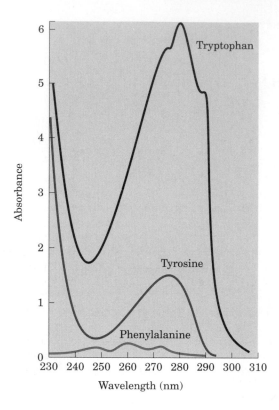

Absorbance

Wavelength (nm)

Figure 5–7 Comparison of the light absorbance spectra of the aromatic amino acids at pH 6.0. The amino acids are present in equimolar amounts (10^{-3} M) under identical conditions. The light absorbance of tryptophan is as much as fourfold higher than that of tyrosine. Phenylalanine absorbs less light than either tryptophan or tyrosine. Note that the absorbance maximum for tryptophan and tyrosine occurs near a wavelength of 280 nm.

Aromatic R Groups **Phenylalanine, tyrosine,** and **tryptophan,** with their aromatic side chains (Fig. 5–6), are relatively nonpolar (hydrophobic). All can participate in hydrophobic interactions, which are particularly strong when the aromatic groups are stacked on one another. The hydroxyl group of tyrosine can form hydrogen bonds, and it acts as an important functional group in the activity of some enzymes. Tyrosine and tryptophan are significantly more polar than phenylalanine because of the tyrosine hydroxyl group and the nitrogen of the tryptophan indole ring.

Tryptophan and tyrosine, and to a lesser extent phenylalanine, absorb ultraviolet light (Fig. 5–7 and Box 5–1). This accounts for the characteristic strong absorbance of light by proteins at a wavelength of 280 nm, and is a property exploited by researchers in the characterization of proteins.

Polar, Uncharged R Groups The R groups of these amino acids (Fig. 5–6) are more soluble in water, or hydrophilic, than those of the nonpolar amino acids, because they contain functional groups that form hydrogen bonds with water. This class of amino acids includes **serine, threonine, cysteine, methionine, asparagine,** and **glutamine.** The polarity of serine and threonine is contributed by their hydroxyl groups; that of cysteine and methionine by their sulfur atom; and that of asparagine and glutamine by their amide groups.

Asparagine and glutamine are the amides of two other amino acids also found in proteins, aspartate and glutamate, respectively, to which asparagine and glutamine are easily hydrolyzed by acid or base. Cysteine has an R group (a thiol group) that is approximately as acidic as the hydroxyl group of tyrosine. Cysteine requires special mention for another reason. It is readily oxidized to form a covalently linked dimeric amino acid called **cystine,** in which two cysteine molecules are joined by a disulfide bridge. Disulfide bridges of this kind occur in many proteins, stabilizing their structures.

Negatively Charged (Acidic) R Groups The two amino acids having R groups with a net negative charge at pH 7.0 are **aspartate** and **glutamate,** each with a second carboxyl group (Fig. 5–6). These amino acids are the parent compounds of asparagine and glutamine, respectively.

Positively Charged (Basic) R Groups The amino acids in which the R groups have a net positive charge at pH 7.0 are **lysine,** which has a second amino group at the ϵ position on its aliphatic chain; **arginine,** which has a positively charged guanidino group; and **histidine,** containing an imidazole group (Fig. 5–6). Histidine is the only standard amino acid having a side chain with a pK_a near neutrality.

COO⁻
H₃N⁺—C—H
CH₂—SH
Cysteine

COO⁻
H₃N⁺—C—H
HS—CH₂
Cysteine

2H ↕ 2H

COO⁻   COO⁻
H₃N⁺—C—H H₃N⁺—C—H
CH₂—S—S—CH₂
Cystine

BOX 5-1   **Absorption of Light by Molecules: The Lambert–Beer Law**

Measurement of light absorption is an important tool for analysis of many biological molecules. The fraction of the incident light absorbed by a solution at a given wavelength is related to the thickness of the absorbing layer (path length) and the concentration of the absorbing species. These two relationships are combined into the Lambert–Beer law, given in integrated form as

$$\log \frac{I_0}{I} = \epsilon c l$$

where I_0 is the intensity of the incident light, I is the intensity of the transmitted light, ϵ is the molar absorption coefficient (in units of liters per mole-centimeter), c the concentration of the absorbing species (in moles per liter), and l the path length of the light-absorbing sample (in centimeters). The Lambert–Beer law assumes that the incident light is parallel and monochromatic and that the solvent and solute molecules are randomly

oriented. The expression $\log (I_0/I)$ is called the absorbance, designated A.

It is important to note that each millimeter path length of absorbing solution in a 1.0 cm cell absorbs not a constant amount but a constant fraction of the incident light. However, with an absorbing layer of fixed path length, *the absorbance A is directly proportional to the concentration of the absorbing solute.*

The molar absorption coefficient varies with the nature of the absorbing compound, the solvent, the wavelength, and also with pH if the light-absorbing species is in equilibrium with another species having a different spectrum through gain or loss of protons.

In practice, absorbance measurements are usually made on a set of standard solutions of known concentration at a fixed wavelength. A sample of unknown concentration can then be compared with the resulting standard curve, as shown in Figure 1.

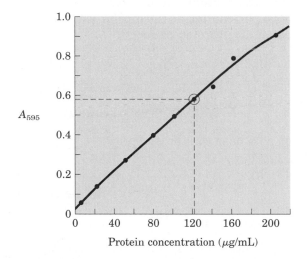

Figure 1 Eight standard solutions containing known amounts of protein and one sample containing an unknown amount of protein were reacted with the Bradford reagent. This reagent contains a dye that shifts its absorption maximum to 595 nm when it binds amino acid residues. The A_{595} (absorbance at 595 nm) of the standard samples was plotted against the protein concentration to create the standard curve, shown here. The A_{595} of the unknown sample, 0.58, corresponds to a protein concentration of 122 μg/mL.

Cells Also Contain Nonstandard Amino Acids

In addition to the 20 standard amino acids that are common in all proteins, other amino acids have been found as components of only certain types of proteins (Fig. 5–8a). Each of these is derived from one of the 20 standard amino acids, in a modification reaction that occurs after the standard amino acid has been inserted into a protein. Among the nonstandard amino acids are **4-hydroxyproline,** a derivative of proline, and **5-hydroxylysine;** the former is found in plant cell-wall proteins, and both are found in the fibrous protein collagen of connective tissues. ***N*-Methyllysine** is found in myosin, a contractile protein of muscle. Another important nonstandard amino acid is

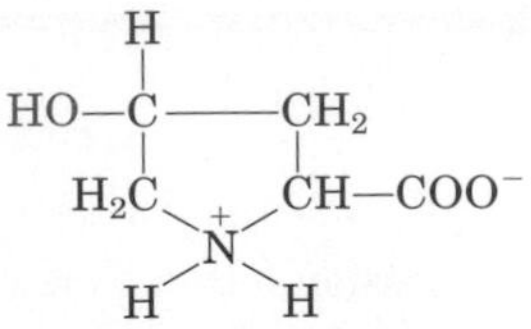

4-Hydroxyproline

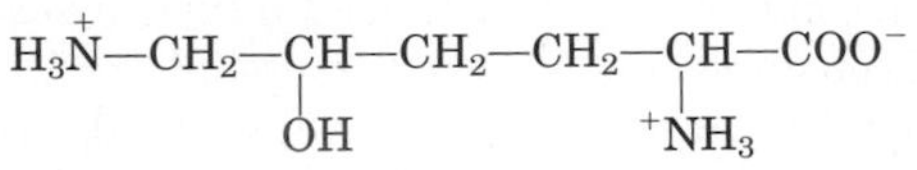

5-Hydroxylysine

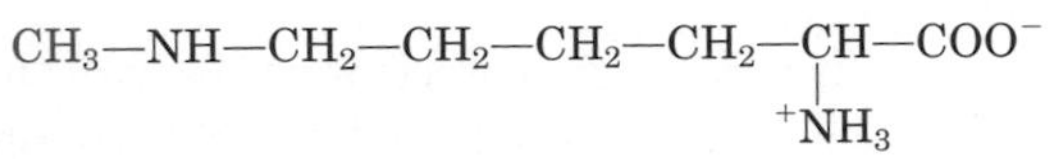

$H_3\overset{+}{N}-CH_2-CH-CH_2-CH_2-CH-COO^-$ with OH and $^+NH_3$

6-N-Methyllysine

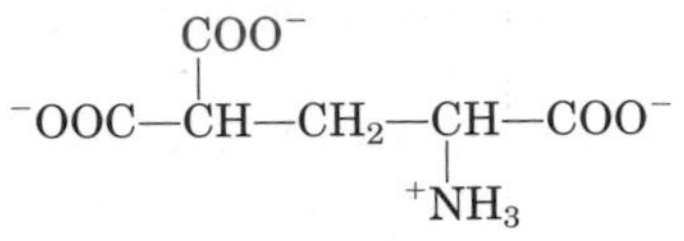

$CH_3-NH-CH_2-CH_2-CH_2-CH_2-CH-COO^-$ with $^+NH_3$

γ-Carboxyglutamate

$^-OOC-CH-CH_2-CH-COO^-$ with COO$^-$ and $^+NH_3$

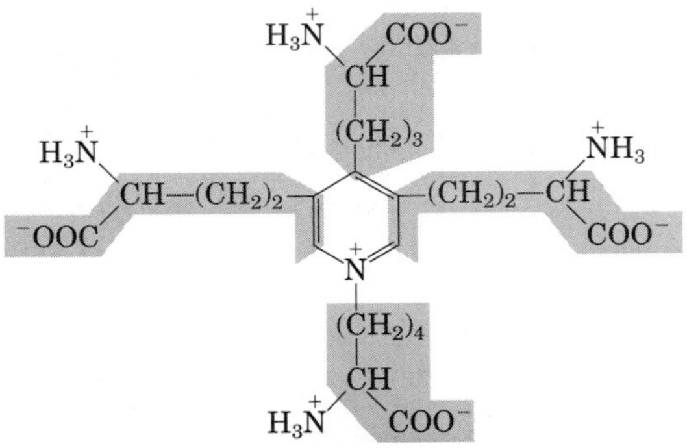

Desmosine

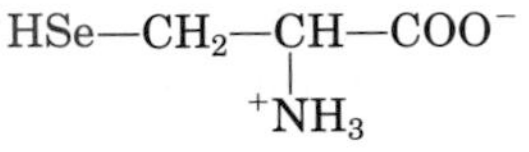

Selenocysteine

$HSe-CH_2-CH-COO^-$ with $^+NH_3$

(a)

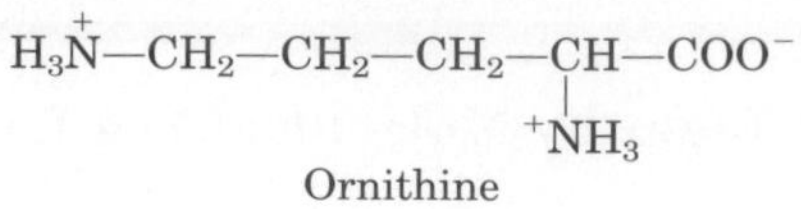

Ornithine

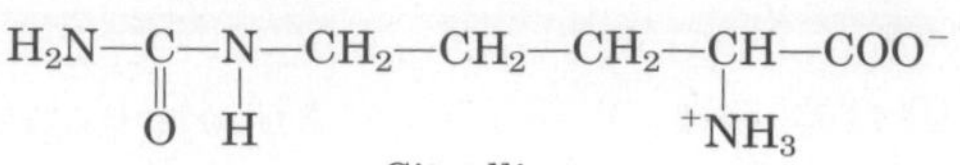

$H_3\overset{+}{N}-CH_2-CH_2-CH_2-CH-COO^-$ with $^+NH_3$

Citrulline

$H_2N-\underset{O}{\overset{}{C}}-\underset{H}{\overset{}{N}}-CH_2-CH_2-CH_2-CH-COO^-$ with $^+NH_3$

(b)

Figure 5–8 (a) Some nonstandard amino acids found in proteins; all are derived from standard amino acids. The extra functional groups are shown in red. Desmosine is formed from four residues of lysine, whose carbon backbones are shaded in gray. Selenocysteine is derived from serine. **(b)** Ornithine and citrulline are intermediates in the biosynthesis of arginine and in the urea cycle. Note that two systems are used to number carbons in the naming of these amino acids. The α, β, γ system used for γ-carboxyglutamate begins at the α carbon (see Fig. 5–2) and extends into the R group. The α-carboxyl group is not included. In contrast, the numbering system used to identify the modified carbon in 4-hydroxyproline, 5-hydroxylysine, and 6-N-methyllysine includes the α-carboxyl carbon, which is designated carbon 1 (or C-1).

γ-carboxyglutamate, found in the blood-clotting protein prothrombin as well as in certain other proteins that bind Ca^{2+} in their biological function. More complicated is the nonstandard amino acid **desmosine,** a derivative of four separate lysine residues, found in the fibrous protein elastin. **Selenocysteine** contains selenium rather than the oxygen of serine, and is found in glutathione peroxidase and a few other proteins.

Some 300 additional amino acids have been found in cells and have a variety of functions but are not substituents of proteins. **Ornithine** and **citrulline** (Fig. 5–8b) deserve special note because they are key intermediates in the biosynthesis of arginine and in the urea cycle. These pathways are described in Chapters 21 and 17, respectively.

Amino Acids Can Act as Acids and as Bases

When a crystalline amino acid, such as alanine, is dissolved in water, it exists in solution as the dipolar ion, or zwitterion, which can act either as an acid (proton donor):

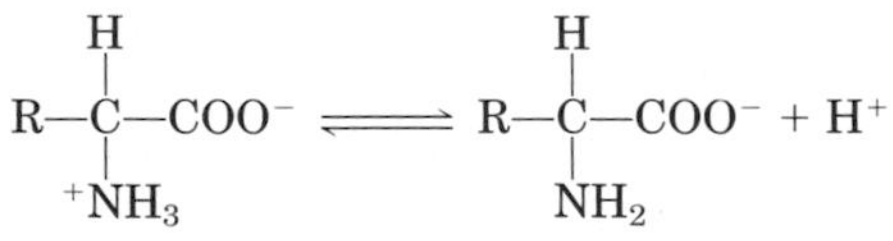

$$R-\underset{^+NH_3}{\overset{H}{\underset{|}{\overset{|}{C}}}}-COO^- \rightleftharpoons R-\underset{NH_2}{\overset{H}{\underset{|}{\overset{|}{C}}}}-COO^- + H^+$$

or as a base (proton acceptor):

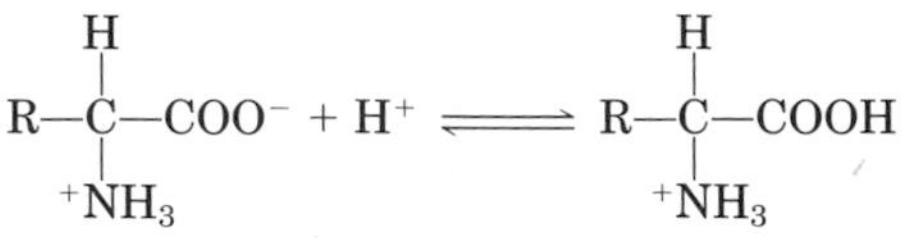

$$R-\underset{^+NH_3}{\overset{H}{\underset{|}{\overset{|}{C}}}}-COO^- + H^+ \rightleftharpoons R-\underset{^+NH_3}{\overset{H}{\underset{|}{\overset{|}{C}}}}-COOH$$

Substances having this dual nature are **amphoteric** and are often called **ampholytes,** from "amphoteric electrolytes." A simple monoamino monocarboxylic α-amino acid, such as alanine, is actually a diprotic acid when it is fully protonated, that is, when both its carboxyl group and amino group have accepted protons. In this form it has two groups that can ionize to yield protons, as indicated in the following equation:

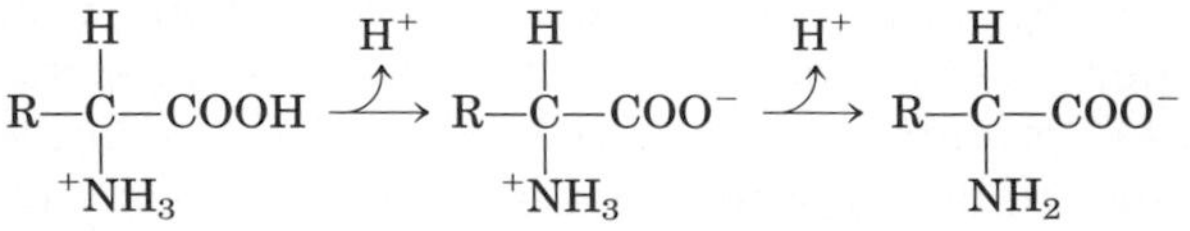

$$R-\underset{^+NH_3}{\overset{H}{\underset{|}{\overset{|}{C}}}}-COOH \xrightarrow{H^+} R-\underset{^+NH_3}{\overset{H}{\underset{|}{\overset{|}{C}}}}-COO^- \xrightarrow{H^+} R-\underset{NH_2}{\overset{H}{\underset{|}{\overset{|}{C}}}}-COO^-$$

Amino Acids Have Characteristic Titration Curves

Titration involves the gradual addition or removal of protons. Figure 5–9 shows the titration curve of the diprotic form of glycine. Each molecule of added base results in the net removal of one proton from one molecule of amino acid. The plot has two distinct stages, each corresponding to the removal of one proton from glycine. Each of the two stages resembles in shape the titration curve of a monoprotic acid, such as acetic acid (see Fig. 4–10), and can be analyzed in the same way. At very low pH, the predominant ionic species of glycine is $^+H_3N—CH_2—COOH$, the fully protonated form. At the midpoint in the first stage of the titration, in which the —COOH group of glycine loses its proton, equimolar concentrations of proton-donor ($^+H_3N—CH_2—COOH$) and proton-acceptor ($^+H_3N—CH_2—COO^-$) species are present. At the midpoint of a titration (see Fig. 4–11), the pH is equal to the pK_a of the protonated group being titrated. For glycine, the pH at the midpoint is 2.34, thus its —COOH group has a pK_a of 2.34. [Recall that pH and pK_a are simply convenient notations for proton concentration and the equilibrium constant for ionization, respectively (Chapter 4). The pK_a is a measure of the tendency of a group to give up a proton, with that tendency decreasing tenfold as the pK_a increases by one unit.] As the titration proceeds, another important point is reached at pH 5.97. Here there is a point of inflection, at which removal of the first proton is essentially complete, and removal of the second has just begun. At this pH the glycine is present largely as the dipolar ion $^+H_3N—CH_2—COO^-$. We shall return to the significance of this inflection point in the titration curve shortly.

The second stage of the titration corresponds to the removal of a proton from the —NH_3^+ group of glycine. The pH at the midpoint of this stage is 9.60, equal to the pK_a for the —NH_3^+ group. The titration is complete at a pH of about 12, at which point the predominant form of glycine is $H_2N—CH_2—COO^-$.

From the titration curve of glycine we can derive several important pieces of information. First, it gives a quantitative measure of the pK_a of each of the two ionizing groups, 2.34 for the —COOH group and 9.60 for the —NH_3^+ group. Note that the carboxyl group of glycine is over 100 times more acidic (more easily ionized) than the carboxyl group of acetic acid, which has a pK_a of 4.76. This effect is caused by the nearby positively charged amino group on the α-carbon atom, as described in Figure 5–10.

The second piece of information given by the titration curve of glycine (Fig. 5–9) is that this amino acid has *two* regions of buffering power (see Fig. 4–12). One of these is the relatively flat portion of the curve centered about the first pK_a of 2.34, indicating that glycine is a good buffer near this pH. The other buffering zone extends for ~1.2 pH units centered around pH 9.60. Note also that glycine is not a good buffer at the pH of intracellular fluid or blood, about 7.4.

The Henderson–Hasselbalch equation (Chapter 4) can be used to calculate the proportions of proton-donor and proton-acceptor species of glycine required to make a buffer at a given pH within the buffering ranges of glycine; it also makes it possible to solve other kinds of buffer problems involving amino acids (see Box 4–2).

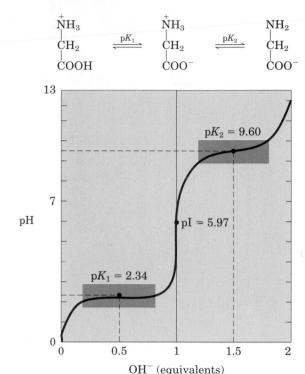

Figure 5–9 The titration curve of 0.1 M glycine at 25 °C. The ionic species predominating at key points in the titration are shown above the graph. The shaded boxes, centered about $pK_1 = 2.34$ and $pK_2 = 9.60$, indicate the regions of greatest buffering power.

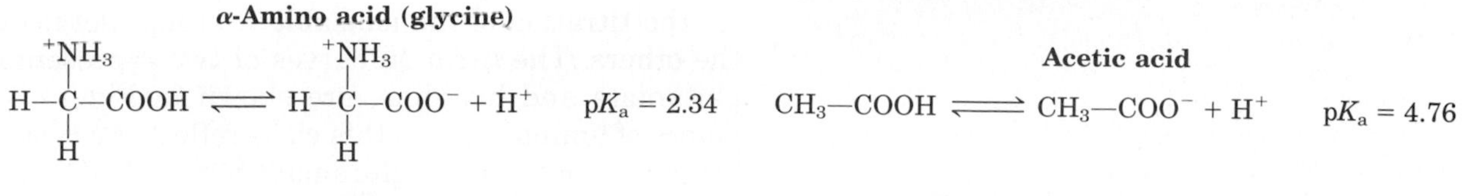

Figure 5–10 (a) Interactions between the α-amino and α-carboxyl groups in an α-amino acid. The nearby positive charge of the —NH_3^+ group makes ionization of the carboxyl group more likely (i.e., lowers the pK_a for —COOH). This is due to a stabilizing interaction between opposite charges on the zwitterion and a repulsive interaction between the positive charges of the amino group and the departing proton. **(b)** The normal pK_a for a carboxyl group is approximately 4.76, as for acetic acid.

α-Amino acid (glycine)

$^+NH_3$ ... $^+NH_3$
H—C—COOH ⇌ H—C—COO$^-$ + H$^+$ $pK_a = 2.34$
H ... H

(a)

Acetic acid

$CH_3—COOH$ ⇌ $CH_3—COO^- + H^+$ $pK_a = 4.76$

(b)

The Titration Curve Predicts the Electric Charge of Amino Acids

Another important piece of information derived from the titration curve of an amino acid is the relationship between its net electric charge and the pH of the solution. At pH 5.97, the point of inflection between the two stages in its titration curve, glycine is present as its dipolar form, fully ionized but with no *net* electric charge (Fig. 5–9). This characteristic pH is called the **isoelectric point** or **isoelectric pH,** designated **pI** or pH_I. For an amino acid such as glycine, which has no ionizable group in the side chain, the isoelectric point is the arithmetic mean of the two pK_a values:

$$pI = \frac{1}{2}(pK_1 + pK_2)$$

which in the case of glycine is

$$pI = \frac{1}{2}(2.34 + 9.60) = 5.97$$

As is evident in Figure 5–9, glycine has a net negative charge at any pH above its pI and will thus move toward the positive electrode (the anode) when placed in an electric field. At any pH below its pI, glycine has a net positive charge and will move toward the negative electrode, the cathode. The farther the pH of a glycine solution is from its isoelectric point, the greater the net electric charge of the population of glycine molecules. At pH 1.0, for example, glycine exists entirely as the form $^+H_3N—CH_2—COOH$, with a net positive charge of 1.0. At pH 2.34, where there is an equal mixture of $^+H_3N—CH_2—COOH$ and $^+H_3N—CH_2—COO^-$, the average or net positive charge is 0.5. The sign and the magnitude of the net charge of any amino acid at any pH can be predicted in the same way.

This information has practical importance. For a solution containing a mixture of amino acids, the different amino acids can be separated on the basis of the direction and relative rate of their migration when placed in an electric field at a known pH.

Amino Acids Differ in Their Acid–Base Properties

The shared properties of many amino acids permit some simplifying generalizations about the acid–base behavior of different classes of amino acids.

All amino acids with a single α-amino group, a single α-carboxyl group, and an R group that does not ionize have titration curves resembling that of glycine (Fig. 5–9). This group of amino acids is characterized by having very similar, although not identical, values for pK_1 (the pK of the —COOH group) in the range of 1.8 to 2.4 and for pK_2 (of the —NH_3^+ group) in the range of 8.8 to 11.0 (Table 5–1).

Amino acids with an ionizable R group (Table 5–1) have more complex titration curves with *three* stages corresponding to the three possible ionization steps; thus they have three pK_a values. The third stage for the titration of the ionizable R group merges to some extent with the others. The titration curves of two representatives of this group, glutamate and histidine, are shown in Figure 5–11. The isoelectric points of amino acids in this class reflect the type of ionizing R groups present. For example, glutamate has a pI of 3.22, considerably lower than that of glycine. This is a result of the presence of two carboxyl

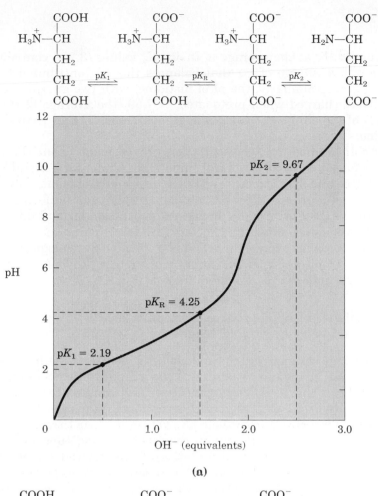

(a)

Figure 5–11 The titration curves of **(a)** glutamate and **(b)** histidine. The pK_a of the R group is designated pK_R.

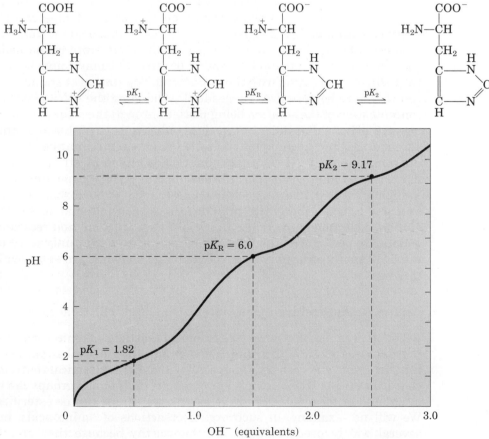

(b)

groups which, at the average of their pK_a values (3.22), contribute a net negative charge of -1 that balances the $+1$ contributed by the amino group. Similarly, the pI of histidine, with two groups that are positively charged when protonated, is 7.59 (the average of the pK_a values of the amino and imidazole groups), much higher than that of glycine.

Another important generalization can be made about the acid–base behavior of the 20 standard amino acids. Under the general condition of free and open exposure to the aqueous environment, only histidine has an R group ($pK_a = 6.0$) providing significant buffering power near the neutral pH usually found in the intracellular and intercellular fluids of most animals and bacteria. All the other amino acids have pK_a values too far away from pH 7 to be effective physiological buffers (Table 5–1), although in the interior of proteins the pK_a values of amino acid side chains are often altered.

Ion-Exchange Chromatography Separates Amino Acids by Electric Charge

Ion-exchange chromatography is the most widely used method for separating, identifying, and quantifying the amounts of each amino acid in a mixture. This technique primarily exploits differences in the sign and magnitude of the net electric charges of amino acids at a given pH, which are predictable from their pK_a values or their titration curves.

The chromatographic column consists of a long tube filled with particles of a synthetic resin containing fixed charged groups; those with fixed anionic groups are called **cation-exchange** resins and those with fixed cationic groups, **anion-exchange** resins. A simple form of ion-exchange chromatography on a cation-exchange resin is described in Figure 5–12. The affinity of each amino acid for the resin is affected by pH (which determines the ionization state of the molecule) and the concentration of other salt ions that may compete with the resin by associating with the amino acid. Separation of amino acids can therefore be optimized by gradually changing the pH and/or salt concentration of the solution being passed through the column so as to create a pH or salt gradient. A modern enhancement of this and other chromatographic techniques is called **high-performance liquid chromatography** (HPLC). This takes advantage of stronger resin material and improved apparatus designed to permit chromatography at high pressures, allowing better separations in a much shorter time. For amino acids, the entire procedure has been automated, so that elution, collection of fractions, analysis of each fraction, and recording of data are performed automatically in an **amino-acid analyzer.** Figure 5–13 shows a chromatogram of an amino acid mixture analyzed in this way.

Amino Acids Undergo Characteristic Chemical Reactions

As for all organic compounds, the chemical reactions of amino acids are those characteristic of their functional groups. Because all amino acids contain amino and carboxyl groups, all will undergo chemical reactions characteristic for these groups. For example, their amino groups can be acetylated or formylated, and their carboxyl groups can be esterified. We will not examine all such organic reactions of amino acids, but several widely used reactions are noteworthy because they greatly simplify the detection, measurement, and identification of amino acids.

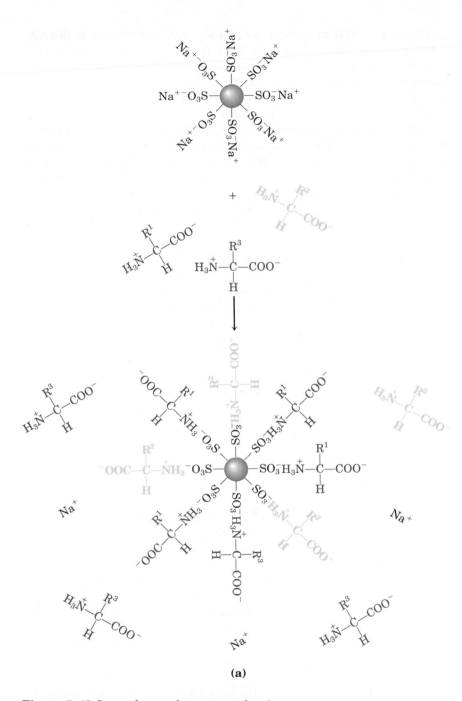

(a)

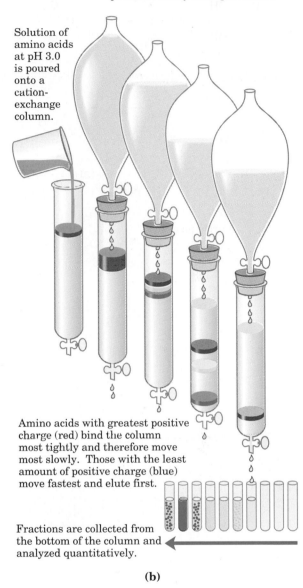

Reservoir of buffer allows sample to percolate slowly through column.

Solution of amino acids at pH 3.0 is poured onto a cation-exchange column.

Amino acids with greatest positive charge (red) bind the column most tightly and therefore move most slowly. Those with the least amount of positive charge (blue) move fastest and elute first.

Fractions are collected from the bottom of the column and analyzed quantitatively.

(b)

Figure 5–12 Ion-exchange chromatography. An example of a cation-exchange resin is presented. **(a)** Negatively charged sulfonate groups ($-SO_3^-$) on the resin surface attract and bind cations, such as H^+, Na^+, or cationic forms of amino acids. **(b)** An acidic solution (pH 3.0) of the amino acid mixture is poured on a column packed with resin and allowed to percolate through slowly. At pH 3.0 the amino acids are largely cations with net positive charges, but they differ in the pK_a values of their R groups, and hence in the extent to which they are ionized and in their tendency to bind to the anionic resin. As a result, they move through the column at different rates.

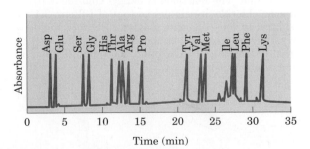

Figure 5–13 Automatically recorded high-performance liquid chromatographic analysis of amino acids on a cation-exchange resin. The area under each peak on the chromatogram is proportional to the amount of each amino acid in the mixture.

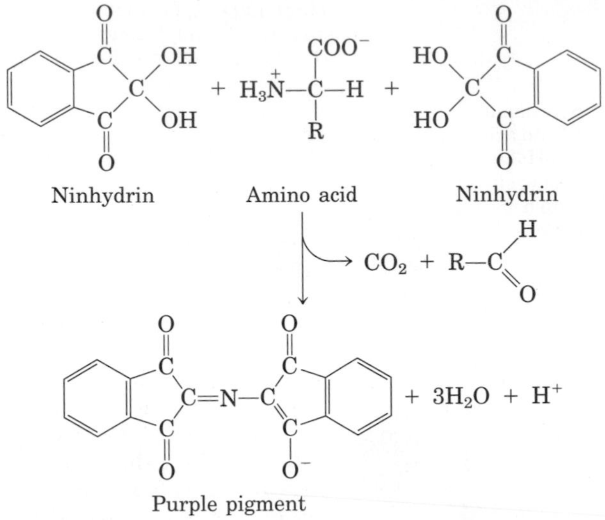

Ninhydrin   Amino acid   Ninhydrin

$\rightarrow CO_2 + R-C\begin{smallmatrix}H\\\\O\end{smallmatrix}$

$+ 3H_2O + H^+$

Purple pigment

One of the most important, technically and historically, is the ninhydrin reaction, which has been used for many years to detect and quantify microgram amounts of amino acids. When amino acids are heated with excess ninhydrin, all those having a free α-amino group yield a purple product. Proline, in which the α-amino group is substituted (forming an imino group), yields a yellow product. Under appropriate conditions the intensity of color produced (optical absorbance of the solution; see Box 5–1) is proportional to the amino acid concentration. Comparing the absorbance to that of appropriate standard solutions is an accurate and technically simple method for measuring amino acid concentration.

Several other convenient reagents are available that react with the α-amino group to form colored or fluorescent derivatives. Unlike ninhydrin, these have the advantage that the intact R group of the amino acid remains part of the product, so that derivatives of different amino acids can be distinguished. Fluorescamine reacts rapidly with amino acids and provides great sensitivity, yielding a highly fluorescent derivative that permits the detection of nanogram quantities of amino acids (Fig. 5–14). Dabsyl chloride, dansyl chloride, and 1-fluoro-2,4-dinitrobenzene (Fig. 5–14) yield derivatives that are stable under harsh conditions such as those used in the hydrolysis of proteins.

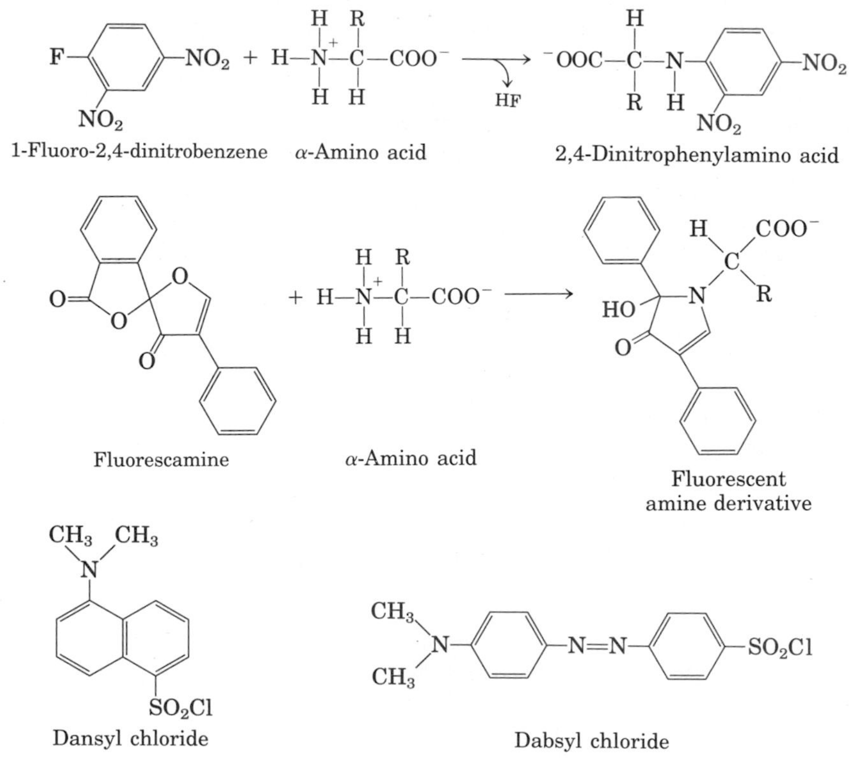

1-Fluoro-2,4-dinitrobenzene α-Amino acid   2,4-Dinitrophenylamino acid

Fluorescamine α-Amino acid   Fluorescent amine derivative

Figure 5–14 Reagents that react with the α-amino group of amino acids. The reactions producing 2,4-dinitrophenyl and fluorescamine derivatives are illustrated. The reactions of dansyl chloride and dabsyl chloride are similar to that of 1-fluoro-2,4-dinitrobenzene (Sanger's reagent). Because the derivatives of these reagents absorb light, they greatly facilitate the detection and quantification of the amino acids.

Dansyl chloride

Dabsyl chloride

Peptides

We now turn to polymers of amino acids, the **peptides.** Biologically occurring peptides range in size from small molecules containing only two or three amino acids to macromolecules containing thousands of amino acids. The focus here is on the structure and chemical properties of the smaller peptides, providing a prelude to the discussion of the large peptides called proteins in the next two chapters.

Peptides Are Chains of Amino Acids

Two amino acid molecules can be covalently joined through a substituted amide linkage, termed a **peptide bond,** to yield a dipeptide. Such a linkage is formed by removal of the elements of water from the α-carboxyl group of one amino acid and the α-amino group of another (Fig. 5–15). Peptide-bond formation is an example of a condensation reaction, a common class of reaction in living cells. Note that as shown in Figure 5–15, this reaction has an equilibrium that favors reactants rather than products. To make the reaction thermodynamically more favorable, the carboxyl group must be chemically modified or activated so that the hydroxyl group can be more readily eliminated. A chemical approach to this problem is outlined at the end of this chapter (see Box 5–2). The biological approach to peptide bond formation is a major topic of Chapter 26.

Three amino acids can be joined by two peptide bonds to form a tripeptide; similarly, amino acids can be linked to form tetrapeptides and pentapeptides. When a few amino acids are joined in this fashion, the structure is called an **oligopeptide.** When many amino acids are joined, the product is called a **polypeptide.** Proteins may have thousands of amino acid units. Although the terms "protein" and "polypeptide" are sometimes used interchangeably, molecules referred to as polypeptides generally have molecular weights below 10,000.

Figure 5–16 shows the structure of a pentapeptide. The amino acid units in a peptide are often called **residues** (each has lost a hydrogen atom from its amino group and a hydroxyl moiety from its carboxyl group). The amino acid residue at that end of a peptide having a free α-amino group is the **amino-terminal** (or N-terminal) residue; the residue at the other end, which has a free carboxyl group, is the **carboxyl-terminal** (C-terminal) residue. By convention, short peptides are named from the sequence of their constituent amino acids, beginning at the left with the amino-terminal residue and proceeding toward the carboxyl terminus at the right (Fig. 5–16).

Although hydrolysis of peptide bonds is an exergonic reaction, it occurs slowly because of its high activation energy. As a result, the peptide bonds in proteins are quite stable under most intracellular conditions.

The peptide bond is the single most important covalent bond linking amino acids in peptides and proteins. The only other type of covalent bond that occurs frequently enough to deserve special mention here is the disulfide bond sometimes formed between two cysteine residues. Disulfide bonds play a special role in the structure of many proteins, particularly those that function extracellularly, such as the hormone insulin and the immunoglobulins or antibodies.

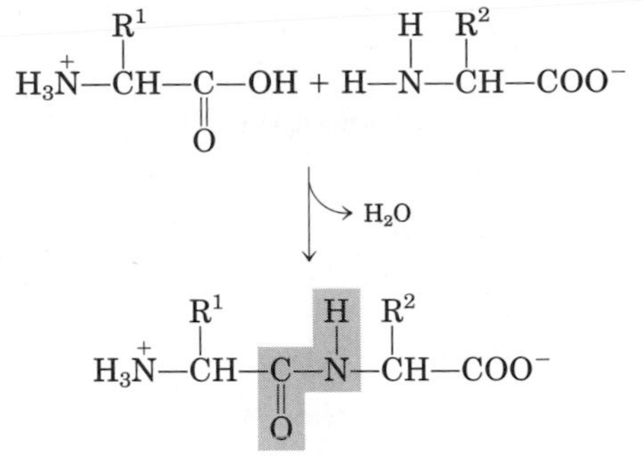

Figure 5–15 Formation of a peptide bond (shaded in gray) in a dipeptide. This is a condensation reaction. The α-amino group of amino acid 2 acts as a nucleophile (see Table 3–6) to displace the hydroxyl group of amino acid 1 (red). Amino groups are good nucleophiles, but the hydroxyl group is a poor leaving group and is not readily displaced. At physiological pH the reaction as shown does not occur to any appreciable extent. Peptide bond formation is endergonic, with a free-energy change of about +21 kJ/mol.

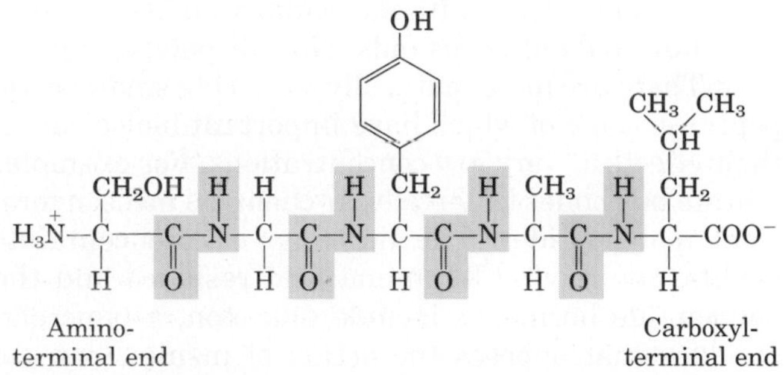

Amino-terminal end   Carboxyl-terminal end

Figure 5–16 Structure of the pentapeptide serylglycyltyrosinylalanylleucine, or Ser–Gly–Tyr–Ala–Leu. Peptides are named beginning with the amino-terminal residue, which by convention is placed at the left. The peptide bonds are shown shaded in gray, the R groups in red.

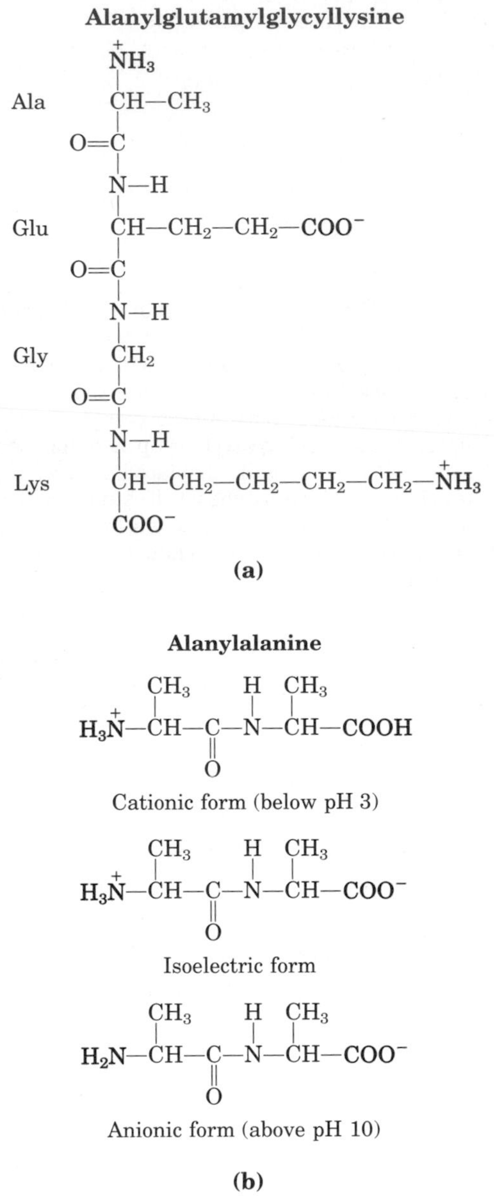

Alanylglutamylglycyllysine

$$\overset{+}{\text{NH}_3}$$

Ala $\text{CH}-\text{CH}_3$

$\text{O}=\text{C}$

$\text{N}-\text{H}$

Glu $\text{CH}-\text{CH}_2-\text{CH}_2-\text{COO}^-$

$\text{O}=\text{C}$

$\text{N}-\text{H}$

Gly CH_2

$\text{O}=\text{C}$

$\text{N}-\text{H}$

Lys $\text{CH}-\text{CH}_2-\text{CH}_2-\text{CH}_2-\text{CH}_2-\overset{+}{\text{NH}_3}$

COO^-

(a)

Alanylalanine

$$\text{H}_3\overset{+}{\text{N}}-\overset{\overset{\displaystyle\text{CH}_3}{|}}{\text{CH}}-\overset{\overset{\displaystyle}{|}}{\underset{\underset{\displaystyle\text{O}}{\|}}{\text{C}}}-\overset{\overset{\displaystyle\text{H}}{|}}{\text{N}}-\overset{\overset{\displaystyle\text{CH}_3}{|}}{\text{CH}}-\text{COOH}$$

Cationic form (below pH 3)

$$\text{H}_3\overset{+}{\text{N}}-\overset{\overset{\displaystyle\text{CH}_3}{|}}{\text{CH}}-\overset{\underset{\underset{\displaystyle\text{O}}{\|}}{}}{\text{C}}-\overset{\overset{\displaystyle\text{H}}{|}}{\text{N}}-\overset{\overset{\displaystyle\text{CH}_3}{|}}{\text{CH}}-\text{COO}^-$$

Isoelectric form

$$\text{H}_2\text{N}-\overset{\overset{\displaystyle\text{CH}_3}{|}}{\text{CH}}-\overset{\underset{\underset{\displaystyle\text{O}}{\|}}{}}{\text{C}}-\overset{\overset{\displaystyle\text{H}}{|}}{\text{N}}-\overset{\overset{\displaystyle\text{CH}_3}{|}}{\text{CH}}-\text{COO}^-$$

Anionic form (above pH 10)

(b)

Figure 5–17 Ionization and electric charge of peptides. The groups ionized at pH 7.0 are in red. **(a)** A tetrapeptide with two ionizable R groups. **(b)** The cationic, isoelectric, and anionic forms of a dipeptide lacking ionizable R groups.

Peptides Can Be Distinguished by Their Ionization Behavior

Peptides contain only one free α-amino group and one free α-carboxyl group (Fig. 5–17). These groups ionize as they do in simple amino acids, although the ionization constants are different because the oppositely charged group is absent from the α carbon. The α-amino and α-carboxyl groups of all other constituent amino acids are covalently joined in the form of peptide bonds, which do not ionize and thus do not contribute to the total acid–base behavior of peptides. However, the R groups of some amino acids can ionize (Table 5–1), and in a peptide these contribute to the overall acid–base properties (Fig. 5–17). Thus the acid–base behavior of a peptide can be predicted from its single free α-amino and α-carboxyl groups and the nature and number of its ionizable R groups. Like free amino acids, peptides have characteristic titration curves and a characteristic isoelectric pH at which they do not move in an electric field. These properties are exploited in some of the techniques used to separate peptides and proteins (Chapter 6).

Peptides Undergo Characteristic Chemical Reactions

Like other organic molecules, peptides undergo chemical reactions that are characteristic of their functional groups: the free amino and carboxyl groups and the R groups.

Peptide bonds can be hydrolyzed by boiling with either strong acid (typically 6 M HCl) or base to yield the constituent amino acids.

$$^+\text{H}_3\text{N}-\overset{\overset{\displaystyle\text{H}}{|}}{\underset{\underset{\displaystyle\text{R}^1}{|}}{\text{C}}}-\overset{\underset{\underset{\displaystyle\text{O}}{\|}}{}}{\text{C}}-\overset{\overset{\displaystyle\text{H}}{|}}{\text{N}}-\overset{\overset{\displaystyle\text{H}}{|}}{\underset{\underset{\displaystyle\text{R}^2}{|}}{\text{C}}}-\text{COO}^- \xrightarrow{\text{H}_2\text{O}} {}^+\text{H}_3\text{N}-\overset{\overset{\displaystyle\text{H}}{|}}{\underset{\underset{\displaystyle\text{R}^1}{|}}{\text{C}}}-\text{COO}^- + {}^+\text{H}_3\text{N}-\overset{\overset{\displaystyle\text{H}}{|}}{\underset{\underset{\displaystyle\text{R}^2}{|}}{\text{C}}}-\text{COO}^-$$

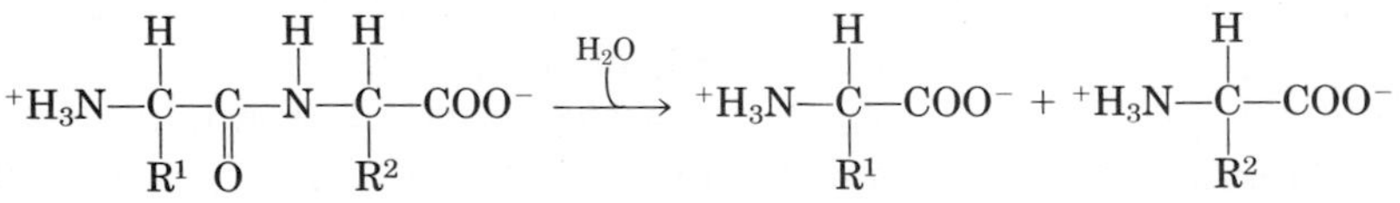

Hydrolysis of peptide bonds in this manner is a necessary step in determining the amino acid composition of proteins. The reagents shown in Figure 5–14 label only free amino groups: those of the amino-terminal residue and the R groups of any lysines present. If dabsyl chloride, dansyl chloride, or 1-fluoro-2,4-dinitrobenzene is used before acid hydrolysis of the peptide, the amino-terminal residue can be separated and identified (Fig. 5–18).

Peptide bonds can also be hydrolyzed by certain enzymes called **proteases.** Proteolytic (protein-cleaving) enzymes are found in all cells and tissues, where they degrade unneeded or damaged proteins or aid in the digestion of food.

Some Small Polypeptides Have Biological Activity

Much of the material in the chapters to follow will revolve around the activities of proteins with molecular weights measured in the tens and even hundreds of thousands. Not all polypeptides are so large, however. There are many naturally occurring small polypeptides and oligopeptides, some of which have important biological activities and exert their effects at very low concentrations. For example, a number of vertebrate hormones (intercellular chemical messengers) (Chapter 22) are small polypeptides. The hormone insulin contains two polypeptide chains, one having 30 amino acid residues and the other 21. Other polypeptide hormones include glucagon, a pancreatic hormone of 29 residues that opposes the action of insulin, and corticotropin, a 39-

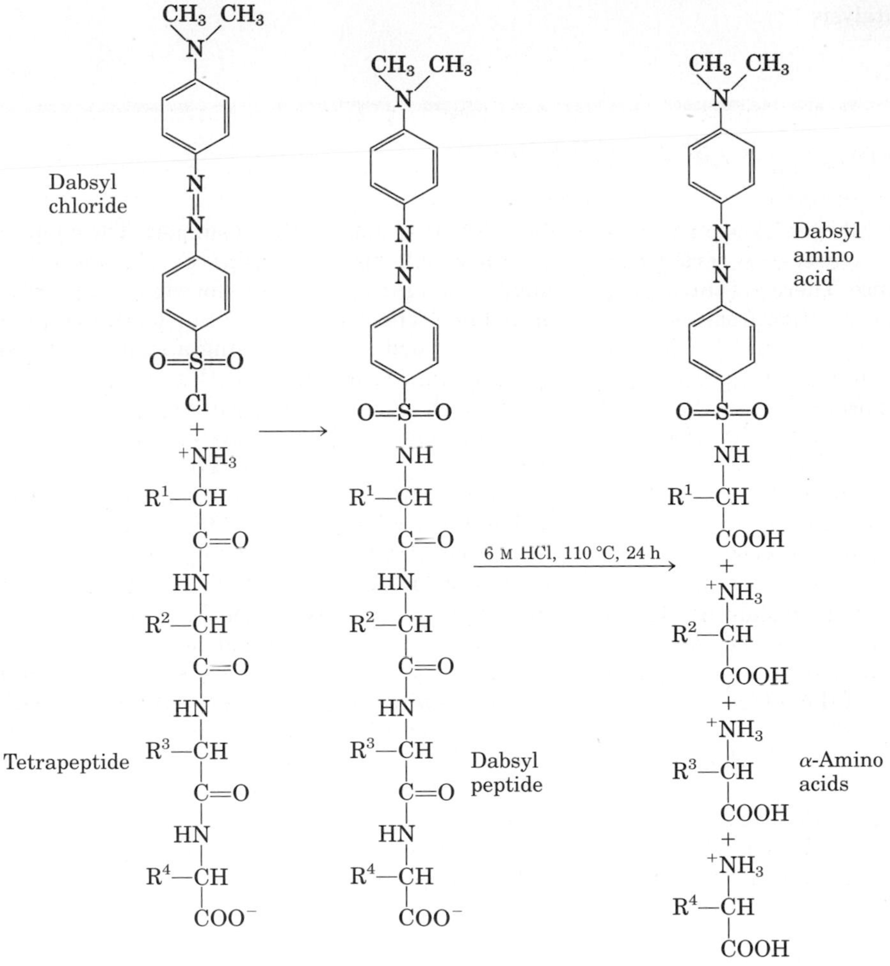

Arg–Pro–Pro–Gly–Phe–Ser–Pro–Phe–Arg

(a)

(b)

Pyroglutamate　　His　　Prolinamide

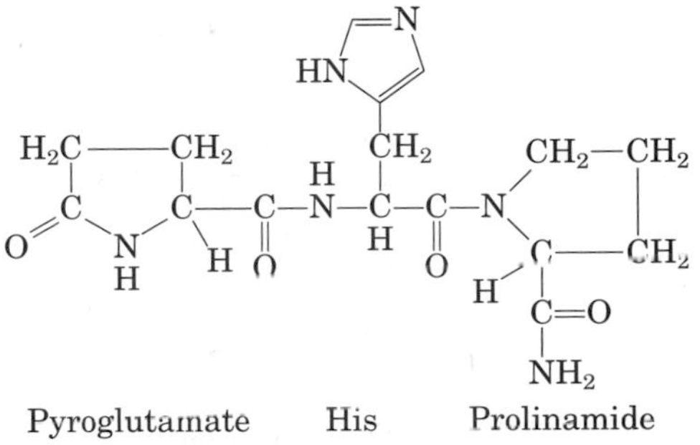

(c)

Tyr–Gly–Gly–Phe–Met

Tyr–Gly–Gly–Phe–Leu

(d)

D-Phe → L-Leu → L-Orn → L-Val → L-Pro

L-Pro ← L-Val ← L-Orn ← L-Leu ← D-Phe

(e)

Figure 5–19 Some naturally occurring peptides with intense biological activity. The amino-terminal residues are at the left end. **(a)** Bradykinin, a hormonelike peptide that inhibits inflammatory reactions. **(b)** Oxytocin, formed by the posterior pituitary gland. The shaded portion is a residue of glycinamide (H_2N—CH_2—$CONH_2$). **(c)** Thyrotropin-releasing factor, formed by the hypothalamus. **(d)** Two enkephalins, brain peptides that affect the perception of pain. **(e)** Gramicidin S, an antibiotic produced by the bacterium *Bacillus brevis*. The arrows indicate the direction from the amino toward the carboxyl end of each residue. The peptide has no termini because it is circular. Orn is the symbol for ornithine, an amino acid that generally does not occur in proteins. Note that gramicidin S contains two residues of a D-amino acid (D-phenylalanine).

residue hormone of the anterior pituitary gland that stimulates the adrenal cortex.

Some biologically important peptides have only a few amino acid residues. That small peptides can have large biological effects is readily illustrated by the activity of the commercially synthesized dipeptide, L-aspartylphenylalanyl methyl ester. This compound is an artificial sweetener better known as aspartame or NutraSweet®:

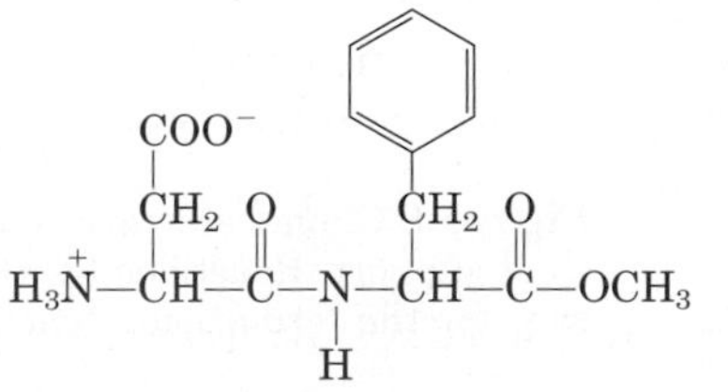

L-Aspartyl-L-phenylalanyl methyl ester
(aspartame)

Among naturally occurring small peptides are hormones such as oxytocin (nine amino acid residues), which is secreted by the posterior pituitary and stimulates uterine contractions; bradykinin (nine residues), which inhibits inflammation of tissues; and thyrotropin-releasing factor (three residues), which is formed in the hypothalamus and stimulates the release of another hormone, thyrotropin, from the anterior pituitary gland (Fig. 5–19). Also noteworthy among short peptides are the enkephalins, compounds formed in the central nervous system

BOX 5–2   Chemical Synthesis of Peptides and Small Proteins

Many peptides are potentially useful as pharmacological reagents, and their synthesis is of considerable commercial importance. There are three ways to obtain a peptide: (1) purification from tissue, a task often made difficult by the vanishingly low concentrations of some peptides; (2) genetic engineering; or (3) direct chemical synthesis. Powerful techniques now make direct chemical synthesis an attractive option in many cases. In addition to commercial applications, the synthesis of specific peptide portions of larger proteins is an increasingly important tool for the study of protein structure and function.

The complexity of proteins makes the traditional synthetic approaches of organic chemistry impractical for peptides with more than four or five amino acids. One problem is the difficulty of purifying the product after each step, because the chemical properties of the peptide change each time a new amino acid is added.

The major breakthrough in this technology was provided by R. Bruce Merrifield. His innovation involved synthesizing a peptide while keeping it attached at one end to a solid support. The support is an insoluble polymer (resin) contained within a column, similar to that used for chromatographic procedures. The peptide is built up on this support one amino acid at a time using a standard set of reactions in a repeating cycle (Fig. 1).

The technology for chemical peptide synthesis has been automated, and several commercial instruments are now available. The most important limitation of the process involves the efficiency of each amino acid addition, as can be seen by calculating the overall yields of peptides of various lengths when the yield for addition of each new amino acid is 96.0 versus 99.8% (Table 1). The chemistry has been optimized to permit the synthesis of proteins 100 amino acids long in about 4 days in reasonable yield. A very similar approach is used to synthesize nucleic acids (Fig. 12–38). It is worth noting that this technology, impressive as it is, still pales when compared with biological processes. The same 100 amino-acid protein would be synthesized with exquisite fidelity in about 5 seconds in a bacterial cell.

Table 1 Effect of stepwise yield on overall yields in peptide synthesis

Number of residues in the final polypeptide	Overall yields of final peptide (%) when the yield of each step is:	
	96.0%	99.8%
11	66	98
21	44	96
31	29	94
51	13	90
100	1.7	82

Figure 1 Chemical synthesis of a peptide on a solid support. Reactions ② through ④ are necessary for the formation of each peptide bond.

that bind to receptors in certain cells of the brain and induce analgesia (deadening of pain sensations). Enkephalins represent one of the body's own mechanisms for control of pain. The enkephalin receptors also bind morphine, heroin, and other addicting opiate drugs (although these are not peptides). Some extremely toxic mushroom poisons, such as amanitin, are also peptides, as are many antibiotics.

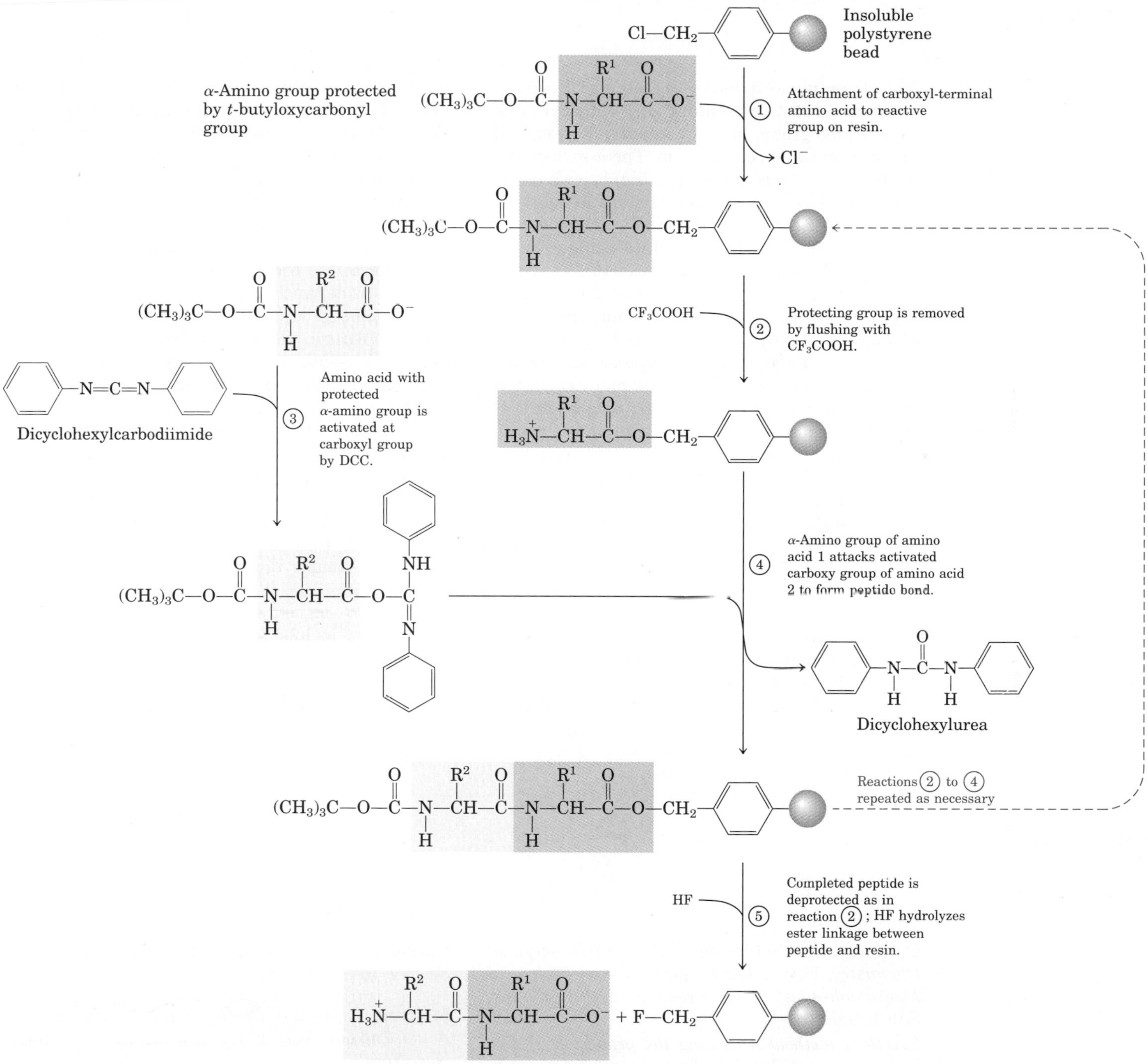

α-Amino group protected by *t*-butyloxycarbonyl group

Insoluble polystyrene bead

1 Attachment of carboxyl-terminal amino acid to reactive group on resin.

Dicyclohexylcarbodiimide

2 Protecting group is removed by flushing with CF_3COOH.

3 Amino acid with protected α-amino group is activated at carboxyl group by DCC.

4 α-Amino group of amino acid 1 attacks activated carboxy group of amino acid 2 to form peptide bond.

Dicyclohexylurea

Reactions 2 to 4 repeated as necessary

5 Completed peptide is deprotected as in reaction 2; HF hydrolyzes ester linkage between peptide and resin.

A growing number of small peptides are proving to be important commercially as pharmaceutical reagents. Unfortunately, they are often present in exceedingly small amounts and hence are hard to purify. For these and other reasons, the chemical synthesis of peptides has become one of the major technologies associated with biochemistry (Box 5–2).

Summary

The 20 amino acids commonly found as hydrolysis products of proteins contain an α-carboxyl group, an α-amino group, and a distinctive R group substituted on the α-carbon atom. The α-carbon atom of the amino acids (except glycine) is asymmetric, and thus amino acids can exist in at least two stereoisomeric forms. Only the L stereoisomers, which are related to the absolute configuration of L-glyceraldehyde, are found in proteins. The amino acids are classified on the basis of the polarity of their R groups. The nonpolar, aliphatic class includes alanine, glycine, isoleucine, leucine, proline, and valine. Phenylalanine, tryptophan, and tyrosine have aromatic side chains and are also relatively hydrophobic. The polar, uncharged class includes asparagine, cysteine, glutamine, methionine, serine, and threonine. The negatively charged (acidic) amino acids are aspartate and glutamate; the positively charge (basic) ones are arginine, histidine, and lysine. There are also a large number of nonstandard amino acids that occur in some proteins (as a result of the modification of standard amino acids) or as free metabolites in cells.

Monoamino monocarboxylic amino acids are diprotic acids ($^+H_3NCH(R)COOH$) at low pH. As the pH is raised to about 6, near the isoelectric point, the proton is lost from the carboxyl group to form the dipolar or zwitterionic species $^+H_3NCH(R)COO^-$, which is electrically neutral. Further increase in pH causes loss of the second proton, to yield the ionic species $H_2NCH(R)COO^-$. Amino acids with ionizable R groups may exist in additional ionic species, depending on the pH and the pK_a of the R group. Thus amino acids vary in their acid–base properties. Amino acids form colored derivatives with ninhydrin. Other colored or fluorescent derivatives are formed in reactions of the α-amino group of amino acids with fluorescamine, dansyl chloride, dabsyl chloride, and 1-fluoro-2,4-dinitrobenzene. Complex mixtures of amino acids can be separated and identified by ion-exchange chromatography or HPLC.

Amino acids can be joined covalently through peptide bonds to form peptides, which can also be formed by incomplete hydrolysis of polypeptides. The acid–base behavior and chemical reactions of a peptide are functions of its amino-terminal amino group, its carboxyl-terminal carboxyl group, and its R groups. Peptides can be hydrolyzed to yield free amino acids. Some peptides occur free in cells and tissues and have specific biological functions. These include some hormones and antibiotics, as well as other peptides with powerful biological activity.

Further Reading

General

Cantor, C.R. & Schimmel, P.R. (1980) *Biophysical Chemistry,* Part I: *The Conformation of Biological Macromolecules,* W.H. Freeman and Company, San Francisco.
Excellent textbook outlining the properties of biological macromolecules and their monomeric subunits.

Creighton, T.E. (1984) *Proteins: Structures and Molecular Properties,* W.H. Freeman and Company, New York.
Very useful general source.

Dickerson, R.E. & Geis, I. (1983) *Proteins: Structure, Function, and Evolution,* 2nd edn, The Benjamin/Cummings Publishing Company, Menlo Park, CA.
Beautifully illustrated and interesting account.

Amino Acids

Corrigan, J.T. (1969) D-Amino acids in animals. *Science* **164,** 142–148.

Meister, A. (1965) *Biochemistry of the Amino Acids,* 2nd edn, Vols. 1 and 2, Academic Press, Inc., New York.
Encyclopedic treatment of the properties, occurrence, and metabolism of amino acids.

Montgomery, R. & Swenson, C.A. (1976) *Quantitative Problems in the Biochemical Sciences,* 2nd edn, W.H. Freeman and Company, New York.

Segel, I.H. (1976) *Biochemical Calculations,* 2nd edn, John Wiley & Sons, New York.

Peptides

Haschemeyer, R. & Haschemeyer, A.H. (1973) *Proteins: A Guide to Study by Physical and Chemical Methods,* John Wiley & Sons, New York.

Merrifield, B. (1986) Solid phase synthesis. *Science* **232,** 341–347.

Smith, L.M. (1988) Automated synthesis and sequence analysis of biological macromolecules. *Analyt. Chem.* **60,** 381A–390A.

Problems

1. *Absolute Configuration of Citrulline* Is citrulline isolated from watermelons (shown below) a D- or L-amino acid? Explain.

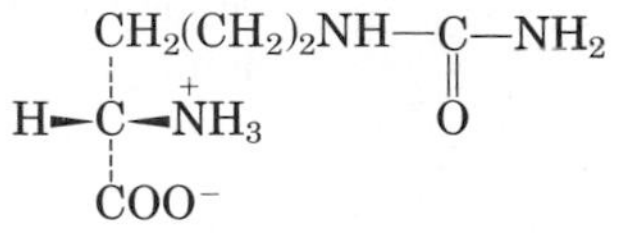

$$CH_2(CH_2)_2NH-C-NH_2$$

$$H-\overset{+}{C}-\overset{}{NH_3} \quad \overset{\|}{O}$$

$$COO^-$$

2. *Relation between the Structures and Chemical Properties of the Amino Acids* The structures and chemical properties of the amino acids are crucial to understanding how proteins carry out their biological functions. The structures of the side chains of 16 amino acids are given below. Name the amino acid that contains each structure and match the R group with the most appropriate description of its properties, (a) to (m). Some of the descriptions may be used more than once.

(a) Small polar R group containing a hydroxyl group; this amino acid is important in the active site of some enzymes.

(b) Provides the least amount of steric hindrance.

(c) R group has $pK_a \approx 10.5$, making it positively charged at physiological pH.

(d) Sulfur-containing R group; neutral at any pH.

(e) Aromatic R group, hydrophobic in nature and neutral at any pH.

(f) Saturated hydrocarbon, important in hydrophobic interactions.

(g) The only amino acid having an ionizing R group with a pK_a near 7; it is an important group in the active site of some enzymes.

(h) The only amino acid having a substituted α-amino group; it influences protein folding by forcing a bend in the chain.

(i) R group has a pK_a near 4 and thus is negatively charged at pH 7.

(j) An aromatic R group capable of forming hydrogen bonds; it has a pK_a near 10.

(k) Forms disulfide cross-links between polypeptide chains; the pK_a of its functional group is about 10.

(l) R group with $pK_a \approx 12$, making it positively charged at physiological pH.

(m) When this polar but uncharged R group is hydrolyzed, the amino acid is converted into another amino acid having a negatively charged R group at pH near 7.

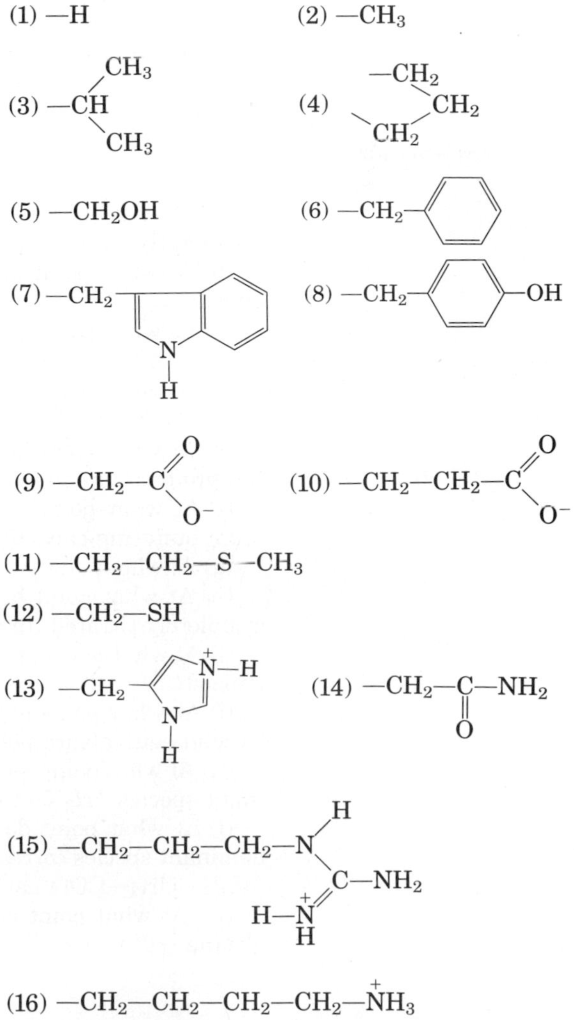

(1) $-H$

(2) $-CH_3$

(3) $-CH\begin{smallmatrix}CH_3\\CH_3\end{smallmatrix}$

(4) $\begin{smallmatrix}-CH_2\\ \ \ \ \ \ \ \ CH_2\\CH_2\end{smallmatrix}$

(5) $-CH_2OH$

(6) $-CH_2-\!\!\bigcirc$

(7) $-CH_2-$ (indole)

(8) $-CH_2-\!\!\bigcirc\!-OH$

(9) $-CH_2-C\begin{smallmatrix}O\\ \\O^-\end{smallmatrix}$

(10) $-CH_2-CH_2-C\begin{smallmatrix}O\\ \\O^-\end{smallmatrix}$

(11) $-CH_2-CH_2-S-CH_3$

(12) $-CH_2-SH$

(13) $-CH_2-$ (imidazole)

(14) $-CH_2-C-NH_2$ (with $\|O$)

(15) $-CH_2-CH_2-CH_2-N\begin{smallmatrix}H\\ \ \ \ \ \ C-NH_2\\ \ \ \ \overset{+}{\|}\\H-N\\ \ \ \ H\end{smallmatrix}$

(16) $-CH_2-CH_2-CH_2-CH_2-\overset{+}{N}H_3$

3. *Relationship between the Titration Curve and the Acid–Base Properties of Glycine* A 100 mL solution of 0.1 M glycine at pH 1.72 was titrated with 2 M NaOH solution. During the titration, the pH was monitored and the results were plotted in the graph shown. The key points in the titration are designated I to V on the graph. For each of the statements below, *identify* the appropriate key point in the titration and *justify* your choice.

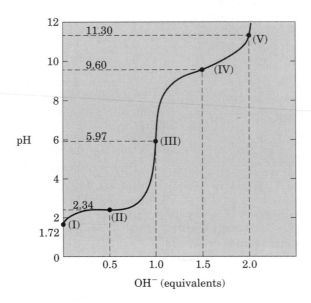

(a) At what point will glycine be present predominantly as the species $^+H_3N—CH_2—COOH$?

(b) At what point is the *average* net charge of glycine $+\frac{1}{2}$?

(c) At what point is the amino group of half of the molecules ionized?

(d) At what point is the pH equal to the pK_a of the carboxyl group?

(e) At what point is the pH equal to the pK_a of the protonated amino group?

(f) At what points does glycine have its maximum buffering capacity?

(g) At what point is the *average* net charge zero?

(h) At what point has the carboxyl group been completely titrated (first equivalence point)?

(i) At what point are half of the carboxyl groups ionized?

(j) At what point is glycine completely titrated (second equivalence point)?

(k) At what point is the structure of the predominant species $^+H_3N—CH_2—COO^-$?

(l) At what point do the structures of the predominant species correspond to a 50:50 mixture of $^+H_3N—CH_2—COO^-$ and $H_2N—CH_2—COO^-$?

(m) At what point is the *average* net charge of glycine -1?

(n) At what point do the structures of the predominant species consist of a 50:50 mixture of $^+H_3N—CH_2—COOH$ and $^+H_3N—CH_2—COO^-$?

(o) What point corresponds to the isoelectric point?

(p) At what point is the *average* net charge on glycine $-\frac{1}{2}$?

(q) What point represents the end of the titration?

(r) If one wanted to use glycine as an efficient buffer, which points would represent the *worst* pH regions for buffering power?

(s) At what point in the titration is the predominant species $H_2N—CH_2—COO^-$?

4. *How Much Alanine Is Present as the Completely Uncharged Species?* At a pH equal to the isoelectric point, the *net* charge on alanine is zero. Two structures can be drawn that have a net charge of zero (zwitterionic and uncharged forms), but the predominant form of alanine at its pI is zwitterionic.

$$H_3\overset{+}{N}—\underset{H}{\overset{CH_3}{C}}—\overset{O}{\underset{O^-}{C}} \qquad H_2N—\underset{H}{\overset{CH_3}{C}}—\overset{O}{\underset{OH}{C}}$$

Zwitterionic   Uncharged

(a) Explain why the form of alanine at its pI is zwitterionic rather than completely uncharged.

(b) Estimate the fraction of alanine present at its pI as the completely uncharged form. Justify your assumptions.

5. *Ionization State of Amino Acids* Each ionizable group of an amino acid can exist in one of two states, charged or neutral. The electric charge on the functional group is determined by the relationship between its pK_a and the pH of the solution. This relationship is described by the Henderson–Hasselbalch equation.

(a) Histidine has three ionizable functional groups. Write the relevant equilibrium equations for its three ionizations and assign the proper pK_a for each ionization. Draw the structure of histidine in each ionization state. What is the net charge on the histidine molecule in each ionization state?

(b) Draw the structures of the predominant ionization state of histidine at pH 1, 4, 8, and 12. Note that the ionization state can be approximated by treating each ionizable group independently.

(c) What is the net charge of histidine at pH 1, 4, 8, and 12? For each pH, will histidine migrate toward the anode (+) or cathode (−) when placed in an electric field?

6. *Preparation of a Glycine Buffer* Glycine is commonly used as a buffer. Preparation of a 0.1 M glycine buffer starts with 0.1 M solutions of glycine hydrochloride ($HOOC—CH_2—NH_3^+Cl^-$) and glycine ($^-$OOC—CH$_2$—NH$_3^+$), two commercially available forms of glycine. What volumes of these two solutions must be mixed to prepare 1 L of 0.1 M glycine buffer having a pH of 3.2? (Hint: See Box 4–2)

7. *Separation of Amino Acids by Ion-Exchange Chromatography* Mixtures of amino acids are analyzed by first separating the mixture into its components through ion-exchange chromatography. On a cation-exchange resin containing sulfonate groups (see Fig. 5–12), the amino acids flow down the column at different rates because of two factors that retard their movement: (1) ionic attraction between the —SO$_3^-$ residues on the column and positively charged functional groups on the amino acids and (2) hydrophobic interaction between amino acid side chains and the strongly hydrophobic backbone of the polystyrene resin. For each pair of amino acids listed, determine which member will be eluted first from an ion-exchange column by a pH 7.0 buffer.

   (a) Asp and Lys

   (b) Arg and Met

   (c) Glu and Val

   (d) Gly and Leu

   (e) Ser and Ala

8. *Naming the Stereoisomers of Isoleucine* The structure of the amino acid isoleucine is:

$$
\begin{array}{c}
\text{COO}^- \\
|\\
\text{H}_3\overset{+}{\text{N}}\text{—C—H} \\
|\\
\text{H—C—CH}_3 \\
|\\
\text{CH}_2 \\
|\\
\text{CH}_3
\end{array}
$$

   (a) How many chiral centers does it have?

   (b) How many optical isomers?

   (c) Draw perspective formulas for all the optical isomers of isoleucine.

9. *Comparison of the pK$_a$ Values of an Amino Acid and Its Peptides* The titration curve of the amino acid alanine shows the ionization of two functional groups with pK$_a$ values of 2.34 and 9.69, corresponding to the ionization of the carboxyl and the protonated amino groups, respectively. The titration of di-, tri-, and larger oligopeptides of alanine also shows the ionization of only two functional groups, although the experimental pK$_a$ values are different. The trend in pK$_a$ values is summarized in the table.

Amino acid or peptide	pK$_1$	pK$_2$
Ala	2.34	9.69
Ala–Ala	3.12	8.30
Ala–Ala–Ala	3.39	8.03
Ala–(Ala)$_n$–Ala, $n \geq 4$	3.42	7.94

   (a) Draw the structure of Ala–Ala–Ala. Identify the functional groups associated with pK$_1$ and pK$_2$.

   (b) The value of pK$_1$ *increases* in going from Ala to an Ala oligopeptide. Provide an explanation for this trend.

   (c) The value of pK$_2$ *decreases* in going from Ala to an Ala oligopeptide. Provide an explanation for this trend.

10. *Peptide Synthesis* In the synthesis of polypeptides on solid supports, the α-amino group of each new amino acid is "protected" by a *t*-butyloxycarbonyl group (see Box 5–2). What would happen if this protecting group were not present?

CHAPTER

An Introduction to Proteins

Almost everything that occurs in the cell involves one or more proteins. Proteins provide structure, catalyze cellular reactions, and carry out a myriad of other tasks. Their central place in the cell is reflected in the fact that genetic information is ultimately expressed as protein. For each protein there is a segment of DNA (a gene; see Chapters 12 and 23) that encodes information specifying its sequence of amino acids. There are thousands of different kinds of proteins in a typical cell, each encoded by a gene and each performing a specific function. Proteins are among the most abundant biological macromolecules and are also extremely versatile in their functions.

The chapter begins with a discussion of some of the general properties of proteins. This is followed by a short summary of some common techniques used to purify and study proteins. Finally, we will examine the **primary structure** of protein molecules: the covalent backbone structure and the sequence of amino acid residues. One goal is to discover the relationships between amino acid sequence and biological function.

Properties of Proteins

An understanding of these important macromolecules must begin with the fundamentals. What do proteins do? How big are they? What forms or shapes do they take? What are their chemical properties? The answers serve as an orientation to much that follows.

Proteins Have Many Different Biological Functions

We can classify proteins according to their biological roles.

Enzymes The most varied and most highly specialized proteins are those with catalytic activity—the enzymes. Virtually all the chemical reactions of organic biomolecules in cells are catalyzed by enzymes. Many thousands of different enzymes, each capable of catalyzing a different kind of chemical reaction, have been discovered in different organisms (Fig. 6–1a).

Transport Proteins Transport proteins in blood plasma bind and carry specific molecules or ions from one organ to another. Hemoglobin of erythrocytes (Fig. 6–1b) binds oxygen as the blood passes through the lungs, carries it to the peripheral tissues, and there releases it to participate in the energy-yielding oxidation of nutrients. The blood

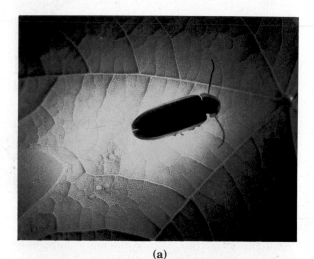

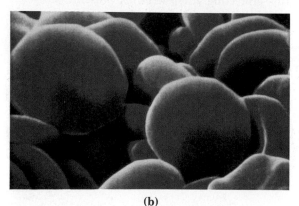

Figure 6–1 Functions of proteins. **(a)** The light produced by fireflies is the result of a light-producing reaction involving luciferin and ATP that is catalyzed by the enzyme luciferase (see Box 13–3). **(b)** Erythrocytes contain large amounts of the oxygen-transporting protein hemoglobin. **(c)** The white color of milk is derived primarily from the protein casein. **(d)** The movement of cilia in protozoans depends on the action of the protein dynein. **(e)** The protein fibroin is the major structural component of spider webs. **(f)** Castor beans contain a highly toxic protein called ricin. **(g)** Cancerous tumors are often made up of cells that have defects involving one or more of the proteins that regulate cell division.

plasma contains lipoproteins, which carry lipids from the liver to other organs. Other kinds of transport proteins are present in the plasma membranes and intracellular membranes of all organisms; these are adapted to bind glucose, amino acids, or other substances and transport them across the membrane.

(c)

Nutrient and Storage Proteins The seeds of many plants store nutrient proteins required for the growth of the germinating seedling. Particularly well-studied examples are the seed proteins of wheat, corn, and rice. Ovalbumin, the major protein of egg white, and casein, the major protein of milk, are other examples of nutrient proteins (Fig. 6–1c). The ferritin found in some bacteria and in plant and animal tissues stores iron.

Contractile or Motile Proteins Some proteins endow cells and organisms with the ability to contract, to change shape, or to move about. Actin and myosin function in the contractile system of skeletal muscle and also in many nonmuscle cells. Tubulin is the protein from which microtubules are built. Microtubules act in concert with the protein dynein in flagella and cilia (Fig. 6–1d) to propel cells.

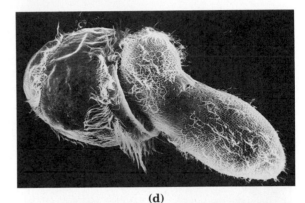

(d)

Structural Proteins Many proteins serve as supporting filaments, cables, or sheets, to give biological structures strength or protection. The major component of tendons and cartilage is the fibrous protein collagen, which has very high tensile strength. Leather is almost pure collagen. Ligaments contain elastin, a structural protein capable of stretching in two dimensions. Hair, fingernails, and feathers consist largely of the tough, insoluble protein keratin. The major component of silk fibers and spider webs is fibroin (Fig. 6–1e). The wing hinges of some insects are made of resilin, which has nearly perfect elastic properties.

Defense Proteins Many proteins defend organisms against invasion by other species or protect them from injury. The immunoglobulins or antibodies, specialized proteins made by the lymphocytes of vertebrates, can recognize and precipitate or neutralize invading bacteria, viruses, or foreign proteins from another species. Fibrinogen and thrombin are blood-clotting proteins that prevent loss of blood when the vascular system is injured. Snake venoms, bacterial toxins, and toxic plant proteins, such as ricin, also appear to have defensive functions (Fig. 6–1f). Some of these, including fibrinogen, thrombin, and some venoms, are also enzymes.

(e)   **(f)**

Regulatory Proteins Some proteins help regulate cellular or physiological activity. Among them are many hormones. Examples include insulin, which regulates sugar metabolism, and the growth hormone of the pituitary. The cellular response to many hormonal signals is often mediated by a class of GTP-binding proteins called G proteins (GTP is closely related to ATP, with guanine replacing the adenine portion of the molecule; see Figs. 1–12 and 3–16b.) Other regulatory proteins bind to DNA and regulate the biosynthesis of enzymes and RNA molecules involved in cell division in both prokaryotes and eukaryotes (Fig. 6–1g).

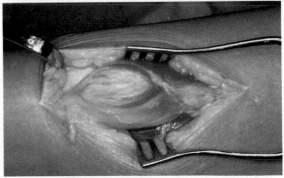

(g)

Other Proteins There are numerous other proteins whose functions are rather exotic and not easily classified. Monellin, a protein of an African plant, has an intensely sweet taste. It is being studied as a nonfattening, nontoxic food sweetener for human use. The blood plasma of some Antarctic fish contains antifreeze proteins, which protect their blood from freezing.

It is extraordinary that all these proteins, with their very different properties and functions, are made from the same group of 20 amino acids.

Proteins Are Very Large Molecules

How long are the polypeptide chains in proteins? Table 6–1 shows that human cytochrome c has 104 amino acid residues linked in a single chain; bovine chymotrypsinogen has 245 amino acid residues. Probably near the upper limit of size is the protein apolipoprotein B, a cholesterol-transport protein with 4,636 amino acid residues in a single polypeptide chain of molecular weight 513,000. Most naturally occurring polypeptides contain less than 2,000 amino acid residues.

Table 6–1 Molecular data on some proteins

	Molecular weight	Number of residues	Number of polypeptide chains
Insulin (bovine)	5,733	51	2
Cytochrome c (human)	13,000	104	1
Ribonuclease A (bovine pancreas)	13,700	124	1
Lysozyme (egg white)	13,930	129	1
Myoglobin (equine heart)	16,890	153	1
Chymotrypsin (bovine pancreas)	21,600	241	3
Chymotrypsinogen (bovine)	22,000	245	1
Hemoglobin (human)	64,500	574	4
Serum albumin (human)	68,500	~550	1
Hexokinase (yeast)	102,000	~800	2
Immunoglobulin G (human)	145,000	~1,320	4
RNA polymerase (*E. coli*)	450,000	~4,100	5
Apolipoprotein B (human)	513,000	4,636	1
Glutamate dehydrogenase (bovine liver)	1,000,000	~8,300	~40

Some proteins consist of a single polypeptide chain, but others, called **multisubunit** proteins, have two or more (Table 6–1). The individual polypeptide chains in a multisubunit protein may be identical or different. If at least some are identical, the protein is sometimes called an **oligomeric** protein and the subunits themselves are referred to as **protomers.** The enzyme ribonuclease has one polypeptide chain. Hemoglobin has four: two identical α chains and two identical β chains, all four held together by noncovalent interactions.

The molecular weights of proteins, which can be determined by various physicochemical methods, may range from little more than 10,000 for small proteins such as cytochrome c (104 residues), to more than 10^6 for proteins with very long polypeptide chains or those with several subunits. The molecular weights of some typical proteins are given in Table 6–1. No simple generalizations can be made about the molecular weights of proteins in relation to their function.

One can calculate the approximate number of amino acid residues in a simple protein containing no other chemical group by dividing its molecular weight by 110. Although the average molecular weight of the

20 standard amino acids is about 138, the smaller amino acids predominate in most proteins; when weighted for the proportions in which the various amino acids occur in proteins (see Table 5–1), the average molecular weight is nearer to 128. Because a molecule of water (M_r 18) is removed to create each peptide bond, the average molecular weight of an amino acid residue in a protein is about $128 - 18 = 110$. Table 6–1 shows the number of amino acid residues in several proteins.

Proteins Have Characteristic Amino Acid Compositions

As is true for simple peptides, hydrolysis of proteins with acid or base yields a mixture of free α-amino acids. When completely hydrolyzed, each type of protein yields a characteristic proportion or mixture of the different amino acids. Table 6–2 shows the composition of the amino acid mixtures obtained on complete hydrolysis of human cytochrome c and of bovine chymotrypsinogen, the inactive precursor of the digestive enzyme chymotrypsin. These two proteins, with very different functions, also differ significantly in the relative numbers of each kind of amino acid they contain. The 20 amino acids almost never occur in equal amounts in proteins. Some amino acids may occur only once per molecule or not at all in a given type of protein; others may occur in large numbers.

Some Proteins Contain Chemical Groups Other Than Amino Acids

Many proteins, such as the enzymes ribonuclease and chymotrypsinogen, contain only amino acids and no other chemical groups; these are considered simple proteins. However, some proteins contain chemical components in addition to amino acids; these are called **conjugated proteins.** The non–amino acid part of a conjugated protein is usually called its **prosthetic group.** Conjugated proteins are classified on the basis of the chemical nature of their prosthetic groups (Table 6–3); for example, **lipoproteins** contain lipids, **glycoproteins** contain sugar groups, and **metalloproteins** contain a specific metal. A number of proteins contain more than one prosthetic group. Usually the prosthetic group plays an important role in the protein's biological function.

Table 6–2 Amino acid composition of two proteins

Amino acid	Number of residues per molecule of protein	
	Human cytochrome c	Bovine chymotrypsinogen
Ala	6	22
Arg	2	4
Asn	5	15
Asp	3	8
Cys	2	10
Gln	2	10
Glu	8	5
Gly	13	23
His	3	2
Ile	8	10
Leu	6	19
Lys	18	14
Met	3	2
Phe	3	6
Pro	4	9
Ser	2	28
Thr	7	23
Trp	1	8
Tyr	5	4
Val	3	23
Total	104	245

Table 6–3 Conjugated proteins

Class	Prosthetic group	Example
Lipoproteins	Lipids	β_1-Lipoprotein of blood
Glycoproteins	Carbohydrates	Immunoglobulin G
Phosphoproteins	Phosphate groups	Casein of milk
Hemoproteins	Heme (iron porphyrin)	Hemoglobin
Flavoproteins	Flavin nucleotides	Succinate dehydrogenase
Metalloproteins	Iron	Ferritin
	Zinc	Alcohol dehydrogenase
	Calcium	Calmodulin
	Molybdenum	Dinitrogenase
	Copper	Plastocyanin

Working with Proteins

The aggregate biochemical picture of protein structure and function is derived from the study of many individual proteins. To study a protein in any detail it must be separated from all other proteins in a cell, and techniques must be available to determine its properties. The necessary methods come from protein chemistry, a discipline as old as biochemistry itself and one that retains a central position in biochemical research. Modern techniques are providing ever newer experimental insights into the critical relationship between the structure of a protein and its function.

Proteins Can Be Separated and Purified

Cells contain thousands of different kinds of proteins. A pure preparation of a given protein is essential before its properties, amino acid composition, and sequence can be determined. How, then, can one protein be purified?

Methods for separating proteins take advantage of properties such as charge, size, and solubility, which vary from one protein to the next. Because many proteins bind to other biomolecules, proteins can also be separated on the basis of their binding properties. The source of a protein is generally tissue or microbial cells. The cells must be broken open and the protein released into a solution called a **crude extract.** If necessary, differential centrifugation can be used to prepare subcellular fractions or to isolate organelles (see Fig. 2–24). Once the extract or organelle preparation is ready, a variety of methods are available for separation of proteins. Ion-exchange chromatography (see Fig. 5–12) can be used to separate proteins with different charges in much the same way that it separates amino acids. Other chromatographic methods take advantage of differences in size, binding affinity, and solubility (Fig. 6–2). Nonchromatographic methods include the selective precipitation of proteins with salt, acid, or high temperatures.

The approach to the purification of a "new" protein, one not previously isolated, is guided both by established precedents and common sense. In most cases, several different methods must be used sequentially to completely purify a protein. The choice of method is somewhat empirical, and many protocols may be tried before the most effective is determined. Trial and error can often be minimized by using purification procedures developed for similar proteins as a guide. Published purification protocols are available for many thousands of proteins. Common sense dictates that inexpensive procedures be used first, when the total volume and number of contaminants is greatest. Chromatographic methods are often impractical at early stages because the amount of chromatographic medium needed increases with sample size. As each purification step is completed, the sample size generally becomes smaller (Table 6–4) and more sophisticated (and expensive) chromatographic procedures can be applied.

Individual Proteins Can Be Quantified

In order to purify a protein, it is essential to have an assay to detect and quantify that protein in the presence of many other proteins. Often, purification must proceed in the absence of any information about the size and physical properties of the protein, or the fraction of the total protein mass it represents in the extract.

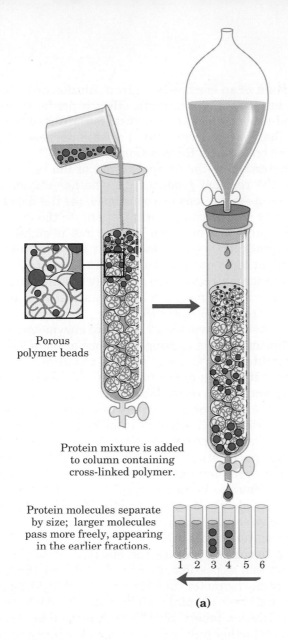

Porous polymer beads

Protein mixture is added to column containing cross-linked polymer.

Protein molecules separate by size; larger molecules pass more freely, appearing in the earlier fractions.

1 2 3 4 5 6

(a)

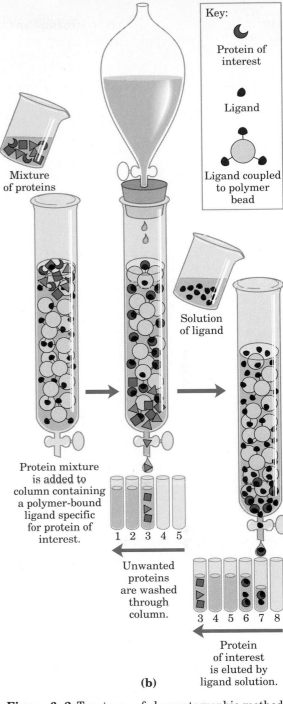

Key:

Protein of interest

Ligand

Ligand coupled to polymer bead

Mixture of proteins

Solution of ligand

Protein mixture is added to column containing a polymer-bound ligand specific for protein of interest.

1 2 3 4 5

Unwanted proteins are washed through column.

3 4 5 6 7 8

Protein of interest is eluted by ligand solution.

(b)

Figure 6–2 Two types of chromatographic methods used in protein purification. **(a)** Size-exclusion chromatography; also called gel filtration. This method separates proteins according to size. The column contains a cross-linked polymer with pores of selected size. Larger proteins migrate faster than smaller ones, because they are too large to enter the pores in the beads and hence take a more direct route through the column. The smaller proteins enter the pores and are slowed by the more labyrinthian path they take through the column. **(b)** Affinity chromatography separates proteins by their binding specificities. The proteins retained on the column are those that bind specifically to a ligand cross-linked to the beads. (In biochemistry, the term "ligand" is used to refer to a group or molecule that is bound.) After nonspecific proteins are washed through the column, the bound protein of particular interest is eluted by a solution containing free ligand.

Table 6–4 A purification table for a hypothetical enzyme*

Procedure or step	Fraction volume (ml)	Total protein (mg)	Activity (units)	Specific activity (units/mg)
1. Crude cellular extract	1,400	10,000	100,000	10
2. Precipitation	280	3,000	96,000	32
3. Ion-exchange chromatography	90	400	80,000	200
4. Size-exclusion chromatography	80	100	60,000	600
5. Affinity chromatography	6	3	45,000	15,000

* All data represent the status of the sample *after* the procedure indicated in the first column has been carried out.

The amount of an enzyme in a given solution or tissue extract can be assayed in terms of the catalytic effect it produces, that is, the *increase* in the rate at which its substrate is converted to reaction products when the enzyme is present. For this purpose one must know (1) the overall equation of the reaction catalyzed, (2) an analytical procedure for determining the disappearance of the substrate or the appearance of the reaction products, (3) whether the enzyme requires cofactors such as metal ions or coenzymes, (4) the dependence of the enzyme activity on substrate concentration, (5) the optimum pH, and (6) a temperature zone in which the enzyme is stable and has high activity. Enzymes are usually assayed at their optimum pH and at some convenient temperature within the range 25 to 38 °C. Also, very high substrate concentrations are generally required so that the initial reaction rate, which is measured experimentally, is proportional to enzyme concentration (Chapter 8).

By international agreement, 1.0 unit of enzyme activity is defined as the amount of enzyme causing transformation of 1.0 μmol of substrate per minute at 25 °C under optimal conditions of measurement. The term **activity** refers to the total units of enzyme in the solution. The **specific activity** is the number of enzyme units per milligram of protein (Fig. 6–3). The specific activity is a measure of enzyme purity: it increases during purification of an enzyme and becomes maximal and constant when the enzyme is pure (Table 6–4).

After each purification step, the activity of the preparation (in units) is assayed, the total amount of protein is determined independently, and their ratio gives the specific activity. Activity and total protein generally decrease with each step. Activity decreases because some loss always occurs due to inactivation or nonideal interactions with chromatographic materials or other molecules in the solution. Total protein decreases because the objective is to remove as much nonspecific protein as possible. In a successful step, the loss of nonspecific protein is much greater than the loss of activity; therefore, specific activity increases even as total activity falls. The data are then assembled in a purification table (Table 6–4). A protein is generally considered pure when further purification steps fail to increase specific activity, and when only a single protein species can be detected (by methods to be described later).

For proteins that are not enzymes, other quantification methods are required. Transport proteins can be assayed by their binding to the molecule they transport, and hormones and toxins by the biological effect they produce; for example, growth hormones will stimulate the growth of certain cultured cells. Some structural proteins represent such a large fraction of a tissue mass that they can be readily extracted and purified without an assay. The approaches are as varied as the proteins themselves.

Figure 6–3 Activity versus specific activity. The difference between these two terms can be illustrated by considering two jars of marbles. The jars contain the same number of red marbles (representing an unknown protein), but different amounts of marbles of other colors. If the marbles are taken to represent proteins, both jars contain the same *activity* of the protein represented by the red marbles. The second jar, however, has the higher *specific activity* because here the red marbles represent a much higher fraction of the total.

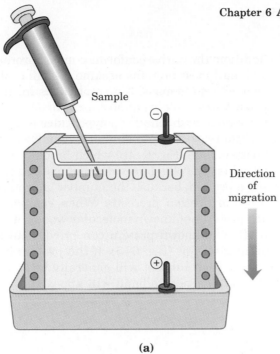

(a)

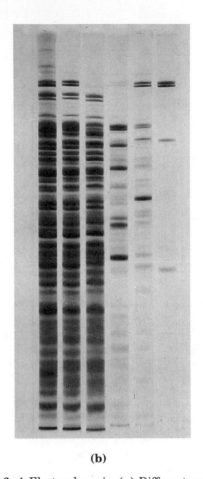

(b)

Proteins Can Be Characterized by Electrophoresis

In addition to chromatography, another important set of methods is available for the separation of proteins, based on the migration of charged proteins in an electric field, a process called **electrophoresis.** These procedures are not often used to purify proteins in large amounts because simpler alternative methods are usually available and electrophoretic methods often inactivate proteins. Electrophoresis is, however, especially useful as an analytical method. Its advantage is that proteins can be visualized as well as separated, permitting a researcher to estimate quickly the number of proteins in a mixture or the degree of purity of a particular protein preparation. Also, electrophoresis allows determination of crucial properties of a protein such as its isoelectric point and approximate molecular weight.

In electrophoresis, the force moving the macromolecule (nucleic acids as well as proteins are separated this way) is the electrical potential, E. The electrophoretic mobility of the molecule, μ, is the ratio of the velocity of the particle, V, to the electrical potential. Electrophoretic mobility is also equal to the net charge of the molecule, Z, divided by the frictional coefficient, f. Thus:

$$\mu = \frac{V}{E} = \frac{Z}{f}$$

Electrophoresis of proteins is generally carried out in gels made up of the cross-linked polymer polyacrylamide (Fig. 6–4). The polyacrylamide gel acts as a molecular sieve, slowing the migration of proteins approximately in proportion to their mass, or molecular weight.

An electrophoretic method commonly used for estimation of purity and molecular weight makes use of the detergent **sodium dodecyl sulfate** (SDS). SDS binds to most proteins (probably by hydrophobic interactions; see Chapter 4) in amounts roughly proportional to the molecular weight of the protein, about one molecule of SDS for every two amino acid residues. The bound SDS contributes a large net negative charge, rendering the intrinsic charge of the protein insignificant.

Figure 6–4 Electrophoresis. **(a)** Different samples are loaded in wells or depressions at the top of the polyacrylamide gel. The proteins move into the gel when an electric field is applied. The gel minimizes convection currents caused by small temperature gradients, and it minimizes protein movements other than those induced by the electric field. **(b)** Proteins can be visualized after electrophoresis by treating the gel with a stain such as Coomassie blue, which binds to the proteins but not to the gel itself. Each band on the gel represents a different protein (or protein subunit); smaller proteins are found nearer the bottom of the gel. This gel illustrates the purification of the enzyme RNA polymerase from the bacterium *E. coli.* The first lane shows the proteins present in the crude cellular extract. Successive lanes show the proteins present after each purification step. The purified protein contains four subunits, as seen in the last lane on the right.

$$Na^+ \ ^-O-\overset{\displaystyle O}{\underset{\displaystyle O}{\overset{\|}{\underset{\|}{S}}}}-O-(CH_2)_{11}CH_3$$

Sodium dodecyl sulfate
(SDS)

In addition, the native conformation of a protein is altered when SDS is bound, and most proteins assume a similar shape, and thus a similar ratio of charge to mass. Electrophoresis in the presence of SDS therefore separates proteins almost exclusively on the basis of mass (molecular weight), with smaller polypeptides migrating more rapidly. After electrophoresis, the proteins are visualized by adding a dye such as Coomassie blue (Fig. 6–4b) which binds to proteins but not to the gel itself. This type of gel provides one method to monitor progress in isolating a protein, because the number of protein bands should decrease as the purification proceeds. When compared with the positions to which proteins of known molecular weight migrate in the gel, the position of an unknown protein can provide an excellent measure of its molecular weight (Fig. 6–5). If the protein has two or more different subunits, each subunit will generally be separated by the SDS treatment, and a separate band will appear for each.

Figure 6–5 Estimating the molecular weight of a protein. The electrophoretic mobility of a protein on an SDS polyacrylamide gel is related to its molecular weight, M_r. **(a)** Standard proteins of known molecular weight are subjected to electrophoresis (lane 1). These marker proteins can be used to estimate the M_r of an unknown protein (lane 2). **(b)** A plot of log M_r of the marker proteins versus relative migration during electrophoresis allows the M_r of the unknown protein to be read from the graph.

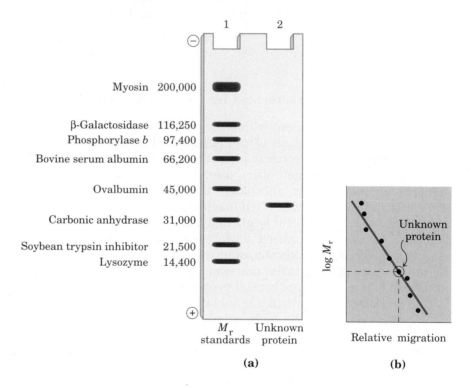

Table 6–5 The isoelectric points of some proteins

	pI
Pepsin	~1.0
Egg albumin	4.6
Serum albumin	4.9
Urease	5.0
β-Lactoglobulin	5.2
Hemoglobin	6.8
Myoglobin	7.0
Chymotrypsinogen	9.5
Cytochrome c	10.7
Lysozyme	11.0

Isoelectric focusing is a procedure used to determine the isoelectric point (pI) of a protein (Fig. 6–6). A pH gradient is established by allowing a mixture of low molecular weight organic acids and bases (ampholytes; see p. 118) to distribute themselves in an electric field generated across the gel. When a protein mixture is applied, each protein migrates until it reaches the pH that matches its pI. Proteins with different isoelectric points are thus distributed differently throughout the gel (Table 6–5).

Combining these two electrophoretic methods in two-dimensional gels permits the resolution of complex mixtures of proteins (Fig. 6–7). This is a more sensitive analytical method than either isoelectric focusing or SDS electrophoresis alone. Two-dimensional electrophoresis separates proteins of identical molecular weight that differ in pI, or proteins with similar pI values but different molecular weights.

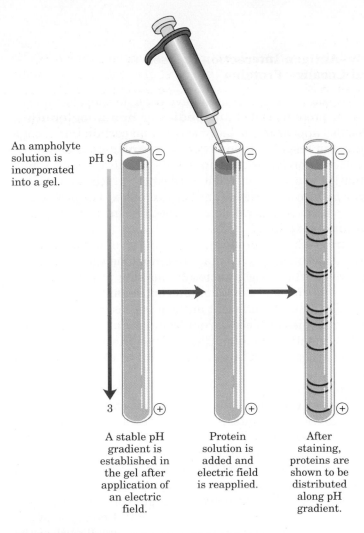

An ampholyte solution is incorporated into a gel.

pH 9

3

A stable pH gradient is established in the gel after application of an electric field.

Protein solution is added and electric field is reapplied.

After staining, proteins are shown to be distributed along pH gradient.

Figure 6–6 Isoelectric focusing. This technique separates proteins according to their isoelectric points. A stable pH gradient is established in the gel by the addition of appropriate ampholytes. A protein mixture is placed in a well on the gel. With an applied electric field, proteins enter the gel and migrate until each reaches a pH equivalent to its pI. Remember that the net charge of a protein is zero when pH = pI.

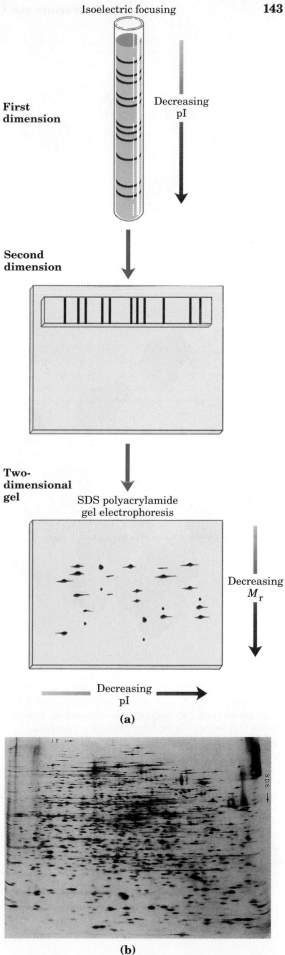

First dimension

Decreasing pI

Second dimension

Two-dimensional gel

SDS polyacrylamide gel electrophoresis

Decreasing M_r

Decreasing pI

(a)

(b)

Figure 6–7 Two-dimensional electrophoresis. **(a)** Proteins are first separated by isoelectric focusing. The gel is then laid horizontally on a second gel, and the proteins are separated by SDS polyacrylamide gel electrophoresis. In this two-dimensional gel, horizontal separation reflects differences in pI; vertical separation reflects differences in molecular weight. **(b)** More than 1,000 different proteins from *E. coli* can be resolved using this technique.

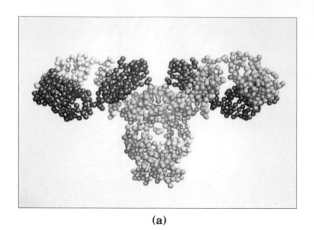

(a)

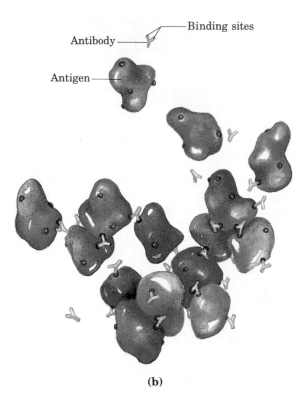

Binding sites

Antibody

Antigen

(b)

Figure 6–8 The immune response and the action of antibodies. **(a)** A molecule of immunoglobulin G (IgG) consists of two polypeptides known as heavy chains (white and light blue) and two known as light chains (purple and dark blue). Immunoglobulins are glycoproteins and contain bound carbohydrate (yellow). **(b)** Each antigen evokes a specific set of antibodies, which will recognize and combine only with that antigen or closely related molecules. (Antibody-binding sites are shown as red areas on the antigen.) The Y-shaped antibodies each have two binding sites for the antigen, and can precipitate the antigen by forming an insoluble, latticelike aggregate.

The Antibody–Antigen Interaction Is Used to Quantify and Localize Proteins

Several sensitive analytical procedures have been developed from the study of a class of proteins called **antibodies** or **immunoglobulins.** Antibody molecules appear in the blood serum and certain tissues of a vertebrate animal in response to injection of an **antigen,** a protein or other macromolecule foreign to that individual. Each foreign protein elicits the formation of a set of different antibodies, which can combine with the antigen to form an antigen–antibody complex. The production of antibodies is part of a general defense mechanism in vertebrates called the **immune response.**

Antibodies are Y-shaped proteins consisting of four polypeptide chains. They have two binding sites that are complementary to specific structural features of the antigen molecule, making possible the formation of a three-dimensional lattice of alternating antigen and antibody molecules (Fig. 6–8). If sufficient antigen is present in a sample, the addition of antibodies or blood serum from an immunized animal will result in the formation of a quantifiable precipitate. No such precipitate is formed when serum of an unimmunized animal is mixed with the antigen.

Antibodies are highly specific for the foreign proteins or other macromolecules that evoke their formation. It is this specificity that makes them valuable analytical reagents. A rabbit antibody formed to horse serum albumin, for example, will combine with the latter but will not usually combine with other horse proteins, such as horse hemoglobin.

Two types of antibody preparations are in use: **polyclonal** and **monoclonal.** Polyclonal antibodies are those produced by many different types (or populations) of antibody-producing cells in an animal immunized with an antigen (in this case a protein). Each type of cell produces an antibody that binds only to a specific, small part of the antigen protein. Consequently, polyclonal preparations contain a mixture of antibodies that recognize different parts of the protein. Monoclonal antibodies, in contrast, are synthesized by a population of identical cells (a **clone**) grown in cell culture. These antibodies are homogeneous, all recognizing the same specific part of the protein. The techniques for producing monoclonal antibodies were worked out by Georges Köhler and Cesar Milstein.

Antibodies are so exquisitely specific that they can in some cases distinguish between two proteins differing by only a single amino acid.

Georges Köhler   Cesar Milstein

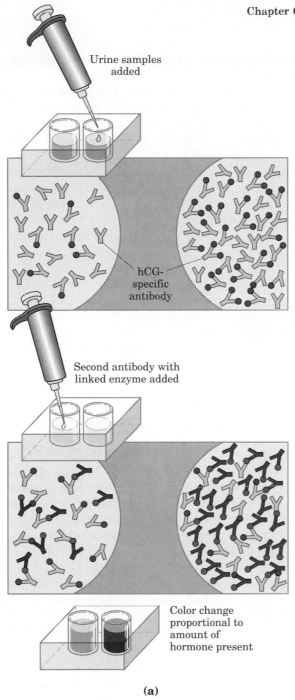

(a)

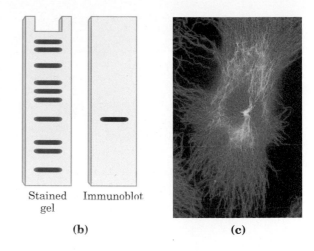

Stained   Immunoblot
gel

(b)   **(c)**

Figure 6–9 Analytical methods based on the interaction of antibodies with antigen. **(a)** An enzyme-linked immunosorbent assay (ELISA) used in testing for human pregnancy. Human chorionic gonadotropin (hCG), a hormone produced by the placenta, is detectable in maternal urine a few days after conception. In the ELISA, an antibody specific for hCG is attached to the bottom of a well in a plastic tray, to which a few drops of urine are added. If any hCG is present, it will bind hCG to the antibodies. The tube is then washed, and a second antibody (also specific for hCG) is added. This second antibody is linked to an enzyme that catalyzes the conversion of a colorless compound to a colored one; the amount of colored compound produced provides a sensitive measure of the amount of hormone present. The ELISA has been adapted for use in determining the amount of specific proteins in tissue samples, in blood, or in urine.

(b) Immunoblot (or Western blot) technique. Proteins are separated by electrophoresis, then antibodies are used to determine the presence and size of the proteins. After separation, the proteins are transferred electrophoretically from an SDS polyacrylamide gel to a special paper (which makes them more accessible). Specific, labeled antibody is added, then the paper is washed to remove unbound antibody. The label can be a radioactive element, a fluorescent compound, or an enzyme as in the ELISA. The position of the labeled antibody defines the M_r of the detected protein. All of the proteins are seen in the stained gel; only the protein bound to the antibody is seen in the immunoblot.

(c) In immunocytochemistry, labeled antibodies are introduced into cells to reveal the subcellular location of a specific protein. Here, fluorescently labeled antibodies and a fluorescence microscope have been used to locate tubulin filaments in a human fibroblast.

When a mixture of proteins is added to a chromatography column in which the antibody is covalently attached to a resin, the antibody will specifically bind its target protein and retain it on the column while other proteins are washed through. The target protein can then be eluted from the resin by a salt solution or some other agent. This can be a powerful tool for protein purification.

A variety of other analytical techniques rely on antibodies. In each case the antibody is attached to a radioactive label or some other reagent to make it easy to detect. The antibody binds the target protein, and the label reveals its presence in a solution or its location in a gel or even a living cell. Several variations of this procedure are illustrated in Figure 6–9. We shall examine some other aspects of antibodies in chapters to follow; they are of extreme importance in medicine and also tell much about the structure of proteins and the action of genes.

The Covalent Structure of Proteins

All proteins in all species, regardless of their function or biological activity, are built from the same set of 20 amino acids (Chapter 5). What is it, then, that makes one protein an enzyme, another a hormone, another a structural protein, and still another an antibody? How do they differ chemically? Quite simply, proteins differ from each other because each has a distinctive number and *sequence* of amino acid residues. The amino acids are the alphabet of protein structure; they can be arranged in an almost infinite number of sequences to make an almost infinite number of different proteins. A specific sequence of amino acids folds up into a unique three-dimensional structure, and this structure in turn determines the function of the protein.

The amino acid sequence of a protein, or its **primary structure,** can be very informative to a biochemist. No other property so clearly distinguishes one protein from another. This now becomes the focus of the remainder of the chapter. We first consider empirical clues that amino acid sequence and protein function are closely linked, then describe how amino acid sequence is determined, and finally outline the many uses to which this information can be put.

The Function of a Protein Depends on Its Amino Acid Sequence

The bacterium *E. coli* produces about 3,000 different proteins. A human being produces 50,000 to 100,000 different proteins. In both cases, each separate type of protein has a unique structure and this structure confers a unique function. Each separate type of protein also has a unique amino acid sequence. Intuition suggests that the amino acid sequence must play a fundamental role in determining the three-dimensional structure of the protein, and ultimately its function, but is this expectation correct? A quick survey of proteins and how they vary in amino acid sequence provides a number of empirical clues that help substantiate the important relationship between amino acid sequence and biological function. First, as we have already noted, proteins with different functions always have different amino acid sequences. Second, more than 1,400 human genetic diseases have been traced to the production of defective proteins (Table 6–6). Perhaps a third of these proteins are defective because of a single change in the amino acid sequence; hence, if the primary structure is altered, the function of the protein may also be changed. Finally, on comparing proteins with similar functions from different species, we find that these proteins often have similar amino acid sequences. An extreme case is ubiquitin, a 76 amino acid protein involved in regulating the degradation of other proteins. The amino acid sequence of ubiquitin is identical in species as disparate as fruit flies and humans.

Is the amino acid sequence absolutely fixed, or invariant, for a particular protein? No; some flexibility is possible. An estimated 20 to 30% of the proteins in humans are **polymorphic,** having amino acid sequence variants in the human population. Many of these variations in sequence have little or no effect on the function of the protein. Furthermore, proteins that carry out a broadly similar function in distantly related species often differ greatly in overall size and amino acid sequence. An example is DNA polymerase, the primary enzyme involved in DNA synthesis. The DNA polymerase of a bacterium is very different in much of its sequence from that of a mouse cell.

Table 6–6 A sampling of genetic diseases linked to loss or defect of a single enzyme or protein

Disease	Physiological effects	Affected enzyme or protein
Cystic fibrosis	Abnormal secretion in lungs, pancreas, sweat glands; chronic pulmonary disease generally leading to death in children or young adults	Chloride channel
Lesch–Nyhan syndrome	Neurological defects, self-mutilation, mental retardation	Hypoxanthine-guanine phosphoribosyl transferase
Immunodeficiency disease	Severe loss of immune response	Purine nucleoside phosphorylase
Immunodeficiency disease	Severe loss of immune response (children must live in a sterile bubble)	Adenosine deaminase
Gaucher's disease	Erosion of bones, hip joints; sometimes brain damage	Glucocerebrosidase
Gout, primary	Overproduction of uric acid resulting in recurring attacks of acute arthritis	Phosphoribosyl pyrophosphate synthetase
Rickets, vitamin D-dependent	Short stature, convulsions	25-Hydroxycholecalciferol-1-hydroxylase
Familial hypercholesterolemia	Atherosclerosis resulting from elevated cholesterol levels in blood; sometimes early death from heart failure	Low-density lipoprotein receptor
Tay-Sachs disease	Motor weakness, mental deterioration, death by age 3 yr	Hexosaminidase-A
Sickle-cell anemia	Pain, swelling in hands and feet; can lead to sudden severe pain in bones or joints and death	Hemoglobin

The amino acid sequence of a protein is inextricably linked to its function. Proteins often contain crucial substructures within their amino acid sequence that are essential to their biological functions. The amino acid sequence in other regions might vary considerably without affecting these functions. The fraction of the sequence that is critical varies from protein to protein, complicating the task of relating sequence to structure, and structure to function. Before we can consider this problem further, however, we must examine how sequence information is obtained.

A chain

```
  +
 NH₃
  |
 Gly
  |
 Ile
  |
 Val
  |
 Glu
  |
5 Gln
  |
 Cys
  |
 Cys ─S──S─ (B chain Cys)
  |
 Ala
  |
 Ser
  |
10 Val
  |
 Cys
  |
 Ser
  |
 Leu
  |
 Tyr
  |
15 Gln
  |
 Leu
  |
 Glu
  |
 Asn
  |
 Tyr
  |
20 Cys ─S─ (B chain Cys)
  |
 Asn
  |
 COO⁻
```

B chain

```
  +
 NH₃
  |
 Phe
  |
 Val
  |
 Asn
  |
 Gln
  |
5 His
  |
 Leu
  |
 Cys
  |
 Gly
  |
 Ser
  |
10 His
  |
 Leu
  |
 Val
  |
 Glu
  |
 Ala
  |
 Leu
  |
 Tyr
  |
 Leu
  |
 Val
  |
 Cys
  |
 Gly
  |
 Glu
  |
 Arg
  |
 Gly
  |
 Phe
  |
25 Phe
  |
 Tyr
  |
 Thr
  |
 Pro
  |
 Lys
  |
30 Ala
  |
 COO⁻
```

Figure 6–10 The amino acid sequence of the two chains of bovine insulin, which are joined by disulfide cross-linkages. The A chain is identical in human, pig, dog, rabbit, and sperm whale insulins. The B chains of the cow, pig, dog, goat, and horse are identical. Such identities between similar proteins of different species are discussed later in this chapter.

The Amino Acid Sequence of Polypeptide Chains Can Be Determined

Two major discoveries in 1953 ushered in the modern era of biochemistry. In that year James D. Watson and Francis Crick deduced the double-helical structure of DNA and proposed a structural basis for the precise replication of DNA (Chapter 12). Implicit in their proposal was the idea that the sequence of nucleotide units in DNA bears encoded genetic information. In that same year, Frederick Sanger worked out the sequence of amino acids in the polypeptide chains of the hormone insulin (Fig. 6–10), surprising many researchers who had long thought that elucidation of the amino acid sequence of a polypeptide would be a hopelessly difficult task. These achievements together suggested that the nucleotide sequence of DNA and the amino acid sequence of proteins were somehow related. Within just over a decade, the nucleotide code that determines the amino acid sequence of protein molecules had been revealed (Chapter 26).

Today the amino acid sequences of thousands of different proteins from many species are known, determined using principles first developed by Sanger. These methods are still in use, although with many variations and improvements in detail.

Short Polypeptides Are Sequenced Using Automated Procedures

Three procedures are used in the determination of the sequence of a polypeptide chain (Fig. 6–11). The first is to hydrolyze it and determine its amino acid composition (Fig. 6–11a). This information is often valuable in later steps, and can also be useful in itself. Because amino acid composition differs from one protein to the next, it can serve as a kind of fingerprint. It can be used, for example, to help determine whether proteins isolated by different laboratories are the same or different.

Often, the next step is to identify the amino-terminal amino acid residue (Fig. 6–11b). For this purpose Sanger developed the reagent 1-fluoro-2,4-dinitrobenzene (FDNB; see Fig. 5–14). Other reagents used to label the amino-terminal residue are dansyl chloride and dabsyl chloride (see Figs. 5–14 and 5–18). The dansyl derivative is highly fluorescent and can be detected and measured in much lower concentrations than dinitrophenyl derivatives. The dabsyl derivative is intensely colored and also provides greater sensitivity than the dinitrophenyl compounds. These methods destroy the polypeptide and their utility is therefore limited to identification of the amino-terminal residue.

Frederick Sanger

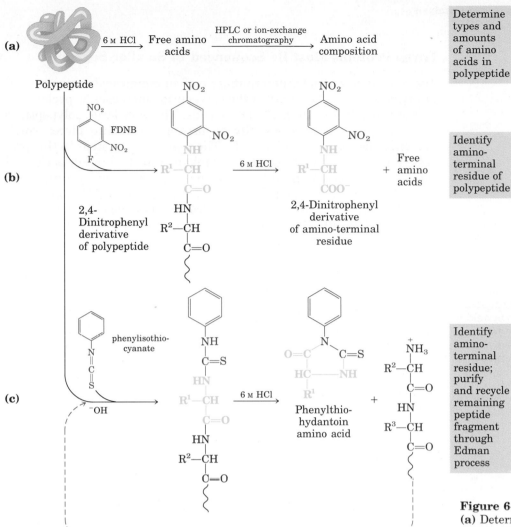

(a) Polypeptide → 6 M HCl → Free amino acids → HPLC or ion-exchange chromatography → Amino acid composition

Determine types and amounts of amino acids in polypeptide

(b) FDNB → 2,4-Dinitrophenyl derivative of polypeptide → 6 M HCl → 2,4-Dinitrophenyl derivative of amino-terminal residue + Free amino acids

Identify amino-terminal residue of polypeptide

(c) phenylisothiocyanate → → 6 M HCl → Phenylthiohydantoin amino acid +

Identify amino-terminal residue; purify and recycle remaining peptide fragment through Edman process

Figure 6–11 Steps in sequencing a polypeptide. **(a)** Determination of amino acid composition and **(b)** identification of the amino-terminal residue are the first steps for many polypeptides. Sanger's method for identifying the amino-terminal residue is shown here. The Edman degradation procedure **(c)** reveals the entire sequence of a peptide. For shorter peptides, this method alone readily yields the entire sequence, and steps **(a)** and **(b)** are often omitted. The latter procedures are useful in the case of larger polypeptides, which are often fragmented into smaller peptides for sequencing (see Fig. 6–13).

To sequence the entire polypeptide, a chemical method devised by Pehr Edman is usually employed. The **Edman degradation** procedure labels and removes only the amino-terminal residue from a peptide, leaving all other peptide bonds intact (Fig. 6–11c). The peptide is reacted with phenylisothiocyanate, and the amino-terminal residue is ultimately removed as a phenylthiohydantoin derivative. After removal and identification of the amino-terminal residue, the *new* amino-terminal residue so exposed can be labeled, removed, and identified by repeating the same series of reactions. This procedure is repeated until the entire sequence is determined. Refinements of each step permit the sequencing of up to 50 amino acid residues in a large peptide.

The many individual steps and the careful bookkeeping required in the determination of the amino acid sequence of long polypeptide chains are usually carried out by programmed and automated analyzers. The Edman degradation is carried out on a programmed machine, called a **sequenator**, which mixes reagents in the proper proportions, separates the products, identifies them, and records the results. Such instruments have greatly reduced the time and labor required to determine the amino acid sequence of polypeptides. These methods are extremely sensitive. Often, less than a microgram of protein is sufficient to determine its complete amino acid sequence.

Large Proteins Must Be Sequenced in Smaller Segments

The overall accuracy for determination of an amino acid sequence generally declines as the length of the polypeptide increases, especially for polypeptides longer than 50 amino acids. The very large polypeptides found in proteins must usually be broken down into pieces small enough to be sequenced efficiently. There are several steps in this process. First, any disulfide bonds are broken, and the protein is cleaved into a set of specific fragments by chemical or enzymatic methods. Each fragment is then purified, and sequenced by the Edman procedure. Finally, the order in which the fragments appear in the original protein is determined and disulfide bonds (if any) are located.

Figure 6–12 Breaking disulfide bonds in proteins. The two common methods are illustrated. Oxidation of cystine with performic acid produces two cysteic acid residues. Reduction by dithiothreitol to form cysteine residues must be followed by further modification of the reactive —SH groups to prevent reformation of the disulfide bond. Acetylation by iodoacetate serves this purpose.

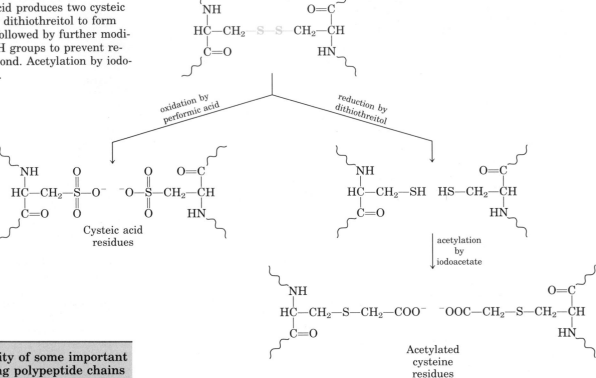

Table 6–7 The specificity of some important methods for fragmenting polypeptide chains

Treatment*	Cleavage points†
Trypsin	Lys, Arg (C)
Submaxillarus protease	Arg (C)
Chymotrypsin	Phe, Trp, Tyr (C)
Staphylococcus aureus V8 protease	Asp, Glu (C)
Asp-*N*-protease	Asp, Glu (N)
Pepsin	Phe, Trp, Tyr (N)
Cyanogen bromide	Met (C)

* All of the enzymes or reagents listed are available from commercial sources.

† Residues furnishing the primary recognition point for the protease; peptide bond cleavage occurs either on the carbonyl (C) or amino (N) side of the indicated group of amino acids.

Breaking Disulfide Bonds Disulfide bonds interfere with the sequencing procedure. A cystine residue (p. 116) that has one of its peptide bonds cleaved by the Edman procedure will remain attached to the polypeptide. Disulfide bonds also interfere with the enzymatic or chemical cleavage of the polypeptide (described below). Two approaches to irreversible breakage of disulfide bonds are outlined in Figure 6–12.

Cleaving the Polypeptide Chain Several methods can be used for fragmenting the polypeptide chain. These involve a set of enzymes (proteases) and chemical reagents that cleave peptide chains adjacent to specific amino acid residues (Table 6–7). The digestive enzyme trypsin, for example, catalyzes the hydrolysis of only those peptide bonds in

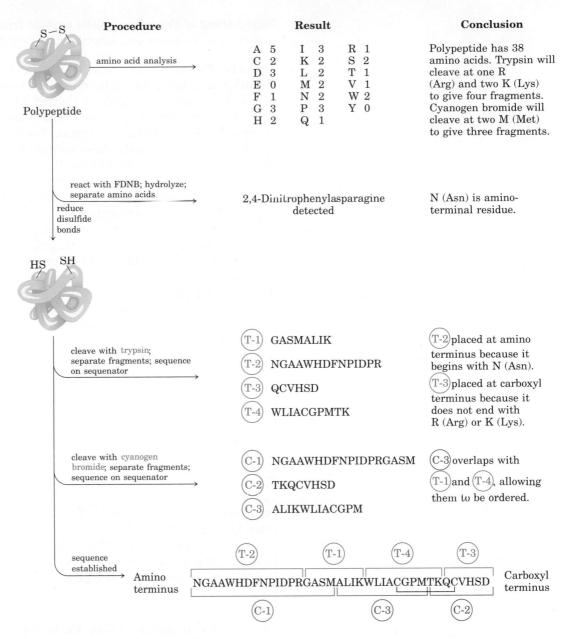

Figure 6–13 Fragmenting proteins prior to sequencing, and placing peptide fragments in their proper order with overlaps. The one-letter abbreviations for amino acids are given in Table 5–1. In this example, there are only two Cys residues, thus one possibility for location of the disulfide bridge (black bracket). In polypeptides with three or more Cys residues, disulfide bridges can be located as described in the text.

which the carbonyl group is contributed by either a Lys or an Arg residue, regardless of the length or amino acid sequence of the chain. The number of smaller peptides produced by trypsin cleavage can thus be predicted from the total number of Lys or Arg residues in the original polypeptide (Fig. 6–13). A polypeptide with five Lys and/or Arg residues will usually yield six smaller peptides on cleavage with trypsin. Moreover, all except one of these will have a carboxyl-terminal Lys or Arg. The fragments produced by trypsin action are separated by chromatographic or electrophoretic methods.

Sequencing of Peptides   All the peptide fragments resulting from the action of trypsin are sequenced separately by the Edman procedure.

Ordering Peptide Fragments   The order of these trypsin fragments in the original polypeptide chain must now be determined. Another sample of the intact polypeptide is cleaved into small fragments using a different enzyme or reagent, one that cleaves peptide bonds at points other than those cleaved by trypsin. For example, the reagent cyanogen bromide cleaves only those peptide bonds in which the carbonyl group is contributed by Met (Table 6–7). The fragments resulting from this new procedure are then separated and sequenced as before.

The amino acid sequences of each fragment obtained by the two cleavage procedures are examined, with the objective of finding peptides from the second procedure whose sequences establish continuity, because of overlaps, between the fragments obtained by the first cleavage procedure (Fig. 6–13). Overlapping peptides obtained from the second fragmentation yield the correct order of the peptide fragments produced in the first. Moreover, the two sets of fragments can be compared for possible errors in determining the amino acid sequence of each fragment. If the amino-terminal amino acid has been identified before the original cleavage of the protein, this information can be used to establish which fragment is derived from the amino terminus.

If the second cleavage procedure fails to establish continuity between all peptides from the first cleavage, a third or even a fourth cleavage method must be used to obtain a set of peptides that can provide the necessary overlap(s). A variety of proteolytic enzymes with different specificities are available (Table 6–7).

Locating Disulfide Bonds   After sequencing is completed, locating the disulfide bonds requires an additional step. A sample of the protein is again cleaved with a reagent such as trypsin, this time without first breaking the disulfide bonds. When the resulting peptides are separated by electrophoresis and compared with the original set of peptides generated by trypsin, two of the original peptides will be missing and a new, larger peptide will appear. The two missing peptides represent the regions of the intact polypeptide that are linked by a disulfide bond.

Amino Acid Sequences Can Be Deduced from DNA Sequences

The approach outlined above is not the only way to obtain amino acid sequences. The development of rapid DNA sequencing methods (Chapter 12), the elucidation of the genetic code (Chapter 26), and the development of techniques for the isolation of genes (Chapter 28) make it possible to deduce the sequence of a polypeptide by determining the sequence of nucleotides in its gene (Fig. 6–14). The two techniques are complementary. When the gene is available, sequencing the DNA can be faster and more accurate than sequencing the protein. If the gene has not been isolated, direct sequencing of peptides is necessary, and this can provide information (e.g., the location of disulfide bonds) not available in a DNA sequence. In addition, a knowledge of the amino acid sequence can greatly facilitate the isolation of the corresponding gene (Chapter 28).

Amino acid
sequence (protein)   Gln–Tyr–Pro–Thr–Ile–Trp

DNA sequence (gene)   CAGTATCCTACGATTTGG

Figure 6–14 Correspondence of DNA and amino acid sequences. Each amino acid is encoded by a specific sequence of three nucleotides (triplet) in DNA. The genetic code is described in detail in Chapter 26.

Amino Acid Sequences Provide Important Biochemical Information

The sequence of amino acids in a protein can offer insights into its three-dimensional structure and its function, cellular location, and evolution. Most of these insights are derived by searching for similarities with other known sequences. Thousands of sequences are known and available in computerized data bases. The comparison of a newly obtained sequence with this large bank of stored sequences often reveals relationships both surprising and enlightening.

The relationship between amino acid sequence and three-dimensional structure, and between structure and function, is not understood in detail. However, a growing number of protein families are being revealed that have at least some shared structural and functional features that can be readily identified on the basis of amino acid sequence similarities alone. For example, there are four major families of proteases, several families of naturally occurring protease inhibitors, a large number of closely related protein kinases, and a similar large number of related protein phosphatases. Individual proteins are generally assigned to families by the degree of similarity in amino acid sequence (identical to other members of the family across 30% or more of the sequence), and proteins in these families generally share at least some structural and functional characteristics. Some families are defined, however, by identities involving only a few amino acids that are critical to a certain function. Many membrane-bound protein receptors share important structural features and have similar amino acid sequences, even though the extracellular molecules they bind are quite different. Even the immunoglobulin family includes a host of extracellular and cell-surface proteins in addition to antibodies.

The similarities may involve the entire protein or may be confined to relatively small segments of it. A number of similar substructures (domains) occur in many functionally unrelated proteins. An example is a 40 to 45 amino acid sequence called the EGF (epidermal growth factor) domain that makes up part of the structure of urokinase, the low-density lipoprotein receptor, several proteins involved in blood clotting, and many others. These domains often fold up into structural configurations that have an unusual degree of stability or that are specialized for a certain environment. Evolutionary relationships can also be inferred from the structural and functional similarities within protein families.

Certain amino acid sequences often serve as signals that determine the cellular location, chemical modification, and half-life of a protein. Special signal sequences, usually at the amino terminus, are used to target certain proteins for export from the cell, while other proteins are distributed to the nucleus, the cell surface, the cytosol, and other cellular locations. Other sequences act as attachment sites for prosthetic groups, such as glycosyl groups in glycoproteins and lipids in lipoproteins. Some of these signals are well characterized, and are easily recognized if they occur in the sequence of a newly discovered protein.

The probability that information about a new protein can be deduced from its primary structure improves constantly with the almost daily addition to the number of published amino acid sequences stored in shared databanks.

Homologous Proteins from Different Species Have Homologous Sequences

Several important conclusions have come from study of the amino acid sequences of homologous proteins from different species. **Homologous proteins** are those that are evolutionarily related. They usually perform the same function in different species; an example is hemoglobin, which has the same oxygen-transport function in different vertebrates. Homologous proteins from different species often have polypeptide chains that are identical or nearly identical in length. Many positions in the amino acid sequence are occupied by the same amino acid in all species and are thus called **invariant residues.** But in other positions there may be considerable variation in the amino acid from one species to another; these are called **variable residues.**

The functional significance of sequence homology can be illustrated by **cytochrome *c*,** an iron-containing mitochondrial protein that transfers electrons during biological oxidations in eukaryotic cells. The polypeptide chain of this protein has a molecular weight of about 13,000 and has about 100 amino acid residues in most species. The amino acid sequences of cytochrome *c* from over 60 different species have been determined, and 27 positions in the chain of amino acid residues are invariant in all species tested (Fig. 6–15), suggesting that they are the most important residues specifying the biological activity of cytochrome *c*. The residues in other positions in the chain exhibit some interspecies variation. There are clear gradations in the number of changes observed in the variable residues. In some positions, all substitutions involve similar amino acid residues (e.g., Arg will replace Lys, both of which are positively charged); these are called **conservative substitutions.** At other positions the substitutions are more random. As we will show in the next chapter, the polypeptide chains of proteins are folded into characteristic and specific conformations and these conformations depend on amino acid sequence. Clearly, the invariant residues are more critical to the structure and function of a protein than the variable ones. Recognizing which amino acids fall into each category is an important step in deciphering the complicated question of how amino acid sequence is translated into a specific three-dimensional structure.

The variable amino acids provide information of another sort. Evolution is sometimes regarded as a theory that is accepted but difficult to test, yet the phylogenetic trees established by taxonomy have been tested and experimentally confirmed through biochemistry. The exam-

Figure 6–15 The amino acid sequence of human cytochrome *c*. Amino acid substitutions found at different positions in the cytochrome *c* of other species are listed below the sequence of the human protein. The amino acids are color-coded to help distinguish conservative and nonconservative substitutions: invariant amino acids are shaded in yellow, conservative amino acid substitutions are shaded in blue, and nonconservative substitutions are unshaded. X is an unusual amino acid, trimethyllysine. The one-letter abbreviations for amino acids are used here (see Table 5–1).

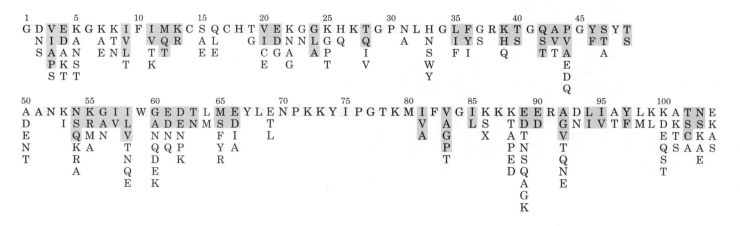

ination of sequences of cytochrome *c* and other homologous proteins has led to an important conclusion: the number of residues that differ in homologous proteins from any two species is in proportion to the phylogenetic difference between those species. For example, 48 amino acid residues differ in the cytochrome *c* molecules of the horse and of yeast, which are very widely separated species, whereas only two residues differ in the cytochrome *c* of the much more closely related duck and chicken. In fact, the cytochrome *c* molecule has identical amino acid sequences in the chicken and the turkey, and in the pig, cow, and sheep. Information on the number of residue differences between homologous proteins of different species allows the construction of evolutionary maps that show the origin and sequence of development of different animals and plants during the evolution of species (Fig. 6–16). The relationships established by taxonomy and biochemistry agree well.

Figure 6–16 Main branches of the evolutionary tree constructed from the number of amino acid differences between cytochrome *c* molecules of different species. The numbers represent the number of residues by which the cytochrome *c* of a given line of organism differs from its ancestors.

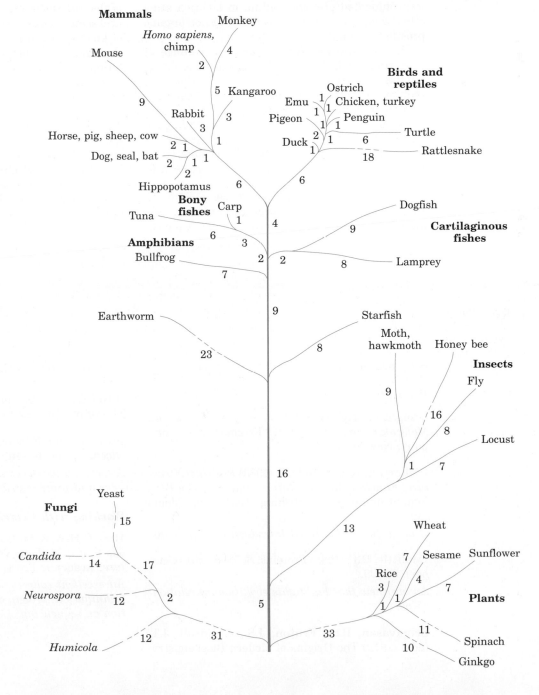

Summary

Cells generally contain thousands of different proteins, each with a different function or biological activity. These functions include enzymatic catalysis, molecular transport, nutrition, cell or organismal motility, structural roles, organismal defense, regulation, and many others. Proteins consist of very long polypeptide chains having from 100 to over 2,000 amino acid residues joined by peptide linkages. Some proteins have several polypeptide chains, which are then referred to as subunits. Simple proteins yield only amino acids on hydrolysis; conjugated proteins contain in addition some other component, such as a metal ion or organic prosthetic group.

Proteins are purified by taking advantage of properties in which they differ, such as size, shape, binding affinities, charge, etc. Purification also requires a method for quantifying or assaying a particular protein in the presence of others. Proteins can be both separated and visualized by electrophoretic methods. Antibodies that specifically bind a certain protein can be used to detect and locate that protein in a solution, a gel, or even in the interior of a cell.

All proteins are made from the same set of 20 amino acids. Their differences in function result from differences in the composition and sequence of their amino acids. The amino acid sequences of polypeptide chains can be established by fragmenting them into smaller pieces using several specific reagents, and determining the amino acid sequence of each fragment by the Edman degradation procedure. The sequencing of suitably sized peptide fragments has been automated. The peptide fragments are then placed in the correct order by finding sequence overlaps between fragments generated by different methods. Protein sequences can also be deduced from the nucleotide sequence of the corresponding gene in the DNA. The amino acid sequence can be compared with the thousands of known sequences, often revealing insights into the structure, function, cellular location, and evolution of the protein.

Homologous proteins from different species show sequence homology: certain positions in the polypeptide chains contain the same amino acids, regardless of the species. In other positions the amino acids may differ. The invariant residues are evidently essential to the function of the protein. The degree of similarity between amino acid sequences of homologous proteins from different species correlates with the evolutionary relationship of the species.

Further Reading

See Chapter 5 for additional useful references.

Properties of Proteins

Creighton, T.E. (1984) *Proteins: Structures and Molecular Properties,* W.H. Freeman and Company, New York.

Dickerson, R.E. & Geis, I. (1983) *Proteins: Structure, Function, and Evolution,* 2nd edn, The Benjamin/Cummings Publishing Company, Menlo Park, CA.
A beautifully illustrated introduction to proteins.

Doolittle, R.F. (1985) Proteins. *Sci. Am.* **253** (October), 88–99.
An overview that highlights evolutionary relationships.

Srinavasan, P.R., Fruton, J.S., & Edsall, J.T. (eds) (1979) The Origins of Modern Biochemistry: A Retrospective on Proteins. *Ann. N.Y. Acad. Sci.* **325.**
A collection of very interesting articles on the history of protein research.

Structure and Function of Proteins. (1989) *Trends Biochem. Sci.* **14** (July).
A special issue devoted to reviews on protein chemistry and protein structure.

Working with Proteins

Hirs, C.H.W. & Timasheff, S.N. (eds) (1983) *Methods in Enzymology,* Vol. 91, Part I: *Enzyme Structure,* Academic Press, Inc., New York.
An excellent collection of authoritative articles on techniques in protein chemistry. Includes information on sequencing.

Kornberg, A. (1990) Why purify proteins? In *Methods in Enzymology*, Vol. 182: *Guide to Protein Purification* (Deutscher, M.P., ed), pp. 1–5, Academic Press, Inc., New York.

The critical role of classical biochemical methods in a new age.

O'Farrell, P.H. (1975) High resolution two-dimensional analysis of proteins. *J. Biol. Chem.* **250,** 4007–4021.

An interesting attempt to count all the proteins in the E. coli *cell.*

Plummer, David T. (1987) *An Introduction to Practical Biochemistry,* 3rd edn, McGraw-Hill, London.

Good descriptions of many techniques for beginning students.

Scopes, R. (1987) *Protein Purification: Principles and Practice,* 2nd edn, Springer Verlag, New York.

Tonegawa, S. (1985) The molecules of the immune system. *Sci. Am.* **253** (October), 122–131.

The Covalent Structure of Proteins

Dickerson, R.E. (1972) The structure and history of an ancient protein. *Sci. Am.* **226** (April), 58–72.

A nice summary of information gleaned from interspecies comparisons of cytochrome c *sequences.*

Doolittle, R. (1981) Similar amino acid sequences: chance or common ancestry. *Science* **214,** 149–159.

A good discussion of what can be learned by comparing amino acid sequences.

Hunkapiller, M.W., Strickler, J.E., & Wilson, K.J. (1984) Contemporary methodology for protein structure determination. *Science* **226,** 304–311.

Reidhaar-Olson, J.F. & Sauer, R.T. (1988) Combinatorial cassette mutagenesis as a probe of the informational content of protein sequences. *Science* **241,** 53–57.

A systematic study of possible amino acid substitutions in a short segment of one protein.

Wilson, A.C. (1985) The molecular basis of evolution. *Sci. Am.* **253** (October), 164–173.

Problems

1. *How Many β-Galactosidase Molecules Are Present in an* E. coli *Cell?* E. coli is a rod-shaped bacterium 2 μm long and 1 μm in diameter. When grown on lactose (a sugar found in milk), the bacterium synthesizes the enzyme β-galactosidase (M_r 450,000), which catalyzes the breakdown of lactose. The average density of the bacterial cell is 1.2 g/mL, and 14% of its total mass is soluble protein, of which 1.0% is β-galactosidase. Calculate the number of β-galactosidase molecules in an *E. coli* cell grown on lactose.

2. *The Number of Tryptophan Residues in Bovine Serum Albumin* A quantitative amino acid analysis reveals that bovine serum albumin contains 0.58% by weight of tryptophan, which has a molecular weight of 204.

(a) Calculate the minimum molecular weight of bovine serum albumin (i.e., assuming there is only one tryptophan residue per protein molecule).

(b) Gel filtration of bovine serum albumin gives a molecular weight estimate of about 70,000. How many tryptophan residues are present in a molecule of serum albumin?

3. *The Molecular Weight of Ribonuclease* Lysine makes up 10.5% of the weight of ribonuclease. Calculate the minimum molecular weight of ribonuclease. The ribonuclease molecule contains ten lysine residues. Calculate the molecular weight of ribonuclease.

4. *The Size of Proteins* What is the approximate molecular weight of a protein containing 682 amino acids in a single polypeptide chain?

5. *Net Electric Charge of Peptides* A peptide isolated from the brain has the sequence

Glu–His–Trp–Ser–Tyr–Gly–Leu–Arg–Pro–Gly

Determine the net charge on the molecule at pH 3. What is the net charge at pH 5.5? At pH 8? At pH 11? Estimate the pI for this peptide. (Use pK_a values for side chains and terminal amino and carboxyl groups as given in Table 5–1.)

6. *The Isoelectric Point of Pepsin* Pepsin of gastric juice (pH ≈ 1.5) has a pI of about 1, much lower than that of other proteins (see Table 6–5). What functional groups must be present in relatively large numbers to give pepsin such a low pI? What amino acids can contribute such groups?

7. *The Isoelectric Point of Histones* Histones are proteins of eukaryotic cell nuclei. They are tightly bound to deoxyribonucleic acid (DNA), which has many phosphate groups. The pI of histones is very high, about 10.8. What amino acids must be present in relatively large numbers in histones? In what way do these residues contribute to the strong binding of histones to DNA?

8. *Solubility of Polypeptides* One method for separating polypeptides makes use of their differential solubilities. The solubility of large polypeptides in water depends upon the relative polarity of their R groups, particularly on the number of ionized groups: the more ionized groups there are, the more soluble the polypeptide. Which of each pair of polypeptides below is more soluble at the indicated pH?

(a) $(Gly)_{20}$ or $(Glu)_{20}$ at pH 7.0

(b) $(Lys-Ala)_3$ or $(Phe-Met)_3$ at pH 7.0

(c) $(Ala-Ser-Gly)_5$ or $(Asn-Ser-His)_5$ at pH 6.0

(d) $(Ala-Asp-Gly)_5$ or $(Asn-Ser-His)_5$ at pH 3.0

9. *Purification of an Enzyme* A biochemist discovers and purifies a new enzyme, generating the purification table below:

Procedure	Total protein (mg)	Activity (units)
1. Crude extract	20,000	4,000,000
2. Precipitation (salt)	5,000	3,000,000
3. Precipitation (pH)	4,000	1,000,000
4. Ion-exchange chromatography	200	800,000
5. Affinity chromatography	50	750,000
6. Size-exclusion chromatography	45	675,000

(a) From the information given in the table, calculate the specific activity of the enzyme solution after each purification procedure.

(b) Which of the purification procedures used for this enzyme is most effective (i.e., gives the greatest increase in purity)?

(c) Which of the purification procedures is least effective?

(d) Is there any indication in this table that the enzyme is now pure? What else could be done to estimate the purity of the enzyme preparation?

10. *Fragmentation of a Polypeptide Chain by Proteolytic Enzymes* Trypsin and chymotrypsin are specific enzymes that catalyze the hydrolysis of polypeptides at specific locations (Table 6–7). The sequence of the B chain of insulin is shown below. Note that the cystine cross-linkage between the A

and B chains has been cleaved through the action of performic acid (see Fig. 6–12).

Phe–Val–Asn–Gln–His–Leu–$CysSO_3^-$–Gly–
Ser–His–Leu–Val–Glu–Ala–Leu–Tyr–Leu–
Val–$CysSO_3^-$–Gly–Glu–Arg–Gly–Phe–Phe–
Tyr–Thr–Pro–Lys–Ala

Indicate the points in the B chain that are cleaved by (a) trypsin and (b) chymotrypsin. Note that these proteases will not remove single amino acids from either end of a polypeptide chain.

11. *Sequence Determination of the Brain Peptide Leucine Enkephalin* A group of peptides that influence nerve transmission in certain parts of the brain has been isolated from normal brain tissue. These peptides are known as opioids, because they bind to specific receptors that bind opiate drugs, such as morphine and naloxone. Opioids thus mimic some of the properties of opiates. Some researchers consider these peptides to be the brain's own pain killers. Using the information below, determine the amino acid sequence of the opioid leucine enkephalin. Explain how your structure is consistent with each piece of information.

(a) Complete hydrolysis by 1 M HCl at 110 °C followed by amino acid analysis indicated the presence of Gly, Leu, Phe, and Tyr, in a 2:1:1:1 molar ratio.

(b) Treatment of the peptide with 1-fluoro-2,4-dinitrobenzene followed by complete hydrolysis and chromatography indicated the presence of the 2,4-dinitrophenyl derivative of tyrosine. No free tyrosine could be found.

(c) Complete digestion of the peptide with pepsin followed by chromatography yielded a dipeptide containing Phe and Leu, plus a tripeptide containing Tyr and Gly in a 1:2 ratio.

12. *Structure of a Peptide Antibiotic from* Bacillus brevis Extracts from the bacterium *Bacillus brevis* contain a peptide with antibiotic properties. Such peptide antibiotics form complexes with metal ions and apparently disrupt ion transport across the cell membrane, killing certain bacterial species. The structure of the peptide has been determined from the following observations.

(a) Complete acid hydrolysis of the peptide followed by amino acid analysis yielded equimolar amounts of Leu, Orn, Phe, Pro, and Val. Orn is ornithine, an amino acid not present in proteins but present in some peptides. It has the structure

$$H_3\overset{+}{N}-CH_2-CH_2-CH_2-\underset{\overset{|}{+}NH_3}{\overset{\overset{H}{|}}{C}}-COO^-$$

(b) The molecular weight of the peptide was estimated as about 1,200.

(c) When treated with the enzyme carboxypeptidase, the peptide failed to undergo hydrolysis.

(d) Treatment of the intact peptide with 1-fluoro-2,4-dinitrobenzene, followed by complete hydrolysis and chromatography, yielded only free amino acids and the following derivative:

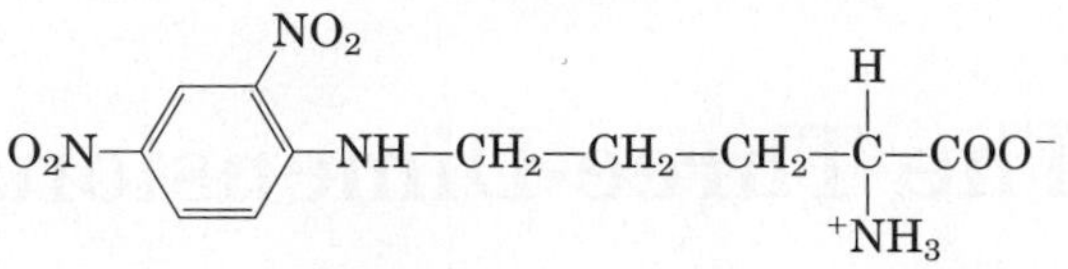

(Hint: Note that the 2,4-dinitrophenyl derivative involves the amino group of a side chain rather than the α-amino group.)

(e) Partial hydrolysis of the peptide followed by chromatographic separation and sequence analysis yielded the di- and tripeptides below (the amino-terminal amino acid is always at the left):

Leu–Phe   Phe–Pro   Orn–Leu   Val–Orn
Val–Orn–Leu   Phe–Pro–Val   Pro–Val–Orn

Given the above information, deduce the amino acid sequence of the peptide antibiotic. Show your reasoning. When you have arrived at a structure, go back and demonstrate that it is consistent with *each* experimental observation.

The Three-Dimensional Structure of Proteins

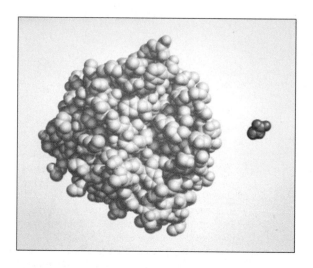

Figure 7–1 The structure of the enzyme chymotrypsin, a globular protein. A molecule of glycine (blue) is shown for size comparison.

The covalent backbone of proteins is made up of hundreds of individual bonds. If free rotation were possible around even a fraction of these bonds, proteins could assume an almost infinite number of three-dimensional structures. Each protein has a specific chemical or structural function, however, strongly suggesting that each protein has a unique three-dimensional structure (Fig. 7–1). The simple fact that proteins can be crystallized provides strong evidence that this is the case. The ordered arrays of molecules in a crystal can generally form only if the molecular units making up the crystal are identical. The enzyme urease (M_r 483,000) was the first protein crystallized, by James Sumner in 1926. This accomplishment demonstrated dramatically that even very large proteins are discrete chemical entities with unique structures, and it revolutionized thinking about proteins.

In this chapter, we will explore the three-dimensional structure of proteins, emphasizing several principles. First, the three-dimensional structure of a protein is determined by its amino acid sequence. Second, the function of a protein depends upon its three-dimensional structure. Third, the three-dimensional structure of a protein is unique, or nearly so. Fourth, the most important forces stabilizing the specific three-dimensional structure maintained by a given protein are noncovalent interactions. Finally, even though the structure of proteins is complicated, several common patterns can be recognized.

The relationship between the amino acid sequence and the three-dimensional structure of a protein is an intricate puzzle that has yet to be solved in detail. Polypeptides with very different amino acid sequences sometimes assume similar structures, and similar amino acid sequences sometimes yield very different structures. To find and understand patterns in this biochemical labyrinth requires a renewed appreciation for fundamental principles of chemistry and physics.

Overview of Protein Structure

The spatial arrangement of atoms in a protein is called a **conformation.** The term conformation refers to a structural state that can, without breaking any covalent bonds, interconvert with other structural states. A change in conformation could occur, for example, by rotation about single bonds. Of the innumerable conformations that are theoretically possible in a protein containing hundreds of single bonds, one generally predominates. This is usually the conformation that is thermodynamically the most stable, having the lowest Gibbs' free energy (G). Proteins in their functional conformation are called **native** proteins.

What principles determine the most stable conformation of a protein? Although protein structures can seem hopelessly complex, close inspection reveals recurring structural patterns. The patterns involve different levels of structural complexity, and we now turn to a biochemical convention that serves as a framework for much of what follows in this chapter.

There Are Four Levels of Architecture in Proteins

Conceptually, protein structure can be considered at four levels (Fig. 7–2). **Primary structure** includes all the covalent bonds between amino acids and is normally defined by the sequence of peptide-bonded amino acids and locations of disulfide bonds. The relative spatial arrangement of the linked amino acids is unspecified.

Polypeptide chains are not free to take up any three-dimensional structure at random. Steric constraints and many weak interactions stipulate that some arrangements will be more stable than others. **Secondary structure** refers to regular, recurring arrangements in space of adjacent amino acid residues in a polypeptide chain. There are a few common types of secondary structure, the most prominent being the α helix and the β conformation. **Tertiary structure** refers to the spatial relationship among all amino acids in a polypeptide; it is the complete three-dimensional structure of the polypeptide. The boundary between secondary and tertiary structure is not always clear. Several different types of secondary structure are often found within the three-dimensional structure of a large protein. Proteins with several polypeptide chains have one more level of structure: **quaternary structure,** which refers to the spatial relationship of the polypeptides, or subunits, within the protein.

Figure 7–2 Levels of structure in proteins. The *primary structure* consists of a sequence of amino acids linked together by covalent peptide bonds, and includes any disulfide bonds. The resulting polypeptide can be coiled into an α helix, one form of *secondary structure*. The helix is a part of the *tertiary structure* of the folded polypeptide, which is itself one of the subunits that make up the *quaternary structure* of the multimeric protein, in this case hemoglobin.

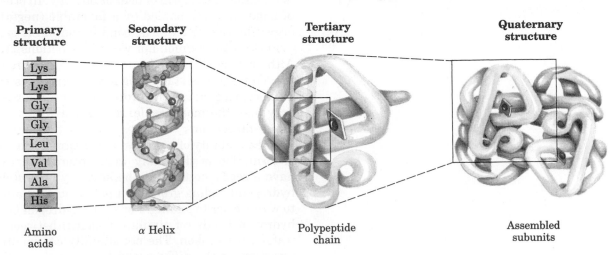

Primary structure	Secondary structure	Tertiary structure	Quaternary structure

Amino acids α Helix   Polypeptide chain   Assembled subunits

Continued advances in the understanding of protein structure, folding, and evolution have made it necessary to define two additional structural levels intermediate between secondary and tertiary structure. A stable clustering of several elements of secondary structure is sometimes referred to as **supersecondary structure.** The term is used to describe particularly stable arrangements that occur in many

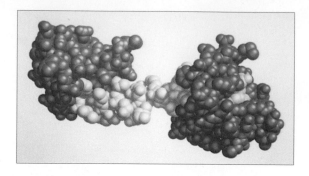

Figure 7–3 The different structural domains in the polypeptide troponin C, a calcium-binding protein associated with muscle. The separate calcium-binding domains, indicated in blue and purple, are connected by a long α helix, shown in white.

different proteins and sometimes many times in a single protein. A somewhat higher level of structure is the **domain.** This refers to a compact region, including perhaps 40 to 400 amino acids, that is a distinct structural unit within a larger polypeptide chain. A polypeptide that is folded into a dumbbell-like shape might be considered to have two domains, one at either end. Many domains fold independently into thermodynamically stable structures. A large polypeptide chain can contain several domains that often are readily distinguishable within the overall structure (Fig. 7–3). In some cases the individual domains have separate functions. As we will see, important patterns exist at each of these levels of structure that provide clues to understanding the overall structure of large proteins.

A Protein's Conformation Is Stabilized Largely by Weak Interactions

The native conformation of a protein is only marginally stable; the difference in free energy between the folded and unfolded states in typical proteins under physiological conditions is in the range of only 20 to 65 kJ/mol. A given polypeptide chain can theoretically assume countless different conformations, and as a result the unfolded state of a protein is characterized by a high degree of conformational entropy. This entropy, and the hydrogen-bonding interactions of many groups in the polypeptide chain with solvent (water), tend to maintain the unfolded state. The chemical interactions that counteract these effects and stabilize the native conformation include disulfide bonds and the weak (noncovalent) interactions described in Chapter 4: hydrogen bonds, and hydrophobic, ionic, and van der Waals interactions. An appreciation of the role of these weak interactions is especially important to understanding how polypeptide chains fold into specific secondary, tertiary, and quaternary structures.

Every time a bond is formed between two atoms, some free energy is released in the form of heat or entropy. In other words, the formation of bonds is accompanied by a favorable (negative) change in free energy. The ΔG for covalent bond formation is generally in the range of -200 to -460 kJ/mol. For weak interactions, $\Delta G = -4$ to -30 kJ/mol. Although covalent bonds are clearly much stronger, weak interactions predominate as a stabilizing force in protein structure because of their number. In general, the protein conformation with the lowest free energy (i.e., the most stable) is the one with the maximum number of weak interactions.

The stability of a protein is not simply the sum of the free energies of formation of the many weak interactions within it, however. We have already noted that the stability of proteins is marginal. Every hydrogen-bonding group in a polypeptide chain was hydrogen bonded to water prior to folding. For every hydrogen bond formed in a protein, hydrogen bonds (of similar strength) between the same groups and water were broken. The net stability contributed by a given weak interaction, or the *difference* in free energies of the folded and unfolded state, is close to zero. We must therefore explain why the native conformation of a protein is favored. The contribution of weak interactions to protein stability can be understood in terms of the properties of water (Chapter 4). Pure water contains a network of hydrogen-bonded water molecules. No other molecule has the hydrogen-bonding potential of water, and other molecules present in an aqueous solution will disrupt the hydrogen bonding of water to some extent. Optimizing the hydrogen bonding of water around a hydrophobic molecule results in the formation of a highly structured shell or solvation layer of water in the immediate vicinity, resulting in an unfavorable decrease in the entropy of water. The association among hydrophobic or nonpolar groups results in a decrease in this structured solvation layer, or a favorable increase in entropy. As described in Chapter 4, this entropy term is the major thermodynamic driving force for the association of hydrophobic groups in aqueous solution, and hydrophobic amino acid side chains therefore tend to be clustered in a protein's interior, away from water.

The formation of hydrogen bonds and ionic interactions in a protein is also driven largely by this same entropic effect. Polar groups can generally form hydrogen bonds with water and hence are soluble in water. However, the number of hydrogen bonds per unit mass is generally greater for pure water than for any other liquid or solution, and there are limits to the solubility of even the most polar molecules because of the net decrease in hydrogen bonding that occurs when they are present. Therefore, a solvation shell of structured water will also form to some extent around polar molecules. Even though the energy of formation of an intramolecular hydrogen bond or ionic interaction between two polar groups in a macromolecule is largely canceled out by the elimination of such interactions between the same groups and water, the release of structured water when the intramolecular interaction is formed provides an entropic driving force for folding. Most of the net change in free energy that occurs when weak interactions are formed within a protein is therefore derived from the increase in entropy in the surrounding aqueous solution.

Of the different types of weak interactions, hydrophobic interactions are particularly important in stabilizing a protein conformation; the interior of a protein is generally a densely packed core of hydrophobic amino acid side chains. It is also important that any polar or charged groups in the protein interior have suitable partners for hydrogen bonding or ionic interactions. One hydrogen bond makes only a small apparent contribution to the stability of a native structure, but the presence of a single hydrogen-bonding group without a partner in the hydrophobic core of a protein can be so *destabilizing* that conformations containing such a group are often thermodynamically untenable.

Most of the structural patterns outlined in this chapter reflect these two simple rules: (1) hydrophobic residues must be buried in the protein interior and away from water, and (2) the number of hydrogen bonds must be maximized. Insoluble proteins and proteins within membranes (Chapter 10) follow somewhat different rules because of their function or their environment, but weak interactions are still critical structural elements.

Protein Secondary Structure

Several types of secondary structure are particularly stable and occur widely in proteins. The most prominent are the α helix and β conformations described below. Using fundamental chemical principles and a few experimental observations, Linus Pauling and Robert Corey predicted the existence of these secondary structures in 1951, several years before the first complete protein structure was elucidated.

Linus Pauling

Robert Corey
1897–1971

In considering secondary structure, it is useful to classify proteins into two major groups: fibrous proteins, having polypeptide chains arranged in long strands or sheets, and globular proteins, with polypeptide chains folded into a spherical or globular shape. Fibrous proteins play important structural roles in the anatomy and physiology of vertebrates, providing external protection, support, shape, and form. They may constitute one-half or more of the total body protein in larger animals. Most enzymes and peptide hormones are globular proteins. Globular proteins tend to be structurally complex, often containing several types of secondary structure; fibrous proteins usually consist largely of a single type of secondary structure. Because of this structural simplicity, certain fibrous proteins played a key role in the development of the modern understanding of protein structure and provide particularly clear examples of the relationship between structure and function; they are considered in some detail after the general discussion of secondary structure.

The Peptide Bond Is Rigid and Planar

Pauling and Corey began their work on protein structure in the late 1930s by first focusing on the structure of the peptide bond. The α carbons of adjacent amino acids are separated by three covalent bonds, arranged C_α—C—N—C_α. X-ray diffraction studies of crystals of amino acids and of simple dipeptides and tripeptides demonstrated that the amide C—N bond in a peptide is somewhat shorter than the C—N bond in a simple amine and that the atoms associated with the bond are coplanar. This indicated a resonance or partial sharing of two pairs of electrons between the carbonyl oxygen and the amide nitrogen (Fig.

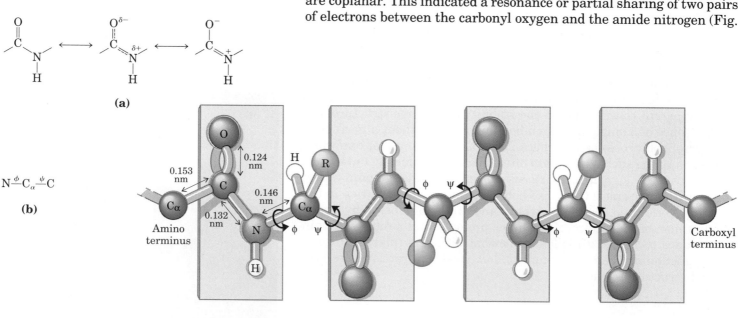

Figure 7–4 **(a)** The planar peptide group. Each peptide bond has some double-bond character due to resonance and cannot rotate. The carbonyl oxygen has a partial negative charge and the amide nitrogen a partial positive charge, setting up a small electric dipole. Note that the oxygen and hydrogen atoms in the plane are on opposite sides of the C—N bond. This is the trans configuration. Virtually all peptide bonds in proteins occur in this configuration, although an exception is noted in Fig. 7–10. **(b)** Three bonds separate sequential C_α carbons in a polypeptide chain. The N—C_α and C_α—C bonds can rotate, with bond angles designated ϕ and ψ, respectively. **(c)** Limited rotation can occur around two of the three types of bonds in a polypeptide chain. The C—N bonds in the planar peptide groups (shaded in blue), which make up one-third of all the backbone bonds, are not free to

7–4a). The oxygen has a partial negative charge and the nitrogen a partial positive charge, setting up a small electric dipole. The four atoms of the peptide group lie in a single plane, in such a way that the oxygen atom of the carbonyl group and the hydrogen atom of the amide nitrogen are trans to each other. From these studies Pauling and Corey concluded that the amide C—N bonds are unable to rotate freely because of their partial double-bond character. The backbone of a polypeptide chain can thus be pictured as a series of rigid planes separated by substituted methylene groups, —CH(R)— (Fig. 7–4c). The rigid peptide bonds limit the number of conformations that can be assumed by a polypeptide chain.

Rotation is permitted about the N—C_α and the C_α—C bonds. By convention the bond angles resulting from rotations are labeled ϕ (phi) for the N—C_α bond and ψ (psi) for the C_α—C bond. Again by convention, both ϕ and ψ are defined as 0° in the conformation in which the two peptide bonds connected to a single α carbon are in the same plane, as shown in Figure 7–4d. In principle, ϕ and ψ can have any value between −180° and +180°, but many values of ϕ and ψ are prohibited by steric interference between atoms in the polypeptide backbone and amino acid side chains. The conformation in which ϕ and ψ are both 0° is prohibited for this reason; this is used merely as a reference point for describing the angles of rotation.

Every possible secondary structure is described completely by the two bond angles ϕ and ψ that are repeated at each residue. Allowed values for ϕ and ψ can be shown graphically by simply plotting ψ versus ϕ, an arrangement known as a **Ramachandran plot.** The Ramachandran plot in Figure 7–5 shows the conformations permitted for most amino acid residues.

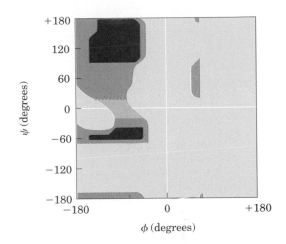

Figure 7–5 A Ramachandran plot. The theoretically allowed conformations of peptides are shown, defined by the values of ϕ and ψ. The shaded areas reflect conformations that can be take up by all amino acids (dark shading) or all except valine and isoleucine (medium shading); the lightest shading reflects conformations that are somewhat unstable but are found in some protein structures.

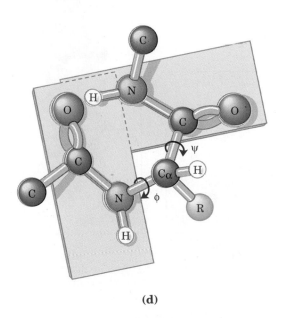

(d)

rotate. Other single bonds in the backbone may also be rotationally hindered, depending on the size and charge of the R groups. **(d)** By convention, ϕ and ψ are both defined as 0° when the two peptide bonds flanking an α carbon are in the same plane. In a protein, this conformation is prohibited by steric overlap between a carbonyl oxygen and an α-amino hydrogen atom.

The α Helix Is a Common Protein Secondary Structure

Pauling and Corey were aware of the importance of hydrogen bonds in orienting polar chemical groups such as the —C=O and —N—H groups of the peptide bond. They also had the experimental results of William Astbury, who in the 1930s had conducted pioneering x-ray studies of proteins. Astbury demonstrated that the protein that makes up hair and wool (the fibrous protein α-keratin) has a regular structure that repeats every 0.54 nm. With this information and their data on the peptide bond, and with the help of precisely constructed models, Pauling and Corey set out to determine the likely conformations of protein molecules.

The simplest arrangement the polypeptide chain could assume with its rigid peptide bonds (but with the other single bonds free to rotate) is a helical structure, which Pauling and Corey called the **α helix** (Fig. 7–6). In this structure the polypeptide backbone is tightly wound around the long axis of the molecule, and the R groups of the amino acid residues protrude outward from the helical backbone. The repeating unit is a single turn of the helix, which extends about 0.56 nm along the long axis, corresponding closely to the periodicity

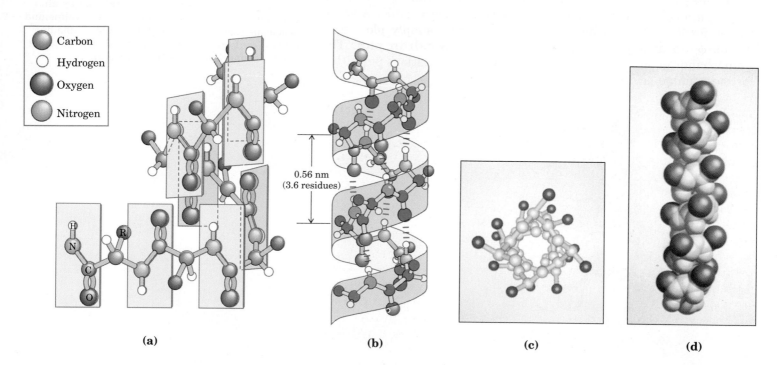

(a)   (b)   (c)   (d)

Figure 7–6 Four models of the α helix, showing different aspects of its structure. **(a)** Formation of a right-handed α helix. The planes of the rigid peptide bonds are parallel to the long axis of the helix. **(b)** Ball-and-stick model of a right-handed α helix, showing the intrachain hydrogen bonds. The repeat unit is a single turn of the helix, 3.6 residues. **(c)** The α helix as viewed from one end, looking down the longitudinal axis. Note the positions of the R groups, represented by red spheres. **(d)** A space-filling model of the α helix.

BOX 7–1   **Knowing the Right Hand from the Left**

There is a simple method for determining the handedness of a helical structure, whether right-handed or left-handed. Make fists of your two hands with thumbs outstretched and pointing away from you. Looking at your right hand, think of a helix spiraling away in the direction indicated by your right thumb, and the spiral occurring in the direction in which the other four fingers are curled as shown (clockwise). The resulting helix is right-handed. Repeating the process with your left hand will produce an image of a left-handed helix, which rotates in the counterclockwise direction as it spirals away from you.

Astbury observed on x-ray analysis of hair keratin. The amino acid residues in an α helix have conformations with $\psi = -45°$ to $-50°$ and $\phi = -60°$, and each helical turn includes 3.6 amino acids. The twisting of the helix has a right-handed sense (Box 7–1) in the most common form of the α helix, although a very few left-handed variants have been observed.

The α helix is one of two prominent types of secondary structure in proteins. It is the predominant structure in α-keratins. In globular proteins, about one-fourth of all amino acid residues are found in α helices, the fraction varying greatly from one protein to the next.

Why does such a helix form more readily than many other possible conformations? The answer is, in part, that it makes optimal use of internal hydrogen bonds. The structure is stabilized by a hydrogen bond between the hydrogen atom attached to the electronegative nitrogen atom of each peptide linkage and the electronegative carbonyl oxygen atom of the fourth amino acid on the amino-terminal side of it in the helix (Fig. 7–6b). Every peptide bond of the chain participates in such hydrogen bonding. Each successive coil of the α helix is held to the adjacent coils by several hydrogen bonds, which in summation give the entire structure considerable stability.

Further model-building experiments have shown that an α helix can form with either L- or D-amino acids. However, the residues must all be of one stereoisomer; a D-amino acid will disrupt any regular structure consisting of L-amino acids, and vice versa. Naturally occurring L-amino acids can form either right- or left-handed helices, but, with rare exceptions, only right-handed helices are found in proteins.

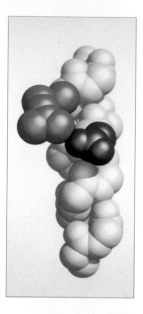

Figure 7–7 Interactions between R groups of amino acids three residues apart in an α helix. An ionic interaction between Asp[100] and Arg[103] in an α-helical region of the protein troponin C is shown in this space-filling model. The polypeptide backbone (carbons, α-amino nitrogens, and α-carbonyl oxygens) is shown in white for a helix segment about 12 amino acids long. The only side chains shown are the interacting Asp and Arg residues, with the aspartate in red and the arginine in blue. The side chain interaction illustrated occurs within the white connecting helix in Fig. 7–3.

Amino Acid Sequence Affects α Helix Stability

Not all polypeptides can form a stable α helix. Additional interactions occur between amino acid side chains that can stabilize or destabilize this structure. For example, if a polypeptide chain has many Glu residues in a long block, this segment of the chain will not form an α helix at pH 7.0. The negatively charged carboxyl groups of adjacent Glu residues repel each other so strongly that they overcome the stabilizing influence of hydrogen bonds on the α helix. For the same reason, if there are many adjacent Lys and/or Arg residues, with positively charged R groups at pH 7.0, they will also repel each other and prevent formation of the α helix. The bulk and shape of certain R groups can also destabilize the α helix or prevent its formation. For example, Asn, Ser, Thr, and Leu residues tend to prevent formation of the α helix if they occur close together in the chain.

The twist of an α helix ensures that critical interactions occur between an amino acid side chain and the side chain three (and sometimes four) residues away on either side of it (Fig. 7–7). Positively charged amino acids are often found three residues away from negatively charged amino acids, permitting the formation of an ionic interaction. Two aromatic amino acids are often similarly spaced, resulting in a hydrophobic interaction.

A minor constraint on the formation of the α helix is the presence of Pro residues. In proline the nitrogen atom is part of a rigid ring (Fig. 5–6), and rotation about the N—C_α bond is not possible. In addition, the nitrogen atom of a Pro residue in peptide linkage has no substituent hydrogen-to-hydrogen bond with other residues. For these reasons, proline is only rarely found within an α helix.

A final factor affecting the stability of an α helix is the identity of the amino acids located near the ends of the α-helical segment of a polypeptide. A small electric dipole exists in each peptide bond (see Fig. 7–4). These dipoles add across the hydrogen bonds in the helix so that the net dipole increases as helix length increases (Fig. 7–8). The four amino acids at either end of the helix do not participate fully in the helix hydrogen bonds. The partial positive and negative charges of the helix dipole actually reside on the peptide amino and carbonyl groups near the amino-terminal and carboxyl-terminal ends of the helix, respectively. For this reason, negatively charged amino acids are often found near the amino terminus of the helical segment, where they have a stabilizing interaction with the positive charge of the helix dipole; a positively charged amino acid at the amino-terminal end is destabilizing. The opposite is true at the carboxyl-terminal end of the helical segment.

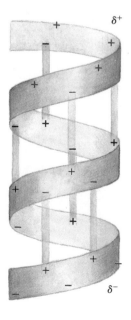

Figure 7–8 The electric dipole of a peptide bond (Fig. 7–4a) is transmitted along an α-helical segment through the intrachain hydrogen bonds, resulting in an overall helix dipole. In this illustration, the amino and carbonyl constituents of each peptide bond are indicated by + and − symbols, respectively. Unbonded amino and carbonyl constituents in the peptide bonds near either end of the α-helical region are shown in red.

Thus there are five different kinds of constraints that affect the stability of an α helix: (1) the electrostatic repulsion (or attraction) between amino acid residues with charged R groups, (2) the bulkiness of adjacent R groups, (3) the interactions between amino acid side chains spaced three (or four) residues apart, (4) the occurrence of Pro residues, and (5) the interaction between amino acids at the ends of the helix and the electric dipole inherent to this structure.

The β Conformation Organizes Polypeptide Chains into Sheets

Pauling and Corey predicted a second type of repetitive structure, the **β conformation.** This is the more extended conformation of the poly- peptide chains, as seen in the silk protein fibroin (a member of a class of fibrous proteins called β-keratins), and its structure has been con- firmed by x-ray analysis. In the β conformation, which like the α helix is common in proteins, the backbone of the polypeptide chain is ex- tended into a zigzag rather than helical structure (Fig. 7–9). In fibroin the zigzag polypeptide chains are arranged side by side to form a struc- ture resembling a series of pleats; such a structure is called a β pleated sheet. In the β conformation the hydrogen bonds can be either intra- chain, or interchain between the peptide linkages of adjacent polypep- tide chains. All the peptide linkages of β-keratin participate in inter- chain hydrogen bonding. The R groups of adjacent amino acids protrude in opposite directions from the zigzag structure, creating an alternating pattern as seen in the side view (Fig. 7–9c).

Figure 7–9 The β conformation of polypeptide chains. Views show the R groups extending out from the β pleated sheet and emphasize the pleated sheet described by the planes of the peptide bonds. Hydrogen-bond cross-links between adjacent chains are also shown. **(a)** Antiparallel β sheets, in which the amino-terminal to carboxyl-terminal orientation of adjacent chains (arrows) is inverse. **(b)** Parallel β sheets. **(c)** Silk fibers are made up of the protein fibroin. Its structure consists of layers of antiparal- lel β sheets rich in Ala (purple) and Gly (yellow) residues. The small side chains interdigitate and allow close packing of each layered sheet, as shown in this side view.

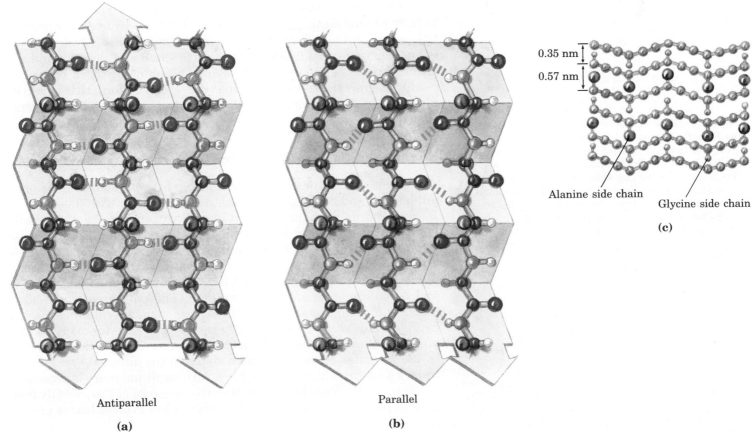

0.35 nm

0.57 nm

Alanine side chain

Glycine side chain

(c)

Antiparallel

(a)

Parallel

(b)

The adjacent polypeptide chains in a β pleated sheet can be either parallel (having the same amino-to-carboxyl polypeptide orientation) or antiparallel (having the opposite amino-to-carboxyl orientation). The structures are similar, although the repeat period is shorter for the parallel conformation (0.65 nm, as opposed to 0.7 nm for antiparallel).

In some structural situations there are limitations to the kinds of amino acids that can occur in the β structure. When two or more pleated sheets are layered closely together within a protein, the R groups of the amino acid residues on the contact surfaces must be relatively small. β-Keratins such as silk fibroin and the protein of spider webs have a very high content of Gly and Ala residues, those with the smallest R groups. Indeed, in silk fibroin Gly and Ala alternate over large parts of the sequence (Fig. 7–9c).

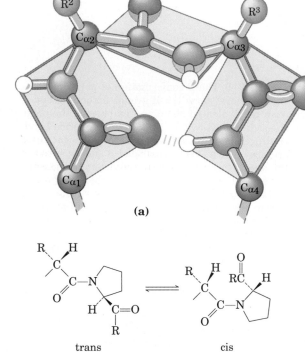

(a)

(b)

trans   cis

Figure 7–10 Structure of a β turn or β bend. **(a)** Note the hydrogen bond between the peptide groups of the first and fourth residues involved in the bend. **(b)** The trans and cis isomers of a peptide bond involving the imino nitrogen of proline. Over 99.95% of the peptide bonds between amino acid residues other than Pro are in the trans configuration. About 6% of the peptide bonds involving the imino nitrogen of proline, however, are in the cis configuration, and many of these occur at β turns.

Other Secondary Structures Occur in Some Proteins

The α helix and the β conformation are the major repetitive secondary structures easily recognized in a wide variety of proteins. Other repetitive structures exist, often in only one or a few specialized proteins. An example is the collagen helix (see Fig. 7–14). One other type of secondary structure is common enough to deserve special mention. This is a β bend or β turn (Fig. 7–10), often found where a polypeptide chain abruptly reverses direction. (These turns often connect the ends of two adjacent segments of an antiparallel β pleated sheet, hence the name.) The structure is a tight turn (~180°) involving four amino acids. The peptide groups flanking the first amino acid are hydrogen bonded to the peptide groups flanking the fourth. Gly and Pro residues often occur in β turns, the former because it is small and flexible; and the latter because peptide bonds involving the imino nitrogen of proline readily assume the cis configuration (Fig. 7–10b), a form that is particularly amenable to a tight turn. β Turns are often found near the surface of a protein.

Secondary Structure Is Affected by Several Factors

The α helix and β conformation are stable because steric repulsion is minimized and hydrogen bonding is maximized. As shown by a Ramachandran plot, these structures fall within a range of sterically allowed structures that is relatively restricted. Values of ϕ and ψ for common secondary structures are shown in Figure 7–11. Most values of ϕ and ψ for amino acid residues, taken from known protein structures, fall into the expected regions, with high concentrations near the α helix and β conformation values as expected. The only amino acid often found in a conformation outside these regions is glycine. Because its hydrogen side chain is small, a Gly residue can take up many conformations that are sterically forbidden for other amino acids.

Some amino acids are accommodated in the different types of secondary structures better than others. An overall summary is presented in Figure 7–12. Some biases, such as the presence of Pro and Gly residues in β turns, can be explained readily; other evident biases are not understood.

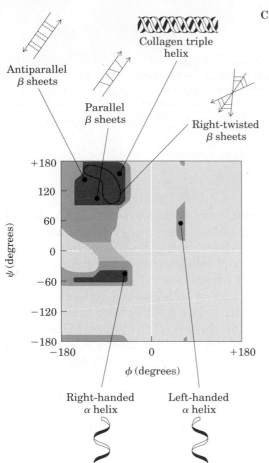

Figure 7–11 A Ramachandran plot. The values of ϕ and ψ for the various secondary structures are overlaid on the plot from Fig. 7–5.

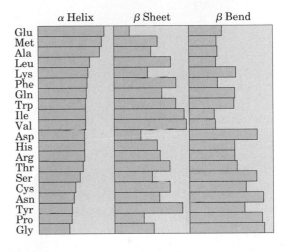

Figure 7–12 Relative probabilities that a given amino acid will occur in the three common types of secondary structure.

Fibrous Proteins Are Adapted for a Structural Function

α-Keratin, collagen, and elastin provide clear examples of the relationship between protein structure and biological function (Table 7–1). These proteins share properties that give strength and/or elasticity to structures in which they occur. They have relatively simple structures, and all are insoluble in water, a property conferred by a high concentration of hydrophobic amino acids both in the interior of the protein and on the surface. These proteins represent an exception to the rule that hydrophobic groups must be buried. The hydrophobic core of the molecule therefore contributes less to structural stability, and covalent bonds assume an especially important role.

Table 7–1 Secondary structures and properties of fibrous proteins

Structure	Characteristics	Examples of occurrence
α Helix, cross-linked by disulfide bonds	Tough, insoluble protective structures of varying hardness and flexibility	α-Keratin of hair, feathers, and nails
β Conformation	Soft, flexible filaments	Fibroin of silk
Collagen triple helix	High tensile strength, without stretch	Collagen of tendons, bone matrix
Elastin chains cross-linked by desmosine and lysinonorleucine	Two-way stretch with elasticity	Elastin of ligaments

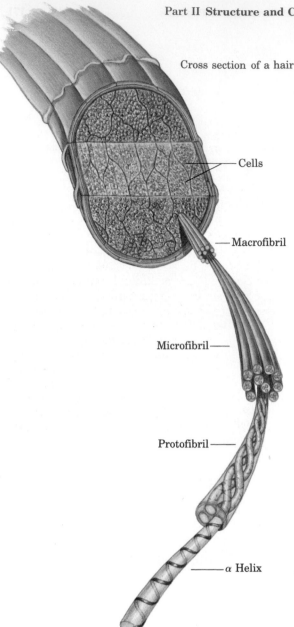

Cross section of a hair

Cells

Macrofibril

Microfibril

Protofibril

α Helix

Figure 7–13 The structure of hair and hair α-keratin. Three α-helical chains form a supercoiled three-stranded "rope" or protofibril, and 11 protofibrils constitute a hair microfibril. The α helices in each protofibril are linked by disulfide bonds, which are more abundant in hard keratins such as tortoise shell. Macrofibrils (bundles of microfibrils) pass through and around the cells in each hair.

α-Keratin and **collagen** have evolved for strength. In vertebrates, α-keratins constitute almost the entire dry weight of hair, wool, feathers, nails, claws, quills, scales, horns, hooves, tortoise shell, and much of the outer layer of skin. Collagen is found in connective tissue such as tendons, cartilage, the organic matrix of bones, and the cornea of the eye. The polypeptide chains of both proteins have simple helical structures. The α-keratin helix is the right-handed α helix found in many other proteins (Fig. 7–13). However, the collagen helix is unique. It is left-handed (see Box 7–1) and has three amino acid residues per turn (Fig. 7–14). In both α-keratin and collagen, a few amino acids predominate. α-Keratin is rich in the hydrophobic residues Phe, Ile, Val, Met, and Ala. Collagen is 35% Gly, 11% Ala, and 21% Pro and Hyp (hydroxyproline; see Fig. 5–8). The unusual amino acid content of collagen is imposed by structural constraints unique to the collagen helix. The amino acid sequence in collagen is generally a repeating tripeptide unit, Gly–X–Pro or Gly–X–Hyp, where X can be any amino acid. The food product gelatin is derived from collagen. Although it is protein, it has little nutritional value because collagen lacks significant amounts of many amino acids that are essential in the human diet.

In both α-keratin and collagen, strength is amplified by wrapping three helical strands together in a superhelix, much the way strings are twisted to make a strong rope (Figs. 7–13, 7–14). In keratin this ropelike structure is called a protofibril. In both proteins the helical path of the supertwists is opposite in sense to the twisting of the individual polypeptide helices, a conformation that permits the closet pos-

Figure 7–14 Structure of collagen. The collagen helix is a repeating secondary structure unique to this protein. **(a)** The repeating tripeptide sequence Gly–X–Pro or Gly–X–Hyp adopts a left-handed helical structure with three residues per turn. The repeating sequence used to generate this model is Gly–Pro–Hyp. **(b)** Space-filling model of the collagen helix shown in **(a)**. **(c)** Three of these helices wrap around one another with a right-handed twist. The resulting three-stranded molecule is referred to as tropocollagen (see Fig. 7–15). **(d)** The three-stranded collagen superhelix shown from one end, in a ball-and-stick representation. Glycine residues are shown in red. Glycine, because of its small size, is required at the tight junction where the three chains are in contact.

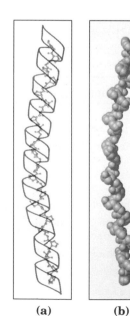
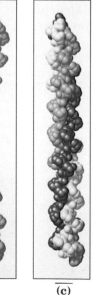
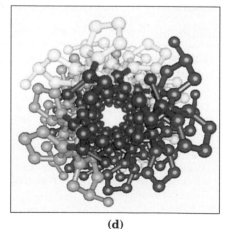

(a)   (b)   (c)   (d)

BOX 7–2   **Permanent Waving Is Biochemical Engineering**

α-Keratins exposed to moist heat can be stretched into the β conformation, but on cooling revert to the α-helical conformation spontaneously. This is because the R groups of α-keratins are larger on average than those of β-keratins and thus are not compatible with a stable β conformation. This characteristic of α-keratins, as well as their content of disulfide cross-linkages, is the basis of permanent waving. The hair to be waved is first bent around a form of appropriate shape. A solution of a reducing agent, usually a compound containing a thiol or sulfhydryl group (—SH), is then applied with heat. The reducing agent cleaves the disulfide cross-linkages by reducing each cystine to two cysteine residues, one in each adjacent chain. The moist heat breaks hydrogen bonds and causes the α-helical structure of the polypeptide chains to uncoil and stretch. After a time the reducing solution is removed, and an oxidizing agent is added to establish *new* disulfide bonds between pairs of Cys residues of adjacent polypeptide chains, but not the same pairs that existed before the treatment. On washing and cooling the hair, the polypeptide chains revert to their α-helical conformation. The hair fibers now curl in the desired fashion because new disulfide cross-linkages have been formed where they will exert some torsion or twist on the bundles of α-helical coils in the hair fibers.

sible packing of the three polypeptide chains. The superhelical twisting is left-handed in α-keratin (Fig. 7–13) and right-handed in collagen (Fig. 7–14). The tight wrapping of the collagen triple helix provides great tensile strength with no capacity to stretch. Collagen fibers can support up to 10,000 times their own weight and are said to have greater tensile strength than a steel wire of equal cross section.

The strength of these structures is also enhanced by covalent cross-links between polypeptide chains within the triple-helical "ropes" and between adjacent ones. In α-keratin, the cross-links are contributed by disulfide bonds (Box 7–2). In the hardest and toughest α-keratins, such as those of tortoise shells and rhinoceros horns, up to 18% of the residues are cysteines involved in disulfide bonds. The arrangement of α-keratin to form a hair fiber is shown in Figure 7–13. In collagen, the cross-links are contributed by an unusual type of covalent link between two Lys residues that creates a nonstandard amino acid residue called lysinonorleucine, found only in certain fibrous proteins.

| Polypeptide chain | Lys residue minus ε-amino group (norleucine) | Lys residue | Polypeptide chain |

Lysinonorleucine

Figure 7–15 The structure of collagen fibers. Tropocollagen (M_r 300,000) is a rod-shaped molecule, about 300 nm long and only 1.5 nm thick. The three helically intertwined polypeptides are of equal length, each having about 1,000 amino acid residues. In some collagens all three chains are identical in amino acid sequence, but in others two chains are identical and the third differs. The heads of adjacent molecules are staggered, and the alignment of the head groups of every fourth molecule produces characteristic cross-striations 64 nm apart that are evident in an electron micrograph.

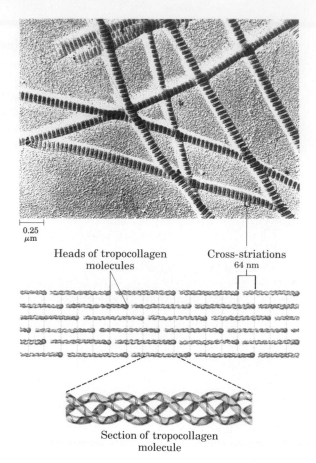

0.25 μm

Heads of tropocollagen molecules

Cross-striations 64 nm

Section of tropocollagen molecule

Collagen fibrils consist of recurring three-stranded polypeptide units called tropocollagen, arranged head to tail in parallel bundles (Fig. 7–15). The rigid, brittle character of the connective tissue in older people is the result of an accumulation of covalent cross-links in collagen as we age.

Human genetic defects involving collagen illustrate the close relationship between amino acid sequence and three-dimensional structure in this protein. Osteogenesis imperfecta results in abnormal bone formation in human babies. Ehlers-Danlos syndrome is characterized by loose joints. Both can be lethal and both result from the substitution of a Cys or Ser residue, respectively, for a Gly (a different Gly residue in each case) in the amino acid sequence of collagen. These seemingly small substitutions have a catastrophic effect on collagen function because they disrupt the Gly–X–Pro repeat that gives collagen its unique helical structure.

Elastic connective tissue contains the fibrous protein **elastin,** which resembles collagen in some of its properties but is very different in others. The polypeptide subunit of elastin fibrils is tropoelastin (M_r 72,000), containing about 800 amino acid residues. Like collagen, it is rich in Gly and Ala residues. Tropoelastin differs from tropocollagen in having many Lys but few Pro residues; it forms a special type of helix, different from the α helix and the collagen helix. Tropoelastin consists of lengths of helix rich in Gly residues separated by short regions containing Lys and Ala residues. The helical portions stretch on applying tension but revert to their original length when tension is released.

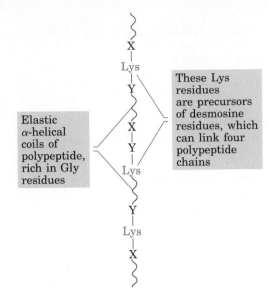

Figure 7–16 Tropoelastin molecules and their linkage to form a network of polypeptide chains in elastin. Elastin consists of tropoelastin molecules cross-linked to give two-dimensional or three-dimensional elasticity. In addition to desmosine residues (in red), which can link two, three, or four tropoelastin molecules, as shown, elastin contains other kinds of cross-linkages, such as lysinonorleucine, also designated in red.

The regions containing Lys residues form covalent cross-links. Four Lys side chains come together and are enzymatically converted into desmosine (see Fig. 5–8) and a related compound, isodesmosine; these amino acids are found only in elastin. Lysinonorleucine (p. 173) also occurs in elastin. These nonstandard amino acids are capable of joining tropoelastin chains into arrays that can be stretched reversibly in all directions (Fig. 7–16).

Protein Tertiary Structure

Although fibrous proteins generally have only one type of secondary structure, globular proteins can incorporate several types of secondary structure in the same molecule. Globular proteins—including enzymes, transport proteins, some peptide hormones, and immunoglobulins—are folded structures much more compact than α or β conformations (as shown for serum albumin in Figure 7–17).

The three-dimensional arrangement of all atoms in a protein is referred to as the tertiary structure, and this now becomes our focus. Whereas the secondary structure of polypeptide chains is determined by the *short-range* structural relationship of amino acid residues, tertiary structure is conferred by *longer-range* aspects of amino acid sequence. Amino acids that are far apart in the polypeptide sequence and reside in different types of secondary structure may interact when the

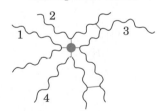

Cross-linking two chains

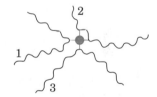

Linking three chains

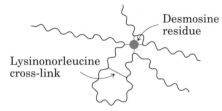

Linking four chains

β Conformation
200 x 0.5 nm

α Helix
90 x 1.1 nm

Native globular form
13 x 3 nm

Figure 7–17 Bovine serum albumin (M_r 64,500) has 584 residues in a single chain. Shown above are the approximate dimensions its single polypeptide chain would have if it occurred entirely in extended β conformation or as an α helix. Also shown (left) is the actual size of native serum albumin in its native globular form, as determined by physicochemical measurements; the polypeptide chain must be very compactly folded to fit into these dimensions.

protein is folded. The formation of bends in the polypeptide chain during folding and the direction and angle of these bends are determined by the number and location of specific bend-producing amino acids, such as Pro, Thr, Ser, and Gly residues. Moreover, loops of the highly folded polypeptide chain are held in their characteristic tertiary positions by different kinds of weak-bonding interactions (and sometimes by covalent bonds such as disulfide cross-links) between R groups of adjacent loops.

We will now consider how secondary structures contribute to the tertiary folding of a polypeptide chain in a globular protein, and how this structure is stabilized by weak interactions, in particular by hydrophobic interactions involving nonpolar amino acid side chains in the tightly packed core of the protein.

X-Ray Analysis of Myoglobin Revealed Its Tertiary Structure

The breakthrough in understanding globular protein structure came from x-ray diffraction studies of the protein myoglobin carried out by John Kendrew and his colleagues in the 1950s (Box 7–3). Myoglobin is a relatively small (M_r 16,700), oxygen-binding protein of muscle cells that functions in the storage and transport of oxygen for mitochondrial oxidation of cell nutrients. Myoglobin contains a single polypeptide chain of 153 amino acid residues of known sequence and a single iron-porphyrin, or **heme**, group (Fig. 7–18), identical to that of hemoglobin, the oxygen-binding protein of erythrocytes. The heme group is responsible for the deep red-brown color of both myoglobin and hemoglobin. Myoglobin is particularly abundant in the muscles of diving mammals such as the whale, seal, and porpoise, whose muscles are so rich in this protein that they are brown. Storage of oxygen by muscle myoglobin permits these animals to remain submerged for long periods of time.

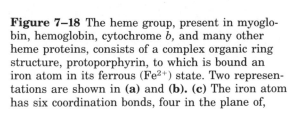

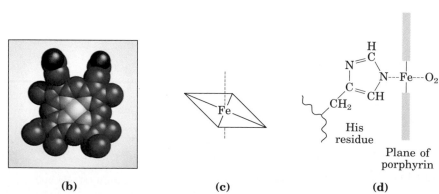

(a)   (b)   (c)   (d)

Figure 7–18 The heme group, present in myoglobin, hemoglobin, cytochrome b, and many other heme proteins, consists of a complex organic ring structure, protoporphyrin, to which is bound an iron atom in its ferrous (Fe^{2+}) state. Two representations are shown in **(a)** and **(b)**. **(c)** The iron atom has six coordination bonds, four in the plane of, and bonded to, the flat porphyrin molecule and two perpendicular to it. **(d)** In myoglobin and hemoglobin, one of the perpendicular coordination bonds is bound to a nitrogen atom of a His residue. The other is "open" and serves as the binding site for an O_2 molecule, as shown here in the edge view.

BOX 7–3   **X-Ray Diffraction**

The spacing of atoms in a crystal lattice can be determined by measuring the angles and the intensities at which a beam of x rays of a given wavelength is diffracted by the electron shells around the atoms. For example, x-ray analysis of sodium chloride crystals shows that Na^+ and Cl^- ions are arranged in a simple cubic lattice. The spacing of the different kinds of atoms in complex organic molecules, even very large ones such as proteins, can also be analyzed by x-ray diffraction methods. However, this is far more difficult than for simple salt crystals because the very large number of atoms in a protein molecule yields thousands of diffraction spots that must be analyzed by computer.

The process may be understood at an elementary level by considering how images are generated in a light microscope. Light from a point source is focused on an object. The light waves are scattered by the object, and these scattered waves are recombined by a series of lenses to generate an enlarged image of the object. The limit to the size of an object whose structure can be determined by such a system (i.e., its resolving power) is determined by the wavelength of the light. Objects smaller than half the wavelength of the incident light cannot be resolved. This is why x rays, with wavelengths in the range of a few tenths of a nanometer (often measured in angstroms, Å; 1 Å=0.1 nm), must be used for proteins. There are no lenses that can recombine x rays to form an image; the pattern of diffracted light is collected directly and converted into an image by computer analysis.

Operationally, there are several steps in x-ray structural analysis. The amount of information obtained depends on the degree of structural order in the sample. Some important structural parameters were obtained from early studies of the diffraction patterns of the fibrous proteins that occur in fairly regular arrays in hair and wool. More detailed three-dimensional structural information, however, requires a highly ordered crystal of a protein. Protein crystallization is something of an empirical science, and the structures of many important proteins are not yet known simply because they have proven difficult to crystallize. Once a crystal is obtained, it is placed in an x-ray beam between the x-ray source and a detector. A regular array of spots called reflections (Fig. 1) is generated by precessional motion of the crystal. The spots represent reflections of the x-ray beam, and each atom in a molecule makes a contribution to each spot. The overall pattern of spots is related to the structure of the protein through a mathematical device called a Fourier transform. The intensity of each spot is measured from the positions and intensities of the spots in several of these diffraction patterns, and the precise three-dimensional structure of the protein is calculated.

John Kendrew found that the x-ray diffraction pattern of crystalline myoglobin from muscles of the sperm whale is very complex, with nearly 25,000 reflections. Computer analysis of these reflections took place in stages. The resolution improved at each stage, until in 1959 the positions of virtually all the atoms in the protein could be determined. The amino acid sequence deduced from the structure agreed with that obtained by chemical analysis. The structures of hundreds of proteins have since been determined to a similar level of resolution, many of them much more complex than myoglobin.

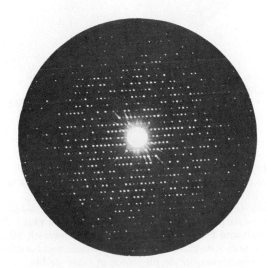

Figure 1 Photograph of the x-ray diffraction pattern of crystalline sperm whale myoglobin.

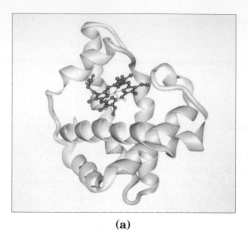

(a)

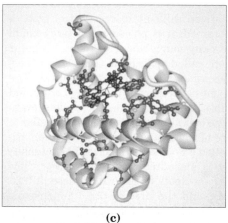

(c)

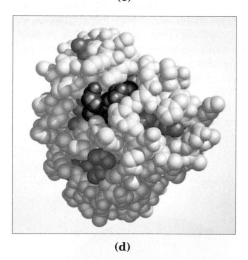

(d)

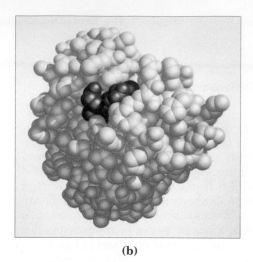

(b)

Figure 7–19 Tertiary structure of sperm whale myoglobin. The orientation of the protein is the same in all panels; the heme group is shown in red. **(a)** The polypeptide backbone, shown in a ribbon representation of a type introduced by Jane Richardson; this highlights regions of secondary structure. The α-helical regions in myoglobin are evident. Amino acid side chains are not shown. **(b)** A space-filling model, showing that the heme group is largely buried. All amino acid side chains are included. **(c)** A ribbon representation, including side chains (purple) for the hydrophobic residues Leu, Ile, Val, and Phe. **(d)** A space-filling model with all amino acid side chains. The hydrophobic residues are again shown in purple; most are not visible because they are buried in the interior of the protein.

Figure 7–19 shows several structural representations of myoglobin, illustrating how the polypeptide chain is folded in three dimensions—its tertiary structure. The backbone of the myoglobin molecule is made up of eight relatively straight segments of α helix interrupted by bends. The longest α helix has 23 amino acid residues and the shortest only seven; all are right-handed. More than 70% of the amino acids in the myoglobin molecule are in these α-helical regions. X-ray analysis also revealed the precise position of each of the R groups, which occupy nearly all the open space between the folded loops.

Other important conclusions were drawn from the structure of myoglobin. The positioning of amino acid side chains reflects a structure that derives much of its stability from hydrophobic interactions. Most of the hydrophobic R groups are in the interior of the myoglobin molecule, hidden from exposure to water. All but two of the polar R groups are located on the outer surface of the molecule, and all of them are hydrated. The myoglobin molecule is so compact that in its interior there is room for only four molecules of water. This dense hydrophobic core is typical of globular proteins. The fraction of space occupied by atoms in an organic liquid is 0.25 to 0.35; in a typical solid the fraction is 0.75. In a protein the fraction is 0.72 to 0.76, very comparable to that in a solid. In this closely packed environment weak interactions strengthen and reinforce each other. For example, the nonpolar side chains in the core are so close together that short-range van der Waals interactions make a significant contribution to stabilizing hydrophobic interactions. By contrast, in an oil droplet suspended in water, the van der Waals interactions are minimal and the cohesiveness of the droplet is based almost exclusively on entropy.

The structure of myoglobin both confirmed some expectations and introduced some new elements of secondary structure. As predicted by Pauling and Corey, all the peptide bonds are in the planar trans configuration. The α helices in myoglobin provided the first direct experimental evidence for the existence of this type of secondary structure. Each of the four Pro residues of myoglobin occurs at a bend (recall that the rigid R group of proline is largely incompatible with α-helical structure). Other bends contain Ser, Thr, and Asn residues, which are among the amino acids that tend to be incompatible with α-helical structure if they are in close proximity (p. 168).

The flat heme group rests in a crevice, or pocket, in the myoglobin molecule. The iron atom in the center of the heme group has two bonding (coordination) positions perpendicular to the plane of the heme.

One of these is bound to the R group of the His residue at position 93; the other is the site to which an O_2 molecule is bound. Within this pocket, the accessibility of the heme group to solvent is highly restricted. This is important for function because free heme groups in an oxygenated solution are rapidly oxidized from the ferrous (Fe^{2+}) form, which is active in the reversible binding of O_2, to the ferric (Fe^{3+}) form, which does not bind O_2.

Proteins Differ in Tertiary Structure

With the elucidation of the tertiary structures of hundreds of other globular proteins by x-ray analysis, it is clear that myoglobin represents only one of many ways in which a polypeptide chain can be folded. In Figure 7–20 the structures of cytochrome c, lysozyme, and ribonuclease are compared. All have different amino acid sequences and different tertiary structures, reflecting differences in function. Like myoglobin, cytochrome c is a small heme protein (M_r 12,400) containing a single polypeptide chain of about 100 residues and a single heme group, which in this case is covalently attached to the polypeptide. It functions as a component of the respiratory chain of mitochondria (Chapter 18). X-ray analysis of cytochrome c (Fig. 7–20) shows that only about 40% of the polypeptide is in α-helical segments, compared with almost 80% of the myoglobin chain. The rest of the cytochrome c chain contains bends, turns, and irregularly coiled and extended segments. Thus, cytochrome c and myoglobin differ markedly in structure, even though both are small heme proteins.

Figure 7–20 The three-dimensional structures of three small proteins: cytochrome c, lysozyme, and ribonuclease. For lysozyme and ribonuclease the active site of the enzyme faces the viewer. Key functional groups (the heme in cytochrome c, and amino acid side chains in the active site of lysozyme and ribonuclease) are shown in red; disulfide bonds are shown in yellow. Two representations of each protein are shown: a space-filling model and a ribbon representation. In the ribbon depictions, the β structures are represented by flat arrows and the α helices by spiral ribbons; the orientation in each case is the same as that of the space-filling model, to facilitate comparison.

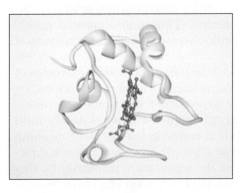
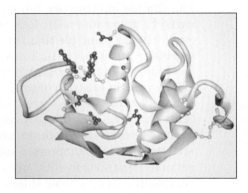
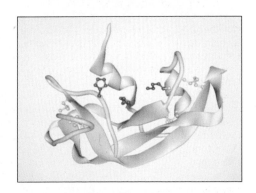
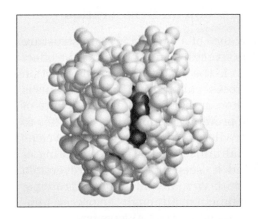
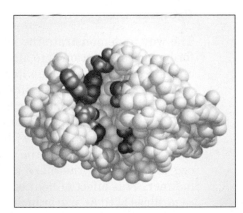
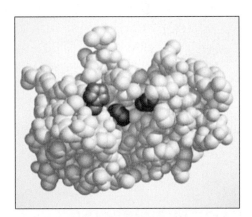

Cytochrome c   **Lysozyme**   **Ribonuclease**

Table 7–2 Approximate amounts of α helix and β conformation in some single-chain proteins*

Protein (total residues)	Residues (%)	
	α Helix	β Conformation
Myoglobin (153)	78	0
Cytochrome c (104)	39	0
Lysozyme (129)	40	12
Ribonuclease (124)	26	35
Chymotrypsin (247)	14	45
Carboxy-peptidase (307)	38	17

Source: Data from Cantor, C.R. & Schimmel, P.R. (1980) *Biophysical Chemistry,* Part I: *The Conformation of Biological Macromolecules,* p. 100, W.H. Freeman and Company, New York.

* Portions of the polypeptide chains that are not accounted for by α helix or β conformation consist of bends and irregularly coiled or extended stretches. Segments of α helix and β conformation sometimes deviate slightly from their normal dimensions and geometry.

Lysozyme (M_r 14,600) is an enzyme in egg white and human tears that catalyzes the hydrolytic cleavage of polysaccharides in the protective cell walls of some families of bacteria. Lysozyme is so named because it can lyse, or dissolve, bacterial cell walls and thus serve as a bactericidal agent. Like cytochrome c, about 40% of its 129 amino acid residues are in α-helical segments, but the arrangement is different and some β structure is also present. Four disulfide bonds contribute stability to this structure. The α helices line a long crevice in the side of the molecule (Fig. 7–20), called the active site, which is the site of substrate binding and action. The bacterial polysaccharide that is the substrate for lysozyme fits into this crevice.

Ribonuclease, another small globular protein (M_r 13,700), is an enzyme secreted by the pancreas into the small intestine, where it catalyzes the hydrolysis of certain bonds in the ribonucleic acids present in ingested food. Its tertiary structure, determined by x-ray analysis, shows that little of its 124 amino acid polypeptide chain is in α-helical conformation, but it contains many segments in the β conformation. Like lysozyme, ribonuclease has four disulfide bonds between loops of the polypeptide chain (Fig. 7–20).

Table 7–2 shows the relative percentages of α helix and β conformation among several small, single-chain, globular proteins. Each of these proteins has a distinct structure, adapted for its particular biological function. These proteins do share several important properties, however. Each is folded compactly, and in each case the hydrophobic amino acid side chains are oriented toward the interior (away from water) and the hydrophilic side chains are on the surface. These specific structures are also stabilized by a multitude of hydrogen bonds and some ionic interactions.

Proteins Lose Structure and Function on Denaturation

The way to demonstrate the importance of a specific protein structure for biological function is to alter the structure and determine the effect on function. One extreme alteration is the total loss or randomization of three-dimensional structure, a process called **denaturation.** This is the familiar process that occurs when an egg is cooked. The white of the egg, which contains the soluble protein egg albumin, coagulates to a white solid on heating. It will not redissolve on cooling to yield a clear solution of protein as in the original unheated egg white. Heating of egg albumin has therefore changed it, seemingly in an irreversible manner. This effect of heat occurs with virtually all globular proteins, regardless of their size or biological function, although the precise temperature at which it occurs may vary and it is not always irreversible. The change in structure brought about by denaturation is almost invariably associated with loss of function. This is an expected consequence of the principle that the specific three-dimensional structure of a protein is critical to its function.

Proteins can be denatured not only by heat, but also by extremes of pH, by certain miscible organic solvents such as alcohol or acetone, by certain solutes such as urea, or by exposure of the protein to detergents. Each of these denaturing agents represents a relatively mild treatment in the sense that no covalent bonds in the polypeptide chain are broken. Boiling a protein solution disrupts a variety of weak interactions. Organic solvents, urea, and detergents act primarily by disrupting the hydrophobic interactions that make up the stable core of globular proteins; extremes of pH alter the net charge on the protein, causing electrostatic repulsion and disruption of some hydrogen bonding. Remember that the native structure of most proteins is only marginally stable. It is not necessary to disrupt *all* of the stabilizing weak interactions to reduce the thermodynamic stability to a level that is insufficient to keep the protein conformation intact.

Amino Acid Sequence Determines Tertiary Structure

The most important proof that the tertiary structure of a globular protein is determined by its amino acid sequence came from experiments showing that denaturation of some proteins is reversible. Some globular proteins denatured by heat, extremes of pH, or denaturing reagents will regain their native structure and their biological activity, a process called **renaturation,** if they are returned to conditions in which the native conformation is stable.

A classic example is the denaturation and renaturation of ribonuclease. Purified ribonuclease can be completely denatured by exposure to a concentrated urea solution in the presence of a reducing agent. The reducing agent cleaves the four disulfide bonds to yield eight Cys residues, and the urea disrupts the stabilizing hydrophobic interactions, thus freeing the entire polypeptide from its folded conformation. Under these conditions the enzyme loses its catalytic activity and undergoes complete unfolding to a randomly coiled form (Fig. 7–21). When the urea and the reducing agent are removed, the randomly coiled, denatured ribonuclease spontaneously refolds into its correct tertiary structure, with full restoration of its catalytic activity (Fig. 7–21). The refolding of ribonuclease is so accurate that the four intrachain disulfide bonds are reformed in the same positions in the renatured molecule as in the native ribonuclease. In theory, the eight Cys residues could have recombined at random to form up to four disulfide bonds in 105 different ways. This classic experiment, carried out by Christian Anfinsen in the 1950s, proves that the amino acid sequence of the polypeptide chain of proteins contains all the information required to fold the chain into its native, three-dimensional structure.

The study of homologous proteins has strengthened this conclusion. We have seen that in a series of homologous proteins, such as cytochrome *c,* from different species, the amino acid residues at certain positions in the sequence are invariant, whereas at other positions the amino acids may vary (see Fig. 6–15). This is also true for myoglobins isolated from different species of whales, from the seal, and from some terrestrial vertebrates. The similarity of the tertiary structures and amino acid sequences of myoglobins from different sources led to the conclusion that the amino acid sequence of myoglobin somehow must determine its three-dimensional folding pattern, an idea substantiated by the similar structures found by x-ray analysis of myoglobins from different species. Other sets of homologous proteins also show this relationship; in each case there are sequence homologies as well as similar tertiary structures.

Many of the invariant amino acid residues of homologous proteins appear to occur at critical points along the polypeptide chain. Some are found at or near bends in the chain, others at cross-linking points between loops in the tertiary structure, such as Cys residues involved in disulfide bonds. Still others occur at the catalytic sites of enzymes or at the binding sites for prosthetic groups, such as the heme group of cytochrome *c.*

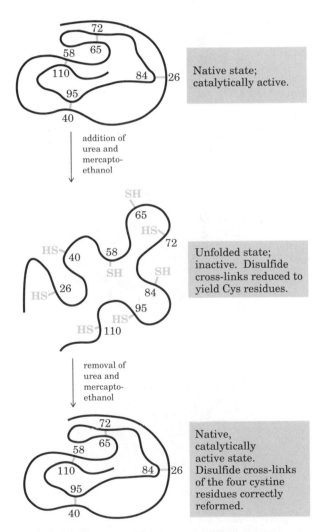

Figure 7–21 Renaturation of unfolded, denatured ribonuclease, with reestablishment of correct disulfide cross-links. Urea is added to denature ribonuclease, and mercaptoethanol ($HOCH_2CH_2SH$) to reduce and thus cleave the disulfide bonds of the four cystine residues to yield eight cysteine residues.

Looking at naturally occurring amino acid substitutions has an important limitation. Any change that abolishes the function of an essential protein (e.g., a change in an invariant residue) usually results in death of the organism very early in development. This severe form of natural selection eliminates many potentially informative changes from study. Fortunately, biochemists have devised methods to specifically alter amino acid sequences in the laboratory and examine the effects of these changes on protein structure and function. These methods are derived from recombinant DNA technology (Chapter 28) and rely on altering the genetic material encoding the protein. By this process, called **site-directed mutagenesis,** specific amino acid sequences can be changed by deleting, adding, rearranging, or substituting amino acid residues. The catalytic roles of certain amino acids lining the active sites of enzymes such as triose phosphate isomerase and chymotrypsin have been elucidated by substituting different amino acids in their place. The importance of certain amino acids in protein folding and structure is being addressed in the same way.

Tertiary Structures Are Not Rigid

Although the native tertiary conformation of a globular protein is the thermodynamically most stable form its polypeptide chain can assume, this conformation must not be regarded as absolutely rigid. Globular proteins have a certain amount of flexibility in their backbones and undergo short-range internal fluctuations. Many globular proteins also undergo small conformational changes in the course of their biological function. In many instances, these changes are associated with the binding of a ligand. The term **ligand** in this context refers to a specific molecule that is bound by a protein (from Latin, *ligare,* "to tie" or "bind"). For example, the hemoglobin molecule, which we shall examine later in this chapter, has one conformation when oxygen is bound, and another when the oxygen is released. Many enzyme molecules also undergo a conformational change on binding their substrates, a process that is part of their catalytic action (Chapter 8).

Polypeptides Fold Rapidly by a Stepwise Process

In living cells, proteins are made from amino acids at a very high rate. For example, *Escherichia coli* cells can make a complete, biologically active protein molecule containing 100 amino acid residues in about 5 s at 37 °C. Yet calculations show that at least 10^{50} yr would be required for a polypeptide chain of 100 amino acid residues to fold itself spontaneously by a random process in which it tries out all possible conformations around every single bond in its backbone until it finds its native, biologically active form. Thus protein folding cannot be a completely random, trial-and-error process. There simply must be shortcuts.

The folding pathway of a large polypeptide chain is unquestionably complicated, and the principles that guide this process have not yet been worked out in detail. For several proteins, however, there is evidence that folding proceeds through several discrete intermediates, and that some of the earliest steps involve local folding of regions of secondary structure. In one model (Fig. 7–22), the process is envisioned as hierarchical, following the levels of structure outlined at the beginning of this chapter. Local secondary structures would form first, followed by longer-range interactions between, say, two α helices with compatible amino acid side chains, a process continuing until folding

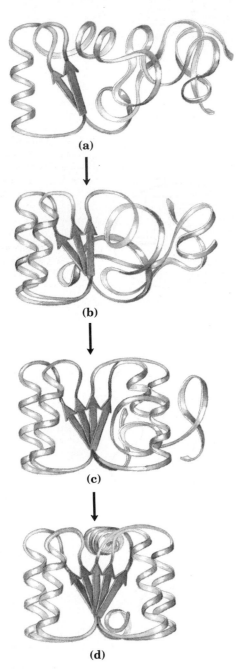

Figure 7–22 A possible protein-folding pathway. **(a)** Protein folding often begins with spontaneous formation of a structural nucleus consisting of a few particularly stable regions of secondary structure. **(b)** As other regions adopt secondary structure, they are stabilized by long-range interactions with the structural nucleus. **(c)** The folding process continues until most of the polypeptide has assumed regular secondary structure. **(d)** The final structure generally represents the most thermodynamically stable conformation.

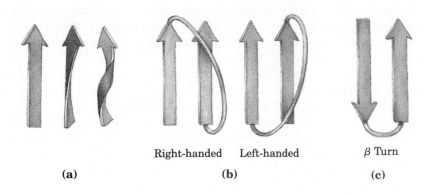

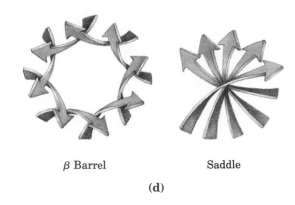

	Right-handed	Left-handed		β Turn		β Barrel	Saddle
(a)		**(b)**		**(c)**		**(d)**	

Figure 7–23 Extended β chains of amino acids tend to twist in a right-handed sense because the slightly twisted conformation is more stable than the linear conformation **(a)**. This influences the conformation of the polypeptide segments that connect two β strands, and also the stable conformations assumed by several adjacent β strands. **(b)** Connections between parallel β chains are right-handed. **(c)** The β turn is a common connector between antiparallel β chains. **(d)** The tendency for right-handed twisting is seen in two particularly stable arrangements of adjacent β chains: the β barrel and the saddle; these structures form the stable core of many proteins.

was complete. In an alternative model, folding is initiated by a spontaneous collapse of the polypeptide into a compact state mediated by hydrophobic interactions among nonpolar residues. The state resulting from this "hydrophobic collapse" may have a high content of secondary structure, but many amino acid side chains are not entirely fixed. Either or both models (and perhaps others) may apply to a given protein.

A number of structural constraints help to guide the interaction of regions of secondary structure. The most common patterns are sometimes referred to as supersecondary structures. A prominent one is a tendency for extended β conformations to twist in a right-handed sense (Fig. 7–23a). This influences both the arrangement of β sheets relative to one another and the path of the polypeptide segment connecting two β strands. Two parallel β strands, for example, must be connected by a crossover strand (Fig. 7–23b). In principle, this crossover could have a right- or left-handed conformation, but only the right-handed form is found in proteins. The twisting of β sheets also leads to a characteristic twisting of the structure formed when many sheets are put together. Two examples of resulting structures are the β barrel and saddle shapes (Fig. 7–23d), which form the core of many larger structures.

Weak-bonding interactions represent the ultimate thermodynamic constraint on the interaction of different regions of secondary structure. The R groups of amino acids project outward from α-helical and β structures, and thus the need to bury hydrophobic residues means that water-soluble proteins must have more than one layer of secondary structure. One simple structural method for burying hydrophobic residues is a supersecondary structural unit called a βαβ loop (Fig. 7–24), a structure often repeated multiple times in larger proteins. More elaborate structures are domains made up of facing β sheets (with hydrophobic residues sandwiched between), and β sheets covered on one side with several α helices, as described later.

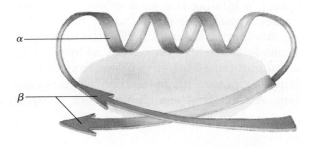

α

β

Figure 7–24 The βαβ loop. The shaded region denotes the area where stabilizing hydrophobic interactions occur.

It becomes more difficult to bury hydrophobic residues in smaller structures, and the number of potential weak interactions available for stabilization decreases. For this reason, smaller proteins are often held together with a number of covalent bonds, principally disulfide linkages. Recall the multiple disulfide bonds in the small proteins insulin (see Fig. 6–10) and ribonuclease (Fig. 7–21). Other types of covalent bonds also occur. The heme group in cytochrome *c,* for example, is covalently linked to the protein on two sides, providing a significant stabilization of the entire protein structure.

Not all proteins fold spontaneously as they are synthesized in the cell. Proteins that facilitate the folding of other proteins have been found in a wide variety of cells. These are called **polypeptide chain binding proteins** or molecular chaperones. Several of these proteins can bind to polypeptide chains, preventing nonspecific aggregation of weak-bonding side chains. They guide the folding of some polypeptides, as well as the assembly of multiple polypeptides into larger structures. Dissociation of polypeptide chain binding proteins from polypeptides is often coupled to ATP hydrolysis. One family of such proteins has structures that are highly conserved in organisms ranging from bacteria to mammals. These proteins (M_r 70,000), as well as several other families of polypeptide chain binding proteins, were originally identified as "heat shock" proteins because they are induced in many cells when heat stress is applied, and apparently help stabilize other proteins.

Some proteins have also been found that promote polypeptide folding by catalyzing processes that otherwise would limit the rate of folding, such as the reversible formation of disulfide bonds or proline isomerization (the interconversion of the cis and trans isomers of peptide bonds involving the imino nitrogen of proline; see Fig. 7–10).

There Are a Few Common Tertiary Structural Patterns

Following the folding patterns outlined above and others yet to be discovered, a newly synthesized polypeptide chain quickly assumes its most stable tertiary structure. Although each protein has a unique structure, several patterns of tertiary structure seem to occur repeatedly in proteins that differ greatly in biological function and amino acid sequence (Fig. 7–25). This may reflect an unusual degree of stability and/or functional flexibility conferred by these particular tertiary structures. It also demonstrates that biological function is determined not only by the overall three-dimensional shape of the protein, but also by the arrangement of amino acids within that shape.

One structural motif is made up of eight β strands arranged in a circle with each β strand connected to its neighbor by an α helix. The β regions are arranged in the barrel structure described in Figure 7–23, and they influence the overall tertiary structure, giving rise to the name α/β barrel (Fig. 7–25a). This structure is found in many enzymes; a binding site for a cofactor or substrate is often found in a pocket formed near an end of the barrel.

Another structural motif is the four-helix bundle (Fig. 7–25b), in which four α helices are connected by three peptide loops. The helices are slightly tilted to form a pocket in the middle, which often contains a binding site for a metal or other cofactors essential for biological function. A somewhat similar structure in which seven helices are ar-

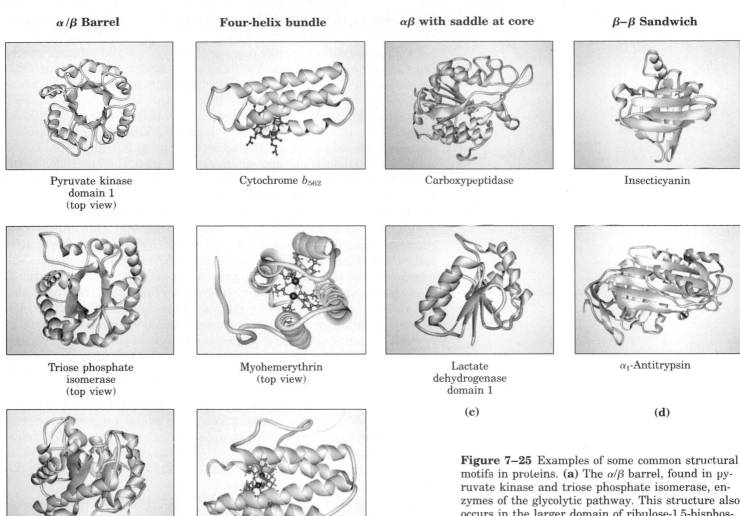

α/β Barrel

Pyruvate kinase
domain 1
(top view)

Triose phosphate
isomerase
(top view)

Triose phosphate
isomerase
(side view)

(a)

Four-helix bundle

Cytochrome b_{562}

Myohemerythrin
(top view)

Myohemerythrin
(side view)

(b)

αβ with saddle at core

Carboxypeptidase

Lactate
dehydrogenase
domain 1

(c)

β–β Sandwich

Insecticyanin

α_1-Antitrypsin

(d)

Figure 7–25 Examples of some common structural motifs in proteins. **(a)** The α/β barrel, found in pyruvate kinase and triose phosphate isomerase, enzymes of the glycolytic pathway. This structure also occurs in the larger domain of ribulose-1,5-bisphosphate carboxylase/oxygenase (known also as rubisco), an enzyme essential to the fixation of CO_2 by plants; in glycolate oxidase, an enzyme in photorespiration; and in a number of other unrelated proteins. **(b)** The four-helix bundle, shown here in cytochrome b_{562} and myohemerythrin. A dinuclear iron center and coordinating amino acids in myohemerythrin are shown in orange. Myohemerythrin is a nonheme oxygen-transporting protein found in certain worms and mollusks. The four-helix bundle is also found in apoferritin and the tobacco mosaic virus coat-protein. Apoferritin is a widespread protein involved in iron transport and storage. **(c)** αβ with saddle at core, in carboxypeptidase, a protein-hydrolyzing (proteolytic) enzyme, and lactate dehydrogenase, a glycolytic enzyme. **(d)** β–β Sandwich. In the protein insecticyanin of moths, the hydrophobic pocket binds biliverdin, a colored substance that plays a role in camouflage. α_1-Antitrypsin is a naturally occurring inhibitor of the proteolytic enzyme trypsin.

ranged in a barrel-like motif is found in some membrane proteins (see Fig. 10–10). The seven helices often surround a channel that spans the membrane.

A third motif has a β sheet in the "saddle" conformation forming a stable core, often surrounded by a number of α-helical regions (Fig. 7–25c). Structures of this kind are found in many enzymes. The location of the substrate binding site varies, determined by the placement of the α helices and other variable structural elements.

One final motif makes use of a sandwich of β sheets, layered so that the strands of the sheets form a quiltlike cross-hatching when viewed from above (Fig. 7–25d). This creates a hydrophobic pocket between the β sheets that is often a binding site for a planar hydrophobic molecule.

Protein Quaternary Structure

Some proteins contain two or more separate polypeptide chains or subunits, which may be identical or different in structure. One of the best-known examples of a multisubunit protein is hemoglobin, the oxygen-carrying protein of erythrocytes. Among the larger, more complex multisubunit proteins are the enzyme RNA polymerase of *E. coli*, responsible for initiation and synthesis of RNA chains; the enzyme aspartate transcarbamoylase (12 chains; see Fig. 8–26), important in the synthesis of nucleotides; and, as an extreme case, the enormous pyruvate dehydrogenase complex of mitochondria, which is a cluster of three enzymes containing a total of 102 polypeptide chains.

The arrangement of proteins and protein subunits in three-dimensional complexes constitutes quaternary structure. The interactions between subunits are stabilized and guided by the same forces that stabilize tertiary structure: multiple noncovalent interactions. The association of polypeptide chains can serve a variety of functions. Many multisubunit proteins serve regulatory functions; their activities are altered by the binding of certain small molecules. Interactions between subunits can permit very large changes in enzyme activity in response to small changes in the concentration of substrate or regulatory molecules (Chapter 8). In other cases, separate subunits can take on separate but related functions. Entire metabolic pathways are often organized by the association of a supramolecular complex of enzymes, permitting an efficient channeling of pathway intermediates from one enzyme to the next. Other associations, such as the histones in a nucleosome or the coat proteins of a virus, serve primarily structural roles. Large assemblies sometimes reflect complex functions. One obvious example is the complicated structure of ribosomes (see Fig. 26–12), which carry out protein synthesis.

X-ray and other analytical methods for structure determination become more difficult as the size and number of subunits in a protein increases. Nevertheless, sufficient data are already available to yield some very important information about the structure and function of multisubunit proteins.

X-Ray Analysis Revealed the Complete Structure of Hemoglobin

The first oligomeric protein to be subjected to x-ray analysis was hemoglobin (M_r 64,500), which contains four polypeptide chains and four heme prosthetic groups, in which the iron atoms are in the ferrous (Fe^{2+}) state. The protein portion, called globin, consists of two α chains (141 residues each) and two β chains (146 residues each). Note that α and β do not refer to secondary structures in this case. Because hemoglobin is four times as large as myoglobin, much more time and effort were required to solve its three-dimensional structure, finally achieved by Max Perutz, John Kendrew, and their colleagues in 1959.

The hemoglobin molecule is roughly spherical, with a diameter of about 5.5 nm. The α and β chains contain several segments of α helix separated by bends, with a tertiary structure very similar to that of the single polypeptide of myoglobin. In fact, there are 27 invariant amino acid residues in these three polypeptide chains, and closely related amino acids at 40 additional positions, indicating that these polypeptides (myoglobin and the α and β chains of hemoglobin) are evolutionarily related. The four polypeptide chains in hemoglobin fit together in an approximately tetrahederal arrangement (Fig. 7–26).

John Kendrew

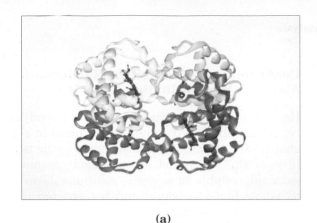

(a)

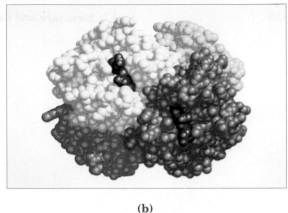

(b)

One heme is bound to each polypeptide chain of hemoglobin. The oxygen-binding sites are rather far apart given the size of the molecule, about 2.5 nm from one another. Each heme is partially buried in a pocket lined with hydrophobic amino acid side chains. It is bound to its polypeptide chain through a coordination bond of the iron atom to the R group of a His residue (see Fig. 7–18). The sixth coordination bond of the iron atom of each heme is available to bind O_2.

Closer examination of the quaternary structure of hemoglobin, with the help of molecular models, shows that although there are few contacts between the two α chains or between the two β chains, there are many contact points between the α and β chains. These contact points consist largely of hydrophobic side chains of amino acid residues, but also include ionic interactions involving the carboxyl-terminal residues of the four subunits.

Naturally occurring changes in the amino acid sequence of hemoglobin provide some useful insights into the relationship between structure and function in proteins. More than 300 genetic variants of hemoglobin are known to occur in the human population. Most of these variations are single amino acid changes that have only minor structural or functional effects. An exception is a substitution of valine for glutamate at position 6 of the β chain. This residue is on the outer surface of the molecule, and the change produces a "sticky" hydrophobic spot on the surface that results in abnormal quaternary association of hemoglobin. When oxygen concentrations are below a critical level, the subunits polymerize into linear arrays of fibers that distort cell shape. The result is a sickling of erythrocytes (Fig. 7–27), the cause of sickle-cell anemia.

Figure 7–26 The three-dimensional (quaternary) structure of deoxyhemoglobin, revealed by x-ray diffraction analysis, showing how the four subunits are packed together. **(a)** A ribbon representation. **(b)** A space-filling model. The α subunits are shown in white and light blue; the β subunits are shown in pink and purple. Note that the heme groups, shown in red, are relatively far apart.

Figure 7–27 Scanning electron micrographs of **(a)** normal and **(b)** sickled human erythrocytes. The sickled cells are fragile, and their breakdown causes anemia.

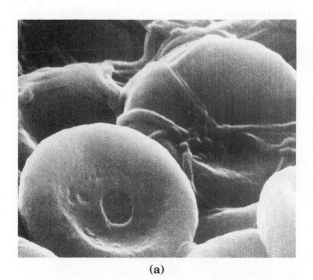

(a)

(b)

Conformational Changes in Hemoglobin Alter Its Oxygen-Binding Capacity

Hemoglobin is an instructive model for studying the function of many regulatory oligomeric proteins. The blood in a human being must carry about 600 L of oxygen from the lungs to the tissues every day, but very little of this is carried by the blood plasma because oxygen is only sparingly soluble in aqueous solutions. Nearly all the oxygen carried by whole blood is bound and transported by the hemoglobin of the erythrocytes. Normal human erythrocytes are small (6 to 9 μm), biconcave disks (Fig. 7–27a). They have no nucleus, mitochondria, endoplasmic reticulum, or other organelles. The hemoglobin of the erythrocytes in arterial blood passing from the lungs to the peripheral tissues is about 96% saturated with oxygen. In the venous blood returning to the heart, the hemoglobin is only about 64% saturated. Thus blood passing through a tissue releases about one-third of the oxygen it carries.

The special properties of the hemoglobin molecule that make it such an effective oxygen carrier are best understood by comparing the O_2-binding or O_2-saturation curves of myoglobin and hemoglobin (Fig. 7–28). These show the percentage of O_2-binding sites of hemoglobin or myoglobin that are occupied by O_2 molecules when solutions of these proteins are in equilibrium with different partial pressures of oxygen in the gas phase. (The partial pressure of oxygen, abbreviated pO_2, is the pressure contributed by oxygen to the overall pressure of a mixture of gases, and is directly related to the concentration of oxygen in the mixture.)

From its saturation curve, it is clear that myoglobin has a very high affinity for oxygen (Fig. 7–28). Furthermore, the O_2-saturation curve of myoglobin is a simple hyperbolic curve, as might be expected from the mass action of oxygen on the equilibrium myoglobin + $O_2 \rightleftharpoons$ oxymyoglobin. In contrast, the oxygen affinity of each of the four O_2-binding sites of deoxyhemoglobin is much lower, and the O_2-saturation curve of hemoglobin is sigmoid (S-shaped) (Fig. 7–28). This shape indicates that whereas the affinity of hemoglobin for binding the first O_2 molecule (to any of the four sites) is relatively low, the second, third, and fourth O_2 molecules are bound with a very much higher affinity. This accounts for the steeply rising portion of the sigmoid curve. The increase in the affinity of hemoglobin for oxygen after the first O_2 molecule is bound is almost 500-fold. Thus the oxygen affinity of each heme–polypeptide subunit of hemoglobin depends on whether O_2 is bound to neighboring subunits. The conversion of deoxyhemoglobin to oxyhemoglobin requires the disruption of ionic interactions involving the carboxyl-terminal residues of the four subunits, interactions that con-

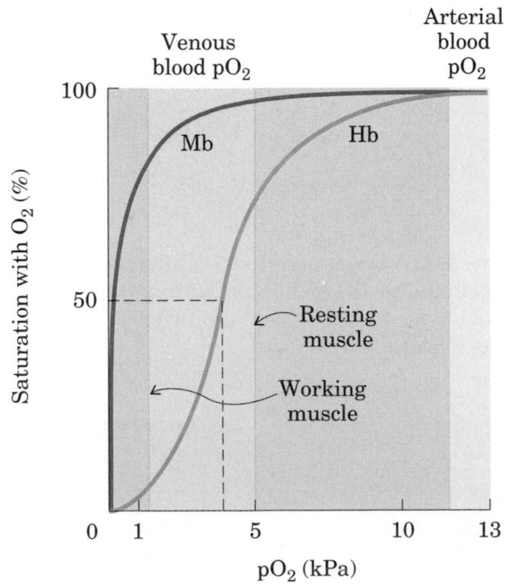

Figure 7–28 The oxygen-binding curves of myoglobin (Mb) and hemoglobin (Hb). Myoglobin has a much greater affinity for oxygen than does hemoglobin. It is 50% saturated at oxygen partial pressures (pO_2) of only 0.15 to 0.30 kPa, whereas hemoglobin requires a pO_2 of about 3.5 kPa for 50% saturation. Note that although both hemoglobin and myoglobin are more than 95% saturated at the pO_2 in arterial blood leaving the lungs (~13 kPa), hemoglobin is only about 75% saturated in resting muscle, where the pO_2 is about 5 kPa, and only 10% saturated in working muscle, where the pO_2 is only about 1.5 kPa. Thus hemoglobin can release its oxygen very effectively in muscle and other peripheral tissues. Myoglobin, on the other hand, is still about 80% saturated at a pO_2 of 1.5 kPa, and therefore unloads very little oxygen even at very low pO_2. Thus the sigmoid O_2-saturation curve of hemoglobin is a molecular adaptation for its transport function in erythrocytes, assuring the binding and release of oxygen in the appropriate tissues.

strain the overall structure in a low-affinity state. The increase in affinity for successive O_2 molecules reflects the fact that more of these ionic interactions must be broken for binding the first O_2 than for binding later ones.

Once the first heme–polypeptide subunit binds an O_2 molecule, it communicates this information to the remaining subunits through interactions at the subunit interfaces. The subunits respond by greatly increasing their oxygen affinity. This involves a change in the conformation of hemoglobin that occurs when oxygen binds (Fig. 7–29). Such communication among the four heme–polypeptide subunits of hemoglobin is the result of cooperative interactions among the subunits. Because binding of one O_2 molecule increases the probability that further O_2 molecules will be bound by the remaining subunits, hemoglobin is said to have **positive cooperativity.** Sigmoid binding curves, like that of hemoglobin for oxygen, are characteristic of positive cooperative binding. Cooperative oxygen binding does not occur with myoglobin, which has only one heme group within a single polypeptide chain and thus can bind only one O_2 molecule; its saturation curve is therefore hyperbolic. The multiple subunits of hemoglobin and the interactions between these subunits result in a fundamental difference between the O_2-binding actions of myoglobin and hemoglobin.

Positive cooperativity is not the only result of subunit interactions in oligomeric proteins. Some oligomeric proteins show **negative cooperativity:** binding of one ligand molecule *decreases* the probability that further ligand molecules will be bound. These and additional regulatory mechanisms used by these proteins are considered in Chapter 8.

Hemoglobin Binds Oxygen in the Lungs and Releases It in Peripheral Tissues

In the lungs the pO_2 in the air spaces is about 13 kPa; at this pressure hemoglobin is about 96% saturated with oxygen. However, in the cells of a working muscle the pO_2 is only about 1.5 kPa because muscle cells use oxygen at a high rate and thus lower its local concentration. As the blood passes through the muscle capillaries, oxygen is released from the nearly saturated hemoglobin in the erythrocytes into the blood plasma and thence into the muscle cells. As is evident from the O_2-saturation curve in Figure 7–28, hemoglobin releases about a third of its bound oxygen as it passes through the muscle capillaries, so that when it leaves the muscle, it is only about 64% saturated. When the blood returns to the lungs, where the pO_2 is much higher (13 kPa), the hemoglobin quickly binds more oxygen until it is 96% saturated again.

Now suppose that the hemoglobin in the erythrocyte were replaced by myoglobin. We see from the hyperbolic O_2-saturation curve of myoglobin (Fig. 7–28) that only 1 or 2% of the bound oxygen can be released from myoglobin as the pO_2 decreases from 13 kPa in the lungs to 3 kPa in the muscle. Myoglobin therefore is not very well adapted for carrying oxygen from the lungs to the tissues, because it has a much higher affinity for oxygen and releases very little of it at the pO_2 in muscles and other peripheral tissues. However, in its true biological function *within* muscle cells, which is to store oxygen and make it available to the mitochondria, myoglobin is in fact much better suited than hemoglobin, because its very high affinity for oxygen at low pO_2 enables it to bind and store oxygen effectively. Thus hemoglobin and myoglobin are specialized and adapted for different kinds of O_2-binding functions.

Figure 7–29 Conformational changes induced in hemoglobin when oxygen binds. (The oxygen-bound form is shown at bottom.) There are multiple structural changes, some not visible here; most of the changes are subtle. The α and β subunits are colored as in Fig. 7–26.

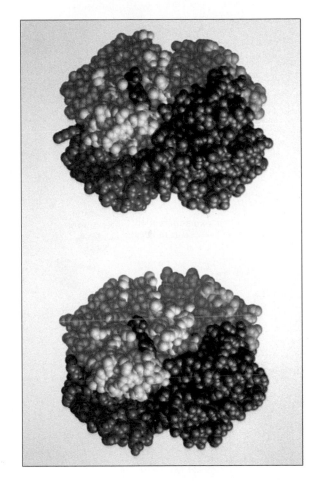

There Are Limits to the Size of Proteins

The relatively large size of proteins reflects their functions. The function of an enzyme, for example, requires a protein large enough to form a specifically structured pocket to bind its substrate. The size of proteins has limits, however, imposed by the genetic coding capacity of nucleic acids and the accuracy of the protein biosynthetic process. The use of many copies of one or a few proteins to make a large enclosing structure is important for viruses because this strategy conserves genetic material. Remember that there is a linear correspondence between the sequence of a gene in nucleic acid and the amino acid sequence of the protein for which it codes (see Fig. 6–14). The nucleic acids of viruses are much too small to encode the information required for a protein shell made of a single polypeptide. By using many copies of much smaller proteins for the virus coat, a much shorter nucleic acid is needed for the protein subunits, and this nucleic acid can be efficiently used over and over again. Cells also use large protein complexes in muscle, cilia, the cytoskeleton, and other structures. It is simply more efficient to make many copies of a small protein than one copy of a very large one. The second factor limiting the size of proteins is the error frequency during protein biosynthesis. This error frequency is low but can become significant for very large proteins. Simply put, the potential for incorporating a "wrong" amino acid in a protein is greater for a large protein than a small one.

Some Proteins Form Supramolecular Complexes

The same principles that govern the stability of secondary, tertiary, and quaternary structure in proteins guide the formation of very large protein complexes. These function, for example, as biological engines (muscle and cilia), large structural enclosures (virus coats), cellular skeletons (actin and tubulin filaments), DNA-packaging complexes (chromatin), and machines for protein synthesis (ribosomes). In many cases the complex consists of a small number of distinct proteins, specialized so that they spontaneously polymerize to form large structures.

Muscle provides an example of a supramolecular complex of multiple copies of a limited number of proteins. The contractile force of muscle is generated by the interaction of two proteins, actin and myosin (Chapter 2). Myosin is a long, rodlike molecule (M_r 540,000) consisting of six polypeptide chains, two so-called heavy chains ($M_r \sim 230,000$) and four light chains ($M_r \sim 20,000$) (Fig. 7–30a). The two heavy chains have long α-helical tails that twist around each other in a right-handed fashion. The large head domain, at one end of each heavy chain, interacts with actin and contains a catalytic site for ATP hydrolysis. Many myosin molecules assemble together to form the **thick filaments** of skeletal muscle (Fig. 7–31).

The other protein, actin, is a polymer of the globular protein G-actin (M_r 42,000); two such polymers coil around each other in a right-handed helix to form a **thin filament** (Fig. 7–30b). The interaction between actin and myosin is dynamic; contacts consist of multiple weak interactions that are strong enough to provide a stable association but weak enough to allow dissociation when needed. Hydrolysis of ATP in the myosin head is coupled to a series of conformational changes that bring about muscle contraction (Fig. 7–32). A similar engine involving an interaction between tubulin and dynein brings about the motion of cilia (Chapter 2).

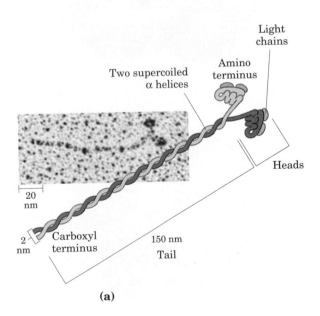

Light chains

Amino terminus

Two supercoiled α helices

Heads

20 nm

2 nm

Carboxyl terminus

150 nm
Tail

(a)

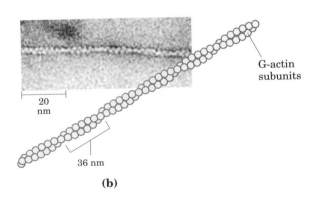

G-actin subunits

20 nm

36 nm

(b)

Figure 7–30 Myosin and actin, the two filamentous proteins of contractile systems. **(a)** The myosin molecule has a long tail consisting of two supercoiled α-helical polypeptide chains (heavy chains). The head of each heavy chain is associated with two light chains and is an enzyme capable of hydrolyzing ATP. **(b)** A representation of an F-actin fiber, which consists of two chains of G-actin subunits coiled about each other to form a filament.

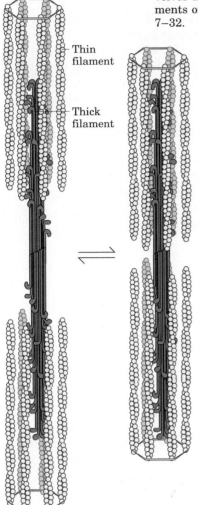

Figure 7–31 The thick and thin filaments of muscle. Many myosin molecules assemble in a bundle to form a thick filament. Muscle contraction involves the sliding of thick filaments past thin filaments of actin, by a mechanism described in Fig. 7–32.

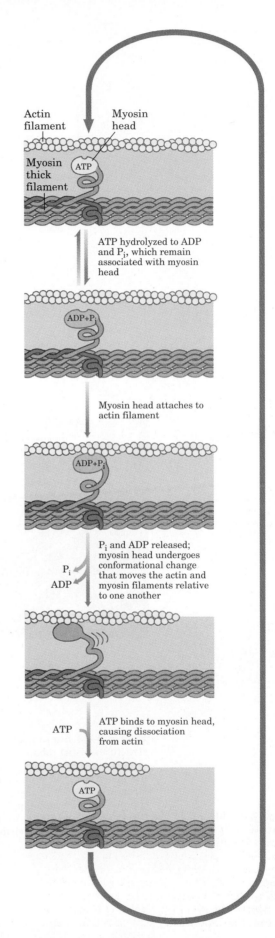

Figure 7–32 The sliding of the thick and thin filaments in muscle involves an interaction between actin and myosin mediated by ATP hydrolysis (ATP $\rightleftharpoons$ ADP + P_i, where P_i is inorganic phosphate, or PO_4^{3-}). Conformational changes in the myosin head that are coupled to stages in the ATP hydrolytic cycle cause myosin to successively dissociate from one actin subunit and then associate with another farther along the actin filament. In this way the myosin heads "walk" along the thin filaments, and draw the thin filament array into the thick filament array.

The protein structures in virus coats (called capsids) generally function simply as enclosures. In many cases capsids are made up of one or a few proteins that assemble spontaneously around a viral DNA or RNA molecule. Two types of viral structures are shown in Figure 7–33. The tobacco mosaic virus is a right-handed helical filament with 2,130 copies of a single protein that interact to form a cylinder enclosing the RNA genome. Another common structure for virus coats is the icosahedron, a regular 12-cornered polyhedron having 20 equilateral triangular faces. Two examples are poliovirus and human rhinovirus 14 (a common cold virus), each made up of 60 protein units (Fig. 7–33). Each protein unit consists of single copies of four different polypeptide chains, three of which are accessible at the outer surface. The resulting shell encloses the genetic material (RNA) of the virus.

The primary forces guiding the assembly of even these very large structures are the weak noncovalent interactions that have dominated this discussion. Each protein has several surfaces that are complementary to surfaces in adjacent protein subunits. Each protein is most stable only when it is part of the larger structure.

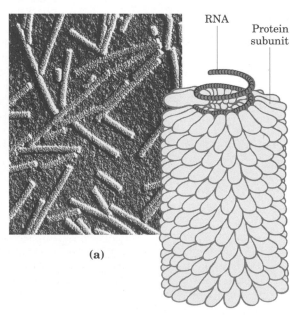

RNA

Protein subunit

(a)

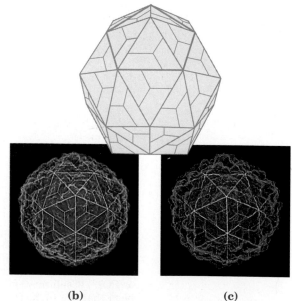

(b)   (c)

Figure 7–33 Supramolecular complexes. The coat structures of **(a)** tobacco mosaic virus, **(b)** rhinovirus 14 (a human cold virus), and **(c)** poliovirus. In the latter two, the proteins assemble into a stable structure called an icosahedron (illustrated above the photos). The rod-shaped tobacco mosaic virus is 300 nm long and 18 nm in diameter. Both rhinovirus 14 and poliovirus have diameters of about 30 nm.

Summary

Every protein has a unique three-dimensional structure that reflects its function, a structure stabilized by multiple weak interactions. Hydrophobic interactions provide the major contribution to stabilizing the globular form of most soluble proteins; hydrogen bonds and ionic interactions are optimized in the specific structure that is thermodynamically most stable.

There are four generally recognized levels of protein structure. Primary structure refers to the amino acid sequence and the location of disulfide bonds. Secondary structure refers to the spatial relationship of adjacent amino acids. Tertiary structure is the three-dimensional conformation of an entire polypeptide chain. Quaternary structure involves the spatial relationship of multiple polypeptide chains (e.g., enzyme subunits) that are tightly associated.

The nature of the bonds in the polypeptide chain places constraints on structure. The peptide bond is characterized by a partial double-bond character that keeps the entire amide group in a rigid planar configuration. The $N—C_\alpha$ and $C_\alpha—C$ bonds can rotate with bond angles ψ and ψ, respectively. Secondary structure can be defined completely by these two bond angles.

There are two general classes of proteins: fibrous and globular. Fibrous proteins, which serve mainly structural roles, have simple repeating structures and provided excellent models for the early studies of protein structure. Two major types of secondary structure were predicted by model building based on information obtained from fibrous proteins: the α helix and the β conformation. Both are characterized by optimal hydrogen bonding between amide nitrogens and carbonyl oxygens in the peptide backbone. The stability of these structures within a protein is influenced by their amino acid content and by the relative placement of amino acids in the sequence. Another nonrepeating type of secondary structure common in proteins is the β bend.

In fibrous proteins such as keratin and collagen, a single type of secondary structure predominates. The polypeptide chains are supertwisted into ropes and then combined in larger bundles to provide strength. The structure of elastin permits stretching.

Globular proteins have more complicated tertiary structures, often containing several types of secondary structure in the same polypeptide chain. The first globular protein structure to be determined, using x-ray diffraction methods, was that of myoglobin. This structure confirmed that a predicted secondary structure (α helix) occurs in proteins; that hydrophobic amino acids are located in the protein interior; and that globular proteins are compact. Subsequent research on protein structure has reinforced these conclusions while demonstrating that different proteins often differ in tertiary structure.

The three-dimensional structure of proteins can be destroyed by treatments that disrupt weak interactions, a process called denaturation. Denaturation destroys protein function, demonstrating a relationship between structure and function. Some denatured proteins (e.g., ribonuclease) can renature spontaneously to give active protein, showing that the tertiary structure of a protein is determined by its amino acid sequence.

The folding of globular proteins is believed to begin with local formation of regions of secondary structure, followed by interactions of these regions and adjustments to reach the final tertiary structure. Sometimes regions of a polypeptide chain, called domains, fold up separately and can have separate functions. The final structure and the steps taken to reach it are influenced by the need to bury hydrophobic amino acid side chains in the protein interior away from water, the tendency of a polypeptide chain to twist in a right-handed sense, and the need to maximize hydrogen bonds and ionic interactions. These constraints give rise to structural patterns such as the $\beta\alpha\beta$ fold and twisted β pleated sheets. Even at the level of tertiary structure, some common patterns are found in proteins that have no known functional relationship.

Quaternary structure refers to the interaction between the subunits of oligomeric proteins or large protein assemblies. The best-studied oligomeric protein is hemoglobin. The four subunits of hemoglobin exhibit cooperative interactions on oxygen binding. Binding of oxygen to one subunit facilitates oxygen binding to the next, giving rise to a sigmoid binding curve. These effects are mediated by subunit–subunit interactions and subunit conformational changes. Very large protein structures consisting of many copies of one or a few different proteins are referred to as supramolecular complexes. These are found in cellular skeletal structures, muscle and other types of cellular "engines," and virus coats.

Further Reading

General

Anfinsen, C.B. (1973) Principles that govern the folding of protein chains. *Science* **181,** 223–230.

The author reviews his classic work on ribonuclease.

Cantor, C.R. & Schimmel, P.R. (1980) *Biophysical Chemistry,* Part I: *The Conformation of Biological Macromolecules,* W.H. Freeman and Company, New York.

Evolution of Catalytic Function. (1987) *Cold Spring Harb. Symp. Quant. Biol.* **52.**

A source of excellent articles on many topics, including protein structure, folding, and function.

Creighton, T.E. (1984) *Proteins: Structures and Molecular Principles,* W.H. Freeman and Company, New York.

Oxender, D.L. (ed) (1987) *Protein Structure, Folding, and Design 2,* UCLA Symposium New Series, Vol. 69, Alan R. Liss, Inc., New York.

Summary papers from a major symposium on the title subject.

Structure and Function of Proteins. (1989) *Trends Biochem. Sci.* **14** (July).

A special issue devoted to reviews on protein chemistry and protein structure. Includes good summaries of protein folding, protein structure prediction, and many other topics.

Secondary, Tertiary, and Quaternary Structure

Dickerson, R.E. & Geis, I. (1982) *Hemoglobin: Structure, Function, Evolution, and Pathology,* The Benjamin/Cummings Publishing Company, Menlo Park, CA.

Ingram, V.M. (1957) Gene mutations in human haemoglobin: the chemical difference between normal and sickle cell haemoglobin. *Nature* **180,** 326–328.

Discovery of the amino acid replacement in sickle-cell hemoglobin (hemoglobin S).

Kendrew, J.C. (1961) The three-dimensional structure of a protein molecule. *Sci. Am.* **205** (December), 96–111.

Describes how the structure of myoglobin was determined and what was learned from it.

Kim, P.S. & Baldwin, R.L. (1990) Intermediates in the folding reactions of small proteins. *Annu. Rev. Biochem.* **59,** 631–660.

Koshland, D.E., Jr. (1973) Protein shape and biological control. *Sci. Am.* **229** (October), 52–64.

A discussion of the importance of flexibility in protein structures.

McPherson, A. (1989) Macromolecular crystals, *Sci. Am.,* **260** (March), 62–69.

Describes how macromolecules such as proteins are crystallized.

Pace, C.N. (1990) Conformational stability of globular proteins. *Trends Biochem. Sci.* **15,** 14–17.

Perutz, M.F. (1978) Hemoglobin structure and respiratory transport. *Sci. Am.* **239** (December), 92–125.

Richards, F.M. (1991) The protein folding problem. *Sci. Am.* **264** (January), 54–63.

Richardson, J.S. (1981) The anatomy and taxonomy of protein structure. *Adv. Prot. Chem.* **34,** 167–339.

An outstanding summary of protein structural patterns and principles; the author originated the very useful "ribbon" representations of protein structure that are used in many places in this chapter.

Rothman, J.E. (1989) Polypeptide chain binding proteins: catalysts of protein folding and related processes in cells. *Cell* **59,** 591–601.

Shortle, D. (1989) Probing the determinants of protein folding and stability with amino acid substitutions. *J. Biol. Chem.* **264,** 5315–5318.

Problems

1. *Properties of the Peptide Bond* In x-ray studies of crystalline peptides Linus Pauling and Robert Corey found that the C—N bond in the peptide link is intermediate in length (0.132 nm) between a typical C—N single bond (0.149 nm) and a C=N double bond (0.127 nm). They also found that the peptide bond is planar (all four atoms attached to the C—N group are located in the same plane) and that the two α-carbon atoms attached to the C—N are always trans to each other (on opposite sides of the peptide bond):

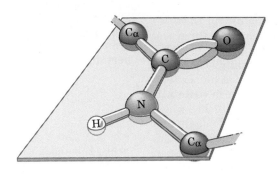

(a) What does the length of the C—N bond in the peptide linkage indicate about its strength and its bond order, i.e., whether it is single, double, or triple?

(b) In light of your answer to part (a), provide an explanation for the observation that such a C—N bond is intermediate in length between a double and single bond.

(c) What do the observations of Pauling and Corey tell us about the ease of rotation about the C—N peptide bond?

2. *Early Observations on the Structure of Wool* William Astbury discovered that the x-ray pattern of wool shows a repeating structural unit spaced about 0.54 nm along the direction of the wool fiber. When he steamed and stretched the wool, the x-ray pattern showed a new repeating structural unit at a spacing of 0.70 nm. Steaming and stretching the wool and then letting it shrink gave an x-ray pattern consistent with the original spacing of about 0.54 nm. Although these observations provided important clues to the molecular structure of wool, Astbury was unable to interpret them at the time. Given our current understanding of the structure of wool, interpret Astbury's observations.

3. *Rate of Synthesis of Hair α-Keratin* In human dimensions, the growth of hair is a relatively slow process, occurring at a rate of 15 to 20 cm/yr. All this growth is concentrated at the base of the hair fiber, where α-keratin filaments are synthesized inside living epidermal cells and assembled into ropelike structures (see Fig. 7–13). The fundamental structural element of α-keratin is the α helix, which has 3.6 amino acid residues per turn and a rise of 0.56 nm per turn (see Fig. 7–6). Assuming that the biosynthesis of α-helical keratin chains is the rate-limiting factor in the growth of hair, calculate the rate at which peptide bonds of α-keratin chains must be synthesized (peptide bonds per second) to account for the observed yearly growth of hair.

4. *The Effect of pH on the Conformations of Polyglutamate and Polylysine* The unfolding of the α helix of a polypeptide to a randomly coiled conformation is accompanied by a large decrease in a property called its specific rotation, a measure of a solution's capacity to rotate plane-polarized light. Polyglutamate, a polypeptide made up of only L-Glu residues, has the α-helical conformation at pH 3. However, when the pH is raised to 7, there is a large decrease in the specific rotation of the solution. Similarly, polylysine (L-Lys residues) is an α helix at pH 10, but when the pH is lowered to 7, the specific rotation also decreases, as shown by the following graph.

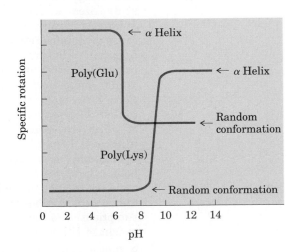

What is the explanation for the effect of the pH changes on the conformations of poly(Glu) and poly(Lys)? Why does the transition occur over such a narrow range of pH?

5. *The Disulfide-Bond Content Determines the Mechanical Properties of Many Proteins* A number of natural proteins are very rich in disulfide bonds, and their mechanical properties (tensile strength, viscosity, hardness, etc.) are correlated with the degree of disulfide bonding. For example, glutenin, a wheat protein rich in disulfide bonds, is responsible for the cohesive and elastic character of dough made from wheat flour. Similarly, the hard, tough nature of tortoise shell is due to the extensive disulfide bonding in its α-keratin. What is the molecular basis for the correlation between disulfide-bond content and mechanical properties of the protein?

6. *Why Does Wool Shrink?* When wool sweaters or socks are washed in hot water and/or dried in an electric dryer, they shrink. From what you know of α-keratin structure, how can you account for this? Silk, on the other hand, does not shrink under the same conditions. Explain.

7. *Heat Stability of Proteins Containing Disulfide Bonds* Most globular proteins are denatured and lose their activity when briefly heated to 65 °C. Globular proteins that contain multiple disulfide bonds often must be heated longer at higher temperatures to denature them. One such protein is bovine pancreatic trypsin inhibitor (BPTI), which has 58 amino acid residues in a single chain and contains three disulfide bonds. On cooling a solution of denatured BPTI, the activity of the protein is restored. Can you suggest a molecular basis for this property?

8. *Bacteriorhodopsin in Purple Membrane Proteins* Under the proper environmental conditions, the salt-loving bacterium *Halobacterium halobium* synthesizes a membrane protein (M_r 26,000) known as bacteriorhodopsin, which is purple because it contains retinal. Molecules of this protein aggregate into "purple patches" in the cell membrane. Bacteriorhodopsin acts as a light-activated proton pump that provides energy for cell functions. X-ray analysis of this protein reveals that it consists of seven parallel α-helical segments, each of which traverses the bacterial cell membrane (thickness 4.5 nm). Calculate the minimum number of amino acids necessary for one segment of α helix to traverse the membrane completely. Estimate the fraction of the bacteriorhodopsin protein that occurs in α-helical form. Justify all your assumptions. (Use an average amino acid residue weight of 110.)

9. *Biosynthesis of Collagen* Collagen, the most abundant protein in mammals, has an unusual amino acid composition. Unlike most other proteins, collagen is very rich in proline and hydroxyproline (see p. 172). Hydroxyproline is not one of the 20 standard amino acids, and its incorporation in collagen could occur by one of two routes: (1) proline is hydroxylated by enzymes *before* incorporation into collagen or (2) a Pro residue is hydroxylated *after* incorporation into collagen. To differentiate between these two possibilities, the following experiments were performed. When [^{14}C]proline was administered to a rat and the collagen from the tail isolated, the newly synthesized tail collagen was found to be radioactive. If, however, [^{14}C]hydroxyproline was administered to a rat, no radioactivity was observed in the newly synthesized collagen. How do these experiments differentiate between the two possible mechanisms for introducing hydroxyproline into collagen?

10. *Pathogenic Action of Bacteria That Cause Gas Gangrene* The highly pathogenic anaerobic bacterium *Clostridium perfringens* is responsible for gas gangrene, a condition in which animal tissue structure is destroyed. This bacterium secretes an enzyme that efficiently catalyzes the hydrolysis of the peptide bond indicated in red in the sequence:

$$-X-Gly-Pro-Y- \xrightarrow{H_2O}$$
$$-X-COO^- + H_3\overset{+}{N}-Gly-Pro-Y-$$

where X and Y are any of the 20 standard amino acids. How does the secretion of this enzyme contribute to the invasiveness of this bacterium in human tissues? Why does this enzyme not affect the bacterium itself?

11. *Formation of Bends and Intrachain Cross-Linkages in Polypeptide Chains* In the following polypeptide, where might bends or turns occur? Where might intrachain disulfide cross-linkages be formed?

```
 1     2     3     4     5     6     7     8     9    10
Ile–Ala –His –Thr–Tyr–Gly –Pro–Phe–Glu–Ala –

11    12    13    14    15    16    17    18    19    20
Ala–Met–Cys–Lys–Trp–Glu–Ala–Gln–Pro–Asp–

21    22    23    24    25    26    27    28
Gly–Met–Glu–Cys–Ala–Phe–His–Arg–
```

12. *Location of Specific Amino Acids in Globular Proteins* X-ray analysis of the tertiary structure of myoglobin and other small, single-chain globular proteins has led to some generalizations about how the polypeptide chains of soluble proteins fold. With these generalizations in mind, indicate the probable location, whether in the interior or on the external surface, of the following amino acid residues in native globular proteins: Asp, Leu, Ser, Val, Gln, Lys. Explain your reasoning.

13. *The Number of Polypeptide Chains in an Oligomeric Protein* A sample (660 mg) of an oligomeric protein of M_r 132,000 was treated with an excess of 1-fluoro-2,4-dinitrobenzene under slightly alkaline conditions until the chemical reaction was complete. The peptide bonds of the protein were then completely hydrolyzed by heating it with concentrated HCl. The hydrolysate was found to contain 5.5 mg of the following compound:

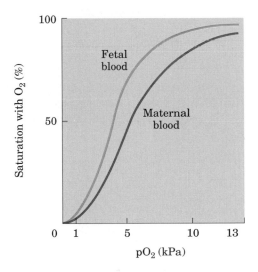

However, 2,4-dinitrophenyl derivatives of the α-amino groups of other amino acids could not be found.

(a) Explain why this information can be used to determine the number of polypeptide chains in an oligomeric protein.

(b) Calculate the number of polypeptide chains in this protein.

14. *Molecular Weight of Hemoglobin* The first indication that proteins have molecular weights greatly exceeding those of the (then known) organic compounds was obtained over 100 years ago. For example, it was known at that time that hemoglobin contains 0.34% by weight of iron.

(a) From this information determine the minimum molecular weight of hemoglobin.

(b) Subsequent experiments indicated that the true molecular weight of hemoglobin is 64,500. What information did this provide about the number of iron atoms in hemoglobin?

15. *Comparison of Fetal and Maternal Hemoglobin* Studies of oxygen transport in pregnant mammals have shown that the O_2-saturation curves of fetal and maternal blood are markedly different when measured under the same conditions. Fetal erythrocytes contain a structural variant of hemoglobin, hemoglobin F, consisting of two α and two γ subunits $(\alpha_2\gamma_2)$, whereas maternal erythrocytes contain the usual hemoglobin A $(\alpha_2\beta_2)$.

(a) Which hemoglobin has a higher affinity for oxygen under physiological conditions, hemoglobin A or hemoglobin F? Explain.

(b) What is the physiological significance of the different oxygen affinities? Explain.

Enzymes

We now come to the most remarkable and highly specialized proteins, the enzymes. Enzymes are the reaction catalysts of biological systems. They have extraordinary catalytic power, often far greater than that of synthetic catalysts. They have a high degree of specificity for their substrates, they accelerate specific chemical reactions, and they function in aqueous solutions under very mild conditions of temperature and pH. Few nonbiological catalysts show all these properties.

Enzymes are one of the keys to understanding how cells survive and proliferate. Acting in organized sequences, they catalyze the hundreds of stepwise reactions in metabolic pathways by which nutrient molecules are degraded, chemical energy is conserved and transformed, and biological macromolecules are made from simple precursors. Some of the many enzymes participating in metabolism are regulatory enzymes, which can respond to various metabolic signals by changing their catalytic activity accordingly. Through the action of regulatory enzymes, enzyme systems are highly coordinated to yield a harmonious interplay among the many different metabolic activities necessary to sustain life.

The study of enzymes also has immense practical importance. In some diseases, especially inheritable genetic disorders, there may be a deficiency or even a total absence of one or more enzymes in the tissues (see Table 6–6). Abnormal conditions can also be caused by the excessive activity of a specific enzyme. Measurements of the activity of certain enzymes in the blood plasma, erythrocytes, or tissue samples are important in diagnosing disease. Enzymes have become important practical tools, not only in medicine but also in the chemical industry, in food processing, and in agriculture. Enzymes play a part even in everyday activities in the home such as food preparation and cleaning.

The chapter begins with descriptions of the properties of enzymes and the principles underlying their catalytic power. Following is an introduction to enzyme kinetics, a discipline that provides much of the framework for any discussion of enzymes. Specific examples of enzyme mechanisms are then provided, illustrating principles introduced earlier in the chapter. We will end with a discussion of regulatory enzymes.

An Introduction to Enzymes

Much of the history of biochemistry is the history of enzyme research. Biological catalysis was first recognized and described in the early 1800s, in studies of the digestion of meat by secretions of the stomach and the conversion of starch into sugar by saliva and various plant extracts. In the 1850s Louis Pasteur concluded that fermentation of sugar into alcohol by yeast is catalyzed by "ferments." He postulated that these ferments, later named **enzymes,** are inseparable from the structure of living yeast cells, a view that prevailed for many years. The discovery by Eduard Buchner in 1897 that yeast extracts can ferment sugar to alcohol proved that the enzymes involved in fermentation can function when removed from the structure of living cells. This encouraged biochemists to attempt the isolation of many different enzymes and to examine their catalytic properties.

James Sumner's isolation and crystallization of urease in 1926 provided a breakthrough in early studies of the properties of specific enzymes. Sumner found that the urease crystals consisted entirely of protein and postulated that all enzymes are proteins. Lacking other examples, this idea remained controversial for some time. Only later in the 1930s, after John Northrop and his colleagues crystallized pepsin and trypsin and found them also to be proteins, was Sumner's conclusion widely accepted. During this period, J.B.S. Haldane wrote a treatise entitled "Enzymes." Even though the molecular nature of enzymes was not yet fully appreciated, this book contained the remarkable suggestion that weak-bonding interactions between an enzyme and its substrate might be used to distort the substrate and catalyze the reaction. This insight lies at the heart of our current understanding of enzymatic catalysis. The latter part of the twentieth century has seen intensive research on the enzymes catalyzing the reactions of cell metabolism. This has led to the purification of thousands of enzymes (Fig. 8–1), elucidation of the structure and chemical mechanism of hundreds of these, and a general understanding of how enzymes work.

James Sumner
1887–1955

J.B.S. Haldane
1892–1964

Most Enzymes Are Proteins

With the exception of a small group of catalytic RNA molecules (Chapter 25), all enzymes are proteins. Their catalytic activity depends upon the integrity of their native protein conformation. If an enzyme is denatured or dissociated into subunits, catalytic activity is usually lost. If an enzyme is broken down into its component amino acids, its catalytic activity is always destroyed. Thus the primary, secondary, tertiary, and quaternary structures of protein enzymes are essential to their catalytic activity.

Enzymes, like other proteins, have molecular weights ranging from about 12,000 to over 1 million. Some enzymes require no chemical groups other than their amino acid residues for activity. Others require an additional chemical component called a **cofactor.** The cofactor may be either one or more inorganic ions, such as Fe^{2+}, Mg^{2+}, Mn^{2+}, or Zn^{2+} (Table 8–1), or a complex organic or metalloorganic molecule called a **coenzyme** (Table 8–2). Some enzymes require *both* a coenzyme and one or more metal ions for activity. A coenzyme or metal ion that is covalently bound to the enzyme protein is called a **prosthetic group.** A complete, catalytically active enzyme together with its coenzyme and/or metal ions is called a **holoenzyme.** The protein part of such an enzyme is called the **apoenzyme** or **apoprotein.** Coenzymes function as transient carriers of specific functional groups (Table 8–2). Many vitamins, organic nutrients required in small amounts in the diet, are precursors of coenzymes. Coenzymes will be considered in more detail as they are encountered in the discussion of metabolic pathways in Part III of this book.

Figure 8–1 Crystals of pyruvate kinase, an enzyme of the glycolytic pathway. The protein in a crystal is generally characterized by a high degree of purity and structural homogeneity.

Table 8–1 Some enzymes containing or requiring inorganic elements as cofactors

Fe^{2+} or Fe^{3+}	Cytochrome oxidase Catalase Peroxidase
Cu^{2+}	Cytochrome oxidase
Zn^{2+}	Carbonic anhydrase Alcohol dehydrogenase
Mg^{2+}	Hexokinase Glucose-6-phosphatase Pyruvate kinase
Mn^{2+}	Arginase Ribonucleotide reductase
K^+	Pyruvate kinase
Ni^{2+}	Urease
Mo	Dinitrogenase
Se	Glutathione peroxidase

Table 8–2 Some coenzymes serving as transient carriers of specific atoms or functional groups*

Coenzyme	Examples of some chemical groups transferred	Dietary precursor in mammals
Thiamine pyrophosphate	Aldehydes	Thiamin (vitamin B_1)
Flavin adenine dinucleotide	Electrons	Riboflavin (vitamin B_2)
Nicotinamide adenine dinucleotide	Hydride ion ($:H^-$)	Nicotinic acid (niacin)
Coenzyme A	Acyl groups	Pantothenic acid, plus other molecules
Pyridoxal phosphate	Amino groups	Pyridoxine (vitamin B_6)
5′-Deoxyadenosyl-cobalamine (coenzyme B_{12})	H atoms and alkyl groups	Vitamin B_{12}
Biocytin	CO_2	Biotin
Tetrahydrofolate	One-carbon groups	Folate
Lipoate acid	Electrons and acyl groups	Not required in diet

* The structure and mode of action of these coenzymes are described in Part III of this book.

Finally, some enzymes are modified by phosphorylation, glycosylation, and other processes. Many of these alterations are involved in the regulation of enzyme activity.

Enzymes Are Classified by the Reactions They Catalyze

Many enzymes have been named by adding the suffix "-ase" to the name of their substrate or to a word or phrase describing their activity. Thus urease catalyzes hydrolysis of urea, and DNA polymerase catalyzes the synthesis of DNA. Other enzymes, such as pepsin and trypsin, have names that do not denote their substrates. Sometimes the same enzyme has two or more names, or two different enzymes have the same name. Because of such ambiguities, and the ever-increasing number of newly discovered enzymes, a system for naming and classifying enzymes has been adopted by international agreement. This system places all enzymes in six major classes, each with subclasses, based on the type of reaction catalyzed (Table 8–3). Each enzyme is assigned a four-digit classification number and a systematic name, which identifies the reaction catalyzed. As an example, the formal systematic name of the enzyme catalyzing the reaction

$$\text{ATP} + \text{D-glucose} \longrightarrow \text{ADP} + \text{D-glucose-6-phosphate}$$

is ATP:glucose phosphotransferase, which indicates that it catalyzes the transfer of a phosphate group from ATP to glucose. Its enzyme classification number (E.C. number) is 2.7.1.1; the first digit (2) denotes the class name (transferase) (see Table 8–3); the second digit (7), the

Table 8–3 International classification of enzymes*

No.	Class	Type of reaction catalyzed
1	Oxidoreductases	Transfer of electrons (hydride ions or H atoms)
2	Transferases	Group-transfer reactions
3	Hydrolases	Hydrolysis reactions (transfer of functional groups to water)
4	Lyases	Addition of groups to double bonds, or formation of double bonds by removal of groups
5	Isomerases	Transfer of groups within molecules to yield isomeric forms
6	Ligases	Formation of C—C, C—S, C—O, and C—N bonds by condensation reactions coupled to ATP cleavage

* Most enzymes catalyze the transfer of electrons, atoms, or functional groups. They are therefore classified, given code numbers, and assigned names according to the type of transfer reaction, the group donor, and the group acceptor.

subclass (phosphotransferase); the third digit (1), phosphotransferases with a hydroxyl group as acceptor; and the fourth digit (1), D-glucose as the phosphate-group acceptor. When the systematic name of an enzyme is long or cumbersome, a trivial name may be used—in this case hexokinase.

A complete list and description of the thousands of known enzymes would be well beyond the scope of this book. This chapter is instead devoted primarily to principles and properties common to all enzymes.

How Enzymes Work

The enzymatic catalysis of reactions is essential to living systems. Under biologically relevant conditions, uncatalyzed reactions tend to be slow. Most biological molecules are quite stable in the neutral-pH, mild-temperature, aqueous environment found inside cells. Many common reactions in biochemistry involve chemical events that are unfavorable or unlikely in the cellular environment, such as the transient formation of unstable charged intermediates or the collision of two or more molecules in the precise orientation required for reaction. Reactions required to digest food, send nerve signals, or contract muscle simply do not occur at a useful rate without catalysis.

An enzyme circumvents these problems by providing a specific environment within which a given reaction is energetically more favorable. The distinguishing feature of an enzyme-catalyzed reaction is that it occurs within the confines of a pocket on the enzyme called the **active site** (Fig. 8–2). The molecule that is bound by the active site and acted upon by the enzyme is called the **substrate.** The enzyme–substrate complex is central to the action of enzymes, and it is the starting point for mathematical treatments defining the kinetic behavior of enzyme-catalyzed reactions and for theoretical descriptions of enzyme mechanisms.

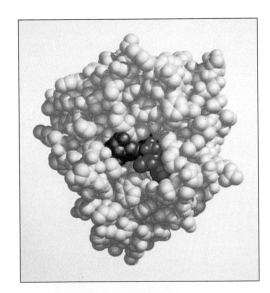

Figure 8–2 Binding of a substrate to an enzyme at the active site. The enzyme chymotrypsin is shown, bound to a substrate (in blue). Some key active-site amino acids are shown in red.

Enzymes Affect Reaction Rates, Not Equilibria

A tour through an enzyme-catalyzed reaction serves to introduce some important concepts and definitions.

A simple enzymatic reaction might be written

$$E + S \rightleftharpoons ES \rightleftharpoons EP \rightleftharpoons E + P \tag{8–1}$$

where E, S, and P represent the enzyme, substrate, and product, respectively. ES and EP are complexes of the enzyme with the substrate and with the product, respectively.

To understand catalysis, we must first appreciate the important distinction between reaction equilibria (discussed in Chapter 4) and reaction rates. The function of a catalyst is to increase the *rate* of a reaction. Catalysts do not affect reaction *equilibria*. Any reaction, such as $S \rightleftharpoons P$, can be described by a reaction coordinate diagram (Fig. 8–3). This is a picture of the energetic course of the reaction. As introduced in Chapters 1 and 3, energy in biological systems is described in terms of free energy, G. In the coordinate diagram, the free energy of the system is plotted against the progress of the reaction (reaction coordinate). In its normal stable form or **ground state,** any molecule (such as S or P) contains a characteristic amount of free energy. To describe the free-energy changes for reactions, chemists define a standard set of conditions (temperature 298 K; partial pressure of gases each 1 atm or 101.3 kPa; concentration of solutes each 1 M), and express the free-energy change for this reacting system as $\Delta G°$, the **standard free-energy change.** Because biochemical systems commonly involve H^+ concentrations far from 1 M, biochemists define a constant $\Delta G°'$, the standard free-energy change *at pH 7.0,* which we will employ throughout the book. A more complete definition of $\Delta G°'$ is given in Chapter 13.

The equilibrium between S and P reflects the difference in the free energy of their ground states. In the example shown in Figure 8–3, the free energy of the ground state of P is lower than that of S, so $\Delta G°'$ for the reaction is negative and the equilibrium favors P. This equilibrium is *not* affected by any catalyst.

A favorable equilibrium, however, does *not* mean that the $S \rightarrow P$ conversion is fast. The rate of a reaction is dependent on an entirely different parameter. There is an energetic barrier between S and P that represents the energy required for alignment of reacting groups, formation of transient unstable charges, bond rearrangements, and other transformations required for the reaction to occur in either direction. This is illustrated by the energetic "hill" in Figures 8–3 and 8–4. To undergo reaction, the molecules must overcome this barrier and therefore must be raised to a higher energy level. At the top of the energy hill is a point at which decay to the S or P state is equally probable (it is downhill either way). This is called the **transition state.** The transition state is not a chemical species with any significant stability and should not be confused with a reaction intermediate. It is simply a fleeting molecular moment in which events such as bond breakage, bond formation, and charge development have proceeded to the precise point at which a collapse to either substrate or product is equally likely. The difference between the energy levels of the ground state and the transition state is called the **activation energy** ($\Delta G^{\ddagger}$). The rate of a reaction reflects this activation energy; a higher activation energy corresponds to a slower reaction. Reaction rates can be increased by raising the temperature, thereby increasing the number of molecules with sufficient energy to overcome this energy barrier. Alternatively

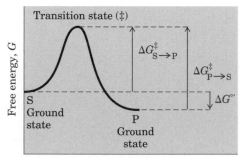

Figure 8–3 Reaction coordinate diagram for a chemical reaction. The free energy of the system is plotted against the progress of the reaction. A diagram of this kind is a description of the energetic course of the reaction, and the horizontal axis (reaction coordinate) reflects the progressive chemical changes (e.g., bond breakage or formation) as S is converted to P. The S and P symbols mark the free energies of the substrate and product ground states. The transition state is indicated by the symbol ‡. The activation energies, $\Delta G^{\ddagger}$, for the $S \rightarrow P$ and $P \rightarrow S$ reactions are indicated. $\Delta G°'$ is the overall standard free-energy change in going from S to P.

the activation energy can be lowered by adding a catalyst (Fig. 8–4). *Catalysts enhance reaction rates by lowering activation energies.*

Enzymes are no exception to the rule that catalysts do not affect reaction equilibria. The bidirectional arrows in Equation 8–1 make this point: any enzyme that catalyzes the reaction S → P also catalyzes the reaction P → S. Its only role is to accelerate the interconversion of S and P. The enzyme is not used up in the process, and the equilibrium point is unaffected. However, the reaction reaches equilibrium much faster when the appropriate enzyme is present because the rate of the reaction is increased.

This general principle can be illustrated by considering the reaction of glucose and O_2 to form CO_2 and H_2O. This reaction has a very large and negative $\Delta G^{\circ\prime}$, and at equilibrium the amount of glucose present is negligible. Glucose, however, is a stable compound, and it can be combined in a container with O_2 almost indefinitely without reacting. Its stability reflects a high activation energy for reaction. In cells, glucose is broken down in the presence of O_2 to CO_2 and H_2O in a pathway of reactions catalyzed by enzymes. These enzymes not only accelerate the reactions, they organize and control them so that much of the energy released in this process is recovered in other forms and made available to the cell for other tasks. This is the primary energy-yielding pathway for cells (Chapters 14 and 18), and these enzymes allow it to occur on a time scale that is useful to the cells.

In practice, any reaction may have several steps involving the formation and decay of transient chemical species called **reaction intermediates**. When the S ⇌ P reaction is catalyzed by an enzyme, the ES and EP complexes are intermediates (Eqn 8–1); they occupy valleys in the reaction coordinate diagram (Fig. 8–4). When several steps occur in a reaction, the overall rate is determined by the step (or steps) with the highest activation energy; this is called the **rate-limiting step.** In a simple case the rate-limiting step is the highest-energy point in the diagram for interconversion of S and P (Fig. 8–4). In practice, the rate-limiting step can vary with reaction conditions, and for many enzymes several steps may have similar activation energies, which means they are all partially rate-limiting.

As described in Chapter 1, activation energies are energetic barriers to chemical reactions; these barriers are crucial to life itself. The stability of a molecule increases with the height of its activation barrier. Without such energetic barriers, complex macromolecules would revert spontaneously to much simpler molecular forms. The complex and highly ordered structures and metabolic processes in every cell could not exist. Enzymes have evolved to lower activation energies *selectively* for reactions that are needed for cell survival.

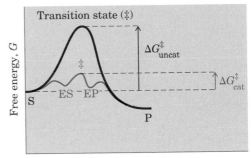

Figure 8–4 Reaction coordinate diagram comparing the enzyme-catalyzed and uncatalyzed reactions S → P. The ES and EP intermediates occupy minima in the energetic progress curve of the enzyme-catalyzed reaction. The terms $\Delta G^{\ddagger}_{uncat}$ and $\Delta G^{\ddagger}_{cat}$ correspond to the activation energies for the uncatalyzed and catalyzed reactions, respectively. The activation energy for the overall process is lower when the enzyme catalyzes the reaction.

Reaction Rates and Equilibria Have Precise Thermodynamic Definitions

Reaction *equilibria* are inextricably linked to $\Delta G^{\circ\prime}$ and reaction *rates* are linked to $\Delta G^{\ddagger}$. A basic introduction to these thermodynamic relationships is the next step in understanding how enzymes work.

As introduced in Chapter 4, an equilibrium such as S ⇌ P is described by an **equilibrium constant,** K_{eq}. Under the standard conditions used to compare biochemical processes, an equilibrium constant is denoted K_{eq}':

$$K_{eq}' = \frac{[P]}{[S]} \qquad (8\text{–}2)$$

From thermodynamics, the relationship between K_{eq}' and $\Delta G^{\circ\prime}$ can be described by the expression

$$\Delta G^{\circ\prime} = -RT \ln K_{eq}' \qquad (8\text{–}3)$$

where R is the gas constant (8.315 J/mol $\cdot$ K) and T is the absolute temperature (298 K). This expression will be developed and discussed in more detail in Chapter 13. The important point here is that the equilibrium constant is a direct reflection of the overall standard free-energy change in the reaction (Table 8–4). A large negative value for $\Delta G^{\circ\prime}$ reflects a favorable reaction equilibrium, but as already noted this does not mean the reaction will proceed at a rapid rate.

The rate of any reaction is determined by the concentration of the reactant (or reactants) and by a **rate constant,** usually denoted by the symbol k. For the unimolecular reaction S $\rightarrow$ P, the rate or velocity of the reaction, V, representing the amount of S that has reacted per unit time, is expressed by a **rate law:**

$$V = k[\text{S}] \qquad (8\text{–}4)$$

In this reaction, the rate depends only on the concentration of S. This is called a first-order reaction. The factor k is a proportionality constant that reflects the probability of reaction under a given set of conditions (pH, temperature, etc.). Here, k is a first-order rate constant and has units of reciprocal time (e.g., s^{-1}). If a first-order reaction has a rate constant k of 0.03 s^{-1}, this may be interpreted (qualitatively) to mean that 3% of the available S will be converted to P in 1 s. A reaction with a rate constant of 2,000 s^{-1} will be over in a small fraction of a second. If the reaction rate depends on the concentration of two different compounds, or if two molecules of the same compound react, the reaction is second order and k is a second-order rate constant (with the units $\text{M}^{-1}\text{s}^{-1}$). The rate law has the form

$$V = k[\text{S}_1][\text{S}_2] \qquad (8\text{–}5)$$

From transition-state theory, an expression can be derived that relates the magnitude of a rate constant to the activation energy:

$$k = \frac{\mathbf{k}\,T}{h}\mathrm{e}^{-\Delta G^{\ddagger}/RT} \qquad (8\text{–}6)$$

where $\mathbf{k}$ is the Boltzmann constant and h is Planck's constant. The important point here is that the relationship between the rate constant, k, and the activation energy, $\Delta G^{\ddagger}$, is inverse and exponential. In simplified terms, this is the basis for the statement that a lower activation energy means a higher reaction rate, and vice versa.

Now we turn from *what* enzymes do to *how* they do it.

A Few Principles Explain the Catalytic Power and Specificity of Enzymes

Enzymes are extraordinary catalysts. The rate enhancements brought about by enzymes are often in the range of 7 to 14 orders of magnitude (Table 8–5). Enzymes are also very specific, readily discriminating between substrates with quite similar structures. How can these enormous and highly selective rate enhancements be explained? Where does the energy come from to provide a dramatic lowering of the activation energies for specific reactions?

Table 8–4 The relationship between K_{eq}' and $\Delta G^{\circ\prime}$ (see Eqn 8–3)

K_{eq}'	$\Delta G^{\circ\prime}$ (kJ/mol)
10^{-6}	34.2
10^{-5}	28.5
10^{-4}	22.8
10^{-3}	17.1
10^{-2}	11.4
10^{-1}	5.7
1	0.0
10^{1}	−5.7
10^{2}	−11.4
10^{3}	−17.1

Table 8–5 Some rate enhancements produced by enzymes

Carbonic anhydrase	10^{7}
Phosphoglucomutase	10^{12}
Succinyl-CoA transferase	10^{13}
Urease	10^{14}

Part of the explanation for enzyme action lies in well-studied chemical reactions that take place between a substrate and enzyme functional groups (specific amino acid side chains, metal ions, and co-enzymes). Catalytic functional groups on enzymes can interact transiently with a substrate and activate it for reaction. In many cases, these groups lower the activation energy (and thereby accelerate the reaction) by providing a lower-energy reaction path. Common types of enzymatic catalysis are outlined later in this chapter.

Catalytic functional groups, however, are not the only contributor to enzymatic catalysis. The energy required to lower activation energies is generally derived from weak, noncovalent interactions between the substrate and the enzyme. The factor that really sets enzymes apart from most nonenzymatic catalysts is the formation of a specific ES complex. The interaction between substrate and enzyme in this complex is mediated by the same forces that stabilize protein structure, including hydrogen bonds and hydrophobic, ionic, and van der Waals interactions (Chapter 7). Formation of each weak interaction in the ES complex is accompanied by a small release of free energy that provides a degree of stability to the interaction. The energy derived from enzyme–substrate interaction is called **binding energy.** Its significance extends beyond a simple stabilization of the enzyme–substrate interaction. *Binding energy is the major source of free energy used by enzymes to lower the activation energies of reactions.*

Two fundamental and interrelated principles provide a general explanation for how enzymes work. First, the catalytic power of enzymes is ultimately derived from the free energy released in forming the multiple weak bonds and interactions that occur between an enzyme and its substrate. This binding energy provides specificity as well as catalysis. Second, weak interactions are optimized in the reaction transition state; enzyme active sites are complementary not to the substrates per se, but to the transition states of the reactions they catalyze. These themes are critical to an understanding of enzymes, and they now become the primary focus of the chapter.

Weak Interactions between Enzyme and Substrate Are Optimized in the Transition State

How does an enzyme use binding energy to lower the activation energy for reaction? Formation of the ES complex is not the explanation in itself, although some of the earliest considerations of enzyme mechanisms began with this idea. Studies on enzyme specificity carried out by Emil Fischer led him to propose, in 1894, that enzymes were structurally complementary to their substrates, so that they fit together like a "lock and key" (Fig. 8–5).

This elegant idea, that a specific (exclusive) interaction between two biological molecules is mediated by molecular surfaces with complementary shapes, has greatly influenced the development of biochemistry, and lies at the heart of many biochemical processes. However, the "lock and key" hypothesis can be misleading when applied to the question of enzymatic catalysis. An enzyme completely complementary to its substrate would be a very poor enzyme. Consider an imaginary reaction, the breaking of a metal stick. The uncatalyzed reaction is shown in Figure 8–6a. We will examine two imaginary enzymes to catalyze this reaction, both of which employ magnetic forces as a paradigm for the binding energy used by real enzymes. We first

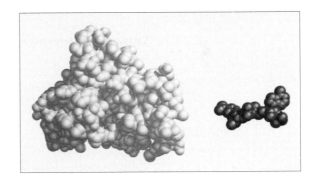

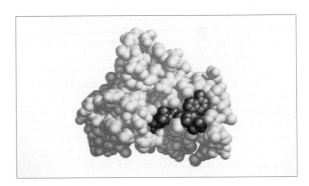

Figure 8–5 Complementary shapes of a substrate and its binding site on an enzyme. The enzyme dihydrofolate reductase is shown with its substrate, NADP⁺ (red), unbound (top) and bound (bottom). Part of a tetrahydrofolate molecule (yellow), also bound to the enzyme, is visible. The NADP⁺ binds to a pocket that is complementary to it in shape and ionic properties. Emil Fischer proposed that enzymes and their substrates have shapes that closely complement each other, like a lock and key. This idea can readily be extended to the interactions of other types of proteins with ligands or other proteins. In reality, the complementarity is rarely perfect, and the interaction of a protein with a ligand often involves changes in the conformation of one or both molecules. This *lack* of perfect complementarity between an enzyme and its substrate (not evident in this figure) is important to enzymatic catalysis.

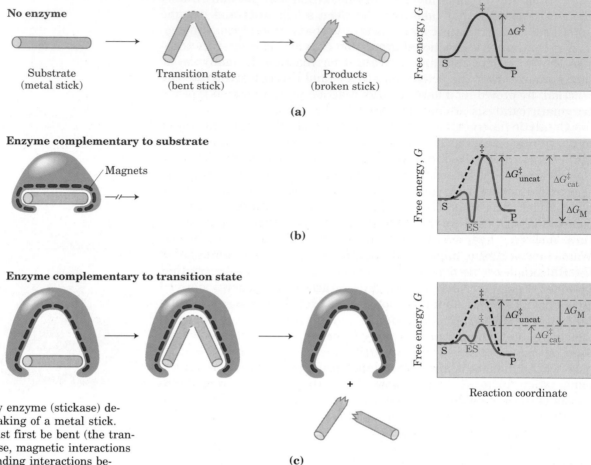

Figure 8–6 An imaginary enzyme (stickase) designed to catalyze the breaking of a metal stick. **(a)** To break, the stick must first be bent (the transition state). In the stickase, magnetic interactions take the place of weak-bonding interactions between enzyme and substrate. **(b)** An enzyme with a magnet-lined pocket complementary in structure to the stick (the substrate) will stabilize this substrate. Bending will be impeded by the magnetic attraction between stick and stickase. **(c)** An enzyme complementary to the reaction transition state will help to destabilize the stick, resulting in catalysis of the reaction. The magnetic interactions provide energy that compensates for the increase in free energy required to bend the stick. Reaction coordinate diagrams show the energetic consequences of complementarity to substrate versus complementarity to transition state. The term ΔG_M represents the energy contributed by the magnetic interactions between the stick and stickase. When the enzyme is complementary to the substrate, as in **(b)**, the ES complex is more stable and has less free energy in the ground state than substrate alone. The result is an *increase* in the activation energy. For simplicity, the EP complexes are not shown.

design an enzyme perfectly complementary to the substrate (Fig. 8–6b). The active site of this "stickase" enzyme is a pocket lined with magnets. To react (break), the stick must reach the transition state of the reaction. The stick fits so tightly in the active site that it cannot bend, because bending of the stick would eliminate some of the magnetic interactions between stick and enzyme. Such an enzyme *impedes* the reaction, stabilizing the substrate instead. In a reaction coordinate diagram (Fig. 8–6), this kind of ES complex would correspond to an energy well from which it would be difficult for the substrate to escape. Such an enzyme would be useless.

The modern notion of enzymatic catalysis was first proposed by Haldane in 1930, and elaborated by Linus Pauling in 1946. In order to catalyze reactions, an enzyme must be complementary to the *reaction transition state*. This means that the optimal interactions (through weak bonding) between substrate and enzyme can occur only in the transition state. Figure 8–6c demonstrates how such an enzyme can work. The metal stick binds, but only a few magnetic interactions are used in forming the ES complex. The bound substrate must still undergo the increase in free energy needed to reach the transition state. Now, however, the increase in free energy required to draw the stick into a bent and partially broken conformation is offset or "paid for" by the magnetic interactions that form between the enzyme and substrate in the transition state. Many of these interactions involve parts of the stick that are distant from the point of breakage; thus interactions between the stickase and nonreacting parts of the stick provide some of the energy needed to catalyze stick breakage. This "energy payment" translates into a lower net activation energy and a faster reaction rate.

Real enzymes work on an analogous principle. Some weak interactions are formed in the ES complex, but the full complement of possible weak interactions between substrate and enzyme are formed only when the substrate reaches the transition state. The free energy (binding energy) released by the formation of these interactions partially offsets the energy required to get to the top of the energy hill. The summation of the unfavorable (positive) $\Delta G^{\ddagger}$ and the favorable (negative) binding energy (ΔG_B) results in a lower *net* activation energy (Fig. 8–7). Even on the enzyme, the transition state represents a brief point in time that the substrate spends atop an energy hill. The enzyme-catalyzed reaction is much faster than the uncatalyzed process, however, because the hill is much smaller. The important principle is that weak-bonding interactions between the enzyme and the substrate provide the major driving force for enzymatic catalysis. The groups on the substrate that are involved in these weak interactions can be at some distance from the bonds that are broken or changed. The weak interactions that are formed only in the transition state are those that make the primary contribution to catalysis.

The requirement for multiple weak interactions to drive catalysis is one reason why enzymes (and some coenzymes) are so large. The enzyme must provide functional groups for ionic interactions, hydrogen bonds, and other interactions, and also precisely position these groups so that binding energy is optimized in the transition state.

Enzymes Use Binding Energy to Provide Reaction Specificity and Catalysis

Can binding energy account for the huge rate accelerations brought about by enzymes? Yes. As a point of reference, Equation 8–6 allows us to calculate that about 5.7 kJ/mol of free energy is required to accelerate a first-order reaction by a factor of ten under conditions commonly found in cells. The energy available from formation of a single weak interaction is generally estimated to be 4 to 30 kJ/mol. The overall energy available from formation of a number of such interactions can lower activation energies by the 60 to 80 kJ/mol required to explain the large rate enhancements observed for many enzymes.

The same binding energy that provides energy for catalysis also makes the enzyme specific. **Specificity** refers to the ability of an enzyme to discriminate between two competing substrates. Conceptually, this idea is easy to distinguish from the idea of catalysis. Catalysis and specificity are much more difficult to distinguish experimentally because they arise from the same phenomenon. If an enzyme active site has functional groups arranged optimally to form a variety of weak interactions with a given substrate in the transition state, the enzyme will not be able to interact as well with any other substrate. For example, if the normal substrate has a hydroxyl group that forms a specific hydrogen bond with a Glu residue on the enzyme, any molecule lacking that particular hydroxyl group will generally be a poorer substrate for the enzyme. In addition, any molecule with an extra functional group for which the enzyme has no pocket or binding site is likely to be excluded from the enzyme. In general, *specificity* is also derived from the formation of multiple weak interactions between the enzyme and many or all parts of its specific substrate molecule.

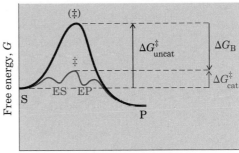

Figure 8–7 The role of binding energy in catalysis. To lower the activation energy for a reaction, the system must acquire an amount of energy equivalent to the amount by which $\Delta G^{\ddagger}$ is lowered. This energy comes largely from binding energy (ΔG_B) contributed by formation of weak noncovalent interactions between substrate and enzyme in the transition state. The role of ΔG_B is analogous to that of ΔG_M in Fig. 8–6.

The general principles outlined above can be illustrated by a variety of recognized catalytic mechanisms. These mechanisms are not mutually exclusive, and a given enzyme will often incorporate several in its own complete mechanism of action. It is often difficult to quantify the contribution of any one catalytic mechanism to the rate and/or specificity of an enzyme-catalyzed reaction.

Binding energy is the dominant driving force in several mechanisms, and these can be the major, and sometimes the only, contribution to catalysis. This can be illustrated by considering what needs to occur for a reaction to take place. Prominent physical and thermodynamic barriers to reaction include (1) entropy, the relative motion of two molecules in solution; (2) the solvated shell of hydrogen-bonded water that surrounds and helps to stabilize most biomolecules in aqueous solution; (3) the electronic or structural distortion of substrates that must occur in many reactions; and (4) the need to achieve proper alignment of appropriate catalytic functional groups on the enzyme. Binding energy can be used to overcome all of these barriers.

A large reduction in the relative motions of two substrates that are to react, or **entropy reduction,** is one of the obvious benefits of binding them to an enzyme. Binding energy holds the substrates in the proper orientation to react—a major contribution to catalysis because productive collisions between molecules in solution can be exceedingly rare. Substrates can be precisely aligned on the enzyme. A multitude of weak interactions between each substrate and strategically located groups on the enzyme clamp the substrate molecules into the proper positions. Studies have shown that constraining the motion of two reactants can produce rate enhancements of as much as 10^8 M (a rate equivalent to that expected if the reactants were present at the impossibly high concentration of 100,000,000 M).

Formation of weak bonds between substrate and enzyme also results in **desolvation** of the substrate. Enzyme–substrate interactions replace most or all of the hydrogen bonds that may exist between the substrate and water in solution.

Binding energy involving weak interactions formed only in the reaction transition state helps to compensate thermodynamically for any **strain** or distortion that the substrate must undergo to react. Distortion of the substrate in the transition state may be electrostatic or structural.

The enzyme itself may undergo a change in conformation when the substrate binds, induced again by multiple weak interactions with the substrate. This is referred to as **induced fit,** a mechanism postulated by Daniel Koshland in 1958. Induced fit may serve to bring specific functional groups on the enzyme into the proper orientation to catalyze the reaction. The conformational change may also permit formation of additional weak-bonding interactions in the transition state. In either case the new conformation may have enhanced catalytic properties.

Specific Catalytic Groups Contribute to Catalysis

Once a substrate is bound, additional modes of catalysis can be employed by an enzyme to aid bond cleavage and formation, using properly positioned catalytic functional groups. Among the best characterized mechanisms are **general acid–base catalysis** and **covalent catalysis.** These are distinct from mechanisms based on binding energy because they generally involve *covalent* interaction with a substrate, or group transfer to or from a substrate.

General Acid–Base Catalysis Many biochemical reactions involve the formation of unstable charged intermediates that tend to break down rapidly to their constituent reactant species, thus failing to undergo reaction (Fig. 8–8). Charged intermediates can often be stabilized (and the reaction thereby catalyzed) by transferring protons to or from the substrate or intermediate to form a species that breaks down to products more readily than to reactants. The proton transfers can involve the constituents of water alone or may involve other weak proton donors or acceptors. Catalysis that simply involves the H^+ (H_3O^+) or OH^- ions present in water is referred to as **specific acid or base catalysis.** If protons are transferred between the intermediate and water faster than the intermediate breaks down to reactants, the intermediate will effectively be stabilized every time it forms. No additional catalysis mediated by other proton acceptors or donors will occur. In many cases, however, water is not enough. The term general acid–base catalysis refers to proton transfers mediated by other classes of molecules. It is observed in aqueous solutions only when the unstable reaction intermediate breaks down to reactants faster than

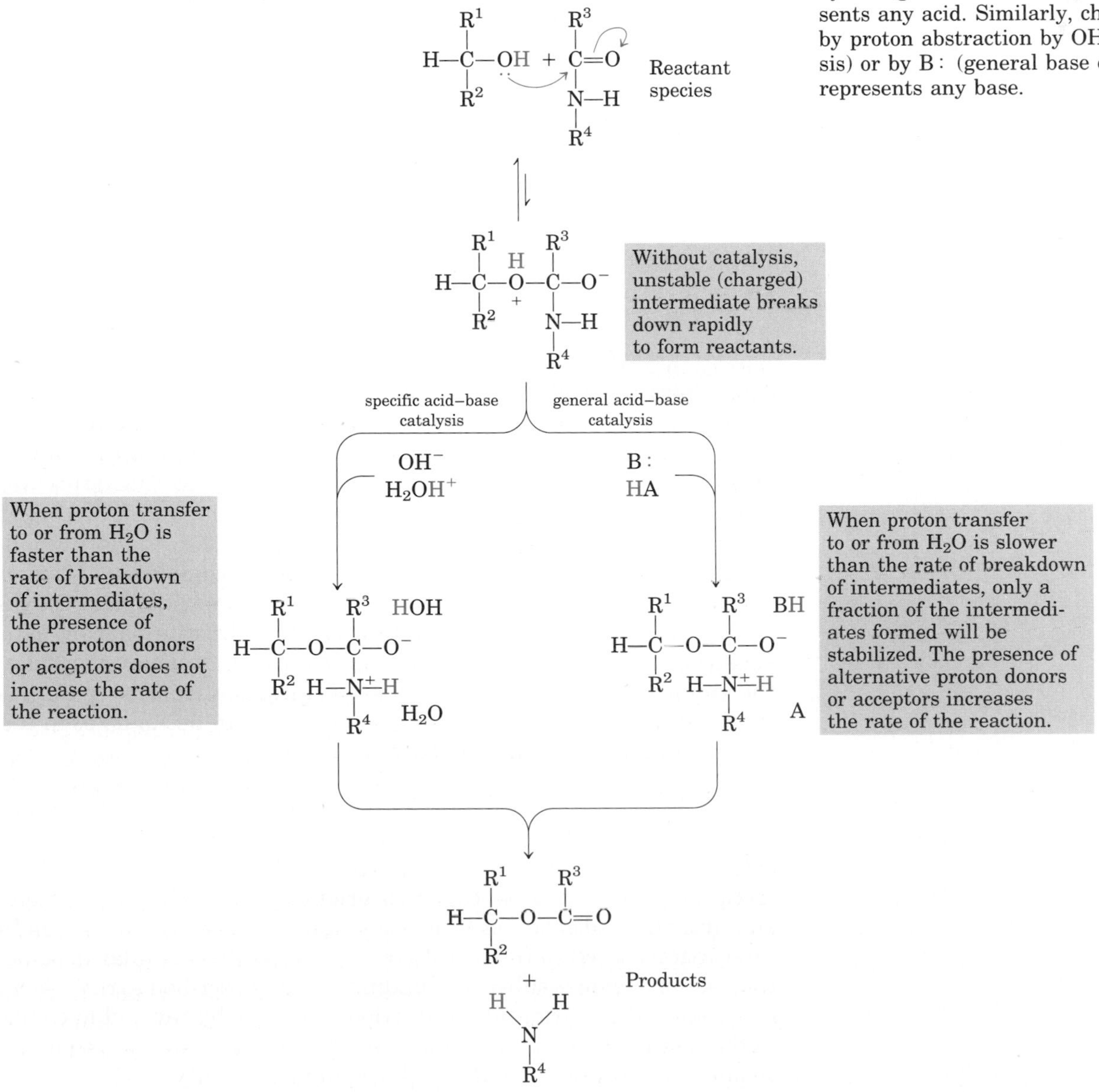

Figure 8–8 Unfavorable charge development during cleavage of an amide. This type of reaction is catalyzed by chymotrypsin and other proteases. Charge development can be circumvented by donation of a proton by H_3O^+ (specific acid catalysis) or by HA (general acid catalysis), where HA represents any acid. Similarly, charge can be neutralized by proton abstraction by OH^- (specific base catalysis) or by B: (general base catalysis), where B: represents any base.

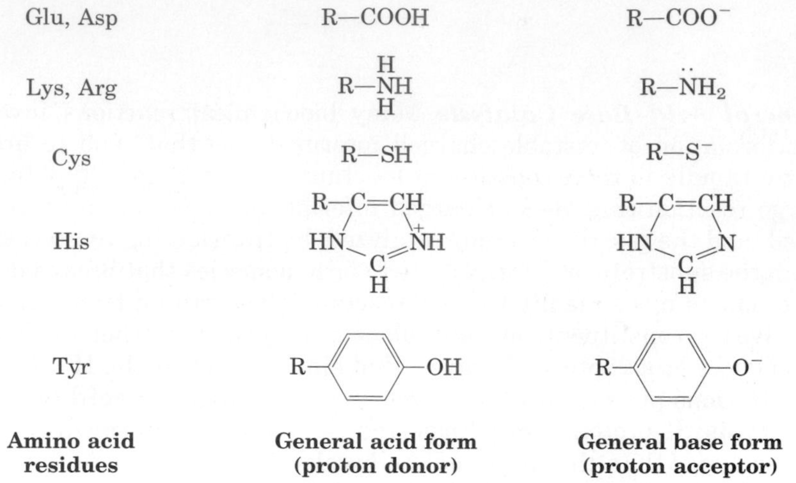

Figure 8–9 Many organic reactions are promoted by proton donors (general acids) or proton acceptors (general bases). The active sites of some enzymes contain amino acid functional groups, such as those shown here, that can participate in the catalytic process as proton donors or proton acceptors.

Amino acid residues	General acid form (proton donor)	General base form (proton acceptor)
Glu, Asp	R—COOH	R—COO⁻
Lys, Arg	R—NH (with H above and below)	R—N̈H₂
Cys	R—SH	R—S⁻
His	R—C=CH, HN—C(H)=N̈H⁺	R—C=CH, HN—C(H)=N
Tyr	R—⬡—OH	R—⬡—O⁻

the rate of proton transfer to or from water. A variety of weak organic acids can supplement water as proton donors in this situation, or weak organic bases can serve as proton acceptors. A number of amino acid side chains can similarly act as proton donors and acceptors (Fig. 8–9). These groups can be precisely positioned in an enzyme active site to allow proton transfers, providing rate enhancements on the order of 10^2 to 10^5.

Covalent Catalysis This involves the formation of a transient covalent bond between the enzyme and substrate. Consider the hydrolysis of a bond between groups A and B:

$$A{-}B \xrightarrow{\text{H}_2\text{O}} A + B$$

In the presence of a covalent catalyst (an enzyme with a nucleophilic group X:) the reaction becomes

$$A{-}B + X: \longrightarrow A{-}X + B \xrightarrow{\text{H}_2\text{O}} A + X: + B$$

This alters the pathway of the reaction and results in catalysis only when the new pathway has a lower activation energy than the uncatalyzed pathway. Both of the new steps must be faster than the uncatalyzed reaction. A number of amino acid side chains (including all of those in Fig. 8–9), as well as the functional groups of some enzyme cofactors, serve as nucleophiles on some enzymes in the formation of covalent bonds with substrates. These covalent complexes always undergo further reaction to regenerate the free enzyme. The covalent bond formed between the enzyme and the substrate can activate a substrate for further reaction in a manner that is usually specific to the group or coenzyme involved. The chemical contribution to catalysis provided by individual coenzymes is described in detail as each coenzyme is encountered in Part III of this book.

Metal Ion Catalysis Metals, whether tightly bound to the enzyme or taken up from solution along with the substrate, can participate in catalysis in several ways. Ionic interactions between an enzyme-bound metal and the substrate can help orient a substrate for reaction or stabilize charged reaction transition states. This use of weak-bonding interactions between the metal and the substrate is similar to some of the uses of enzyme–substrate binding energy described earlier. Metals can also mediate oxidation–reduction reactions by reversible changes in the metal ion's oxidation state. Nearly a third of all known enzymes require one or more metal ions for catalytic activity.

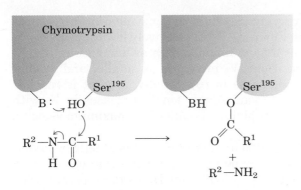

Figure 8–10 The first step in the reaction catalyzed by chymotrypsin, also called the acylation step. The hydroxyl group of Ser[195] is the nucleophile in a reaction aided by general base catalysis (the base is the side chain of His[57]). The chymotrypsin reaction is described in more detail in Fig. 8–19.

A combination of several catalytic strategies is usually employed by an enzyme to bring about a rate enhancement. A good example of the use of both covalent catalysis and general acid–base catalysis occurs in chymotrypsin. The first step in the reaction catalyzed by chymotrypsin is the cleavage of a peptide bond. This is accompanied by formation of a covalent linkage between a Ser residue on the enzyme and part of the substrate; this reaction is enhanced by general base catalysis by other groups on the enzyme (Fig. 8–10). The chymotrypsin reaction is described in more detail later in this chapter.

Enzyme Kinetics as an Approach to Understanding Mechanism

Multiple approaches are commonly used to study the mechanism of action of purified enzymes. A knowledge of the three-dimensional structure of a protein provides important information. The value of structural information is greatly enhanced by classical protein chemistry and modern methods of site-directed mutagenesis (changing the amino acid sequence of a protein in a defined way by genetic engineering; see Chapter 28) that permit enzymologists to examine the role of individual amino acids in structure and enzyme action. However, the *rate* of the catalyzed reaction can also reveal much about the enzyme. The study of reaction rates and how they change in response to changes in experimental parameters is known as **kinetics.** This is the oldest approach to understanding enzyme mechanism, and one that remains most important today. The following is a basic introduction to the kinetics of enzyme-catalyzed reactions. The more advanced student may wish to consult the texts and articles cited at the end of this chapter.

Substrate Concentration Affects the Rate of Enzyme-Catalyzed Reactions

A discussion of kinetics must begin with some fundamental concepts. One of the key factors affecting the rate of a reaction catalyzed by a purified enzyme in vitro is the amount of substrate present, [S]. But studying the effects of substrate concentration is complicated by the fact that [S] changes during the course of a reaction as substrate is converted to product. One simplifying approach in a kinetic experiment is to measure the initial rate (or initial velocity), designated V_0, when [S] is generally much greater than the concentration of enzyme. Then, if the time is sufficiently short following the start of a reaction, changes in [S] are negligible, and [S] can be regarded as a constant.

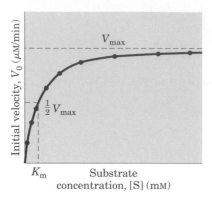

Figure 8–11 Effect of substrate concentration on the initial velocity of an enzyme-catalyzed reaction. V_{max} can only be approximated from such a plot, because V_0 will approach but never quite reach V_{max}. The substrate concentration at which V_0 is half maximal is K_m, the Michaelis–Menten constant. The concentration of enzyme E in an experiment such as this is generally so low that $[S] \gg [E]$ even when $[S]$ is described as low or relatively low. The units given are typical for enzyme-catalyzed reactions and are presented only to help illustrate the meaning of V_0 and $[S]$. (Note that the curve describes *part* of a rectangular hyperbola, with one asymptote at V_{max}. If the curve were continued below $[S] = 0$, it would approach a vertical asymptote at $[S] = -K_m$.)

Leonor Michaelis
1875–1949

Maud Menten
1879–1960

The effect on V_0 of varying $[S]$ when the enzyme concentration is held constant is shown in Figure 8–11. At relatively low concentrations of substrate, V_0 increases almost linearly with an increase in $[S]$. At higher substrate concentrations, V_0 increases by smaller and smaller amounts in response to increases in $[S]$. Finally, a point is reached beyond which there are only vanishingly small increases in V_0 with increasing $[S]$ (Fig. 8–11). This plateau is called the maximum velocity, V_{max}.

The ES complex is the key to understanding this kinetic behavior, just as it represented a starting point for the discussion of catalysis. The kinetic pattern in Figure 8–11 led Victor Henri to propose in 1903 that an enzyme combines with its substrate molecule to form the ES complex as a necessary step in enzyme catalysis. This idea was expanded into a general theory of enzyme action, particularly by Leonor Michaelis and Maud Menten in 1913. They postulated that the enzyme first combines reversibly with its substrate to form an enzyme–substrate complex in a relatively fast reversible step:

$$E + S \underset{k_{-1}}{\overset{k_1}{\rightleftharpoons}} ES \qquad (8–7)$$

The ES complex then breaks down in a slower second step to yield the free enzyme and the reaction product P:

$$ES \underset{k_{-2}}{\overset{k_2}{\rightleftharpoons}} E + P \qquad (8–8)$$

In this model the second reaction (Eqn 8–8) is slower and therefore limits the rate of the overall reaction. It follows that the overall rate of the enzyme-catalyzed reaction must be proportional to the concentration of the species that reacts in the second step, that is, ES.

At any given instant in an enzyme-catalyzed reaction, the enzyme exists in two forms, the free or uncombined form E and the combined form ES. At low $[S]$, most of the enzyme will be in the uncombined form E. Here, the rate will be proportional to $[S]$ because the equilibrium of Equation 8–7 will be pushed toward formation of more ES as $[S]$ is increased. The maximum initial rate of the catalyzed reaction (V_{max}) is observed when virtually all of the enzyme is present as the ES complex and the concentration of E is vanishingly small. Under these conditions, the enzyme is "saturated" with its substrate, so that further increases in $[S]$ have no effect on rate. This condition will exist when $[S]$ is sufficiently high that essentially all the free enzyme will have been converted into the ES form. After the ES complex breaks down to yield the product P, the enzyme is free to catalyze another reaction. The saturation effect is a distinguishing characteristic of enzyme catalysts and is responsible for the plateau observed in Figure 8–11.

When the enzyme is first mixed with a large excess of substrate, there is an initial period called the **pre-steady state** during which the concentration of the ES complex builds up. The pre-steady state is usually too short to be easily observed. The reaction quickly achieves a **steady state** in which [ES] (and the concentration of any other intermediates) remains approximately constant over time. The measured V_0 generally reflects the steady state even though V_0 is limited to early times in the course of the reaction. Michaelis and Menten concerned themselves with the steady-state rate, and this type of analysis is referred to as **steady-state kinetics**.

The Relationship between Substrate Concentration and Enzymatic Reaction Rate Can Be Expressed Quantitatively

Figure 8–11 shows the relationship between [S] and V_0 for an enzymatic reaction. The curve expressing this relationship has the same general shape for most enzymes (it approaches a rectangular hyperbola). The hyperbolic shape of this curve can be expressed algebraically by the Michaelis–Menten equation, derived by these workers starting from their basic hypothesis that the rate-limiting step in enzymatic reactions is the breakdown of the ES complex to form the product and the free enzyme.

The important terms are [S], V_0, V_{max}, and a constant called the Michaelis–Menten constant or K_m. All of these terms are readily measured experimentally.

Here we shall develop the basic logic and the algebraic steps in a modern derivation of the Michaelis–Menten equation. The derivation starts with the two basic reactions involved in the formation and breakdown of ES (Eqns 8–7 and 8–8). At early times in the reaction, the concentration of the product [P] is negligible and the simplifying assumption is made that k_{-2} can be ignored. The overall reaction then reduces to

$$\text{E} + \text{S} \xrightleftharpoons[k_{-1}]{k_1} \text{ES} \xrightarrow{k_2} \text{E} + \text{P} \qquad (8-9)$$

V_0 is determined by the breakdown of ES to give product, which is determined by [ES]:

$$V_0 = k_2[\text{ES}] \qquad (8-10)$$

As [ES] in Equation 8–10 is not easily measured experimentally, we must begin by finding an alternative expression for [ES]. First, we will introduce the term [E_t], representing the total enzyme concentration (the sum of the free and substrate-bound enzyme). Free or unbound enzyme can then be represented by [E_t] − [ES]. Also, because [S] is ordinarily far greater than [E_t], the amount of substrate bound by the enzyme at any given time is negligible compared with the total [S]. With these in mind, the following steps will lead us to an expression for V_0 in terms of parameters that are easily measured.

Step 1. The rates of formation and breakdown of ES are determined by the steps governed by the rate constants k_1 (formation) and $k_{-1} + k_2$ (breakdown), according to the expressions

$$\text{Rate of ES formation} = k_1([\text{E}_t] - [\text{ES}])[\text{S}] \qquad (8-11)$$

$$\text{Rate of ES breakdown} = k_{-1}[\text{ES}] + k_2[\text{ES}] \qquad (8-12)$$

Step 2. An important assumption is now made that the initial rate of reaction reflects a steady state in which [ES] is constant, i.e., the rate of formation of ES is equal to its rate of breakdown. This is called the steady-state assumption. The expressions in Equations 8–11 and 8–12 can be equated at the steady state, giving

$$k_1([\text{E}_t] - [\text{ES}])[\text{S}] = k_{-1}[\text{ES}] + k_2[\text{ES}] \qquad (8-13)$$

Step 3. A series of algebraic steps is now taken to solve Equation 8–13 for [ES]. The left side is multiplied out and the right side is simplified to give

$$k_1[\text{E}_t][\text{S}] - k_1[\text{ES}][\text{S}] = (k_{-1} + k_2)[\text{ES}] \qquad (8-14)$$

Adding the term $k_1[\text{ES}][\text{S}]$ to both sides of the equation and simplifying gives

$$k_1[\text{E}_t][\text{S}] = (k_1[\text{S}] + k_{-1} + k_2)[\text{ES}] \qquad (8\text{--}15)$$

Solving this equation for [ES] gives

$$[\text{ES}] = \frac{k_1[\text{E}_t][\text{S}]}{k_1[\text{S}] + k_{-1} + k_2} \qquad (8\text{--}16)$$

This can now be simplified further, in such a way as to combine the rate constants into one expression:

$$[\text{ES}] = \frac{[\text{E}_t][\text{S}]}{[\text{S}] + (k_2 + k_{-1})/k_1} \qquad (8\text{--}17)$$

The term $(k_2 + k_{-1})/k_1$ is defined as the **Michaelis–Menten constant, K_m.** Substituting this into Equation 8–17 simplifies the expression to

$$[\text{ES}] = \frac{[\text{E}_t][\text{S}]}{K_m + [\text{S}]} \qquad (8\text{--}18)$$

Step 4. V_0 can now be expressed in terms of [ES]. Equation 8–18 is used to substitute for [ES] in Equation 8–10, giving

$$V_0 = \frac{k_2[\text{E}_t][\text{S}]}{K_m + [\text{S}]} \qquad (8\text{--}19)$$

This equation can be further simplified. Because the maximum velocity will occur when the enzyme is saturated and $[\text{ES}] = [\text{E}_t]$, V_{max} can be defined as $k_2[\text{E}_t]$. Substituting this in Equation 8–19 gives

$$V_0 = \frac{V_{max}[\text{S}]}{K_m + [\text{S}]} \qquad (8\text{--}20)$$

This is the **Michaelis–Menten equation,** the rate equation for a one-substrate, enzyme-catalyzed reaction. It is a statement of the quantitative relationship between the initial velocity V_0, the maximum initial velocity V_{max}, and the initial substrate concentration [S], all related through the Michaelis–Menten constant K_m. Does the equation fit the facts? Yes; we can confirm this by considering the limiting situations where [S] is very high or very low, as shown in Figure 8–12.

An important numerical relationship emerges from the Michaelis–Menten equation in the special case when V_0 is exactly one-half V_{max} (Fig. 8–12). Then

$$\frac{V_{max}}{2} = \frac{V_{max}[\text{S}]}{K_m + [\text{S}]} \qquad (8\text{--}21)$$

On dividing by V_{max}, we obtain

$$\frac{1}{2} = \frac{[\text{S}]}{K_m + [\text{S}]} \qquad (8\text{--}22)$$

Solving for K_m, we get $K_m + [\text{S}] = 2[\text{S}]$, or

$$K_m = [\text{S}], \quad \text{when } V_0 = \tfrac{1}{2}V_{max} \qquad (8\text{--}23)$$

This represents a very useful, practical definition of K_m: K_m is equivalent to that substrate concentration at which V_0 is one-half V_{max}. Note that K_m has units of molarity.

The Michaelis–Menten equation (8–20) can be algebraically transformed into forms that are useful in the practical determination of K_m and V_{max} (Box 8–1) and, as we will describe later, in the analysis of inhibitor action (see Box 8–2).

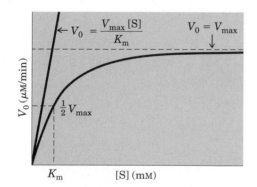

Figure 8–12 Dependence of initial velocity on substrate concentration, showing the kinetic parameters that define the limits of the curve at high and low [S]. At low [S], $K_m \gg [\text{S}]$, and the [S] term in the denominator of the Michaelis–Menten equation (Eqn 8–20) becomes insignificant; the equation simplifies to $V_0 = V_{max}[\text{S}]/K_m$, and V_0 exhibits a linear dependence on [S], as observed. At high [S], where $[\text{S}] \gg K_m$, the K_m term in the denominator of the Michaelis–Menten equation becomes insignificant, and the equation simplifies to $V_0 = V_{max}$; this is consistent with the plateau observed at high [S]. The Michaelis–Menten equation is therefore consistent with the observed dependence of V_0 on [S], with the shape of the curve defined by the terms V_{max}/K_m at low [S] and V_{max} at high [S].

BOX 8–1   **Transformations of the Michaelis–Menten Equation: The Double-Reciprocal Plot**

The Michaelis–Menten equation:

$$V_0 = \frac{V_{max}[S]}{K_m + [S]}$$

can be algebraically transformed into forms that are more useful in plotting experimental data. One common transformation is derived simply by taking the reciprocal of both sides of the Michaelis–Menten equation to give

$$\frac{1}{V_0} = \frac{K_m + [S]}{V_{max}[S]}$$

Separating the components of the numerator on the right side of the equation gives

$$\frac{1}{V_0} = \frac{K_m}{V_{max}[S]} + \frac{[S]}{V_{max}[S]}$$

which simplifies to

$$\frac{1}{V_0} = \frac{K_m}{V_{max}} \frac{1}{[S]} + \frac{1}{V_{max}}$$

This equation is a transform of the Michaelis–Menten equation called the **Lineweaver–Burk equation.** For enzymes obeying the Michaelis–Menten relationship, a plot of $1/V_0$ versus $1/[S]$ (the "double-reciprocal" of the V_0-versus-[S] plot we have been using to this point) yields a straight line (Fig. 1). This line will have a slope of K_m/V_{max}, an intercept of $1/V_{max}$ on the $1/V_0$ axis, and an intercept of $-1/K_m$ on the $1/[S]$ axis. The double-reciprocal presentation, also called a Lineweaver–Burk plot, has the great advantage of allowing a more accurate determination of V_{max}, which can only be *approximated* from a simple plot of V_0 versus [S] (see Fig. 8–12).

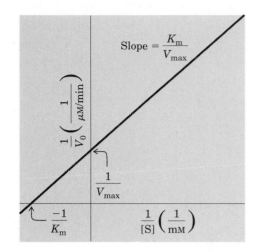

Figure 1
A double-reciprocal, or Lineweaver–Burk, plot.

Other transformations of the Michaelis–Menten equation have been derived and used. Each has some particular advantage in analyzing enzyme kinetic data.

The double-reciprocal plot of enzyme reaction rates is very useful in distinguishing between certain types of enzymatic reaction mechanisms (see Fig. 8–14) and in analyzing enzyme inhibition (see Box 8–2).

The Meaning of V_{max} and K_m Is Unique for Each Enzyme

It is important to distinguish between the Michaelis–Menten equation and the specific kinetic mechanism upon which it was originally based. The equation describes the kinetic behavior of a great many enzymes, and all enzymes that exhibit a hyperbolic dependence of V_0 on [S] are said to follow **Michaelis–Menten kinetics.** The practical rule that $K_m = [S]$ when $V_0 = \frac{1}{2}V_{max}$ (Eqn 8–23) holds for all enzymes that follow Michaelis–Menten kinetics (the major exceptions to Michaelis–Menten kinetics are the regulatory enzymes, discussed at the end of this chapter). However, this equation does not depend on the relatively simple two-step reaction mechanism proposed by Michaelis and Menten (Eqn 8–9). Many enzymes that follow Michaelis–Menten kinetics have quite different reaction mechanisms, and enzymes that catalyze reactions with six or eight identifiable steps will often exhibit the same steady-state kinetic behavior. Even though Equation 8–23 holds true for many enzymes, both the magnitude and the real meaning of V_{max} and K_m can change from one enzyme to the next. This is an important limitation of the steady-state approach to enzyme kinetics. V_{max} and K_m are parameters that can be obtained experimentally for any given enzyme, but by themselves they provide little information about the number, rates, or chemical nature of discrete steps in the reaction. Steady-state kinetics nevertheless represents the standard language by which the catalytic efficiencies of enzymes are characterized and compared. We now turn to the application and interpretation of the terms V_{max} and K_m.

A simple graphical method for obtaining an approximate value for K_m is shown in Figure 8–12. A more convenient procedure, using a **double-reciprocal plot,** is presented in Box 8–1. The K_m can vary greatly from enzyme to enzyme, and even for different substrates of the same enzyme (Table 8–6). The term is sometimes used (inappropriately) as an indication of the affinity of an enzyme for its substrate.

Table 8–6 K_m for some enzymes

Enzyme	Substrate	K_m (mM)
Catalase	H_2O_2	25
Hexokinase (brain)	ATP	0.4
	D-Glucose	0.05
	D-Fructose	1.5
Carbonic anhydrase	HCO_3^-	9
Chymotrypsin	Glycyltyrosinylglycine	108
	N-Benzoyltyrosinamide	2.5
β-Galactosidase	D-Lactose	4.0
Threonine dehydratase	L-Threonine	5.0

The actual meaning of K_m depends on specific aspects of the reaction mechanism such as the number and relative rates of the individual steps of the reaction. Here we will consider reactions with two steps. On page 214 K_m is defined by the expression

$$K_m = \frac{k_2 + k_{-1}}{k_1} \tag{8–24}$$

For the Michaelis–Menten reaction, k_2 is rate-limiting; thus $k_2 \ll k_{-1}$ and K_m reduces to k_{-1}/k_1, which is defined as the **dissociation constant, K_S,** for the ES complex. Where these conditions hold, K_m does represent a measure of the affinity of the enzyme for the substrate in the ES complex. However, this scenario does not apply to all enzymes. Sometimes $k_2 \gg k_{-1}$, and then $K_m = k_2/k_1$. In other cases, k_2 and k_{-1} are comparable, and K_m remains a more complex function of all three rate constants (Eqn 8–24). These situations were first analyzed by Haldane along with George E. Briggs in 1925. The Michaelis–Menten equation and the characteristic saturation behavior of the enzyme still apply, but K_m cannot be considered a simple measure of substrate affinity. Even more common are cases in which the reaction goes through multiple steps after formation of the ES complex; K_m can then become a very complex function of many rate constants.

V_{max} also varies greatly from one enzyme to the next. If an enzyme reacts by the two-step Michaelis–Menten mechanism, V_{max} is equivalent to $k_2[E_t]$, where k_2 is the rate-limiting step. However, the number of reaction steps and the identity of the rate-limiting step(s) can vary from enzyme to enzyme. For example, consider the quite common situation where product release, $EP \rightarrow E + P$, is rate-limiting:

$$E + S \underset{k_{-1}}{\overset{k_1}{\rightleftharpoons}} ES \underset{k_{-2}}{\overset{k_2}{\rightleftharpoons}} EP \xrightarrow{k_3} E + P \qquad (8\text{–}25)$$

In this case, most of the enzyme is in the EP form at saturation, and $V_{max} = k_3[E_t]$. It is useful to define a more general rate constant, k_{cat}, to describe the limiting rate of any enzyme-catalyzed reaction at saturation. If there are several steps in the reaction, and one is clearly rate-limiting, k_{cat} is equivalent to the rate constant for that limiting step. For the Michaelis–Menten reaction, $k_{cat} = k_2$. For the reaction of Equation 8–25, $k_{cat} = k_3$. When several steps are partially rate-limiting, k_{cat} can become a complex function of several of the rate constants that define each individual reaction step. In the Michaelis–Menten equation, $k_{cat} = V_{max}/[E_t]$, and Equation 8–19 becomes

$$V_0 = \frac{k_{cat}[E_t][S]}{K_m + [S]} \qquad (8\text{–}26)$$

The constant k_{cat} is a first-order rate constant with units of reciprocal time, and is also called the **turnover number.** It is equivalent to the number of substrate molecules converted to product in a given unit of time on a single enzyme molecule when the enzyme is saturated with substrate. The turnover numbers of several enzymes are given in Table 8–7.

Table 8–7 Turnover numbers* (k_{cat}) of some enzymes		
Enzyme	Substrate	k_{cat} (s^{-1})
Catalase	H_2O_2	40,000,000
Carbonic anhydrase	HCO_3^-	400,000
Acetylcholinesterase	Acetylcholine	140,000
β-Lactamase	Benzylpenicillin	2,000
Fumarase	Fumarate	800
RecA protein (ATPase)	ATP	0.4

* Number of substrate molecules transformed per second per molecule of enzyme.

The kinetic parameters k_{cat} and K_m are generally useful for the study and comparison of different enzymes, whether their reaction mechanisms are simple or complex. Each enzyme has optimum values of k_{cat} and K_m that reflect the cellular environment, the concentration of substrate normally encountered in vivo by the enzyme, and the chemistry of the reaction being catalyzed.

Comparison of the catalytic efficiency of different enzymes requires the selection of a suitable parameter. The constant k_{cat} is not entirely satisfactory. Two enzymes catalyzing different reactions may have the same k_{cat} (turnover number), yet the rates of the uncatalyzed reactions may be different and thus the rate enhancement brought about by the enzymes may differ greatly. Also, k_{cat} reflects the properties of an enzyme when it is saturated with substrate, and is less useful at low [S]. The constant K_m is also unsatisfactory by itself. As shown by Equation 8–23, K_m must have some relationship to the normal [S] found in the cell. An enzyme that acts on a substrate present at a very low concentration in the cell will tend to have a lower K_m than an enzyme that acts on a substrate that is normally abundant.

The most useful parameter for a discussion of catalytic efficiency is one that includes both k_{cat} and K_m. When [S] $\ll K_m$, Equation 8–26 reduces to the form

$$V_0 = \frac{k_{cat}}{K_m}[E_t][S] \tag{8–27}$$

V_0 in this case depends on the concentration of two reactants, E_t and S; therefore this is a second-order rate law and the constant k_{cat}/K_m is a second-order rate constant. The factor k_{cat}/K_m is generally the best kinetic parameter to use in comparisons of catalytic efficiency. There is an upper limit to k_{cat}/K_m, imposed by the rate at which E and S can diffuse together in an aqueous solution. This diffusion-controlled limit is 10^8 to 10^9 $M^{-1}s^{-1}$, and many enzymes have a value of k_{cat}/K_m near this range (Table 8–8).

Table 8–8 Enzymes for which k_{cat}/K_m is close to the diffusion-controlled limit (10^8 to 10^9 $M^{-1}s^{-1}$)

Enzyme	Substrate	k_{cat} (s^{-1})	K_m (M)	k_{cat}/K_m $(M^{-1}s^{-1})$
Acetylcholinesterase	Acetylcholine	1.4×10^4	9×10^{-5}	1.6×10^8
Carbonic anhydrase	CO_2	1×10^6	1.2×10^{-2}	8.3×10^7
	HCO_3^-	4×10^5	2.6×10^{-2}	1.5×10^7
Catalase	H_2O_2	4×10^7	1.1	4×10^7
Crotonase	Crotonyl-CoA	5.7×10^3	2×10^{-5}	2.8×10^8
Fumarase	Fumarate	8×10^2	5×10^{-6}	1.6×10^8
	Malate	9×10^2	2.5×10^{-5}	3.6×10^7
Triose phosphate isomerase	Glyceraldehyde-3-phosphate	4.3×10^3	4.7×10^{-4}	2.4×10^8
β-Lactamase	Benzylpenicillin	2.0×10^3	2×10^{-5}	1×10^8

Source: From Fersht, A. (1985) *Enzyme Structure and Mechanism*, p. 152, W.H. Freeman and Company, New York.

Many Enzymes Catalyze Reactions Involving Two or More Substrates

We have seen how [S] affects the rate of a simple enzyme reaction (S → P) in which there is only one substrate molecule. In many enzymatic reactions, however, two (or even more) different substrate molecules bind to the enzyme and participate in the reaction. For example, in the reaction catalyzed by hexokinase, ATP and glucose are the substrate molecules, and ADP and glucose-6-phosphate the products:

ATP + glucose ⟶ ADP + glucose-6-phosphate

The rates of such bisubstrate reactions can also be analyzed by the Michaelis–Menten approach. Hexokinase has a characteristic K_m for each of its two substrates (Table 8–6).

Enzyme reaction involving a ternary complex

Random order

$$E \underset{\searrow ES_2 \nearrow}{\overset{\nearrow ES_1 \searrow}{}} ES_1S_2 \longrightarrow E + P_1 + P_2$$

Ordered

$$E + S_1 \rightleftharpoons ES_1 \overset{S_2}{\rightleftharpoons} ES_1S_2 \longrightarrow E + P_1 + P_2$$

(a)

Enzyme reaction in which no ternary complex is formed

$$E + S_1 \rightleftharpoons ES_1 \rightleftharpoons E'P_1 \overset{P_1}{\rightleftharpoons} E' \overset{S_2}{\rightleftharpoons} E'S_2 \longrightarrow E + P_2$$

(b)

Figure 8–13 Common mechanisms for enzyme-catalyzed bisubstrate reactions. In **(a)** the enzyme and both substrates come together to form a ternary complex. In ordered binding, substrate 1 must be bound before substrate 2 can bind productively. In **(b)** an enzyme–substrate complex forms, a product leaves the complex, the altered enzyme forms a second complex with another substrate molecule, and the second product leaves, regenerating the enzyme. Substrate 1 may transfer a functional group to the enzyme (forming E'), which is subsequently transferred to substrate 2. This is a ping-pong or double-displacement mechanism.

Enzymatic reactions in which there are two substrates (bisubstrate reactions) usually involve transfer of an atom or a functional group from one substrate to the other. Such reactions proceed by one of several different pathways. In some cases, both substrates are bound to the enzyme at the same time at some point in the course of the reaction, forming a ternary complex (Fig. 8–13a). Such a complex can be formed by substrates binding in a random sequence or in a specific order. No ternary complex is formed when the first substrate is converted to product and dissociates before the second substrate binds. An example of this is the ping-pong or double-displacement mechanism (Fig. 8–13b). Steady-state kinetics can often help distinguish among these possibilities (Fig. 8–14).

Pre-Steady State Kinetics Can Provide Evidence for Specific Reaction Steps

We have introduced kinetics as a set of methods used to study the steps in an enzymatic reaction, but have also outlined the limitations of the most common kinetic parameters in providing such information. The two most important experimental parameters provided by steady-state kinetics are k_{cat} and k_{cat}/K_m. Variation in these parameters with changes in pH or temperature can sometimes provide additional information about steps in a reaction pathway. In the case of bisubstrate reactions, steady-state kinetics can help determine whether a ternary complex is formed during the reaction (Fig. 8–14). A more complete picture generally requires more sophisticated kinetic methods that go beyond the scope of an introductory text. Here, we briefly introduce one of the most important kinetic approaches for studying reaction mechanisms, pre-steady state kinetics.

A complete description of an enzyme-catalyzed reaction requires direct measurement of the rates of individual reaction steps, for example the measurement of the association of enzyme and substrate to form the ES complex. It is during the pre-steady state that the rates of many reaction steps can be measured independently. Reaction conditions are adjusted to facilitate the measurement of events that occur during the reaction of a single substrate molecule. Because the pre-steady state phase of a reaction is generally very short, this often requires specialized techniques for very rapid mixing and sampling. One objective is to gain a complete and quantitative picture of the energetic course of a reaction. As we have already noted, reaction rates and equilibria are related to the changes in free energy that occur during the

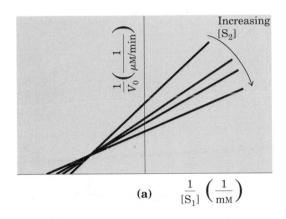

(a) $\dfrac{1}{[S_1]}\left(\dfrac{1}{mM}\right)$

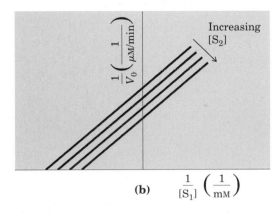

(b) $\dfrac{1}{[S_1]}\left(\dfrac{1}{mM}\right)$

Figure 8–14 Steady-state kinetic analysis of bisubstrate reactions. In these double-reciprocal plots (see Box 8–1), the concentration of substrate 1 is varied while the concentration of substrate 2 is held constant. This is repeated for several values of [S₂], generating several separate lines. The lines intersect if a ternary complex is formed in the reaction **(a)**, but are parallel if the reaction goes through a ping-pong or double-displacement pathway **(b)**.

reaction. Measuring the rate of individual reaction steps defines how energy is used by a specific enzyme, which represents an important component of the overall reaction mechanism. In a number of cases it has proven possible to measure the rates of every individual step in a multistep enzymatic reaction. Some examples of the application of pre-steady state kinetics are included in the descriptions of specific enzymes later in this chapter.

Enzymes Are Subject to Reversible and Irreversible Inhibition

Enzymes catalyze virtually every process in the cell, and it should not be surprising that enzyme inhibitors are among the most important pharmaceutical agents known. For example, aspirin (acetylsalicylate) inhibits the enzyme that catalyzes the first step in the synthesis of prostaglandins, compounds involved in many processes including some that produce pain. The study of enzyme inhibitors also has provided valuable information about enzyme mechanisms and has helped define some metabolic pathways. There are two broad classes of enzyme inhibitors: reversible and irreversible.

One common type of reversible inhibition is called competitive (Fig. 8–15). A **competitive inhibitor** competes with the substrate for the active site of an enzyme, but a reaction usually does not occur once the inhibitor (I) is bound. While the inhibitor occupies the active site it prevents binding by the substrate. Competitive inhibitors are often compounds that resemble the substrate and combine with the enzyme to form an EI complex (Fig. 8–15). This type of inhibition can be analyzed quantitatively by steady-state kinetics (Box 8–2). Because the inhibitor binds reversibly to the enzyme, the competition can be biased to favor the substrate simply by adding more substrate. When enough substrate is present the probability that an inhibitor molecule will bind is minimized, and the reaction exhibits a normal V_{max}. However, the [S] at which $V_0 = \frac{1}{2}V_{max}$, the K_m, will increase in the presence of inhibitor. This effect on the apparent K_m and the absence of an effect on V_{max} is diagnostic of competitive inhibition, and is readily revealed in a double-reciprocal plot (Box 8–2). The equilibrium constant for inhibitor binding, K_I, can be obtained from the same plot.

Competitive inhibition is used therapeutically to treat patients who have ingested methanol, a solvent found in gas-line antifreeze. Methanol is converted to formaldehyde by the action of the enzyme alcohol dehydrogenase. Formaldehyde damages many tissues, and blindness is a common result because the eyes are particularly sensitive. Ethanol competes effectively with methanol as a substrate for alcohol dehydrogenase. The therapy for methanol poisoning is intravenous infusion of ethanol, which slows the formation of formaldehyde sufficiently so that most of the methanol can be excreted harmlessly in the urine.

Two other types of reversible inhibition, noncompetitive and uncompetitive, are often defined in terms of one-substrate enzymes but in practice are only observed with enzymes having two or more substrates. A **noncompetitive inhibitor** is one that binds to a site distinct from that which binds the substrate (Fig. 8–15); inhibitor binding does not block substrate binding (or vice versa). The enzyme is inactivated when inhibitor is bound, whether or not substrate is also present. The inhibitor effectively lowers the concentration of active enzyme and hence lowers the apparent V_{max} ($V_{max} = k_{cat} [E_t]$). There is often

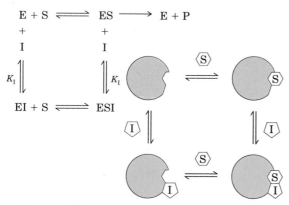

Noncompetitive inhibition

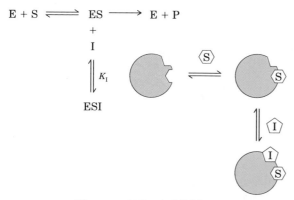

Uncompetitive inhibition

Figure 8–15 Three types of reversible inhibition. Competitive inhibitors bind to the enzyme's active site. Noncompetitive inhibitors generally bind at a separate site. Uncompetitive inhibitors also bind at a separate site, but they bind only to the ES complex. K_I is the equilibrium constant for inhibitor binding.

BOX 8–2   Kinetic Tests for Determining Inhibition Mechanisms

The double-reciprocal plot (see Box 8–1) offers an easy way of determining whether an enzyme inhibitor is competitive or noncompetitive. Two sets of rate experiments are carried out, in both of which the enzyme concentration is held constant. In the first set, [S] is also held constant, permitting measurement of the effect of increasing inhibitor concentration [I] on the initial rate V_0 (not shown). In the second set, [I] is held constant but [S] is varied. In the double-reciprocal plot $1/V_0$ is plotted versus $1/[S]$.

Figure 1 shows a set of double-reciprocal plots obtained in the absence of the inhibitor and with two different concentrations of a competitive inhibitor. Increasing [I] results in the production of a family of lines with a common intercept on the $1/V_0$ axis but with different slopes. Because the intercept on the $1/V_0$ axis is equal to $1/V_{max}$, we can see that V_{max} is unchanged by the presence of a competitive inhibitor. That is, regardless of the concentration of a competitive inhibitor, there is always some high substrate concentration that will displace the inhibitor from the enzyme's active site.

In noncompetitive inhibition, similar plots of the rate data give the family of lines shown in Figure 2, having a common intercept on the $1/[S]$ axis. This indicates that K_m for the substrate is not altered by a noncompetitive inhibitor, but V_{max} decreases.

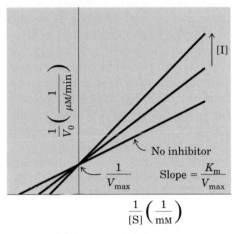

Figure 1 Competitive inhibition.

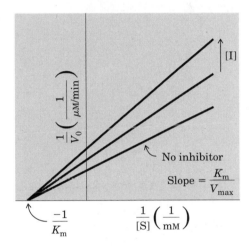

Figure 2 Noncompetitive inhibition.

little or no effect on K_m. These characteristic effects of a noncompetitive inhibitor are further analyzed in Box 8–2. An **uncompetitive inhibitor** (Fig. 8–15) also binds at a site distinct from the substrate. However, an uncompetitive inhibitor will bind only to the ES complex. (The noncompetitive inhibitor binds to *either* free enzyme or the ES complex.)

With these definitions in mind, consider a bisubstrate enzyme with separate binding sites within the active site for two substrates, S_1 and S_2, and suppose an inhibitor (I) binds to the site for S_2. If S_1 and S_2 normally bind to the enzyme independently (in random order), I may act as a competitive inhibitor of S_2. However, since I binds at a site distinct from the site for S_1, but will exclude S_2 and thereby block the reaction of S_1, I may act as a noncompetitive inhibitor of S_1. Alternatively, if S_1 normally binds to the enzyme before S_2 (ordered binding), then I may bind only to the ES_1 complex and act as an uncompetitive inhibitor of S_1. These are only a few of the scenarios that can be encountered with reversible inhibition of bisubstrate enzymes, and the effects of these inhibitors can provide much information about reaction mechanisms.

Figure 8–16 Reaction of chymotrypsin with diisopropylfluorophosphate (DIFP). This reaction led to the discovery that Ser[195] is the key active-site serine. DIFP also acts as a poison nerve gas because it irreversibly inactivates the enzyme acetylcholinesterase by a mechanism similar to that shown here. Acetylcholinesterase cleaves the neurotransmitter acetylcholine, an essential step in normal functioning of the nervous system.

Irreversible inhibitors are those that combine with or destroy a functional group on the enzyme that is essential for its activity. Formation of a covalent link between an irreversible inhibitor and an enzyme is common. Irreversible inhibitors are very useful in studying reaction mechanisms. Amino acids with key catalytic functions in the active site can sometimes be identified by determining which amino acid is covalently linked to an inhibitor after the enzyme is inactivated. An example is shown in Figure 8–16.

A very special class of irreversible inhibitors are the **suicide inhibitors.** These compounds are relatively unreactive until they bind to the active site of a specific enzyme. A suicide inhibitor is designed to carry out the first few chemical steps of the normal enzyme reaction. Instead of being transformed into the normal product, however, the inhibitor is converted to a very reactive compound that combines irreversibly with the enzyme. These are also called **mechanism-based inactivators,** because they utilize the normal enzyme reaction mechanism to inactivate the enzyme. These inhibitors play a central role in the modern approach to obtaining new pharmaceutical agents, a process called rational drug design. Because the inhibitor is designed to be specific for a single enzyme and is unreactive until within that enzyme's active site, drugs based on this approach are often very effective and have few side effects (see Box 21–1).

Enzyme Activity Is Affected by pH

Enzymes have an optimum pH or pH range in which their activity is maximal (Fig. 8–17); at higher or lower pH their activity decreases. This is not surprising because some amino acid side chains act as weak acids and bases that perform critical functions in the enzyme active site. The change in ionization state (titration) of groups in the active site is a common reason for the activity change, but it is not the only one. The group being titrated might instead affect some critical aspect of the protein structure. Removing a proton from a His residue outside the active site might, for example, eliminate an ionic interaction essential for stabilization of the active conformation of the enzyme. Less common are cases in which the group being titrated is on the substrate.

The pH range over which activity changes can provide a clue as to what amino acid is involved (see Table 5–1). A change in enzyme activity near pH 7.0, for example, often reflects titration of a His residue. The effects of pH must be interpreted with some caution, however. In the closely packed environment of a protein, the pK of amino acid side chains can change significantly. For example, a nearby positive charge can lower the pK of a Lys residue, and a nearby negative charge can increase its pK. Such effects sometimes result in a pK that is perturbed by 2 or more pH units from its normal value. One Lys residue in the enzyme acetoacetate decarboxylase has a pK of 6.6 (10.5 is normal) due to electrostatic effects of nearby positive charges.

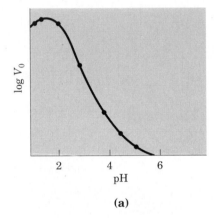

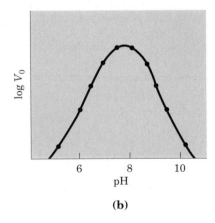

Figure 8–17 pH-activity profiles of two enzymes. Such curves are constructed from measurements of initial velocities when the reaction is carried out in buffers of different pH. The pH optimum for the activity of an enzyme generally reflects the cellular environment in which it is normally found. **(a)** Pepsin, which hydrolyzes certain peptide bonds of proteins during digestion in the stomach, has a pH optimum of about 1.6. The pH of gastric juice is between 1 and 2. **(b)** Glucose-6-phosphatase of hepatocytes, with a pH optimum of about 7.8, is responsible for releasing glucose into the blood. The normal pH of the cytosol of hepatocytes is about 7.2.

Examples of Enzymatic Reactions

This chapter has focused on the general principles of catalysis and an introduction to some of the kinetic parameters used to describe enzyme action. Principles and kinetics are combined in Box 8–3, which describes some of the evidence that reinforces the notion that binding energy and transition-state complementarity are central to enzymatic catalysis. We now turn to several examples of specific enzyme reaction mechanisms.

An understanding of the complete mechanism of action of a purified enzyme requires a knowledge of (1) the temporal sequence in which enzyme-bound reaction intermediates occur, (2) the structure of each intermediate and transition state, (3) the rates of interconversion between intermediates, (4) the structural relationship of the enzyme with each intermediate, and (5) the energetic contributions of all reacting and interacting groups with respect to intermediate complexes and transition states. There is probably no enzyme for which current understanding meets this standard exactly. Many decades of research, however, have produced mechanistic information about hundreds of enzymes, and in some cases this information is highly detailed.

Reaction Mechanisms Illustrate Principles

Mechanisms are presented for three enzymes: chymotrypsin, hexokinase, and tyrosyl-tRNA synthetase. These are chosen not necessarily because they are the best-understood enzymes or cover all possible classes of enzyme chemistry, but because they help to illustrate some general principles outlined in this chapter. The discussion concentrates on selected principles, along with some key experiments that have helped to bring them into focus. Much mechanistic detail and experimental evidence is omitted, and in no instance do the mechanisms described below provide a complete explanation for the catalytic rate enhancements brought about by these enzymes.

Chymotrypsin This enzyme is a protease (M_r 25,000) specific for peptide bonds adjacent to aromatic amino acid residues (see Table 6–7). The three-dimensional structure of chymotrypsin is shown in Figure 8–18, with functional groups in the active site emphasized. This enzyme reaction illustrates the principle of transition-state stabilization by an enzyme, and also provides a classic example of the use of general acid–base catalysis and covalent catalysis (Fig. 8–19, p. 226).

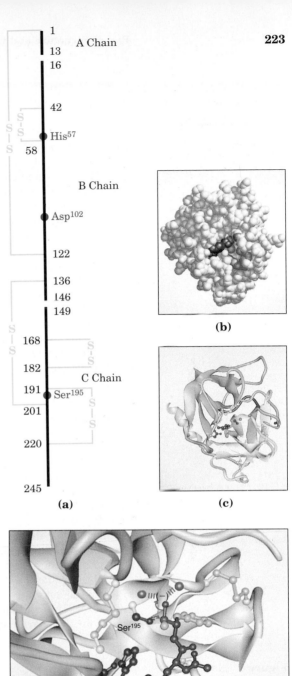

(b)

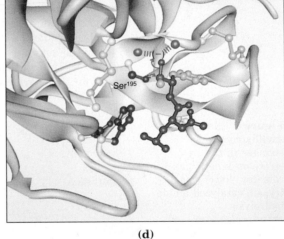

(a)   **(c)**

(d)

Figure 8–18 The structure of chymotrypsin. **(a)** A representation of primary structure, showing disulfide bonds and the location of key amino acids. Note that the protein consists of three polypeptide chains. The active-site amino acids are found grouped together in the three-dimensional structure. **(b)** A space-filling model of chymotrypsin. The pocket in which the aromatic amino acid side chain is bound is shown in green. The amide nitrogens of Gly[193] and Ser[195] in the polypeptide backbone make up the oxyanion hole (see Fig. 8–19), and are shown in orange. The side chains of other key active site residues, including Ser[195], His[57], and Asp[102] are shown in red, and are explained in Fig. 8–19. **(c)** The polypeptide backbone of chymotrypsin shown as a ribbon structure; disulfide bonds are shown in yellow. **(d)** A close-up of the chymotrypsin active site with a substrate bound. Ser[195] attacks the carbonyl group of the substrate (shown in purple); the developing negative charge on the oxygen is stabilized by the oxyanion hole (amide nitrogens shown in orange), as explained in Fig. 8–19. In the substrate, the aromatic amino acid side chain and the amide nitrogen of the peptide bond to be cleaved are shown in light blue.

BOX 8–3   Evidence for Enzyme–Transition State Complementarity

The transition state of a reaction is difficult to study, because by definition it has no finite lifetime. To understand enzymatic catalysis, however, we must dissect the interaction between the enzyme and this ephemeral moment in the course of a reaction. The idea that an enzyme is complementary to the transition state is virtually a requirement for catalysis, because the energy hill upon which the transition state sits is what the enzyme must lower if catalysis is to occur. How can we obtain evidence that the idea of enzyme–transition state complementarity is really correct? Fortunately, there are a variety of approaches, old and new, to this problem. Each has provided compelling evidence in support of this general principle of enzyme action.

Structure–Activity Correlations

If enzymes are complementary to reaction transition states, then some functional groups in the substrate and in the enzyme must interact preferentially with the transition state rather than the ES complex. Altering these groups should have little effect on formation of the ES complex, and hence should not affect kinetic parameters (K_S, or sometimes K_m if $K_S = K_m$) that reflect the $E + S \rightleftharpoons ES$ equilibrium. Changing the same groups, however, should have a large effect on the overall rate (k_{cat} or k_{cat}/K_m) of the reaction, because the bound substrate lacks potential binding interactions needed to lower the activation energy.

An excellent example is seen in a series of substrates for the enzyme chymotrypsin (Fig. 1). Chymotrypsin normally catalyzes the hydrolysis of peptide bonds next to aromatic amino acids, and the substrates shown in Figure 1 are convenient smaller models for the natural substrates (long polypeptides and proteins; see Chapter 6). The additional chemical groups added in going from A to B to C are shaded in red. Note that the interaction between the enzyme and these added functional groups has a minimal effect on K_m (which is taken here as a reflection of K_S), but a large, positive effect on k_{cat} and k_{cat}/K_m. This is what we would expect if the interaction occurred only in the transition state. Chymotrypsin is described in more detail beginning on page 223.

A complementary experimental approach to this problem is to modify the enzyme, eliminating certain enzyme–substrate interactions, by replacing specific amino acids through site-directed mutagenesis (Chapters 7 and 28). A good example is found in tyrosyl-tRNA synthetase (p. 227).

Transition-State Analogs

Even though transition states cannot be observed directly, chemists can often predict the approximate structure of a transition state based on accumulated knowledge about reaction mechanisms. The transition state by definition is transient and so unstable that direct measurement of the binding interaction between this species and the enzyme is impossible. In some cases, however, stable molecules can be designed that resemble transition states. These are called **transition-state analogs.** In principle, they should bind to an enzyme

Figure 1 Effects of small structural changes in the substrate on kinetic parameters for chymotrypsin-catalyzed amide hydrolysis.

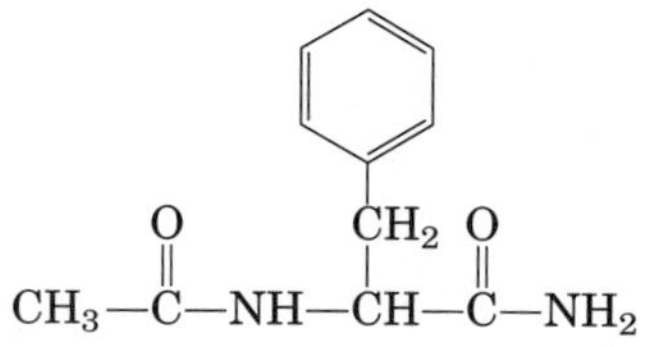

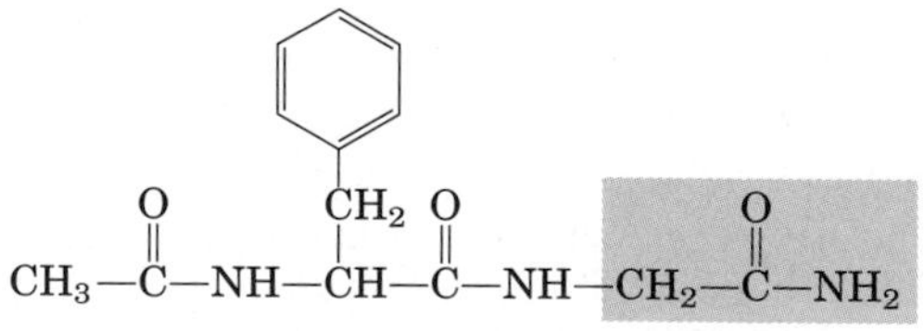

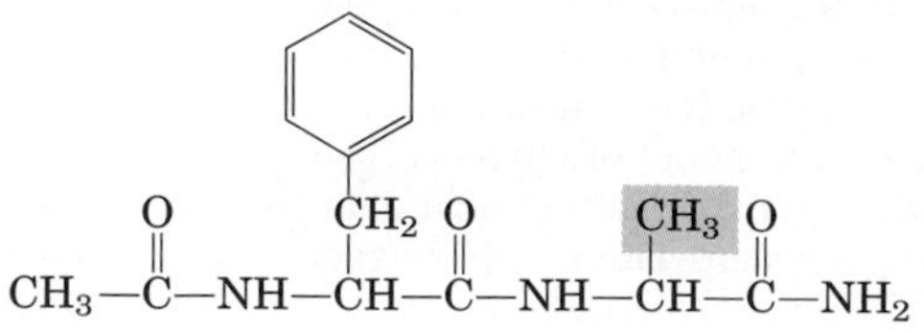

	k_{cat} (s^{-1})	K_m (mM)	k_{cat}/K_m ($M^{-1}\,s^{-1}$)
Substrate A	0.06	31	2
Substrate B	0.14	15	10
Substrate C	2.8	25	114

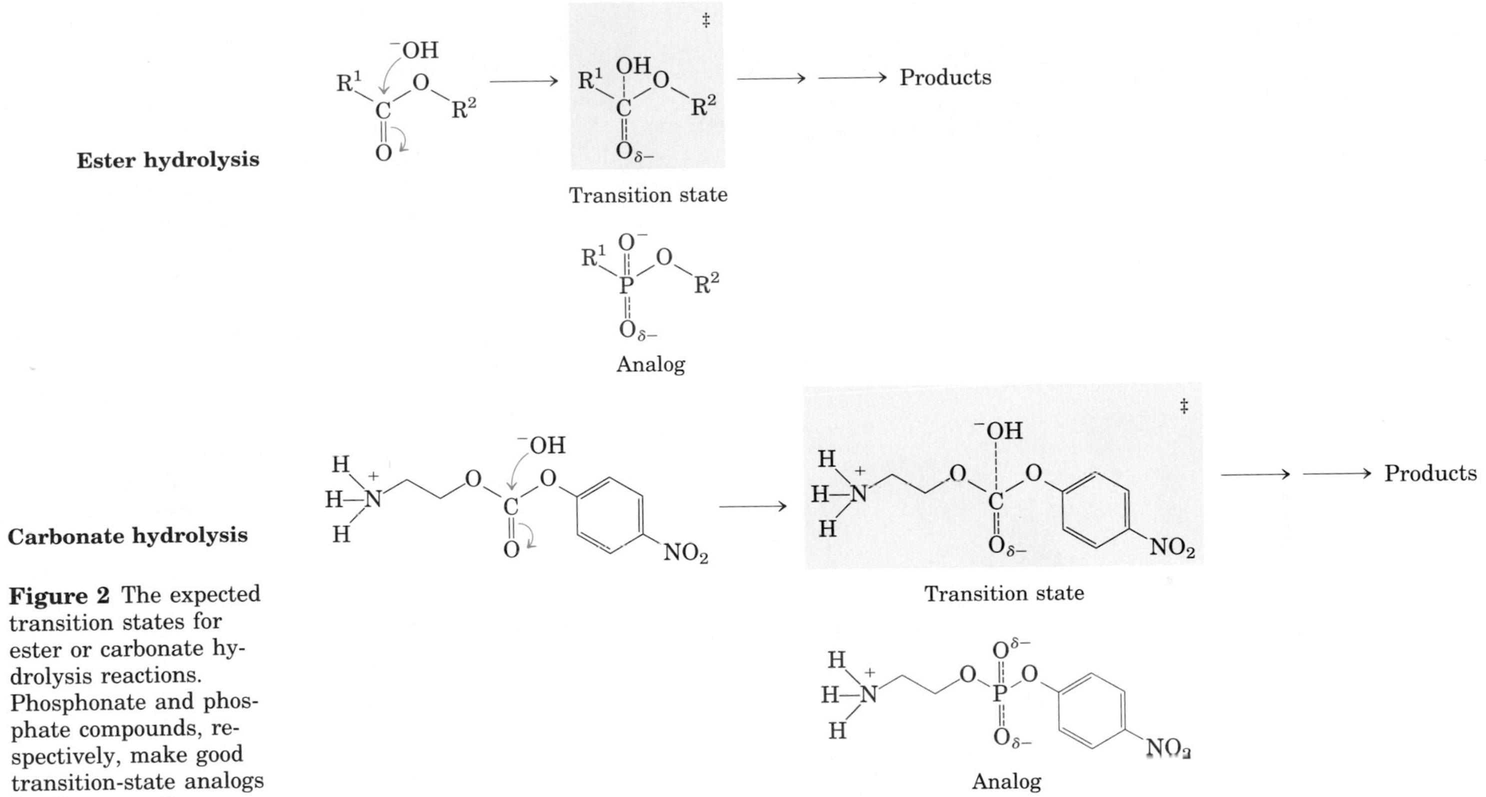

Ester hydrolysis

Transition state

Analog

Carbonate hydrolysis

Transition state

Analog

Figure 2 The expected transition states for ester or carbonate hydrolysis reactions. Phosphonate and phosphate compounds, respectively, make good transition-state analogs for these reactions.

more tightly than the substrate binds in the ES complex, because they should fit in the active site better (i.e., form more weak interactions) than the substrate itself. The idea of transition-state analogs was suggested by Pauling in the 1940s, and it has been used for a number of enzymes. These experiments have the limitation that a transition-state analog can never mimic a transition state perfectly. Analogs have been found, however, that bind an enzyme 10^2 to 10^6 times more tightly than the normal substrate, providing good evidence that enzyme active sites are indeed complementary to transition states.

Catalytic Antibodies

If a transition-state analog can be designed for the reaction S → P, then an antibody that binds tightly to the transition-state analog might catalyze S → P. Antibodies (immunoglobulins; see Fig. 6–8) are key components of the immune response. A molecule or chemical group that is bound tightly and specifically by a given antibody is referred to as an **antigen.** When a transition-state analog is used as an antigen to stimulate the production of antibodies, the antibodies that bind it are potential catalysts of the corresponding reaction. This approach, first suggested by William P. Jencks in 1969, has become practical with the development of

laboratory techniques to produce antibodies that are all identical and bind one specific antigen (these are known as monoclonal antibodies; see Chapter 6).

Pioneering work in the laboratories of Richard Lerner and Peter Schultz has resulted in the isolation of a number of monoclonal antibodies that catalyze the hydrolysis of esters or carbonates (Fig. 2). In these reactions, the attack by water (OH$^-$) on the carbonyl carbon produces a tetrahedral transition state in which a partial negative charge has developed on the carbonyl oxygen. Phosphonate compounds mimic the structure and charge distribution of this transition state in ester hydrolysis, making them good transition-state analogs; phosphate compounds are used for carbonate reactions. Antibodies that bind the phosphonate or phosphate tightly have been found to catalyze the corresponding ester or carbonate hydrolysis reaction by factors of 10^3 to 10^4. Structural analyses of a few of these catalytic antibodies have shown that the catalytic amino acid side chains are arranged where they could interact with the substrate only in the transition state. These studies provide additional evidence for enzyme–transition state complementarity and suggest that new classes of antibody catalysts might be developed for research and industry.

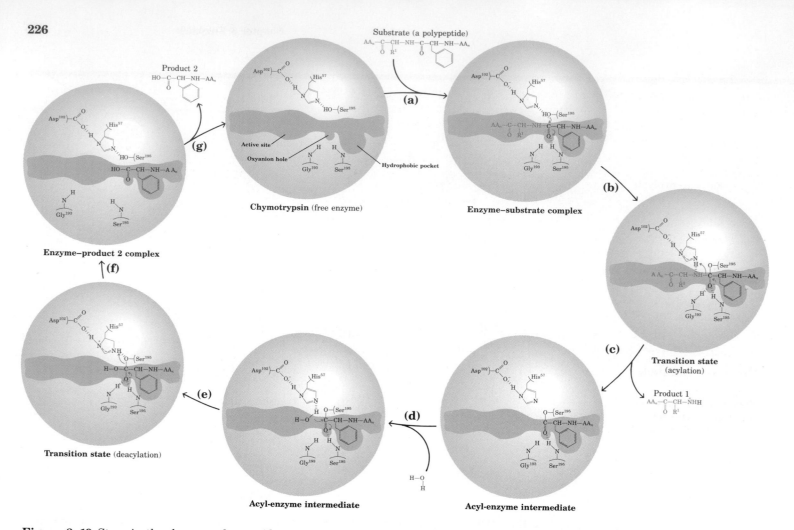

Figure 8–19 Steps in the cleavage of a peptide bond by chymotrypsin. The substrate (a polypeptide or protein) is bound at the active site. The peptide bond to be cleaved is positioned by the binding of the adjacent hydrophobic amino acid side chain (a Phe residue in this example) in a special hydrophobic pocket on the enzyme, as shown. The reaction consists of two phases: **(a)** to **(c)** formation of a covalent acyl-enzyme intermediate coupled to cleavage of the peptide bond (the acylation phase) and **(d)** to **(g)**, deacylation to regenerate the free enzyme (the deacylation phase). In both phases, the carbonyl oxygen of the substrate acquires a negative charge in the transition state. The charge is stabilized by a hydrogen bond to the amide nitrogens of Gly^{193} and Ser^{195}; the hydrogen bond to Gly^{193} forms only in the transition state. Deacylation is essentially the reverse of acylation, with water serving in place of the amine component of the substrate. The His and Asp residues cooperate in a catalytic triad, providing general base catalysis of steps **(b)** and **(e)** and general acid catalysis of steps **(c)** and **(f)**.

The enzyme reaction of chymotrypsin has two major phases: acylation, in which the peptide bond is cleaved and an ester linkage is formed between the peptide carbonyl carbon and the enzyme; and deacylation, in which the ester linkage is hydrolyzed and the enzyme regenerated. The nucleophile in the acylation phase is the oxygen of Ser^{195}. A serine hydroxyl is normally protonated at neutral pH, but in the enzyme Ser^{195} is hydrogen-bonded to His^{57}, which is further hydrogen-bonded to Asp^{102}. These three amino acids are often referred to as a catalytic triad. As the serine oxygen attacks the carbonyl carbon of a peptide bond, the hydrogen-bonded His^{57} functions as a general base to abstract the serine proton, and the negatively charged Asp^{102} stabilizes the positive charge that forms on the His residue. This prevents the development of a very unstable positive charge on the serine hydroxyl and increases its nucleophilicity. His^{57} can also act as a proton donor to protonate the amino group in the displaced portion of the substrate (the leaving group). A similar set of proton transfers occurs in the deacylation step (Fig. 8–19).

As the serine oxygen attacks the carbonyl group in the substrate, a transition state is reached in which the carbonyl oxygen acquires a negative charge. This charge is formed within a pocket on the enzyme called the oxyanion hole, and it is stabilized by hydrogen bonds contributed by the amide nitrogens of two peptide bonds in the protein backbone. One of these hydrogen bonds occurs only in the transition state and thereby reduces the energy required to reach the transition state. This represents an example of the use of binding energy in catalysis. The importance of binding energy in catalysis by chymotrypsin is discussed further in Box 8–3.

The first evidence for a covalent acyl-enzyme intermediate came from a classic application of pre-steady state kinetics. In addition to its action on polypeptides, chymotrypsin will catalyze the hydrolysis of small ester and amide compounds. These reactions are much slower because less binding energy is available with these substrates, but they are easier to study. Studies by B.S. Hartley and B.A. Kilby found that the hydrolysis of p-nitrophenylacetate by chymotrypsin, as measured by release of p-nitrophenol, proceeded with a rapid burst before leveling off to a slower rate (Fig. 8–20). By extrapolating back to zero time, they concluded that the burst phase corresponded to just under one molecule of p-nitrophenol released for every enzyme molecule present. They suggested that this reflected a rapid acylation of all the enzyme molecules (with release of p-nitrophenol) but that subsequent turnover of the enzyme was limited in rate by a slow deacylation step. Similar results have been obtained with many enzymes.

Hexokinase This is a bisubstrate enzyme (M_r 100,000), catalyzing the interconversion of glucose and ATP with glucose-6-phosphate and ADP. The hydroxyl at position 6 of the glucose molecule (to which the γ-phosphate of ATP is transferred) is similar in chemical reactivity to water, and water freely enters the enzyme active site. Yet hexokinase discriminates between glucose and water, with glucose favored by a factor of 10^6.

Mg · ATP + Glucose $\rightleftharpoons$ (hexokinase) Mg · ADP + Glucose-6-phosphate

Hexokinase can discriminate between glucose and water because of a conformational change in the enzyme that occurs when the correct substrates are bound (Fig. 8–21). The enzyme thus provides a good example of induced fit. When glucose is not present, the enzyme is in an inactive conformation with the active-site amino acid side chains out of position for reaction. When glucose (but not water) and ATP bind, the binding energy derived from this interaction induces a change to the catalytically active enzyme conformation.

Tyrosyl-tRNA Synthetase This enzyme (M_r 95,000) catalyzes the attachment of tyrosine to an RNA molecule called a transfer RNA, activating the amino acid to form a precursor for protein synthesis (described in Chapter 26). The reaction proceeds in two phases:

$$\text{Enz} + \text{Tyr} + \text{ATP} \rightleftharpoons \text{Enz} \cdot \text{Tyr-AMP} + \text{PP}_i$$

$$\text{Enz} \cdot \text{Tyr-AMP} + \text{tRNA} \rightleftharpoons \text{Tyr-tRNA} + \text{Enz} + \text{AMP}$$

(PP_i is the abbreviation for inorganic pyrophosphate. P_i, used later in this chapter, is the abbreviation for inorganic phosphate.) Kinetic studies have shown that ATP and tyrosine bind to the enzyme in random order. The tyrosyl-AMP intermediate is not released by the enzyme, and is sufficiently stable to allow study of the first reaction phase in isolation. The following discussion focuses on this phase.

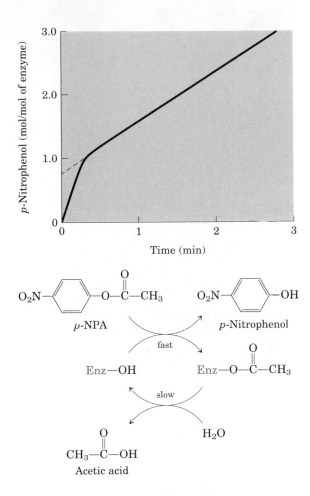

Figure 8–20 Observed kinetics of the hydrolysis of p-nitrophenylacetate (p-NPA) by chymotrypsin as measured by release of p-nitrophenol (a colored product). A rapid release (burst) of an amount of p-nitrophenol nearly stoichiometric with the amount of enzyme present is observed. This reflects the fast acylation phase of the reaction. The subsequent rate is slower because enzyme turnover is limited by the rate of the slower deacylation phase.

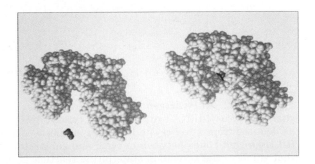

Figure 8–21 The conformational change induced in hexokinase by the binding of a substrate (D-glucose, shown in blue).

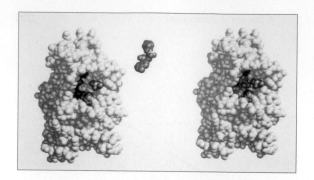

Figure 8–22 The structure of tyrosyl-tRNA synthetase. At left, the enzyme is shown without substrate bound. Active site residues that hydrogen-bond to the tyrosyl–AMP intermediate (Fig. 8–23a) are shown in red, and two residues (Thr⁴⁰ and His⁴⁵) that contribute hydrogen bonds in the reaction transition state (Fig. 8–23b) are shown in orange. The amino acid side chains of Lys⁸² and Arg⁸⁶, which cover part of the active site, are not shown in order to expose more of the active site residues. At right, the same view is shown with the tyrosyl–AMP intermediate (shown in blue, with a phosphorus atom in yellow) bound.

The structure of tyrosyl-tRNA synthetase in its complex with tyrosyl-AMP is shown in Figure 8–22. This structure indicates a number of potential hydrogen bonds between enzyme and substrate (Fig. 8–23). Alan Fersht and colleagues have used this information to create, by site-directed mutagenesis (Chapter 28), a series of mutant enzymes lacking one or another of the amino acid side chains contributing to these hydrogen bonds.

Hydrogen bonds formed in the ES complex affect K_S, which can be directly measured for this enzyme. Hydrogen bonds formed only in the transition state affect k_{cat}. In several cases, substitution of a non-hydrogen-bonding amino acid at key positions in the active site affects

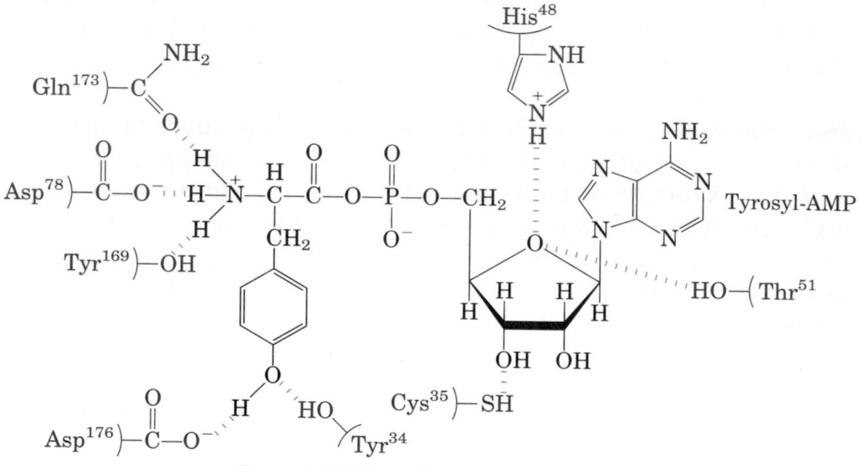

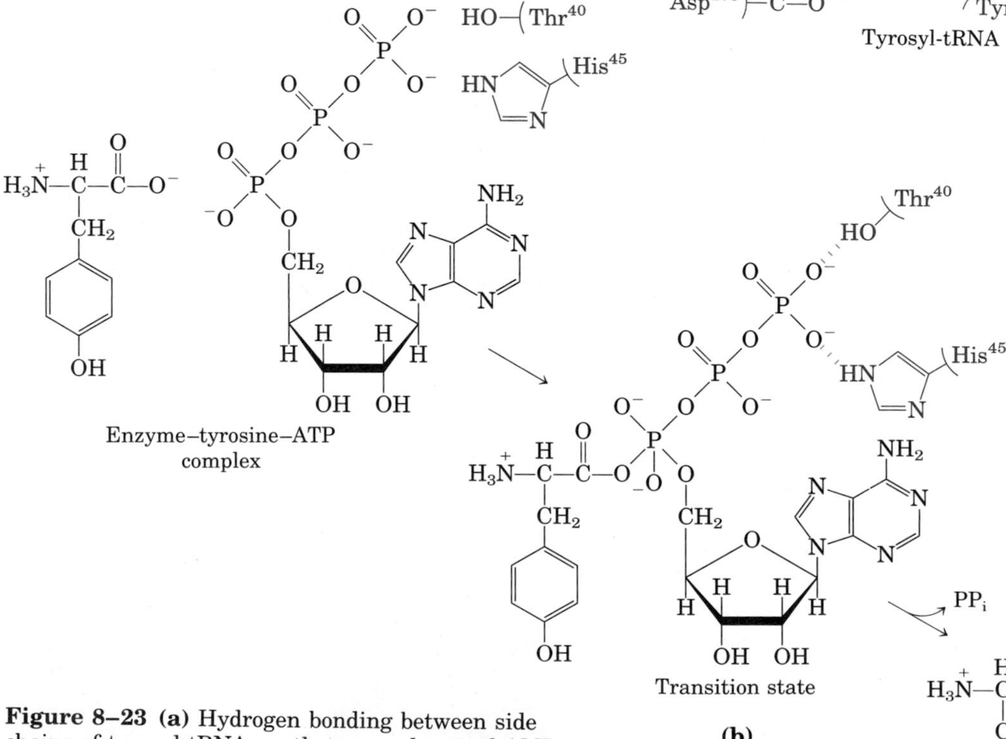

Tyrosyl-tRNA synthetase

(a)

Enzyme–tyrosine–ATP complex

Transition state

(b)

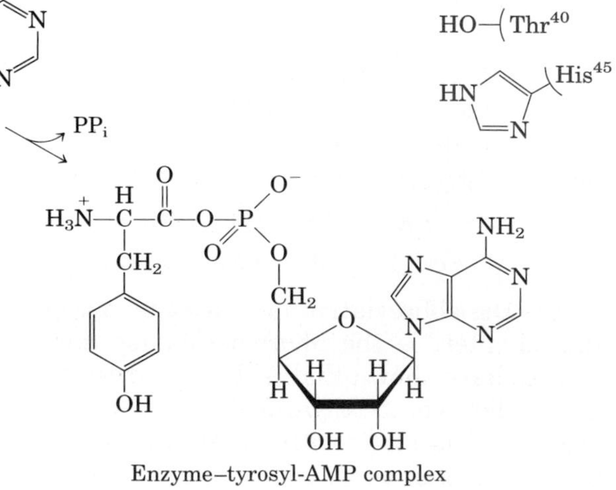

Enzyme–tyrosyl-AMP complex

Figure 8–23 (a) Hydrogen bonding between side chains of tyrosyl-tRNA synthetase and tyrosyl-AMP, as deduced from x-ray crystallographic studies. **(b)** Hydrogen bonds between enzyme and substrate that stabilize the transition state in the reaction leading to formation of tyrosyl-AMP on tyrosyl-tRNA synthetase. Amino acids that contribute hydrogen bonds primarily in the transition state are shown in red.

k_{cat} but does not affect K_S. For example, substitution of an alanine for Thr^{40} and a glycine for His^{45} has little effect on the observed K_S for ATP. However, these substitutions lower the k_{cat} for the reaction by a factor of 300,000. In other words, this altered enzyme can still bind its substrates to form the ES complex, but once bound, the substrates do not react as rapidly because the enzyme can no longer form two essential hydrogen bonds that normally help to lower the activation energy required to reach the transition state (Fig. 8–23).

Regulatory Enzymes

We now turn to a special class of enzymes that represent exceptions to some of the rules outlined so far in this chapter. In cell metabolism, groups of enzymes work together in sequential pathways to carry out a given metabolic process, such as the multireaction conversion of glucose into lactate in skeletal muscle or the multireaction synthesis of an amino acid from simpler precursors in a bacterial cell. In such enzyme systems, the reaction product of the first enzyme becomes the substrate of the next, and so on (Figure 8–24).

Most of the enzymes in each system follow kinetic patterns already described. In each enzyme system, however, there is at least one enzyme that sets the rate of the overall sequence because it catalyzes the slowest or rate-limiting reaction. These **regulatory enzymes** exhibit increased or decreased catalytic activity in response to certain signals. By the action of such regulatory enzymes, the rate of each metabolic sequence is constantly adjusted to meet changes in the cell's demands for energy and for biomolecules required in cell growth and repair. In most multienzyme systems the first enzyme of the sequence is a regulatory enzyme. Catalyzing even the first few reactions of a pathway that leads to an unneeded product diverts energy and metabolites from more important processes. An excellent place to regulate a metabolic pathway, therefore, is at the point of commitment to the pathway. The other enzymes in the sequence are usually present in amounts providing a large excess of catalytic activity; they can promote their reactions only as fast as their substrates are made available from preceding reactions.

The activity of regulatory enzymes is modulated through various types of signal molecules, which are generally small metabolites or cofactors. There are two major classes of regulatory enzymes in metabolic pathways. **Allosteric enzymes** function through reversible, noncovalent binding of a regulatory metabolite called a modulator. The term allosteric derives from Greek *allos*, "other," and *stereos*, "solid" or "shape." Allosteric enzymes are those having "other shapes" or conformations induced by the binding of modulators. The second class includes enzymes regulated by reversible covalent modification. Both classes of regulatory enzymes tend to have multiple subunits, and in some cases the regulatory site(s) and the active site are on separate subunits.

There are at least two other mechanisms by which enzyme activity is regulated. Some enzymes are stimulated or inhibited by separate control proteins that bind to them and affect their activity. Others are activated by proteolytic cleavage, which unlike the other mechanisms is irreversible. Important examples of both these mechanisms are found in physiological processes such as digestion, blood clotting, hormone action, and vision.

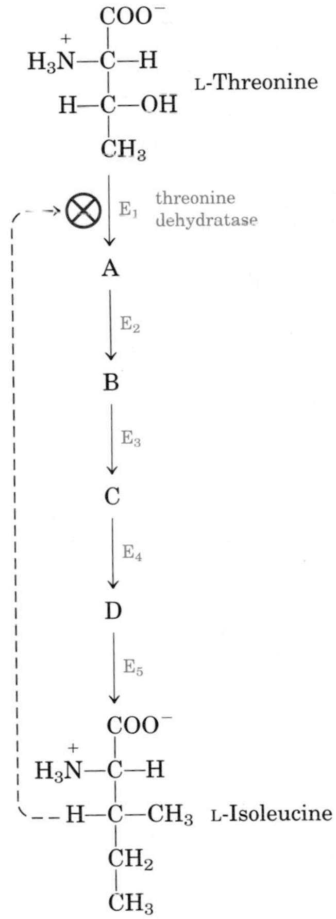

Figure 8–24 Feedback inhibition of the conversion of L-threonine into L-isoleucine, catalyzed by a sequence of five enzymes (E_1 to E_5). Threonine dehydratase (E_1) is specifically inhibited allosterically by L-isoleucine, the end product of the sequence, but not by any of the four intermediates (A to D). Such inhibition is indicated by the dashed feedback line and the $\otimes$ symbol at the threonine dehydratase reaction arrow.

No single rule governs the occurrence of different types of regulation in different systems. To a degree, allosteric (noncovalent) regulation may permit fine-tuning of metabolic pathways that are required continuously but at different levels of activity as cellular conditions change. Regulation by covalent modification tends to be all-or-none. However, both types of regulation are observed in a number of regulatory enzymes.

Allosteric Enzymes Are Regulated by Noncovalent Binding of Modulators

In some multienzyme systems the regulatory enzyme is specifically inhibited by the end product of the pathway, whenever the end product increases in excess of the cell's needs. When the regulatory enzyme reaction is slowed, all subsequent enzymes operate at reduced rates because their substrates are depleted by mass action. The rate of production of the pathway's end product is thereby brought into balance with the cell's needs. This type of regulation is called **feedback inhibition.** Buildup of the pathway's end product ultimately slows the entire pathway.

One of the first discovered examples of such allosteric feedback inhibition was the bacterial enzyme system that catalyzes the conversion of L-threonine into L-isoleucine (Fig. 8–24). In this system, the first enzyme, threonine dehydratase, is inhibited by isoleucine, the product of the last reaction of the series. Isoleucine is quite specific as an inhibitor. No other intermediate in this sequence of reactions inhibits threonine dehydratase, nor is any other enzyme in the sequence inhibited by isoleucine. Isoleucine binds not to the active site, but to another specific site on the enzyme molecule, the regulatory site. This binding is noncovalent and thus readily reversible; if the isoleucine concentration decreases, the rate of threonine dehydratase activity increases. Thus threonine dehydratase activity responds rapidly and reversibly to fluctuations in the concentration of isoleucine in the cell.

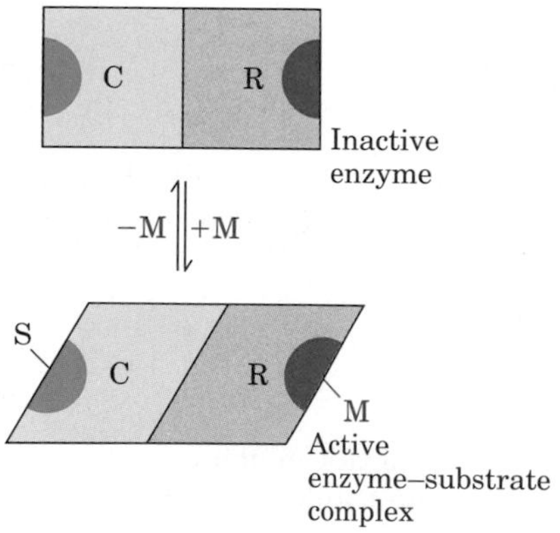

Allosteric Enzymes Are Exceptions to Many General Rules

The modulators for allosteric enzymes may be either inhibitory or stimulatory. An activator is often the substrate itself, and regulatory enzymes for which substrate and modulator are identical are called **homotropic.** When the modulator is a molecule other than the substrate the enzyme is **heterotropic.** Some enzymes have two or more modulators.

As already noted, the properties of allosteric enzymes are significantly different from those of simple nonregulatory enzymes discussed earlier in this chapter. Some of the differences are structural. In addition to active or catalytic sites, allosteric enzymes generally have one or more regulatory or allosteric sites for binding the modulator (Fig. 8–25). Just as an enzyme's active site is specific for its substrate, the allosteric site is specific for its modulator. Enzymes with several modulators generally have different specific binding sites for each. In homotropic enzymes the active site and regulatory site are the same.

Allosteric enzymes are also generally larger and more complex than simple enzymes. Most of them have two or more polypeptide chains or subunits. Aspartate transcarbamoylase, which catalyzes the first reaction in the biosynthesis of pyrimidine nucleotides (Chapter 21), has 12 polypeptide chains organized into catalytic and regulatory

Figure 8–25 Schematic model of the subunit interactions in an allosteric enzyme, and interactions with inhibitors and activators. In many allosteric enzymes the substrate binding site and the modulator binding site(s) are on different subunits, the catalytic (C) and regulatory (R) subunits, respectively. Binding of the positive modulator (M) to its specific site on the regulatory subunit is communicated to the catalytic subunit through a conformational change. This change renders the catalytic subunit active and capable of binding the substrate (S) with higher affinity. On dislocation of the modulator from the regulatory subunit, the enzyme reverts to its inactive or less active form.

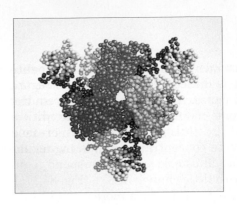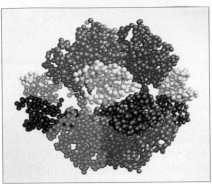

Figure 8–26 The three-dimensional subunit architecture of the regulatory enzyme aspartate transcarbamoylase; two different views. This allosteric regulatory enzyme has two catalytic clusters, each with three catalytic polypeptide chains, and three regulatory clusters, each with two regulatory polypeptide chains. The catalytic polypeptides in each cluster are shown in shades of blue and purple. Binding sites for allosteric modulators are found on the regulatory subunits (shown in white and red). Modulator binding produces large changes in enzyme conformation and activity. The role of this enzyme in nucleotide synthesis, and details of its regulation, will be discussed in Chapter 21.

subunits. Figure 8–26 shows the quaternary structure of this enzyme, deduced from x-ray analysis.

Other differences between nonregulated enzymes and allosteric enzymes involve kinetic properties. Allosteric enzymes show relationships between V_0 and [S] that differ from normal Michaelis–Menten behavior. They do exhibit saturation with the substrate when [S] is sufficiently high, but for some allosteric enzymes, when V_0 is plotted against [S] (Fig. 8–27) a sigmoid saturation curve results, rather than the hyperbolic curve shown by nonregulatory enzymes. Although we can find a value of [S] on the sigmoid saturation curve at which V_0 is half-maximal, we cannot refer to it with the designation K_m because the enzyme does not follow the hyperbolic Michaelis–Menten relationship. Instead the symbol $[S]_{0.5}$ or $K_{0.5}$ is often used to represent the substrate concentration giving half-maximal velocity of the reaction catalyzed by an allosteric enzyme (Fig. 8–27).

Sigmoid kinetic behavior generally reflects cooperative interactions between multiple protein subunits. In other words, changes in the structure of one subunit are translated into structural changes in adjacent subunits, an effect that is mediated by noncovalent interactions at the subunit–subunit interface. The principles are similar to those discussed for cooperativity in oxygen binding to the nonenzyme protein hemoglobin (p. 188). Homotropic allosteric enzymes generally have multiple subunits. In many cases the same binding site on each subunit functions as both the active site and the regulatory site. The substrate can function as a positive modulator (an activator) because the subunits act cooperatively. The binding of one molecule of the substrate to one binding site alters the enzyme's conformation and greatly enhances the binding of subsequent substrate molecules. This accounts for the sigmoid rather than hyperbolic increase in V_0 with increasing [S].

Figure 8–27 Substrate-activity curves for representative allosteric enzymes. Three examples of complex responses given by allosteric enzymes to their modulators. **(a)** The sigmoid curve given by a homotropic enzyme, in which the substrate also serves as a positive (stimulatory) modulator. Note that a relatively small increase in [S] in the steep part of the curve can cause a very large increase in V_0. Note also the resemblance to the oxygen-saturation curve of hemoglobin (see Fig. 7–28). **(b)** The effects of a positive modulator $\oplus$, a negative modulator $\ominus$, and no modulator $\circledcirc$ on an allosteric enzyme in which $K_{0.5}$ is modulated without a change in V_{max}. **(c)** A less common type of modulation, in which V_{max} is modulated with $K_{0.5}$ nearly constant.

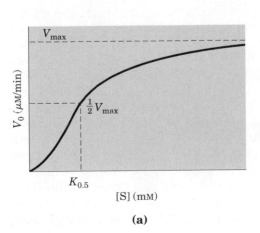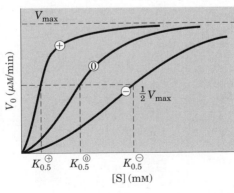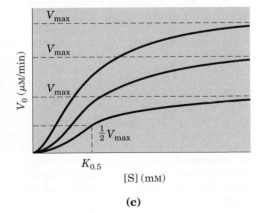

(a)   (b)   (c)

With heterotropic enzymes, in which the modulator is a metabolite other than the substrate itself, it is difficult to generalize about the shape of the substrate-saturation curve. An activator may cause the substrate-saturation curve to become more nearly hyperbolic, with a decrease in $K_{0.5}$ but no change in V_{max}, thus resulting in an increased reaction velocity at a fixed substrate concentration (V_0 is higher for any value of [S]) (Fig. 8–27b). Other allosteric enzymes respond to an activator by an increase in V_{max}, with little change in $K_{0.5}$ (Fig. 8–27c). A negative modulator (an inhibitor) may produce a *more* sigmoid substrate-saturation curve, with an increase in $K_{0.5}$ (Fig. 8–27b). Allosteric enzymes therefore show different kinds of responses in their substrate-activity curves because some have inhibitory modulators, some have activating modulators, and some have both.

Two Models Explain the Kinetic Behavior of Allosteric Enzymes

The sigmoidal dependence of V_0 on [S] reflects subunit cooperativity, and has inspired two models to explain these cooperative interactions.

In the first model (the symmetry model), proposed by Jacques Monod and colleagues in 1965, an allosteric enzyme can exist in only two conformations, active and inactive (Fig. 8–28a). All subunits are in the active form or all are inactive. Every substrate molecule that binds increases the probability of a transition from the inactive to the active state.

In the second model (the sequential model) (Fig. 8–28b), proposed by Koshland in 1966, there are still two conformations, but subunits can undergo the conformational change individually. Binding of substrate increases the probability of the conformational change. A conformational change in one subunit makes a similar change in an adjacent subunit, as well as the binding of a second substrate molecule, more likely. There are more potential intermediate states in this model than in the symmetry model. The two models are not mutually exclusive; the symmetry model may be viewed as the "all-or-none" limiting case of the sequential model. The precise mechanism of allosteric interaction has not been established. Different allosteric enzymes may have different mechanisms for cooperative interactions.

Figure 8–28 Two general models for the interconversion of inactive and active forms of allosteric enzymes. Four subunits are shown because the model was originally proposed for the oxygen-carrying protein hemoglobin. In the symmetry, or all-or-none, model **(a)** all the subunits are postulated to be in the same conformation, either all ◯ (low affinity or inactive) or all ☐ (high affinity or active). Depending on the equilibrium, K_1, between ◯ and ☐ forms, the binding of one or more substrate (S) molecules will pull the equilibrium toward the ☐ form. Subunits with bound S are shaded. A possible pathway is given by the gray shading. In the sequential model **(b)** each individual subunit can be in either the ◯ or ☐ form. A very large number of conformations is thus possible, but the shaded pathway (diagonal arrows) is the most probable route.

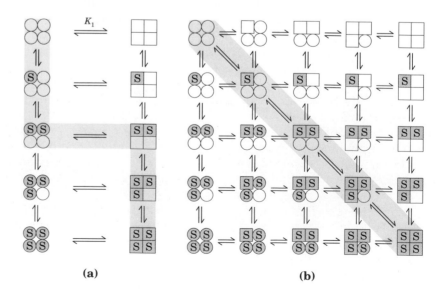

(a)   (b)

Other Mechanisms of Enzyme Regulation

In another important class of regulatory enzymes activity is modulated by covalent modification of the enzyme molecule. Modifying groups include phosphate, adenosine monophosphate, uridine monophosphate, adenosine diphosphate ribose, and methyl groups. These are generally covalently linked to and removed from the regulatory enzyme by separate enzymes (some examples are given in Box 8–4).

An important example of regulation by covalent modification is glycogen phosphorylase (M_r 94,500) of muscle and liver (Chapter 14), which catalyzes the reaction

$$(\text{Glucose})_n + \text{P}_i \longrightarrow (\text{glucose})_{n-1} + \text{glucose-1-phosphate}$$

Glycogen   Shortened
                                                  glycogen
                                                  chain

The glucose-1-phosphate so formed can then be broken down into lactate in muscle or converted to free glucose in the liver. Glycogen phosphorylase occurs in two forms: the active form phosphorylase a and the relatively inactive form phosphorylase b (Fig. 8–29). Phosphorylase a has two subunits, each with a specific Ser residue that is phosphorylated at its hydroxyl group. These serine phosphate residues are required for maximal activity of the enzyme. The phosphate groups can be hydrolytically removed from phosphorylase a by a separate enzyme called phosphorylase phosphatase:

$$\text{Phosphorylase } a + 2\text{H}_2\text{O} \longrightarrow \text{phosphorylase } b + 2\text{P}_i$$

(More active)   (Less active)

In this reaction phosphorylase a is converted into phosphorylase b by the cleavage of two serine–phosphate covalent bonds.

Phosphorylase b can in turn be reactivated—covalently transformed back into active phosphorylase a—by another enzyme, phosphorylase kinase, which catalyzes the transfer of phosphate groups from ATP to the hydroxyl groups of the specific Ser residues in phosphorylase b:

$$2\text{ATP} + \text{phosphorylase } b \longrightarrow 2\text{ADP} + \text{phosphorylase } a$$

(Less active)   (More active)

The breakdown of glycogen in skeletal muscles and the liver is regulated by variations in the ratio of the two forms of the enzyme. The a and b forms of phosphorylase differ in their quaternary structure; the active site undergoes changes in structure and, consequently, changes in catalytic activity as the two forms are interconverted.

Some of the more complex regulatory enzymes are located at particularly crucial points in metabolism, so that they respond to multiple regulatory metabolites through both allosteric and covalent modification. Glycogen phosphorylase is an example. Although its primary regulation is through covalent modification, it is also modulated in a noncovalent, allosteric manner by AMP, which is an activator of phosphorylase b, and several other molecules that are inhibitors.

Glutamine synthetase of *E. coli,* one of the most complex regulatory enzymes known, provides examples of regulation by allostery, reversible covalent modification, and regulating proteins. It has at least eight allosteric modulators. The glutamine synthetase system is described in more detail in Chapter 21.

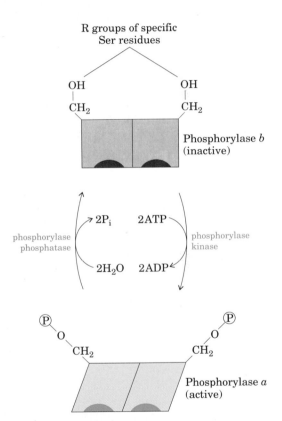

Figure 8–29 Regulation of glycogen phosphorylase activity by covalent modification. In the active form of the enzyme, phosphorylase a, specific Ser residues, one on each subunit, are in the phosphorylated state. Phosphorylase a is converted into phosphorylase b, which is relatively inactive, by enzymatic loss of these phosphate groups, promoted by phosphorylase phosphatase. Phosphorylase b can be reactivated to form phosphorylase a by the action of phosphorylase kinase.

BOX 8–4   **Regulation of Protein Activity by Reversible Covalent Modification**

A variety of chemical groups are used in reversible covalent modification of regulatory proteins to produce activity changes (Fig. 1). An example of phosphorylation (glycogen phosphorylase) is given in the text. An excellent example of methylation involves the methyl-accepting chemotaxis protein of bacteria. This protein is part of a system that permits a bacterium to swim toward an attractant in solution (such as a sugar) and away from repellent chemicals. The methylating agent is S-adenosylmethionine (adoMet), described in Chapter 17. ADP-ribosylation is an especially interesting reaction observed in only a few proteins. ADP-ribose is derived from nicotinamide adenine dinucleotide (see Fig. 12–41). This type of modification occurs for dinitrogenase reductase, resulting in the regulation of the important process of biological nitrogen fixation. In addition, both diphtheria toxin and cholera toxin are enzymes that catalyze the ADP-ribosylation (and inactivation) of key cellular enzymes or proteins. Diphtheria toxin acts on and inhibits elongation factor 2, a protein involved in protein biosynthesis. Cholera toxin acts on a specific G protein (Chapter 22) leading ultimately to several physiological responses including a massive loss of body fluids and sometimes death.

Figure 1 Some well-studied examples of enzyme modification reactions.

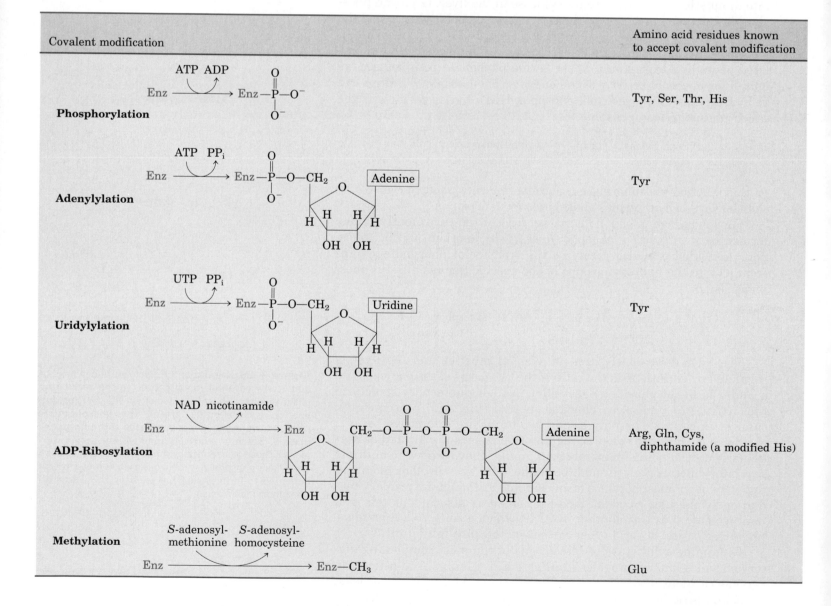

Covalent modification	Amino acid residues known to accept covalent modification
Phosphorylation	Tyr, Ser, Thr, His
Adenylylation	Tyr
Uridylylation	Tyr
ADP-Ribosylation	Arg, Gln, Cys, diphthamide (a modified His)
Methylation	Glu

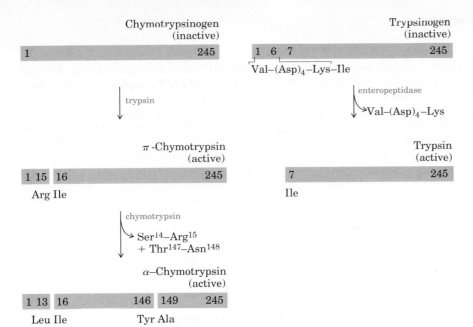

Figure 8–30 Activation of the zymogens of chymotrypsin and trypsin by proteolytic cleavage. The bars represent the primary sequence of the polypeptide chains. Amino acids at the termini of the polypeptide fragments generated by cleavage are indicated below the bars. The numbers represent the positions of the amino acids in the primary sequence of chymotrypsinogen or trypsinogen. (The amino-terminal amino acid is number 1.)

Activation of an enzyme by proteolytic cleavage is a somewhat different type of regulatory mechanism. An inactive precursor of the enzyme, called a **zymogen,** is cleaved to form the active enzyme. Many proteolytic enzymes (proteases) of the stomach and pancreas are regulated this way. Chymotrypsin and trypsin are initially synthesized as chymotrypsinogen and trypsinogen, respectively (Fig. 8–30). Specific cleavage causes conformational changes that expose the enzyme active site. Because this type of activation is irreversible, other mechanisms are needed to inactivate these enzymes. Proteolytic enzymes are inactivated by inhibitor proteins that bind very tightly to the enzyme active site. Pancreatic trypsin inhibitor (M_r 6,000) binds to and inhibits trypsin; α_1-antiproteinase (M_r 53,000) primarily inhibits elastase. An insufficiency of α_1-antiproteinase, believed to be caused by exposure to cigarette smoke, leads to lung damage and the condition known as emphysema.

Other examples of zymogen activation occur in hormones, connective tissue, and the blood-clotting system. The hormone insulin is produced by cleavage of proinsulin, collagen is initially synthesized as a soluble precursor called procollagen, and blood clotting is mediated by a complicated cascade of zymogen activations.

Summary

Virtually every biochemical reaction is catalyzed by enzymes. With the exception of a few catalytic RNAs, all known enzymes are proteins. Enzymes are extraordinarily effective catalysts, commonly producing reaction rate enhancements of 10^7 to 10^{14}. To be active, some enzymes require a chemical cofactor, which can be loosely or tightly bound. Each enzyme is classified according to the specific reaction it catalyzes.

Enzyme-catalyzed reactions are characterized by the formation of a complex between substrate and enzyme (an ES complex). The binding occurs in a pocket on the enzyme called the active site. The function of enzymes and other catalysts is to lower the activation energy for the reaction and thereby enhance the reaction rate. The equilibrium of a reaction is unaffected by the enzyme.

The energy used for enzymatic rate enhancements is derived from weak interactions (hydrogen bonds and van der Waals, hydrophobic, and ionic interactions) between the substrate and enzyme. The enzyme active site is structured so that many of these weak interactions occur only in the reaction transition state, thus stabilizing the transition state. The energy available from the numerous weak interactions between enzyme and substrate (the binding energy) is substantial and can generally account for observed rate enhancements. The need for multiple interactions is one reason for the large size of enzymes. Binding energy can be used to lower substrate entropy, to strain the substrate, or to cause a conformational change in the enzyme (induced fit). This same binding energy accounts for the exquisite specificity exhibited by enzymes for their substrates. Other catalytic mechanisms include general acid–base catalysis and covalent catalysis. Details of the reaction mechanisms have been worked out for many enzymes.

Kinetics is an important method for the study of enzyme mechanisms. Most enzymes have some common kinetic properties. As the concentration of the substrate is increased, the catalytic activity of a fixed concentration of an enzyme will increase in a hyperbolic fashion to approach a characteristic maximum rate V_{max}, at which essentially all the enzyme is in the form of the ES complex. The substrate concentration giving one-half V_{max} is the Michaelis–Menten constant K_m, which is characteristic for each enzyme acting on a given substrate. The Michaelis–Menten equation

$$V_0 = \frac{V_{max}[S]}{K_m + [S]}$$

relates the initial velocity of an enzymatic reaction to the substrate concentration and V_{max} through the constant K_m. Both K_m and V_{max} can be measured; they have different meanings for different enzymes. The limiting rate of an enzyme-catalyzed reaction at saturation is described by the constant k_{cat}, also called the turnover number. The ratio k_{cat}/K_m provides a good measure of catalytic efficiency. The Michaelis–Menten equation is also applicable to bisubstrate reactions, which occur by either ternary complex or double-displacement (ping-pong) pathways. Each enzyme has an optimum pH, as well as a characteristic specificity for the substrates on which it acts.

Enzymes can be inactivated by irreversible modification of a functional group essential for catalytic activity. They can also be reversibly inhibited, competitively or noncompetitively. Competitive inhibitors compete reversibly with the substrate for binding to the active site, but they are not transformed by the enzyme. Noncompetitive inhibitors bind to some other site on the free enzyme or to the ES complex.

Some enzymes regulate the rate of metabolic pathways in cells. In feedback inhibition, the end product of a pathway inhibits the first enzyme of that pathway. The activity of some regulatory enzymes, called allosteric enzymes, is adjusted by reversible, noncovalent binding of a specific modulator to a regulatory or allosteric site. Such modulators may be inhibitory or stimulatory and may be either the substrate itself or some other metabolite. The kinetic behavior of allosteric enzymes reflects cooperative interactions among the enzyme subunits. Other regulatory enzymes are modulated by covalent modification of a specific functional group necessary for activity, or by proteolytic cleavage of a zymogen.

Further Reading

General

Evolution of Catalytic Function. (1987) *Cold Spring Harb. Symp. Quant. Biol.* **52.**
A collection of excellent papers on many topics discussed in this chapter.

Fersht, A. (1985) *Enzyme Structure and Mechanism,* W.H. Freeman and Company, New York.
A clearly written, concise introduction. More advanced.

Friedmann, H. (ed) (1981) *Benchmark Papers in Biochemistry,* Vol. 1: *Enzymes,* Hutchinson Ross Publishing Company, Stroudsburg, PA.
A collection of classic papers in enzyme chemistry, with historical commentaries by the editor. Extremely interesting.

Jencks, W.P. (1987) *Catalysis in Chemistry and Enzymology,* Dover Publications, Inc., New York.
A new printing of an outstanding book on the subject. More advanced.

Principles of Catalysis

Hansen, D.E. and Raines, R.T. (1990) Binding energy and enzymatic catalysis. *J. Chem. Educ.* **67,** 483–489.
A good place for the beginning student to acquire a better understanding of principles.

Jencks, W.P. (1975) Binding energy, specificity, and enzymic catalysis: the Circe effect. *Adv. Enzymol.* **43,** 219–410.

Kraut, J. (1988) How do enzymes work? *Science* **242,** 533–540.

Lerner, R.A., Benkovic, S.J., & Schulz, P.G. (1991) At the crossroads of chemistry and immunology: catalytic antibodies. *Science* **252,** 659–667.

Schultz, P.G. (1988) The interplay between chemistry and biology in the design of enzymatic catalysts. *Science* **240,** 426–433.

Kinetics

Cleland, W.W. (1977) Determining the chemical mechanisms of enzyme-catalyzed reactions by kinetic studies. *Adv. Enzymol.* **45,** 273–387.

Raines, R. T. & Hansen, D.E. (1988) An intuitive approach to steady-state kinetics. *J. Chem. Educ.* **65,** 757–759.

Segel, I.H. (1975) *Enzyme Kinetics: Behavior and Analysis of Rapid Equilibrium and Steady State Enzyme Systems,* John Wiley & Sons, Inc., New York.
A more advanced treatment.

Enzyme Examples

Anderson, C.M., Zucker, F.H., & Steitz, T.A. (1979) Space-filling models of kinase clefts and conformation changes. *Science* **204,** 375–380.
Structure of hexokinase and other enzymes utilizing ATP.

Fersht, A.R. (1987) Dissection of the structure and activity of the tyrosyl-tRNA synthetase by site-directed mutagenesis. *Biochemistry* **26,** 8031–8037.

Warshel, A., Naray-Szabo, G., Sussman, F., & Hwang, J.-K. (1989) How do serine proteases really work? *Biochemistry* **28,** 3629–3637.

Regulatory Enzymes

Dische, Z. (1976) The discovery of feedback inhibition. *Trends Biochem. Sci.* **1,** 269–270.

Koshland, D.E., Jr. & Neet, K.E. (1968) The catalytic and regulatory properties of enzymes. *Annu. Rev. Biochem.* **37,** 359–410.

Monod, J., Changeux, J.-P. & Jacob, F. (1963) Allosteric proteins and cellular control systems. *J. Mol. Biol.* **6,** 306–329.
A classic paper introducing the concept of allosteric regulation.

Problems

1. *Keeping the Sweet Taste of Corn* The sweet taste of freshly picked corn is due to the high level of sugar in the kernels. Store-bought corn (several days after picking) is not as sweet, because about 50% of the free sugar of corn is converted into starch within one day of picking. To preserve the sweetness of fresh corn, the husked ears are immersed in boiling water for a few minutes ("blanched") and then cooled in cold water. Corn processed in this way and stored in a freezer maintains its sweetness. What is the biochemical basis for this procedure?

2. *Intracellular Concentration of Enzymes* To ap-proximate the actual concentration of enzymes in a bacterial cell, assume that the cell contains 1,000 different enzymes in solution in the cytosol, and that each protein has a molecular weight of 100,000, and that all 1,000 enzymes are present in equal concentrations. Assume that the bacterial cell is a cylinder (diameter 1 μm, height 2.0 μm). If the cytosol (specific gravity 1.20) is 20% soluble protein by weight, and if the soluble protein consists entirely of different enzymes, calculate the *average* molar concentration of each enzyme in this hypothetical cell.

3. *Rate Enhancement by Urease*   The enzyme urease enhances the rate of urea hydrolysis at pH 8.0 and 20 °C by a factor of 10^{14}. If a given quantity of urease can completely hydrolyze a given quantity of urea in 5 min at 20 °C and pH 8.0, how long will it take for this amount of urea to be hydrolyzed under the same conditions in the absence of urease? Assume that both reactions take place in sterile systems so that bacteria cannot attack the urea.

4. *Requirements of Active Sites in Enzymes*   The active site of an enzyme usually consists of a pocket on the enzyme surface lined with the amino acid side chains necessary to bind the substrate and catalyze its chemical transformation. Carboxypeptidase, which sequentially removes the carboxyl-terminal amino acid residues from its peptide substrates, consists of a single chain of 307 amino acids. The two essential catalytic groups in the active site are furnished by Arg^{145} and Glu^{270}.

(a) If the carboxypeptidase chain were a perfect α helix, how far apart (in nanometers) would Arg^{145} and Glu^{270} be? (*Hint:* See Fig. 7–6.)

(b) Explain how it is that these two amino acids, so distantly separated in the sequence, can catalyze a reaction occurring in the space of a few tenths of a nanometer.

(c) If only these two catalytic groups are involved in the mechanism of hydrolysis, why is it necessary for the enzyme to contain such a large number of amino acid residues?

5. *Quantitative Assay for Lactate Dehydrogenase* The muscle enzyme lactate dehydrogenase catalyzes the reaction

$$CH_3-\overset{\overset{\displaystyle O}{\|}}{C}-COO^- + NADH + H^+ \longrightarrow$$
Pyruvate

$$CH_3-\overset{\overset{\displaystyle OH}{|}}{\underset{\underset{\displaystyle H}{|}}{C}}-COO^- + NAD^+$$
Lactate

NADH and NAD^+ are the reduced and oxidized forms, respectively, of the coenzyme NAD. Solutions of NADH, but *not* NAD^+, absorb light at 340 nm. This property is used to determine the concentration of NADH in solution by measuring spectrophotometrically the amount of light absorbed at 340 nm by the solution. Explain how these properties of NADH can be used to design a quantitative assay for lactate dehydrogenase.

6. *Estimation of V_{max} and K_m by Inspection*   Although graphical methods are available for accurate determination of the values of V_{max} and K_m of an enzyme-catalyzed reaction (see Box 8–1), these values can be quickly estimated by inspecting values of V_0 at increasing [S]. Estimate the approxi-

mate value of V_{max} and K_m for the enzyme-catalyzed reaction for which the following data were obtained:

[S] (M)	$V_0 (\mu M/min)$
2.5×10^{-6}	28
4.0×10^{-6}	40
1×10^{-5}	70
2×10^{-5}	95
4×10^{-5}	112
1×10^{-4}	128
2×10^{-3}	139
1×10^{-2}	140

7. *Relation between Reaction Velocity and Substrate Concentration: Michaelis–Menten Equation*

(a) At what substrate concentration will an enzyme having a k_{cat} of 30 s^{-1} and a K_m of 0.005 M show one-quarter of its maximum rate?

(b) Determine the fraction of V_{max} that would be found in each case when $[S] = \frac{1}{2}K_m$, $2K_m$, and $10K_m$.

8. *Graphical Analysis of V_{max} and K_m Values* The following experimental data were collected during a study of the catalytic activity of an intestinal peptidase capable of hydrolyzing the dipeptide glycylglycine:

$$\text{Glycylglycine} + H_2O \longrightarrow 2 \text{ glycine}$$

[S] (mM)	Product formed ($\mu mol/min$)
1.5	0.21
2.0	0.24
3.0	0.28
4.0	0.33
8.0	0.40
16.0	0.45

From these data determine by graphical analysis (see Box 8–1) the values of K_m and V_{max} for this enzyme preparation and substrate.

9. *The Turnover Number of Carbonic Anhydrase* Carbonic anhydrase of erythrocytes (M_r 30,000) is among the most active of known enzymes. It catalyzes the reversible hydration of CO_2:

$$H_2O + CO_2 \rightleftharpoons H_2CO_3$$

which is important in the transport of CO_2 from the tissues to the lungs.

(a) If 10 μg of pure carbonic anhydrase catalyzes the hydration of 0.30 g of CO_2 in 1 min at 37 °C under optimal conditions, what is the turnover number (k_{cat}) of carbonic anhydrase (in units of min^{-1})?

(b) From the answer in (a), calculate the activation energy of the enzyme-catalyzed reaction (in kJ/mol).

(c) If carbonic anhydrase provides a rate enhancement of 10^7, what is the activation energy for the uncatalyzed reaction?

10. *Irreversible Inhibition of an Enzyme* Many enzymes are inhibited irreversibly by heavy-metal ions such as Hg^{2+}, Cu^{2+}, or Ag^+, which can react with essential sulfhydryl groups to form mercaptides:

$$Enz—SH + Ag^+ \longrightarrow Enz—S—Ag + H^+$$

The affinity of Ag^+ for sulfhydryl groups is so great that Ag^+ can be used to titrate —SH groups quantitatively. To 10 mL of a solution containing 1.0 mg/mL of a pure enzyme was added just enough $AgNO_3$ to completely inactivate the enzyme. A total of 0.342 μmol of $AgNO_3$ was required. Calculate the *minimum* molecular weight of the enzyme. Why does the value obtained in this way give only the minimum molecular weight?

11. *Protection of an Enzyme against Denaturation by Heat* When enzyme solutions are heated, there is a progressive loss of catalytic activity with time. This loss is the result of the unfolding of the native enzyme molecule to a randomly coiled conformation, because of its increased thermal energy. A solution of the enzyme hexokinase incubated at 45 °C lost 50% of its activity in 12 min, but when hexokinase was incubated at 45 °C in the presence of a very large concentration of one of its substrates, it lost only 3% of its activity. Explain why thermal denaturation of hexokinase was retarded in the presence of one of its substrates.

12. *Clinical Application of Differential Enzyme Inhibition* Human blood serum contains a class of enzymes known as acid phosphatases, which hydrolyze biological phosphate esters under slightly acidic conditions (pH 5.0):

$$R—O—\underset{\underset{O}{\|}}{\overset{\overset{O^-}{|}}{P}}—O^- + H_2O \longrightarrow R—OH + HO—\underset{\underset{O}{\|}}{\overset{\overset{O^-}{|}}{P}}—O^-$$

Acid phosphatases are produced by erythrocytes, the liver, kidney, spleen, and prostate gland. The enzyme from the prostate gland is clinically important because an increased activity in the blood is frequently an indication of cancer of the prostate gland. The phosphatase from the prostate gland is strongly inhibited by the tartrate ion, but acid phosphatases from other tissues are not. How can this information be used to develop a specific procedure for measuring the activity of the acid phosphatase of the prostate gland in human blood serum?

13. *Inhibition of Carbonic Anhydrase by Acetazolamide* Carbonic anhydrase is strongly inhibited by the drug acetazolamide, which is used as a diuretic (increases the production of urine) and to treat glaucoma (reduces excessively high pressure within the eyeball). Carbonic anhydrase plays an important role in these and other secretory processes, because it participates in regulating the pH and bicarbonate content of a number of body fluids. The experimental curve of reaction velocity (given here as percentage of V_{max}) versus [S] for the carbonic anhydrase reaction is illustrated below (upper curve). When the experiment is repeated in the presence of acetazolamide, the lower curve is obtained. From an inspection of the curves and your knowledge of the kinetic properties of competitive and noncompetitive enzyme inhibitors, determine the nature of the inhibition by acetazolamide. Explain.

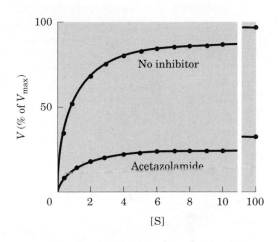

14. *pH Optimum of Lysozyme* The enzymatic activity of lysozyme is optimal at pH 5.2.

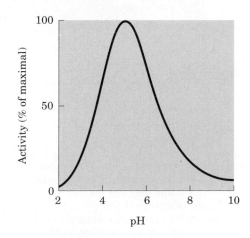

The active site of lysozyme contains two amino acid residues essential for catalysis: Glu^{35} and Asp^{52}. The pK_a values of the carboxyl side chains of these two residues are 5.9 and 4.5, respectively. What is the ionization state (protonated or deprotonated) of each residue at the pH optimum of lysozyme? How can the ionization states of these two amino acid residues explain the pH-activity profile of lysozyme shown above?

Lipids

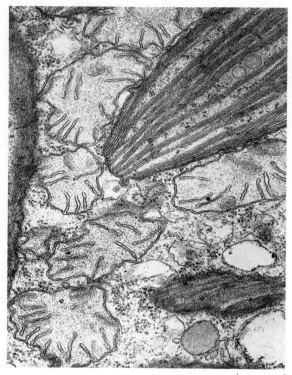

0.4 μm

Lipids play an important role in cell structure and function. In this electron micrograph of the cytoplasm of the photosynthetic alga *Euglena,* the lipid-containing membranes of a chloroplast (upper right) and several mitochondria (surrounding the chloroplast and lower left) are visible. Two lipid droplets, stores of chemical energy, can be seen in the chloroplast. The gray oval structure at the lower right is a lipid-filled inclusion in the cytoplasm.

Biological lipids are a chemically diverse group of compounds, the common and defining feature of which is their insolubility in water. The biological functions of the lipids are equally diverse. Fats and oils are the principal stored forms of energy in many organisms, and phospholipids and sterols make up about half the mass of biological membranes. Other lipids, although present in relatively small quantities, play crucial roles as enzyme cofactors, electron carriers, light-absorbing pigments, hydrophobic anchors, emulsifying agents, hormones, and intracellular messengers. This chapter introduces representative lipids of each type, with emphasis on their chemical structure and physical properties.

Storage Lipids

The fats and oils used almost universally as stored forms of energy in living organisms are highly reduced compounds, derivatives of **fatty acids.** The fatty acids are hydrocarbon derivatives, at about the same low oxidation state (that is, as highly reduced) as the hydrocarbons in fossil fuels. The complete oxidation of fatty acids (to CO_2 and H_2O) in cells, like the explosive oxidation of fossil fuels in internal combustion engines, is highly exergonic.

We will introduce here the structure and nomenclature of the fatty acids most commonly found in living organisms. Two types of fatty acid-containing compounds, triacylglycerols and waxes, are described to illustrate the diversity of structure and physical properties in this family of compounds.

Fatty Acids Are Hydrocarbon Derivatives

Fatty acids are carboxylic acids with hydrocarbon chains of 4 to 36 carbons. In some fatty acids, this chain is fully saturated (contains no double bonds) and unbranched; others contain one or more double bonds (Table 9–1). A few contain three-carbon rings or hydroxyl groups. A simplified nomenclature for these compounds specifies the chain length and number of double bonds, separated by a colon; the 16-carbon saturated palmitic acid is abbreviated 16:0, and the 18-carbon oleic acid, with one double bond, is 18:1. The positions of any double bonds are specified by superscript numbers following Δ (delta); a 20-carbon fatty acid with one double bond between C-9 and C-10 (C-1

Table 9–1 Some naturally occurring fatty acids

Carbon skeleton	Structure*	Systematic name†	Common name (derivation)	Melting point (°C)	Solubility at 30 °C (mg/g solvent)	
					Water	Benzene
12:0	$CH_3(CH_2)_{10}COOH$	*n*-Dodecanoic acid	Lauric acid (Latin *laurus*, laurel plant)	44.2	0.063	2,600
14:0	$CH_3(CH_2)_{12}COOH$	*n*-Tetradecanoic acid	Myristic acid (Latin *Myristica*, nutmeg genus)	53.9	0.024	874
16:0	$CH_3(CH_2)_{14}COOH$	*n*-Hexadecanoic acid	Palmitic acid (Greek *palma*, palm tree)	63.1	0.0083	348
18:0	$CH_3(CH_2)_{16}COOH$	*n*-Octadecanoic acid	Stearic acid (Greek *stear*, hard fat)	69.6	0.0034	124
20:0	$CH_3(CH_2)_{18}COOH$	*n*-Eicosanoic acid	Arachidic acid (Latin *Arachis*, legume genus)	76.5		
24:0	$CH_3(CH_2)_{22}COOH$	*n*-Tetracosanoic acid	Lignoceric acid (Latin *lignum*, wood + *cera*, wax)	86.0		
16:1(Δ^9)	$CH_3(CH_2)_5CH{=}CH(CH_2)_7COOH$		Palmitoleic acid	−0.5		
18:1(Δ^9)	$CH_3(CH_2)_7CH{=}CH(CH_2)_7COOH$		Oleic acid (Greek *oleum*, oil)	13.4		
18:2($\Delta^{9,12}$)	$CH_3(CH_2)_4CH{=}CHCH_2CH{=}CH(CH_2)_7COOH$		α-Linoleic acid (Greek *linon*, flax)	−5		
18:3($\Delta^{9,12,15}$)	$CH_3CH_2CH{=}CHCH_2CH{=}CHCH_2CH{=}CH(CH_2)_7COOH$		Linolenic acid	−11		
20:4($\Delta^{5,8,11,14}$)	$CH_3(CH_2)_4CH{=}CHCH_2CH{=}CHCH_2CH{=}CHCH_2CH{=}CH(CH_2)_3COOH$		Arachidonic acid	−49.5		

* All acids are shown in their un-ionized form. At pH 7, all free fatty acids have an ionized carboxylate. Note that numbering of carbon atoms begins at the carboxyl group carbon.

† The prefix *n*- indicates the "normal" unbranched structure. For instance, "dodecanoic" simply indicates 12 carbon atoms, which could be arranged in a variety of branched forms. Thus "*n*-dodecanoic" specifies the linear, unbranched form.

being the carboxyl carbon), and another between C-12 and C-13, is designated 20:2($\Delta^{9,12}$), for example. The most commonly occurring fatty acids have even numbers of carbon atoms in an unbranched chain of 12 to 24 carbons (Table 9–1). As we shall see in Chapter 20, the even number of carbons results from the mode of synthesis of these compounds, which involves condensation of acetate (two-carbon) units.

The position of double bonds is also regular; in most monounsaturated fatty acids the double bond is between C-9 and C-10 (Δ^9), and the other double bonds of polyunsaturated fatty acids are generally Δ^{12} and Δ^{15} (Table 9–1). The double bonds of polyunsaturated fatty acids are almost never conjugated (alternating single and double bonds, as in —CH=CH—CH=CH—), but are separated by a methylene group (—CH=CH—CH_2—CH=CH—). The double bonds of almost all naturally occurring unsaturated fatty acids are in the cis configuration.

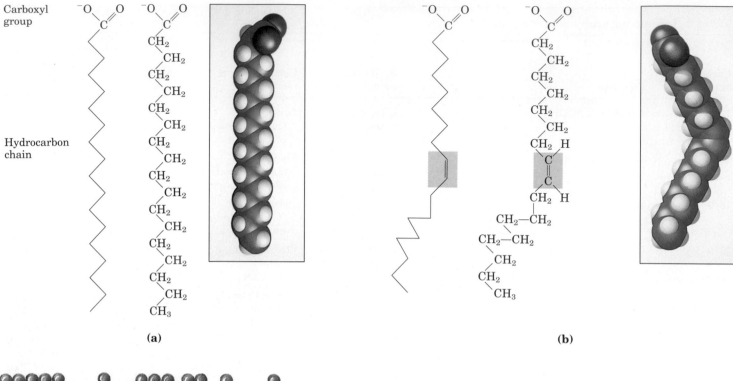

Carboxyl group

Hydrocarbon chain

(a)

(b)

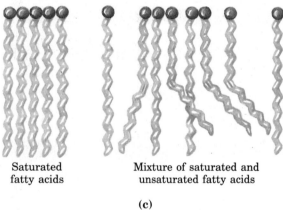

Saturated fatty acids

Mixture of saturated and unsaturated fatty acids

(c)

Figure 9–1 The packing of fatty acids depends on their degree of saturation. **(a)** Stearic acid (stearate at pH 7) is shown in its usual extended conformation. **(b)** The cis double bond (shaded) in oleic acid (oleate) does not permit rotation and introduces a rigid bend in the hydrocarbon tail. All the other bonds are free to rotate. **(c)** Fully saturated fatty acids in the extended form pack into nearly crystalline arrays, stabilized by many hydrophobic interactions. The presence of one or more cis double bonds interferes with this tight packing, and results in less stable aggregates.

The physical properties of the fatty acids, and of compounds that contain them, are largely determined by the length and degree of unsaturation of the hydrocarbon chain. The nonpolar hydrocarbon chain accounts for the poor solubility of fatty acids in water. Lauric acid (12:0, M_r 200), for example, has a solubility of 0.063 mg/g of water—much less than that of glucose (M_r 180), which is 1,100 mg/g of water. The longer the fatty acyl chain and the fewer the double bonds, the lower the solubility in water (Table 9–1). The carboxylic acid group is polar (and ionized at neutral pH), and accounts for the slight solubility of short-chain fatty acids in water.

The melting points of fatty acids and of compounds that contain them are also strongly influenced by the length and degree of unsaturation of the hydrocarbon chain (Table 9–1). At room temperature (25 °C), the saturated fatty acids from 12:0 to 24:0 have a waxy consistency, whereas unsaturated fatty acids of these lengths are oily liquids. In the fully saturated compounds, free rotation around each of the carbon–carbon bonds gives the hydrocarbon chain great flexibility; the most stable conformation is this fully extended form (Fig. 9–1a), in which the steric hindrance of neighboring atoms is minimized. These molecules can pack together tightly in nearly crystalline arrays, with atoms all along their lengths in van der Waals contact with the atoms of neighboring molecules (Fig. 9–1c). A cis double bond forces a kink in the hydrocarbon chain (Fig. 9–1b). Fatty acids with one or several such kinks cannot pack together as tightly as fully saturated fatty acids (Fig. 9–1c), and their interactions with each other are therefore weaker. Because it takes less thermal energy to disorder these poorly ordered arrays of unsaturated fatty acids, they have lower melting points than saturated fatty acids of the same chain length (Table 9–1).

In vertebrate animals, free fatty acids (having a free carboxylate group) circulate in the blood bound to a protein carrier, serum albumin. However, fatty acids are present mostly as carboxylic acid deriva-

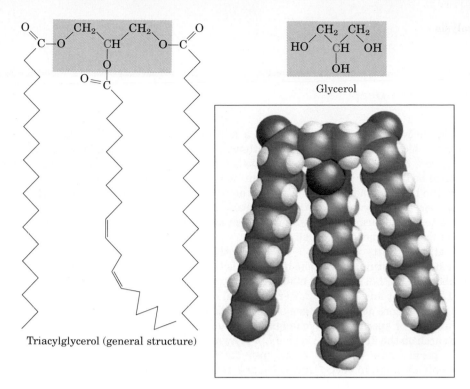

Glycerol

Triacylglycerol (general structure)

Figure 9–2 Glycerol and triacylglycerols. The triacylglycerol shown here has identical fatty acids (palmitate, 18:0) in positions 1 and 3. When there are two different fatty acids in positions 1 and 3 of the glycerol, C-2 (in red) of glycerol (shaded) becomes a chiral center (see Fig. 3–9). Biological triacylglycerols have the L configuration.

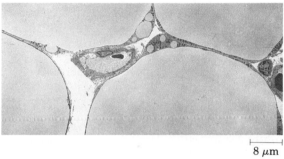

8 μm

(a)

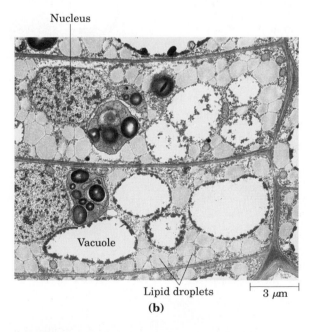

Nucleus

Vacuole

Lipid droplets   3 μm

(b)

Figure 9–3 Fat stores in cells. **(a)** Cross-section of four guinea pig adipocytes, showing huge fat droplets that virtually fill the cells. Also visible are several capillaries in cross-section. **(b)** Two cambial cells from the underground stem of the plant *Isoetes muricata*, a quillwort. In winter, these cells store fats as lipid droplets.

tives such as esters or amides. Lacking the charged carboxylate group, these fatty acid derivatives are generally even less soluble in water than are the free carboxylic acids.

Triacylglycerols Are Fatty Acid Esters of Glycerol

The simplest lipids constructed from fatty acids are the **triacylglycerols,** also referred to as triglycerides, fats, or neutral fats. Triacylglycerols are composed of three fatty acids each in ester linkage with a single glycerol (Fig. 9–2). Those containing the same kind of fatty acid in all three positions are called simple triacylglycerols, and are named after the fatty acid they contain. Simple triacylglycerols of 16:0, 18:0, and 18:1, for example, are tristearin, tripalmitin, and triolein, respectively. Mixed triacylglycerols contain two or more different fatty acids; to name these compounds unambiguously, the name and position of each fatty acid must be specified.

Because the polar hydroxyls of glycerol and the polar carboxylates of the fatty acids are bound in ester linkages, triacylglycerols are nonpolar, hydrophobic molecules, essentially insoluble in water. This explains why oil–water mixtures (oil-and-vinegar salad dressing, for example) have two phases. Because lipids have lower specific gravities than water, the oil floats on the aqueous phase.

Triacylglycerols Provide Stored Energy and Insulation

In most eukaryotic cells, triacylglycerols form a separate phase of microscopic, oily droplets in the aqueous cytosol, serving as depots of metabolic fuel. Specialized cells in vertebrate animals, called adipocytes, or fat cells, store large amounts of triacylglycerols as fat droplets, which nearly fill the cell (Fig. 9–3). Triacylglycerols are also stored in the seeds of many types of plants, providing energy and biosynthetic precursors when seed germination occurs.

BOX 9-1   Sperm Whales: Fatheads of the Deep

Studies of sperm whales have uncovered another way in which triacylglycerols are biologically useful. The sperm whale's head is very large, accounting for over one-third of its total body weight (Fig. 1). About 90% of the weight of the head is made up of the spermaceti organ, a blubbery mass that contains up to 18,000 kg (about 4 tons) of spermaceti oil, a mixture of triacylglycerols and waxes containing an abundance of unsaturated fatty acids. This mixture is liquid at the normal resting body temperature of the whale, about 37 °C, but it begins to crystallize at about 31 °C and becomes solid when the temperature drops several more degrees.

The probable biological function of spermaceti oil has been deduced from research on the anatomy and feeding behavior of the sperm whale. These mammals feed almost exclusively on squid in very deep water. In their feeding dives they descend 1,000 m or more; the record dive is 3,000 m (almost 2 miles). At these depths the sperm whale has no competitors for the very plentiful squid. The sperm whale rests quietly, waiting for schools of squid to pass. For a marine animal to remain at a given depth, without a constant swimming effort, it must have the same density as the surrounding water. The sperm whale can change its buoyancy to match the density of its surroundings—from the tropical ocean surface to great depths where the water is much colder and thus has a greater density.

The key to the sperm whale's ability to change its buoyancy is the freezing point of spermaceti oil. When the temperature of liquid spermaceti oil is lowered several degrees during a deep dive, it congeals or crystallizes and becomes more dense, thus changing the buoyancy of the whale to match the density of seawater. Various physiological mechanisms promote rapid cooling of the oil during a dive. During the return to the surface, the congealed spermaceti oil is warmed again and melted, decreasing its density to match that of the surface water. Thus we see in the sperm whale a remarkable anatomical and biochemical adaptation, perfected by evolution. The triacylglycerols synthesized by the sperm whale contain fatty acids of the necessary chain length and degree of unsaturation to give the spermaceti oil the proper melting point for the animal's diving habits.

Unfortunately for the sperm whale population, spermaceti oil is commercially valuable as a lubricant. Several centuries of intensive hunting of these mammals have depleted the world's population of sperm whales.

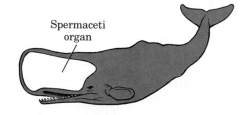

Spermaceti organ

Figure 1 Silhouette of a sperm whale, showing the spermaceti organ, a huge enlargement of the snout that lies above the upper jaw.

As stored fuels, triacylglycerols have two significant advantages over polysaccharides such as glycogen and starch. The carbon atoms of fatty acids are more reduced than those of sugars, and oxidation of triacylglycerols yields more than twice as much energy, gram for gram, as that of carbohydrates. Furthermore, because triacylglycerols are hydrophobic and therefore unhydrated, the organism that carries fat as fuel does not have to carry the extra weight of water of hydration that is associated with stored polysaccharides. In humans, fat tissue, which is composed primarily of adipocytes, occurs under the skin, in the abdominal cavity, and in the mammary glands. Obese people may have 15 or 20 kg of triacylglycerols deposited in their adipocytes, sufficient to supply energy needs for months. In contrast, the human body can store less than a day's energy supply in the form of glycogen. Carbohydrates such as glucose and glycogen do offer certain advantages as quick sources of metabolic energy, one of which is their ready solubility in water.

In some animals, triacylglycerols stored under the skin serve not only as energy stores but as insulation against very low temperatures. Seals, walruses, penguins, and other warm-blooded polar animals are amply padded with triacylglycerols. In hibernating animals (bears, for example) the huge fat reserves accumulated before hibernation also serve as energy stores (see Box 16–1). The low density of triacylglycerols is the basis for another remarkable function of these compounds. In sperm whales, a store of triacylglycerols allows the animals to match the buoyancy of their bodies to that of their surroundings during deep dives in cold water (Box 9–1).

Many Foods Contain Triacylglycerols

Most natural fats, such as those in vegetable oils, dairy products, and animal fat, are complex mixtures of simple and mixed triacylglycerols. These contain a variety of fatty acids differing in chain length and degree of saturation (Table 9–2). Vegetable oils such as corn and olive oil are composed largely of triacylglycerols with unsaturated fatty acids, and thus are liquids at room temperature. They are converted industrially into solid fats by catalytic hydrogenation, which reduces some of their double bonds to single bonds. Triacylglycerols containing only saturated fatty acids, such as tristearin, the major component of beef fat, are white, greasy solids at room temperature.

Table 9–2 Fatty acid composition of three natural food fats*

	State at room temperature (25 °C)	Fatty acids (%)[†]				
		Saturated				Unsaturated
		C_4–C_{12}	C_{14}	C_{16}	C_{18}	C_{16} + C_{18}
Olive oil	Liquid	<2	<2	13	3	80
Butter	Solid (soft)	11	10	26	11	40
Beef fat	Solid (hard)	<2	<2	29	21	46

* These fats consist of mixtures of triacylglycerols, differing in their fatty acid composition and thus in their melting points.

[†] Values are given as percentage of total fatty acids.

When lipid-rich foods are exposed too long to the oxygen in air, they may spoil and become rancid. The unpleasant taste and smell associated with rancidity result from the oxidative cleavage of the double bonds in unsaturated fatty acids to produce aldehydes and carboxylic acids of shorter chain length and therefore higher volatility.

Hydrolysis of Triacylglycerols Produces Soaps

The ester linkages of triacylglycerols are susceptible to hydrolysis by either acid or alkali. Heating animal fats with NaOH or KOH produces glycerol and the Na^+ or K^+ salts of the fatty acids, known as soaps (Fig. 9–4). The usefulness of soaps is in their ability to solubilize or disperse water-insoluble materials by forming microscopic aggregates (micelles). When used in "hard" water (having high concentrations of Ca^{2+} and Mg^{2+}), soaps are converted into their insoluble calcium or magnesium salts, forming a residue. Synthetic detergents such as sodium dodecylsulfate (SDS; see p. 141) are less prone to precipitation in hard water, and have largely replaced natural soaps in many industrial applications.

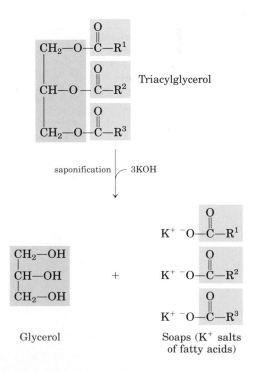

Figure 9–4 Triacylglycerol breakdown by alkaline hydrolysis: the process of saponification. R^1, R^2, R^3 represent long alkyl chains. Household soap is made by hydrolyzing a mixture of triacylglycerols (animal fat, for example) with KOH. The K^+ salts of the fatty acids are collected, washed free of KOH, and pressed into cakes.

At neutral pH, a variety of **lipases** catalyze the enzymatic hydrolysis of triacylglycerols. Lipases in the intestine aid in the digestion and absorption of dietary fats. Adipocytes and germinating seeds contain lipases that break down stored triacylglycerols, releasing fatty acids for export to other tissues where they are required as fuel.

Waxes Serve as Energy Stores and Water-Impermeable Coatings

Biological waxes are esters of long-chain saturated and unsaturated fatty acids (having 14 to 36 carbon atoms) with long-chain alcohols (having 16 to 30 carbon atoms) (Fig. 9–5). Their melting points (60 to 100 °C) are generally higher than those of triacylglycerols. In marine organisms that constitute the plankton, waxes are the chief storage form of metabolic fuel.

Waxes also serve a diversity of other functions in nature, related to their water-repellent properties and their firm consistency. Certain skin glands of vertebrates secrete waxes to protect the hair and skin and to keep them pliable, lubricated, and waterproof. Birds, particularly waterfowl, secrete waxes from their preen glands to make their feathers water-repellent. The shiny leaves of holly, rhododendrons, poison ivy, and many tropical plants are coated with a layer of waxes, which protects against parasites and prevents excessive evaporation of water.

$$CH_3(CH_2)_{14}-\overset{\overset{\displaystyle O}{\|}}{C}-O-CH_2-(CH_2)_{28}-CH_3$$

Palmitic acid   1-Triacontanol

(a)

Figure 9–5 (a) Triacontanylpalmitate, the major component of beeswax. It is an ester of palmitic acid with the alcohol triacontanol. **(b)** A honeycomb, constructed of beeswax, is firm at 25 °C and completely impervious to water. The term "wax" originates in the Old English word *weax,* meaning "the material of the honeycomb."

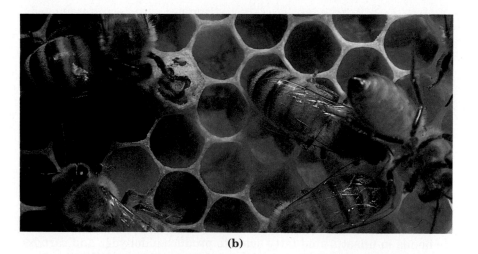

(b)

Biological waxes find a variety of applications in the pharmaceutical, cosmetic, and other industries. Lanolin (from lamb's wool), beeswax (Fig. 9–5), carnauba wax (from a Brazilian palm tree), and spermaceti oil (from whales) are widely used in the manufacture of lotions, ointments, and polishes.

Structural Lipids in Membranes

The central architectural feature of biological membranes is a double layer of lipids, which constitutes a barrier to the passage of polar molecules and ions. Membrane lipids are amphipathic; the orientation of their hydrophobic and hydrophilic regions directs their packing into membrane bilayers. Three general types of membrane lipids will be described: glycerophospholipids, in which the hydrophobic regions are composed of two fatty acids joined to glycerol; sphingolipids, in which a single fatty acid is joined to a fatty amine, sphingosine; and sterols, compounds characterized by a rigid system of four fused hydrocarbon rings. The hydrophilic moieties in these amphipathic compounds may be as simple as a single —OH group at one end of the sterol ring system, or they may be more complex. Glycerophospholipids and sphingolipids contain polar or charged alcohols at their polar ends; some also contain phosphate groups (Fig. 9–6). Within these three classes of membrane lipids, enormous diversity results from various combinations of fatty acid "tails" and polar "heads." We describe here a representative sample of the types of membrane lipids found in living organisms. The arrangement of these lipids in membranes, and their structural and functional roles therein, are considered in the next chapter.

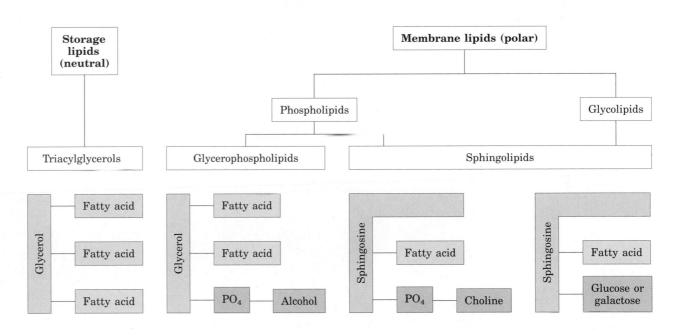

Figure 9–6 The principal classes of storage and membrane lipids. All of the classes shown here have either glycerol or sphingosine as the backbone. A third class of membrane lipids, the sterols, is described later (see Fig. 9–13).

Glycerophospholipids Are Derivatives of Phosphatidic Acid

Membranes contain several classes of lipids in which two fatty acids are ester-linked to glycerol at C-1 and C-2, and a highly polar or charged (and therefore hydrophilic) head group is attached to C-3 (Fig. 9–6). The most abundant of these polar lipids in most membranes are the **glycerophospholipids,** sometimes called phosphoglycerides (Fig. 9–7). In glycerophospholipids, a polar alcohol is joined to C-3 of glycerol through a phosphodiester bond. All glycerophospholipids are derivatives of phosphatidic acid (Fig. 9–7) and are named for their polar head groups (phosphatidylcholine and phosphatidylethanolamine, for example). All have a negative charge on the phosphate group at pH 7.0. The head-group alcohol may also contribute one or more charges at pH near 7.

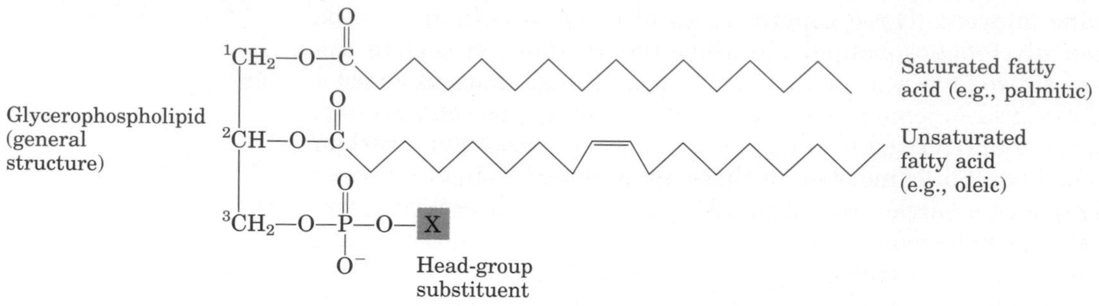

Name of X	Formula of X	Name of glycerophospholipid	Net charge (at pH 7)
—	$-H$	Phosphatidic acid	−1
Ethanolamine	$-CH_2-CH_2-\overset{+}{N}H_3$	Phosphatidylethanolamine	0
Choline	$-CH_2-CH_2-\overset{+}{N}(CH_3)_3$	Phosphatidylcholine	0
Serine	$-CH_2-CH(COO^-)-\overset{+}{N}H_3$	Phosphatidylserine	−1
Glycerol	$-CH_2-CH(OH)-CH_2-OH$	Phosphatidylglycerol	−1
Inositol	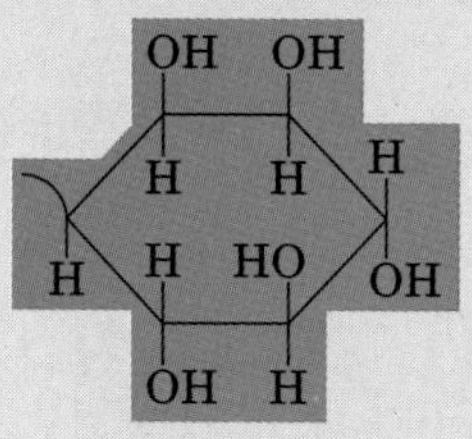	Phosphatidylinositol	−1
Phosphatidyl glycerol	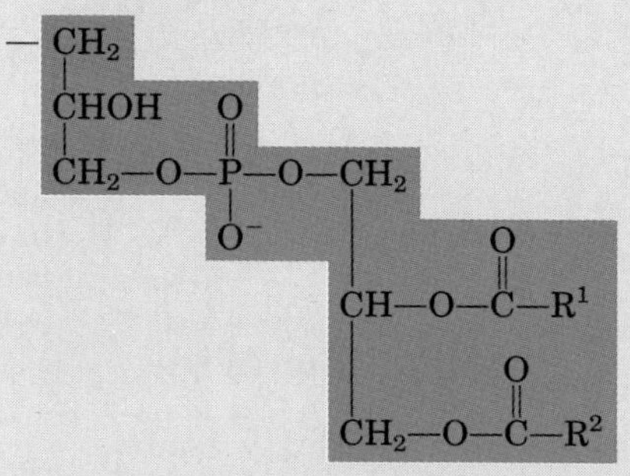	Cardiolipin	−2

Figure 9–7 The common glycerophospholipids are diacylglycerols linked to head-group alcohols through a phosphodiester bond. Phosphatidic acid is the parent compound, a phosphomonoester. Each derivative is named for the head-group alcohol (X), with the prefix "phosphatidyl." In cardiolipin, two phosphatidic acids share a single glycerol.

The fatty acids in glycerophospholipids can be any of a wide variety. They are different in different species, in different tissues of the same species, and in different types of glycerophospholipids in the same cell or tissue. In general, glycerophospholipids contain a saturated fatty acid at C-1 and an unsaturated fatty acid at C-2, and the fatty acyl groups are commonly 16 or 18 carbons long—but there are many exceptions.

Some Phospholipids Have Ether-Linked Fatty Acids

Some animal tissues and some unicellular organisms are rich in ether lipids, in which one of the two acyl chains is attached to glycerol in ether, rather than ester, linkage. The ether-linked chain may be saturated, as in the alkyl ether lipids, or may contain a double bond between C-1 and C-2, as in **plasmalogens** (Fig. 9–8). Vertebrate heart tissue is uniquely enriched in ether lipids; about half of the heart phospholipids are plasmalogens. The membranes of halophilic bacteria, of ciliated protists, and of certain invertebrates also contain high proportions of ether lipids. Their functional significance in these membranes is unknown; perhaps they confer resistance to phospholipases that cleave ester-linked fatty acids from membrane lipids. At least one ether lipid, **platelet-activating factor** (Fig. 9–8), is an important hormone. It is released from white blood cells called basophils and stimulates platelet aggregation and the release of serotonin from platelets. It exerts a variety of effects on liver, smooth muscle, heart, uterine, and lung tissues, and plays an important role in inflammation and the allergic response.

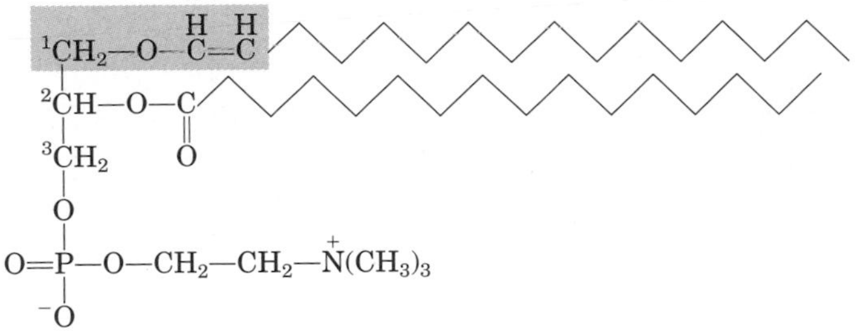

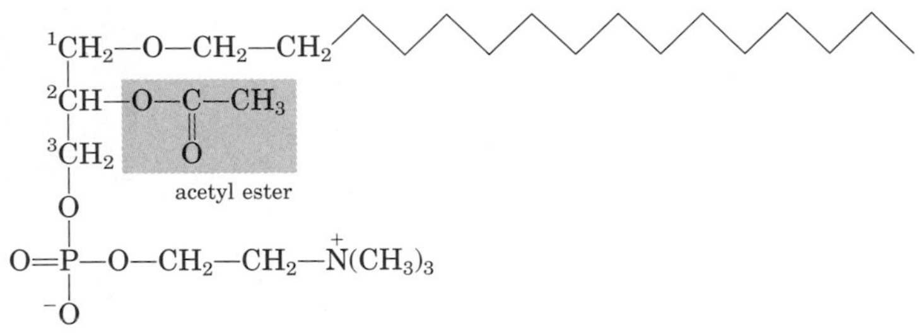

Plasmalogen

Platelet-activating factor

Figure 9–8 Plasmalogens and platelet-activating factor. Plasmalogens have one *ether*-linked alkenyl chain where most glycerophospholipids have an *ester*-linked fatty acid (compare Fig. 9–7). Platelet-activating factor has a long ether-linked alkyl chain at C-1 of glycerol, but C-2 is ester-linked to a very short fatty acid (acetic acid), which makes the compound much more water-soluble than most glycerophospholipids and plasmalogens. The head group alcohol is choline in plasmalogens and platelet-activating factor.

Sphingolipids Are Derivatives of Sphingosine

Sphingolipids, the second large class of membrane lipids, also have a polar head and two nonpolar tails, but unlike glycerophospholipids they contain no glycerol. Sphingolipids are composed of one molecule of the long-chain amino alcohol sphingosine (4-sphingenine) or one of its derivatives, one molecule of a long-chain fatty acid, a polar head alcohol, and sometimes phosphoric acid in diester linkage at the polar head group (Fig. 9–9).

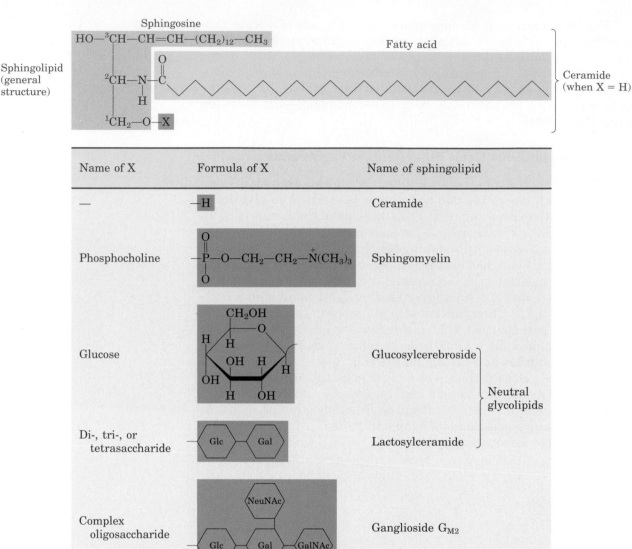

Figure 9–9 Sphingolipids. The first three carbons at the polar end of sphingosine are analogous to the three carbons of glycerol in glycerophospholipids. In ceramide, the parent compound for this group, the amino group at C-2 bears a fatty acid in amide linkage. Individual sphingolipids differ in the polar head group (X) attached at C-1. The fatty acid components of sphingolipids are usually saturated or monounsaturated, and contain 16, 18, 22, or 24 carbon atoms. Gangliosides have very complex oligosaccharide head groups. These compounds are given identifying symbols (e.g., G_{M1}, G_{M2}) that indicate the structure of the head group. At least 15 different classes of gangliosides have been found in higher animals. Standard symbols for sugars are used in this figure: Glc, D-glucose; Gal, D-galactose; GalNAc, N-acetyl-D-galactosamine; NeuNAc, N-acetylneuraminic acid (sialic acid).

Carbons C-1, C-2, and C-3 of the sphingosine molecule bear functional groups (—OH, —NH$_2$, —OH) that are structurally homologous with the three hydroxyl groups of glycerol in glycerophospholipids. When a fatty acid is attached in amide linkage to the —NH$_2$, the resulting compound is a **ceramide** (Fig. 9–9), which is structurally similar to a diacylglycerol. Ceramide is the fundamental structural unit common to all sphingolipids.

There are three subclasses of sphingolipids, all derivatives of ceramide, but differing in their head groups: sphingomyelins, neutral (uncharged) glycolipids, and gangliosides (Fig. 9–9). **Sphingomyelins** contain phosphocholine or phosphoethanolamine as their polar head group, and are therefore classified as phospholipids, together with glycerophospholipids. Indeed, sphingomyelins resemble phosphatidylcholines in their general properties and three-dimensional structure, and in having no net charge on their head groups (Fig. 9–10). Sphingomyelins are present in plasma membranes of animal cells; the myelin sheath which surrounds and insulates the axons of myelinated neurons is a good source of sphingomyelins, and gives them their name.

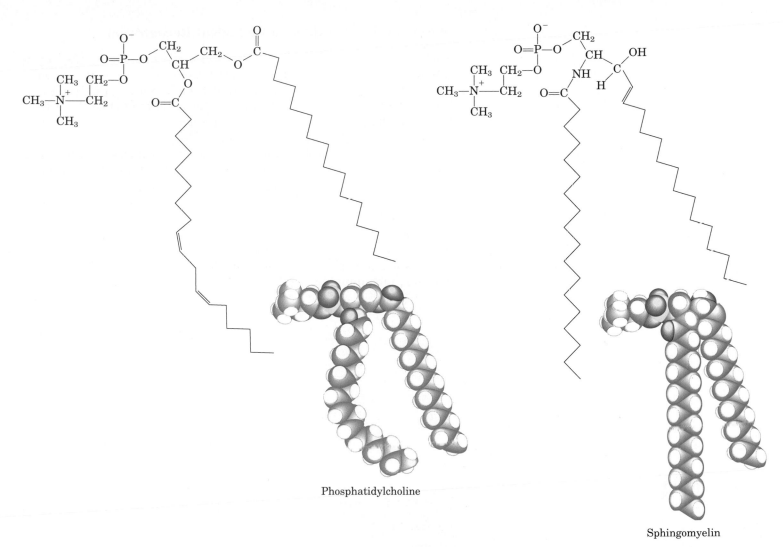

Phosphatidylcholine

Sphingomyelin

Figure 9–10 The similarities in shape and in molecular structure of phosphatidylcholine (a glycerophospholipid) and sphingomyelin (a sphingolipid) are clear when their space-filling and structural formulas are drawn as here.

Neutral glycolipids and gangliosides have one or more sugars in their head group, connected directly to the —OH at C-1 of the ceramide moiety; they do not contain phosphate. These sugar-containing sphingolipids are sometimes called **glycosphingolipids.** Neutral glycolipids contain one to six (sometimes more) sugar units, which may be D-glucose, D-galactose, or *N*-acetyl-D-galactosamine (Fig. 9–9). These glycosphingolipids occur largely in the outer face of the plasma membrane. **Cerebrosides** have a single sugar linked to ceramide (Fig. 9–9); those with galactose are characteristically found in the plasma membranes of cells in neural tissue, and those with glucose, in the plasma membranes of cells in nonneural tissues.

Gangliosides, the most complex sphingolipids (Fig. 9–9), contain very large polar heads made up of several sugar units. One or more of the terminal sugar units of gangliosides is *N*-acetylneuraminic acid, also called sialic acid, which has a negative charge at pH 7. Gangliosides make up about 6% of the membrane lipids in the gray matter of the human brain, and they are present in lesser amounts in the membranes of most nonneural animal tissues.

N-Acetylneuraminic acid
(sialic acid)

Sphingolipids Are Sites of Biological Recognition

When the sphingolipids were discovered a century ago by the physician-chemist Johann Thudicum, their biological role seemed as enigmatic as the Sphinx, for which he named them. Sphingolipids are now known to be involved in various recognition events at the cell surface. For example, glycosphingolipids are the determinants of the human blood groups A, B, and O (Fig. 9–11). The ganglioside G_{M1}, which doubtless plays some role of value to the animal cell that contains it, is the point of attachment of cholera toxin as it attacks an animal cell, a case of coevolution of a host cell and its pathogenic parasite. The membranes of the human nervous system contain at least 15 different gangliosides for which no function is yet known. However, it is clearly important that the synthesis and breakdown of these compounds be tightly regulated; derangements in the metabolism of cerebrosides and gangliosides underlie the devastating effects of several human genetic diseases, including Tay-Sachs and Niemann-Pick diseases (Box 9–2).

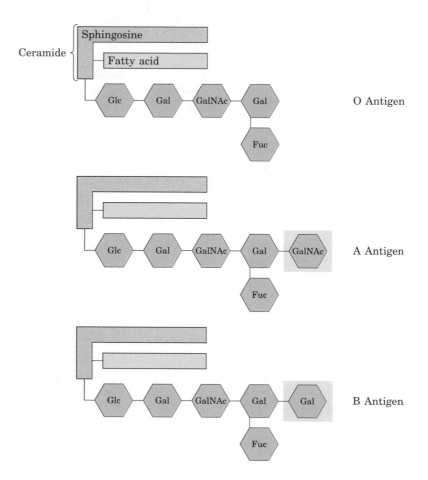

Figure 9–11 The human blood groups (O, A, B) are determined in part by the sugar head groups in these glycosphingolipids. The same three types of complex sugar groups are also found attached to certain blood proteins of individuals of blood types O, A, and B, respectively. The symbol Fuc represents the sugar fucose.

BOX 9-2   Some Inherited Human Diseases Resulting from Abnormal Accumulations of Membrane Lipids

The polar lipids of membranes undergo constant metabolic turnover, the rate of their synthesis normally being counterbalanced by an equal rate of breakdown. The breakdown of lipids is promoted by hydrolytic enzymes, each capable of hydrolyzing a specific covalent bond. For example, the degradation of phosphatidylcholine, a major membrane lipid, takes place by the action of several different phospholipases (see Fig. 9–12).

The metabolism of membrane sphingolipids, including sphingomyelin, cerebrosides, and gangliosides, is prone to genetic defects of enzymes involved in their degradation. When they are synthesized at a normal rate but their degradation is impaired, sphingolipids or their partial breakdown products accumulate in the tissues. For example, in Niemann-Pick disease, sphingomyelin accumulates in the brain, spleen, and liver. The disease first becomes evident in infants, causing mental retardation and early death. Niemann-Pick disease is caused by a rare genetic defect in the hydrolytic enzyme sphingomyelinase, which cleaves phosphocholine from sphingomyelin.

Much more common is Tay-Sachs disease, in which a specific ganglioside accumulates in the brain and spleen owing to the lack of the lysosomal enzyme hexosaminidase A, a degradative enzyme that normally hydrolyzes a specific bond between an N-acetyl-D-galactosamine and a D-galactose residue in the polar head of the ganglioside (see Fig. 9–9). As a result, the partially degraded gangliosides accumulate, causing degeneration of the nervous system. The symptoms of Tay-Sachs disease are progressive retardation in development, paralysis, blindness, and death by the age of 3 or 4 yr.

Tay-Sachs disease is rare in the population at large (1 in 300,000 births) but has a very high incidence (1 in 3,600 births) in Ashkenazic Jews (those of Eastern European extraction), who make up more than 90% of the Jewish population of the United States. One in 28 Ashkenazic Jews carries the defective gene in recessive form, which means that when both parents are carriers, there is a one in four probability that a child will develop Tay-Sachs disease. Genetic counseling of parents has become important in averting the occurrence of this disease. Tests have been devised to determine the presence of the recessive gene in prospective parents. These tests involve measuring the level of hexosaminidase A in skin cells. Carriers of the defective gene have a reduced (but for these individuals, functional) level of the enzyme. Tests of the

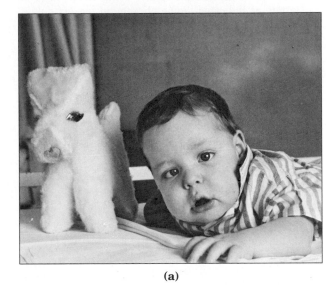

(a)

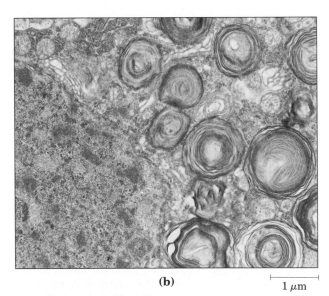

(b) $1\ \mu m$

Figure 1 **(a)** A 1-year-old infant with Tay-Sachs disease. **(b)** Electron micrograph of a portion of an affected brain cell, showing the abnormal ganglioside deposits in the lysosomes.

fetus can also be made during pregnancy by taking a sample of amniotic fluid, the fluid surrounding the growing fetus, in a process known as amniocentesis. The activity of hexosaminidase A can be measured in fetal cells contained in this fluid.

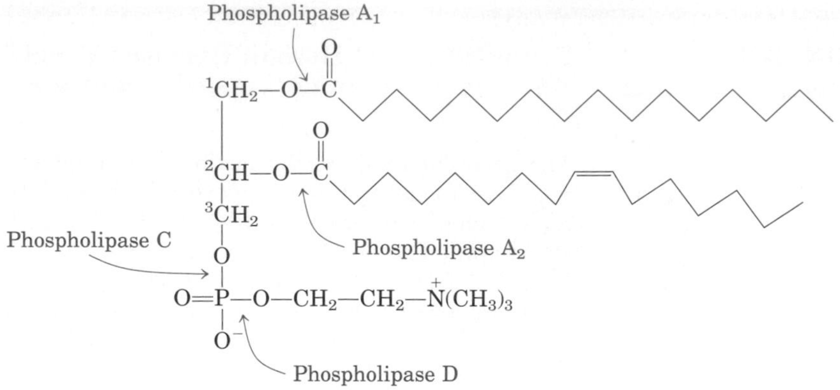

Figure 9–12 The specificities of phospholipases. Phospholipases A_1 and A_2 hydrolyze the ester bonds of intact glycerophospholipids at C-1 and C-2 of glycerol, respectively. Phospholipases C and D each split one of the phosphodiester bonds in the head group, as indicated. Some phospholipases act only on one type of glycerophospholipid, such as phosphatidylinositol or phosphatidylcholine; others are less specific. When one of the fatty acids has been removed by a type-A phospholipase, the second fatty acid is cleaved from the molecule by a lysophospholipase.

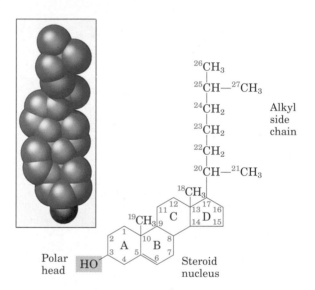

Figure 9–13 Cholesterol. To simplify reference to derivatives of the steroid nucleus, the rings are labeled A through D and the carbon atoms are numbered (in blue) as shown. The hydroxyl group on C-3 represents the polar head group. For storage and transport of the sterol, this hydroxyl group condenses with a fatty acid to form a sterol ester.

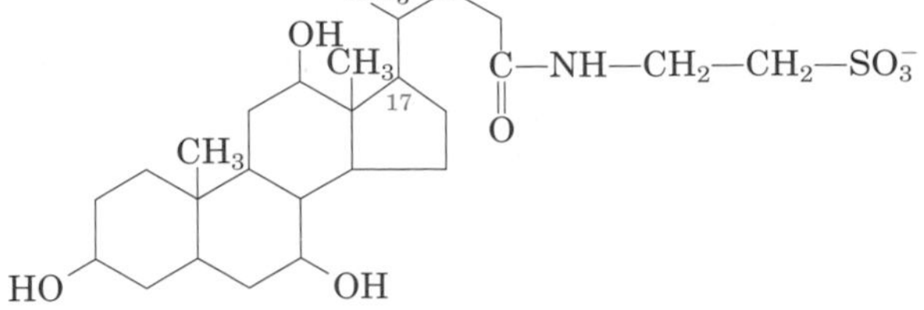

Taurocholic acid
(a bile acid)

Specific Phospholipases Degrade Membrane Phospholipids

Most cells continually degrade and replace their membrane lipids. For each of the bonds in a glycerophospholipid, there is a specific hydrolytic enzyme (Fig. 9–12). Phospholipases of the A type remove one of the two fatty acids, producing a lysophospholipid; these esterases do not attack the ether link in plasmalogens. Lysophospholipases remove the remaining fatty acid.

Phospholipid breakdown is part of at least two signaling processes in animal cells. Extracellular signals (certain hormones, for example) activate a phospholipase C that specifically cleaves phosphatidylinositols, releasing diacylglycerol and inositol phosphates, which serve as intracellular signals. Other extracellular stimuli activate a phospholipase A that releases arachidonic acid from membrane lipids; arachidonate serves as a precursor in the synthesis of one of the eicosanoids that act as intracellular messengers. These messenger roles for lipids are discussed later in this chapter.

Sterols Have Four Fused Hydrocarbon Rings

Sterols are structural lipids present in the membranes of most eukaryotic cells. Their characteristic structure is the steroid nucleus consisting of four fused rings, three with six carbons and one with five (Fig. 9–13). The steroid nucleus is almost planar, and relatively rigid; the fused rings do not allow rotation about C—C bonds. **Cholesterol,** the major sterol in animal tissues, is amphipathic, with a polar head group (the hydroxyl group at C-3) and a nonpolar hydrocarbon body (the steroid nucleus and the hydrocarbon side chain at C-17) about as long as a 16-carbon fatty acid in its extended form. Similar sterols are found in other eukaryotes: stigmasterol in plants and ergosterol in fungi, for example. With rare exceptions, bacteria lack sterols. The sterols of all species are synthesized from simple five-carbon isoprene subunits (as are the fat-soluble vitamins, quinones, and dolichols described below).

In addition to their roles as membrane constituents, the sterols serve as precursors for a variety of products with specific biological activities. Bile acids, in which the side chain at C-17 is hydrophilic, act as detergents in the intestine, emulsifying dietary fats to make them more readily accessible to digestive lipases. A variety of steroid hormones (described below) are also produced from cholesterol by oxidation of the side chain at C-17.

On receiving the Nobel Prize in 1985 for their work on cholesterol metabolism, Michael Brown and Joseph Goldstein recounted in their lecture the extraordinary history of cholesterol:

> Cholesterol is the most highly decorated small molecule in biology. Thirteen Nobel Prizes have been awarded to scientists who devoted major parts of their careers to cholesterol. Ever since it was isolated from gallstones in 1784, cholesterol has exerted an almost hypnotic fascination for scientists from the most diverse areas of science and medicine.

We shall return to cholesterol later, to consider its role in biological membranes, its remarkable biosynthetic pathway, and its role as precursor to the steroid hormones.

Amphipathic Lipids Aggregate

We have noted that glycerophospholipids, sphingolipids, and sterols are virtually insoluble in water. When mixed with water, these amphipathic compounds form microscopic lipid aggregates in a phase separate from their aqueous surroundings. Lipid molecules cluster together with their hydrophobic moieties in contact with each other and their hydrophilic groups interacting with the surrounding water. Recall that lipid clustering reduces the amount of hydrophobic surface exposed to water and thus minimizes the number of molecules in the shell of ordered water at the lipid–water interface (see Fig. 4–7), resulting in an increase in entropy. Hydrophobic interactions among lipid molecules provide the thermodynamic driving force for the formation and maintenance of these structures.

Depending on the precise conditions and the nature of the lipids used, three types of lipid aggregates can form when amphipathic lipids are mixed with water (Fig. 9–14). **Micelles** are relatively small, spherical structures involving a few dozen to a few thousand molecules arranged so that their hydrophobic regions aggregate in the interior, excluding water, and their hydrophilic head groups are at the surface, in contact with water. Micelle formation is favored when the cross-sectional area of the head group is greater than that of the acyl side chain(s) (Fig. 9–14a), as it is in free fatty acids, lysophospholipids (which lack one fatty acid), and the detergent SDS.

Figure 9–14 Amphipathic lipid aggregates that form in water. **(a)** In spherical micelles, the hydrophobic chains of the fatty acids are sequestered at the core of the sphere. There is virtually no water in the hydrophobic interior of the micelle. **(b)** In a bilayer, all acyl side chains except those at the edges of the sheet are protected from interaction with water. **(c)** When an extensive two-dimensional bilayer folds on itself, it forms a liposome, a three-dimensional hollow vesicle enclosing an aqueous cavity.

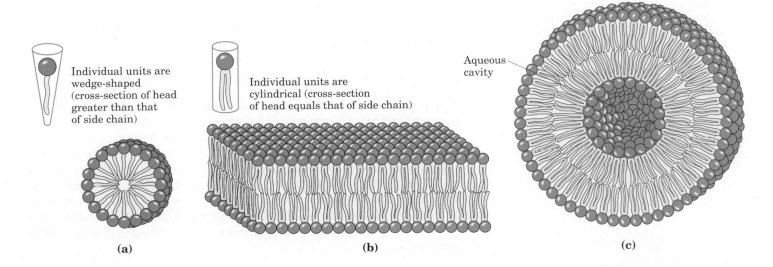

Individual units are wedge-shaped (cross-section of head greater than that of side chain)

Individual units are cylindrical (cross-section of head equals that of side chain)

Aqueous cavity

(a)   (b)   (c)

A second type of lipid aggregate in water is the **bilayer,** in which two lipid monolayers combine to form a two-dimensional sheet. Bilayer formation occurs most readily when the cross-sectional areas of the head group and side chain(s) are similar (Fig. 9–14b), as in glycerophospholipids and sphingolipids. The hydrophobic portions in each monolayer interact, excluding water. The hydrophilic head groups interact with water at the two surfaces of the bilayer.

The third type of lipid aggregate is formed when a lipid bilayer folds back on itself to form a hollow sphere called a **liposome** or vesicle (Fig. 9–14c). By forming vesicles, bilayer sheets lose their hydrophobic edge regions, achieving maximal stability in their aqueous environment. These bilayer vesicles enclose water, creating a separate aqueous compartment. It is likely that the first living cells resembled liposomes, their aqueous contents segregated from the rest of the world by a hydrophobic shell. We shall see in the next chapter that lipid bilayers are fundamental to the structure of all biological membranes.

Lipids with Specific Biological Activities

The two classes of lipids considered thus far (storage lipids and structural lipids) are major cellular components; membrane lipids represent 5 to 10% of the dry mass of most cells, and storage lipids, more than 50% of the mass of an adipocyte. With some important exceptions, these lipids play a *passive* role in the cell; fuels are acted on by oxidative enzymes, and lipid membranes form impermeable barriers that separate cellular compartments. Another group of lipids, although relatively minor cellular components on a mass basis, have specific and essential biological activities. These include hundreds of steroids—compounds that share the four-ring steroid nucleus but are more polar than cholesterol—and large numbers of isoprenoids, which are synthesized from five-carbon precursors related to isoprene:

$$\text{Isoprene} \qquad CH_2{=}\overset{\overset{\displaystyle CH_3}{|}}{C}{-}CH{=}CH_2$$

The isoprenoids include vitamins A, D, E, and K, first recognized as fatty materials essential to the normal growth of animals, and numerous biological pigments. Other "active" lipids serve as essential cofactors for enzymes, as electron carriers, or as intracellular signals. To illustrate the range of their structures and biological activities we will briefly describe a few of these compounds. In later chapters, their synthesis and biological roles will be considered in more detail.

Steroid Hormones Carry Messages between Tissues

The major groups of steroid hormones are the male and female sex hormones and the hormones of the adrenal cortex, cortisol and aldosterone (Fig. 9–15). All of these hormones contain an intact steroid nucleus. They are produced in one tissue and carried in the bloodstream to target tissues, where they bind to highly specific receptor proteins and trigger changes in gene expression and metabolism. Because of the very high affinity of receptor for hormone, very low concentrations of hormone (as low as 10^{-9} M) suffice to produce the effect on target tissues. These hormones and their actions are described in more detail in Chapter 22.

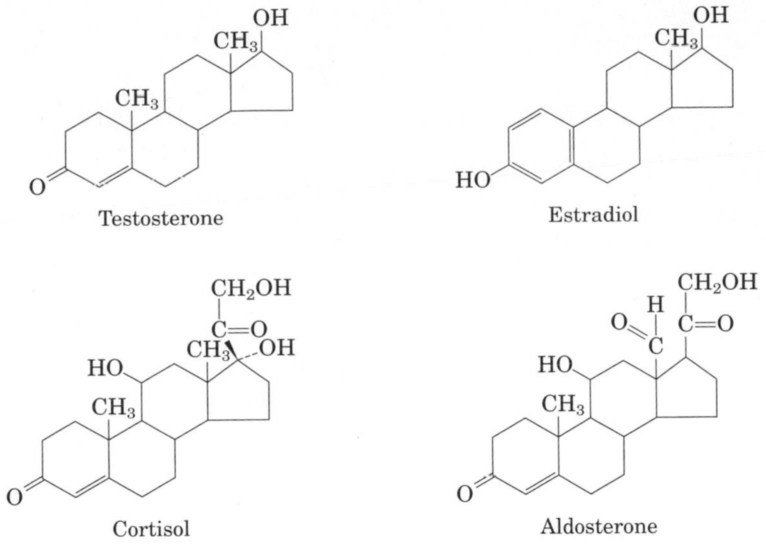

Testosterone

Estradiol

Cortisol

Aldosterone

Figure 9–15 Steroids derived from cholesterol. Testosterone, the male sex hormone, is produced in the testes. Estradiol, one of the female hormones, is produced in the ovaries and placenta. Cortisol and aldosterone are hormones produced in the cortex of the adrenal gland; they regulate glucose metabolism and salt excretion, respectively.

Hydrolysis of Phosphatidylinositol Produces Intracellular Messengers

Phosphatidylinositol and its phosphorylated derivatives (Fig. 9–16) are components of the plasma membranes of all eukaryotic cells. They serve as a reservoir of messenger molecules that are released inside the cell when certain extracellular signals interact with specific receptors in the plasma membrane. For example, when the hormone vasopressin binds to receptor molecules in the plasma membranes of cells in the kidney and the blood vessels, a specific phospholipase in the membrane is activated. This phospholipase breaks the bond between glycerol and phosphate in phosphatidylinositol-4,5-bisphosphate (Fig. 9–16), releasing two products: inositol-1,4,5-trisphosphate and diacylglycerol. Inositol-1,4,5-triphosphate causes the release of Ca^{2+} sequestered in membrane-bounded compartments of the cell, triggering the activation of a variety of Ca^{2+}-dependent enzymes and hormonal responses. Diacylglycerol binds to and activates an enzyme, protein kinase C, that transfers phosphate groups from ATP to several cytosolic proteins, thereby altering their enzymatic activities.

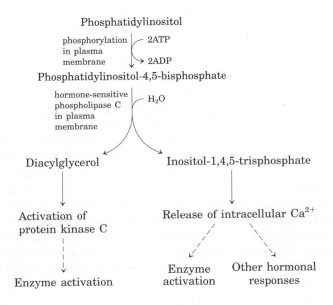

Figure 9–16 Phosphatidylinositol-4,5-bisphosphate, formed in the plasma membrane by phosphorylation of phosphatidylinositol, is hydrolyzed by a specific phospholipase C in response to hormonal signals. Both of the products of hydrolysis act as intracellular messengers.

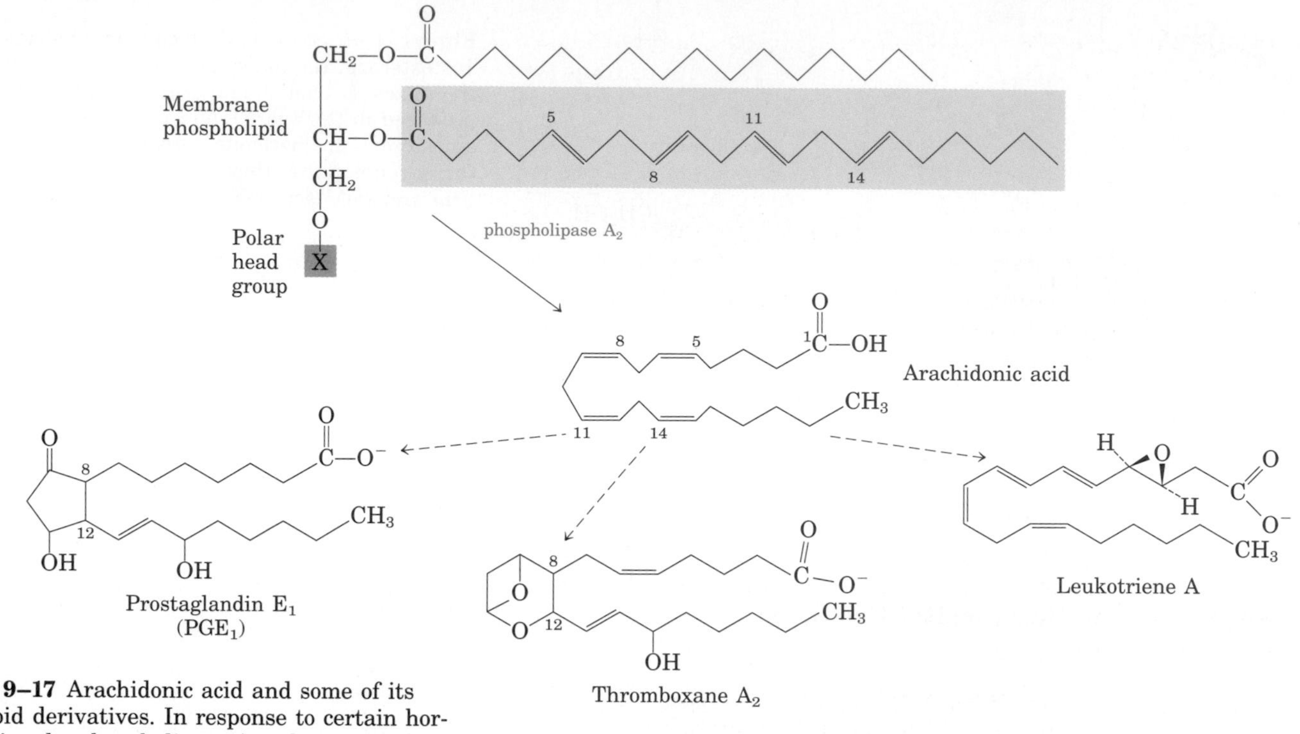

Figure 9–17 Arachidonic acid and some of its eicosanoid derivatives. In response to certain hormonal signals, phospholipase A_2 releases arachidonic acid (arachidonate at pH 7) from membrane phospholipids; arachidonic acid then serves as a precursor to various eicosanoids. These include prostaglandins such as PGE_1, in which carbon atoms 8 and 12 of arachidonic acid are joined to form the characteristic five-membered ring; thromboxane A_2, in which carbons 8 and 12 are joined and an oxygen atom is added to form the six-membered ring; and leukotriene A, containing a series of three conjugated double bonds. Aspirin and ibuprofen block the formation of prostaglandins and thromboxanes from arachidonic acid.

Eicosanoids Are Potent Biological Effectors

Eicosanoids (Fig. 9–17) are fatty acid derivatives with a variety of extremely potent hormonelike actions on various tissues of vertebrate animals. Unlike hormones, they are not transported between tissues in the blood, but act on the tissue in which they are produced. This family of compounds is known to be involved in reproductive function; in the inflammation, fever, and pain associated with injury or disease; in the formation of blood clots and the regulation of blood pressure; in gastric acid secretion; and in a variety of other processes important in human health or disease. More roles for the eicosanoids doubtless remain to be discovered.

Eicosanoids are all derived from the 20-carbon polyunsaturated fatty acid arachidonic acid, $20:4(\Delta^{5,8,11,14})$ (Fig. 9–17), from which they take their general name (Greek *eikosi*, "twenty"). There are three classes of eicosanoids: prostaglandins, thromboxanes, and leukotrienes. Various eicosanoids are produced in different cell types by different synthetic pathways, and have different target cells and biological actions.

The **prostaglandins** (PG) (Fig. 9–17) all contain a five-membered ring of carbon atoms originally part of the chain of arachidonic acid. They derive their name from the tissue in which they were first recognized (the prostate gland). Two groups were originally defined: PGE, for ether-soluble, and PGF, for phosphate buffer-soluble (*fosfat* in Swedish). Each contains numerous subtypes, named PGE_1, PGE_2, etc. Prostaglandins are now known to act in many tissues by regulating the synthesis of the intracellular messenger molecule 3′,5′-cyclic AMP (cAMP). Because cAMP mediates the action of many hormones, the prostaglandins affect a wide range of cellular and tissue functions. Some prostaglandins stimulate contraction of the smooth muscle of the uterus during labor or menstruation. Others affect blood flow to specific organs, the wake–sleep cycle, and the responsiveness of certain tissues to hormones such as epinephrine and glucagon. Prostaglandins in a third group elevate body temperature (producing fever) and cause inflammation, resulting in pain.

The **thromboxanes,** first isolated from blood platelets (also known as thrombocytes), have a six-membered ring containing an ether (Fig. 9–17). They are produced by platelets and act in formation of blood clots and the reduction of blood flow to the site of a clot.

Leukotrienes, found first in leukocytes, contain three conjugated double bonds (Fig. 9–17). They are powerful biological signals; for example, they induce contraction of the muscle lining the airways to the lung. Overproduction of leukotrienes causes asthmatic attacks. The strong contraction of the smooth muscles of the lung that occurs during anaphylactic shock is part of the potentially fatal allergic reaction in individuals hypersensitive to bee stings, penicillin, or various other agents.

Vitamins A, D, E, and K Are Fat-Soluble

During the first third of this century, a major focus of research in physiological chemistry was the identification of **vitamins**—compounds essential to the health of humans and other vertebrate animals that cannot by synthesized by these animals and must therefore be obtained in the diet. Early nutritional studies identified two general classes of such compounds: those soluble in nonpolar organic solvents (fat-soluble vitamins) and those that could be extracted from foods with aqueous solvents (water-soluble vitamins). Eventually the fat-soluble group was resolved into the four vitamins A, D, E, and K, all of which are isoprenoid compounds. Isoprenoids are synthesized by the condensation of isoprene units.

Vitamin A (retinol) (Fig. 9–18) is a pigment essential to vision. It was first recognized as an essential nutritional factor for laboratory animals, and was later isolated from fish liver oils. Vitamin A itself

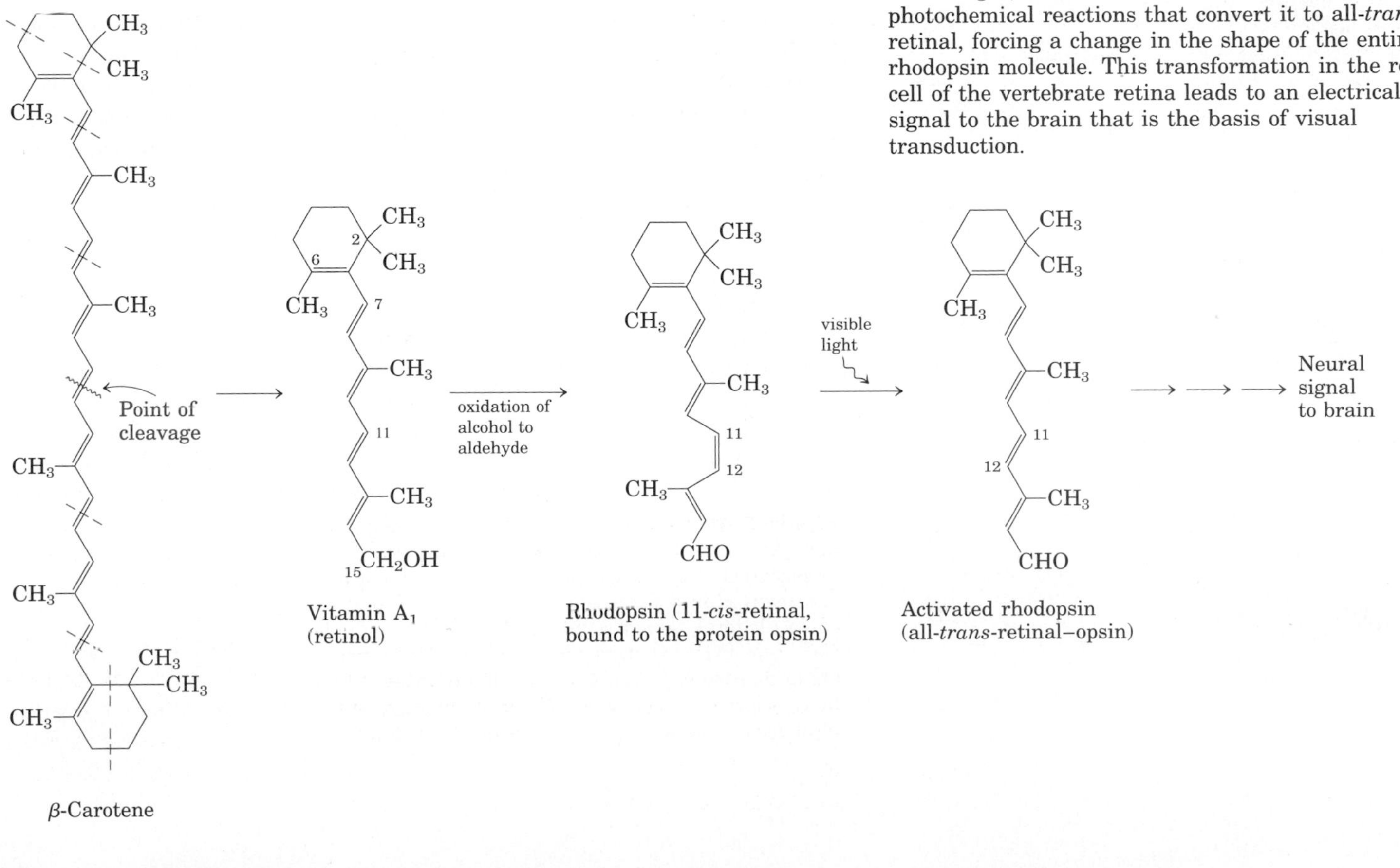

Figure 9–18 Vitamin A_1 and its precursor, β-carotene. The isoprene structural units are set off by dashed red lines. Cleavage of β-carotene yields two molecules of vitamin A_1 (retinol). Oxidation at C-15 converts retinol to the aldehyde, retinal. Rhodopsin, a visual pigment widely employed in nature, consists of retinal and the protein opsin. In the dark, retinal of rhodopsin is in the 11-cis form. When a rhodopsin molecule is excited with visible light, the 11-*cis*-retinal undergoes a series of photochemical reactions that convert it to all-*trans*-retinal, forcing a change in the shape of the entire rhodopsin molecule. This transformation in the rod cell of the vertebrate retina leads to an electrical signal to the brain that is the basis of visual transduction.

β-Carotene

Vitamin A_1 (retinol)

Rhodopsin (11-*cis*-retinal, bound to the protein opsin)

Activated rhodopsin (all-*trans*-retinal–opsin)

visible light

oxidation of alcohol to aldehyde

Point of cleavage

Neural signal to brain

does not occur in plants, but many plants contain carotenoids, light-absorbing pigments that can be enzymatically converted into vitamin A by most animals. Figure 9–18 shows, for example, how vitamin A can be formed by cleavage of β-carotene, the pigment that gives carrots, sweet potatoes, and other yellow vegetables their characteristic color. Deficiency of vitamin A leads to a variety of symptoms in humans and experimental animals, which include dry skin, xerophthalmia (dry eyes), dry mucous membranes, retarded development and growth, sterility in male animals, and night blindness, an early symptom commonly used in the medical diagnosis of vitamin A deficiency.

Vitamin D is a derivative of cholesterol and the precursor to a hormone essential in calcium and phosphate metabolism in vertebrate animals. Vitamin D_3, also called cholecalciferol, is normally formed in the skin in a photochemical reaction driven by the ultraviolet component of sunlight (Figure 9–19). It is also abundant in fish liver oils, and is added to commercial milk as a nutritional supplement. Vitamin D_3 itself is not biologically active, but it is the precursor of 1,25-dihydroxycholecalciferol, a potent hormone that regulates the uptake of calcium in the intestine and the balance of release and deposition of bone calcium and phosphate. Deficiency of vitamin D leads to defective bone formation, resulting in the disease rickets.

Vitamin E (Fig. 9–20) is the collective name for a group of closely related lipids called tocopherols, all of which contain a substituted aromatic ring and a long hydrocarbon side chain. Tocopherols are found in hens' eggs and vegetable oils, and are especially abundant in wheat germ. Deficiency of vitamin E is very rare in humans, but when laboratory animals are fed diets depleted of vitamin E, they develop scaly skin, muscular weakness and wasting, and sterility. Tocopherols can undergo oxidation–reduction reactions on the aromatic ring. The vitamin activity of tocopherols likely results from their ability to prevent oxidative damage to the lipids of cellular membranes. Recall the reactions of unsaturated fatty acids with oxygen that cause rancidity in foods. If such reactions were to occur in living cells, the resulting defects in membrane function might cause cell death. Tocopherols react with and destroy the most reactive forms of oxygen, protecting unsaturated fatty acids from oxidation. Tocopherols are used commercially to retard spoilage of certain foods.

Vitamin K is a lipid cofactor required for normal blood clotting. Vitamin K_1 (phylloquinone; Fig. 9–20) is found in green plant leaves,

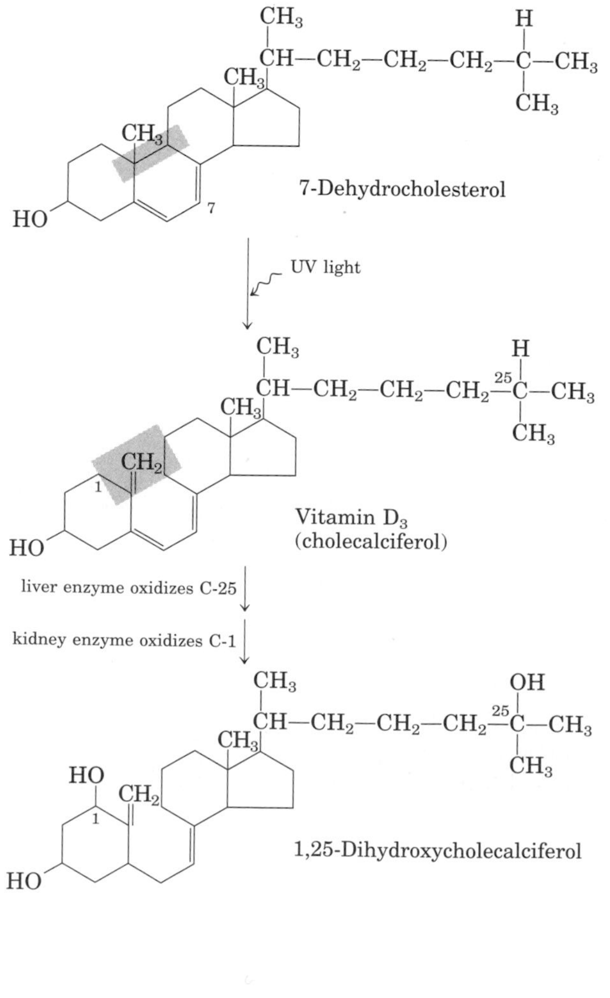

Figure 9–19 Vitamin D_3 production and metabolism. Vitamin D_3 is produced by irradiation of 7-dehydrocholesterol in the skin, and in the kidney is converted into the active hormone, 1,25-dihydroxycholecalciferol, which regulates the metabolism of Ca^{2+} and PO_4^{3-}. Dietary vitamin D prevents rickets, a disease once common in cold climates, where heavy clothing blocked the UV component of sunlight necessary to vitamin D_3 production in skin.

Vitamin E: an antioxidant

Vitamin K_1: a blood-clotting cofactor (phylloquinone)

Warfarin: a blood anticoagulant

Ubiquinone: a mitochondrial electron carrier (coenzyme Q) ($n = 4$–8)

Plastoquinone: a chloroplast electron carrier ($n = 4$–8)

Dolichol: a sugar carrier ($n = 9$–22)

Figure 9–20 Some other biologically active isoprenoid compounds or derivatives. Note that the values of n exclude the first and last isoprene unit in each isoprenoid side chain as represented here. Warfarin does not occur naturally. It is an analog of vitamin K that lacks an isoprenoid side chain.

and a related form, vitamin K_2 (menaquinone), is formed by bacteria residing in the animal intestine. The vitamin acts in the formation of prothrombin, a blood plasma protein essential in blood-clot formation. Prothrombin is a proteolytic enzyme that splits specific peptide bonds in the blood protein fibrinogen, converting it to fibrin, the insoluble, fibrous protein that holds blood clots together. Deficiency of vitamin K results in slowed blood clotting, which can be fatal to a wounded animal. Henrik Dam and Edward A. Doisy are given credit for having independently discovered the antihemorrhaghic action of vitamin K.

Warfarin (Fig. 9–20) is a synthetic analog of vitamin K, which acts as a competitive inhibitor of prothrombin formation. It is extremely poisonous to rats, causing death by internal bleeding. Ironically, this potent rodenticide is also a valuable anticoagulant drug for the treatment of human patients in whom excessive blood clotting is dangerous— surgical patients and victims of coronary thrombosis.

Henrik Dam
1895–1976

Edward A. Doisy
1893–1986

Lipid Quinones Carry Electrons

Ubiquinone and plastoquinone (Fig. 9–20), also isoprenoid derivatives, function as electron carriers in the production of ATP in mitochondria and chloroplasts. In most mammalian tissues, ubiquinone (also called coenzyme Q) has ten isoprene units. Plastoquinone is the plant equivalent of ubiquinone. In their roles as electron carriers, both ubiquinone and plastoquinone can accept either one or two electrons and either one or two protons to be reduced, as shown in Figure 18–2.

Dolichols Form Activated, Hydrophobic Sugar Derivatives

During the assembly of the complex carbohydrates of bacterial cell walls, and during the addition of polysaccharide units to certain proteins (glycoproteins) in eukaryotes, the sugar units to be added are chemically activated by attachment to **dolichols** (Fig. 9–20), another group of isoprenoids. Dolichols from animals have between 17 and 21 isoprene units (85 to 105 carbon atoms), bacterial dolichols have 11 units, and those of plants and fungi have 14 to 24 isoprene units. These very hydrophobic compounds have strong hydrophobic interactions with membrane lipids, anchoring the attached sugars to the membrane where they participate in sugar-transfer reactions.

Resolution and Analysis of Lipids

In exploring the role of lipids in a biological process, it is often useful to know which lipids are present, and in what proportions. Because lipids are insoluble in water, their extraction from tissues and subsequent fractionation require the use of organic solvents and some techniques not commonly used in the purification of water-soluble molecules such as proteins and carbohydrates. In general, complex mixtures of lipids are separated by differences in their polarity or solubility in nonpolar solvents. Lipids that contain ester- or amide-linked fatty acids can be hydrolyzed (saponified) by treatment with acid or alkali, to yield their component parts for analysis.

Lipid Extraction Requires Organic Solvents

Neutral lipids (triacylglycerols, waxes, pigments, etc.) are readily extracted from tissues with ethyl ether, chloroform, or benzene, solvents in which lipid clustering driven by hydrophobic interactions does not occur. Membrane lipids are more effectively extracted by more polar organic solvents, such as ethanol or methanol, which reduce the hydrophobic interactions among lipid molecules but also weaken the hydrogen bonds and electrostatic interactions that bind membrane lipids to membrane proteins. A commonly used extractant is a mixture of chloroform, methanol, and water, initially in proportions that are miscible, producing a single phase (1:2:0.8, v/v/v). After homogenizing tissue in this solvent to extract all lipids, more water is added to the resulting extract, and it separates into two phases, methanol/water (top phase) and chloroform (bottom phase). The lipids remain in the chloroform, and more polar molecules (proteins, sugars) partition into the polar phase of methanol/water (Fig. 9–21).

Figure 9–21 Some common procedures used in the extraction, separation, and identification of cellular lipids. **(a)** Tissue is homogenized in a chloroform/methanol/water mixture, which on addition of water and removal of unextractable sediment by centrifugation yields two phases. Different types of extracted lipids in the chloroform phase may be separated by **(b)** adsorption chromatography on a column of silica gel, through which solvents of increasing polarity are passed, or **(c)** thin-layer chromatography (TLC), in which lipids are carried by a rising solvent front, less polar lipids traveling farther than more polar or charged lipids. TLC with appropriate solvents also can be used to separate individual lipid species from a single class; for example, the charged lipids phosphatidylserine, phosphatidylglycerol, and phosphatidylinositol are easily separated by TLC. For the determination of fatty acid composition, a lipid fraction containing ester-linked fatty acids is **(d)** transesterified in a warm aqueous solution of NaOH and methanol, producing a mixture of fatty acyl methyl esters, which are then **(e)** separated on the basis of chain length and degree of saturation by gas-liquid chromatography. Precise determination of molecular mass, by mass spectroscopy (not shown), allows unambiguous identification of individual lipids. The lipid is ionized and volatilized by heat and the resulting molecular ion is passed through an electromagnetic field, which deflects ions to a degree dependent on their size. By comparison with standard ions of known molecular mass, the mass of the unknown molecular ion is determined with such great accuracy that the structure of the lipid can be deduced.

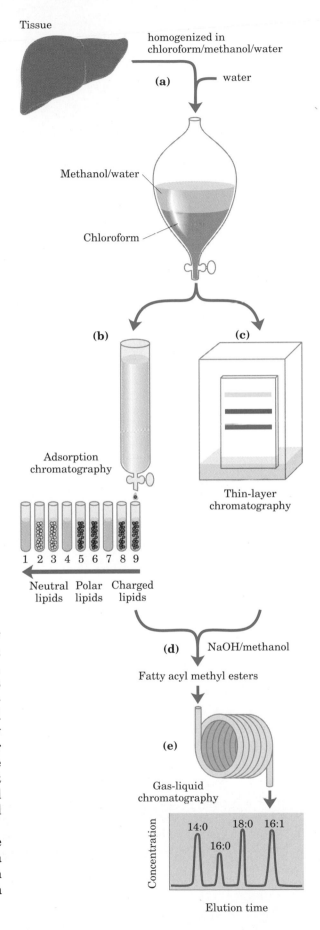

Adsorption Chromatography Separates Lipids of Different Polarity

The complex mixture of tissue lipids can be fractionated further by chromatographic procedures based on the different polarities of each class of lipid. In adsorption chromatography (Fig. 9–21), an insoluble, polar material such as silica gel (a form of silicic acid, Si(OH)$_4$), is packed into a long, thin glass column, and the lipid mixture (in chloroform solution) is applied to the top of the column. The polar lipids bind tightly to the polar silicic acid, but the neutral lipids pass directly through the column and emerge in the first chloroform wash. The polar lipids are then eluted, in order of increasing polarity, by washing the column with solvents of progressively higher polarity. Uncharged but polar lipids (cerebrosides, for example) are eluted with acetone, and very polar or charged lipids (such as glycerophospholipids) are eluted with methanol.

Thin-layer chromatography on silicic acid (Fig. 9–21) employs the same principle. A thin layer of silica gel (silicic acid) is spread onto a glass plate, to which it adheres. A small sample of lipids dissolved in chloroform is applied near one edge of the plate, which is dipped in a shallow container of an organic solvent within a closed chamber saturated with the solvent vapor. As the solvent rises on the plate by capillary action, it carries lipids with it. The less polar lipids move farthest, as they have less tendency to bind to the polar silicic acid. The lipids can be detected after their separation by spraying the plate with a dye (rhodamine), which fluoresces when associated with lipids, or by exposing the plate to iodine fumes. Iodine reacts with the double bonds in fatty acids, giving the lipids that contain them a yellow or brown color. For subsequent analysis, regions containing separated lipids can be scraped from the plate and the lipids recovered by extraction with an organic solvent.

Gas-Liquid Chromatography Resolves Mixtures of Volatile Lipid Derivatives

Gas-liquid chromatography separates volatile components of a mixture according to their relative tendencies to dissolve in the inert material packed in the chromatography column, and to volatilize and move through the column, carried by a current of an inert gas such as helium. Some lipids are naturally volatile, but most must first be derivatized to increase their volatility (that is, lower their boiling point). For the analysis of the fatty acids present in a sample of phospholipids, the lipids are first heated in a methanol/HCl or methanol/NaOH mixture, which converts fatty acids esterified to glycerol into their methyl esters (transesterification). These fatty acyl methyl esters are then loaded onto the gas-liquid chromatography column, and the column is heated to volatilize the compounds. Those fatty acyl esters most soluble in the column material partition into (dissolve in) that material; those less soluble are carried by the stream of helium and emerge first from the column (Fig. 9–21). The order of elution depends on the nature of the solid adsorbant in the column, and on the boiling point of the components of the lipid mixture. Using these techniques, mixtures of fatty acids with various chain lengths and various degrees of unsaturation can be completely resolved.

Specific Hydrolysis Aids in Determination of Lipid Structure

Certain classes of lipids are susceptible to degradation under specific conditions. For example, all ester-linked fatty acids in triacylglycerols, phospholipids, and sterol esters are released by mild acid or alkaline treatment, and somewhat harsher hydrolysis conditions release amide-bound fatty acids from sphingolipids. Enzymes that specifically hydrolyze certain lipids are also useful in the determination of lipid structure. Phospholipases A, C, and D (see Fig. 9–12) each split specific bonds in phospholipids and yield products with characteristic solubilities and chromatographic behaviors. Phospholipase C, for example, releases a water-soluble phosphoryl alcohol (phosphocholine from phosphatidylcholine) and a chloroform-soluble diacylglycerol, each of which can be characterized separately to determine the structure of the intact phospholipid. The combination of specific hydrolysis with characterization of the products by thin-layer chromatography or gas-liquid chromatography often allows determination of the structure of a lipid. To establish unambiguously the length of a hydrocarbon chain, or the position of double bonds, mass spectral analysis of lipids or their volatile derivatives is invaluable.

Summary

Lipids are water-insoluble components of cells that can be extracted by nonpolar solvents. Some lipids serve as structural components of membranes and others as storage forms of fuel. Fatty acids, which provide the hydrocarbon components of lipids, usually have an even number (12 to 24) of carbon atoms and may be saturated or unsaturated; unsaturated fatty acids have double bonds in the cis configuration. In most unsaturated fatty acids, one double bond is at the Δ^9 position (between C-9 and C-10).

Triacylglycerols contain three fatty acid molecules esterified to the three hydroxyl groups of glycerol. Simple triacylglycerols contain only one type of fatty acid; mixed triacylglycerols contain at least two different types. Triacylglycerols are primarily storage fats; they are present in many types of foods.

The polar lipids, which have polar heads and nonpolar tails, are major components of membranes. The most abundant are the glycerophospholipids, which contain two fatty acid molecules esterified to two hydroxyl groups of glycerol, and a second alcohol, the head group, esterified to the third hydroxyl of glycerol via a phosphodiester bond. Glycerophospholipids differ in the structure of the head group; common glycerophospholipids are phosphatidylethanolamine and phosphatidylcholine. The polar heads of the glycerophospholipids carry electric charges at pH near 7. The sphingolipids, also membrane components, contain sphingosine, a long-chain aliphatic amino alcohol, but no glycerol. Sphingomyelin possesses, in addition to phosphoric acid and choline, two long hydrocarbon chains, one contributed by a fatty acid and the other by sphingosine. Two other classes of sphingolipids are neutral glycolipids and gangliosides, which contain various sugar components.

Cholesterol, a sterol, is a precursor of many steroids and is also an important component of plasma membranes of animal cells. All polar lipids are amphipathic; they have polar or charged heads and nonpolar hydrocarbon tails. They spontaneously form micelles, bilayers, and liposomes, stabilized by hydrophobic interactions.

Some types of lipids, although present in relatively small quantities, play critical roles as cofactors or signals. Steroid hormones are derived from sterols. Phosphatidylinositol is hydrolyzed to yield two intracellular messengers, diacylglycerol and inositol trisphosphate. Prostaglandins, thromboxanes, and leukotrienes are extremely potent hormonelike molecules derived from arachidonic acid. Vitamins A, D, E, and K are fat-soluble compounds made up of isoprene units. All play essential roles in the metabolism or physiology of animals. Vitamin A furnishes the visual pigment of the vertebrate eye. Vitamin D is parent to a hormone that regulates calcium and phosphate metabolism. Vitamin E probably functions in the protection of membrane lipids from oxidative damage, and vitamin K is essential in the blood-clotting process. Ubiquinones and plastoquinones, also isoprenoid derivatives, function as electron carriers in animals and plants, respectively. Dolichols activate and anchor sugars on cellular membranes for use in the synthesis of certain complex carbohydrates and glycoproteins.

In the determination of lipid composition, lipids are extracted from tissues with organic solvents and separated by thin-layer or gas-liquid chromatography. Individual lipids are identified by their chromatographic behavior, their susceptibility to hydrolysis by specific enzymes, or by mass spectral determination of their molecular masses.

Further Reading

General

Gurr, M.I. & Harwood, J.L. (1990) *Lipid Biochemistry. An Introduction*, 4th edn, Chapman & Hall, London.
A good general resource on lipid structure and metabolism, at the intermediate level.

Harwood, J.L. & Russell, N.J. (1984) *Lipids in Plants and Microbes*, George Allen & Unwin, Ltd., London.
Short, clear descriptions of lipid types, their distribution, metabolism, and function in plants and microbes; intermediate level.

Mead, J.F., Alfin-Slater, R.B., Howton, D.R., & Popjak, G. (1986) *Lipids: Chemistry, Biochemistry and Nutrition,* Plenum Press, New York.

An intermediate level textbook on chemical, metabolic, and nutritional aspects of lipids.

Vance, D.E. & Vance, J.E. (eds) (1991) *Biochemistry of Lipids, Lipoproteins and Membranes,* New Comprehensive Biochemistry, Vol. 20, Elsevier Science Publishing Co., Inc., New York.

An excellent collection of reviews on various aspects of lipid structure, biosynthesis, and function. Particularly germane are the chapters by K. Bloch (Cholesterol: evolution of structure and function); P.R. Cullis & M.J. Hope (Physical properties and functional roles of lipids in membranes); C.C. Sweeley (Sphingolipids); and W.L. Smith, P. Borgeat, & F.A. Fitzpatrick (The eicosanoids: cyclooxygenase, lipoxygenase, and epoxygenase pathways).

Structural Lipids in Membranes

Hakamori, S. (1986) Glycosphingolipids. *Sci. Am.* **254** (May), 44–53.

Ostro, M.J. (1987) Liposomes. *Sci. Am.* **256** (January), 102–111.

Sastry, P.S. (1985) Lipids of nervous tissue: composition and metabolism. *Prog. Lipid Res.* **24,** 69–176.

Spector, A.A. & Yorek, M.A. (1985) Membrane lipid composition and cellular function. *J. Lipid Res.* **26,** 1015–1035.

Lipids with Specific Biological Activities

Chojnacki, T. & Dallner, G. (1988) The biological role of dolichol. *Biochem. J.* **251,** 1–9.

Fisher, S.K., Heacock, A.M., & Agranoff, B.W. (1992) Inositol lipids and signal transduction in the nervous system: an update. *J. Neurochem.* **58,** 18–38.

A good discussion of the growing number of processes known to be controlled by metabolites of inositol-containing membrane lipids.

Machlin, L.J. & Bendich, A. (1987) Free radical tissue damage: protective role of antioxidant nutrients. *FASEB J.* **1,** 441–445.

Brief discussion of tocopherols as antioxidants and their role in preventing damage by oxygen free radicals.

Shimizu, T. & Wolfe, L.S. (1990) Arachidonic acid cascade and signal transduction. *J. Neurochem.* **55,** 1–15.

Role of arachidonic acid (20 : 4) as precursor to the eicosanoids.

Snyder, F., Lee, T.-C., & Blank, M.L. (1989) Platelet-activating factor and related ether lipid mediators: biological activities, metabolism, and regulation. *Ann. N. Y. Acad. Sci.* **568,** 35–43.

Vermeer, C. (1990) γ-Carboxyglutamate-containing proteins and the vitamin K-dependent carboxylase. *Biochem. J.* **266,** 625–636.

Biochemical basis for the requirement of vitamin K in blood clotting, and the importance of carboxylation in the synthesis of the blood-clotting protein thrombin.

Viitala, J. & Järnefelt, J. (1985) The red cell surface revisited. *Trends Biochem. Sci.* **10,** 392–395.

Includes discussion of the human A, B, and O blood type determinants.

Resolution and Analysis of Lipids

Kates, M. (1986) *Techniques of Lipidology: Isolation, Analysis and Identification of Lipids,* 2nd edn, Laboratory Techniques in Biochemistry and Molecular Biology, Vol. 3, Part 2 (Burdon, R.H. & van Knippenberg, P.H., eds), Elsevier Science Publishing Co., Inc., New York.

Problems

1. *Melting Points of Fatty Acids* The melting points of a series of 18-carbon fatty acids are stearic acid, 69.6 °C; oleic acid, 13.4 °C; linoleic acid, −5 °C; and linolenic acid, −11 °C. What structural aspect of these 18-carbon fatty acids can be correlated with the melting point? Provide a molecular explanation for the trend in melting points.

2. *Spoilage of Cooking Fats* Some fats used in cooking, such as olive oil, spoil rapidly upon exposure to air at room temperature, whereas others, such as solid shortening, remain unchanged. Why?

3. *Preparation of Béarnaise Sauce* During the preparation of béarnaise sauce, egg yolks are incorporated into melted butter to stabilize the sauce and avoid separation. The stabilizing agent in the egg yolks is lecithin (phosphatidylcholine). Suggest why this works.

4. *Hydrolysis of Lipids* Name the products of mild hydrolysis of the following lipids with dilute NaOH:

    (a) 1-stearoyl-2,3-dipalmitoylglycerol

    (b) 1-palmitoyl-2-oleoylphosphatidylcholine

5. *Number of Detergent Molecules per Micelle* When a small amount of sodium dodecyl sulfate $(Na^+CH_3(CH_2)_{11}OSO_3^-)$ is dissolved in water, the detergent ions go into solution as monomeric species. As more detergent is added, a point is reached (the critical micelle concentration) at which the monomers associate to form micelles. The critical micelle concentration of SDS is 8.2 mM. An examination of the micelles shows that they have an average particle weight (the sum of the molecular weights of the constituent monomers) of 18,000. Calculate the number of detergent molecules in the average micelle.

6. *Hydrophobic and Hydrophilic Components of Membrane Lipids* A common structural feature of membrane lipid molecules is their amphipathic nature. For example, in phosphatidylcholine, the two fatty acid chains are hydrophobic and the phosphocholine head group is hydrophilic. For each of the following membrane lipids, name the components that serve as the hydrophobic and hydrophilic units:

    (a) phosphatidylethanolamine

    (b) sphingomyelin

    (c) galactosylcerebroside

    (d) ganglioside

    (e) cholesterol

7. *Properties of Lipids and Lipid Bilayers* Lipid bilayers formed between two aqueous phases have this important property: they form two-dimensional sheets, the edges of which close upon each other, and undergo self-sealing to form liposomes.

    (a) What properties of lipids are responsible for this property of bilayers? Explain.

    (b) What are the biological consequences of this property with regard to the structure of biological membranes?

8. *Chromatographic Separation of Lipids* A mixture of the following lipids is applied to a silica gel column, and the column is then washed with progressively more polar solvents. The mixture consists of: phosphatidylserine, cholesteryl palmitate (a sterol ester), phosphatidylethanolamine, phosphatidylcholine, sphingomyelin, palmitic acid, *n*-tetradecanol, triacylglycerol, and cholesterol. In what order do you expect the lipids to elute from the column?

9. *Storage of Fat-Soluble Vitamins* In contrast to water-soluble vitamins, which must be a part of our daily diet, fat-soluble vitamins can be stored in the body in amounts sufficient for many months. Suggest an explanation for this difference based on solubilities.

10. *Alkali Lability of Triacylglycerols* A common procedure for cleaning the grease trap in a sink is to add a product that contains sodium hydroxide. Explain why this works.

11. *Dependence of Melting Point on Fatty Acid Unsaturation* Draw all of the possible triacylglycerols that you could construct from glycerol, palmitic acid, and oleic acid. Rank them in order of increasing melting point.

12. *Operational Definition of Lipids* How is the definition of "lipid" different from the definitions of other types of biomolecules that we have considered, such as amino acids, nucleic acids, and proteins?

13. *Effect of Polarity on Solubility* Rank, in order of increasing solubility in water, a triacylglycerol, a diacylglycerol, and a monoacylglycerol, all containing only palmitic acid.

14. *Intracellular Messengers from Phosphatidylinositols* When the hormone vasopressin stimulates cleavage of phosphatidylinositol-4,5-bisphosphate by hormone-sensitive phospholipase C, two products are formed. Compare their properties and solubilities in water, and predict whether either would be expected to diffuse readily through the cytosol.

15. *Identification of Unknown Lipids* Johann Thudichum, who practiced medicine in London about 100 years ago, also dabbled in lipid chemistry in his spare time. He isolated a variety of lipids from neural tissue, and characterized and named many of them. His carefully sealed and labeled vials of isolated lipids were rediscovered many years later. How would you confirm, using techniques available to you but not to him, that the vials he labeled "sphingomyelin" and "cerebroside" actually contain these compounds?

16. *Analysis of Choline-Containing Phospholipids* How would you distinguish sphingomyelin from phosphatidylcholine by chemical, physical, or enzymatic tests?

CHAPTER

Biological Membranes and Transport

The first living cell probably came into being when a membrane formed, separating that cell's precious contents from the rest of the universe. Membranes define the external boundary of cells and regulate the molecular traffic across that boundary; they divide the internal space into discrete compartments to segregate processes and components (Fig. 10–1); they organize complex reaction sequences; and they are central to both biological energy conservation and cell-to-cell communication. The biological activities of membranes flow from their remarkable physical properties. Membranes are tough but flexible, self-sealing, and selectively permeable to polar solutes. Their flexibility permits the shape changes that accompany cell growth and movement (such as amoeboid movement). Their ability to seal over temporary breaks in their continuity allows two membranes to fuse, as in exocytosis, or a single membrane-enclosed compartment to undergo fission, yielding two sealed compartments, as in endocytosis or cell division, without creating gross leaks through the cell surface. Because membranes are selectively permeable, they retain certain compounds and ions within cells and within specific cellular compartments, and exclude others.

Membranes are not merely passive barriers. They include an array of proteins specialized for promoting or catalyzing a variety of molecular events. Pumps move specific organic solutes and inorganic ions across the membrane against a concentration gradient; energy transducers convert one form of energy into another; receptors on the plasma membrane sense extracellular signals, converting them into molecular changes within the cell.

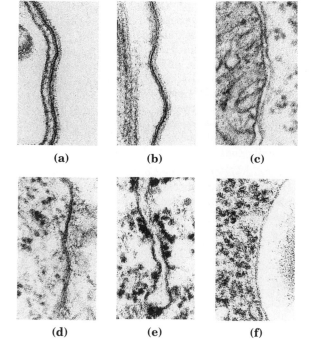

(a)   **(b)**   **(c)**

(d)   **(e)**   **(f)**

Figure 10–1 Viewed in cross section, all intracellular membranes share a characteristic trilaminar appearance. The protozoan *Paramecium* contains a variety of specialized membrane-bounded organelles. When a thin section of a *Paramecium* is stained with osmium tetroxide to highlight membranes, each of the membranes appears as a three-layer structure, 5 to 8 nm thick. The trilaminar images consist of two electron-dense layers on the inner and outer surfaces separated by a less dense central region. At left are high-magnification views of the membranes of **(a)** a cell body (plasma and alveolar membranes tightly apposed), **(b)** a cilium, **(c)** a mitochondrion, **(d)** a digestive vacuole, **(e)** the endoplasmic reticulum, and **(f)** a secretory vesicle.

Membranes are composed of just two layers of molecules, and are therefore very thin; they can be thought of as essentially two-dimensional. A large number of cellular processes are associated with membranes (such as the synthesis of lipids and certain proteins, and the energy transductions in mitochondria and chloroplasts). Because intermolecular collisions are far more probable in this two-dimensional space than in three-dimensional space, the efficiency of certain enzyme-catalyzed pathways organized within a two-dimensional membrane is vastly increased.

In this chapter we first describe the composition of cellular membranes and their chemical architecture—the physical structure that underlies their biological functions. We then turn to membrane transport, the protein-mediated transmembrane passage of solutes. In later chapters we will discuss the role of membranes in energy transduction, lipid synthesis, signal transduction, and protein synthesis.

The Molecular Constituents of Membranes

One approach to understanding membrane function is to study membrane composition—to determine, for example, which components are commonly present in membranes and which are unique to membranes with specific functions. Knowledge of composition is also invaluable in studies of membrane structure, as any viable model for membrane structure must conform to and explain the known composition. Before describing membrane structure and function, we therefore consider the molecular components of membranes.

Proteins and polar lipids account for almost all of the mass of biological membranes; the small amount of carbohydrate present is generally part of glycoproteins or glycolipids. The relative proportions of protein and lipid differ in different membranes (Table 10–1), reflecting the diversity of biological roles. The myelin sheath, which serves as a passive electrical insulator wrapped around certain neurons, consists primarily of lipids, but the membranes of bacteria, mitochondria, and chloroplasts, in which many enzyme-catalyzed metabolic processes take place, contain more protein than lipid.

Table 10–1 Major components of plasma membranes of different species*

	Protein (%)	Phospholipid (%)	Other lipids	Sterol (%)	Sterol type
Mouse liver	45	27	—	25	Cholesterol
Corn leaf	47	26	Galactolipids	7	Sitosterol
Yeast	52	7	Triacylglycerols Steryl esters	4	Ergosterol
Paramecium (ciliate protist)	56	40	—	4	Stigmasterol
E. coli	75	25	—	0	—

* Values are given as weight percentages.

Each Membrane Has a Characteristic Lipid Composition

For studies of membrane composition, it is essential first to isolate the membrane of interest. When eukaryotic cells are subjected to mechani-

Table 10–2 Lipid composition of organelle membranes of a rat liver cell*

	Chol	PC	PE	PS	PI	PG	CL	SM
Plasma membrane	30	18	11	9	4	0	0	14
Golgi complex	8	40	15	4	6	0	0	10
Smooth endoplasmic reticulum	10	50	21	0	7	0	2	12
Rough endoplasmic reticulum	6	55	16	3	8	0	0	3
Nuclear membrane	10	55	20	3	7	0	0	3
Lysosomal membrane	14	25	13	0	7	0	5	24
Mitochondrial membrane								
Inner	3	45	24	1	6	2	18	3
Outer	5	45	23	2	13	3	4	5

* Values are given as weight percentages. Chol designates cholesterol; PC, phosphatidylcholine; PE, phosphatidylethanolamine; PS, phosphatidylserine; PI, phosphatidylinositol; PG, phosphatidylglycerol; CL, cardiolipin; SM, sphingomyelin.

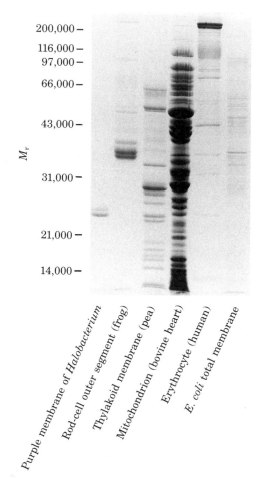

Figure 10–2 Membranes with specialized functions differ in protein composition, as revealed by electrophoretic separation on a polyacrylamide gel in the presence of the detergent SDS (p. 141). The purple membrane of *Halobacterium* and the rod-cell outer segment membrane are very rich in bacteriorhodopsin and rhodopsin, respectively. The myelin sheath also contains relatively few kinds of proteins. The other membranes shown have more complex functions, reflected in a wider variety of membrane proteins.

cal shear, their plasma membranes are torn and fragmented, releasing cytosolic components and membrane-bounded organelles: mitochondria, chloroplasts, lysosomes, nuclei, and others. The plasma membrane fragments and intact organelles can be isolated by centrifugal techniques described in Chapter 2 (see Fig. 2–24).

Chemical analysis of membranes isolated from various sources reveals certain common properties. Membrane lipid composition is characteristic for each kingdom, each species, each tissue, and each organelle within a given cell type (Table 10–2). Cells clearly have mechanisms to control the kinds and amounts of membrane lipids synthesized and to target specific lipids to particular organelles. These distinct combinations doubtless confer advantages on cells and organisms during evolution, but in most cases the functional significance of these characteristic lipid compositions remains to be discovered.

Membranes with Different Functions Have Different Proteins

The protein composition of membranes from different sources (Fig. 10–2) varies even more widely than their lipid composition, reflecting functional specialization. The outer segment of the rod cells of the vertebrate retina is highly specialized for the reception of light; more than 90% of its membrane protein is the light-absorbing protein rhodopsin (see Fig. 9–18). The less-specialized plasma membrane of the erythrocyte has about 20 prominent proteins as well as dozens of minor ones; many of these serve as transporters, each responsible for moving a specific solute across the membrane. The inner (plasma) membrane of *E. coli* contains hundreds of different proteins, various transporters, as well as many enzymes involved in energy-conserving metabolism, lipid synthesis, protein export, and cell division. The outer membrane of *E. coli* has a different function (protection) and a different set of proteins. Some membrane proteins have more or less complex arrays of covalently bound carbohydrates, which may make up from 1 to 70% of the total mass of these glycoproteins. In the rhodopsin of the vertebrate eye, a single hexasaccharide makes up 4% of the mass; in glycophorin, a glycoprotein of the plasma membrane of erythrocytes, 60% of the mass consists of complex polysaccharide units covalently attached to specific amino acid residues. Ser, Thr, and Asn residues are often the points of attachment (see Fig. 11–23). In general, plasma

Figure 10–3 Covalently attached lipids of several types anchor membrane proteins to the lipid bilayer. The farnesyl side chain is an isoprenoid (p. 256).

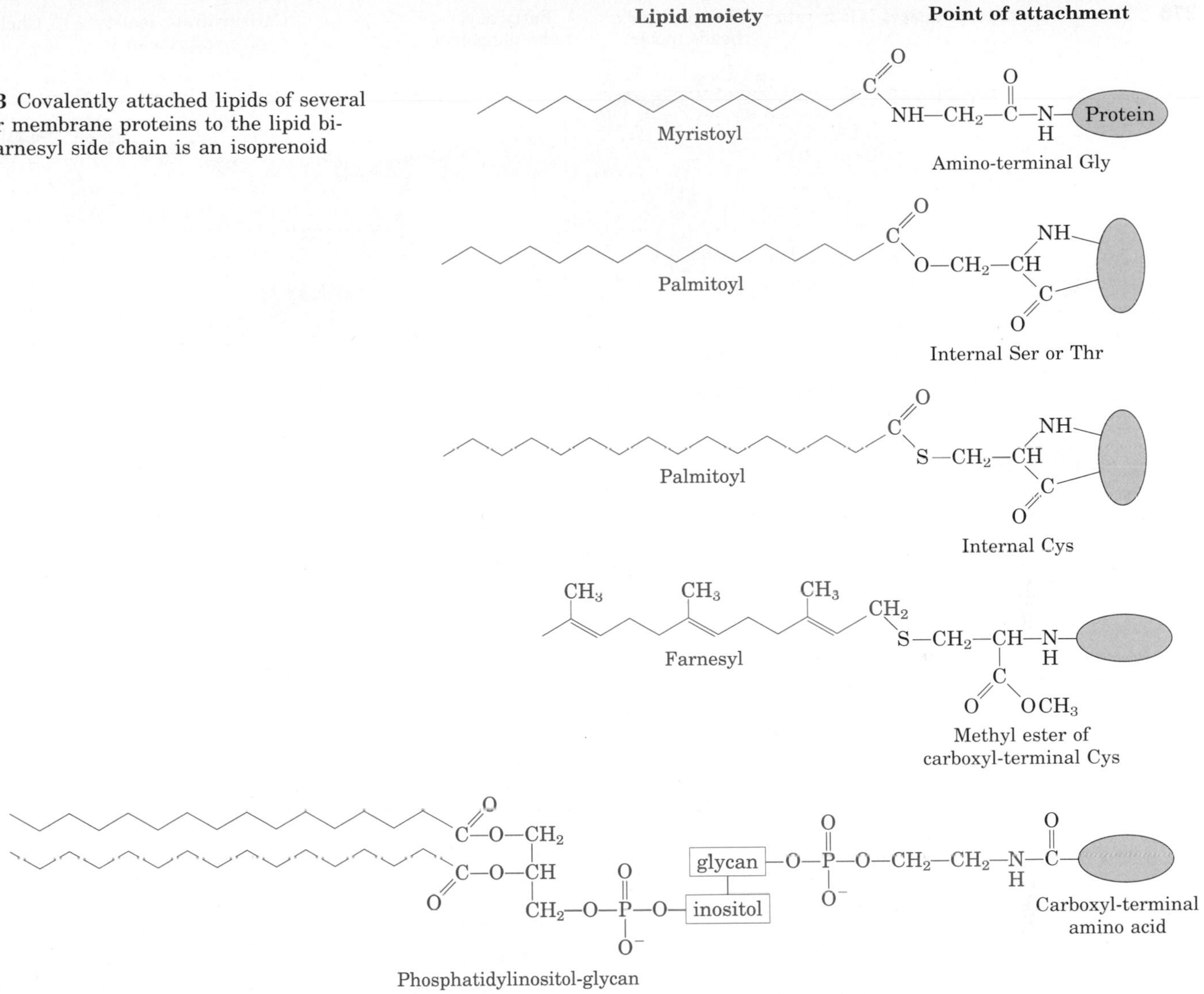

membranes contain many glycoproteins, but intracellular membranes such as those of mitochondria and chloroplasts rarely contain covalently bound carbohydrates. The sugar moieties of surface glycoproteins influence the protein folding, transport to the cell surface, and receptor functions of these glycoproteins.

Certain membrane proteins are covalently attached to one or more lipids, which probably serve as hydrophobic anchors, holding the proteins to the membrane. The lipid moiety on some membrane proteins is a fatty acid, attached in amide or ester linkage; other proteins have a long-chain isoprenoid covalently attached, and others are joined through a complex polysaccharide (a glycan; see Chapter 11) to a molecule of phosphatidylinositol (Fig. 10–3).

The Supramolecular Architecture of Membranes

All biological membranes share certain fundamental properties. They are impermeable to most polar or charged solutes, but permeable to nonpolar compounds; are 5 to 8 nm thick; appear trilaminar (three-layered) when viewed in cross section with the electron microscope (see Fig. 10–1). The combined evidence from electron microscopy, chemical composition, and physical studies of permeability and of the motion of

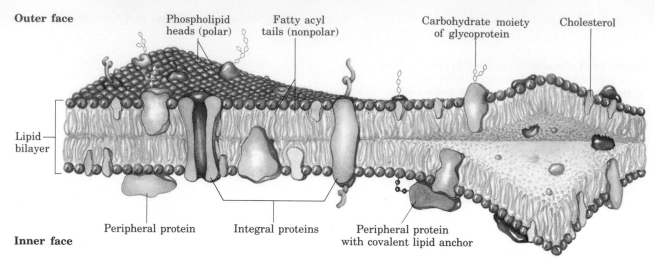

Outer face

Phospholipid heads (polar)

Fatty acyl tails (nonpolar)

Carbohydrate moiety of glycoprotein

Cholesterol

Lipid bilayer

Peripheral protein

Inner face

Integral proteins

Peripheral protein with covalent lipid anchor

Figure 10–4 The fluid mosaic model for membrane structure. The fatty acyl chains in the interior of the membrane form a fluid, hydrophobic region. Integral membrane proteins float in this sea of lipid, held by hydrophobic interactions with their nonpolar amino acid side chains. Both proteins and lipids are free to move laterally in the plane of the bilayer, but movement of either from one face of the bilayer to the other is restricted. The carbohydrate moieties attached to some proteins and lipids of the plasma membrane are invariably exposed on the extracellular face of the membrane.

individual protein and lipid molecules within membranes supports the **fluid mosaic model** for the structure of biological membranes (Fig. 10–4). Amphipathic phospholipids and sterols form a lipid bilayer, with the nonpolar regions of lipids facing each other at the core of the bilayer and their polar head groups facing outward. In this lipid bilayer, globular proteins are embedded at irregular intervals, held by hydrophobic interactions between the membrane lipids and hydrophobic domains in the proteins. Some proteins protrude from one or the other face of the membrane; some span its entire width. The orientation of proteins in the bilayer is asymmetric, giving the membrane "sidedness"; the protein domains exposed on one side of the bilayer are different from those exposed on the other side, reflecting functional asymmetry. The individual lipid and protein subunits in a membrane form a fluid mosaic; its pattern, unlike a mosaic of ceramic tile and mortar, is free to change constantly. The membrane mosaic is fluid because the interactions among lipids, and between lipids and proteins, are noncovalent, leaving individual lipid and protein molecules free to move laterally in the plane of the membrane.

We will now look at some of these features of the fluid mosaic model in more detail, and consider the experimental evidence that supports it.

A Lipid Bilayer Is the Basic Structural Element

We saw in Chapter 9 that lipids, when suspended in water, spontaneously form bilayer structures that are stabilized by hydrophobic interactions (see Fig. 9–14). The thickness of biological membranes (5 to 8 nm, measured by electron microscopy) is about that expected for a lipid bilayer 3 nm thick with proteins protruding on each side. X-ray diffraction by membranes shows the distribution of electron density expected for a bilayer structure. Liposomes (lipid vesicles) formed in the laboratory show the same relative impermeability to polar solutes as is seen in biological membranes (although the latter are permeable to solutes for which they have specific transporters). In short, all evidence indicates that biological membranes are constructed of lipid bilayers.

Membrane lipids are asymmetric in their distribution on the two faces of the bilayer, although the asymmetry, unlike that of membrane proteins, is not absolute. In the plasma membrane, for example, certain lipids are typically found primarily in the outer face of the bilayer, and others in the inner (cytoplasmic) face (Fig. 10–5).

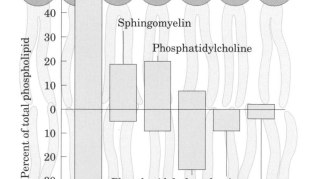

Outer face of bilayer

Total phospholipid

Sphingomyelin

Phosphatidylcholine

Percent of total phospholipid

Phosphatidylethanolamine

Phosphatidylserine

Phosphatidylinositol

Inner face of bilayer

Figure 10–5 The distribution of specific erythrocyte membrane lipids between the inner and outer face is asymmetric.

Membrane Lipids Are in Constant Motion

Although the lipid bilayer structure itself is stable, the individual phospholipid and sterol molecules have great freedom of motion within the plane of the membrane (Fig. 10–6). They diffuse laterally so fast that an individual lipid molecule can circumnavigate an erythrocyte in a few seconds. The interior of the bilayer is also fluid; individual hydrocarbon chains of fatty acids are in constant motion produced by rotation about the carbon–carbon bonds of the long acyl side chains.

The degree of fluidity depends on lipid composition and temperature. At low temperature, relatively little lipid motion occurs and the bilayer exists as a nearly crystalline (paracrystalline) array. Above a temperature that is characteristic for each membrane, lipids can undergo rapid motion. The temperature of the transition from paracrystalline solid to fluid depends upon the lipid composition of the membrane. Saturated fatty acids pack well into a paracrystalline array, but the kinks in unsaturated fatty acids (see Fig. 9–1) interfere with this packing, preventing the formation of a paracrystalline solid state. The higher the proportion of saturated fatty acids, the higher is the solid-to-fluid transition temperature of the membrane.

The sterol content of a membrane also is an important determinant of this transition temperature. The rigid planar structure of the steroid nucleus, inserted between fatty acyl side chains, has two effects on fluidity: below the temperature of the solid-to-fluid transition, sterol insertion prevents the highly ordered packing of fatty acyl chains, and thus fluidizes the membrane. Above the thermal transition point, the rigid ring system of the sterol reduces the freedom of neighboring fatty acyl chains to move by rotation about carbon–carbon bonds, and thus reduces the fluidity in the core of the bilayer. Sterols therefore tend to moderate the extremes of solidity and fluidity of the membranes that contain them.

Both microorganisms and cultured animal cells regulate their lipid composition so as to achieve a constant fluidity under various growth conditions. For example, when cultured at low temperatures, bacteria synthesize more unsaturated fatty acids and fewer saturated ones than when cultured at higher temperatures (Table 10–3). As a result of this adjustment in lipid composition, membranes of bacteria cultured at high or low temperature have about the same degree of fluidity.

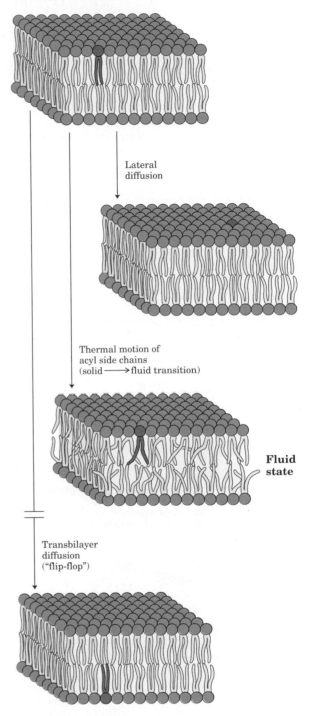

Paracrystalline state (solid)

Lateral diffusion

Thermal motion of acyl side chains (solid ⟶ fluid transition)

Fluid state

Transbilayer diffusion ("flip-flop")

Table 10–3 Fatty acid composition of *E. coli* cells cultured at different temperatures

Fatty acid	Percentage of total fatty acids*			
	10 °C	20 °C	30 °C	40 °C
Myristic (14:0)	4	4	4	8
Palmitic (16:0)	18	25	29	48
Palmitoleic (16:1)	26	24	23	9
Oleic (18:1)	38	34	30	12
Hydroxymyristic	13	10	10	8
Ratio of unsaturated: saturated†	2.9	2.0	1.6	0.38

Source: Data from Marr, A.G. & Ingraham, J.L. (1962) Effect of temperature on the composition of fatty acids in *Escherichia coli. J. Bacteriol.* **84,** 1260.

* The exact fatty acid composition depends not only on growth temperature, but also on growth stage and growth medium composition.

† The total percentage of 16:1 plus 18:1 divided by total percentage of 14:0 plus 16:0. Hydroxymyristic acid was omitted from this calculation.

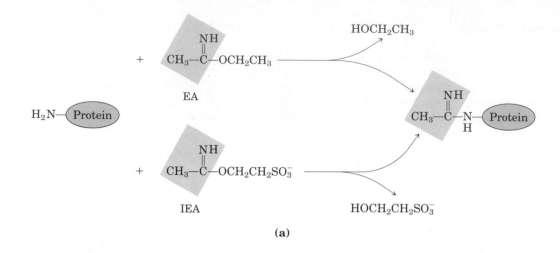

(a)

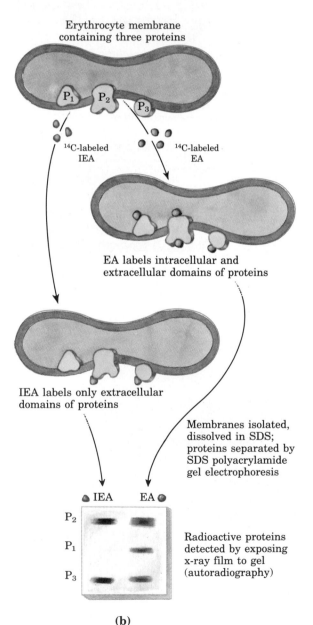

Erythrocyte membrane
containing three proteins

^{14}C-labeled
IEA

^{14}C-labeled
EA

EA labels intracellular and
extracellular domains of proteins

IEA labels only extracellular
domains of proteins

Membranes isolated,
dissolved in SDS;
proteins separated by
SDS polyacrylamide
gel electrophoresis

IEA   EA

P$_2$

P$_1$

P$_3$

Radioactive proteins
detected by exposing
x-ray film to gel
(autoradiography)

(b)

Although lateral migration of membrane components and thermal flexing of the acyl chains clearly occurs, a third kind of motion is restricted: transbilayer or "flip-flop" diffusion, the movement of a lipid from one face of the bilayer to the other (Fig. 10–6). For such motion to occur, the lipid head group, which is polar and may be charged, must leave its aqueous environment and move into the hydrophobic interior of the bilayer, a process with a large, positive free-energy change (highly endergonic). During synthesis of the bacterial plasma membrane, for example, phospholipids produced on the inside surface of the membrane must undergo flip-flop diffusion to enter the outer face of the bilayer. There are proteins that facilitate flip-flop diffusion, providing a transmembrane path that is energetically more favorable.

Membrane Proteins Penetrate and Span the Lipid Bilayer

When individual protein molecules and multiprotein complexes in freeze-fractured biological membranes are visualized with the electron microscope (Box 10–1), some proteins appear on only one face of the membrane; others span the full thickness of the bilayer, and protrude from both inner and outer membrane surfaces. Among the latter are some proteins that conduct solutes or signals across the membrane.

Membrane protein localization has also been investigated with reagents that react with protein side chains but cannot cross membranes (Fig. 10–7). The human erythrocyte is convenient for such studies, because the plasma membrane is the only membrane present.

Figure 10–7 Experiments to determine the transmembrane arrangement of membrane proteins. **(a)** Both ethylacetimidate (EA) and isoethionylacetimidate (IEA) react with free amino groups in proteins, but only the ethyl derivative (EA) diffuses freely through the membrane. **(b)** Comparison of the labeling patterns with the two reagents reveals whether a given protein is exposed only on the outer surface or only on the inner surface. Proteins labeled by both reagents, but more heavily by the permeant reagent, are exposed on both sides, and thus span the membrane.

B O X   10–1   **Electron Microscopy of Membranes**

Combined with different staining procedures and tissue-preparation methods, electron microscopy has revealed important details of membrane structure. Shown here are three different aspects of the erythrocyte membrane, visualized by electron microscopy after three different ways of preparing the cells for examination.

Figure 1 is a **transmission electron micrograph** of a section through the erythrocyte membrane, showing the two dense lines visible after osmium tetroxide staining of cells, corresponding to the outer and inner polar layers of the membrane-lipid head groups. The clear zone between the lines is the hydrophobic portion of the lipid bilayer, which contains the nonpolar fatty acyl tails.

Figure 2 shows the glycocalyx on the outer surface of the erythrocyte, visualized by a special staining procedure. This "fuzzy coat," which consists of hydrophilic oligosaccharide groups of membrane glycoproteins and glycolipids, is over 100 nm thick, more than ten times the thickness of the lipid bilayer itself.

A view of the inside of the erythrocyte membrane (illustrated in Figure 3), is produced by the **freeze-fracture** method. In this procedure the cells are frozen and the frozen block is shattered or split. The fracture lines sometimes split a membrane along a plane between the two lipid layers. The exposed surface is coated with a very thin layer of carbon, then visualized with the electron microscope (Fig. 4). The inside surface of one lipid layer forms the smooth background; the clusters of globular bodies are molecules of integral membrane proteins.

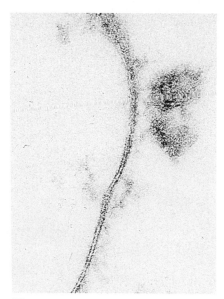

Figure 1

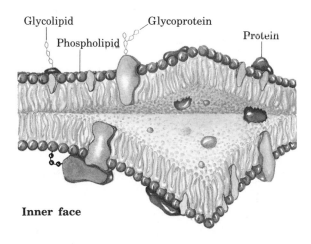

Outer face

Glycolipid   Glycoprotein

Phospholipid   Protein

Inner face

Figure 3

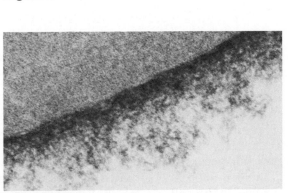

Figure 2

Figure 4

Experiments like those described in Figure 10–7 show that the glycophorin molecule spans the erythrocyte membrane. Its amino-terminal domain (bearing the carbohydrate) is on the outer surface of the erythrocyte, and its carboxyl terminus protrudes on the inside of the cell. The amino-terminal and carboxyl-terminal domains contain many polar or charged amino acid residues, and are therefore quite hydrophilic. However, a long segment in the center of the protein contains mainly hydrophobic amino acid residues. These findings suggest the transmembrane arrangement of glycophorin shown in Figure 10–8.

Figure 10–8 Transbilayer disposition of glycophorin in the erythrocyte. One hydrophilic domain, containing all the sugar residues, is on the outer surface, and another hydrophilic domain protrudes on the inner surface. The red hexagons represent a tetrasaccharide (containing two NeuNAc, Gal, and GalNAc) O-linked to a serine or threonine residue; the blue hexagon represents an oligosaccharide chain N-linked to an asparagine residue. A segment of relatively hydrophobic residues forms an α helix that traverses the membrane bilayer.

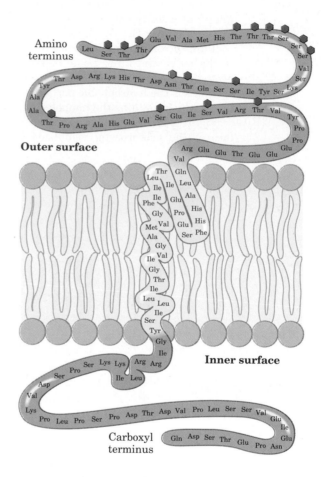

Membrane Proteins Are Oriented Asymmetrically

One further fact may be deduced from the results of the experiments with glycophorin: it does not move from one face of the bilayer to the other; its disposition in the membrane is asymmetric. Similar studies of other membrane proteins show that each has a specific orientation in the bilayer, and that protein reorientation by flip-flop diffusion occurs seldom, if ever. Furthermore, glycoproteins of the plasma membrane are invariably situated with their sugar residues on the outer surface of the cell. As we shall see, the asymmetric arrangement of membrane proteins results in functional asymmetry; all the molecules of a given ion pump, for example, have the same orientation and therefore all pump in the same direction.

Integral Membrane Proteins Are Insoluble in Water

Membrane proteins may be divided operationally into two groups: **integral** (intrinsic) **proteins,** which are very firmly bound to the membrane, and **peripheral** (extrinsic) **proteins,** which are bound more loosely, or reversibly. Peripheral membrane proteins can be released from membranes by relatively mild treatments (Fig. 10–9), and once released from the membrane they are generally water soluble. In contrast, the release of integral proteins from membranes requires the action of agents (detergents, organic solvents, or denaturants) that interfere with hydrophobic interactions. Even after integral proteins have been solubilized, removal of the detergent may cause the protein to precipitate as an insoluble aggregate. The insolubility of integral membrane proteins results from the presence of domains rich in hydrophobic amino acids; hydrophobic interactions between the protein and the lipids of the membrane account for the firm attachment of the protein.

Some Integral Proteins Have Hydrophobic Transmembrane Anchors

Integral membrane proteins generally have domains rich in hydrophobic amino acids. In some proteins, there is a single hydrophobic sequence in the middle of the protein (as in glycophorin) or at the amino or carboxyl terminus. Other membrane proteins have multiple hydrophobic sequences, each long enough to span the lipid bilayer when in the α-helical conformation.

One of the best studied membrane-spanning proteins, bacteriorhodopsin, contains seven very hydrophobic internal sequences, and crosses the lipid bilayer seven times. Bacteriorhodopsin is a light-driven proton pump that is densely packed in regular arrays in the purple membrane of the bacterium *Halobacterium halobium*. When these arrays are viewed with the electron microscope from several angles, the resulting images allow a three-dimensional reconstruction of the bacteriorhodopsin molecule (Fig. 10–10). Seven α-helical segments, each traversing the lipid bilayer, are connected by nonhelical loops at the inner or outer face of the membrane. The amino acid sequence of bacteriorhodopsin has seven segments with about 20 hydrophobic residues, each segment just long enough to make a helix that spans the lipid bilayer. Hydrophobic interactions between the nonpolar amino acids and the acyl side chains of the membrane lipids firmly anchor the protein in the membrane, providing a transmembrane pathway for proton translocation.

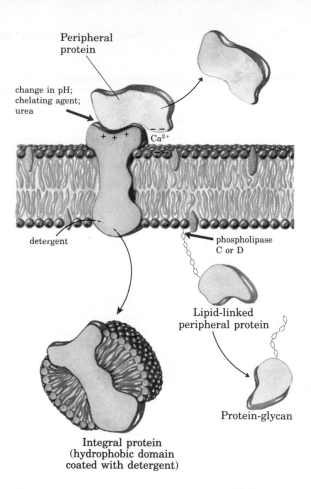

Figure 10–9 Membrane proteins can be distinguished by the conditions required to release them from the membrane. Most peripheral proteins can be released by changes in pH or ionic strength, removal of Ca^{2+} by a chelating agent, or addition of urea, which breaks hydrogen bonds. Peripheral proteins covalently attached to a membrane lipid, such as through a phosphatidylinositol–glycan anchor (see Fig. 10–3), are released by phospholipase C or D. Integral proteins can be extracted with detergents, which disrupt the hydrophobic interactions with the lipid bilayer, forming micelles with individual protein molecules.

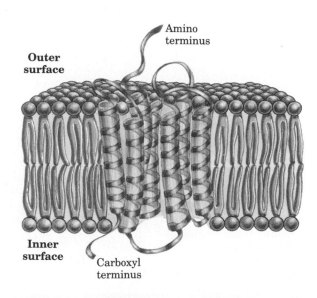

Figure 10–10 The single polypeptide chain of bacteriorhodopsin folds into seven hydrophobic α helices, each of which traverses the lipid bilayer and is roughly perpendicular to the plane of the membrane. The seven transmembrane helices are clustered, and the space around and between them is filled with the acyl chains of membrane lipids.

BOX 10–2   Predicting the Topology of Membrane Proteins

Table 1 Residue hydrophobicity

Amino acid	Free energy of transfer (kJ/mol)
Ile	3.1
Phe	2.5
Val	2.3
Leu	2.2
Trp	1.5
Met	1.1
Ala	1.0
Gly	0.67
Cys	0.17
Tyr	0.08
Pro	−0.29
Thr	−0.75
Ser	−1.1
His	−1.7
Glu	−2.6
Asn	−2.7
Gln	−2.9
Asp	−3.0
Lys	−4.6
Arg	−7.5

Source: From Eisenberg, D., et al. (1982) Hydrophobic moments in protein structure. *Faraday Symp. Chem. Soc.* **17,** 109–120.

It is generally much easier to determine the amino acid sequence of a membrane protein (by sequencing the protein itself or its gene) than to determine its three-dimensional structure. Consequently, only a few three-dimensional structures are known, but hundreds of sequences are available for membrane proteins. The sequences of most integral proteins contain one or more regions rich in hydrophobic residues and long enough to span the 3 nm thick lipid bilayer. An α-helical peptide of 20 residues is just long enough to span the bilayer (the length per residue is 0.15 nm). Because a polypeptide chain surrounded by lipids has no water molecules with which to form hydrogen bonds, it will tend to fold into α helices or β sheets, in which intrachain hydrogen bonding is maximized. If the side chains of all amino acids in a helix are nonpolar, hydrophobic interactions with the surrounding lipids further stabilize the helices.

Several simple methods of analyzing amino acid sequences have been found to yield reasonably accurate predictions of secondary structure for transmembrane proteins. The relative polarity of each of the 20 amino acids has been determined experimentally by measuring the free-energy change of moving a given residue from a hydrophobic solvent into water. This free energy of transfer ranges from very exergonic for charged or polar residues to very endergonic for amino acids with aromatic or aliphatic hydrocarbon side chains (Table 1). To estimate the overall hydrophobicity of a sequence of amino acids, one sums the free energies of transfer for those residues, obtaining a **hydropathy index** for that region. To search a sequence for potential membrane-spanning segments, one calculates the hydropathy index for successive segments of a given size (a "window," which may be from 7 to 20 residues). For a window of 7 residues, the indexes for residues 1 to 7, 2 to 8, 3 to 9, and so on, are plotted as in Figure 1. A region of about 20 residues of high hydropathy index is presumed to be a transmembrane segment. When the sequences of membrane proteins of known three-dimensional structure are scanned in this way, a reasonably good correspondence is found between predicted and known membrane-spanning segments. Hydropathy analysis predicts a single hydrophobic helix for glycophorin (Fig. 1a), five for the M subunit of the photosynthetic reaction center protein (Fig. 1b), seven transmembrane segments for bacteriorhodopsin (Fig. 1c), and twelve segments for the chloride–bicarbonate exchanger (Fig. 1d).

Many of the transport proteins described in this chapter are believed, on the basis of their amino acid sequences and hydropathy plots, to have multiple membrane-spanning helical regions. These assignments of topology should be considered tentative until confirmed by direct structural determination.

Figure 1 Plots of hydropathy index against residue number for four integral membrane proteins.

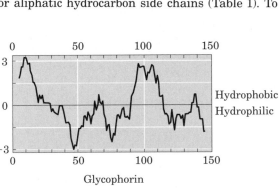

Glycophorin

(a)

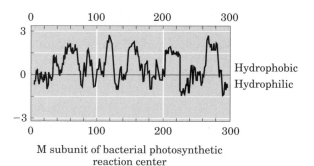

M subunit of bacterial photosynthetic reaction center

(b)

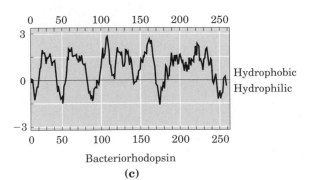

Bacteriorhodopsin

(c)

Chloride–bicarbonate exchange protein

(d)

This pattern of seven hydrophobic membrane-spanning helices has proven to be a common motif in membrane structure, seen in at least ten other membrane proteins, all involved in signal reception. Although no information is yet available on the three-dimensional structures of these proteins, it seems likely that they will prove to be structurally similar to bacteriorhodopsin. The presence of long hydrophobic regions along the amino acid sequence of a membrane protein is generally taken as evidence that such sequences traverse the lipid bilayer, acting as hydrophobic anchors or forming transmembrane channels; virtually all integral membrane proteins have at least one such sequence (Box 10–2). When sequence information yields predictions consistent with chemical studies of protein localization (such as those described above for glycophorin and bacteriorhodopsin), the assumption that hydrophobic regions correspond to membrane-spanning domains is better justified.

The Structure of a Crystalline Integral Membrane Protein Has Been Determined

The same techniques that have allowed determination of the three-dimensional structures of many soluble proteins can in principle be applied to membrane proteins. However, very few membrane proteins have been crystallized; they tend instead to form amorphous aggregates. One instructive exception is the photosynthetic reaction center from a purple bacterium (Fig. 10–11). The protein has four subunits, three of which contain α-helical segments that span the membrane. These segments are rich in nonpolar amino acids, and their hydrophobic side chains are oriented toward the outside of the protein, interacting with the hydrocarbon side chains of membrane lipids. The architecture of the reaction center protein is therefore the inverse of that seen in most water-soluble proteins, which have their hydrophobic residues buried within the protein core and their hydrophilic residues on the surface available for polar interactions with water (recall the structures of myoglobin and hemoglobin, for example). The structures of only a few membrane proteins are known, but the hydrophobic exterior of the reaction center protein seems to be typical of integral membrane proteins.

Figure 10–11 Three-dimensional structure of the photosynthetic reaction center of a purple bacterium, *Rhodopseudomonas viridis*. This was the first integral membrane protein to have its atomic structure determined by x-ray diffraction methods. Eleven α-helical segments from three of the four subunits span the lipid bilayer, forming a cylinder 4.5 nm long, with hydrophobic residues on the exterior, interacting with lipids of the bilayer. In the ribbon representation at left, the residues that are part of the transmembrane helices are shown in purple. The high density of nonpolar residues in the region of the bilayer is illustrated in the space-filling model on the right, in which the four very hydrophobic residues (Phe, Val, Ile, Leu) are shown in purple. In both views, the prosthetic groups (light-absorbing pigments and electron carriers; see Fig. 18–47) are shown in yellow.

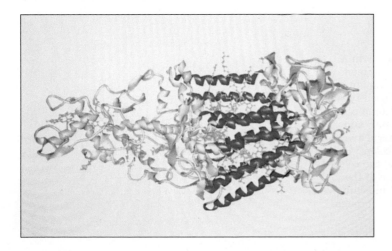

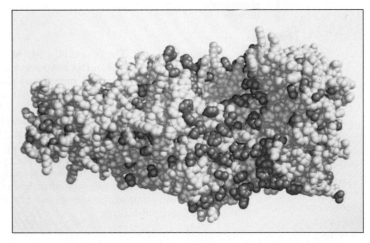

Peripheral Proteins Associate Reversibly with the Membrane

Many peripheral proteins are held to the membrane by electrostatic interactions and hydrogen bonding with the hydrophilic domains of integral membrane proteins, and perhaps with the polar head groups of membrane lipids. They can be released by relatively mild treatments that interfere with electrostatic interactions or break hydrogen bonds (see Fig. 10–9). These peripheral proteins may serve as regulators of membrane-bound enzymes, or as tethers that connect integral membrane proteins to intracellular structures or limit the mobility of certain membrane proteins.

Lipids attached covalently to certain membrane proteins (see Fig. 10–3) anchor these proteins to the lipid bilayer by hydrophobic interactions. Proteins thus held can be released from the membrane by the breakage of a single bond; the action of phospholipase C or D, for example, frees the membrane protein from the hydrophobic portion of a phosphatidylinositol "anchor" (see Fig. 10–9). It seems likely that this type of quick-release mechanism gives cells the capacity to change their membrane surface architecture rapidly, or to alter the subcellular localization of proteins that shuttle between membrane and cytosol.

Although these proteins with lipid anchors resemble integral membrane proteins in that they can be solubilized by detergent treatment, they are generally considered peripheral membrane proteins on the basis of their other properties: their association with the membrane is often weak and reversible, they do not contain long hydrophobic sequences, and once solubilized (by phospholipase action, for example), they behave like typical soluble proteins.

Membrane Proteins Diffuse Laterally in the Bilayer

Many membrane proteins behave as though they were afloat in a sea of lipids. We noted earlier that membrane lipids are free to diffuse laterally in the plane of the bilayer, and are in constant motion. The experiment diagrammed in Figure 10–12 shows that this is also true of some membrane proteins. Other experimental techniques confirm that many but not all membrane proteins undergo rapid lateral diffusion, but there are many exceptions to this generalization. Some membrane proteins associate with adjacent membrane proteins to form large aggregates ("patches") on the surface of a cell or organelle, in which indi-

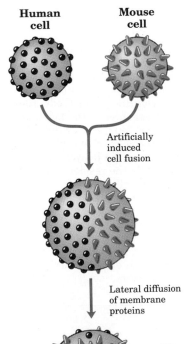

Human cell   **Mouse cell**

Artificially induced cell fusion

Lateral diffusion of membrane proteins

Figure 10–12 The fusion of a mouse cell with a human cell results in the randomization of membrane proteins from the mouse and the human cell. After fusion of the cells, the location of each type of membrane protein is determined by staining cells with species-specific antibodies. Anti-mouse and anti-human antibodies are specifically tagged with molecules that fluoresce with different colors. Observed with the fluorescence microscope, the colored antibodies are seen to mix on the surface of the hybrid cell within minutes after fusion, indicating rapid diffusion of the membrane proteins throughout the lipid bilayer.

Figure 10–13 The chloride–bicarbonate exchange protein of the erythrocyte spans the membrane and is tethered to the cytoskeletal protein spectrin by ankyrin, limiting lateral mobility. Ankyrin contains a covalently bound palmitoyl side chain (Fig. 10–3), which may hold to the membrane. Spectrin is a long, filamentous protein that forms a network attached to the cytoplasmic face of the membrane, thereby stabilizing it against deformation in shape.

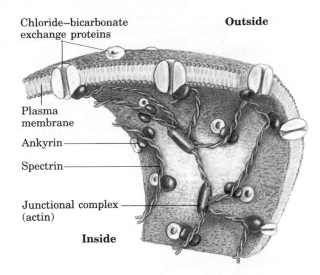

vidual protein molecules do not move relative to one another. Acetylcholine receptors (p. 292) form dense patches at synapses. Other membrane proteins are anchored to internal structures that prevent their free diffusion in the membrane bilayer. In the erythrocyte membrane, both glycophorin and the chloride–bicarbonate exchanger (p. 286) are tethered from the inside to a filamentous cytoskeletal protein, spectrin (Fig. 10–13).

Membrane Fusion Is Central to Many Biological Processes

Although membranes are stable, they are by no means static. The fluid mosaic structure is dynamic and flexible enough to allow fusion of two membranes. Within the endomembrane system described in Chapter 2 (see Fig. 2–10) there is constant reorganization of the membranous compartments, as small vesicles bud from the Golgi complex carrying newly synthesized lipids and proteins to other organelles and to the plasma membrane. Exocytosis, endocytosis, fusion of egg and sperm cells, and cell division all involve membrane reorganization in which the fundamental operation is fusion of two membrane segments without loss of continuity (Fig. 10–14).

To fuse, two membranes must first approach each other within molecular distances (a few nanometers). Much evidence suggests that an increase in intracellular Ca^{2+} concentration is the signal for certain fusion events such as exocytosis. **Annexins** are a family of proteins located just beneath the plasma membrane. They bind avidly to the head groups of phospholipids in bilayers, but only in the presence of Ca^{2+}. Some annexins also associate with specific intracellular vesicles fated for exocytosis. These proteins cause clumping of liposomes in vitro, presumably by cross-linking lipid molecules of two different vesi-

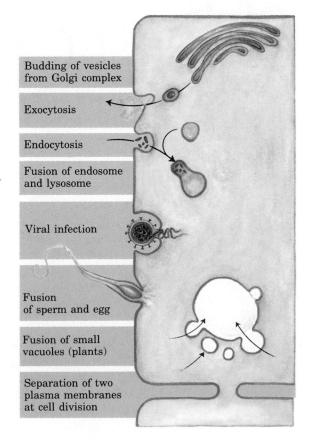

Figure 10–14 Membrane fusion is central to a variety of cellular processes, involving both organelles and the plasma membrane.

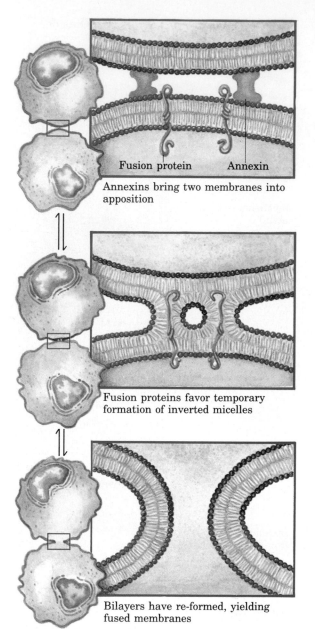

Annexins bring two membranes into apposition

Fusion protein   Annexin

Figure 10–15 One plausible model for membrane fusion. The inverted micelles shown here are only one kind of nonbilayer structure that might be assumed by membrane lipids. When the lipids revert to bilayer structures, either the original structure (top) or the fusion product (bottom) can form. The fusion proteins may act by favoring the formation of the nonbilayer intermediate.

Fusion proteins favor temporary formation of inverted micelles

Bilayers have re-formed, yielding fused membranes

cles. In one simple model of membrane fusion (Fig. 10–15), annexins hold two membranes in close and stable apposition in the first step of fusion.

Another family of proteins believed to act in fusion is the **fusion proteins,** which are typified by the integral membrane protein HA, essential for the entry of the influenza virus into host cells (Fig. 10–14). (Note that these "fusion proteins" are unrelated to the products of two fused genes, also called fusion proteins, discussed in Chapter 28.) Like most other fusion proteins, the HA protein contains two regions rich in nonpolar amino acids, one a typical membrane-spanning domain, the other (the "fusion peptide") rich in Ala and Gly residues. The fusion protein may bridge the two membranes, with its membrane-spanning domain in one membrane and the fusion peptide inserted into the other (Fig. 10–15).

Fusion proteins may bring about transient distortions of the bilayer structure in the region of fusion. Physical studies of pure phospholipids in vitro have shown that several nonbilayer structures, such as inverted micelles, can form under some circumstances. In the model illustrated in Figure 10–15, an alternative structure exists in equilibrium with the bilayer, and represents the transition structure between unfused and fused membranes. Fusion proteins are believed to favor the formation of the transition structure, thus easing the phospholipid reorganization that results in fusion. In addition to annexins and fusion proteins, several, perhaps many, other proteins are probably involved; the machinery for fusion may be much more complex than implied by the simple model described here.

Solute Transport across Membranes

Every living cell must acquire from its surroundings the raw materials for biosynthesis and for energy production, and must release to its environment the byproducts of metabolism. The plasma membrane contains proteins that specifically recognize and carry into the cell such necessities as sugars, amino acids, and inorganic ions. In some cases, these components are brought into the cell against a concentration gradient—"pumped" in. Certain other species are pumped out, to keep their cytosolic concentrations lower than those in the surrounding medium. With few exceptions, the traffic of small molecules across the plasma membrane occurs by protein-mediated processes, via transmembrane channels, carriers, or pumps. Within the eukaryotic cell, different compartments have different concentrations of metabolic intermediates and products, and these, too, must move across intracellular membranes in tightly regulated, protein-mediated processes. Table 10–4 summarizes the properties of membrane transport systems.

Table 10–4 Summary of transport types

Type of transport	Protein carrier?	Saturable with substrate?	Produces concentration gradient?	Energy-dependent?	Energy source (if any)	Examples
Simple diffusion	No	No	No	No	—	H_2O, O_2, N_2, CH_4
Passive transport (facilitated diffusion)	Yes	Yes	No	No	—	Glucose permease of erythrocytes
Active transport						
Primary	Yes	Yes	Yes	Yes	ATP, light, substrate oxidation	H^+ ATPase (plant plasma membrane); Na^+K^+ ATPase (animal plasma membrane)
Secondary	Yes	Yes	Yes	Yes	Ion gradient	Amino acids and sugars (Na^+-driven; intestine); lactose (H^+-driven; bacteria)
Ion channels	Yes	No	No	No*	—	Na^+ channel of acetylcholine receptor (plasma membrane of neuron)

* Although the mechanism of transport via ion channels is not directly energy dependent, the direction of ion flow is determined by the transmembrane differences in electrochemical potential. Ions always move *down* their electrochemical gradient through ion channels.

Passive Transport Is Downhill Diffusion Facilitated by Membrane Proteins

When two aqueous compartments containing unequal concentrations of a soluble compound or ion are separated by a permeable divider, the solute moves by **simple diffusion** from the region of higher concentration, through the divider, to the region of lower concentration, until the two compartments have equal solute concentrations (Fig. 10–16). This behavior of solutes is in accord with the second law of thermodynamics: molecules will tend spontaneously to assume the distribution of greatest randomness, i.e., entropy will increase.

In living organisms, simple diffusion is impeded by selectively permeable barriers—the membranes that separate intracellular compartments and surround cells. To pass through the bilayer, a polar or charged solute must give up its interactions with the water molecules in its hydration shell, then diffuse about 3 nm through a solvent in which it is poorly soluble (the central region of the lipid bilayer), before reaching the other side and regaining its water of hydration (Fig. 10–17). The energy used to strip away the hydration shell and move a polar compound from water into lipid is regained as the compound leaves the membrane on the other side and is rehydrated. However, the intermediate stage of transmembrane passage represents a high-

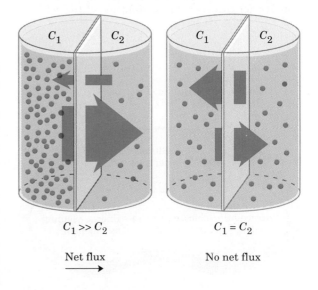

$C_1 \gg C_2$ $C_1 = C_2$

Net flux   No net flux
⟶

Figure 10–16 The rate of net movement of a solute across a permeable membrane depends upon the size of the concentration gradient. C_1 and C_2 are the solute concentrations on the left and right sides of the membrane.

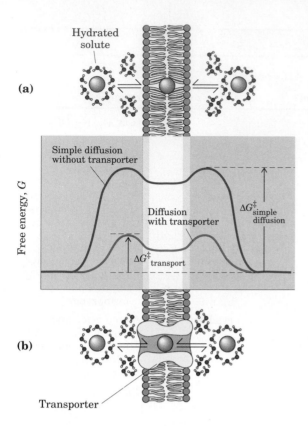

Figure 10–17 Energy changes that occur as a solute in aqueous solution passes through the lipid bilayer of a biological membrane. **(a)** In simple diffusion, the removal of the hydration shell is highly endergonic, and the energy of activation ($\Delta G^{\ddagger}$) for diffusion through the bilayer is very high. **(b)** A transporter protein—by forming noncovalent interactions with the dehydrated solute to replace its hydrogen bonds with water, and by providing a hydrophilic transmembrane passageway—reduces the $\Delta G^{\ddagger}$ for transmembrane diffusion of the solute.

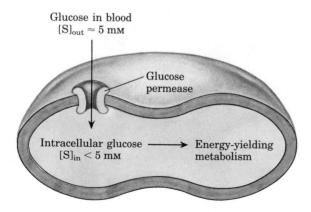

Figure 10–18 The glucose permease of erythrocytes facilitates the passage of glucose into the cell, down its concentration gradient. The 12 transmembrane segments of the permease form a hydrophilic path through the hydrophobic center of the membrane.

energy state comparable to the transition state in an enzyme-catalyzed chemical reaction. In both cases, an activation barrier must be overcome to reach the intermediate stage (Fig. 10–17; compare with Fig. 8–4). The energy of activation for translocation of a polar solute across the bilayer is so large that pure lipid bilayers are virtually impermeable to polar and charged species over the periods of time important to cells.

Water itself is an exception to this generalization. Although polar, it diffuses rapidly across biological membranes by mechanisms not fully understood. When the *solute* concentrations on two sides of a membrane are very different, there is a concentration gradient of *solvent* (water) molecules, and this osmotic imbalance results in the transmembrane flux of water until the osmotic strength equalizes on both sides of the membrane. A few biologically important gases also cross membranes by simple diffusion: molecular oxygen (O_2), nitrogen (N_2), and methane (CH_4), all of which are relatively nonpolar.

Transmembrane passage of polar compounds and ions is made possible by membrane proteins that lower the activation energy for transport by providing an alternative path for specific solutes through the lipid bilayer. Proteins that bring about this **facilitated diffusion** or **passive transport** are not enzymes in the usual sense; their "substrates" are moved from one compartment to another, but are not chemically altered. Membrane proteins that speed the movement of a solute across a membrane by facilitating diffusion are called **transporters** or **permeases**.

The kind of detailed structural information obtained for many soluble enzymes by x-ray crystallography is not yet available for most membrane transporters; as a group, these proteins are both difficult to purify and difficult to crystallize. However, from studies of the specificity and kinetics of transporters it is clear that their action is closely analogous to that of enzymes. Like enzymes, transporters bind their substrates through many weak, noncovalent interactions and with stereochemical specificity. The negative free-energy change that occurs with these weak interactions, $\Delta G_{binding}$, counterbalances the positive free-energy change that accompanies loss of the water of hydration from the substrate, $\Delta G_{dehydration}$, thereby lowering the activation energy, $\Delta G^{\ddagger}$, for transmembrane passage (Fig. 10–17). Transporter proteins span the lipid bilayer at least once, and usually several times, forming a transmembrane channel lined with hydrophilic amino acid side chains. The channel provides an alternative path for its specific substrate to move across the lipid bilayer, without having to dissolve in it, further lowering $\Delta G^{\ddagger}$ for transmembrane diffusion. The result is an increase of orders of magnitude in the rate of transmembrane passage of the substrate.

The Glucose Permease of Erythrocytes Mediates Passive Transport

Energy-yielding metabolism in the erythrocyte depends on a constant supply of glucose from the blood plasma, where its concentration is maintained at about 5 mM. Glucose enters the erythrocyte by facilitated diffusion via a specific glucose permease (Fig. 10–18). This integral membrane protein (M_r 45,000) has 12 hydrophobic segments, and probably spans the membrane 12 times. It allows glucose entry into the cell at a rate about 50,000 times greater than its unaided diffusion through a lipid bilayer. Because glucose transport into erythrocytes is a typical example of passive transport, we will look at it in some detail.

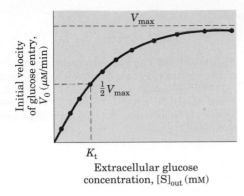

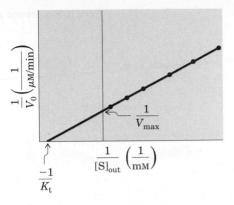

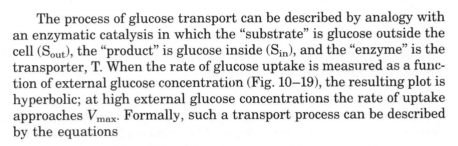

Extracellular glucose
concentration, $[S]_{out}$ (mM)

Figure 10–19 The initial rate of glucose entry into an erythrocyte depends upon the initial concentration of glucose on the outside, $[S]_{out}$. The kinetics of facilitated diffusion are analogous to the kinetics of an enzyme-catalyzed reaction. Compare these plots with Fig. 8–11, and Fig. 1 in Box 8-1. Note that K_t is analogous to K_m, the Michaelis–Menten constant.

The process of glucose transport can be described by analogy with an enzymatic catalysis in which the "substrate" is glucose outside the cell (S_{out}), the "product" is glucose inside (S_{in}), and the "enzyme" is the transporter, T. When the rate of glucose uptake is measured as a function of external glucose concentration (Fig. 10–19), the resulting plot is hyperbolic; at high external glucose concentrations the rate of uptake approaches V_{max}. Formally, such a transport process can be described by the equations

$$S_{out} + T \underset{k_{-1}}{\overset{k_1}{\rightleftharpoons}} S_{out} \cdot T \underset{k_{-2}}{\overset{k_2}{\rightleftharpoons}} S_{in} \cdot T \underset{k_{-3}}{\overset{k_3}{\rightleftharpoons}} S_{in} + T$$

in which k_1, k_{-1}, etc., are the forward and reverse rate constants for each step. The first step is the binding of glucose to a stereospecific site on the transporter protein on the exterior surface of the membrane; step 2 is the transmembrane passage of the substrate; and step 3 is the release of the substrate (product), now on the inner surface of the membrane, from the transporter into the cytoplasm.

The rate equations for this process can be derived exactly as for enzyme-catalyzed reactions (Chapter 8), yielding an expression analogous to the Michaelis–Menten equation:

$$V_0 = \frac{V_{max}[S]_{out}}{K_t + [S]_{out}}$$

in which V_0 is the initial velocity of accumulation of glucose inside the cell when its concentration in the surrounding medium is $[S]_{out}$, and K_t ($K_{transport}$) is a constant, analogous to the Michaelis–Menten constant, a combination of rate constants characteristic of each transport system. This equation describes the *initial* velocity—the rate observed when $[S]_{in} = 0$.

Because no chemical bonds are made or broken in the conversion of S_{out} into S_{in}, neither "substrate" nor "product" is intrinsically more stable, and the process of entry is therefore fully reversible. As $[S]_{in}$ approaches $[S]_{out}$, the rates of entry and exit become equal. Such a system is therefore incapable of accumulating the substrate (glucose) within cells at concentrations above that in the surrounding medium; it simply achieves equilibration of glucose on the two sides of the membrane at a much higher rate than would occur in the absence of a specific transporter. The glucose transporter is specific for D-glucose, for which the measured K_t is 1.5 mM. For the close analogs D-mannose and D-galactose, which differ only in the position of one hydroxyl group, the values of K_t are 20 and 30 mM, respectively, and for L-glucose, K_t exceeds 3,000 mM! (Recall that a high K_t generally reflects a low affinity of transporter for substrate.) The glucose transporter of the erythrocyte therefore shows the three hallmarks of passive transport: high rates of diffusion down a concentration gradient, saturability, and specificity.

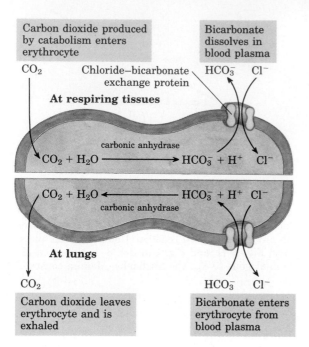

Carbon dioxide produced by catabolism enters erythrocyte

CO_2

Chloride–bicarbonate exchange protein

At respiring tissues

$CO_2 + H_2O \xrightarrow{\text{carbonic anhydrase}} HCO_3^- + H^+ \quad Cl^-$

Bicarbonate dissolves in blood plasma

$HCO_3^- \quad Cl^-$

$CO_2 + H_2O \xleftarrow{\hspace{1cm}} HCO_3^- + H^+ \quad Cl^-$

carbonic anhydrase

At lungs

CO_2

$HCO_3^- \quad Cl^-$

Carbon dioxide leaves erythrocyte and is exhaled

Bicarbonate enters erythrocyte from blood plasma

Figure 10–20 The chloride–bicarbonate exchanger of the erythrocyte membrane allows the entry and exit of HCO_3^- without changes in the transmembrane electrical potential. The role of this shuttle system is to increase the CO_2-carrying capacity of the blood.

Chloride and Bicarbonate Are Cotransported across the Erythrocyte Membrane

The erythrocyte contains another facilitated diffusion system, an anion exchanger, which is essential in CO_2 transport from tissues such as muscle and liver to the lungs. Waste CO_2 released from respiring tissues into the blood plasma enters the erythrocyte, where it is converted into bicarbonate (HCO_3^-) by the enzyme carbonic anhydrase (Fig. 10–20). The HCO_3^- reenters the blood plasma for transport to the lungs. Because HCO_3^- is much more soluble in blood plasma than is CO_2, this roundabout route increases the blood's capacity to carry carbon dioxide from the tissues to the lungs. In the lungs, HCO_3^- reenters the erythrocyte and is converted to CO_2, which is eventually exhaled. For this shuttle to be effective, very rapid movement of HCO_3^- across the erythrocyte membrane is required.

The **chloride–bicarbonate exchanger,** also called the **anion exchange protein,** or (for historical reasons) band 3, increases the permeability of the erythrocyte membrane to HCO_3^- by a factor of more than a million. Like the glucose transporter, it is an integral membrane protein that probably spans the membrane 12 times. Unlike the glucose transporter, this protein mediates a bidirectional exchange; for each HCO_3^- ion that moves in one direction, one Cl^- ion must move in the opposite direction (Fig. 10–20). The result of this paired movement of two monovalent anions is no net change in the charge or electrical potential across the erythrocyte membrane; the process is not electrogenic. The coupling of Cl^- and HCO_3^- movement is obligatory; in the absence of chloride, bicarbonate transport stops. In this respect, the anion exchanger resembles many other systems that simultaneously carry two solutes across a membrane, all of which are called **cotransport systems.** When, as in this case, the two substrates move in opposite directions, the process is **antiport.** In **symport,** two substrates are moved simultaneously in the same direction (Fig. 10–21). Transporters that carry only one substrate, such as the glucose permease, are sometimes called **uniport** systems.

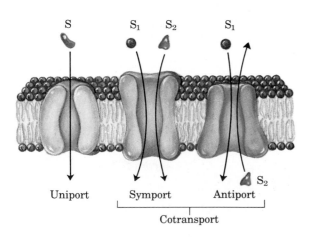

S

$S_1 \quad S_2$

S_1

S_2

Uniport   Symport   Antiport

Cotransport

Figure 10–21 The three general classes of transport systems differ in the number of solutes (substrates) transported and the direction in which each is transported. Examples of all three types of transporters are discussed in the text. Note that this classification tells us nothing about whether these are energy-requiring (active transport) or energy-independent (passive transport) processes.

Active Transport Results in Solute Movement against a Concentration Gradient

In passive transport, the transported species always moves down its concentration gradient, and no net accumulation occurs. **Active transport,** by contrast, results in the accumulation of a solute on one side of a membrane. Active transport is thermodynamically unfavorable (endergonic), and occurs only when coupled (directly or indirectly) to an exergonic process such as the absorption of sunlight, an oxidation reaction, the breakdown of ATP, or the concomitant flow of some other chemical species down its concentration gradient. In primary active transport, solute accumulation is coupled *directly* to an exergonic reaction (e.g., conversion of ATP to ADP + P_i). Secondary active transport occurs when endergonic (uphill) transport of one solute is coupled to the exergonic (downhill) flow of a different solute that was originally pumped uphill by primary active transport.

The amount of energy needed for the transport of a solute against a gradient can easily be calculated from the initial concentration gradient. The general equation for the free-energy change in the chemical process that converts S into P is

$$\Delta G = \Delta G^{\circ\prime} + RT \ln [P]/[S] \qquad (10\text{–}1)$$

where R is the gas constant 8.315 J/mol $\cdot$ K and T is the absolute temperature. When the "reaction" is simply transport of a solute from a region where its concentration is C_1 to another region where its concentration is C_2, no bonds are made or broken and the standard free-energy change, $\Delta G^{\circ\prime}$, equals zero. The free-energy change for transport, ΔG_t, is then

$$\Delta G_t = RT \ln (C_2/C_1) \qquad (10\text{–}2)$$

For a tenfold gradient, the cost of moving 1 mol of an uncharged solute across the membrane separating two compartments at 25 °C is therefore

$$\Delta G_t = (8.315 \text{ J/mol} \cdot \text{K})(298 \text{ K})(\ln 10/1) = 5{,}705 \text{ J/mol}$$

or 5.7 kJ/mol. Equation 10–2 holds for all uncharged solutes. When the solute is an ion, its movement without an accompanying counterion results in the endergonic separation of positive and negative charges. The energetic cost of moving an ion therefore depends on the difference of electrical potential across the membrane as well as the difference in the chemical concentrations (that is, the **electrochemical potential**):

$$\Delta G_t = RT \ln (C_2/C_1) + Z\mathcal{F}\Delta\psi \qquad (10\text{–}3)$$

where Z is the charge on the ion, $\mathcal{F}$ is the Faraday constant (96,480 J/V $\cdot$ mol), and $\Delta\psi$ is the transmembrane electrical potential (in volts). Eukaryotic cells typically have electrical potentials across their plasma membranes of the order of 0.05 to 0.1 V, so the second term of Equation 10–3 can be a significant contribution to the total free-energy change for transporting an ion. Most cells maintain ion gradients larger than tenfold across their plasma or intracellular membranes, and for many cells and tissues, active transport is therefore a major energy-consuming process.

The mechanism of active transport is of fundamental importance in biology. As we shall see in Chapter 18, the formation of ATP in mitochondria and chloroplasts occurs by a mechanism that is essentially ATP-driven ion transport operating in reverse. The energy made available by the spontaneous flow of protons across a membrane is calculable from Equation 10–3; remember that for flow *down* a concentration gradient, the sign of ΔG is opposite to that for transport *against* the gradient.

Figure 10–22 The Na^+K^+ ATPase is primarily responsible for setting and maintaining the intracellular concentrations of Na^+ and K^+ and for generating the transmembrane electrical potential, which it does by moving 3 Na^+ out of the cell for every 2 K^+ it moves in. The electrical potential is central to electrical signaling in neurons, and the gradient of Na^+ is used to drive uphill cotransport of various solutes in a variety of cell types.

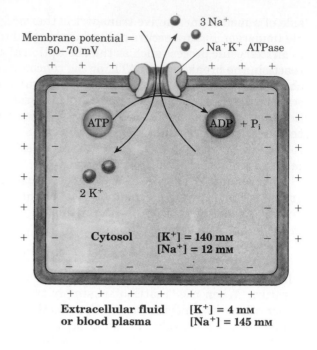

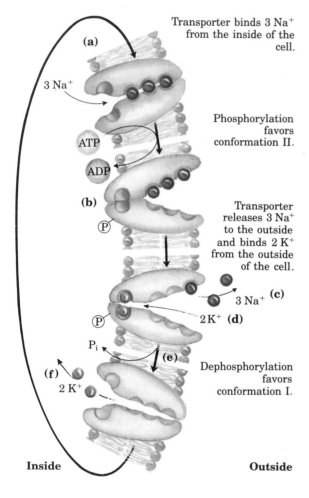

Figure 10–23 Postulated mechanism of Na^+ and K^+ transport by the Na^+K^+ ATPase. The process begins with the binding of three Na^+ to high-affinity sites on the large subunit of the transport protein on the inner surface of the membrane **(a).** This same part of the large subunit also has the ATP-binding site. Phosphorylation of the transporter changes its conformation **(b)** and decreases its affinity for Na^+, leading to Na^+ release on the outer surface **(c).** Next, K^+ on the outside binds to high-affinity sides on the extracellular portion of the large subunit **(d),** the enzyme is dephosphorylated, reducing its affinity for K^+ **(e),** and K^+ is discharged on the inside **(f).** The transport protein is now ready for another cycle of Na^+ and K^+ pumping.

Active Cotransport of Na^+ and K^+ Is Energized by ATP

Virtually every animal cell maintains a lower concentration of Na^+ and a higher concentration of K^+ than is found in its surrounding medium (in vertebrates, extracellular fluid or the blood plasma) (Fig. 10–22). This imbalance is established and maintained by a primary active transport system in the plasma membrane, involving the enzyme **Na^+K^+ ATPase,** which couples breakdown of ATP to the simultaneous movement of both Na^+ and K^+ against their concentration gradients. For each molecule of ATP converted to ADP and P_i, this transporter moves two K^+ ions inward and three Na^+ ions outward, across the plasma membrane. The Na^+K^+ ATPase is an integral membrane protein with two subunits ($M_r \sim 50{,}000$ and $\sim 110{,}000$), both of which span the membrane.

The detailed mechanism by which ATP hydrolysis is coupled to transport remains to be established, but a current working model (Fig. 10–23) supposes that the ATPase cycles between two conformations: conformation II, a phosphorylated form (designated $P–Enz_{II}$) with high affinity for K^+ and low affinity for Na^+, and conformation I, a dephosphorylated form (Enz_I) with high affinity for Na^+ and low affinity for K^+. The conversion of ATP to ADP and P_i occurs in two steps catalyzed by the enzyme:

(1) formation of phosphoenzyme:
$$ATP + Enz_I \longrightarrow ADP + P–Enz_{II}$$

(2) hydrolysis of phosphoenzyme:
$$P–Enz_{II} + H_2O \longrightarrow Enz_I + P_i$$

which sum to the hydrolysis of ATP: $ATP + H_2O \longrightarrow ADP + P_i$.

Because three Na^+ ions move outward for every two K^+ ions that move inward, the process is **electrogenic**—it creates a net separation of charge across the membrane, making the inside of the cell negative relative to the outside. The resulting transmembrane potential of -50

to -70 mV (inside negative relative to outside) is essential to the conduction of action potentials in neurons, and is also characteristic of most nonneuronal animal cells. The activity of this Na^+K^+ ATPase in extruding Na^+ and accumulating K^+ is an essential cell function; about 25% of the energy-yielding metabolism of a human at rest goes to support the Na^+K^+ ATPase.

Ouabain (pronounced 'wä-bān), a steroid derivative extracted from the seeds of an African shrub, is a potent and specific inhibitor of the Na^+K^+ ATPase. Ouabain is a powerful poison used to tip hunting arrows; its name is derived from *waba yo,* meaning "arrow poison."

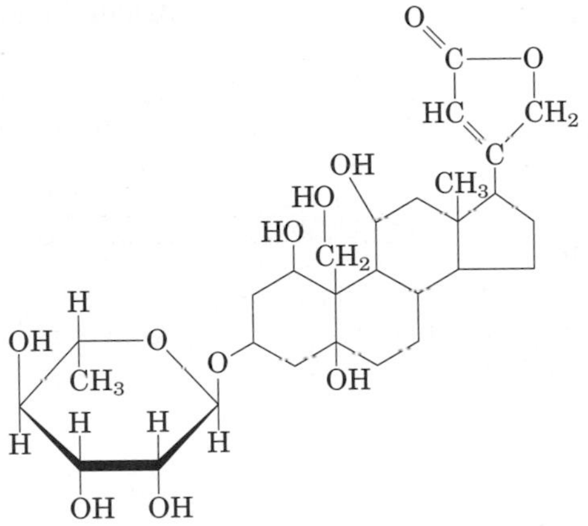

Ouabain

There Are Three General Types of Transport ATPases

The Na^+K^+ ATPase is the prototype for a class of transporters (Table 10–5), all of which are reversibly phosphorylated as part of the transport cycle—thus the name, **P-type ATPase.** All P-type transport ATPases share amino acid sequence homology, especially near the Asp residue that undergoes phosphorylation, and all are sensitive to inhibition by the phosphate analog **vanadate.** Each is an integral membrane protein having multiple membrane-spanning regions. P-type transporters are very widely distributed. In higher plants, a P-type H^+ ATPase pumps protons out of the cell, establishing a difference of as much as 2 pH units and 250 mV across the plasma membrane. For each proton transported, one ATP is consumed. A similar P-type ATPase is responsible for pumping protons from the bread mold *Neurospora,* and for pumping H^+ and K^+ across the plasma membranes of cells that line the mammalian stomach, acidifying its contents (Table 10–5).

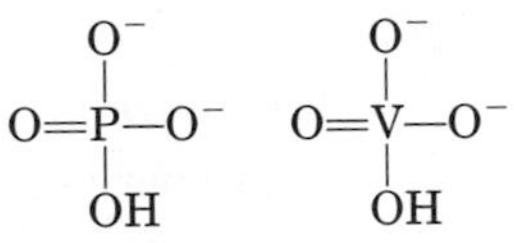

A distinctly different class of transport ATPases is responsible for acidifying intracellular compartments in many organisms. Within the vacuoles of higher plants and of fungi, for example, the pH is maintained well below that of the surrounding cytoplasm by the action of **V-type ATPase–proton pumps.** V-type (for vacuole) ATPases are also responsible for the acidification of lysosomes, endosomes, the Golgi complex, and secretory vesicles in animal cells. Unlike P-type ATPases, these proton-pumping ATPases (Table 10–5) do not undergo cyclic phosphorylation and dephosphorylation, and are not inhibited by vanadate or ouabain. The mechanism by which they couple ATP hydrolysis to the concentrative transport of protons is not yet known.

Phosphate   Vanadate

Table 10–5 Three classes of ion transport ATPases

Transported ion(s)	Organism	Type of membrane	Role of the ATPase
P-type ATPases			
Na^+K^+	Higher eukaryotes	Plasma	Maintains low $[Na^+]$, high $[K^+]$ inside cell; creates transmembrane electrical potential
H^+K^+	Acid-secreting cells of mammals	Plasma	Acidifies contents of stomach
H^+	Fungi *(Neurospora)*	Plasma	Creates low pH in compartment; activating proteases and other hydrolytic enzymes
H^+	Higher plants	Plasma	
Ca^{2+}	Higher eukaryotes	Plasma	Maintains low $[Ca^{2+}]$ in cytosol
Ca^{2+}	Muscle cells of animals	Sarcoplasmic reticulum (endoplasmic reticulum)	Sequesters intracellular Ca^{2+}, keeping cytosolic $[Ca^{2+}]$ low
V-type ATPases			
H^+	Animals	Lysosomal, endosomal, secretory vesicles	Creates low pH in compartment, activating proteases and other hydrolytic enzymes
H^+	Higher plants	Vacuolar	
H^+	Fungi	Vacuolar	
F-type ATPases			
H^+	Eukaryotes	Inner mitochondrial	Catalyzes formation of ATP from $ADP + P_i$
H^+	Higher plants	Thylakoid	
H^+	Prokaryotes	Plasma	

A third family of ATP-splitting proton pumps plays the central role in energy-conserving reactions in bacteria, mitochondria, and chloroplasts. This group of related enzymes, the **F-type ATPases** (Table 10–5), will be discussed when we describe ATP formation in mitochondria and chloroplasts (Chapter 18). (The *F* in their name originated in their identification as energy-coupling *f*actors.) They catalyze the reversible transmembrane passage of protons, driven by ATP hydrolysis. Flow of protons across the membrane *down* their concentration gradient is accompanied by ATP synthesis from ADP and P_i, the reversal of ATP hydrolysis. In this role, the F-type ATPase is more appropriately named **ATP synthase.** In some cases, the formation of the proton gradient in energy-conserving processes is driven by an energy source other than ATP, such as substrate oxidation or sunlight, as we will discuss in Chapter 18.

Ion Gradients Provide the Energy for Secondary Active Transport

The ion gradients formed by primary transport of Na^+ or H^+ driven by light, oxidation, or ATP hydrolysis can themselves provide the driving force for the cotransport of other solutes (Fig. 10–24). Many cells contain transport systems that couple the spontaneous, downhill flow of H^+ or Na^+ to the simultaneous uphill pumping of another ion, sugar, or amino acid (Table 10–6). The galactoside permease of *E. coli* allows the accumulation of the disaccharide lactose to levels 100 times that in the surrounding growth medium (Fig. 10–25). *E. coli* normally has a proton gradient across its plasma membrane, produced by energy-yielding metabolism; protons tend spontaneously to flow back into the cell, down this gradient. The lipid bilayer is impermeable to protons, but the galactoside permease provides a route for proton reentry, and lactose is simultaneously carried into the cell on the symporter protein (permease). The endergonic accumulation is thereby coupled to the exergonic flow of protons; the total free-energy change for the coupled process is negative.

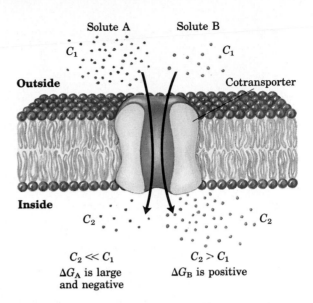

Figure 10–24 In secondary active transport, a single cotransporter couples the flow of one solute (solute 1; e.g., H^+ or Na^+) *down* its concentration gradient to the pumping of a second solute (solute 2; e.g., lactose or glucose) *against* its concentration gradient. When $\Delta G_1 + \Delta G_2 < 0$, cotransport occurs spontaneously. A specific example of this general process is shown in Fig. 10–25.

Table 10–6 Cotransport systems driven by gradients of Na^+ or H^+

Organism or tissue	Transported solute (symport or antiport)	Cotransported solute
E. coli	Lactose (symport)	H^+
	Proline (symport)	H^+
	Dicarboxylic acids (symport)	H^+
Intestine, kidney	Glucose (symport)	Na^+
	Amino acids (symport)	Na^+
Vertebrate cells (many types)	Ca^{2+} (antiport)	Na^+
Higher plants	K^+ (antiport)	H^+
Fungi (*Neurospora*)	K^+ (antiport)	H^+

Figure 10–25 Lactose uptake in *E. coli*. **(a)** The primary transport of H^+ out of the cell, driven by the oxidation of a variety of fuels, establishes a proton gradient. Secondary active transport of lactose into the cell involves symport of H^+ and lactose by the galactoside permease. The uptake of lactose against its concentration gradient is entirely dependent on this inflow of H^+. **(b)** When the energy-yielding oxidation reactions are blocked by cyanide (CN^-), there is an efflux of lactose from the cell, and no further accumulation occurs. The broken line represents the concentration of lactose in the surrounding medium. When active transport is blocked by cyanide, the galactoside permease allows equilibration of lactose inside and outside the cell (passive transport).

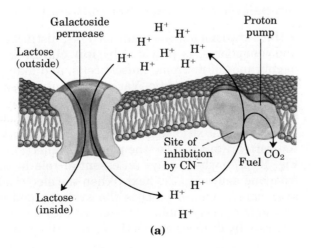

(a)

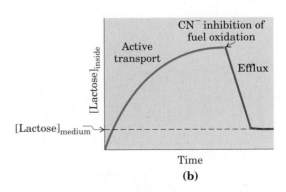

(b)

In intestinal epithelial cells, glucose and certain amino acids are accumulated by symport with Na^+, using the Na^+ gradient established by the Na^+K^+ ATPase. Most cells of vertebrate animals also have an antiport system that simultaneously pumps one Ca^{2+} ion out of a cell and allows three Na^+ ions in, thereby maintaining the very low intracellular Ca^{2+} concentration necessary for normal function. The role of Na^+ in symport and antiport systems such as these requires the continued outward pumping of Na^+ to maintain the transmembrane Na^+ gradient. Clearly, the Na^+K^+ ATPase is a central element in many cotransport processes.

Because of the essential role of ion gradients in active transport and energy conservation, natural products and drugs that collapse the ion gradients across cellular membranes are poisons, and may serve as antibiotics. Valinomycin is a small, cyclic peptide that folds around K^+ and neutralizes its positive charge (Fig. 10–26). The peptide then acts as a shuttle, carrying K^+ across membranes down its concentration gradient and deflating that gradient. Compounds that shuttle ions across membranes in this way are called **ionophores,** literally "ion-bearers." Both valinomycin and monensin (a Na^+-carrying ionophore) are antibiotics; they kill microbial cells by disrupting secondary transport processes and energy-conserving reactions.

Figure 10–26 Valinomycin, a peptide ionophore that binds K^+. The oxygen (red) atoms that bind K^+ (the green atom at the center) are part of a central hydrophilic cavity. Hydrophobic amino acid side chains (yellow) coat the outside of the molecule. Because the exterior of the K^+-valinomycin complex is hydrophobic, it readily diffuses through membranes, carrying K^+ down its concentration gradient. The resulting dissipation of the transmembrane ion gradient kills cells, making valinomycin a potent antibiotic.

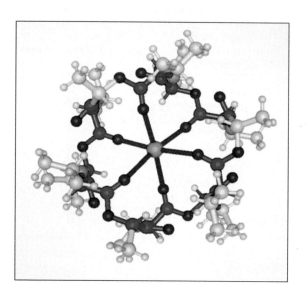

Ion-Selective Channels Act in Signal Transductions

A third transmembrane path for ions, distinct from both transporters and ionophores, is provided by **ion channels,** found in the plasma membranes of neurons, muscle cells, and many other cells, both prokaryotic and eukaryotic. Various stimuli cause rapid changes in the electrical potential across the plasma membranes of neurons and muscle cells, the result of the rapid opening and closing of ion channels. One of the best-studied ion channels is the **acetylcholine receptor** of the vertebrate synapse (the point of connection between two neurons; Fig. 10–27), which plays an essential role in the passage of a signal from one neuron to the next. When the electrical signal carried by the presynaptic neuron reaches the synaptic end of the cell, the neurotransmitter acetylcholine is released into the synaptic cleft. Acetylcholine rapidly diffuses across the cleft to the postsynaptic neuron, where

$$CH_3-C\underset{O-CH_2-CH_2-\overset{\pm}{N}-CH_3}{\overset{O}{\parallel}}$$

Acetylcholine

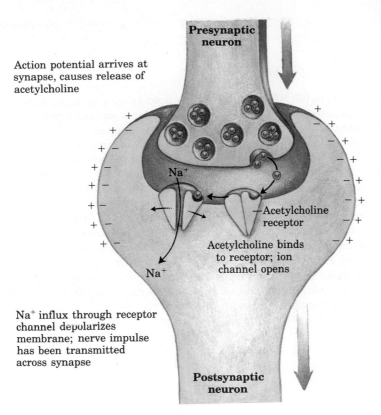

Action potential arrives at synapse, causes release of acetylcholine

Na$^+$

Na$^+$

Acetylcholine receptor

Acetylcholine binds to receptor; ion channel opens

Na$^+$ influx through receptor channel depolarizes membrane; nerve impulse has been transmitted across synapse

Presynaptic neuron

Postsynaptic neuron

Figure 10–27 Communication occurs across the synaptic cleft between adjoining neurons as acetylcholine released by the presynaptic cell diffuses to specific receptors on the postsynaptic cell. The binding of acetylcholine changes the conformation of the receptor, and results in the opening of a receptor-associated ion channel. Na$^+$ flows in, down its concentration gradient, carrying positive charge and reducing the membrane electrical potential (depolarizing the cell). Depolarization initiates an electrical signal (action potential) that sweeps through the postsynaptic neuron at very high speed and is conducted to the next synapse, where these events may be repeated.

it binds to high-affinity sites on the acetylcholine receptor. This binding induces a change in receptor structure, opening a transmembrane channel in the receptor protein. Cations in the extracellular fluid, present at higher concentration than in the cytosol of the postsynaptic neuron, flow through the opened channel into the cell, down their concentration gradient, thereby depolarizing (decreasing the transmembrane electrochemical gradient of) the postsynaptic cell. The acetylcholine receptor allows Na$^+$ and K$^+$ to pass with equal ease, but other cations and all anions are unable to pass through it.

The rate of Na$^+$ movement through the acetylcholine receptor ion channel is linear with respect to extracellular Na$^+$ concentration; the process is not saturable in the way that transporter-catalyzed translocation is saturable with substrate (see Fig. 10–19). This kinetic property distinguishes ion channels from passive transporters. Na$^+$ movement through the channel is very fast—almost at the rate expected for unhindered diffusion of the ion. Under physiological conditions of ion concentrations and membrane potential, about 2×10^7 Na$^+$ ions can pass through a single channel in 1 s. A comparison of this figure with the turnover numbers for typical enzymes, which are in the range of 10 to 10^5 s^{-1}, shows the efficiency of transmembrane diffusion through an ion channel. In short, the ion channel of the acetylcholine receptor behaves as though it provided a hydrophilic pore through the lipid bilayer through which an ion of the right size, charge, and geometry can diffuse very rapidly down its electrochemical gradient. This receptor/channel is typical of many ion channels in cells that produce or respond to electrical signals: it has a "gate" that opens in response to stimulation by acetylcholine, and an intrinsic timing mechanism that closes the gate after a split second. Thus the acetylcholine signal is transient—an essential feature of electrical signal conduction.

Summary

Biological membranes are central to life. They define cellular boundaries, divide cells into discrete compartments, organize complex reaction sequences, and act in signal reception and energy transformations. Membranes are composed of lipids and proteins in varying combinations that are specific to each species, cell type, and organelle. The fluid mosaic model describes certain features common to all biological membranes. The lipid bilayer is the basic structural unit. Fatty acyl chains of phospholipids and the steroid nucleus of sterols are oriented toward the interior of the bilayer; their hydrophobic interactions stabilize the bilayer but allow the structure to be flexible. Lipids and most proteins are free to diffuse laterally within the membrane, and the hydrophobic moieties of the lipids undergo rapid thermal motion, making the interior of the bilayer fluid. Fluidity is affected by temperature, fatty acid composition, and sterol content. Cells strive to maintain a constant fluidity when external circumstances change.

Peripheral membrane proteins are loosely associated with the membrane through electrostatic interactions and hydrogen bonds or by covalently attached lipid anchors. Integral membrane proteins associate with the lipid bilayer by hydrophobic interactions with their nonpolar amino acid side chains, which are oriented toward the outside of the protein molecule. Some membrane proteins span the lipid bilayer several times, with hydrophobic sequences of about 20 amino acids, each capable of forming a transmembrane α helix. Such hydrophobic sequences can be detected and used to predict the structure and transmembrane disposition of these proteins. The lipids and proteins of the membrane are inserted into the bilayer with specific sidedness; the membrane is structurally and functionally asymmetric. Many membrane proteins contain covalently attached polysaccharides of various degrees of complexity. Plasma membrane glycoproteins are always oriented with the carbohydrate-bearing domain on the extracellular surface. Annexins and fusion proteins mediate the fusion of two membranes, which accompanies processes such as endocytosis and exocytosis.

The lipid bilayer is impermeable to polar substances. Water is an important exception; it is able to diffuse passively across the bilayer. Other polar species cross biological membranes only by way of specific membrane proteins. Ion channels provide hydrophilic pores through which select ions can diffuse, moving down their electrical or chemical concentration gradients.

The movement of many ions and compounds across cellular membranes is catalyzed by specific transport proteins (transporters), which, like enzymes, show saturation and substrate specificity. Transport via these systems may be passive (down the electrochemical gradient, hence independent of metabolic energy) such as glucose transport into erythrocytes, or active (against the gradient, and dependent on metabolic energy). The energy input for active transport may come from light, oxidation reactions, ATP hydrolysis, or cotransport of some other solute. Some transporters carry out symport, the simultaneous passage of two species in the same direction; others mediate antiport, in which two species move in opposite directions, but simultaneously. An example of antiport is the chloride–bicarbonate exchanger of erythrocytes. In animal cells, the differences in cytosolic and extracellular concentrations of Na^+ and K^+ are established and maintained by active transport via the Na^+K^+ ATPase, and the resulting Na^+ gradient is used as an energy source by a variety of symport and antiport systems.

There are three general types of ion-pumping ATPases. P-type ATPases undergo reversible phosphorylation during their catalytic cycle, and are inhibited by the phosphate analog vanadate. V-type ATPases produce gradients of protons across the membranes of a variety of intracellular organelles, including plant vacuoles. F-type proton pumps (ATP synthases) are central to energy-conserving mechanisms in mitochondria and chloroplasts. Ion-selective channels such as the acetylcholine receptor act in the passage of electrical signals in neurons, muscle cells, and other cells sensitive to a variety of stimuli.

Further Reading

General

Finean, J.B., Coleman, R., & Mitchell, R.H. (1984) *Membranes and Their Cellular Functions,* 3rd edn, Blackwell Scientific Publications, Oxford.
An introductory text on membrane composition, structure, and function.

Harold, F.M. (1986) *The Vital Force: A Study of Bioenergetics*, W.H. Freeman and Company, New York.

Chapters 9 and 10 of this excellent book concern the energetics and mechanisms of transport, and Chapter 4 is a discussion of the energy of ion gradients.

Jain, M.K. (1988) *Introduction to Biological Membranes*, 2nd edn, John Wiley & Sons, Inc., New York.

A textbook of membranology, longer and more advanced than Finean et al.

Martonosi, A.N. (ed) (1985) *The Enzymes of Biological Membranes*, 2nd edn, Plenum Press, New York.

This four-volume set has 61 individual reviews covering many of the topics in this chapter, including the electron microscopy of membranes, protein–lipid interactions in membranes, the energetics of active transport, and the Na^+K^+ ATPase.

Stein, W.D. (1986) *Transport and Diffusion Across Cell Membranes*, Academic Press, Inc., New York.

This excellent textbook on biological transport covers all the transport systems described in this chapter, at a more advanced level.

Tanford, C. (1980) *The Hydrophobic Effect: Formation of Micelles and Biological Membranes*, 2nd edn, John Wiley & Sons, Inc., New York.

A description of the forces that stabilize micelles, bilayers, and membranes, in rigorous physical–chemical terms.

Molecular Constituents of Membranes

Bretscher, M.S. (1985) The molecules of the cell membrane. *Sci. Am.* **253** (October), 100–109.

Supramolecular Architecture of Membranes

Fasman, G.D. & Gilbert, W.A. (1990) The prediction of transmembrane protein sequences and their conformation: an evaluation. *Trends Biochem. Sci.* **15**, 89–92.

Short but clear survey of several methods of predicting transmembrane helices from sequence (hydropathy plots).

McIlhinney, R.A.J. (1990) The fats of life: the importance and function of protein acylation. *Trends Biochem. Sci.* **15**, 387–391.

A short but good summary of the kinds of proteins that contain covalently bound lipid, and the functional significance of lipid attachment.

Rothman, J.E. & Lenard, J. (1977) Membrane asymmetry. *Science* **195**, 743–753.

Excellent summary of the experimental evidence that both lipids and proteins of membranes are asymmetrically disposed in the bilayer.

Singer, S.J. & Nicolson, G.L. (1972) The fluid mosaic model of the structure of cell membranes. *Science* **175**, 720–731.

The classic statement of the model.

Unwin, N. & Henderson, R. (1984) The structure of proteins in biological membranes. *Sci. Am.* **250** (February), 78–94.

Describes the technical problems and solutions in determining the structure of integral membrane proteins, with emphasis on bacteriorhodopsin.

Solute Transport across Membranes

Forgac, M. (1989) Structure and function of vacuolar class of ATP-driven proton pumps. *Physiol. Rev.* **69**, 765–796.

Hille, B. (1988) Ionic channels: molecular pores of excitable membranes. *Harvey Lect.* **82**, 47–69.

Jennings, M.L. (1989) Structure and function of the red blood cell anion transport protein. *Annu. Rev. Biophys. Biophys. Chem.* **18**, 397–430.

Detailed, advanced treatment of this transporter.

Kaback, H.R. (1989) Molecular biology of active transport: from membrane to molecule to mechanism. *Harvey Lect.* **83**, 77–105.

Description of the galactoside permease of E. coli.

Lienhard, F.E. Slot, J.W. James, D.E. & Mueckler, M.M. (1992) How cells absorb glucose. *Sci. Am.* **266** (January), 86–91.

Introductory level description of the glucose transporter and its regulation by insulin.

Lodish, H.F. (1988) Anion-exchange and glucose transport proteins: structure, function and distribution. *Harvey Lect.* **82**, 19–46.

Numa, S. (1989) A molecular view of neurotransmitter receptors and ionic channels. *Harvey Lect.* **83**, 121–165.

Pedersen, P.L. & Carafoli, E. (1987) Ion motive ATPases. I. Ubiquity, properties, and significance to cell function. *Trends Biochem. Sci.* **12**, 146–150. II. Energy coupling and work output. *Trends Biochem. Sci.* **12**, 186–189.

White, J.M. (1990) Viral and cellular membrane fusion proteins. *Annu. Rev. Physiol.* **52**, 675–697.

Problems

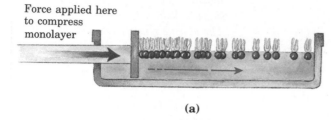

Force applied here
to compress
monolayer

(a)

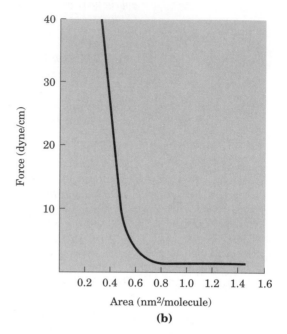

(b)

1. *Determining the Cross-Sectional Area of a Lipid Molecule* When phospholipids are layered gently onto the surface of water, they orient at the air–water interface with their head groups in the water and their hydrophobic tails in the air. The experimental apparatus pictured above **(a)** pushes these lipids together by reducing the surface area available to them. By measuring the force necessary to push the lipids together, it is possible to determine when the molecules are packed tightly together in a continuous monolayer; when that area is approached, the pressure needed to further reduce the surface area increases sharply **(b)**. How would you use such an experimental apparatus to determine the average area occupied by a single lipid molecule in a lipid monolayer?

2. *Evidence for Lipid Bilayer* In 1925, E. Gorter and F. Grendel used an apparatus like that described in Problem 1 to determine the surface area of a lipid monolayer formed by lipids extracted from erythrocytes of several animal species. They used a microscope to measure the dimensions of individual cells, from which they calculated the average surface area of one erythrocyte. They obtained the data shown below. Were these investigators justified in concluding that "chromocytes [erythrocytes] are covered by a layer of fatty substances that is two molecules thick" (i.e., a lipid bilayer)?

Animal	Volume of packed cells (mL)	Number of cells (per mm³)	Total surface area of lipid monolayer from cells (m²)	Total surface area of one erythrocyte (μm²)
Dog	40	8,000,000	62	98
Sheep	10	9,900,000	2.95	29.8
Human	1	4,740,000	0.47	99.4

Source: Data from Gorter, E. & Grendel, F. (1925) On bimolecular layers of lipoids on the chromocytes of the blood. *J. Exp. Med.* **41,** 439–443.

3. *Length of a Fatty Acid Molecule* The carbon–carbon bond distance for single-bonded carbons such as those in a saturated fatty acyl chain is about 0.15 nm. Estimate the length of a single molecule of palmitic acid in its fully extended form. If two molecules of palmitic acid were laid end to end, how would their total length compare with the thickness of the lipid bilayer in a biological membrane?

4. *Temperature Dependence of Lateral Diffusion* The experiment described in Figure 10–12 was done at 37 °C. If, instead, the whole experiment were carried out at 10 °C, what effect would you predict on the rate of cell–cell fusion, and the rate of membrane protein mixing? Why?

5. *Synthesis of Gastric Juice: Energetics* Gastric juice (pH 1.5) is produced by pumping HCl from blood plasma (pH 7.4) into the stomach. Calculate the amount of free energy required to concentrate the H^+ in 1 L of gastric juice at 37 °C. Under cellular conditions, how many moles of ATP must be hydrolyzed to provide this amount of free energy? (The free-energy change for ATP hydrolysis under cellular conditions is about −58 kJ/mol, as we will explain in Chapter 13.)

6. *Energetics of the Na^+K^+ ATPase* The concentration of Na^+ inside a vertebrate cell is about 12 mM, and the cell is bathed in blood plasma containing about 145 mM Na^+. For a typical cell with a transmembrane potential of −0.07 V (inside negative relative to outside), what is the free-energy change for transporting 1 mol of Na^+ out of the cell at 37 °C?

7. *Action of Ouabain on Kidney Tissue* Ouabain specifically inhibits the Na^+K^+ ATPase activity of animal tissues but is not known to inhibit any other enzyme. When ouabain is added in graded concentrations to thin slices of living kidney tissue, it inhibits oxygen consumption by 66%. Explain the basis of this observation. What does it tell us about the use of respiratory energy by kidney tissue?

8. *Membrane Protein Topology* The receptor for the hormone epinephrine in animal cells is an integral membrane protein (M_r 64,000) that is believed to span the membrane seven times. Show that a protein of this size is capable of spanning the membrane seven times. If you were given the amino acid sequence of this protein, how would you go about predicting which regions of the protein form the membrane-spanning helices?

9. *Energetics of Symport* Suppose that you determined experimentally that a cellular transport system for glucose, driven by symport of Na^+, could accumulated glucose to concentrations 25 times greater than in the external medium, while the external $[Na^+]$ was only ten times greater than the intracellular $[Na^+]$. Is this a violation of the laws of thermodynamics? If not, how do you explain this observation?

10. *Location of a Membrane Protein* An unknown membrane protein, X, can be extracted from disrupted erythrocyte membranes into a concentrated salt solution. Isolated X can be cleaved into fragments by proteolytic enzymes. But treatment of erythrocytes, first with proteolytic enzymes, followed by disruption and extraction of membrane components, yields intact X. In contrast, treatment of erythrocyte "ghosts" (which consist of only membranes, produced by disrupting the cells and wash-

ing out the hemoglobin) with proteolytic enzymes, followed by disruption and extraction, yields extensively fragmented X. What do these experiments indicate about the location of X in the plasma membrane? On the basis of this information, do the properties of X resemble those of glycophorin or those of ankyrin?

11. *Membrane Self-Sealing* Cell membranes are self-sealing—if they are punctured or disrupted mechanically, they quickly and automatically reseal. What properties of membranes are responsible for this important feature?

12. *Lipid Melting Temperatures* Membrane lipids in tissue samples obtained from different parts of the leg of a reindeer show different fatty acid compositions. Membrane lipids from tissue near the hooves contain a larger proportion of unsaturated fatty acids than lipids from tissue in the upper part of the leg. What is the significance of this observation?

13. *Flip-Flop Diffusion* The inner face of the human erythrocyte membrane consists predominantly of phosphatidylethanolamine and phosphatidylserine. The outer face consists predominantly of phosphatidylcholine and sphingomyelin. Although the phospholipid components of the membrane can diffuse in the fluid bilayer, this sidedness is preserved all all times. How?

14. *Membrane Permeability* At pH 7, tryptophan crosses a lipid bilayer membrane about 1,000 times more slowly than does the closely related substance indole:

Suggest an explanation for this observation.

CHAPTER

Carbohydrates

Carbohydrates are the most abundant biomolecules on earth. Each year, photosynthesis by plants and algae converts more than 100 billion metric tons of CO_2 and H_2O into cellulose and other plant products. Certain carbohydrates (sugar and starch) are a staple of the human diet in most parts of the world, and the oxidation of carbohydrates is the central energy-yielding pathway in most nonphotosynthetic cells. Insoluble carbohydrate polymers serve as structural and protective elements in the cell walls of bacteria and plants and in the connective tissues and cell coats of animals. Other carbohydrate polymers lubricate skeletal joints and provide adhesion between cells. Complex carbohydrate polymers, covalently attached to proteins or lipids, act as signals that determine the intracellular location or the metabolic fate of these glycoconjugates. This chapter introduces the major classes of carbohydrates and glycoconjugates, and provides a few examples of their many structural and functional roles.

Carbohydrates are polyhydroxy aldehydes or ketones, or substances that yield such compounds on hydrolysis. Most substances of this class have empirical formulas suggesting that they are carbon "hydrates," in which the ratio of C:H:O is 1:2:1. For example, the empirical formula of glucose is $C_6H_{12}O_6$, which can also be written $(CH_2O)_6$ or $C_6(H_2O)_6$. Although many common carbohydrates conform to the empirical formula $(CH_2O)_n$, others do not; some carbohydrates also contain nitrogen, phosphorus, or sulfur.

There are three major size classes of carbohydrates: monosaccharides, oligosaccharides, and polysaccharides (the word "saccharide" is derived from the Greek *sakkharon,* meaning "sugar"). **Monosaccharides,** or simple sugars, consist of a single polyhydroxy aldehyde or ketone unit. The most abundant monosaccharide in nature is the six-carbon sugar D-glucose.

Oligosaccharides consist of short chains of monosaccharide units joined together by characteristic glycosidic linkages. The most abundant are the **disaccharides,** with two monosaccharide units. Typical is sucrose, or cane sugar, which consists of the six-carbon sugars D-glucose and D-fructose joined covalently. All common monosaccharides and disaccharides have names ending with the suffix "-ose." Most oligosaccharides having three or more units do not occur as free entities but are joined to nonsugar molecules (lipids or proteins) in hybrid structures (glycoconjugates).

Polysaccharides consist of long chains having hundreds or thousands of monosaccharide units. Some polysaccharides, such as cellulose, occur in linear chains, whereas others, such as glycogen, have branched chains. The most abundant polysaccharides, starch and cellulose made by plants, consist of recurring units of D-glucose, but they differ in the type of glycosidic linkage.

Monosaccharides and Disaccharides

The simplest of the carbohydrates are the monosaccharides, the subunits from which disaccharides, oligosaccharides, and polysaccharides are constructed. Monosaccharides are either aldehydes or ketones, with one or more hydroxyl groups; the six-carbon monosaccharides glucose and fructose have five hydroxyl groups. The carbon atoms to which hydroxyl groups are attached are often chiral centers, and stereoisomerism is common among monosaccharides.

We begin by describing the families of monosaccharides with backbones of three to seven carbons—their structure and stereoisomeric forms, and the means of representing their three-dimensional structures on paper. We then discuss several special chemical reactions of the carbonyl groups of monosaccharides. One such reaction, the addition of a hydroxyl group from the same molecule, produces cyclic forms of five- and six-carbon sugars and creates a new chiral center, adding further stereochemical complexity to this class of compounds. The nomenclature for unambiguously specifying the configuration about each carbon atom in a cyclic form and the means of representing these structures on paper are therefore described in some detail; these will be useful later in our discussion of the metabolism of monosaccharides.

We will encounter a variety of derivatives of the simple monosaccharides in later chapters, and some of the most common of these are introduced here.

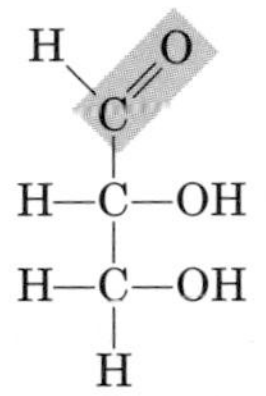

There Are Two Families of Monosaccharides

Monosaccharides are colorless, crystalline solids that are freely soluble in water but insoluble in nonpolar solvents. Most have a sweet taste. The backbone of monosaccharides is an unbranched carbon chain in which all the carbon atoms are linked by single bonds. One of the carbon atoms is double-bonded to an oxygen atom to form a carbonyl group; each of the other carbon atoms has a hydroxyl group. If the carbonyl group is at an end of the carbon chain, the monosaccharide is an aldehyde and is called an **aldose;** if the carbonyl group is at any other position, the monosaccharide is a ketone and is called a **ketose.** The simplest monosaccharides are the two three-carbon trioses: glyceraldehyde, an aldose, and dihydroxyacetone, a ketose.

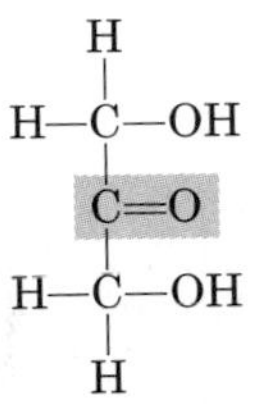

Monosaccharides with four, five, six, and seven carbon atoms in their backbones are called, respectively, tetroses, pentoses, hexoses, and heptoses. There are aldoses and ketoses of each of these chain lengths: aldotetroses and ketotetroses, aldopentoses and ketopentoses, and so on. The hexoses, which include the aldohexose D-glucose and the ketohexose D-fructose (Fig. 11–1), are the most common monosaccharides in nature. The aldopentoses D-ribose and 2-deoxy-D-ribose (Fig. 11–1) are components of nucleotides and nucleic acids (Chapter 12).

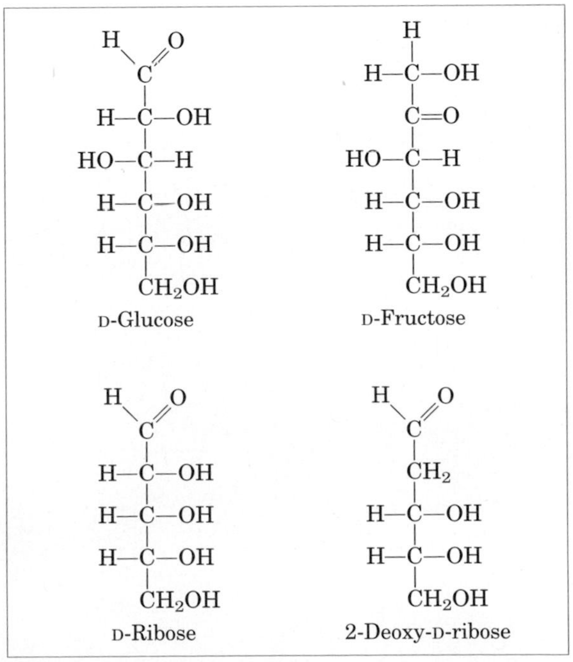

Glyceraldehyde,
an aldose

Dihydroxyacetone,
a ketose

D-Glucose

D-Fructose

D-Ribose

2-Deoxy-D-ribose

Figure 11–1 Two common hexoses, and the pentose components of nucleic acids. D-Ribose is a component of ribonucleic acid (RNA), and 2-deoxy-D-ribose is a component of deoxyribonucleic acid (DNA).

Monosaccharides Have Asymmetric Centers

All the monosaccharides except dihydroxyacetone contain one or more asymmetric (chiral) carbon atoms and thus occur in optically active isomeric forms (Chapter 3). The simplest aldose, glyceraldehyde, contains a chiral center (the middle carbon atom) and therefore has two different optical isomers, or enantiomers (Fig. 11–2; see also Fig. 3–9). By convention, one of these two forms is designated the D isomer of glyceraldehyde; the other is the L isomer. To represent three-dimensional sugar structures on paper, we often use **Fischer projection formulas** (Fig. 11–2).

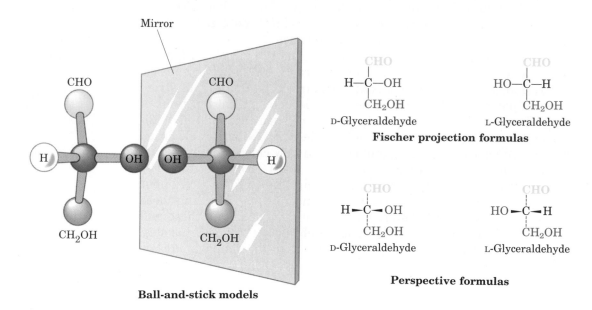

Fischer projection formulas

Perspective formulas

Ball-and-stick models

Figure 11–2 Three ways to represent the two stereoisomers of glyceraldehyde, which are mirror images of each other. Ball-and-stick models show the actual configuration of molecules. The conventions for projection and perspective formulas were explained in Fig. 5–3.

In general, a molecule with n chiral centers can have 2^n stereoisomers. Glyceraldehyde has $2^1 = 2$; the aldohexoses, with four chiral centers, have $2^4 = 16$ stereoisomers. The stereoisomers of monosaccharides of each carbon chain length can be divided into two groups, which differ in the configuration about the chiral center most distant from the carbonyl carbon; those with the same configuration at this reference carbon as that of D-glyceraldehyde are designated D isomers, and those with the configuration of L-glyceraldehyde are L isomers. When the hydroxyl group on the reference carbon is on the right in the projection formula, the sugar is the D isomer; when on the left, the L isomer. Of the 16 possible aldohexoses, 8 are D forms and 8 are L. Most of the hexoses found in living organisms are D isomers.

Figure 11–3a shows the structures of all the stereoisomers of aldotrioses, aldotetroses, aldopentoses, and aldohexoses of the D series; Figure 11–3b shows the D-ketoses. The carbon atoms of a sugar are numbered beginning at the end of the chain nearest the carbonyl group. Each of the eight D-aldohexoses, which differ in the stereochemistry at C-2, C-3, and C-4 (Fig. 11–3a), has its own name: D-glucose, D-galactose, D-mannose, etc. Ketoses are commonly designated by inserting *ul* into the name of the corresponding aldose; for example, D-ribulose is the ketopentose corresponding to the aldopentose D-ribose. However, a few ketoses are named otherwise, such as fructose (from Latin *fructus,* meaning "fruit"; fruits are a good source of this sugar).

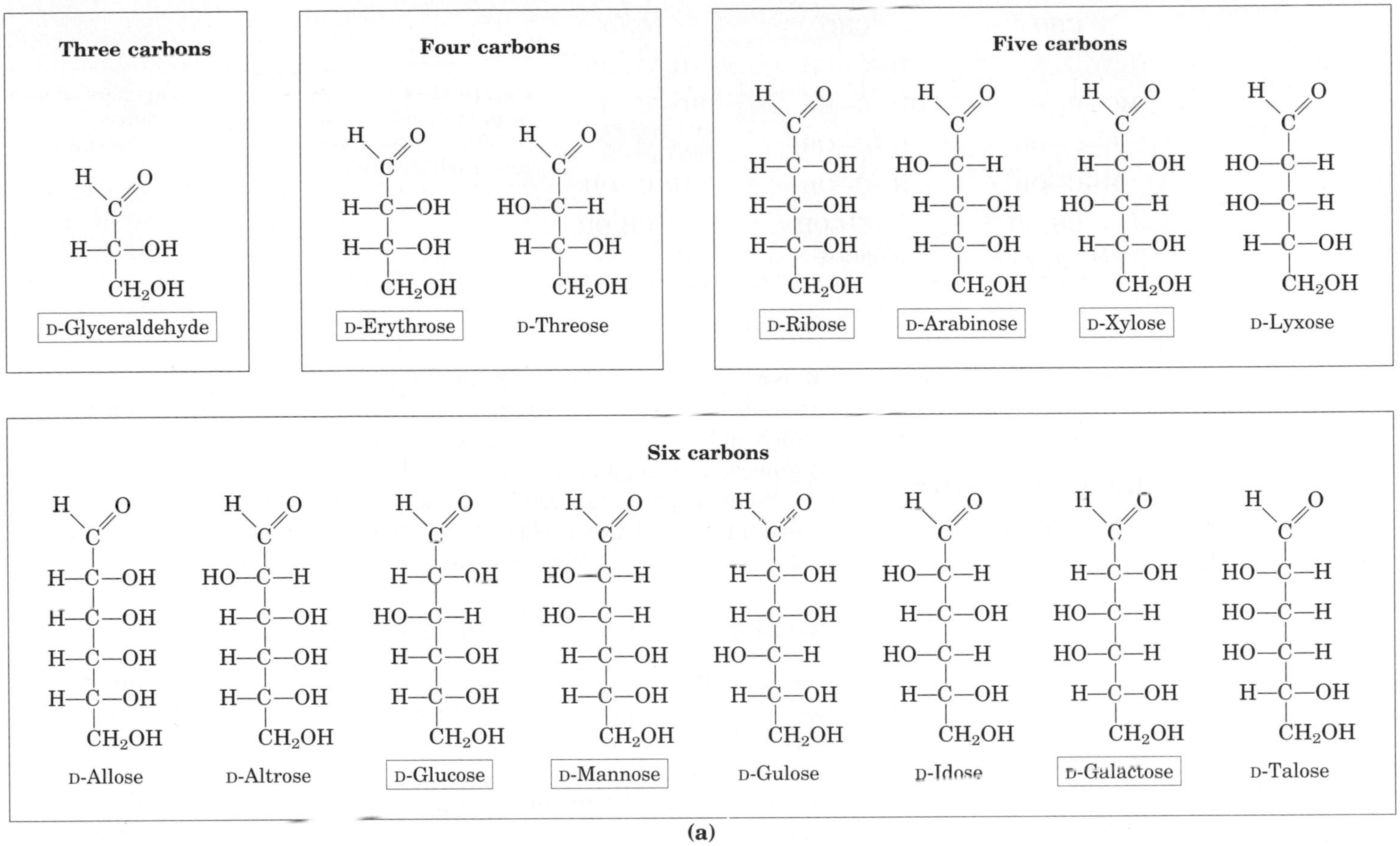

(a)

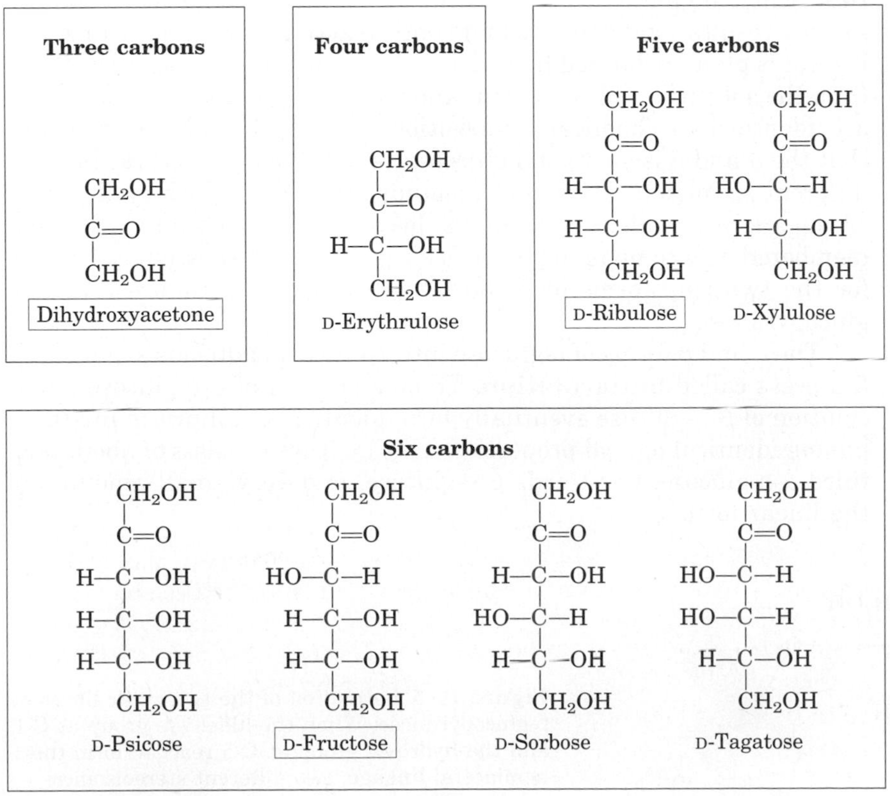

Figure 11–3 The family of **(a)** D-aldoses and **(b)** D-ketoses having from three to six carbon atoms, shown as projection formulas. The carbon atoms in red are chiral centers. In all of these D isomers, the chiral carbon *most distant from the carbonyl carbon* has the same configuration as the chiral carbon in D-glyceraldehyde. The sugars named in boxes are the most abundant in nature; we shall encounter these again in this and later chapters.

(b)

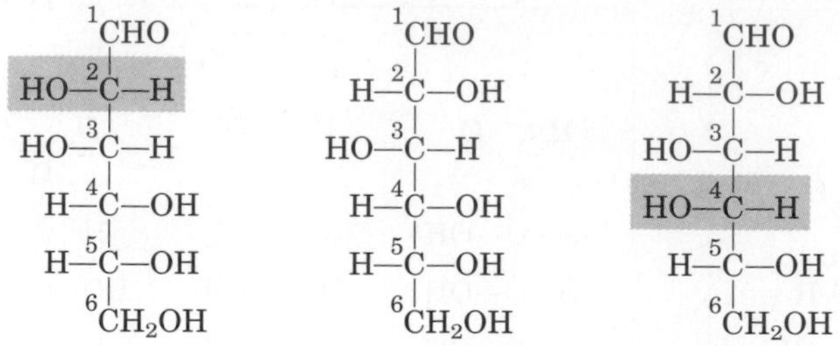

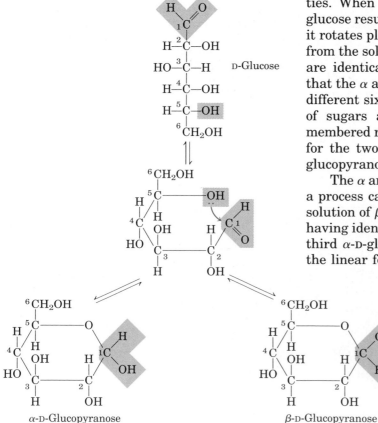

Figure 11–4 D-Glucose and its two epimers, shown as projection formulas. Each epimer differs from D-glucose in the configuration at only one chiral center (shaded red).

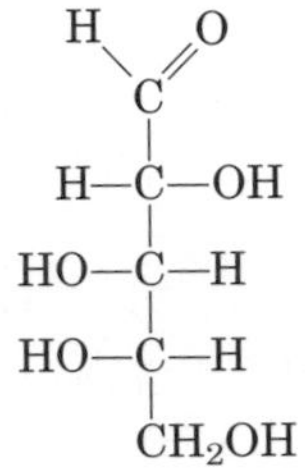

D-Mannose (epimer at C-2)   D-Glucose   D-Galactose (epimer at C-4)

L-Arabinose

When two sugars differ only in the configuration around one carbon atom, they are called **epimers** of each other; D-glucose and D-mannose, which differ only in the stereochemistry at C-2, are epimers, as are D-glucose and D-galactose (at C-4) (Fig. 11–4).

Some sugars do occur naturally in their L form; examples are L-arabinose and the L isomers of some sugar derivatives (discussed below) that are common components of glycoproteins.

The Common Monosaccharides Occur in Cyclic Forms

We have thus far represented the structures of various aldoses and ketoses as straight-chain forms (Figs. 11–3, 11–4). In fact, monosaccharides with five or more carbon atoms in the backbone usually occur in aqueous solution as cyclic (ring) structures, in which the carbonyl group has formed a covalent bond with the oxygen of a hydroxyl group along the chain. One indication that D-glucose has a ring structure is that it has two crystalline forms with slightly different optical properties. When D-glucose is crystallized from water, a form called α-D-glucose results, which differs in its optical activity (the degree to which it rotates plane-polarized light) from the form of D-glucose crystallized from the solvent pyridine, a form known as β-D-glucose. The two forms are identical in chemical composition. Chemical evidence indicates that the α and β isomers of D-glucose are not linear structures but two different six-membered ring compounds (Fig. 11–5). Such cyclic forms of sugars are called **pyranoses** because they resemble the six-membered ring compound pyran (see Fig. 11–7). The systematic names for the two ring forms of D-glucose are α-D-glucopyranose and β-D-glucopyranose.

The α and β forms of D-glucose interconvert in aqueous solution, by a process called **mutarotation.** Thus a solution of α-D-glucose and a solution of β-D-glucose eventually form identical equilibrium mixtures having identical optical properties. This mixture consists of about one-third α-D-glucose, two-thirds β-D-glucose, and very small amounts of the linear form.

D-Glucose

α-D-Glucopyranose   β-D-Glucopyranose

Figure 11–5 Formation of the two cyclic forms of D-glucopyranose. When the aldehyde group at C-1 and the hydroxyl group at C-5 react to form the hemiacetal linkage, two different stereoisomers (anomers), designated α and β, may be formed at C-1.

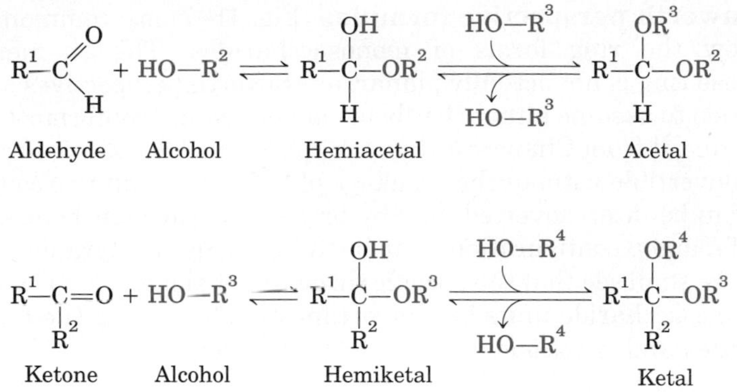

Figure 11–6 An aldehyde or ketone can react with an alcohol in a 1:1 ratio to yield a hemiacetal or hemiketal, respectively, creating a new chiral center at the carbonyl carbon. Addition of a second alcohol molecule produces an acetal or ketal.

The formation of pyranose rings in D-glucose is the result of a general reaction between aldehydes and alcohols to form derivatives called **hemiacetals** (Fig. 11–6), which contain an additional asymmetric carbon atom and thus can exist in two stereoisomeric forms. D-Glucopyranose is an intramolecular hemiacetal, in which the free hydroxyl group at C-5 has reacted with the aldehydic C-1, rendering the latter asymmetric and producing the α and β stereoisomers of D-glucose (Fig. 11–5). Isomeric forms of monosaccharides that differ from each other only in their configuration about the hemiacetal carbon atom, such as α-D-glucopyranose and β-D-glucopyranose (or about the hemiketal carbon, as described below) are called **anomers.** The hemiacetal or carbonyl carbon atom is called the **anomeric carbon.** Only aldoses having five or more carbon atoms can form pyranose rings. Aldohexoses also exist in cyclic forms having five-membered rings, which, because they resemble the five-membered ring compound furan, are called **furanoses.** However, the six-membered aldopyranose ring is much more stable than the aldofuranose ring and predominates in aldohexose solutions.

Ketohexoses also occur in α and β anomeric forms. In these compounds the hydroxyl group on C-5 (or C-6) reacts with the keto group at C-2, forming a furanose (or pyranose) ring containing a **hemiketal** linkage (Fig. 11–6). D-Fructose forms two anomers (Fig. 11–7); the more common form is β-D-fructofuranose.

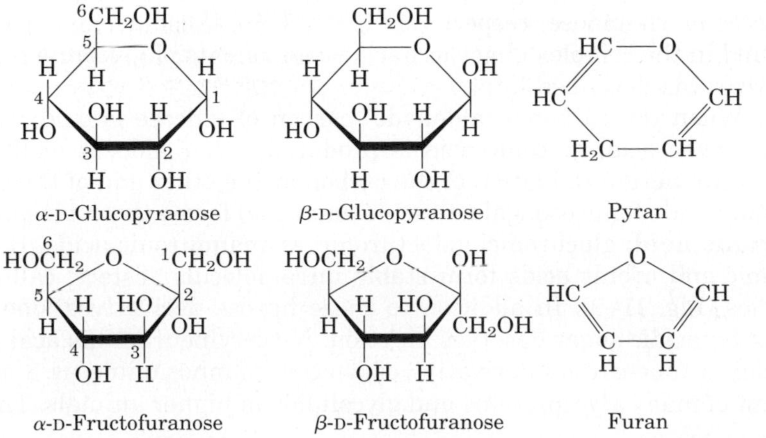

α-D-Glucopyranose $\qquad$ β-D-Glucopyranose $\qquad$ Pyran

α-D-Fructofuranose $\qquad$ β-D-Fructofuranose $\qquad$ Furan

Figure 11–7 Haworth perspective formulas of the pyranose forms of D-glucose and the furanose forms of D-fructose. The edges of the ring nearest the reader are represented by bold lines. Note that hydroxyl groups below the plane of the ring in Haworth perspectives are at the right side of a Fischer projection (compare with Fig. 11–5). Pyran and furan are shown for comparison.

Haworth perspective formulas (Fig. 11–7) are commonly used to show the ring forms of monosaccharides. The six-membered pyranose ring is not actually planar, as Haworth perspectives suggest, but tends to assume either the "boat" or the "chair" conformation (Fig. 11–8). Recall from Chapter 3 that two *conformations* of a molecule are interconvertible without the breakage of any bonds, but two *configurations* can be interconverted only by breaking a covalent bond—in the case of α and β configurations, the bond involving the pyranose oxygen atom. We shall see that the specific three-dimensional conformations of the monosaccharide units are important in determining the biological properties and functions of some polysaccharides.

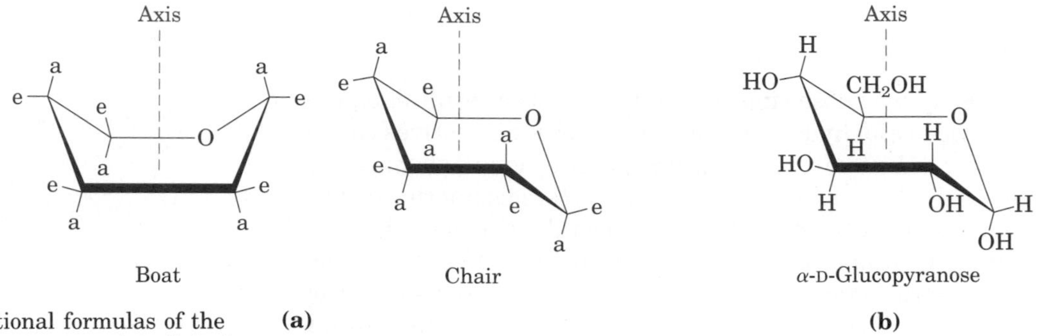

Boat   Chair   α-D-Glucopyranose

(a)   (b)

Figure 11–8 (a) Conformational formulas of the boat and chair forms of the pyranose ring. Substituents on the ring carbons may be either axial (a), projecting almost parallel with the vertical axis through the ring, or equatorial (e), projecting roughly perpendicular to this axis. Generally, substituents in the equatorial positions are less sterically hindered by neighboring substituents, and conformations with their bulky substituents in equatorial positions are favored. The boat conformation is uncommon except in derivatives with very bulky substituents. **(b)** The conformational formula of α-D-glucopyranose, which has the chair conformation.

Organisms Contain a Variety of Hexose Derivatives

In addition to simple hexoses such as glucose, galactose, and mannose, there are a number of derivatives in which a hydroxyl group in the parent compound is replaced with another substituent, or a carbon atom is oxidized to a carboxylic acid (Fig. 11–9). We have already encountered some of these derivatives as components of glycolipids (see Figs. 9–9, 9–11) and glycoproteins (see Fig. 10–8). In glucosamine, galactosamine, and mannosamine, the hydroxyl at C-2 of the parent compound is replaced with an amino group. The amino group may be condensed with acetic acid, as in *N*-acetylglucosamine (Fig. 11–9). This glucosamine derivative is part of many structural polymers, including those of the bacterial cell wall. Bacterial cell walls also contain another derivative of glucosamine in which the three-carbon carboxylic acid lactic acid is ether-linked to the oxygen at C-3 of *N*-acetylglucosamine to form *N*-acetylmuramic acid (Fig. 11–9). The substitution of a hydrogen for the hydroxyl group at C-6 of galactose or mannose produces fucose or rhamnose, respectively (Fig. 11–9); these deoxy sugars are found in the complex oligosaccharide components of glycoproteins and glycolipids described later.

When the carbonyl (aldehyde) carbon of glucose is oxidized to a carboxylic acid, gluconic acid is produced; other aldoses yield other **aldonic acids.** Oxidation of the carbon at the other end of the carbon chain (C-6 of glucose, galactose, or mannose) forms the corresponding **uronic acid;** glucuronic, galacturonic, or mannuronic acids. Both aldonic and uronic acids form stable intramolecular esters, called lactones (Fig. 11–9). In addition to these hexose derivatives, one nine-carbon acidic sugar deserves mention. *N*-acetylneuraminic acid (sialic acid), a nine-carbon derivative of *N*-acetylmannosamine, is a component of many glycoproteins and glycolipids in higher animals. The car-

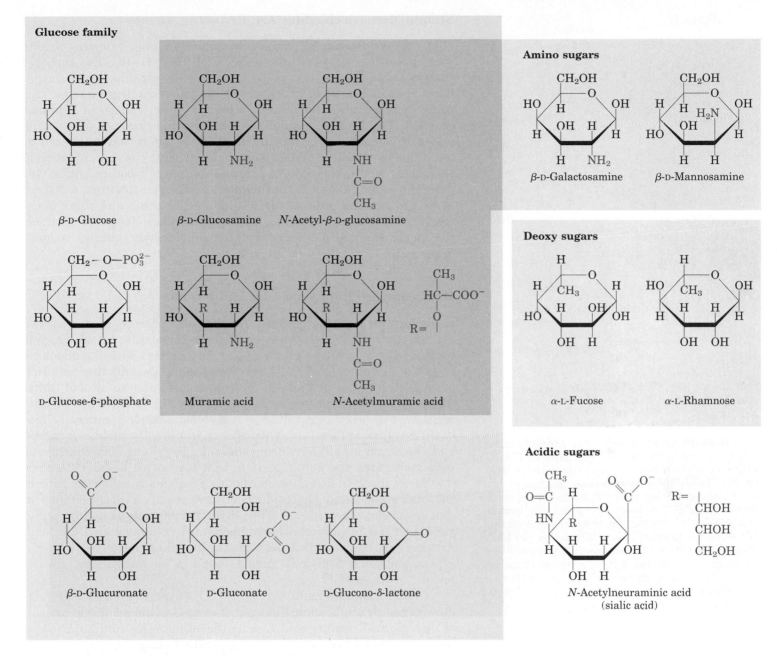

Glucose family

β-D-Glucose

β-D-Glucosamine

N-Acetyl-β-D-glucosamine

D-Glucose-6-phosphate

Muramic acid

N-Acetylmuramic acid

β-D-Glucuronate

D-Gluconate

D-Glucono-δ-lactone

Amino sugars

β-D-Galactosamine

β-D-Mannosamine

Deoxy sugars

α-L-Fucose

α-L-Rhamnose

Acidic sugars

N-Acetylneuraminic acid
(sialic acid)

boxylic acid groups of these sugar derivatives are ionized at pH 7, and the compounds are therefore correctly named as carboxylates— glucuronate, galacturonate, etc.

In the synthesis and metabolism of carbohydrates, the intermediates are very often not the sugars themselves, but their phosphorylated derivatives. Condensation of phosphoric acid with one of the hydroxyl groups of a sugar forms a phosphate ester, as in glucose-6-phosphate (Fig. 11–9). Sugar phosphates are relatively stable at neutral pH, and bear a negative charge. One effect of sugar phosphorylation within cells is to prevent the diffusion of the sugar out of the cell; highly charged molecules do not, in general, cross biological membranes without specific transport systems (Chapter 10). Phosphorylation also activates sugars for subsequent chemical transformation. Several important phosphorylated derivatives of sugars will be discussed in the next chapter.

Figure 11–9 Some hexose derivatives important in biology. In amino sugars, an —NH₂ group replaces one of the —OH groups in the parent hexose. Substitution of —H for —OH produces a deoxy sugar. Note that the deoxy sugars shown here occur as the L isomers. The acidic sugars contain a carboxylate group, which confers a negative charge at neutral pH. D-Glucono-δ-lactone results from formation of an ester linkage between the C-1 carboxylate and the C-5 (also known as the δ carbon) hydroxyl group of gluconate.

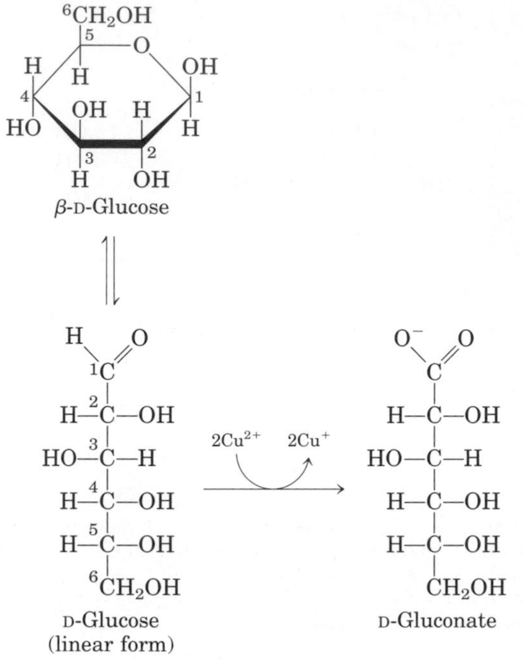

(a)

$$\text{D-Glucose} + O_2 \xrightarrow{\text{glucose oxidase}} \text{D-Gluconate} + H_2O_2$$

(b)

Figure 11–10 (a) Oxidation of the anomeric carbon of glucose and other sugars is the basis for Fehling's reaction; cuprous ion (Cu^+) produced in this reaction forms a red cuprous oxide precipitate. In the hemiacetal (ring) form, C-1 of glucose cannot be oxidized by Cu^{2+}. However, the open chain form is in equilibrium with the ring form, and eventually the oxidation reaction goes to completion. **(b)** Blood glucose concentration is commonly determined by measuring the amount of H_2O_2 produced in the reaction catalyzed by glucose oxidase. The H_2O_2 reacts with certain dyes to produce an easily detected color change.

Simple Monosaccharides Are Reducing Agents

Monosaccharides can be oxidized by relatively mild oxidizing agents such as ferric (Fe^{3+}) or cupric (Cu^{2+}) ion (Fig. 11–10). The carbonyl carbon is oxidized to a carboxylic acid. Glucose and other sugars capable of reducing ferric or cupric ion are called **reducing sugars.** This property is useful in the analysis of sugars; it is the basis of Fehling's reaction (Fig. 11–10a), a qualitative test for the presence of reducing sugar. By measuring the amount of oxidizing agent that is reduced by a solution of a sugar, it is also possible to estimate the concentration of that sugar. For many years, the glucose content of blood and urine was determined in this way in the diagnosis of diabetes mellitus, a disease in which the blood glucose level is abnormally high and there is excessive urinary excretion of glucose. Now, more sensitive methods for measuring blood glucose employ an enzyme, glucose oxidase (Fig. 11–10b).

Disaccharides Contain a Glycosidic Bond

Disaccharides such as maltose, lactose, and sucrose consist of two monosaccharides joined covalently by an **O-glycosidic bond,** which is formed when a hydroxyl group on one sugar reacts with the anomeric carbon on the other (Fig. 11–11). This reaction represents the formation of an acetal from a hemiacetal (glucopyranose) and an alcohol (a hydroxyl group of a second sugar molecule) (see Fig. 11–6). When an anomeric carbon becomes involved in a glycosidic bond, it cannot be oxidized by cupric or ferric ion. The sugar containing the anomeric carbon atom no longer acts as a reducing sugar. In describing disaccharides or polysaccharides, the end of a chain that has a free anomeric carbon (i.e., is not involved in a glycosidic bond) is commonly called the **reducing end** of the chain. Glycosidic bonds are readily hydrolyzed by acid (but resist cleavage by base). Thus disaccharides can be hydrolyzed to yield their free monosaccharide components by boiling with dilute acid. Another type of glycosidic bond joins the anomeric carbon of a sugar to a nitrogen atom; *N*-glycosyl bonds, found in all nucleotides, are described in Chapter 12.

The disaccharide maltose (Fig. 11–11) contains two D-glucose residues joined by a glycosidic linkage between C-1 (the anomeric carbon) of one glucose residue and C-4 of the other. The configuration of the anomeric carbon atom in the glycosidic linkage between the two D-glucose residues is α. Although the anomeric carbon atom involved in the glycosidic bond is not available for reaction with cupric or ferric ion, maltose is nonetheless a reducing sugar; the free anomeric carbon (of the second glucose residue) can be oxidized (Fig. 11–11). The two glucose units in maltose are therefore not identical chemically. The second glucose residue is capable of existing in α- and β-pyranose forms; the β form is shown in Figure 11–11.

To name disaccharides such as maltose unambiguously, and especially to name more complex oligosaccharides, several rules are followed. First, the compound is written with its nonreducing end to the left. An *O* precedes the name of the first (left) monosaccharide unit, as a reminder that the sugar–sugar linkage is through an oxygen atom. The configuration at the anomeric carbon joining the first (left) monosaccharide unit to the second is then given (α or β). To distinguish five- and six-membered ring structures, "furanosyl" or "pyranosyl" is inserted into the name of each monosaccharide residue. The two carbon

α-D-Glucose   β-D-Glucose

hydrolysis condensation
H_2O ⇌ H_2O

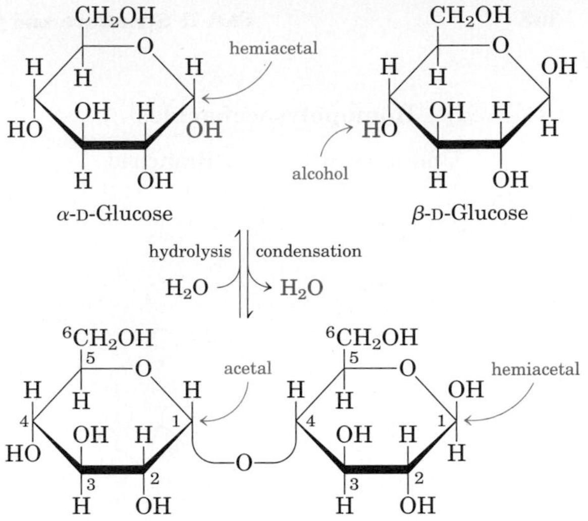

Maltose
(O-α-D-glucopyranosyl-(1→4)-β-D-glucopyranose)

atoms joined by the glycosidic bond are then indicated in parentheses, with an arrow connecting the two numbers; for example, (1→4) shows that C-1 of the first sugar residue is joined to C-4 of the second. If there is a third residue, the second glycosidic bond is described next, by the same conventions. To shorten the description of a complex polysaccharide, three-letter abbreviations for each monosaccharide (Table 11–1) are often used. Following this convention for naming oligosaccharides, maltose is therefore named O-α-D-glucopyranosyl-(1→4)-β-D-glucopyranose. Because most sugars encountered in this book are the D enantiomers, and the pyranose form of hexoses predominates, we will generally use a shortened version of the formal name of such compounds, giving the configuration of the anomeric carbon and naming the carbons joined by the glycosidic bond. In this abbreviated nomenclature, maltose is Glc(α1→4)Glc.

Figure 11–11 A disaccharide is formed from two monosaccharides (here, two molecules of D-glucose) when an alcoholic —OH of one glucose molecule (right) condenses with the intramolecular hemiacetal of the other glucose molecule (left), with the elimination of H_2O and formation of a glycosidic bond. The reversal of this reaction is hydrolysis—attack by H_2O on the glycosidic bond. The maltose molecule retains a reducing hemiacetal at the C-1 not involved in the glycosidic bond.

Table 11–1 Abbreviations for common monosaccharides and their derivatives

Abequose	Abe	Gluconic acid	GlcA
Arabinose	Ara	Glucuronic acid	GlcUA
Fructose	Fru	Galactosamine	GalN
Fucose	Fuc	Glucosamine	GlcN
Galactose	Gal	N-Acetylgalactosamine	GalNAc
Glucose	Glc	N-Acetylglucosamine	GlcNAc
Mannose	Man	Muramic acid	Mur
Rhamnose	Rha	N-Acetylmuramic acid	MurNAc
Ribose	Rib	N-Acetylneuraminic acid	NeuNAc
Xylose	Xyl	(sialic acid)	

The disaccharide lactose (Fig. 11–12), which yields D-galactose and D-glucose on hydrolysis, occurs only in milk. The anomeric carbon of the glucose residue is available for oxidation, and thus lactose is a reducing disaccharide. Its abbreviated name is Gal(β1→4)Glc. Sucrose (table sugar) is a disaccharide of glucose and fructose. It is formed by plants but not by higher animals. In contrast to maltose and lactose, sucrose contains no free anomeric carbon atom; the anomeric carbons of both monosaccharide units are involved in the glycosidic bond (Fig. 11–12). Sucrose is therefore not a reducing sugar, and has no reducing end. Its abbreviated name is given correctly as either Glc(α1→2)Fru or Fru(β2→1)Glc. Sucrose is a major intermediate product of photosynthesis; in many plants it is the principal form in which sugar is transported from the leaves to other portions of plants via their vascular systems. Trehalose (Glc(α1→α1)Glc; Fig. 11–12) is a disaccharide of D-glucose and, like sucrose, is a nonreducing sugar; the anomeric carbons of both glucose moieties are involved in the glycosidic bond. Trehalose is a major constituent of the circulating fluid (hemolymph) of insects, in which it serves as an energy storage compound.

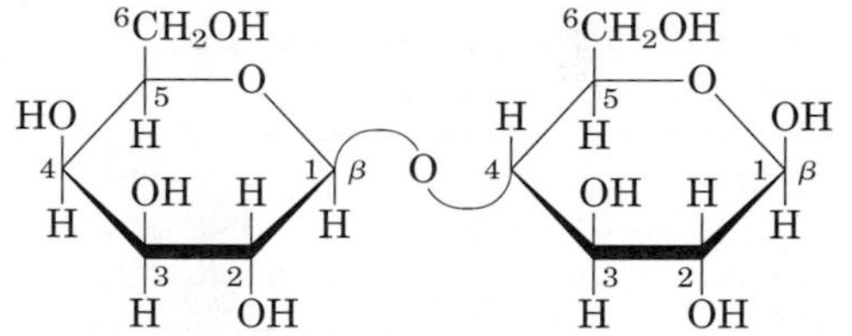

Lactose (β form)
(O-β-D-galactopyranosyl-(1→4)-β-D-glucopyranose)
(Gal(β1→4)Glc)

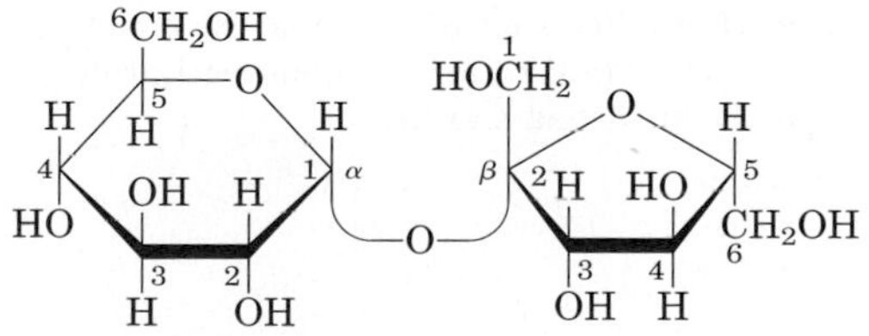

Sucrose
(O-α-D-glucopyranosyl-(1→2)-β-D-fructofuranoside)
(Glc(α1→2)Fru)

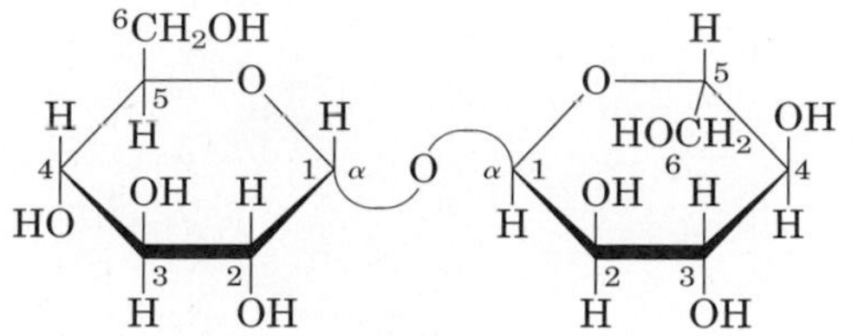

Trehalose
(O-α-D-glucopyranosyl-(1→1)-α-D-glucopyranose)
(Glc(α1→α1)Glc)

Figure 11–12 Common disaccharides, shown as Haworth perspectives. The full systematic names and abbreviations are in parentheses.

307

Homopolysaccharides

Unbranched   Branched

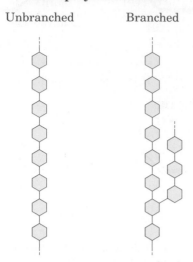

Heteropolysaccharides

Two monomer   Multiple
types,   monomers,
unbranched   branched

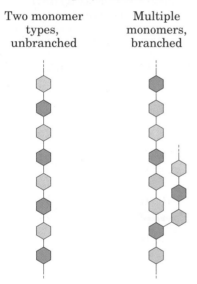

Figure 11–13 Polysaccharides may be composed of one, two, or several different monosaccharides, in straight or branched chains.

Polysaccharides and Proteoglycans

Most of the carbohydrates found in nature occur as polysaccharides, polymers of high molecular weight. Polysaccharides, also called **glycans,** differ from each other in the identity of their recurring monosaccharide units, in the length of their chains, in the types of bonds linking the units, and in the degree of branching. **Homopolysaccharides** contain only a single type of monomeric unit; **heteropolysaccharides** contain two or more different kinds of monomeric units (Fig. 11–13). Some homopolysaccharides serve as storage forms of monosaccharides used as fuels; starch and glycogen are homopolysaccharides of this type. Other homopolysaccharides (cellulose and chitin, for example) serve as structural elements in plant cell walls and animal exoskeletons. Heteropolysaccharides provide extracellular support for organisms of all kingdoms. The rigid layer of the bacterial cell envelope (the peptidoglycan) is a heteropolysaccharide built from two alternating monosaccharide units. In animal tissues, the extracellular space is occupied by several types of heteropolysaccharides, which form a matrix that holds individual cells together and provides protection, shape, and support to cells, tissues, or organs. Hyaluronic acid, one of the polymers that accounts for the toughness and flexibility of cartilage and tendon, typifies this group of extracellular polysaccharides. Other heteropolysaccharides, sometimes in very large aggregates with proteins (proteoglycans), account for the high viscosity and lubricating properties of some extracellular secretions.

Unlike proteins, polysaccharides generally do not have definite molecular weights. This difference is a consequence of the mechanisms of assembly of the two types of polymers. Proteins are synthesized on a template (messenger RNA) of defined sequence and length, by enzymes that copy the template exactly. For polysaccharide synthesis, there is no template; rather, the program for polysaccharide synthesis is intrinsic to the enzymes that catalyze the polymerization of monomeric units. For each type of monosaccharide to be added to the growing polymer there is a separate enzyme, and each enzyme acts only when the enzyme that inserts the preceding subunit has acted. The alternating action of several enzymes produces a polymer with a precisely repeating sequence, but the exact length varies from molecule to molecule, within a general size class. The mechanisms that set the upper size limits are unknown.

Starch and Glycogen Are Stored Fuels

The most important storage polysaccharides in nature are starch in plant cells and glycogen in animal cells. Both polysaccharides occur intracellularly as large clusters or granules (Fig. 11–14). Starch and glycogen molecules are heavily hydrated because they have many exposed hydroxyl groups available to hydrogen bond with water. Most plant cells have the ability to form starch, but it is especially abundant in tubers, such as potatoes, and in seeds, such as corn.

Starch contains two types of glucose polymer, amylose and amylopectin. The former consists of long, unbranched chains of D-glucose units connected by ($\alpha1{\rightarrow}4$) linkages (Fig. 11–15a). Such chains vary in molecular weight from a few thousand to 500,000. Amylopectin also has a high molecular weight (up to 1 million) but is highly branched (Fig. 11–15b). The glycosidic linkages joining successive glucose residues in amylopectin chains are ($\alpha1{\rightarrow}4$), but the branch points, occurring every 24 to 30 residues, are ($\alpha1{\rightarrow}6$) linkages (Fig. 11–15c).

Starch granules

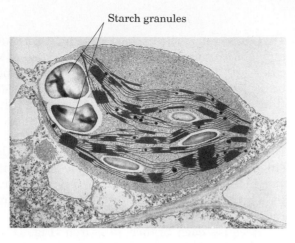

(a)

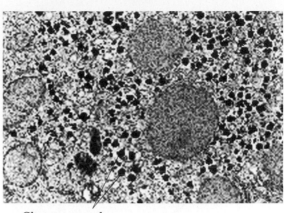

Glycogen granules

(b)

Figure 11–14 Electron micrographs of starch and glycogen granules. **(a)** Large starch granules in a single chloroplast. Starch is made from D-glucose formed photosynthetically.

(b) Glycogen granules in a hepatocyte. These granules are much smaller ($\sim$0.1 μm) than the starch granules ($\sim$1.0 μm).

(a)

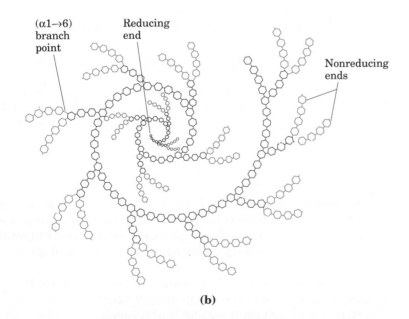

(b)

(c)

Figure 11–15 Amylose and amylopectin, the polysaccharides of starch. **(a)** Amylose, a linear polymer of D-glucose units in ($\alpha1\rightarrow4$) linkage. Each polymer chain can contain several thousand glucose residues. **(b)** Amylopectin. Each hexagon represents one glucose residue. The colored hexagons represent residues of the outer branches, which are removed enzymatically one at a time during the intracellular

mobilization of starch for energy production. This diagram shows only a very small portion of a very long molecule. Glycogen has a similar structure but is more highly branched and more compact. **(c)** Structure of an ($\alpha1\rightarrow6$) branch point. During starch breakdown, a separate enzyme is required to break the ($\alpha1\rightarrow6$) linkage.

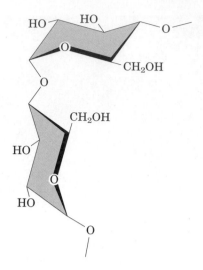

(α1→4)-linked D-glucose units

(a)

(b)

Figure 11–16 The structure of starch (amylose). **(a)** In the most stable conformation of adjacent rigid chairs, the polysaccharide chain is curved, rather than linear as in cellulose (see Fig. 11–17). **(b)** Scale drawing of a segment of amylose. The (α1→4) linkages of amylose, amylopectin, and glycogen cause these polymers to assume a tightly coiled helical structure. This compact structure produces the dense granules of stored starch or glycogen seen in many cells (Fig. 11–14).

Glycogen is the main storage polysaccharide of animal cells. Like amylopectin, glycogen is a polymer of (α1→4)-linked subunits of glucose, with (α1→6)-linked branches, but glycogen is more extensively branched (branches occur every 8 to 12 residues) and more compact than starch. Glycogen is especially abundant in the liver, where it may constitute as much as 7% of the wet weight; it is also present in skeletal muscle. In hepatocytes glycogen is found in large granules (Fig. 11–14), which are themselves clusters of smaller granules composed of single, highly branched glycogen molecules with an average molecular weight of several million. Such glycogen granules also contain, in tightly bound form, the enzymes responsible for the synthesis and degradation of glycogen.

Because each branch in starch (Fig. 11–15b) and glycogen ends with a nonreducing sugar (one without a free anomeric carbon), these polymers have as many nonreducing ends as they have branches, but only one reducing end. When starch or glycogen is used as an energy source, glucose units are removed one at a time from the nonreducing ends. Because of the branching of amylopectin and glycogen, degradative enzymes (which act at nonreducing ends) can work simultaneously at many ends, speeding the conversion of the polymer to monosaccharides.

Why not store glucose in its monomeric form? Liver and skeletal muscle contain glycogen equivalent to several percent of their wet weight, in an essentially insoluble form that contributes very little to the osmotic strength of the cytosol. If the cytosol were a 2% glucose solution (about 0.1 M), the osmolarity of the cell would be threateningly elevated. Furthermore, with an intracellular glucose concentration of 0.1 M and an external concentration of about 5 mM (in a mammal), the free-energy change for glucose uptake would be prohibitively large (recall Eqn 10–2).

The three-dimensional structure of starch is shown in Figure 11–16, and is compared with the structure of cellulose below.

Cellulose and Chitin Are Structural Homopolysaccharides

Cellulose, a fibrous, tough, water-insoluble substance, is found in the cell walls of plants, particularly in stalks, stems, trunks, and all the woody portions of plant tissues. Cellulose constitutes much of the mass of wood, and cotton is almost pure cellulose. Because cellulose is a linear, unbranched homopolysaccharide of 10,000 to 15,000 D-glucose units, it resembles amylose and the main chains of glycogen. But there is a very important difference: in cellulose the glucose residues have the β configuration (Fig. 11–17a), whereas in amylose, amylopectin, and glycogen the glucose is in the α configuration. The glucose residues in cellulose are linked by (β1→4) glycosidic bonds. This difference gives cellulose and amylose very different three-dimensional structures and physical properties.

The three-dimensional structure of carbohydrate-containing macromolecules can be understood using the same principles that explain the structure of polypeptides and nucleic acids: subunits with a more-or-less rigid structure dictated by covalent bonds form three-dimensional macromolecular structures that are stabilized by weak interactions. Because polysaccharides have so many hydroxyl groups, hydrogen bonding has an especially important influence on their structures. Polymers of β-D-glucose, such as cellulose, can be represented as a series of rigid pyranose rings in the chair conformation, connected by

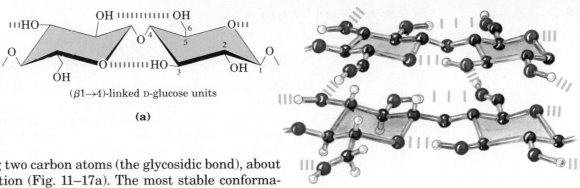

$(\beta1{\to}4)$-linked D-glucose units

(a)

(b)

Figure 11–17 The structure of cellulose. **(a)** Part of a cellulose chain; the D-glucose units are in $(\beta1{\to}4)$ linkage. The rigid chair structures can rotate relative to one another. **(b)** Scale drawing of segments of two parallel cellulose chains, showing the actual conformation of the D-glucose residues and the hydrogen-bond cross-links. In the hexose unit at lower left, all hydrogen atoms are shown; in the other three hexose units all hydrogens attached to carbon have been omitted for clarity, as they do not participate in hydrogen bonding.

an oxygen atom bridging two carbon atoms (the glycosidic bond), about which there is free rotation (Fig. 11–17a). The most stable conformation for the polymer is that in which each chair is turned 180° relative to the preceding subunit, yielding a straight, extended chain. Several chains lying side by side can form the stabilizing network of inter- and intrachain hydrogen bonds shown in Figure 11–17b, resulting in straight, stable supramolecular fibers of great tensile strength. The tensile strength of cellulose has made it a useful substance to civilizations for millenia. Many manufactured products, including paper, cardboard, rayon, insulating tiles, and other packing and building materials, are derived from cellulose.

In contrast to the straight fibers produced by $(\beta1{\to}4)$-linked polymers such as cellulose, the most favorable conformation for $(\alpha1{\to}4)$-linked polymers of D-glucose, such as starch and glycogen, is a tightly coiled helical structure stabilized by hydrogen bonds (Fig. 11–16).

Glycogen and starch ingested in the diet are hydrolyzed by α-amylases, enzymes in saliva and intestinal juice that break $(\alpha1{\to}4)$ glycosidic bonds between glucose units. Cellulose cannot be used by most animals as a source of stored fuel, because the $(\beta1{\to}4)$ linkages of cellulose are not hydrolyzed by α-amylases. Termites readily digest cellulose (and therefore wood), but only because their intestinal tract harbors a symbiotic microorganism, *Trichonympha*, which secretes cellulase, an enzyme that hydrolyzes $(\beta1{\to}4)$ linkages between glucose units. Wood-rot fungi and bacteria also produce cellulase. The only vertebrates able to use cellulose as food are cattle and other ruminant animals (sheep, goats, camels, giraffes). The extra stomachs (rumens) of these animals teem with bacteria and protists that secrete cellulase.

Chitin is a linear homopolysaccharide composed of N-acetyl-D-glucosamine residues in β linkage (Fig. 11–18). The only chemical difference from cellulose is the replacement of a hydroxyl group at C-2 with an acetylated amino group. Chitin forms extended fibers similar to those of cellulose, and like cellulose is indigestible by vertebrate animals. Chitin is the principal component of the hard exoskeletons of nearly a million species of arthropods—insects, lobsters, and crabs, for example—and is probably the second most abundant polysaccharide, next to cellulose, in nature.

Figure 11–18 A short segment of chitin, a homopolymer of N-acetyl-D-glucosamine units in $(\beta1{\to}4)$ linkage.

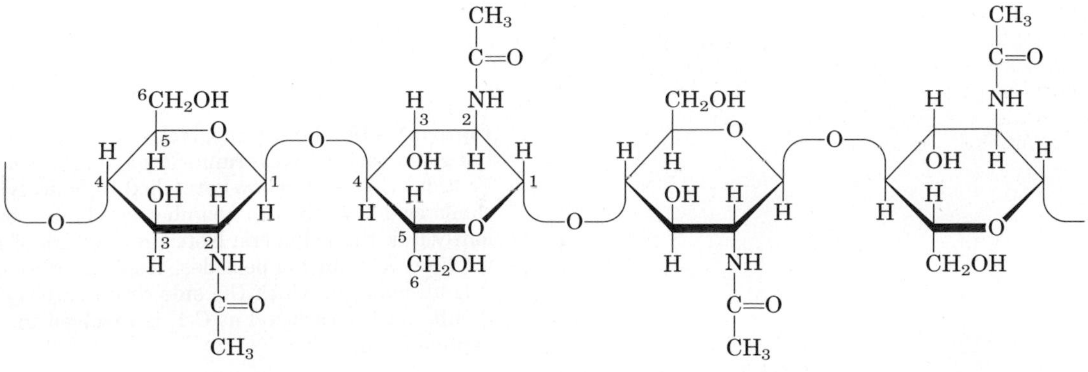

311

The Bacterial Cell Wall Contains a Heteropolysaccharide

The rigid component of bacterial cell walls is a heteropolymer of alternating ($\beta1\rightarrow4$)-linked N-acetylglucosamine and N-acetylmuramic acid units (Fig. 11–19). Many such linear polymers lie side by side in the cell wall, cross-linked by short peptides, the exact structure of which depends on the bacterial species (Fig. 11–19). The cross-linked **peptidoglycan** is degraded by the enzyme lysozyme, which hydrolyzes the glycosidic bond between N-acetylglucosamine and N-acetylmuramic acid, killing bacterial cells. Lysozyme is present in tears, presumably a defense against bacterial infections of the eye. It is also produced by certain bacterial viruses to ensure their release from the host bacteria, an essential step of the viral infection cycle.

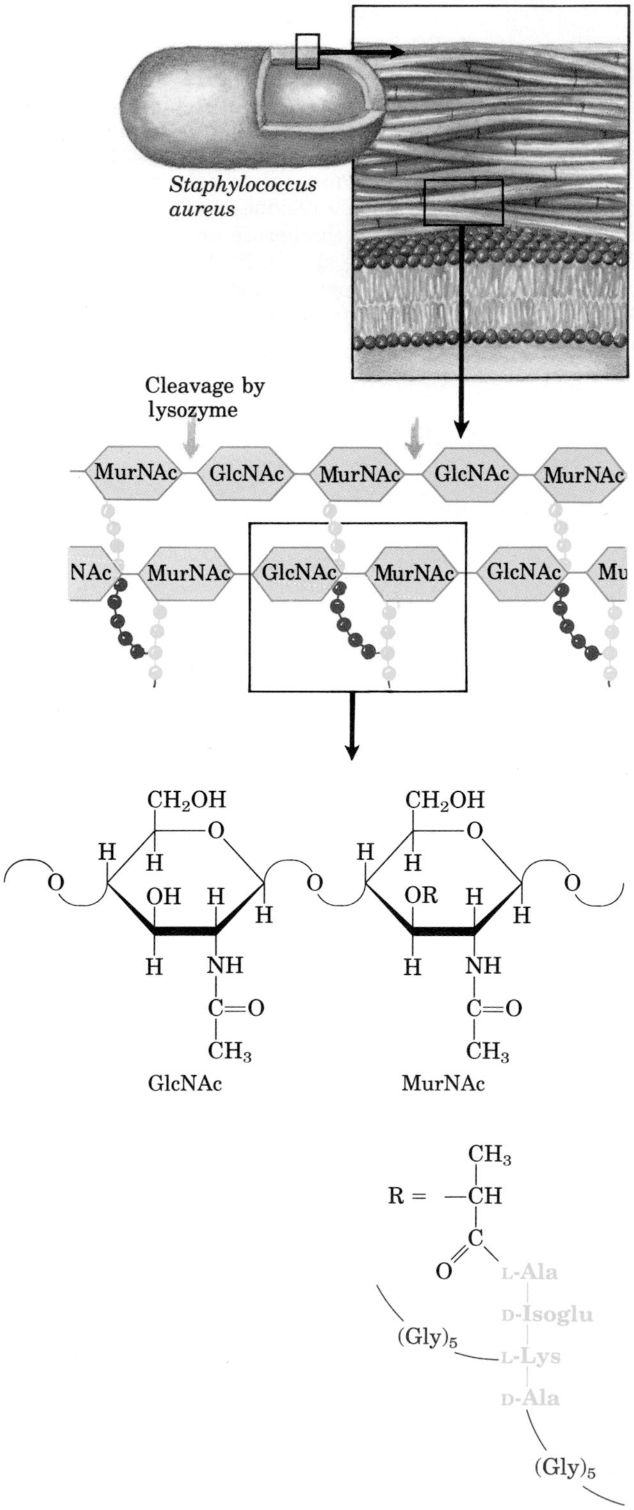

Glycosaminoglycans and Proteoglycans Are Components of the Extracellular Matrix

The extracellular space in animal tissues is filled with a gel-like material, the **extracellular matrix,** also called ground substance, which holds the cells of a tissue together and provides a porous pathway for the diffusion of nutrients and oxygen to individual cells. The extracellular matrix is composed of an interlocking meshwork of heteropolysaccharides and fibrous proteins. The heteropolysaccharides, called **glycosaminoglycans,** are a family of linear polymers composed of repeating disaccharide units (Fig. 11–20). One of the two monosaccharides is always either N-acetylglucosamine or N-acetylgalactosamine; the other is in most cases a uronic acid, usually glucuronic acid. In some glycosaminoglycans, one or more of the hydroxyls of the amino sugar is esterified with sulfate. The combination of these sulfate groups and the carboxylate groups of the uronic acid residues gives the glycosaminoglycans a very high density of negative charge. To minimize the repulsive forces among neighboring charged groups, these molecules assume an extended conformation in solution. One consequence of this extended conformation is the very high viscosity of solutions of these long, thin molecules. Glycosaminoglycans are attached to extracellular proteins to form **proteoglycans** (discussed below), enormous aggregates in which the polysaccharide makes up most of the mass, often 95% or more.

The glycosaminoglycan **hyaluronic acid** (hyaluronate at physiological pH) of the extracellular matrix of animal tissues contains alternating units of D-glucuronic acid and N-acetylglucosamine (Fig. 11–20). Hyaluronates have molecular weights greater than 1 million; they form clear, highly viscous solutions, which serve as lubricants in

Figure 11–19 The peptidoglycan of the cell wall of the gram-positive bacterium *Staphylococcus aureus.* Peptides (red and yellow) attached to N-acetylmuramic acid units in two neighboring chains covalently link the polymers. Note the mixture of L and D amino acids in the peptides. Isoglu represents isoglutamate, in which the side chain carboxyl group, not the carboxyl at C-1, is involved in a peptide bond.

Glycosaminoglycan	Repeating disaccharide	Number of disaccharides per chain

Figure 11–20 Some of the common heteropolysaccharide (glycosaminoglycan) components of extracellular matrix. The ionized carboxylate and sulfate groups give these polymers their characteristic high negative charge.

Hyaluronate — GlcUA ($\beta1\rightarrow3$) GlcNAc ($\beta1\rightarrow4$) — ~50,000

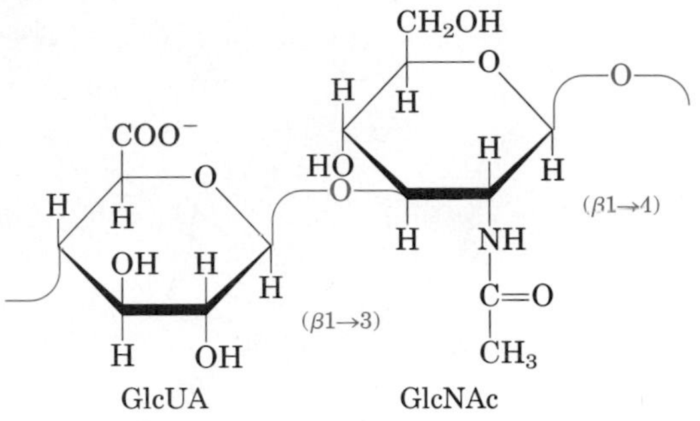

Chondroitin sulfate — GlcUA ($\beta1\rightarrow3$) GalNAcSO$_4^-$ ($\beta1\rightarrow4$) — 20–60

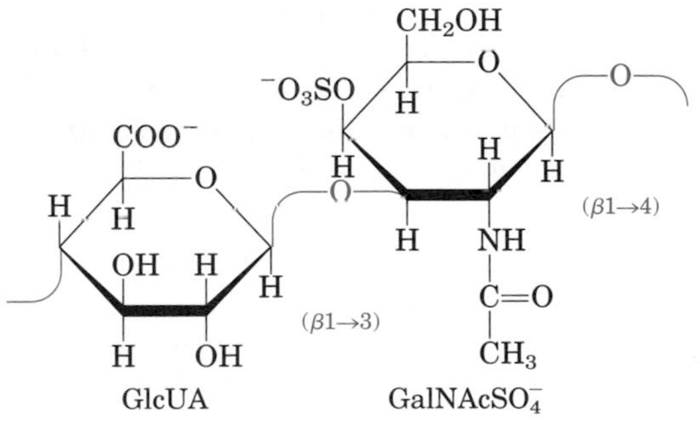

Keratan sulfate — Gal ($\beta1\rightarrow4$) GlcNAcSO$_4^-$ ($\beta1\rightarrow3$) — ~25

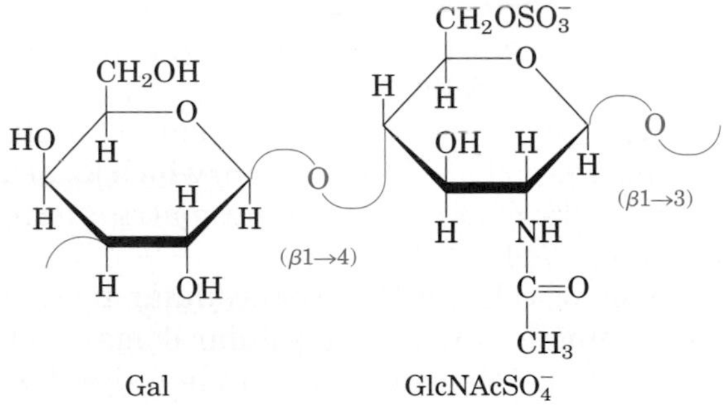

the synovial fluid of joints, and give the vitreous humor of the vertebrate eye its jellylike consistency. Hyaluronate is also a central component of the extracellular matrix of cartilage and tendons, to which it contributes tensile strength and elasticity. Hyaluronidase, an enzyme secreted by some pathogenic (disease-causing) bacteria, can hydrolyze the glycosidic linkages of hyaluronate, rendering tissues more susceptible to invasion by the bacteria. A similar enzyme in sperm hydrolyzes an outer glycosaminoglycan coat around the ovum of many organisms, allowing sperm penetration.

Proteoglycans (Fig. 11–21) are composed of a very long strand of hyaluronate to which numerous molecules of core protein are bound noncovalently, at about 40 nm intervals. Each core protein is bound covalently to many shorter glycosaminoglycan molecules, such as chondroitin sulfate, keratan sulfate (Fig. 11–20), heparan sulfate, and dermatan sulfate. The covalent attachments between glycosaminoglycans and core protein are glycosidic bonds between sugar residues and

Figure 11-21 A proteoglycan aggregate. A single long molecule of hyaluronate is associated noncovalently with many molecules of core protein, each containing covalently bound chondroitin sulfate and keratan sulfate.

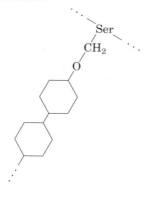

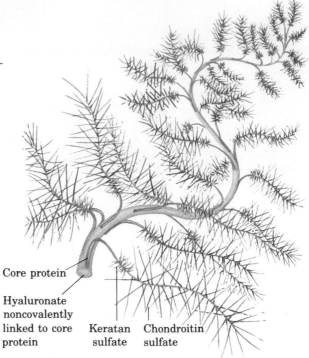

Core protein

Hyaluronate
noncovalently
linked to core
protein   Keratan   Chondroitin
            sulfate   sulfate

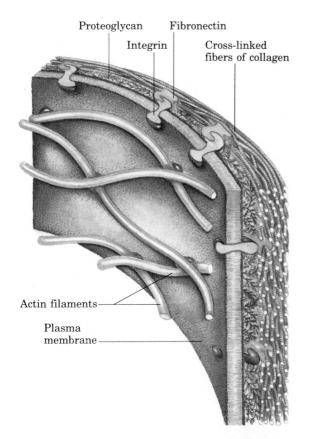

Proteoglycan   Fibronectin

Integrin   Cross-linked
            fibers of collagen

Actin filaments

Plasma
membrane

the hydroxyl groups of Ser residues in the protein. A typical proteoglycan in human cartilage contains about 150 polysaccharide chains (each of $M_r \sim 20{,}000$) covalently bound as side chains to each core protein. When a hundred or more of these "decorated" core proteins bind a single, extended molecule of hyaluronate, the resulting proteoglycan and its associated water of hydration occupy a volume about equal to that of an entire bacterial cell!

Interwoven with these enormous extracellular proteoglycans are fibrous proteins such as collagen and elastin, which form a cross-linked meshwork that gives the whole extracellular matrix strength and resiliency. (See also Table 7-1.)

The attachment of cells to the extracellular meshwork involves several families of proteins. The extracellular domains of certain integral membrane proteins (integrins) have binding sites for another family of adhesion proteins (including fibronectin and laminin), that bind to proteoglycans (Fig. 11-22). It is likely that this complex system of binding proteins serves not merely to anchor cells to the extracellular matrix, but also to direct the migration of cells in developing tissue along paths determined by the organization of the extracellular matrix.

Table 11-2 summarizes the composition, properties, and occurrence of the polysaccharides described in this section.

Figure 11-22 The association between cells and the proteoglycan of extracellular matrix is mediated by a membrane protein (integrin) and an extracellular protein (fibronectin in this example) with binding sites for both integrin and the proteoglycan. Note the close association of collagen fibers with the fibronectin and proteoglycan.

Table 11–2 Summary: Structure and role of some polysaccharides and glycoconjugates

Polymer	Type*	Repeating unit	Size (number of monosac-charide units)	Roles
Starch Amylose	Homo-	($\alpha1\rightarrow4$)Glc, linear	A few thousand to 500,000	Energy storage: in plants
Amylopectin	Homo-	($\alpha1\rightarrow4$)Glc, with ($\alpha1\rightarrow6$)Glc branches every 24–30 residues	Up to 10^6	Energy storage: in plants
Glycogen	Homo-	($\alpha1\rightarrow4$)Glc, with ($\alpha1\rightarrow6$)Glc branches every 8–12 residues	Varies (several million)	Energy storage: in bacteria, animal cells
Cellulose	Homo-	($\beta1\rightarrow4$)Glc	Up to 15,000	Structural: gives rigidity, strength to plant cell walls
Chitin	Homo-	($\beta1\rightarrow4$)GlcNAc	Very large	Structural: gives rigidity, strength to exoskeletons of insects, spiders, crustaceans
Peptidoglycan	Hetero-, with peptides attached	MurNAc($\beta1\rightarrow4$)GlcNAc	Very large	Structural: gives rigidity, strength to bacterial cell envelope
Glycosaminoglycan (hyaluronate)	Hetero-, acidic	GlcUA($\beta1\rightarrow3$)GlcNAc	Varies ($>10^6$)	Structural: extracellular matrix in skin, connective tissue; viscosity, lubrication in joints of vertebrates
Proteoglycans	Hetero-, with protein attached; largely carbohydrate	Uronic acid ($\beta1\rightarrow3$)-linked with sulfated hexosamine	Varies	Structural: resilience, viscosity, lubrication in joints of vertebrates

* Each polymer is classified as a homopolysaccharide (homo-) or heteropolysaccharide (hetero-).

Glycoproteins and Glycolipids

We saw in Chapter 10 that many membrane proteins and certain classes of membrane lipids have more or less complex arrays of covalently attached oligosaccharides; these are **glycoproteins** and **glycolipids.** Most proteins that are secreted by eukaryotic cells are also glycoproteins. The biological advantage in the addition of oligosaccharides to proteins or lipids is not fully understood. The very hydrophilic clusters of carbohydrate alter the polarity and solubility of the proteins or lipids with which they are conjugated. Oligosaccharide chains attached to newly synthesized proteins in the Golgi complex may also influence the sequence of polypeptide-folding events that lead to the tertiary structure of the protein (see Fig. 7–22). Steric interactions between peptide and oligosaccharide may preclude one folding route and favor another. When numerous negatively charged oligosaccharide chains are clustered in a single region of a protein, the charge repulsion among them favors the formation of extended, rodlike structure in that region. The bulkiness and negative charge of oligosaccharide chains also protect some proteins from attack by proteolytic enzymes.

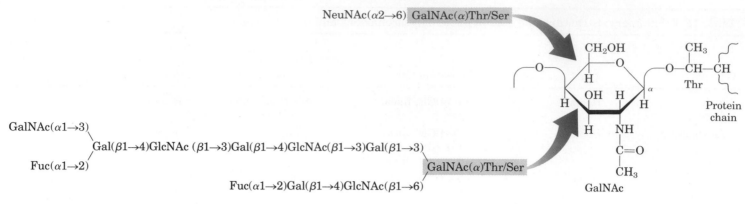

GalNAc($\alpha1\rightarrow3$)

Gal($\beta1\rightarrow4$)GlcNAc ($\beta1\rightarrow3$)Gal($\beta1\rightarrow4$)GlcNAc($\beta1\rightarrow3$)Gal($\beta1\rightarrow3$)

Fuc($\alpha1\rightarrow2$)

Fuc($\alpha1\rightarrow2$)Gal($\beta1\rightarrow4$)GlcNAc($\beta1\rightarrow6$)

O-linked to Ser or Thr residue in protein

(a)

Figure 11–23 Some common N-linked and O-linked oligosaccharide chains in glycoproteins, and two types of glycosidic linkage between protein and oligosaccharide. **(a)** The O-glycosidic bond to the hydroxyl group of Ser or Thr side chains. **(b)** The N-glycosidic bond to the nitrogen of the Asn side chain. The protein-linked ends of the oligosaccharides (shaded in red) are shown in detail on the right.

Beyond these global physical effects on protein structure, there are more specific biological effects of oligosaccharide chains in glycoproteins and glycolipids. We have noted earlier the difference between the information-rich linear sequences of nucleic acids and proteins and the monotonous regularity of homopolysaccharides such as cellulose (see Fig. 3–15). The oligosaccharides attached to glycoproteins and glycolipids are generally not monotonous, but are enormously rich in structural information. Consider the oligosaccharide chains in Figure 11–23, typical of those found in many glycoproteins. The most complex of those shown contains 14 monosaccharide units, of four different kinds, variously linked (1→2), (1→3), (1→4), (1→6), (2→3), and (2→6), some with the α and some with the β configuration. The number of possible permutations and combinations of monosaccharide types and glycosidic linkages in an oligosaccharide this size is astronomical. Each of the oligosaccharides in Figure 11–23 therefore presents a unique face, recognizable by the enzymes and receptors that interact with it.

Because the analysis of oligosaccharide structure is much more difficult than the determination of the linear sequence of bases in nucleic acids, or of amino acid residues in proteins, the oligosaccharide structures of relatively few glycoproteins are known. From the few known structures, it is already clear that a given protein may have several different types of oligosaccharides attached at different positions, and that different glycoproteins have different oligosaccharides. Cells apparently use complex oligosaccharides to encode information about how a protein will fold, where in the cell it will be located, and whether it will be recognized by other proteins. We present here a few examples to illustrate this point.

The Oligosaccharides of Glycoproteins Have Biological Functions

The carbohydrate chains covalently attached to glycoproteins are generally oligosaccharides of much lower molecular weight than the glycosaminoglycans discussed above. The carbohydrate portion commonly constitutes from 1% to about 70% of a glycoprotein by weight, and never 99% as in the proteoglycans. Some glycoproteins have only one or a few carbohydrate groups; others have numerous oligosaccharide

NeuNAc($\alpha2\rightarrow3$)Gal($\beta1\rightarrow4$)GlcNAc($\beta1\rightarrow2$)Man($\alpha1\rightarrow6$)

NeuNAc($\alpha2\rightarrow6$)Gal($\beta1\rightarrow4$)GlcNAc($\beta1\rightarrow2$)   Man($\beta1\rightarrow4$)GlcNAc($\beta1\rightarrow4$) GlcNAc(β)Asn

Man($\alpha1\rightarrow3$)

NeuNAc($\alpha2\rightarrow3$)Gal($\beta1\rightarrow4$)GlcNAc($\beta1\rightarrow4$)

GlcNAc

N-linked to Asn residue in protein

(b)

side chains, which may be linear or branched (Fig. 11–23). Many of the proteins of plasma membranes are glycoproteins, with their oligosaccharide moieties invariably located on the external surface of the membrane. One of the best-characterized membrane glycoproteins is glycophorin of the erythrocyte membrane (see Fig. 10–8), which contains 60% carbohydrate by weight in the form of 16 oligosaccharide chains (totaling 60 to 70 monosaccharide units) covalently attached to residues near the amino terminus of the polypeptide chain. Fifteen of the oligosaccharide units are O-linked to Ser or Thr side chains, and one is N-linked to an Asn residue, two basic types of linkages in glycoproteins (Fig. 11–23; see also Fig. 10–3). Many soluble glycoproteins are also known, including certain carrier proteins and immunoglobulins (antibodies) in the blood of vertebrates, and many of the proteins contained within lysosomes.

The sialic acid (NeuNAc) residues (see Fig. 11–9) found at the ends of the oligosaccharide chains of many soluble glycoproteins (Fig. 11–23) carry a message that determines whether a given protein will continue to circulate in the bloodstream or be removed by the liver. For example, ceruloplasmin is a copper-transporting glycoprotein in the blood of humans and other vertebrates. It has several oligosaccharide chains that end in sialic acid. When these terminal sialic acid units are lost, ceruloplasmin rapidly disappears from the blood. The plasma membrane of hepatocytes has specific binding sites for glycoproteins lacking sialic acid, known as asialoglycoprotein receptors. Glycoproteins bound by these receptors are taken up by the hepatocytes and degraded in lysosomes. Ceruloplasmin is only one of many sialoglycoproteins whose removal from the bloodstream is triggered by the loss of sialic acid units.

Removal of sialic acid is probably one of the ways in which the body marks "old" proteins for destruction and replacement. A similar mechanism is apparently responsible for removing old erythrocytes from the circulation of mammals. Newly synthesized erythrocytes have several membrane glycoproteins with oligosaccharide chains that end in sialic acid. When the sialic acid residues are removed experimentally (by withdrawing blood, treating it with sialidase in vitro, and reintroducing it into the bloodstream), the erythrocytes disappear from the bloodstream within a few hours, whereas cells with intact oligosaccharides continue to circulate for days.

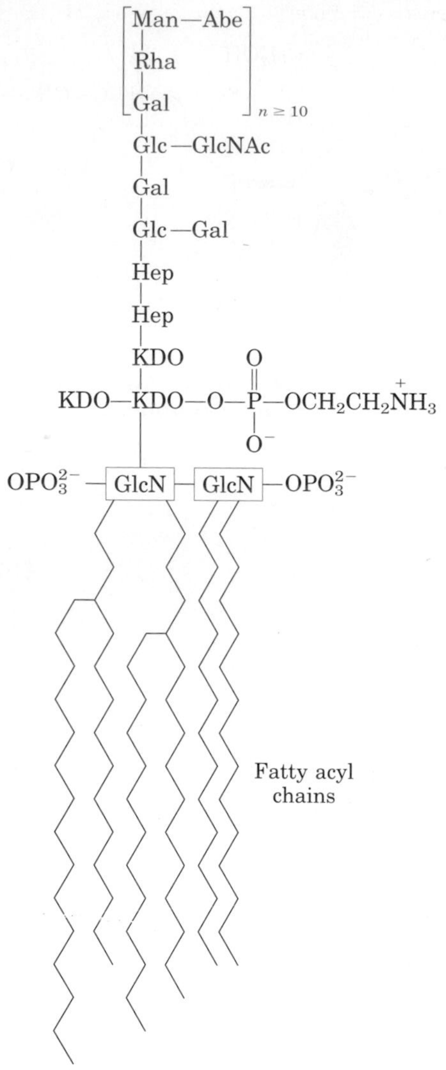

$$
\begin{bmatrix} \text{Man—Abe} \\ | \\ \text{Rha} \\ | \\ \text{Gal} \end{bmatrix}_{n \ge 10}
$$

Glc—GlcNAc
|
Gal
|
Glc—Gal
|
Hep
|
Hep
|
KDO   O
             ‖
KDO—KDO—O—P—OCH₂CH₂NH₃⁺
             |
             O⁻

OPO₃²⁻—⎡GlcN⎤—⎡GlcN⎤—OPO₃²⁻

Fatty acyl
chains

Figure 11–24 The lipopolysaccharide of the outer membrane of *Salmonella typhimurium.* KDO represents 2-keto-3-deoxyoctanoic acid; Hep, L-glycero-D-mannoheptose; Abe, abequose (a 2,3,6-deoxyhexose).

In some cases, attachment of a particular oligosaccharide to a newly synthesized protein targets that protein for a specific intracellular organelle, or for export (by secretion) or placement on the outer surface of a cell. For example, the addition (in the Golgi complex) of mannose-6-phosphate units to the end of the oligosaccharide chains of certain degradative enzymes targets them for transport to lysomes; this targeting is described in detail in Chapter 26. It is probable that many more recognition functions of carbohydrates in glycoproteins remain to be discovered.

Glycolipids and Lipopolysaccharides Are Membrane Components

Glycoproteins and proteoglycans are not the only cellular components that bear complex oligosaccharide chains; some lipids, too, contain covalently bound oligosaccharide chains. In gangliosides (see Fig. 9–9), the polar head group is a complex oligosaccharide containing sialic acid and other monosaccharide units. **Lipopolysaccharides** are major components of the outer membrane of gram-negative bacteria such as *E. coli* and *Salmonella typhimurium*. The lipopolysaccharides of *S. typhimurium* contain six fatty acids bound to two glucosamine residues, one of which is the point of attachment for a complex oligosaccharide (Fig. 11–24). Lipopolysaccharides are the dominant surface feature of gram-negative bacteria; they are prime targets of the antibodies produced by the immune system in response to bacterial infection. The lipopolysaccharide of some bacteria is toxic to humans and other animals; for example, it is responsible for the dangerously lowered blood pressure that occurs with toxic shock syndrome in *Staphylococcus aureus* infections of humans.

Analysis of Carbohydrates

Mixtures of carbohydrates can be resolved into their individual components (Fig. 11–25) by many of the same techniques useful in protein and amino acid separation: differential centrifugation (see Fig. 2–24), ion-exchange chromatography (see Fig. 5–12), and gel filtration (see Fig. 6–2). Hydrolysis in strong acid yields a mixture of monosaccharides, which, after conversion to suitable volatile derivatives, may be separated, identified, and quantified by gas-liquid chromatography (see p. 264) to yield the overall composition of the polymer. For simple, linear polymers such as amylose, the position of the glycosidic bond between monosaccharides is determined by treating the intact polysaccharide with methyl iodide to convert all free hydroxyls to acid-stable methyl ethers (Fig. 11–25). When the methylated polysaccharide is hydrolyzed in acid, the only free hydroxyls present in the monosaccharides produced are those that were involved in glycosidic bonds. To determine the stereochemistry at the anomeric carbon, the intact polymer is tested for sensitivity to purified glycosidases known to hydrolyze only α- or only β-glycosides. Total structure determination for complex heteropolysaccharides is much more difficult. Stepwise degradation with highly specific glycosidases, followed by isolation and identification of the products, is often helpful. Mass spectral analysis and high-resolution nuclear magnetic resonance (NMR) spectroscopy are extremely powerful analytic tools for carbohydrates, especially when combined with selective degradation by glycosidases.

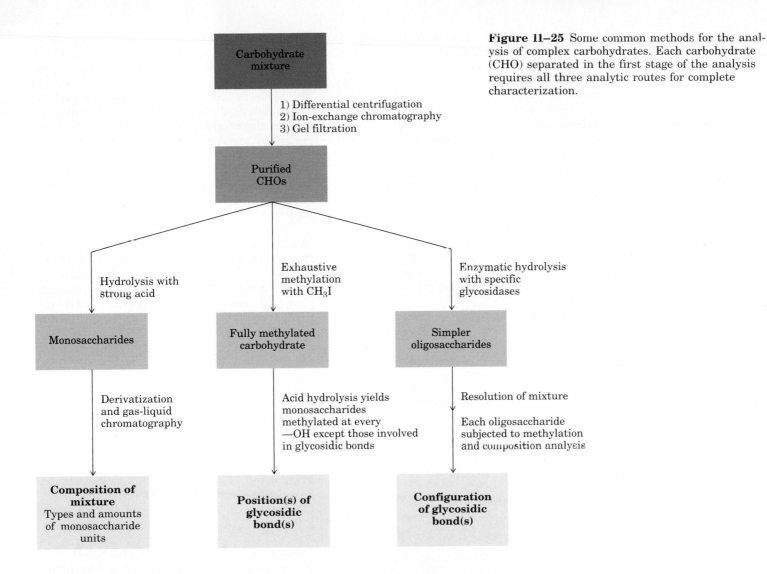

Figure 11–25 Some common methods for the analysis of complex carbohydrates. Each carbohydrate (CHO) separated in the first stage of the analysis requires all three analytic routes for complete characterization.

Summary

Carbohydrates are polyhydroxy aldehydes or ketones having the empirical formula $(CH_2O)_n$. They are classified as monosaccharides or sugars (aldoses or ketoses); oligosaccharides (several monosaccharide units); and polysaccharides (large linear or branched molecules containing many monosaccharide units). Monosaccharides or simple sugars have at least one asymmetric carbon atom and thus exist in stereoisomeric forms. Most common, naturally occurring sugars, such as ribose, glucose, fructose, and mannose, are of the D series. Simple sugars having five or more carbon atoms may exist in the form of closed-ring hemiacetals or hemiketals, either furanoses (five-membered ring) or pyranoses (six-membered ring). Furanoses and pyranoses occur in anomeric α and β forms, which are interconverted in the process of mutarotation.

Sugars with free, oxidizable anomeric carbons are called reducing sugars. Many derivatives of the simple sugars are found in living cells, including amino sugars and their acetylated derivatives, aldonic acids, and uronic acids. The hydroxyl groups of monosaccharide units can also form phosphate and sulfate esters. Disaccharides consist of two monosaccharides joined by a glycosidic bond.

Polysaccharides (glycans) contain many monosaccharide units in glycosidic linkage. Some function as storage forms of carbohydrate. The most important storage polysaccharides are starch and glycogen, high molecular weight, branched polymers of glucose having (α1→4) linkages in the main chains and (α1→6) linkages at the branch points. Other polysaccharides play a structural role in cell walls. Cellulose, the structural polysaccharide of plants, has D-glucose units in (β1→4) linkage. Chitin, the structural polysaccharide of insect exoskeletons, is a linear polymer of *N*-acetylglucosamine, joined in (β1→4) linkages. The rigid porous walls of bacterial cells contain peptidoglycans, linear polysaccharides of alternating *N*-acetylmuramic acid and *N*-acetylglucosamine units, cross-linked by short peptide chains. The extracellular matrix surrounding cells in animal tissues contains very large aggregates of heteropolysaccharides (glycosaminoglycans) and proteins, called proteoglycans. Among the glycosaminoglycans in proteoglycans are hyaluronate, a high molecular weight polymer of alternating D-glucuronic acid and *N*-acetyl-D-glucosamine, and a variety of shorter, very acidic heteropolysaccharides covalently bound to a core protein. Polysaccharides make up most of the mass of proteoglycans.

Glycoproteins contain one or more sugar residues, but most of their mass is amino acid resi-dues. Many cell-surface or extracellular proteins are glycoproteins, as are most secreted proteins. The carbohydrate moieties of glycoproteins influence the physical structure of the proteins, and also serve as biological labels, marking proteins with different oligosaccharides for different fates. Certain oligosaccharides tag a glycoprotein for secretion or insertion into the plasma membrane; others signal transfer to lysosomes. Sialoglycoproteins in the blood that lose their sialic acid residues are targeted for removal and destruction. Glycolipids and lipopolysaccharides are carbohydrate conjugates at the outer surface of cell membranes.

The structure of oligosaccharides and polysaccharides is investigated by a combination of specific enzymatic hydrolysis to determine stereochemistry and produce simple fragments for further analysis, methylation analysis to locate the glycosidic bonds, and high-resolution NMR spectroscopy to establish sequences and confirm configurations.

Further Reading

General

Aspinall, G.O. (ed) (1982, 1983, 1985) *The Polysaccharides,* Vols. 1–3, Academic Press, Inc., New York.

Binkley, R.W. (1988) *Modern Carbohydrate Chemistry,* Marcel Dekker, Inc., San Diego, CA.
A comprehensive and up-to-date survey.

El Khadem, H.S. (1988) *Carbohydrate Chemistry: Monosaccharides and Their Oligomers,* Academic Press, Inc., New York.

Ginsburg, V. & Robbins, P. (eds) (1981, 1984) *Biology of Carbohydrates,* Vols. 1 and 2, Wiley Interscience, New York.
A collection of excellent reviews of carbohydrates and glycoproteins.

Morrison, R.T. & Boyd, R.N. (1987) *Organic Chemistry,* 5th edn, Allyn & Bacon, Inc., Boston.
Chapters 38 and 39 cover the structure, stereochemistry, nomenclature, and chemical reactions of carbohydrates.

Pigman, W. & Horton, D. (eds) (1970, 1972, 1980) *The Carbohydrates: Chemistry and Biochemistry,* Vols. IA, IB, IIA, and IIB, Academic Press, Inc., New York.
Articles on several aspects of carbohydrate chemistry.

Preis, J. (ed) (1980) *Carbohydrates: Structure and Function.* Vol. 3 of *The Biochemistry of Plants: A Comprehensive Treatise* (Stumpf, P.K. & Conn, E.E., eds), Academic Press, Inc., New York.

Jackson, R.L., Busch, S.J., & Cardin, A.D. (1991) Glycosaminoglycans: molecular properties, protein interactions, and role in physiological processes. *Physiol. Rev.* **71,** 481–539.
An advanced review of the chemistry and biology of glycosaminoglycans.

Polysaccharides and Proteoglycans

Buck, C.A. & Horwitz, A.F. (1987) Cell surface receptors for extracellular matrix molecules. *Annu. Rev. Cell Biol.* **3,** 179–205.

Caplan, A.I. (1984) Cartilage. *Sci. Am.* **251** (October), 84–94.
A lucid introductory account of proteoglycan structure and function.

Carney, S.L. & Muir, H. (1988) The structure and function of cartilage proteoglycans. *Physiol. Rev.* **68,** 858–910.
A more advanced review.

Fransson, L.-Å. (1987) Structure and function of cell-associated proteoglycans. *Trends Biochem. Sci.* **12,** 406–411.

Ruoslahti, E. (1988) Structure and biology of proteoglycans. *Annu. Rev. Cell Biol.* **4,** 229–255.
A lengthy and advanced review of recent developments.

Ruoslahti, E. (1989) Proteoglycans in cell regulation. *J. Biol. Chem.* **264,** 13369–13372.
A short, clear review.

Sharon, N. (1975) *Complex Carbohydrates: Their Chemistry, Biosynthesis, and Functions,* Addison-Wesley Publishing Co., Inc., Reading, MA.

Varner, J.E. & Lin, L.-S. (1989) Plant cell wall architecture. *Cell* **56,** 231–239.
A good introductory review.

Glycoproteins and Glycolipids

Hynes, R.O. (1987) Integrins: a family of cell surface receptors. *Cell* **48,** 549–554.

Jentoft, N. (1990) Why are proteins *O*-glycosylated? *Trends Biochem. Sci.* **15,** 291–294.

Lennarz, W.J. (ed) (1980) *The Biochemistry of Glycoproteins and Proteoglycans,* Plenum Press, New York.
An advanced-level text.

Paulson, J.C. (1989) Glycoproteins: what are the sugar chains for? *Trends Biochem. Sci.* **14,** 272–276.
An introductory account, with emphasis on function.

Schauer, R. (1985) Sialic acids and their role as biological masks. *Trends Biochem. Sci.* **10,** 357–360.

Steer, C.J. & Ashwell, G. (1986) Hepatic membrane receptors for glycoproteins. *Prog. Liver Dis.* **8,** 99–123.
A review of asialoglycoprotein receptors.

Analysis of Carbohydrates

Biermann, C.J. & McGinnis, G.D. (eds) (1989) *Analysis of Carbohydrates by GLC and MS,* CRC Press, Inc., Boca Raton, FL.
Describes both theory and practice of carbohydrate isolation and identification. Chapter 1 is an excellent introduction to the subject.

McCleary, B.V. & Matheson, N.K. (1986) Enzymic analysis of polysaccharide structure. *Adv. Carbohydr. Chem. Biochem.* **44,** 147–276.
On the use of purified enzymes in analysis of structure and stereochemistry.

Sweeley, C.C. & Nunez, H.A. (1985) Structural analysis of glycoconjugates by mass spectrometry and nuclear magnetic resonance spectroscopy. *Annu. Rev. Biochem.* **54,** 765–801.
An advanced review of recent progress.

Vliegenthart, J.F.G., Dorland, L., & van Halbeek, H. (1983) High-resolution, [1]H-nuclear magnetic resonance spectroscopy as a tool in the structural analysis of carbohydrates related to glycoproteins. *Adv. Carbohydr. Chem. Biochem.* **41,** 209–374.
An excellent and extensive technical review.

Problems

1. *Interconversion of D-Galactose Forms* A solution of one stereoisomer of a given monosaccharide will rotate plane-polarized light to the left (counterclockwise) and is called the levorotatory isomer, designated (−); the other stereoisomer will rotate plane-polarized light to the same extent but to the right (clockwise) and is called the dextrorotatory isomer, designated (+). An equimolar mixture of the (+) and (−) forms will not rotate plane-polarized light.

The optical activity of a stereoisomer is expressed quantitatively by its *optical rotation,* the number of degrees by which plane-polarized light is rotated on passage through a given path length of a solution of the compound at a given concentration. The *specific rotation* $[\alpha]_D^{25°C}$ of an optically active compound is defined thus:

$$[\alpha]_D^{25°C} = \frac{\text{observed optical rotation (°)}}{\text{length of optical path (dm)} \times \text{concentration (g/mL)}}$$

The temperature and the wavelength of the light employed (usually the D line of sodium, 589 nm) must be specified in the definition.

A freshly prepared solution of the α form of D-galactose (1 g/mL in a 10 cm cell) shows an optical rotation of +150.7°. When the solution is allowed to stand for a prolonged period of time the observed rotation gradually decreases and reaches an equilibrium value of +80.2°. In contrast, a freshly prepared solution (1 g/mL) of the β form shows an optical rotation of only +52.8°. Moreover, when the solution is allowed to stand for several hours, the rotation increases to an equilibrium value of +80.2°, identical to the equilibrium value reached by α-D-galactose.

(a) Draw the Haworth perspective formulas of the α and β forms of galactose. What feature distinguishes the two forms?

(b) Why does the optical rotation of a freshly prepared solution of the α form gradually decrease with time? Why do solutions of the α and β forms (at equal concentrations) reach the same optical rotation at equilibrium?

(c) Calculate the percentage composition of the two forms of galactose at equilibrium.

2. *Invertase "Inverts" Sucrose* The hydrolysis of sucrose (specific rotation +66.5°) yields an equimolar mixture of D-glucose (specific rotation +52.5°) and D-fructose (specific rotation −92°).

(a) Suggest a convenient way to determine the rate of hydrolysis of sucrose by an enzyme preparation extracted from the lining of the small intestine.

(b) Explain why an equimolar mixture of D-glucose and D-fructose formed by hydrolysis of sucrose is called invert sugar in the food industry.

(c) The enzyme invertase (its preferred name is now sucrase) is allowed to act on a solution of sucrose until the optical rotation of the solution becomes zero. What fraction of the sucrose has been hydrolyzed?

3. *Manufacture of Liquid-Filled Chocolates* The manufacture of chocolates containing a liquid center is an interesting application of enzyme engineering. The flavored liquid center consists largely of an aqueous solution of sugars rich in fructose to provide sweetness. The technical dilemma is the following: the chocolate coating must be prepared by pouring hot melted chocolate over a solid (or almost solid) core, yet the final product must have a liquid, fructose-rich center. Suggest a way to solve this problem. (Hint: The solubility of sucrose is much lower than the solubility of a mixture of glucose and fructose.)

4. *Anomers of Sucrose?* Although lactose exists in two anomeric forms, no anomeric forms of sucrose have been reported. Why?

5. *Growth Rate of Bamboo* The stems of bamboo, a tropical grass, can grow at the phenomenal rate of 0.3 m/d under optimal conditions. Given that the stems are composed almost entirely of cellulose fibers oriented in the direction of growth, calculate the number of sugar residues per second that must be added enzymatically to growing cellulose chains to account for the growth rate. Each D-glucose unit in the cellulose molecule is about 0.45 nm long.

6. *Enzymatic Digestibility of Cellulose and Starch* Both cellulose and α-amylose consist of (1→4)-linked D-glucose units and can be extensively hydrated. Despite this similarity, a person on a diet consisting predominantly of α-amylose (starch) will gain weight, whereas a person on a diet of cellulose (wood) will starve. Why?

7. *Physical Properties of Cellulose and Glycogen* The practically pure cellulose obtained from the seed threads of the plant genus *Gossypium* (cotton) is tough, fibrous, and completely insoluble in water. In contrast, glycogen obtained from muscle or liver disperses readily in hot water to make a turbid solution. Although they have markedly different physical properties, both substances are composed of (1→4)-linked D-glucose polymers of comparable molecular weight. What features of their structures cause these two polysaccharides to differ in their physical properties? Explain the biological advantages of their respective properties.

8. *Glycogen as Energy Storage: How Long Can a Game Bird Fly?* Since ancient times it had been observed that certain game birds, such as grouse, quail, and pheasants, are easily fatigued. The Greek historian Xenophon wrote: "The bustards, on the other hand, can be caught if one is quick in starting them up, for they will fly only a short distance, like partridges, and soon tire; and their flesh is delicious." The flight muscles of game birds rely almost entirely on the metabolic breakdown of glucose-1-phosphate for the necessary energy, in the form of ATP (see Chapter 14). In game birds, glucose-1-phosphate is formed by the breakdown of stored muscle glycogen, catalyzed by the enzyme glycogen phosphorylase. The rate of ATP production is limited by the rate at which glycogen can be broken down. During a "panic flight," the game bird's rate of glycogen breakdown is quite high, approximately 120 μmol/min of glucose-1-phosphate produced per gram of fresh tissue. Given that the flight muscles usually contain about 0.35% glycogen by weight, calculate how long a game bird can fly.

9. *Determination of the Extent of Branching in Amylopectin* The extent of branching (number of ($\alpha 1 \rightarrow 6$) glycosidic bonds) in amylopectin can be determined by the following procedure. A weighed sample of amylopectin is exhaustively treated with a methylating agent (methyl iodide) that replaces all the hydrogens on the sugar hydroxyls with methyl groups, converting —OH to —OCH$_3$. All the glycosidic bonds in the treated sample are then hydrolyzed with aqueous acid. The amount of 2,3-dimethylglucose in the hydrolyzed sample is determined.

2,3-Dimethylglucose

(a) Explain the basis of this procedure for determining the number of ($\alpha 1 \rightarrow 6$) branch points in amylopectin. What happens to the unbranched glucose residues in amylopectin during the methylation and hydrolysis procedure?

(b) A 258 mg sample of amylopectin treated as described above yielded 12.4 mg of 2,3-dimethylglucose. Determine what percentage of the glucose residues in amylopectin contain an ($\alpha 1 \rightarrow 6$) branch.

10. *Structure Determination of a Polysaccharide* A polysaccharide of unknown structure was isolated, subjected to exhaustive methylation, and hydrolyzed. Analysis of the products revealed three methylated sugars: 2,3,4-tri-*O*-methyl-D-glucose, 2,4-di-*O*-methyl-D-glucose, and 2,3,4,6-tetra-*O*-methyl-D-glucose, in the ratio 20:1:1. What is the structure of the polysaccharide?

11. *Empirical Formula Determination* An unknown substance containing only C, H, and O was isolated from goose liver. A 0.423 g sample produced 0.620 g of CO_2 and 0.254 g of H_2O after complete combustion in excess oxygen. Is the empirical formula of this substance consistent with its being a carbohydrate? Explain.

12. *Reaction with Fehling's Reagent* A sample of disaccharide is either lactose or sucrose. No reddish precipitate forms in Fehling's reaction, unless the compound is first warmed in dilute acid. Is it lactose or sucrose? Explain.

13. *Glucose Oxidase in Determination of Blood Glucose* The enzyme glucose oxidase isolated from the mold *Penicillium notatum* catalyzes the oxidation of β-D-glucose to D-glucono-δ-lactone. This enzyme is highly specific for the β anomer of glucose and does not affect the α anomer. In spite of this specificity, the reaction catalyzed by glucose oxidase is commonly used in a clinical assay for total blood glucose i.e., solutions consisting of a mixture of β- and α-D-glucose. How is this possible? Aside from allowing the detection of smaller quantities of glucose, what advantage does glucose oxidase offer over Fehling's reaction for the determination of blood glucose?

14. *Volume of Chondroitin Sulfate in Solution* One of the critical functions of chondroitin sulfate is to act as a lubricant in skeletal joints by creating a gel-like medium that is resilient to friction and shock. This function appears to be related to a distinctive property of chondroitin sulfate: the volume occupied by the molecule is much greater in solution than in the dehydrated solid. Why is the volume occupied by the molecule so much larger in solution?

15. *Information Content of Oligosaccharides* The carbohydrate portion of some glycoproteins may serve as a cellular recognition site. In order to perform this function, the oligosaccharide moiety of glycoproteins must have the potential to occur in a large variety of forms. Which can produce a larger variety of structures: oligopeptides composed of five different amino acid residues or oligosaccharides composed of five different monosaccharide residues? Explain.

CHAPTER

12

Nucleotides and Nucleic Acids

The final classes of biomolecules to be considered, the nucleotides and molecules derived from them, represent a clear case in which last is not least. Nucleotides themselves participate in a plethora of crucial supporting roles in cell metabolism, and polymers of nucleotides, the nucleic acids, provide the script for everything that occurs in a cell.

Nucleotides are energy-rich compounds that drive metabolic processes (primarily biosyntheses) in all cells. They also serve as chemical signals, key links in cellular systems that respond to hormones and other extracellular stimuli, and are structural components of a number of enzyme cofactors and metabolic intermediates.

The nucleic acids, deoxyribonucleic acid (DNA) and ribonucleic acid (RNA), are the molecular repositories for genetic information. The structure of every protein, and ultimately of every cell constituent, is a product of information programmed into the nucleotide sequence of a cell's nucleic acids.

This chapter provides an overview of the nucleotides and nucleic acids found in most cells. The metabolism of nucleotides is discussed in Chapter 21, and a more detailed examination of the function of nucleic acids is the focus of Part IV of this text.

Some Basics

The amino acid sequence of every protein and the nucleotide sequence of every RNA molecule in a cell are specified by that cell's DNA. The necessary protein or RNA sequence information is found in corresponding nucleotide sequences in the DNA. A segment of DNA that contains the information required for the synthesis of a functional biological product (protein or RNA) is referred to as a **gene.** A cell typically has many thousands of genes, and DNA molecules, not surprisingly, tend to be very large. The storage of biological information is the only known function of DNA.

Several classes of RNAs are found in cells, each with a distinct function. **Ribosomal RNAs** (rRNA) are structural components of ribosomes, the large complexes that carry out the synthesis of proteins. **Messenger RNAs** (mRNA) are nucleic acids that carry the information from one or a few genes to the ribosome, where the corresponding proteins can be synthesized. **Transfer RNAs** (tRNA) are adapter molecules that faithfully translate the information in mRNA into a specific sequence of amino acids. In addition to these major classes there are a wide variety of special-function RNAs, described in depth in Part IV. We introduce here the chemical structures of nucleotides and nucleic acids.

Nucleotides Have Characteristic Bases and Pentoses

Nucleotides have three characteristic components: (1) a nitrogenous base, (2) a pentose, and (3) a phosphate (Fig. 12–1a). The nitrogenous bases are derivatives of two parent compounds, **pyrimidine** and **purine** (Fig. 12–1b). The bases and pentoses found in the common nucleotides are heterocyclic compounds. The carbon and nitrogen atoms in the parent structures are conventionally numbered to facilitate naming and identification of the many derivative compounds. The convention for the pentose ring follows rules outlined in Chapter 11, but in the pentoses of nucleotides the carbon numbers are given a prime (′) designation (Fig. 12–1a) to distinguish them from the numbered atoms of the nitrogenous bases.

The base is joined covalently (at N-1 of pyrimidines and N-9 of purines) in an *N*-glycosyl linkage to the 1′ carbon of the pentose, and the phosphate is esterified to the 5′ carbon. The *N*-glycosyl bond is formed by removal of the elements of water (a hydroxyl group from the pentose and hydrogen from the base), as in *O*-glycosyl bond formation (see Fig. 11–11). Without the phosphate group, the molecule is called a **nucleoside.**

DNA and RNA both contain two major purine bases, **adenine** (A) and **guanine** (G). DNA and RNA also contain two major pyrimidines; in both types of nucleic acid one of these is **cytosine** (C). The single important difference between the bases of DNA and those of RNA is the nature of the second major pyrimidine: **thymine** (T) in DNA and **uracil** (U) in RNA. Only rarely does thymine occur in RNA or uracil in DNA. The structures of the five major bases are shown in Figure 12–2, and the nomenclature of their corresponding nucleotides and nucleosides is summarized in Table 12–1.

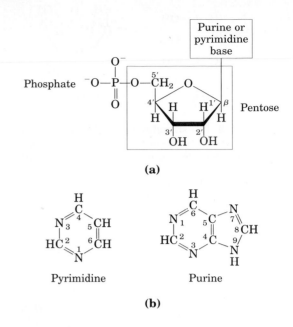

Figure 12–1 **(a)** The general structure of nucleotides, showing the numbering convention for the pentose. The structure shown is that of a ribonucleotide. In deoxyribonucleotides the —OH group on the 2′ carbon (in red) is replaced with —H. **(b)** The parent compounds of the pyrimidine and purine bases of nucleotides and nucleic acids, showing the numbering conventions for the ring structures.

Table 12–1 Nucleotide and nucleic acid nomenclature

Base	Nucleoside*	Nucleotide*	Nucleic acid
Purines			
Adenine	Adenosine	Adenylate	RNA
	Deoxyadenosine	Deoxyadenylate	DNA
Guanine	Guanosine	Guanylate	RNA
	Deoxyguanosine	Deoxyguanylate	DNA
Pyrimidines			
Cytosine	Cytidine	Cytidylate	RNA
	Deoxycytidine	Deoxycytidylate	DNA
Thymine	Thymidine or deoxythymidine	Thymidylate or deoxythymidylate	DNA
Uracil	Uridine	Uridylate	RNA

* *Nucleoside* and *nucleotide* are generic terms that include both ribo- and deoxyribo- forms. Note that here ribonucleosides and ribonucleotides are designated simply as nucleosides and nucleotides (e.g., riboadenosine as adenosine) and deoxyribonucleosides and deoxyribonucleotides as deoxynucleosides and deoxynucleotides (e.g., deoxyriboadenosine as deoxyadenosine). Both forms of naming are acceptable, but the shortened names are more commonly used.

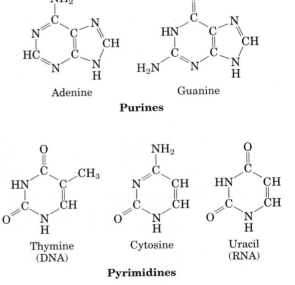

Figure 12–2 The major purine and pyrimidine bases of nucleic acids. Some of the common names of these bases reflect the circumstances of their discovery. Guanine, for example, was first isolated from guano (bird manure), and thymine was first isolated from thymus tissue.

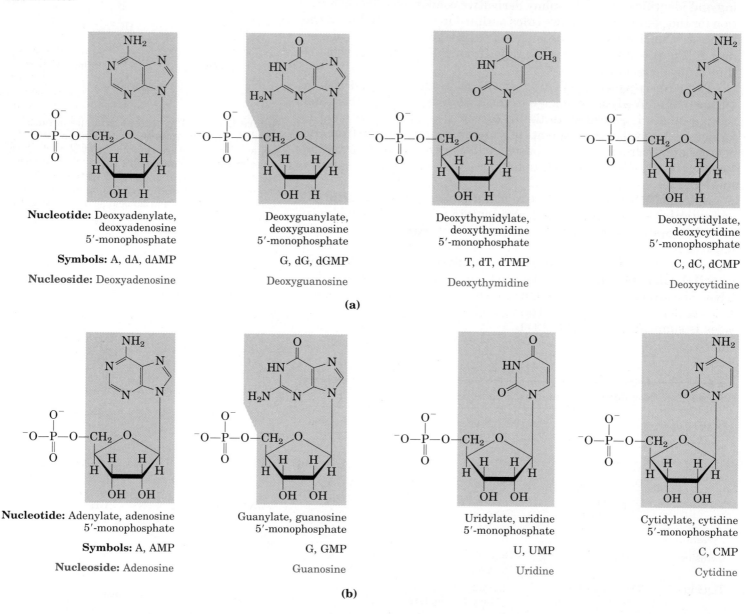

Figure 12–3 The straight-chain (aldehyde) and ring (β-furanose) forms of ribose. When ribose is free in solution, the two forms are in equilibrium. RNA contains only the ring form, β-D-ribofuranose. Deoxyribose undergoes a similar interconversion in solution, but in DNA exists solely as β-2′-deoxy-D-ribofuranose.

Two kinds of pentoses are found in nucleic acids. The recurring deoxyribonucleotide units of DNA contain 2′-deoxy-D-ribose, and the ribonucleotide units of RNA contain D-ribose. In nucleotides, both types of pentoses are in their β-furanose (closed five-member ring) form (Fig. 12–3).

Figure 12–4 gives the structures and names of the four major **deoxyribonucleotides** (deoxyribonucleoside 5′-monophosphates), the structural units of DNAs, and the four major **ribonucleotides** (ribonucleoside 5′-monophosphates), the structural units of RNAs. Specific long sequences of A, T, G, and C nucleotides in DNA encode the genetic information. Although nucleotides bearing one of these major bases are most common, both DNA and RNA also contain some minor bases

Nucleotide: Deoxyadenylate, deoxyadenosine 5′-monophosphate

Symbols: A, dA, dAMP

Nucleoside: Deoxyadenosine

Deoxyguanylate, deoxyguanosine 5′-monophosphate

G, dG, dGMP

Deoxyguanosine

Deoxythymidylate, deoxythymidine 5′-monophosphate

T, dT, dTMP

Deoxythymidine

Deoxycytidylate, deoxycytidine 5′-monophosphate

C, dC, dCMP

Deoxycytidine

(a)

Nucleotide: Adenylate, adenosine 5′-monophosphate

Symbols: A, AMP

Nucleoside: Adenosine

Guanylate, guanosine 5′-monophosphate

G, GMP

Guanosine

Uridylate, uridine 5′-monophosphate

U, UMP

Uridine

Cytidylate, cytidine 5′-monophosphate

C, CMP

Cytidine

(b)

Figure 12–4 (a) The deoxyribonucleotide units of DNA in free form at pH 7.0. In DNA they are usually symbolized as A, G, T, and C, and sometimes as dA, dG, dT, and dC. In their free form these nucleotides are commonly abbreviated dAMP, dGMP, dTMP, and dCMP. **(b)** The ribonucleotide units of RNAs. All abbreviations assume that the phosphate group is at the 5′ position. The nucleoside portion of each molecule is shaded in red. In this and the following illustrations, the ring carbons are not shown in the purine and pyrimidine bases, as is also the convention for the pentoses.

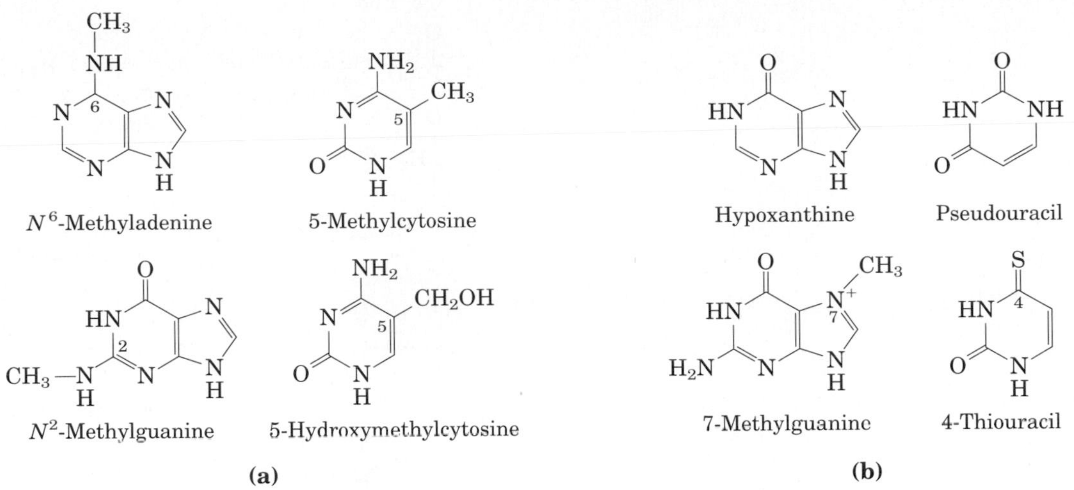

N^6-Methyladenine 5-Methylcytosine

N^2-Methylguanine 5-Hydroxymethylcytosine

(a)

Hypoxanthine Pseudouracil

7-Methylguanine 4-Thiouracil

(b)

Figure 12–5 Some minor purine and pyrimidine bases. **(a)** Minor bases found in DNA. 5-Methylcytosine occurs in the DNA of animals and higher plants, N^6-methyladenine in bacterial DNA, and 5-hydroxymethylcytosine in bacteria infected with certain bacteriophages. **(b)** Some minor bases of tRNAs. Note that pseudouracil is identical to uracil; the distinction is the point of attachment to the ribose—uracil is attached through N-1, the normal attachment point for pyrimidines, and pseudouracil is attached through C-5.

(Fig. 12–5). In DNA the most common of these are methylated forms of the major bases, but in some viral DNAs certain bases may be hydroxymethylated or glucosylated. Such altered or unusual bases in DNA molecules are in many cases specific signals for regulating or protecting the genetic information. Minor bases of many types are also found in RNAs, especially in tRNA.

The nomenclature used for the minor bases can be confusing. As indicated in Figures 12–4 and 12–5, many of the minor bases (such as hypoxanthine) have common names, just as the major bases do. For substituted forms of these bases, when the substitution involves an atom in the purine or pyrimidine rings, the usual convention (used here) is simply to indicate the ring position of the substitution by its number (e.g., 5-methylcytosine, 7-methylguanine, and 5 hydroxymethylcytosine in Fig. 12–5). The type of atom to which the substituent is attached (N, C, O, etc.) is not identified. The convention changes when the substituted atom is exocyclic, in which case the type of atom is identified and the ring position to which it is attached is denoted with a superscript. The amino group attached to C-6 in adenine becomes N^6; similarly, the carbonyl oxygen and the amino group at C-6 and C-2 of guanine become O^6 and N^2, respectively. Examples of bases substituted on exocyclic atoms are N^6-methyladenine, and N^2-methylguanine, as shown in Figure 12–5.

Cells also contain nucleotides with phosphate groups in positions other than on the 5′ carbon (Fig. 12–6). **Ribonucleoside 2′,3′-cyclic phosphates** are intermediates and **ribonucleoside 3′-phosphates** are end products of the hydrolysis of RNA by certain ribonucleases. Another variation is represented by adenosine 3′,5′-cyclic monophosphate (cAMP) and guanosine 3′,5′-cyclic monophosphate (cGMP), considered at the end of this chapter.

Figure 12–6 Some adenosine monophosphates. Adenosine 2′-monophosphate, 3′-monophosphate, and 2′,3′-cyclic monophosphate are intermediates in the alkaline hydrolysis of RNA.

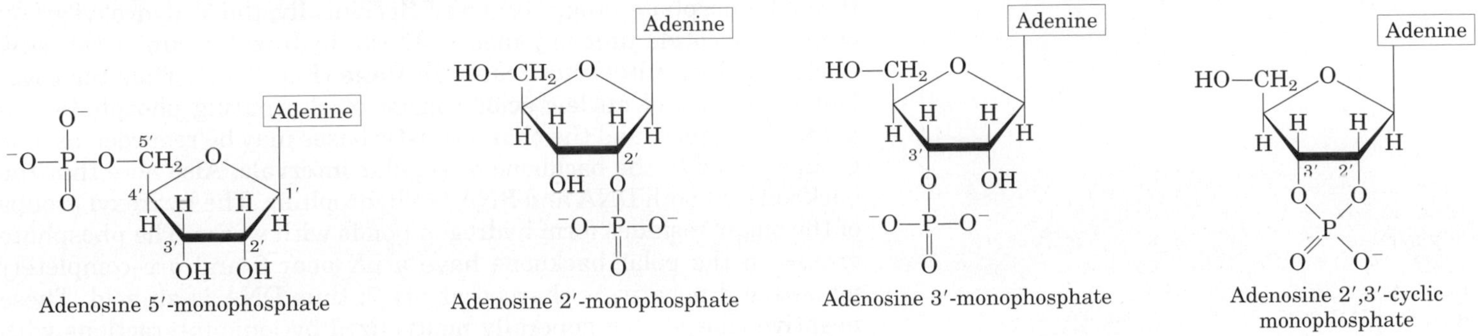

Adenosine 5′-monophosphate

Adenosine 2′-monophosphate

Adenosine 3′-monophosphate

Adenosine 2′,3′-cyclic monophosphate

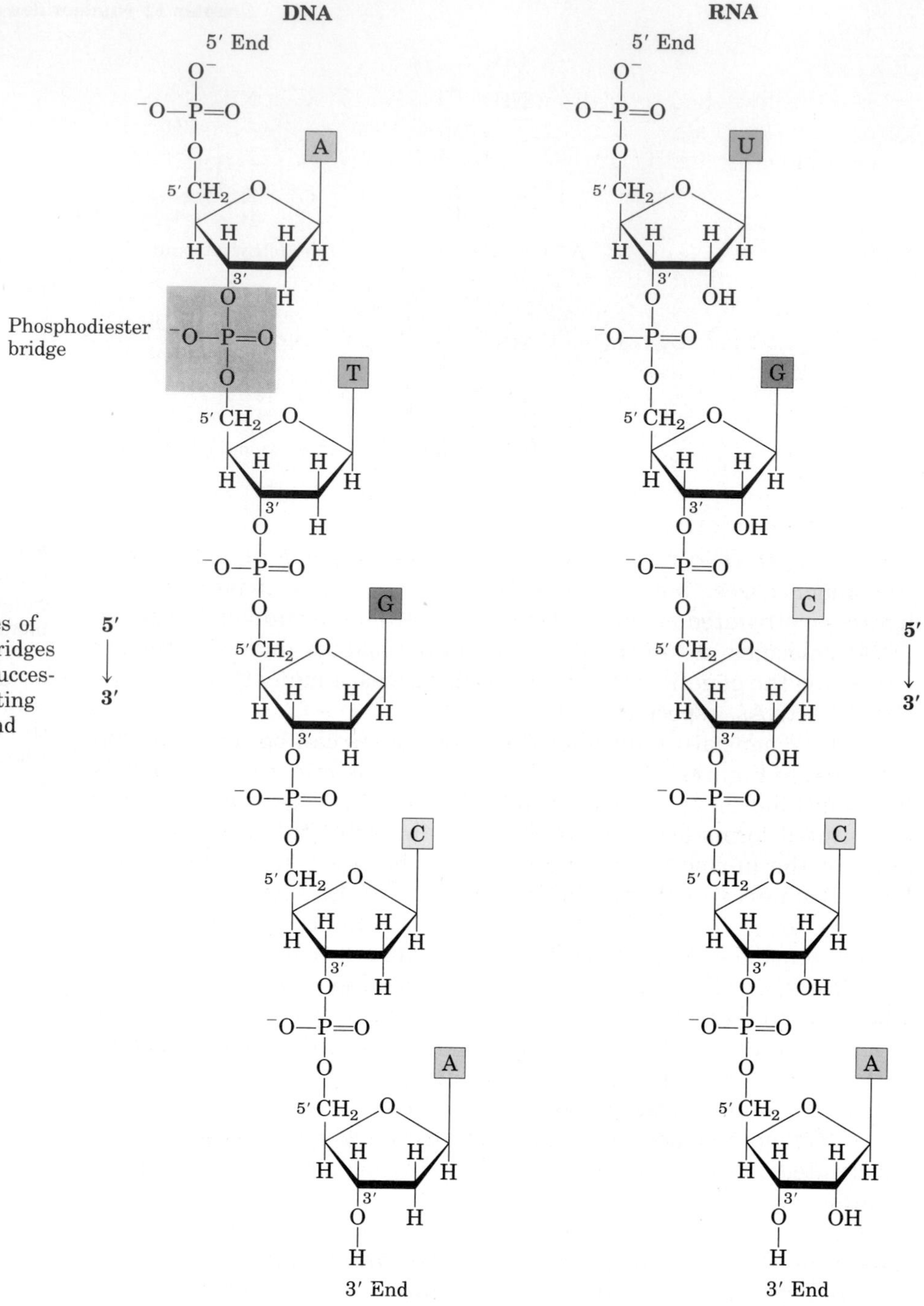

Figure 12–7 The covalent backbone structures of DNA and RNA, showing the phosphodiester bridges (one of which is shaded in the DNA) linking successive nucleotide units. The backbone of alternating pentose and phosphate groups of both DNA and RNA is highly polar.

Phosphodiester Bonds Link Successive Nucleotides in Nucleic Acids

The successive nucleotides of both DNA and RNA are covalently linked through phosphate-group "bridges." Specifically, the 5'-hydroxyl group of one nucleotide unit is joined to the 3'-hydroxyl group of the next nucleotide by a **phosphodiester linkage** (Fig. 12–7). Thus the covalent backbones of nucleic acids consist of alternating phosphate and pentose residues, and the characteristic bases may be regarded as side groups joined to the backbone at regular intervals. Also note that the backbones of both DNA and RNA are hydrophilic. The hydroxyl groups of the sugar residues form hydrogen bonds with water. The phosphate groups in the polar backbone have a pK near 0 and are completely ionized and negatively charged at pH 7; thus DNA is an acid. These negative charges are generally neutralized by ionic interactions with positive charges on proteins, metal ions, and polyamines.

All the phosphodiester linkages in DNA and RNA strands have the same orientation along the chain (Fig. 12–7), giving each linear nucleic acid strand a specific polarity and distinct 5' and 3' ends. By definition, the **5' end** lacks a nucleotide at the 5' position, and the **3' end** lacks a nucleotide at the 3' position (Fig. 12–7). Other groups (most often one or more phosphates) may be present on one or both ends.

The covalent backbone of DNA and RNA is subject to slow, nonenzymatic hydrolysis of the phosphodiester bonds. In the test tube, RNA is hydrolyzed rapidly under alkaline conditions, but DNA is not; the 2'-hydroxyl groups in RNA (absent in DNA) are directly involved in the process. Cyclic 2',3'-monophosphates are the first products of the action of alkali on RNA, and are rapidly hydrolyzed further to yield a mixture of 2'- and 3'-nucleoside monophosphates (Fig. 12–8).

The nucleotide sequences of nucleic acids can be represented schematically, as illustrated (at right) by a segment of DNA having five nucleotide units. The phosphate groups are symbolized by Ⓟ and each deoxyribose by a vertical line. The carbons in the deoxyribose are represented from 1' at the top to 5' at the bottom of the vertical line (even though the sugar is always in its closed-ring β-furanose form in nucleic acids). The connecting lines between nucleotides (through Ⓟ) are drawn diagonally from the middle (3') of the deoxyribose of one nucleotide to the bottom (5') of the next. By convention, the structure of a single strand of nucleic acid is always written with the 5' end at the left and the 3' end at the right; i.e., in the 5'→3' direction. Some simpler representations of the pentadeoxyribonucleotide illustrated are pA-C-G-T-A$_{OH}$, pApCpGpTpA, and pACGTA. A short nucleic acid is referred to as an **oligonucleotide.** The definition of "short" is somewhat arbitrary, but the term oligonucleotide is often used for polymers containing 50 or fewer nucleotides. A longer nucleic acid is called a **polynucleotide.**

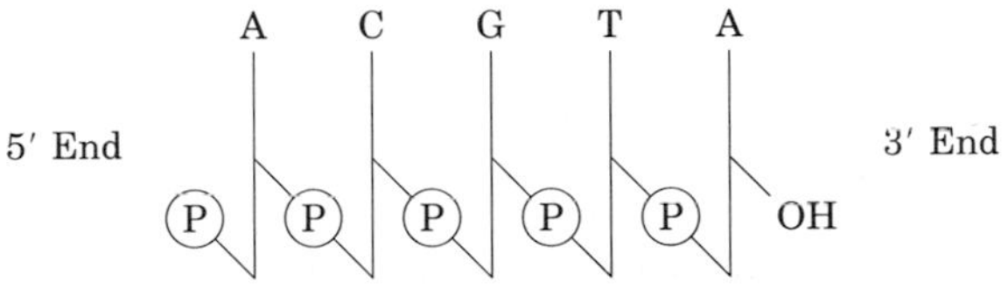

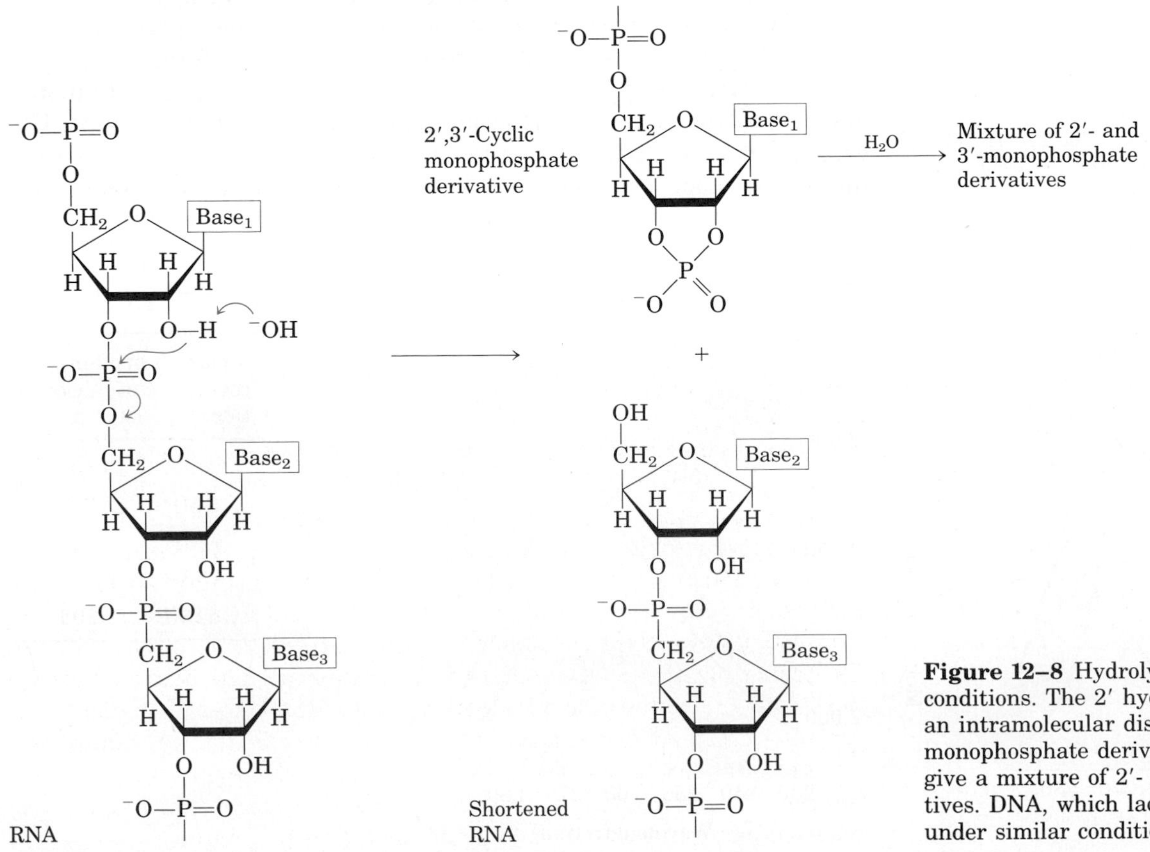

Figure 12–8 Hydrolysis of RNA under alkaline conditions. The 2' hydroxyl acts as a nucleophile in an intramolecular displacement, the 2',3'-cyclic monophosphate derivative is further hydrolyzed to give a mixture of 2'- and 3'-monophosphate derivatives. DNA, which lacks 2' hydroxyls, is stable under similar conditions.

The Properties of Nucleotide Bases Affect the Structure of Nucleic Acids

The bases have a variety of chemical properties that affect the structure, and ultimately the function, of nucleic acids. Free pyrimidines and purines are weakly basic compounds, and are thus called bases. The purines and pyrimidines common in DNA and RNA are highly conjugated molecules (see Fig. 12–2). This property has important effects on the structure, electron distribution, and light absorption of nucleic acids. Resonance involving many atoms in the ring gives most of the bonds a partially double-bonded character. One result is that pyrimidines are planar molecules; purines are very nearly planar, with a slight pucker. Free pyrimidine and purine bases may exist in two or more tautomeric forms depending upon the pH. Uracil, for example, occurs in lactam, lactim, and double lactim forms (Fig. 12–9). The structures of the purines and pyrimidines shown in Figure 12–2 are the tautomers predominating at pH 7.0. Again as a result of resonance, all of the bases absorb UV light, and nucleic acids are characterized by a strong absorption at wavelengths near 260 nm (Fig. 12–10).

The purines and pyrimidines are also hydrophobic and relatively insoluble in water at the near neutral pH of the cell. At acidic or alkaline pH the purines and pyrimidines become charged, and their solubility in water increases. Hydrophobic stacking interactions in which two or more bases are positioned with the planes of their rings parallel (similar to a stack of coins) represent one of two important modes of interaction between two bases. The stacking involves a combination of van der Waals and dipole–dipole interactions between the bases. These base-stacking interactions help to minimize contact with water and are very important in stabilizing the three-dimensional structure of nucleic acids, as described later. The close interaction between stacked bases in a nucleic acid has the effect of decreasing the absorption of UV light relative to a solution with the same concentration of free nucleotides. This is called the **hypochromic effect.**

The most important functional groups of pyrimidines and purines are ring nitrogens, carbonyl groups, and exocyclic amino groups. Hydrogen bonds involving the amino and carbonyl groups are the second important mode of interaction between bases. Hydrogen bonds be-

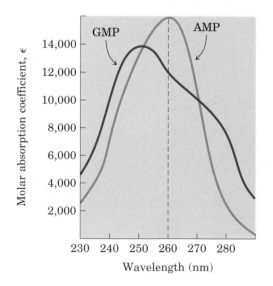

Figure 12–9 Tautomeric forms of uracil. At pH 7.0, the lactam form predominates; the other forms become more prominent as pH decreases.

Figure 12–10 The absorption spectra of the common nucleotides and their molar absorption coefficients at 260 nm and pH 7.0 (ϵ_{260}). The spectra of the corresponding ribonucleotides and deoxyribonucleotides, as well as the nucleosides, are essentially identical. When mixtures of nucleotides are present, the wavelength at 260 nm (dashed vertical lines) is used for measurements.

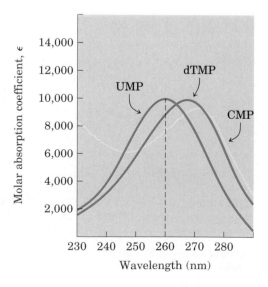

Molar absorption coefficient of nucleotides, ϵ_{260} ($M^{-1}cm^{-1}$)	
AMP	15,400
GMP	11,700
CMP	7,500
UMP	9,900
dTMP	9,200

Figure 12–11 Hydrogen-bonding patterns in the base pairs defined by Watson and Crick.

tween bases permit a complementary association of two and occasionally three strands of nucleic acid. The most important hydrogen-bonding patterns are those defined by James Watson and Francis Crick in 1953, in which A bonds specifically to T (or U) and G bonds to C (Fig. 12–11). These two types of base pairs predominate in double-stranded DNA and RNA, and the tautomers shown in Figure 12–2 are responsible for these patterns. This specific pairing of bases permits the duplication of genetic information by the synthesis of nucleic acid strands that are complementary to existing strands, as we shall discuss later in this chapter.

Nucleic Acid Structure

The discovery of the structure of DNA by Watson and Crick in 1953 was a momentous event in science, an event that gave rise to entirely new disciplines and influenced the course of many others. Our present understanding of the storage and utilization of a cell's genetic information is based on work made possible by this discovery. Although information pathways are not treated in detail until Part IV of this book, the outline of these pathways presented in Chapters 1 and 3 is now a prerequisite for discussion of any area of biochemistry. Here, we concern ourselves with DNA structure itself, events that led to its discovery, and more recent refinements in our understanding. RNA structure will also be introduced.

As in the case of protein structure (Chapters 6 and 7), it is sometimes useful to describe nucleic acid structure in terms of hierarchical levels of complexity (primary, secondary, tertiary). The primary structure of a nucleic acid is its covalent structure and nucleotide sequence. Any regular, stable structure taken up by some or all of the nucleotides in a nucleic acid can be referred to as secondary structure. All of the structures considered in the following pages of this chapter fall under the heading of secondary structure. The complex folding of large chromosomes within the bacterial nucleoid and eukaryotic chromatin is generally considered tertiary structure; this is considered in Chapter 23.

James Watson

Francis Crick

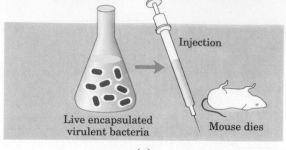

(a)

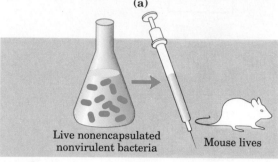

(b)

Live encapsulated
virulent bacteria

Mouse dies

Live nonencapsulated
nonvirulent bacteria

Mouse lives

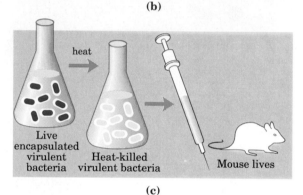

heat

Live
encapsulated
virulent
bacteria

Heat-killed
virulent bacteria

Mouse lives

(c)

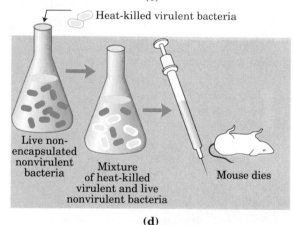

Heat-killed virulent bacteria

Live non-
encapsulated
nonvirulent
bacteria

Mixture
of heat-killed
virulent and live
nonvirulent bacteria

Mouse dies

(d)

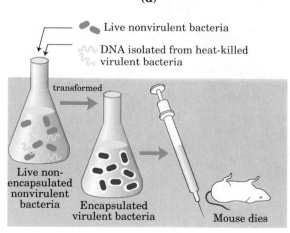

Live nonvirulent bacteria

DNA isolated from heat-killed
virulent bacteria

transformed

Live non-
encapsulated
nonvirulent
bacteria

Encapsulated
virulent bacteria

Mouse dies

(e)

DNA Stores Genetic Information

The biochemical investigation of DNA began with Friedrich Miescher, who carried out the first systematic chemical studies of cell nuclei. In 1868 Miescher isolated a phosphorus-containing substance, which he called "nuclein," from the nuclei of pus cells (leukocytes) obtained from discarded surgical bandages. He found nuclein to consist of an acidic portion, which we know today as DNA, and a basic portion, protein. Miescher later found a similar acidic substance in the heads of salmon sperm cells. Although he partially purified the nucleic acid and studied its properties, the covalent (primary) structure of DNA (as shown in Fig. 12–7) did not become known with certainty until the late 1940s.

Miescher and many others suspected that nuclein or nucleic acid was associated in some way with cell inheritance, but the first direct evidence that DNA is the bearer of genetic information came in 1944 through a discovery made by Oswald T. Avery, Colin MacLeod, and Maclyn McCarty. These investigators found that DNA extracted from a virulent (disease-causing) strain of the bacterium *Streptococcus pneumoniae,* also known as pneumococcus, genetically transformed a nonvirulent strain of this organism into a virulent form (Fig. 12–12). Avery and his colleagues concluded that the DNA extracted from the virulent strain carried the inheritable genetic message for virulence. Not everyone accepted these conclusions, because traces of protein impurities present in the DNA could have been the actual carrier of the genetic information. This possibility was soon eliminated by the finding that treatment of the DNA with proteolytic enzymes did not destroy the transforming activity, but treatment with deoxyribonucleases (DNA-hydrolyzing enzymes) did.

A second important experiment provided independent evidence that DNA carries genetic information. In 1952 Alfred D. Hershey and Martha Chase used radioactive phosphorus (^{32}P) and radioactive sulfur (^{35}S) tracers to show that when the bacterial virus (bacteriophage) T2 infects its host cell, *E. coli,* it is the phosphorus-containing DNA of

Figure 12–12 The Avery–MacLeod–McCarty experiment. When injected into mice, the encapsulated strain of pneumococcus **(a)** is lethal, whereas the nonencapsulated strain **(b)** is harmless, as is the heat-killed encapsulated strain **(c).** Earlier research by the bacteriologist Frederick Griffith had shown that adding heat-killed virulent bacteria (which alone are harmless to mice) to a live nonvirulent strain permanently transformed the latter into lethal, virulent, encapsulated bacteria **(d).** He concluded that a transforming factor in the heat-killed virulent bacteria had gained entrance into the live nonvirulent bacteria and rendered them virulent and encapsulated.

Avery and his colleagues identified the Griffith transforming factor as DNA. **(e)** They extracted the DNA from heat-killed virulent pneumococci, removing the protein as completely as possible, and added this DNA to nonvirulent bacteria. The nonvirulent pneumococci were permanently transformed into a virulent strain. The DNA evidently gained entrance into the nonvirulent bacteria, and the genes for virulence and capsule formation became incorporated into the chromosomes of the nonvirulent bacteria. All subsequent generations of these bacteria were therefore virulent and encapsulated.

Figure 12–13 Summary of the Hershey–Chase experiment. Two batches of isotopically labeled bacteriophage particles were prepared. One was labeled with ^{32}P in the phosphate groups of the DNA and the other with ^{35}S in the sulfur-containing amino acids of the protein coats (capsids). (Note that DNA contains no sulfur, and viral protein no phosphorus.) The two batches of labeled phage were then added to separate suspensions of unlabeled bacteria. Each suspension of phage-infected cells was agitated in a blender to shear the viral capsids from the bacteria. The bacteria and empty viral coats (ghosts) were then separated by centrifugation. The cells infected with the ^{32}P-labeled phage were found to contain ^{32}P, indicating that the labeled viral DNA had entered the cells, and the viral ghosts contained no radioactivity. The cells infected with ^{35}S-labeled phage were found to have no radioactivity after blender treatment, but the viral ghosts contained ^{35}S. Progeny virus particles were produced in both batches of bacteria some time after the viral coats were removed, thus the genetic message for their replication had been introduced by viral DNA, not by viral protein.

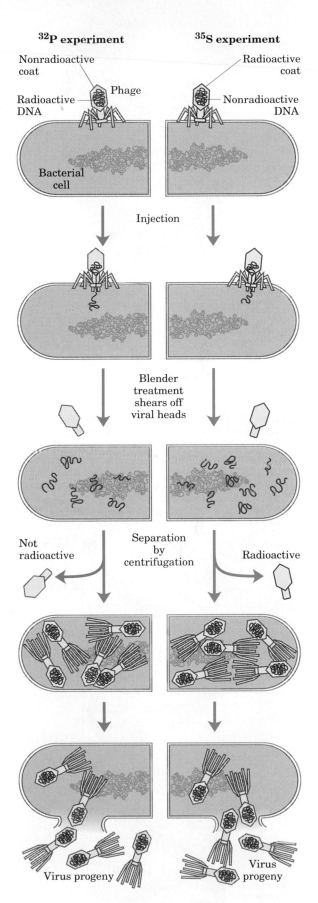

the viral particle, not the sulfur-containing protein of the viral coat, that actually enters the host cell and furnishes the genetic information for viral replication (Fig. 12–13).

These important early experiments and many other lines of evidence have shown that DNA is definitely the exclusive chromosomal component bearing the genetic information of living cells.

DNAs Have Distinctive Base Compositions

A most important clue to the structure of DNA came from the work of Erwin Chargaff and his colleagues in the late 1940s. They found that the four nucleotide bases in DNA occur in different ratios in the DNAs of different organisms and that the amounts of certain bases are closely related. These data, collected from DNAs of a great many different species, led Chargaff to the following conclusions:

1. The base composition of DNA generally varies from one species to another.

2. DNA specimens isolated from different tissues of the same species have the same base composition.

3. The base composition of DNA in a given species does not change with the organism's age, nutritional state, or changing environment.

4. In *all* DNAs, regardless of the species, the number of adenine residues is equal to the number of thymine residues (that is, A = T), and the number of guanine residues is equal to the number of cytosine residues (G = C). From these relationships it follows that the sum of the purine residues equals the sum of the pyrimidine residues; that is, A + G = T + C.

These quantitative relationships, sometimes called "Chargaff's rules," were confirmed by many subsequent researchers. They were a key to establishing the three-dimensional structure of DNA and yielded clues to how genetic information is encoded in DNA and passed from one generation to the next.

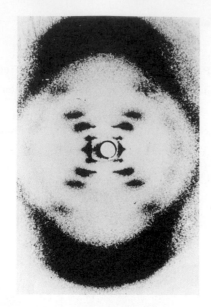

Figure 12–14 The x-ray diffraction pattern of DNA. The spots forming a cross in the center denote a helical structure. The heavy bands at the top and bottom correspond to the recurring bases.

Figure 12–15 The Watson–Crick model for the structure of DNA. The original model proposed that there are 10 base pairs or 3.4 nm per turn of the helix. Subsequent measurements have shown that there are 10.5 base pairs or 3.6 nm per turn. (a) Schematic representation, showing dimensions of the helix. (b) Line model showing the backbone and stacking of the bases. (c) Space-filling model.

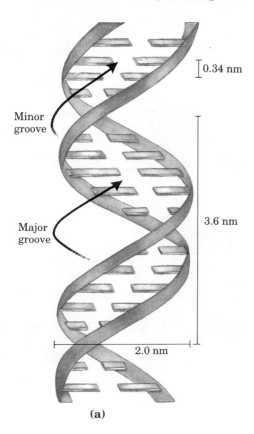

0.34 nm

Minor groove

Major groove

3.6 nm

2.0 nm

(a)

DNA Is a Double Helix

To shed more light on the structure of DNA, Rosalind Franklin and Maurice Wilkins used the powerful method of x-ray diffraction (see Box 7–3) to analyze DNA crystals. They showed in the early 1950s that DNA produces a characteristic x-ray diffraction pattern (Fig. 12–14). From this pattern it was deduced that DNA polymers are helical with two periodicities along their long axis, a primary one of 0.34 nm and a secondary one of 3.4 nm. The pattern also indicated that the molecule contains two strands, a clue that was crucial to determining the structure. The problem then was to formulate a three-dimensional model of the DNA molecule that could account not only for the x-ray diffraction data but also for the specific A = T and G = C base equivalences discovered by Chargaff and for the other chemical properties of DNA.

In 1953 Watson and Crick postulated a three-dimensional model of DNA structure that accounted for all of the available data (Fig. 12–15). It consists of two helical DNA chains coiled around the same axis to form a right-handed double helix (see Box 7–1 for an explanation of the right- or left-handed sense of a helical structure). The hydrophilic backbones of alternating deoxyribose and negatively charged phosphate groups are on the outside of the double helix, facing the surrounding water. The purine and pyrimidine bases of both strands are stacked inside the double helix, with their hydrophobic and nearly planar ring structures very close together and perpendicular to the long axis of the helix. The spatial relationship between these strands creates a **major groove** and **minor groove** between the two strands. Each base of one strand is paired in the same plane with a base of the other strand. Watson and Crick found that the hydrogen-bonded base pairs illustrated in Figure 12–11 are those that fit best within the structure, providing a rationale for Chargaff's rules. It is important to note that three hydrogen bonds can form between G and C, symbolized G≡C, but only two can form between A and T, symbolized A=T. Other pairings of bases tend (to varying degrees) to destabilize the double-helical structure.

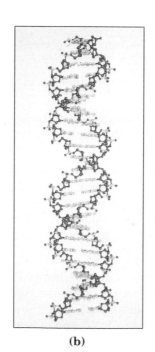

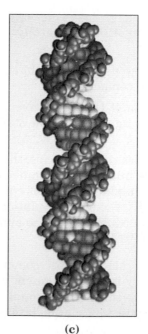

(b)   **(c)**

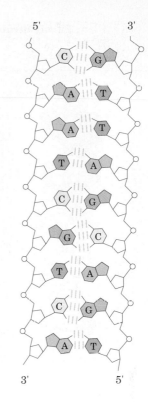

Figure 12–16 Schematic drawing of complementary antiparallel strands of DNA following the pairing rules proposed by Watson and Crick. The base-paired antiparallel strands differ in base composition: the left strand has the composition A_3 T_2 G_1 C_3; the right, A_2 T_3 G_3 C_1. They also differ in sequence when each chain is read in the $5' \rightarrow 3'$ direction. Note the base equivalences: A = T and G = C. In this and following illustrations, the hydrogen bonds between base pairs are often represented by sets of blue lines.

In the Watson–Crick structure, the two chains or strands of the helix are **antiparallel;** their $5',3'$-phosphodiester bonds run in opposite directions. Later work with DNA polymerases (Chapter 24) provided experimental evidence, confirmed by x-ray crystallography, that the strands are indeed antiparallel.

To account for the periodicities observed in the x-ray diffraction pattern, Watson and Crick used molecular models to show that the vertically stacked bases inside the double helix would be 0.34 nm apart and that the secondary repeat distance of about 3.4 nm could be accounted for by the presence of 10 (now 10.5) nucleotide residues in each complete turn of the double helix (Fig. 12–15a). As can be seen in Figure 12–16, the two antiparallel polynucleotide chains of double-helical DNA are not identical in either base sequence or composition. Instead they are **complementary** to each other. Wherever adenine appears in one chain, thymine is found in the other; similarly, wherever guanine is found in one chain, cytosine is found in the other.

The DNA double helix or duplex is held together by two sets of forces, as described earlier: hydrogen bonding between complementary base pairs (Fig. 12–11) and base-stacking interactions. The specificity that maintains a given base sequence in each DNA strand is contributed entirely by the hydrogen bonding between base pairs. The base-stacking interactions, which are largely nonspecific with respect to the identity of the stacked bases, make the major contribution to the stability of the double helix.

The important features of the double-helical model of DNA structure are supported by much chemical and biological evidence. Moreover, the model immediately suggested a mechanism for the transmission of genetic information. The essential feature of the model is the complementarity of the two DNA strands. Making a copy of this structure (replication) could logically proceed by (1) separating the two strands and (2) synthesizing a complementary strand for each by joining nucleotides in a sequence specified by the base-paring rules stated above. Each preexisting strand could function as a template to guide the synthesis of the complementary strand (Fig. 12–17). These expectations have been experimentally confirmed, and this discovery was a revolution in our understanding of DNA metabolism.

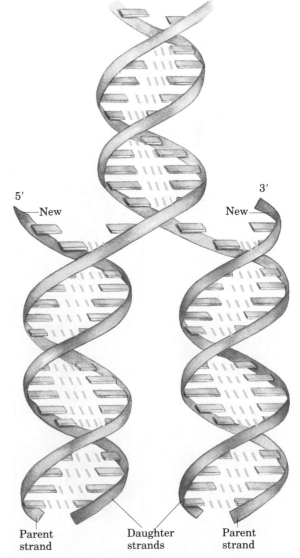

Figure 12–17 Replication of DNA as suggested by Watson and Crick. The parent strands become separated, and each forms the template for biosynthesis of a complementary daughter strand (in red).

Parent strand

Daughter strands

Parent strand

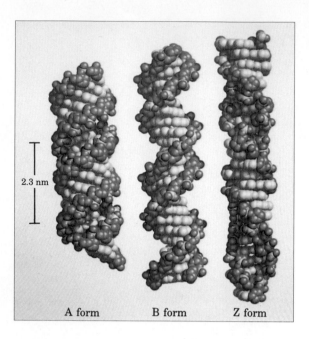

2.3 nm

A form   B form   Z form

Figure 12–18 Comparison of the A, B, and Z forms of DNA. There are 24 base pairs in each of the structures shown.

DNA Can Occur in Different Structural Forms

DNA is a remarkably flexible molecule. Considerable rotation is possible around a number of bonds in the sugar–phosphate backbone, and thermal fluctuation can produce bending, stretching, and unpairing (melting) in the structure. Many significant deviations from the Watson–Crick DNA structure are found in cellular DNA, and some or all of these may play important roles in DNA metabolism. These structural variations generally do not affect the key properties of DNA defined by Watson and Crick: strand complementarity, antiparallel strands, and the requirement for A=T and G≡C base pairs.

The Watson–Crick structure is also referred to as B-form DNA. The B form is the most stable structure for a random-sequence DNA molecule under physiological conditions, and is therefore the standard point of reference in any study of the properties of DNA. Two DNA structural variants that have been well characterized in crystal structures are the A and Z forms (Fig. 12–18). The A form is favored in many solutions that are relatively devoid of water. The DNA is still arranged in a right-handed double helix, but the rise per base pair is 0.23 nm and the number of base pairs per helical turn is 11, relative to the 0.34 nm rise and 10.5 base pairs per turn found in B-DNA. For a given DNA molecule, the A form will be shorter and have a greater diameter than the B form. The reagents used to promote crystallization of DNA tend to dehydrate it, and this leads to a tendency for many DNAs to crystallize in the A form.

Z-form DNA is a more radical departure from the B structure; the most obvious distinction is the left-handed helical rotation. There are 12 base pairs per helical turn, with a rise of 0.38 nm per base pair. The DNA backbone takes on a zig-zag appearance. Certain nucleotide sequences fold up into left-handed Z helices more readily than do others. Prominent examples are sequences in which pyrimidines alternate with purines, especially alternating C and G or 5-methyl-C and G. Whether A-form DNA actually occurs in cells is uncertain, but there is evidence for some short stretches (tracts) of Z-DNA in both prokaryotes and eukaryotes. These Z-DNA tracts may play an as yet undefined role in the regulation of the expression of some genes or in genetic recombination.

Certain DNA Sequences Adopt Unusual Structures

A number of other sequence-dependent structural variations have been detected that may serve locally important functions in DNA metabolism. For example, some sequences cause bends in the DNA helix. Bends are produced whenever four or more adenine residues appear sequentially in one of the two strands (Fig. 12–19). Six adenines in a row produce a bend of about 18°. The bending observed with this and other sequences may be important in the binding of some proteins to DNA.

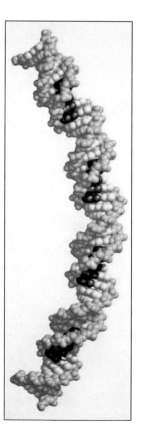

Figure 12–19 A model for the bending of DNA produced by poly(A) tracts. The bend here is produced by four (dA)₅ tracts, separated by five base pairs. The adenine bases are shown in red.

Figure 12–20 Palindromes and mirror repeats. Palindromes are defined in nucleic acids as sequences with twofold symmetry. In order to superimpose one repeat (shaded sequence) on the other, it must be rotated 180° around the horizontal axis and then again about the vertical axis, as shown by the colored arrows. A mirror repeat, on the other hand, has a symmetric sequence on each strand. Superimposing one repeat on the other requires only a single 180° rotation about the vertical axis.

Palindrome

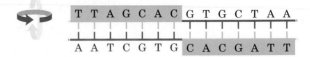

Mirror repeat

A rather common type of sequence found in DNA is a **palindrome.** A palindrome is a word, phrase, or sentence that is spelled identically reading forward or backward; two examples are ROTATOR and NURSES RUN. The term is applied to regions of DNA in which there are **inverted repetitions** of base sequence with twofold symmetry occurring over two strands of DNA (Fig. 12–20). Such sequences are self-complementary within each of the strands and therefore have the potential to form hairpin or cruciform (cross-shaped) structures (Fig. 12–21). When the inverted sequence occurs within each individual strand of the DNA, the sequence is called a **mirror repeat.** Mirror repeats do not have complementary sequences within the same strand and cannot form hairpin or cruciform structures. Sequences of these types are found in virtually every large DNA molecule and can involve a few or up to thousands of base pairs. It is not known how many palindromes actually occur as cruciforms in cells, although the existence of at least some cruciform structures has been demonstrated in vivo in *E. coli.* Self-complementary sequences cause isolated single strands of DNA to fold up in solution into complex structures containing multiple hairpins.

A particularly unusual DNA structure, known as H-DNA, is found in polypyrimidine/polypurine tracts that also incorporate a mirror repeat within the sequence. One simple example is a long stretch of alternating T and C residues, as shown in Figure 12–22. A novel feature of H-DNA is the pairing and interwinding of *three* strands of DNA to form a triple helix. Triple-helical DNA forms spontaneously only within long sequences containing only pyrimidines (or only purines) in one strand. Two of the three strands in the H-DNA triple helix (Fig. 12–22c, d) contain pyrimidines and the third contains purines.

These structural variations are interesting because there is a tendency for many of them to appear at sites where important events in DNA metabolism (replication, recombination, transcription) are initi-

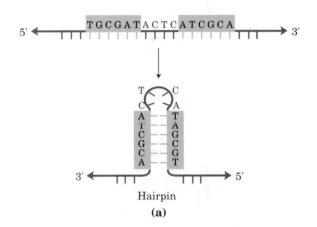

Hairpin

(a)

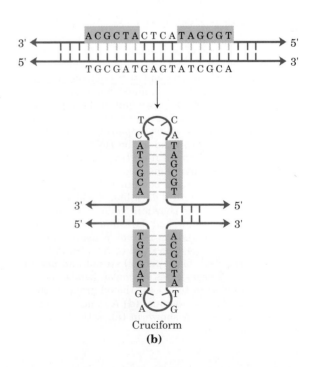

Cruciform

(b)

Figure 12–21 Hairpins and cruciforms. Palindromic DNA (or RNA) sequences can form alternative structures with intrastrand base pairing. When only a single DNA (or RNA) strand is involved it is called a hairpin **(a).** When both strands of a duplex DNA are involved, the structure is called a cruciform **(b).** Blue shading highlights asymmetric sequences that can pair alternatively with a complementary sequence in the same or opposite strand.

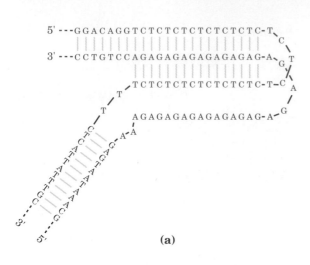

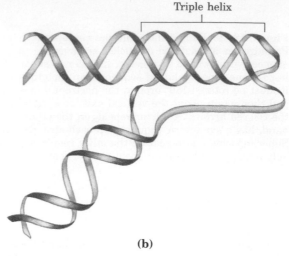

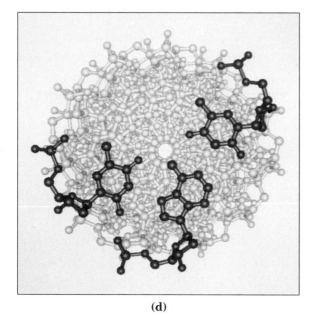

Figure 12–22 H-DNA. A sequence of alternating T and C residues can be considered a mirror repeat centered about one of the central T or C residues **(a).** These sequences form an unusual structure in which the strands in one half of the mirror repeat are separated, and the pyrimidine-containing strand folds back on the other half of the repeat to form a triple helix **(b).** The purine strand (alternating A and G residues) is left unpaired. This structure produces a sharp bend in the DNA. **(c)** A triple-helical DNA formed from two pyrimidine strands (polydeoxythymidine, shown with gray and light blue backbones) and one purine strand (polydeoxyadenine, with a dark blue backbone). Phosphorus atoms are shown in yellow. In this structure the light blue and dark blue strands are antiparallel and paired via normal Watson–Crick base pairing patterns. The third (gray) strand is parallel to the dark blue (purine) strand and paired through non–Watson–Crick hydrogen bonds, including one between the C-4 carbonyl group of thymine and the N-7 of adenine. **(d)** An end view of the triple-helical DNA shown in **(c),** with the base triplet at one end.

ated or regulated. For example, the sites recognized by many sequence-specific DNA-binding proteins (Chapter 27) are arranged as palindromes, and sequences that can form H-DNA are found within regions involved in the regulation of expression of a number of genes in eukaryotes. Much work is still required to define these structures and determine their functional significance.

Messenger RNAs Code for Polypeptide Chains

We now turn our attention briefly from DNA structure to the expression of the genetic information contained in DNA. RNA, the second major form of nucleic acid in cells, plays the role of intermediary in converting this information into a functional protein.

In eukaryotes DNA is largely confined to the nucleus, whereas protein synthesis occurs on ribosomes in the cytoplasm. Therefore some molecule other than DNA must carry the genetic message for protein synthesis from the nucleus to the cytoplasm. As early as the 1950s, RNA was considered the logical candidate: RNA is found in both the nucleus and cytoplasm, and the onset of protein synthesis is accompanied by an increase in the amount of RNA in the cytoplasm and an increase in its rate of turnover. These and other observations led several researchers to suggest that RNA carries genetic information from DNA to the protein biosynthetic machinery of the ribosome. In 1961, Francois Jacob and Jacques Monod presented a unified (and essentially correct) picture of many aspects of this process. They proposed the name messenger RNA (mRNA) for that portion of the total cell RNA carrying the genetic information from DNA to the ribosomes, where the messengers provide the templates for specifying amino acid sequences in polypeptide chains. Although mRNAs from different genes can vary greatly in length, the mRNAs from a particular gene will generally have a defined size. The process of forming mRNA on a DNA template is known as transcription.

Figure 12–23 Schematic diagram of monocistronic **(a)** and polycistronic **(b)** mRNAs of prokaryotes. The polycistronic transcript shown here contains three genes. Noncoding RNA separates the genes.

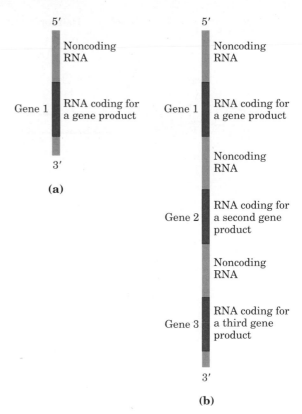

In prokaryotes a single mRNA molecule may code for one or several polypeptide chains. If it carries the code for only one polypeptide, the mRNA is **monocistronic;** if it codes for two or more different polypeptides, the mRNA is **polycistronic.** In eukaryotes, most mRNAs are monocistronic. (The term **cistron,** for purposes of this discussion, refers to a gene. The term itself has historical roots in the science of genetics, and its formal genetic definition is beyond the scope of this text.) The minimum length of an mRNA is set by the length of the polypeptide chain for which it codes. For example, a polypeptide chain of 100 amino acid residues requires an RNA coding sequence of at least 300 nucleotides, because each amino acid is coded by a nucleotide triplet (Chapter 26). However, mRNAs transcribed from DNA are always somewhat longer than needed simply to specify the code for the polypeptide sequence(s). The additional noncoding RNA includes sequences that regulate protein synthesis (Chapter 26). Figure 12–23 summarizes the general structure of prokaryotic mRNAs.

Many RNAs Have More Complex Structures

Messenger RNA is only one of several classes of cellular RNA. Transfer RNAs serve as adapter molecules in protein synthesis; covalently linked to an amino acid at one end, they pair with the mRNA in such a way that the amino acids are joined in the correct sequence. Ribosomal RNAs are structural components of ribosomes. There is also a wide variety of special-function RNAs. All of these are considered in detail in Chapter 25.

Regardless of the class of RNA being synthesized, the product of transcription is always a single strand of RNA. The single-stranded nature of these molecules does not mean their structure is random. The single strands tend to take up a right-handed helical conformation that is dominated by base-stacking interactions (Fig. 12–24). The stacking interactions are stronger between two purines than between a purine and a pyrimidine or between two pyrimidines. The purine–purine interaction is so strong that a pyrimidine separating two purines will often be displaced from the stacking pattern so that the purines can interact. Any self-complementary sequences in the molecule will lead to more complex and specific structures. RNA can base-pair with complementary strands of either RNA or DNA. The standard base-pairing rules are identical to those for DNA: guanine pairs with cytosine and adenine pairs with uracil (or thymine). One difference is that one unusual base pairing—between guanine and uracil—is fairly common between two strands of RNA; see Fig. 12–26. The paired strands in RNA or RNA–DNA are antiparallel, as in DNA.

Figure 12–24 Typical right-handed stacking pattern found in a single strand of RNA. The bases are shown in white, the ribose rings in green, and the phosphate atoms in yellow. The right-handed twist of the backbone is evident.

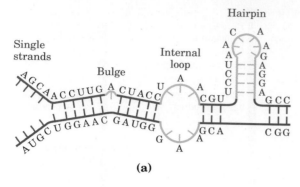

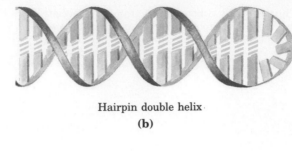

Hairpin double helix
(b)

(a)

Figure 12–25 (a) Types of secondary structure found in some RNAs. The paired regions generally have an A-form right-handed helix, as shown for a hairpin **(b)**.

Unlike the double helix of DNA, there is no simple, regular secondary structure that forms a reference point for RNA structure. The three-dimensional structures of many RNAs, like those of proteins, are complex and unique. Weak interactions, especially base-stacking (hydrophobic) interactions, again play a major role in stabilizing structures. Where complementary sequences are present, the predominant double-stranded structure is an A-form right-handed double helix. Z-form helices have been made in the laboratory (under very high-salt or high-temperature conditions). The B form of RNA has not been observed. Breaks in the regular A-form helix caused by mismatched or unmatched bases in one or both strands are common, and result in bulges or internal loops (Fig. 12–25). Hairpin loops form between nearby self-complementary sequences in the RNA strand (Fig. 12–25). The potential for base-paired helical structures in many RNAs is extensive (Fig. 12–26), and the resulting hairpins can be considered the most common type of secondary structure in RNA. Certain short base sequences, such as UUCG, are often found at the ends of RNA hairpins and are known to form particularly tight and stable loops. Such sequences may play an important role in nucleating the folding of an RNA molecule into its precise three-dimensional structure. Important additional structural contributions are made by hydrogen bonds that are not part of standard Watson–Crick base pairs. For example, the 2′-hydroxyl group of ribose can form a hydrogen bond with other groups, and a variety of nonstandard base-pairing patterns are also observed. Some of these properties are evident in the structure of the phenylalanine transfer RNA of yeast (Fig. 12–27).

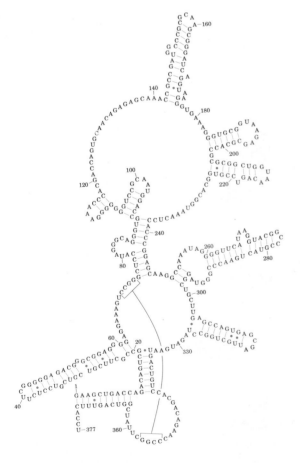

Figure 12–26 Possible secondary structure of the M1 RNA component of the enzyme RNase P of *E. coli,* showing many hairpins. RNase P also contains a protein component (not shown). This enzyme functions in the processing of transfer RNAs,

as described in Chapter 25. Brackets indicate additional complementary sequences that may be paired in the three-dimensional structure. The dots (·) indicate non–Watson–Crick G≡U base pairs, as shown above.

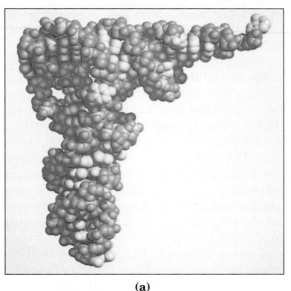

(a)

**Figure 12–27 Phenylalanine tRNA of yeast.
(a)** Three-dimensional structure. **(b)** Some unusual base-pairing patterns. Note also the involvement of a phosphodiester bond oxygen in one hydrogen-bonding arrangement, and the involvement of a 2′-hydroxyl group in another (both in red).

(b)

The analysis of RNA structure and its relationship to function is an emerging field of inquiry that has many of the same complexities as the analysis of protein structure. The importance of understanding RNA structure grows as we become aware of an increasing number of functions of RNA molecules.

Nucleic Acid Chemistry

To understand how nucleic acids function, we must understand their chemical properties as well as their structures. DNA functions well as a repository of genetic information in part because of its inherent stability. The chemical transformations that do occur are generally very slow in the absence of an enzyme catalyst. The long-term storage of information without alteration is so important to a cell, however, that even very slow reactions that alter DNA structure can be physiologically significant. Processes such as carcinogenesis and aging may be intimately linked to slowly accumulating, irreversible alterations of DNA. Nondestructive alterations, such as the strand separation that must precede DNA replication or transcription, are also important. In addition to providing these insights into physiological processes, our understanding of nucleic acid chemistry has given us a powerful array of technologies that have applications in molecular biology, medicine, and forensic science. We now examine the chemical properties of DNA and some of these technologies.

Double-Helical DNA and RNA Can Be Denatured

Solutions of carefully isolated, native DNA are highly viscous at pH 7.0 and room temperature (20 to 25 °C). When such a solution is subjected to extremes of pH or to temperatures above 80 to 90 °C, its viscosity decreases sharply, indicating that the DNA has undergone a physical change. Just as heat and extremes of pH cause denaturation of globular proteins, so too will they cause denaturation or melting of double-helical DNA. This involves disruption of the hydrogen bonds between the paired bases and the hydrophobic interactions between the stacked bases. As a result, the double helix unwinds to form two single strands,

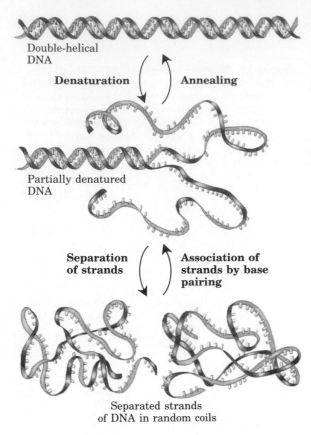

Figure 12–28 Stages in the reversible denaturation and annealing (renaturation) of DNA.

Double-helical DNA

Denaturation — **Annealing**

Partially denatured DNA

Separation of strands — **Association of strands by base pairing**

Separated strands of DNA in random coils

Figure 12–29 (a) The denaturation or melting curve of two DNA specimens. The temperature at the midpoint of the transition (t_m) is the melting point; it depends on pH and ionic strength, and on the size and base composition of the DNA. **(b)** Relationship between t_m and the G≡C content of a DNA, in a solution containing 0.15 M NaCl and 0.015 M sodium citrate.

completely separate from each other along the entire length, or part of the length (partial denaturation), of the molecule. No covalent bonds in the DNA are broken (Fig. 12–28).

Renaturation of DNA is a rapid one-step process as long as a double-helical segment of a dozen or more residues still unites the two strands. When the temperature or pH is returned to the biological range, the unwound segments of the two strands spontaneously rewind or **anneal** to yield the intact duplex (Fig. 12–28). However, if the two strands are completely separated, renaturation occurs in two steps. The first step is relatively slow, because the two strands must first "find" each other by random collisions and form a short segment of complementary double helix. The second step is much faster: the remaining unpaired bases successively come into register as base pairs, and the two strands "zipper" themselves together to form the double helix.

Viral or bacterial DNA molecules in solution denature at characteristic temperatures when they are heated slowly (Fig. 12–29). The transition from double-stranded DNA to the single-stranded, denatured form can be detected by an increase in the absorption of UV light (the hyperchromic effect) or a decrease in the viscosity of the DNA solution. Each species of DNA has a characteristic denaturation temperature or melting point: the higher its content of G≡C base pairs, the higher the melting point of the DNA. This is because G≡C base pairs, with three hydrogen bonds, are more stable and require more heat energy to dissociate than A=T base pairs. Careful determination of the melting point of a DNA specimen, under fixed conditions of pH and ionic strength, can yield an estimate of its base composition. If denaturation conditions are carefully controlled, regions that are rich in A=T base pairs will specifically denature while most of the DNA remains double-stranded. Such denatured regions can be visualized with electron microscopy (Fig. 12–30). Strand separation of DNA *must* occur in vivo during processes such as DNA replication and transcription. As we will see, the DNA sites where these processes are initiated are often rich in A=T base pairs.

Double-stranded nucleic acids with two RNA strands or with one RNA strand and one DNA strand (RNA–DNA hybrids) can also be denatured. Notably, RNA duplexes are more stable than DNA du-

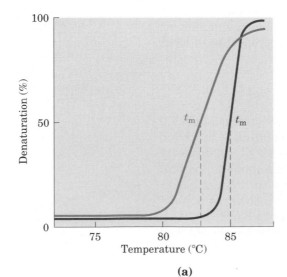

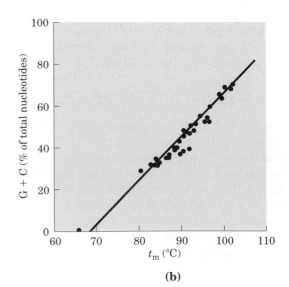

(a)

(b)

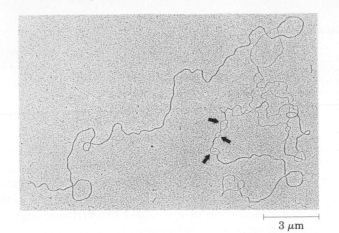

Figure 12–30 Electron micrograph of partially denatured DNA. Few structural details are evident. The shadowing method used to visualize the DNA increases its diameter approximately fivefold and obliterates the details of the helix. However, length measurements can be obtained, and single-stranded regions are readily distinguished from double-stranded regions. The arrows point to some single-stranded bubbles in the DNA, where denaturation has occurred. The regions that denature are highly reproducible and are rich in A=T base pairs.

3 μm

plexes. At neutral pH, a double-helical RNA will often denature at temperatures 20 °C or more higher than a DNA molecule with a comparable sequence. The stability of an RNA–DNA hybrid is generally intermediate between that of RNA and that of DNA. The physical basis for these differences in stability is not known.

Nucleic Acids from Different Species Can Form Hybrids

The capability of two complementary DNA strands to pair with one another can be used to detect similar DNA sequences in two different species or within the genome of a single species. If duplex DNAs isolated from human cells and from mouse cells are completely denatured by heating, then mixed and kept at 65 °C for many hours, much of the DNA will anneal. Most of the mouse DNA strands anneal with complementary mouse DNA strands to form mouse duplex DNA; similarly, many of the human DNA strands anneal with complementary human DNA strands. However, some strands of the mouse DNA will associate with human DNA strands to yield **hybrid duplexes,** in which segments of the mouse DNA strand form base-paired regions with segments of the human DNA strand (Fig. 12–31). This reflects the fact that different organisms have some common evolutionary heritage; they generally have some proteins and RNAs with similar functions and, often, similar structures. In many cases, the DNA encoding these proteins and RNAs will have similar (homologous) sequences. The closer the evolutionary relationship between the species, the more extensively will their DNAs hybridize. For example, human DNA hybridizes much more extensively with mouse DNA than with DNA from yeast.

The hybridization of DNA strands from different sources forms the basis of a powerful set of techniques essential to the modern practice of molecular genetics. It is possible to detect a specific DNA sequence or gene in the presence of many other sequences if one already has an appropriate complementary DNA strand (usually labeled in some way) to hybridize with it (Chapter 28). The complementary DNA can be from a different species or from the same species; in some cases it is synthesized in the laboratory, using techniques described later in this chapter. Hybridization techniques can be varied to detect a specific RNA rather than DNA. The isolation and identification of specific genes and RNAs relies on these techniques, and new applications of this technology are making it possible to accurately identify an individual on the basis of a single hair left at the scene of a crime or predict the onset of some diseases in an individual decades before symptoms appear (see Box 28–1).

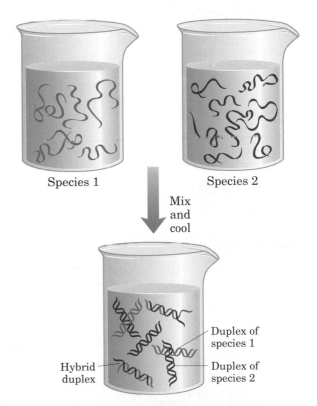

Species 1   Species 2

Mix and cool

Hybrid duplex — Duplex of species 1 — Duplex of species 2

Figure 12–31 Principle of the hybridization test. Two DNAs from different species are completely denatured by heating. When mixed and slowly cooled, complementary DNA strands of each species will associate and anneal to form normal duplexes. If the different DNAs have significant sequence homology, they will tend to form partial duplexes or hybrids with each other: the greater the sequence homology between the two DNAs, the greater the number of hybrids formed. Hybrid formation can be measured by different procedures, such as chromatography or isopycnic centrifugation. Usually one of the DNAs is labeled with a radioactive isotope to simplify the measurements.

Deamination

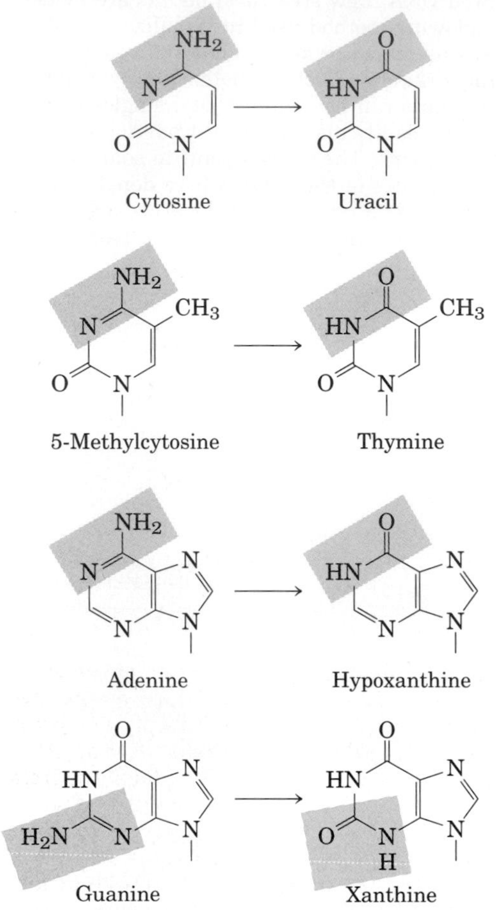

Cytosine → Uracil

5-Methylcytosine → Thymine

Adenine → Hypoxanthine

Guanine → Xanthine

(a)

Nucleotides and Nucleic Acids Undergo Nonenzymatic Transformations

Purines and pyrimidines, along with the nucleotides of which they are a part, undergo a number of reactions involving spontaneous alteration of their covalent structure. These reactions are generally *very slow,* but they are physiologically significant because of the cell's very low tolerance for alterations in genetic information. Alterations in DNA structure that lead to permanent changes in the genetic information encoded therein are called **mutations,** and much evidence suggests an intimate link between the accumulation of mutations and the processes of aging and cancer.

Several bases undergo spontaneous loss of their exocyclic amino groups (deamination) (Fig. 12–32a). For example, under conditions found in a typical cell, deamination of cytosine (in DNA) to uracil will occur in about one of every 10^7 cytosines in 24 h. This corresponds to about 100 spontaneous events per day in an average mammalian cell. Deamination of adenine and guanine is about 100 times slower.

The slow cytosine deamination reaction seems innocuous enough, but it is almost certainly the reason why DNA contains thymine rather than uracil. The product of cytosine deamination (uracil) is readily recognized as foreign in DNA and is removed by a repair system (Chapter 24). If DNA normally contained uracil, recognition of uracils resulting from cytosine deamination would be more difficult, and unrepaired uracils would lead to permanent sequence changes as they were paired with adenines during replication. Cytosine deamination would gradually lead to a decrease in G≡C base pairs and an increase in A=U base pairs in the DNA of all cells. Over the millennia, the cytosine deamination reaction could eliminate G≡C base pairs and the genetic code that depends on them. Establishing thymine as one of the four bases in DNA may well have been one of the key turning points in evolution, making the long-term storage of genetic information possible.

Depurination

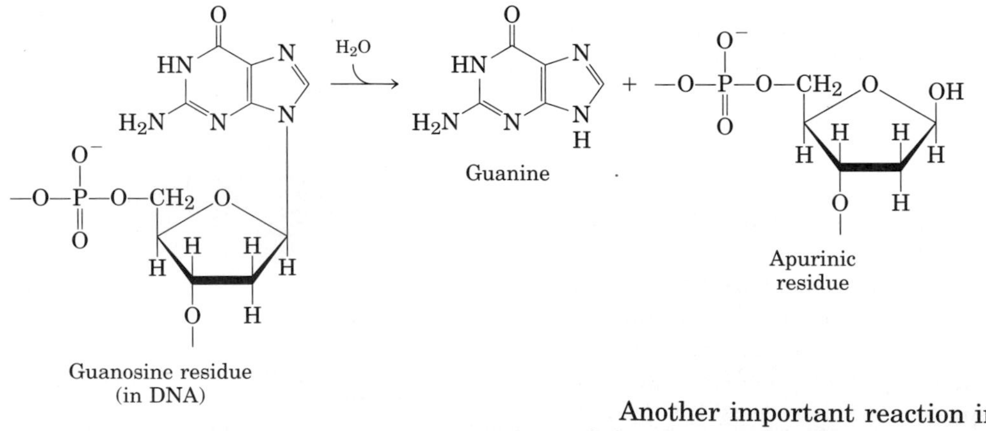

Guanosine residue
(in DNA)

Guanine + Apurinic residue

(b)

Figure 12–32 Some well-characterized reactions of nucleotides. **(a)** Deamination reactions. Only the base is shown. **(b)** Depurination, in which a purine is lost by hydrolysis of the *N*-glycosyl bond. The deoxyribose remaining after depurination is readily converted from the β-furanose to the aldehyde form (see Fig. 12–3).

Another important reaction in deoxynucleotides is the hydrolysis of the glycosyl bond between the base and the pentose (Fig. 12–32b). This occurs much faster for purines than for pyrimidines. In DNA as many as one in 10^5 purines (10,000 per mammalian cell) are lost every 24 h under typical cellular conditions. Depurination of ribonucleotides and RNA is much slower and generally is not considered physiologically significant. In the test tube, loss of purines can be accelerated by dilute acid. Incubation of DNA at pH 3 causes selective removal of the purine bases, resulting in a derivative called **apurinic acid.**

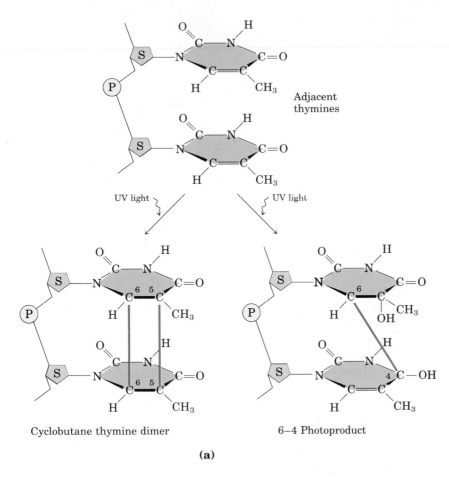

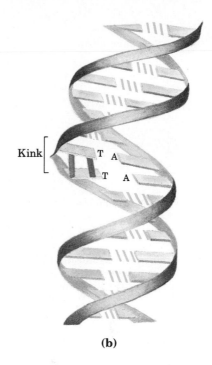

(b)

Figure 12–33 Formation of thymine dimers induced by UV light. **(a)** The reaction on the left results in the formation of a cyclobutyl ring involving C-5 and C-6 of each thymine residue, just as UV light will induce the formation of cyclobutane from two molecules of ethylene. An alternative light-induced reaction between adjacent thymines results in a linkage between C-6 of one residue and C-4 of its neighbor, as shown on the right. A bend or kink is introduced into the DNA on formation of a cyclobutane thymine dimer **(b)**.

Other reactions are promoted by certain types of radiation. In the laboratory, UV light will induce the condensation of two ethylene groups to form a cyclobutane ring. In the cell, the same reaction occurs between adjacent pyrimidine bases in nucleic acids to form cyclobutane pyrimidine dimers. This happens most frequently between adjacent thymine residues on the same DNA strand (Fig. 12–33). A second type of pyrimidine dimer formed during UV irradiation, called a 6–4 photoproduct, is also shown in Figure 12–33. Ionizing radiation (x rays and gamma rays) can cause ring opening and fragmentation of bases as well as breaks in the covalent backbone of nucleic acids.

Virtually all forms of life are exposed to energy-rich radiation capable of causing chemical changes in DNA. UV radiation (having wavelengths of 200 to 400 nm), which makes up a significant portion of the solar spectrum, can cause pyrimidine dimer formation and other chemical changes in the DNA of bacteria and of human skin cells. There is a constant field of ionizing radiation around us in the form of cosmic rays, which can penetrate deep into the earth, as well as radiation emitted from radioactive elements, such as radium, plutonium, uranium, radon, ^{14}C, and ^{3}H. X rays used in medical or dental examinations and in radiation therapy of cancer and other diseases are another form of ionizing radiation. It is estimated that UV and ionizing radiations are responsible for about 10% of all DNA damage caused by nonbiological agents.

DNA also may be damaged by reactive chemicals introduced into the environment as products of industrial activity. Such products may not be injurious per se, but may be metabolized by cells into forms that are. There are three major classes of such reactive chemical agents

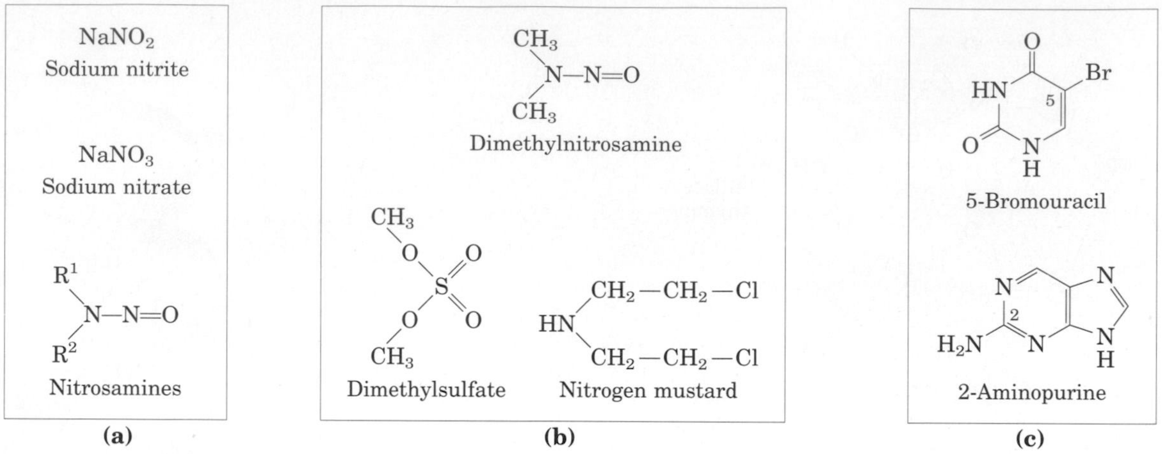

Figure 12–34 Chemical agents that cause some types of DNA damage. **(a)** Nitrous acid precursors. **(b)** Alkylating agents. **(c)** Base analogs.

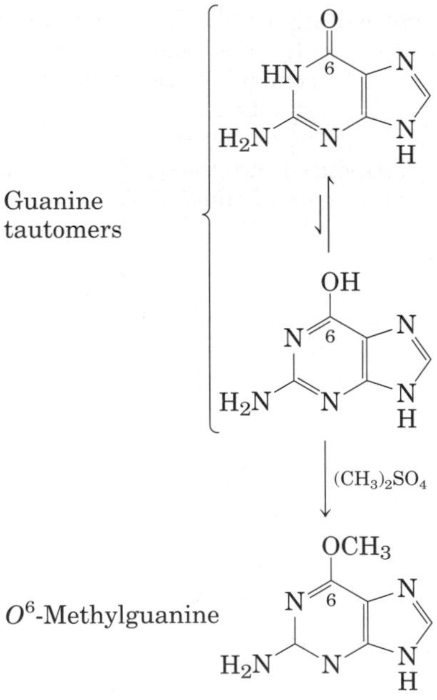

(Fig. 12–34): (1) deaminating agents, particularly nitrous acid (HNO_2) or compounds that can be metabolized to nitrous acid or nitrites, (2) alkylating agents, and (3) compounds that can simulate or mimic the normal bases present in DNA.

Nitrous acid, formed from organic precursors such as nitrosamines and from nitrite and nitrate salts, is a potent reagent that accelerates the deamination of bases described above. Bisulfite has similar effects. Both agents are used as preservatives in processed foods to prevent the growth of toxic bacteria. They do not appear to significantly increase cancer risks when used in this way, perhaps because they are used in small amounts and their contribution to the overall levels of DNA damage is minor. (The potential health risk from food spoilage if these preservatives were not used is much greater.)

Alkylating agents can alter certain bases of DNA. For example, the highly reactive chemical dimethylsulfate (Fig. 12–34b) can methylate a guanine residue to yield O^6-methylguanine, which is unable to base-pair with cytosine. Many similar reactions are brought about by alkylating agents normally present in cells, such as S-adenosylmethionine (see Fig. 17–20) and other compounds.

Possibly the most important source of mutagenic alterations in DNA is oxidative damage. Excited-oxygen species such as hydrogen peroxide, hydroxyl radicals, and superoxide radicals arise during irradiation or as a byproduct of aerobic metabolism. Cells possess an elaborate defense system to destroy these reactive species, including enzymes such as catalase and superoxide dismutase. A fraction of these oxidants inevitably escapes cellular defenses, however, and damage to DNA involves a large, complex group of reactions ranging from oxidation of sugar and base moieties to breaking strands. Accurate estimates for the extent of this damage are not yet available, but it is clear that each day the DNA in each human cell is subject to thousands of damaging oxidative reactions.

This is merely a sampling of the best-understood reactions. Many carcinogenic compounds present in food, water, or air exert their cancer-causing effects by modifying bases in DNA. In the cell, the integrity of DNA as a polymer is nevertheless maintained better than that of either RNA or protein, because DNA is the only macromolecule having biochemical repair systems. These repair processes (described in Chapter 24) greatly lessen the impact of damage to DNA.

DNA Is Often Methylated

Certain nucleotide bases in DNA molecules are often enzymatically methylated. Adenine and cytosine are methylated more often than guanine and thymine. Methylation of these bases is not random but is generally confined to certain sequences or regions of a DNA molecule. In some cases the function of methylation is well understood; in others the function is still unclear. All known DNA methylases use S-adenosylmethionine as a methyl group donor. In $E.\ coli$ there are two prominent methylation systems. One serves as part of a cellular defense mechanism that helps to distinguish the cell's own DNA from foreign DNA (restriction modification, described in Chapter 28). The other system methylates adenine to N^6-methyladenine (see Fig. 12–5a) within the sequence (5′)GATC(3′). This is mediated by an enzyme called the Dam methylase, which functions as part of a system that repairs mismatched base pairs formed occasionally during DNA replication (Chapter 24).

In eukaryotic cells, about 5% of cytosine residues are methylated to form 5-methylcytosine (see Fig. 12–5a). Methylation is most common at CpG sequences, producing methyl-CpG symmetrically on both strands of the DNA. The extent of methylation of CpG sequences varies in different regions of large eukaryotic DNA molecules, and is often inversely related to the degree of gene expression. These methylations have structural as well as regulatory significance. The presence of 5-methylcytosine in an alternating CpG sequence markedly increases the tendency for that sequence to take up the Z conformation.

Long DNA Sequences Can Be Determined

In its capacity as a repository of information, the most important property of a DNA molecule is its nucleotide sequence. Until the late 1970s, obtaining the sequence of a nucleic acid containing even five or ten nucleotides was difficult and very laborious. The development of two new techniques in 1977, one by Alan Maxam and Walter Gilbert and the other by Frederick Sanger, has made it possible to sequence ever larger DNA molecules with an ease unimagined just a few decades ago. The techniques depend upon an improved understanding of nucleotide chemistry and DNA metabolism, and on electrophoretic methods that allow the separation of DNA strands differing in size by only one nucleotide. Electrophoresis of DNA is similar to the electrophoresis of proteins (see Fig. 6–4). Polyacrylamide is often used as the gel matrix for short DNAs (up to a few hundred nucleotides). Agarose is generally used as the gel matrix for separating longer DNAs.

In both Sanger (dideoxy) and Maxam–Gilbert sequencing, the general principle is to reduce the DNA to be sequenced to four sets of labeled fragments. The reaction producing each set is base-specific, so that the lengths of the fragments correspond to positions in the DNA sequence where a certain base occurs. For example, for an oligonucleotide with the sequence pAATCGACT, a reaction that produces only fragments ending in C will generate fragments four and seven nucleotides long, whereas a reaction producing fragments ending in G will produce only a five-nucleotide fragment. The fragment sizes correspond to the relative positions of C and G residues in the sequence. When the sets of fragments corresponding to each of the four bases are electrophoretically separated side by side, they produce a ladder of bands from which the sequence can be read directly (Figs. 12–35, 12–

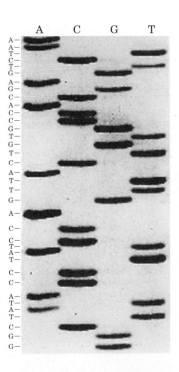

Figure 12–35 Section of an autoradiogram produced by the method developed by Sanger and colleagues. Side-by-side electrophoresis of the DNA fragments generated by each dideoxynucleotide generates a ladder of bands. Each band on the film corresponds to a population of DNA fragments of a specific length produced in the sequencing reactions (see Fig. 12–36). The identity of the base at each position in the sequence is determined from the lane in which a band is observed; the order of the bands read from the bottom of the gel corresponds to the DNA sequence.

Figure 12–36 DNA sequencing by the Sanger (dideoxy) method. This method makes use of the mechanism of DNA synthesis by DNA polymerases (Chapter 24). DNA polymerases require both a primer, to which nucleotides are added, and a template strand to guide selection of each new nucleotide **(a)**. The 3′-hydroxyl group of the primer reacts with the incoming deoxynucleoside triphosphate (dNTP), forming a new phosphodiester bond. The Sanger sequencing procedure uses dideoxynucleoside triphosphate (ddNTP) analogs **(b)** to interrupt DNA synthesis. When the dNTP is replaced by the ddNTP, strand elongation is halted after the analog is added because it lacks the 3′-hydroxyl group needed for the next step. The DNA to be sequenced is used as the template strand, and a short primer (usually radioactively labeled) is annealed to it **(c)**. By adding small amounts of a single ddNTP, for example ddCTP, to an otherwise normal reaction system, the synthesized strands will be prematurely terminated at locations where dC normally occurs. Because there is much more dCTP than ddCTP, there is only a small chance that the analog will be incorporated whenever a dC is to be added, but there is generally enough ddCTP that each new strand has a high probability of acquiring one ddC at some point during synthesis. The result is a solution containing fragments representing each C residue in the sequence. The size of the fragments, separated by electrophoresis, reveals the location of C residues in the sequence. This procedure is repeated separately for each of the four ddNTPs, and the sequence can be read directly from an autoradiogram of the gel **(c)**. Because shorter DNA fragments migrate faster, the fragments near the bottom represent the nucleotide positions closest to the primer (the 5′ end), and the sequence is read from bottom to top. Note that the sequence obtained is that of the strand *complementary* to the strand being analyzed. An actual sequencing gel is shown in Fig. 12–35.

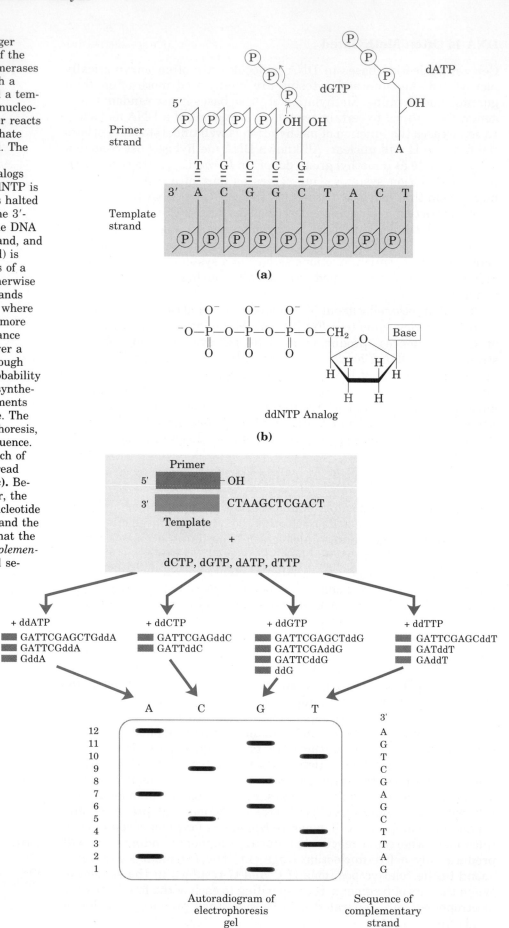

36). The Sanger method (Fig. 12–36) is in more widespread use because it has proven to be technically easier. It involves the enzymatic synthesis of a DNA strand complementary to the strand to be analyzed.

DNA sequencing is now automated, using a variation of Sanger's sequencing method in which the primer used for each reaction is labeled with a differently colored fluorescent tag (Fig. 12–37). This technology allows sequences of thousands of nucleotides to be obtained in a few hours, and very large DNA-sequencing projects are being contemplated. The most ambitious of these, now underway, is the Human Genome Initiative, in which all of the 3 billion base pairs of DNA in a human cell will be sequenced.

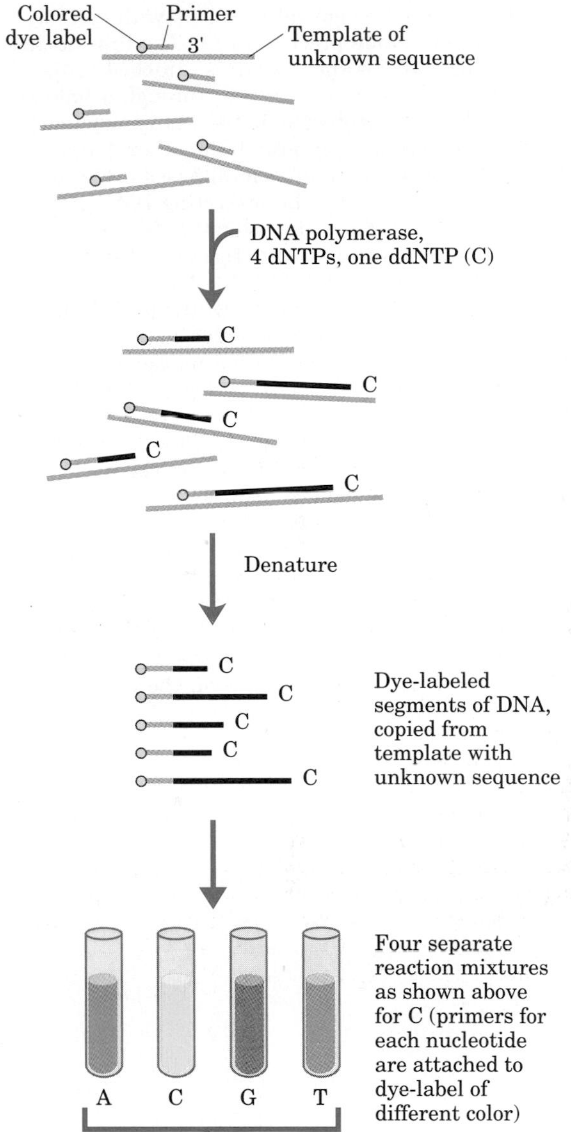

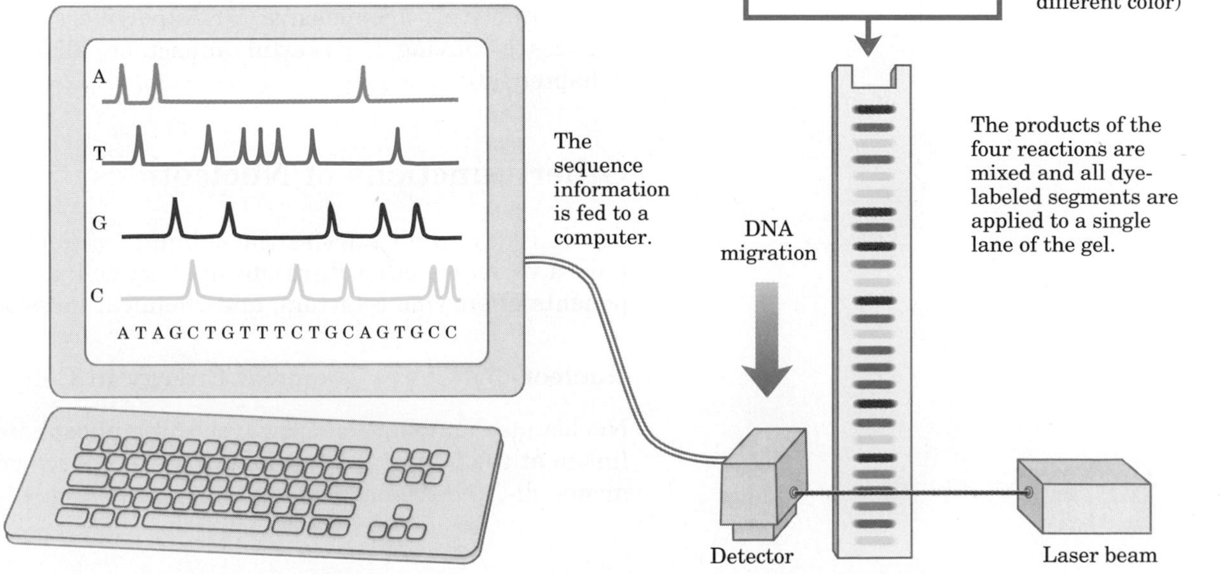

Figure 12–37 A prototype strategy for automating DNA sequencing reactions. The short oligonucleotides used as a primer for DNA synthesis in the Sanger method can be linked to a fluorescent molecule that gives the DNA strand a color. If each nucleotide is assigned a different color, the nucleotide on the end of each fragment can be identified by color. The dideoxy method is used with a different ddNTP added to each of the four tubes according to the color assignments. The resulting colored DNA fragments are mixed and then separated by size in a single electrophoretic gel lane. The fragments of a given length migrate through the gel in a peak, and the color associated with each successive peak is detected using a laser beam. The DNA sequence is read by determining the sequence of colors in the peaks as they pass the detector, and this information is fed directly to a computer.

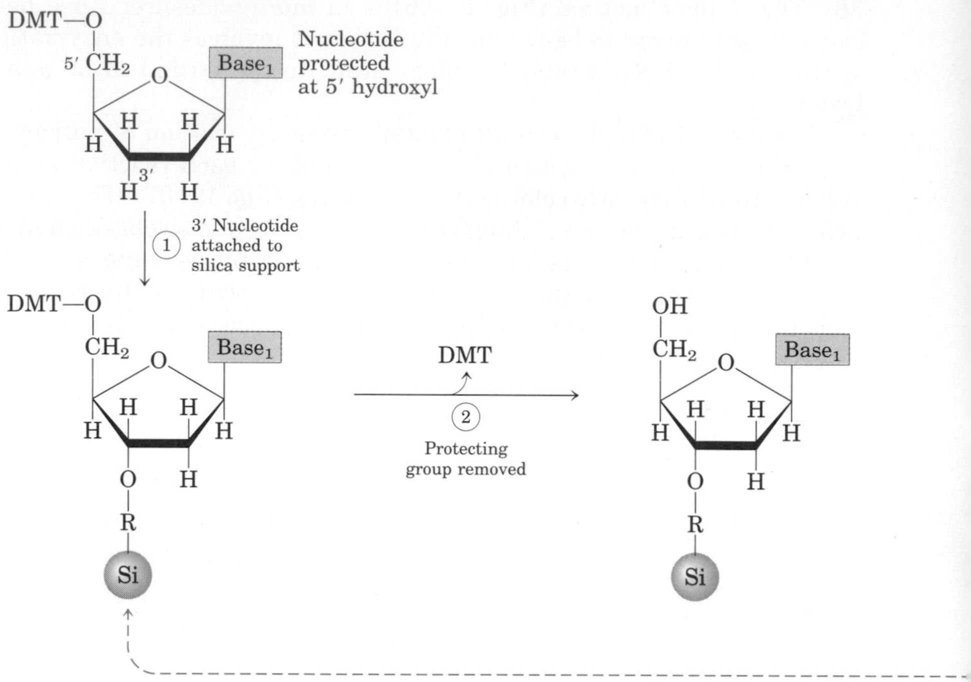

Figure 12–38 Automated synthesis of DNA is conceptually similar to the solid-state synthesis of polypeptides. The desired oligonucleotide is built up on a solid support (silica) one nucleotide at a time in a repeated series of chemical reactions with suitably protected nucleotide precursors. ① The first nucleotide (which will be the 3′ end) is attached to the silica support at the 3′ hydroxyl (through a linking group, R), and is protected at the 5′ hydroxyl with an acid-labile protecting group (dimethoxytrityl, DMT). The reactive groups on all bases are also chemically blocked. ② The protecting DMT group is removed by washing the column with acid (the DMT group is colored, so this reaction can be followed spectrophotometrically). ③ The next nucleotide is activated and reacted with the bound nucleotide to form a 5′-3′ linkage, which in ④ is oxidized with iodine to produce a phosphotriester linkage. (One of the phosphate oxygens is methylated.) Reactions ② through ④ are repeated until all nucleotides are added. At each step, excess nucleotide is removed before addition of the next nucleotide. In ⑤ and ⑥ the remaining blocking groups on the bases and the methyl groups on the phosphates are removed, and in ⑦ the oligonucleotide is separated from the solid support and purified. The chemistry of RNA synthesis has lagged far behind the procedures for DNA synthesis because of difficulties in protecting the 2′ hydroxyl of ribose without adverse effects on the reactivity of the 3′ hydroxyl.

The Chemical Synthesis of DNA Has Been Automated

Another technology that has paved the way for many biochemical advances is the chemical synthesis of oligonucleotides with any chosen sequence. The chemical methods for synthesizing nucleic acids were developed primarily by H. Gobind Khorana in the 1970s. Refinement and automation of these methods has made it possible to synthesize DNA strands rapidly and accurately. The synthesis is carried out with the growing strand attached to a solid support (Fig. 12–38), using principles similar to those used by Merrifield in peptide synthesis (see Box 5–2). The efficiency of each addition step is very high, allowing the routine laboratory synthesis of polymers of 70 or 80 nucleotides. In some laboratories much longer strands are synthesized. The availability of relatively inexpensive DNA polymers with predesigned sequences is having a powerful impact on all areas of biochemistry (Chapter 28).

Other Functions of Nucleotides

In addition to their roles as the subunits of nucleic acids, nucleotides have a variety of other functions in every cell: as energy carriers, components of enzyme cofactors, and chemical messengers.

Nucleotides Carry Chemical Energy in Cells

Nucleotides may have one, two, or three phosphate groups covalently linked at the 5′ hydroxyl of ribose. These are referred to as nucleoside mono-, di-, and triphosphates, respectively (Fig. 12–39). Starting from

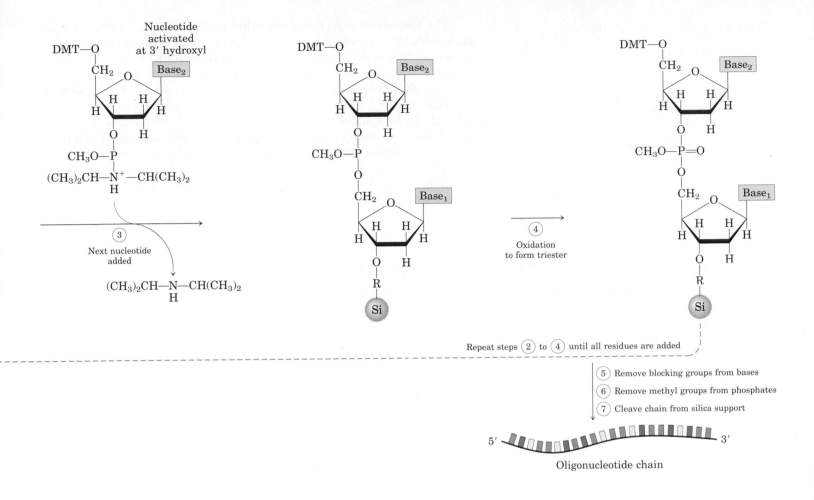

the ribose, the three phosphates are generally labeled α, β, and γ. Nucleoside triphosphates are used as a source of chemical energy to drive a wide variety of biochemical reactions. ATP is by far the most widely used, but UTP, GTP, and CTP are used in specific reactions. Nucleoside triphosphates also serve as the activated precursors of DNA and RNA synthesis, as will be described in Chapters 24 and 25.

Figure 12–39 General structure of nucleoside 5'-mono-, 5'-di-, and 5'-triphosphates (NMPs, NDPs, and NTPs) and their standard abbreviations. In the deoxyribonucleoside phosphates (dNMPs, dNDPs, and dNTPs) the pentose is 2'-deoxy-D-ribose.

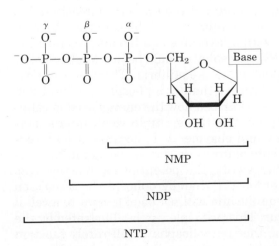

Abbreviations of ribonucleoside 5'-phosphates			
Base	Mono-	Di-	Tri-
Adenine	AMP	ADP	ATP
Guanine	GMP	GDP	GTP
Cytosine	CMP	CDP	CTP
Uracil	UMP	UDP	UTP

Abbreviations of deoxyribonucleoside 5'-phosphates			
Base	Mono-	Di-	Tri-
Adenine	dAMP	dADP	dATP
Guanine	dGMP	dGDP	dGTP
Cytosine	dCMP	dCDP	dCTP
Thymine	dTMP	dTDP	dTTP

The hydrolysis of ATP and the other nucleoside triphosphates is an energy-yielding reaction because of the chemistry of the triphosphate structure. The bond between the ribose and the α phosphate is an ester linkage. The α–β and β–γ linkages are phosphoric acid anhydrides (Fig. 12–40). Hydrolysis of the ester linkage yields about 14 kJ/mol, whereas hydrolysis of each of the anhydride bonds yields about 30 kJ/mol. In biosynthesis, ATP hydrolysis often drives less favorable metabolic reactions (i.e., those with $\Delta G^{\circ\prime} > 0$). When coupled to a reaction with a positive free-energy change, ATP hydrolysis shifts the equilibrium of the overall process to favor product formation (recall the relationship between equilibrium and free-energy change described in Chapter 8).

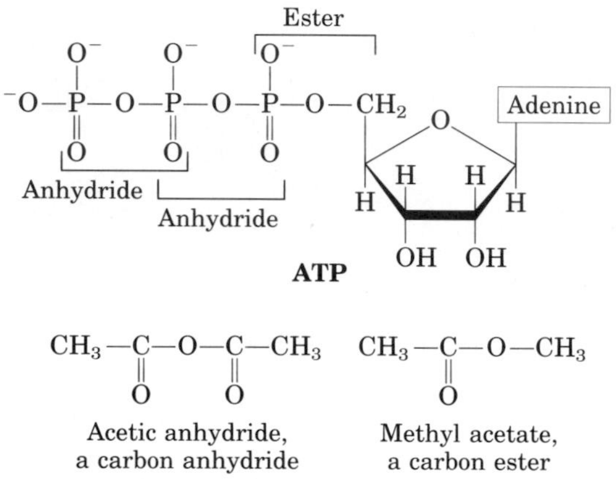

Figure 12–40 The phosphate ester and phosphoric acid anhydride bonds of ATP. Hydrolysis of an anhydride bond yields more energy than hydrolysis of the ester. A carbon anhydride and ester are shown for comparison.

It is appropriate to ask why ATP serves as the primary carrier of energy in the cell. The chemical energy potential of pyrophosphate (~33 kJ/mol), a much simpler molecule, is almost identical to that of ATP because pyrophosphate also contains a phosphoric acid anhydride. Pyrophosphate would be so much easier to synthesize than ATP that the selection of ATP at first seems to contradict evolutionary logic.

The explanation can be found in the fundamental energetic principles governing every chemical reaction. In promoting chemically unfavorable reactions such as those in many biosynthetic processes, the cell must deal with both the *equilibrium* and the *rate* of the reaction. We have seen that an unfavorable equilibrium can be overcome by coupling such a reaction to one with a favorable equilibrium, such as the hydrolysis of an anhydride. Pyrophosphate would be just as effective as ATP in its potential effects on reaction equilibria. Therefore, the advantage to cells in using ATP rather than pyrophosphate must lie in reaction rates. In Chapter 8 we described how the energy used in catalysis is derived from binding energy, the multiple weak interactions that occur between substrate and enzyme. ATP, because of its larger structure, clearly can contribute many more of these weak interactions than pyrophosphate. In other words, the potential for reaction *rate* enhancement is much greater for ATP than pyrophosphate. A reaction with a favorable energetic equilibrium will not be of benefit to a cell if it takes several years to occur. This principle can be illustrated by the simple empirical observation that pyrophosphate will rarely function in an enzymatic reaction requiring ATP, even though it should fit into any enzyme active site that can accommodate ATP.

Nucleotides Are Components of Many Enzyme Cofactors

A variety of enzyme cofactors serving a wide range of chemical functions include adenosine as part of their structure (Fig. 12–41). They are unrelated structurally except for the presence of adenosine. In none of these cofactors does the adenosine portion participate directly in the primary function, but removal of adenosine from these structures generally results in a drastic reduction of their activities. For example, removal of the adenosine nucleotide (3'-P-ADP; see Fig. 12–41) from acetoacetyl-CoA reduces its reactivity as a substrate for β-ketoacyl-CoA transferase (an enzyme of lipid metabolism) by a factor of 10^6. Although the reason for this requirement for adenosine has not been examined in detail, it must involve the binding energy between enzyme and substrate (or cofactor) that is used both in catalysis and to stabilize the initial ES complex (Chapter 8). In the case of CoA transferase, the nucleotide appears to be a binding "handle" that helps to

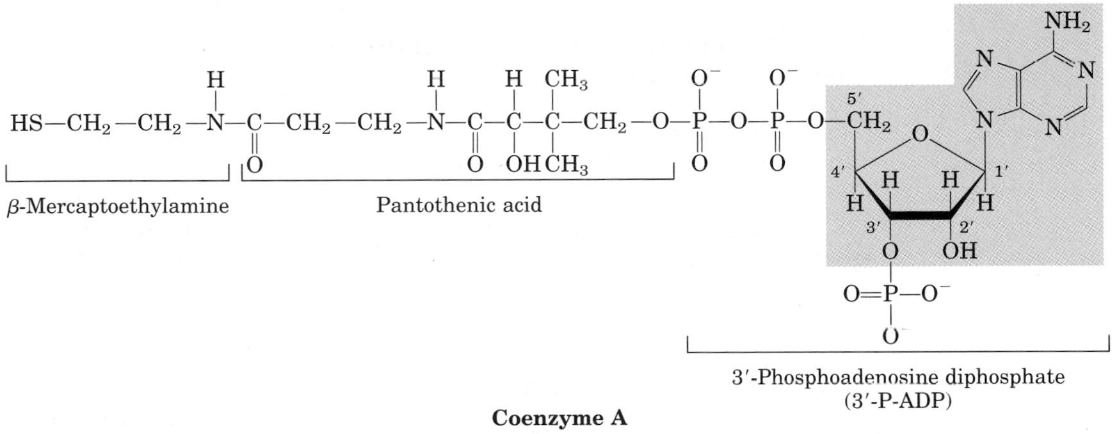

Figure 12–41 Enzyme cofactors and coenzymes incorporating adenosine in their structure. The adenosine portion is shaded in red. Coenzyme A functions in acyl group transfer reactions; NAD^+ participates in hydride transfers; FAD, the active form of vitamin B_2 (riboflavin), participates in electron transfers. Another coenzyme incorporating adenosine in its structure is 5'-deoxyadenosyl-cobalamin, the active form of vitamin B_{12} (see Box 16–2). This coenzyme is involved in intramolecular group transfers between adjacent carbons.

Coenzyme A

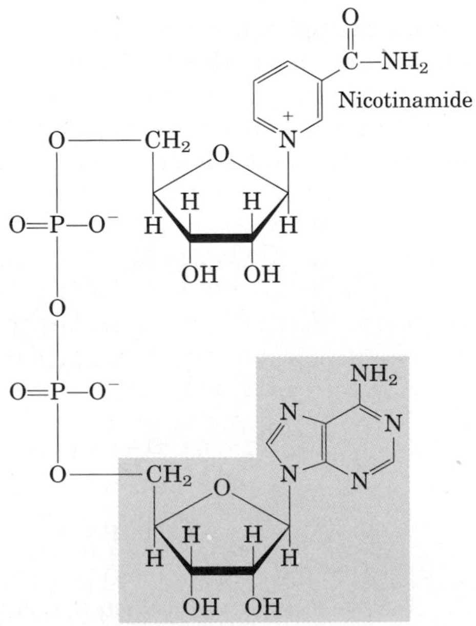

Nicotinamide adenine dinucleotide (NAD^+)

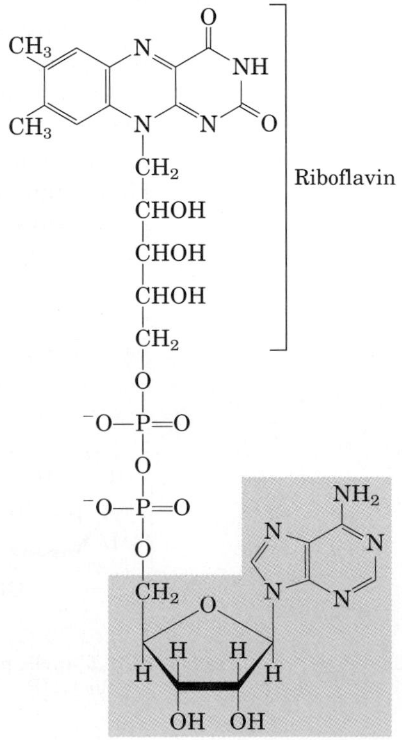

Flavin adenine dinucleotide (FAD)

pull the substrate into the active site. Similar roles may be found for the nucleoside portion of other nucleotide cofactors.

Now we may ask why adenosine, rather than some other large molecule, is used in these structures. The answer here may involve a kind of evolutionary economy. Adenosine is certainly not unique in the amount of potential binding energy it can contribute. The importance of adenosine probably lies not so much in some special chemical characteristic, but rather that an advantage existed in making one compound a standard. Once ATP became the standard source of chemical energy, systems developed to synthesize ATP more efficiently than the other nucleotides; because it is abundant, it becomes the logical choice for incorporation into a wide variety of structures. The economy extends to protein structure. A protein domain that binds adenosine can be used in a wide variety of different enzymes. Such a structure, called a **nucleotide-binding fold,** is found in many enzymes that bind ATP and nucleotide cofactors.

Some Nucleotides Are Intermediates in Cellular Communication

Cells respond to their environment by taking cues from hormones or other chemical signals in the surrounding medium. The interaction of these extracellular chemical signals (first messengers) with receptors on the cell surface often leads to the production of **second messengers** inside the cell, which in turn lead to adaptive changes in the cell interior (Chapter 22). Often, the second messenger is a nucleotide.

One of the most common second messengers is the nucleotide **adenosine 3′,5′-cyclic monophosphate (cyclic AMP, or cAMP),** formed from ATP in a reaction catalyzed by adenylate cyclase, associated with the inner face of the plasma membrane. Cyclic AMP serves regulatory functions in virtually every cell outside the plant kingdom, and these are described in detail in Chapter 22. Guanosine 3′,5′-cyclic monophosphate (cGMP) occurs in many cells and also has regulatory functions.

Another regulatory nucleotide, ppGpp, is produced in bacteria in response to the slowdown in protein synthesis that occurs during amino acid starvation. This nucleotide inhibits the synthesis of the rRNA and tRNA molecules (Chapter 27) needed for protein synthesis, preventing the unnecessary production of nucleic acids.

Adenosine 3′,5′-cyclic monophosphate
(cyclic AMP; cAMP)

Guanosine 3′,5′-cyclic monophosphate
(cyclic GMP; cGMP)

Guanosine 3′-diphosphate,5′-diphosphate
(guanosine tetraphosphate)
(ppGpp)

Summary

Nucleotides serve a diverse set of important functions in cells. As subunits of nucleic acids they carry genetic information. They are also the primary carriers of chemical energy in cells, structural components of many enzyme cofactors, and cellular second messengers.

A nucleotide consists of a nitrogenous base (purine or pyrimidine), a pentose sugar, and one or more phosphate groups. Nucleic acids are polymers of nucleotides, linked together by phosphodiester bridges between the 5' hydroxyl of one pentose and the 3' hydroxyl of the next. There are two types of nucleic acid: RNA and DNA. The nucleotides in RNA contain ribose, and the common pyrimidine bases are uracil and cytosine. In DNA, the nucleotides contain 2'-deoxyribose, and the pyrimidine bases are generally thymine and cytosine. The primary purines are adenosine and guanine in both RNA and DNA.

Many lines of evidence show that DNA bears genetic information. In particular, the Avery–MacLeod–McCarty experiment showed that DNA isolated from one strain of a bacterium can enter and transform the cells of another strain, endowing it with some of the inheritable characteristics of the donor. The Hershey–Chase experiment showed that the DNA of a bacterial virus, but not its protein coat, carries the genetic message for replication of the virus in the host cell.

From x-ray diffraction studies of DNA fibers and the base equivalences in DNA discovered by Chargaff (A = T and G = C), Watson and Crick postulated that native DNA consists of two antiparallel chains in a right-handed double-helical arrangement. Complementary base pairs, A=T and G≡C, are formed by hydrogen bonding within the helix, and the hydrophilic sugar–phosphate backbones are located on the outside. The base pairs are stacked perpendicular to the long axis, 0.34 nm apart; there are about 10 base pairs in each complete turn of the double helix.

DNA can exist in several structural forms. Two variations from the Watson–Crick B-form DNA, the A and Z forms, have been characterized in DNA crystal structures. The A-form helix is shorter and of greater diameter than a B-form helix with the same sequence. The Z form is a left-handed helix. Some sequence-dependent structural variations cause bends in the DNA. DNA strands with self-complementary inverted repeats can form hairpin or cruciform structures. Polypyrimidine tracts arranged in mirror repeats can take up a triple-helical structure called H-DNA.

Messenger RNA is the vehicle by which genetic information is transferred to ribosomes for protein synthesis. Transfer RNA and ribosomal RNA are also involved in protein synthesis. RNA can be structurally complex, with single RNA strands often folded into hairpins, double-stranded regions, and complex loops.

Native DNA undergoes reversible unwinding and separation (melting) of strands on heating or at extremes of pH. Because G≡C base pairs are more stable than A=T pairs, the melting point of DNAs rich in G≡C pairs is higher than that of DNAs rich in A=T pairs. Denatured single-stranded DNAs from two species can form a hybrid duplex, the degree of hybridization depending on the extent of sequence homology. Hybridization is the basis for important techniques used to study and isolate specific genes and RNAs.

DNA is a relatively stable polymer. Very slow, spontaneous reactions such as deamination of certain bases, hydrolysis of base–sugar N-glycosidic bonds, formation of pyrimidine dimers (radiation damage), and oxidative damage are important because of the very low tolerance of cells for changes in genetic material. DNA sequences can be determined and DNA polymers synthesized using simple protocols involving chemical and enzymatic methods.

ATP is the central carrier of chemical energy in cells, probably reflecting the requirement for binding energy in catalysis. The presence of adenosine in the structure of a variety of enzyme cofactors may also be related to binding energy requirements. Cyclic AMP is a common second messenger produced in response to hormones and other chemical signals. It is formed from ATP in a reaction catalyzed by adenylate cyclase.

Further Reading

General

Friedberg, E.C. (1985) *DNA Repair,* W.H. Freeman and Company, New York.
A good source for more information on the chemistry of nucleotides and nucleic acids.

Kornberg, A. & Baker, T.A. (1991) *DNA Replication,* 2nd edn, W.H. Freeman and Company, New York.
The best place to start for learning more about DNA structure.

Saenger, W. (1984) *Principles of Nucleic Acid Structure,* Springer-Verlag, New York.
A more detailed treatment.

Watson, J.D., Hopkins, N.H., Roberts, J.W., Steitz, J.A., & Weiner, A.M. (1987) *Molecular Biology of the Gene,* 4th edn, The Benjamin/Cummings Publishing Company, Menlo Park, CA.
Excellent general reference.

Variations in DNA Structure

Dickerson, R.E. (1983) The DNA helix and how it is read. *Sci. Am.* **249** (December), 94–111.

Htun, H. & Dahlberg, J.E. (1989) Topology and formation of triple-stranded H-DNA. *Science* **243,** 1571–1576.

Rich, A., Nordheim, A., & Wang, A.H.-J. (1984) The chemistry and biology of left-handed Z-DNA. *Annu. Rev. Biochem.* **53,** 791–846.

Wells, R.D. & Harvey, S.C. (eds) (1988) *Unusual DNA Structures,* Springer-Verlag, New York.

Wells, R.D. (1988) Unusual DNA structures. *J. Biol. Chem.* **263,** 1095–1098.
Minireview; a concise summary.

ATP As Energy Carrier

Jencks, W.P. (1987) Economics of enzyme catalysis. *Cold Spring Harb. Symp. Quant. Biol.* **52,** 65–73.
A relatively short article, full of insights.

Historical

Olby, R. (1974) *The Path to the Double Helix,* University of Washington Press, Seattle.

Sayre, A. (1978) *Rosalind Franklin and DNA,* W.W. Norton & Co., Inc., New York.

Watson, J.D. (1968) *The Double Helix,* Atheneum Publishers, New York.
A personal account of the human aspects of the discovery.

Problems

1. *Determination of Protein Concentration by UV Absorption in a Solution Containing Nucleic Acids* The concentration of protein or nucleic acid in solutions containing both can be estimated by using their light absorption properties. Proteins have a strong absorption centered at a wavelength of 280 nm, whereas nucleic acids absorb most strongly at 260 nm. When both proteins and nucleic acids are present in a solution, their respective concentrations can be estimated by measuring the absorbance (A) of the solution at 280 nm and 260 nm and using the table at the top of page 357. $R_{280/260}$ is the ratio of the absorbance at 280 and 260 nm. The table indicates the percentage of total mass that is nucleic acid, and provides a factor, F, to correct the A_{280} reading and give a more accurate protein estimate. The protein concentration (in mg/ml) is equal to $F \times A_{280}$ (assuming the cuvette is 1 cm wide). What are the protein and nucleic acid concentration if $A_{280} = 0.69$ and $A_{260} = 0.94$?

$R_{280/260}$	Proportion of nucleic acid (%)	F
1.75	0.00	1.116
1.63	0.25	1.081
1.52	0.50	1.054
1.40	0.75	1.023
1.36	1.00	0.994
1.30	1.25	0.970
1.25	1.50	0.944
1.16	2.00	0.899
1.09	2.50	0.852
1.03	3.00	0.814
0.979	3.50	0.776
0.939	4.00	0.743
0.874	5.00	0.682
0.846	5.50	0.656
0.822	6.00	0.632
0.804	6.50	0.607
0.784	7.00	0.585
0.767	7.50	0.565
0.753	8.00	0.545
0.730	9.00	0.508
0.705	10.00	0.478
0.671	12.00	0.422
0.644	14.00	0.377
0.615	17.00	0.322
0.595	20.00	0.278

2. *Nucleotide Structure* What positions in a purine ring have the potential to form hydrogen bonds, but are not involved in the hydrogen bonds of Watson–Crick base pairs?

3. *Base Sequence of Complementary DNA Strands* Write the base sequence of the complementary strand of double-helical DNA in which one strand has the sequence (5′)ATGCCCGTATGCATTC(3′).

4. *DNA of the Human Body* Calculate the weight in grams of a double-helical DNA molecule stretching from the earth to the moon (~320,000 km). The DNA double helix weighs about 1×10^{-18} g per 1,000 nucleotide pairs; each base pair extends 0.34 nm. For an interesting comparison, your body contains about 0.5 g of DNA!

5. *DNA Bending* Assume that a poly(A) tract five base pairs long produces a bend of about 20°. Calculate the total (net) bend produced in the DNA if the center base pairs (the third of five) of two successive $(dA)_5$ tracts are located (a) 10 or (b) 15 base pairs apart. Assume that there are 10 base pairs per turn in the DNA double helix.

6. *Distinction between DNA Structure and RNA Structure* Hairpins may form at palindromic sequences in single strands of either RNA or DNA. How is the helical structure of a hairpin in RNA different from that of a hairpin in DNA?

7. *Nucleotide Chemistry* In the cells of many eukaryotic organisms, there are highly specialized systems that specifically repair G–T mismatches in DNA. The mismatch is repaired to form a G≡C base pair (not A=T). This G–T mismatch repair system occurs in addition to a more general system that repairs virtually all mismatches. Can you think of a reason why cells require a specialized system to repair G–T mismatches?

8. *Nucleic Acid Structure* Explain why there is an increase in the absorption of UV light (hyperchromic effect) when double-stranded DNA is denatured.

9. *Base Pairing in DNA* In samples of DNA isolated from two unidentified species of bacteria, adenine makes up 32 and 17%, respectively, of the total bases. What relative proportions of adenine, guanine, thymine, and cytosine would you expect to find in the two DNA samples? What assumptions have you made? One of these bacteria was isolated from a hot spring (64 °C). Which DNA came from this thermophilic bacterium? What is the basis for your answer?

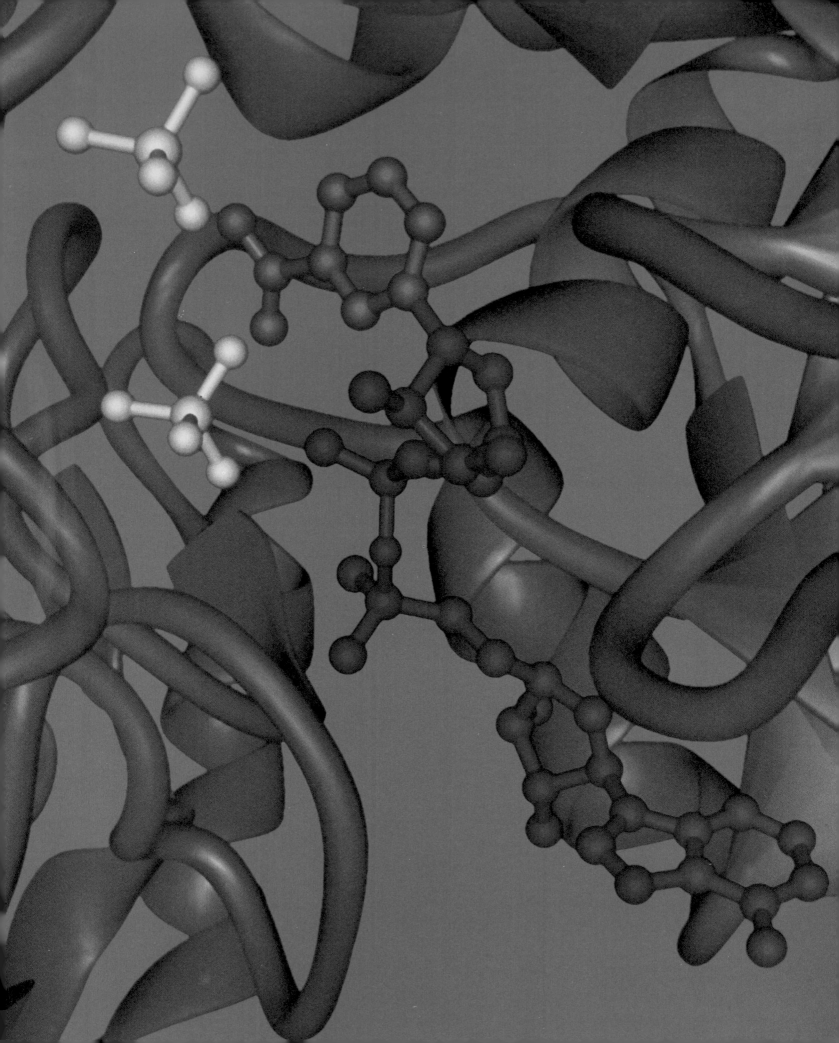

Bioenergetics and Metabolism

Metabolism is a highly coordinated and directed cell activity, in which many multienzyme systems cooperate to accomplish four functions: (1) to obtain chemical energy by capturing solar energy or by degrading energy-rich nutrients from the environment, (2) to convert nutrient molecules into the cell's own characteristic molecules, including macromolecular precursors, (3) to polymerize monomeric precursors into proteins, nucleic acids, lipids, polysaccharides, and other cell components, and (4) to synthesize and degrade biomolecules required in specialized cellular functions.

Although metabolism embraces hundreds of different enzyme-catalyzed reactions, the central metabolic pathways—our major concern—are few in number and are remarkably similar in all forms of life. Living organisms can be divided into two large groups according to the chemical form in which they obtain carbon from the environment. **Autotrophs** (such as photosynthetic bacteria and higher plants) can use carbon dioxide from the atmosphere as their sole source of carbon, from which they construct all their carbon-containing biomolecules (see Fig. 2–4). Some autotrophic organisms, such as cyanobacteria, can also use atmospheric nitrogen to generate all their nitrogenous components. **Heterotrophs** cannot use atmospheric carbon dioxide and must obtain carbon from their environment in the form of relatively complex organic molecules, such as glucose. The cells of higher animals and most microorganisms are heterotrophic. Autotrophic cells are relatively self-sufficient, whereas heterotrophic cells, with their requirements for carbon in more complex forms, must subsist on the products of other cells.

Many autotrophic organisms are photosynthetic and obtain their energy from sunlight, whereas heterotrophic cells obtain their energy from the degradation of organic nutrients made by autotrophs. In our biosphere, autotrophs and heterotrophs live together in a vast, interdependent cycle in which autotrophic organisms use atmospheric CO_2 to build their organic biomolecules, some of them generating oxygen from H_2O in the process. Heterotrophs in turn use the organic products of autotrophs as nutrients and return CO_2 to the atmosphere. The oxidation reactions that produce CO_2 also consume O_2, converting it to H_2O. Thus carbon, oxygen, and water are constantly cycled between the heterotrophic and autotrophic worlds, solar energy ultimately providing the driving force for this massive process (Fig. 1).

Facing page: The active site of glyceraldehyde-3-phosphate dehydrogenase, with the bound cofactor nicotinamide adenine dinucleotide (NAD) shown in red. This enzyme catalyzes the oxidation of glyceraldehyde-3-phosphate to 1,3-bisphosphoglycerate, a step in glycolysis, a central pathway in glucose metabolism. This is the earliest known example of an enzymatic reaction in which the energy released by electron transfer (oxidation) drives the formation of a high-energy phosphate compound.

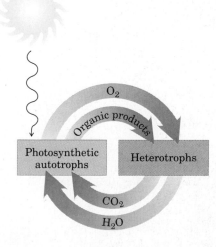

Figure 1 The cycling of carbon dioxide and oxygen between the autotrophic (photosynthetic) and the heterotrophic domains in the biosphere. The flow of mass through this cycle is enormous; about 4×10^{11} metric tons of carbon are turned over in the biosphere annually.

All living organisms also require a source of nitrogen, which is necessary for the synthesis of amino acids, nucleotides, and other compounds. Plants are generally able to use either ammonia or soluble nitrates as their sole source of nitrogen, but vertebrate animals must obtain some nitrogen in the form of amino acids or other organic compounds. Only a few organisms—the cyanobacteria and a few species of soil bacteria that live symbiotically on the roots of certain plants (legumes)—are capable of converting ("fixing") atmospheric nitrogen (N_2) into ammonia. Other microbial organisms (nitrifying bacteria) carry out the oxidation of ammonia to nitrites and nitrates. Thus, in addition to the global carbon and oxygen cycle (Fig. 1), a nitrogen cycle operates in the biosphere in which huge amounts of nitrogen undergo cycling and turnover (Fig. 2). The cycling of carbon, oxygen, and nitrogen, which involves many species of living organisms, depends on a proper balance between the activities of the producers (autotrophs) and consumers (heterotrophs) in our biosphere.

Figure 2 The cycling of nitrogen in the biosphere. Gaseous nitrogen (N_2) makes up 80% of our atmosphere.

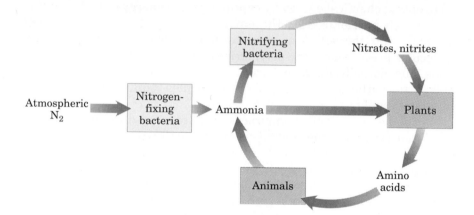

These great cycles of matter are driven by an enormous flow of energy through the biosphere, which begins with the capture of solar energy by photosynthetic organisms and its use to generate energy-rich carbohydrates and other organic nutrients; these nutrients are then used as energy sources by heterotrophic organisms. In the metabolic processes of each organism participating in these cycles, and in all energy-requiring activities, there is a loss of useful energy (free energy) and an inevitable increase in the amount of unavailable energy as heat and entropy. In contrast to the cycling of matter, therefore, energy flows one-way through the biosphere; useful energy can never be regenerated in living organisms from energy dissipated as heat and entropy. Carbon, oxygen, and nitrogen recycle continuously, but energy is constantly transformed into unusable forms.

Metabolism, the sum of all of the chemical transformations that occur in a cell or organism, occurs in a series of enzyme-catalyzed reactions that constitute metabolic pathways. Each of the consecutive steps in such a pathway brings about a small, specific chemical change, usually the removal, transfer, or addition of a specific atom, functional group, or molecule. In this sequence of steps (the **pathway**), a precursor is converted into a product through a series of metabolic intermediates (**metabolites**). The term intermediary metabolism is often applied to the combined activities of all of the metabolic pathways that interconvert precursors, metabolites, and products of low molecular weight (not including macromolecules).

Catabolism is the degradative phase of metabolism, in which organic nutrient molecules (carbohydrates, fats, and proteins) are converted into smaller, simpler end products (e.g., lactic acid, CO_2, NH_3). Catabolic pathways release free energy, some of which is conserved in the formation of ATP and reduced electron carriers (NADH and NADPH). In **anabolism,** also called biosynthesis, small, simple precursors are built up into larger and more complex molecules, including lipids, polysaccharides, proteins, and nucleic acids. Anabolic reactions require the input of energy, generally in the forms of the free energy of hydrolysis of ATP and the reducing power of NADH and NADPH (Fig. 3).

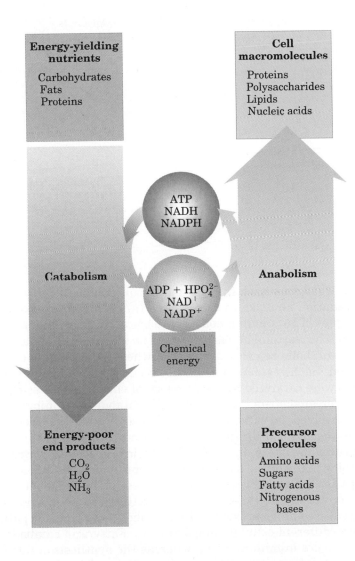

Figure 3 Energy relationships between catabolic and anabolic pathways. Catabolic pathways deliver chemical energy in the form of ATP, NADH, and NADPH. These are used in anabolic pathways to convert small precursor molecules into cell macromolecules.

Metabolic pathways are sometimes linear and sometimes branched, yielding several useful end products from a single precursor or converting several starting materials into a single product. In general, catabolic pathways are convergent and anabolic pathways divergent (Fig. 4). Some pathways are even cyclic: one of the starting components of the pathway is regenerated in the series of reactions that converts another starting component into a product. We shall see examples of each type of pathway in the following chapters.

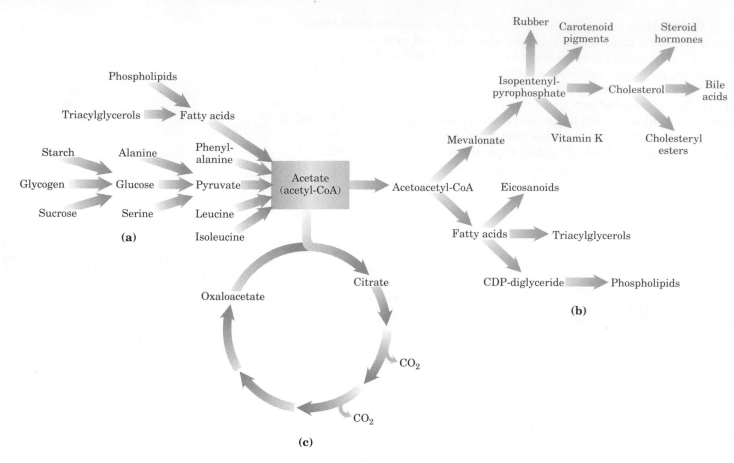

Figure 4 Three types of nonlinear metabolic pathways: **(a)** converging, catabolic; **(b)** diverging, anabolic; and **(c)** a cyclic pathway, in which one of the starting materials (oxaloacetate) is regenerated and reenters the pathway. Acetate, a key metabolic intermediate, can be produced by the breakdown of a variety of fuels **(a)**, can serve as the precursor for the biosynthesis of an array of products **(b)**, or can be consumed in the catabolic pathway known as the citric acid cycle **(c)**.

Most organisms have the enzymatic equipment to carry out both the degradation and the synthesis of certain compounds (fatty acids, for example). The simultaneous synthesis and degradation of fatty acids would be wasteful and is prevented by separately regulating anabolic and catabolic reaction sequences: when one occurs, the other is suppressed. Such regulation could not occur if anabolic and catabolic pathways were catalyzed by the same set of enzymes, operating in one direction for anabolism, the opposite for catabolism. Inhibition of an enzyme involved in catabolism would also inhibit the reaction sequence in the anabolic direction. Catabolic and anabolic pathways that connect the same two end points (a fatty acid and acetate, for example) may employ many of the same enzymes, but invariably at least one of the steps is catalyzed by different enzymes in the catabolic and the anabolic directions, and these enzymes are the sites of separate regulation. It is also common for such paired catabolic and anabolic pathways to occur in different cellular compartments. Fatty acid catabolism, for example, occurs in mitochondria, whereas the synthesis of fatty acids takes place in the cytosol. The concentrations of intermediates, enzymes, and regulators can be maintained at different levels in different compartments, further contributing to the separate regulation of catabolic and anabolic reaction sequences. These devices for separation of anabolic and catabolic processes will be of particular interest in our discussions of metabolism.

Metabolic pathways are regulated at three levels. The first and most immediately responsive form of regulation is through the action of allosteric enzymes, which are capable of changing their catalytic activity in response to stimulatory or inhibitory modulators (p. 230). We shall meet examples of allosteric regulation throughout the following chapters. Metabolic control is exerted at a second level in higher organisms by hormonal regulation. Hormones are chemical messengers released by one tissue that stimulate or inhibit some process in another tissue. Hormones serve to coordinate the metabolic activities of different tissues, and their actions and effects are generally on a somewhat longer time scale than those of allosteric effectors. The third level of metabolic regulation is control of the rate of a metabolic step by regulating the concentration of its enzyme in the cell. The concentration of an enzyme at any given time is the result of a balance between its rate of synthesis and its rate of degradation, both of which are subject to regulation on a time scale of minutes to hours.

The number of metabolic transformations that occur in a typical cell can seem overwhelming to a beginning student. Fortunately, there are recurring patterns in the metabolic pathways that make learning easier. Certain types of reactions occur in many different metabolic pathways but always employ the same coenzyme(s) and the same general mechanism. Many of the coenzymes are derived from vitamins (see Table 8–2), compounds essential in the diets of animals. The coenzymes are critical to the reaction mechanisms in which they participate. Once you have learned the general mechanism of a reaction, including the role of the cofactor, the recurring pattern in a variety of metabolic pathways will be easily recognizable. In the chapters that follow, we will usually discuss the general mechanism for each of these reactions when we first encounter the cofactor in its typical role.

In the first half of Part III we consider the major catabolic pathways by which cells obtain energy from the oxidation of various fuels: first, the central pathways of hexose conversion to triose (Chapter 14) and triose oxidation to carbon dioxide (Chapter 15); then the pathways of fatty acid oxidation (Chapter 16) and amino acid oxidation (Chapter 17). Chapter 18 is the pivotal point of our discussion of metabolism; it concerns chemiosmotic energy coupling, the universal mechanism in which a transmembrane electrochemical potential, produced either by substrate oxidation or by light absorption, drives the synthesis of ATP.

The second half of this part describes the major anabolic pathways by which cells use ATP to produce carbohydrates (Chapter 19), lipids (Chapter 20), and amino acids and nucleotides (Chapter 21) from simpler precursors. Finally, in Chapter 22 we step back from the details of the metabolic pathways and consider how those pathways are regulated and integrated in mammals by hormonal mechanisms.

We begin our study of intermediary metabolism with an introduction to bioenergetics (Chapter 13). But before we begin, a final word. Try not to forget that the myriad reactions described on these pages take place in, and play crucial roles in, living organisms. Ask of each reaction and of each pathway, "What is accomplished for the cell or the organism by this reaction or pathway? How does this pathway interconnect with the other pathways occurring simultaneously in the same cell to produce the energy and products required for cell maintenance and growth? How do the multilayered regulatory mechanisms cooperate to balance metabolic and energetic inputs and outputs, achieving the dynamic steady state of life?" Learned with this perspective, metabolism provides fascinating and revealing insights into life.

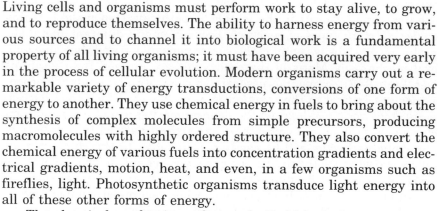

CHAPTER

13

Principles of Bioenergetics

Living cells and organisms must perform work to stay alive, to grow, and to reproduce themselves. The ability to harness energy from various sources and to channel it into biological work is a fundamental property of all living organisms; it must have been acquired very early in the process of cellular evolution. Modern organisms carry out a remarkable variety of energy transductions, conversions of one form of energy to another. They use chemical energy in fuels to bring about the synthesis of complex molecules from simple precursors, producing macromolecules with highly ordered structure. They also convert the chemical energy of various fuels into concentration gradients and electrical gradients, motion, heat, and even, in a few organisms such as fireflies, light. Photosynthetic organisms transduce light energy into all of these other forms of energy.

The chemical mechanisms that underlie biological energy transductions have fascinated and challenged biologists for centuries. Antoine Lavoisier, before he lost his head in the French Revolution, recognized that animals somehow transform chemical fuels (foods) into heat and that this process of respiration is essential to life. He observed that

> . . . in general, respiration is nothing but a slow combustion of carbon and hydrogen, which is entirely similar to that which occurs in a lighted lamp or candle, and that, from this point of view, animals that respire are true combustible bodies that burn and consume themselves. . . . One may say that this analogy between combustion and respiration has not escaped the notice of the poets, or rather the philosophers of antiquity, and which they had expounded and interpreted. This fire stolen from heaven, this torch of Prometheus, does not only represent an ingenious and poetic idea, it is a faithful picture of the operations of nature, at least for animals that breathe; one may therefore say, with the ancients, that the torch of life lights itself at the moment the infant breathes for the first time, and it does not extinguish itself except at death.*

Antoine Lavoisier
1743–1794

In this century, biochemical studies have revealed much of the chemistry of energy transductions in living organisms. Biological energy transductions obey the same physical laws that govern all other natural processes. It is therefore essential for a student of biochemistry to understand these laws and the ways in which they apply to the flow of energy in the biosphere. In this chapter we first review the laws of

---

* From a memoir by Armand Seguin and Antoine Lavoisier, dated 1789, quoted in Lavoisier, A. (1862) *Oeuvres de Lavoisier*, Imprimerie Impériale, Paris.

thermodynamics and the quantitative relationships among free energy, enthalpy, and entropy. We then describe the special role of ATP in biological energy exchanges. Finally, we consider the importance of oxidation–reduction reactions in living cells, the energetics of such electron transfer reactions, and the electron carriers commonly employed as cofactors of the enzymes that catalyze these reactions.

Bioenergetics and Thermodynamics

Bioenergetics is the quantitative study of the energy transductions that occur in living cells and of the nature and function of the chemical processes underlying these transductions. Although many of the principles of thermodynamics have been introduced in earlier chapters and may be familiar to you, it is worth reviewing the quantitative aspects of these principles.

Biological Energy Transformations Follow the Laws of Thermodynamics

Many quantitative observations made by physicists and chemists on the interconversion of different forms of energy led to the formulation, in the nineteenth century, of two fundamental laws of thermodynamics. The first law is the principle of the conservation of energy: *in any physical or chemical change, the total amount of energy in the universe remains constant, although the form of the energy may change.* The second law of thermodynamics, which can be stated in several forms, says that the universe always tends toward more and more disorder: *in all natural processes, the entropy of the universe increases.*

Living organisms consist of collections of molecules much more highly organized than the surrounding materials from which they are constructed, and they maintain and produce order, seemingly oblivious to the second law of thermodynamics. Living organisms do not violate the second law; they operate strictly within it. To discuss the application of the second law to biological systems, we must first define those systems and the universe in which they occur. The reacting system is the collection of matter that is undergoing a particular chemical or physical process; it may be an organism, a cell, or two reacting compounds. The reacting system and its surroundings together constitute the universe. Some chemical or physical processes can be made to take place in isolated or closed systems, in which no material or energy is exchanged with the surroundings. Living cells and organisms are open systems, which exchange both material and energy with their surroundings; living systems are never at equilibrium with their surroundings.

We have defined earlier in this text three thermodynamic quantities that describe the energy changes occurring in a chemical reaction. Gibbs free energy (G) expresses the amount of energy capable of doing work during a reaction at constant temperature and pressure (p. 8). When a reaction proceeds with the release of free energy (i.e., when the system changes so as to possess less free energy), the free-energy change, ΔG, has a negative sign and the reaction is said to be exergonic. In endergonic reactions, the system gains free energy and ΔG is positive. Enthalpy, H, is the heat content of the reacting system. It reflects the number and kinds of chemical bonds in the reactants and

"Now, in the *second* law of thermodynamics . . . "

BOX 13-1   Entropy: The Advantages of Being Disorganized

The term entropy, which literally means "a change within," was first used in 1851 by Rudolf Clausius, one of the promulgators of the second law. A rigorous quantitative definition of entropy involves statistical and probability considerations. However, its nature can be illustrated qualitatively by three simple examples, each of which shows one aspect of entropy. The key descriptors of entropy are *randomness* or *disorder,* manifested in different ways.

Case 1: The Teakettle and the Randomization of Heat

We know that steam generated from boiling water can do useful work. But suppose we turn off the burner under a teakettle full of water at 100 °C (the "system") in the kitchen (the "surroundings") and allow it to cool. As it cools, no work will be done, but heat will pass from the teakettle to the surroundings, raising the temperature of the surroundings (the kitchen) by an infinitesimally small amount until complete equilibrium is attained. At this point all parts of the teakettle and the kitchen will be at precisely the same temperature. The free energy that was once concentrated in the teakettle of hot water at 100 °C, *potentially* capable of doing work, has disappeared. Its equivalent in heat energy is still present in the teakettle + kitchen (i.e., the "universe") but has become completely randomized throughout. This energy is no longer available to do work because there is no temperature differential within the kitchen. Moreover, the increase in entropy of the kitchen (the surround-ings) is irreversible. We know from everyday experience that heat will never spontaneously pass back from the kitchen into the teakettle to raise the temperature of the water to 100 °C again.

Case 2: The Oxidation of Glucose

Entropy is a state or condition not only of energy but also of matter. Aerobic organisms extract free energy from glucose obtained from their surroundings. To extract this energy they oxidize the glucose with molecular oxygen, also obtained from the surroundings. The end products of the oxidative metabolism of glucose are CO_2 and H_2O, which are returned to the surroundings. In this process the surroundings undergo an increase in entropy, whereas the organism itself remains in a steady state and undergoes no change in its internal order. Although some of the entropy arises from the dissipation of heat, entropy also arises from another kind of disorder, illustrated by the equation for the oxidation of glucose by living organisms, which we can write as

$$C_6H_{12}O_6 + 6O_2 \longrightarrow 6CO_2 + 6H_2O$$

or represent schematically as

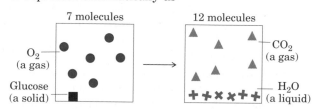

Table 13-1 Some physical constants and units frequently used in thermodynamics

Boltzmann constant, $\mathbf{k} = 1.381 \times 10^{-23}$ J/K
Avogadro's number, $N = 6.022 \times 10^{23}$ mol^{-1}
Faraday constant, $\mathscr{F} = 96,480$ J/V·mol
Gas constant, $R = 8.315$ J/mol·K
          ($= 1.987$ cal/mol·K)

Units of ΔG and ΔH are J/mol (or cal/mol)
Units of ΔS are J/mol·K (or cal/mol·K)
          1 cal = 4.184 J

Units of absolute temperature, T, are degrees Kelvin, K
          25 °C = 298 K
     At 25 °C, $RT = 2.479$ kJ/mol
          ($= 0.592$ kcal/mol)
          $\ln x = 2.303 \log x$

products. When a chemical reaction releases heat, it is said to be exothermic; the heat content of the products is less than that of the reactants and ΔH has a negative value. Reacting systems that take up heat from their surroundings are endothermic and have positive values of ΔH (p. 66). Entropy, S, is a quantitative expression for the randomness or disorder in a system (Box 13-1). When the products of a reaction are less complex and more disordered than the reactants, the reaction is said to proceed with a gain in entropy (p. 72). The units of ΔG and ΔH are joules/mole or calories/mole (recall that 1 cal equals 4.18 J); units of entropy are joules/mole·degree Kelvin (J/mol·K) (Table 13-1).

Under the conditions existing in biological systems (at constant temperature and pressure), changes in free energy, enthalpy, and entropy are related to each other quantitatively by the equation

$$\Delta G = \Delta H - T\,\Delta S \qquad (13\text{-}1)$$

in which ΔG is the change in Gibbs free energy of the reacting system,

The atoms contained in 1 molecule of glucose plus 6 molecules of oxygen, a total of 7 molecules, are more randomly dispersed by the oxidation reaction and are now present in a total of 12 molecules ($6CO_2 + 6H_2O$).

Whenever a chemical reaction proceeds so that there is an increase in the number of molecules—or when a solid substance, such as glucose, is converted into liquid or gaseous products, which have more freedom to move or fill space than a solid—there is an increase in molecular disorder and thus an increase in entropy.

Case 3: Information and Entropy

The following short passage from *Julius Caesar,* Act IV, Scene 3, is spoken by Brutus, when he realizes that he must face Mark Antony's army. It is an information-rich nonrandom arrangement of 125 letters of the English alphabet:

> There is a tide in the affairs of men,
> Which, taken at the flood, leads on to fortune;
> Omitted, all the voyage of their life
> Is bound in shallows and in miseries.

In addition to what this quotation says overtly, it has many hidden meanings. It not only reflects a complex sequence of events in the play, it also echoes the play's ideas on conflict, ambition, and the demands of leadership. Permeated with Shakespeare's understanding of human nature, it is very rich in information.

However, if the 125 letters making up this quotation were allowed to fall into a completely random, chaotic pattern, as shown in the following box, they would have no meaning whatsoever.

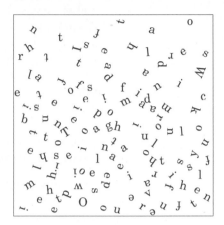

In this form the 125 letters would contain little or no information, but would be very rich in entropy. Such considerations have led to the conclusion that information is a form of energy; information has been called "negative entropy." In fact, the branch of mathematics called information theory, which is basic to the programming logic of computers, is closely related to thermodynamic theory. Living organisms are highly ordered, nonrandom structures, immensely rich in information and thus entropy-poor.

ΔH is the change in enthalpy of the system, T is the absolute temperature, and ΔS is the change in entropy of the reacting system. By convention ΔS has a positive sign when entropy increases and ΔH has a negative sign when heat is released by the system to its surroundings. Either of these conditions, which are typical of favorable processes, will tend to make ΔG negative. In fact, ΔG of a spontaneously reacting system is always negative.

The second law of thermodynamics states that the entropy *of the universe* increases during all chemical and physical processes, but it does not require that the entropy increase take place *in the reacting system* itself. The order produced within cells as they grow and divide is more than compensated for by the disorder they create in their surroundings in the course of growth and division (Box 13–1, case 2). In short, living organisms preserve their internal order by taking from the surroundings free energy in the form of nutrients or sunlight, and returning to their surroundings an equal amount of energy as heat and entropy.

Cells Require Sources of Free Energy

Cells are isothermal systems—they function at essentially constant temperature (and at constant pressure). Heat flow is not a source of energy for cells because heat can do work only as it passes from a zone or object at one temperature to a zone or object at a lower temperature. The energy that cells can and must use is free energy, described by the Gibbs free-energy function G, which allows prediction of the direction of chemical reactions, their exact equilibrium position, and the amount of work they can in theory perform at constant temperature and pressure. Heterotrophic cells acquire free energy from nutrient molecules, and photosynthetic cells acquire it from absorbed solar radiation. Both kinds of cells transform this free energy into ATP and other energy-rich compounds, capable of providing energy for biological work at constant temperature.

Standard Free-Energy Change Is Directly Related to the Equilibrium Constant

The composition of a reacting system (a mixture of chemical reactants and products) will tend to continue changing until equilibrium is reached. At the equilibrium concentration of reactants and products, the rates of the forward and reverse reactions are exactly equal and no further net change occurs in the system. The concentrations of reactants and products *at equilibrium* define the equilibrium constant (p. 90). In the general reaction $a\text{A} + b\text{B} \rightleftharpoons c\text{C} + d\text{D}$, where a, b, c, and d are the number of molecules of A, B, C, and D participating, the equilibrium constant is given by

$$K_{eq} = \frac{[\text{C}]^c[\text{D}]^d}{[\text{A}]^a[\text{B}]^b} \tag{13–2}$$

where [A], [B], [C], and [D] are the molar concentrations of the reaction components at the point of equilibrium.

When a reacting system is not at equilibrium, the tendency to move toward equilibrium represents a driving force, the magnitude of which can be expressed as the free-energy change for the reaction, ΔG. Under standard conditions (298 K (25 °C)), when reactants and products are initially present at 1 M concentrations or, for gases, at partial pressures of 101.3 kPa (1 atm), the force driving the system toward equilibrium is defined as the standard free-energy change, $\Delta G°$. By this definition, the standard state for reactions that involve hydrogen ions is $[\text{H}^+] = 1$ M, or pH is 0. Most biochemical reactions occur in well-buffered aqueous solutions near pH 7; both the pH and the concentration of water (55.5 M) are essentially constant. For convenience of calculations, biochemists therefore define a slightly different standard state, in which the concentration of H^+ is 10^{-7} M (pH is 7) and that of water is 55.5 M. Physical constants based on this biochemical standard state are written with a prime (e.g., $\Delta G°'$ and K'_{eq}) to distinguish them from the constants used by chemists and physicists. Under this convention, when H_2O or H^+ are reactants or products, their concentrations are not included in equations such as Equation 13–2, but are instead incorporated into the constants $\Delta G°'$ and K'_{eq}.

Just as K'_{eq} is a physical constant characteristic for each reaction, so too is $\Delta G°'$ a constant. As we noted in Chapter 8 (p. 204), there is a simple relationship between K'_{eq} and $\Delta G°'$:

$$\Delta G°' = -RT \ln K'_{eq}$$

The standard free-energy change of a chemical reaction is simply an alternative mathematical way of expressing its equilibrium constant. Table 13–2 shows the relationship between $\Delta G^{\circ\prime}$ and K'_{eq}. If the equilibrium constant for a given chemical reaction is 1.0, the standard free-energy change of that reaction is 0.0 (the natural logarithm of 1.0 is zero). If K'_{eq} of a reaction is greater than 1.0, its $\Delta G^{\circ\prime}$ is negative. If K'_{eq} is less than 1.0, $\Delta G^{\circ\prime}$ is positive. Because the relationship between $\Delta G^{\circ\prime}$ and K'_{eq} is exponential, relatively small changes in $\Delta G^{\circ\prime}$ correspond to large changes in K'_{eq}.

It may be helpful to think of the standard free-energy change in another way. $\Delta G^{\circ\prime}$ is the difference between the free-energy content of the products and the free-energy content of the reactants under standard conditions. When $\Delta G^{\circ\prime}$ is negative, the products contain less free energy than the reactants. The reaction will therefore proceed spontaneously to form the products under standard conditions, because all chemical reactions tend to go in the direction that results in a decrease in the free energy of the system. A positive value of $\Delta G^{\circ\prime}$ means that the products of the reaction contain more free energy than the reactants. The reaction will therefore tend to go in the reverse direction if we start with 1.0 M concentrations of all components. Table 13–3 summarizes these points.

Table 13–2 Relationship between the equilibrium constants of chemical reactions and their standard free-energy changes	
K'_{eq}	$\Delta G^{\circ\prime}$ (kJ/mol)
0.001	17.1
0.01	11.4
0.1	5.7
1.0	0.0
10.0	−5.7
100.0	−11.4
1,000.0	−17.1

Table 13–3 Relationships among K'_{eq}, $\Delta G^{\circ\prime}$, and the direction of chemical reactions under standard conditions

When K'_{eq} is	$\Delta G^{\circ\prime}$ is	Starting with 1 M components the reaction
>1.0	Negative	Proceeds forward
1.0	Zero	Is at equilibrium
<1.0	Positive	Proceeds in reverse

As an example, let us make a simple calculation of the standard free-energy change of the reaction catalyzed by the enzyme phosphoglucomutase:

$$\text{Glucose-1-phosphate} \rightleftharpoons \text{glucose-6-phosphate}$$

Chemical analysis shows that whether we start with, say, 20 mM glucose-1-phosphate (but no glucose-6-phosphate) in the presence of phosphoglucomutase, or with 20 mM glucose-6-phosphate, the final equilibrium mixture in either case will contain 1 mM glucose-1-phosphate and 19 mM glucose-6-phosphate at 25 °C and pH 7.0. (Remember that enzymes do not affect the point of equilibrium of a reaction; they merely hasten its attainment.) From these data we can calculate the equilibrium constant:

$$K'_{eq} = \frac{[\text{glucose-6-phosphate}]}{[\text{glucose-1-phosphate}]} = \frac{19\ \text{mM}}{1\ \text{mM}} = 19$$

From this value of K'_{eq} we can calculate the standard free-energy change:

$$\Delta G^{\circ\prime} = -RT \ln K'_{eq}$$
$$= -(8.315\ \text{J/mol} \cdot \text{K})(298\ \text{K})(\ln 19)$$
$$= -7{,}296\ \text{J/mol} = -7.3\ \text{kJ/mol}$$

Because the standard free-energy change is negative, when the reaction starts with 1.0 M glucose-1-phosphate and 1.0 M glucose-6-phosphate, the conversion of glucose-1-phosphate into glucose-6-phosphate proceeds with a loss (release) of free energy. For the reverse reaction (the conversion of glucose-6-phosphate to glucose-1-phosphate), $\Delta G^{\circ\prime}$ has the same magnitude but the opposite sign.

Table 13–4 gives the standard free-energy changes for several representative chemical reactions. Note that hydrolysis of simple esters, amides, peptides, and glycosides, as well as rearrangements and eliminations, proceed with relatively small standard free-energy changes, whereas hydrolysis of acid anhydrides occurs with relatively large decreases in standard free energy. The oxidation of organic compounds to CO_2 and H_2O proceeds with especially large decreases in standard free energy. However, standard free-energy changes such as those in Table 13–4 tell how much free energy is available from a reaction under *standard conditions*. To describe the energy released under the conditions that exist within cells, an expression for the *actual* free-energy change is essential.

Table 13–4 Standard free-energy changes of some chemical reactions at pH 7.0 and 25 °C (298 K)

	$\Delta G^{\circ\prime}$	
Reaction type	(kJ/mol)	(kcal/mol)*
Hydrolysis reactions		
Acid anhydrides		
Acetic anhydride + $H_2O \longrightarrow$ 2 acetate	−91.1	−21.8
ATP + $H_2O \longrightarrow$ ADP + P_i	−30.5	−7.3
Esters		
Ethyl acetate + $H_2O \longrightarrow$ ethanol + acetate	−19.6	−4.7
Glucose-6-phosphate + $H_2O \longrightarrow$ glucose + P_i	−13.8	−3.3
Amides and peptides		
Glutamine + $H_2O \longrightarrow$ glutamate + NH_4^+	−14.2	−3.4
Glycylglycine + $H_2O \longrightarrow$ 2 glycine	−9.2	−2.2
Glycosides		
Maltose + $H_2O \longrightarrow$ 2 glucose	−15.5	−3.7
Lactose + $H_2O \longrightarrow$ glucose + galactose	−15.9	−3.8
Rearrangements		
Glucose-1-phosphate $\longrightarrow$ glucose-6-phosphate	−7.3	−1.74
Fructose-6-phosphate $\longrightarrow$ glucose-6-phosphate	−1.7	−0.40
Elimination of water		
Malate $\longrightarrow$ fumarate + H_2O	3.1	0.75
Oxidations with molecular oxygen		
Glucose + $6O_2 \longrightarrow 6CO_2 + 6H_2O$	−2,840	−686
Palmitic acid + $23O_2 \longrightarrow 16CO_2 + 16H_2O$	−9,770	−2,338

* Although joules and kilojoules are the standard units of energy and are used throughout this text, biochemists sometimes express $\Delta G^{\circ\prime}$ values in kilocalories per mole. We have therefore included values in both kilojoules and kilocalories in this table and in Table 13–5. To convert kilojoules to kilocalories, divide the number of kilojoules by 4.184.

The Actual Free-Energy Change Depends on the Reactant and Product Concentrations

We must be careful to distinguish between two different quantities, the free-energy change, ΔG, and the standard free-energy change, $\Delta G^{\circ\prime}$. Each chemical reaction has a characteristic standard free-energy change, which may be positive, negative, or zero, depending on the equilibrium constant of the reaction. The standard free-energy change tells us in which direction and how far a given reaction will go to reach equilibrium *when the initial concentration of each component is 1.0 M, the pH is 7.0, and the temperature is 25 °C*. Thus $\Delta G^{\circ\prime}$ is a constant: it has a characteristic, unchanging value for a given reaction. But the *actual* free-energy change, ΔG, of a given chemical reaction is a function of the concentrations and of the temperature actually prevailing during the reaction, which are not necessarily the standard conditions as defined above. Moreover, the ΔG of any reaction proceeding spontaneously toward its equilibrium is always negative, becomes less negative as the reaction proceeds, and is zero at the point of equilibrium, indicating that no more work can be done by the reaction.

ΔG and $\Delta G^{\circ\prime}$ for any reaction A + B $\rightleftharpoons$ C + D are related by the equation

$$\Delta G = \Delta G^{\circ\prime} + RT \ln \frac{[C][D]}{[A][B]} \qquad (13\text{-}3)$$

in which the terms in red are those *actually prevailing* in the system under observation. The concentration terms in this equation express the effects commonly called mass action. As an example, let us suppose that the reaction A + B $\rightleftharpoons$ C + D is taking place at the standard conditions of temperature (25 °C) and pressure (101.3 kPa) but that the concentrations of A, B, C, and D are *not* equal and that none of them is present at the standard concentration of 1.0 M. To determine the actual free-energy change, ΔG, that will occur under these nonstandard conditions of concentration as the reaction proceeds from left to right, we simply put in the *actual* concentrations of A, B, C, and D; the values of R, T, and $\Delta G^{\circ\prime}$ are the standard values. ΔG will be negative and will approach zero as the reaction proceeds because the actual concentrations of A and B will be getting smaller and the concentrations of C and D will be getting larger. Notice that when a reaction is at equilibrium, where there is no force driving the reaction in either direction and ΔG is equal to zero, Equation 13–3 reduces to

$$0 = \Delta G^{\circ\prime} + RT \ln \frac{[C]_{eq}[D]_{eq}}{[A]_{eq}[B]_{eq}}$$

or

$$\Delta G^{\circ\prime} = -RT \ln K'_{eq}$$

the equation that, as we noted above (p. 368), relates the standard free-energy change and the equilibrium constant.

Even a reaction for which $\Delta G^{\circ\prime}$ is positive can go in the forward direction, *if ΔG is negative*. This is possible if the term $RT \ln$ ([products]/[reactants]) in Equation 13–3 is negative and has a larger absolute value than $\Delta G^{\circ\prime}$. For example, the immediate removal of the products of a reaction can keep the ratio [products]/[reactants] well below 1, giving the term $RT \ln$ ([products]/[reactants]) a large, negative value.

$\Delta G^{\circ\prime}$ and ΔG are expressions of the *maximum* amount of free energy that a given reaction can *theoretically* deliver. This amount of energy could be realized only if there were a perfectly efficient device available to trap or harness it. Given that no such device is available, the amount of work done by the reaction at constant temperature and pressure is always less than the theoretical amount.

It is also essential to understand that some reactions that are thermodynamically favorable (i.e., for which ΔG is large and negative) nevertheless do not occur at measurable rates. For example, firewood can be converted into CO_2 and H_2O by combustion in a reaction that is very favorable thermodynamically. Nevertheless, firewood is stable for years, because the activation energy (see Fig. 8–4) for its combustion is higher than that provided by room temperature. If the necessary activation energy is provided (with a lighted match, for example), combustion will begin, converting the wood to the more stable products CO_2 and H_2O and releasing energy as heat and light.

In living cells, reactions that would be extremely slow if uncatalyzed are caused to occur, not by supplying additional heat but by lowering the activation energy with an enzyme (see Fig. 8–4). *The free-energy change ΔG for a reaction is independent of the pathway by which the reaction occurs;* it depends only on the nature and concentration of the initial reactants and the final products. An enzyme provides an alternative reaction pathway with a lower activation energy, so that at room temperature a large fraction of the substrate molecules have enough thermal energy to overcome the activation barrier, and the reaction rate increases dramatically. Enzymes cannot change equilibrium constants; but they can and do increase the rate at which a reaction proceeds in the direction dictated by thermodynamics.

Standard Free-Energy Changes Are Additive

In the case of two sequential chemical reactions, $A \rightleftharpoons B$ and $B \rightleftharpoons C$, each reaction has its own equilibrium constant and each has its characteristic standard free-energy change, $\Delta G_1^{\circ\prime}$ and $\Delta G_2^{\circ\prime}$. As the two reactions are sequential, B cancels out and the overall reaction is $A \rightleftharpoons C$. Reaction $A \rightleftharpoons C$ will also have its own equilibrium constant and thus will also have its own standard free-energy change, $\Delta G_{total}^{\circ\prime}$. *The $\Delta G^{\circ\prime}$ values of sequential chemical reactions are additive.* For the overall reaction $A \rightleftharpoons C$, $\Delta G_{total}^{\circ\prime}$ is the algebraic sum of the individual standard free-energy changes, $\Delta G_1^{\circ\prime}$ and $\Delta G_2^{\circ\prime}$, of the two separate reactions: $\Delta G_{total}^{\circ\prime} = \Delta G_1^{\circ\prime} + \Delta G_2^{\circ\prime}$. This principle of bioenergetics explains how a thermodynamically unfavorable (endergonic) reaction can be driven in the forward direction by coupling it to a highly exergonic reaction through a common intermediate. For example, the synthesis of glucose-6-phosphate is the first step in the utilization of glucose by many organisms:

$$\text{Glucose} + P_i \longrightarrow \text{glucose-6-phosphate} + H_2O \qquad \Delta G^{\circ\prime} = 13.8 \text{ kJ/mol}$$

The positive value of $\Delta G^{\circ\prime}$ predicts that under standard conditions the reaction will tend not to proceed spontaneously in the direction written. Another cellular reaction, the hydrolysis of ATP to ADP and P_i, is very exergonic:

$$\text{ATP} + H_2O \longrightarrow \text{ADP} + P_i \qquad \Delta G^{\circ\prime} = -30.5 \text{ kJ/mol}$$

These two reactions share the common intermediates P_i and H_2O and may be expressed as sequential reactions:

(1)   Glucose + P_i $\longrightarrow$ glucose-6-phosphate + H_2O
(2)   ATP + H_2O $\longrightarrow$ ADP + P_i

Sum: ATP + glucose $\longrightarrow$ ADP + glucose-6-phosphate

The overall standard free-energy change is obtained by adding the $\Delta G^{\circ\prime}$ values for individual reactions:

$$\Delta G^{\circ\prime} = +13.8 \text{ kJ/mol} + (-30.5 \text{ kJ/mol}) = -16.7 \text{ kJ/mol}$$

The overall reaction is exergonic. In this case, energy stored in the bonds of ATP is used to drive the synthesis of glucose-6-phosphate, a product whose formation from glucose and phosphate is endergonic. The *pathway* of glucose-6-phosphate formation by phosphate transfer from ATP is different from reactions (1) and (2) above, but the net result is the same as the sum of the two reactions. In thermodynamic calculations, all that matters is the initial and final states; the route between them is immaterial.

We have said that $\Delta G^{\circ\prime}$ is a way of expressing the equilibrium constant for a reaction. For reaction (1) above,

$$K'_{eq_1} = \frac{\text{[glucose-6-phosphate]}}{\text{[glucose][P}_i\text{]}} = 3.9 \times 10^{-3}\ \text{M}^{-1}$$

Notice that H_2O is not included in this expression. The equilibrium constant for the hydrolysis of ATP is

$$K'_{eq_2} = \frac{\text{[ADP][P}_i\text{]}}{\text{[ATP]}} = 2 \times 10^5\ \text{M}$$

The equilibrium constant for the two coupled reactions is

$$K'_{eq_3} = \frac{\text{[glucose-6-phosphate][ADP][P}_i\text{]}}{\text{[glucose][P}_i\text{][ATP]}}$$

$$= (K'_{eq_1})(K'_{eq_2}) = (3.9 \times 10^{-3}\ \text{M}^{-1})(2.0 \times 10^5\ \text{M})$$

$$= 7.8 \times 10^2$$

By coupling ATP hydrolysis to glucose-6-phosphate synthesis, the K_{eq} for formation of glucose-6-phosphate has been raised by a factor of about 2×10^5.

This strategy is employed by all living cells in the synthesis of metabolic intermediates and cellular components. Obviously, the strategy only works if compounds such as ATP are continuously available. In the following chapters we consider several of the most important cellular pathways for producing ATP.

Phosphate Group Transfers and ATP

Having developed some fundamental principles of energy changes in chemical systems, we can now examine the energy cycle in cells and the special role of ATP in linking catabolism and anabolism (see Fig. 1–13). Heterotrophic cells obtain free energy in a chemical form by the catabolism of nutrient molecules and use that energy to make ATP from ADP and P_i. ATP then donates some of its chemical energy to endergonic processes such as the synthesis of metabolic intermediates and macromolecules from smaller precursors, transport of substances across membranes against concentration gradients, and mechanical motion. This donation of energy from ATP generally involves the covalent participation of ATP in the reaction that is to be driven, with the result that ATP is converted to ADP and P_i or to AMP and $2P_i$. We discuss here the chemical basis for the large free-energy changes that accompany hydrolysis of ATP and other high-energy phosphate compounds, and show that most cases of energy donation by ATP involve group transfer, not simple hydrolysis of ATP. To illustrate the range of energy transductions in which ATP provides energy, we consider the synthesis of information-rich macromolecules, the transport of solutes across membranes, and motion produced by muscle contraction.

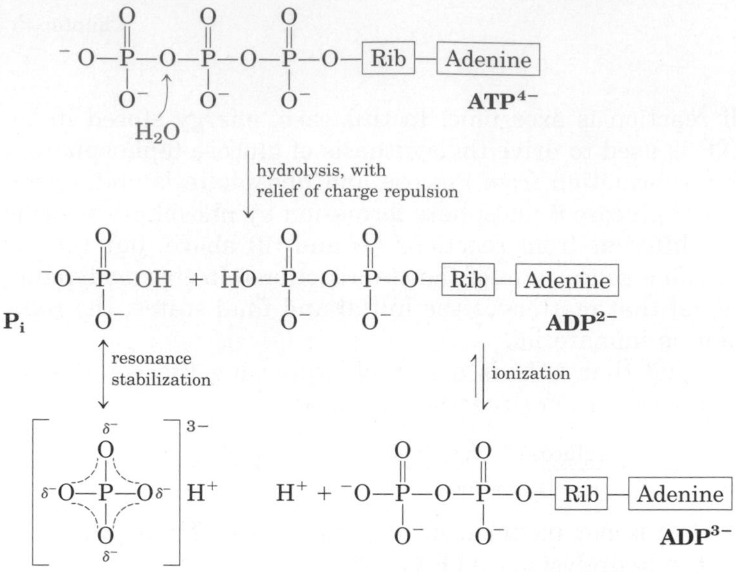

Figure 13–1 The chemical basis for the large free-energy change associated with ATP hydrolysis. (1) Electrostatic repulsion among the four negative charges on ATP is relieved by charge separation after hydrolysis. (2) Inorganic phosphate (P_i) released by hydrolysis is stabilized by formation of a resonance hybrid (left), in which each of the four P—O bonds has the same degree of double-bond character and the hydrogen ion is not permanently associated with any one of the oxygens. (3) The other direct product of hydrolysis, ADP^{2-}, also immediately ionizes (right), releasing a proton into a medium of very low [H^+] (pH 7). A fourth factor (not shown) that favors ATP hydrolysis is the greater degree of solvation (hydration) of the products P_i and ADP relative to ATP, which further stabilizes the products relative to the reactants.

$$ATP^{4-} + H_2O \longrightarrow ADP^{3-} + P_i^{2-} + H^+$$
$$\Delta G^{\circ\prime} = -30.5 \text{ kJ/mol}$$

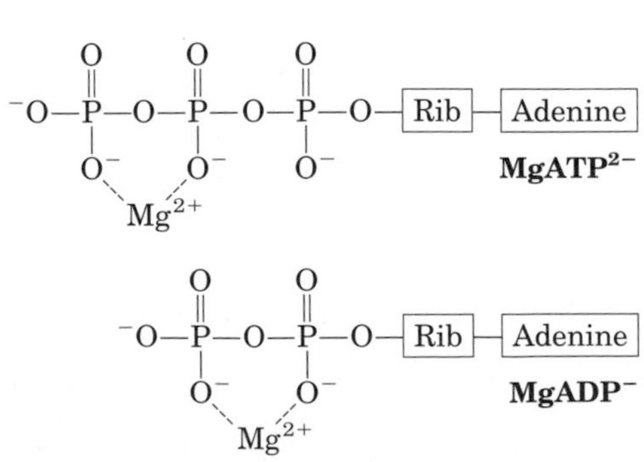

Figure 13–2 Formation of Mg^{2+} complexes partially shields the negative charges and influences the conformation of the phosphate groups in nucleotides such as ATP and ADP.

The Free-Energy Change for ATP Hydrolysis Is Large and Negative

Figure 13–1 summarizes the chemical basis for the relatively large, negative, standard free energy of hydrolysis of ATP. The hydrolytic cleavage of the terminal phosphoric acid anhydride (phosphoanhydride) bond in ATP separates off one of the three negatively charged phosphates and thus relieves some of the electrostatic repulsion in ATP; the P_i (HPO_4^{2-}) released by hydrolysis is stabilized by the formation of several resonance forms not possible in ATP; and ADP^{2-}, the other direct product of hydrolysis, immediately ionizes, releasing H^+ into a medium of very low [H^+]($\sim 10^{-7}$ M). The low concentration of the direct products favors, by mass action, the hydrolysis reaction.

Although its hydrolysis is highly exergonic ($\Delta G^{\circ\prime} = -30.5$ kJ/mol), ATP is kinetically stable toward nonenzymatic breakdown at pH 7 because the activation energy for ATP hydrolysis is relatively high. Rapid cleavage of the phosphoric acid anhydride bonds occurs only when catalyzed by an enzyme.

Although the $\Delta G^{\circ\prime}$ for ATP hydrolysis is -30.5 kJ/mol under standard conditions, the *actual* free energy of hydrolysis (ΔG) of ATP in living cells is very different. This is because the concentrations of ATP, ADP, and P_i in living cells are not identical and are much lower than the standard 1.0 M concentrations (Table 13–5). Furthermore, the cytosol contains Mg^{2+}, which binds to ATP and ADP (Fig. 13–2). In most enzymatic reactions that involve ATP as phosphoryl donor, the true substrate is $MgATP^{2-}$ and the relevant $\Delta G^{\circ\prime}$ is that for $MgATP^{2-}$ hydrolysis. Box 13–2 shows how ΔG for ATP hydrolysis in the intact erythrocyte can be calculated from the data in Table 13–4. ΔG for ATP hydrolysis in intact cells, usually designated ΔG_p, is much more negative than $\Delta G^{\circ\prime}$; in most cells ΔG_p ranges from -50 to -65 kJ/mol. ΔG_p is often called the **phosphorylation potential.** In the following discussion we use the standard free-energy change for ATP hydrolysis, because this allows convenient comparison with the energetics of other cellular reactions for which the actual free-energy changes within cells are not known with certainty.

BOX 13–2   **The Free Energy of Hydrolysis of ATP within Cells:**
The Real Cost of Doing Metabolic Business

The standard free energy of hydrolysis of ATP has the value -30.5 kJ/mol. In the cell, however, the concentrations of ATP, ADP, and P_i are not only unequal but are also much lower than the standard 1 M concentrations (see Table 13–5). Moreover, the pH inside cells may differ somewhat from the standard pH of 7.0. Thus the *actual* free energy of hydrolysis of ATP under intracellular conditions (ΔG_p) differs from the standard free-energy change, $\Delta G^{\circ\prime}$. We can easily calculate ΔG_p. For example, in human erythrocytes the concentrations of ATP, ADP, and P_i are 2.25, 0.25, and 1.65 mM, respectively (Table 13–5). Let us assume for simplicity that the pH is 7.0 and the temperature is 25 °C, the standard pH and temperature. The actual free energy of hydrolysis of ATP in the erythrocyte under these conditions is given by the relationship

$$\Delta G = \Delta G^{\circ\prime} + RT \ln \frac{[\text{ADP}][\text{P}_i]}{[\text{ATP}]}$$

Substituting the appropriate values we obtain

$$\Delta G = -30{,}500 \text{ J/mol} + (8.315 \text{ J/mol} \cdot \text{K})(298 \text{ K})$$
$$\ln \frac{(2.50 \times 10^{-4})(1.65 \times 10^{-3})}{2.25 \times 10^{-3}}$$

$$= -30{,}500 \text{ J/mol} + (2{,}480 \text{ J/mol}) \ln (1.83 \times 10^{-4})$$

$$= -30{,}500 \text{ J/mol} - 21{,}300 \text{ J/mol} = -51{,}800 \text{ J/mol}$$

$$= -51.8 \text{ kJ/mol}$$

Thus ΔG_p, the actual free-energy change for ATP hydrolysis in the intact erythrocyte (-51.8 kJ/mol), is much larger than the standard free-energy change (-30.5 kJ/mol). By the same token, the free energy required to *synthesize* ATP from ADP and P_i under the conditions prevailing in the erythrocyte would be 51.8 kJ/mol.

Because the concentrations of ATP, ADP, and P_i may differ from one cell type to another (Table 13–5), ΔG_p for ATP hydrolysis likewise differs. Moreover, in any given cell ΔG_p can vary from time to time, depending on the metabolic conditions in the cell and how they influence the concentrations of ATP, ADP, P_i, and H^+ (pH). We can calculate the actual free-energy change for any given metabolic reaction as it occurs in the cell, providing we know the concentrations of all the reactants and products of the reaction and other factors (such as pH, temperature, and the concentration of Mg^{2+}) that may affect the equilibrium constant and thus the free-energy change.

Table 13–5 Adenine nucleotide, inorganic phosphate, and phosphocreatine concentrations in some cells*

	Concentration (mM)				
	ATP	ADP	AMP	P_i	PCr
Rat hepatocyte	3.38	1.32	0.29	4.8	0
Rat myocyte	8.05	0.93	0.04	8.05	28
Human erythrocyte	2.25	0.25	0.02	1.65	0
Rat neuron	2.59	0.73	0.06	2.72	4.7
E. coli cell	7.90	1.04	0.82	7.9	0

* For erythrocytes the concentrations are those of the cytosol (human erythrocytes lack a nucleus and mitochondria). In the other types of cells the data are for the entire cell contents, although the cytosol and the mitochondria have very different concentrations of ADP. Phosphocreatine (PCr) is discussed later in this chapter.

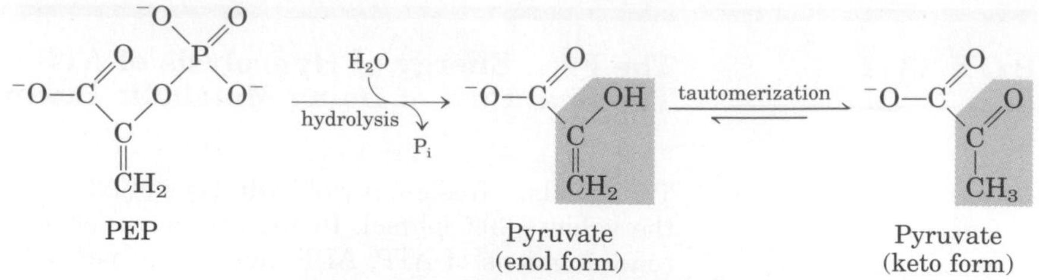

Figure 13–3 Hydrolysis of phosphoenolpyruvate (PEP), catalyzed by pyruvate kinase, is followed by spontaneous tautomerization of the product. Tautomerization is not possible in PEP, and thus the product of hydrolysis is stabilized relative to the reactant. Resonance stabilization of P_i also occurs, as shown in Fig. 13–1.

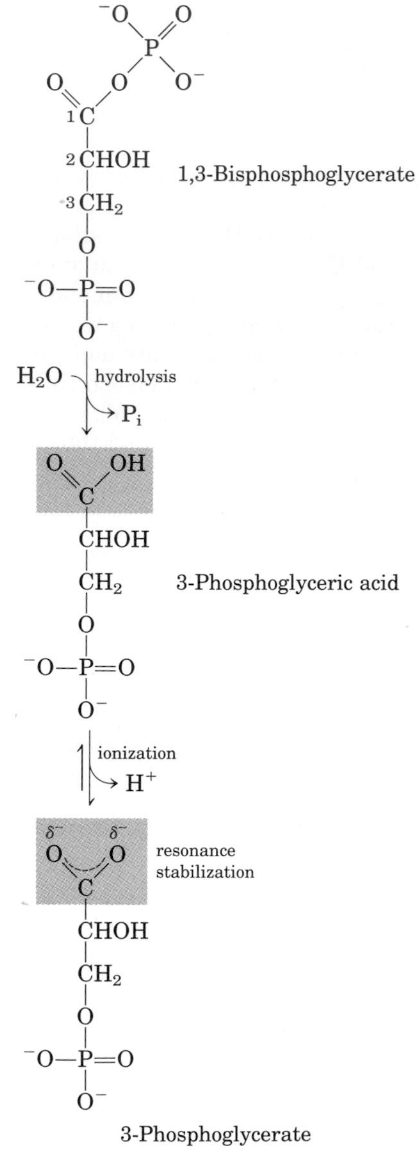

PEP   Pyruvate (enol form)   Pyruvate (keto form)

$$PEP^{3-} + H_2O \longrightarrow pyruvate^- + P_i^{2-}$$
$$\Delta G^{\circ\prime} = -61.9 \text{ kJ/mol}$$

1,3-Bisphosphoglycerate

3-Phosphoglyceric acid

3-Phosphoglycerate

$$1,3\text{-Bisphosphoglycerate}^{4-} + H_2O \longrightarrow$$
$$3\text{-phosphoglycerate}^{3-} + P_i^{2-} + H^+$$
$$\Delta G^{\circ\prime} = 49.3 \text{ kJ/mol}$$

Figure 13–4 Hydrolysis of 1,3-bisphosphoglycerate. The direct product of hydrolysis is 3-phosphoglyceric acid, with an undissociated carboxylic acid group, but dissociation occurs immediately. This ionization and the resonance structures it makes possible stabilize the product relative to the reactants. Resonance stabilization of P_i further contributes to the free-energy change.

Other Phosphorylated Compounds and Thioesters Also Have Large Free Energies of Hydrolysis

Phosphoenolpyruvate (Fig. 13–3) contains a phosphate ester bond that can undergo hydrolysis to yield the enol form of pyruvate, which immediately tautomerizes to the more stable keto form. Because the product of hydrolysis can exist in either of two tautomeric forms (enol and keto), whereas the reactant has only one form (enol), the product is stabilized relative to the reactant. This is the main reason for the high standard free energy of hydrolysis of phosphoenolpyruvate: $\Delta G^{\circ\prime} = -61.9$ kJ/mol.

Another three-carbon compound, 1,3-bisphosphoglycerate (Fig. 13–4), contains an anhydride bond between the carboxyl group at C-1 and phosphoric acid. Hydrolysis of this acyl phosphate is accompanied by a large, negative, standard free-energy change ($\Delta G^{\circ\prime} = -49.3$ kJ/mol), which can be rationalized in terms of the structure of reactant and products. When H_2O is added across the anhydride bond, one of the direct products (3-phosphoglyceric acid) can immediately lose a proton. Removal of this direct product favors the forward reaction and results in the formation of the carboxylate ion (3-phosphoglycerate), which has two equally probable resonance forms (Fig. 13–4).

In phosphocreatine (Fig. 13–5), the P—N bond can be hydrolyzed to generate free creatine and P_i. As in the previous cases, the release of P_i favors the forward reaction. Creatine can exist in two resonance forms, and this resonance stabilization of the product favors the forward reaction. The standard free-energy change in this reaction is large, about -43 kJ/mol.

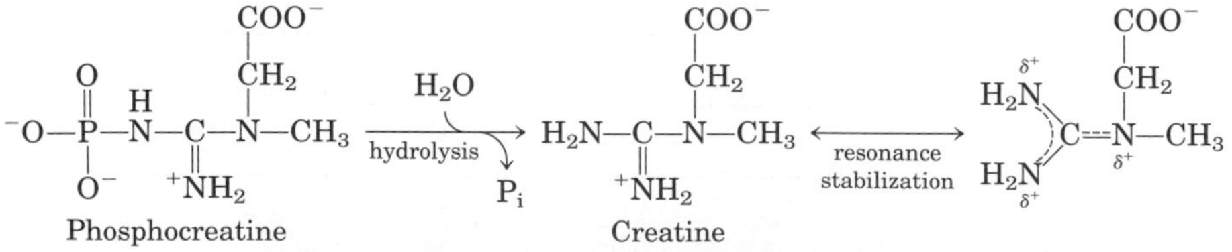

Phosphocreatine   Creatine

$$Phosphocreatine^{2-} + H_2O \longrightarrow creatine + P_i^{2-}$$
$$\Delta G^{\circ\prime} = -43.0 \text{ kJ/mol}$$

Figure 13–5 Hydrolysis of phosphocreatine. Breakage of the P—N bond in phosphocreatine produces creatine, which forms a resonance hybrid and is thus stabilized. The other product, P_i, is also resonance stabilized.

Table 13–6 Standard free energies of hydrolysis of some phosphorylated compounds and acetyl-coenzyme A

	$\Delta G^{\circ\prime}$	
	(kJ/mol)	(kcal/mol)
Phosphoenolpyruvate	−61.9	−14.8
1,3-bisphosphoglycerate ($\longrightarrow$ 3-phosphoglycerate + P_i)	−49.3	−11.8
Phosphocreatine	−43.0	−10.3
ADP ($\longrightarrow$ AMP + P_i)	−30.5	−7.3
ATP ($\longrightarrow$ ADP + P_i)		
ATP ($\longrightarrow$ AMP + PP_i)	−32.2	−7.7
AMP ($\longrightarrow$ adenosine + P_i)	−14.2	−3.4
PP_i ($\longrightarrow$ 2P_i)	−33.4	−8.0
Glucose-1-phosphate	−20.9	−5.0
Fructose-6-phosphate	−15.9	−3.8
Glucose-6-phosphate	−13.8	−3.3
Glycerol-1-phosphate	−9.2	−2.2
Acetyl-CoA	−31.4	−7.5

Source: Data mostly from Jencks, W.P. (1976) In *Handbook of Biochemistry and Molecular Biology,* 3rd edn (Fasman, G.D., ed), *Physical and Chemical Data,* Vol. I, pp. 296–304, CRC Press, Cleveland, OH.

In all of the reactions that liberate P_i, the several resonance forms available to P_i (Fig. 13–1) stabilize this product relative to the reactant, further contributing to a negative free-energy change for the hydrolysis reactions. Table 13–6 lists the standard free energies of hydrolysis for a number of phosphorylated compounds.

Thioesters are compounds that do not release P_i on hydrolysis but nevertheless have large, negative, standard free energies of hydrolysis. Acetyl-coenzyme A (Fig. 13–6) is a thioester that we will encounter repeatedly in later chapters. There is no resonance stabilization in thioesters comparable to that in oxygen esters (Fig. 13–7); consequently, the difference in free energy between the thioester and its hydrolysis products, which *are* resonance-stabilized, is greater than that for comparable oxygen esters. In both cases, hydrolysis of the ester generates a carboxylic acid, which can ionize and assume two resonance forms as described above for acyl phosphates. The free energy of hydrolysis for acetyl-CoA is large and negative, about −31 kJ/mol.

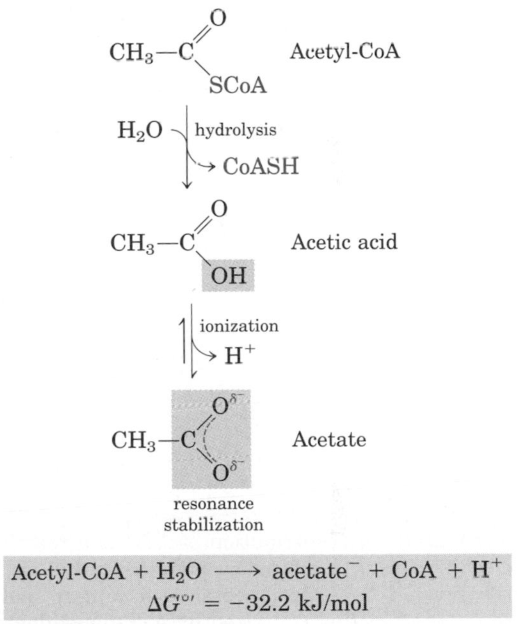

Figure 13–6 Hydrolysis of acetyl-coenzyme A, a thioester with a large, negative, free energy of hydrolysis. Thioesters contain a sulfur atom in the position where an oxygen atom is present in oxygen esters. (The complete structure of coenzyme A is shown in Fig. 12–41.)

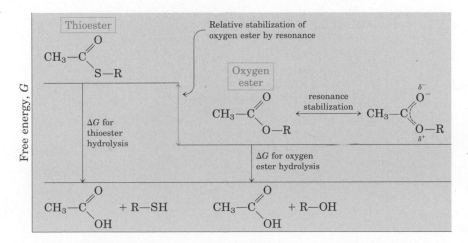

Figure 13–7 The free energy of hydrolysis of thioesters is large relative to that of oxygen esters. The *products* of both types of hydrolysis reactions have about the same free-energy content (G), but the thioester has a higher free-energy content than the oxygen ester. Orbital overlap between the O and C atoms allows resonance stabilization in oxygen esters, but orbital overlap between S and C is poorer and little resonance stabilization occurs.

To summarize, compounds with large, negative, standard free energies of hydrolysis give products that are more stable than the reactants because of one or more of the following: (1) the bond strain in reactants due to electrostatic repulsion is relieved by charge separation, as in the case of ATP (described earlier), (2) the products are stabilized by ionization, as in the case of ATP, acyl phosphates, and thioesters, (3) the products are stabilized by isomerization (tautomerization), as for phosphoenolpyruvate, and/or (4) the products are stabilized by resonance, as for creatine from phosphocreatine, the carboxylate ion from acyl phosphates and thioesters, and phosphate (P_i) from all of the phosphorylated compounds.

ATP Provides Energy by Group Transfers, Not by Simple Hydrolysis

Throughout this book we will refer to reactions or processes for which ATP supplies energy, and the contribution of ATP to these reactions will commonly be indicated as in Figure 13–8a, with a single arrow showing the conversion of ATP into ADP and P_i, or of ATP into AMP and PP_i (pyrophosphate). When written this way, these reactions of ATP appear to be simple hydrolysis reactions in which water displaces either P_i or PP_i, and one is tempted to say that an ATP-dependent reaction is "driven by the hydrolysis of ATP." This is *not* the case. ATP hydrolysis per se usually accomplishes nothing but the liberation of heat, which cannot drive a chemical process in an isothermal system.

Single reaction arrows such as those in Figure 13–8a almost invariably represent two-step processes (Fig. 13–8b) in which part of the ATP molecule, either a phosphate group or the adenylate moiety (AMP), is first transferred to a substrate molecule or to an amino acid residue in an enzyme, becoming covalently attached to and raising the free-energy content of the substrate or enzyme. In the second step, the phosphate-containing moiety transferred in the first step is displaced,

Figure 13–8 The contribution of ATP to a reaction is often shown with a single arrow **(a),** but is almost always a two-step process, such as that shown here for the reaction catalyzed by ATP-dependent glutamine synthetase **(b).**

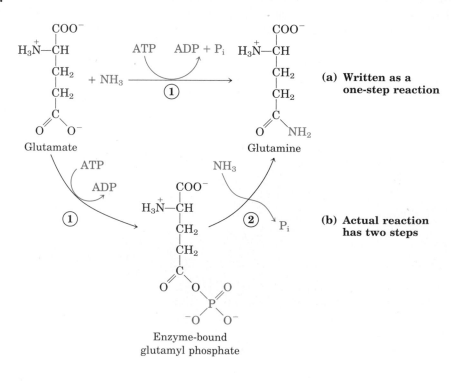

(a) **Written as a one-step reaction**

(b) **Actual reaction has two steps**

generating either P_i or AMP. Thus ATP participates in the enzyme-catalyzed reaction to which it contributes free energy. There is one important class of exceptions to this generalization: those processes in which noncovalent binding of ATP (or of GTP), followed by its hydrolysis to ADP and P_i, provides the energy to cycle a protein between two conformations, producing mechanical motion, as in muscle contraction or in the movement of enzymes along DNA (discussed below).

The phosphate compounds found in living organisms can be divided arbitrarily into two groups, based on their standard free energies of hydrolysis (Fig. 13–9). "High-energy" compounds have a $\Delta G^{\circ\prime}$ of hydrolysis more negative than -25 kJ/mol; "low-energy" compounds have a less negative $\Delta G^{\circ\prime}$. ATP, for which $\Delta G^{\circ\prime}$ of hydrolysis is -30.5 kJ/mol (-7.3 kcal/mol), is a high-energy compound; glucose-6-phosphate, with a standard free energy of hydrolysis of -13.8 kJ/mol (-3.3 kcal/mol), is a low-energy compound.

The term "high-energy phosphate bond," although long used by biochemists, is incorrect and misleading, as it wrongly suggests that the bond itself contains the energy. In fact, the breaking of chemical bonds requires an *input* of energy. The free energy released by hydrolysis of phosphate compounds thus does not come from the specific bond that is broken but results from the products of the reaction having a smaller free-energy content than the reactants. For simplicity, we will sometimes use the term "high-energy phosphate compound" when referring to ATP or other phosphate compounds with a large, negative, standard free energy of hydrolysis.

From the additivity of free-energy changes of sequential reactions, one can see that the synthesis of any phosphorylated compound can be accomplished by coupling it to the breakdown of another phosphorylated compound with a more negative free energy of hydrolysis (Fig. 13–9). One can therefore describe phosphorylated compounds as having a high or low **phosphate group transfer potential.** The phosphate group transfer potential of phosphoenolpyruvate is very high, that of ATP is high, and that of glucose-6-phosphate is low.

Figure 13–9 Flow of phosphate groups, represented by Ⓟ, from high-energy phosphate donors via ATP to acceptor molecules (such as glucose and glycerol) to form their low-energy phosphate derivatives. This flow of phosphate groups, which is catalyzed by enzymes called kinases, proceeds with an overall loss of free energy under intracellular conditions. Hydrolysis of low-energy phosphate compounds releases P_i, which has an even lower group transfer potential.

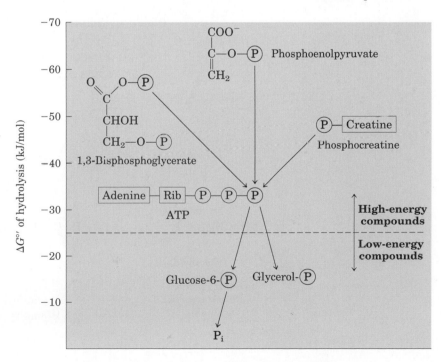

Much of catabolism is directed toward the synthesis of high-energy phosphate compounds, but their formation is not an end in itself; it is the means of activating a very wide variety of compounds for further chemical transformation. The transfer of a phosphate group to a compound effectively puts free energy into that compound, so that it has more free energy to give up during subsequent metabolic transformations. We described above how the synthesis of glucose-6-phosphate is accomplished by phosphate group transfer from ATP. We shall see in the next chapter that this phosphorylation of glucose activates or "primes" the glucose for catabolic reactions that occur in nearly every living cell.

In some reactions that involve ATP, both of its terminal phosphate groups are released in one piece as PP_i. Simultaneously, the remainder of the ATP molecule (adenylate) is joined to another compound, which is thereby activated. For example, the first step in the activation of a fatty acid either for energy-yielding oxidation (Chapter 16) or for use in the synthesis of more complex lipids (Chapter 20) is its attachment to the carrier coenzyme A (Fig. 13–10). The direct condensation of a fatty acid with coenzyme A is endergonic, but the formation of fatty acyl–CoA is made exergonic by coupling it to the net breakdown, in two steps, of ATP.

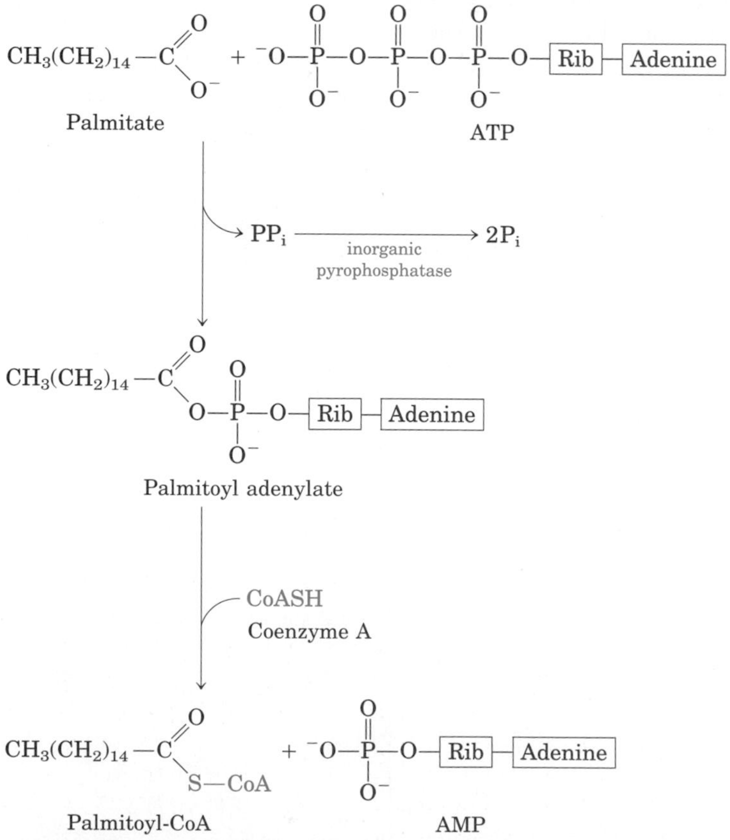

Figure 13–10 Both phosphoric acid anhydride bonds in ATP are eventually broken in the formation of palmitoyl-coenzyme A. In the first step of the reaction, ATP donates adenylate (AMP), forming the fatty acyl adenylate and releasing PP_i. The pyrophosphate is subsequently hydrolyzed by inorganic pyrophosphatase. The "energized" fatty acyl group is then transferred to coenzyme A.

Overall reaction:
Palmitate + ATP + CoASH $\longrightarrow$ palmitoyl-CoA + AMP + $2P_i$
$\Delta G^{\circ\prime} = -32.5$ kJ/mol

In the first step, adenylate (AMP) is transferred from ATP to the carboxyl group of the fatty acid, forming a mixed anhydride (fatty acyl adenylate) and liberating PP$_i$. In the second step, the thiol group of coenzyme A displaces the adenylate group and forms a thioester with the fatty acid. The sum of these two reactions is the exergonic hydrolysis of ATP to AMP and PP$_i$ ($\Delta G^{\circ\prime} = -32.2$ kJ/mol) and the endergonic formation of fatty acyl–CoA ($\Delta G^{\circ\prime} = 31.4$ kJ/mol).

The formation of fatty acyl–CoA is made energetically favorable by a third step, in which the PP$_i$ formed in the first step is hydrolyzed by the ubiquitous enzyme **inorganic pyrophosphatase** to yield 2P$_i$:

$$\text{PP}_i + \text{H}_2\text{O} \longrightarrow 2\text{P}_i \qquad \Delta G^{\circ\prime} = -33.4 \text{ kJ/mol}$$

Thus, in the activation of a fatty acid, both of the phosphoric acid anhydride bonds of ATP are broken. The resulting $\Delta G^{\circ\prime}$ is the sum of the $\Delta G^{\circ\prime}$ values for the breakage of these bonds:

$$\text{ATP} + 2\text{H}_2\text{O} \longrightarrow \text{AMP} + 2\text{P}_i \qquad \Delta G^{\circ\prime} = -65.6 \text{ kJ/mol}$$

The activation of amino acids before their polymerization into proteins (Chapter 26) is accomplished by an analogous set of reactions. An aminoacyl adenylate is first formed from the amino acid and ATP, with the elimination of PP$_i$. The adenylate group is then displaced by a transfer RNA, which is thereby joined to the amino acid. In this case, too, the PP$_i$ formed in the first step is hydrolyzed by inorganic pyrophosphatase. An unusual use of the cleavage of ATP to AMP and PP$_i$ occurs in the firefly, which uses ATP as an energy source to produce light flashes (Box 13–3, p. 382).

The AMP produced in adenylate transfers is returned to the ATP cycle by the action of **adenylate kinase,** which catalyzes the reversible reaction

$$\text{ATP} + \text{AMP} \underset{\text{Mg}^{2+}}{\rightleftharpoons} \text{ADP} + \text{ADP} \qquad \Delta G^{\circ\prime} \approx 0$$

The ADP so formed can be phosphorylated to ATP, using reactions described in detail in later chapters.

Assembly of Informational Macromolecules Requires Energy

When simple precursors are assembled into high molecular weight polymers with defined sequences (DNA, RNA, proteins), as described in detail in Part IV, energy is required both for the condensation of monomeric units and for the creation of *ordered* sequences. The precursors for DNA and RNA synthesis are nucleoside triphosphates, and polymerization is accompanied by cleavage of the phosphoric acid anhydride linkage between the α- and β-phosphates, with the release of PP$_i$ (Fig. 13–11). The moieties transferred to the growing polymer in these polymerization reactions are adenylate (AMP), guanylate (GMP), cytidylate (CMP), or uridylate (UMP) for RNA synthesis, and their deoxy analogs for DNA synthesis. We have seen that the activation of amino acids for protein synthesis involves the donation of adenylate groups from ATP, and we shall see later that the formation of peptide bonds on the ribosome is also accompanied by GTP hydrolysis (Chapter 26). In all of these cases, the exergonic breakdown of a nucleoside triphosphate is coupled to the endergonic process of synthesizing a polymer of a specific sequence.

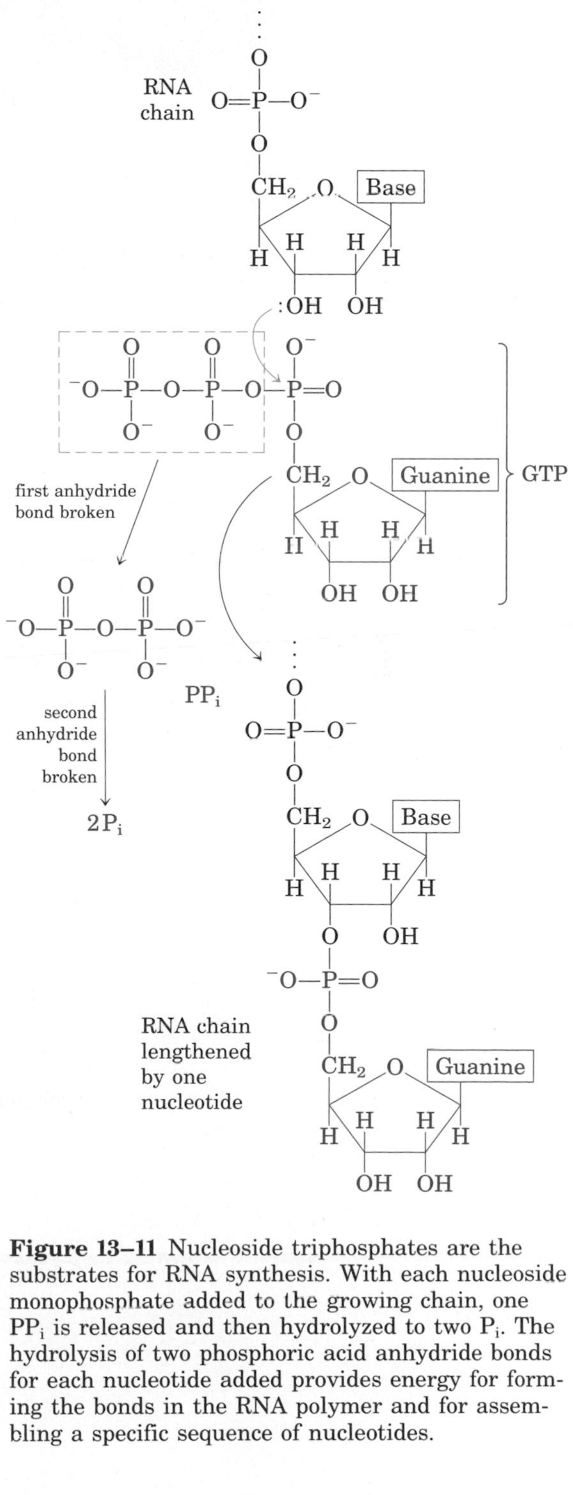

Figure 13–11 Nucleoside triphosphates are the substrates for RNA synthesis. With each nucleoside monophosphate added to the growing chain, one PP$_i$ is released and then hydrolyzed to two P$_i$. The hydrolysis of two phosphoric acid anhydride bonds for each nucleotide added provides energy for forming the bonds in the RNA polymer and for assembling a specific sequence of nucleotides.

BOX 13–3   Firefly Flashes: Glowing Reports of ATP

Figure 1 The firefly, a beetle of the Lampyridae family.

Many fungi, marine microorganisms, jellyfish, and crustaceans as well as the firefly (Fig. 1) are capable of generating bioluminescence, which requires considerable amounts of energy. In the firefly, ATP is used in a set of reactions that converts chemical energy into light energy. From many thousands of firefly lanterns collected by children in and around Baltimore, William McElroy and his colleagues at The Johns Hopkins University isolated the principal biochemical components involved, luciferin (Fig. 2), a complex carboxylic acid, and luciferase, an enzyme. The generation of a light flash requires activation of luciferin by an enzymatic reaction with ATP in which a pyrophosphate cleavage of ATP occurs, to form luciferyl adenylate (Fig. 2). This compound is then acted upon by molecular oxygen and luciferase to bring about the oxidative decarboxylation of the luciferin to yield oxyluciferin. This reaction, which has intermediate steps, is accompanied by emission of light (Fig. 2). The color of the light flash differs with firefly species and appears to be determined by differences in the structure of the luciferase. Luciferin is then regenerated from oxyluciferin in a subsequent series of reactions. Other bioluminescent organisms use other types of enzymatic reactions to generate light.

In the laboratory, pure firefly luciferin and luciferase are used to measure minute quantities of ATP by the intensity of the light flash produced. As little as a few picomoles (10^{-12} mol) of ATP can be measured in this way.

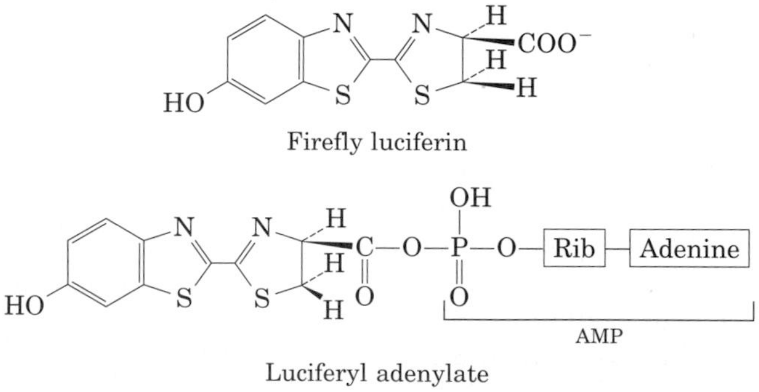

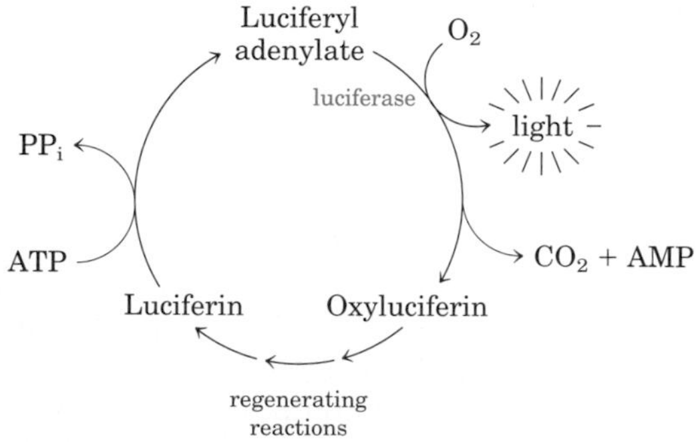

Firefly luciferin

Luciferyl adenylate

Figure 2 Important components in firefly bioluminescence, and the firefly bioluminescence cycle.

ATP Energizes Active Transport across Membranes

ATP can supply the energy for transporting an ion or a molecule across a membrane into another aqueous compartment where its concentration is higher. Recall from Chapter 10 that the free-energy change (ΔG_t) for the transport of a nonionic solute from one compartment to another is given by

$$\Delta G_t = RT \ln (C_2/C_1) \qquad (13\text{--}4)$$

where C_1 is the molar concentration of the solute in the compartment from which the ion or molecule moves and C_2 is its molar concentration in the compartment into which it moves. When a proton or other charged species moves across a membrane without a counterion, the separation of electrical charge requires extra electrical work beyond the osmotic work against a concentration gradient. The extra electrical work is $Z\mathcal{F}\Delta\psi$, where Z is the (unitless) electrical charge of the transported species, $\Delta\psi$ is the transmembrane electrical potential (in volts), and $\mathcal{F}$ is the Faraday constant (96.48 kJ/V · mol). The total energy cost of moving a charged species against an electrochemical gradient is

$$\Delta G_t = RT \ln (C_2/C_1) + Z\mathcal{F}\Delta\psi \qquad (13\text{--}5)$$

Transport processes are major consumers of energy; in tissues such as human kidney and brain, as much as two-thirds of the energy consumed at rest is used to pump Na^+ and K^+ across plasma membranes via the Na^+K^+ ATPase. Na^+ and K^+ transport is driven by cyclic phosphorylation and dephosphorylation of the transporter protein, with ATP as the phosphate donor (see Fig. 10–23). Na^+-dependent phosphorylation of the Na^+K^+ ATPase forces a change in the protein's conformation, and K^+-dependent dephosphorylation favors return to the original conformation. Each cycle in the transport process results in the conversion of ATP to ADP and P_i, and it is the free-energy change of ATP hydrolysis that drives the pumping of Na^+ and K^+. In animal cells, the net hydrolysis of one ATP is accompanied by the outward transport of three Na^+ ions and the uptake of two K^+ ions.

ATP Is the Energy Source for Muscle Contraction

In the contractile system of skeletal muscle cells, myosin and actin are specialized to transduce the chemical energy of ATP into motion. ATP binds tightly but noncovalently to the head portion of one conformation of myosin, holding the protein in that conformation. When myosin (which is also an ATPase) catalyzes the hydrolysis of its bound ATP, the ADP and P_i produced dissociate from the protein, allowing it to relax into a second conformation until another molecule of ATP binds (Fig. 13–12). The binding and subsequent hydrolysis of ATP thus provide the energy that forces cyclic changes in the conformation of the myosin head. The change in conformation of many individual myosin molecules results in the sliding of myosin fibrils along actin filaments (see Fig. 7–32), which translates into macroscopic contraction of the muscle fiber.

This production of mechanical motion at the expense of ATP is one of the few cases in which ATP hydrolysis per se, and not group transfer from ATP, is the source of the chemical energy in a coupled process. The energy-dependent reactions catalyzed by helicases, RecA protein, and some topoisomerases (Chapter 24) and by certain GTP-binding proteins (Chapter 22) also involve direct hydrolysis of phosphoric acid anhydride bonds.

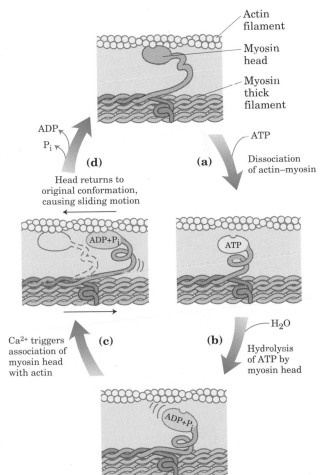

Figure 13–12 ATP hydrolysis drives the cross-bridge cycle during the sliding motion of actin–myosin complexes in muscle. This proposed mechanism begins with each myosin head bound to an actin filament. Binding of ATP to myosin **(a)** causes dissociation of the actin–myosin cross-bridge. ATP hydrolysis **(b)** leaves myosin with bound ADP and P_i, which favors a different conformation of the myosin head. In this conformation, the myosin head binds to an adjacent actin filament **(c)** when elevated cytosolic Ca^{2+} signals contraction. This cross-bridge formation induces the release of bound ADP and P_i **(d)**, which provides the free energy for a conformational change in the myosin head; the head tilts, forcing the thin (actin) filament to slide relative to the thick (myosin) filament, producing contraction. ATP then binds to the myosin head to dissociate the cross-bridge and start another cycle. Each cycle occurs in about 1 msec.

Biological Oxidation–Reduction Reactions

The transfer of phosphate groups is one of the central features of metabolism. Metabolic electron transfer reactions are also of crucial importance. These oxidation–reduction reactions involve the loss of electrons by one chemical species, which is thereby oxidized, and the gain by another, which is reduced. The flow of electrons in oxidation–reduction reactions is responsible, directly or indirectly, for all of the work done by living organisms. In nonphotosynthetic organisms, the source of electrons is reduced compounds (food); in photosynthetic organisms, the initial electron donor is a chemical species excited by the absorption of light. The path of electron flow in metabolism is complex. Electrons move from various metabolic intermediates to specialized electron carriers in enzyme-catalyzed reactions. Those carriers in turn donate electrons to acceptors with higher electron affinities, with the release of energy. Cells contain a variety of molecular energy transducers, which convert the energy of electron flow into useful work.

We begin our discussion with a description of the general types of metabolic reactions that involve electron transfers. After considering the theoretical and experimental basis for measuring energy changes in oxidation reactions in terms of electromotive force, we will discuss the relationship between this force, expressed in volts, and the free-energy change, expressed in joules. We conclude by introducing the structures and oxidation–reduction chemistry of the most common of the specialized electron carriers, which we shall meet repeatedly in later chapters.

The Flow of Electrons Can Do Biological Work

The conversion of electron flow to biological work requires molecular transducers, analogous to the electric motors that convert electron flow through macroscopic circuits into mechanical motion. The analogy between a circuit connecting a battery with an electric motor and the submicroscopic electron circuits in cells is instructive.

In the macroscopic circuit (Fig. 13–13a), the source of electrons is a battery containing two chemical species that differ in affinity for electrons. The electrical wires provide a pathway for electron flow from the chemical species at one pole of the battery, through the motor, to the chemical species at the other pole of the battery. Because the two chemical species differ in their affinity for electrons, electrons flow spontaneously through the circuit, driven by a force proportional to the difference in electron affinity, the electromotive force. The electromotive force (typically a few volts) can accomplish work if an appropriate energy transducer such as a motor is placed in the circuit. The motor can be coupled to a variety of mechanical devices to accomplish work.

In an analogous biological "circuit" (Fig. 13–13b), the source of electrons is a relatively reduced compound such as glucose. As glucose is enzymatically oxidized, electrons are released and flow spontaneously through a series of electron carrier intermediates to another chemical species with a high affinity for electrons, such as O_2. Electron flow is spontaneous and exergonic, because O_2 has a higher affinity for electrons than do the intermediates that donate electrons. The resulting electromotive force provides energy to molecular transducers that do biological work. In the mitochondrion, for example, membrane-bound transducers couple electron flow to the production of a transmembrane

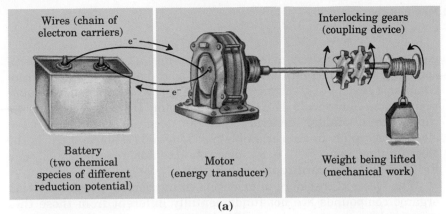

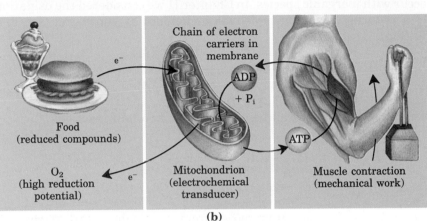

Figure 13–13 The analogy between macroscopic **(a)** and microscopic **(b)** electrical circuits. In both circuits, the energy of electron flow is harnessed to do work.

pH difference, accomplishing osmotic and electrical work. The proton gradient thus formed has potential energy, sometimes called proton-motive force by analogy with electromotive force. Another molecular transducer in the mitochondrial membrane uses the proton-motive force to do chemical work: ATP is synthesized from ADP and P_i as protons flow spontaneously across the membrane. Similarly, membrane-localized transducers in *E. coli* convert electromotive to proton-motive force, which is then used to power flagellar motion.

The principles of electrochemistry that govern energy changes in the circuit with a motor and battery apply with equal validity to the microscopic processes accompanying electron flow in living cells. We turn now to a review of those principles.

Oxidation–Reductions Can Be Described as Half-Reactions

Although oxidation and reduction must occur together, it is convenient when describing electron transfers to consider the two halves of an oxidation–reduction reaction separately. For example, the oxidation of ferrous ion by cupric ion:

$$Fe^{2+} + Cu^{2+} \rightleftharpoons Fe^{3+} + Cu^{+}$$

can be described in terms of two half-reactions:

$$(1) \quad Fe^{2+} \rightleftharpoons Fe^{3+} + e^{-}$$

$$(2) \quad Cu^{2+} + e^{-} \rightleftharpoons Cu^{+}$$

The electron-donating molecule in an oxidation–reduction reaction is called the reducing agent or reductant; the electron-accepting molecule is the oxidizing agent or oxidant. A given agent, such as an iron cation in the ferrous (Fe^{2+}) and the ferric (Fe^{3+}) state, functions as a conjugate reductant–oxidant pair (redox pair), just as an acid and corresponding base function as a conjugate acid–base pair. Recall from Chapter 4 that in acid–base reactions we can write the general equation: proton donor $\rightleftharpoons$ H^+ + proton acceptor. In redox reactions we can write a similar general equation: electron donor $\rightleftharpoons$ e^- + electron acceptor. In the reversible half-reaction (1) above, Fe^{2+} is the electron donor and Fe^{3+} is the electron acceptor; together, Fe^{2+} and Fe^{3+} constitute a **conjugate redox pair.**

The electron transfers in oxidation–reduction reactions involving organic compounds are not fundamentally different from those that occur with inorganic species. In Chapter 11 we considered the oxidation of a reducing sugar (an aldehyde or ketone) by cupric ion (see Fig. 11–10a):

$$R-C\underset{H}{\overset{O}{\big\backslash}} + 4OH^- + 2Cu^{2+} \rightleftharpoons R-C\underset{OH}{\overset{O}{\big\backslash}} + Cu_2O + 2H_2O$$

This overall reaction can be expressed as two half-reactions:

$$(1) \quad R-C\underset{H}{\overset{O}{\big\backslash}} + 2OH^- \rightleftharpoons R-C\underset{OH}{\overset{O}{\big\backslash}} + 2e^- + H_2O$$

$$(2) \quad 2Cu^{2+} + 2e^- + 2OH^- \rightleftharpoons Cu_2O + H_2O$$

Because two electrons are removed from the aldehyde carbon, the second half-reaction (the one-electron reduction of cupric to cuprous ion) must be doubled to balance the overall equation.

Biological Oxidations Often Involve Dehydrogenation

Carbon occurs in living cells in five different oxidation states (Fig. 13–14). In the most reduced compounds carbon atoms are rich in electrons and in hydrogen, whereas in the more highly oxidized compounds a carbon atom is bonded to more oxygen and to less hydrogen. In the oxidation of ethane to ethanol (Fig. 13–14), the compound does not lose a hydrogen but one of the carbon atoms does; the hydrogen of the —OH group is, of course, not bonded directly to carbon. In the series of compounds shown in Figure 13–14, oxidation of a carbon atom is synonymous with its dehydrogenation. When a carbon atom shares an electron pair with another atom such as oxygen, the sharing is unequal, in favor of the more electronegative atom (oxygen). Thus oxidation has the effect of removing electrons from the carbon atom.

Not all biological oxidation–reduction reactions involve oxygen and carbon. For example, the conversion of molecular nitrogen into ammonia, $6H^+ + 6e^- + N_2 \longrightarrow 2NH_3$, represents a reduction of the nitrogen atoms.

Electrons are transferred from one molecule to another in one of four different ways:

1. They may be transferred directly as *electrons*. For example, the Fe^{2+}/Fe^{3+} redox pair can transfer an electron to the Cu^+/Cu^{2+} redox pair:

$$Fe^{2+} + Cu^{2+} \rightleftharpoons Fe^{3+} + Cu^+$$

2. Electrons may be transferred in the form of *hydrogen atoms*. Recall that a hydrogen atom consists of a proton (H^+) and a single electron (e^-). In this case we can write the general equation

$$AH_2 \rightleftharpoons A + 2e^- + 2H^+$$

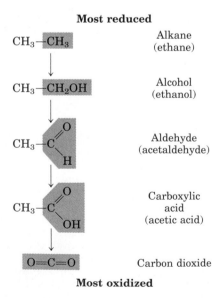

Most reduced

$CH_3{-}CH_3$ — Alkane (ethane)

$CH_3{-}CH_2OH$ — Alcohol (ethanol)

$CH_3{-}C\underset{H}{\overset{O}{\big\backslash}}$ — Aldehyde (acetaldehyde)

$CH_3{-}C\underset{OH}{\overset{O}{\big\backslash}}$ — Carboxylic acid (acetic acid)

$O{=}C{=}O$ — Carbon dioxide

Most oxidized

Figure 13–14 The oxidation states of carbon. Each of the arrows indicates an oxidation reaction; all except the last reaction are oxidations brought about by dehydrogenation.

where AH$_2$ acts as the hydrogen (or electron) donor. AH$_2$ and A together constitute a conjugate redox pair, which can reduce another compound B by transfer of hydrogen atoms:

$$AH_2 + B \rightleftharpoons A + BH_2$$

3. Electrons may be transferred from an electron donor to an acceptor in the form of a *hydride ion* ($:H^-$), which includes two electrons, as in the case of NAD-linked dehydrogenases described below.

4. Electron transfer also takes place when there is a direct combination of an organic reductant with *oxygen,* to give a product in which the oxygen is covalently incorporated, as in the oxidation of a hydrocarbon to an alcohol:

$$R\text{—}CH_3 + \tfrac{1}{2}O_2 \longrightarrow R\text{—}CH_2\text{—}OH$$

In this reaction the hydrocarbon is the electron donor and the oxygen atom is the electron acceptor.

All four types of electron transfer occur in cells. The neutral term **reducing equivalent** is commonly used to designate a single electron equivalent participating in an oxidation–reduction reaction, no matter whether this equivalent be in the form of an electron per se, a hydrogen atom, or a hydride ion, or whether the electron transfer takes place in a reaction with oxygen to yield an oxygenated product. Because biological fuel molecules usually undergo enzymatic dehydrogenation to lose *two* reducing equivalents at a time, and because each oxygen atom can accept two reducing equivalents, biochemists by convention refer to the unit of biological oxidations as two reducing equivalents passing from substrate to oxygen.

Reduction Potentials Measure Affinity for Electrons

When two conjugate redox pairs are present together in solution, electron transfer from the electron donor of one pair to the electron acceptor of the other may occur spontaneously. The tendency of such a reaction to occur depends upon the relative affinity of the electron acceptor of each redox pair for electrons. The **standard reduction potential,** E_0, a measure of this affinity, is determined in an experiment such as that described in Figure 13–15. Electrochemists have chosen as a standard of reference the half-reaction

$$H^+ + e^- \longrightarrow \tfrac{1}{2}H_2$$

The electrode at which this half-reaction occurs is arbitrarily assigned a standard reduction potential of 0.00 V. When this hydrogen electrode is connected through an external circuit to another half-cell in which the oxidized and reduced species are both present at standard concentrations (each solute at 1 M, each gas at 1 atm), electrons will tend to flow through the external circuit from the half-cell of lower standard reduction potential to the half-cell of higher standard reduction potential. By convention, the half-cell with the stronger tendency to acquire electrons is assigned a positive value of E_0 (in volts).

The reduction potential of a half-cell depends not only upon the chemical species present but also upon their activities, approximated by their concentrations. About a century ago, Walther Nernst derived an equation that relates standard reduction potential (E_0) to reduction

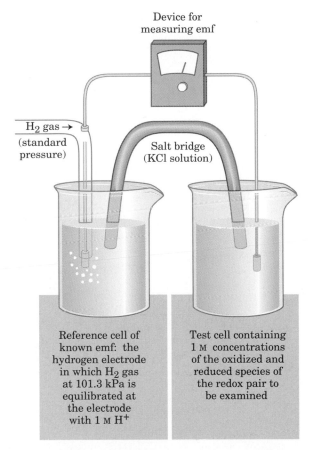

Device for measuring emf

H$_2$ gas →
(standard pressure)

Salt bridge
(KCl solution)

Reference cell of known emf: the hydrogen electrode in which H$_2$ gas at 101.3 kPa is equilibrated at the electrode with 1 M H$^+$

Test cell containing 1 M concentrations of the oxidized and reduced species of the redox pair to be examined

Figure 13–15 Measurement of the standard reduction potential (E_0') of a redox pair. Electrons flow from the test electrode to the reference electrode, or vice versa. The ultimate reference half-cell is the hydrogen electrode, as shown here. The arbitrary electromotive force (emf) of this electrode is 0.00 V. At pH 7, E_0' for the hydrogen electrode is −0.414 V. The direction of electron flow depends upon the relative electron "pressure" or potential of the two cells. A salt bridge containing a saturated KCl solution provides a path for counter-ion movement between the test cell and the reference cell. From the observed emf and the known emf of the reference cell, the emf of the test cell containing the redox pair is obtained. The cell that gains electrons has, by convention, the more positive reduction potential.

potential (E) at any concentration of oxidized and reduced species in the cell:

$$E = E_0 + \frac{RT}{n\mathcal{F}} \ln \frac{[\text{electron acceptor}]}{[\text{electron donor}]} \qquad (13\text{–}6)$$

where R and T have their usual meanings (Table 13–1), n is the number of electrons transferred per molecule, and $\mathcal{F}$ is the Faraday constant, 96.48 kJ/V · mol. At 298 K (25 °C), this expression reduces to:

$$E = E_0 + \frac{0.026\ \text{V}}{n} \ln \frac{[\text{electron acceptor}]}{[\text{electron donor}]} \qquad (13\text{–}7)$$

Many half-reactions of interest to biochemists involve protons. As in the definition of $\Delta G^{\circ\prime}$, biochemists define the standard state for oxidation–reduction reactions as pH 7 and express reduction potential as E_0', the standard reduction potential at pH 7. The values for standard reduction potentials given in Table 13–7 and used throughout this book are for E_0' and are therefore only valid for calculations involving systems at neutral pH. Each value represents the potential difference when the conjugate redox pair at 1 M concentrations at pH 7 is connected with the standard (pH 0) hydrogen electrode. Notice in Table 13–7 that when the conjugate pair $2H^+/H_2$ at pH 7 is connected with the standard hydrogen electrode (pH 0), electrons tend to flow from the pH 7 cell to the standard (pH 0) cell; the measured $\Delta E_0'$ for the $2H^+/H_2$ pair is -0.414 V.

Table 13–7 Standard reduction potentials of some biologically important half-reactions (at 25 °C, pH 7)

Half-reaction	$E_0'(\text{V})$
$\frac{1}{2}O_2 + 2H^+ + 2e^- \longrightarrow H_2O$	0.816
$Fe^{3+} + e^- \longrightarrow Fe^{2+}$	0.771
$NO_3^- + 2H^+ + 2e^- \longrightarrow NO_2^- + H_2O$	0.421
Cytochrome f (Fe^{3+}) + $e^- \longrightarrow$ cytochrome f (Fe^{2+})	0.365
$Fe(CN)_6^{3-}$ (ferricyanide) + $e^- \longrightarrow Fe(CN)_6^{4-}$	0.36
$O_2 + 2H^+ + 2e^- \longrightarrow H_2O_2$	0.295
Cytochrome a (Fe^{3+}) + $e^- \longrightarrow$ cytochrome a (Fe^{2+})	0.29
Cytochrome c (Fe^{3+}) + $e^- \longrightarrow$ cytochrome c (Fe^{2+})	0.254
Cytochrome c_1 (Fe^{3+}) + $e^- \longrightarrow$ cytochrome c_1 (Fe^{2+})	0.22
Ubiquinone + $2H^+ + 2e^- \longrightarrow$ ubiquinol + H_2	0.045
Cytochrome b (Fe^{3+}) + $e^- \longrightarrow$ cytochrome b (Fe^{2+})	0.077
Fumarate^{2-} + $2H^+ + 2e^- \longrightarrow$ succinate^{2-}	0.031
$2H^+ + 2e^- \longrightarrow H_2$ (at standard conditions, pH 0)	0.000
Crotonyl-CoA + $2H^+ + 2e^- \longrightarrow$ butyryl-CoA	-0.015
Oxaloacetate^{2-} + $2H^+ + 2e^- \longrightarrow$ malate^{2-}	-0.166
Pyruvate$^-$ + $2H^+ + 2e^- \longrightarrow$ lactate$^-$	-0.185
Acetaldehyde + $2H^+ + 2e^- \longrightarrow$ ethanol	-0.197
FAD + $2H^+ + 2e^- \longrightarrow$ FADH$_2$	-0.219
Glutathione + $2H^+ + 2e^- \longrightarrow$ 2 reduced glutathione	-0.23
S + $2H^+ + 2e^- \longrightarrow H_2S$	-0.243
Lipoic acid + $2H^+ + 2e^- \longrightarrow$ dihydrolipoic acid	-0.29
NAD$^+$ + H$^+$ + $2e^- \longrightarrow$ NADH	-0.320
NADP$^+$ + H$^+$ + $2e^- \longrightarrow$ NADPH	-0.324
Acetoacetate + $2H^+ + 2e^- \longrightarrow \beta$-hydroxybutyrate	-0.346
α-Ketoglutarate + CO_2 + $2H^+ + 2e^- \longrightarrow$ isocitrate	-0.38
$2H^+ + 2e^- \longrightarrow H_2$ (at pH 7)	-0.414
Ferredoxin (Fe^{3+}) + $e^- \longrightarrow$ ferredoxin (Fe^{2+}) (spinach)	-0.432

Data mostly from Loach, P.A. (1976) In *Handbook of Biochemistry and Molecular Biology,* 3rd edn (Fasman, G.D., ed), *Physical and Chemical Data,* Vol. I, pp. 122–130, CRC Press, Cleveland, OH.

Standard Reduction Potentials Allow the Calculation of Free-Energy Change

The usefulness of reduction potentials stems from the fact that when E has been determined for any two half-cells, relative to the standard hydrogen electrode, their reduction potentials relative to each other are also known. One can therefore predict the direction in which electrons will tend to flow when these two half-cells are connected through an external circuit, or when the components of the two half-cells are present together in the same solution. Electrons will tend to flow to the half-cell with the more positive E, and the strength of that tendency is proportional to the difference in reduction potentials, ΔE.

The energy made available to do work by this spontaneous electron flow (the free-energy change for the oxidation–reduction reaction) is proportional to ΔE:

$$\Delta G = -n\mathcal{F}\Delta E, \quad \text{or} \quad \Delta G^{\circ\prime} = -n\mathcal{F}\Delta E_0' \qquad (13\text{–}8)$$

Here n represents the number of electrons transferred in the reaction. With this equation it is possible to calculate the free-energy change for any oxidation–reduction reaction from the values of E_0' (found in a table of reduction potentials) and the concentrations of the species involved in the reaction.

Consider the reaction in which acetaldehyde is reduced by the biological electron carrier NADH:

$$\text{Acetaldehyde} + \text{NADH} + \text{H}^+ \longrightarrow \text{ethanol} + \text{NAD}^+$$

The relevant half-reactions and their E_0' values (Table 13–7) are:

(1) Acetaldehyde + 2H$^+$ + 2e$^-$ ⟶ ethanol $E_0' = -0.197$ V
(2) NAD$^+$ + 2H$^+$ + 2e$^-$ ⟶ NADH + H$^+$ $E_0' = -0.320$ V

For the overall reaction, $\Delta E_0' = -0.197 \text{ V} - (-0.320 \text{ V}) = 0.123 \text{ V}$, and n is 2. Therefore, $\Delta G^{\circ\prime} = -n\mathcal{F}\Delta E_0' = -2(96.5 \text{ kJ/V} \cdot \text{mol})(0.123 \text{ V}) = -23.7 \text{ kJ/mol}$. This is the free-energy change for the oxidation–reduction reaction when acetaldehyde, ethanol, NAD$^+$, and NADH are all present at 1 M concentrations. If, instead, acetaldehyde and NADH were present at 1 M, but ethanol and NAD$^+$ were present at 0.1 M, the value for ΔG would be calculated as follows. First, the values of E for both reductants are determined (Eqn 13–7):

$$E_{\text{acetaldehyde}} = E_0' + \frac{RT}{n\mathcal{F}} \ln \frac{[\text{acetaldehyde}]}{[\text{ethanol}]}$$

$$= -0.197 \text{ V} + \frac{0.026 \text{ V}}{2} \ln \frac{1.0}{0.1} = -0.167 \text{ V}$$

$$E_{\text{NADH}} = E_0' + \frac{RT}{n\mathcal{F}} \ln \frac{[\text{NAD}^+]}{[\text{NADH}]}$$

$$= -0.320 \text{ V} + \frac{0.026 \text{ V}}{2} \ln \frac{0.1}{1.0} = -0.350 \text{ V}$$

Then ΔE is used to calculate ΔG (Eqn 13–8):

$$\Delta E = -0.167 \text{ V} - (-0.350) \text{ V} = 0.183 \text{ V}$$

$$\Delta G = -n\mathcal{F}\Delta E$$

$$= -2(96.5 \text{ kJ/V} \cdot \text{mol})(0.183 \text{ V})$$

$$= -35.3 \text{ kJ/mol}$$

It is thus possible to calculate the free-energy change for any biological oxidation at any concentrations of the redox pairs.

Cells Oxidize Glucose to Carbon Dioxide in Steps Involving Specialized Electron Carriers

In many organisms, the oxidation of glucose supplies energy for the production of ATP. For the complete oxidation of glucose:

$$C_6H_{12}O_6 + 6O_2 \longrightarrow 6CO_2 + 6H_2O$$

$\Delta G^{\circ\prime}$ is $-2,840$ kJ/mol. This is a much larger change in free energy than that occurring during ATP synthesis (50 to 60 kJ/mol; see Box 13–2). Cells do not convert glucose to CO_2 in a single, very energetic reaction, but rather in a series of reactions, some of which are oxidations. The free-energy change of these oxidation steps is larger than, but of the same order of magnitude as, that required for ATP synthesis from ADP. Electrons removed in these oxidation steps are transferred to coenzymes specialized for carrying electrons, such as NAD^+ and FAD, which are described below.

A Few Types of Cofactors and Proteins Serve as Universal Electron Carriers

Most cells have enzymes to catalyze the oxidation of hundreds of different compounds. These enzymes channel electrons from their substrates into a few types of universal electron carriers. The nucleotides NAD^+, $NADP^+$, FMN, and FAD are water-soluble cofactors that undergo reversible oxidation and reduction in many of the electron transfer reactions of metabolism. Their reduction in catabolic processes results in the conservation of free energy released by substrate oxidation. The nucleotides NAD^+ and $NADP^+$ move readily from one enzyme to another, but the flavin nucleotides FMN and FAD are very tightly bound to the enzymes, called flavoproteins, for which they serve as prosthetic groups. Lipid-soluble quinones such as ubiquinone and plastoquinone act in the nonaqueous environment of membranes, accepting electrons and conserving free energy. Iron–sulfur proteins and cytochromes are proteins with tightly bound prosthetic groups that undergo reversible oxidation and reduction; they, too, serve as electron carriers in many oxidation–reduction reactions. Some of these proteins are soluble, but others are peripheral or integral membrane proteins (p. 277). We will describe some chemical features of nucleotide cofactors and of certain enzymes (dehydrogenases and flavoproteins) that use them. The oxidation–reduction chemistry of quinones, iron-sulfur proteins, and cytochromes will be discussed in Chapter 18.

NADH and NADPH Act with Dehydrogenases as Soluble Electron Carriers

Nicotinamide adenine dinucleotide (NAD^+ in its oxidized form) and its close analog nicotinamide adenine dinucleotide phosphate ($NADP^+$) are composed of two nucleotides joined through their phosphate groups by a phosphoric acid anhydride bond (Fig. 13–16). Because their nicotinamide ring resembles pyridine, these compounds are sometimes called pyridine nucleotides. The vitamin niacin provides the nicotinamide moiety for the synthesis of the pyridine nucleotides.

Both coenzymes undergo reversible reduction of the nicotinamide ring (Fig. 13–16). As a substrate molecule undergoes oxidation (dehydrogenation), giving up two hydrogen atoms, the oxidized form of the nucleotide (NAD^+ or $NADP^+$) accepts a hydride ion ($:H^-$, the equiva-

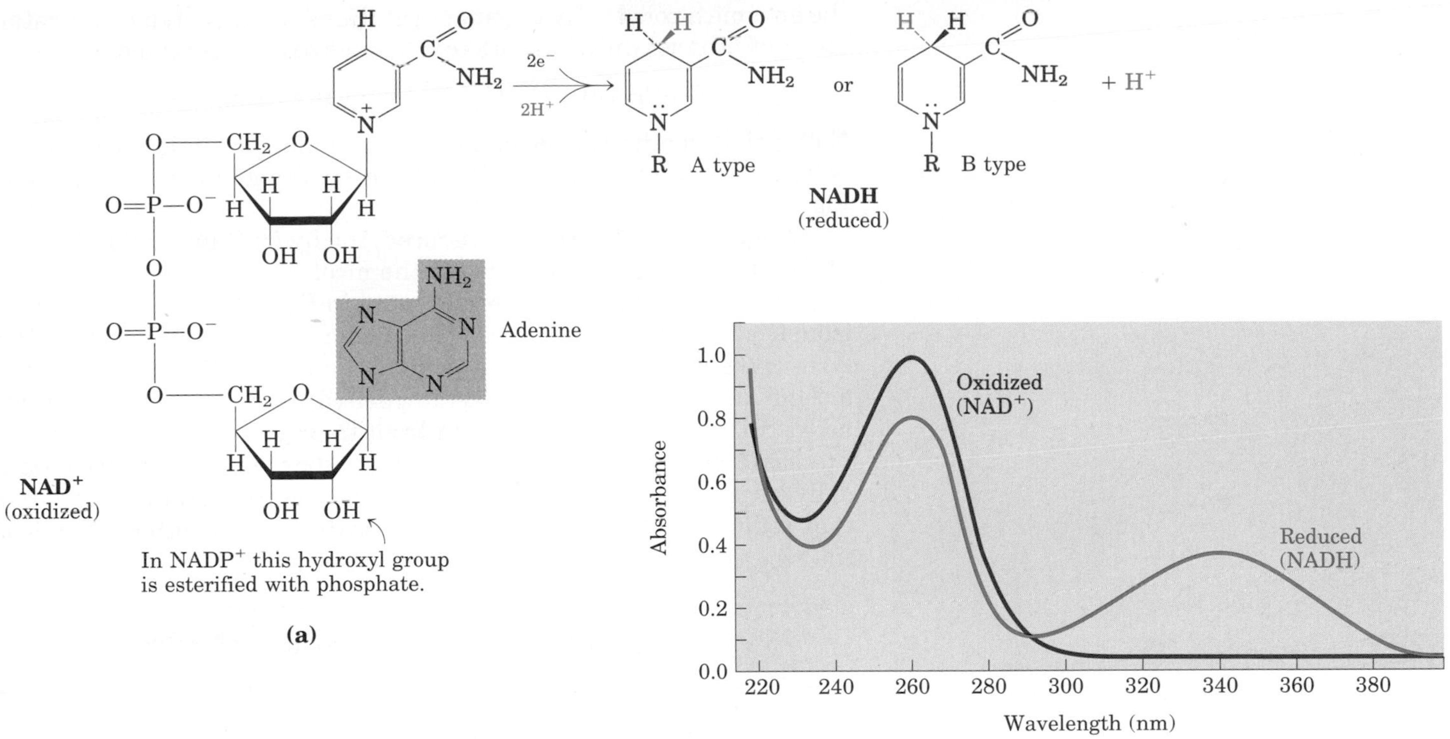

(a)

(b)

lent of a proton and two electrons) and is transformed into the reduced form (NADH or NADPH). The second H^+ removed from the substrate is released to the aqueous solvent. The half-reaction for each nucleotide is therefore

$$NAD^+ + 2e^- + 2H^+ \longrightarrow NADH + H^+$$
$$NADP^+ + 2e^- + 2H^+ \longrightarrow NADPH + H^+$$

In the abbreviations NADH and NADPH, the H denotes this added hydride ion, and the loss of the positive charge when H^- is added to the oxidized form is also made clear. To refer to one of these nucleotides without specifying its oxidation state, we will use NAD or NADP.

The total concentration of $NAD^+ + NADH$ in most tissues is about 10^{-5} M; that of $NADP^+ + NADP$ is about 10 times lower. In many cells and tissues, the ratio of NAD^+ (oxidized) to NADH (reduced) is high, favoring hydride transfer *to* NAD^+ to form NADH; by contrast, NADPH (reduced) is generally present in greater amounts than its oxidized form, $NADP^+$, favoring hydride transfer *from* NADPH. This reflects the specialized metabolic roles of the two cofactors: NAD^+ generally functions in catabolic oxidations, and NADPH is the usual cofactor in anabolic reductions. A few enzymes will use either cofactor, but most show a strong preference for one cofactor over the other. This functional specialization allows a cell to maintain two distinct pools of electron carriers in the same cellular compartment.

More than 200 enzymes are known to catalyze reactions in which NAD^+ or $NADP^+$ accepts a hydride ion from some reduced substrate, or NADH or NADPH donates a hydride ion to an oxidized substrate. The general reactions are

$$AH_2 + NAD^+ \rightleftharpoons A + NADH + H^+$$
$$AH_2 + NADP^+ \rightleftharpoons A + NADPH + H^+$$

where AH_2 is the reduced substrate and A the oxidized substrate. The general name for enzymes of this type is **oxidoreductase** (see Table 8–3); they are also commonly called **dehydrogenases.** For example,

Figure 13–16 (a) Nicotinamide adenine dinucleotide (NAD^+) and its phosphorylated analog $NADP^+$ undergo reduction to NADH or NADPH, accepting a hydride ion (two electrons and one proton) from an oxidizable substrate. The hydride ion may be added to either the front (A type) or the back (B type) of the planar nicotinamide ring (see Table 13–8). **(b)** The UV absorption spectra of NAD^+ and NADH. Reduction of the nicotinamide ring produces a new, broad absorption band with a maximum at 340 nm. The production of NADH during an enzyme-catalyzed oxidation can be conveniently followed by observing the appearance of the absorbance at 340 nm.

the enzyme alcohol dehydrogenase catalyzes the first step in the catabolism of ethanol, in which ethanol is oxidized to acetaldehyde:

$$CH_3CH_2OH + NAD^+ \rightleftharpoons CH_3CHO + NADH + H^+$$

Notice that in ethanol, one of the carbon atoms has undergone the loss of hydrogen and has been oxidized from an alcohol to an aldehyde (see Fig. 13–14).

When NAD^+ or $NADP^+$ is reduced, the hydride ion could in principle be transferred to either side of the nicotinamide ring: the front (A type) or the back (B type) as represented in Figure 13–16. Studies with isotopically labeled substrates have shown that a given enzyme catalyzes one or the other type of transfer, but not both. For example, yeast alcohol dehydrogenase and lactate dehydrogenase from vertebrate heart both transfer a hydride ion from their respective substrates to the same side of the nicotinamide ring; they are classed as type A dehydrogenases to distinguish them from another group of enzymes that reduce NAD^+ by transferring the hydride to the other (B) side of the ring (Fig. 13–16; Table 13–8).

Table 13–8 Dehydrogenases that employ NAD^+ or $NADP^+$ as cofactors

Enzyme	Cofactor	Stereochemical specificity for nicotinamide ring (A or B)
Isocitrate dehydrogenase	NAD^+	A
α-Ketoglutarate dehydrogenase	NAD^+	B
Glucose-6-phosphate dehydrogenase	$NADP^+$	B
Malate dehydrogenase	NAD^+	A
Glutamate dehydrogenase	NAD^+ or $NADP^+$	B
Glyceraldehyde-3-phosphate dehydrogenase	NAD^+	B
Lactate dehydrogenase	NAD^+	A
Alcohol dehydrogenase	NAD^+	A

The association between a given dehydrogenase and NAD or NADP is relatively loose; the cofactor readily diffuses from the surface of one enzyme to that of another, acting as a water-soluble carrier of electrons from one metabolite to another. For example, in the production of alcohol during fermentation of glucose by yeast cells, a hydride ion is removed from glyceraldehyde-3-phosphate by one enzyme (glyceraldehyde-3-phosphate dehydrogenase) and transferred to NAD^+. The NADH thereby produced then leaves the enzyme surface and diffuses to another enzyme, alcohol dehydrogenase, which transfers a hydride ion from NADH to acetaldehyde, producing ethanol:

(1)   Glyceraldehyde-3-phosphate + NAD^+ $\longrightarrow$ 3-phosphoglycerate
$+ NADH + H^+$

(2)   Acetaldehyde + NADH + H^+ $\longrightarrow$ ethanol + NAD^+

Sum: Glyceraldehyde-3-phosphate + acetaldehyde $\longrightarrow$
3-phosphoglycerate + ethanol

Notice that in the overall reaction there is no net production or consumption of NAD^+ or NADH; the cofactors function catalytically, being recycled repeatedly without a net change in the concentration of NAD^+ + NADH.

Flavoproteins Contain Flavin Nucleotides

Flavoproteins (Table 13–9) are enzymes that catalyze oxidation–reduction reactions using either flavin mononucleotide (FMN) or flavin adenine dinucleotide (FAD) as cofactor (Fig. 13–17). These cofactors are derived from the vitamin riboflavin. The fused ring structure of flavin nucleotides (the isoalloxazine ring) undergoes reversible reduction, accepting either one or two electrons in the form of hydrogen atoms (electron plus proton) from a reduced substrate; the reduced forms are abbreviated $FADH_2$ and $FMNH_2$. When a fully oxidized flavin nucleotide accepts only one electron (one hydrogen atom), the semiquinone form of the isoalloxazine ring (Fig. 13–17) is produced. Because flavoproteins can participate in either one- or two-electron transfers, this class of proteins is involved in a greater diversity of reactions than the NAD-linked dehydrogenases. As with nicotinamide coenzymes (Fig. 13–16), the reduction of flavin nucleotides is accompanied by a change in a major absorption band. This change can often be used in assaying a reaction involving a flavoprotein.

The flavin nucleotide in most flavoproteins is bound rather tightly and, in some enzymes such as succinate dehydrogenase, covalently. Such tightly bound cofactors are properly called prosthetic groups. They do not carry electrons by diffusing away from one enzyme and to the next; rather, they provide a means by which the flavoprotein can temporarily hold electrons while it catalyzes electron transfer from a reduced substrate to an electron acceptor. One important feature of the flavoproteins is the variability in standard reduction potential (E_0') of the bound flavin nucleotide; tight association between the enzyme and prosthetic group confers on the flavin ring a reduction potential typical of the specific flavoprotein, sometimes quite different from that of the

Table 13–9 Some enzymes (flavoproteins) that employ flavin coenzymes

Enzyme	Flavin nucleotide
Fatty acyl–CoA dehydrogenase	FAD
Dihydrolipoyl dehydrogenase	FAD
Succinate dehydrogenase	FAD
α-Glycerophosphate dehydrogenase	FAD
NADH dehydrogenase	FMN
Glycolate dehydrogenase	FMN

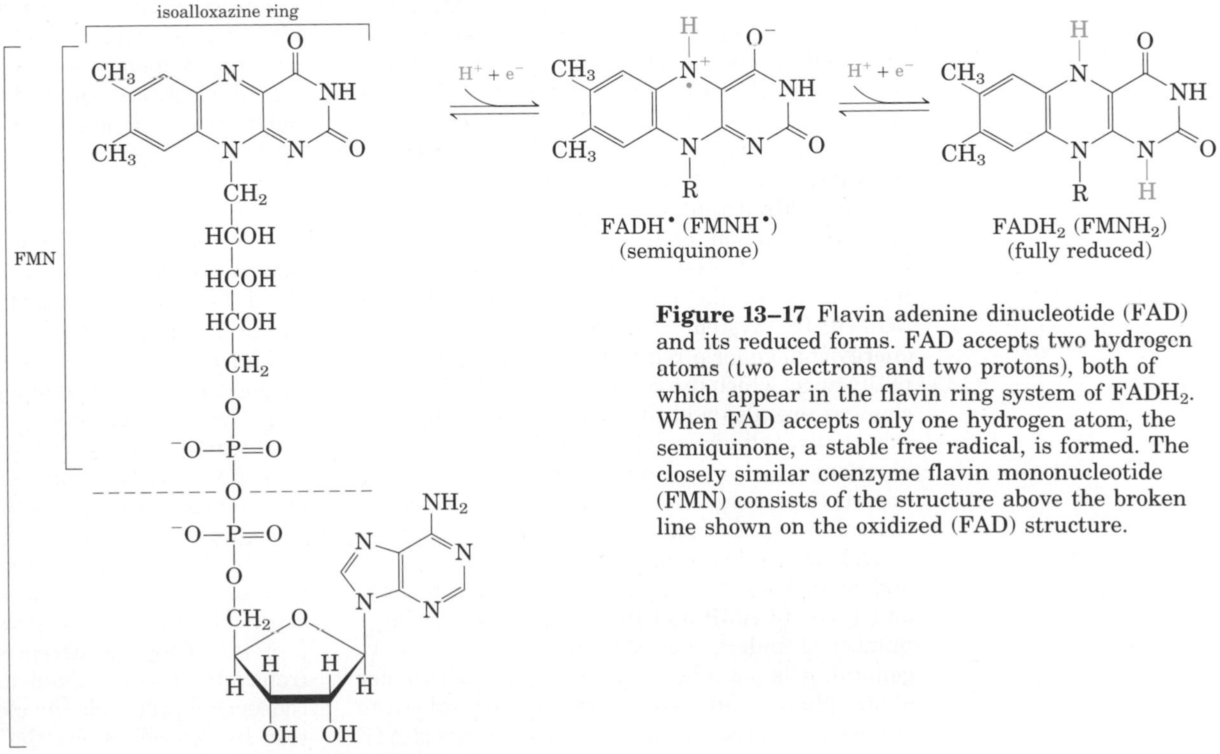

Flavin adenine dinucleotide (FAD) and flavin mononucleotide (FMN)

Figure 13–17 Flavin adenine dinucleotide (FAD) and its reduced forms. FAD accepts two hydrogen atoms (two electrons and two protons), both of which appear in the flavin ring system of $FADH_2$. When FAD accepts only one hydrogen atom, the semiquinone, a stable free radical, is formed. The closely similar coenzyme flavin mononucleotide (FMN) consists of the structure above the broken line shown on the oxidized (FAD) structure.

free flavin nucleotide. Flavoproteins are often very complex; some have, in addition to a flavin nucleotide, tightly bound inorganic ions (iron or molybdenum, for example) capable of participating in electron transfers.

Summary

Living cells constantly perform work and thus require energy for the maintenance of highly organized structures, for the synthesis of cellular components, for movement, for the generation of electrical currents, for the production of light, and for many other processes. Bioenergetics is the quantitative study of energy relationships and energy conversions in biological systems. Biological energy transformations obey the laws of thermodynamics. All chemical reactions are influenced by two forces: the tendency to achieve the most stable bonding state (for which enthalpy, H, is a useful expression) and the tendency to achieve the highest degree of randomness, expressed as entropy, S. The net driving force in a reaction is ΔG, the free-energy change, which represents the net effect of these two factors: $\Delta G = \Delta H - T \Delta S$. Cells require sources of free energy to perform work.

The standard free-energy change, $\Delta G^{\circ\prime}$, is a physical constant characteristic for a given reaction, and can be calculated from the equilibrium constant for the reaction: $\Delta G^{\circ\prime} = -RT \ln K'_{eq}$. The actual free-energy change, ΔG, is a variable, which depends on $\Delta G^{\circ\prime}$ and on the concentrations of reactants and products: $\Delta G = \Delta G^{\circ\prime} + RT \ln$ ([products]/[reactants]). When ΔG is large and negative, the reaction tends to go in the forward direction; when it is large and positive, the reaction tends to go in the reverse direction; and when $\Delta G = 0$, the system is at equilibrium. The free-energy change for a reaction is independent of the pathway by which the reaction occurs. Free-energy changes are also additive; the net chemical reaction that results from the successive occurrence of reactions sharing a common intermediate has an overall free-energy change that is the sum of the ΔG values for the individual reactions.

ATP is the chemical link between catabolism and anabolism. Its exergonic conversion to ADP and P_i, or to AMP and PP_i, is coupled to a large number of endergonic reactions and processes. In general, it is not ATP hydrolysis, but the transfer of phosphate or adenylate from ATP to a substrate or enzyme molecule that couples the energy of ATP breakdown to endergonic transformations of substrates. By these group transfer reactions ATP provides the energy for anabolic reactions, including the synthesis of informational molecules, and for the transport of molecules and ions across membranes against concentration and electrical potential gradients. Muscle contraction is one of several exceptions to this generalization; ATP hydrolysis drives the conformational changes in myosin that produce contraction in muscle.

Cells contain other metabolites with large, negative, free energies of hydrolysis, including phosphoenolpyruvate, 1,3-bisphosphoglycerate, and phosphocreatine. These high-energy compounds, like ATP, have a high phosphate group transfer potential; they are good donors of the phosphate group. Thioesters also have high free energies of hydrolysis.

Biological oxidation–reduction reactions can be described in terms of two half-reactions, each with a characteristic standard reduction potential, $E_0{}'$. When two electrochemical half-cells, each containing the components of a half-reaction, are connected, electrons tend to flow to the half-cell with the higher reduction potential. The strength of this tendency is proportional to the difference between the two reduction potentials (ΔE), and is a function of the concentrations of oxidized and reduced species. The standard free-energy change for an oxidation–reduction reaction is directly proportional to the difference in standard reduction potentials of the two half-cells: $\Delta G^{\circ\prime} = -n\mathcal{F}\Delta E_0'$.

Many biological oxidation reactions are dehydrogenations in which one or two hydrogen atoms (electron and proton) are transferred from a substrate to a hydrogen acceptor. Oxidation–reduction reactions in cells involve specialized electron carrier cofactors. NAD and NADP are the freely diffusible cofactors of many dehydrogenases of cells. Both cofactors accept two electrons and one proton. FAD and FMN, the flavin nucleotides, serve as tightly bound prosthetic groups of flavoproteins. They can accept either one or two electrons. In many organisms, a central energy-conserving process is the stepwise oxidation of glucose to CO_2, in which the energy of oxidation is conserved in ATP as electrons are passed to O_2.

Further Reading

Bioenergetics and Thermodynamics

Atkins, P.W. (1984) *The Second Law,* Scientific American Books, Inc., New York.
A well-illustrated and elementary discussion of the second law and its implications.

Blum, H.F. (1968) *Time's Arrow and Evolution,* 3rd edn, Princeton University Press, Princeton, NJ.

Cantor, C.R. & Schimmel, P.R. (1980) *Biophysical Chemistry,* W.H. Freeman and Company, San Francisco.
This and the next two books are outstanding advanced treatments of thermodynamics.

Dickerson, R.E. (1969) *Molecular Thermodynamics,* W.A. Benjamin, Inc., Menlo Park, CA.

Edsall, J.T. & Gutfreund, H. (1983) *Biothermodynamics: The Study of Biochemical Processes at Equilibrium,* John Wiley & Sons, Inc., New York.

Ingraham, L.L. & Pardee, A.B. (1967) Free energy and entropy in metabolism. In *Metabolic Pathways,* 3rd edn, Vol. I (Greenberg, D.M., ed), pp. 1–46, Academic Press, Inc., New York.

Klotz, I.M. (1967) *Energy Changes in Biochemical Reactions,* Academic Press, Inc., New York.
Brief and nonmathematical introduction to thermodynamics for biochemists, with many illustrative examples.

Morowitz, H.J. (1970) *Entropy for Biologists: An Introduction to Thermodynamics,* Academic Press, Inc., New York.
A good introduction to thermodynamics in biology, not limited to a discussion of entropy.

Rothman, T. (1989) *Science à la Mode,* Princeton University Press, Princeton, NJ.
Chapter 4, "The Evolution of Entropy," is an excellent discussion of entropy in biology.

van Holde, K.E. (1985) *Physical Biochemistry,* 2nd edn, Prentice-Hall, Inc., Englewood Cliffs, NJ.
Chapters 1 through 3 cover the thermodynamic concepts discussed in this chapter.

Phosphate Group Transfers and ATP

Alberty, R.A. (1969) Standard Gibbs free energy, enthalpy and entropy changes as a function of pH and pMg for several reactions involving adenosine phosphates. *J. Biol. Chem.* **244,** 3290–3302.
This research paper documents the strong dependence of the free energy of ATP hydrolysis on the concentrations of H^+ and Mg^{2+}.

Bock, R.M. (1960) Adenine nucleotides and properties of pyrophosphate compounds. In *The Enzymes,* 2nd edn, Vol. 2 (Boyer, P.D., Lardy, H., & Myrbäck, K., eds), pp. 3–38, Academic Press, Inc., New York.

Bridger, W.A. & Henderson, J.F. (1983) *Cell ATP,* John Wiley & Sons, Inc., New York.
The chemistry of ATP, the role of ATP in metabolic regulation, and the catabolic and anabolic roles of ATP.

Hanson, R.W. (1989) The role of ATP in metabolism. *Biochem. Educ.* **17,** 86–92.
Excellent summary of the chemistry and biology of ATP.

Harold, F.M. (1986) *The Vital Force: A Study of Bioenergetics,* W.H. Freeman and Company, New York.
A beautifully clear discussion of thermodynamics in biological processes.

Jencks, W.P. (1990) How does ATP make work? *Chemtracts—Biochem. Mol. Biol.* **1,** 1–13.
A clear and sophisticated description of ATP energy transductions in ion transport, muscle contraction, oxidative phosphorylation, and photophosphorylation.

Kalckar, H.M. (1969) *Biological Phosphorylations: Development of Concepts,* Prentice-Hall, Inc., Englewood Cliffs, NJ.
An historical account by one of the central participants in the study of biological phosphorylations.

Lipmann, F. (1941) Metabolic generation and utilization of phosphate bond energy. *Adv. Enzymol.* **11,** 99–162.
The classic description of the role of high-energy phosphate compounds in biology.

Pullman, B. & Pullman, A. (1960) Electronic structure of energy-rich phosphates. *Radiat. Res.* Suppl. 2, pp. 160–181.
An advanced discussion of the chemistry of ATP and other "energy-rich" compounds.

Westheimer, F.H. (1987) Why nature chose phosphates. *Science* **235,** 1173–1178.
A chemist's description of the unique suitability of phosphate esters and anhydrides for metabolic transformations.

Biological Oxidation–Reduction Reactions

Dolphin, D., Avramovic, O., & Poulson, R. (eds) (1987) *Pyridine Nucleotide Coenzymes: Chemical, Biochemical, and Medical Aspects*, John Wiley & Sons, Inc., New York.

An excellent two-volume collection of authoritative reviews. Among the most useful of these are the chapters by Kaplan, Westheimer, Veech, and Ohno and Ushio.

Latimer, W.M. (1952) *Oxidation Potentials*, 2nd edn, Prentice-Hall, Inc., New York.

Montgomery, R. & Swenson, C.A. (1976) *Quantitative Problems in the Biochemical Sciences*, 2nd edn, W.H. Freeman and Company, San Francisco.

Segel, I.H. (1976) *Biochemical Calculations*, 2nd edn, John Wiley & Sons, Inc., New York.

Problems

1. *Entropy Changes during Egg Development* Consider an ecosystem consisting of an egg in an incubator. The white and yolk of the egg contain proteins, carbohydrates, and lipids. If fertilized, the egg is transformed from a single cell to a complex organism. Discuss this irreversible process in terms of the entropy changes in the system, surroundings, and universe. Be sure that you first clearly define the system and surroundings.

2. *Calculation of $\Delta G^{\circ\prime}$ from Equilibrium Constants* Calculate the standard free-energy changes of the following metabolically important enzyme-catalyzed reactions at 25 °C and pH 7.0 from the equilibrium constants given.

(a) Glutamate + oxaloacetate $\xrightleftharpoons[\text{aminotransferase}]{\text{aspartate}}$

aspartate + α-ketoglutarate $K'_{eq} = 6.8$

(b) Dihydroxyacetone phosphate $\xrightleftharpoons[\text{isomerase}]{\text{triose phosphate}}$

glyceraldehyde-3-phosphate $K'_{eq} = 0.0475$

(c) Fructose-6-phosphate + ATP $\xrightleftharpoons{\text{phosphofructokinase}}$

fructose-1,6-bisphosphate + ADP $K'_{eq} = 254$

3. *Calculation of Equilibrium Constants from $\Delta G^{\circ\prime}$* Calculate the equilibrium constants K'_{eq} for each of the following reactions at pH 7.0 and 25 °C, using the $\Delta G^{\circ\prime}$ values of Table 13–4:

(a) Glucose-6-phosphate + H_2O $\xrightleftharpoons[\text{6-phosphatase}]{\text{glucose-}}$

glucose + P_i

(b) Lactose + H_2O $\xrightleftharpoons{\beta\text{-galactosidase}}$

glucose + galactose

(c) Malate $\xrightleftharpoons{\text{fumarase}}$ fumarate + H_2O

4. *Experimental Determination of K'_{eq} and $\Delta G^{\circ\prime}$* If a 0.1 M solution of glucose-1-phosphate is incubated with a catalytic amount of phosphoglucomutase, the glucose-1-phosphate is transformed to glucose-6-phosphate until equilibrium is established. The equilibrium concentrations are

Glucose-1-phosphate $\rightleftharpoons$ glucose-6-phosphate
4.5×10^{-3} M 9.6×10^{-2} M

Calculate K'_{eq} and $\Delta G^{\circ\prime}$ for this reaction at 25 °C.

5. *Experimental Determination of $\Delta G^{\circ\prime}$ for ATP Hydrolysis* A direct measurement of the standard free-energy change associated with the hydrolysis of ATP is technically demanding because the minute amount of ATP remaining at equilibrium is difficult to measure accurately. The value of $\Delta G^{\circ\prime}$ can be calculated indirectly, however, from the equilibrium constants of two other enzymatic reactions having less favorable equilibrium constants:

Glucose-6-phosphate + H_2O $\longrightarrow$ glucose + P_i
$K'_{eq} = 270$

ATP + glucose $\longrightarrow$ ADP + glucose-6-phosphate
$K'_{eq} = 890$

Using this information, calculate the standard free energy of hydrolysis of ATP. Assume a temperature of 25 °C.

6. *Difference between $\Delta G^{\circ\prime}$ and ΔG* Consider the following interconversion, which occurs in glycolysis (Chapter 14):

Fructose-6-phosphate $\rightleftharpoons$ glucose-6-phosphate
$K'_{eq} = 1.97$

(a) What is $\Delta G^{\circ\prime}$ for the reaction (assuming that the temperature is 25 °C)?

(b) If the concentration of fructose-6-phosphate is adjusted to 1.5 M and that of glucose-6-phosphate is adjusted to 0.5 M, what is ΔG?

(c) Why are $\Delta G^{\circ\prime}$ and ΔG different?

7. *Dependence of ΔG on pH* The free energy released by the hydrolysis of ATP under standard conditions at pH 7.0 is -30.5 kJ/mol. If ATP is hydrolyzed under standard conditions but at pH 5.0, is more or less free energy released? Why?

8. *The $\Delta G^{\circ\prime}$ for Coupled Reactions* Glucose-1-phosphate is converted into fructose-6-phosphate in two successive reactions:

Glucose-1-phosphate $\longrightarrow$ glucose-6-phosphate
Glucose-6-phosphate $\longrightarrow$ fructose-6-phosphate

Using the $\Delta G^{\circ\prime}$ values in Table 13–4, calculate the equilibrium constant, K'_{eq}, for the sum of the two reactions at 25 °C:

Glucose-1-phosphate $\longrightarrow$ fructose-6-phosphate

9. *Strategy for Overcoming an Unfavorable Reaction: ATP-Dependent Chemical Coupling* The phosphorylation of glucose to glucose-6-phosphate is the initial step in the catabolism of glucose. The direct phosphorylation of glucose by P_i is described by the equation

Glucose + P_i $\longrightarrow$ glucose-6-phosphate + H_2O
$$\Delta G^{\circ\prime} = 13.8 \text{ kJ/mol}$$

(a) Calculate the equilibrium constant for the above reaction. In the rat hepatocyte the physiological concentrations of glucose and P_i are maintained at approximately 4.8 mM. What is the equilibrium concentration of glucose-6-phosphate obtained by the direct phosphorylation of glucose by P_i? Does this route represent a reasonable metabolic route for the catabolism of glucose? Explain.

(b) In principle, at least, one way to increase the concentration of glucose-6-phosphate is to drive the equilibrium reaction to the right by increasing the intracellular concentrations of glucose and P_i. Assuming a fixed concentration of P_i at 4.8 mM, how high would the intracellular concentration of glucose have to be to have an equilibrium concentration of glucose-6-phosphate of 250 μM (normal physiological concentration)? Would this route be a physiologically reasonable approach, given that the maximum solubility of glucose is less than 1 M?

(c) The phosphorylation of glucose in the cell is coupled to the hydrolysis of ATP; that is, part of the free energy of ATP hydrolysis is utilized to effect the endergonic phosphorylation of glucose:

(1)   Glucose + P_i $\longrightarrow$ glucose-6-phosphate + H_2O
$$\Delta G^{\circ\prime} = 13.8 \text{ kJ/mol}$$
(2)   ATP + H_2O $\longrightarrow$ ADP + P_i
$$\Delta G^{\circ\prime} = -30.5 \text{ kJ/mol}$$

Sum: Glucose + ATP $\longrightarrow$ glucose-6-phosphate + ADP

Calculate K'_{eq} for the overall reaction. When the ATP-dependent phosphorylation of glucose is carried out, what concentration of glucose is needed to achieve a 250 μM intracellular concentration of glucose-6-phosphate when the concentrations of ATP and ADP are 3.38 and 1.32 mM, respectively?

Does this coupling process provide a feasible route, at least in principle, for the phosphorylation of glucose as it occurs in the cell? Explain.

(d) Although coupling ATP hydrolysis to glucose phosphorylation makes thermodynamic sense, how this coupling is to take place has not been specified. Given that coupling requires a common intermediate, one conceivable route is to use ATP hydrolysis to raise the intracellular concentration of P_i and thus drive the unfavorable phosphorylation of glucose by P_i. Is this a reasonable route? Explain.

(e) The ATP-coupled phosphorylation of glucose is catalyzed in the hepatocyte by the enzyme glucokinase. This enzyme binds ATP and glucose to form a glucose–ATP–enzyme complex, and the phosphate is transferred directly from ATP to glucose. Explain the advantages of this route.

10. *Calculations of $\Delta G^{\circ\prime}$ for ATP-Coupled Reactions* From data in Table 13–6 calculate the $\Delta G^{\circ\prime}$ value for the reactions

(a) Phosphocreatine + ADP $\longrightarrow$
creatine + ATP

(b) ATP + fructose $\longrightarrow$
ADP + fructose-6-phosphate

11. *Coupling ATP Cleavage to an Unfavorable Reaction* This problem explores the consequences of coupling ATP hydrolysis under physiological conditions to a thermodynamically unfavorable biochemical reaction. Because we want to explore these consequences in stages, we shall consider the hypothetical transformation, X $\longrightarrow$ Y, a reaction for which $\Delta G^{\circ\prime} = 20$ kJ/mol.

(a) What is the ratio $[Y]/[X]$ at equilibrium?

(b) Suppose X and Y participate in a sequence of reactions during which ATP is hydrolyzed to ADP and P_i. The overall reaction is

X + ATP + H_2O $\longrightarrow$ Y + ADP + P_i

Calculate $[Y]/[X]$ for this reaction at equilibrium. Assume for the purposes of this calculation that the concentrations of ATP, ADP, and P_i are all 1 M when the reaction is at equilibrium.

(c) We know that [ATP], [ADP], and [P_i] are *not* 1 M under physiological conditions. Calculate the ratio $[Y]/[X]$ for the ATP-coupled reaction when the values of [ATP], [ADP], and [P_i] are those found in rat myocytes (Table 13–5).

12. *Calculations of ΔG at Physiological Concentrations* Calculate the physiological ΔG (not $\Delta G^{\circ\prime}$) for the reaction

Phosphocreatine + ADP $\longrightarrow$ creatine + ATP

at 25 °C as it occurs in the cytosol of neurons, in which phosphocreatine is present at 4.7 mM, creatine at 1.0 mM, ADP at 0.20 mM, and ATP at 2.6 mM.

13. *Free Energy Required for ATP Synthesis under Physiological Conditions* In the cytosol of rat hepatocytes, the mass-action ratio is

$$\frac{[ATP]}{[ADP][P_i]} = 5.33 \times 10^2 \text{ M}^{-1}$$

Calculate the free energy required to synthesize ATP in the rat hepatocyte.

14. *Daily ATP Utilization by Human Adults*

(a) A total of 30.5 kJ/mol of free energy is needed to synthesize ATP from ADP and P_i when the reactants and products are at 1 M concentration (standard state). Because the actual physiological concentrations of ATP, ADP, and P_i are not 1 M, the free energy required to synthesize ATP under physiological conditions is different from $\Delta G^{\circ\prime}$. Calculate the free energy required to synthesize ATP in the human hepatocyte when the physiological concentrations of ATP, ADP, and P_i are 3.5, 1.50, and 5.0 mM, respectively.

(b) A normal 68 kg (150 lb) adult requires a caloric intake of 2,000 kcal (8,360 kJ) of food per day (24 h). This food is metabolized and the free energy used to synthesize ATP, which is then utilized to do the body's daily chemical and mechanical work. Assuming that the efficiency of converting food energy into ATP is 50%, calculate the weight of ATP utilized by a human adult in a 24 h period. What percentage of the body weight does this represent?

(c) Although adults synthesize large amounts of ATP daily, their body weight, structure, and composition do not change significantly during this period. Explain this apparent contradiction.

15. *ATP Reserve in Muscle Tissue* The ATP concentration in muscle tissue (approximately 70% water) is about 8.0 mM. During strenuous activity each gram of muscle tissue uses ATP at the rate of 300 μmol/min for contraction.

(a) How long would the reserve of ATP last during a 100 meter dash?

(b) The phosphocreatine level in muscle is about 40.0 mM. How does this help extend the reserve of muscle ATP?

(c) Given the size of the reserve ATP pool, how can a person run a marathon?

16. *Rates of Turnover of γ- and β-Phosphates of ATP* If a small amount of ATP labeled with radioactive phosphorus in the terminal position, [γ-^{32}P]ATP, is added to a yeast extract, about half of the ^{32}P activity is found in P_i within a few minutes, but the concentration of ATP remains unchanged. Explain. If the same experiment is carried out using ATP labeled with ^{32}P in the central position, [β-^{32}P]ATP, the ^{32}P does not appear in P_i within the same number of minutes. Why?

17. *Cleavage of ATP to AMP and PP$_i$ during Metabolism* The synthesis of the activated form of acetate (acetyl-CoA) is carried out in an ATP-dependent process:

Acetate + CoA + ATP $\longrightarrow$

acetyl-CoA + AMP + PP$_i$

(a) The $\Delta G^{\circ\prime}$ for the hydrolysis of acetyl-CoA to acetate and CoA is -32.2 kJ/mol and that for hydrolysis of ATP to AMP and PP$_i$ is -30.5 kJ/mol. Calculate $\Delta G^{\circ\prime}$ for the ATP-dependent synthesis of acetyl-CoA.

(b) Almost all cells contain the enzyme inorganic pyrophosphatase, which catalyzes the hydrolysis of PP$_i$ to P_i. What effect does the presence of this enzyme have on the synthesis of acetyl-CoA? Explain.

18. *Are All Metabolic Reactions at Equilibrium?*

(a) Phosphoenolpyruvate is one of the two phosphate donors in the synthesis of ATP during glycolysis. In human erythrocytes, the steady-state concentration of ATP is 2.24 mM, that of ADP is 0.25 mM, and that of pyruvate is 0.051 mM. Calculate the concentration of phosphoenolpyruvate at 25 °C, assuming that the pyruvate kinase reaction (Fig. 13–3) is at equilibrium in the cell.

(b) The physiological concentration of phosphoenolpyruvate in human erythrocytes is 0.023 mM. Compare this with the value obtained in (a). What is the significance of this difference? Explain.

19. *Standard Reduction Potentials* The standard reduction potential, E_0', of any redox pair is defined for the half-cell reaction:

Oxidizing agent + n electrons $\rightleftharpoons$

reducing agent

The E_0' values for the NAD$^+$/NADH and pyruvate/lactate conjugate redox pairs are -0.32 and -0.19 V, respectively.

(a) Which conjugate pair has the greater tendency to lose electrons? Explain.

(b) Which is the stronger oxidizing agent? Explain.

(c) If we begin with 1 M concentrations of each reactant and product at pH 7, in which direction will the following reaction proceed?

Pyruvate + NADH + H$^+$ $\rightleftharpoons$ lactate + NAD$^+$

(d) What is the standard free-energy change ($\Delta G^{\circ\prime}$) at 25 °C for this reaction?

(e) What is the equilibrium constant (K'_{eq}) for this reaction?

20. *Energy Span of the Respiratory Chain* Electron transfer in the mitochondrial respiratory chain may be represented by the net reaction equation

NADH + H$^+$ + $\frac{1}{2}$O$_2$ $\rightleftharpoons$ H$_2$O + NAD$^+$

(a) Calculate the value of $\Delta E_0'$ for the net reaction of mitochondrial electron transfer.

(b) Calculate $\Delta G^{\circ\prime}$ for this reaction.

(c) How many ATP molecules can *theoretically* be generated by this reaction if the standard free energy of ATP synthesis is 30.5 kJ/mol?

21. *Dependence of Electromotive Force on Concentrations* Calculate the electromotive force (in volts) registered by an electrode immersed in a solution containing the following mixtures of NAD^+ and NADH at pH 7.0 and 25 °C, with reference to a half-cell of $E_0' = 0.00$ V.

(a) 1.0 mM NAD^+ and 10 mM NADH

(b) 1.0 mM NAD^+ and 1.0 mM NADH

(c) 10 mM NAD^+ and 1.0 mM NADH

22. *Electron Affinity of Compounds* List the following substances in order of increasing tendency to accept electrons: (a) α-ketoglutarate + CO_2 (yielding isocitrate), (b) oxaloacetate, (c) O_2, (d) $NADP^+$.

23. *Direction of Oxidation–Reduction Reactions* Which of the following reactions would be expected to proceed in the direction shown under standard conditions, assuming that the appropriate enzymes are present to catalyze them?

(a) Malate + NAD^+ $\longrightarrow$
$$\text{oxaloacetate} + \text{NADH} + H^+$$

(b) Acetoacetate + NADH + H^+ $\longrightarrow$
$$\beta\text{-hydroxybutyrate} + NAD^+$$

(c) Pyruvate + NADH + H^+ $\longrightarrow$
$$\text{lactate} + NAD^+$$

(d) Pyruvate + β-hydroxybutyrate $\longrightarrow$
$$\text{lactate} + \text{acetoacetate}$$

(e) Malate + pyruvate $\longrightarrow$
$$\text{oxaloacetate} + \text{lactate}$$

(f) Acetaldehyde + succinate $\longrightarrow$
$$\text{ethanol} + \text{fumarate}$$

Glycolysis and the Catabolism of Hexoses

Having examined the organizing principles of cell metabolism and bioenergetics, we are ready to see how the chemical energy stored in glucose and other fuel molecules is released to perform biological work. D-Glucose is the major fuel of most organisms and occupies a central position in metabolism. It is relatively rich in potential energy; its complete oxidation to carbon dioxide and water proceeds with a standard free-energy change of −2,840 kJ/mol. By storing glucose as a high molecular weight polymer, a cell can stockpile large quantities of hexose units while maintaining a relatively low cytosolic osmolarity. When the cell's energy demands suddenly increase, glucose can be released quickly from these intracellular storage polymers.

Glucose is not only an excellent fuel, it is also a remarkably versatile precursor, capable of supplying a huge array of metabolic intermediates, the necessary starting materials for biosynthetic reactions. *E. coli* can obtain from glucose the carbon skeletons for every one of the amino acids, nucleotides, coenzymes, fatty acids, and other metabolic intermediates needed for growth. A study of the numerous metabolic fates of glucose would encompass hundreds or thousands of transformations. In the higher plants and animals glucose has three major fates: it may be stored (as a polysaccharide or as sucrose), oxidized to a three-carbon compound (pyruvate) via glycolysis, or oxidized to pentoses via the pentose phosphate (phosphogluconate) pathway (Fig. 14–1).

This chapter begins with a description of the individual reactions that constitute the glycolytic pathway and of the enzymes that catalyze them. We then consider fermentation, the operation of the glycolytic pathway under anaerobic conditions. The sources of glucose units for glycolysis are diverse, and we next describe pathways that bring carbon into glycolysis from hexoses other than glucose and from disaccharides and polysaccharides. Like all metabolic pathways, glycolysis is under tight regulation. We discuss the general principles of metabolic regulation, then illustrate these principles with the glycolytic pathway. The chapter concludes with a brief description of two other catabolic pathways that begin with glucose: one leading to pentoses, the other to glucuronate and ascorbic acid (vitamin C).

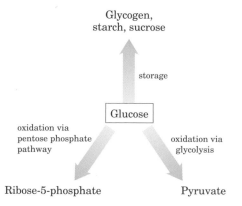

Figure 14–1 Major pathways of glucose utilization in cells of higher plants and animals. Although not the only possible fates for glucose, these three pathways are the most significant in terms of the amount of glucose that flows through them in most cells.

Glycolysis

In **glycolysis** (from the Greek *glykys,* meaning "sweet," and *lysis,* meaning "splitting") a molecule of glucose is degraded in a series of enzyme-catalyzed reactions to yield two molecules of pyruvate. During the sequential reactions of glycolysis some of the free energy released from glucose is conserved in the form of ATP. Glycolysis was the first metabolic pathway to be elucidated and is probably the best understood. From the discovery by Eduard Buchner (in 1897) of fermentation in broken extracts of yeast cells until the clear recognition by Fritz Lipmann and Herman Kalckar (in 1941) of the metabolic role of high-energy compounds such as ATP in metabolism, the reactions of glycolysis in extracts of yeast and muscle were central to biochemical research. The development of methods of enzyme purification, the discovery and recognition of the importance of cofactors such as NAD, and the discovery of the pivotal role in metabolism of phosphorylated compounds all came out of studies of glycolysis. By now, all of the enzymes of glycolysis have been purified from many organisms and thoroughly studied, and the three-dimensional structures of all of the glycolytic enzymes are known from x-ray crystallographic studies.

Fritz Lipmann
1899–1986

Glycolysis is an almost universal central pathway of glucose catabolism. It is the pathway through which the largest flux of carbon occurs in most cells. In certain mammalian tissues and cell types (erythrocytes, renal medulla, brain, and sperm, for example), glucose is the sole or major source of metabolic energy through glycolysis. Some plant tissues that are modified for the storage of starch (such as potato tubers) and some plants adapted to growth in areas regularly inundated by water (watercress, for example) derive most of their energy from glycolysis; many types of anaerobic microorganisms are entirely dependent on glycolysis.

Fermentation is a general term denoting the *anaerobic* degradation of glucose or other organic nutrients into various products (characteristic for different organisms) to obtain energy in the form of ATP. Because living organisms first arose in an atmosphere lacking oxygen, anaerobic breakdown of glucose is probably the most ancient biological mechanism for obtaining energy from organic fuel molecules. In the course of evolution this reaction sequence has been completely conserved; the glycolytic enzymes of vertebrate animals are closely similar, in amino acid sequence and three-dimensional structure, to their homologs in yeast and spinach. The process of glycolysis differs from one species to another only in the details of its regulation and in the subsequent metabolic fate of the pyruvate formed. The thermodynamic principles and the types of regulatory mechanisms in glycolysis are found in all pathways of cell metabolism. A study of glycolysis can serve as a model of many aspects of the pathways discussed later in this book.

Herman Kalckar
1908–1991

Before examining each step of the pathway in some detail, we will take a look at glycolysis as a whole.

An Overview: Glycolysis Has Two Phases

The breakdown of the six-carbon glucose into two molecules of the three-carbon pyruvate occurs in ten steps, the first five of which constitute the *preparatory phase* (Fig. 14–2a). In these reactions glucose is first phosphorylated at the hydroxyl group on C-6 (step ①). The D-glucose-6-phosphate thus formed is converted to D-fructose-6-phosphate (step ②), which is again phosphorylated, this time at C-1, to yield D-fructose-1,6-bisphosphate (step ③). For both phosphorylations, ATP is the phosphate donor. As all of the sugar derivatives that occur in the glycolytic pathway are the D isomers, we will omit the D designation except when emphasizing stereochemistry.

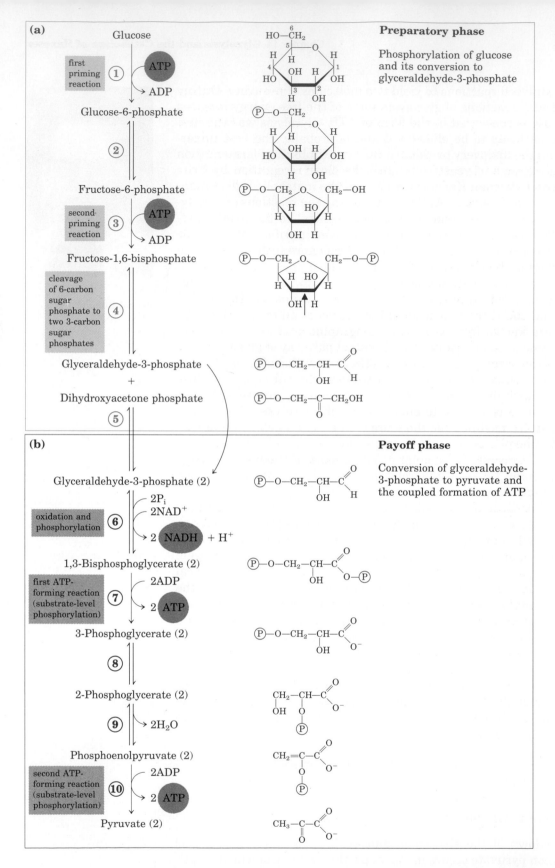

(a) Glucose

Preparatory phase

Phosphorylation of glucose and its conversion to glyceraldehyde-3-phosphate

① first priming reaction — ATP → ADP

Glucose-6-phosphate

②

Fructose-6-phosphate

③ second priming reaction — ATP → ADP

Fructose-1,6-bisphosphate

④ cleavage of 6-carbon sugar phosphate to two 3-carbon sugar phosphates

Glyceraldehyde-3-phosphate
+
Dihydroxyacetone phosphate

⑤

(b)

Payoff phase

Conversion of glyceraldehyde-3-phosphate to pyruvate and the coupled formation of ATP

Glyceraldehyde-3-phosphate (2)

⑥ oxidation and phosphorylation — $2P_i$, $2NAD^+$ → 2 NADH + H^+

1,3-Bisphosphoglycerate (2)

⑦ first ATP-forming reaction (substrate-level phosphorylation) — 2ADP → 2 ATP

3-Phosphoglycerate (2)

⑧

2-Phosphoglycerate (2)

⑨ → $2H_2O$

Phosphoenolpyruvate (2)

⑩ second ATP-forming reaction (substrate-level phosphorylation) — 2ADP → 2 ATP

Pyruvate (2)

Figure 14–2 The two phases of glycolysis. For each molecule of glucose that passes through the preparatory phase **(a)**, two molecules of glyceraldehyde-3-phosphate are formed; both pass through the payoff phase **(b)**. Pyruvate is the end product of the second phase under aerobic conditions, but under anaerobic conditions pyruvate is reduced to lactate to regenerate NAD^+. For each glucose molecule, two ATP are consumed in the preparatory phase and four ATP are produced in the payoff phase, giving a net yield of two molecules of ATP per one of glucose converted to pyruvate. The number beside each reaction step corresponds to its numbered heading in the text discussion. Keep in mind that each phosphate group, represented here as ⓅP, has two negative charges ($-PO_3^{2-}$).

Fructose-1,6-bisphosphate is next split to yield two three-carbon molecules, dihydroxyacetone phosphate and glyceraldehyde-3-phosphate (step ④); this is the "lysis" step that gives the process its name. The dihydroxyacetone phosphate is isomerized to a second molecule of glyceraldehyde-3-phosphate (step ⑤); this ends the first phase of glycolysis. Note that two molecules of ATP must be invested to activate, or prime, the glucose molecule for its cleavage into two three-carbon pieces; later there will be a good return on this investment. To sum up: in the preparatory phase of glycolysis the energy of ATP is invested, raising the free-energy content of the intermediates, and the carbon chains of all the metabolized hexoses are converted into a common product, glyceraldehyde-3-phosphate.

The energetic gain comes in the *payoff phase* of glycolysis (Fig. 14–2b). Each molecule of glyceraldehyde-3-phosphate is oxidized and phosphorylated by inorganic phosphate (*not* by ATP) to form 1,3-bisphosphoglycerate (step ⑥). Energy is released as the two molecules of 1,3-bisphosphoglycerate are converted into two molecules of pyruvate (steps ⑦ through ⑩). Much of this energy is conserved by the coupled phosphorylation of four molecules of ADP to ATP. The net yield is two molecules of ATP per molecule of glucose used, because two molecules of ATP were invested in the preparatory phase of glycolysis. Energy is also conserved in the payoff phase in the formation of two molecules of NADH per molecule of glucose.

In the sequential reactions of glycolysis, three types of chemical transformation are particularly noteworthy: (1) the degradation of the carbon skeleton of glucose to yield pyruvate, (2) the phosphorylation of ADP to ATP by high-energy phosphate compounds formed during glycolysis, and (3) the transfer of hydrogen atoms or electrons to NAD^+, forming NADH. The fate of the product, pyruvate, depends on the cell type and the metabolic circumstances.

Fate of Pyruvate   Three alternative catabolic routes are taken by the pyruvate formed by glycolysis. In aerobic organisms or tissues, under aerobic conditions, glycolysis constitutes only the first stage in the complete degradation of glucose (Fig. 14–3). Pyruvate is oxidized, with loss of its carboxyl group as CO_2, to yield the acetyl group of acetyl-coenzyme A, which is then oxidized completely to CO_2 by the citric acid cycle (Chapter 15). The electrons from these oxidations are passed to O_2 through a chain of carriers in the mitochondrion, forming H_2O. The energy from the electron transfer reactions drives the synthesis of ATP in the mitochondrion (Chapter 18).

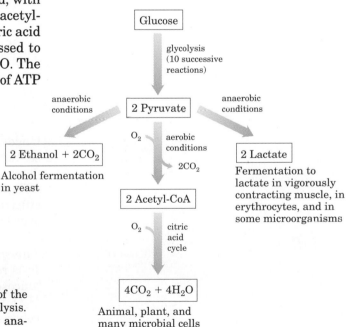

Figure 14–3 Three possible catabolic fates of the pyruvate formed in the payoff phase of glycolysis. Pyruvate also serves as a precursor in many anabolic reactions, not shown here.

The second route for pyruvate metabolism is its reduction to lactate via **lactic acid fermentation.** When a tissue such as vigorously contracting skeletal muscle must function anaerobically, the pyruvate cannot be oxidized further for lack of oxygen. Under these conditions pyruvate is reduced to lactate. Certain tissues and cell types (retina, brain, erythrocytes) convert glucose to lactate even under aerobic conditions. Lactate (the dissociated form of lactic acid) is also the product of glycolysis under anaerobic conditions in microorganisms that carry out the lactic acid fermentation (Fig. 14–3).

The third major route for catabolism of pyruvate leads to ethanol. In some plant tissues and in certain invertebrates, protists, and microorganisms such as brewer's yeast, pyruvate is converted anaerobically into ethanol and CO_2, a process called **alcohol** (or **ethanol**) **fermentation** (Fig. 14–3).

The focus of this chapter is catabolism, but pyruvate has other, anabolic, fates. It can, for example, provide the carbon skeleton for the synthesis of the amino acid alanine. We shall return to these anabolic reactions of pyruvate in later chapters.

ATP Formation Coupled to Glycolysis During glycolysis some of the energy of the glucose molecule is conserved in the form of ATP, while much remains in the product, pyruvate. The overall equation for glycolysis is

$$\text{Glucose} + 2\text{NAD}^+ + 2\text{ADP} + 2\text{P}_i \longrightarrow$$
$$2 \text{ pyruvate} + 2\text{NADH} + 2\text{H}^+ + 2\text{ATP} + 2\text{H}_2\text{O} \quad (14\text{–}1)$$

For each molecule of glucose degraded to pyruvate, two molecules of ATP are generated from ADP and P_i. We can now resolve the equation of glycolysis into two processes: (1) the conversion of glucose into pyruvate, which is exergonic:

$$\text{Glucose} + 2\text{NAD}^+ \longrightarrow 2 \text{ pyruvate} + 2\text{NADH} + 2\text{H}^+ \quad (14\text{–}2)$$
$$\Delta G_1^{\circ\prime} = -146 \text{ kJ/mol}$$

and (2) the formation of ATP from ADP and P_i, which is endergonic:

$$2\text{ADP} + 2\text{P}_i \longrightarrow 2\text{ATP} + 2\text{H}_2\text{O} \quad (14\text{–}3)$$
$$\Delta G_2^{\circ\prime} = 2(30.5 \text{ kJ/mol}) = 61 \text{ kJ/mol}$$

If we now write the sum of Equations 14–2 and 14–3, we can also determine the overall standard free-energy change of glycolysis (Eqn 14–1), including ATP formation, as the algebraic sum, $\Delta G_s^{\circ\prime}$, of $\Delta G_1^{\circ\prime}$ and $\Delta G_2^{\circ\prime}$:

$$\Delta G_s^{\circ\prime} = \Delta G_1^{\circ\prime} + \Delta G_2^{\circ\prime} = -146 \text{ kJ/mol} + 61 \text{ kJ/mol} = -85 \text{ kJ/mol}$$

Under either standard or intracellular conditions, glycolysis is an essentially irreversible process, driven to completion by this large net decrease in free energy. At the actual intracellular concentrations of ATP, ADP, and P_i (see Table 13–5) and of glucose and pyruvate, the efficiency of recovery of the energy of glycolysis in the form of ATP is over 60%.

Energy Remaining in the Pyruvate Produced by Glycolysis Glycolysis releases only a small fraction of the total available energy of the glucose molecule. When glucose is oxidized completely to CO_2 and H_2O, the total standard free-energy change is $-2,840$ kJ/mol. The glycolytic degradation of glucose to two molecules of pyruvate ($\Delta G^{\circ\prime} =$

−146 kJ/mol) therefore yields only (146/2,840)100 = 5.2% of the total energy that can be released by complete oxidation. The two molecules of pyruvate formed by glycolysis still contain most of the biologically available energy of the glucose molecule, energy that can be extracted by oxidative reactions in the citric acid cycle.

Importance of Phosphorylated Intermediates Each of the nine glycolytic intermediates between glucose and pyruvate is phosphorylated (Fig. 14–2). The phosphate groups appear to have three functions.

1. The phosphate groups are ionized at pH 7, thus giving each of the intermediates of glycolysis a net negative charge. Because the plasma membrane is impermeable to molecules that are charged, the phosphorylated intermediates cannot diffuse out of the cell. After the initial phosphorylation, the cell does not have to spend further energy in retaining phosphorylated intermediates despite the large difference between the intracellular and extracellular concentrations of these compounds.

2. Phosphate groups are essential components in the enzymatic conservation of metabolic energy. Energy released in the breakage of phosphoric acid anhydride bonds (such as those in ATP) is partially conserved in the formation of phosphate esters such as glucose-6-phosphate. High-energy phosphate compounds formed in glycolysis (1,3-bisphosphoglycerate and phosphoenol pyruvate) donate phosphate groups to ADP to form ATP.

3. Binding of phosphate groups to the active sites of enzymes provides binding energy that contributes to lowering the activation energy and increasing the specificity of enzyme-catalyzed reactions (Chapter 8). The phosphate groups of ADP, ATP, and the glycolytic intermediates form complexes with Mg^{2+}, and the substrate binding sites of many of the glycolytic enzymes are specific for these Mg^{2+} complexes. Nearly all the glycolytic enzymes require Mg^{2+} for activity.

The Preparatory Phase of Glycolysis Requires ATP

The preparatory phase of glycolysis requires the investment of two molecules of ATP and results in cleavage of the hexose chain into two triose phosphates. The realization that *phosphorylated* hexoses were intermediates in glycolysis came slowly and serendipitously. In 1906, Arthur Harden and William Young sought to test their hypothesis that inhibitors of proteolytic enzymes would stabilize the glucose-fermenting enzymes in yeast extract. They added blood serum (known to contain inhibitors of proteolytic enzymes) to yeast extracts and observed the predicted stimulation of glucose metabolism. However, in a control experiment intended to show that boiling the serum destroyed the stimulatory activity, they discovered that boiled serum was just as effective at stimulating glycolysis. Careful examination of the contents of the boiled serum revealed that inorganic phosphate was responsible for the stimulation. Harden and Young soon discovered that glucose added to their yeast extract was converted into a hexose bisphosphate (the "Harden–Young ester," eventually identified as fructose-1,6-bisphosphate). This was the beginning of a long series of investigations of the role of organic esters of phosphate in biochemistry, which has led to our current understanding of the central role of phosphate group transfer in biology.

Arthur Harden
1865–1940

William Young
1878–1942

① *Phosphorylation of Glucose* In the first step of glycolysis, glucose is primed for subsequent reactions by its phosphorylation at C-6 to yield **glucose-6-phosphate;** ATP is the phosphate donor:

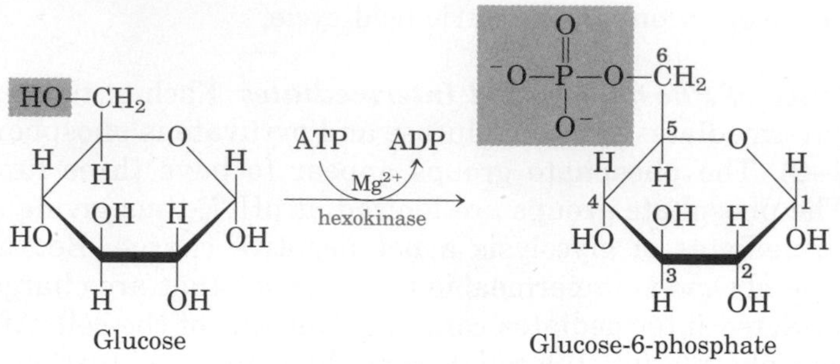

Glucose

Glucose-6-phosphate

$$\Delta G^{\circ\prime} = -16.7 \text{ kJ/mol}$$

This reaction, which is irreversible under intracellular conditions, is catalyzed by **hexokinase.** The common name **kinase** is applied to enzymes that catalyze the transfer of the terminal phosphate group from ATP to some acceptor—a hexose, in the case of hexokinase. Kinases are a subclass of transferases (see Table 8–3).

Hexokinase catalyzes the phosphorylation not only of D-glucose but also of certain other common hexoses, such as D-fructose and D-mannose. Hexokinase, like many other kinases, requires Mg^{2+} for its activity, because the true substrate of the enzyme is not ATP^{4-} but the $MgATP^{2-}$ complex (see Fig. 13–2). Detailed studies of the hexokinase of yeast show that the enzyme undergoes a profound change in its shape, an induced fit, when it binds the hexose molecule (see Fig. 8–21). Hexokinase is universally present in cells of all types. Hepatocytes also contain a form of hexokinase called hexokinase D or glucokinase, which is more specific for glucose and differs from other forms of hexokinase in kinetic and regulatory properties (p. 432).

② *Conversion of Glucose-6-Phosphate to Fructose-6-Phosphate* **Phosphohexose isomerase (phosphoglucose isomerase)** catalyzes the reversible isomerization of glucose-6-phosphate, an aldose, to yield **fructose-6-phosphate,** a ketose:

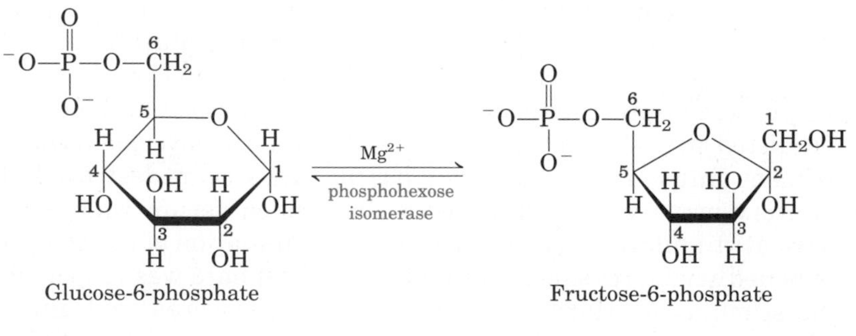

Glucose-6-phosphate

Fructose-6-phosphate

$$\Delta G^{\circ\prime} = 1.7 \text{ kJ/mol}$$

This reaction proceeds readily in either direction, as is predicted from the relatively small change in standard free energy. Phosphohexose isomerase also requires Mg^{2+} and is specific for glucose-6-phosphate and fructose-6-phosphate.

③ **Phosphorylation of Fructose-6-Phosphate to Fructose-1,6-Bisphosphate** In the second of the two priming reactions of glycolysis, **phosphofructokinase-1** catalyzes the transfer of a phosphate group from ATP to fructose-6-phosphate to yield **fructose-1,6-bisphosphate:**

Fructose-6-phosphate   Fructose-1,6-bisphosphate

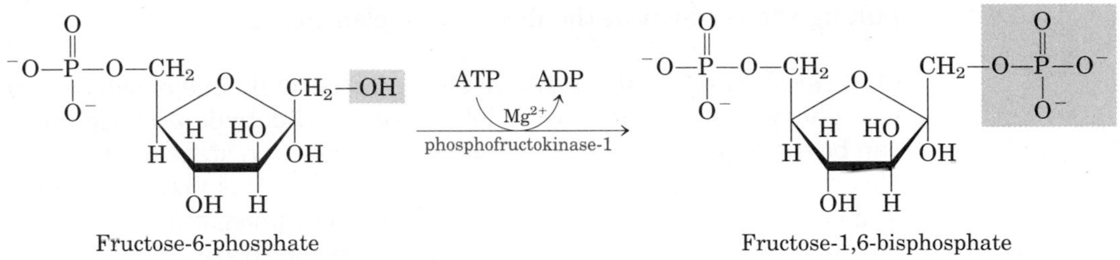

$$\Delta G^{\circ\prime} = -14.2 \text{ kJ/mol}$$

The reaction is essentially irreversible under cellular conditions. This enzyme is called phosphofructokinase-1 (PFK-1) to distinguish it from a second enzyme (PFK-2; Chapter 19) that catalyzes the formation of fructose-2,6-bisphosphate from fructose-6-phosphate.

In some bacteria and protists, and in most or all plants, there is a phosphofructokinase that uses pyrophosphate (PP_i), not ATP, as the phosphate group donor in the synthesis of fructose-1,6-bisphosphate:

$$\text{Fructose-6-phosphate} + PP_i \xrightarrow{Mg^{2+}} \text{fructose-1,6-bisphosphate} + P_i$$
$$\Delta G^{\circ\prime} = -14 \text{ kJ/mol}$$

Phosphofructokinase-1, like hexokinase, is a regulatory enzyme (Chapter 8), one of the most complex known. It is the major point of regulation in glycolysis. The activity of PFK-1 is increased whenever the ATP supply of the cell becomes depleted or when there is an excess of ATP breakdown products, ADP and AMP, particularly the latter. The enzyme is inhibited whenever the cell has ample ATP and when it is well supplied by other fuels such as fatty acids. Fructose-2,6-bisphosphate, structurally similar to the product of this reaction but not an intermediate in glycolysis, is a potent stimulator of both the ATP-dependent and the PP_i-dependent enzymes. The regulation of this step in glycolysis is discussed in greater detail later in the chapter.

④ *Cleavage of Fructose-1,6-Bisphosphate* The enzyme **fructose-1,6-bisphosphate aldolase,** often simply called **aldolase,** catalyzes a reversible aldol condensation. Fructose-1,6-bisphosphate is cleaved to yield two different triose phosphates, **glyceraldehyde-3-phosphate,** an aldose, and **dihydroxyacetone phosphate,** a ketose:

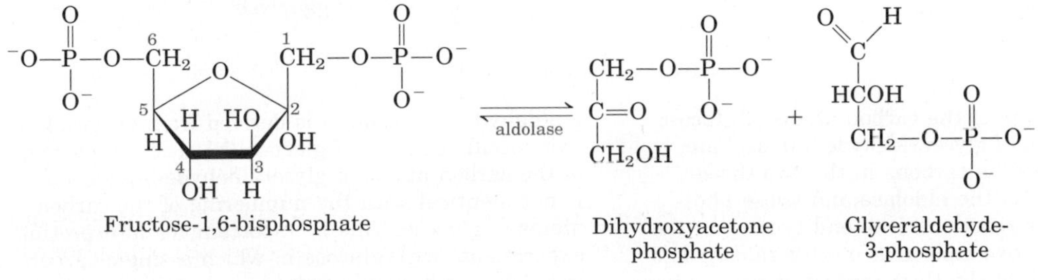

Fructose-1,6-bisphosphate   Dihydroxyacetone   Glyceraldehyde-
                                        phosphate   3-phosphate

$$\Delta G^{\circ\prime} = 23.8 \text{ kJ/mol}$$

The aldolase of vertebrate animal tissues does not require a divalent cation, but in many microorganisms aldolase is a Zn^{2+}-containing enzyme. Although the aldolase reaction has a strongly positive standard free-energy change in the direction of cleavage, in cells it can proceed readily in either direction. During glycolysis the reaction products (two triose phosphates) are removed quickly by the next two steps, pulling the reaction in the direction of cleavage.

⑤ *Interconversion of the Triose Phosphates* Only one of the two triose phosphates formed by aldolase—glyceraldehyde-3-phosphate—can be directly degraded in the subsequent reaction steps of glycolysis. However, the other product, dihydroxyacetone phosphate, is rapidly and reversibly converted into glyceraldehyde-3-phosphate by the fifth enzyme of the glycolytic sequence, **triose phosphate isomerase:**

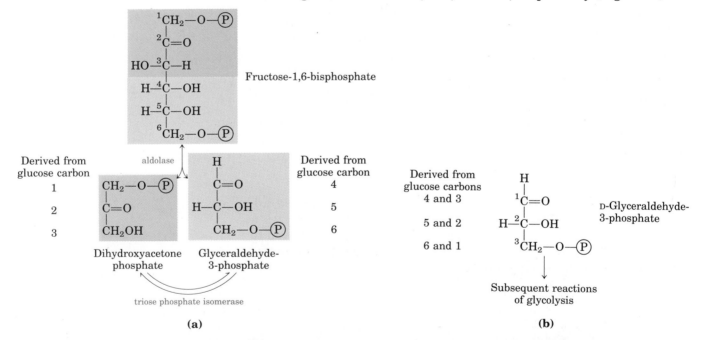

$$\Delta G^{\circ\prime} = 7.5 \text{ kJ/mol}$$

By this reaction C-1, C-2, and C-3 of the starting glucose now become indistinguishable from C-6, C-5, and C-4, respectively (Fig. 14–4).

Figure 14–4 Fate of the carbon atoms of glucose in the formation of glyceraldehyde-3-phosphate. (a) The origin of the carbons in the two three-carbon products of the aldolase and triose phosphate isomerase reactions. The end product of the two reactions is two molecules of glyceraldehyde-3-phosphate. Each of the three carbon atoms of glyc- eraldehyde-3-phosphate is derived from either of two specific carbons of glucose (b). The numbering of the carbon atoms of glyceraldehyde-3-phosphate is not identical with the numbering of the carbon atoms of glucose. This is important for interpreting experiments with glucose in which a single carbon is labeled with a radioisotope.

This reaction completes the preparatory phase of glycolysis, in which the hexose molecule has been phosphorylated at C-1 and C-6 and then cleaved to form, ultimately, two molecules of glyceraldehyde-3-phosphate. Other hexoses, such as D-fructose, D-mannose, and D-galactose, are also convertible into glyceraldehyde-3-phosphate, as we shall see later.

The Payoff Phase of Glycolysis Produces ATP

The payoff phase of glycolysis (Fig. 14–2b) includes the energy-conserving phosphorylation steps in which some of the free energy of the glucose molecule is conserved in the form of ATP. Remember that one molecule of glucose yields two molecules of glyceraldehyde-3-phosphate; both halves of the glucose molecule follow the same pathway in the second phase of glycolysis. The conversion of two molecules of glyceraldehyde-3-phosphate into two of pyruvate is accompanied by the formation of four molecules of ATP from ADP. However, the net yield of ATP per molecule of glucose degraded is only two, because two molecules of ATP were invested in the preparatory phase of glycolysis to phosphorylate the two ends of the hexose molecule.

⑥ *Oxidation of Glyceraldehyde-3-Phosphate to 1,3-Bisphosphoglycerate* The first step in the payoff phase of glycolysis is the conversion of glyceraldehyde-3-phosphate to **1,3-bisphosphoglycerate,** catalyzed by **glyceraldehyde-3-phosphate dehydrogenase:**

$$\Delta G^{\circ\prime} = 6.3 \text{ kJ/mol}$$

This is the first of the two energy-conserving reactions of glycolysis that eventually lead to the formation of ATP. The aldehyde group of glyceraldehyde-3-phosphate is dehydrogenated, not to a free carboxyl group, as one might expect, but to a carboxylic acid anhydride with phosphoric acid. This type of anhydride, called an **acyl phosphate,** has a very high standard free energy of hydrolysis ($\Delta G^{\circ\prime} = -49.3$ kJ/mol; see Fig. 13–4 and Table 13–6). Much of the free energy of oxidation of the aldehyde group of glyceraldehyde-3-phosphate is conserved by formation of the acyl phosphate group at C-1 of 1,3-bisphosphoglycerate.

The acceptor of hydrogen in the glyceraldehyde-3-phosphate dehydrogenase reaction is the coenzyme NAD$^+$ (see Fig. 13–16), the oxidized form of nicotinamide adenine dinucleotide. The reduction of NAD$^+$ proceeds by the enzymatic transfer of a hydride ion ($:H^-$) from the aldehyde group of glyceraldehyde-3-phosphate to the nicotinamide ring of NAD$^+$, to yield the reduced coenzyme NADH. The other hydrogen atom of the substrate molecule appears in solution as H$^+$ (p. 391).

Oxidation of glyceraldehyde-3-phosphate involves an intermediate in which the substrate is covalently bound to the enzyme (Fig. 14–5a). The aldehyde group of glyceraldehyde-3-phosphate first reacts with the —SH group of an essential Cys residue in the active site of the enzyme. This reaction is homologous with the formation of a hemiacetal (see Fig. 11–6), but in this case the product is a *thio*hemiacetal. The discovery that glyceraldehyde-3-phosphate dehydrogenase is inhibited by iodoacetate (Fig. 14–5b) was important in the history of research on glycolysis; the addition of this enzyme inhibitor to crude extracts of yeast or muscle caused the accumulation of the hexose phosphates produced in glycolysis, allowing their isolation and identification.

The NADH formed in this step of glycolysis must be reoxidized to NAD^+. Cells contain limited amounts of NAD^+, and glycolysis would soon come to a halt for lack of NAD^+ were the NADH not reoxidized. The reactions in which NAD^+ is regenerated anaerobically are described in detail later, in connection with the alternative fates of pyruvate.

Figure 14–5 (a) A more detailed representation of the glyceraldehyde-3-phosphate dehydrogenase reaction. In step ①, a covalent thiohemiacetal linkage forms between the substrate and the sulfhydryl group of a Cys residue in the enzyme's active site. This enzyme–substrate intermediate is oxidized by NAD^+ (step ②), also bound to the active site, converting it into a covalent acyl-enzyme intermediate, a thioester. The enzyme-bound NADH is reoxidized by free NAD^+ (step ③). The bond between the acyl group and the thiol group of the enzyme has a very high standard free energy of hydrolysis. In step ④, the thioester bond undergoes phosphorolysis (attack by P_i), releasing the free enzyme and an acyl phosphate product (1,3-bisphosphoglycerate), the formation of which conserves much of the free energy liberated during oxidation of the aldehyde. **(b)** Iodoacetate is a potent inhibitor of glyceraldehyde-3-phosphate dehydrogenase because it forms a covalent derivative of the essential —SH group of the enzyme active site, rendering it inactive.

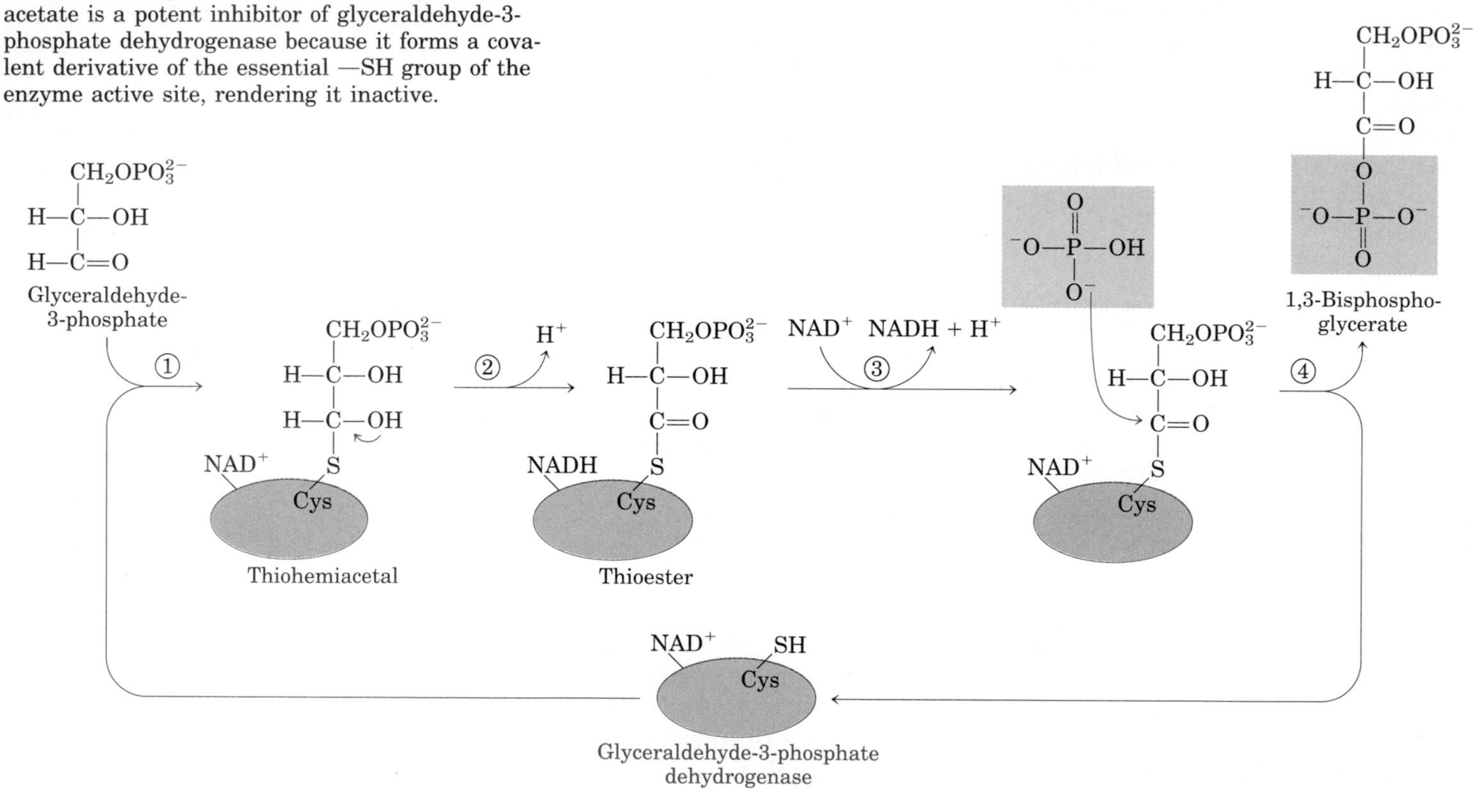

(a)

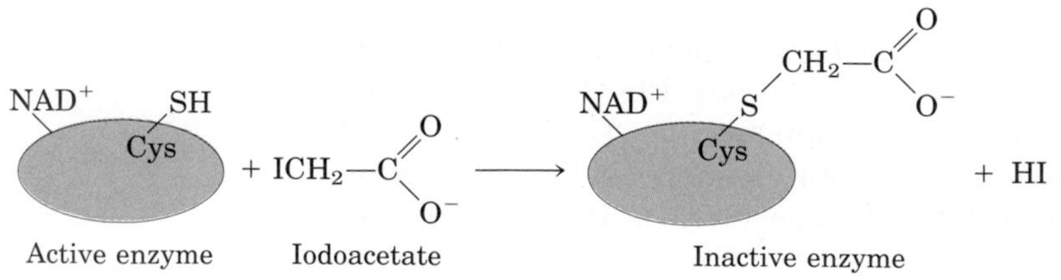

(b)

⑦ Transfer of Phosphate from 1,3-Bisphosphoglycerate to ADP The enzyme **phosphoglycerate kinase** transfers the high-energy phosphate group from the carboxyl group of 1,3-bisphosphoglycerate to ADP, forming ATP and **3-phosphoglycerate:**

| 1,3-Bisphosphoglycerate | ADP | 3-Phosphoglycerate | ATP |

$$\Delta G^{\circ\prime} = -18.5 \text{ kJ/mol}$$

This and the preceding reaction of glycolysis together constitute an energy-coupling process. In these two reactions (steps ⑥ and ⑦), 1,3-bisphosphoglycerate is the common intermediate; it is formed in the first reaction (which is endergonic), and its acyl phosphate group is transferred to ADP to form ATP in the second reaction (which is strongly exergonic). The sum of these two sequential reactions is

Glyceraldehyde-3-phosphate + ADP + P_i + NAD$^+$ ⇌
$$\text{3-phosphoglycerate + ATP + NADH + H}^+$$
$$\Delta G^{\circ\prime} = -12.5 \text{ kJ/mol}$$

Thus the overall reaction is exergonic.

Recall from Chapter 13 that the actual free-energy change, ΔG, is determined by the standard free-energy change, $\Delta G^{\circ\prime}$, and the mass-action ratio, which is the ratio [products]/[reactants] (See Eqn 13–3, p. 371). For the first of these two reactions (step ⑥)

$$\Delta G = \Delta G^{\circ\prime} + RT \ln \frac{[\text{1,3-bisphosphoglycerate}][\text{NADH}]}{[\text{glyceraldehyde-3-phosphate}][P_i][\text{NAD}^+]}$$

Notice that [H$^+$] is not included in the mass-action ratio for this reaction. In biochemical calculations [H$^+$] is assumed to be a constant (10^{-7} M), and this constant is included in the definition of $\Delta G^{\circ\prime}$ (Chapter 13).

The second reaction (step ⑦), by consuming the 1,3-bisphosphoglycerate produced in the first, reduces the concentration of 1,3-bisphosphoglycerate and thereby reduces the mass-action ratio for the overall process. When this ratio is less than 1.0, its natural logarithm has a negative sign. If the mass-action ratio is very small, the contribution of the logarithmic term can make ΔG strongly negative. This is simply another way of showing that the two reactions are coupled through a shared intermediate.

The outcome of these two coupled reactions, both reversible under cellular conditions, is that the energy released on oxidation of an aldehyde to a carboxylate group is conserved by the coupled formation of ATP from ADP and P_i. The formation of ATP by phosphate group transfer from a substrate such as 1,3-bisphosphoglycerate is referred to as a **substrate-level phosphorylation.** We shall later contrast substrate-level phosphorylation with respiration-linked phosphorylation (oxidative phosphorylation), which occurs in mitochondria.

⑧ *Conversion of 3-Phosphoglycerate to 2-Phosphoglycerate* The enzyme **phosphoglycerate mutase** catalyzes a reversible shift of the phosphate group between C-2 and C-3 of glycerate. Mg^{2+} is essential for this reaction:

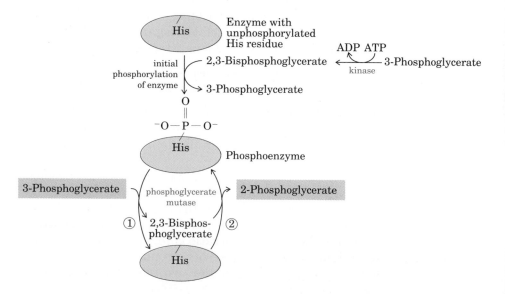

3-Phosphoglycerate   2-Phosphoglycerate

$$\Delta G^{\circ\prime} = 4.4 \text{ kJ/mol}$$

The reaction occurs in two steps (Fig. 14–6). A phosphate group initially attached to a His residue in the active site of the enzyme is transferred to the hydroxyl group at C-2 of 3-phosphoglycerate, forming 2,3-bisphosphoglycerate. The phosphate at C-3 of 2,3-bisphosphoglycerate is then transferred to the same His residue of the enzyme, producing 2-phosphoglycerate and regenerating the phosphorylated enzyme. Because the enzyme is initially phosphorylated by phosphate transfer from 2,3-bisphosphoglycerate, this compound functions as a cofactor; it is required in small quantities to initiate the catalytic cycle, and is continuously regenerated by that cycle.

Figure 14–6 Mechanism of the phosphoglycerate mutase reaction. The enzyme is initially phosphorylated on a His residue by transfer of a phosphate group from 2,3-bisphosphoglycerate. In step ① of the catalytic reaction, the phosphoenzyme transfers its phosphate group to 3-phosphoglycerate, forming 2,3-bisphosphoglycerate. In step ② the phosphate group at C-3 of 2,3-bisphosphoglycerate is transferred to the same His residue on the enzyme, producing 2-phosphoglycerate and regenerating the phosphoenzyme. The 2,3-bisphosphoglycerate required initially to phosphorylate the enzyme is formed from 3-phosphoglycerate by a specific ATP-dependent kinase; it is then regenerated in step ① of each catalytic cycle.

Essentially the same mechanism is employed by the enzyme phosphoglucomutase, described below, in the conversion of glucose-1-phosphate into glucose-6-phosphate. In that reaction, glucose-1,6-bisphosphate serves as the essential cofactor. The general name **mutase** is often given to enzymes that catalyze the transfer of a functional group from one position to another on the same molecule. Mutases are a subclass of **isomerases,** enzymes that interconvert stereoisomers or structural or positional isomers (see Table 8–3).

⑨ Dehydration of 2-Phosphoglycerate to Phosphoenolpyruvate The second glycolytic reaction that generates a compound with high phosphate group transfer potential is catalyzed by **enolase.** This enzyme promotes reversible removal of a molecule of water from 2-phosphoglycerate to yield **phosphoenolpyruvate:**

2-Phosphoglycerate   Phosphoenolpyruvate $\Delta G^{\circ\prime} = 7.5$ kJ/mol

Despite the relatively small standard free-energy change in this reaction, there is a very large difference in the standard free energy of hydrolysis of the phosphate groups of the reactant and product. That of 2-phosphoglycerate (a low-energy phosphate compound) is -17.6 kJ/mol and that of phosphoenolpyruvate (a super high-energy phosphate compound) is -61.9 kJ/mol (see Fig. 13–3 and Table 13–6). Although 2-phosphoglycerate and phosphoenolpyruvate contain nearly the same *total* amount of energy, the loss of the water molecule from 2-phosphoglycerate causes a redistribution of energy within the molecule; the standard free-energy change accompanying hydrolysis of the phosphate group is much greater for phosphoenolpyruvate than for 2-phosphoglycerate.

⑩ Transfer of the Phosphate Group from Phosphoenolpyruvate to ADP The last step in glycolysis is the transfer of the phosphate group from phosphoenolpyruvate to ADP, catalyzed by **pyruvate kinase:**

Phosphoenolpyruvate   ADP   Pyruvate $\Delta G^{\circ\prime} = -31.4$ kJ/mol

ATP

In this reaction, a substrate-level phosphorylation, the product **pyruvate** first appears in its enol form. However, the enol form tautomerizes rapidly and nonenzymatically to yield the keto form of pyruvate, the form that predominates at pH 7. The overall reaction has a large, negative standard free-energy change, due in large part to the spontaneous conversion of the enol form of pyruvate into the keto form (see Fig. 13–3). The $\Delta G^{\circ\prime}$ of phosphoenolpyruvate hydrolysis is -61.9 kJ/mol; about half of this energy is conserved in the formation of the phosphoric acid anhydride bond of ATP ($\Delta G^{\circ\prime} = -30.5$ kJ/mol) and the rest (-31.4 kJ/mol) constitutes a large driving force pushing the reaction toward ATP synthesis. The pyruvate kinase reaction is essentially irreversible under intracellular conditions. Pyruvate kinase requires K^+ and either Mg^{2+} or Mn^{2+}. It is an important site of regulation, as described later.

Pyruvate   Pyruvate
(enol form)   (keto form)

The Overall Balance Sheet Shows a Net Gain of ATP

We can now construct a balance sheet for glycolysis to account for (1) the fate of the carbon skeleton of glucose, (2) the input of P_i and ADP and the output of ATP, and (3) the pathway of electrons in the oxidation–reduction reactions. The left-hand side of the following equation shows all the inputs of ATP, NAD^+, ADP, and P_i (consult Fig. 14–2), and the right-hand side shows all the outputs (keep in mind that each molecule of glucose yields two molecules of glyceraldehyde-3-phosphate):

$$\text{Glucose} + 2\text{ATP} + 2\text{NAD}^+ + 4\text{ADP} + 2P_i \longrightarrow$$
$$2 \text{ pyruvate} + 2\text{ADP} + 2\text{NADH} + 2\text{H}^+ + 4\text{ATP} + 2\text{H}_2\text{O}$$

If we cancel out common terms on both sides of the equation, we get the overall equation for glycolysis under aerobic conditions:

$$\text{Glucose} + 2\text{NAD}^+ + 2\text{ADP} + 2P_i \longrightarrow$$
$$2 \text{ pyruvate} + 2\text{NADH} + 2\text{H}^+ + 2\text{ATP} + 2\text{H}_2\text{O}$$

The two molecules of NADH formed by glycolysis in the cytosol are, under aerobic conditions, reoxidized to NAD^+ by transfer of their electrons to the respiratory chain, which in eukaryotic cells is located in the mitochondria. Here these electrons are ultimately passed to O_2:

$$2\text{NADH} + 2\text{H}^+ + O_2 \longrightarrow 2\text{NAD}^+ + 2\text{H}_2\text{O}$$

Electron transfer from NADH to O_2 in mitochondria provides the energy for synthesis of ATP by respiration-linked phosphorylation (Chapter 18).

In the overall process, one molecule of glucose is converted into two molecules of pyruvate (the pathway of carbon). Two molecules of ADP and two of P_i are converted into two molecules of ATP (the pathway of phosphate groups). Four electrons (two hydride ions) are transferred from two molecules of glyceraldehyde-3-phosphate to two of NAD^+ (the pathway of electrons).

Intermediates Are Channeled between Glycolytic Enzymes

Although the enzymes of glycolysis usually are described as soluble components of the cytosol, there is growing evidence that within the cell these enzymes exist as multienzyme complexes. The classic approach of enzymology—the purification of individual proteins from extracts of broken cells—was applied with great success to the enzymes of glycolysis; we have noted that each of the enzymes has been purified to homogeneity. However, the first casualty of cell breakage is higher-level organization within a cell—the noncovalent and reversible interaction of one protein with another, or of an enzyme with some structural component such as a membrane, microtubule, or microfilament. When cells are broken open, their contents, including enzymes, undergo dilution by a factor of a hundred or a thousand (Fig. 14–7).

When the purified enzymes of glycolysis are combined in vitro at relatively high concentrations, they form specific, functional aggregates, which may reflect their true state inside cells. Several types of evidence suggest that such complexes act in cells to ensure efficient passage of the product of one enzyme to the next enzyme in the pathway, for which that product serves as substrate. Kinetic evidence for the **channeling** of 1,3-bisphosphoglycerate from glyceraldehyde-3-

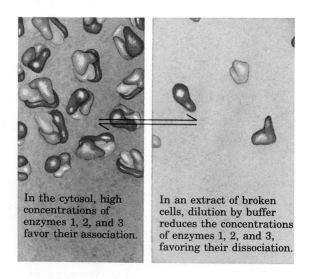

In the cytosol, high concentrations of enzymes 1, 2, and 3 favor their association.

In an extract of broken cells, dilution by buffer reduces the concentrations of enzymes 1, 2, and 3, favoring their dissociation.

Figure 14–7 Dilution of a solution containing a noncovalent protein complex favors dissociation of the complex into its constituents.

phosphate dehydrogenase to phosphoglycerate kinase without entering solution (Fig. 14–8) is corroborated by physical evidence that these two enzymes form stable, noncovalent complexes. There is similar evidence for channeling of intermediates between other glycolytic enzymes, such as glyceraldeyde-3-phosphate from aldolase to glyceraldehyde-3-phosphate dehydrogenase.

Furthermore, certain glycolytic enzymes form specific noncovalent complexes with structural components of the cell, which may serve to organize reaction sequences and assure efficient transfer of intermediates between cellular compartments. Certain glycolytic enzymes bind to microtubules or to actin microfilaments (see Fig. 2–18), bringing those enzymes into close association and holding them in a specific region of the cytoplasm. Hexokinase binds specifically to the outer membrane of mitochondria. This association may allow ATP produced within the mitochondrion to move directly to the catalytic site of hexokinase without entering, and being diluted by, the cytosol. There is strong evidence for substrate channeling through multienzyme complexes in other metabolic pathways, and it seems likely that many enzymes now thought of as "soluble" actually function in the cell as highly organized complexes that channel intermediates.

Glycolysis Is under Tight Regulation

During his studies of the fermentation of glucose by yeast, Louis Pasteur discovered that both the rate and the total amount of glucose consumption were many times greater under anaerobic conditions than under aerobic conditions. Later studies of muscle showed the same large difference in the rate of glycolysis under anaerobic and aerobic conditions. The biochemical basis of this "Pasteur effect" is now clear. The ATP yield from glycolysis under anaerobic conditions (2 ATP per molecule of glucose) is much smaller than that from the complete oxidation of glucose to CO_2 under aerobic conditions (36 or 38 ATP per glucose molecule; see Chapter 18). About 18 times as much glucose must therefore be consumed anaerobically as aerobically to yield the same amount of ATP.

The flux of glucose through the glycolytic pathway is regulated to achieve constant ATP levels (as well as adequate supplies of glycolytic intermediates that serve biosynthetic roles). The required adjustment in the rate of glycolysis is achieved by the regulation of two glycolytic enzymes: phosphofructokinase-1 and pyruvate kinase. Both enzymes are regulated allosterically by second-to-second fluctuations in the concentration of certain key metabolites that reflect the cellular balance between ATP production and consumption. We return to a more detailed discussion of the regulation of glycolysis later in the chapter.

Fates of Pyruvate under Aerobic and Anaerobic Conditions

Pyruvate, the product of glycolysis, represents an important junction point in carbohydrate catabolism (Fig. 14–3). Under aerobic conditions pyruvate is oxidized to acetate, which enters the citric acid cycle (Chapter 15) and is oxidized to CO_2 and H_2O. The NADH formed by the

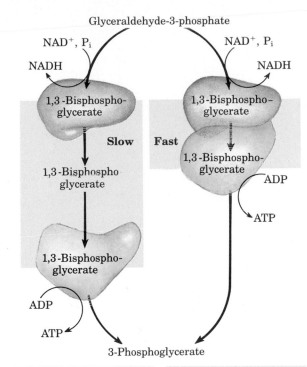

| Sequential action of two separate enzymes: the product of the first enzyme (1,3-bisphosphoglycerate) diffuses to the second enzyme. | Substrate channeling through a functional complex of two enzymes: the intermediate (1,3-bisphosphoglycerate) is never released to the solvent. |

Figure 14–8 Channeling of a substrate between two enzymes in the glycolytic pathway. When glyceraldehyde-3-phosphate dehydrogenase (blue) and 3-phosphoglycerate kinase (yellow) are combined in vitro, they catalyze the two-step conversion of glyceraldehyde-3-phosphate to 3-phosphoglycerate (Figs. 14–5, 14–6) at a rate greater than the rate at which the first step is catalyzed in the presence of the first enzyme only. Apparently the transfer of 1,3-bisphosphoglycerate from the surface of the dehydrogenase to that of the kinase is faster than the dissociation of 1,3-bisphosphoglycerate from the dehydrogenase into the surrounding medium (which occurs in the absence of the kinase). Physical studies show that the two enzymes can form a stable complex, as is required for substrate channeling between them.

dehydrogenation of glyceraldehyde-3-phosphate is reoxidized to NAD^+ by passage of its electrons to O_2 in the process of mitochondrial respiration (Chapter 18). However, under anaerobic conditions (as in very active skeletal muscles, in submerged plants, or in lactic acid bacteria, for example), NADH generated by glycolysis cannot be reoxidized by O_2. Failure to regenerate NAD^+ would leave the cell with no electron acceptor for the oxidation of glyceraldehyde-3-phosphate, and the energy-yielding reactions of glycolysis would stop. NAD^+ must therefore be regenerated by some other reaction.

The earliest cells to arise during evolution lived in an atmosphere almost devoid of oxygen and had to develop strategies for carrying out glycolysis under anaerobic conditions. Most modern organisms have retained the ability to continually regenerate NAD^+ during anaerobic glycolysis by transferring electrons from NADH to form a reduced end product such as lactate or ethanol.

Pyruvate Is the Terminal Electron Acceptor in Lactic Acid Fermentation

When animal tissues cannot be supplied with sufficient oxygen to support aerobic oxidation of the pyruvate and NADH produced in glycolysis, NAD^+ is regenerated from NADH by the reduction of pyruvate to **lactate.** Certain other tissues and cell types (retina, brain, erythrocytes) also produce lactate from glucose under aerobic conditions; lactate is a major product of erythrocyte metabolism. The reduction of pyruvate is catalyzed by **lactate dehydrogenase,** which forms the L isomer of lactic acid (lactate at pH 7). The overall equilibrium of this reaction strongly favors lactate formation, as shown by the large negative standard free-energy change. In glycolysis, dehydrogenation of the two molecules of glyceraldehyde-3-phosphate derived from each molecule of glucose converts two molecules of NAD^+ to two of NADH. Because the reduction of two molecules of pyruvate to two of lactate regenerates two molecules of NAD^+, the overall process is balanced and can continue indefinitely: one molecule of glucose is converted to two of lactate, with the generation of two ATP molecules from one of glucose, and NAD^+ and NADH are continuously interconverted with no net gain or loss in the amount of either.

Although there are two oxidation–reduction steps as glucose is converted into lactate, there is no net change in the oxidation state of carbon; in glucose ($C_6H_{12}O_6$) and lactic acid ($C_3H_6O_3$), the H : C ratio is the same. Nevertheless, some of the energy of the glucose molecule has been extracted by its conversion to lactate, enough to give a net yield of two molecules of ATP for every one of glucose consumed. The lactate formed by active muscles of vertebrate animals can be recycled; it is carried in the blood to the liver where it is converted into glucose during the recovery from strenuous muscle activity (Box 14–1).

Many microorganisms ferment glucose and other hexoses to lactate. Certain lactobacilli and streptococci, for example, ferment the lactose in milk to lactic acid. The dissociation of lactic acid to lactate and H^+ in the fermentation mixture lowers the pH, denaturing casein and other milk proteins and causing them to precipitate. Under the correct conditions, the resultant curdling produces cheese or yogurt, depending on which microorganism is involved.

Pyruvate

lactate dehydrogenase

NADH + H⁺

NAD⁺

Lactate

$\Delta G^{\circ\prime} = -25.1$ kJ/mol

BOX 14–1   Glycolysis without Oxygen: Alligators and Coelacanths

Most vertebrates are essentially aerobic organisms; they first convert glucose into pyruvate by glycolysis and then oxidize the pyruvate completely to CO_2 and H_2O using molecular oxygen. Anaerobic catabolism of glucose (fermentation to lactate) occurs in most vertebrates, including human beings, during short bursts of extreme muscular activity, for example in a 100 m sprint, during which oxygen cannot be carried to the muscles fast enough to oxidize pyruvate for generating ATP. Instead, the muscles use their stored glycogen as fuel to generate ATP by fermentation, with lactate as the end product. In a sprint, the lactate in the blood builds up to high concentrations. It is slowly converted back into glucose by gluconeogenesis in the liver in the subsequent rest or recovery period, during which oxygen is consumed at a gradually diminishing rate until the breathing rate returns to normal. The excess oxygen consumed in the recovery period represents the repayment of the **oxygen debt.** This is the amount of oxygen required to supply ATP for gluconeogenesis during recovery respiration, in order to regenerate the glycogen "borrowed" from liver and muscle to carry out intense muscular activity in the sprint. The cycle of reactions that includes glucose conversion to lactate in muscle and lactate conversion to glucose in liver is called the **Cori cycle,** for Carl and Gerty Cori, whose studies in the 1930s and 1940s clarified the pathway and its role.

The circulatory systems of most small vertebrates can carry oxygen to their muscles fast enough to avoid having to use muscle glycogen anaerobically. For example, migrating birds often fly great distances at high speeds without rest and without incurring an oxygen debt. Many running animals of moderate size also have an essentially aerobic metabolism in their skeletal muscle. However in larger animals, including humans, the circulatory system cannot completely sustain aerobic metabolism in skeletal muscles during long bursts of muscular activity. Such animals generally are slow-moving under normal circumstances and engage in intense muscular activity only in the gravest emergencies, because such bursts of activity require long recovery periods to repay the oxygen debt.

Alligators and crocodiles, for example, are normally sluggish and torpid. Yet when provoked these animals are capable of lightning-fast charges and dangerous lashings of their powerful tails. Such intense bursts of activity are short and must be followed by long periods of recovery. The fast emergency movements require lactate fermentation to generate ATP in skeletal muscles. Because

the stores of muscle glycogen are not large, they are rapidly expended in intense muscular activity. Moreover, in such bursts of action, lactate reaches very high concentration in muscles and extracellular fluid. Whereas a trained athlete can recover from a 100 m sprint in 30 min or less, an alligator may require many hours of rest and extra oxygen consumption to clear the excess lactate from its blood and regenerate muscle glycogen.

Other large animals, such as the elephant and rhinoceros, have similar metabolic problems, as do diving mammals such as whales and seals. Dinosaurs and other huge, now-extinct animals probably had to depend on lactate fermentation to supply energy for muscular activity, followed by very long recovery periods during which they were vulnerable to attack by smaller predators better able to use oxygen and thus better adapted to continuous, sustained muscular activity.

Deep-sea explorations have revealed many species of marine life at great ocean depths, where the oxygen concentration is near zero. For example, the primitive coelacanth, a large fish recovered from depths of 4,000 m or more off the coast of South Africa, has been found to have an essentially anaerobic metabolism in virtually all its tissues. It converts carbohydrates by anaerobic mechanisms into lactate and other products, most of which must be excreted. Some marine vertebrates ferment glucose to ethanol and CO_2 in order to obtain energy in the form of ATP.

BOX 14–2   **Brewing Beer**

Beer is made by alcohol fermentation of the carbo-hydrates present in cereal grains (seeds) such as barley, but these carbohydrates, largely polysac-charides, are not available to the glycolytic en-zymes in yeast cells until they have been degraded to disaccharides and monosaccharides. The barley must first undergo a process called malting. The cereal seeds are allowed to germinate until they form the hydrolytic enzymes required to break down the polysaccharides of their cell walls and the starch and other polysaccharide food reserves. Germination is then stopped by controlled heating, before further growth of the seedlings occurs. The product is malt, which now contains enzymes such as α-amylase and maltase, capable of breaking down starch to maltose, glucose, and other simple sugars. The malt also contains enzymes specific for the β linkages of cellulose and other cell-wall poly-saccharides of the barley husks, which must be broken down in order to allow α-amylase to act on the starch within the seeds.

In the next step the brewer prepares the wort, the nutrient medium required for the subsequent fermentation by yeast cells. The malt is mixed with water and then mashed or crushed. This allows the enzymes formed in the malting process to act on the cereal polysaccharides to form maltose, glu-cose, and other simple sugars, which are soluble in the aqueous medium. The remaining cell matter is then separated, and the liquid wort is boiled with hops, to give flavor. The wort is cooled and then aerated.

Now the yeast cells are added. In the aerobic wort the yeast grows and reproduces very rapidly, using energy obtained from some of the sugars in the wort. In this phase no alcohol is formed be-cause the yeast, being amply supplied with oxygen, oxidizes the pyruvate formed by glycolysis to CO_2 and H_2O via the citric acid cycle. When all the dis-solved oxygen in the vat of wort has been con-sumed, the yeast cells switch to anaerobic metabo-lism of the sugar. From this point on, the yeast ferments the sugars of the wort into ethanol and CO_2. The fermentation process is controlled in part by the concentration of the ethanol formed, by the pH, and by the amount of remaining sugar. After the fermentation has been stopped, the cells are removed, and the "raw" beer is ready for final processing.

In the final steps of brewing, the amount of foam or head on the beer, which results from dis-solved proteins, is adjusted. Normally this is con-trolled by the action of proteolytic enzymes that appear in the malting process. If these enzymes act on the beer proteins too long, the beer will have very little head and will be flat; if they do not act long enough, the beer will not be clear when it is cold. Sometimes proteolytic enzymes from other sources are added to control the head.

Ethanol Is the Reduced Product in Alcohol Fermentation

Yeast and other microorganisms ferment glucose to ethanol and CO_2, rather than to lactate. Glucose is converted to pyruvate by glycolysis, and the pyruvate is converted to ethanol and CO_2 in a two-step pro-cess. In the first step, pyruvate undergoes decarboxylation in the irre-versible reaction catalyzed by **pyruvate decarboxylase.** This reac-tion is a simple decarboxylation and does not involve the net oxidation of pyruvate. Pyruvate decarboxylase requires Mg^{2+} and has a tightly bound coenzyme, thiamine pyrophosphate, discussed in more detail below.

In the second step, acetaldehyde is reduced to ethanol, with NADH derived from glyceraldehyde-3-phosphate dehydrogenation furnishing the reducing power, through the action of **alcohol dehydrogenase.** Ethanol and CO_2, instead of lactate, are thus the end products of alco-hol fermentation. The overall equation of alcohol fermentation is

$$\text{Glucose} + 2ADP + 2P_i \longrightarrow 2 \text{ ethanol} + 2CO_2 + 2ATP + 2H_2O$$

As in lactic acid fermentation, there is no net change in the ratio of hydrogen to carbon atoms when glucose (H : C ratio = 12/6 = 2) is fermented to two ethanol and two CO_2 (combined $H:C$ ratio $= 12/6 = 2$). In all fermentations, the $H:C$ ratio of the reactants and products remains the same.

Pyruvate decarboxylase is characteristically present in brewer's and baker's yeast and in all other organisms that promote alcohol fermentation, including some plants. The CO_2 produced by pyruvate decarboxylation in brewer's yeast is responsible for the characteristic carbonation of champagne. The ancient art of brewing beer involves a number of enzymatic processes in addition to the reactions of alcohol fermentation (Box 14–2). In baking, CO_2 released by pyruvate decarboxylase when yeast is mixed with a fermentable sugar causes dough to rise. The enzyme is absent in the tissues of vertebrate animals and in other organisms, such as the lactic acid bacteria, that carry out lactic acid fermentation.

Alcohol dehydrogenase is present in many organisms that metabolize alcohol, including humans. In human liver it brings about the *oxidation* of ethanol, either ingested or produced by intestinal microorganisms, with the concomitant reduction of NAD^+ to NADH.

Thiamine Pyrophosphate Carries "Active Aldehyde" Groups

The pyruvate decarboxylase reaction in alcohol fermentation represents our first encounter with **thiamine pyrophosphate** (TPP) (Fig. 14–9), a coenzyme derived from vitamin B_1. The absence of vitamin B_1 in the human diet leads to the condition known as beriberi, characterized by an accumulation of body fluids (swelling), pain, paralysis, and ultimately death.

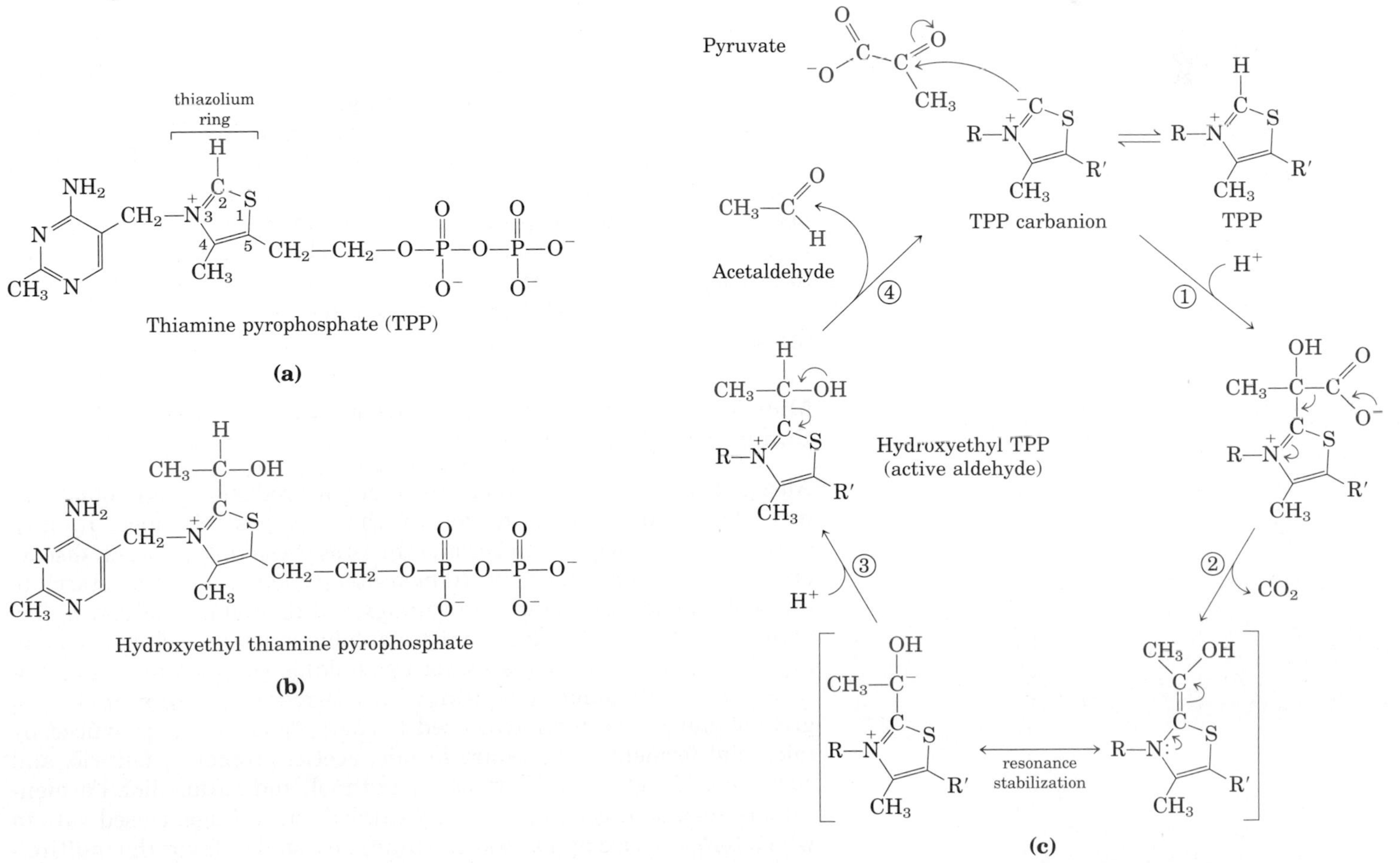

Figure 14–9 (a) Thiamine pyrophosphate (TPP), the coenzyme form of vitamin B_1 (thiamin). The reactive carbon atom in the thiazolium ring is shown in red. In the reaction catalyzed by pyruvate decarboxylase, two of the three carbons of pyruvate are carried transiently on TPP in the form of hydroxyethyl thiamine pyrophosphate **(b).** This "active acetaldehyde" group (in red) is subsequently released as acetaldehyde. **(c)** The cleavage of a carbon–carbon bond often leaves behind a free electron pair or carbanion on one of the products. The strong tendency of a carbanion to form a new bond generally renders a carbanion intermediate unstable. The thiazolium ring of TPP stabilizes carbanion intermediates by providing an electrophilic (electron-deficient) structure into which the carbanion electrons can be delocalized by resonance. Structures with this property are often called "electron sinks," and they play a role in many biochemical reactions. This principle is illustrated here for the reaction catalyzed by pyruvate decarboxylase. In step ①, the TPP carbanion acts as a nucleophile, adding to the carbonyl group of pyruvate. In step ②, a carbanion is formed following decarboxylation. The thiazolium ring acts as an electron sink, stabilizing the carbanion by resonance. After protonation (step ③), the reaction product acetaldehyde is released (step ④).

Thiamine pyrophosphate (TPP)

(a)

Hydroxyethyl thiamine pyrophosphate

(b)

Pyruvate

TPP carbanion

TPP

Acetaldehyde

Hydroxyethyl TPP (active aldehyde)

resonance stabilization

(c)

Table 14–1 Some reactions in which thiamine pyrophosphate is an essential cofactor

Enzyme	Pathway	Bond cleaved	Bond formed
Pyruvate decarboxylase	Alcohol fermentation	$R^1-\overset{\displaystyle O}{\underset{}{C}}-\overset{\displaystyle O}{\underset{O^-}{C}}$	$R^1-\overset{\displaystyle O}{\underset{H}{C}}$
Pyruvate dehydrogenase	Synthesis of acetyl-CoA		
α-Ketoglutarate dehydrogenase	Citric acid cycle	$R^2-\overset{\displaystyle O}{\underset{}{C}}-\overset{\displaystyle O}{\underset{O^-}{C}}$	$R^2-\overset{\displaystyle O}{\underset{S\text{-CoA}}{C}}$
Transketolase	Carbon-fixation reactions of photosynthesis	$R^3-\overset{\displaystyle O}{\underset{}{C}}-\overset{OH}{\underset{H}{C}}-R^4$	$R^3-\overset{\displaystyle O}{\underset{}{C}}-\overset{OH}{\underset{H}{C}}-R^5$
Acetolactate synthetase	Valine, leucine biosynthesis	$R^6-\overset{\displaystyle O}{\underset{}{C}}-\overset{\displaystyle O}{\underset{O^-}{C}}$	$R^6-\overset{\displaystyle O}{\underset{}{C}}-\overset{OH}{\underset{C}{C}}-$, $\overset{O}{\underset{O^-}{C}}$

Thiamine pyrophosphate plays an important role in the cleavage of bonds adjacent to a carbonyl group (such as the decarboxylation of α-keto acids) and in chemical rearrangements involving transfer of an activated aldehyde group from one carbon atom to another (Table 14–1). The functional part of thiamine pyrophosphate is the thiazolium ring (Fig. 14–9a). The proton at C-2 of the ring is relatively acidic, and loss of this acidic proton produces a carbanion that is the active species in TPP-dependent reactions (Fig. 14–9c). This carbanion readily adds to carbonyl groups, and the thiazolium ring is thereby positioned to act as an "electron sink" that greatly facilitates reactions such as the decarboxylation catalyzed by pyruvate decarboxylase.

Microbial Fermentations Yield Other End Products of Commercial Value

Although lactate and ethanol are common products of microbial fermentations, they are by no means the only possible ones. In 1910 Chaim Weizmann (later to become the first president of Israel) discovered that a bacterium, *Clostridium acetobutyricum,* ferments starch to butanol and acetone. This discovery opened the field of industrial fermentations, in which some readily available material rich in carbohydrate (corn starch or molasses, for example) is supplied to a pure culture of a specific microorganism, which ferments it into a product of greater value. The methanol used to make "gasohol" is produced by microbial fermentation, as are formic, acetic, propionic, butyric, and succinic acids, glycerol, isopropanol, butanol, and butanediol. Fermentations such as these are generally carried out in huge, closed vats in which temperature and access to air are adjusted to favor the multiplication of the desired microorganism and to exclude contaminating organisms (Fig. 14–10). The beauty of industrial fermentations is that complicated, multistep chemical transformations are carried out in high yields and with few side products by chemical factories that reproduce themselves—microbial cells. In some cases it is possible to immobilize the cells in an inert support, to pass the starting material continuously through a bed of immobilized cells, and to collect the desired product in the effluent: an engineer's dream!

Feeder Pathways for Glycolysis

In addition to glucose, many other carbohydrates ultimately enter the glycolytic pathway to undergo energy-yielding degradation. The most significant are the storage polysaccharides glycogen and starch, the disaccharides maltose, lactose, trehalose, and sucrose, and the monosaccharides fructose, mannose, and galactose. We shall now consider the pathways by which these carbohydrates can enter glycolysis.

Glycogen and Starch Are Degraded by Phosphorolysis

The glucose units of the outer branches of glycogen and starch gain entrance into the glycolytic pathway through the sequential action of two enzymes: glycogen phosphorylase (or the similar starch phosphorylase in plants) and phosphoglucomutase. Glycogen phosphorylase catalyzes the reaction in which an ($\alpha 1 \rightarrow 4$) glycosidic linkage joining two glucose residues in glycogen undergoes attack by inorganic phosphate, removing the terminal glucose residue as **α-D-glucose-1-phosphate** (Fig. 14–11). This *phosphorolysis* reaction that occurs during intracellular mobilization of glycogen stores is different from the *hydrolysis* of glycosidic bonds by amylase during intestinal degradation of glycogen or starch; in phosphorolysis, some of the energy of the glycosidic bond is preserved in the formation of the phosphate ester, glucose-1-phosphate.

Figure 14–10 An industrial-scale fermentation. Microorganisms are cultured in a sterilizable vessel containing thousands of liters of growth medium made up of an inexpensive carbon-and-energy source under carefully controlled conditions, including low oxygen concentration and constant temperature. After centrifugal separation of the cells from the growth medium, the valuable products of the fermentation are recovered from the cells or the supernatant fluid.

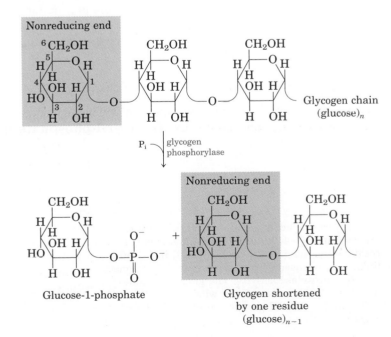

Glycogen chain
(glucose)$_n$

P$_i$ — glycogen phosphorylase

Glucose-1-phosphate   +   Nonreducing end

Glycogen shortened
by one residue
(glucose)$_{n-1}$

Figure 14–11 Removal of a terminal glucose residue from the nonreducing end of a glycogen chain by the action of glycogen phosphorylase. This process is repetitive, removing successive glucose residues until it reaches the fourth glucose unit from a branch point (see Fig. 14–12). Amylopectin is degraded in a similar fashion by starch phosphorylase.

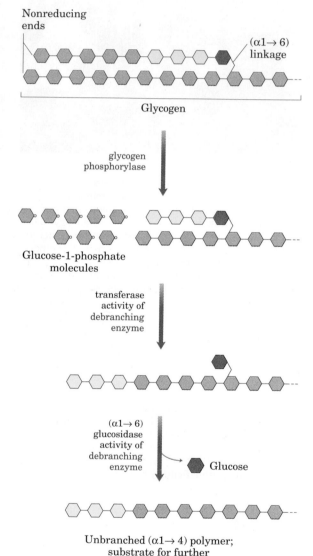

Pyridoxal phosphate

Nonreducing ends

$(\alpha 1 \rightarrow 6)$ linkage

Glycogen

glycogen phosphorylase

Glucose-1-phosphate molecules

transferase activity of debranching enzyme

$(\alpha 1 \rightarrow 6)$ glucosidase activity of debranching enzyme

Glucose

Unbranched $(\alpha 1 \rightarrow 4)$ polymer; substrate for further phosphorylase action

Pyridoxal phosphate is an essential cofactor in the glycogen phosphorylase reaction; its phosphate group acts as a general acid catalyst, promoting attack by P_i on the glycosidic bond. A quite different role of pyridoxal phosphate as a cofactor in amino acid metabolism will be described in detail in Chapter 17.

Glycogen phosphorylase (or starch phosphorylase) acts repetitively on the nonreducing ends of glycogen (or amylopectin) branches until it reaches a point four glucose residues away from an $(\alpha 1 \rightarrow 6)$ branch point (see Fig. 11–15). Here the action of glycogen or starch phosphorylase stops. Further degradation can occur only after the action of a **"debranching enzyme," oligo $(\alpha 1 \rightarrow 6)$ to $(\alpha 1 \rightarrow 4)$ glucantransferase,** which catalyzes two successive reactions that remove branches (Fig. 14–12).

Glucose-1-phosphate, the end product of the glycogen and starch phosphorylase reactions, is converted into glucose-6-phosphate by **phosphoglucomutase,** which catalyzes the reversible reaction

$$\text{Glucose-1-phosphate} \rightleftharpoons \text{glucose-6-phosphate}$$

Phosphoglucomutase requires as a cofactor **glucose-1,6-bisphosphate;** its role is analogous to that of 2,3-bisphosphoglycerate in the reaction catalyzed by phosphoglycerate mutase (Fig. 14–6). Phosphoglucomutase, like phosphoglycerate mutase, cycles between a phosphorylated and nonphosphorylated form. In phosphoglucomutase, however, it is the hydroxyl group of a Ser residue in the active site that is transiently phosphorylated in the catalytic cycle.

Other Monosaccharides Can Enter the Glycolytic Pathway

In most organisms, hexoses other than glucose can undergo glycolysis after conversion to a phosphorylated derivative. D-Fructose, present in free form in many fruits and formed by hydrolysis of sucrose in the small intestine, can be phosphorylated by hexokinase, which acts on a number of different hexoses:

$$\text{Fructose} + \text{ATP} \xrightarrow{Mg^{2+}} \text{fructose-6-phosphate} + \text{ADP}$$

In the muscles and kidney of vertebrates this is a major pathway. In the liver, however, fructose gains entry into glycolysis by a different pathway. The liver enzyme **fructokinase** catalyzes the phosphorylation of fructose, not at C-6, but at C-1:

$$\text{Fructose} + \text{ATP} \xrightarrow{Mg^{2+}} \text{fructose-1-phosphate} + \text{ADP}$$

The fructose-1-phosphate is then cleaved to form glyceraldehyde and dihydroxyacetone phosphate by **fructose-1-phosphate aldolase.**

Figure 14–12 Glycogen breakdown near $(\alpha 1 \rightarrow 6)$ branch points. Following the sequential removal of terminal glucose residues by glycogen phosphorylase (Fig. 14–11), glucose residues near a branch are removed in a two-step process that requires the action of a bifunctional "debranching enzyme." First, the transferase activity of this enzyme shifts a block of three glucose residues from the branch to a nearby nonreducing end, to which they are reattached in $(\alpha 1 \rightarrow 4)$ linkage. Then the single glucose residue remaining at the branch point, in $(\alpha 1 \rightarrow 6)$ linkage, is released as free glucose by the enzyme's $(\alpha 1 \rightarrow 6)$ glucosidase activity. The glucose residues are shown in shorthand form, which omits the —H, —OH, and —CH$_2$OH groups from the pyranose rings.

Fructose-6-phosphate

$\overset{\text{fructose-1-phosphate}}{\underset{\text{aldolase}}{\rightleftharpoons}}$

Dihydroxyacetone phosphate

+

Glyceraldehyde

Dihydroxyacetone phosphate is converted into glyceraldehyde-3-phosphate by the glycolytic enzyme triose phosphate isomerase. Glyceraldehyde is phosphorylated by ATP and **triose kinase** to glyceraldehyde-3-phosphate:

$$\text{Glyceraldehyde} + \text{ATP} \xrightarrow{\text{Mg}^{2+}} \text{glyceraldehyde-3-phosphate} + \text{ADP}$$

Thus both products of fructose hydrolysis enter the glycolytic pathway as glyceraldehyde-3-phosphate.

D-Galactose, derived by hydrolysis of the disaccharide lactose (milk sugar), is first phosphorylated at C-1 at the expense of ATP by the enzyme **galactokinase:**

$$\text{Galactose} + \text{ATP} \longrightarrow \text{galactose-1-phosphate} + \text{ADP}$$

The galactose-1-phosphate is then converted into its epimer at C-4, glucose-1-phosphate, by a set of reactions in which **uridine diphosphate** (UDP) functions as a coenzymelike carrier of hexose groups (Fig. 14–13).

There are several human genetic diseases in which galactose metabolism is affected. In the most common form of **galactosemia,** the enzyme UDP-glucose:galactose-1-phosphate uridylyltransferase (Fig. 14–13) is genetically defective, preventing the overall conversion of galactose into glucose. Other forms of galactosemia result when either galactokinase or UDP-glucose-4-epimerase is genetically defective.

D-Mannose, which arises from the digestion of various polysaccharides and glycoproteins present in foods, can be phosphorylated at C-6 by hexokinase:

$$\text{Mannose} + \text{ATP} \xrightarrow{\text{Mg}^{2+}} \text{mannose-6-phosphate} + \text{ADP}$$

Mannose-6-phosphate is then isomerized by the action of **phosphomannose isomerase,** to yield fructose-6-phosphate, an intermediate of glycolysis.

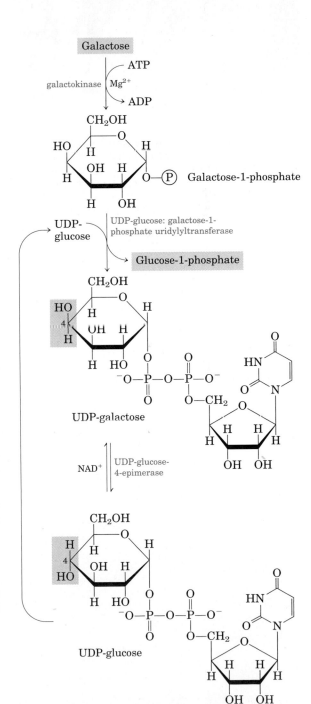

Figure 14–13 Pathway of the conversion of D-galactose into D-glucose. The conversion proceeds through a sugar–nucleotide derivative, UDP-galactose, which is formed when galactose-1-phosphate displaces glucose-1-phosphate from UDP-glucose. UDP-galactose is then converted by UDP-glucose-4-epimerase to UDP-glucose. The UDP-glucose is recycled through another round of the same reaction. The net effect of this cycle is the conversion of galactose-1-phosphate to glucose-1-phosphate; there is no net production or consumption of UDP-galactose or UDP-glucose.

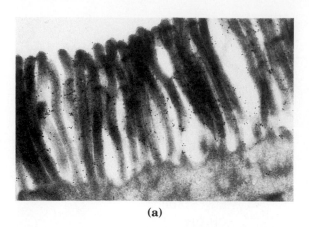

(a)

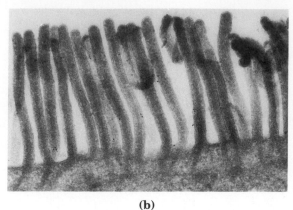

(b)

Figure 14–14 Lactase, a disaccharidase of the intestinal epithelium, can be detected by treating a thin section of intestinal tissue with an antibody that specifically binds to the enzyme. The antibodies are made visible in the electron microscope by attaching to them tiny colloidal particles of gold, which appear as black (electron-dense) dots in electron micrographs. **(a)** Tissue from an adult who has retained high levels of lactase. Microvilli are heavily labeled with antibodies that detect lactase. **(b)** Intestinal microvilli in tissue from an adult with lactose intolerance are much less heavily labeled with antibodies against lactase.

Dietary Disaccharides Are Hydrolyzed to Monosaccharides

Disaccharides cannot directly enter the glycolytic pathway; indeed they cannot enter cells without first being hydrolyzed to monosaccharides extracellularly. In vertebrates, ingested disaccharides must first be hydrolyzed by enzymes attached to the outer surface of the epithelial cells lining the small intestine (Fig. 14–14), to yield their monosaccharide units:

$$\text{Maltose} + H_2O \xrightarrow{\text{maltase}} 2\ \text{D-glucose}$$

$$\text{Lactose} + H_2O \xrightarrow{\text{lactase}} \text{D-galactose} + \text{D-glucose}$$

$$\text{Sucrose} + H_2O \xrightarrow{\text{sucrase}} \text{D-fructose} + \text{D-glucose}$$

$$\text{Trehalose} + H_2O \xrightarrow{\text{trehalase}} 2\ \text{D-glucose}$$

The monosaccharides so formed are transported into the cells lining the intestine, from which they pass into the blood and are carried to the liver. There they are phosphorylated and funneled into the glycolytic sequence as described above.

Lactose intolerance is a condition, common among adults of most human races except Northern Europeans and some Africans, in which the ingestion of milk or other foods containing lactose leads to abdominal cramps and diarrhea. Lactose intolerance is due to the disappearance after childhood of most or all of the lactase activity of the intestinal cells (Fig. 14–14b), so that lactose cannot be completely digested and absorbed. Lactose not absorbed in the small intestine is converted by bacteria in the large intestine into toxic products that cause the symptoms of the condition. In those parts of the world where lactose intolerance is prevalent, milk is simply not used as a food by adults. Milk products digested with lactase are commercially available in some countries as an alternative to excluding milk products from the diet. In certain diseases of humans, several or all of the intestinal disaccharidases are missing because of genetic defects or dietary factors, resulting in digestive disturbances triggered by disaccharides in the diet (Fig. 14–14b). Altering the diet to reduce disaccharide content sometimes alleviates the symptoms of these defects.

Figure 14–15 summarizes the feeder pathways that funnel hexoses, disaccharides, and polysaccharides into the central glycolytic pathway.

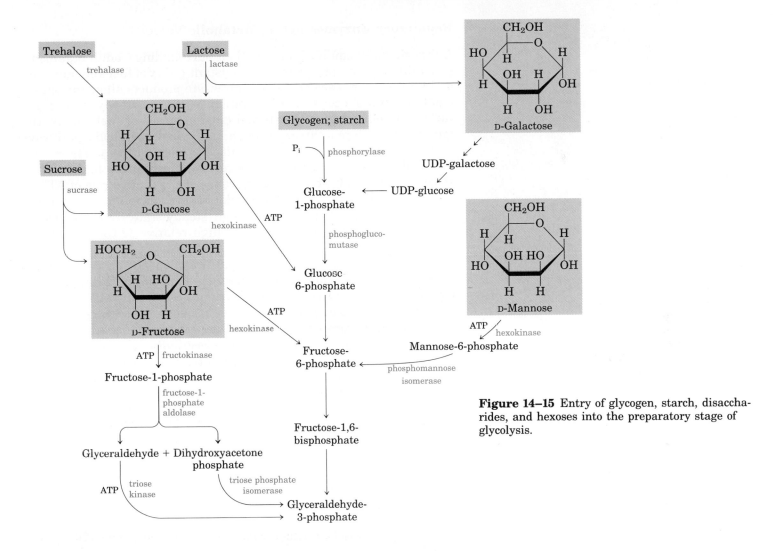

Figure 14–15 Entry of glycogen, starch, disaccharides, and hexoses into the preparatory stage of glycolysis.

Regulation of Carbohydrate Catabolism

Carbohydrate catabolism provides ATP as well as precursors for a variety of biosynthetic processes. It is crucial to a cell to maintain a sufficient concentration of ATP, at a nearly constant level, regardless of which fuel is used to produce ATP and regardless of the rate at which ATP is consumed. An organism that undergoes a change in circumstances, such as increased muscular activity, decreased availability of oxygen, or decreased dietary intake of carbohydrate, must alter its catabolic patterns to change the flow of carbohydrate fuel, whether from stored reserves or from extracellular sources, through glycolysis. These changes in catabolic patterns are accomplished by the regulation of key enzymes in the catabolic pathways. In glycolysis in muscle and liver tissue, four enzymes play a regulatory role: glycogen phosphorylase, hexokinase, phosphofructokinase-1, and pyruvate kinase. Our discussion of the regulation of glycolysis necessarily involves some details of the reciprocally regulated process of glucose synthesis (gluconeogenesis), which is more fully discussed in Chapter 19.

Before describing the regulation of glucose catabolism, we will consider some general principles that apply to the regulation of all biochemical pathways.

Regulatory Enzymes Act as Metabolic Valves

Although not at equilibrium with their surroundings, adult organisms generally exist in a steady state. A constant influx of fuel and nutrients and a constant release of energy and waste products allow the organism to maintain a constant composition. When the steady state is disturbed by some change in external circumstances or fuel supply, the temporarily altered fluxes through individual metabolic pathways trigger regulatory mechanisms intrinsic to each pathway. The net effect of all of these adjustments is to return the organism to the steady state—to achieve **homeostasis.** Because of the central role of ATP in cellular activities, evolution has produced catabolic enzymes with regulatory properties that ensure a high steady-state concentration of ATP, "high" in this context meaning high relative to the breakdown products ADP and AMP.

The flux through a biochemical pathway depends on the activities of the enzymes that catalyze each reaction. For some of the enzymes in a pathway such as glycolysis, the reaction is essentially at equilibrium within the cell; the activity of such an enzyme is sufficiently high that the substrate is converted to product as fast as the substrate is supplied. The flux through this step is essentially *substrate-limited*—determined by the instantaneous concentration of the substrate.

Other cellular reactions are far from equilibrium. In the glycolytic pathway, the equilibrium constant (K'_{eq}) for the reaction catalyzed by phosphofructokinase-1 is about 250, but the mass action ratio [fructose-1,6-bisphosphate][ADP]/[fructose-6-phosphate][ATP] in a typical cell in the steady state is about 0.04. (The intracellular concentrations of some glycolytic enzymes and reactants are given in Table 14–2.) The

Table 14–2 Cytosolic concentrations of enzymes and metabolites of the glycolytic pathway in skeletal muscle

Enzyme	Concentration (μM)	Metabolite	Concentration (μM)
Aldolase	810	Glucose-6-phosphate	3,900
Triose phosphate isomerase	220	Fructose-6-phosphate	1,500
Glyceraldehyde-3-phosphate dehydrogenase	1,400	Fructose-1,6-bisphosphate	80
		Dihydroxyacetone phosphate	160
Phosphoglycerate kinase	130	Glyceraldehyde-3-phosphate	80
Phosphoglycerate mutase	240	1,3-Bisphosphoglycerate	50
Enolase	540	3-Phosphoglycerate	200
Pyruvate kinase	170	2-Phosphoglycerate	20
Lactate dehydrogenase	300	Phosphenolpyruvate	65
Phosphoglucomutase	32	Pyruvate	380
		Lactate	3,700
		ATP	8,000
		ADP	600
		P_i	8,000
		NAD$^+$	540
		NADH	50

Source: From Srivastava, D.K. & Bernhard, S.A. (1987) Biophysical chemistry of metabolic reaction sequences in concentrated solution and in the cell. *Annu. Rev. Biophys. Biophys. Chem.* **16,** 175–204.

reaction is so far from equilibrium because the rate of conversion of fructose-6-phosphate to fructose-1,6-bisphosphate is limited by the activity of PFK-1. Increased production of fructose-6-phosphate by the preceding enzymes in the glycolytic pathway does not increase the flux through this step, but instead leads to the accumulation of the substrate, fructose-6-phosphate. Thus PFK-1 functions as a valve, regulating the flow of carbon through glycolysis; increasing the activity of this enzyme (by allosteric activation, for example) increases the overall flux through the pathway. Metabolite flux through this pathway is determined not by mass action (by substrate and product concentrations) but by how far this enzymatic valve is "opened."

In every metabolic pathway there is at least one reaction that, in the cell, is far from equilibrium because of the relatively low activity of the enzyme that catalyzes it (Fig. 14–16). The rate of this reaction is not limited by substrate availability, but only by the activity of the enzyme. The reaction is therefore said to be *enzyme-limited*, and because its rate limits the rate of the whole reaction sequence, the step is called the *rate-limiting step* in the pathway. In general, these rate-limiting steps are very exergonic reactions and are therefore essentially irreversible under cellular conditions. *Enzymes that catalyze these exergonic, rate-limiting steps are commonly the targets of metabolic regulation.* In addition to very rapid allosteric enzyme regulation within individual cells, multicellular organisms use hormonal signals to coordinate the metabolic activities of different tissues and organs (Chapter 22). Hormone action alters the activities of key enzymes, often within seconds or minutes. When external circumstances change on a longer time scale, as when a human's diet shifts from primarily fat to primarily carbohydrate, adjustments in the flux through specific pathways are brought about by changes in the number of molecules of specific regulatory enzymes. This is accomplished by changing the relative rates of synthesis and degradation of the enzymes (Chapters 26 and 27).

Many regulatory enzymes are situated at critical branch points in metabolism; their activities determine the allocation of a metabolite to each of the several pathways through which it might pass. For example, glucose-6-phosphate can be metabolized either by glycolysis or by the pentose phosphate pathway (described later in this chapter). The first enzyme unique to each of these pathways (phosphofructokinase-1 and glucose-6-phosphate dehydrogenase, respectively) catalyzes the "committed" step for its pathway. Both are regulatory enzymes, which respond to a variety of allosteric regulators that signal the need for the products of each pathway.

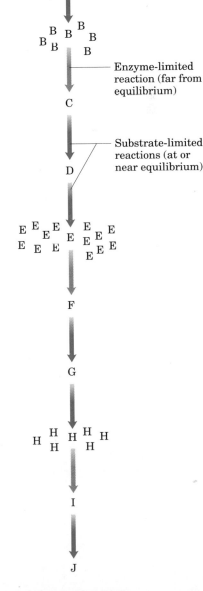

Enzyme-limited reaction (far from equilibrium)

Substrate-limited reactions (at or near equilibrium)

Figure 14–16 Regulation of the flux through multistep pathways occurs at steps that are enzyme-limited. At each of these steps (orange arrows), which are generally exergonic, the substrate is not in equilibrium with the product because the enzyme-catalyzed reaction is relatively slow. The substrate for this reaction tends to accumulate, just as river water accumulates behind a dam. The slowest of these enzyme-limited steps is the rate-limiting step in the overall process. The reactions between enzyme-limited steps are faster, and for each of these reactions (blue arrows) the substrate and product are essentially at their equilibrium concentrations; these reactions are substrate-limited.

Cells commonly have the enzymatic capacity to carry out both the catabolism of some complex molecule into a simpler product and the anabolic conversion of that product back into the starting molecule. Glycolysis degrades glucose to pyruvate; gluconeogenesis converts pyruvate to glucose. Paired catabolic and anabolic pathways often employ many of the same enzymes—those that catalyze readily reversible reactions. Phosphoglycerate mutase, for example, acts in both glycolysis and gluconeogenesis. However, paired pathways almost invariably employ at least one enzyme in the catabolic direction that is different from the enzyme catalyzing the reverse reaction in the anabolic direction. These distinctive enzymes are the points of regulation of the two opposing pathways. The reactions catalyzed by these path-specific enzymes are generally exergonic reactions, irreversible under cellular conditions, and out of equilibrium in the steady state; they are enzyme-limited, not substrate-limited. Having separate enzymes for catabolic and anabolic pathways allows separate regulation of the flux in each direction, avoiding the wasteful "futile cycling" that would result if the breakdown and energy-consuming resynthesis of a compound were allowed to proceed simultaneously.

Regulation of Glucose Metabolism Is Different in Muscle and Liver

The regulatory enzymes that control the rate of breakdown of carbohydrates via glycolysis illustrate these general principles of metabolic regulation. Glucose catabolism is doubtless regulated in all organisms, but the regulatory mechanisms have been studied especially well in vertebrate muscle and liver.

In muscle the end served by glycolysis is ATP production, and the rate of glycolysis increases as muscle contracts more vigorously or more frequently, demanding more ATP. The liver has a different role in whole-body metabolism, and glucose metabolism in the liver is correspondingly different. The liver serves to keep a constant level of glucose in the blood, producing and exporting glucose when the tissues demand it, and importing and storing glucose when it is provided in excess in the diet.

Glycogen Phosphorylase of Muscle Is Regulated Allosterically and Hormonally

In skeletal muscle cells (myocytes), the mobilization of stored glycogen to provide fuel for glycolysis is brought about by glycogen phosphorylase, which degrades glycogen to glucose-1-phosphate (Fig. 14–11). The case of glycogen phosphorylase is an especially instructive example of enzyme regulation. It was the first enzyme shown to be allosterically regulated and the first shown to be controlled by reversible phosphorylation. It is also one of only a few allosteric enzymes for which the detailed three-dimensional structures of the active and inactive forms are known from x-ray crystallographic studies.

In skeletal muscle, glycogen phosphorylase occurs in two forms: a catalytically active form, **phosphorylase a,** and a usually inactive form, **phosphorylase b** (Fig. 14–17); the latter predominates in resting muscle. The rate of glycogen breakdown in muscle depends in part on the ratio of phosphorylase a (active) to phosphorylase b (less active), which is adjusted by the action of hormones such as epinephrine. Phosphorylase a consists of two identical subunits (M_r 94,500), in each of

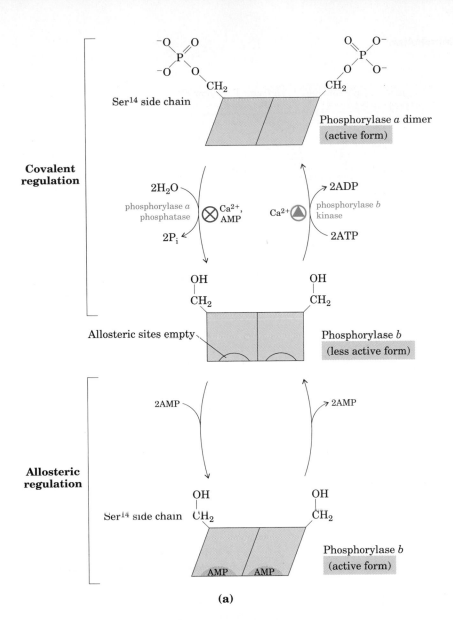

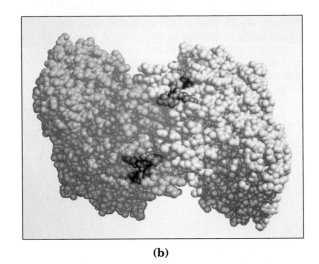

(b)

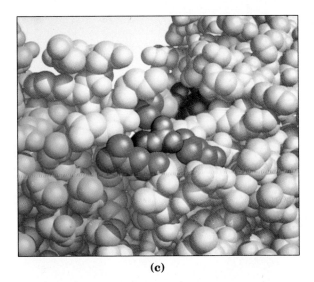

(c)

(a)

Figure 14–17 Covalent and allosteric regulation of glycogen phosphorylase in muscle. **(a)** The enzyme has two identical subunits, each of which can be phosphorylated by phosphorylase b kinase at Ser^{14} to give phosphorylase a, a reaction promoted by Ca^{2+}. Phosphorylase a phosphatase, also called phosphoprotein phosphatase-1, removes these phosphate groups, inactivating the enzyme. Phosphorylase b can also be activated by noncovalent binding of AMP at its allosteric sites. Conformational changes in the enzyme are indicated schematically. Liver glycogen phosphorylase undergoes similar a and b interconversions, but has different regulatory mechanisms. **(b)** The three-dimensional structure of the enzyme from muscle. The two subunits (gray and blue) of the glycogen phosphorylase a dimer,

showing the location of the phosphates (orange) attached to the Ser^{14} residues (red) in each. In phosphorylase b, the amino-terminal peptide containing Ser^{14} is disordered. However, with the attachment of the negatively charged phosphate group at Ser^{14} this peptide folds toward several nearby (positively charged) Arg residues (pink), forcing compensatory changes in regions distant from Ser^{14} and activating the enzyme. AMP, the allosteric activator of phosphorylase b, binds very near Ser^{14}. On the back side of the enzyme is a deep channel that admits the substrate glycogen to the active site, which is 3.3 nm away from the allosteric site. **(c)** A close-up view of the region around the phospho-Ser residue; note its proximity to the interface between dimers.

which the Ser residue at position 14 is phosphorylated. Phosphorylase b is structurally identical except that the Ser^{14} residues are not phosphorylated. Phosphorylase a is converted into the less active phosphorylase b by dephosphorylation, catalyzed by **phosphorylase a phosphatase** (Fig. 14–17). Phosphorylase b is converted back into phosphorylase a by the enzyme **phosphorylase b kinase,** which catalyzes phosphate transfer from ATP to Ser^{14}.

Figure 14–18 Hormonal regulation of glycogen phosphorylase in muscle and liver. A cascade of enzymatic activations leads to activation of glycogen phosphorylase by epinephrine in muscle and by glucagon in liver. When catalysts activate catalysts large amplifications of the initial signal result.

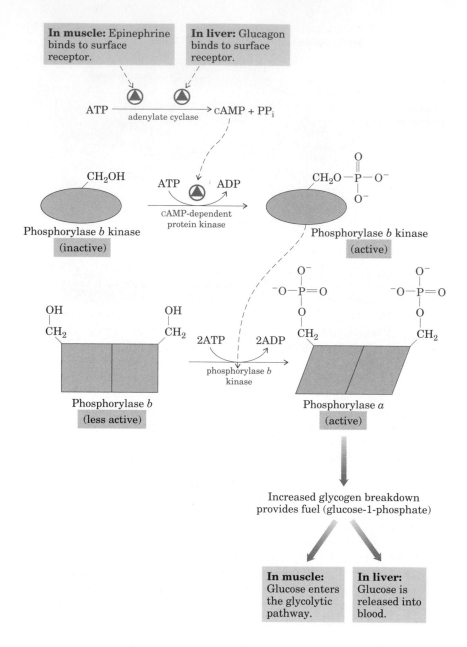

In muscle: Epinephrine binds to surface receptor.

In liver: Glucagon binds to surface receptor.

ATP $\xrightarrow{\text{adenylate cyclase}}$ cAMP + PP$_i$

Phosphorylase b kinase (inactive)

$\xrightarrow[\text{cAMP-dependent protein kinase}]{\text{ATP} \quad \text{ADP}}$

Phosphorylase b kinase (active)

Phosphorylase b (less active)

$\xrightarrow[\text{phosphorylase } b \text{ kinase}]{\text{2ATP} \quad \text{2ADP}}$

Phosphorylase a (active)

Increased glycogen breakdown provides fuel (glucose-1-phosphate)

In muscle: Glucose enters the glycolytic pathway.

In liver: Glucose is released into blood.

Hormones ultimately regulate the interconversion of phosphorylase a and b by regulating the activities of phosphorylase a phosphatase and phosphorylase b kinase. **Epinephrine** is released into the blood by the adrenal gland when an animal is suddenly confronted by a situation that requires vigorous muscular activity. Epinephrine is a signal to skeletal muscle to turn on the processes that lead to production of ATP, which will be needed for muscle contraction. Glycogen phosphorylase is activated to provide glucose-1-phosphate to be fed into the glycolytic pathway. By the cascade of events shown in Figure 14–18, the binding of epinephrine to its specific receptor in the plasma membrane of a muscle cell activates phosphorylase b kinase and inactivates phosphorylase a phosphatase, tipping the balance toward formation of the active (a) form of glycogen phosphorylase. The cascade of activations allows one molecule of hormone to cause activation of many molecules of target enzyme (glycogen phosphorylase).

When the emergency is over, release of epinephrine ceases, the phosphorylase b kinase reverts to its original, lower activity, and the ratio of phosphorylase a to phosphorylase b returns to that in resting muscle.

Superimposed on the hormonal control is faster, allosteric regulation of glycogen phosphorylase b by ATP and AMP. Phosphorylase b, the relatively inactive form, is activated by its allosteric effector AMP (Fig. 14–17), which increases in concentration in muscle during the ATP breakdown accompanying contraction. The stimulation of phosphorylase b by AMP can be prevented by high concentrations of ATP, which blocks the AMP binding site. The activity of phosphorylase b thus reflects the ratio of AMP to ATP. Phosphorylase a, which is not stimulated by AMP, is sometimes referred to as the AMP-independent form, and phosphorylase b as the AMP-dependent form.

In resting muscle nearly all the phosphorylase is in the b form, which is inactive because ATP is present at a much higher concentration than AMP. Vigorous muscular activity increases the AMP : ATP ratio, very rapidly activating (in milliseconds) phosphorylase b by allosteric means. On a longer time scale (seconds to minutes) hormone-triggered phosphorylation of phosphorylase b converts it into phosphorylase a, the activity of which is independent of the AMP : ATP ratio.

There is yet a third type of control on glycogen phosphorylase in skeletal muscle. Calcium, the intracellular signal for muscle contraction, is also an allosteric activator of phosphorylase b kinase. When a transient rise in intracellular Ca^{2+} triggers muscle contraction, it also accelerates conversion of phosphorylase b to the more active phosphorylase a (Fig. 14–17).

Liver Glycogen Phosphorylase Is Regulated by Hormones and Blood Glucose

The glycogen phosphorylase of liver is similar to that of muscle; it too is a dimer of identical subunits, and it undergoes phosphorylation and dephosphorylation on Ser^{14}, interconverting the b and a forms. However, its regulatory properties are slightly different from those of the muscle enzyme, reflecting the different role of glycogen breakdown in liver. Liver glycogen serves as a reservoir that releases glucose into the blood when blood glucose levels fall below the normal level (4 to 5 mM). Glucose-1-phosphate formed by liver phosphorylase is converted (as in muscle) into glucose-6-phosphate by the action of phosphoglucomutase (p. 422). Then **glucose-6-phosphatase,** an enzyme present in liver but not in muscle, removes the phosphate:

$$\text{Glucose-6-phosphate} + H_2O \longrightarrow \text{glucose} + P_i$$

When the blood glucose level is low, the free glucose produced from glycogen in the liver by these reactions is released into the bloodstream and carried to tissues (such as the brain) that require it as a fuel. (For most tissues, glucose is only one of several equally useful fuels.)

Glycogen phosphorylase of liver, like that of muscle, is under hormonal control. **Glucagon** is a hormone released by the pancreas when blood glucose levels fall below normal. When glucagon binds to its receptor in the plasma membrane of a hepatocyte, a cascade of events essentially similar to that in muscle (Fig. 14–18) results in the conversion of phosphorylase b to phosphorylase a, increasing the rate of glycogen breakdown and thereby increasing the rate of glucose release into the blood.

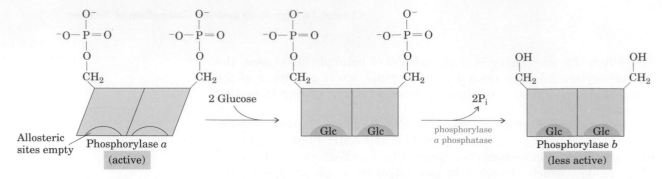

Figure 14–19 Glycogen phosphorylase as a glucose sensor. Glucose binding to an allosteric site in liver glycogen phosphorylase a induces a conformational change that exposes the phosphorylated Ser[14] residues to the action of phosphorylase a phosphatase, which converts phosphorylase a to b, reducing its activity in response to high blood glucose.

Liver glycogen phosphorylase, like that of muscle, is subject to allosteric regulation, but in this case the allosteric regulator is glucose, not AMP. When the concentration of glucose in the blood rises, glucose enters hepatocytes and binds to the regulatory site of glycogen phosphorylase a, causing a conformational change that exposes the phosphorylated Ser[14] residues to dephosphorylation by phosphorylase a phosphatase (Fig. 14–19). In this way, glycogen phosphorylase a acts as the glucose sensor of liver, slowing the breakdown of glycogen whenever the level of blood glucose is high.

Hexokinase Is Allosterically Inhibited by Its Product

Hexokinase, which catalyzes the entry of free glucose into the glycolytic pathway, is another regulatory enzyme. The hexokinase of myocytes has a high affinity for glucose (it is half-saturated at about 0.1 mM). Glucose entering myocytes from the blood (in which the glucose concentration is 4 to 5 mM) produces an intracellular glucose concentration high enough to saturate hexokinase, so that it normally acts at its maximal rate. Muscle hexokinase is allosterically inhibited by its product, glucose-6-phosphate. Whenever the concentration of glucose-6-phosphate in the cell rises above its normal level, hexokinase is temporarily and reversibly inhibited, bringing the rate of glucose-6-phosphate formation into balance with the rate of its utilization and reestablishing the steady state.

Mammals have several forms of hexokinase, all of which catalyze the conversion of glucose into glucose-6-phosphate. Different proteins that catalyze the same reaction are called **isozymes** (Box 14–3). The predominant hexokinase isozyme in liver is hexokinase D, also called **glucokinase,** which differs in two important respects from the hexokinase isozymes in muscle.

First, the glucose concentration at which glucokinase is half-saturated (about 10 mM) is higher than the usual concentration of glucose in the blood. Because the concentration of glucose in liver is maintained at a level close to that in the blood by an efficient glucose transporter, this property of glucokinase allows its direct regulation by the level of blood glucose. When the glucose concentration in the blood is high, as it is after a meal rich in carbohydrates, excess blood glucose is transported into hepatocytes, where glucokinase converts it into glucose-6-phosphate.

Second, glucokinase is inhibited not by its reaction product glucose-6-phosphate but by its isomer, fructose-6-phosphate, which is always in equilibrium with glucose-6-phosphate because of the action of phosphoglucose isomerase. The partial inhibition of glucokinase by fructose-6-phosphate is mediated by an additional protein, the **regulator protein.** This regulator protein also has affinity for fructose-1-phosphate, which competes with fructose-6-phosphate and cancels its inhibitory effect on glucokinase. Because fructose-1-phosphate is pres-

BOX 14–3   **Isozymes: Different Proteins That Catalyze the Same Reaction**

The several forms of hexokinase found in mammalian tissues are but one example of a common situation in which the same reaction is catalyzed by two or more different molecular forms of an enzyme. These multiple forms, called isozymes or isoenzymes, may occur in the same species, in the same tissue, or even in the same cell. The different forms of the enzyme generally differ in kinetic or regulatory properties, in the form of cofactor they use (NADH or NADPH for dehydrogenase isozymes, for example) or in their subcellular distribution (soluble or membrane-bound). Isozymes commonly have similar, but not identical, amino acid sequences, and in many cases they clearly share an evolutionary origin.

One of the first enzymes found to have isozymes was lactate dehydrogenase (LDH) (p. 416). LDH occurs in vertebrate tissues as at least five different isozymes separable by electrophoresis. All LDH isozymes contain four polypeptide chains (each of M_r 33,500), but the five isozymes contain different ratios of two kinds of polypeptides that differ in composition and sequence. The A chains (also designated M for muscle) and the B chains (also designated H for heart) are encoded by two different genes. In skeletal muscle the predominant isozyme contains four A chains, and in heart the predominant isozyme contains four B chains. LDH isozymes in other tissues are a mixture of the five possible forms, which may be designated A_4, A_3B, A_2B_2, AB_3, and B_4. The different LDH isozymes have significantly different values of V_{max} and K_m, particularly for pyruvate. The properties of LDH isozyme A_4 favor rapid reduction of very low concentrations of pyruvate to lactate in skeletal muscle, whereas those of isozyme B_4 tend to favor rapid oxidation of lactate to pyruvate in the heart.

The distribution of different isozyme forms of a given enzyme reflects at least four factors:

1. The differing metabolic patterns in different organs. The two forms of glycogen phosphorylase found in skeletal muscle and in liver differ in their regulatory properties, reflecting the different roles of glycogen breakdown in these two tissues, as described in the text.

2. The different locations and metabolic roles of a given enzyme within one type of cell. The isocitrate dehydrogenase isozymes of the cytosol and the mitochondrion are an example (Chapter 15).

3. The differentiation and development of adult tissues from their embryonic or fetal forms. For example, the fetal liver has a characteristic isozyme distribution of LDH, which changes as the organ undergoes differentiation to its adult form. An interesting discovery is that some of the enzymes of glucose catabolism in malignant (cancer) cells occur as their fetal, not adult, isozymes.

4. The fine-tuning of metabolic rates through the different responses of isozyme forms to allosteric modulators. Hexokinase D (glucokinase) of liver and the hexokinase isozymes found in other tissues differ in their sensitivity to inhibition by their product, glucose-6-phosphate (p. 432).

---

ent in liver only when there is fructose in the blood, this property of the regulator protein explains the observation that ingested fructose stimulates the phosphorylation of glucose in the liver.

The pancreatic β cells, which are responsible for the release of insulin when blood glucose levels rise above normal, also contain glucokinase and the inhibitory regulator protein.

Pyruvate Kinase Is Inhibited by ATP

In vertebrates there are at least three isozymes of pyruvate kinase, differing somewhat in their tissue distribution and in their response to modulators. High concentrations of ATP inhibit pyruvate kinase allosterically, by decreasing the affinity of the enzyme for its substrate phosphoenolpyruvate (PEP). The level of PEP normally found in cells is not high enough to saturate the enzyme, and the reaction rate will accordingly be low at normal PEP concentrations.

Pyruvate kinase is also inhibited by acetyl-CoA and by long-chain fatty acids, both important fuels for the citric acid cycle. (Recall that acetyl-CoA (acetate) is produced by the catabolism of fats and amino acids, as well as by glucose catabolism; see Fig. 4a, p. 362.) Because the citric acid cycle is a major source of energy for ATP production, the availability of these other fuels reduces the dependence on glycolysis for ATP.

Thus, whenever the cell has a high concentration of ATP, or whenever ample fuels are already available for energy-yielding respiration, glycolysis is inhibited by the slowed action of pyruvate kinase. When the ATP concentration falls, the affinity of pyruvate kinase for PEP increases, enabling the enzyme to catalyze ATP synthesis even though the concentration of PEP is relatively low. The result is a high steady-state concentration of ATP.

Phosphofructokinase-1 Is under Complex Allosteric Regulation

Glucose-6-phosphate can flow either into glycolysis or through one of the secondary oxidative pathways described later in this chapter. The irreversible reaction catalyzed by PFK-1 is the step that commits a cell to the passage of glucose through glycolysis. In addition to the binding sites for its substrates, fructose-6-phosphate and ATP, this complex enzyme has several regulatory sites where allosteric activators or inhibitors bind.

ATP is not only the substrate for PFK-1, but also the end product of the glycolytic pathway. When high ATP levels signal that the cell is producing ATP faster than it is consuming it, ATP inhibits PFK-1 by binding to an allosteric site and lowering the affinity of the enzyme for its substrate fructose-6-phosphate (Fig. 14–20). ADP and AMP, which rise in concentration when the consumption of ATP outpaces its production, act allosterically to relieve this inhibition by ATP. These effects combine to produce higher enzyme activity when fructose-6-phosphate, ADP, or AMP builds up, and lower activity when ATP accumulates.

Citrate (the ionized form of citric acid), a key intermediate in the aerobic oxidation of pyruvate (Chapter 15), also serves as an allosteric regulator of PFK-1; high citrate concentration increases the inhibitory effect of ATP, further reducing the flow of glucose through glycolysis. In this case, as in several others to be encountered later, citrate serves as an intracellular signal that the cell's needs for energy-yielding metabolism and for biosynthetic intermediates are being met.

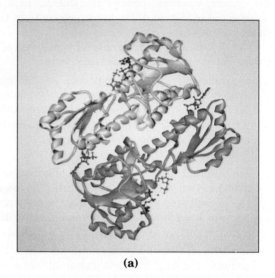

(a)

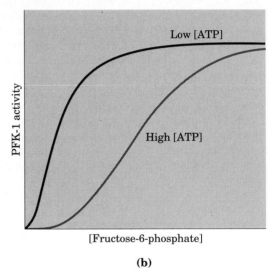

(b)

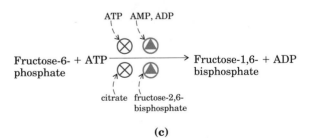

(c)

Figure 14–20 (a) A ribbon diagram of *E. coli* phosphofructokinase-1, a tetramer of identical subunits. Each subunit has its own catalytic site and its own binding sites for the allosteric activators. **(b)** Allosteric regulation of muscle PFK-1 by ATP, shown by a substrate-activity curve. At low concentrations of ATP the $K_{0.5}$ (p. 231) for fructose-6-phosphate is relatively low, enabling the enzyme to function at a high rate at relatively low concentrations of fructose-6-phosphate. At high ATP, $K_{0.5}$ for fructose-6-phosphate is greatly increased, as indicated by the sigmoid relationship between substrate concentration and enzyme activity. **(c)** A summary of the regulators affecting PFK-1 activity.

The most significant allosteric regulator of PFK-1 is fructose-2,6-bisphosphate, which, as noted earlier, strongly activates the enzyme. The concentration of fructose-2,6-bisphosphate in liver decreases in response to the hormone glucagon, slowing glycolysis and stimulating glucose synthesis in liver.

Glycolysis and Gluconeogenesis Are Coordinately Regulated

Most organisms can synthesize glucose from simpler precursors such as pyruvate or lactate. In mammals this process, called **gluconeogenesis,** occurs primarily in the liver, and its role is to provide glucose for export to other tissues when other sources of glucose are exhausted. Gluconeogenesis employs most of the same enzymes that act in glycolysis, but it is not simply the reversal of glycolysis. Seven of the glycolytic reactions are freely reversible, and the enzymes that catalyze these reactions also function in gluconeogenesis. Three reactions of glycolysis are so exergonic as to be essentially irreversible: those catalyzed by hexokinase, phosphofructokinase-1, and pyruvate kinase. Detours around each of these irreversible steps are employed in gluconeogenesis. For example, in gluconeogenesis the conversion of fructose-1,6-bisphosphate to fructose-6-phosphate is catalyzed by **fructose-1,6-bisphosphatase (FBPase-1)** (Fig. 14–21).

To prevent futile cycling in which glucose is simultaneously degraded by glycolysis and resynthesized by gluconeogenesis (Chapter 19), the enzymes unique to each pathway are reciprocally regulated by common allosteric effectors. Fructose-2,6-bisphosphate, a potent activator of liver PFK-1 and therefore of glycolysis, also inhibits FBPase-1, thereby slowing gluconeogenesis.

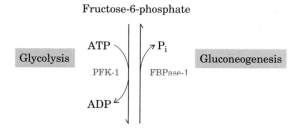

Figure 14–21 The reaction in gluconeogenesis that bypasses the irreversible phosphofructokinase-1 reaction in glycolysis. The conversion of fructose-1,6-bisphosphate to fructose-6-phosphate is catalyzed by fructose-1,6-bisphosphatase (called FBPase-1 to distinguish it from a similar enzyme described in Chapter 19).

Fructose-2,6-bisphosphate

Glucagon, the hormone that signals low blood sugar, lowers the level of fructose-2,6-bisphosphate in liver, slowing the consumption of glucose by glycolysis and stimulating the production of glucose for export by gluconeogenesis. We will return to a more complete discussion of this coordinate regulation in Chapter 19, when we have discussed gluconeogenesis in more detail.

Fructose-2,6-bisphosphate is found in all animals, in fungi, and in some plants, but not in bacteria. It stimulates all known animal PFK-1 activities as well as PFK-1 from yeast. In plants, fructose-2,6-bisphosphate also regulates carbohydrate metabolism, but by mechanisms not identical with those in liver; plants do not, of course, have glucagon. Fructose-2,6-bisphosphate inhibits the PP_i-dependent phosphofructokinase of plants that is responsible for fructose-1,6-bisphosphate formation in glycolysis (p. 407), but does not inhibit the ATP-dependent PFK-1 of plants. Plant PFK-1 is, however, strongly inhibited by phosphoenolpyruvate, a glycolytic intermediate downstream from fructose-1,6-bisphosphate.

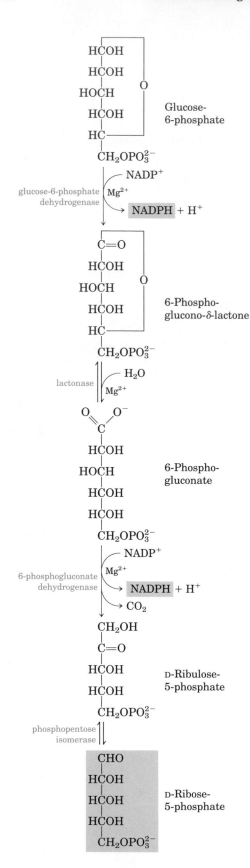

HCOH
HCOH
HOCH   O
HCOH
HC

CH₂OPO₃²⁻   Glucose-
6-phosphate

glucose-6-phosphate
dehydrogenase

NADP⁺
Mg²⁺
NADPH + H⁺

C=O
HCOH
HOCH   O   6-Phospho-
HCOH   glucono-δ-lactone
HC

CH₂OPO₃²⁻

lactonase

H₂O
Mg²⁺

O   O⁻
  C
HCOH
HOCH   6-Phospho-
HCOH   gluconate
HCOH
HCOH

CH₂OPO₃²⁻

6-phosphogluconate
dehydrogenase

NADP⁺
Mg²⁺
NADPH + H⁺
CO₂

CH₂OH
C=O
HCOH   D-Ribulose-
HCOH   5-phosphate

CH₂OPO₃²⁻

phosphopentose
isomerase

CHO
HCOH
HCOH   D-Ribose-
HCOH   5-phosphate
HCOH

CH₂OPO₃²⁻

Figure 14–22 The oxidative reactions of the pentose phosphate pathway, leading to D-ribose-5-phosphate and producing NADPH.

Secondary Pathways of Glucose Oxidation

In animal tissues, most of the glucose consumed is catabolized via glycolysis to pyruvate. Most of the pyruvate in turn is oxidized via the citric acid cycle. The main function of glucose catabolism by this route is to generate ATP. There are, however, other catabolic pathways taken by glucose that lead to specialized products needed by the cell, and these pathways constitute part of the *secondary metabolism* of glucose. Two such pathways produce pentose phosphates and uronic and ascorbic acids.

Oxidative Decarboxylation Yields Pentose Phosphates and NADPH

The **pentose phosphate pathway,** also called the **phosphogluconate pathway** (Fig. 14–22), produces NADPH and ribose-5-phosphate. Recall that NADPH is a carrier of chemical energy in the form of reducing power (Chapter 13). In mammals this function is especially prominent in tissues actively carrying out the biosynthesis of fatty acids and steroids from small precursors, particularly the mammary gland, adipose tissue, the adrenal cortex, and the liver. The biosynthesis of fatty acids requires reducing power in the form of NADPH to reduce the double bonds and carbonyl groups of intermediates in this process (Chapter 20). Other tissues less active in synthesizing fatty acids, such as skeletal muscle, are virtually lacking in the pentose phosphate pathway. A second function of the pentose phosphate pathway is to generate essential pentoses, particularly D-ribose, used in the biosynthesis of nucleic acids (Chapter 21).

The first reaction of the pentose phosphate pathway is the enzymatic dehydrogenation of glucose-6-phosphate by **glucose-6-phosphate dehydrogenase** to form **6-phosphoglucono-δ-lactone,** an intramolecular ester, which is hydrolyzed to the free acid **6-phosphogluconate** by a specific **lactonase** (Fig. 14–22). NADP⁺ is the electron acceptor, and the overall equilibrium lies far in the direction of formation of NADPH. In the next step 6-phosphogluconate undergoes dehydrogenation and decarboxylation by **6-phosphogluconate dehydrogenase** to form the ketopentose D-**ribulose-5-phosphate,** a reaction that generates a second molecule of NADPH. **Phosphopentose isomerase** then converts ribulose-5-phosphate into its aldose isomer D-**ribose-5-phosphate.** In some tissues, the pentose phosphate pathway ends at this point, and its overall equation is then written

$$\text{Glucose-6-phosphate} + 2\text{NADP}^+ + \text{H}_2\text{O} \longrightarrow$$
$$\text{ribose-5-phosphate} + \text{CO}_2 + 2\text{NADPH} + 2\text{H}^+$$

The net result is the production of NADPH for reductive biosynthetic reactions and the production of ribose-5-phosphate as a precursor for nucleotide synthesis.

In tissues that require primarily NADPH rather than ribose-5-phosphate, pentose phosphates are recycled into glucose-6-phosphate in a series of reactions (Fig. 14–23) that will be examined in more detail in Chapter 19. First, ribulose-5-phosphate is epimerized to xylulose-5-phosphate. Then, in a series of rearrangements of the carbon skeletons of sugar phosphate intermediates, six five-carbon sugar phosphates are converted into five six-carbon sugar phosphates (Fig. 14–23b), completing the cycle and allowing continued oxidation of glucose-6-phosphate with the production of NADPH.

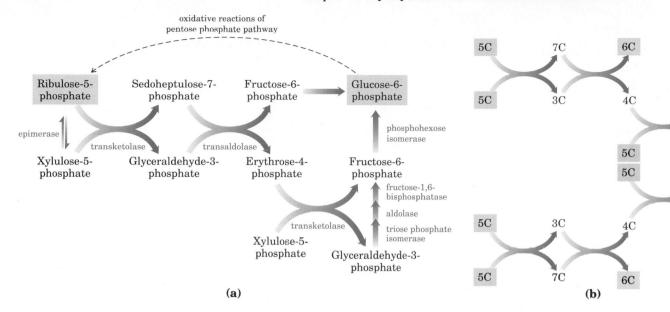

(a)

(b)

Figure 14–23 (a) The nonoxidative reactions of the pentose phosphate pathway convert pentose phosphates back into hexose phosphates, allowing the oxidative reactions (see Fig. 14–22) to continue. The enzymes transaldolase and transketolase (discussed in more detail in Chapter 19) are specific to this pathway; the other enzymes also serve in the glycolytic or gluconeogenic pathways. **(b)** A simplified schematic diagram showing the pathway leading from six pentoses (5C) to five hexoses (6C). Note that this involves two sets of the interconversions shown in **(a)**.

In the nonoxidative part of the pentose phosphate pathway (Fig. 14–23a), transketolase, a thiamine pyrophosphate–dependent enzyme, catalyzes the transfer of a two-carbon fragment (C-1 and C-2) of xylulose-5-phosphate to ribose-5-phosphate, forming the seven-carbon product sedoheptulose-7-phosphate; the remaining three-carbon fragment of xylulose is glyceraldehyde-3-phosphate. (The detailed mechanism for transketolase is shown in Fig. 19–25.) Transaldolase then catalyzes a reaction similar to the aldolase reaction in glycolysis: a three-carbon fragment is removed from sedoheptulose-7-phosphate and condensed with glyceraldehyde-3-phosphate, forming fructose-6-phosphate; the remaining four-carbon fragment of sedoheptulose is erythrose-4-phosphate. Now transaldolase acts again, forming fructose-6-phosphate and glyceraldehyde-3-phosphate from erythrose-4-phosphate and xylulose-5-phosphate. Two molecules of glyceraldehyde-3-phosphate formed by two iterations of these reactions can be converted into fructose-1,6-bisphosphate (Fig. 14–23b). The cycle is then complete: six pentose phosphates have been converted back into five hexose phosphates.

All of the reactions of the nonoxidative part of the pentose phosphate pathway are readily reversible, and thus also provide a means of converting hexose phosphates into pentose phosphates. As we shall see in Chapter 19, this is essential in the fixation of CO_2 by photosynthetic plants.

Glucose Is Converted to Glucuronic Acid and Ascorbic Acid

Another secondary pathway for glucose leads to two specialized products: **D-glucuronate,** important in the detoxification and excretion of foreign organic compounds, and **L-ascorbic acid** or **vitamin C.** Although the amount of glucose diverted into this secondary pathway is very small compared with the large amounts of glucose proceeding through glycolysis and the citric acid cycle, the products are vital to the organism.

In this pathway (Fig. 14–24) glucose-1-phosphate is first converted into **UDP-glucose** by reaction with UTP. The glucose portion of UDP-glucose is then dehydrogenated to yield **UDP-glucuronate,** another example (see also Fig. 14–13) of the use of UDP derivatives as intermediates in the enzymatic transformations of sugars.

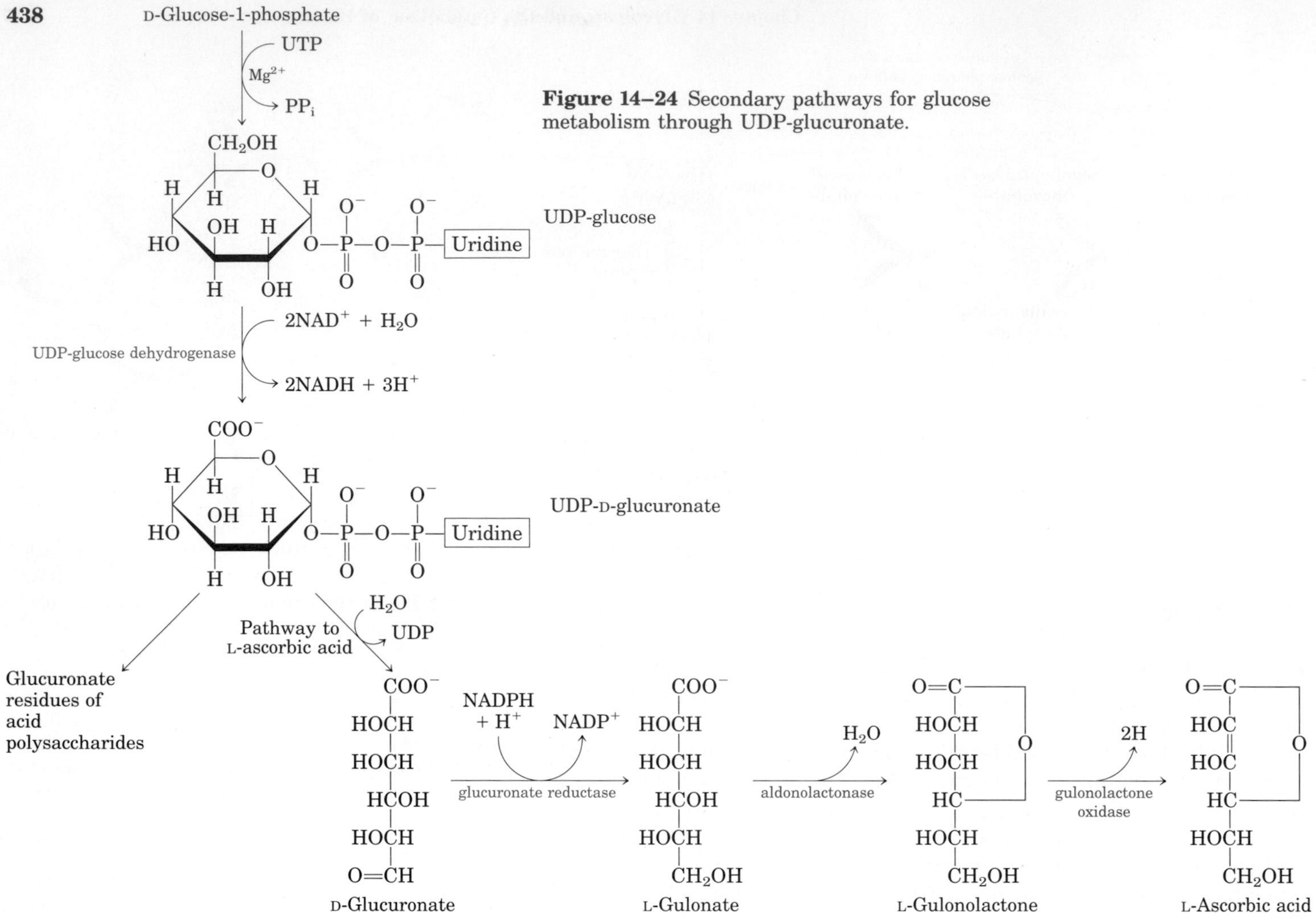

D-Glucose-1-phosphate

UTP

Mg^{2+}

PP$_i$

CH$_2$OH

OH H

HO

H OH

H

O—P—O—P—Uridine UDP-glucose

O$^-$ O$^-$

O O

2NAD$^+$ + H$_2$O

UDP-glucose dehydrogenase

2NADH + 3H$^+$

COO$^-$

H H H

HO

H OH

H

O—P—O—P—Uridine UDP-D-glucuronate

O$^-$ O$^-$

O O

Glucuronate residues of acid polysaccharides

H$_2$O

Pathway to L-ascorbic acid

UDP

COO$^-$

HOCH

HOCH

HCOH

HOCH

O=CH

D-Glucuronate

NADPH + H$^+$ NADP$^+$

glucuronate reductase

COO$^-$

HOCH

HOCH

HCOH

HOCH

CH$_2$OH

L-Gulonate

H$_2$O

aldonolactonase

O=C

HOCH

HOCH O

HC

HOCH

CH$_2$OH

L-Gulonolactone

2H

gulonolactone oxidase

O=C

HOC

HOC O

HC

HOCH

CH$_2$OH

L-Ascorbic acid

Figure 14–24 Secondary pathways for glucose metabolism through UDP-glucuronate.

UDP-glucuronate is the glucuronosyl donor used by a family of detoxifying enzymes that act on a variety of relatively nonpolar drugs, environmental toxins, and carcinogens. The conjugation of these compounds with glucuronate (**glucuronidation**) converts them into much more polar derivatives that are more easily cleared from the blood by the kidneys and excreted in the urine. For example, the sedative drug phenobarbital, the anti-AIDS drug AZT, and the hydroxylated form of the carcinogen benzo[a]pyrene (3-hydroxybenzo[a]pyrene) all undergo glucuronidation catalyzed by UDP-glucuronosyl transferases in the human liver (Fig. 14–25). Chronic exposure to the drug or toxin induces increased synthesis of the enzyme specific for that compound, increasing tolerance for the drug or resistance to the toxin. UDP-glucuronate is also the precursor of the glucuronate residues of such acidic polysaccharides as hyaluronate and chondroitin sulfate (see Fig. 11–20).

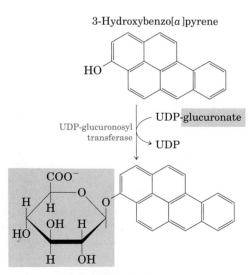

3-Hydroxybenzo[a]pyrene

HO

UDP-glucuronosyl transferase

UDP-glucuronate

UDP

COO$^-$

H H

OH H

HO

H OH

O

O

Hydroxybenzo[a]pyrene glucuronoside (water-soluble)

Figure 14–25 Detoxification of 3-hydroxybenzo[a]pyrene, a toxic component of tobacco smoke. Glucuronidation by transfer of glucuronate from UDP-glucuronate converts the nonpolar toxin to a polar compound more easily removed by the kidneys.

D-Glucuronate is an intermediate in the conversion of D-glucose into L-ascorbic acid (Fig. 14–24). It is reduced by NADPH to the six-carbon sugar acid **L-gulonate,** which is converted into its lactone. L-Gulonolactone then undergoes dehydrogenation by the flavoprotein **gulonolactone oxidase** to yield L-ascorbic acid. Some animal species, including humans, guinea pigs, monkeys, some birds, and some fish lack the enzyme gulonolactone oxidase and are unable to synthesize ascorbic acid; they require it ready-made in the diet (as vitamin C).

Humans who do not obtain enough vitamin C in the diet develop the serious disease scurvy, in which the synthesis of connective tissue containing collagen is defective. The symptoms of scurvy include swollen and bleeding gums with loosened teeth, stiffness and soreness of joints, bleeding under the skin, and slow wound healing. For centuries the disease was very common among sailors on long sea voyages, during which no fresh fruit was available, and in 1753 the Scottish naval surgeon James Lind showed that scurvy was prevented and cured by ingestion of citrus juice. In 1932 the antiscurvy vitamin C was isolated from lemon juice and named ascorbic acid (from the Latin *scorbutus,* meaning "scurvy").

Summary

Glycolysis is a universal metabolic pathway for the catabolism of glucose to pyruvate accompanied by the formation of ATP. The process is catalyzed by ten cytosolic enzymes, and all of the intermediates are phosphorylated compounds. In the preparatory phase of glycolysis, ATP is invested to convert glucose to the phosphorylated intermediate fructose-1,6-bisphosphate, then the carbon–carbon bond between C-3 and C-4 is broken to yield two molecules of triose phosphate. In the payoff phase of glycolysis, each of the two molecules of glyceraldehyde-3-phosphate derived from glucose undergoes oxidation at C-1; the energy of this oxidation reaction is conserved in the formation of NADH and an acyl phosphate bond in 1,3-bisphosphoglycerate. This compound has a high phosphate group transfer potential, and in a substrate-level phosphorylation catalyzed by phosphoglycerate kinase the phosphate group is transferred to ADP, forming ATP and 3-phosphoglycerate. Rearrangement of the atoms in 3-phosphoglycerate with the loss of H_2O gives rise to phosphoenolpyruvate, another compound with high phosphate group transfer potential. Phosphoenolpyruvate donates a phosphate group to ADP to form ATP in the second substrate-level phosphorylation; the other product of this reaction is pyruvate, the end product of the payoff phase of glycolysis. The overall equation for glycolysis is

$$\text{Glucose} + 2NAD^+ + 2ADP + 2P_i \longrightarrow$$
$$2 \text{ pyruvate} + 2NADH + 2H^+ + 2ATP + 2H_2O$$

There is a net gain of two ATP.

The NADH formed in glycolysis must be recycled to regenerate NAD^+, which is required as electron acceptor in the first step of the payoff phase of glycolysis. Under aerobic conditions, electrons pass from NADH to O_2 through a chain of electron carriers in the process of mitochondrial respiration. Under anaerobic conditions, many organisms regenerate NAD^+ by transferring electrons from NADH to pyruvate, forming lactate. This process occurs in vertebrate muscle during intense muscular activity, when energy demand outstrips the ability to deliver O_2 to the muscles. Other organisms, such as yeast, regenerate NAD^+ by reducing pyruvate to ethanol and CO_2. In these anaerobic processes, called fermentations, no *net* oxidation or reduction of the carbons of glucose occurs. A variety of alcohols and organic acids are produced commercially by exploiting the ability of microorganisms to ferment glucose to these products.

Glycogen and starch, polymeric storage forms of glucose, enter glycolysis in a two-step process that begins with phosphorolytic cleavage of a glucose residue from an end of the polymer, forming glucose-1-phosphate. This is catalyzed by glycogen (or starch) phosphorylase. Phosphoglucomutase then converts glucose-1-phosphate to glucose-6-phosphate, the first intermediate in glycolysis. Ingested disaccharides are converted into monosaccharides in the animal intestine by specific hydrolytic enzymes on the outer surface of intestinal epithelial cells; the monosaccharides are then taken up and transported to the liver or other tissues.

A variety of D-hexoses, including fructose, mannose, and galactose, can be funneled into glycolysis. Each is first phosphorylated by a kinase that uses ATP as phosphate donor, then converted into either glucose-6-phosphate or fructose-6-phosphate (the second intermediate in glycolysis). The conversion of galactose-1-phosphate to glucose-1-phosphate involves two intermediates that are nucleotide derivatives: UDP-galactose and UDP-glucose. Genetic defects in the enzymes that catalyze conversion of galactose into glucose-1-phosphate result in galactosemia, a serious human disease.

In a metabolically active cell at steady state, intermediates are formed and consumed at equal rates. Of paramount importance to a cell is the maintenance of a high steady-state concentration of ATP. When some perturbation alters the rate of formation or consumption of an intermediate or product such as ATP, compensating changes in the activities of the relevant enzymes bring the system back into the steady state. These changes in enzyme activity are achieved by allosteric regulation or covalent modification (such as phosphorylation) of the enzymes, often triggered by hormonal signals.

In multistep processes such as glycolysis, certain of the enzyme-catalyzed reactions are essentially at equilibrium in the steady state; the rates of these reactions rise and fall with substrate concentration, and they are said to be substrate-limited. Other reactions are out of equilibrium; their rates are too slow to produce instant equilibration of substrate and product, and these reactions are said to be enzyme-limited. The enzyme-limited reactions in a multistep process are often highly exergonic and therefore practically irreversible, and the enzymes that catalyze these reactions are commonly the points at which flux through the pathway is regulated. In the glycolytic pathway from glycogen to pyruvate, the regulated steps include those catalyzed by glycogen phosphorylase, hexokinase, phosphofructokinase-1 (PFK-1), and pyruvate kinase; all are exergonic, enzyme-limited reactions.

Glycogen phosphorylase of vertebrate muscle is activated by phosphorylation at the Ser^{14} residue of each subunit, catalyzed by phosphorylase b kinase, which is itself activated by a cascade of regulatory events triggered by the hormone epinephrine. Inactivation of glycogen phosphorylase results from the action of a specific phosphatase that removes the phosphate groups from the Ser^{14} residues; this enzyme, too, is under hormonal regulation. The glycogen phosphorylase of liver is also regulated by phosphorylation and dephosphorylation, but the details of its regulation differ from those of the muscle enzyme, reflecting the different roles of muscle and liver in the metabolism of glucose. Liver serves as a buffer against changes in blood glucose concentrations; it releases glucose from stored glycogen when the hormone glucagon signals that blood glucose is too low. The dephosphorylation of the Ser^{14} residues in the liver enzyme is stimulated when glucose binds to an allosteric site on the phosphorylase. When intracellular glucose rises, signaling that there is sufficient glucose in the blood, the glycogen phosphorylase is dephosphorylated and thus inactivated, slowing the mobilization of free glucose from liver glycogen.

Hexokinase is inhibited by high concentrations of its product, glucose-6-phosphate; thus, when the product of the reaction accumulates, its rate of production is lowered. Pyruvate kinase is likewise allosterically inhibited by one of its products, ATP.

Gluconeogenesis is a multistep process in which pyruvate is converted into glucose. Some of the enzymatic steps in gluconeogenesis are catalyzed by the same enzymes used in glycolysis; these are the substrate-limited reactions in both processes. The glycolytic reactions catalyzed by hexokinase, PFK-1, and pyruvate kinase are essentially irreversible. In gluconeogenesis, these three reactions are bypassed by reactions catalyzed by different enzymes. To prevent futile cycling—the simultaneous production and consumption of a cellular component (glucose in this case)—the enzyme-limited reactions of glycolysis and gluconeogenesis are subject to reciprocal allosteric control; when the glycolytic reactions are stimulated, the gluconeogenic reactions are inhibited, and vice versa. Fructose-2,6-bisphosphate is an allosteric activator of PFK-1 (glycolysis) and an allosteric inhibitor of fructose-1,6-bisphosphatase (gluconeogenesis). The hormone glucagon triggers a series of enzymatic changes that cause reduction of the level of the regulator fructose-2,6-bisphosphate. The result is a slowing of glycolysis and an increased rate of gluconeogenesis.

Glucose has catabolic fates other than glycolysis. The pentose phosphate pathway results in oxidation and decarboxylation at the C-1 position of glucose, producing NADPH and pentose phosphates; NADPH provides reducing power for biosynthetic reactions, and pentose phosphates are essential components of nucleotides and nucleic acids. Other oxidative pathways transform glucose into glucuronic acid and ascorbic acid (vitamin C). Glucuronidation converts certain nonpolar toxins into polar derivatives that can be excreted. Humans cannot synthesize ascorbic acid; the lack of this vitamin in the human diet leads to the disease scurvy.

Further Reading

General

Dennis, D.T. (1987) *The Biochemistry of Energy Utilization in Plants,* Chapman & Hall, New York.
A short, excellent presentation of energy-conserving reactions in higher plants, written at about the level of this textbook; it includes a very good chapter (Chapter 6) on glycolysis and the pentose phosphate pathway.

Fruton, J.S. (1972) *Molecules and Life: Historical Essays on the Interplay of Chemistry and Biology,* Wiley-Interscience, New York.
This text includes a detailed historical account of research on glycolysis.

Hochachka, P.W. (1980) *Living Without Oxygen: Closed and Open Systems in Hypoxia Tolerance,* Harvard University Press, Cambridge, MA.
Comparative biochemistry and physiology of glycolysis in different organisms under anaerobic conditions.

Glycolysis

Phillips, D., Blake, C.C.F., & Watson, H.C. (eds) (1981) The Enzymes of Glycolysis: Structure, Activity and Evolution. *Phil. Trans. R. Soc. Lond. [Biol.]* **293,** 1–214.
A collection of excellent reviews on the enzymes of glycolysis, written at a level challenging but comprehensible to a beginning student of biochemistry.

Physiological Significance of Metabolite Channeling. (1991) *J. Theoret. Biol.* **152,** 1–140.
A special issue containing a collection of 29 short papers on all aspects of metabolite channeling.

Srivastava, D.K. & Bernhard, S.A. (1987) Biophysical chemistry of metabolic reaction sequences in concentrated enzyme solution and in the cell. *Annu. Rev. Biophys. Biophys. Chem.* **16,** 175–204.
A detailed consideration of the evidence for protein–protein interactions and substrate channeling in concentrated protein solutions, with examples from the glycolytic pathway.

Regulation of Carbohydrate Metabolism

Barford, D., Hu, S.-H., & Johnson, L.N. (1991) Structural mechanism for glycogen phosphorylase control by phosphorylation and AMP. *J. Mol. Biol.* **218,** 233–260.
Clear discussion of the regulatory changes in the structure of glycogen phosphorylase, based on the structures (from x-ray diffraction studies) of the active and inactive forms of the enzyme.

Hue, L. & Rider, M.H. (1987) Role of fructose 2,6-bisphosphate in the control of glycolysis in mammalian tissues. *Biochem. J.* **245,** 313–324.

Ochs, R.S., Hanson, R.W., & Hall, J. (eds) (1985) *Metabolic Regulation,* Elsevier Science Publishing Co., Inc., New York.
A collection of short essays first published in Trends in Biochemical Sciences, *better known as* TIBS.

Pilkis, S.J. (ed) (1990) *Fructose-2,6-bisphosphate,* CRC Press, Inc., Boca Raton, FL.
An excellent collection of reviews of the discovery of fructose-2,6-bisphosphate and its role in the regulation of carbohydrate metabolism in vertebrate animals, higher plants, and eukaryotic microbial organisms.

Pilkis, S.J., & Claus, T.H. (1991) Hepatic gluconeogenesis/glycolysis: regulation and structure/function relationships of substrate cycle enzymes. *Annu. Rev. Nutr.* **11,** 465–515.
A review of the hormonal regulation of the enzymes of these pathways; advanced level.

Turner, J.F. & Turner, D.H. (1980) The regulation of glycolysis and the pentose phosphate pathway. In *Biochemistry of Plants: A Comprehensive Treatise* (Stumpf, P.K. & Conn, E.E., eds), Vol. 2: *Metabolism and Respiration* (Davies, D.D., ed), pp. 279–316, Academic Press, New York.

Secondary Pathways of Glucose Oxidation

Chayen, J., Howat, D.W., & Bitensky, L. (1986) Cellular biochemistry of glucose 6-phosphate and 6-phosphogluconate dehydrogenase activities. *Cell Biochem. Funct.* **4,** 249–253.

Tephyl, T.R. & Burchell, B. (1990) UDP-glucuronosyl transferases: a family of detoxifying enzymes. *Trends Pharmacol. Sci.* **11,** 276–279.

Wood, T. (1986) Physiological functions of the pentose phosphate pathway. *Cell Biochem. Funct.* **4,** 241–247.

Problems

1. *Equation for the Preparatory Phase of Glycolysis*
Write balanced equations for all of the reactions in the catabolism of D-glucose to two molecules of D-glyceraldehyde-3-phosphate (the preparatory phase of glycolysis). For each equation write the standard free-energy change. Then write the overall or net equation for the preparatory phase of glycolysis, including the net standard free-energy change.

2. *The Payoff Phase of Glycolysis in Skeletal Muscle* In working skeletal muscle under anaerobic conditions, glyceraldehyde-3-phosphate is converted into pyruvate (the payoff phase of glycolysis), and the pyruvate is reduced to lactate. Write balanced equations for all of the reactions in this process, with the standard free-energy change for each. Then write the overall or net equation for the payoff phase of glycolysis (with lactate as the end product), including the net standard free-energy change.

3. *Pathway of Atoms in Fermentation* A "pulse-chase" experiment using ^{14}C-labeled carbon sources is carried out on a yeast extract maintained under strictly anaerobic conditions to produce ethanol. The experiment consists of incubating a small amount of ^{14}C-labeled substrate (the pulse) with the yeast extract just long enough for each intermediate in the pathway to become labeled. The label is then "chased" through the pathway by the addition of excess unlabeled glucose. The "chase" effectively prevents any further entry of labeled glucose into the pathway.

(a) If [1-^{14}C] glucose (glucose labeled at C-1 with ^{14}C) is used as a substrate, what is the location of ^{14}C in the product ethanol? Explain.

(b) Where would ^{14}C have to be located in the starting glucose molecule in order to assure that all the ^{14}C activity were liberated as $^{14}CO_2$ during fermentation to ethanol? Explain.

4. *Equivalence of Triose Phosphates* ^{14}C-Labeled glyceraldehyde-3-phosphate was added to a yeast extract. After a short time, fructose-1,6-bisphosphate labeled with ^{14}C at C-3 and C-4 was isolated. What was the location of the ^{14}C label in the starting glyceraldehyde-3-phosphate? Where did the second ^{14}C label in fructose-1,6-bisphosphate come from? Explain.

5. *Glycolysis Shortcut* Suppose you discovered a mutant yeast whose glycolytic pathway was shorter because of the presence of a new enzyme catalyzing the reaction

Glyceraldehyde-3-phosphate + H_2O $\longrightarrow$ 3-phosphoglycerate

NAD$^+$ NADH + H$^+$

Although this mutant enzyme shortens glycolysis by one step, how would it affect anaerobic ATP production? Aerobic ATP production?

6. *Role of Lactate Dehydrogenase* During strenuous activity, muscle tissue demands vast quantities of ATP compared with resting tissue. In rabbit leg muscle or turkey flight muscle, this ATP is produced almost exclusively by lactate fermentation. ATP is produced in the payoff phase of glycolysis by two enzymatic reactions, promoted by phosphoglycerate kinase and pyruvate kinase. Suppose skeletal muscle were devoid of lactate dehydrogenase. Could it carry out strenuous physical activity; that is, could it generate ATP at a high rate by glycolysis? Explain. Remember that the lactate dehydrogenase reaction does not involve ATP. A clear understanding of the answer to this question is essential for comprehension of the glycolytic pathway.

7. *Free-Energy Change for Triose Phosphate Oxidation* The oxidation of glyceraldehyde-3-phosphate to 1,3-bisphosphoglycerate, catalyzed by glyceraldehyde-3-phosphate dehydrogenase, proceeds with an unfavorable equilibrium constant ($K'_{eq} = 0.08$; $\Delta G^{\circ\prime} = +6.3$ kJ/mol). Despite this unfavorable equilibrium, the flow through this point in the pathway proceeds smoothly. How does the cell overcome the unfavorable equilibrium?

8. *Arsenate Poisoning* Arsenate is structurally and chemically similar to phosphate (P_i), and many enzymes that require phosphate will also use arsenate. Organic compounds of arsenate are less stable than analogous phosphate compounds, however. For example, acyl arsenates decompose rapidly by hydrolysis in the absence of catalysts:

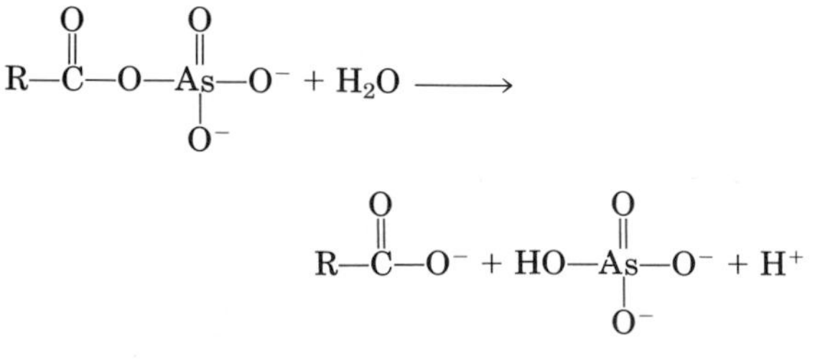

On the other hand, acyl *phosphates,* such as 1,3-bisphosphoglycerate, are more stable and are transformed in cells by enzymatic action.

(a) Predict the effect on the net reaction catalyzed by glyceraldehyde-3-phosphate dehydrogenase if phosphate were replaced by arsenate.

(b) What would be the consequence to an organism if arsenate were substituted for phosphate? Arsenate is very toxic to most organisms. Explain why.

9. *Requirement for Phosphate in Alcohol Fermentation* In 1906 Harden and Young carried out a series of classic studies on the fermentation of glucose to ethanol and CO_2 by extracts of brewer's yeast and made the following observations. (1) Inorganic phosphate was essential to fermentation; when the supply of phosphate was exhausted, fermentation ceased before all the glucose was used. (2) During fermentation under these conditions, ethanol, CO_2, and a biphosphorylated hexose accumulated. (3) When arsenate was substituted for phosphate, no biphosphorylated hexose accumulated, but the fermentation proceeded until all the glucose was converted into ethanol and CO_2.

(a) Why does fermentation cease when the supply of phosphate is exhausted?

(b) Why do ethanol and CO_2 accumulate? Is the conversion of pyruvate into ethanol and CO_2 essential? Why? Identify the biphosphorylated hexose that accumulates. Why does it accumulate?

(c) Why does the substitution of arsenate for phosphate prevent the accumulation of the biphosphorylated hexose yet allow the fermentation to ethanol and CO_2 to go to completion? (See Problem 8.)

10. *Intracellular Concentration of Free Glucose* The concentration of glucose in human blood plasma is maintained at about 5 mM. The concentration of free glucose inside muscle cells is much lower. Why is the concentration so low in the cell? What happens to the glucose upon entry into the cell?

11. *Metabolism of Glycerol* Glycerol (see below) obtained from the breakdown of fat is metabolized by being converted into dihydroxyacetone phosphate, an intermediate in glycolysis, in two enzyme-catalyzed reactions. Propose a reaction sequence for the metabolism of glycerol. On which known enzyme-catalyzed reactions is your proposal based? Write the net equation for the conversion of glycerol to pyruvate based on your scheme.

$$HOCH_2-\underset{\underset{H}{|}}{\overset{\overset{OH}{|}}{C}}-CH_2OH$$

Glycerol

12. *Measurement of Intracellular Metabolite Concentrations* Measuring the concentrations of metabolic intermediates in the living cell presents a difficult experimental problem. Because cellular enzymes rapidly catalyze metabolic interconversions, a common problem associated with perturbing the cell experimentally is that the measured concentrations of metabolites reflect not the physiological concentrations but the equilibrium concentrations. Hence, a reliable experimental technique requires all enzyme-catalyzed reactions to be instantaneously stopped in the intact tissue, so that the metabolic intermediates do not undergo

change. This objective is accomplished by rapidly compressing the tissue between large aluminum plates cooled with liquid nitrogen ($-190°C$), a process called **freeze-clamping.** After freezing, which stops enzyme action instantly, the tissue is powdered and the enzymes are inactivated by precipitation with perchloric acid. The precipitate is removed by centrifugation, and the clear supernatant extract is analyzed for metabolites. To calculate the actual intracellular concentration of the metabolite in the cell, the intracellular volume is determined from the total water content of the tissue and a measurement of the extracellular volume.

The actual intracellular concentrations of the substrates and products involved in the phosphorylation of fructose-6-phosphate by the enzyme phosphofructokinase-1 in isolated rat heart tissue are given in the table below.

Metabolite	Apparent concentration (mM)*
Fructose-6-phosphate	0.087
Fructose-1,6-bisphosphate	0.022
ATP	11.42
ADP	1.32

Source: From Williamson, J.R. (1965) Glycolytic control mechanisms I. Inhibition of glycolysis by acetate and pyruvate in the isolated, perfused rat heart. *J. Biol. Chem.* **240,** 2308–2321.
*Calculated as μmol/mL of intracellular water.

(a) Using the information in the table, calculate the mass-action ratio, [fructose-1,6-bisphosphate][ADP]/[fructose-6-phosphate][ATP], for the phosphofructokinase-1 reaction under physiological conditions.

(b) Given that $\Delta G°'$ for the PFK-1 reaction is -14.2 kJ/mol, calculate the equilibrium constant for this reaction.

(c) Compare the values of the mass-action ratio and K'_{eq}. Is the physiological reaction at equilibrium? Explain. What does this experiment say about the role of PFK-1 as a regulatory enzyme?

13. *Pasteur Effect* The regulated steps of glycolysis in intact cells are identified by studying the catabolism of glucose in whole tissues or organs. For example, the consumption of glucose by heart muscle can be measured by artificially circulating blood through an isolated intact heart and measuring the concentration of glucose before and after the blood passes through the heart. If the circulating blood is deoxygenated, heart muscle consumes glucose at a steady rate. When oxygen is added to the blood, the rate of glucose consumption drops dramatically, then continues at the new, lower rate. Why?

14. *Regulation of Phosphofructokinase-1* The effect of ATP on the allosteric enzyme PFK-1 is shown below. For a given concentration of fructose-6-phosphate, the PFK-1 activity increases with increasing concentrations of ATP, but a point is reached beyond which increasing concentrations of ATP cause inhibition of the enzyme.

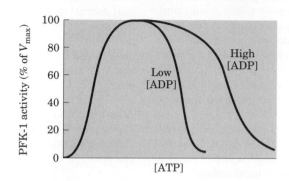

(a) Explain how ATP can be both a substrate and an inhibitor of PFK-1. How is the enzyme regulated by ATP?

(b) In what ways is glycolysis regulated by ATP levels?

(c) The inhibition of PFK-1 by ATP is diminished when the ADP concentration is high, as shown in the illustration above. How can this observation be explained?

15. *Enzyme Activity and Physiological Function* The V_{max} of the enzyme glycogen phosphorylase from skeletal muscle is much larger than the V_{max} of the same enzyme from liver tissue.

(a) What is the physiological function of glycogen phosphorylase in skeletal muscle? In liver tissue?

(b) Why does the V_{max} of the muscle enzyme need to be larger than that of the liver enzyme?

16. *Glycogen Phosphorylase Equilibrium* Glycogen phosphorylase catalyzes the removal of glucose from glycogen. Given that $\Delta G^{\circ\prime}$ for this reaction is 3.1 kJ/mol, calculate the ratio of [P_i] to [glucose-1-phosphate] when this reaction is at equilibrium. (Hint: The removal of glucose units from glycogen does not change the glycogen concentration.) The measured ratio of [P_i] to [glucose-1-phosphate] in muscle cells under physiological conditions is more than 100 to 1. What does this indicate about the direction of metabolite flow through the glycogen phosphorylase reaction? Why are the equilibrium and physiological ratios different? What is the possible significance of this difference?

17. *Regulation of Glycogen Phosphorylase* In muscle tissue, the rate of conversion of glycogen to glucose-6-phosphate is determined by the ratio of phosphorylase *a* (active) to phosphorylase *b* (less active). Determine what happens to the rate of glycogen breakdown if a muscle preparation containing glycogen phosphorylase is treated with (a) phosphorylase *b* kinase and ATP; (b) phosphorylase *a* phosphatase; (c) epinephrine.

18. *Glycogen Breakdown in Rabbit Muscle* The intracellular use of glucose and glycogen is tightly regulated at four points. In order to compare the regulation of glycolysis when oxygen is plentiful and when it is depleted, consider the utilization of glucose and glycogen by rabbit leg muscle in two physiological settings: a resting rabbit, whose leg-muscle ATP demands are low, and a rabbit who has just sighted its mortal enemy, the coyote, and dashes into its burrow at full speed. For each setting, determine the relative levels (high, intermediate, or low) of AMP, ATP, citrate, and acetyl-CoA and how these levels affect the flow of metabolites through glycolysis by regulating specific enzymes. In periods of stress, rabbit leg muscle produces much of its ATP by anaerobic glycolysis (lactate fermentation) and very little by oxidation of acetyl-CoA derived from fat breakdown.

19. *Glycogen Breakdown in Migrating Birds* Unlike the rabbit with its short dash, migratory birds require energy for extended periods of time. For example, ducks generally fly several thousand miles during their annual migration. The flight muscles of migratory birds have a high oxidative capacity and obtain the necessary ATP through the oxidation of acetyl-CoA (obtained from fats) via the citric acid cycle. Compare the regulation of muscle glycolysis during short-term intense activity, as in the fleeing rabbit, and during extended activity, as in the migrating duck. Why must the regulation in these two settings be different?

20. *Enzyme Defects in Carbohydrate Metabolism* Summaries of four clinical case studies follow. For each case determine which enzyme is defective and designate the appropriate treatment, from the lists provided. Justify your choices. Answer the questions contained in each case study.

Case A The patient develops vomiting and diarrhea shortly after milk ingestion. A lactose tolerance test is administered. (The patient ingests a standard amount of lactose, and the blood-plasma glucose and galactose concentrations are measured at intervals. In normal individuals the levels increase to a maximum in about 1 h and then recede.) The patient's blood glucose and galactose concentrations do not rise but remain constant. Explain why the blood glucose and galactose increase and then decrease in normal individuals. Why do they fail to rise in the patient?

Case B The patient develops vomiting and diarrhea after ingestion of milk. His blood is found to have a low concentration of glucose but a much higher than normal concentration of reducing sugars. The urine gives a positive test for galactose. Why is the reducing-sugar concentration in the blood high? Why does galactose appear in the urine?

Case C The patient complains of painful muscle cramps when performing strenuous physical exercise but is otherwise normal. A muscle biopsy indicates that muscle glycogen concentration is much higher than in normal individuals. Why does glycogen accumulate?

Case D The patient is lethargic, her liver is enlarged, and a biopsy of the liver shows large amounts of excess glycogen. She also has a lower than normal level of blood glucose. Account for the low blood glucose concentration in this patient.

Defective Enzyme
(a) Muscle phosphofructokinase-1
(b) Phosphomannose isomerase
(c) Galactose-1-phosphate uridylyltransferase
(d) Liver glycogen phosphorylase
(e) Triose kinase
(f) Lactase in intestinal mucosa
(g) Maltase in intestinal mucosa

Treatment
1. Jogging 5 km each day
2. Fat-free diet
3. Low-lactose diet
4. Avoiding strenuous exercise
5. Large doses of niacin (the precursor of NAD$^+$)
6. Frequent and regular feedings

21. *Severity of Clinical Symptoms Due to Enzyme Deficiency* The clinical symptoms of the two forms of galactosemia involving the deficiency of galactokinase and galactose-1-phosphate uridylyltransferase show radically different severity. Although both deficiencies produce gastric discomfort upon milk ingestion, the deficiency of the latter enzyme leads to liver, kidney, spleen, and brain dysfunction and eventual death. What products accumulate in the blood and tissues with each enzyme deficiency? Estimate the relative toxicities of these products from the above information.

22. *Preparation of [γ-^{32}P]ATP* Highly radioactive ATP labeled with ^{32}P in the γ position (terminal phosphate) is used extensively in metabolic studies. In one such procedure, investigators prepared [γ-^{32}P]ATP by incubating the following components:

 1 L 50 mM pH 8.0 buffer

 10 mmol MgCl$_2$

 2 mmol reducing agent (to inhibit disulfide bond formation)

 0.4 mmol 3-phosphoglycerate

 0.05 mmol NAD$^+$ (*not* NADH)

 0.2 mmol ATP (not radioactive, free of ADP)

 0.4 mg glyceraldyde-3-phosphate dehydrogenase

 0.2 mg phosphoglycerate kinase

 small amount of ^{32}P-labeled sodium phosphate

After the mixture was incubated for 1 h, the ATP was recovered by chromatography. Almost all the ^{32}P was found in the γ position of the ATP. How does this procedure work? Explain the role of all the components except the buffer and the reducing agent.

The Citric Acid Cycle

We saw in Chapter 14 how cells obtain energy (ATP) by fermentation, breaking down glucose in the absence of oxygen. However, most eukaryotic cells and many bacteria normally are aerobic and oxidize their organic fuels completely to CO_2 and H_2O. Under these conditions the pyruvate formed in the glycolytic breakdown of glucose is not reduced to lactate, ethanol, or some other fermentation product, as occurs under anaerobic conditions, but instead is oxidized to CO_2 and H_2O in the aerobic phase of catabolism, **respiration.** In the broader physiological or macroscopic sense, respiration refers to a multicellular organism's uptake of O_2 from its environment and release of CO_2, but biochemists and cell biologists use the term in a microscopic sense to refer to the molecular processes involved in O_2 consumption and CO_2 formation by cells. This latter process may be more precisely termed **cellular respiration.**

Cellular respiration occurs in three major stages (Fig. 15–1). In the first stage, organic fuel molecules—glucose, fatty acids, and some amino acids—are oxidized to yield two-carbon fragments in the form of the acetyl group of acetyl-coenzyme A (acetyl-CoA). In the second stage, these acetyl groups are fed into the citric acid cycle, which enzymatically oxidizes them to CO_2. The energy released by oxidation is conserved in the reduced electron carriers NADH and $FADH_2$. In the third stage of respiration, these reduced cofactors are themselves oxidized, giving up protons (H^+) and electrons. The electrons are transferred along a chain of electron-carrying molecules, known as the respiratory chain, to O_2, which they reduce to form H_2O. During this process of electron transfer much energy is released and conserved in the form of ATP, in the process called oxidative phosphorylation. Respiration is more complex than glycolysis and is believed to have evolved much later, after the appearance of cyanobacteria, which added oxygen—the electron acceptor of respiration—to the earth's atmosphere.

We will discuss the third stage of cellular respiration (electron transfer and oxidative phosphorylation) in Chapter 18. In this chapter we examine the complete oxidation of pyruvate and the **citric acid cycle,** also called the **tricarboxylic acid cycle** or the **Krebs cycle** (after its discoverer). We consider first the cycle reactions and the enzymes that catalyze them. Because intermediates of the citric acid cycle are often used as biosynthetic precursors, some means of replenishing these intermediates is essential to the continued operation of the cycle; we discuss several such replenishing reactions. The mecha-

Figure 15–1 Catabolism of proteins, fats, and carbohydrates occurs in the three stages of cellular respiration. Stage 1: Oxidation of fatty acids, glucose, and some amino acids yields acetyl-CoA. Stage 2: Oxidation of acetyl groups via the citric acid cycle includes four steps in which electrons are abstracted. Stage 3: Electrons carried by NADH and $FADH_2$ are funneled into a chain of mitochondrial (or plasma membrane–bound, in bacteria) electron carriers—the respiratory chain—ultimately reducing O_2 to H_2O. This electron flow drives the synthesis of ATP, in the process of oxidative phosphorylation.

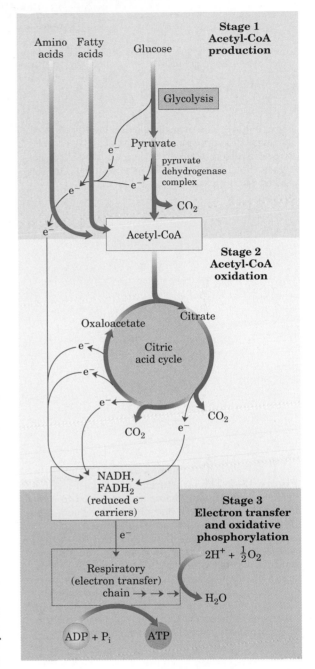

nisms that regulate the flux of material through the citric acid cycle are then considered. Finally, we describe a metabolic sequence, the glyoxylate pathway, that employs some of the same enzymes and reactions that occur in the citric acid cycle to convert acetate into oxaloacetate, a precursor of glucose.

Production of Acetate

In aerobic organisms, glucose and other sugars, fatty acids, and most of the amino acids are ultimately oxidized to CO_2 and H_2O via the citric acid cycle. Before they can enter the cycle, the carbon skeletons of sugars and fatty acids must be degraded to the acetyl group of acetyl-CoA, the form in which the citric acid cycle accepts most of its fuel input. Many amino acid carbons also enter the cycle this way, although several amino acids are degraded to other intermediates of the cycle. In Chapters 16 and 17, respectively, we shall see how fatty acids and amino acids enter the citric acid cycle. Here we consider how pyruvate, derived from glucose by glycolysis, is oxidized to yield acetyl-CoA and CO_2 by a structured cluster of three enzymes, the **pyruvate dehydrogenase complex,** located in the mitochondria of eukaryotic cells and in the cytosol of prokaryotes.

A careful examination of this enzyme complex can be rewarding in several respects. The pyruvate dehydrogenase complex is a classic example, very well studied, of a multienzyme complex in which a series of chemical intermediates remain bound to the surface of the enzyme molecules as the substrate is transformed into the final product. Five cofactors participate in this key reaction mechanism, all of which are coenzymes derived from vitamins. The regulation of this enzyme complex also illustrates how a combination of covalent modification and allosteric regulation results in precisely regulated flux through a metabolic step. Finally, the pyruvate dehydrogenase complex is the prototype for two other important enzyme complexes that we will encounter: α-ketoglutarate dehydrogenase (of the citric acid cycle) and the branched-chain α-ketoacid dehydrogenase involved in the oxidative degradation of several amino acids (Chapter 17). The remarkable similarity in the protein structure, cofactor requirements, and reaction mechanisms of these three complexes doubtless reflects a common evolutionary origin.

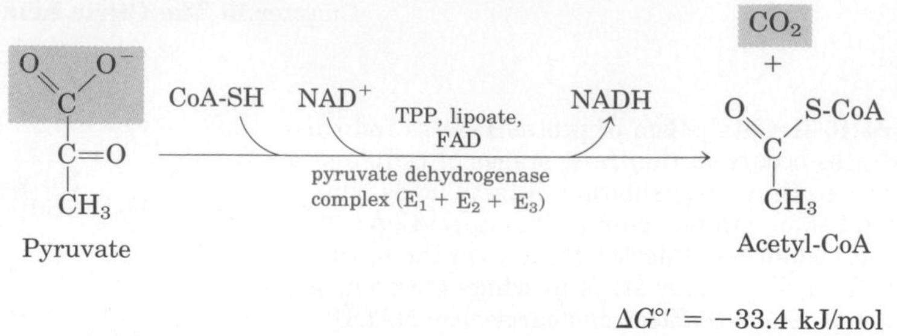

$$\Delta G^{\circ\prime} = -33.4 \text{ kJ/mol}$$

Figure 15–2 The overall reaction catalyzed by the pyruvate dehydrogenase complex. The five coenzymes involved in this reaction, and the three enzymes that make up the enzyme complex, are discussed in the text.

Pyruvate Is Oxidized to Acetyl-CoA and CO_2

The overall reaction catalyzed by the pyruvate dehydrogenase complex is **oxidative decarboxylation,** an irreversible oxidation process in which the carboxyl group is removed from pyruvate as a molecule of CO_2 and the two remaining carbons become the acetyl group of acetyl-CoA (Fig. 15–2). The NADH formed in this reaction gives up a hydride ion with its two electrons ($:H^-$) to the respiratory chain (Fig. 15–1), which carries the electrons to oxygen or, in anaerobic microorganisms, to an alternative electron acceptor such as sulfur. This electron transfer to oxygen ultimately generates three molecules of ATP per pair of electrons.

The irreversibility of the pyruvate dehydrogenase reaction has been proved by isotopic labeling experiments: radioactively labeled CO_2 cannot be reattached to acetyl-CoA to yield pyruvate labeled in the carboxyl group.

The Pyruvate Dehydrogenase Complex Requires Five Coenzymes

The combined dehydrogenation and decarboxylation of pyruvate to acetyl-CoA (Fig. 15–2) involves the sequential action of three different enzymes, as well as five different coenzymes or prosthetic groups— thiamine pyrophosphate (TPP), flavin adenine dinucleotide (FAD), coenzyme A (CoA), nicotinamide adenine dinucleotide (NAD), and lipoate. Four different vitamins required in human nutrition are vital components of this system: thiamin (in TPP), riboflavin (in FAD), niacin (in NAD), and pantothenate (in coenzyme A).

We have already described the roles of FAD and NAD as electron carriers (Chapter 13), and we have encountered TPP as the coenzyme of the enzyme that catalyzes decarboxylation of pyruvate to acetaldehyde in alcohol fermentation (see Fig. 14–9). Pantothenate, present in all living organisms, is an essential component of coenzyme A (Fig.

Figure 15–3 The structure of coenzyme A. A hydroxyl group of pantothenic acid is joined to a modified ADP moiety by a phosphate ester bond, and its carboxyl group is attached to β-mercaptoethylamine in amide linkage. The hydroxyl group at the 3' position of the ADP moiety has a phosphate group not present in ADP itself. The —SH group of the mercaptoethylamine moiety forms a thioester with acetate in acetyl-CoA (lower left). Coenzyme A is abbreviated as CoA, and acetyl-coenzyme A as acetyl-CoA; the reactive —SH group is generally shown only in chemical structures.

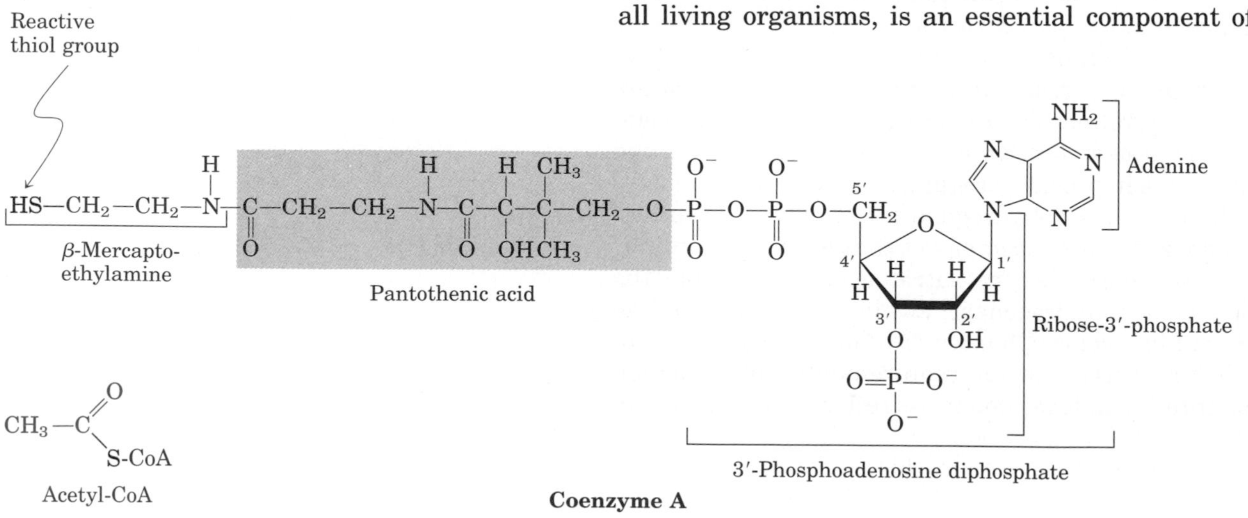

15–3). Coenzyme A has a reactive thiol (—SH) group that is critical to its role as an acyl carrier in a number of metabolic reactions; acyl groups become covalently linked to this thiol group, forming **thioesters.** Because of their relatively high free energy of hydrolysis (see Figs. 13–6, 13–7), thioesters have a high acyl group transfer potential, donating their acyl groups to a variety of acceptor molecules. The acyl group attached to coenzyme A may thus be thought of as activated for group transfer.

The fifth cofactor for the pyruvate dehydrogenase reaction, **lipoate** (Fig. 15–4), has two thiol groups, both essential to its role as cofactor. In the reduced form of lipoate both sulfur atoms are present as —SH groups, but oxidation produces a disulfide (—S—S—) bond, similar to that between two Cys residues in a protein. Because of this capacity to undergo oxidation–reduction reactions, lipoate can serve both as an electron carrier and as an acyl carrier; both functions are important in the action of the pyruvate dehydrogenase complex.

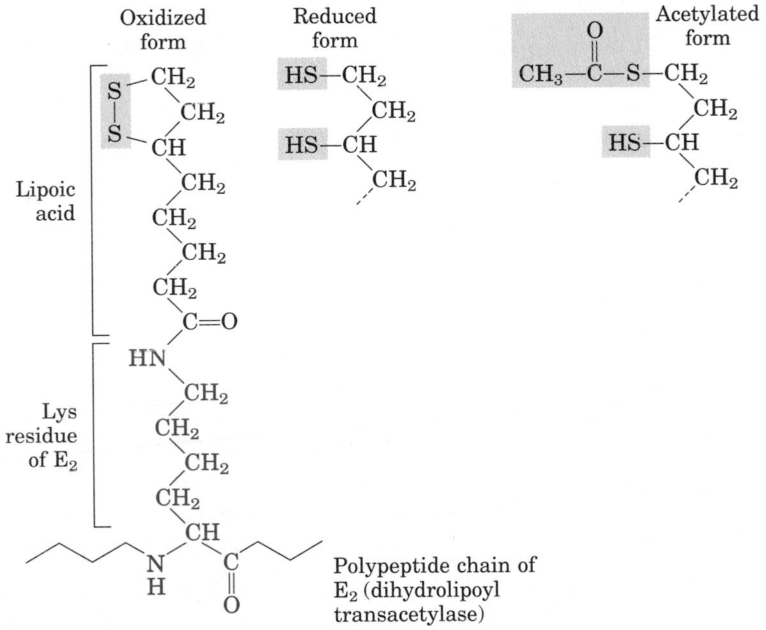

Figure 15-4 Lipoic acid in amide linkage with the side chain of a Lys residue is the prosthetic group of dihydrolipoyl transacetylase (E_2). The lipoyl group occurs in oxidized (disulfide) or reduced (dithiol) form and can act as a carrier of both hydrogen and an acetyl (or other acyl) group.

The Pyruvate Dehydrogenase Complex Consists of Three Distinct Enzymes

The pyruvate dehydrogenase complex consists of multiple copies of each of the three enzymes **pyruvate dehydrogenase** (E_1), **dihydrolipoyl transacetylase** (E_2), and **dihydrolipoyl dehydrogenase** (E_3) (Table 15–1). The number of copies of each subunit, and therefore the size of the complex, varies from one organism to another. The pyruvate dehydrogenase complex isolated from *E. coli* ($M_r > 4.5 \times 10^6$) is about 45 nm in diameter, slightly larger than a ribosome, and can be visualized with the electron microscope (Fig. 15–5a). The "core" of the cluster, to which the other enzymes are attached, is dihydrolipoyl transacetylase (E_2). In the complex from *E. coli*, 24 copies of this polypeptide chain, each containing three molecules of covalently bound lipoate, constitute this core. In the complex from mammals, there are 60 copies of E_2 and six of a related protein, protein X, which also contains covalently bound lipoate. The attachment of lipoate to the ends of

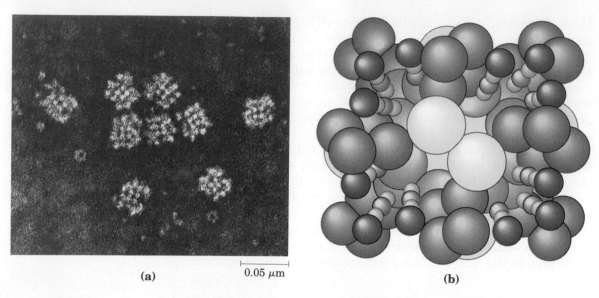

(a)
0.05 μm
(b)

Figure 15–5 (a) Electron micrograph of the pyruvate dehydrogenase complex isolated from *E. coli,* showing its subunit structure. **(b)** Interpretive model of the organization of the mammalian pyruvate dehydrogenase complex. The 24 E_2 subunits are depicted as having an inner catalytic domain (green) with an attached flexible lipoyllysyl domain (red) and an E_1/E_3 binding domain (blue) joined by linker segments (gray). The complex also contains 24 dimeric E_1 components (orange) and 6 dimeric E_3 components (yellow). Note that the mammalian complex is larger and has more subunits than the *E. coli* complex.

Lys side chains in E_2 produces lipoyllysyl groups (Fig. 15–4)—long, flexible arms that can carry acetyl groups from one active site to another in the pyruvate dehydrogenase complex. Bound to the core of E_2 molecules are 12 copies of pyruvate dehydrogenase (E_1), each composed of two identical subunits, and six copies of dihydrolipoyl dehydrogenase (E_3), each also composed of two identical subunits. Pyruvate dehydrogenase (E_1) contains bound TPP, and dihydrolipoyl dehydrogenase (E_3) contains bound FAD. Two regulatory proteins that are also part of the pyruvate dehydrogenase complex (a protein kinase and a phosphoprotein phosphatase) will be discussed below.

Table 15–1 Subunit composition of the *E. coli* pyruvate dehydrogenase complex

Enzyme	Coenzyme(s)	Molecular weight of subunit	Number of subunits per complex
Pyruvate dehydrogenase (E_1)	TPP	96,000	24
Dihydrolipoyl transacetylase (E_2)	Lipoate, CoA	65,000–70,000	24
Dihydrolipoyl dehydrogenase (E_3)	FAD, NAD	56,000	12

Source: Modified from Eley, M.H., Namihira, G., Hamilton, L., Munk, P., & Reed, L.J. (1972) α-Ketoacid dehydrogenases. XVIII: subunit composition of the *E. coli* pyruvate dehydrogenase complex. *Arch. Biochem. Biophys.* **152,** 655–669.

Intermediates Remain Bound to the Enzyme Surface

Figure 15–6 shows schematically how the pyruvate dehydrogenase complex carries out the five consecutive reactions in the decarboxylation and dehydrogenation of pyruvate. Step ① is essentially identical to the reaction catalyzed by pyruvate decarboxylase (see Fig. 14–9c); C-1 of pyruvate is released as CO_2, and C-2, which in pyruvate has the oxidation state of an aldehyde, is attached to TPP as a hydroxyethyl group. In step ② this group is oxidized to a carboxylic acid (acetate).

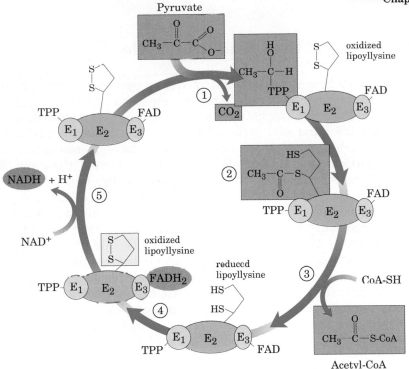

Figure 15–6 Steps in the oxidative decarboxylation of pyruvate to acetyl-CoA by the pyruvate dehydrogenase complex. The fate of pyruvate is traced in pink. In step ① pyruvate reacts with the bound thiamine pyrophosphate (TPP) of pyruvate dehydrogenase (E_1), undergoing decarboxylation to form the hydroxyethyl derivative. Pyruvate dehydrogenase also carries out step ②, the transfer of two electrons and the acetyl group from TPP to the oxidized form of the lipoyllysyl group of the core enzyme, dihydrolipoyl transacetylase (E_2), to form the acetyl thioester of the reduced lipoyl group. Step ③ is a transesterification in which the —SH group of CoA replaces the —SH group of E_2 to yield acetyl-CoA and the fully reduced (dithiol) form of the lipoyl group. In step ④ dihydrolipoyl dehydrogenase (E_3) promotes transfer of two hydrogen atoms from the reduced lipoyl groups of E_2 to the FAD prosthetic group of E_3, restoring the oxidized form of the lipoyllysyl group of E_2 (shaded yellow). In step ⑤ the reduced $FADH_2$ group on E_3 transfers a hydride ion to NAD^+, forming NADH. The enzyme complex is now ready for another catalytic cycle. The structures of TPP and its hydroxyethyl derivative are shown in Fig. 14–9. E_1, pyruvate dehydrogenase; E_2, dihydrolipoyl transacetylase; E_3, dihydrolipoyl dehydrogenase.

The two electrons removed in the oxidation reaction reduce the —S—S— of a lipoyl group on E_2 to two thiol (—SH) groups. The acetate produced in this oxidation–reduction reaction is first esterified to one of the lipoyl —SH groups, then transesterified to CoA to form acetyl-CoA (step ③). Thus the energy of oxidation drives the formation of a high-energy thioester of acetate. The remainder of the reactions catalyzed by the pyruvate dehydrogenase complex (steps ④ and ⑤) are electron transfers necessary to regenerate the disulfide form of the lipoyl group of E_2 to prepare the enzyme complex for another round of oxidation. The electrons removed from the hydroxyethyl group derived from pyruvate eventually appear in NADH after passing through FAD.

Central to the process are the swinging lipoyllysyl arms of E_2, which pass the two electrons and the acetyl group derived from pyruvate from E_1 to E_3. All these enzymes and coenzymes are clustered, allowing the intermediates to react quickly without ever diffusing away from the surface of the enzyme complex. The five-reaction sequence shown in Figure 15–6 is another example of substrate channeling, as described in Chapter 14 (see Fig. 14–8).

As one might predict, mutations in the genes for pyruvate dehydrogenase subunits, or thiamin deficiency in the diet, have severe consequences. Thiamin-deficient animals are unable to oxidize pyruvate normally, which is of particular importance in the brain; this organ usually obtains all its energy from the aerobic oxidation of glucose, and pyruvate oxidation is therefore vital. Beriberi, a disease that results from dietary deficiency of thiamin, is characterized by loss of neural function. This disease occurs primarily in populations that depend on white (polished) rice for most of their food; white rice lacks the hulls in which most of the thiamin of rice is found. People who habitually consume large amounts of alcohol can also develop thiamin deficiency; much of their dietary intake is in the form of the "empty (vitamin-free) calories" of distilled spirits. An elevated level of pyruvate in the blood is often an indicator of defects in pyruvate oxidation due to one of these causes.

Reactions of the Citric Acid Cycle

Having seen how acetyl-CoA is formed from pyruvate, we are ready to examine the citric acid cycle. An overview of how the cycle functions will be helpful. First we note a fundamental difference between glycolysis and the citric acid cycle. Glycolysis takes place by a linear sequence of enzyme-catalyzed steps, whereas the sequence of reactions in the citric acid cycle is cyclical. To begin a turn of the cycle (Fig. 15–7), acetyl-CoA donates its acetyl group to the four-carbon compound oxaloacetate to form the six-carbon citrate. Citrate is then transformed into isocitrate, also a six-carbon molecule, which is dehydrogenated with loss of CO_2 to yield the five-carbon compound α-ketoglutarate. The latter undergoes loss of CO_2 and ultimately yields the four-carbon compound succinate and a second molecule of CO_2. Succinate is then enzymatically converted in three steps into the four-carbon oxaloacetate, with which the cycle began; thus, oxaloacetate is ready to react with another molecule of acetyl-CoA to start a second turn. In each turn of the cycle, one acetyl group (two carbons) enters as acetyl-CoA and two molecules of CO_2 leave. In each turn, one molecule of oxaloacetate is used to form citrate but—after a series of reactions—the oxaloacetate is regenerated. Therefore no net removal of oxaloacetate occurs; one molecule of oxaloacetate can theoretically suffice to bring about oxidation of an infinite number of acetyl groups. Four of the eight steps in this process are oxidations, in which the energy of oxidation is conserved, with high efficiency, in the formation of reduced cofactors (NADH and $FADH_2$).

After discussing the individual reactions of the citric acid cycle in detail, we will briefly consider the experiments that led to its discovery, and the evolutionary origins of the cycle.

Although the citric acid cycle is central to energy-yielding metabolism, its role is not limited to energy conservation. Four- and five-carbon intermediates of the cycle serve as biosynthetic precursors for a wide variety of products. To replace intermediates removed for this purpose, cells employ anaplerotic (replenishing) reactions, which are briefly described.

Eugene Kennedy and Albert Lehninger showed in 1948 that in eukaryotes the entire set of reactions of the citric acid cycle takes place in the mitochondria. Isolated mitochondria were found to contain not only all the enzymes and coenzymes required for the citric acid cycle but also all the enzymes and proteins necessary for the last stage of respiration, namely, electron transfer and ATP synthesis by oxidative phosphorylation. As we shall see in later chapters, mitochondria also contain the enzymes that catalyze the oxidation of fatty acids to acetyl-CoA and the oxidative degradation of amino acids to acetyl-CoA or, for some amino acids, α-ketoglutarate, succinyl-CoA, or oxaloacetate. Thus in nonphotosynthetic eukaryotes the mitochondrion is the site of most energy-yielding oxidative reactions and of the synthesis of ATP coupled to those reactions. In photosynthetic eukaryotes, mitochondria are the major site of ATP production in the dark, but in daylight chloroplasts produce most of the ATP. Most prokaryotes contain the enzymes of the citric acid cycle in their cytosol, and their plasma membrane plays a role analogous to that of the inner mitochondrial membrane in ATP synthesis (Chapter 18).

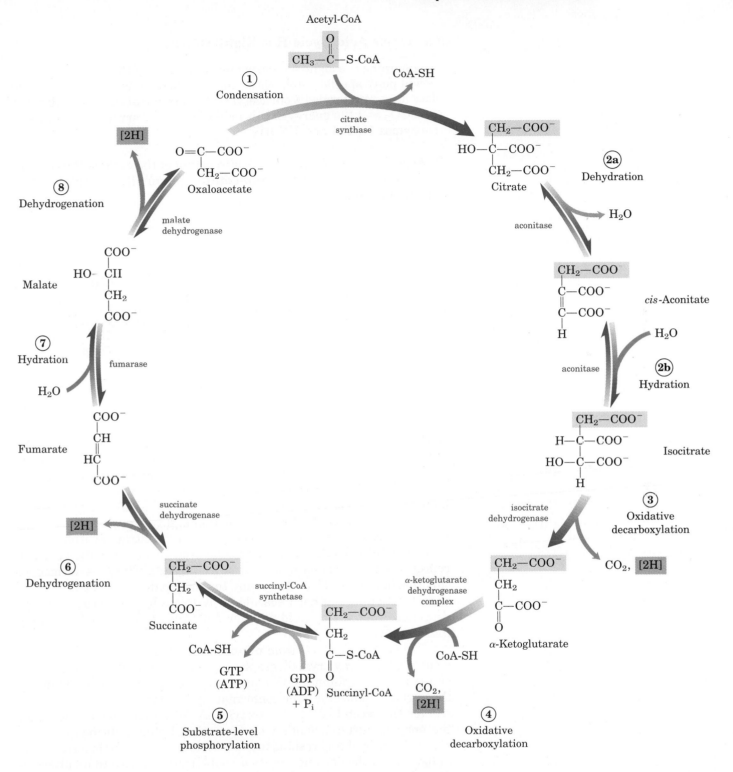

Figure 15–7 The reactions of the citric acid cycle. The carbon atoms shaded in red are those originally derived from the acetate of acetyl-CoA in the first turn of the cycle. These carbons are *not* the ones released as CO_2 in the first turn. Note that in fumarate, the two-carbon group derived from acetate can no longer be specifically denoted; because succinate and fumarate are symmetric molecules, C-1 and C-2 are indistinguishable from C-4 and C-3. The number beside each reaction step corresponds to a numbered heading in the text. Steps ①, ③, and ④ are essentially irreversible in the cell; all of the others are reversible.

The Citric Acid Cycle Has Eight Steps

In examining the eight successive reaction steps of the citric acid cycle, we will place special emphasis on the chemical transformations taking place as citrate formed from acetyl-CoA and oxaloacetate is oxidized to yield CO_2 and the energy of this oxidation is conserved in the reduced coenzymes NADH and $FADH_2$.

① *Formation of Citrate* The first reaction of the cycle is the condensation of acetyl-CoA with **oxaloacetate** to form **citrate,** catalyzed by **citrate synthase:**

$$\Delta G^{\circ\prime} = -32.2 \text{ kJ/mol}$$

In this reaction the methyl carbon of the acetyl group is joined to the carbonyl group (C-2) of oxaloacetate. Citroyl-CoA is a transient intermediate. It is formed on the active site of the enzyme and rapidly undergoes hydrolysis to yield free CoA and citrate, which are then released from the active site. The hydrolysis of the high-energy thioester intermediate makes the forward reaction highly exergonic. The large, negative free-energy change associated with the citrate synthase reaction is essential to the operation of the cycle, because of the very low concentration of oxaloacetate normally present. The CoA formed in this reaction is recycled; it is ready to participate in the oxidative decarboxylation of another molecule of pyruvate by the pyruvate dehydrogenase complex to yield another molecule of acetyl-CoA for entry into the cycle.

The citrate synthase from mitochondria has been crystallized and visualized by x-ray crystallography in the presence and absence of its substrates and inhibitors. Oxaloacetate, the first substrate to bind to the enzyme, induces a large conformational change, creating a binding site for the second substrate, acetyl-CoA. When citroyl-CoA forms on the enzyme surface, another conformational change brings the side chain of a crucial Asp residue into position to cleave the thioester. This induced fit of the enzyme first to its substrate and then to its intermediate decreases the likelihood of premature and unproductive cleavage of the thioester bond of acetyl-CoA.

② *Formation of Isocitrate via* **cis**-*Aconitate* The enzyme **aconitase** (more formally, **aconitate hydratase**) catalyzes the reversible transformation of citrate to **isocitrate,** through the intermediary formation of the tricarboxylic acid *cis*-**aconitate,** which normally does not dissociate from the active site. Aconitase can promote the reversible addition of H_2O to the double bond of enzyme-bound *cis*-aconitate in two different ways, one leading to citrate and the other to isocitrate:

$$\Delta G^{\circ\prime} = 13.3 \text{ kJ/mol}$$

Although the equilibrium mixture at pH 7.4 and 25 °C contains less than 10% isocitrate, in the cell the reaction is pulled to the right because isocitrate is rapidly consumed in the subsequent step of the cycle, lowering its steady-state concentration. Aconitase contains an **iron–sulfur center** (Fig. 15–8), which acts both in the binding of the substrate at the active site and in catalysis of the addition or removal of H_2O.

③ *Oxidation of Isocitrate to α-Ketoglutarate and CO_2* In the next step **isocitrate dehydrogenase** catalyzes oxidative decarboxylation of isocitrate to form **α-ketoglutarate**:

$$\Delta G^{\circ\prime} = -20.9 \text{ kJ/mol}$$

There are two different forms of isocitrate dehydrogenase (isozymes; see Box 14–3), one requiring NAD^+ as electron acceptor and the other requiring $NADP^+$. The overall reactions catalyzed by the two isozymes are otherwise identical. The NAD-dependent enzyme is found in the mitochondrial matrix and serves in the citric acid cycle to produce α-ketoglutarate. The NADP-dependent isozyme is found in both the mitochondrial matrix and the cytosol. It may function primarily in the generation of NADPH, which is essential for reductive anabolic reactions.

④ *Oxidation of α-Ketoglutarate to Succinyl-CoA and CO_2* The next step is another oxidative decarboxylation, in which α-ketoglutarate is converted to **succinyl-CoA** and CO_2 by the action of the **α-ketoglutarate dehydrogenase complex;** NAD^+ serves as electron acceptor:

$$\Delta G^{\circ\prime} = -33.5 \text{ kJ/mol}$$

This reaction is virtually identical to the pyruvate dehydrogenase reaction discussed above; both involve the oxidation of an α-keto acid

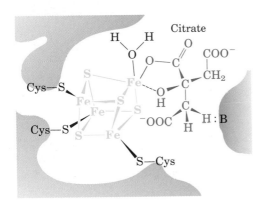

Figure 15–8 The iron–sulfur center in aconitase acts in substrate binding and catalysis. Three Cys residues of the enzyme bind three iron atoms in the iron–sulfur center (yellow); the fourth iron is bound to one of the carboxyl groups of citrate (blue). :B represents a basic residue on the enzyme that helps to position the citrate in the active site.

with loss of the carboxyl group as CO_2. The energy of oxidation of α-ketoglutarate is conserved in the formation of the thioester bond of succinyl-CoA. In both structure and function the α-ketoglutarate dehydrogenase complex closely resembles the pyruvate dehydrogenase complex. It includes three enzymes, analogous to E_1, E_2, and E_3 of the pyruvate dehydrogenase complex, as well as enzyme-bound TPP, bound lipoate, FAD, NAD, and coenzyme A. Although E_1 of the α-ketoglutarate dehydrogenase complex is structurally similar to E_1 of pyruvate dehydrogenase, their amino acid sequences are different; the E_1 components specifically bind either α-ketoglutarate or pyruvate, conferring substrate specificity upon their respective enzyme complexes. The subunits of E_3 for the two complexes are virtually identical. The E_2 components of the two complexes are very similar; both have covalently bound lipoyl moieties. It is a near certainty that the proteins of these two multienzyme complexes share a common evolutionary origin.

(5) *Conversion of Succinyl-CoA to Succinate* Succinyl-CoA, like acetyl-CoA, has a strongly negative free energy of hydrolysis of its thioester bond ($\Delta G^{\circ\prime} \approx -36$ kJ/mol). In the next step of the citric acid cycle, energy released in the breakage of this bond is used to drive the synthesis of a phosphoanhydride bond in either GTP or ATP, and **succinate** is also formed in the process:

$$\Delta G^{\circ\prime} = -2.9 \text{ kJ/mol}$$

The enzyme that catalyzes this reversible reaction is called either **succinyl-CoA synthetase** or **succinic thiokinase;** both names indicate the participation of a nucleoside triphosphate in the reaction (Box 15–1). It was long thought that the enzyme from animal tissues used GDP exclusively, whereas that from plants and bacteria used predominantly ADP. It now appears that some animal cells contain two isozymes, one specific for ADP and the other for GDP.

This energy-conserving reaction involves an intermediate step in which the enzyme molecule itself becomes phosphorylated at a His residue in the active site (Fig. 15–9). This phosphate group, which has a high group transfer potential, is transferred to ADP (or GDP) to form ATP (or GTP). The coupled formation of ATP (or GTP) at the expense of the energy released by the oxidative decarboxylation of α-ketoglutarate is another example of a substrate-level phosphorylation, like the synthesis of ATP coupled to the oxidation of glyceraldehyde-3-phosphate in glycolysis (p. 411). These reactions are called substrate-level phosphorylations to distinguish them from oxidative or respiration-linked phosphorylation. Substrate-level phosphorylations involve soluble enzymes and chemical intermediates such as the phosphohistidine residue in succinyl-CoA synthetase, or 1,3-bisphosphoglycerate in the glycolytic pathway. Respiration-linked phosphorylation, on the other hand, involves membrane-bound enzymes and transmembrane gradients of protons (Chapter 18).

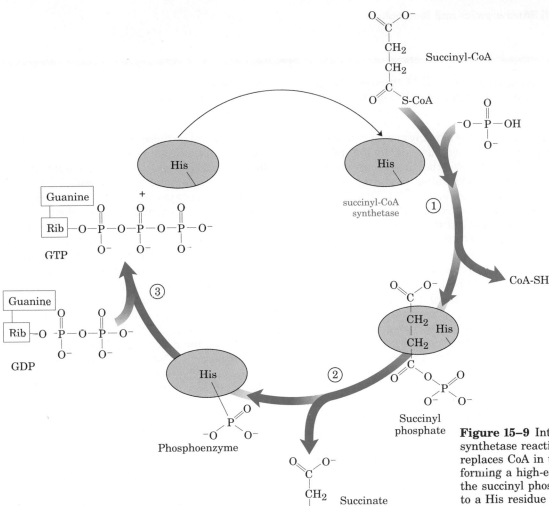

Figure 15–9 Intermediates in the succinyl-CoA synthetase reaction. In step ① a phosphate group replaces CoA in the enzyme-bound succinyl-CoA, forming a high-energy acyl phosphate. In step ② the succinyl phosphate donates its phosphate group to a His residue on the enzyme, forming the high-energy phosphohistidyl enzyme. In step ③ the phosphate group is transferred from the His residue to the terminal phosphate of GDP, forming GTP.

The GTP formed by succinyl-CoA synthetase may donate its terminal phosphate group to ADP to form ATP, by the reversible action of **nucleoside diphosphate kinase:**

$$GTP + ADP \underset{Mg^{2+}}{\rightleftharpoons} GDP + ATP \qquad \Delta G^{\circ\prime} = 0 \text{ kJ/mol}$$

Thus the net result of the activity of either isozyme of succinyl-CoA synthetase is the conservation of energy as ATP. There is no change in free energy for the nucleoside diphosphate kinase reaction; ATP and GTP are energetically equivalent.

⑥ *Oxidation of Succinate to Fumarate* The succinate formed from succinyl-CoA is oxidized to **fumarate** by the flavoprotein **succinate dehydrogenase** (right). In eukaryotes, succinate dehydrogenase is tightly bound to the inner mitochondrial membrane (in prokaryotes, to the plasma membrane); it is the only enzyme of the citric acid cycle that is membrane-bound. The enzyme from beef heart mitochondria contains three different iron–sulfur clusters as well as one molecule of covalently bound FAD. Electrons pass from succinate through the FAD and iron–sulfur centers before entering the chain of electron carriers in the mitochondrial inner membrane (or the plasma membrane of bacteria). Electron flow from succinate through these carriers to the final electron acceptor, O_2, is coupled to the synthesis of two ATP mole-

BOX 15–1

Synthases, Synthetases, Ligases, Lyases, and Kinases: A Refresher on Enzyme Nomenclature

Citrate synthase is one of many enzymes that catalyze condensation reactions, yielding a product more chemically complex than its precursors. **Synthases** catalyze condensation reactions in which no nucleoside triphosphate (ATP, GTP, etc.) is required as an energy source. **Synthetases** also catalyze condensations, but this name is reserved for enzymes that *do* use ATP or another nucleoside triphosphate as a source of energy for the synthetic reaction. Succinyl-CoA synthetase is such an enzyme. **Ligase** (from the Latin *ligare,* meaning "to tie together") is the general name applied to those enzymes that catalyze condensation reactions in which two atoms are joined using the energy of ATP or another energy source. DNA ligase, for example, closes breaks in DNA molecules, using energy supplied by either ATP or NAD^+; it is widely used in joining DNA pieces for genetic engineering (Chapter 28). Ligases are not to be confused with **lyases,** enzymes that catalyze cleavages (or, in the reverse directions, additions) in which electronic rearrangements occur. The pyruvate dehydrogenase complex, which oxidatively cleaves CO_2 from pyruvate, is in this large class of enzymes. The name **kinase** is applied to those enzymes that transfer a phosphate group from a nucleoside triphosphate such as ATP to some acceptor molecule—a sugar (as in hexokinase and glucokinase), a protein (as in glycogen phosphorylase kinase), another nucleotide (as in nucleoside diphosphate kinase), or a metabolic intermediate such as oxaloacetate (as in PEP carboxykinase).

Unfortunately, these descriptions of enzyme types overlap, and many enzymes are commonly called by two or more names. Succinyl-CoA synthetase, for example, is also called succinate thiokinase; the enzyme is clearly both a synthetase in the citric acid cycle and a kinase when acting in the direction of succinyl-CoA synthesis. This raises another source of confusion in the naming of enzymes: an enzyme is sometimes discovered by the use of an assay of the conversion of, say, A to B and is named for that reaction, but later it is found to function in the cell primarily in converting B to A. Commonly, the first name continues to be used, although the metabolic role of the enzyme would clearly be better described by naming it for the reverse reaction. The enzyme pyruvate kinase (which acts in glycolysis) illustrates this situation (p. 413). To a beginner in biochemistry, this duplication in nomenclature can be bewildering. International committees have made heroic efforts to systematize the nomenclature of enzymes (see Table 8–3 for a brief summary of the system), but often the systematic names are so long and cumbersome that they are not frequently used in biochemical conversation.

We have tried throughout this book to use the name most commonly used by working biochemists, and to point out cases in which a given enzyme has more than one widely used name. For an authoritative treatment of systematic enzyme nomenclature, see: Webb, E. (1984) *Enzyme Nomenclature,* Academic Press, Inc., Orlando, FL.

Malonate   Succinate

cules per pair of electrons (respiration-linked phosphorylation; Chapter 18). Malonate, an analog of succinate, is a strong competitive inhibitor of succinate dehydrogenase and therefore blocks the citric acid cycle.

⑦ *Hydration of Fumarate to Produce Malate* The reversible hydration of fumarate to L-**malate** is catalyzed by **fumarase (fumarate hydratase):**

Fumarate   L-Malate

$$\Delta G^{\circ\prime} = -3.8 \ \text{kJ/mol}$$

This enzyme is highly stereospecific; it catalyzes hydration of the trans double bond of fumarate but does not act on maleate, the cis isomer of fumarate. In the reverse direction (from L-malate to fumarate), fumarase is equally stereospecific: D-malate is not a substrate.

⑧ *Oxidation of Malate to Oxaloacetate* In the last reaction of the citric acid cycle, NAD-linked L-**malate dehydrogenase** catalyzes the oxidation of L-malate to oxaloacetate:

$$\Delta G^{\circ\prime} = 29.7 \text{ kJ/mol}$$

The equilibrium of this reaction lies far to the left under standard thermodynamic conditions. However, in intact cells oxaloacetate is continually removed by the highly exergonic citrate synthase reaction (p. 454). This keeps the concentration of oxaloacetate in the cell extremely low ($< 10^{-6}$ M), pulling the malate dehydrogenase reaction toward oxaloacetate formation.

The Energy of Oxidations in the Cycle Is Efficiently Conserved

We have now covered one complete turn of the citric acid cycle (Fig. 15–10). An acetyl group, containing two carbon atoms, was fed into the cycle by combining with oxaloacetate. Two carbon atoms emerged from the cycle as CO_2 with the oxidation of isocitrate and α-ketoglutarate, and at the end of the cycle a molecule of oxaloacetate was regenerated.

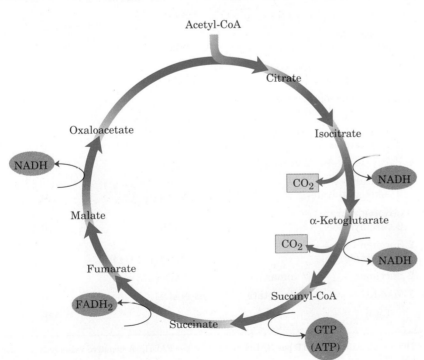

Figure 15–10 Each turn of the citric acid cycle produces three NADH and one $FADH_2$, as well as one GTP (or ATP). Two CO_2 are released in oxidative decarboxylation reactions. (Here and in several following figures, all cycle reactions are shown in one direction only. Keep in mind that most of the reactions are actually reversible, as shown in Fig. 15–7.)

Note that the two carbon atoms appearing as CO_2 are not the same two carbons that entered in the form of the acetyl group; additional turns around the cycle are required before the carbon atoms that entered as an acetyl group finally appear as CO_2 (Fig. 15–7).

Although the citric acid cycle itself directly generates only one molecule of ATP per turn (in the conversion of succinyl-CoA to succinate), the four oxidation steps in the cycle provide a large flow of electrons into the respiratory chain and thus eventually lead to formation of a large number of ATP molecules during oxidative phosphorylation.

We saw in the previous chapter that the energy yield from glycolysis (in which one molecule of glucose is converted into two of pyruvate) is two ATP molecules from one of glucose. When both pyruvate molecules are completely oxidized to yield six CO_2 molecules in the reactions catalyzed by the pyruvate dehydrogenase complex and the enzymes of the citric acid cycle, and the electrons are transferred to O_2 via the respiratory chain, as many as 38 ATP are obtained per glucose (Table 15–2). In round numbers, this represents the conservation of 38×30.5 kJ/mol = 1,160 kJ/mol, or 40% of the theoretical maximum of 2,840 kJ/mol available from the complete oxidation of glucose. These calculations employ the standard free-energy changes; when corrected for the actual free energy required to form ATP within cells (Box 13–2; p. 375), the calculated efficiency of the process is even greater.

Table 15–2 The stoichiometry of coenzyme reduction and ATP formation in the aerobic oxidation of a molecule of glucose via glycolysis, the pyruvate dehydrogenase reaction, and the citric acid cycle

Reaction	Number of ATP or reduced coenzymes directly formed	Number of ATP ultimately formed*
Glucose ⟶ glucose-6-phosphate	−1 ATP	−1
Fructose-6-phosphate ⟶ fructose-1,6-bisphosphate	−1 ATP	−1
2 Glyceraldehyde-3-phosphate ⟶ 2 1,3-bisphosphoglycerate	2 NADH	6
2 1,3-Bisphosphoglycerate ⟶ 2 3-phosphoglycerate	2 ATP	2
2 Phosphoenolpyruvate ⟶ 2 pyruvate	2 ATP	2
2 Pyruvate ⟶ 2 acetyl-CoA	2 NADH	6
2 Isocitrate ⟶ 2 α-ketoglutarate	2 NADH	6
2 α-Ketoglutarate ⟶ 2 succinyl-CoA	2 NADH	6
2 Succinyl-CoA ⟶ 2 succinate	2 ATP (or 2 GTP)	2
2 Succinate ⟶ 2 fumarate	2 $FADH_2$	4
2 Malate ⟶ 2 oxaloacetate	2 NADH	6
Total		38

* This is calculated as 3 ATP per NADH and 2 ATP per $FADH_2$. A negative value indicates consumption.

How Was the Cyclic Nature of the Pathway Established?

The historical road that led to the understanding of this cyclic metabolic pathway is instructive; it illustrates the general strategies used by biochemists since early in this century to unravel complex metabolic relationships. The citric acid cycle was first postulated as the pathway of pyruvate oxidation in animal tissues by Hans Krebs, in 1937. The idea of the cycle came to him during a study of the effect of the anions of various organic acids on the rate of oxygen consumption during pyruvate oxidation by suspensions of minced pigeon-breast muscle. This muscle, used in flight, has a very high rate of respiration and was thus especially appropriate for the study of oxidative activity. Earlier investigators, particularly Albert Szent-Györgyi, had found that certain four-carbon dicarboxylic acids known to be present in animal tissues—succinate, fumarate, malate, and oxaloacetate—stimulate the consumption of oxygen by muscle. Krebs confirmed this observation and found that they also stimulate the oxidation of pyruvate. Moreover, he found that oxidation of pyruvate by muscle is also stimulated by the six-carbon tricarboxylic acids citrate, *cis*-aconitate, and isocitrate, and the five-carbon α-ketoglutarate. No other naturally occurring organic acids that were tested possessed such activity. The stimulatory effect of the active acids was remarkable: the addition of even a small quantity of any one of them could promote the oxidation of many times that amount of pyruvate.

Hans Krebs
1900–1981

The second important observation made by Krebs was that malonate (p. 458), a close analog of succinate and a competitive inhibitor of succinate dehydrogenase, inhibits the aerobic utilization of pyruvate by muscle suspensions, *regardless of which active organic acid is added*. This indicated that succinate and succinate dehydrogenase must be essential components in the enzymatic reactions involved in the oxidation of pyruvate. Krebs further found that when malonate is used to inhibit the aerobic utilization of pyruvate by a suspension of muscle tissue, there is an accumulation of citrate, α-ketoglutarate, and succinate in the suspending medium. This suggested that citrate and α-ketoglutarate are normally precursors of succinate.

From these basic observations and other evidence, Krebs concluded that the active tricarboxylic and dicarboxylic acids listed above can be arranged in a logical chemical sequence. Because the incubation of pyruvate and oxaloacetate with ground muscle tissue resulted in accumulation of citrate in the medium, Krebs reasoned that this sequence functions in a circular rather than linear manner: its beginning and end are linked together. For the missing link that closes the circle he proposed the reaction

$$\text{Pyruvate} + \text{oxaloacetate} \longrightarrow \text{citrate} + CO_2$$

(Notice that in this detail Krebs was originally incorrect.)

From these simple experiments and logical reasoning, Krebs postulated what he called the citric acid cycle as the main pathway for oxidation of carbohydrate in muscle. The citric acid cycle is also called the tricarboxylic acid cycle because for some years after Krebs postulated the cycle it was uncertain whether citrate or some other tricarboxylic acid such as isocitrate was the first product formed by reaction of pyruvate and oxaloacetate. In the years since its discovery, the citric acid cycle has been found to function not only in muscles but in virtually all tissues of aerobic animals and plants and in many aerobic microorganisms.

The citric acid cycle was first postulated from experiments carried out on suspensions of minced muscle tissue. Subsequently its details were worked out by study of the highly purified enzymes of the cycle. One might ask whether these enzymes really function in a cycle in intact living cells and whether the rate of the cycle is high enough to account for the overall rate of glucose oxidation in animal tissues. These questions have been studied by the use of isotopically labeled metabolites, such as pyruvate or acetate, in which the isotopes ^{13}C or ^{14}C are used to mark a given carbon atom in the molecule. Many stringent experiments with the isotope tracer technique have confirmed that the citric acid cycle does take place in living cells, and at a high rate.

Some of the earliest experiments with isotopes produced an unexpected result, however, which aroused considerable controversy about the pathway and mechanism of the citric acid cycle. In fact, these experiments at first seemed to show that citrate was not the first tricarboxylic acid to be formed. Box 15–2 gives some details of this episode in the scientific history of the cycle.

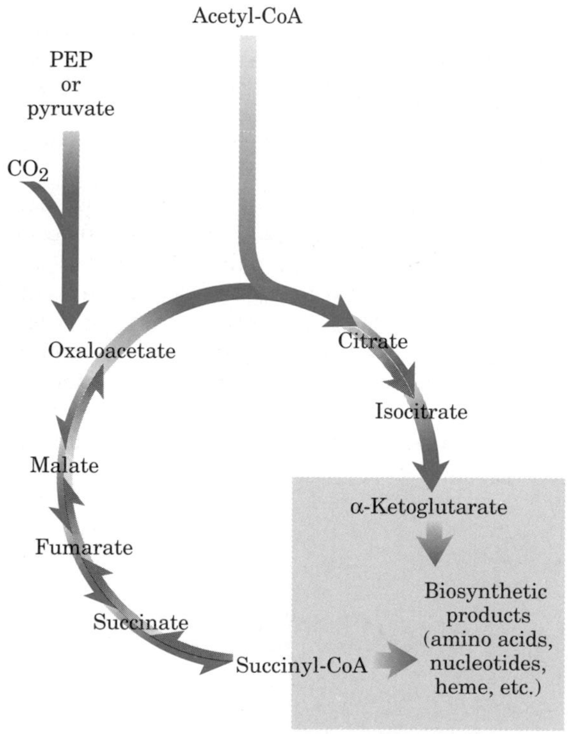

Why Is the Oxidation of Acetate So Complicated?

This eight-step, cyclic process for oxidation of the simple two-carbon acetyl groups to CO_2 may seem unnecessarily cumbersome and not in keeping with the principle of maximum economy in the molecular logic of living cells. The role of the citric acid cycle is not confined to the oxidation of acetate, however; this pathway is the hub of intermediary metabolism. Four- and five-carbon end products of many catabolic processes are fed into the cycle to serve as fuels. Oxaloacetate and α-ketoglutarate are, for example, produced from asparate and glutamate, respectively, when proteins in the diet are degraded. Under other circumstances, intermediates are drawn out of the cycle to be used as precursors in a variety of biosynthetic pathways.

The pathway used in modern organisms is the product of evolution, much of which occurred before the advent of aerobic organisms. It does not necessarily represent the *shortest* pathway from acetate to CO_2, but is rather the pathway that has conferred the greatest selective advantage on its possessors throughout evolution. Early anaerobes very probably used some of the reactions of the citric acid cycle in linear biosynthetic processes. In fact, there are modern anaerobic microorganisms in which an incomplete citric acid cycle serves as a source, not of energy, but of biosynthetic precursors (Fig. 15–11). These organisms use the first three reactions of the citric acid cycle to make α-ketoglutarate, but they lack α-ketoglutarate dehydrogenase and therefore cannot carry out the complete set of citric acid cycle reactions. They do have the four enzymes that catalyze the reversible conversion of oxaloacetate into succinyl-CoA (Fig. 15–11), and these anaerobes make malate, fumarate, succinate, and succinyl-CoA from oxaloacetate in a reversal of the "normal" (oxidative) direction of flow through the cycle.

With the evolution of cyanobacteria that produced O_2 from water, the earth's atmosphere became aerobic and there was selective pressure on organisms to develop aerobic metabolism, which, as we have seen, is much more efficient than anaerobic fermentation.

Figure 15–11 The noncyclic reactions (blue arrows) that provide biosynthetic precursors in anaerobically growing bacteria. These cells lack α-ketoglutarate dehydrogenase and therefore cannot carry out the complete citric acid cycle, which normally follows the direction shown with gray arrows. α-Ketoglutarate and succinyl-CoA serve as precursors in a variety of biosynthetic reactions (see Fig. 15–12).

BOX 15–2   **Is Citric Acid the First Tricarboxylic Acid Formed in the Cycle?**

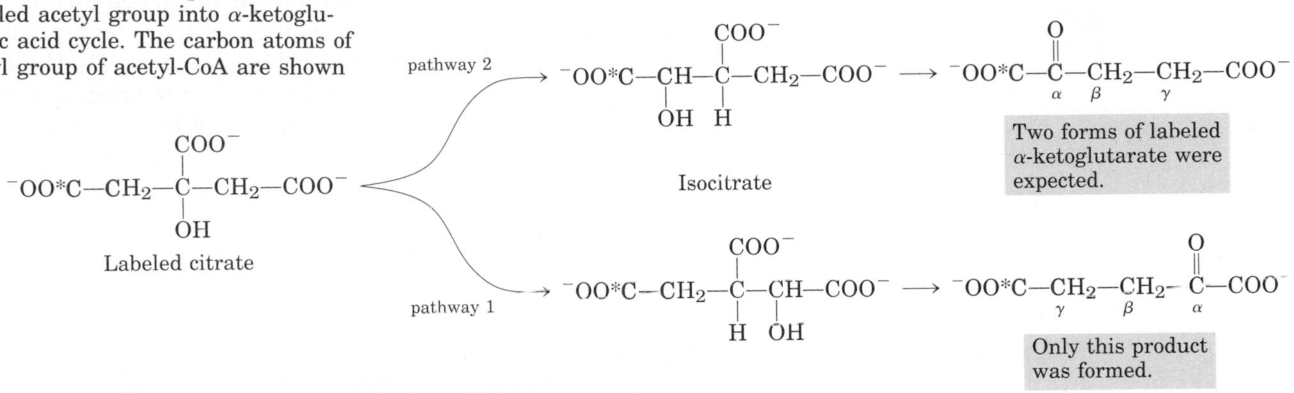

Figure 1 Incorporation of the isotopic carbon (asterisk) of the labeled acetyl group into α-ketoglutarate by the citric acid cycle. The carbon atoms of the entering acetyl group of acetyl-CoA are shown in red.

pathway 2 $\longrightarrow$ $^-OO^*C-CH-\underset{OH\ \ \ H}{\overset{|\ \ \ \ |}{C}}-CH_2-COO^-$ $\longrightarrow$ $^-OO^*C-\underset{\alpha}{\overset{O}{C}}-\underset{\beta}{CH_2}-\underset{\gamma}{CH_2}-COO^-$

Two forms of labeled α-ketoglutarate were expected.

$^-OO^*C-CH_2-\underset{OH}{\overset{COO^-}{C}}-CH_2-COO^-$

Labeled citrate

Isocitrate

pathway 1 $\longrightarrow$ $^-OO^*C-CH_2-\underset{H\ \ \ OH}{\overset{COO^-}{C}}-CH-COO^-$ $\longrightarrow$ $^-OO^*C-\underset{\gamma}{CH_2}-\underset{\beta}{CH_2}-\underset{\alpha}{\overset{O}{C}}-COO^-$

Only this product was formed.

When the heavy-carbon isotope ^{13}C and the radioactive carbon isotopes ^{11}C and ^{14}C became available, they were very soon put to use to trace the pathway of carbon atoms through the citric acid cycle. In one such experiment, which initiated the controversy over the role of citric acid, acetate labeled in the carboxyl group (designated [1-^{14}C]acetate) was incubated aerobically with an animal tissue preparation. Acetate is enzymatically converted into acetyl-CoA in animal tissues, and the pathway of the labeled carboxyl carbon of the acetyl group in the cycle reactions could be traced. α-Ketoglutarate was isolated from the tissue after incubation, then degraded by known chemical reactions to establish the position(s) of the isotopic carbon derived from carboxyl-labeled acetate. Condensation of unlabeled oxaloacetate with carboxyl-labeled acetate would be expected to produce citrate labeled in one of the two primary carboxyl groups (Fig. 1). Because citrate is a sym-

metric molecule, with no asymmetric carbon, its two terminal carboxyl groups are chemically indistinguishable. Therefore, half of the labeled citrate molecules were expected to yield α-ketoglutarate labeled in the α-carboxyl group and the other half to yield α-ketoglutarate labeled in the γ-carboxyl group; that is, the α-ketoglutarate isolated should have been a mixture of molecules labeled in both carboxyl groups.

Contrary to this expectation, the labeled α-ketoglutarate isolated from the tissue suspension contained the isotope only in the γ-carboxyl group (Fig. 1). It was concluded that citrate itself, or any other symmetric molecule, could not possibly be an intermediate in the pathway from acetate to α-ketoglutarate. Hence an asymmetric tricarboxylic acid, presumably *cis*-aconitate or isocitrate, had to be the first condensation product formed from acetate and oxaloacetate.

In 1948, however, Alexander Ogston pointed out that although citrate has no chiral center (see Fig. 3–9), it has the *potential* of reacting asymmetrically if the enzyme acting upon it has an active site that is asymmetric. He suggested that the active site of aconitase, the enzyme acting on the newly formed citrate, may have three points to which the citrate molecule must be bound and that the citrate molecule must undergo a specific three-point attachment to these binding points. As seen in Figure 2, the binding of citrate to the three points can happen in only one way, and this would account for the formation of only one type of labeled α-ketoglutarate. Organic molecules such as citrate that have no chiral center, but are potentially capable of reacting asymmetrically with an asymmetric active site, are now called **prochiral** molecules.

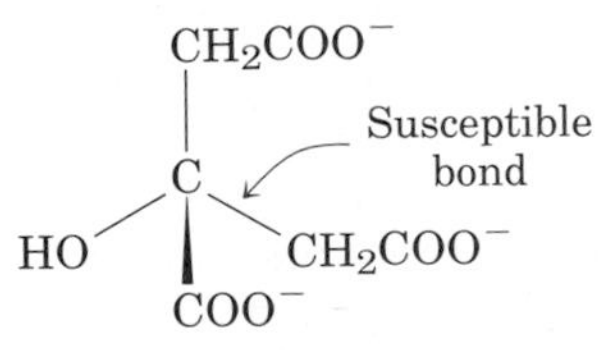

$\underset{COO^-}{\overset{\underset{\textstyle CH_2COO^-}{|}}{\underset{|}{\overset{|}{C}}}}$ HO — C — CH$_2$COO$^-$

Susceptible bond

(a)

$\underset{Y}{\overset{Z}{\underset{|}{\overset{|}{C}}}}$ X $\diagdown$ $\diagup$ Z

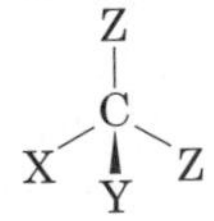

(b)

This bond cannot be positioned correctly and is not attacked.

This bond can be positioned correctly and is attacked.

$\underset{Y}{\overset{Z}{\underset{|}{\overset{|}{C}}}}$ X $\diagdown$ $\diagup$ Z

X'   Z'
  Y'

Active site has complementary binding points.

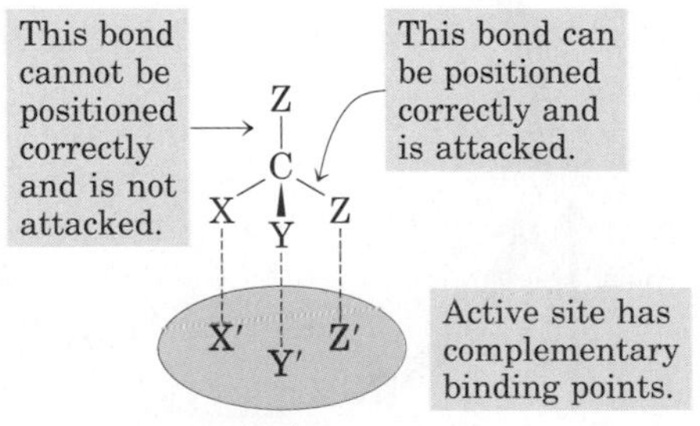

(c)

Figure 2 The prochiral nature of citrate. **(a)** Structure of citrate. **(b)** Schematic representation of citrate: X = —OH, Y = —COO$^-$, Z = —CH$_2$COO$^-$. **(c)** Correct complementary fit of citrate to the binding site on aconitase. There is only one way in which the three specified groups of citrate can fit on the three binding sites. Thus only one of the two —CH$_2$COO$^-$ groups is bound by aconitase.

Citric Acid Cycle Components Are Important Biosynthetic Intermediates

In aerobic organisms, the citric acid cycle is an **amphibolic pathway** (i.e., it serves in both catabolic and anabolic processes). It not only functions in the oxidative catabolism of carbohydrates, fatty acids, and amino acids, but also provides precursors for many biosynthetic pathways (Fig. 15–12), as in anaerobic ancestors. By the action of several important auxiliary enzymes, certain intermediates of the citric acid cycle, particularly α-ketoglutarate and oxaloacetate, can be removed from the cycle to serve as precursors of amino acids. Aspartate and glutamate have the same carbon skeletons as oxaloacetate and α-ketoglutarate, respectively, and are synthesized from them by simple transamination (Chapter 21). Through aspartate and glutamate the carbons of oxaloacetate and α-ketoglutarate are used to build other amino acids as well as purine and pyrimidine nucleotides. We will see in Chapter 19 how oxaloacetate is converted into glucose in the process of gluconeogenesis. Succinyl-CoA is a central intermediate in the synthesis of the porphyrin ring of heme groups, which serve as oxygen carriers (in hemoglobin and myoglobin) and electron carriers (in cytochromes).

Given the number of biosynthetic products derived from citric acid cycle intermediates, this cycle clearly serves a critical role apart from its function in energy-yielding metabolism.

Figure 15–12 Intermediates of the citric acid cycle are drawn off as precursors in many biosynthetic pathways, yielding the products in the shaded areas. Shown in red are four anaplerotic reactions that replenish depleted intermediates of the citric acid cycle (see Table 15–3).

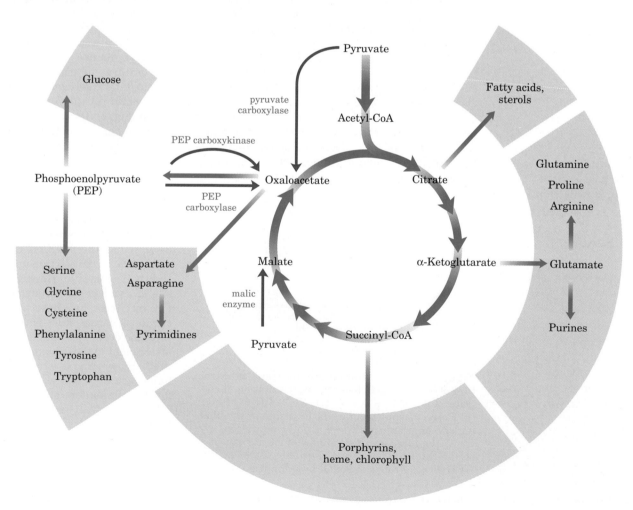

Anaplerotic Reactions Replenish Citric Acid Cycle Intermediates

When intermediates of the citric acid cycle are removed to serve as biosynthetic precursors, the resulting decrease in the concentration of these intermediates would be expected to slow the flux through the citric acid cycle. However, the intermediates can be replenished by **anaplerotic reactions** (Fig. 15–12; Table 15–3). Under normal circumstances the reactions by which the cycle intermediates are drained away and those by which they are replenished are in dynamic balance, so that the concentrations of the citric acid cycle intermediates remain almost constant.

Table 15–3 Anaplerotic reactions

Reaction	Tissue(s)/organism(s)
Pyruvate + HCO_3^- + ATP $\xrightleftharpoons{\text{pyruvate carboxylase}}$ oxaloacetate + ADP + P_i	Liver, kidney
Phosphoenolpyruvate + CO_2 + GDP $\xrightleftharpoons{\text{PEP carboxykinase}}$ oxaloacetate + GTP	Heart, skeletal muscle
Phosphoenolpyruvate + HCO_3^- $\xrightleftharpoons{\text{PEP carboxylase}}$ oxaloacetate + P_i	Higher plants, yeast, bacteria
Pyruvate + HCO_3^- + NAD(P)H $\xrightleftharpoons{\text{malic enzyme}}$ malate + NAD(P)$^+$	Widely distributed in eukaryotes and prokaryotes

In animal tissues one important anaplerotic reaction is the reversible carboxylation of pyruvate by CO_2 to form oxaloacetate, catalyzed by **pyruvate carboxylase** (Table 15–3). The three other anaplerotic reactions shown in Table 15–3 also serve, in various tissues and organisms, to convert either pyruvate or phosphoenolpyruvate to oxaloacetate. When the citric acid cycle is deficient in oxaloacetate or any of the other intermediates, pyruvate is carboxylated to produce more oxaloacetate. The enzymatic addition of a carboxyl group to the pyruvate molecule requires energy, which is supplied by ATP. Because the standard free-energy change of the overall reaction is very small, we can conclude that the free energy required to attach a carboxyl group to pyruvate is about equal to the free energy available from ATP. The carboxylation of pyruvate also requires the vitamin **biotin** (Fig. 15–13a), which is the prosthetic group of pyruvate carboxylase.

The pyruvate carboxylase reaction is the most important anaplerotic reaction in the liver and kidney of mammals. Other anaplerotic reactions are more important in other tissues and organisms (Table 15–3).

Pyruvate carboxylase is a regulatory enzyme and is virtually inactive in the absence of acetyl-CoA, its positive allosteric modulator. Whenever acetyl-CoA, which is the fuel for the citric acid cycle, is present in excess, it stimulates the pyruvate carboxylase reaction to produce more oxaloacetate, enabling the cycle to use more acetyl-CoA in the citrate synthase reaction.

The other anaplerotic reactions are also regulated to keep the level of intermediates high enough to support the activity of the citric acid cycle. Phosphoenolpyruvate carboxylase, for example, is activated by the glycolytic intermediate fructose-1,6-bisphosphate, the level of which rises when the citric acid cycle operates too slowly to process the pyruvate generated by glycolysis.

Figure 15–13 The role of biotin in carboxylation reactions. **(a)** Biotin. **(b)** The role of biotin in the reaction catalyzed by pyruvate carboxylase. Bicarbonate is first converted to CO_2 in an ATP-dependent reaction; this conversion of HCO_3^- to CO_2 is necessitated by the absence of significant amounts of free CO_2 in solution at neutral pH (see Table 4–3). The CO_2 thus formed at the active site is then readily added to biotin (step 1). In the second step, the carboxyl group is transferred to pyruvate to form oxaloacetate. Other biotin-dependent carboxylation reactions occur by very similar mechanisms.

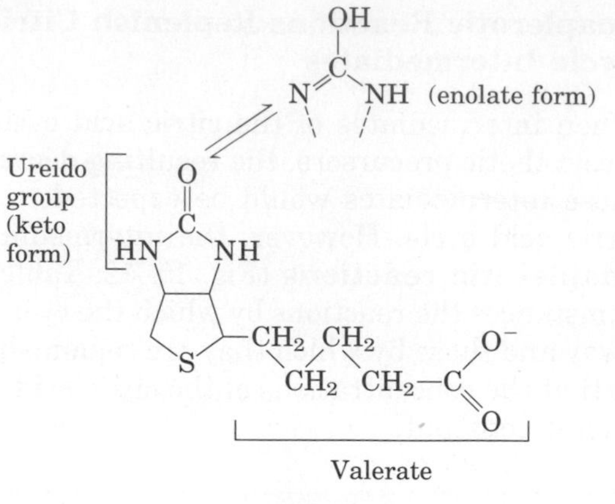

(a)

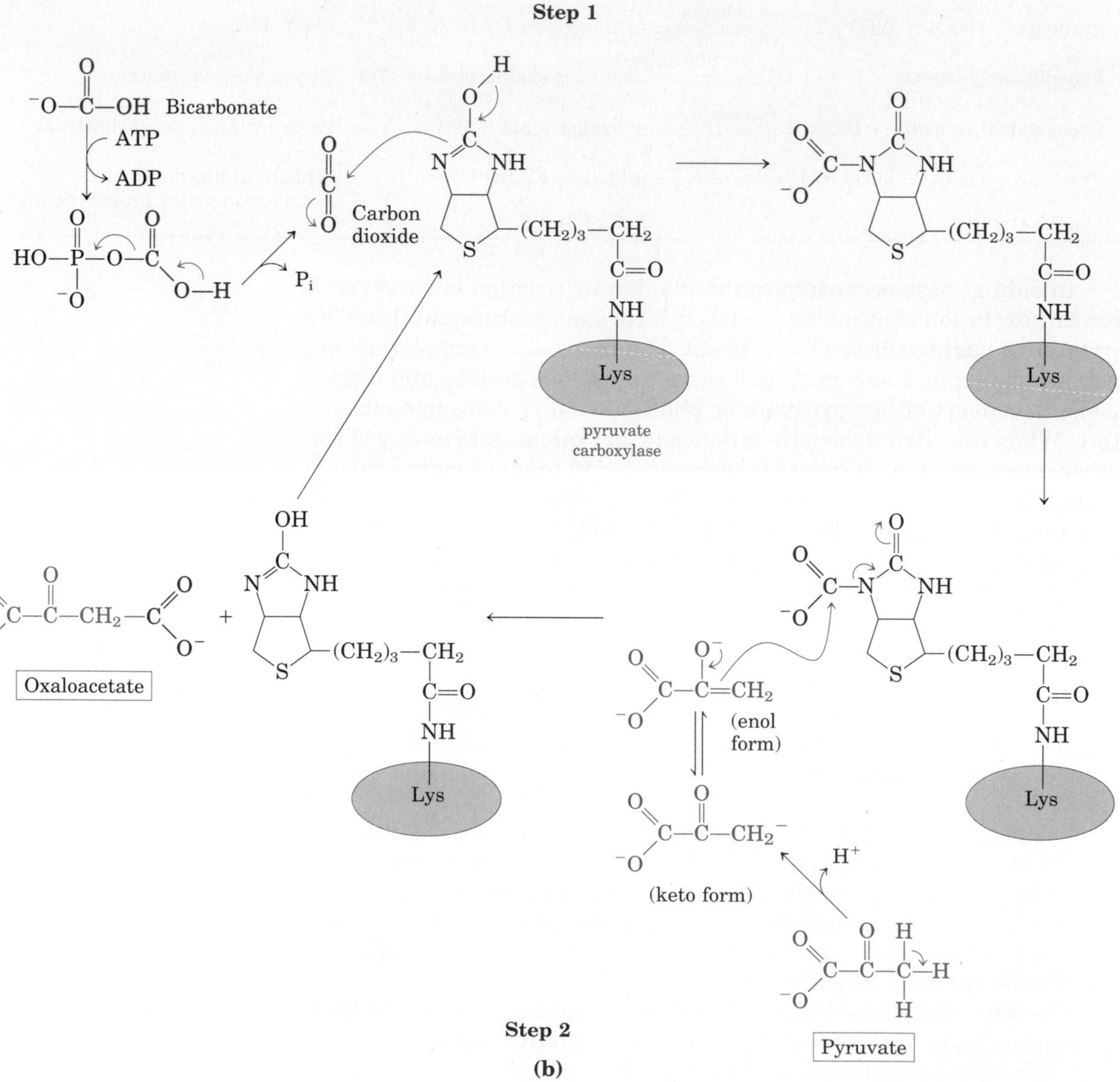

(b)

Biotin Carries CO$_2$ Groups

Biotin plays a key role in many carboxylation reactions. This vitamin is a specialized carrier of one-carbon groups in their most oxidized form: CO$_2$. (The transfer of one-carbon groups in more reduced forms is mediated by other cofactors, notably tetrahydrofolate and S-adenosyl-methionine, as described in Chapter 17.) Carboxyl groups are attached to biotin at the ureido group within the biotin ring system (Fig. 15–13a).

Pyruvate carboxylase is composed of four identical subunits, each containing a molecule of biotin covalently attached through an amide linkage between its valerate side chain and the ϵ-amino group of a specific Lys residue in the enzyme active site (Fig. 15–13b); this biotinyllysine is called **biocytin.** Carboxylation of pyruvate proceeds in two steps; first, a carboxyl group derived from HCO$_3^-$ is attached to biotin, then the carboxyl group is transferred to pyruvate to form oxaloacetate. These two steps occur at separate active sites; the long flexible arm of biocytin permits the transfer of activated carboxyl groups from the first active site to the second, much as the long lipoyllysyl arm of E$_2$ functions in the pyruvate dehydrogenase complex.

Biotin is a vitamin required in the human diet; it is abundant in many foods and is synthesized by intestinal bacteria. Deficiency diseases are rare and are generally observed only when large quantities of raw eggs are consumed. Egg whites contain a large amount of the protein **avidin** (M_r 70,000), which binds to biotin very tightly and prevents its absorption in the intestine. The avidin in egg whites may be a defense mechanism, inhibiting the growth of bacteria. When eggs are cooked, avidin is denatured (and thereby inactivated) along with all other egg white proteins.

Regulation of the Citric Acid Cycle

We saw in Chapter 14 that key enzymes in metabolic pathways are regulated by allosteric effectors and by covalent modification, to assure production of intermediates and products at the rates required to keep the cell in a stable steady state and to avoid wasteful overproduction of intermediates. The flow of carbon atoms from pyruvate into and through the citric acid cycle is under tight regulation at two levels: the conversion of pyruvate into acetyl-CoA, the starting material for the cycle (the pyruvate dehydrogenase complex reaction), and the entry of acetyl-CoA into the cycle (the citrate synthase reaction). Because pyruvate is not the sole source of acetyl-CoA (most cells can obtain acetyl-CoA by the oxidation of fatty acids and certain amino acids), the availability of intermediates from these other pathways is also important in the regulation of pyruvate oxidation and of the citric acid cycle. The cycle is also regulated at the isocitrate dehydrogenase and α-ketoglutarate dehydrogenase reactions.

The Production of Acetyl-CoA by the Pyruvate Dehydrogenase Complex Is Regulated

The pyruvate dehydrogenase complex of vertebrates is regulated both allosterically and by covalent modification. The complex is strongly inhibited by ATP, as well as by acetyl-CoA and NADH, the products of

Figure 15–14 Regulation of metabolite flow from pyruvate through the citric acid cycle. The pyruvate dehydrogenase complex is allosterically inhibited at high [ATP]/[ADP], [NADH]/[NAD$^+$], and [acetyl-CoA]/[CoA] ratios, all of which indicate the energy-sufficient metabolic state. When these ratios decrease, allosteric activation of pyruvate oxidation results. The rate of flow through the citric acid cycle can be limited by the availability of the substrates oxaloacetate and acetyl-CoA or by the depletion of NAD$^+$ by its conversion to NADH, which slows the three oxidation steps for which NAD$^+$ is the cofactor. Feedback inhibition by succinyl-CoA, citrate, and ATP also slows the cycle by inhibiting early steps. In muscle tissue, Ca^{2+} signals contraction and stimulates energy-yielding metabolism to replace the ATP consumed by contraction.

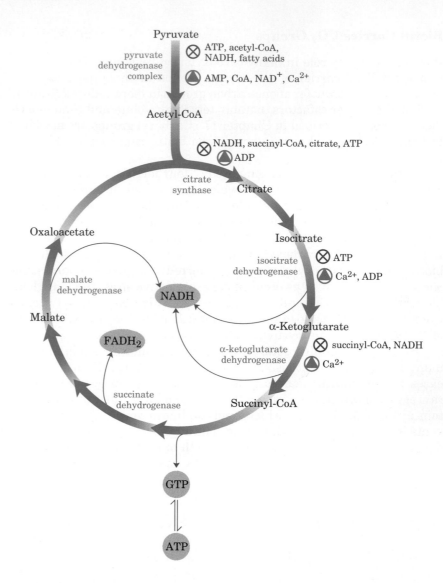

the reaction (Fig. 15–14). The allosteric inhibition of pyruvate oxidation is greatly enhanced when long-chain fatty acids are available. AMP, CoA, and NAD$^+$, all of which accumulate when too little acetate flows into the citric acid cycle, allosterically activate the pyruvate dehydrogenase complex. Thus this enzyme activity is turned off when ample fuel is available in the form of fatty acids and acetyl-CoA and when the cell's ATP concentration and [NADH]/[NAD$^+$] ratio are high, and turned on when energy demands are high and greater flux of acetyl-CoA into the citric acid cycle is required.

In the pyruvate dehydrogenase complex of vertebrates, these allosteric regulation mechanisms are complemented by a second level of regulation, covalent protein modification. The enzyme complex is inhibited by the reversible phosphorylation of a specific Ser residue on one of the two subunits of E$_1$. As noted earlier, in addition to the enzymes E$_1$, E$_2$, and E$_3$, the pyruvate dehydrogenase complex contains two regulatory proteins, the sole purpose of which is to regulate the activity of the complex. A specific protein kinase phosphorylates and thereby inactivates E$_1$, and a specific phosphoprotein phosphatase removes the phosphate group by hydrolysis, thereby activating E$_1$. The kinase is allosterically activated by ATP: when ATP levels are high (reflecting a sufficient supply of energy), the pyruvate dehydrogenase complex is inactivated by phosphorylation of E_1. When the ATP level declines, kinase activity decreases and phosphatase action removes the phosphates from E_1, activating the complex.

The pyruvate dehydrogenase complex of plants, which is found in the mitochondrial matrix and within plastids (p. 39), is strongly inhibited by NADH, which may be its primary regulator. There is also evidence for inactivation of the plant mitochondrial enzyme by reversible phosphorylation. The pyruvate dehydrogenase complex of *E. coli* is under allosteric regulation similar to that of the vertebrate enzyme, but the regulation by phosphorylation apparently does not occur with the bacterial enzyme.

Three Enzymes of the Citric Acid Cycle Are Regulated

The flow of metabolites through the citric acid cycle is under stringent, but not complex, regulation. Three factors govern the rate of flux through the cycle: substrate availability, inhibition by accumulating products, and allosteric feedback inhibition of early enzymes by later intermediates in the cycle.

There are three strongly exergonic steps in the cycle, those catalyzed by citrate synthase, isocitrate dehydrogenase, and α-ketoglutarate dehydrogenase (Fig. 15–14). Each can become the rate-limiting step under some circumstances. The availability of the substrates for citrate synthase (acetyl-CoA and oxaloacetate) varies with the metabolic circumstances and sometimes limits the rate of citrate formation. NADH, a product of the oxidation of isocitrate and α-ketoglutarate, accumulates under some conditions, and when the [NADH]/[NAD$^+$] ratio becomes large, both dehydrogenase reactions are severely inhibited by mass action. Similarly, the malate dehydrogenase reaction is essentially at equilibrium in the cell (i.e., it is substrate limited; see Fig. 14–16), and when [NADH]/[NAD$^+$] is large, the concentration of oxaloacetate is low, slowing the first step in the cycle. Product accumulation inhibits all three of the limiting steps of the cycle: succinyl-CoA inhibits α-ketoglutarate dehydrogenase (and also citrate synthase); citrate blocks citrate synthase; and the end product, ATP, inhibits both citrate synthase and isocitrate dehydrogenase (Fig. 15–14). The inhibition of citrate synthase by ATP is relieved by ADP, an allosteric activator of this enzyme. Calcium ions, which in vertebrate muscle are the signal for contraction and the concomitant increased demand for ATP, activate both isocitrate dehydrogenase and α-ketoglutarate dehydrogenase, as well as the pyruvate dehydrogenase complex. In short, the concentrations of substrates and intermediates of the citric acid cycle set the flux through this pathway at a rate that provides optimal concentrations of ATP and NADH.

Under normal conditions the rates of glycolysis and of the citric acid cycle are integrated so that only as much glucose is metabolized to pyruvate as is needed to supply the citric acid cycle with its fuel, the acetyl groups of acetyl-CoA. Pyruvate, lactate, and acetyl-CoA are normally maintained at steady-state concentrations. The rate of glycolysis is matched to the rate of the citric acid cycle not only by its inhibition by high levels of ATP and NADH, which are common components of both the glycolytic and respiratory stages of glucose oxidation, but also by the concentration of citrate. Citrate, the product of the first step of the citric acid cycle, serves as an important allosteric inhibitor of the phosphorylation of fructose-6-phosphate by phosphofructokinase-1 in the glycolytic pathway (p. 434).

The Glyoxylate Cycle

The conversion of phosphoenolpyruvate to pyruvate (p. 413) and of pyruvate to acetyl-CoA (Fig. 15–2) are so exergonic as to be essentially irreversible. If a cell cannot convert acetate into phosphoenolpyruvate, acetate cannot serve as the starting material for the gluconeogenic pathway that leads from phosphoenolpyruvate to glucose (Chapter 19). Without this capacity, a cell or organism is unable to convert fuels that are degraded to acetate (fatty acids and certain amino acids) into carbohydrates.

As we saw in our discussion of anaplerotic reactions (Table 15–3), phosphoenolpyruvate can be synthesized from oxaloacetate in the reversible reaction catalyzed by PEP carboxykinase:

$$\text{Oxaloacetate} + \text{GTP} \rightleftharpoons \text{phosphoenolpyruvate} + CO_2 + \text{GDP}$$

In Chapter 19 we will see how phosphoenolpyruvate is converted to glucose by the gluconeogenic pathway.

Because carbon atoms from acetate molecules that enter the citric acid cycle appear eight steps later in oxaloacetate, it might appear that operation of the citric acid cycle could generate oxaloacetate from acetate, and thus generate phosphoenolpyruvate for gluconeogenesis. However, examination of the stoichiometry of the cycle reveals that there is no *net* conversion of acetate into oxaloacetate via the cycle; for every two carbons that enter the cycle as acetyl-CoA, two leave as CO_2.

In plants, in certain invertebrates, and in some microorganisms such as *E. coli* and yeast, acetate can serve both as an energy-rich fuel and as a source of phosphoenolpyruvate for carbohydrate synthesis. These organisms have a pathway, the **glyoxylate cycle,** that allows the net conversion of acetate to oxaloacetate. In these organisms, some enzymes of the citric acid cycle operate in two modes: (1) they can function in the citric acid cycle for the oxidation of acetyl-CoA to CO_2, as it occurs in most tissues, and (2) they can operate as part of a specialized modification, the glyoxylate cycle (Fig. 15–15). The glyoxylate cycle may have evolved before, and given rise to, the citric acid cycle. The overall reaction equation of the glyoxylate cycle, which may also be regarded as an anaplerotic pathway, is

$$2 \text{ Acetyl-CoA} + NAD^+ + 2H_2O \longrightarrow \text{succinate} + 2\text{CoA} + \text{NADH} + H^+$$

The Glyoxylate Cycle Is a Variation of the Citric Acid Cycle

In the glyoxylate cycle, acetyl-CoA condenses with oxaloacetate to form citrate exactly as in the citric acid cycle. The breakdown of isocitrate does not occur via the isocitrate dehydrogenase reaction, however, but through a cleavage catalyzed by the enzyme **isocitrate lyase,** to form succinate and **glyoxylate.** The glyoxylate then condenses with acetyl-CoA to yield malate in a reaction catalyzed by **malate synthase.** The malate is subsequently oxidized to oxaloacetate, which can condense with another molecule of acetyl-CoA to start another turn of the cycle (Fig. 15–15). In each turn of the glyoxylate cycle, two molecules of acetyl-CoA enter and there is a net synthesis of one molecule of succinate, available for biosynthetic purposes. The succinate may be converted through fumarate and malate into oxaloacetate, which can then be converted into phosphoenolpyruvate by the PEP carboxykinase reaction described above. Phosphoenolpyruvate can then serve as a precursor of glucose in gluconeogenesis.

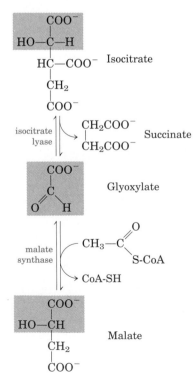

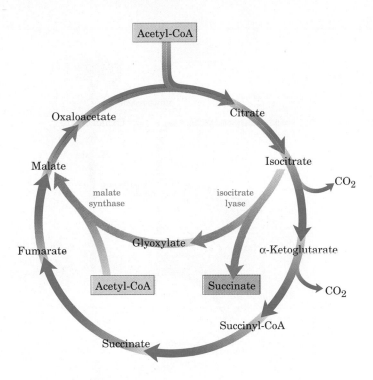

Figure 15–15 The glyoxylate cycle and its relationship to the citric acid cycle. The orange reaction arrows represent the glyoxylate cycle, and the blue arrows, the citric acid cycle. Notice that the glyoxylate cycle bypasses the two decarboxylation steps of the citric acid cycle, and that two molecules of acetyl-CoA enter the glyoxylate cycle during each turn, but only one enters the citric acid cycle. The glyoxylate cycle was elucidated by Hans Kornberg and Neil Madsen in the laboratory of Hans Krebs.

In plants, the enzymes of the glyoxylate cycle are sequestered in membrane-bounded organelles called glyoxysomes (Fig. 15–16); those enzymes common to the citric acid and glyoxylate cycles have two isozymes, one specific to mitochondria, the other to glyoxysomes. Glyoxysomes are not present in all plant tissues at all times. They develop in lipid-rich seeds during germination, before the developing plants acquire the ability to make glucose by photosynthesis. In addition to glyoxylate cycle enzymes, glyoxysomes also contain all of the enzymes needed for the degradation of fatty acids stored in seed oils (Chapter 16). Acetyl-CoA formed from lipids is converted into malate via the glyoxylate cycle, and the malate serves as a source of oxaloacetate (through the malate dehydrogenase reaction) for gluconeogenesis. Germinating plants are therefore able to convert the carbon of seed lipids into glucose.

Vertebrate animals do not have the enzymes specific to the glyoxylate cycle (isocitrate lyase and malate synthase) and therefore cannot bring about the net synthesis of glucose from lipids.

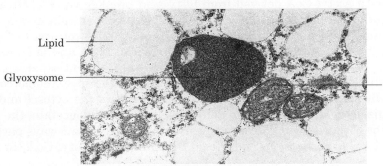

Figure 15–16 Electron micrograph of a germinating cucumber seed, showing a glyoxysome, mitochondria, and surrounding lipid bodies.

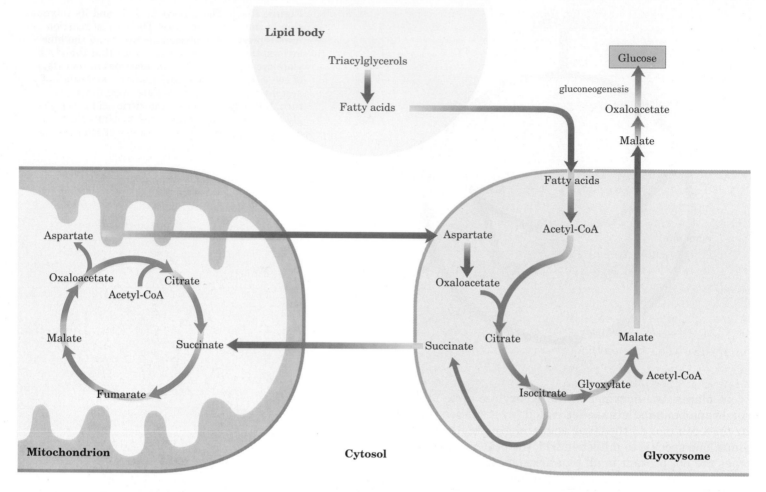

Figure 15–17 The reactions of the glyoxylate cycle (in glyoxysomes) proceed simultaneously with, and mesh with, those of the citric acid cycle (in mitochondria), as intermediates pass through the cytosol between these compartments. The reactions involved in the oxidation of fatty acids to acetyl-CoA and the conversion of oxaloacetate to aspartate will be discussed in Chapters 16 and 21, respectively.

The Citric Acid and Glyoxylate Cycles Are Coordinately Regulated

In germinating plant seeds, the enzymatic transformations of dicarboxylic and tricarboxylic acids occur in three intracellular compartments: mitochondria, glyoxysomes, and the cytosol. There is a continuous interchange of intermediates among these compartments (Fig. 15–17).

Aspartate carries the carbon skeleton of oxaloacetate from the citric acid cycle (in mitochondria) to the glyoxysome, where it condenses with acetyl-CoA derived from fatty acid breakdown. The citrate thus formed is converted to isocitrate by aconitase, then split into glyoxylate and succinate by isocitrate lyase. The succinate returns to the mitochondrion, where it reenters the citric acid cycle and is transformed into oxaloacetate, which can again be exported (via aspartate) to the glyoxysome. The glyoxylate formed within the glyoxysome combines with acetyl-CoA to yield malate, which enters the cytosol and is oxidized (by cytosolic malate dehydrogenase) to oxaloacetate, the precursor of glucose via gluconeogenesis. Four distinct pathways participate in these conversions: fatty acid breakdown to acetyl-CoA (in glyoxysomes), the glyoxylate cycle (in glyoxysomes), the citric acid cycle (in mitochondria), and gluconeogenesis (in the cytosol).

The sharing of common intermediates requires that these pathways be regulated and coordinated. Isocitrate is a crucial intermediate, standing at the branch point between the glyoxylate and citric acid cycles (Fig. 15–18). Isocitrate dehydrogenase is regulated by covalent modification: a specific protein kinase phosphorylates and thereby inactivates the dehydrogenase. Inactivation of isocitrate dehydrogenase shunts isocitrate to the glyoxylate cycle, where it begins the synthetic route toward glucose. A phosphoprotein phosphatase removes the phosphate group from isocitrate dehydrogenase, reactivating the enzyme and sending more isocitrate through the energy-yielding citric acid cycle. The regulatory protein kinase and phosphoprotein phosphatase are separate enzymatic activities, but both reside in the same polypeptide.

Some bacteria, including *E. coli*, have the full complement of enzymes for the glyoxylate and citric acid cycles in the cytosol. *E. coli* can therefore grow with acetate as its sole source of carbon and energy. The phosphatase activity that causes activation of isocitrate dehydrogenase is stimulated by intermediates of the citric acid cycle and of glycolysis, and by indicators of reduced cellular energy supply (Table 15–4; Fig. 15–18). The same metabolites *inhibit* the protein kinase activity of the bifunctional enzyme. Thus, the accumulation of intermediates of the central energy-yielding pathways, or energy depletion, results in the activation of isocitrate dehydrogenase. When the concentration of these regulators falls, signaling enough flux through the energy-yielding citric acid cycle, isocitrate dehydrogenase is inactivated by the protein kinase.

Table 15–4 Allosteric effectors of the regulatory kinase/phosphatase protein of isocitrate dehydrogenase

Citric acid cycle intermediates	Glycolytic intermediates	Cofactors that indicate energy depletion
Citrate	Phosphoenolpyruvate*	AMP*
Isocitrate*	Pyruvate*	ADP*
α-Ketoglutarate*	3-Phosphoglycerate*	NADP+
Oxaloacetate*	Fructose-6-phosphate	

All compounds shown *inhibit* the kinase activity. Compounds with * *stimulate* the phosphatase activity. The overall result is activation of isocitrate dehydrogenase and thus of the citric acid cycle.

The same intermediates of the glycolytic and citric acid cycles that lead to activation of isocitrate dehydrogenase are allosteric inhibitors of isocitrate lyase. When energy-yielding metabolism is sufficiently fast to keep the concentrations of intermediates of glycolysis and the citric acid cycle low, isocitrate dehydrogenase is inactivated, the inhibition of isocitrate lyase is relieved, and isocitrate flows into the glyoxylate pathway, where it is used in the biosynthesis of carbohydrates, amino acids, and other cellular components.

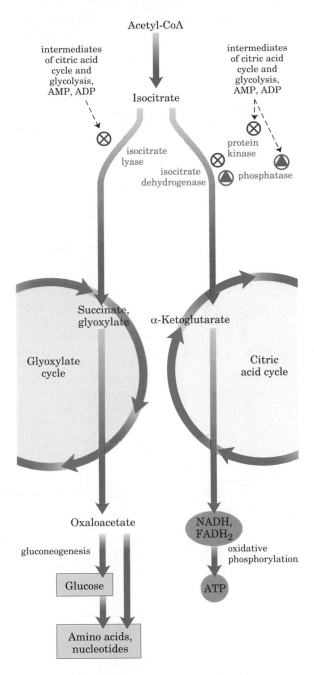

Figure 15–18 Regulation of isocitrate dehydrogenase activity determines the partitioning of isocitrate between the glyoxylate cycle and the citric acid cycle. When isocitrate dehydrogenase is inactivated by phosphorylation (by a specific protein kinase), isocitrate is directed into biosynthetic reactions via the glyoxylate cycle; when the enzyme is activated by dephosphorylation (by a specific phosphatase), isocitrate enters the citric acid cycle, and ATP production results.

Summary

Cellular respiration occurs in three stages: (1) the oxidative formation of acetyl-CoA from pyruvate, fatty acids, and some amino acids, (2) the degradation of acetyl residues by the citric acid cycle to yield CO_2 and electrons, and (3) the transfer of electrons to molecular oxygen, coupled to the phosphorylation of ADP to ATP. The oxidative catabolism of glucose yields much more energy than the fermentation pathways.

Pyruvate, the end product of glycolysis, undergoes dehydrogenation and decarboxylation by the pyruvate dehydrogenase complex, which contains three sequentially acting enzymes and requires five coenzymes, to yield acetyl-CoA and CO_2. The acetyl-CoA enters the citric acid cycle, which occurs in the mitochondria of eukaryotes and in the cytosol of prokaryotes. Citrate synthase catalyzes the condensation of acetyl-CoA with oxaloacetate to form citrate. Aconitase catalyzes the reversible formation of isocitrate from citrate; isocitrate is then oxidized to α-ketoglutarate by isocitrate dehydrogenase in a reaction that also yields CO_2. The α-ketoglutarate undergoes another dehydrogenation and decarboxylation to succinyl-CoA and CO_2. Succinyl-CoA reacts with ADP (or GDP) and P_i to form free succinate and ATP (or GTP), in a substrate-level phosphorylation. The succinate is then oxidized to fumarate by succinate dehydrogenase, an FAD-linked enzyme that is part of the inner membrane of the mitochondrion (or of the plasma membrane in bacteria). Fumarate is reversibly hydrated by fumarase to L-malate, which is oxidized by NAD-linked L-malate dehydrogenase to regenerate a molecule of oxaloacetate. The latter can now combine with another molecule of acetyl-CoA and start another turn of the cycle.

Isotopic tracer experiments with carbon-labeled fuel molecules or intermediates have established that the citric acid cycle is the major pathway of carbohydrate oxidation in aerobic cells. The pyruvate dehydrogenase complex of vertebrates is inhibited by the allosteric effectors NADH, ATP, and acetyl-CoA. The enzyme complex is also inhibited by reversible phosphorylation catalyzed by a protein kinase and phosphatase that are part of the complex. The overall rate of the cycle is controlled by the rate of conversion of pyruvate to acetyl-CoA and by the flux through three enzymes of the cycle: citrate synthase, isocitrate dehydrogenase, and α-ketoglutarate dehydrogenase. These fluxes are largely determined by the concentrations of substrates and products; the end products ATP and NADH are inhibitory.

Citric acid cycle intermediates are also used as precursors in biosynthesis of amino acids and other biomolecules. The cycle intermediates are then replenished by anaplerotic reactions catalyzed by pyruvate carboxylase, PEP carboxykinase, PEP carboxylase, or malic enzyme. In the germinating seeds of some plants, and in certain microorganisms that can live on acetate as sole carbon source for the synthesis of carbohydrate, a variation of the citric acid cycle, the glyoxylate cycle, comes into play. The process involves two additional enzymes: isocitrate lyase and malate synthase, located within glyoxysomes. This cycle makes possible the *net* formation of succinate, oxaloacetate, and other cycle intermediates from acetyl-CoA. Oxaloacetate thus formed can be used to synthesize glucose via gluconeogenesis. Vertebrates lack the glyoxylate cycle and cannot synthesize glucose from acetate. In organisms with both the citric acid cycle and the glyoxylate cycle, the partitioning of isocitrate between the two pathways is controlled at the level of isocitrate dehydrogenase. This enzyme is subject to regulation by reversible phosphorylation.

Further Reading

General

Gottschalk, G. (1986) *Bacterial Metabolism,* 2nd edn, Springer-Verlag, New York.
An excellent account of the metabolic diversity of bacteria and the richness of their energy-generating pathways.

Kay, J. & Weitzman, P.D.J. (eds) (1987) *Krebs' Citric Acid Cycle: Half a Century and Still Turn-* ing, Biochemical Society Symposium 54, The Biochemical Society, London.
A multiauthor book on the citric acid cycle, including molecular genetics, regulatory mechanisms, variations on the cycle in microorganisms from unusual ecological niches, and evolution of the pathway. Especially relevant are the chapters by H. Cest (Evolutionary roots of the citric acid cycle in

prokaryotes), W.H. Holms *(Control of flux through the citric acid cycle and the glyoxylate bypass in* Escherichia coli*), and* R.N. Perham et al. *(α-Keto acid dehydrogenase complexes).*

Production of Acetate (Pyruvate Dehydrogenase Complex)

Patel, M.S. & Roche, T.E. (1990) Molecular biology and biochemistry of pyruvate dehydrogenase complexes. *FASEB J.* **4,** 3224–3233.

Recent results from the cloning of the enzymes of the complex are described, and the enzymes from a number of species are compared in composition, structure, and action. The protein X of vertebrate pyruvate dehydrogenase is included in this excellent review.

Reed, L.J. & Hackert, M.L. (1990) Structure-function relationships in dihydrolipoamide acyltransferases. *J. Biol. Chem.* **265,** 8971–8974.

A short, clear review of the structure of the dihydrolipoyl transacetylase of both the pyruvate dehydrogenase complex and α-ketoglutarate dehydrogenase, including a comparison of the structures of bacterial, yeast, and human versions of the protein.

Roche, T.E. & Patel, M.S. (eds) (1990) *Alpha-Keto Acid Dehydrogenase Complexes: Organization, Regulation, and Biomedical Ramifications. Ann. N.Y. Acad. Sci.* **573.**

This volume contains about 60 papers covering all aspects of the enzyme group that includes the pyruvate dehydrogenase complex and α-ketoglutarate dehydrogenase.

Citric Acid Cycle Enzymes

Knowles, J. (1989) The mechanism of biotin-dependent enzymes. *Annu. Rev. Biochem.* **58,** 195–221.

Krebs, H.A. & Johnson, W.A. (1937) The role of citric acid in intermediate metabolism in animal tissues. *Enzymologia* **4,** 148–156.

One of the classic papers on the citric acid cycle.

Singer, T.P. & Johnson, M.K. (1985) The prosthetic groups of succinate dehydrogenase: 30 years from discovery to identification. *FEBS Lett.* **190,** 189–198.

A description of the structure and role of the iron–sulfur centers in this enzyme.

Weigand, G. & Remington, S.J. (1986) Citrate synthase: structure, control, and mechanism. *Annu. Rev. Biophys. Biophys. Chem.* **15,** 97–117.

Regulation of the Citric Acid Cycle

Hansford, R.G. (1980) Control of mitochondrial substrate oxidation. *Curr. Top. Bioenerget.* **10,** 217–278.

A detailed review of the regulation of the citric acid cycle.

Kaplan, N.O. (1985) The role of pyridine nucleotides in regulating cellular metabolism. *Curr. Top. Cell. Regul.* **26,** 371–381.

An excellent general discussion of the importance of the [NADH]/[NAD$^+$] ratio in cellular regulation.

Reed, L.J., Damuni, Z., & Merryfield, M.L. (1985) Regulation of mammalian pyruvate and branched-chain α-keto acid dehydrogenase complexes by phosphorylation-dephosphorylation. *Curr. Top. Cell. Regul.* **27,** 41–49.

Glyoxylate Cycle

Holms, W.H. (1986) The central metabolic pathways of *Escherichia coli:* relationship between flux and control at a branch point, efficiency of conversion to biomass, and excretion of acetate. *Curr. Top. Cell. Regul.* **28,** 69–106.

Problems

1. *Balance Sheet for the Citric Acid Cycle* The citric acid cycle uses eight enzymes to catabolize acetyl-CoA: citrate synthase, aconitase, isocitrate dehydrogenase, α-ketoglutarate dehydrogenase, succinyl-CoA synthetase, succinate dehydrogenase, fumarase, and malate dehydrogenase.

(a) Write a balanced equation for the reaction catalyzed by each enzyme.

(b) What cofactor(s) are required by each enzyme reaction?

(c) For each enzyme determine which of the following describes the type of reaction catalyzed: condensation (carbon–carbon bond formation); dehydration (loss of water); hydration (addition of water); decarboxylation (loss of CO_2); oxidation–reduction; substrate-level phosphorylation; isomerization.

(d) Write a balanced net equation for the catabolism of acetyl-CoA to CO_2.

$$CH_3-\underset{\underset{\displaystyle}{\overset{\displaystyle O}{\|}}}{C}-H + H-H \underset{\text{oxidation}}{\overset{\text{reduction}}{\rightleftharpoons}} \left[CH_3-\underset{\underset{\displaystyle H}{|}}{\overset{\displaystyle O \leftarrow H}{C \leftarrow H}} \right] \underset{\text{oxidation}}{\overset{\text{reduction}}{\rightleftharpoons}} CH_3-\underset{\underset{\displaystyle H}{|}}{\overset{\displaystyle O-H}{C}}-H \qquad (1)$$

Acetaldehyde   Ethanol

$$CH_3-\underset{\displaystyle O^-}{\overset{\displaystyle O}{\|}}{C} + H^+ + H-H \underset{\text{oxidation}}{\overset{\text{reduction}}{\rightleftharpoons}} \left[CH_3-C \underset{\underset{\displaystyle H}{\overset{H}{\diagdown}}}{\overset{\displaystyle O}{\diagup}}\times \underset{\displaystyle H^+}{O^-} \right] \underset{\text{oxidation}}{\overset{\text{reduction}}{\rightleftharpoons}} CH_3-\underset{}{\overset{\displaystyle O}{\|}}{C}-H + \underset{\displaystyle H}{\overset{\displaystyle H}{O}} \qquad (2)$$

Acetate   Acetaldehyde

2. *Recognizing Oxidation and Reduction Reactions in Metabolism* The biochemical strategy of living organisms is the stepwise oxidation of organic compounds to carbon dioxide and water. By properly coupling these reactions, a major part of the energy produced in oxidation is conserved in the form of ATP. It is important to be able to recognize oxidation–reduction processes in metabolism. The reduction of an organic molecule results from the hydrogenation of a double bond (Eqn 1 above) or of a single bond with accompanying cleavage (Eqn 2). Conversely, the oxidation of an organic molecule results from dehydrogenation. In biochemical redox reactions (see Problem 3) the coenzymes NAD and FAD function to dehydrogenate/hydrogenate organic molecules in the presence of the proper enzymes.

For each of the following metabolic transformations, determine whether oxidation or reduction has occurred. Balance each transformation by inserting H—H, and H_2O where necessary.

(a) $CH_3-OH \longrightarrow H-\underset{\displaystyle}{\overset{\displaystyle O}{\|}}{C}-H$

    Methanol   Formaldehyde

(b) $H-\overset{\displaystyle O}{\|}C-H \longrightarrow H-\underset{\displaystyle O^-}{\overset{\displaystyle O}{\|}}C + H^+$

    Formaldehyde   Formate

(c) $O=C=O \longrightarrow H-\underset{\displaystyle O^-}{\overset{\displaystyle O}{\|}}C + H^+$

    Carbon dioxide   Formate

(d) $\underset{\displaystyle H}{\overset{\displaystyle OH}{CH_2}}-\underset{\displaystyle H}{\overset{\displaystyle OH}{C}}-\underset{\displaystyle O^-}{\overset{\displaystyle O}{C}} + H^+ \longrightarrow \underset{\displaystyle H}{\overset{\displaystyle OH}{CH_2}}-\underset{\displaystyle H}{\overset{\displaystyle OH}{C}}-\underset{\displaystyle H}{\overset{\displaystyle O}{C}}$

    Glycerate   Glyceraldehyde

(e) $\underset{\displaystyle H}{\overset{\displaystyle OH}{CH_2}}-\overset{\displaystyle OH}{C}-\overset{\displaystyle OH}{CH_2} \longrightarrow \overset{\displaystyle OH}{CH_2}-\overset{\displaystyle O}{C}-\overset{\displaystyle OH}{CH_2}$

    Glycerol   Dihydroxyacetone

(f)

    Toluene   Benzoate

(g) $\underset{\displaystyle O^-}{\overset{\displaystyle O}{C}}-CH_2-CH_2-\overset{\displaystyle O}{C}\underset{\displaystyle O}{} \longrightarrow$

    Succinate   Fumarate

(h) $CH_3-\underset{\underset{\displaystyle O}{\overset{\displaystyle |}{C-O^-}}}{\overset{\displaystyle O}{\|}}C \longrightarrow CH_3-\underset{\displaystyle O^-}{\overset{\displaystyle O}{C}} + CO_2$

    Pyruvate   Acetate

3. *Nicotinamide Coenzymes as Reversible Redox Carriers* The nicotinamide coenzymes (see Fig. 13–16) can undergo reversible oxidation–reduction reactions with specific substrates in the presence of the appropriate dehydrogenase. The nicotinamide ring is the portion of the coenzyme involved in the redox reaction; the remaining portion of the coenzyme serves as a binding group recognized by the dehydrogenase protein. Formally, NADH + H^+ serves as the hydrogen source (H—H), as described in Problem 2. Whenever the coenzyme is oxidized, a substrate must be simultaneously reduced:

Substrate	+ NADH + H^+	$\rightleftharpoons$ product	+ NAD$^+$
Oxidized	Reduced	Reduced	Oxidized

For each of the following reactions, determine whether the substrate has been oxidized or reduced or is unchanged in oxidation state (see Problem 2). For substrates that have undergone a redox change, balance the reaction with the necessary amount of NAD$^+$, NADH, H$^+$, and H_2O. The objective is to recognize when a redox coenzyme is necessary in a metabolic reaction.

(a) $CH_3CH_2OH \longrightarrow CH_3-C\overset{O}{\underset{H}{\big\langle}}$

Ethanol   Acetaldehyde

(b) $^{2-}O_3PO-CH_2-\overset{OH}{\underset{H}{\overset{|}{C}}}-C\overset{O}{\underset{OPO_3^{2-}}{\big\langle}} \longrightarrow {}^{2-}O_3PO-CH_2-\overset{OH}{\underset{H}{\overset{|}{C}}}-C\overset{O}{\underset{H}{\big\langle}} + HPO_4^{2-}$

1,3-Bisphosphoglycerate   Glyceraldehyde-3-phosphate

(c) $CH_3-\overset{O}{\overset{||}{C}}-C\overset{O^-}{\underset{O}{\big\langle}} \longrightarrow CH_3-C\overset{O^-}{\underset{H}{\big\langle}} + CO_2$

Pyruvate   Acetaldehyde

(d) $CH_3-\overset{O}{\overset{||}{C}}-C\overset{O^-}{\underset{O}{\big\langle}} \longrightarrow CH_3-C\overset{O}{\underset{O^-}{\big\langle}} + CO_2$

Pyruvate   Acetate

(e) $^-OOC-CH_2-\overset{O}{\overset{||}{C}}-COO^- \longrightarrow {}^-OOC-CH_2-\overset{OH}{\underset{H}{\overset{|}{C}}}-COO^-$

Oxaloacetate   Malate

(f) $CH_3-\overset{O}{\overset{||}{C}}-CH_2-C\overset{O}{\underset{O}{\big\langle}} + H^+ \longrightarrow CH_3-\overset{O}{\overset{||}{C}}-CH_3 + CO_2$

Acetoacetate   Acetone

4. *Stimulation of Oxygen Consumption by Oxaloacetate and Malate* In the early 1930s, Albert Szent-Györgyi reported the interesting observation that the addition of small amounts of oxaloacetate or malate to suspensions of minced pigeon-breast muscle stimulated the oxygen consumption of the preparation. Surprisingly, when the amount of oxygen consumed was measured, it was about seven times more than the amount necessary to oxidize the added oxaloacetate or malate completely to carbon dioxide and water.

(a) Why does the addition of oxaloacetate or malate stimulate oxygen consumption?

(b) Why is the amount of oxygen consumed so much greater than the amount necessary to oxidize the added oxaloacetate or malate completely?

5. *The Number of Molecules of Oxaloacetate in a Mitochondrion* In the last reaction of the citric acid cycle, malate is dehydrogenated to regenerate the oxaloacetate necessary for the entry of acetyl-CoA via the citrate synthase reaction:

$$\text{L-Malate} + NAD^+ \longrightarrow \text{oxaloacetate} + NADH + H^+$$
$$\Delta G^{\circ\prime} = 30 \text{ kJ/mol}$$

(a) Calculate the equilibrium constant for the reaction at 25 °C.

(b) Because $\Delta G^{\circ\prime}$ assumes a standard pH of 7, the equilibrium constant obtained in (a) corresponds to

$$K'_{eq} = \frac{[\text{oxaloacetate}][\text{NADH}]}{[\text{L-malate}][\text{NAD}^+]}$$

The measured concentration of L-malate in rat liver mitochondria is about 0.20 mM when $[NAD^+]/[NADH]$ is 10. Calculate the concentration of oxaloacetate at pH 7 in these mitochondria.

(c) Rat liver mitochondria are roughly spherical, with a diameter of about 2 μm. To appreciate the magnitude of the oxaloacetate concentration in mitochondria, calculate the number of oxaloacetate molecules in a single rat liver mitochondrion.

6. *Respiration Studies in Isolated Mitochondria* Cellular respiration can be studied using isolated mitochondria and measuring their oxygen consumption under different conditions. If 0.01 M sodium malonate is added to actively respiring mitochondria using pyruvate as a fuel source, respiration soon stops and a metabolic intermediate accumulates.

(a) What is the structure of the accumulated intermediate?

(b) Explain why it accumulates.

(c) Explain why oxygen consumption stops.

(d) Aside from removing malonate, how can the inhibition of respiration by malonate be overcome? Explain.

7. *Labeling Studies in Isolated Mitochondria* The metabolic pathways of organic compounds have often been delineated by using a radioactively labeled substrate and following the fate of the label.

(a) How can you determine whether glucose added to a suspension of isolated mitochondria is metabolized to CO_2 and H_2O?

(b) Suppose you add [3-^{14}C]pyruvate (labeled in the methyl position) to the mitochondria. After one turn of the citric acid cycle, what is the location of the ^{14}C in the oxaloacetate? Explain by tracing the ^{14}C label through the pathway.

(c) How many turns of the citric acid cycle must the ^{14}C go through before all the isotope is released as $^{14}CO_2$? Explain.

8. *[1-^{14}C]Glucose Catabolism* If an actively respiring bacterial culture is briefly incubated with [1-^{14}C]glucose and the glycolytic and citric acid cycle intermediates are isolated, where is the ^{14}C in each of the intermediates listed below? Consider only the initial incorporation of ^{14}C into these molecules, in the first pass of labeled glucose through the pathways.

(a) Fructose-1,6-bisphosphate
(b) Glyceraldehyde-3-phosphate
(c) Phosphoenolpyruvate
(d) Acetyl-CoA
(e) Citrate
(f) α-Ketoglutarate
(g) Oxaloacetate

9. *Synthesis of Oxaloacetate by the Citric Acid Cycle* Oxaloacetate is formed in the last step of the citric acid cycle by the NAD^+-dependent oxidation of L-malate. Can a net synthesis of oxaloacetate take place from acetyl-CoA using only the enzymes and cofactors of the citric acid cycle, without depleting the intermediates of the cycle? Explain. How is the oxaloacetate lost from the cycle (to biosynthetic reactions) replenished?

10. *Mode of Action of the Rodenticide Fluoroacetate* Fluoroacetate, prepared commercially for rodent control, is also produced naturally by a South African plant. After entering a cell, fluoroacetate is converted into fluoroacetyl-CoA in a reaction catalyzed by the enzyme acetate thiokinase:

$$F—CH_2COO^- + CoA\text{-}SH + ATP \longrightarrow$$
$$F—CH_2\underset{\underset{O}{\|}}{C}—S\text{-}CoA + AMP + PP_i$$

The toxic effect of fluoroacetate was studied in a metabolic experiment on intact isolated rat heart. After the heart was perfused with 0.22 mM fluoroacetate, the measured rate of glucose uptake and glycolysis decreased and glucose-6-phosphate and fructose-6-phosphate accumulated. An examination of the citric acid cycle intermediates indicated that their concentrations were below normal except for citrate, which had a concentration 10 times higher than normal.

(a) Where does the block in the citric acid cycle occur? What causes citrate to accumulate and the other cycle intermediates to be depleted?

(b) Fluoroacetyl-CoA is enzymatically transformed in the citric acid cycle. What is the structure of the metabolic end product of fluoroacetate? Why does it block the citric acid cycle? How might the inhibition be overcome?

(c) Why do glucose uptake and glycolysis decrease in the heart upon fluoroacetate perfusion? Why do hexose monophosphates accumulate?

(d) Why is fluoroacetate poisoning fatal?

11. *Net Synthesis of α-Ketoglutarate* α-Ketoglutarate plays a central role in the biosynthesis of several amino acids. Write a series of known enzymatic reactions that result in the net synthesis of α-ketoglutarate from pyruvate. Your proposed sequence must not involve the net consumption of other citric acid cycle intermediates. Write the overall reaction for your proposed sequence and identify the source of each reactant.

12. *Regulation of Citrate Synthase* In the presence of saturating amounts of oxaloacetate, the activity of citrate synthase from pig heart tissue shows a sigmoid dependence on the concentration of acetyl-CoA, as shown below. When succinyl-CoA is added, the curve shifts to the right and becomes even more sigmoid.

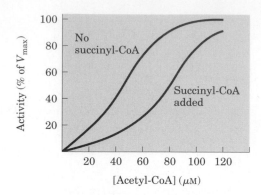

On the basis of these observations, explain how succinyl-CoA regulates the activity of citrate synthase (Hint: See Fig. 8–27). Why is succinyl-CoA an appropriate signal for regulation of the citric acid cycle? How does the regulation of citrate synthase control the rate of cellular respiration in pig heart tissue?

13. *Regulation of Pyruvate Carboxylase* The carboxylation of pyruvate by pyruvate carboxylase occurs at a very low rate unless acetyl-CoA, a positive allosteric modulator, is present. If you have just completed a meal rich in fatty acids (triacylglycerols) but low in carbohydrates (glucose), how does this regulatory property shut down the oxidation of glucose to CO_2 and H_2O but increase the oxidation of acetyl-CoA derived from fatty acids?

14. *Relationship between Respiration and the Citric Acid Cycle* Although oxygen does not participate directly in the citric acid cycle, the cycle operates only when O_2 is present. Why?

15. *Thermodynamics of Citrate Synthase Reaction In Vivo* Citrate is formed by the citrate synthase–catalyzed condensation of acetyl-CoA with oxaloacetate:

$$\text{Oxaloacetate} + \text{acetyl-CoA} + H_2O \longrightarrow$$
$$\text{citrate} + \text{CoA} + H^+$$

In rat heart mitochondria at pH 7.0 and 25 °C, the concentrations of reactants and products are: oxaloacetate, 1 μM; acetyl-CoA, 1 μM; citrate, 220 μM; and CoA, 65 μM. On the basis of these concentrations and the value of the standard free-energy change for the citrate synthase reaction (-32.2 kJ/mol), determine the direction of metabolite flow through the citrate synthase reaction in the cell. Explain.

16. *Reactions of the Pyruvate Dehydrogenase Complex* Two of the steps in the oxidative decarboxylation of pyruvate (steps ④ and ⑤, Fig. 15–6) do not involve any of the three carbons of pyruvate yet are essential to the operation of the pyruvate dehydrogenase complex. Explain.

Oxidation of Fatty Acids

The oxidation of long-chain fatty acids to acetyl-CoA is a central energy-yielding pathway in animals, many protists, and some bacteria. The electrons removed during fatty acid oxidation pass through the mitochondrial respiratory chain, driving ATP synthesis, and the acetyl-CoA produced from the fatty acids may be completely oxidized to CO_2 via the citric acid cycle, resulting in further energy conservation. In some organisms, acetyl-CoA produced by fatty acid oxidation has alternative fates. In vertebrate animals, acetyl-CoA may be converted in the liver into ketone bodies—water-soluble fuels exported to the brain and other tissues when glucose is not available. In higher plants, acetyl-CoA from fatty acid oxidation serves primarily as a biosynthetic precursor, and only secondarily as fuel. Although the biological role of fatty acid oxidation differs from organism to organism, the mechanism is essentially the same. This chapter is centered on the four-step process, called **β oxidation,** by which fatty acids are converted into acetyl-CoA.

In Chapter 9 we described the properties of triacylglycerols (also called triglycerides or neutral fats) that make them especially suitable as storage fuels. The long alkyl chains of their constituent fatty acids are essentially hydrocarbons, highly reduced structures with an energy of complete oxidation ($\sim$38 kJ/g) more than twice that for the same weight of carbohydrate or protein. Because of their hydrophobicity and extreme insolubility in water, triacylglycerols are segregated into lipid droplets, which do not raise the osmolarity of the cytosol and, unlike polysaccharides, do not contain extra weight as water of solvation. The relative chemical inertness of triacylglycerols allows their intracellular storage in large quantity without the risk of undesired chemical reactions with other cellular constituents.

The same properties that make triacylglycerols good storage compounds present problems in their role as fuels. Because of their insolubility in water, ingested triacylglycerols must be emulsified before they can be digested by water-soluble enzymes in the intestine, and triacylglycerols absorbed in the intestine or mobilized from storage tissues must be carried in the blood by proteins that counteract their insolubility. The relative stability of the C—C bonds in a fatty acid is overcome by activation of the carboxyl group at C-1 by attachment to coenzyme A, which allows stepwise oxidation of the fatty acyl group at the C-3 position. This latter carbon atom is also called the *beta* (β) carbon in common nomenclature, from which the oxidation of fatty acids gets its common name: β oxidation.

We begin this chapter with a brief discussion of the sources of fatty acids and the routes by which they are carried to the site of their oxidation, with special emphasis on the case of vertebrate animals. The chemical steps of fatty acid oxidation in mitochondria are then described. Three stages in this process can be distinguished: the oxidation of long-chain fatty acids to two-carbon fragments, in the form of acetyl-CoA; the oxidation of acetyl-CoA to CO_2 via the citric acid cycle (Chapter 15); and the transfer of electrons from reduced electron carriers to the mitochondrial respiratory chain (Chapter 18). Our emphasis in this chapter is on the first of these stages. We consider the simple case in which a fully saturated fatty acid with an even number of carbon atoms is degraded to acetyl-CoA, then we look briefly at the extra transformations necessary for the degradation of unsaturated fatty acids and of fatty acids with an odd number of carbons. Finally, we discuss variations on the β-oxidation theme that occur in specialized organelles—peroxisomes and glyoxysomes. The chapter concludes with the description of an alternative fate for the acetyl-CoA formed by β oxidation in vertebrates: the production of ketone bodies in the liver.

Digestion, Mobilization, and Transport of Fatty Acids

Cells that derive energy from the oxidation of fatty acids may obtain those fatty acids from three sources: fats in the diet, fats stored in cells as lipid droplets, and (in animals) fats newly synthesized in one organ for export to another. Some organisms use all three sources under various circumstances, whereas others obtain fatty acids from only one or two of these sources. Vertebrates, for example, obtain fats in the diet, mobilize fats stored in specialized tissue (adipose tissue), and convert excess dietary carbohydrates to fats in the liver for export to other tissues. On the average, 40% or more of the daily energy requirement of humans in highly industrialized countries is supplied by dietary triacylglycerols (although most nutritional guidelines recommend that no more than 30% of the daily caloric intake be from fats). Triacylglycerols provide more than half the energy requirements of some organs, particularly the liver, heart, and resting skeletal muscle. Stored triacylglycerols are virtually the sole source of energy in hibernating animals and migrating birds. Protists obtain fats by consuming organisms lower in the food chain, and some also store fats in cytosolic lipid droplets. Higher plants mobilize fats stored in seeds during the process of germination, but do not otherwise depend on fats for energy.

Dietary Fats Are Absorbed in the Small Intestine

Before ingested triacylglycerols can be absorbed through the intestinal wall, they must be converted from insoluble macroscopic fat particles into finely dispersed microscopic micelles. Bile salts such as taurocholic acid are synthesized from cholesterol in the liver, stored in the gallbladder, and released into the small intestine after ingestion of a fatty meal. These amphipathic compounds act as biological detergents, converting dietary fats into mixed micelles of bile salts and triacylglycerols (Fig. 16–1, step ①). Micelle formation enormously increases the fraction of lipid molecules accessible to the action of

Taurocholic acid

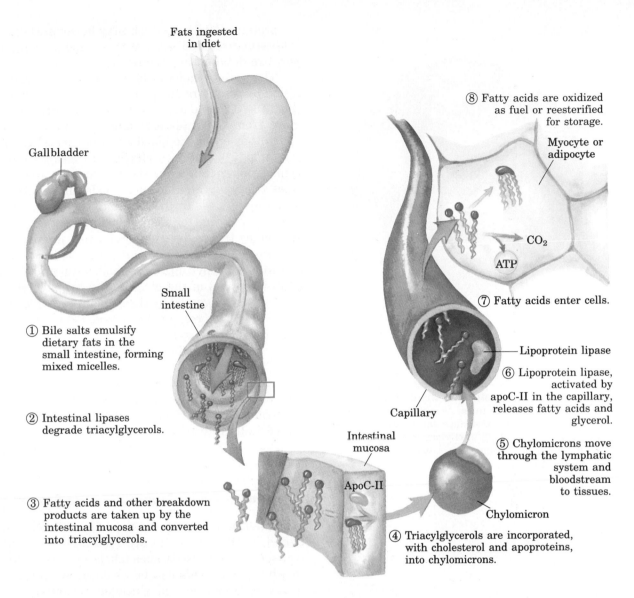

Fats ingested
in diet

Gallblader

① Bile salts emulsify
dietary fats in the
small intestine, forming
mixed micelles.

Small
intestine

② Intestinal lipases
degrade triacylglycerols.

③ Fatty acids and other breakdown
products are taken up by the
intestinal mucosa and converted
into triacylglycerols.

Intestinal
mucosa

ApoC-II

④ Triacylglycerols are incorporated,
with cholesterol and apoproteins,
into chylomicrons.

Chylomicron

⑤ Chylomicrons move
through the lymphatic
system and
bloodstream
to tissues.

Capillary

⑥ Lipoprotein lipase,
activated by
apoC-II in the capillary,
releases fatty acids and
glycerol.

Lipoprotein lipase

⑦ Fatty acids enter cells.

⑧ Fatty acids are oxidized
as fuel or reesterified
for storage.

Myocyte or
adipocyte

CO_2

ATP

Figure 16–1 Uptake of dietary lipid in the intestine of a vertebrate animal, and delivery of fatty acids to muscle and adipose tissues. The eight steps are discussed in the text.

water-soluble lipases in the intestine, and lipase action converts triacylglycerols into monoacylglycerols (monoglycerides) and diacylglycerols (diglycerides), free fatty acids, and glycerol (step ②). These products of lipase action diffuse into the epithelial cells lining the intestinal surface (the intestinal mucosa) (step ③), where they are reconverted to triacylglycerols and packaged with dietary cholesterol and specific proteins into lipoprotein aggregates called **chylomicrons** (Fig. 16–2; see also Fig. 16–1, step ④).

Apolipoproteins are lipid-binding proteins in the blood, responsible for the transport of triacylglycerols, phospholipids, cholesterol, and cholesteryl esters between organs. Apolipoproteins ("apo" designates the protein in its lipid-free form) combine with various lipids to form several classes of lipoprotein particles, spherical aggregates with hydrophobic lipids at the core and hydrophilic protein side chains and lipid head groups at the surface. Various combinations of lipid and protein produce particles of different densities, ranging from chylomicrons and very low-density lipoproteins (VLDL) to very high-density

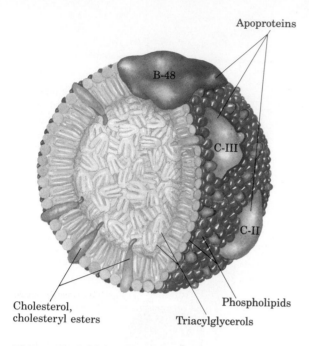

Apoproteins

B-48

C-III

C-II

Cholesterol,
cholesteryl esters

Phospholipids

Triacylglycerols

Figure 16–2 Molecular structure of a chylomicron. The surface is covered with a layer of phospholipids, with head groups facing the aqueous phase. Triacylglycerols sequestered in the interior make up more than 80% of the mass. Several apoproteins that protrude from the surface act as signals in the uptake and metabolism of chylomicron contents. The diameter of chylomicrons ranges from about 100 to about 500 nm.

lipoproteins (VHDL), which may be separated by ultracentrifugation. The structures and roles of these lipoprotein particles in lipid transport are detailed in Chapter 20.

The protein moieties of lipoproteins act as points of specific recognition by receptors on cell surfaces. In lipid uptake from the intestine (Fig. 16–1), chylomicrons, which contain apoprotein C-II (apoC-II), move from the intestinal mucosa into the lymphatic system, from which they enter the blood and are carried to muscle and adipose tissue (step ⑤). In the capillaries of these tissues, the extracellular enzyme **lipoprotein lipase** is activated by apoC-II. This enzyme hydrolyzes triacylglycerols to fatty acids and glycerol (step ⑥), which are taken up by cells in the target tissues (step ⑦). In muscle, the fatty acids are oxidized for energy; in adipose tissue, they are reesterified for storage as triacylglycerols (step ⑧).

The remnants of chylomicrons, depleted of most of their triacylglycerols but still containing cholesterol and the apoproteins apoE and apoB-48, travel in the blood to the liver, where they are taken up by endocytosis, triggered by their apoproteins. Triacylglycerols that enter the liver by this route may be oxidized to provide energy or to provide precursors for the synthesis of ketone bodies, as described later in this chapter. When the diet contains more fatty acids than are needed immediately for fuel or as precursors, they are converted into triacylglycerols in the liver, and the triacylglycerols are packaged with specific apolipoproteins into VLDLs. VLDLs are transported in the blood from the liver to adipose tissues, where the triacylglycerols are removed and stored in lipid droplets within adipocytes.

Hormones Trigger Mobilization of Stored Triacylglycerols

When hormones signal the need for metabolic energy, triacylglycerols stored in adipose tissue are mobilized (brought out of storage) and transported to those tissues (skeletal muscle, heart, and renal cortex) in which fatty acids can be oxidized for energy production. The hormones epinephrine and glucagon, secreted in response to low blood glucose levels, activate adenylate cyclase in the adipocyte plasma membrane (Fig. 16–3), raising the intracellular concentration of cAMP (see Fig. 14–18). A cAMP-dependent protein kinase, in turn, phosphorylates and thereby activates **hormone-sensitive triacylglycerol lipase,** which catalyzes hydrolysis of the ester linkages of triacylglycerols. The fatty acids thus released diffuse from the adipocyte into the blood, where they bind to the blood protein **serum albumin.** This protein (M_r 62,000), which constitutes about half of the total serum protein, binds as many as 10 fatty acids per protein monomer by noncovalent interactions. Bound to this soluble protein, the otherwise insoluble fatty acids are carried to tissues such as skeletal muscle, heart, and renal cortex. Here, fatty acids dissociate from albumin and diffuse into the cytosol of the cells in which they will serve as fuel.

About 95% of the biologically available energy of triacylglycerols resides in their three long-chain fatty acids; only 5% is contributed by the glycerol moiety. The glycerol released by lipase action is phosphorylated by **glycerol kinase** (Fig. 16–4), and the resulting glycerol-3-phosphate is oxidized to dihydroxyacetone phosphate. The glycolytic enzyme triose phosphate isomerase converts this compound to glyceraldehyde-3-phosphate, which is oxidized via glycolysis.

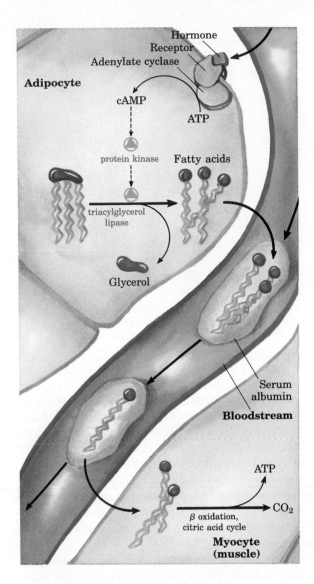

Figure 16–3 Mobilization of triacylglycerols stored in adipose tissue. Low levels of glucose in the blood trigger the mobilization of triacylglycerols through the action of epinephrine and glucagon on the adipocyte adenylate cyclase. The subsequent steps in mobilization are described in the text.

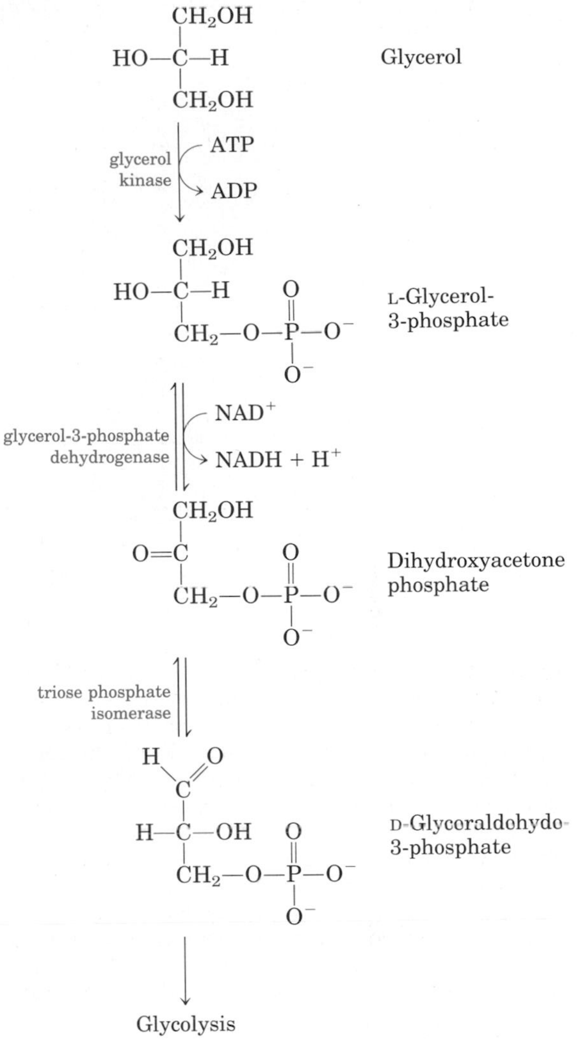

Figure 16–4 Pathway by which glycerol derived from triacylglycerols enters glycolysis.

Fatty Acids Are Activated and Transported into Mitochondria

The enzymes of fatty acid oxidation in animal cells are located in the mitochondrial matrix, as demonstrated in 1948 by Eugene P. Kennedy and Albert Lehninger. The free fatty acids that enter the cytosol from the blood cannot pass directly through the mitochondrial membranes, but must first undergo a series of three enzymatic reactions. The first is catalyzed by a family of isozymes present in the outer mitochondrial membrane, **acyl-CoA synthetases,** which promote the general reaction

$$\text{Fatty acid} + \text{CoA} + \text{ATP} \Longleftrightarrow \text{fatty acyl–CoA} + \text{AMP} + \text{PP}_i$$

484

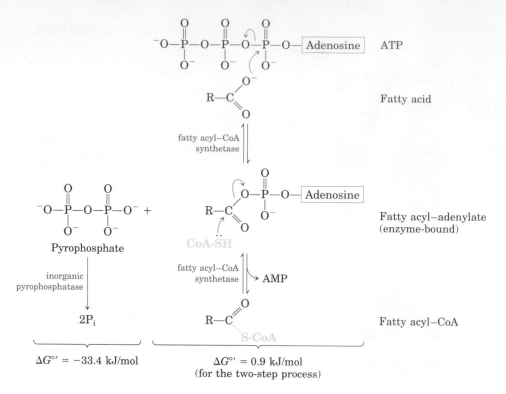

Figure 16–5 The reactions catalyzed by acyl-CoA synthetase and inorganic pyrophosphatase. Fatty acid activation by the formation of the fatty acyl–CoA derivative occurs in two steps. First, the carboxylate ion displaces the outer two (β and γ) phosphates of ATP to form a fatty acyl–adenylate, the mixed anhydride of a carboxylic acid and a phosphoric acid. The other product is PP_i, an excellent leaving group that is immediately hydrolyzed to two P_i, pulling the reaction in the forward direction. Coenzyme A carries out nucleophilic attack on the mixed anhydride, displacing AMP and forming the thioester fatty acyl–CoA. The overall reaction is highly exergonic.

The different acyl-CoA synthetase isozymes act on fatty acids of short, intermediate, and long carbon chains, respectively. Acyl-CoA synthetase catalyzes the formation of a thioester linkage between the fatty acid carboxyl group and the thiol group of coenzyme A to yield a **fatty acyl–CoA;** simultaneously, ATP undergoes cleavage to AMP and PP_i. Recall our description of this reaction in Chapter 13 to illustrate how the free energy released by cleavage of phosphoric acid anhydride bonds in ATP could be coupled to the formation of a high-energy compound (see Fig. 13–10). The reaction occurs in two steps, and involves a fatty acyl–adenylate intermediate (Fig. 16–5).

Fatty acyl–CoAs, like acetyl-CoA, are high-energy compounds; their hydrolysis to free fatty acid and CoA has a large, negative standard free-energy change ($\Delta G^{\circ\prime} \approx -31$ kJ/mol). The formation of fatty acyl–CoAs is made more favorable by the hydrolysis of *two* high-energy bonds in ATP; the pyrophosphate formed in the activation reaction is immediately hydrolyzed by a second enzyme, inorganic pyrophosphatase (Fig. 16–5), which pulls the preceding activation reaction in the direction of the formation of fatty acyl–CoA. The overall reaction is

$$\text{Fatty acid} + \text{CoA} + \text{ATP} \longrightarrow \text{fatty acyl–CoA} + \text{AMP} + 2P_i \qquad (16\text{–}1)$$
$$\Delta G^{\circ\prime} = -32.5 \text{ kJ/mol}$$

Fatty acyl–CoA esters formed in the outer mitochondrial membrane do not cross the inner mitochondrial membrane intact. Instead, the fatty acyl group is transiently attached to the hydroxyl group of **carnitine** and the fatty acyl–carnitine is carried across the inner mitochondrial membrane by a specific transporter (Fig. 16–6). In this second enzymatic reaction required for fatty acid movement into mitochondria, **carnitine acyltransferase I,** present on the outer face of the inner membrane, catalyzes transesterification of the fatty acyl group from coenzyme A to carnitine. The fatty acyl–carnitine ester crosses the inner mitochondrial membrane into the matrix by facilitated diffusion through the **acyl-carnitine/carnitine transporter.**

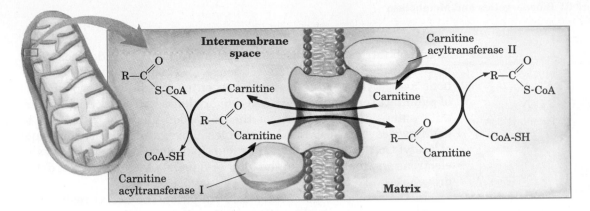

Figure 16–6 Fatty acid entry into mitochondria via the acyl-carnitine/carnitine transporter. After its formation at the outer surface of the inner mitochondrial membrane, fatty acyl–carnitine moves into the matrix by facilitated diffusion through the transporter. In the matrix, the acyl group is transferred back to CoA, freeing carnitine to return to the intermembrane space via the same transporter. The acyltransferase I and II enzymes are bound to the outer and inner surfaces, respectively, of the mitochondrial inner membrane. This entry process is the rate-limiting step for oxidation of fatty acids in mitochondria, as discussed later in this chapter.

In the third and final step of the entry process, the fatty acyl group is enzymatically transferred from carnitine to intramitochondrial coenzyme A by **carnitine acyltransferase II.** This isozyme is located on the inner face of the inner mitochondrial membrane, where it regenerates fatty acyl–CoA and releases it, along with free carnitine, into the matrix (Fig. 16–6). Carnitine reenters the space between the inner and outer mitochondrial membranes via the acyl-carnitine/carnitine transporter.

This three-step process for transferring fatty acids into the mitochondrion has the effect of separating the cytosolic and mitochondrial pools of coenzyme A, which have different functions. The mitochondrial pool of coenzyme A is largely used in oxidative degradation of pyruvate, fatty acids, and some amino acids, whereas the cytosolic pool of coenzyme A is used in the biosynthesis of fatty acids (Chapter 20).

Once inside the mitochondrion, the fatty acyl–CoA is ready for the oxidation of its fatty acid component by a set of enzymes in the mitochondrial matrix.

β Oxidation

Mitochondrial oxidation of fatty acids takes place in three stages (Fig. 16–7). In the first stage—β oxidation—the fatty acids undergo oxidative removal of successive two-carbon units in the form of acetyl-CoA, starting from the carboxyl end of the fatty acyl chain. For example, the

Figure 16–7 Stages of fatty acid oxidation. Stage 1: A long-chain fatty acid is oxidized to yield acetyl residues in the form of acetyl-CoA. Stage 2: The acetyl residues are oxidized to CO_2 via the citric acid cycle. Stage 3: Electrons derived from the oxidations of stages 1 and 2 are passed to O_2 via the mitochondrial respiratory chain, providing the energy for ATP synthesis by oxidative phosphorylation.

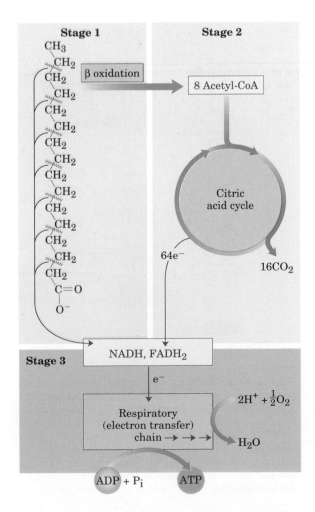

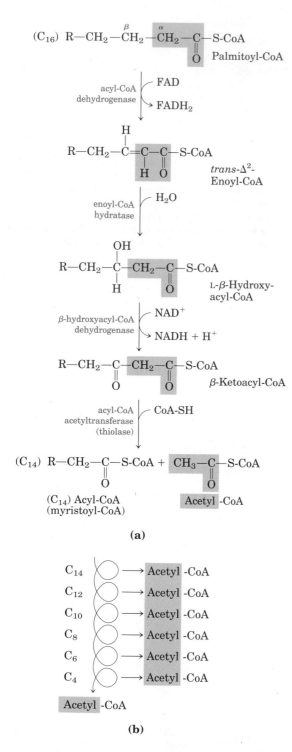

Figure 16–8 The fatty acid oxidation (β-oxidation) pathway. **(a)** In each pass through this sequence, one acetyl residue (shaded in red) is removed in the form of acetyl-CoA from the carboxyl end of palmitate (C_{16}), which enters as palmitoyl-CoA. **(b)** Six more passes through the pathway yield seven more molecules of acetyl-CoA, the seventh arising from the last two carbon atoms of the 16-carbon chain. Eight molecules of acetyl-CoA are formed in all.

16-carbon fatty acid palmitic acid (palmitate at pH 7) undergoes seven passes through this oxidative sequence, in each pass losing two carbons as acetyl-CoA. At the end of seven cycles the last two carbons of palmitate (originally C-15 and C-16) are left as acetyl-CoA. The overall result is the conversion of the 16-carbon chain of palmitate to eight two-carbon acetyl-CoA molecules. Formation of each molecule of acetyl-CoA requires removal of four hydrogen atoms (two pairs of electrons and four H^+) from the fatty acyl moiety by the action of dehydrogenases.

In the second stage of fatty acid oxidation the acetyl residues of acetyl-CoA are oxidized to CO_2 via the citric acid cycle, which also takes place in the mitochondrial matrix. Acetyl-CoA derived from fatty acid oxidation thus enters a final common pathway of oxidation along with acetyl-CoA derived from glucose via glycolysis and pyruvate oxidation (see Fig. 15–1).

The first two stages of fatty acid oxidation produce the reduced electron carriers NADH and $FADH_2$, which in the third stage donate electrons to the mitochondrial respiratory chain, through which the electrons are carried to oxygen (Fig. 16–7). Coupled to this flow of electrons is the phosphorylation of ADP to ATP, to be described in Chapter 18. Thus energy released by fatty acid oxidation is conserved as ATP.

We will now look in more detail at the first stage of fatty acid oxidation, for the simple case of a saturated chain with an even number of carbons, and for the slightly more complicated cases of unsaturated and odd-number chains. We then consider the regulation of fatty acid oxidation, and the β-oxidative processes occurring in organelles other than mitochondria.

β Oxidation of Saturated Fatty Acids Has Four Basic Steps

Four enzyme-catalyzed reactions are involved in the first stage of fatty acid oxidation (Fig. 16–8a). First, dehydrogenation produces a double bond between the α and β carbon atoms (C-2 and C-3), yielding a **trans-Δ²-enoyl-CoA.** The symbol Δ^2 designates the position of the double bond. (It may be helpful to review fatty acid nomenclature, described on p. 240.) The new double bond has the trans configuration; recall that naturally occurring unsaturated fatty acids normally have their double bonds in the cis configuration. We shall consider the significance of this difference later.

The enzyme responsible for this first step, **acyl-CoA dehydrogenase,** includes FAD as a prosthetic group. The electrons removed from the fatty acyl–CoA are transferred to the FAD, and the reduced form of the dehydrogenase then immediately donates its electrons to an electron carrier, the **electron-transferring flavoprotein** (ETFP). ETFP, an integral protein of the inner mitochondrial membrane, is one of the electron carriers of the mitochondrial respiratory chain (Fig. 16–9). The transfer of a pair of electrons from the $FADH_2$ of acyl-CoA dehydrogenase to O_2 via the respiratory chain provides the energy for the synthesis of two ATP molecules.

The oxidation catalyzed by acyl-CoA dehydrogenase is analogous to succinate dehydrogenation in the citric acid cycle (p. 457); in both reactions the enzyme is bound to the inner membrane, a double bond is introduced into a carboxylic acid between the α and β carbons, FAD is the electron acceptor, and electrons from the reaction ultimately enter the respiratory chain and are carried to O_2 with the concomitant synthesis of two ATP molecules per electron pair.

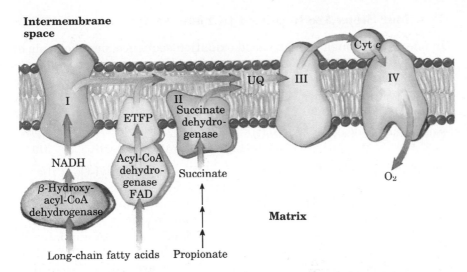

Intermembrane space

Matrix

NADH

β-Hydroxy-acyl-CoA dehydrogenase

Long-chain fatty acids

ETFP

Acyl-CoA dehydro-genase FAD

Propionate

UQ

II Succinate dehydro-genase

Succinate

III

Cyt *c*

IV

O$_2$

Figure 16–9 Electrons removed from fatty acids during β oxidation pass into the mitochondrial respiratory chain and eventually to O$_2$. The structures I through IV are enzyme complexes that catalyze portions of the electron transfer to oxygen. Fatty acyl–CoA dehydrogenase feeds electrons into an electron-transferring flavoprotein (ETFP) containing an iron–sulfur center, which in turn reduces a lipid-soluble electron carrier, ubiquinone (UQ, or coenzyme Q). β-Hydroxyacyl-CoA dehydrogenase transfers electrons to NAD$^+$, and the resulting NADH is reoxidized by NADH dehydrogenase (Complex I of the respiratory chain). Propionate produced from odd-chain fatty acids is converted to succinate. Succinate dehydrogenase, which acts in the citric acid cycle (p. 457), feeds electrons into the respiratory chain at Complex II. Cytochrome *c* (cyt *c*) is a soluble electron carrier that transfers electrons between Complexes III and IV. All of these transfers are described in detail in Chapter 18.

In the second step of the fatty acid oxidation cycle (Fig. 16–8a), water is added to the double bond of the *trans*-Δ²-enoyl-CoA to form the L stereoisomer of **β-hydroxyacyl-CoA** (also designated **3-hydroxyacyl-CoA**). This reaction, catalyzed by **enoyl-CoA hydratase,** is formally analogous to the fumarase reaction in the citric acid cycle, in which H$_2$O adds across an α–β double bond (p. 458).

In the third step, the L-β-hydroxyacyl-CoA is dehydrogenated to form **β-ketoacyl-CoA** by the action of **β-hydroxyacyl-CoA dehydrogenase** (Fig. 16–8a); NAD$^+$ is the electron acceptor. This enzyme is absolutely specific for the L stereoisomer. The NADH formed in this reaction donates its electrons to **NADH dehydrogenase (Complex I),** an electron carrier of the respiratory chain (Fig. 16–9). Three ATP molecules are generated from ADP per pair of electrons passing from NADH to O$_2$ via the respiratory chain. The reaction catalyzed by β-hydroxyacyl-CoA dehydrogenase is closely analogous to the malate dehydrogenase reaction of the citric acid cycle (p. 459).

The fourth and last step of the fatty acid oxidation cycle is catalyzed by **acyl-CoA acetyltransferase** (more commonly called **thiolase**), which promotes reaction of β-ketoacyl-CoA with a molecule of free coenzyme A to split off the carboxyl-terminal two-carbon fragment of the original fatty acid as acetyl-CoA. The other product is the coenzyme A thioester of the original fatty acid, now shortened by two carbon atoms (Fig. 16–8a). This reaction is called thiolysis, by analogy with the process of hydrolysis, because the β-ketoacyl-CoA is cleaved by reaction with the thiol group of coenzyme A.

The carbon–carbon single bond that connects methylene (—CH$_2$—) groups in fatty acids is relatively stable. The β-oxidation sequence represents an elegant solution to the problem of breaking these bonds. The first three reactions of β oxidation have the effect of creating a much less stable C—C bond, in which one of the carbon atoms (the α carbon, C-2) is bonded to *two* carbonyl carbons. The ketone function on the β carbon (C-3) makes it a good point for nucleophilic attack by —SH of coenzyme A, catalyzed by thiolase. The acidity of the α carbon makes the terminal —CH$_2$—CO—S-CoA a good leaving group, facilitating breakage of the α–β bond.

The Four Steps Are Repeated to Yield Acetyl-CoA and ATP

In one pass through the fatty acid oxidation sequence, one molecule of acetyl-CoA, two pairs of electrons, and four H^+ ions are removed from the long-chain fatty acyl–CoA, to shorten it by two carbon atoms. The equation for one pass, beginning with the coenzyme A ester of our example, palmitate, is

$$\text{Palmitoyl-CoA} + \text{CoA} + \text{FAD} + \text{NAD}^+ + H_2O \longrightarrow$$
$$\text{myristoyl-CoA} + \text{acetyl-CoA} + \text{FADH}_2 + \text{NADH} + H^+ \quad (16\text{-}2)$$

Following removal of one acetyl-CoA unit from palmitoyl-CoA, the coenzyme A thioester of the shortened fatty acid remains, in this case the 14-carbon myristate. The myristoyl-CoA can now enter the β-oxidation sequence and go through another set of four reactions, exactly analogous to the first, to yield a second molecule of acetyl-CoA and lauroyl-CoA, the coenzyme A thioester of the 12-carbon laurate. Altogether, seven passes through the β-oxidation sequence are required to oxidize one molecule of palmitoyl-CoA to eight molecules of acetyl-CoA (Fig. 16–8b). The overall equation is

$$\text{Palmitoyl-CoA} + 7\text{CoA} + 7\text{FAD} + 7\text{NAD}^+ + 7H_2O \longrightarrow$$
$$8 \text{ acetyl-CoA} + 7\text{FADH}_2 + 7\text{NADH} + 7H^+ \quad (16\text{-}3)$$

Each molecule of FADH_2 formed during oxidation of the fatty acid donates a pair of electrons to ETFP of the respiratory chain (Fig. 16–9); two molecules of ATP are generated during the ensuing transfer of the electron pair to O_2 and the coupled oxidative phosphorylations. Similarly, each molecule of NADH formed delivers a pair of electrons to the mitochondrial NADH dehydrogenase; the subsequent transfer of each pair of electrons to O_2 results in formation of three molecules of ATP. Thus five molecules of ATP are formed for each two-carbon unit removed in one pass through the sequence as it occurs in animal tissues, such as the liver or heart. Note that water is also produced in this process. Condensation of ADP and P_i releases one H_2O for each ATP formed, and transfer of electrons from NADH or FADH_2 to O_2 yields one H_2O per electron pair. Reduction of O_2 by NADH also consumes one H^+ per NADH: $\text{NADH} + H^+ + \frac{1}{2}O_2 \longrightarrow \text{NAD}^+ + H_2O$. In hibernating animals, fatty acid oxidation provides metabolic energy, heat, and water—all essential for survival of an animal that neither eats nor drinks for long periods (Box 16–1).

The overall equation for the oxidation of palmitoyl-CoA to eight molecules of acetyl-CoA, including the electron transfers and oxidative phosphorylation, is

$$\text{Palmitoyl-CoA} + 7\text{CoA} + 7O_2 + 35P_i + 35\text{ADP} \longrightarrow$$
$$8 \text{ acetyl-CoA} + 35\text{ATP} + 42H_2O \quad (16\text{-}4)$$

Acetyl-CoA Can Be Further Oxidized via the Citric Acid Cycle

The acetyl-CoA produced from the oxidation of fatty acids can be oxidized to CO_2 and H_2O by the citric acid cycle. The following equation represents the balance sheet for the second stage in the oxidation of our example, palmitoyl-CoA, together with the coupled phosphorylations of the third stage:

$$8 \text{ Acetyl-CoA} + 16O_2 + 96P_i + 96\text{ADP} \longrightarrow$$
$$8\text{CoA} + 96\text{ATP} + 104H_2O + 16CO_2 \quad (16\text{-}5)$$

BOX 16–1   **Fat Bears Carry On β Oxidation in Their Sleep**

Many animals depend on fat stores for energy during hibernation or dormancy, during migratory periods, and in other situations involving radical metabolic adjustments (as in the case of the camel, which can obtain its water supply from the oxidation of fat).

One of the most pronounced adjustments of fat metabolism occurs in the hibernation of the grizzly bear (Fig. 1). Bears go into a continuous state of dormancy for periods as long as seven months without arousal. Unlike most other hibernating species, the bear maintains its body temperature between 32 and 35 °C, nearly the normal level. Although the bear in this state expends about 6,000 kcal/day (25,000 kJ/day), it does not eat, drink, urinate, or defecate for months at a time. When accidentally aroused, the bear is almost immediately alert and ready to defend itself.

Experimental studies have shown that the bear uses body fat as its sole fuel during hibernation. The oxidation of fat yields sufficient energy for maintaining body temperature, for active synthesis of amino acids and proteins, and for other energy-requiring activities, such as membrane transport. Fat oxidation also releases large amounts of water (p. 488), which replenishes water loss during breathing. In addition, degradation of triacylglycerols yields glycerol, which, following its enzymatic phosphorylation to glycerol-3-phosphate and oxidation to dihydroxyacetone phosphate, is converted into blood glucose. Urea formed during the degradation of amino acids is reabsorbed and recycled by the bear, the amino groups being used to make new amino acids for maintaining body proteins.

Bears store an enormous amount of body fat in preparation for their long hibernation periods. Normally, an adult grizzly bear consumes about 9,000 kcal/day during the late spring and summer. But as winter approaches bears will feed 20 hours a day and consume up to 20,000 kcal, in response to seasonal changes in hormone secretion. Large amounts of body triacylglycerols are formed from the huge amounts of carbohydrate consumed during the fattening-up period. Other hibernating species, including the tiny dormouse, also accumulate large amounts of body fat. The camel, although not a hibernator, can synthesize and store triacylglycerols in large amounts in its hump, a metabolic source of both energy and water under desert conditions.

Figure 1 A grizzly bear prepares its hibernation nest, near the McNeil River in Canada.

Combining Equations 16–4 and 16–5, we obtain the overall equation for the complete oxidation of palmitoyl-CoA to carbon dioxide and water:

$$\text{Palmitoyl-CoA} + 23O_2 + 131P_i + 131ADP \longrightarrow$$
$$\text{CoA} + 131ATP + 16CO_2 + 146H_2O \quad (16\text{–}6)$$

Because the activation of palmitate to palmitoyl-CoA consumes two ATP equivalents (p. 484), the net gain per molecule of palmitate is 129 ATP. Table 16–1 summarizes the yields of NADH, $FADH_2$, and ATP in the successive steps of fatty acid oxidation. The standard free-energy change for the oxidation of palmitate to $CO_2 + H_2O$ is about 9,800 kJ/mol. Under standard conditions, $30.5 \times 129 = 3,940$ kJ/mol (about 40% of the theoretical maximum) is recovered as the phosphate bond energy of ATP. However, when the free-energy changes are calculated from actual concentrations of reactants and products under intracellular conditions (see Box 13–2), the free-energy recovery is over 80%; the energy conservation is remarkably efficient.

Table 16–1 Yield of ATP during oxidation of one molecule of palmitoyl-CoA to CO_2 and H_2O

Enzyme catalyzing oxidation step	Number of NADH or $FADH_2$ formed	Number of ATP ultimately formed
Acyl-CoA dehydrogenase	7 $FADH_2$	14
β-Hydroxyacyl-CoA dehydrogenase	7 NADH	21
Isocitrate dehydrogenase	8 NADH	24
α-Ketoglutarate dehydrogenase	8 NADH	24
Succinyl-CoA synthetase		8*
Succinate dehydrogenase	8 $FADH_2$	16
Malate dehydrogenase	8 NADH	24
Total		131

* GTP produced directly in this step yields ATP in the reaction catalyzed by nucleoside diphosphate kinase (p. 457).

Oxidation of Unsaturated Fatty Acids Requires Two Additional Reactions

The fatty acid oxidation sequence just described is typical when the incoming fatty acid is saturated (having only single bonds in its carbon chain). However, most of the fatty acids in the triacylglycerols and phospholipids of animals and plants are unsaturated, having one or more double bonds. These bonds are in the cis configuration and cannot be acted upon by enoyl-CoA hydratase, the enzyme catalyzing the addition of H_2O to the trans double bond of the Δ^2-enoyl-CoA generated during β oxidation. However, by the action of two auxiliary enzymes, the fatty acid oxidation sequence described above can also break down the common unsaturated fatty acids. The action of these two enzymes, one an isomerase and the other a reductase, will be illustrated by two examples.

First, let us follow the oxidation of oleate, an abundant 18-carbon monounsaturated fatty acid with a cis double bond between C-9 and C-10 (denoted Δ^9). Oleate is converted into oleoyl-CoA (Fig. 16–10),

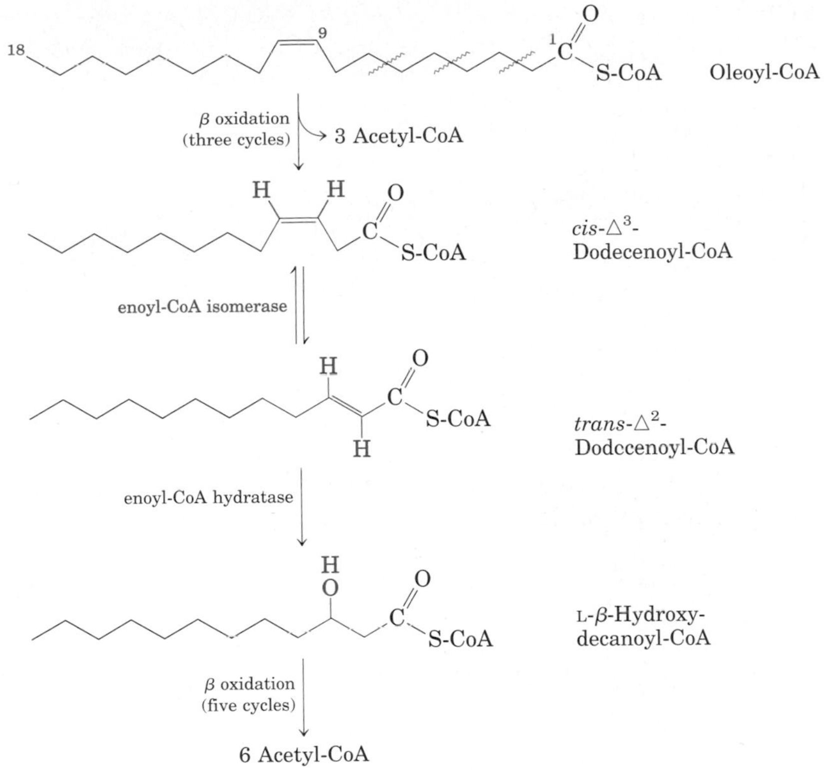

Figure 16–10 The oxidation of a monounsaturated fatty acyl–CoA, such as oleoyl-CoA (Δ^9), requires an additional enzyme, enoyl-CoA isomerase. This enzyme repositions the double bond, converting the cis isomer to a trans isomer, a normal intermediate in β oxidation.

which is transported through the mitochondrial membrane as oleoylcarnitine and then converted back into oleoyl-CoA in the matrix (Fig. 16–6). Oleoyl-CoA then undergoes three passes through the fatty acid oxidation cycle to yield three molecules of acetyl-CoA and the coenzyme A ester of a Δ^3, 12-carbon unsaturated fatty acid, *cis*-Δ^3-dodecenoyl-CoA (Fig. 16–10). This product cannot be acted upon by the next enzyme of the β-oxidation pathway, enoyl-CoA hydratase, which acts only on trans double bonds. However, by the action of the auxiliary enzyme, **enoyl-CoA isomerase,** the *cis*-Δ^3-enoyl-CoA is isomerized to yield the *trans*-Δ^2-enoyl-CoA, which is converted by enoyl-CoA hydratase into the corresponding L-β-hydroxyacyl-CoA (*trans*-Δ^2-dodecenoyl-CoA). This intermediate is now acted upon by the remaining enzymes of β oxidation to yield acetyl-CoA and a 10-carbon saturated fatty acid as its coenzyme A ester (decanoyl-CoA). The latter undergoes four more passes through the pathway to yield altogether nine acetyl-CoAs from one molecule of the 18-carbon oleate.

The other auxiliary enzyme (a reductase) is required for oxidation of polyunsaturated fatty acids. As an example, we take the 18-carbon linoleate, which has a *cis*-Δ^9,*cis*-Δ^{12} configuration (Fig. 16–11). Linoleoyl-CoA undergoes three passes through the standard β-oxidation sequence to yield three molecules of acetyl-CoA and the coenzyme A ester of a 12-carbon unsaturated fatty acid with a *cis*-Δ^3,*cis*-Δ^6 configuration. This intermediate cannot be used by the enzymes of the β-oxidation pathway; its double bonds are in the wrong position and have the wrong configuration (cis, not trans). However, the combined action of enoyl-CoA isomerase and **2,4-dienoyl-CoA reductase** (Fig. 16–11) allows reentry of this intermediate into the normal β-oxidation pathway and its degradation to six acetyl-CoAs. The overall result is conversion of linoleate to nine molecules of acetyl-CoA.

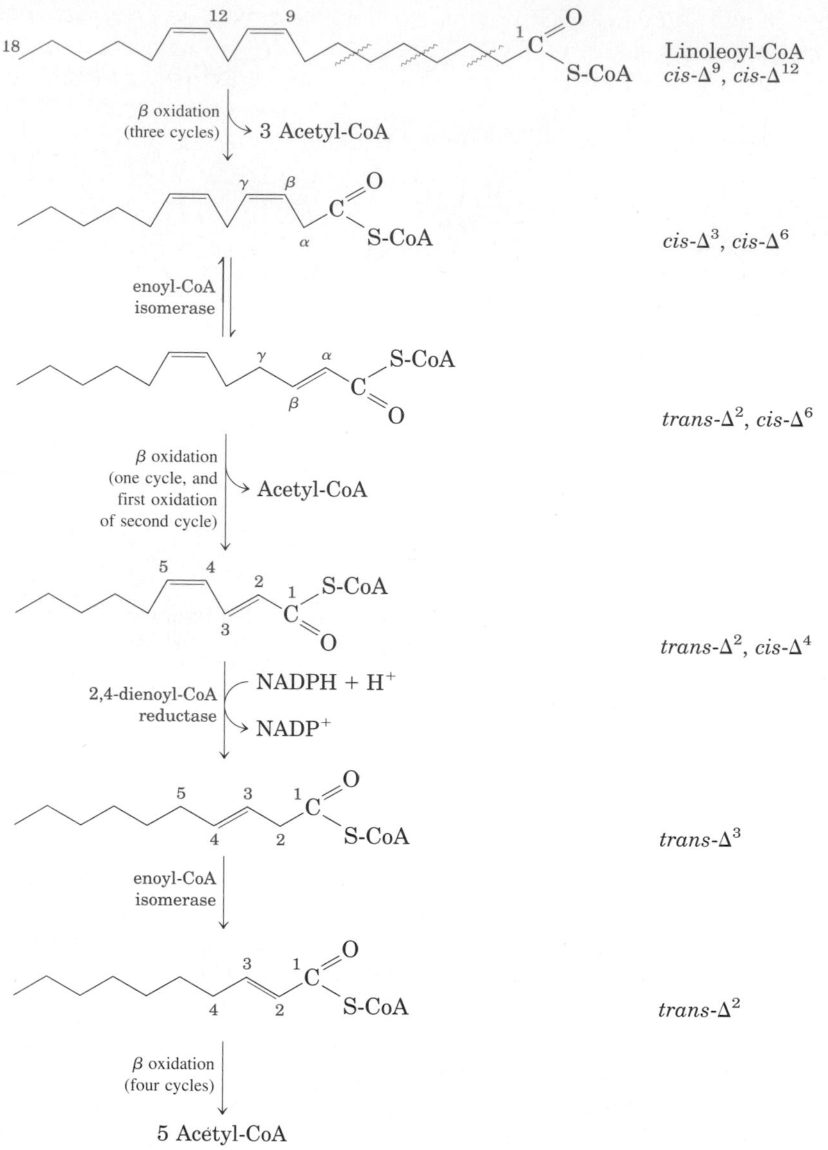

Figure 16–11 Oxidation of polyunsaturated fatty acids requires a second auxiliary enzyme in addition to enoyl-CoA isomerase: NADPH-dependent 2,4-dienoyl-CoA reductase. The combined action of these two enzymes converts a *trans-*Δ^2,*cis-*Δ^4-dienoyl-CoA intermediate into the *trans-*Δ^2-enoyl-CoA substrate necessary for β oxidation.

Oxidation of Odd-Chain Fatty Acids Requires Three Extra Reactions

Although most naturally occurring lipids contain fatty acids with an even number of carbon atoms, fatty acids with an odd number of carbons are found in significant amounts in the lipids of many plants and some marine organisms. Small quantities of the three-carbon **propionate** are added as a mold inhibitor to some breads and cereals, and thus propionate enters the human diet. Moreover, cattle and other ruminant animals form large amounts of propionate during fermentation of carbohydrates in the rumen. The propionate so formed is absorbed into the blood and oxidized by the liver and other tissues.

Long-chain odd-carbon fatty acids are oxidized by the same pathway as the even-carbon acids, beginning at the carboxyl end of the chain. However, the substrate for the last pass through the β-oxidation sequence is a fatty acyl–CoA in which the fatty acid has five carbon

$$CH_3—CH_2—COO^-$$
Propionate

atoms. When this is oxidized and ultimately cleaved, the products are acctyl-CoA and **propionyl-CoA.** The acetyl-CoA is of course oxidized via the citric acid cycle, but propionyl-CoA takes a rather unusual enzymatic pathway, involving three enzymes. Propionyl-CoA is carboxylated to form the D stereoisomer of **methylmalonyl-CoA** (Fig. 16–12) by **propionyl-CoA carboxylase,** which contains the cofactor biotin. In this enzymatic reaction, as in the pyruvate carboxylase reaction (see Fig. 15–13), CO_2 (or its hydrated ion, HCO_3^-) is activated by attachment to biotin before its transfer to the propionate moiety. The formation of the carboxybiotin intermediate requires energy, which is provided by the cleavage of ATP to AMP and PP_i.

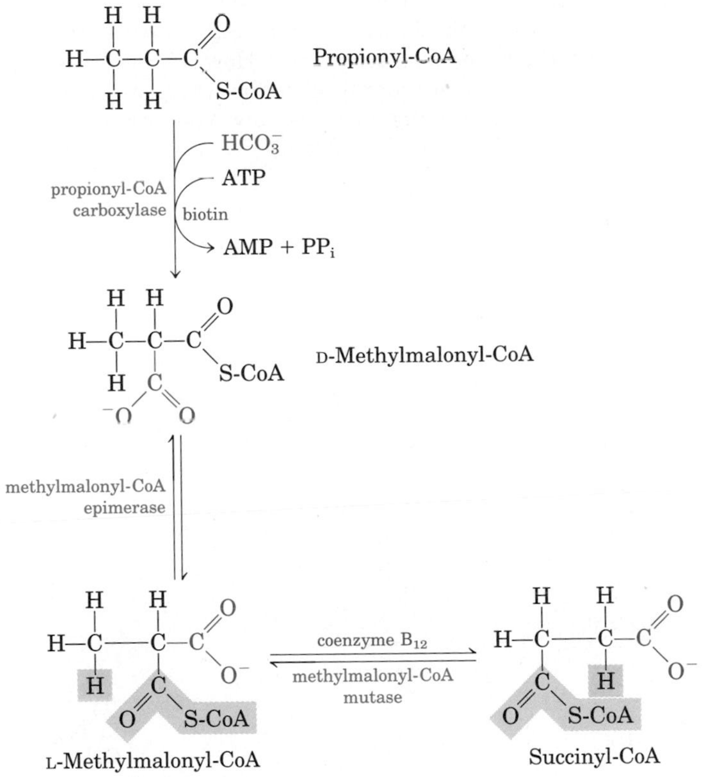

Figure 16–12 Strategy for the oxidation of propionyl-CoA, involving the carboxylation of propionyl-CoA to D-methylmalonyl-CoA and conversion of the latter to succinyl-CoA. This conversion requires epimerization of D- to L-methylmalonyl-CoA, followed by a remarkable reaction in which substituents on adjacent carbon atoms exchange positions; Box 16–2 describes the role of coenzyme B_{12} in this exchange reaction.

The D-methylmalonyl-CoA thus formed is enzymatically epimerized to its L stereoisomer by the action of **methylmalonyl-CoA epimerase** (Fig. 16–12). The L-methylmalonyl-CoA undergoes an intramolecular rearrangement to form succinyl-CoA, which can enter the citric acid cycle. This rearrangement is catalyzed by **methylmalonyl-CoA mutase,** which requires as its coenzyme **deoxyadenosylcobalamin,** or **coenzyme B_{12},** derived from vitamin B_{12} (cobalamin) (Box 16–2).

BOX 16–2   Coenzyme B_{12}: A Radical Solution to a Perplexing Problem

In the methylmalonyl-CoA mutase reaction (see Fig. 16–12), the group —CO—S-CoA at C-2 of the original propionate exchanges position with a hydrogen atom at C-3 of the original propionate (Fig. 1a). Coenzyme B_{12} is the cofactor for this reaction, as it is for almost all enzymes that catalyze reactions of this general type (Fig. 1b). These coenzyme B_{12}–dependent reactions are among the very few enzymatic reactions in biology in which there is an exchange of an alkyl or substituted alkyl group (X) with a hydrogen atom of an adjacent carbon, *with no mixing of the hydrogen atom transferred with the hydrogen of the solvent,* H_2O. How is it possible for the hydrogen atom to move between two carbons without mixing with the enormous excess of hydrogen atoms in the solvent?

Coenzyme B_{12} is the cofactor form of vitamin B_{12}, which is unique among all the vitamins in that it contains not only a complex organic molecule but also an essential trace element, cobalt. The complex **corrin ring system** of vitamin B_{12} (colored blue in Fig. 2), to which cobalt (as Co^{3+}) is coordinated, is chemically related to the porphyrin ring system of heme and heme proteins (see Fig. 7–18). A fifth coordination position of cobalt is filled by a nucleotide we have not encountered before, dimethylbenzimidazole ribonucleotide (yellow), bound covalently by its 5′-phosphate group to one of the side chains of the corrin ring through aminoisopropanol.

Vitamin B_{12} as usually isolated is called **cyanocobalamin** because it contains a cyano group

Figure 1

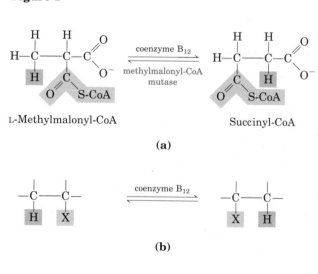

L-Methylmalonyl-CoA   Succinyl-CoA

(a)

(b)

Figure 2

5′-Deoxy-adenosine

Corrin ring system

Amino-isopropanol

Dimethyl-benzimidazole ribonucleotide

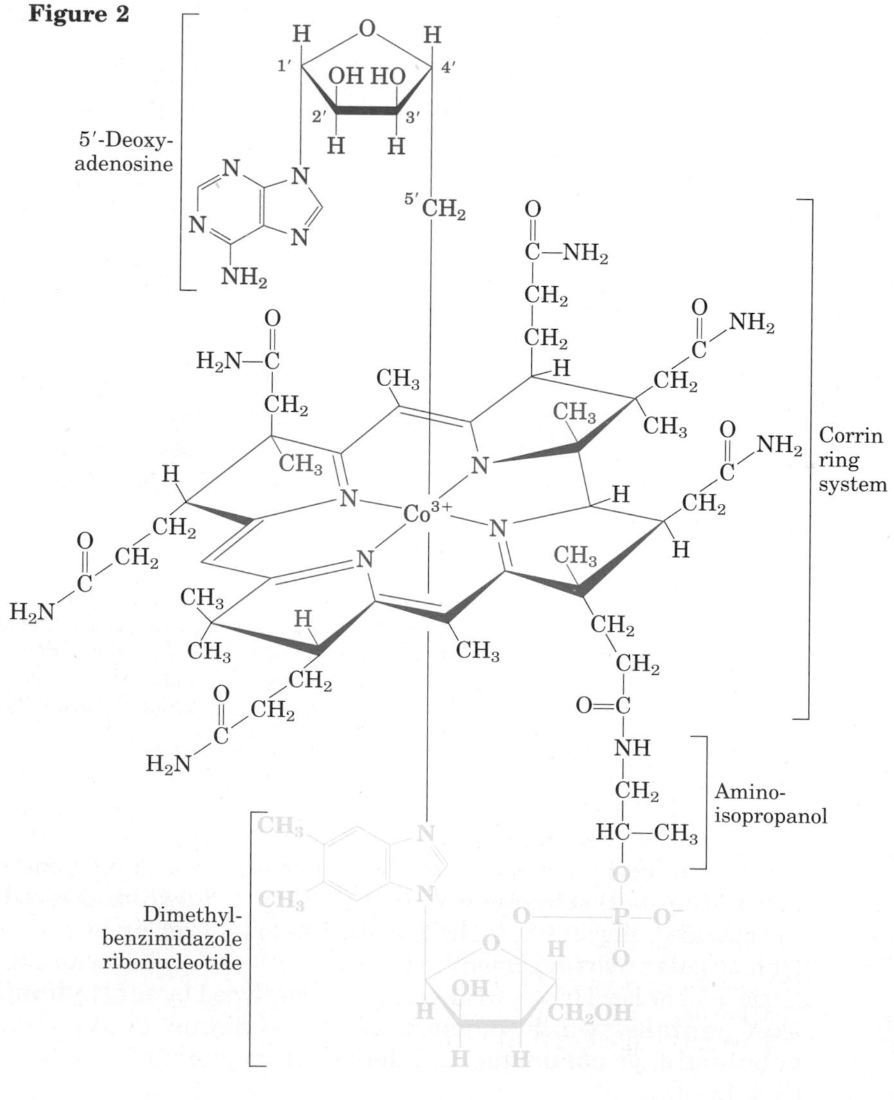

Figure 3

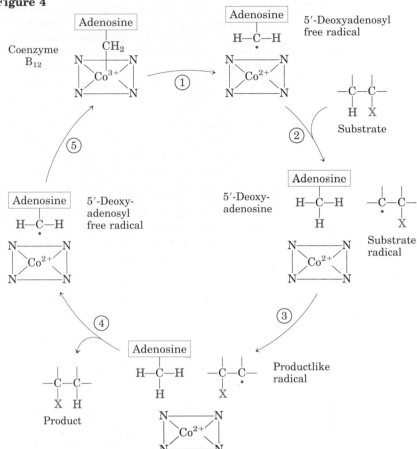

Figure 4

(picked up during purification) attached to cobalt in the sixth coordination position. In **5′-deoxyadenosylcobalamin,** the cofactor for methylmalonyl-CoA mutase, the cyano group is replaced by the **5′-deoxyadenosyl** group (red in Fig. 2), covalently bound through C-5′ to the cobalt. The three-dimensional structure of the cofactor was determined by x-ray crystallography by Dorothy Crowfoot Hodgkin in 1956.

The formation of this complex cofactor (Fig. 3) is one of only two known cases in which triphosphate is cleaved from ATP; the other case is the formation of S-adenosylmethionine from ATP and methionine (see Fig. 17–20).

The key to understanding how coenzyme B_{12} catalyzes hydrogen exchange lies in the properties of the covalent bond between cobalt and C-5′ of the deoxyadenosyl group (Fig. 2). This is a relatively weak bond; its bond dissociation energy is about 110 kJ/mol, compared with 348 kJ/mol for a typical C—C bond or 414 kJ/mol for a C—H bond. Merely illuminating the compound with visible light is

enough to break this bond. (This extreme photolability probably accounts for the fact that plants do not contain vitamin B_{12}.) Dissociation produces a 5′-deoxyadenosyl radical and the Co^{2+} form of the vitamin. The chemical function of 5′-deoxyadenosylcobalamin is to generate free radicals in this way, initiating a series of transformations such as that illustrated in Figure 4, a postulated mechanism for the reaction catalyzed by methylmalonyl-CoA mutase and a number of other coenzyme B_{12}–dependent transformations.

The enzyme first breaks the Co—C bond in the cofactor, leaving the coenzyme in its Co^{2+} form and producing the 5′-deoxyadenosyl free radical (step ①). This radical now abstracts a hydrogen atom from the substrate, converting the substrate to a radical and producing 5′-deoxyadenosine (step ②). Rearrangement of the substrate radical (step ③) yields another radical, in which the migrating

Dorothy Crowfoot Hodgkin

Continued

group X (—CO—S-CoA for methylmalonyl-CoA mutase) has moved to the adjacent carbon to form a productlike radical. The hydrogen atom initially abstracted from the substrate is now part of the CH_3— group of 5'-deoxyadenosine; one of the hydrogens from this same CH_3— group (it can be the same one originally abstracted) is returned to the productlike radical, generating the product and regenerating the deoxyadenosyl free radical (step ④). Finally, the bond re-forms between cobalt and the ˙CH_2— group of the deoxyadenosyl radical (step ⑤), destroying the free radical and regenerating the cofactor in its Co^{3+} form, ready to undergo another catalytic cycle.

In this postulated mechanism, the migrating hydrogen atom never exists as a free species and is thus never free to exchange with the hydrogen of surrounding water molecules.

Vitamin B_{12} deficiency results in serious disease. Vitamin B_{12} is not made by either plants or animals and can be synthesized by only a few species of microorganisms. It is required in only minute amounts, about 3 μg/day, by healthy people, but the severe disease **pernicious anemia** results from failure to absorb vitamin B_{12} efficiently from the intestine, where it is synthesized by intestinal bacteria or obtained from digestion of meat in the diet. A glycoprotein essential to vitamin B_{12} absorption, called **intrinsic factor,** is not produced in sufficient quantity in individuals with this disease. The pathology in pernicious anemia includes reduced production of erythrocytes, reduced levels of hemoglobin, and severe, progressive impairment of the central nervous system. Administration of large doses of vitamin B_{12} alleviates these symptoms in at least some cases.

Fatty Acid Oxidation Is Tightly Regulated

In the liver, fatty acyl–CoAs formed in the cytosol have two major pathways open to them: (1) β oxidation by enzymes in the mitochondria or (2) conversion into triacylglycerols and phospholipids by enzymes in the cytosol. The pathway taken depends upon the rate of transfer of long-chain fatty acyl–CoAs into the mitochondria. The three-step process by which fatty acyl groups are carried from cytosolic fatty acyl–CoA into the mitochondrial matrix (Fig. 16–6) is rate-limiting for fatty acid oxidation. Once fatty acyl groups have entered the mitochondria, they are committed to oxidation to acetyl-CoA.

Malonyl-CoA, the first intermediate in the cytosolic biosynthesis of long-chain fatty acids from acetyl-CoA (Chapter 20), increases in concentration whenever the animal is well supplied with carbohydrate; excess glucose that cannot be oxidized or stored as glycogen is converted in the cytosol into fatty acids for storage as triacylglycerol. The inhibition of carnitine acyltransferase I by malonyl-CoA assures that the oxidation of fatty acids is inhibited whenever the liver is amply supplied with glucose as fuel and is actively making triacylglycerols from excess glucose.

Two of the enzymes of β oxidation are also regulated by metabolites that signal energy sufficiency. When the [NADH]/[NAD$^+$] ratio is high, β-hydroxyacyl-CoA dehydrogenase is inhibited; in addition, high concentrations of acetyl-CoA inhibit thiolase.

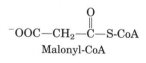

Malonyl-CoA

Peroxisomes Also Carry Out β Oxidation

Although the major site of fatty acid oxidation in animal cells is the mitochondrial matrix, other compartments in certain cells also contain enzymes capable of oxidizing fatty acids to acetyl-CoA, by a pathway similar to, but not identical with, that in mitochondria. Peroxisomes are membrane-enclosed cellular compartments (see p. 38) in animals and plants, where hydrogen peroxide is produced by fatty acid oxidation and then destroyed enzymatically. As in the oxidation of fatty acids in mitochondria, the intermediates are coenzyme A derivatives, and the process consists of four steps (Fig. 16–13): (1) dehydrogenation; (2) addition of water to the resulting double bond; (3) oxidation of the β-hydroxyacyl-CoA to a ketone, and (4) thiolytic cleavage by coenzyme A. The difference between the peroxisomal and mitochondrial pathways is in the first step. In peroxisomes, the flavoprotein dehydrogenase that introduces the double bond passes electrons directly to O_2, producing H_2O_2 (Fig. 16–13). This strong and potentially damaging oxidant is immediately cleaved to H_2O and O_2 by **catalase.** By contrast, in mitochondria the electrons removed in the first oxidation step pass through the respiratory chain to O_2, and H_2O is the product, a process accompanied by ATP synthesis. In peroxisomes, the energy released in the first oxidative step of fatty acid breakdown is dissipated as heat.

High concentrations of fats in the diet result in increased synthesis of the enzymes of peroxisomal β oxidation in mammalian liver. Liver peroxisomes do not contain the enzymes of the citric acid cycle and cannot catalyze the oxidation of acetyl-CoA to CO_2. Instead, the acetate produced by fatty acid oxidation is exported from peroxisomes. Presumably some of this acetate enters mitochondria and is oxidized there.

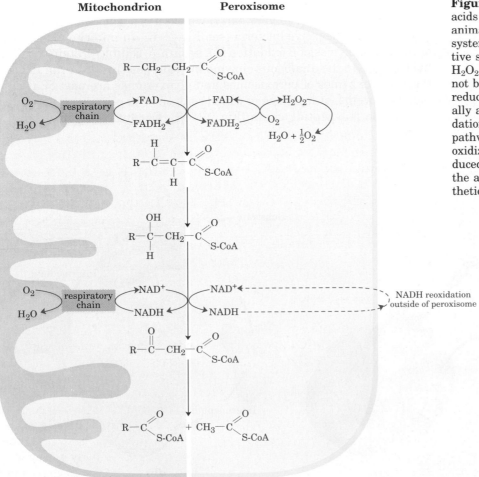

Mitochondrion　　　　**Peroxisome**

Figure 16–13 Comparison of β oxidation of fatty acids as it occurs in animal mitochondria and in animal and plant peroxisomes. The peroxisomal system differs in two respects: (1) in the first oxidative step electrons pass directly to O_2, generating H_2O_2, and (2) the NADH formed in β oxidation cannot be reoxidized, and the peroxisome must export reducing equivalents to the cytosol. (These eventually are passed on to mitochondria.) Fatty acid oxidation in glyoxysomes occurs by the peroxisomal pathway. In mitochondria, acetyl-CoA is further oxidized via the citric acid cycle. Acetyl-CoA produced by peroxisomes and glyoxysomes is exported; the acetate from glyoxysomes serves as a biosynthetic precursor (see Fig. 16–14).

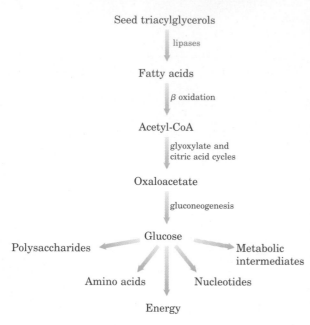

Figure 16–14 The role of β oxidation in the conversion of seed triacylglycerols into glucose in germinating seeds.

Plant Peroxisomes and Glyoxysomes Use Acetyl-CoA from β Oxidation as a Biosynthetic Precursor

Fatty acid oxidation in plants occurs in the peroxisomes of leaf tissue and the glyoxysomes of germinating seeds. Plant peroxisomes and glyoxysomes are similar in structure and function. Glyoxysomes occur only during seed germination, and may be considered specialized peroxisomes.

In plants, the biological role of β oxidation in peroxisomes and glyoxysomes is clear: it provides biosynthetic precursors from stored lipids. The β-oxidation pathway is not an important source of metabolic energy in plants; in fact, plant mitochondria do not contain the enzymes of β oxidation. During germination, triacylglycerols stored in seeds are converted into glucose and a wide variety of essential metabolites (Fig. 16–14). Fatty acids released from triacylglycerols are activated to their coenzyme A derivatives and oxidized in glyoxysomes by the same four-step process that occurs in peroxisomes (Fig. 16–13). The acetyl-CoA produced is converted via the glyoxylate cycle (Chapter 15) to four-carbon precursors for gluconeogenesis (Chapter 19). Glyoxysomes, like peroxisomes, contain high concentrations of catalase, which converts the H_2O_2 produced by β oxidation to H_2O and O_2.

The β-Oxidation Enzymes Have Diverged during Evolution

Although the β-oxidation reaction sequence in mitochondria is essentially the same as that in peroxisomes and glyoxysomes, the mitochondrial enzymes differ significantly from their isozymes in those compartments. These differences apparently reflect an evolutionary divergence that occurred very early, with the separation of gram-positive and gram-negative bacteria (see Fig. 2–6). In mitochondria, each of the four enzymes of β oxidation is a separate, soluble protein, similar in structure to the analogous enzyme in gram-positive bacteria. In contrast, the enzymes of peroxisomes and glyoxysomes are part of a complex of proteins, at least one of which contains two enzymatic activities in a single polypeptide chain (Fig. 16–15). Enoyl-CoA hydratase and

Figure 16–15 A schematic diagram of the structure of the enzymes of β oxidation in gram-negative **(a)** and gram-positive **(b)** bacteria. The complex of four enzyme activities, two of which are part of a single polypeptide chain, is typical of gram-negative bacteria and is also found in the peroxisomal and glyoxysomal β-oxidation systems **(a).** The four enzymes of β oxidation in mitochondria are separate entities, similar to those of gram-positive bacteria **(b).** Enz₁, acyl-CoA dehydrogenase; Enz₂, enoyl-CoA hydratase; Enz₃, ʟ-β-hydroxyacyl-CoA dehydrogenase; Enz₄, thiolase.

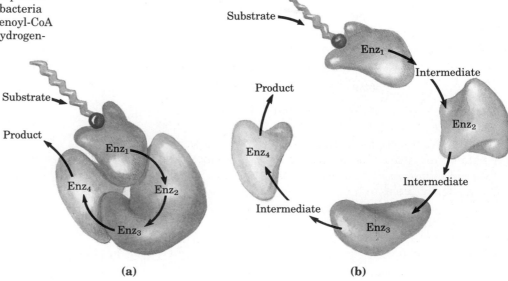

L-β-hydroxyacyl-CoA dehydrogenase activities both reside in a single, monomeric protein (M_r ~150,000), closely similar to the bifunctional protein of the gram-negative bacterium *E. coli*. The evolutionary selective value of retaining both types of β-oxidation system in the same organism is not yet apparent.

Ketone Bodies

In human beings and most other mammals, acetyl-CoA formed in the liver during oxidation of fatty acids may enter the citric acid cycle (stage 2 of Fig. 16–7) or it may be converted to the "ketone bodies" **acetoacetate, D-β-hydroxybutyrate,** and **acetone** for export to other tissues. (The term "bodies" is a historical artifact; these compounds are soluble in blood and urine.) Acetone, produced in smaller quantities than the other ketone bodies, is exhaled. Acetoacetate and D-β-hydroxybutyrate are transported by the blood to the extrahepatic tissues, where they are oxidized via the citric acid cycle to provide much of the energy required by tissues such as skeletal and heart muscle and the renal cortex. The brain, which normally prefers glucose as a fuel, can adapt to the use of acetoacetate or D-β-hydroxybutyrate under starvation conditions, when glucose is unavailable.

A major determinant of the pathway taken by acetyl-CoA in liver mitochondria is the availability of oxaloacetate to initiate entry of acetyl-CoA into the citric acid cycle. Under some circumstances (such as starvation) oxaloacetate is drawn out of the citric acid cycle for use in synthesizing glucose. When the oxaloacetate concentration is very low, little acetyl-CoA enters the cycle, and ketone body formation is favored. The production and export of ketone bodies from the liver to extrahepatic tissues allows continued oxidation of fatty acids in the liver when acetyl-CoA is not being oxidized via the citric acid cycle. Overproduction of ketone bodies can occur in conditions of severe starvation and in uncontrolled diabetes.

$$CH_3-\underset{\underset{O}{\|}}{C}-CH_2-\underset{\underset{O^-}{}}{C}\overset{O}{}$$

Acetoacetate

$$CH_3-\underset{\underset{H}{|}}{\overset{\overset{OH}{|}}{C}}-CH_2-\underset{\underset{O^-}{}}{C}\overset{O}{}$$

D-β-Hydroxybutyrate

$$CH_3-\underset{\underset{O}{\|}}{C}-CH_3$$

Acetone

Ketone Bodies Formed in the Liver Are Exported to Other Organs

The first step in formation of acetoacetate in the liver (Fig. 16–16) is the enzymatic condensation of two molecules of acetyl-CoA, catalyzed by thiolase; this is simply the reversal of the last step of β oxidation. The acetoacetyl-CoA then condenses with acetyl-CoA to form **β-hydroxy-β-methylglutaryl-CoA** (HMG-CoA), which is cleaved to free acetoacetate and acetyl-CoA.

The free acetoacetate so produced is reversibly reduced by **D-β-hydroxybutyrate dehydrogenase,** a mitochondrial enzyme, to D-β-hydroxybutyrate (Fig. 16–16). This enzyme is specific for the D stereoisomer; it does not act on L-β-hydroxyacyl-CoAs and is not to be confused with L-β-hydroxyacyl-CoA dehydrogenase, which acts in the β-oxidation pathway. In healthy people, acetone is formed in very small amounts from acetoacetate by the loss of a carboxyl group. Acetoacetate is easily decarboxylated; the carboxyl group may be lost spontaneously or by the action of **acetoacetate decarboxylase** (Fig. 16–16). Because untreated diabetics produce large quantities of acetoacetate, their blood contains significant amounts of acetone, which is toxic. Acetone is volatile and imparts a characteristic odor to the breath, which is sometimes useful in diagnosing the severity of the disease.

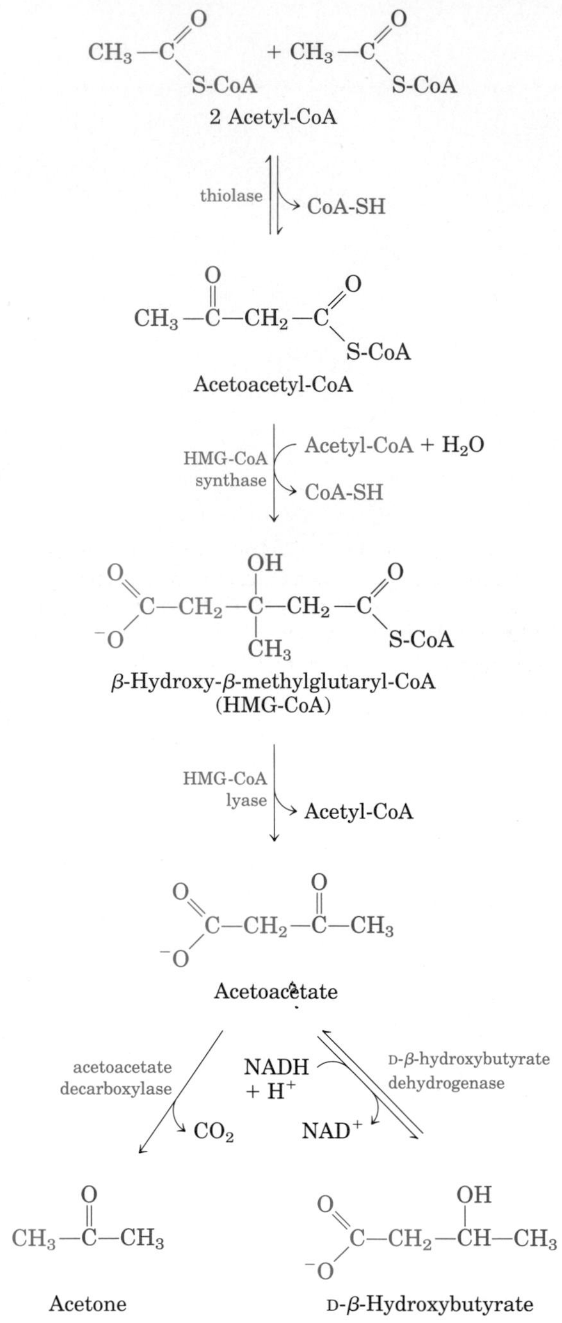

Figure 16–16 Formation of ketone bodies from acetyl-CoA. Under circumstances that cause acetyl-CoA accumulation (starvation or untreated diabetes, for example), thiolase catalyzes the condensation of two acetyl-CoA molecules to acetoacetyl-CoA, the parent of the three ketone bodies. These reactions all occur within the mitochondrial matrix. The six-carbon compound β-hydroxy-β-methylglutaryl-CoA (HMG-CoA) is also an intermediate of sterol biosynthesis, but the enzyme that forms HMG-CoA in that pathway is cytosolic. HMG-CoA lyase is present in the mitochondrial matrix but not in the cytosol.

Extrahepatic Tissues Use Ketone Bodies as Fuels

In the extrahepatic tissues D-β-hydroxybutyrate is oxidized to acetoacetate by D-β-hydroxybutyrate dehydrogenase (Fig. 16–17). Acetoacetate is activated to form its coenzyme A ester by transfer of CoA from succinyl-CoA, an intermediate of the citric acid cycle (see Fig. 15–7), in a reaction catalyzed by **β-ketoacyl-CoA transferase.** The acetoacetyl-CoA is then cleaved by thiolase to yield two acetyl-CoAs, which enter the citric acid cycle.

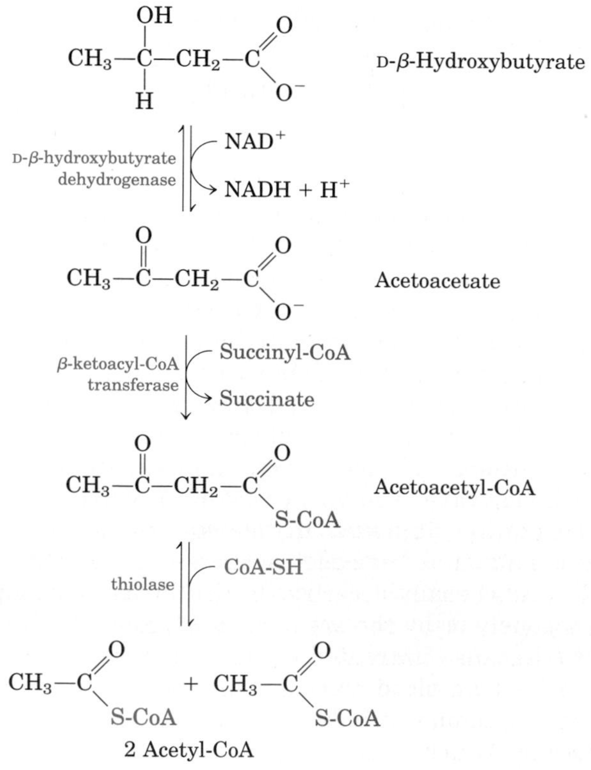

Figure 16–17 Hydroxybutyrate as a fuel. D-β-Hydroxybutyrate synthesized in the liver passes into the blood and thus to other tissues, where it is converted to acetyl-CoA for energy production. It is first oxidized to acetoacetate, which is activated with coenzyme A donated from succinyl-CoA, then split by thiolase.

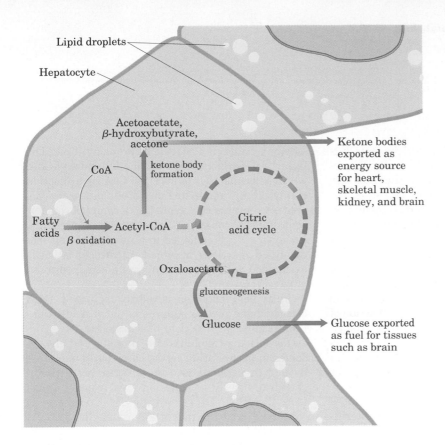

Figure 16–18 Ketone body formation and export from the liver. Conditions that increase gluconeogenesis (diabetes, fasting) slow the citric acid cycle (by drawing off oxaloacetate) and enhance the conversion of acetyl-CoA to acetoacetate. The released coenzyme A allows continued β oxidation of fatty acids.

Ketone Bodies Are Overproduced in Diabetes and during Starvation

The production and export of ketone bodies from the liver allows continued oxidation of fatty acids with only minimal oxidation of acetyl-CoA in the liver (Fig. 16–18). When, for example, intermediates of the citric acid cycle are being used for glucose synthesis via gluconeogenesis, oxidation of citric acid cycle intermediates slows, and so does acetyl-CoA oxidation. Moreover, the liver contains a limited amount of coenzyme A, and when most of it is tied up in acetyl-CoA, β oxidation of fatty acids slows for lack of the free coenzyme. The production and export of ketone bodies frees coenzyme A, allowing continued fatty acid oxidation.

Severe starvation or untreated diabetes mellitus leads to overproduction of ketone bodies, with several associated medical problems. During starvation, gluconeogenesis depletes citric acid cycle intermediates, diverting acetyl-CoA to ketone body production (Fig. 16–18). In untreated diabetes, insulin is present in insufficient quantity, and the extrahepatic tissues cannot take up glucose efficiently from the blood (Chapter 22). To raise the blood glucose level, gluconeogenesis in the liver accelerates, as does fatty acid oxidation in liver and muscle, with the result that ketone bodies are produced beyond the capacity of extrahepatic tissues to oxidize them. The rise in blood levels of acetoacetate and D-β-hydroxybutyrate lowers the blood pH, causing the condition known as **acidosis.** Extreme acidosis can lead to coma and in some cases death. Ketone bodies in the blood and urine of untreated diabetics may reach extraordinary levels (Table 16–2); this condition is **ketosis.** In individuals on very low-calorie diets, fats stored in adipose tissue become the major energy source. The levels of ketone bodies in the blood and urine should be monitored to avoid the dangers of acidosis and ketosis (ketoacidosis).

Table 16–2 Ketone body accumulation in diabetic ketosis

	Urinary excretion (mg/24 h)	Blood concentration (mg/100 mL)
Normal	$\leq$125	<3
Extreme ketosis (untreated diabetic)	5,000	90

Summary

The fatty acid components of triacylglycerols furnish a large fraction of the oxidative energy in animals. Triacylglycerols ingested in the diet are emulsified in the small intestine by bile salts, hydrolyzed by intestinal lipases, absorbed by intestinal epithelial cells and reconverted into triacylglycerols, then formed into chylomicrons by combination with specific apolipoproteins. Chylomicrons deliver triacylglycerols to tissues, where lipoprotein lipase releases free fatty acids for entry into cells. Triacylglycerols stored in adipose tissue of vertebrate animals are mobilized by the action of hormones through a hormone-sensitive triacylglycerol lipase. The fatty acids released by this enzyme bind to serum albumin and are carried in the blood to the heart, skeletal muscle, and other tissues that use fatty acids for fuel.

Once inside cells, free fatty acids are activated at the outer mitochondrial membrane by esterification with coenzyme A to form fatty acyl–CoA thioesters. These are converted into fatty acyl–carnitine esters, which are carried by a specific transporter across the inner mitochondrial membrane into the matrix, where fatty acyl–CoA esters are formed again. All subsequent steps in the oxidation of fatty acids take place in the form of their coenzyme A thioesters, within the mitochondrial matrix.

In the first stage of fatty acid β oxidation, four reactions are required to remove each acetyl-CoA unit from the carboxyl end of saturated fatty acyl–CoAs: (1) dehydrogenation of the α and β carbons (C-2 and C-3) by FAD-linked acyl-CoA dehydrogenases, (2) hydration of the resulting $trans$-Δ^2 double bond by enoyl-CoA hydratase, (3) dehydrogenation of the resulting L-β-hydroxyacyl-CoA by NAD-linked β-hydroxyacyl-CoA dehydrogenase, and (4) CoA-requiring cleavage by thiolase of the resulting β-ketoacyl-CoA to form acetyl-CoA and the coenzyme A thioester of the original fatty acid, shortened by two carbons. The shortened fatty acyl–CoA can then reenter the sequence, with loss of another acetyl-CoA. For example, the 16-carbon palmitate yields altogether eight molecules of acetyl-CoA, which in the second stage of fatty acid oxidation can be oxidized to CO_2 via the citric acid cycle. A large fraction of the theoretical yield of free energy from fatty acid oxidation is recovered as ATP by oxidative phosphorylation, the third and final stage of the oxidative pathway.

Oxidation of unsaturated fatty acids requires the action of two additional enzymes: enoyl-CoA isomerase and 2,4-dienoyl-CoA reductase. Odd-carbon fatty acids are oxidized by the same pathway but yield one molecule of propionyl-CoA. The latter is carboxylated to methylmalonyl-CoA, which is isomerized to succinyl-CoA by a reaction catalyzed by methylmalonyl-CoA mutase. This enzyme requires coenzyme B_{12}, a complex cofactor containing a cobalt ion in a corrin ring system. Coenzyme B_{12} is involved in a number of enzyme-catalyzed reactions in which a hydrogen atom is exchanged with a functional group attached to an adjacent carbon.

Fatty acid oxidation is tightly regulated. High carbohydrate intake suppresses fatty acid oxidation in favor of fatty acid biosynthesis.

Peroxisomes of plants and animals, and glyoxysomes in germinating seeds, carry out β oxidation by four steps similar to those occurring in mitochondria. The first oxidation step transfers electrons directly to O_2, generating H_2O_2; no energy is conserved, and the potentially damaging H_2O_2 is destroyed by catalase. In glyoxysomes, β oxidation serves to convert stored lipids into four-carbon compounds (via the glyoxylate cycle); these compounds are precursors of a variety of intermediates and products required during seed germination.

The ketone bodies acetoacetate, D-β-hydroxybutyrate, and acetone are formed in the liver and are carried to other tissues, where they serve as fuel molecules, being oxidized to acetyl-CoA and thus entering the citric acid cycle. The overproduction of ketone bodies in uncontrolled diabetes or severe starvation can lead to acidosis or ketosis.

Further Reading

General

Boyer, P.D. (1983) *The Enzymes,* 3rd edn, Vol. 16: *Lipid Enzymology,* Academic Press, Inc., San Diego, CA.

Gurr, M.I. & Harwood, J.L. (1991) *Lipid Biochemistry: An Introduction,* 4th edn, Chapman & Hall, London.

Numa, S. (ed) (1984) *Fatty Acid Metabolism and Its Regulation,* New Comprehensive Biochemistry, Vol. 7 (Neuberger, A. & van Deenen, L.L.M., series eds), Elsevier Biomedical Press, Amsterdam.
An excellent collection of articles on the enzymes of prokaryotes and eukaryotes and their regulation.

β Oxidation

Galliard, T. (1980) Degradation of acyl lipids: hydrolytic and oxidative enzymes. In *The Biochemistry of Plants,* Vol. 4 (Stumpf, P.K., ed), pp. 85–116, Academic Press, Inc., San Diego, CA.
A description of the enzymes of β oxidation in plants.

Greville, G.D. & Tubbs, P.K. (1968) The catabolism of long-chain fatty acids in mammalian tissues. *Essays Biochem.* **4,** 155–212.
An early review, but basic to more recent developments.

Harwood, J.L. (1988) Fatty acid metabolism. *Annu. Rev. Plant Physiol. Plant Mol. Biol.* **39,** 101–138.

Kindl, H. (1984) Lipid degradation in higher plants. In *Fatty Acid Metabolism and Its Regulation* (Numa, S., ed), pp. 181–204, Elsevier Biomedical Press, Amsterdam.

Schulz, H. (1985) Oxidation of fatty acids. In *Biochemistry of Lipids and Membranes* (Vance, D.E. & Vance, J.E., eds), pp. 116–142, The Benjamin/Cummings Publishing Company, Menlo Park, CA.

Schulz, H. & Kunau, W.-H. (1987) Beta-oxidation of unsaturated fatty acids: a revised pathway. *Trends Biochem. Sci.* **12,** 403–406.

Wang, C.S., Hartsuck, J., & McConathy, W.J. (1992) Structure and functional properties of lipoprotein lipase. *Biochim. Biophys. Acta* **1123,** 1–17.
Advanced-level discussion of the enzyme that releases fatty acids from lipoproteins in the capillaries of muscle and adipose tissue.

Ketone Bodies

Foster, D.W. & McGarry, J.D. (1983) The metabolic derangements and treatment of diabetic ketoacidosis. *N. Engl. J. Med.* **309,** 159–169.

McGarry, J.D. & Foster, D.W. (1980) Regulation of hepatic fatty acid oxidation and ketone body production. *Annu. Rev. Biochem.* **49,** 395–420.

Robinson, A.M. & Williamson, D.H. (1980) Physiological roles of ketone bodies as substrates and signals in mammalian tissues. *Physiol. Rev.* **60,** 143–187.

Problems

1. *Energy in Triacylglycerols* On a per-carbon basis, where does the largest amount of biologically available energy in triacylglycerols reside: in the fatty acid portions or the glycerol portion? Indicate how knowledge of the chemical structure of triacylglycerols provides the answer.

2. *Fuel Reserves in Adipose Tissue* Triacylglycerols have the highest energy content of the major nutrients.

(a) If 15% of the body mass of a 70 kg adult consists of triacylglycerols, calculate the total available fuel reserve, in both kilojoules and kilocalories, in the form of triacylglycerols. Recall that 1.00 kcal = 4.18 kJ, and that 1.0 kcal = 1.0 nutritional Calorie.

(b) If the basal energy requirement is approximately 8,400 kJ/day (2,000 kcal/day) how long could this person survive if the oxidation of fatty acids stored as triacylglycerols were the only source of energy?

(c) What would be the weight loss per day in pounds under such starvation conditions (1 lb = 0.454 kg)?

3. *Common Reaction Steps in the Fatty Acid Oxidation Cycle and Citric Acid Cycle* Cells often follow the same enzyme reaction pattern for bringing about analogous metabolic reactions. For example, the steps in the oxidation of pyruvate and α-ketoglutarate to acetyl-CoA and succinyl-CoA, although catalyzed by different enzymes, are very similar. The first stage in the oxidation of fatty acids follows a reaction sequence closely resembling one in the citric acid cycle. Show by equations the analogous reaction sequences in the two pathways.

4. *The Chemistry of the Acyl-CoA Synthetase Reaction* Fatty acids are converted into their coenzyme A esters by the reversible reaction catalyzed by acyl-CoA synthetase:

$$R-COO^- + ATP + CoA \rightleftharpoons$$

$$\underset{\substack{|| \\ O}}{R-C}-CoA + AMP + PP_i$$

(a) The enzyme-bound intermediate in this reaction has been identified as the mixed anhydride of the fatty acid and adenosine monophosphate (AMP), acyl-AMP:

Write two equations corresponding to the two steps involved in the reaction catalyzed by acyl-CoA synthetase.

(b) The reaction above is readily reversible, with an equilibrium constant near 1. How can this reaction be made to favor formation of fatty acyl–CoA?

5. *Oxidation of Tritiated Palmitate* Palmitate uniformly labeled with tritium (^3H) to a specific activity of 2.48×10^8 counts per minute (cpm) per micromole of palmitate is added to a mitochondrial preparation that oxidizes it to acetyl-CoA. The acetyl-CoA is isolated and hydrolyzed to acetate. The specific activity of the isolated acetate is 1.00×10^7 cpm per micromole. Is this result consistent with the β-oxidation pathway? Explain. What is the final fate of the removed tritium?

6. *Compartmentation in β Oxidation* Free palmitate is activated to its coenzyme A derivative (palmitoyl-CoA) in the cytosol before it can be oxidized in the mitochondrion. If palmitate and [^{14}C]coenzyme A are added to a liver homogenate, palmitoyl-CoA isolated from the cytosolic fraction is radioactive, but that isolated from the mitochondrial fraction is not. Explain.

7. *Effect of Carnitine Deficiency* A patient developed a condition characterized by progressive muscular weakness and aching muscle cramps. These symptoms were aggravated by fasting, exercise, and a high-fat diet. The homogenate of a muscle specimen from the patient oxidized added oleate more slowly than did control homogenates of muscle specimens from healthy individuals. When carnitine was added to the patient's muscle homogenate, the rate of oleate oxidation equaled that in the control homogenates. The patient was diagnosed as having a carnitine deficiency.

(a) Why did added carnitine increase the rate of oleate oxidation in the patient's muscle homogenate?

(b) Why were the symptoms aggravated by fasting, exercise, and a high-fat diet?

(c) Suggest two possible reasons for the deficiency of muscle carnitine in the patient.

8. *Fatty Acids as a Source of Water* Contrary to legend, camels do not store water in their humps, which actually consist of a large fat deposit. How can these fat deposits serve as a source of water? Calculate the amount of water (in liters) that can be produced by the camel from 1 kg (0.45 lb) of fat. Assume for simplicity that the fat consists entirely of tripalmitoylglycerol.

9. *Petroleum as a Microbial Food Source* Some microorganisms of the genera *Nocardia* and *Pseudomonas* can grow in an environment where hydrocarbons are the only food source. These bacteria oxidize straight-chain aliphatic hydrocarbons, for example, octane, to their corresponding carboxylic acids:

$$CH_3(CH_2)_6CH_3 + NAD^+ + O_2 \longrightarrow$$
$$CH_3(CH_2)_6COOH + NADH + H^+$$

How can these bacteria be used to clean up oil spills?

10. *Metabolism of a Straight-Chain Phenylated Fatty Acid* A crystalline metabolite was isolated from the urine of a rabbit that had been fed a straight-chain fatty acid containing a terminal phenyl group:

$$\text{—CH}_2\text{—(CH}_2)_n\text{—COO}^-$$

A 302 mg sample of the metabolite in aqueous solution was completely neutralized by adding 22.2 mL of 0.1 M NaOH.

(a) What is the probable molecular weight and structure of the metabolite?

(b) Did the straight-chain fatty acid fed to the rabbit contain an even or an odd number of methylene (—CH$_2$—) groups (i.e., is n even or odd)? Explain.

11. *Fatty Acid Oxidation in Diabetics* When the acetyl-CoA produced during β oxidation in the liver exceeds the capacity of the citric acid cycle, the excess acetyl-CoA reacts to form the ketone bodies acetoacetate, D-β-hydroxybutyrate, and acetone. This condition exists in cases of severe diabetes because the patient's tissues cannot use glucose; they oxidize large amounts of fatty acids instead. Although acetyl-CoA is not toxic, the mitochondrion must divert the acetyl-CoA to ketone bodies. Why? How does this diversion solve the problem?

12. *Consequences of a High-Fat Diet with No Carbohydrates* Suppose you had to subsist on a diet of whale and seal blubber with little or no carbohydrate.

(a) What would be the effect of carbohydrate deprivation on the utilization of fats for energy?

(b) If your diet were totally devoid of carbohydrate, would it be better to consume odd- or even-numbered fatty acids? Explain.

13. *Formation of Acetyl-CoA from Fatty Acid Precursors* Write a balanced net equation for the formation of acetyl-CoA from the following substances, including all activation steps:

(a) Myristoyl-CoA

(b) Stearate

(c) D-β-Hydroxybutyrate

14. *Pathway of Labeled Atoms during Fatty Acid Oxidation* [9-^{14}C]Palmitate is oxidized under conditions in which the citric acid cycle is operating. What will be the location of ^{14}C in (a) acetyl-CoA,

(b) citrate, and (c) butyryl-CoA? Assume only one turn of the citric acid cycle.

15. *Net Equation for Complete Oxidation of D-β-Hydroxybutyrate* Write the net equation for the complete oxidation of D-β-hydroxybutyrate in the kidney. Include any required activation steps and all oxidative phosphorylations.

16. *Role of FAD as Electron Acceptor* Acyl-CoA dehydrogenase uses enzyme-bound FAD as a prosthetic group to dehydrogenate the α and β carbons of fatty acyl–CoA. What is the advantage of using FAD as an electron acceptor rather than NAD$^+$? Explain in terms of the standard reduction potentials for the Enz-FAD/FADH$_2$ ($E_o' = -0.219$ V) and NAD$^+$/NADH ($E_o' = -0.320$ V) half-reactions.

17. *β Oxidation of Arachidic Acid* How many turns of the fatty acid oxidation cycle are required to oxidize arachidic acid (see Table 9–1) completely to acetyl-CoA?

18. *Sources of H$_2$O Produced in β Oxidation* The complete oxidation of palmitate to carbon dioxide and water is represented by the overall equation

Palmitate + 23O$_2$ + 129P$_i$ + 129ADP $\longrightarrow$
$$16CO_2 + 129ATP + 145H_2O$$

The 145 H$_2$O molecules come from two separate reactions. What are they, and how many H$_2$O molecules are produced in each?

19. *Fate of Labeled Propionate* If [3-^{14}C]propionate (^{14}C in the methyl group) is added to a liver homogenate, ^{14}C-labeled oxaloacetate is rapidly produced. Draw a flow chart for the pathway by which propionate is transformed to oxaloacetate and indicate the location of the ^{14}C in oxaloacetate.

20. *Biological Importance of Cobalt* Cattle, deer, sheep, and other ruminant animals produce large amounts of propionate in the rumen through the bacterial fermentation of ingested plant matter. This propionate is the principal source of glucose for the animals via the route

Propionate $\longrightarrow$ oxaloacetate $\longrightarrow$ glucose

In some areas of the world, notably Australia, ruminant animals sometimes show symptoms of anemia with concomitant loss of appetite and retarded growth. These symptoms are the result of the animals' inability to transform propionate to oxaloacetate, which is due to a cobalt deficiency caused by very low cobalt levels in the soil. Explain.

Amino Acid Oxidation and the Production of Urea

Amino acids, derived largely from protein in the diet or from degradation of intracellular proteins, are the final class of biomolecules whose oxidation makes a significant contribution to the generation of metabolic energy. The fraction of metabolic energy derived from amino acids varies greatly with the type of organism and with the metabolic situation in which an organism finds itself. Carnivores, immediately following a meal, may obtain up to 90% of their energy requirements from amino acid oxidation. Herbivores may obtain only a small fraction of their energy needs from this source. Most microorganisms can scavenge amino acids from their environment if they are available; these can be oxidized as fuel when required by metabolic conditions. Photosynthetic plants, on the other hand, rarely, if ever, oxidize amino acids to provide energy. Instead, they convert CO_2 and H_2O into the carbohydrate that is used almost exclusively as an energy source. The amounts of amino acids in plant tissues are carefully regulated to just meet the requirements for biosynthesis of proteins, nucleic acids, and a few other molecules needed to support growth. Amino acid catabolism does occur in plants, but it is generally concerned with the production of metabolites for other biosynthetic pathways.

In animals, amino acids can undergo oxidative degradation in three different metabolic circumstances. (1) During the normal synthesis and degradation of cellular proteins (protein turnover; Chapter 26) some of the amino acids released during protein breakdown will undergo oxidative degradation if they are not needed for new protein synthesis. (2) When a diet is rich in protein, and amino acids are ingested in excess of the body's needs for protein synthesis, the surplus may be catabolized; amino acids cannot be stored. (3) During starvation or in diabetes mellitus, when carbohydrates are either unavailable or not properly utilized, body proteins are called upon as fuel. Under these different circumstances, amino acids lose their amino groups, and the α-keto acids so formed may undergo oxidation to CO_2 and H_2O. In addition, and often equally important, the carbon skeletons of the amino acids provide three- and four-carbon units that can be converted to glucose, which in turn can fuel the functions of the brain, muscle, and other tissues.

Amino acid degradative pathways are quite similar in most organisms. The focus of this chapter is on vertebrates, because amino acid catabolism has received the most attention in these organisms. As is the case for sugar and fatty acid catabolic pathways, the processes of amino acid degradation converge on the central catabolic pathways for carbon metabolism. The carbon skeletons of the amino acids generally find their way to the citric acid cycle, and from there they are either oxidized to produce chemical energy or funneled into gluconeogenesis. In some cases the reaction pathways closely parallel steps in the catabolism of fatty acids (Chapter 16).

However, one major factor distinguishes amino acid degradation from the catabolic processes described to this point: every amino acid contains an amino group. Every degradative pathway therefore passes through a key step in which the α-amino group is separated from the carbon skeleton and shunted into the specialized pathways for amino group metabolism (Fig. 17–1). This biochemical fork in the road is the point around which this chapter is organized. We deal first with amino group metabolism and nitrogen excretion, then with the fate of the carbon skeletons derived from the amino acids.

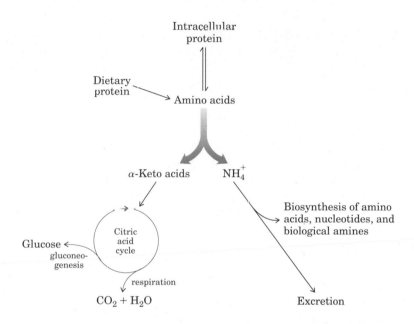

Figure 17–1 Overview of the catabolism of amino acids. The separate paths taken by the carbon skeleton and the amino groups are emphasized by the orange divergent arrow.

Metabolic Fates of Amino Groups

Nitrogen ranks fourth, behind carbon, hydrogen, and oxygen, in its contribution to the mass of living cells. Atmospheric nitrogen, N_2, is abundant but is too inert for use in most biochemical processes. Because only a few microorganisms can convert N_2 to biologically useful forms such as NH_3, amino groups are used with great economy in biological systems.

An overview of the catabolism of ammonia and amino groups in vertebrates is provided in Figure 17–2 (p. 508). Amino acids derived from dietary proteins are the source of most amino groups. Most of the amino acids are metabolized in the liver. Some of the ammonia that is generated is recycled and used in a variety of biosynthetic processes; the excess is either excreted directly or converted to uric acid or urea for excretion, depending on the organism. Excess ammonia generated in other (extrahepatic) tissues is transported to the liver (in the form of amino groups, as described below) for conversion to the appropriate excreted form. With these reactions we encounter the coenzyme pyridoxal phosphate, the functional form of vitamin B_6 and a coenzyme of major importance in nitrogen metabolism.

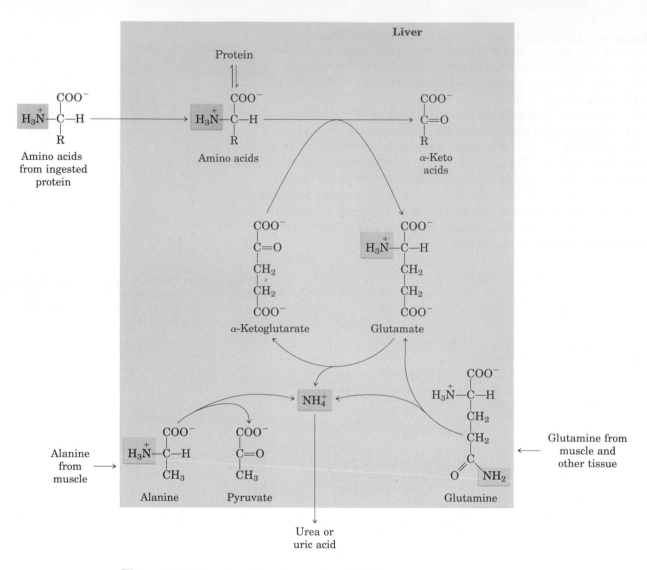

Figure 17–2 Overview of amino group catabolism in the vertebrate liver (shaded). Excess NH_4^+ is excreted as urea or uric acid.

The amino acids glutamate and glutamine play especially critical roles in these pathways (Fig. 17–2). Amino groups from amino acids are generally first transferred to α-ketoglutarate in the cytosol of liver cells (hepatocytes) to form glutamate. Glutamate is then transported into the mitochondria; only here is the amino group removed to form NH_4^+. Excess ammonia generated in most other tissues is converted to the amide nitrogen of glutamine, then transported to liver mitochondria. In most tissues, one or both of these amino acids are found in elevated concentrations relative to other amino acids.

In muscle, excess amino groups are generally transferred to pyruvate to form alanine. Alanine is another important molecule in the transport of amino groups, conveying them from muscle to the liver.

We begin with a discussion of the breakdown of dietary proteins to amino acids, then turn to a general description of the metabolic fates of amino groups.

Dietary Protein Is Enzymatically Degraded to Amino Acids

In humans, the degradation of ingested proteins into their constituent amino acids occurs in the gastrointestinal tract. Entry of protein into the stomach stimulates the gastric mucosa to secrete the hormone **gastrin,** which in turn stimulates the secretion of hydrochloric acid by the parietal cells of the gastric glands (Fig. 17–3a) and pepsinogen by the chief cells. The acidity of gastric juice (pH 1.5 to 2.5) acts as an antiseptic and kills most bacteria and other foreign cells. Globular proteins denature at low pH, rendering their internal peptide bonds more accessible to enzymatic hydrolysis. **Pepsinogen** (M_r 40,000), an inactive precursor or zymogen (p. 235), is converted into active pepsin in the gastric juice by the enzymatic action of pepsin itself. In this process, 42 amino acid residues are removed from the amino-terminal end of the polypeptide chain. The portion of the molecule that remains intact is enzymatically active **pepsin** (M_r 33,000). In the stomach, pepsin hydrolyzes ingested proteins at peptide bonds on the amino-terminal side of the aromatic amino acid residues Tyr, Phe, and Trp (see Table 6–7), cleaving long polypeptide chains into a mixture of smaller peptides.

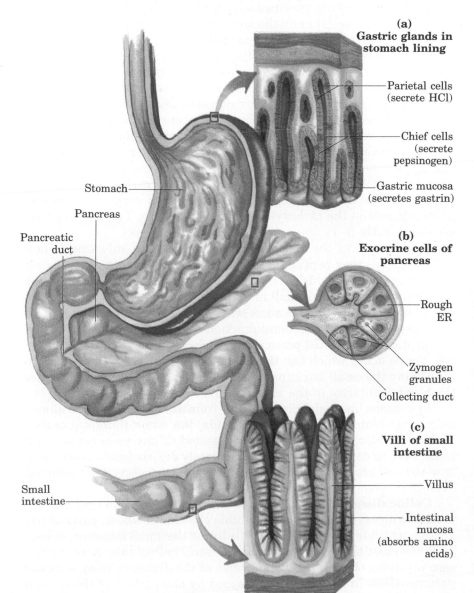

(a)
Gastric glands in stomach lining

Parietal cells (secrete HCl)

Chief cells (secrete pepsinogen)

Stomach

Gastric mucosa (secretes gastrin)

Pancreas

Pancreatic duct

(b)
Exocrine cells of pancreas

Rough ER

Zymogen granules

Collecting duct

(c)
Villi of small intestine

Small intestine

Villus

Intestinal mucosa (absorbs amino acids)

Figure 17–3 A portion of the human digestive tract. **(a)** Gastric glands in the stomach lining. The parietal cells and chief cells secrete their products in response to the hormone gastrin. **(b)** Exocrine cells of the pancreas. The cytoplasm is completely filled with rough endoplasmic reticulum, on which the ribosomes synthesize the polypeptide chains of the zymogens of many digestive enzymes. The zymogens are concentrated in condensing vesicles, ultimately forming mature zymogen granules. When the cell is stimulated, the plasma membrane fuses with the membrane around the zymogen granules, and the zymogens are released into the lumen of the collecting duct by exocytosis. The collecting ducts ultimately lead to the pancreatic duct and thence to the small intestine. **(c)** Villi of the small intestine. Amino acids are absorbed through the epithelial cell layer (intestinal mucosa) and enter the capillaries. Recall that the products of lipid hydrolysis in the small intestine enter the lymphatic system following absorption by the intestinal mucosa (Fig. 16–1).

As the acidic stomach contents pass into the small intestine, the low pH triggers the secretion of the hormone **secretin** into the blood. Secretin stimulates the pancreas to secrete bicarbonate into the small intestine to neutralize the gastric HCl, increasing the pH abruptly to about pH 7. The digestion of proteins continues in the small intestine. The entry of amino acids into the upper part of the intestine (duodenum) releases the hormone **cholecystokinin,** which stimulates secretion of several pancreatic enzymes, whose activity optima occur at pH 7 to 8. Three of these, **trypsin, chymotrypsin,** and **carboxypeptidase,** are made by the exocrine cells of the pancreas (Fig. 17–3b) as their respective enzymatically inactive zymogens, **trypsinogen, chymotrypsinogen,** and **procarboxypeptidase.**

Synthesis of these enzymes as inactive precursors protects the exocrine cells from destructive proteolytic attack. The pancreas protects itself against self-digestion in another way—by making a specific inhibitor, itself a protein, called pancreatic trypsin inhibitor (p. 235). Free trypsin can activate not only trypsinogen but also three other digestive zymogens: chymotrypsinogen, procarboxypeptidase, and **proelastase;** trypsin inhibitor effectively prevents premature production of free proteolytic enzymes within the pancreatic cells.

After trypsinogen enters the small intestine, it is converted into its active form, trypsin, by **enteropeptidase,** a specialized proteolytic enzyme secreted by intestinal cells. Once some free trypsin has been formed, it also can catalyze the conversion of trypsinogen into trypsin (see Fig. 8–30). Trypsin, as noted above, can convert chymotrypsinogen and procarboxypeptidase into chymotrypsin and carboxypeptidase.

Trypsin and chymotrypsin thus hydrolyze into smaller peptides the peptides resulting from the action of pepsin in the stomach. This stage of protein digestion is accomplished very efficiently because pepsin, trypsin, and chymotrypsin have different amino acid specificities. Trypsin hydrolyzes those peptide bonds whose carbonyl groups are contributed by Lys and Arg residues, and chymotrypsin hydrolyzes peptide bonds on the carboxyl-terminal side of Phe, Tyr, and Trp residues (see Table 6–7).

Degradation of the short peptides in the small intestine is now completed by other peptidases. The first is carboxypeptidase, a zinc-containing enzyme, which removes successive carboxyl-terminal residues from peptides. The small intestine also secretes an **aminopeptidase,** which can hydrolyze successive amino-terminal residues from short peptides. By the sequential action of these proteolytic enzymes and peptidases, ingested proteins are hydrolyzed to yield a mixture of free amino acids, which can then be transported across the epithelial cells lining the small intestine (Fig. 17–3c). The free amino acids enter the blood capillaries in the villi and are transported to the liver.

In humans, most globular proteins from animal sources are almost completely hydrolyzed into amino acids, but some fibrous proteins, such as keratin, are only partially digested. Many proteins of plant foods, such as cereal grains, are incompletely digested because the protein part of grains of seeds is surrounded by indigestible cellulose husks.

Celiac disease is a condition in which the intestinal enzymes are unable to digest certain water-insoluble proteins of wheat, particularly gliadin, which is injurious to the cells lining the small intestine. Wheat products must therefore be avoided by such individuals. Another disease involving the proteolytic enzymes of the digestive tract is **acute pancreatitis.** In this condition, caused by obstruction of the normal pathway of secretion of pancreatic juice into the intestine, the zymogens of the proteolytic enzymes are converted into their catalytically active forms prematurely, *inside* the pancreatic cells. As a result these powerful enzymes attack the pancreatic tissue itself, causing a painful and serious destruction of the organ, which can be fatal.

Pyridoxal Phosphate Facilitates the Transfer of α-Amino Groups to Glutamate

The α-amino groups of the 20 L-amino acids commonly found in proteins are removed during the oxidative degradation of the amino acids. If not reused for synthesis of new amino acids or other nitrogenous products, these amino groups are channeled into a single excretory end product (Fig. 17–4). Many aquatic organisms simply release ammonia as NH_4^+ into the surrounding medium. Most terrestrial vertebrates first convert the ammonia into urea (humans, other mammals, and adult amphibians) or uric acid (birds, reptiles).

The removal of the α-amino groups, the first step in the catabolism of most of the L-amino acids, is promoted by enzymes called **aminotransferases** or **transaminases**. In these **transamination** reactions, the α-amino group is transferred to the α-carbon atom of α-ketoglutarate, leaving behind the corresponding α-keto acid analog of the amino acid (Fig. 17–5). There is no net deamination (i.e., loss of amino groups) in such reactions because the α-ketoglutarate becomes aminated as the α-amino acid is deaminated. The effect of transamination reactions is to collect the amino groups from many different amino acids in the form of only one, namely, L-glutamate. The glutamate channels amino groups either into biosynthetic pathways or into a final sequence of reactions by which nitrogenous waste products are formed and then excreted.

Cells contain several different aminotransferases, many specific for α-ketoglutarate as the amino group acceptor. The aminotransferases differ in their specificity for the other substrate, the L-amino acid that donates the amino group, and are named for the amino group donor (Fig. 17–5b). The reactions catalyzed by the aminotransferases are freely reversible, having an equilibrium constant of about 1.0 ($\Delta G^{\circ\prime} \simeq 0$ kJ/mol).

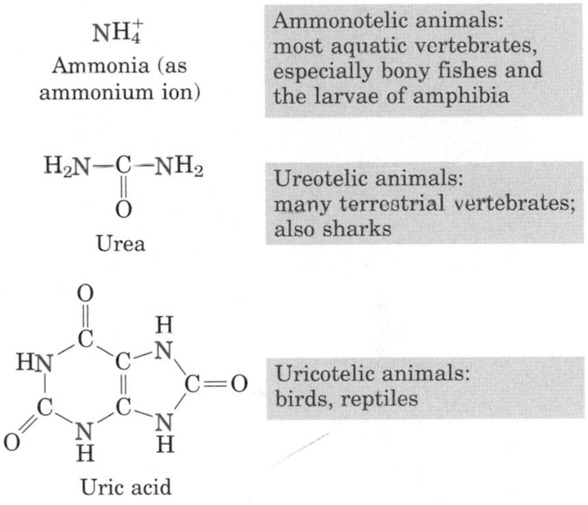

Figure 17–4 Excretory forms of amino group nitrogen in different forms of life. Notice that the carbon atoms of urea and uric acid are at a high oxidation state; the organism discards carbon only after having obtained most of its available energy of oxidation.

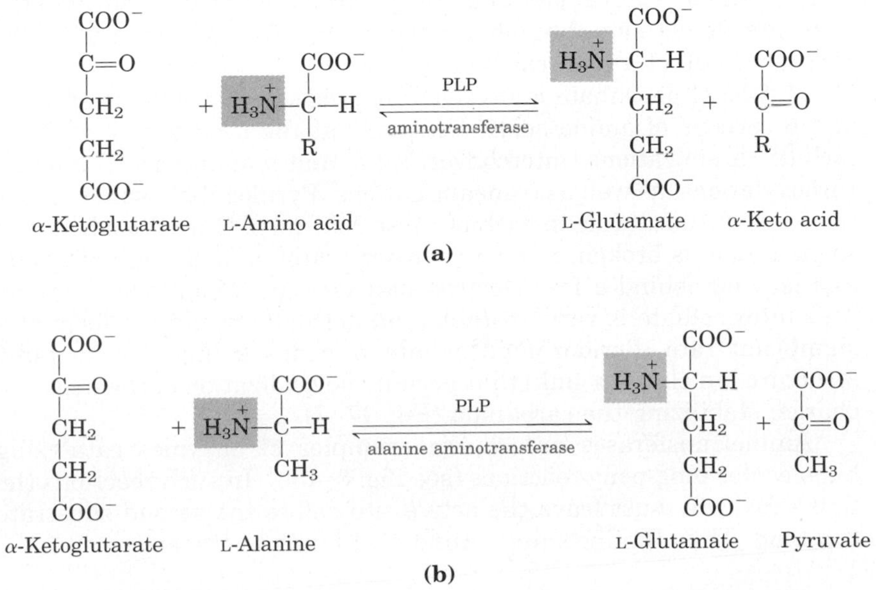

Figure 17–5 (a) The aminotransferase reaction (transamination). In many aminotransferase reactions, α-ketoglutarate is the amino group acceptor. All aminotransferases have pyridoxal phosphate (PLP) as cofactor. (b) The reaction of alanine aminotransferase is shown as an example.

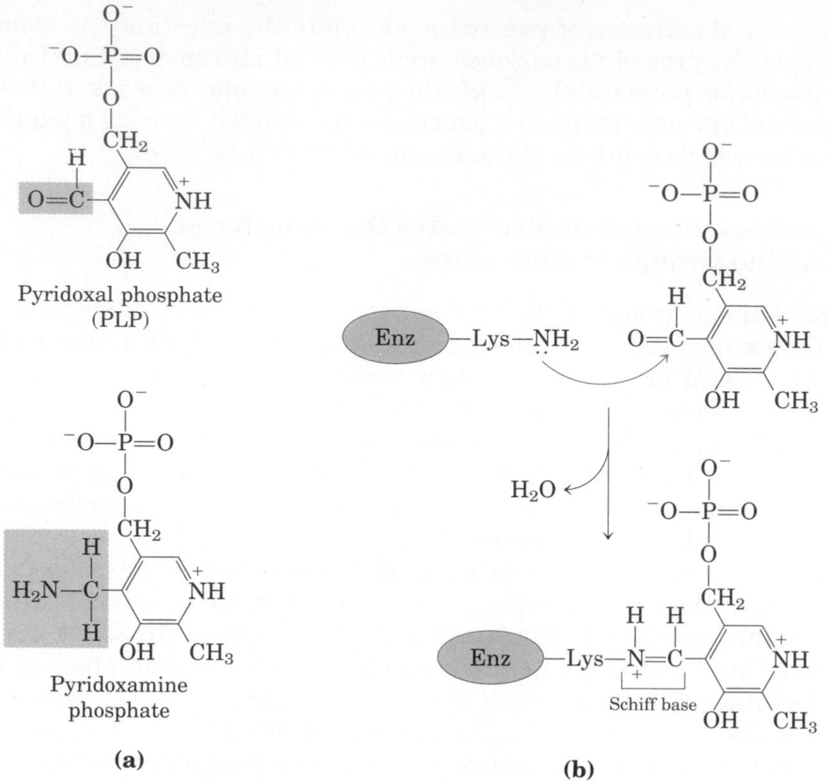

Figure 17–6 The prosthetic group of aminotransferases. **(a)** Pyridoxal phosphate (PLP) and its aminated form pyridoxamine phosphate are the tightly bound coenzymes of aminotransferases. The functional groups involved in their action are shaded in red. Pyridoxal phosphate is bound to the enzyme both through strong noncovalent interactions and through formation of a Schiff-base linkage involving a Lys residue at the active site **(b)**.

All aminotransferases share a common prosthetic group and a common reaction mechanism. The prosthetic group is **pyridoxal phosphate** (PLP), the coenzyme form of pyridoxine or vitamin B_6. Pyridoxal phosphate was briefly introduced in Chapter 14 (p. 422) as a cofactor in the glycogen phosphorylase reaction. Its role in that reaction, however, is not representative of its normal coenzyme function. Its more typical functions occur in the metabolism of molecules with amino groups.

Pyridoxal phosphate functions as an intermediate carrier of amino groups at the active site of aminotransferases. It undergoes reversible transformations between its aldehyde form, pyridoxal phosphate, which can accept an amino group, and its aminated form, pyridoxamine phosphate, which can donate its amino group to an α-keto acid (Fig. 17–6a). Pyridoxal phosphate is generally bound covalently to the enzyme's active site through an imine (Schiff-base) linkage to the ϵ-amino group of a Lys residue (Fig. 17–6b).

Pyridoxal phosphate is involved in a variety of reactions at the α and β carbons of amino acids. Reactions at the α carbon (Fig. 17–7) include racemizations (interconverting L- and D-amino acids) and decarboxylations, as well as transaminations. Pyridoxal phosphate plays the same chemical role in each of these reactions. One of the bonds to the α carbon is broken, removing either a proton or a carboxyl group and leaving behind a free electron pair on the carbon (a carbanion). This intermediate is very unstable and normally would not form at a significant rate. Pyridoxal phosphate provides a highly conjugated structure (an electron sink) that permits delocalization of the negative charge, stabilizing the carbanion (Fig. 17–7).

Aminotransferases are classic examples of enzymes catalyzing bimolecular ping-pong reactions (see Fig. 8–13b). In such reactions the first substrate must leave the active site before the second substrate can bind. Thus the incoming amino acid binds to the active site, do-

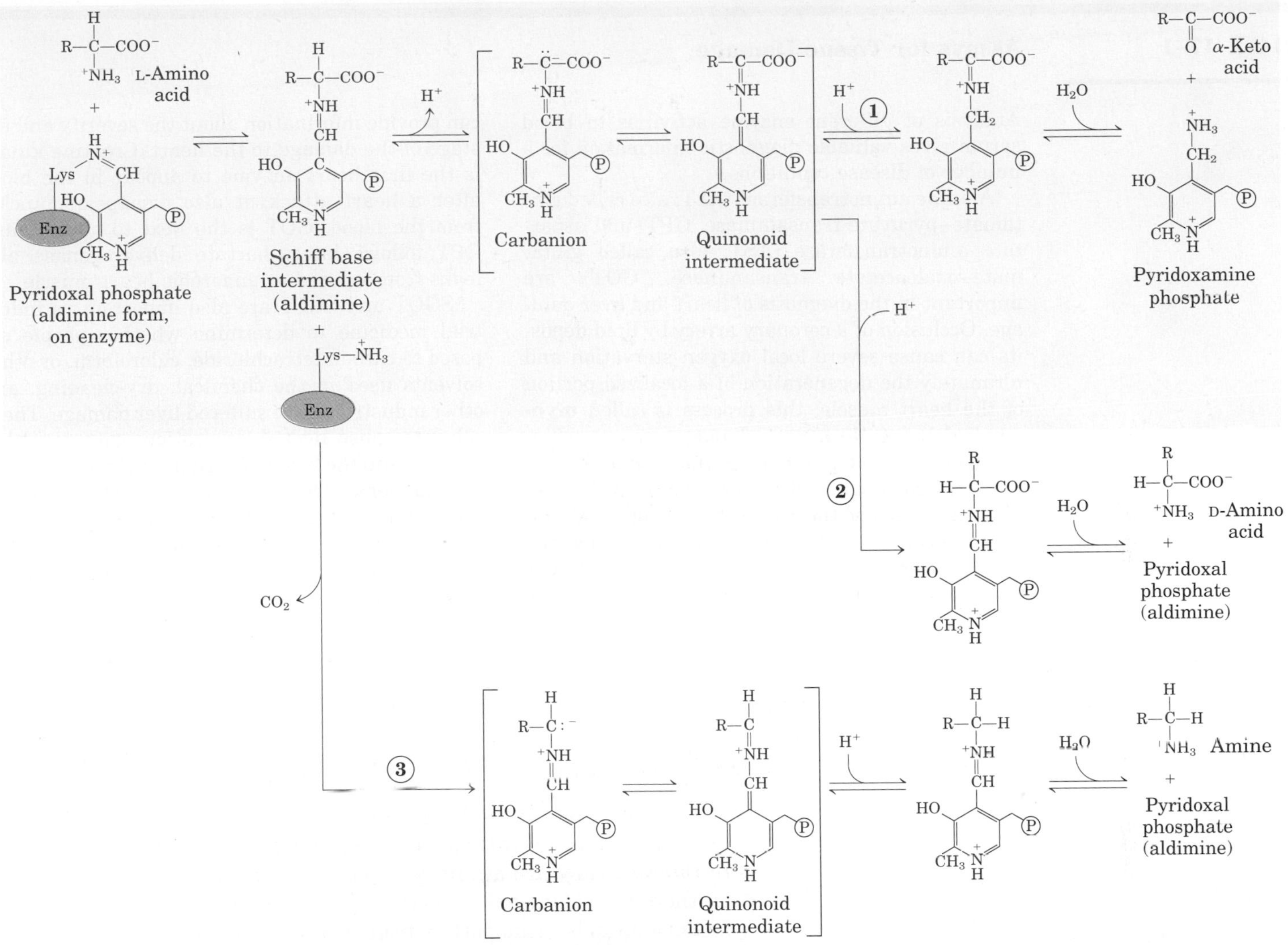

Figure 17–7 Some of the amino acid transformations facilitated by pyridoxal phosphate. Pyridoxal phosphate is generally bound to the enzyme by means of a Schiff base (see Fig. 17–6b). Reactions begin with formation of a new Schiff base (aldimine) between the α-amino group of the amino acid and PLP, which substitutes for the enzyme–PLP linkage. The amino acid then can have three alternative fates, each involving formation of a carbanion: ① transamination, ② racemization, or ③ decarboxylation. The Schiff base formed between PLP and the amino acid is in conjugation with the pyridine ring, which acts as an electron sink, permitting delocalization of the negative charge of the carbanion (as shown within the brackets). A quinonoid intermediate is involved in all of the reactions. The transamination route is especially important in the pathways described in this chapter. The highlighted transamination pathway (shown left to right) represents only part of the reaction catalyzed by aminotransferases. To complete the process, a second α-keto acid replaces the one that is released and is converted to an amino acid in a reversal of the reaction (right to left).

nates its amino group to pyridoxal phosphate, and departs in the form of an α-keto acid. Then the incoming α-keto acid is bound, accepts the amino group from pyridoxamine phosphate, and departs in the form of an amino acid.

The measurement of alanine aminotransferase and aspartate aminotransferase levels in blood serum is an important diagnostic procedure in medicine, used as an indicator of heart damage and to monitor recovery from the damage (Box 17–1).

BOX 17–1   Assays for Tissue Damage

Analysis of different enzyme activities in blood serum gives valuable diagnostic information for a number of disease conditions.

Alanine aminotransferase (ALT; also called glutamate–pyruvate transaminase, GPT) and aspartate aminotransferase (AST; also called glutamate–oxaloacetate transaminase, GOT) are important in the diagnosis of heart and liver damage. Occlusion of a coronary artery by lipid deposits can cause severe local oxygen starvation and ultimately the degeneration of a localized portion of the heart muscle; this process is called **myocardial infarction.** Such damage causes aminotransferases, among other enzymes, to leak from the injured heart cells into the bloodstream. Measurements of the concentration in the blood serum of these two aminotransferases by the SGPT and SGOT tests (S for serum) and of another heart enzyme, **creatine kinase** (the SCK test),

can provide information about the severity and the stage of the damage to the heart. Creatine kinase is the first heart enzyme to appear in the blood after a heart attack; it also disappears quickly from the blood. GOT is the next to appear, and GPT follows later. Lactate dehydrogenase also leaks from injured or anaerobic heart muscle.

SGOT and SGPT are also important in industrial medicine to determine whether people exposed to carbon tetrachloride, chloroform, or other solvents used in the chemical, dry-cleaning, and other industries have suffered liver damage. These solvents cause liver degeneration, with resulting leakage into the blood of various enzymes from the injured hepatocytes. Aminotransferases, because they are very active in liver and their activity can be detected in very small amounts, are most useful in the monitoring of people exposed to such industrial chemicals.

Ammonia Is Formed from Glutamate

We have seen that, in the liver, amino groups are removed from many of the α-amino acids by transamination with α-ketoglutarate to form L-glutamate. How are amino groups removed from glutamate to prepare them for excretion?

Glutamate is transported from the cytosol to the mitochondria, where it undergoes **oxidative deamination** catalyzed by **L-glutamate dehydrogenase** (M_r 330,000). This enzyme, which is present only in the mitochondrial matrix, requires NAD^+ (or $NADP^+$) as the acceptor of the reducing equivalents (Fig. 17–8). The combined action of the aminotransferases and glutamate dehydrogenase is referred to as **transdeamination.** A few amino acids bypass the transdeamination pathway and undergo direct oxidative deamination. The fate of the NH_4^+ produced by either of these processes is discussed in detail later.

As might be expected from its central role in amino group metabolism, glutamate dehydrogenase is a complex allosteric enzyme. The

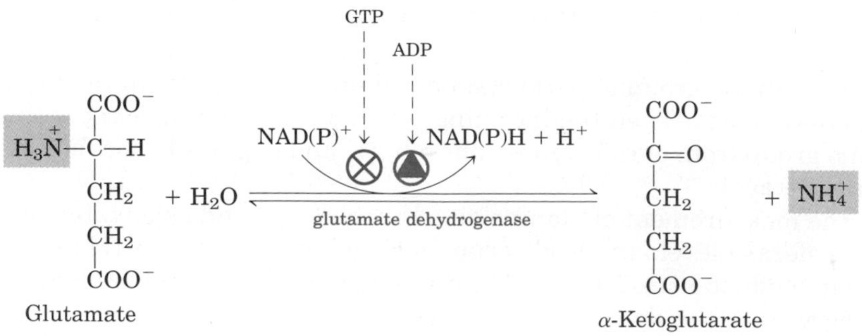

Figure 17–8 The reaction catalyzed by glutamate dehydrogenase. This enzyme can employ either NAD^+ or $NADP^+$ as cofactor, and is allosterically regulated by GTP and ADP.

enzyme molecule consists of six identical subunits. It is influenced by the positive modulator ADP and by the negative modulator GTP, a product of the succinyl-CoA synthetase reaction in the citric acid cycle (p. 456). Whenever a hepatocyte needs fuel for the citric acid cycle, glutamate dehydrogenase activity increases, making α-ketoglutarate available for the citric acid cycle and releasing NH_4^+ for excretion. On the other hand, whenever GTP accumulates in the mitochondria as a result of high citric acid cycle activity, oxidative deamination of glutamate is inhibited.

Glutamine Carries Ammonia to the Liver

Ammonia is quite toxic to animal tissues (we examine some possible reasons for this toxicity later). In most animals excess ammonia is converted into a nontoxic compound before export from extrahepatic tissues into the blood and thence to the liver or kidneys. Glutamate, which is so critical to intracellular amino group metabolism, is supplanted by L-**glutamine** for this transport function. In many tissues, including the brain, ammonia is enzymatically combined with glutamate to yield glutamine by the action of **glutamine synthetase.** This reaction requires ATP and occurs in two steps. In the first step, glutamate and ATP react to form ADP and a γ-glutamyl phosphate intermediate, which reacts with ammonia to produce glutamine and inorganic phosphate. We will encounter glutamine synthetase again in Chapter 21 when we consider nitrogen metabolism in microorganisms, where this enzyme serves as a portal for the entry of fixed nitrogen into biological systems. Glutamine is a nontoxic, neutral compound that can readily pass through cell membranes, whereas glutamate, which bears a net negative charge, cannot. In most land animals glutamine is carried in the blood to the liver. As is the case for the amino group of glutamate, the amide nitrogen is released as ammonia only within liver mitochondria, where the enzyme **glutaminase** converts glutamine to glutamate and NH_4^+.

Glutamine is a major transport form of ammonia; it is normally present in blood in much higher concentrations than other amino acids. In addition to its role in the transport of amino groups, glutamine serves as a source of amino groups in a variety of biosynthetic reactions (Chapter 21).

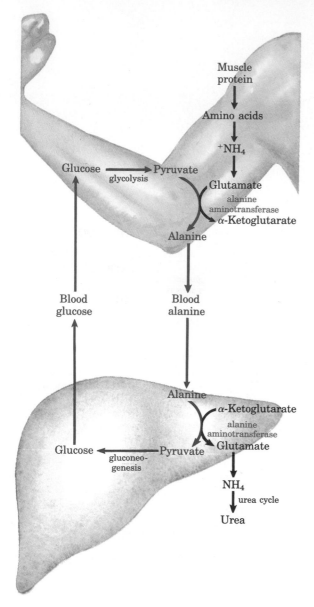

Figure 17–9 The glucose–alanine cycle. Alanine serves as a carrier of ammonia equivalents and of the carbon skeleton of pyruvate from muscle to liver. The ammonia is excreted, and the pyruvate is used to produce glucose, which is returned to the muscle.

Alanine Carries Ammonia from Muscles to the Liver

Alanine also plays a special role in transporting amino groups to the liver in a nontoxic form, by the **glucose–alanine cycle** (Fig. 17–9). In muscle and certain other tissues that degrade amino acids for fuel, amino groups are collected in glutamate by transamination (Fig. 17–2). Glutamate may then be converted to glutamine for transport to the liver, or it may transfer its α-amino group to pyruvate, a readily available product of muscle glycolysis, by the action of **alanine aminotransferase** (Fig. 17–9). The alanine, with no net charge at pH near 7, passes into the blood and is carried to the liver. As with glutamine, excess nitrogen carried to the liver as alanine is eventually delivered as ammonia in the mitochondria. In a reversal of the alanine aminotransferase reaction described above, alanine transfers its amino group to α-ketoglutarate, forming glutamate in the cytosol. Some of this glutamate is transported into the mitochondria and acted on by glutamate dehydrogenase, releasing NH_4^+ (Fig. 17–8). Alternatively, transamination with oxaloacetate moves amino groups from glutamate to aspartate, another nitrogen donor in urea synthesis.

The use of alanine to transport ammonia from hard-working skeletal muscles to the liver is another example of the intrinsic economy of living organisms. Vigorously contracting skeletal muscles operate anaerobically, producing not only ammonia from protein breakdown but also large amounts of pyruvate from glycolysis. Both these products must find their way to the liver—ammonia to be converted into urea for excretion and pyruvate to be rebuilt into glucose and returned to the muscles. Animals thus solve two problems with one cycle: they move the carbon atoms of pyruvate, as well as excess ammonia, from muscle to liver as alanine. In the liver, alanine yields pyruvate, the starting material for gluconeogenesis, and releases NH_4^+ for urea synthesis. The energetic burden of gluconeogenesis is thus imposed on the liver rather than the muscle, so that the available ATP in the muscle can be devoted to muscle contraction.

Ammonia Is Toxic to Animals

The catabolic production of ammonia poses a serious biochemical problem because ammonia is very toxic. The molecular basis for this toxicity is not entirely understood. The terminal stages of ammonia intoxication in humans are characterized by the onset of a comatose state and other effects on the brain, so that much of the research and speculation has focused on this tissue. The major toxic effects of ammonia in brain probably involve changes in cellular pH and the depletion of certain citric acid cycle intermediates.

The protonated form of ammonia (ammonium ion) is a weak acid, and the unprotonated form is a strong base:

$$NH_4^+ \underset{pK' = 9.5}{\rightleftharpoons} NH_3 + H^+$$

Most of the ammonia generated in catabolism is present as NH_4^+ at neutral pH. Although many of the reactions that produce ammonia, such as the glutamate dehydrogenase reaction, yield NH_4^+, a few reactions, such as that of adenosine deaminase (Chapter 21), produce NH_3. Excessive amounts of NH_3 cause alkalization of cellular fluids, which has complex effects on cellular metabolism.

Ridding the cytosol of excess ammonia involves reductive amination of α-ketoglutarate to form glutamate by glutamate dehydrogenase (the reverse of the reaction described earlier) and conversion of glutamate to glutamine by glutamine synthetase. Both of these enzymes occur in high levels in the brain, although the second probably represents the more important pathway for removal of ammonia. The first reaction depletes cellular NADH and α-ketoglutarate required for ATP production in the cell. The second reaction depletes ATP itself. Overall, NH_3 may interfere with the very high levels of ATP production required to maintain brain function.

Depletion of glutamate in the glutamine synthetase reaction may have additional effects on the brain. Glutamate, and the compound γ-aminobutyrate (GABA) that is derived from it (Chapter 21), are both important neurotransmitters; the sensitivity of the brain to ammonia may well reflect a depletion of neurotransmitters as well as changes in cellular pH and ATP metabolism.

As we close this discussion of amino group metabolism, note that we have described several processes that deposit excess ammonia in the mitochondria of hepatocytes (Fig. 17–2). We now turn to a discussion of the fate of that ammonia.

Nitrogen Excretion and the Urea Cycle

As already noted (Fig. 17–4), most aquatic species, such as the bony fishes, excrete amino nitrogen as ammonia and are thus called **ammonotelic** animals; most terrestrial animals excrete amino nitrogen in the form of urea and are thus **ureotelic;** and birds and reptiles excrete amino nitrogen as uric acid and are called **uricotelic.** Plants recycle virtually all amino groups, and nitrogen excretion occurs only under very unusual circumstances. There is no general pathway for nitrogen excretion in plants.

In ureotelic organisms, the ammonia in the mitochondria of hepatocytes is converted to urea via the **urea cycle.** This pathway was discovered in 1932 by Hans Krebs (who later also discovered the citric acid cycle) and a medical student associate, Kurt Henseleit. Urea production occurs almost exclusively in the liver, and it represents the fate of most of the ammonia that is channeled there. This pathway now becomes the focus of our discussion.

Urea Is Formed in the Liver

Using thin slices of liver suspended in a buffered aerobic medium, Krebs and Henseleit found that the rate of urea formation from ammonia was greatly accelerated by adding any one of three α-amino acids: ornithine, citrulline, or arginine. Each of these three compounds stimulated urea synthesis to a far greater extent than any of the other common nitrogenous compounds tested, and their structures suggested that they might be related in a sequence.

From these and other facts Krebs and Henseleit deduced that a cyclic process occurs (Fig. 17–10), in which ornithine plays a role resembling that of oxaloacetate in the citric acid cycle. A molecule of ornithine combines with one molecule of ammonia and one of CO_2 to form citrulline. A second amino group is added to citrulline to form arginine, which is then hydrolyzed to yield urea, with regeneration of ornithine. Ureotelic animals have large amounts of the enzyme arginase in the liver. This enzyme catalyzes the irreversible hydrolysis of arginine to urea and ornithine. The ornithine is then ready for the next turn of the urea cycle. The urea is passed via the bloodstream to the kidneys and is excreted into the urine.

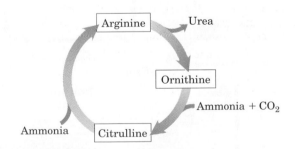

Figure 17–10 The urea cycle. The three amino acids found by Krebs and Henseleit to stimulate urea formation from ammonia in liver slices are boxed. As shown, ornithine and citrulline can serve as successive precursors of arginine. Note that citrulline and ornithine are nonstandard amino acids that are not found in proteins.

Figure 17–11 The urea cycle and the reactions that feed amino groups into it. Note that the enzymes catalyzing these reactions (named in the text) are distributed between the mitochondrial matrix and the cytosol. One amino group enters the urea cycle from carbamoyl phosphate (step ①), formed in the matrix; the other (entering at step ②) is derived from aspartate, also formed in the matrix via transamination of oxaloacetate and glutamate in a reaction catalyzed by aspartate aminotransferase. The urea cycle itself consists of four steps: ① Formation of citrulline from ornithine and carbamoyl phosphate. Citrulline passes into the cytosol. ② Formation of argininosuccinate through a citrullyl-AMP intermediate. ③ Formation of arginine from argininosuccinate. This reaction releases fumarate, which enters the citric acid cycle. ④ Formation of urea. The arginase reaction also regenerates the starting compound in the cycle, ornithine. The pathways by which NH_4^+ arrives in the mitochondrial matrix are discussed earlier in the text.

The Production of Urea from Ammonia Involves Five Enzymatic Steps

The urea cycle begins inside the mitochondria of hepatocytes, but three of the steps occur in the cytosol; the cycle thus spans two cellular compartments (Fig. 17–11). The first amino group to enter the urea cycle is derived from ammonia inside the mitochondria, arising by the multiple pathways described above. Some ammonia also arrives at the liver via the portal vein from the intestine, where it is produced by bacterial oxidation of amino acids. Whatever its source, the NH_4^+ generated in liver mitochondria is immediately used, together with HCO_3^- produced by mitochondrial respiration, to form **carbamoyl phosphate** in the matrix (Fig. 17–12; see also Fig. 17–11). This ATP-dependent reaction is catalyzed by the enzyme **carbamoyl phosphate synthetase I.** The mitochondrial form of the enzyme is distinct from the cytosolic (II) form, which has a separate function in pyrimidine biosynthesis (Chapter 21). Carbamoyl phosphate synthetase I is a regulatory enzyme; it requires N-acetylglutamate as a positive modulator (see below). Carbamoyl phosphate may be regarded as an activated carbamoyl group donor.

The carbamoyl phosphate now enters the urea cycle, which entails four enzymatic steps. Carbamoyl phosphate donates its carbamoyl group to **ornithine** to form **citrulline** and release P_i (Fig. 17–11, step

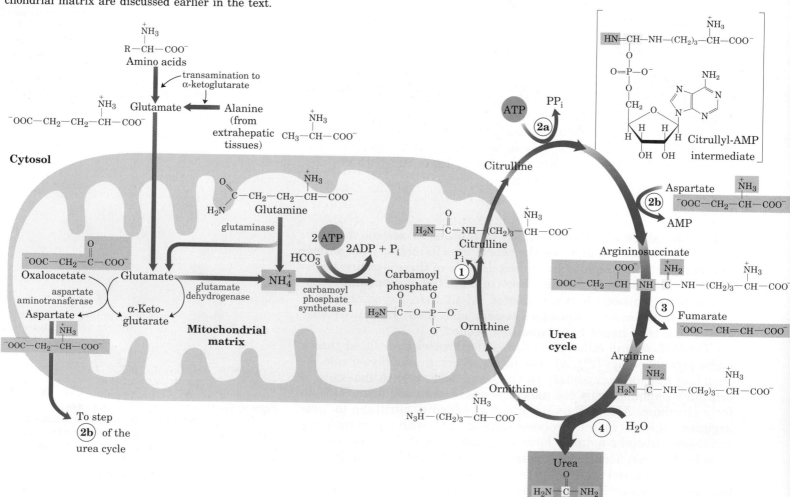

(1)) in a reaction catalyzed by **ornithine transcarbamoylase.** The citrulline is released from the mitochondrion into the cytosol.

The second amino group is introduced from aspartate (generated in the mitochondria by transamination (Fig. 17–11) and transported to the cytosol) by a condensation reaction between the amino group of aspartate and the ureido (carbonyl) group of citrulline to form **argininosuccinate** (step (2)). This reaction, catalyzed by **argininosuccinate synthetase** of the cytosol, requires ATP and proceeds through a citrullyl-AMP intermediate. The argininosuccinate is then reversibly cleaved by **argininosuccinate lyase** to form free arginine and fumarate (step (3)), which enters the pool of citric acid cycle intermediates. In the last reaction of the urea cycle the cytosolic enzyme **arginase** cleaves arginine to yield **urea** and ornithine (step (4)). Ornithine is thus regenerated and can be transported into the mitochondrion to initiate another round of the urea cycle.

As we noted in Chapter 14, enzymes of many metabolic pathways are not randomly distributed within cellular compartments, but instead are clustered (p. 414). The product of one enzyme is often channeled directly to the next enzyme in the pathway. In the urea cycle, mitochondrial and cytosolic enzymes appear to be clustered in this way. The citrulline transported out of the mitochondria is not diluted into the general pool of metabolites in the cytosol. Instead, each molecule of citrulline is passed directly into the active site of a molecule of argininosuccinate synthetase. This channeling continues for argininosuccinate, arginine, and ornithine. Only the urea is released into the general pool within the cytosol.

The Citric Acid and Urea Cycles Are Linked

The fumarate produced in the argininosuccinate lyase reaction is also an intermediate of the citric acid cycle. Fumarate enters the mitochondria, where the combined activities of fumarase (fumarate hydratase) (p. 458) and malate dehydrogenase (p. 459) transform fumarate into oxaloacetate (Fig. 17–13). Aspartate, which acts as a nitrogen donor in the urea cycle reaction catalyzed by argininosuccinate synthetase in the cytosol, is formed from oxaloacetate by transamination from glutamate; the other product of this transamination is α-ketoglutarate, another intermediate of the citric acid cycle. Because the reactions of the urea and citric acid cycles are inextricably intertwined, together they have been called the "Krebs bicycle."

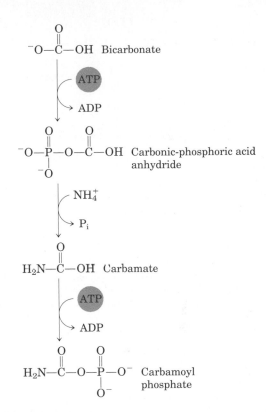

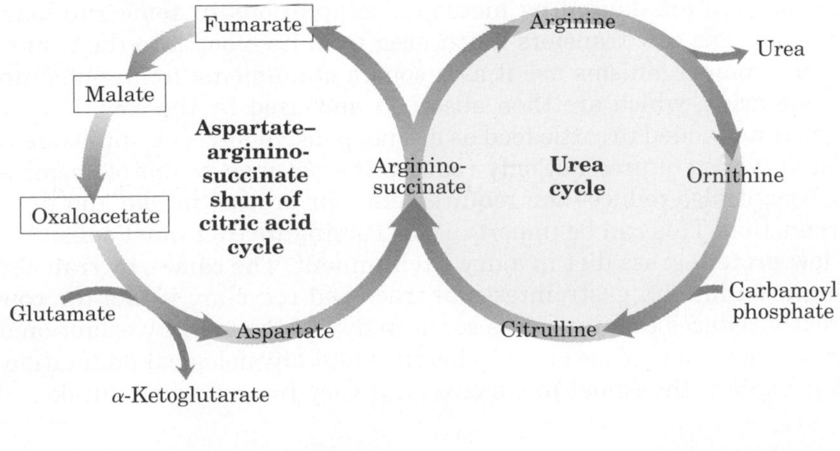

Figure 17–12 The reaction catalyzed by carbamoyl phosphate synthetase I. The formation of carbamoyl phosphate in the mitochondrial matrix is strongly stimulated by the allosteric effector *N*-acetylglutamate (see Fig. 17–14). Note that the terminal phosphate groups of two molecules of ATP are used to form one molecule of carbamoyl phosphate: two activation steps occur in the carbamoyl phosphate synthetase I reaction.

Figure 17–13 The "Krebs bicycle," composed of the urea cycle on the right, which meshes with the aspartate–argininosuccinate shunt of the citric acid cycle on the left. Fumarate produced in the cytosol by argininosuccinate lyase of the urea cycle enters the citric acid cycle in the mitochondrion and is converted in several steps to oxaloacetate. Oxaloacetate accepts an amino group from glutamate by transamination, and the aspartate thus formed leaves the mitochondrion and donates its amino group to the urea cycle in the argininosuccinate synthetase reaction. Intermediates in the citric acid cycle are boxed.

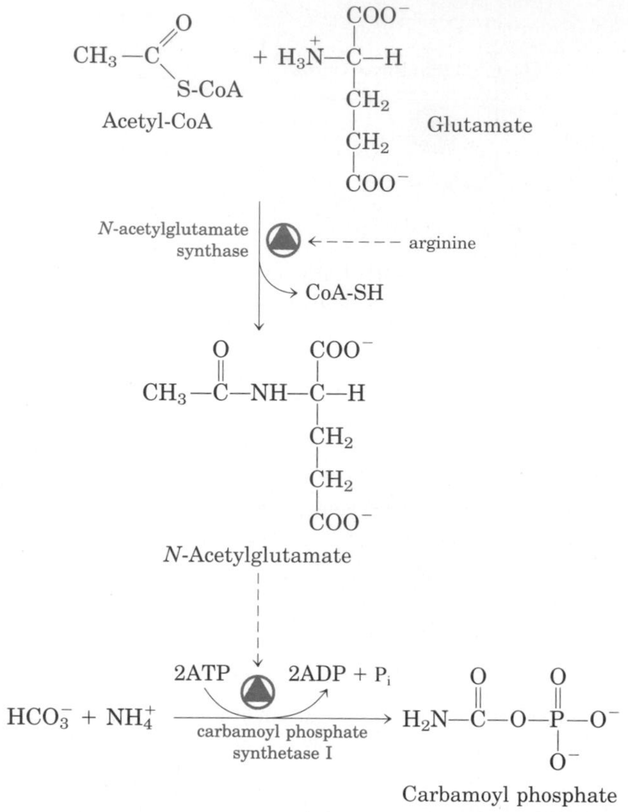

Figure 17–14 Synthesis of N-acetylglutamate, the allosteric activator of carbamoyl phosphate synthetase I, is stimulated by high concentrations of arginine. Increasing arginine levels signal the need for more flux through the urea cycle.

The Activity of the Urea Cycle Is Regulated

The flux of nitrogen through the urea cycle varies with the composition of the diet. When the diet is primarily protein, the use of the carbon skeletons of amino acids for fuel results in the production of much urea from the excess amino groups. During severe starvation, when breakdown of muscle protein supplies much of the metabolic fuel, urea production also increases substantially, for the same reason.

These changes in demand for urea cycle activity are met in the long term by regulation of the rates of synthesis of the urea cycle enzymes and carbamoyl phosphate synthetase I in the liver. All five enzymes are synthesized at higher rates during starvation or in animals on very high-protein diets than in well-fed animals on diets containing primarily carbohydrates and fats. Animals on protein-free diets produce even lower levels of the urea cycle enzymes.

On a shorter time scale, allosteric regulation of at least one key enzyme is involved in adjusting flux through the cycle. The first enzyme in the pathway, carbamoyl phosphate synthetase I, is allosterically activated by **N-acetylglutamate,** which is synthesized from acetyl-CoA and glutamate (Fig. 17–14). **N-Acetylglutamate synthase** is in turn activated by arginine, a urea cycle intermediate that accumulates when urea production is too slow to accommodate the ammonia produced by amino acid catabolism.

The Urea Cycle Is Energetically Expensive

The urea cycle brings together two amino groups and HCO_3^- to form a molecule of urea, which diffuses from the liver into the bloodstream, thence to be excreted into the urine by the kidneys. The overall equation of the urea cycle is

$$2NH_4^+ + HCO_3^- + 3ATP^{4-} + H_2O \longrightarrow$$
$$\text{urea} + 2ADP^{3-} + 4P_i^{2-} + AMP^{2-} + 5H^+$$

The synthesis of one molecule of urea requires four high-energy phosphate groups. Two ATPs are required to make carbamoyl phosphate, and one ATP is required to make argininosuccinate. In the latter reaction, however, the ATP undergoes a pyrophosphate cleavage to AMP and pyrophosphate, which may be hydrolyzed to yield two P_i.

It has been estimated that, because of the necessity of excreting nitrogen as urea instead of ammonia, ureotelic animals lose about 15% of the energy of the amino acids from which the urea was derived. This loss is counterbalanced by metabolic adaptations in some ruminant animals. The cow transfers much urea from its blood into the rumen, where microorganisms use it as a source of ammonia to manufacture amino acids, which are then absorbed and used by the cow. Urea is sometimes added to cattle feed as an inexpensive nitrogen supplement. The recycling of urea not only reduces the net investment of chemical energy, it also reduces the requirements for protein intake and urine production. This can be important for ruminants that must subsist on a low-protein grass diet in a dry environment. The camel, by transferring urea into its gastrointestinal tract and recycling it like the cow, greatly reduces the water loss connected with the urinary excretion of urea. This is one of several biochemical and physiological adaptations that enables the camel to survive on a very limited water intake.

Genetic Defects in the Urea Cycle Can Be Life-Threatening

People with genetic defects in any enzyme involved in the formation of urea have an impaired ability to convert ammonia to urea. They cannot tolerate a protein-rich diet because amino acids ingested in excess of the minimum daily requirements for protein synthesis would be deaminated in the liver, producing free ammonia in the blood. As we have seen, ammonia is toxic and causes mental disorders, retarded development, and, in high amounts, coma and death. Humans, however, are incapable of synthesizing half of the 20 standard amino acids, and these **essential amino acids** (Table 17–1) must be provided in the diet. Patients with defects in the urea cycle are often treated by substituting in the diet the α-keto acid analogs of the essential amino acids, which are the indispensable parts of the amino acids. The α-keto acid analogs can then accept amino groups from excess nonessential amino acids by aminotransferase action (Fig. 17–15). In this way the essential amino acids are made available for biosynthesis, and nonessential amino acids are kept from delivering their amino groups to the blood in the form of ammonia.

Table 17–1 Nonessential and essential amino acids for humans and the albino rat

Nonessential	Essential
Alanine	Arginine*
Asparagine	Histidine
Aspartate	Isoleucine
Cysteine	Leucine
Glutamate	Lysine
Glutamine	Methionine
Glycine	Phenylalanine
Proline	Threonine
Serine	Tryptophan
Tyrosine	Valine

* Essential in young, growing animals but not in adults.

Habitat Determines the Molecular Pathway for Nitrogen Excretion

Urea synthesis is not the only, or even the most common, pathway among organisms for excreting ammonia. The basis for differences in the molecular form in which amino groups are excreted lies in the anatomy and physiology of different organisms in relation to their usual habitat. Bacteria and free-living protozoa simply release ammonia to their aqueous environment, in which it is diluted and thus made harmless. In the bony fishes (ammonotelic animals), ammonia is rapidly cleared from the blood at the gills by the large volume of water passing through these respiratory structures. Although quite sensitive to NH_3, fish are relatively tolerant of NH_4^+. Liver is also the primary site of amino acid catabolism in fish, and NH_4^+ produced by transdeamination is simply released from the liver into the blood for transport to the gills and excretion. The bony fishes thus do not require a complex urinary system to excrete ammonia.

Organisms that excrete ammonia could not survive in an environment in which water is limited. The evolution of terrestrial species depended upon mutations that conferred the ability to convert ammonia to nontoxic substances that could be excreted in a small volume of water. Two main methods of excreting nitrogen have evolved: conversion to either urea or uric acid.

The importance of the habitat in excretion of amino nitrogen is illustrated by the change in the pathway of nitrogen excretion as the tadpole undergoes metamorphosis into the adult frog. Tadpoles are entirely aquatic and excrete amino nitrogen as ammonia through their gills. The tadpole liver lacks the necessary enzymes to make urea, but during metamorphosis it begins synthesizing these enzymes and loses the ability to excrete ammonia. In the adult frog, which is more terrestrial in habit, amino nitrogen is excreted almost entirely as urea.

In birds and reptiles, availability of water is an especially important consideration. Excretion of urea into urine requires simultaneous excretion of a relatively large volume of water; the weight of the required water would impede flight in birds, and reptiles living in arid environments must conserve water. Instead, these animals convert

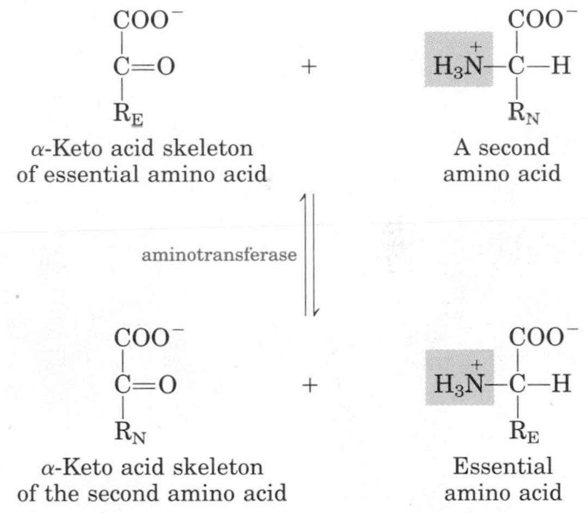

Figure 17–15 The essential amino acids (those with carbon skeletons that cannot be synthesized by animals and must be obtained in the diet) can be synthesized from the corresponding α-keto acids by transamination. The dietary requirement for essential amino acids can therefore be met by the α-keto acid skeletons. R_E and R_N represent R groups of essential and nonessential amino acids, respectively.

Figure 17–16 A view on San Lorenzo Island, one of the guano islands off the coast of Peru. Hundreds of thousands of "gooney" birds nest on these islands, and over the centuries, enormous clifflike deposits of guano, which are largely solid uric acid, have built up. Guano is a valuable fertilizer because of its nitrogen content.

amino nitrogen into uric acid (Fig. 17–4), a relatively insoluble compound that is extracted as a semisolid mass of uric acid crystals with the feces. For the advantage of excreting amino nitrogen in the form of solid uric acid, birds and reptiles must carry out considerable metabolic work; uric acid is a purine (see Fig. 12–1), and the biosynthesis of uric acid is a complex energy-requiring process that is part of the purine catabolic pathway (Chapter 21).

On many islands off the coast of South America, which serve as immense rookeries for sea birds, uric acid is deposited in enormous amounts (Fig. 17–16). These huge guano deposits are used as fertilizer, thus returning organic nitrogen to the soil, to be used again for the synthesis of amino acids by plants and soil microorganisms (Chapter 21).

Pathways of Amino Acid Degradation

There are 20 standard amino acids in proteins, with a variety of carbon skeletons. Correspondingly, there are 20 different catabolic pathways for amino acid degradation. In humans, these pathways taken together normally account for only 10 to 15% of the body's energy production. Therefore, the individual amino acid degradative pathways are not nearly as active as glycolysis and fatty acid oxidation. In addition, the activity of the catabolic pathways can vary greatly from one amino acid to the next, depending upon the balance between requirements for biosynthetic processes and the amounts of a given amino acid available. For this reason, we shall not examine them all in detail. The 20 catabolic pathways converge to form only five products, all of which enter the citric acid cycle. From here the carbons can be diverted to gluconeogenesis or ketogenesis, or they can be completely oxidized to CO_2 and H_2O (Fig. 17–17).

All or part of the carbon skeletons of ten of the amino acids are ultimately broken down to yield acetyl-CoA. Five amino acids are converted into α-ketoglutarate, four into succinyl-CoA, two into fumarate, and two into oxaloacetate. The individual pathways for the 20 amino acids will be summarized by means of flow diagrams, each leading to a specific point of entry into the citric acid cycle. In these diagrams the amino acid carbon atoms that enter the citric acid cycle are shown in color. Note that some amino acids appear more than once, reflecting the fact that different parts of their carbon skeletons have different fates. Some of the enzymatic reactions in these pathways that are particularly noteworthy for their mechanisms or their medical significance will be singled out for special discussion.

Several Enzyme Cofactors Play Important Roles in Amino Acid Catabolism

A variety of interesting chemical rearrangements are found among the amino acid catabolic pathways. Before examining the pathways themselves, it is useful to note classes of reactions that recur and to introduce the enzymatic cofactors required. We have already considered one important class, the transamination reactions requiring pyridoxal phosphate. Another common type of reaction in amino acid catabolism is a one-carbon transfer. One-carbon transfers usually involve one of three cofactors: biotin, tetrahydrofolate, or S-adenosylmethionine (Fig. 17–18). These cofactors are used to transfer one-carbon groups in dif-

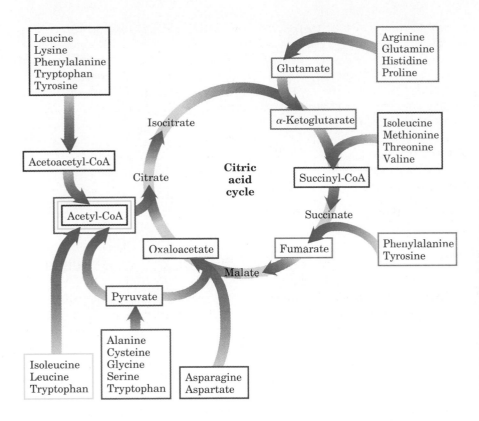

Figure 17–17 A summary of the points of entry of the standard amino acids into the citric acid cycle. (The boxes around the amino acids are color-matched to the end-products (shaded) of the catabolic pathways, here and in figures throughout the rest of this chapter.) Some amino acids are listed more than once; these are broken down to yield different fragments, each of which enters the citric acid cycle at a different point. This scheme represents the major catabolic pathways in vertebrate animals, but there are minor variations from organism to organism. Threonine, for instance, is degraded into acetyl-CoA via pyruvate in some organisms via a pathway illustrated in Fig. 17–22.

ferent oxidation states. The most oxidized state of carbon, CO_2, is transferred by biotin (see Fig. 15–13). The remaining two cofactors are especially important in amino acid and nucleotide metabolism. Tetrahydrofolate is generally involved in transfers of one-carbon groups in the intermediate oxidation states, and S-adenosylmethionine in transfers of methyl groups, the most reduced state of carbon.

Tetrahydrofolate (H_4 folate) consists of a substituted pteridine, p-aminobenzoate, and glutamate linked together as in Figure 17–18. This cofactor is synthesized in bacteria and is a vitamin for mammals.

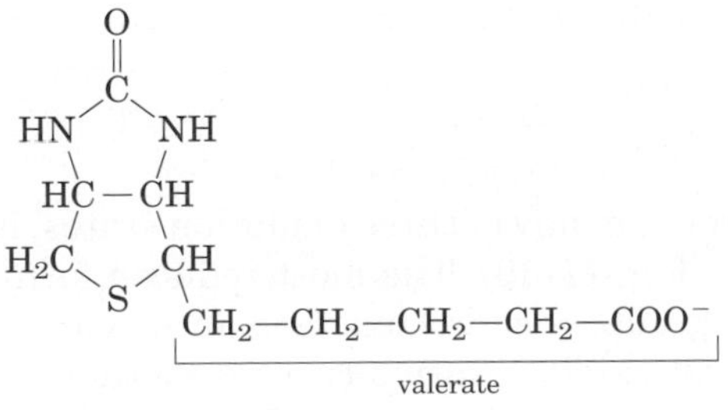

Biotin

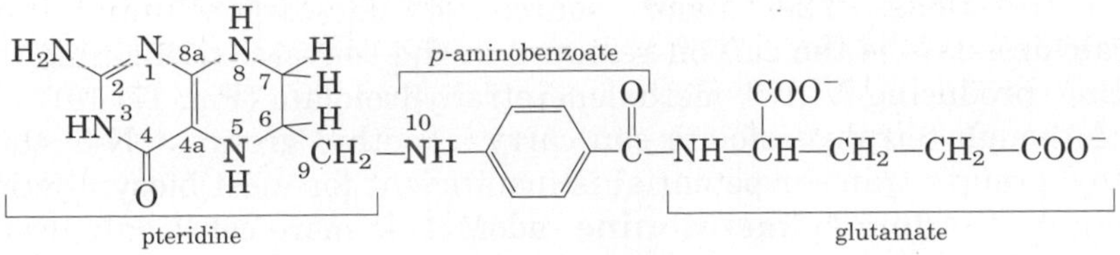

Tetrahydrofolate (H_4 folate)

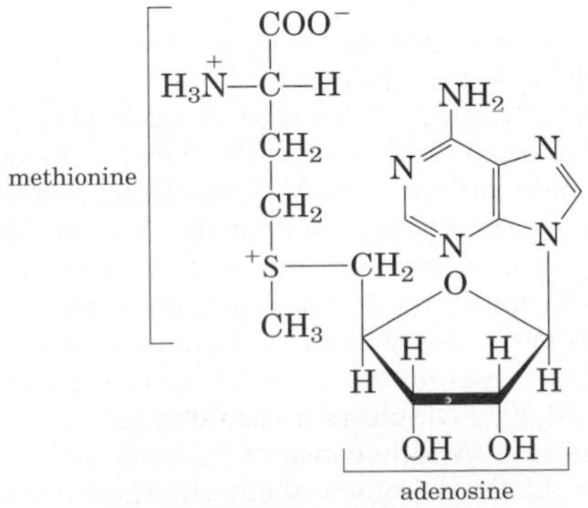

S-Adenosylmethionine (adoMet)

Figure 17–18 The structures of enzyme cofactors important in one-carbon transfer reactions. The nitrogen atoms to which one-carbon groups are attached in tetrahydrofolate are shown in blue.

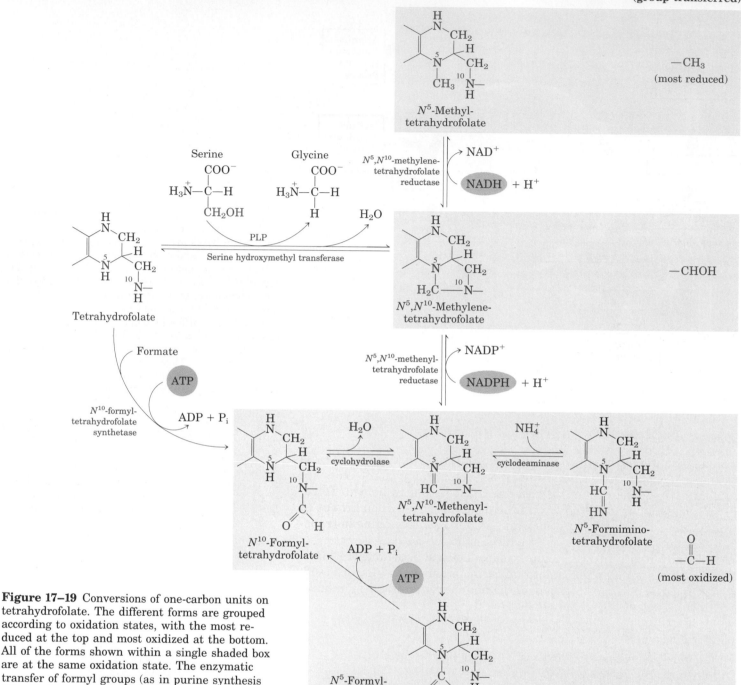

Figure 17–19 Conversions of one-carbon units on tetrahydrofolate. The different forms are grouped according to oxidation states, with the most reduced at the top and most oxidized at the bottom. All of the forms shown within a single shaded box are at the same oxidation state. The enzymatic transfer of formyl groups (as in purine synthesis (Fig. 21–27) and formation of formylmethionine in prokaryotes (p. 917)) generally uses N^{10}-formyltetrahydrofolate rather than N^5-formyltetrahydrofolate. The latter species is significantly more stable, and hence is not as good a donor of formyl groups. Over time, the equilibria of the reactions that interconnect these species will favor formation of N^5-formyltetrahydrofolate. The N^5 species must be converted to N^{10}-formyltetrahydrofolate in a reaction that requires ATP because of its unfavorable equilibrium. Little is known about the mechanism of this reaction. Note that N^5-formiminotetrahydrofolate is derived from histidine in a pathway shown in Fig. 17–29.

The one-carbon group, in any of three oxidation states, is bonded to N-5 or N-10 or to both (Fig. 17–19). The most reduced form of the cofactor carries a methyl group, a more oxidized form carries a methylene group, and the most oxidized forms carry a methenyl, formyl, or formimino group. The different forms of tetrahydrofolate are interconvertible and serve as donors of one-carbon units in a variety of biosynthetic reactions. The major source of one-carbon units for tetrahydrofolate is the carbon removed in the conversion of serine to glycine, producing N^5,N^{10}-methylenetetrahydrofolate (Fig. 17–19).

Although tetrahydrofolate can carry a methyl group at N-5, the methyl group's transfer potential is insufficient for most biosynthetic reactions. **S-Adenosylmethionine** (adoMet) is more commonly used for methyl group transfers. It is synthesized from ATP and methionine

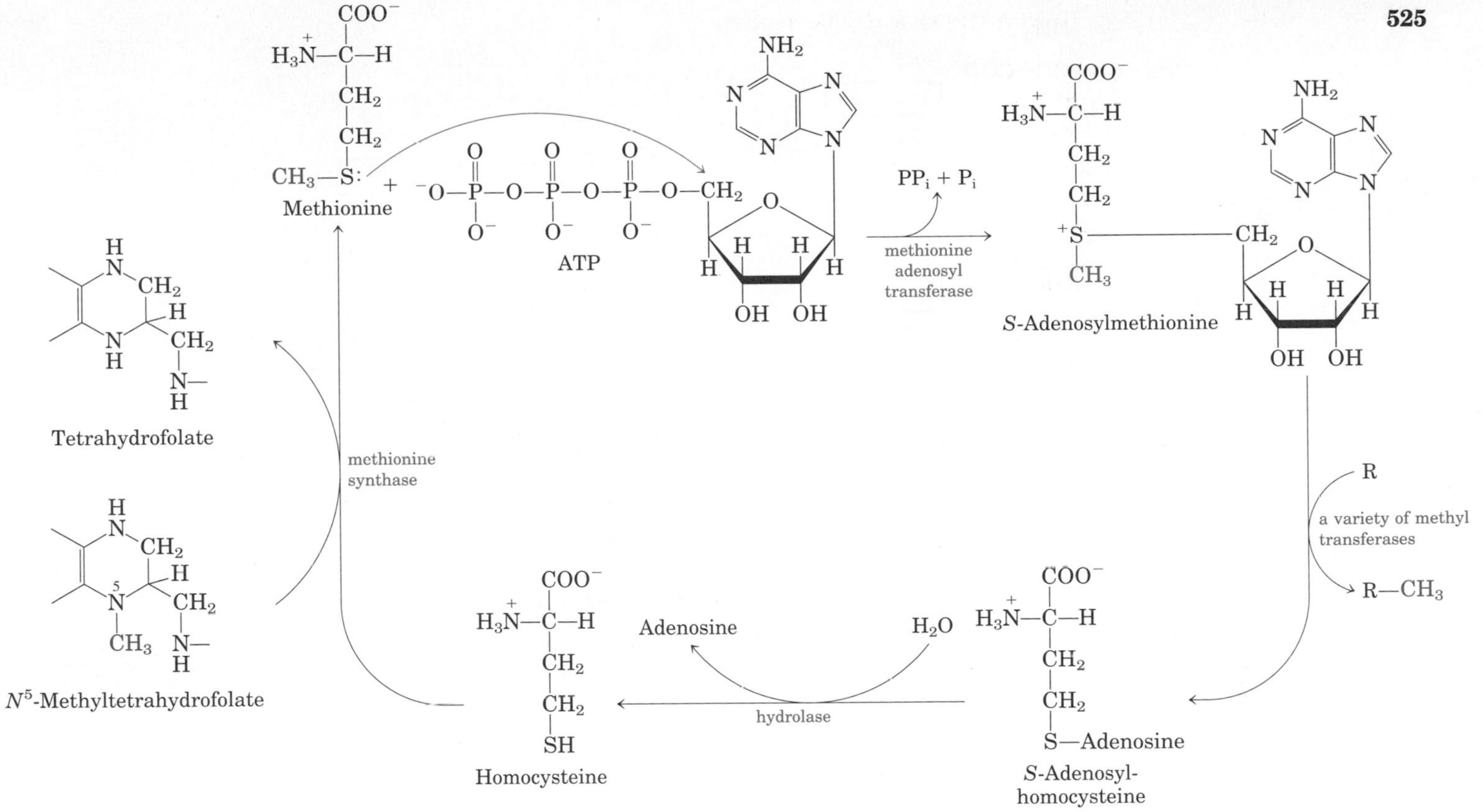

Figure 17–20 Synthesis of methionine and S-adenosylmethionine as part of an activated methyl cycle. The methyl group donor in the methionine synthase reaction is methylcobalamin in some organisms. S-Adenosylmethionine, which has a positively charged sulfur (and is thus a sulfonium ion), is a powerful methylating agent in a number of biosynthetic reactions. The methyl group acceptor is designated R.

by the action of **methionine adenosyl transferase** (Fig. 17–20). This reaction is unusual in that the nucleophilic sulfur atom of methionine attacks at the 5' carbon of the ribose moiety of ATP, releasing triphosphate, rather than attacking at one of the phosphorus atoms. The triphosphate is cleaved to P_i and PP_i on the enzyme, and the PP_i is later cleaved by inorganic pyrophosphatase, so that three high-energy bonds are expended in this reaction. The only other reaction known in which triphosphate is displaced from ATP occurs in the synthesis of coenzyme B_{12} (see Box 16–2, Fig. 3).

S-Adenosylmethionine is a potent alkylating agent by virtue of its destabilizing sulfonium ion. The methyl group is subject to attack by nucleophiles and is about 1,000 times more reactive than the methyl group of N^5-methyltetrahydrofolate.

Transfer of a methyl group from S-adenosylmethionine to an acceptor yields **S-adenosylhomocysteine,** which is subsequently broken down to homocysteine and adenosine (Fig. 17–20). Methionine is regenerated by the transfer of a methyl group to homocysteine in a reaction catalyzed by **methionine synthase.** One form of this enzyme is common in bacteria and uses N^5-methyltetrahydrofolate as a methyl donor. Another form that occurs in bacteria and mammals uses methylcobalamin derived from coenzyme B_{12}. This reaction and the rearrangement of L-methylmalonyl-CoA to succinyl-CoA (Box 16–2, Fig. 1a) are the only coenzyme B_{12}–dependent reactions known in mammals. Methionine is reconverted to S-adenosylmethionine to complete an activated methyl cycle (Fig. 17–20).

Tetrahydrobiopterin is another cofactor introduced in these pathways, but it is not involved in one-carbon transfers. Tetrahydrobiopterin is structurally related to the flavin coenzymes, and it participates in biological oxidation reactions. It belongs to a widespread class of biological compounds called pterins (Fig. 17–21), and we will consider its mode of action when we discuss phenylalanine degradation.

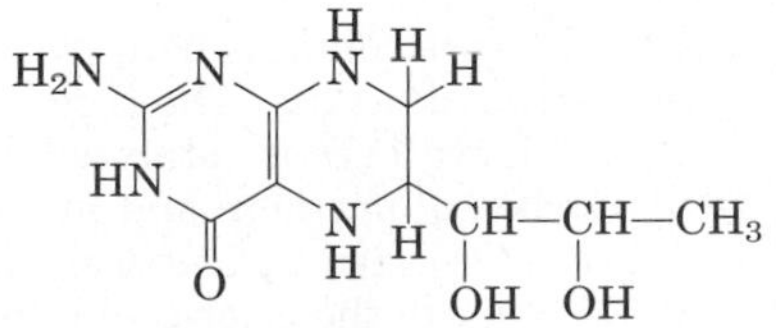

Figure 17–21 Tetrahydrobiopterin and its parent compound, pterin. Tetrahydrobiopterin is a cofactor for the enzyme phenylalanine hydroxylase.

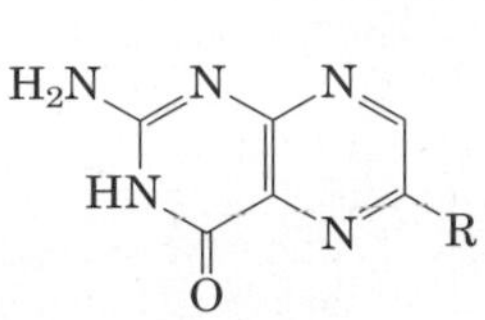

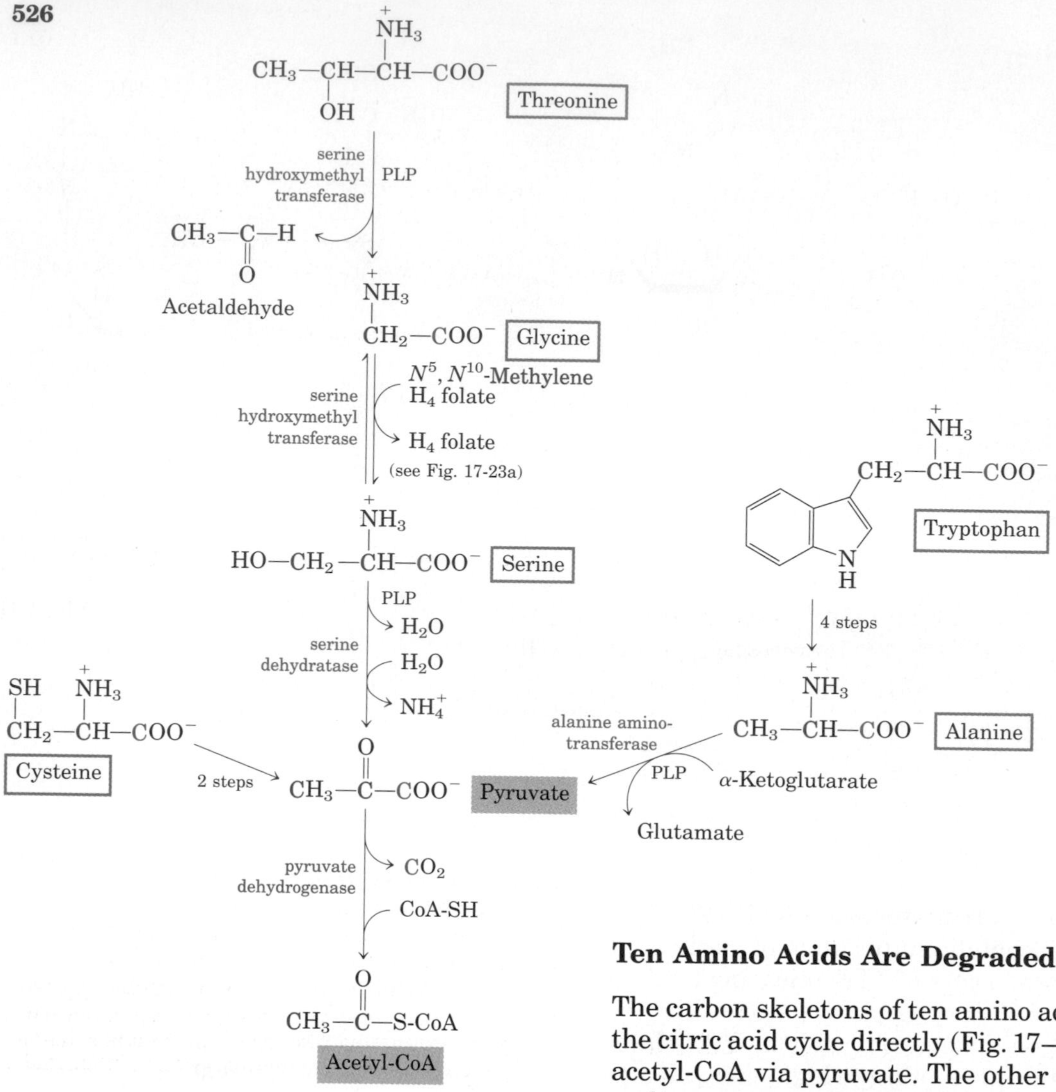

Figure 17–22 Outline of the catabolic pathways for alanine, glycine, serine, cysteine, tryptophan, and threonine. The fate of the indole group of tryptophan is shown in Fig. 17–24. Details of the glycine-to-serine conversion, and a second fate for glycine, are shown in Fig. 17–23. In some organisms (including humans) threonine is degraded to succinyl-CoA by another pathway (Fig. 17–30). There are several pathways for cysteine degradation, all of which lead to pyruvate. The enzyme serine hydroxymethyl transferase contains both pyridoxal phosphate and tetrahydrofolate. The threonine cleavage reaction shown here is catalyzed by the same enzyme. Carbon atoms here and in subsequent figures are color-coded as necessary to trace their fates, in addition to the color-coding for pathways described in Fig. 17–17.

Ten Amino Acids Are Degraded to Acetyl-CoA

The carbon skeletons of ten amino acids yield acetyl-CoA, which enters the citric acid cycle directly (Fig. 17–17). Five of the ten are degraded to acetyl-CoA via pyruvate. The other five are converted into acetyl-CoA and/or acetoacetyl-CoA, which is then cleaved to form acetyl-CoA.

The five amino acids entering via pyruvate are **alanine, glycine, serine, cysteine,** and **tryptophan** (Fig. 17–22). In some organisms **threonine** is also degraded to form acetyl-CoA, as shown in Figure 17–22; in humans it is degraded to succinyl-CoA, as described later. Alanine yields pyruvate directly on transamination with α-ketoglutarate, and the side chain of tryptophan is cleaved to yield alanine and thus pyruvate. Cysteine is converted to pyruvate in two steps, one to remove the sulfur atom, the other a transamination. Serine is converted to pyruvate by serine dehydratase. Both the β-hydroxyl and the α-amino groups of serine are removed in this single PLP-dependent reaction (an analogous reaction with threonine is shown in Fig. 17–30). Glycine has two pathways. It can be converted into serine by enzymatic addition of a hydroxymethyl group (Fig. 17–23a). This reaction, catalyzed by **serine hydroxymethyl transferase,** requires the coenzymes tetrahydrofolate and pyridoxal phosphate. The second pathway for glycine, which predominates in animals, involves its oxidative cleavage into CO_2, NH_4^+, and a methylene group ($-CH_2-$) (Fig. 17–23b). This readily reversible reaction, catalyzed by **glycine synthase,** also requires tetrahydrofolate, which accepts the methylene group. In this oxidative cleavage pathway the two carbon atoms of glycine do not enter the citric acid cycle. One is lost as CO_2, and the other becomes the methylene group of N^5,N^{10}-methylene-tetrahydrofolate (Fig. 17–19), which is used as a one-carbon group donor in certain biosynthetic pathways.

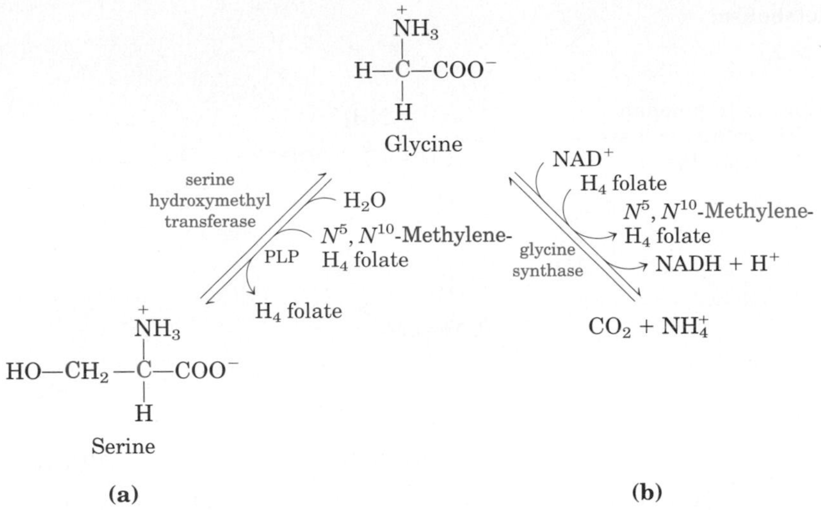

Figure 17–23 Two metabolic fates of glycine: **(a)** conversion to serine and **(b)** breakdown to CO_2 and ammonia. The cofactor tetrahydrofolate carries one-carbon units in both of these reactions. The structure of H_4 folate is shown in Fig. 17–18, and its role as a cofactor in one-carbon transfers in Fig. 17–19.

Portions of the carbon skeleton of six amino acids—**tryptophan, lysine, phenylalanine, tyrosine, leucine,** and **isoleucine**—yield acetyl-CoA and/or acetoacetyl-CoA; the latter is then converted into acetyl-CoA (Fig. 17–24). Some of the final steps in the degradative pathways for leucine, lysine, and tryptophan resemble steps in the oxidation of fatty acids. The breakdown of two of these six amino acids deserves special mention.

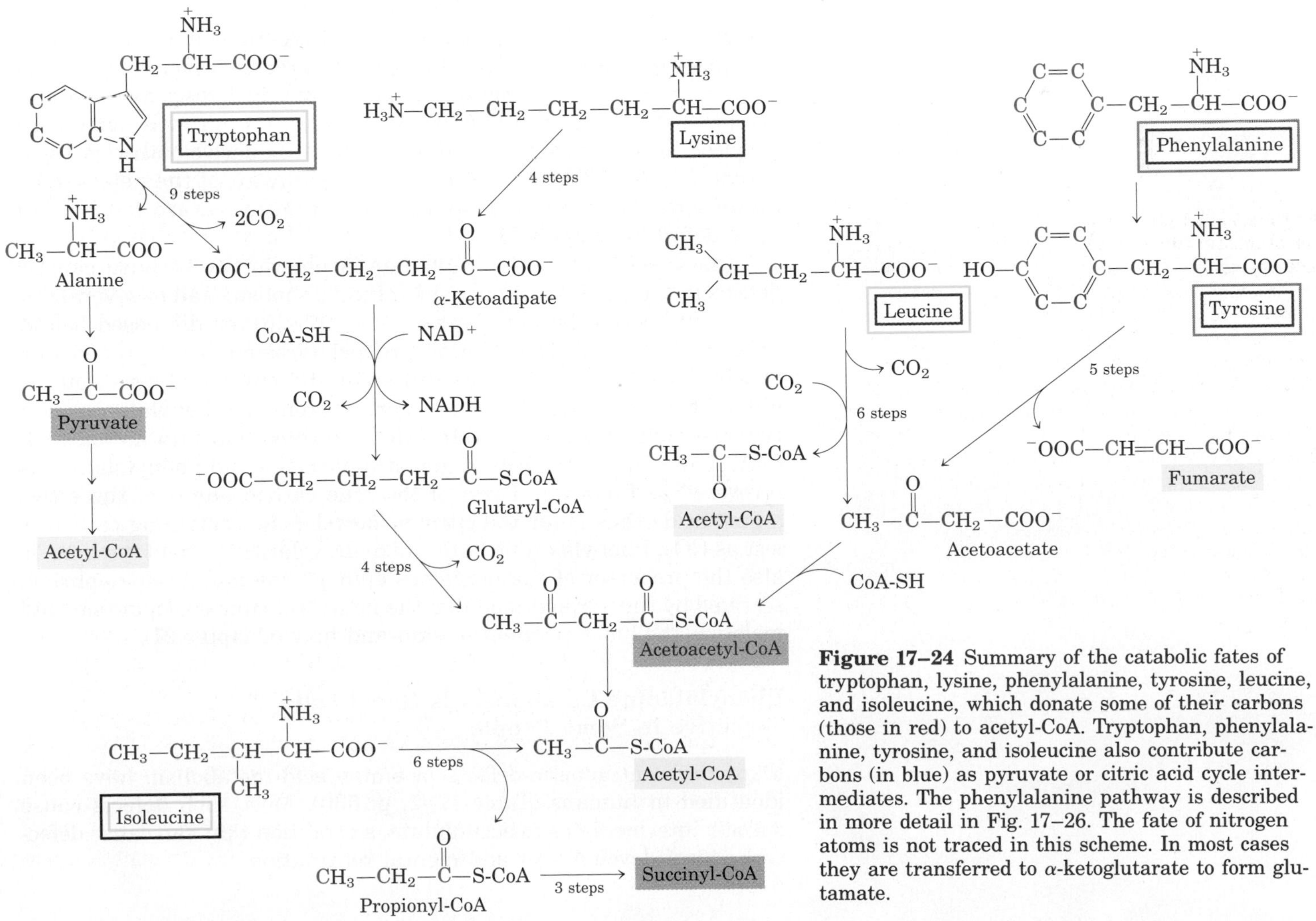

Figure 17–24 Summary of the catabolic fates of tryptophan, lysine, phenylalanine, tyrosine, leucine, and isoleucine, which donate some of their carbons (those in red) to acetyl-CoA. Tryptophan, phenylalanine, tyrosine, and isoleucine also contribute carbons (in blue) as pyruvate or citric acid cycle intermediates. The phenylalanine pathway is described in more detail in Fig. 17–26. The fate of nitrogen atoms is not traced in this scheme. In most cases they are transferred to α-ketoglutarate to form glutamate.

Figure 17–25 The aromatic rings of tryptophan are precursors of nicotinate, indoleacetate, and serotonin. Colored atoms are used to trace the source of the ring atoms in nicotinate.

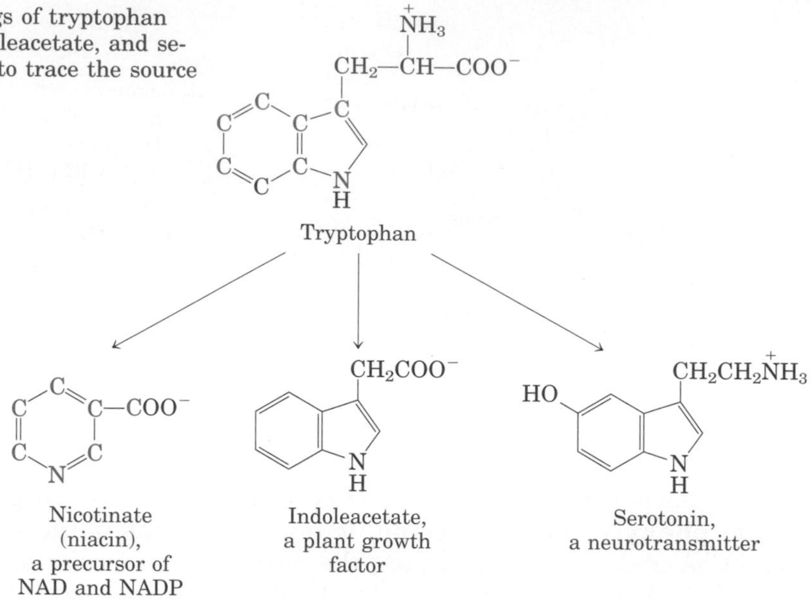

The dehydration of tryptophan is the most complex of all the pathways of amino acid catabolism in animal tissues; portions of tryptophan (six carbons total) yield acetyl-CoA by two different pathways, one via pyruvate and one via acetoacetyl-CoA. Some of the intermediates in tryptophan catabolism are required precursors for biosynthesis of other important biomolecules (Fig. 17–25), including nicotinate, a precursor of NAD and NADP. In plants, the growth factor indoleacetate is derived from tryptophan by an oxidative pathway. Tryptophan is also the parent, by a different pathway, of the neurotransmitter serotonin. Some of these biosynthetic pathways are described in more detail in Chapter 21.

The breakdown of phenylalanine is noteworthy because genetic defects in the enzymes of phenylalanine catabolism lead to several different inheritable human diseases (Fig. 17–26), as discussed below. Phenylalanine and its oxidation product tyrosine are degraded into two fragments, each of which can enter the citric acid cycle, but at different points. Four of the nine carbon atoms of phenylalanine and tyrosine yield free acetoacetate, which is converted into acetoacetyl-CoA. A second four-carbon fragment of tyrosine and phenylalanine is recovered as fumarate. Eight of the nine carbon atoms of these two amino acids thus enter the citric acid cycle; the remaining carbon is lost as CO_2. Phenylalanine, after its hydroxylation to yield tyrosine, is also the precursor of the hormones epinephrine and norepinephrine, secreted by the adrenal medulla, the neurotransmitter dopamine, and melanin, the black pigment of skin and hair (Chapter 21).

Phenylalanine Catabolism Is Genetically Defective in Some People

Many different genetic defects in amino acid metabolism have been identified in humans (Table 17–2, p. 530). Most such defects cause specific intermediates to accumulate, a condition that can cause defective neural development and mental retardation.

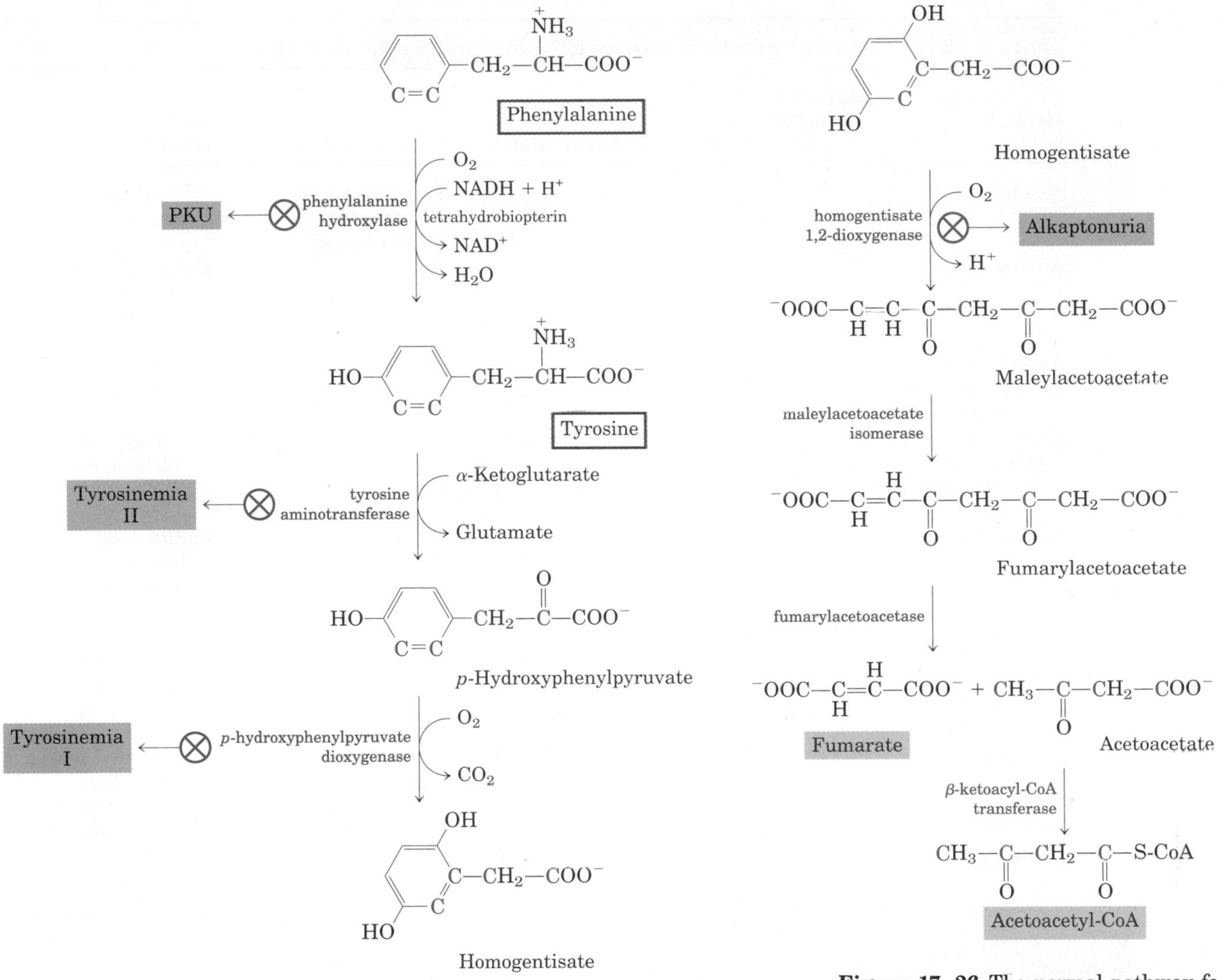

Figure 17–26 The normal pathway for conversion of phenylalanine and tyrosine into acetoacetyl-CoA and fumarate in humans. Genetic defects in each of the first four enzymes in this pathway are known to cause inheritable human diseases (shaded in red).

The first enzyme in the catabolic pathway for phenylalanine (Fig. 17–26), **phenylalanine hydroxylase,** catalyzes the hydroxylation of phenylalanine to tyrosine. A genetic defect in phenylalanine hydroxylase is responsible for the disease **phenylketonuria** (PKU). Phenylketonuria is the most common cause of elevated levels of phenylalanine (hyperphenylalaninemia). Phenylalanine hydroxylase inserts one of the two oxygen atoms of O_2 into phenylalanine to form the hydroxyl group of tyrosine; the other oxygen atom is reduced to H_2O by the NADH also required in the reaction. This is one of a general class of reactions catalyzed by enzymes called **mixed-function oxidases** (see Box 20–1), all of which catalyze simultaneous hydroxylation of a substrate by O_2 and reduction of the other oxygen atom of O_2 to H_2O. Phenylalanine hydroxylase requires a cofactor, **tetrahydrobiopterin,** which carries electrons from NADH to O_2 in the hydroxylation of phenylalanine. During the hydroxylation reaction the coenzyme is oxidized to dihydrobiopterin (Fig. 17–27). It is subsequently reduced again by the enzyme **dihydrobiopterin reductase** in a reaction that requires NADH.

Table 17–2 Some human genetic disorders affecting amino acid catabolism

Medical condition	Approximate incidence (per 100,000 births)	Defective process	Defective enzyme	Symptoms and effects
Albinism	3	Melanin synthesis from tyrosine	Tyrosine 3-mono-oxygenase (tyrosinase)	Lack of pigmentation; white hair, pink skin
Alkaptonuria	0.4	Tyrosine degradation	Homogentisate 1,2-dioxygenase	Dark pigment in urine; late-developing arthritis
Argininemia	<0.5	Urea synthesis	Arginase	Mental retardation
Argininosuccinic acidemia	1.5	Urea synthesis	Argininosuc-cinate lyase	Vomiting, convulsions
Carbamoyl phosphate synthetase I deficiency	>0.5	Urea synthesis	Carbamoyl phosphate synthetase I	Lethargy, convulsions, early death
Homocystinuria	0.5	Methionine degradation	Cystathione β-synthase	Faulty bone development, mental retardation
Maple syrup urine disease (branched-chain ketoaciduria)	0.4	Isoleucine, leucine, and valine degradation	Branched-chain α-keto acid dehydrogenase complex	Vomiting, convulsions, mental retardation, early death
Methylmalonic acidemia	<0.5	Conversion of propionyl-CoA to succinyl-CoA	Methylmalonyl-CoA mutase	Vomiting, convulsions, mental retardation, early death
Phenylke-tonuria	8	Conversion of phenylalanine to tyrosine	Phenylalanine hydroxylase	Neonatal vomiting; mental retardation

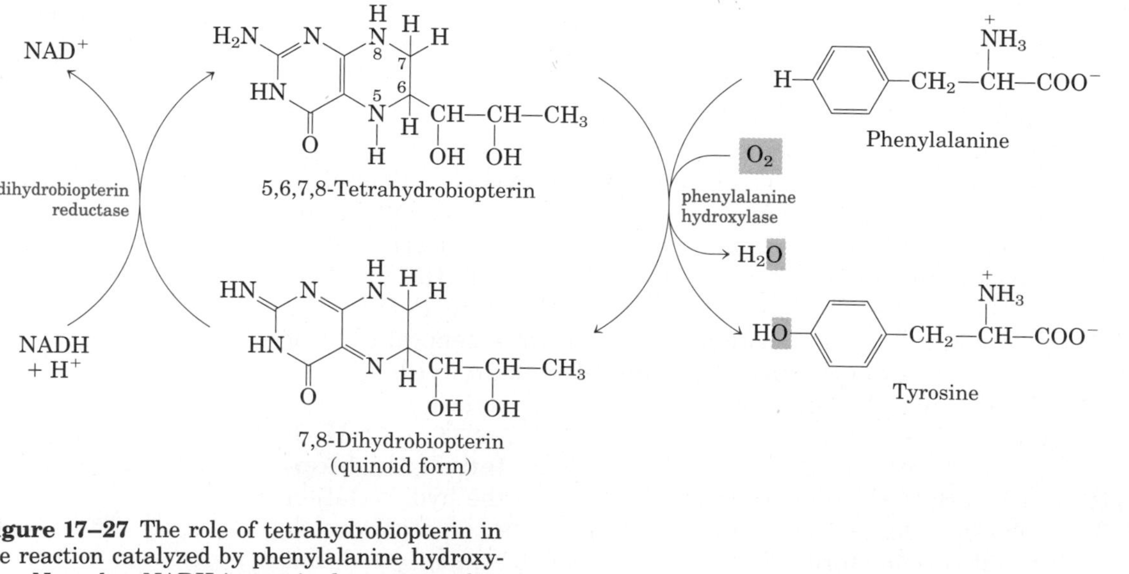

Figure 17–27 The role of tetrahydrobiopterin in the reaction catalyzed by phenylalanine hydroxylase. Note that NADH is required to restore the reduced form of the coenzyme.

When phenylalanine hydroxylase is genetically defective, a secondary pathway of phenylalanine metabolism, normally little used, comes into play. In this minor pathway phenylalanine undergoes transamination with pyruvate to yield **phenylpyruvate** (Fig. 17–28). Phenylalanine and phenylpyruvate accumulate in the blood and tissues and are excreted in the urine: hence the name of the condition, phenylketonuria. Much of the phenylpyruvate is either decarboxylated to produce phenylacetate or reduced to form phenyllactate. Phenylacetate imparts a characteristic odor to the urine that has been used by nurses to detect PKU in infants. The accumulation of phenylalanine or its metabolites in early life impairs the normal development of the brain, causing severe mental retardation. Excess phenylalanine may compete with other amino acids for transport across the blood–brain barrier, resulting in a depletion of some required metabolites.

Phenylketonuria was among the first human genetic defects of metabolism discovered. When this condition is recognized early enough in infancy, mental retardation can largely be prevented by rigid dietary control. The diet must supply just enough phenylalanine and tyrosine to meet the needs for protein synthesis. Consumption of foods that are rich in protein must be curtailed, and warnings are listed on foods artificially sweetened with aspartame (a dipeptide of the methyl ester of phenylalanine and aspartate; see Fig. 3–13). Natural proteins, such as casein of milk, must first be hydrolyzed and much of the phenylalanine removed to provide an appropriate diet for phenylketonurics, at least through childhood.

Phenylketonuria can also be caused by a defect in the enzyme that catalyzes the regeneration of tetrahydrobiopterin. The treatment in this case is more complex than providing a diet that restricts intake of phenylalanine and tyrosine. Because tetrahydrobiopterin is also required for the formation of L-3,4-dihydroxyphenylalanine (L-dopa) and 5-hydroxytryptophan (essential precursors of the neurotransmitters norepinephrine and serotonin, respectively), these compounds must be supplied in the diet. Supplying tetrahydrobiopterin in the diet is insufficient because it is unstable and does not cross the blood–brain barrier.

Screening newborns for genetic diseases can be highly cost-effective, especially in the case of PKU. The tests are relatively inexpensive, and the detection and early treatment of PKU in infants (eight to ten cases per 100,000 individuals) saves millions of dollars each year that would otherwise be spent on institutionalized care and special programs to address the mental retardation. The human emotional trauma avoided by these simple tests is, of course, inestimable.

People who inherit a genetic defect in another of the enzymes in the phenylalanine catabolic pathway, **homogentisate dioxygenase,** are more fortunate than those with PKU. The defect results in no serious ill effects (although these individuals excrete large amounts of homogentisate, which is rapidly oxidized and turns the urine black), but it has historical importance. This defect, called **alkaptonuria,** was studied by Archibald Garrod in the early 1900s. Garrod discovered that the condition was inherited, and he could trace it to the absence of a single enzyme. Garrod was the first to make a connection between an inheritable trait and an enzyme, a major advance on the path that ultimately led to our current understanding of genes and the information pathways described in Part IV of this book.

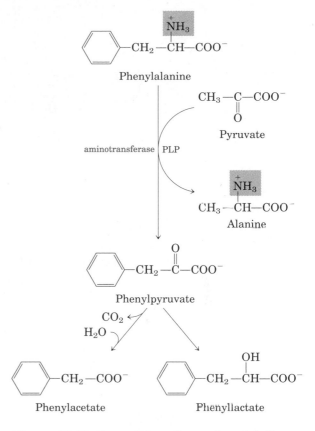

Figure 17–28 Alternative pathways for catabolism of phenylalanine in phenylketonurics. Phenylpyruvate accumulates in the tissues, blood, and urine. Phenylacetate and phenyllactate can also be found in the urine.

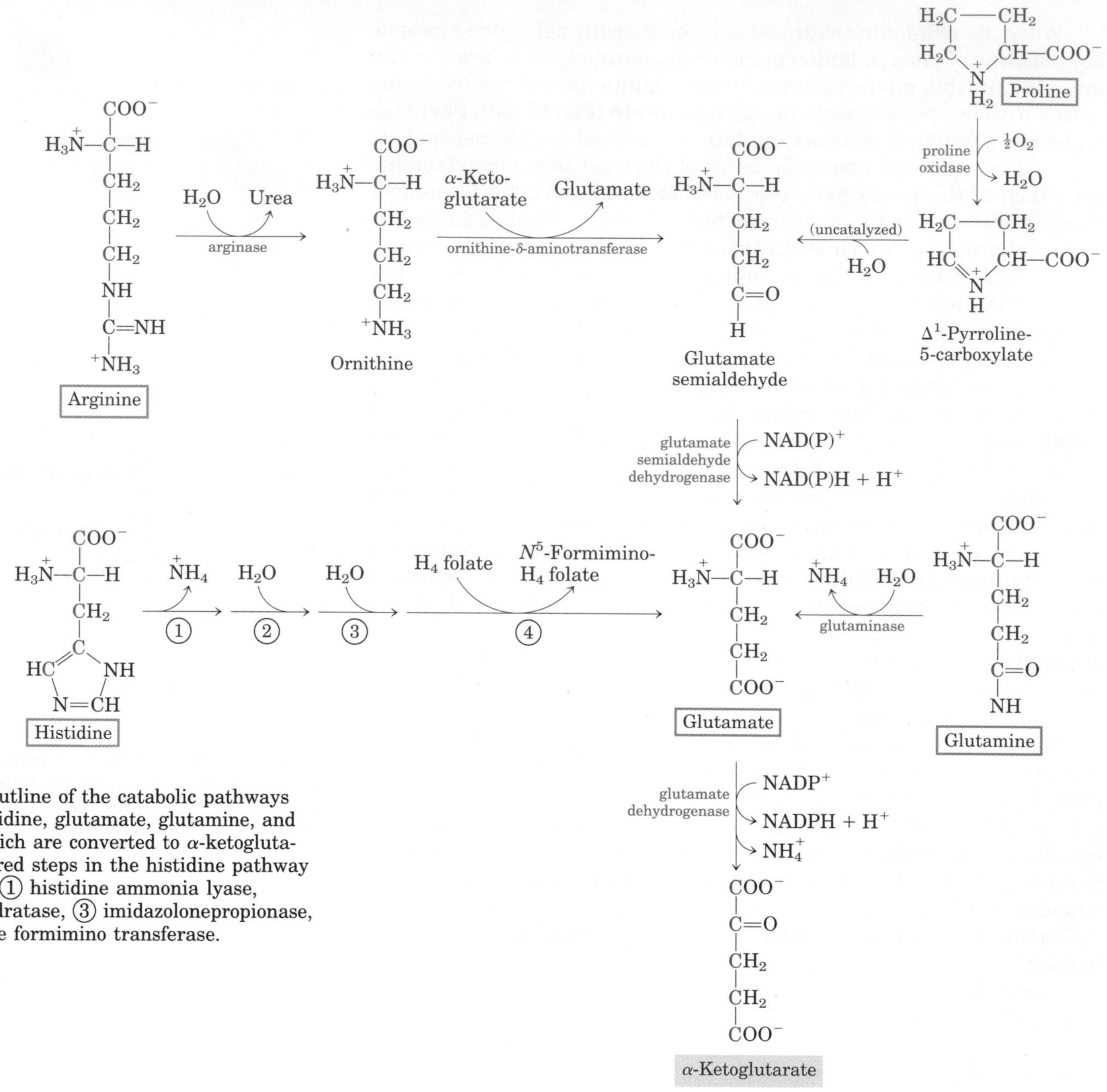

Figure 17–29 Outline of the catabolic pathways for arginine, histidine, glutamate, glutamine, and proline, all of which are converted to α-ketoglutarate. The numbered steps in the histidine pathway are catalyzed by ① histidine ammonia lyase, ② urocanate hydratase, ③ imidazolonepropionase, and ④ glutamate formimino transferase.

Five Amino Acids Are Converted into α-Ketoglutarate

The carbon skeletons of five amino acids (arginine, histidine, glutamate, glutamine, and proline) enter the citric acid cycle via α-ketoglutarate (Fig. 17–29). **Proline, glutamate,** and **glutamine** have five-carbon skeletons. The cyclic structure of proline is opened by oxidation of the carbon most distant from the carboxyl group to create a Schiff base and hydrolysis of the Schiff base to a linear semialdehyde (glutamate semialdehyde). This is further oxidized at the same carbon to produce glutamate. The action of glutaminase, or any of several reactions in which glutamine donates its amide nitrogen to some acceptor, converts glutamine to glutamate. Transamination or deamination of glutamate produces the citric acid cycle intermediate α-ketoglutarate.

Arginine and **histidine** contain five adjacent carbons and a sixth carbon attached through a nitrogen atom. The catabolic conversion of these two amino acids to glutamate is therefore slightly more complex than the path from proline or glutamine to glutamate (Fig. 17–29). Arginine is converted to the five-carbon skeleton of ornithine in the urea cycle (see Fig. 17–11), and the ornithine is transaminated to glutamate semialdehyde. The conversion of histidine to the five-carbon glutamate occurs in a multistep pathway; the extra carbon is removed in a step that employs tetrahydrofolate as a cofactor (Fig. 17–29).

Four Amino Acids Are Converted into Succinyl-CoA

The carbon skeletons of methionine, isoleucine, threonine, and valine are degraded by pathways that yield succinyl-CoA (Fig. 17–30), an intermediate of the citric acid cycle. **Methionine** donates its methyl group to one of several possible acceptors through S-adenosylmethionine, and three of the four remaining atoms of its carbon skeleton are converted into those of propionate as propionyl-CoA.

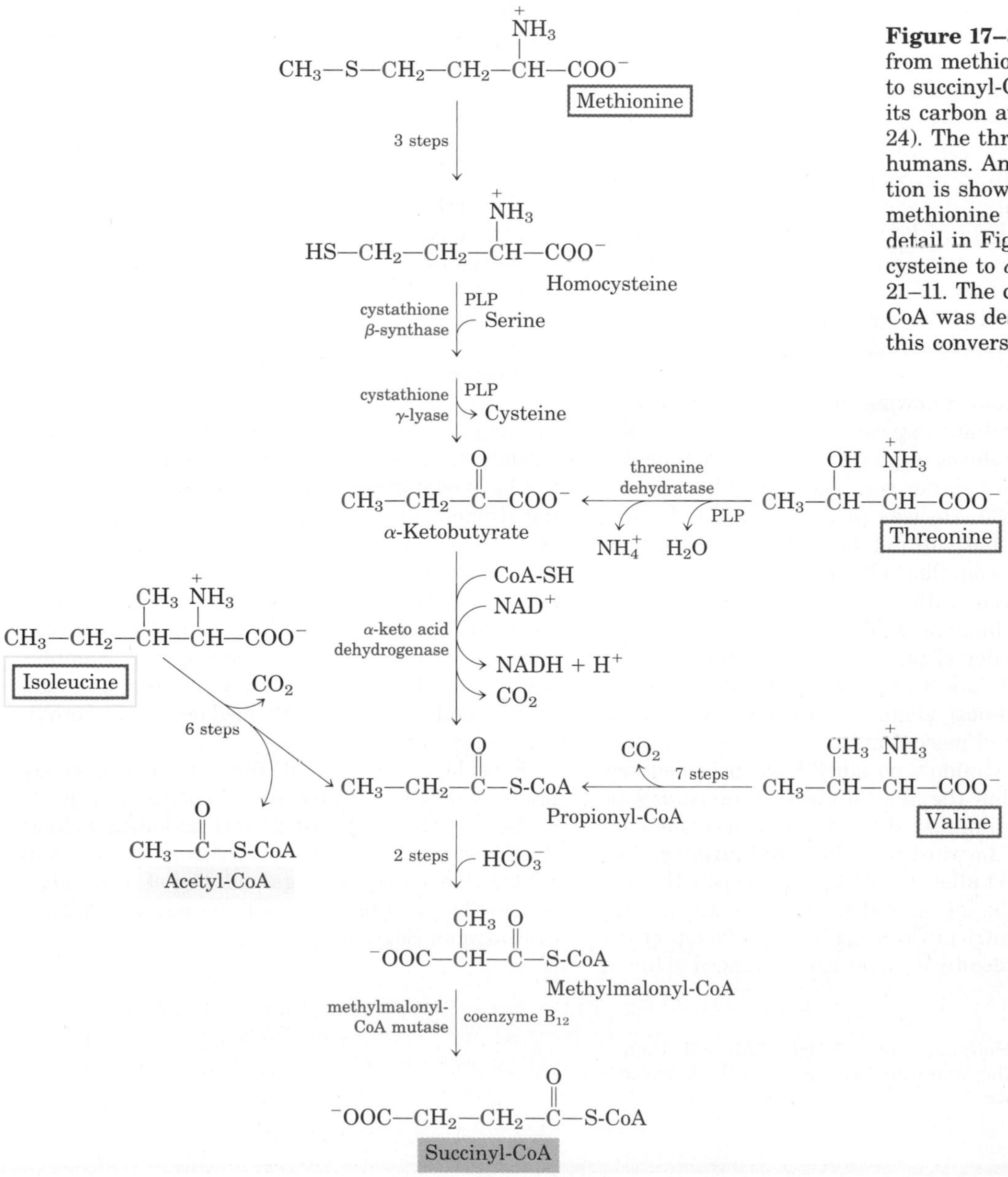

Figure 17–30 Outline of the catabolic pathways from methionine, isoleucine, threonine, and valine to succinyl-CoA. Isoleucine also contributes two of its carbon atoms to acetyl-CoA (see also Fig. 17–24). The threonine pathway shown here occurs in humans. Another pathway for threonine degradation is shown in Fig. 17–22. The pathway from methionine to homocysteine is described in more detail in Fig. 17–20. The steps that convert homocysteine to α-ketobutyrate are illustrated in Fig. 21–11. The conversion of propionyl-CoA to succinyl-CoA was described in Chapter 16; the last step in this conversion requires coenzyme B_{12}.

BOX 17–2   Scientific Sleuths Solve a Murder Mystery

Truth can sometimes be stranger than fiction—or at least as strange as a made-for-TV movie. Take, for example, the case of Patricia Stallings. Convicted of the murder of her infant son, she was sentenced to life in prison—but was later found innocent, thanks to the medical sleuthing of three persistent researchers.

The story began in the summer of 1989 when Stallings brought her three-month-old son, Ryan, to the emergency room of Cardinal Glennon Children's Hospital in St. Louis. The child had labored breathing, uncontrollable vomiting, and gastric distress. According to the attending physician, a toxicologist, the child's symptoms indicated that he had been poisoned with ethylene glycol, an ingredient of antifreeze, a conclusion apparently confirmed by analysis by a commercial lab.

After he recovered, the child was placed in a foster home, and Stallings and her husband, David, were allowed to see him in supervised visits. But when the infant became ill, and subsequently died, after a visit in which Stallings had been briefly left alone with him, she was charged with first-degree murder and held without bail. At the time, the evidence seemed compelling as both the commercial lab and the hospital lab found large amounts of ethylene glycol in the boy's blood and traces of it in a bottle of milk Stallings had fed her son during the visit.

But without knowing it, Stallings had performed a brilliant experiment. While in custody, she learned she was pregnant; she subsequently gave birth to another son, David Stallings Jr., in February 1990. He was placed immediately in a foster home, but within two weeks he started having symptoms similar to Ryan's. David was eventually diagnosed with a rare metabolic disorder called methylmalonic acidemia (MMA). A recessive genetic disorder of amino acid metabolism, MMA affects about 1 in 48,000 newborns and presents symptoms almost identical with those caused by ethylene glycol poisoning.

Stallings couldn't possibly have poisoned her second son, but the Missouri state prosecutor's office was not impressed by the new developments and pressed forward with her trial anyway. The court wouldn't allow the MMA diagnosis of the second child to be introduced as evidence, and in January 1991 Patricia Stallings was convicted of assault with a deadly weapon and sentenced to life in prison.

Fortunately for Stallings, however, William Sly, chairman of the Department of Biochemistry and Molecular Biology, and James Shoemaker, head of a metabolic screening lab, both at St. Louis University, got interested in her case when they heard about it from a television broadcast. Shoemaker performed his own analysis of Ryan's blood and didn't detect ethylene glycol. He and Sly then contacted Piero Rinaldo, a metabolic disease expert at Yale University School of Medicine whose lab is equipped to diagnose MMA from blood samples.

When Rinaldo analyzed Ryan's blood serum, he found high concentrations of methylmalonic acid, a breakdown product of the branched chain amino acids isoleucine and valine, which accumulates in MMA patients because the enzyme that should convert it to the next product in the metabolic pathway is defective. And particularly telling, he says, the child's blood and urine contained massive amounts of ketones, another metabolic consequence of the disease. Like Shoemaker, he did not find any ethylene glycol in a sample of the baby's bodily fluids. The bottle couldn't be tested, since it had mysteriously disappeared. Rinaldo's analyses convinced him that Ryan had died from MMA, but how to account for the results from two labs, indicating that the boy had ethylene glycol in his blood? Could they both be wrong?

When Rinaldo obtained the lab reports, what he saw was, he says, "scary." One lab said that Ryan Stallings' blood contained ethylene glycol, even though the blood sample analysis did not match the lab's own profile for a known sample containing ethylene glycol. "This was not just a matter of questionable interpretation. The quality of their analysis was unacceptable," Rinaldo says. And the second laboratory? According to Rinaldo, that lab detected an abnormal component in Ryan's blood and just "assumed it was ethylene glycol." Samples from the bottle had produced nothing unusual, says Rinaldo, yet the lab claimed evidence of ethylene glycol in that, too.

Rinaldo presented his findings to the case's prosecutor, George McElroy, who called a press conference the very next day. "I no longer believe the laboratory data," he told reporters. Having concluded that Ryan Stallings had died of MMA after all, McElroy dismissed all charges against Patricia Stallings on September 20, 1991.

*By Michelle Hoffman (1991). *Science* **253,** 931. Copyright 1991 by the American Association for the Advancement of Science.

Isoleucine undergoes transamination, followed by oxidative decarboxylation of the resulting α-keto acid. The remaining five-carbon skeleton derived from isoleucine undergoes further oxidation, yielding acetyl-CoA and propionyl-CoA. In human tissues, **threonine** is also converted to propionyl-CoA. Propionyl-CoA derived from these three amino acids is converted to succinyl-CoA by the pathway described in Chapter 16 for propionate derived from the oxidation of fatty acids of uneven chain length: carboxylation to methylmalonyl-CoA, epimerization of the methylmalonyl-CoA, and finally its conversion to succinyl-CoA by the coenzyme B_{12}–dependent enzyme methylmalonyl-CoA mutase (see Fig. 16–12). A rare genetic defect that results in the absence of methylmalonyl-CoA mutase causes a serious genetic disease called methylmalonyl acidemia (Table 17–2, Box 17–2).

The oxidation of **valine** follows a path similar to those for the other three amino acids in this group (Fig. 17–30). After transamination and decarboxylation of valine, a series of oxidation reactions converts the remaining four carbons into methylmalonyl-CoA, which is transformed into succinyl-CoA. Some parts of the valine and isoleucine degradative pathways closely parallel steps in fatty acid degradation (Chapter 16).

Branched-Chain Amino Acids Are Not Degraded in the Liver

Although much of the catabolism of amino acids occurs in liver, the three amino acids with branched side chains (leucine, isoleucine, and valine) are oxidized as fuels primarily in muscle, adipose, kidney, and brain tissue. These extrahepatic tissues contain a single aminotransferase not present in liver that acts on all three branched-chain amino acids to produce the corresponding α-keto acids (Fig. 17–31).

Figure 17–31 The three branched-chain amino acids (valine, isoleucine, and leucine) share the first two enzymes in their catabolic pathways, which occur in extrahepatic tissues. The second enzyme, the branched-chain α-keto acid dehydrogenase complex, is defective in people with maple syrup urine disease. This dehydrogenase complex is analogous to the pyruvate and α-ketoglutarate dehydrogenases, and requires the same five cofactors (some not shown here).

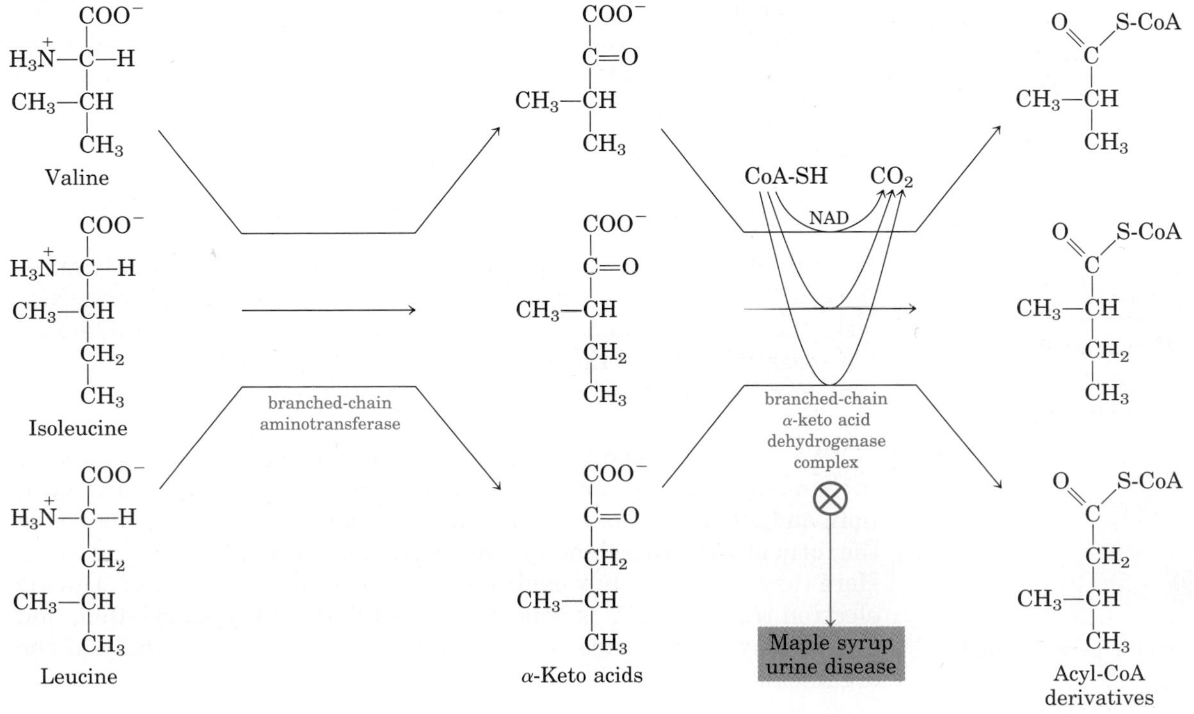

There is a relatively rare human genetic disease in which these three α-keto acids accumulate in the blood and "spill over" into the urine. This condition, which unless treated results in abnormal development of the brain, mental retardation, and death in early infancy, is called **maple syrup urine disease** because of the characteristic odor imparted to the urine by the α-keto acids. It is treated by rigid control over the diet to limit intake of valine, isoleucine, and leucine to the minimum required to permit normal growth. All three α-keto acids derived from the branched-chain amino acids are acted on by a single enzyme, which is defective in patients with maple syrup urine disease. This explains why all three α-keto acids accumulate along with the corresponding amino acids (especially leucine) in affected individuals. This enzyme, the **branched-chain α-keto acid dehydrogenase complex,** catalyzes oxidative decarboxylation of each of the three α-keto acids, releasing the carboxyl group as CO_2 and producing the acyl-CoA derivative (Fig. 17–31).

This decarboxylation reaction is formally analogous to two others encountered in Chapter 15: the oxidation of pyruvate to acetyl-CoA by the pyruvate dehydrogenase complex (p. 450) and the oxidation of α-ketoglutarate to succinyl-CoA in the citric acid cycle by the α-ketoglutarate dehydrogenase complex (p. 455). In fact, all three enzymes are closely homologous in structure, and the reaction mechanism is essentially the same for all. Five cofactors (thiamine pyrophosphate, FAD, NAD, lipoate, and coenzyme A) participate, and the three proteins in each of the complexes catalyze homologous reactions. This is clearly a case in which enzymatic machinery that evolved to catalyze one reaction was "borrowed" by gene duplication and further evolved to catalyze similar reactions in other pathways.

The branched-chain α-keto acid dehydrogenase complex of the rat is regulated by covalent modification in response to the content of branched-chain amino acids in the diet. When there is little or no excess dietary intake of branched-chain amino acids, the enzyme complex is phosphorylated and thereby inactivated by a protein kinase. Addition of excess branched-chain amino acids to the diet results in dephosphorylation and consequent activation of the enzyme. Recall that the pyruvate dehydrogenase complex is subject to similar regulation by phosphorylation and dephosphorylation (p. 468).

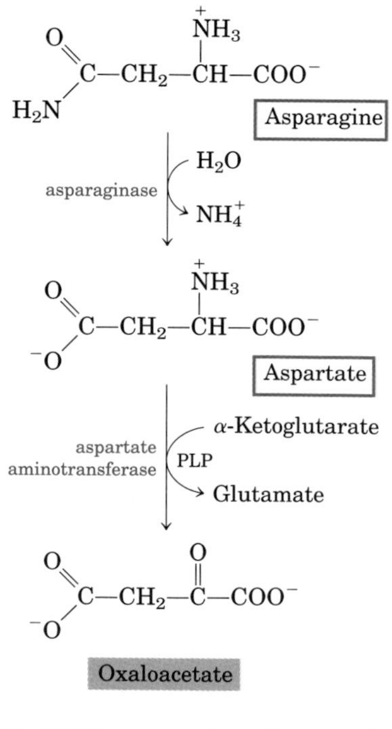

Asparagine and Aspartate Are Degraded to Oxaloacetate

The carbon skeletons of **asparagine** and **aspartate** ultimately enter the citric acid cycle via oxaloacetate. The enzyme **asparaginase** catalyzes the hydrolysis of asparagine to yield aspartate, which undergoes a transamination reaction with α-ketoglutarate to yield glutamate and oxaloacetate (Fig. 17–32). The latter enters the citric acid cycle.

We have now seen how the 20 different amino acids, after loss of their nitrogen atoms, are degraded by dehydrogenation, decarboxylation, and other reactions to yield portions of their carbon backbones in the form of five central metabolites that can enter the citric acid cycle. Here they are completely oxidized to carbon dioxide and water. During electron transfer, ATP is generated by oxidative phosphorylation, and in this way amino acids contribute to the total energy supply of the organism.

Figure 17–32 The conversion of asparagine and aspartate to oxaloacetate.

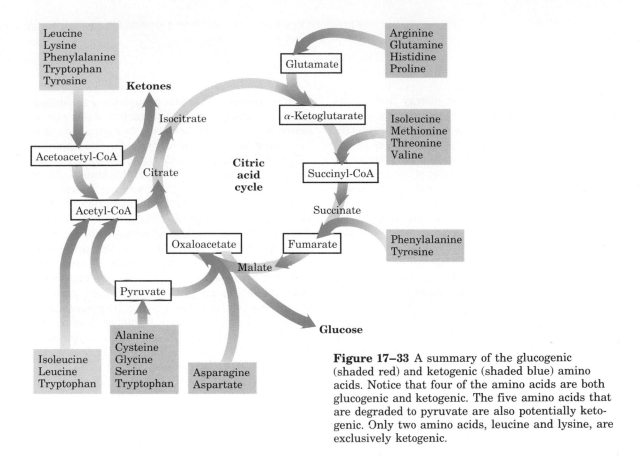

Figure 17–33 A summary of the glucogenic (shaded red) and ketogenic (shaded blue) amino acids. Notice that four of the amino acids are both glucogenic and ketogenic. The five amino acids that are degraded to pyruvate are also potentially ketogenic. Only two amino acids, leucine and lysine, are exclusively ketogenic.

Some Amino Acids Can Be Converted to Glucose, Others to Ketones

Some carbon atoms from six of the amino acids (those that are degraded to acetoacetyl-CoA and/or acetyl-CoA: tryptophan, phenylalanine, tyrosine, isoleucine, leucine, and lysine) can yield ketones in the liver, by conversion of acetoacetyl-CoA into acetone and β-hydroxybutyrate (see Fig. 16–16). These amino acids are **ketogenic** (Fig. 17–33). Their ability to form ketones is particularly evident in untreated diabetes mellitus, in which large amounts of ketones are produced by the liver, not only from fatty acids but from the ketogenic amino acids. Degradation of leucine, an exclusively ketogenic amino acid that is very common in proteins, makes a substantial contribution to ketosis during starvation.

The amino acids that can be converted into pyruvate, α-ketoglutarate, succinyl-CoA, fumarate, and oxaloacetate can be converted into glucose and glycogen by pathways described in Chapter 19. They are called **glucogenic** amino acids. The division between ketogenic and glucogenic amino acids is not sharp; four amino acids (tryptophan, phenylalanine, tyrosine, and isoleucine) are both ketogenic and glucogenic. Some of the amino acids that can be converted into pyruvate, particularly alanine, cysteine, and serine, can also potentially form acetoacetate via acetyl-CoA, especially in severe starvation and untreated diabetes mellitus (Chapter 16).

Summary

A small fraction of oxidative energy in humans comes from the catabolism of amino acids. Amino acids are derived from the normal breakdown (recycling) of cellular proteins, degradation of ingested proteins, or breakdown of body proteins in lieu of other fuel sources during starvation or in untreated diabetes mellitus. Ingested proteins are degraded in the stomach and small intestine by proteases. Most proteases are initially synthesized as inactive zymogens, which are activated in the stomach or intestine by proteolytic removal of parts of their polypeptide chains. An early step in the catabolism of amino acids is the separation of the amino group from the carbon skeleton. In most cases, the amino group is transferred to α-ketoglutarate to form glutamate. This type of reaction is called a transamination and requires the coenzyme pyridoxal phosphate. Glutamate is transported to liver mitochondria, where an amino group is liberated as ammonia (NH_4^+) by the enzyme glutamate dehydrogenase. Ammonia formed in other tissues is transported to liver mitochondria as the amide nitrogen of glutamine or as the amino group of alanine. Most of the alanine is generated in muscle and transported in the blood to the liver. After deamination the resulting pyruvate is converted to glucose, which is transported back to muscle as part of the glucose–alanine cycle.

Ammonia is highly toxic to animal tissues. Ammonotelic animals (bony fishes, tadpoles) excrete amino nitrogen from their gills as ammonia. Ureotelic animals (adult terrestrial amphibians and all mammals) excrete amino nitrogen as urea, formed in the liver by the urea cycle. Arginine is the immediate precursor of urea. Arginase hydrolyzes arginine to yield urea and ornithine, and arginine is resynthesized in the urea cycle. Ornithine is converted to citrulline at the expense of carbamoyl phosphate, and an amino group is transferred to citrulline from aspartate, re-forming arginine. Ornithine is regenerated in each turn of the cycle. Several of the intermediates and byproducts of the urea cycle are also intermediates in the citric acid cycle, and the two cycles are thus interconnected. The activity of the urea cycle is regulated at the levels of enzyme synthesis and allosteric regulation of the enzyme that forms carbamoyl phosphate. Uricotelic animals (birds and reptiles) excrete amino nitrogen in semisolid form as uric acid, a derivative of purine. The mode of nitrogen excretion is determined by habitat. The formation of the nontoxic urea and of solid uric acid has a high ATP cost. Genetic defects in enzymes of the urea cycle can be compensated for by dietary regulation.

After removal of amino groups by transamination to α-ketoglutarate, the carbon skeletons of amino acids undergo oxidation to compounds that can enter the citric acid cycle for oxidation to CO_2 and H_2O. In these pathways, the cofactors tetrahydrofolate and S-adenosylmethionine facilitate one-carbon transfer reactions, and the cofactor tetrahydrobiopterin facilitates the oxidation of phenylalanine catalyzed by phenylalanine hydroxylase. There are five intermediates through which carbon skeletons of amino acids enter the citric acid cycle: (1) acetyl-CoA, (2) α-ketoglutarate, (3) succinyl-CoA, (4) fumarate, and (5) oxaloacetate. The amino acids producing acetyl-CoA are divided into two groups. Alanine, cysteine, glycine, tryptophan, and serine yield acetyl-CoA via pyruvate; leucine, lysine, phenylalanine, tyrosine, and tryptophan yield acetyl-CoA via acetoacetyl-CoA. Isoleucine, leucine, and tryptophan also form acetyl-CoA directly. Proline, histidine, arginine, glutamine, and glutamate enter the citric acid cycle via α-ketoglutarate; threonine, methionine, isoleucine, and valine enter via succinyl-CoA; four carbon atoms of phenylalanine and tyrosine enter via fumarate; and asparagine and aspartate enter via oxaloacetate. The branched-chain amino acids (leucine, isoleucine, and valine), unlike the other amino acids, are degraded in extrahepatic tissues. A number of serious human diseases can be traced to genetic defects in specific enzymes in the pathways of amino acid catabolism.

Some amino acids can be converted to ketone bodies; some can be converted to glucose.

Further Reading

General

Bender, D.A. (1985) *Amino Acid Metabolism,* 2nd edn, Wiley-Interscience, Inc., New York.

Campbell, J.W. (1991) Excretory nitrogen metabolism. In *Environmental and Metabolic Animal Physiology,* 4th edn (Prosser, C.L., ed), pp. 277–324, John Wiley & Sons, Inc., New York.

Mazelis, M. (1980) Amino acid catabolism. In *The Biochemistry of Plants: A Comprehensive Treatise* (Stumpf, P.K. & Conn, E.E., eds), Vol. 5: *Amino Acids and Derivatives* (Miflin, B.J., ed), pp. 541–567, Academic Press, Inc., New York.
A discussion of the various fates of amino acids in plants.

Mehler, A. (1992) Amino acid metabolism I: general pathways; Amino acid metabolism II: metabolism of the individual amino acids. In *Textbook of Biochemistry with Clinical Correlations,* 3rd edn (Devlin, T.M., ed), pp 475–528, Wiley-Liss, New York.

Powers-Lee, S.G. & Meister, A. (1988) Urea synthesis and ammonia metabolism. In *The Liver: Biology and Pathobiology,* 2nd edn (Arias, I.M., Jakoby, W.B., Popper, H., Schachter, D., & Shafritz, D.A., eds), pp. 317–329, Raven Press, New York.

Walsh, C. (1979) *Enzymatic Reaction Mechanisms,* W.H. Freeman and Company, San Francisco.
A good source for in-depth discussion of the classes of enzymatic reaction mechanisms described in the chapter.

Amino Group Metabolism

Christen, P. & Metzler, D.E. (1985) *Transaminases,* Wiley-Interscience, Inc., New York.

The Urea Cycle

Holmes, F.L. (1980) Hans Krebs and the discovery of the ornithine cycle. *Fed. Proc.* **39,** 216–225.
A medical historian reconstructs the events leading to the discovery of the urea cycle.

Kirsch, J.F., Eichele, G., Ford, G.C., Vincent, M.G., Jansonius, J.N., Gehring, H., & Christen, P. (1984) Mechanism of action of aspartate aminotransferase proposed on the basis of its spatial structure. *J. Mol. Biol.* **174,** 497–525.

Disorders of Amino Acid Degradation

Ledley, F.D., Levy, H.L., & Woo, S.L.C. (1986) Molecular analysis of the inheritance of phenylketonuria and mild hyperphenylalaninemia in families with both disorders. *N. Engl. J Med.* **314,** 1276 1280.

Nyhan, W.L. (1984) *Abnormalities in Amino Acid Metabolism in Clinical Medicine,* Appleton-Century-Crofts, Norwalk, CT.

Scriver, C.R., Kaufman, S., & Woo, S.L.C. (1988) Mendelian hyperphenylalaninemia. *Annu. Rev. Genet.* **22,** 301–321.

Stanbury, J.B., Wyngaarden, J.B., Fredrickson, D.S, Goldstein, J.L., & Brown, M.S. (eds) (1983) *The Metabolic Basis of Inherited Disease,* 5th edn, Part 3: *Disorders of Amino Acid Metabolism,* McGraw-Hill Book Company, New York.

Problems

1. *Products of Amino Acid Transamination* Draw the structure and give the name of the α-keto acid resulting when the following amino acids undergo transamination with α-ketoglutarate:

(a) Aspartate   (c) Alanine
(b) Glutamate   (d) Phenylalanine

2. *Measurement of the Alanine Aminotransferase Reaction Rate* The activity (reaction rate) of alanine aminotransferase is usually measured by including an excess of pure lactate dehydrogenase and NADH in the reaction system. The rate of alanine disappearance is equal to the rate of NADH disappearance measured spectrophotometrically. Explain how this assay works.

3. *Distribution of Amino Nitrogen* If your diet is rich in alanine but deficient in aspartate, will you show signs of aspartate deficiency? Explain.

4. *A Genetic Defect in Amino Acid Metabolism: A Case History* A two-year-old child was brought to the hospital. His mother indicated that he vomited frequently, especially after feedings. The child's weight and physical development were below normal. His hair, although dark, contained patches of white. A urine sample treated with ferric chloride ($FeCl_3$) gave a green color characteristic of the presence of phenylpyruvate. Quantitative analysis of urine samples gave the results shown in the table below.

Substance	Concentration in patient's urine (mM)	Normal concentration in urine (mM)
Phenylalanine	7.0	0.01
Phenylpyruvate	4.8	0
Phenyllactate	10.3	0

(a) Suggest which enzyme might be deficient. Propose a treatment for this condition.

(b) Why does phenylalanine appear in the urine in large amounts?

(c) What is the source of phenylpyruvate and phenyllactate? Why does this pathway (normally not functional) come into play when the concentration of phenylalanine rises?

(d) Why does the patient's hair contain patches of white?

5. *Role of Cobalamin in Amino Acid Catabolism* Pernicious anemia is caused by impaired absorption of vitamin B_{12}. What is the effect of this impairment on the catabolism of amino acids? Are all amino acids affected equally? (Hint: See Box 16–2.)

6. *Lactate versus Alanine as Metabolic Fuel: The Cost of Nitrogen Removal* The three carbons in lactate and alanine have identical states of oxidation, and animals can use either carbon source as a metabolic fuel. Compare the net ATP yield (moles of ATP per mole of substrate) for the complete oxidation (to CO_2 and H_2O) of lactate versus alanine when the cost of nitrogen excretion as urea is included.

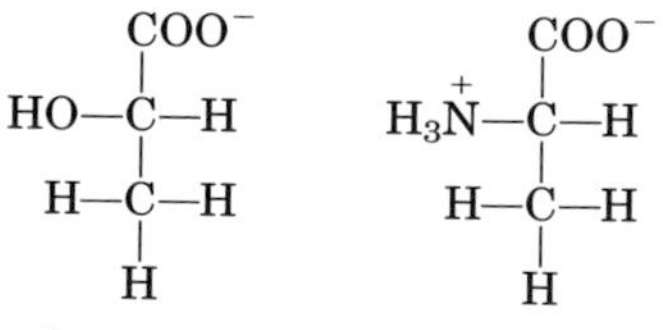

Lactate   Alanine

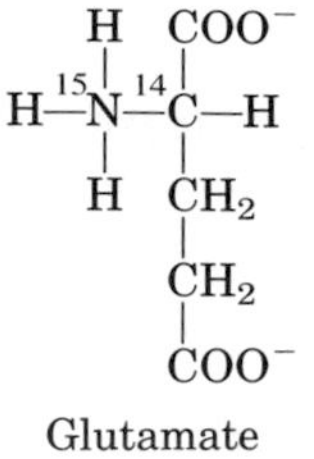

Glutamate

7. *Pathway of Carbon and Nitrogen in Glutamate Metabolism* When [2-^{14}C,^{15}N]glutamate undergoes oxidative degradation in the liver of a rat, in which atoms of the following metabolites will each isotope be found?

(a) Urea
(b) Succinate
(c) Arginine
(d) Citrulline
(e) Ornithine
(f) Aspartate

8. *Chemical Strategy of Isoleucine Catabolism* Isoleucine is degraded by a series of six steps to propionyl-CoA and acetyl-CoA:

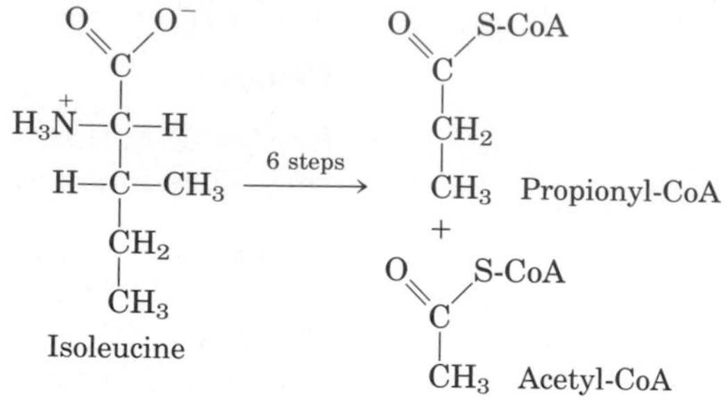

(a) The chemical process of isoleucine degradation consists of strategies analogous to those found in the citric acid cycle and the β oxidation of fatty acids. The intermediates involved in isoleucine degradation (I to V) shown below are not in the proper order. Use your knowledge and understanding of the citric acid cycle and β-oxidation pathway to arrange the intermediates into the proper metabolic sequence for isoleucine degradation.

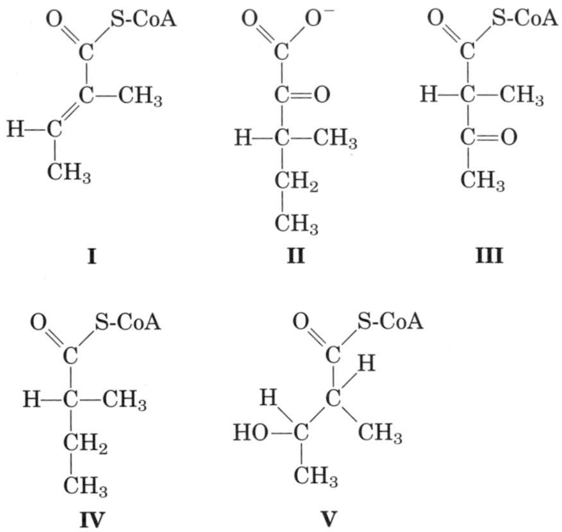

(b) For each step proposed above, describe the chemical process, provide an analogous example from the citric acid cycle or β-oxidation pathway, and indicate any necessary cofactors.

9. *Ammonia Intoxication Resulting from an Arginine-Deficient Diet* In a study conducted some years ago, cats were fasted overnight then given a single meal complete in amino acids but without arginine. Within 2 h, blood ammonia levels increased from a normal level of 18 μg/L to 140 μg/L, and the cats showed the clinical symptoms of ammonia toxicity. A control group fed a complete amino acid diet or an amino acid diet in which arginine was replaced by ornithine showed no unusual clinical symptoms.

(a) What was the role of fasting in the experiment?

(b) What caused the ammonia levels to rise? Why did the absence of arginine lead to ammonia toxicity? Is arginine an essential amino acid in cats? Why or why not?

(c) Why can ornithine be substituted for arginine?

10. *Oxidation of Glutamate* Write a series of balanced equations and the net reaction describing the oxidation of 2 mol of glutamate to 2 mol of α-ketoglutarate plus 1 mol of excreted urea.

11. *The Role of Pyridoxal Phosphate in Glycine Metabolism* The enzyme serine hydroxymethyl transferase (Fig. 17–23) requires a pyridoxal phosphate cofactor. Propose a mechanism for this reaction that explains the requirement. (Hint: See Fig. 17–7.)

12. *Parallel Pathways for Amino Acid and Fatty Acid Degradation* The carbon skeleton of leucine is degraded by a series of reactions (at right) closely analogous to those of the citric acid cycle and fatty acid oxidation. For each reaction, indicate its type, provide an analogous example from the citric acid cycle or β-oxidation pathways, and indicate any necessary cofactors.

13. *Transamination and the Urea Cycle* Aspartate aminotransferase has the highest activity of all the mammalian liver aminotransferases. Why?

14. *The Case against the Liquid Protein Diet* A weight-reducing diet heavily promoted some years ago required the daily intake of "liquid protein" (soup of hydrolyzed gelatin), water, and an assortment of vitamins. All other food and drink were to be avoided. People on this diet typically lost 10 to 14 lb in the first week.

(a) Opponents argued that the weight loss was almost entirely water and would be regained almost immediately when a normal diet was resumed. What is the biochemical basis for the opponents' argument?

(b) A number of people on this diet died. What are some of the dangers inherent in the diet and how can they lead to death?

15. *Alanine and Glutamine in the Blood* Blood plasma contains all the amino acids required for the synthesis of body proteins, but they are not present in equal concentrations. Two amino acids, alanine and glutamine, are present in much higher concentrations in normal human blood plasma than any of the other amino acids. Suggest possible reasons for their abundance.

Problem 12

Oxidative Phosphorylation and Photophosphorylation

Oxidative phosphorylation (ATP synthesis driven by electron transfer to oxygen) and photophosphorylation (ATP synthesis driven by light) are arguably the two most important energy transductions in the biosphere. These two processes together account for most of the ATP synthesized by aerobic organisms. Oxidative phosphorylation is the culmination of energy-yielding metabolism in aerobic organisms. All the enzymatic steps in the oxidative degradation of carbohydrates, fats, and amino acids in aerobic cells converge at this final stage of cellular respiration, in which electrons flow from catabolic intermediates to O_2, yielding energy for the generation of ATP from ADP and P_i. Photophosphorylation is the means by which photosynthetic organisms capture the energy of sunlight, the ultimate source of energy in the biosphere.

In eukaryotes, oxidative phosphorylation occurs in mitochondria; photophosphorylation occurs in chloroplasts. Oxidative phosphorylation involves the *reduction* of O_2 to H_2O with electrons donated by NADH and $FADH_2$, and occurs equally well in light or darkness. Photophosphorylation involves the *oxidation* of H_2O to O_2, with $NADP^+$ as electron acceptor, and it is absolutely dependent on light. These two highly efficient energy-conserving processes occur by fundamentally similar mechanisms.

Our current understanding of ATP synthesis in mitochondria and chloroplasts is based on a hypothesis, introduced by Peter Mitchell in 1961, in which transmembrane differences in proton concentration are central to energy transduction. This **chemiosmotic theory** has been accepted as one of the great unifying principles of twentieth century biology. It provides insight into the processes of oxidative phosphorylation and photophosphorylation, and into such apparently disparate energy transductions as active transport across membranes and the motion of bacterial flagella. Many biochemical details of these processes remain unsolved, but the chemiosmotic model described in this chapter provides the intellectual framework for investigating those details.

There are three fundamental similarities between oxidative phosphorylation and photophosphorylation. (1) Both processes involve the flow of electrons through a chain of redox intermediates, membrane-bound carriers that include quinones, cytochromes, and iron–sulfur proteins. (2) The free energy made available by this "downhill" (exergonic) electron flow is coupled to the "uphill" transport of protons across a proton-impermeable membrane, conserving some of the free energy of oxidation of metabolic fuels as a transmembrane electrochemical potential (p. 287). (3) The transmembrane flow of protons down their concentration gradient through specific protein channels provides the free energy for synthesis of ATP.

In this chapter we first consider the process of oxidative phosphorylation. We begin with descriptions of the components of the electron transfer chain in mitochondria, the sequence in which these carriers act, and their organization into large functional complexes in the mitochondrial inner membrane. We then look at the chemiosmotic mechanism by which electron transfer is used to drive ATP synthesis, and the means by which this process is regulated in coordination with other energy-yielding pathways. The evolutionary origins of mitochondria, touched upon in Chapter 2, are further considered.

With this understanding of mitochondrial oxidative phosphorylation, we turn to photophosphorylation. Light-absorbing pigments in the membranes of chloroplasts and photosynthetic bacteria transfer the energy of absorbed light to reaction centers where electron flow is initiated. Electron flow occurs through a series of carriers and, as in mitochondria, this flow drives ATP synthesis by the chemiosmotic mechanism.

Albert L. Lehninger
1917–1986

Mitochondrial Electron Flow

The discovery in 1948 by Eugene Kennedy and Albert Lehninger that mitochondria are the site of oxidative phosphorylation in eukaryotes marked the beginning of the modern phase of studies of biological energy transductions.

Mitochondria are organelles of eukaryotic cells, believed to have arisen during evolution when aerobic bacteria capable of oxidative phosphorylation took up symbiotic residence within a primitive, anaerobic, eukaryotic host cell (see Fig. 2–17). Mitochondria, like gram-negative bacteria, have two membranes (Fig. 18–1). The outer mitochondrial membrane is readily permeable to small molecules and ions; transmembrane channels composed of the protein porin allow most molecules of molecular weight less than 5,000 to pass easily. The inner membrane is impermeable to most small molecules and ions, including protons (H^+); the only species that cross the inner membrane are those for which there are specific transporter proteins. The inner membrane bears the components of the respiratory chain and the enzyme complex responsible for ATP synthesis.

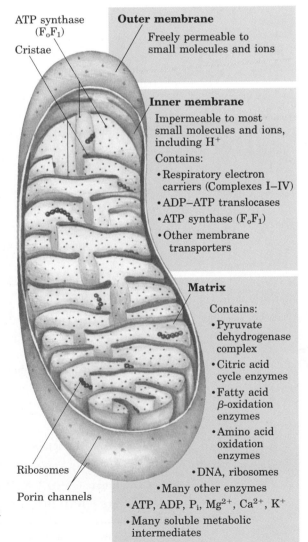

Figure 18–1 Biochemical anatomy of a mitochondrion. The convolutions (cristae) of the inner membrane give it a very large surface area. The inner membrane of a single liver mitochondrion may have over 10,000 sets of electron transfer systems (respiratory chains) and ATP synthase molecules, distributed over the whole surface of the inner membrane. Heart mitochondria, which have very profuse cristae and thus a much larger area of inner membrane, contain over three times as many sets of electron transfer systems as liver mitochondria. The mitochondrial pool of coenzymes and intermediates is functionally separate from the cytosolic pool. The mitochondria of invertebrates, plants, and microbial eukaryotes are similar to those shown here, although there is much variation in size, shape, and degree of convolution of the inner membrane. See Chapter 2 for other details of mitochondrial structure.

ATP synthase (F_oF_1)

Cristae

Outer membrane

Freely permeable to small molecules and ions

Inner membrane

Impermeable to most small molecules and ions, including H^+

Contains:
• Respiratory electron carriers (Complexes I–IV)
• ADP–ATP translocases
• ATP synthase (F_oF_1)
• Other membrane transporters

Matrix

Contains:
• Pyruvate dehydrogenase complex
• Citric acid cycle enzymes
• Fatty acid β-oxidation enzymes
• Amino acid oxidation enzymes
• DNA, ribosomes
• Many other enzymes
• ATP, ADP, P_i, Mg^{2+}, Ca^{2+}, K^+
• Many soluble metabolic intermediates

Ribosomes

Porin channels

Recall that the mitochondrial matrix, the space enclosed by the inner membrane, contains the pyruvate dehydrogenase complex and the enzymes of the citric acid cycle, the fatty acid β-oxidation pathway, and the pathways of amino acid oxidation—all of the pathways of fuel oxidation except glycolysis, which occurs in the cytosol. Because the inner membrane is selectively permeable, it segregates the intermediates and enzymes of cytosolic metabolic pathways from those of metabolic processes occurring in the matrix. Specific transporters carry pyruvate, fatty acids, and amino acids or their α-keto derivatives into the matrix for access to the machinery of the citric acid cycle. Similarly, ADP and P_i are specifically transported into the matrix as the newly synthesized ATP is transported out.

We will discuss here in some detail the electron-carrying components of the mitochondrial respiratory chain.

Electrons Are Funneled into Universal Electron Carriers

Most of the electrons entering the mitochondrial respiratory chain arise from the action of dehydrogenases that collect electrons from the oxidative reactions of the pyruvate dehydrogenase complex, the citric acid cycle, the β-oxidation pathway, and the oxidative steps of amino acid catabolism and funnel them as electron pairs into the respiratory chain. These dehydrogenases use either pyridine nucleotides (NAD or NADP; Table 18–1) or flavin nucleotides (FMN or FAD) as electron acceptors.

All of the pyridine nucleotide–linked dehydrogenases catalyze reversible reactions of the following general types:

$$\text{Reduced substrate} + \text{NAD}^+ \rightleftharpoons \text{oxidized substrate} + \text{NADH} + \text{H}^+$$

$$\text{Reduced substrate} + \text{NADP}^+ \rightleftharpoons \text{oxidized substrate} + \text{NADPH} + \text{H}^+$$

Most dehydrogenases are specific for NAD^+ as electron acceptor (Table 18–1), but some, such as glucose-6-phosphate dehydrogenase

---

Table 18–1 Some important reactions catalyzed by NAD(P)H-linked dehydrogenases

Reaction[*]	Location[†]
NAD-linked	
α-Ketoglutarate + CoA + NAD$^+$ $\rightleftharpoons$ succinyl-CoA + CO$_2$ + NADH + H$^+$	M
L-Malate + NAD$^+$ $\rightleftharpoons$ oxaloacetate + NADH + H$^+$	M and C
Pyruvate + CoA + NAD$^+$ $\rightleftharpoons$ acetyl-CoA + CO$_2$ + NADH + H$^+$	M
Glyceraldehyde-3-phosphate + P$_i$ + NAD$^+$ $\rightleftharpoons$ 1,3-bisphosphoglycerate + NADH + H$^+$	C
Lactate + NAD$^+$ $\rightleftharpoons$ pyruvate + NADH + H$^+$	C
β-Hydroxyacyl-CoA + NAD$^+$ $\rightleftharpoons$ β-ketoacyl-CoA + NADH + H$^+$	M
NADP-linked	
Glucose-6-phosphate + NADP$^+$ $\rightleftharpoons$ 6-phosphogluconate + NADPH + H$^+$	C
NAD- or NADP-linked	
L-Glutamate + H$_2$O + NAD(P)$^+$ $\rightleftharpoons$ α-ketoglutarate + NH$_4^+$ + NAD(P)H	M
Isocitrate + NAD(P)$^+$ $\rightleftharpoons$ α-ketoglutarate + CO$_2$ + NAD(P)H + H$^+$	M and C

[*] All of these reactions and their enzymes have been discussed in Chapters 14 through 17.

[†] M designates mitochondria; C, cytosol.

(see Fig. 14–22), require $NADP^+$. A few, such as glutamate dehydrogenase, can react with either NAD^+ or $NADP^+$. Some pyridine nucleotide–linked dehydrogenases are located in the cytosol, some in the mitochondria, and still others have two isozymes, one mitochondrial and the other cytosolic.

As was described in Chapter 13, the NAD-linked dehydrogenases remove two hydrogen atoms from their substrates. One of these is transferred as a hydride ion ($:H^-$) to the NAD^+; the other appears as H^+ in the medium (see Fig. 13–16). NAD^+ can also collect reducing equivalents from substrates acted upon by NADP-linked dehydrogenases. This is made possible by **pyridine nucleotide transhydrogenase,** which catalyzes the reaction

$$NADPH + NAD^+ \rightleftharpoons NADP^+ + NADH$$

NADH and NADPH are water-soluble electron carriers that associate *reversibly* with dehydrogenases. NADH acts as a diffusible carrier, transporting the electrons derived from catabolic reactions to their point of entry into the respiratory chain, the NADH dehydrogenase complex described below. NADPH is a diffusible carrier that supplies electrons to anabolic reactions.

Flavoproteins contain a very tightly, sometimes covalently, bound flavin nucleotide, either FMN or FAD (see Fig. 13–17). The oxidized flavin nucleotide can accept either one electron (yielding the semiquinone form) or two (yielding $FADH_2$ or $FMNH_2$). The standard reduction potential of a flavin nucleotide, unlike that of pyridine nucleotides, depends on the protein with which it is associated. Local interactions with functional groups in the protein distort the electron orbitals in the flavin ring, changing the relative stabilities of oxidized and reduced forms. The relevant standard reduction potential is therefore that of the particular flavoprotein, not that of isolated FAD or FMN, and the flavin nucleotide should be considered part of the flavoprotein's active site, not as a reactant or product in the electron transfer reaction. Because flavoproteins can participate in either one- or two-electron transfers, they can serve as intermediates between reactions in which two electrons are donated (as in dehydrogenations) and those in which only one electron is accepted (as in the reduction of a quinone to a hydroquinone, described below).

Several Types of Electron Carriers Act in the Respiratory Chain

The mitochondrial respiratory chain consists of a series of electron carriers, most of which are integral membrane proteins, with prosthetic groups capable of accepting and donating either one or two electrons. Each component of the chain can accept electrons from the preceding carrier and transfer them to the following one, in a specific sequence. We noted earlier (Chapter 13) that there are four types of electron transfers in biological systems: (1) direct transfer of electrons, as in the reduction of Fe^{3+} to Fe^{2+}; (2) transfer as a hydrogen atom ($H^+ + e^-$); (3) transfer as a hydride ion ($:H^-$), which bears two electrons; and (4) direct combination of an organic reductant with oxygen. Each of the first three types occurs in the respiratory chain. Some of the reactions in this sequence are one-electron transfers, and others involve the transfer of pairs of electrons. Whatever the form, the term reducing equivalent is used to designate a single electron equivalent that is transferred in an oxidation–reduction reaction.

Ubiquinone (UQ)
(fully oxidized)

$H^+ + e^-$

Semiquinone radical
(UQH$^\bullet$)

$H^+ + e^-$

Ubiquinol (UQH$_2$)
(fully reduced)

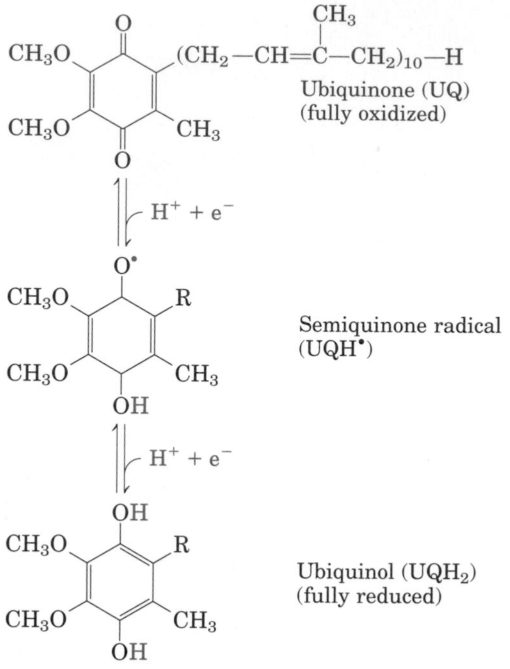

Figure 18–2 Ubiquinone (UQ, or coenzyme Q), a respiratory chain electron carrier. Complete reduction of ubiquinone requires two electrons and two protons, and occurs in two steps through the semiquinone radical intermediate. The same chemistry is involved in the reduction of plastoquinone (a photosynthetic electron carrier) in chloroplasts and menaquinone (a respiratory chain carrier) in bacteria.

In addition to NAD and the flavoproteins described above, three other types of electron-carrying groups function in the respiratory chain: a hydrophobic benzoquinone (ubiquinone) and two different types of iron-containing proteins (cytochromes and iron–sulfur proteins).

Ubiquinone (also called **coenzyme Q,** or simply UQ) is a fat-soluble benzoquinone with a very long isoprenoid side chain (Fig. 18–2). The closely related compounds plastoquinone (found in plant chloroplasts) and menaquinone (found in bacteria) play roles analogous to that of ubiquinone; all carry electrons in membrane-associated electron transfer chains. Ubiquinone can accept one electron to become the semiquinone radical UQH$^\bullet$ or two electrons to form ubiquinol (UQH$_2$) (Fig. 18–2) and, like flavoprotein carriers, it is therefore able to act at the junction between a two-electron donor and a one-electron acceptor. Because ubiquinone is both small and hydrophobic, it is freely diffusible within the lipid bilayer of the inner mitochondrial membrane, and can shuttle reducing equivalents between other, less mobile, electron carriers in the membrane.

The **cytochromes** are iron-containing electron transfer proteins of the mitochondrial inner membrane, the thylakoid membranes of chloroplasts, and the plasma membrane of bacteria. The characteristic strong colors of cytochromes are produced by the heme prosthetic group (Fig. 18–3).

There are three classes of cytochromes distinguished by differences in their light-absorption spectra and designated a, b, and c. Each type of cytochrome in its reduced (Fe^{2+}) state has three absorption bands in the visible range (Fig. 18–4). The longest-wavelength band (the α band) is near 600 nm in type a cytochromes, near 560 nm in type b, and near 550 in type c. To distinguish among closely related cyto-

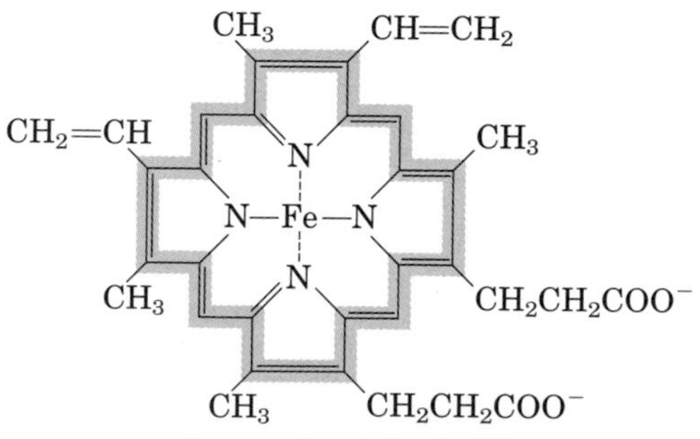

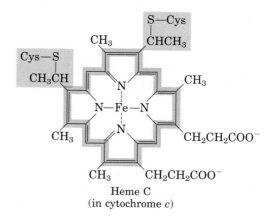

Iron protoporphyrin IX
(in b-type cytochromes)

Heme C
(in cytochrome c)

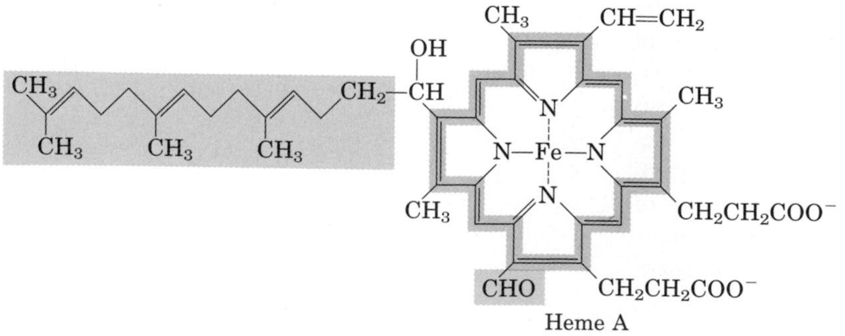

Heme A
(in a-type cytochromes)

Figure 18–3 The prosthetic groups of cytochromes have four five-membered, nitrogen-containing rings in a cyclic structure called a porphyrin. The four nitrogen atoms are coordinated with a central Fe ion that can be either Fe^{2+} or Fe^{3+}. Iron protoporphyrin IX is found in b-type cytochromes and in hemoglobin and myoglobin (see Fig. 7–18). Heme C is bound covalently to the protein of cytochrome c through thioether bonds to two Cys residues. Heme A, found in the a-type cytochromes, has a long isoprenoid tail attached to one of the five-membered rings. The conjugated double bond system (shaded in red) of the porphyrin ring accounts for the absorption of visible light by these hemes.

chromes of one type, their exact absorption maximum is sometimes used in their names, as in cytochrome b_{562} (the three-dimensional structure of this protein was shown in Fig. 7–25).

The heme groups of a and b cytochromes are tightly, but not covalently, bound to their associated proteins; heme groups of c-type cytochromes are covalently attached (through Cys residues; Fig. 18–3). As with the flavoproteins, the standard reduction potential of the iron atom in the heme of a cytochrome depends heavily on its interaction with protein side chains and is therefore different for each cytochrome. The cytochromes of type a and b and some of type c are integral membrane proteins. One striking exception is cytochrome c of mitochondria, a soluble protein that associates through electrostatic interactions with the outer surface of the mitochondrial inner membrane.

The ubiquitous occurrence of cytochrome c in aerobic organisms, together with its small size (104 amino acid residues), has allowed the determination of its amino acid sequence in many species from every phylum. The degree of sequence similarity in cytochrome c has been used as a measure of the evolutionary distances that separate species (see Fig. 6–16).

In some iron-containing electron transfer proteins, the **iron–sulfur proteins,** the iron is present not in heme (as it is in cytochromes) but in association with inorganic sulfur atoms and/or the sulfur atoms of Cys residues in the protein. These iron–sulfur (Fe–S) centers range from simple structures with a single Fe atom coordinated to four Cys residues through the sulfur in their side chains to more complex Fe–S centers with two or four Fe atoms (Fig. 18–5).

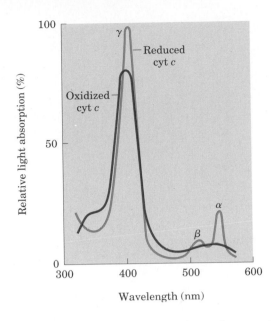

Figure 18–4 Absorption spectra of cytochrome c in its oxidized (red) and reduced (blue) forms. Also labeled are the characteristic α, β, and γ bands of the reduced form.

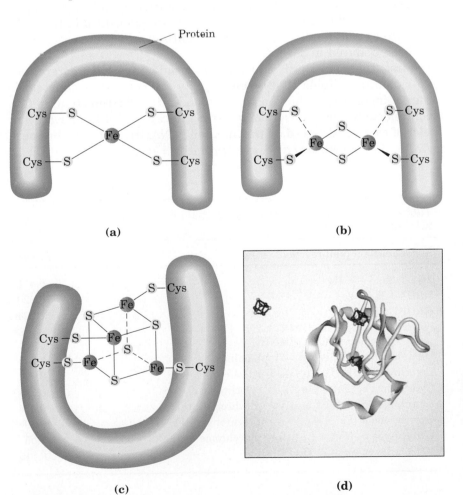

(a)

(b)

(c)

(d)

Figure 18–5 The Fe–S centers of iron–sulfur proteins may be as simple as in **(a)**, with a single Fe ion surrounded by S atoms of four Cys residues. Other centers include both inorganic and Cys S atoms, as in the 2Fe–2S **(b)** or 4Fe–4S centers **(c)**. The ferredoxin in **(d)**, from the cyanobacterium *Spirulina platensis*, has two 2Fe–2S centers. (Note that only the inorganic S atoms are counted in these designations. For example, in the 2Fe–2S center **(b)**, each Fe ion is actually surrounded by four S atoms.) The exact standard reduction potential of the iron in these centers depends on the type of center and the details of its interaction with its associated protein.

These proteins all participate in one-electron transfers, in which one of the Fe atoms is oxidized or reduced. With some exceptions, iron–sulfur proteins have low standard reduction potentials—they are good electron donors.

Many of the iron–sulfur proteins absorb visible light in the region of 400 to 460 nm, and in this region absorption decreases by about 50% when the proteins are reduced. Although the visible absorption spectrum can be used as a measure of oxidation state for purified proteins, this is not possible in more complex systems such as mitochondrial membranes, where the absorption is masked by the many other pigments present. However, the Fe atom(s) in iron–sulfur proteins are paramagnetic (i.e., they possess electrons with unpaired spins) and can therefore be detected by electron paramagnetic resonance (epr) spectroscopy. The epr signal, which is only observable at temperatures well below 0 °C, is the best measure of the presence, and the oxidation state, of a given iron–sulfur protein. The difficulty of detecting and identifying iron–sulfur proteins at room temperature has seriously complicated studies of their function in electron transfer, but it is clear that a number of iron–sulfur proteins play crucial roles in mitochondria (and chloroplasts).

Mitochondrial Electron Carriers Function in Serially Ordered Complexes

In the overall reaction catalyzed by the mitochondrial respiratory chain, electrons move from NADH, succinate, or some other primary electron donor through flavoproteins, ubiquinone, iron–sulfur proteins, and cytochromes (nearly all of which are embedded in the inner membrane), and finally to O_2. The sequence in which the carriers act has been deduced in several ways.

First, their standard reduction potentials have been determined experimentally (Table 18–2). One expects the carriers to function in order of increasing reduction potential, because electrons tend to flow spontaneously from carriers of lower E_0' to carriers of higher E_0'. The order of carriers deduced by this method is NADH, UQ, cytochrome b, cytochrome c_1, cytochrome c, cytochrome $a + a_3$.

Table 18–2 Standard reduction potentials for respiratory chain and related electron carriers

Redox reaction (half-reaction)	E_0' (V)
$2H^+ + 2e^- \longrightarrow H_2$	−0.414
$NAD^+ + H^+ + 2e^- \longrightarrow NADH$	−0.320
$NADP^+ + H^+ + 2e^- \longrightarrow NADPH$	−0.324
NADH dehydrogenase (FMN) $+ 2H^+ + 2e^- \longrightarrow$ NADH dehydrogenase ($FMNH_2$)	−0.30
Ubiquinone $+ 2H^+ + 2e^- \longrightarrow$ ubiquinol	0.045
Cytochrome b (Fe^{3+}) $+ e^- \longrightarrow$ cytochrome b (Fe^{2+})	0.077
Cytochrome c_1 (Fe^{3+}) $+ e^- \longrightarrow$ cytochrome c_1 (Fe^{2+})	0.22
Cytochrome c (Fe^{3+}) $+ e^- \longrightarrow$ cytochrome c (Fe^{2+})	0.254
Cytochrome a (Fe^{3+}) $+ e^- \longrightarrow$ cytochrome a (Fe^{2+})	0.29
Cytochrome a_3 (Fe^{3+}) $+ e^- \longrightarrow$ cytochrome a_3 (Fe^{2+})	0.55
$\frac{1}{2}O_2 + 2H^+ + 2e^- \longrightarrow H_2O$	0.816

Second, when the entire chain of carriers is reduced experimentally by providing an electron source but no electron acceptor (no O_2), and then O_2 is suddenly introduced into the system, the rate at which each electron carrier becomes oxidized (measured spectroscopically) shows the order in which the carriers function (Fig. 18–6). The carrier nearest O_2 (at the end of the chain) gives up its electrons first, the second carrier from the end is oxidized next, and so on. Such experiments have confirmed the sequence deduced from standard reduction potentials.

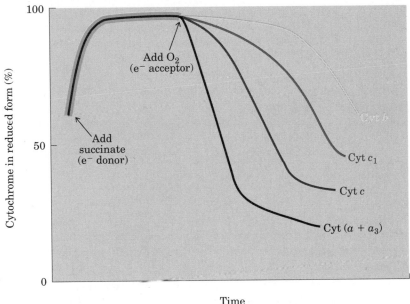

Deduced sequence: succinate $\rightarrow$ cyt b $\rightarrow$ cyt c_1 $\rightarrow$ cyt c $\rightarrow$ cyt $(a + a_3)$ $\rightarrow$ O_2

Figure 18–6 The sequence of electron carriers can be determined by the kinetics of their oxidation. Isolated mitochondria are incubated with a source of electrons (succinate in this case) but without O_2. Electrons from succinate enter the respiratory chain (through $FADH_2$), reducing each of the electron carriers almost completely. Using rapid and sensitive spectrophotometric techniques, the rate of oxidation of each carrier is determined immediately after introducing O_2 into the system. The carriers closest to O_2 (cytochromes a and a_3) are oxidized first; the most distant carrier (cytochrome b) is oxidized last.

Third, agents that inhibit the flow of electrons through the chain have been used in combination with measurement of the degree of oxidation of each carrier (Fig. 18–7). In the presence of O_2 and an electron donor, carriers that function before the inhibited step are expected to become fully reduced, and those that function after the block should be completely oxidized. By using several inhibitors that block earlier or later in the chain, the entire sequence has been deduced; it is the same as predicted from the first two approaches.

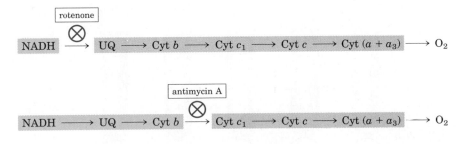

Figure 18–7 Determination of the sequence of electron carriers by the effects of inhibitors of electron transfer on the oxidation state of each carrier. In the presence of an electron donor and O_2, each inhibitor causes a characteristic pattern of oxidized/reduced carriers: those before the block become reduced (blue), and those after the block become oxidized (red).

Fourth, gentle treatment of the inner mitochondrial membrane with detergents allows the resolution of four electron-carrier complexes, each representing a fraction of the entire respiratory chain (Fig. 18–8). Each of the four separated complexes has its own unique composition (Table 18–3), and each is capable of catalyzing electron transfer through a portion of the chain. Complexes I and II catalyze electron transfer to ubiquinone from two different electron donors: NADH (Complex I) and succinate (Complex II). Complex III carries electrons from ubiquinone to cytochrome c, and Complex IV completes the sequence by transferring electrons from cytochrome c to O_2 (Fig. 18–8).

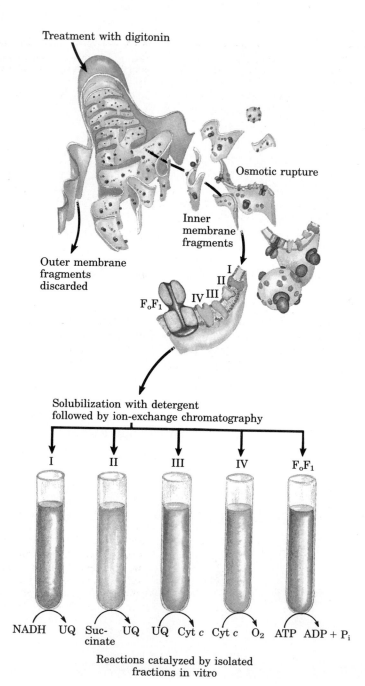

Figure 18–8 Resolution of functional complexes of the respiratory chain. The outer mitochondrial membrane is first removed by treatment with the detergent digitonin. Fragments of inner membrane are then obtained by osmotic rupture of the mitochondria, and the fragments are gently dissolved in a second detergent. The resulting mixture of inner membrane proteins is resolved by ion-exchange chromatography into different complexes (I through IV) of the respiratory chain, each with its unique protein composition (see Table 18–3), and the enzyme ATP synthase (sometimes called Complex V). The isolated Complexes I through IV catalyze transfers between donors (NADH and succinate), intermediate carriers (UQ and cytochrome c), and O_2, as shown. ATP synthase, which is composed of peripheral (F_1) and integral (F_o) proteins (discussed later in this chapter), has only ATP-hydrolyzing (ATPase), not ATP-synthesizing, activity in vitro.

Table 18–3 Protein components of the mitochondrial electron transfer chain

Enzyme complex[*]	Mass (kDa)	Number of subunits	Prosthetic group(s)
I NADH dehydrogenase	850	>25	FMN, Fe–S
II Succinate dehydrogenase	140	4	FAD, Fe–S
III Ubiquinone-cytochrome c oxidoreductase	250	10	Hemes, Fe–S
Cytochrome c	13	1	Heme
IV Cytochrome oxidase	160	6–13	Hemes; Cu_A, Cu_B

Sources: DePierre, J.W. & Ernster, L. (1977) Enzyme topology of intracellular membranes. *Annu. Rev. Biochem.* **46,** 201–262; Hatefi, Y. (1985) The mitochondrial electron transport and oxidative phosphorylation system. *Annu. Rev. Biochem.* **54,** 1015–1069.

[*] Note that cytochrome c is not part of an enzyme complex, but moves between Complexes III and IV as a freely soluble protein.

Complex I: NADH to Ubiquinone Complex I, also called the **NADH dehydrogenase complex,** is a huge flavoprotein complex containing more than 25 polypeptide chains (Table 18–3). The entire complex is embedded in the inner mitochondrial membrane, oriented with its NADH-binding site facing the matrix such that it can interact with NADH produced by any of the several matrix dehydrogenases. The overall reaction catalyzed by Complex I is

$$NADH + H^+ + UQ \rightleftharpoons NAD^+ + UQH_2$$

in which oxidized ubiquinone (UQ) accepts a hydride ion (two electrons and one proton) from NADH and a proton from the solvent water in the matrix. The enzyme complex first transfers a pair of reducing equivalents from NADH to its prosthetic group, FMN (Fig. 18–9). The complex also contains seven Fe–S centers of at least two different types, through which electrons pass on their way from FMN to ubiquinone. Amytal (a barbiturate drug), rotenone (a plant product commonly used as an insecticide), and the antibiotic piericidin A all inhibit electron flow from these Fe–S centers to ubiquinone (Table 18–4).

Figure 18–9 Path of electrons from NADH, succinate, fatty acyl–CoA, and glycerol-3-phosphate to ubiquinone (UQ). Electrons from NADH pass through a flavoprotein to a series of iron–sulfur proteins (in Complex I) and then to UQ. Electrons from succinate pass through a flavoprotein and several Fe–S centers (in Complex II) on the way to UQ. Glycerol-3-phosphate donates electrons to a flavoprotein (glycerol-3-phosphate dehydrogenase) on the outer face of the inner mitochondrial membrane, from which they pass through Fe–S centers to UQ. Acyl-CoA dehydrogenase (the first enzyme in β oxidation) transfers electrons to electron-transferring flavoprotein (ETFP), from which they pass via ETFP-ubiquinone oxidoreductase to UQ.

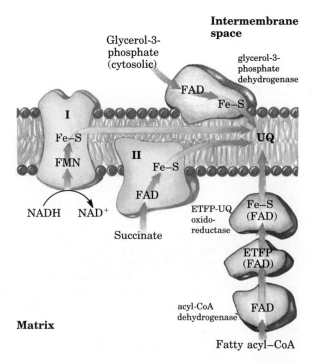

Table 18–4 Some agents that interfere with oxidative phosphorylation or photophosphorylation

Type of interference	Compound	Target/mode of action
Inhibition of electron transfer	Cyanide / Carbon monoxide	Inhibit cytochrome oxidase
	Antimycin A	Blocks electron transfer from cytochrome b to cytochrome c_1
	Rotenone / Amytal / Piericidin A	Prevent electron transfer from Fe–S center to ubiquinone
	DCMU	Competes with Q_B for binding site in photosystem II
Inhibition of ATP synthase	Oligomycin / Venturicidin	Inhibit F_1 and CF_1
	Dicyclohexyl-carbodiimide (DCCD)	Blocks proton flow through F_o and CF_o
Uncoupling of phosphorylation from electron transfer	Carbonyl-cyanide phenylhydrazone / Dinitrophenol	Hydrophobic proton carriers
	Valinomycin	K^+ ionophore
	Uncoupling protein (thermogenin)	Forms proton-conducting pores in inner membrane of brown fat mitochondria
Inhibition of ATP–ADP exchange	Atractyloside	Inhibits adenine nucleotide translocase

Ubiquinol (UQH_2, the fully reduced form; Fig. 18–2) diffuses in the membrane from Complex I to Complex III, where it is oxidized to UQ. The flow of electrons through Complex I to ubiquinone to Complex III is accompanied by the movement of protons from the mitochondrial matrix to the outer (cytosolic) side of the inner mitochondrial membrane (the intermembrane space), as described below.

Complex II: Succinate to Ubiquinone We encountered Complex II in Chapter 15 under a different name: **succinate dehydrogenase;** it is the only membrane-bound enzyme in the citric acid cycle (p. 457). Although smaller and simpler than Complex I, it contains two types of prosthetic groups and at least four different proteins (Table 18–3). One protein has a covalently bound FAD and an Fe–S center with four Fe atoms; a second iron–sulfur protein is also present. Electrons are believed to pass from succinate to FAD, then through the Fe–S centers to ubiquinone.

Other substrates for mitochondrial dehydrogenases also pass electrons into the respiratory chain at the level of ubiquinone, but not through Complex II. The first step in the β oxidation of fatty acyl–CoA, catalyzed by the flavoprotein **acyl-CoA dehydrogenase** (p. 486), involves transfer of electrons from the substrate to the FAD of the dehydrogenase, then to electron-transferring flavoprotein (ETFP), which in turn passes its electrons to **ETFP-ubiquinone oxidoreductase** (Fig. 18–9). This reductase, an iron–sulfur protein that also contains a bound flavin nucleotide, passes electrons into the respiratory chain by reducing ubiquinone in the inner mitochondrial membrane.

In Chapter 16 we noted that glycerol released in the degradation of triacylglycerols is phosphorylated, then converted into dihydroxyacetone phosphate by **glycerol-3-phosphate dehydrogenase** (see Fig. 16–4). This enzyme is a flavoprotein located on the outer face of the inner mitochondrial membrane, and like succinate dehydrogenase and acyl-CoA dehydrogenase it channels electrons into the respiratory chain by reducing ubiquinone (Fig. 18–9). The important role of glycerol-3-phosphate dehydrogenase in shuttling reducing equivalents from cytosolic NADH into the mitochondrial matrix is described later (see Fig. 18–26).

Complex III: Ubiquinone to Cytochrome c Complex III, also called **cytochrome bc_1 complex** or **ubiquinone-cytochrome** c **oxidoreductase,** contains cytochromes b_{562} and b_{566}, cytochrome c_1, an iron–sulfur protein, and at least six other protein subunits (Table 18–3). These proteins are asymmetrically disposed in the inner mitochondrial membrane; cytochrome b spans the membrane, and both cytochrome c_1 and the iron–sulfur protein are on the outer surface. The switch between the two-electron carrier ubiquinone and the one-electron carriers (cytochromes b_{562}, b_{566}, c_1, and c) is accomplished in a series of reactions called the Q cycle (Fig. 18–10). Although the path of electron flow through this segment of the respiratory chain is complicated, the net effect of the transfer is simple: UQH_2 is oxidized to UQ and cytochrome c is reduced.

Complex III functions as a proton pump; as a result of the asymmetric orientation of the complex, protons produced when UQH_2 is oxidized to UQ are released to the intermembrane space, producing a transmembrane difference of proton concentration—a proton gradient. The importance of this proton gradient to mitochondrial ATP synthesis will soon become clear.

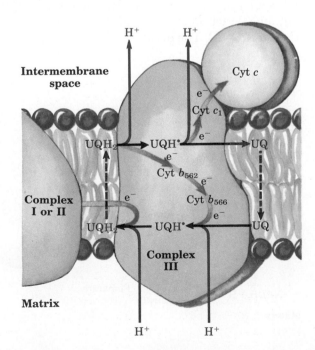

Figure 18–10 The path of electrons through Complex III probably involves a "Q cycle" such as that shown here (blue arrows). The broken arrows represent diffusion of UQH_2 (ubiquinol) or its oxidized form UQ across the membrane. Notice that the electron transfers between cytochromes and ubiquinone are one-electron reactions, producing the semiquinone radical as an intermediate (see Fig. 18–2). The net effects of the reactions here are (1) movement of electrons from UQH_2 to cytochrome c, and (2) movement of protons from the inside (matrix) to the outer (cytosolic) side of the inner membrane (the intermembrane space). (For simplicity the oxidation–reduction cycles of the individual cytochromes are not shown here.)

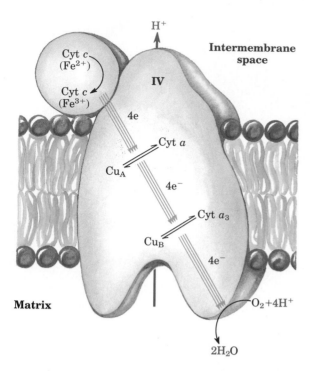

Figure 18–11 Path of electrons through Complex IV. Cu_A (Cu^{2+}) and cyt a (Fe^{2+}) form one bimetallic redox center capable of accepting two electrons; Cu_B and cyt a_3 constitute a second two-electron redox center. The detailed path of electron flow between cyt c and O_2 is not known with certainty; apparently electrons first move from cyt c to Cu_A or cyt a, which are in rapid redox equilibrium with each other. This bimetallic center then donates electrons to Cu_B and cyt a_3, also in redox equilibrium, which in turn donate the electrons that reduce O_2 to H_2O. The four protons used in the reduction of O_2 to H_2O are taken up from the matrix side of the inner mitochondrial membrane. Consequently, cytochrome oxidase pumps protons out of the matrix as electrons are transferred to O_2.

Complex IV: Reduction of O_2 Complex IV, also called **cytochrome oxidase,** contains cytochromes a and a_3. These cytochromes consist of two heme groups bound to different regions of the same large protein that are therefore spectrally and functionally distinct. Cytochrome oxidase also contains two copper ions, Cu_A and Cu_B, that are crucial to the transfer of electrons to O_2. This complex enzyme has evolved to carry out the four-electron reduction of O_2 (Fig. 18–11) without generating incompletely reduced intermediates such as hydrogen peroxide or hydroxyl free radicals—very reactive species that would damage cellular components.

The flow of electrons from cytochrome c to O_2 through Complex IV causes net movement of protons from the matrix to the intermembrane space; Complex IV functions as a proton pump that contributes to the proton-motive force.

Electron Transfer to O_2 Is Highly Exergonic

The energetics of electron transfer (oxidation–reduction) reactions were described in Chapter 13. In oxidative phosphorylation, two electrons pass from NADH through the respiratory chain to molecular oxygen:

$$NADH + H^+ + \tfrac{1}{2}O_2 \longrightarrow H_2O + NAD^+$$

For the redox pair $NAD^+/NADH$, E_0' is -0.320 V, and for the pair O_2/H_2O, E_0' is 0.816 V. The $\Delta E_0'$ for this reaction is therefore $+1.14$ V, and the standard free-energy change (p. 389) is

$$\Delta G^{\circ\prime} = -n\mathcal{F}\Delta E_0'$$
$$= (-2)(96.5 \text{ kJ/V} \cdot \text{mol})(1.14 \text{ V})$$
$$= -220 \text{ kJ/mol}$$

A similar calculation for succinate oxidation shows that electron transfer from succinate (E_0' for fumarate/succinate = 0.031 V) to O_2 has a smaller, but still negative, standard free-energy change of about -152 kJ/mol.

In the mitochondrion, the combined action of Complexes I, III, and IV results in the transfer of electrons from NADH to O_2; Complexes II, III, and IV act together to catalyze electron transfer from succinate to O_2 (Fig. 18–12). The actual free-energy changes in respiring mitochon-

Figure 18–12 Summary of the flow of electrons and protons through the four complexes of the respiratory chain. Electrons reach UQ via Complexes I and II. UQH_2 serves as a mobile carrier of electrons and protons. It passes electrons to Complex III, which passes them to another mobile connecting link, cytochrome c. Complex IV transfers electrons from reduced cytochrome c to O_2. Electron flow through Complexes I, III, and IV is accompanied by proton flow from the matrix to the intermembrane space. Recall that electrons from fatty acid β oxidation can also enter the respiratory chain through UQ (see Figs. 16–9 and 18–9).

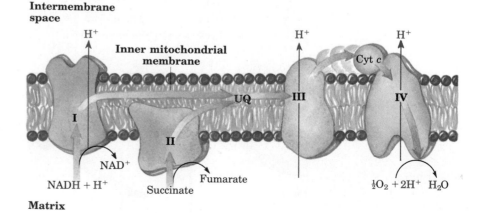

dria are not the same as the standard free-energy changes, because reactants are not present at 1 M concentrations. However, it is clear from experimental measurements that under cellular conditions, the mitochondrial oxidation of NADH or succinate releases more free energy than the 51.8 kJ/mol required to drive the synthesis of ATP from ADP and P_i (see Box 13–2).

ATP Synthesis Coupled to Respiratory Electron Flow

We now turn to the most fundamental question about mitochondrial oxidative phosphorylation: how does the flow of electrons through the respiratory chain channel energy into the synthesis of ATP? We have seen that electron transfer through the respiratory chain releases more than enough free energy to form ATP. Mitochondrial oxidative phosphorylation therefore poses no thermodynamic problem. However, one cannot deduce from thermodynamic considerations the chemical mechanism by which energy released in one exergonic reaction (the oxidation of NADH by O_2) is channeled into a second, endergonic, reaction (the condensation of ADP and P_i). To describe the process of oxidative phosphorylation completely, we need to identify the physical and chemical changes that result from electron flow and cause ADP phosphorylation—the mechanism that *couples* oxidation with phosphorylation.

We begin our discussion by considering the stoichiometry of oxidation and phosphorylation in isolated mitochondria and the evidence for obligatory coupling of the two processes. The chemiosmotic interpretation of oxidative phosphorylation is then presented, with the major lines of evidence that support it. The enzyme ATP synthase, which is directly responsible for ATP synthesis, is the equivalent of an F-type ATP-dependent proton pump working in reverse; the flow of protons down their electrochemical gradient through this "pump" drives the condensation of P_i and ADP. We describe also the membrane transport systems that move substrates, products, and reducing equivalents between the cytosol and the mitochondrial matrix. Having looked in detail at the coupling of ATP synthesis to electron flow, we will see that in the mitochondria of some tissues the two processes are deliberately "uncoupled" to produce heat.

We conclude with a summary of the overall regulation of ATP-producing processes in the cell, and a look at two further interesting aspects of mitochondria: the mitochondrial genome (and the effects of mutations therein) and the likely evolutionary origins of these organelles.

Phosphorylation of ADP Is Coupled to Electron Transfer

When isolated mitochondria are suspended in a buffer containing ADP, P_i, and an oxidizable substrate such as succinate, three easily measured processes occur: (1) the substrate is oxidized (succinate yields fumarate), (2) O_2 is consumed (respiration occurs), and (3) ATP is synthesized. Careful experimental measurements of the stoichiometry of electron transfer to O_2 and the associated synthesis of ATP show that with NADH as electron donor, mitochondria synthesize nearly 3.0 ATP per pair of electrons passed to O_2, and with succinate nearly 2.0 ATP

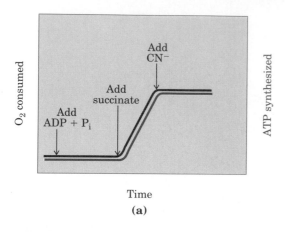

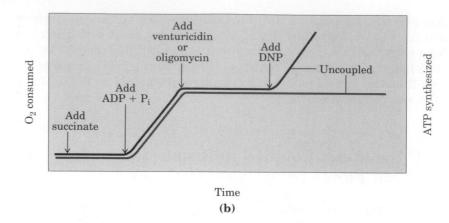

Figure 18–13 Electron transfer to O_2 is tightly coupled to ATP synthesis in mitochondria, as is demonstrated in these experiments. Mitochondria are suspended in a buffered medium, and an O_2 electrode is used to monitor O_2 consumption. At intervals, samples are removed and assayed for the presence of ATP. **(a)** The addition of ADP and P_i alone results in little or no increase in either respiration (O_2 consumption; black) or ATP synthesis (red). When succinate is added, respiration begins immediately and ATP is synthesized. The addition of cyanide (CN^-), which blocks electron transfer between cytochrome oxidase and O_2, inhibits both respiration and ATP synthesis. **(b)** Mitochondria provided with succinate respire and synthesize ATP only when ADP and P_i are added. Subsequent addition of venturicidin or oligomycin, inhibitors of ATP synthase, blocks both ATP synthesis and respiration. Dinitrophenol (DNP) allows respiration to continue without ATP synthesis; DNP acts as an uncoupler.

per electron pair. Oxygen consumption and ATP synthesis are dependent upon substrate oxidation, as can be seen in the experiments diagrammed in Figure 18–13.

Because the energy of substrate oxidation drives ATP synthesis in mitochondria, it is not surprising that inhibitors of the passage of electrons to O_2 (e.g., cyanide ion, carbon monoxide, and antimycin A) block ATP synthesis (Fig. 18–13a). It is perhaps not so obvious that the converse is true: inhibition of ATP synthesis blocks electron transfer in intact mitochondria. This obligatory coupling can be demonstrated in isolated mitochondria by providing O_2 and oxidizable substrates, but not ADP (Fig. 18–13b). Under these conditions, no ATP synthesis can occur, and electron transfer to O_2 is also strikingly reduced. Coupling of oxidation and phosphorylation can also be demonstrated using oligomycin or venturicidin, toxic antibiotics that bind to the ATP synthase in mitochondria. These compounds are potent inhibitors of both ATP synthesis *and* the transfer of electrons through the chain of carriers to O_2 (Fig. 18–13b). Because oligomycin is known not to interact directly with the electron carriers but only with ATP synthase, it follows that electron transfer and ATP synthesis are obligatorily coupled; neither reaction occurs without the other.

There are, however, certain conditions and reagents that uncouple oxidation from phosphorylation. When intact mitochondria are disrupted by treatment with detergent or physical shear, the resulting membrane fragments are still capable of catalyzing electron transfer from succinate or NADH to O_2, but no ATP synthesis is coupled to this respiration. Certain chemical compounds also cause uncoupling (Fig. 18–13b), without disrupting mitochondrial structure. The chemical uncouplers (Table 18–4) include 2,4-dinitrophenol (DNP) and a group of compounds related to carbonylcyanide phenylhydrazone (Fig. 18–14). All of these uncouplers are weak acids with hydrophobic properties. Ionophores (p. 560) also uncouple oxidative phosphorylation. These agents bind to inorganic ions and surround them with hydrophobic moieties; the ionophore–metal ion complexes pass easily through membranes. We shall see later how the chemiosmotic theory accounts for the action of uncouplers.

OH → O⁻

2,4-Dinitrophenol
(DNP)

Carbonylcyanide-*m*-chlorophenylhydrazone
(CCCP)

Figure 18–14 Two chemical uncouplers of oxidative phosphorylation. Both have a dissociable proton and are very hydrophobic. They act by carrying protons across the inner mitochondrial membrane, dissipating the proton gradient. Both also uncouple photophosphorylation (p. 585).

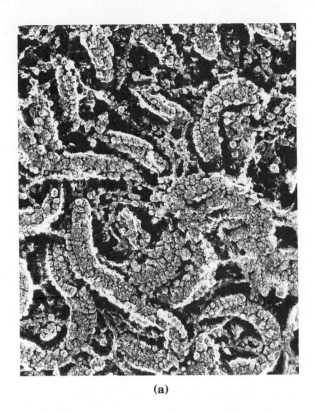

(a)

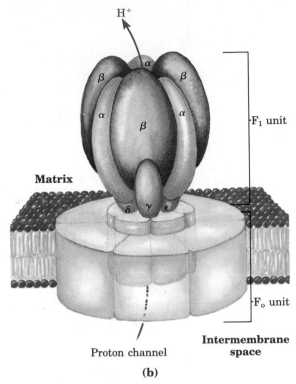

(b)

ATP Synthase Is a Large Membrane–Protein Complex

ATP synthase, the ATP-synthesizing enzyme complex of the inner mitochondrial membrane, has two major components (or factors), F_1 and F_o (Fig. 18–15). The subscript letter o in F_o denotes that it is the portion of the ATP synthase that confers sensitivity to oligomycin, a potent inhibitor of this enzyme complex and thus of oxidative phosphorylation (Table 18–4).

F_1 was first extracted from the mitochondrial inner membrane and purified by Efraim Racker and his colleagues in the early 1960s. Isolated F_1 cannot synthesize ATP from ADP and P_i; because it can catalyze the reverse reaction—hydrolysis of ATP—the enzyme was originally called F_1ATPase. When F_1 is carefully extracted from inside-out vesicles prepared from the inner mitochondrial membrane (Fig. 18–16, p. 558), the vesicles still contain intact respiratory chains and can catalyze electron transfer, but cannot make ATP. When a preparation of isolated F_1 is added back to such depleted vesicles, their capacity to couple electron transfer and ATP synthesis is restored. Membrane reconstitution experiments of this kind, pioneered by Racker, opened new doors to research on membrane structure and function.

F_1, which in all aerobic organisms consists of six subunits, contains several binding sites for ATP and ADP, including the catalytic site for ATP synthesis. It is a peripheral membrane protein complex, held to the membrane by its interaction with F_o, an integral membrane protein complex of four different polypeptides that forms a transmembrane channel through which protons can cross the membrane. High-resolution electron micrographs of the isolated F_oF_1 complex show the knoblike F_1 head, a stalk, and a base piece (F_o), which normally extends across the inner membrane (Fig. 18–15b).

Figure 18–15 The ATP synthase complex from mitochondria. **(a)** Electron micrographs showing the knoblike protrusions from the mitochondrial inner membrane. **(b)** Schematic diagram showing the likely organization of the subunits to form the proton-conducting F_o portion, and the ATP-synthesizing F_1 unit. The F_1 complex consists of three α, three β, and one each of γ, δ, and ϵ subunits, as discussed later in the chapter (p. 562).

Efraim Racker
1913–1991

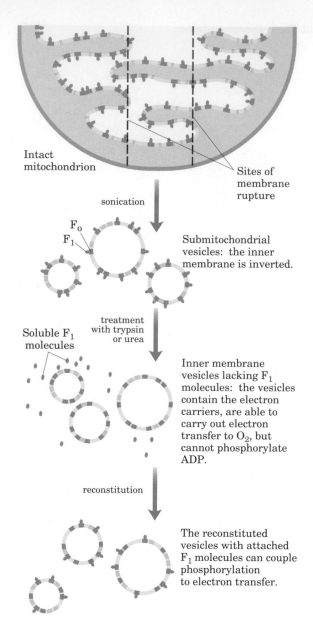

Peter Mitchell
1920–1992

Figure 18–16 Formation of inner-membrane vesicles by sonic treatment of mitochondria, and the reconstitution of oxidative phosphorylation by combining separated fractions of the membrane.

The complete F_oF_1 complex, like isolated F_1, can hydrolyze ATP to ADP and P_i, but its biological function is to catalyze the condensation of ADP and P_i to form ATP. The F_oF_1 complex is therefore more appropriately called ATP synthase.

ATP Synthase Is Related to ATP-Dependent Proton Pumps

The structure and catalytic activity of ATP synthase from mitochondria show that it belongs to a larger class of enzymes that we have already encountered: the F-type ATPases (see Table 10–5). These protein complexes share the F_oF_1 structure, and there is sequence homology between the subunits of the F-type ATPases and those of mitochondrial ATP synthase. F-type ATPases are ATP-dependent proton pumps; they use the energy released by ATP hydrolysis to move protons across membranes against a concentration gradient, thereby creating a difference of pH and electrical charge (an electrochemical potential) across the membrane. The F-type ATPases must have evolved very early; they are found in the membranes of eubacteria and archaebacteria and in the mitochondria of all aerobic eukaryotes. Chloroplasts also contain an ATP synthase that is an F-type ATPase, which acts in light-driven ATP synthesis, as we shall see shortly.

The Chemiosmotic Model: A Proton Gradient Couples Electron Flow and Phosphorylation

How does the oxidation of substrates via electron transfer through the respiratory chain cooperate with the ATP synthase to bring about phosphorylation of ADP to ATP? Early investigators of mitochondrial oxidative phosphorylation had before them a well-documented example of an oxidation coupled with a phosphorylation and involving a high-energy chemical intermediate: the glyceraldehyde-3-phosphate dehydrogenase reaction. In this glycolytic step, glyceraldehyde-3-phosphate is oxidized and simultaneously converted to 1,3-bisphosphoglycerate, a compound with a high-energy group at the site of the oxidation (see Fig. 14–5). ATP is formed when 1,3-bisphosphoglycerate transfers its activated P_i to ADP. The formal similarity between this oxidative phosphorylation and that which takes place in mitochondria led naturally to the chemical coupling hypothesis, according to which some high-energy chemical intermediate is the direct product of electron transfer through one of three coupling sites along the chain of mitochondrial electron carriers. The energy in this putative chemical intermediate would then be used to drive the synthesis of ATP. An enormous amount of effort was expended in the search for such a chemical intermediate; none was found.

In the early 1960s Peter Mitchell suggested a new paradigm that has become central to current thinking and research on biological energy transductions. Mitchell's chemiosmotic hypothesis accounted for the experimental observations and made certain novel and experimentally testable predictions. Having survived more than two decades of vigorous testing and refinement, the hypothesis has become a generally accepted theory that is applied not only to mitochondrial oxidative phosphorylation, but to a wide array of other energy transductions, including light-driven ATP formation in photosynthetic organisms.

As postulated in the **chemiosmotic theory** (Fig. 18–17), the transfer of electrons along the respiratory chain is accompanied by outward pumping of protons across the inner mitochondrial mem-

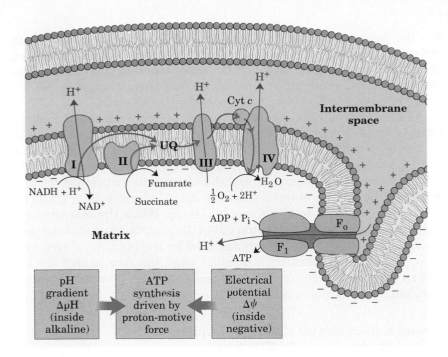

Figure 18–17 In this simple version of the chemiosmotic theory applied to mitochondria, electrons from NADH and other oxidizable substrates pass through a chain of carriers (cytochromes, etc.) arranged asymmetrically in the membrane. Electron flow is accompanied by proton transfer across the mitochondrial membrane, producing both a chemical (ΔpH) and an electrical ($\Delta\psi$) gradient. The inner mitochondrial membrane is impermeable to protons; protons can reenter the matrix only through proton-specific channels (F_o). The proton-motive force that drives protons back into the matrix provides the energy for ATP synthesis, catalyzed by the F_1 complex associated with F_o.

brane, which results in a transmembrane difference in proton concentration (a proton gradient) and thus in pH; the matrix becomes alkaline relative to the cytosolic side of the membrane. The electrochemical energy inherent in this difference in proton concentration and separation of charge, the **proton-motive force,** represents a conservation of part of the energy of oxidation. The proton-motive force is subsequently used to drive the synthesis of ATP catalyzed by F_1 as protons flow passively back into the matrix through proton pores formed by F_o. To emphasize this crucial role of the proton-motive force, the equation for ATP synthesis is sometimes written

$$\text{ADP} + P_i + \{\text{H}^+\}_{out} \rightleftharpoons \text{ATP} + H_2O + \{\text{H}^+\}_{in} \qquad (18\text{--}1)$$

In the more general case, the electrochemical energy of a transmembrane gradient of any charged species is seen to be interconvertible with the energy of chemical bonds. The energy stored in such a gradient has two components, as implied in the word "electrochemical." One is the chemical potential energy due to the difference in concentration of a chemical species in the two regions separated by the membrane; the other is the electrical potential energy that results from the separation of charge when an ion moves across the membrane without its counterion (Fig. 18–17).

We showed in Chapter 10 that the free-energy change for the creation of an electrochemical gradient by an ion pump (Fig. 18–18) is

$$\Delta G = RT \ln (C_2/C_1) + Z\mathcal{F}\Delta\psi \qquad (18\text{--}2)$$

where C_2/C_1 is the concentration ratio for the ion that moves, Z is the absolute value of its electrical charge (1 for a proton), and $\Delta\psi$ is the transmembrane difference in electrical potential, measured in volts.

For the case of protons at 25 °C

$$\ln (C_2/C_1) = 2.3(\log [\text{H}^+]_{out} - \log [\text{H}^+]_{in}) = 2.3(\text{pH}_{in} - \text{pH}_{out}) = 2.3 \,\Delta\text{pH}$$

and Equation 18–2 reduces to

$$\Delta G = 2.3RT \,\Delta\text{pH} + \mathcal{F}\Delta\psi$$

$$\Delta G = (5.70 \text{ kJ/mol})\Delta\text{pH} + (96.5 \text{ kJ/V} \cdot \text{mol})\Delta\psi \qquad (18\text{--}3)$$

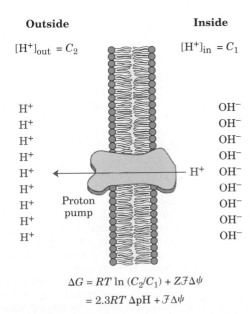

Outside   Inside

$[\text{H}^+]_{out} = C_2$ $[\text{H}^+]_{in} = C_1$

$$\Delta G = RT \ln (C_2/C_1) + Z\mathcal{F}\Delta\psi$$
$$= 2.3RT \,\Delta\text{pH} + \mathcal{F}\Delta\psi$$

Figure 18–18 The inner mitochondrial membrane separates two compartments of different pH, producing differences in both chemical concentration (ΔpH) and charge distribution, creating an electrical potential difference ($\Delta\psi$). The net effect of these differences is the proton-motive force (ΔG), which can be calculated as shown. This is explained more fully in the text.

For the pumping of ions against an electrochemical gradient, both terms on the right-hand side of this equation have positive values, and ΔG is therefore positive. When protons flow spontaneously *down* their electrochemical gradient, an amount of free energy equal to ΔG becomes available to do work; it is the proton-motive force.

Chemiosmotic theory readily explains the dependence of electron transfer on ATP synthesis in mitochondria (Fig. 18–17). When ATP synthesis is blocked (with oligomycin, for example), protons cannot flow into the matrix through the F_oF_1 complex. With no path for the return of protons to the matrix and the continued extrusion of protons driven by the activity of the respiratory chain, a large proton gradient, hence a large proton-motive force, builds up. When the cost (free energy) of pumping one more proton out of the matrix against this gradient equals or exceeds the energy released by the transfer of electrons from NADH to O_2, electron flow must stop; the free energy for the overall process of electron flow coupled to proton pumping becomes zero, and equilibrium is attained.

Strong Experimental Evidence Implicates the Proton-Motive Force in ATP Synthesis

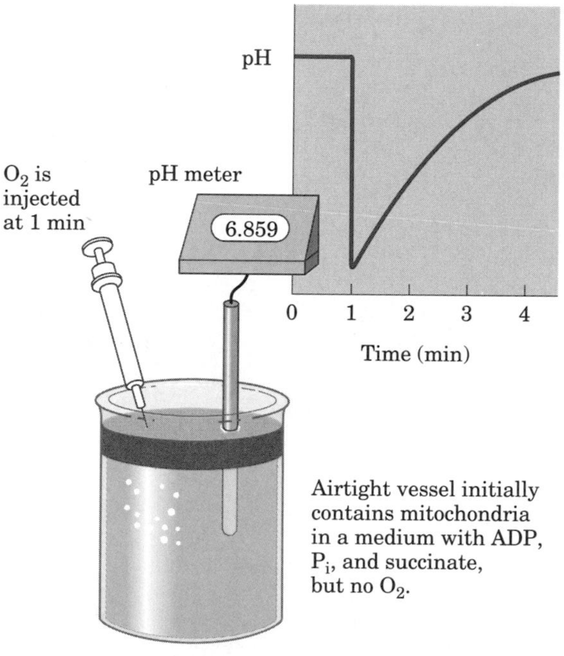

The transmembrane proton gradient predicted by chemiosmotic theory has been experimentally measured. When intact mitochondria are suspended with an oxidizable substrate such as succinate in a lightly buffered medium, the addition of O_2 results in acidification of the suspending medium (Fig. 18–19). The stoichiometry of proton pumping can be estimated from the pH changes, after correcting for buffering effects. The measurement is difficult, and there is no universal agreement on the stoichiometry; about ten protons are pumped out for each pair of electrons transferred from NADH to O_2. The result is a steady state with a difference of about one pH unit between the matrix and the medium, inside (matrix) alkaline. Indirect measurements of the transmembrane electrical potential in respiring mitochondria yield a value of 0.1 to 0.2 V, inside negative. Substituting these values into the equation for proton-motive force (Eqn 18–3) gives a value of 15 to 25 kJ/mol, the free-energy change for the transmembrane movement of one equivalent of protons back into the mitochondrial matrix. This free-energy change is large enough to account for the synthesis of one ATP when two to three protons flow into the matrix.

Uncouplers are hydrophobic weak acids (Fig. 18–14). Their hydrophobicity allows them to diffuse readily across mitochondrial membranes. After entering the mitochondrial matrix in the protonated form, they can release a proton (dissociate), thus dissipating the proton gradient. **Ionophores** uncouple electron transfer from oxidative phosphorylation by creating electrical short circuits across the mitochondrial membrane (Fig. 18–20). The toxic ionophore valinomycin forms a lipid-soluble complex with K^+, which is abundant in the cytosol (see Fig. 10–26). Whereas K^+ penetrates the mitochondrial inner membrane only very slowly, the K^+–valinomycin complex readily passes through. The influx of positive ions neutralizes the excess of negative charge inside the matrix, diminishing the electrical component of the proton-motive force. Valinomycin slows mitochondrial ATP synthesis without blocking electron transfer to O_2 (Table 18–4).

If the role of electron transfer in mitochondrial ATP synthesis is simply to pump protons to create the electrochemical potential of the proton-motive force, an artificially created proton gradient should be able to replace electron transfer in driving ATP synthesis. This predic-

pH meter

6.859

O_2 is injected at 1 min

pH

Time (min)

Airtight vessel initially contains mitochondria in a medium with ADP, P_i, and succinate, but no O_2.

Figure 18–19 Isolated mitochondria are suspended in a medium containing ADP, P_i, and succinate, but initially no O_2. When a small amount of O_2 is injected into the reaction mixture, succinate oxidation and electron transfer to O_2 begin immediately. A pH electrode registers a sudden decrease in the pH of the medium, indicating that protons are moving out of the mitochondria. As the injected O_2 is consumed, protons slowly leak back into the mitochondria, and the external pH returns to the initial level. From the known amount of O_2 added and the measured pH change, one can in principle calculate the number of protons extruded per molecule of O_2 consumed.

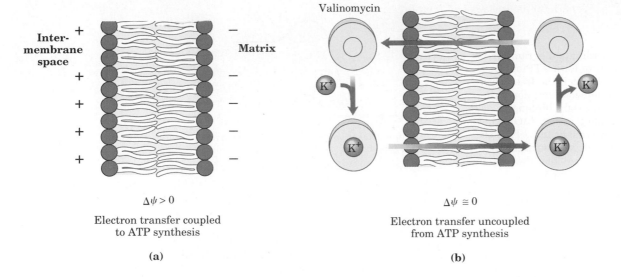

$$\Delta\psi > 0$$

Electron transfer coupled
to ATP synthesis

(a)

$$\Delta\psi \cong 0$$

Electron transfer uncoupled
from ATP synthesis

(b)

Figure 18–20 Ionophores such as valinomycin uncouple oxidative phosphorylation by dissipating ion gradients across the inner mitochondrial membrane, eliminating the contribution of $\Delta\psi$ to the proton-motive force. **(a)** A transmembrane electrical potential ($\Delta\psi$) exists because of the unequal distribution of protons on either side. **(b)** Valinomycin combines reversibly with K^+ ions to form a membrane-permeable complex that diffuses across the inner membrane and releases K^+ on the inside. This movement of charge reduces the value of $\Delta\psi$, and uncouples electron transfer from ATP synthesis.

tion of the chemiosmotic theory has been experimentally tested and confirmed (Fig. 18–21). Mitochondria manipulated so as to impose a difference of proton concentration and a separation of charge across the inner membrane (Fig. 18–21) carry out the synthesis of ATP *in the absence of an oxidizable substrate;* the proton-motive force alone suffices to drive ATP synthesis.

Chemiosmotic theory also accounts for a third condition that uncouples oxidation from phosphorylation—mechanical disruption of the mitochondrial membrane. Without an intact membrane there can be no proton gradient, hence no energy conservation and no ATP synthesis.

The Detailed Mechanism of ATP Formation Remains Elusive

Although it is clear that a transmembrane proton gradient provides the energy for ATP synthesis, it is not clear how this energy is transmitted to the ATP synthase. This enzyme catalyzes the conversion of enzyme-bound ADP and P_i into bound ATP even in the absence of a proton gradient. Remarkably, it appears that the reaction ADP + $P_i \rightleftharpoons$ ATP + H_2O is readily reversible—that the free-energy change

Figure 18–21 An artificially imposed electrochemical gradient can drive ATP synthesis in the absence of an oxidizable substrate as electron donor. In this two-step experiment, isolated mitochondria are first incubated in a pH 9 buffer containing 0.1 M KCl **(a)**. The slow leakage of buffer and KCl into the mitochondria eventually brings the matrix into equilibrium with the surrounding medium. No oxidizable substrates are present. **(b)** In the second step, mitochondria are separated from the pH 9 buffer and resuspended in pH 7 buffer containing valinomycin but no KCl. The change in buffer creates a difference of two pH units. The outward flow of K^+, carried (without a counterion) down its concentration gradient by valinomycin, creates a charge imbalance across the membrane (matrix negative). The sum of the chemical potential provided by the pH difference and the electrical potential provided by the separation of charges is a proton-motive force large enough to support ATP synthesis in the absence of an oxidizable substrate.

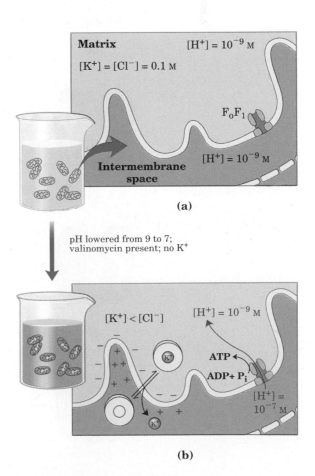

(a)

pH lowered from 9 to 7;
valinomycin present; no K^+

(b)

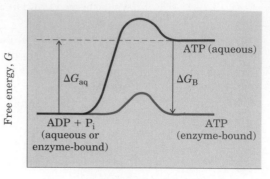

Figure 18–22 Reaction coordinate diagram for the condensation of ADP and P_i to form ATP on the surface of ATP synthase. Although the free-energy change for this reaction in aqueous solution (ΔG_{aq}) is large and positive, the very tight binding of ATP to the enzyme provides binding energy (ΔG_B) that brings the free energy of enzyme-bound ATP close to that of ADP + P_i. On the enzyme surface, the reaction is therefore readily reversible; the equilibrium constant is believed to be near 1.

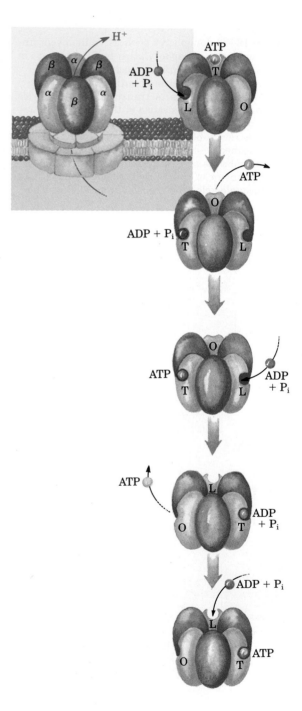

for ATP synthesis on the enzyme surface is close to zero. However, the ATP formed in this reaction remains tightly bound to the active site, preventing further catalysis at that site. The proton-motive force apparently supplies the energy needed to force dissociation of the tightly bound ATP from the enzyme, allowing another catalytic cycle to begin.

How is it possible that ATP synthesis, known to be strongly endergonic in aqueous solution ($\Delta G^{\circ\prime} = 30.5$ kJ/mol), is readily reversible on the enzyme surface? The very tight noncovalent binding of enzyme to ATP may supply enough binding energy (Chapter 8) to make bound ATP about as stable as its hydrolysis products (Fig. 18–22). Alternatively, the reaction may occur in a very hydrophobic pocket in the enzyme interior, where the energetics of hydrolysis in water do not apply.

Each F_1 complex has the subunit composition $\alpha_3\beta_3\gamma\delta\epsilon$ (Fig. 18–15). A tight binding site for ATP, apparently identical to the catalytic site, is located on each β subunit, or perhaps between each β and its associated α subunit. Every F_1 complex therefore has three ATP-synthesizing sites, which interact with F_o through the single copies of γ, δ, and ϵ subunits.

On the basis of detailed kinetic and binding studies of the reactions catalyzed by F_oF_1, Paul Boyer has suggested a mechanism (Fig. 18–23) in which the three active sites on F_1 alternate in catalyzing ATP synthesis. The limiting step in the process is the release of newly synthesized ATP from the enzyme. A conformational transition driven by the proton-motive force reduces the enzyme's affinity for ATP.

Figure 18–23 The "binding change" model for ATP synthase action. The enzyme has three equivalent adenine nucleotide binding sites, one for each pair of α and β subunits. At any given moment, one of these sites is in the T (tight-binding) conformation, a second is in the L (loose-binding) conformation, and a third is in the O (open; very loose-binding) conformation. At the beginning of a catalytic cycle, the T site is occupied by ATP, and ADP and P_i bind loosely to the L site. The proton-motive force causes, perhaps by the flow of protons through the F_o channel, a cooperative conformational change in which the T site is converted to an O site, and ATP dissociates from it; the L site is converted to a T site, where ADP and P_i quickly condense to form ATP; and the O site becomes an L site, where ADP and P_i loosely bind. The experimental results require that at least two of the three catalytic sites alternate in activity; ATP cannot be released from one site unless and until ADP and P_i are bound at the other.

The transition induced by the proton-motive force may be envisioned as a 120° rotation of the $\alpha_3\beta_3$ portion of F_1 (Fig. 18–23), placing one of the three α–β pairs in a special position relative to the proton channel of F_o. However, no physical rotation of F_1 has been demonstrated, and the model does not require it; the conformational change that interconverts the three types of sites may be an allosteric transition, in which changes at one α–β pair force compensating changes in the other two pairs.

What makes the mechanism of this enzyme particularly difficult to solve is the *vectorial* nature of the process it catalyzes. Somehow, the enzyme must sense not merely a certain concentration of protons, but a difference of proton concentration in two regions of space. Determination of the detailed structure of F_1 by x-ray crystallography should shed light on the mechanism of its action.

The Proton-Motive Force Energizes Active Transport

The primary role of electron transfer in mitochondria is to furnish energy for the synthesis of ATP during oxidative phosphorylation, but this energy serves also to drive several transport processes essential to oxidative phosphorylation. We have seen that the inner mitochondrial membrane is generally impermeable to charged species, but two specific systems in the inner mitochondrial membrane transport ADP and P_i into the matrix and ATP out to the cytosol (Fig. 18–24). The **adenine nucleotide translocase,** which extends across the inner membrane, binds ADP^{3-} on the outside (cytosolic) surface of the inner membrane and transports it inward in exchange for an ATP^{4-} molecule simultaneously transported outward (see Fig. 13–1 for the ionic forms of ATP and ADP). Because this antiporter moves four negative charges out for each three moved in, its activity is favored by the transmembrane electrochemical gradient, which gives the matrix a net negative charge; the proton-motive force drives ATP–ADP exchange.

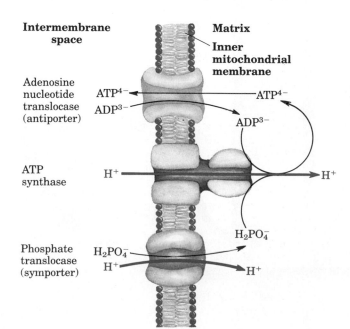

Intermembrane space   **Matrix**

Inner mitochondrial membrane

Adenosine nucleotide translocase (antiporter) ATP^{4-} ATP^{4-} ADP^{3-} ADP^{3-}

ATP synthase H^+ H^+

$H_2PO_4^-$

Phosphate translocase (symporter) $H_2PO_4^-$ H^+ H^+

Figure 18–24 Transport systems of the mitochondrial inner membrane carry ADP and P_i into the matrix and allow the newly synthesized ATP to leave. The ATP–ADP translocase is an antiporter; the same protein moves ADP into the matrix and ATP out. The effect of replacing ATP^{4-} with ADP^{3-} is the net efflux of one negative charge, which is favored because the matrix is electrically negative relative to the outside. At pH 7, P_i is present as both HPO_4^{2-} and $H_2PO_4^-$; the transport system that carries P_i into the matrix is specific for $H_2PO_4^-$. There is no net flow of charge during symport of $H_2PO_4^-$ and H^+, but the relatively low proton concentration in the matrix favors the inward movement of H^+. Thus the proton-motive force is responsible both for providing the energy for ATP synthesis by ATP synthase and for transporting substrates (ADP and P_i) in and product (ATP) out of the mitochondrial matrix.

Adenine nucleotide translocase is specifically inhibited by atractyloside, a toxic glycoside formed by a species of thistle; for centuries it has been known that grazing cattle are poisoned when they ingest this plant. If the transport of ADP into and ATP out of the mitochondria is inhibited, cytosolic ATP cannot be regenerated from ADP, explaining the toxicity of atractyloside (Table 18–4).

A second membrane transport system essential to oxidative phosphorylation is the **phosphate translocase,** which promotes symport of one $H_2PO_4^-$ and one H^+ into the matrix. This transport process, too, is favored by the transmembrane proton gradient (Fig. 18–24).

Shuttle Systems Are Required for Mitochondrial Oxidation of Cytosolic NADH

The NADH dehydrogenase of the inner mitochondrial membrane of animal cells can accept electrons only from NADH in the matrix. Given that the inner membrane is not permeable to cytosolic NADH, how can the NADH generated by glycolysis outside mitochondria be reoxidized to NAD^+ by O_2 via the respiratory chain? Special shuttle systems carry

Figure 18–25 The malate–aspartate shuttle for transporting reducing equivalents from cytosolic NADH into the mitochondrial matrix. ① NADH in the cytosol (intermembrane space) passes two reducing equivalants to oxaloacetate, producing malate. ② Malate is transported across the inner membrane by the malate–α-ketoglutarate transporter. ③ In the matrix, malate passes two reducing equivalents to NAD^+; the resulting matrix NADH is oxidized by the mitochondrial respiratory chain. The oxaloacetate formed from malate cannot pass directly into the cytosol. It is first transaminated to form aspartate ④, which can leave via the glutamate–aspartate transporter ⑤. Oxaloacetate is regenerated in the cytosol ⑥, completing the cycle.

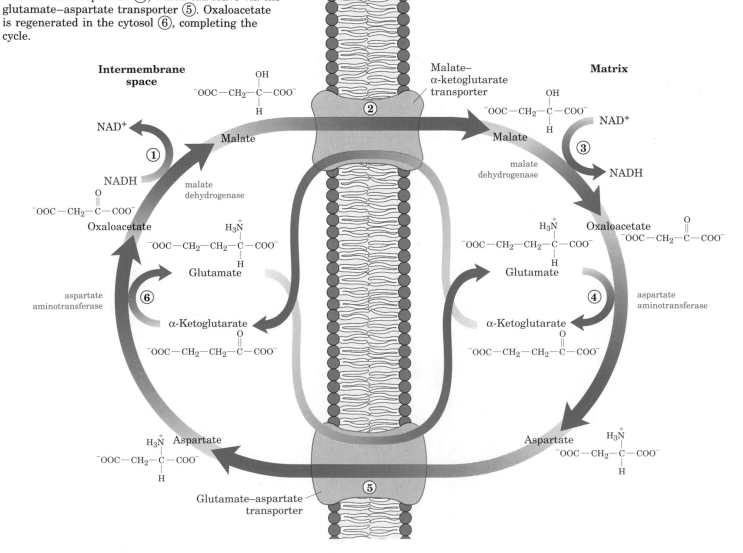

reducing equivalents from cytosolic NADH into mitochondria by an indirect route. The most active NADH shuttle, which functions in liver, kidney, and heart mitochondria, is the **malate–aspartate shuttle** (Fig. 18–25). The reducing equivalents of cytosolic NADH are first transferred to cytosolic oxaloacetate to yield malate by the action of cytosolic malate dehydrogenase. The malate thus formed passes through the inner membrane into the matrix via the malate–α-ketoglutarate transport system. Within the matrix the reducing equivalents are passed by the action of matrix malate dehydrogenase to matrix NAD^+, forming NADH; this NADH can then pass electrons directly to the respiratory chain in the inner membrane. Three molecules of ATP are generated as this pair of electrons passes to O_2. Cytosolic oxaloacetate must be regenerated via transamination reactions (see Fig. 17–5) and the activity of membrane transporters (Fig. 18–25) to start another cycle of the shuttle.

In skeletal muscle and brain, another type of NADH shuttle, the **glycerol-3-phosphate shuttle,** occurs (Fig. 18–26). It differs from the malate–aspartate shuttle in that it delivers the reducing equivalents from NADH into Complex III, not Complex I (Fig. 18–9), providing only enough energy to synthesize two ATP molecules per pair of electrons.

The mitochondria of higher plants have an externally oriented NADH dehydrogenase that is able to transfer electrons directly from cytosolic NADH into the respiratory chain.

Oxidative Phosphorylation Produces Most of the ATP Made in Aerobic Cells

The complete oxidation of a molecule of glucose to CO_2 yields two ATP and two NADH from glycolysis in the cytosol (Chapter 14); two NADH from pyruvate oxidation in the mitochondrial matrix (Chapter 15); and two ATP, six NADH, and two $FADH_2$ from citric acid cycle reactions in the matrix (Chapter 15). Each NADH produced in the matrix yields three ATP from mitochondrial oxidative phosphorylation, and for each $FADH_2$, two ATP are generated. Cytosolic NADH, after shuttling into the matrix, yields two or three ATP, depending on which shuttle is used. The total yield of glucose oxidation is therefore 36 or 38 ATP per glucose (Table 18–5).

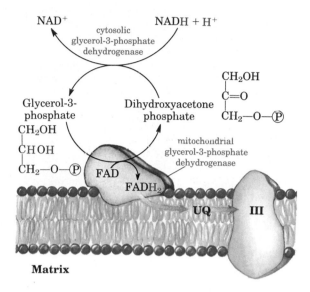

Figure 18–26 The glycerol-3-phosphate shuttle, an alternative means of moving reducing equivalents from the cytosol to the mitochondrial matrix. Dihydroxyacetone phosphate in the cytosol accepts two reducing equivalents from cytosolic NADH in a reaction catalyzed by cytosolic glycerol-3-phosphate dehydrogenase. A membrane-bound isozyme of glycerol-3-phosphate dehydrogenase, located on the outer face of the inner membrane, transfers two reducing equivalents from glycerol-3-phosphate in the intermembrane space to ubiquinone. Note that this shuttle does not involve membrane transport systems.

Table 18–5 ATP yield from complete oxidation of glucose

Process	Direct product	Final ATP
Glycolysis	2 NADH (cytosolic)	4 or 6*
	2 ATP	2
Pyruvate oxidation (two per glucose)	2 NADH (mitochondrial matrix)	6
Acetyl-CoA oxidation (two per glucose)	6 NADH (mitochondrial matrix)	18
	2 $FADH_2$	4
	2 ATP or 2 GTP	2
Total yield per molecule of glucose		36 or 38

* The number depends on which shuttle system is used to transfer reducing equivalents into the mitochondrial matrix.

Table 18–6 ATP yield from complete oxidation of palmitoyl-CoA*

Process	Direct product	Final ATP
β Oxidation	7 NADH	21
	7 FADH$_2$	14
	8 Acetyl-CoA	
Citric acid cycle	24 NADH	72
	8 FADH$_2$	16
	8 ATP or 8 GTP	8
Total yield per molecule of palmitoyl-CoA		131

* The activation of palmitate to palmitoyl-CoA costs two ATP.

By comparison, glycolysis under anaerobic conditions (lactate fermentation) yields only two ATP per glucose. Clearly, the evolution of oxidative phosphorylation provided a tremendous increase in the energetic efficiency of catabolism.

The oxidation of fatty acids and amino acids also takes place within the mitochondrial matrix, and the FADH$_2$ and NADH produced by these oxidative pathways also serve as electron donors for oxidative phosphorylation. Complete oxidation of the 16-carbon saturated fatty acid palmitate to CO_2 produces eight acetyl-CoAs, seven FADH$_2$, and seven NADH (Chapter 16). The oxidation of each acetyl-CoA via the citric acid cycle produces three NADH, one FADH$_2$, and one ATP (GTP). The gain in ATP is therefore 131 ATP (Table 18–6); but, because the activation of palmitate to palmitoyl-CoA costs two ATP equivalents, the net gain is 129 ATP per palmitate. A similar calculation may be made for the ATP yield upon oxidation of each of the amino acids (Chapter 17). The important conclusion here is that aerobic oxidative pathways that result in electron transfer to O_2 accompanied by oxidative phosphorylation account for the vast majority of the ATP produced in catabolism.

Oxidative Phosphorylation Is Regulated by Cellular Energy Needs

The rate of respiration (O_2 consumption) in mitochondria is under tight regulation; it is generally limited by the availability of ADP as a substrate for phosphorylation. As we saw in Figure 18–13b, the respiration rate in isolated mitochondria is low in the absence of ADP and increases strikingly with the addition of ADP; this phenomenon is part of the definition of coupling of oxidation and phosphorylation. The dependence of the rate of O_2 consumption on the concentration of the P_i acceptor ADP, called **acceptor control** of respiration, can be dramatic. In some animal tissues the **acceptor control ratio,** the ratio of the maximal rate of ADP-induced O_2 consumption to the basal rate in the absence of ADP, is at least 10.

The intracellular concentration of ADP is one measure of the energy status of cells. Another, related measure is the **mass-action ratio** of the ATP–ADP system: $[ATP]/([ADP][P_i])$. Normally this ratio is very high, so that the ATP–ADP system is almost fully phosphorylated. When the rate of some energy-requiring process in cells (protein synthesis, for example) increases, there is an increased rate of breakdown of ATP to ADP and P_i, lowering the mass-action ratio. With more ADP available for oxidative phosphorylation, the rate of respiration increases, causing regeneration of ATP. This continues until the mass-action ratio returns to its normal high level, at which point respiration slows again. The rate of oxidation of cell fuels is regulated with such sensitivity and precision that the ratio $[ATP]/([ADP][P_i])$ fluctuates only slightly in most tissues, even during extreme variations in energy demand. In short, ATP is formed only as fast as it is used in energy-requiring cell activities.

Uncoupled Mitochondria in Brown Fat Produce Heat

There is a remarkable and instructive exception to the general rule that respiration slows when the ATP supply is adequate. In most mammals, including humans, newborns have a type of adipose tissue called

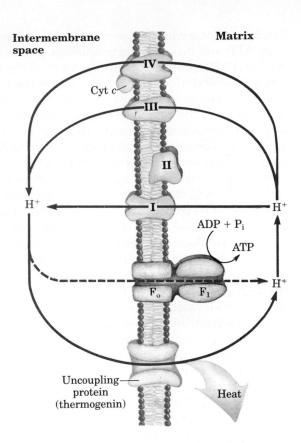

Figure 18–27 The uncoupling protein (thermogenin) of brown fat mitochondria, by providing an alternative route for protons to reenter the mitochondrial matrix, causes the energy conserved by proton pumping to be dissipated as heat.

brown fat in which fuel oxidation serves not to produce ATP, but to generate heat to keep the newborn warm. Hibernating mammals (grizzly bears, for example) also have large amounts of brown fat. This specialized adipose tissue, located at the back of the neck of human infants, is brown because of the presence of large numbers of mitochondria and thus large amounts of cytochromes, whose heme groups are strong absorbers of visible light.

The mitochondria of brown fat oxidize fuels (particularly fatty acids) normally, passing electrons through the respiratory chain to O_2. This electron transfer is accompanied by proton pumping out of the matrix, as in other mitochondria. The mitochondria of brown fat, however, have a unique protein in their inner membrane: **thermogenin,** also called the **uncoupling protein** (UCP) (Table 18–4). This protein, an integral membrane protein, provides a path for protons to return to the matrix without passing through the F_oF_1 complex (Fig. 18–27). As a result of this short-circuiting of protons, the energy of oxidation is not conserved by ATP formation but is dissipated as heat, which contributes to maintaining the body temperature. For the hairless newborn infant, maintaining body heat is an important use of metabolic energy. Hibernating animals depend on uncoupled mitochondria of brown fat to generate heat during their long winter period of dormancy (see Box 16–1).

ATP-Producing Pathways Are Coordinately Regulated

The major catabolic pathways (glycolysis, the citric acid cycle, fatty acid and amino acid oxidation, and oxidative phosphorylation) have interlocking and concerted regulatory mechanisms that allow them to

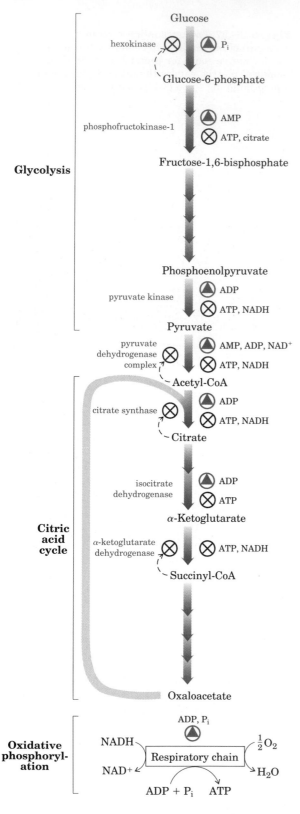

function together in an economical and self-regulating manner to produce ATP and biosynthetic precursors. The relative concentrations of ATP and ADP control not only the rates of electron transfer and oxidative phosphorylation but also the rates of the citric acid cycle, pyruvate oxidation, and glycolysis (Fig. 18–28). Whenever there is an increased drain on ATP, the rate of electron transfer and oxidative phosphorylation increases. Simultaneously, the rate of pyruvate oxidation via the citric acid cycle increases, thus increasing the flow of electrons into the respiratory chain. These events can in turn evoke an increase in the rate of glycolysis, increasing the rate of pyruvate formation. When the conversion of ADP to ATP lowers the ADP concentration, acceptor control will slow electron transfer and thus oxidative phosphorylation. Glycolysis and the citric acid cycle will also slow, because ATP is an allosteric inhibitor of phosphofructokinase-1 (see Fig. 14–20) and of pyruvate dehydrogenase (see Fig. 15–14).

Phosphofructokinase-1 is inhibited not only by ATP but by citrate, the first intermediate of the citric acid cycle. When both ATP and citrate are elevated they produce a concerted allosteric inhibition that is greater than the sum of their individual effects.

In many types of tumor cells, this interlocking coordination appears to be defective: glycolysis proceeds at a higher rate than required by the citric acid cycle. As a result, cancer cells use far more blood glucose than normal cells, but cannot oxidize the excess pyruvate formed by rapid glycolysis, even in the presence of O_2. To reoxidize cytoplasmic NADH, most of the pyruvate is reduced to lactate (Chapter 14), which passes from the cells into the blood. The high glycolytic rate may result in part from smaller numbers of mitochondria in tumor cells. In addition, some tumor cells overproduce an isozyme of hexokinase that associates with the cytosolic face of the mitochondrial inner membrane and is insensitive to feedback inhibition by glucose-6-phosphate (p. 432). This enzyme may monopolize the ATP produced in mitochondria, using it to produce glucose-6-phosphate and committing the cell to continue glycolysis.

Mutations in Mitochondrial Genes Cause Human Disease

Mitochondria contain their own genome, a circular, double-stranded DNA molecule. There are 37 genes (16,569 base pairs) in the human mitochondrial chromosome (Fig. 18–29), including 13 that encode proteins of the respiratory chain (Table 18–7); the remaining genes code for ribosomal and transfer RNA molecules essential to the protein-synthesizing machinery of mitochondria. Many of the mitochondrial proteins are encoded in nuclear genes, synthesized on cytoplasmic ribosomes, then imported and assembled within mitochondria.

Figure 18–28 Interlocking regulation of glycolysis, pyruvate oxidation, the citric acid cycle, and oxidative phosphorylation by the relative concentrations of ATP, ADP, and AMP, and by NADH. High [ATP] (or low [ADP] and [AMP]) produces low rates of glycolysis, pyruvate oxidation, acetate oxidation via the citric acid cycle, and oxidative phosphorylation. All four pathways are accelerated when there is an increase in the rate of ATP utilization and increased formation of ADP, AMP, and P_i. Interlocking of glycolysis and the citric acid cycle by citrate, which inhibits glycolysis, supplements the action of the adenine nucleotide system. In addition, increased levels of NADH and acetyl-CoA also inhibit the oxidation of pyruvate to acetyl-CoA, and high [NADH]/[NAD$^+$] ratios inhibit the dehydrogenase reactions of the citric acid cycle (p. 469).

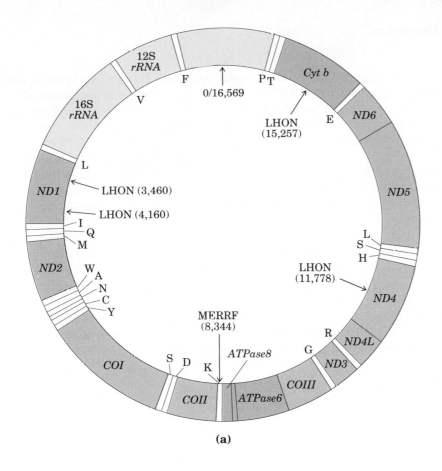

(a)

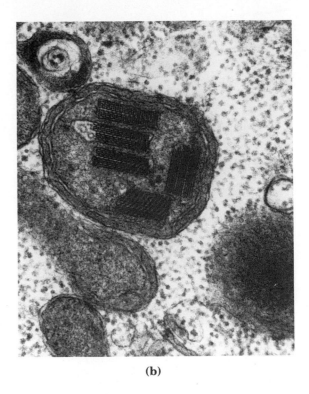

(b)

Figure 18–29 (a) Map of human mitochondrial DNA, showing the genes that encode proteins of Complex I, the NADH dehydrogenase (*ND1* to *ND6*); the cytochrome *b* of Complex III (*Cyt b*); the subunits of cytochrome oxidase (Complex IV) (*COI* to *COIII*); and two subunits of ATP synthase (*ATPase6* and *ATPase8*). The colors of the genes correspond to the colors of the complexes shown in Fig. 18–8. Also shown are the genes for ribosomal RNAs (*rRNA*) and for a number of mitochondrion-specific transfer RNAs. Transfer RNA specificity is indicated by the one-letter codes for amino acids. Arrows indicate the positions where alterations of base sequence (mutations) are known to cause Leber's hereditary optic neuropathy (LHON) and myoclonic epilepsy and ragged-red fiber disease (MERRF). The numbers in parentheses indicate the position of the altered nucleotides; nucleotide number 1 is at the top of the circle. **(b)** Electron micrograph of an abnormal mitochondrion from the muscle of an individual with MERRF, showing the paracrystalline protein inclusions sometimes present in the mutant mitochondria.

Table 18–7 Respiratory proteins encoded in the human mitochondrial chromosome

Complex	Total number of subunits	Number of subunits encoded in mitochondrial DNA
I NADH dehydrogenase	>25	7
II Succinate dehydrogenase	4	0
III Ubiquinone-cytochrome *c* oxidoreductase	9	1
IV Cytochrome oxidase	13	3
V ATP synthase	12	2

Mutations in the mitochondrial DNA are known to cause the human disease called **Leber's hereditary optic neuropathy** (LHON). This rare genetic disease affects the central nervous system, including the optic nerves, causing bilateral loss of vision, of rapid onset, in early adulthood. The disease is invariably inherited from the female parent, a consequence of the fact that all of the mitochondria of the developing embryo are derived from the mother. The unfertilized egg contains many mitochondria, and the few mitochondria in the much smaller sperm cell do not enter the egg at the time of fertilization. Mitochondria arise only from the division of preexisting mitochondria.

A single base change in the mitochondrial gene *ND4* (Fig. 18–29a) changes an Arg residue to a His residue in one of the proteins of Complex I, and the result is mitochondria partially defective in electron transfer from NADH to ubiquinone. Although these mitochondria are able to produce some ATP by electron transfer from succinate, they apparently cannot supply ATP in sufficient quantity to support the very active metabolism of neurons. One result is damage to the optic nerve, leading to blindness.

A single base change in the mitochondrial gene for cytochrome b, a component of Complex III, also produces LHON, demonstrating that the pathology of the disease results from a general reduction of mitochondrial function, not just from a defect in electron transfer through Complex I.

Another serious human genetic disease, **myoclonic epilepsy and ragged-red fiber disease** (MERRF) is caused by a mutation in the mitochondrial gene that encodes a transfer RNA (Fig. 18–29). This disease, characterized by uncontrollable muscular jerking, apparently results from defective production of several of the proteins synthesized using mitochondrial transfer RNAs. Skeletal muscle fibers of individuals with MERRF have abnormally shaped mitochondria that sometimes contain paracrystalline structures (Fig. 18–29b).

Mitochondria Probably Evolved from Endosymbiotic Bacteria

The fact that mitochondria contain their own DNA, ribosomes, and transfer RNAs supports the theory of the endosymbiotic origin of mitochondria (see Fig. 2–17). This theory supposes that the first organisms capable of aerobic metabolism, including respiration-linked ATP production, were prokaryotes. Primitive eukaryotes that lived anaerobically (by fermentation) acquired the ability to carry out oxidative phosphorylation when they established a symbiotic relationship with bacteria living in their cytosol. After much evolution and the movement of many bacterial genes into the nucleus of the "host" eukaryote, the endosymbiotic bacteria eventually became mitochondria.

This theory presumes that early free-living prokaryotes had the enzymatic machinery for oxidative phosphorylation, and it predicts that their modern prokaryotic descendents have respiratory chains closely similar to those of modern eukaryotes. They do.

Aerobic bacteria carry out NAD-linked electron transfer from substrates to O_2, coupled to the phosphorylation of cytosolic ADP. The dehydrogenases are located in the bacterial cytosol, but the electron carriers of the respiratory chain are in the plasma membrane. The electron carriers are similar to those of mitochondria, act in the same sequence (Fig. 18–30), and translocate protons outward across the plasma membrane concomitantly with electron transfer to O_2. We noted earlier that bacteria such as *E. coli* have F_oF_1 complexes in their plasma membranes; the F_1 portions protrude into the cytosol and cata-

Figure 18–30 Respiratory chain in the inner membrane of *E. coli*. Electrons from NADH pass to menaquinone (MQ), at which point the chain branches. The upper path is dominant in cells grown under normal aerobic conditions, but when O_2 is the limiting factor in growth, the lower route predominates.

$$\text{NADH} \longrightarrow \text{MQ} \begin{array}{c} \nearrow \text{Cyt } b_{562} \longrightarrow \text{Cyt } o \longrightarrow \text{Cyt } c \longrightarrow O_2 \\ \\ \searrow \text{Cyt } b_{558} \longrightarrow \text{Cyt } b_{595} \longrightarrow \text{Cyt } d \longrightarrow O_2 \end{array}$$

lyze ATP synthesis from ADP and P_i as protons flow back into the cell through proton channels formed by F_o.

Certain bacterial transport systems bring about uptake of extracellular nutrients (lactose, for example) against a concentration gradient, in symport with protons (see Fig. 10–25). The respiration-linked transmembrane proton extrusion provides the driving force for this uptake. The rotary motion of bacterial flagella, which move cells through their surroundings, is provided by "proton turbines," molecular rotary motors driven not by ATP but directly by the transmembrane electrochemical potential generated by respiration-linked proton pumping (Fig. 18–31). It appears likely that the chemiosmotic mechanism evolved early (before the emergence of eukaryotes). The protonmotive force can clearly be used to power processes other than ATP synthesis.

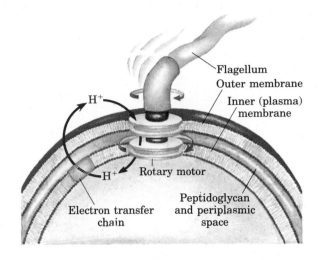

Figure 18–31 Rotation of bacterial flagella by proton-motive force. The shaft and rings at the base of the flagellum make up a rotary motor that has been called a "proton turbine." Protons ejected by electron transfer flow back into the cell through the "turbine," causing rotation of the shaft of the flagellum. This motion differs fundamentally from the motion of muscle or of eukaryotic flagella and cilia, for which ATP hydrolysis is the energy source.

Photosynthesis: Harvesting Light Energy

We now turn to another reaction sequence in which the flow of electrons is coupled to the synthesis of ATP: light-driven phosphorylation. The capture of solar energy by photosynthetic organisms and its conversion into the chemical energy of reduced organic compounds is the ultimate source of nearly all biological energy. Photosynthetic and heterotrophic organisms live in a balanced steady state in the biosphere (Fig. 18–32). Photosynthetic organisms trap solar energy and form ATP and NADPH, which they use as energy sources to make carbohydrates and other organic components from CO_2 and H_2O; simultaneously, they release O_2 into the atmosphere. Aerobic heterotrophs (we humans, for example) use the O_2 so formed to degrade the energy-rich organic products of photosynthesis to CO_2 and H_2O, generating ATP for their own activities. The CO_2 formed by respiration in heterotrophs returns to the atmosphere, to be used again by photosynthetic organisms. Solar energy thus provides the driving force for the continuous cycling of atmospheric CO_2 and O_2 through the biosphere and provides the reduced substrates (fuels), such as glucose, on which nonphotosynthetic organisms depend.

Enormous amounts of energy are stored as products of photosynthesis. Each year at least 10^{17} kJ of free energy from sunlight is captured and used for biosynthesis by photosynthetic organisms. This is more than ten times the fossil-fuel energy used each year by people the world over. Even fossil fuels (coal, oil, and natural gas) are the products of photosynthesis that took place millions of years ago. Because of our global dependence upon solar energy, past and present, for both energy and food, discovering the mechanism of photosynthesis is a central goal of biochemical research.

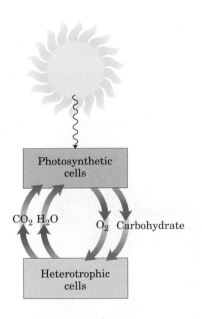

Figure 18–32 Solar energy is the ultimate source of all biological energy. Photosynthetic organisms use the energy of sunlight to manufacture glucose and other organic cell products, which heterotrophic cells use as energy and carbon sources.

The overall equation of photosynthesis describes an oxidation–reduction reaction in which H_2O donates electrons (as hydrogen) for the reduction of CO_2 to carbohydrate (CH_2O):

$$CO_2 + H_2O \xrightarrow{\text{light}} O_2 + (CH_2O)$$

Unlike NADH (the hydrogen donor in oxidative phosphorylation), H_2O is a poor electron donor; its standard reduction potential is 0.82 V, compared with -0.32 V for NADH. The central difference between photophosphorylation and oxidative phosphorylation is that the latter process begins with a good electron donor, whereas the former requires the input of energy in the form of light to *create* a good electron donor. Except for this crucial difference, the two processes are remarkably similar. In photophosphorylation, electrons flow through a series of membrane-bound carriers including cytochromes, quinones, and iron–sulfur proteins, while protons are pumped across a membrane to create an electrochemical potential. This potential is the driving force for ATP synthesis from ADP and P_i by a membrane-bound ATP synthase complex closely similar to that which functions in oxidative phosphorylation.

Photosynthesis encompasses two processes: the **light reactions,** which occur only when plants are illuminated, and the **carbon fixation reactions,** or so-called dark reactions, which occur in both light and darkness (Fig. 18–33). In the light reactions, chlorophyll and other pigments of the photosynthetic cells absorb light energy and conserve it in chemical form as the two energy-rich products ATP and NADPH; simultaneously, O_2 is evolved. In the carbon fixation reactions, ATP and NADPH are used to reduce CO_2 to form glucose and other organic products. The formation of O_2, which occurs only in the light, and the reduction of CO_2, which does not require light, thus are distinct and separate processes. In this chapter we are concerned only with the light reactions; the reduction of CO_2 is described in Chapter 19.

In photosynthetic eukaryotic cells, both the light and carbon fixation reactions take place in the chloroplasts (Fig. 18–34). When solar energy is not available, mitochondria in the plant cell generate ATP by oxidizing carbohydrates originally produced in chloroplasts in the light.

Chloroplasts may assume many different shapes in different species, and they usually have a much larger volume than mitochondria.

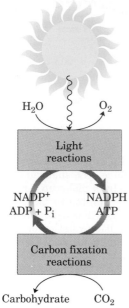

H_2O O_2

Light reactions

NADP⁺   NADPH
ADP + P_i ATP

Carbon fixation reactions

Carbohydrate CO_2

Figure 18–33 The light reactions generate energy-rich NADPH and ATP at the expense of solar energy. These products are used in the carbon fixation reactions, which occur in light or darkness, to reduce CO_2 to form trioses and more complex compounds (such as glucose) derived from trioses.

Figure 18–34 Schematic diagram **(a)** and electron micrograph **(b)** of a single chloroplast at high magnification.

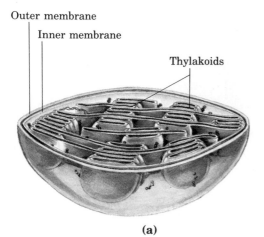

Outer membrane
Inner membrane
Thylakoids

(a)

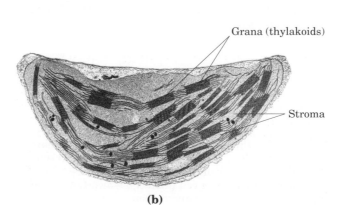

Grana (thylakoids)

Stroma

(b)

They are surrounded by a continuous outer membrane, which, like the outer mitochondrial membrane, is permeable to small molecules and ions. An inner membrane system encloses the internal compartment, in which there are many flattened, membrane-surrounded vesicles or sacs, called **thylakoids,** which are usually arranged in stacks called **grana.** The thylakoid membranes are separate from the inner chloroplast membrane. Embedded in the thylakoid membranes are the photosynthetic pigments and all the enzymes required for the primary light reactions. The fluid in the compartment surrounding the thylakoids, the **stroma,** contains most of the enzymes required for the carbon fixation reactions, in which CO_2 is reduced to form triose phosphates and, from them, glucose.

Our discussion here focuses on the nature of the light-absorbing systems and their roles in photosynthesis.

Light Produces Electron Flow in Chloroplasts

How do the pigment molecules of the thylakoid membranes transduce absorbed light energy into chemical energy? The key to answering this question came from a discovery made in 1937 by Robert Hill, a pioneer in photosynthesis research. He found that when leaf extracts containing chloroplasts were supplemented with a nonbiological hydrogen acceptor and then illuminated, evolution of O_2 and simultaneous reduction of the hydrogen acceptor took place, according to an equation now known as the **Hill reaction:**

$$2H_2O + 2A \xrightarrow{\text{light}} 2AH_2 + O_2$$

where A is the artificial hydrogen acceptor. One of the nonbiological hydrogen acceptors used by Hill was the dye 2,6-dichlorophenolindophenol, now called a **Hill reagent,** which in its oxidized form (A) is blue and in its reduced form (AH_2) is colorless. When the leaf extract supplemented with the dye was illuminated, the blue dye became colorless and O_2 was evolved. In the dark neither O_2 evolution nor dye reduction took place. This was the first specific clue to how absorbed light energy is converted into chemical energy: it causes electrons to flow from H_2O to an electron acceptor. Moreover, Hill found that CO_2 was not required for this reaction, nor was it reduced to a stable form under these conditions. He therefore concluded that O_2 evolution can be dissociated from CO_2 reduction. Several years later it was found that $NADP^+$ is the biological electron acceptor in chloroplasts, according to the equation

$$2H_2O + 2NADP^+ \xrightarrow{\text{light}} 2NADPH + 2H^+ + O_2$$

This equation shows an important distinction between mitochondrial oxidative phosphorylation and the analogous process in chloroplasts: in chloroplasts electrons flow from H_2O to $NADP^+$, whereas in mitochondrial respiration electrons flow in the opposite direction, from NADH or NADPH to O_2, with the release of free energy. Because light-induced electron flow in chloroplasts is in the reverse or "uphill" direction, from H_2O to $NADP^+$, it cannot occur without the input of free energy; this energy comes from light. To understand how this occurs, we must first consider the effects of light absorption on molecular structure.

Oxidized form (blue) Reduced form (colorless)

Dichlorophenolindophenol

Absorption of Light Excites Molecules

Visible light is electromagnetic radiation of wavelengths 400 to 700 nm, a small part of the electromagnetic spectrum (Fig. 18–35), ranging from violet to red. The energy of a single **photon** (a quantum of light) is greater at the violet end of the spectrum than at the red end. The energy in a "mole" of photons (one einstein; 6×10^{23} photons) is 170 to 300 kJ, about an order of magnitude more than the energy required to synthesize a mole of ATP from ADP and P_i (about 30 kJ under standard conditions). The energy of an einstein in the infrared or microwave regions of the spectrum is too small to be useful in the kinds of photochemical events that occur in photosynthesis. UV light and x rays, on the other hand, have so much energy that they damage proteins and nucleic acids and induce mutations that are often lethal.

Figure 18–35 The spectrum of electromagnetic radiation, and the energy of photons in the visible range of the spectrum. One einstein is 6×10^{23} photons.

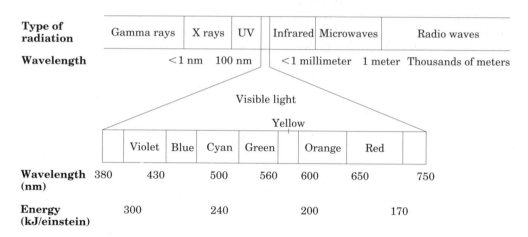

The ability of a molecule to absorb light depends upon the arrangement of electrons around the atomic nuclei in its structure. When a photon is absorbed, an electron is lifted to a higher energy level. This happens on an all-or-none basis; to be absorbed, the photon must contain a quantity of energy (a **quantum**) that exactly matches the energy of the electronic transition. A molecule that has absorbed a photon is in an **excited state,** which is generally unstable. The electrons lifted into higher-energy orbitals usually return rapidly to their normal lower-energy orbitals; the excited molecule reverts (decays) to the stable **ground state,** giving up the absorbed quantum as light or heat or using it to do chemical work. The light emitted upon decay of excited molecules, called **fluorescence,** is always of a longer wavelength (lower energy) than the absorbed light. Excitation of molecules by light and their fluorescent decay are extremely fast processes, occurring in about 10^{-15} and 10^{-12} s, respectively. The initial events in photosynthesis are therefore very rapid.

Chlorophylls Absorb Light Energy for Photosynthesis

The most important light-absorbing pigments in the thylakoid membranes are the **chlorophylls,** green pigments with polycyclic, planar structures resembling the protoporphyrin of hemoglobin (see Fig. 7–18), except that Mg^{2+}, not Fe^{2+}, occupies the central position (Fig. 18–36). Chlorophyll *a*, present in the chloroplasts of all green plant cells, contains four substituted pyrrole rings, one of which (ring IV) is reduced, and a fifth ring that is not a pyrrole. All chlorophylls have a long **phytol** side chain, esterified to a carboxyl-group substituent in ring IV. The four inward-oriented nitrogen atoms of chlorophyll *a* are coordinated with the Mg^{2+}.

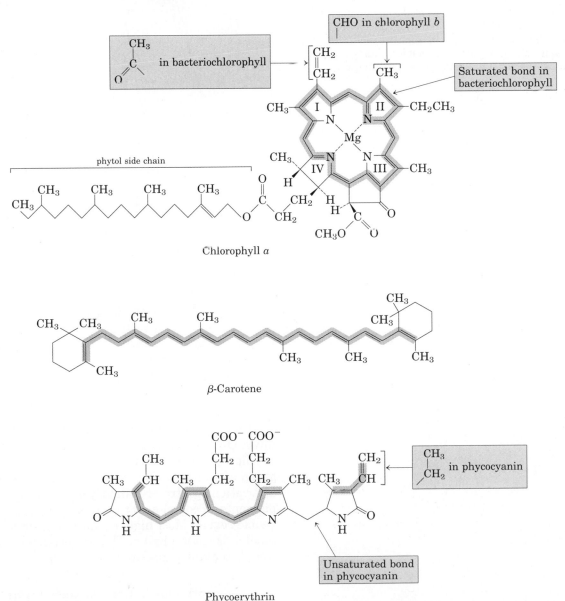

Chlorophyll *a*

β-Carotene

Phycoerythrin

Figure 18–36 Structures of the primary photopigments chlorophylls *a* and *b* and bacteriochlorophyll, and of the accessory pigments β-carotene (a carotenoid) and phycoerythrin and phycocyanin (phycobilins). The areas shaded pink represent conjugated systems (alternating single and double bonds), which largely account for the absorption of visible light.

The heterocyclic five-ring system that surrounds the Mg^{2+} has an extended polyene structure, with alternating single and double bonds. Such polyenes characteristically show strong absorption in the visible region of the spectrum (Fig. 18–37); the chlorophylls have unusually high molar absorption coefficients (see Box 5–1) and are therefore particularly well-suited for absorbing visible light during photosynthesis.

Figure 18–37 Absorption of visible light by photopigments shown in Fig. 18–36. Plants are green because their pigments absorb light from the red and violet regions of the spectrum, leaving primarily green light to be reflected or transmitted. Compare the absorption spectra of the pigments with the spectrum of sunlight reaching the earth's surface; the combination of chlorophylls (chl a and chl b) and accessory pigments enables plants to harvest most of the energy available from sunlight.

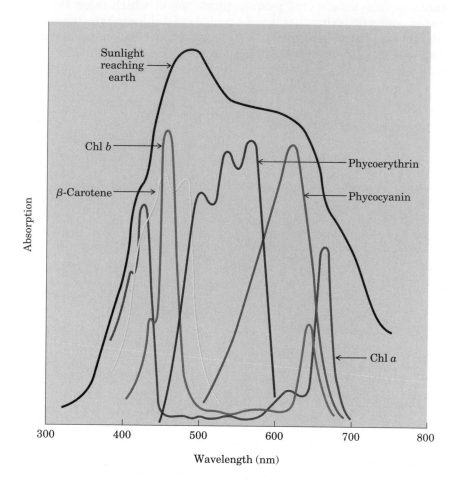

Chloroplasts of higher plants always contain two types of chlorophyll. One is invariably chlorophyll a, and the second in many species is chlorophyll b, which has an aldehyde group instead of a methyl group attached to ring II (Fig. 18–36). Although both are green, their absorption spectra are slightly different (Fig. 18–37), allowing the two pigments to complement each other's range of light absorption in the visible region. Most higher plants contain about twice as much chlorophyll a as chlorophyll b. The bacterial chlorophylls differ only slightly from the plant pigments (Fig. 18–36).

Accessory Pigments Also Absorb Light

In addition to chlorophylls, the thylakoid membranes contain secondary light-absorbing pigments, together called the **accessory pigments,** the carotenoids and phycobilins. **Carotenoids** may be yellow, red, or purple. The most important are **β-carotene** (Fig. 18–36), a red-orange isoprenoid compound that is the precursor of vitamin A in animals, and the yellow carotenoid **xanthophyll.** The carotenoid pigments absorb light at wavelengths other than those absorbed by the chlorophylls (Fig. 18–37) and thus are supplementary light receptors. **Phycobilins** are linear tetrapyrroles that have the extended polyene system found in chlorophylls, but not their cyclic structure or central Mg^{2+}. Examples are phycoerythrin and phycocyanin (Fig. 18–36).

The relative amounts of the chlorophylls and the accessory pigments are characteristic for different plant species. It is variation in the proportions of these pigments that is responsible for the range of colors of photosynthetic organisms, which vary from the deep blue-green of spruce needles, to the greener green of maple leaves, to the red, brown, or even purple color of different species of multicellular algae and the leaves of some decorative plants.

Experimental determination of the effectiveness of light of different colors in promoting photosynthesis yields an **action spectrum** (Fig. 18–38), often useful in identifying the pigment primarily responsible for a biological effect of light. By capturing light in a region of the spectrum not used by other plants, a photosynthetic organism can claim its unique ecological niche. For example, the phycobilins, present only in red algae and cyanobacteria, absorb in the region 520 to 630 nm, allowing these organisms to live in niches where light of lower or higher wavelength has been filtered out by the pigments of other organisms living in the water above them, or by the water itself.

Chlorophyll Funnels Absorbed Energy to Reaction Centers

The light-absorbing pigments of thylakoid membranes are arranged in functional sets or arrays called **photosystems.** In spinach chloroplasts each photosystem contains about 200 molecules of chlorophylls and about 50 molecules of carotenoids. The clusters can absorb light over the entire visible spectrum but especially well between 400 to 500 nm and 600 to 700 nm (Fig. 18–37). All the pigment molecules in a photosystem can absorb photons, but only a few can transduce the light energy into chemical energy. A transducing pigment consists of several chlorophyll molecules combined with a protein complex also containing tightly bound quinones; this complex is called a **photochemical reaction center.** The other pigment molecules in a photosystem are called **light-harvesting** or **antenna molecules.** They function to absorb light energy and transmit it at a very high rate to the reaction center where the photochemical reactions occur (Fig. 18–39, p. 578), which will be described in detail later.

The chlorophyll molecules in thylakoid membranes are bound to integral membrane proteins (chlorophyll *a/b*-binding, or CAB, proteins) that orient the chlorophyll relative to the plane of the membrane and confer light absorption properties that are subtly different from those of free chlorophyll. When isolated chlorophyll molecules in vitro are excited by light, the absorbed energy is quickly released as fluorescence and heat, but when chlorophyll in intact spinach leaves is excited by visible light (Fig. 18–40 (step ①), p. 579), very little fluorescence is observed. Instead, a direct transfer of energy from the excited chlorophyll (an antenna chlorophyll) to a neighboring chlorophyll molecule occurs, exciting the second molecule and allowing the first to return to its ground state (step ②). This **resonance energy transfer** is repeated to a third, fourth, or subsequent neighbor, until the chlorophyll at the photochemical reaction center becomes excited (step ③). In this special chlorophyll molecule, an electron is promoted by excitation to a higher-energy orbital. This electron then passes to a nearby electron

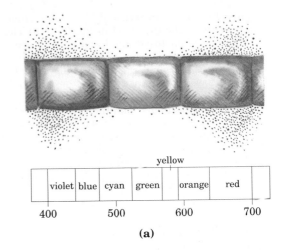

(a)

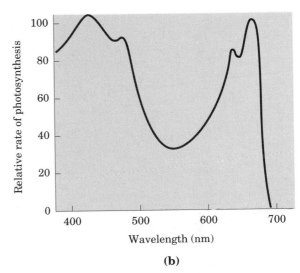

(b)

Figure 18–38 Two ways to determine the action spectrum for photosynthesis. **(a)** The results of a classic experiment done by T.W. Englemann in 1882 to determine what wavelength of light was most effective in supporting photosynthesis. Englemann placed a filamentous, photosynthetic alga on a microscope stage and illuminated it with light from a prism, so that cells in one part of the filament received mainly blue light, another yellow, another red. To determine which cells carried out photosynthesis most actively, bacteria known to migrate toward regions of high O_2 concentration were also placed on the microscope slide. The distribution of bacteria showed highest O_2 levels (produced by photosynthesis) in the regions illuminated with violet and red light. **(b)** A similar experiment using modern techniques for the measurement of O_2 production yields the same result. An action spectrum describes the relative rate of photosynthesis for illumination with a constant number of photons of different wavelengths. Such an action spectrum is useful because it suggests (by comparison with absorption spectra such as those in Fig. 18–37) which pigments are able to channel energy into photosynthesis.

Figure 18–39 Organization of the photosystem components in the thylakoid membrane. **(a)** The distribution of photosystems I and II, ATP synthase, and the cytochrome *bf* complex in the thylakoid membranes is not random. Photosystem I and ATP synthase are almost completely excluded from the regions with tightly stacked membranes, whereas photosystem II and the cytochrome *bf* complex are enriched in these regions of tight packing. This separation of photosystems I and II prevents energy absorbed by photosystem II from being transferred directly to photosystem I, and also places photosystem I in the regions most accessible to NADP⁺ from the stroma. **(b)** An enlargement of a photosystem showing the reaction center, antenna chlorophylls, and accessory pigments. Cytochrome *bf* and ATP synthase of chloroplasts are described later in this chapter.

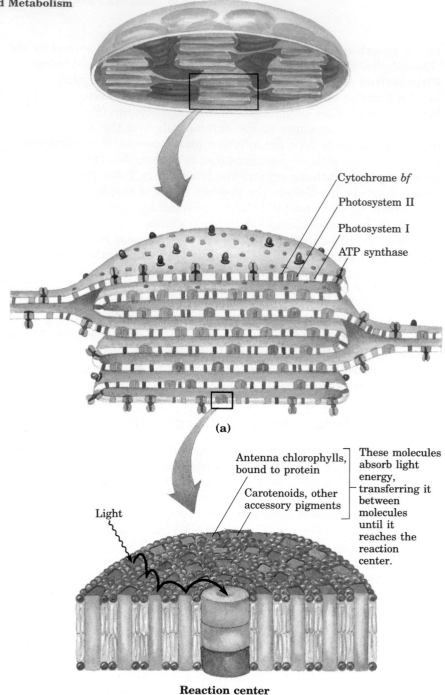

Cytochrome *bf*

Photosystem II

Photosystem I

ATP synthase

(a)

Antenna chlorophylls, bound to protein

Carotenoids, other accessory pigments

These molecules absorb light energy, transferring it between molecules until it reaches the reaction center.

Light

Reaction center
Photochemical reaction here converts the energy of a photon into a separation of charge, initiating electron flow.

(b)

acceptor that is part of the electron transfer chain of the chloroplast, leaving the excited chlorophyll molecule with an empty orbital (an "electron hole") (step ④). The electron acceptor thus acquires a negative charge. The electron lost by the reaction-center chlorophyll is replaced by an electron from a neighboring electron donor molecule (step ⑤), which becomes positively charged. In this way, *excitation by light causes electric charge separation and initiates an oxidation–reduction chain.* Coupled to the light-dependent electron flow along this chain are processes that generate ATP and NADPH.

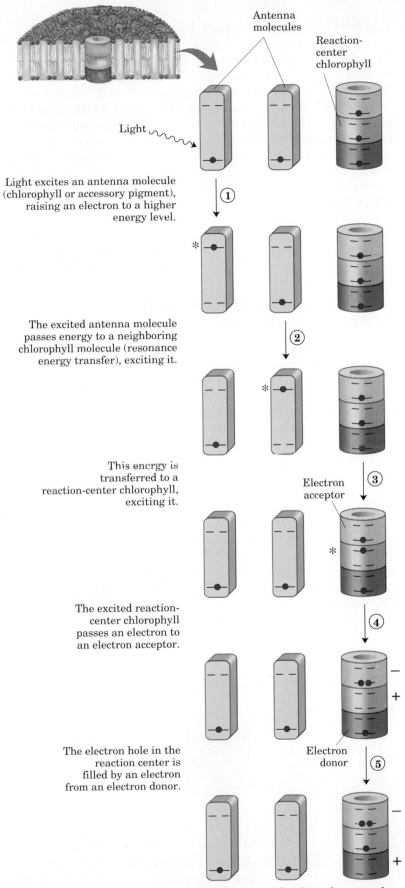

Antenna molecules

Reaction-center chlorophyll

Light

Light excites an antenna molecule (chlorophyll or accessory pigment), raising an electron to a higher energy level.

The excited antenna molecule passes energy to a neighboring chlorophyll molecule (resonance energy transfer), exciting it.

This energy is transferred to a reaction-center chlorophyll, exciting it.

Electron acceptor

The excited reaction-center chlorophyll passes an electron to an electron acceptor.

The electron hole in the reaction center is filled by an electron from an electron donor.

Electron donor

The absorption of a photon has caused separation of charge in the reaction center.

Figure 18–40 A generalized scheme showing the conversion of energy from an absorbed photon into separation of charges at the photosystem reaction center. The steps are further described in the text. Note that step ② may be repeated a number of times between successive antenna molecules until a reaction-center chlorophyll is reached. The asterisk (*) represents the excited state of an antenna molecule.

Light-Driven Electron Flow

Thylakoid membranes have two different kinds of photosystems, each with its own type of photochemical reaction center and a set of antenna molecules. The two systems have distinct and complementary functions. **Photosystem I** has a reaction center designated **P700** and a high ratio of chlorophyll a to chlorophyll b. **Photosystem II,** with its reaction center **P680,** contains roughly equal amounts of chlorophyll a and b and may also contain a third type, chlorophyll c. The thylakoid membranes of a single spinach chloroplast have many hundreds of each kind of photosystem. All O_2-evolving photosynthetic cells—those of higher plants, algae, and cyanobacteria—contain both photosystems I and II; all other species of photosynthetic bacteria, which do not evolve O_2, contain only photosystem I.

It is between photosystems I and II that light-driven electron flow occurs, producing NADPH and a transmembrane proton gradient.

Light Absorption by Photosystem II Initiates Charge Separation

How can light energy captured by chloroplasts induce electrons to flow energetically "uphill"? Excitation briefly creates a chemical species of very low standard reduction potential—an excellent electron donor. Excited P680, designated P680*, within picoseconds transfers an electron to **pheophytin** (a chlorophyll-like accessory pigment lacking Mg^{2+}), giving it a negative charge (designated Ph$^-$) (Fig. 18–41a). With the loss of its electron, P680* is transformed into a radical cation, designated P680$^+$. Thus excitation, in creating Ph$^-$ and P680$^+$, causes charge separation. Ph$^-$ very rapidly passes its extra electron to a protein-bound **plastoquinone,** Q_A, which in turn passes its electron to another, more loosely bound quinone, Q_B (Fig. 18–42; see also Fig. 18–44). When Q_B has acquired two electrons in two such transfers from Q_A and two protons from the solvent water, it is in its fully reduced quinol form, Q_BH_2. This molecule dissociates from its protein and diffuses away from the photochemical reaction center, carrying in its chemical bonds some of the energy of the photons that originally excited P680. The overall reaction initiated by light in photosystem II is therefore

$$4\ P680 + 4H^+ + 2Q_B + light\ (4\ photons) \longrightarrow 4\ P680^+ + 2Q_BH_2 \quad (18\text{–}4)$$

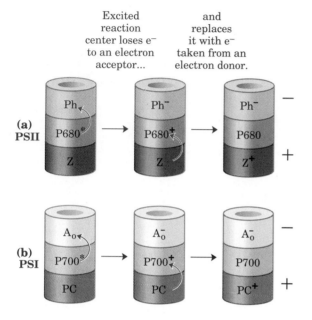

**(a)
PSII**

**(b)
PSI**

Figure 18–41 Photochemical events following excitation of photosystems by light absorption. (The steps shown here are equivalent to steps ④ and ⑤ in Fig. 18–40.) **(a)** Photosystem II (PSII). Z represents a Tyr residue in the D1 protein of PSII; Ph, pheophytin. **(b)** Photosystem I (PSI). A_o is a chlorophyll molecule near the reaction center of PSI; it accepts an electron from P700 to become the powerful reducing agent A_o^-. PC, plastocyanin.

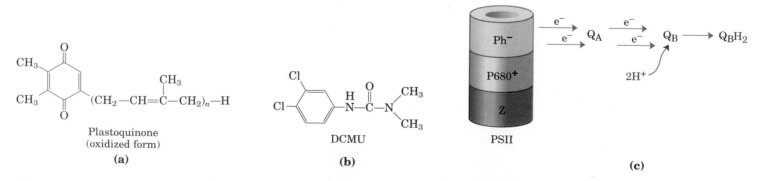

Figure 18–42 (a) Plastoquinone (Q_A). **(b)** The herbicide DCMU (3-(3,4-dichlorophenyl)-1,1-dimethylurea), which displaces Q_B from its binding site in photosystem II and blocks electron transfer from photosystem II to photosystem I. **(c)** The role of Q_A and Q_B in transferring electrons away from photosystem II. Q_BH_2 carries some of the energy of light absorbed by PSII.

Eventually, the electrons in Q_BH_2 are transferred through a chain of membrane-bound carriers to $NADP^+$, reducing it to NADPH and releasing H^+ (as described later). The potent herbicide DCMU (Fig. 18–42) competes with Q_B for the Q_B binding site in photosystem II, thus blocking photosynthetic electron transfer (Table 18–4).

In the meantime, $P680^+$ must acquire an electron to return to its ground state in preparation for the capture of another photon of energy (Fig. 18–41a). In principle, the required electron might come from any number of organic or inorganic compounds. Photosynthetic bacteria can use a variety of electron donors for this purpose—acetate, succinate, malate, or sulfide—depending on what is available in a particular ecological niche. About 3 billion years ago, evolution of primitive photosynthetic bacteria (the progenitors of the modern cyanobacteria) produced a photosystem capable of taking electrons from a donor that is always available—water. In this process two water molecules are split, yielding four electrons, four protons, and molecular oxygen: $2H_2O \longrightarrow 4H^+ + 4e^- + O_2$. A single photon of visible light does not possess enough energy to break the bonds in water; four photons are required in this photolytic cleavage reaction.

The four electrons abstracted from water do not pass directly to $P680^+$, which can only accept one electron at a time. Instead, a remarkable molecular device, the **water-splitting complex,** passes four electrons one at a time to $P680^+$. The immediate electron donor to $P680^+$ is a Tyr residue (often represented by the symbol Z) in protein D1 of the photosystem II reaction center:

$$4\ P680^+ + 4Z \longrightarrow 4\ P680 + 4Z^+ \tag{18–5}$$

This Tyr residue regains its missing electron by oxidizing a cluster of four manganese ions in the water-splitting complex. With each single-electron transfer, this Mn cluster becomes more oxidized; four single-electron transfers, each corresponding to the absorption of one photon, produce a charge of +4 on the Mn complex (Fig. 18–43):

$$4Z^+ + [\text{Mn complex}]^0 \longrightarrow 4Z + [\text{Mn complex}]^{4+} \tag{18–6}$$

In this state, the Mn complex can take four electrons from a pair of water molecules, releasing $4H^+$ and O_2:

$$[\text{Mn complex}]^{4+} + 2H_2O \longrightarrow [\text{Mn complex}]^0 + 4H^+ + O_2 \tag{18–7}$$

The sum of Equations 18–4 through 18–7 is

$$2H_2O + 2Q_B + 4\ \text{photons} \longrightarrow O_2 + 2Q_BH_2 \tag{18–8}$$

The water-splitting activity is an integral part of the photosystem II reaction center, and it has proved exceptionally difficult to purify. The detailed structure of the Mn cluster is not yet known. Manganese can exist in stable oxidation states from +2 to +7, so a cluster of four Mn ions can certainly donate or accept four electrons; the chemical details of this process, however, remain to be clarified.

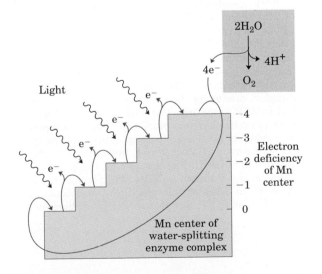

Figure 18–43 The four-step process that produces a four-electron oxidizing agent, believed to be a complex of several Mn ions, in the water-splitting complex of photosystem II. The sequential absorption of four photons, each causing the loss of one electron from the Mn center, produces an oxidizing agent that can take four electrons from two molecules of water, producing O_2. The electrons lost from the Mn center pass one at a time to a Tyr residue (Z^+) in a reaction-center protein.

Light Absorption by Photosystem I Creates a Powerful Reducing Agent

The photochemical events that follow excitation of photosystem I (P700) are formally similar to those in photosystem II (following the general scheme of Figure 18–40). Light is first captured by any one of about 200 chlorophyll a and b molecules, or additional accessory pigments, that serve as antennae, and the absorbed energy moves to P700 by resonance energy transfer. The excited reaction center P700* loses an electron to an acceptor, A_0 (believed to be a special form of chlorophyll, functionally analogous to the pheophytin of photosystem II), creating A_0^- and P700$^+$ (Fig. 18–41b); again, excitation has resulted in charge separation at the photochemical reaction center. P700$^+$ is a strong oxidizing agent, which quickly acquires an electron from **plastocyanin,** a soluble Cu-containing electron transfer protein.

A_0^- is an exceptionally strong reducing agent, which passes its electron through a chain of carriers that leads to NADP$^+$ (Fig. 18–44). First, **phylloquinone** (A_1) accepts an electron from A_0^- and passes it on to an iron–sulfur protein. From here, the electron moves to **ferredoxin** (Fd), another iron–sulfur protein loosely associated with the thylakoid membrane. Spinach ferredoxin (M_r 10,700), which has been isolated and crystallized, contains a 2Fe–2S center (Fig. 18–5). The Fe atoms of ferredoxin transfer electrons via one-electron Fe^{2+} to Fe^{3+} valence changes.

Figure 18–44 The integration of photosystems I and II. This "Z scheme" shows the pathway of electron transfer from H_2O (lower left) to NADP$^+$ (upper right) in noncyclic photosynthesis. The position on the vertical scale of each electron carrier reflects its standard reduction potential. To raise the energy of electrons derived from H_2O to the energy level required to reduce NADP$^+$ to NADPH, each electron must be "lifted" twice (heavy arrows) by photons absorbed in photosystems I and II. One photon is required per electron boosted in each photosystem. After each excitation, the high-energy electrons flow "downhill" via the carrier chains shown. Protons move across the thylakoid membrane during the water-splitting reaction and during electron transfer through the cytochrome bf complex, producing the proton gradient that is central to ATP formation. The dashed arrow is the path of cyclic electron transfer, in which only photosystem I is involved; electrons return via the cyclic pathway to photosystem I, instead of reducing NADP$^+$ to NADPH. Ph, pheophytin; Q_A, plastoquinone; Q_B, a second quinone; PC, plastocyanin; A_o, electron acceptor chlorophyll; A_1, phylloquinone; Fd, ferredoxin; FP, ferredoxin-NADP$^+$ oxidoreductase.

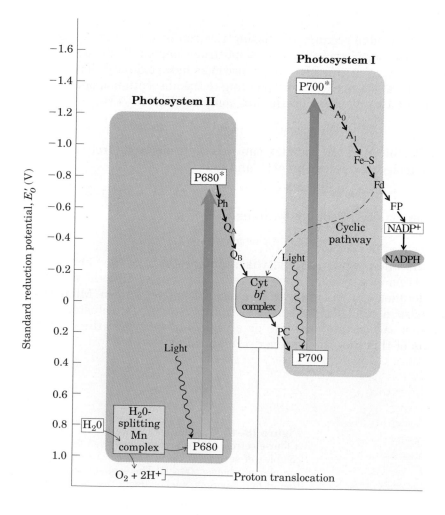

The fourth electron carrier in the chain is a flavoprotein called **ferredoxin-NADP$^+$ oxidoreductase.** It transfers electrons from reduced ferredoxin (Fd$^{2+}_{red}$) to NADP$^+$, reducing the latter to NADPH:

$$2Fd^{2+}_{red} + 2H^+ + NADP^+ \longrightarrow 2Fd^{3+}_{ox} + NADPH + H^+$$

Photosystems I and II Cooperate to Carry Electrons from H$_2$O to NADP$^+$

Early studies by Robert Emerson established that maximum rates of photosynthesis, measured as O$_2$ evolution, required light of at least two wavelengths, now known to excite photosystems I and II. The two photosystems must function together in the O$_2$-evolving light reactions of photosynthesis. The diagram in Figure 18–44, often called the **Z scheme** because of its overall form, outlines the pathway of electron flow between the two photosystems as well as the energy relationships in the light reactions.

When photons are absorbed by photosystem I, electrons are expelled from the reaction center and flow down a chain of electron carriers to NADP$^+$ to reduce it to NADPH. P700$^+$, transiently electron-deficient, accepts an electron expelled by illumination of photosystem II, which arrives via a second connecting chain of electron carriers. This leaves an "electron hole" in photosystem II, which is filled by electrons from H$_2$O. We have described how water is split to yield: (1) electrons, which are donated to the electron-deficient photosystem II; (2) H$^+$ ions (protons), which are released inside the thylakoid lumen; and (3) O$_2$, which is released into the gas phase. The Z scheme thus describes the complete route by which electrons flow from H$_2$O to NADP$^+$ according to the equation

$$2H_2O + 2NADP^+ + 8 \text{ photons} \longrightarrow O_2 + 2NADPH + 2H^+$$

For each electron transferred from H$_2$O to NADP$^+$, two photons are absorbed, one by each photosystem. To form one molecule of O$_2$, which requires transfer of four electrons from two H$_2$O to two NADP$^+$, a total of eight photons must be absorbed, four by each photosystem.

The Cytochrome *bf* Complex Links Photosystems II and I

Electrons stored in Q$_B$H$_2$ as a result of the excitation of P680 in photosystem II (Fig. 18–42c) are carried to P700 of photosystem I via an assembly of several integral membrane proteins known as the **cytochrome *bf* complex** and the soluble protein plastocyanin (Fig. 18–44). The purified cytochrome *bf* complex contains a *b*-type cytochrome with two heme groups (cytochrome b_{563}), an iron–sulfur protein (M_r 20,000), and cytochrome *f* (named for the Latin *frons*, meaning "leaf"), also called cytochrome c_{552}. Electrons flow through the cytochrome *bf* complex from Q$_B$H$_2$ to cytochrome *f*; the detailed path is uncertain. Cytochrome *f* passes its electron to plastocyanin, the donor for P700 reduction (Figs. 18–41b, 18–44).

The cytochrome *bf* complex of plants is remarkably similar to the cytochrome bc_1 complex (Complex III) of the mitochondrial electron transfer chain, and it carries out a similar function. Both complexes convey electrons from a reduced quinone, a mobile, lipid-soluble carrier of two electrons (UQ in mitochondria, Q$_B$ in chloroplasts), to a water-soluble protein that carries one electron (cytochrome *c* in mitochondria, plastocyanin in chloroplasts). As in mitochondria, the function of

584

Figure 18–45 Proton and electron circuits in chloroplast thylakoids. Electrons (blue arrows) move from H_2O through photosystem II, the intermediate chain of carriers, photosystem I, and finally to $NADP^+$. Protons (red arrows) are pumped into the thylakoid lumen by the flow of electrons through the chain of carriers between photosystem II and photosystem I, and reenter the stroma through proton channels formed by the F_o portion of the ATP synthase, designated CF_o in the chloroplast enzyme. The F_1 subunit (CF_1) catalyzes synthesis of ATP.

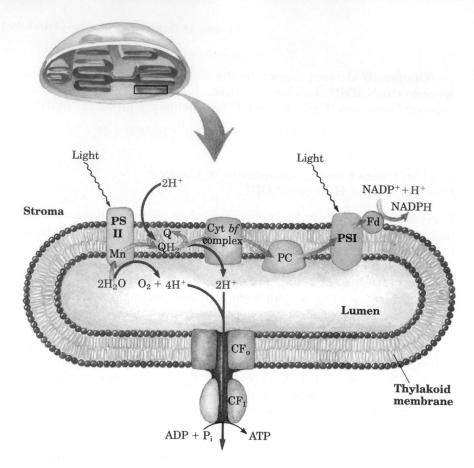

this complex involves a Q cycle (see Fig. 18–10) in which electrons pass from Q_BH_2 to cytochrome b one at a time. As in the mitochondrial Complex III, this cycle results in the pumping of protons across the membrane; in chloroplasts, the direction of proton movement is from the stromal compartment to the thylakoid lumen. The result is the production of a proton gradient across the thylakoid membrane as electrons pass from photosystem II to photosystem I (Fig. 18–45). Because the volume of the flattened thylakoid lumen is small, the influx of a small number of protons has a relatively large effect on lumenal pH. The measured difference in pH between the stroma (pH 8) and the thylakoid lumen (pH 4.5) represents a 3,000-fold difference in proton concentration—a powerful driving force for ATP synthesis.

Coupling ATP Synthesis to Light-Driven Electron Flow

We have now seen how one of the two energy-rich products formed in the light reactions, NADPH, is generated by photosynthetic electron transfer from H_2O to $NADP^+$. What about the other energy-rich product, ATP?

In 1954 Daniel Arnon and his colleagues discovered that ATP is generated from ADP and P_i during photosynthetic electron transfer in illuminated spinach chloroplasts. Support for these findings came from the work of Albert Frenkel who detected light-dependent ATP production in membranous pigment-containing structures called **chromatophores,** derived from photosynthetic bacteria. They concluded that some of the light energy captured by the photosynthetic systems of these organisms is transformed into the phosphate bond energy of ATP. This process is called **photophosphorylation** or **photosyn-**

Daniel Arnon

thetic phosphorylation, to distinguish it from oxidative phosphorylation in respiring mitochondria.

Recall that oxidative phosphorylation of ADP to ATP in mitochondria occurs at the expense of the free energy released as high-energy electrons flow downhill along the electron transfer chain from substrates to O_2. In a similar way, photophosphorylation of ADP to ATP is coupled to the energy released as high-energy electrons flow down the photosynthetic electron transfer chain from excited photosystem II to the electron-deficient photosystem I. The direct effect of electron flow is the formation of a proton gradient, which then provides the energy for ATP synthesis by an ATP synthase.

ATP synthesis in chloroplasts can be coupled to two types of electron flow—cyclic and noncyclic—as we shall see. We will also turn our attention to photophosphorylation in organisms other than green plants, and to the possible bacterial origins of chloroplasts. The chapter concludes with the development of an overall equation for photosynthesis in plants.

A Proton Gradient Couples Electron Flow and Phosphorylation

Several properties of photosynthetic electron transfer and photophosphorylation in chloroplasts show a role for a proton gradient as in mitochondrial oxidative phosphorylation: (1) the reaction centers, electron carriers, and ATP-forming enzymes are located in a membrane—the thylakoid membrane; (2) photophosphorylation requires intact thylakoid membranes; (3) the thylakoid membrane is impermeable to protons; (4) photophosphorylation can be uncoupled from electron flow by reagents that promote the passage of protons through the thylakoid membrane; (5) photophosphorylation can be blocked by venturicidin and similar agents that inhibit the formation of ATP from ADP and P_i by ATP synthase of mitochondria (see Fig. 18–13); and (6) ATP synthesis is catalyzed by F_oF_1 complexes, located on the outer surface of the thylakoid membranes, that are very similar in structure and function to the F_oF_1 complexes of mitochondria.

Electron-transferring molecules in the connecting chain between photosystem II and photosystem I are oriented asymmetrically in the thylakoid membrane, so that photoinduced electron flow results in the net movement of protons across the membrane, from the *outside* of the thylakoid membrane to the inner compartment (Fig. 18–45).

In 1966 André Jagendorf showed that a pH gradient across the thylakoid membrane (alkaline outside) could furnish the driving force to generate ATP. Jagendorf's early observations provided some of the most important experimental evidence in support of Mitchell's chemiosmotic hypothesis. In the dark, he soaked chloroplasts in a pH 4 buffer, which slowly penetrated into the inner compartment of the thylakoids, lowering their internal pH. He added ADP and P_i to the dark suspension of chloroplasts and then suddenly raised the pH of the outer medium to 8, momentarily creating a large pH gradient across the membrane. As protons moved out of the thylakoids into the medium, ATP was generated from ADP and P_i. Because the formation of ATP occurred in the dark (with no input of energy from light), this experiment showed that a pH gradient across the membrane is a high-energy state that can, as in mitochondrial oxidative phosphorylation, mediate the transduction of energy from electron transfer into the chemical energy of ATP.

André Jagendorf

The stoichiometry for this process (protons transported per electron) is not well established. Electron transfer between photosystems II and I through the cytochrome *bf* complex contributes to the proton gradient an amount of proton-motive force roughly equivalent to one to two ATP formed per pair of electrons.

The ATP Synthase of Chloroplasts Is Like That of Mitochondria

The enzyme responsible for ATP synthesis in chloroplasts is a large complex with two functional components, CF_o and CF_1 (the C denoting chloroplast origin). CF_o is a transmembrane proton pore composed of several integral membrane proteins and is homologous with mitochondrial F_o. CF_1 is a peripheral membrane protein complex very similar in subunit composition, structure, and function to mitochondrial F_1 (Table 18–8). Together these proteins constitute the ATP synthase of chloroplasts. (Bacteria also contain ATP synthases remarkably similar in structure and function to those of chloroplasts and mitochondria (Table 18–8).)

Table 18–8 Equivalent subunits in ATP synthase of mitochondria, chloroplasts, and bacteria (*E. coli*)[*]

Portion of ATP synthase	Mitochondria		Chloroplasts		*E. coli*	
	Subunit	Number	Subunit	Number	Subunit	Number
F_1	α	3	α	3	α	3
	β	3	β	3	β	3
	γ	1	γ	1	γ	1
	OSCP	1	δ	1	δ	1
	δ	1	ϵ	1	ϵ	1
	ϵ	1	—	—	—	—
F_o	a	1	a (IV)	2	a	1
	b	1	b and b′ (I and II)	1 + 1	b	2
	c	6–12	c (III)	6–12	c	10–12

Source: Data primarily from Walker, J.E., Lutter, R., Dupuis, A., & Runswick, M.J. (1991) Identification of the subunits of F_1F_o-ATPase from bovine heart mitochondria. *Biochemistry* **30**, 5369–5378.

[*] Subunits on the same horizontal line are structurally related and are believed to be functionally homologous. Chloroplasts contain two nonidentical subunits, b and b′, that are together the homologs of mitochondrial b; *E. coli* has two identical b subunits. OSCP is oligomycin sensitivity-conferring protein. Alternative nomenclatures for the four chloroplast F_o subunits are shown in parentheses. In addition to the subunits shown here, the mitochondrial enzyme complex contains at least five more proteins with no apparent homologs in the chloroplast or bacterium: F6, inhibitor protein, A6L, d, and e. These subunits are released as soluble proteins when F_1 and F_o are dissociated, but they are believed to be part of the intact, functioning complex in mitochondria.

Electron microscopy of sectioned chloroplasts shows ATP synthase complexes as knoblike projections on the *outside* (stromal) surface of thylakoid membranes; these correspond to the ATP synthase complexes seen to project on the *inside* (matrix) surface of the inner mitochondrial membrane (see Fig. 18–15). Thus both the orientation of the ATP synthase and the direction of proton pumping in chloroplasts are opposite to those in mitochondria. In both cases, the F_1 portion of ATP synthase is located on the more alkaline side of the membrane through which protons flow down their concentration gradient; the direction of proton flow relative to F_1 is the same in both cases (Fig. 18–46).

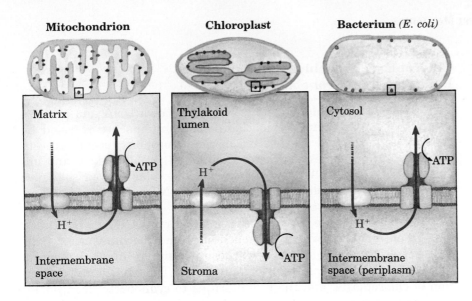

Mitochondrion

Chloroplast

Bacterium (*E. coli*)

Matrix

ATP

H⁺

Intermembrane space

Thylakoid lumen

H⁺

Stroma

ATP

Cytosol

ATP

H⁺

Intermembrane space (periplasm)

Figure 18–46 Comparison of the topology of proton movement and ATP synthase orientation in the membranes of mitochondria, chloroplasts, and the bacterium *E. coli*. In each case, the orientation of the proton gradient relative to the ATP synthase is the same.

The mechanism of chloroplast ATP synthase is also believed to be essentially identical to that of its mitochondrial analog; ADP and P_i readily condense to form ATP on the enzyme surface, but the release of this enzyme-bound ATP requires a proton-motive force (see Fig. 18–23). As for the mitochondrial ATP synthase, the details of this mechanism remain to be determined.

Cyclic Electron Flow Produces ATP but Not NADPH or O_2

There is an alternative path of light-induced electron flow that allows chloroplasts to vary the ratio of NADPH and ATP formed during illuminations; this is called **cyclic electron flow** to differentiate it from the normally unidirectional or **noncyclic electron flow** that proceeds from H_2O to $NADP^+$, as we have discussed thus far. Cyclic electron flow involves only photosystem I (Fig. 18–44). Electrons passed from P700 to ferredoxin do not continue to $NADP^+$, but move back through the cytochrome *bf* complex to plastocyanin. Plastocyanin donates electrons to P700, the illumination of which promotes electron transfer to ferredoxin. Thus illumination of photosystem I can cause electrons to cycle continuously out of the reaction center of photosystem I and back into it, each electron being propelled around the cycle by the energy yielded by absorption of one photon. Cyclic electron flow is not accompanied by net formation of NADPH or the evolution of O_2. However, it is accompanied by proton pumping and by the phosphorylation of ADP to ATP, referred to as **cyclic photophosphorylation.** The overall reaction equation for cyclic electron flow and photophosphorylation is simply

$$ADP + P_i \xrightarrow{\text{light}} ATP + H_2O$$

Cyclic electron flow and photophosphorylation are believed to occur when the plant cell is already amply supplied with reducing power in the form of NADPH but requires additional ATP for other metabolic needs. By regulating the partitioning of electrons between $NADP^+$ reduction and cyclic photophosphorylation, a plant adjusts the ratio of NADPH and ATP produced in the light reactions to match the needs for these products in the carbon fixation reactions and in other energy-requiring processes.

Chloroplasts Probably Evolved from Endosymbiotic Cyanobacteria

Like mitochondria, chloroplasts contain their own DNA and protein-synthesizing machinery. Some chloroplast proteins are encoded by chloroplast genes and synthesized in the chloroplast; others are encoded by nuclear genes, synthesized outside the chloroplast, and imported. When plant cells grow and divide, chloroplasts give rise to new chloroplasts by division, during which their DNA is replicated and divided between daughter chloroplasts.

Cyanobacteria are photosynthetic prokaryotes (formerly called blue-green algae) in which the machinery and mechanism for light capture, electron flow, and ATP synthesis are similar in many respects to those in the chloroplasts of higher plants. The bacteriumlike division of chloroplasts, and their similarities to cyanobacteria in photosynthetic components and mechanisms, support the hypothesis that chloroplasts arose during evolution from endosymbiotic prokaryotes that also gave rise to modern cyanobacteria (see Fig. 2–17). Studies of photosynthesis in lower eukaryotes and prokaryotes may therefore yield insight into photosynthetic mechanisms that is generalizable to the higher plants.

Diverse Photosynthetic Organisms Use Hydrogen Donors Other Than H_2O

Photosynthesis occurs not only in green plants but in lower eukaryotic organisms such as algae, euglenoids, dinoflagellates, and diatoms, and also in certain prokaryotes. The photosynthetic prokaryotes include the cyanobacteria, the green sulfur bacteria found in mountain lakes, the purple bacteria common in the ocean, and the purple sulfur bacteria of sulfur springs. The cyanobacteria, found in both fresh and salt waters, are perhaps the most versatile of photosynthetic organisms. Because they can also fix atmospheric nitrogen (Chapter 21), cyanobacteria are among the most self-sufficient organisms in the biosphere. At least half of the photosynthetic activity on earth occurs in the many different microorganisms that constitute the phytoplankton in oceans, rivers, and lakes.

With the exception of the cyanobacteria, which have an O_2-producing photosynthetic system resembling that of plants, photosynthetic bacteria do not produce O_2. Many are obligate anaerobes that cannot tolerate O_2. Some photosynthetic bacteria use inorganic compounds as hydrogen donors. For example, the green sulfur bacteria use hydrogen sulfide, according to the equation

$$2H_2S + CO_2 \xrightarrow{\text{light}} (CH_2O) + H_2O + 2S$$

These bacteria, instead of giving off molecular O_2, produce elemental sulfur as the oxidation product of H_2S. Other photosynthetic bacteria use organic compounds as hydrogen donors, for example, lactate:

$$2 \text{ Lactate} + CO_2 \xrightarrow{\text{light}} (CH_2O) + H_2O + 2 \text{ pyruvate}$$

Plant and bacterial photosynthesis are fundamentally similar processes despite the differences in the hydrogen donors they employ. This similarity becomes obvious when the equation of photosynthesis is written in a more general form:

$$2H_2D + CO_2 \xrightarrow{\text{light}} (CH_2O) + H_2O + 2D$$

in which H_2D symbolizes a hydrogen donor and D is its oxidized form. H_2D thus may be water, hydrogen sulfide, lactate, or some other organic compound, depending upon the species.

The Structure of a Bacterial Photosystem Reaction Center Has Been Determined

The three-dimensional structure of the photoreaction center of a photosynthetic bacterium, *Rhodopseudomonas viridis*, which is analogous in some ways to photosystem II in higher plants, is known from x-ray crystallographic studies (Fig. 18–47). This structure sheds light on how phototransduction takes place in the bacterium, and presumably in higher plants as well.

The bacterial reaction center has four types of proteins: a *c*-type cytochrome, two subunits (L and M) associated with bacteriochlorophyll, and a fourth protein, the H subunit. A single reaction center contains four hemes (cytochromes), four bacteriochlorophylls that are similar to the chlorophylls of chloroplasts, two bacteriopheophytins, one inorganic Fe, and a quinone. From a variety of physical studies, the extremely rapid sequence of events shown in Figure 18–47b has been deduced.

A pair of closely spaced bacteriochlorophylls (the "special pair") constitutes the site of the initial photochemistry in the bacterial reaction center. The "electron hole" that develops in the chlorophyll is filled with electrons from the *c*-type cytochrome, a role played by H_2O in photosystem II of higher plants.

The bacteriochlorophylls, bacteriopheophytins, and quinone are held rigidly in a fixed orientation relative to each other by the proteins of the reaction center. The photochemical reactions among these components therefore take place in a virtually solid state, accounting for the high efficiency and rapidity of the reactions; nothing is left to chance collision or dependent upon random diffusion.

Figure 18–47 The purple sulfur bacterium *Rhodopseudomonas viridis* performs light-driven ATP synthesis using machinery closely similar to photosystem II of higher plants, although the bacterium has no analog of photosystem I. (a) Superimposed on the structure of the reaction center proteins (see Fig. 10–11) are the prosthetic groups that participate in the photochemical events. There are four molecules of bacteriochlorophyll, two of which (the "special pair," dark blue) constitute the site of the first photochemical changes after light absorption. The other two bacteriochlorophylls (orange) are called accessory pigments; their role in the photochemical events is not well understood. Two bacterial pheophytins (light blue) lie beneath the bacteriochlorophylls, and beneath them, two bacterial quinones, Q_A and Q_B (green), separated by a non-heme iron atom (red). Shown at the top of the figure are four heme groups (red) associated with *c*-type cytochromes of the reaction center. (b) The sequence of events that occurs upon excitation of the special pair of bacteriochlorophylls and the time scale of the electron transfers (in parentheses). The excited special pair passes an electron to bacteriopheophytin ①, from which the electron moves rapidly to the tightly bound quinone Q_A ②. This quinone much more slowly passes electrons to the diffusible quinone Q_B through the non-heme iron atom ③. Meanwhile, the "electron hole" in the special pair is filled by an electron from one of the hemes of the four *c*-type cytochromes ④.

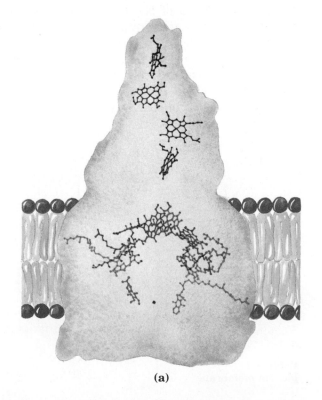

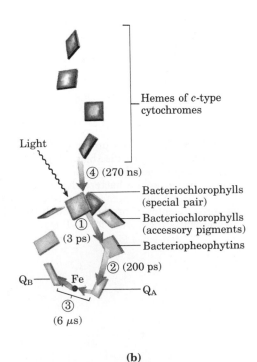

Hemes of *c*-type cytochromes

Light

④ (270 ns)

Bacteriochlorophylls (special pair)

① Bacteriochlorophylls (accessory pigments)

(3 ps) Bacteriopheophytins

② (200 ps)

Q_B — Fe — Q_A

③ (6 µs)

(a)   (b)

Although the reaction centers of plants have not yet been seen in such detail, the similarities between the bacterial and eukaryotic reaction centers suggest that the principles deduced from studies of the geometry and mechanism of the bacterial reaction center will also apply to the reaction centers of plants.

Salt-Loving Bacteria Use Light Energy to Make ATP

The halophilic ("salt-loving") bacterium *Halobacterium halobium* conserves energy derived from absorbed sunlight by an interesting variation on the principle employed by true photosynthetic organisms. These unusual bacteria live only in brine ponds and salt lakes (Great Salt Lake and the Dead Sea, for example), where the high salt concentration results from water loss by evaporation; indeed, they cannot live in NaCl concentrations lower than 3 M. Halobacteria are aerobes and normally use O_2 to oxidize organic fuel molecules. However, the solubility of O_2 is so low in brine ponds, in which the NaCl concentration may exceed 4 M, that these bacteria must sometimes call on another source of energy, namely sunlight. The plasma membrane of *H. halobium* contains patches of light-absorbing pigments, called **purple patches.** These patches are made up of closely packed molecules of the protein **bacteriorhodopsin** (see Fig. 10–10), which contains retinal (vitamin A aldehyde; see Fig. 9–18) as a prosthetic group. When the cells are illuminated, the bacteriorhodopsin molecules are excited by an absorbed photon. As the excited molecules revert to their initial ground state, an induced conformational change results in the release of protons outside the cell, forming an acid-outside pH gradient across the plasma membrane. Protons tend to diffuse back into the cell through an ATP synthase complex in the membrane, very similar to that of mitochondria and of chloroplasts, supplying the energy for ATP synthesis (Fig. 18–48). Thus halobacteria can use light to supplement the ATP synthesized by oxidative phosphorylation with the O_2 that is available. However, halobacteria do not evolve O_2, nor do they carry out photoreduction of $NADP^+$; their phototransducing machinery is therefore much simpler than that of cyanobacteria or higher plants.

Bacteriorhodopsin, with only 247 amino acid residues, is the simplest light-driven proton pump known. The determination of its molecular structure should yield important insights into light-dependent energy transduction and the action of the proton pumps that function in respiration and photosynthesis.

Figure 18–48 Light-driven proton currents in *Halobacterium halobium* provide the proton-motive force for ATP synthesis. Illumination of the phototransducing protein bacteriorhodopsin results in outward proton movement, generating a proton-motive force. Reentry of protons through the F_oF_1 complex provides the energy for ATP synthesis.

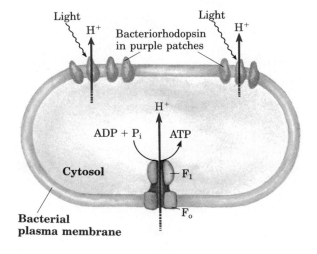

Photosynthesis Uses the Energy in Light Very Efficiently

The standard free-energy change for the synthesis of glucose from CO_2 and H_2O during photosynthesis by the reaction

$$6CO_2 + 6H_2O \xrightarrow{\text{light}} C_6H_{12}O_6 + 6O_2 \qquad (18\text{--}9)$$

is 2,840 kJ/mol. (Recall that oxidation of glucose by the reverse of this equation proceeds with a *decrease* of 2,840 kJ/mol.) Now let us compare this energy requirement with the energy yielded by the light reactions of plant photosynthesis.

Recall that two photons must be absorbed, one by each photosystem, to cause flow of one electron from H_2O to $NADP^+$. To generate one molecule of O_2, four electrons must be transferred. Therefore, production of six molecules of O_2 (as in Eqn 18–9) requires the absorption and use of 48 photons: (2 photons/e^-)(4e^-/O_2)(6O_2) = 48 photons. Because the energy of one einstein may range from 300 kJ at 400 nm to about 170 kJ at 700 nm (Fig. 18–35), anywhere from 8,160 to 14,400 kJ (depending upon the wavelength of the absorbed light) is required under standard conditions to make 1 mol of glucose "costing" 2,840 kJ.

In the next chapter we turn to a consideration of the carbon fixation reactions by which photosynthetic organisms use the ATP and NADPH produced in the light reactions to carry out the reduction of CO_2 to carbohydrates.

Summary

Chemiosmotic theory provides the intellectual framework for understanding many biological energy transductions, including the processes of oxidative phosphorylation in mitochondria and photophosphorylation in chloroplasts. The mechanism of energy coupling is similar in both cases. The conservation of free energy involves the passage of electrons through a chain of membrane-bound oxidation–reduction (redox) carriers and the concomitant pumping of protons across the membrane, producing an electrochemical gradient, the proton-motive force. This force drives the synthesis of ATP by membrane-bound enzyme complexes through which protons flow back across the membrane, down their electrochemical gradient. Proton-motive force also drives other energy-requiring processes of cells.

In mitochondria, H atoms removed from substrates by the action of NAD-linked dehydrogenases donate their electrons to the respiratory (electron transfer) chain, which transfers them to molecular O_2, reducing it to H_2O. Shuttle systems convey reducing equivalents from cytosolic NADH to mitochondrial NADH. Reducing equivalents from all NAD-linked dehydrogenations are transferred to mitochondrial NADH dehydrogenase (Complex I), which contains FMN as its prosthetic group. They are then passed via a series of Fe–S centers to ubiquinone, which transfers the electrons to cytochrome b, the first carrier in Complex III. In this complex, electrons pass through two b-type cytochromes and cytochrome c_1 before reaching an Fe–S center. The Fe–S center passes electrons, one at a time, through cytochrome c and into Complex IV, cytochrome oxidase. This copper-containing enzyme, which also contains cytochromes a and a_3, accumulates electrons, then passes them to O_2, reducing it to H_2O.

There are alternative paths of entry of electrons into this chain of carriers. Succinate, for example, is oxidized by succinate dehydrogenase (Complex II), which contains a flavoprotein (with FAD) that passes electrons through several Fe–S centers and into the chain at the level of ubiquinone. Electrons derived from the oxidation of fatty acids pass into ubiquinone via the electron-transferring flavoprotein (ETFP).

The flow of electrons through Complexes I, III, and IV results in the pumping of protons across the mitochondrial inner membrane, making the matrix alkaline relative to the extramitochondrial space. This proton gradient provides the energy (proton-motive force) for ATP synthesis from ADP and P_i by an inner-membrane protein complex, ATP synthase, also called F_oF_1 ATPase. The details of this ATP-synthesizing mechanism are still under investigation. Bacteria carry out oxidative phosphorylation by essentially the same mechanism, using electron carriers and an ATP synthase in the plasma membrane. Oxidative phosphorylation produces most of the ATP required by aerobic cells; it is regulated by cellular energy demands. In brown fat tissue, which is specialized for the production of metabolic heat, electron transfer is uncoupled from ATP synthesis; the energy of fatty acid oxidation is therefore dissipated as heat.

Photophosphorylation in the chloroplasts of green plants and in cyanobacteria also involves electron flow through a series of membrane-bound carriers. In the light reactions of plants, the absorption of a photon excites chlorophyll molecules and other (accessory) pigments that funnel the energy into reaction centers in the thylakoid membranes of chloroplasts. At the reaction centers, photoexcitation results in a charge separation that produces one chemical species that is a good electron donor (reducing agent) and another that is a good electron acceptor. In chloroplasts there are two different photoreaction centers, which function together. Photosystem I passes electrons from its excited reaction center, P700, through a series of carriers to ferredoxin, which then reduces $NADP^+$ to NADPH. The reaction center, P680, of photosystem II passes electrons to plastoquinone, reducing it to the quinol form. The electrons lost from P680 are replaced by electrons abstracted from H_2O (hydrogen donors other than H_2O are used in other organisms). This light-driven splitting of H_2O is catalyzed by a Mn-containing protein complex; O_2 is produced. Reduced plastoquinone carries electrons from photosystem II to the cytochrome bf complex; these electrons pass to the soluble protein plastocyanin, and then to P700 to replace those lost during its photoexcitation. Electron flow through the cytochrome bf complex is accompanied by proton pumping across the thylakoid membrane, and the proton-motive force thus created drives ATP synthesis by a CF_oCF_1 complex closely similar to the F_oF_1 complex of mitochondria. This flow of electrons through photosystems II and I thus produces both NADPH and ATP. A second type of electron flow (cyclic flow) produces ATP only.

Both mitochondria and chloroplasts contain their own genomes and are believed to have originated from prokaryotic endosymbionts of early eukaryotic cells. Oxidative phosphorylation in aerobic bacteria and photophosphorylation in photosynthetic bacteria are closely similar, in machinery and mechanism, to the homologous processes in mitochondria and chloroplasts.

Further Reading

History and Background

Arnon, D.I. (1984) The discovery of photosynthetic phosphorylation. *Trends Biochem. Sci.* **9,** 258–262.

Harold, F.M. (1986) *The Vital Force: A Study in Bioenergetics,* W.H. Freeman and Company, New York.
A very readable synthesis of the principles of bioenergetics and their application to energy transductions.

Kalckar, H.M. (1991) 50 years of biological research—from oxidative phosphorylation to energy requiring transport regulation. *Annu. Rev. Biochem.* **60,** 1–37.
A delightful autobiographical account by one of the pioneers in the field.

Keilin, D. (1966) *The History of Cell Respiration and Cytochrome,* Cambridge University Press, London.
An authoritative and absorbing account of the discovery of cytochromes and of their roles in respiration, written by the discoverer of cytochromes.

Lehninger, A.L. (1964) *The Mitochondrion: Molecular Basis of Structure and Function,* The Benjamin Co., Inc., New York.
A classic description of early work on mitochondria.

Mitchell, P. (1979) Keilin's respiratory chain concept and its chemiosmotic consequences. *Science* **206,** 1148–1159.
The author's Nobel lecture, outlining the evolution of the chemiosmotic hypothesis.

Skulachev, V.P. (1992) The laws of cell energetics. *Eur. J. Biochem.* **208,** 203–209.

On the interconvertibility of ATP and ion gradients.

Slater, E.C. (1987) The mechanism of the conservation of energy of biological oxidations. *Eur. J. Biochem.* **166,** 489–504.

A clear and critical account of the evolution of the chemiosmotic model.

Staehelin, L.A. & Arntzen, C.J. (eds) (1986) *Photosynthesis III: Photosynthetic Membranes and Light Harvesting Systems,* Encyclopedia of Plant Physiology, Vol. 19, Springer-Verlag, Berlin.

Authoritative reviews of many aspects of photosynthesis.

Respiratory Electron Flow

Babcock, G.T. & Wickström, M. (1992) Oxygen activation and the conservation of energy in cell respiration. *Nature* **356,** 301–309.

An advanced discussion of the reduction of water and pumping of protons by cytochrome oxidase.

Douce, R. & Neuburger, M. (1989) The uniqueness of plant mitochondria. *Annu. Rev. Plant Physiol. Plant Mol. Biol.* **40,** 371–414.

A focus on the features of plant mitochondria that distinguish them from mitochondria of animal cells.

Hinkle, P.C. & McCarty, R.E. (1978) How cells make ATP. *Sci. Am.* **238** (March), 104–123.

Although not recent, this is an excellent, readable, and well-illustrated description of oxidative phosphorylation.

Lehninger, A.L., Reynafarje, B., Alexandre, A., & Villalobo, A. (1980) Respiration-coupled H^+ ejection by mitochondria. *Ann. N. Y. Acad. Sci.* **341,** 585–592.

The methods and problems in measuring proton efflux stoichiometry.

Malmström, B.G. (1989) The mechanism of proton translocation in respiration and photosynthesis. *FEBS Lett.* **250,** 9–21.

Comparative review of the electron-transferring complexes of mitochondria and chloroplasts.

Trumpower, B.L. (1990) The protonmotive Q cycle: energy transduction by coupling of proton translocation to electron transfer by the cytochrome bc_1 complex. *J. Biol. Chem.* **265,** 11409–11412.

Short, clear description of the Q cycle and electron flow through Complex III.

Coupling ATP Synthesis to Respiratory Electron Flow

Boyer, P.D. (1989) A perspective of the binding change mechanism for ATP synthesis. *FASEB J.* **3,** 2164–2178.

An article on the historical development and current state of the binding-change model, by its principal architect.

Futai, M., Noumi, T., & Maeda, M. (1987) Molecular biological studies on structure and mechanism of proton translocating ATPase (H^+-ATPase, F_oF_1). *Adv. Biophys.* **23,** 1–37.

Insight into the mechanism of ATP synthase from studies of the genes that encode its subunits.

Pedersen, P.L. & Carafoli, E. (1987) Ion motive ATPases. I. Ubiquity, properties, and significance to cell function. *Trends Biochem. Sci.* **12,** 145–150. II. Energy coupling and work output. *Trends Biochem. Sci.* **12,** 186–189.

Two short reviews that place ATP synthase within the family of ATP-dependent proton pumps; include their general mechanisms.

Penefsky, H.S. & Cross, R.L. (1991) Structure and mechanism of F_oF_1-type ATP synthases and ATPases. *Adv. Enzymol. Relat. Areas Mol. Bio.* **64,** 173–214.

An advanced discussion.

Ricquier, D., Casteilla, L., & Bouillaud, F. (1991) Molecular studies of the uncoupling protein. *FASEB J.* **5,** 2237–2242.

A discussion of the protein and its role in thermogenesis.

Senior, A.E. (1988) ATP synthesis by oxidative phosphorylation. *Physiol. Rev.* **68,** 177–231.

An advanced but very clear review, with an emphasis on the mechanism of ATP synthase.

Regulation of Mitochondrial Oxidative Phosphorylation

Brand, M.D. & Murphy, M.P. (1987) Control of electron flux through the respiratory chain in mitochondria and cells. *Biol. Rev. Cambridge Phil. Soc.* **62,** 141–193.

An advanced description of respiratory control.

Harris, D.A. & Das, A.M. (1991) Control of mitochondrial ATP synthesis in the heart. *Biochem. J.* **280,** 561–573.

An advanced discussion of regulation of the ATP synthase by Ca^{2+} and other factors.

Photosynthesis: Harvesting Light Energy

Green, B.R., Pichersky, E., & Kloppstech, K. (1991) Chlorophyll *a/b*-binding proteins: an extended family. *Trends Biochem. Sci.* **16**, 181–186.

An intermediate-level description of the proteins that orient chlorophyll molecules in chloroplasts.

Huber, R. (1990) A structural basis of light energy and electron transfer in biology. *Eur. J. Biochem.* **187**, 283–305.

The author's Nobel lecture, describing the physics and chemistry of phototransductions. An exceptionally clear and well-illustrated discussion, based on crystallographic studies of reaction centers.

Zuber, H. (1986) Structure of light-harvesting antenna complexes of photosynthetic bacteria, cyanobacteria and red algae. *Trends Biochem. Sci.* **11**, 414–419.

Light-Driven Electron Flow

Andréasson, L.-E. & Vänngård, T. (1988) Electron transport in photosystems I and II. *Annu. Rev. Plant Physiol. Plant Mol. Biol.* **39**, 379–411.

An advanced description of the path of electron flow in chloroplasts, studied with spectroscopic techniques.

Blankenship, R.E. & Prince, R.C. (1985) Excited-state redox potentials and the Z scheme of photosynthesis. *Trends Biochem. Sci.* **10**, 382–383.

A concise and lucid statement of the redox properties of excited states.

Deisenhofer, J. & Michel, H. (1991) Structures of bacterial photosynthetic reaction centers. *Annu. Rev. Cell Biol.* **7**, 1–23.

The structure of the reaction center of purple bacteria, and implications for the function of bacterial and plant reaction centers.

Glazer, A.N. & Melis, A. (1987) Photochemical reaction centers: structure, organization, and function. *Annu. Rev. Plant Physiol.* **38**, 11–45.

An advanced description of the structure and function of reaction centers of green plants, cyanobacteria, and purple and green bacteria.

Golbeck, J.H. (1992) Structure and function of photosystem I. *Annu. Rev. Plant Physiol. Plant Mol. Biol.* **43**, 293–324.

Govindjee & Coleman, W.J. (1990) How plants make oxygen. *Sci. Am.* **262** (February), 50–58.

An exceptionally clear account of the water-splitting activity of photosystem II.

Nitschke, W. & Rutherford, A.W. (1991) Photosynthetic reaction centres: variations on a common structural theme? *Trends Biochem. Sci.* **16**, 241–245.

A comparison of the structure and function of photosystems I and II and the reaction centers of several photosynthetic bacteria.

Coupling ATP Synthesis to Light-Driven Electron Flow

Cramer, W.A., Widger, W.R., Herrmann, R.G., & Trebst, A. (1985) Topography and function of thylakoid membrane proteins. *Trends Biochem. Sci.* **10**, 125–129.

Jagendorf, A.T. (1967) Acid-base transitions and phosphorylation by chloroplasts. *Fed. Proc.* **26**, 1361–1369.

The classic experiment establishing the ability of a proton gradient to drive ATP synthesis in the dark.

Youvan, D.C. & Marrs, B.L. (1987) Molecular mechanisms of photosynthesis. *Sci. Am.* **256** (June), 42–48.

An excellent description of the chemical basis for light reactions.

Problems

1. *Oxidation–Reduction Reactions* The NADH dehydrogenase complex of the mitochondrial respiratory chain promotes the following series of oxidation–reduction reactions, in which Fe^{3+} and Fe^{2+} represent the iron in iron–sulfur centers, UQ is ubiquinone, UQH_2 is ubiquinol, and E is the enzyme:

$$
\begin{array}{ll}
(1) & NADH + H^+ + E\text{–}FMN \rightarrow NAD^+ + E\text{–}FMNH_2 \\
(2) & E\text{–}FMNH_2 + 2Fe^{3+} \rightarrow E\text{–}FMN + 2Fe^{2+} + 2H^+ \\
(3) & 2Fe^{2+} + 2H^+ + UQ \rightarrow 2Fe^{3+} + UQH_2 \\
\hline
\textit{Sum:} & NADH + H^+ + UQ \rightarrow NAD^+ + UQH_2
\end{array}
$$

For each of the three reactions catalyzed by the NADH dehydrogenase complex, identify (a) the electron donor, (b) the electron acceptor, (c) the conjugate redox pair, (d) the reducing agent, and (e) the oxidizing agent.

2. *Standard Reduction Potentials* The standard reduction potential of any redox couple is defined for the half-cell reaction (or half-reaction):

Oxidizing agent + n electrons $\longrightarrow$
$$\text{reducing agent}$$

The standard reduction potentials of the NAD^+/ NADH and pyruvate/lactate redox pairs are -0.320 and -0.185 V, respectively.

(a) Which redox pair has the greater tendency to lose electrons? Explain.

(b) Which is the stronger oxidizing agent? Explain.

(c) Beginning with 1 M concentrations of each reactant and product at pH 7, in which direction will the following reaction proceed?

$$\text{Pyruvate} + \text{NADH} + \text{H}^+ \rightleftharpoons \text{lactate} + \text{NAD}^+$$

(d) What is the standard free-energy change, $\Delta G^{\circ\prime}$, at 25 °C for this reaction?

(e) What is the equilibrium constant for this reaction at 25 °C?

3. *Energy Span of the Respiratory Chain* Electron transfer in the mitochondrial respiratory chain may be represented by the net reaction equation

$$\text{NADH} + \text{H}^+ + \tfrac{1}{2}\text{O}_2 \rightleftharpoons \text{H}_2\text{O} + \text{NAD}^+$$

(a) Calculate the value of the change in standard reduction potential, $\Delta E_0'$, for the net reaction of mitochondrial electron transfer.

(b) Calculate the standard free-energy change, $\Delta G^{\circ\prime}$, for this reaction.

(c) How many ATP molecules could *theoretically* be generated per molecule of NADH oxidized by this reaction, given a standard free energy of ATP synthesis of 30.5 kJ/mol?

(d) How many ATP molecules could be synthesized under typical cellular conditions (see Box 13–2)?

4. *Use of FAD Rather Than NAD^+ in the Oxidation of Succinate* All the dehydrogenation steps in glycolysis and the citric acid cycle use NAD^+ (E_0' for NAD^+/NADH $= -0.32$ V) as the electron acceptor except succinate dehydrogenase, which uses covalently bound FAD (E_0' for FAD/FADH$_2$ in this enzyme $= 0.05$ V). Why is FAD a more appropriate electron acceptor than NAD^+ in the dehydrogenation of succinate? Give a possible explanation based on a comparison of the E_0' values of the fumarate/succinate pair ($E_0' = 0.03$), the NAD^+/NADH pair, and the succinate dehydrogenase FAD/FADH$_2$ pair.

5. *Degree of Reduction of Electron Carriers in the Respiratory Chain* The degree of reduction of each electron carrier in the respiratory chain is determined by the conditions existing in the mitochondrion. For example, when the supply of NADH and

O_2 is abundant, the steady-state degree of reduction of the carriers decreases as electrons pass from the substrate to O_2. When electron transfer is blocked, the carriers before the block become more reduced while those beyond the block become more oxidized (Fig. 18–7). For each of the conditions below, predict the state of oxidation of each carrier in the respiratory chain (ubiquinone and cytochromes b, c_1, c, and $a + a_3$).

(a) Abundant supply of NADH and O_2 but cyanide added

(b) Abundant supply of NADH but O_2 exhausted

(c) Abundant supply of O_2 but NADH exhausted

(d) Abundant supply of NADH and O_2

6. *The Effect of Rotenone and Antimycin A on Electron Transfer* Rotenone, a toxic natural product from plants, strongly inhibits NADH dehydrogenase of insect and fish mitochondria. Antimycin A, a toxic antibiotic, strongly inhibits the oxidation of ubiquinol.

(a) Explain why rotenone ingestion is lethal to some insect and fish species.

(b) Explain why antimycin A is a poison.

(c) Assuming that rotenone and antimycin A are equally effective in blocking their respective sites in the electron transfer chain, which would be a more potent poison? Explain.

7. *Uncouplers of Oxidative Phosphorylation* In normal mitochondria the rate of electron transfer is tightly coupled to the demand for ATP. Thus when the rate of utilization of ATP is relatively low, the rate of electron transfer is also low. Conversely, when ATP is demanded at a high rate, electron transfer is rapid. Under such conditions of tight coupling, the number of ATP molecules produced per atom of oxygen consumed when NADH is the electron donor—known as the P/O ratio—is close to 3.

(a) Predict the effect of a relatively low and a relatively high concentration of an uncoupling agent on the rate of electron transfer and the P/O ratio.

(b) The ingestion of uncouplers causes profuse sweating and an increase in body temperature. Explain this phenomenon in molecular terms. What happens to the P/O ratio in the presence of uncouplers?

(c) The uncoupler 2,4-dinitrophenol was once prescribed as a weight-reducing drug. How can this agent, in principle, serve as a reducing aid? Such uncoupling agents are no longer prescribed because some deaths occurred following their use. Why can the ingestion of uncouplers lead to death?

8. *Mode of Action of Dicyclohexylcarbodiimide (DCCD)* When DCCD is added to a suspension of tightly coupled, actively respiring mitochondria, the rate of electron transfer (measured by O_2 consumption) and the rate of ATP production dramatically decrease. If a solution of 2,4-dinitrophenol is now added to the inhibited mitochondrial preparation, O_2 consumption returns to normal but ATP production remains inhibited.

(a) What process in electron transfer or oxidative phosphorylation is affected by DCCD?

(b) Why does DCCD affect the O_2 consumption of mitochondria? Explain the effect of 2,4-dinitrophenol on the inhibited mitochondrial preparation.

(c) Which of the following inhibitors does DCCD most resemble in its action: antimycin A, rotenone, or oligomycin?

9. The Malate–α-Ketoglutarate Transport System of Mitochondria   The inner mitochondrial membrane transport system that promotes the transport of malate and α-ketoglutarate across the membrane (Fig. 18–25) is inhibited by n-butylmalonate. Suppose n-butylmalonate is added to an aerobic suspension of kidney cells using glucose exclusively as fuel. Predict the effect of this inhibitor on

(a) Glycolysis

(b) Oxygen consumption

(c) Lactate formation

(d) ATP synthesis

10. The Pasteur Effect   When O_2 is added to an anaerobic suspension of cells using glucose at a high rate, the rate of glucose consumption declines dramatically as the added O_2 is consumed. In addition, the accumulation of lactate ceases. This effect, first observed by Louis Pasteur in the 1860s, is characteristic of most cells capable of both aerobic and anaerobic utilization of glucose.

(a) Why does the accumulation of lactate cease after O_2 is added?

(b) Why does the presence of O_2 decrease the rate of glucose consumption?

(c) How does the onset of O_2 consumption slow down the rate of glucose consumption? Explain in terms of specific enzymes.

11. How Many Protons in a Mitochondrion?   Electron transfer functions to translocate protons from the mitochondrial matrix to the external medium to establish a pH gradient across the inner membrane, the outside more acidic than the inside. The tendency of protons to diffuse from the outside into the matrix, where [H^+] is lower, is the driving force for ATP synthesis via the ATP synthase. During oxidative phosphorylation by a suspension of mitochondria in a medium of pH 7.4, the internal pH of the matrix has been measured as 7.7.

(a) Calculate [H^+] in the external medium and in the matrix under these conditions.

(b) What is the outside:inside ratio of [H^+]? Comment on the energy inherent in this concentration. (Hint: See p. 383, Eqn 13–5.)

(c) Calculate the number of protons in a respiring liver mitochondrion, assuming its inner matrix compartment is a sphere of diameter 1.5 μm.

(d) From these data would you think the pH gradient alone is sufficiently great to generate ATP?

(e) If not, can you suggest how the necessary energy for synthesis of ATP arises?

12. Rate of ATP Turnover in Rat Heart Muscle   Rat heart muscle operating aerobically fills more than 90% of its ATP needs by oxidative phosphorylation. This tissue consumes O_2 at the rate of 10 μmol/min·g of tissue, with glucose as the fuel source.

(a) Calculate the rate at which this tissue consumes glucose and produces ATP.

(b) If the steady-state concentration of ATP in rat heart muscle is 5 μmol/g of tissue, calculate the time required (in seconds) to completely turn over the cellular pool of ATP. What does this result indicate about the need for tight regulation of ATP production? (Note: Concentrations are expressed as micromoles per gram of muscle tissue because the tissue is mostly water.)

13. Rate of ATP Breakdown in Flight Muscle   ATP production in the flight muscles of the fly *Lucilia sericata* results almost exclusively from oxidative phosphorylation. During flight, 187 ml of O_2/h·g of fly body weight is needed to maintain an ATP concentration of 7 μmol/g of flight muscle. Assuming that the flight muscles represent 20% of the weight of the fly, calculate the rate at which the flight-muscle ATP pool turns over. How long would the reservoir of ATP last in the absence of oxidative phosphorylation? Assume that reducing equivalents are transferred by the glycerol-3-phosphate shuttle and that O_2 is at 25 °C and 101.3 kPa (1 atm). (Note: Concentrations are expressed in micromoles per gram of flight muscle.)

14. Transmembrane Movement of Reducing Equivalents   Under aerobic conditions, extramitochondrial NADH must be oxidized by the mitochondrial electron transfer chain. Consider a preparation of rat hepatocytes containing mitochondria and all the enzymes of the cytosol. If [4–^{3}H]NADH is introduced, radioactivity appears quickly in the mitochondrial matrix. However, if [7–^{14}C]NADH is introduced, no radioactivity appears in the matrix. What do these observations tell us about the oxidation of extramitochondrial NADH by the electron transfer chain?

[4-^{3}H]NADH   [7-^{14}C]NADH

15. *Photochemical Efficiency of Light at Different Wavelengths* The rate of photosynthesis, measured by O_2 production, is higher when a green plant is illuminated with light of wavelength 680 nm than with light of 700 nm. However, illumination by a combination of light of 680 nm and 700 nm gives a higher rate of photosynthesis than light of either wavelength alone. Explain.

16. *Role of H_2S in Some Photosynthetic Bacteria* Illuminated purple sulfur bacteria carry out photosynthesis in the presence of H_2O and $^{14}CO_2$, but only if H_2S is added and O_2 is absent. During the course of photosynthesis, measured by formation of [^{14}C]glucose, H_2S is converted into elemental sulfur, but no O_2 is evolved. What is the role of the conversion of H_2S into sulfur? Why is no O_2 evolved?

17. *Boosting the Reducing Power of Photosystem I by Light Absorption* When photosystem I absorbs red light at 700 nm, the standard reduction potential of P700 changes from 0.4 to about -1.2 V. What fraction of the absorbed light is trapped in the form of reducing power?

18. *Mode of Action of the Herbicide DCMU* When chloroplasts are treated with 3-(3,4-dichlorophenyl)-1,1-dimethylurea (DCMU, or Diuron), a potent herbicide, O_2 evolution and photophosphorylation cease. Oxygen evolution but not photophosphorylation can be restored by the addition of an external electron acceptor, or Hill reagent. How does this herbicide act as a weed killer? Suggest a location for the inhibitory site of this herbicide in the scheme shown in Figure 18–44. Explain.

19. *Bioenergetics of Photophosphorylation* The steady-state concentrations of ATP, ADP, and P_i in isolated spinach chloroplasts under full illumination at pH 7.0 are 120, 6, and 700 μM, respectively.

(a) What is the free-energy requirement for the synthesis of 1 mol of ATP under these conditions?

(b) The energy for ATP synthesis is furnished by light-induced electron transfer in the chloroplasts. What is the minimum voltage drop necessary during the transfer of a pair of electrons to synthesize ATP under these conditions? (You may need to refer to p. 389, Eqn 13–8.)

20. *Equilibrium Constant for Water-Splitting Reactions* The coenzyme $NADP^+$ is the terminal electron acceptor in chloroplasts, according to the reaction

$$2H_2O + 2NADP^+ \longrightarrow 2NADPH + 2H^+ + O_2$$

Use the information in Table 18–2 to calculate the equilibrium constant at 25 °C for this reaction. (The relationship between K'_{eq} and $\Delta G^{\circ\prime}$ is discussed on p. 368.) How can the chloroplast overcome this unfavorable equilibrium?

21. *Energetics of Phototransduction* During photosynthesis, eight photons of light must be absorbed (four by each photosystem) for every O_2 molecule produced:

$$2H_2O + 2NADP^+ + 8 \text{ photons} \longrightarrow$$
$$2NADPH + 2H^+ + O_2$$

Assuming that these photons have a wavelength of 700 nm (red) and that the absorption and utilization of light energy are 100% efficient, calculate the free-energy change for the process.

22. *Electron Transfer to a Hill Reagent* Isolated spinach chloroplasts evolve O_2 when illuminated in the presence of potassium ferricyanide (the Hill reagent), according to the equation

$$2H_2O + 4Fe^{3+} \longrightarrow O_2 + 4H^+ + 4Fe^{2+}$$

where Fe^{3+} represents ferricyanide and Fe^{2+}, ferrocyanide. Is NADPH produced in this process? Explain.

23. *How Often Does a Chlorophyll Molecule Absorb a Photon?* The amount of chlorophyll a (M_r 892) in a spinach leaf is about 20 $\mu g/cm^2$ of leaf. In noonday sunlight (average energy 5.4 $J/cm^2 \cdot min$), the leaf absorbs about 50% of the radiation. How often does a single chlorophyll molecule absorb a photon? If the average lifetime of an excited chlorophyll molecule in vivo is 1 ns, what fraction of chlorophyll molecules are excited at any one time?

24. *Effect of Monochromatic Light on Electron Flow* The extent to which an electron carrier is oxidized or reduced during photosynthetic electron transfer can sometimes be observed directly with a spectrophotometer. When chloroplasts are illuminated with 700 nm light, cytochrome f, plastocyanin, and plastoquinone are oxidized. When chloroplasts are illuminated with 680 nm light, however, these electron carriers are reduced. Explain.

25. *Function of Cyclic Photophosphorylation* When the [NADPH]/[NADP$^+$] ratio in chloroplasts is high, photophosphorylation is predominantly cyclic (Fig. 18–44). Is O_2 evolved during cyclic photophosphorylation? Explain. Can the chloroplast produce NADPH this way? What is the main function of cyclic photophosphorylation?

CHAPTER

Carbohydrate Biosynthesis

We have now reached a turning point in the study of cellular metabolism. The preceding chapters of Part III have described how the major foodstuffs—carbohydrates, fatty acids, and amino acids—are degraded via converging *catabolic* pathways to enter the citric acid cycle and yield their electrons to the respiratory chain. The exergonic flow of electrons to oxygen is coupled to the endergonic synthesis of ATP. We now turn to *anabolic* pathways, which use chemical energy in the form of ATP and NADH or NADPH to synthesize cell components from simple precursor molecules. Anabolic pathways are generally reductive rather than oxidative. Catabolism and anabolism proceed simultaneously in a dynamic steady state, so that the energy-yielding degradation of cell components is counterbalanced by biosynthetic processes, which create and maintain the intricate orderliness of living cells.

Several organizing principles of biosynthesis deserve emphasis at the outset. The first principle is that the pathway taken in the synthesis of a biomolecule is usually different from the pathway taken in its degradation. Although the two opposing pathways may share many reversible reactions, there is always at least one enzymatic step that is unique to each pathway. If the reactions of catabolism and anabolism were catalyzed by the same set of enzymes acting reversibly, the flow of carbon through these pathways would be dictated exclusively by mass action (p. 371), not by the cell's changing needs for energy, precursors, or macromolecules.

Second, corresponding anabolic and catabolic pathways are controlled by different regulatory enzymes. These opposing pathways are regulated in a coordinated, reciprocal manner, so that stimulation of the biosynthetic pathway is accompanied by inhibition of the corresponding degradative pathway, and vice versa. Biosynthetic pathways are usually regulated at their initial steps, so that the cell avoids wasting precursors to make unneeded intermediates; intrinsic economy prevails in the molecular logic of living cells.

Third, energy-requiring biosynthetic processes are coupled to the energy-yielding breakdown of ATP in such a way that the overall process is essentially irreversible in vivo. Thus the total amount of ATP (and NAD(P)H) energy used in a given biosynthetic pathway always exceeds the minimum amount of free energy required to convert the precursor into the biosynthetic product. The resulting large, negative, free-energy change for the overall process assures that it will occur even when the concentrations of precursors are relatively low.

This chapter provides many opportunities for elaboration of the three principles outlined above, as we describe pathways for carbohydrate biosynthesis. The chapter is divided into four parts. First we consider gluconeogenesis, the ubiquitous pathway for synthesis of glucose. We then describe how glucose is converted into a variety of polysaccharides: glycogen in animals and many microorganisms, starch and sucrose in plants. At this point the focus shifts entirely to plants. The third topic is the incorporation of CO_2 into more complex molecules (CO_2 fixation), a process that represents the ultimate source of reduced carbon compounds for all organisms. The chapter ends with a discussion of the regulation of carbohydrate metabolism in plants. The overall regulation of carbohydrate metabolism in mammals is covered separately in Chapter 22.

Gluconeogenesis

We begin our survey of biosynthetic processes with the central pathway that leads to the formation of different carbohydrates from noncarbohydrate precursors in animal tissues (Fig. 19–1). The biosynthesis of glucose is an absolute necessity in all mammals, because the brain and nervous system, as well as the kidney medulla, testes, erythrocytes, and embryonic tissues, require glucose from the blood as their sole or major fuel source. The human brain alone requires over 120 g of

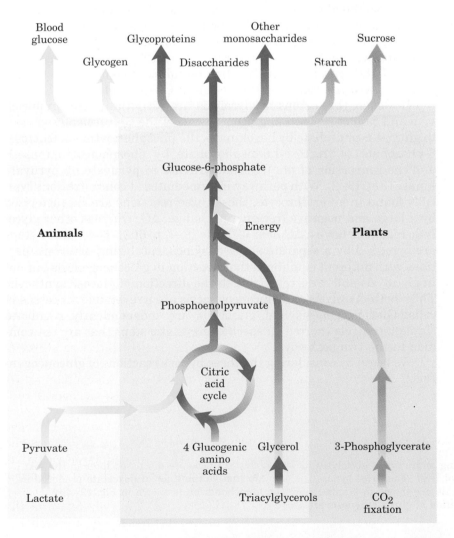

Figure 19–1 The pathway from phosphoenolpyruvate to glucose-6-phosphate is common to the biosynthetic conversion of many different precursors into carbohydrates in animals and plants.

glucose per day. Mammalian cells constantly make glucose directly from simpler precursors, such as pyruvate and lactate, and then pass the glucose into the blood. Other important carbohydrates, including glycogen, are also made from simple precursors (Fig. 19–1).

The formation of glucose from nonhexose precursors is called **gluconeogenesis** ("formation of new sugar"). Gluconeogenesis is a universal pathway, found in all animals, plants, fungi, and microorganisms. The reactions are the same in every case. The important precursors of glucose in animals are lactate, pyruvate, glycerol, and most of the amino acids (Fig. 19–1). In higher animals gluconeogenesis occurs largely in the liver and to a much smaller extent in kidney cortex. In plant seedlings, stored fats and proteins are converted into the disaccharide sucrose for transport throughout the developing plant. Glucose and its derivatives are precursors in the synthesis of plant cell walls, nucleotides and coenzymes, and a variety of other essential metabolites. Many microorganisms are able to grow on simple organic compounds such as acetate, lactate, and propionate, which they convert to glucose by gluconeogenesis.

Although the reactions of gluconeogenesis are the same in all organisms, the metabolic context and the regulation of the pathway differ from organism to organism, and from tissue to tissue. In this chapter we focus first on gluconeogenesis as it occurs in the mammalian liver. Later, its role and regulation in plants will be considered.

Just as the glycolytic conversion of glucose into pyruvate is a central pathway for catabolism of carbohydrates, the conversion of pyruvate into glucose is a central pathway in gluconeogenesis. These pathways are not identical, although they share several steps (Fig. 19–2). Seven of the ten enzymatic reactions of gluconeogenesis are the reverse of glycolytic reactions (discussed in Chapter 14).

However, three steps in glycolysis are essentially irreversible in vivo and cannot be used in gluconeogenesis: the conversion of glucose to glucose-6-phosphate by hexokinase, the phosphorylation of fructose-6-phosphate to fructose-1,6-bisphosphate by phosphofructokinase-1, and the conversion of phosphoenolpyruvate to pyruvate by pyruvate kinase (Fig. 19–2). With pathway intermediates at concentrations typically found in an erythrocyte, these three reactions are characterized by a large and negative free-energy change, ΔG, whereas other glycolytic reactions have a ΔG near 0 (Table 19–1, p. 602). These three steps are bypassed by a separate set of enzymes, catalyzing different reactions with different equilibria; they function in gluconeogenesis but not in glycolysis and are irreversible in the direction of glucose synthesis. Thus, both glycolysis and gluconeogenesis are irreversible processes in cells. Gluconeogenesis and glycolysis are independently regulated through controls exerted on specific enzymatic steps that are not common to the two pathways.

We begin by considering the three bypass reactions of gluconeogenesis.

Figure 19–2 The opposing pathways of glycolysis and gluconeogenesis in rat liver. The three bypass reactions of gluconeogenesis are shown in orange. Two major sites of regulation of gluconeogenesis are also shown; these are discussed later in the text. An alternative route for oxaloacetate produced in the mitochondrion is shown in Fig. 19–4.

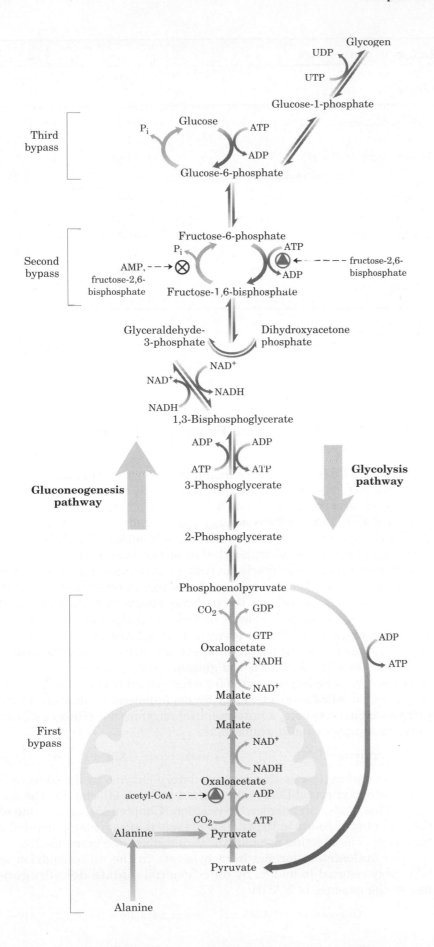

Table 19–1 Free-energy changes of glycolytic reactions in erythrocytes*

Glycolytic reaction step	$\Delta G^{\circ\prime}$ (kJ/mol)	ΔG (kJ/mol)
① Glucose + ATP $\longrightarrow$ glucose-6-phosphate + ADP + H$^+$	−16.7	−33.4
② Glucose-6-phosphate $\rightleftharpoons$ fructose-6-phosphate	1.7	−2.5
③ Fructose-6-phosphate + ATP $\longrightarrow$ fructose-1,6-bisphosphate + ADP + H$^+$	−14.2	−22.2
④ Fructose-1,6-bisphosphate $\rightleftharpoons$ dihydroxyacetone phosphate + glyceraldehyde-3-phosphate	23.8	−1.25
⑤ Dihydroxyacetone phosphate $\rightleftharpoons$ glyceraldehyde-3-phosphate	7.5	2.5
⑥ Glyceraldehyde-3-phosphate + P$_i$ + NAD$^+$ $\rightleftharpoons$ 1,3-bisphosphoglycerate + NADH + H$^+$	6.3	−1.7
⑦ 1,3-Bisphosphoglycerate + ADP $\rightleftharpoons$ 3-phosphoglycerate + ATP	−18.8	1.25
⑧ 3-Phosphoglycerate $\rightleftharpoons$ 2-phosphoglycerate	4.4	0.8
⑨ 2-Phosphoglycerate $\rightleftharpoons$ phosphoenolpyruvate + H$_2$O	7.5	−3.3
⑩ Phosphoenolpyruvate + ADP + H$^+$ $\longrightarrow$ pyruvate + ATP	−31.4	−16.7

* $\Delta G^{\circ\prime}$ is the standard free-energy change, as defined in Chapter 13. ΔG is the free-energy change calculated from the actual concentrations of glycolytic intermediates present under physiological conditions in erythrocytes (at pH 7.0). The glycolytic reactions bypassed in gluconeogenesis are shown in red.

Conversion of Pyruvate into Phosphoenolpyruvate Requires a Bypass

The first of the bypass reactions in gluconeogenesis is the conversion of pyruvate into phosphoenolpyruvate (Fig. 19–2). This reaction cannot occur by reversal of the pyruvate kinase reaction (p. 413), which has a large, negative, standard free-energy change and has been found to be irreversible under the conditions that exist in intact cells (Table 19–1, step ⑩). Instead, the phosphorylation of pyruvate is achieved by a roundabout sequence of reactions that in mammals and some other organisms requires the cooperation of enzymes in both the cytosol and mitochondria. As we will see, the pathway shown in Figure 19–2 and described in more detail below is one of two paths from pyruvate to phosphoenolpyruvate; it is the predominant one when pyruvate or alanine is the gluconeogenic precursor. A second pathway, described later, predominates when lactate is the gluconeogenic precursor.

First, pyruvate is transported from the cytosol to the mitochondria, or generated within mitochondria by deamination of alanine. Then, **pyruvate carboxylase,** a mitochondrial enzyme that requires biotin, converts the pyruvate to oxaloacetate:

$$\text{Pyruvate} + \text{HCO}_3^- + \text{ATP} \longrightarrow \text{oxaloacetate} + \text{ADP} + \text{P}_i + \text{H}^+ \quad (19\text{–}1)$$

Pyruvate carboxylase is the first regulatory enzyme in the gluconeogenic pathway; acetyl-CoA is a required positive effector for the enzyme. This is also an anaplerotic reaction (Chapter 15), and as one of the major entry points in the citric acid cycle it affects many metabolic pathways. The reaction mechanism is described in Figure 15–13b.

The oxaloacetate formed from pyruvate in the mitochondrion is reversibly reduced to malate by mitochondrial **malate dehydrogenase** at the expense of NADH:

$$\text{Oxaloacetate} + \text{NADH} + \text{H}^+ \rightleftharpoons \text{L-malate} + \text{NAD}^+ \quad (19\text{–}2)$$

The malate then leaves the mitochondrion via the malate–α-ketoglutarate transporter in the inner mitochondrial membrane (see Fig. 18–25). In the cytosol, malate is reoxidized to oxaloacetate, with the production of cytosolic NADH:

$$\text{Malate} + \text{NAD}^+ \longrightarrow \text{oxaloacetate} + \text{NADH} + \text{H}^+ \qquad (19\text{–}3)$$

The oxaloacetate is then converted to phosphoenolpyruvate (PEP) by **phosphoenolpyruvate carboxykinase** in a Mg^{2+}-dependent reaction in which GTP serves as the phosphate donor (Fig. 19–3):

$$\text{Oxaloacetate} + \text{GTP} \rightleftharpoons \text{phosphoenolpyruvate} + \text{CO}_2 + \text{GDP} \quad (19\text{–}4)$$

This reaction is reversible under intracellular conditions.

The overall equation for this set of bypass reactions, the sum of Equations 19–1 through 19–4, is

$$\text{Pyruvate} + \text{ATP} + \text{GTP} + \text{HCO}_3^- \longrightarrow$$
$$\text{phosphoenolpyruvate} + \text{ADP} + \text{GDP} + \text{P}_i + \text{H}^+ + \text{CO}_2 \quad (19\text{–}5)$$
$$\Delta G^{\circ\prime} = 0.9 \text{ kJ/mol}$$

Two high-energy phosphate groups (one from ATP and one from GTP), each yielding 30.5 kJ/mol under standard conditions, must be expended to phosphorylate one molecule of pyruvate to PEP, thus requiring an input of 61 kJ/mol under standard conditions. In contrast, when PEP is converted into pyruvate during glycolysis, only one ATP is generated from ADP. Although the standard free-energy change ($\Delta G^{\circ\prime}$) of the net reaction leading to PEP is 0.9 kJ/mol, the actual free-energy change (ΔG), calculated from measured cellular concentrations of intermediates, is very strongly negative ($\Delta G = -25$ kJ/mol); this results from the ready consumption of PEP in other reactions, such that its concentration remains low. The reaction is thus irreversible in the cell.

Note that the CO_2 lost in the PEP carboxykinase reaction is the same molecule that is added to pyruvate in the pyruvate carboxylase step (Fig. 19–3). This carboxylation–decarboxylation sequence represents a way of "activating" pyruvate in that the decarboxylation of oxaloacetate facilitates PEP formation. In Chapter 20 we will see that a similar carboxylation step is used to activate acetyl-CoA for fatty acid biosynthesis.

The path of these reactions through the mitochondria is not coincidental. The [NADH]/[NAD$^+$] ratio in the cytosol is 8×10^{-4}, about 10^5 times lower than in mitochondria. Because cytosolic NADH is consumed in gluconeogenesis (in the conversion of 1,3-bisphosphoglycerate to glyceraldehyde-3-phosphate; Fig. 19–2), glucose biosynthesis cannot proceed unless NADH is made available. The transport of malate from the mitochondrion to the cytosol and its reconversion there to oxaloacetate has the effect of moving reducing equivalents in the form of NADH to the cytosol, where they are scarce. This path from pyruvate to PEP therefore provides an important balance between NADH produced and consumed in the cytosol during gluconeogenesis.

A second and shorter pyruvate $\longrightarrow$ PEP bypass predominates when lactate is the gluconeogenic precursor (Fig. 19–4). This pathway makes use of lactate produced by glycolysis in erythrocytes or muscle, for example, and it is particularly important in larger vertebrates after vigorous exercise (see Box 14–1). The conversion of lactate to pyruvate in the hepatocyte cytosol yields NADH, and the export of malate from mitochondria is no longer necessary. After the pyruvate produced by the lactate dehydrogenase reaction is transported into the mitochon-

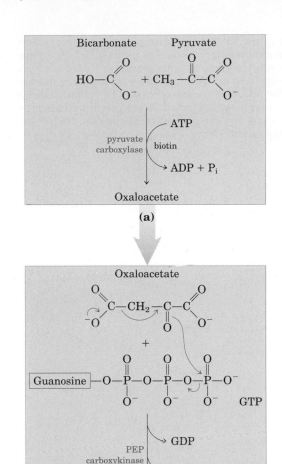

Figure 19–3 Synthesis of phosphoenolpyruvate from pyruvate. **(a)** Pyruvate is converted to oxaloacetate in a biotin-requiring reaction catalyzed by pyruvate carboxylase. **(b)** Oxaloacetate is converted into phosphoenolpyruvate by PEP carboxykinase. The CO_2 that was fixed in the pyruvate carboxylase reaction is lost here as CO_2. The decarboxylation leads to a rearrangement of electrons that facilitates the attack of the carbonyl oxygen of the pyruvate moiety on the γ phosphate of GTP.

Figure 19–4 Alternative paths from pyruvate to phosphoenolpyruvate. The paths differ depending upon the gluconeogenic precursor (lactate or pyruvate) and are determined by cytosolic requirements for NADH in gluconeogenesis. The abbreviated path occurring when lactate is the precursor is made possible by the generation of cytosolic NADH in the lactate dehydrogenase reaction (p. 416).

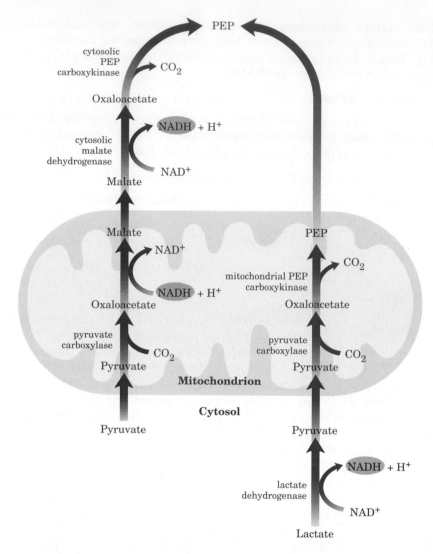

drion, it is converted to oxaloacetate by pyruvate carboxylase as described above. This oxaloacetate, however, is converted directly to PEP by a mitochondrial form of PEP carboxykinase. The PEP is then transported out of the mitochondrion and continues on the gluconeogenic path. The mitochondrial and cytosolic forms of PEP carboxykinase are encoded by separate nuclear genes, providing an example of two distinct enzymes catalyzing the same reaction but having different cellular locations and metabolic roles.

Conversion of Fructose-1,6-Bisphosphate into Fructose-6-Phosphate Is the Second Bypass

The second reaction of the catabolic glycolytic sequence that cannot participate in the anabolic process of gluconeogenesis is the phosphorylation of fructose-6-phosphate by phosphofructokinase-1 (Table 19–1, Step ③). Because this reaction is irreversible in intact cells, the generation of fructose-6-phosphate from fructose-1,6-bisphosphate (Fig. 19–2) is catalyzed by a different enzyme, Mg^{2+}-dependent **fructose-1,6-bisphosphatase,** which promotes the essentially irreversible hydrolysis of the C-1 phosphate:

Fructose-1,6-bisphosphate + H_2O $\longrightarrow$ fructose-6-phosphate + P_i

$$\Delta G^{\circ\prime} = -16.3 \text{ kJ/mol}$$

Conversion of Glucose-6-Phosphate into Free Glucose Is the Third Bypass

The third bypass is the final reaction of gluconeogenesis, the dephosphorylation of glucose-6-phosphate to yield free glucose (Fig. 19–2). Because the hexokinase reaction of glycolysis is irreversible (Table 19–1, step ①), the hydrolytic reaction is catalyzed by another enzyme, **glucose-6-phosphatase:**

$$\text{Glucose-6-phosphate} + H_2O \longrightarrow \text{glucose} + P_i \qquad \Delta G^{\circ\prime} = -13.8 \text{ kJ/mol}$$

This Mg^{2+}-dependent enzyme is found in the endoplasmic reticulum of hepatocytes. Glucose-6-phosphatase is not present in muscles or in the brain, and gluconeogenesis does not occur in these tissues. Instead, glucose produced by gluconeogenesis in the liver or ingested in the diet is delivered to brain and muscle through the bloodstream.

Gluconeogenesis Is Energetically Costly

The sum of the biosynthetic reactions leading from pyruvate to free blood glucose (Table 19–2) is

$$2 \text{ Pyruvate} + 4ATP + 2GTP + 2NADH + 4H_2O \longrightarrow$$
$$\text{glucose} + 4ADP + 2GDP + 6P_i + 2NAD^+ + 2H^+$$

For each molecule of glucose formed from pyruvate, six high-energy phosphate groups are required, four from ATP and two from GTP. In addition, two molecules of NADH are required for the reduction of two molecules of 1,3-bisphosphoglycerate. This equation is clearly not the simple reverse of the equation for the conversion of glucose into pyruvate by glycolysis, which yields only two molecules of ATP:

$$\text{Glucose} + 2ADP + 2P_i + 2NAD^+ \longrightarrow$$
$$2 \text{ pyruvate} + 2ATP + 2NADH + 2H^+ + 2H_2O$$

Table 19–2 Sequential reactions in gluconeogenesis starting from pyruvate[*]

Pyruvate + HCO_3^- + ATP $\longrightarrow$ oxaloacetate + ADP + P_i + H^+	×2
Oxaloacetate + GTP $\rightleftharpoons$ phosphoenolpyruvate + CO_2 + GDP	×2
Phosphoenolpyruvate + H_2O $\rightleftharpoons$ 2-phosphoglycerate	×2
2-Phosphoglycerate $\rightleftharpoons$ 3-phosphoglycerate	×2
3-Phosphoglycerate + ATP $\rightleftharpoons$ 1,3-bisphosphoglycerate + ADP + H^+	×2
1,3-Bisphosphoglycerate + NADH + H^+ $\rightleftharpoons$ glyceraldehyde-3-phosphate + NAD^+ + P_i	×2
Glyceraldehyde-3-phosphate $\rightleftharpoons$ dihydroxyacetone phosphate	
Glyceraldehyde-3-phosphate + dihydroxyacetone phosphate $\rightleftharpoons$ fructose-1,6-bisphosphate	
Fructose-1,6-bisphosphate + H_2O $\longrightarrow$ fructose-6-phosphate + P_i	
Fructose-6-phosphate $\rightleftharpoons$ glucose-6-phosphate	
Glucose-6-phosphate + H_2O $\rightleftharpoons$ glucose + P_i	

Sum: 2 Pyruvate + 4ATP + 2GTP + 2NADH + 4H_2O $\longrightarrow$ glucose + 4ADP + 2GDP + 6P_i + 2NAD^+ + 2H^+

[*] The bypass reactions are in red; all other reactions are reversible steps of glycolysis. The figures at the right indicate that the reaction is to be counted twice, because two three-carbon precursors are required to make a molecule of glucose. Note that the reactions required to replace the cytosolic NADH consumed in the glyceraldehyde-3-phosphate dehydrogenase reaction (the conversion of lactate to pyruvate in the cytosol or the transport of reducing equivalents from the mitochondria to the cytosol in the form of malate) are not considered in this summary.

Thus the synthesis of glucose from pyruvate is a relatively costly process. Much of this high energy cost is necessary to ensure that gluconeogenesis is irreversible. Under intracellular conditions, the overall free-energy change of glycolysis is at least -63 kJ/mol. Under the same conditions the overall free-energy change of gluconeogenesis from pyruvate is also highly negative. Thus glycolysis and gluconeogenesis are both essentially irreversible processes under intracellular conditions.

Citric Acid Cycle Intermediates and Many Amino Acids Are Glucogenic

The biosynthetic pathway to glucose described above allows the net synthesis of glucose not only from pyruvate but also from the citric acid cycle intermediates citrate, isocitrate, α-ketoglutarate, succinate, fumarate, and malate. All may undergo oxidation in the citric acid cycle to yield oxaloacetate. However, only three carbon atoms of oxaloacetate are converted into glucose; the fourth is released as CO_2 in the conversion of oxaloacetate to phosphoenolpyruvate by PEP carboxykinase (Fig. 19–3).

In Chapter 17 we showed that some or all of the carbon atoms of many of the amino acids derived from proteins are ultimately converted by mammals into either pyruvate or certain intermediates of the citric acid cycle. Such amino acids can therefore undergo net conversion into glucose and are called **glucogenic** amino acids (Table 19–3). Alanine and glutamine make especially important contributions in that they are the principal molecules used to transport amino groups from extrahepatic tissues to the liver. After removal of their amino groups in liver mitochondria, the carbon skeletons remaining (pyruvate and α-ketoglutarate, respectively) are readily funneled into gluconeogenesis.

In contrast, there is no net conversion of even-carbon fatty acids into glucose by mammals because such fatty acids yield only acetyl-CoA on oxidative cleavage. Acetyl-CoA cannot be used as a precursor of glucose by mammals. The pyruvate dehydrogenase reaction is irreversible under intracellular conditions, and no other pathway exists by which acetyl-CoA can be converted into pyruvate. For every two carbons of acetate that enter the citric acid cycle, two carbons are lost as CO_2 (Fig. 19–5); therefore, there can be no net conversion of acetate to oxaloacetate or pyruvate. However, fatty acids make an important contribution to gluconeogenesis in another way. Oxidation of fatty acids delivered to the liver during starvation provides much of the ATP and NADH needed to drive gluconeogenesis energetically.

Futile Cycles in Carbohydrate Metabolism Consume ATP

At the three points where a glycolytic reaction is bypassed by a different type of enzymatic reaction in gluconeogenesis, simultaneous operation of both pathways would be wasteful. For example, phosphofructokinase-1 and fructose-1,6-bisphosphatase catalyze opposing reactions:

$$\text{ATP} + \text{fructose-6-phosphate} \longrightarrow \text{ADP} + \text{fructose-1,6-bisphosphate} + \text{H}^+$$
$$\text{Fructose-1,6-bisphosphate} + \text{H}_2\text{O} \longrightarrow \text{fructose-6-phosphate} + \text{P}_i$$

The sum of these two reactions is

$$\text{ATP} + \text{H}_2\text{O} \longrightarrow \text{ADP} + \text{P}_i + \text{H}^+ + \text{heat}$$

an energy-wasting reaction resulting in the hydrolysis of ATP without

Table 19–3 Glucogenic amino acids, grouped by site of entry[*]

Pyruvate	*Succinyl-CoA*
Alanine	Valine
Serine	Threonine
Cysteine	Methionine
Glycine	[†]Isoleucine
[†]Tryptophan	
	Fumarate
α-Ketoglutarate	[†]Phenylalanine
Glutamate	[†]Tyrosine
Glutamine	
Proline	*Oxaloacetate*
Arginine	Asparagine
Histidine	Aspartate

[*] These amino acids are precursors of blood glucose or liver glycogen because they can be converted into pyruvate or citric acid cycle intermediates. Only leucine and lysine are totally unable to furnish carbon for net glucose synthesis.

[†] These amino acids are also ketogenic (Chapter 17).

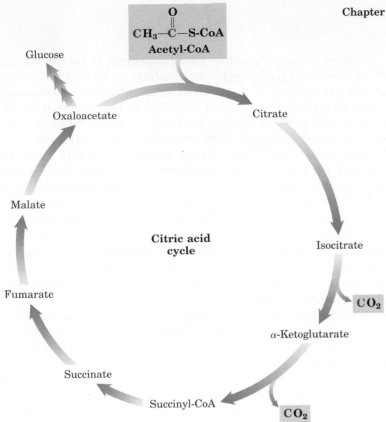

Figure 19–5 Fatty acids with even-numbered chains of carbon atoms cannot be a source of carbon for the net synthesis of glucose in animals and microorganisms. Fatty acids are catabolized to acetyl-CoA, which enters the citric acid cycle. For every two carbons entering the cycle as acetyl-CoA, two carbons are lost as CO_2, so there is no net production of oxaloacetate to support glucose biosynthesis by this path. However, fatty acid oxidation does provide extraordinary amounts of energy (in the form of NADH, ATP, and GTP) to support gluconeogenesis (Chapter 16). Amino acids that are degraded to acetyl-CoA are also not glucogenic (see Fig. 17–33), for the reason illustrated here.

any net metabolic work being done. Clearly, if these two reactions were allowed to proceed simultaneously at a high rate in the same cell, a large amount of chemical energy would be dissipated as heat. Such an ATP-degrading cycle is called a **futile cycle.** A similar futile cycle could occur with the other two sets of bypass reactions.

Under normal circumstances futile cycles probably do not take place at a significant rate, because they are prevented by reciprocal regulatory mechanisms (as discussed below). However, futile cycling sometimes occurs physiologically to produce heat. For example, in cold weather bumblebees cannot fly until they have warmed their muscles to about 30 °C by futile cycling of fructose-6-phosphate and fructose-1,6-bisphosphate and the consequent heat-generating hydrolysis of ATP.

Gluconeogenesis and Glycolysis Are Reciprocally Regulated

To assure that futile cycling does not occur under normal circumstances, gluconeogenesis and glycolysis are regulated separately and reciprocally. The first control point occurs in the reactions catalyzed by the pyruvate dehydrogenase complex of glycolysis and pyruvate carboxylase of gluconeogenesis (Fig. 19–6); the latter enzyme requires the positive allosteric modulator acetyl-CoA for its activity. As a consequence, the biosynthesis of glucose from pyruvate is promoted only when excess mitochondrial acetyl-CoA builds up beyond the immediate needs of the cell for the citric acid cycle. When the cell's energetic needs are being met, oxidative phosphorylation slows, NADH accumulation inhibits the citric acid cycle, and acetyl-CoA accumulates. The increased concentration of acetyl-CoA inhibits the pyruvate dehydrogenase complex, slowing the formation of acetyl-CoA from pyruvate, and stimulates gluconeogenesis by activating pyruvate carboxylase. This allows excess pyruvate to be converted to glucose.

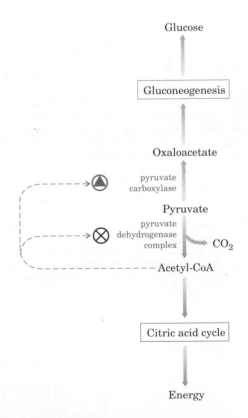

Figure 19–6 Two alternative fates for pyruvate: conversion to glucose and glycogen via gluconeogenesis, or oxidation to acetyl-CoA for energy production. The first enzyme in each path is regulated allosterically; acetyl-CoA stimulates the activity of pyruvate carboxylase and inhibits the activity of the pyruvate dehydrogenase complex.

The second control point in gluconeogenesis is the reaction catalyzed by fructose-1,6-bisphosphatase, which is strongly inhibited by AMP. The corresponding glycolytic enzyme, phosphofructokinase-1, is stimulated by AMP and ADP but inhibited by citrate and ATP, so that these opposing steps in the two pathways are regulated in a coordinated or reciprocal manner. When sufficient concentrations of acetyl-CoA or of the product of acetyl-CoA condensation with oxaloacetate (citrate) are present, or when a high proportion of the cell's adenylate is in the form of ATP, gluconeogenesis is favored, thus promoting formation of glucose and its storage as glycogen (described later).

The special role of liver in maintaining a constant blood glucose level requires additional regulatory mechanisms to coordinate glucose production and consumption. When the blood glucose level decreases, the hormone glucagon signals the liver to produce and release more glucose. One source of glucose is glycogen stored in the liver; another source is gluconeogenesis.

The hormonal regulation of glycolysis and gluconeogenesis in liver is mediated by **fructose-2,6-bisphosphate,** an allosteric effector for the enzymes phosphofructokinase-1 (PFK-1) and fructose-1,6-bisphosphatase (FBPase-1) (Fig. 19–7). When fructose-2,6-bisphosphate binds to its allosteric site on PFK-1, it increases that enzyme's affinity for its substrate fructose-6-phosphate and reduces its affinity for the allosteric inhibitors ATP and citrate. Fructose-2,6-bisphosphate therefore activates PFK-1 and stimulates glycolysis in liver. Fructose-2,6-bisphosphate also inhibits FBPase-1, thereby slowing gluconeogenesis.

Figure 19–7 The effect of fructose-2,6-bisphosphate (F-2,6-BP) on the enzymatic activities of phosphofructokinase-1 (PFK-1, a glycolytic enzyme) and fructose-1,6-bisphosphatase-1 (FBPase-1, an enzyme of gluconeogenesis). **(a)** PFK-1 activity in the absence of fructose-2,6-bisphosphate (blue curve) is half-maximal when the concentration of the substrate fructose-6-phosphate is 2 mM (that is, $K_{0.5} = 2$ mM; recall from Chapter 8 that $K_{0.5}$ or K_m is equivalent to the substrate concentration at which half-maximal enzyme activity occurs). When 0.13 μM fructose-2,6-bisphosphate is present (red curve), the $K_{0.5}$ for fructose-6-phosphate is only 0.08 mM. Thus fructose-2,6-bisphosphate activates PFK-1 by increasing its apparent affinity (p. 216) for fructose-6-phosphate. **(b)** FBPase-1 activity is inhibited by as little as 1 μM fructose-2,6-bisphosphate and is strongly inhibited by 25 μM. In the absence of this inhibitor (blue curve) the $K_{0.5}$ for the substrate fructose-1,6-bisphosphate is 5 μM, but in the presence of 25 μM fructose-2,6-bisphosphate (red curve) the $K_{0.5}$ is >70 μM. Fructose-2,6-bisphosphate also makes this enzyme more sensitive to inhibition by another allosteric regulator, AMP.

Fructose-2,6-bisphosphate

Although structurally related to fructose-1,6-bisphosphate, fructose-2,6-bisphosphate is clearly not an intermediate in gluconeogenesis or glycolysis; it is a *regulator* whose cellular level reflects the level of glucagon in the blood (see below), which in turn varies with the blood glucose level. Because fructose-2,6-bisphosphate is effective at low concentrations (in the micromolar range), very little fructose is diverted from metabolic pathways for its synthesis.

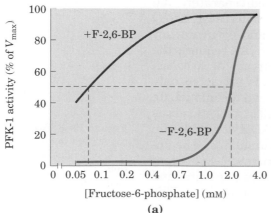

(a)

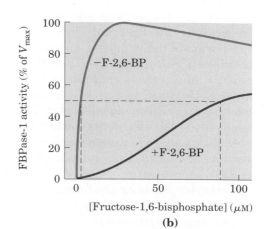

(b)

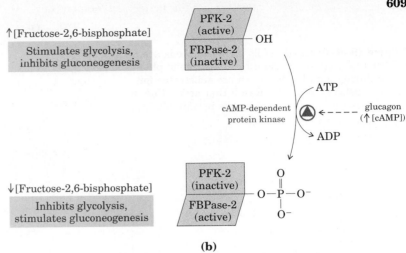

Figure 19–8 **(a)** The cellular concentration of the regulator fructose-2,6-bisphosphate is determined by the rates of its synthesis by phosphofructokinase-2 (PFK-2) and breakdown by fructose-2,6-bisphosphatase (FBPase-2). **(b)** Both of these enzymes are part of the same polypeptide chain, and both are regulated, in a reciprocal fashion, by glucagon. Here and elsewhere arrows are used to indicate increasing ($\uparrow$) and decreasing ($\downarrow$) levels of metabolites.

The cellular concentration of fructose-2,6-bisphosphate is set by the relative rates of its formation and breakdown. Fructose-2,6-bisphosphate is formed by phosphorylation of fructose-6-phosphate, catalyzed by **phosphofructokinase-2** (PFK-2), and is broken down by **fructose-2,6-bisphosphatase** (FBPase-2) (Fig. 19–8). (Note that these enzymes are distinct from PFK-1 and FBPase-1, which catalyze the formation and breakdown, respectively, of fructose-1,6-bisphosphate.) PFK-2 and FBPase-2 are two distinct enzymatic activities, but both are part of a single, bifunctional protein. The balance of these two activities in the liver, and therefore the cellular level of fructose-2,6-bisphosphate, is regulated by **glucagon.** Glucagon stimulates adenylate cyclase, an enzyme that synthesizes 3′,5′-cyclic AMP (cAMP) from ATP (see Fig. 14–18). Cyclic AMP in turn stimulates a cAMP-dependent protein kinase, which transfers a phosphate group from ATP to the bifunctional protein PFK-2/FBPase-2 (Fig. 19–8). Phosphorylation of this protein enhances its FBPase-2 activity and inhibits PFK-2. Glucagon thereby lowers the cellular level of fructose-2,6-bisphosphate, inhibiting glycolysis and stimulating gluconeogenesis. The resulting production of more glucose enables the liver to replenish blood glucose in response to glucagon.

Gluconeogenesis Converts Fats and Proteins to Glucose in Germinating Seeds

Many plants store lipids and proteins in their seeds, to be used as sources of energy and biosynthetic precursors during germination, before photosynthetic mechanisms can supply both. Active gluconeogenesis occurs in germinating seeds, providing glucose for the synthesis of polysaccharides and many metabolites derived from hexoses, and for the production of sucrose (the transport form of carbon in plants).

Unlike animals, plants can convert acetyl-CoA derived from fatty acid oxidation into glucose. Triacylglycerols are converted into glucose by the combined action of the β-oxidation pathway (Chapter 16), the glyoxylate cycle (Chapter 15), and gluconeogenesis. Hydrolysis of storage triacylglycerols produces glycerol-3-phosphate, which can enter the gluconeogenic pathway after its oxidation to dihydroxyacetone phosphate (Fig. 19–9). Fatty acids, the other product of lipolysis, are activated, then oxidized to acetyl-CoA in glyoxysomes by isozymes of the β-oxidation pathway (Fig. 19–10). The separation of this process from the citric acid cycle, the enzymes of which are in mitochondria but not glyoxysomes, prevents the further oxidation of acetyl-CoA to CO_2.

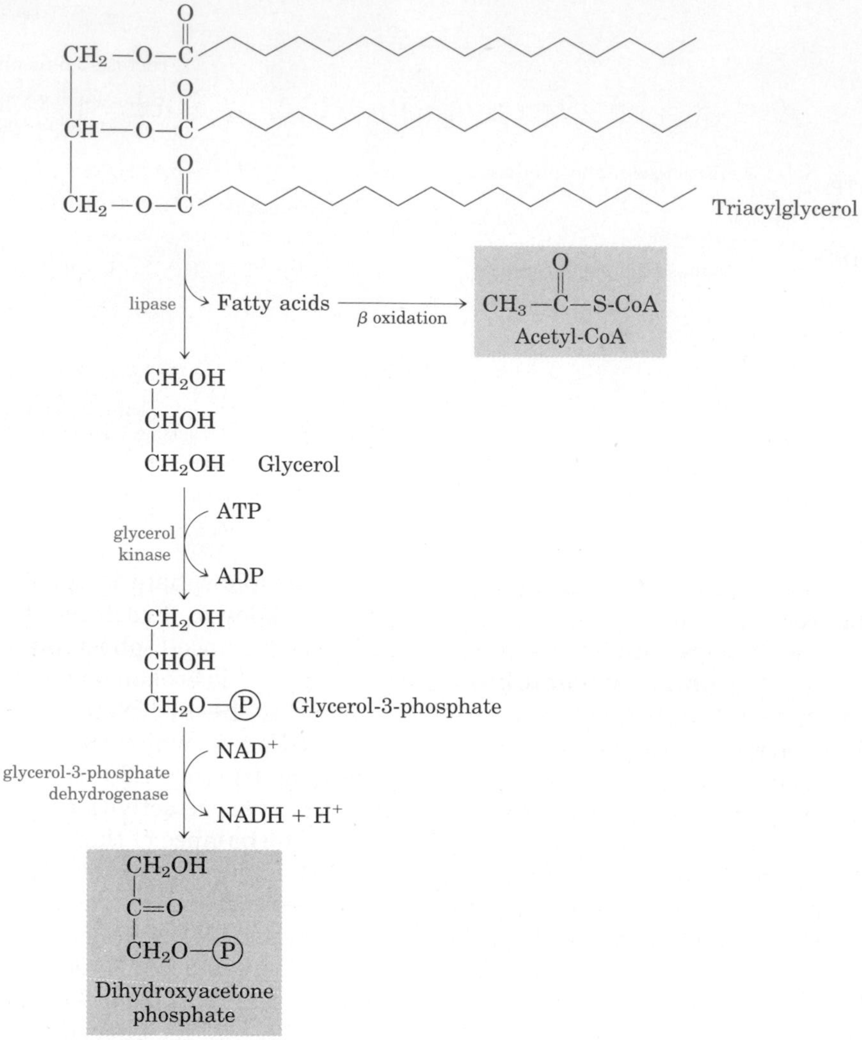

Figure 19–9 Triacylglycerols stored in seeds are oxidized to acetyl-CoA and dihydroxyacetone phosphate during germination; both are substrates for gluconeogenesis in plants. Recall that acetyl-CoA is not a substrate for gluconeogenesis in animals (Fig. 19–5).

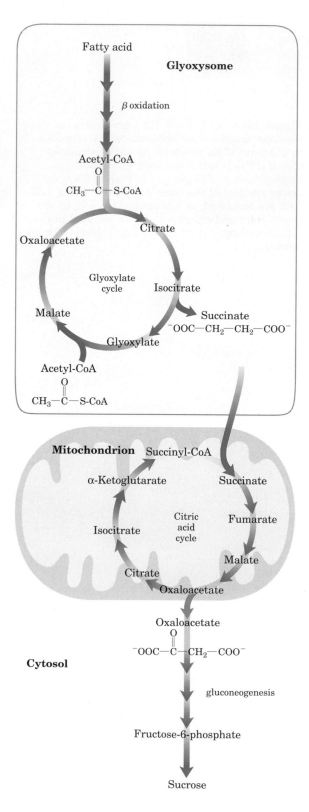

Instead, the acetyl-CoA is converted, via the glyoxylate cycle, into succinate. Succinate passes to the mitochondrial matrix, where it is converted by citric acid cycle enzymes into oxaloacetate, which passes out of the mitochondrion and into the cytosol. Cytosolic oxaloacetate is converted by gluconeogenesis to fructose-6-phosphate, the precursor to sucrose. Thus, the integration of reaction sequences occurring in three subcellular compartments is required for the production of fructose-6-phosphate or sucrose from stored lipids. About 75% of the carbon in the stored lipids of seeds is converted into carbohydrate by this means; the other 25% is lost as CO_2 in the conversion of oxaloacetate to phosphoenolpyruvate.

Amino acids derived from the breakdown of stored seed proteins also yield precursors for gluconeogenesis. They are deaminated and oxidized, via pathways discussed in Chapter 17, to succinyl-CoA, pyruvate, oxaloacetate, fumarate, and α-ketoglutarate—all good starting materials for gluconeogenesis.

Figure 19–10 The conversion of stored fatty acids to sucrose in germinating seeds begins in glyoxysomes, which produce succinate and export it to mitochondria. There it is converted to oxaloacetate by enzymes of the citric acid cycle. Oxaloacetate enters the cytosol and serves as the starting material for gluconeogenesis and the synthesis of sucrose, the transported sugar in plants.

Biosynthesis of Glycogen, Starch, and Sucrose

In a wide range of organisms, excess glucose is converted into polymeric forms for storage and transport. The principal storage forms of glucose are glycogen in vertebrates and many microorganisms, and starch in plants. In vertebrates, glucose itself is generally transported in the blood, but the transport form in plants is sucrose, or its galactosylated derivatives.

Although the intermediates in glycolysis and gluconeogenesis are sugar phosphates, many of the reactions in which hexoses are transformed or polymerized involve a different type of activating group, a nucleotide bound to the anomeric hydroxyl of the sugar through a phosphate ester linkage. **Sugar nucleotides** are the substrates for polymerization into disaccharides, glycogen, starch, cellulose, and more complex extracellular polysaccharides. They are also key intermediates in the production of aminohexoses and deoxyhexoses found in some of these polysaccharides. The role of sugar nucleotides (specifically **UDP-glucose**) in the biosynthesis of glycogen and many other carbohydrate derivatives was discovered by Luis Leloir.

The suitability of sugar nucleotides for biosynthetic reactions stems from several properties:

1. Their formation by the condensation of a nucleoside triphosphate with a hexose phosphate splits one high-energy bond and releases PP_i, which is further hydrolyzed by inorganic pyrophosphatase; there is a net cleavage of two high-energy bonds (Fig. 19–11). The resulting large, negative, free-energy change drives the synthetic reaction and reflects a strategy common to many biological polymerization reactions.

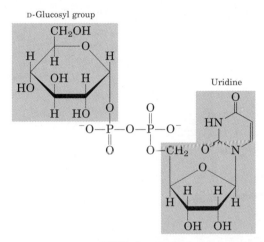

UDP-glucose

Net reaction: Sugar phosphate + NTP $\longrightarrow$ NDP-sugar + 2P$_i$

Figure 19–11 Formation of a sugar nucleotide by condensation of a nucleoside triphosphate (NTP) with a sugar phosphate. The negatively charged oxygen on the sugar phosphate serves as a nucleophile, attacking at the α phosphate in the nucleoside triphosphate and displacing pyrophosphate. The reaction is pulled in the forward direction by the formation and subsequent hydrolysis of PP_i by the enzyme inorganic pyrophosphatase.

Luis Leloir

2. Although the chemical transformations of sugar nucleotides do not involve the atoms of the nucleotide itself, the sugar nucleotide molecule offers many potential groups for noncovalent interactions with enzymes; the free energy of binding contributes significantly to the catalytic activity of the enzyme (Chapter 8; see also p. 353).

3. Like phosphate, the nucleotidyl group is an excellent leaving group, activating the sugar carbon to which it is attached so as to facilitate nucleophilic attack.

4. By "tagging" some hexoses with nucleotidyl groups, cells may set them aside for one purpose (glycogen synthesis, for example) in a pool separate from hexose phosphates earmarked for another purpose (such as glycolysis).

UDP-Glucose Is the Substrate for Glycogen Synthesis

In animals and some microorganisms, excess glucose available from carbohydrates in the diet or from gluconeogenesis is stored as glycogen. Glycogen synthesis occurs in virtually all animal tissues but is especially prominent in the liver and skeletal muscles. In the liver, glycogen serves as a reservoir of glucose, readily converted into blood glucose for distribution to other tissues, whereas in muscle, glycogen is broken down via glycolysis to provide ATP energy for muscle contraction. The starting point for synthesis of glycogen is **glucose-6-phosphate.** This can be derived from free glucose by the **hexokinase** (in liver; **glucokinase** in muscle) reaction:

$$\text{D-Glucose} + \text{ATP} \longrightarrow \text{D-glucose-6-phosphate} + \text{ADP}$$

However, much of the glucose ingested during a meal takes a more roundabout path to glycogen. It is first taken up by erythrocytes in the bloodstream and converted to lactate glycolytically; the lactate is then taken up by the liver and converted to glucose-6-phosphate by gluconeogenesis.

To initiate glycogen synthesis, the glucose-6-phosphate is reversibly converted into **glucose-1-phosphate** by **phosphoglucomutase:**

$$\text{Glucose-6-phosphate} \rightleftharpoons \text{glucose-1-phosphate}$$

The formation of UDP-glucose by the action of **UDP-glucose pyrophosphorylase** is a key reaction in glycogen biosynthesis:

$$\text{Glucose-1-phosphate} + \text{UTP} \longrightarrow \text{UDP-glucose} + \text{PP}_i$$

This reaction proceeds in the direction of UDP-glucose formation because pyrophosphate is rapidly hydrolyzed to orthophosphate by inorganic pyrophosphatase ($\Delta G^{\circ\prime} = -25$ kJ/mol) (Fig. 19–11).

UDP-glucose, as we have already noted (see Fig. 14–13), is an intermediate in the conversion of galactose into glucose. UDP-glucose is the immediate donor of glucose residues in the enzymatic formation of glycogen by the action of **glycogen synthase,** which promotes the transfer of the glucosyl residue from UDP-glucose to a nonreducing end of the branched glycogen molecule (Fig. 19–12). The overall equilibrium of the path from glucose-6-phosphate to lengthened glycogen greatly favors synthesis of glycogen. Glycogen synthase requires as a primer an ($\alpha1\rightarrow4$) polyglucose chain or branch having at least four glucose residues.

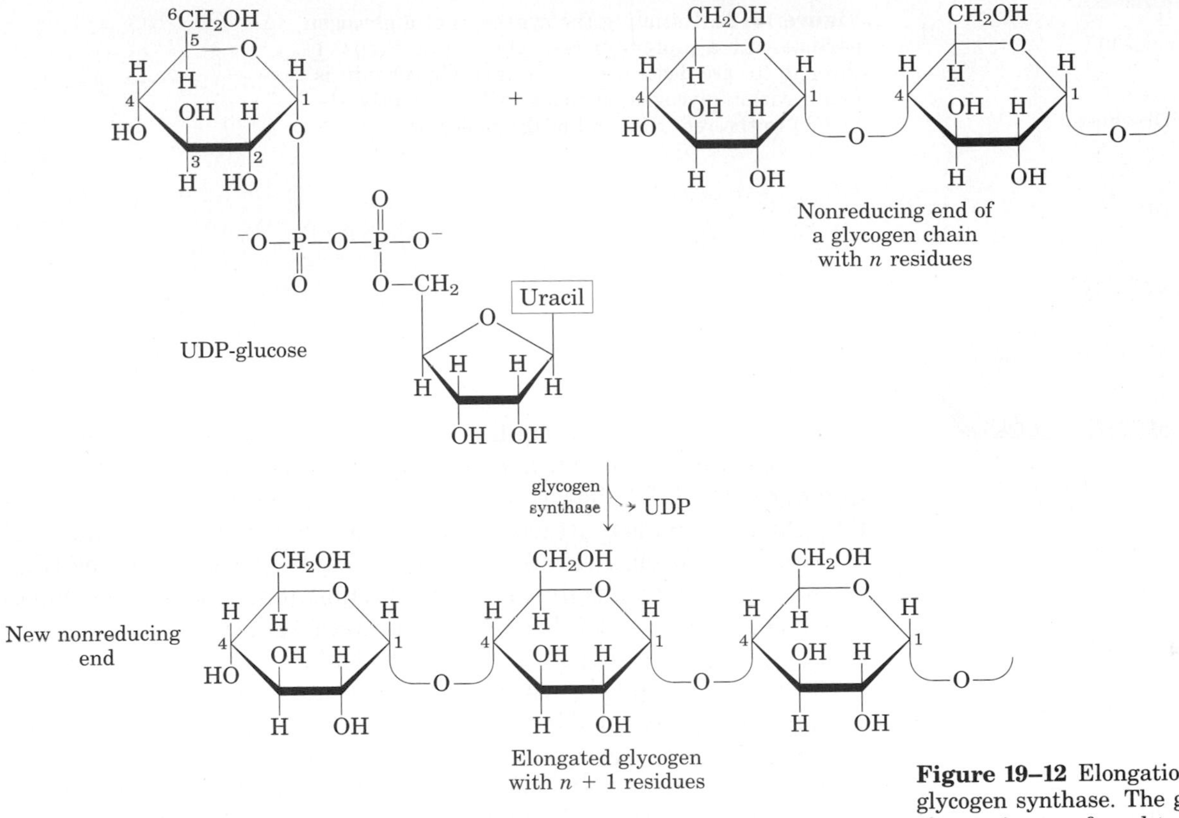

Figure 19–12 Elongation of a glycogen chain by glycogen synthase. The glucosyl residue of UDP-glucose is transferred to the nonreducing end of a glycogen branch (see Fig. 11–15) to make a new ($\alpha1\rightarrow4$) linkage.

Glycogen synthase cannot make the ($\alpha1\rightarrow6$) bonds found at the branch points of glycogen (see Fig. 11–15); instead, these are formed by a glycogen-branching enzyme, **amylo ($1\rightarrow4$) to ($1\rightarrow6$) transglycosylase** or **glycosyl-($4\rightarrow6$)-transferase.** Glycosyl-($4\rightarrow6$)-transferase catalyzes transfer of a terminal fragment of six or seven glucosyl residues from the nonreducing end of a glycogen branch having at least eleven residues to the C-6 hydroxyl group of a glucose residue of the same or another glycogen chain at a more interior point, thus creating a new branch (Fig. 19–13). Further glucosyl residues may be added to the new branch by glycogen synthase. The biological effect of branching is to make the glycogen molecule more soluble and to increase the number of nonreducing ends, thus making the glycogen more reactive to both glycogen phosphorylase and glycogen synthase.

Figure 19–13 The glycogen-branching enzyme glycosyl-($4\rightarrow6$)-transferase (or amylo ($1\rightarrow4$) to ($1\rightarrow6$) transglycosylase) forms a new branch point during glycogen synthesis.

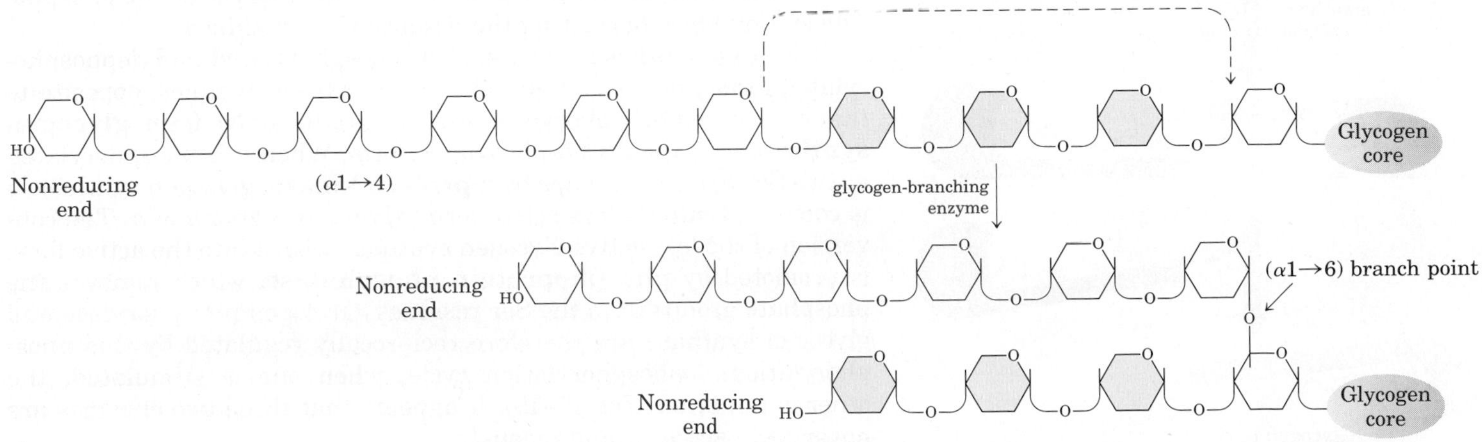

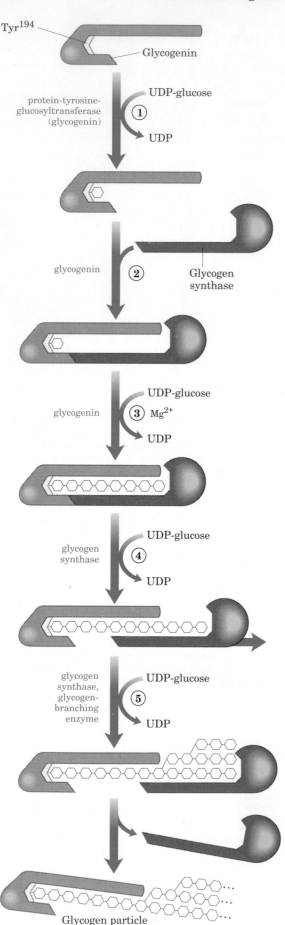

Figure 19–14 Initiating the synthesis of a glycogen particle with a protein primer, glycogenin. Steps ① through ⑤ are described in the text. Glycogenin is found within glycogen particles, still covalently attached to the reducing end of the molecule.

If glycogen synthase requires a primer, how is a new glycogen molecule initiated? The answer lies in an intriguing protein called **glycogenin** (M_r 37,284), which itself acts as the primer to which the first glucose residue is attached and also as the catalyst for synthesis of a nascent glycogen molecule with up to eight glucose residues (Fig. 19–14). The first step is the covalent attachment of a glucose residue to Tyr[194] of glycogenin, catalyzed by the protein's glucosyltransferase activity (step ①). Glycogenin then forms a tight 1:1 complex with glycogen synthase (step ②), within which the next few steps occur. The glucan chain is extended by the sequential addition of up to seven more glucose residues (step ③). Each new residue is derived from UDP-glucose, and the reactions are autocatalytic (mediated by the glucosyltransferase of glycogenin). At this point, glycogen synthase takes over, extending the glycogen chain and dissociating from glycogenin (step ④). The combined action of glycogen synthase and the branching enzyme (step ⑤) completes the glycogen particle. Glycogenin remains buried within the particle, covalently attached to one end.

Glycogen Synthase and Glycogen Phosphorylase Are Reciprocally Regulated

Earlier we saw that the breakdown of glycogen is regulated by both covalent and allosteric modulation of glycogen phosphorylase (see Fig. 14–17). Phosphorylase a, the active form, which contains essential phosphorylated Ser residues, is dephosphorylated by phosphorylase a phosphatase to yield phosphorylase b, the relatively inactive form, which can be stimulated by AMP, its allosteric modulator. Phosphorylase b kinase can convert phosphorylase b back into the active phosphorylase a by phosphorylating the essential Ser residues.

Glycogen synthase also occurs in phosphorylated and dephosphorylated forms, but it is regulated in a reciprocal manner, opposite to that of glycogen phosphorylase (Fig. 19–15). Its active form, **glycogen synthase a,** is the dephosphorylated form. When it is phosphorylated at two Ser hydroxyl groups by a **protein kinase,** glycogen synthase a is converted into its less active form **glycogen synthase b.** The conversion of the less active glycogen synthase b back into the active form is promoted by **phosphoprotein phosphatase,** which removes the phosphate groups from the Ser residues. Glycogen phosphorylase and glycogen synthase are therefore reciprocally regulated by this phosphorylation–dephosphorylation cycle; when one is stimulated, the other is inhibited (Fig. 19–15). It appears that these two enzymes are never fully active simultaneously.

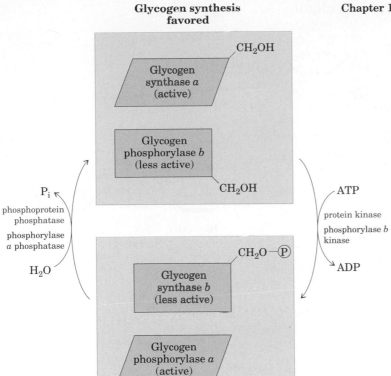

Glycogen synthesis favored

Glycogen breakdown favored

Figure 19–15 Reciprocal regulation of glycogen synthase and glycogen phosphorylase by phosphorylation and dephosphorylation. In both enzymes the site of phosphorylation is a Ser residue (represented here by —CH₂OH). The enzymes of glycogen synthesis are shown in red; those of glycogen breakdown in black.

The balance between glycogen synthesis and breakdown in liver is controlled by the hormones glucagon and **insulin** (Table 19–4). These hormones, by regulating the level of cAMP in their target tissues, determine the ratio of active to less active forms of glycogen phosphorylase and glycogen synthase. The hormones also regulate the concentration of fructose-2,6-bisphosphate and thereby the balance between gluconeogenesis and glycolysis. **Epinephrine** has effects similar to those of glucagon, but its target is primarily muscle whereas glucagon's primary action is on liver. The regulatory role of these hormones and the details of their action on target tissue will be considered further in Chapter 22.

Table 19–4 The effects of glucagon, epinephrine, and insulin on carbohydrate metabolism in mammals

	Glucagon	Epinephrine	Insulin
Source	Pancreatic α cells	Adrenal medulla	Pancreatic β cells
Primary target	Liver	Muscle > liver	Muscle, liver, adipose
Effects on			
[cAMP]	↑	↑	↓
[Fructose-2,6-bisphosphate]	↓	↓	↑
Gluconeogenesis	↑	↑	↓
Glycogen breakdown (glycolysis)	↑	↑	↓

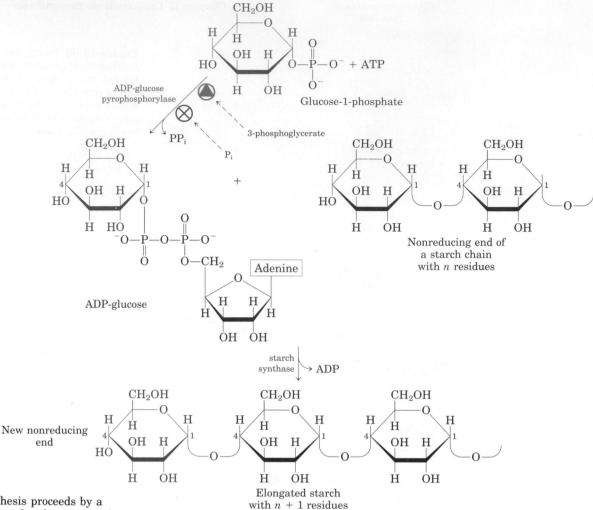

Figure 19–16 Starch synthesis proceeds by a mechanism analogous to that for glycogen synthesis (see Fig. 19–12), except that the activated substrate is ADP-glucose. Glucose is transferred to the nonreducing end of an existing starch molecule, in (α1→4) linkage. ADP-glucose pyrophosphorylase is a regulatory enzyme (p. 633).

ADP-Glucose Is the Substrate for Starch Synthesis in Plants

Starch, like glycogen, is a high molecular weight polymer of D-glucose in (α1→4) linkage (see Fig. 11–15). Starch synthesis occurs in chloroplasts. The mechanism of hexose polymerization in starch synthesis is essentially similar to that in glycogen synthesis. An activated nucleotide sugar (**ADP-glucose** in this case) is formed by condensation of glucose-1-phosphate with ATP. **Starch synthase** then transfers glucose residues from ADP-glucose to the nonreducing end of preexisting starch molecules that act as primers (Fig. 19–16). The reaction involves displacement of the ADP of ADP-glucose by the attacking 4′ hydroxyl of the primer, forming the characteristic (α1→4) linkages of starch. The starch fraction called amylose is unbranched, but amylopectin has numerous (α1→6)-linked branches like those of glycogen, from which it differs in molecular weight and extent of branching (Chapter 11). Chloroplasts contain a branching enzyme similar to that involved in glycogen synthesis (Fig. 19–13); it introduces the (α1→6) branches of amylopectin.

With the hydrolysis by inorganic pyrophosphatase of PP_i produced during ADP-glucose synthesis, the overall reaction for starch formation from glucose-1-phosphate is

$$\text{Starch}_n + \text{glucose-1-phosphate} + \text{ATP} \longrightarrow \text{starch}_{n+1} + \text{ADP} + 2P_i$$
$$\Delta G^{\circ\prime} = -50 \text{ kJ/mol}$$

Starch synthesis is regulated at the level of ADP-glucose formation (Fig. 19–16), as discussed later in this chapter.

Figure 19–17 Sucrose is synthesized from activated glucose (UDP-glucose) and fructose-6-phosphate. The sucrose-6-phosphate formed in the first step accumulates when the final product, sucrose, inhibits the enzyme sucrose-6-phosphate phosphatase. The first step is also inhibited by its product, sucrose-6-phosphate. UDP-glucose and fructose-6-phosphate are synthesized from triose phosphates in the cytosol of plant cells by pathways identical to those shown in Figs. 19–11 and 19–22.

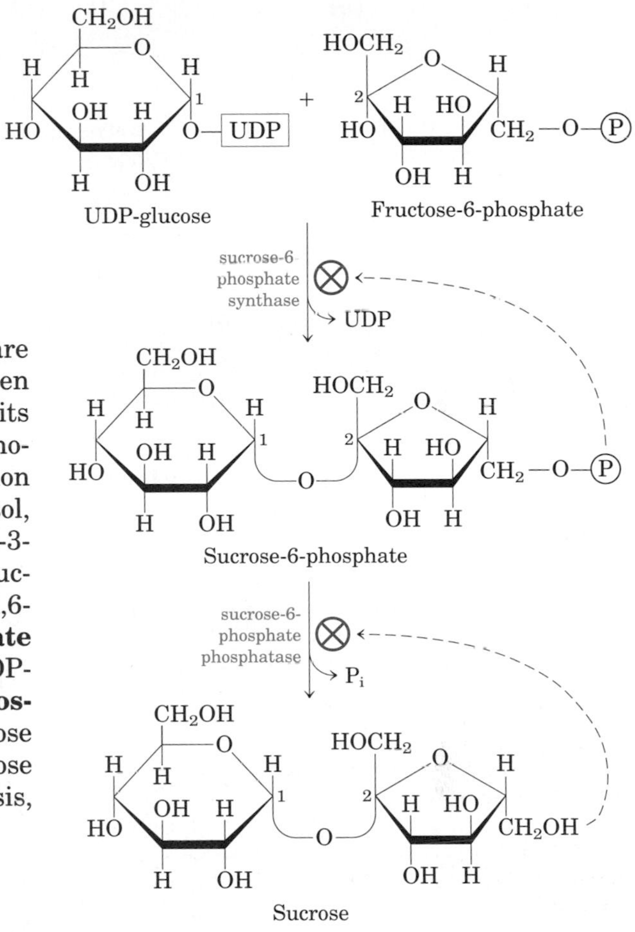

UDP-glucose   Fructose-6-phosphate

Sucrose-6-phosphate

Sucrose

UDP-Glucose Is the Substrate for Sucrose Synthesis in Plants

Most of the triose phosphates generated by CO_2 fixation in plants are converted into sucrose (Fig. 19–17) or starch. Sucrose may have been selected during evolution as the transport form of carbon because its unusual linkage, which joins the anomeric C-1 of glucose and the anomeric C-2 of fructose, is not hydrolyzed by amylases or other common carbohydrate-cleaving enzymes. Sucrose is synthesized in the cytosol, beginning with dihydroxyacetone phosphate and glyceraldehyde-3-phosphate exported from the chloroplast. After condensation to fructose-1,6-bisphosphate (by aldolase), hydrolysis by fructose-1,6-bisphosphatase yields fructose-6-phosphate. **Sucrose-6-phosphate synthase** catalyzes the reaction of fructose-6-phosphate with UDP-glucose to form **sucrose-6-phosphate.** Finally, **sucrose-6-phosphate phosphatase** removes the phosphate group, making sucrose available for export from the cell to other tissues of the plant. Sucrose synthesis is regulated and closely coordinated with starch synthesis, as we shall see.

Lactose Synthesis Is Regulated in a Unique Way

The synthesis of lactose in lactating mammary gland occurs by a mechanism similar to that for glycogen synthesis. However, the regulation of lactose synthesis is unusual. Most vertebrate tissues contain the enzyme **galactosyl transferase** (Fig. 19–18a), which promotes the transfer of an activated galactose residue in UDP-galactose to the monosaccharide N-acetylglucosamine:

UDP-D-galactose + N-acetyl-D-glucosamine $\longrightarrow$
  D-galactosyl-N-acetyl-D-glucosamine + UDP

This reaction has no role in lactose synthesis; it is a step in the biosynthesis of the carbohydrate portion of galactose-containing glycoproteins in animal tissues.

In the lactating mammary gland, however, this same enzyme participates in lactose synthesis (Fig. 19–18b). Galactosyl transferase, which is very active with N-acetylglucosamine but only feebly active with glucose as galactosyl acceptor, is present in most tissues, including the mammary gland, as noted above. Immediately after a female gives birth, the specificity of galactosyl transferase in the mammary gland changes: it now transfers the galactosyl group of UDP-galactose to glucose at a very high rate, thus making lactose:

UDP-D-galactose + D-glucose $\longrightarrow$ D-lactose + UDP

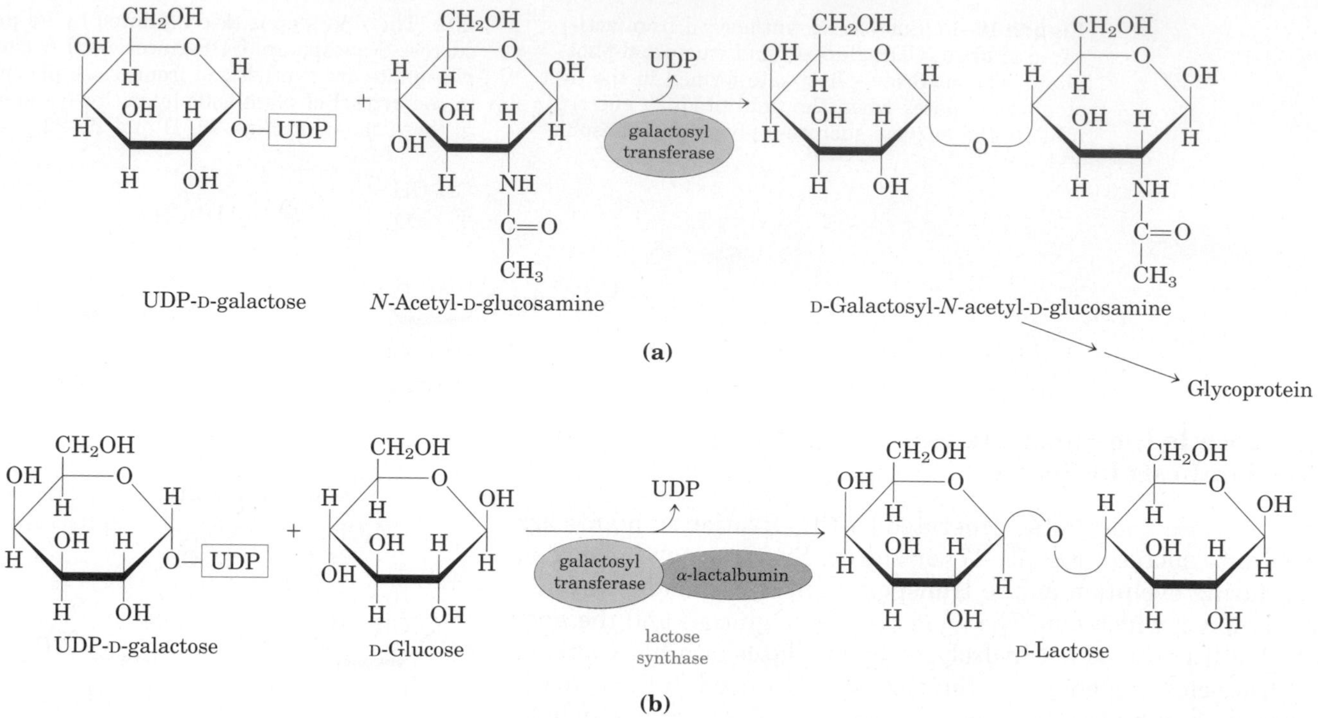

(a)

(b)

Figure 19–18 Two distinct reactions are catalyzed by galactosyl transferase, depending on whether the protein α-lactalbumin, produced only in lactating mammary gland, is present. **(a)** The reaction in nonlactating tissues; **(b)** the reaction in lactating mammary gland.

This "new" enzyme is called **lactose synthase** (Fig. 19–18b).

The change in specificity of galactosyl transferase is caused by the synthesis of **α-lactalbumin** (M_r 13,500), a milk protein, whose function was long unknown. α-Lactalbumin has been found to be a specificity-modifying subunit; its synthesis in the mammary gland, which is regulated by the hormones promoting lactation, leads to the formation of an **α-lactalbumin–galactosyl transferase** complex, that is, lactose synthase.

Photosynthetic Carbohydrate Synthesis

The synthesis of carbohydrates in animal cells always employs precursors having at least three carbons, all at an oxidation state lower than that of carbon in CO_2. Photosynthetic organisms, by contrast, can make carbohydrates from CO_2 and water. They synthesize glucose, sucrose, and other carbohydrates by reducing CO_2 at the expense of energy furnished by the ATP and NADPH generated in photosynthetic electron transfer. This process represents a fundamental difference between autotrophic (phototrophic or chemotrophic) and heterotrophic organisms. Autotrophs can use CO_2 as the sole source of all the carbon atoms required for biosynthesis not only of cellulose and starch, but also of the lipids and proteins and all of the many other organic components of plant cells. By contrast, heterotrophic organisms in general are unable to bring about the net reduction of CO_2 to form "new" glucose in any significant amounts. Carbon dioxide can be taken up by animal tissues, as in the pyruvate carboxylase reaction during gluconeogenesis, but the CO_2 molecule incorporated into oxaloacetate is lost in a subsequent reaction step (Fig. 19–3). Similarly, the CO_2 taken up by acetyl-CoA carboxylase during fatty acid synthesis in animal tissues (Chapter 20) or by carbamoyl phosphate synthetase I during urea formation (Chapter 17) is lost in later steps.

Green plants contain in their chloroplasts unique enzymatic machinery to catalyze the conversion of CO_2 into simple (reduced) organic compounds, a process called **CO_2 fixation,** or **carbon fixation.** Plants convert these simple products of photosynthesis into more complex biomolecules, including sugars, polysaccharides, and metabolites derived from them, using metabolic pathways similar to those in animals. The reactions that result in CO_2 fixation make up a cyclic pathway in which key intermediates are constantly regenerated. The pathway was elucidated in the early 1950s by Melvin Calvin and coworkers, and is often called the **Calvin cycle.**

Melvin Calvin

Carbon Dioxide Fixation Occurs in Three Stages

The first stage in the fixation of CO_2 into organic linkage (Fig. 19–19) is its condensation with a five-carbon acceptor, **ribulose-1,5-bisphosphate,** to form two molecules of 3-phosphoglycerate. (Note that Figure 19–19 shows the number of molecules reacting to give net formation of one molecule of triose—this takes three molecules of CO_2.) In the second stage the 3-phosphoglycerate is reduced to glyceraldehyde-3-phosphate: three molecules of CO_2 are fixed to three molecules of ribulose-1,5-bisphosphate to form six molecules of glyceraldehyde-3-phosphate (18 carbons). One molecule of this triose phosphate (three carbons) can either be used for energy production via glycolysis and the citric acid cycle, or condensed to hexose phosphates to be used in the synthesis of starch or sucrose. In the third stage, five of the six molecules of glyceraldehyde-3-phosphate (15 carbons) are used to regenerate three molecules of ribulose-1,5-bisphosphate, the starting material.

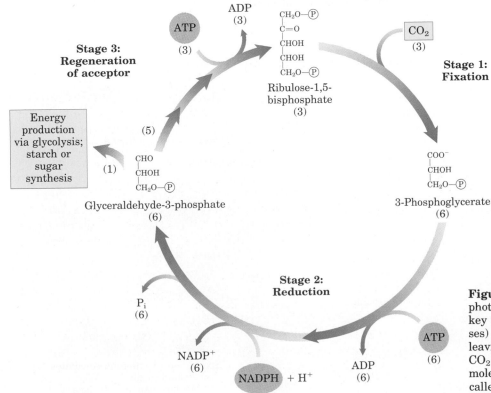

Figure 19–19 The three stages of CO_2 fixation in photosynthetic organisms. Stoichiometries of three key intermediates are shown (numbers in parentheses) so that the fate of carbon atoms entering and leaving the cycle is apparent. As shown here, three CO_2 are fixed to permit the net synthesis of one molecule of glyceraldehyde-3-phosphate. This is called the photosynthetic carbon reduction cycle or the Calvin cycle.

Thus the overall process is cyclical and allows the continuous conversion of CO_2 into triose and hexose phosphates. Fructose-6-phosphate is a key intermediate in stage 3. The pathway from hexose phosphate to pentose bisphosphate involves the same reactions used in animal cells for the conversion of pentose phosphates to hexose phosphates during the operation of the pentose phosphate pathway, an alternative route for glucose oxidation (Fig. 14–22). In photosynthetic fixation of CO_2, this pathway operates in the opposite direction, converting hexose phosphates into pentose phosphates.

Stage 1: Fixation of CO_2 into 3-Phosphoglycerate   An important clue to the nature of the CO_2 fixation mechanisms in photosynthetic organisms first came in the late 1940s when Calvin and his associates illuminated a suspension of green algae in the presence of radioactive carbon dioxide ($^{14}CO_2$) for only a few seconds. They quickly killed the cells, extracted their contents, and with the help of chromatographic methods searched for the metabolites in which the labeled carbon first appeared. The first compound that became labeled was **3-phosphoglycerate,** with the ^{14}C predominantly located in the carboxyl carbon atom. This atom does not become labeled rapidly in animal tissues in the presence of radioactive CO_2. These experiments strongly suggested that 3-phosphoglycerate is an early intermediate in photosynthesis. The enzyme in green leaf extracts that catalyzes incorporation of CO_2 into organic form is **ribulose-1,5-bisphosphate carboxylase** or **RuBP carboxylase/oxygenase** (often called **rubisco** for short). (The enzyme's oxygenase activity is discussed later in this chapter.) As a carboxylase, rubisco catalyzes the covalent attachment of CO_2 to the five-carbon sugar ribulose-1,5-bisphosphate and the cleavage of the unstable six-carbon intermediate to form two molecules of 3-phosphoglycerate, one of which bears the new carbon introduced as CO_2 in its carboxyl group (Fig. 19–20). The rubisco of plants (M_r 550,000) has a

Figure 19–20 The first stage of CO_2 fixation is the reaction catalyzed by ribulose-1,5-bisphosphate carboxylase (rubisco). The carboxylated reaction intermediate is believed to be the enzyme-bound six-carbon β-keto acid shown here, which is hydrolyzed and released from the enzyme surface as two identical three-carbon products, one of which contains the carbon atom from CO_2 (red).

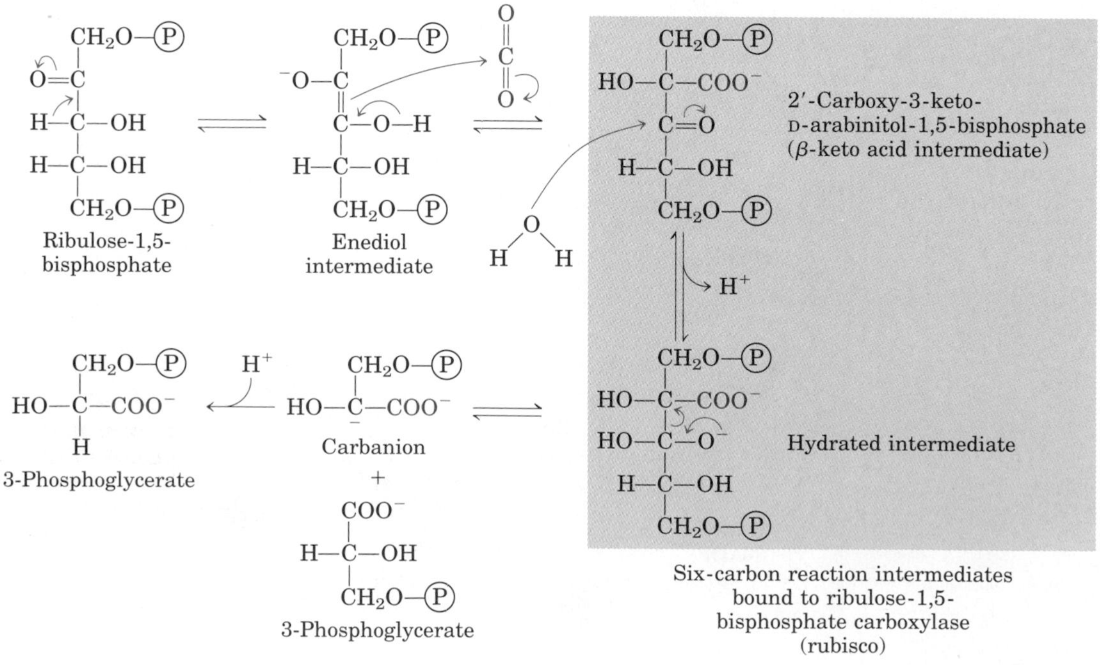

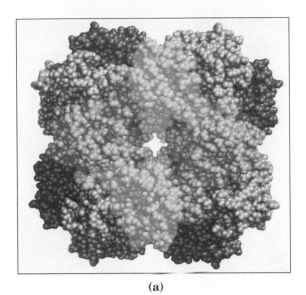

(a)

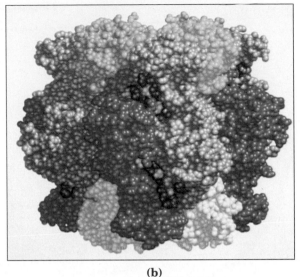

(b)

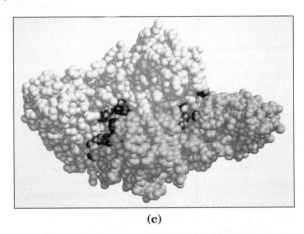

(c)

complex structure (Fig. 19–21a, b). There are eight large subunits (each of M_r 56,000), each containing an active site, and eight small subunits (each of M_r 14,000), whose function is not well understood. The subunit structure of rubisco of photosynthetic bacteria is quite different, with two subunits that resemble the large subunits of the plant enzyme in many respects (Fig. 19–21c). The plant enzyme is located in the chloroplast stroma, where it makes up about 50% of the total chloroplast protein. Rubisco, which does not occur in animals, is the most abundant enzyme in the biosphere; it is the key enzyme in the production of biomass from CO_2.

Rubisco is subject to regulation by the covalent addition of a molecule of CO_2 to the ϵ-amino group of a certain Lys residue to form a carbamate (Fig. 19–21). This carbamate binds to a Mg^{2+} ion; the resulting complex is essential for activity, as we shall see later (see Fig. 19–30) and may function directly in catalysis. Note that this is *not* the same molecule of CO_2 that is attached to ribulose-1,5-bisphosphate in the reaction catalyzed by this enzyme.

Stage 2: Conversion of 3-Phosphoglycerate to Glyceraldehyde-3-Phosphate 3-Phosphoglycerate is converted into glyceraldehyde-3-phosphate in two steps that are essentially the reversal of the corresponding steps in glycolysis, with one exception: the nucleotide cofactor for the reduction of 1,3-bisphosphoglycerate is NADPH, not NADH (Fig. 19–22, p. 622). The chloroplast stroma contains the full complement of glycolytic enzymes. These stromal enzymes are isozymes (see Box 14–3) of those found in the cytosol; both sets of enzymes catalyze the same reactions, but they are products of different genes.

In the first step of the sequence, **3-phosphoglycerate kinase** in the stroma catalyzes the transfer of phosphate from ATP to 3-phosphoglycerate, yielding 1,3-bisphosphoglycerate (Fig. 19–22). Then NADPH donates electrons in a reduction catalyzed by **glyceraldehyde-3-phosphate dehydrogenase,** producing glyceraldehyde-3-phosphate. In addition to its role as an intermediate in CO_2 fixation, glyceraldehyde-3-phosphate has several possible fates in the plant cell. It may be oxidized via glycolysis for energy production or used for hexose synthesis (Fig. 19–22).

Figure 19–21 Structure and function of ribulose-1,5-bisphosphate carboxylase (rubisco). **(a)** Top view and **(b)** side view of a space-filling model of rubisco from tobacco, based on x-ray diffraction analysis of the crystalline enzyme. There are eight large subunits (shown in gray and dark blue) and eight small ones (white and shades of light blue), tightly packed into a structure of M_r 550,000. Rubisco is present at a concentration of about 250 mg/mL in the chloroplast stroma, corresponding to an extraordinarily high concentration of active sites (~4 mM). Active-site amino acid residues are shown in red. Sulfate molecules bound to the active site in the crystal structure (these are not normal substrates) are in yellow. **(c)** Space-filling model of rubisco from the bacterium *Rhodospirillum rubrum.* The subunits are shown in white and light blue. Amino acid residues at the active site are in red. A Lys residue at the active site, which is carboxylated to form a carbamate in the active enzyme (see Fig. 19–30), is shown in yellow.

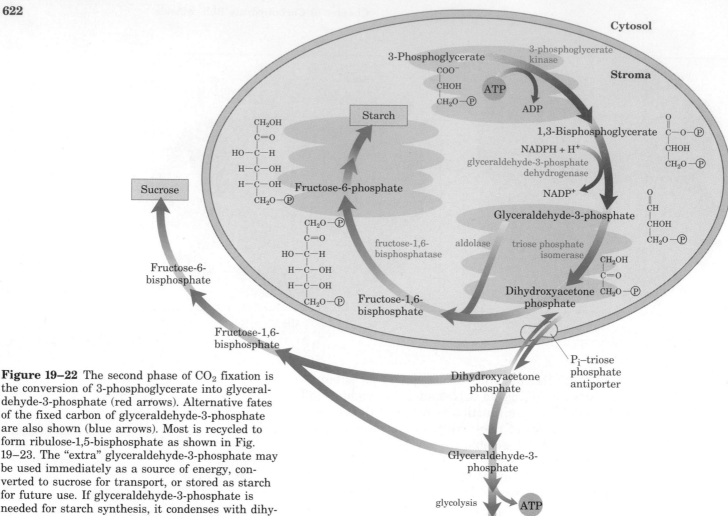

Figure 19–22 The second phase of CO_2 fixation is the conversion of 3-phosphoglycerate into glyceraldehyde-3-phosphate (red arrows). Alternative fates of the fixed carbon of glyceraldehyde-3-phosphate are also shown (blue arrows). Most is recycled to form ribulose-1,5-bisphosphate as shown in Fig. 19–23. The "extra" glyceraldehyde-3-phosphate may be used immediately as a source of energy, converted to sucrose for transport, or stored as starch for future use. If glyceraldehyde-3-phosphate is needed for starch synthesis, it condenses with dihydroxyacetone phosphate in the stroma and is converted to fructose-6-phosphate, a precursor of starch. In other situations it is converted to dihydroxyacetone phosphate, which leaves the chloroplast via a specific transporter (see Fig. 19–28). In the cytosol, dihydroxyacetone phosphate can be degraded via glycolysis to provide energy, or used to form fructose-6-phosphate and hence sucrose.

Stage 3: Regeneration of Ribulose-1,5-Bisphosphate from Triose Phosphates As we have seen, the first reaction in the fixation of CO_2 into triose phosphates consumes ribulose-1,5-bisphosphate. For continuous flow of CO_2 into carbohydrate, ribulose-1,5-bisphosphate must be constantly regenerated. Plant cells solve this problem with a series of reactions that, together with stages 1 and 2 discussed above, form a cyclic pathway (Fig. 19–23). By this pathway, the product of the first reaction (3-phosphoglycerate) passes through a series of transformations that eventually lead to the regeneration of the starting material, ribulose-1,5-bisphosphate.

The regeneration of ribulose-1,5-bisphosphate involves rearrangements of the carbon skeletons of glyceraldehyde-3-phosphate and dihydroxyacetone phosphate produced in the first two stages of carbon fixation. The intermediates in the pathway include three-, four-, five-, six-, and seven-carbon sugars. In the following discussion, all step numbers refer to Figure 19–23.

Figure 19–23 (a) The third stage of CO_2 fixation consists of the remaining set of reactions of the Calvin cycle, in which ribulose-1,5-bisphosphate is regenerated from the triose phosphates. The starting materials are the triose phosphates: glyceraldehyde-3-phosphate and dihydroxyacetone phosphate. Reactions catalyzed by aldolase (step ②) and transketolase (steps ①, ④, and ⑤) produce pentose phosphates, all of which are eventually converted to ribulose-1,5-bisphosphate. The two-carbon group carried by TPP in steps ① and ④ (TPP-2C) is a ketol group: $CH_2OH—CO—$. The individual reactions are described in the text. (b) A simplified schematic diagram showing the interconversions of triose phosphates (three-carbon (3C) compounds) into pentose phosphates (five-carbon (5C) compounds).

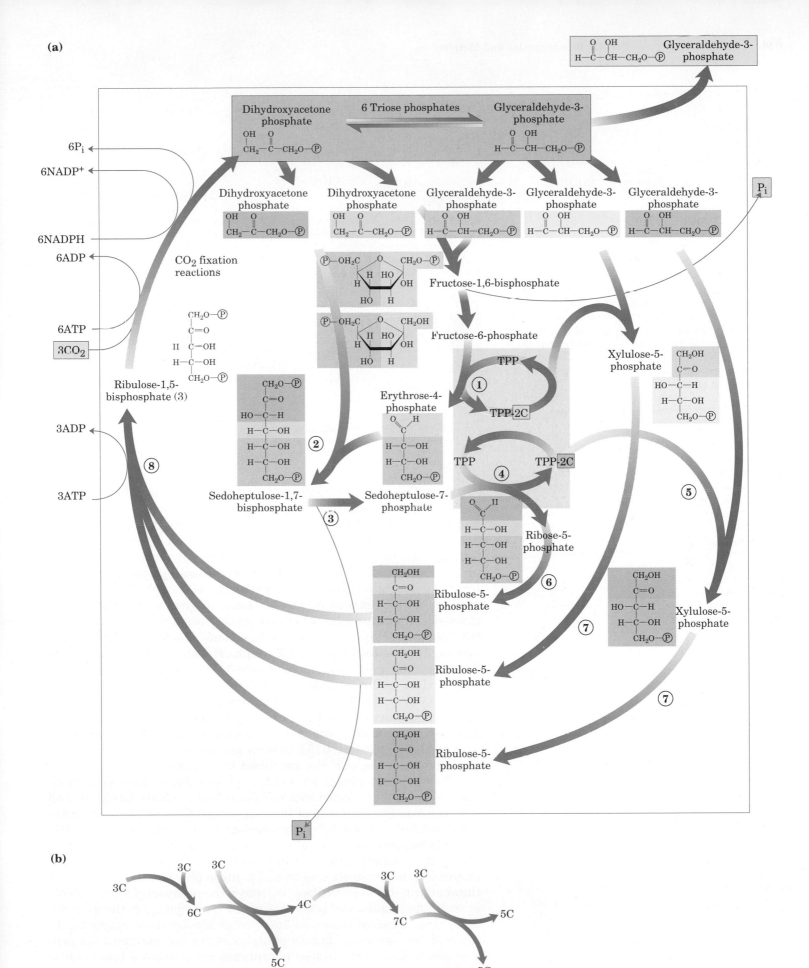

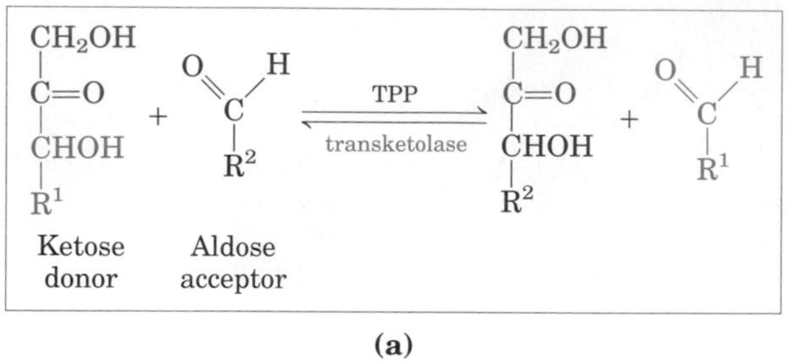

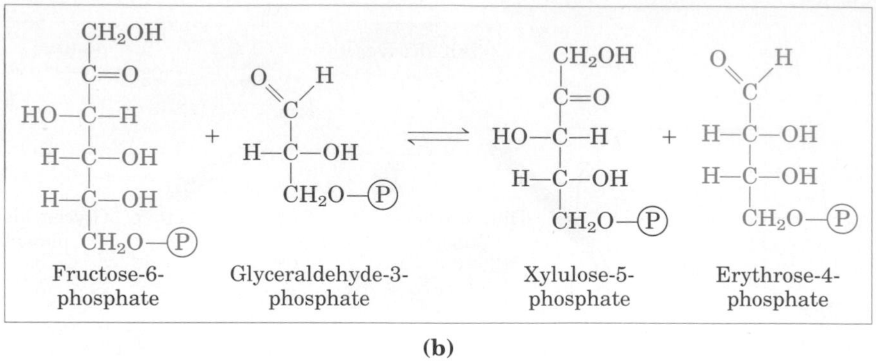

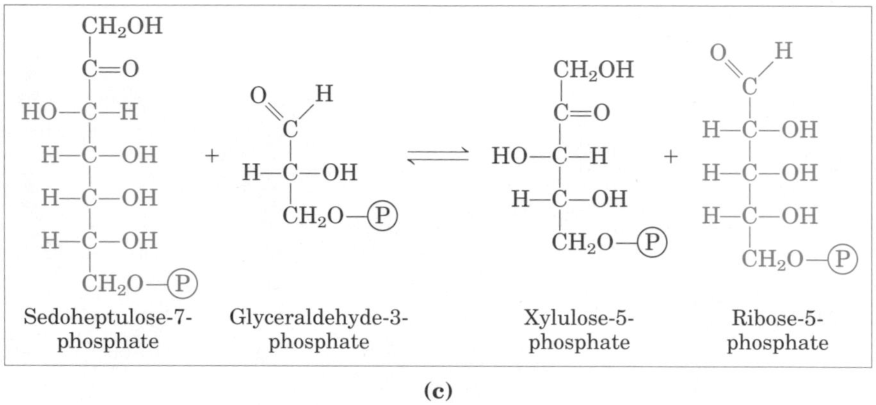

(a)

(b)

Figure 19–24 Transketolase catalyzes two very similar reactions in the Calvin cycle. **(a)** The general reaction catalyzed by transketolase is the transfer of a two-carbon group, carried temporarily on enzyme-bound TPP, from a ketose donor to an aldose acceptor. **(b)** Conversion of a hexose and a triose to a pentose and a four-carbon sugar by transketolase action (step ① of Fig. 19–23). **(c)** Conversion of seven-carbon and three-carbon skeletons to two pentoses by transketolase (steps ④ and ⑤ of Fig. 19–23).

(c)

Step ① is catalyzed by the enzyme **transketolase,** which contains thiamine pyrophosphate (TPP) as its prosthetic group (see Fig. 14–9a) and requires Mg^{2+}. Transketolase catalyzes the reversible transfer of a ketol (CH_2OH—CO—) group from a ketose phosphate donor, fructose-6-phosphate, to an aldose phosphate acceptor, glyceraldehyde-3-phosphate (Fig. 19–24b). The products are **xylulose-5-phosphate** (a pentose) and the four-carbon sugar **erythrose-4-phosphate.**

Step ② is promoted by an aldolase similar to that which acts in glycolysis; it catalyzes the reversible condensation of an aldehyde, erythrose-4-phosphate, with dihydroxyacetone phosphate, yielding the seven-carbon **sedoheptulose-1,7-bisphosphate.** After removal of the C-1 phosphate group by sedoheptulose-1,7-bisphosphatase (step ③), the product sedoheptulose-7-phosphate is split by transketolase (step ④) (Fig. 19–24c) into a pentose phosphate (**ribose-5-phosphate**) and a two-carbon fragment, carried on TPP (Fig. 19–25). This two-carbon fragment is condensed with the three carbons of glyceraldehyde-3-phosphate (step ⑤) to form another molecule of xylulose-5-phosphate in a reaction also catalyzed by transketolase.

The pentose phosphates—ribose-5-phosphate and xylulose-5-phosphate—are converted into **ribulose-5-phosphate** (steps ⑥ and ⑦), which in the final step of the cycle (step ⑧) is phosphorylated to ribulose-1,5-bisphosphate by ribulose-5-phosphate kinase (Fig. 19–26).

This pathway is essentially the reversal of the oxidative pentose phosphate pathway described in Chapter 14, and employs the same enzymes to interconvert hexose and pentose phosphates. To highlight the similarity between the two pathways, the pathway of Figure 19–23 is sometimes called the reductive pentose phosphate pathway, even though no chemical reduction steps occur among these reactions. In the oxidative sequence, hexose phosphates are regenerated from pentose phosphates; here, hexose phosphates are converted into pentose phosphates.

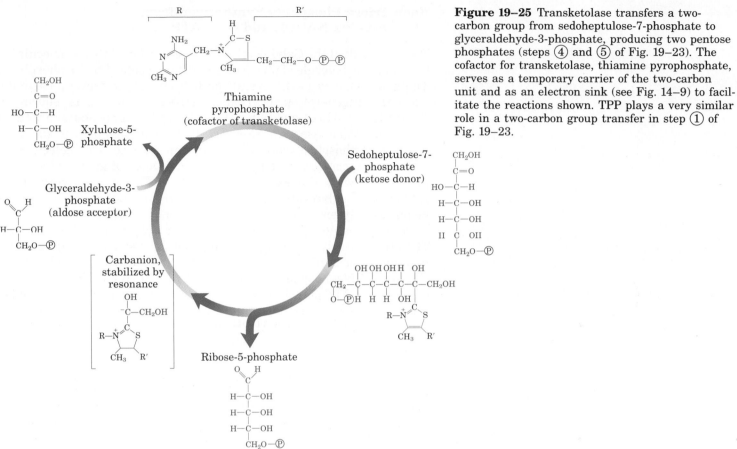

Figure 19–25 Transketolase transfers a two-carbon group from sedoheptulose-7-phosphate to glyceraldehyde-3-phosphate, producing two pentose phosphates (steps ④ and ⑤ of Fig. 19–23). The cofactor for transketolase, thiamine pyrophosphate, serves as a temporary carrier of the two-carbon unit and as an electron sink (see Fig. 14–9) to facilitate the reactions shown. TPP plays a very similar role in a two-carbon group transfer in step ① of Fig. 19–23.

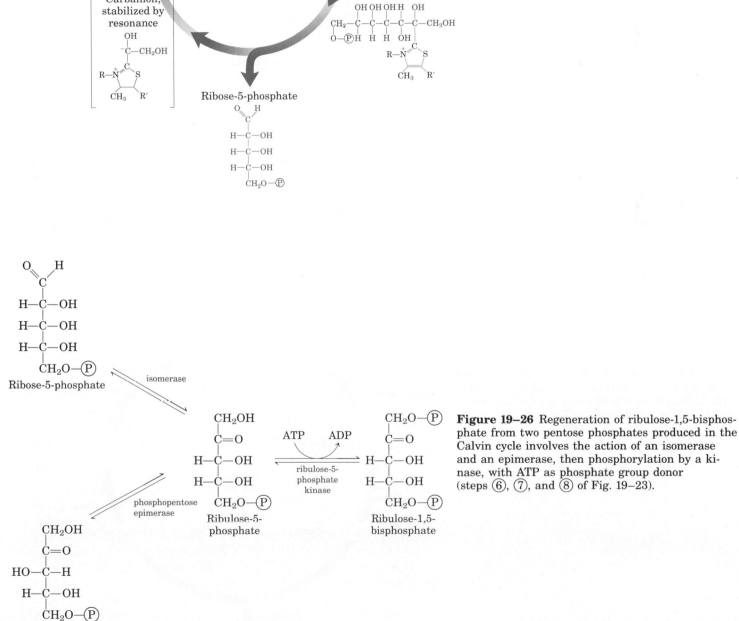

Figure 19–26 Regeneration of ribulose-1,5-bisphosphate from two pentose phosphates produced in the Calvin cycle involves the action of an isomerase and an epimerase, then phosphorylation by a kinase, with ATP as phosphate group donor (steps ⑥, ⑦, and ⑧ of Fig. 19–23).

Each Triose Phosphate Synthesized from CO$_2$ Costs Six NADPH and Nine ATP

The net result of the Calvin cycle is the conversion of three molecules of CO$_2$ and one molecule of phosphate into a molecule of triose phosphate. The stoichiometry of the overall path from CO$_2$ to triose phosphate, with the regeneration of ribulose-1,5-bisphosphate, is shown in Figure 19–27. Three molecules of ribulose-1,5-bisphosphate (a total of 15 carbons) condense with three CO$_2$ (three carbons) to form six molecules of 3-phosphoglycerate (18 carbons). These six molecules of 3-phosphoglycerate are reduced to six molecules of glyceraldehyde-3-phosphate, with the expenditure of six ATP (in the synthesis of 1,3-bisphosphoglycerate) and six NADPH (in the reduction of 1,3-bisphosphoglycerate to glyceraldehyde-3-phosphate). *One of these molecules of glyceraldehyde-3-phosphate is the net product of the process.* The other five glyceraldehyde-3-phosphate molecules (15 carbons) are rearranged in steps ① to ⑧ of Figure 19–23 to form three molecules of ribulose-1,5-bisphosphate (15 carbons). The last step in this conversion requires one ATP per ribulose-1,5-bisphosphate, or a total of three ATP. Thus, for every molecule of triose phosphate produced by photosynthetic CO$_2$ fixation, six NADPH and nine ATP are required.

The source of ATP and NADPH for these reactions is the light-driven reactions of photophosphorylation (Chapter 18). Of the nine ATP molecules converted to ADP and phosphate in the generation of a molecule of triose phosphate, eight of the phosphates are released as P$_i$ and combined with eight ADP to regenerate ATP. The ninth phosphate is incorporated into the triose phosphate itself. To convert the ninth ADP to ATP, a molecule of P$_i$ must be imported from the cytosol, as we will see later.

In the dark, the production of ATP and NADPH by photophosphorylation ceases, and the incorporation of CO$_2$ into triose phosphate (by

Figure 19–27 The stoichiometry of CO$_2$ fixation via the Calvin cycle. For every three CO$_2$ molecules fixed, one molecule of triose phosphate (glyceraldehyde-3-phosphate) is produced and nine ATP and six NADPH are consumed.

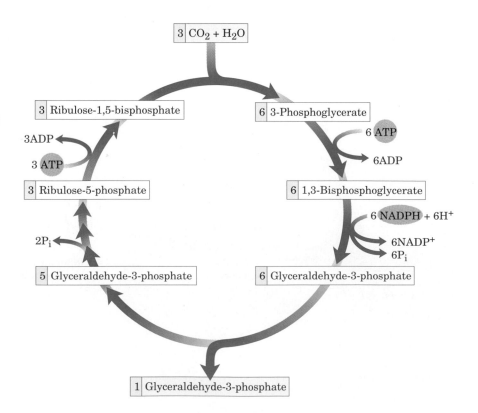

the so-called "dark reactions") also stops. The "dark reactions" of photosynthesis are so named to distinguish them from the *primary* light-driven reactions of electron transfer to NADP$^+$ and synthesis of ATP, described in Chapter 18. They do not, in fact, occur at significant rates in the dark in photosynthetic organisms (although they do in chemotrophic organisms). Therefore, these reactions are more appropriately called the **carbon fixation reactions** of photosynthesis.

Fixed carbon generated in the chloroplast is also stored there in significant amounts. Within the chloroplast stroma are all the enzymes necessary to convert the triose phosphates produced by CO_2 fixation (glyceraldehyde-3-phosphate and dihydroxyacetone phosphate) into starch, which is stored in the chloroplast as insoluble granules. Aldolase condenses the trioses to fructose-1,6-bisphosphate, fructose-1,6-bisphosphatase produces fructose-6-phosphate, phosphohexose isomerase yields glucose-6-phosphate, and phosphoglucomutase produces glucose-1-phosphate, the starting material for starch synthesis.

All the reactions of the Calvin cycle except the first, catalyzed by rubisco, and the last, catalyzed by ribulose-5-phosphate kinase, also take place in animal tissues. For lack of these two enzymes animals cannot carry out net conversion of CO_2 into glucose.

A Transport System Exports Triose Phosphates and Imports Phosphate

The inner chloroplast membrane is impermeable to most phosphorylated compounds, including fructose-6-phosphate, glucose-6-phosphate, and fructose-1,6-bisphosphate. There is, however, a specific transporter (antiporter) that catalyzes the one-for-one exchange of P_i with triose phosphates, either dihydroxyacetone phosphate or 3-phosphoglycerate (Fig. 19–28; see also Fig. 19–22). This antiporter simultaneously moves triose phosphate out of the chloroplast to the cytosol and P_i into the chloroplast, where it is used in photophosphorylation.

Without this antiport system, CO_2 fixation in the chloroplast would quickly come to a halt. The net transport of triose phosphates

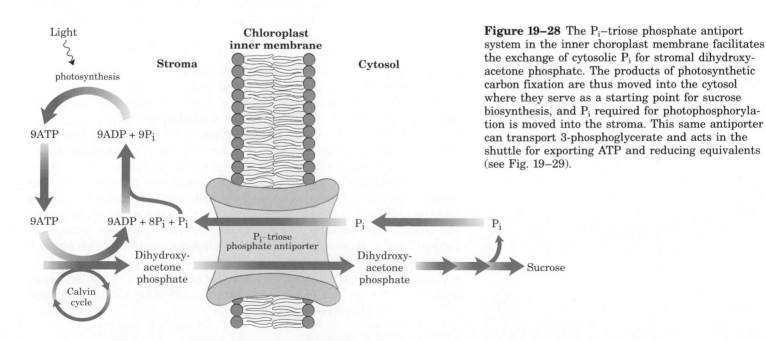

Figure 19–28 The P_i–triose phosphate antiport system in the inner choroplast membrane facilitates the exchange of cytosolic P_i for stromal dihydroxyacetone phosphate. The products of photosynthetic carbon fixation are thus moved into the cytosol where they serve as a starting point for sucrose biosynthesis, and P_i required for photophosphorylation is moved into the stroma. This same antiporter can transport 3-phosphoglycerate and acts in the shuttle for exporting ATP and reducing equivalents (see Fig. 19–29).

Figure 19–29 The P_i–triose phosphate antiporter in the inner chloroplast membrane allows exit of dihydroxyacetone phosphate, which is converted to glyceraldehyde-3-phosphate in the cytosol. The activity of cytosolic glyceraldehyde-3-phosphate dehydrogenase and phosphoglycerate kinase then produces NADH and ATP; the 3-phosphoglycerate also produced reenters the chloroplast via the antiporter and is reduced to dihydroxyacetone phosphate in the stroma, completing a cycle that effectively moves ATP and reducing equivalents (NADPH/NADH) out of the chloroplast.

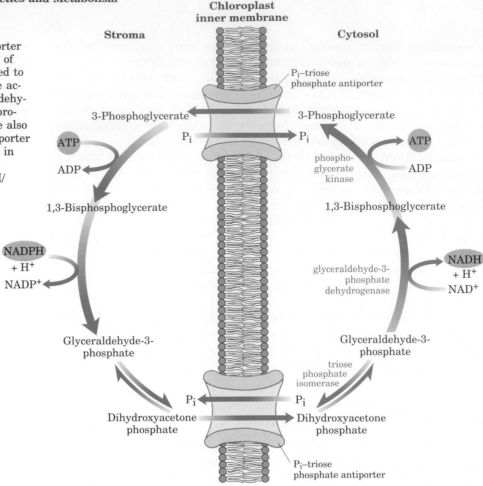

out of the chloroplast serves the important function of removing the triose phosphate products of carbon fixation. In the cytosol, the triose phosphates are converted to sucrose by the pathways illustrated in Figures 19–22 and 19–17. Sucrose synthesis in the cytosol and starch synthesis in the chloroplast are the major pathways by which the excess triose phosphates are "harvested." The last step of sucrose synthesis yields one molecule of free P_i (Fig. 19–17). This is transported back into the chloroplast and used in the synthesis of ATP, effectively replacing the molecule of P_i that is used to generate triose phosphate, as described above. For every molecule of triose phosphate removed from the chloroplast, one P_i is transported into the chloroplast. If this exchange were blocked, triose phosphate synthesis would quickly deplete the available P_i in the chloroplast and prevent further CO_2 fixation.

This P_i–triose phosphate antiport system serves one additional function. ATP and reducing power are needed in the cytosol for a variety of synthetic and energy-requiring reactions. These requirements are met to an as yet undetermined degree by the mitochondria. A second potential source of energy is the ATP and NADPH generated in the stroma during the light reactions of photosynthesis; however, ATP and NADPH do not cross the chloroplast membrane. The antiport system has the indirect effect of moving ATP and reducing equivalents across the chloroplast membrane (Fig. 19–29). Dihydroxyacetone phosphate formed in the stroma by CO_2 fixation is transported to the cytosol, where it is converted by glycolytic enzymes to 3-phosphoglycerate, generating ATP and NADH. 3-Phosphoglycerate reenters the chloroplast, completing the cycle. The net effect is transport of NADPH/NADH and ATP from the chloroplast to the cytosol.

Regulation of Carbohydrate Metabolism in Plants

Carbohydrate metabolism in plant cells is more complex than in animal cells or nonphotosynthetic microorganisms. In addition to the universal pathways of glycolysis and gluconeogenesis, plants have the unique reaction sequences for CO_2 reduction to triose phosphates and the associated reductive pentose phosphate pathway—all of which must be coordinately regulated to avoid wasteful futile cycling and to ensure proper allocation of carbon to energy production and synthesis of starch and sucrose. One level of coordination is achieved by light activation of certain enzymes associated with the carbon-fixing reactions of photosynthesis. Some enzymes are activated by changes in pH that result from light-induced proton movements; some enzymes are activated by reduction of disulfide bonds involved in their catalytic activity, using electrons flowing from photosystem I; other enzymes are subject to more conventional allosteric regulation by one or more metabolic intermediates. The compartmentation of metabolic sequences in organelles also contributes to the regulation of carbohydrate metabolism.

We will end with a discussion of another reaction catalyzed by rubisco, the condensation of O_2 with ribulose-1,5-bisphosphate. This is the starting point for a process called photorespiration, which greatly affects the efficiency of carbon fixation in plants.

Rubisco Is Subject to Both Positive and Negative Regulation

As the site where photosynthetic CO_2 fixation is initiated, rubisco is a prime target for regulation. One type of regulation involves the carbamylation of a Lys residue (Fig. 19–30a, b). At high CO_2 levels this occurs nonenzymatically. However, the substrate for this enzyme, ribulose-1,5-bisphosphate, inhibits carbamylation, and this effect is almost complete at physiological CO_2 concentrations. An enzyme called **rubisco activase** overcomes this inhibition and promotes an ATP-dependent activation of rubisco that results in carbamylation. The mechanism of activation is unknown, and it is not clear why ATP is required.

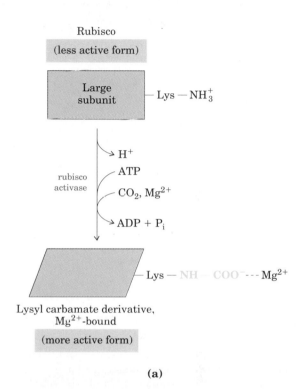

(a)

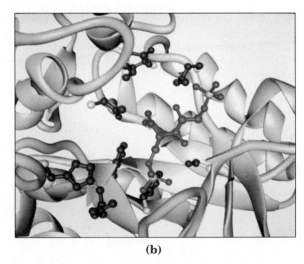

(b)

Figure 19–30 Regulation of rubisco. **(a)** Activation of rubisco by formation of a carbamate derivative of a Lys residue at the active site. The reaction is catalyzed by the enzyme rubisco activase. **(b)** The active site of rubisco from the bacterium *Rhodospirillum rubrum*. The substrate-binding site is occupied here by an inhibitor, 2-carboxy-ᴅ-arabinitol-1,5-bisphosphate (blue), that was cocrystallized with the enzyme. The amino acid side chains that interact with the bound inhibitor are shown in red. **(c)** The naturally occurring transition-state analog 2-carboxyarabinitol-1-phosphate, compared here with the β-keto acid intermediate (see Fig. 19–20) of the rubisco reaction. This analog is quite similar to the inhibitor shown bound to the enzyme in **(b)**.

(c)

Rubisco is also inhibited by 2-carboxyarabinitol-1-phosphate, a naturally occurring transition-state analog (see Box 8–3) with a structure similar to that of the β-keto acid intermediate of the rubisco reaction (Figs. 19–20, 19–30c). This compound is synthesized in the dark by some plants to depress rubisco activity, and it is sometimes called the "nocturnal inhibitor." It is broken down when light returns, permitting a reactivation of rubisco.

Certain Enzymes of the Calvin Cycle Are Indirectly Activated by Light

The reductive fixation of CO_2 requires ATP and NADPH, and their stromal concentrations increase when chloroplasts are illuminated (Fig. 19–31). The light-induced transport of protons across the thylakoid membrane (Chapter 18) also makes the stromal compartment alkaline and is accompanied by a flow of Mg^{2+} out of the thylakoid compartment into the stroma. Several stromal enzymes have evolved to take advantage of these light-dependent conditions that signal the availability of ATP and NADPH; they have pH or Mg^{2+} optima that are better suited to alkaline conditions and high $[Mg^{2+}]$. Activation of rubisco by formation of the lysyl carbamate is faster at alkaline pH, and high stromal $[Mg^{2+}]$ favors formation of the active Mg^{2+} complex.

Figure 19–31 ATP and NADPH produced by the light reactions are essential substrates for the reduction of CO_2; their availability limits the rate of CO_2 fixation. The photosynthetic reactions that produce ATP and NADPH are accompanied by movement of protons (red) from the stroma into the thylakoid, creating alkaline conditions in the stroma. Mg^{2+} ions pass from the thylakoid into the stroma, increasing the stromal $[Mg^{2+}]$.

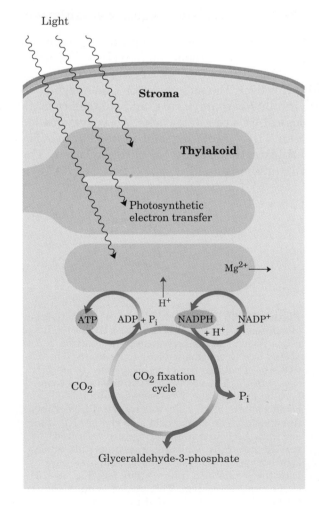

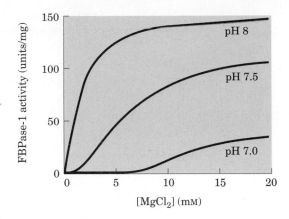

Figure 19–32 Activation of spinach chloroplast fructose-1,6-bisphosphatase (FBPase-1) by pH and Mg^{2+}. The combination of high pH and high $[Mg^{2+}]$ in the stroma, both of which are produced by illumination, strongly stimulates the activity of this gluconeogenic enzyme.

Fructose-1,6-bisphosphatase requires Mg^{2+} and is very dependent upon pH (Fig. 19–32). Its activity increases by a factor of more than 100 when the pH and $[Mg^{2+}]$ rise during chloroplast illumination.

Three other enzymes essential to the Calvin cycle's operation are subject to another type of regulation by light. Ribulose-5-phosphate kinase, fructose-1,6-bisphosphatase, and sedoheptulose-1,7-bisphosphatase can exist in either of two forms, differing in the oxidation state of Cys residues essential to their catalytic activity. When these Cys residues are oxidized as disulfide bonds, the enzymes are inactive; this is the normal situation in the dark. With illumination, electrons flow from photosystem I to ferredoxin (see Fig. 18–44), which passes electrons to a small, soluble, disulfide-containing protein called **thioredoxin** (Fig. 19–33). Thioredoxin donates electrons for the reduction of the disulfide bridges of these light-activated enzymes and is then reactivated in a disulfide-exchange reaction catalyzed by **thioredoxin reductase.**

Gluconeogenesis and Glycolysis Are Reciprocally Regulated in Plants

The possibility of futile cycling by the simultaneous operation of glycolysis and gluconeogenesis exists in plants as in animals. Plant cells (the chloroplast stroma and cytosol) have all the enzymes of glycolysis, and during dark periods glycolytic breakdown of starch is a major source of energy. During periods of illumination, photosynthetic plant cells produce the triose phosphate intermediates common to glycolysis and gluconeogenesis, and convert these into hexoses, sucrose, and starch as we have just seen. Futile cycling through these two paths is prevented in plants by the regulation of key enzymes of each cytosolic pathway by fructose-2,6-bisphosphate, whose concentration reflects the level of photosynthetic activity.

The concentration of fructose-2,6-bisphosphate varies inversely with the rate of photosynthesis in higher plants (Fig. 19–34). The enzyme phosphofructokinase-2, responsible for fructose-2,6-bisphosphate synthesis, is inhibited by dihydroxyacetone phosphate or 3-phosphoglycerate and stimulated by P_i. During active photosynthesis, dihydroxyacetone phosphate is produced and P_i is consumed, resulting in inhibition of PFK-2 and lowered concentrations of fructose-2,6-bisphosphate.

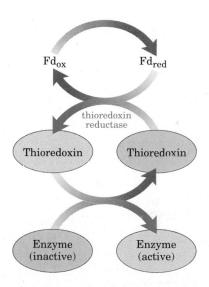

Figure 19–33 Light activation of several enzymes of the Calvin cycle is mediated by thioredoxin, a small, disulfide-containing protein. In the light, thioredoxin is reduced by electrons from ferredoxin (Fd) (blue arrows), then reduces critical disulfides of sedoheptulose-1,7-bisphosphatase, fructose-1,6-bisphosphatase, and ribulose-5-phosphate kinase, thereby activating these enzymes.

Figure 19–34 The concentration of the allosteric regulator fructose-2,6-bisphosphate in plant cells is regulated by the products of photosynthetic carbon fixation and by P_i. Dihydroxyacetone phosphate and 3-phosphoglycerate produced by CO_2 fixation inhibit phosphofructokinase-2 (PFK-2), the enzyme that synthesizes the regulator; P_i stimulates that enzyme. The concentration of the regulator is therefore inversely proportional to the rate of photosynthesis. In the dark, the concentration of fructose-2,6-bisphosphate increases and stimulates the glycolytic enzyme PP_i-dependent phosphofructokinase-1 (PFK-1), while inhibiting the gluconeogenic enzyme fructose-1,6-bisphosphatase (FBPase-1). When photosynthesis is active (in the light), the concentration of the regulator drops and the synthesis of fructose-6-phosphate and sucrose is favored. FBPase-2 represents fructose-2,6-bisphosphatase.

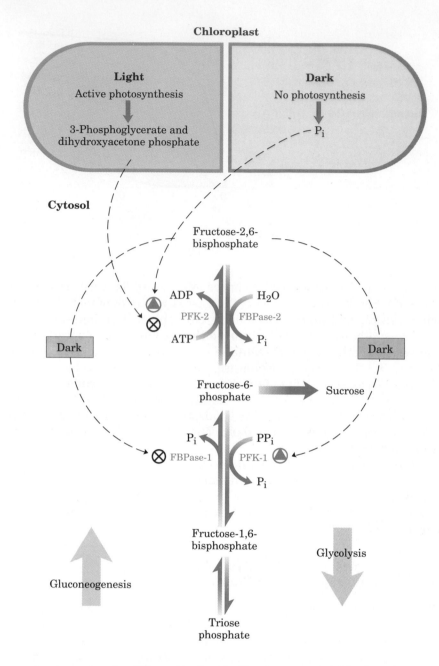

As in animals (Figs. 19–7, 19–8), fructose-2,6-bisphosphate in plants is an activator of glycolysis and an inhibitor of gluconeogenesis (Fig. 19–34). It slows gluconeogenesis by inhibiting the cytosolic fructose-1,6-bisphosphatase, which catalyzes a rate-limiting step in the synthesis of fructose-6-phosphate. It stimulates glycolysis by activating the PP_i-dependent form of phosphofructokinase-1 (p. 435); this enzyme, a critical control point in glycolysis, is virtually inactive in the absence of fructose-2,6-bisphosphate.

Thus, when photosynthesis ceases, the concentration of fructose-2,6-bisphosphate rises, stimulating glycolysis and inhibiting gluconeogenesis; glycolysis (combined with mitochondrial oxidative phosphorylation) provides energy in the dark. In the light, the fructose-2,6-bisphosphate concentration drops, glycolysis is inhibited, gluconeogenesis is turned on, and the energy and precursors produced by photosynthesis are used to make hexoses to be transported as sucrose or stored as starch.

Sucrose and Starch Synthesis Are Coordinately Regulated

In most plant cells, triose phosphates produced during CO_2 fixation are converted primarily to sucrose and starch. The balance between these processes is tightly regulated, and both processes must be coordinated with the rate of carbon fixation. Communication and coordination between sucrose synthesis in the cytosol and carbon fixation and starch synthesis in the chloroplast are mediated by the P_i–triose phosphate antiport system. If sucrose synthesis is too rapid, the transport of the excess P_i into the chloroplast will result in the removal of too much triose phosphate; this will have a deleterious effect on the rate of carbon fixation because five of every six triose phosphate molecules produced in the Calvin cycle are needed to regenerate ribulose-1,5-bisphosphate and complete the cycle. If sucrose synthesis is too slow, insufficient P_i will be made available to the chloroplast for triose phosphate synthesis.

Sucrose synthesis is regulated primarily at three steps, those catalyzed by fructose-1,6-bisphosphatase, sucrose-6-phosphate synthase, and sucrose phosphate phosphatase. When light first strikes a leaf in the morning, triose phosphate levels in the cytosol (derived from carbon fixation in the chloroplast) rise. The triose phosphates inhibit the activity of phosphofructokinase-2, and the levels of fructose-2,6-bisphosphate fall. This releases the inhibition of fructose-1,6-bisphosphatase, permitting the synthesis of fructose-6-phosphate and thus other hexose phosphates, including glucose-6-phosphate. Glucose-6-phosphate is an allosteric activator of sucrose-6-phosphate synthase. Sucrose phosphate phosphatase catalyzes the final step in sucrose synthesis as its substrate becomes available.

The key regulatory enzyme in starch synthesis is ADP-glucose pyrophosphorylase (Fig. 19–16), which is indirectly activated as cytosolic sucrose levels rise. Sucrose-6-phosphate synthase and sucrose phosphate phosphatase are both inhibited to some degree by sucrose. This inhibition becomes effective as sucrose levels increase, leading to an increase in the cytosolic pool of hexose phosphates that effectively sequesters much of the available phosphate in a form that cannot be transported back to the chloroplast. The levels of free P_i in the cytosol, and the resulting decrease in the activity of the P_i–triose phosphate antiporter, lead to a decrease in P_i and an increase in three-carbon compounds in the chloroplast. ADP-glucose pyrophosphorylase is inhibited by P_i and stimulated by 3-phosphoglycerate, so that one effect of the increase in cytosolic sucrose concentration is an increase in starch synthesis in the chloroplast. In the steady state, all of these activities are balanced so that sucrose and starch synthesis each consume about 50% of the triose phosphate produced during carbon fixation, and the rates of both processes are regulated so that precursors and intermediates required for carbon fixation are maintained at optimal levels.

Condensation of O_2 with Ribulose-1,5-Bisphosphate Initiates Photorespiration

Rubisco is not absolutely specific for CO_2 as a substrate; O_2 competes with CO_2 at the active site, and rubisco catalyzes the condensation of O_2 with ribulose-1,5-bisphosphate to form one molecule of 3-phosphoglycerate and one of **phosphoglycolate** (Fig. 19–35). This is the enzyme's oxygenase activity, evident in its name: RuBP carboxylase/

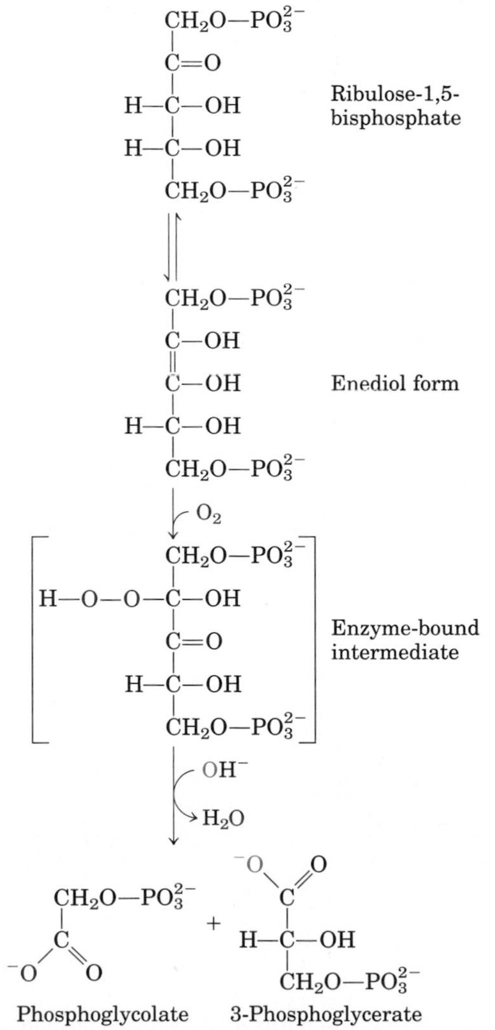

Figure 19–35 The oxygenase activity of ribulose-1,5-bisphosphate carboxylase/oxygenase (rubisco) results in the incorporation of O_2, not CO_2, into ribulose-1,5-bisphosphate. The unstable intermediate thus formed splits into phosphoglycolate, which is recycled as described in Fig. 19–36, and 3-phosphoglycerate, which can reenter the Calvin cycle.

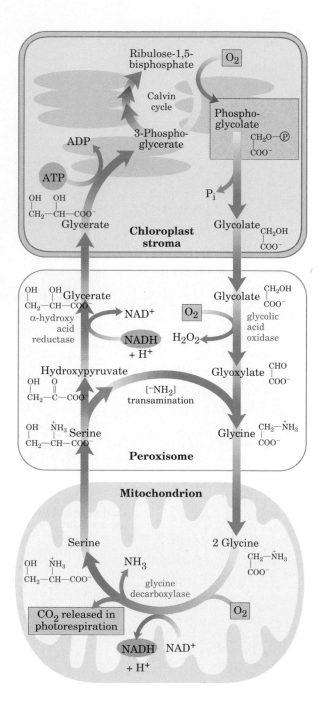

Figure 19–36 The pathway by which phosphoglycolate (shaded red) formed during photorespiration is salvaged by conversion into serine and thus 3-phosphoglycerate. This path is long and involves three cellular compartments. Glycolate formed by dephosphorylation of phosphoglycolate in chloroplasts is transaminated to glycine in peroxisomes. In mitochondria, two glycine molecules condense to form serine and the CO_2 released during photorespiration (shaded green). This reaction is catalyzed by glycine decarboxylase, an enzyme present at very high levels in the mitochondria of C_3 plants (see text above). The serine is converted to glycerate in peroxisomes, then reenters the chloroplasts to be phosphorylated, rejoining the Calvin cycle. Oxygen is consumed during photorespiration, in three steps (shaded blue).

oxygenase. The metabolic function of this reaction is not clear. It results in no fixation of carbon and appears to be a net liability to the cell in which it occurs; salvaging the carbons from phosphoglycolate, which is not a useful metabolite, uses cellular energy. The condensation of O_2 with ribulose-1,5-bisphosphate occurs concurrently with CO_2 fixation, with the latter predominating by a factor of about three.

The salvage pathway (Fig. 19–36) involves the conversion of two molecules of phosphoglycolate into a molecule of serine (which has three carbons) and a molecule of CO_2. The oxygenase activity of rubisco combined with the salvage pathway consumes O_2 and produces CO_2, a process called **photorespiration.** Unlike mitochondrial respiration, however, this process does not conserve energy.

Apparently the evolution of rubisco produced an active site not able to discriminate well between CO_2 and O_2, perhaps because much evolution occurred before O_2 was an important component of the atmosphere. The K_m for CO_2 is about 20 μM, and for O_2 is about 200 μM. (A solution that is in equilibrium with air at room temperature contains about 10 μM CO_2 and 250 μM O_2.) The modern atmosphere contains about 20% O_2 and only 0.04% CO_2, proportions that allow O_2 "fixation" by rubisco to constitute a significant waste of energy. During photosynthesis CO_2 is consumed in the fixation reactions, altering the CO_2 to O_2 ratio in the air spaces around the leaf in favor of O_2. In addition, the affinity of rubisco for CO_2 decreases with increasing temperature, exacerbating the tendency of the enzyme to catalyze the wasteful oxygenase reaction. Photorespiration may inhibit net biomass formation as much as 50%. As we shall see, this has led to adaptations in the process by which carbon fixation takes place, particularly in plants that live in warm climates.

Some Plants Have a Mechanism to Prevent Photorespiration

Most plants in the tropics, as well as temperate-zone crop plants native to the tropics, such as corn, sugar cane, and sorghum, have evolved a mechanism to circumvent the problem of wasteful photorespiration. The ultimate step of CO_2 fixation into a three-carbon product, 3-phosphoglycerate, is preceded by several steps, one of which is a preliminary fixation of CO_2 into a compound with four carbon atoms. These plants are referred to as **C_4 plants.** Plants in which the *first step* in carbon fixation is reaction of CO_2 with ribulose-1,5-bisphosphate to form 3-phosphoglycerate—as we have described thus far—are called **C_3 plants.**

C_4 plants, which typically grow in high light intensity and temperatures, have several important characteristics: high photosynthetic rates, high growth rates, low photorespiration rates, low rates of water loss, and an unusual leaf structure. Photosynthesis in the leaves of C_4 plants involves two cell types: mesophyll and bundle-sheath cells (Fig. 19–37). There are three known patterns of C_4 metabolism. The best-understood pathway, worked out in the 1960s by two plant biochemists, Marshall Hatch and Rodger Slack, is described here.

In plants of tropical origin the first intermediate in which $^{14}CO_2$ is fixed is not 3-phosphoglycerate but **oxaloacetate,** a four-carbon compound. This reaction, which occurs in leaf mesophyll cells (Fig. 19–37), is catalyzed by **phosphoenolpyruvate carboxylase:**

$$\text{Phosphoenolpyruvate} + HCO_3^- \longrightarrow \text{oxaloacetate} + P_i$$

This enzyme does not occur in animal tissues and is not to be confused with phosphoenolpyruvate carboxykinase (p. 603), which catalyzes the gluconeogenic reaction

$$\text{Oxaloacetate} + GTP \rightleftharpoons \text{phosphoenolpyruvate} + CO_2 + GDP$$

The oxaloacetate formed in the mesophyll cells is reduced to malate at the expense of NADPH:

$$\text{Oxaloacetate} + NADPH + H^+ \longrightarrow \text{L-malate} + NADP^+$$

Alternatively, oxaloacetate may be converted to aspartate by transamination:

$$\text{Oxaloacetate} + \alpha\text{-amino acid} \longrightarrow \text{L-aspartate} + \alpha\text{-keto acid}$$

The malate or aspartate formed in the mesophyll cells, which contains the fixed CO_2, is transferred into the neighboring bundle-sheath cells via special junctions (plasmodesmata; p. 49) between the cells (Fig. 19–37). In the bundle-sheath cells the malate is oxidized and decarboxylated to yield pyruvate and CO_2 by the action of **malic enzyme:**

$$\text{L-Malate} + NADP^+ \longrightarrow \text{pyruvate} + CO_2 + NADPH + H^+$$

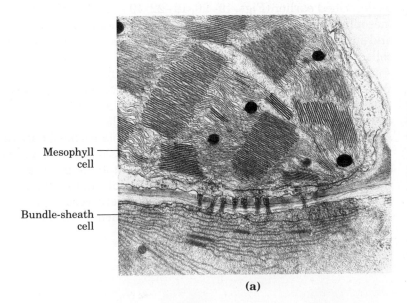

(a)

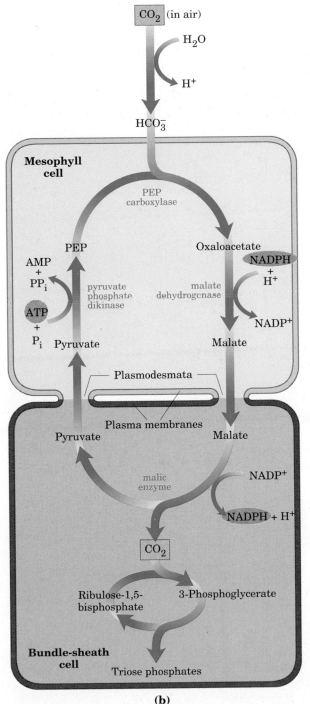

(b)

Figure 19–37 (a) An electron micrograph showing chloroplasts of connected mesophyll (above) and bundle-sheath (below) cells. The bundle-sheath cell contains starch granules. The plasmodesmata linking the two cells are evident. **(b)** The Hatch–Slack pathway of CO_2 fixation, via a four-carbon intermediate. This pathway prevails in plants of tropical origin (C_4 plants).

In plants that employ aspartate as the carrier of CO_2, aspartate is first transaminated to form oxaloacetate then reduced to malate in bundle-sheath cells before the release of CO_2 by malic enzyme. The free CO_2 formed in the bundle-sheath cells is the same CO_2 molecule that was originally fixed into oxaloacetate in the mesophyll cells.

In the bundle-sheath cells the CO_2 arising from the decarboxylation of malate is fixed again (Fig. 19–37), this time by rubisco, in exactly the same reaction that occurs in C_3 plants, leading to incorporation of CO_2 into the C-1 of 3-phosphoglycerate. The pyruvate formed by the decarboxylation of malate in the bundle-sheath cells is transferred back to the mesophyll cells, where it is converted into PEP by an unusual enzymatic reaction, catalyzed by the enzyme **pyruvate phosphate dikinase:**

$$\text{Pyruvate} + \text{ATP} + P_i \longrightarrow \text{phosphoenolpyruvate} + \text{AMP} + PP_i$$

This enzyme is called a dikinase because two different molecules are simultaneously phosphorylated by one molecule of ATP: pyruvate is phosphorylated to PEP, and phosphate is phosphorylated to pyrophosphate. The pyrophosphate is subsequently hydrolyzed to phosphate, so two high-energy phosphate groups of ATP are used in regenerating PEP. The PEP is now ready to fix another molecule of CO_2 in the mesophyll cell.

The PEP carboxylase of mesophyll cells has a high affinity for HCO_3^- and can fix CO_2 more efficiently than can rubisco. Unlike rubisco, PEP carboxylase does not use O_2 as an alternative substrate, so there is no competition between CO_2 and O_2 with this enzyme. This reaction serves to fix and concentrate CO_2 in the form of malate. Release of CO_2 from malate in the bundle-sheath cells yields a sufficiently high local concentration of CO_2 for rubisco to function near its maximal rate, with relatively little of the competitive side reaction with O_2.

Once CO_2 is fixed into 3-phosphoglycerate in the bundle-sheath cells, all the other reactions of the C_3 or Calvin cycle take place exactly as described earlier (Figs. 19–20, 19–22, 19–23). Thus in C_4 plants the mesophyll cells carry out CO_2 fixation by the C_4 pathway, but starch and sucrose biosynthesis occur by the C_3 pathway in the bundle-sheath cells.

The pathway of CO_2 fixation in C_4 plants has a greater energy cost than in C_3 plants. For each molecule of CO_2 fixed in the C_4 pathway, a molecule of PEP must be regenerated at the expense of two high-energy phosphate groups of ATP. Thus C_4 plants need a total of five ATPs to fix one molecule of CO_2, whereas C_3 plants need only three. As the temperature increases (and the affinity of rubisco for CO_2 decreases, as noted above), a point is reached at about 28 to 30 °C where the gain in efficiency from the elimination of photorespiration in C_4 plants more than compensates for this energetic cost. C_4 plants (crabgrass, for example) outgrow most C_3 plants during the summer, as any experienced gardener can attest!

Summary

Gluconeogenesis is the formation of carbohydrate from noncarbohydrate precursors, the most important of which are pyruvate, lactate, and alanine. In vertebrates, gluconeogenesis in the liver and kidney provides glucose for use by the brain, muscle, and erythrocytes. Like all biosynthetic pathways, gluconeogenesis proceeds by an enzymatic route that differs from the corresponding catabolic pathway, is independently regulated, and requires ATP. The biosynthetic pathway from pyruvate to glucose occurs in all organisms. It employs seven of the glycolytic enzymes, which function reversibly. Three irreversible steps in the glycolytic pathway cannot be used in gluconeogenesis in the cell, and these are bypassed by reactions catalyzed by nonglycolytic enzymes: conversion of pyruvate into phosphoenolpyruvate via oxaloacetate, involving several enzymes and two high-energy phosphate groups; dephosphorylation of fructose-1,6-bisphosphate by fructose-1,6-bisphosphatase; and dephosphorylation of glucose-6-phosphate by glucose-6-phosphatase. The path from pyruvate to phosphoenolpyruvate varies somewhat depending upon whether lactate or pyruvate itself serves as the gluconeogenic precursor. Formation of one molecule of glucose from pyruvate requires four molecules of ATP and two of GTP. Three carbon atoms of each of the citric acid cycle intermediates and some or all carbons of many of the amino acids are convertible into glucose.

Gluconeogenesis in the liver is regulated at two major points: (1) the carboxylation of pyruvate by pyruvate carboxylase, which is stimulated by the allosteric effector acetyl-CoA, and (2) the dephosphorylation of fructose-1,6-bisphosphate by fructose-1,6-bisphosphatase, which is inhibited by fructose-2,6-bisphosphate and AMP and stimulated by citrate. Fructose-2,6-bisphosphate also stimulates the glycolytic enzyme phosphofructokinase-1 and is crucial to the balance between gluconeogenesis and glycolysis. The levels of fructose-2,6-bisphosphate are hormonally regulated in animals. Reciprocal regulation of gluconeogenesis and glycolysis prevents futile cycling with its accompanying loss of ATP energy.

Unlike animals, plants can convert acetyl-CoA derived from fatty acid oxidation into glucose. They do so by a combination of the glyoxylate and citric acid cycles and gluconeogenic enzymes, in reactions compartmented among the glyoxysomes, mitochondria, and cytosol.

Glycogen synthesis also proceeds via a pathway different from its breakdown. It requires conversion of glucose-1-phosphate into UDP-glucose, a sugar nucleotide. Sugar phosphates are activated and earmarked for a particular synthetic path by ester linkage of a nucleoside diphosphate to the anomeric carbon of the sugar. Glycogen synthase adds glucose units from UDP-glucose to the nonreducing end of the growing glycogen chain, forming ($\alpha1\rightarrow4$) links. A branching enzyme, glycosyl-($4\rightarrow6$)-transferase, is necessary to add ($\alpha1\rightarrow6$) branch points. The initiation of glycogen synthesis requires a primer protein called glycogenin. The synthesis and breakdown of glycogen are reciprocally regulated by hormone-dependent phosphorylation of glycogen synthase (inactivating it) and of glycogen phosphorylase (activating it).

In plants, triose phosphates can be condensed to hexose phosphates and polymerized to starch for storage within the chloroplast. Starch synthase catalyzes the addition of glucose units from ADP-glucose to starch by a mechanism similar to that of glycogen synthase. Alternatively, triose phosphates can pass into the cytosol via an antiporter and serve as precursors for sucrose synthesis. Sucrose-6-phosphate synthase, which condenses UDP-glucose with fructose-6-phosphate, is inhibited when sucrose accumulates in the cytosol. Starch synthesis is stimulated by sucrose accumulation.

Lactose synthesis in the lactating mammary gland is brought about by the α-lactalbumin–galactosyl transferase (lactose synthase) enzyme complex, using UDP-galactose and glucose as substrates. The α-lactalbumin serves as a specificity-modifying subunit, whose formation is regulated by hormones promoting lactation.

In plant cells photosynthesis takes place in chloroplasts. In the CO_2-fixing reactions of photosynthesis (the Calvin cycle), ATP and NADPH are used to reduce CO_2 to form triose phosphates. The reactions required for CO_2 fixation occur in three stages: the fixation reaction itself, catalyzed by the stromal enzyme ribulose-1,5-bisphosphate carboxylase/oxygenase (rubisco); the reduction of the resulting 3-phosphoglycerate to glyceraldehyde-3-phosphate, which can be used in the synthesis of hexoses or in glycolysis; and the regeneration of ribulose-1,5-bisphosphate from triose phosphates.

Rubisco condenses CO_2 with the five-carbon acceptor ribulose-1,5-bisphosphate, then hydrolyzes the resulting hexose into two molecules of 3-phosphoglycerate. Stromal isozymes of the glycolytic enzymes, acting in the "reverse" direction, catalyze reduction of 3-phosphoglycerate to glyceraldehyde-3-phosphate; each molecule reduced requires one ATP and one NADPH. Finally, stromal enzymes including transketolase and aldolase rearrange the carbon skeletons of triose phosphates, generating a series of intermediates of three, four, five, six, and seven carbons and yielding pentose phosphates. The pentose phosphates are converted to ribulose-5-phosphate, then phosphorylated to ribulose-1,5-bisphosphate to complete the Calvin cycle. The energetic cost of fixing three CO_2 into triose phosphate is nine ATP and six NADPH.

An antiport system in the inner chloroplast membrane exchanges P_i in the cytosol for 3-phosphoglycerate or dihydroxyacetone phosphate produced by CO_2 fixation in the stroma. Dihydroxyacetone phosphate oxidation in the cytosol generates ATP and NADH, moving ATP and reducing equivalents from the chloroplast to the cytosol.

Rubisco is regulated by covalent modification and by a natural transition-state analog. Other enzymes of the Calvin cycle are inhibited by light-induced processes. Gluconeogenesis and glycolysis are regulated in plants by fructose-2,6-bisphosphate, the level of which varies inversely with the rate of photosynthesis: as the photosynthetic rate increases, fructose-2,6-bisphosphate levels fall and gluconeogenesis is activated.

Photorespiration wastes photosynthetic energy in C_3 plants by forming and oxidizing phosphoglycolate, a product of the oxygenation of ribulose-1,5-bisphosphate by rubisco. In C_4 plants a pathway exists to avoid photorespiration; CO_2 is first fixed in mesophyll cells into a four-carbon compound, which passes into bundle-sheath cells and releases CO_2 in high concentrations. This CO_2 is fixed in the bundle-sheath cells by rubisco, and the remaining reactions of the Calvin cycle occur as in C_3 plants.

Further Reading

Gluconeogenesis

Hers, H.G. & Hue, L. (1983) Gluconeogenesis and related aspects of glycolysis. *Annu. Rev. Biochem.* **52,** 617–653.

Hue, L. (1987) Gluconeogenesis and its regulation. *Diabetes Metab. Rev.* **3,** 111–126.

Pilkis, S.J., El-Maghrabi, M.R., & Claus, T.H. (1988) Hormonal regulation of hepatic gluconeogenesis and glycolysis. *Annu. Rev. Biochem.* **57,** 755–783.

Polysaccharide Synthesis

Akazawa, T. & Okamoto, K. (1980) Biosynthesis and metabolism of sucrose. In *The Biochemistry of Plants: A Comprehensive Treatise,* Vol. 3: *Carbohydrates: Structure and Function* (Preis, J., ed), pp. 199–220, Academic Press, Inc., New York.

Ap Rees, T. (1980) Integration of pathways of synthesis and degradation of hexose phosphates. In *The Biochemistry of Plants: A Comprehensive Treatise,* Vol. 3: *Carbohydrates: Structure and Function* (Preis, J., ed), pp. 1–42, Academic Press, Inc., New York.

Beck, E. & Ziegler, P. (1989) Biosynthesis and degradation of starch in higher plants. *Annu. Rev. Plant Physiol. Plant Mol. Biol.* **40,** 95–117.

Feingold, D.S. & Avigad, G. (1980) Sugar nucleotide transformations in plants. In *The Biochemistry of Plants: A Comprehensive Treatise,* Vol. 3: *Carbohydrates: Structure and Function* (Preis, J., ed), pp. 101–170, Academic Press, Inc., New York.

Geddes, R. (1986) Glycogen: a metabolic viewpoint. *Biosci. Rep.* **6,** 415–428.

Leloir, L.F. (1971) Two decades of research on the biosynthesis of saccharides. *Science* **172,** 1299–1303.

Leloir's Nobel address, including a discussion of the role of sugar nucleotides in metabolism.

Nuttall, F.Q., Gilboe, D.P., Gannon, M.C., Niewoehner, C.B., & Tan, A.W.H. (1988) Regulation of glycogen synthesis in the liver. *Am. J. Med.* **85, Supplement 5A,** 77–85.

Preis, J. & Levi, C. (1980) Starch biosynthesis and degradation. In *The Biochemistry of Plants: A Comprehensive Treatise,* Vol. 3: *Carbohydrates: Structure and Function* (Preis, J., ed), pp. 371–423, Academic Press, Inc., New York.

Smythe, C. & Cohen, P. (1991) The discovery of glycogenin and the priming mechanism for glycogen biogenesis. *Eur. J. Biochem.* **200,** 625–631.

Carbon Dioxide Fixation

Andersson, I., Knight, S., Schneider, G., Lindqvist, Y., Lundqvist, T., Brändén, C.-I., & Lorimer, G.H. (1989) Crystal structure of the active site of ribulose-bisphosphate carboxylase. *Nature* **337,** 229–234.

Edwards, G.E. & Huber, S.C. (1981) The C_4 pathway. In *The Biochemistry of Plants: A Comprehensive Treatise,* Vol. 8: *Photosynthesis* (Hatch, M.D. & Boardman, N.K., eds), pp. 273–281, Academic Press, Inc., New York.

Halliwell, B. (1984) *Chloroplast Metabolism: The Structure and Function of Chloroplasts in Green Leaf Cells,* Clarendon Press, Oxford.

Hoober, J.K. (1984) *Chloroplasts,* Plenum Press, New York.

Horecker, B.L. (1976) Unravelling the pentose phosphate pathway. In *Reflections on Biochemistry* (Kornberg, A., Cornudella, L., Horecker, B.L., & Oro, J., eds), pp. 65–72, Pergamon Press, Inc., Oxford.

Huber, S.C. (1986) Fructose 2,6-bisphosphate as a regulatory metabolite in plants. *Annu. Rev. Plant Physiol.* **37,** 233–246.

Husic, D.W., Husic, H.D., & Tolbert, N.E. (1987) The oxidative photosynthetic carbon cycle or C_2 cycle. *CRC Crit. Rev. Plant Sci.* **5,** 45–100.

Lorimer, G.H. & Andrews, T.J. (1981) The C_2 chemo- and photorespiratory carbon oxidation cycle. In *The Biochemistry of Plants: A Comprehensive Treatise,* Vol. 8: *Photosynthesis* (Hatch, M.D. & Boardman, N.K., eds), pp. 329–374, Academic Press, Inc., New York.

Miziorko, H.M. & Lorimer, G.H. (1982) Ribulose-1,5-bisphosphate carboxylase-oxygenase. *Annu. Rev. Biochem.* **52,** 507–535.

Portis, A.R., Jr. (1990) Rubisco activase. *Biochim. Biophys. Acta* **1015,** 15–28.

Robinson, S.P. & Walker, D.A. (1981) Photosynthetic carbon reduction cycle. In *The Biochemistry of Plants: A Comprehensive Treatise,* Vol. 8: *Photosynthesis* (Hatch, M.D. & Boardman, N.K., eds), pp. 193–236, Academic Press, Inc., New York.

Schneider, G., Lindqvist, Y., Brändén, C.-I., & Lorimer, G. (1986) Three-dimensional structure of ribulose-1,5-bisphosphate carboxylase/oxygenase from *Rhodospirillum rubrum* at 2.9 Å resolution. *EMBO J.* **5,** 3409–3415.

Wood, T. (1985) *The Pentose Phosphate Pathway,* Academic Press, Inc., Orlando, FL.

Woodrow, I.E. & Berry, J.A. (1988) Enzymatic regulation of photosynthetic CO_2 fixation in C_3 plants. *Annu. Rev. Plant Physiol. Plant Mol. Biol.* **39,** 533–594.

Problems

1. *Role of Oxidative Phosphorylation in Gluconeogenesis* Is it possible to obtain a net synthesis of glucose from pyruvate if the citric acid cycle and oxidative phosphorylation are totally inhibited?

2. *Pathway of Atoms in Gluconeogenesis* A liver extract capable of carrying out all the normal metabolic reactions of the liver is briefly incubated in separate experiments with the following ^{14}C-labeled precursors:

(a) [^{14}C]Bicarbonate, $HO—^{14}C \overset{O^-}{\underset{O}{}}$

(b) [1-^{14}C]Pyruvate, $CH_3—\underset{O}{C}—^{14}COO^-$

Trace the pathway of each precursor through gluconeogenesis. Indicate the location of ^{14}C in all intermediates of the process and in the product, glucose.

3. *Pathway of CO_2 in Gluconeogenesis* In the first bypass step in gluconeogenesis, the conversion of pyruvate to phosphoenolpyruvate, pyruvate is carboxylated by pyruvate carboxylase to oxaloacetate and is subsequently decarboxylated by PEP carboxykinase to yield phosphoenolpyruvate. The observation that the addition of CO_2 is directly followed by the loss of CO_2 suggests that ^{14}C of $^{14}CO_2$ would not be incorporated into PEP, glucose, or any of the intermediates in gluconeogenesis. However, it has been found that if rat liver slices synthesize glucose in the presence of $^{14}CO_2$, ^{14}C slowly appears in PEP and eventually appears in C-3 and C-4 of glucose. How does the ^{14}C label get into PEP and glucose? (Hint: During gluconeogenesis in the presence of $^{14}CO_2$, several of the four-carbon citric acid cycle intermediates also become labeled.)

4. *Regulation of Fructose-1,6-Bisphosphatase and Phosphofructokinase-1* What are the effects of increasing concentrations of ATP and AMP on the catalytic activities of fructose-1,6-bisphosphatase and phosphofructokinase-1? What are the consequences of these effects on the relative flow of metabolites through gluconeogenesis and glycolysis?

5. *Glucogenic Substrates* A common procedure for determining the effectiveness of compounds as precursors of glucose in mammals is to fast the animal until the liver glycogen stores are depleted and then administer the substrate in question. A substrate that leads to a *net* increase in liver glycogen is termed glucogenic because it must first be converted to glucose-6-phosphate. Show by means of known enzymatic reactions which of the following substances are glucogenic:

(a) $^-OOC—CH_2—CH_2—COO^-$
Succinate

(b)
$$\underset{\text{Glycerol}}{CH_2-\overset{\overset{\displaystyle OH}{|}}{\underset{\underset{\displaystyle H}{|}}{C}}-CH_2}$$
with OH groups

(c)
$$\underset{\text{Acetyl-CoA}}{CH_3-\overset{\overset{\displaystyle O}{\|}}{C}-S\text{-CoA}}$$

(d)
$$\underset{\text{Pyruvate}}{CH_3-\overset{\overset{\displaystyle O}{\|}}{C}-COO^-}$$

(e)
$$\underset{\text{Butyrate}}{CH_3-CH_2-CH_2-COO^-}$$

6. *Blood Lactate Levels during Vigorous Exercise* The concentration of lactate in blood plasma before, during, and after a 400 m sprint are shown below.

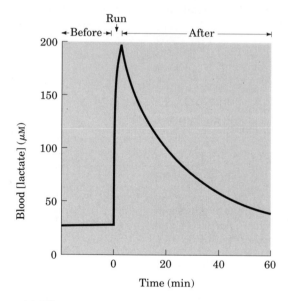

(a) What causes the rapid rise in lactate concentration?

(b) What causes the decline in lactate concentration after completion of the run? Why does the decline occur more slowly than the increase?

(c) Why is the concentration of lactate not zero during the resting state?

7. *Excess O_2 Uptake during Gluconeogenesis* Lactate absorbed by the liver is converted to glucose. This process requires the input of 6 mol of ATP for every mole of glucose produced. The extent of this process in rat liver slices can be monitored by administering [^{14}C]lactate and measuring the amount of [^{14}C]glucose produced. Because the stoichiometry of O_2 consumption and ATP production is known (Chapter 18), we can predict the extra O_2 consumption above the normal rate when a given amount of lactate is administered. The extra amount of O_2 necessary for the synthesis of glucose from lactate, however, when actually measured is always higher than predicted by known stoichiometric relationships. Suggest a possible explanation for this observation.

8. *At What Point Is Glycogen Synthesis Regulated?* Explain how the two following observations identify the point of regulation in the synthesis of glycogen in skeletal muscle:

(a) The measured activity of glycogen synthase in resting muscle, expressed in micromoles of UDP-glucose used per gram per minute, is lower than the activity of phosphoglucomutase or UDP-glucose pyrophosphorylase, each measured in terms of micromoles of substrate transformed per gram per minute.

(b) Stimulation of glycogen synthesis leads to a small decrease in the concentrations of glucose-6-phosphate and glucose-1-phosphate, a large decrease in the concentration of UDP-glucose, but a substantial increase in the concentration of UDP.

9. *What Is the Cost of Storing Glucose as Glycogen?* Write the sequence of steps and the net reaction required to calculate the cost in number of ATPs of converting cytosolic glucose-6-phosphate into glycogen and back into glucose-6-phosphate. What fraction of the maximum number of ATPs that are available from complete catabolism of glucose-6-phosphate to CO_2 and H_2O does this cost represent?

10. *Identification of a Defective Enzyme in Carbohydrate Metabolism* A sample of liver tissue was obtained post mortem from the body of a patient believed to be genetically deficient in one of the enzymes of carbohydrate metabolism. A homogenate of the liver sample had the following characteristics: (1) it degraded glycogen to glucose-6-phosphate, (2) it was unable to make glycogen from any sugar or to utilize galactose as an energy source, and (3) it synthesized glucose-6-phosphate from lactate. Which of the following three enzymes was deficient?

(a) Glycogen phosphorylase

(b) Fructose-1,6-bisphosphatase

(c) UDP-glucose pyrophosphorylase

Give reasons for your choice.

11. *Ketosis in Sheep* The udder of a ewe uses almost 80% of the total glucose synthesized by the animal. The glucose is used for milk production, principally in the synthesis of lactose and of glycerol-3-phosphate, used in the formation of milk triacylglycerols. During the winter when food quality is poor, milk production decreases and the ewes sometimes develop ketosis, that is, increased levels of plasma ketone bodies. Why do these changes occur? A standard treatment for this condition is the administration of large doses of propionate (which is readily converted to succinyl-CoA in ruminants). How does this treatment work?

12. *Adaptation to Galactosemia* Galactosemia is a pathological condition in which there is deficient utilization of galactose derived from lactose in the diet. One form of this disease is due to the absence of the enzyme UDP-glucose: galactose-1-phosphate uridylyltransferase. If an individual survives the disease in early life, some capacity to metabolize ingested galactose may develop in later life, because of increased production of the enzyme UDP-galactose pyrophosphorylase, which catalyzes the reaction

$$\text{Galactose-1-phosphate} + \text{UTP} \longrightarrow \text{UDP-galactose} + PP_i$$

How does the presence of this enzyme increase the capacity of such individuals to metabolize galactose?

13. *Phases of Photosynthesis* When a suspension of green algae is illuminated in the absence of CO_2 and then incubated with $^{14}CO_2$ in the dark, $^{14}CO_2$ is converted into [^{14}C]glucose for a brief time. What is the significance of this observation with regard to the CO_2 fixation process and how is it related to the light reactions of photosynthesis? Why does the conversion of $^{14}CO_2$ into [^{14}C]glucose stop after a brief time?

14. *Identification of Key Intermediates in CO_2 Fixation* Calvin and his colleagues used the unicellular green alga *Chlorella* to study the carbon fixation reactions of photosynthesis. In their experiments $^{14}CO_2$ was incubated with illuminated suspensions of algae under different conditions. They followed the time course of appearance of ^{14}C in two compounds, X and Y, under two sets of conditions.

(a) Illuminated *Chlorella* were grown on unlabeled CO_2; then the lights were turned off, and $^{14}CO_2$ was added (vertical dashed line in graph **a** below). Under these conditions X was the first compound to become labeled with ^{14}C. Compound Y was unlabeled.

(b) Illuminated *Chlorella* cells were grown in $^{14}CO_2$. Illumination was continued until all the $^{14}CO_2$ had disappeared (vertical dashed line in graph **b** below). Under these conditions compound X became labeled quickly but lost its radioactivity with time, whereas compound Y became more radioactive with time.

Suggest the identities of X and Y based on your understanding of the Calvin cycle.

15. *Pathway of CO_2 Fixation in Maize* If a maize (corn) plant is illuminated in the presence of $^{14}CO_2$, after about 1 s more than 90% of all the radioactivity incorporated in the leaves is found in the C-4 atoms of malate, aspartate, and oxaloacetate. Only after 60 s does ^{14}C appear in the C-1 atom of 3-phosphoglycerate. Explain.

16. *Chemistry of Malic Enzyme: Variation on a Theme* Malic enzyme, found in the bundle-sheath cells of C_4 plants, carries out a reaction that has a counterpart in the citric acid cycle. What is the analogous reaction? Explain.

17. *Sucrose and Dental Caries* The most prevalent infection in humans worldwide is dental caries, which stems from the colonization and destruction of tooth enamel by a variety of acidifying microorganisms. These organisms synthesize and live within a water-insoluble network of dextrans, called dental plaque, composed of ($\alpha 1\rightarrow 6$)-linked polymers of glucose with many ($\alpha 1\rightarrow 3$) branch points. Polymerization of dextran requires dietary sucrose, and the reaction is catalyzed by a bacterial enzyme, dextran–sucrose glucosyltranferase.

(a) Write the overall reaction for dextran polymerization.

(b) In addition to providing a substrate for the formation of dental plaque, how does dietary sucrose also provide oral bacteria with an abundant source of metabolic energy?

18. *Regulation of Carbohydrate Synthesis in Plants* Sucrose synthesis occurs in the cytosol, and starch synthesis occurs in the chloroplast stroma; yet both reactions are intricately balanced.

(a) What factors shift the reactions in favor of starch synthesis?

(b) What factors shift the reactions to favor sucrose synthesis?

(c) Given that these two synthetic pathways occur in separate cellular compartments, what enables the two processes to influence each other?

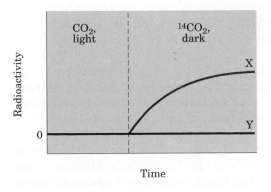

(a)

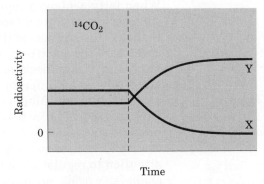

(b)

Lipid Biosynthesis

Lipids play a variety of cellular roles, including some only recently recognized. They are the principal form of stored energy in most organisms, as well as major constituents of cell membranes. Specialized lipids serve as pigments (retinal), cofactors (vitamin K), detergents (bile salts), transporters (dolichols), hormones (vitamin D derivatives, sex hormones), extracellular and intracellular messengers (eicosanoids and derivatives of phosphatidylinositol), and anchors for membrane proteins (covalently attached fatty acids, prenyl groups, and phosphatidylinositol). The ability to synthesize a variety of lipids is therefore essential to all organisms. This chapter describes biosynthetic pathways for some of the principal lipids present in most cells, illustrating the strategies employed in assembling these water-insoluble products from simple, water-soluble precursors such as acetate. Like other biosynthetic pathways, these reaction sequences are endergonic and reductive. They use ATP as a source of metabolic energy and a reduced electron carrier (usually NADPH) as a reductant.

We first describe the biosynthesis of fatty acids, the major components of both triacylglycerols and phospholipids. Then we examine the assembly of fatty acids into triacylglycerols and into the simpler types of membrane phospholipids. Finally, we consider the synthesis of cholesterol, a component of some membranes and the precursor of such steroid products as the bile acids, sex hormones, and adrenal cortical hormones.

Biosynthesis of Fatty Acids and Eicosanoids

When fatty acid oxidation was found to occur by oxidative removal of successive two-carbon (acetyl-CoA) units (see Fig. 16–8), biochemists thought that the biosynthesis of fatty acids might proceed by simple reversal of the same enzymatic steps used in their oxidation. However, fatty acid biosynthesis and breakdown occur by different pathways, are catalyzed by different sets of enzymes, and take place in different parts of the cell. Moreover, a three-carbon intermediate, **malonyl-CoA,** participates in the biosynthesis of fatty acids but not in their breakdown.

We focus first on the pathway of fatty acid synthesis, then turn our attention to regulation of the pathway and to the biosynthesis of long-chain fatty acids, unsaturated fatty acids, and their eicosanoid derivatives.

Malonyl-CoA

Malonyl-CoA Is Formed from Acetyl-CoA and Bicarbonate

The irreversible formation of malonyl-CoA from acetyl-CoA is catalyzed by **acetyl-CoA carboxylase** (Fig. 20–1). Acetyl-CoA carboxylase contains biotin as its prosthetic group, covalently bound in amide linkage to the ε-amino group of a Lys residue on one of the three subunits of the enzyme molecule. The two-step reaction is very similar to other biotin-dependent carboxylation reactions, such as those catalyzed by pyruvate carboxylase (see Fig. 15–13) and propionyl-CoA carboxylase (see Fig. 16–12). The carboxyl group, derived from bicarbonate (HCO_3^-), is first transferred to biotin in an ATP-dependent reaction. The biotinyl group serves as a temporary carrier of CO_2, transferring it to acetyl-CoA in the second step to yield malonyl-CoA.

Acetyl-CoA carboxylase from bacteria has three separate polypeptide subunits, as shown in Figure 20–1. In higher plants and animals, all three activities are part of a single multifunctional polypeptide.

Figure 20–1 The acetyl-CoA carboxylase reaction. Acetyl-CoA carboxylase has three functional regions: biotin carrier protein (gray); biotin carboxylase, which activates CO_2 by attaching it to a nitrogen in the biotin ring in an ATP-dependent reaction (see Fig. 15–13b); and transcarboxylase, which transfers activated CO_2 from biotin to acetyl-CoA, producing malonyl-CoA. The long, flexible biotin arm carries the activated CO_2 from the biotin carboxylase region to the transcarboxylase active site, as shown in the diagrams below the reaction arrows. The active enzyme in each case is shaded in blue.

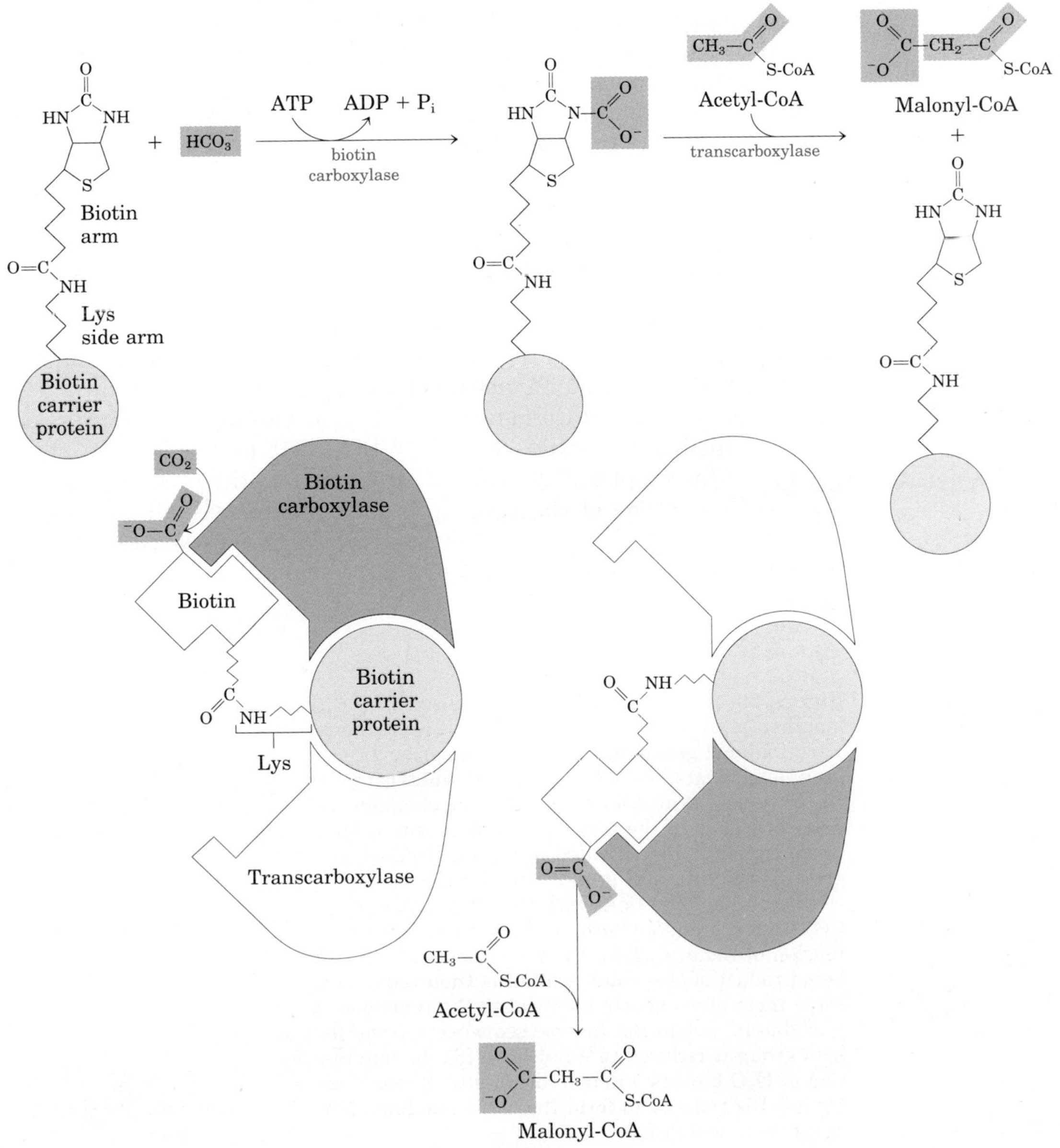

Malonyl group

Acetyl group
(first acyl group)

Fatty acid synthase

condensation ① → CO_2

reduction ② NADPH + H^+ → $NADP^+$

dehydration ③ → H_2O

reduction ④ NADPH + H^+ → $NADP^+$

Saturated acyl group,
lengthened by two carbons

The Biosynthesis of Fatty Acids Proceeds by a Distinctive Pathway

The fundamental reaction sequence by which the long chains of carbon atoms in fatty acids are assembled consists of four steps (Fig. 20–2). The saturated acyl group produced during this set of reactions is recycled to become the substrate in another condensation with an activated malonyl group. With each passage through the cycle, the fatty acyl chain is extended by two carbons. When the chain length reaches 16, the product (palmitate, 16:0; see Table 9–1) leaves the cycle. The methyl and carboxyl carbon atoms of the acetyl group become C-16 and C-15, respectively, of the palmitate (Fig. 20–3); the rest of the carbon atoms are derived from malonyl-CoA.

Both the electron carrier cofactor and the activating groups in the reductive anabolic sequence are different from those that act in the oxidative catabolic process. Recall that in β oxidation, NAD^+ and FAD serve as electron acceptors, and the activating group is the thiol (—SH) group of coenzyme A (see Fig. 16–8). By contrast, the reducing agent in the synthetic sequence is NADPH, and the activating groups are two different enzyme-bound —SH groups, to be described below.

All of the reactions in the synthetic process are catalyzed by a multienzyme complex, the **fatty acid synthase.** The detailed structure of this multienzyme complex and its location in the cell differ from one species to another, but the reaction sequence is identical in all organisms.

The Fatty Acid Synthase Complex Has Seven Different Active Sites

The fatty acid synthase system from *E. coli* consists of seven separate polypeptides that are tightly associated in a single, organized complex (Table 20–1). The proteins act together to catalyze the formation of fatty acids from acetyl-CoA and malonyl-CoA. Throughout the process, the intermediates remain covalently attached to one of two thiol groups of the complex. One point of attachment is the —SH group of a Cys residue in one of the seven proteins (β-ketoacyl-ACP synthase, described below); the other is the —SH group of acyl carrier protein, with which the acyl intermediates of fatty acid synthesis form a thioester.

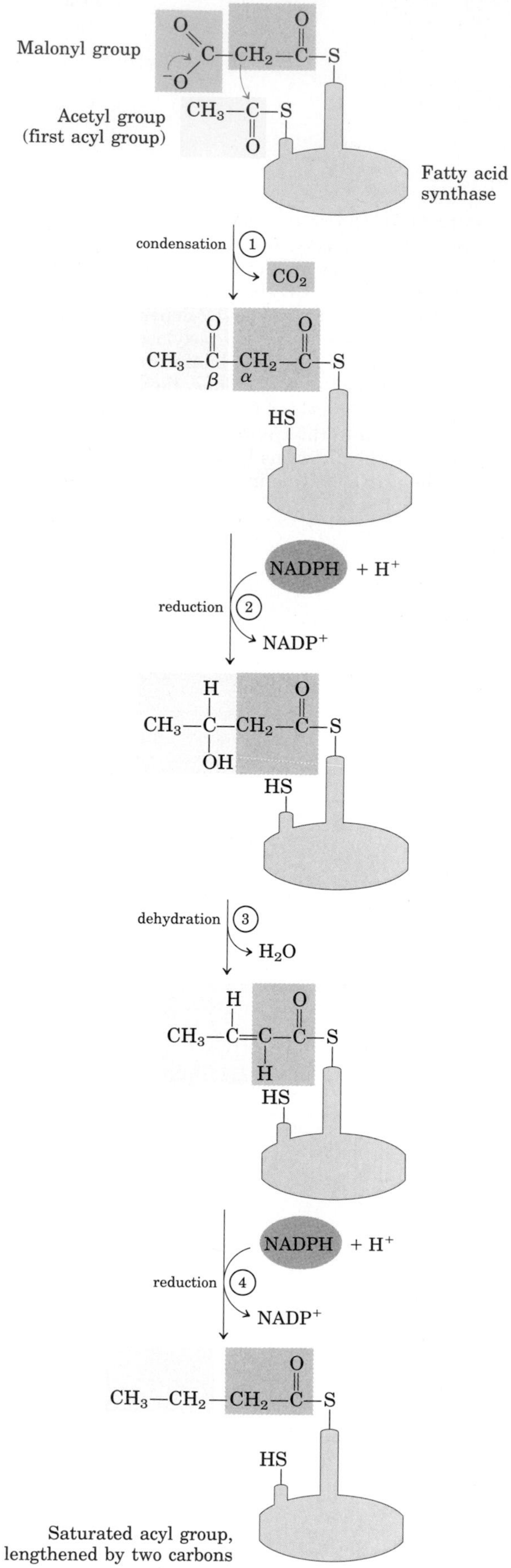

Figure 20–2 The four-step sequence used to lengthen a growing fatty acyl chain by two carbons. Each malonyl group and acetyl (or longer acyl) group is activated by a thioester that links it to the fatty acid synthase, a multienzyme complex described later in the text. ① The first step is the condensation of an activated acyl group (an acetyl group is the first acyl group) and two carbons derived from malonyl-CoA, with the elimination of CO_2 from the malonyl group; the net effect is extension of the acyl chain by two carbons. The β-keto product of this condensation is then reduced in three more steps nearly identical to the reactions of β oxidation, but in the reverse sequence: ② the β-keto group is reduced to an alcohol, ③ the elimination of H_2O creates a double bond, and ④ the double bond is reduced to form the corresponding saturated fatty acyl group.

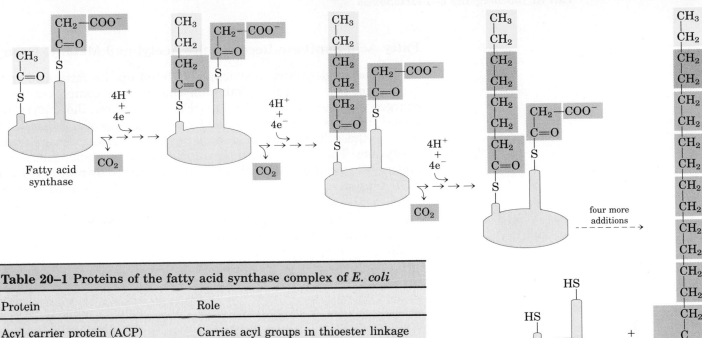

Table 20–1 Proteins of the fatty acid synthase complex of *E. coli*

Protein	Role
Acyl carrier protein (ACP)	Carries acyl groups in thioester linkage
Acetyl-CoA–ACP transacetylase (AT)	Transfers acyl group from CoA to Cys residue of KS
Malonyl-CoA–ACP transferase (MT)	Transfers malonyl group from CoA to ACP
β-Ketoacyl-ACP synthase (KS)	Condenses acyl and malonyl groups
β-Ketoacyl-ACP reductase (KR)	Reduces β-keto group to β-hydroxy group
β-Hydroxyacyl-ACP dehydratase (HD)	Removes H_2O from β-hydroxyacyl-ACP, creating double bond
Enoyl-ACP reductase (ER)	Reduces double bond, forming saturated acyl-ACP

Figure 20–3 The overall process of palmitate synthesis. The fatty acyl chain grows by two-carbon units donated by activated malonate, with loss of CO_2. After each two-carbon addition, reductions convert the growing chain to a saturated fatty acid of four, then six, then eight carbons, and so on. The final product is palmitate (16:0).

Acyl carrier protein (ACP) of *E. coli* (Table 20–1) is a small protein (M_r 8,860) containing the prosthetic group **4′-phosphopantetheine** (Fig. 20–4), an intermediate in the synthesis of coenzyme A (see Fig. 12–41). The thioester that links ACP to the fatty acyl group has a high free energy of hydrolysis, and the energy released when this bond is broken helps to make the first reaction in fatty acid synthesis, condensation, thermodynamically favorable. The 4′-phosphopantetheine prosthetic group of ACP is believed to serve as a flexible arm, tethering the growing fatty acyl chain to the surface of the fatty acid synthase complex and carrying the reaction intermediates from one enzyme active site to the next.

Although the details of enzyme structure differ in prokaryotes such as *E. coli* and in eukaryotes, the four-step process of fatty acid synthesis is the same in all organisms. We first describe the process as it occurs in *E. coli,* then consider the differences in enzyme structure among other organisms.

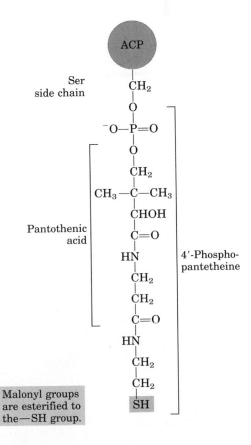

Figure 20–4 Acyl carrier protein (ACP). The prosthetic group is 4′-phosphopantetheine, which is covalently attached to the hydroxyl group of a Ser residue in ACP. Phosphopantetheine contains the B vitamin pantothenate, also found in the coenzyme A molecule. Its —SH group is the site of entry of malonyl groups during fatty acid synthesis.

Fatty Acid Synthase Receives the Acetyl and Malonyl Groups

Before the condensation reactions that build up the fatty acid chain can begin, the two thiol groups on the enzyme complex must be charged with the correct acyl groups (Fig. 20–5). First, the acetyl group of acetyl-CoA is transferred to the Cys —SH group of the β-ketoacyl-ACP synthase. This reaction is catalyzed by **acetyl-CoA–ACP transacetylase.** The second reaction, transfer of the malonyl group from malonyl-CoA to the —SH group of ACP, is catalyzed by **malonyl-CoA–ACP transferase,** also part of the complex. In the charged synthase complex, the acetyl and malonyl groups are very close to each other and are activated for the chain-lengthening process, which consists of the four steps outlined earlier. These steps are now considered in some detail.

① *Condensation* The first step in the formation of a fatty acid chain is condensation of the activated acetyl and malonyl groups to form an acetoacetyl group bound to ACP through the phosphopantetheine —SH group: **acetoacetyl-ACP;** simultaneously, a molecule of CO_2 is produced (Fig. 20–5). In this reaction, catalyzed by **β-ketoacyl-ACP synthase,** the acetyl group is transferred from the Cys —SH group of this enzyme to the malonyl group on the —SH of ACP, becoming the methyl-terminal two-carbon unit of the new acetoacetyl group.

The carbon atom in the CO_2 formed in this reaction is the same carbon atom that was originally introduced into malonyl-CoA from HCO_3^- by the acetyl-CoA carboxylase reaction (Fig. 20–1). Thus CO_2 is only transiently in covalent linkage during fatty acid biosynthesis; it is removed as each two-carbon unit is inserted.

Why do cells go to the trouble of adding CO_2 to make a malonyl group from an acetyl group, only to lose CO_2 again during the formation of acetoacetate? Remember that in the β oxidation of fatty acids, cleavage of the bond between two acyl groups (the cleavage of an acetyl unit from the acyl chain) is highly exergonic. Therefore, the simple condensation of two acyl groups (of two acetyl-CoA molecules, for example) is endergonic. The condensation reaction is made thermodynamically favorable by the involvement of activated malonyl, rather than acetyl, groups. The methylene carbon (C-2) of the malonyl group, sandwiched between carbonyl and carboxyl carbons, is an especially good nucleophile. In the condensation step (Fig. 20–5), decarboxylation of the malonyl group facilitates the nucleophilic attack of this methylene carbon on the thioester linking the acetyl group to β-ketoacyl-ACP synthase, displacing the enzyme's —SH group. Coupling the condensation to the decarboxylation of the malonyl group renders the overall process highly exergonic. Recall that a similar carboxylation–decarboxylation sequence facilitates the formation of phosphoenolpyruvate from pyruvate in gluconeogenesis (see Fig. 19–3).

By using activated malonyl groups in the synthesis of fatty acids and activated acetate in their degradation, the cell manages to make both processes favorable, although one is effectively the reversal of the other. The extra energy required to make fatty acid synthesis favorable is provided by the ATP used to synthesize malonyl-CoA from acetyl-CoA and HCO_3^- (Fig. 20–1).

② *Reduction of the Carbonyl Group* The acetoacetyl-ACP formed in the condensation step next undergoes reduction of the carbonyl group

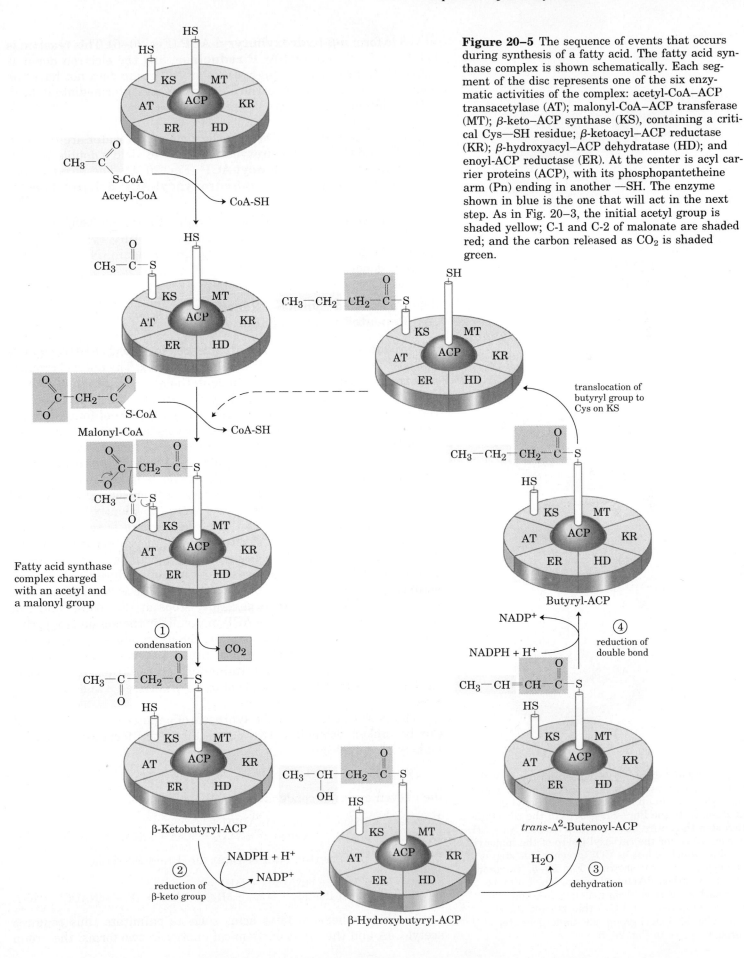

Figure 20–5 The sequence of events that occurs during synthesis of a fatty acid. The fatty acid synthase complex is shown schematically. Each segment of the disc represents one of the six enzymatic activities of the complex: acetyl-CoA–ACP transacetylase (AT); malonyl-CoA–ACP transferase (MT); β-keto–ACP synthase (KS), containing a critical Cys—SH residue; β-ketoacyl–ACP reductase (KR); β-hydroxyacyl–ACP dehydratase (HD); and enoyl-ACP reductase (ER). At the center is acyl carrier proteins (ACP), with its phosphopantetheine arm (Pn) ending in another —SH. The enzyme shown in blue is the one that will act in the next step. As in Fig. 20–3, the initial acetyl group is shaded yellow; C-1 and C-2 of malonate are shaded red; and the carbon released as CO_2 is shaded green.

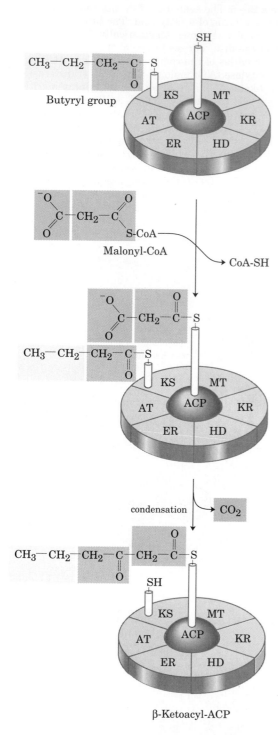

Figure 20–6 Beginning of the second round of the fatty acid synthesis cycle. The butyryl group is on the Cys —SH group. The incoming malonyl group is attached to the Pn —SH group. In the condensation step the entire butyryl group on the Cys —SH is exchanged for the carboxyl group of the malonyl residue, which is lost as CO_2 (green). This step is analogous with that shown in Fig. 20–5. The product, a six-carbon β-ketoacyl group, now contains four carbons derived from malonyl-CoA and two derived from the acetyl-CoA that started the reaction. The β-ketoacyl group now undergoes steps ② through ④, as in Fig. 20–5.

at C-3 to form D-β-hydroxybutyryl-ACP (Fig. 20–5). This reaction is catalyzed by **β-ketoacyl-ACP reductase**, and the electron donor is NADPH. Notice that the D-β-hydroxybutyryl group does not have the same stereoisomeric form as the L-β-hydroxyacyl intermediate in fatty acid oxidation (see Fig. 16–8).

③ *Dehydration* In the third step, the elements of water are removed from C-2 and C-3 of D-β-hydroxybutyryl-ACP to yield a double bond in the product, ***trans*-Δ^2-butenoyl-ACP** (Fig. 20–5). The enzyme that catalyzes this dehydration is **β-hydroxyacyl-ACP dehydratase**.

④ *Reduction of the Double Bond* Finally, the double bond of *trans*-Δ^2-butenoyl-ACP is reduced (saturated) to form **butyryl-ACP** by the action of **enoyl-ACP reductase** (Fig. 20–5); again, NADPH is the electron donor.

The Fatty Acid Synthase Reactions Are Repeated to Form Palmitate

The production of the four-carbon, saturated fatty acyl–ACP completes one pass through the fatty acid synthase complex. The butyryl group is now transferred from the phosphopantetheine —SH group of ACP to the Cys —SH group of β-ketoacyl-ACP synthase, which initially bore the acetyl group (Fig. 20–5). To start the next cycle of four reactions that lengthens the chain by two more carbons, another malonyl group is linked to the now unoccupied phosphopantetheine —SH group of ACP (Fig. 20–6). Condensation occurs as the butyryl group, acting exactly as did the acetyl group in the first cycle, is linked to two carbons of the malonyl-ACP group with concurrent loss of CO_2. The product of this condensation is a six-carbon acyl group, covalently bound to the phosphopantetheine —SH group. Its β-keto group is reduced in the next three steps of the synthase cycle to yield the six-carbon saturated acyl group, exactly as in the first round of reactions.

Seven cycles of condensation and reduction produce the 16-carbon saturated palmitoyl group, still bound to ACP. For reasons not well understood, chain elongation generally stops at this point, and free palmitate is released from the ACP molecule by the action of a hydrolytic activity in the synthase complex. Small amounts of longer fatty acids such as stearate (18:0) are also formed. In certain plants (coconut and palm, for example) chain termination occurs earlier; up to 90% of the fatty acids in the oils of these plants are between 8 and 14 carbons long.

The overall reaction for the synthesis of palmitate from acetyl-CoA can be broken down into two parts. First, the formation of seven malonyl-CoA molecules:

$$7 \text{ Acetyl-CoA} + 7CO_2 + 7ATP \longrightarrow 7 \text{ malonyl-CoA} + 7ADP + 7P_i \quad (20\text{–}1)$$

then seven cycles of condensation and reduction:

$$\text{Acetyl-CoA} + 7 \text{ malonyl-CoA} + 14NADPH + 14H^+ \longrightarrow$$
$$\text{palmitate} + 7CO_2 + 8CoA + 14NADP^+ + 6H_2O \quad (20\text{–}2)$$

The overall process (the sum of Eqns 20–1 and 20–2) is

$$8 \text{ Acetyl-CoA} + 7ATP + 14NADPH + 14H^+ \longrightarrow$$
$$\text{palmitate} + 8CoA + 6H_2O + 7ADP + 7P_i + 14NADP^+ \quad (20\text{–}3)$$

The biosynthesis of fatty acids such as palmitate thus requires acetyl-CoA and the input of chemical energy in two forms: the group transfer potential of ATP and the reducing power of NADPH. The ATP is required to attach CO_2 to acetyl-CoA to make malonyl-CoA; the NADPH is required to reduce the double bonds. We shall return to the sources of acetyl-CoA and NADPH soon, but let us first consider the structure of the remarkable enzyme complex that catalyzes the synthesis of fatty acids.

The Fatty Acid Synthase of Some Organisms Is Composed of Multifunctional Proteins

We noted earlier that the seven active sites for fatty acid synthesis (six enzymes and ACP) reside in seven separate polypeptides in the fatty acid synthase of *E. coli;* the same is true of the enzyme complex from higher plants (Fig. 20–7). In these complexes each enzyme is positioned with its active site near that of the preceding and succeeding enzymes of the sequence. The flexible pantetheine arm of ACP can reach all of the active sites, and it carries the growing fatty acyl chain from one site to the next; the intermediates are not released from the enzyme complex until the finished product is obtained. As in the cases we have encountered in earlier chapters, this channeling of intermediates from one active site to the next increases the efficiency of the overall process.

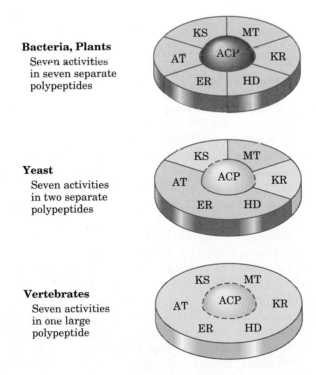

Bacteria, Plants
Seven activities in seven separate polypeptides

Yeast
Seven activities in two separate polypeptides

Vertebrates
Seven activities in one large polypeptide

Figure 20–7 The fatty acid synthase from bacteria and plants is a complex of seven different polypeptides. In yeast all seven activities reside in only two polypeptides, and in vertebrates, in a single large polypeptide.

The fatty acid synthases of yeast and of vertebrates are also multienzyme complexes, but their integration is even more complete than in *E. coli* and plants. In yeast, the seven distinct active sites reside in only two large, multifunctional polypeptides, and in vertebrates, a single large polypeptide (M_r 240,000) contains all seven enzymatic activities as well as a hydrolytic activity that cleaves the fatty acid from the ACP-like part of the enzyme complex. The active form of this multifunctional protein is a dimer (M_r 480,000).

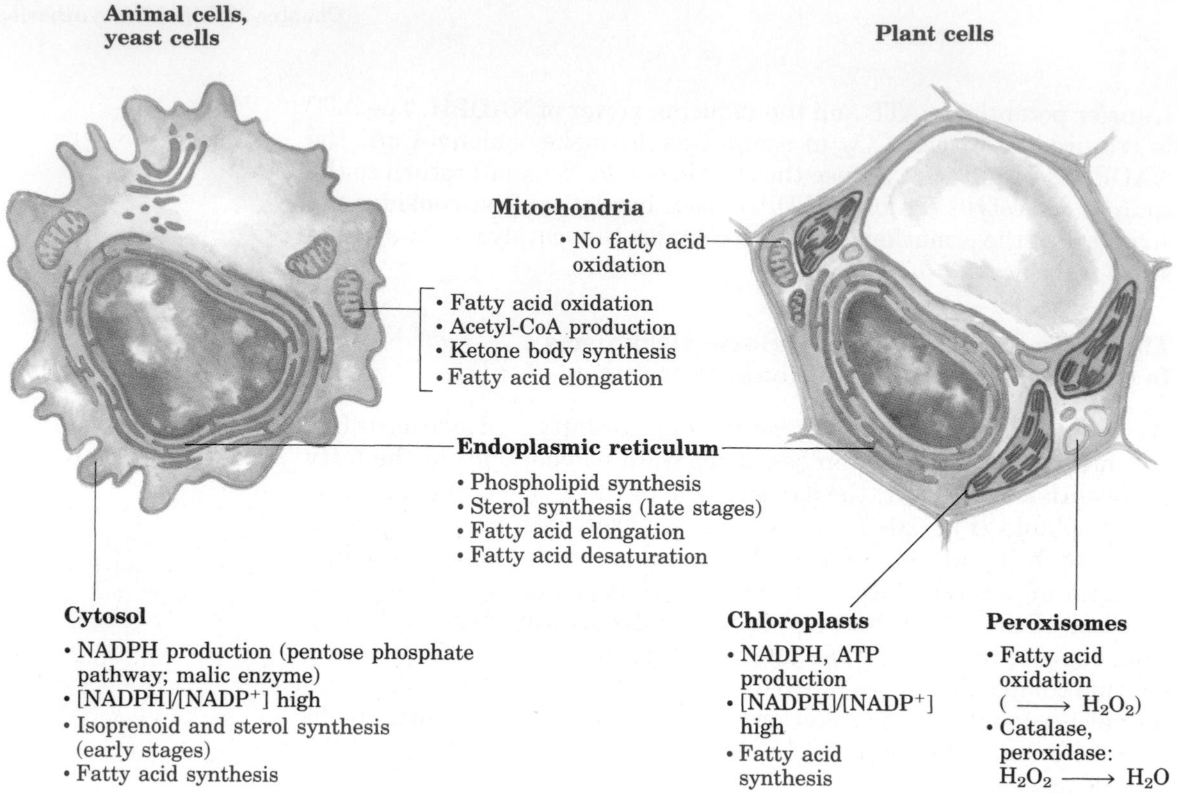

Animal cells, yeast cells

Plant cells

Mitochondria
- No fatty acid oxidation

- Fatty acid oxidation
- Acetyl-CoA production
- Ketone body synthesis
- Fatty acid elongation

Endoplasmic reticulum
- Phospholipid synthesis
- Sterol synthesis (late stages)
- Fatty acid elongation
- Fatty acid desaturation

Cytosol
- NADPH production (pentose phosphate pathway; malic enzyme)
- [NADPH]/[NADP$^+$] high
- Isoprenoid and sterol synthesis (early stages)
- Fatty acid synthesis

Chloroplasts
- NADPH, ATP production
- [NADPH]/[NADP$^+$] high
- Fatty acid synthesis

Peroxisomes
- Fatty acid oxidation ($\longrightarrow H_2O_2$)
- Catalase, peroxidase: $H_2O_2 \longrightarrow H_2O$

Figure 20–8 Subcellular localization of lipid metabolism in yeast and in vertebrate animal cells differs from that in higher plants. Fatty acid synthesis takes place in the compartment in which NADPH is available for reductive synthesis (i.e., where the [NADPH]/[NADP$^+$] ratio is high). Processes in red are covered in this chapter.

Fatty Acid Synthesis Occurs in the Cytosol of Many Organisms but in the Chloroplasts of Plants

In mammals, the fatty acid synthase complex is found exclusively in the cytosol (Fig. 20–8), as are the biosynthetic enzymes for nucleotides, amino acids, and glucose. This location segregates synthetic processes from degradative reactions, many of which take place in the mitochondrial matrix. There is a corresponding segregation of electron-carrying cofactors for anabolism (generally a reductive process) and those for catabolism (generally oxidative). Usually, NADPH is the electron carrier for anabolic reactions, and NAD$^+$ serves in catabolic reactions. In hepatocytes, the ratio [NADPH]/[NADP$^+$] is very high (about 75) in the cytosol, furnishing a strongly reducing environment for the reductive synthesis of fatty acids and other biomolecules. Because the cytosolic [NADH]/[NAD$^+$] ratio is much smaller (only about 8×10^{-4}), the NAD$^+$-dependent oxidative catabolism of glucose can occur in the same compartment, at the same time, as fatty acid synthesis. The [NADH]/[NAD$^+$] ratio within the mitochondrion is much higher than in the cytosol because of the flow of electrons into NAD$^+$ from the oxidation of fatty acids, amino acids, pyruvate, and acetyl-CoA. This high [NADH]/[NAD$^+$] ratio favors the reduction of oxygen via the respiratory chain.

In adipocytes cytosolic NADPH is largely generated by **malic enzyme** (Fig. 20–9). (We encountered an NAD-linked malic enzyme in the carbon fixation pathway of C$_4$ plants (see Fig. 19–37); this enzyme is unrelated in function.) The pyruvate produced in this reaction reenters the mitochondrion. In hepatocytes and in the mammary gland of lactating animals, the NADPH required for fatty acid biosynthesis is supplied primarily by the reactions of the pentose phosphate pathway (Chapter 14).

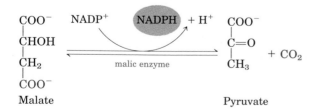

Figure 20–9 Production of NADPH by the malic enzyme.

In the photosynthetic cells of plants, fatty acid synthesis occurs not in the cytosol, but in the chloroplast stroma (Fig. 20–8). This location makes sense when we recall that NADPH is produced in chloroplasts by the light reactions of photosynthesis (Fig. 20–10). Again, the resulting high [NADPH]/[NADP$^+$] ratio provides the reducing environment that favors reductive anabolic processes such as fatty acid synthesis.

$$H_2O + NADP^+ \xrightarrow{\text{light}} \tfrac{1}{2}O_2 + NADPH + H^+$$

Figure 20–10 Production of NADPH by photosynthesis.

Acetate Is Shuttled out of Mitochondria as Citrate

In nonphotosynthetic eukaryotes, nearly all the acetyl-CoA used in fatty acid synthesis is formed in mitochondria from pyruvate oxidation and from the catabolism of the carbon skeletons of amino acids. Acetyl-CoA arising from the oxidation of fatty acids does not represent a significant source of acetyl-CoA for fatty acid biosynthesis in animals because the two pathways are regulated reciprocally, as described below. Because the mitochondrial inner membrane is impermeable to acetyl-CoA, an indirect shuttle transfers acetyl group equivalents across the inner membrane (Fig. 20–11). Intramitochondrial acetyl-CoA first re-

Figure 20–11 The acetyl group shuttle for transfer of acetyl groups from mitochondria to the cytosol for fatty acid synthesis. (The outer mitochondrial membrane is freely permeable to all of these compounds.) Acetyl groups pass out of the mitochondrion as citrate; in the cytosol they are delivered as acetyl-CoA for fatty acid synthesis. Malate returns to the mitochondrial matrix, where it is converted to oxaloacetate. An alternative fate for cytosolic malate is oxidation by malic enzyme to generate cytosolic NADPH; the pyruvate produced returns to the mitochondrial matrix.

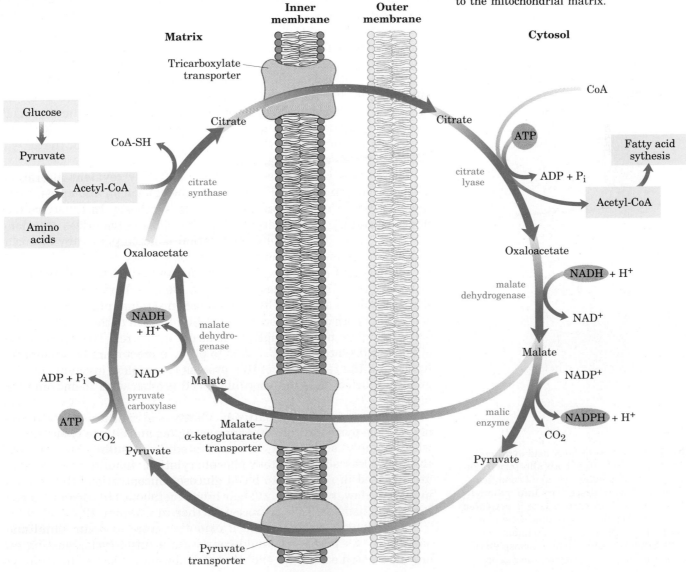

acts with oxaloacetate to form citrate, in the citric acid cycle reaction catalyzed by **citrate synthase** (see Fig. 15–7). Citrate then passes into the cytosol through the mitochondrial inner membrane on the **tricarboxylate transporter.** In the cytosol, citrate cleavage by **citrate lyase** regenerates acetyl-CoA; this reaction is driven by the investment of energy from ATP. Oxaloacetate cannot return to the matrix directly; there is no transporter for it. Instead, oxaloacetate is reduced by cytosolic malate dehydrogenase to malate, which returns to the mitochondrial matrix on the malate–α-ketoglutarate transporter in exchange for citrate, and is reoxidized to oxaloacetate to complete the shuttle. Alternatively, the malate produced in the cytosol is used to generate cytosolic NADPH through the activity of malic enzyme, as described above.

Plants have another means of acquiring acetyl-CoA for fatty acid synthesis. They produce acetyl-CoA from pyruvate using a stromal isozyme of **pyruvate dehydrogenase** (see Fig. 15–2).

Fatty Acid Biosynthesis Is Tightly Regulated

When a cell or organism has more than enough metabolic fuel available to meet its energetic needs, the excess is generally converted to fatty acids and stored as lipids such as triacylglycerols. The reaction catalyzed by acetyl-CoA carboxylase is the rate-limiting step in the biosynthesis of fatty acids, and this enzyme is an important site of regulation. In vertebrates, palmitoyl-CoA, the principal product of fatty acid synthesis, acts as a feedback inhibitor of the enzyme, and citrate is an allosteric activator (Fig. 20–12a). When there is an increase in the concentrations of mitochondrial acetyl-CoA and of ATP, citrate is transported out of the mitochondria and becomes both the precursor of cytosolic acetyl-CoA and an allosteric signal for the activation of acetyl-CoA carboxylase.

Acetyl-CoA carboxylase is also regulated by covalent alteration. Phosphorylation triggered by the hormones glucagon and epinephrine inactivates it, thereby slowing fatty acid synthesis. In its active (dephosphorylated) form, acetyl-CoA carboxylase polymerizes into long filaments (Fig. 20–12b); phosphorylation is accompanied by dissociation into monomeric subunits and loss of activity.

The acetyl-CoA carboxylase from plants and bacteria is not regulated by citrate or by a phosphorylation–dephosphorylation cycle. The plant enzyme is activated by an increase in stromal pH and Mg^{2+} concentration, both of which occur upon illumination of the plant (p. 630). Bacteria do not use triacylglycerols as energy stores. The primary role of fatty acid synthesis in *E. coli* is to provide precursors for membrane lipids, and the regulation of this process is complex, involving certain guanine nucleotides that coordinate cell growth with membrane formation.

Other enzymes in the pathway of fatty acid synthesis are also regulated. The pyruvate dehydrogenase complex and citrate lyase, both of which supply acetyl-CoA, are activated by insulin (Fig. 20–12a) through a cascade of protein phosphorylation. Insulin and glucagon are released in response to blood glucose concentrations that are too high or too low, respectively. Their broad metabolic effects and molecular mechanisms will be discussed further in Chapter 22.

If fatty acid synthesis and β oxidation were to occur simultaneously, the two processes would constitute a futile cycle, wasting energy. We noted earlier (p. 496) that β oxidation is blocked by malonyl-

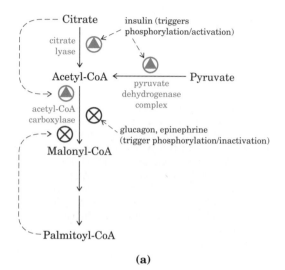

(a)

(b)

Figure 20–12 Regulation of fatty acid synthesis. **(a)** In the cells of vertebrates, both allosteric regulation and hormone-dependent covalent modification influence the flow of precursors into malonyl-CoA. In plants, acetyl-CoA carboxylase is activated by the changes in $[Mg^{2+}]$ and pH that accompany illumination (not shown here). **(b)** Filaments of acetyl-CoA carboxylase (the active, dephosphorylated form) as seen with the electron microscope.

CoA, which inhibits carnitine acyltransferase I. Thus during fatty acid synthesis, the production of the first intermediate, malonyl-CoA, shuts down β oxidation at the level of a transport system in the mitochondrial inner membrane. This control mechanism illustrates another advantage to a cell of segregating synthetic and degradative pathways in different cellular compartments.

Long-Chain Fatty Acids Are Synthesized from Palmitate

Palmitate, the principal product of the fatty acid synthase system in animal cells, is the precursor of other long-chain fatty acids (Fig. 20–13). It may be lengthened to form stearate (18:0) or even longer saturated fatty acids by further additions of acetyl groups, through the action of **fatty acid elongation systems** present in the smooth endoplasmic reticulum and the mitochondria. The more active elongation system of the endoplasmic reticulum extends the 16-carbon chain of palmitoyl-CoA by two carbons, forming stearoyl-CoA. Although different enzyme systems are involved, and coenzyme A rather than ACP is the acyl carrier directly involved in the reaction, the mechanism of elongation is otherwise identical with that employed in palmitate synthesis: donation of two carbons by malonyl-ACP, followed by reduction, dehydration, and reduction to the saturated 18-carbon product, stearoyl-CoA.

Some Fatty Acids Are Desaturated

Palmitate and stearate serve as precursors of the two most common monounsaturated fatty acids of animal tissues: palmitoleate, $16:1(\Delta^9)$, and oleate, $18:1(\Delta^9)$ (Fig. 20–13). Each of these fatty acids has a single cis double bond in the Δ^9 position (between C-9 and C-10). The double bond is introduced into the fatty acid chain by an oxidative reaction catalyzed by **fatty acyl–CoA desaturase** (Fig. 20–14). This enzyme

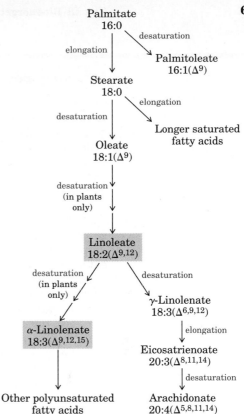

Figure 20–13 Routes of synthesis of other fatty acids. Palmitate is the precursor of stearate and longer-chain saturated fatty acids, as well as the monounsaturated acids, palmitoleate and oleate. Mammals cannot convert oleate into linoleate or α-linolenate (shaded red), which are therefore required in the diet as essential fatty acids. Conversion of linoleate into other polyunsaturated fatty acids and eicosanoids is outlined. Unsaturated fatty acids are symbolized by indicating the number of carbons and the number and position of the double bonds, as in Table 9–1.

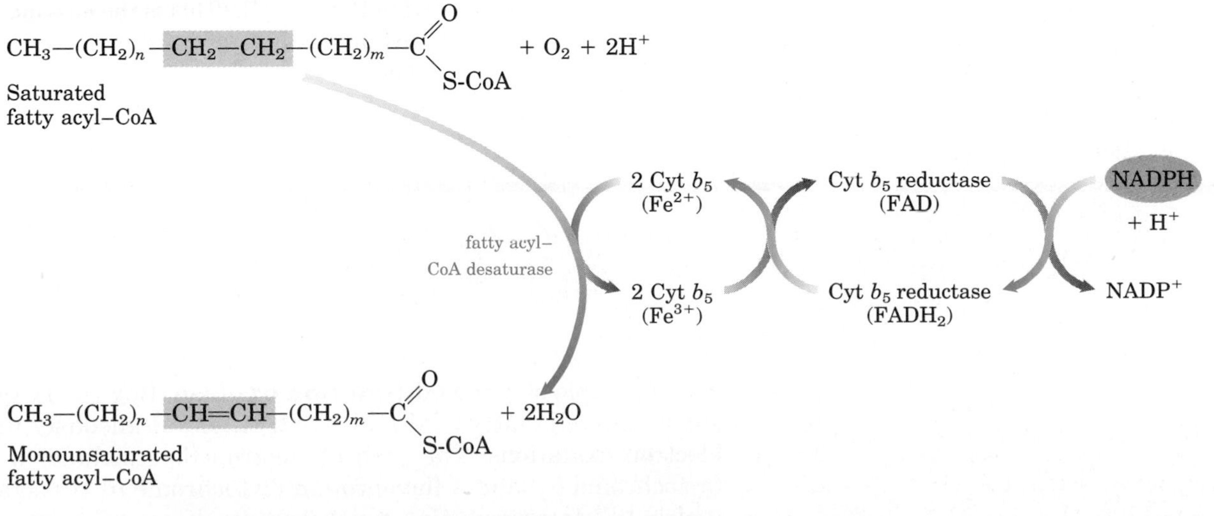

Figure 20–14 The pathway of electron transfer (blue arrows) in the desaturation of fatty acids by a mixed-function oxidase in vertebrate animals. Two different substrates—a fatty acyl–CoA and NADPH—undergo oxidation by molecular oxygen. These reactions occur on the lumenal face of the smooth endoplasmic reticulum. A similar pathway, but with different electron carriers, occurs in plants.

BOX 20–1   Mixed-Function Oxidases, Oxygenases, and Cytochrome P-450

In this chapter we encounter several enzymes that carry out oxidation–reduction reactions in which molecular oxygen is a participant. The reaction that introduces a double bond into a fatty acyl chain (Fig. 20–14) is such a reaction.

The nomenclature for enzymes that catalyze reactions of this general type is often confusing to students, as is the mechanism. **Oxidase** is the general name for enzymes that catalyze oxidations in which molecular oxygen is the electron acceptor but oxygen atoms do not appear in the oxidized product (but there is an exception to this "rule," as we shall see!). The enzyme that creates a double bond in fatty acyl–CoA during the oxidation of fatty acids in peroxisomes (see Fig. 16–13) is an oxidase of this type, as is the cytochrome oxidase of the mitochondrial electron transfer chain (see Fig. 18–11). In the first case, the transfer of two electrons to H_2O produces hydrogen peroxide, H_2O_2; in the second, two electrons reduce $\frac{1}{2}O_2$ to H_2O. Many, but not all, oxidases are flavoproteins.

Oxygenases catalyze oxidative reactions in which oxygen atoms *are* directly incorporated into the substrate molecule, forming a new hydroxyl or carboxyl group, for example. **Dioxygenases** catalyze reactions in which both of the oxygen atoms of O_2 are incorporated into the organic substrate molecule. An example of a dioxygenase is tryptophan 2,3-dioxygenase, which catalyzes the opening of the five-membered ring of tryptophan in the catabolism of this amino acid:

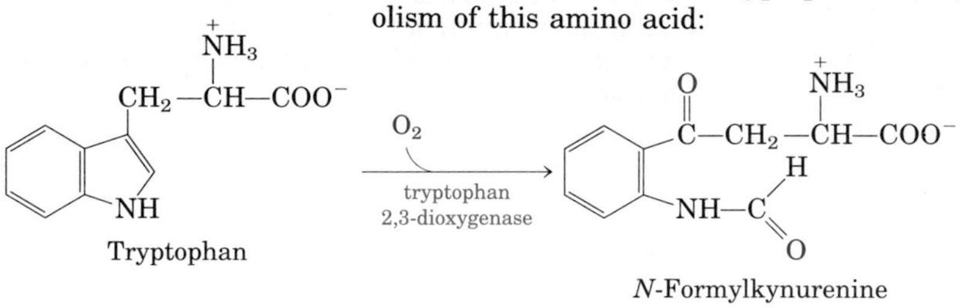

When this reaction occurs in the presence of $^{18}O_2$, the isotopic oxygen atoms are found in the two carbonyl groups of the product (shown in red).

Monooxygenases, which are more abundant and more complex in their action, catalyze reactions in which only one of the two oxygen atoms of O_2 is incorporated into the organic substrate, the other being reduced to H_2O. Monooxygenases require two substrates to serve as reductants of the two oxygen atoms of O_2. The main substrate accepts one of the two oxygen atoms, and a cosubstrate furnishes hydrogen atoms to reduce the other oxygen atom to H_2O. The general reaction equation for monooxygenases is

$$AH + BH_2 + O\!-\!O \longrightarrow A\!-\!OH + B + H_2O$$

where AH is the main substrate and BH_2 the cosubstrate. Because most monooxygenases catalyze reactions in which the main substrate becomes hydroxylated, they are also called **hydroxylases.** They are also sometimes called **mixed-function oxidases** or **mixed-function oxygenases,** to indicate that they oxidize two different substrates simultaneously. (Note here the use of "oxidase"—a deviation from the general meaning of this term.)

There are different classes of monooxygenases, depending upon the nature of the cosubstrate. Some use reduced flavin nucleotides ($FMNH_2$ or $FADH_2$), others use NADH or NADPH, and still others use α-ketoglutarate as the cosubstrate. The enzyme that hydroxylates the phenyl ring of phenylalanine to give tyrosine is a monooxygenase for which tetrahydrobiopterin serves as cosubstrate (see Fig. 17–27). (This is the enzyme that is defective in the human genetic disease phenylketonuria.)

Tryptophan

N-Formylkynurenine

is an example of a **mixed-function oxidase** (Box 20–1). Two different substrates, the fatty acid and NADPH, simultaneously undergo two-electron oxidations. The path of electron flow includes a cytochrome (cytochrome b_5) and a flavoprotein (cytochrome b_5 reductase), both of which, like fatty acyl–CoA desaturase itself, are present in the smooth endoplasmic reticulum.

Mammalian hepatocytes can readily introduce double bonds at the Δ^9 position of fatty acids but cannot introduce additional double bonds in the fatty acid chain between C-10 and the methyl-terminal end.

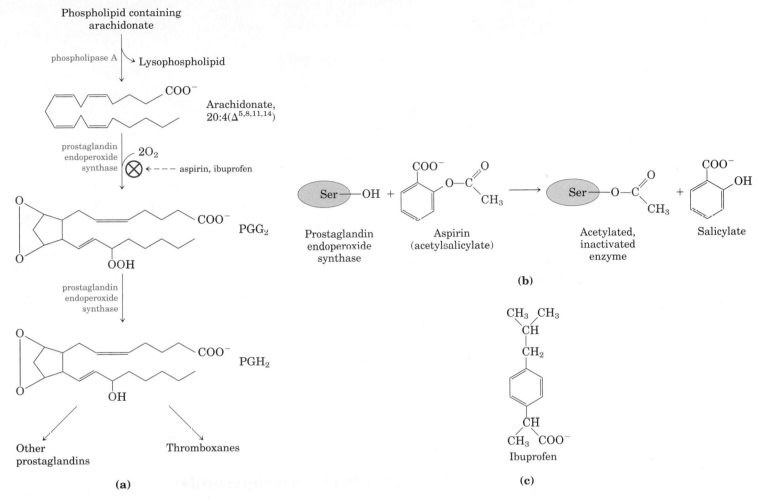

(a)

(b)

(c)

Figure 20–16 The "cyclic" pathway from arachidonate to prostaglandins and thromboxanes. **(a)** After release of arachidonate, prostaglandin endoperoxide synthase catalyzes the first two reactions, producing PGH_2, the precursor of other prostaglandins and thromboxanes. **(b)** Aspirin inhibits prostaglandin endoperoxide synthase by acetylating an essential Ser residue on the enzyme. Ibuprofen **(c)** also inhibits this step, probably by mimicking the structure of the substrate or an intermediate in the reaction.

Thromboxane synthase present in blood platelets (thrombocytes) converts PGH_2 into thromboxane A_2, from which other **thromboxanes** are derived (Fig. 20–16a). Thromboxanes induce blood vessel constriction and platelet aggregation, early steps in blood clotting. Low doses of aspirin, taken regularly, are believed to reduce the probability of heart attacks and strokes by reducing thromboxane production.

Thromboxanes, like prostaglandins, contain a ring of five or six atoms, and the pathway that leads from arachidonate to these two classes of compounds is sometimes called the "cyclic" pathway, to distinguish it from the "linear" pathway that leads from arachidonate to the **leukotrienes,** which are linear (Fig. 20–17, p. 658). Leukotriene synthesis begins with the action of several lipoxygenases that catalyze the incorporation of molecular oxygen into arachidonate. These enzymes, found in leukocytes and in heart, brain, lung, and spleen, are mixed-function oxidases that use cytochrome P-450 (Box 20–1). The various leukotrienes differ in the position of the peroxide that is introduced by these lipoxygenases. This linear pathway from arachidonate, unlike the cyclic pathway, is not inhibited by aspirin or the other nonsteroidal antiinflammatory drugs.

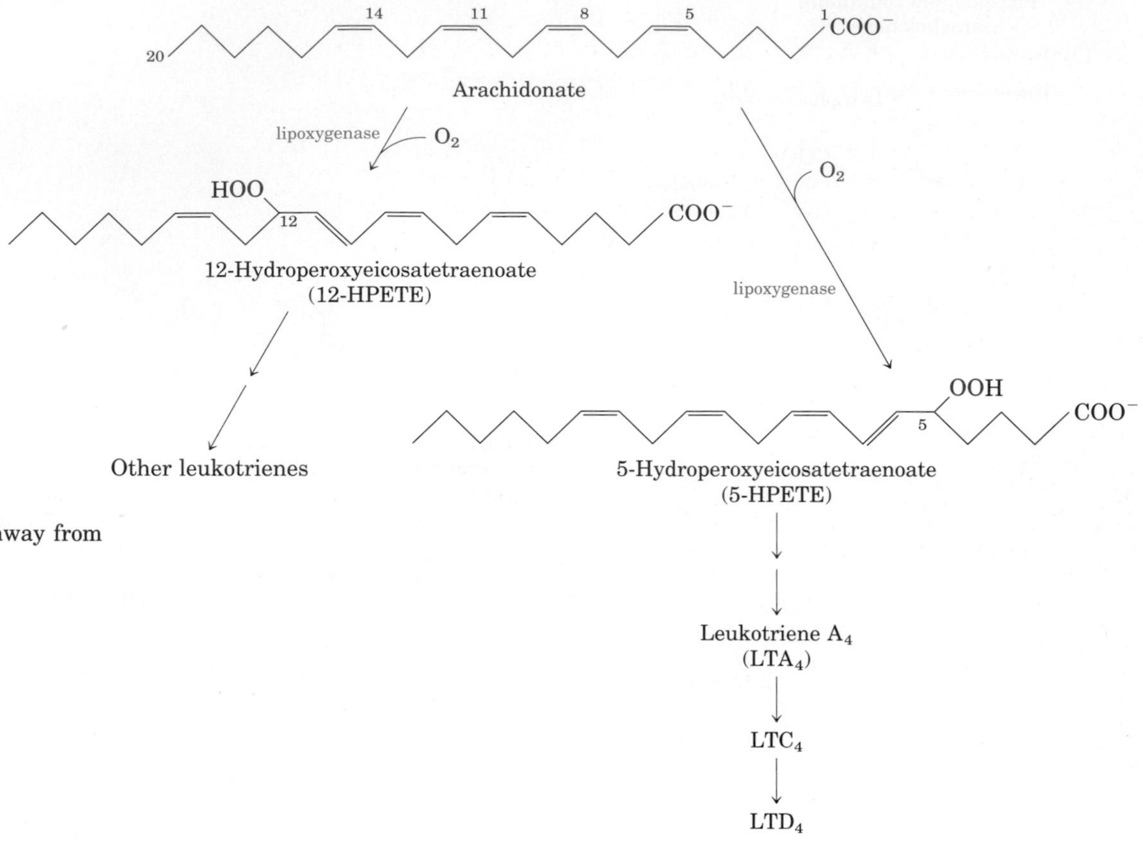

Figure 20–17 The "linear" pathway from arachidonate to leukotrienes.

Biosynthesis of Triacylglycerols

Most of the fatty acids synthesized or ingested by an organism have one of two fates: incorporation into triacylglycerols for the storage of metabolic energy or incorporation into the phospholipid components of membranes. The partitioning between these alternative fates depends on the requirements of the organism. During rapid growth, the synthesis of new membranes requires membrane phospholipid synthesis; organisms that have a plentiful supply of food but are not actively growing shunt most of their fatty acids into storage fats. The pathways to storage fats and several classes of membrane phospholipids begin at the same point: the formation of fatty acyl esters of glycerol. First we discuss the route to triacylglycerols and its regulation.

Triacylglycerols and Glycerophospholipids Are Synthesized from Common Precursors

Animals can synthesize and store large quantities of triacylglycerols, to be used later as fuel (see Box 16–1). In humans only a few hundred grams of glycogen can be stored in the liver and muscles, barely enough to supply the body's energy needs for 12 hours. In contrast, the total amount of stored triacylglycerol in a 70 kg man of average build is about 15 kg, enough to supply his basal energy needs for as long as 12 weeks (see Table 22–5). Whenever carbohydrate is ingested in excess of the capacity to store glycogen, it is converted into triacylglycerols and stored in adipose tissue. Plants also manufacture triacylglycerols as an energy-rich fuel, stored especially in fruits, nuts, and seeds.

Triacylglycerols and glycerophospholipids such as phosphatidylethanolamine share two precursors (fatty acyl–CoAs and glycerol-3-phosphate) and several enzymatic steps in their biosynthesis in animal tissues. Glycerol-3-phosphate can be formed in two ways (Fig. 20–18). It can arise from dihydroxyacetone phosphate generated during glycolysis by the action of the cytosolic NAD-linked **glycerol-3-phosphate dehydrogenase,** and in liver and kidney it is also formed from glycerol by the action of **glycerol kinase.** The other precursors of triacylglycerols are fatty acyl–CoAs, formed from fatty acids by **acyl-CoA synthetases** (Fig. 20–18), the same enzymes responsible for the activation of fatty acids for β oxidation (Chapter 16).

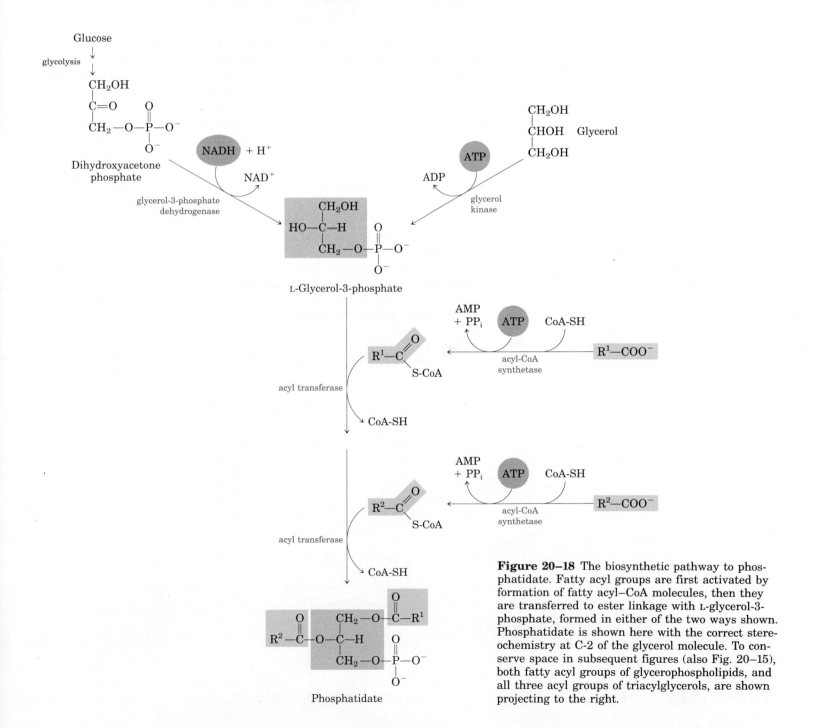

Figure 20–18 The biosynthetic pathway to phosphatidate. Fatty acyl groups are first activated by formation of fatty acyl–CoA molecules, then they are transferred to ester linkage with L-glycerol-3-phosphate, formed in either of the two ways shown. Phosphatidate is shown here with the correct stereochemistry at C-2 of the glycerol molecule. To conserve space in subsequent figures (also Fig. 20–15), both fatty acyl groups of glycerophospholipids, and all three acyl groups of triacylglycerols, are shown projecting to the right.

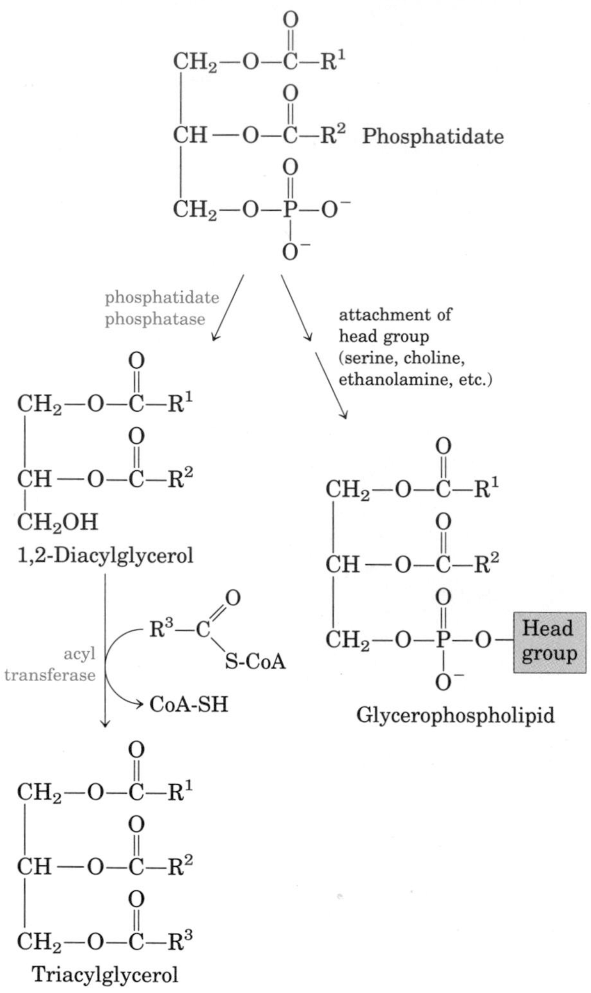

The first stage in the biosynthesis of triacylglycerols is the acylation of the two free hydroxyl groups of glycerol-3-phosphate by two molecules of fatty acyl–CoA to yield **diacylglycerol-3-phosphate,** more commonly called **phosphatidate** (Fig. 20–18). Phosphatidate occurs in only trace amounts in cells, but is a central intermediate in lipid biosynthesis; it can be converted either to a triacylglycerol or to a glycerophospholipid. In the pathway to triacylglycerols, phosphatidate is hydrolyzed by **phosphatidate phosphatase** to form a 1,2-diacylglycerol (Fig. 20–19). Diacylglycerols are then converted into triacylglycerols by transesterification with a third fatty acyl–CoA.

Triacylglycerol Biosynthesis in Animals Is Regulated by Hormones

In humans, the amount of body fat stays relatively constant over long periods, although there may be minor short-term changes as the caloric intake fluctuates. However, if carbohydrate, fat, or protein is consumed in amounts exceeding energy needs, the excess is stored in the form of triacylglycerols. The fat stored in this way can be drawn upon for energy and enables the body to withstand periods of fasting.

The biosynthesis and degradation of triacylglycerols are regulated reciprocally, with the favored path depending upon the metabolic resources and requirements of the moment. The rate of triacylglycerol biosynthesis is profoundly altered by the action of several hormones. Insulin, for example, promotes the conversion of carbohydrate into triacylglycerols (Fig. 20–20). People with severe diabetes mellitus, due to failure of insulin secretion or action, not only are unable to use glucose properly but also fail to synthesize fatty acids from carbohydrates or amino acids. They show increased rates of fat oxidation and ketone body formation (Chapter 16). As a consequence they lose weight. Triacylglycerol metabolism is also influenced by glucagon (Chapter 22), and by pituitary growth hormone and adrenal cortical hormones.

Figure 20–19 Phosphatidate is the precursor of both triacylglycerols and glycerophospholipids. The mechanisms for head group attachment in phospholipid synthesis are described later.

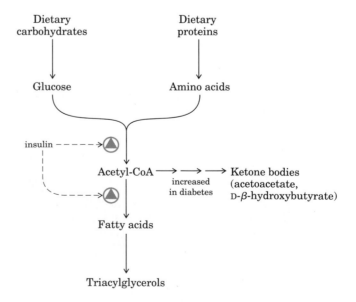

Figure 20–20 Insulin stimulates conversion of dietary carbohydrates and proteins into fat. In individuals with untreated diabetes mellitus, acetyl-CoA from catabolism of carbohydrates and proteins is instead shunted to ketone body production, because of the lack of insulin.

Biosynthesis of Membrane Phospholipids

In Chapter 9 we introduced two major classes of membrane phospholipids: glycerophospholipids and sphingolipids. We noted that many different phospholipid species can be constructed by combining various fatty acids and polar head groups with the glycerol or sphingosine backbones (see Figs. 9–7, 9–9). Although the number of different end products of phospholipid biosynthesis is very large, all of these diverse products are synthesized according to a few basic patterns. We will describe the biosynthesis of selected membrane lipids to illustrate these patterns. In general, the assembly of phospholipids from simple precursors requires (1) synthesis of the backbone molecule (glycerol or sphingosine); (2) attachment of fatty acid(s) to the backbone, in ester or amide linkage; (3) addition of a hydrophilic head group, joined to the backbone through a phosphodiester linkage; and in some cases, (4) alteration or exchange of the head group to yield the final phospholipid product.

   In eukaryotic cells, phospholipid synthesis occurs primarily at the surface of the smooth endoplasmic reticulum. Some newly synthesized phospholipids remain in that membrane, but most are destined for other cellular locations. The process by which water-insoluble phospholipids move from the site of their synthesis to the point of their eventual function is not fully understood, but we will conclude by discussing some mechanisms that have emerged in recent years.

There Are Two Strategies for Attaching Head Groups

The first steps of glycerophospholipid synthesis are shared with the pathway to triacylglycerols (Fig. 20–19): two fatty acyl groups are esterified to C-1 and C-2 of L-glycerol-3-phosphate to form phosphatidate. Commonly but not invariably, the fatty acid at C-1 is saturated and that at C-2 is unsaturated. A second route to phosphatidate is the phosphorylation of a diacylglycerol by a specific kinase.

   The polar head group of glycerophospholipids is attached through a phosphodiester bond, in which each of two alcoholic hydroxyls (one on the polar head group and one on C-3 of glycerol) forms an ester with phosphoric acid (Fig. 20–21). In the biosynthetic process, one of the hydroxyls is first activated by attachment of a nucleotide, cytidine diphosphate (CDP). Cytidine monophosphate (CMP) is then displaced in a nucleophilic attack by the other hydroxyl (Fig. 20–22, p. 662). The CDP is attached either to the diacylglycerol, forming in effect an activated phosphatidate, **CDP-diacylglycerol** (strategy 1), or to the hydroxyl of the head group (strategy 2). The central importance of cytidine nucleotides in lipid biosynthesis was discovered by Eugene P. Kennedy in the early 1960s.

Eugene P. Kennedy

Figure 20–21 The phospholipid head group is attached to a diacylglycerol by a phosphodiester bond, formed when phosphoric acid condenses with two alcohols, eliminating two molecules of H_2O.

Figure 20–22 Two general strategies for forming the phosphodiester bond of phospholipids. In both cases CDP supplies the phosphate group of the phosphodiester bond.

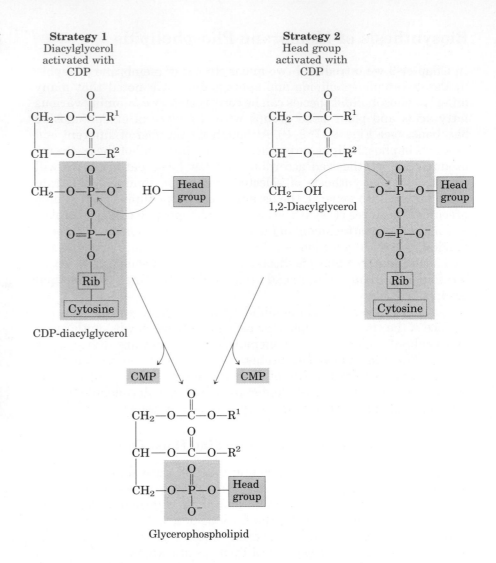

Strategy 1
Diacylglycerol activated with CDP

CDP-diacylglycerol

Strategy 2
Head group activated with CDP

1,2-Diacylglycerol

CMP   CMP

Glycerophospholipid

Phospholipid Synthesis in *E. coli* Employs CDP-Diacylglycerol

The first strategy for head group attachment is illustrated by the synthesis of phosphatidylserine, phosphatidylethanolamine, and phosphatidylglycerol in *E. coli*. The diacylglycerol is activated by condensation of phosphatidate with CTP to form CDP-diacylglycerol, with the elimination of pyrophosphate (Fig. 20–23). Displacement of CMP through nucleophilic attack by the hydroxyl group of serine or by the C-1 hydroxyl of glycerol-3-phosphate yields **phosphatidylserine** or phosphatidylglycerol-3-phosphate, respectively. The latter is processed further by cleavage of the phosphate monoester (with release of P_i) to yield **phosphatidylglycerol.**

Phosphatidylserine and phosphatidylglycerol can both serve as precursors of other membrane lipids in bacteria (Fig. 20–23). Decarboxylation of the serine moiety in phosphatidylserine by phosphatidylserine decarboxylase yields **phosphatidylethanolamine.** In *E. coli,* condensation of two molecules of phosphatidylglycerol, with the elimination of one glycerol, yields **cardiolipin,** in which two diacylglycerols are joined through a common head group.

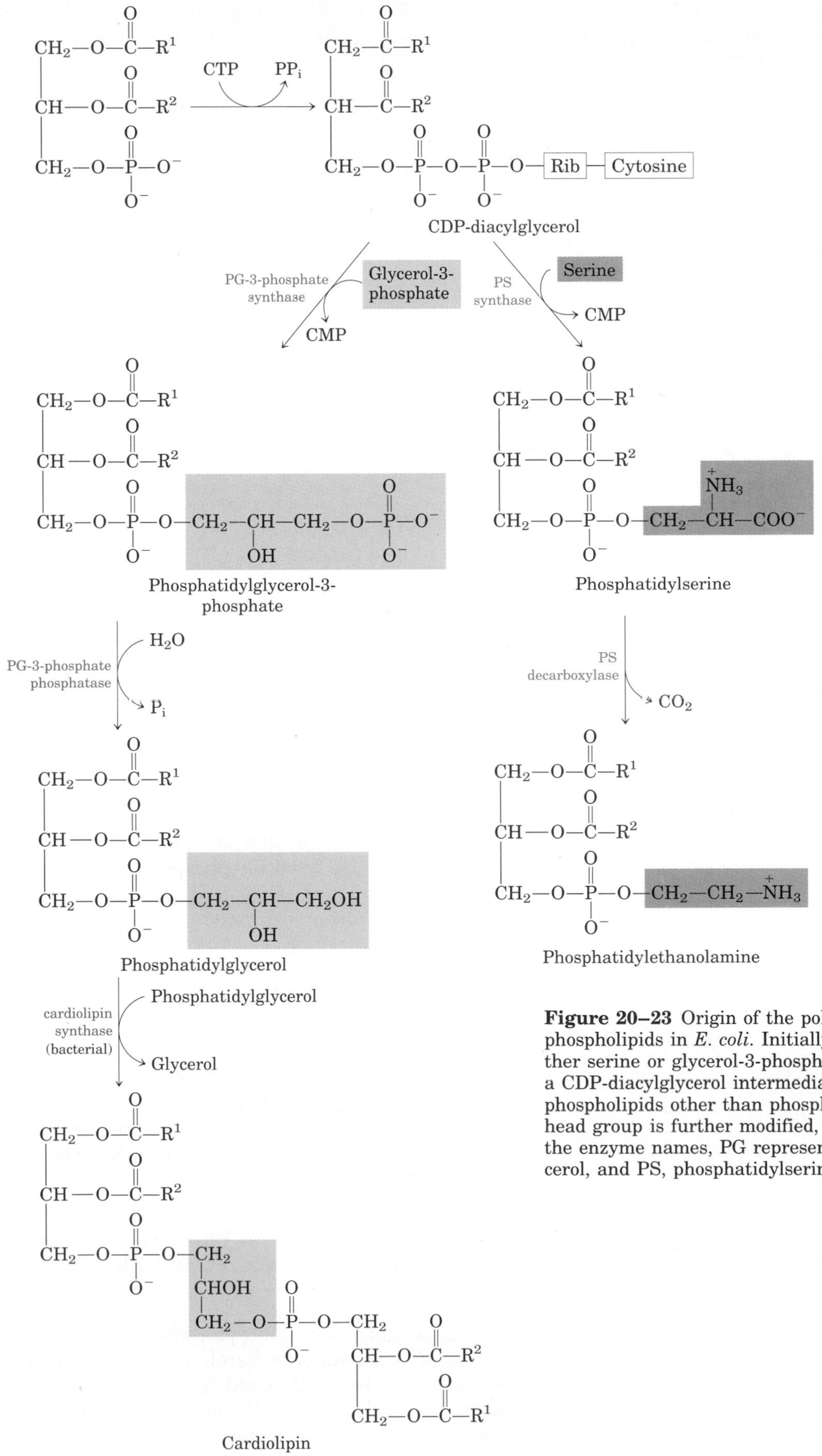

Figure 20–23 Origin of the polar head groups of phospholipids in *E. coli*. Initially, a head group (either serine or glycerol-3-phosphate) is attached via a CDP-diacylglycerol intermediate (strategy 1). For phospholipids other than phosphatidylserine, the head group is further modified, as shown here. In the enzyme names, PG represents phosphatidylglycerol, and PS, phosphatidylserine.

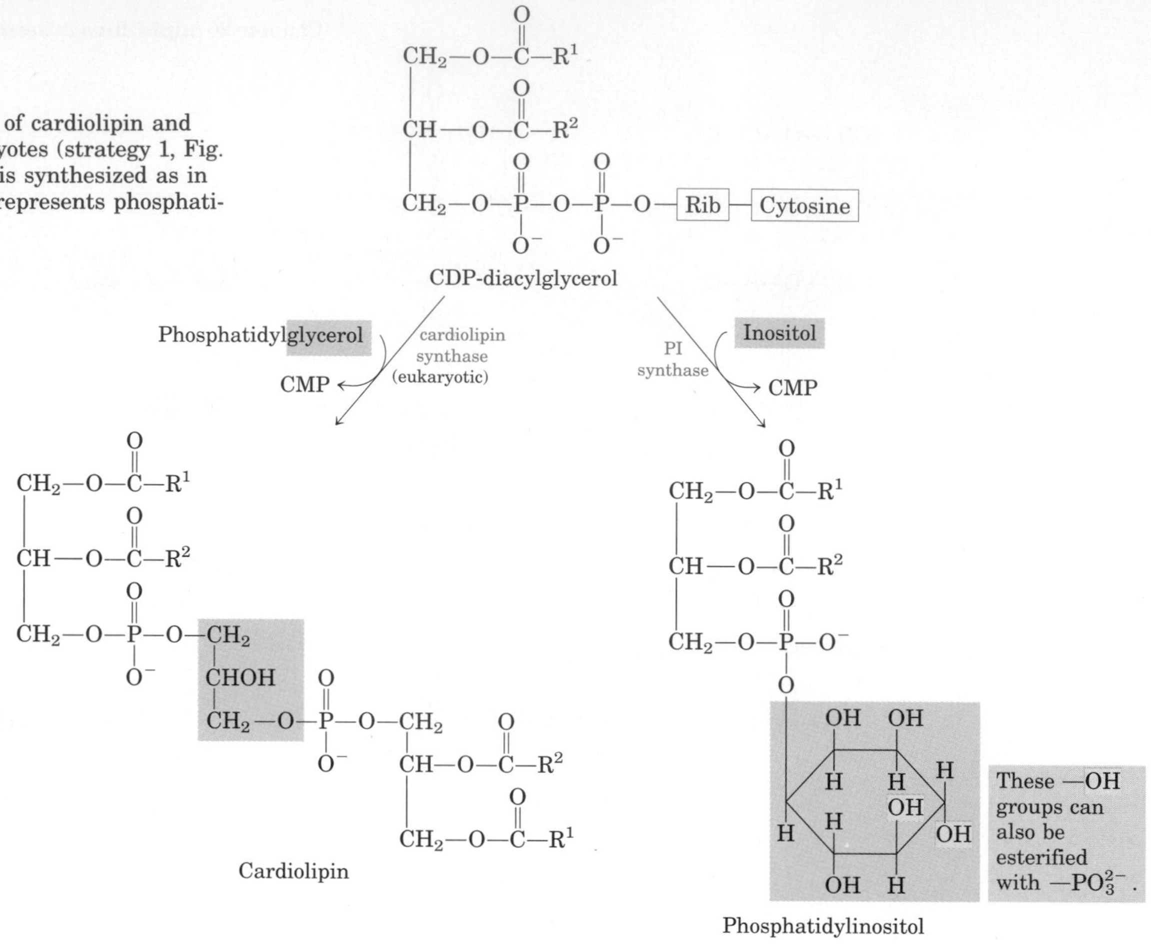

Figure 20–24 The synthesis of cardiolipin and phosphatidylinositol in eukaryotes (strategy 1, Fig. 20–22). Phosphatidylglycerol is synthesized as in bacteria (see Fig. 20–23). PI represents phosphatidylinositol.

Eukaryotes Synthesize Acidic Phospholipids from CDP-Diacylglycerol

In eukaryotes, phosphatidylglycerol, cardiolipin, and the phosphatidylinositols (all acidic phospholipids; see Fig. 9–7) are synthesized by the same strategy used for phospholipid synthesis in bacteria. Phosphatidylglycerol is made exactly as in bacteria. Cardiolipin synthesis in eukaryotes differs slightly: phosphatidylglycerol condenses with CDP-diacylglycerol (Fig. 20–24), not another molecule of phosphatidylglycerol as in *E. coli* (Fig. 20–23).

Phosphatidylinositol is synthesized by condensation of CDP-diacylglycerol with inositol (Fig. 20–24). Specific **phosphatidylinositol kinases** then convert phosphatidylinositol into its phosphorylated derivatives (see Fig. 9–16). Phosphatidylinositol and its phosphorylated products in the plasma membrane play a central role in signal transduction in eukaryotes, as we noted in Chapter 9. These signal transduction mechanisms will be described in more detail in Chapter 22.

Eukaryotic Pathways to Phosphatidylserine, Phosphatidylethanolamine, and Phosphatidylcholine Are Interrelated

In yeast as in bacteria, phosphatidylserine can be produced by condensation of CDP-diacylglycerol and serine, and phosphatidylethanolamine can be synthesized from phosphatidylserine in the reaction catalyzed by phosphatidylserine decarboxylase (Fig. 20–25). An alternative route to phosphatidylserine is a head group exchange reaction,

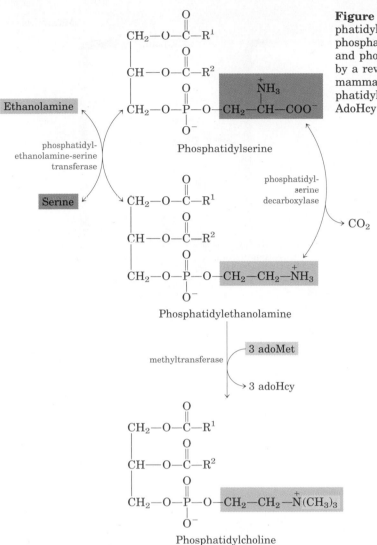

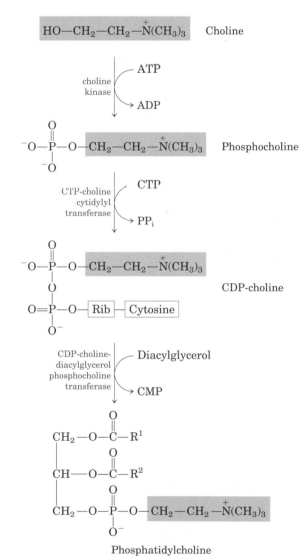

Figure 20–25 The "salvage" pathway from phosphatidylserine to phosphatidylethanolamine and phosphatidylcholine in yeast. Phosphatidylserine and phosphatidylethanolamine are interconverted by a reversible head group exchange reaction. In mammals, phosphatidylserine is derived from phosphatidylethanolamine by a reversal of this reaction. AdoHcy represents *S*-adenosylhomocysteine.

in which free serine displaces ethanolamine. Phosphatidylethanolamine may also be converted to **phosphatidylcholine** (lecithin) by the addition of three methyl groups to its amino group. All three methylation reactions are catalyzed by a single enzyme (a methyltransferase) with *S*-adenosylmethionine as the methyl group donor (see Fig. 17–20).

In mammals, phosphatidylserine is not synthesized from CDP-diacylglycerol; instead, it is derived from phosphatidylethanolamine via the head group exchange reaction shown in Figure 20–25. In mammals, synthesis of all nitrogen-containing phospholipids occurs by strategy 2 of Figure 20–22: phosphorylation and activation of the head group followed by condensation with diacylglycerol. For example, choline is reused ("salvaged") by being phosphorylated then converted into CDP-choline by condensation with CTP. A diacylglycerol displaces CMP from CDP-choline, producing phosphatidylcholine (Fig. 20–26).

Figure 20–26 The pathway for phosphatidylcholine synthesis from choline in mammals (strategy 2, Fig. 20–22). The same strategy is used for salvaging ethanolamine in phosphatidylethanolamine synthesis.

Figure 20–27 Summary of the pathways to phosphatidylcholine and phosphatidylethanolamine. Note that the conversion of phosphatidylethanolamine to phosphatidylcholine in mammals occurs only in the liver.

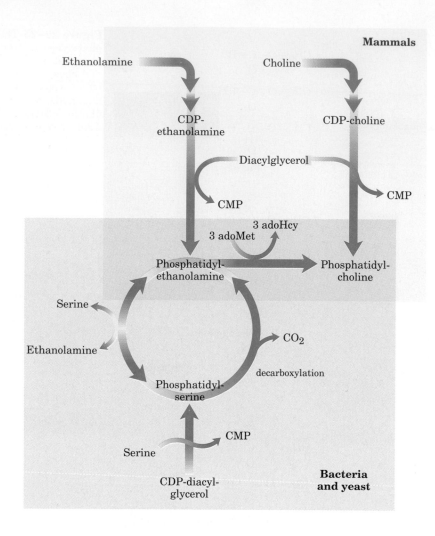

An analogous salvage pathway converts ethanolamine obtained in the diet into phosphatidylethanolamine. In the liver, phosphatidylcholine is also produced by methylation of phosphatidylethanolamine using S-adenosylmethionine, as described above. In all other tissues, however, phosphatidylcholine is produced only by condensation of diacylglycerol and CDP-choline. The pathways to phosphatidylcholine and phosphatidylethanolamine in various organisms are summarized in Figure 20–27.

Plasmalogen Synthesis Requires Formation of an Ether-Linked Fatty Alcohol

The biosynthetic pathway to ether lipids, including **plasmalogens** and the **platelet-activating factor** (see Fig. 9–8), involves the displacement of an esterified fatty acyl group by a long-chain alcohol to form the ether linkage (Fig. 20–28). Head group attachment follows, by mechanisms essentially like those for the common ester-linked phospholipids. Finally, the characteristic double bond of plasmalogens is introduced by the action of a mixed-function oxidase similar to that responsible for desaturation of fatty acids (Fig. 20–14).

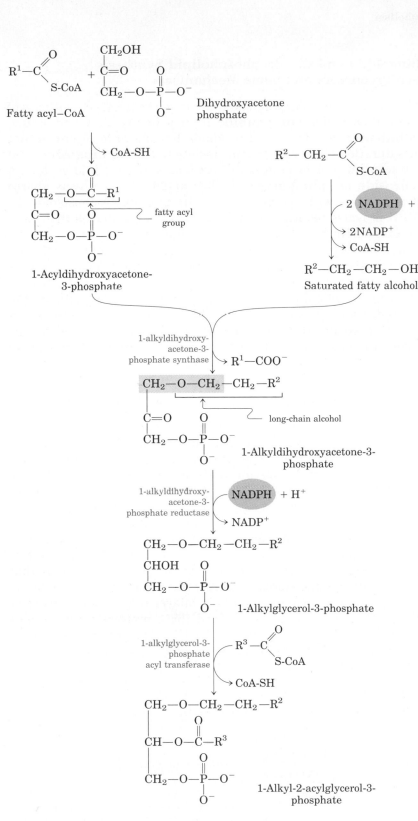

Figure 20–28 Synthesis of the ether linkage (shaded in red) in ether lipids, and the formation of plasmalogens. The intermediate 1-alkyl-2-acylglycerol-3-phosphate is the ether analog of phosphatidate. Mechanisms for attaching head groups to ether lipids are essentially the same as for their ester-linked analogs. The characteristic double bond of plasmalogens (shaded in blue) is introduced in a final step by a mixed-function oxidase system similar to that shown in Fig. 20–14.

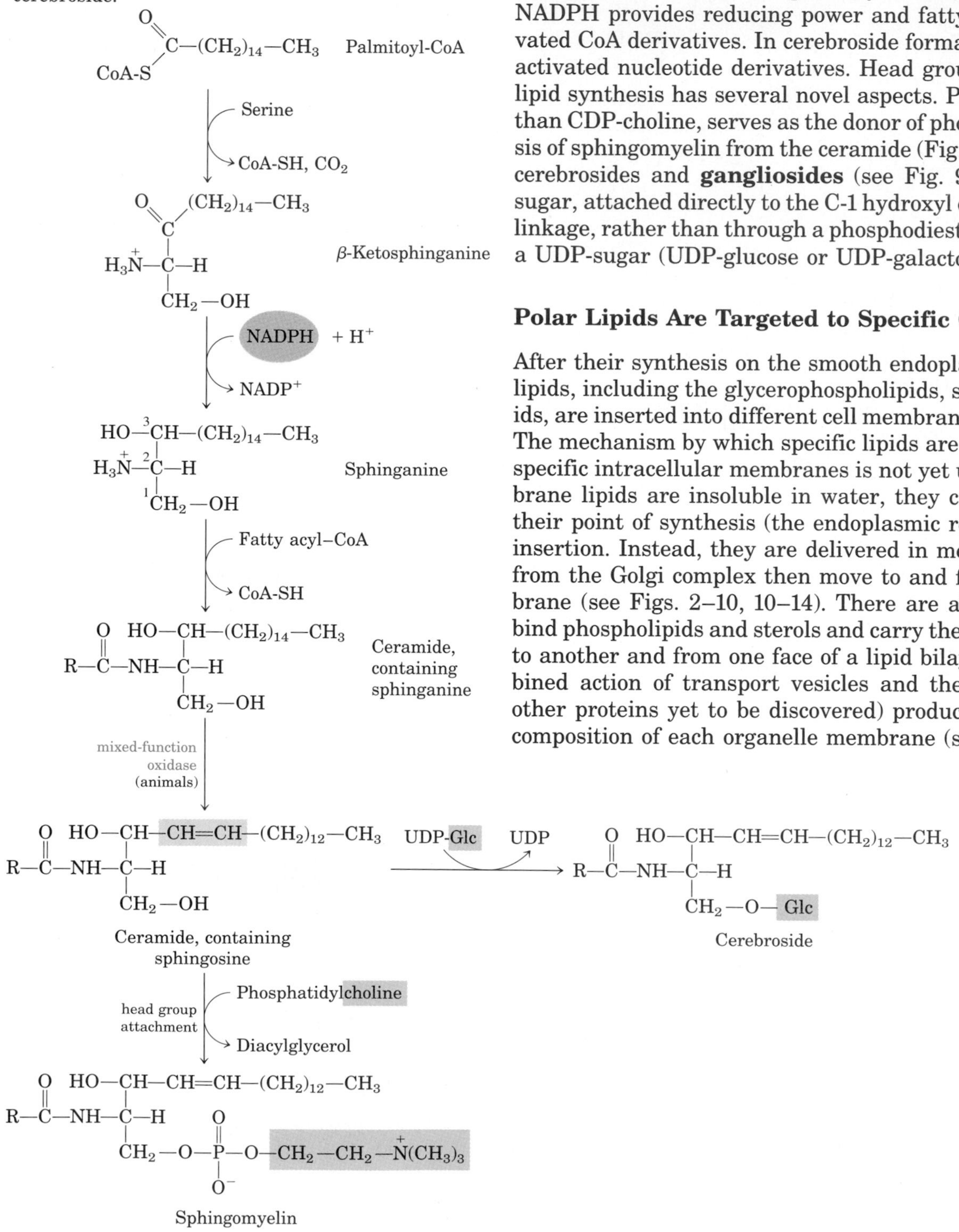

Sphingolipid and Glycerophospholipid Synthesis Share Precursors and Some Mechanisms

The biosynthesis of sphingolipids occurs in four stages: (1) synthesis of the 18-carbon amine **sphinganine** from palmitoyl-CoA and serine; (2) attachment of a fatty acid in amide linkage to form **ceramide;** (3) desaturation of the sphinganine moiety to form **sphingosine;** and (4) attachment of a head group to produce a sphingolipid such as a **cerebroside** or **sphingomyelin** (Fig. 20–29). The pathway shares several features with the pathways leading to glycerophospholipids: NADPH provides reducing power and fatty acids enter as their activated CoA derivatives. In cerebroside formation, sugars enter as their activated nucleotide derivatives. Head group attachment in sphingolipid synthesis has several novel aspects. Phosphatidylcholine, rather than CDP-choline, serves as the donor of phosphocholine in the synthesis of sphingomyelin from the ceramide (Fig. 20–29). In glycolipids, the cerebrosides and **gangliosides** (see Fig. 9–9), the head group is a sugar, attached directly to the C-1 hydroxyl of sphingosine in glycosidic linkage, rather than through a phosphodiester bond; the sugar donor is a UDP-sugar (UDP-glucose or UDP-galactose).

Polar Lipids Are Targeted to Specific Cell Membranes

After their synthesis on the smooth endoplasmic reticulum, the polar lipids, including the glycerophospholipids, sphingolipids, and glycolipids, are inserted into different cell membranes in different proportions. The mechanism by which specific lipids are targeted for insertion into specific intracellular membranes is not yet understood. Because membrane lipids are insoluble in water, they cannot simply diffuse from their point of synthesis (the endoplasmic reticulum) to their point of insertion. Instead, they are delivered in membrane vesicles that bud from the Golgi complex then move to and fuse with the target membrane (see Figs. 2–10, 10–14). There are also cytosolic proteins that bind phospholipids and sterols and carry them from one cell membrane to another and from one face of a lipid bilayer to the other. The combined action of transport vesicles and these proteins (and perhaps other proteins yet to be discovered) produces the characteristic lipid composition of each organelle membrane (see Table 10–2).

Figure 20–29 Biosynthesis of sphingolipids. First, the condensation of palmitate and serine yields sphinganine, which is then acylated to form a ceramide. In animals, a double bond (shaded in red) is then created by a mixed-function oxidase, before the final addition of a head group: phosphatidylcholine, to form sphingomyelin; or glucose, to form a cerebroside.

Biosynthesis of Cholesterol, Steroids, and Isoprenoids

Cholesterol is doubtless the most publicized lipid in nature, because of the strong correlation between high levels of cholesterol in the blood and the incidence of diseases of the cardiovascular system in humans. Less well-advertised is the critical role of cholesterol in the structure of many membranes and as a precursor of steroid hormones and bile acids. Cholesterol is an essential molecule in many animals, including humans. It is not required in the mammalian diet because the liver can synthesize it from simple precursors.

Although the structure of this 27-carbon compound suggests complexity in its biosynthesis, all of its carbon atoms are provided by a single precursor—acetate (Fig. 20–30). The biosynthetic pathway to cholesterol is instructive in several respects. The study of this pathway has led to an understanding of the transport of cholesterol and other lipids between organs, of the process by which cholesterol enters cells (receptor-mediated endocytosis), of the means by which intracellular cholesterol production is influenced by dietary cholesterol, and of how failure to regulate cholesterol production affects health. Finally, the **isoprene** units that are key intermediates in the pathway from acetate to cholesterol are precursors to many other natural lipids, and the mechanisms by which isoprene units are polymerized are similar in all of these pathways.

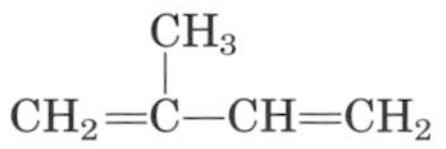

We begin with an account of the major steps in the biosynthesis of cholesterol from acetate, then discuss the transport of cholesterol in the blood, its uptake by cells, and the regulation of cholesterol synthesis in normal individuals and in those with defects in cholesterol uptake or transport. We also consider other cellular components derived from cholesterol, such as bile acids and steroid hormones. Finally, the biosynthetic pathways to some of the many compounds derived from isoprene units, which share early steps with the pathway to cholesterol, are outlined to illustrate the extraordinary versatility of isoprenoid condensations in biosynthesis.

$$CH_2{=}\underset{\underset{CH_3}{|}}{C}{-}CH{=}CH_2$$

Isoprene

Cholesterol Is Made from Acetyl-CoA in Four Stages

Cholesterol, like long-chain fatty acids, is made from acetyl-CoA, but the assembly plan is quite different in the two cases. In early experiments animals were fed acetate labeled with ^{14}C in either the methyl carbon or the carboxyl carbon. The pattern of labeling in the cholesterol isolated from the two groups of animals (Fig. 20–30) provided the blueprint for working out the enzymatic steps in cholesterol biosynthesis.

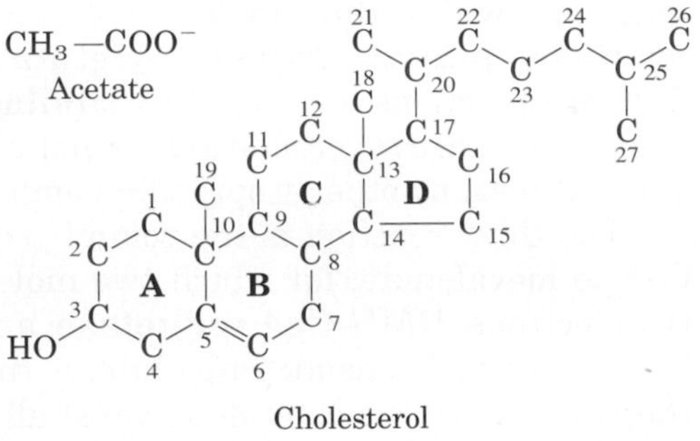

Figure 20–30 The origin of the carbon atoms of cholesterol, deduced from tracer experiments with acetate labeled in the methyl carbon (black) or the carboxyl carbon (red). The individual rings in the fused-ring system are designated A through D.

Figure 20–31 A summary of cholesterol biosynthesis, showing the four stages discussed in the text. The isoprene units in squalene are set off by red dashed lines.

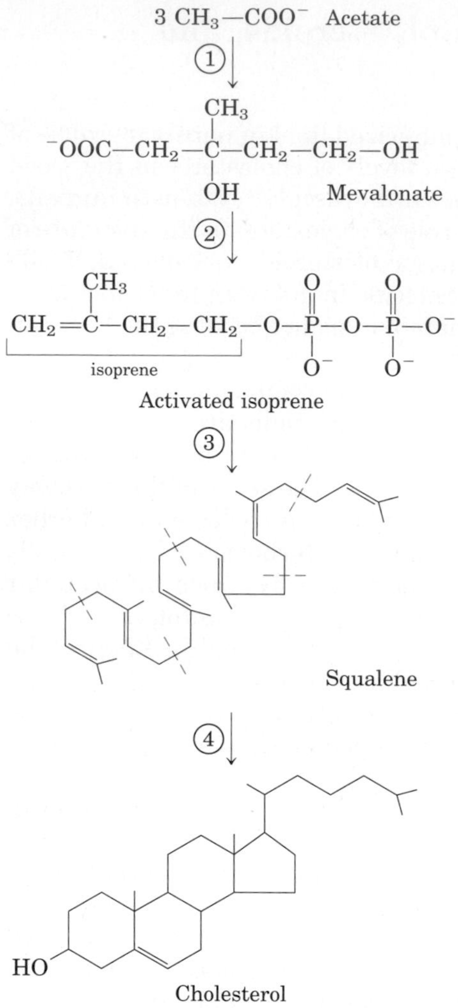

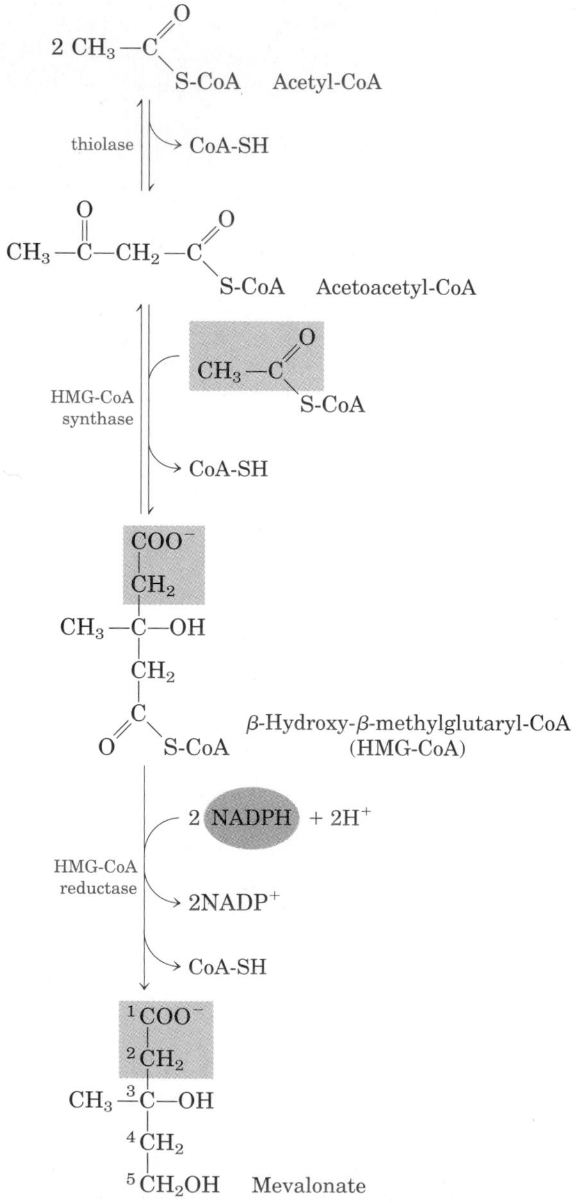

Figure 20–32 Formation of mevalonate from acetyl-CoA. The origin of C-1 and C-2 of mevalonate from acetyl-CoA is shown in red.

The process occurs in four stages (Fig. 20–31). In stage ① the three acetate units condense to form a six-carbon intermediate, mevalonate. Stage ② involves the conversion of mevalonate into activated isoprene units, and stage ③ the polymerization of six 5-carbon isoprene units to form the 30-carbon linear structure of squalene. Finally (stage ④), the cyclization of squalene forms the four rings of the steroid nucleus, and a further series of changes (oxidations, removal or migration of methyl groups) leads to the final product, cholesterol.

① **Synthesis of Mevalonate from Acetate** The first stage in cholesterol biosynthesis leads to the intermediate **mevalonate** (Fig. 20–32). Two molecules of acetyl-CoA condense, forming acetoacetyl-CoA, which condenses with a third molecule of acetyl-CoA to yield the six-carbon compound **β-hydroxy-β-methylglutaryl-CoA (HMG-CoA).** These first two reactions, catalyzed by **thiolase** and **HMG-CoA synthase,** respectively, are reversible and do not commit the cell to the synthesis of cholesterol or other isoprenoid compounds.

The third reaction is the committed step: the reduction of HMG-CoA to mevalonate, for which two molecules of NADPH each donate two electrons. **HMG-CoA reductase,** an integral membrane protein of the smooth endoplasmic reticulum, is the major point of regulation on the pathway to cholesterol, as we shall see.

② *Conversion of Mevalonate to Two Activated Isoprenes* In the next stage of cholesterol synthesis, three phosphate groups are transferred from three ATP molecules to mevalonate (Fig. 20–33). The phosphate attached to the C-3 hydroxyl group of mevalonate in the intermediate 3-phospho-5-pyrophosphomevalonate is a good leaving group; in the next step this phosphate and the nearby carboxyl group both leave, producing a double bond in the five-carbon product, **Δ³-isopentenyl pyrophosphate.** This is the first of the two activated isoprenes central to cholesterol formation. Isomerization of Δ³-isopentenyl pyrophosphate yields the second activated isoprene, **dimethylallyl pyrophosphate** (Fig. 20–33).

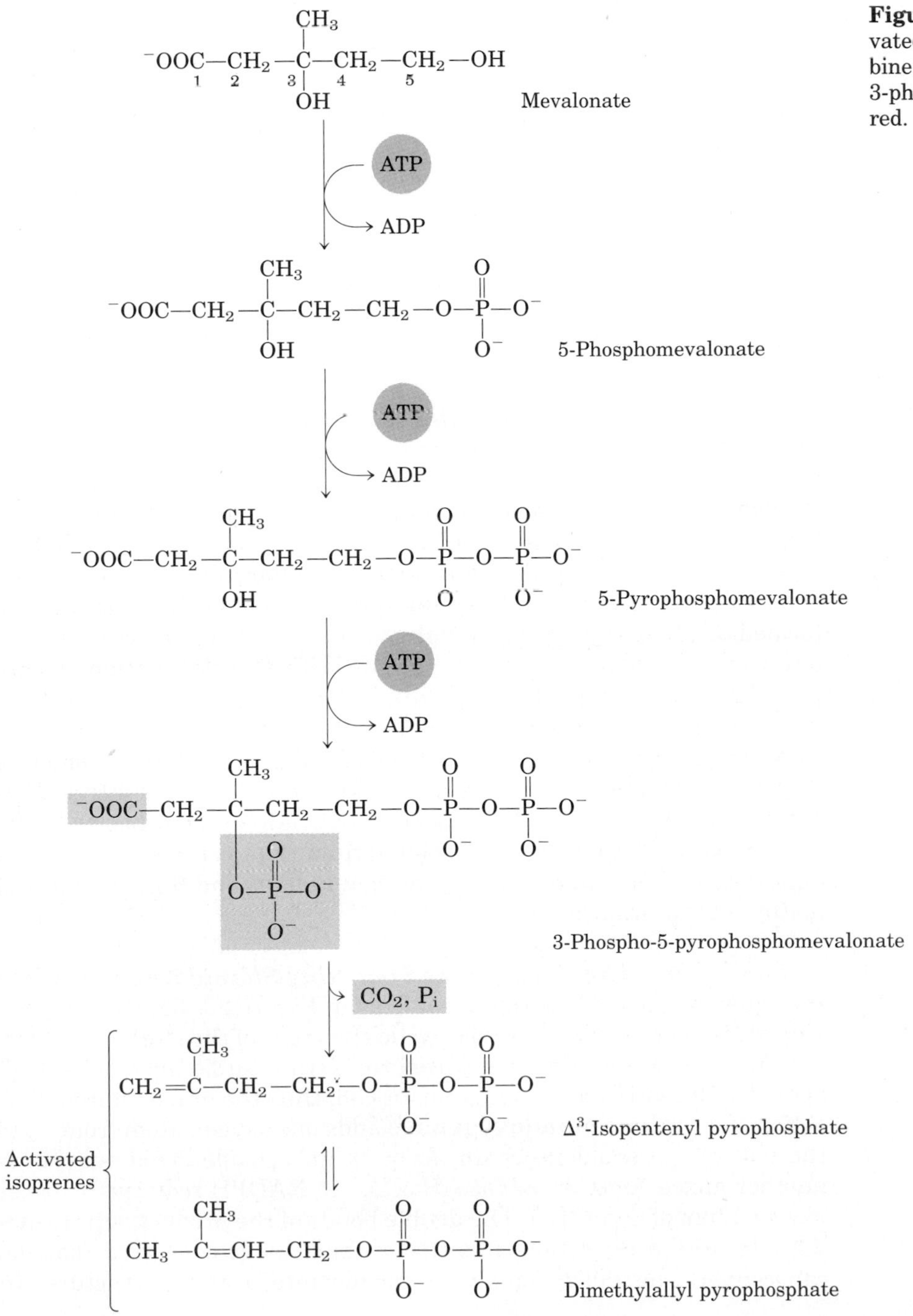

Figure 20–33 Conversion of mevalonate into activated isoprene units. Six of these units will combine to form squalene. The leaving groups of 3-phospho-5-pyrophosphomevalonate are shaded in red.

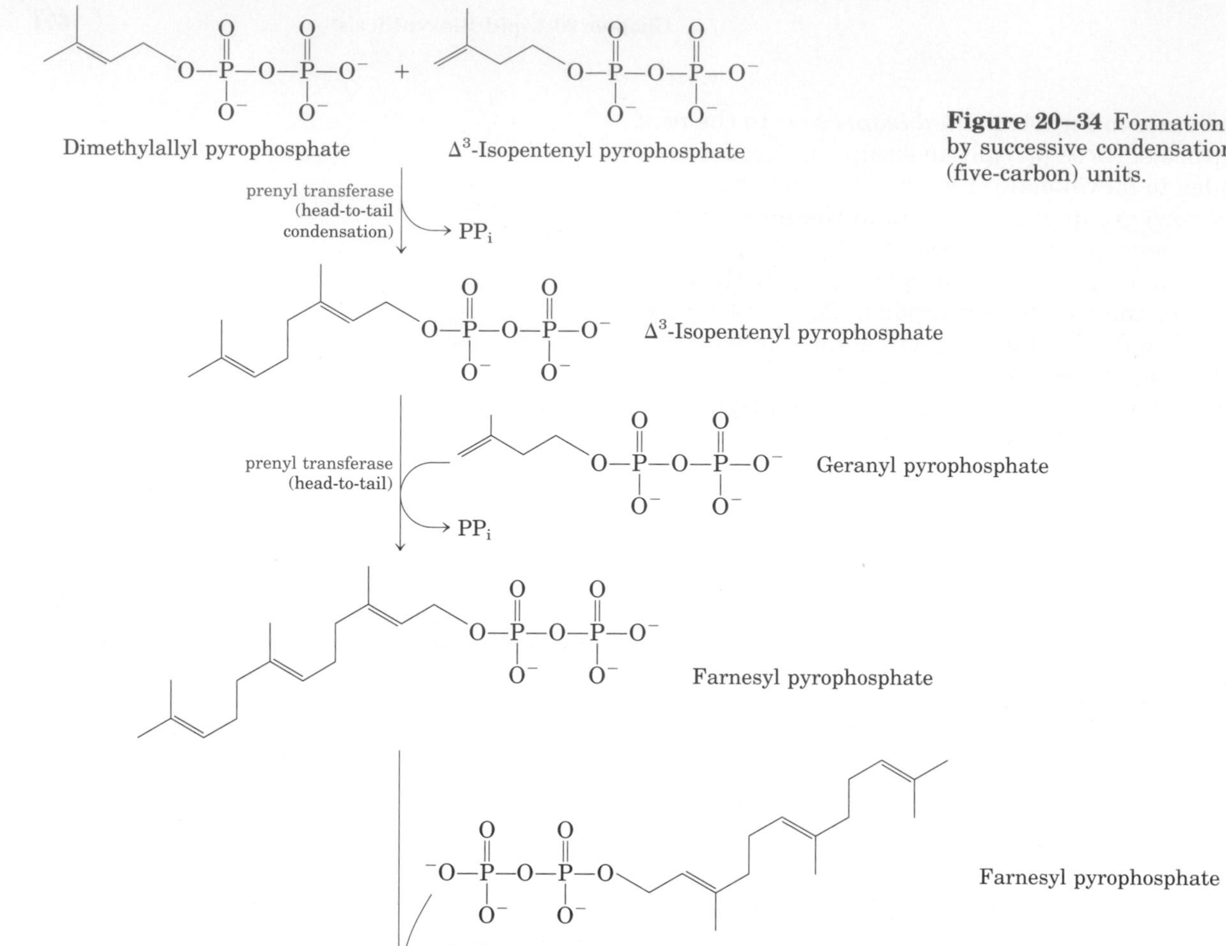

Dimethylallyl pyrophosphate

Δ^3-Isopentenyl pyrophosphate

Figure 20–34 Formation of squalene (30 carbons) by successive condensations of activated isoprene (five-carbon) units.

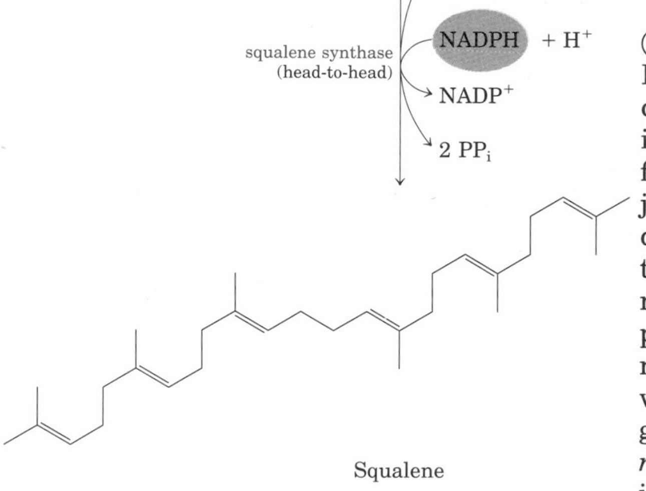

prenyl transferase (head-to-tail condensation) → PP$_i$

Δ^3-Isopentenyl pyrophosphate

prenyl transferase (head-to-tail) → PP$_i$

Geranyl pyrophosphate

Farnesyl pyrophosphate

Farnesyl pyrophosphate

squalene synthase (head-to-head)

NADPH + H$^+$

NADP$^+$

2 PP$_i$

Squalene

③ **Condensation of Six Activated Isoprene Units to Form Squalene** Isopentenyl pyrophosphate and dimethylallyl pyrophosphate now undergo a "head-to-tail" condensation in which one pyrophosphate group is displaced and a 10-carbon chain, **geranyl pyrophosphate,** is formed (Fig. 20–34). (The "head" is the end to which pyrophosphate is joined.) Geranyl pyrophosphate undergoes another head-to-tail condensation with isopentenyl pyrophosphate, yielding the 15-carbon intermediate **farnesyl pyrophosphate.** Finally, two molecules of farnesyl pyrophosphate join head to head, with the elimination of both pyrophosphate groups, forming **squalene** (Fig. 20–34). The common names of these compounds derive from the sources from which they were first isolated. Geraniol, a component of rose oil, has the smell of geraniums, and farnesol is a scent found in the flowers of a tree, *Farnese acacia*. Many natural scents of plant origin are synthesized from isoprene units. Squalene, first isolated from the liver of sharks (genus *Squalus*), has 30 carbons, 24 in the main chain and 6 in the form of methyl group branches.

④ **Conversion of Squalene to the Four-Ring Steroid Nucleus** When the squalene molecule is represented as in Figure 20–35, the relationship of its linear structure to the cyclic structure of the sterols is apparent. All of the sterols have four fused rings (the steroid nucleus) and all are alcohols, with a hydroxyl group at C-3; thus the name "sterol." The action of **squalene monooxygenase** adds one oxygen atom from O$_2$ to the end of the squalene chain, forming an epoxide. This enzyme is another mixed-function oxidase (Box 20–1); NADPH reduces the other oxygen atom of O$_2$ to H$_2$O. The double bonds of the product, **squalene-2,3-epoxide,** are positioned so that a remarkable concerted reaction can convert the linear squalene epoxide into a cyclic structure. In

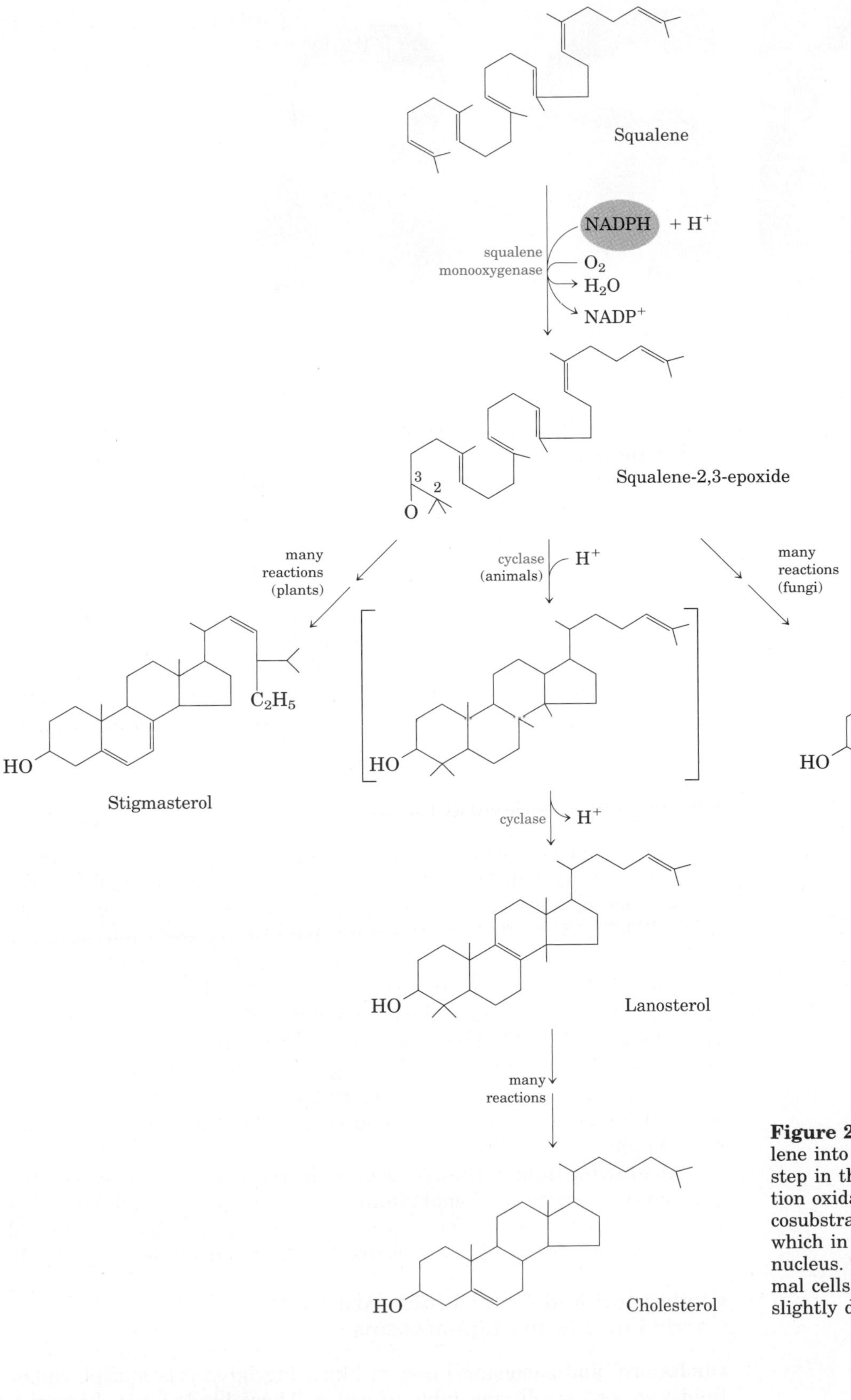

Figure 20–35 Ring closure converts linear squalene into the condensed steroid nucleus. The first step in this sequence is catalyzed by a mixed-function oxidase (a monooxygenase), for which the cosubstrate is NADPH. The product is an epoxide, which in the next step is cyclized to the steroid nucleus. The final product of these reactions in animal cells is cholesterol, but in other organisms, slightly different sterols are produced.

Konrad Bloch

Feodor Lynen
1911–1979

John Cornforth

George Popják

animal cells, this cyclization results in the formation of **lanosterol,** which contains the four rings characteristic of the steroid nucleus. Lanosterol is finally converted into cholesterol in a series of about 20 reactions, including the migration of some methyl groups and the removal of others. Elucidation of this extraordinary biosynthetic pathway, one of the most complex known, was accomplished by Konrad Bloch, Feodor Lynen, John Cornforth, and George Popják in the late 1950s.

Cholesterol is the sterol characteristic of animal cells, but plants, fungi, and protists make other, closely related sterols instead of cholesterol, using the same synthetic pathway as far as squalene-2,3-epoxide. At this point the synthetic pathways diverge slightly, yielding other sterols: stigmasterol in many plants and ergosterol in fungi, for example (Fig. 20–35).

Cholesterol Has Several Fates

Most of the cholesterol synthesis in vertebrates takes place in the liver. A small fraction of the cholesterol made there is incorporated into the membranes of hepatocytes, but most of it is exported in one of two forms: bile acids or cholesteryl esters. **Bile acids** and their salts are relatively hydrophilic cholesterol derivatives that are synthesized in the liver and aid in lipid digestion (p. 480). **Cholesteryl esters** are formed in the liver through the action of **acyl-CoA–cholesterol acyl transferase (ACAT).** This enzyme catalyzes the transfer of a fatty acid from coenzyme A to the hydroxyl group of cholesterol (Fig. 20–36), converting the cholesterol into a more hydrophobic form. Cholesteryl esters are stored in the liver or transported to other tissues that use cholesterol.

All growing animal tissues need cholesterol for membrane synthesis, and some organs (adrenal gland and gonads, for example) use cholesterol as a precursor for steroid hormone production (discussed later). Cholesterol is also a precursor of vitamin D (see Fig. 9–19).

Cholesterol and Other Lipids Are Carried on Plasma Lipoproteins

Cholesterol and cholesteryl esters, like triacylglycerols and phospholipids, are essentially insoluble in water. These lipids must, however, be moved from the tissue of origin (liver, where they are synthesized, or intestine, where they are absorbed) to the tissues in which they will be stored or consumed. They are carried in the blood plasma from one

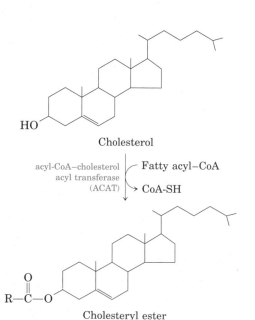

HO

Cholesterol

acyl-CoA–cholesterol   Fatty acyl–CoA
acyl transferase
(ACAT)   CoA-SH

R–C–O

Cholesteryl ester

Figure 20–36 Synthesis of cholesteryl esters converts cholesterol into an even more hydrophobic form for storage and transport.

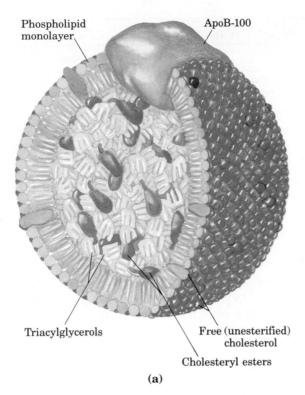

Phospholipid monolayer

ApoB-100

Triacylglycerols

Free (unesterified) cholesterol

Cholesteryl esters

(a)

Figure 20–37 (a) Structure of a low-density lipo-protein (LDL). Apolipoprotein B-100 (apoB-100) is one of the largest single polypeptide chains known, with 4,636 amino acid residues (M_r 513,000). (b) Four classes of lipoproteins visualized in the electron microscope after negative staining. From top to bottom: chylomicrons (50–200 nm in diameter); VLDL (28–70 nm); LDL (20–25 nm); and HDL (8–11 nm). For properties of lipoproteins, see Table 20–2.

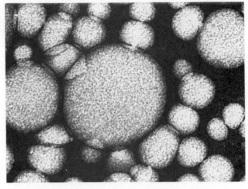

Chylomicrons

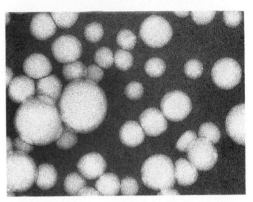

VLDL

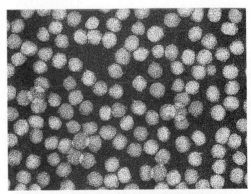

LDL

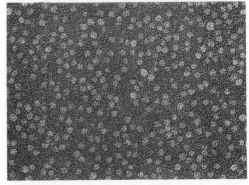

HDL   **(b)**

tissue to another as **plasma lipoproteins,** molecular aggregates of specific carrier proteins called **apolipoproteins** with various combinations of phospholipids, cholesterol, cholesteryl esters, and triacylglycerols.

Apolipoproteins ("apo" designates the protein in its lipid-free form) combine with lipids to form several classes of lipoprotein particles, spherical aggregates with hydrophobic lipids at the core and the hydrophilic side chains of protein amino acids at the surface (Fig. 20–37a). Differing combinations of lipid and protein produce particles of different densities, ranging from very low-density lipoproteins (VLDL) to very high-density lipoproteins (VHDL), which may be separated by ultracentrifugation (Table 20–2, p. 676) and visualized by electron microscopy (Fig. 20–37b).

Each class of lipoprotein has a specific function, determined by its point of synthesis, lipid composition, and apolipoprotein content. At least nine different apolipoproteins are found in the lipoproteins of human plasma (Table 20–3); they can be distinguished by their size, their reactions with specific antibodies, and their characteristic distribution in the lipoprotein classes. These protein components act as signals, targeting lipoproteins to specific tissues or activating enzymes that act on the lipoproteins.

Table 20–2 Major classes of human plasma lipoproteins: some properties

Lipoprotein	Density (g/mL)	Composition (wt %)				
		Protein	Free cholesterol	Cholesteryl esters	Phospholipids	Triacylglycerols
Chylomicrons	<1.006	2	1	3	9	85
VLDL	0.95–1.006	10	7	12	18	50
LDL	1.006–1.063	23	8	37	20	10
HDL	1.063–1.210	55	2	15	24	4

Source: Modified from Kritchevsky, D. (1986) Atherosclerosis and nutrition. *Nutr. Int.* **2,** 290–297.

We discussed **chylomicrons** in Chapter 16, in connection with the movement of dietary triacylglycerols from the intestine to other tissues. They are the largest of the lipoproteins and the least dense, containing a high proportion of triacylglycerols (see Fig. 16–2). Chylomicrons are synthesized in the smooth endoplasmic reticulum of epithelial cells that line the small intestine, then move through the lymphatic system, entering the bloodstream through the subclavian artery. The apolipoproteins of chylomicrons include apoB-48 (unique to this class of lipoproteins), apoE, and apoC-II (Table 20–3). ApoC-II activates lipoprotein lipase in the capillaries of adipose, heart, skeletal muscle, and lactating mammary tissues, allowing the release of free fatty acids to these tissues. Chylomicrons thus carry fatty acids obtained in the diet to the tissue in which they will be consumed or stored as fuel. The remnants of chylomicrons, depleted of most of their triacylglycerols but still containing cholesterol, apoE, and apoB-48, move through the bloodstream to the liver, where they are taken up and recycled.

Table 20–3 Apolipoproteins of the human plasma lipoproteins

Apolipoprotein	Molecular weight	Lipoprotein association	Function (if known)
ApoA-I	28,331	HDL	Activates LCAT
ApoA-II	17,380	HDL	
ApoB-48	240,000	Chylomicrons	
ApoB-100	513,000	VLDL, LDL	Binds to LDL receptor
ApoC-I	7,000	VLDL, HDL	
ApoC-II	8,837	Chylomicrons, VLDL, HDL	Activates lipoprotein lipase
ApoC-III	8,751	Chylomicrons, VLDL, HDL	Inhibits lipoprotein lipase
ApoD	32,500	HDL	
ApoE	34,145	Chylomicrons, VLDL, HDL	Triggers clearance of VLDL and chylomicron remnants

Source: Modified from Vance, D.E. & Vance, J.E. (eds) (1985) *Biochemistry of Lipids and Membranes.* The Benjamin/Cummings Publishing Company, Menlo Park, CA.

When the diet contains more fatty acids than are needed immediately as fuel, they are converted into triacylglycerols in the liver and packaged with specific apolipoproteins into **very low-density lipoprotein, VLDL.** Excess carbohydrate in the diet can also be converted into triacylglycerols in the liver and exported as VLDLs. In addition to triacylglycerols, VLDLs contain some cholesterol and cholesteryl esters, as well as apoB-100, apoC-I, apoC-II, apoC-III, and apo-E (Table 20–3). These lipoproteins are transported in the blood from the liver to adipose tissue, where activation of lipoprotein lipase by apoC-II causes the release of free fatty acids from the triacylglycerols of the VLDL. Adipocytes take up these fatty acids, resynthesize triacylglycerols from them, and store the products in intracellular lipid droplets.

The loss of triacylglycerols converts VLDL to **low-density lipoprotein, LDL** (Table 20–2). Very rich in cholesterol and cholesteryl esters and containing apoB-100 as their major apoprotein, LDLs carry cholesterol to peripheral tissues (those other than the liver) that have specific surface receptors that recognize apoB-100. These receptors mediate the uptake of cholesterol and cholesteryl esters in a process described below.

The fourth major lipoprotein type, **high-density lipoprotein, HDL,** is synthesized in the liver as small, protein-rich particles containing relatively little cholesterol and cholesteryl esters. HDLs contain apoC-I and apoC-II, among other apolipoproteins (Table 20–3), as well as the enzyme **lecithin-cholesterol acyl transferase** (LCAT), which catalyzes the formation of cholesteryl esters from lecithin (phosphatidylcholine) and cholesterol (Fig. 20–38). After release into the bloodstream, the nascent (newly synthesized) HDL collects cholesteryl esters from other circulating lipoproteins. Chylomicrons and VLDLs, after the removal of their triacylglycerols by lipoprotein lipase, are rich in cholesterol and phosphatidylcholine. LCAT on the surface of nascent HDL converts this phosphatidylcholine and cholesterol to cholesteryl

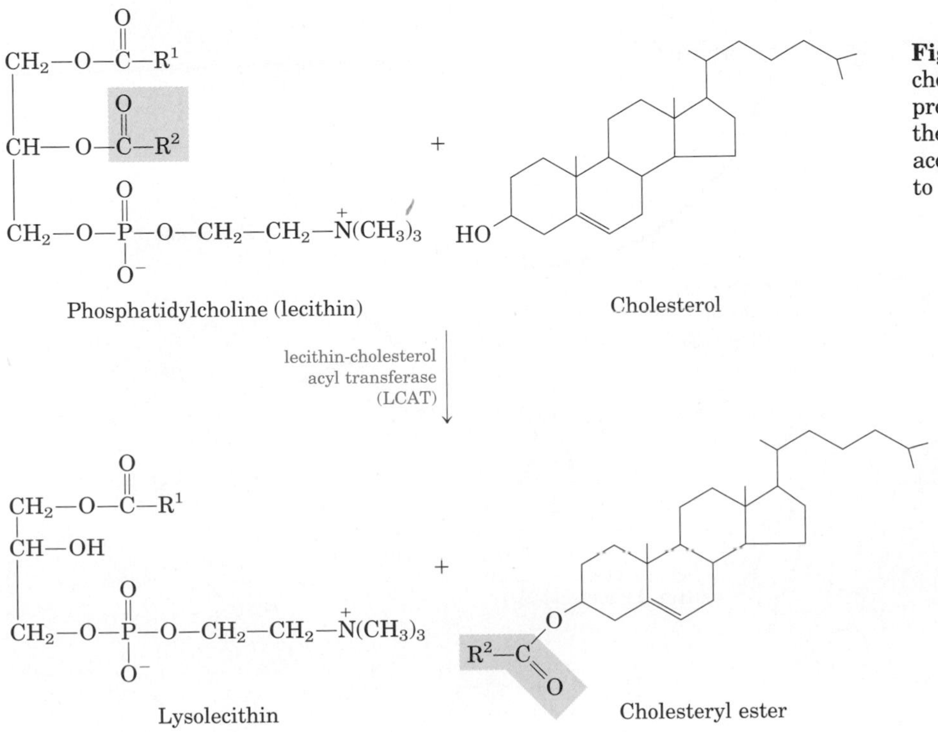

Figure 20–38 The reaction catalyzed by lecithin-cholesterol acyl transferase (LCAT). This enzyme is present on the surface of HDL and is stimulated by the HDL component apoA-I. The cholesteryl esters accumulate within nascent HDLs, converting them to mature HDLs.

esters, which enter the interior of the nascent HDL, converting it from a flat disc to a sphere—a mature HDL. This cholesterol-rich lipoprotein now returns to the liver, where the cholesterol is unloaded. Some of this cholesterol is converted into bile salts.

Cholesteryl Esters Enter Cells by Receptor-Mediated Endocytosis

Each LDL particle circulating in the bloodstream contains apoB-100, which as noted above is recognized by specific surface receptor proteins, **LDL receptors,** on cells that need to take up cholesterol. The binding of LDL to an LDL receptor initiates endocytosis (see Fig. 2–10), which brings the LDL and its associated receptor into the cell within an endosome (Fig. 20–39). This endosome eventually fuses with a lysosome, which contains enzymes that hydrolyze the cholesteryl es-

Figure 20–39 Uptake of cholesterol by receptor-mediated endocytosis. Endocytosis is also described in Chapter 2 (p. 32).

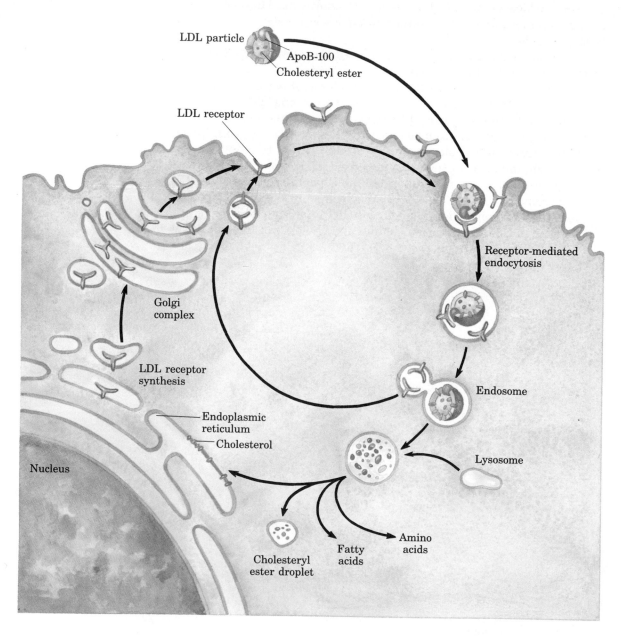

ters, releasing cholesterol and fatty acid into the cytosol. The apoB-100 of LDL is also degraded to amino acids, which are released to the cytosol, but the LDL receptor escapes degradation and returns to the cell surface, where it can again function in LDL uptake. This pathway for the transport of cholesterol in blood and its **receptor-mediated endocytosis** by target tissues was elucidated by Michael Brown and Joseph Goldstein.

Cholesterol entering cells by this path may be incorporated into membranes or may be reesterified by ACAT (Fig. 20–36) for storage within cytosolic lipid droplets. The accumulation of excess intracellular cholesterol is prevented by reducing the rate of cholesterol synthesis when sufficient cholesterol is available from LDL in the blood.

Michael Brown and Joseph Goldstein

Cholesterol Biosynthesis Is Regulated by Several Factors

Cholesterol synthesis is a complex and energy-expensive process, and it is clearly advantageous to an organism to be able to regulate the synthesis of cholesterol so as to complement the intake of cholesterol in the diet. In mammals, cholesterol production is regulated by intracellular cholesterol concentration and by the hormones glucagon and insulin. The rate-limiting step in the pathway to cholesterol is the conversion of β-hydroxy-β-methylglutaryl-CoA (HMG-CoA) into mevalonate (Fig. 20–32), and the enzyme that catalyzes this reaction, HMG-CoA reductase, is a complex regulatory enzyme whose activity is modulated over a 100-fold range. It is allosterically inhibited by as yet unidentified derivatives of cholesterol and of the key intermediate mevalonate (Fig. 20–40). HMG-CoA reductase is also hormonally regulated. The enzyme exists in phosphorylated (inactive) and dephosphorylated (active) forms. Glucagon stimulates phosphorylation (inactivation), and insulin promotes dephosphorylation, activating the enzyme and favoring cholesterol synthesis.

In addition to its immediate inhibition of existing HMG-CoA reductase, high intracellular cholesterol also slows the synthesis of new molecules of the enzyme. Furthermore, high intracellular concentrations of cholesterol also activate ACAT (Fig. 20–40), increasing esterification of cholesterol for storage. Finally, high intracellular cholesterol causes reduced production of the LDL receptor, slowing the uptake of cholesterol from the blood.

Unregulated cholesterol production can lead to serious disease. When the sum of the cholesterol synthesized and obtained in the diet exceeds the amount required for the synthesis of membranes, bile salts, and steroids, pathological accumulations of cholesterol in blood vessels (atherosclerotic plaques) can develop in humans, resulting in obstruction of blood vessels (**atherosclerosis**). Heart failure from occluded coronary arteries is a leading cause of death in industrialized societies. Atherosclerosis is linked to high levels of cholesterol in the blood, and particularly to high levels of LDL-bound cholesterol; there is a *negative* correlation between HDL levels and arterial disease.

In the human genetic disease known as familial hypercholesterolemia, blood levels of cholesterol are extremely high, and afflicted individuals develop severe atherosclerosis in childhood. The LDL receptor is defective in these individuals, and the receptor-mediated uptake of cholesterol carried by LDL does not occur. Consequently, cholesterol obtained in the diet is not cleared from the blood; it accumulates and contributes to the formation of atherosclerotic plaques. Endogenous

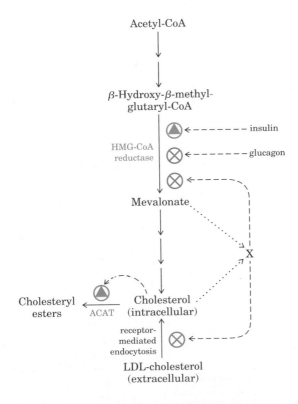

Figure 20–40 Regulation of cholesterol biosynthesis balances synthesis with dietary uptake. Glucagon acts by promoting phosphorylation of HMG-CoA reductase, insulin by promoting dephosphorylation. X represents unidentified metabolites of cholesterol and mevalonate, or other unidentified second messengers.

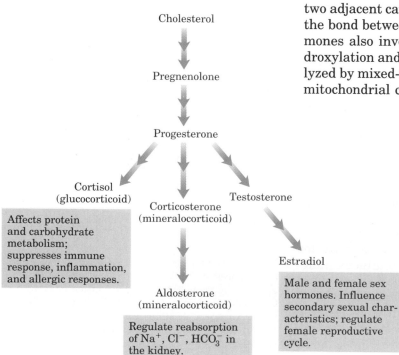

HO   O

H— in compactin → CH₃

Lovastatin

cholesterol synthesis continues even in the presence of excessive cholesterol in the blood, because the extracellular cholesterol cannot enter the cytosol to regulate intracellular synthesis. Two natural products derived from fungi, **lovastatin** and **compactin,** have shown promise in treating patients with familial hypercholesterolemia. Both are competitive inhibitors of HMG-CoA reductase and thus inhibit cholesterol synthesis. Lovastatin treatment lowers serum cholesterol by as much as 30% in individuals who carry one copy of the gene for familial hypercholesterolemia. When combined with an edible resin that binds bile acids and prevents their reabsorption from the intestine, the drug is even more effective.

Steroid Hormones Are Formed by Side Chain Cleavage and Oxidation

All steroid hormones in humans are derived from cholesterol (Fig. 20–41). Two classes of steroid hormones are synthesized in the cortex of the adrenal gland: **mineralocorticoids,** which control the reabsorption of inorganic ions (Na^+, Cl^-, and HCO_3^-) by the kidney, and **glucocorticoids,** which help regulate gluconeogenesis and also reduce the inflammatory response. The sex hormones are produced in male and female gonads and the placenta. They include **androgens** (e.g., testosterone) and **estrogens** (e.g., estradiol), which influence the development of secondary sexual characteristics in males and females, respectively, and **progesterone,** which regulates the reproductive cycle in females. The steroid hormones are effective at very low concentrations, and they are therefore synthesized in relatively small quantities. In comparison with the bile salts, their production consumes relatively little cholesterol.

The synthesis of these hormones requires removal of some or all of the carbons in the "side chain" that projects from C-17 of the D ring of cholesterol. Side chain removal takes place in the mitochondria of tissues that make steroid hormones. It involves first the hydroxylation of two adjacent carbons in the side chain (C-20 and C-22) then cleavage of the bond between them (Fig. 20–42). Formation of the individual hormones also involves the introduction of oxygen atoms. All of the hydroxylation and oxygenation reactions in steroid biosynthesis are catalyzed by mixed-function oxidases (Box 20–1) that use NADPH, O_2, and mitochondrial cytochrome P-450.

Cholesterol

↓

Pregnenolone

↓

Progesterone

Cortisol (glucocorticoid)   Corticosterone (mineralocorticoid)   Testosterone

Affects protein and carbohydrate metabolism; suppresses immune response, inflammation, and allergic responses.

Estradiol

Aldosterone (mineralocorticoid)

Male and female sex hormones. Influence secondary sexual characteristics; regulate female reproductive cycle.

Regulate reabsorption of Na^+, Cl^-, HCO_3^- in the kidney.

Figure 20–41 Some steroid hormones derived from cholesterol. The structures of some of these compounds are shown in Fig. 9–15.

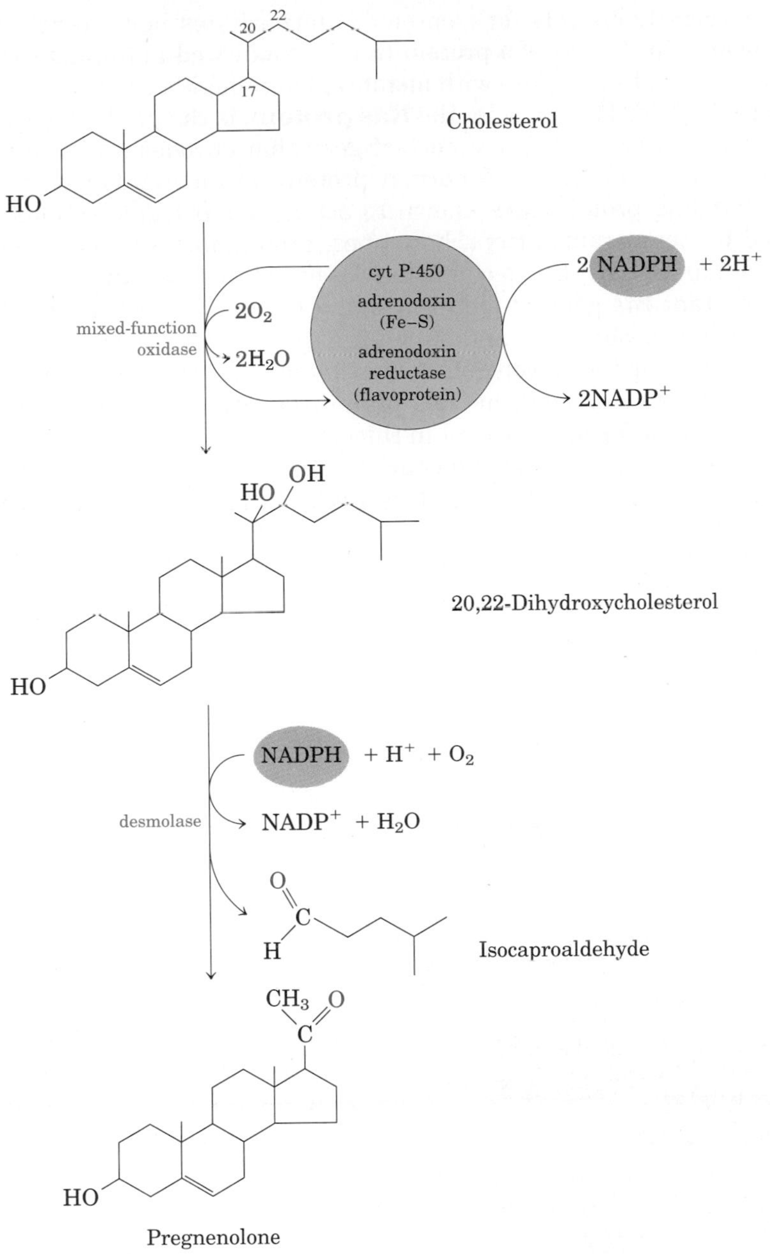

Cholesterol

mixed-function
oxidase

$2O_2$

$2H_2O$

cyt P-450
adrenodoxin
(Fe–S)
adrenodoxin
reductase
(flavoprotein)

2 NADPH $+ 2H^+$

$2NADP^+$

OH

HO

20,22-Dihydroxycholesterol

HO

desmolase

NADPH $+ H^+ + O_2$

$NADP^+ + H_2O$

Isocaproaldehyde

CH_3

HO

Pregnenolone

Figure 20–42 Side chain cleavage in the synthesis of steroid hormones involves oxidation of adjacent carbons. Cytochrome P-450 acts as electron carrier in this mixed-function oxidase system, which also requires the electron-transferring proteins adrenodoxin and adrenodoxin reductase. This side chain–cleaving system is found in mitochondria of the adrenal cortex, where active steroid production occurs. Pregnenolone is the precursor of all other steroid hormones (see Fig. 20–41).

Intermediates in Cholesterol Biosynthesis Have Many Alternative Fates

In addition to its role as an intermediate in cholesterol biosynthesis, isopentenyl pyrophosphate is the activated precursor of a huge array of biomolecules with diverse biological roles (Fig. 20–43). They include vitamins A, E, and K; plant pigments such as carotene and the phytol chain of chlorophyll; natural rubber; many essential oils, such as the fragrant principles of lemon oil, eucalyptus, and musk; insect juvenile hormone, which controls metamorphosis; dolichols, which serve as lipid-soluble carriers in complex polysaccharide synthesis; and ubiquinone and plastoquinone, electron carriers in mitochondria and chloroplasts.

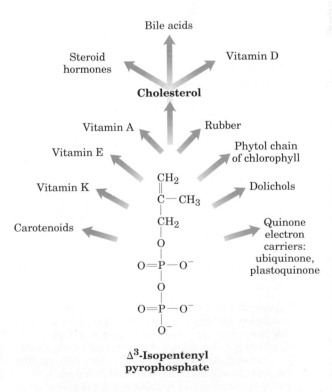

Δ^3-**Isopentenyl pyrophosphate**

Figure 20–43 An overview of isoprenoid biosynthesis. The structures of most of the end products shown here are given in Chapter 9.

A remarkable role for isoprenyl intermediates has recently been discovered in studies of a protein that is implicated in human cancers and is known to associate with membranes through a covalently bound isoprenyl lipid. This protein, the **Ras protein,** is the product of the *ras* gene, a mutant version of a normal gene that encodes a GTP-binding protein (Chapter 22). The normal protein and a number of related GTP-binding proteins are known to act in signal transductions triggered by neurotransmitters, hormones, growth factors, and other extracellular signals, by mechanisms described in detail in Chapter 22. The mutant *ras* gene is found in many humans with cancers of the lung, colon, or pancreas, and the mutant gene product is believed to be responsible for the uncontrolled division of the cancerous cells.

After its synthesis, the Ras protein is covalently altered by the attachment of farnesyl alcohol in thioether linkage with a Cys residue located four residues from the carboxyl terminus of the protein (Fig. 20–44). The farnesyl donor in this **prenylation** reaction is farnesyl

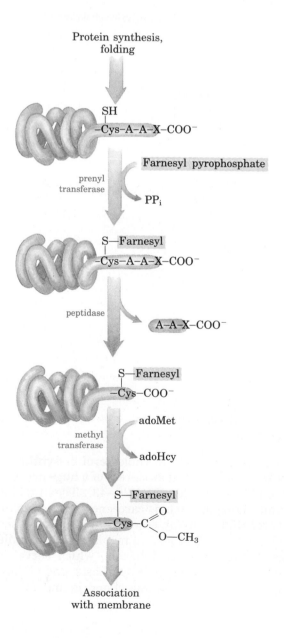

Figure 20–44 Prenylation of proteins leads to membrane association. Protein targeted for prenylation has a carboxyl-terminal sequence of Cys–A–A–X, where A is an aliphatic amino acid residue and X is the carboxyl terminus. If X is Ser, Met, or Gln, the protein will be farnesylated; if X is Leu, a geranylgeranyl group is attached. After prenylation, the three terminal residues are cleaved and the new carboxyl-terminal Cys is methylated, with *S*-adenosylmethionine as the methyl donor.

pyrophosphate (Fig. 20–34). When prenylation is prevented, the *ras* gene product does not cause uncontrolled cell division; the cancer-causing activity of the Ras protein depends on the presence of the farnesyl group. It appears that the prenylation somehow targets the Ras protein for association with the plasma membrane and that without this association the protein does not function.

Other proteins also undergo prenylation; a number of proteins in humans and a variety of other organisms have covalently attached isoprenyl derivatives that earmark them for membrane association. In some of these, the attached lipid is the 15-carbon farnesyl group; others have the 20-carbon geranylgeranyl group. Different enzymes attach the two types of lipids, and it is possible that the prenylation reactions target proteins to different membranes, depending upon which lipid is attached. Clearly, specific inhibitors of the prenylation of the Ras protein would be of interest as possible therapeutic agents for cancers caused by mutation in the *ras* gene. These protein prenylation reactions represent another important role for the isoprene derivatives formed on the pathway to cholesterol.

Summary

Long-chain saturated fatty acids are synthesized from acetyl-CoA by a cytosolic complex of six enzymes plus acyl carrier protein (ACP), which contains phosphopantetheine as its prosthetic group. The fatty acid synthase, which in some organisms consists of multifunctional polypeptides, contains two types of —SH groups (one furnished by the phosphopantetheine of ACP and the other by a Cys residue of the enzyme α-ketoacyl-ACP synthase) that function as carriers of the fatty acyl intermediates. Malonyl-ACP, formed from acetyl-CoA (shuttled out of mitochondria) and CO_2, condenses with an acetyl bound to the Cys —SH to yield acetoacetyl-ACP with release of CO_2. Reduction to the D-β-hydroxy derivative and its dehydration to the *trans*-Δ^2-unsaturated acyl-ACP is followed by reduction to butyryl-ACP. For both reduction steps, NADPH is the electron donor. Six more molecules of malonyl-ACP react successively at the carboxyl end of the growing fatty acid chain to form palmitoyl-ACP, the end product of the fatty acid synthase reaction. Free palmitate is released by hydrolysis. Fatty acid synthesis is regulated at the level of malonyl-CoA formation.

Palmitate may be elongated to yield the 18-carbon stearate. Palmitate and stearate in turn can be desaturated to yield palmitoleate and oleate, respectively, by the action of mixed-function oxidases. Mammals cannot make linoleate and must obtain it from plant sources. Mammals convert exogenous linoleate into arachidonate, the parent compound of a family of very potent hormonelike eicosanoids (prostaglandins, thromboxanes, and leukotrienes).

Triacylglycerols are formed by reaction of two molecules of fatty acyl—CoA with glycerol-3-phosphate to form phosphatidate, which is dephosphorylated to a diacylglycerol then acylated by a third molecule of fatty acyl—CoA to yield a triacylglycerol. This process is hormonally regulated. Triacylglycerols are carried in the blood in chylomicrons. Diacylglycerols are also the major precursors of glycerophospholipids. In bacteria, phosphatidylserine is formed by the condensation of serine with CDP-diacylglycerol, and decarboxylation of phosphatidylserine produces phosphatidylethanolamine. Phosphatidylglycerol is formed by condensation of CDP-diacylglycerol with glycerol-3-phosphate followed by removal of the phosphate in monoester linkage. Yeasts use similar pathways in the synthesis of phosphatidylserine, phosphatidylethanolamine, and phosphatidylglycerol; phosphatidylcholine is formed by methylation of phosphatidylethanolamine. Mammalian cells have somewhat different pathways for synthesizing phosphatidylcholine and phosphatidylethanolamine. The head group alcohol (choline or ethanolamine) is activated as the CDP-derivative, then condensed with diacylglycerol. Phosphatidylserine is derived only from phosphatidylethanolamine. The synthesis of plasmalogens involves formation of their characteristic double bond by a mixed-function oxidase. The head groups of sphingolipids are attached by unique mechanisms. Phospholipids are moved to their intracellular destinations by transport vesicles or specific proteins.

Cholesterol is formed from acetyl-CoA in a complex series of reactions through the intermediates β-hydroxy-β-methylglutaryl-CoA, mevalonate, and two activated isoprenes, dimethylallyl pyrophosphate and isopentenyl pyrophosphate. Condensation of isoprene units produces the noncyclic squalene, which is cyclized to yield the steroid ring system and side chain. Cholesterol synthesis is inhibited by elevated intracellular cholesterol. Cholesterol and cholesteryl esters are carried in the blood as plasma lipoproteins. Very low-density lipoprotein (VLDL) carries cholesterol, cholesteryl esters, and triacylglycerols from the liver to other tissues, where the triacylglycerols are degraded by lipoprotein lipase, converting VLDL to low-density lipoprotein (LDL). The LDL, rich in cholesterol and its esters, is taken up by receptor-mediated endocytosis, in which the apolipoprotein B-100 of LDL is recognized by LDL receptors in the plasma membrane. High-density lipoprotein (HDL) serves to remove cholesterol from the blood, carrying it to the liver. Dietary conditions or genetic defects in cholesterol metabolism may lead to atherosclerosis and heart disease.

The steroid hormones (glucocorticoids, mineralocorticoids, and sex hormones) are produced from cholesterol by alteration of the side chain and the introduction of oxygen atoms into the steroid ring system. In addition to cholesterol, a very wide variety of isoprenoid compounds are derived from mevalonate through condensations of isopentenyl pyrophosphate and dimethylallyl pyrophosphate. Prenylation of certain proteins targets them for association with cell membranes and is essential for their biological activity.

Further Reading

The general references in Chapters 9 and 16 will also be useful.

General

Gotto, A.M., Jr. (ed) (1987) *Plasma Lipoproteins,* New Comprehensive Biochemistry, Vol. 14 (Neuberger, A. & van Deenen, L.L.M., series eds), Elsevier Biomedical Press, Amsterdam.

Twelve reviews covering the structure, synthesis, and metabolism of lipoproteins, regulation of cholesterol synthesis, and the enzymes LCAT and lipoprotein lipase.

Hawthorne, J.N. & Ansell, G.B. (eds) (1982) *Phospholipids,* New Comprehensive Biochemistry, Vol. 4 (Neuberger, A. & van Deenen, L.L.M., series eds), Elsevier Biomedical Press, Amsterdam.

A collection of reviews that includes excellent coverage of the biosynthetic pathways to glycerophospholipids and sphingolipids, phospholipid transfer proteins, and bilayer assembly.

Mead, J.F., Alfin–Slater, R.B., Howton, D.R., & Popják, G. (1986) *Lipids: Chemistry, Biochemistry, and Nutrition,* Plenum Press, New York.

Chapters 8 (fatty acid synthesis), 9 (desaturation of fatty acids), 12 (digestion and absorption), 15 (cholesterol synthesis), 17 (glycerophospholipid metabolism), and 18 (sphingolipid metabolism) are especially germane to the topics in this chapter.

Numa, S. (ed) (1984) *Fatty Acid Metabolism and Its Regulation,* New Comprehensive Biochemistry, Vol. 7 (Neuberger, A. & van Deenen, L.L.M., series eds), Elsevier Biomedical Press, Amsterdam.

An extremely helpful collection of reviews; Chapters 1 (acetyl-CoA carboxylase), 2 (bacterial and animal fatty acid synthase), 4 (fatty acid desaturation), and 6 (fatty acid synthesis in plants) are particularly relevant.

Biosynthesis of Fatty Acids and Eicosanoids

Capdevila, J.H., Falck, J.R., & Estabrook, R.W. (1992) Cytochrome P450 and the arachidonate cascade. *FASEB J.* **6,** 731–736.

This issue of the FASEB J. *contains 20 articles on the structure and function of various cytochrome P-450s.*

Harwood, J.L. (1988) Fatty acid metabolism. *Annu. Rev. Plant Physiol. Plant Mol. Biol.* **39,** 101–138.

A good account of the compartmentation and enzymology of fatty acid synthesis in plants; also discusses the desaturase systems.

Kim, K.-H., López–Casillas, F., Bai, D.H., Luo, X., & Pape, M.E. (1989) Role of reversible phosphorylation of acetyl-CoA carboxylase in long-chain fatty acid synthesis. *FASEB J.* **3,** 2250–2256.

An advanced discussion of hormonal regulation of this enzyme by covalent alteration.

Lands, W.E.M. (1991) Biosynthesis of prostaglandins. *Annu. Rev. Nutr.* **11,** 41–60.

Discussion of the nutritional requirement for unsaturated fatty acids and recent biochemical work on pathways from arachidonate to prostaglandins; advanced level.

McCarthy, A.D. & Hardie, D.G. (1984) Fatty acid synthase—an example of protein evolution by gene fusion. *Trends Biochem. Sci.* **9,** 60–63.

A short account of the structure of the proteins of fatty acid synthase in bacteria, yeast, and vertebrates and the likely evolutionary route to their formation.

Wakil, S.J., Stoops, J.K., & Joshi, V.C. (1983) Fatty acid synthesis and its regulation. *Annu. Rev. Biochem.* **52,** 537–579.

An advanced discussion.

Biosynthesis of Membrane Phospholipids

Bishop, W.R. & Bell, R.M. (1988) Assembly of phospholipids into cellular membranes: biosynthesis, transmembrane movement and intracellular translocation. *Annu. Rev. Cell. Biol.* **4,** 579–610.

The enzymology and cell biology of phospholipid synthesis and targeting; advanced level.

Browse, J. & Somerville, C. (1991) Glycerolipid synthesis: biochemistry and regulation. *Annu. Rev. Plant Physiol. Plant Mol. Biol.* **42,** 467–506.

A detailed review of the pathways to glycerol-containing phospholipids in higher plants.

Kennedy, E.P. (1962) The metabolism and function of complex lipids. *Harvey Lectures* **57,** 143–171.

A classic description of the role of cytidine nucleotides in phospholipid synthesis.

Kent, C., Carman, G.M., Spence, M.W., & Dowhan, W. (1991) Regulation of eukaryotic phospholipid metabolism. *FASEB J.* **5,** 2258–2266.

The genetics and biochemistry of phospholipid metabolism in yeast, and the factors that regulate synthesis and intracellular transport of phospholipids.

Raetz, C.R.H. & Dowhan, W. (1990) Biosynthesis and function of phospholipids in *Escherichia coli.* *J. Biol. Chem.* **265,** 1235–1238.

A brief review of bacterial biosynthesis of phospholipids and lipopolysaccharides.

Wirtz, K.W.A. (1991) Phospholipid transfer proteins. *Annu. Rev. Biochem.* **60,** 73–99.

Discussion of the proteins that are believed responsible for transport of newly synthesized phospholipids from their sites of formation to their intracellular targets; advanced level.

Biosynthesis of Cholesterol, Steroids, and Isoprenoids

Benveniste, P. (1986) Sterol biosynthesis. *Annu. Rev. Plant Physiol.* **37,** 275–308.

A detailed review of sterol synthesis, with emphasis on the differences between the path to cholesterol and that to the plant sterols.

Bloch, K. (1965) The biological synthesis of cholesterol. *Science* **150,** 19–28.

The author's Nobel address; a classic description of cholesterol synthesis in animals.

Brown, M.S. & Goldstein, J.L. (1984) How LDL receptors influence cholesterol and atherosclerosis. *Sci. Am.* **251** (November), 58–66.

An introduction to the role of low-density lipoproteins in cholesterol transport in health and disease.

Glickman, R.M. & Sabesin, S.M. (1988) Lipoprotein metabolism. In *The Liver: Biology and Pathobiology,* 2nd edn (Arias, I.M., Jakoby, W.B., Popper, H., Schachter, D., & Shafritz, D.A., eds), pp. 331–354, Raven Press, New York.

A very useful description of the structure, composition, synthesis, and roles of plasma lipoproteins.

Goldstein, J.L. & Brown, M.S. (1990) Regulation of the mevalonate pathway. *Nature* **343,** 425–430.

The allosteric and covalent regulation of the enzymes of the mevalonate pathway; includes a short discussion of the prenylation of Ras and other proteins.

Kleinig, H. (1989) The role of plastids in isoprenoid biosynthesis. *Annu. Rev. Plant Physiol. Plant Mol. Biol.* **40,** 39–59.

The emphasis is on the unique features of isoprenoid synthesis in plants.

Myant, N.B. (1990) *Cholesterol Metabolism, LDL, and the LDL Receptor.* Academic Press, Inc., New York.

This advanced book covers the genetics, biochemistry, and cell biology of cholesterol synthesis and uptake in healthy individuals and in patients with familial hypercholesterolemia.

Rine, J. & Kim, S.-H. (1990) A role for isoprenoid lipids in the localization and function of an oncoprotein. *New Biol.* **2,** 219–226.

A discussion of the isoprenylation of the Ras protein.

Problems

1. *Pathway of Carbon in Fatty Acid Synthesis* Using your knowledge of fatty acid biosynthesis, provide an explanation for the following experimental observations:

(a) The addition of uniformly labeled [^{14}C]acetyl-CoA to a soluble liver fraction yields palmitate uniformly labeled with ^{14}C.

(b) However, the addition of a *trace* of uniformly labeled [^{14}C]acetyl-CoA in the presence of an excess of unlabeled malonyl-CoA to a soluble liver fraction yields palmitate labeled with ^{14}C only in C-15 and C-16.

2. *Synthesis of Fatty Acids from Glucose* After a person has consumed large amounts of sucrose, the glucose and fructose that exceed caloric requirements are transformed to fatty acids for triacylglycerol synthesis. This fatty acid synthesis consumes acetyl-CoA, ATP, and NADPH. How are these substances produced from glucose?

3. *Net Equation of Fatty Acid Synthesis* Write the net equation for the biosynthesis of palmitate in rat liver, starting from mitochondrial acetyl-CoA and cytosolic NADPH, ATP, and CO_2.

4. *Pathway of Hydrogen in Fatty Acid Synthesis* Consider a preparation that contains all the enzymes and cofactors necessary for fatty acid biosynthesis from added acetyl-CoA and malonyl-CoA.

(a) If [2-^{2}H]acetyl-CoA (labeled with deuterium, the heavy isotope of hydrogen):

$$\begin{array}{c} ^2H \\ | \\ ^2H-C-C \\ | \quad\quad \diagdown \\ ^2H \quad\quad S\text{-}CoA \end{array} \quad \overset{O}{\diagup}$$

and an excess of unlabeled malonyl-CoA are added as substrates, how many deuterium atoms are incorporated into every molecule of palmitate? What are their locations? Explain.

(b) If unlabeled acetyl-CoA and [2-^{2}H]malonyl-CoA:

$$\begin{array}{c} ^2H \\ | \\ ^-OOC-C-C \\ | \quad\quad \diagdown \\ ^2H \quad\quad S\text{-}CoA \end{array} \quad \overset{O}{\diagup}$$

are added as substrates, how many deuterium atoms are incorporated into every molecule of palmitate? What are their locations? Explain.

5. *Energetics of β-Ketoacyl-ACP Synthase* In the condensation reaction catalyzed by β-ketoacyl-ACP synthase (Fig. 20–5), a four-carbon unit is synthesized by the combination of a two-carbon unit and a three-carbon unit, with the release of CO_2. What is the thermodynamic advantage of this process over one that simply combines two two-carbon units?

6. *Modulation of Acetyl-CoA Carboxylase* Acetyl-CoA carboxylase is the principal regulation point in the biosynthesis of fatty acids. Some of the properties of the enzyme are described below:

(a) The addition of citrate or isocitrate raises the V_{max} of the enzyme by as much as a factor of 10.

(b) The enzyme exists in two interconvertible forms that differ markedly in their activities:

Protomer (inactive) $\rightleftharpoons$
    filamentous polymer (active)

Citrate and isocitrate bind preferentially to the filamentous form, and palmitoyl-CoA binds preferentially to the protomer.

Explain how these properties are consistent with the regulatory role of acetyl-CoA carboxylase in the biosynthesis of fatty acids.

7. *Shuttling of Acetyl Groups across the Inner Mitochondrial Membrane* The acetyl group of acetyl-CoA, produced by the oxidative decarboxylation of pyruvate in the mitochondrion, is transferred to the cytosol by the acetyl group shuttle outlined in Figure 20–11.

(a) Write the overall equation for the transfer of one acetyl group from the mitochondrion to the cytosol.

(b) What is the cost of this process in ATPs per acetyl group?

(c) In Chapter 16 we encountered an acyl group shuttle in the transfer of fatty acyl–CoA from the cytosol to the mitochondrion in preparation for β oxidation (see Fig. 16–6). One result of that shuttle was separation of the mitochondrial and cytosolic pools of CoA. Does the acetyl group shuttle also accomplish this?

8. *Oxygen Requirement for Desaturases* The biosynthesis of palmitoleate (Fig. 20–14), a common unsaturated fatty acid with a cis double bond in the Δ^9 position, uses palmitate as a precursor. Can this be carried out under strictly anaerobic conditions? Explain.

9. *Energy Cost of Triacylglycerol Synthesis* Use a net equation for the biosynthesis of tripalmitoylglycerol (tripalmitin) from glycerol and palmitate to show how many ATPs are required per molecule of tripalmitin formed.

10. *Turnover of Triacylglycerols in Adipose Tissue* When [^{14}C]glucose is added to the balanced diet of adult rats, there is no increase in the total amount of stored triacylglycerols, but the triacylglycerols become labeled with ^{14}C. Explain.

11. *Energy Cost of Phosphatidylcholine Synthesis* Write the sequence of steps and the net reaction for the biosynthesis of phosphatidylcholine by the salvage pathway from oleate, palmitate, dihydroxyacetone phosphate, and choline. Starting from these precursors, what is the cost in number of ATPs of the synthesis of phosphatidylcholine by the salvage pathway?

12. *Salvage Pathway for Synthesis of Phosphatidylcholine* A young rat maintained on a diet deficient in methionine fails to thrive unless choline is included in the diet. Explain.

13. *Synthesis of Isopentenyl Pyrophosphate* If 2-[^{14}C]acetyl-CoA is added to a rat liver homogenate that is synthesizing cholesterol, where will the ^{14}C label appear in Δ^3-isopentenyl pyrophosphate, the activated form of an isoprene unit?

14. *HMG-CoA in Ketone Body Synthesis* The rate-limiting step in the early stages of cholesterol biosynthesis is the conversion of β-hydroxy-β-methylglutaryl-CoA to mevalonate, catalyzed by HMG-CoA reductase (Fig. 20–32). The liver of a fasting animal has decreased reductase activity. When the flow through this reaction is reduced, what is the effect on the formation of ketone bodies from acetyl-CoA? How does this explain increased ketosis during fasting? (Hint: See Figure 16–16.)

15. *Activated Donors in Lipid Synthesis* In the biosynthesis of complex lipids, components are assembled by transfer of the appropriate group from an activated donor. For example, the activated donor of acetyl groups is acetyl-CoA. For each of the following groups, give the form of the activated donor: (a) phosphate; (b) D-glucosyl; (c) phosphoethanolamine; (d) D-galactosyl; (e) fatty acyl; (f) methyl; (g) the two-carbon group in fatty acid biosynthesis; (h) Δ^3-isopentenyl.

Biosynthesis of Amino Acids, Nucleotides, and Related Molecules

Nitrogen ranks behind only carbon, hydrogen, and oxygen in its contribution to the mass of living systems, as noted in Chapters 3 and 17. Most of this nitrogen is bound up in amino acids and nucleotides. We discussed the catabolism of amino acids in Chapter 17. All other aspects of the metabolism of nitrogen-containing compounds will be addressed in this chapter: first the biosynthesis of amino acids and other classes of molecules derived from them, then nucleotide metabolism.

There are several reasons for discussing the biosynthetic pathways leading to amino acids and nucleotides together; the most obvious are that both classes of molecules contain nitrogen (which arises from common biological sources) and that they are the precursors of proteins and nucleic acids. Perhaps more germane to a discussion of metabolism, however, is the simple fact that the two sets of pathways are extensively intertwined. Several key intermediates are shared in the biosynthetic pathways for nucleotides and some amino acids. Certain amino acids or parts of amino acids are incorporated into the structure of purines and pyrimidines, and in one case part of a purine ring is incorporated into the structure of an amino acid (histidine). Both sets of pathways also share much common chemistry, in particular a preponderance of reactions involving transfer of nitrogen or one-carbon groups.

The pathways to be described in the following pages can be intimidating to the beginning biochemistry student. Their apparent complexity arises not so much from the chemistry itself, which in many cases is well understood, but from the sheer number of steps and the structural complexity of many of the intermediates. They are best approached by maintaining a focus on metabolic principles already discussed, key intermediates and precursors, and common classes of reactions that recur. Even a cursory look at the chemistry can be rewarding, for some of the most unusual chemical transformations to be found in biological systems occur in these pathways. For instance, prominent examples of the rare biological use of the metals molybdenum, selenium, and vanadium are found here. The effort also offers a practical dividend, especially for students of human or veterinary medicine. Many genetic diseases of humans and animals have been traced to an absence of one or more of the enzymes of these pathways. Many pharmaceuticals used to combat infectious diseases are inhibitors of enzymes in these pathways, as are many of the most important agents in cancer chemotherapy.

Regulation is the final theme of this chapter. Because each of the amino acids and nucleotides is required in relatively small amounts, the metabolic flow through most of these pathways is not nearly as great as the biosynthetic flow leading to carbohydrate or fat in animal tissues. But, because the different amino acids and nucleotides must be made in the correct ratios and at the right time for protein and nucleic acid synthesis, their biosynthetic pathways must be accurately regulated and coordinated with each other. As discussed several times in earlier chapters, pathways can be regulated by changes in either the activity or the amounts of the enzymes involved. The pathways presented in this chapter provide some of the best-understood examples of the regulation of enzyme activity. A general discussion of the regulation of the amounts of different enzymes in a cell (that is, of their synthesis and degradation) can be found in Chapter 27.

Overview of Nitrogen Metabolism

The biosynthetic pathways to the amino acids and nucleotides share a requirement for nitrogen, but soluble, biologically useful nitrogen compounds are generally scarce in natural environments. For this reason ammonia, amino acids, and nucleotides are used economically by most organisms. Indeed, we will see that free amino acids, purines, and pyrimidines, formed during metabolic turnover, are often salvaged and reused. We will now examine the pathways by which nitrogen from the environment is introduced into biological systems.

The Nitrogen Cycle Maintains a Pool of Biologically Available Nitrogen

The most abundant form of nitrogen is present in air, which is four-fifths molecular nitrogen (N_2). However, only a relatively few species can convert atmospheric nitrogen into forms useful to living organisms; therefore the metabolic processes of different organisms function in an interdependent manner to salvage and reuse biologically available nitrogen in a vast **nitrogen cycle** (Fig. 21–1). The first step in the

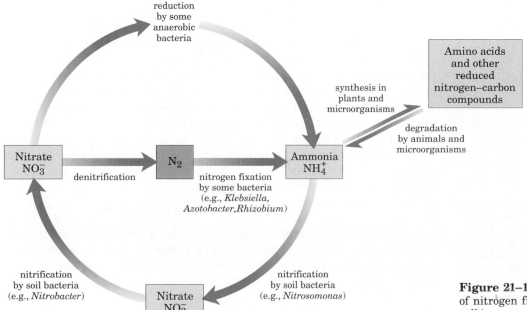

Figure 21–1 The nitrogen cycle. The total amount of nitrogen fixed annually in the biosphere exceeds 10^{11} kg.

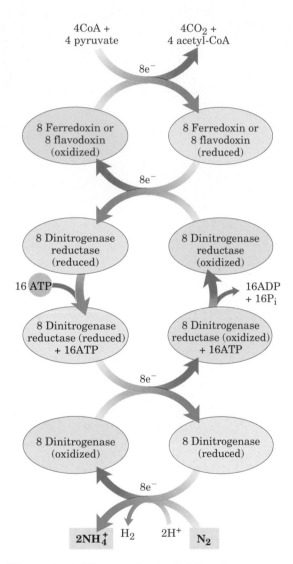

Figure 21–2 Nitrogen fixation by the nitrogenase complex. Electrons are transferred from pyruvate to dinitrogenase via ferredoxin (or flavodoxin) and dinitrogenase reductase. Dinitrogenase is reduced one electron at a time by dinitrogenase reductase, and must be reduced by at least six electrons to fix one molecule of N_2. An additional two electrons (thus a total of eight) are used to reduce two H^+ to H_2 in a process that obligatorily accompanies nitrogen fixation in anaerobes. The subunit structures and metal cofactors of the dinitrogenase reductase and dinitrogenase proteins are described in the text.

nitrogen cycle is the **fixation** (reduction) of atmospheric nitrogen by nitrogen-fixing bacteria to yield ammonia (NH_3 or NH_4^+). Although ammonia can be used by most living organisms, soil bacteria that derive their energy by oxidizing ammonia to nitrite (NO_2^-) and ultimately nitrate (NO_3^-) are so abundant and active that nearly all ammonia reaching the soil ultimately becomes oxidized to nitrate. This process is known as **nitrification.** Plants and many bacteria can readily reduce nitrate to ammonia by the action of nitrate reductases. Ammonia so formed can be built into amino acids by plants, which are then used by animals as a source of both nonessential and essential amino acids to build animal proteins. When organisms die, the microbial degradation of their proteins returns ammonia to the soil, where nitrifying bacteria convert it into nitrite and nitrate again. A balance is maintained between fixed nitrogen and atmospheric nitrogen by bacteria that convert nitrate to N_2 under anaerobic conditions. In this process, called **denitrification** (Fig. 21–1), these soil bacteria use NO_3^- rather than O_2 as the ultimate electron acceptor in a series of reactions that (like oxidative phosphorylation; Chapter 18) generates a transmembrane proton gradient that is used to synthesize ATP.

Now let us examine the process of nitrogen fixation, the first step in the nitrogen cycle.

Nitrogen Is Fixed by Enzymes of the Nitrogenase Complex

Only a relatively few species of microorganisms, all of them prokaryotes, can fix atmospheric nitrogen. The cyanobacteria, which inhabit soils and fresh and salt waters, as well as other kinds of free-living soil bacteria, such as *Azotobacter* species, are capable of fixing atmospheric nitrogen. Other nitrogen-fixing bacteria live as **symbionts** in the root nodules of leguminous plants. The first important product of nitrogen fixation in all of these organisms is ammonia, which can be used by other organisms, either directly or after its conversion into other soluble compounds, such as nitrites, nitrates, or amino acids.

The reduction of nitrogen to ammonia is an exergonic reaction:

$$N_2 + 3H_2 \longrightarrow 2NH_3 \qquad \Delta G^{\circ\prime} = -33.5 \text{ kJ/mol}$$

The N≡N triple bond, however, is very stable, with a bond energy of 942 kJ/mol. Nitrogen fixation therefore has an extremely high activation energy, and atmospheric nitrogen is almost chemically inert under normal conditions. Ammonia is produced industrially by the Haber process (named for Fritz Haber, who invented it in 1910), which uses temperatures of 400 to 500 °C and pressures of tens of thousands of kilopascals (several hundred atmospheres) of N_2 and H_2 to provide the necessary activation energy. Biological nitrogen fixation must occur at 0.8 atm of nitrogen, and the high activation barrier is overcome, at least in part, by the binding and hydrolysis of ATP (described below). The overall reaction can be written

$$N_2 + 10H^+ + 8e^- + 16ATP \longrightarrow 2NH_4^+ + 16ADP + 16P_i + H_2$$

Biological nitrogen fixation is carried out by a highly conserved complex of proteins called the **nitrogenase complex** (Fig. 21–2). The two key components of this complex are **dinitrogenase reductase** and **dinitrogenase.** Dinitrogenase reductase (M_r 60,000) is a dimer of two identical subunits (shown at right). It contains a single Fe_4–S_4 redox center (see Fig. 18–5) and can be oxidized and reduced by one electron. It also has two binding sites for ATP. Dinitrogenase is a tetramer with two copies of two different subunits (combined M_r 240,000). Dinitrogenase contains both iron and molybdenum, and its redox centers have a total of 2 Mo, 32 Fe, and 30 S per tetramer. About half of the Fe and S is present as four Fe_4–S_4 centers. The remainder is present as part of a novel iron–molybdenum cofactor of unknown structure. A form of nitrogenase that contains vanadium rather than molybdenum has been detected, and both types of nitrogenase systems can be produced by some bacterial species. The vanadium enzyme may be the primary nitrogen fixation system under some environmental conditions, but it has not been well characterized.

Nitrogen fixation is carried out by a highly reduced form of dinitrogenase, and it requires eight electrons: six for the reduction of N_2 and two to produce one molecule of H_2 as an obligate part of the reaction mechanism. Dinitrogenase is reduced by the transfer of electrons from dinitrogenase reductase (Fig. 21–2). Dinitrogenase has two binding sites for the reductase, and the required eight electrons are transferred to dinitrogenase one at a time, with the reduced reductase binding and the oxidized reductase dissociating from dinitrogenase in a cycle. This cycle requires the hydrolysis of ATP by the reductase. The immediate source of electrons to reduce dinitrogenase reductase varies, with reduced ferredoxin (p. 582; see also Fig. 18–5), reduced flavodoxin, and perhaps other sources playing a role in some systems. In at least one instance, the ultimate source of electrons is pyruvate (Fig. 21–2).

The role of ATP in this process is interesting in that it appears to be catalytic rather than thermodynamic. Remember that ATP can contribute not only chemical energy, through the hydrolysis of one or more of its phosphodiester bonds, but also binding energy (pp. 205 and 353) through noncovalent interactions that can be used to lower the activation energy. In the reaction carried out by dinitrogenase reductase, both ATP binding and ATP hydrolysis bring about protein conformational changes that evidently help overcome the high activation energy of nitrogen fixation. ATP binding to the reductase shifts the reduction potential (E_0') of this protein from -250 to -400 mV, an enhancement of its reducing power that is required to transfer electrons to dinitrogenase. Two ATP molecules are then hydrolyzed during the actual transfer of each electron from dinitrogenase reductase to dinitrogenase.

Another important characteristic of the nitrogenase complex is an extreme lability when oxygen is present. The reductase is inactivated in air, with a half-life of 30 s. The dinitrogenase has a half-life of 10 min in air. Free-living bacteria that fix nitrogen avoid or solve this problem in a variety of ways. Some exist only anaerobically or repress nitrogenase synthesis when oxygen is present. Some aerobic bacteria, such as *Azotobacter vinelandii,* partially uncouple electron transport from ATP synthesis so that oxygen is burned off as rapidly as it enters the cell (Chapter 18). When fixing nitrogen, cultures of these bacteria actually warm up as a result of their efforts to remove oxygen. The nitrogen-fixing cyanobacteria use still another approach. One of every nine cells differentiates into a heterocyst, a cell specialized for nitrogen fixation, with thick walls to prevent oxygen from entering.

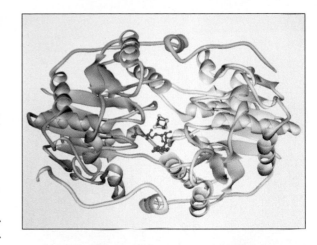

Ribbon diagram of the structure of dinitrogenase reductase. The two subunits are shown in gray and light blue. A bound ADP is shown in dark blue. Iron and sulfur atoms in the Fe_4–S_4 complex are shown in red and yellow, respectively.

(a)

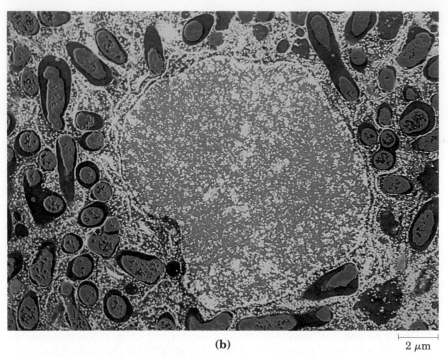

(b) $2\ \mu m$

Figure 21–3 (a) Nitrogen-fixing nodules on the roots of bird's-foot trefoil, a legume. **(b)** Electron micrograph of a thin section through a pea root nodule. Symbiotic nitrogen-fixing bacteria (bacteroids, shown in red) live inside the nodule cells, surrounded by the peribacteroid membrane (blue). Bacteroids produce the enzyme nitrogenase, which converts atmospheric nitrogen (N_2) into ammonium (NH_4^+); without the bacteroids, the plant is unable to utilize N_2. The root cells provide some factors essential for nitrogen fixation, particularly leghemoglobin, which has a very high affinity for binding oxygen. Oxygen is highly inhibitory to nitrogenase. (The cell nucleus is shown in yellow/green. The infected plant cell also contains other organelles, not visible in this micrograph, that are normally found in plant cells.)

The symbiotic relationship between leguminous plants and the nitrogen-fixing bacteria in their root nodules (Fig. 21–3) solves both the energetic requirements of the reaction and the oxygen lability of the enzymes. The energy required for nitrogen fixation was probably the evolutionary driving force for this association of plants with bacteria. The bacteria in root nodules have access to a large reservoir of energy in the form of the abundant carbohydrate made available by the plant. Because of this energy source, the bacteria in root nodules may fix hundreds of times more nitrogen than their free-living cousins under conditions generally encountered in soils. To solve the oxygen-toxicity problem, the bacteria in root nodules are bathed in a solution of an oxygen-binding protein called **leghemoglobin.** This protein is produced by the plant (although the heme may be contributed by the bacteria). Leghemoglobin efficiently delivers oxygen to the electron transfer system of the bacteria, and it binds all of the oxygen so that it cannot interfere with nitrogen fixation. The efficiency of the symbiosis between plants and bacteria is evident in the enrichment of soil nitrogen brought about by leguminous plants. This enrichment is the basis of the crop rotation methods used by many farmers, in which plantings of nonleguminous plants (such as corn) that extract fixed nitrogen from the soil are alternated every few years with planting of legumes such as alfalfa, peas, or clover.

Nitrogen fixation is the subject of intense study because of its immense practical importance. The expense of producing ammonia industrially for use in fertilizers increases with the cost of energy supplies, and this has led to efforts to develop recombinant or transgenic organisms that can fix nitrogen. Recombinant DNA techniques are being used to transfer the DNA that encodes nitrogenase and related enzymes into non–nitrogen-fixing bacteria and plants (Chapter 28). Success in these efforts will depend on overcoming the problem of oxygen toxicity in any cell producing nitrogenase.

Ammonia Is Incorporated into Biomolecules through Glutamate and Glutamine

Reduced nitrogen in the form of NH_4^+ can be assimilated, first into amino acids and then into other nitrogen-containing biomolecules. Two amino acids, **glutamate** and **glutamine,** provide the critical entry point. Recall that these same two amino acids play central roles in amino acid catabolism (Chapter 17). The amino groups of most other amino acids are derived from glutamate via transamination reactions (the reverse of the reaction shown in Fig. 17–5a). The amide nitrogen of glutamine is the source of amino groups in a wide range of biosynthetic processes. In most types of cells (and intercellular fluids in higher organisms), one or both of these amino acids is present at elevated concentrations, sometimes of an order of magnitude or more higher than those of other amino acids. In *E. coli* so much glutamate is required that it is one of the primary solutes in the cell. Its concentration is regulated and varied, not only in response to nitrogen requirements, but also to keep the interior of the cell in osmotic balance with the external medium.

The biosynthetic pathways to glutamate and glutamine are simple and appear to be similar in all forms of life. The most important pathway for the assimilation of NH_4^+ into glutamate requires two reactions. First, glutamate and NH_4^+ react to yield glutamine by the action of **glutamine synthetase,** which has a high affinity for NH_4^+ and is found in all organisms:

$$\text{Glutamate} + NH_4^+ + \text{ATP} \longrightarrow \text{glutamine} + \text{ADP} + P_i + H^+$$

Recall that this reaction takes place in two steps, with enzyme-bound γ-glutamyl phosphate as an intermediate (p. 515):

(1) $\text{Glutamate} + \text{ATP} \rightleftharpoons \gamma\text{-glutamyl phosphate} + \text{ADP}$
(2) $\gamma\text{-Glutamyl phosphate} + NH_4^+ \rightleftharpoons \text{glutamine} + P_i + H^+$
_____

Sum: $\text{Glutamate} + NH_4^+ + \text{ATP} \rightleftharpoons \text{glutamine} + \text{ADP} + P_i + H^+$

In addition to its importance for NH_4^+ assimilation in bacteria, this is a central reaction in amino acid metabolism in mammals; it is the main pathway for converting toxic free ammonia into the nontoxic glutamine for transport in the blood (Chapter 17).

In bacteria, glutamate is then produced by the action of the enzyme **glutamate synthase.** This enzyme catalyzes the reductive amination of α-ketoglutarate, an intermediate of the citric acid cycle, using glutamine as nitrogen donor.

$$\alpha\text{-Ketoglutarate} + \text{glutamine} + \text{NADPH} + H^+ \longrightarrow 2 \text{ glutamate} + \text{NADP}^+$$

The net reaction of these two enzymes (glutamate synthase and glutamine synthetase) in bacteria is

$$\alpha\text{-Ketoglutarate} + NH_4^+ + \text{NADPH} + \text{ATP} \longrightarrow$$
$$\text{L-glutamate} + \text{NADP}^+ + \text{ADP} + P_i$$

Thus there is a net synthesis of one molecule of glutamate.

In animals, glutamate synthase is not known to occur; glutamate is maintained at high levels by processes such as the transamination of α-ketoglutarate during amino acid catabolism (Chapter 17).

Glutamate can also be formed from α-ketoglutarate and NH_4^+ by the action of **L-glutamate dehydrogenase,** present in all organisms.

The required reducing power is furnished by NADPH:

$$\alpha\text{-Ketoglutarate} + NH_4^+ + NADPH \rightleftharpoons \text{L-glutamate} + NADP^+ + H_2O$$

We encountered this reaction in the catabolism of amino acids (Chapter 17). In eukaryotic cells, L-glutamate dehydrogenase is located in the mitochondrial matrix. The equilibrium for the reaction favors reactants, and the K_m for NH_4^+ (~1 mM) is so high that this reaction probably makes only a modest contribution to NH_4^+ assimilation. (Recall that the glutamate dehydrogenase reaction, in reverse, is a primary source of NH_4^+ destined for the urea cycle.) Soil bacteria and plants rarely encounter sufficiently high NH_4^+ concentrations for this reaction to make a significant contribution to glutamate levels, and generally rely on the two-enzyme pathway outlined above.

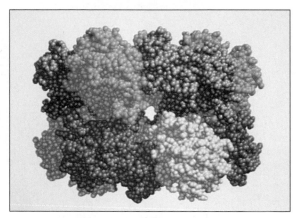

(a)

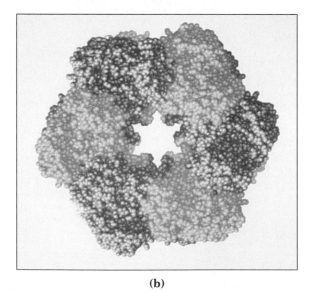

(b)

Figure 21–4 Subunit structure of glutamine synthetase as determined by x-ray diffraction. **(a)** Side view. The subunits are identical; they are differently colored to illustrate packing and placement. **(b)** Top view. The red atoms visible in each subunit are manganese ions bound in the enzyme's active sites.

Glutamine Synthetase Is a Primary Regulatory Point in Nitrogen Metabolism

Glutamine synthetase in bacteria is one of the most complex regulatory enzymes known—not surprising in light of its central role as the entry point for reduced nitrogen in metabolism. It is subject to both allosteric regulation and control by covalent modification. The enzyme has 12 identical subunits (Fig. 21–4). At least six end products of glutamine metabolism plus alanine and glycine are allosteric inhibitors of the enzyme (Fig. 21–5), and each subunit (M_r 50,000) has binding sites for all eight inhibitors as well as an active site for catalysis. Each inhibitor alone gives only partial inhibition. The effects of the different inhibitors, however, are more than additive, and all eight together virtually shut down the enzyme. This control mechanism provides a minute-by-minute adjustment of the supply of glutamine to the metabolic processes that require it.

Superimposed on the allosteric regulation is inhibition by adenylylation of (addition of AMP to) Tyr^{397} (Fig. 21–6a), which is located near the enzyme's active site. This covalent modification increases the enzyme's sensitivity to the allosteric inhibitors, and the enzyme's activity decreases as more of the 12 subunits are adenylylated. Both adenylylation and deadenylylation are promoted by the enzyme **adenylyl transferase,** part of a complex enzymatic cascade that responds to levels of glutamine, α-ketoglutarate, ATP, and P_i (Fig. 21–6b). The activity of adenylyl transferase is modulated by binding to a regulatory protein called P_{II}. The effect of P_{II}, in turn, is regulated by covalent modification (uridylylation), again at a Tyr residue. The adenylyl transferase complex with P_{II}–UMP stimulates deadenylylation, whereas the same complex with deuridylylated P_{II} stimulates adenylylation of glutamine synthetase. The uridylylation and deuridylylation of P_{II} is brought about by a single enzyme, **uridylyl transferase,** with both uridylylation and deuridylylation activities. Uridylylation is stimulated by α-ketoglutarate and ATP but inhibited by glutamine and P_i. The deuridylylation activity is not regulated.

The net result of this complex mechanism is a decrease in glutamine synthetase activity when glutamine levels are high and an increase in activity when glutamine levels are low and the α-ketoglutarate and ATP substrates are available.

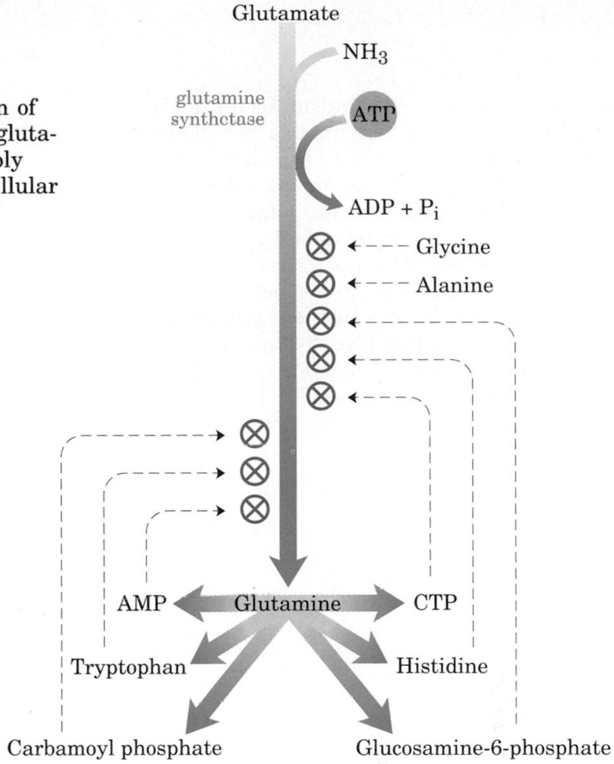

Figure 21–5 Cumulative allosteric regulation of glutamine synthetase by six end products of glutamine metabolism. Alanine and glycine probably serve as indicators of the general status of cellular amino acid metabolism.

Figure 21–6 Second level of regulation of glutamine synthetase: covalent modifications. **(a)** Structure of an adenylylated Tyr residue. **(b)** Cascade leading to adenylylation (inactivation) of glutamine synthetase. AT represents adenylyl transferase; UT, uridylyl transferase. The details of this cascade are discussed in the text.

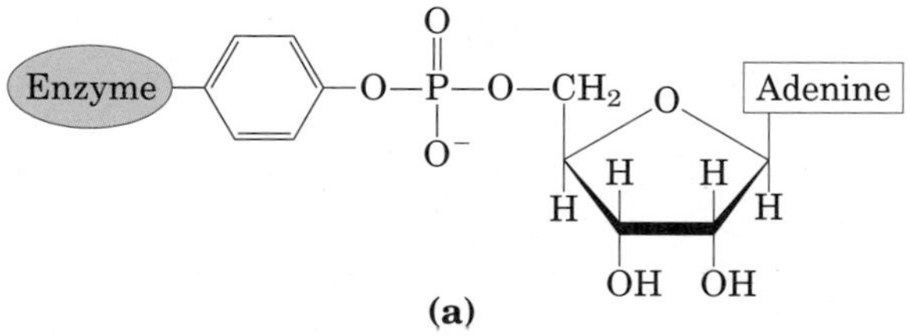

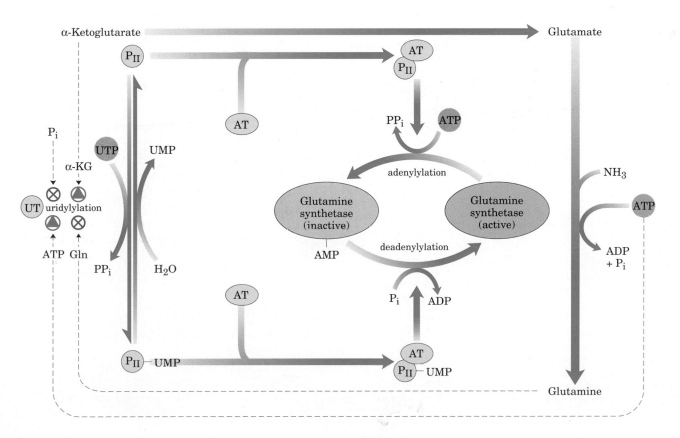

Figure 21–7 Proposed mechanism for glutamine amidotransferases. Each enzyme has two domains. The glutamine-binding domain has a number of structural elements conserved among many of these enzymes, including a Cys residue required for activity. The NH$_3$-acceptor (second substrate) domain varies. The γ-amido nitrogen of glutamine (red) is released in the form of NH$_3$ in a reaction that probably involves formation of a covalent glutamyl–enzyme intermediate. Two types of amino acceptors are shown. X represents an activating group, typically a phosphate derived from ATP, that facilitates displacement of a hydroxyl group from R—OH by NH$_3$.

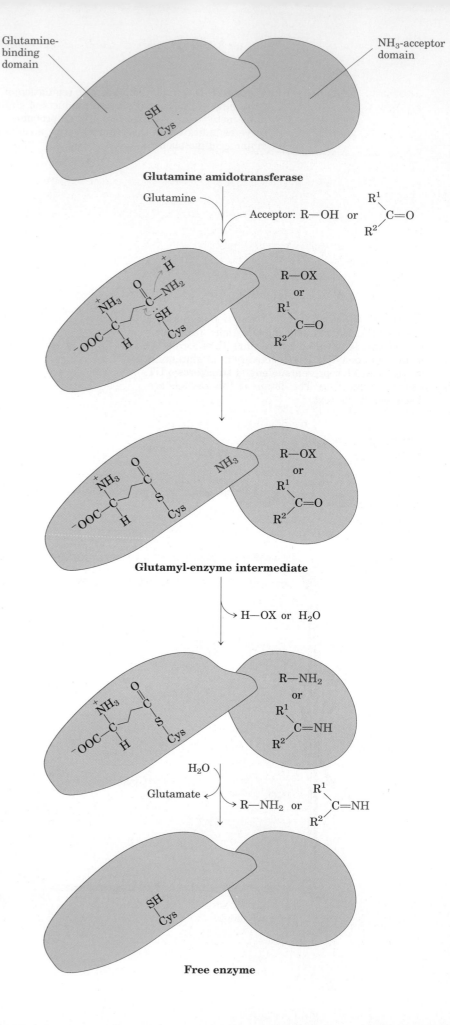

Several Classes of Reactions Play Special Roles in the Biosynthesis of Amino Acids and Nucleotides

The pathways described in this chapter offer examples of a variety of interesting chemical rearrangements. Several of these recur and deserve special note before we discuss the pathways themselves. These are (1) the transamination reactions and other rearrangements promoted by enzymes containing pyridoxal phosphate, (2) the transfer of one-carbon groups using either tetrahydrofolate or *S*-adenosylmethionine as a cofactor, and (3) the transfer of amino groups derived from the amide nitrogen of glutamine.

Pyridoxal phosphate (PLP), tetrahydrofolate (H$_4$ folate), and *S*-adenosylmethionine (adoMet) were described in some detail in Chapter 17; reactions promoted by these enzymatic cofactors were described in Figures 17–7, 17–19, and 17–20, respectively. Here we will focus on amido group transfer from glutamine.

There are over a dozen known biosynthetic reactions in which glutamine is the major physiological source of ammonia, and most of these appear in the pathways outlined in this chapter. As a class, the enzymes catalyzing these reactions are called **glutamine amidotransferases,** and all have two structural domains. One domain binds glutamine and the other binds the second substrate, which serves as amino group acceptor (Fig. 21–7, opposite). In the reaction, a conserved Cys residue in the glutamine-binding domain is believed to act as a nucleophile, cleaving the amide bond of glutamine and forming a covalent glutamyl–enzyme intermediate. The NH$_3$ produced in this reaction remains at the active site and reacts with the second substrate to form the aminated product. The covalent intermediate is hydrolyzed to form the free enzyme and glutamate. If the second substrate must be activated, ATP is generally used to generate an acyl phosphate intermediate (represented as R—OX in Fig. 21–7). The enzyme glutaminase is similar but has no second substrate, and this reaction simply yields NH$_4^+$ and glutamate (p. 515).

Biosynthesis of Amino Acids

All amino acids are derived from intermediates in glycolysis, the citric acid cycle, or the pentose phosphate pathway (Fig. 21–8). Nitrogen enters these pathways by way of glutamate and glutamine. Some pathways are simple, others are not. Ten of the amino acids are only one or a few enzymatic steps removed from their precursors. The pathways for others, such as the aromatic amino acids, are more complex.

Different organisms vary greatly in their ability to synthesize the 20 amino acids. Whereas most bacteria and plants can synthesize all 20, mammals can synthesize only about half of them (see Table 17–1).

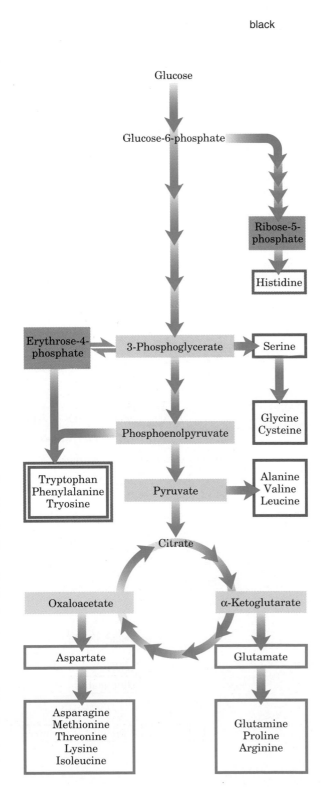

Figure 21–8 Overview of amino acid biosynthesis. Precursors from glycolysis (red), the citric acid cycle (blue), and the pentose phosphate pathway (purple) are shaded, and the amino acids derived from them are boxed in the corresponding colors. [The same device—color-matching precursors with pathway end products—will be used in illustrations of the individual pathways (Figs. 21–9 through 21–17).]

Those that are synthesized in mammals are generally those with simple pathways. These are called the **nonessential amino acids** to denote the fact that they are not needed in the diet. The remainder, the **essential amino acids,** must be obtained from food. Unless otherwise indicated, the pathways presented below are those operative in bacteria.

A useful way to organize the amino acid biosynthetic pathways is to group them into families corresponding to the metabolic precursor of each amino acid (Table 21–1). This approach is used in the detailed descriptions of these pathways presented below.

Table 21–1 Amino acid biosynthetic families, grouped by metabolic precursor

α-Ketoglutarate	*Oxaloacetate*	*Phosphoenolpyruvate and erythrose-4-phosphate*
Glutamate	Aspartate	
Glutamine	Asparagine	Tryptophan*
Proline	Methionine*	Phenylalanine*
Arginine†	Threonine*	Tyrosine
	Lysine*	
3-Phosphoglycerate	Isoleucine*	*Ribose-5-phosphate*
Serine		Histidine*
Glycine	*Pyruvate*	
Cysteine	Alanine	
	Valine*	
	Leucine*	

* Essential amino acids.

† Essential in young animals.

O
‖
⁻O—P—O—CH₂   O   H
| H H
O⁻   O   O
| ‖   ‖
H H   H O—P—O—P—O⁻
| | |
OH OH   O⁻   O⁻

5-Phosphoribosyl-1-pyrophosphate
(PRPP)

In addition to these precursors, there is a notable intermediate that recurs in several pathways: **phosphoribosyl pyrophosphate** (PRPP). PRPP is synthesized from ribose-5-phosphate derived from the pentose phosphate pathway (see Fig. 14–22), in a reaction catalyzed by **ribose phosphate pyrophosphokinase:**

Ribose-5-phosphate + ATP $\rightleftharpoons$ 5-phosphoribosyl-1-pyrophosphate + AMP

Ribose phosphate pyrophosphokinase is allosterically regulated by many of the biomolecules for which PRPP is a precursor. PRPP is an intermediate in tryptophan and histidine biosynthesis, with the ribose ring contributing several of its carbons to the final structure of these amino acids. It is also of fundamental importance in the biosynthesis of nucleotides, as we shall see later in this chapter.

α-Ketoglutarate Gives Rise to Glutamate, Glutamine, Proline, and Arginine

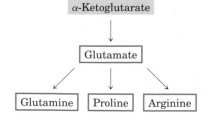

The biosynthesis of glutamate and glutamine was described earlier in this chapter. The formation of **proline,** a cyclized derivative of glutamate, is shown in Figure 21–9. In the first reaction, ATP reacts with the γ-carboxyl group of glutamate to form an acyl phosphate, which is reduced by NADPH to form glutamate γ-semialdehyde. This intermediate is then cyclized and reduced further to yield proline.

Arginine is synthesized from glutamate via ornithine and the urea cycle (Chapter 17). Ornithine could also be synthesized from glutamate γ-semialdehyde by transamination, but the cyclization of the semialdehyde that occurs in the proline pathway is a rapid spontane-

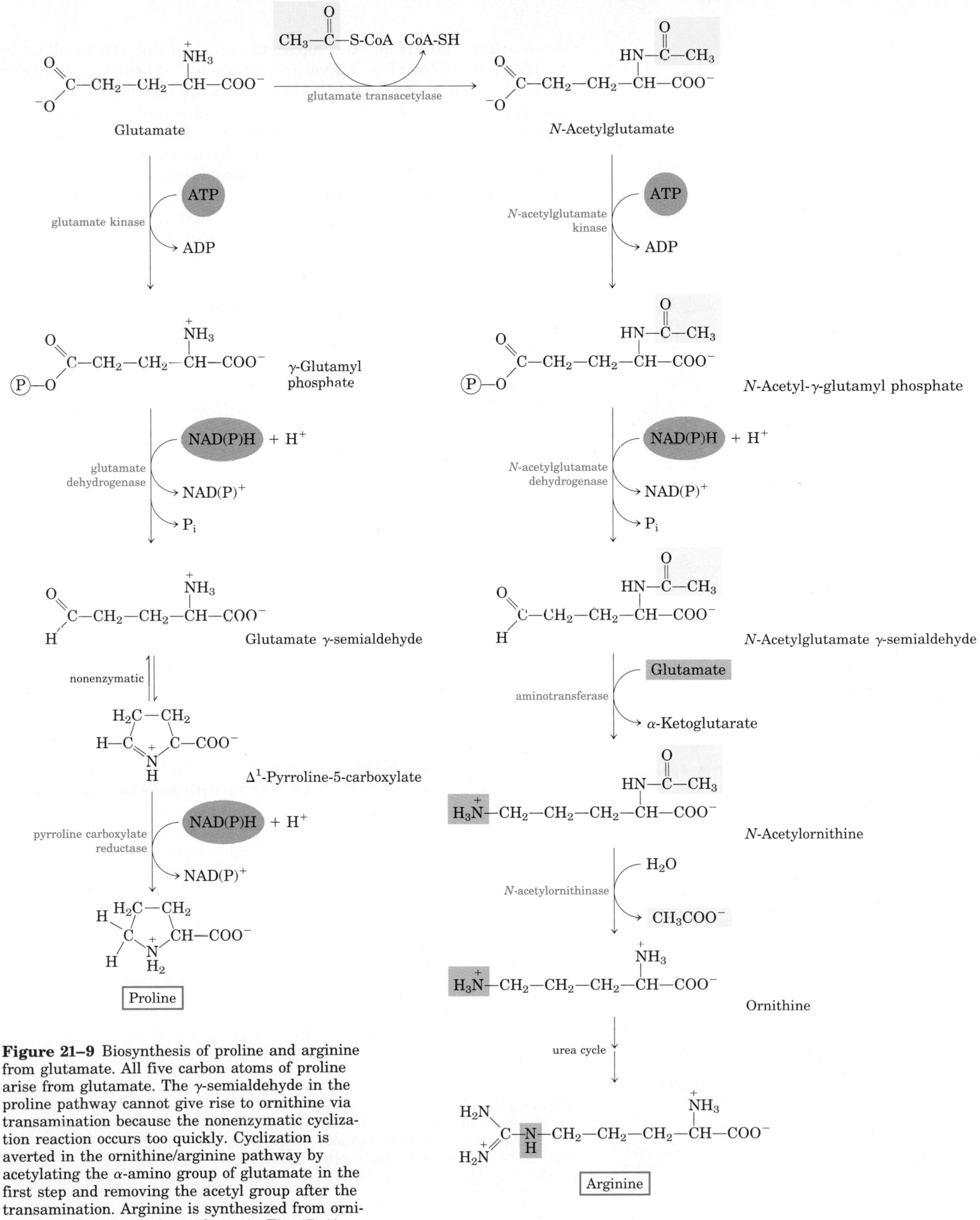

Figure 21–9 Biosynthesis of proline and arginine from glutamate. All five carbon atoms of proline arise from glutamate. The γ-semialdehyde in the proline pathway cannot give rise to ornithine via transamination because the nonenzymatic cyclization reaction occurs too quickly. Cyclization is averted in the ornithine/arginine pathway by acetylating the α-amino group of glutamate in the first step and removing the acetyl group after the transamination. Arginine is synthesized from ornithine via the urea cycle as shown in Fig. 17–11.

ous reaction that precludes a sufficient supply of this intermediate for ornithine synthesis. The biosynthetic pathway for ornithine therefore parallels some steps of the proline pathway, but includes two additional steps to chemically block the amino group of glutamate γ-semialdehyde and prevent cyclization (Fig. 21–9). At the outset the α-amino group of glutamate is blocked by acetylation in a reaction involving acetyl-CoA, and after the transamination step the acetyl group is removed to yield ornithine. Most of the arginine formed in mammals is cleaved to form urea, a process that depletes the available arginine and makes it an essential amino acid in young animals that require higher amounts of amino acids for growth.

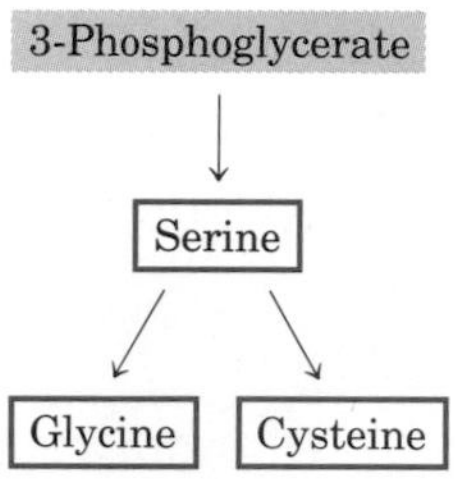

Serine, Glycine, and Cysteine Are Derived from 3-Phosphoglycerate

The major pathway for the formation of **serine** is shown in Figure 21–10. In the first step the hydroxyl group of 3-phosphoglycerate is oxidized by NAD^+ to yield 3-phosphohydroxypyruvate. Transamination from glutamate yields 3-phosphoserine, which undergoes hydrolysis by phosphoserine phosphatase to yield free serine.

The three-carbon amino acid serine is the precursor of the two-carbon **glycine** through removal of one carbon atom by **serine hydroxymethyl transferase** (Fig. 21–10). Tetrahydrofolate is the acceptor of the β-carbon atom of serine during its cleavage to yield glycine. This carbon atom forms a methylene bridge between N-5 and N-10 of tetrahydrofolate to yield N^5,N^{10}-methylenetetrahydrofolate (see Fig. 17–19). The overall reaction, which is reversible, also requires pyridoxal phosphate.

In the liver of vertebrates, glycine can be made by another route (the reverse of the reaction shown in Fig. 17–23b), catalyzed by the enzyme **glycine synthase:**

$$CO_2 + NH_4^+ + NADH + H^+ + N^5,N^{10}\text{-methylenetetrahydrofolate} \rightleftharpoons$$
$$\text{glycine} + NAD^+ + \text{tetrahydrofolate}$$

In mammals, **cysteine** is made from two other amino acids: methionine furnishes the sulfur atom and serine furnishes the carbon skeleton. In a series of reactions the —OH group of serine is replaced by an —SH group derived from methionine to form cysteine. In the first reaction methionine is converted into S-adenosylmethionine (see Fig. 17–20). After the enzymatic transfer of the methyl group to any of a number of different acceptors, S-adenosylhomocysteine, the demethylated product, is hydrolyzed to free homocysteine. Homocysteine next

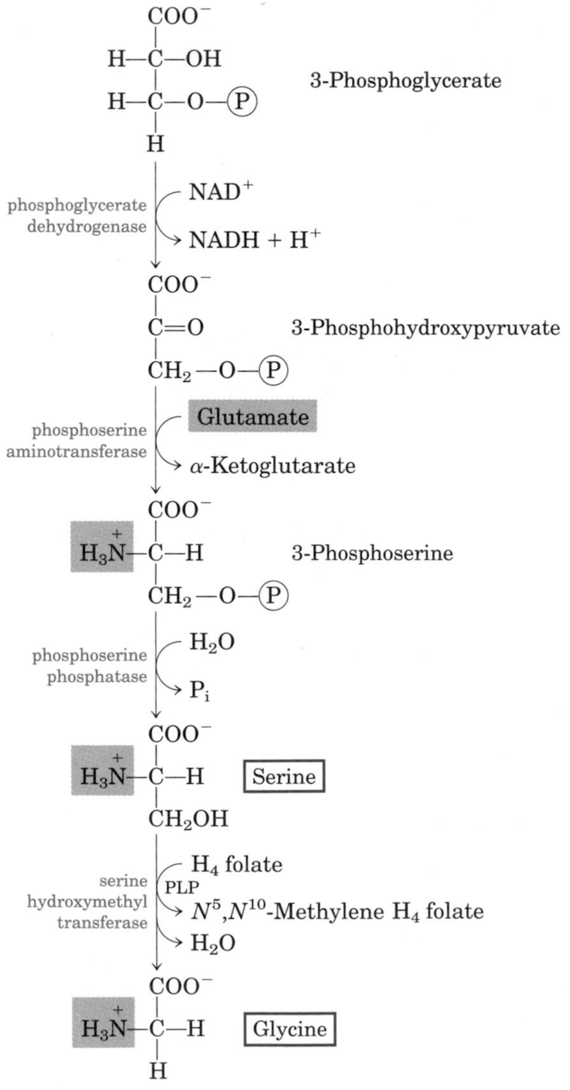

Figure 21–10 Biosynthesis of serine from 3-phosphoglycerate and the subsequent conversion of serine into glycine. Glycine is also made from CO_2 and NH_4^+ by the action of glycine synthase, which uses N^5,N^{10}-methylenetetrahydrofolate as methyl group donor (see text).

$$^-OOC-CH-CH_2-CH_2-SH \ + \ HOCH_2-\overset{\overset{+}{N}H_3}{\underset{}{CH}}-COO^-$$

$$\underset{^+NH_3}{}$$

Homocysteine   Serine

cystathionine-β-synthase | PLP
→ H₂O

$$^-OOC-CH-CH_2-CH_2-S-CH_2-\overset{\overset{+}{N}H_3}{\underset{}{CH}}-COO^-$$

Cystathionine

cystathionine-γ-lyase | H₂O
PLP
→ NH₄⁺

$$^-OOC-\overset{}{\underset{\underset{O}{\parallel}}{C}} \ \ CH_2-CH_3 \ + \ HS-CH_2-\overset{\overset{+}{N}H_3}{\underset{}{CH}}-COO^-$$

α-Ketobutyrate   Cysteine

Figure 21–11 Biosynthesis of cysteine from homocysteine and serine.

reacts with serine in a reaction catalyzed by **cystathionine-β-synthase** to yield cystathionine (Fig. 21–11). In the last step **cystathionine-γ-lyase,** a PLP-requiring enzyme, catalyzes the removal of ammonia and cleavage of cystathionine to yield free cysteine.

Three Nonessential and Six Essential Amino Acids Are Synthesized from Oxaloacetate and Pyruvate

Alanine and **aspartate** are synthesized from pyruvate and oxaloacetate, respectively, by transamination from glutamate. **Asparagine** is then synthesized by amidation of aspartate, with glutamine donating the NH_4^+. These amino acids are nonessential and their simple biosynthetic pathways are found in all organisms.

The amino acids methionine, threonine, lysine, isoleucine, valine, and leucine are essential amino acids. The biosynthetic pathways for these amino acids are complex and interconnected. In some cases there are significant differences in the pathways present in bacteria, fungi, and plants. The bacterial pathways are outlined in Figure 21–12 (pp. 702–703).

Aspartate gives rise to **methionine, threonine,** and **lysine.** Branch points occur at aspartate-β-semialdehyde, an intermediate in all three pathways, and at homoserine, a precursor of threonine and methionine. Threonine, in turn, is one of the precursors of isoleucine. The **valine** and **isoleucine** pathways share four enzymes. Pyruvate gives rise to valine and isoleucine in pathways that begin with the condensation of two carbons of pyruvate (in the form of hydroxyethyl thiamine pyrophosphate; see Fig. 14–9) with another molecule of pyruvate (valine path) or with α-ketobutyrate (isoleucine path). The α-ketobutyrate is derived from threonine in a reaction that requires pyridoxal phosphate. An intermediate in the valine pathway, α-ketoisovalerate, is the starting point for a four-step branch pathway leading to **leucine.**

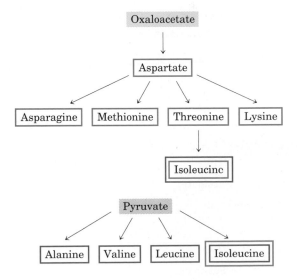

702

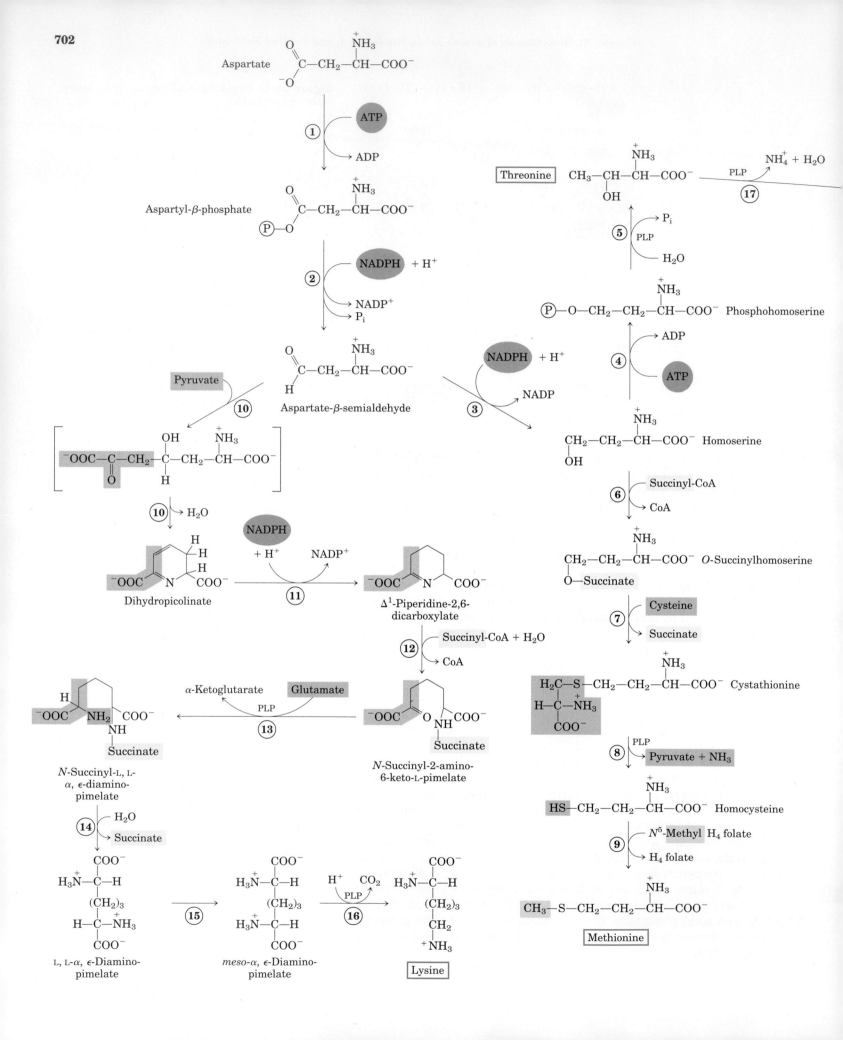

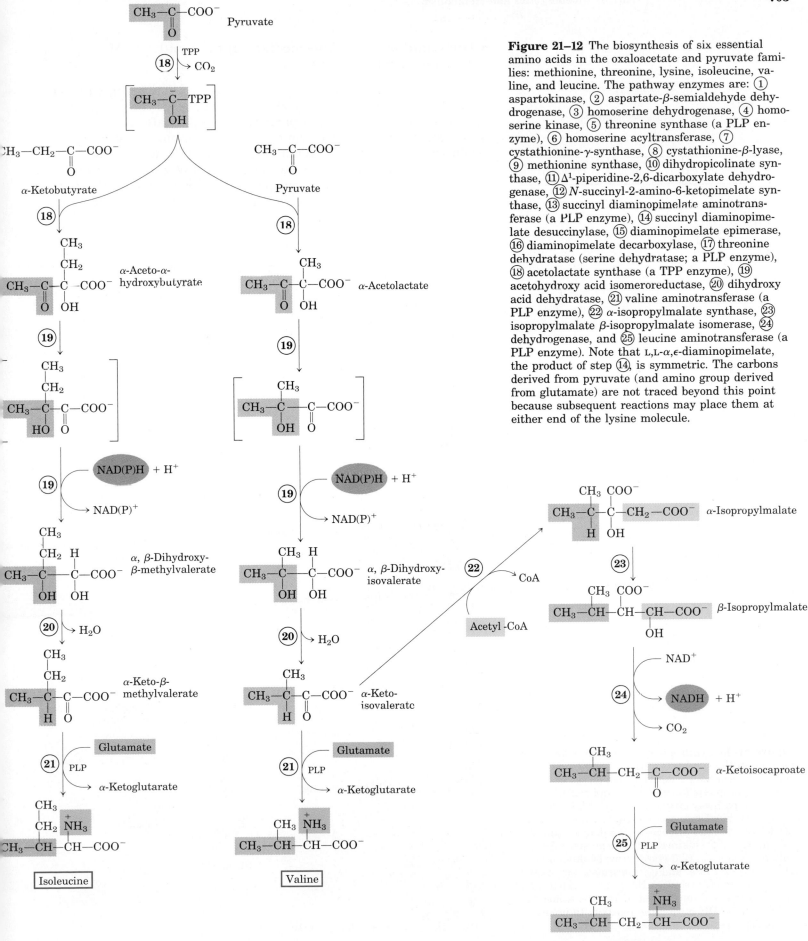

Figure 21–12 The biosynthesis of six essential amino acids in the oxaloacetate and pyruvate families: methionine, threonine, lysine, isoleucine, valine, and leucine. The pathway enzymes are: ① aspartokinase, ② aspartate-β-semialdehyde dehydrogenase, ③ homoserine dehydrogenase, ④ homoserine kinase, ⑤ threonine synthase (a PLP enzyme), ⑥ homoserine acyltransferase, ⑦ cystathionine-γ-synthase, ⑧ cystathionine-β-lyase, ⑨ methionine synthase, ⑩ dihydropicolinate synthase, ⑪ Δ¹-piperidine-2,6-dicarboxylate dehydrogenase, ⑫ N-succinyl-2-amino-6-ketopimelate synthase, ⑬ succinyl diaminopimelate aminotransferase (a PLP enzyme), ⑭ succinyl diaminopimelate desuccinylase, ⑮ diaminopimelate epimerase, ⑯ diaminopimelate decarboxylase, ⑰ threonine dehydratase (serine dehydratase; a PLP enzyme), ⑱ acetolactate synthase (a TPP enzyme), ⑲ acetohydroxy acid isomeroreductase, ⑳ dihydroxy acid dehydratase, ㉑ valine aminotransferase (a PLP enzyme), ㉒ α-isopropylmalate synthase, ㉓ isopropylmalate β-isopropylmalate isomerase, ㉔ dehydrogenase, and ㉕ leucine aminotransferase (a PLP enzyme). Note that L,L-α,ε-diaminopimelate, the product of step ⑭, is symmetric. The carbons derived from pyruvate (and amino group derived from glutamate) are not traced beyond this point because subsequent reactions may place them at either end of the lysine molecule.

Chorismate Is a Key Intermediate in the Synthesis of Tryptophan, Phenylalanine, and Tyrosine

Tryptophan, phenylalanine, and tyrosine are synthesized in bacteria by well-understood pathways that share a number of early steps. The first four steps result in the production of shikimate, in which the seven carbons are derived from erythrose-4-phosphate and phosphoenolpyruvate (Fig. 21–13). Shikimate is converted to chorismate in three more steps that include the addition of three more carbons from another molecule of phosphoenolpyruvate. Chorismate is the first branch point, with one branch leading to tryptophan and the other to phenylalanine and tyrosine.

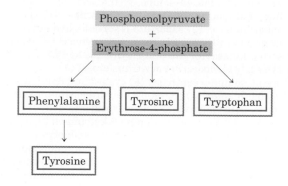

Figure 21–13 Synthesis of chorismate, a key intermediate in the synthesis of the aromatic amino acids. All carbons are derived from either erythrose-4-phosphate (purple) or phosphoenolpyruvate (red). The pathway enzymes are: ① 2-keto-3-deoxy-D-arabinoheptulosonate-7-phosphate synthase, ② dehydroquinate synthase, ③ 3-dehydroquinate dehydratase, ④ shikimate dehydrogenase, ⑤ shikimate kinase, ⑥ 3-enoylpyruvylshikimate-5-phosphate synthase, and ⑦ chorismate synthase. Note that step ② requires NAD^+ as a cofactor, and NAD^+ is released unchanged. It may be transiently reduced to NADH during the reaction, to produce an oxidized reaction intermediate.

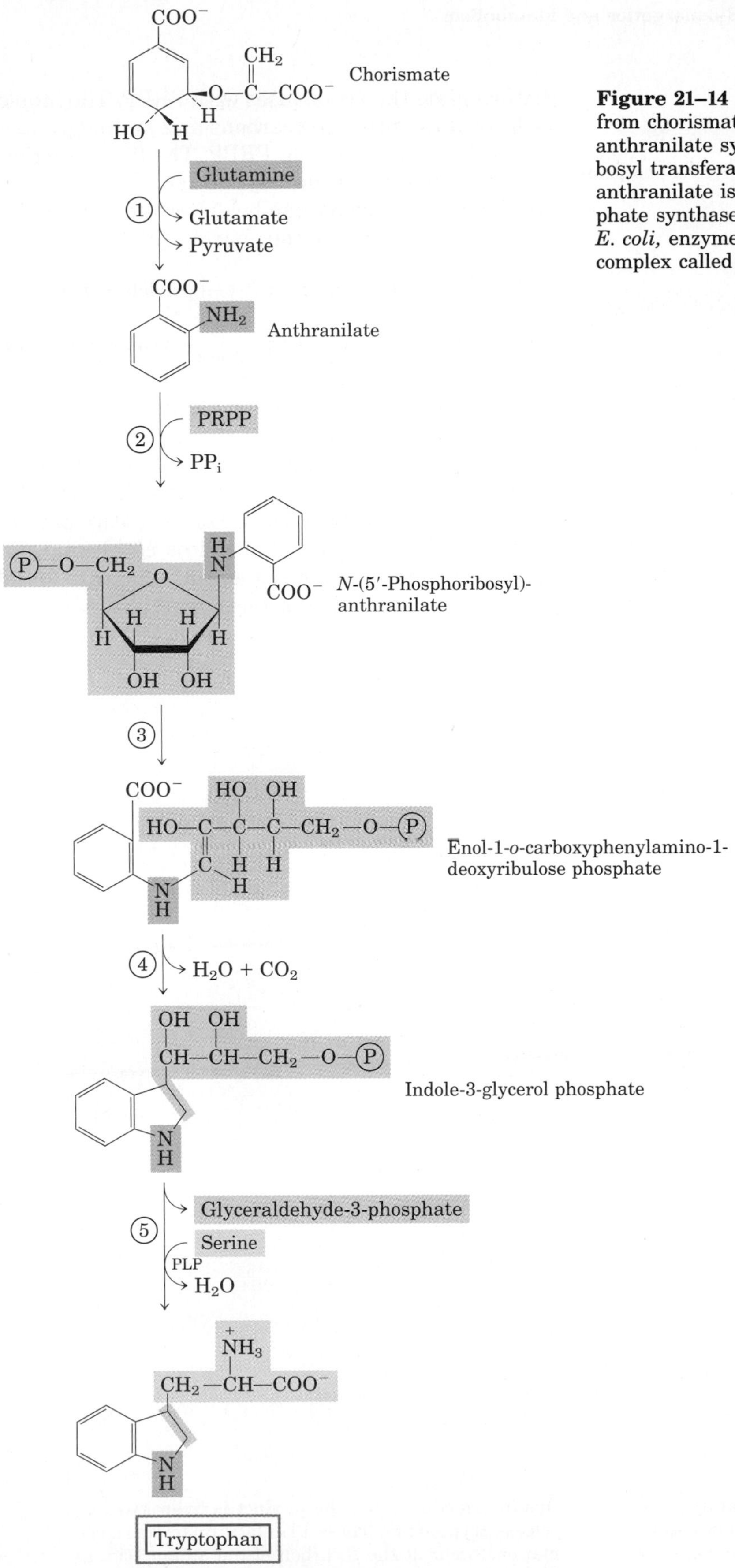

Figure 21–14 Biosynthesis of tryptophan from chorismate. The pathway enzymes are: ① anthranilate synthase, ② anthranilate phosphoribosyl transferase, ③ *N*-(5′-phosphoribosyl)-anthranilate isomerase, ④ indole-3-glycerol phosphate synthase, and ⑤ tryptophan synthase. In *E. coli,* enzymes ① and ② are subunits of a single complex called anthranilate synthase.

On the **tryptophan** branch (Fig. 21–14), chorismate is converted first to anthranilate. In this reaction, glutamine donates a nitrogen that ultimately becomes part of the completed indole ring.

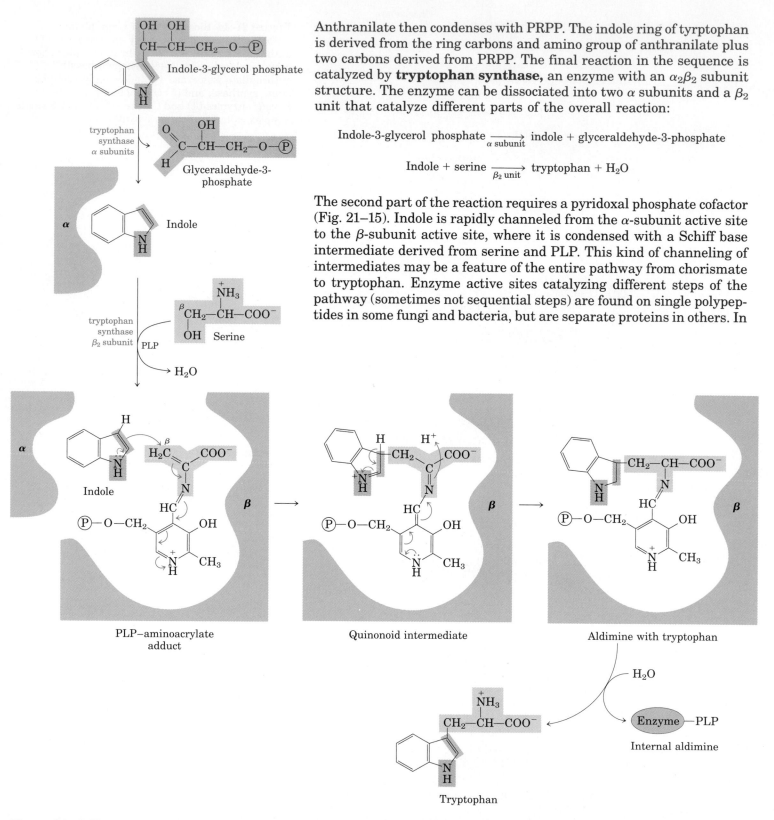

Anthranilate then condenses with PRPP. The indole ring of tyrptophan is derived from the ring carbons and amino group of anthranilate plus two carbons derived from PRPP. The final reaction in the sequence is catalyzed by **tryptophan synthase,** an enzyme with an $\alpha_2\beta_2$ subunit structure. The enzyme can be dissociated into two α subunits and a β_2 unit that catalyze different parts of the overall reaction:

$$\text{Indole-3-glycerol phosphate} \xrightarrow[\alpha \text{ subunit}]{} \text{indole} + \text{glyceraldehyde-3-phosphate}$$

$$\text{Indole} + \text{serine} \xrightarrow[\beta_2 \text{ unit}]{} \text{tryptophan} + \text{H}_2\text{O}$$

The second part of the reaction requires a pyridoxal phosphate cofactor (Fig. 21–15). Indole is rapidly channeled from the α-subunit active site to the β-subunit active site, where it is condensed with a Schiff base intermediate derived from serine and PLP. This kind of channeling of intermediates may be a feature of the entire pathway from chorismate to tryptophan. Enzyme active sites catalyzing different steps of the pathway (sometimes not sequential steps) are found on single polypeptides in some fungi and bacteria, but are separate proteins in others. In

Figure 21–15 Tryptophan synthase catalyzes a multistep reaction with different types of chemical rearrangements. First there is an aldol cleavage to form indole and release glyceraldehyde-3-phosphate. This reaction does not require PLP. The next step is the dehydration of serine to form a PLP–aminoacrylate intermediate. Indole condenses with this intermediate, and the product is hydrolyzed to release tryptophan. These PLP-facilitated transformations occur at the β carbon of the amino acid, as opposed to the α-carbon reactions described in Fig. 17–7. The β carbon of serine is attached to the indole ring system.

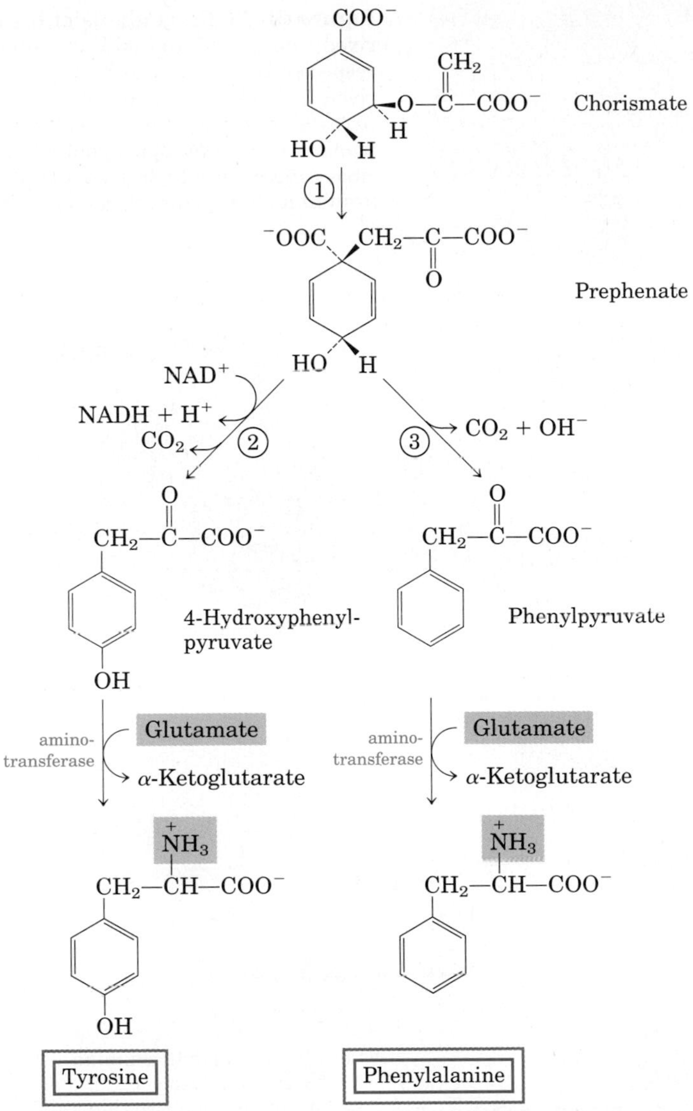

Figure 21–16 Biosynthesis of phenylalanine and tyrosine from chorismate. The enzymes are: ① chorismate mutase, ② prephenate dehydrogenase, and ③ prephenate dehydratase.

addition, the activity of some of these enzymes requires a noncovalent association with other enzymes of the pathway. These observations suggest that all are parts of a large multienzyme complex in both prokaryotes and eukaryotes. Although such complexes are generally not preserved intact when the enzymes are isolated using traditional biochemical methods, evidence for the existence of multienzyme complexes in cells is accumulating for a number of metabolic pathways.

Phenylalanine and **tyrosine** are synthesized from chorismate in plants and microorganisms via simpler pathways using the common intermediate prephenate (Fig. 21–16). The paths branch at prephenate, and the final step in both cases is transamination with glutamate as amino group donor.

Tyrosine can also be made by animals directly from phenylalanine via hydroxylation at C-4 of the phenyl group by **phenylalanine hydroxylase,** which also participates in the degradation of phenylalanine (see Figs. 17–26, 17–27). Tyrosine is considered a nonessential amino acid only because it can be synthesized from the essential amino acid phenylalanine.

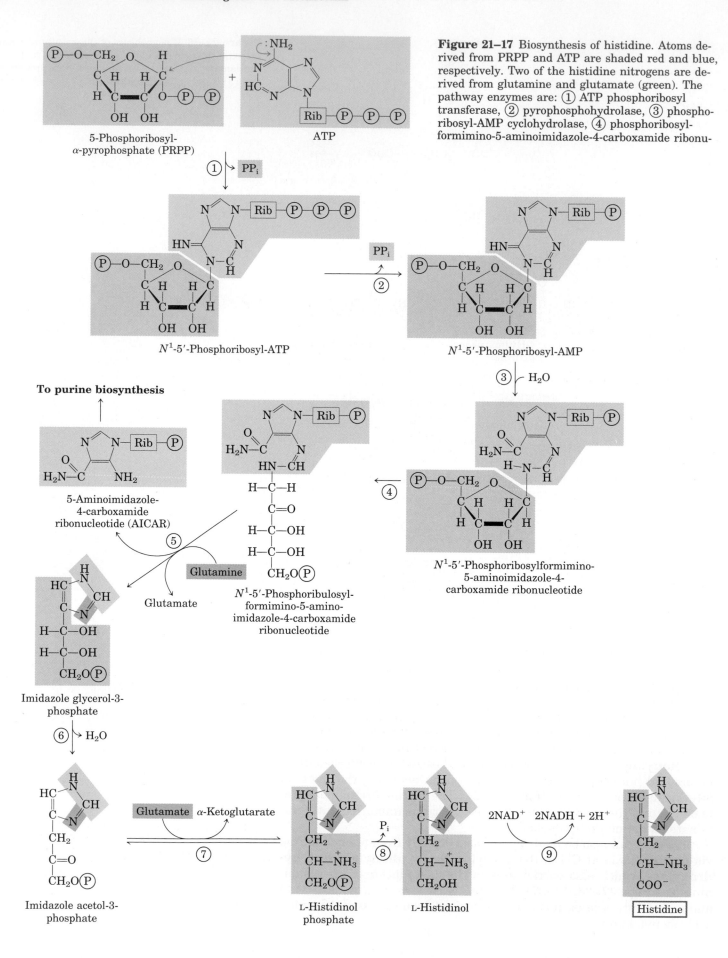

Figure 21–17 Biosynthesis of histidine. Atoms derived from PRPP and ATP are shaded red and blue, respectively. Two of the histidine nitrogens are derived from glutamine and glutamate (green). The pathway enzymes are: ① ATP phosphoribosyl transferase, ② pyrophosphohydrolase, ③ phosphoribosyl-AMP cyclohydrolase, ④ phosphoribosyl-formimino-5-aminoimidazole-4-carboxamide ribonu-

cleotide isomerase, ⑤ glutamine amidotransferase, ⑥ imidazole glycerol-3-phosphate dehydratase, ⑦ L-histidinol phosphate aminotransferase, ⑧ histidinol phosphate phosphatase, and ⑨ histidinol dehydrogenase. Note that the derivative of ATP remaining after step ⑤ is an intermediate in purine biosynthesis (Fig. 21–27), so that ATP is rapidly regenerated.

Histidine Biosynthesis Uses Precursors of Purine Biosynthesis

The **histidine** biosynthetic pathway in all plants and bacteria is novel in several respects. Histidine is derived from three precursors (Fig. 21–17): PRPP contributes five carbons, the purine ring of ATP contributes a nitrogen and a carbon, and the second ring nitrogen comes from glutamine. The key steps are the condensation of ATP and PRPP (N-1 of the purine ring becomes linked to the activated C-1 of the ribose in PRPP) (step ① in Fig. 21–17), purine ring opening that ultimately leaves N-1 and C-2 linked to the ribose (step ③), and formation of the imidazole ring in a reaction during which glutamine donates a nitrogen (step ⑤). The use of ATP as a metabolite rather than a high-energy cofactor is unusual, but not wasteful because it dovetails with the purine biosynthetic pathway. The remnant of ATP that is released after the transfer of N-1 and C-2 is 5-aminoimidazole-4-carboxamide ribonucleotide, an intermediate in the biosynthesis of purines (see Fig. 21–27) that can rapidly be recycled to ATP.

Amino Acid Biosynthesis Is under Allosteric Regulation

The most responsive manner in which amino acid synthesis is controlled is through feedback inhibition of the first reaction in the biosynthetic sequence by its final end product. The first reaction of such a sequence, which is usually irreversible, is catalyzed by an allosteric enzyme. As an example, Figure 21–18 shows the allosteric regulation of the synthesis of isoleucine from threonine, discussed earlier (Fig. 21–12). The end product, isoleucine, is a negative modulator of the first reaction in the sequence. Such allosteric or noncovalent modulation of amino acid synthesis is responsive on a minute-to-minute basis in bacteria.

Allosteric regulation can be considerably more complex. An example is the remarkable set of allosteric controls exerted on the activity of glutamine synthetase of *E. coli* (Fig. 21–5). Six products of glutamine metabolism in *E. coli* are now known to serve as negative feedback modulators of the activity of glutamine synthetase, and the overall effects of these and other modulators are more than additive. This kind of regulation is called **concerted inhibition.**

Because the 20 amino acids must be made in the correct proportions for protein synthesis, cells have developed ways not only of controlling the rate of synthesis of individual amino acids but also of coordinating their formation. Such coordination is especially well

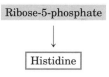

Ribose-5-phosphate

↓

Histidine

$$\overset{+}{N}H_3$$
$$CH_3\!-\!CH\!-\!CH\!-\!COO^-\quad\text{Threonine}$$
$$\underset{OH}{|}$$

→ ⊗ | threonine dehydratase

$$\overset{O}{\underset{\|}{}}$$
$$CH_3\!-\!CH_2\!-\!C\!-\!COO^-\quad\alpha\text{-Ketobutyrate}$$

↓

↓

↓

↓

$$\overset{+}{N}H_3$$
$$CH_3\!-\!CH_2\!-\!CH\!-\!CH\!-\!COO^-\quad\text{Isoleucine}$$
$$\underset{CH_3}{|}$$

Figure 21–18 The first reaction in the pathway leading from threonine to isoleucine is inhibited by the end product, isoleucine. This was one of the first examples of allosteric feedback inhibition to be discovered. The steps from α-ketobutyrate to isoleucine correspond to steps ⑱ through ㉑ in Fig. 21–12.

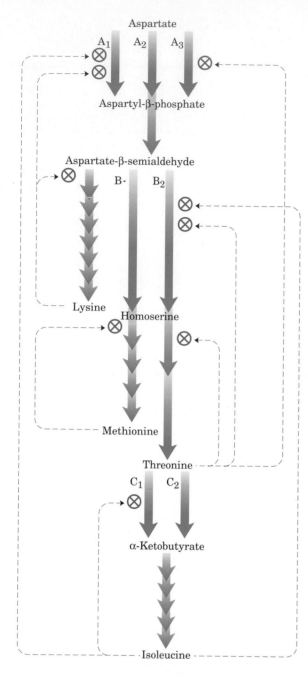

Figure 21–19 Interlocking network of regulatory mechanisms in the biosynthesis of several amino acids derived from aspartate in *E. coli*. Three enzymes (A, B, C) are shown that have either two or three isozyme forms, indicated by numerical subscripts. In each case one of the isozymes (A_2, B_1, and C_2) is shown as having no allosteric regulation; these are regulated by varying the amounts synthesized at the genetic level. Synthesis of isozymes A_2 and B_1 is repressed when methionine levels are high, and synthesis of isozyme C_2 is repressed when isoleucine levels are high. This type of genetic regulation is described in Chapter 27. Enzyme A is aspartokinase; B, homoserine dehydrogenase; C, threonine dehydratase.

developed in fast-growing bacterial cells. Figure 21–19 shows how *E. coli* cells coordinate the synthesis of lysine, methionine, threonine, and isoleucine, all made from aspartate. Several important types of inhibition patterns are evident. The step from aspartate to aspartyl-β-phosphate is catalyzed by three isozyme forms (see Box 14–3), each of which can be independently controlled by different modulators. This **enzyme multiplicity** prevents one biosynthetic end product from shutting down key steps in a pathway when other products of the same pathway are required. The steps from aspartate-β-semialdehyde to homoserine and from threonine to α-ketobutyrate (Fig. 21–12) are also catalyzed by dual, independently controlled isozymes. One of the isozymes for the conversion of aspartate to aspartyl-β-phosphate can be allosterically inhibited by two different modulators, lysine and isoleucine, whose action is more than additive. This is another example of concerted inhibition. The sequence from aspartate to isoleucine shows multiple, overlapping negative feedback inhibition; for example, isoleucine inhibits the conversion of threonine to α-ketobutyrate (as described above), and threonine inhibits its own formation at three points: from homoserine, from aspartate-β-semialdehyde, and from aspartate (steps ④, ③, and ① in Fig. 21–12). This overall action is called **sequential feedback inhibition.**

Molecules Derived from Amino Acids

In addition to their role as the building blocks of proteins, amino acids are precursors of many specialized biomolecules, including hormones, coenzymes, nucleotides, alkaloids, cell-wall polymers, porphyrins, antibiotics, pigments, and neurotransmitters—all of which serve essential biological roles. A number of pathways in which amino acids serve as precursors for other biomolecules will be described here.

Glycine Is a Precursor of Porphyrins

The biosynthesis of **porphyrins,** for which glycine is a major precursor, is our first example because of the central importance of the porphyrin nucleus in heme proteins such as hemoglobin and the cytochromes, and the Mg^{2+}-containing porphyrin derivative, chlorophyll. The porphyrins are constructed from four molecules of the monopyrrole derivative **porphobilinogen** (Fig. 21–20). In the first reaction, glycine reacts with succinyl-CoA to yield α-amino-β-ketoadipate, which is then decarboxylated to give δ-aminolevulinate. Two molecules of δ-aminolevulinate condense to form porphobilinogen, and four molecules of porphobilinogen come together to form **protoporphyrin,** through a series of complex enzymatic reactions. The iron atom is incorporated after the protoporphyrin has been assembled. Porphyrin biosynthesis is regulated by the concentration of the heme protein product, such as hemoglobin, which can serve as a feedback inhibitor of early steps in porphyrin synthesis.

In humans, genetic defects of certain enzymes in the biosynthetic pathway from glycine to porphyrins lead to the accumulation of specific porphyrin precursors in erythrocytes, in body fluids, and in the liver. These genetic diseases are known as **porphyrias.** In one of the porphyrias, which affects mainly erythrocytes, there is an accumulation of uroporphyrinogen I, an abnormal isomer of a precursor of protoporphyrin. It stains the urine red and causes the teeth to fluoresce

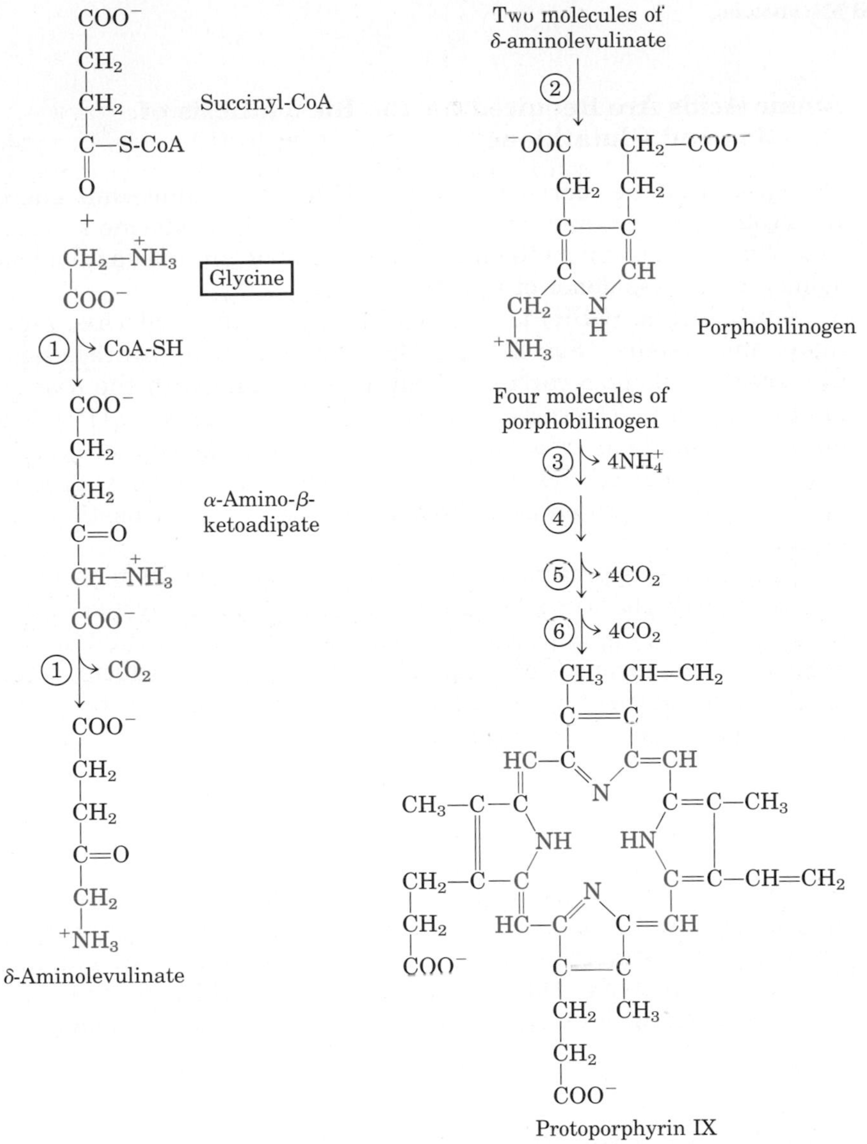

Figure 21–20 Biosynthesis of protoporphyrin IX, the porphyrin of hemoglobin and myoglobin. The atoms furnished by glycine are shown in red. The remaining carbon atoms are derived from the succinyl group of succinyl-CoA. Pathway enzymes are: ① δ-aminolevulinate synthase, ② porphobilinogen synthase, ③ uroporphyrinogen synthase, ④ uroporphyrinogen III cosynthase, ⑤ uroporphyrinogen decarboxylase, and ⑥ coproporphyrinogen oxidase.

strongly in ultraviolet light and the skin to show abnormal sensitivity to sunlight. Because insufficient heme is synthesized, patients with this disease are anemic, shy away from sunlight, and have a propensity to drink blood. This condition may have given rise to the vampire myths in medieval folk legend. Another type of porphyria causes accumulation of porphobilinogen in the liver, as well as intermittent neurological and behavioral aberrations.

Degradation of Heme Yields Bile Pigments

The iron–porphyrin or heme group of hemoglobin, released from dying erythrocytes in the spleen, is degraded to yield free Fe^{3+} and ultimately **bilirubin,** a linear (open) tetrapyrrole derivative. Bilirubin binds to serum albumin and is transported to the liver, where it is transformed into the bile pigment bilirubin glucuronide, which is sufficiently water soluble to be secreted with other components of bile into the small intestine. Impaired liver function or blocked bile secretion causes bilirubin to leak into the blood, resulting in a yellowing of the skin and eyeballs, a general condition called jaundice. Determination of bilirubin concentration in the blood is useful in diagnosing underlying liver disease.

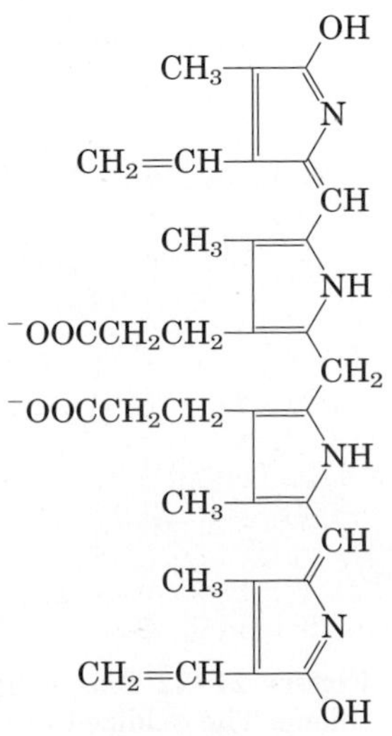

Bilirubin

Figure 21–21 Biosynthesis of creatine and phosphocreatine. Creatine is made from three amino acids: glycine, arginine, and methionine. This pathway shows the versatility of amino acids as precursors in the biosynthesis of other nitrogenous biomolecules.

Amino Acids Are Required for the Biosynthesis of Creatine and Glutathione

Phosphocreatine, derived from **creatine,** is an important energy reservoir in skeletal muscle. Creatine is derived from glycine and arginine (Fig. 21–21), and methionine plays an important role (as *S*-adenosylmethionine) as donor of a methyl group.

Glutathione (GSH) is a tripeptide derived from glycine, glutamate, and cysteine (Fig. 21–22). The first step in its synthesis is a condensation of the γ-carboxyl group of glutamate with the α-amino group of cysteine. The carboxyl group is first activated by ATP to form an acyl phosphate intermediate, which is then attacked by the cysteine amino group. The second step is similar, with the α-carboxyl group of cysteine activated to an acyl phosphate to permit condensation with glycine.

Glutathione is present in virtually all cells, often at high levels, and can be thought of as a kind of redox buffer. It probably helps maintain the sulfhydryl groups of proteins in the reduced state and the iron of heme in the ferrous (Fe^{2+}) state, and it serves as a reducing agent for glutaredoxin (see Fig. 21–32). Its redox function can also be used in removing toxic peroxides that form in the course of growth and metabolism under aerobic conditions:

$$2\,GSH + R{-}O{-}O{-}H \longrightarrow GSSG + H_2O + R{-}OH$$

This reaction is catalyzed by **glutathione peroxidase,** a remarkable enzyme in that it contains a covalently bound selenium (Se) atom in the form of selenocysteine (see Fig. 5–8). The selenium is essential for the enzyme's activity. The oxidized form of glutathione (GSSG) contains two molecules of glutathione linked by a disulfide bond (Fig. 21–22).

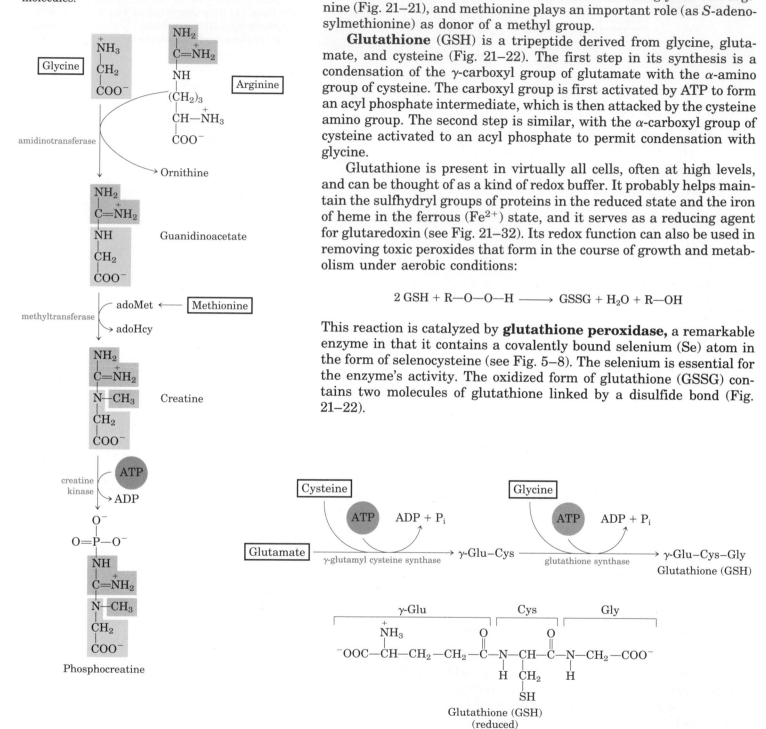

Figure 21–22 Biosynthesis and structure of glutathione. The oxidized form of glutathione is also shown.

D-Amino Acids Are Found Primarily in Bacteria

Although D-amino acids do not generally occur in proteins, they do serve some special functions in the structure of bacterial cell walls and peptide antibiotics. The peptidoglycans (see Fig. 11–19) of bacteria contain both D-alanine and D-glutamate. D-Amino acids arise directly from the L isomers by the action of amino acid racemases, which have pyridoxal phosphate as a required cofactor (see Fig. 17–7). Amino acid racemization is uniquely important to bacterial metabolism, and enzymes such as alanine racemase represent prime targets for pharmaceutical agents. One such agent, **L-fluoroalanine,** is being tested as an antibacterial drug. Another, **cycloserine,** is already used to treat urinary tract infections and tuberculosis. Both inhibitors also affect some other PLP-requiring enzymes.

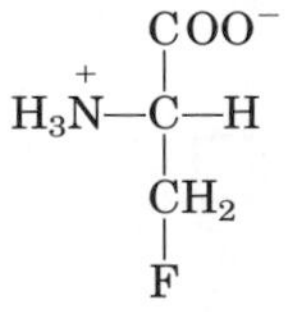

L-Fluoroalanine

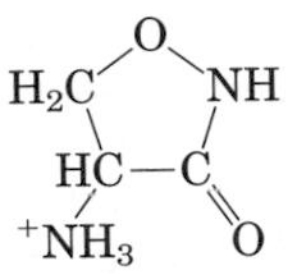

Cycloserine

Aromatic Amino Acids Are Precursors of Many Plant Substances

Phenylalanine, tyrosine, and tryptophan are converted into a variety of important compounds in plants. The rigid polymer **lignin** is derived from phenylalanine and tyrosine. It is second only to cellulose in abundance in plant tissues. The structure of lignin is complex and not well understood. Phenylalanine and tyrosine also give rise to many commercially significant natural products, including tannins that inhibit oxidation in wines; alkaloids such as morphine that have potent physiological effects; and flavor components of products such as cinnamon oil, nutmeg, cloves, vanilla, and cayenne pepper.

Tryptophan gives rise to the plant growth hormone, indole-3-acetate or **auxin** (Fig. 21–23). This molecule has been implicated in the regulation of a wide range of biological processes in plant cells.

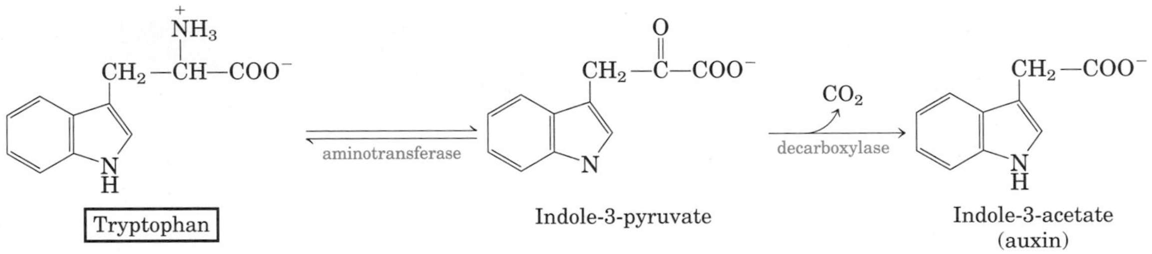

Figure 21–23 Biosynthesis of indole-3-acetate (auxin).

Amino Acids Are Converted to Biological Amines by Decarboxylation

Many important neurotransmitters are primary or secondary amines derived from amino acids in simple pathways. In addition, some polyamines that are complexed with DNA are derived from the amino acid ornithine. A common denominator of many of these pathways is amino acid decarboxylation, another reaction involving pyridoxal phosphate (see Fig. 17–7).

The synthesis of some neurotransmitters is illustrated in Figure 21–24. Tyrosine gives rise to a family of catecholamines that includes **dopamine, norepinephrine,** and **epinephrine.** Levels of catecholamines are correlated with (among other things) changes in blood pressure in animals. The neurological disorder Parkinson's disease is associated with an underproduction of dopamine, and it has been treated

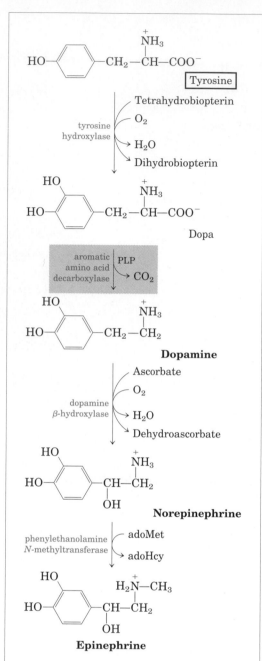

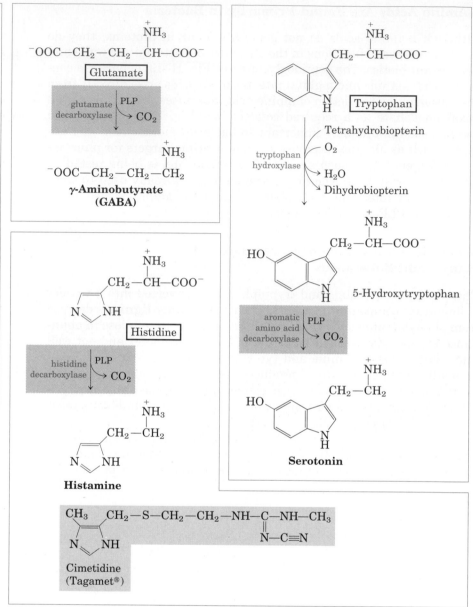

Figure 21–24 Some neurotransmitters derived from amino acids. The key biosynthetic step is the same in each case: a PLP-dependent decarboxylation (shaded in red). Cimetidine (shaded beige), a histamine analog, is used to treat duodenal ulcers.

by administering L-dopa. An overproduction of dopamine in the brain is associated with psychological disorders such as schizophrenia. Glutamate decarboxylation gives rise to **γ-aminobutyrate** (GABA), an inhibitory neurotransmitter. Its underproduction is associated with epileptic seizures. GABA is used pharmacologically in the treatment of epilepsy and hypertension. Another important neurotransmitter, **serotonin,** is derived from tryptophan in a two-step pathway.

Histidine is decarboxylated to form **histamine,** a powerful vasodilator present in animal tissues. Histamine is released in large amounts as part of the allergic response and it also stimulates acid secretion in the stomach. A growing array of pharmaceutical agents are being designed to interfere with either the synthesis or action of histamine. A prominent example is the histamine receptor antagonist **cimetidine,** also known as Tagamet®. Cimetidine is a structural analog of histamine; it promotes healing of duodenal ulcers by inhibiting secretion of gastric acid.

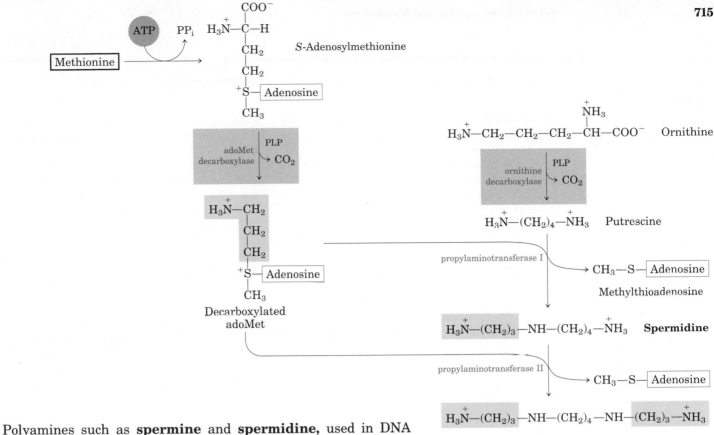

Polyamines such as **spermine** and **spermidine,** used in DNA packaging, are derived from methionine and ornithine by the pathway in Figure 21–25. The first step is the decarboxylation of ornithine, a component of the urea cycle and a precursor of arginine (Fig. 21–9). **Ornithine decarboxylase** is a PLP-requiring enzyme and is the target of several powerful inhibitors developed commercially as pharmaceutical agents (Box 21–1, pp. 716–717).

Figure 21–25 Biosynthesis of spermidine and spermine. The PLP-dependent decarboxylation steps are shaded. In these reactions, *S*-adenosylmethionine (in its decarboxylated form) acts as a source of propylamino groups.

Biosynthesis and Degradation of Nucleotides

As discussed in Chapter 12, nucleotides play a variety of important roles in all cells. First, they are precursors of DNA and RNA. Second, ATP and to some extent GTP are essential carriers of chemical energy. Third, nucleotides are components of the cofactors NAD, FAD, *S*-adenosylmethionine, and coenzyme A, as well as of activated biosynthetic intermediates such as UDP-glucose and CDP-diacylglycerol. Some, such as cAMP and cGMP, are also cellular second messengers.

There are two types of pathways leading to nucleotides: the **de novo pathways** and the **salvage pathways.** De novo synthesis of nucleotides begins with their metabolic precursors: amino acids, ribose-5-phosphate, CO_2, and NH_3. Salvage pathways recycle the free bases and nucleosides released from nucleic acid breakdown. Both types of pathways are important in cellular metabolism.

The de novo pathways for purine and pyrimidine biosynthesis appear to be present in identical form in nearly all living organisms. Notably, the free bases guanine, adenine, thymine, cytidine, and uracil are *not* intermediates in these pathways; that is, the bases are not synthesized and then attached to ribose, as might be expected. The purine ring structure is built up one or a few atoms at a time, attached to the ribose throughout the process. The pyrimidine ring is synthesized in the form of **orotate,** attached to ribose, and then converted into the common pyrimidine nucleotides used in nucleic acid synthesis.

BOX 21–1   Curing African Sleeping Sickness with a Biochemical Trojan Horse

African sleeping sickness is caused by protists (single-celled eukaryotes) called trypanosomes (Fig. 1). Until recently, this disease (also called African trypanosomiasis) was virtually incurable. This and related diseases are medically and economically important in many developing nations. Vaccines are ineffective because this parasite has a novel mechanism to evade the host immune system. The cell coat is covered with a single protein, to which the immune system responds. Every so often, however, a few individual cells switch to a new protein coat not recognized by the immune system [this occurs by a process of genetic recombination, as discussed in Chapter 27 (see Table 27–2)]. This process of "changing coats" can occur perhaps hundreds of times. The result is a cyclic chronic infection. The patient develops a fever, which subsides as the immune system beats back the first infection. The cells that have changed coats, however, become the seed for a second infec-

tion, and the fever reappears. This cycle can go on for weeks, and the weakened patient eventually dies.

Some modern approaches to treating this disease have been developed, based on an understanding of enzymology and metabolism. In at least one case, this involves pharmaceutical agents designed as mechanism-based enzyme inactivators (suicide inhibitors; see Chapter 8). A vulnerable point in the metabolism of this organism was found in the pathway for polyamine biosynthesis. The polyamines spermine and spermidine are used in DNA packaging, and they are required in large amounts in rapidly dividing cells. The first step in their synthesis is catalyzed by the enzyme ornithine decarboxylase, a PLP-requiring enzyme (see Fig. 21–25). In mammalian cells, ornithine decarboxylase is turned over rapidly; that is, it is degraded and new enzyme is synthesized continuously to replace it. For reasons not well

Figure 1 *Trypanosoma brucei rhodesiense,* the causative agent of African sleeping sickness.

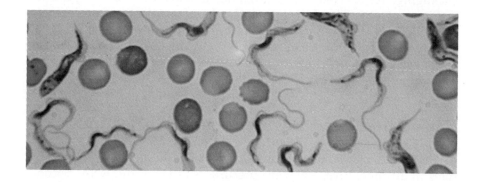

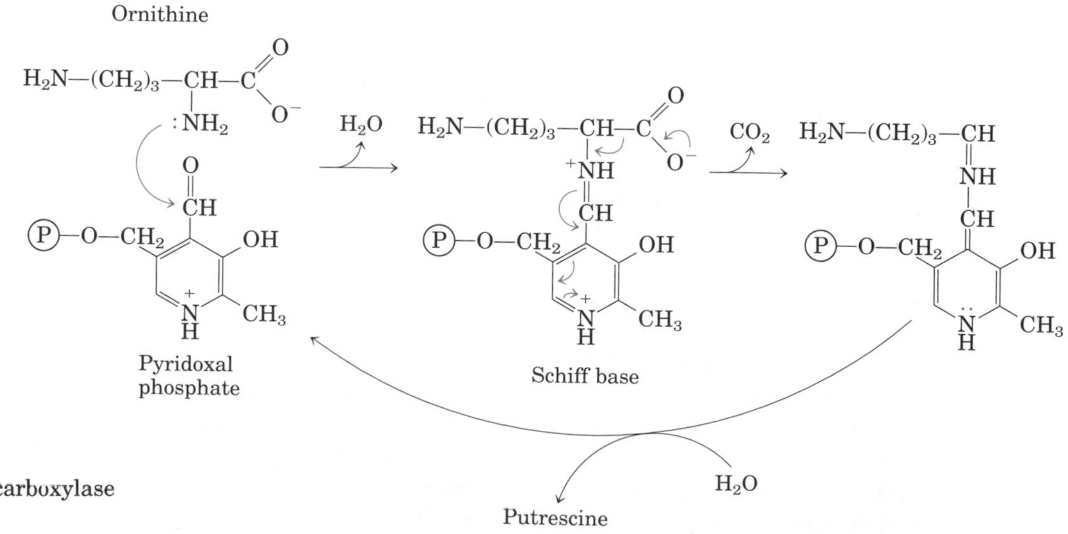

Figure 2 Mechanism of ornithine decarboxylase reaction.

understood, the trypanosome enzyme is stable and not readily replaced by new synthesis. An inhibitor of ornithine decarboxylase that binds permanently to the enzyme would thus have little effect on mammalian cells, which could rapidly replace the enzyme by new synthesis, but the inactivated enzyme of the parasites would not be replaced and reproduction would be inhibited.

The first few steps of the normal reaction catalyzed by ornithine decarboxylase, as determined experimentally, are shown in Figure 2. The direction of flow of electron pairs is denoted by blue arrows. Once CO_2 is released, the electron movement is reversed and putrescine is ultimately released (see Fig. 21–25). Based on this mechanism, several suicide inhibitors have been designed for this enzyme. One of these is difluoromethylornithine (DFMO). DFMO is relatively inert in solution. When it binds to ornithine decarboxylase, however, the enzyme is quickly inactivated (Fig. 3).

This inhibitor provides an alternative electron sink in the form of two strategically placed fluorine atoms, which are excellent leaving groups. Instead of electrons moving into the ring structure of PLP, the reaction results in displacement of a fluorine atom. Nucleophilic amino acid side chains (represented by B:) at the enzyme's active site may then react with the highly reactive PLP–inhibitor adduct, forming a covalent complex in an essentially irreversible reaction. In this way the inhibitor makes use of the enzyme's own reaction mechanisms to kill it. DFMO has proven highly effective against African sleeping sickness in clinical trials in Africa.

Approaches such as this show great promise for treating a wide range of diseases. The ability to design drugs based on enzyme mechanism and structure is replacing the more traditional trial-and-error method for producing new drugs.

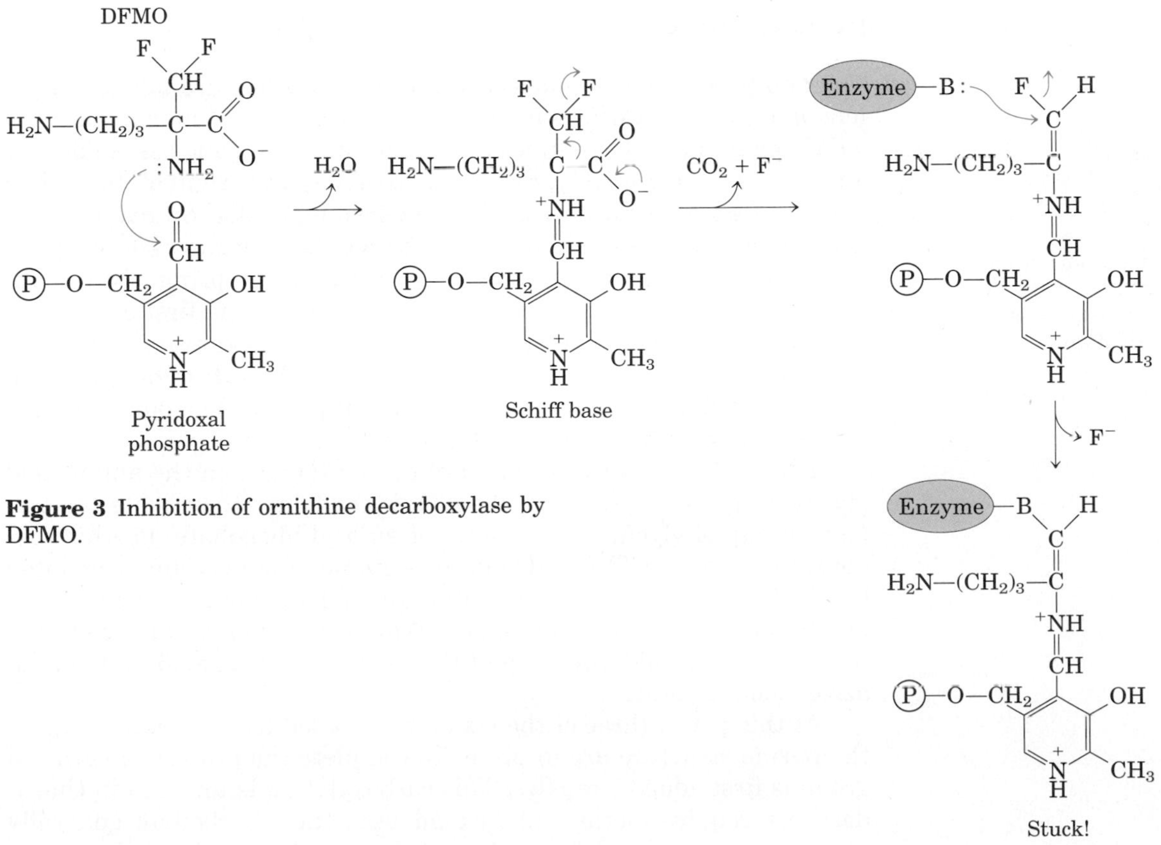

Figure 3 Inhibition of ornithine decarboxylase by DFMO.

Although the free bases are not intermediates in the de novo pathways, they *are* intermediates in some of the salvage pathways.

Several important precursors are shared by the de novo pathways for pyrimidines and purines. PRPP is important in both, and here the structure of ribose is retained in the product nucleotide, in contrast to its fate in the tryptophan and histidine pathways discussed earlier. An amino acid is an important precursor in each pathway: glycine in the case of purines and aspartate for pyrimidines. Glutamine again is the most important source of amino groups, playing this role in five different steps in these pathways. Aspartate is also used twice in the purine pathways as the source of an amino group.

Two other features deserve mention. First, there is evidence, especially in the purine pathway, that the enzymes used are present as large, multienzyme complexes in the cell, a recurring theme in our discussion of metabolism. Second, the pools of nucleotides in cells (exclusive of ATP) are quite small, perhaps 1% or less of the amounts required to synthesize the cellular DNA. Therefore, nucleotide synthesis must continue during nucleic acid synthesis and in some cases may limit the rates of DNA replication and transcription. The importance of these processes in dividing cells has made agents that inhibit nucleotide synthesis particularly important to modern medicine.

We examine here the biosynthetic pathways of purine and pyrimidine nucleotides and their regulation, the formation of the deoxynucleotides, and the degradation of purines and pyrimidines to uric acid and urea. We end with a discussion of chemotherapeutic agents that affect nucleotide synthesis.

De Novo Purine Synthesis Begins with PRPP

The two parent purine nucleotides of nucleic acids are adenosine 5′-monophosphate (AMP; adenylate) and guanosine 5′-monophosphate (GMP; guanylate). These nucleotides contain the purine bases adenine and guanine, respectively. Figure 21–26 shows the origin of the carbon and nitrogen atoms of the purine ring system, as determined by John Buchanan using isotopic tracer experiments in birds. The detailed pathway of purine biosynthesis was worked out primarily by Buchanan and G. Robert Greenberg in the 1950s. In the first committed step of the pathway, an amino group donated by glutamine is attached at C-1 of PRPP (Fig. 21–27). The resulting **5-phosphoribosylamine** is highly unstable, with a half-life of 30 s at pH 7.5. The purine ring is subsequently built up on this structure.

The next step is the addition of three atoms from the amino acid glycine (Fig. 21–27, step ②). An ATP is consumed to activate the carboxyl group of glycine (in the form of an acyl phosphate) for this condensation reaction. The added glycine amino group is then formylated by N^{10}-formyltetrahydrofolate (step ③), and a nitrogen is contributed by glutamine (step ④), before dehydration and ring closure yield the five-membered imidazole ring of the purine nucleus, as 5-aminoimidazole ribonucleotide (step ⑤).

At this point, three of the six atoms needed for the second ring in the purine structure are in place. To complete the process, a carboxyl group is first added (step ⑥). This carboxylation is unusual in that it does not require biotin, but instead uses the bicarbonate generally present in aqueous solutions. Aspartate then donates its amino group to the imidazole ring in two steps (⑦ and ⑧): formation of an amide bond is followed by elimination of the carbon skeleton of aspartate (as

Figure 21–26 Origin of the ring atoms of purines, determined from isotopic experiments with ^{14}C- or ^{15}N-labeled precursors. The formate is supplied in the form of N^{10}-formyltetrahydrofolate.

John Buchanan

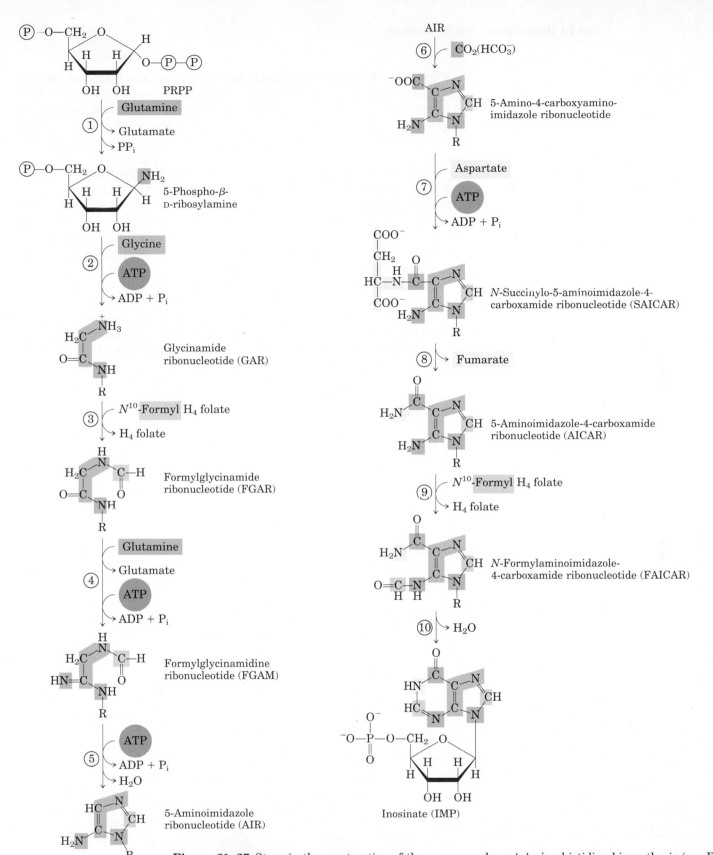

Figure 21–27 Steps in the construction of the purine ring of inosinate. Each addition to the purine ring is shaded to match Fig. 21–26. After step ②, R symbolizes the 5-phospho-D-ribosyl group on which the purine ring is built. Formation of 5-phosphoribosylamine (step ①) is the first committed step in purine synthesis. Note that the product of step ⑧ is 5-aminoimidazole-4-carboxamide ribonucleotide (AICAR), the remnant of ATP that is released during histidine biosynthesis (see Fig. 21–17, step ⑤). Abbreviations are given for most intermediates to simplify the naming of the pathway enzymes. The enzymes are: ① glutamine-PRPP amidotransferase, ② GAR synthetase, ③ GAR transformylase, ④ FGAR amidotransferase, ⑤ FGAM cyclase (AIR synthetase), ⑥ AIR carboxylase, ⑦ SAICAR synthetase, ⑧ SAICAR lyase, ⑨ AICAR transformylase, and ⑩ IMP synthase.

fumarate). Recall that aspartate plays an analogous role in two steps of the urea cycle (see Fig. 17–11). The final carbon is contributed by N^{10}-formyltetrahydrofolate (step ⑨), and a second ring closure takes place to yield the second of the two fused rings of the purine nucleus (step ⑩). The first intermediate to have a complete purine ring is **inosinate** (IMP).

As in the tryptophan and histidine biosynthetic pathways, the enzymes in the pathway leading to IMP appear to exist as large multienzyme complexes in the cell. Once again, evidence comes from the existence of single polypeptides with several functions, some of which catalyze several nonsequential steps in the pathway. In eukaryotic cells ranging from yeast to fruit flies to chickens, steps ②, ③, and ⑤ in Figure 21–27 are catalyzed by such a multifunctional protein. Additional multifunctional proteins catalyze steps ⑥ and ⑦ and steps ⑨ and ⑩. In bacteria, these activities are found on separate proteins, but a large noncovalent complex may exist there as well. The channeling of reaction intermediates from one enzyme to the next permitted by these complexes is probably especially important in the case of unstable intermediates such as 5-phosphoribosylamine.

The conversion of inosinate to adenylate (Fig. 21–28) requires the insertion of an amino group derived from aspartate; this takes place by a series of two reactions similar to those used to introduce N-1 of the purine ring (Fig. 21–27, steps ⑦ and ⑧). A key difference is that GTP is used in place of ATP as the source of the high-energy phosphate in synthesizing adenylosuccinate. Guanylate is formed by the oxidation of inosinate at C-2 using NAD^+, followed by the addition of an amino group derived from glutamine. ATP is cleaved to AMP and PP$_i$ in the final step (Fig. 21–28).

Figure 21–28 Synthesis of AMP and GMP from IMP. The enzymes are: ① adenylosuccinate synthetase, ② adenylosuccinate lyase, ③ IMP dehydrogenase, and ④ XMP-glutamine amidotransferase.

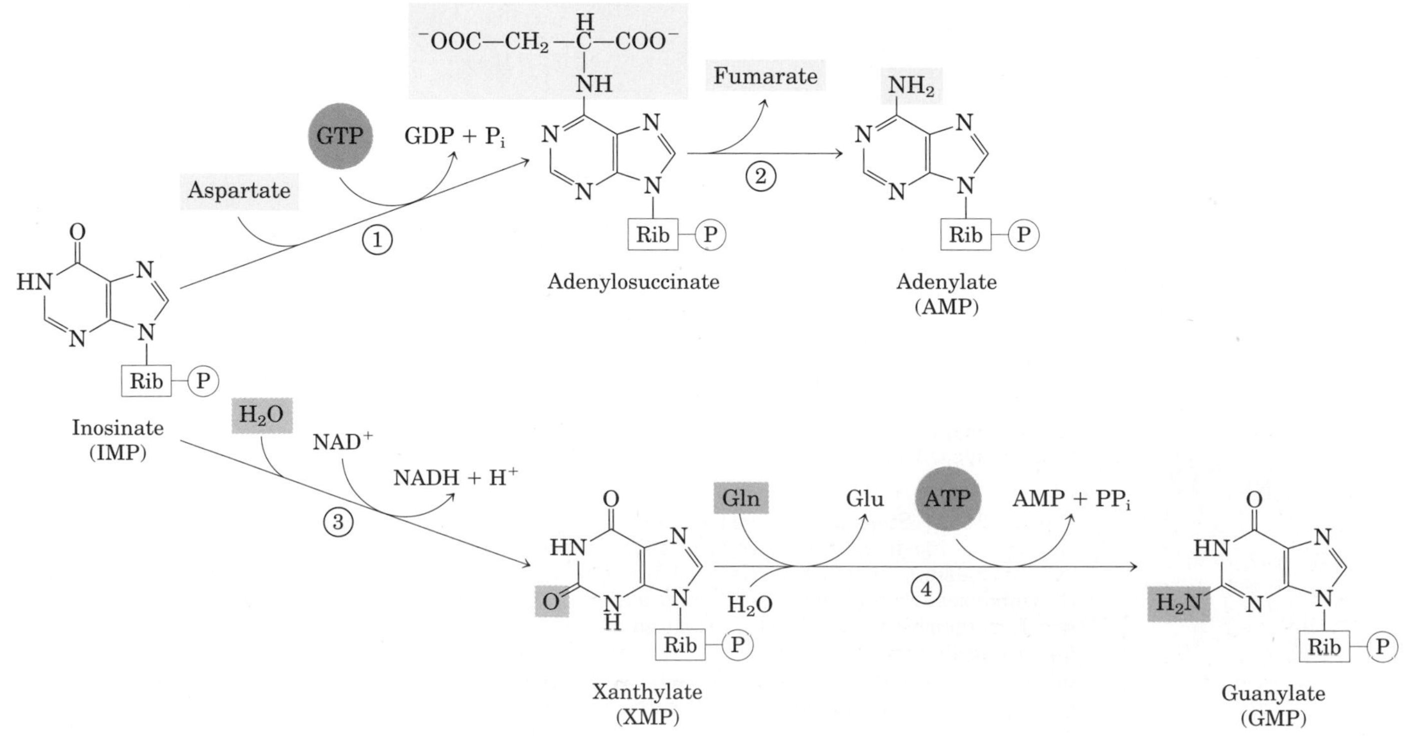

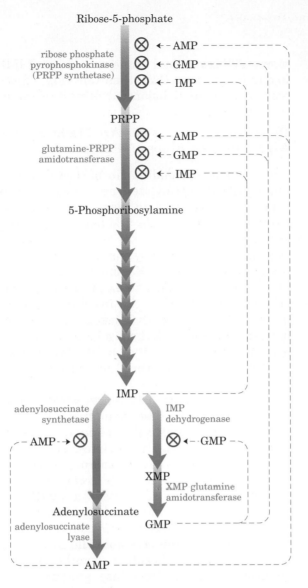

Ribose-5-phosphate

ribose phosphate
pyrophosphokinase
(PRPP synthetase)

⊗ ← AMP
⊗ ← GMP
⊗ ← IMP

PRPP

glutamine-PRPP
amidotransferase

⊗ ← AMP
⊗ ← GMP
⊗ ← IMP

5-Phosphoribosylamine

IMP

adenylosuccinate
synthetase

IMP
dehydrogenase

– AMP – → ⊗   ⊗ ← GMP –

XMP

XMP glutamine
amidotransferase

Adenylosuccinate

GMP

adenylosuccinate
lyase

AMP

Figure 21–29 Feedback control mechanisms in the biosynthesis of adenine and guanine nucleotides in *E. coli.* Regulation of these pathways varies in other organisms.

Purine Nucleotide Biosynthesis Is Regulated by Feedback Control

Three major feedback mechanisms cooperate in regulating the overall rate of de novo purine nucleotide synthesis and the relative rates of formation of the two end products, adenylate and guanylate (Fig. 21–29). The first of these control mechanisms is exerted on the first reaction that is unique to purine synthesis—the transfer of an amino group to PRPP to form 5-phosphoribosylamine. This reaction is catalyzed by the allosteric enzyme glutamine-PRPP amidotransferase, which is inhibited by the end products IMP, AMP, and GMP. These same nucleotides inhibit the synthesis of PRPP from ribose phosphate by ribose phosphate pyrophosphokinase. AMP and GMP act synergistically in this inhibition. Thus, whenever either AMP or GMP accumulates to excess, the first step in its biosynthesis from PRPP is partially inhibited.

In the second control mechanism, exerted at a later stage, an excess of GMP in the cell inhibits formation of xanthanylate from inosinate by IMP dehydrogenase, without affecting the formation of AMP (Fig. 21–29). Conversely, an accumulation of adenylate results in inhibition of formation of adenylosuccinate by adenylosuccinate synthetase, without affecting the biosynthesis of GMP. In the third mecha-

Aspartate

aspartate
trans-
carbamoylase

Carbamoyl
phosphate

P_i

N-Carbamoylaspartate

dihydroorotase

H_2O

L-Dihydroorotate

dihydroorotate
dehydrogenase

NAD^+

$NADH + H^+$

Orotate

orotate
phosphoribosyl
transferase

PRPP

PP_i

Orotidylate

orotidylate
decarboxylase

CO_2

Uridylate (UMP)

kinases

2 ATP

2ADP

Uridylate
5'-triphosphate (UTP)

cytidylate
synthetase

Gln

Glu

ATP

ADP + P_i

Cytidine 5'-triphosphate (CTP)

nism, GTP is required in the conversion of IMP to AMP, whereas ATP is required to form GMP from IMP (Fig. 21–28), a reciprocal arrangement that tends to balance synthesis of the two ribonucleotides.

Pyrimidine Nucleotides Are Made from Aspartate and Ribose-5-Phosphate

The common pyrimidine ribonucleotides are cytidine 5'-monophosphate (CMP; cytidylate) and uridine 5'-monophosphate (UMP; uridylate), which contain the pyrimidines cytosine and uracil, respectively. Pyrimidine nucleotide biosynthesis (Fig. 21–30) proceeds in a somewhat different manner from purine nucleotide synthesis; in this case the six-membered pyrimidine ring is made first and then attached to ribose-5-phosphate. Required in this process is carbamoyl phosphate, also an intermediate in the urea cycle (see Fig. 17–11). However, as we noted in Chapter 17, in animals the carbamoyl phosphate required in urea synthesis is made in the mitochondria by a mitochondrial enzyme, carbamoyl phosphate synthetase I, whereas the carbamoyl phosphate required in pyrimidine biosynthesis is made in the cytosol by a different form of the enzyme, **carbamoyl phosphate synthetase II.**

Carbamoyl phosphate reacts with aspartate to yield N-carbamoylaspartate in the first committed step of pyrimidine biosynthesis. This reaction is catalyzed by **aspartate transcarbamoylase.** In bacteria, this step is highly regulated, and bacterial aspartate transcarbamoylase is one of the most thoroughly studied allosteric enzymes, as discussed again below. By removal of water from N-carbamoylaspartate, a reaction catalyzed by **dihydroorotase,** the pyrimidine ring is closed to form L-dihydroorotate. This compound is oxidized to yield the pyrimidine derivative orotate, a reaction in which NAD^+ is the ultimate electron acceptor. The first three enzymes in this pathway, carbamoyl phosphate synthetase II, aspartate transcarbamoylase, and dihydroorotase, are part of a single trifunctional protein in eukaryotes. The protein, which is known by the acronym CAD, contains three identical polypeptide chains (each of M_r 230,000), each of which has active sites for all three reactions. This suggests that large, multienzyme complexes may be the rule in this pathway as elsewhere.

Once orotate is formed, the ribose-5-phosphate side chain, provided once again by PRPP, is attached to orotate to yield orotidylate. Orotidylate is then decarboxylated to yield uridylate, which is phosphorylated to UTP. CTP is formed from UTP by the action of **cytidylate synthetase** (Fig. 21–30). This reaction occurs by way of an acyl phosphate intermediate (consuming one ATP), and the nitrogen donor is glutamine in animals or NH_4^+ in some bacteria.

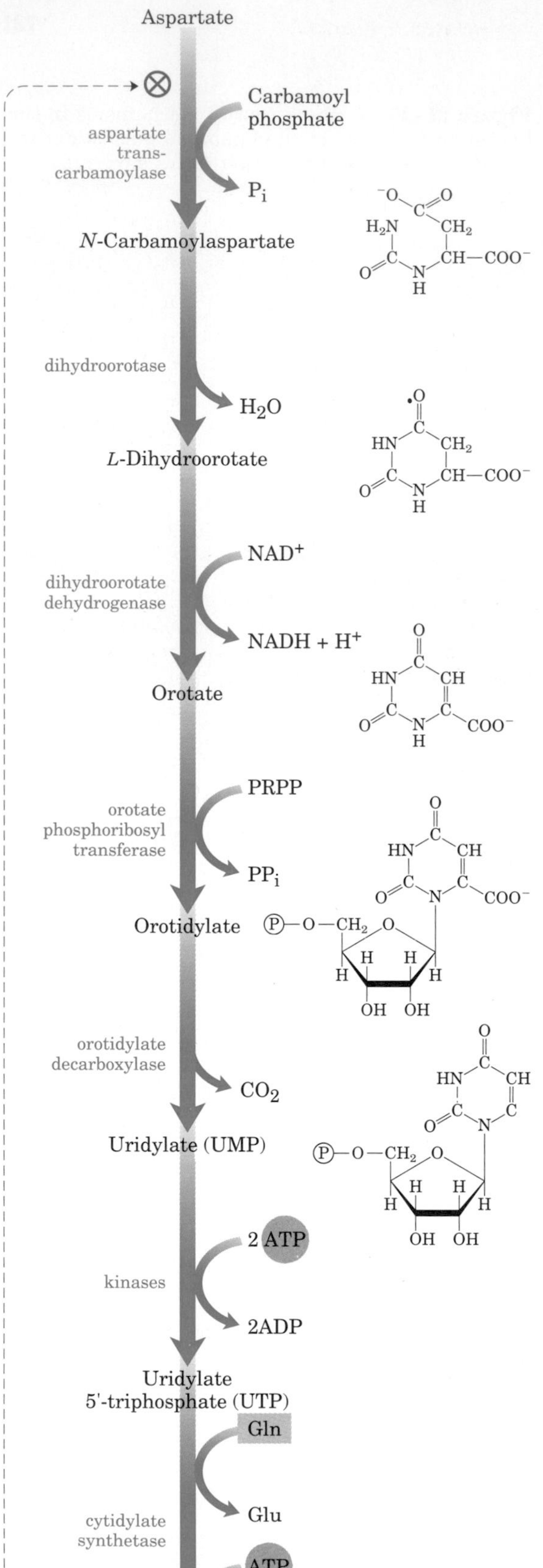

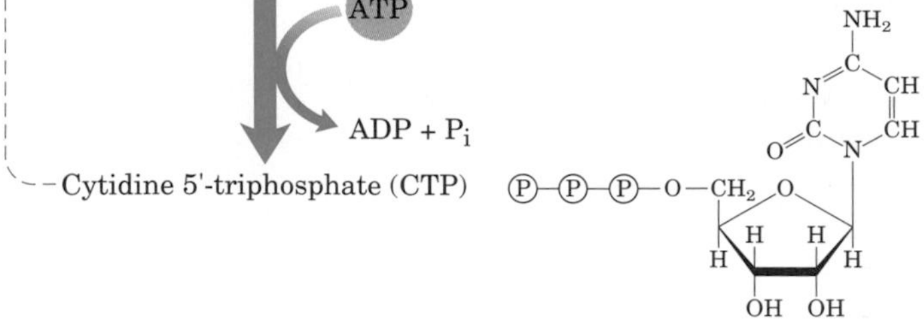

Figure 21–30 Biosynthesis of the pyrimidine nucleotides UTP and CTP via orotidylate. The ribose-5-phosphate is added to the completed pyrimidine ring by orotate phosphoribosyl transferase.

Pyrimidine Nucleotide Biosynthesis Is Regulated by Feedback Inhibition

The regulation of the rate of pyrimidine nucleotide synthesis in bacteria occurs in large part through the enzyme aspartate transcarbamoylase (ATCase), which catalyzes the first reaction in the sequence. This enzyme is inhibited by CTP, the end product of this sequence of reactions (Fig. 21–30). The bacterial ATCase molecule consists of six catalytic subunits and six regulatory subunits (see Fig. 8–26). The catalytic subunits bind the substrate molecules, and the allosteric subunits bind the allosteric inhibitor CTP. The entire ATCase molecule, as well as its subunits, exists in two conformations, active and inactive. When the regulatory subunits are empty, the enzyme is maximally active. However, when CTP accumulates it is bound by the regulatory subunits, causing a change in their conformation. This change is transmitted to the catalytic subunits, which then also shift to an inactive conformation. The presence of ATP prevents the changes induced by CTP. Figure 21–31 shows the effects of the allosteric regulators on the activity of ATCase.

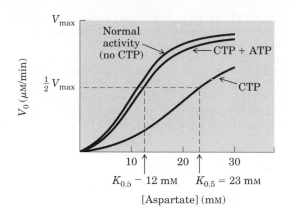

Figure 21–31 Effect of the allosteric modulators CTP and ATP on the rate of conversion of aspartate into N-carbamoylaspartate by aspartate transcarbamoylase. The addition of 0.8 mM CTP, the allosteric inhibitor of ATCase, increases the $K_{0.5}$ for aspartate (lower curve). ATP at 0.6 mM fully reverses this effect (middle curve).

Nucleoside Monophosphates Are Converted to Nucleoside Triphosphates

Nucleotides are generally used in biosynthesis in the form of nucleoside triphosphates. The conversion pathways are common to all cells. The phosphorylation of AMP to ADP is promoted by **adenylate kinase,** in the reaction

$$\text{ATP} + \text{AMP} \rightleftharpoons 2\text{ADP}$$

The ADP so formed is then phosphorylated to ATP by the glycolytic enzymes or through oxidative phosphorylation.

ATP also brings about the formation of other nucleoside diphosphates by the action of a class of enzymes called **nucleoside monophosphate kinases.** These enzymes, which are generally specific for a particular base but nonspecific as to whether the sugar is ribose or deoxyribose, catalyze the reaction

$$\text{ATP} + \text{NMP} \rightleftharpoons \text{ADP} + \text{NDP}$$

The efficient cellular systems for rephosphorylation of ADP to ATP tend to pull this reaction in the direction of products.

Nucleoside diphosphates are converted to triphosphates by the action of a ubiquitous enzyme, **nucleoside diphosphate kinase,** which catalyzes the reaction

$$\text{NTP}_D + \text{NDP}_A \rightleftharpoons \text{NDP}_D + \text{NTP}_A$$

This enzyme is notable in that it is nonspecific for the base (purines or pyrimidines) or for ribose or deoxyribose. This nonspecificity applies to both phosphate acceptor (A) and donor (D), although the donor is almost invariably ATP because it is present in high concentrations as a respiratory product under aerobic conditions.

Ribonucleotides Are the Precursors of the Deoxyribonucleotides

Deoxyribonucleotides, the building blocks of DNA, are derived from the corresponding ribonucleotides by reactions in which the 2′-carbon

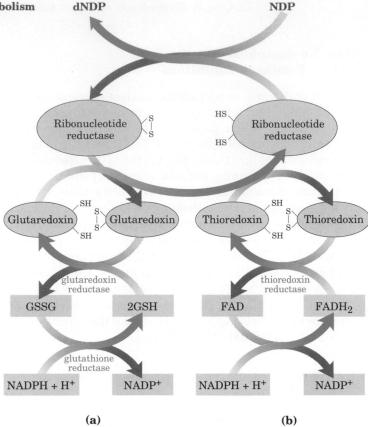

Figure 21–32 Reduction of ribonucleotides by ribonucleotide reductase. Electrons are transmitted (blue arrows) to the enzyme from NADPH via either (a) glutaredoxin or (b) thioredoxin. The sulfide groups in glutaredoxin reductase are contributed by two molecules of bound glutathione (GSH). Note that thioredoxin reductase is a flavoenzyme, with FAD as prosthetic group. Oxidized glutathione is denoted GSSG (Fig. 21–22).

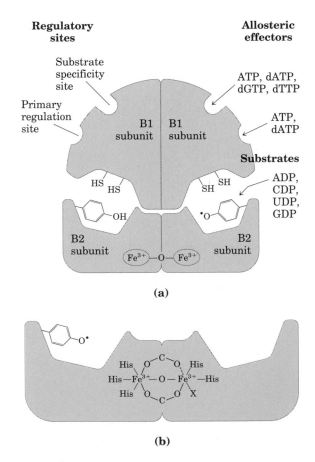

Figure 21–33 (a) Model for the structure of ribonucleotide reductase. The results of effector binding to the two types of regulatory sites are shown in Fig. 21–35. (b) The postulated structure of the binuclear iron cofactor and the tyrosyl radical.

atom of the D-ribose portion of the ribonucleotide is directly reduced to form the 2′-deoxy derivative. The substrates for this reaction, catalyzed by the enzyme **ribonucleotide reductase,** are ribonucleoside diphosphates. In this way, for example, adenosine diphosphate (ADP) is reduced to form 2′-deoxyadenosine diphosphate (dADP), and GDP is reduced to dGDP.

The reduction of the D-ribose portion of the ribonucleoside diphosphates to 2′-deoxy-D-ribose requires a pair of hydrogen atoms, which are ultimately donated by NADPH via an intermediate hydrogen-carrying protein, **thioredoxin.** This protein has pairs of —SH groups that carry hydrogen atoms from NADPH to the ribonucleoside diphosphate. The oxidized or disulfide form of thioredoxin is reduced by NADPH in a reaction catalyzed by **thioredoxin reductase** (Fig. 21–32). The reduced thioredoxin is then used by ribonucleotide reductase to reduce the nucleoside diphosphates (NDPs) to deoxyribonucleoside diphosphates (dNDPs). A second source of reducing equivalents for ribonucleotide reductase is glutathione (GSH), which serves as the reductant for a protein closely related to thioredoxin called **glutaredoxin.** Reduced glutaredoxin then transfers the reducing power of glutathione to ribonucleotide reductase (Fig. 21–32).

Ribonucleotide reductase is notable in that its reaction mechanism provides the best-characterized example of the involvement of free radicals in biochemical transformations. As such, it is a prototype for reactions involving radical intermediates, once thought to be rare in biological systems. The enzyme in *E. coli* and most eukaryotes is a dimer of two subunits, B1 and B2 (Fig. 21–33). The B1 subunit contains two kinds of regulatory effector-binding sites as described below. The two active sites of the enzyme are formed at the interface between the B1 and B2 subunits. At each active site, B1 contributes two thiol groups required for activity and B2 contributes a stable tyrosyl radical. The B2 subunit also has a binuclear iron cofactor that helps generate and stabilize the tyrosyl radicals (Fig. 21–33).

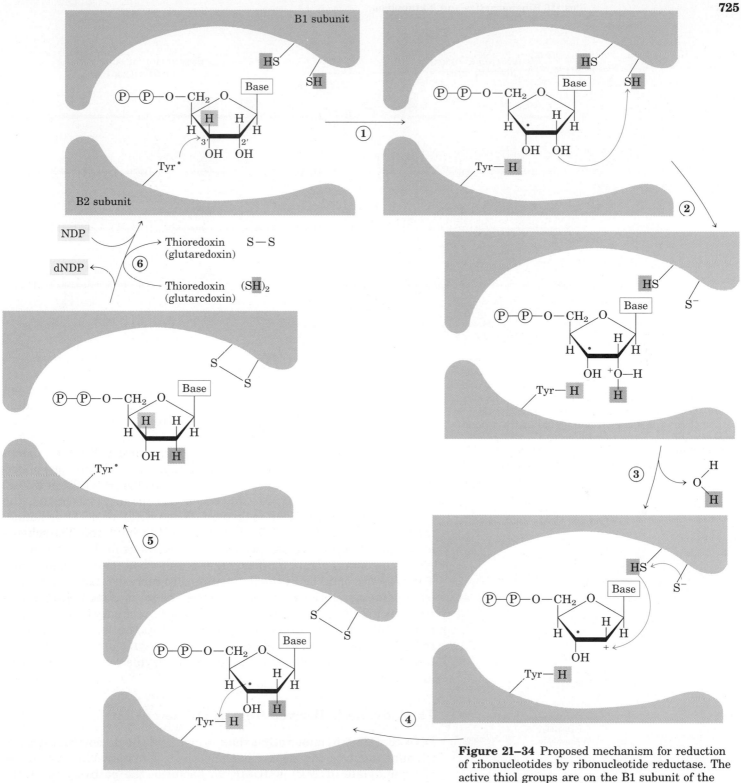

Figure 21–34 Proposed mechanism for reduction of ribonucleotides by ribonucleotide reductase. The active thiol groups are on the B1 subunit of the enzyme; the tyrosyl radical is on the B2 subunit (see Fig. 21–33). Steps (1) through (6) are described in the text.

A likely mechanism for ribonucleotide reductase is illustrated in Figure 21–34. The 3′-ribonucleotide radical formed in step (1) helps stabilize the cation subsequently formed at the 2′ carbon after the loss of H_2O (steps (2) and (3)). Two one-electron transfers accompanied by oxidation of the dithiol of B1 reduce the radical cation and regenerate the 3′-ribonucleotide radical (step (4)). Step (5) is the reverse of step (1), regenerating the tyrosyl radical and forming the deoxy product. The oxidized dithiol of B1 is reduced by thioredoxin or glutaredoxin to complete the cycle (step (6)).

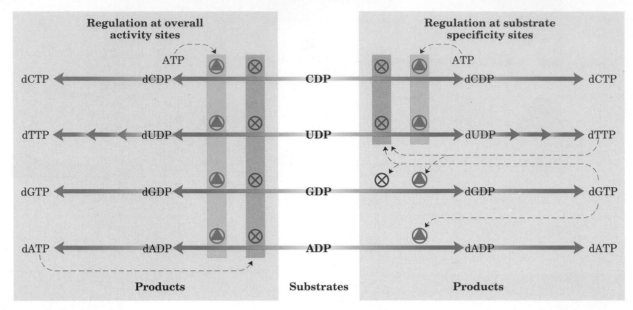

Figure 21–35 Regulation of ribonucleotide reductase by deoxynucleoside triphosphates. The overall activity of the enzyme is affected by binding at one type of regulatory site (shown on the left). The substrate specificity of the enzyme is affected by the nature of the effector molecule bound at the second type of regulatory site (shown on the right). Inhibition or stimulation of the enzyme's activity with the four different substrates is indicated. The pathway from dUDP to dTTP is described later (see Fig. 21–36).

The regulation of ribonucleotide reductase is unusual in that not only its *activity* but its *substrate specificity* is regulated by the binding of effector molecules. There are two types of regulatory sites on each B1 subunit (Fig. 21–33). One type affects overall enzyme activity and binds either ATP, which activates the enzyme, or dATP, which inactivates it. The second type of regulatory site alters substrate specificity in response to the effector molecule (either ATP, dATP, dTTP, or dGTP) that is bound there (Fig. 21–35). When ATP or dATP is bound, the reduction of UDP and CDP is favored. When dTTP or dGTP is bound, the reduction of GDP and ADP, respectively, is stimulated. The scheme is designed to provide a balanced pool of precursors for DNA synthesis. ATP is a general signal for biosynthesis and ribonucleotide reduction. The presence of dATP in small amounts also increases the reduction of pyrimidine nucleotides. An oversupply of the pyrimidine dNTPs is signaled by high levels of dTTP, which shifts the specificity to favor reduction of GDP. High levels of dGTP, in turn, shift the specificity to ADP reduction, and high levels of dATP shut the enzyme down. These effectors are thought to induce several distinct enzyme conformations with altered specificities.

Thymidylate Is Derived from dCDP and dUMP

DNA contains thymine rather than uracil, and the de novo pathway to thymine involves only deoxyribonucleotides. The immediate precursor of thymidylate (dTMP) is dUMP. In bacteria, the pathway to dUMP begins with formation of dUTP, either by deamination of dCTP or by phosphorylation of dUDP (Fig. 21–36). The dUTP is converted to dUMP by a dUTPase. This latter reaction must be efficient to keep dUTP pools low and prevent the incorporation of uridylate into DNA.

The conversion of dUMP to dTMP is catalyzed by the enzyme **thymidylate synthase.** In this reaction a one-carbon unit is transferred from N^5,N^{10}-methylenetetrahydrofolate to dUMP at the hydroxymethyl (—CH$_2$OH) oxidation level (see Fig. 17–19), then reduced to a methyl group (Fig. 21–37). The reduction comes at the expense of oxidation of tetrahydrofolate to dihydrofolate and is unusual in reactions

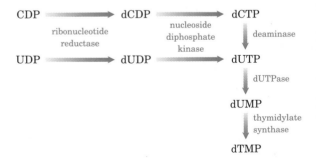

Figure 21–36 Origin of thymidylate (dTMP). The pathways are shown beginning with the reaction catalyzed by ribonucleotide reductase. Details of the thymidylate synthase reaction are shown in Fig. 21–37.

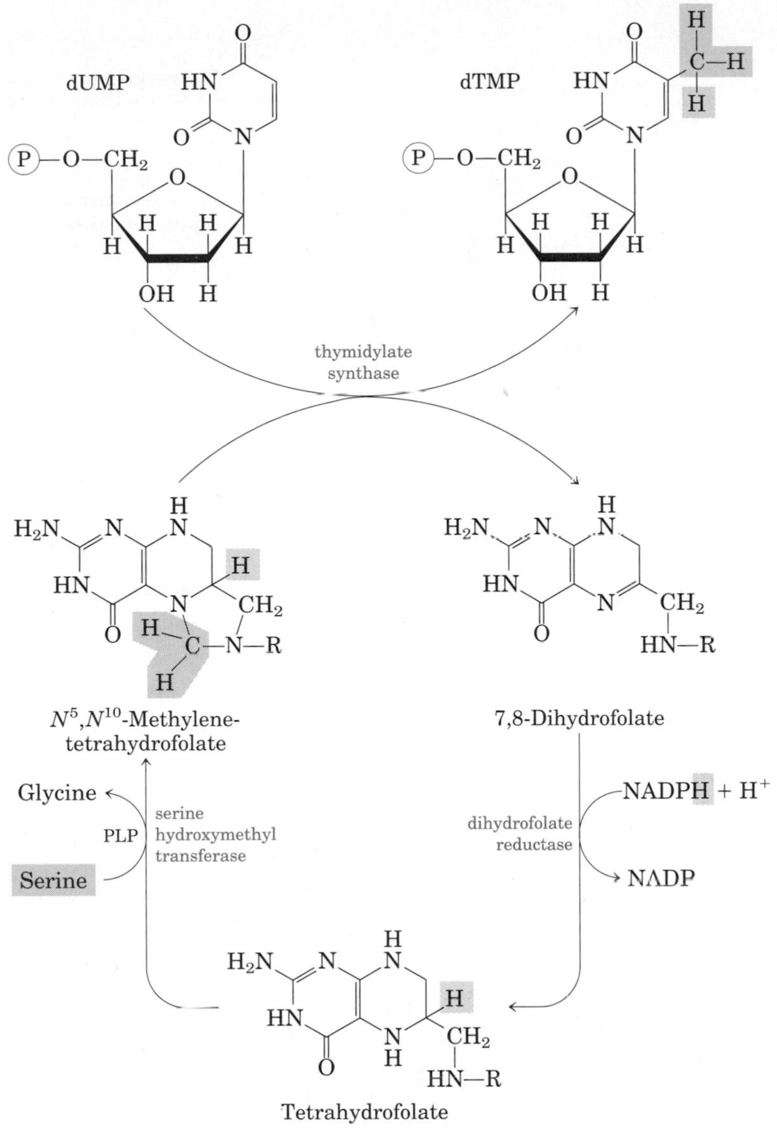

Figure 21–37 Conversion of dUMP to dTMP by thymidylate synthase and dihydrofolate reductase. Serine hydroxymethyl transferase is required for regeneration of the N^5,N^{10}-methylene form of H_4 folate. In the synthesis of dTMP, all three hydrogens of the added methyl group are derived from the N^5,N^{10}-methylenetetrahydrofolate, as shown in red and gray.

using tetrahydrofolate as a cofactor. (Details of this reaction are shown in Fig. 21–43.) Dihydrofolate is reduced again to tetrahydrofolate by the enzyme **dihydrofolate reductase.** This regeneration of tetrahydrofolate is essential for the many processes that depend on this form of the coenzyme. In at least one protist, thymidylate synthase and dihydrofolate reductase are combined in a single polypeptide as a bifunctional protein.

Degradation of Purines and Pyrimidines Leads to Uric Acid and Urea, Respectively

Purine nucleotides are degraded by a pathway (Fig. 21–38) in which the phosphate group is lost by the action of **5′-nucleotidase.** Adenylate yields adenosine, which is then deaminated to inosine by **adenosine deaminase.** Inosine is hydrolyzed to yield its purine base hypoxanthine and D-ribose. Hypoxanthine is oxidized successively to xanthine and then uric acid by **xanthine oxidase,** a flavoenzyme that contains an atom of molybdenum and four iron–sulfur centers (see Fig. 18–5) in its prosthetic group. Molecular oxygen is the electron acceptor in this complex reaction.

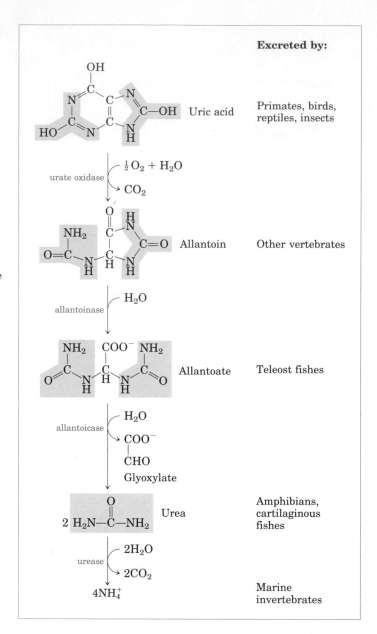

Figure 21–38 Purine nucleotide catabolism. Note that in primates, much more nitrogen is excreted as urea via the urea cycle (Chapter 17) than as uric acid from purine degradation. Similarly, in fish much more nitrogen is excreted as NH_4^+ than as the urea produced by the pathway shown here.

GMP catabolism also yields uric acid as end product. GMP is first hydrolyzed to yield the nucleoside guanosine, which is then cleaved to free guanine. Guanine undergoes hydrolytic removal of its amino group to yield xanthine, which is converted into uric acid by xanthine oxidase (Fig. 21–38).

Uric acid is the excreted end product of purine catabolism in primates, birds, and some other animals. However, in many other vertebrates uric acid is degraded further to the excretory product **allantoin,** by the action of **urate oxidase.** In other organisms the pathway is further extended, as shown in Figure 21–38. The rate of uric acid excretion by the normal adult human is about 0.6 g/24 h, arising in part from ingested purines and in part from the turnover of the purine nucleotides of nucleic acids.

The pathways for degradation of pyrimidines generally lead to urea. Thymine, for example, is degraded to methylmalonyl semialdehyde (Fig. 21–39), which is an intermediate in the valine degradation pathway. It is further degraded to methylmalonyl-CoA and then to succinyl-CoA (Fig. 17–30).

Genetic aberrations in human purine metabolism have been found, some with serious consequences. For example, **adenosine deaminase deficiency** leads to severe immunodeficiency diseases in humans. The T lymphocytes and B lymphocytes crucial to the immune system do not develop properly. A lack of adenosine deaminase leads to a 100-fold increase in the concentration of dATP, a strong negative effector of ribonucleotide reductase (Fig. 21–35). The increase in dATP leads to the general deficiency in the levels of other dNTPs that is observed in T lymphocytes. The basis for B-lymphocyte toxicity is less clear. Patients with adenosine deaminase deficiencies lack an effective immune system and do not survive unless they are kept in a sterile "bubble" environment.

Purine and Pyrimidine Bases Are Recycled by Salvage Pathways

Free purine and pyrimidine bases are constantly formed in cells during the metabolic degradation of nucleotides by pathways described above. However, free purines formed on degradation of purine nucleotides are in large part salvaged and used again to make nucleotides. This occurs by a pathway that is quite different from the de novo biosynthesis of purines described earlier, in which the purine ring system is assembled step by step on ribose-5-phosphate in a long series of reactions. The salvage pathways are much simpler. One of the primary salvage pathways consists of a single reaction catalyzed by **adenosine phosphoribosyltransferase,** in which free adenine reacts with PRPP to yield the corresponding adenine nucleotide:

$$\text{Adenine} + \text{PRPP} \longrightarrow \text{AMP} + \text{PP}_i$$

Free guanine and hypoxanthine (the deamination product of adenine; see Fig. 21–38) are salvaged in the same way by a different enzyme, **hypoxanthine-guanine phosphoribosyltransferase.** A similar salvage pathway exists for pyrimidine bases in microorganisms, but pyrimidines are not salvaged in significant amounts in mammals.

The genetic lack of hypoxanthine-guanine phosphoribosyltransferase activity, seen almost exclusively in male children, results in a bizarre set of symptoms, called **Lesch–Nyhan syndrome.** Children with this genetic disorder, which becomes manifest by the age of 2 years, are mentally retarded and badly coordinated. In addition, they are extremely hostile and show compulsive self-destructive tendencies: they mutilate themselves by biting off their fingers, toes, and lips.

Lesch–Nyhan syndrome illustrates the importance of the salvage pathways. Hypoxanthine and guanine arise continually as breakdown products of nucleic acids. A lack of the crucial salvage enzyme hypoxanthine-guanine phosphoribosyltransferase results in a rise in PRPP levels, which leads to a general increase in de novo purine synthesis. Overproduction of purines leads to high levels of uric acid production, and goutlike damage to tissue occurs (see below). The brain is especially dependent on the salvage pathways, and this may account for the central nervous system damage that occurs in children with Lesch–Nyhan syndrome. This syndrome, and the immunodeficiency disease resulting from a lack of adenosine deaminase, are among the targets of early trials in human gene therapy (see Box 28–2).

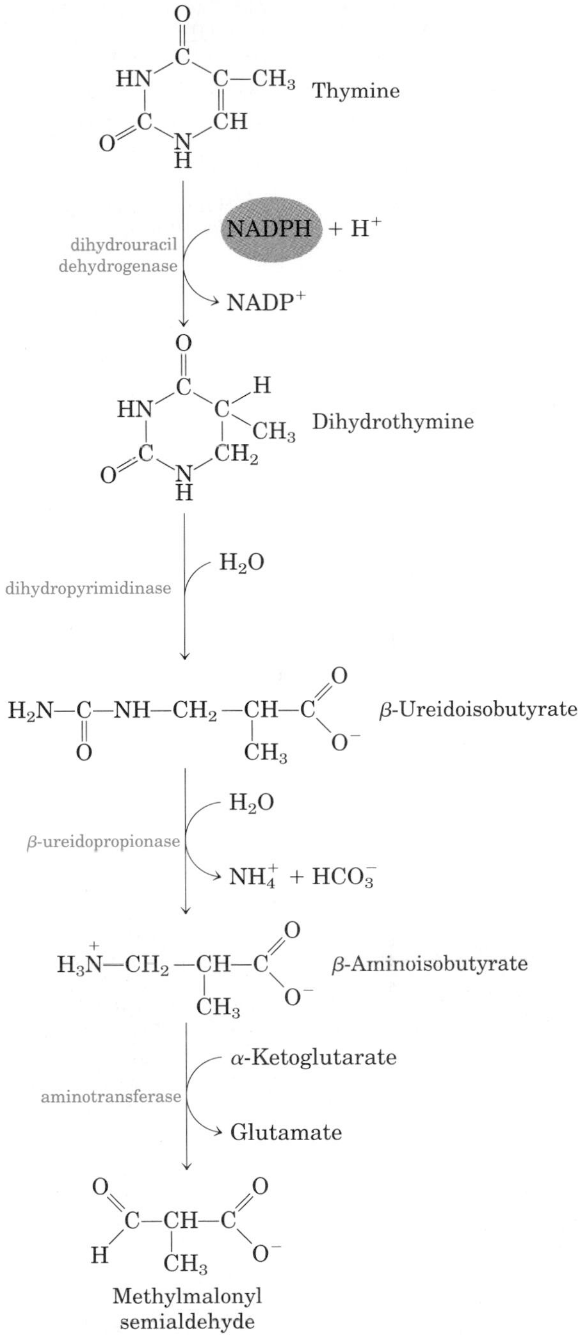

Figure 21–39 Degradative pathway for thymine.

Allopurinol   Hypoxanthine
                     (enol form)

Figure 21–40 Allopurinol, an inhibitor of xanthine oxidase. Only a slight alteration (shaded) in the structure of the substrate hypoxanthine yields a medically effective enzyme inhibitor. Allopurinol is an example of a useful drug that was designed to be a competitive inhibitor.

Gertrude Elion

George Hitchings

Overproduction of Uric Acid Causes Gout

The disease gout, long erroneously thought to be due to "high living," is a disease of the joints, usually in males, caused by an elevated concentration of uric acid in the blood and tissues. The joints become inflamed, painful, and arthritic, owing to the abnormal deposition of crystals of sodium urate. The kidneys are also affected, because excess uric acid is deposited in the kidney tubules. The precise cause of gout is not known, but it is suspected to be due to a genetic deficiency of one or another enzyme concerned in purine metabolism.

Gout can be effectively treated by a combination of nutritional and drug therapies. Foods especially rich in nucleotides and nucleic acids, such as liver or glandular products, are withheld from the diet. In addition, major improvement follows use of the drug **allopurinol** (Fig. 21–40), an inhibitor of xanthine oxidase, the enzyme responsible for converting purines into uric acid. When xanthine oxidase is inhibited, the excreted products of purine metabolism are xanthine and hypoxanthine, which are more soluble in water than uric acid and less likely to form crystalline deposits. Allopurinol was developed by Gertrude Elion and George Hitchings, who also developed acyclovir, used to treat AIDS, and other purine analogs used in cancer chemotherapy.

Many Chemotherapeutic Agents Target Enzymes in the Nucleotide Biosynthetic Pathways

Cancer cells grow more rapidly than the cells of most normal tissues, and thus they have greater requirements for nucleotides as precursors to DNA and RNA synthesis. Consequently, cancer cells are generally more sensitive to inhibitors of nucleotide biosynthesis than are normal cells. A growing array of important chemotherapeutic agents act by inhibiting one or more enzymes in these pathways. We will examine several well-studied examples that both illustrate productive approaches to treatment of cancer and facilitate an understanding of how these enzymes work.

The first set of examples includes compounds that inhibit glutamine amidotransferases. Recall that glutamine acts as a nitrogen donor in at least half a dozen separate reactions in nucleotide biosynthesis. The binding sites for glutamine and the mechanism by which NH_4^+ is extracted are quite similar in many of these enzymes. Most are strongly inhibited by glutamine analogs such as **azaserine** and **acivicin** (Fig. 21–41). Azaserine, characterized by John Buchanan in the 1950s, was one of the first examples of a mechanism-based enzyme inactivator (suicide inhibitor; see p. 222 and Box 21–1). Acivicin shows promise as a cancer chemotherapeutic agent.

Other useful targets for pharmaceutical agents are the enzymes thymidylate synthase and dihydrofolate reductase (Fig. 21–42a). These enzymes provide the only cellular pathway for synthesis of thymine. One inhibitor that acts on thymidylate synthase, **fluorouracil** (Fig. 21–42b), is an important chemotherapeutic agent. Fluorouracil itself is not the inhibitor. In the cell, salvage pathways convert it to the deoxynucleoside monophosphate FdUMP, which then binds and inactivates the enzyme. The mechanism of action of FdUMP (Fig. 21–43, p. 732) represents a classic example of mechanism-based enzyme inactivation. Another prominent chemotherapeutic agent, **methotrexate** (Fig. 21–42b) is an inhibitor of dihydrofolate reductase. Methotrexate is a folate analog and acts as a competitive inhibitor. The enzyme binds methotrexate about 100 times better than dihydrofolate. **Aminopterin** also inhibits dihydrofolate reductase.

The medical potential of inhibitors of nucleotide biosynthesis is not limited to cancer treatment. All fast-growing cells (including bacteria and protists) are potential targets. Parasitic protists, such as the trypanosomes that cause African sleeping sickness (African trypanosomiasis) lack pathways for de novo nucleotide biosynthesis and are particularly sensitive to agents that interfere with their scavenging of nucleotides from the surrounding environment via salvage pathways. Allopurinol (Fig. 21–40) and a number of related purine analogs have shown promise for the treatment of African trypanosomiasis and related afflictions. (See Box 21–1 for another approach to the treatment of African trypanosomiasis that has been made possible by advances in our understanding of metabolism and enzyme mechanism.)

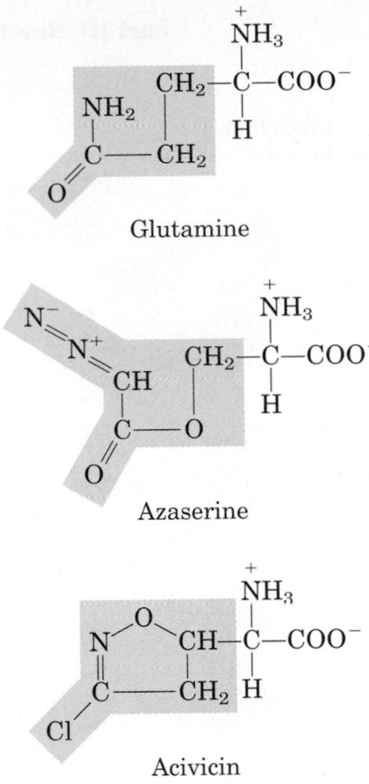

Figure 21–41 Glutamine and two analogs, azaserine and acivicin, that inhibit glutamine amidotransferases and thus interfere in a number of amino acid and nucleotide biosynthetic pathways.

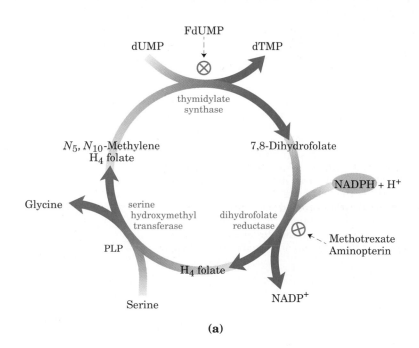

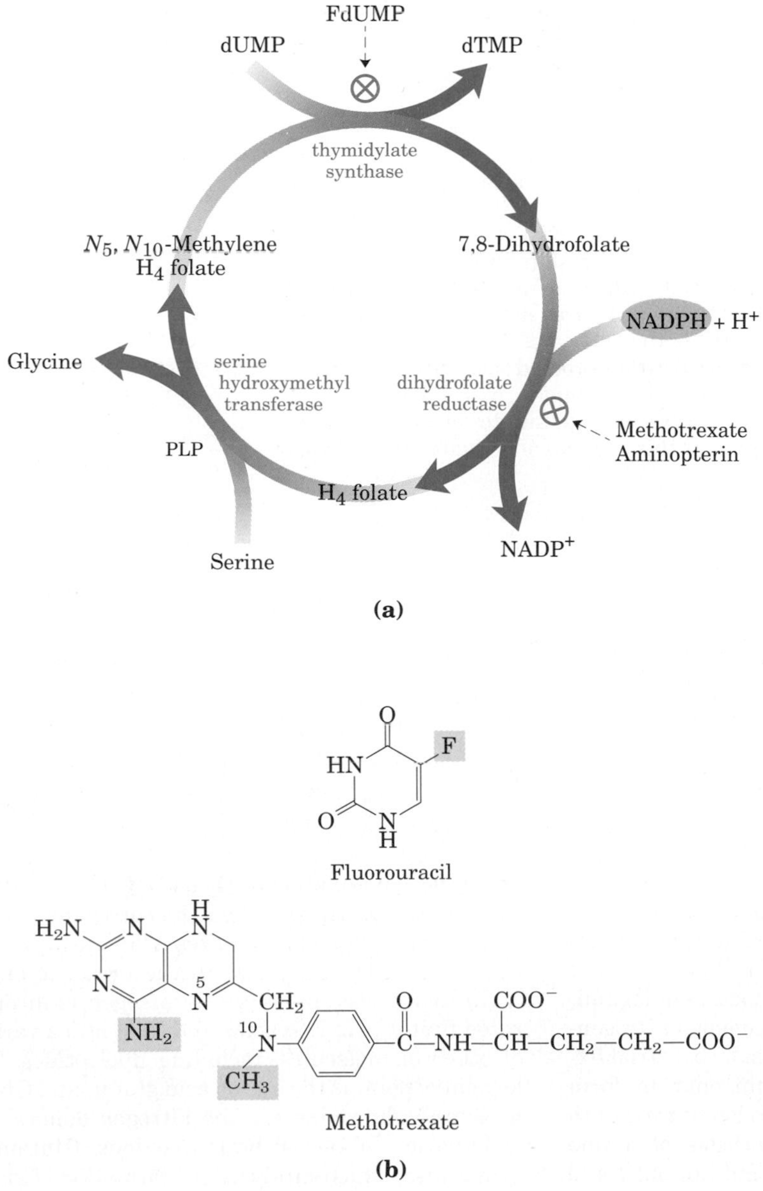

(a)

(b)

Figure 21–42 Thymidylate synthesis and folate metabolism as targets of chemotherapy. **(a)** During thymidylate synthesis, N^5,N^{10}-methylenetetrahydrofolate is converted to 7,8-dihydrofolate. The N^5,N^{10}-methylenetetrahydrofolate is regenerated in two steps. The details of this reaction are shown in Fig. 21–37. The resulting folate cycle is a major target of several chemotherapeutic agents. **(b)** The important chemotherapeutic agents fluorouracil and methotrexate. Fluorouracil is converted to FdUMP by the cell and inhibits thymidylate synthase. Methotrexate is a structural analog of tetrahydrofolate that inhibits dihydrofolate reductase. The shaded amino and methyl groups on methotrexate replace a carbonyl oxygen and a proton, respectively, in hydrofolate (see Fig. 21–37). Another important folate analog, aminopterin, is identical to methotrexate except that it lacks the shaded methyl group.

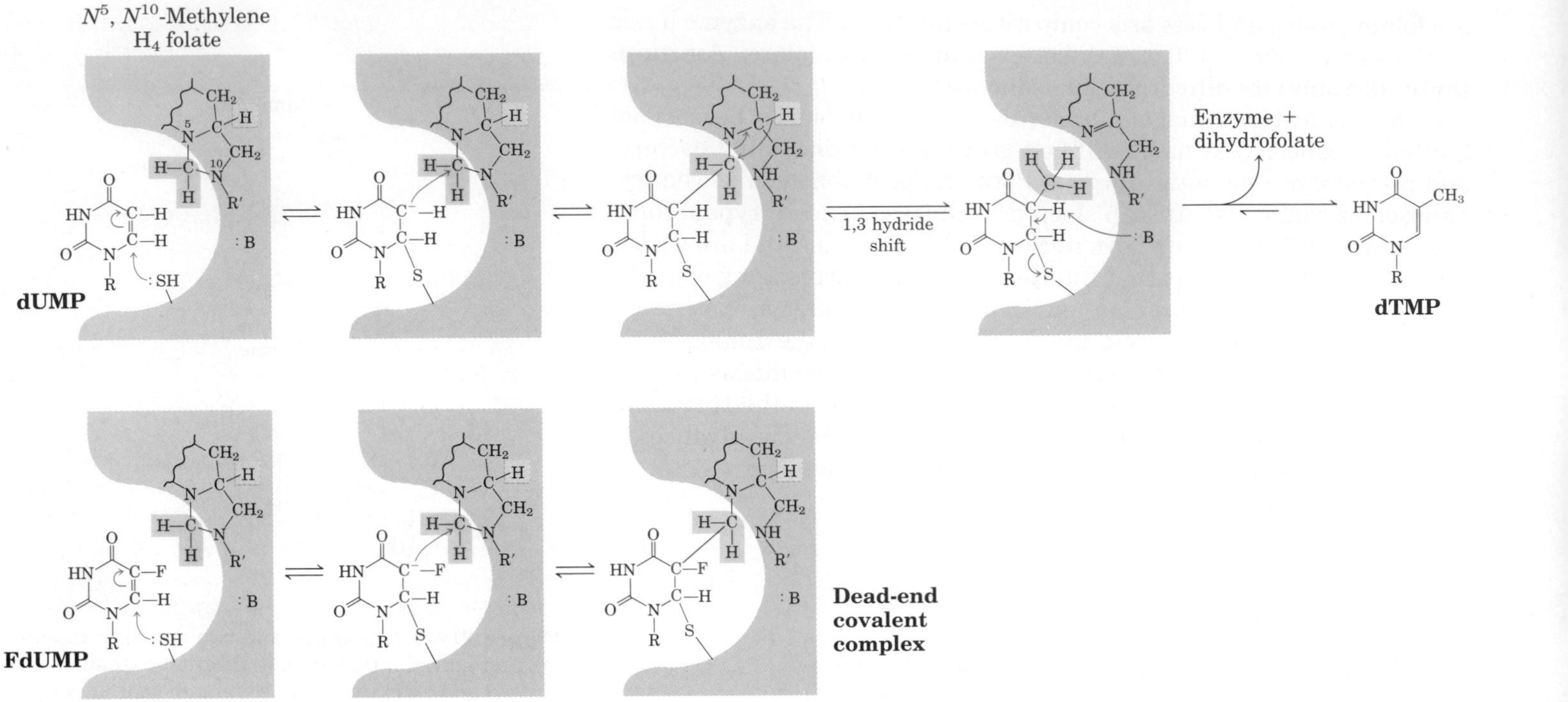

N^5, N^{10}-Methylene
H_4 folate

Figure 21–43 Mechanism of the conversion of dUMP to dTMP catalyzed by thymidylate synthase, and its inhibition by FdUMP. The nucleophilic sulfhydryl group contributed by the enzyme and the ring atoms of dUMP taking part in the reaction are shown in red; :B denotes an amino acid side chain that acts as a base to abstract a proton in the last step. The hydrogens derived from the methylene group of N^5,N^{10}-methylenetetrahydrofolate are shaded in gray. A novel feature of the reaction mechanism of thymidylate synthase (top) is a 1,3 hydride shift, which moves a hydride ion (shaded in red) from C-6 of H_4 folate to the methyl group of thymidine. This is the third step in the reaction as depicted here, and it results in the oxidation of tetrahydrofolate to dihydrofolate. It is this hydride shift that apparently does not occur when the analog FdUMP is the substrate (below). The first two steps of the reaction proceed normally, but result in a stable covalent complex, with FdUMP linked covalently to the enzyme and to tetrahydrofolate. Formation of this complex inactivates the enzyme.

Summary

The molecular nitrogen that makes up 80% of the earth's atmosphere is unavailable to living organisms until it is reduced. Fixation of atmospheric N_2 takes place in certain free-living soil bacteria and in symbiotic bacteria in the root nodules of leguminous plants, by the action of the complex nitrogenase system. Formation of ammonia by bacterial fixation of N_2, nitrification of ammonia to form nitrate by soil organisms, conversion of nitrate to ammonia by higher plants, synthesis of amino acids from ammonia by plants and animals, and conversion of nitrate to N_2 by some soil bacteria in the process of denitrification constitute the nitrogen cycle. The fixation of N_2 as NH_3 is carried out by a protein complex called the nitrogenase complex, in a reaction that requires ATP. The nitrogenase complex is very labile in the presence of O_2.

In living systems, reduced nitrogen is incorporated first into amino acids and then into a variety of other biomolecules, including nucleotides. The key entry point is the amino acid glutamate. Glutamate and glutamine are the nitrogen donors in a wide variety of biosynthetic reactions. Glutamine synthetase, which catalyzes the formation of glutamine from glutamate, is a key regulatory enzyme of nitrogen metabolism.

The amino acid and nucleotide biosynthetic pathways make repeated use of the biological cofactors pyridoxal phosphate, tetrahydrofolate, and *S*-adenosylmethionine. Pyridoxal phosphate is required for transamination reactions involving glutamate and for a number of other amino acid transformations. One-carbon transfers are carried out using *S*-adenosylmethionine (at the —CH$_3$ oxidation level) and tetrahydrofolate (usually at the —CHO and —CH$_2$OH oxidation levels). Enzymes called glutamine amidotransferases are used in reactions that incorporate nitrogen derived from glutamine.

Mammals (e.g., humans and the albino rat) can synthesize 10 of the 20 amino acids of proteins. The remainder, which are required in the diet (essential amino acids), can be synthesized by plants and bacteria. Among the nonessential amino acids, glutamate is formed by reductive amination of α-ketoglutarate and is the precursor of glutamine, proline, and arginine. Alanine and aspartate (and thus asparagine) are formed from pyruvate and oxaloacetate, respectively, by transamination. The carbon chain of serine is derived from 3-phosphoglycerate. Serine is a precursor of glycine; the β-carbon atom of serine is transferred to tetrahydrofolate. Cysteine is formed from methionine and serine by a series of reactions in which *S*-adenosylmethionine and cystathionine are intermediates. The aromatic amino acids (phenylalanine, tyrosine, and tryptophan) are formed via a pathway in which the intermediate chorismate occupies a key branch point. Phosphoribosyl pyrophosphate is a precursor of tryptophan and histidine, both essential amino acids. The biosynthetic pathway to histidine is interconnected with the purine synthetic pathway. Tyrosine can also be formed by hydroxylation of phenylalanine, an essential amino acid. The pathways for biosynthesis of the other essential amino acids in bacteria and plants are complex. The amino acid biosynthetic pathways are subject to allosteric end-product inhibition; the regulatory enzyme is usually the first in the sequence. The regulation of these synthetic pathways is coordinated.

Many other important biomolecules are derived from amino acids. Glycine is a precursor of porphyrins; porphyrins, in turn, are degraded to form bile pigments. Glycine and arginine give rise to creatine and phosphocreatine. Glutathione, a tripeptide, is an important cellular reducing agent. D-Amino acids are synthesized from L-amino acids in bacteria in racemization reactions requiring pyridoxal phosphate. The PLP-dependent decarboxylation of certain amino acids yields some important biological amines, including neurotransmitters. The aromatic amino acids are precursors of a number of plant substances.

The purine ring system is built up in a step-by-step fashion on 5-phosphoribosylamine. The amino acids glutamine, glycine, and aspartate furnish all the nitrogen atoms of purines. Two ring-closure steps ensue to form the purine nucleus. Pyrimidines are synthesized from carbamoyl phosphate and aspartate. Ribose-5-phosphate is then attached to yield the pyrimidine ribonucleotides. Purine and pyrimidine biosynthetic pathways are regulated by feedback inhibition. Nucleoside monophosphates are converted to their triphosphates by enzymatic phosphorylation reactions. Ribonucleotides are converted to deoxyribonucleotides by the action of ribonucleotide reductase, an enzyme with novel mechanistic and regulatory characteristics. The thymine nucleotides are derived from the deoxyribonucleotides dCDP and dUMP. Uric acid and urea are the end products of purine and pyrimidine degradation. Free purines can be salvaged and rebuilt into nucleotides by a separate pathway. Genetic deficiencies in certain salvage enzymes cause serious genetic diseases such as Lesch–Nyhan syndrome and severe immunodeficiency disease. Another genetic deficiency results in the accumulation of uric acid crystals in the joints, causing gout. The enzymes of the nucleotide biosynthetic pathways are targets for an array of chemotherapeutic agents used to treat cancer and other diseases.

Further Reading

Nitrogen Fixation

Burris, R.H. (1991) Nitrogenases. *J. Biol. Chem.* **266,** 9339–9342.

A short and well-written summary.

Orme-Johnson, W.H. (1985) Molecular basis of biological nitrogen fixation. *Annu. Rev. Biophys. Biophys. Chem.* **14,** 419–459.

Shah, V.K., Ugalde, R.A., Imperial, J., & Brill, W.J. (1984) Molybdenum in nitrogenase. *Annu. Rev. Biochem.* **53,** 231–257.

Pathways of Amino Acid Biosynthesis

Bender, D.A. (1985) *Amino Acid Metabolism,* 2nd edn, Wiley-Interscience, New York.

Cooper, A.J.L. (1983) Biochemistry of sulfur-containing amino acids. *Annu. Rev. Biochem.* **52,** 187–222.

Cunningham, E.B. (1978) *Biochemistry: Mechanisms of Metabolism,* McGraw-Hill Book Company, New York.
Excellent description of the enzymatic steps.

Umbarger, H.E. (1978) Amino acid biosynthesis and its regulation. *Annu. Rev. Biochem.* **47,** 533–606.
Definitive review by a pioneer in research on the regulation of these pathways.

Walsh, C. (1979) *Enzymatic Reaction Mechanisms,* W.H. Freeman and Company, New York.
This book includes excellent accounts of reaction mechanisms, including one-carbon metabolism and pyridoxal phosphate enzymes.

Compounds Derived from Amino Acids

Granick, S. & Beale, S.I. (1978) Hemes, chlorophylls, and related compounds: biosynthesis and metabolic regulation. *Adv. Enzymol.* **46,** 33–203.
A good review.

Meister, A. & Anderson, M.E. (1983) Glutathione. *Annu. Rev. Biochem.* **52,** 711–760.

Stadtman, T.C. (1991) Biosynthesis and function of selenocysteine-containing enzymes. *J. Biol. Chem.* **266,** 16257–16260.
A compact review of selenium biochemistry.

Stocker, R. Yamamoto, Y., McDonagh, A.F., Glazer, A.N., & Ames, B.N. (1987) Bilirubin is an antioxidant of possible physiologic importance. *Science* **235,** 1043–1046.

Nucleotide Biosynthesis

Benkovic, S.J. (1980) On the mechanism of action of folate- and biopterin-requiring enzymes. *Annu. Rev. Biochem.* **49,** 227–251.

Blakley, R.L. & Benkovic, S.J. (1985) *Folates and Pterins,* Vol. 2: *Chemistry and Biochemistry of Pterins,* Wiley-Interscience, New York.

Daubner, S.C., Schrimsher, J.L., Schendel, F.J., Young, M., Henikoff, S., Patterson, D., Stubbe, J., & Benkovic, S.J. (1985) A multifunctional protein possessing glycinamide ribonucleotide synthetase, glycinamide ribonucleotide transformylase, and aminoimidazole ribonucleotide synthetase activities in *de novo* purine biosynthesis. *Biochemistry* **24,** 7059–7062.

Hardy, L.W., Finer-Moore, J.S., Montfort, W.R., Jones, M.O., Santi, D.V., & Stroud, R.M. (1987) Atomic structure of thymidylate synthase: target for rational drug design. *Science* **235,** 448–455.

Holmgren, A. (1985) Thioredoxin. *Annu. Rev. Biochem.* **54,** 237–271.

Jones, M.E. (1980) Pyrimidine nucleotide biosynthesis in animals: genes, enzymes, and regulation of UMP biosynthesis. *Annu. Rev. Biochem.* **49,** 253–279.

Kornberg, A. & Baker, T.A. (1991) *DNA Replication,* 2nd edn, W.H. Freeman and Company, New York.
This text includes an up-to-date summary of nucleotide biosynthesis.

Lee, L., Kelly, R.E., Pastra-Landis, S.C., & Evans, D.R. (1985) Oligomeric structure of the multifunctional protein CAD that initiates pyrimidine biosynthesis in mammalian cells. *Proc. Natl. Acad. Sci. USA* **82,** 6802–6806.

Reichard, P. & Ehrenberg, A. (1983) Ribonucleotide reductase—a radical enzyme. *Science* **221,** 514–519.

Stubbe, J. (1989) Protein radical development in biological catalysis? *Annu. Rev. Biochem.* **58,** 257–285.
A discussion of free radical mechanisms in ribonucleotide reductase and some other enzymes.

Stubbe, J. (1990) Ribonucleotide reductases: amazing and confusing. *J. Biol. Chem.* **265,** 5329–5332.

Villafranca, J.E., Howell, E.E., Voet, D.H., Strobel, M.S., Ogden, R.C., Abelson, J.N., & Kraut, J. (1983) Directed mutagenesis of dihydrofolate reductase. *Science* **222,** 782–788.
Structural studies on this important enzyme.

Genetic Diseases

Scriver, C.R., Beandet, A.L., Sly, W.S., & Valle, D. (eds) (1989) *The Metabolic Basis of Inherited Disease,* 6th edn, McGraw-Hill Information Services Company, Health Sciences Division, New York.
This book has good chapters on disorders of amino acid, porphyrin, and heme metabolism. See also the chapters on inborn errors of purine and pyrimidine metabolism.

Problems

1. *Cofactors for One-Carbon Transfer Reactions* Most one-carbon transfers are promoted by one of three cofactors: biotin, tetrahydrofolate, or S-adenosylmethionine (Chapter 17). S-Adenosylmethionine is used as a methyl group donor in most reactions; the transfer potential of the methyl group in N^5-methyltetrahydrofolate is insufficient for most biosynthetic reactions. However, one example of the use of N^5-methyltetrahydrofolate in a methyl group transfer occurs in the methionine synthase reaction (step ⑨ of Fig. 21–12), and methionine is the immediate precursor of S-adenosylmethionine (see Fig. 17–20). Explain how the methyl group of S-adenosylmethionine can be derived from N^5-methyltetrahydrofolate, even though the transfer potential of the methyl group in N^5-methyltetrahydrofolate is 10^3 times *lower* than that in S-adenosylmethionine.

2. *Defect in Phenylalanine Hydroxylase and Diet* Tyrosine is normally a nonessential amino acid, but individuals with a genetic defect in phenylalanine hydroxylase require tyrosine in their diet for normal growth. Explain.

3. *Equation for the Synthesis of Aspartate from Glucose* Write the net equation for the synthesis of the nonessential amino acid aspartate from glucose, carbon dioxide, and ammonia.

4. *Inhibition of Nucleotide Synthesis by Azaserine* The diazo compound O-(2-diazoacetyl)-L-serine, known also as azaserine (Fig. 21–41), is a powerful inhibitor of those enzymes that transfer ammonia from glutamine to an acceptor (amidotransferases) during biosynthesis. If growing cells are treated with azaserine, what intermediates in nucleotide biosynthesis would you expect to accumulate? Explain.

5. *Nucleotide Biosynthesis in Amino Acid Auxotrophic Bacteria* Although normal *E. coli* cells can synthesize all the amino acids, some mutants, called amino acid auxotrophs, are unable to synthesize specific amino acids and require the addition of that amino acid to the culture medium for optimal growth. In addition to their role in protein synthesis, specific amino acids are also required in the biosynthesis of other nitrogenous cell products. Consider the three amino acid auxotrophs that are unable to synthesize glycine, glutamine, and aspartate, respectively. For each mutant what nitrogenous cell products other than proteins would fail to be synthesized?

6. *Inhibitors of Nucleotide Biosynthesis* Suggest mechanisms for the inhibition of (a) alanine racemase by L-fluoroalanine and (b) glutamine amidotransferases by azaserine.

7. *Nucleotides Are Poor Sources of Energy* In most organisms, nucleotides are not employed as energy-yielding fuels. What observations support this conclusion? Why are nucleotides relatively poor sources of energy in mammals?

8. *Mode of Action of Sulfa Drugs* Some bacteria require the inclusion of p-aminobenzoate in the culture medium for normal growth. The growth of such bacteria is severely inhibited by the addition of sulfanilamide, one of the earliest antibacterial sulfa drugs. Moreover, in its presence, 5'-aminoimidazole-4-carboxamide ribonucleotide (AICAR; see Fig. 21–27) accumulates in the culture medium. Both effects are reversed by the addition of excess p-aminobenzoate.

p-Aminobenzoate   Sulfanilamide

(a) What is the role of p-aminobenzoate? (Hint: See Fig. 17–18).

(b) Why does AICAR accumulate in the presence of sulfanilamide?

(c) Why is the inhibition and accumulation reversed by the addition of excess p-aminobenzoate?

9. *Treatment of Gout* Allopurinol (Fig. 21–40), an inhibitor of xanthine oxidase, is used to treat chronic gout. Explain the biochemical basis for this treatment. Patients treated with allopurinol sometimes develop xanthine stones in the kidneys, although the incidence of kidney damage is much lower than in untreated gout. Explain this observation in light of the following solubilities in urine: uric acid, 0.15 g/L; xanthine, 0.05 g/L; and hypoxanthine, 1.4 g/L.

10. *ATP Consumption by Root Nodules in Legumes* The bacteria residing in the root nodules of the pea plant consume more than 20% of all the ATP produced by the plant. Suggest a reason why these bacteria consume so much ATP.

11. *Pathway of Carbon in Pyrimidine Biosynthesis* What are the locations of ^{14}C in the orotate molecule present in cells grown on a small amount of uniformly labeled [^{14}C]succinate? Explain.

Integration and Hormonal Regulation of Mammalian Metabolism

In Chapters 13 through 21 we have discussed metabolism at the level of the individual cell, emphasizing those pathways common to almost all cells, prokaryotic and eukaryotic. We have seen how metabolic processes in single cells are regulated at the level of individual enzymes by substrate availability, by allosteric mechanisms, and/or by phosphorylation or other covalent modifications of the enzyme molecules.

To appreciate fully the significance of individual metabolic pathways and their regulation, we must view these pathways in the context of the whole organism. An essential characteristic of multicellular organisms is cell differentiation and division of labor. In addition to the central pathways of energy-yielding metabolism that occur in all cells, the tissues and organs of complex organisms such as humans have specialized functions and thus characteristic fuel requirements and patterns of metabolism. Hormonal signals integrate and coordinate the metabolic activities of different tissues and bring about the optimal allocation of fuels and precursors to each organ. In this chapter our focus is on mammals, and we deal with two distinct but interrelated themes: (1) the specialized metabolism of several major organs and tissues and the integration of metabolism in the whole organism, and (2) the structure and action of the hormones that regulate these metabolic processes.

We begin by examining the distribution of nutrients to various tissues and organs—in which liver plays a central role—and the metabolic cooperation among these tissues and organs. After reviewing the major classes of hormones, we consider the means by which they coordinate the diverse metabolic activities of the organism. The chapter concludes with a description of the fundamental mechanisms by which hormones and neurotransmitters interact with cellular receptors to alter and integrate metabolism.

Tissue-Specific Metabolism: The Division of Labor

Each tissue and organ of the human body has a specialized function that is reflected in its anatomy and its metabolic activity. Skeletal muscle, for example, uses metabolic energy to produce motion; adipose tissue stores and releases fats, which serve as fuel throughout the body; the brain pumps ions to produce electrical signals. The liver plays a central processing and distributing role in metabolism and furnishes all the other organs and tissues with a proper mix of nutrients via the bloodstream. The functional centrality of the liver is indicated by the common reference to all other tissues and organs as "extrahepatic" or "peripheral." We therefore begin our discussion of the division of metabolic labor by considering the transformations of carbohydrates, amino acids, and fats in the mammalian liver. This is followed by brief descriptions of the major metabolic functions of adipose tissue, muscle, the brain, and the tissue that interconnects all others: the blood.

The Liver Processes and Distributes Nutrients

During digestion in the gastrointestinal tract of mammals, the three major classes of nutrients (carbohydrates, proteins, and lipids) undergo enzymatic hydrolysis into their monomeric subunits. This breakdown is necessary because the epithelial cells lining the intestinal lumen are able to absorb only relatively small molecules. Many of the fatty acids and monoacylglycerols released by digestion in the intestine are reconverted within these epithelial cells into triacylglycerols (Chapter 16).

After being absorbed, most of the sugars and amino acids and some triacylglycerols pass to the blood and are taken up by hepatocytes in the liver; the remaining triacylglycerols take a different path via the lymphatic system and enter adipose tissue. Hepatocytes transform the nutrients obtained from the diet into the fuels and precursors required by each of the tissues, and export them in the blood. The kinds and amounts of nutrients supplied to the liver vary with several factors, including the diet and the time interval between meals. The demand of the extrahepatic tissues for fuels and precursors varies among organs and with the activity of the organism. To meet these changing circumstances, the liver has remarkable metabolic flexibility. For example, when the diet is rich in protein, hepatocytes contain high levels of enzymes for amino acid catabolism and gluconeogenesis. Within hours after a shift to a high-carbohydrate diet, the levels of these enzymes drop and the synthesis of enzymes essential to carbohydrate metabolism begins. Other tissues also adjust their metabolism to the prevailing conditions, but none is as adaptable as the liver, and none is so central to the organism's overall metabolic activities. What follows is a survey of the possible fates of sugars, amino acids, and lipids that enter the liver from the bloodstream. To help you recall the metabolic transformations discussed here, Table 22–1 (p. 738) shows the major pathways and processes to which we will refer and the chapter in which each pathway is discussed in detail.

Sugars Glucose entering the liver is phosphorylated by glucokinase to yield glucose-6-phosphate. Fructose, galactose, and mannose, absorbed from the small intestine, are also converted into glucose-6-phosphate by enzymatic pathways examined earlier. Glucose-6-phosphate is at the crossroads of carbohydrate metabolism in the liver. It may take any of five major metabolic routes (Fig. 22–1), depending on the current metabolic needs of the organism. By the action of various allosterically regulated enzymes, and through hormonal regulation of enzyme synthesis and activity, the flow of glucose is directed into one or more of these pathways in the liver.

① Glucose-6-phosphate is dephosphorylated by glucose-6-phosphatase to yield free glucose (p. 605), which is exported to replenish blood glucose. Export is the pathway of choice when the amount of glucose-6-phosphate is limited, because the blood glucose concentra-

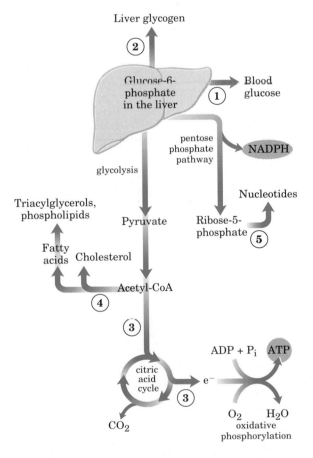

Figure 22–1 Metabolic pathways for glucose-6-phosphate in the liver. Here and in the following figures, anabolic pathways are shown leading upward, catabolic pathways leading downward, and distribution to other organs horizontally; the numbered processes correspond to descriptions in the text.

Table 22–1 Pathways of carbohydrate, amino acid, and fat metabolism discussed in earlier chapters

Citric acid cycle: acetyl-CoA $\longrightarrow$ $2CO_2$	Chapter 15
Oxidative phosphorylation: ATP synthesis	Chapter 18
Carbohydrate catabolism	
Glycogenolysis: glycogen $\longrightarrow$ glucose-1-phosphate $\longrightarrow$ blood glucose	
Hexose entry into glycolysis: fructose, mannose, galactose $\longrightarrow$ glucose-6-phosphate	Chapter 14
Glycolysis: glucose $\longrightarrow$ pyruvate	
Pyruvate dehydrogenase reaction: pyruvate $\longrightarrow$ acetyl-CoA	Chapter 15
Lactic acid fermentation: glycogen $\longrightarrow$ lactate + 2ATP	
Pentose phosphate pathway: glucose-6-phosphate $\longrightarrow$ pentose phosphates + NADPH	Chapter 14
Carbohydrate anabolism	
Gluconeogenesis: citric acid cycle intermediates $\longrightarrow$ glucose	Chapter 19
Glucose–alanine cycle: glucose $\longrightarrow$ pyruvate $\longrightarrow$ alanine $\longrightarrow$ glucose	Chapter 17
Glycogen synthesis: glucose-6-phosphate $\longrightarrow$ glucose-1-phosphate $\longrightarrow$ glycogen	Chapter 19
Amino acid and nucleotide metabolism	
Amino acid degradation: amino acids $\longrightarrow$ acetyl-CoA, citric acid cycle intermediates	Chapter 17
Amino acid synthesis	Chapter 21
Urea cycle: NH_3 $\longrightarrow$ urea	
Glucose–alanine cycle: alanine $\longrightarrow$ glucose	Chapter 17
Nucleotide synthesis: amino acids $\longrightarrow$ purines, pyrimidines	
Hormone synthesis (and synthesis of other nitrogenous compounds)	Chapter 19
Fat catabolism	
β Oxidation of fatty acids: fatty acid $\longrightarrow$ acetyl-CoA	
Oxidation of ketone bodies: β-hydroxybutyrate $\longrightarrow$ acetyl-CoA $\longrightarrow$ CO_2	Chapter 16
Fat anabolism	
Fatty acid synthesis: acetyl-CoA $\longrightarrow$ fatty acids	
Triacylglycerol synthesis: acetyl-CoA $\longrightarrow$ fatty acids $\longrightarrow$ triacylglycerol	Chapter 20
Ketone body formation: acetyl-CoA $\longrightarrow$ acetoacetate, β-hydroxybutyrate	Chapter 16
Synthesis of cholesterol and its derivatives: acetyl-CoA $\longrightarrow$ cholesterol; cholesterol $\longrightarrow$ bile salts, cholesteryl esters	
Phospholipid synthesis: fatty acids $\longrightarrow$ phospholipids	Chapter 20

tion must be kept sufficiently high (4 mM) to provide adequate energy for the brain and other tissues. ② Glucose-6-phosphate not immediately needed to form blood glucose is converted into liver glycogen. ③ Glucose-6-phosphate may be oxidized for energy production via glycolysis, decarboxylation of pyruvate (by the pyruvate dehydrogenase reaction), and the citric acid cycle. The ensuing electron transfer and oxidative phosphorylation yield ATP. (Normally, however, fatty acids are the preferred fuel for energy production in hepatocytes.) ④ Excess glucose-6-phosphate not used to make blood glucose or liver glycogen is degraded via glycolysis and the pyruvate dehydrogenase reaction into acetyl-CoA, which serves as the precursor for the synthesis of lipids: fatty acids, which are incorporated into triacylglycerols and phospholipids, and cholesterol. Much of the lipid synthesized in the liver is exported to other tissues, carried there by blood lipoproteins. ⑤ Finally, glucose-6-phosphate is the substrate for the pentose phosphate pathway, yielding both reducing power (NADPH), needed for the biosynthesis of fatty acids and cholesterol, and D-ribose-5-phosphate, a precursor in nucleotide biosynthesis.

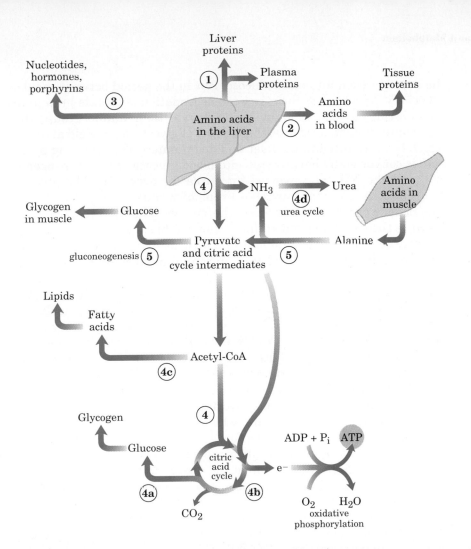

Figure 22-2 Metabolism of amino acids in the liver.

Amino Acids Amino acids that enter the liver have several important metabolic routes (Fig. 22–2). ① They act as precursors for protein synthesis in hepatocytes, a process discussed in Chapter 26. The liver constantly renews its own proteins, which have a very high turnover rate, with an average half-life of only a few days. The liver is also the site of biosynthesis of most of the plasma proteins of the blood. ② Alternatively, amino acids may pass from the liver into the blood and thus to other organs, to be used as precursors in the synthesis of tissue proteins. ③ Certain amino acids are precursors in the biosynthesis of nucleotides, hormones, and other nitrogenous compounds in the liver and other tissues.

④ Amino acids not needed for biosynthesis of proteins and other molecules in the liver or elsewhere are deaminated and degraded to yield acetyl-CoA and citric acid cycle intermediates. Citric acid cycle intermediates so formed may be converted into glucose and glycogen via the gluconeogenic pathway (④a). Acetyl-CoA may be oxidized via the citric acid cycle for ATP energy (④b), or it may be converted into lipids for storage (④c). The ammonia released on degradation of amino acids is converted by hepatocytes into the excretory product, urea (④d).

Finally, the liver participates in the metabolism of amino acids arriving intermittently from the peripheral tissues. The blood is adequately supplied with glucose just after the digestion and absorption of dietary carbohydrate or, between meals, by the conversion of some of the liver glycogen into blood glucose. But in the period between meals, especially if prolonged, there is some degradation of muscle protein to amino acids ⑤. These amino acids donate their amino groups (by transamination) to pyruvate, the product of glycolysis, to yield alanine, which is transported to the liver and deaminated. The resulting pyruvate is converted by hepatocytes into blood glucose (via gluconeogenesis), and the NH_3 is converted into urea for excretion. The glucose returns to the skeletal muscles to replenish muscle glycogen stores. One benefit of this cyclic process, the glucose–alanine cycle (see Fig. 17–9), is the smoothing out of fluctuations in blood glucose in the periods between meals. The amino acid deficit incurred in the muscles is made up after the next meal from incoming dietary amino acids.

Lipids The fatty acid components of the lipids entering hepatocytes also have several different pathways (Fig. 22–3). ① Fatty acids are converted into liver lipids. ② Under most circumstances, fatty acids are the major oxidative fuel in the liver. Free fatty acids may be activated and oxidized to yield acetyl-CoA and NADH. The acetyl-CoA is further oxidized via the citric acid cycle to yield ATP by oxidative phosphorylation. ③ Excess acetyl-CoA released on oxidation of fatty acids and not required by the liver is converted into the ketone bodies, acetoacetate and D-β-hydroxybutyrate, which are circulated in the blood to peripheral tissues, to be used as fuel for the citric acid cycle. The ketone bodies may be regarded as a transport form of acetyl groups. They can supply a significant fraction of the energy in some peripheral tissues, up to one-third in the heart, and 60 to 70% in the brain during prolonged fasting. ④ Some of the acetyl-CoA derived from fatty acids (and from glucose) is used for the biosynthesis of cholesterol, which is required for membrane biosynthesis. Cholesterol is also the precursor of all steroid hormones and of the bile salts, which are essential for the digestion and absorption of lipids.

The final two metabolic fates of lipids involve specialized mechanisms for the transport of insoluble lipids in the blood. ⑤ Fatty acids are converted to the phospholipids and triacylglycerols of the plasma lipoproteins, which carry lipids to adipose (fat) tissue for storage as triacylglycerols. Cholesterol and cholesteryl esters are also transported as lipoproteins. ⑥ Some free fatty acids become bound to serum albumin and are carried in the blood to the heart and skeletal muscles, which absorb and oxidize free fatty acids as a major fuel. Serum albumin is the most abundant plasma protein; one molecule of serum albumin can carry up to 10 molecules of free fatty acid, releasing them at the consuming tissue where they are taken up by passive diffusion.

Thus, the liver serves as the body's distribution center: exporting nutrients in the correct proportions to the other organs, smoothing out fluctuations in metabolism caused by the intermittent nature of food intake, and processing excess amino groups into urea and other products to be disposed of by the kidneys.

In addition to the processing and distribution of carbohydrates, fats, and amino acids, the liver is also active in the enzymatic detoxification of foreign organic compounds, such as drugs, food additives, preservatives, and other possibly harmful agents with no food value. Detoxification usually involves the cytochrome P-450–dependent hydroxylation of relatively insoluble organic compounds to make them sufficiently soluble for further breakdown and excretion (see Box 20–1).

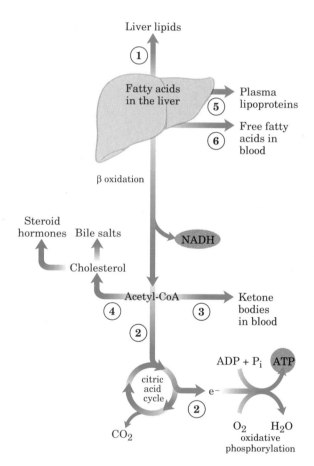

Figure 22–3 Metabolism of fatty acids in the liver.

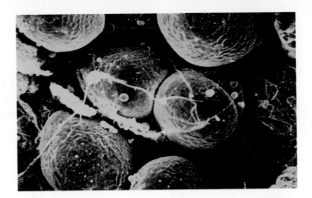

Figure 22–4 Scanning electron micrograph of human adipocytes ($\times 440$ magnification). Capillaries and collagen fibers form a supporting network around adipocytes in fat tissues. Almost the entire volume of the cells is filled with fat droplets, which are very active metabolically.

Adipose Tissue Stores and Supplies Fatty Acids

Adipose tissue, which consists of adipocytes (fat cells) (Fig. 22–4), is amorphous and widely distributed in the body: under the skin, around the deep blood vessels, and in the abdominal cavity. It typically makes up about 15% of the mass of a young adult human, with approximately 65% of this mass being in the form of triacylglycerols. Adipocytes are metabolically very active, responding quickly to hormonal stimuli in a metabolic interplay with the liver, skeletal muscles, and the heart.

Like other cell types in the body, adipocytes have an active glycolytic metabolism, use the citric acid cycle to oxidize pyruvate and fatty acids, and carry out mitochondrial oxidative phosphorylation. During periods of high carbohydrate intake, adipose tissue can convert glucose via pyruvate and acetyl-CoA into fatty acids, from which triacylglycerols are made and stored as large fat globules. In humans, however, most fatty acid synthesis occurs in hepatocytes, not in adipocytes. Adipocytes store triacylglycerols arriving from the liver (carried in the blood as VLDLs) and from the intestinal tract, particularly after meals rich in fat.

When fuel is needed, triacylglycerols stored in adipose tissue are hydrolyzed by lipases within the adipocytes to release free fatty acids, which may then be delivered via the bloodstream to skeletal muscles and the heart. The release of fatty acids from adipocytes is greatly accelerated by the hormone epinephrine, which stimulates the conversion of the inactive form of triacylglycerol lipase into its active form (see Fig. 16–3). Insulin counterbalances this effect of epinephrine, decreasing the activity of triacylglycerol lipase.

Humans and many other animals, particularly those that hibernate, have adipose tissue called brown fat, which is specialized to generate heat rather than ATP during the oxidation of fatty acids (see Fig. 18–27).

Muscle Uses ATP for Mechanical Work

Skeletal muscle accounts for over 50% of the total O_2 consumption in a resting human being and up to 90% during very active muscular work. Metabolism in skeletal muscle is primarily specialized to generate ATP as the immediate source of energy. Moreover, skeletal muscle is adapted to do its mechanical work in an intermittent fashion, on demand. Sometimes skeletal muscles must deliver much work in a short time, as in a 100 m sprint; at other times more extended work is required, as in running a marathon or giving birth.

742

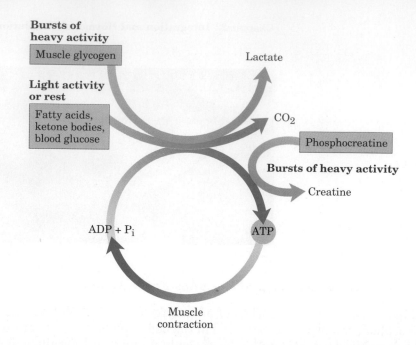

Figure 22–5 Energy sources for muscle contraction: fuels used for ATP synthesis during bursts of heavy activity and during light activity or rest. ATP can be obtained rapidly from phosphocreatine.

Skeletal muscles can use free fatty acids, ketone bodies, or glucose as fuel, depending on the degree of muscular activity (Fig. 22–5). In resting muscle the primary fuels are free fatty acids from adipose tissue and ketone bodies from the liver. These are oxidized and degraded to yield acetyl-CoA, which enters the citric acid cycle for oxidation to CO_2. The ensuing transfer of electrons to O_2 provides the energy for ATP synthesis by oxidative phosphorylation. Moderately active muscles use blood glucose in addition to fatty acids and ketone bodies. The glucose is phosphorylated, then degraded by glycolysis to pyruvate, which is converted to acetyl-CoA and oxidized via the citric acid cycle. However, in maximally active muscles, the demand for ATP is so great that the blood flow cannot provide O_2 and fuels fast enough to produce the necessary ATP by aerobic respiration alone. Under these conditions, the stored muscle glycogen is broken down to lactate by fermentation, with a yield of two molecules of ATP per glucose unit degraded. Lactic acid fermentation thus provides extra ATP energy quickly, supplementing the basal ATP production resulting from the aerobic oxidation of other fuels via the citric acid cycle. The use of blood glucose and muscle glycogen as emergency fuels for muscular activity is greatly enhanced by the secretion of epinephrine, which stimulates the formation of blood glucose from glycogen in the liver and the breakdown of glycogen in muscle tissue. Skeletal muscle does not contain glucose-6-phosphatase and cannot convert glucose-6-phosphate to free glucose for export to other tissues. Consequently, muscle glycogen is completely dedicated to providing energy in the muscle, via glycolytic breakdown.

Because skeletal muscles store relatively little glycogen (about 1% of their total weight), there is an upper limit to the amount of glycolytic energy available during all-out exertion. Moreover, the accumulation of lactate and the consequent decrease in pH that occurs in maximally active muscles reduces their efficiency.

After a period of intense muscular activity, heavy breathing continues for some time. Much of the O_2 thus obtained is used for the production of ATP by oxidative phosphorylation in the liver. This ATP is used for gluconeogenesis from lactate, carried in the blood from the muscles to the liver. The glucose thus formed returns to the muscles to replenish their glycogen, completing the Cori cycle (Fig. 22–6; see also Box 14–1).

Figure 22–6 Metabolic cooperation between skeletal muscles and the liver. During extremely active muscular work, skeletal muscle uses glycogen as its energy source, via glycolysis. During recovery, some of the lactate formed in the muscles is transported to the liver and used to form glucose, which is released to the blood and returned to the muscles to replenish their glycogen stores. This pathway (glucose $\longrightarrow$ lactate $\longrightarrow$ glucose) constitutes the Cori cycle.

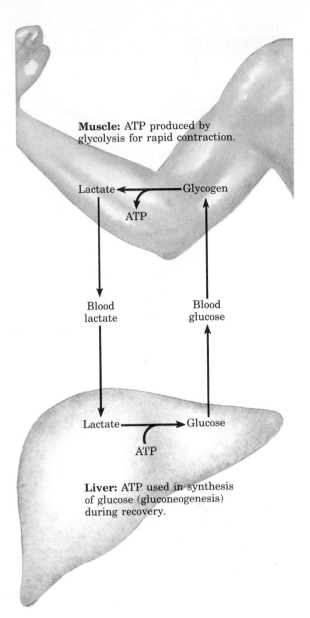

Muscle: ATP produced by glycolysis for rapid contraction.

Liver: ATP used in synthesis of glucose (gluconeogenesis) during recovery.

Skeletal muscles contain considerable amounts of phosphocreatine, which can rapidly regenerate ATP from ADP by the creatine kinase reaction. During periods of active contraction and glycolysis, this reaction proceeds predominantly in the direction of ATP synthesis (Fig. 22–5), but during recovery from exertion, the same enzyme is used to resynthesize phosphocreatine from creatine at the expense of ATP.

$$\text{ATP} + \underset{\text{Creatine}}{\begin{array}{c} \text{NH}_2 \\ | \\ \overset{+}{\text{C}}{=}\text{NH}_2 \\ | \\ \text{CH}_3{-}\text{N} \\ | \\ \text{CH}_2 \\ | \\ \text{COO}^- \end{array}} \underset{\text{kinase}}{\overset{\text{creatine}}{\rightleftharpoons}} \underset{\text{Phosphocreatine}}{\begin{array}{c} \text{O}^- \\ | \\ ^-\text{O}{-}\text{P}{=}\text{O} \\ | \\ \text{N}{-}\text{H} \\ | \\ \overset{+}{\text{C}}{=}\text{NH}_2 \\ | \\ \text{CH}_3{-}\text{N} \\ | \\ \text{CH}_2 \\ | \\ \text{COO}^- \end{array}} + \text{ADP}$$

Heart muscle differs from skeletal muscle in that it is continuously active in a regular rhythm of contraction and relaxation. In contrast to skeletal muscle, the heart has a completely aerobic metabolism at all times. Mitochondria are much more abundant in heart muscle than in skeletal muscle; they make up almost half the volume of the cells (Fig. 22–7). The heart uses as fuel a mixture of glucose, free fatty acids, and ketone bodies arriving from the blood. These fuels are oxidized via the citric acid cycle to deliver the energy required to generate ATP by oxidative phosphorylation. Like skeletal muscle, heart muscle does not store lipids or glycogen in large amounts. Small amounts of reserve energy are stored in the form of phosphocreatine. Because the heart is normally aerobic and obtains its energy from oxidative phosphorylation, the failure of O_2 to reach a portion of the heart muscle when the blood vessels are blocked by lipid deposits (atherosclerosis) or blood clots (coronary thrombosis) can cause this region of the heart muscle to die, a process known as myocardial infarction, more commonly called a heart attack.

Figure 22–7 Electron micrograph of heart muscle, showing the profuse mitochondria in which pyruvate, fatty acids, and ketone bodies are oxidized to drive ATP synthesis. This steady aerobic metabolism allows the human heart to pump blood at a rate of 5 quarts per minute, or 75 gallons per hour, or 18 million barrels in a 70 year lifetime.

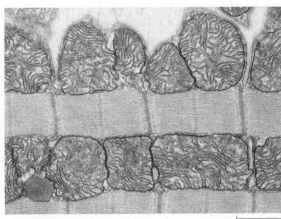

1 μm

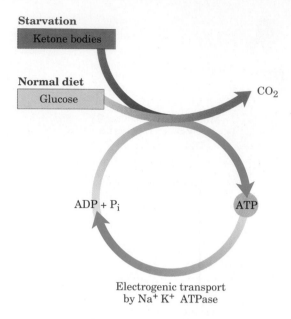

Figure 22–8 Energy sources in the brain vary with nutritional state. The ketone body used by the brain is β-hydroxybutyrate.

The Brain Uses Energy for Transmission of Impulses

The metabolism of the brain is remarkable in several respects. First, the brain of adult mammals normally uses only glucose as fuel (Fig. 22–8). Second, the brain has a very active respiratory metabolism; it uses almost 20% of the total O_2 consumed by a resting human adult. The use of O_2 by the brain is fairly constant in rate and does not change significantly during active thought or sleep. Because the brain contains very little glycogen, it is continuously dependent on incoming glucose from the blood. If the blood glucose should fall significantly below a certain critical level for even a short period of time, severe and sometimes irreversible changes in brain function may occur.

Although the brain cannot directly use free fatty acids or lipids from the blood as fuels, it can, when necessary, use D-β-hydroxybutyrate (a ketone body) formed from fatty acids in hepatocytes. The capacity of the brain to oxidize β-hydroxybutyrate via acetyl-CoA becomes important during prolonged fasting or starvation, after essentially all the liver glycogen has been depleted, because it allows the brain to use body fat as a source of energy. The use of β-hydroxybutyrate by the brain during severe starvation also spares muscle proteins, which become the ultimate source of glucose for the brain (via gluconeogenesis) during severe starvation.

Glucose is oxidized by the glycolytic pathway and the citric acid cycle, providing almost all of the ATP used by the brain. ATP energy is required to create and maintain an electrical potential across the plasma membrane of neurons (Fig. 22–8). The plasma membrane contains an ATP-driven antiporter, the Na^+K^+ ATPase, which simultaneously pumps K^+ ions into and Na^+ ions out of the neuron (see Fig. 10–22). Because three Na^+ ions are transported out and only two K^+ ions are transported in for each molecule of ATP hydrolyzed, the Na^+K^+ ATPase is electrogenic—it generates an electrical potential difference across the neuronal membrane, with the inside negative relative to the outside. This transmembrane potential changes transiently as an electrical signal (action potential) sweeps from one end of a neuron to the other, as we will see later in this chapter (see Fig. 22–34). Action potentials are the chief method of information transfer in the nervous system.

Blood Carries Oxygen, Metabolites, and Hormones

The blood flows through and connects all of the tissues, mediating the metabolic interactions among them. It transports nutrients from the small intestine to the liver, and from the liver and adipose tissue to other organs; it also transports waste products from the tissues to the kidneys for excretion. Oxygen moves in the blood from the lungs to the tissues, and CO_2 generated by tissue respiration returns in the blood to the lungs for exhalation. Blood also carries hormonal signals from one tissue to another. In its role as signal carrier, the circulatory system resembles the nervous system; both serve to regulate and integrate the activities of different organs.

The average adult human has 5 to 6 L of blood. Almost half of this volume is occupied by three types of blood cells (Fig. 22–9): **erythrocytes** (red cells), filled with hemoglobin and specialized for carrying O_2 and CO_2; much smaller numbers of **leukocytes** (white cells) of several types, central to the immune system that defends against infections; and **platelets,** which help to mediate the blood clotting that prevents

Figure 22–9 The composition of blood. Whole blood is separated into blood plasma and cells by centrifugation. About 10% of blood plasma is solutes, of which about 10% consists of inorganic salts, 20% small organic molecules, and 70% plasma proteins. The major dissolved components are shown. Blood contains many other substances, often in trace amounts, including other metabolites, enzymes, hormones, vitamins, trace elements, and bile pigments. Measurements of the concentrations of components in blood plasma are important in the diagnosis and treatment of disease.

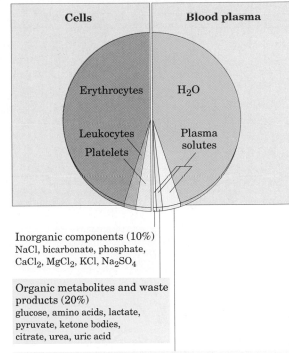

Inorganic components (10%)
NaCl, bicarbonate, phosphate, $CaCl_2$, $MgCl_2$, KCl, Na_2SO_4

Organic metabolites and waste products (20%)
glucose, amino acids, lactate, pyruvate, ketone bodies, citrate, urea, uric acid

Plasma proteins (70%)
Major plasma proteins: serum albumin, very low-density lipoproteins (VLDL), low-density lipoproteins (LDL), high-density lipoproteins (HDL), immunoglobulins (hundreds of kinds), fibrinogen, prothrombin, many specialized transport proteins such as transferrin

loss of blood after injury. The liquid portion is the **blood plasma,** which is 90% water and 10% solutes. The plasma is very complex in chemical composition; in it are dissolved or suspended a large variety of proteins, lipoproteins, nutrients, metabolites, waste products, inorganic ions, and hormones. Over 70% of the plasma solids are **plasma proteins** (Fig. 22–9). Major plasma proteins include immunoglobulins (circulating antibodies), serum albumin, apolipoproteins involved in the transport of lipids (as VLDL, LDL, HDL), transferrin (for iron transport), and blood-clotting proteins such as fibrinogen and prothrombin.

The ions and low molecular weight solutes in the blood plasma are not fixed components, but are in constant flux between blood and various tissues. Dietary uptake of inorganic ions is, in general, counterbalanced by their excretion in the urine. For many of the components of blood, something near a dynamic steady state is achieved; the concentration of the component changes little, although a continual flux occurs from the digestive tract, through the blood, and to the urine. For example, almost regardless of the dietary intake of Na^+, K^+, and Ca^{2+}, the plasma levels of these ions remain close to 140, 5, and 2.5 mM, respectively. Any significant departure from these values can result in serious illness or death. The kidneys play an especially important role in maintaining the ion balance, serving as a selective filter that allows waste products and excess ions to pass from the blood to the urine while preventing the loss of essential nutrients and ions.

The concentration of glucose dissolved in the plasma is also subject to tight regulation. We have noted the requirement of the brain for glucose and the role of the liver in maintaining the glucose concentration near the normal level of 80 mg/100 mL of blood (about 4.5 mM). When blood glucose in a human drops to half this value (the hypoglycemic condition), the person experiences discomfort and mental confusion (Fig. 22–10); further reductions lead to coma, convulsions, and in extreme hypoglycemia, death. Maintaining the normal concentration of glucose in the blood is therefore a very high priority of the organism, and a variety of regulatory mechanisms have evolved to achieve that end. Among the most important regulators of blood glucose are the hormones insulin, glucagon, and epinephrine. Before considering their specific action, we turn now to a general discussion of hormones.

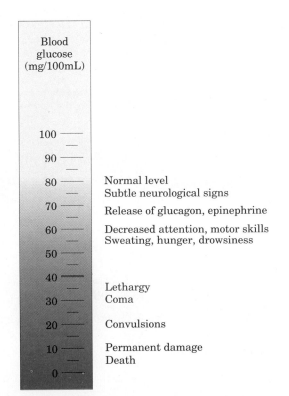

Blood glucose (mg/100mL)

100 —
90 —
80 — Normal level / Subtle neurological signs
70 — Release of glucagon, epinephrine
60 — Decreased attention, motor skills / Sweating, hunger, drowsiness
50 —
40 — Lethargy
30 — Coma
20 — Convulsions
10 — Permanent damage / Death
0 —

Figure 22–10 Physiological effects of low blood glucose in humans. Blood glucose levels of 40 mg/100 mL and below constitute severe hypoglycemia.

Hormones: Communication among Cells and Tissues

The coordination of metabolism in the separate organs of mammals is achieved by hormonal and neuronal signaling. Individual cells in one tissue sense a change in the organism's circumstances and respond by secreting an extracellular chemical messenger. Endocrine cells secrete hormones; neurons secrete neurotransmitters. In each case, the extracellular messenger passes to another cell where it binds to a specific receptor molecule and triggers a change in the activity of the second cell. In neuronal signaling (Fig. 22–11a), the chemical messenger (neurotransmitter; acetylcholine, for example) may travel only a fraction of a micrometer, across the synaptic cleft to the next neuron in a chain. In contrast, hormones are carried rapidly in the blood between distant organs and tissues; they may travel a meter or more before encountering their target cell (Fig. 22–11b). Except for this anatomical difference, the chemical signaling in the neural and endocrine systems is remarkably similar in mechanism. Even some of the chemical messengers are common to both systems. Epinephrine and norepinephrine, for example, serve as neurotransmitters in certain synapses of the brain and smooth muscle and also as hormones regulating fuel metabolism in the liver and in muscle. Although the neural and endocrine systems were traditionally treated as separate entities, it has become clear that in the regulation of metabolism they merge into a single **neuroendocrine system.** In the following discussion of cellular signaling, we emphasize hormone action and the endocrine system, drawing illustrations from our previous discussions of fuel metabolism. However, most of the fundamental mechanisms described here also occur in neurotransmitter action in the neural system.

The word **hormone** is derived from the Greek verb *horman,* meaning "to stir up or excite." The concept of hormones as internal signals is not new; the physiologist Claude Bernard used the term "internal secretion" in 1855 to distinguish between components secreted into the bloodstream and "external secretions" such as sweat and tears. Ernest

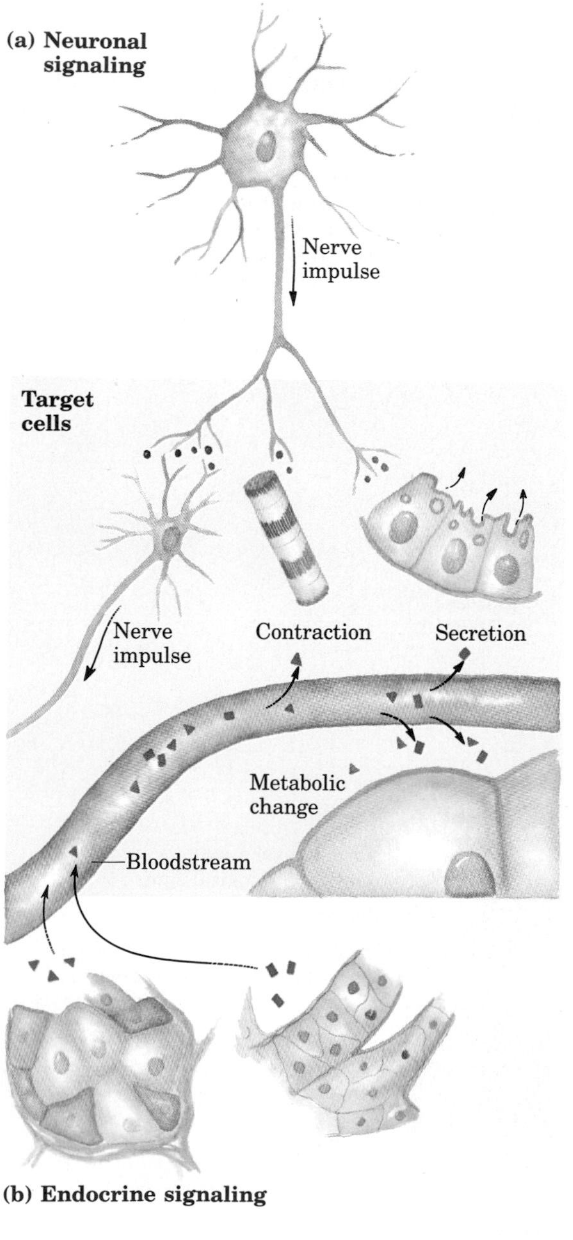

(a) Neuronal signaling

Nerve impulse

Target cells

Nerve impulse   Contraction   Secretion

Metabolic change

Bloodstream

(b) Endocrine signaling

Figure 22–11 Signaling by the neural and endocrine systems. **(a)** In neuronal signaling electrical signals (nerve impulses) originate in the cell body and are carried very rapidly over long distances to the axon tip, where neurotransmitters are released and diffuse to the target cell. The target cell, which may be another neuron, a myocyte, or a secretory cell, is only a fraction of a micrometer or a few micrometers away from the site of neurotransmitter release. **(b)** In the endocrine system hormones are secreted by the producing cell into the bloodstream, which carries them throughout the body to target tissues that may be more than a meter away from the secreting cell. Both neurotransmitters and hormones interact with specific receptors on or in the target cell, triggering responses.

Henry Starling introduced the term hormone in 1905 in a famous lecture "The Chemical Correlation of the Functions of the Body." We now know that hormones control not only different aspects of metabolism but also many other functions: cell and tissue growth, heart rate, blood pressure, kidney function, motility of the gastrointestinal tract, secretion of digestive enzymes and of other hormones, lactation, and the activity of the reproductive systems.

We introduce here the major classes of hormones, their tissues of origin, and their general properties.

Hormones Are Chemically Diverse, Biologically Potent Molecules

There are three chemically distinct classes of hormones: peptides, amines, and steroids (Table 22–2). A fourth group of extracellular signals, the eicosanoids, are hormonelike in their actions, but act locally.

Claude Bernard
1813–1878

Ernest Henry Starling
1866–1927

Table 22–2 The classes of hormones and hormonelike compounds, with some examples

Hormone	Secreting organ/tissue/cells	Function or activity
Peptide hormones		
Thyrotropin-releasing hormone (TRH)	Hypothalamus	Stimulates thyrotropin release from anterior pituitary
Corticotropin (adrenocorticotropic hormone, ACTH)	Anterior pituitary	Stimulates synthesis of adrenocortical steroids in adrenal cortex
Vasopressin (antidiuretic hormone, ADH)	Posterior pituitary	Increases blood pressure, promotes water reabsorption by kidney
Insulin	Pancreas	Stimulates glucose uptake and utilization
Glucagon	Pancreas	Stimulates glucose production by liver
Amine hormones		
Epinephrine (adrenaline)	Adrenal medulla	Controls responses to stress, increases heart rate
Thyroxine (thyroid hormone)	Thyroid	Stimulates metabolism in many tissues
Steroid hormones		
Cortisol	Adrenal cortex	Limits glucose utilization, increases blood glucose
Aldosterone	Adrenal cortex	Regulates sodium retention and blood pressure
β-Estradiol	Ovary	Regulates activity in female reproductive tissues
Testosterone	Testis	Regulates activity in male reproductive tissues
Progesterone	Corpus luteum (in ovary)	Regulates activity in female reproductive organs during menstrual cycle and pregnancy
Eicosanoids (hormonelike)		
Prostaglandins	Most tissues	Trigger smooth muscle contraction; fever; inflammation
Leukotrienes	Leukocytes (white blood cells), spleen, others	Cause bronchial constriction; involved in hypersensitivity reactions
Thromboxanes	Platelets and other tissues	Regulate blood clotting; vasoconstriction; platelet aggregation

The **peptide hormones,** which may have from 3 to over 200 amino acid residues, include all of the hormones of the hypothalamus and pituitary and the pancreatic hormones insulin, glucagon, and somatostatin. The **amine hormones,** low molecular weight compounds derived from the amino acid tyrosine, include water-soluble epinephrine and norepinephrine of the adrenal medulla and the less water-soluble thyroid hormones. The **steroid hormones,** which are fat-soluble, include the adrenal cortical hormones, hormone forms of vitamin D, and the androgens and estrogens (the male and female sex hormones) (see Fig. 9–15). They move through the bloodstream bound to specific carrier proteins. **Eicosanoids** are derivatives of the 20-carbon polyunsaturated fatty acid arachidonate (see Fig. 9–17). All three subclasses of eicosanoids (prostaglandins, leukotrienes, and thromboxanes) are unstable and insoluble in water; these signaling molecules generally do not move far from the tissue that produced them, and they act primarily on cells very near their point of release.

Tyr–Gly–Gly–Phe–Leu
Leu-enkephalin, a peptide hormone

Epinephrine, an amine hormone

Testosterone, a steroid hormone

Prostaglandin E$_1$, an eicosanoid

Hormones normally occur in very low concentrations in the blood, in the micromolar (10^{-6} M) to picomolar (10^{-12} M) range; this may be contrasted with the normal concentration of glucose in the blood, which is in the millimolar range (about 4×10^{-3} M). For this reason, hormones have been very difficult to isolate, identify, and measure accurately. The exceedingly sensitive technique of radioimmunoassay developed by Rosalyn Yalow and Solomon A. Berson revolutionized hormone research by making possible the quantitative and specific measurement of many hormones in minute concentrations. A newer variation of this technique, called an enzyme-linked immunosorbent assay (ELISA), is illustrated in Figure 6–9.

When a given hormone is secreted, its concentration in the blood rises, sometimes by orders of magnitude. When secretion stops, the hormone concentration quickly returns to the resting level. Hormones have a short existence in the blood, often only minutes; once their presence is no longer required they are quickly inactivated enzymatically.

Some hormones yield immediate physiological or biochemical responses. Seconds after epinephrine is secreted into the bloodstream by the adrenal medulla, the liver responds by pouring glucose into the blood. By contrast, the thyroid hormones and the estrogens promote maximal responses in their target tissues only after hours or even days. These differences in response time correspond to a difference in mode of action (Fig. 22–12). In general, the fast-acting hormones lead to a change in the activity of one or more preexisting enzyme(s) in the cell, by allosteric mechanisms or by covalent modification of the enzyme(s). The slower-acting hormones generally alter gene expression, resulting in the synthesis of more or fewer copies of the regulated protein(s).

Figure 22–12 Two general mechanisms of hormone action. The peptide and amine hormones are faster acting than steroid and thyroid hormones.

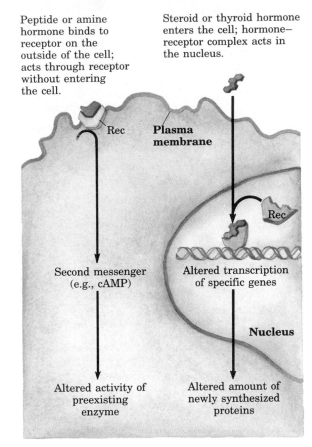

Peptide or amine hormone binds to receptor on the outside of the cell; acts through receptor without entering the cell.

Steroid or thyroid hormone enters the cell; hormone–receptor complex acts in the nucleus.

All hormones act through specific receptors present in hormone-sensitive target cells, to which hormones bind with high specificity and high affinity. Each cell type has its own combination of hormone receptors, defining the range of its hormone responsiveness. Two cell types with the same receptor may have different intracellular targets of hormone action and thus may respond differently to the same hormone.

The water-soluble peptide and amine hormones do not penetrate cell membranes readily; their receptors are located on the outer surface of the target cells (Fig. 22–12). The lipid-soluble steroid and thyroid hormones readily pass through the plasma membrane of their target cells; their receptors are specific proteins located in the nucleus. Upon hormone binding to a plasma membrane receptor, the receptor protein undergoes a conformational change analogous to that produced in an allosteric enzyme by effector binding. In its altered form, the receptor either produces or causes production of an intracellular messenger molecule, often called the **second messenger.** We have already encountered one second messenger, adenosine 3′,5′-cyclic monophosphate (cAMP), in our discussion of the regulation of glycogen synthesis and breakdown (see Fig. 14–18; Table 19–4). The second messenger conveys the signal from the hormone receptor to some enzyme or molecular system in the cell, which then responds. The second messenger either regulates a specific enzymatic reaction or changes the rate at which a specific gene or set of genes is translated into protein(s). In the case of steroid and thyroid hormones, the hormone–receptor complex itself carries the message; it alters the expression of specific genes.

Hormones Function in a Complex Hierarchy

Now let us briefly examine the major endocrine systems of the human body and some of their functional interrelationships. Figure 22–13 shows the anatomical location of the major endocrine glands important in the regulation of metabolism in humans. The word endocrine (from the Greek *endon,* meaning "within," and *krinein,* "to release") means that the secretions of such glands are internal, that is, released into the blood. [Exocrine glands secrete their products (tears, sweat, digestive enzymes) "outward," through ducts that lead to the body surface or the intestinal lumen.] Figure 22–14 (p. 750) is a schematic master plan of the regulatory relationships among the endocrine glands and their target tissues in humans.

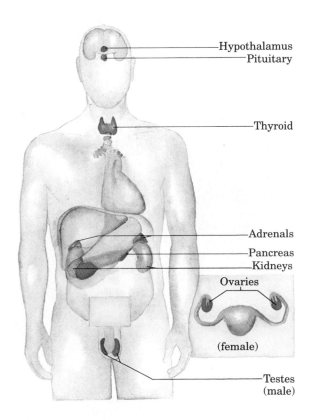

Figure 22–13 The major endocrine glands (shaded red).

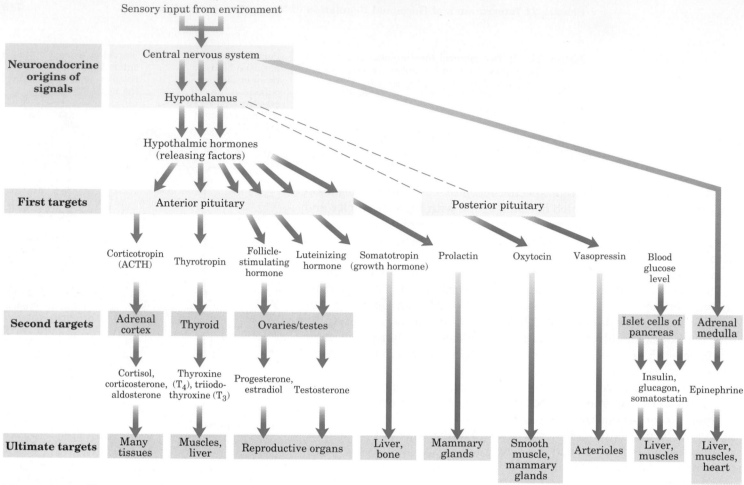

Roger Guillemin

Andrew Schally

Figure 22–14 The major endocrine systems and their target tissues. Signals originating in the central nervous system (top) are passed via a series of relays to the ultimate target tissues (bottom). In addition to the systems shown, the thymus and pineal glands, as well as groups of cells in the gastrointestinal tract, also secrete hormones.

The **hypothalamus,** a specialized portion of the brain (Fig. 22–15), is the coordination center of the endocrine system; it receives and integrates messages from the central nervous system. In response to these messages the hypothalamus produces a number of regulatory hormones, which pass to the **anterior pituitary gland,** located just below the hypothalamus. Some hypothalamic hormones ("releasing factors") stimulate the anterior pituitary to secrete a given hormone; others are inhibitory. Once stimulated, the anterior pituitary secretes hormones into the blood to be carried to the next rank of endocrine glands, which includes the **adrenal cortex,** the **thyroid gland,** the **ovary** and **testis,** and the endocrine cells of the **pancreas.** These glands in turn are stimulated to secrete their specific hormones, which are carried by the blood to hormone receptors on or in the cells of the target tissues.

The **posterior pituitary** contains the axonal endings of many neurons that originate in the hypothalamus. In these neurons, two short peptide hormones, oxytocin and vasopressin (Fig. 22–16), are formed from longer precursor peptides. These peptide hormones move down the hypothalamic axons to the nerve endings in the pituitary, where they are stored in secretory granules. Oxytocin (M_r 1,007) acts on the smooth muscles of the uterus and mammary gland, causing uterine contractions during labor and promoting milk release during lactation. Vasopressin (also called antidiuretic hormone, ADH; M_r 1,040) increases water reabsorption in the kidney and promotes the constriction of blood vessels, thereby increasing blood pressure.

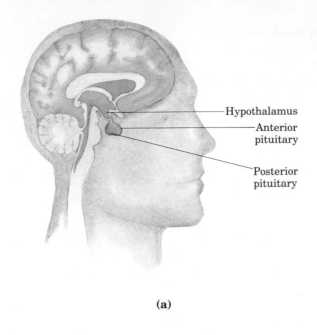

(a)

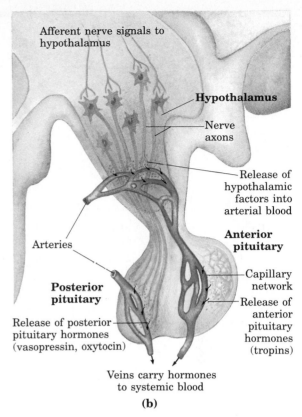

Afferent nerve signals to hypothalamus

Hypothalamus

Nerve axons

Release of hypothalamic factors into arterial blood

Anterior pituitary

Arteries

Capillary network

Release of anterior pituitary hormones (tropins)

Posterior pituitary

Release of posterior pituitary hormones (vasopressin, oxytocin)

Veins carry hormones to systemic blood

(b)

The final link in this system is the intracellular mechanism triggered by the hormone receptor: either a second messenger that carries the message from the hormone receptor to the specific cell structure or enzyme that is the ultimate target, or the alteration of gene expression by a hormone–receptor complex bound to DNA. Thus each endocrine system resembles a set of relays, carrying messages through several steps from the central nervous system to a specific effector molecule in the target cells.

The hypothalamus functions at the top of the hierarchy of many hormone-producing tissues (Fig. 22–14). It receives neural input from diverse regions of the brain and feedback signals from hormones circulating in the blood. These signals are integrated in the hypothalamus, which responds by releasing appropriate hormones to the next tissue in the cascade, the pituitary. The hormones secreted by the hypothalamus are relatively short peptides (see, for example, Fig. 5–19c), produced in very small quantities. A number of these were first isolated and characterized by Roger Guillemin and Andrew Schally.

The hypothalamic hormones pass directly to the nearby pituitary gland through special blood vessels and neurons that connect the two glands (Fig. 22–15b). The pituitary gland has two functionally distinct parts. The **anterior pituitary** responds to hypothalamic hormones carried in the blood, by producing six **tropic hormones** or **tropins** (from the Greek *tropos,* meaning "turn"), relatively long polypeptides

Figure 22–15 (a) Location of the hypothalamus and pituitary gland. **(b)** Details of the hypothalamus–pituitary system. Signals arriving from connecting neurons stimulate the hypothalamus to secrete hormones destined for the anterior pituitary into a special blood vessel, which carries the hormones directly to a capillary network in the anterior pituitary. In response to each hypothalamic hormone, the anterior pituitary releases its appropriate hormone into the general circulation. Posterior pituitary hormones are made in neurons arising in the hypothalamus, transported in axons to nerve endings in the posterior pituitary, and stored there until released into the blood in response to a neuronal signal.

Figure 22–16 Two hormones of the posterior pituitary gland. The carboxyl-terminal residues are glycinamide (—NH—CH₂—CONH₂); amidation of the carboxyl terminus is common in short peptide hormones. These two hormones, identical in all but two residues (shaded) have very different biological effects.

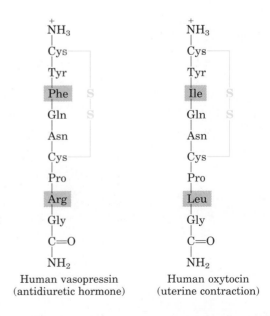

Human vasopressin (antidiuretic hormone)

Human oxytocin (uterine contraction)

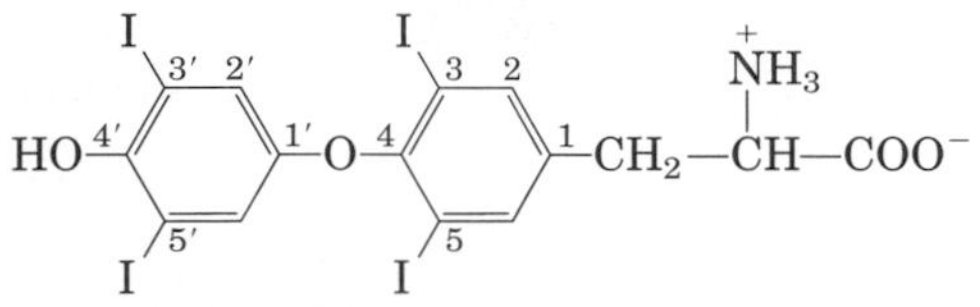

Thyroxine (L-3,5,3',5'-tetraiodothyronine, or T_4)

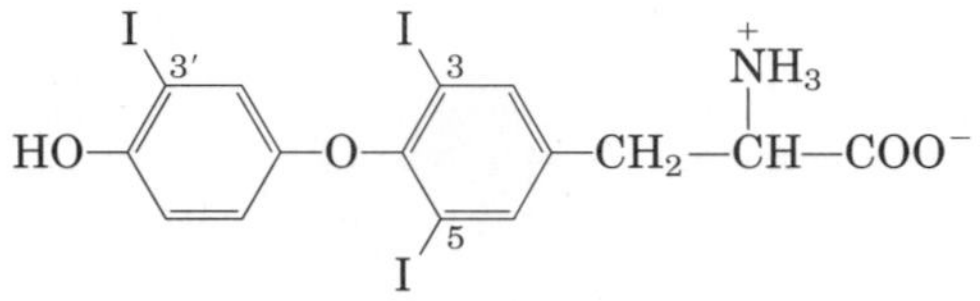

Triiodothyronine (L-3,5,3'-triiodothyronine, or T_3)

Figure 22–17 The thyroid hormones T_4 and T_3. Both are derived from L-tyrosine. Synthesis begins with the iodination of certain Tyr residues in the protein thyroglobulin (M_r 650,000). Further modifications of these residues yield T_4 and T_3, which remain bound to thyroglobulin until released by proteolytic enzymes.

that activate the next rank of endocrine glands (Fig. 22–14). Adrenocorticotropic hormone (ACTH, also called corticotropin; M_r 4,500) stimulates the adrenal cortex; thyroid-stimulating hormone (TSH, also called thyrotropin; M_r 28,000) acts on the thyroid gland; follicle-stimulating hormone (FSH; M_r 34,000) and luteinizing hormone (LH; M_r 20,500) act on the gonads; and growth hormone (GH, also called somatotropin; M_r 21,500) stimulates the liver to produce several growth factors.

Thyroid Hormones The thyroid hormones are released when the hypothalamus secretes thyrotropin-releasing hormone, which stimulates the anterior pituitary to release thyrotropin, which in turn stimulates the thyroid gland to secrete its two characteristic hormones: L-thyroxine (T_4) and L-triiodothyronine (T_3) (Fig. 22–17). Small amounts of T_4 and T_3 stimulate energy-yielding metabolism, especially in liver and muscle. These hormones bind to a specific intracellular receptor protein; the hormone–receptor complex activates certain genes encoding energy-related enzymes, increasing their synthesis and thus increasing the basal metabolic rate of the animal.

The **basal metabolic rate** (BMR) is a measure of the rate of O_2 consumption by an individual at complete rest, 12 hours after a meal. Measurement of the BMR is useful in the diagnosis of thyroid malfunction. Hyperthyroid individuals (those who oversecrete thyroid hormones) have an elevated BMR; hypothyroidism is characterized by a lowered BMR. The level of protein-bound iodine (T_3 and T_4 bound to their carrier proteins) in the blood also is a useful measure of thyroid function.

Steroid Hormones The major steroid hormones are the adrenocortical hormones, the sex hormones (androgens and estrogens), and vitamin D–derived hormones. These hormones are lipid-soluble and readily pass through plasma membranes into the cytosol of target cells. Here they combine with specific intracellular receptor proteins, and these complexes, like the thyroid hormone–receptor complexes, act in the nucleus, causing certain genes to be expressed (Fig. 22–12). Most steroid hormone receptors are localized in the nucleus; others may move from the cytosol to the nucleus only when bound to the hormone. Adrenocortical hormones are produced by cells in the outer portion (cortex) of the adrenal glands, which are located just atop the kidneys ("adrenal"). When an animal is under stress, the hypothalamus secretes corticotropin-releasing hormone, which stimulates the anterior pituitary to release corticotropin into the blood. Corticotropin in turn signals the adrenal cortex to produce its characteristic **corticosteroid hormones,** including cortisol, corticosterone, and aldosterone (Fig. 22–14). Over 50 corticosteroid hormones of two general types are produced in the adrenal cortex: **glucocorticoids** and **mineralocorticoids.** Glucocorticoids affect primarily the metabolism of carbohydrates, and mineralocorticoids regulate the concentrations of electrolytes in the blood.

The androgens (testosterone) and the estrogens (such as estradiol; see Fig. 9–15) are synthesized in the testes and ovaries, respectively (Fig. 22–14). They affect sexual development, sexual behavior, and a variety of other reproductive and nonreproductive functions. Steroid hormones produced from vitamin D by enzymes in the liver and kidneys (see Fig. 9–19) regulate the uptake and metabolism of Ca^{2+} and phosphate, including the formation and mobilization of calcium phosphate in bone.

Tyrosine
(the parent compound)

Norepinephrine

3,4-Dihydroxyphenylalanine
(dopa)

Epinephrine

Dopamine

Catechol
(1,2-dihydroxybenzene)

Figure 22–18 The catecholamine hormones. They are formed from tyrosine (top) and are derivatives of catechol (bottom). The abbreviation "dopa" is derived from the German name of the compound, *dioxyphenylalanin.*

Amine Hormones The water-soluble hormones epinephrine, norepinephrine, dopa, and dopamine are in a class of amines called **catecholamines,** derivatives of catechol (Fig. 22–18). **Epinephrine** (adrenaline) and **norepinephrine** (noradrenaline) are closely related hormones, made and secreted by the inner portion (medulla) of the adrenal glands in response to signals from the central nervous system (Fig. 22–14). Normally the epinephrine level in blood is only about 10^{-10} M, but sensory stimuli that alarm the animal and galvanize it for action lead to the release of epinephrine from the adrenal medulla and a 1,000-fold increased concentration in the blood within seconds or minutes. The catecholamines are also made in the brain and other neural tissue, where they function as neurotransmitters.

Peptide Hormones: Insulin, Glucagon, and Somatostatin The pancreas has two major biochemical functions: exocrine cells produce digestive enzymes for secretion into the intestine, and endocrine cells produce and secrete peptide hormones that regulate fuel metabolism throughout the body. The peptide hormones insulin, glucagon, and somatostatin are produced by clusters of specialized cells called the islets of Langerhans (Fig. 22–19). Each islet cell type produces a single hormone: α cells produce glucagon; β cells, insulin; and δ cells, somatostatin.

Figure 22–19 The endocrine system of the pancreas. In addition to the exocrine or acinar cells (see Fig. 17–3b), which secrete digestive enzymes in the form of zymogens, the pancreas contains endocrine tissue consisting of the islets of Langerhans. The islets contain several different types of cells, each of which excretes a specific polypeptide hormone. **(a)** Schematic drawing of an islet, showing the α, β, and δ cells (also known as A, B, and D cells, respectively). **(b)** Micrograph of a portion of an islet of Langerhans of human pancreas.

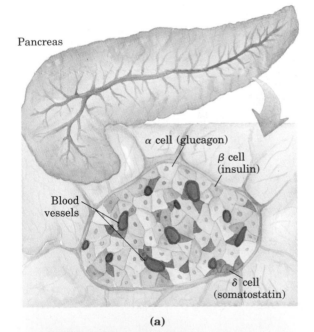

Pancreas

α cell (glucagon)

β cell (insulin)

Blood vessels

δ cell (somatostatin)

(a)

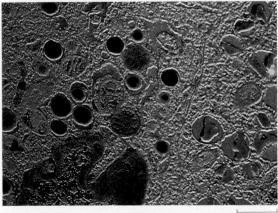

(b)　　　　0.6 μm

Figure 22–20 Insulin formation. **(a)** Mature insulin is formed from its larger precursor preproinsulin by proteolytic processing. Removal of 23 amino acids (the signal sequence) at the amino terminus of preproinsulin and formation of three disulfide bonds produces proinsulin. Further proteolytic cuts remove the C peptide, leaving mature insulin, composed of A and B chains. **(b)** Space-filling and **(c)** ribbon models of porcine insulin. The amino acid sequence of bovine insulin is shown in Fig. 6–10.

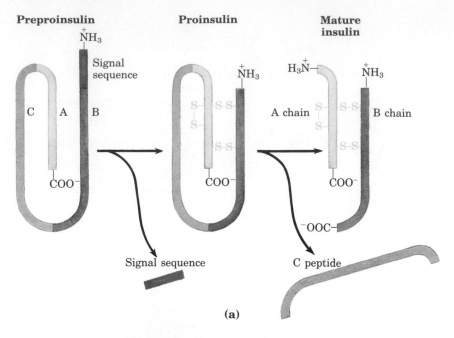

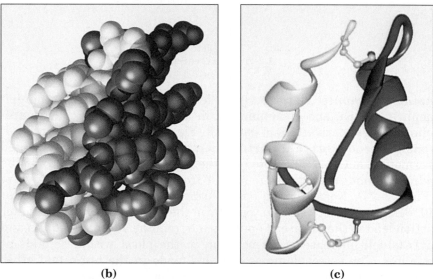

(a)

(b)   (c)

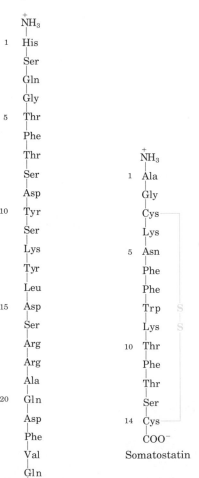

Figure 22–21 The primary structures of glucagon and somatostatin. Glucagon is formed by proteolytic processing similar to that shown for insulin in Fig. 22–20.

Insulin is a small protein (M_r 5,700) with two polypeptide chains, A and B, joined by two disulfide bonds. It is synthesized in the pancreatic β cells as an inactive single-chain precursor, preproinsulin (Fig. 22–20), with an amino-terminal "signal sequence" that directs its passage into secretory vesicles. (Signal sequences are discussed in Chapter 26; see Fig. 26–35.) Proteolytic removal of the signal sequence and formation of three disulfide bonds produces proinsulin, which is stored in secretory granules in the β cells. When elevated blood glucose triggers insulin secretion, proinsulin is converted into active insulin by specific peptidases, which cleave two peptide bonds to form the mature insulin molecule (Fig. 22–20).

Glucagon (Fig. 22–21) is a single polypeptide chain of 29 amino acid residues, and like insulin is derived from larger precursors (preproglucagon and proglucagon) by precise proteolytic cleavages. **Somatostatin,** also a polypeptide hormone (Fig. 22–21), inhibits the secretion of insulin and glucagon by the pancreas. Somatostatin is produced and secreted not only by pancreatic δ cells, but also by the hypothalamus and certain intestinal cells.

Hormonal Regulation of Fuel Metabolism

Our discussions of metabolic regulation and hormone action now come together as we return to the hormonal regulation of blood glucose level. The minute-by-minute adjustments that keep the blood glucose level near 4.5 mM involve the combined actions of insulin, glucagon, and epinephrine on metabolic processes in many body tissues, but especially in liver, muscle, and adipose tissue. Insulin signals these tissues that the blood glucose concentration is higher than necessary; as a result, the excess glucose is taken up from the blood into cells and converted to storage compounds, glycogen and triacylglycerols. Glucagon carries the message that blood glucose is too low, and the tissues respond by producing glucose through glycogen breakdown and gluconeogenesis and by oxidizing fats to reduce the use of glucose. Epinephrine is released into the blood to prepare the muscles, lungs, and heart for a burst of activity. Insulin, glucagon, and epinephrine are the primary determinants of the metabolic activities of muscle, liver, and adipose tissue.

Epinephrine Signals Impending Activity

When an animal is confronted with a stressful situation that requires increased activity—fighting or fleeing, in the extreme case—neuronal signals from the brain trigger the release of epinephrine and norepinephrine from the adrenal medulla. Both hormones increase the rate and strength of the heartbeat and raise the blood pressure, thereby increasing the flow of O_2 and fuels to the tissues, and dilate the respiratory passages, facilitating the uptake of O_2 (Table 22–3).

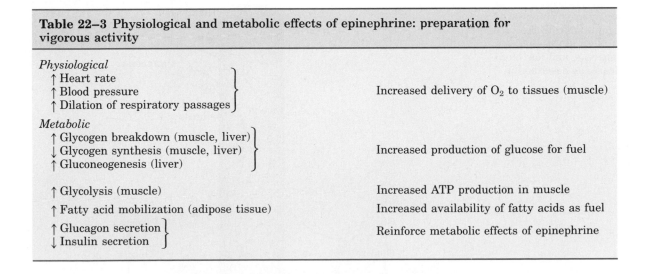

Table 22–3 Physiological and metabolic effects of epinephrine: preparation for vigorous activity

Physiological
↑ Heart rate
↑ Blood pressure   Increased delivery of O_2 to tissues (muscle)
↑ Dilation of respiratory passages

Metabolic
↑ Glycogen breakdown (muscle, liver)
↓ Glycogen synthesis (muscle, liver)   Increased production of glucose for fuel
↑ Gluconeogenesis (liver)

↑ Glycolysis (muscle)   Increased ATP production in muscle

↑ Fatty acid mobilization (adipose tissue)   Increased availability of fatty acids as fuel

↑ Glucagon secretion
↓ Insulin secretion   Reinforce metabolic effects of epinephrine

In its effects on metabolism, epinephrine acts primarily on muscle, adipose tissue, and liver. It activates glycogen phosphorylase and inactivates glycogen synthase (by cAMP-dependent phosphorylation of the enzymes; see Fig. 14–18 and p. 615), thus stimulating the conversion of liver glycogen into blood glucose, the fuel for anaerobic muscular work. Epinephrine also promotes the anaerobic breakdown of the glycogen of

skeletal muscle into lactate by fermentation, thus stimulating glycolytic ATP formation. The stimulation of glycolysis is accomplished by
raising the concentration of fructose-2,6-bisphosphate, a potent allosteric activator of the key glycolytic enzyme phosphofructokinase-1
(see Figs. 19–7, 19–8). Epinephrine also stimulates fat mobilization in
adipose tissue, activating (by cAMP-dependent phosphorylation) the
triacylglycerol lipase (see Fig. 16–3). Finally, epinephrine stimulates
the secretion of glucagon and inhibits the secretion of insulin, reinforc-
ing its effect of mobilizing fuels and inhibiting fuel storage.

Glucagon Signals Low Blood Glucose

Even in the absence of significant physical activity or stress, several
hours after the intake of dietary carbohydrate, blood glucose levels fall
to below 4.5 mM because of the continued oxidation of glucose by the
brain and other tissues. Lowered blood glucose triggers secretion of
glucagon and decreases insulin release (Fig. 22–22). Glucagon causes
an increase in blood glucose concentration in two ways (Table 22–4).
Like epinephrine, glucagon stimulates the net breakdown of liver glycogen by activating glycogen phosphorylase and inactivating glycogen
synthase; both effects are the result of phosphorylation of the regulated enzymes, triggered by cAMP. But, unlike epinephrine, glucagon
inhibits glucose breakdown by glycolysis in the liver and stimulates
glucose synthesis by gluconeogenesis. Both of these effects result from

Figure 22–22 Regulation of blood glucose by insulin and glucagon. Blue arrows indicate processes
stimulated by insulin; red arrows indicate processes
stimulated by glycogen. High blood glucose results
in insulin secretion by the pancreas, and low blood
glucose leads to glucagon release, as described in
the text.

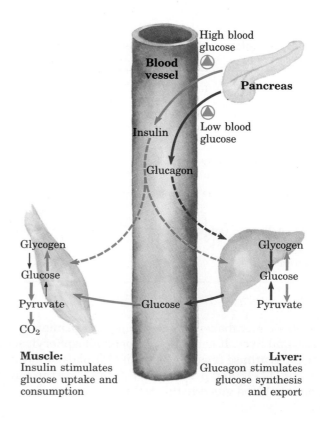

lowering the level of fructose-2,6-bisphosphate, an allosteric inhibitor of the gluconeogenic enzyme fructose-1,6-bisphosphatase (FBPase-1) and an activator of phosphofructokinase-1. Recall that the fructose-2,6-bisphosphate level is ultimately controlled by a cAMP-dependent protein phosphorylation reaction (see Fig. 19–8). Glucagon also inhibits the glycolytic enzyme pyruvate kinase (by promoting its cAMP-dependent phosphorylation), thus blocking the conversion of phosphoenolpyruvate to pyruvate and preventing oxidation of pyruvate via the citric acid cycle; the resulting accumulation of phosphoenolpyruvate favors gluconeogenesis.

Table 22–4 Effects of glucagon on blood glucose: production and release of glucose by the liver

Metabolic effect	Effect on glucose metabolism	Target enzyme
↑ Glycogen breakdown (liver)	Glycogen $\longrightarrow$ glucose	↑ Glycogen phosphorylase
↓ Glycogen synthesis (liver)	Less glucose stored as glycogen	↓ Glycogen synthase
↓ Glycolysis (liver)	Less glucose used as fuel in liver	↓ Phosphofructokinase-1
↑ Gluconeogenesis (liver)	Amino acids Glycerol $\Big\} \longrightarrow$ glucose Oxaloacetate	↑ Fructose-1,6-bisphosphatase ↓ Pyruvate kinase
↑ Fatty acid mobilization (adipose tissue)	Less glucose used as fuel by liver, muscle	↑ Triacylglycerol lipase

By stimulating liver glycogen breakdown, preventing glucose utilization in the liver by glycolysis, and promoting gluconeogenesis, glucagon enables the liver to export glucose to the blood, restoring blood glucose to its normal level (Fig. 22–22).

Although its primary target is the liver, glucagon (like epinephrine) also affects adipose tissue, activating triacylglycerol lipase by causing its cAMP-dependent phosphorylation. This lipase liberates free fatty acids, which are exported to the liver and other tissues as fuel, thus sparing glucose for the brain. The net effect of glucagon is therefore to stimulate glucose synthesis and release by the liver and to cause the mobilization of fatty acids from adipose tissue, to be used instead of glucose as fuel for tissues other than the brain (Table 22–4). All of these affects of glucagon are mediated by cAMP-dependent protein phosphorylation.

During Starvation, Metabolism Shifts to Provide Fuel for the Brain

The fuel reserves of a normal adult human are of three types: glycogen stored in the liver and in muscle in relatively small quantities; larger quantities of triacylglycerols in adipose tissues; and tissue proteins, which can be degraded when necessary to provide fuel (Table 22–5).

Figure 22–23 shows the changes in fuel metabolism during starvation. After an overnight fast, almost all of the liver glycogen and most of the muscle glycogen have been depleted. Within 24 hours, the blood glucose concentration begins to fall, insulin secretion slows, and glucagon secretion is stimulated. These hormonal signals result in the mobilization of triacylglycerols, which become the primary fuels for muscle and liver. To provide glucose for the brain, the liver degrades certain proteins (those most expendable in an organism not ingesting food). Their amino groups are converted into urea in the liver; the urea is exported via the bloodstream to the kidney and is excreted. Also in the liver, the carbon skeletons of glucogenic amino acids (see Table 19–3) are converted into pyruvate or intermediates of the citric acid cycle. These intermediates, as well as the glycerol derived from triacylglycerols in adipose tissue, provide the starting materials for gluconeogenesis in the liver, yielding glucose for the brain.

Table 22–5 Available metabolic fuels in a normal 70 kg man and in an obese man at the beginning of a fast

Type of fuel	Weight (kg)	Caloric equivalent (thousands of kcal (kJ))	Estimated survival time (months)*
Normal 70 kg man:			
Triacylglycerols (adipose tissue)	15	141 (589)	
Proteins (mainly muscle)	6	24 (100)	
Glycogen (muscle, liver)	0.225	0.90 (3.8)	
Circulating fuels (glucose, fatty acids, triacylglycerols, etc.)	0.023	0.10 (0.42)	
Total		166 (694)	3
Obese man:			
Triacylglycerols (adipose tissue)	80	752 (3,140)	
Proteins (mainly muscle)	8	32 (134)	
Glycogen (muscle, liver)	0.23	0.92 (3.8)	
Circulating fuels	0.025	0.11 (0.46)	
Total		785 (3,280)	14

* Survival time is calculated on the assumption of a basal energy expenditure of 1,800 kcal/day.

Eventually the use of citric acid cycle intermediates for gluconeogenesis depletes oxaloacetate, preventing the entry of acetyl-CoA into the cycle (Fig. 22–23). Acetyl-CoA produced by fatty acid oxidation accumulates, favoring the formation of acetoacetyl-CoA and ketone bodies in the liver. After a few days of fasting, the levels of ketone bodies in the blood rise as these fuels are exported from the liver to heart and skeletal muscle and the brain, which use them instead of glucose.

The triacylglycerols stored in the adipose tissue of an adult of normal weight provide enough fuel to maintain a basal rate of metabolism for about three months; a very obese adult has enough stored fuel to endure a fast of more than a year (Table 22–5). However, such a fast would be extremely dangerous; it would almost certainly lead to severe overproduction of ketone bodies (described below), and perhaps to death. When fat reserves are gone, the degradation of essential proteins begins; this leads to loss of heart and liver function, and death.

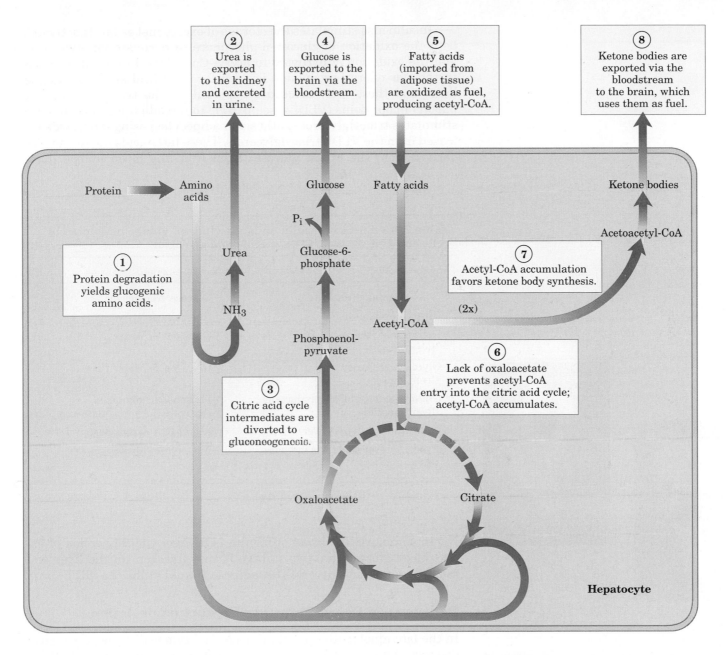

Figure 22–23 Fuel metabolism in the liver during prolonged starvation. After the depletion of stored carbohydrates, proteins become an important source of glucose, produced from glucogenic amino acids by gluconeogenesis (steps ① through ④). Fatty acids imported from adipose tissue are converted into ketone bodies for export to the brain (steps ⑤ through ⑧). The broken arrows represent reactions through which there is reduced flux during starvation.

Insulin Signals High Blood Glucose

When glucose enters the bloodstream from the intestine after a carbohydrate-rich meal, the resulting increase in blood glucose causes increased secretion of insulin and decreased secretion of glucagon (Fig. 22–22). Insulin stimulates glucose uptake by muscle tissue (Table 22–6), where the glucose is converted to glucose-6-phosphate. Insulin also activates glycogen synthase and inactivates glycogen phosphorylase, so that much of the glucose-6-phosphate is channeled into glycogen. As a consequence of accelerated uptake of glucose from the blood, the blood glucose concentration falls to the normal level, slowing the rate of insulin release from the pancreas. Thus there is a closely adjusted feedback relationship between the rate of insulin secretion and the blood glucose concentration. The effect of this regulation is to hold the blood glucose concentration nearly constant in the face of large fluctuations in the dietary intake of glucose.

Insulin also stimulates the storage of excess fuel as fat. It activates both the oxidation of glucose-6-phosphate to pyruvate via glycolysis and the oxidation of pyruvate to acetyl-CoA. Acetyl-CoA not oxidized further for energy production is used for fatty acid synthesis in the liver, and these fatty acids are exported as the triacylglycerols of plasma lipoproteins (VLDLs; see p. 677) to the adipose tissue. Insulin stimulates triacylglycerol synthesis in adipocytes, using fatty acids released from the VLDL triacylglycerols. These fatty acids are ultimately derived from the excess glucose taken from the blood by the liver.

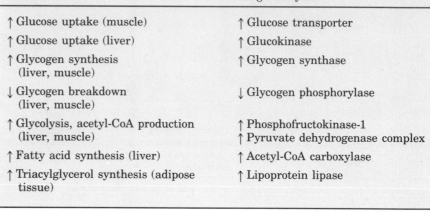

Table 22–6 Effect of insulin on blood glucose: uptake of glucose by cells and storage as triacylglycerols and glycogen

Metabolic effect	Target enzyme
↑ Glucose uptake (muscle)	↑ Glucose transporter
↑ Glucose uptake (liver)	↑ Glucokinase
↑ Glycogen synthesis (liver, muscle)	↑ Glycogen synthase
↓ Glycogen breakdown (liver, muscle)	↓ Glycogen phosphorylase
↑ Glycolysis, acetyl-CoA production (liver, muscle)	↑ Phosphofructokinase-1 ↑ Pyruvate dehydrogenase complex
↑ Fatty acid synthesis (liver)	↑ Acetyl-CoA carboxylase
↑ Triacylglycerol synthesis (adipose tissue)	↑ Lipoprotein lipase

J.B. Collip
1892–1965

J.J.R. Macleod
1876–1935

Charles Best
1899–1978

Frederick G. Banting
1891–1941

In summary, the effect of insulin is to favor the conversion of excess blood glucose into two storage forms: glycogen (in the liver and muscle) and triacylglycerols (in adipose tissue) (Table 22–6).

Diabetes Is a Defect in Insulin Production or Action

In the late nineteenth century, the surgical removal of the pancreas of dogs was found to cause a condition closely resembling human diabetes mellitus. Injection of extracts of normal pancreas into these dogs alleviated the diabetic symptoms. The active factor present in pancreatic extracts, insulin ("islet substance"), was finally isolated in pure form in 1922 by Banting, Best, Collip, and Macleod. Insulin quickly came into use in the treatment of human diabetes and has become one of the most important therapeutic agents known to medicine; it has prolonged countless lives.

Diabetes mellitus, caused by a deficiency in the secretion or action of insulin, is a relatively common disease: nearly 5% of the United States population shows some degree of abnormality in glucose metabolism indicative of diabetes or a tendency toward it. Diabetes mellitus is really a group of diseases in which the regulatory activity of insulin may be defective in different ways. Moreover, several other hormones can influence the metabolism of glucose. There are two major clinical classes of the disease: insulin-dependent diabetes mellitus (IDDM) and non–insulin-dependent diabetes mellitus (NIDDM).

In the former, the disease begins early in life and quickly becomes severe. The latter is slow to develop, milder, and often goes unrecognized. IDDM requires insulin therapy and careful, lifelong control of the balance between glucose intake and insulin dose. Characteristic symptoms of diabetes are excessive thirst and frequent urination (polyuria), leading to the intake of large volumes of water (polydipsia). These changes are due to the excretion of large amounts of glucose in the urine, a condition known as **glucosuria.** The term diabetes mellitus means "excessive excretion of sweet urine."

Another characteristic metabolic change resulting from the defect in insulin action in diabetes is excessive but incomplete oxidation of fatty acids in the liver, resulting in an overproduction of the ketone bodies acetoacetate and β-hydroxybutyrate, which cannot be used by the extrahepatic tissues as fast as they are made in the liver. In addition to β-hydroxybutyrate and acetoacetate, the blood of diabetics also contains acetone, which results from the spontaneous decarboxylation of acetoacetate:

$$CH_3-\overset{\displaystyle O}{\overset{\|}{C}}-CH_2-COO^- + H_2O \longrightarrow CH_3-\overset{\displaystyle O}{\overset{\|}{C}}-CH_3 + HCO_3^-$$

$$\text{Acetoacetate} \qquad\qquad\qquad\qquad \text{Acetone}$$

Acetone is volatile and is exhaled, giving the breath of an untreated diabetic a characteristic odor sometimes mistaken for ethanol. A diabetic experiencing mental confusion because of high blood glucose is occasionally misdiagnosed as intoxicated, an error that can be fatal. The overproduction of ketone bodies, called **ketosis,** results in their appearance in greatly increased concentrations in the blood (ketonemia) and urine (ketonuria) (see Table 16–2).

The oxidation of triacylglycerols to form ketone bodies produces carboxylic acids, which ionize, releasing protons. In uncontrolled diabetes this can overwhelm the capacity of the bicarbonate buffering system of blood and produce a lowering of blood pH called **acidosis,** a potentially life-threatening condition.

Biochemical measurements on the blood and urine are essential in the diagnosis and treatment of diabetes, which causes profound changes in metabolism. A sensitive diagnostic criterion is provided by the **glucose-tolerance test.** After a night without food, the patient drinks a test dose of 100 g of glucose dissolved in a glass of water. The blood glucose concentration is measured before the test dose and at 30 min intervals for several hours thereafter. A normal individual assimilates the glucose readily, the blood glucose rising to no more than about 9 or 10 mM; little or no glucose appears in the urine. Diabetic individuals show a marked deficiency in assimilating the test dose of glucose. The blood glucose level increases far above the kidney threshold, which is about 10 mM, causing glucose to appear in the urine.

Molecular Mechanisms of Signal Transduction

The number of known hormones, and the number of physiological and biochemical effects attributable to hormones, continue to grow, with no end in sight. Fortunately for the student of biochemistry, most of these hormones act through a few fundamentally similar mechanisms. We first consider one of the best-understood hormone mechanisms—

involving cAMP as the second messenger—which mediates the cellular response to epinephrine. We then describe examples of several other fundamental hormone mechanisms, involving different second messengers (cGMP, diacylglycerols, an inositol trisphosphate, Ca^{2+}), a protein–tyrosine kinase activity, and ligand- and voltage-activated ion channels. Similar mechanisms mediate the action of growth factors and of some oncogenes. The phosphorylation and dephosphorylation of specific proteins are shown to be central to these mechanisms. Finally, we describe how steroid hormones function through the regulation of gene activity.

Receptors for Epinephrine Trigger Cyclic AMP Production

The current understanding of the mechanism of epinephrine (and glucagon) action originated in the work of Earl W. Sutherland, Jr., and his colleagues in the early 1950s. These investigators showed that epinephrine stimulates the activity of glycogen phosphorylase, which promotes the breakdown of glycogen to glucose-1-phosphate, the rate-limiting step in the conversion of glycogen to glucose. Sutherland's laboratory identified **adenosine 3′,5′-cyclic monophosphate (cyclic AMP or cAMP)** as the intracellular messenger produced in response to extracellular epinephrine. Figure 22–24 schematizes the multistep path from the initial stimulus to the elevation of blood glucose. Several of these steps amplify the effect of hormone binding to the receptor, so that a single molecule of hormone can change the catalytic activity of thousands of enzyme molecules.

Eventually, five proteins essential to the epinephrine response were identified and purified (Fig. 22–25): (1) a hormone receptor in the plasma membrane; (2) the enzyme adenylate cyclase, which catalyzes cAMP formation; (3) G_s protein, which shuttles between the receptor and adenylate cyclase, activating the cyclase when hormone is bound to the receptor; (4) a cAMP-dependent protein kinase, which phosphorylates target enzymes within the cell, altering their activities; and (5) cyclic nucleotide phosphodiesterase, which degrades cAMP and thereby terminates the intracellular signal.

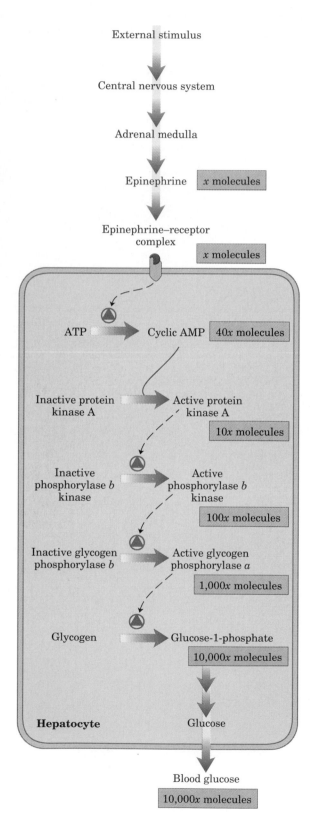

Figure 22–24 Epinephrine triggers a series of reactions in hepatocytes in which catalysts activate catalysts, resulting in great amplification of the signal. Binding of a small number of molecules of epinephrine to specific receptors on the cell surface activates adenylate cyclase. For the sake of illustration, we have shown 40 molecules of cAMP produced by each molecule of adenylate cyclase. These 40 cAMP molecules activate 10 molecules of protein kinase, each of which in turn activates 10 molecules of the next enzyme in the cascade. The amplifications shown here for each step are probably gross underestimates.

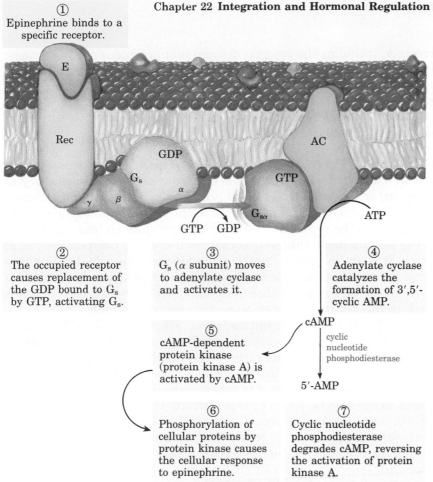

① Epinephrine binds to a specific receptor.

② The occupied receptor causes replacement of the GDP bound to G_s by GTP, activating G_s.

③ G_s (α subunit) moves to adenylate cyclase and activates it.

④ Adenylate cyclase catalyzes the formation of 3′,5′-cyclic AMP.

⑤ cAMP-dependent protein kinase (protein kinase A) is activated by cAMP.

⑥ Phosphorylation of cellular proteins by protein kinase causes the cellular response to epinephrine.

⑦ Cyclic nucleotide phosphodiesterase degrades cAMP, reversing the activation of protein kinase A.

Figure 22–25 The mechanism that couples binding of epinephrine (E) to its receptor (Rec) with the activation of adenylate cyclase (AC). The seven steps are further discussed in the text. The same adenylate cyclase molecule in the plasma membrane may be regulated by a stimulatory G protein, G_s, as shown or an inhibitory G protein, G_i (not shown). G_s and G_i are under the influence of different hormones. Hormones that induce GTP binding to G_i cause *inhibition* of adenylate cyclase, resulting in lower cellular levels of cAMP.

Earl W. Sutherland, Jr. 1915–1974

The Epinephrine–β-Adrenergic Receptor Complex The action of epinephrine begins with the binding of the hormone to a protein receptor in the plasma membrane of a hormone-sensitive cell, a hepatocyte or myocyte (Fig. 22–25, step ①). The binding is tight but noncovalent, like the binding of an allosteric effector to an allosterically regulated enzyme. The binding site on the receptor is stereospecific and will accommodate only the natural hormone ligand or molecules with a closely similar three-dimensional geometry. Structural analogs that bind to a receptor and mimic the effects of its natural ligand are called **agonists; antagonists** are analogs that bind without triggering the normal effect, and thereby block the effects of agonists.

Adrenergic receptors (the term "adrenergic" reflects the alternative name for epinephrine: *adren*aline) are of four general types, defined by subtle differences in their affinities and responses to a group of agonists and antagonists. The four types (α_1, α_2, β_1, β_2) are found in different target tissues and mediate different responses to epinephrine. Here we will focus on the **β-adrenergic receptors** found in muscle, liver, and adipose tissue. These receptors mediate the changes in fuel metabolism described above, including the increased breakdown of glycogen and fat.

β-Adrenergic receptors are integral membrane proteins with amino acid sequences that contain seven hydrophobic regions of 20 to 28 residues, suggesting that the protein traverses the lipid bilayer seven times (see Box 10–2). The binding site for epinephrine is on the outer face of the plasma membrane; the hormone causes an intracellular change without itself crossing the plasma membrane. The binding of epinephrine apparently promotes a conformational change in the receptor, including the receptor domain that protrudes on the cytosolic face of the membrane. The first stage of hormone action is therefore comparable with the action of an allosteric effector on an allosterically regulated enzyme. The structural change in the intracellular domain of the receptor allows its interaction with the second protein in the signal transduction pathway, a GTP-binding protein.

GTP-Binding Protein and Adenylate Cyclase The next element in the signal-transducing pathway is a protein called a **stimulatory G protein,** or **G_s,** located on the cytosolic face of the plasma membrane (Fig. 22–25). (G_s takes its name from the fact that, when bound to *G*TP, it *s*timulates the production of cAMP by adenylate cyclase, an enzyme of the plasma membrane.) G_s is composed of three polypeptides, α, β, and γ. It is one of a large family of guanosine nucleotide–binding proteins that mediate a wide variety of signal transductions, including those triggered by many other hormones (some of which are discussed later in this chapter) as well as certain sensory stimuli.

G_s can exist in either of two forms. When its nucleotide-binding site (on the α subunit) is occupied by GTP, G_s is active and can interact with and activate adenylate cyclase. With GDP bound to the site, G_s is inactive and incapable of activating adenylate cyclase. Binding of epinephrine causes the receptor to catalyze the displacement of the GDP bound to inactive G_s by GTP; this converts G_s to its active form (Fig. 22–25, step ②). As this occurs, the β and γ subunits dissociate from the α subunit; $G_{s\alpha}$, with its bound GTP, then moves in the plane of the membrane from the receptor to a nearby molecule of adenylate cyclase (step ③).

Adenylate cyclase is an integral protein of the plasma membrane, with its active site on the cytosolic face (Fig. 22–25, step ④). The association of active $G_{s\alpha}$ with adenylate cyclase converts the cyclase to its catalytically active form; the enzyme catalyzes the production of cAMP from ATP, raising the cytosolic level of this second messenger.

Adenosine 3′,5′-cyclic monophosphate (cAMP)

Activation of adenylate cyclase by $G_{s\alpha}$ is self-limiting; $G_{s\alpha}$ has a weak GTPase activity and turns itself off by converting its bound GTP to GDP (Fig. 22–26). The now inactive $G_{s\alpha}$ dissociates from adenylate cyclase, thereby inactivating it. After $G_{s\alpha}$ reassociates with the β and γ subunits, G_s again becomes available for interaction with hormone-bound receptor.

Signal transduction through adenylate cyclase involves two steps in sequence that amplify the original hormone signal (Fig. 22–24).

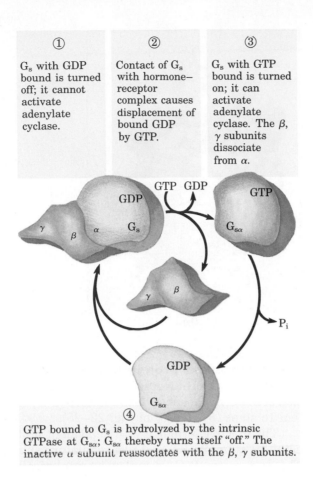

① G_s with GDP bound is turned off; it cannot activate adenylate cyclase.

② Contact of G_s with hormone–receptor complex causes displacement of bound GDP by GTP.

③ G_s with GTP bound is turned on; it can activate adenylate cyclase. The β, γ subunits dissociate from α.

④ GTP bound to G_s is hydrolyzed by the intrinsic GTPase at $G_{s\alpha}$; $G_{s\alpha}$ thereby turns itself "off." The inactive α subunit reassociates with the β, γ subunits.

Figure 22–26 The protein G_s acts as a self-inactivating switch.

First, *one hormone molecule* bound to one receptor catalytically activates *several G_s molecules*. Second, by activating a molecule of adenylate cyclase, *one active $G_{s\alpha}$ molecule* leads to the catalytic synthesis of *many molecules of cAMP*. The net effect of this cascade is a very significant amplification of the hormonal signal, which accounts for the very low concentration of epinephrine (and of other hormones) required for activity.

Cyclic AMP, the intracellular second messenger in this system, is short-lived; it is quickly degraded by **cyclic nucleotide phosphodiesterase** to 5′-AMP (Fig. 22–25, step ⑦; Fig. 22–27), which is not active as a second messenger. The intracellular signal therefore persists only as long as the hormone receptor remains occupied by epinephrine. Methyl xanthines such as theophylline (a component of tea) inhibit the phosphodiesterase, potentiating the action of agents that act through adenylate cyclase.

Most signal receptors also have some mechanism for reducing their sensitivity to a chronically present signal. In the case of the β-adrenergic receptor, this desensitization involves phosphorylation of receptor molecules by a specific protein kinase. In other cases, desensitization occurs by removal of receptors from the cell surface.

Cyclic AMP–Dependent Protein Kinase We have seen that one effect of epinephrine is to activate glycogen phosphorylase b. Recall that this conversion is promoted by the enzyme phosphorylase b kinase, which catalyzes the phosphorylation of two specific Ser residues in phosphorylase b, converting it into phosphorylase a (see Figs. 14–17, 14–18).

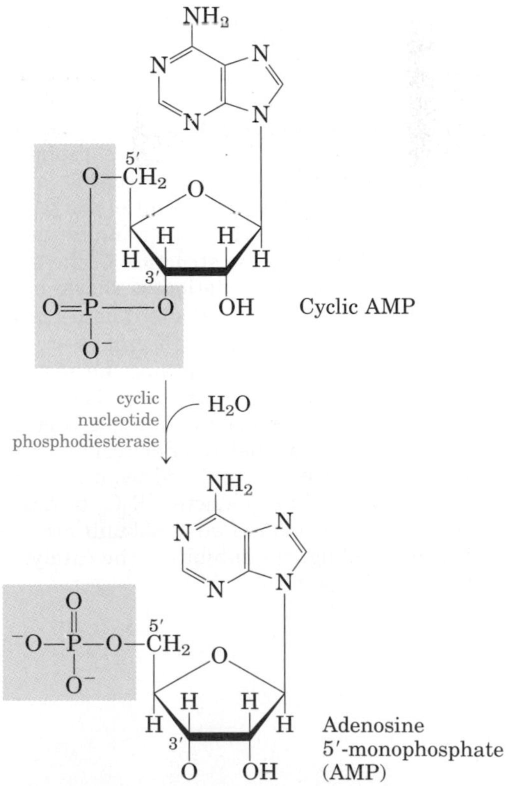

Cyclic AMP

cyclic nucleotide phosphodiesterase — H_2O

Adenosine 5′-monophosphate (AMP)

Figure 22–27 The degradation of cAMP by cyclic nucleotide phosphodiesterase counterbalances the synthesis of cAMP from ATP by adenylate cyclase. In many tissues phosphodiesterase is stimulated by Ca^{2+}, an effect mediated by the regulatory Ca^{2+}-binding protein calmodulin (see Fig. 22–33).

Inactive

Regulatory subunits: empty cAMP sites

Catalytic subunits: substrate binding sites blocked by autoinhibitory domains of R subunits

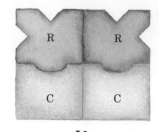

4 cAMP   4 cAMP

Regulatory subunits: autoinhibitory domains buried

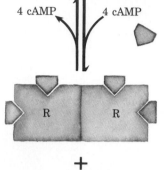

+

Active

Catalytic subunits: open substrate binding sites

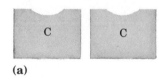

(a)

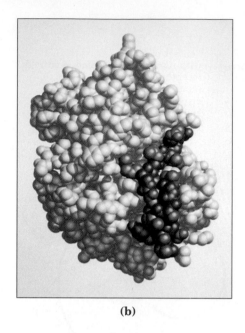

(b)

Figure 22–28 (a) Activation of cAMP-dependent protein kinase. Cyclic AMP activates the protein kinase by causing dissociation of the catalytic (C) subunits from the inhibitory regulatory (R) subunits. This allows phosphorylation and activation of phosphorylase b kinase, which in turn phosphorylates and activates glycogen phosphorylase.
(b) Structure of one catalytic subunit of the cAMP-dependent protein kinase. A potent inhibitor peptide (PKI), which mimics the structure of the normal substrates for phosphorylation, is shown here occupying the substrate binding site. The peptide backbone of PKI is shown in red; R groups are shown in blue. This inhibitor contains the sequence Arg–Arg–Gln–Ala–Ile, which corresponds to the consensus sequence recognized by protein kinase A (see Table 22–9), except that the Ser residue phosphorylated in substrates is replaced by an Ala residue in the inhibitor. In the inactive R_2C_2 tetramer, the autoinhibitory domain of an R subunit occupies the substrate binding site, inhibiting the catalytic activity of the C subunit.

Cyclic AMP does not affect phosphorylase b kinase directly. Rather, **cAMP-dependent protein kinase,** also called **protein kinase A,** which is allosterically activated by cAMP, catalyzes the phosphorylation of inactive phosphorylase b kinase to yield the active form (equivalent to steps ⑤ and ⑥ in Fig. 22–25).

The inactive form of cAMP-dependent protein kinase contains two catalytic subunits (C) and two regulatory subunits (R) (Fig. 22–28a). The tetrameric R_2C_2 complex is catalytically inactive, because an autoinhibitory domain of each R subunit occupies the substrate binding site of each C subunit. When cAMP binds to two sites on each of the two R subunits, the R subunits undergo a conformational change, and the R_2C_2 complex dissociates to yield two free, catalytically active C subunits.

Phosphorylase b kinase is not the only target of cAMP-dependent protein kinase. This protein kinase can phosphorylate the hydroxyl group of Thr and Ser residues (Fig. 22–29) in a number of other important enzymes in different kinds of target cells, thereby altering their catalytic activities (Table 22–7). Although the proteins regulated by cAMP-dependent phosphorylation have diverse catalytic activities, they share a region of sequence homology around the Ser or Thr residue that undergoes phosphorylation. This "consensus sequence" marks these proteins for regulation by cAMP; it is the region recognized by the substrate binding site of the cAMP-dependent protein kinase. The autoinhibitory domain of the regulatory subunit of the protein kinase also contains a consensus sequence. Determination of the three-dimensional structure of this enzyme by x-ray crystallography (Fig. 22–28b) has shown how this autoinhibitor domain can occupy the substrate binding site of a C subunit and prevent access to protein substrates (such as phosphorylase b kinase) that are targets for phosphorylation, and thus for regulation by this enzyme.

Biochemists now appreciate that the details of epinephrine action represent a particular example of a more general theme. As we will see, hormone signals often lead, via pathways similar to that shown in Figures 22–24 and 22–25, to the phosphorylation of certain target enzymes by one or more protein kinases. As noted above, the cAMP-

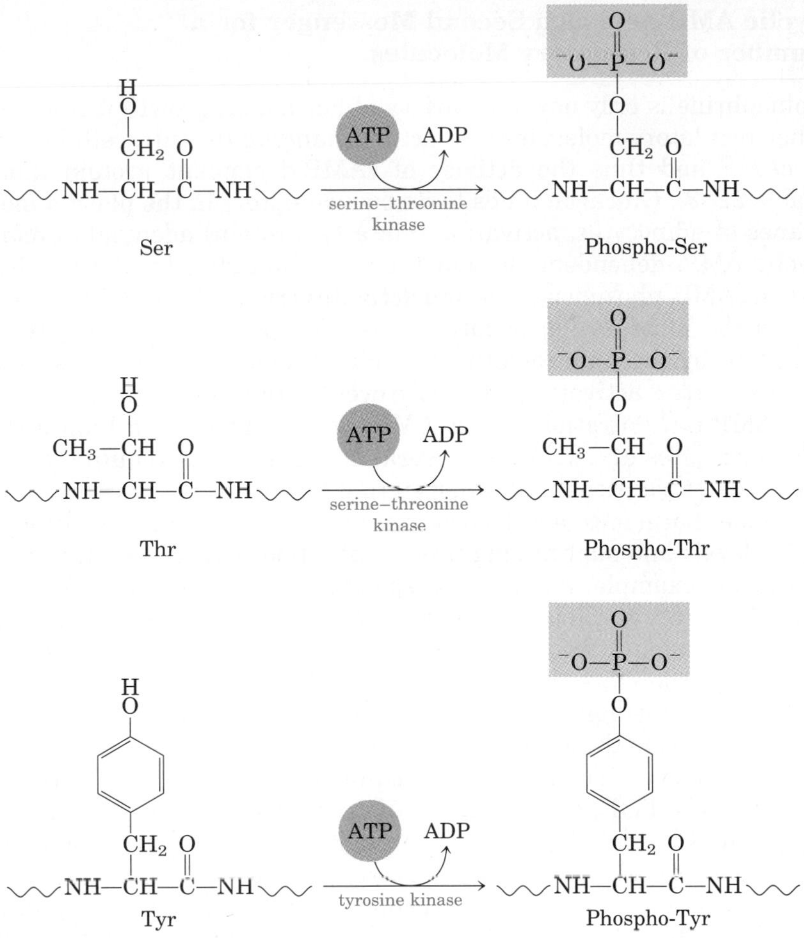

Figure 22–29 The reactions catalyzed by protein kinases involve phosphate group transfer to the hydroxyl in the side chain of a Ser, Thr, or Tyr residue. One large class of protein kinases, typified by cAMP-dependent protein kinase, phosphorylate only Ser or Thr residues and thus are called serine–threonine kinases. Another class of protein kinases, typified by the insulin receptor, act only on Tyr residues. In each case, the addition of the highly charged and bulky phosphate group alters the conformation of the phosphorylated protein, bringing about a change in its activity or in its kinetic properties.

dependent protein kinase phosphorylates Ser and Thr residues in target proteins. Other protein kinases phosphorylate Tyr residues (Fig. 22–29). Note that the target of phosphorylation by cAMP-dependent protein kinase being considered here—phosphorylase *b* kinase—is another protein kinase. This pattern, in which phosphorylation of one protein kinase leads to phosphorylation of another in a cascade of reactions that amplify the initial signal, is common in intracellular signal transduction pathways. We will encounter it again later in this chapter.

Table 22–7 Some enzymes regulated by cAMP-dependent phosphorylation

Enzyme	Pathway	Reference
Glycogen synthase	Glycogen synthesis	Fig. 19–15
Phosphorylase *b* kinase	Glycogen breakdown	Fig. 14–18
Acetyl-CoA carboxylase	Fatty acid synthesis	Fig. 20–12
Pyruvate dehydrogenase complex	Pyruvate oxidation to acetyl-CoA	Page 468
Triacylglycerol lipase	Triacylglycerol mobilization/fatty acid oxidation	Fig. 16–3
Phosphofructokinase-2/ fructose-2,6-bisphosphatase	Glycolysis/gluconeogenesis	Fig. 19–8

Table 22–8 Some hormones that use cAMP as second messenger

Glucagon
Epinephrine
Corticotropin (ACTH)
Parathyroid hormone
Thyroid-stimulating hormone
Follicle-stimulating hormone
Luteinizing hormone

Cyclic AMP Acts as a Second Messenger for a Number of Regulatory Molecules

Epinephrine is only one of a variety of hormones, growth factors, and other regulatory molecules that act by changing the intracellular level of cAMP and thus the activity of cAMP-dependent protein kinase (Table 22–8). Glucagon binds to its own receptors in the plasma membranes of adipocytes, activating (via a G_s protein) adenylate cyclase. Cyclic AMP–dependent protein kinase, stimulated by the resulting rise in cAMP, phosphorylates and activates triacylglycerol lipase, leading to the mobilization of fatty acids. The peptide hormone ACTH, produced by the anterior pituitary, binds to specific receptors in the adrenal cortex, activating adenylate cyclase and raising the intracellular cAMP concentration. Cyclic AMP–dependent protein kinase then phosphorylates and activates several of the enzymes required for the synthesis of cortisone and other steroid hormones.

Some hormones act by *inhibiting* adenylate cyclase, *lowering* cAMP levels and suppressing protein phosphorylation. When somatostatin, for example, binds to its specific receptor, an **inhibitory G protein,** or G_i, which is structurally homologous to G_s, is activated. G_i inhibits adenylate cyclase and lowers the concentration of cAMP. Somatostatin therefore counterbalances the effects of glucagon. In adipose tissue, prostaglandin E_1 (PGE_1) (see Fig. 9–17) inhibits adenylate cyclase, lowering the cAMP concentration and slowing the mobilization of lipid reserves triggered by epinephrine and glucagon. In certain other tissues, PGE_1 stimulates cAMP synthesis because its receptors are coupled to adenylate cyclase through a stimulatory G protein, G_s. In tissues with α_2-adrenergic receptors, epinephrine lowers the cAMP concentration, because the α_2 receptors are coupled to adenylate cyclase through an inhibitory G protein. In short, an extracellular signal such as epinephrine or PGE_1 can have quite different effects on different tissues or cell types, depending upon (1) the type of receptors, (2) the type of G protein (G_s or G_i) with which the receptor is coupled, and (3) the set of enzymes susceptible to phosphorylation by cAMP-dependent protein kinase in each cell type.

Cyclic GMP Also Acts as a Second Messenger

Another cyclic nucleotide, **guanosine 3′,5′-cyclic monophosphate (cyclic GMP or cGMP),** functions as a second messenger in certain cells, including those of the intestinal lining, heart, blood vessels, brain, and the collecting ducts of the kidneys. The message carried by cGMP varies with the tissue in which it acts: in the kidney and intestine it leads to changes in ion transport and water retention; in cardiac (smooth) muscle it signals relaxation; in brain it may be involved both in development and in adult brain function.

At least two isozymes of **guanylate cyclase** produce cGMP from GTP in a reaction analogous to that catalyzed by adenylate cyclase: GTP $\longrightarrow$ cGMP + PP_i. One of the isozymes is an integral protein of the plasma membrane, with the hormone receptor domain on the outer face and the cGMP-forming domain on the cytosolic face (Fig. 22–30). In mammals, this guanylate cyclase is activated by the binding of the hormone **atrial natriuretic factor** (ANF), which is released by cells in the atrium of the heart when increased blood volume stretches the atrium. Carried in the blood to the kidney, ANF activates guanylate cyclase in cells of the collecting ducts, and the resulting rise in cGMP triggers increased renal excretion of Na^+ and, consequently, of water.

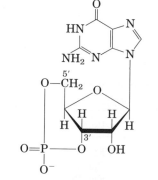

Guanosine 3′,5′-cyclic monophosphate (cGMP)

Water loss reduces the blood volume, countering the stimulus that initially led to ANF secretion. Vascular smooth muscle also has an ANF receptor–guanylate cyclase; upon binding to the receptor, ANF causes vessel relaxation (vasodilation), which reduces blood pressure.

A similar receptor–guanylate cyclase in the plasma membrane of intestinal epithelial cells is activated by a heat-stable bacterial endotoxin (a small peptide) produced by *E. coli* (Fig. 22–30). The resulting elevation in cGMP causes decreased reabsorption of water by the intestinal epithelium, producing the diarrhea characteristic of this toxin's action.

A second and distinctly different isozyme of guanylate cyclase is a cytosolic protein with a tightly associated heme group (Fig. 22–30). This enzyme is activated by its natural ligand nitric oxide (NO), and by several nitrovasodilators—compounds such as nitroglycerin and nitroprusside—used in the treatment of heart disease. The nitrovasodilators spontaneously break down, yielding NO.

$$CH_2-O-NO_2$$
$$CH-O-NO_2$$
$$CH_2-O-NO_2$$
Nitroglycerin

Nitric oxide is produced from arginine by a Ca^{2+}-dependent mixed-function oxidase (see Box 20–1), **NO synthase,** present in many mammalian tissues. NO diffuses from its cell of origin into nearby cells, where it binds to the heme group of guanylate cyclase and activates that enzyme to produce cGMP. In heart, cGMP brings about less forceful contractions by stimulating the ion pump(s) that maintain a low cytosolic Ca^{2+} concentration. This relaxation of the heart muscle is the same response brought about by nitroglycerin tablets taken to relieve angina, the pain caused by contraction of a heart deprived of O_2 because of blocked coronary arteries.

Nitric oxide is unstable and its action is brief; within seconds of its formation, NO undergoes oxidation to nitrite or nitrate. Because it is only slowly converted to NO, nitroglycerin produces long-lasting relaxation of cardiac muscle.

Most of the actions of cGMP are believed to be mediated by **cGMP-dependent protein kinase,** also called **protein kinase G.** This enzyme is widely distributed among eukaryotic organisms; certain mammalian tissues, including smooth muscle and brain, are enriched for the enzyme. It contains both catalytic and regulatory domains on a single polypeptide ($M_r \sim 80,000$). The catalytic domain contains sequences homologous with those of the C subunit of cAMP-dependent protein kinase (protein kinase A), and the regulatory domain resembles the R subunit of the cAMP-dependent enzyme (Fig. 22–28). Binding of cGMP to protein kinase G forces an autoinhibitory domain out of the substrate binding site, allowing the enzyme to phosphorylate proteins that contain Ser or Thr residues surrounded by the appropriate consensus sequence. Protein kinases A and G recognize different consensus sequences and therefore regulate different proteins.

The Insulin Receptor Is a Tyrosine-Specific Protein Kinase

The receptor for insulin is itself a protein kinase, which transfers a phosphate group from ATP to the hydroxyl group of Tyr residues (not

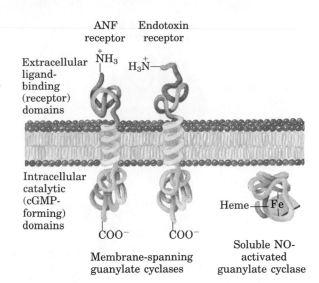

Figure 22–30 Two types (isozymes) of guanylate cyclase involved in signal transduction. The first type exists in two similar membrane-spanning forms that are activated by their extracellular ligands: atrial natriuretic factor (ANF) (receptors occurring in the cells of the collecting ducts of the kidney and in smooth muscle cells of blood vessels) and bacterial endotoxin (receptors in intestinal epithelial cells). The second type is a soluble form that is activated by intracellular nitric oxide (NO); this form is found in many tissues, including smooth muscle of the heart and blood vessels.

NH₂ ... structures (Arginine → Citrulline + NO via NO synthase)

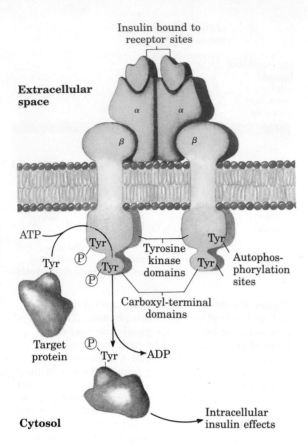

Insulin bound to receptor sites

Extracellular space

α α

β β

ATP

Tyr Tyr Tyrosine kinase domains Tyr

Tyr P Tyr Tyr Autophosphorylation sites

P

Carboxyl-terminal domains

Target protein P Tyr → ADP

Cytosol

Intracellular insulin effects

Figure 22–31 The insulin receptor consists of two α chains located on the outer face of the plasma membrane and two β chains that traverse the membrane and protrude on the cytosolic face. Binding of insulin to the α chains triggers autophosphorylation of Tyr residues in the carboxyl-terminal domain of the β subunits, which allows the tyrosine kinase domain to catalyze phosphorylation of other target proteins.

Inositol-1,4,5-trisphosphate

Ser or Thr; Fig. 22–29). The insulin receptor has two identical α chains that protrude from the outer face of the plasma membrane, and two transmembrane β subunits, with their carboxyl termini on the cytosolic face (Fig. 22–31). The α chains contain the insulin-binding domain, and the β chains have the tyrosine kinase domain.

Insulin binding to the α chains activates the tyrosine kinase activity of the β chains. The enzyme first phosphorylates itself on critical Tyr residues in the β chain (Fig. 22–31), and this autophosphorylation activates the enzyme to phosphorylate other proteins of the membrane or cytosol. Although the detailed sequence of events that follows stimulation by insulin has not been fully established, it appears likely that the binding of insulin to its receptor starts a cascade of protein phosphorylations in which the insulin receptor (a tyrosine kinase) activates a second protein kinase, which may then activate a third, serine or threonine kinase. Eventually, phosphorylation of Ser or Thr residues alters the activity of one or more enzymes crucial to some aspect of cellular function; insulin has hit its target(s).

Individuals with "insulin-resistant" diabetes (NIDDM; p. 760) secrete insulin normally, but their tissues do not respond to their own insulin or to injected insulin. In some of these people there is a mutation in the tyrosine kinase domain of the insulin receptor. Insulin binds normally to the mutant receptor, but the tyrosine kinase is inactive and the downstream consequences of insulin binding do not occur.

The insulin receptor is the prototype of a variety of other hormone and growth-factor receptors, all of which resemble it in structure and have tyrosine kinase activity. The receptors for epidermal growth factor (EGF) and platelet-derived growth factor (PDGF), for example, show structural and sequence homologies with the insulin receptor, and both receptors have a tyrosine kinase activity in their intracellular, carboxyl-terminal domains.

Two Second Messengers Are Derived from Phosphatidylinositols

A third class of signal receptors are coupled, through a G protein, to a plasma membrane **phospholipase C** specific for the plasma membrane lipid phosphatidylinositol-4,5-bisphosphate (see Fig. 9–16). This hormone-sensitive enzyme catalyzes the formation of two potent second messengers: **diacylglycerol** and **inositol-1,4,5-trisphosphate** (Fig. 22–32). The list of hormones known to act through this transduction mechanism is growing rapidly; it includes, for example, vasopressin acting on hepatocytes and thyrotropin-releasing hormone acting on pituitary cells. When these hormones bind to their specific receptors in the plasma membrane, the hormone–receptor complex catalyzes GTP–GDP exchange on an associated G protein, G_p, activating it exactly as the adrenergic (epinephrine) receptor activates G_s (Fig. 22–25). The activated G_p in turn activates a specific membrane-bound phospholipase C, which produces the two second messengers by hydrolysis of phosphatidylinositol-4,5-bisphosphate in the plasma membrane.

Diacylglycerol serves as second messenger by activating a membrane-bound, Ca^{2+}-dependent enzyme, **protein kinase C** (C for calcium). Protein kinase C phosphorylates Ser or Thr residues of specific target proteins, changing their catalytic activities. There are a number of protein kinase C isozymes, each with a characteristic tissue distribution and characteristic sensitivity to activation by Ca^{2+} and diacylglycerol.

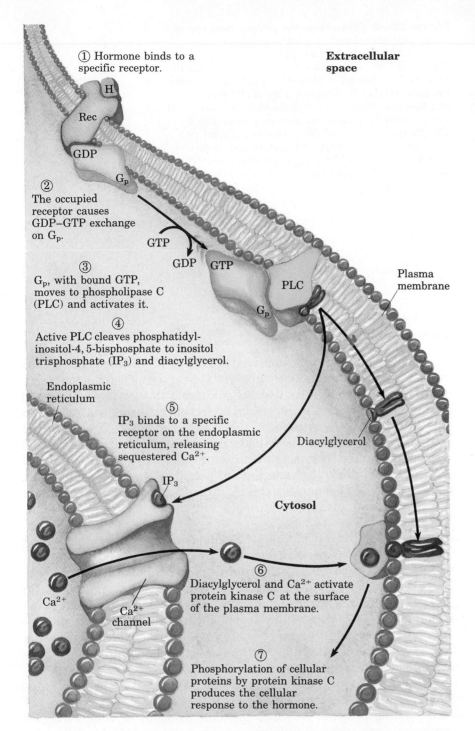

① Hormone binds to a specific receptor.

Extracellular space

② The occupied receptor causes GDP–GTP exchange on G_p.

③ G_p, with bound GTP, moves to phospholipase C (PLC) and activates it.

④ Active PLC cleaves phosphatidyl-inositol-4, 5-bisphosphate to inositol trisphosphate (IP_3) and diacylglycerol.

Endoplasmic reticulum

⑤ IP_3 binds to a specific receptor on the endoplasmic reticulum, releasing sequestered Ca^{2+}.

IP_3

GDP

G_p

GTP

GDP   GTP

PLC

G_p

Plasma membrane

Diacylglycerol

Cytosol

Ca^{2+}

Ca^{2+} channel

⑥ Diacylglycerol and Ca^{2+} activate protein kinase C at the surface of the plasma membrane.

⑦ Phosphorylation of cellular proteins by protein kinase C produces the cellular response to the hormone.

Figure 22–32 Two intracellular second messengers are produced in the hormone-sensitive phosphatidylinositol system: inositol-1,4,5-trisphosphate (IP_3) and diacylglycerol. Both contribute to the activation of protein kinase C; IP_3, by raising cytosolic [Ca^{2+}], also activates other Ca^{2+}-dependent enzymes. Thus Ca^{2+} also acts as a second messenger. H represents the hormone; Rec, receptor; PLC, phospholipase C.

The water-soluble product derived from phospholipase C action, inositol-1,4,5-trisphosphate, diffuses from the plasma membrane to the endoplasmic reticulum, where it binds to specific receptors and causes Ca^{2+} channels within the reticulum to open, releasing sequestered Ca^{2+} into the cytosol (Fig. 22–32). The cytosolic Ca^{2+} concentration rises more than 100-fold, from below 10^{-8} M to about 10^{-6} M. The activation of protein kinase C is but one of many effects of this increase in cytosolic Ca^{2+}.

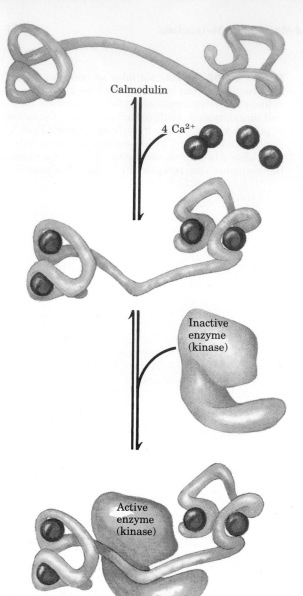

Calmodulin

4 Ca²⁺

Inactive enzyme (kinase)

Active enzyme (kinase)

(a)

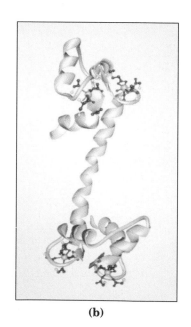

(b)

Figure 22–33 Calmodulin, the protein mediator of many Ca²⁺-stimulated enzymatic reactions, contains four high-affinity Ca²⁺-binding sites. **(a)** The binding of Ca²⁺ induces a conformational change in calmodulin, allowing it to interact productively with the proteins that it regulates. One of the many enzymes regulated by calmodulin and Ca²⁺ is Ca²⁺/calmodulin-dependent protein kinase, which phosphorylates Ser and Thr residues in target proteins. **(b)** A ribbon model of the structure of calmodulin, determined by x-ray crystallography. The four Ca²⁺-binding sites (red) are shown occupied by Ca²⁺ (yellow).

Calcium Is a Second Messenger in Many Signal Transductions

In hormone-sensitive cells, neurons, muscle cells, and many other cells that respond to extracellular signals, Ca^{2+} serves as a second messenger to trigger intracellular responses. Among the processes triggered by Ca^{2+} are exocytosis in nerve and endocrine cells and contraction in muscle. Normally the cytosolic $[Ca^{2+}]$ is kept very low ($<10^{-7}$ M) by the action of Ca^{2+} pumps in the endoplasmic reticulum, mitochondria, and plasma membrane. Hormonal, neural, or other stimuli cause influx of Ca^{2+} into the cell through specific Ca^{2+} channels in the plasma membrane or release of sequestered Ca^{2+} from the endoplasmic reticulum or mitochondria, raising the cytosolic $[Ca^{2+}]$ and triggering the cellular response.

One way in which Ca^{2+} triggers cellular responses is by activating a variety of Ca^{2+}-dependent enzymes, including yet another protein kinase, the **Ca²⁺/calmodulin-dependent protein kinase.** The regulatory subunit of this enzyme is a Ca^{2+}-binding protein, calmodulin (Fig. 22–33). When intracellular $[Ca^{2+}]$ increases in response to some stimulus, the Ca^{2+}/calmodulin-dependent protein kinase phosphorylates and thus regulates a number of target enzymes. Phosphorylase b kinase, which is activated by Ca^{2+}, also has calmodulin as one of its subunits, as does the NO synthase described above.

Calmodulin (M_r 17,000) is an acidic protein with four high-affinity Ca^{2+}-binding sites (Fig. 22–33). It is a member of a large family of Ca^{2+}-binding proteins that also includes troponin C, which triggers skeletal muscle contraction in response to increased $[Ca^{2+}]$. When intracellular $[Ca^{2+}]$ rises to about 10^{-6} M (1 μM), the binding of Ca^{2+} drives a conformational change in calmodulin. In its Ca^{2+}-bound state, calmodulin associates with a variety of proteins and modulates their activities (thus the name "calmodulin").

One isozyme of the cyclic nucleotide phosphodiesterase that degrades cAMP (Fig. 22–27) is a Ca^{2+}/calmodulin-dependent enzyme. The influence of one second messenger (Ca^{2+}) on the level of another (cAMP) is typical of many transduction systems; there is crosstalk and feedback among the several systems of a cell, producing a further complexity in signaling that in most cases remains incompletely understood.

Ion Channels Are Gated by Ligands and by Membrane Potential

In a fourth class of signal transducers, receptors are coupled directly or indirectly to ion channels in the plasma membrane. The best-understood example of such a receptor is the **nicotinic acetylcholine receptor,** which responds to the neurotransmitter acetylcholine. It is found in the postsynaptic cells in certain nerve synapses (Fig. 22–34) and in the junction between a muscle fiber and the neuron that controls it. The acetylcholine receptor complex (M_r 250,000) is composed of four different polypeptide chains, one of which is present in two copies. The transmembrane arrangement of these five chains provides a hydrophilic channel through which ions can traverse the lipid bilayer. When acetylcholine released from the presynaptic nerve ending binds

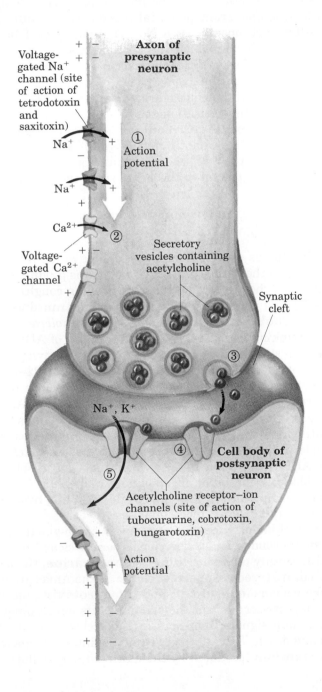

Figure 22–34 Role of voltage-gated and ligand-gated ion channels in passage of an electrical signal between two neurons. Initially, the plasma membrane of the presynaptic neuron is polarized, with the inside negative; this results from the action of the electrogenic Na^+K^+ ATPase, which pumps three Na^+ outward for every two K^+ pumped into the neuron (see Fig. 10–22). ① A stimulus to this neuron causes an action potential to move downward along its axon (white arrow). The opening of one voltage-gated Na^+ channel allows Na^+ entry, and the resulting local depolarization causes the adjacent Na^+ channel to open, and so on. The directionality of movement of the action potential is ensured by the brief refractory period that follows the opening of each voltage-gated Na^+ channel. ② When this wave of depolarization reaches the axon tip, voltage-gated Ca^{2+} channels open, allowing Ca^{2+} entry into the presynaptic neuron. ③ The resulting increase in internal [Ca^{2+}] triggers exocytosis of the neurotransmitter acetylcholine into the space between the neurons (synaptic cleft). ④ Acetylcholine binds to its specific receptor in the plasma membrane of the cell body of the postsynaptic neuron, causing the ligand-gated ion channel that is part of the receptor to open. ⑤ Extracellular Na^+ and K^+ enter through this channel, depolarizing the postsynaptic cell. The electrical signal has thus passed to the postsynaptic cell, and will move along its axon to a third neuron by this same sequence of events. The effects of the toxins shown in parentheses are discussed on p. 774.

Figure labels: Voltage-gated Na^+ channel (site of action of tetrodotoxin and saxitoxin); Na^+; Na^+; Ca^{2+}; Voltage-gated Ca^{2+} channel; Axon of presynaptic neuron; ① Action potential; ② ; Secretory vesicles containing acetylcholine; Synaptic cleft; ③ ; Na^+, K^+; ④ ; ⑤ ; Cell body of postsynaptic neuron; Acetylcholine receptor–ion channels (site of action of tubocurarine, cobrotoxin, bungarotoxin); Action potential

$$CH_3-C \overset{O}{\underset{O-CH_2-CH_2-\overset{+}{N}(CH_3)_3}{\big\langle}}$$

Acetylcholine

to its receptor in the postsynaptic cell (Fig. 22–34), the receptor–ion channel opens, allowing transmembrane passage of Na^+ and K^+ ions (pp. 292–293). The receptor is therefore referred to as a **ligand-gated ion channel.** The resulting depolarization of the postsynaptic membrane triggers muscle contraction or initiates an action potential in the postsynaptic neuron.

The action potential is a wave of transient depolarization that sweeps the neuron from the site of the initial stimulus (in the cell body of the neuron), along the long, thin cytoplasmic extension (axon), to the next synapse. Essential to this signaling mechanism are several types of "voltage-gated" ion channels in the plasma membrane of the neuron. These channels, formed by transmembrane proteins, open and close in response to changes in the transmembrane electrical potential. Along the entire length of the axon are **voltage-gated Na^+ channels** (Fig. 22–34), which are closed when the membrane is polarized, but open briefly when the membrane potential is reduced (i.e., during depolarization). After each opening of a Na^+ channel there follows a brief refractory period during which the channel cannot open again, and thus a unidirectional wave of depolarization sweeps from the nerve cell body toward the end of the axon.

At the distal tip of the neuron are **voltage-gated Ca^{2+} channels.** When the wave of depolarization reaches these channels they open, letting Ca^{2+} enter from the extracellular space and triggering acetylcholine release into the synaptic cleft (Fig. 22–34). Acetylcholine diffuses to the postsynaptic cell, where it binds to acetylcholine receptors; thus the message is passed to the next cell in the circuit.

Toxins, Oncogenes, and Tumor Promoters Interfere with Signal Transductions

Biochemical studies of signal transductions have led to an improved understanding of the pathological effects of toxins produced by the bacteria that cause cholera and pertussis (whooping cough). Both toxins are enzymes that interfere with normal signal transductions in the host animal. **Cholera toxin,** secreted by *Vibrio cholerae* found in contaminated drinking water, catalyzes the transfer of ADP-ribose from NAD^+ to the α subunit of G_s, blocking its GTPase activity (Fig. 22–26) and thereby rendering it permanently activated (Fig. 22–35). This results in continuous activation of the adenylate cyclase of intestinal epithelial cells, and the resultant high concentration of cAMP triggers continual secretion of Cl^-, HCO_3^-, and water into the intestinal lumen. The resulting dehydration and electrolyte loss are the major pathologies in cholera. The **pertussis toxin** produced by *Bordetella pertussis* catalyzes ADP-ribosylation of G_i, preventing GDP displacement by GTP and blocking inhibition of adenylate cyclase by G_i; this defect produces the symptoms of whooping cough, including hypersensitivity to histamines and lowered blood glucose.

The critical importance of ligand- and voltage-gated ion channels in nerve signal conduction as described above is clear from the effects of several naturally occurring toxins. **Tubocurarine,** the active component of curare (used as an arrow poison in the Amazon), and toxins from snake venoms (**cobrotoxin** and **bungarotoxin**), block the acetylcholine receptor or prevent the opening of its ion channel (Fig. 22–34). By blocking signals from nerves to muscles, these toxins cause paralysis and death. **Tetrodotoxin** (from the internal organs of puffer fish) and **saxitoxin** (produced by the marine dinoflagellate that occa-

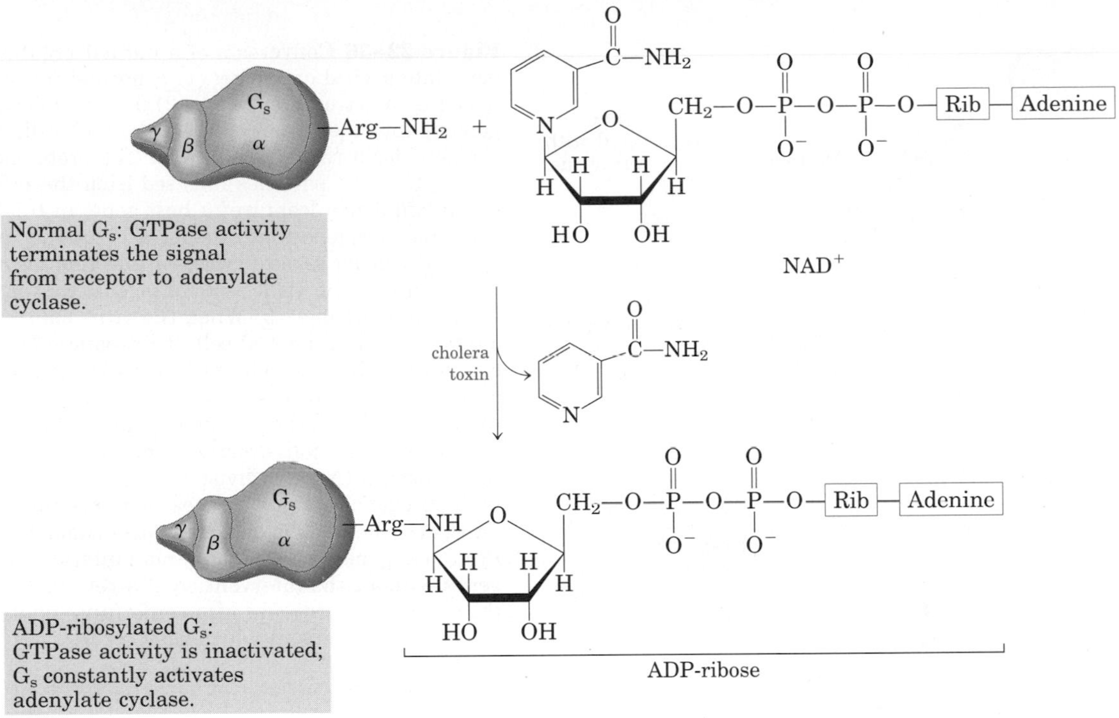

Normal G$_s$: GTPase activity terminates the signal from receptor to adenylate cyclase.

ADP-ribosylated G$_s$: GTPase activity is inactivated; G$_s$ constantly activates adenylate cyclase.

NAD$^+$

cholera toxin

ADP-ribose

Figure 22–35 The toxins produced by the bacteria that cause cholera and whooping cough (pertussis) are enzymes that catalyze transfer of the ADP-ribose moiety of NAD$^+$ to an Arg residue of G proteins: G$_s$ in the case of cholera (as shown here) and G$_i$ in whooping cough. The G proteins thus modified fail to respond to normal hormonal stimuli. The pathology of both diseases results from defective regulation of adenylate cyclase and overproduction of cAMP.

sionally causes "red tides") are also deadly poisons, which block neurotransmission by preventing the opening of Na$^+$ channels.

Tumors and cancer are the result of uncontrolled cell division. Normally, cell division is highly regulated by a family of **growth factors,** proteins that cause resting cells to undergo cell division and, in some cases, differentiation. Some growth factors are cell type–specific, stimulating division of only those cells with appropriate receptors; other growth factors are more general in their effects. Among the well-studied growth factors are epidermal growth factor (EGF), nerve growth factor (NGF), fibroblast growth factor (FGF), platelet-derived growth factor (PDGF), erythropoietin, and a family of proteins called lymphokines, which includes interleukins (IL-1, IL-2, etc.) and interferon γ. There are also extracellular factors that antagonize the effects of growth factors, slowing or preventing cell division; transforming growth factor β (TGFβ) and tumor necrosis factor (TNF) are such factors.

These extracellular signals act through cell-surface receptors very similar to those for hormones, and by similar mechanisms: the production of intracellular second messengers, protein phosphorylation, and ultimately, alteration of gene expression.

It is becoming clear that many types of cancer are the result of abnormal signal-transducing proteins, which lead to continual production of the signal for cell division. The mutated genes that encode these defective signaling proteins are **oncogenes.** (Oncogenes, and gene function in general, are discussed in Chapter 25.) Oncogenes were originally discovered in tumor-causing viruses, then later found to be closely similar to or derived from genes present in the animal host cells. Most likely, these viral genes originated from normal host genes (proto-oncogenes) that encode growth-regulating proteins. During certain types of viral infections, these DNA sequences can be copied by the virus and incorporated into its genome (Fig. 22–36). At some point during the cycle of viral infection, the gene can become defective as a

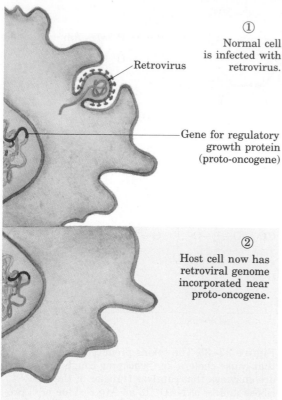

① Normal cell is infected with retrovirus.

Retrovirus

Gene for regulatory growth protein (proto-oncogene)

② Host cell now has retroviral genome incorporated near proto-oncogene.

Figure 22–36 Conversion of a normal regulatory gene into a viral oncogene. ① A normal cell is infected by a retrovirus, which ② inserts its own genome into the chromosome of the host cell, near the gene for a regulatory protein (the proto-oncogene). ③ Virus particles released from the infected cell infrequently "capture" a host gene, in this case the proto-oncogene that encodes a regulatory protein. ④ During several cycles of infection, a mutation occurs in the viral proto-oncogene, converting it into an oncogene. ⑤ When the virus subsequently infects a normal cell, it introduces the oncogene into the host-cell DNA. Transcription of the oncogene leads to the production of a defective regulatory protein that continuously gives the signal for host-cell division, overriding normal mechanisms for limiting cell division. Host cells infected with oncogene-carrying viruses therefore undergo unregulated cell division—they form tumors. Proto-oncogenes can also undergo mutation to oncogenes without the intervention of a retrovirus; these cellular oncogenes also confer unregulated growth on the cells in which they occur.

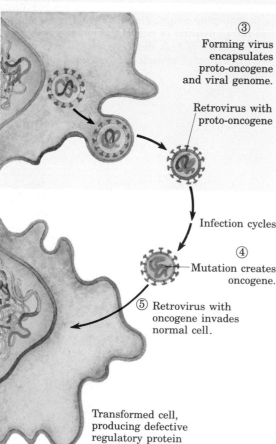

③ Forming virus encapsulates proto-oncogene and viral genome.

Retrovirus with proto-oncogene

Infection cycles

④ Mutation creates oncogene.

⑤ Retrovirus with oncogene invades normal cell.

Transformed cell, producing defective regulatory protein

result of truncation or some other mutation. During a subsequent infection, when this viral oncogene is expressed in its host cell, the abnormal protein product interferes with normal regulation of cell growth, and the unregulated growth can result in a tumor. Oncogenes can also arise from proto-oncogenes without viral involvement. Chromosomal rearrangements, chemical agents, radiation, or other factors can cause mutations in the genes that encode signal-transducing proteins. The resulting oncogenes express defective proteins and defective signaling, once again leading to tumor growth.

Many viral oncogenes encode unregulated tyrosine kinase activities, and in some cases the oncogene product is nearly identical to a normal animal-cell receptor, but with the normal signal-binding site defective or missing. For example, the *erb*B oncogene product, a protein called ErbB, is essentially identical to the normal receptor for epidermal growth factor, except that ErbB lacks the domain that normally binds EGF (Fig. 22–37, p. 777). The *erb*B2 oncogene is commonly associated with adenocarcinomas (cancers) of the breast, stomach, and ovary.

Other signal-transducing proteins with oncogene analogs are the GTP-binding (G) proteins. One well-characterized oncogene, *ras,* encodes a protein with normal GTP binding but no GTPase activity. When the Ras protein (p. 682) is produced in an animal cell, it remains always in the activated form, regardless of the signals coming through normal receptors. Again, the result is unregulated growth—cancer. Mutations in *ras* are associated with 30 to 50% of lung and colon carcinomas and over 90% of pancreatic carcinomas.

The action of a group of compounds known as **tumor promoters** can also be understood in the light of what we know of signal transduction. The best understood of these compounds, phorbol esters, are chemically synthesized compounds that are potent activators of protein kinase C. They apparently mimic cellular diacylglycerol as second messengers (Fig. 22–32), but unlike naturally occurring diacylglycerols they are not rapidly metabolized. By permanently activating protein kinase C, these synthetic tumor promoters interfere with the normal regulation of cell growth and division.

Protein Phosphorylation and Dephosphorylation Are Central to Cellular Control

One common denominator in signal transductions—whether they involve adenylate cyclase, a transmembrane receptor–tyrosine kinase, phospholipase C, or an ion channel—is the eventual regulation of the activity of a protein kinase. We have seen examples of kinases activated by cAMP, insulin, Ca^{2+}/calmodulin, Ca^{2+}/diacylglycerol, and by phosphorylation catalyzed by another protein kinase. The number of known protein kinases has grown remarkably since their discovery by Edwin G. Krebs and Edmond H. Fischer in 1959. Hundreds of different protein kinases, each with its own specific activator and its own specific protein target(s), may be present in eukaryotic cells. Although many other types of covalent modifications are known to occur on proteins it is clear that phosphorylations make up the vast majority of known regulatory modifications of proteins.

The addition of a phosphate group to a Ser, Thr, or Tyr residue introduces a bulky, highly charged group into a region that was only moderately polar. When the modified side chain is located in a region of the protein critical to its three-dimensional structure, phosphorylation can be expected to have dramatic effects on protein conformation and thus on the catalytic activity of the protein. As a result of evolution, the kinase-phosphorylated Ser, Thr, and/or Tyr residues of regulated proteins occur within common structural motifs (consensus sequences) that are recognized by their specific protein kinases (Table 22–9).

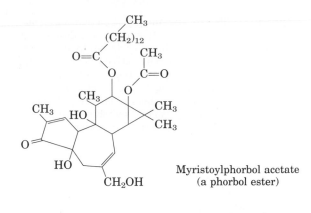

Myristoylphorbol acetate
(a phorbol ester)

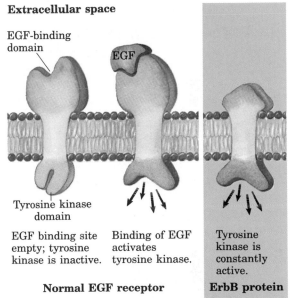

Figure 22–37 The product of the *erb*B oncogene (the ErbB protein) is a truncated version of the normal receptor for epidermal growth factor (EGF). Its intracellular domain has the structure normally induced by EGF binding, but the protein lacks the extracellular binding site for EGF. Unregulated by EGF, ErbB continuously signals cell division.

Table 22–9 Consensus sequences for protein kinases

Protein kinase	Consensus sequence*
Protein kinase A	–X–R–(R/K)–X–(S/T)–X–
Protein kinase G	–X–(R/K)$_{2-3}$–X–(S/T)–X–
Protein kinase C	–X–(R/K$_{1-3}$, X$_{0-2}$)–(S/T)–(X$_{0-2}$, R/K$_{1-3}$)–X–
Ca^{2+}/calmodulin kinase II	–X–R–X–X–(S/T)–X–
Phosphorylase *b* kinase	–K–R–K–Q–I–(S/T)–V–R–
Insulin receptor kinase	–T–R–D–I–Y–E–T–D–Y–Y–R–K–
EGF receptor kinase	–T–A–E–N–A–E–Y–L–R–V–A–P–

Source: Data from Kemp, B.E. & Pearson, R.B. (1990) Protein kinase recognition sequence motifs. *Trends Biochem. Sci.* **15,** 342–346; and Kennelly, P.J. & Krebs, E.G. (1991) Consensus sequences as substrate specificity determinants for protein kinases and protein phosphatases. *J. Biol. Chem.* **266,** 15555–15558.

* (S/T) and Y are the Ser (or Thr) and Tyr residues that are phosphorylated. X is a less essential residue; any of several amino acids may be at this position. Essential residues are indicated by their one-letter abbreviations (see Table 5–1). The notation –(R/K$_{1-3}$, X$_{0-2}$)– means that at this position there are from one to three amino acids, which can be R (Arg) or K (Lys), as well as zero to two of any amino acids, in any sequence (the comma indicates that no sequence is implied).

Figure 22–38 The enzyme glycogen synthase contains at least nine separate sites in five designated regions susceptible to phosphorylation by one of the cellular protein kinases. The activity of this enzyme is therefore capable of modulation in response to a variety of second messengers produced in response to different extracellular signals. Thus regulation is a matter not of binary (on/off) switching but of finely tuned modulation of the activity over a wide range.

Glycogen synthase molecule

Kinase	Glycogen synthase sites phosphorylated	Degree of synthase inactivation
cAMP-dependent protein kinase	1A, 1B, 2, 4	+
cGMP-dependent protein kinase	1A, 1B, 2	+
Phosphorylase *b* kinase	2	+
Ca²⁺/calmodulin-dependent kinase	1B, 2	+
Glycogen synthase kinase 3	3A, 3B, 3C	+ + +
Glycogen synthase kinase 4	2	+
Casein kinase II	5	0
Casein kinase I	At least 9 sites	+ + + +
Protein kinase C	1A	+

Not all cases of regulation by phosphorylation are as simple as those we have described. Some proteins have consensus sequences recognized by several different protein kinases, each of which can phosphorylate the protein and alter its enzymatic activity. For example, glycogen synthase is inactivated by cAMP-dependent phosphorylation of specific Ser residues, and is also modulated by at least four other protein kinases that phosphorylate four other sites in the protein (Fig. 22–38). Some of the phosphorylations inhibit the enzyme more than others, and some combinations of phosphorylation are cumulative. The result of all of these regulations is the potential for extremely subtle modulation of the activity of glycogen synthase, allowing very finely tuned responses to varying metabolic circumstances.

The end effect of epinephrine's interaction with the β-adrenergic receptor is the phosphorylation of several cellular enzymes, including glycogen synthase and glycogen phosphorylase. To serve as an effective regulatory mechanism, this phosphorylation must be reversible, allowing the regulated enzymes to return to their prestimulus level when the hormonal signal stops. In muscle, for example, the enzyme phosphoprotein phosphatase-1 dephosphorylates glycogen phosphorylase, phosphorylase *b* kinase, and glycogen synthase (see Figs. 14–17, 19–15), reversing the effects of cAMP on the activities of these enzymes. This enzyme (sometimes called phosphorylase *a* phosphatase, synthase phosphatase, or kinase phosphatase to indicate its substrate specificity) is regulated by another protein, **phosphoprotein phosphatase inhibitor.** This inhibitor, when phosphorylated by protein kinase A, inhibits phosphoprotein phosphatase-1. A rise in the concentration of cAMP therefore stimulates phosphorylation of certain regulated proteins such as glycogen phosphorylase and also slows dephosphorylation of these proteins, prolonging the effect of phosphorylation.

Cells contain a family of phosphoprotein phosphatases that hydrolyze specific phosphoserine, phosphothreonine, and phosphotyrosine esters, releasing P_i. Although this class of enzymes is not yet as thoroughly studied as the protein kinases, it is very likely that these phosphatases will turn out to be just as important as the protein kinases in regulating cellular processes and metabolism. The known phosphoprotein phosphatases show substrate specificity, acting on only a subset of phosphoproteins, and they are in some cases regulated by a second messenger or an extracellular signal. Some protein phosphatases are transmembrane proteins of the plasma membrane, with extracellular receptorlike domains and intracellular phosphatase domains; they may well prove to be regulated by extracellular signals in a fashion similar to regulation of the tyrosine kinase of the insulin receptor. The complexity and the subtlety of the regulatory mechanisms achieved by evolution strain the imagination, and the experimental challenges of discovering the full range of regulatory mechanisms remain to be met.

Steroid and Thyroid Hormones Act in the Nucleus to Change Gene Expression

The mechanism by which steroid and thyroid hormones exert their effects is fundamentally different from that for the other types of hormones. Steroid hormones (estrogen, progesterone, and cortisol, for example), too hydrophobic to dissolve readily in the blood, are carried on specific carrier proteins from the point of their release to their target tissues. In the target tissue, these hormones pass through the plasma membrane by simple diffusion and bind to specific receptor proteins in the nucleus (Fig. 22–39). The hormone–receptor complexes act by

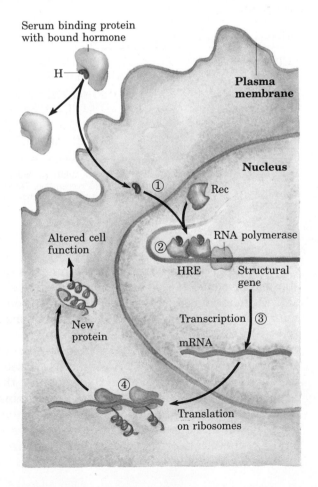

Figure 22–39 The general mechanism by which steroid and thyroid hormones, retinoids, and vitamin D act to regulate gene expression. ① Hormone (H) carried to the target tissue on serum binding proteins diffuses across the plasma membrane and binds to its specific receptor protein (Rec) in the nucleus. ② Hormone binding changes the conformation of the receptor, allowing it to form dimers in the nucleus with other hormone–receptor complexes of the same type and to bind to specific regulatory regions, hormone response elements (HREs), in the DNA adjacent to specific genes. ③ This binding somehow facilitates transcription of the adjacent gene(s) by RNA polymerase (Chapter 25), increasing the rate of messenger RNA formation and ④ bringing about new synthesis of the hormone-regulated gene product. The changed level of the newly synthesized protein produces the cellular response to the hormone. The details of protein synthesis are discussed in Chapter 26.

binding to highly specific DNA sequences called **hormone response elements** (HREs) (Fig. 22–39) and altering gene expression. Hormone binding triggers changes in the conformation of the receptor proteins so that they become capable of interacting with specific transcription factors (Chapter 27). The bound hormone–receptor complex can either enhance or suppress the expression (transcription into messenger RNA; Chapter 25) of specific genes adjacent to HREs, and thus the synthesis of the genes' protein products (Chapter 26).

The DNA sequences (HREs) to which hormone–receptor complexes bind are similar in length and arrangement, but different in sequence, for the various steroid hormones. The HRE sequences recognized by a given receptor are very similar but not identical; for each receptor there is a "consensus sequence" (Table 22–10), which the hormone–receptor complex binds at least as well as it binds the natural HREs. Each HRE consensus sequence consists of two six-nucleotide sequences, either contiguous or separated by three nucleotides. The two hexameric sequences occur either in tandem or in a palindromic arrangement (Fig. 12–20). The hormone–receptor complex binds to the DNA as a dimer, with each monomer recognizing one of the six-nucleotide sequences. The ability of a given hormone to alter the expression of a specific gene depends upon the HRE element's exact sequence and on its position relative to the gene and the number of HREs associated with the gene.

Table 22–10 Consensus sequences of some hormone response elements

Hormone	Sequence of DNA (both strands)*	
Glucocorticoid	(5') AGAACAXXXTGTTCT (3')	(strand 1)
	(3') TCTTGTXXXACAAGA (5')	(strand 2)
Estrogen	(5') AGGTCAXXXTGACCT (3')	(strand 1)
	(3') TCCAGTXXXACTGGA (5')	(strand 2)
Thyroid	(5') AGGTCATGACCT (3')	(strand 1)
	(3') TCCAGTACTGGA (5')	(strand 2)

Source: Data from Schwabe, J.W.R. & Rhodes, D. (1991) Beyond zinc fingers: steroid hormone receptors have a novel structural motif for DNA recognition. *Trends Biochem. Sci.* **16,** 291–296; and Fuller, P.J. (1991) The steroid receptor superfamily: mechanisms of diversity. *FASEB J.* **5,** 3092–3099.

* X represents any nucleotide.

Comparison of the amino acid sequences of receptors for several steroid hormones as well as receptors for thyroid hormone, vitamin D, and retinoids has revealed several highly conserved sequences and some regions in which the sequences differ considerably with receptor type (Fig. 22–40). (Retinoids are compounds related to retinoate, the carboxylate form of vitamin A_1 (see Fig. 9–18), which have hormonelike actions on some cell types.) A centrally located sequence of 66 to 68 residues is very similar in all of the receptors; this is the DNA-binding region, which resembles regions of other proteins known to bind DNA. All of these DNA-binding regions share the "zinc finger" structure (see Fig. 27–12), a sequence containing eight Cys residues that provide binding sites for two Zn^{2+} ions, which stabilize the DNA-binding domain.

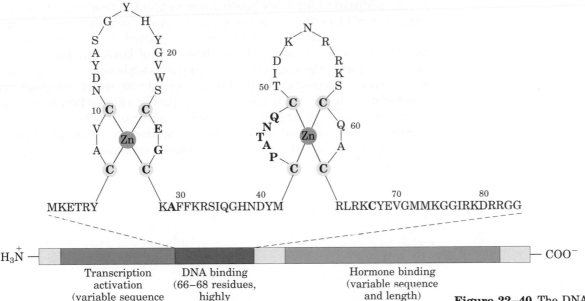

MKETRY KAFFKRSIQGHNDYM RLRKCYEVGMMKGGIRKDRRGG

$H_3\overset{+}{N}$ — COO⁻

Transcription activation (variable sequence and length)

DNA binding (66–68 residues, highly conserved)

Hormone binding (variable sequence and length)

Figure 22–40 The DNA-binding domain common to a number of steroid hormone receptor proteins. These proteins have a binding site for the hormone, a DNA-binding domain, and a region that activates the transcription of the regulated gene. The DNA-binding region is highly conserved. The sequence shown here (see Table 5–1 for amino acid abbreviations) is that for the estrogen receptor, but the residues in bold type are common to all such receptors. Eight critical Cys residues bind to two Zn^{2+} ions that stabilize the "zinc finger" structure shared with many other DNA-binding proteins (see Fig. 22–41). The regulation of gene expression is described in more detail in Chapter 27.

The region of the hormone receptor responsible for hormone binding (the ligand-binding region, always at the carboxyl terminus) is quite different in different members of the hormone receptor family. The glucocorticoid receptor is only 30% homologous with the estrogen receptor and 17% homologous with the thyroid hormone receptor. In the vitamin D receptor, the ligand-binding region consists of only 25 residues, whereas it has 603 residues in the mineralocorticoid receptor. The different sequences are reflected in different specificities for hormone binding. Mutations that change one amino acid residue in this region result in loss of responsiveness to a specific hormone; some humans unable to respond to cortisol, testosterone, vitamin D, or thyroxine have been shown to have such mutations in the corresponding hormone receptor.

The specificity of the ligand-binding site is exploited in the use of a drug, **tamoxifen,** in the treatment of breast cancer in humans. In some types of breast cancer, division of the cancerous cells depends on the continued presence of the hormone estrogen. Tamoxifen competes with estrogen in binding to the estrogen receptor, but the tamoxifen–receptor complex is inactive in gene regulation. Consequently, tamoxifen administration after surgery or chemotherapy for this type of breast cancer slows or stops the growth of remaining cancerous cells, prolonging the life of the patient.

Another steroid analog, the drug **RU486,** is used in the very early termination of pregnancy. An antagonist of the hormone progesterone, RU486 binds to the progesterone receptor and blocks hormone actions essential to the implantation of the fertilized ovum in the uterus. As of 1992, RU486 had not been approved for use in the United States.

The ability of a given steroid or thyroid hormone to act on a specific cell type depends not only on whether the receptor for that hormone is synthesized by the cell, but also on whether the cell contains enzymes that metabolize the hormone. Some hormones (testosterone, thyroxine, vitamin D) are enzymatically converted into more active derivatives within the target cell; others, such as cortisol, are converted to an inactive form in some cells, making these cells resistant to that hormone.

Tamoxifen

RU486 (mifepristone)

In addition to the DNA-binding and ligand-binding regions, steroid receptors also have two domains that interact (in a way not fully understood) with elements of the transcriptional (RNA-synthesizing) machinery in the nucleus. The combination of DNA binding and this interaction with the transcriptional apparatus allows the steroid hormone–receptor complex to modulate the rate at which proteins are produced from a specific gene. The relatively slow action of steroid hormones (hours or days are required for their full effect) is a consequence of their mode of action; time is required for RNA synthesis in the nucleus and for the subsequent protein synthesis.

Summary

In mammals there is a division of metabolic labor among specialized tissues and organs. Coordination of the body's diverse metabolic activities is accomplished by hormonal signals that circulate in the blood. The liver is the central distributing and processing organ for nutrients. Sugars and amino acids produced in digestion cross the intestinal epithelium and enter the blood, which carries them to the liver. Some triacylglycerols derived from ingested lipids also make their way to the liver, where the constituent fatty acids are used in a variety of processes. Glucose-6-phosphate is the key intermediate in carbohydrate metabolism. It may be polymerized into glycogen, dephosphorylated to blood glucose, or converted to fatty acids via acetyl-CoA. It may undergo degradation by glycolysis and the citric acid cycle to yield ATP energy or by the pentose phosphate pathway to yield pentoses and NADPH. Amino acids are used to synthesize liver and plasma proteins, or their carbon skeletons may be converted into glucose and glycogen by gluconeogenesis; the ammonia formed by their deamination is converted into urea. Fatty acids may be converted by the liver into other triacylglycerols, cholesterol, or plasma lipoproteins for transport to and storage in adipose tissue. They may also be oxidized to yield ATP, and to form ketone bodies to be circulated to other tissues.

Skeletal muscle is specialized to produce ATP for mechanical work. During strenuous muscular activity, glycogen is the ultimate fuel and is fermented into lactate, supplying ATP. During recovery the lactate is reconverted (through gluconeogenesis) to glycogen and glucose in the liver. Phosphocreatine is an immediate source of ATP during active contraction. Heart muscle obtains all of its ATP from oxidative phosphorylation. The brain uses only glucose and β-hydroxybutyrate as fuels, the latter being important during fasting or starvation. The brain uses most of its ATP energy for the active transport of Na$^+$ and K$^+$ and the

maintenance of the electrical potential of neuronal membranes. The blood links all of the organs, carrying nutrients, waste products, and hormonal signals between them.

Hormones are chemical messengers (peptides, amines, or steroids) secreted by certain tissues into the blood, serving to regulate the activity of other tissues. They act in a hierarchy of functions. Nerve impulses stimulate the hypothalamus to send specific hormones to the pituitary gland, stimulating (or inhibiting) the release of tropic hormones. The anterior pituitary hormones in turn stimulate other endocrine glands (thyroid, adrenals, pancreas) to secrete their characteristic hormones, which in turn stimulate specific target tissues.

The concentration of glucose in the blood is hormonally regulated. Fluctuations in blood glucose (which is normally about 80 mg/100 mL or 4.5 mM) due to dietary uptake or vigorous exercise are counterbalanced by a variety of hormonally triggered changes in the metabolism of several organs. Epinephrine prepares the body for increased activity by mobilizing blood glucose from glycogen and other precursors. Low blood glucose results in the release of glucagon, which stimulates glucose release from liver glycogen and shifts the fuel metabolism in liver and muscle to fatty acids, sparing glucose for use by the brain. In prolonged fasting, triacylglycerols become the principal fuels; the liver converts the fatty acids to ketone bodies for export to other tissues, including the brain. High blood glucose elicits the release of insulin, which speeds the uptake of glucose by tissues and favors the storage of fuels as glycogen and triacylglycerols. In untreated diabetes, insulin is either not produced or is not recognized by the tissues, and the utilization of blood glucose is compromised. When blood glucose levels are high, glucose is excreted intact into the urine. Tissues then depend upon fatty acids for fuel (producing ketone bodies) and degrade cellular proteins to make glucose from their glucogenic amino acids. Untreated diabetes is characterized by high glucose levels in the blood and urine and the production and excretion of ketone bodies.

Hormones act through a small number of fundamentally similar mechanisms. Epinephrine binds to specific β-adrenergic receptors on the outer face of hepatocytes and myocytes. A stimulatory GTP-binding protein (G_s) mediates between the adrenergic receptor and adenylate cyclase on the inner face of the plasma membrane. When the adrenergic receptor is occupied, adenylate cyclase is activated and converts ATP to cAMP (the second messenger), which then activates the cAMP-dependent protein kinase. This protein kinase phosphorylates and activates inactive phosphorylase b kinase, which in a subsequent step phosphorylates and activates glycogen phosphorylase. Cyclic nucleotide phosphodiesterase terminates the signal by converting cAMP to AMP. The cAMP-dependent protein kinase also phosphorylates and regulates a number of other enzymes present in target tissues. (Glucagon acts by an essentially similar mechanism except that the tissue distribution of glucagon receptors is different; this hormone acts primarily on the liver.) This cascade of events, in which a single molecule of hormone activates a catalyst that in turn activates another catalyst and so on, results in large signal amplification; this is characteristic of all hormone-activated systems. Cyclic GMP acts as the second messenger for other hormones, by a similar mechanism.

Protein phosphorylation is a universal mechanism for rapid and reversible enzyme regulation. To reverse the effects of signal-stimulated protein kinases, cells contain a variety of phosphatases. These enzymes, too, are subject to regulation by extracellular and intracellular signals.

The insulin receptor represents a second signal-transducing mechanism. The receptor is an integral protein of the plasma membrane. Binding of insulin to its extracellular domain activates a tyrosine-specific protein kinase in the receptor's cytosolic domain. This kinase activates several protein kinases by phosphorylating specific Tyr residues.

The phosphorylated protein kinases bring about changes in metabolism by phosphorylating additional key enzymes, altering their enzymatic activities.

A third general class of hormone mechanisms involves the coupling of hormone receptors, via another group of GTP-binding proteins, to a phospholipase C of the plasma membrane. Hormone binding activates this enzyme, which hydrolyzes inositol-containing phospholipids in the plasma membrane. This generates two second messengers: diacylglycerol, which activates protein kinase C, and inositol-1,4,5-trisphosphate (IP_3), which causes the release of Ca^{2+} sequestered in the endoplasmic reticulum. Ca^{2+} is a common second messenger in hormone-sensitive cells and in neural signaling; it alters the enzymatic activities of specific protein kinases. Calmodulin is a small Ca^{2+}-binding subunit of a number of Ca^{2+}-dependent enzymes.

The fourth general transduction mechanism triggered by hormones is the opening of hormone-sensitive ion channels. The nicotinic acetylcholine receptor is a ligand-gated ion channel, which, when occupied by acetylcholine, allows transmembrane passage of Na^+ and K^+ ions and consequent depolarization of the target cell. A wave of depolarization sweeps along nerves through the action of voltage-gated Na^+ and Ca^{2+} ion channels, triggering neurotransmitter release.

A variety of pathological conditions are associated with defects in signal-transduction mechanisms. Some bacterial toxins interfere with signal transductions. Oncogenes in a cell's DNA permit uncontrolled cell division, possibly through formation of defective signal-transducing proteins that are insensitive to modulation by growth factors or hormonal signals. Tumor promoters also interfere with cell regulation and growth.

Steroid hormones enter cells and bind to specific receptor proteins. The hormone–receptor complex binds specific regions of nuclear DNA called hormone response elements and regulates the expression of nearby genes. Tamoxifen and RU486 are drugs that act as steroid hormone antagonists.

Further Reading

General Background and History

Molecular Biology of Signal Transduction. (1988) *Cold Spring Harb. Symp. Quant. Biol.* **53.**
This entire volume is filled with short research and review papers on a wide variety of signal-transducing systems, from bacteria to humans.

Nishizuka, Y., Tanaka, C., & Endo, M. (eds) (1990) *The Biology and Medicine of Signal Transduction,* Adv. Second Messenger Phosphoprotein Res., **24.**
A collection of papers on receptor–transducer systems and the medical effects of defective signal transducers.

Sutherland, E.W. (1972) Studies on the mechanisms of hormone action. *Science* **177,** 401–408.
The author's Nobel lecture, describing the classic experiments on cAMP.

Wilson, J.D. & Foster, D.W. (eds) (1992) *Williams Textbook of Endocrinology,* 8th edn, W.B. Saunders Company, Philadelphia.
Especially relevant are Chapter 1, an introduction to hormonal regulation; Chapter 3, on the mechanism of action of steroid hormones; and Chapter 4, on the mechanisms of hormones that act at the cell surface.

Yalow, R.S. (1978) Radioimmunoassay: a probe for the fine structure of biologic systems. *Science* **200,** 1236–1245.
A history of the development of radioimmunoassays; the author's Nobel lecture.

Tissue-Specific Metabolism: Division of Labor

Arias, I.M., Jakoby, W.B., Popper, H., Schachter, D., & Shafritz, D.A. (eds) (1988) *The Liver: Biology and Pathobiology,* 2nd edn, Raven Press, New York.
An advanced-level text; includes chapters on the metabolism of carbohydrates, fats, and proteins in the liver.

Hormones: Communication among Cells and Tissues

Crapo, L. (1985) *Hormones: The Messengers of Life,* W.H. Freeman and Company, New York.
A short, entertaining account of the history and recent state of hormone research.

Snyder, S.H. (1985) The molecular basis of communication between cells. *Sci. Am.* **253** (October), 132–141.
An introductory-level discussion of the human endocrine system.

Hormonal Regulation of Fuel Metabolism

Harris, R.A. & Crabb, D.W. (1992) Metabolic interrelationships. In *Textbook of Biochemistry with Clinical Correlations,* 3rd edn (Devlin, T.M., ed), pp. 576–606, John Wiley & Sons, Inc., New York.
A description of the metabolic interplay among human tissues during normal metabolism, and the effect on tissue-specific energy metabolism of the stresses of exercise, lactation, diabetes, and renal disease.

Pilkis, S.J. & Claus, T.H. (1991) Hepatic gluconeogenesis/glycolysis: regulation and structure/function relationships of substrate cycle enzymes. *Annu. Rev. Nutr.* **11,** 465–515.

A review at the advanced level.

Roach, P.J. (1990) Control of glycogen synthase by hierarchal protein phosphorylation. *FASEB J.* **4,** 2961–2968.
Phosphorylation of one enzyme at several positions by several different protein kinases can produce finely graded changes in enzyme activity.

Molecular Mechanisms of Signal Transduction

Aaronson, S.A. (1991) Growth factors and cancer. *Science* **254,** 1146–1153.
A clear description of defects in the signal-transducing mechanisms that regulate cell division, which result from mutations in the genes for growth-factor receptors.

Becker, A.B. & Roth, R.A. (1990) Insulin receptor structure and function in normal and pathological conditions. *Annu. Rev. Med.* **41,** 99–115.
A brief description of the structure of the receptor and its gene, and a discussion of the clinical syndromes associated with receptor defects.

Berridge, M.J. (1985) The molecular basis of communication within the cell. *Sci. Am.* **253** (October), 142–152.
An introduction to the transductions mediated by adenylate cyclase, guanylate cyclase, and phospholipase C.

Berridge, M.J. & Irvine, R.F. (1989) Inositol phosphates and cell signalling. *Nature* **341,** 197–205.
Not the latest, but one of the best descriptions of the role of inositol phospholipids in signal transduction.

Brent, G.A., Moore, D.D., & Larsen, P.R. (1991) Thyroid hormone regulation of gene expression. *Annu. Rev. Physiol.* **53,** 17–36.
An advanced discussion.

Collins, S., Lohse, M.J., O'Dowd, B., Caron, M.G., & Lefkowitz, R.J. (1991) Structure and regulation of G protein-coupled receptors: the β_2-adrenergic receptor as a model. *Vitam. Horm.* **46,** 1–39.
An advanced discussion.

Fisher, S.K., Heacock, A.M., & Agranoff, B.W. (1992) Inositol lipids and signal transduction in the nervous system: an update. *J. Neurochem.* **58,** 18–38.
A review of inositol phospholipids in signaling, including a good description of the various phosphorylated derivatives of inositol and their functions as second messengers; advanced level.

Gilman, A.G. (1989) G proteins and regulation of adenylyl cyclase. *JAMA* **262,** 1819–1825.

Hille, B. (1991) *Ionic Channels of Excitable Membranes,* 2nd edn, Sinauer Associates, Sunderland, MA.

Very broad coverage, at an intermediate level.

Hollenberg, M.D. (1991) Structure-activity relationships for transmembrane signaling: the receptor's turn. *FASEB J.* **5,** 178–186.

A description of how information about the amino acid sequences of receptors, derived from cloning receptor genes, can be used to discover structural bases for receptor interactions with ligands, G proteins, and other elements of a transducing system.

Kennelly, P.J. & Krebs, E.G. (1991) Consensus sequences as substrate specificity determinants for protein kinases and protein phosphatases. *J. Biol. Chem.* **266,** 15555–15558.

A concise summary of the sequence specificity of protein kinases.

Krebs, E.G. (1989) Role of the cyclic AMP–dependent protein kinase in signal transduction. *JAMA* **262,** 1815–1818.

A clear account of the research on protein kinase A and its history.

Linder, M.E. & Gilman, A.G. (1992) G proteins. *Sci. Am.* **267** (July), 56–65.

An introductory level description of the discovery and functions of GTP-binding proteins.

O'Malley, B.W., Tsai, S.Y., Bagchi, M., Weigel, N.L., Schrader, W.T., & Tsai, M.-J. (1991) Molecular mechanism of action of a steroid hormone receptor. *Recent Prog. Horm. Res.* **47,** 1–26.

A brief history of the discovery of steroid hormone receptors and their genes, and a review of the effects of the hormone–receptor complex on mRNA and protein synthesis in vitro.

Rasmussen, H. (1989) The cycling of calcium as an intracellular messenger. *Sci. Am.* **261** (October), 66–73.

An introduction to the role of Ca^{2+} as a second messenger.

Snyder, S.H. & Bredt, D.S. (1992) Biological roles of nitric oxide. *Sci. Am.* **266** (May), 68–77.

An intermediate-level review of the role of NO as a second messenger.

Taylor, S.S., Buechler, J.A., & Yonemoto, W. (1990) cAMP-dependent protein kinase: framework for a diverse family of regulatory enzymes. *Annu. Rev. Biochem.* **59,** 971–1005.

An advanced review of the structure and function of protein kinase A and a comparison of its activation mechanism and catalytic mechanism with those of other protein kinases.

Ulmann, A., Teutsch, G., & Philibert, D. (1990) RU 486. *Sci. Am.* **262** (June), 42–48.

The effects of this steroid antagonist, the "morning-after pill," on the female reproductive system; an introduction.

Ullrich, A. & Schlessinger, J. (1990) Signal transduction by receptors with tyrosine kinase activity. *Cell* **61,** 203–212.

A review of the common structural and functional features of receptors in the insulin receptor family.

Problems

1. *ATP and Phosphocreatine as Sources of Energy for Muscle* In contracting skeletal muscle, the concentration of phosphocreatine drops while the concentration of ATP remains fairly constant. Explain how this happens.

In a classic experiment, Robert Davies found that if the muscle is first treated with 1-fluoro-2,4-dinitrobenzene (see Fig. 5–14), the concentration of ATP in the muscle declines rapidly, whereas the concentration of phosphocreatine remains unchanged during a series of contractions. Suggest an explanation.

2. *Metabolism of Glutamate in the Brain* Glutamate in the blood flowing into the brain is transformed into glutamine, which appears in the blood leaving the brain. What is accomplished by this metabolic conversion? How does it take place? Actually, the brain can generate more glutamine than can be made from the glutamate entering in the blood. How does this extra glutamine arise? (Hint: You may want to review amino acid catabolism in Chapter 17. Recall that NH_3 is very toxic to the brain.)

3. *Absence of Glycerol Kinase in Adipose Tissue* Glycerol-3-phosphate is a key intermediate in the biosynthesis of triacylglycerols. Adipocytes, which are specialized for the synthesis and degradation of triacylglycerols, cannot directly use glycerol because they lack glycerol kinase, which catalyzes the reaction

$$\text{Glycerol} + \text{ATP} \longrightarrow \text{glycerol-3-phosphate} + \text{ADP}$$

How does adipose tissue obtain the glycerol-3-phosphate necessary for triacylglycerol synthesis? Explain.

4. *Hyperglycemia in Patients with Acute Pancreatitis* Patients with acute pancreatitis are treated by withholding protein from the diet and by intravenous administration of glucose–saline solution. What is the biochemical basis for these measures? Patients undergoing this treatment commonly experience hyperglycemia. Why?

5. *Oxygen Consumption during Exercise* A sedentary adult consumes about 0.05 L of O_2 during a 10 s period. A sprinter, running a 100 m race, consumes about 1 L of O_2 during the same time period. After finishing the race, the sprinter will continue to breathe at an elevated but declining rate for some minutes, consuming an extra 4 L of O_2 above the amount consumed by the sedentary individual.

(a) Why do the O_2 needs increase dramatically during the sprint?

(b) Why do the O_2 demands remain high after the sprint is completed?

6. *Thiamin Deficiency and Brain Function* Individuals with thiamin deficiency display a number of characteristic neurological signs: loss of reflexes, anxiety, and mental confusion. Suggest a reason why thiamin deficiency is manifested by changes in brain function.

7. *Significance of Hormone Concentration* Under normal conditions, the human adrenal medulla secretes epinephrine ($C_9H_{13}NO_3$) at a rate sufficient to maintain a concentration of 10^{-10} M in the circulating blood. To appreciate what that concentration means, calculate the diameter of a round swimming pool, with a water depth of 2 m, that would be needed to dissolve 1 g (about 1 teaspoon) of epinephrine to a concentration equal to that in blood.

8. *Regulation of Hormone Levels in the Blood* The half-life of most hormones in the blood is relatively short. For example, if radioactively labeled insulin is injected into an animal, one can determine that within 30 min half the hormone has disappeared from the blood.

(a) What is the importance of the relatively rapid inactivation of circulating hormones?

(b) In view of this rapid inactivation, how can the circulating hormone level be kept constant under normal conditions?

(c) In what ways can the organism make possible rapid changes in the level of circulating hormones?

9. *Water-Soluble versus Lipid-Soluble Hormones* On the basis of their physical properties, hormones fall into one of two categories: those that are very soluble in water but relatively insoluble in lipids (e.g., epinephrine) and those that are relatively insoluble in water but highly soluble in lipids (e.g., steroid hormones). In their role as regulators of cellular activity, most water-soluble hormones do not penetrate into the interior of their target cells. The lipid-soluble hormones, by contrast, do penetrate into their target cells and ultimately act in the nucleus. What is the correlation between solubility, the location of receptors, and the mode of action of the two classes of hormones?

10. *Hormone Experiments in Cell-Free Systems* In the 1950s, Earl Sutherland and his colleagues carried out pioneering experiments to elucidate the mechanism of action of epinephrine and glucagon. In the light of our current understanding of hormone action as described in this chapter, interpret each of the experiments described below. Identify the components and indicate the significance of the results.

(a) The addition of epinephrine to a homogenate or broken-cell preparation of normal liver resulted in an increase in the activity of glycogen phosphorylase. However, if the homogenate was first centrifuged at a high speed and epinephrine or glucagon was added to the clear supernatant fraction containing phosphorylase, no increase in phosphorylase activity was observed.

(b) When the particulate fraction sedimented from a liver homogenate by centrifugation was separated and treated with epinephrine, a new substance was produced. This substance was isolated and purified. Unlike epinephrine, this substance activated glycogen phosphorylase when added to the clear supernatant fraction of the homogenate.

(c) The substance obtained from the particulate fraction was heat-stable; that is, heat treatment did not prevent its capacity to activate phosphorylase. (Hint: Would this be the case if the substance were a protein?) The substance appeared nearly identical to a compound obtained when pure ATP was treated with barium hydroxide. (Figure 12–6 will be helpful.)

11. *Effect of Dibutyryl-cAMP versus cAMP on Intact Cells* The physiological effects of the hormone epinephrine should in principle be mimicked by the addition of cAMP to the target cells. In practice, the addition of cAMP to intact target cells elicits only a minimal physiological response. Why? When the structurally related derivative dibutyryl-cAMP (shown below) is added to intact cells, the expected physiological responses can readily be seen. Explain the basis for the difference in cellular response to these two substances. Dibutyryl cAMP is a widely used derivative in studies of cAMP function.

Dibutyryl-cAMP

12. *Effect of Cholera Toxin on Adenylate Cyclase* The gram-negative bacterium *Vibrio cholerae* produces a protein, cholera toxin (M_r 90,000), responsible for the characteristic symptoms of cholera: extensive loss of body water and Na^+ through continuous, debilitating diarrhea. If body fluids and Na^+ are not replaced, severe dehydration will occur; untreated, the disease is often fatal. When the cholera toxin gains access to the human intestinal tract it binds tightly to specific sites in the plasma membrane of the epithelial cells lining the small intestine, causing adenylate cyclase to undergo activation that persists for hours or days.

(a) What is the effect of cholera toxin on the level of cAMP in the intestinal cells?

(b) Based on the information above, can you suggest how cAMP normally functions in intestinal epithelial cells?

(c) Suggest a possible treatment for cholera.

13. *Metabolic Differences in Muscle and Liver in a "Fight or Flight" Situation* During a "fight or flight" situation, the release of epinephrine promotes glycogen breakdown in the liver, heart, and skeletal muscle. The end product of glycogen breakdown in the liver is glucose. In contrast, the end product in skeletal muscle is pyruvate.

(a) Why are different products of glycogen breakdown observed in the two tissues?

(b) What is the advantage to the organism during a "fight or flight" condition of having these specific glycogen breakdown routes?

14. *Excessive Amounts of Insulin Secretion: Hyperinsulinism* Certain malignant tumors of the pancreas cause excessive production of insulin by the β cells. Affected individuals exhibit shaking and trembling, weakness and fatigue, sweating, and hunger. If this condition is prolonged, brain damage occurs.

(a) What is the effect of hyperinsulinism on the metabolism of carbohydrate, amino acids, and lipids by the liver?

(b) What are the causes of the observed symptoms? Suggest why this condition, if prolonged, leads to brain damage.

15. *Thermogenesis Caused by Thyroid Hormones* Thyroid hormones are intimately involved in regulating the basal metabolic rate. Liver tissue of animals given excess thyroxine shows an increased rate of O_2 consumption and increased heat output (thermogenesis), but the ATP concentration in the tissue is normal. Different explanations have been offered for the thermogenic effect of thyroxine. One is that excess thyroid hormone causes uncoupling of oxidative phosphorylation in mitochondria. How could such an effect account for the observations? Another explanation suggests that the thermogenesis is due to an increased rate of ATP utilization by the thyroid-stimulated tissue. Is this a reasonable explanation? Why?

16. *Function of Prohormones* What are the possible advantages in the synthesis of hormones as prohormones or preprohormones?

17. *Action of Aminophylline* Aminophylline, a purine derivative resembling theophylline of tea, is often administered together with epinephrine to individuals with acute asthma. What is the purpose and biochemical basis for this treatment?

Information Pathways

The fourth and final part of this book considers biochemical questions raised by the genetic continuity and the evolution of living organisms. What is the molecular nature of the genetic material? How is genetic information transmitted with such fidelity? How is it ultimately translated into the amino acid sequence of protein molecules?

The fundamental unit of information in living systems is the **gene.** A gene is defined biochemically as that segment of DNA (or in a few cases RNA) that encodes the information required to produce a functional biological product. This product is most often a protein, and much of the material in the chapters to follow concerns genes that encode proteins. However, a gene product can also be one of several classes of RNA molecules. The storage and metabolism of these informational units now becomes the focal point of our discussion.

Modern biochemical research on gene structure and function has brought to biology a revolution comparable to that evoked over 100 years ago by Darwin's theory on the origin of species. An understanding of how information is stored and used in cells has brought penetrating new insights into some of the most fundamental problems concerning the structure and function of cells. Moreover, it has led to a more comprehensive conceptual framework for the science of biochemistry.

Today's knowledge of information pathways has arisen from the convergence of three different disciplines: genetics, physics, and biochemistry. The contributions of these three fields are epitomized by the discovery that opened the modern era of genetic biochemistry: the double-helical structure of DNA, as postulated by James Watson and Francis Crick in 1953 (see Fig. 12–15). Genetic theory contributed the concept of coding by genes. Physics made possible the determination of molecular structure by x-ray diffraction analysis. Biochemistry revealed the chemical composition of DNA. The great impact of the Watson–Crick hypothesis was largely due to its ability to account for a wide range of results derived from these varied sources.

A vastly improved understanding of DNA structure inevitably led to questions about its function. The structure itself suggested how DNA might be copied so that the information contained therein could be transmitted from one generation to the next. Understanding how the information in DNA was converted into functional proteins became possible through the discovery of messenger RNA and transfer RNA and the solution of the genetic code. These and other major advances led to the central dogma of molecular genetics, which defines three major processes in the cellular utilization of genetic information. The first is **replication,** the copying of parental DNA to form daughter

Facing page: The two β subunits of *E. coli* DNA polymerase III bound to DNA. The subunits, shown as gray ribbon structures form a circle around the DNA, tethering the DNA polymerase III (which has at least 9 other subunits) to the DNA. This permits the enzyme to synthesize long stretches of DNA without dissociation. The complex set of operations by which macromolecules containing information are faithfully synthesized requires a great many enzymes, of which this is just part of one.

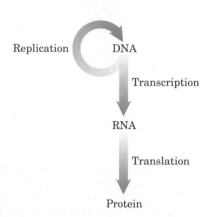

Replication   DNA

Transcription

RNA

Translation

Protein

The central dogma of molecular genetics, showing the general pathways of information flow via the processes of replication, transcription, and translation. The term "dogma" is a misnomer here. It was introduced by Francis Crick at a time when little evidence supported these ideas. The "dogma" is now a well-established principle.

DNA molecules having identical nucleotide sequences. The second is **transcription,** the process by which parts of the coded genetic message in DNA are copied precisely in the form of RNA. The third is **translation,** in which the genetic message coded in messenger RNA is translated on the ribosomes into a protein with a specific sequence of amino acids.

Part IV is devoted to an explanation of these and related processes. First (Chapter 23) we will examine the structure, topology, and packaging of chromosomes and genes. The processes that make up the central dogma will be elaborated in Chapters 24 through 26. Then, as we have done for biosynthetic pathways, we will turn to regulation and examine how the expression of genetic information is controlled (Chapter 27).

A major theme running through these chapters is the added complexity encountered in the biosynthesis of a macromolecule when that macromolecule contains information. Assembling nucleic acids and proteins with the correct sequences of nucleotides and amino acids, respectively, represents nothing less than preserving the faithful expression of the template upon which life itself is based. The formation of phosphodiester bonds in DNA or peptide bonds in proteins might be expected to be a trivial feat for cells, given the arsenal of enzymatic and chemical tools described in Part III of this book. Nevertheless, the framework of patterns and rules established in the examination of metabolic pathways must be enlarged considerably when information is added to the equation. Forming *specific* bonds and preventing sequence errors in these polymers has an enormous impact on the thermodynamics, chemistry, and enzymology of the synthetic processes. For example, formation of a peptide bond should require an input of only about 21 kJ, and relatively simple enzymes that catalyze comparable reactions are known. To synthesize the correct peptide bond between two specific amino acids at a given point in a protein, however, the cell invests about 125 kJ in chemical energy and makes use of the combined activities of over 200 RNA molecules, enzymes, and specialized proteins. Information is expensive.

The dynamic interaction between nucleic acids and proteins is another central theme of Part IV. With the important exception of a few catalytic RNA molecules (discussed in Chapter 25), the processes that make up the pathways of cellular information flow are catalyzed and regulated by proteins. An understanding of these enzymes and proteins can have practical as well as intellectual rewards because they form the basis of the development of recombinant DNA technology. This technology is making possible the prenatal diagnosis of genetic disease; the production of a wide range of potent new pharmaceutical agents; the sequencing of the entire human genome; the introduction of new traits into bacteria, plants, and animals for industry and agriculture; human gene therapy; and many other advances. We finish our tour of the information pathways, and indeed the entire book, in Chapter 28 with a look at this technology and its implications for the future.

Genes and Chromosomes

CHAPTER

23

Every cell of a multicellular organism generally contains the same genetic material. One has only to look at a human being to marvel at the wealth of information contained in each human cell. It should come as no surprise that the DNA molecules containing the cellular genes are by far the largest macromolecules in cells. They are commonly packaged into structures called **chromosomes.** Most bacteria and viruses have a single chromosome; eukaryotes usually have many. A single chromosome typically contains thousands of individual genes. The sum of all the genes and intergenic DNA on all the different chromosomes of a cell is referred to as the cellular **genome.**

Measurements carried out in the 1950s indicated that the largest DNAs had molecular weights of 10^6 or less, equivalent to about 15,000 base pairs. But with improved methods for isolation of native DNAs, their molecular weights were found to be much higher. Today we know that native DNA molecules, such as those from *E. coli* cells, are so large that they are easily broken by mechanical shear forces, and therefore are not readily isolated in intact form.

The size of DNA molecules represents an interesting biological problem in itself. Chromosomal DNAs are often many orders of magnitude longer than the biological packages (cells or viruses) that contain them (Fig. 23–1). In this chapter we move from the secondary structure of DNA considered in Chapter 12 to the extraordinary degree of organization required for the tertiary packaging of DNA into chromosomes. First we examine the size of viral DNAs and cellular chromosomes and the organization of genes and other sequences within them. We then turn to the discipline of DNA topology to give formal definition to the twisting and coiling of DNA molecules. Finally, we consider the protein–DNA interactions that organize chromosomes into compact structures.

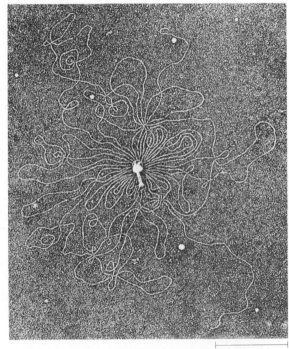

$\vdash\!\!-\!\!-\!\!-\!\!\dashv$ 0.5 μm

Figure 23–1 Electron micrograph of bacteriophage T2 surrounded by its single, linear molecule of DNA. The DNA was released by lysing the bacteriophage in distilled water and allowing the DNA to spread on the water surface.

The Size and Sequence Structure of DNA Molecules

We begin with a survey of the DNA molecules of viruses and of cells, both prokaryotic and eukaryotic. Chromosomes contain, in addition to genes, special-function sequences that aid in the packaging and segregation of chromosomes to daughter cells at cell division. The structure of chromosomes will be examined, with a focus on the various types of DNA sequences found within them.

Viral DNA Molecules Are Small

Viruses generally require considerably less genetic information than cells, because they rely on many functions of a host cell to reproduce themselves. Viral genomes can be made up of either RNA or DNA. Almost all plant viruses and some bacterial and animal viruses contain RNA. RNA viruses tend to have particularly small genomes. The genomes of DNA viruses, in contrast, span a wide range of sizes (Table 23–1). From the molecular weight of a double-stranded (duplex) viral DNA it is possible to calculate its **contour length** (its helix length), given that each nucleotide pair has an average molecular weight of about 650 and there is one nucleotide pair for every 0.36 nm of the duplex (see Fig. 12–15). Note that the DNA found in some viruses is single-stranded rather than double-stranded.

Table 23–1 The DNA and particle sizes of some bacterial viruses

Virus	Viral particle weight ($\times 10^6$)	Long dimension of particle (nm)	Number of base pairs
ϕX174 (duplex form)	6	25	5,386
T7	38	78	39,936
λ (lambda)	50	190	48,502
T2, T4	220	210	182,000*

* The complete base sequence of T2 and T4 DNA is not known; this is an approximation.

Many viral DNAs have covalently linked ends and are therefore circular (in the sense of an endless belt, rather than a perfect round) during at least part of their life cycle. During viral replication within a host cell, specific types of viral DNA called **replicative forms** may appear; for example, linear DNAs often become circular and all single-stranded DNAs become double-stranded.

A typical medium-sized DNA virus is bacteriophage λ (lambda) of *E. coli*. In its replicative form inside cells, its DNA is a circular double helix. Double-stranded λ DNA contains 48,502 base pairs and has a contour length of 17.5 μm. Bacteriophage ϕX174 is a much smaller DNA virus; the DNA in a ϕX174 viral particle is a single-stranded circle. Its double-stranded replicative form contains 5,386 base pairs. Another important point about viral DNAs will be echoed in sections to follow: their contour lengths are much greater than the long dimensions of the viral particles in which they are found. The DNA of bacteriophage T2, for example, is about 3,500 times longer than the viral particle itself (Fig. 23–1).

Bacteria Contain Chromosomes and Extrachromosomal DNA

Bacteria contain much more DNA than the DNA viruses. For example, a single *E. coli* cell contains almost 200 times as much DNA as a bacteriophage λ particle. The DNA in an *E. coli* cell is a single, covalently closed double-stranded circular molecule. It contains about 4.7×10^6 base pairs and has a contour length of about 1.7 mm, some 850 times the length of an *E. coli* cell (Fig. 23–2). Again, the DNA molecule must have a tightly compacted tertiary structure.

In addition to the very large, circular DNA chromosome found in the nucleoid, many species of bacteria contain one or more small, circular DNA molecules that are free in the cytosol. These extrachromosomal elements are called **plasmids** (Fig. 23–3). Many plasmids are only a few thousand base pairs long, but some contain over 10^5 base pairs. Plasmids carry genetic information and undergo replication to yield daughter plasmids, which pass into the daughter cells at cell division. Ordinarily, plasmids exist separately, detached from the chromosomal DNA. A few classes of plasmid DNAs are sometimes inserted into the chromosomal DNA and later excised in a precise manner by means of specialized recombination processes.

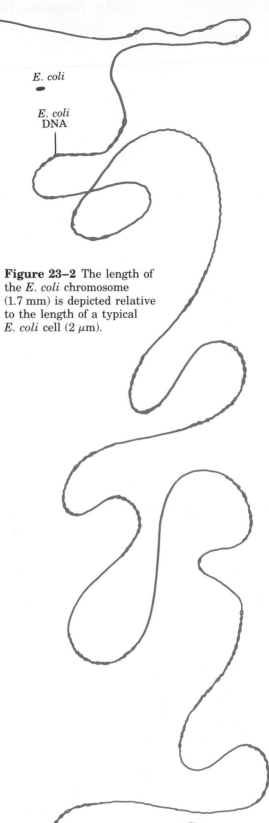

Figure 23–2 The length of the *E. coli* chromosome (1.7 mm) is depicted relative to the length of a typical *E. coli* cell (2 μm).

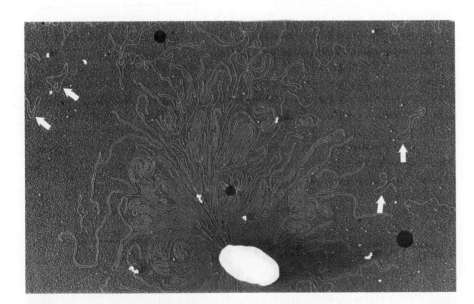

Figure 23–3 Electron micrograph of DNA from a lysed *E. coli* cell. Several small, circular plasmid DNAs are indicated by arrows. The black spots and white specks are artifacts of the preparation.

Plasmids have been found in yeast and other fungi as well as in bacteria. In many cases plasmids confer no obvious advantage on their host, and their sole function appears to be self-propagation. However, some plasmids carry genes that make a host bacterium resistant to antibacterial agents. For example, plasmids carrying the gene for the enzyme β-lactamase confer resistance to β-lactam antibiotics such as penicillin and amoxicillin. Plasmids also may pass from an antibiotic-resistant cell to an antibiotic-sensitive cell of the same or another bacterial species, thus rendering the latter resistant. The extensive use of antibiotics has served as a strong selective force for the spread of these plasmids in disease-causing bacteria, creating multiply resistant bacterial strains, particularly in hospital settings. Physicians are becoming reluctant to prescribe antibiotics unless a bacterial infection is confirmed. For similar reasons, the widespread use of antibiotics in animal feeds is being curbed.

Plasmids are useful models for the study of many processes in DNA metabolism. They are relatively small DNA molecules and hence can quite easily be isolated intact from bacterial and yeast cells. Plasmids have also become a central component of the modern technologies associated with the isolation and cloning of genes. Genes from a variety of species can be inserted into isolated plasmids, and the modified plasmid can then be reintroduced into its normal host cell. Such a plasmid will be replicated and transcribed, and may also cause the host cell to make the proteins coded by the foreign gene, even though it is not part of the normal genome of the cell. Chapter 28 describes how such **recombinant DNAs** are made.

Eukaryotic Cells Contain More DNA than Prokaryotes

An individual cell of a yeast, one of the simplest eukaryotes, has four times more DNA than an *E. coli* cell. Cells of *Drosophila,* the fruit fly used in classical genetic studies, have more than 25 times as much DNA as *E. coli* cells. Each cell of human beings and many other mammals has about 600 times as much DNA as *E. coli,* and the cells of many plants and amphibians have an even greater amount. Note that the nuclear DNA molecules of eukaryotic cells are linear, not circular.

The total contour length of all the DNA in a *single* human cell is about 2 m, compared with 1.7 mm for *E. coli* DNA. In the approximately 10^{14} cells of the adult human body, the total length of all the DNA would be about 2×10^{13} m or 2×10^{10} km. Compare this with the circumference of the earth (4×10^4 km) or the distance between the earth and the sun (1.5×10^8 km). Once again it becomes clear that DNA packaging in cells must involve an extraordinary degree of organization and compaction.

Microscopic observation of nuclei in dividing eukaryotic cells has shown that the genetic material is subdivided into chromosomes, their diploid number depending upon the species of organism (Table 23–2). Human cells, for example, have 46 chromosomes. Each chromosome of a eukaryotic cell, such as that shown in Figure 23–4a, can contain a single, very large, duplex DNA molecule, which may be from 4 to 100 times larger than that of an *E. coli* cell. For example, the DNA of one of the smaller human chromosomes has a contour length of about 30 mm, almost 15 times longer than the DNA of *E. coli*. The DNA molecules in the 24 different types of chromosomes of human cells (22 + X + Y) vary in length over a 25-fold range. Each different chromosome in eukaryotes carries a characteristic set of genes.

Table 23–2 Normal chromosome number in different organisms*

Bacteria	1
Fruit fly	8
Red clover	14
Garden pea	14
Yeast	16
Honeybee	16
Corn	20
Frog	26
Hydra	30
Fox	34
Cat	38
Mouse	40
Rat	42
Rabbit	44
Human	46
Chicken	78

* For all eukaryotic organisms listed, the diploid chromosome number is shown.

Figure 23–4 Eukaryotic chromosomes. **(a)** A chromosome from a human cell. **(b)** A complete set of chromosomes from a leukocyte from one of the authors. There are 46 chromosomes in every human somatic cell.

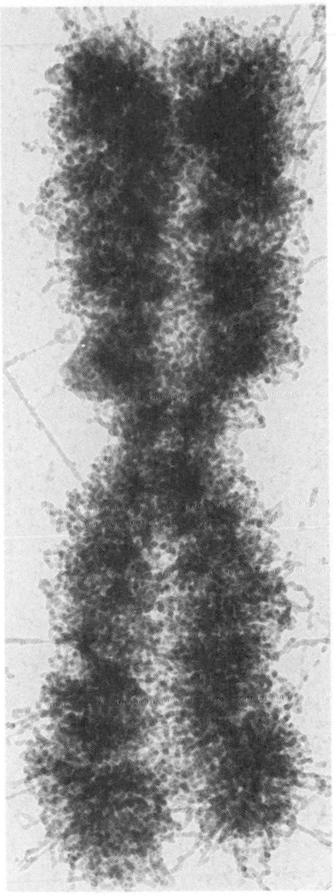

Organelles of Eukaryotic Cells Also Contain DNA

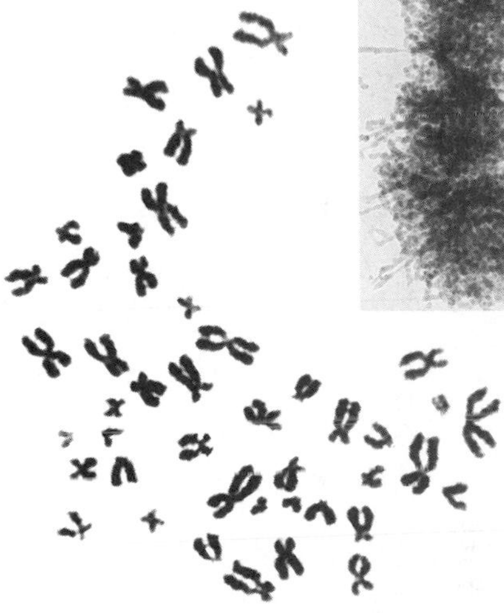

In addition to the DNA in the nucleus of eukaryotic cells, very small amounts of DNA, differing in base sequence from nuclear DNA, are present within the mitochondria. Chloroplasts of photosynthetic cells also contain DNA. Usually less than 0.1% of all the cell DNA is present in the mitochondria in typical somatic cells, but in fertilized and dividing egg cells, where the mitochondria are much more numerous, the total amount of mitochondrial DNA is correspondingly larger. Mitochondrial DNA (mDNA) is a very small molecule compared with the nuclear chromosomes. In animal cells it contains less than 20,000 base pairs (16,569 base pairs in human mDNA) and occurs as a circular duplex. Chloroplast DNA molecules also exist as circular duplexes and are considerably larger than those of mitochondria.

The evolutionary origin of mitochondrial and chloroplast DNAs has been the subject of much speculation. A widely accepted view is that they are vestiges of the chromosomes of ancient bacteria that gained access to the cytoplasm of host cells and became the precursors of these organelles (see Fig. 2–17). Mitochondrial DNA codes for the mitochondrial tRNAs and rRNAs and for a few mitochondrial proteins. More than 95% of mitochondrial proteins are encoded by nuclear DNA. Mitochondria and chloroplasts divide when the cell divides (Fig. 23–5). Before and during division of these organelles their DNA is replicated and the daughter DNA molecules pass into the daughter organelles.

(a)

(b)

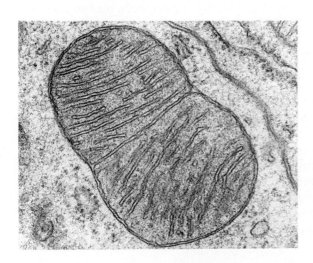

Figure 23–5 A dividing mitochondrion. Many mitochondrial proteins and RNAs are encoded by the mitochondrial DNA (not visible here), which is replicated each time the mitochondrion divides.

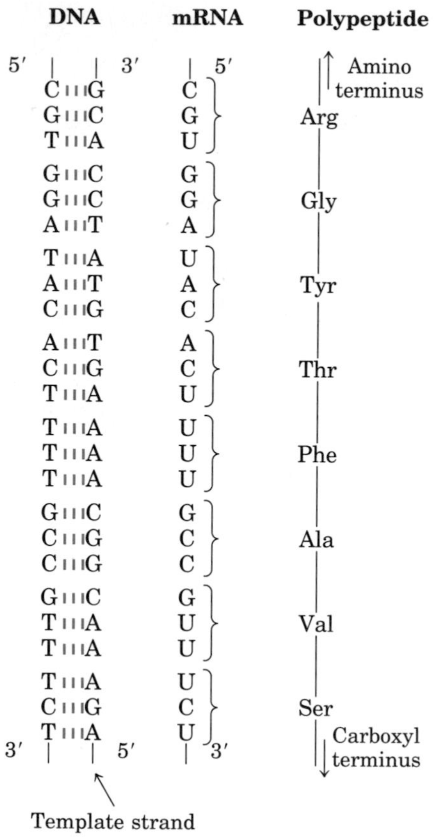

DNA   **mRNA**   **Polypeptide**

```
 5' |   | 3'    | 5'        ↑ Amino
    C···G        C  ⎫        | terminus
    G···C        G  ⎬ Arg
    T···A        U  ⎭
    G···C        G  ⎫
    G···C        G  ⎬ Gly
    A···T        A  ⎭
    T···A        U  ⎫
    A···T        A  ⎬ Tyr
    C···G        C  ⎭
    A···T        A  ⎫
    C···G        C  ⎬ Thr
    T···A        U  ⎭
    T···A        U  ⎫
    T···A        U  ⎬ Phe
    T···A        U  ⎭
    G···C        G  ⎫
    C···G        C  ⎬ Ala
    C···G        C  ⎭
    G···C        G  ⎫
    T···A        U  ⎬ Val
    T···A        U  ⎭
    T···A        U  ⎫
    C···G        C  ⎬ Ser
    T···A        U  ⎭   ↓ Carboxyl
 3' |   | 5'    | 3'        terminus
```

Template strand

Figure 23–6 Colinearity of the nucleotide sequences of DNA, mRNA, and the amino acid sequence of polypeptide chains. The triplets of nucleotide units in DNA determine the sequence of amino acids in proteins through the intermediary formation of mRNA, which has nucleotide triplets (codons) complementary to those of the DNA. Only one of the DNA strands, the template strand, serves as a template for mRNA synthesis.

Genes Are Segments of DNA That Code for Polypeptide Chains and RNAs

Our present understanding of the gene has evolved considerably over the last century. A gene is defined in the classical biological sense as a portion of a chromosome that determines or affects a single character or **phenotype** (visible property), for example, eye color. But there is also a molecular definition, first proposed by George Beadle and Edward Tatum in 1940. They exposed spores of the mold *Neurospora crassa* to x rays and other agents that damage DNA and sometimes cause alterations in the DNA sequence (**mutations).** Some mutants were found to be deficient in one or another specific enzyme, resulting in the failure of a metabolic pathway. This observation led Beadle and Tatum to conclude that a gene is a segment of the genetic material that determines or codes for one enzyme: the **one gene–one enzyme** hypothesis. Later this concept was broadened to **one gene–one protein,** because some genes code for proteins that are not enzymes.

The present biochemical definition of a gene is somewhat more precise. Recall that many proteins have multiple polypeptide chains (Chapter 6). In some multichain proteins, all the polypeptide chains are identical, in which case they can all be encoded by the same gene. Others have two or more different kinds of polypeptide chains, each with a distinctive amino acid sequence. Hemoglobin A, the major adult hemoglobin of humans, for example, has two kinds of polypeptide chains, α and β chains, which differ in amino acid sequence and are encoded by two different genes. Thus the gene–protein relationship is more accurately described by the phrase "one gene–one polypeptide."

However, not all genes are ultimately expressed in the form of polypeptide chains. Some genes code for the different kinds of RNAs such as tRNAs and rRNAs (Chapters 12 and 25). Genes that code for either polypeptides or RNAs are known as **structural genes:** they encode the primary sequence of some final gene product, such as an enzyme or a stable RNA. DNA also contains other segments or sequences that have a purely regulatory function. **Regulatory sequences** provide signals that may denote the beginning and end of structural genes, or participate in turning on or off the transcription of structural genes, or function as initiation points for replication or recombination (Chapter 27).

The minimum overall size of genes can be estimated directly. As will be described in detail in Chapter 26, each amino acid of a polypeptide chain is coded by a sequence of three consecutive nucleotides in a single strand of DNA (Fig. 23–6). Because there are no signals for "commas" in the genetic code, the coding triplets of DNA are generally arranged sequentially, corresponding to the sequence of amino acids in the polypeptide for which it codes. Figure 23–6 shows the principle of the coding relationships between DNA, RNA, and proteins. A single polypeptide chain may have anywhere from about fifty to several thousand amino acid residues in a specific sequence, thus a gene coding for the biosynthesis of a polypeptide chain must have, correspondingly, at least 150 to 6,000 or more base pairs. For an average polypeptide chain of 350 amino acid residues, this would correspond to 1,050 base pairs. We will see later that many genes in eukaryotes and a few in prokaryotes are interrupted by noncoding DNA segments called introns, and can therefore be considerably longer than the simple calculations outlined above would suggest.

There Are Many Genes in a Single Chromosome

How many genes are in a single chromosome? We can give an approximate answer to this question in the case of *E. coli*. If the average gene is 1,050 base pairs long, the 4.7 million base pairs in the *E. coli* chromosome could accommodate about 4,400 genes. The products of over 1,000 *E. coli* genes have already been characterized, and the number is increasing. A growing fraction of the *E. coli* chromosome has been sequenced, and the number of genes it contains will be known with some precision when this effort is completed.

Eukaryotic Chromosomes Are Very Complex

Bacteria usually have only one chromosome per cell, and in nearly all cases each chromosome contains only one copy of any given gene. A very few genes, such as those for rRNAs, are repeated several times. Regulatory and structural gene sequences account for much of the DNA in prokaryotes. Moreover, almost every gene is precisely colinear with the amino acid sequence (or RNA sequence) for which it codes (Fig. 23–6).

The organization of genes in eukaryotic DNA is structurally and functionally much more complex, and the study of eukaryotic chromosome structure has yielded many surprises. Tests made of the extent to which segments of mouse DNA occur in multiple copies had an unexpected outcome. About 10% of mouse DNA consists of short lengths of less than 10 base pairs that are repeated millions of times per cell. These are called **highly repetitive** segments. Another 20% of mouse DNA was found to occur in lengths up to a few hundred base pairs that are repeated at least 1,000 times, designated **moderately repetitive.** The remainder, some 70% of the DNA, consists of unique segments and segments that are repeated only a few times.

Some of the repetitive DNA may simply be "junk DNA," vestiges of evolutionary sidetracks. At least some of it has functional significance, however. The most highly repeated sequences are called **satellite DNA** because their base compositions are generally unusual, permitting their separation from the rest of the DNA when fragmented cellular DNA samples are centrifuged in cesium chloride density gradients. Satellite DNA is not believed to encode proteins or RNAs. Much of the highly repetitive DNA is associated with two important structures in eukaryotic chromosomes—centromeres and telomeres.

Each chromosome has a single **centromere,** which functions as an attachment point for proteins that link the chromosome to the microtubules of the mitotic spindle (see Fig. 2–14). This attachment is essential for the ordered segregation of chromosomes to daughter cells during cell division. The centromeres of yeast chromosomes have been isolated and studied (Fig. 23–7). The sequences essential to centromere function are about 130 base pairs long and are very rich in A=T pairs. The centromeres of higher eukaryotes are much larger. In higher eukaryotes (but not in yeast), satellite DNA is generally found in the centromeric region and consists of thousands of tandem (side-by-side and in the same orientation) copies of one or a few short sequences. Characterized satellite sequences are generally 5 to 10 base pairs long. The precise role of satellite DNA in centromere function is not yet understood.

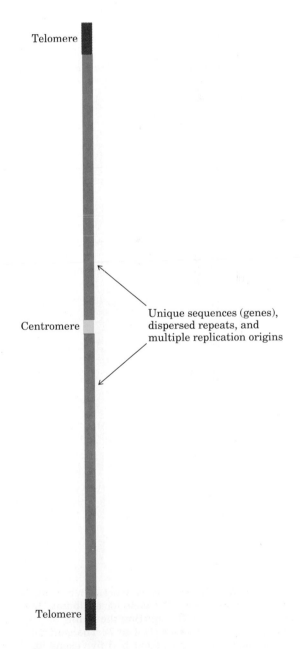

Telomere

Centromere

Unique sequences (genes), dispersed repeats, and multiple replication origins

Telomere

Figure 23–7 Important structural features of a yeast chromosome.

Telomeres are sequences located at the ends of the linear eukaryotic chromosomes, which help stabilize them. The best-characterized telomeres are those of simpler eukaryotes. Yeast telomeres end with about 100 base pairs of imprecisely repeated sequences of the form

$$(5')(T_xG_y)_n$$
$$(3')(A_xC_y)_n$$

where x and y generally fall in the range of 1 to 4. The ends of a linear DNA molecule cannot be replicated by the cellular replication machinery (which may be one reason why bacterial DNA molecules are circular). The repeated sequences in telomeres are added to chromosome ends by special enzymes, one of which is telomerase, which will be discussed in more detail in Chapter 25. What controls the number of repeats in a telomere is not known. The telomere repeats are a very unusual DNA structure.

Efforts have begun to construct artificial chromosomes as a means of better understanding the functional significance of many structural features of eukaryotic chromosomes. A reasonably stable, artificial, linear chromosome requires only three components: a centromere, telomeres at the ends, and sequences that direct the initiation of DNA replication.

Most moderately repetitive DNA consists of 150 to 300 base-pair repeats scattered throughout the genome of higher eukaryotes. Some of these repeats have been characterized. A number of them have some of the structural properties of transposable elements, sequences that move about the genome at very low frequency (Chapter 24). In humans, one class of these repeats (about 300 base pairs long) is called the *Alu* family, so named because their sequence generally includes one copy of the recognition sequence for the restriction endonuclease *Alu*I. (Restriction endonucleases are described in Chapter 28.) Hundreds of thousands of *Alu* repeats occur in the human genome, comprising 1 to 3% of the total DNA. They apparently were derived from a gene for 7SL RNA, a component of a complex called the signal-recognition particle (SRP, Chapter 26) that functions in protein synthesis. The *Alu* repeats, however, lack parts of the 7SL RNA gene sequence and do not produce functional 7SL RNAs. When *Alu* repeats are grouped with other classes of repeats with similar sizes and sequence structures, they make up 5 to 10% of the DNA in the human genome. No function for this DNA is known.

The unique sequences in eukaryotic chromosomes include most of the genes. There are an estimated 100,000 different genes in the human genome.

Many Eukaryotic Genes Contain Intervening Nontranscribed Sequences (Introns)

Many, if not most, eukaryotic genes have a distinctive and puzzling structural feature: their nucleotide sequences contain one or more intervening segments of DNA that do not code for the amino acid sequence of the polypeptide product. These nontranslated inserts interrupt the otherwise precisely colinear relationship between the nucleotide sequence of the gene and the amino acid sequence of the polypeptide it encodes (Fig. 23–8). Such nontranslated DNA segments in genes are called **intervening sequences,** or **introns,** and the coding segments are called **exons.** A well-known example is the gene coding for the single polypeptide chain of the avian egg protein ovalbumin.

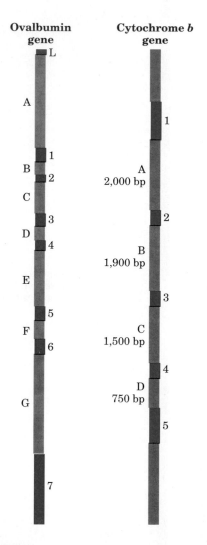

Figure 23–8 Intervening sequences, or introns, in two eukaryotic genes. The gene for ovalbumin has seven introns (A to G), splitting the coding sequences into eight exons (L, 1 to 7). The gene for cytochrome *b* has four introns and five exons. In both cases, more DNA is devoted to introns than to exons. The number of base pairs (bp) in the introns of the cytochrome *b* gene is shown.

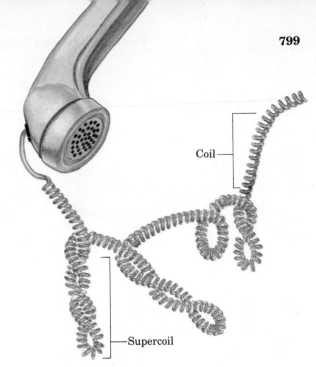

Figure 23–9 Supercoils. A typical phone cord is a coil. A phone cord twisted as shown is a supercoil. The illustration is especially appropriate, because an examination of the twisting of phone cords helped lead Jerome Vinograd and colleagues to the insight that many properties of small, circular DNAs could be explained by supercoiling. They first detected DNA supercoiling in small, circular viral DNAs in 1965.

As can be seen in Figure 23–8, the introns of this particular gene are much longer than the exons; altogether the introns make up 85% of the DNA of this gene. Most eukaryotic genes examined thus far appear to contain introns that vary in number, position, and the fraction of the total length of the gene they occupy. For example, the serum albumin gene contains 6 introns, the gene for the protein conalbumin of the chicken egg contains 17 introns, and a collagen gene has been found to have over 50 introns. Genes for histones provide an example of a family of genes that appear to have no introns. Only a few prokaryotic genes contain introns. In most cases the function of introns is not clear.

DNA Supercoiling

From the examples given above, it is clear that cellular DNA must be very tightly compacted just to fit into the cell. This implies a high degree of structural organization. It is not enough just to fold the DNA into a small space, however. The packaging must permit access to the information in the DNA for processes such as replication and transcription. Before considering how this is accomplished, we must examine an important property of DNA structure that we have not yet considered—DNA supercoiling.

The term "supercoiling" means literally the coiling of a coil. A telephone cord for example, is typically a coiled wire. The twisted path often taken by that wire as it goes from the base of the phone to the receiver generally describes a supercoil (Fig. 23–9). DNA is coiled in the form of a double helix. Let us define an axis about which both strands of the DNA coil. A bending or twisting of that axis upon itself (Fig. 23–10) is referred to as **DNA supercoiling.** As detailed below, DNA supercoiling is generally a manifestation of structural strain. Conversely, if there is no net bending of the DNA axis upon itself, the DNA is said to be in a **relaxed** state.

It is probably apparent that DNA compaction must involve some form of supercoiling. Perhaps less apparent is the fact that replicating or transcribing DNA also must induce some degree of supercoiling.

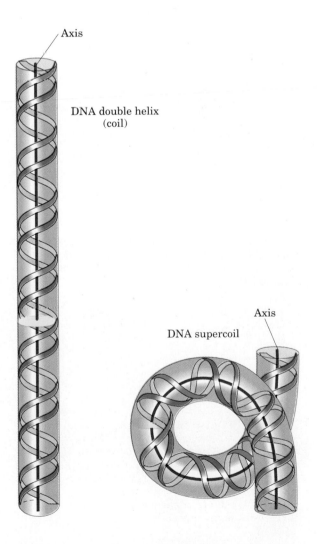

Figure 23–10 Supercoiling of DNA. Supercoiling is the twisting of the DNA axis upon itself.

Figure 23–11 Supercoiling induced by separating the strands of a helical structure. Twist two linear strands of rubber band into a right-handed double-helix as shown. Fix the left end by having a friend hold onto it. If the two strands are pulled apart at the right end, the resulting strain will produce supercoiling as shown.

Replication and transcription both require a transient separation of the strands of DNA, and this is not a simple process in a DNA structure in which the two strands are helically interwound. Figure 23–11 illustrates this point.

That supercoiling must occur in cellular DNA would seem almost trivial were it not for one additional fact: many circular DNA molecules remain highly supercoiled even after they are purified from protein and other cellular components. Supercoiling is an important and intrinsic aspect of DNA tertiary structure that is ubiquitous in cellular DNAs and highly regulated by each cell.

A number of quantifiable properties of supercoiling have been established, the study of which has provided many insights into DNA structure and function. This work has drawn heavily on concepts derived from a branch of mathematics called topology, the study of properties of an object that do not change under continuous deformations. In the case of DNA, a topological property is one that is not affected by twisting and turning of the DNA axis and can only be changed by breakage and rejoining of the DNA backbone. We now turn to an examination of the fundamental properties of supercoiling and the physical origin of the phenomenon itself.

Most Cellular DNA Is Underwound

To understand supercoiling we must now focus on the properties of small, circular DNAs such as plasmids and the DNAs derived from many small DNA viruses. When these DNAs contain no breaks in either strand, they are called **closed-circular DNAs.** If the DNA making up a closed-circular molecule conforms closely to the B-form structure (see Fig. 12–15), with one turn of the double helix for each 10.5 base pairs, the DNA will be relaxed rather than supercoiled (Fig. 23–12). Supercoiling is not a random process and does not occur unless the DNA is subject to some form of structural strain. When purified, however, closed-circular DNAs are rarely relaxed regardless of their biological origin. Furthermore, the degree of supercoiling tends to be well defined and characteristic of DNAs derived from a given cellular source. These facts suggest that the DNA structure is strained in some way to induce the supercoiling, and that the degree of strain introduced is regulated by the cell.

Figure 23–12 Electron micrographs of relaxed and supercoiled plasmid DNAs. The molecule at the left is relaxed, and the degree of supercoiling increases from left to right.

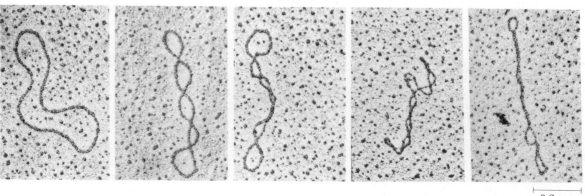

0.2 μm

In almost every instance, the strain is a result of an **underwinding** of the DNA in the closed circle. In other words, there are *fewer* helical turns in the DNA than would be expected for the B-form structure. The effect of underwinding is illustrated in Figure 23–13 for an 84 base pair segment of a circular DNA. If the DNA were relaxed, this segment would contain eight double-helical turns, or one for every 10.5 base pairs. If one of these turns is removed, there will be 84/7 or about 12.0 base pairs per turn rather than the 10.5 found in B-DNA. This is a deviation from the most stable DNA form, and the molecule is thermodynamically strained as a result. The strain can be accommodated in one of two ways. First, the two strands can simply separate over the distance corresponding to one turn of B-DNA—10.5 base pairs (Fig. 23–13). Alternatively, the DNA can form a supercoil. When the axis of the DNA is twisted on itself in a certain manner, neighboring base pairs in underwound DNA can stack in positions that more closely approximate those they would assume in B-DNA.

Every cell actively underwinds its DNA with the aid of enzymatic processes to be described below. The resulting strained state of the DNA represents a form of stored energy. In isolated closed-circular DNA, strain introduced by underwinding generally is accommodated by supercoiling rather than strand separation, because twisting the axis of the DNA usually requires less energy than breaking the hydrogen bonds that stabilize paired bases. As we shall see below, however, the underwinding of DNA in vivo makes it easier to separate DNA strands and thereby gain access to the information they contain. Facilitating strand separation is one important reason for maintaining DNA in an underwound state.

The underwound state can be maintained only if the DNA is a closed circle or if it is bound and stabilized by proteins such that the strands are not free to rotate about each other. If there is a break in one of the strands of a protein-free circular DNA, free rotation at that point will cause the underwound DNA to revert spontaneously to the relaxed state. In a closed-circular DNA, however, the number of helical turns present is fixed and cannot be changed without at least transiently breaking one of the DNA strands. The number of helical turns in DNA is quantifiable and leads to a more precise description of supercoiling.

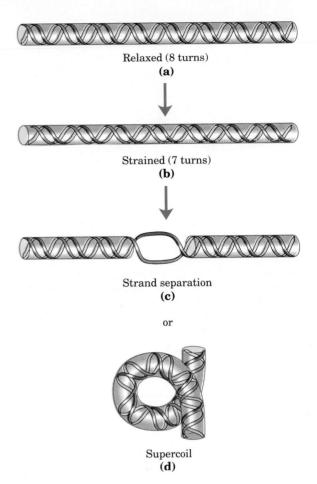

Figure 23–13 The effects of DNA underwinding. **(a)** A segment of DNA, 84 base pairs long, in its relaxed form with eight helical turns. **(b)** Removal of one turn induces structural strain that can be accommodated by **(c)** strand separation over 10.5 base pairs or by **(d)** formation of a supercoil.

DNA Underwinding Is Defined by Topological Linking Number

The branch of mathematics called topology provides a number of ideas that are useful in this discussion. Perhaps foremost among these is the concept of **linking number.** The linking number of a DNA molecule rigorously specifies the number of helical turns in a closed-circular DNA, in the absence of any supercoiling. Linking number is a topological property because it does not vary when double-stranded DNA is twisted or deformed in any way, as long as both DNA strands remain intact.

The concept of linking number (Lk) is illustrated in Figure 23–14. We begin by separating the two strands of a double-stranded circular DNA. If these two strands are linked as shown in Figure 23–14a, they are effectively joined by what can be described as a topological bond.

$Lk = 1$
(a)

$Lk = 6$
(b)

Figure 23–14 Linking number, Lk. The molecule in **(a)** has a linking number of 1. The molecule in **(b)** has a linking number of 6. One of the strands in **(b)** is kept untwisted for illustrative purposes to define the border of an imaginary surface (shaded blue). The number of times the twisting strand penetrates this surface provides one definition of linking number.

Even if all hydrogen bonds and base-stacking interactions are abolished such that the strands are not in physical contact, this topological bond will still link the two strands. If one of the circular strands is thought of as the boundary of an imaginary surface (much as a soap film might span the space framed by a circular wire), the linking number can be defined rigorously as the number of times the second strand pierces this surface. For the molecule in Figure 23–14a $Lk = 1$; for that in Figure 23–14b $Lk = 6$. The linking number for a closed-circular DNA is always an integer. By convention, if the links between two DNA strands are arranged so that the strands are interwound in a right-handed helix, the linking number is defined as positive $(+)$. Conversely, for strands interwound as a left-handed helix the linking number is negative $(-)$. Given that left-handed Z-DNA occurs only rarely, negative linking numbers are not encountered in studies of DNA for all practical purposes.

We can now extend these ideas to a closed-circular DNA with 210 base pairs (Fig. 23–15). For a closed-circular DNA molecule that is relaxed, the linking number is simply the number of base pairs divided by 10.5; in this case, $Lk = 20$. For a circular DNA molecule to have a topological property such as linking number, neither strand may contain a break. If there is a break in either strand, it is possible in principle to unravel the strands and separate them completely (Fig. 23–15b). Clearly, no topological bond exists in this case, and Lk is undefined.

We can now describe DNA underwinding in terms of changes in the linking number. The linking number in relaxed DNA is used as a reference and called Lk_0. In the molecule shown in Figure 23–15a, $Lk_0 = 20$; if two turns are removed from this molecule, Lk will equal 18. The change can be described by the equation

$$\Delta Lk = Lk - Lk_0 = 18 - 20 = -2$$

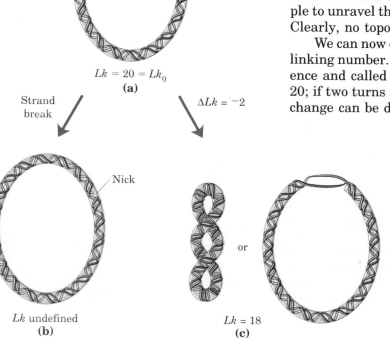

$Lk = 20 = Lk_0$
(a)

Strand break

$\Delta Lk = -2$

Nick

Lk undefined
(b)

or

$Lk = 18$
(c)

Figure 23–15 Linking number applied to closed-circular DNA molecules. A 210 base pair circular DNA is shown in three forms: **(a)** relaxed, $Lk = 20$; **(b)** relaxed with a nick (break) in one strand, Lk undefined; **(c)** underwound by two turns, $Lk = 18$. The underwound molecule can occur as a super-coiled (left) or strand-separated (right) structure.

It is often convenient to express the change in linking number in terms of a length-independent quantity called the **specific linking difference** (σ), which is a measure of the turns removed relative to those present in relaxed DNA. The term σ is also called the superhelical density and is defined as

$$\sigma = \frac{\Delta Lk}{Lk_0}$$

In the example in Figure 23–15c, $\sigma = -0.10$, which means that 10% of the helical turns present in the DNA (in its B form) have been removed. The degree of underwinding in cellular DNAs generally falls into the range of 5 to 7%; that is, $\sigma = -0.05$ to -0.07. The negative sign of σ denotes that the change in linking number comes about as a result of underwinding the DNA. The supercoiling induced by underwinding is therefore defined as negative supercoiling. Conversely, under some conditions DNA can be overwound, and the resulting supercoiling is defined as positive. Note that the twisting path taken by the axis of the DNA helix when the DNA is underwound (negative supercoiling) is the mirror image of that taken when the DNA is overwound (positive supercoiling) (Fig. 23–16). Supercoiling is not a random process; the path of the supercoiling is largely prescribed by the torsional strain imparted to the DNA by decreasing or increasing the linking number relative to B-DNA.

The linking number can be changed by ±1 by breaking one DNA strand, rotating one of the ends 360° about the unbroken strand, and rejoining the broken ends. This change has no effect on the number of base pairs, or indeed on the number of atoms in the circular DNA molecule. Two forms of a given circular DNA that differ only in a topological property such as linking number are referred to as **topoisomers.**

Linking number can be broken down into two structural components called writhe (W_r) and twist (T_w) (Fig. 23–17). These are more difficult to describe intuitively than linking number, but to a first approximation W_r may be thought of as a measure of the coiling of the helix axis and T_w as determining the local twisting or spatial relationship of neighboring base pairs. When a change in linking number occurs, some of the resulting strain is usually compensated by writhe (supercoiling) and some by changes in twist, giving rise to the equation

$$Lk = T_w + W_r$$

Twist and writhe are geometric rather than topological properties, because they may be changed by deformation of a closed-circular DNA molecule. In addition, T_w and W_r need not be integers.

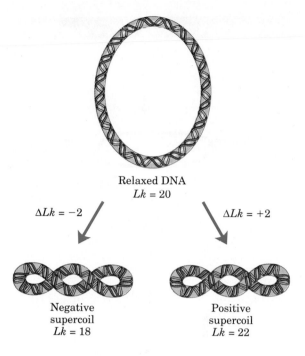

Relaxed DNA
$Lk = 20$

$\Delta Lk = -2$

$\Delta Lk = +2$

Negative supercoil
$Lk = 18$

Positive supercoil
$Lk = 22$

Figure 23–16 For the relaxed DNA molecule of Figure 23–15a, underwinding or overwinding by two helical turns ($Lk = 18$ or 22) will produce negative or positive supercoiling as shown. Note that the twisting of the DNA axis is opposite in sign in the two cases.

Straight ribbon (relaxed DNA)
(a)

Large writhe, small change in twist
(b)

Zero writhe, large change in twist
(c)

Figure 23–17 A ribbon model for illustrating twist and writhe. The ribbon in **(a)** represents the axis of a relaxed DNA molecule. Strain introduced by twisting the ribbon (underwinding the DNA) can be manifested as a change in writhe **(b)** or a change in twist **(c)**. Changes in linking number are usually accompanied by changes in both writhe and twist.

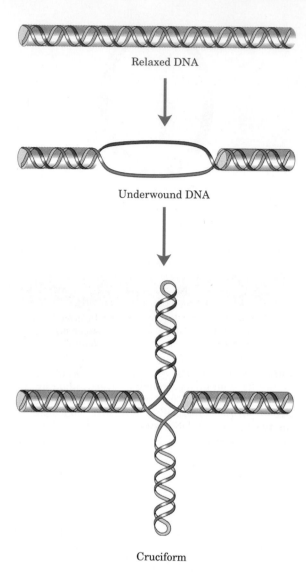

Figure 23–18 DNA underwinding promotes cruciform structures. In relaxed DNA, cruciforms seldom occur because the linear DNA accommodates more paired bases than does the cruciform structure. Underwinding the DNA facilitates the partial strand separation needed to promote cruciform formation at appropriate sequences (palindromes).

The concepts outlined above can be summarized by considering the supercoiling of a typical bacterial plasmid DNA. Plasmids are generally closed-circular DNA molecules. Because DNA is a right-handed helix, a plasmid will have a positive linking number. When the DNA is relaxed, the linking number or Lk_0 is simply the number of base pairs divided by 10.5. A typical plasmid, however, is generally underwound in the cell. Therefore, Lk is less than Lk_0, σ is negative, and the plasmid is negatively supercoiled. Typically for a bacterial plasmid, $\sigma = -0.05$ to -0.07.

Underwinding DNA facilitates a number of structural changes in the molecule. Strand separation occurs more readily in underwound DNA. This is critical to the processes of replication and transcription, and represents a major reason why DNA is maintained in an underwound state. Other structural changes are of less physiological importance but help illustrate the effects of underwinding. A cruciform (see Fig. 12–21) generally contains a few unpaired bases, and DNA underwinding helps to maintain the required strand separation (Fig. 23–18). In addition, underwinding a right-handed DNA helix facilitates the formation of short regions of left-handed Z-DNA, where the DNA sequence is consistent with Z-DNA formation (Chapter 12).

Topoisomerases Catalyze Changes in the Linking Number of DNA

In every cell, DNA supercoiling is a precisely regulated process that influences many aspects of DNA metabolism. Not surprisingly, there are enzymes in every cell whose sole purpose is to underwind and/or relax DNA. The enzymes that increase or decrease the extent of DNA underwinding are called **topoisomerases,** and the property of DNA they affect is the linking number. These enzymes play an especially important role in processes such as replication and DNA packaging. There are two classes of topoisomerases. Type 1 topoisomerases act by transiently breaking one of the two DNA strands, rotating one of the ends about the unbroken strand, and rejoining the broken ends; they change Lk in increments of 1. Type 2 topoisomerases break both DNA strands and change Lk in increments of 2.

The effects of these enzymes can be demonstrated using agarose gel electrophoresis (Fig. 23–19). A population of identical plasmid DNAs with the same linking number will migrate as a discrete band during electrophoresis. Topoisomers with Lk values differing by as little as 1 can be separated by this method. In this way changes in linking number induced by topoisomerases can readily be observed.

There are at least four different topoisomerases in *E. coli,* distinguished by Roman numerals I through IV. The type 1 topoisomerases (topoisomerases I and III) generally relax DNA by removing negative supercoils (they increase Lk). One bacterial type 2 enzyme, called topoisomerase II or, alternatively, DNA gyrase, can introduce negative supercoils (decrease Lk). It uses the energy of ATP and a surprising mechanism to accomplish this (Fig. 23–20). The superhelical density of bacterial DNA is balanced by regulation of the net activity of topoisomerases I and II.

Eukaryotic cells also have type 1 and type 2 topoisomerases; in most eukaryotes there is one known example of each type, called topoisomerase I and II, respectively. The type 2 enzymes in eukaryotic cells cannot underwind DNA (introduce negative supercoils), although both types can relax both positive and negative supercoils. We will consider one probable origin of negative supercoils in eukaryotic cells in our discussion of chromatin.

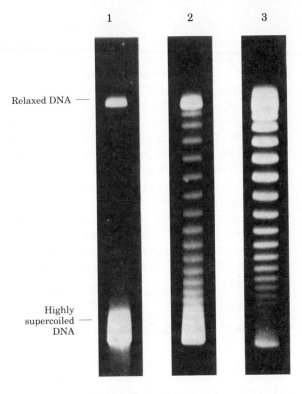

Figure 23–19 Circular DNA molecules that differ in linking number can be separated by gel electrophoresis. All the DNA molecules shown here have the same number of base pairs. Because supercoiled DNA molecules are more compact, they migrate more rapidly in a gel than the corresponding relaxed molecules. Gels such as those shown here separate topoisomers only over a limited range of superhelical density, so that highly supercoiled DNA migrates in a single band (lane 1) even though many different topoisomers may be present. Lanes 2 and 3 illustrate the effect of treating the supercoiled DNA with a type I topoisomerase (the DNA in lane 3 was treated for a longer time than that in lane 2).

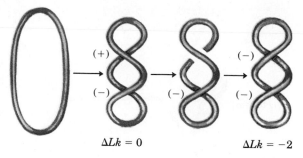

Figure 23–20 *E. coli* topoisomerase II (DNA gyrase) alters the linking number of circular DNA molecules by an unusual mechanism. Two regions of a DNA molecule are overlaid in a specific configuration in the bound complex (a positive (+) node). A compensating (−) node forms spontaneously elsewhere in the DNA molecule. As shown, both strands of one DNA segment are broken, the other segment is passed through the break, and the break is then resealed. The product now contains two minus nodes, and a comparison with Fig. 23–16 shows that the DNA now contains two negative supercoils. The change in structure reflects a change in Lk of -2.

DNA Compaction Requires a Special Form of Supercoiling

Supercoiled DNA molecules are remarkably uniform in many respects; the characteristic form is illustrated in Figure 23–21. Supercoils are right-handed in a negatively supercoiled DNA molecule (Fig. 23–16). Supercoiled DNA also tends to be extended and narrow rather than compacted, and it often exhibits multiple branches. At superhelical densities normally encountered in cells, the length of the supercoil axis, including branches, is about 40% the length of the DNA itself. This type of supercoiling is referred to as plectonemic (from the Greek *plektos,* "twisted," and *nema,* "thread") supercoiling.

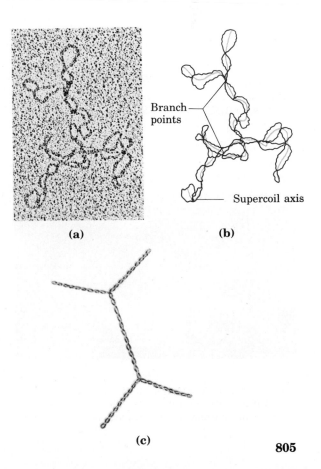

Figure 23–21 (a) An electron micrograph of plectonemically supercoiled plasmid DNA with (b) an interpretation of the observed structure. The blue lines define the axis of the supercoil. Note the branching of this molecule. (c) An idealized representation of this structure.

805

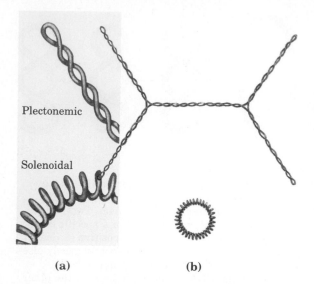

(a)   **(b)**

Figure 23–22 (a) Plectonemic and solenoidal forms of supercoiling. Solenoidal negative supercoiling takes the form of tight left-handed turns about an imaginary tubelike structure. The two forms are readily interconverted, although the solenoidal form is generally not observed unless certain proteins are bound to the DNA. **(b)** Plectonemic and solenoidal supercoiling of the same DNA molecule, drawn to scale. Note that solenoidal supercoiling provides a much greater degree of compaction.

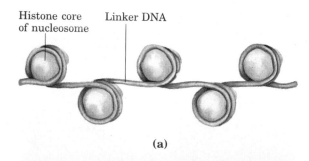

(a)

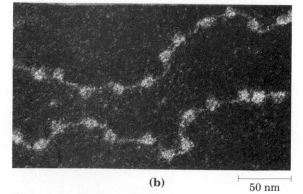

(b)   50 nm

Figure 23–23 Regularly spaced nucleosomes, consisting of histone complexes bound to DNA. **(a)** Schematic illustration; **(b)** electron micrograph.

Although plectonemic coiling is the form observed in underwound DNAs in solution, it does not give the compaction required to package DNA in the cell. A second form of supercoiling, called **solenoidal supercoiling** (Fig. 23–22), can be adopted by an underwound DNA. Instead of the extended right-handed supercoils characteristic of the plectonemic form, solenoidal supercoiling involves tighter, left-handed turns. The structure is similar to that taken up by a garden hose neatly wrapped on a reel. Although their structures are dramatically different, plectonemic and solenoidal supercoiling represent two forms of negative supercoiling that can be taken up by the same underwound DNA. The two forms are readily interconvertible. Although the plectonemic form is more stable in solution, the solenoidal form can be stabilized by protein binding and is the form found in chromatin. It provides a much greater degree of compaction (Fig. 23–22b). Solenoidal supercoiling explains how underwinding contributes to actual DNA compaction.

Chromatin and Nucleoid Structure

The term chromosome today refers to the nucleic acid molecule that is the repository of the genetic information of a virus, a bacterium, a eukaryotic cell, or an organelle. But the word chromosome was originally used in another sense, to refer to the densely colored bodies in eukaryotic nuclei that can be visualized with the light microscope after the cells are stained with a dye. Eukaryotic chromosomes, in the original sense of the word, appear as sharply defined bodies in the nucleus during the period just before and during mitosis, the process of nuclear division in somatic cells (see Fig. 2–14). In nondividing eukaryotic cells, the chromosomal material, called **chromatin,** is amorphous and appears to be randomly dispersed throughout the nucleus. But when the cells prepare to divide, the chromatin condenses and assembles itself into a species-specific number of well-defined chromosomes (see Fig. 23–4).

Chromatin has been isolated and analyzed. It consists of fibers that contain protein and DNA in approximately equal masses, plus a small amount of RNA. The DNA in the chromatin is very tightly associated with proteins called **histones,** which package and order the DNA into structural units called **nucleosomes** (Fig. 23–23). Also found in chromatin are many nonhistone proteins, some of which regulate the expression of specific genes (Chapter 27). Beginning with nucleosomes, eukaryotic chromosomal DNA is packaged into a succession of higher-order structures that ultimately yield the compact chromosome seen with the light microscope. We now turn to a description of this structure in eukaryotes, and compare the DNA packaging in bacterial cells.

Histones Are Small, Basic Proteins

Found in the chromatin of all eukaryotic cells, histones have molecular weights of between 11,000 and 21,000 and are very rich in the basic amino acids arginine and lysine (together these make up about one-fourth of the amino acid residues). Five major classes of histones are found in all eukaryotic cells, differing in molecular weight and amino acid composition (Table 23–3). The H3 histones are nearly identical in amino acid sequence in all eukaryotes, as are the H4 histones, suggesting strict conservation of their functions. Comparing the 102 amino acid H4 histones, for example, only two differences are found in the H4 molecules of peas and cows, and only eight differences in those of humans and yeast. Histones H1, H2A, and H2B show a lesser degree of sequence homology between eukaryotic species.

Each of the histones can exist in different forms because certain amino acid side chains are enzymatically modified by methylation, ADP-ribosylation, phosphorylation, or acetylation. Such modifications change the histone molecules' net electric charge, shape, and other properties, but the functional significance of the changes is not well understood.

Table 23–3 Histones

Histone	Molecular weight	Number of amino acid residues	Content of basic amino acids (as % of total)	
			Lys	Arg
H1*	21,130	223	29.5	1.3
H2A*	13,960	129	10.9	9.3
H2B*	13,774	125	16.0	6.4
H3	15,273	135	9.6	13.3
H4	11,236	102	10.8	13.7

* The size of these histones varies somewhat from species to species. The numbers given here are for bovine histones.

Nucleosomes Are the Fundamental Organizational Units in Chromatin

The eukaryotic chromosome depicted in Figure 23–4 represents the compaction of a DNA molecule about 10^5 μm long into a cell nucleus that is typically 5 to 10 μm in diameter. This compaction involves several layers of highly organized folding. Subjecting chromosomes to treatments that partially unfold them reveals a structure in which the DNA is bound tightly to beads of protein that are often regularly spaced (Fig. 23–23). The "beads" in this "beads-on-a-string" arrangement are complexes of histones and DNA called nucleosomes. They are the fundamental units of organization upon which the higher-order packing of chromatin is built. Each nucleosome contains eight histone molecules, two copies each of H2A, H2B, H3, and H4. The spacing of the nucleosome beads along the DNA defines a repeating unit typically of about 200 base pairs, of which 146 base pairs are bound tightly around the histone core and the remainder serve as a linker between nucleosomes. Histone H1 is not part of the nucleosome core, but it is generally bound to the linker DNA. When chromatin is treated with enzymes that digest DNA, the linker DNA is degraded, releasing nucleosome particles. Each particle contains 146 base pairs of bound DNA that are protected from digestion. Nucleosomes obtained in this way have been crystallized and studied by x-ray diffraction analysis. This has revealed a particle made up of the eight histone molecules, with the DNA wrapped around it in the form of a left-handed solenoidal supercoil (Fig. 23–24).

A close inspection of this structure can explain why eukaryotic DNA is underwound even though eukaryotic cells lack enzymes that underwind DNA. Recall that the solenoidal wrapping of DNA seen in nucleosomes is one form taken up by underwound (negatively super-

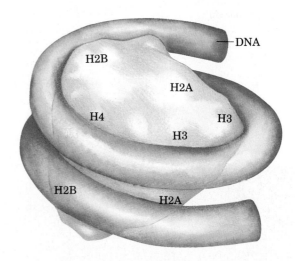

Figure 23–24 Structure of DNA (146 base pairs) wrapped around a nucleosome core. The DNA is bound in a left-handed solenoidal supercoil that circumnavigates the histone complex 1.8 times.

Figure 23–25 (a) Chromatin assembly on relaxed, closed-circular DNA. (b) Binding of a histone core to form a nucleosome will induce one negative supercoil; but in the absence of any strand breaks, a positive supercoil must also form elsewhere in the DNA ($\Delta Lk = 0$). (c) Relaxation of this positive supercoil by cellular topoisomerases leaves one net negative supercoil ($\Delta Lk = -1$).

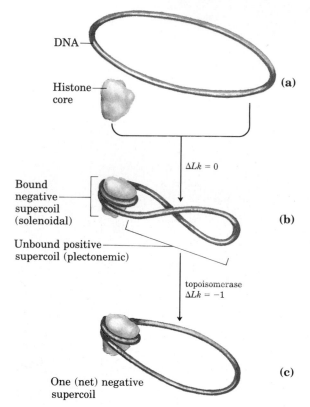

DNA

Histone core

(a)

$\Delta Lk = 0$

Bound negative supercoil (solenoidal)

Unbound positive supercoil (plectonemic)

(b)

topoisomerase
$\Delta Lk = -1$

One (net) negative supercoil

(c)

coiled) DNA. Wrapping DNA tightly around histone cores in nucleosome particles (Fig. 23–24) requires the removal of about one helical turn in the DNA to accommodate the tight turns. When the protein core of a nucleosome binds in vitro to a relaxed, closed-circular DNA, the binding will introduce a negative supercoil. This binding process does not break the DNA or change the linking number, however, so that formation of a negative solenoidal supercoil must be accompanied by a compensatory unbound positive supercoil elsewhere in the DNA (Fig. 23–25). The eukaryotic topoisomerases, unlike the bacterial DNA gyrase, cannot underwind DNA but they can relax positive supercoils. Relaxing the unbound positive supercoil leaves the negative supercoil fixed by virtue of nucleosome binding, and results in a net decrease in linking number. Not surprisingly, topoisomerases have proved necessary for assembling chromatin from histones and intact circular DNA in a test tube.

Another factor important in the binding of DNA to histones in nucleosomes is the sequence of the bound DNA. The histone cores do not bind randomly to the DNA, but nucleosomes tend to position themselves at certain locations. This positioning is not understood in all cases, but part of the explanation appears to be that nucleosomes form where A=T base pairs are abundant wherever the minor groove of the DNA helix (see p. 334) contacts the nucleosome core (Fig. 23–26). The tight wrapping of the DNA around the protein core requires compression of the minor groove at these points, and a cluster of two or three A=T base pairs makes this compression easier.

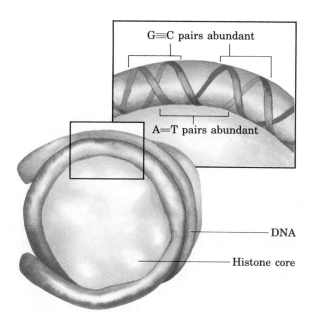

G≡C pairs abundant

A=T pairs abundant

DNA

Histone core

Figure 23–26 The positioning of a nucleosome to make optimal use of A=T base pairs where the histone core is in contact with the minor groove of the DNA.

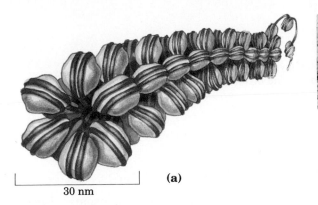

(a)

30 nm

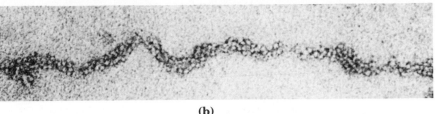

(b)

Figure 23–27 Probable structure of the 30 nm fiber, a higher-order organization of nucleosomes. **(a)** A schematic illustration of nucleosome packing in the fiber; **(b)** an electron micrograph of this structure.

Nucleosomes Are Packed into Successively Higher-Order Structures

Wrapping DNA about a nucleosome core compacts it about sevenfold. The total compaction in a chromosome is greater than 10,000-fold, which in itself provides ample evidence for even higher orders of structural organization. In chromosomes isolated by very gentle methods, nucleosomes themselves appear to be organized to form a structure called simply a 30 nm fiber (Fig. 23–27). Packing requires one molecule of histone H1 per nucleosome, although it is unclear where the H1 is bound. Organization into 30 nm fibers does not extend over the entire chromosome, but is punctuated by regions that are bound by sequence-specific (nonhistone) DNA-binding proteins. The structure observed also appears to depend on the transcriptional activity of the particular region of DNA. Regions containing genes that are being transcribed are apparently in a less-ordered state that contains little, if any, histone H1.

The 30 nm fibers provide an approximately 100-fold compaction of the DNA. The next level of folding is not yet understood, but it appears that certain regions of the DNA associate with a nuclear scaffold (Fig. 23–28). The scaffold-associated regions are separated by loops of DNA with perhaps 20,000 to 100,000 base pairs. The DNA in these loops may contain a set of related genes. For example, in *Drosophila,* complete sets of histone-coding genes seem to be clustered together in loops that are bounded by scaffold attachment sites (Fig. 23–29). The scaffold itself appears to contain several proteins, notably large amounts of histone H1 and topoisomerase II. The presence of topoisomerase II further emphasizes the important relationship between DNA underwinding and chromatin assembly. Evidence exists for additional layers of organization in eukaryotic chromosomes, each enhancing the degree of compaction multiplicatively. One model for this is illustrated in Figure 23–30. The principle is straightforward: DNA compaction in eukaryotic chromosomes is likely to involve coils upon coils upon coils. . .

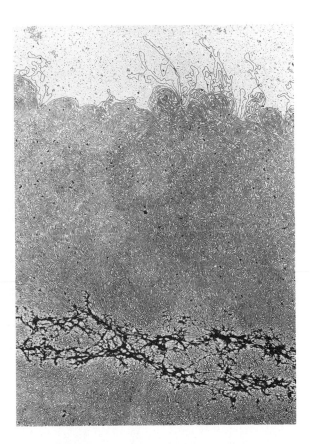

Figure 23–28 A partially unraveled human chromosome, revealing numerous loops of DNA attached to a scaffoldlike structure.

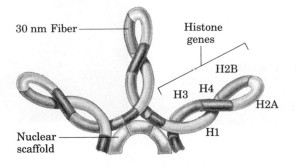

30 nm Fiber

Histone genes

H2B

H3 H4

H2A

H1

Nuclear scaffold

Figure 23–29 A schematic illustration of loops of chromosomal DNA attached to a nuclear scaffold. The DNA in the loops is packaged as 30 nm fibers, so that the loops represent the next level of organization. Within loops there are often groups of genes with related functions. Complete sets of histone-coding genes, as shown here, appear to be clustered in loops of this kind. Unlike most genes, all of the histone genes occur in multiple copies in the genomes of many eukaryotes.

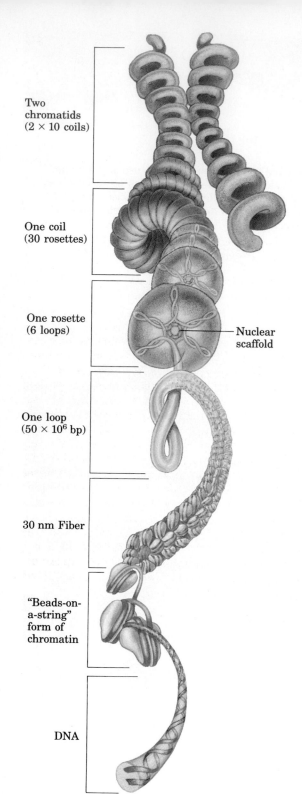

Two chromatids
(2 × 10 coils)

One coil
(30 rosettes)

One rosette
(6 loops)

Nuclear scaffold

One loop
(50 × 10⁶ bp)

30 nm Fiber

"Beads-on-a-string" form of chromatin

DNA

Figure 23–30 A model for layers of organization in a eukaryotic chromosome. The layers take the form of coils upon coils.

Bacterial DNA Is Also Highly Organized

We now turn briefly to the structure of bacterial chromosomes. Bacterial DNA is compacted in a structure called the **nucleoid,** which occupies a large fraction of the bacterial cell's volume (Fig. 23–31). The DNA of bacterial cells appears to be attached at one or more points to the inner surface of the plasma membrane. Much less is known about the structure of the nucleoid than of eukaryotic chromatin. In *E. coli,* a scaffoldlike structure appears to exist that organizes the circular chromosome into a series of looped domains, as described above for chromatin. The local organization provided by nucleosomes in eukaryotes does not seem to be duplicated by any comparable structure in bacterial DNA. Histonelike proteins are abundant in *E. coli,* and the best-characterized example is a protein with two subunits called HU (M_r 19,000). However, these proteins bind and dissociate on a time scale of minutes, and no regular, stable structure has been found. The bacterial chromosome is a relatively dynamic structure, possibly reflecting a requirement for more ready access to the genetic information it contains. The bacterial cell division cycle can be as short as 15 min, whereas a typical eukaryotic cell may not divide for many months. In addition, structural genes account for a much greater fraction of prokaryotic DNA, and high rates of cellular metabolism in bacteria mean that a much higher proportion of the DNA is being transcribed or replicated at a given time than in most eukaryotic cells.

With this overview of the complexity of DNA structure, we are now ready to turn to a discussion of DNA metabolism.

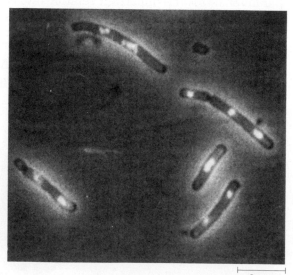

Figure 23–31 *E. coli* cells. The DNA is stained with a dye that fluoresces when exposed to UV light. The light area defines the nucleoid. Note that some cells have replicated their DNA but have not yet undergone cell division, and hence have multiple nucleoids.

2 μm

Summary

The DNA molecules in chromosomes are the largest macromolecules in cells. Many smaller DNAs also occur in cells, in the form of viral DNAs, plasmids, and (in eukaryotes) mitochondrial or chloroplast DNAs. Many DNAs, especially those in bacteria, mitochondria, and chloroplasts, are circular. Viral and chromosomal DNAs have one major feature in common: they are generally much longer than the viral particles or cells in which they are packaged. The total DNA content of a eukaryotic cell is much greater than that of a bacterial cell.

Genes are segments of a chromosome that contain the information for a functional polypeptide or RNA molecule. In addition to these structural genes, chromosomes contain a variety of regulatory sequences involved in replication, transcription, and other processes. In eukaryotic chromosomes, there are two important special-function repetitive DNA sequences: centromeres, which are attachment points for the mitotic spindle, and telomeres, which occur at the ends of the linear chromosomes. Many genes in eukaryotic cells, and occasionally in bacteria, are interrupted by noncoding sequences called introns. The coding segments separated by introns are called exons.

Most cellular DNAs are supercoiled. Supercoiling is a manifestation of structural strain imparted by the underwinding of the DNA molecule. Underwinding is a decrease in the total number of helical turns in the DNA relative to the relaxed or B form. To maintain an underwound state, DNA must be a closed circle or be bound with protein. Supercoils resulting from underwinding are defined as negative supercoils. Underwinding is quantified by a topological parameter called linking number, Lk. The linking number of a relaxed, closed-circular DNA is used as a reference (Lk_0) and is equal to the number of base pairs divided by 10.5. Underwinding is measured in terms of the specific linking difference or σ, which equals $(Lk - Lk_0)/Lk_0$. For cellular DNAs, σ typically equals -0.05 to -0.07, which means that approximately 5 to 7% of the helical turns in the DNA have been removed. DNA underwinding facilitates strand separation for processes such as transcription or replication. The plectonemic supercoils in negatively supercoiled DNA in solution are right-handed, and the overall structure is narrow and extended. An alternative form called solenoidal supercoiling provides a much greater degree of compaction, and this form predominates in the cell.

DNAs that differ only in their linking number are called topoisomers. The enzymes that underwind and/or relax DNA are called topoisomerases, and they act by catalyzing changes in linking number. There are two classes, type 1 and type 2, which change Lk in increments of 1 or 2, respectively. In a bacterial cell, the superhelical density of the DNA represents a regulated balance between the activities of topoisomerases that increase and decrease linking number.

In the chromatin of eukaryotic cells, the fundamental unit of organization is the nucleosome, which consists of DNA and a protein particle containing eight histones, two copies each of histones H2A, H2B, H3, and H4. The segment of DNA (about 146 base pairs) wrapped around the protein core is in the form of a left-handed solenoidal supercoil. Nucleosomes are organized into 30 nm fibers, and the fibers themselves are extensively folded to provide the 10,000-fold compaction required to fit a typical eukaryotic chromosome into a cell nucleus. The higher-order folding involves attachment to a nuclear scaffold that contains large amounts of histone H1 and topoisomerase II. Bacterial chromosomes are also extensively compacted into a structure called a nucleoid, but the chromosome appears to be much more dynamic and irregular in structure than eukaryotic chromatin, reflecting the shorter cell cycle and very active metabolism of a bacterial cell.

Further Reading

General

Alberts, B., Bray, D., Lewis, J., Raff, M., Roberts, K., & Watson, J.D. (1989) *Molecular Biology of the Cell,* 2nd edn, Garland Publishing, Inc., New York.
An excellent general reference.

Kornberg, A. & Baker, T.A. (1991) *DNA Replication,* 2nd edn, W.H. Freeman and Company, New York.
A good place to start for further information on the structure and function of DNA.

Singer, M. & Berg, P. (1991) *Genes and Genomes: A Changing Perspective,* University Science Books, Mill Valley, CA.
An up-to-date discussion of genes, chromosome structure, and many other topics.

Genes and Chromosomes

Blackburn, E.H. (1990) Telomeres: structure and synthesis. *J. Biol. Chem.* **265,** 5919–5921.

Jelinek, W.R. & Schmid, C.W. (1982) Repetitive sequences in eukaryotic DNA and their expression. *Annu. Rev. Biochem.* **51,** 813–844.

Murray, A.W. & Szostak, J.W. (1987) Artificial chromosomes. *Sci. Am.* **257** (November), 62–68.

Novick, R.P. (1980) Plasmids. *Sci. Am.* **243** (December), 102–127.

Sharp, P.A. (1985) On the origin of RNA splicing and introns. *Cell* **42,** 397–400.

Ullu, E. & Tschudi, C. (1984) *Alu* sequences are processed 7SL RNA genes. *Nature* **312,** 171–172.

Supercoiling and Topoisomerases

Bauer, W.R., Crick, F.H.C., & White, J.H. (1980) Supercoiled DNA. *Sci. Am.* **243** (July), 118–133.

Boles, T.C., White, J.H., & Cozzarelli, N.R. (1990) Structure of plectonemically supercoiled DNA. *J. Mol. Biol.* **213,** 931–951.
A study that defines several fundamental features of supercoiled DNA.

Cozzarelli, N.R., Boles, T.C., & White, J.H. (1990) Primer on the topology and geometry of DNA supercoiling. In *DNA Topology and Its Biological Effects* (Cozzarelli, N.R. & Wang, J.C., eds), pp. 139–184, Cold Spring Harbor Laboratory Press, Cold Spring Harbor, NY.
This provides a more advanced and thorough discussion.

Lebowitz, J. (1990) Through the looking glass: the discovery of supercoiled DNA. *Trends Biochem. Sci.* **15,** 202–207.
A short and interesting historical note.

Liu, L.F. (1989) DNA topoisomerase poisons as antitumor drugs. *Annu. Rev. Biochem.* **58,** 351–375.
A review of eukaryotic topoisomerases and the use of topoisomerase inhibitors in cancer chemotherapy.

Wang, J.C. (1985) DNA topoisomerases. *Annu. Rev. Biochem.* **54,** 665–697.

Wang, J.C. (1991) DNA topoisomerases: Why so many? *J. Biol. Chem.* **266,** 6659–6662.
A good short summary of topoisomerase functions.

Chromatin and Nucleosomes

Filipski, J., Leblanc, J., Youdale, T., Sikorska, M., & Walker, P.R. (1990) Periodicity of DNA folding in higher order chromatin structures. *EMBO J.* **9,** 1319–1327.

Kornberg, R.D. (1974) Chromatin structure: a repeating unit of histones and DNA. *Science* **184,** 868–871.
The classic paper that introduced the subunit model for chromatin.

Richmond, T.J., Finch, J.T., Rushton, B., Rhodes, D., & Klug, A. (1984) Structure of the nucleosome core particle at 7Å resolution. *Nature* **311,** 532–537.

van Holde, K.E. (1989) *Chromatin,* Springer-Verlag, New York.

Problems

1. *How Long Is the Ribonuclease Gene?* What is the minimum number of nucleotide pairs in the gene for pancreatic ribonuclease (124 amino acids long)? Suggest a reason why the number of nucleotide pairs in the gene might be much larger than your answer.

2. *Packaging of DNA in a Virus* The DNA of bacteriophage T2 has a molecular weight of 120×10^6. The head of the T2 phage is about 210 nm long. Assuming the molecular weight of a nucleotide pair is 650, calculate the length of T2 DNA and compare it with the length of the T2 head. Your answer will show the necessity of very compact packaging of DNA in viruses (see Fig. 23–1).

3. *The DNA of Phage M13* Bacteriophage M13 DNA has the following base composition: A, 23%; T, 36%; G, 21%; C, 20%. What does this information tell us about the DNA of this phage?

4. *Base Composition of ϕX174 DNA* Bacteriophage ϕX174 DNA occurs in two forms, single-stranded in the isolated virion and double-stranded during viral replication in the host cell. Would you expect them to have the same base composition? Give your reasons.

5. *Size of Eukaryotic Genes* An enzyme present in rat liver has a polypeptide chain of 192 amino acid residues. It is coded for by a gene having 1,440 base pairs. Explain the relationship between the number of amino acid residues in this enzyme and the number of nucleotide pairs in its gene.

6. *DNA Supercoiling* A covalently closed circular DNA molecule has an *Lk* of 500 when it is relaxed. Approximately how many base pairs are in this DNA? How will the linking number be altered (increase, decrease, no change, become undefined) if (a) a protein complex is bound to form a nucleosome, (b) one DNA strand is broken, (c) DNA gyrase is added with ATP, or (d) the double helix is denatured (base pairs are separated) by heat?

7. *DNA Structure* Explain how the underwinding of a B-DNA helix might facilitate or stabilize the formation of Z-DNA.

8. *Chromatin* One of the important early pieces of evidence that helped define the structure of the nucleosome is illustrated by the agarose gel shown below, in which the thick bands represent DNA. It was generated by treating chromatin briefly with an enzyme that degrades DNA, then removing all protein and subjecting the purified DNA to electrophoresis. Numbers at the side of the gel denote the position to which a linear DNA of the indicated size (in base pairs) would migrate. What does this gel tell you about chromatin structure? Why are the DNA bands thick and spread out rather than sharp?

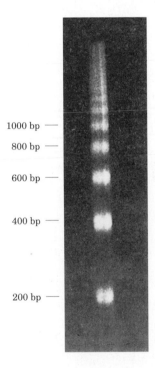

1000 bp —
800 bp —
600 bp —
400 bp —
200 bp —

DNA Metabolism

As the repository of genetic information, DNA occupies a unique and central place among biological macromolecules. The nucleotide sequences of DNA ultimately describe the primary structures of all cellular RNAs and proteins, and through enzymes can indirectly affect the synthesis of all other cellular constituents, determining the size, shape, and function of every living thing.

The structure of DNA is a marvelous device for the stable storage of genetic information. The phrase "stable storage," however, conveys a static and incomplete picture of the biochemical role of DNA in the cell. A proper description of DNA function must also explain how that information is transmitted from one generation of cells to the next. The term "DNA metabolism" can be used to describe the process by which faithful copies of DNA molecules are made (replication), along with the processes that affect the structure of the information within (repair and recombination). Together they are the focus of this chapter.

Perhaps more than any other factor, it is the requirement for an exquisite degree of accuracy that shapes these processes. At the level of joining one nucleotide to the next, the chemistry of DNA replication is simple and elegant, almost deceptively so. But as we will see, the synthesis of all macromolecules that contain information involves complex devices to ensure that the information is transmitted intact. If left uncorrected, errors in DNA synthesis can have dire consequences because they are essentially permanent. The enzymes that synthesize DNA must copy DNA molecules that often contain millions of bases, and they do so with great fidelity and speed. They must also act on a DNA substrate that is highly compacted and bound with other proteins. The enzymes that catalyze the formation of phosphodiester bonds are therefore only part of an elaborate system involving myriad proteins and enzymes.

The importance of maintaining the integrity of the information stored in DNA is underscored when the discussion turns to repair. As detailed in Chapter 12, DNA is susceptible to many types of damaging reactions. Though generally slow, they are nevertheless significant because of the very low biological tolerance for changes in DNA sequence. DNA is the only macromolecule for which repair systems exist, and their number, diversity, and complexity reflect the wide range of insults to which a DNA molecule is subject.

The processes by which genetic information is rearranged, collectively called recombination, seem to belie the principles just established. If the integrity of the genetic information is paramount, why rearrange it? One explanation seems to be the need for maintaining a level of genetic diversity by providing new combinations of alleles, the alternative forms of a single gene. Even without this explanation, however, recombination is not really so renegade a set of processes. Most recombination events are conservative in the sense that genetic information is neither lost nor gained. Indeed, with a closer look at a recombination event, one often finds a DNA repair or gene regulation process in disguise.

Special emphasis is given in this chapter to the enzymes that catalyze these processes. They are well worth getting acquainted with if for no other reason than their everyday use as reagents in a wide range of modern biochemical technologies. Because many of the seminal discoveries in DNA metabolism have been made with *E. coli*, the well-understood enzymes obtained from this bacterium are generally used here to illustrate the ground rules. A quick look at the relevant genes on the *E. coli* genetic map (Fig. 24–1) provides just a hint of what is to come.

Figure 24–1 A map of the *E. coli* chromosome, showing the relative positions of genes encoding some of the proteins important in DNA metabolism. The number of known genes involved provides a hint of the complexity of these processes. The numbers 0 to 100 denote a genetic measurement called minutes, with each minute corresponding to about 40,000 base pairs. The acronyms consisting of three lowercase letters generally reflect some aspect of the gene's function. These include *mut, mut*agenesis; *dna,* DNA replication; *pol,* DNA *pol*ymerase; *rpo, R*NA *po*lymerase; *uvr, UV*-resistance; *rec, re*combination; *ter, ter*mination of replication; *ori, ori*gin of replication; *dam, D*NA *a*denine *m*ethylation; *lig,* DNA *lig*ase; *cou, cou*mermycin resistance; and *nal, nal*idixic acid resistance (coumermycin and nalidixic acid inhibit DNA replication by binding to the subunits of DNA gyrase encoded by these genes).

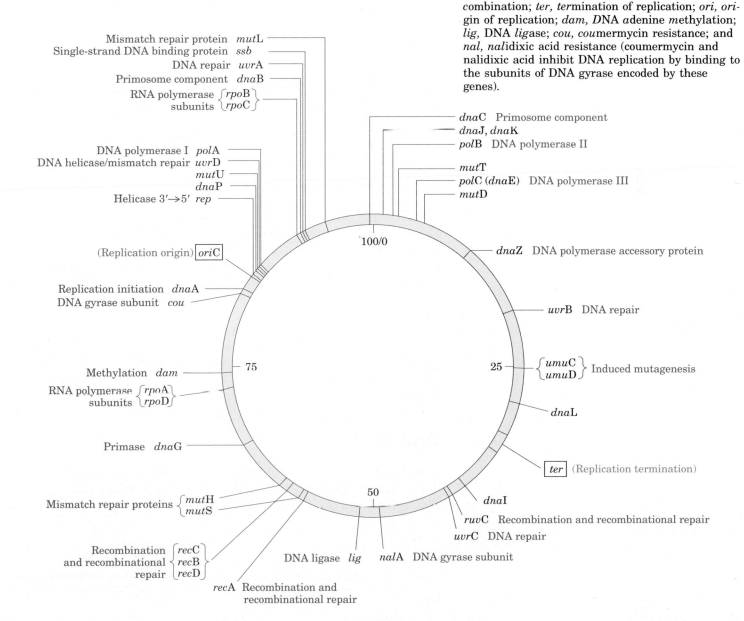

Before moving on to replication, we must entertain two short digressions. The first concerns the use of acronyms in naming genes and proteins. Bacterial genetics is a powerful tool that has facilitated much of the work described in this chapter. Bacterial genes that affect a given cellular process such as replication often have been identified before the roles of their protein products were understood. By convention, acronyms used to identify bacterial (and sometimes eukaryotic) genes are generally three lowercase, italicized letters that reflect function, such as *dna, uvr,* or *rec* for genes that affect *DNA* replication, *r*esistance to the damaging effects of *UV* radiation, or *rec*ombination, respectively. In the case of multiple genes that affect the same process, the designation A, B, C, etc., is added, usually reflecting the temporal order of gene discovery rather than a reaction sequence. In most cases, the protein product of each gene is ultimately isolated and characterized. Sometimes the product is identified as a previously isolated protein. The *dna*E gene, for example, was found to encode the polymerizing subunit of DNA polymerase III; consequently, the *dna*E gene was renamed *pol*C to reflect that function more clearly. In many cases the protein product has turned out to be novel, with an activity not easily described by a simple enzyme name. In a practice that can be confusing, these proteins often retain the name of their genes; for example, the products of the *dna*A and *rec*A genes are simply called the DnaA and RecA proteins, respectively. Many examples of this practice are found in this chapter. Here we use the convention that names in italics refer to genes or important DNA sequences, and roman type is used when the name refers to a protein.*

The second digression is needed to introduce enzymes that degrade DNA rather than synthesize it, because directed DNA degradation plays a significant role in all of the processes described in this chapter. These enzymes are called **nucleases** or, alternatively, **DNases** if they are specific for DNA. Every cell contains several different nucleases, and these fall into two broad classes: exonucleases and endonucleases. **Exonucleases** degrade DNA from one end of the molecule. Many are specific for degradation in either the $5' \rightarrow 3'$ or $3' \rightarrow 5'$ direction; that is, they remove nucleotides specifically from the 5' or 3' end, respectively, of one strand of a double-stranded nucleic acid (see Fig. 12–7). **Endonucleases** act in the interior of nucleic acids, reducing them to smaller and smaller fragments. A few exonucleases and endonucleases degrade only single-stranded DNA. There are also a few important classes of endonucleases that cleave only at specific nucleotide sequences (e.g., the restriction endonucleases considered in Chapter 28). Many types of nucleases will be encountered in this and subsequent chapters.

DNA Replication

Long before the structure of DNA became known, scientists had wondered first at the ability of organisms to create reasonable copies of themselves, and later at the ability of cells to produce many identical

---

* For eukaryotic proteins, these naming conventions are somewhat different and vary sufficiently from one organism to the next that no single convention can be presented here.

copies of large and complex macromolecules. Speculation about these problems centered around the concept of a **template.** The molecular template had to be a surface upon which molecules could be lined up in a specific order and joined to create a macromolecule with a unique structure and function.

The process of DNA replication provided the first biological example of the use of a molecular template to guide the synthesis of a macromolecule. The 1940s brought the revelation that DNA was the genetic molecule, but not until James Watson and Francis Crick deduced its structure did it become clear how DNA could act as a template for the replication and transmission of genetic information. *One strand is the complement of the other.* The strict base-pairing rules mean that the use of one strand as a template will result in another strand with a predictable, complementary sequence.

The fundamental properties of the DNA replication process and the mechanisms used by the enzymes that catalyze it have proven to be essentially identical in all organisms. This mechanistic unity will be a major theme as we proceed from general properties of the replication process to *E. coli* replication enzymes and finally to replication in eukaryotes.

DNA Replication Is Governed by a Set of Fundamental Rules

DNA Replication Is Semiconservative If each DNA strand serves as a template for the synthesis of a new strand, two new DNA molecules will result, each with one new strand and one old strand. This is called **semiconservative replication.**

The hypothesis of semiconservative replication was proposed by Watson and Crick soon after publication of their paper on the structure of DNA; the theory was proven in ingeniously designed experiments by Matthew Meselson and Franklin Stahl in 1957 (Fig. 24–2). Meselson and Stahl grew *E. coli* cells for many generations in a medium in which the sole nitrogen source (NH_4Cl) contained ^{15}N, the "heavy" isotope of nitrogen, instead of the normal, more abundant "light" isotope ^{14}N. The DNA isolated from these cells had a density about 1% greater than that of normal [^{14}N]DNA. Although this is only a small difference, a mixture of heavy [^{15}N]DNA and light [^{14}N]DNA can be separated by centrifugation to equilibrium in a cesium chloride density gradient.

The *E. coli* cells grown in the ^{15}N medium were transferred to a fresh medium containing only the ^{14}N isotope, where they were allowed to grow until the cell population had just doubled. The DNA isolated from these first-generation cells formed a single band in the CsCl gradient at a position indicating that the double-helical DNAs of the daughter cells were hybrids containing one new ^{14}N strand and one parental ^{15}N strand (Fig. 24–2).

This result argued against conservative replication, an alternative hypothesis in which one progeny DNA molecule would consist of two newly synthesized DNA strands and the other would contain the two parental strands; this would never yield hybrid DNA molecules in the Meselson–Stahl experiment. The semiconservative replication hypothesis was further supported in the next step of the experiment. Cells were allowed to double in number again in the ^{14}N medium, and the isolated DNA product of this second cycle of replication exhibited *two* bands, one having a density equal to that of light DNA and the other having the density of the hybrid DNA observed after the first cell doubling.

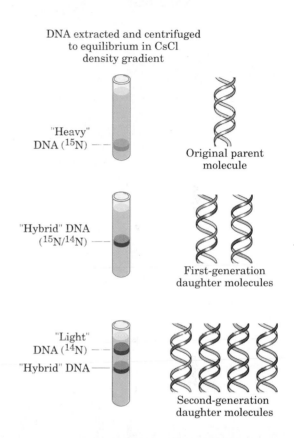

DNA extracted and centrifuged to equilibrium in CsCl density gradient

"Heavy" DNA (^{15}N) — Original parent molecule

"Hybrid" DNA ($^{15}N/^{14}N$) — First-generation daughter molecules

"Light" DNA (^{14}N) — "Hybrid" DNA — Second-generation daughter molecules

Figure 24–2 The Meselson–Stahl experiment was designed to distinguish between two alternative DNA replication mechanisms. Cells were grown for many generations in a medium containing only heavy nitrogen, ^{15}N, so that all the nitrogen in the DNA was ^{15}N. The cells were then transferred to a medium containing only light nitrogen, ^{14}N, and the density of the DNA was monitored closely for the next two cell generations. Cellular DNA was isolated after the first and second generations and centrifuged to equilibrium in a CsCl density gradient. The [^{15}N]DNA (shown in blue) came to equilibrium at a lower position in the CsCl gradient than [^{14}N]DNA (shown in red). Hybrid DNA equilibrated in an intermediate position. If DNA replication were conservative, each of the two heavy strands of parental DNA would be replicated to yield the original heavy duplex DNA and a DNA duplex containing two new light strands. Continuation of conservative replication would yield in the next generation one heavy DNA and three light DNAs but no hybrid DNAs. The Meselson–Stahl experiment, however, showed that replication is semiconservative, resulting in two daughter duplexes each containing one parental heavy strand and one new light strand. The next generation yielded two hybrid DNAs and two light DNAs.

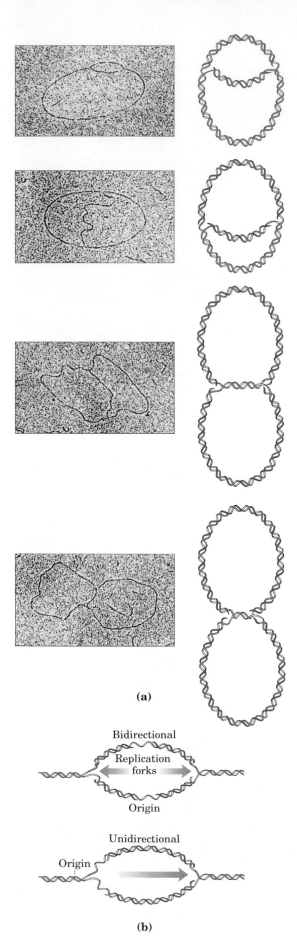

(a)

Bidirectional

Replication forks

Origin

Unidirectional

Origin

(b)

Figure 24–3 Replication of a circular chromosome produces a structure resembling the Greek letter theta (θ). **(a)** Labeling with tritium (^{3}H) shows that both strands are replicated at the same time (new strands shown in red). The electron micrographs illustrate the replication of a circular *E. coli* plasmid as visualized by autoradiography. **(b)** Addition of ^{3}H for a short period just before the reaction is stopped allows a distinction to be made between unidirectional and bidirectional replication, by determining whether label (in red) is found at one or both replication forks seen in autoradiograms. This technique has revealed bidirectional replication in *E. coli, B. subtilis,* and other bacteria. **(c)** Autoradiogram of a replicating *E. coli* chromosome taken from a culture grown for two generations in [^{3}H]thymidine.

Replication Begins at an Origin and Usually Proceeds Bidirectionally

A host of questions now arises. Are the parental DNA strands completely unwound before each is replicated? Does replication begin at random places or at a unique point? After initiation at any point in the DNA, does replication proceed in one direction or both? An early indication that replication is a highly coordinated process in which the parental strands are unwound and replicated simultaneously was provided by John Cairns using the technique of autoradiography. He made the DNA of *E. coli* cells radioactive by growing them in a medium containing thymidine labeled with tritium (^{3}H). When the DNA was carefully isolated, spread, and overlaid with a photographic emulsion, and left for several weeks, the radioactive thymidine residues generated "tracks" of silver grains in the emulsion, producing an image of the DNA molecule. These tracks revealed that the intact chromosome of *E. coli* is a single giant circle, 1.7 mm long (see Fig. 23–2). Radioactive DNA isolated from cells during replication showed an extra radioactive loop (Fig. 24–3). The amount of radioactivity in the loop relative to the remainder of the DNA led Cairns to conclude that the loop in the DNA was the result of the formation of two radioactive daughter strands, each complementary to a parent strand. One or both ends of the loop are dynamic points, termed **replication forks,** where parental DNA is being unwound and the separated strands quickly repli-

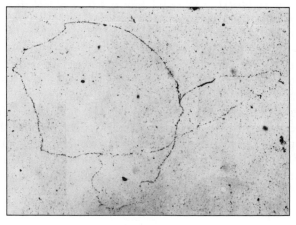

(c)

cated. This demonstrated that both DNA strands are replicated simultaneously, and a variation of this experiment (Fig. 24–3b) indicated that replication of bacterial chromosomes is bidirectional: both ends of the loop have active replication forks.

To determine whether the loops originated at a unique point in the DNA, landmarks were needed in the DNA "string." These were provided by a technique called **denaturation mapping,** developed by Ross Inman and colleagues. Using the 48,502 base pair chromosome from bacteriophage λ, Inman showed that DNA could be selectively denatured at sequences unusually rich in A=T base pairs. This generates a reproducible pattern of single-stranded bubbles (see Fig. 12–30). When isolated DNAs containing replication loops are partially denatured in this way, the progress of the replication forks can be measured and mapped using the denatured regions as points of reference. The technique revealed that the replication loops always initiate at a unique point, called an **origin.** In addition, this work reinforced the earlier observation that replication is usually bidirectional. For circular DNA molecules, the two replication forks meet at a point on the side of the circle opposite to the origin.

DNA Synthesis Proceeds in a 5′→3′ Direction and Is Semidiscontinuous A new strand of DNA is always synthesized in the 5′→3′ direction (the 5′ and 3′ ends of a DNA strand are defined as shown in Figure 12–7). Because the two DNA strands are antiparallel, the strand acting as template is being read from its 3′ end toward its 5′ end.

If synthesis always proceeds in the 5′→3′ direction, how can both strands be synthesized simultaneously? If both were synthesized continuously as the replication fork moved, one would have to undergo 3′→5′ synthesis. This problem was resolved by Reiji Okazaki and colleagues in the 1960s. Okazaki found that one of the new DNA strands is synthesized in short pieces, now called **Okazaki fragments.** This work ultimately led to the conclusion that one strand is synthesized continuously and the other discontinuously (Fig. 24–4). The continuous or **leading strand** is the one in which 5′→3′ synthesis proceeds in the same direction as replication fork movement. The discontinuous or **lagging strand** is the one in which 5′→3′ synthesis proceeds in the direction opposite to the direction of fork movement. Okazaki fragments range in length from a few hundred to a few thousand nucleotides, depending on the cell type.

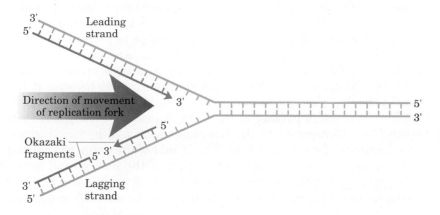

Figure 24–4 A new DNA strand (red) is always synthesized in the 5′→3′ direction. The template is copied in the opposite direction: 3′→5′. The strand that is continuously synthesized (in the direction taken by the replication fork) is the leading strand. The other strand, the lagging strand, is synthesized discontinuously in short pieces (Okazaki fragments) in a direction opposite to the direction of replication fork movement. The Okazaki fragments are then spliced together by DNA ligase. In bacteria the Okazaki fragments are about 1,000 to 2,000 nucleotides long. In eukaryotic cells they are 150 to 200 nucleotides long.

Arthur Kornberg

DNA Is Synthesized by DNA Polymerases

The search for an enzyme that could synthesize DNA was initiated in 1955 by Arthur Kornberg and colleagues. This work led to the purification and characterization of DNA polymerase from *E. coli* cells, a single-polypeptide enzyme now called **DNA polymerase I** (M_r 103,000). Much later, it was found that *E. coli* contains at least two other distinct DNA polymerases, which will be described below.

Detailed studies of DNA polymerase I revealed features of the DNA synthetic process that have proven to be common to all DNA polymerases. The fundamental reaction is a nucleophilic attack by the 3′-hydroxyl group of the nucleotide at the 3′ end of the growing strand on the 5′-α-phosphorus of the incoming deoxynucleoside 5′-triphosphate (Fig. 24–5). Inorganic pyrophosphate is released in the reaction. The general reaction equation is

$$\underset{\text{DNA}}{(\text{dNMP})_n} + \text{dNTP} \longrightarrow \underset{\substack{\text{Lengthened}\\\text{DNA}}}{(\text{dNMP})_{n+1}} + \text{PP}_i \qquad (24\text{–}1)$$

where dNMP and dNTP are deoxynucleoside 5′-monophosphate and 5′-triphosphate, respectively.

Figure 24–5 Elongation of a DNA chain. A single unpaired strand is required to act as template, and a primer strand is needed to provide a free 3′ end to which new nucleotide units are added. Each incoming nucleotide is selected by virtue of base pairing to the appropriate nucleotide in the template strand.

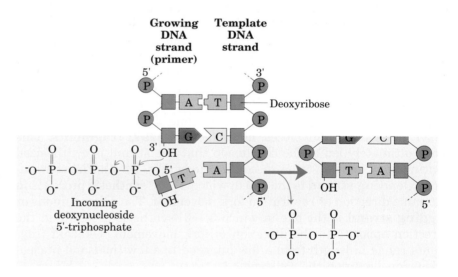

Early work on DNA polymerase I led to the definition of two central requirements for DNA polymerization. First, all DNA polymerases require a **template** (Fig. 24–5). The polymerization reaction is guided by a template DNA strand according to the base-pairing rules predicted by Watson and Crick: where a guanine is present in the template, a cytosine is added to the new strand, and so on. This was a particularly important discovery, not only because it provided a chemical basis for semiconservative DNA replication, but because it represented the first example of the use of a template to guide a biosynthetic reaction. Second, a **primer** is required. A primer is a segment of new strand (complementary to the template) with a 3′-hydroxyl group to which nucleotides can be added. The 3′ end of the primer is called the **primer terminus.** In other words, part of the new strand must already be in place; the polymerase can only add nucleotides to a preexisting strand. This has proven to be the case for all DNA polymerases, and this discovery provided an interesting wrinkle in the DNA replication story. No DNA-synthesizing enzyme can initiate synthesis of a new DNA strand. As we will see later in this chapter, enzymes that synthesize RNA do have the capability of initiating synthesis, and as a consequence, primers are often oligonucleotides of RNA.

After a nucleotide is added to a growing DNA strand, the DNA polymerase must either dissociate or move along the template and add another nucleotide. Dissociation and reassociation of the polymerase can limit the overall reaction rate, thus the rate generally increases if a polymerase adds additional nucleotides without dissociating from the template. The number of nucleotides added, on average, before a polymerase dissociates is defined as its **processivity.** DNA polymerases vary greatly in processivity, with some adding just a few nucleotides and others adding many thousands before dissociation occurs.

Polymerization Is a Thermodynamically Favorable Reaction

Throughout this book we have emphasized the importance of noncovalent as well as covalent interactions in biochemical processes. A discussion of the energetics of the polymerization reaction can be deceptive if only the covalent bonds are considered. The rearrangement of covalent bonds is straightforward: one phosphoric anhydride bond (in the dNTP) is hydrolyzed and one phosphodiester bond (in the DNA) is formed. This results in a slightly positive (unfavorable) change in standard free energy ($\Delta G^{\circ\prime} \simeq 2$ kJ/mol) for the overall reaction shown in Equation 24–1. Hydrolysis of the pyrophosphate to two molecules of inorganic phosphate by the pyrophosphatases present in all cells yields a $\Delta G^{\circ\prime}$ of -30 kJ/mol, and by coupling these two reactions the cell can provide a strong thermodynamic pull in the direction of polymerization, with a net $\Delta G^{\circ\prime}$ of -28 kJ/mol. This is important to the cell, but in this case it is not the whole story. If this calculation were complete, polymerases would tend to catalyze DNA degradation in the absence of pyrophosphate hydrolysis. Purified DNA polymerases, however, carry out polymerization very efficiently in vitro in the absence of pyrophosphatases. The explanation of this seeming paradox now is clear: noncovalent interactions not considered in the calculation above make an important thermodynamic contribution to the polymerization reaction. Every new nucleotide added to the growing chain is held there not just by the new phosphodiester bond but also by hydrogen bonds to its partner in the template and base-stacking interactions with the adjacent nucleotide in the same chain (p. 330). The additional energy released by these multiple weak interactions helps drive the reaction in the direction of polymerization.

DNA Polymerases Are Very Accurate

Replication *must* proceed with a very high degree of fidelity. In *E. coli*, a mistake is made only once for every 10^9 to 10^{10} nucleotides added. For the *E. coli* chromosome of about 4.7×10^6 base pairs, this means that an error will be made only once per 1,000 to 10,000 replications. During polymerization, discrimination between correct and incorrect nucleotides relies upon the hydrogen bonds that specify the correct pairing between complementary bases. Incorrect bases will not form the correct hydrogen bonds and can be rejected before the phosphodiester bond is formed. The accuracy of the polymerization reaction itself,

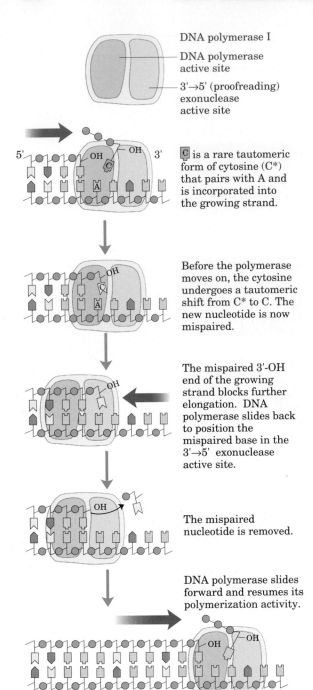

DNA polymerase I

DNA polymerase active site

3'→5' (proofreading) exonuclease active site

5'   OH   OH   3'

C is a rare tautomeric form of cytosine (C*) that pairs with A and is incorporated into the growing strand.

Before the polymerase moves on, the cytosine undergoes a tautomeric shift from C* to C. The new nucleotide is now mispaired.

The mispaired 3'-OH end of the growing strand blocks further elongation. DNA polymerase slides back to position the mispaired base in the 3'→5' exonuclease active site.

The mispaired nucleotide is removed.

DNA polymerase slides forward and resumes its polymerization activity.

Figure 24–6 An example of error correction by the 3'→5' exonuclease activity of DNA polymerase I. Structural analysis has located the exonuclease activity ahead of the polymerization activity as the enzyme is oriented in its movement along the DNA. A mismatched base (here, a C–A mismatch) impedes translocation of the enzyme to the next site. Sliding backward, the enzyme corrects the mistake with its 3'→5' exonuclease activity, then resumes its polymerase activity in the 5'→3' direction.

however, is insufficient to account for the high degree of fidelity in replication. Careful measurements in vitro have shown that DNA polymerases insert one incorrect nucleotide for every 10^4 to 10^5 correct ones. These mistakes sometimes occur because a base is briefly in an unusual tautomeric form (see Fig. 12–9), allowing it to hydrogen-bond with an incorrect partner. The error rate is reduced further in vivo by additional enzymatic mechanisms.

One mechanism intrinsic to virtually all DNA polymerases is a separate 3'→5' exonuclease activity that serves to double-check each nucleotide after it is added. This nuclease activity permits the enzyme to remove a nucleotide just added and is highly specific for mismatched base pairs (Fig. 24–6). If the wrong nucleotide has been added, translocation of the polymerase to the position where the next nucleotide is to be added is inhibited. The 3'→5' exonuclease activity removes the mispaired nucleotide, and the polymerase begins again. This activity, called **proofreading,** is not simply the reverse of the polymerization reaction, because pyrophosphate is not involved. The polymerizing and proofreading activities of a DNA polymerase can be measured separately. Such measurements have shown that proofreading improves the inherent accuracy of the polymerization reaction by 10^2- to 10^3-fold.

The discrimination between correct and incorrect bases during proofreading depends on the same base-pairing interactions that are used during polymerization. This strategy of enhancing fidelity by using complementary noncovalent interactions for discrimination twice in successive steps is common in the synthesis of information-containing molecules. A similar strategy is used to ensure the fidelity of protein synthesis (Chapter 26).

Overall, a DNA polymerase makes about one error for every 10^6 to 10^8 bases added. The measured accuracy of replication in *E. coli* cells, however, is still higher. The remaining degree of accuracy is accounted for by a separate enzyme system that repairs mismatched base pairs remaining after replication. This process, called mismatch repair, is described with other DNA repair processes later in this chapter.

E. coli Has at Least Three DNA Polymerases

More than 90% of the DNA polymerase activity in *E. coli* extracts can be accounted for by DNA polymerase I. Nevertheless, almost immediately after the isolation of this enzyme in 1955, evidence began to accumulate that it is not suited for replication of the large *E. coli* chromosome. First, the rate at which nucleotides are added by this enzyme (600 nucleotides/min) is too slow, by a factor of 20 or more, to account for observed rates of fork movement in the bacterial cell. Second, DNA polymerase I has a relatively low processivity; only about 50 nucleotides are added before the enzyme dissociates. Third, genetic studies have shown that many genes, and therefore many proteins, are involved in replication: DNA polymerase I clearly does not act alone. Finally, and most important, in 1969 John Cairns isolated a bacterial strain in which the gene for DNA polymerase I was altered, inactivating the enzyme. This strain was nevertheless viable!

A search for other DNA polymerases led to the discovery of *E. coli* **DNA polymerase II** and **DNA polymerase III** in the early 1970s. DNA polymerase II appears to have a highly specialized DNA repair function (described later in this chapter). DNA polymerase III is the primary replication enzyme in *E. coli*. Properties of the three DNA polymerases are compared in Table 24–1. DNA polymerase III is a much more complex enzyme than polymerase I. It is a multimeric enzyme with at least ten different subunits (Table 24–2). Notably, the polymerization and proofreading activities of DNA polymerase III are located in separate subunits. The β subunit of this complex enzyme has been crystallized. Its structure is depicted in Fig. 24–7.

Table 24–1 Comparison of DNA polymerases of *E. coli*

| | DNA polymerase | | |
	I	II	III
Structural gene*	*pol*A	*pol*B (*dna*A)	*pol*C (*dna*E)
Subunits	1	≥4	≥10
M_r	103,000	88,000[†]	~900,000
3′→5′ Exonuclease (proofreading)	Yes	Yes	Yes
5′→3′ Exonuclease	Yes	No	No
Polymerization rate (nucleotides/s)	16–20	~7	250–1000
Processivity (nucleotides added before dissociation)	3–200	≥10,000	≥500,000

* For enzymes with more than one subunit, the gene listed encodes the subunit with polymerization activity. Gene names in parentheses represent earlier designations of the same gene (p. 815). The acronym *din* stands for *d*amage *in*ducible; *din* genes were originally identified as those that were induced as part of the SOS response to heavy DNA damage, as described later in this chapter.

[†] Polymerization subunit only. DNA polymerase II shares several subunits with DNA polymerase III, including the β, γ, and δ subunits (see Table 24–2) and possibly others.

Table 24–2 Subunits of DNA polymerase III of *E. coli*

Subunit	M_r	Gene	Function*	
α	132,000	*pol*C (*dna*E)	Polymerization activity	Core subunits
ϵ	27,000	*dna*Q (*mut*D)	3′→5′ Proofreading exonuclease	
θ	10,000	*hol*E		
τ	71,000	*dna*X	Stable template binding; core enzyme dimerization	
γ	52,000	*dna*X[†]	Enhanced processivity	
δ	35,000	*hol*A	Enhanced processivity	
δ'	33,000	*hol*B		
χ	15,000	*hol*C		
ψ	12,000	*hol*D		
β	37,000	*dna*N	ATPase required for optimal processivity	

* Where no function is listed, the molecular role of the subunit is not entirely clear.

[†] The γ subunit is encoded by a portion of the gene for the τ subunit, such that the amino-terminal 80% of the τ subunit has the same amino acid sequence as the γ subunit. The γ subunit is generated by a translational frameshifting mechanism (see Box 26–1) that leads to premature translational termination.

(a)

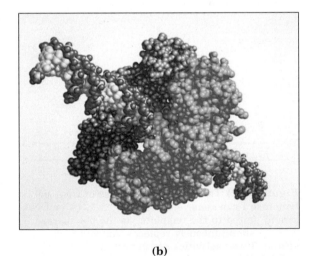

(b)

Figure 24–7 The two β subunits of *E. coli* polymerase III (shown in gray and light blue) form a circular clamp that surrounds DNA, shown in (a) from above as a ribbon structure and in (b) from the side as a space-filling model. The clamp slides along the DNA, enhancing the processivity of the polymerase by preventing its dissociation. The γ and δ subunits of DNA polymerase III facilitate the binding of the β subunits to DNA.

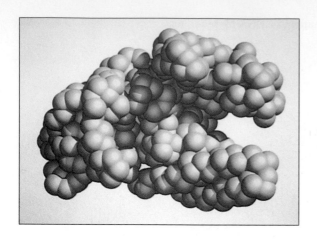

Figure 24–8 The Klenow fragment of *E. coli* DNA polymerase I, produced by proteolytic treatment of the polymerase, includes the polymerization activity of the enzyme. The horizontal groove evident on this face of the protein is the likely binding site for DNA.

DNA polymerase I is far from irrelevant, however. This enzyme serves a host of "clean-up" functions during replication, recombination, and repair, as discussed later in the chapter. These special functions are enhanced by an additional enzymatic activity of DNA polymerase I, a 5′→3′ exonuclease activity. This activity is distinct from the 3′→5′ proofreading exonuclease and is located in a distinct structural domain that can be separated from the enzyme by mild protease treatment. When the 5′→3′ exonuclease domain is removed, the remaining fragment (M_r 68,000) retains the polymerization and proofreading activities, and is called the **large** or **Klenow fragment.** The structure of the Klenow fragment has been determined, and it is this fragment of DNA polymerase I that is depicted in Figure 24–8. The 5′→3′ exonuclease activity of intact DNA polymerase I permits it to extend DNA strands even if the template is already paired to an existing strand of nucleic acid (Fig. 24–9). Using this activity, DNA polymerase I can degrade or displace a segment of DNA (or RNA) paired to the template and replace it with newly synthesized DNA. Most other DNA polymerases, including DNA polymerase III, lack a 5′→3′ exonuclease activity.

DNA Replication Requires Many Enzymes and Protein Factors

We now know that replication in *E. coli* requires not just a single DNA polymerase but 20 or more different enzymes and proteins, each performing a specific task. Although not yet obtained as a physical entity, the entire complex has been called the **DNA replicase system** or the **replisome.** The enzymatic complexity of replication reflects the requirements imposed on the process by the structure of DNA. We will introduce some of the major classes of replication enzymes by considering the problems that they overcome.

To gain access to the DNA strands that are to act as templates the two parent strands must be separated. This is generally accomplished by enzymes called **helicases,** which move along the DNA and separate the strands using chemical energy from ATP. Strand separation creates topological stress in the helical DNA structure, which is relieved by the action of **topoisomerases** (Chapter 23). The separated strands are stabilized by **DNA-binding proteins.** Primers must be present or synthesized before DNA polymerases can synthesize DNA. The primers are generally short segments of RNA laid down by enzymes called **primases.** Ultimately, the RNA primers must be removed and replaced by DNA. In *E. coli,* this is one of the many functions of DNA polymerase I. After removal of the RNA segments and filling in of the gap with DNA, there remain points in the DNA backbone where a phosphodiester bond is broken. These breaks, called nicks, must be sealed by enzymes called **DNA ligases.** All of these processes must be coordinated and regulated. The interplay of these and other enzymes has been best characterized in the *E. coli* system.

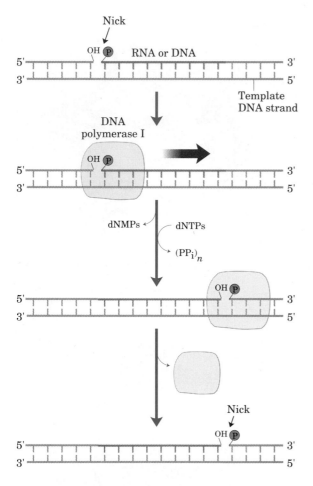

Figure 24–9 The 5′→3′ exonuclease of DNA polymerase I can remove or degrade an RNA or DNA strand paired to the template, as the polymerase activity simultaneously replaces the degraded strand. These activities are important for the role of DNA polymerase I in DNA repair and in removal of RNA primers during replication, as described later in this chapter. The strand of nucleic acid (DNA or RNA) to be removed is shown in green; the replacement strand is shown in red. A nick (a phosphodiester bond broken to leave a free 3′ OH and 5′ phosphate) is found where DNA synthesis starts. After synthesis, a nick remains where DNA polymerase I dissociates. This action of polymerase I has effectively extended the nontemplate DNA strand and moved the nick down the DNA, a process that is sometimes called nick translation.

Replication of the *E. coli* Chromosome Proceeds in Stages

The synthesis of a DNA molecule can be divided into three stages: initiation, elongation, and termination. These are distinguished by differences in the reactions taking place and in the enzymes required. In the next two chapters we will see that the synthesis of the other major biological polymers, RNAs and proteins, can be similarly broken down into the same three stages, each with unique characteristics. The events described below reflect information derived from in vitro experiments using purified *E. coli* proteins.

Initiation The *E. coli* replication origin, called *oriC*, consists of 245 base pairs, many of which are highly conserved among bacteria. The general arrangement of the conserved sequences is illustrated in Figure 24–10. The key sequences for this discussion are two series of short repeats; three repeats of a 13 base pair sequence and four repeats of a 9 base pair sequence.

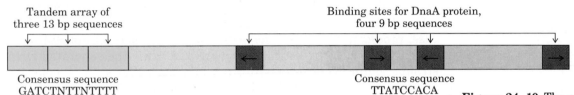

Tandem array of three 13 bp sequences — Consensus sequence GATCTNTTNTTTT

Binding sites for DnaA protein, four 9 bp sequences — Consensus sequence TTATCCACA

Figure 24–10 The arrangement of sequences in the *E. coli* replication origin, called *oriC*. The repeated sequences are shaded in color. The term "consensus sequence" is used to describe a repeated sequence that varies somewhat from one copy to the next; it depicts the most common nucleotide residues found at each position in the sequence (N represents any of the four nucleotides). Individual copies of the repeated sequence may differ from the consensus at one or several positions. The arrows indicate the orientations of the nucleotide sequences.

At least eight different enzymes or proteins (summarized in Table 24–3) participate in the initiation phase of replication. They open the DNA helix at the origin and establish a prepriming complex that sets the stage for subsequent reactions. The key component in the initiation

Table 24–3 Proteins required to initiate replication at the *E. coli* origin

Protein	M_r	Number of subunits	Function
DnaA protein	50,000	1	Opens duplex at specific sites in origin
DnaB protein (helicase)	300,000	6*	Unwinds DNA
DnaC protein	29,000	1	Required for DnaB binding at origin
HU	19,000	2	Histonelike protein; stimulates initiation
Primase (DnaG protein)	60,000	1	Synthesizes RNA primers
SSB	75,600	4*	Binds single-stranded DNA
RNA polymerase	454,000	6	Facilitates DnaA activity
DNA topoisomerase II (gyrase)	400,000	4	Relieves torsional strain generated by DNA unwinding

* Subunits in these cases are identical.

Figure 24–11 A model for initiation of replication at the *E. coli* origin, *ori*C. **(a)** About 20 DnaA protein molecules, each with a bound ATP, bind at the four 9 base pair repeats. The DNA is wrapped around this complex. **(b)** The three 13 base pair repeats are then denatured sequentially to give the open complex. **(c)** The DnaB protein binds to the open complex, with the aid of DnaC protein, and the DnaB helicase activity further unwinds the DNA in preparation for priming and DNA synthesis.

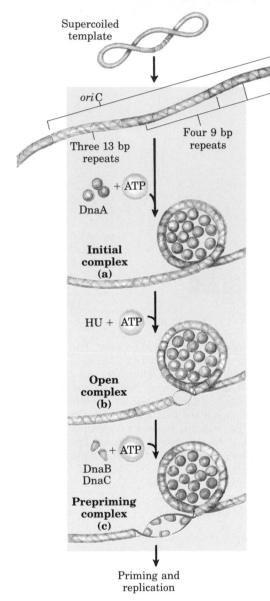

Supercoiled
template

ori C

Three 13 bp
repeats

Four 9 bp
repeats

DnaA   + ATP

**Initial
complex
(a)**

HU + ATP

**Open
complex
(b)**

DnaB
DnaC   + ATP

**Prepriming
complex
(c)**

Priming and
replication

process is the DnaA protein (Fig. 24–11). A complex of about 20 DnaA protein molecules binds to the four 9 base pair repeats in the origin. In a reaction that requires ATP and is facilitated by the bacterial histone-like protein HU, the DnaA protein recognizes and successively denatures the DNA in the region of the three 13 base pair repeats, which are rich in A=T pairs. The DnaB protein then binds to this region in a reaction that requires the DnaC protein. The DnaB protein is a helicase that unwinds the DNA bidirectionally, creating two potential replication forks. If the *E. coli* single-strand DNA-binding protein (SSB) and DNA gyrase (DNA topoisomerase II) are added to this reaction in vitro, thousands of base pairs are rapidly unwound by the DnaB helicase, proceeding out from the origin. Multiple molecules of SSB bind cooperatively to single-stranded DNA, stabilizing the separated DNA strands and preventing renaturation. Gyrase relieves the topological stress created by the DnaB helicase reaction. When additional replication proteins are added as described below, the DNA unwinding mediated by DnaB protein is coupled to replication.

DNA replication must be precisely regulated so that it occurs once and only once in each cell cycle. Initiation is the only phase of replication that is regulated, but the mechanism is not yet well understood. Biochemical studies have provided a few insights. The DnaA protein hydrolyzes its tightly bound ATP slowly (about 1 hour) to form an inactive DnaA–ADP complex. Reactivating this complex (replacing ADP with ATP) is facilitated by an interaction between DnaA protein and acidic phospholipids in the bacterial plasma membrane. Initiation at inappropriate times is prevented by the presence of the inactive DnaA-ADP complex, by the binding of a protein called IciA (*i*nhibitor of *c*hromosomal *i*nitiation) to the 13 base pair repeats, and perhaps by other factors. Deciphering the complex interactions in this regulatory network remains an active area of research.

Elongation The elongation phase of replication consists of two seemingly similar operations that are mechanistically quite distinct: leading strand synthesis and lagging strand synthesis. Several enzymes at the replication fork are important to the synthesis of both strands. DNA helicases unwind the parental DNA. DNA topoisomerases relieve the topological stress induced by the helicases, and SSB stabilizes the separated strands. In other respects, synthesis of DNA in the two strands is sharply different. We will begin with leading strand synthesis, the more straightforward of the two.

Leading strand synthesis begins with the synthesis by primase of a short (10 to 60 nucleotide) RNA primer at the replication origin. Deoxyribonucleotides are then added to this primer by DNA polymerase III. Once begun, leading strand synthesis proceeds continuously, keeping pace with the replication fork (Fig. 24–12).

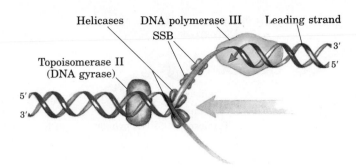

Helicases   DNA polymerase III   Leading strand
SSB
Topoisomerase II
(DNA gyrase)

Figure 24–12 Synthesis of the leading strand. DNA polymerase III keeps pace with the replication fork. Helicases separate the two DNA strands at the fork, molecules of SSB bind to and stabilize the separated strands, and DNA topoisomerase II acts to relieve torsional stress generated by the helicases.

Lagging strand synthesis, which must be accomplished in short fragments (Okazaki fragments) synthesized in the direction opposite to fork movement, is a more intricate problem. It is solved by a protein machine that incorporates several specialized proteins in addition to polymerase III. Each fragment must have its own RNA primer, synthesized by primase, and positioning of the primers must be controlled and coordinated with fork movement. The regulatory apparatus for lagging strand synthesis is a traveling protein machine called a **primosome,** which consists of seven different proteins including the DnaB protein, DnaC protein, and primase mentioned above (Table 24–4). The primosome moves along the lagging strand template in the $5' \rightarrow 3'$ direction, keeping pace with the replication fork. As it moves, the primosome at intervals compels primase to synthesize a short (10 to 60) residue RNA primer to which DNA is then added by DNA poly-

Table 24–4 *E. coli* **proteins at the replication fork**

Protein	M_r	Number of subunits	Function
SSB	75,600	4	Binding to single-stranded DNA
Protein i (DnaT protein)	66,000	3	Primosome constituent
Protein n	28,000	2	Primosome assembly and function
Protein n'	76,000	1	Primosome constituent
Protein n"	17,000	1	Primosome constituent
DnaC protein	29,000	1	Primosome constituent
DnaB protein (helicase)	300,000	6	DNA unwinding; primosome constituent
Primase (DnaG protein)	60,000	1	RNA primer synthesis; primosome constituent
DNA polymerase III	900,000	2×10	Processive chain elongation
DNA polymerase I	103,000	1	Filling of gaps, excision of primers
DNA ligase	74,000	1	Ligation
DNA topoisomerase II (gyrase)	400,000	4	Supercoiling
Rep (helicase)	65,000	1	Unwinding
DNA helicase II	75,000	1	Unwinding
DNA topoisomerase I	100,000	4	Relaxing negative supercoils

* Modified from Kornberg, A. (1982) *Supplement to DNA Replication,* Table S11–2, W.H. Freeman and Company, New York.

Figure 24–13 Synthesis of Okazaki fragments. The multiprotein primosome complex travels in the same direction as the replication fork. **(a)** At intervals, primase synthesizes an RNA primer for a new Okazaki fragment. Note that this synthesis formally proceeds in the direction opposite to fork movement. **(b)** Each primer is extended by DNA polymerase III. **(c)** DNA synthesis continues until the primer of the previously added Okazaki fragment is encountered. (Helicases, DNA topoisomerase II, and SSB have the functions outlined in Fig. 24–12.)

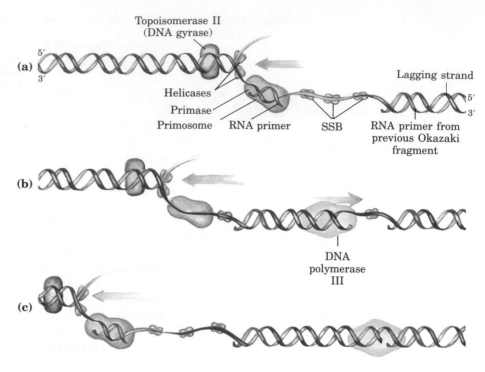

merase III (Fig. 24–13). Note that the direction of the synthetic reactions of primase and polymerase III is opposite to the direction of primosome movement. When the new Okazaki fragment is complete, the RNA primer is removed by DNA polymerase I (using its $5'{\rightarrow}3'$ exonuclease activity) and is replaced with DNA by the same enzyme. The remaining nick is sealed by DNA ligase (Fig. 24–14). The proteins acting at the replication fork are summarized in Table 24–4.

DNA ligase catalyzes the formation of a phosphodiester bond between a $3'$ hydroxyl at the end of one DNA strand and a $5'$ phosphate at the end of another strand. In *E. coli* the phosphate must be activated using NAD^+ (ATP is used in some organisms) to supply the required chemical energy. The reaction pathway, as established by I. Robert Lehman and colleagues, is shown in Figure 24–15. The use by the ligase of *E. coli* of the nucleotide NAD^+—a cofactor that normally functions in hydride transfer reactions (see Fig. 13–16)—as the source of the AMP activating group is unusual. DNA ligase is another enzyme of DNA metabolism that has become an important reagent in recombinant DNA experiments (Chapter 28).

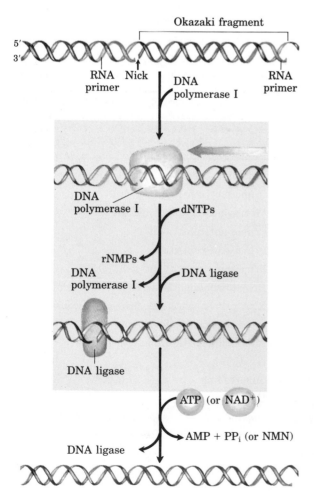

Figure 24–14 Removal of RNA primers in the lagging strand. The RNA primer is removed by the $5'{\rightarrow}3'$ exonuclease activity of DNA polymerase I, and it is replaced with DNA by the same enzyme. The remaining nick is sealed by DNA ligase. The role of NAD^+ is shown in Fig. 24–15.

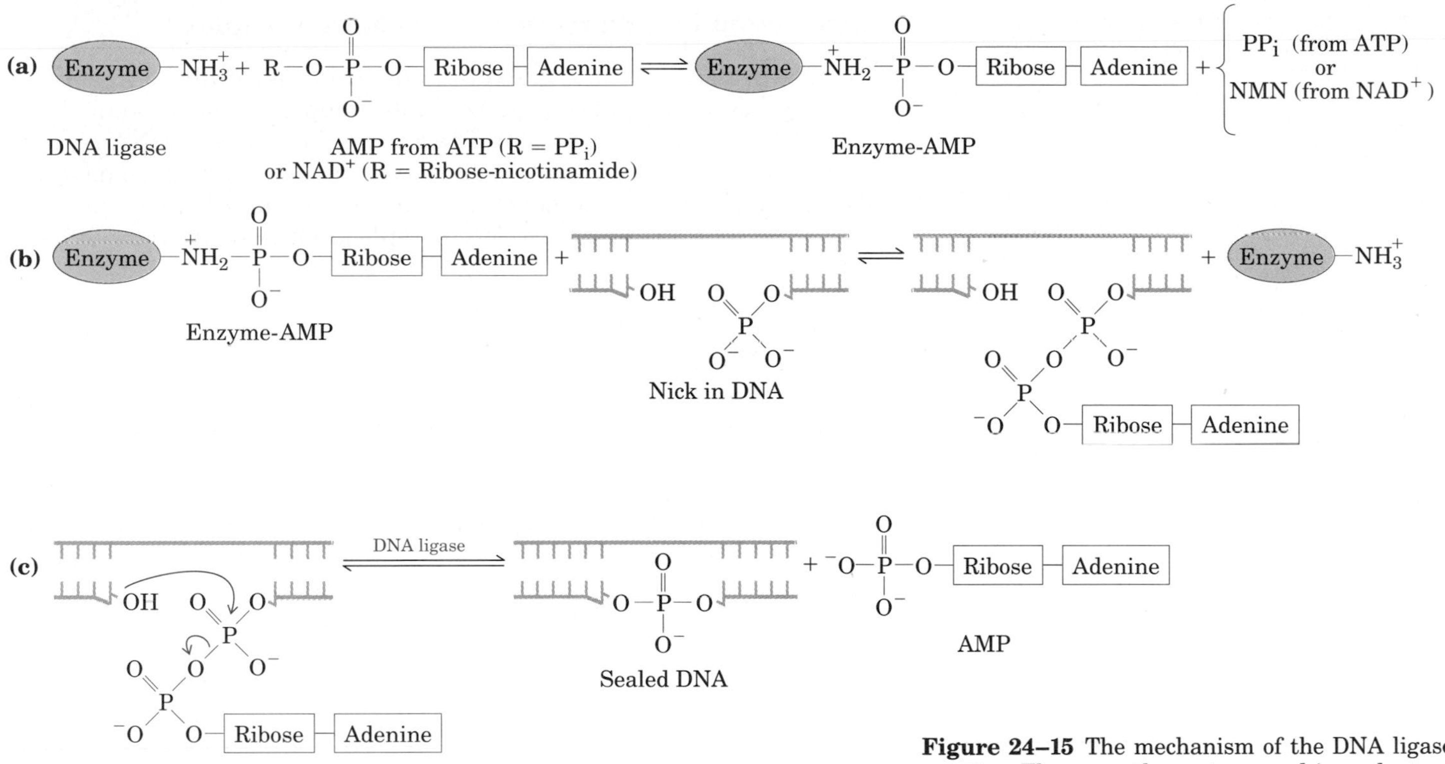

(a) Enzyme—NH$_3^+$ + R—O—P—O—Ribose—Adenine $\rightleftharpoons$ Enzyme—NH$_2$—P—O—Ribose—Adenine + PP$_i$ (from ATP) or NMN (from NAD$^+$)

DNA ligase   AMP from ATP (R = PP$_i$)   Enzyme-AMP
                    or NAD$^+$ (R = Ribose-nicotinamide)

(b) Enzyme—NH$_2$—P—O—Ribose—Adenine + [Nick in DNA] $\rightleftharpoons$ [DNA] + Enzyme—NH$_3^+$

Enzyme-AMP

Nick in DNA

(c) [DNA with AMP] $\xrightarrow{\text{DNA ligase}}$ [Sealed DNA] + $^-$O—P—O—Ribose—Adenine

Sealed DNA   AMP

Figure 24–15 The mechanism of the DNA ligase reaction. There are three steps, and in each step one phosphodiester bond is formed at the expense of another. Steps **(a)** and **(b)** lead to activation of the 5′ phosphate in the nick. An AMP group is transferred first to a Lys residue on the enzyme and then to the 5′ phosphate in the nick. **(c)** The 3′-OH group then attacks this phosphate and displaces AMP, leading to the formation of a phosphodiester bond to seal the nick. The AMP is derived from NAD$^+$ in the case of *E. coli* DNA ligase. The DNA ligases isolated from a number of other prokaryotic and eukaryotic sources use ATP rather than NAD$^+$, and release pyrophosphate rather than nicotinamide mononucleotide (NMN) in step **(a)**.

In *E. coli*, synthesis of the leading and lagging strands may actually be coupled as shown in Figure 24–16. This can be accomplished by looping the lagging strand template so that synthesis can be carried out concurrently on both strands by a single dimeric polymerase III acting in concert with the primosome and all of the other proteins at the replication fork (Table 24–4).

Termination Eventually, the two replication forks meet at the other side of the circular *E. coli* chromosome. Very little is known about this stage of the reaction, though the action of a type 2 topoisomerase called DNA topoisomerase IV appears to be necessary for final separation of the two completed circular DNA molecules. Nor is much understood about the process of partitioning the two DNA molecules into daughter cells at division.

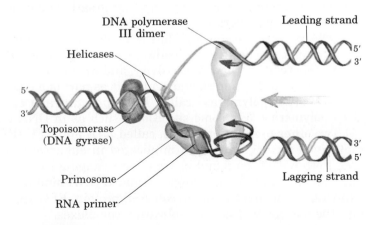

Figure 24–16 Coupling the synthesis of leading and lagging strands with a dimeric DNA polymerase III. The template for the lagging strand is looped tightly so that the direction of synthesis has the same orientation for both strands. As polymerization proceeds, the loop grows until the previous Okazaki fragment is encountered. Here, the polymerase synthesizing the lagging strand must dissociate and reinitiate at a new primer and with a new tight loop. This must be coordinated to keep pace with that part of the polymerase synthesizing the leading strand.

Replication in Eukaryotic Cells Is More Complex

The DNA molecules in eukaryotic cells are considerably larger than those in bacteria and are organized into complex nucleoprotein structures (chromatin) (Chapter 23). The essential features of DNA replication are the same in eukaryotes and prokaryotes. However, some interesting variations on the general principles discussed above promise new insights into the regulation of replication and its link with the cell cycle.

Origins of replication, called *a*utonomously *r*eplicating *s*equences (ARS), have been identified and studied in yeast. ARS elements span regions of about 300 base pairs and contain several conserved sequences that are essential for ARS function. There are about 400 ARS elements in yeast, with most chromosomes having several. Proteins that specifically bind the ARS region have been identified in yeast, although their functions are not yet understood.

The rate of replication fork movement in eukaryotes (~50 nucleotides/s) is only one-tenth that observed in *E. coli*. At this rate, replication of an average human chromosome proceeding from a single origin would take more than 500 hours. Instead, replication of human chromosomes proceeds bidirectionally from multiple origins spaced 30,000 to 300,000 base pairs apart. With the exception of the ARS elements of yeast, the structure of the origins of replication in eukaryotes is not known. Because eukaryotic chromosomes are almost uniformly much larger than bacterial chromosomes, the presence of multiple origins on a eukaryotic chromosome is probably a general rule.

As in bacteria, there are several types of DNA polymerases in eukaryotic cells. Some have been linked to special functions such as the replication of the DNA in mitochondria. The replication of nuclear chromosomes involves an enzyme called **DNA polymerase α,** in association with another polymerase called **DNA polymerase δ.** DNA polymerase α is typically a four-subunit enzyme with similar structure and properties in all eukaryotic cells. One of the subunits has a primase activity. The largest subunit ($M_r \sim 180,000$) contains the polymerization activity. DNA polymerase δ has two subunits. This enzyme exhibits a very interesting association with and stimulation by a protein called *p*roliferating *c*ell *n*uclear *a*ntigen (PCNA; M_r 29,000) found in large amounts in the nuclei of proliferating cells. The PCNA from yeast will function with DNA polymerase δ from calf thymus, and the calf thymus PCNA with yeast DNA polymerase δ, suggesting a conservation of the structure and function of these key components of the cell division apparatus in all eukaryotic cells. PCNA appears to have a function analogous to the β subunit of *E. coli* DNA polymerase III (see Fig. 24–7), forming a circular clamp that greatly enhances the processivity of DNA polymerase δ.

DNA polymerase δ, which has a 3'→5' proofreading exonuclease activity, appears to carry out leading strand synthesis. DNA polymerase α has a relatively low processivity, and with its associated primase it may carry out lagging strand synthesis as part of a eukaryotic replisome. Another polymerase called **DNA polymerase ϵ,** may replace DNA polymerase δ in some situations, such as in DNA repair.

Two other protein complexes, called RFA and RFC (RF stands for replication factor), have been implicated in eukaryotic DNA replication. Both have been found in organisms ranging from yeast to mammals. RFA is a eukaryotic single-stranded DNA-binding protein, with a function equivalent to the *E. coli* SSB protein. RFC appears to facilitate the assembly of active replication complexes.

DNA Repair

A cell generally has only one or two sets of genomic DNA, and whereas proteins and RNA molecules can, if damaged, be quickly replaced using information encoded in the DNA, the DNA molecules themselves are irreplaceable. Maintaining the integrity of the information contained in DNA is therefore a cellular imperative, and an elaborate set of DNA repair systems is present in every cell. As described in Chapter 12, DNA can be damaged by a variety of processes, some spontaneous, some catalyzed by environmental agents. In addition, replication can occasionally leave mispaired bases. The chemistry of DNA damage is diverse and complex. Not surprisingly, the cellular response includes a wide range of enzymatic systems that catalyze some of the most interesting chemical transformations to be found in DNA metabolism. We will first examine the effects of alterations in DNA sequence and then turn to specific repair systems.

Mutations Are Linked to Cancer

There is no better way of illustrating the importance of DNA repair than to consider the effects of unrepaired DNA damage (a lesion). The most serious outcome is a change in the base sequence of the DNA, which if replicated and transmitted to future cell generations becomes permanent. Such permanent changes in the nucleotide sequence of DNA are called **mutations.** Mutations can range from the replacement of one base pair with another (substitution mutation) to the addition or deletion of one or more base pairs (insertion or deletion mutations). If the mutation affects nonessential DNA or if it has a negligible effect on the function of a gene, it is called a **silent mutation.** Favorable mutations that confer some advantage to the cell in which they occur are rare, although the frequency is sufficient to provide the variation necessary for natural selection and thus evolution. The majority of mutations, however, are deleterious to the cell.

In mammals there is a strong correlation between the accumulation of mutations and cancer, as seen from a simple test for mutagenic compounds developed by Bruce Ames (Fig. 24–17). The Ames test mea-

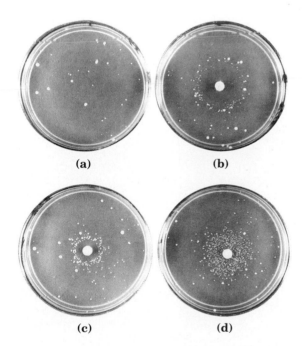

(a)   **(b)**

(c)   **(d)**

Figure 24–17 The Ames test for carcinogens, based on their mutagenicity. *Salmonella typhimurium* cells having a mutation that inactivates an enzyme of the histidine biosynthetic pathway are plated on a histidine-free medium. Most of the cells are unable to grow. **(a)** The few small colonies of histidine-less *S. typhimurium* that do grow on a histidine-free medium are the result of spontaneous back-mutations. To each of three identical nutrient plates **(b), (c),** and **(d)** inoculated with an equal number of cells has been added a disk of filter paper containing progressively lower concentrations of a mutagen, which greatly increases the rate of back-mutation and hence the number of colonies. In the clear area around the filter paper, the concentration of mutagen is so high that it is lethal to the cells. As the mutagen diffuses outward, away from the filter paper, it is diluted to sublethal concentrations that promote back-mutation. Mutagens are compared on the basis of the increased mutation rate they produce. Because many compounds undergo a variety of chemical transformations when they enter a cell, compounds are sometimes tested for mutagenicity after incubating them with a liver extract. A number of compounds have been found to be mutagenic only after this treatment.

sures the potential of a given chemical compound to promote certain easily detected mutations in a specialized bacterial strain. Few of the chemicals that we might encounter day to day score as mutagens in this test. However, of the compounds known to be carcinogenic from extensive animal trials, more than 90% are also found to be mutagenic in the Ames test. Because of the strong correlation between mutagenesis and carcinogenesis, the Ames test for mutagens is widely used as a rapid and inexpensive screen for potential carcinogens.

Defects in genes that encode DNA repair enzymes can have catastrophic effects. A rare human genetic disease called xeroderma pigmentosum is caused by a defect in the multienzyme process by which pyrimidine dimers and other bulky lesions in DNA are repaired. Many of these lesions are induced by UV light (see Fig. 12–33), and patients with this disease rapidly develop multiple skin cancers if exposed to sunlight.

The genome of a typical mammalian cell accumulates many thousands of lesions in a 24-hour period. However, as a result of DNA repair, less than one lesion in 1,000 becomes a mutation. DNA is a relatively stable molecule, but without repair systems the cumulative effect of many infrequent but damaging reactions would make life impossible.

All Cells Have Multiple DNA Repair Systems

The number and diversity of repair systems reflect the importance of DNA repair to cell survival and the diverse sources of DNA damage. For some common types of lesions there is even a built-in redundancy, with several distinct systems available to repair them (e.g., pyrimidine dimers; Table 24–5). As a complementary theme, it is worth noting that many DNA repair processes appear to be extraordinarily ineffi-

Table 24–5 Types of DNA repair systems in *E. coli*

System	Enzymes/proteins	Type of damage
Mismatch repair	Dam methylase MutH, MutL, MutS proteins DNA helicase II SSB DNA polymerase III Exonuclease I DNA ligase	Mismatches
Base-excision repair	DNA glycosylases AP endonucleases DNA polymerase I DNA ligase	Abnormal bases (uracil, hypoxanthine, xanthine); alkylated bases; pyrimidine dimers in some other organisms
Nucleotide-excision repair	ABC excinuclease DNA polymerase I DNA ligase	DNA lesions that cause large structural changes, e.g., pyrimidine dimers
Direct repair	DNA photolyases O^6-Methylguanine-DNA methyltransferase	Pyrimidine dimers O^6-Methylguanine

cient in an energetic sense. This represents an exception to the pattern observed in the metabolic pathways, as described in Part III, where every ATP is generally accounted for and used optimally. When the integrity of the genetic information is at stake, the amount of chemical energy invested in a repair process seems to be almost irrelevant.

DNA repair is possible largely because the DNA molecule consists of two complementary strands. DNA damage in one strand can be removed and accurately replaced by following the template instructions in the undamaged complementary strand.

We now turn to a consideration of the principal classes of repair systems, beginning with those that repair the rare nucleotide mismatches that are left behind by replication.

Mismatch Repair The correction of mismatches after replication in *E. coli* improves the overall fidelity of the replication process by a factor of 10^2 to 10^3. The mismatches are nearly always corrected to correspond to the information in the template strand, thus the system must somehow discriminate between the template and the newly synthesized strand. The cell accomplishes this discrimination by tagging the old (template) DNA with methyl groups to distinguish it from newly synthesized strands. The mismatch repair system of *E. coli* includes at least nine protein components (Table 24–5) that function either in strand discrimination or in the repair process itself.

Strand discrimination is based on the action of an enzyme called the Dam methylase, which methylates DNA at the N^6 position of all adenines that occur within (5')GATC sequences. Immediately after replication there is a short lag (a few seconds or minutes) during which the template strand is methylated but the newly synthesized strand is not yet methylated (Fig. 24–18). It is this transient undermethylation of GATC sequences in the newly synthesized strand that permits strand discrimination. Replication mismatches in the vicinity of a GATC sequence are then repaired according to the information in the methylated parent (template) strand. If both strands are methylated at a GATC sequence, little repair occurs. If neither strand is methylated, repair occurs but does not favor either strand. This system, sometimes referred to as methyl-directed mismatch repair, correctly repairs mismatches as much as 1,000 base pairs distant from a partially or hemimethylated GATC sequence.

The mechanism by which mismatch corrections are directed by relatively distant GATC sequences is not completely understood, but the proteins involved have been purified from *E. coli* cells and the reaction has been reconstituted in vitro. This work has inspired the model illus-

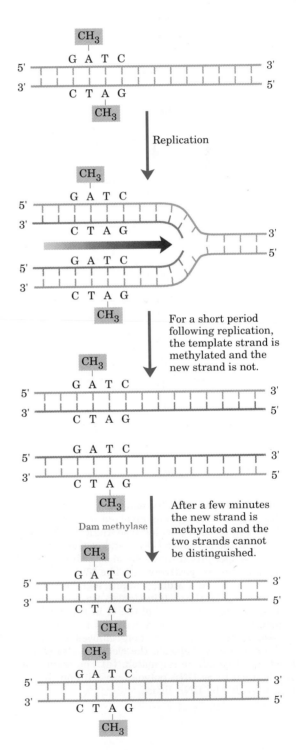

Figure 24–18 Methylation of DNA strands can serve to distinguish parental (template) strands from newly synthesized strands in *E. coli* DNA, a function which is critical to mismatch repair (Fig. 24–19). The methylation occurs at N^6 of adenines (Fig. 12–5a) in (5')GATC sequences. This sequence is a palindrome (see Fig. 12–20) and thus is present in opposite orientations on both strands.

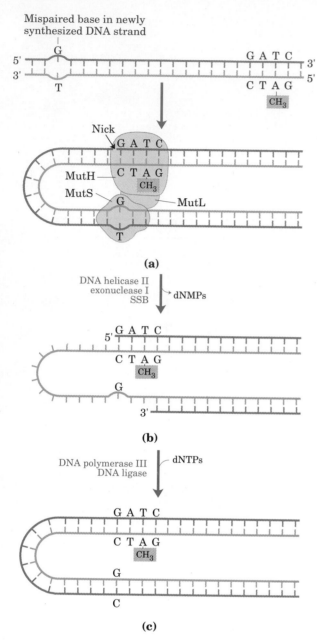

Mispaired base in newly
synthesized DNA strand

(a)

DNA helicase II
exonuclease I → dNMPs
SSB

(b)

DNA polymerase III → dNTPs
DNA ligase

(c)

Figure 24–19 A model for methyl-directed mismatch repair. The proteins involved in this process in *E. coli* have been purified (see Table 24–5). Recognition of the sequence GATC and of the mismatch are specialized functions of the MutH and MutS proteins, respectively. **(a)** The MutL protein links the MutH and MutS proteins together in a complex. The MutH protein cleaves the unmethylated strand on the 5' side of the G in the GATC sequence. **(b)** The combined action of DNA helicase II, exonuclease I, and SSB then removes a segment of the new strand between the cleavage site and a point just beyond the mismatch. **(c)** The resulting gap is filled in by DNA polymerase III, and the nick is sealed by DNA ligase.

trated in Figure 24–19. The MutS, MutH, and MutL proteins play key roles in the process. The MutS protein binds to a wide range of mismatched base pairs. The MutH protein binds to GATC sequences. MutL may be an interface protein, linking the MutS and MutH proteins in a complex. If only one of the two strands is methylated at the GATC sequence and a mismatched base pair exists nearby (within ~1,000 base pairs), the MutH protein acts as a site-specific endonuclease, cleaving the unmethylated strand on the 5' side of the G in GATC, thereby marking the strand for repair. Further steps in the pathway depend upon where the mismatch is located relative to this cleavage site. When the mismatch is on the 5' side of the cleavage site, evidence suggests that the unmethylated strand is unwound and degraded in the 3'→5' direction from the cleavage site through the mismatch, and replaced with new DNA. This process requires the combined action of DNA helicase II, SSB, exonuclease I (which degrades only single-stranded DNA in the 3'→5' direction), DNA polymerase III, and DNA ligase (Fig. 24–19). The pathway for repair of mismatches on the 3' side of the cleavage site is similar, except that exonuclease VII (which degrades single-stranded DNA, 5'→3' or 3'→5') or RecJ protein (an exonuclease that degrades single-stranded DNA 5'→3') replaces exonuclease I.

Energetically, mismatch repair is a particularly expensive process. The mismatch may be 1,000 base pairs or more from the GATC sequence, and the degradation and replacement of a strand segment of this length represents an enormous investment in activated deoxynucleotide precursors to repair a single DNA mismatch. This once again illustrates the importance of DNA repair to the cell.

All mismatches are recognized and repaired by this system, but not equally well. Those that stand out are G–T mismatches, which are generally repaired more efficiently than the others, and C–C mismatches, which are repaired poorly.

Base-Excision Repair Every cell has a class of enzymes called **DNA glycosylases** that recognize particularly common DNA lesions (such as the products of cytosine and adenine deamination; see Fig. 12–32a) and remove the affected base by cleaving the *N*-glycosyl bond. This creates an apurinic or apyrimidinic site in the DNA, both commonly referred to as abasic or **AP sites.** Each DNA glycosylase is generally specific for one type of lesion.

An example common to most cells is uracil glycosylase, which removes from DNA the uracil that results from spontaneous deamination of cytosine. This glycosylase is of necessity very specific; it does not remove uracil residues from RNA, nor does it remove thymine residues from DNA. The problem posed by cytosine deamination suggests a reason for the long-puzzling fact that DNA contains thymine instead of uracil (p. 344).

Other DNA glycosylases recognize and remove hypoxanthine (arising from adenine deamination) and alkylated bases such as 3-methyladenine and 7-methylguanine. Glycosylases that recognize other lesions, including pyrimidine dimers, have been identified. Remember that AP sites also arise from the slow, spontaneous hydrolysis of the *N*-glycosyl bonds in DNA (see Fig. 12–32b).

Once an AP site has been formed, another group of enzymes must repair it. The repair is *not* made by simply inserting a new base and re-forming the *N*-glycosyl bond. Instead, the deoxyribose 5'-phosphate left behind is removed and replaced with a new nucleotide. This pro-

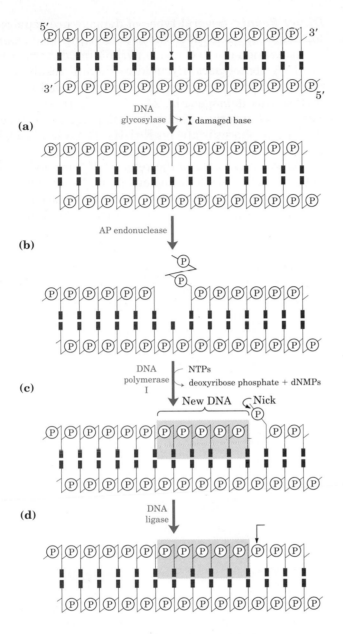

(a)

(b)

(c)

(d)

Figure 24–20 Repair by the base-excision repair pathway. **(a)** DNA glycosylase recognizes a damaged base and cleaves between the base and deoxyribose in the backbone. **(b)** An AP endonuclease cleaves the phosphodiester backbone near the AP site. **(c)** DNA polymerase I initiates repair synthesis from the free 3′ OH at the nick, removing a portion of the damaged strand (with its 5′→3′ exonuclease activity) and replacing it with undamaged DNA. **(d)** The nick remaining after DNA polymerase I has dissociated is sealed by DNA ligase.

cess begins with enzymes called **AP endonucleases,** which cut the DNA strand containing the AP site. The position of the incision relative to the AP site (5′ or 3′) varies with different AP endonucleases. A segment of DNA including the AP site is then removed, the DNA is replaced by the action of DNA polymerase I, and the remaining nick is sealed by DNA ligase (Fig. 24–20).

Nucleotide-Excision Repair   DNA lesions that cause large distortions in the helical structure of DNA generally are repaired by the nucleotide-excision system. In *E. coli* the key enzyme is made up of three subunits, products of the *uvr*A, *uvr*B, and *uvr*C genes, and is called the ABC excinuclease (M_r 246,000). This enzyme recognizes many types of lesions, including cyclobutane pyrimidine dimers, 6–4 photoproducts (see Fig. 12–33), and several other types of base adducts. The ABC excinuclease's nucleolytic activity is novel in the sense that two cuts are made in the DNA (Fig. 24–21). The term "excinuclease" is meant to distinguish this activity from that of standard endonucleases.

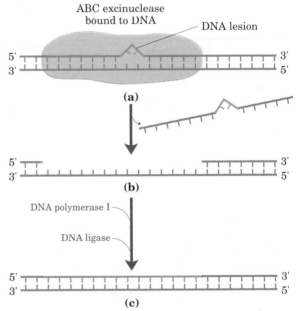

(a)

(b)

(c)

Figure 24–21 The mechanism of nucleotide-excision repair in *E. coli.* **(a)** A specialized nuclease (the ABC excinuclease; see Table 24–5) binds to DNA at the site of a bulky lesion and cleaves the damaged strand at the eighth phosphodiester bond on the 5′ side of the lesion and at the fourth or fifth phosphodiester bond on the 3′ side. **(b)** The excinuclease then removes the resulting 12 to 13 base pair oligonucleotide that spans the damaged base. **(c)** The resulting gap is filled in by DNA polymerase I and sealed by ligase.

Direct Repair Several types of damage are repaired without removing a base or nucleotide. The best-characterized example is direct photoreactivation of cyclobutane pyrimidine dimers, a reaction promoted by DNA photolyases. Pyrimidine dimers result from a light-induced reaction, and photolyases use energy derived from absorbed light to reverse this damage (Fig. 24–22). Photolyases generally contain two cofactors that serve as light-absorbing agents, or chromophores. One of the chromophores is always $FADH_2$. The other is a folate in *E. coli* and yeast.

Another example is the repair of O^6-methylguanine, which forms in the presence of alkylating agents and is a common and highly mutagenic lesion. It tends to pair with thymine rather than cytosine during replication, and therefore causes $G\equiv C$ to $A=T$ mutations (Fig. 24–23).

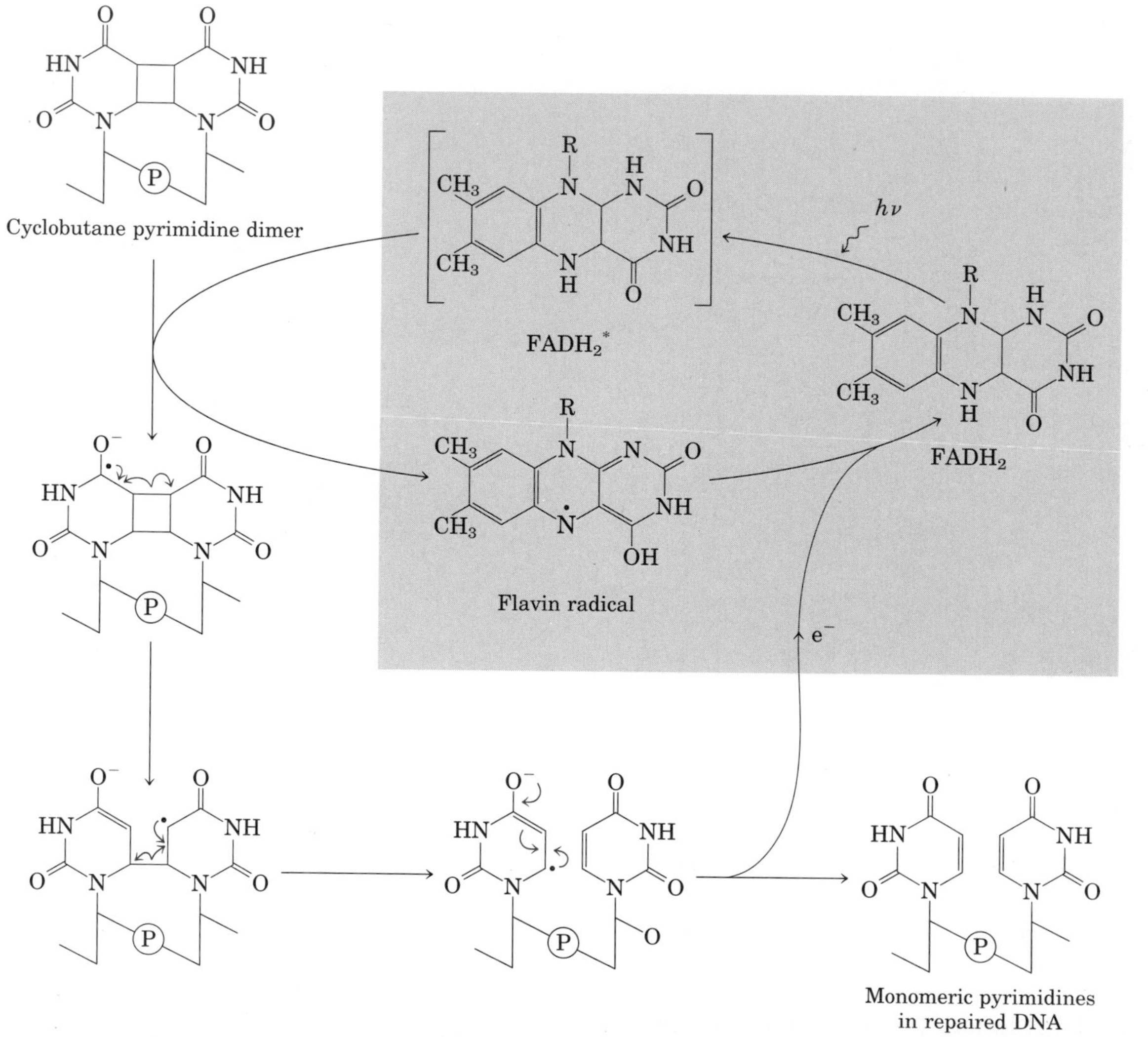

Figure 24–22 Repair of pyrimidine dimers with photolyase. Shown here is a simplified representation of a pyrimidine dimer (see Fig. 12–33). Energy derived from absorbed light is used to reverse the photoreaction that caused the lesion. The two chromophores in *E. coli* photolyase (M_r 54,000), 5,10-methenyltetrahydrofolate and $FADH_2$, complement each other in terms of the light wavelengths at which they absorb efficiently. Most of the photoreactivating light energy is absorbed by the folate and transferred to $FADH_2$; some is absorbed directly by $FADH_2$. The resulting excited form of $FADH_2$ [$FADH_2$*] transfers an electron to the pyrimidine dimer, regenerating $FADH_2$. The resulting pyrimidine dimer species (which contains a free radical) is unstable and breaks down to form the monomeric pyrimidines.

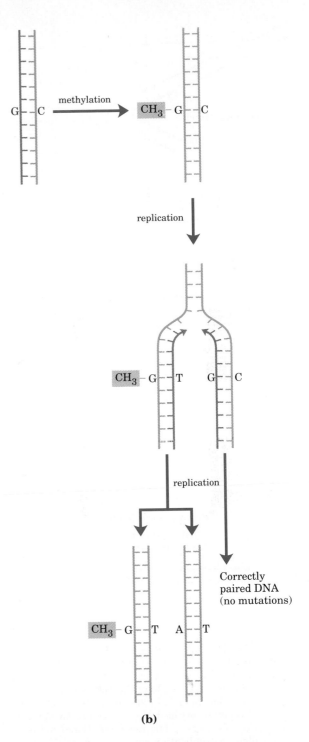

Direct repair of O^6-methylguanine is carried out by O^6-methylguanine-DNA methyltransferase, which catalyzes the transfer of the methyl group of O^6-methylguanine to a specific Cys residue on the same protein. This methyltransferase is not strictly an enzyme, because a single methyl transfer event inactivates the protein. The consumption of an entire protein molecule to correct a single damaged base is another vivid illustration of the central importance of maintaining the integrity of cellular DNA.

When DNA Damage Is Extensive, Repair Becomes Error-Prone

Up to this point, our discussion has focused on the accurate repair of the relatively rare DNA lesions that occur daily in any cell. However, in *E. coli,* when the chromosome is subjected to heavy damage through exposure to UV light or a DNA-damaging reagent, DNA repair becomes significantly less accurate and a high mutation rate is observed. This is referred to as **error-prone repair,** a distinct and unusual pathway.

Given the energetic investment made to maintain the structural and sequence integrity of cellular DNA, it may seem incongruous that mechanisms exist to *increase* mutation rates. As is often the case in biochemistry, however, an examination of an apparent exception to a

Figure 24–23 An example of how DNA damage results in mutations. The methylation product O^6-methylguanine pairs with thymine rather than cytosine **(a).** If not repaired, this leads to a G≡C to A=T mutation after replication **(b).**

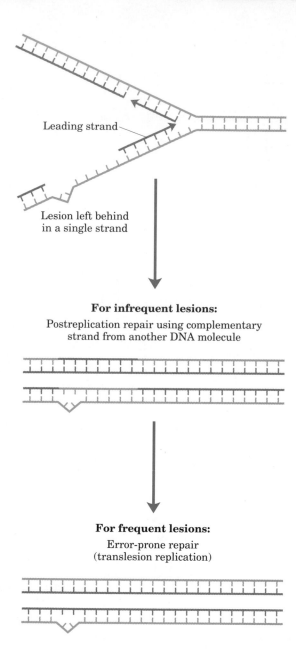

Leading strand

Lesion left behind
in a single strand

For infrequent lesions:
Postreplication repair using complementary
strand from another DNA molecule

For frequent lesions:
Error-prone repair
(translesion replication)

Figure 24–24 DNA damage and its effect on DNA replication. If an unrepaired lesion is encountered at the replication fork, replication generally stops and is resumed farther along the chromosome. The lesion is left behind in an unreplicated, single-stranded segment of the DNA. There are two possible avenues for repair. The recombinational pathway, called postreplication repair, is described in Fig. 24–34. When lesions are so numerous that normal replication is inhibited, a second repair mechanism operates. The specialized system uses DNA polymerase II and can replicate over many types of lesions. This is called error-prone repair because mutations often result.

general rule can throw light on the rule itself. In this instance we must examine the complex interrelationships among repair, replication, and recombination. In *E. coli,* normal DNA replication with DNA polymerase III cannot proceed past many types of DNA lesions. Under normal circumstances, most lesions are repaired before the replication complex arrives. The occasional unrepaired lesion blocks replication, but replication begins again beyond the site of the lesion (Fig. 24–24) and the lesion itself can eventually be repaired with the aid of recombination processes (postreplication repair) described later in this chapter. Higher levels of DNA damage, however, effectively bring normal DNA replication to a halt and trigger a stress response in the cell involving a regulated increase (induction) in the levels of a number of proteins. This is called, appropriately enough, the **SOS response.** Some of the proteins induced, such as the UvrA and UvrB proteins, have roles in DNA repair (Table 24–6). A number of the induced proteins, however, are part of a specialized replication system that can replicate past the DNA lesions that block DNA polymerase III. Because proper base pairing is often impossible at the site of a lesion, this translesion replication is error-prone. The resulting increase in mutagenesis does not contradict the general principle that replication accuracy is important—the resulting mutations actually kill many cells. This is the biological price that is paid, however, to overcome the general barrier to replication and permit at least a few mutant cells to survive.

Table 24–6 Genes induced as part of the SOS response

Gene name	Role in DNA repair
Genes of known function	
*pol*B (*din*A)	Encodes polymerization subunit of DNA polymerase II, required for error-prone repair
*uvr*A *uvr*B *uvr*C	Encode ABC excinuclease
*umu*C *umu*D	Encode proteins required for error-prone repair
*sul*A	Encodes protein that inhibits cell division, possibly to allow time for DNA repair
*rec*A	Encodes RecA protein required for error-prone repair and recombinational repair
Genes involved in DNA metabolism, but role in DNA repair unknown	
ssb	Encodes single-strand binding protein (SSB)
*uvr*D	Encodes DNA helicase II (DNA-unwinding protein)
*him*A	Encodes subunit of integration host factor, involved in site-specific recombination, replication, transposition, regulation of expression of a number of genes
*rec*N	Involved in recombinational repair
Genes of unknown function	
*din*B	
*din*D	
*din*F	

Translesion replication brings us back to a discussion of *E. coli* DNA polymerase II. This polymerase is induced as part of the SOS response and, unlike DNA polymerase III, it is capable of replication past lesions such as AP sites. This enzyme has some of the same subunits as DNA polymerase III, and at least some of these protein subunits are synthesized in larger amounts as part of the SOS response. In addition to this unusual polymerase activity, error-prone repair requires the activities of the UmuC, UmuD, and RecA proteins, although their precise molecular functions in mutagenesis are not understood.

The RecA protein merits some additional discussion because it has several distinct functions (besides mutagenesis) in the bacterial cell. RecA protein is involved in recombination and in the regulation of the SOS response, and in these cases its molecular function is well characterized. The regulation of the SOS response is described in Chapter 27. We now turn to a discussion of genetic recombination.

DNA Recombination

The rearrangement of genetic information in and among DNA molecules encompasses a variety of processes that are collectively placed under the heading of genetic recombination. An understanding of how DNA rearrangements occur is finding practical application as scientists explore new methods for altering the genomes of a variety of organisms (Chapter 28).

Genetic recombination events fall into at least three general classes. **Homologous genetic recombination** involves genetic exchanges between any two DNA molecules (or segments of the same molecule) that share an extended region with homologous sequences. The actual sequence of bases in the DNA is irrelevant as long as the sequences in the two DNAs are similar. **Site-specific recombination** differs in that these exchanges occur only at a defined DNA sequence. **DNA transposition** is distinct in that it usually involves a short segment of DNA with the remarkable capacity to move from one location in a chromosome to another. These "hopping genes" were first observed in maize in the 1950s by Barbara McClintock. In addition to these well-characterized classes, there is a wide range of unusual rearrangements for which no mechanism or purpose has been proposed. We will focus only on the first three classes noted above.

Any discussion of the mechanics of recombination must always include unusual DNA structures. In homologous genetic recombination, the two DNA molecules interact and align their similar sequences at some stage in the reaction. This alignment process may involve the formation of novel DNA intermediates in which three or possibly even four strands are interwound. (Recall the three-stranded structure of H-DNA; see Fig. 12–22.) Branched DNA structures are also found as recombination intermediates. The exchange of information between two large, helical macromolecules often involves a complex interweaving of strands.

The functions of genetic recombination systems are as varied as their mechanisms. The maintenance of genetic diversity, specialized DNA repair systems, the regulation of expression of certain genes, and programmed genetic rearrangements during development represent some of the recognized roles for genetic recombination events. To illustrate these functions, we must first describe the recombination reactions themselves.

Barbara McClintock
1902–1992

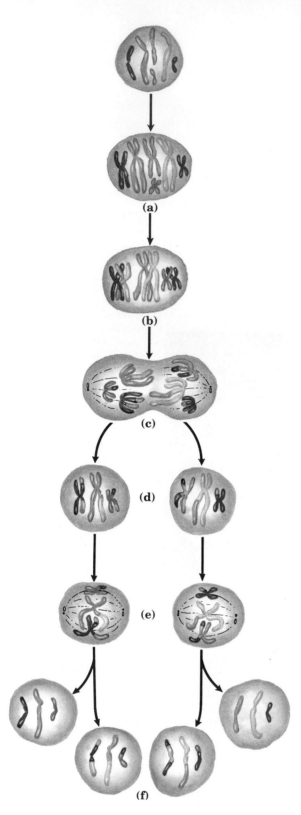

(a)

(b)

(c)

(d)

(e)

(f)

Figure 24–25 Meiosis in eukaryotic germ-line cells. **(a)** The chromosomes of a germ-line cell (six chromosomes; three homologous pairs) are replicated, except for centromeres. While the product DNA molecules remain attached at their centromeres, they are called chromatids (sometimes, "sister chromatids"). **(b)** In prophase I, just prior to the first meiotic division, the three homologous sets of chromatids are aligned to form tetrads, held together by covalent links at homologous junctions (chiasmata). Crossovers occur within the chiasmata (see Fig. 24–26). **(c)** Homologous pairs separate toward opposite poles of the cell. **(d)** The first meiotic division produces two daughter cells, each with three pairs of chromatids. **(e)** The homologous pairs align in the center of the cell in preparation for separation of the chromatids (now chromosomes). **(f)** The second meiotic division produces daughter cells with three chromosomes, half the number of the germ-line cell. The chromosomes have resorted and recombined.

Homologous Genetic Recombination Has Multiple Functions

Homologous genetic recombination (also called general recombination) is tightly linked to cell division in eukaryotes. The process occurs with the highest frequency during **meiosis,** the process in which a germ-line cell with two matching sets of chromosomes (a diploid cell) divides to produce a set of gametes—sperm cells or ova in higher eukaryotes—each gamete having only one member of each chromosome pair (haploid cells). The process of meiosis is illustrated in Figure 24–25. In outline, meiosis begins with replication of the DNA in the germ-line cell so that each DNA molecule is present in four copies. The cell then goes through two meiotic cell divisions that reduce the DNA content to the haploid level in each of four daughter cells.

After the DNA is replicated during prophase I (prophase of the first meiotic division), the resulting DNA copies remain associated at their centromeres and are referred to as sister chromatids. Each set of four homologous DNA molecules is therefore arranged as two pairs of chromatids. Genetic information is exchanged between the closely associated homologous chromatids at this stage of meiosis by means of homologous genetic recombination. This process involves a breakage and rejoining of DNA. The exchange is also called crossing over, and can be observed cytologically (Fig. 24–26). Crossing over links the two pairs of sister chromatids together at points called chiasmata (singular, chiasma). This effectively links together all four homologous chromatids, and this linkage is essential to the proper segregation of chromosomes in the subsequent meiotic cell divisions. To a first approximation, recombination, or crossing over, can occur with equal probability at almost any point along the length of two homologous chromosomes. The frequency of recombination in a region separating two points on a chromosome is therefore proportional to the distance between the points.

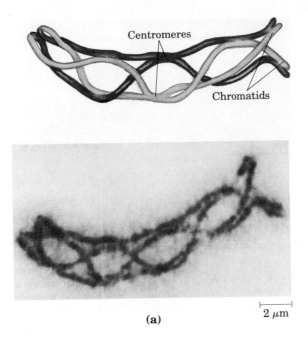

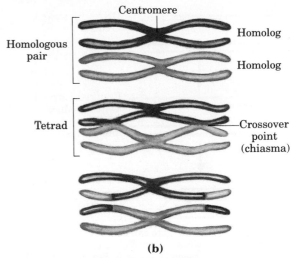

(a)   2 µm

(b)

Figure 24–26 Crossing over. **(a)** The homologous chromosomes of a grasshopper are shown during prophase I of meiosis. Multiple points of joining (chiasmata) are evident between the two homologous pairs of chromatids. These chiasmata are the physical manifestation of prior homologous recombination (crossing over) events. **(b)** Crossing over often results in an exchange of genetic material.

This fact has been used by geneticists for many decades to map the relative positions and distances between genes; homologous recombination is therefore the molecular process that underpins much of the classical application of the science of genetics.

In bacteria, which do not of course undergo meiosis, homologous genetic recombination occurs in processes such as conjugation, a mating in which chromosomal DNA is transferred between two closely linked bacterial cells, or it can occur within a single cell between the two homologous chromosomes present during or immediately after replication.

This type of recombination serves at least three identifiable functions: (1) it contributes to genetic diversity in a population; (2) it provides in eukaryotes a transient physical link between chromatids that is apparently critical to the orderly segregation of chromosomes to the daughter cells in the first meiotic cell division; and (3) it contributes to the repair of several types of DNA damage.

The first and second functions are often of most interest to scientists studying genes, and homologous recombination is often described as a source of genetic diversity. However, the DNA repair function is almost certainly the most important role in the cell. DNA repair as described thus far is predicated on the fact that a DNA lesion in one strand can be accurately repaired because the genetic information is preserved in an undamaged complementary strand. In certain types of lesions, such as double-strand breaks, double-strand cross-links, or lesions left behind in single strands during replication (Fig. 24–27), the complementary strand is itself damaged or absent. When this occurs, the information required for accurate DNA repair must come from a separate, homologous chromosome, and the repair involves homologous recombination. These kinds of lesions commonly result from ionizing radiation and oxidative reactions, and their repair is critical to the production of viable gametes in eukaryotes and to the everyday existence of bacteria. Repair that is mediated by homologous genetic recombination is simply called **recombinational repair;** it is discussed in detail later in this chapter.

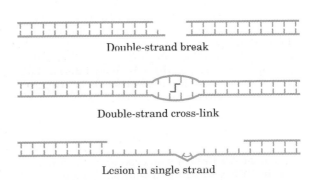

Double-strand break

Double-strand cross-link

Lesion in single strand

Figure 24–27 Types of DNA damage that require recombinational repair. In each case the damage to one strand cannot be repaired by mechanisms described earlier in this chapter because the complementary strand required to direct accurate repair is damaged or absent.

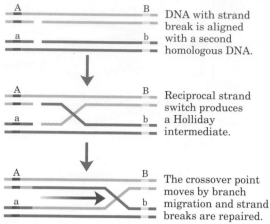

DNA with strand break is aligned with a second homologous DNA.

Reciprocal strand switch produces a Holliday intermediate.

The crossover point moves by branch migration and strand breaks are repaired.

The Holliday intermediate can be cleaved (or resolved) in two ways, producing two possible sets of products. Below, the orientation of the Holliday intermediate is changed to clarify differences in the two cleavage patterns:

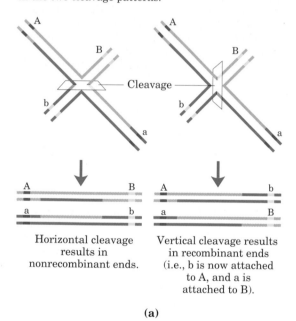

Horizontal cleavage results in nonrecombinant ends.

Vertical cleavage results in recombinant ends (i.e., b is now attached to A, and a is attached to B).

(a)

Figure 24–28 (a) The Holliday model for homologous genetic recombination. Two genes on the homologous chromosomes are indicated by the regions in red and blue. Each chromosome has different alleles of these genes, as indicated by uppercase and lowercase letters. Note which alleles are linked in the four final products. **(b)** A Holliday intermediate formed between two bacterial plasmids in vivo, as seen with the electron microscope.

Figure 24–29 Branch migration occurs within a branched DNA structure in which at least one strand is partially paired with each of two complementary strands. The branch "migrates" when a base pair to one of the two complementary strands is broken and replaced with a base pair to the second strand. In the absence of an enzyme to direct it, this process can move the branch spontaneously in either direction.

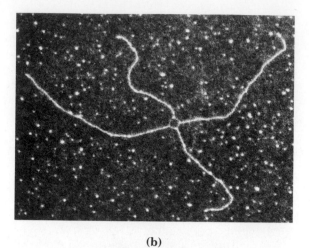

(b)

An important contribution to understanding homologous recombination is a model proposed by Robin Holliday in 1964, a version of which is presented in Figure 24–28. There are four key features of this model: (1) homologous DNAs are aligned by an unspecified mechanism; (2) one strand of each DNA is broken and joined to the other to form a crossover structure called a **Holliday intermediate;** (3) the region in which strands from different DNA molecules are paired, called **heteroduplex DNA,** is extended by **branch migration** (Fig. 24–29); and (4) two strands of the Holliday intermediate are cleaved and the breaks are repaired to form recombinant products. Homologous recombination can vary in many details from one species to another, but most of these steps are generally present in some form. Holliday intermediates have been observed in vivo in bacteria and in bacteriophage DNA (Fig. 24–28b). Note that there are two ways to cleave or "resolve" the Holliday intermediate so that the process is conservative, that is, so that the two products contain the same genes linked in the same linear order as in the substrates. If cleaved one way, the DNA flanking the heteroduplex region is recombined; if cleaved the other way, the flanking DNA is not recombined (Fig. 24–28a). Both outcomes are observed in vivo in both eukaryotes and prokaryotes.

Homologous recombination as illustrated in Figure 24–28 is a very elaborate process with subtle molecular consequences. To understand how this process affects genetic diversity, it is important to note that *homologous* does not necessarily mean *identical*. The two homologous chromosomes that are recombined may contain the same linear array

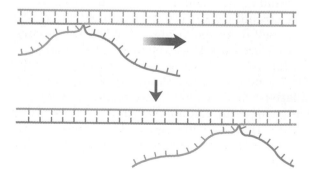

of genes, but each chromosome may have slightly different base sequences in some of these genes. In a human, for example, one chromosome may contain the normal gene for hemoglobin while the other contains a hemoglobin gene with the sickle-cell mutation. The differences may represent no more than a change in a base pair or two among millions of identical base pairs. Although homologous recombination does not change the linear array of genes, it can determine which of the different versions (or alleles) of the genes are linked together on a single chromosome (Fig. 24–28).

Recombination Requires Specific Enzymes

Enzymes have been isolated from both prokaryotes and eukaryotes that promote one or more steps of homologous recombination. Again, progress in both identifying and understanding these enzymes has been greatest in *E. coli*. Important recombination enzymes are encoded by the *rec*A, B, C, and D genes, and by the *ruv*C gene. The *rec*B, C, and D genes encode the RecBCD enzyme, which can initiate recombination by unwinding DNA and occasionally cleaving one strand. The RecA protein promotes all the central steps in the process: the pairing of two DNAs, formation of Holliday intermediates, and branch migration as described below. A novel class of nucleases that specifically cleave Holliday intermediates have also been isolated from bacteria and yeast. These nucleases are often called resolvases; the *E. coli* resolvase is the RuvC protein.

The RecBCD enzyme binds to linear DNA at one end and uses the energy of ATP to travel along the helix, unwinding the DNA ahead and rewinding it behind (Fig. 24–30). Rewinding is slower than unwinding so that a single-stranded bubble is gradually formed and enlarged. The single strands in the bubble are cut when the enzyme encounters a certain sequence called *chi,* (5′)GCTGGTGG(3′). There are about 1,000 of these sequences in the *E. coli* genome, and they have the effect of increasing the frequency of recombination in the regions where they occur. Sequences that enhance recombination frequency have also been identified in several other organisms.

The RecA protein is unusual among proteins involved in DNA metabolism in that its active form is an ordered, helical filament that assembles cooperatively on DNA and can involve thousands of RecA monomers (Fig. 24–31). Formation of this filament normally occurs on single-stranded DNA such as that produced by the RecBCD enzyme. The filament will also form on a duplex DNA with a single-stranded gap, in which case the first RecA monomers bind to the single-stranded DNA in the gap and then filament assembly rapidly envelops the neighboring duplex.

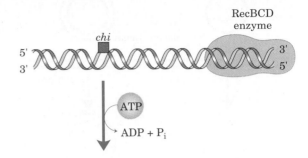

Helicase activity of enzyme produces single-stranded bubbles.

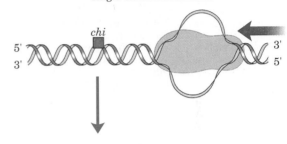

On reaching a *chi* sequence, nuclease activity cleaves the adjacent single strand.

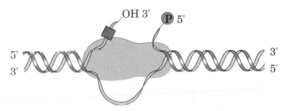

Figure 24–30 Helicase and nuclease activities of the RecBCD enzyme. Unwinding of DNA ahead of the moving enzyme and slower rewinding behind create single-stranded bubbles. One strand is cleaved when the enzyme encounters a *chi* sequence. Movement of the enzyme requires ATP hydrolysis. This enzyme is believed to help initiate homologous genetic recombination in *E. coli*.

Figure 24–31 (a) Nucleoprotein filament of RecA protein on single-stranded DNA, as seen with the electron microscope. The striations make evident the right-handed helical structure of the filament. **(b)** A computer enhancement of the structure seen with the electron microscope.

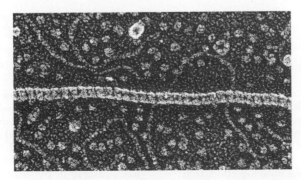

(a)

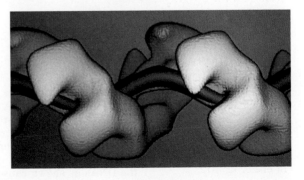

(b)

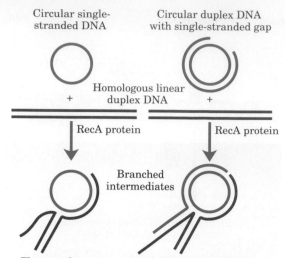

Circular single-
stranded DNA

Circular duplex DNA
with single-stranded gap

Homologous linear
duplex DNA

+   +

RecA protein   RecA protein

Branched
intermediates

The complementary strand of the linear DNA is
paired with a circular single strand. The other
linear strand is displaced (left) or paired with its
complement in the circular duplex to yield a
Holliday structure (right).

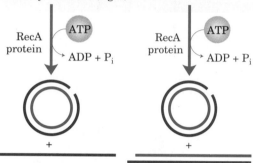

RecA
protein   ATP   RecA
protein   ATP

ADP + P$_i$ ADP + P$_i$

+   +

Continued branch migration yields a circular duplex
with a nick and a displaced linear strand (left), or a
partially single-stranded linear duplex (right).

Figure 24–32 DNA strand-exchange reactions promoted by RecA protein in vitro. Strand exchange
involves the separation of one strand of a duplex
DNA from its complement and transfer to an alternative complementary strand to form a new duplex
(heteroduplex) DNA. The transfer is gradual, and a
branched intermediate is formed. Converting the
intermediate to products involves a RecA protein–
facilitated branch migration. The reaction can involve three strands (left), or a reciprocal exchange
can occur between two homologous duplexes (four
strands in all) (right). In the case of four strands,
the branched intermediate is a Holliday structure.
RecA protein promotes these reactions with the
energy of ATP hydrolysis.

A useful in vitro paradigm for the recombination activities of this
RecA filament is a reaction called DNA strand exchange (Fig. 24–32).
The DNA within the filament is aligned with a second duplex DNA,
and strands are exchanged between the two DNAs to create heteroduplex DNA. The exchange occurs at a rate of about 3 to 6 base pairs/s
and progresses in a unique direction, 5′→3′ relative to the single-
stranded DNA within the filament. As shown in Figure 24–32, this
reaction can involve either three or four strands, and in the latter case
a Holliday structure is an intermediate in the process.

A more complete sequence of events as presented in Figure 24–33
introduces two additional features of a RecA protein–mediated three-
strand exchange reaction. First, the alignment of the two DNAs may
involve the formation of an unusual DNA structure in which three
strands are interwound. The details of the structure are unknown.
Second, because DNA is a helical structure, strand exchange requires
an ordered rotation of the two aligned DNAs. This brings about a spooling action that moves the branch point along the helix. ATP is hydrolyzed by RecA protein as this reaction proceeds.

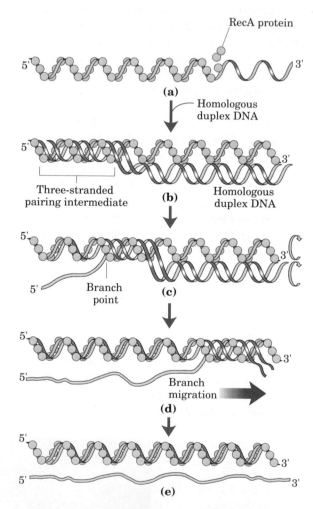

Figure 24–33 Model for RecA protein–mediated
DNA strand exchange. A three-strand reaction is
shown. **(a)** In the first step, RecA protein forms a
filament on the single-stranded DNA. **(b)** A homologous duplex wraps around this complex, forming a
three-stranded pairing intermediate. **(c)** Rotation of
the DNAs as shown causes a spooling effect that

moves the three-stranded region from left to right.
Within the three-stranded region, one of the
strands of the incoming duplex switches pairing
partners, and its original complement is displaced.
As rotation continues **(d, e),** the displaced strand is
eventually separated entirely. ATP is hydrolyzed by
RecA protein in the course of this strand exchange.

Once a Holliday intermediate has been formed, enzymes involved in completing recombination include topoisomerases, a resolvase, other nucleases, DNA polymerase I or III, and DNA ligase. The RuvC protein (M_r 20,000) of *E. coli* cleaves Holliday intermediates in the manner depicted in Figure 24–28a. Many details of the reactions carried out by the recombination enzymes and the coordination of these reactions in the cell are not yet understood.

Homologous Recombination Is an Important Pathway for DNA Repair

Recombination provides an avenue for accurate DNA repair when the necessary sequence information is not available from a strand paired with the damaged strand (see Fig. 24–27). To illustrate the role of recombination in DNA repair, we will examine the fate of lesions encountered during normal replication and left behind unreplicated in single-stranded DNA (see Figs. 24–24, 24–27). Repair of these lesions is called postreplication repair, and in *E. coli* this process requires RecA protein.

A plausible pathway for postreplication repair is presented in Figure 24–34. A lesion in an unpaired DNA strand cannot be excised, because this would leave breaks in both DNA strands, an outcome that could be lethal to the cell. To prevent chromosomal breakage and allow for repair, the region containing the lesion must acquire a complementary strand. The recombination pathway makes use of the homologous DNA on the other leg of the replication fork. A RecA protein–mediated strand-exchange reaction transfers an undamaged complementary strand from the homologous DNA, converting the region containing the lesion into heteroduplex DNA. A notable property of RecA protein–mediated DNA strand exchange is that it proceeds efficiently past most DNA lesions with the aid of energy supplied by ATP hydrolysis. Once the lesion is made part of a duplex, the damage can be readily repaired. The repair of lesions of this type is clearly a major function of the homologous recombination system of every cell.

Site-Specific Recombination Results in Precise DNA Rearrangements

We now turn from general recombination, which can involve any two homologous sequences, to a very different type of recombination that is limited to specific sequences. Site-specific recombination reactions occur in virtually every cell, but their functions are specialized and vary greatly from one species to the next. These functions include the regulation of expression of certain genes, the promotion of programmed DNA rearrangements that occur during development in many organisms, and DNA rearrangements tied to the replication cycle of some viral and plasmid DNAs, as illustrated later. A site-specific recombination system consists of an enzyme called a recombinase and a short (20 to 200 base pairs, depending on the system) unique DNA sequence where the recombinase acts (the recombination site). Some systems also include one or more auxiliary proteins that regulate the timing or outcome of the reaction.

From in vitro studies of more than a dozen site-specific recombination systems, some principles have emerged. The fundamental reaction pathway for many systems is illustrated in Figure 24–35. A recombinase recognizes and binds to each of two recombination sites on dif-

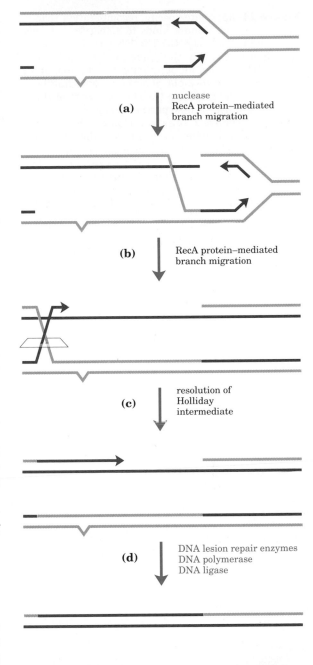

Figure 24–34 Model for the role of RecA protein in postreplication repair. **(a)** A region of single-stranded DNA containing a lesion remains unreplicated. **(b)** A RecA protein–mediated strand exchange transfers a complementary strand from the homologous DNA. **(c)** A RecA protein–mediated branch migration results in formation of a Holliday intermediate, which is then cleaved. **(d)** The lesion can now be repaired, and the transferred strand can be replaced by DNA polymerase and ligase activities.

Figure 24–35 A site-specific recombination reaction. **(a)** A recombinase binds to a specific sequence or recombination site. **(b)** The DNA is cleaved at specific points within the sequence. The nucleophile is the OH group of an active-site Tyr residue, and the product is a covalent phosphotyrosine link between protein and DNA. **(c)** The two parts of the site are rejoined to new partners. This first cleavage/exchange results in a Holliday intermediate. **(d)** An isomerization step moves the crossover to the point where a second cleavage/exchange **(e, f, g)** completes the reaction by a reversal of the first three steps. The original sequence of the recombination site is regenerated, but the DNA flanking the site on either side is recombined. Areas in red denote sequences identical in the two DNAs. Note that the size of the crossover region has been exaggerated in this drawing.

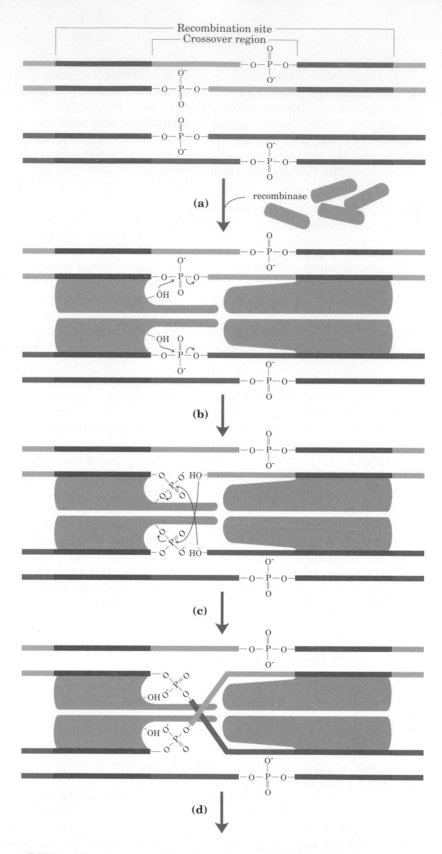

ferent DNA molecules or within the same DNA. One DNA strand in each site is cleaved at a specific point within the site, and the recombinase becomes covalently linked to the DNA at the cleavage site through a phosphotyrosine (sometimes phosphoserine) bond. The transient protein–DNA linkage preserves the phosphodiester bond lost in cleaving the DNA, and high-energy cofactors such as ATP are unnecessary in subsequent steps. The cleaved DNA strands are rejoined to new partners, with new phosphodiester bonds created at the expense of the

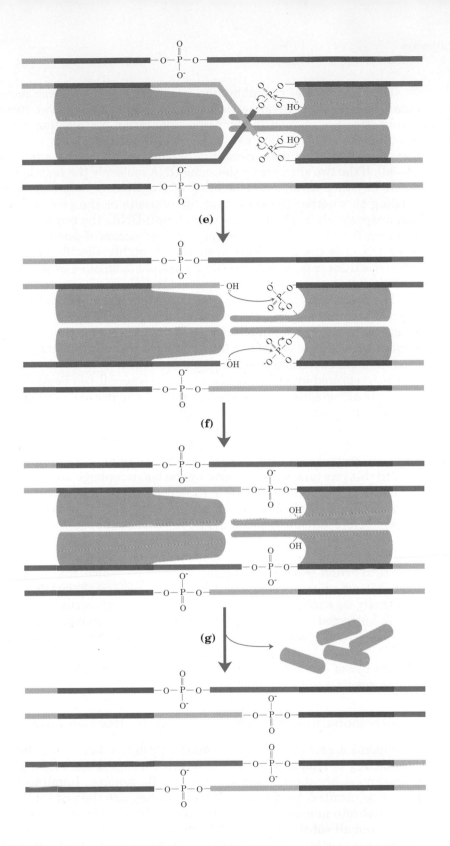

protein–DNA linkage. The result of this initial breakage and rejoining process is a Holliday intermediate. To complete the reaction, the process must be repeated at a second point within each of the two recombination sites. In some systems both strands of each recombination site may be cut concurrently and rejoined to new partners without the Holliday intermediate. The exchange in each case is reciprocal and precise so that the recombination sites are regenerated after the reaction. In summary, the recombinase can be viewed as a site-specific endonuclease and ligase in one package.

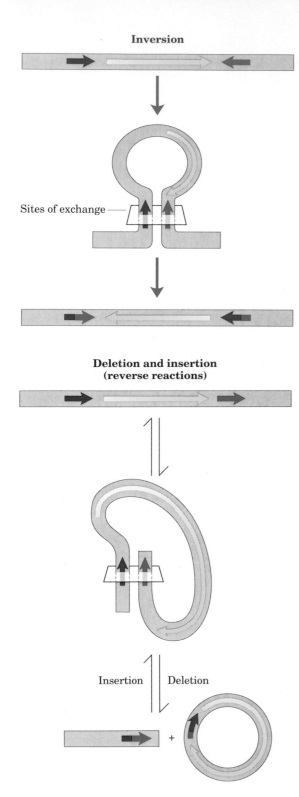

Inversion

Sites of exchange

**Deletion and insertion
(reverse reactions)**

Insertion   Deletion

+

Figure 24–36 Possible outcome of site-specific recombination, depending on location and orientation of recombination sites (red and blue) in a double-stranded DNA molecule. Inversion and deletion and insertion are illustrated. Orientation here refers to the order of nucleotides in the recombination site, not the 5′→3′ direction.

The sequences of recombination sites recognized by these recombinases are partially asymmetric (nonpalindromic), and the two recombining sites are aligned in the same orientation for reaction by the recombinase. The reaction can have several outcomes, depending on the relative location and orientation of the recombination sites (Fig. 24–36). If the two sites are on the same DNA molecule the reaction will result in either inversion or deletion of the DNA between them, depending on whether the sites have the opposite or the same orientation, respectively. If the sites are on different DNAs the recombination is intermolecular, and an insertion reaction occurs if one or both of these DNAs is circular. Some systems are highly specific for one of these reactions (e.g., inversions) and will not act on sites in the wrong relative orientation.

The first site-specific recombination system identified and studied in vitro was that encoded by the bacteriophage λ. When λ phage DNA enters an *E. coli* cell, a complex series of regulatory events ensues that commits the DNA to one of two fates: either it is replicated and used to produce more bacteriophages (in which case the host cell is destroyed), or it is integrated into the host chromosome where it can be replicated passively along with the host chromosome for many cell generations. Integration is accomplished by a phage-encoded recombinase called the λ integrase, acting at recombination sites (attachment sites in the bacteriophage λ system) on the phage and bacterial DNAs called *att*P and *att*B, respectively (Fig. 24–37). Several auxiliary proteins also are used in this reaction, some encoded by the bacteriophage and others by the bacterial host cell. Note that a site-specific recombination reaction (Fig. 24–35) is chemically symmetric in terms of the chemical bonds present before and after, and it should have an equilibrium constant of 1.0. A major function of the auxiliary proteins in λ integration is to alter this equilibrium by permitting integration and/or preventing the reverse reaction (excision). The mechanism by which this is accomplished is not understood in detail. When the bacteriophage DNA must eventually be excised from the chromosome (which occurs when the cell is subjected to a variety of environmental stresses), the site-specific excision reaction uses a different set of auxiliary proteins (Fig. 24–37).

The use of site-specific recombination to regulate gene expression will be considered in Chapter 27.

Immunoglobulin Genes Are Assembled by Recombination

An important example of a programmed recombination event that occurs during development is the generation of immunoglobulin genes from gene segments that are separate in the genome. Immunoglobulins (or antibodies), produced by B lymphocytes, are the foot soldiers of the vertebrate immune system—the molecules that bind to infectious agents and all substances foreign to the organism. A mammal such as a human is capable of producing many millions of different antibodies with distinct binding specificities. However, the human genome contains only about 100,000 genes. Recombination allows an organism to produce an extraordinary diversity of antibodies from a relatively small amount of DNA-coding capacity.

Vertebrates generally produce multiple classes of immunoglobulins. To illustrate how antibody diversity is generated, we will focus on the immunoglobulin G (IgG) class from humans. Immunoglobulins consist of two heavy and two light polypeptide chains (Fig. 24–38a).

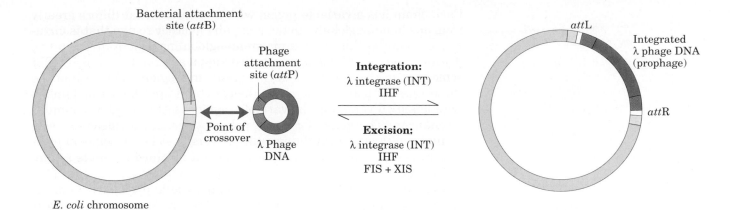

E. coli chromosome

Figure 24–37 The integration and excision of bacteriophage λ DNA at the chromosomal target site. The attachment site on the λ phage DNA (*att*P) shares only 15 base pairs of complete homology with the bacterial site (*att*B) in the region of the crossover. The reaction generates two new attachment sites (*att*R and *att*L) flanking the integrated phage DNA. The recombinase is the λ integrase or INT protein. Integration and excision use different attachment sites and different auxiliary proteins. Excision uses the proteins XIS, encoded by the bacteriophage, and FIS, encoded by the bacterium. Both reactions require the protein IHF (*integration host factor*), encoded by the bacterium.

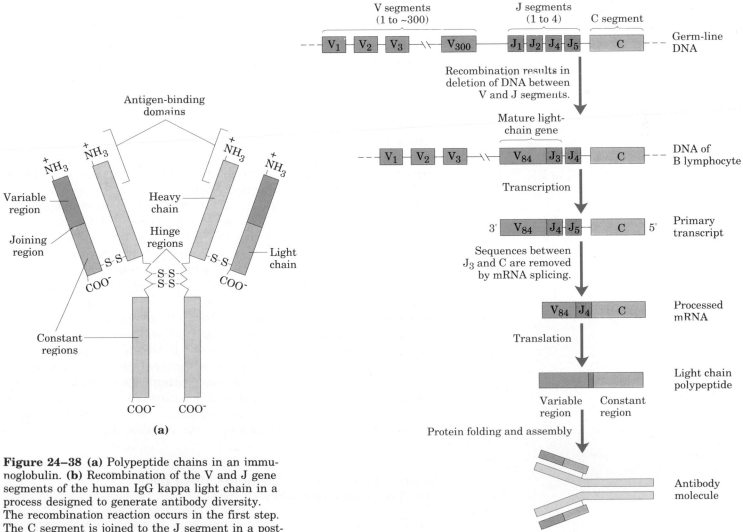

Figure 24–38 (a) Polypeptide chains in an immunoglobulin. (b) Recombination of the V and J gene segments of the human IgG kappa light chain in a process designed to generate antibody diversity. The recombination reaction occurs in the first step. The C segment is joined to the J segment in a post-transcriptional splicing reaction (Chapter 25).

Each chain has a variable region with a sequence that differs greatly from one immunoglobulin to the next, and another region that is virtually constant within a class of immunoglobulins. There are also two distinct families of light chains, called kappa and lambda, which differ somewhat in the sequences of their constant regions. For each of the three types of polypeptide chain (heavy chain, and kappa or lambda light chain), diversity in the variable regions is generated by a similar mechanism. The genes for these polypeptides are divided into segments, and clusters containing multiple versions of each segment exist in the genome. One version of each segment is joined to create a complete gene.

The organization of the DNA encoding the kappa light chains of human IgG and the process by which a mature kappa light chain is generated are shown in Figure 24–38b. In undifferentiated cells, the coding information for this polypeptide chain is separated into three segments. The V (*variable*) segment encodes the first 95 amino acid residues of the variable region, the J (*joining*) segment encodes the remaining 12 amino acid residues of the variable region, and the C segment encodes the *constant* region. There are about 300 different V segments, 4 different J segments, and 1 C segment. As a stem cell in the bone marrow differentiates to form a mature B lymphocyte, one V and one J are brought together by site-specific recombination. This is effectively a programmed DNA deletion event, and the intervening DNA is discarded. There are $300 \times 4 = 1,200$ possible combinations. The recombination process is not as precise as the site-specific recombination described earlier, and some additional variation occurs in the sequence at the V–J junction that adds a factor of at least 2.5 to the total variation possible, so that about $2.5 \times 1,200 = 3,000$ different V–J combinations can be generated. The final joining of this V–J combination to the C region is accomplished by an RNA-splicing reaction after transcription (Fig. 24–38b). RNA splicing will be described in the next chapter. The genes for the heavy chains and lambda light chains are formed similarly. For heavy chains, there are more gene segments and more than 5,000 possible combinations. Because any heavy chain can combine with any light chain to generate an immunoglobulin, there are at least $3,000 \times 5,000$ or 1.5×10^7 possible IgGs. Additional diversity is generated because the V sequences are subject to high mutation rates (of unknown mechanism) during B-lymphocyte differentiation. Each mature B lymphocyte produces only one type of antibody, but the range of antibodies produced by different cells is clearly enormous. The enzymes that catalyze these gene rearrangements have not been isolated, but sequences critical to the V–J joining process that are presumably recognized by these enzymes have been identified.

This recombination process helps to illustrate the principle that recombination does not destroy the integrity of the genetic material that the replication and repair processes attempt to maintain. Here we see a precisely orchestrated process that occurs only in specialized cells (germ-line DNA is not affected) and enables the organism to make much more efficient use of its genetic information resource.

Transposable Genetic Elements Move from One Location to Another

Finally, we consider the recombination of transposable elements or **transposons.** Transposons are segments of DNA, found in virtually

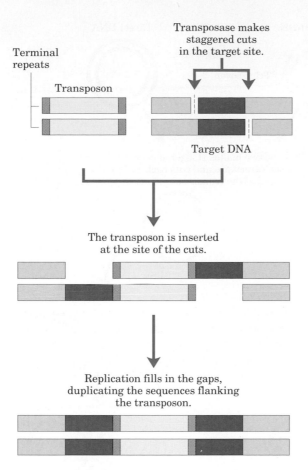

Terminal repeats

Transposon

Transposase makes staggered cuts in the target site.

Target DNA

The transposon is inserted at the site of the cuts.

Replication fills in the gaps, duplicating the sequences flanking the transposon.

Figure 24–39 Duplication of the DNA sequence at a target site when a transposon is inserted. The duplicated sequences are shown in red. These sequences are generally only a few base pairs long, and their size (relative to that of a typical transposon) is greatly exaggerated in this drawing.

all cells, that move or "hop" from one place on a chromosome (the donor site) to another on the same or a different chromosome (the target site). No homology is usually required for the movement, called transposition, to occur. The new location is chosen more or less randomly. Because insertion of a transposon in an essential gene could kill the cell, the events are tightly regulated and occur very infrequently (perhaps once in a million cell divisions).

There are two classes of transposons in bacteria. **Insertion sequences** (simple transposons) contain only the sequences required for their transposition and the genes for proteins (transposases) that promote the process. **Complex transposons** contain one or more genes besides those needed for transposition. These additional genes often confer resistance to antibiotics. The spread of antibiotic-resistance elements through disease-causing bacterial populations, mediated in part by transposition, is rendering some antibiotics ineffectual (p. 796).

Bacterial transposons vary in structure, but most have short repeats at the two ends of the element that serve as binding sites for the transposase. When transposition occurs, a short sequence at the target site (5 to 10 base pairs) is duplicated to form an additional short repeat flanking each end of the inserted transposon (Fig. 24–39). These short terminal repeats reflect the cutting mechanism used to insert a transposon into the DNA at a new location.

There are two general pathways for transposition in bacteria (Fig. 24–40). In direct or simple transposition, cuts are made on each side of the transposon to excise it, and the transposon moves to a new location, leaving a double-stranded break in the DNA from which it came. At the target site, a staggered cut is made, the transposon is spliced into the break, and some DNA replication is needed to duplicate the target site

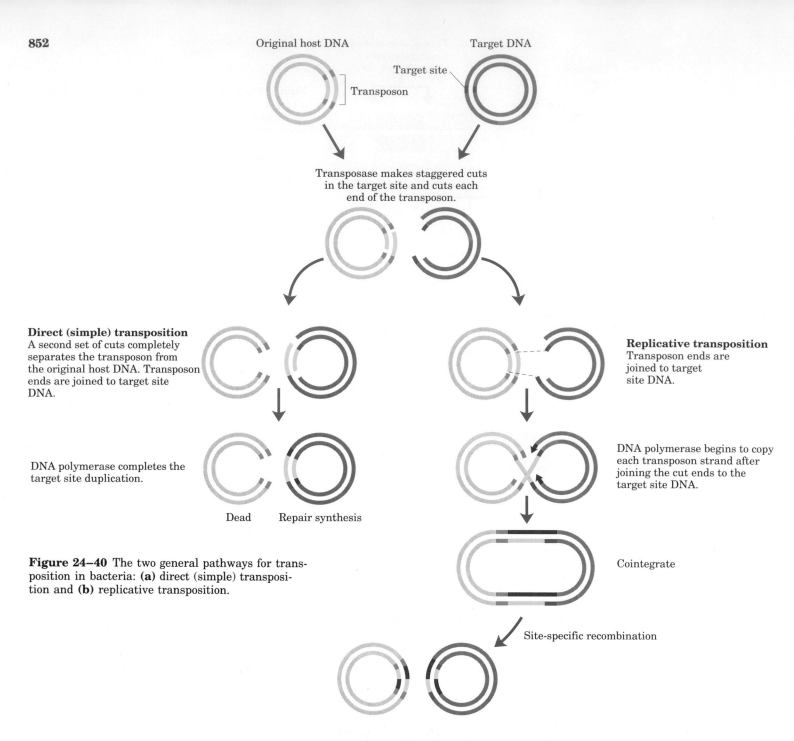

Original host DNA

Target DNA

Target site

Transposon

Transposase makes staggered cuts in the target site and cuts each end of the transposon.

Direct (simple) transposition
A second set of cuts completely separates the transposon from the original host DNA. Transposon ends are joined to target site DNA.

Replicative transposition
Transposon ends are joined to target site DNA.

DNA polymerase completes the target site duplication.

DNA polymerase begins to copy each transposon strand after joining the cut ends to the target site DNA.

Dead   Repair synthesis

Cointegrate

Figure 24–40 The two general pathways for transposition in bacteria: **(a)** direct (simple) transposition and **(b)** replicative transposition.

Site-specific recombination

sequence (Fig. 24–40a). In replicative transposition, the entire transposon is replicated so that a copy is left behind in its original donor location (Fig. 24–40b). An intermediate in the latter reaction is a **cointegrate,** in which the donor region is covalently linked to DNA at the target site. Two complete copies of the transposon are present in this intermediate, arranged in the same relative orientation in the DNA. This intermediate is converted to products in some well-characterized transposons by site-specific recombination in which specialized recombinases promote the required deletion reaction.

Transposons are also found in eukaryotes. These are structurally similar to the bacterial transposons, and some utilize similar transposition mechanisms. However, in some cases the mechanism of transposition is quite different and appears to involve an RNA intermediate. These transposons will be described in the next chapter, in which we leave DNA metabolism and move to a discussion of RNA.

Summary

The integrity of the structure and nucleotide sequence of DNA is of utmost importance to the cell. This is reflected in the complexity and redundancy of the enzyme systems that participate in DNA replication, repair, and recombination.

Replication of DNA occurs with very high fidelity and within a designated time period in the cell cycle. Replication is semiconservative, with each strand acting as a template for a new daughter strand. The reaction starts at a sequence in the DNA called the origin, and usually proceeds bidirectionally from that point. DNA is synthesized in the $5' \rightarrow 3'$ direction by DNA polymerases. At the replication fork, the leading strand is synthesized continuously and in the same direction as replication fork movement. The lagging strand is synthesized discontinuously. The fidelity of DNA replication is maintained by (1) base selection by the polymerase, (2) a $3' \rightarrow 5'$ proofreading exonuclease activity that is part of most DNA polymerases, and (3) a specific repair system that repairs any mismatches left behind after replication.

Most cells have several DNA polymerases. In *E. coli,* DNA polymerase III is the primary replication enzyme. DNA polymerase I is responsible for special functions during replication, recombination, and repair. DNA polymerase II has a specialized replication activity that allows it to replicate past DNA lesions in error-prone DNA repair. Replication of the *E. coli* chromosome involves many enzymes and protein factors organized into complexes. Initiation of replication requires binding of DnaA protein to the origin, strand separation, and the entry of the DnaB and DnaC proteins to set up two replication forks. The action of DnaA is associated with the *E. coli* membrane and is regulated by the action of acidic phospholipids. Initiation is the only phase of replication that is regulated. The process of elongation has different requirements for each strand. DNA strands are separated by helicases, and the resulting topological strain is relieved by topoisomerases. Single-strand DNA binding proteins stabilize the separated strands. In synthesis of the lagging strand, the primosome protein complex moves with the fork and regulates the synthesis of RNA primers by primase. Synthesis of the leading and lagging strands by DNA polymerase III may be coupled. RNA primers are removed and replaced with DNA by DNA polymerase I, and nicks are sealed by DNA ligase.

A similar pattern of replication occurs in eukaryotic cells, but eukaryotic chromosomes have multiple replication origins. Several eukaryotic DNA polymerases have been identified.

Every cell also has multiple and sometimes redundant systems for DNA repair. Mismatch repair in *E. coli* is directed by transient undermethylation of (5')GATC sequences on the newly synthesized strand after replication. Other systems recognize and repair damage caused by environmental agents such as radiation and alkylating agents, and damage caused by spontaneous reactions of nucleotides. Some repair systems recognize and excise only damaged or incorrect bases (e.g., uracil), leaving an AP (apurinic or apyrimidinic) site in the DNA. This is repaired by excising and replacing the segment of DNA containing the AP site. Other excision repair systems recognize and remove pyrimidine dimers and other modified nucleotides. Some types of DNA damage can also be repaired by direct reversal of the reaction causing the damage: pyrimidine dimers are directly converted to monomeric pyrimidines by photolyase, and the methyl group in O^6-methylguanine is removed by a specific methyltransferase. Error-prone repair is a specialized and mutagenic replication process observed when DNA damage is so heavy that the need for some replication outweighs the need to avoid errors.

DNA sequences are rearranged in recombination reactions. Homologous genetic recombination occurs between any two DNAs that share sequence homology. This reaction takes place in meiosis (in eukaryotes) and is one of the processes that creates genetic diversity. Homologous recombination also is needed for repair of some types of DNA damage. A Holliday intermediate in which a crossover has occurred between the strands of two homologous DNAs is formed during the process. In *E. coli,* the RecA protein promotes formation of Holliday intermediates and branch migration to extend heteroduplex DNA.

Site-specific recombination occurs only at specific target sequences and can also involve a Holliday intermediate. The recombinases cleave the DNA at specific points and ligate the strands to new partners. This type of recombination is found in virtually all cells, and its many functions include DNA integration and regulation of gene expression. In vertebrates, a programmed recombination reaction related to site-specific recombination is used to join immunoglobulin gene segments to form immunoglobulin genes during B-lymphocyte differentiation. Some small segments of DNA, called transposons, are capable of moving from one point in a chromosome to another point in the same or another chromosome. These elements are found in virtually all cells.

Further Reading

General

Kornberg, A. & Baker, T.A. (1991) *DNA Replication,* 2nd edn, W.H. Freeman and Company, New York.
An excellent primary source.

Kucherlapati, R. & Smith, G.R. (eds) (1988) *Genetic Recombination,* American Society of Microbiology, Washington, DC.
Excellent reviews on a wide assortment of recombination topics.

Richardson, C.C. & Lehman, I.R. (eds) (1990) *Molecular Mechanisms in DNA Replication and Recombination,* Alan R. Liss, Inc., New York.
A collection of a papers from a major symposium on the topic.

Replication

Bramhill, D. & Kornberg, A. (1988) A model for initiation at origins of DNA replication. *Cell* **54,** 915–918.

Burgers, P.M.J. (1989) Eukaryotic DNA polymerases α and δ: conserved properties and interactions, from yeast to mammalian cells. *Prog. Nucleic Acid Res. Mol. Biol.* **37,** 235–280.

Campbell, J. (1988) Eukaryotic DNA replication: yeast bares its ARSs. *Trends Biochem. Sci.* **13,** 212–217.

Echols, H. & Goodman, M.F. (1991) Fidelity mechanisms in DNA replication. *Annu. Rev. Biochem.* **60,** 477–511.

Marians, K.J. (1992) Prokaryotic DNA replication. *Annu. Rev. Biochem.* **61,** 673–719.

McHenry, C.S. (1991) DNA polymerase III holoenzyme. *J. Biol. Chem.* **266,** 19127–19130.

Radman, M. & Wagner, R. (1988) The high fidelity of DNA duplication. *Sci. Am.* **259** (August), 40–46.

Wang, T.S.-F. (1991) Eukaryotic DNA polymerases. *Annu. Rev. Biochem.* **60,** 513–552.

Repair

Friedberg, E.C. (1985) *DNA Repair,* W.H. Freeman and Company, New York.

Modrich, P. (1991) Mechanisms and biological effects of mismatch repair. *Annu. Rev. Genet.* **25,** 229–253.

Sancar, A. & Sancar, G.B. (1988) DNA repair enzymes. *Annu. Rev. Biochem.* **57,** 29–67.

Recombination

Berg, D.E. & Howe, M.M. (eds) (1989) *Mobile DNA,* American Society for Microbiology, Washington, DC.
Reviews covering many topics related to transposition.

Cox, M.M. & Lehman, I.R. (1987) Enzymes of genetic recombination. *Annu. Rev. Biochem.* **56,** 229–262.

Craig, N.L. (1988) The mechanism of conservative site-specific recombination. *Annu. Rev. Genet.* **22,** 77–105.

Landy, A. (1989) Dynamic, structural, and regulatory aspects of λ site-specific recombination. *Annu. Rev. Biochem.* **58,** 913–949.
A thorough description of the protein–DNA complexes involved in this reaction.

Mizuuchi, K. (1992) Transpositional recombination: insights from Mu and other elements. *Annu. Rev. Biochem.* **61,** 1011–1051.

Radding, C.M. (1991) Helical interactions in homologous pairing and strand exchange driven by RecA protein. *J. Biol. Chem.* **266,** 5355–5358.
A good, short summary.

Roca, A.I. & Cox, M.M. (1990) The RecA protein: structure and function. *Crit. Rev. Biochem. Mol. Biol.* **25,** 415–456.

Taylor, A.F. (1992) Movement and resolution of Holliday junctions by enzymes from *E. coli. Cell* **69,** 1063–1065.

Problems

1. *Conclusions from the Meselson–Stahl Experiment* The Meselson–Stahl experiment proved that DNA undergoes semiconservative replication in *E. coli*. In the "dispersive" model of DNA replication, the parent DNA strands are cleaved into pieces of random size and are then joined with pieces of the newly replicated DNA to yield daughter duplexes in which, in the Meselson–Stahl experiment, both strands would contain random segments of both heavy and light DNA. Explain how the results of the Meselson–Stahl experiment ruled out such a model.

2. *Number of Turns in the* E. coli *Chromosome* How many turns must be unwound during replication of the *E. coli* chromosome? The chromosome contains about 4.7×10^6 base pairs.

3. *Replication Time in* E. coli From the data in this chapter, how long would it take to replicate the *E. coli* chromosome at 37 °C, if two replication forks start from the origin? Under some conditions *E. coli* cells can divide every 20 min. Can you suggest how this is possible?

4. *Base Composition of DNAs Made from Single-Stranded Templates* Determine the base composition you might expect in the total DNA synthesized by DNA polymerase on templates provided by an equimolar mixture of the two complementary strands of circular bacteriophage ϕX174 DNA. The base composition of one strand is A, 24.7%; G, 24.1%; C, 18.5%; and T, 32.7%. What assumption is necessary to answer this problem?

5. *Okazaki Fragments* In the replication of the *E. coli* chromosome, about how many Okazaki fragments would be formed? What factors guarantee that the numerous Okazaki fragments are assembled in the correct order in the new DNA?

6. *Leading and Lagging Strands* List and compare the precursors and enzymes needed to make the leading versus lagging strands during DNA replication in *E. coli*.

7. *Fidelity of Replication of DNA* What factors participate in ensuring the fidelity of replication during the synthesis of the leading strand of a new DNA? Would you expect the lagging strand to be made with the same fidelity as the leading strand? Give reasons for your answers.

8. *DNA Repair Mechanisms* Vertebrate and plant cells often methylate cytosine in DNA to form 5-methylcytosine (see Fig. 12–5a). In these same cells, there is a specialized repair system that recognizes G–T mismatches and repairs them to G≡C base pairs. Rationalize this repair system in terms of the presence of 5-methylcytosine in the DNA.

9. *Holliday Intermediates* How are the Holliday intermediates formed in homologous genetic recombination and in site-specific recombination different?

10. *DNA Recombination* A circular DNA molecule is converted to two smaller circles by an enzyme or enzymes in a crude cellular extract. What types of recombination could account for this reaction, and what else must you know to determine which type it is?

RNA Metabolism

The expression of the genetic information contained in a segment of DNA always involves the generation of a molecule of RNA. At first glance, a strand of RNA may seem quite similar to a strand of DNA, differing only in the hydroxyl group at the 2′ position and the substitution of uracil for thymine. As we will see, however, these small differences confer on RNA the potential for much greater structural diversity than DNA, a diversity that allows RNA to assume a variety of cellular functions. RNA molecules not only carry and express genetic information, they can also act as catalysts.

RNA is the only macromolecule known to have both informational and catalytic functions, leading to much speculation that it may have been the essential chemical intermediate in the development of life on this planet. The discovery of catalytic RNAs has changed the very definition of the word "enzyme." Many RNAs are also complexed with proteins, forming complicated biochemical machines with a wide variety of functions.

With the exception of the RNA genomes of certain viruses, all RNA molecules are derived from information permanently stored in DNA. In a process called **transcription,** an enzyme system converts the genetic information of a segment of DNA into an RNA strand with a base sequence complementary to one of the DNA strands. Three major kinds of RNA are produced. **Messenger RNA (mRNA)** carries the sequences that encode the amino acid sequence of one or more polypeptides specified by a gene or set of genes in the chromosomes. **Transfer RNA (tRNA)** is an adapter that reads the information encoded in the mRNA and transfers the appropriate amino acid to the growing polypeptide chain during protein synthesis. **Ribosomal RNA (rRNA)** molecules associate with proteins to form the intricate protein synthetic machine, the ribosome. In addition, there are many specialized RNAs with regulatory or catalytic functions.

Replication and transcription differ in one important respect. During replication the entire chromosome is copied to yield daughter DNAs identical to the parent DNA, whereas transcription is selective: only particular genes or groups of genes are transcribed at any one time. The transcription of DNA can therefore be regulated so that only genetic information needed by the cell at a particular moment is transcribed. Specific regulatory sequences indicate the beginning and end of the segments of DNA to be transcribed, as well as which DNA strand is to be used as template. Regulation also involves a variety of proteins that will be described in more detail in Chapter 27.

In this chapter we begin by describing the synthesis of RNA on a DNA template, a process similar in many respects to DNA synthesis. We then turn to postsynthetic processing and turnover of RNA molecules. Many of the specialized functions of RNA will be encountered in this discussion of the posttranscriptional reactions. Indeed, the substrates for RNA enzymes are generally other RNA molecules. We conclude the chapter with an examination of systems in which RNA rather than DNA serves as a template for the transfer of genetic information. Here, the information pathways are expanded and come full circle, and template-directed nucleic acid synthesis is revealed as a process with standard rules that apply regardless of whether the template or product is RNA or DNA. This biological interconversion of DNA and RNA as information carriers leads finally to a discussion of the origin of biological information.

DNA-Dependent Synthesis of RNA

We can most usefully begin our discussion of RNA synthesis by comparing it with DNA replication as described in Chapter 24. Transcription is very similar to replication in terms of chemical mechanism, polarity (direction of synthesis), and use of a template. The two processes differ, however, in that transcription does not require a primer, it generally involves only short segments of a DNA molecule, and within those segments only one of the two DNA strands serves as a template. We begin our discussion by introducing the enzymes responsible for transcription.

RNA Is Synthesized by RNA Polymerases

The discovery of DNA polymerase and its dependence on a DNA template encouraged a search for an enzyme that synthesizes an RNA strand complementary to a DNA template. Such an enzyme, capable of forming an RNA polymer from ribonucleoside 5'-triphosphates, was isolated from bacterial extracts in 1959 by four independent research groups. This enzyme, **DNA-directed RNA polymerase,** requires, in addition to a DNA template, all four ribonucleoside 5'-triphosphates (ATP, GTP, UTP, and CTP) as precursors of the nucleotide units of RNA, as well as Mg^{2+}. The purified enzyme also contains Zn^{2+}. The fundamental chemistry of RNA synthesis has much in common with DNA synthesis. RNA polymerase elongates an RNA strand by adding ribonucleotide units to the 3'-hydroxyl end of the RNA chain and thus builds RNA chains in the 5'→3' direction. The 3'-hydroxyl group acts as nucleophile, attacking at the α-phosphate of the incoming ribonucleoside triphosphate (as illustrated for DNA synthesis in Fig. 24–5) and releasing pyrophosphate. The overall reaction is

$$\underset{\text{RNA}}{(NMP)_n} + NTP \longrightarrow \underset{\substack{\text{Lengthened} \\ \text{RNA}}}{(NMP)_{n+1}} + PP_i \qquad (25\text{--}1)$$

RNA polymerase requires DNA for activity and is most active with a double-stranded DNA as template. Only one of the two DNA strands is used as a template, copied in the 3'→5' direction (antiparallel to the new RNA strand) just as in DNA replication. Each nucleotide in the newly formed RNA is selected by Watson–Crick base-pairing interactions; uridylate (U) residues are inserted in the RNA opposite to adenylate residues in the DNA template, adenylate residues are inserted opposite to thymidylate residues. Guanylate and cytidylate residues in DNA specify cytidylate and guanylate, respectively, in the new RNA strand.

Unlike DNA polymerase, RNA polymerase does not require a primer to initiate synthesis. Initiation of RNA synthesis, however, occurs only at specific sequences called promoters (described below). RNA synthesis usually starts with a GTP or ATP residue, whose 5'-triphosphate group is not cleaved to release PP_i but remains intact throughout transcription. During transcription the new RNA strand base-pairs temporarily with the DNA template to form a short length of hybrid RNA–DNA double helix, which is essential to the correct readout of the DNA strand (Fig. 25–1). The RNA in this hybrid duplex "peels off" shortly after its formation.

To enable RNA polymerase to synthesize an RNA strand complementary to one of the DNA strands, the DNA duplex must unwind over a short distance, forming a transcription "bubble." During transcription, the *E. coli* RNA polymerase generally keeps about 17 base pairs unwound, unwinding the DNA ahead and rewinding it behind. Because the DNA is a helix, this process requires considerable rotation of the nucleic acid molecules (Fig. 25–1a). Rotation is restricted in most DNAs by DNA-binding proteins and other structural barriers, and a moving RNA polymerase generates waves of positive supercoils ahead of and negative supercoils behind the point at which transcription is occurring (Fig. 25–1b). This transcription-driven supercoiling of DNA has been observed both in vitro and, in bacteria, in vivo. In the cell, the

Figure 25–1 Transcription by RNA polymerase in *E. coli*. To synthesize an RNA strand complementary to one of two DNA strands, the DNA is transiently unwound. Strand designations are summarized in Table 25–1. **(a)** About 17 base pairs are unwound at any given time. A short RNA–DNA hybrid (about 12 base pairs) is present in the unwound region. The transcription bubble moves from left to right as shown, keeping pace with RNA synthesis. The DNA is unwound ahead and rewound behind as RNA is transcribed. Arrows show the direction in which the DNA and the RNA–DNA hybrid must rotate to permit this process. As the DNA is rewound, the RNA–DNA hybrid is displaced and the RNA strand is extruded. **(b)** Supercoiling of DNA brought about by transcription. Positive supercoils form ahead of the transcription bubble and negative supercoils form behind.

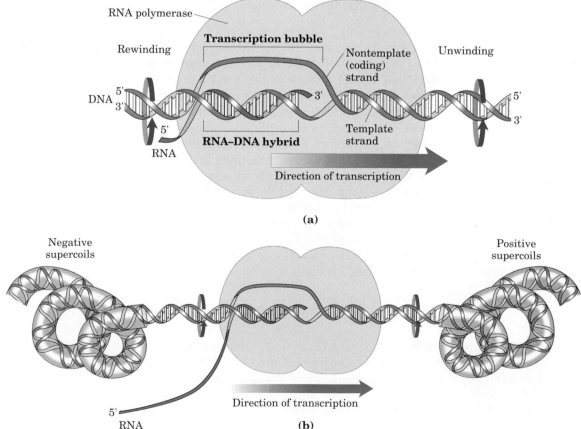

(a)

(b)

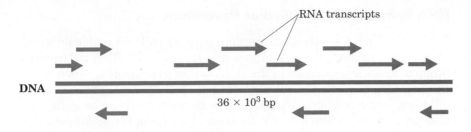

DNA

36×10^3 bp

RNA transcripts

topological problems caused by transcription are relieved through the action of topoisomerases. Once begun, transcription in *E. coli* proceeds at a rate of about 50 nucleotides per second.

The sequences of two complementary DNA strands are different, and the two strands serve different functions in transcription. A variety of designations are used to distinguish the two strands (Table 25–1). The strand that serves as template for RNA synthesis is called the **template strand** or minus (−) strand. In any chromosome, different genes may use different strands as template (Fig. 25–2). The DNA strand complementary to the template is called the **nontemplate strand** or plus (+) strand. It is identical in base sequence with the RNA transcribed from the gene, with U in place of T (Fig. 25–3). The nontemplate strand is also sometimes called the **coding strand,** even though it has no direct function in either transcription or protein synthesis. The regulatory sequences needed for transcription (described later in this chapter) are by convention given as sequences in the nontemplate (or coding or +) strand.

Figure 25–2 The genetic information of the adenovirus is encoded by a double-stranded DNA molecule (36,000 base pairs), both strands of which encode proteins. The information for most proteins is encoded by the top strand (transcribed left to right), but some is encoded by the bottom strand and is transcribed in the opposite direction. Synthesis of mRNAs in adenovirus is actually much more complex than shown here. Many of the mRNAs shown for the upper strand are initially synthesized as one long transcript derived from more than two-thirds of the length of the DNA. The transcript is extensively processed to produce the mRNAs for most of the individual gene products. Adenovirus causes some types of upper respiratory tract infections in some vertebrates.

Table 25–1 Alternative designations for DNA strands in transcription

Template strand	Nontemplate strand
Minus (−) strand	Plus (+) strand
	Coding strand

(5′) C G C T A T A G C G T T T (3′)   DNA nontemplate (+) strand
(3′) G C G A T A T C G C A A A (5′)   DNA template (−) strand

(5′) C G C U A U A G C G U U U (3′)   RNA transcript

Figure 25–3 The two complementary strands of DNA are defined by their function in transcription. The RNA transcript is synthesized on the complementary template (−) strand, and it is identical in sequence (with U in place of T) to the nontemplate (+) or coding strand.

E. coli has a single DNA-directed RNA polymerase that synthesizes all types of RNA. It is a large (M_r 390,000) and complex enzyme, containing five core subunits and a sixth subunit, called σ or σ^{70} (M_r 70,000), that binds transiently to the core and directs the enzyme to specific initiation sites on the DNA (described below). These six subunits constitute the RNA polymerase holoenzyme (Fig. 25–4). RNA polymerases, whether from *E. coli* or other organisms, lack a proofreading 3′→5′ exonuclease activity such as that found in many DNA polymerases. As a result, during transcription about one error is made for every 10^4 to 10^5 ribonucleotides incorporated into RNA. Given that many copies of an RNA are generally produced from a single gene and that all of the RNAs are eventually degraded and replaced, a rare mistake in an RNA molecule is of less consequence to the cell than a mistake in the permanent information stored in DNA.

Figure 25–4 The subunit structure of *E. coli* RNA polymerase. The α (of which there are two), β, β', ω, and σ subunits have molecular weights of 36,500, 151,000, 155,000, 11,000, and 70,000, respectively. The σ subunit is also called σ^{70}. The catalytic site for RNA synthesis is believed to be in the β subunit.

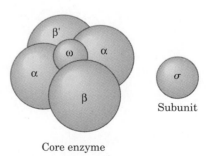

Core enzyme

Subunit

RNA Synthesis Is Initiated at Promoters

Initiation of RNA synthesis at random points in a DNA molecule would be an extraordinarily wasteful process. Instead, the RNA polymerase binds to specific sequences in the DNA called **promoters,** which direct the transcription of adjacent segments of DNA (genes). The sequences adjacent to genes where RNA polymerases must bind can be quite variable, and much research has focused on identifying the sequences that are critical to promoter function. Analysis and comparison of sequences in many different bacterial promoters have revealed similarities in two short sequences located about 10 and 35 base pairs away from the point where RNA synthesis is initiated (Fig. 25–5). By convention the base pair that begins an RNA molecule is given the number +1, so these sequences are commonly called the −10 and −35 regions. The sequences are not identical for all bacterial promoters, but certain nucleotides are found much more often than others at each position. The most common nucleotides form what is called a **consensus sequence** (recall the consensus sequences of *oriC* in the *E. coli* chromosome; see Fig. 24–10). For most promoters in *E. coli* and related bacteria, the consensus sequence for the −10 region (also called the Pribnow box) is (5′)TATAAT(3′), and the consensus sequence at the −35 region is (5′)TTGACA(3′).

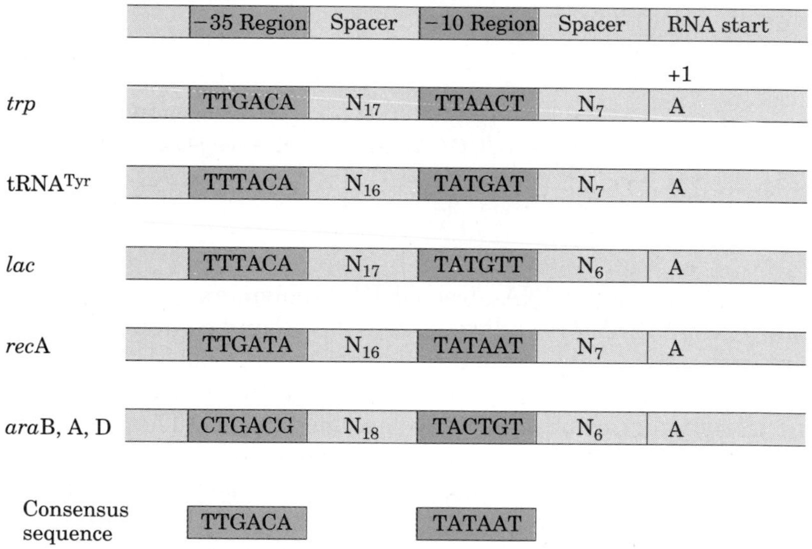

Figure 25–5 The sequences of five *E. coli* promoters. These include promoters for genes involved in *trp*tophan, *lac*tose, and *ara*binose metabolism. The sequences vary from one promoter to the next, but comparisons of many promoters reveal similarities in the −10 and −35 regions. The consensus sequences of the −10 and −35 regions are shown at the bottom. The −10 region is often called the Pribnow box, after David Pribnow, the investigator who first recognized it in 1975. All sequences shown are those of the coding (nontemplate) strand and read 5′→3′, left to right, as is the convention in representations of this kind. The spacer regions contain variable numbers of nucleotides (N). Only the first nucleotide coding the RNA transcript (at position +1) is shown.

	−35 Region	Spacer	−10 Region	Spacer	RNA start
					+1
trp	TTGACA	N_{17}	TTAACT	N_7	A
tRNA^{Tyr}	TTTACA	N_{16}	TATGAT	N_7	A
lac	TTTACA	N_{17}	TATGTT	N_6	A
*rec*A	TTGATA	N_{16}	TATAAT	N_7	A
*ara*B, A, D	CTGACG	N_{18}	TACTGT	N_6	A
Consensus sequence	TTGACA		TATAAT		

Many independent lines of evidence attest to the functional importance of these sequences. Mutations that affect the function of a given promoter usually involve one of the base pairs in the −35 or −10 region. Natural variations in the consensus sequence also affect the efficiency of RNA polymerase binding and transcription initiation. Differences of a few base pairs can decrease the rate of initiation by several orders of magnitude, providing one means by which *E. coli* can modulate the expression of different genes. In addition, specific binding of RNA polymerase to these sequences has been directly demonstrated in vitro (Box 25–1).

BOX 25–1   RNA Polymerase Leaves Its Footprint on a Promoter

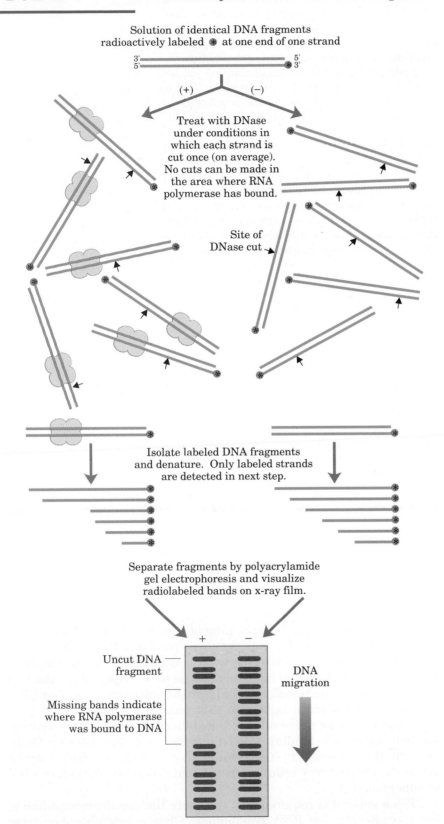

Solution of identical DNA fragments radioactively labeled ✱ at one end of one strand

(+)   (−)

Treat with DNase under conditions in which each strand is cut once (on average). No cuts can be made in the area where RNA polymerase has bound.

Site of DNase cut

Isolate labeled DNA fragments and denature. Only labeled strands are detected in next step.

Separate fragments by polyacrylamide gel electrophoresis and visualize radiolabeled bands on x-ray film.

Uncut DNA fragment

Missing bands indicate where RNA polymerase was bound to DNA

+   −

DNA migration

Footprinting, a technique derived from principles used in DNA sequencing (see Fig. 12–35), is used to identify the specific DNA sequences that are bound by a particular protein. A DNA fragment thought to contain sequences recognized by the DNA binding protein is isolated and radiolabeled at one end of one strand (Fig. 1). Chemical or enzymatic cleavage introduces random breaks in the DNA fragment (averaging about one per molecule). Separation of the labeled cleavage products (broken fragments of various lengths) by high-resolution electrophoresis reveals a "ladder" of radioactive bands. In a separate tube the cleavage procedure is repeated on the original DNA fragment to which the protein is bound. The protein prevents cleavage of the DNA in the region to which it is bound. The second set of cleavage prod-

Figure 1 Footprint analysis of the binding site for RNA polymerase on a DNA fragment. Separate experiments are carried out in the presence (+) and absence (−) of RNA polymerase.

Continued on next page

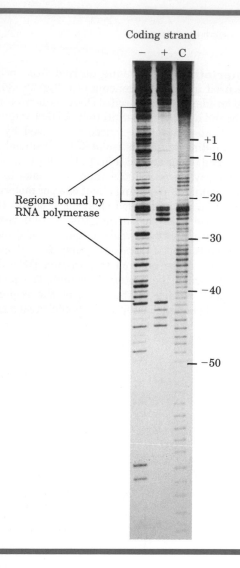

Coding strand

− + C

— +1
— −10

— −20

Regions bound by
RNA polymerase

— −30

— −40

— −50

ucts is subjected to electrophoresis side by side with the products of the original reaction. A hole or "footprint" is revealed in the "ladder" of radioactive bands derived from the protein-containing sample. The hole results from the protection of the DNA by protein binding, and it defines the sequences recognized by the protein. The precise location of this binding site can be determined by directly sequencing (see Fig. 12–35) the original DNA fragment and including the sequencing lanes (not shown here) on the same gel with the footprint. Footprinting results for the binding of RNA polymerase to a DNA fragment containing a promoter are shown in Figure 2. The polymerase covers 50 to 60 base pairs; protection by the bound enzyme is concentrated in the −10 and −35 regions.

Figure 2 Footprinting results of RNA polymerase binding to the *lac* promoter (see Fig. 25–5). In this experiment the 5′ end of the coding strand was radioactively labeled. The C lane is a control in which the labeled DNA fragment is cleaved with a chemical reagent that produces a more uniform banding pattern.

RNA polymerase binds to the promoter in at least two distinguishable steps (Fig. 25–6). The holoenzyme first binds the DNA and migrates to the −35 region, forming what is called the "closed complex." The DNA is then unwound for about 17 base pairs beginning at the −10 region, exposing the template strand at the initiation site. The RNA polymerase binds more tightly to this unwound region, forming an "open complex" (the name reflects the state of the DNA). RNA synthesis then begins. The binding of RNA polymerase to promoters is facilitated by the supercoiling (underwinding) of the DNA, which may be one of the reasons why cellular DNA is maintained in an underwound or supercoiled state.

The σ subunit is required only to ensure the specific recognition of the promoter by the RNA polymerase. Once a few phosphodiester bonds are formed the σ subunit dissociates, leaving the core polymerase to complete synthesis of the RNA molecule.

Figure 25–6 Steps in the initiation of transcription by *E. coli* RNA polymerase. RNA polymerase binding to a promoter requires two steps: formation of the closed and open complexes. Messenger RNA synthesis is almost always initiated with a purine (Pu) nucleotide. N is any nucleoside.

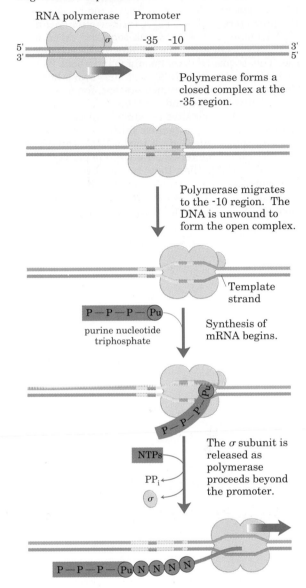

RNA polymerase holoenzyme binds to DNA and migrates to the promoter.

Polymerase forms a closed complex at the -35 region.

Polymerase migrates to the -10 region. The DNA is unwound to form the open complex.

Synthesis of mRNA begins.

The σ subunit is released as polymerase proceeds beyond the promoter.

Some *E. coli* promoters differ greatly from the standard promoters described above, and recognition of these promoters by RNA polymerase is mediated by different σ factors. An example occurs in a set of genes called the heat-shock genes, which are induced (their gene products are made at higher levels) when the cell is under the stress that accompanies an insult such as a sudden temperature jump. RNA polymerase binds to these promoters when its normal σ subunit (designated σ^{70} because it has a molecular weight of 70,000) is replaced with a different σ subunit that is specific for the heat-shock promoters (see Fig. 27–3). This distinct σ subunit has a molecular weight of 32,000 and is therefore called σ^{32}. The use of different σ factors allows the cell to coordinately express sets of genes involved in major changes in cell physiology.

Initiation of Transcription Is Regulated

Under certain conditions and at different developmental stages, the cellular requirements for any given gene product may very greatly. To provide proteins to the cell in the proportions needed, the transcription of each gene is carefully regulated. The variation in affinity of RNA polymerase for promoters due to differences in promoter sequences, as discussed above, is only one level of control. A variety of proteins bind to sequences in and around the promoter and either activate transcription by facilitating RNA polymerase binding or repress transcription by blocking the activity of polymerase. In *E. coli,* an example of a protein that activates transcription is the **catabolite gene activator protein (CAP),** which increases the transcription of genes coding for enzymes that metabolize sugars other than glucose when cells are grown in the absence of glucose. **Repressors,** typified by the Lac repressor, are proteins that block the synthesis of RNA at specific genes. In the case of the Lac repressor, RNA synthesis is blocked at the genes for enzymes involved in lactose metabolism when lactose is unavailable. Because transcription is the first step in a complicated and energy-intensive pathway leading to protein synthesis, much of the regulation of protein levels in both bacterial and eukaryotic cells is directed at transcription initiation. In Chapter 27 we will describe many mechanisms by which this is accomplished.

Eukaryotic Cells Have Three Kinds of RNA Polymerases

The transcriptional machinery in the nucleus of a eukaryotic cell is much more complex than that in bacteria. Eukaryotes have three different RNA polymerases, designated I, II, and III. Each has a specific function and binds to a different promoter sequence. RNA polymerase I (Pol I) is responsible for the synthesis of only one type of RNA, a

Figure 25–7 The consensus sequences of some common elements in promoters used by eukaryotic RNA polymerase II, derived from a comparison of 100 promoters of this type. A transcription factor (TFIID) binds at the A=T-rich sequence called a TATA box, facilitating the binding of the polymerase. This sequence is commonly found about 25 base pairs before the RNA start site. Two other elements are also sometimes present, found somewhere between −110 and −40: the CCAAT box and GC box are binding sites for other transcription factors that affect polymerase function. Other sequences, some quite distant in the DNA, can affect transcription (Chapter 27). Eukaryotic promoters are more variable than their bacterial counterparts, and some RNA polymerase II promoters lack all of the sequences shown. As in Fig. 25–5, the sequences are those in the coding (nontemplate) strand.

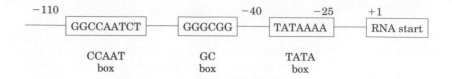

preribosomal RNA transcript that contains the precursor for the 18S, 5.8S, and 28S rRNAs (see Fig. 26–12). Its promoter varies greatly in sequence from one species to another. RNA polymerase II (Pol II) has the central function of synthesizing mRNAs, as well as some special-function RNAs. This enzyme must recognize thousands of promoters, many of which share some key sequence similarities in most eukaryotes (Fig. 25–7). These sequences are generally binding sites for proteins called **transcription factors,** which modulate the binding of RNA polymerase to the promoter. RNA polymerase III (Pol III) makes tRNAs, the 5S rRNA, and some other small specialized RNAs. The promoter recognized by RNA polymerase III is well characterized. Interestingly, some of the sequences required for the regulated initiation of transcription by RNA polymerase III are located within the gene itself, whereas others are found in more conventional locations before the RNA start site (Chapter 27).

Specific Sequences Signal Termination of RNA Synthesis

RNA synthesis proceeds until the RNA polymerase encounters a sequence that triggers its dissociation. This process is not well understood in eukaryotes, and our focus again shifts to bacteria. In *E. coli* there are at least two classes of such termination signals or terminators. One class relies on a protein factor called ρ (rho), and the other is ρ-independent.

The ρ-independent class has two distinguishing features (Fig. 25–8). The first is a region that is transcribed into self-complementary sequences, permitting the formation of a hairpin structure (see Fig. 12–21) centered 15 to 20 nucleotides before the end of the RNA. The second feature is a run of adenylates in the template strand that are transcribed into uridylates at the end of the RNA. It is thought that formation of the hairpin disrupts part of the RNA–DNA hybrid in the transcription complex. The remaining hybrid duplex (oligoribo-U–oligodeoxy-A) contains a particularly unstable combination of bases, and the entire complex simply dissociates.

The ρ-dependent terminators lack the sequence of repeated adenylates in the template but do usually have a short sequence that is transcribed to form a hairpin. RNA polymerase pauses at these sequences, and dissociates if ρ protein is present. The ρ protein has an ATP-dependent RNA–DNA helicase activity and probably disrupts the RNA–DNA hybrid formed during transcription. ATP is hydrolyzed by ρ protein during the termination process, but the detailed mechanism by which the protein acts is not known.

DNA-Directed RNA Polymerase Can Be Selectively Inhibited

The elongation of RNA chains by RNA polymerase in both bacteria and eukaryotes is specifically inhibited by the antibiotic **actinomycin D** (Fig. 25–9). The planar portion of this molecule intercalates (inserts

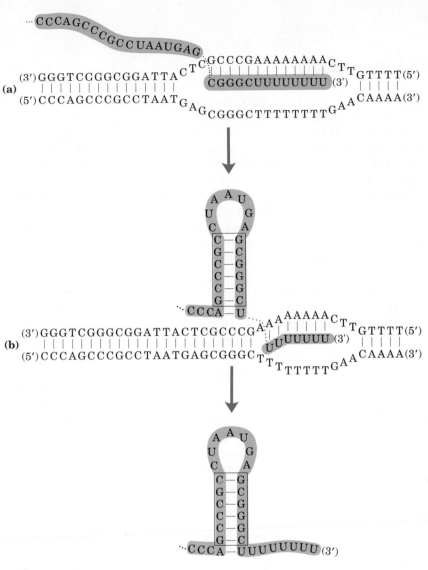

Figure 25–8 A model for ρ-independent termination of transcription in *E. coli*. **(a)** The poly(U) region is synthesized by RNA polymerase. **(b)** Intramolecular pairing of complementary sequences in the RNA forms a hairpin, destroying part of the RNA–DNA hybrid. The remaining A≡U hybrid region is relatively unstable, and **(c)** the RNA dissociates completely.

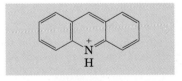

Actinomycin D

Acridine

Figure 25–9 Structure of actinomycin D and acridine, inhibitors of DNA transcription. The shaded portion of actinomycin D is planar and intercalates between two successive G≡C base pairs in duplex DNA. The two cyclic peptide structures of the actinomycin D molecule bind to the minor groove of the double helix. Sarcosine (Sar) is *N*-methylglycine; meVal represents methylvaline. The linkages between sarcosine, L-proline, and D-valine are peptide bonds. Acridine also acts by intercalation in the DNA.

itself) into the double-helical DNA between successive G≡C base pairs, deforming the DNA. This local alteration prevents the movement of the polymerase along the template. In effect, actinomycin D jams the zipper. Because actinomycin D inhibits RNA elongation in intact cells, as well as in cell extracts, it has become very useful for identifying cell processes that depend upon RNA synthesis. **Acridine** inhibits RNA synthesis in a similar fashion (Fig. 25–9).

Rifampicin is an antibiotic inhibitor of RNA synthesis that binds specifically to the β subunit of bacterial RNA polymerases (see Fig. 25–4), preventing the initiation of transcription. A specific inhibitor of RNA synthesis in animal cells is **α-amanitin,** a toxic component of the poisonous mushroom *Amanita phalloides*. It blocks mRNA synthesis by RNA polymerase II and, at higher concentrations, by RNA polymerase III. It does not affect RNA synthesis in bacteria. This mushroom has developed a very effective defense mechanism: a substance that inhibits mRNA formation in organisms that might try to eat it but is evidently harmless to the mushroom's own transcription mechanism.

RNA Processing

Many of the RNA molecules in bacteria and virtually all of the RNA molecules in eukaryotes are processed to some degree after they are synthesized. Many of the most interesting molecular events in RNA metabolism are to be found among these postsynthetic reactions. The study of these processes has revealed that some of them are catalyzed by enzymes made up of RNA rather than protein. The discovery of catalytic RNAs has brought on a revolution in thinking about RNA function and about the origin of life.

A newly synthesized RNA molecule is called a **primary transcript.** Perhaps the most extensive processing of primary transcripts occurs in eukaryotic mRNAs and in tRNAs of both bacteria and eukaryotes. A primary transcript for a eukaryotic mRNA typically contains sequences encompassing one gene. The sequences encoding the polypeptide, however, usually are not contiguous. Instead, in the majority of cases, the coding sequence is interrupted by noncoding tracts called introns; the coding segments are called exons (see the discussion of introns and exons in DNA, p. 798). In a process called **splicing,** the introns are removed from the primary transcript and the exons joined to form a contiguous sequence specifying a functional polypeptide. Eukaryotic mRNAs are also modified at each end. A structure called a cap is added at the 5′ end, and a polymer containing 20 to 250 adenylate residues, poly(A), is added to the 3′ end. These processes are outlined in Figure 25–10 and described in more detail below.

The primary transcripts of most tRNAs (in all organisms) are also processed by the removal of sequences from each end (called cleavage) and sometimes by the removal of introns (splicing). Many bases in tRNAs are also modified; mature tRNAs are replete with unusual bases not found in other nucleic acids.

The ultimate postsynthetic modification reaction is the complete degradation of the RNA. All RNAs eventually meet this fate and are replaced with newly synthesized RNAs. The rate of turnover of RNAs is critical to determining their steady-state level and the rate at which cells can shut down expression of a gene whose product is no longer needed.

Figure 25–10 Formation of the primary transcript and its processing during maturation of the mRNA in a eukaryotic cell. The 5′ cap (in red) is added before synthesis of the primary transcript is complete. Noncoding sequences following the last exon are shown in orange. Splicing may occur either before or after the cleavage and polyadenylation steps. All of the processes represented here take place within the nucleus.

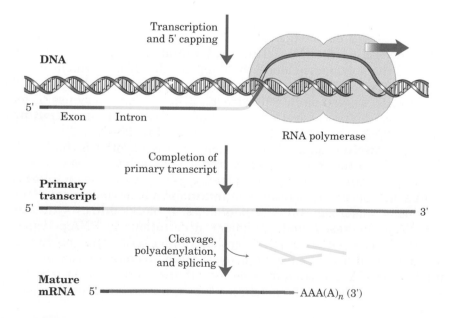

Transcription
and 5′ capping

DNA

5′

Exon   Intron

RNA polymerase

Completion of
primary transcript

**Primary
transcript**

5′   3′

Cleavage,
polyadenylation,
and splicing

**Mature
mRNA**   5′   AAA(A)$_n$ (3′)

The Introns Transcribed into RNA Are Removed by Splicing

In bacteria, a polypeptide chain is generally encoded by a DNA sequence that is colinear with the amino acid sequence, continuing along the DNA template without interruption until the information needed to specify the polypeptide is complete. The notion that all genes are continuous was unexpectedly disproven in 1977 with the discovery that the genes for polypeptides in eukaryotes are often interrupted by the noncoding sequences now called introns. Introns are present in the vast majority of genes in vertebrates; among the few exceptions are the genes that encode certain histones. The occurrence of introns in other eukaryotes is variable. Most genes in the yeast *Saccharomyces cerevisiae* lack introns, although introns are more prevalent in the genes of some other yeast species. Introns are also found in a few prokaryotic genes.

Introns are spliced from the primary transcript, and exons are joined to form a mature, functional RNA. Introns were discovered when mRNA and the DNA from which it was derived were compared using methods such as that illustrated in Figure 25–11. If the DNA containing a gene is completely denatured and then renatured in the presence of the mature RNA derived from the gene, an RNA–DNA hybrid is formed. This kind of experiment revealed DNA sequences that were not present in the RNA and therefore were looped out as in Figure 25–11. Experiments using this and other methods have shown the presence of multiple introns in many genes, with some genes interrupted by introns more than 40 times. In eukaryotic mRNAs most exons are less than 1,000 nucleotides long, with many clustered in the 100 to 200 nucleotide size range. Most exons therefore encode polypeptide chains that are 30 to 50 amino acids long. Introns are much more variable in size (50 to 20,000 nucleotides). Genes of higher eukaryotes, including humans, typically have much more DNA devoted to introns than to exons; it is not uncommon to find genes that are 50,000 to 200,000 nucleotides long and that contain numerous introns.

There are four classes of introns. The first two, called group I and group II, share some key characteristics but differ in the details of their splicing mechanisms. Group I introns are found in some nuclear, mitochondrial, and chloroplast genes coding for rRNAs; group II introns are generally found in the primary transcripts of mitochondrial or chloroplast mRNAs. Both groups share the property that no high-energy cofactors (such as ATP) are required for splicing. Both splicing

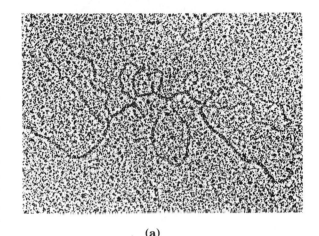

(a)

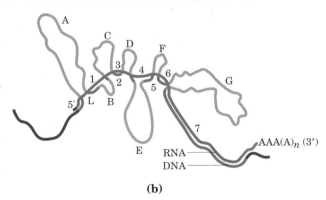

(b)

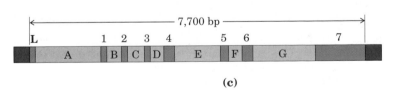

(c)

Figure 25–11 Defining the structure of the chicken ovalbumin gene by hybridization. Mature mRNA was hybridized to denatured DNA containing the ovalbumin gene, and the resulting molecules were visualized with the electron microscope. Some regions of the DNA have no complement in the mRNA because of splicing of the primary transcript. The resulting single-stranded DNA loops are evident in the electron micrograph **(a).** The loops define the locations and sizes of introns. The introns are labeled A to G and the seven exons are numbered in the interpretive drawing **(b).** The poly(A) tail defines the 3' end of the mRNA. The L sequence encodes a signal sequence that targets the protein for export from the cell. **(c)** A linear representation of the ovalbumin gene showing introns and exons.

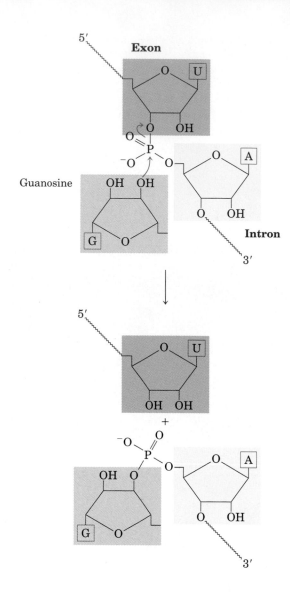

Figure 25–12 A transesterification reaction. This is the first step in the splicing of group I introns. Here, the 3′ OH of a guanosine molecule acts as nucleophile.

Figure 25–13 Splicing mechanism of group I introns. The nucleophile in the first step may be guanosine, GMP, GDP, or GTP.

mechanisms involve two transesterification reaction steps (Fig. 25–12). A 2′- or 3′-hydroxyl group of a ribose makes a nucleophilic attack on a phosphorus, and in each step a new phosphodiester bond is formed at the expense of the old, maintaining an energy balance. Note that these reactions are very similar to the DNA breaking and rejoining reactions promoted by topoisomerases (Chapter 23) and site-specific recombinases (Chapter 24).

The group I splicing reaction requires a guanine nucleoside or nucleotide cofactor. This cofactor is not used as a source of energy; instead, the 3′-hydroxyl group of guanosine is used as a nucleophile in the first step of the splicing pathway. The guanosine 3′-hydroxyl forms a normal 3′,5′-phosphodiester bond with the 5′ end of the intron (Fig. 25–13). The 3′-hydroxyl of the exon that is displaced in this step then acts as a nucleophile in a similar reaction at the 3′ end of the intron. The result is precise excision of the intron and ligation of the exons.

In group II introns the pattern is similar except for the nucleophile in the first step. Instead of an external cofactor, the nucleophile is the 2′-hydroxyl group of an adenylate residue within the intron (Fig. 25–14). An unusual branched lariat structure is formed as an intermediate.

Attempts to identify the enzymes that promote splicing of group I and group II introns produced a major surprise; many of these introns are *self-splicing*—no protein enzymes are involved. This was first revealed in studies of the splicing mechanism of the group I rRNA intron from the ciliated protozoan *Tetrahymena thermophila* by Thomas Cech and colleagues in 1982. These workers proved that no proteins were involved by transcribing *Tetrahymena* DNA (including the intron) in

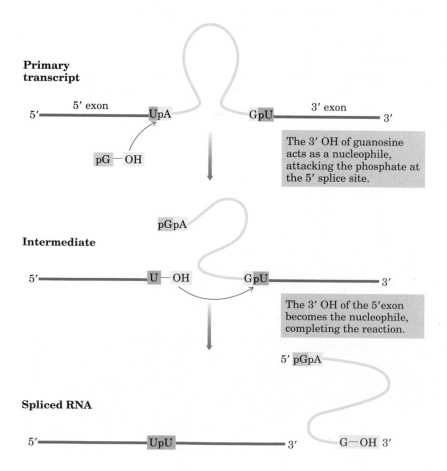

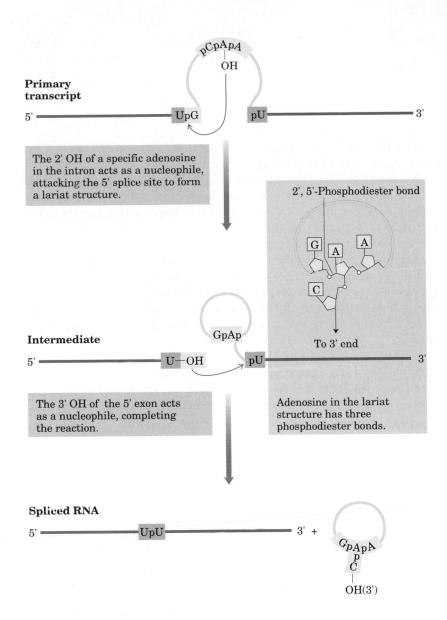

Primary transcript

The 2' OH of a specific adenosine in the intron acts as a nucleophile, attacking the 5' splice site to form a lariat structure.

2', 5'-Phosphodiester bond

To 3' end

Intermediate

The 3' OH of the 5' exon acts as a nucleophile, completing the reaction.

Adenosine in the lariat structure has three phosphodiester bonds.

Spliced RNA

Figure 25–14 Splicing mechanism of group II introns. The chemistry is similar to that of group I intron splicing, except for the nucleophile in the first step and the novel lariatlike intermediate with one branch having a 2′,5′-phosphodiester bond.

vitro using bacterial RNA polymerase. The resulting RNA spliced itself accurately even though it had never been in contact with any enzymes from *Tetrahymena*. The realization that RNAs, as well as proteins, could have catalytic functions was a milestone in thinking about biological systems. RNA catalysts are discussed in more detail later in this chapter.

The third and largest group of introns, found in nuclear mRNA primary transcripts, undergo splicing by the same lariat-formation mechanism as the group II introns. However, they are not self-splicing. Splicing requires the action of specialized RNA–protein complexes containing a class of eukaryotic RNAs called **small nuclear RNAs (snRNAs).** Five snRNAs, U1, U2, U4, U5, and U6, are involved in splicing reactions. They are found in abundance in the nuclei of many eukaryotes, range in size from 106 (U6) to 189 (U2) nucleotides, and are complexed with proteins to form particles called *small nuclear ribonucleoproteins* (snRNPs, often referred to as "snurps"). The RNAs and proteins in snRNPs are highly conserved among vertebrates and insects. Small nuclear RNAs similar to these are also found in yeast and slime molds.

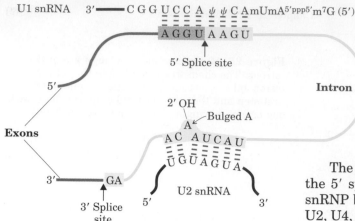

U1 snRNA 3′ ━━━ C G G U C C A ψ ψ C AmUmA⁵′ppp⁵′m⁷G (5′)

AGGUAAGU

5′ Splice site

Intron

2′ OH

A — Bulged A

AC AUCAU

UGUAGUA

U2 snRNA

5′   3′

Exons

5′

3′   GA

3′ Splice site

Figure 25–15 Splicing mechanism in mRNA primary transcripts. The splice sites that mark the intron–exon boundaries of many eukaryotic mRNAs have some conserved sequences. The U1 snRNA has a sequence near its 5′ end complementary to the splice site at the 5′ end of the intron. Base pairing of U1 to this region of the primary transcript helps define the 5′ splice site. ψ represents pseudouridine (see Fig. 25–25), and "m" indicates methylated residues. Base pairing of U2 snRNA to the branch site displaces (bulges) and perhaps activates the adenosine, whose 2′ OH forms the lariat structure through a 2′,5′-phosphodiester bond.

The U1 snRNA has a sequence complementary to sequences near the 5′ splice site of nuclear mRNA introns (Fig. 25–15), and the U1 snRNP binds to this region in the primary transcript. Addition of the U2, U4, U5, and U6 snRNPs leads to formation of a complex called the "spliceosome" within which the actual splicing reaction occurs. ATP is required for assembly of the spliceosome, but there is no reason to believe that the splicing reactions require ATP.

The fourth class of intron, found in certain tRNAs, is distinguished from the group I and II introns in that its splicing requires ATP. In this case, a splicing endonuclease cleaves the phosphodiester bonds at both ends of the intron, and the two exons are joined as shown in Figure 25–16. The joining reaction is similar to the DNA ligase mechanism (see Fig. 24–15).

Introns are not limited to eukaryotes. Although very rare, several genes with introns have now been found in bacteria and bacterial viruses. Bacteriophage T4, for example, has several genes with group I introns. Introns appear to be more common in archaebacteria (p. 25) than in *E. coli*.

Figure 25–16 (Facing page) The splicing of yeast tRNA. This splicing pathway requires a high-energy cofactor (ATP) for the ligation step. **(a)** The intron is first removed by endonuclease-catalyzed cleavage at both ends. **(b)** The 2′,3′-cyclic phosphate on the 5′ exon is cleaved by a cyclic nucleotide phosphodiesterase, leaving a 2′ phosphate. **(c)** The 5′ OH left on the 3′ exon is then activated in two steps. **(d)** The free 3′ hydroxyl of the 5′ exon acts as a nucleophile to displace AMP, joining the two exons with a 3′,5′-phosphodiester bond. **(e)** The 2′ phosphate is removed to yield the final product.

Eukaryotic mRNAs Undergo Additional Processing

In eukaryotes, mature mRNAs have distinctive structural features at both ends. Most have a **5′ cap,** a residue of 7-methylguanosine linked to the 5′-terminal residue of the mRNA through an unusual 5′,5′-triphosphate linkage (Fig. 25–17). At the 3′ end, most eukaryotic mRNAs have a "tail" of 20 to 250 adenylate residues, called the **poly(A) tail.** The functions of the 5′ cap and the 3′ poly(A) tail are only partially known. The 5′ cap binds to a protein and may participate in the binding of the mRNA to the ribosome to initiate translation (Chapter 26). The poly(A) tail also is bound by a specific protein. It is likely that the 5′ cap and poly(A) tail and their associated proteins help protect the mRNA from enzymatic destruction.

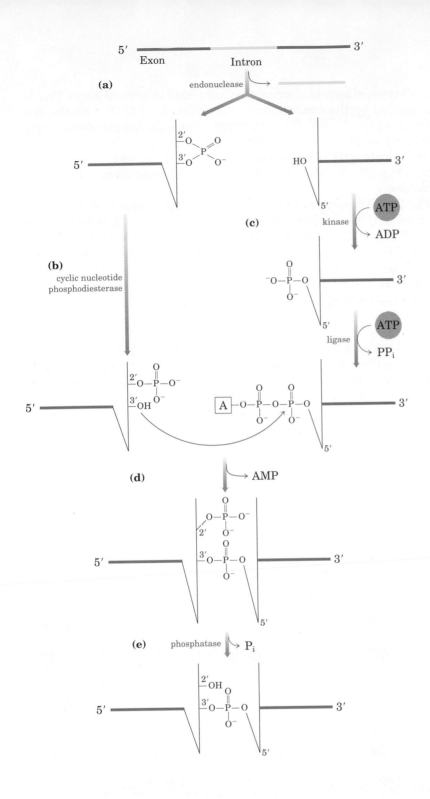

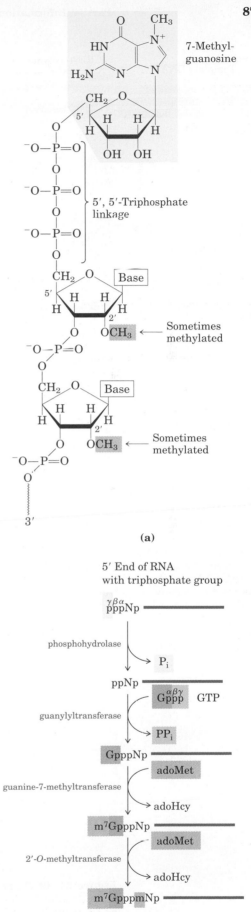

(a)

(b)

Figure 25–17 (Right) The 5′ cap of 7-methyl-guanosine is found on almost all eukaryotic mRNAs. **(a)** 7-Methylguanosine is joined to the 5′ end of the mRNA in a novel 5′,5′-triphosphate linkage. Methyl groups (screened in red) are also often found at the 2′ position of the first and second nucleotides. Yeasts lack the 2′-methyl groups, and the 2′ methyl on the second nucleotide is generally only found in vertebrates. **(b)** Generation of the 5′ cap involves four to five separate steps. adoHcy is an abbreviation for S-adenosylhomocysteine.

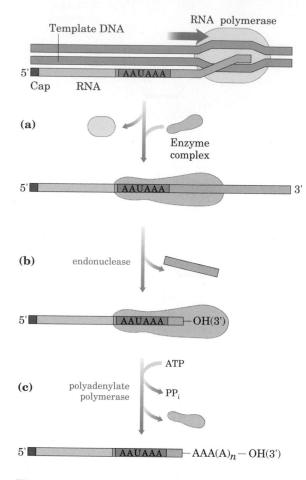

Figure 25–18 Addition of the poly(A) tail to the primary RNA transcript of eukaryotes. RNA polymerase synthesizes RNA beyond the segment of the transcript containing the cleavage signal ((5′)AAUAAA). **(a)** A complex including an endonuclease and polyadenylate polymerase binds to this signal sequence. **(b)** The RNA is cleaved 11 to 30 nucleotides 3′ to AAUAAA, and **(c)** polyadenylate polymerase synthesizes a poly(A) tail of 20 to 250 nucleotides, beginning at the cleavage site.

Both types of terminal structures are added in several steps. The 5′ cap is formed by the condensation of a molecule of GTP with the triphosphate at the 5′ end of the transcript. The guanine is subsequently methylated at N-7, and additional methyl groups often are added at the 2′ hydroxyls of the first and second nucleotides adjacent to the cap (Fig. 25–17). The methyl groups are derived from S-adenosylmethionine.

The poly(A) tail is not simply added to the 3′ end of the primary transcript at the site where transcription terminates. The transcript is extended beyond the site where the poly(A) tail is to be added, then is cleaved at the poly(A) addition site by a specific riboendonuclease (Fig. 25–18). This cleavage generates the free 3′-hydroxyl group that defines the end of the mRNA and to which adenylate residues are immediately added by **polyadenylate polymerase,** catalyzing the reaction

$$\text{RNA} + n\text{ATP} \longrightarrow \text{RNA-(AMP)}_n + n\text{PP}_i$$

where $n = 20$ to 250. This enzyme requires no template but does require the mRNA as a primer. The site where cleavage and poly(A) addition occur is marked in the mRNA by the highly conserved sequence (5′)AAUAAA(3′), situated 11 to 30 nucleotides on the 5′ side of the cleavage site. A complex containing the riboendonuclease, polyadenylate polymerase, and possibly other proteins and one or more snRNAs binds to this sequence and carries out the processing reactions.

The processing of a typical eukaryotic mRNA is summarized in Figure 25–19. In some cases the polypeptide-coding region of the mRNA is also modified. The origin and mechanism of this RNA "editing" are not understood (see Box 26–1).

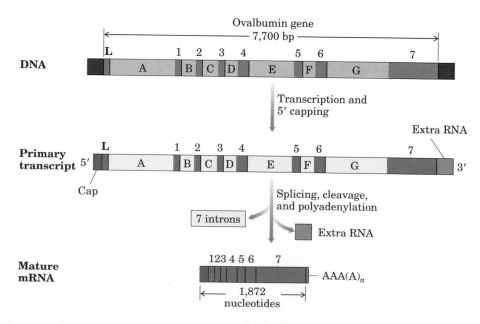

Figure 25–19 Overview of the processing of a eukaryotic mRNA. The ovalbumin gene is again used as an example (see Fig. 25–11). Introns are lettered and exons are numbered. About three-quarters of the RNA is removed during processing. Introns can make up more than 90% of the length of other genes. RNA polymerase II extends the primary transcript well beyond the cleavage and polyadenylation site ("extra RNA"). Termination signals for RNA polymerase II have not been defined.

Multiple Products Are Derived from One Gene by Differential RNA Processing

The transcription of introns consumes energy without apparently returning any benefit to the organism, but evolution would select against interrupted genes if they did not confer some practical advantage to cells. Although these benefits are not yet clear in most cases, an obvious advantage can be seen for transcripts that use splicing to produce multiple gene products.

Primary transcripts for mRNAs fall into two classes. **Simple transcripts** produce only one mature mRNA and one type of polypeptide product. **Complex transcripts,** in contrast, produce two or more *different* mRNAs and polypeptides. In most cases the complex transcripts are still monocistronic, i.e., only one type of mature mRNA and polypeptide is derived from any given transcript molecule at any one time. The primary transcript, however, has the molecular signals for two or more alternative processing pathways so that one of two or more different mRNAs may result depending upon which pathway is chosen. In different cells or at different stages of development, the transcript might be processed to produce different gene products.

Complex transcripts have either more than one site for cleavage and polyadenylation or alternative splicing patterns, or both (Fig. 25–20). If there are two sites for cleavage and polyadenylation, the use of the one closest to the 5′ end will remove more of the primary transcript sequence (Fig. 25–20a). Using this mechanism, immunoglobulin heavy chains are produced that differ at their carboxyl termini; this process is called poly(A) site choice. In fruit flies, using alternative splice sites (Fig. 25–20b), three different forms of the myosin heavy chain are produced at different stages of development. In rats, both

Figure 25–20 Two mechanisms for the differential processing of complex transcripts in eukaryotes: **(a)** multiple sites for cleavage and polyadenylation (here, two poly(A) sites, A_1 and A_2, are shown), and **(b)** alternative splicing patterns (two different 3′ splice sites are shown).

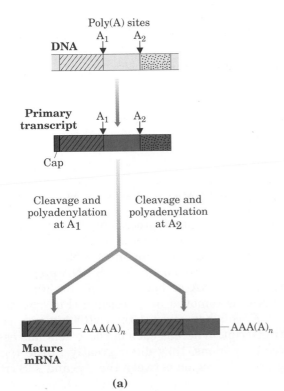

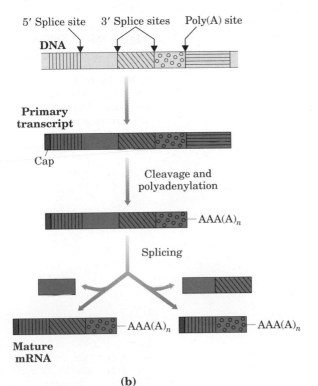

(a)   (b)

mechanisms are used to produce from a common primary transcript the calcium-regulating hormone calcitonin in the thyroid and a different hormone (calcitonin gene–related peptide) in the brain (Fig. 25–21).

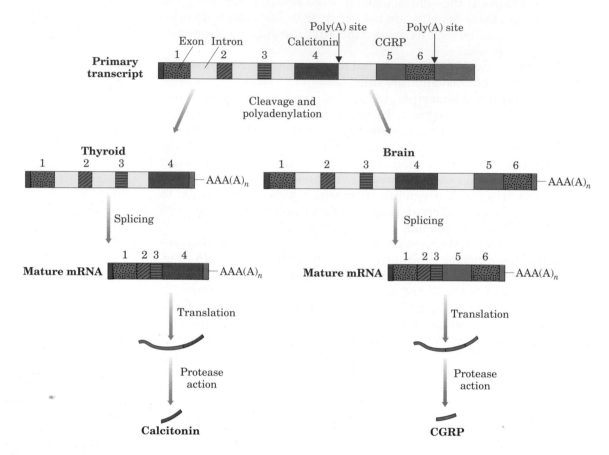

Figure 25–21 Differential processing of the calcitonin gene transcript in rats. The primary transcript has two poly(A) sites; the first predominates in the thyroid, the second in the brain. Splicing in the brain eliminates the calcitonin exon; in the thyroid this exon is retained. The resulting peptides are processed further to yield the final hormone products: calcitonin in the thyroid and calcitonin gene–related peptide (CGRP) in the brain.

Ribosomal RNAs and tRNAs Also Undergo Processing

Posttranscriptional processing is not limited to mRNA. Ribosomal RNAs of both bacterial and eukaryotic cells are made from longer precursors called **preribosomal RNAs.** In bacteria, 16S, 23S, and 5S rRNAs arise from a single 30S RNA precursor having about 6,500 nucleotides. RNA at both ends of the 30S precursor and between the rRNAs is removed in processing (Fig. 25–22).

E. coli has seven sets of rRNA genes, each yielding a precursor transcript. Whereas each of these sets has essentially identical rRNA coding regions, they differ greatly in the regions between the rRNA genes. The region between the 16S and 23S rRNA genes generally con-

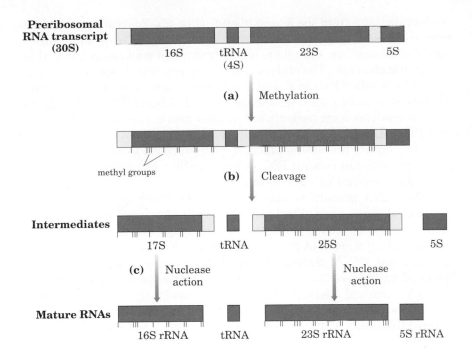

Preribosomal RNA transcript (30S)

16S tRNA (4S) 23S 5S

methyl groups

(a) Methylation

(b) Cleavage

Intermediates

17S tRNA 25S 5S

(c) Nuclease action   Nuclease action

Mature RNAs

16S rRNA tRNA 23S rRNA 5S rRNA

Figure 25–22 Processing of preribosomal RNA transcripts in bacteria. **(a)** Prior to cleavage of the 30S RNA precursor, it is methylated at specific bases. **(b)** Cleavage produces 17S and 25S intermediates, and **(c)** the final 16S and 23S rRNA products are made by the action of specific nucleases. The 5S rRNA arises from the 3′ end of the 30S precursor. From the midsection (and sometimes the 3′ end), one or more tRNAs are formed. There are seven copies of the gene for the preribosomal RNA in *E. coli*, and these differ in the number, location, and identity of tRNAs included in the primary transcript.

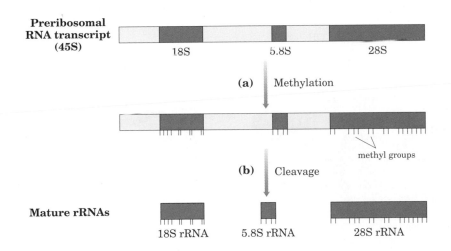

Preribosomal RNA transcript (45S)

18S 5.8S 28S

(a) Methylation

methyl groups

(b) Cleavage

Mature rRNAs

18S rRNA 5.8S rRNA 28S rRNA

Figure 25–23 Processing of preribosomal RNA transcripts in eukaryotes. **(a)** The 45S precursor is methylated at more than 100 of its 14,000 nucleotides, mostly on the 2′-OH groups of ribose units retained in the final products. **(b)** A series of enzymatic cleavages produces the 18S, 5.8S, and 28S rRNAs. The 5S rRNA arises separately.

tains genes for one or two tRNAs, with different tRNAs present in different rRNA precursors. Transfer RNAs are also found on the 3′ side of the 5S rRNA in the precursor transcript.

In eukaryotes, a 45S preribosomal RNA is processed in the nucleolus to form the 18S, 28S, and 5.8S rRNAs characteristic of eukaryotic ribosomes (Fig. 25–23). The 5S rRNA of most eukaryotes is made as a completely separate transcript.

Most cells have 40 to 50 distinct tRNAs. In eukaryotic cells there are multiple copies of many of the tRNA genes. Transfer RNAs are derived from longer RNA precursors by enzymatic removal of extra nucleotide units from the 5' and 3' ends (Fig. 25–24). Introns are occasionally present and must be excised. In some cases two or more different tRNAs occur on a single primary transcript and are separated by enzymatic cleavage. The endonuclease that removes RNA at the 5' end of tRNAs is called RNase P, and the 3' end is processed by one or more nucleases including an exonuclease called RNase D. RNase P is found in all organisms from bacteria to humans, and it contains both protein and RNA. The RNA component is essential for activity, and this RNA can carry out its processing function with precision even in the absence of the protein component. RNase P is another example of a catalytic RNA, as described in more detail below.

The tRNA precursors may undergo two other types of posttranscriptional processing. First, the 3'-terminal trinucleotide CCA(3'), characteristic of tRNAs, is absent in the sequence of some bacterial and all eukaryotic tRNA precursors and is added to the 3' terminus by the enzyme tRNA nucleotidyltransferase; some prokaryotic tRNA genes already have this 3'-terminal sequence encoded by the DNA. It is the 3'-terminal adenylate residue that covalently attaches to a specific amino acid in preparing the amino acid for addition to the growing polypeptide chain on the ribosome (Chapter 26). The tRNA nucleotidyl-transferase is an unusual enzyme in that it promotes the formation of all three phosphodiester bonds at once; it has multiple active sites that separately bind each of the ribonucleoside triphosphate precursors and

Primary transcript

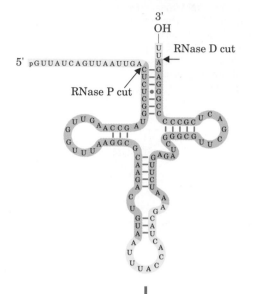

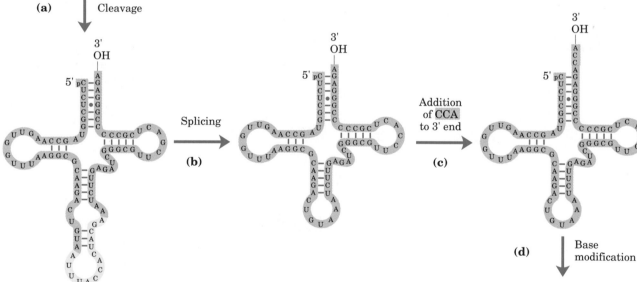

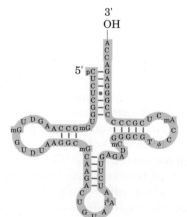

Figure 25–24 Processing of tRNAs in bacteria and eukaryotes. The sequences shown in yellow are removed, **(a)** some by specific ribonucleases and **(b)** some by splicing reactions. **(c)** CCA is added to the 3' end in eukaryotic tRNAs and in those bacterial tRNAs that lack this sequence in the primary transcript; this reaction is catalyzed by tRNA nucleotidyl transferase. **(d)** Finally, specific bases are modified (see Fig. 25–25).

Mature tRNA

catalyze formation of the succession of phosphodiester bonds needed to generate the CCA(3′) sequence. Although a defined sequence of nucleotides is added, the reaction is template independent and thus represents a mechanism distinct from that used by DNA and RNA polymerases.

The final type of tRNA processing is the modification of some of the bases (Fig. 25–25) by methylation, by deamination, or by reduction. Some of these modified bases occur at characteristic positions in all tRNAs (Fig. 25–24).

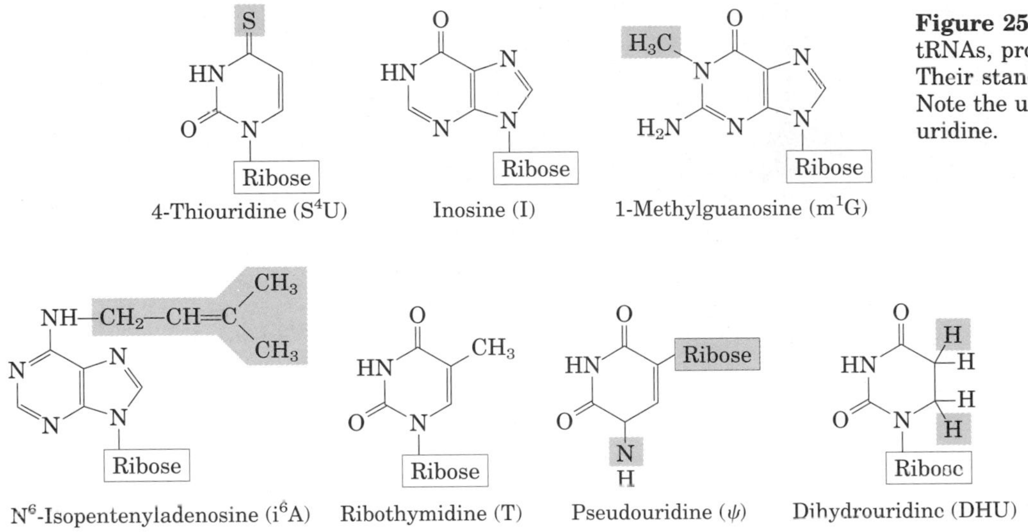

4-Thiouridine (S⁴U)   Inosine (I)   1-Methylguanosine (m¹G)

N⁶-Isopentenyladenosine (i⁶A)   Ribothymidine (T)   Pseudouridine (ψ)   Dihydrouridine (DHU)

Figure 25–25 Some of the modified bases found in tRNAs, produced in posttranscriptional reactions. Their standard symbols are shown in parentheses. Note the unusual ribose attachment point in pseudouridine.

Some Events in RNA Metabolism Are Catalyzed by RNA Enzymes

The study of posttranscriptional processing of RNA molecules has led to one of the most exciting discoveries in modern biochemistry—the existence of RNA enzymes or **ribozymes.** The best characterized ribozymes are the self-splicing group I introns and RNase P. Most of the activities of these ribozymes are based on two fundamental reactions: transesterification (see Fig. 25–12) and phosphodiester bond hydrolysis (cleavage). Where bonds are cleaved by either of these ribozymes, the products have 3′-hydroxyl and 5′-phosphate termini, in contrast to the 5′-hydroxyl and 2′- or 3′-phosphate products formed by random alkaline hydrolysis of RNA. The substrate for ribozymes is often an RNA molecule; sometimes the substrate is itself part of the ribozyme. With an RNA substrate, an RNA catalyst can make use of base-pairing interactions to align the substrate for reaction.

Ribozymes tend to be large. Although the three-dimensional structure of these catalytic RNAs is not known, it is clearly important for function. Activity is lost if the RNA is heated beyond its melting temperature, if denaturing agents are added, or if complementary oligonucleotides are added that can disrupt normal base-pairing patterns. They can also be inactivated if some essential nucleotides are changed. The secondary structure and possible tertiary structure of a self-splicing intron from the 26S rRNA precursor of *Tetrahymena* are shown in Figure 25–26.

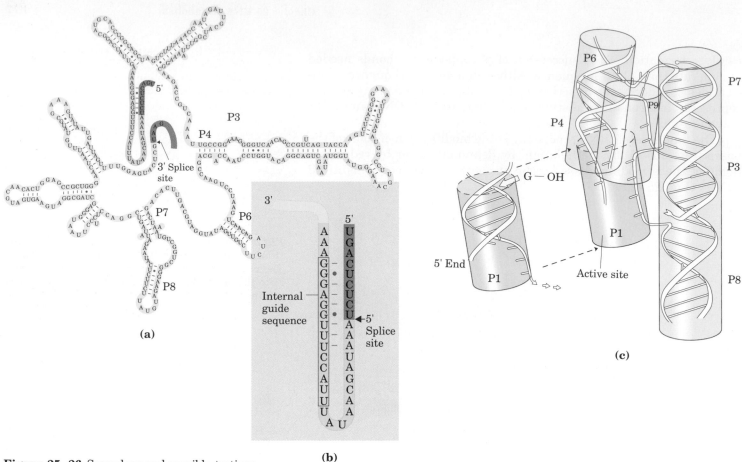

(a)

(b)

(c)

Figure 25–26 Secondary and possible tertiary structure of the self-splicing rRNA intron from *Tetrahymena*. Intron and exon sequences are shaded yellow and green, respectively. **(a)** Some base-paired regions are labeled (P1, P3, etc.) according to an established convention for this RNA molecule. The P1 region, which contains the internal guide sequence, is the location of the 5′ splice site. **(b)** An enlargement of the internal guide sequence of group I introns. This sequence can base-pair with the 5′ splice site to bring about its proper alignment for reaction. The remainder of the intron forms a three-dimensional structure that catalyzes the splicing reaction (see Fig. 25–24b). **(c)** In a proposed tertiary structure, the P3, P4, P6, P7, and P8 regions fold up to form an active site into which P1 can fit.

The self-splicing group I introns have several properties of enzymes other than greatly accelerating the reaction rate. The binding of the guanosine cofactor to the *Tetrahymena* group I rRNA intron (p. 868) is saturable ($K_m \approx 30\ \mu\text{M}$) and can be competitively inhibited by 3′-deoxyguanosine. The intron is also very precise in its excision reaction, largely due to an internal guide sequence that can base-pair with exon sequences near the 5′ splice site (Fig. 25–26b). This helps provide the proper alignment of bonds to be cleaved and rejoined.

Because the intron itself is used up (excised) during the splicing reaction, it may appear that it lacks one key enzymatic property: the ability to catalyze multiple reactions. Closer inspection has shown that after excision, the 414 nucleotide intron from *Tetrahymena* rRNA does in fact act as a true enzyme. A series of intramolecular cyclization/cleavage reactions in the excised intron leads to the loss of 19 nucleotides from its 5′ end. The remaining 395 nucleotide linear RNA, called the L-19 IVS (for *i*nter*v*ening *s*equence *l*acking *19* nucleotides), promotes nucleotidyl transfer reactions in which some oligonucleotides are lengthened at the expense of others (Fig. 25–27). The best substrates are oligonucleotides, for example a synthetic (C)₅ oligomer, which can base-pair with a specific guanylate-rich sequence in L-19 IVS. Enzymatic activity results from a cycle of transesterification reactions mechanistically similar to self-splicing. L-19 IVS, however, is not used up, and each ribozyme processes about 100 substrate molecules per hour. Thus L-19 IVS is a catalyst. It follows Michaelis–Menten kinetics, is specific for RNA oligonucleotides, and is competitively inhibited by (dC)₅. The k_{cat}/K_m is $10^3\ \text{M}^{-1}\,\text{s}^{-1}$, low compared to many enzymes, but the hydrolysis of (C)₅ by this ribozyme is accelerated by a factor of 10^{10} relative to the uncatalyzed reaction. This RNA molecule can clearly be quite effective as an enzyme.

(5′) G̲ A A A U A G C A A U A U | U U A C C U U U G G A G G G | A

**Spliced
rRNA intron**

(385 nucleotides)

G — OH (3′)

19 Nucleotides from 5′ end

(385 nucleotides)

L-19 IVS   (5′) | U U G G A G G G | A ⎯⎯⎯⎯⎯⎯ G — OH (3′)

(a)

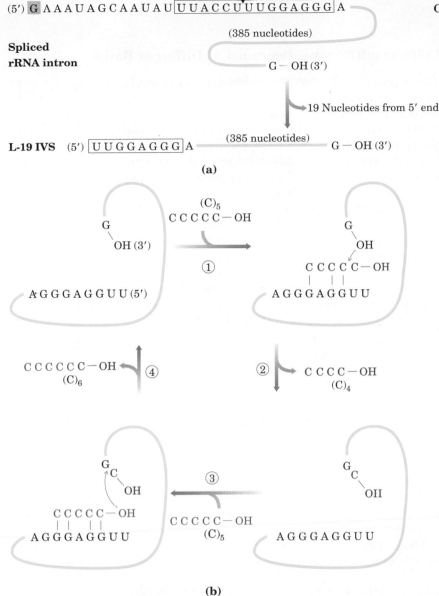

(b)

Figure 25–27 In vitro catalytic activity of L-19 IVS of *Tetrahymena*. **(a)** L-19 IVS is generated by the autocatalytic removal of 19 nucleotides from the 5′ end of the spliced intron. The G residue (shaded in red) added in the first step of the splicing reaction (see Fig. 25–13) is part of the removed sequence. A portion of the internal guide sequence (boxed) remains at the 5′ end of L-19 IVS. **(b)** Lengthening of RNA oligonucleotides catalyzed by L-19 IVS. Some oligonucleotides are lengthened at the expense of others in a cycle of transesterification reactions (steps ① through ④). The 3′ OH of the G residue at the 3′ end of L-19 IVS plays a key role in this cycle of reactions (note that this is *not* the G residue that was added in the splicing reaction). (C)₅ is one of the better substrates because it can base-pair with the guide sequence remaining in the intron. Although this activity is probably irrelevant to the cell, it has important implications for current hypotheses on evolution. This catalytic activity was elucidated by Thomas Cech and coworkers.

The second well-characterized ribozyme is derived from RNase P. The *E. coli* RNase P has an RNA component called the M1 RNA (377 nucleotides) and a protein component (M_r 17,500). In 1983, Sidney Altman, Norman Pace, and coworkers discovered that under some conditions the M1 RNA alone is sufficient for catalysis, cleaving tRNA precursors at the correct position. The protein apparently is needed only to stabilize the RNA or facilitate its function under the particular conditions present in the cell. In spite of some sequence differences, RNase P derived from such diverse organisms as bacteria and humans accurately processes the tRNA precursors from other species.

Other catalytic RNAs are known. Some small RNAs of plant RNA viruses form a structure that promotes a self-cleavage reaction. In one case the specific cleavage activity has been localized to a 19 nucleotide fragment, making it the smallest ribozyme known. The significance of RNA catalysis has been enhanced further by another discovery—that the synthesis of peptide bonds in proteins is catalyzed largely by an RNA component of ribosomes (Chapter 26). The discovery of catalytic RNAs has provided new insights into catalytic function in general and has important implications for the origin of life on this planet, a topic we will discuss at the end of this chapter.

Cellular mRNAs Are Degraded at Different Rates

The expression of genes is regulated at many levels. Perhaps the most important factor in gene expression is the concentration of a specific mRNA in a cell. The concentration of any molecule depends on two factors: its rate of synthesis and its rate of degradation. When synthesis and degradation of an mRNA are balanced, its concentration remains at a constant, steady-state level. An increase in the rate of synthesis or degradation leads to net accumulation or depletion, respectively, of the mRNA.

In eukaryotic cells, the rates of mRNA degradation vary greatly for mRNAs derived from different genes. Gene products needed only briefly may have mRNAs with a half-life measured in minutes or even seconds. Gene products needed constantly by the cell may have mRNAs that are stable for many cell generations. On average, the half-life of an mRNA in a vertebrate cell is about three hours, the mRNA turning over about ten times in each cell generation. In bacterial cells the half-life of mRNAs is only about 1.5 min, but because bacteria divide much faster than eukaryotic cells, bacterial mRNAs also turn over about ten times in each cell cycle.

RNA is degraded by ribonucleases present in all cells. A prominent enzyme in most cells is a $3' \rightarrow 5'$ exoribonuclease; this may represent the primary degradative activity. Stable mRNAs generally have some sequence at or near the $3'$ end that inhibits this enzyme. In bacteria, the hairpin structure present in mRNAs with a ρ-independent terminator (see Fig. 25–8) confers stability. Similar hairpin structures can confer stability on selected regions of a primary transcript, leading to nonuniform degradation of some polycistronic transcripts. In eukaryotic cells, the $3'$ poly(A) tail may be important to the stability of many mRNAs. Removal of this tail (and possibly some proteins that are normally bound to it) leads to rapid degradation of some normally stable mRNAs. These degradative processes ensure that RNAs do not build up in the cell and direct the synthesis of unnecessary proteins.

Polynucleotide Phosphorylase Makes Random RNA-like Polymers

In 1955 Marianne Grunberg-Manago and Severo Ochoa discovered the bacterial enzyme **polynucleotide phosphorylase,** which in vitro catalyzes the reaction

$$(NMP)_n + NDP \longrightarrow \underset{\substack{\text{Lengthened} \\ \text{polynucleotide}}}{(NMP)_{n+1}} + P_i$$

Polynucleotide phosphorylase was the first nucleic acid–synthesizing enzyme found (Arthur Kornberg's discovery of DNA polymerase followed soon thereafter). The reaction catalyzed by polynucleotide phosphorylase differs fundamentally from the other polymerizing activities discussed so far in that it is not template-directed. The enzyme requires the $5'$-diphosphates of ribonucleosides and cannot act on the homologous $5'$-triphosphates or on deoxyribonucleoside $5'$-diphosphates. The RNA polymer formed by polynucleotide phosphorylase contains normal $3',5'$-phosphodiester linkages, which can be hydrolyzed by ribonuclease. The reaction is readily reversible and can be pushed in the direction of breakdown of the polyribonucleotide by increasing the phosphate concentration. The probable function of this enzyme in the cell is the degradation of mRNAs to form nucleoside diphosphates.

Marianne
Grunberg-Manago

Severo Ochoa

Because the polynucleotide phosphorylase reaction does not use a template, it does not form a polymer having a specific base sequence. The reaction proceeds as well with only one of the nucleoside diphosphates as with all four. The base composition of the polymer formed by the enzyme reflects the relative concentrations of the 5'-diphosphate substrates in the medium.

Polynucleotide phosphorylase can be used for the laboratory preparation of many different kinds of RNA polymers with different sequences and frequencies of bases. Such synthetic RNA polymers made it possible to deduce the genetic code for the amino acids (Chapter 26).

RNA-Dependent Synthesis of RNA and DNA

In our discussion of DNA and RNA synthesis up to this point, the role of template strand has been reserved for DNA. However, enzymes that use an RNA template in nucleic acid synthesis are surprisingly widely distributed. With a very important exception, these enzymes play only a modest role in information pathways. The exception is viruses having an RNA genome. These viruses are the source of most RNA-dependent polymerases characterized so far.

The existence of RNA replication requires an elaboration of the information pathways described in the introduction to Part IV of this book (p. 790) (Fig. 25–28). The additional pathways are important, not simply because the enzymes involved are extremely useful in recombinant DNA technology (Chapter 28), but because they have profound implications for any discussion of the nature of self-replicating molecules that may have existed in prebiotic times.

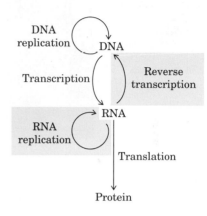

Figure 25–28 Extension of the central dogma to include RNA-dependent synthesis of RNA and DNA.

Reverse Transcriptase Produces DNA from Viral RNA

Certain RNA viruses of animal tissues contain within the viral particle a unique RNA-directed DNA polymerase called **reverse transcriptase.** On infection, the single-stranded RNA viral genome (~10,000 nucleotides in length) and the enzyme enter the host cell, and the reverse transcriptase catalyzes the synthesis of a DNA strand complementary to the viral RNA (Fig. 25–29). The same enzyme degrades the RNA strand in the resulting RNA–DNA hybrid and replaces it with DNA. The duplex DNA so formed often becomes incorporated into the genome of the eukaryotic host cell. Under some conditions such integrated (and dormant) viral genes become activated and transcribed to generate new viruses.

Figure 25–29 Retroviral infection of a mammalian cell and integration of the retrovirus into the host chromosome. The integration has many characteristics of the insertion of transposons in bacteria (Fig. 24–40). For example, a few base pairs of host DNA are duplicated at the site of integration, forming short (4 to 6 base pair) repeats at each end (not shown). Note that on infection of the host cell, also entering from the viral particle are reverse transcriptase and a tRNA base-paired to the viral RNA. The function of this tRNA is described later in the text.

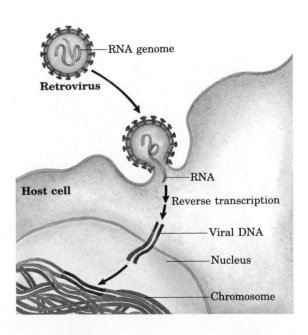

Howard Temin

David Baltimore

The existence of reverse transcriptases in RNA viruses was predicted by Howard Temin in 1962, and the enzymes were ultimately demonstrated to occur in such viruses by Temin and independently by David Baltimore in 1970. Their discovery aroused much attention, particularly because it constituted molecular proof that genetic information can sometimes flow "backward" from RNA to DNA. The RNA viruses containing reverse transcriptases are also known as **retroviruses** (*retro* is the Latin word for "backward").

Retroviruses typically have three genes: *gag* (derived from the historical designation: *group associated antigen*), *pol,* and *env* (Fig. 25–30). The *gag* gene encodes a "polyprotein" that is cleaved into three or four proteins that make up the interior core of the virus particle structure. The protease catalyzing this cleavage is itself part of the polyprotein in many retroviruses. The *pol* gene codes for reverse transcriptase. Reverse transcriptase often has two subunits, α and β; the *pol* gene encodes the β subunit (M_r 90,000), and the α subunit (M_r 65,000) is simply a proteolytic fragment of the β subunit. The *pol* gene product is also a polyprotein that includes the reverse transcriptase and a separate integrase needed for inserting the viral DNA into the host genome. The integrase is separated from the reverse transcriptase by the protease described above. The *env* gene specifies another polyprotein from which the proteins of the viral envelope are derived. At each end of the linear RNAs are long terminal repeat (LTR) sequences a few hundred nucleotides long. These contain sequences required for integration into the host DNA and the regulation of viral gene expression.

Figure 25–30 General structure of an integrated retrovirus genome. The long terminal repeats (LTRs) have sequences needed for the regulation and initiation of transcription. The sequence denoted ψ is required for packaging retroviral RNAs into mature virus particles.

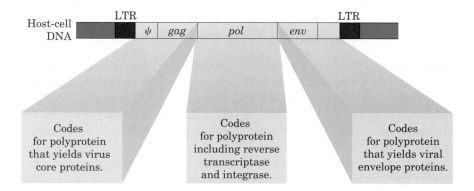

Viral reverse transcriptases contain Zn^{2+}, as do all DNA and RNA polymerases. They are most active with the RNA of their own type of virus but can be used experimentally to make DNA complementary to a variety of RNAs. Reverse transcriptases catalyze three different reactions: (1) RNA-directed DNA synthesis, (2) RNA degradation, and (3) DNA-directed DNA synthesis. The reverse transcriptases require a primer for their initial DNA synthesis. The primer is a tRNA included within the virus particle (obtained during an earlier infection), base-paired at its 3' end with a complementary sequence in the viral RNA. The new DNA strand is synthesized in the $5' \rightarrow 3'$ direction, as in all RNA and DNA polymerase reactions. Reverse transcriptases, like RNA polymerases, do not have $3' \rightarrow 5'$ proofreading exonucleases. They generally have error rates of about one per 20,000 nucleotides added, similar to the expected fidelity of base selection itself. As a result, reverse transcription exhibits a relatively high error rate, and this appears to be a feature of most enzymes that replicate the RNA genomes of these and other RNA viruses. A likely consequence is a faster rate of evolution, and this may be a factor in the frequent appearance of new strains of disease-causing viruses.

Reverse transcriptases have become important reagents in the study of DNA–RNA relationships and in cloning DNA. They make possible the laboratory synthesis of a DNA complementary in base sequence to any RNA template, whether it is mRNA, tRNA, or rRNA. A synthetic DNA prepared in this manner is called a **complementary DNA (cDNA).** Later we shall see how this process is used to clone cellular genes (Chapter 28).

Some Retroviruses Cause Cancer or AIDS

Retroviruses have played an important role in recent advances in the molecular understanding of cancer. Most retroviruses do not kill their host cells. They remain integrated in the cellular DNA and are replicated along with it. Some retroviruses, however, have an additional gene that can cause the cell to become cancerous (i.e., to grow abnormally), and these viruses are classified as RNA tumor viruses. The first retrovirus of this type to be studied was the Rous sarcoma virus (also called avian sarcoma virus; Fig. 25–31), named for Peyton Rous who studied chicken tumors now known to be caused by this virus. The cancer-causing gene in this and other tumor viruses is called an **oncogene.** Since the initial discovery of oncogenes by Harold Varmus and Michael Bishop, more than 40 different such genes have been found in retroviruses.

Cancer results from a malfunction in the normal cellular mechanisms that regulate cell division. A cancerous tumor is made up of cells that are growing out of control because of these malfunctions (see Fig. 6–1g). The oncogenes in retroviruses are derived from normal cellular genes called **proto-oncogenes,** incorporated into the viral genome by rare recombination events. Proto-oncogenes generally encode proteins involved in normal cell division, growth, and development. However, the oncogenes tend to differ slightly in sequence from the proto-oncogenes, and hence the activity of their protein products may differ from that of their cellular analogs. Known oncogenes include genes for tyrosine kinases, growth factors, receptors for growth factors, and G proteins, all of which are known to participate in normal growth-control signals (Chapter 22). Some oncogenes encode nuclear proteins that are involved in gene regulation. Overexpression of oncogene-encoded proteins by the integrated retrovirus contributes to an imbalance that causes the cells to grow out of control. Cancer can be induced without the participation of a virus, by mutations in the cellular proto-oncogenes from which oncogenes are derived. Study of these genes is providing insight into mechanisms controlling normal growth and development, as well as the perturbations in these mechanisms that can lead to cancer.

Figure 25–31 The Rous sarcoma virus genome. The *src* gene encodes a tyrosine-specific protein kinase, one of a class of enzymes known to function in systems that affect cell division, cell–cell interactions, and intercellular communication. The same gene is found in the DNA of normal chickens and in the genomes of many other eukaryotes including humans. When associated with the virus, this oncogene is expressed at abnormally high levels, contributing to unregulated cell division and cancer.

BOX 25–2   Fighting AIDS with Inhibitors of HIV Reverse Transcriptase

A knowledge of the fundamental chemistry of template-directed nucleic acid biosynthesis, combined with modern techniques of molecular biology, has led to a rapid understanding of the life cycle and structure of the human immunodeficiency virus (HIV), the RNA virus that causes AIDS. Just a few years after the isolation of HIV, these advances also resulted in the development of drugs capable of prolonging the lives of those infected by HIV. The first of these drugs to be approved for clinical use was AZT (Fig. 1), a structural analog of deoxythymidine. AZT was first synthesized in 1964 by Jerome P. Horwitz. It failed as an anticancer drug (the purpose for which it was made), but in 1985 it was found to be an effective treatment for AIDS. AZT is taken up by the T lymphocytes, immune system cells that are particularly vulnerable to HIV infection, and converted to AZT triphosphate (AZT triphosphate cannot be given directly because it cannot cross the plasma membrane). The HIV reverse transcriptase has a higher affinity for AZT triphosphate than for dTTP; binding of AZT triphosphate to the enzyme competitively inhibits dTTP binding. In addition, AZT can be added to the 3' end of the growing RNA chain, but because AZT has no 3' hydroxyl the RNA chain is prematurely terminated and viral RNA synthesis quickly grinds to a halt.

The compound is not as toxic to the T lymphocytes themselves, because the cellular DNA polymerases have a lower affinity for AZT triphosphate than for dTTP. AZT is effective at concentrations of 1 to 5 μM, high enough to affect HIV reverse transcription but too low to significantly affect most cellular DNA replication. Unfortunately, the drug appears to be toxic to the bone marrow cells that are the progenitors of erythrocytes, and patients often develop anemia. AZT can increase the survival time of patients with advanced AIDS by about a year and has also been shown to delay the onset of AIDS in individuals in the early stages of HIV infection. Some newer drugs, such as dideoxyinosine (Fig. 1), have a similar mechanism of action.

Figure 1

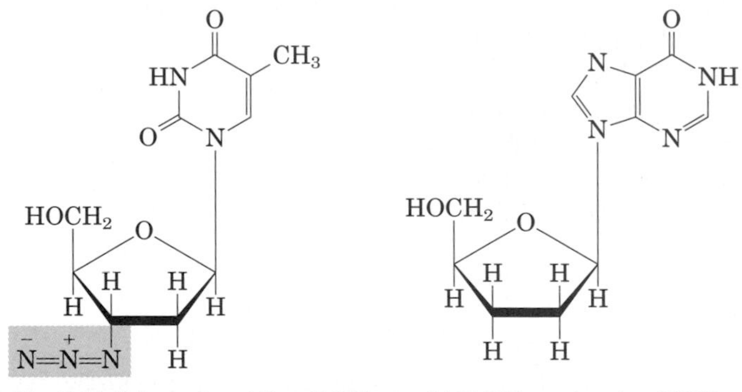

3'-Azido-2',3'-dideoxythymidine (AZT)   2',3'-Dideoxyinosine (DDI)

The human immunodeficiency virus (HIV), the causative agent of *a*cquired *i*mmune *d*eficiency *s*yndrome (AIDS), is also a retrovirus. Identified in 1983, HIV has an RNA genome with standard retroviral genes along with several other unusual genes (Fig. 25–32). The *env* gene in this virus (along with the rest of the genome) undergoes mutation at a very rapid rate, complicating the development of an effective vaccine. The reverse transcriptase of HIV is about tenfold less accurate in replication than other known reverse transcriptases, and this fact is largely responsible for the increased mutation rates in this virus. There are generally one or more errors made every time the viral genome is replicated, so that any two viral RNA molecules are almost never identical. Reverse transcriptase is the target of the drugs most widely used to treat HIV-infected individuals (Box 25–2).

Figure 25–32 The genome of HIV, the virus that causes AIDS. In addition to the typical retroviral genes, there are several small genes with a variety of functions. Some of these genes overlap (see Chapter 26). Alternative splicing mechanisms lead to the production of many different proteins from this small (9.7×10^6 bases) genome.

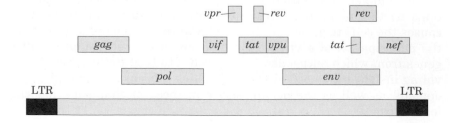

Many Eukaryotic Transposons Are Related to Retroviruses

Some well-characterized transposons from eukaryotes as diverse as yeast and fruit flies have a structure very similar to that of retroviruses and are sometimes called retrotransposons (Fig. 25–33). They have coding regions specifying an enzyme homologous to the retroviral reverse transcriptase, and the coding regions are flanked by LTR sequences. They transpose from one position to another in the genome by means of an RNA intermediate, probably using reverse transcriptase to make a DNA copy of the RNA, followed by integration at a new site. Most transposons in eukaryotes probably use this mechanism for transposition, distinguishing them from the bacterial transposons described in Chapter 24 that move as DNA directly from one chromosomal location to another (see Fig. 24–40). Retrotransposons lack an *env* gene and so cannot form virus particles capable of transferring among cells. They can be thought of as defective viruses trapped in one cell. The relationship between retroviruses and these eukaryotic transposons suggests that reverse transcriptase is an ancient enzyme that predates the evolution of multicellular organisms.

Telomerase Is an Enzyme Resembling Reverse Transcriptase

Telomeres (Chapter 23) are the specialized structures at the ends of linear eukaryotic chromosomes. They generally consist of many tandem copies of a short oligonucleotide sequence, usually of the form T_xG_y in one strand and C_yA_x in the complementary strand, where x and y typically fall in the range of 1 to 4 (see p. 798).

The structure of telomeres poses a particular biological problem. DNA replication requires a primer, but in a linear DNA molecule it is impossible to synthesize an RNA primer starting at the end nucleotide and replace it by the normal mechanisms. Without a special mechanism for replicating the ends, chromosomes would be shortened somewhat in each cell generation. The problem is solved by an enzyme called **telomerase,** which adds telomeres to chromosome ends. Although the existence of this enzyme may not be surprising, the mechanism by which it acts is unprecedented. Telomerase, like some other enzymes described in this chapter, contains both RNA and protein components. The RNA component is about 150 nucleotides long and contains about 1.5 copies of the appropriate C_yA_x telomere repeat. This part of the RNA acts as a template for synthesis of the T_xG_y strand of the telomere. In effect, telomerase is a reverse transcriptase that synthesizes only a segment of DNA that is complementary to an *internal* RNA template.

Telomere synthesis requires a short T_xG_y primer and proceeds in the usual $5' \rightarrow 3'$ direction. Having synthesized one copy of the repeat, the enzyme must be repositioned to resume extension of the telomere. This may occur in an inchworm-like process as outlined in Figure 25–34.

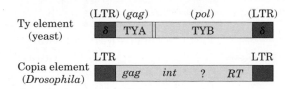

Figure 25–33 Eukaryotic transposons, as exemplified by Ty of the yeast *Saccharomyces* and copia of *Drosophila* (fruit flies), often have a structure similar to retroviruses but lack the *env* gene. The δ sequences are functionally equivalent to retroviral LTRs. In copia, *int* and *RT* are homologous to the integrase and reverse transcriptase segments, respectively, of the *pol* gene.

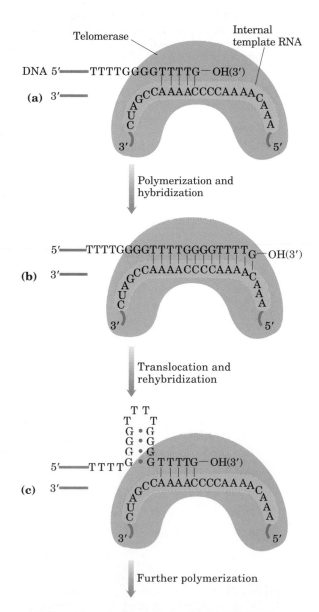

Figure 25–34 Synthesis of the TG strand of a telomere by telomerase: the "inchworm" model. **(a)** Telomerase binds to the TG primer, with base pairing between the primer and the enzyme's internal RNA template. **(b)** The enzyme adds more T and G residues to the primer, then **(c)** shifts to reposition the internal template for addition of more TG repeats. The newly synthesized telomere strand can form a hairpin structure by nonstandard base pairing between G residues.

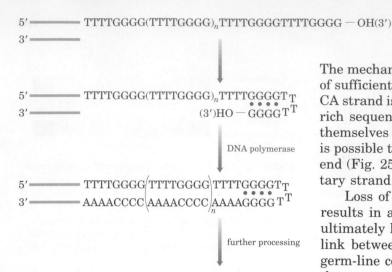

5′ —————— TTTTGGGG(TTTTGGGG)$_n$TTTTGGGGTTTTGGGG — OH(3′)
3′ ——————

5′ —————— TTTTGGGG(TTTTGGGG)$_n$TTTTGGGGT T
3′ ——————   (3′)HO — GGGG T^T

DNA polymerase

5′ —————— TTTTGGGG(TTTTGGGG)TTTTGGGGT T
3′ —————— AAAACCCC(AAAACCCC)$_n$AAAAGGGG T^T

further processing

Figure 25–35 Nonstandard G–G base pairing could permit the end of the telomere TG strand to fold over as shown, creating a primer for synthesis of the complementary strand (shown in red).

The mechanisms by which this process is terminated when a telomere of sufficient length has been synthesized, and how the complementary CA strand is synthesized, are not yet known. One feature of guanylate-rich sequences, however, is that they are capable of folding back on themselves to form non–Watson–Crick G–G base pairs (Fig. 25–34). It is possible that the TG strand folds back on itself in this way near the end (Fig. 25–35), providing a primer for synthesis of the complementary strand.

Loss of telomerase activity in protozoans (such as *Tetrahymena*) results in a gradual shortening of telomeres with each cell division, ultimately leading to the death of the cell line. In humans, a similar link between telomere length and cell death has been observed. In germ-line cells telomere lengths are maintained, but in somatic cells they are not. There is a linear, inverse relationship between the length of telomeres in cultured fibroblasts and the age of the individual from whom the fibroblasts were taken: telomeres in human somatic cells gradually shorten as an individual ages. One inference is that germ-line cells contain telomerase activity but somatic cells do not. Is this gradual shortening of telomeres a key to the aging process? Is our natural life span determined by the length of the telomeres we are born with? Further research in this area should yield some fascinating insights.

Some Viral RNAs Are Replicated by RNA-Directed RNA Polymerase

Some *E. coli* bacteriophages, including f2, MS2, R17, and Qβ, have RNA genomes. The single-stranded RNA chromosomes of these viruses, which also function as mRNAs for the synthesis of viral proteins, are replicated in the host cell by the action of enzymes called **RNA-directed RNA polymerases** or **RNA replicases.** RNA replicase (M_r ~ 210,000) has four subunits. Only one of these subunits (M_r 65,000), the product of the viral replicase gene, is encoded by the viral RNA; this subunit includes the active site for replication. The other three subunits are host proteins normally involved in protein synthesis: the elongation factors Tu (M_r 30,000) and Ts (M_r 45,000) of *E. coli,* which normally function in ferrying amino acyl–tRNAs to the ribosome, and the protein S1, which is normally an integral part of the 30S ribosomal subunit. These host proteins may help the replicase locate and bind to the 3′ ends of viral RNA.

RNA replicase isolated from Qβ-infected *E. coli* cells catalyzes the formation of an RNA complementary to the viral RNA in a reaction similar to that of DNA-directed RNA polymerases (see p. 857). Synthesis of the new RNA strand proceeds in the 5′→3′ direction, and the chemical mechanism is identical to that for all other template-requiring nucleic acid synthetic reactions. RNA replicase requires RNA as template and will not function with DNA. The enzyme lacks a proofreading endonuclease, and the frequency of error is similar to that of RNA polymerase. Unlike the DNA and RNA polymerases, RNA replicases are specific for the RNA of their own virus; the RNAs of the host cell are generally not replicated. This explains how RNA viruses are preferentially replicated in the host cell, which contains many other types of RNA.

RNA Synthesis Offers Important Clues to Biochemical Evolution

The extraordinary complexity and order that distinguish living systems from inanimate ones are only the outward manifestation of more fundamental processes that hold the key to understanding "life." Maintaining the living state requires that *selected* chemical transformations occur very rapidly—in particular those required to make efficient use of energy sources in the environment and to synthesize the elaborate and specialized macromolecules found in a cell. The crucial conditions for life, then, are the existence of powerful and very selective catalysts— the enzymes—and an informational system capable of storing a blueprint for these enzymes and reproducing them generation after generation. The cellular chromosomes do not encode the blueprint for a cell; they encode the structure of the enzymes needed to construct and maintain a cell. But how did this system come into being?

In the 1960s, the unveiling of the structural and functional complexity of RNA led Carl Woese, Francis Crick, and Leslie Orgel to propose that this macromolecule might have catalytic as well as informational functions in the cell. The discovery of catalytic RNAs has taken this remarkable insight from conjecture to reality. The presence of both of these important functions in one macromolecule suggests that self-replicating RNA molecules might exist, or might once have existed; this has a number of important implications for evolution. It is now possible to speculate about two stages of biochemical evolution: an early stage in which RNA molecules first achieved a self-replicating activity, and a later stage when proteins were evolving to their present form.

The existence of catalytic RNAs allows us to envision an early biological world made up entirely or almost entirely of RNA (see Fig. 3–22). Proteins became abundant later when their greater versatility as catalysts proved advantageous. This "RNA world" scenario has been made much more plausible by the discovery that the synthesis of the peptide bonds of proteins is catalyzed by the rRNA component of ribosomes. The first true cells may have contained only RNA, protein, and the smaller molecules needed to form a cell wall and provide metabolic energy. Later, DNA entered the picture to provide a more stable molecular form for long-term information storage. The activities of the L-19 IVS, in particular its catalysis of a crude form of polymerization that is partially dependent on a template (internal guide sequence) and the activities of ribosomal RNA, suggest that RNA molecules can catalyze all the reactions needed to duplicate themselves and synthesize proteins if appropriate precursor molecules (oligonucleotides or perhaps nucleoside triphosphates and activated amino acids) are available in sufficient quantities. There are shortcomings to the RNA → protein → DNA scenario in its simplest form. Precursors must be synthesized, and the jump from L-19 IVS to a true template-dependent RNA polymerase is a large one.

Protein enzymes presumably emerged through a complex series of evolutionary steps that coincided with the development of a genetic code, allowing specific protein sequences that exhibited useful properties to be reproduced. As pathways (and enzymes) for conversion of RNA to DNA, DNA to DNA, and DNA to RNA developed, the superior stability of DNA would have gradually led to its adoption as a long-term information-storage polymer. RNA replicase and reverse tran-

Carl Woese

Francis Crick

Leslie Orgel

scriptase may be modern versions of enzymes that once played important roles in making this transition to the modern DNA-based system.

Once proteins had appeared, they became the focus of natural selection as cells required an ever-increasing array of catalysts and structural components. Introns provide an important clue to one of the mechanisms by which proteins probably evolved. Introns could have two evolutionary origins: they were either (1) present in the earliest genes and then gradually lost from bacteria or (2) inserted into genes (primarily eukaryotic) gradually over evolutionary time. Evidence now favors the first evolutionary pathway. For example, introns are at the same positions in some genes in organisms as disparate as yeast and humans, and the splicing apparatus for mammalian mRNAs will splice yeast genes. It is likely that introns were present, along with a splicing mechanism, in the first cells. In this view, introns helped to assemble early genes from assorted pieces and were gradually lost from bacteria and some yeast species as their genomes became streamlined for rapid cell division. Bacteria, which produce one or more new generations each hour, evolve much more rapidly than humans.

Introns often separate DNA regions (exons) that encode distinct folding domains of a polypeptide. In evolution, the separation of these exons could permit their recombination or shuffling to create new proteins made up of domains of proven stability. Introns provide a large region of DNA where recombination can occur with little chance that it will be detrimental to gene-coding information. Several striking examples of complete domains encoded within a single exon have been found, as well as clear examples of the shuffling of these domains during evolution. Perhaps the best example is an exon that encodes a small domain (~40 amino acids) in the 1,217 amino acid precursor protein from which the peptide hormone epidermal growth factor (EGF) is derived (Fig. 25–36). This domain is also found in several other proteins including the low-density lipoprotein (LDL) receptor and the blood clotting proteins factor IX, factor X, and protein C. The LDL receptor itself is a mosaic of domains derived from other proteins. Of its 18 exons, 13 encode domains homologous to those found in other

Figure 25–36 Examples of exon shuffling during evolution. Exons that encode various protein domains are indicated by colored boxes (the box size is *not* representative of exon size, and the variation in size of the intervening introns is not shown). The leftmost exons encode domains near the amino-terminal end of the protein. A protein domain called the EGF (epidermal growth factor) repeat contains about 40 amino acids and six Cys residues (three disulfide bonds). The exon encoding the EGF repeat (red box) corresponds to 8 of the 24 exons in the gene for the EGF precursor protein. A very similar exon appears three times (of 18 exons) in the gene for the LDL receptor, twice in the gene for blood coagulation factor IX, and once in the gene for the tissue plasminogen activator protein. The genes for the LDL receptor and EGF precursor share a longer region of homology, as shown. Certain exons in the LDL receptor gene (green boxes) correspond to a domain also occurring once in a protein component of the immune system called complement factor C9. In the LDL receptor gene, this repeated sequence occurs seven times; four of these are found in separate exons and three are found in one exon. Two introns (arrows) separating the repeats in this latter exon were probably lost during evolution. Altogether, 13 of the 18 exons found in the LDL receptor gene are also found in genes for other proteins.

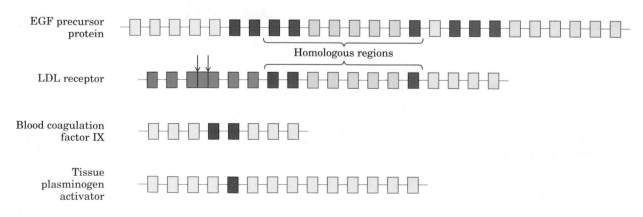

proteins. Many proteins are clearly derived, at least in part, from exon shuffling during evolution. Walter Gilbert and colleagues have suggested that all present-day proteins may have been assembled from as few as 1,000 to 7,000 primordial exons encoding small polypeptides each 30 to 50 amino acids long.

The origin of life still offers a major intellectual challenge. Even though we cannot go back billions of years and observe the events firsthand, many clues to the puzzle lie buried in the fundamental chemistry of living cells.

Walter Gilbert

Summary

Transcription is catalyzed by DNA-directed RNA polymerase, a complex enzyme that synthesizes RNA complementary to a segment of one strand (the template strand) of duplex DNA, starting from ribonucleoside 5'-triphosphates. To initiate transcription, RNA polymerase binds to a DNA site called a promoter. Bacterial RNA polymerase requires a special subunit for recognizing the promoter. As the first committed step in transcription, binding of RNA polymerase to promoters is subject to many forms of regulation. Eukaryotic cells have three different types of RNA polymerases. Transcription stops at specific sequences called terminators. Many copies of an RNA chain can be transcribed simultaneously from a single gene.

Ribosomal RNAs and transfer RNAs are made from longer precursor RNAs that are trimmed by nucleases, and some bases are modified enzymatically to yield the mature RNAs. In eukaryotes, messenger RNAs are also formed from longer precursors. Primary RNA transcripts often contain noncoding regions called introns, which are removed by splicing. Group I introns are found in rRNAs and their excision requires a guanosine cofactor. Some group I and some group II introns are capable of self-splicing; no protein enzymes are required. Nuclear mRNA precursors have a third class of introns that are spliced with the aid of RNA–protein complexes called snRNPs. The fourth class of introns, found in some tRNAs, are the only ones known to be spliced by protein enzymes. Messenger RNAs are also modified by addition of a 7-methylguanosine residue at the 5' end, and cleavage and polyadenylation at the 3' end to form a long poly(A) tail.

The self-splicing introns and the RNA component of RNase P (the enzyme that cleaves the 5' end of tRNA precursors) form a new class of biological catalysts called ribozymes. These have the properties of true enzymes and are effective catalysts. They promote two types of reaction, hydrolytic cleavage and transesterification, using RNA as substrate. Combinations of these reactions are promoted by the excised group I rRNA intron from *Tetrahymena*, resulting in a type of RNA polymerization reaction. The study of these reactions and of introns themselves has provided insights into likely pathways for biochemical evolution.

Polynucleotide phosphorylase can reversibly form RNA-like polymers from ribonucleoside 5'-diphosphates, adding or removing ribonucleotides at the 3'-hydroxyl end of the polymer. It acts in vivo to degrade RNA.

RNA-directed DNA polymerases, also called reverse transcriptases, are produced in animal cells infected by RNA viruses called retroviruses. These enzymes transcribe the viral RNA into DNA. This process can be used experimentally to form complementary DNA. Many eukaryotic transposons are related to retroviruses, and their mechanism of transposition includes an RNA intermediate. The enzyme that synthesizes telomeres, called telomerase, is a specialized reverse transcriptase that contains an internal RNA template.

RNA-directed RNA polymerases, or replicases, are found in bacterial cells infected with certain RNA viruses. They are template-specific for the viral RNA.

The existence of catalytic RNAs and pathways for the interconversion of RNA and DNA has led to speculation that the earliest living things were made up entirely or largely of RNA molecules that served both for information storage and for catalysis of replication.

Further Reading

General

Darnell, J.E., Jr. (1985) RNA. *Sci. Am.* **253** (October), 68–78.

Evolution of Catalytic Function. (1987) *Cold Spring Harb. Symp. Quant. Biol.* **52.**
An excellent source for articles on catalytic RNA, evolution, and many other topics discussed in this chapter.

Jacob, F. & Monod, J. (1961) Genetic regulatory mechanisms in the synthesis of proteins. *J. Mol. Biol.* **3,** 318–356.
A classic article that introduced many important ideas.

Watson, J.D., Hopkins, N.H., Roberts, J.W., Steitz, J.A., & Weiner, A.M. (1987) *Molecular Biology of the Gene,* 4th edn, The Benjamin/Cummings Publishing Company, Menlo Park, CA.

DNA-Directed RNA Synthesis

Conaway, J.W. & Conaway, R.C. (1991) Initiation of eukaryotic messenger RNA synthesis. *J. Biol. Chem.* **266,** 17721–17724.
A good minireview.

McClure, W.R. (1985) Mechanism and control of transcription initiation in prokaryotes. *Annu. Rev. Biochem.* **54,** 171–204.

Platt, T. (1986) Transcription termination and the regulation of gene expression. *Annu. Rev. Biochem.* **55,** 339–372.

Sawadogo, M. & Sentenac, A. (1990) RNA polymerase B (II) and general transcription factors. *Annu. Rev. Biochem.* **59,** 711–754.
A good review of eukaryotic RNA polymerase II.

RNA Processing

Breitbart, R.E., Andreadis, A., & Nadal-Ginard, B. (1987) Alternative splicing: a ubiquitous mechanism for the generation of multiple protein isoforms from single genes. *Annu. Rev. Biochem.* **56,** 467–495.

Cech, T.R. (1986) RNA as an enzyme. *Sci. Am.* **255** (November), 64–75.

Cech, T.R. (1987) The chemistry of self-splicing RNA and RNA enzymes. *Science* **236,** 1532–1539.

Deutscher, M.P. (1990) Ribonucleases, tRNA nucleotidyltransferase, and the 3' processing of tRNA. *Prog. Nucleic Acid Res. Mol. Biol.* **39,** 209–240.
A good overview of tRNA processing reactions.

Green, M.R. (1986) Pre-mRNA splicing. *Annu. Rev. Genet.* **20,** 671–708.

McCorkle, G.M. & Altman, S. (1987) RNAs as catalysts: a new class of enzyme. *J. Chem. Educ.* **64,** 221–226.

Pace, N.R. & Smith, D. (1990) Ribonuclease P: function and variation. *J. Biol. Chem.* **265,** 3587–3590.

Ross, J. (1989) The turnover of messenger RNA. *Sci. Am.* **260** (April), 48–55.

Sharp, P.A. (1987) Splicing of messenger RNA precursors. *Science* **235,** 766–771.

Wahle, E. & Keller, W. (1992) The biochemistry of 3'-end cleavage and polyadenylation of messenger RNA precursors. *Annu. Rev. Biochem.* **61,** 419–440.

Wickens, M. (1990) How the messenger got its tail: addition of poly(A) in the nucleus. *Trends Biochem. Sci.* **15,** 277–281.

RNA-Directed RNA or DNA Synthesis

Belfort, M. (1991) Self-splicing introns in prokaryotes: migrant fossils? *Cell* **64,** 9–11.
A discussion of the evolutionary significance of prokaryotic introns.

Bishop, J.M. (1991) Molecular themes in oncogenesis. *Cell* **64,** 235–248.
A good overview of oncogenes; it introduces a series of more detailed reviews included in the same issue of Cell.

Blackburn, E.H. (1991) Telomeres. *Trends Biochem. Sci.* **16,** 378–381.

Blackburn, E.H. (1992) Telomerases. *Annu. Rev. Biochem.* **61,** 113–129.

Boeke, J.D. (1990) Reverse transcriptase, the end of the chromosome, and the end of life. *Cell* **61,** 193–195.
The possible role of telomerase in regulating the life span of an organism.

Dorit, R.L., Schoenbach, L., & Gilbert, W. (1990) How big is the universe of exons? *Science* **250,** 1377–1382.

Interesting speculation on the origin and function of exons.

Gallo, R.C. & Montagnier, L. (1988) AIDS in 1988. *Sci. Am.* **259** (October), 40–48.

The introductory article to an entire Scientific American *issue devoted to AIDS.*

Kingsman, A.J. & Kingsman, S.M. (1988) Ty: a retroelement moving forward. *Cell* **53,** 333–335.

Describes a well-studied yeast transposon related to retroviruses.

Pace, N.R. (1991) Origin of life—facing up to the physical setting. *Cell* **65,** 531–533.

A discussion of the conditions believed to have existed when life began evolving.

Perlman, P.S. & Butow, R.A. (1989) Mobile introns and intron-encoded proteins. *Science* **246,** 1106–1109.

This article describes a special class of introns capable of colonizing homologous genes that lack introns.

Temin, H.M. (1976) The DNA provirus hypothesis: the establishment and implications of RNA-directed DNA synthesis. *Science* **192,** 1075–1080.

A discussion of the original proposal for reverse transcription in retroviruses.

Varmus, H. (1987) Reverse transcription. *Sci. Am.* **257** (September), 56–64.

Varmus, H.E. (1989) Reverse transcription in bacteria. *Cell* **56,** 721–724.

Problems

1. *RNA Polymerase* How long would it take for the *E. coli* RNA polymerase to synthesize the primary transcript for *E. coli* rRNAs (6500 bases)?

2. *Error Correction by RNA Polymerases* DNA polymerases are capable of editing and error correction, but RNA polymerases do not appear to have this capacity. Given that a single base error in either replication or transcription can lead to an error in protein synthesis, can you give a possible biological explanation for this striking difference?

3. *The Rate of Transcription* From what you know of the rate at which *E. coli* RNA polymerase synthesizes RNA, predict how far the transcription "bubble" formed by RNA polymerase will move along the DNA in 10 s.

4. *RNA Posttranscriptional Processing* Predict the likely effects of a mutation in the sequence (5')AAUAAA in a eukaryotic mRNA transcript.

5. *Coding vs. Template Strands* The RNA genome of phage Qβ is the nontemplate or (+) strand, and when introduced into the cell it functions as an mRNA. Suppose the RNA replicase of phage Qβ synthesized primarily (−) strand RNA and uniquely incorporated it into the virus particles, rather than (+) strands. What would be the fate of the (−) strands when they entered a new cell? What enzyme would such a (−) strand virus need to include in the virus particle to successfully invade a host cell?

6. *The Chemistry of Nucleic Acid Biosynthesis* Describe three properties common to the reactions catalyzed by DNA polymerase, RNA polymerase, reverse transcriptase, and RNA replicase.

7. *RNA Splicing* What is the minimum number of transesterification reactions needed to splice an intron from an mRNA transcript? Why?

8. *Telomerase* Assuming that the RNA component of telomerase is fixed within the protein structure, in what respect might the active site of this enzyme differ from the active site of reverse transcriptases, RNA polymerases, and DNA polymerases? (Hint: The latter three enzymes add one nucleotide at a time.)

9. *RNA Genomes* The RNA viruses have relatively small genomes. For example, the single-stranded RNAs of retroviruses have about 10,000 nucleotides and the Qβ RNA is only 4,220 nucleotides long. Given the properties of reverse transcriptase and RNA replicase described in this chapter, can you suggest a reason for the small size of these viral genomes?

Protein Metabolism

Proteins are the end products of most information pathways. A typical cell requires thousands of different proteins at any given moment. These must be synthesized in response to the cell's current needs, transported (targeted) to the appropriate cellular location, and degraded when the need has passed. The protein synthesis pathway is much better understood than protein targeting or degradation, and coverage in this chapter reflects that fact.

Protein synthesis is the most complex of biosynthetic mechanisms, and understanding it has been one of the greatest challenges in the history of biochemistry. In eukaryotic cells, protein synthesis requires the participation of over 70 different ribosomal proteins; 20 or more enzymes to activate the amino acid precursors; a dozen or more auxiliary enzymes and other specific protein factors for the initiation, elongation, and termination of polypeptides; perhaps 100 additional enzymes for the final processing of different kinds of proteins; and 40 or more kinds of transfer and ribosomal RNAs. Thus almost 300 different macromolecules must cooperate to synthesize polypeptides. Many of these macromolecules are organized into the complex three-dimensional structure of the ribosome to carry out stepwise translocation of the mRNA as the polypeptide is assembled.

To appreciate the central importance of protein synthesis to every cell, it can be enlightening to consider the fraction of cellular resources that are devoted to this process. Protein synthesis can account for up to 90% of the chemical energy used by a cell for all biosynthetic reactions. In *E. coli*, the numbers of different types of proteins and RNA molecules involved in protein synthesis are similar to those in eukaryotic cells. Both prokaryotic and eukaryotic cells contain thousands of copies of each protein and RNA type per cell. When totaled, the 20,000 ribosomes, 100,000 related protein factors and enzymes, and 200,000 tRNAs present in a typical bacterial cell (with a volume of 100 nm^3) can account for more than 35% of the cell's dry weight.

Despite this great complexity, proteins are made at exceedingly high rates. A complete polypeptide chain of 100 residues is synthesized in an *E. coli* cell at 37 °C in about 5 s. The synthesis of the thousands of different proteins in each cell is tightly regulated so that only the required number of molecules of each is made under any given set of metabolic circumstances. To maintain the appropriate mix and concentration of proteins in a cell, the targeting and degradative processes must keep pace with synthesis. Research is gradually unraveling the extraordinary set of biochemical processes that shepherd each protein to its proper location in the cell and selectively degrade proteins no longer required.

Protein Synthesis and the Genetic Code

Three major advances in the 1950s set the stage for our present knowledge of protein biosynthesis. In the early 1950s Paul Zamecnik and his colleagues designed a set of experiments to investigate the question: Where in the cell are proteins synthesized? They injected radioactive amino acids into rats, and at different time intervals after the injection the liver was removed, homogenized, and fractionated by centrifugation. The subcellular fractions were then examined for the presence of radioactive protein. When hours or days were allowed to elapse after injection of the labeled amino acids, *all* the subcellular fractions contained labeled proteins. However, when the liver was removed and fractionated only minutes after injection of the labeled amino acids, labeled protein was found only in a fraction containing small ribonucleoprotein particles. These particles, earlier discovered in animal tissues by electron microscopy, were thus identified as the site of protein synthesis from amino acids; later they were named ribosomes (Fig. 26–1).

Paul Zamecnik

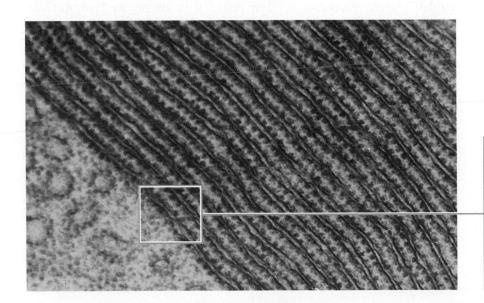

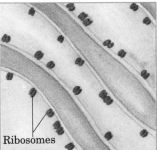

Ribosomes

Figure 26–1 Electron micrograph and schematic drawing of a portion of a pancreatic cell, showing ribosomes attached to the outer (cytosolic) face of the endoplasmic reticulum. The ribosomes are the numerous small dots bordering the parallel layers of membranes.

    The second advance was made by Mahlon Hoagland and Zamecnik; they found that when incubated with ATP and the cytosolic fraction of liver cells, amino acids became "activated." The amino acids were attached to a special form of heat-stable soluble RNA, later called transfer RNA (tRNA), to form **aminoacyl-tRNAs.** The enzymes catalyzing this process are the **aminoacyl-tRNA synthetases.**

    The third major advance occurred when Francis Crick asked: How is the genetic information that is coded in the 4-letter language of nucleic acids translated into the 20-letter language of proteins? Crick reasoned that tRNA must serve the role of an adapter, one part of the

Figure 26–2 Crick's hypothesis of the adapter function of tRNA. Today we know that the amino acid is covalently bound at the 3' end of the tRNA and that a specific nucleotide triplet elsewhere in the tRNA molecule interacts with a specific triplet codon in the mRNA through hydrogen bonding of complementary bases.

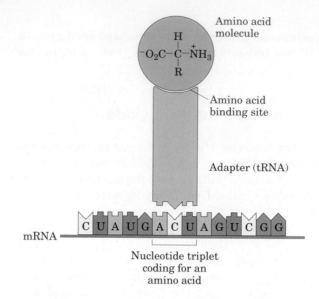

tRNA molecule binding a specific amino acid and some other part of the tRNA recognizing a short nucleotide sequence in the mRNA coding for that amino acid (Fig. 26–2). This idea was soon verified. The tRNA adapter "translates" the nucleotide sequence of an mRNA into the amino acid sequence of a polypeptide. The overall process of mRNA-guided protein synthesis is often referred to simply as **translation.**

These developments soon led to recognition of the major stages of protein synthesis and ultimately to the elucidation of the genetic code words for the amino acids. The nature of this code is the focus of the discussion that follows.

The Genetic Code Has Been Solved

By the 1960s it had long been apparent that at least three nucleotide residues of DNA are required to code for each amino acid. The four code letters of DNA (A, T, G, and C) in groups of two can yield only $4^2 = 16$ different combinations, not sufficient to code for 20 amino acids. But four bases in groups of three can yield $4^3 = 64$ different combinations. Early genetic experiments conclusively proved not only that the genetic code words or **codons** for amino acids are triplets of nucleotides but also that the codons do not overlap and there is no punctuation between codons for successive amino acid residues (Figs. 26–3, 26–4).

Figure 26–3 The triplet, nonoverlapping code. Evidence for the general nature of the genetic code came from many types of experiments, including genetic experiments on the effects of deletion and insertion mutations. Inserting or deleting one base pair (shown here in the mRNA transcript) alters the sequence of triplets in a nonoverlapping code, as shown, and all amino acids coded by the mRNA following the change are affected. Combining insertion and deletion mutations affects some amino acids but eventually restores the correct amino acid sequence. Adding or subtracting three nucleotides (not shown) leaves the remaining triplets intact, providing evidence that a codon has three, rather than four or five, nucleotides. The triplet codons shaded in gray are those transcribed from the original gene; codons shaded in blue are new codons resulting from the insertion or deletion mutations.

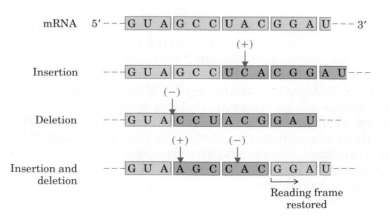

Nonoverlapping   A U A C G A G U C _ _ _ _
      code   1   2   3

Overlapping   A U A C G A G U C
      code   1
                      2
                          3

Figure 26–4 Overlapping versus nonoverlapping codes. In nonoverlapping codes, codons do not share nucleotides. In the example shown, the consecutive codons are numbered. In an overlapping code, some nucleotides in the mRNA are shared by different codons. A triplet code with maximum overlap, with consecutive codons defined by the numbered brackets, will have many nucleotides (such as the third nucleotide here) shared by three different codons. Note that in an overlapping code, the sequence of the first codon limits the possible sequences for the second codon. A nonoverlapping code provides much more flexibility in the sequence of neighboring codons and ultimately in the possible amino acid sequences designated by the code. The code used in all living systems is nonoverlapping.

The amino acid sequence of a protein is therefore defined by a linear sequence of contiguous triplet codons. The first codon in the sequence establishes a **reading frame,** in which a new codon begins every three nucleotide residues. In this scheme there are three possible reading frames for any given DNA sequence, and each will generally give a different sequence of codons (Fig. 26–5). Although it seemed clear that only one reading frame was likely to contain the information required for a given protein, the ultimate questions still loomed: What are the specific three-letter code words for the different amino acids? How could they be identified experimentally?

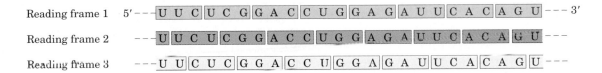

Reading frame 1   5'---UUCUCGGACCUGGAGAUUCACAGU---3'

Reading frame 2   ---UUCUCGGACCUGGAGAUUCACAGU---

Reading frame 3   ---UUCUCGGACCUGGAGAUUCACAGU---

Figure 26–5 In a triplet, nonoverlapping code, all mRNAs have three potential reading frames, shaded here in different colors. Note that the triplets, and hence the amino acids specified, are very different in each reading frame.

    In 1961 Marshall Nirenberg and Heinrich Matthaei reported an observation that provided the first breakthrough. They incubated the synthetic polyribonucleotide polyuridylate (designated poly(U)) with an *E. coli* extract, GTP, and a mixture of the 20 amino acids in 20 different tubes. In each tube a different amino acid was radioactively labeled. Poly(U) can be regarded as an artificial mRNA containing many successive UUU triplets, and it should promote the synthesis of a polypeptide from only one of the 20 different amino acids—that coded by the triplet UUU. A radioactive polypeptide was formed in only one of the 20 tubes, that containing radioactive phenylalanine. Nirenberg and Matthaei therefore concluded that the triplet UUU codes for phenylalanine. The same approach revealed that the synthetic polyribonucleotide polycytidylate or poly(C) codes for formation of a polypeptide containing only proline (polyproline), and polyadenylate or poly(A) codes for polylysine. Thus the triplet CCC must code for proline and the triplet AAA for lysine.
    The synthetic polynucleotides used in such experiments were made by the action of polynucleotide phosphorylase (p. 880), which catalyzes the formation of RNA polymers starting from ADP, UDP, CDP, and GDP. This enzyme requires no template and makes polymers with a base composition that directly reflects the relative concentrations of the nucleoside 5'-diphosphate precursors in the medium. If polynucleotide phosphorylase is presented with UDP, it makes only poly(U). If it is presented with a mixture of five parts of ADP and one of

Marshall Nirenberg

CDP, it will make a polymer in which about five-sixths of the residues are adenylate and one-sixth cytidylate. Such a random polymer is likely to have many triplets of the sequence AAA, lesser numbers of AAC, ACA, and CAA triplets, relatively few ACC, CCA, and CAC triplets, and very few CCC triplets (Table 26–1). With the use of different artificial mRNAs made by polynucleotide phosphorylase from different starting mixtures of ADP, GDP, UDP, and CDP, the base compositions of the triplets coding for almost all the amino acids were soon identified. However, these experiments could not reveal the *sequence* of the bases in each coding triplet.

H. Gobind Khorana

Table 26–1 Incorporation of amino acids into polypeptides in response to random polymers of RNA*

Amino acid	Observed frequency of incorporation (Lys = 100)	Tentative assignment for nucleotide composition† of corresponding codon	Expected frequency of incorporation based on assignment (Lys = 100)
Asparagine	24	$(A)_2C$	20
Glutamine	24	$(A)_2C$	20
Histidine	6	$A(C)_2$	4
Lysine	100	$(A)_3$	100
Proline	7	$A(C)_2, (C)_3$	4.8
Threonine	26	$(A)_2C, A(C)_2$	24

* Presented here is a summary of data from one of the early experiments designed to elucidate the genetic code. An RNA synthesized enzymatically, and containing only A and C residues in a 5:1 ratio, was used to direct polypeptide synthesis. Both the identity and quantity of amino acids incorporated were determined. Based upon the relative abundance of A and C residues in the synthetic RNA, and if the codon AAA (the most likely) is assigned a frequency of 100, there should be three different codons of composition $(A)_2C$, each at a relative frequency of 20; three codons of composition $A(C)_2$, each at a relative frequency of 4.0; and the codon CCC should occur at a relative frequency of 0.8. The CCC assignment here was based on information derived from prior studies with poly(C). Where two tentative codon assignments are made, both are proposed to code for the same amino acid.

† Note that these designations of nucleotide composition contain no information on nucleotide sequence.

In 1964 Nirenberg and Philip Leder achieved another breakthrough. They found that isolated *E. coli* ribosomes will bind a specific aminoacyl-tRNA if the corresponding synthetic polynucleotide messenger is present. For example, ribosomes incubated with poly(U) and phenylalanyl-tRNA[Phe] (or Phe-tRNA[Phe]) will bind both nucleotides, but if the ribosomes are incubated with poly(U) and some other aminoacyl-tRNA, the aminoacyl-tRNA will not be bound because it will not recognize the UUU triplets in poly(U) (Table 26–2). (Note that by convention, the identity of a tRNA is indicated by a superscript and an aminoacylated tRNA is indicated by a hyphenated name. For example, correctly aminoacylated tRNA[Ala] is alanyl-tRNA[Ala] or Ala-tRNA[Ala]. If the tRNA is incorrectly aminoacylated, e.g., with valine, one would have Val-tRNA[Ala].) The shortest polynucleotide that could promote specific binding of Phe-tRNA[Phe] was the trinucleotide UUU. By use of simple trinucleotides of known sequence it was possible to determine which aminoacyl-tRNA bound to each of about 50 of the 64 possible triplet codons. For some codons, either no aminoacyl-tRNAs would bind, or more than one were bound. Another method was needed to complete and confirm the entire genetic code.

Table 26–2 Experiment showing that trinucleotides are sufficient to induce specific binding of aminoacyl-tRNAs to ribosomes

| Trinucleotide | ^{14}C-Labeled aminoacyl-tRNA bound to ribosome* | | |
	Phe-tRNA^{Phe}	Lys-tRNA^{Lys}	Pro-tRNA^{Pro}
UUU	4.6	0	0
AAA	0	7.7	0
CCC	0	0	3.1

Source: Modified from Nirenberg, M. & Leder, P. (1964) RNA code words and protein synthesis. *Science* **145**, 1399.

* The numbers represent factors by which the amount of bound ^{14}C increased when the indicated trinucleotide was present, relative to controls in which no trinucleotide was added.

At about this time, a complementary approach was provided by H. Gobind Khorana, who developed methods to synthesize polyribonucleotides with defined, repeating sequences of two to four bases. The polypeptides produced using these RNAs as messengers had one or a few amino acids in repeating patterns. These patterns, when combined with information from the random polymers used by Nirenberg and colleagues, permitted unambiguous codon assignments. The copolymer $(AC)_n$, for example, has alternating ACA and CAC codons, regardless of the reading frame:

$$A\ C\ A\ |\ C\ A\ C\ |\ A\ C\ A\ |\ C\ A\ C\ |\ A\ C\ A$$

The polypeptide synthesized in response to this polymer was found to have equal amounts of threonine and histidine. Because the experiment described in Table 26–1 revealed a histidine codon with one A and two Cs, CAC must code for histidine and ACA for threonine.

Similarly, an RNA with three bases in a repeating pattern should yield three different types of polypeptide. Each polypeptide would be derived from a different reading frame and would contain a single kind of amino acid. An RNA with four bases in a repeating pattern should yield a single type of polypeptide with a repeating pattern of four amino acids (Table 26–3). Results from all of these experiments with polymers permitted the assignment of 61 of 64 possible codons. The other three were identified as termination codons, in part because they disrupted amino acid coding patterns when included in the sequence of a synthetic RNA polymer (Fig. 26–6; Table 26–3).

Table 26–3 Polypeptides produced in response to synthetic RNA polymers with repeating sequences of three and four bases

Polynucleotide	Polypeptide products
Trinucleotide repeats	
$(UUC)_n$	$(Phe)_n$, $(Ser)_n$, $(Leu)_n$
$(AAG)_n$	$(Lys)_n$, $(Arg)_n$, $(Glu)_n$
$(UUG)_n$	$(Leu)_n$, $(Cys)_n$, $(Val)_n$
$(CCA)_n$	$(Pro)_n$, $(His)_n$, $(Thr)_n$
$(GUA)_n$	$(Val)_n$, $(Ser)_n$, (chain terminator)*
$(UAC)_n$	$(Tyr)_n$, $(Thr)_n$, $(Leu)_n$
$(AUC)_n$	$(Ile)_n$, $(Ser)_n$, $(His)_n$
$(GAU)_n$	$(Asp)_n$, $(Met)_n$, (chain terminator)*
Tetranucleotide repeats	
$(UAUC)_n$	$(Tyr–Leu–Ser–Ile)_n$
$(UUAC)_n$	$(Leu–Leu–Thr–Tyr)_n$
$(GUAA)_n$	Di- and tripeptides*
$(AUAG)_n$	Di- and tripeptides*

* With these polynucleotides, the patterns of amino acid incorporation into polypeptides are affected by the presence of codons that are termination signals for protein biosynthesis. In the repeating three-base sequences, one of the three reading frames includes only termination codons and thus only two homopolypeptides are observed (generated from the remaining two reading frames). In some of the repeating four-base sequences, every fourth codon is a termination codon in every reading frame, so that only short peptides are produced. This is illustrated in Figure 26–6 for $(GUAA)_n$.

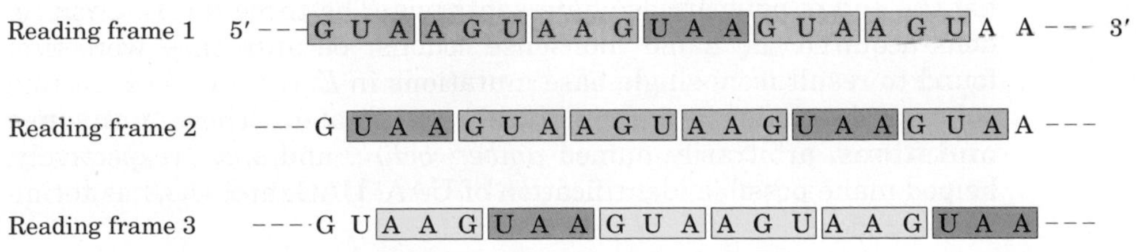

Figure 26–6 The effect of a termination codon incorporated within a repeating tetranucleotide. Dipeptides or tripeptides will be synthesized, depending on where the ribosome initially binds. The three different reading frames are shown in different colors. Termination codons (indicated in red) are encountered every fourth codon in all three reading frames.

Reading frame 1   5′ - - - G U A A G U A A G U A A G U A A G U A A - - - 3′

Reading frame 2   - - - G U A A G U A A G U A A G U A A G U A A - - -

Reading frame 3   - - - G U A A G U A A G U A A G U A A G U A A - - -

With these approaches the base sequences of all the triplet code words for each of the amino acids were established by 1966. Since then, these code words have been verified in many different ways. The complete codon "dictionary" for the amino acids is given in Figure 26–7. The cracking of the genetic code is regarded as the greatest scientific discovery of the 1960s.

Figure 26–7 The "dictionary" of amino acid code words as they occur in mRNAs. The codons are written in the 5′→3′ direction. The third base of each codon, shown in bold type, plays a lesser role in specifying an amino acid than the first two. The three termination codons are shaded in red, and the initiation codon AUG is shaded in green. Note that all the amino acids except methionine and tryptophan have more than one codon. In most cases, codons that specify the same amino acid differ only in the third base.

Second letter of codon

	U		C		A		G	
U	UUU	Phe	UCU	Ser	UAU	Tyr	UGU	Cys
	UUC	Phe	UCC	Ser	UAC	Tyr	UGC	Cys
	UUA	Leu	UCA	Ser	UAA	Stop	UGA	Stop
	UUG	Leu	UCG	Ser	UAG	Stop	UGG	Trp
C	CUU	Leu	CCU	Pro	CAU	His	CGU	Arg
	CUC	Leu	CCC	Pro	CAC	His	CGC	Arg
	CUA	Leu	CCA	Pro	CAA	Gln	CGA	Arg
	CUG	Leu	CCG	Pro	CAG	Gln	CGG	Arg
A	AUU	Ile	ACU	Thr	AAU	Asn	AGU	Ser
	AUC	Ile	ACC	Thr	AAC	Asn	AGC	Ser
	AUA	Ile	ACA	Thr	AAA	Lys	AGA	Arg
	AUG	Met	ACG	Thr	AAG	Lys	AGG	Arg
G	GUU	Val	GCU	Ala	GAU	Asp	GGU	Gly
	GUC	Val	GCC	Ala	GAC	Asp	GGC	Gly
	GUA	Val	GCA	Ala	GAA	Glu	GGA	Gly
	GUG	Val	GCG	Ala	GAG	Glu	GGG	Gly

First letter of codon (5′ end)

The Genetic Code Has Several Important Characteristics

The key to the organization of the genetic information specifying a protein can be found in codons and in the array of codons that constitutes a reading frame. Keep in mind that no punctuation or signal is required to indicate the end of one codon and the beginning of the next. The reading frame must therefore be correctly set at the beginning of the readout of an mRNA molecule and then moved sequentially from one triplet to the next. If the initial reading frame is off by one or two bases, or if the ribosome accidentally skips a nucleotide in the mRNA, all the subsequent codons will be out of register and will lead to formation of a "missense" protein with a garbled amino acid sequence.

Several of the codons serve special functions. The **initiation codon,** AUG, signals the beginning of polypeptide chains. AUG not only is the initiation codon in both prokaryotes and eukaryotes but also codes for Met residues in internal positions of polypeptides. Of the 64 possible nucleotide triplets, three (UAA, UAG, and UGA) do not code for any known amino acids (Fig. 26–7); they are the **termination codons** (also called stop codons or nonsense codons), which normally signal the end of polypeptide chain synthesis. The three termination codons acquired the name "nonsense codons" because they were first found to result from single-base mutations in *E. coli* in which certain polypeptide chains are prematurely terminated. These **nonsense mutations,** arbitrarily named *amber, ochre,* and *opal,* respectively, helped make possible identification of UAA, UAG, and UGA as termination codons.

In a random sequence of nucleotides, one in every 20 codons in each reading frame, on average, will be a termination codon. Where a reading frame exists without a termination codon for 50 or more codons, the region is called an **open reading frame.** Long open reading frames usually correspond to genes that encode proteins. An uninterrupted gene coding for a typical protein with a molecular weight of 60,000 would require an open reading frame with 500 or more codons. See Box 26–1 (p. 900) for some interesting exceptions to this general pattern.

Perhaps the most striking feature of the genetic code is that it is **degenerate,** meaning that a given amino acid may be specified by more than one codon (Table 26–4). Only methionine and tryptophan have single codons. Degenerate does not mean imperfect; the genetic code is unambiguous because no codon specifies more than one amino acid. Note that the degeneracy of the code is not uniform. For example, leucine and serine have six codons, glycine and alanine have four, and glutamate, tyrosine, and histidine have two.

When an amino acid has multiple codons, the difference between the codons usually lies in the third base (at the 3' end). For example, alanine is coded by the triplets GCU, GCC, GCA, and GCG. The codons for nearly all of the amino acids can be symbolized by XY_G^A or XY_C^U. The first two letters of each codon are therefore the primary determinants of specificity. This has some interesting consequences.

Table 26–4 Degeneracy of the genetic code

Amino acid	Number of codons	Amino acid	Number of codons
Ala	4	Leu	6
Arg	6	Lys	2
Asn	2	Met	1
Asp	2	Phe	2
Cys	2	Pro	4
Gln	2	Ser	6
Glu	2	Thr	4
Gly	4	Trp	1
His	2	Tyr	2
Ile	3	Val	4

Wobble Allows Some tRNAs to Recognize More than One Codon

Transfer RNAs recognize codons by base pairing between the mRNA codon and a three-base sequence on the tRNA called the **anticodon.** The two RNAs are paired antiparallel, the first base of the codon (always reading in the 5'→3' direction) pairing with the third base of the anticodon (Fig. 26–8).

One might expect the anticodon triplet of a given tRNA to recognize only one codon triplet through Watson–Crick base pairing, so that there would be a different tRNA for each codon of an amino acid. However, the number of different tRNAs for each amino acid is *not* the same as the number of its codons. Moreover, some of the tRNAs contain the nucleotide inosinate (designated I), which contains the uncommon base hypoxanthine (see Fig. 12–5b). Molecular models show that inosinate can form hydrogen bonds with three different nucleotides, U, C, and A, but these pairings are rather weak compared with the strong hydrogen bonds between the Watson–Crick base pairs G≡C and A=U. In yeast, for example, one tRNA^{Arg} has the anticodon (5')ICG, which can recognize three different arginine codons, (5')CGA, (5')CGU, and (5')CGC. The first two bases of these codons are identical (CG) and form strong Watson–Crick base pairs (blue) with the corresponding bases of the anticodon:

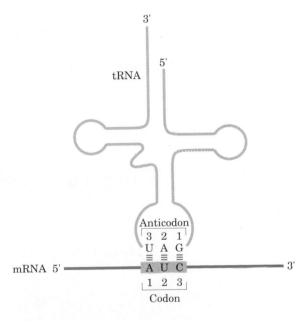

Figure 26–8 The pairing relationship of codon and anticodon. Alignment of the two RNAs is antiparallel. The tRNA is presented in the traditional cloverleaf configuration.

		3 2 1		3 2 1		3 2 1	
Anticodon	(3')	G–C–**I**		G–C–**I**		G–C–**I**	(5')
Codon	(5')	C̄–Ḡ–**A**		C̄–Ḡ–**U**		C̄–Ḡ–**C**	(3')
		1 2 3		1 2 3		1 2 3	

BOX 26–1

Translational Frameshifting and RNA Editing: mRNAs That Change Horses in Midstream

Proteins are synthesized according to a pattern of contiguous triplet codons. Once the reading frame is set, codons are translated in order, without overlap or punctuation, until a termination codon is encountered. Usually, the other two possible reading frames within a gene contain no useful genetic information. However, a few genes are structured so that ribosomes "hiccup" at a certain point in the translation of the mRNA, leading to a change in the reading frame from that point on. In some cases this appears to be a mechanism used to produce two or more related proteins from a single transcript or to regulate the synthesis of a protein.

The best-documented example occurs in the translation of the mRNA for the *gag* and *pol* genes of the Rous sarcoma virus (see Fig. 25–31). The two genes overlap, with *pol* encoded by the reading frame in which each codon is offset to the left by one base pair (−1 reading frame) relative to *gag* (Fig. 1). The product of the *pol* gene (reverse transcriptase; p. 882) is translated initially as a larger *gag–pol* fusion protein using the same mRNA used for the *gag* protein alone. This fusion protein is later trimmed to the mature reverse transcriptase by proteolytic digestion. The large fusion protein is produced by a translational frameshift that occurs in the overlap region and allows the ribosome to bypass the UAG termination codon at the end of the *gag* gene (shown in red in Fig. 1). This frameshift occurs in about 5% of the translation events, so that the *gag–pol* fusion protein, and ultimately reverse transcriptase, is synthesized at the appropriate level for efficient replication of the viral genome—about 20-fold less than the *gag* protein. A similar mechanism is used to produce both the τ

and γ subunits of *E. coli* DNA polymerase III from *dnaX* gene transcripts (see Table 24–2).

An example of the use of this mechanism for regulation occurs in the gene for *E. coli* release factor 2 (RF₂), a protein required for termination of protein synthesis at the termination codons UAA and UGA (described later in this chapter). The 26th codon of the gene for RF₂ is UGA, which would normally halt protein synthesis. The remainder of the gene is in the +1 reading frame (offset one base pair to the right) relative to this UGA codon. Low levels of RF₂ lead to a translational pause at this codon, because UGA is not recognized as a termination codon unless RF₂ binds to it. The absence of RF₂ prevents the termination of protein synthesis at this UGA and allows time for a frameshift so that UGA plus the C that follows it (UGAC) is read as GAC = Asp. Translation then proceeds in the new reading frame to complete synthesis of RF₂. In this way, RF₂ regulates its own synthesis in a feedback loop.

An especially unusual frameshifting mechanism occurs through the editing of mRNAs prior to translation. The genes in mitochondrial DNA that encode the cytochrome oxidase subunit II in some protists do not have open reading frames that correspond precisely to the protein product. Instead, the codons specifying the amino terminus of the protein are in a different reading frame from the codons specifying the carboxyl terminus. The problem is corrected not on the ribosome, but by a posttranscriptional editing process in which four uridines are added to create three new codons and shift the reading frame so that the entire gene can be translated directly, as shown in Figure 2a; the

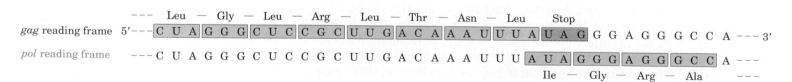

Figure 1 The *gag–pol* overlap region in Rous sarcoma virus.

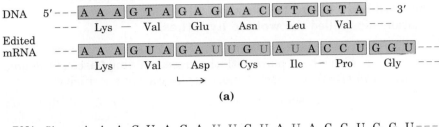

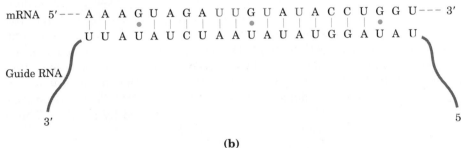

(a)

(b)

Figure 2 RNA editing of the transcript of the cytochrome oxidase subunit II gene from mitochondria of *Tetrahymena brucei*.

added uridine residues are shown in red. Only a small part of the gene (the region affected by editing) is shown. Neither the function nor mechanism of this editing process is understood. A special class of RNA molecules encoded by these mitochondria have been detected that have sequences complementary to the final, edited mRNAs. These appear to act as templates for the editing process and are referred to as guide RNAs (Fig. 2b). Note that the base pairing involves a number of G=U base pairs (symbolized by blue dots), which are common in RNA molecules.

A distinct form of RNA editing occurs in the gene for the apolipoprotein B component of low-density lipoprotein in vertebrates. One form of apo-

lipoprotein B, called apoB-100 (M_r 513,000), is synthesized in the liver. A second form, apoB-48 (M_r 250,000), is synthesized in the intestine. Both are synthesized from an mRNA produced from the gene for apoB-100. A cytosine deaminase enzyme found only in the intestine binds to the mRNA at codon 2,153 (CAA = Gln) and converts the C to a U to introduce the termination codon UAA at this position. The apoB-48 produced in the intestine from the modified mRNA is simply an abbreviated form (corresponding to the amino-terminal half) of apoB-100 (Fig. 3). This reaction permits the synthesis of two different proteins from one gene in a tissue-specific manner.

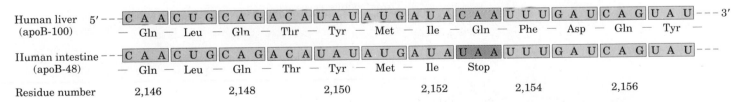

Figure 3 RNA editing of the transcript of the gene for the apolipoprotein B-100 component of low-density lipoprotein.

The third bases of the arginine codons (A, U, and C) form rather weak hydrogen bonds with the I residue at the first position of the anticodon. Examination of these and other codon–anticodon pairings led Crick to conclude that the third base of most codons pairs rather loosely with the corresponding base of its anticodons; to use his picturesque word, the third bases of such codons "wobble." Crick proposed a set of four relationships called the **wobble hypothesis:**

1. The first two bases of a codon in mRNA always form strong Watson–Crick base pairs with the corresponding bases of the anticodon in tRNA and confer most of the coding specificity.

2. The first base of some anticodons (reading in the $5' \rightarrow 3'$ direction; remember that this is paired with the third base of the codon) determines the number of codons read by a given tRNA. When the first base of the anticodon is C or A, binding is specific and only one codon is read by that tRNA. However, when the first base is U or G, binding is less specific and two different codons may be read. When inosinate (I) is the first, or wobble, nucleotide of an anticodon, three different codons can be read by that tRNA. This is the maximum number of codons that can be recognized by a tRNA. These relationships are summarized in Table 26–5.

3. When an amino acid is specified by several different codons, those codons that differ in either of the first two bases require different tRNAs.

4. A minimum of 32 tRNAs are required to translate all 61 codons.

Table 26–5 The wobble base of the anticodon determines how many codons of a given amino acid a tRNA can recognize

In the following, X and Y denote complementary bases capable of strong Watson–Crick base pairing with each other. The bases in the wobble or 3′ position of the codons and 5′ position of the anticodons are shaded in red.

1. One codon recognized:

| Anticodon | $(3')$ X–Y–**C** $(5')$ | $(3')$ X–Y–**A** $(5')$ |
| Codon | $(5')$ Y–X–**G** $(3')$ | $(5')$ Y–X–**U** $(3')$ |

2. Two codons recognized:

| Anticodon | $(3')$ X–Y–**U** $(5')$ | $(3')$ X–Y–**G** $(5')$ |
| Codon | $(5')$ Y–X–$\begin{smallmatrix}A\\G\end{smallmatrix}$ $(3')$ | $(5')$ Y–X–$\begin{smallmatrix}C\\U\end{smallmatrix}$ $(3')$ |

3. Three codons recognized:

| Anticodon | $(3')$ X–Y–**I** $(5')$ |
| Codon | $(5')$ Y–X–$\begin{smallmatrix}A\\U\\C\end{smallmatrix}$ $(3')$ |

What can the reason be for this unexpected complexity of codon–anticodon interactions? In brief, the first two bases of a codon confer most of the codon–anticodon specificity. The wobble (or third) base of the codon contributes to specificity, but because it pairs only loosely with its corresponding base in the anticodon, it permits rapid dissociation of the tRNA from its codon during protein synthesis. If all three bases of mRNA codons engaged in strong Watson–Crick pairing with the three bases of the tRNA anticodons, tRNAs would dissociate too slowly and severely limit the rate of protein synthesis. Codon–anticodon interactions optimize both accuracy *and* speed.

Overlapping Genes in Different Reading Frames Are Found in Some Viral DNAs

Although a given nucleotide sequence can, in principle, be read in any of its three reading frames, most DNA sequences encode a protein product in only one reading frame. In the coding frame there must be no termination codons, and each codon must correspond to the appropriate amino acid. As illustrated in Figure 26–9, the genetic code imposes strict limits on the numbers of amino acids that can be encoded by the codons of reading frame 2 without changing the amino acids specified by reading frame 1. Sometimes one amino acid (and its corresponding codon) may be substituted for another in reading frame 1 and still retain the function of the encoded protein, making it more likely that reading frames 2 or 3 might also encode a useful protein; but even taking these factors into account, the flexibility in other reading frames is very limited.

Figure 26–9 An amino acid sequence specified by one reading frame severely limits the potential amino acids encoded by any other reading frame. **(a)** The codons that can exist in reading frame 1 to produce the indicated amino acid sequence. Most of the permitted nucleotide changes (red) are in the third (wobble) position of each codon. **(b)** At the top are shown the codons that can exist in reading frame 2 without changing the amino acid sequence encoded by reading frame 1. Below are shown the alternative codons that correspond to the alternative mRNA sequences listed in **(a)**. The possible amino acids that can be encoded by reading frame 2 without changing the amino acid sequence encoded by reading frame 1 are in parentheses.

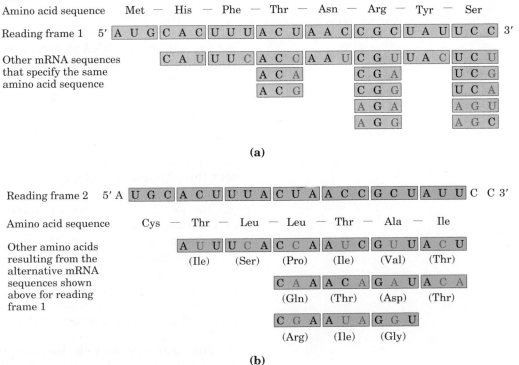

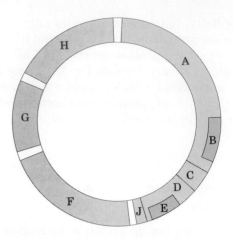

Figure 26–10 Genes within genes. The circular DNA of φX174 contains nine genes (A to J). Gene B lies within the sequence of gene A but uses a different reading frame. Similarly, gene E lies within gene D and also uses a different reading frame (see Fig. 26–11). The unshaded segments are untranslated spacer regions.

Although only one reading frame is generally used to encode a protein and genes do not overlap, there are a few interesting exceptions. In several viruses the same DNA base sequence codes for two different proteins by employing two different reading frames. The discovery of such "genes within genes" arose from the observation that the DNA of bacteriophage φX174, which contains 5,386 nucleotide residues, is not long enough to code for the nine different proteins that are known to be the products of the φX174 DNA genome, unless the genes overlap. The entire nucleotide sequence of the φX174 chromosome was compared with the amino acid sequences of the proteins encoded by the φX174 genes; this indicated several overlapping gene sequences. Figure 26–10 shows that genes B and E are nested within A and D, respectively. There are also five cases (not shown) in which the initiation codon of one gene overlaps the termination codon of the other gene. Figure 26–11 shows how genes D and E share a segment of DNA but use different reading frames; a similar situation exists for genes A and B. The sum of all the nested and overlapping sequences accounts completely for the surprisingly small size of the φX174 genome compared with the number of amino acid residues in the nine proteins for which it codes.

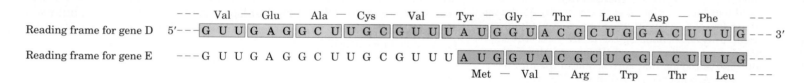

Figure 26–11 Portion of the nucleotide sequence of the mRNA transcript of gene D of φX174 DNA, showing how gene E, which is nested within gene D, is coded by a different reading frame from that used by gene D.

This discovery was quickly followed by similar observations in other viral DNAs, including those of phage λ, the cancer-causing simian virus 40 (SV40), RNA phages such as Qβ and Q17, and phage G4, a close relative of φX174. Phage G4 is remarkable in that at least one codon is shared by *three* different genes. It has been suggested that overlapping genes or genes within genes may be found only in viruses because the fixed, small size of the viral capsid requires economical use of a limited amount of DNA to code for the variety of proteins needed to infect a host cell and replicate within it. Also, because viruses reproduce (and therefore evolve) faster than their host cells, they may represent the ultimate in biological streamlining.

The genetic code is nearly universal. With the intriguing exception of a few minor variations that have been found in mitochondria, some bacteria, and some single-celled eukaryotes (Box 26–2, p. 906), amino acid codons are identical in all species that have been examined. Human beings, *E. coli,* tobacco plants, amphibians, and viruses share the same genetic code. Thus it would appear that all life forms had a common evolutionary ancestor with a single genetic code that has been very well preserved throughout the course of biological evolution.

The genetic code tells us how protein sequence information is stored in nucleic acids and provides some clues about how that information is translated into protein. We now turn to the molecular mechanisms of the translation process.

Protein Synthesis

As we have seen for DNA and RNA, the synthesis of polymeric biomolecules can be separated into initiation, elongation, and termination stages. Protein synthesis is no exception. The activation of amino acid precursors prior to their incorporation into polypeptides and the post-translational processing of the completed polypeptide constitute two important and especially complex additional stages in the synthesis of proteins, and therefore require separate discussion. The cellular components required for each of the five stages in *E. coli* and other bacteria are listed in Table 26–6. The requirements in eukaryotic cells are quite similar. An overview of these stages will provide a useful outline for the discussion that follows.

Table 26–6 Components required for the five major stages in protein synthesis in *E. coli*

Stage	Necessary components
1. Activation of amino acids	20 amino acids 20 aminoacyl-tRNA synthetases 20 or more tRNAs ATP Mg^{2+}
2. Initiation	mRNA *N*-Formylmethionyl-tRNA Initiation codon in mRNA (AUG) 30S ribosomal subunit 50S ribosomal subunit Initiation factors (IF-1, IF-2, IF-3) GTP Mg^{2+}
3. Elongation	Functional 70S ribosome (initiation complex) Aminoacyl-tRNAs specified by codons Elongation factors (EF-Tu, EF-Ts, EF-G) Peptidyl transferase GTP Mg^{2+}
4. Termination and release	Termination codon in mRNA Polypeptide release factors (RF$_1$, RF$_2$, RF$_3$) ATP
5. Folding and processing	Specific enzymes and cofactors for removal of initiating residues and signal sequences, additional proteolytic processing, modification of terminal residues, attachment of phosphate, methyl, carboxyl, carbohydrate, or prosthetic groups

Stage 1: Activation of Amino Acids During this stage, which takes place in the cytosol, not on the ribosomes, each of the 20 amino acids is covalently attached to a specific tRNA at the expense of ATP energy. These reactions are catalyzed by a group of Mg^{2+}-dependent activating enzymes called aminoacyl-tRNA synthetases, each specific for one amino acid and its corresponding tRNAs. Where two or more tRNAs exist for a given amino acid, one aminoacyl-tRNA synthetase generally aminoacylates all of them. Aminoacylated tRNAs are commonly referred to as being "charged."

BOX 26–2   Natural Variations in the Genetic Code

In biochemistry, as in other disciplines, exceptions to general rules can be problematic for educators and frustrating for students. At the same time they teach us that life is complex and inspire us to search for more surprises. Understanding the exceptions can even reinforce the original rule in surprising ways.

It would seem that there is little room for variation in the genetic code. Recall from Chapters 6 and 7 that even a single amino acid substitution can have profoundly deleterious effects on the structure of a protein. Suppose that somewhere there was a bacterial cell in which one of the codons specifying alanine suddenly began specifying arginine; the resulting substitution of arginine for alanine at multiple positions in scores of proteins would unquestionably be lethal. Variations in the code occur in some organisms nonetheless, and they are both interesting and instructive. The very rarity of these variations and the types of variations that occur together provide powerful evidence for a common evolutionary origin of all living things.

The mechanism for altering the code is straightforward: changes must occur in one or more tRNAs, with the obvious target for alterations being the anticodon. This will lead to the systematic insertion of an amino acid at a codon that does not specify that amino acid in the normal code (Fig. 26–7). The genetic code, in effect, is defined by the anticodons on tRNAs (which determine where an amino acid is placed in a growing polypeptide) and by the specificity of the enzymes—aminoacyl-tRNA synthetases—that charge the tRNAs (which determine the identity of the amino acid attached to a given tRNA).

Because of the catastrophic effects most sudden code changes would have on cellular proteins, one might predict that code alterations would occur only in cases where relatively few proteins would be affected. This could happen in small genomes encoding only a few proteins. The biological consequences of a code change could also be limited by restricting changes to the three termination codons, because these do not generally occur within genes (see Box 26–1 for exceptions to *this* rule). A change that converts a termination codon to a codon specifying an amino acid will affect termination in the products of only a subset of genes, and sometimes the effects in those genes will be minor because some genes have multiple (redundant) termination codons. This pattern is in fact observed.

Changes in the genetic code are very rare. Most of the characterized code variations occur in mitochondria, whose genomes encode only 10 to 20 proteins. Mitochondria have their own tRNAs, and the code variations do not affect the much larger cellular genomes. The most common changes in mitochondria, and the only changes observed in cellular genomes, involve termination codons.

In mitochondria, the changes can be viewed as a kind of genomic streamlining. Vertebrate mDNAs have genes that encode 13 proteins, 2 rRNAs, and 22 tRNAs (see Fig. 18–29). An unusual set of wobble rules allows the 22 tRNAs to decode all 64 possible codon triplets, rather than the 32 tRNAs required for the normal code. Four codon families (where the amino acid is determined entirely by the first two nucleotides) are decoded by a single tRNA with a U in the first (or wobble) position in the anticodon. Either the U pairs somehow with all four bases in the third position of the codon, or a "two out of three" mechanism is used in these cases (i.e., no pairing occurs at the third position of the codon). Other tRNAs recognize codons with either A or G in the third position, and yet others recognize U or C, so that virtually all the tRNAs recognize either two or four codons.

In the normal code, only two amino acids are specified by single codons, methionine and tryptophan (Table 26–4). If all mitochondrial tRNAs recognize two codons, then additional codons for Met and Trp might be expected in mitochondria. Hence, the single most common code variation observed is the UGA specification, from "termination" to Trp. A single tRNA^{Trp} can be used to recognize and insert a Trp residue at the codon UGA and the normal Trp codon UGG. Converting AUA from an Ile codon to a Met codon has a similar effect; the normal Met codon is AUG, and a single tRNA can be used for both codons. This turns out to be the second most common mitochondrial code variation. The known coding variations in mitochondria are summarized in Table 1.

Turning to the much rarer changes in the codes for cellular (as distinct from mitochondrial) genomes, we find that the only known variation in a prokaryote is again the use of UGA to encode Trp residues in the simplest free-living cell, *Mycoplasma capricolum*. In eukaryotes, the only known extramitochondrial coding changes occur in a few species of ciliated protists, where the termination codons UAA and UAG both specify glutamine.

Changes in the code need not be absolute—a codon need not always encode the same amino acid. In *E. coli* there are two examples of amino

Table 1 Known variant codon assignments in mitochondria

	Codons*				
	UGA	AUA	AGA AGG	CUN	CCG
Normal code assignment	Stop	Ile	Arg	Leu	Arg
Animals					
Vertebrates	Trp	Met	Stop	+	+
Drosophila	Trp	Met	Ser	+	+
Yeasts					
Saccharomyces					
cerevisiae	Trp	Met	+	Thr	+
Torulopsis					
glabrata	Trp	Met	+	Thr	?
Schizosaccharomyces					
pombe	Trp	+	+	+	+
Filamentous fungi	Trp	+	+	+	+
Trypanosomes	Trp	+	+	+	+
Higher plants	+	+	+	+	Trp
Chlamydomonas					
reinhardtii	?	+	+	+	?

* ? Indicates that the codon has not been observed in the indicated mitochondrial genome; N, any nucleotide; +, the codon has the same meaning as in the normal code.

acids being inserted at positions not specified in the general code. The first is the occasional use of the codon GUG (Val) as an initiating codon. This occurs only for those genes in which the GUG is properly located relative to special translation initiating signals in the mRNA (as discussed later in this chapter) that override the normal coding pattern. Thus, GUG has an altered coding specification only when it is positioned within a certain "context" of other sequences.

The use of contextual signals to alter coding patterns also applies to the second *E. coli* example. A few proteins in all cells (e.g., formate dehydrogenase in bacteria and glutathione peroxidase in mammals) require the element selenium for their activity. It is generally present in the form of the modified amino acid selenocysteine (Fig. 1). Modified amino acids are generally produced in posttranslational reactions (described later in this chapter), but in *E. coli*, selenocysteine is introduced into formate dehydrogenase during translation in response to an in-frame UGA codon. A specialized type of serine tRNA, present at lower levels than other serine tRNAs, recognizes UGA and

no other codons. This tRNA is charged with serine, and the serine is then enzymatically converted to selenocysteine prior to its use on the ribosome. The charged tRNA will not recognize just any UGA codon; instead some contextual signal in the mRNA, still to be identified, permits the tRNA to recognize only those few UGA codons that specify selenocysteine within certain genes. In effect, there are 21 standard amino acids in *E. coli*, and UGA doubles as a codon for termination and (sometimes) for selenocysteine.

These variations tell us that the code is not quite as universal as once believed, but they also tell us that flexibility in the code is severely constrained. It is clear that the variations are derivatives of the general code; no example of a completely different code has ever been found. The variants do not provide evidence for new forms of life, nor do they undermine the concepts of evolution or universality of the genetic code. The limited scope of code variants strengthens the principle that all life on this planet evolved on the basis of a single (very slightly flexible) genetic code.

$$\begin{array}{c} COO^- \\ | \\ H_3\overset{+}{N}-CH \\ | \\ CH_2 \\ | \\ Se \\ | \\ H \end{array}$$

Figure 1 Selenocysteine.

Stage 2: Initiation   Next, the mRNA bearing the code for the polypeptide to be made binds to the smaller of two major ribosomal subunits; this is followed by the binding of the initiating aminoacyl-tRNA and the large ribosomal subunit to form an initiation complex. The initiating aminoacyl-tRNA base-pairs with the mRNA codon AUG that signals the beginning of the polypeptide chain. This process, which requires GTP, is promoted by specific cytosolic proteins called initiation factors.

Stage 3: Elongation   The polypeptide chain is now lengthened by covalent attachment of successive amino acid units, each carried to the ribosome and correctly positioned by its tRNA, which base-pairs to its corresponding codon in the mRNA. Elongation is promoted by cytosolic proteins called elongation factors. The binding of each incoming aminoacyl-tRNA and the movement of the ribosome along the mRNA are facilitated by the hydrolysis of two molecules of GTP for each residue added to the growing polypeptide.

Stage 4: Termination and Release   The completion of the polypeptide chain is signaled by a termination codon in the mRNA. The polypeptide chain is then released from the ribosome, aided by proteins called release factors.

Stage 5: Folding and Processing   In order to achieve its biologically active form the polypeptide must fold into its proper three-dimensional conformation. Before or after folding, the new polypeptide may undergo enzymatic processing to remove one or more amino acids from the amino terminus; to add acetyl, phosphate, methyl, carboxyl, or other groups to certain amino acid residues; to cleave the protein proteolytically; or to attach oligosaccharides or prosthetic groups.

In our expanded discussion of these stages a particular emphasis will be placed on stage 1. The reason is evident on considering the overall goal of the process: to synthesize a polypeptide chain with a defined sequence. To accomplish this task, two fundamental chemical requirements must be met: (1) the carboxyl group of each amino acid must be activated to facilitate formation of a peptide bond (see Fig. 5–15), and (2) a link must be maintained between each new amino acid and the information that encodes it in the mRNA. As we will see, both of these requirements are met by attaching the amino acid to a tRNA, and attaching the right amino acid to the right tRNA is therefore critical to the overall process of protein biosynthesis.

Before examining each stage in detail, we must introduce two key components in protein biosynthesis: the ribosome and tRNAs.

The Ribosome Is a Complex Molecular Machine

Each *E. coli* cell contains 15,000 or more ribosomes, which make up almost a quarter of the dry weight of the cell. Bacterial ribosomes contain about 65% rRNA and about 35% protein. They have a diameter of about 18 nm and a sedimentation coefficient of 70S.

Bacterial ribosomes consist of two subunits of unequal size (Fig. 26–12), the larger having a sedimentation coefficient of 50S and the smaller of 30S. The 50S subunit contains one molecule of 5S rRNA, one molecule of 23S rRNA, and 34 proteins. The 30S subunit contains one molecule of 16S rRNA and 21 proteins. The proteins are designated

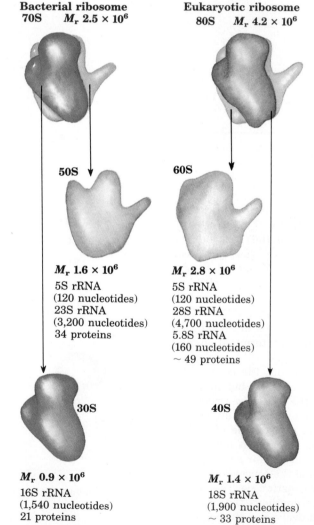

Bacterial ribosome
70S M_r 2.5 × 10⁶

Eukaryotic ribosome
80S M_r 4.2 × 10⁶

50S

60S

M_r 1.6 × 10⁶
5S rRNA
(120 nucleotides)
23S rRNA
(3,200 nucleotides)
34 proteins

M_r 2.8 × 10⁶
5S rRNA
(120 nucleotides)
28S rRNA
(4,700 nucleotides)
5.8S rRNA
(160 nucleotides)
~ 49 proteins

30S

40S

M_r 0.9 × 10⁶
16S rRNA
(1,540 nucleotides)
21 proteins

M_r 1.4 × 10⁶
18S rRNA
(1,900 nucleotides)
~ 33 proteins

Figure 26–12 Components of bacterial and eukaryotic ribosomes. The designation S (Svedberg units) refers to rates of sedimentation in the centrifuge. The S values (sedimentation coefficients) are not necessarily additive when subunits are combined.

Figure 26–13 Predicted folding patterns in *E. coli* 16S and 5S rRNAs, based on maximizing the potential intrastrand base pairing.

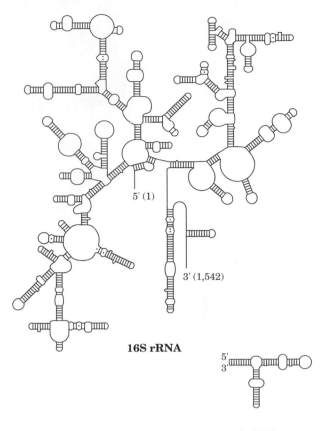

5' (1)

3' (1,542)

16S rRNA

5S rRNA

by numbers. Those in the large 50S subunit are numbered L1 to L34 (L for large) and those in the smaller subunit S1 to S21 (S for small). All the ribosomal proteins of *E. coli* have been isolated and many have been sequenced. Their variety is enormous, with molecular weights from about 6,000 to 75,000.

The sequences of nucleotides in the rRNAs of many organisms have been determined. Each of the three single-stranded rRNAs of *E. coli* has a specific three-dimensional conformation conferred by intrachain base pairing. Figure 26–13 shows a postulated representation of the 16S and 5S rRNAs in a maximally base-paired conformation. The rRNAs appear to serve as a framework to which the ribosomal proteins are bound.

In a method pioneered by Masayasu Nomura, the ribosome can be broken down into its RNA and protein components, then reconstituted in vitro. When the 21 different proteins and the 16S rRNA of the 30S subunit are isolated from *E. coli* and then mixed under appropriate experimental conditions, they spontaneously reassemble to form 30S subunits identical in structure and activity to native 30S subunits. Similarly, the 50S subunit can assemble itself from its 34 proteins and its 5S and 23S rRNAs, providing the 30S subunit is also present. Each of the 55 proteins in the bacterial ribosome is believed to play a role in the synthesis of polypeptides, either as an enzyme or as a structural component in the overall process. However, the detailed function of only a few of the ribosomal proteins is known.

The two ribosomal subunits have irregular shapes. The three-dimensional structures of the 30S and 50S subunits of *E. coli* ribosomes (Fig. 26–12) have been deduced from x-ray diffraction, electron microscopy, and other structural methods. The two oddly shaped subunits fit together in such a way that a cleft is formed through which the mRNA passes as the ribosome moves along it during the translation process and from which the newly formed polypeptide chain emerges (Fig. 26–14).

The ribosomes of eukaryotic cells (other than mitochondrial and chloroplast ribosomes) are substantially larger and more complex than bacterial ribosomes (Fig. 26–12). They have a diameter of about 23 nm and a sedimentation coefficient of about 80S. They also have two subunits, which vary in size between species but on average are 60S and 40S. The rRNAs and most of the proteins of eukaryotic ribosomes have also been isolated. The small subunit contains an 18S rRNA, and the large subunit contains 5S, 5.8S, and 28S rRNAs. Altogether, eukaryotic ribosomes contain over 80 different proteins. (In contrast, the ribosomes of mitochondria and chloroplasts are somewhat smaller and simpler than bacterial ribosomes.)

Masayasu Nomura

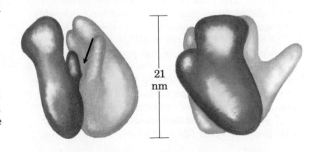

21 nm

Figure 26–14 Two different views of models of an *E. coli* ribosome, showing the relationship between the 30S and 50S subunits. The arrow indicates the cleft between the subunits.

Robert W. Holley

Transfer RNAs Have Characteristic Structural Features

To understand how tRNAs can serve as adapters in translating the language of nucleic acids into the language of proteins, we must first examine their structure in more detail. As shown in Chapter 12, tRNAs are relatively small and consist of a single strand of RNA folded into a precise three-dimensional structure (see Fig. 12–27a). In bacteria and in the cytosol of eukaryotes, tRNAs have between 73 and 93 nucleotide residues, corresponding to molecular weights between 24,000 and 31,000. (Mitochondria contain distinctive tRNAs that are somewhat smaller.) As we have noted earlier in this chapter, there is at least one kind of tRNA for each amino acid; for some amino acids there are two or more specific tRNAs. At least 32 tRNAs are required to recognize all the amino acid codons (some recognize more than one codon), but some cells have many more than 32.

Many tRNAs have been isolated in homogeneous form. In 1965, after several years of work, Robert W. Holley and his colleagues worked out the complete nucleotide sequence of alanine tRNA (tRNA^Ala) from yeast. This, the very first nucleic acid to be sequenced in its entirety, was found to contain 76 nucleotide residues, ten of which have modified bases. Its complete base sequence is shown in Figure 26–15.

Since Holley's pioneering studies, the base sequences of many other tRNAs from various species have been worked out and have revealed many common denominators of structure. Eight or more of the nucleotide residues of all tRNAs have unusual modified bases, many of which are methylated derivatives of the principal bases. Most tRNAs have a guanylate (pG) residue at the 5' end, and all have the trinucleotide sequence CCA(3') at the 3' end. All tRNAs, if written in a form in which there is maximum intrachain base pairing through the allowed

Figure 26–15 The nucleotide sequence of yeast tRNA^Ala as deduced by Holley and his colleagues. The cloverleaf conformation shown here is that in which intrastrand base-pairing is maximal. In addition to A, G, U, and C, the following symbols are used for the modified nucleotides: ψ, pseudouridine; I, inosine; T, ribothymidine; DHU, 5,6-dihydrouridine; m¹I, 1-methylinosine; m¹G, 1-methylguanosine; m²G, N²-dimethylguanosine. The modified bases are shaded in red, and most are illustrated in Fig. 25–25. The blue lines between the parallel sections indicate base pairs. The anticodon is capable of recognizing three codons for alanine (GCA, GCU, and GCC). Other features of tRNA structure are shown in Fig. 26–16. Note the presence of G═U base pairs in both the amino acid arm (top) and the DHU arm (left), signified by a blue dot to indicate a non–Watson–Crick pairing. In RNAs guanosine is often found base-paired with uridine, although the G═U pair is not as stable as the Watson–Crick G≡C pair (Chapter 12).

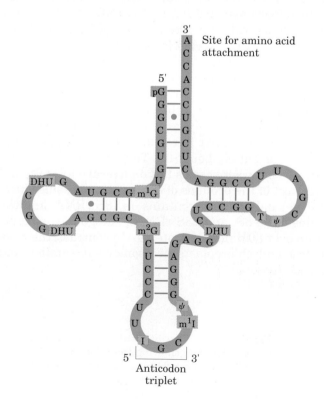

Figure 26–16 General structure of all tRNAs. When drawn with maximum intrachain base pairing, all tRNAs show the cloverleaf structure. The large dots on the backbone represent nucleotide residues, and the blue lines represent base pairings. Characteristic and/or invariant residues common to all tRNAs are shaded in red. Transfer RNAs differ in length, from 73 to 93 nucleotides. Extra nucleotides occur in the extra arm or in the DHU arm. At the end of the anticodon arm is the anticodon loop, which always contains seven unpaired nucleotides. The DHU arm contains up to three DHU residues, depending on the tRNA. In some tRNAs the DHU arm has only three hydrogen-bonded base pairs. In addition to the symbols explained in Fig. 26–15: Pu, purine nucleotide; Py, pyrimidine nucleotide; G*, guanylate *or* 2'-*O*-methylguanylate.

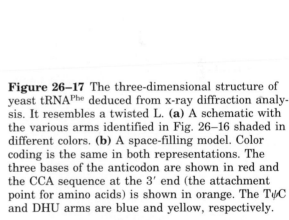

pairs A=U, G≡C, and G=U (see Fig. 12–26), form a cloverleaflike structure with four arms; the longer tRNAs have a short fifth or extra arm (Fig. 26–16; also evident in Fig. 26–15). The actual three-dimensional structure of a tRNA looks more like a twisted L than a cloverleaf (Fig. 26–17).

Two of the arms of a tRNA are critical for the adapter function. The **amino acid** or **AA arm** carries a specific amino acid esterified by its carboxyl group to the 2'- or 3'-hydroxyl group of the adenosine residue at the 3' end of the tRNA. The **anticodon arm** contains the anticodon. The other major arms are the **DHU** or **dihydrouridine arm,** which contains the unusual nucleotide dihydrouridine, and the **TψC arm,** which contains ribothymidine (T), not usually present in RNAs, and pseudouridine (ψ), which has an unusual carbon–carbon bond between the base and pentose (see Fig. 25–25). The functions of the DHU and TψC arms have not yet been determined.

Figure 26–17 The three-dimensional structure of yeast tRNA^Phe deduced from x-ray diffraction analysis. It resembles a twisted L. **(a)** A schematic with the various arms identified in Fig. 26–16 shaded in different colors. **(b)** A space-filling model. Color coding is the same in both representations. The three bases of the anticodon are shown in red and the CCA sequence at the 3' end (the attachment point for amino acids) is shown in orange. The TψC and DHU arms are blue and yellow, respectively.

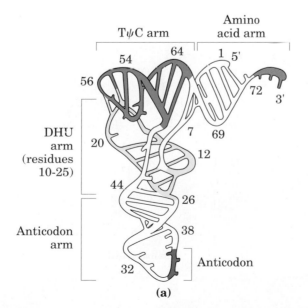

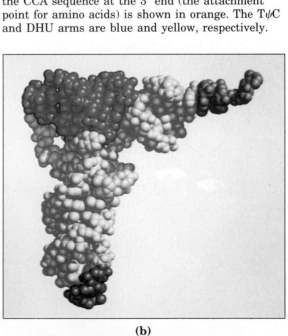

(a)

(b)

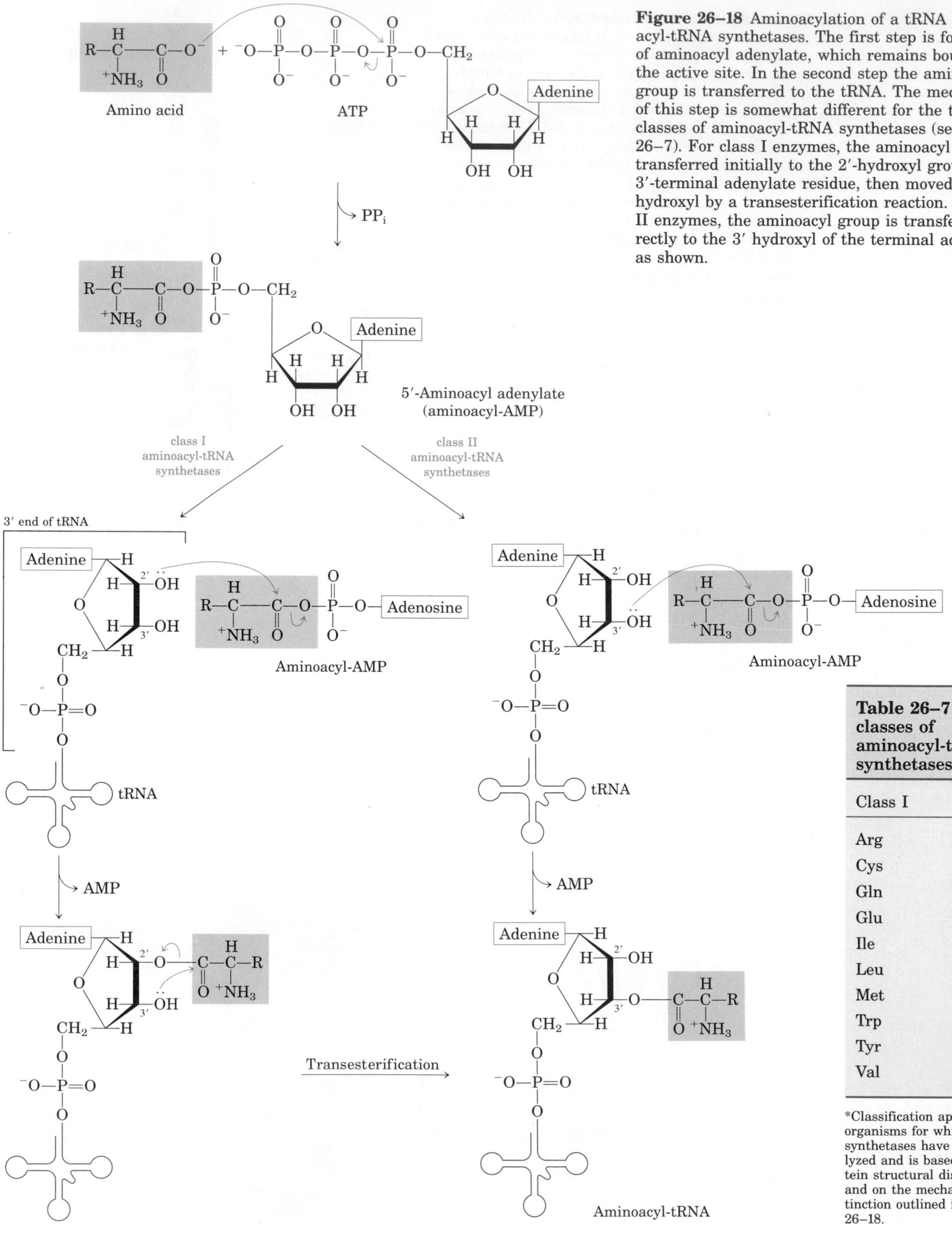

Figure 26–18 Aminoacylation of a tRNA by amino-acyl-tRNA synthetases. The first step is formation of aminoacyl adenylate, which remains bound to the active site. In the second step the aminoacyl group is transferred to the tRNA. The mechanism of this step is somewhat different for the two classes of aminoacyl-tRNA synthetases (see Table 26–7). For class I enzymes, the aminoacyl group is transferred initially to the 2′-hydroxyl group of the 3′-terminal adenylate residue, then moved to the 3′ hydroxyl by a transesterification reaction. For class II enzymes, the aminoacyl group is transferred directly to the 3′ hydroxyl of the terminal adenylate, as shown.

Table 26–7 Two classes of aminoacyl-tRNA synthetases*

Class I	Class II
Arg	Ala
Cys	Asn
Gln	Asp
Glu	Gly
Ile	His
Leu	Lys
Met	Phe
Trp	Pro
Tyr	Ser
Val	Thr

*Classification applies to all organisms for which tRNA synthetases have been analyzed and is based on protein structural distinctions and on the mechanistic distinction outlined in Figure 26–18.

Aminoacyl-tRNA Synthetases Attach the Correct Amino Acids to Their tRNAs

In the first stage of protein synthesis, which takes place in the cytosol, the 20 different amino acids are esterified to their corresponding tRNAs by aminoacyl-tRNA synthetases, each of which is specific for one amino acid and one or more corresponding tRNA. In most organisms there is generally one aminoacyl-tRNA synthetase for each amino acid. As noted earlier, for amino acids that have two or more corresponding tRNAs the same aminoacyl-tRNA synthetase usually aminoacylates all of them. In *E. coli,* the only exception to this rule is lysine, for which there are two aminoacyl-tRNA synthetases. There is only one tRNA^{Lys} in *E. coli,* and the biological rationale for the presence of two Lys-tRNA synthetases is unclear. Nearly all the aminoacyl-tRNA synthetases of *E. coli* have been isolated; all have been sequenced (either the protein itself or its gene), and a number have been crystallized. They have been divided into two classes (Table 26–7) based on distinctions in primary and tertiary structure and on differences in reaction mechanism, as detailed below. The overall reaction catalyzed by these enzymes is

$$\text{Amino acid} + \text{tRNA} + \text{ATP} \underset{}{\overset{\text{Mg}^{2+}}{\rightleftharpoons}} \text{aminoacyl-tRNA} + \text{AMP} + \text{PP}_\text{i}$$

The activation reaction occurs in two separate steps in the enzyme active site. In the first step, an enzyme-bound intermediate, aminoacyl adenylate (aminoacyl-AMP) is formed by reaction of ATP and the amino acid at the active site (Fig. 26–18). In this reaction, the carboxyl group of the amino acid is bound in anhydride linkage with the 5′-phosphate group of the AMP, with displacement of pyrophosphate.

In the second step the aminoacyl group is transferred from enzyme-bound aminoacyl-AMP to its corresponding specific tRNA. As shown in Figure 26–18, the course of this second step depends upon the class to which the enzyme belongs (Table 26–7). The reason for the mechanistic distinction between the two enzyme classes is unknown. The resulting ester linkage between the amino acid and the tRNA (Fig. 26–19) has a high standard free energy of hydrolysis ($\Delta G^{\circ\prime} = -29$ kJ/mol). The pyrophosphate formed in the activation reaction undergoes hydrolysis to phosphate by inorganic pyrophosphatase. Thus *two* high-energy phosphate bonds are ultimately expended for each amino acid molecule activated, rendering the overall reaction for amino acid activation essentially irreversible:

$$\text{Amino acid} + \text{tRNA} + \text{ATP} \overset{\text{Mg}^{2+}}{\longrightarrow} \text{aminoacyl-tRNA} + \text{AMP} + 2\text{P}_\text{i}$$
$$\Delta G^{\circ\prime} \approx -29 \text{ kJ/mol}$$

Some Aminoacyl-tRNA Synthetases Are Capable of Proofreading

The aminoacylation of tRNA accomplishes two things: the activation of an amino acid for peptide bond formation and attachment of the amino acid to an adapter tRNA that directs its placement within a growing polypeptide. As we will see, the identity of the amino acid attached to a tRNA is not checked on the ribosome. Attaching the correct amino acid to each tRNA is therefore essential to the fidelity of protein synthesis as a whole.

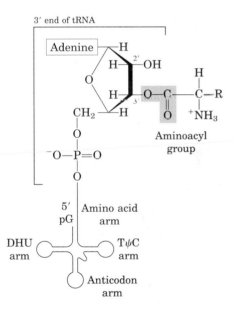

Figure 26–19 General structure of aminoacyl-tRNAs. The aminoacyl group is esterified to the 3′ position of the terminal adenylate residue. The ester linkage that both activates the amino acid and joins it to the tRNA is shaded red.

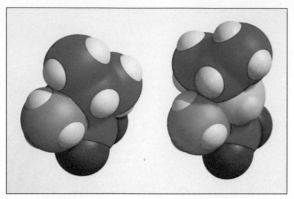

Valine   Isoleucine

The potential for any enzyme to discriminate between two different substrates is limited by the available binding energy that can be derived from enzyme–substrate interactions (Chapter 8). Discrimination between two similar amino acid substrates has been studied in detail in the case of Ile-tRNA^{Ile} synthetase, which faces the molecular problem that valine differs from isoleucine only by one methylene (CH_2) group. For this enzyme, activation of isoleucine (to form Ile-AMP) is favored over valine by a factor of 200, in the range expected given the potential contribution of binding energy from a methylene group. However, valine is incorporated into proteins in positions normally occupied by isoleucine at a frequency of only about 1 in 3,000.

The difference is brought about by a separate proofreading function of Ile-tRNA synthetase; this function is also present in some other aminoacyl-tRNA synthetases. All aminoacyl-AMPs produced by Ile-tRNA synthetase are checked in a second active site on the same enzyme, and incorrect ones are hydrolyzed. This proofreading activity reflects a general principle already seen in the discussion of proofreading by DNA polymerases (p. 822). If available binding interactions involving different groups on two substrates do not provide for a sufficient discrimination between the two on the enzyme, then this available binding energy must be used *twice* (or more) in separate steps requiring discrimination. Forcing the system through two successive "filters" rather than one increases the potential fidelity by a power of 2. In the case of Ile-tRNA synthetase, the first filter is the initial amino acid binding and activation to aminoacyl-AMP. The second filter is the separate active site, which catalyzes deacylation of incorrect aminoacyl-AMPs. The aminoacyl-AMP intermediates remain bound to the enzyme. When tRNA^{Ile} binds to the enzyme, the presence of Ile-AMP leads to aminoacylation of the tRNA. If Val-AMP is present on the enzyme instead, it is hydrolyzed to valine and AMP and the tRNA is not aminoacylated. Because the R group of valine is slightly smaller than that of isoleucine, the Val-AMP fits the hydrolytic (proofreading) site of the Ile-tRNA synthetase, but Ile-AMP does not.

In addition to proofreading after formation of the aminoacyl-AMP intermediate, most aminoacyl-tRNA synthetases are also capable of hydrolyzing the ester linkage between amino acids and tRNAs in aminoacyl-tRNAs. This hydrolysis is greatly accelerated for incorrectly charged tRNAs, providing yet a third filter to enhance the fidelity of the overall process. In contrast, in a few aminoacyl-tRNA synthetases that activate amino acids that have no close structural relatives, little or no proofreading occurs; in these cases the active site can sufficiently discriminate between the proper substrate amino acid and incorrect amino acids.

The overall error rate of protein synthesis (~1 mistake per 10^4 amino acids incorporated) is not nearly as low as for DNA replication, perhaps because a mistake in a protein is erased by destroying the protein and is not passed on to future generations. This degree of fidelity is sufficient to ensure that most proteins contain no mistakes and that the large amount of energy required to synthesize a protein is rarely wasted.

The Interaction between Aminoacyl-tRNA Synthetase and tRNA Constitutes a "Second Genetic Code"

An individual aminoacyl-tRNA synthetase must be specific not only for a single amino acid but for a certain tRNA as well. Discriminating among several dozen tRNAs is just as important for the overall fidelity of protein biosynthesis as is distinguishing among amino acids. The interaction between aminoacyl-tRNA synthetases and tRNAs has been referred to as the "second genetic code," to reflect its critical role in maintaining the accuracy of protein synthesis. The "coding" rules are apparently more complex than those in the "first" code.

Figure 26–20 summarizes what is known about the nucleotides involved in recognition by some or all aminoacyl-tRNA synthetases. Some nucleotides are conserved in all tRNAs and therefore cannot be used for discrimination. Nucleotide positions that are involved in discrimination by the aminoacyl-tRNA synthetases have been identified by the fact that changes at those nucleotides alter the enzyme's substrate specificity. These interactions seem to be concentrated in the amino acid arm and the anticodon arm, but are also located in many other parts of the molecule. The conformation of the tRNA (as opposed to its sequence) can also be important in recognition.

Some aminoacyl-tRNA synthetases recognize the tRNA anticodon itself. Changing the anticodon of one tRNA^Val from UAC to CAU makes this tRNA an excellent substrate for Met-tRNA synthetase. The Val-tRNA synthetase will similarly recognize a modified tRNA^Met in which the anticodon has been changed to UAC. Recognition by aminoacyl-tRNA synthetases of other tRNAs (about half of them, including those for alanine and serine) is affected little or not at all by changes at the anticodon. In some cases ten or more specific nucleotides are involved in recognition of a tRNA by its specific aminoacyl-tRNA synthetase (Fig. 26–20). In contrast, across a range of organisms from bacteria to humans the primary determinant for tRNA recognition by the Ala-tRNA synthetases is a single G=U base pair in the amino acid arm of tRNA^Ala (Fig. 26–21a). A short RNA with as few as seven base pairs arranged in a simple hairpin minihelix is efficiently aminoacylated by the Ala-tRNA synthetase as long as the RNA contains this critical G=U (Fig. 26–21b).

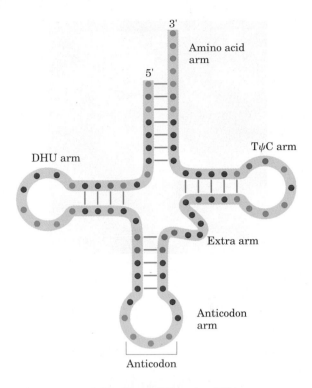

Figure 26–20 Known positions in tRNAs recognized by aminoacyl-tRNA synthetases. Positions in blue are the same in all tRNAs and therefore cannot be used to discriminate one from another. Other positions are known recognition points for one (red) or more (green) tRNA synthetases. Structural features other than sequence are important for recognition by some tRNA synthetases.

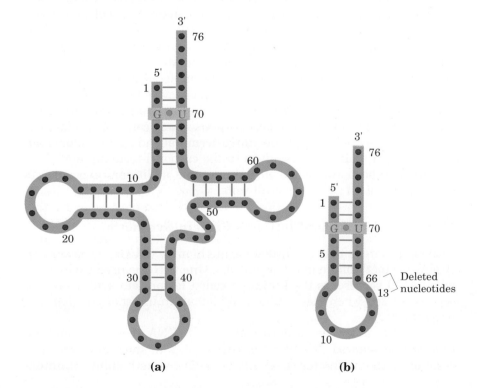

Figure 26–21 (a) The tRNA^Ala structural elements recognized by the Ala-tRNA synthetase are unusually simple. A single G=U base pair (red) is the only element needed for specific binding and aminoacylation. **(b)** A short synthetic RNA minihelix, which has the critical G=U base pair but lacks most of the remaining tRNA structure, is specifically aminoacylated with alanine almost as efficiently as the complete tRNA^Ala.

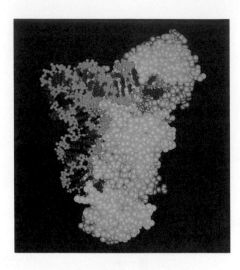

Figure 26–22 Structure of Gln-tRNA synthetase (white) bound to its cognate tRNA^{Gln} (green and red) and ATP. The three phosphate groups of the ATP, shown in yellow, are visible. In this case, bases in both the anticodon arm and the amino acid arm are the key structural features of the tRNA used for recognition by the aminoacyl-tRNA synthetase. Additional contacts between the enzyme and the tRNA revealed in this crystal structure occur along the inside of the L structure of the tRNA (see Fig. 26–17), but many of these involve residues conserved in all tRNAs and may not contribute to discrimination among different tRNAs.

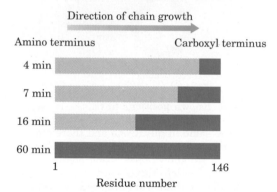

Figure 26–23 Proof that polypeptide chains grow by addition of new amino acid residues to the carboxyl end. The dark red zones show the portions of completed α-globin chains containing radioactive Leu residues at different times after addition of labeled leucine. At 4 min, only a few residues at the carboxyl end of α-globin were labeled. This is because the only *complete* globin chains that contained label after 4 min were those that had nearly completed synthesis at the time the label was added. On longer times of incubation with labeled leucine, successively longer segments of the polypeptide chain contained labeled residues, always in a block at the carboxyl end of the chain. The unlabeled end of the polypeptide (the amino terminus) was thus defined as the initiating end, and the polypeptide chain grows by successive addition of amino acids at the carboxyl end.

A complete understanding of the structural factors guiding these interactions remains an area of very active investigation. The solution of the crystal structures of two aminoacyl-tRNA synthetases (Gln and Asp) complexed with their cognate tRNAs and ATP is an important advance (Fig. 26–22). The relatively simple alanine system described above may be an evolutionary relic of a period when RNA oligonucleotides (ancestors to tRNA) were aminoacylated in a primitive system for protein synthesis.

Polypeptide Synthesis Begins at the Amino-Terminal End

Does polypeptide chain growth begin from the amino-terminal or from the carboxyl-terminal end? The answer came from isotope tracer experiments carried out by Howard Dintzis in 1961. Reticulocytes (immature erythrocytes) that were actively synthesizing hemoglobin were incubated with radioactive leucine. Leucine was chosen because it occurs frequently along both the α- and β-globin chains. Samples of completed α chains were isolated from the reticulocytes at various times after addition of radioactive leucine, and the distribution of radioactivity along the α chain was determined with the expectation that it would be concentrated in the end that was synthesized last. In those globin chains isolated after 60 min of incubation, nearly all the Leu residues were radioactive. However, in completed globin chains that were isolated only a few minutes after radioactive leucine was added, radioactive Leu residues were concentrated at the carboxyl-terminal end (Fig. 26–23). From these observations it was concluded that polypeptide chains are begun at the amino-terminal end and are elongated by sequential addition of residues to the carboxyl-terminal end. This pattern has been confirmed in innumerable additional experiments and applies to all proteins in all cells.

A Specific Amino Acid Initiates Protein Synthesis

Although there is only one codon for methionine (AUG), there are two tRNAs for methionine in all organisms. One tRNA is used exclusively when AUG represents the initiation codon for protein synthesis. The second is used when methionine is added at an internal position in a polypeptide.

In bacteria, the two separate classes of tRNA specific for methionine are designated tRNA^{Met} and tRNA^{fMet}. The starting amino acid residue at the amino-terminal end is *N*-formylmethionine. It enters

N-Formylmethionine

the ribosome as N-formylmethionyl-tRNA^{fMet} (fMet-tRNA^{fMet}), which is formed in two successive reactions. First, methionine is attached to tRNA^{fMet} by the Met-tRNA synthetase:

$$\text{Methionine} + \text{tRNA}^{fMet} + \text{ATP} \longrightarrow \text{Met-tRNA}^{fMet} + \text{AMP} + \text{PP}_i$$

As already noted, there is only one of these enzymes in *E. coli,* and it aminoacylates both tRNA^{fMet} and tRNA^{Met}. Second, a formyl group is transferred to the amino group of the Met residue from N^{10}-formyltetrahydrofolate by a transformylase enzyme:

$$N^{10}\text{-Formyltetrahydrofolate} + \text{Met-tRNA}^{fMet} \longrightarrow$$
$$\text{tetrahydrofolate} + \text{fMet-tRNA}^{fMet}$$

This transformylase is more selective than the Met-tRNA synthetase, and it cannot formylate free methionine or Met residues attached to tRNA^{Met}. Instead, it is specific for Met residues attached to tRNA^{fMet}, presumably recognizing some unique structural feature of that tRNA. The other Met-tRNA species, Met-tRNA^{Met}, is used to insert methionine in interior positions in the polypeptide chain. Blocking of the amino group of methionine by the N-formyl group not only prevents it from entering interior positions but also allows fMet-tRNA^{fMet} to be bound at a specific initiation site on the ribosome that does not accept Met-tRNA^{Met} or any other aminoacyl-tRNA.

In eukaryotic cells, all polypeptides synthesized by cytosolic ribosomes begin with a Met residue (as opposed to fMet), but again a specialized initiating tRNA is used that is distinct from the tRNA^{Met} used at interior positions. In contrast, polypeptides synthesized by the ribosomes in the mitochondria and chloroplasts of eukaryotic cells begin with N-formylmethionine. This and other similarities in the protein-synthesizing machinery of these organelles and bacteria strongly support the view that mitochondria and chloroplasts originated from bacterial ancestors symbiotically incorporated into the precursors of eukaryotic cells at an early stage of evolution (see Fig. 2–17).

We are now left with a puzzle. There is only one codon for methionine, namely (5')AUG. How can this single codon serve to identify both the starting N-formylmethionine (or methionine in the case of eukaryotes) and those Met residues that occur in interior positions in polypeptide chains? The answer will be found in the next section.

Initiation of Polypeptide Synthesis Has Several Steps

We now turn to a detailed examination of the second stage of protein synthesis: **initiation.** The focus here, and in the discussion of elongation and termination to follow, is on protein synthesis in bacteria; the process is not as well understood in eukaryotes. The initiation of polypeptide synthesis in bacteria requires (1) the 30S ribosomal subunit, which contains 16S rRNA, (2) the mRNA coding for the polypeptide to be made, (3) the initiating fMet-tRNA^{fMet}, (4) a set of three proteins called initiation factors (IF-1, IF-2, and IF-3), (5) GTP, (6) the 50S ribosomal subunit, and (7) Mg^{2+}. The formation of the initiation complex takes place in three steps (Fig. 26–24).

In the first step, the 30S ribosomal subunit binds initiation factor 3 (IF-3), which prevents the 30S and 50S subunits from combining prematurely. Binding of the mRNA to the 30S subunit then takes place in such a way that the initiation codon (AUG) binds to a precise location on the 30S subunit (Fig. 26–24).

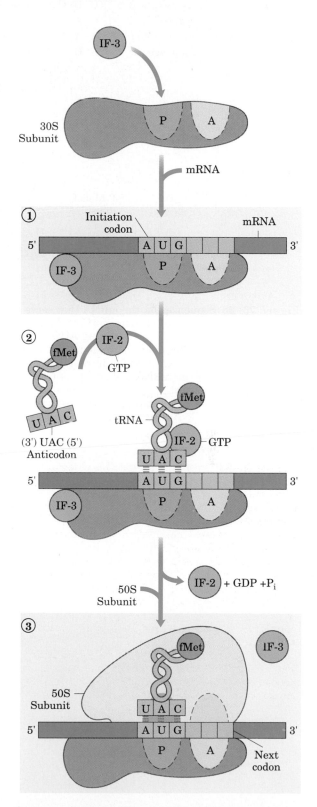

Figure 26–24 Formation of the initiation complex in three steps (described in the text) at the expense of the hydrolysis of GTP to GDP and P_i. IF-2 and IF-3 are initiation factors. P designates the peptidyl site, A the aminoacyl site.

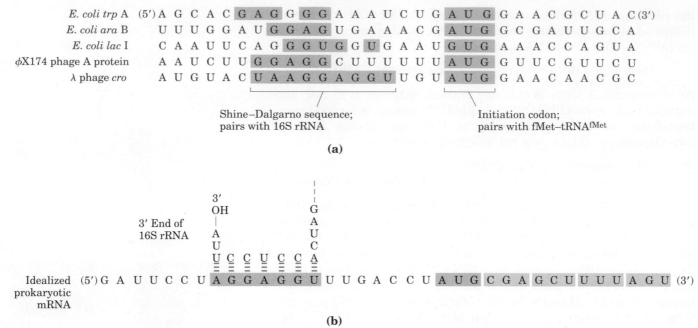

(a)

(b)

Figure 26–25 Sequences on the mRNA that serve as signals for initiation of protein synthesis in prokaryotes. **(a)** Alignment of the initiating AUG (shaded in green) in the P site depends in part on Shine–Dalgarno sequences (shaded in red) upstream. Portions of the mRNA transcripts of five prokaryotic genes are shown. **(b)** The Shine–Dalgarno sequences pair with a sequence near the 3′ end of the 16S rRNA, as shown for an idealized prokaryotic mRNA.

The initiating AUG is guided to the correct position on the 30S subunit by an initiating signal called the **Shine–Dalgarno sequence** in the mRNA, centered 8 to 13 base pairs to the 5′ side of the initiation codon (Fig. 26–25). Generally consisting of four to nine purine residues, the Shine–Dalgarno sequence is recognized by, and base-pairs (antiparallel) with, a complementary pyrimidine-rich sequence near the 3′ end of the 16S rRNA of the 30S subunit. This mRNA–rRNA interaction fixes the mRNA so that the AUG is correctly positioned for initiation of translation. The specific AUG where fMet-tRNA^{fMet} is to be bound is thereby distinguished from interior methionine codons by its proximity to the Shine–Dalgarno sequence in the mRNA.

Ribosomes have two sites that bind aminoacyl-tRNAs, the **aminoacyl** or **A site** and the **peptidyl** or **P site.** Both the 30S and the 50S subunits contribute to the characteristics of each site. The initiating AUG is positioned in the P site, which is the only site to which fMet-tRNA^{fMet} can bind (Fig. 26–24). However, fMet-tRNA^{fMet} is the exception: during the subsequent elongation stage, all other incoming aminoacyl-tRNAs, including the Met-tRNA^{Met} that binds to interior AUGs, bind to the A site. The P site is the site from which the "uncharged" tRNAs leave during elongation.

In the second step of the initiation process (Fig. 26–24), the complex consisting of the 30S subunit, IF-3, and mRNA now forms a still larger complex by binding IF-2, which already is bound to GTP and the initiating fMet-tRNA^{fMet}. The anticodon of this tRNA pairs correctly with the initiation codon in this step.

In the third step, this large complex combines with the 50S ribosomal subunit; simultaneously, the GTP molecule bound to IF-2 is hydrolyzed to GDP and P$_i$ (which are released). IF-3 and IF-2 also depart from the ribosome.

A major difference in protein synthesis between prokaryotes and eukaryotes is the existence of at least nine eukaryotic initiation factors. One of these, called cap binding protein or CBPI, binds to the 5′ cap of mRNA and facilitates formation of a complex between the mRNA and the 40S ribosomal subunit. The mRNA is then scanned to

Table 26–8 Protein factors required for translation initiation in bacteria and eukaryotes

Bacteria		Eukaryotes	
Factor	Function	Factor	Function
IF-1	Stimulates activities of IF-2 and IF-3		
IF-2	Facilitates binding of fMet-tRNA^fMet to 30S ribosomal subunit	eIF2*	Facilitates binding of initiating Met-tRNA^Met to 40S ribosomal subunit
IF-3	Binds to 30S subunit; prevents premature association of 50S subunit	eIF3, eIF4C	First factors to bind 40S subunit; facilitate subsequent steps
		CBPI	Binds to 5′ cap of mRNA
		eIF4A, eIF4B, eIF4F	Bind to mRNA; facilitate scanning of mRNA to locate first AUG
		eIF5	Promotes dissociation of several other initiation factors from 40S subunit as prelude to association of 60S subunit to form 80S initiation complex
		eIF6	Facilitates dissociation of inactive 80S ribosome into 40S and 60S subunits

*Surprisingly, eIF2 appears to be a multifunctional protein. In addition to its role in the initiation of translation, it is also involved in the splicing of mRNA precursors in the nucleus. This finding provides an intriguing link between transcription and translation in eukaryotic cells.

locate the first AUG codon, which signals the beginning of the reading frame. Several additional initiation factors are required in this mRNA scanning reaction, and in assembly of the complete 80S initiation complex in which the initiating Met-tRNA^Met and mRNA are bound and ready for elongation to proceed. The roles of the various bacterial and eukaryotic initiation factors in the overall process are summarized in Table 26–8. The mechanism by which these proteins act remains a very important area of investigation.

In bacteria, the steps in Figure 26–24 result in a functional 70S ribosome called the **initiation complex,** containing the mRNA and the initiating fMet-tRNA^fMet. The correct binding of the fMet-tRNA^fMet to the P site in the complete 70S initiation complex is assured by two points of recognition and attachment: the codon–anticodon interaction involving the initiating AUG fixed in the P site, and binding interactions between the P site and the fMet-tRNA^fMet. The initiation complex is now ready for the elongation steps.

Peptide Bonds Are Formed during the Elongation Stage

The third stage of protein synthesis is elongation, the stepwise addition of amino acids to the polypeptide chain. Again, our discussion focuses on bacteria. Elongation requires (1) the initiation complex described above, (2) the next aminoacyl-tRNA, specified by the next codon in the mRNA, (3) a set of three soluble cytosolic proteins called **elongation factors** (EF-Tu, EF-Ts, and EF-G), and (4) GTP. Three steps take place in the addition of each amino acid residue, and this cycle is repeated as many times as there are residues to be added.

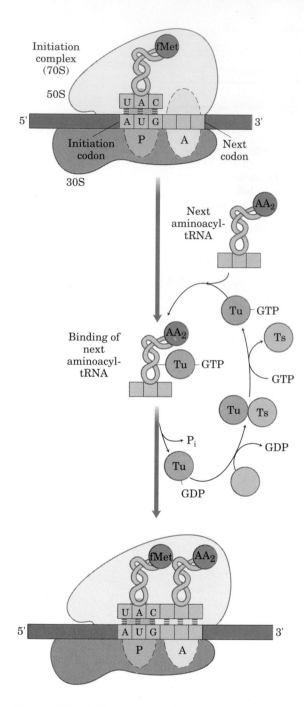

In the first step of the elongation cycle (Fig. 26–26), the next aminoacyl-tRNA is first bound to a complex of EF-Tu containing a molecule of bound GTP. The resulting aminoacyl-tRNA–EF-Tu·GTP complex is then bound to the A site of the 70S initiation complex. The GTP is hydrolyzed, an EF-Tu·GDP complex is released from the 70S ribosome, and an EF-Tu·GTP complex is regenerated (Fig. 26–26).

In the second step, a new peptide bond is formed between the amino acids bound by their tRNAs to the A and P sites on the ribosome (Fig. 26–27). This occurs by the transfer of the initiating *N*-formylmethionyl group from its tRNA to the amino group of the second amino acid now in the A site. The α-amino group of the amino acid in the A site acts as nucleophile, displacing the tRNA in the P site to form the peptide bond. This reaction produces a dipeptidyl-tRNA in the A site and the now "uncharged" (deacylated) tRNA^fMet remains bound to the P site.

The enzymatic activity that catalyzes peptide bond formation has historically been referred to as **peptidyl transferase** and was widely assumed to be intrinsic to one or more of the proteins in the large subunit. In 1992, Harry Noller and his colleagues discovered that this activity was catalyzed not by a protein but by the 23S rRNA, adding another critical biological function for ribozymes. As indicated in Chapter 25, this startling discovery has important implications for our understanding of the evolution of life on this planet.

In the third step of the elongation cycle, called **translocation,** the ribosome moves by the distance of one codon toward the 3′ end of the mRNA. Because the dipeptidyl-tRNA is still attached to the second codon of the mRNA, the movement of the ribosome shifts the dipeptidyl-tRNA from the A site to the P site, and the deacylated tRNA is released from the initial P site back into the cytosol. The third codon of the mRNA is now in the A site and the second codon in the P site. This shift of the ribosome along the mRNA requires EF-G (also called the translocase) and the energy provided by hydrolysis of another molecule of GTP (Fig. 26–28). A change in the three-dimensional conformation of the entire ribosome is believed to take place at this step in order to move the ribosome along the mRNA.

The ribosome, with its attached dipeptidyl-tRNA and mRNA, is now ready for another elongation cycle to attach the third amino acid residue. This process occurs in precisely the same way as the addition of the second. For each amino acid residue added to the chain, two GTPs are hydrolyzed to GDP and P_i. The ribosome moves from codon to codon along the mRNA toward the 3′ end, adding one amino acid residue at a time to the growing chain.

The polypeptide chain always remains attached to the tRNA of the last amino acid to have been inserted. This continued attachment to a tRNA is the chemical glue that makes the entire process work. The ester linkage between the tRNA and the carboxyl terminus of the polypeptide activates the terminal carboxyl group for nucleophilic attack by the incoming amino acid to form a new peptide bond (as in Fig. 26–27). At the same time, this tRNA represents the only link between the growing polypeptide and the information in the mRNA. As the existing ester linkage between the polypeptide and tRNA is broken during peptide bond formation, a new linkage is formed because each new amino acid is itself attached to a tRNA.

Figure 26–26 First step in elongation: the binding of the second aminoacyl-tRNA. The second aminoacyl-tRNA enters bound to EF-Tu (shown as Tu), which also contains bound GTP. Binding of the second aminoacyl-tRNA to the A site in the ribosome is accompanied by hydrolysis of the GTP to GDP and P_i, and an EF-Tu·GDP complex leaves the ribosome. The bound GDP is released when the EF-Tu·GDP complex binds to EF-Ts, and EF-Ts is subsequently released when another molecule of GTP becomes bound to EF-Tu. This recycles EF-Tu and permits it to bind another aminoacyl-tRNA.

In eukaryotes, the elongation cycle is quite similar. Three eukaryotic elongation factors called eEF1α, eEF1βγ, and eEF2 have functions analogous to the bacterial elongation factors EF-Tu, EF-Ts, and EF-G, respectively.

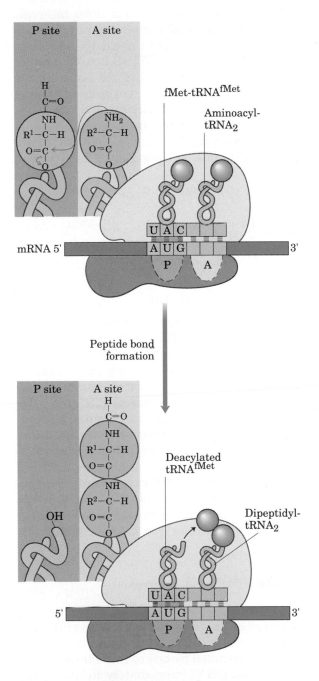

Figure 26–27 Second step in elongation: formation of the first peptide bond in bacterial protein synthesis, catalyzed by the 23S rRNA ribozyme. The *N*-formylmethionyl group is transferred to the amino group of the second aminoacyl-tRNA in the A site, forming a dipeptidyl-tRNA.

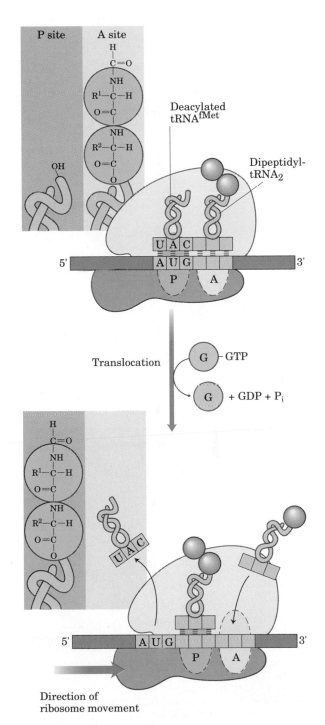

Figure 26–28 Third step in elongation: translocation. The ribosome moves one codon toward the 3′ end of mRNA, using energy provided by hydrolysis of GTP bound to EF-G (translocase; shown as G). The dipeptidyl-tRNA is now in the P site, leaving the A site open for the incoming (third) aminoacyl-tRNA.

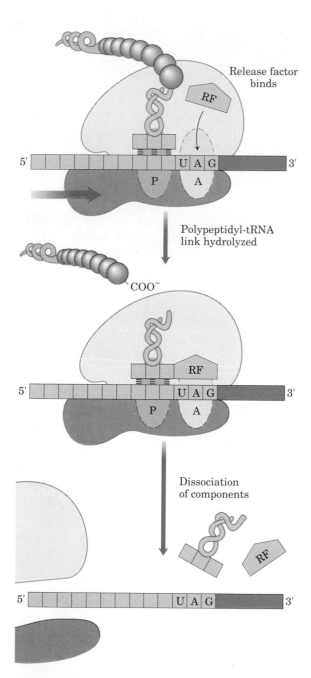

Figure 26–29 The termination of protein synthesis in bacteria in response to a termination codon in the A site. First, a release factor, RF₁ or RF₂ depending on which termination codon is present, binds to the A site. This leads in the second step to hydrolysis of the ester linkage between the nascent polypeptide and the tRNA in the P site, and release of the completed polypeptide. Finally, the mRNA, deacylated tRNA, and release factor leave the ribosome, and the ribosome dissociates into its 30S and 50S subunits.

Proofreading on the Ribosome Is Limited to Codon–Anticodon Interactions

The GTPase activity of EF-Tu makes an important contribution to the rate and fidelity of the overall biosynthetic process. The EF-Tu·GTP complex exists for a few milliseconds, and the EF-Tu·GDP complex also exists for a similar period before it dissociates. Both of these intervals provide an opportunity for the codon–anticodon interactions to be verified (i.e., proofread). Incorrect aminoacyl-tRNAs normally dissociate during one of these periods. If the GTP analog GTPγS is used in place of GTP, hydrolysis is slowed, improving the fidelity but reducing the rate of protein synthesis. The process of protein synthesis (including the characteristics of codon–anticodon pairing already described) has clearly been optimized through evolution to balance the requirements of both speed and fidelity. Improved fidelity might diminish speed, whereas increases in speed would probably compromise fidelity.

Guanosine 5′-O-(3-thiotriphosphate) (GTPγS)

This proofreading mechanism establishes only that the proper codon–anticodon pairing has taken place. The identity of the amino acids attached to tRNAs is not checked at all on the ribosome. This was demonstrated experimentally in 1962 by two research groups led by Fritz Lipmann and Seymour Benzer. They isolated enzymatically formed Cys-tRNA^Cys and then chemically converted it into Ala-tRNA^Cys. This hybrid aminoacyl-tRNA, which carries alanine but contains the anticodon for cysteine, was then incubated with a cell-free system capable of protein synthesis. The newly synthesized polypeptide was found to contain Ala residues in positions that should have been occupied by Cys residues. This important experiment also provided timely proof for Crick's adapter hypothesis. The fact that the amino acids themselves are never checked on the ribosome reinforces the central role of aminoacyl-tRNA synthetases in maintaining the fidelity of protein biosynthesis.

Termination of Polypeptide Synthesis Requires a Special Signal

Elongation continues until the ribosome adds the last amino acid, completing the polypeptide coded by the mRNA. Termination, the fourth stage of polypeptide synthesis, is signaled by one of three termination codons in the mRNA (UAA, UAG, UGA), immediately following the last amino acid codon (Box 26–3).

In bacteria, once a termination codon occupies the ribosomal A site three **termination** or **release factors**, the proteins RF₁, RF₂, and RF₃, contribute to (1) the hydrolysis of the terminal peptidyl-tRNA bond, (2) release of the free polypeptide and the last tRNA, now uncharged, from the P site, and (3) the dissociation of the 70S ribosome into its 30S and 50S subunits, ready to start a new cycle of polypeptide synthesis (Fig. 26–29). RF₁ recognizes the termination codons UAG

BOX 26-3   **Induced Variation in the Genetic Code: Nonsense Suppression**

When a termination codon is introduced in the interior of a gene by mutation, translation is prematurely halted and the incomplete polypeptide chains are often inactive. Such mutations are called nonsense mutations. Restoring the gene to its normal function requires a second mutation that either converts the termination codon to a codon specifying an amino acid or alternatively suppresses the effects of the termination codon. The second class of restorative mutations are called **nonsense suppressors,** and they generally involve mutations in tRNA genes that produce altered (suppressor) tRNAs that can recognize the termination codon and insert an amino acid at that position. Most suppressor tRNAs are created by single base substitutions in the anticodons of minor tRNA species.

Suppressor tRNAs constitute an experimentally induced variation in the genetic code involving the reading of what are usually termination codons, as is the case for many naturally occurring code variations described in Box 26–2. Nonsense suppression does not completely disrupt information transfer in the cell. This is because there are usually several copies of the genes for some tRNAs in any cell; some of these duplicate genes are weakly expressed and account for only a minor part of the cellular pool of a particular tRNA. Suppressor mutations usually involve these "minor" tRNA species, leaving the major tRNA to read its codon normally. For example, there are three identical genes for tRNA^{Tyr} in *E. coli*, each producing a tRNA with the anticodon (5')GUA. One of these is

expressed at relatively high levels and thus represents the major tRNA^{Tyr} species; the other two genes are duplicates transcribed in only small amounts. A change in the anticodon of the tRNA product of one of these duplicate tRNA^{Tyr} genes, from (5')GUA to (5')CUA, produces a minor tRNA^{Tyr} species that will insert tyrosine at UAG stop codons. This insertion of tyrosine at UAG is inefficient, but can permit production of enough useful full-length protein from a gene with a nonsense mutation to allow the cell to live. The major tRNA^{Tyr} maintains the normal genetic code for the majority of the proteins.

The base change in the tRNA that leads to the creation of a suppressor tRNA does not always occur in the anticodon. The suppression of UGA nonsense codons, interestingly, generally involves the tRNA^{Trp} that normally recognizes UGG. The alteration that allows it to read UGA (and insert Trp at these positions) does not occur in the anticodon. Instead, a G → A change at position 24 (in an arm of the tRNA somewhat removed from the anticodon) alters the anticodon pairing so that it can read *both* UGG and UGA. A similar change is found in tRNAs involved in the most common naturally occurring variation in the genetic code (UGA = Trp; see Box 26–2).

Suppression should lead to many abnormally long proteins, but, for reasons that are not entirely clear, this does not always occur. Many details of the molecular events that occur during translation termination and nonsense suppression are not understood.

---

and UAA, and RF$_2$ recognizes UGA and UAA. Either RF$_1$ or RF$_2$ (as appropriate, depending on which codon is present) binds at a termination codon and induces peptidyl transferase to transfer the growing peptide chain to a water molecule rather than to another amino acid. The specific function of RF$_3$ has not been firmly established. In eukaryotes, a single release factor called eRF recognizes all three termination codons.

Fidelity in Protein Synthesis Is Energetically Expensive

The enzymatic formation of each aminoacyl-tRNA used two high-energy phosphate groups. Additional ATPs are used each time incorrectly activated amino acids are hydrolyzed by the deacylation activity of some aminoacyl-tRNA synthetases (p. 914). One molecule of GTP is cleaved to GDP and P$_i$ during the first elongation step, and another GTP is hydrolyzed in the translocation step. Therefore a total of at least four high-energy bonds is ultimately required for the formation of each peptide bond of the completed polypeptide chain.

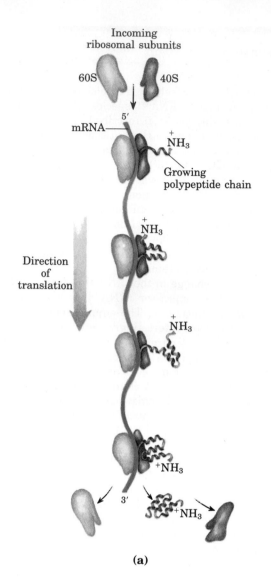

(a)

This represents an exceedingly large thermodynamic "push" in the direction of synthesis: at least $4 \times 30.5 = 122$ kJ/mol of phosphodiester bond energy is required to generate a peptide bond having a standard free energy of hydrolysis of only about -21 kJ/mol. The net free-energy change in peptide-bond synthesis is thus -101 kJ/mol. Although this large energy expenditure may appear wasteful, it is again important to remember that proteins are information-containing polymers. The biochemical problem is not simply the formation of a peptide bond, but the formation of a peptide bond between *specific* amino acids. Each of the high-energy bonds expended in this process plays a role in a step that is critical to maintaining proper alignment between each new codon in the mRNA and the amino acid it encodes at the growing end of the polypeptide. This energy makes possible the nearly perfect fidelity in the biological translation of the genetic message of mRNA into the amino acid sequence of proteins.

Polysomes Allow Rapid Translation of a Single Message

Large clusters of 10 to 100 ribosomes can be isolated from either eukaryotic or bacterial cells that are very active in protein synthesis. Such clusters, called **polysomes,** can be fragmented into individual ribosomes by the action of ribonuclease. Furthermore, a connecting fiber between adjacent ribosomes is visible in electron micrographs (Fig. 26–30). The connecting strand is a single strand of mRNA, being translated simultaneously by many ribosomes, spaced closely together. The simultaneous translation of a single mRNA by many ribosomes allows highly efficient use of the mRNA.

In bacteria there is a very tight coupling between transcription and translation. Messenger RNAs are synthesized in the $5' \rightarrow 3'$ direction and are translated in the same direction. As shown in Figure 26–31, ribosomes begin translating the $5'$ end of the mRNA before transcription is complete. The situation is somewhat different in eukaryotes, where newly transcribed mRNAs must be transferred out of the nucleus before they can be translated.

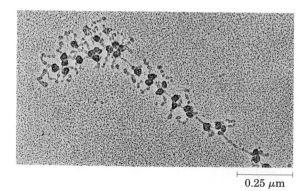

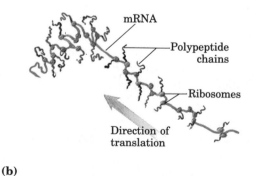

(b)

Figure 26–30 A polysome. **(a)** Four ribosomes are shown translating a eukaryotic mRNA molecule simultaneously, moving from the $5'$ end to the $3'$ end. **(b)** Electron micrograph and explanatory diagram of a polysome from the silk gland of a silk-worm larva. The mRNA is being translated by many ribosomes simultaneously. The polypeptide chains become longer as the ribosomes move toward the $3'$ end of the mRNA. The final product of this process is silk fibroin.

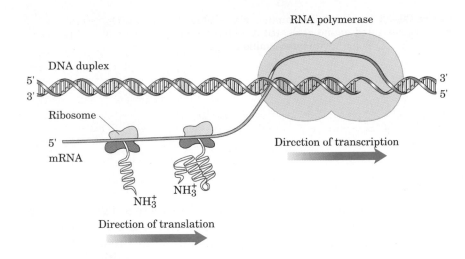

Figure 26–31 The coupling of transcription and translation in bacteria. The mRNA is translated by ribosomes while it is still being transcribed from DNA by RNA polymerase. This is possible because the mRNA in bacteria does not have to be transported from a nucleus to the cytoplasm before encountering ribosomes. In this schematic diagram the ribosomes are depicted as smaller than the RNA polymerase. In reality the ribosomes (M_r 2.5×10^6) are an order of magnitude larger than the RNA polymerase (M_r 3.9×10^5).

Bacterial mRNAs generally exist for only a few minutes (p. 880) before they are degraded by nucleases. Therefore, in order to maintain high rates of protein synthesis, the mRNA for a given protein or set of proteins must be made continuously and translated with maximum efficiency. The short lifetime of mRNAs in bacteria allows synthesis of a protein to cease rapidly when it is no longer needed by the cell.

Polypeptide Chains Undergo Folding and Processing

In the fifth and final step of protein synthesis, the nascent polypeptide chain is folded and processed into its biologically active form. At some point during or after its synthesis, the polypeptide chain spontaneously assumes its native conformation, which permits the maximum number of hydrogen bonds and van der Waals, ionic, and hydrophobic interactions (see Fig. 7–22). In this way, the linear or one-dimensional genetic message in the mRNA is converted into the three-dimensional structure of the protein. Some newly made proteins do not attain their final biologically active conformation until they have been altered by one or more processing reactions called **posttranslational modifications.** Both prokaryotic and eukaryotic posttranslational modifications are considered in what follows.

Amino-Terminal and Carboxyl-Terminal Modifications Initially, all polypeptides begin with a residue of *N*-formylmethionine (in bacteria) or methionine (in eukaryotes). However, the formyl group, the amino-terminal Met residue, and often additional amino-terminal and carboxyl-terminal residues may be removed enzymatically and thus do not appear in the final functional proteins.

In as many as 50% of eukaryotic proteins, the amino group of the amino-terminal residue is acetylated after translation. Carboxyl-terminal residues are also sometimes modified.

Loss of Signal Sequences As we shall see, the 15 to 30 residues at the amino-terminal end of some proteins play a role in directing the protein to its ultimate destination in the cell. Such **signal sequences** are ultimately removed by specific peptidases.

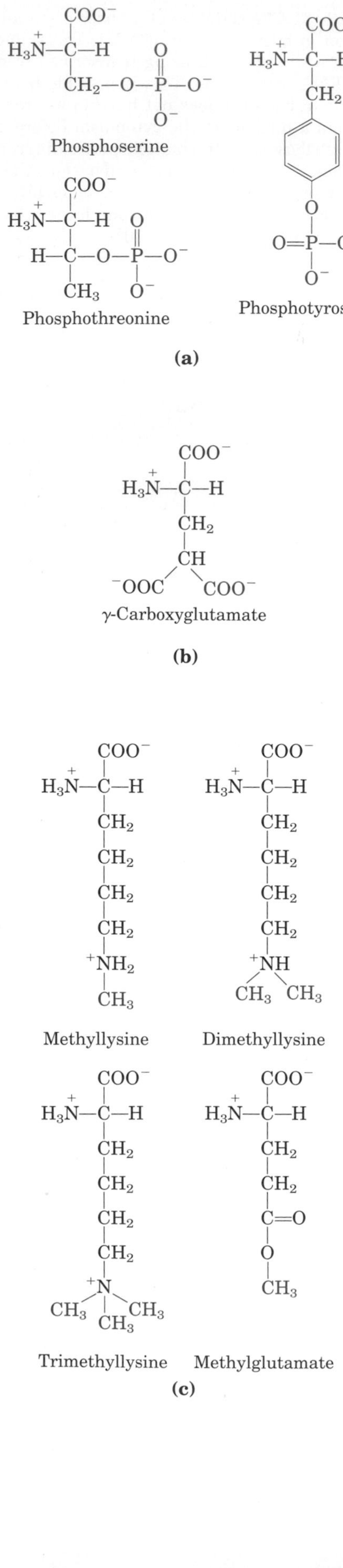

Figure 26–32 Some modified amino acid residues. **(a)** Phosphorylated amino acids. **(b)** A carboxylated amino acid. **(c)** Some methylated amino acids.

(a)

(b)

(c)

Modification of Individual Amino Acids The hydroxyl groups of certain Ser, Thr, and Tyr residues of some proteins are enzymatically phosphorylated by ATP (Fig. 26–32a); the phosphate groups add negative charges to these polypeptides. The functional significance of this modification varies from one protein to the next. For example, the milk protein casein has many phosphoserine groups, which function to bind Ca^{2+}. Given that Ca^{2+} and phosphate, as well as amino acids, are required by suckling young, casein provides three essential nutrients. The phosphorylation and dephosphorylation of the hydroxyl group of certain Ser residues is required to regulate the activity of some enzymes, such as glycogen phosphorylase (see Fig. 14–17). Phosphorylation of specific Tyr residues of some proteins is an important step in the transformation of normal cells into cancer cells (see Fig. 22–37).

Extra carboxyl groups may be added to Asp and Glu residues of some proteins. For example, the blood-clotting protein prothrombin contains a number of γ-carboxyglutamate residues (Fig. 26–32b) in its amino-terminal region, introduced by a vitamin K–requiring enzyme. These groups bind Ca^{2+}, required to initiate the clotting mechanism.

In some proteins certain Lys residues are methylated enzymatically (Fig. 26–32c). Monomethyl- and dimethyllysine residues are present in some muscle proteins and in cytochrome c. The calmodulin of most organisms contains one trimethyllysine residue at a specific position. In other proteins the carboxyl groups of some Glu residues undergo methylation (Fig. 26–32c), which removes their negative charge.

Attachment of Carbohydrate Side Chains The carbohydrate side chains of glycoproteins are attached covalently during or after the synthesis of the polypeptide chain. In some glycoproteins the carbohydrate side chain is attached enzymatically to Asn residues (*N*-linked oligosaccharides), in others to Ser or Thr residues (*O*-linked oligosaccharides; see Fig. 11–23). Many proteins that function extracellularly, as well as the "lubricating" proteoglycans coating mucous membranes, contain oligosaccharide side chains (see Fig. 11–21).

Addition of Isoprenyl Groups A number of eukaryotic proteins are isoprenylated; a thioether bond is formed between the isoprenyl group and a Cys residue of the protein (see Fig. 10–3). The isoprenyl groups are derived from pyrophosphate intermediates of the cholesterol biosynthetic pathway (see Fig. 20–34), such as farnesyl pyrophosphate (Fig. 26–33). Proteins modified in this way include the products of the *ras* oncogenes and proto-oncogenes (Chapter 22), G proteins (Chapter 22), and proteins called lamins, found in the nuclear matrix. In some

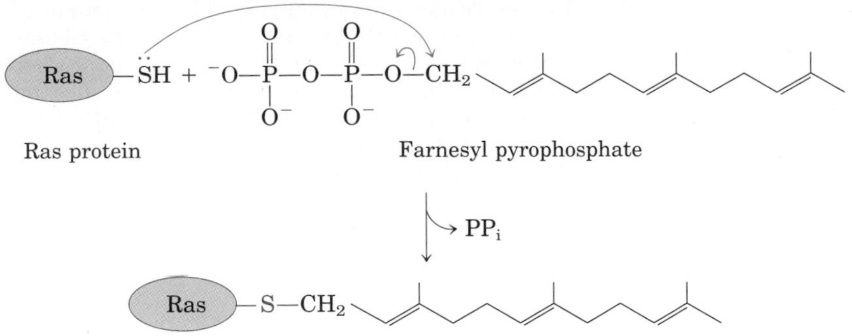

Figure 26–33 Farnesylation of a Cys residue on a protein. The thioether linkage is shown in red. The ras protein is the product of the *ras* oncogene.

cases the isoprenyl group serves to help anchor the protein in a membrane. The transforming (carcinogenic) activity of the *ras* oncogene is lost when isoprenylation is blocked, stimulating great interest in identifying inhibitors of this posttranslational modification pathway for use in cancer chemotherapy.

Addition of Prosthetic Groups Many prokaryotic and eukaryotic proteins require for their activity covalently bound prosthetic groups; these are attached to the polypeptide chain after it leaves the ribosome. Two examples are the covalently bound biotin molecule in acetyl-CoA carboxylase and the heme group of cytochrome *c*.

Proteolytic Processing Many proteins—for example, insulin (see Fig. 22–20), some viral proteins, and proteases such as trypsin and chymotrypsin (see Fig. 8–30)—are initially synthesized as larger, inactive precursor proteins. These precursors are proteolytically trimmed to produce their final, active forms.

Formation of Disulfide Cross-Links Proteins to be exported from eukaryotic cells, after undergoing spontaneous folding into their native conformations, are often covalently cross-linked by the formation of intrachain or interchain disulfide bridges between Cys residues. The cross-links formed in this way help to protect the native conformation of the protein molecule from denaturation in an extracellular environment that can differ greatly from that inside the cell.

Protein Synthesis Is Inhibited by Many Antibiotics and Toxins

Protein synthesis is a central function in cellular physiology, and as such it is the primary target of a wide variety of naturally occurring antibiotics and toxins. Except as noted, these antibiotics inhibit protein synthesis in bacteria. The differences between bacterial and eukaryotic protein synthesis are sufficient that most of these compounds are relatively harmless to eukaryotic cells. Antibiotics are important "biochemical weapons," synthesized by some microorganisms and extremely toxic to others. Antibiotics have become valuable tools in the study of protein synthesis; nearly every step in protein synthesis can be specifically inhibited by one antibiotic or another.

One of the best-understood inhibitory antibiotics is **puromycin,** made by the mold *Streptomyces alboniger*. Puromycin has a structure

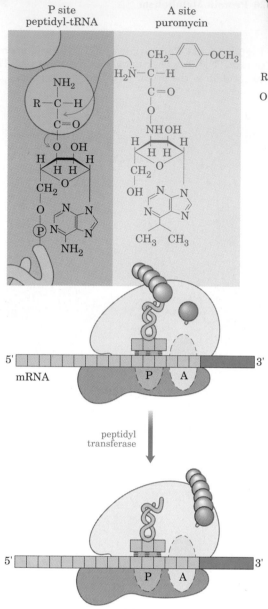

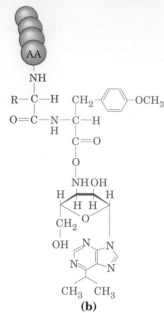

P site
peptidyl-tRNA

A site
puromycin

mRNA

P A

peptidyl transferase

P A

(a)

Figure 26–34 Puromycin resembles the aminoacyl end of a charged tRNA and can bind to the ribosomal A site, where it can participate in peptide bond formation (**a**). The product of this reaction, instead of being translocated to the P site, dissociates from the ribosome, causing premature chain termination. (**b**) Peptidyl puromycin.

(b)

very similar to the 3′ end of an aminoacyl-tRNA (Fig. 26–34). It binds to the A site and participates in all elongation steps up to and including peptide bond formation, producing a peptidyl puromycin. However, puromycin will not bind to the P site, nor does it engage in translocation. It dissociates from the ribosome shortly after it is linked to the carboxyl terminus of the peptide, prematurely terminating synthesis of the polypeptide.

Tetracyclines inhibit protein synthesis in bacteria by blocking the A site on the ribosome, inhibiting binding of aminoacyl-tRNAs. **Chloramphenicol** inhibits protein synthesis by bacterial (and mitochondrial and chloroplast) ribosomes by blocking peptidyl transfer but does not affect cytosolic protein synthesis in eukaryotes. Conversely, **cycloheximide** blocks the peptidyl transferase of 80S eukaryotic ribosomes but not that of 70S bacterial (and mitochondrial and chloroplast) ribosomes. **Streptomycin,** a basic trisaccharide, causes misreading of the genetic code in bacteria at relatively low concentrations and inhibits initiation at higher concentrations.

Several other inhibitors of protein synthesis are notable because of their toxicity to humans and other mammals. **Diphtheria toxin** (M_r 65,000) catalyzes the ADP-ribosylation of a diphthamide (a modified histidine) residue on eukaryotic elongation factor eEF2, thereby inactivating it (see Box 8–4). **Ricin,** an extremely toxic protein of the castor bean, inactivates the 60S subunit of eukaryotic ribosomes.

Tetracycline

Chloramphenicol

Cycloheximide

Streptomycin

Protein Targeting and Degradation

The eukaryotic cell is made up of many structures, compartments, and organelles, each with specific functions requiring distinct sets of proteins and enzymes. The synthesis of almost all these proteins begins on free ribosomes in the cytosol. How are these proteins directed to their final cellular destinations?

The answer to this question is at once complex, fascinating, and unfortunately incomplete. Enough is known, however, to outline many key steps in this process. Proteins destined for secretion, integration in the plasma membrane, or inclusion in lysosomes generally share the first few steps of a transport pathway that begins in the endoplasmic reticulum. Proteins destined for mitochondria, chloroplasts, or the nucleus each use separate mechanisms, and proteins destined for the cytosol simply remain where they are synthesized. The pathways by which proteins are sorted and transported to their proper cellular location are often referred to as **protein targeting** pathways.

The most important element in all of these targeting systems (with the exception of cytosolic and nuclear proteins) is a short amino acid sequence at the amino terminus of a newly synthesized polypeptide called the **signal sequence.** This signal sequence, whose function was first postulated by David Sabatini and Günter Blobel in 1970, directs a protein to its appropriate location in the cell and is removed during transport or when the protein reaches its final destination. In many cases, the targeting capacity of particular signal sequences has been confirmed by fusing the signal sequence from one protein, say protein A, to a different protein B, and showing that the signal directs protein B to the location where protein A is normally found.

The selective degradation of proteins no longer needed in the cell also relies largely on a set of molecular signals embedded in each protein's structure; most of these signals are not yet understood. The final part of this chapter is devoted to the processes of targeting and degradation, with emphasis on the underlying signals and molecular regulation that are so crucial to cellular metabolism. Except where noted, the focus is on eukaryotic cells.

Günter Blobel

Posttranslational Modification of Many Eukaryotic Proteins Begins in the Endoplasmic Reticulum

Perhaps the best-characterized targeting system begins in the endoplasmic reticulum (ER). Most lysosomal, membrane, or secreted proteins have an amino-terminal signal sequence that marks them for translocation into the lumen of the ER. More than 100 signal sequences for proteins in this group have been determined (Fig. 26–35). The se-

Figure 26–35 Amino-terminal signal sequences of some eukaryotic proteins, directing translocation into the endoplasmic reticulum. The hydrophobic core (yellow) is preceded by one or more basic residues (blue). Note the presence of polar and short-side-chain residues immediately preceding the cleavage sites (indicated by red arrows).

		cleavage site
Human influenza virus A	Met Lys Ala Lys Leu Leu Val Leu Leu Tyr Ala Phe Val Ala	Gly Asp Gln --
Human preproinsulin	Met Ala Leu Trp Met Arg Leu Leu Pro Leu Leu Ala Leu Leu Ala Leu Trp Gly Pro Asp Pro Ala Ala Ala	Phe Val --
Bovine growth hormone	Met Met Ala Ala Gly Pro Arg Thr Ser Leu Leu Leu Ala Phe Ala Leu Leu Cys Leu Pro Trp Thr Gln Val Val Gly	Ala Phe --
Bee promellitin	Met Lys Phe Leu Val Asn Val Ala Leu Val Phe Met Val Val Tyr Ile Ser Tyr Ile Tyr Ala	Ala Pro --
Drosophila glue protein	Met Lys Leu Leu Val Val Ala Val Ile Ala Cys Met Leu Ile Gly Phe Ala Asp Pro Ala Ser Gly	Cys Lys --

George Palade

quences vary in length (13 to 36 amino acid residues), but all have (1) a sequence of hydrophobic amino acids, typically 10 to 15 residues long, (2) one or more positively charged amino acid residues, usually near the amino terminus preceding the hydrophobic sequence, and (3) a short sequence at the carboxyl terminus (near the cleavage site) that is relatively polar, with amino acid residues having short side chains (especially Ala) predominating in the positions closest to the cleavage site.

As originally demonstrated by George Palade, proteins with these signal sequences are synthesized on ribosomes attached to the ER. The signal sequence itself is instrumental in directing the ribosome to the ER. The overall pathway summarized in Figure 26–36 begins with the initiation of protein synthesis on free ribosomes. The signal sequence appears early in the synthetic process because it is at the amino terminus. As it leaves the ribosome, this sequence and the ribosome itself are rapidly bound by a large complex called the **signal recognition particle** (SRP). This binding event halts elongation when the peptide is about 70 amino acids long and the signal sequence has emerged completely from the ribosome. The bound SRP directs the ribosome with the incomplete polypeptide to a specific set of SRP receptors in the cytosolic face of the ER. The nascent polypeptide is delivered to a **peptide translocation complex** in the ER, the SRP dissociates from the ribosome, and synthesis of the protein resumes. The translocation complex feeds the growing polypeptide into the lumen of the ER in a reaction that is driven by the energy of ATP. The signal sequence is removed by a signal peptidase within the lumen of the ER. Once the complete protein has been synthesized, the ribosome dissociates from the ER.

In the lumen of the ER, newly synthesized proteins are modified in several ways. In addition to the removal of signal sequences, polypeptide chains fold and disulfide bonds form. Many proteins are also glycosylated.

Figure 26–36 Directing eukaryotic proteins with the appropriate signals to the endoplasmic reticulum: the SRP cycle and nascent polypeptide translocation and cleavage. ① The ribosomal subunits assemble in an initiation complex at the initiation codon and begin protein synthesis. ② If an appropriate signal sequence appears at the amino terminus of the nascent polypeptide, ③ the SRP binds to the ribosome and halts elongation. ④ The ribosome–SRP complex is bound by receptors on the ER, and ⑤ the SRP dissociates and is recycled. ⑥ Protein synthesis resumes, coupled to translocation of the polypeptide chain into the lumen of the ER. ⑦ The signal sequence is cleaved by a signal peptidase within the lumen of the ER. ⑧ The ribosome is recycled.

The SRP is a rod-shaped complex containing a 300 nucleotide RNA (called 7SL-RNA) and six different proteins, with a combined molecular weight of 325,000. One protein subunit of the SRP binds directly to the signal sequence, inhibiting elongation by sterically blocking entry of aminoacyl-tRNAs and inhibiting peptidyl transferase. The SRP receptor is a heterodimer of α (M_r 69,000) and β (M_r 30,000) subunits.

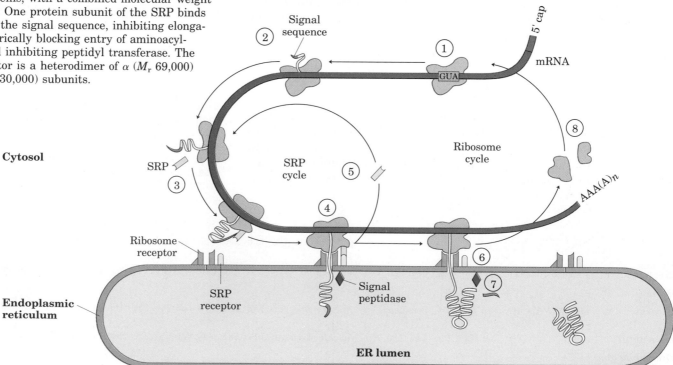

Glycosylation Plays a Key Role in Protein Targeting

Glycosylated proteins, or glycoproteins, often are linked to their oligosaccharides through Asn residues. These *N*-linked oligosaccharides are very diverse (Chapter 11), but the many pathways by which they form all have a common first step. A 14 residue core oligosaccharide (containing two *N*-acetylglucosamine, nine mannose, and three glucose residues) is transferred from a dolichol phosphate donor molecule to certain Asn residues on the proteins.

$$^-O-\overset{\overset{O}{\|}}{\underset{\underset{O^-}{|}}{P}}-O-CH_2-CH_2-\overset{\overset{CH_3}{|}}{\underset{\underset{H}{|}}{C}}-CH_2-\left(CH_2-CH=\overset{\overset{CH_3}{|}}{C}-CH_2\right)_n-CH_2-CH=\overset{\overset{CH_3}{|}}{C}-CH_3$$

Dolichol phosphate
($n = 9$–22)

The core oligosaccharide is built up on the phosphate group of dolichol phosphate (an isoprenoid derivative) by the successive addition of monosaccharide units. Once this core oligosaccharide is complete, it is enzymatically transferred from dolichol phosphate to the protein (Fig. 26–37). The transferase is located on the lumenal face of the ER and thus does not catalyze glycosylation of cytosolic proteins. After the transfer, the core oligosaccharide is trimmed and elaborated in different ways on different proteins, but all *N*-linked oligosaccharides retain a pentasaccharide core derived from the original 14 resi-

Figure 26–37 Synthesis of the core oligosaccharide of glycoproteins. The core oligosaccharide is built up in a series of steps as shown. The first few steps occur on the cytosolic face of the ER. Completion occurs within the lumen of the ER after a translocation step (upper left) in which the incomplete oligosaccharide is moved across the membrane. The mechanism of this translocation is not shown. The synthetic precursors that contribute additional mannose and glucose residues to the growing oligosaccharide in the lumen are themselves dolichol phosphate derivatives. The dolichol —℗—Man and dolichol —℗—Glc are synthesized from dolichol phosphate and GDP-mannose or UDP-glucose, respectively. After it is transferred to the protein, the core oligosaccharide is further modified in the ER and the Golgi complex in pathways that differ for different proteins. The five sugar residues enclosed in a beige screen (lower right) are retained in the final structure of all *N*-linked oligosaccharides. In the first step in the construction of the *N*-linked oligosaccharide moiety of a glycoprotein, the core oligosaccharide is transferred from dolichol phosphate to an Asn residue of the protein within the lumen of the ER. The released dolichol pyrophosphate is recycled.

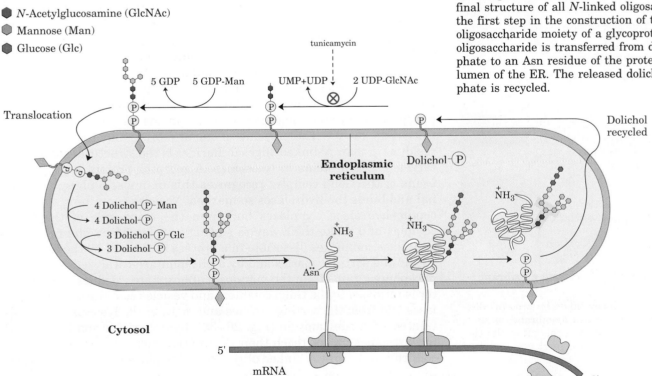

● *N*-Acetylglucosamine (GlcNAc)

◗ Mannose (Man)

⬣ Glucose (Glc)

Figure 26–38 The structure of tunicamycin, an antibiotic produced by *Streptomyces* that mimics UDP–*N*-acetylglucosamine and blocks the first step in the synthesis of the core oligosaccharide of glyco-proteins on dolichol phosphate (see Fig. 26–37). Tunicamycin is actually a family of antibiotics produced by (and isolated as a mixture from) *Streptomyces lysosuperficens*. They all contain uracil, *N*-acetylglucosamine, an 11 carbon aminodialdose called tunicamine, and a fatty acyl side chain. The structure of the fatty acyl side chain varies in the different compounds within the family. In addition to the variation in length of the fatty acyl side chain (indicated in the figure), some homologs lack the isopropyl group at the end and/or α,β-unsaturation.

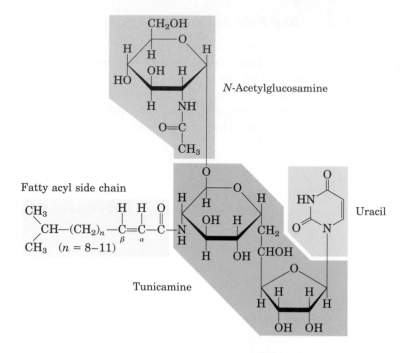

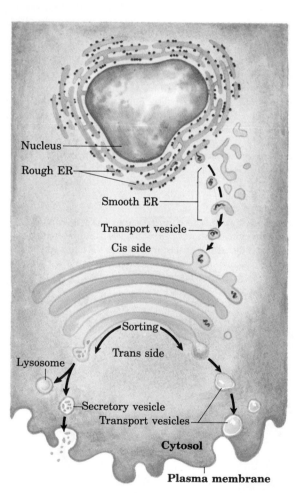

Figure 26–39 The pathway taken by proteins destined for lysosomes, the plasma membrane, or secretion. Proteins are moved from the ER to the cis side of the Golgi complex in transport vesicles. Sorting occurs primarily in the trans side of the Golgi complex.

due oligosaccharide (Fig. 26–37). Several antibiotics interfere with one or more steps in this process. The best-characterized is **tunicamycin** (Fig. 26–38), which blocks the first step.

Proteins are moved from the ER to the Golgi complex in transport vesicles (Fig. 26–39). In the Golgi complex, *O*-linked oligosaccharides are added and *N*-linked oligosaccharides are further modified. By mechanisms only partially understood, proteins are also sorted here and sent to their final destinations (Fig. 26–39). Within the Golgi complex, the processes that segregate proteins destined for the cell exterior from those destined for the plasma membrane or lysosomes must distinguish between proteins on the basis of structural features other than the signal sequence, which was removed in the lumen of the ER.

This sorting process is perhaps best understood in the case of hydrolases destined for transport to lysosomes. Upon arrival in the Golgi complex from the ER, some as yet undetermined feature of the three-dimensional structure of these hydrolases (sometimes called a "signal patch") is recognized by a phosphotransferase that catalyzes the phosphorylation of certain mannose residues in the enzymes' oligosaccharides (Fig. 26–40). The presence of one or more mannose-6-phosphate residues in their *N*-linked oligosaccharides is the structural signal that targets these proteins to lysosomes. A receptor protein in the membrane of the Golgi complex recognizes this mannose-6-phosphate signal and binds the hydrolases so marked. Vesicles containing these receptor–hydrolase complexes bud from the trans side of the Golgi complex and make their way to sorting vesicles. Here, the receptor–hydrolase complexes dissociate in a process facilitated by the lower pH within the sorting vesicles and by a phosphatase-catalyzed removal of phosphate groups from the mannose-6-phosphate residues. The receptor is returned to the Golgi complex, and vesicles containing the hydrolases bud from the sorting vesicles and move to the lysosomes. In cells treated with tunicamycin (Fig. 26–38), hydrolases normally targeted for lysosomes do not reach their destination but are secreted instead, confirming that the *N*-linked oligosaccharide plays a key role in targeting these enzymes to lysosomes.

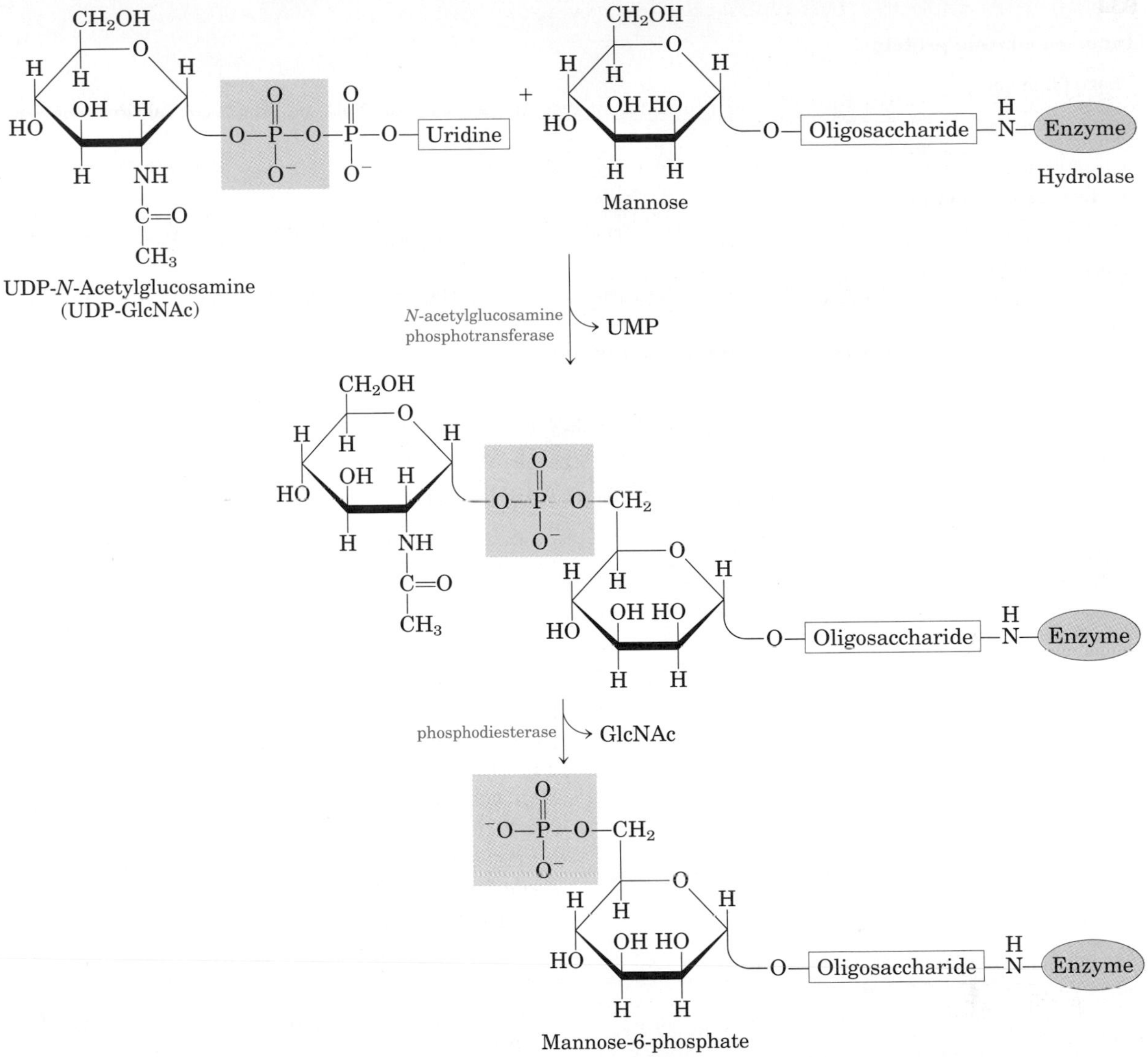

Figure 26–40 The two-step process by which mannose residues on lysosome-targeted enzymes, such as hydrolases, are phosphorylated. *N*-Acetylglucosamine phosphotransferase recognizes some as yet unidentified structural feature of lysosome-destined hydrolases.

For proteins destined for the plasma membrane or for secretion, and for those destined to reside permanently in the ER or the Golgi complex, the signals are less well understood. These targeting pathways are not impeded by tunicamycin, indicating that the signals are not carbohydrates.

Cellular proteins targeted to the mitochondria, chloroplasts, or nucleus use their own distinct signal sequences. For mitochondria and chloroplasts, the signal sequences are again found at the amino terminus of the proteins and are cleaved once the proteins arrive at their final destinations. The signal sequences that target some proteins to the nucleus (an example is the sequence –Pro–Lys–Lys–Lys–Arg–Lys–Val–) are located internally and are not cleaved. These signals permit proteins such as DNA polymerases and RNA polymerases to enter the nucleus rapidly through nuclear pores.

Bacteria Also Use Signal Sequences for Protein Targeting

Bacteria must also target some proteins to the inner or outer membranes, the periplasmic space between the membranes, or the extracellular medium (secretion). This targeting uses signal sequences at the amino terminus of the proteins much like those found on eukaryotic proteins targeted to the ER (Fig. 26–41).

Inner membrane proteins

| | | | | | | | | | | | | | | | | | | | | | | | | cleavage site | |

Phage fd, major coat protein — Met Lys Lys Ser Leu Val Leu Lys Ala Ser Val Ala Val Ala Thr Leu Val Pro Met Leu Ser Phe Ala↓Ala Glu --

Phage fd, minor coat protein — Met Lys Lys Leu Leu Phe Ala Ile Pro Leu Val Val Pro Phe Tyr Ser His Ser↓Ala Glu --

Periplasmic proteins

Alkaline phosphatase — Met Lys Gln Ser Thr Ile Ala Leu Ala Leu Leu Pro Leu Leu Phe Thr Pro Val Thr Lys Ala↓Arg Thr --

Leucine-specific binding protein — Met Lys Ala Asn Ala Lys Thr Ile Ile Ala Gly Met Ile Ala Leu Ala Ile Ser His Thr Ala Met Ala↓Asp Asp --

β-Lactamase of pBR322 — Met Ser Ile Gln His Phe Arg Val Ala Leu Ile Pro Phe Phe Ala Ala Phe Cys Leu Pro Val Phe Ala↓His Pro --

Outer membrane proteins

Lipoprotein — Met Lys Ala Thr Lys Leu Val Leu Gly Ala Val Ile Leu Gly Ser Thr Leu Leu Ala Gly↓Cys Ser --

LamB — Leu Arg Lys Leu Pro Leu Ala Val Ala Val Ala Ala Gly Val Met Ser Ala Gln Ala Met Ala↓Val Asp --

OmpA — Met Met Ile Thr Met Lys Lys Thr Ala Ile Ala Ile Ala Val Ala Leu Ala Gly Phe Ala Thr Val Ala Gln Ala↓Ala Pro --

Figure 26–41 Signal sequences used for targeting to different locations in bacteria. Basic amino acids (blue) near the amino terminus and hydrophobic core amino acids (yellow) are highlighted. The cleavage sites marking the ends of the signal sequences are marked by red arrows. Note that the inner membrane (see Fig. 2–6) is where phage fd coat proteins and DNA are assembled into phage particles.

Some proteins that are translocated through one or more membranes to reach their final destinations must be maintained in a distinct "translocation-competent" conformation until this process is complete. The functional conformation is assumed after translocation, and proteins purified in this final form are often found to be no longer capable of translocation. There is growing evidence that the translocation conformation is stabilized by a specialized set of proteins in all bacterial cells. These bind to the protein to be translocated while it is being synthesized, preventing it from folding into its final three-dimensional structure. In *E. coli,* a protein called trigger factor (M_r 63,000) appears to facilitate the translocation of at least one outer membrane protein through the inner membrane.

Cells Import Proteins by Receptor-Mediated Endocytosis

Some proteins are imported into certain cells from the surrounding medium; these include low-density lipoprotein (LDL), the iron-carrying protein transferrin, peptide hormones, and circulating proteins that are destined to be degraded. These proteins bind to receptors on the outer face of the plasma membrane. The receptors are concentrated in invaginations of the membrane called **coated pits,** which are coated on their cytosolic side with a lattice made up of the protein **clathrin** (Fig. 26–42). Clathrin forms closed polyhedral structures, and as more of the receptors become occupied with target proteins, the clathrin lattice grows until a complete membrane-bounded endocytic vesicle buds off the plasma membrane and moves into the cytoplasm.

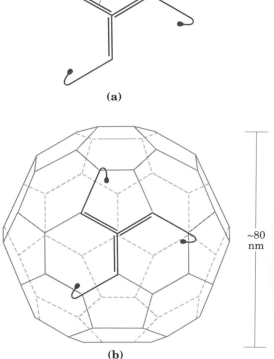

Heavy chain
Light chain

(a)

~80 nm

(b)

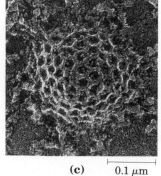

(c) 0.1 μm

Figure 26–42 Clathrin is a trimer of three light (L) chains (M_r 35,000) and three heavy (H) chains (M_r 180,000). **(a)** The (HL)$_3$ clathrin unit is organized as a three-legged structure called a triskelion. **(b)** Triskelions have a propensity to assemble into polyhedral lattices. **(c)** Electron micrograph of a coated pit on the cytosolic face of the plasma membrane of a fibroblast.

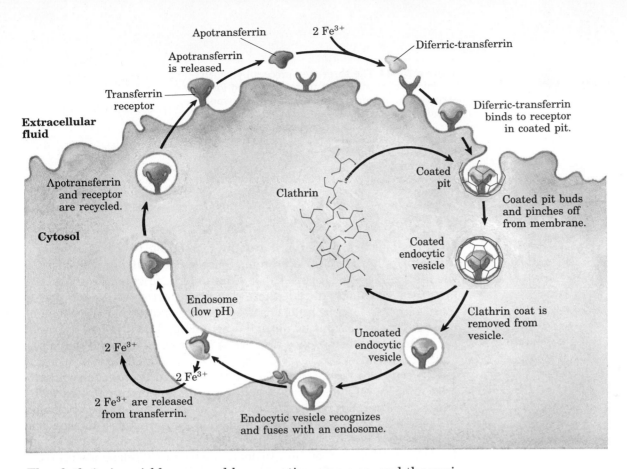

The clathrin is quickly removed by uncoating enzymes, and the vesicles fuse with endosomes. The pH of endosomes is lowered by the activity of V-type ATPases in their membranes (see Table 10–5), producing an environment that facilitates dissociation of receptors from their target proteins. Proteins and receptors then go their separate ways, their fates varying according to the system. Transferrin and its receptor are eventually recycled (Fig. 26–43). Some hormones, growth factors, and immune complexes are degraded along with their receptors after they have elicited the appropriate response. LDL is degraded after the associated cholesterol has been delivered to its destination, but its receptor is recycled (see Fig. 20–39).

Receptor-mediated endocytosis is exploited by some toxins and viruses to gain entry to cells. Diphtheria toxin, cholera toxin, and influenza virus all enter cells this way. HIV, the virus that causes AIDS, also binds to specific receptors on the cell surface and may gain entry by endocytosis. In humans, the receptor that binds HIV, known as CD4, is a glycoprotein found primarily on the surface of immune system cells called helper T cells. CD4 is normally involved in the complex communication between cells of the immune system that is required to execute the immune response.

Figure 26–43 The transferrin cycle transports iron into cells. Diferric-transferrin (transferrin containing two bound Fe^{3+} ions) is bound by receptors in coated pits (top right), which form endocytic vesicles coated with clathrin. Uncoating is catalyzed by ATP-dependent enzymes. This is followed by receptor-mediated fusion of the vesicles with endosomes (bottom). The low pH within the endosome causes dissociation of the Fe^{3+}. At low pH, the receptor retains a high affinity for apotransferrin, which is returned to the cell surface still bound to the receptors. Here the neutral pH lowers the affinity of the receptor for apotransferrin, permitting its dissociation. At neutral pH, the receptor has a high affinity for diferric-transferrin, allowing more molecules of diferric-transferrin to bind, thereby continuing the cycle.

Protein Degradation Is Mediated by Specialized Systems in All Cells

Proteins are constantly being degraded in all cells to prevent the buildup of abnormal or unwanted proteins and to facilitate the recycling of amino acids. Degradation is a selective process. The lifetime of any particular protein is regulated by proteolytic systems specialized for this task, as opposed to proteolytic events that might occur during posttranslational processing. The half-lives of different proteins can vary from half a minute to many hours or even days in eukaryotes.

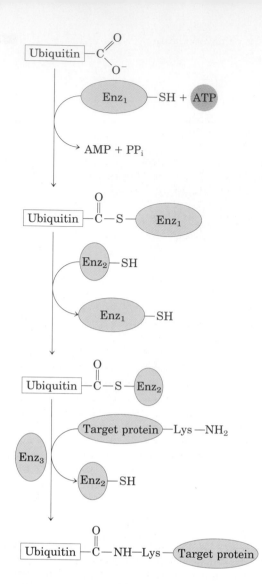

Figure 26–44 The three-step process by which ubiquitin is attached to a protein targeted for destruction in eukaryotes. Two different enzyme–ubiquitin intermediates are involved. The free carboxyl of ubiquitin's carboxyl-terminal Gly residue is ultimately linked through an amide (isopeptide) bond to an ϵ-amino group of a Lys residue of the target protein.

Most proteins are turned over rapidly in relation to the lifetime of a cell, although a few stable proteins (such as hemoglobin) can last for the life span of a cell (about 110 days for an erythrocyte). Proteins that are degraded rapidly include those that are defective because of one or more incorrect amino acids inserted during synthesis or because of damage that occurs during normal functioning. Also targeted for rapid turnover are many enzymes that act at key regulatory points in metabolic pathways.

Defective proteins and those with characteristically short half-lives are generally degraded in both bacteria and eukaryotes by ATP-dependent cytosolic systems. A second system in vertebrates operates in lysosomes and serves to recycle membrane proteins, extracellular proteins, and proteins with characteristically long half-lives.

In *E. coli*, many proteins are degraded by an ATP-dependent protease called La. The ATPase is activated only in the presence of defective proteins or those slated for rapid turnover; two ATP molecules are hydrolyzed for every peptide bond cleaved. The precise molecular function of ATP hydrolysis during peptide-bond cleavage is unclear. Once a protein is reduced to small inactive peptides, other ATP-independent proteases complete the degradation process.

In eukaryotes, the ATP-dependent pathway is quite different. A key component in this system is the 76 amino acid protein **ubiquitin,** so named because of its presence throughout the eukaryotic kingdoms. One of the most highly conserved proteins known, ubiquitin is essentially identical in organisms as different as yeasts and humans. Ubiquitin is covalently linked to proteins slated for destruction via an ATP-dependent pathway involving three separate enzymes (Fig. 26–44). How attachment of one or more molecules of ubiquitin to a protein targets that protein for proteolysis is not yet understood. The ATP-dependent proteolytic system in eukaryotes is a large complex ($M_r \geq 1 \times 10^6$). The mode of action of the protease component of the system and the role of ATP are unknown.

The signals that trigger ubiquitination are also not all understood, but one simple one has been found. The amino-terminal residue (i.e., the residue remaining after removal of methionine and any other proteolytic processing of the amino-terminal end) has a profound influence on the half-lives of many proteins (Table 26–9). These amino-terminal signals have evidently been conserved during billions of years of evolution; the signals are the same in bacterial protein degradation systems and in the human ubiquitination pathway. The degradation of proteins is as important to a cell's survival in a changing environment as is the protein synthetic process, and much remains to be learned about these interesting pathways.

Table 26–9 Relationship between the half-life of a protein and its amino-terminal amino acid

Amino-terminal residue	Half-life*
Stabilizing	
Met, Gly, Ala, Ser, Thr, Val	>20 h
Destabilizing	
Ile, Gln	~30 min
Tyr, Glu	~10 min
Pro	~7 min
Leu, Phe, Asp, Lys	~3 min
Arg	~2 min

Source: Modified from Bachmair, A., Finley, D., & Varshavsky, A. (1986) In vivo half-life of a protein is a function of its amino-terminal residue. *Science* **234,** 179–186.

* Half-lives were measured in yeast for a single protein that was modified so that in each experiment it had a different amino-terminal amino acid residue. (See Chapter 28 for a discussion of techniques used to engineer proteins with altered amino acid sequences.) Half-lives may vary for different proteins and in different organisms, but this general pattern appears to hold for all organisms: amino acids listed here as stabilizing when present at the amino terminus have a stabilizing effect on proteins in all cells.

Summary

Proteins are synthesized with a particular amino acid sequence through the translation of information encoded in messenger RNA by an RNA–protein complex called a ribosome. Amino acids are specified by informational units in the mRNA called codons. Translation requires adapter molecules, the transfer RNAs, which recognize codons and insert amino acids into their appropriate sequential positions in the polypeptide.

The codons for the amino acids consist of specific nucleotide triplets. The base sequences of the codons were deduced from experiments using synthetic mRNAs of known composition and sequence. The genetic code is degenerate: it has multiple code words for nearly all the amino acids. The third position in each codon is much less specific than the first and second and is said to wobble. The standard genetic code words are probably universal in all species, although some minor deviations exist in mitochondria and a few single-celled organisms. The initiating amino acid, N-formylmethionine in bacteria, is coded by AUG. Recognition of a particular AUG as the initiation codon requires a purine-rich initiating signal (the Shine–Dalgarno sequence) on the 5' side of the AUG. The triplets UAA, UAG, and UGA do not code for amino acids but are signals for chain termination. In some viruses two different proteins may be coded by the same nucleotide sequence but translated with different reading frames.

Protein synthesis occurs on the ribosomes. Bacteria have 70S ribosomes, with a large (50S) subunit and a small (30S) subunit. Ribosomes of eukaryotes are significantly larger and contain more proteins than do bacterial ribosomes.

In stage 1 of protein synthesis, amino acids are activated by specific aminoacyl-tRNA synthetases in the cytosol. These enzymes catalyze the formation of aminoacyl-tRNAs, with simultaneous cleavage of ATP to AMP and PP_i. The fidelity of protein synthesis depends to a large extent on the accuracy of this reaction, and some of these enzymes carry out proofreading steps at separate active sites. Transfer RNAs have 73 to 93 nucleotide units, several of which have modified bases. They have an amino acid arm with the terminal sequence CCA(3') to which an amino acid is esterified, an anticodon arm, a TψC arm, and a DHU arm; some tRNAs have a fifth or extra arm. The anticodon nucleotide triplet of tRNA is responsible for the specificity of interaction between the aminoacyl-tRNA and the complementary codon on the mRNA. The growth of polypeptide chains on ribosomes begins with the amino-terminal amino acid and proceeds by successive additions of new residues to the carboxyl-terminal end.

In bacteria, the initiating aminoacyl-tRNA in all proteins is N-formylmethionyl-tRNA^{fMet}. Initiation of protein synthesis (stage 2) involves formation of a complex between the 30S ribosomal subunit, mRNA, GTP, fMet-tRNA^{fMet}, two initiation factors, and the 50S subunit; GTP is hydrolyzed to GDP and P_i. In the subsequent elongation steps (stage 3), GTP and three elongation factors are required for binding the incoming aminoacyl-tRNA to the aminoacyl site on the ribosome. In the first peptidyl transfer reaction, the fMet residue is transferred to the amino group of the incoming aminoacyl-tRNA. Movement of the ribosome along the mRNA then translocates the dipeptidyl-tRNA from the aminoacyl site to the peptidyl site, a process requiring hydrolysis of GTP. After many such elongation cycles, synthesis of the polypeptide chain is terminated (stage 4) with the aid of release factors. A polysome consists of an mRNA molecule to which are attached several or many ribosomes, each independently reading the mRNA and forming a polypeptide. At least four high-energy phosphate bonds are required to generate each peptide bond, an energy investment required to guarantee fidelity of translation. In stage 5 of protein synthesis, polypeptides undergo folding into their active, three-dimensional forms. Many proteins also are further processed by posttranslational modification reactions.

After synthesis, many proteins are directed to particular locations in the cell. One targeting mechanism involves peptide signal sequences generally found at the amino terminus of newly synthesized proteins. In eukaryotes, one class of these signal sequences is recognized and bound by a large protein–RNA complex called the signal recognition particle (SRP). The SRP binds the signal sequence as soon as it appears on the ribosome and transfers the entire ribosome and incomplete polypeptide to the endoplasmic reticulum. Polypeptides with these signal sequences are moved into the lumen of the endoplasmic reticulum as they are synthesized; there they may be modified and moved to the Golgi complex, and then sorted and sent to lysosomes, the plasma membrane, or secretory vesicles. Other known targeting signals include carbohydrates (mannose-6-phosphate targets proteins to lysosomes) and three-dimensional structural features of the proteins called signal patches. Some proteins are imported into the cell by receptor-mediated endocytosis. These receptors are also used by some toxins and viruses to gain entry into cells.

Proteins are eventually degraded by specialized proteolytic systems present in all cells. Defective proteins and those slated for rapid turnover are generally degraded by an ATP-dependent proteolytic system. In eukaryotes, proteins to be broken down by this system are first tagged by linking them to a highly conserved protein called ubiquitin.

Further Reading

General

Hill, W.E., Dahlberg, A., Garrett, R.A., Moore, P.B., Schlessinger, D., & Warner, J.R. (1990) *The Ribosome: Structure, Function, and Evolution,* The American Society for Microbiology, Washington, DC.
Many good articles covering a wide range of topics.

Spirin, A.S. (1986) *Ribosome Structure and Protein Biosynthesis,* The Benjamin/Cummings Publishing Company, Menlo Park, CA.

The Genetic Code

Barrell, B.G., Air, G.M., & Hutchison, C.A., III (1976) Overlapping genes in bacteriophage φX174. *Nature* **264,** 34–41.

Crick, F.H.C. (1966) The genetic code: III. *Sci. Am.* **215** (October), 55–62.
An insightful overview of the genetic code at a time when the code words had just been worked out.

Fox, T.D. (1987) Natural variation in the genetic code. *Annu. Rev. Genet.* **21,** 67–91.

Hatfield, D. & Oroszlan, S. (1990) The *where, what* and *how* of ribosomal frameshifting in retroviral protein synthesis. *Trends Biochem. Sci.* **15,** 186–190.

Nirenberg, M.W. (1963) The genetic code: II. *Sci. Am.* **208** (March), 80–94.
A description of the original experiments.

Stuart, K. (1991) RNA editing in mitochondrial mRNA of trypanosomatids. *Trends Biochem. Sci.* **16,** 68–72.

Weiner, A.M. & Maizels, N. (1990) RNA editing: guided but not templated? *Cell* **61,** 917–920.

Protein Synthesis

Björk, G.R., Ericson, J.U., Gustafsson, C.E.D., Hagervall, T.G., Jönsson, Y.H., & Wikström, P.M. (1987) Transfer RNA modification. *Annu. Rev. Biochem.* **56,** 263–288.

Burbaum, J.J. & Schimmel, P. (1991) Structural relationships and the classification of aminoacyl-tRNA synthetases. *J. Biol. Chem.* **266,** 16965–16968.

Chapeville, F., Lipmann, F., von Ehrenstein, G., Weisblum, B., Ray, W.J., Jr., & Benzer, S. (1962) On the role of soluble ribonucleic acid in coding for amino acids. *Proc. Natl. Acad. Sci. USA* **48,** 1086–1092.
Classic experiments providing proof for Crick's adapter hypothesis and showing that amino acids are not checked after they are linked to tRNAs.

Clarke, S. (1992) Protein isoprenylation and methylation at carboxyl-terminal cysteine residues. *Annu. Rev. Biochem.* **61,** 355–386.

Dahlberg, A.E. (1989) The functional role of ribosomal RNA in protein synthesis. *Cell* **57,** 525–529.

Dintzis, H.M. (1961) Assembly of the peptide chains of hemoglobin. *Proc. Natl. Acad. Sci. USA* **47,** 247–261.
A classic experiment establishing that proteins are assembled beginning at the amino terminus.

Fersht, A. (1985) *Enzyme Structure and Mechanism,* W.H. Freeman and Company, New York.
See Chapter 13 for a discussion of proofreading in aminoacyl-tRNA synthetases.

Gualerzi, C.O. & Pon, C.L. (1990) Initiation of mRNA translation in prokaryotes. *Biochemistry* **29,** 5881–5889.

Lake, J.A. (1981) The ribosome. *Sci. Am.* **245** (August), 84–97.

Maden, B.E.H. (1990) The numerous modified nucleotides in eukaryotic ribosomal RNA. *Prog. Nucleic Acid Res. Mol. Biol.* **39,** 241–303.

Moldave, K. (1985) Eukaryotic protein synthesis. *Annu. Rev. Biochem.* **54,** 1109–1149.

Noller, H.F., Hoffarth, V., & Zimniak, L. (1992) Unusual resistance of peptidyl transferase to protein extraction procedures. *Science* **256,** 1416–1419.

Normanly, J. & Abelson, J. (1989) tRNA identity. *Annu. Rev. Biochem.* **58**, 1029–1049.

Rich, A. & Kim, S.H. (1978) The three-dimensional structure of transfer RNA. *Sci. Am.* **238**, (January), 52–62.

Riis, B., Rattan, S.I.S., Clark, B.F.C., & Merrick, W.C. (1990) Eukaryotic protein elongation factors. *Trends Biochem. Sci.* **15**, 420–424.

Schimmel, P. (1989) Parameters for the molecular recognition of transfer RNAs. *Biochemistry* **28**, 2747–2759.

Protein Targeting and Secretion

Bachmair, A., Finley, D., & Varshavsky, A. (1986) In vivo half-life of a protein is a function of its amino-terminal residue. *Science* **234**, 179–186.

Balch, W.E. (1989) Biochemistry of interorganelle transport: a new frontier in enzymology emerges from versatile in vitro model systems. *J. Biol. Chem.* **264**, 16965–16968.

Dahms, N.M., Lobel, P., & Kornfeld, S. (1989) Mannose 6-phosphate receptors and lysosomal enzyme targeting. *J. Biol. Chem.* **264**, 12115–12118.

Goldstein, J.L., Brown, M.S., Anderson, R.G.W., Russell, D.W., & Schneider, W.J. (1985) Receptor-mediated endocytosis: concepts emerging from the LDL receptor system. *Annu. Rev. Cell Biol.* **1**, 1–39.

Hershko, A. & Ciechanover, A. (1992) The ubiquitin system for protein degradation. *Annu. Rev. Biochem.* **61**, 761–807.

Hurt, E.C. & van Loon, A.P.G.M. (1986) How proteins find mitochondria and intramitochondrial compartments. *Trends Biochem. Sci.* **11**, 204–207.

Mellman, I., Fuchs, R., & Helenius, A. (1986) Acidification of the endocytic and exocytic pathways. *Annu. Rev. Biochem.* **55**, 663–700.

Meyer, D.I. (1988) Preprotein conformation: the year's major theme in translocation studies. *Trends Biochem. Sci.* **13**, 471–474.

Pfeffer, S.R. & Rothman, J.E. (1987) Biosynthetic protein transport and sorting by the endoplasmic reticulum and Golgi. *Annu. Rev. Biochem.* **56**, 829–852.

Pryer, N.K., Wuestehube, L.J. & Schekman, R. (1992) Vesicle-mediated protein sorting. *Annu. Rev. Biochem.* **61**, 471–516.

Randall, L.L. & Hardy, S.J.S. (1984) Export of protein in bacteria. *Microbiol. Rev.* **48**, 290–298.

Rapoport, T.A. (1990) Protein transport across the ER membrane. *Trends Biochem. Sci.* **15**, 355–358.

Rothman, J.E. (1985) The compartmental organization of the Golgi apparatus. *Sci. Am.* **253** (September), 74–89.

Schmidt, G.W. & Mishkind, M.L. (1986) The transport of proteins into chloroplasts. *Annu. Rev. Biochem.* **55**, 879–912.

Silver, P.A. (1991) How proteins enter the nucleus. *Cell* **64**, 489–497.

Ward, W.H.J. (1987) Diphtheria toxin: a novel cytocidal enzyme. *Trends Biochem. Sci.* **12**, 28–31.

Wickner, W.T. & Lodish, H.F. (1985) Multiple mechanisms of protein insertion into and across membranes. *Science* **230**, 400–407.

Problems

1. *Messenger RNA Translation* Predict the amino acid sequences of peptides formed by ribosomes in response to the following mRNAs, assuming that the initial codon is the first three bases in each sequence.

    (a) GGUCAGUCGCUCCUGAUU
    (b) UUGGAUGCGCCAUAAUUUGCU
    (c) CAUGAUGCCUGUUGCUAC
    (d) AUGGACGAA

2. *How Many mRNAs Can Specify One Amino Acid Sequence?* Write all the possible mRNA sequences that can code for the simple tripeptide segment Leu–Met–Tyr. Your answer will give you some idea as to the number of possible mRNAs that can code for one polypeptide.

3. *Can the Base Sequence of an mRNA Be Predicted from the Amino Acid Sequences of Its Polypeptide Product?* A given sequence of bases in an mRNA will code for one and only one sequence of amino acids in a polypeptide, if the reading frame is specified. From a given sequence of amino acid residues in a protein such as cytochrome *c*, can we predict the base sequence of the unique mRNA that coded for it? Give reasons for your answer.

4. *Coding of a Polypeptide by Duplex DNA*   The template strand of a sample of double-helical DNA contains the sequence

(5′)CTTAACACCCCTGACTTCGCGCCGTCG

(a) What is the base sequence of mRNA that can be transcribed from this strand?

(b) What amino acid sequence could be coded by the mRNA base sequence in (a), starting from the 5′ end?

(c) Suppose the other (nontemplate) strand of this DNA sample is transcribed and translated. Will the resulting amino acid sequence be the same as in (b)? Explain the biological significance of your answer.

5. *Methionine Has Only One Codon*   Methionine is one of the two amino acids having only one codon. Yet the single codon for methionine can specify both the initiating residue and interior Met residues of polypeptides synthesized by *E. coli*. Explain exactly how this is possible.

6. *Synthetic mRNAs*   How would you make a polyribonucleotide that could serve as an mRNA coding predominantly for many Phe residues and a small number of Leu and Ser residues? What other amino acid(s) would be coded for by this polyribonucleotide but in smaller amounts?

7. *The Direct Energy Cost of Protein Biosynthesis*   Determine the minimum energy cost, in terms of high-energy phosphate groups expended, required for the biosynthesis of the β-globin chain of hemoglobin (146 residues), starting from a pool including all necessary amino acids, ATP, and GTP. Compare your answer with the direct energy cost of the biosynthesis of a linear glycogen chain of 146 glucose residues in (α1→4) linkage, starting from a pool including glucose, UTP, and ATP (Chapter 19). From your data, what is the *extra* energy cost of imparting the genetic information inherent in the β-globin molecule?

8. *Indirect Costs of Protein Synthesis*   In addition to the direct energy cost for the synthesis of a protein, as developed in Problem 7, there are indirect energy costs—those required for the cell to make the necessary biocatalysts for protein synthesis. Contrast the relative magnitude of the indirect costs to a eukaryotic cell of the biosynthesis of linear (α1→4) glycogen chains versus the indirect costs of the biosynthesis of polypeptides. (Compare the enzymatic machinery used to synthesize proteins and glycogen.)

9. *Predicting Anticodons from Codons*   Most amino acids have more than one codon and will be attached to more than one tRNA, each with a different anticodon. Write all possible anticodons for the four codons for glycine: (5′)GGU, GGC, GGA, and GGG.

(a) From your answer, which of the positions in the anticodons are primary determinants of their codon specificity in the case of glycine?

(b) Which of these anticodon–codon pairings have a wobbly base pair?

(c) In which of the anticodon–codon pairings do all three positions exhibit strong Watson–Crick hydrogen bonding?

10. *The Effect of Single-Base Changes on Amino Acid Sequence*   Much important confirmatory evidence on the genetic code has come from the nature of single-residue changes in the amino acid sequence of mutant proteins. Which of the following single-residue amino acid replacements would be consistent with the genetic code? Which cannot be the result of single-base mutations? Why?

(a) Phe → Leu   (e) Ile → Leu
(b) Lys → Ala   (f) His → Glu
(c) Ala → Thr   (g) Pro → Ser
(d) Phe → Lys

11. *The Basis of the Sickle-Cell Mutation*   In sickle-cell hemoglobin there is a Val residue at position 6 of the β-globin chain, instead of the Glu residue found in this position in normal hemoglobin A. Can you predict what change took place in the DNA codon for glutamate to account for its replacement by valine?

12. *Importance of the "Second Genetic Code"*   Some aminoacyl-tRNA synthetases do not bind the anticodon of their cognate tRNAs but instead use other structural features of the tRNAs to impart binding specificity. The tRNAs for alanine apparently fall into this category. Describe the consequences of a C → G mutation in the third position of the anticodon of tRNA^{Ala}. What other kinds of mutations might have similar effects? Mutations of these kinds are never found in natural populations of any organism. Why? (Hint: Consider what might happen both to individual proteins and to the organism as a whole.)

13. *Maintaining the Fidelity of Protein Synthesis*   The chemical mechanisms used to avoid errors in protein synthesis are different from those used during DNA replication. DNA polymerases utilize a 3′→5′ exonuclease proofreading activity to remove mispaired nucleotides incorrectly inserted into a growing DNA strand. There is no analogous proofreading function on ribosomes; and, in fact, the identity of amino acids attached to incoming tRNAs and added to the growing polypeptide is never checked. A proofreading step that hydrolyzed the last peptide bond formed when an incorrect amino acid was inserted into a growing polypeptide (analogous to the proofreading step of DNA polymerases) would actually be chemically impractical. Why? (Hint: Consider how the link between the growing polypeptide and the mRNA is maintained during the elongation phase of protein synthesis; see Figs. 26–27 and 26–28.)

Regulation of Gene Expression

Of the 4,000 genes in the typical bacterial genome or the estimated 100,000 genes in the human genome, only a fraction are expressed at any given time. Some gene products have functions that mandate their presence in very large amounts. The elongation factors required for protein synthesis, for example, are among the most abundant proteins in bacteria. Other gene products are needed in much smaller amounts; for instance, a cell may contain only a few molecules of the enzymes that repair rare DNA lesions. Requirements for a given gene product may also change with time. The need for enzymes in certain metabolic pathways may wax or wane as food sources change or are depleted. During development in a multicellular eukaryote, some proteins that influence cellular differentiation are present for only a brief time in a small subset of an organism's cells. The specialization of some cells for particular functions can also dramatically affect the need for various gene products, one example being the uniquely high concentration of hemoglobin in erythrocytes.

The regulation of gene expression is a critical component in regulating cellular metabolism and in orchestrating and maintaining the structural and functional differences that exist in cells during development. Given the high energetic cost of protein synthesis, regulation of gene expression is essential if the cell is to make optimal use of available energy.

Regulating the concentration of a cellular protein involves a delicate balance of many processes. There are at least six potential points at which the amount of protein can be regulated (Fig. 27–1): synthesis of the primary RNA transcript, posttranscriptional processing of mRNA, mRNA degradation, protein synthesis (translation), posttranslational modification of proteins, and protein degradation. The concentration of a given protein is controlled by regulatory mechanisms at any or all of these points. Some of these mechanisms have been examined in previous chapters. Posttranscriptional modification of mRNAs by processes such as differential splicing (p. 873) or RNA editing (see Box 26–1) can affect which proteins are produced from an mRNA transcript and in what amounts. A variety of sequences can affect the rate at which an mRNA is degraded (p. 880). Many factors that affect the rate at which an mRNA is translated into a protein, as well as the posttranslational modification and eventual degradation of that protein, were described in Chapter 26.

Our primary focus in this chapter is the regulation of transcription initiation (although some aspects of the regulation of translation will

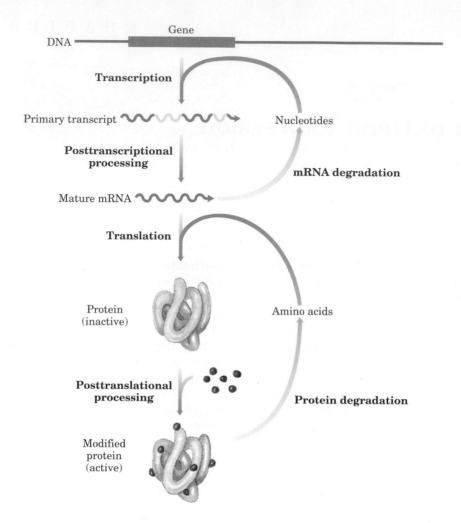

Figure 27–1 Six processes that affect the steady-state concentration of a protein. Each of these processes is a potential point of regulation.

be described). Of all the processes illustrated in Figure 27–1, regulation at the level of transcription initiation is the best documented and may be the most common. At least one important reason is clear: as for all biosynthetic pathways, the most efficient place for regulation is the first reaction in the pathway. In this way, unnecessary biosynthesis can be halted before energy is invested. Transcription initiation also is an excellent point at which to coordinate the regulation of multiple genes whose products have interdependent activities. For example, when DNA is heavily damaged, bacterial cells require a coordinated increase in the levels of many enzymes involved in DNA repair. Perhaps the most sophisticated form of coordination occurs in the complex regulatory circuits that guide the development of multicellular eukaryotes.

In this chapter, we first describe the interactions between proteins and DNA that are the key to transcriptional regulation. Specific proteins that regulate the expression of specific genes will then be discussed, first for prokaryotes and then for eukaryotes. In the course of this discussion we will examine several different mechanisms by which cells regulate gene expression and coordinate the expression of multiple genes.

Gene Regulation: Principles and Proteins

Just as the cellular requirements for different proteins vary, the mechanisms by which their respective genes are regulated also vary. The degree and type of regulation naturally reflect the function of the protein product of the gene. Some gene products are required all the time and their genes are expressed at a more or less constant level in virtually all the cells of a species or organism. Many of the genes for enzymes that catalyze steps in central metabolic pathways such as the citric acid cycle fall into this category. These genes are often referred to as **housekeeping genes.** Constant, seemingly unregulated expression of a gene is called **constitutive** gene expression. The amounts of other gene products rise and fall in response to molecular signals. Gene products that increase in concentration under prescribed molecular circumstances are referred to as inducible, and the process of increasing the expression of the gene is called **induction.** The expression of many genes encoding DNA repair enzymes, for example, is induced in response to high levels of DNA damage. Conversely, gene products that decrease in concentration in response to a molecular signal are referred to as repressible, and the decrease in gene expression is called **repression.** For example, the presence of ample supplies of the amino acid tryptophan leads to repression of the genes for the enzymes catalyzing tryptophan biosynthesis in bacteria.

Transcription is mediated and regulated by protein–DNA interactions. The central component is RNA polymerase, an enzyme described in some detail in Chapter 25. We begin here with a further description of RNA polymerase from the standpoint of regulation, then proceed to a general description of the proteins that modulate the activity of RNA polymerase. Finally we discuss the molecular basis for the recognition of specific DNA sequences by DNA-binding proteins.

The Activity of RNA Polymerase Is Regulated

RNA polymerases bind to DNA and initiate transcription at specific sites in the DNA called promoters (Chapter 25). Promoters generally are found very near the position where RNA synthesis begins on the DNA template. The regulation of transcription initiation is, in effect, regulation of the interaction of RNA polymerase with its promoter.

Promoters vary considerably in their nucleotide sequence, and this affects the binding affinity of RNA polymerases. The binding affinity in turn affects the frequency of transcription initiation. In *E. coli,* some genes are transcribed once each second whereas others are transcribed less than once per cell generation. Much of this variation is accounted for simply by differences in promoter sequences. In the absence of regulatory proteins, differences in the sequences of two promoters may affect the frequency of transcription initiation by factors of 1,000 or more. Recall (see Fig. 25–5) that *E. coli* promoters have a consensus sequence (Fig. 27–2). Promoters that exactly match the consensus se-

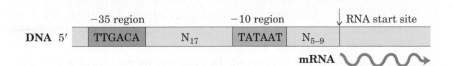

Figure 27–2 Consensus sequence for many *E. coli* promoters. N indicates any nucleotide. Most base substitutions in the −10 and −35 regions have a negative effect on promoter function. (Recall from Chapter 25 that by convention, DNA sequences are shown as they occur on the coding (nontemplate) strand.)

quence generally have the highest affinity for RNA polymerase and the highest frequency of transcription initiation. Mutations that change a consensus base pair to a nonconsensus pair generally decrease promoter function: mutations that change a nonconsensus base pair to a consensus pair usually enhance promoter function.

Although housekeeping genes are expressed constitutively, the proteins they encode are present in widely varying amounts. For these genes the RNA polymerase–promoter interaction is the only factor affecting transcription initiation, and differences in promoter sequences allow the cell to maintain the required level of each housekeeping protein.

Transcription initiation at the promoters of many genes that do not fall in the housekeeping category is further regulated in response to molecular signals. These promoters have a basal rate of transcription initiation (determined by the promoter sequence), superimposed on which is regulation mediated by several types of regulatory proteins. These proteins affect the interaction between RNA polymerase and the promoters.

Transcription Initiation Is Regulated by Proteins Binding to or near Promoters

At least three types of proteins regulate transcription initiation by RNA polymerase: (1) **specificity factors** alter the specificity of RNA polymerase for a given promoter or set of promoters; (2) **repressors** bind to a promoter, blocking access of RNA polymerase to the promoter; (3) **activators** bind near a promoter, enhancing the RNA–promoter interaction.

We encountered prokaryotic specificity factors in Chapter 25, although they were not given that name. The σ subunit (M_r 70,000) called σ^{70} of the E. coli RNA polymerase holoenzyme is a prototypical specificity factor that mediates specific promoter recognition and binding. Under some conditions, notably when the bacteria are subjected to heat stress, σ^{70} is replaced with another specificity factor (M_r 32,000) called σ^{32} (p. 863). When bound to σ^{32}, RNA polymerase does not bind to the standard E. coli promoters (Fig. 27–2), but instead is directed to a specialized set of promoters with the sequence structure shown in Figure 27–3. The promoters control the expression of a set of genes that make up the heat-shock response. Altering the polymerase to direct it to different promoters is one mechanism by which a set of related genes can be coordinately regulated. Other mechanisms will be encountered throughout this chapter.

Figure 27–3 Consensus sequence for promoters that regulate the expression of genes involved in the heat-shock response in E. coli. This system responds to temperature increases as well as some other environmental stresses, and it involves the induction of a set of proteins. Binding of RNA polymerase to heat-shock promoters is mediated by a specialized σ subunit of the enzyme called σ^{32}, which replaces σ^{70}.

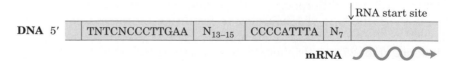

Repressors bind to specific sites in the DNA. In prokaryotes, the binding sites for repressors are called **operators.** Operator sites are generally near and often overlap the promoter so that RNA polymerase binding, or its movement along the DNA after binding, is blocked whenever the repressor is present. Regulation by means of a repressor

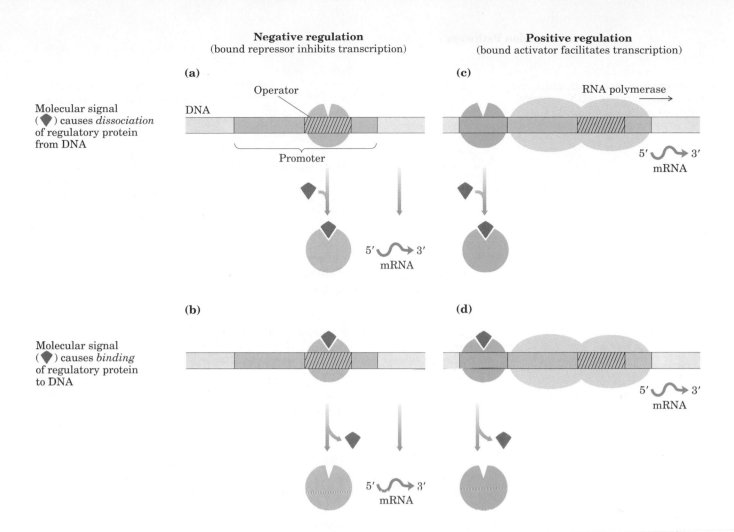

(a)

Molecular signal
(◆) causes *dissociation*
of regulatory protein
from DNA

Operator

DNA

Promoter

RNA polymerase

5′ ～ 3′
mRNA

5′ ～ 3′
mRNA

(b)

Molecular signal
(◆) causes *binding*
of regulatory protein
to DNA

(c)

(d)

5′ ～ 3′
mRNA

5′ ～ 3′
mRNA

protein that binds to DNA and blocks transcription is referred to as **negative regulation.** Repressor binding is regulated by a molecular signal, usually a specific small molecule that binds to and induces a conformational change in the repressor. The interaction between repressor and signal molecule may lead to either an increase or a decrease in transcription. In some cases the conformational change results in dissociation of a DNA-bound repressor from the operator (Fig. 27–4a). Transcription initiation can then proceed unhindered. In other cases the interaction between an inactive repressor and the signal molecule causes the repressor to bind to the operator (Fig. 27–4b).

Activators provide a molecular counterpoint to repressors. Regulation mediated by an activator is called **positive regulation.** Activators bind to sites adjacent to a promoter and enhance the binding and activity of RNA polymerase at that promoter. The binding sites for activators are often found adjacent to promoters that are normally bound weakly or not at all by RNA polymerase. Transcription at these genes is therefore often negligible in the absence of activator. Sometimes the activator is normally bound to DNA and dissociates when it binds to the signal molecule, often a specific small molecule or another protein (Fig. 27–4c). When bound to the DNA, the activator protein facilitates RNA polymerase binding and increases the rate of transcription initiation. In other cases the activator is not bound to the DNA until it also binds to a molecular signal (Fig. 27–4d). Positive regulation is particularly common in eukaryotes, as we shall see. We now turn to a fundamental unit of gene expression, the study of which gave rise to much of our current understanding of the regulation of gene expression.

Figure 27–4 Common patterns of regulation of transcription initiation. Two types of negative regulation are illustrated. **(a)** The repressor (red) is bound to the operator in the absence of the molecular signal; the signal causes dissociation of the repressor to permit transcription. **(b)** The repressor is bound in the presence of the signal; the repressor dissociates and transcription ensues when the signal is removed. Positive regulation is mediated by gene activators. **(c)** The activator (green) binds in the absence of the molecular signal and transcription proceeds; the activator dissociates and transcription is inhibited when the signal is added. **(d)** The activator binds in the presence of the signal; it dissociates only when the signal is removed. Note that "positive" and "negative" regulation are defined by the type of regulatory protein involved. In either case the addition of the molecular signal may increase or decrease transcription, depending on the effect of the signal on the regulatory protein.

Figure 27–5 An operon. Genes A, B, and C are transcribed on one polycistronic mRNA. Typical regulatory sequences include binding sites for proteins that either activate or repress transcription from the promoter.

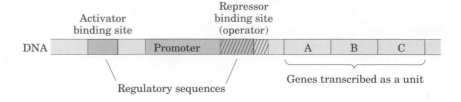

Many Prokaryotic Genes Are Regulated in Units Called Operons

Bacteria have a simple general mechanism for coordinating the regulation of genes whose products are involved in related processes: the genes are clustered on the chromosome and transcribed together. Most prokaryotic mRNAs are polycistronic. The single promoter required to initiate transcription of the cluster is the point where expression of all of the genes is regulated. The gene cluster, the promoter, and additional sequences that function in regulation are together called an **operon** (Fig. 27–5). Operons that include 2 to 6 genes transcribed as a unit are common; some operons contain 20 or more genes.

Many of the principles guiding the regulation of gene expression in bacteria were defined by studies of the regulation of lactose metabolism in *E. coli*. The disaccharide lactose can be used as the sole carbon source for the growth of *E. coli*. In 1960, François Jacob and Jacques Monod published a short paper in the *Proceedings of the French Academy of Sciences* demonstrating that two genes involved in lactose metabolism were coordinately regulated by a genetic element located adjacent to them. The genes were those for β-galactosidase, which cleaves lactose to galactose and glucose, and galactoside permease, which transports lactose into the cell (Fig. 27–6). The terms operon and operator were first introduced in this paper. The operon model that evolved from this and subsequent studies permitted biochemists to think about gene regulation in molecular terms for the first time.

The *lac* Operon Is Subject to Negative Regulation

The model for regulation of the lactose *(lac)* operon deduced from these studies is shown in Figure 27–7; it follows the pattern outlined in Figure 27–4a. In addition to the genes for β-galactosidase (Z) and galactoside permease (Y), the operon includes a gene for thiogalactoside transacetylase (A), whose physiological function is unknown. Each of the three genes is preceded by translational signals (not shown in Fig. 27–7) to guide ribosome binding and protein synthesis (Chapter 26). In

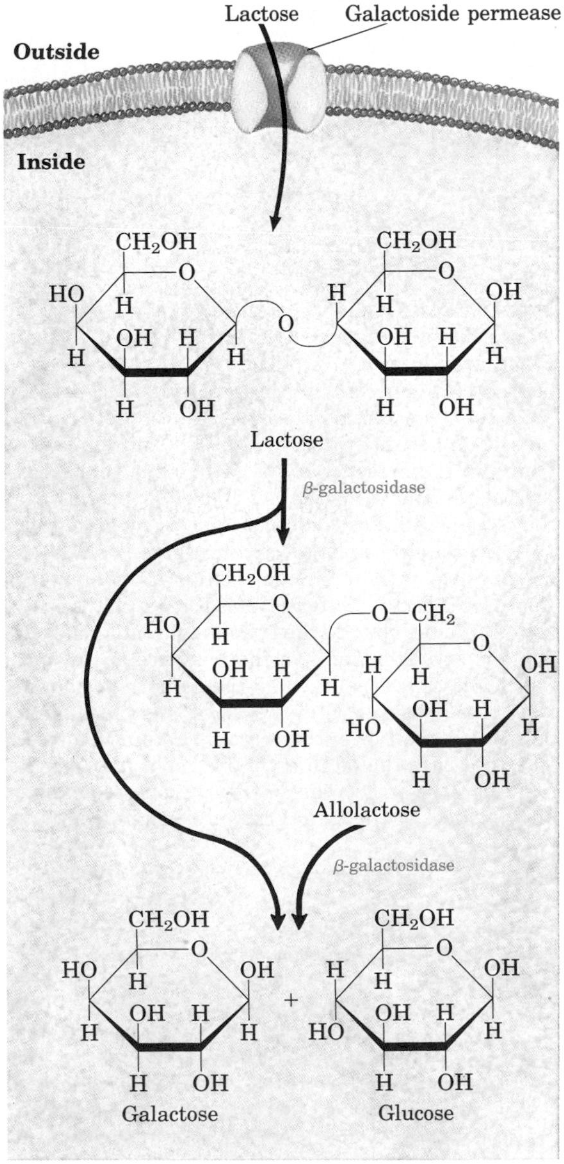

Figure 27–6 The activities of galactoside permease and β-galactosidase in lactose metabolism in *E. coli*. The conversion of lactose to allolactose by transglycosylation is a minor reaction catalyzed by β-galactosidase.

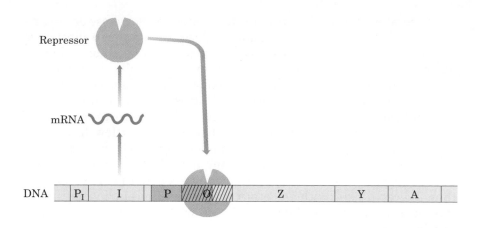

the absence of the substrate lactose, the *lac* operon genes are repressed, and β-galactosidase is present in only a few copies (a few molecules) per cell. Jacob and Monod found that mutations in the operator or in another gene called I led to constitutive synthesis of the *lac* operon gene products. When the I gene was defective, repression could be restored by introducing a functional I gene to the cell on another DNA molecule. This showed that the I gene encoded a diffusible molecule that caused gene repression; the molecule was later shown to be a protein, now called the Lac repressor. Repression is not absolute. Even in the repressed state each cell has a few copies of β-galactosidase and galactoside permease, presumably synthesized on the rare occasions when the repressor briefly dissociates from its DNA binding site (the operator).

When cells are provided with lactose, the *lac* operon is induced. An inducer molecule binds to a specific site on the repressor causing a conformational change in the repressor that results in its dissociation from the operator (Fig. 27–8). The inducer in this system is not lactose itself but an isomer of lactose called allolactose (Fig. 27–6). Lactose entering the *E. coli* cell is converted to allolactose in a reaction catalyzed by the few copies of β-galactosidase in the cell. Allolactose then binds to the Lac repressor. After the repressor dissociates, the *lac* operon genes are expressed and the concentration of β-galactosidase increases by a factor of 1,000.

Jacques Monod

François Jacob

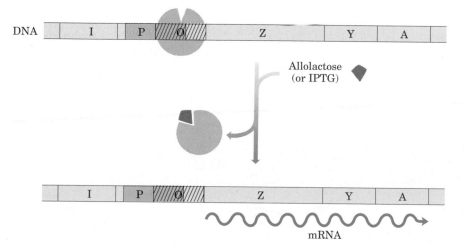

Figure 27–8 Induction of the *lac* operon in response to a molecular signal. Binding of allolactose to the Lac repressor causes a conformational change. The repressor dissociates from the operator, allowing transcription to proceed. Other β-galactosides, such as isopropylthiogalactoside (IPTG), can also act as inducers.

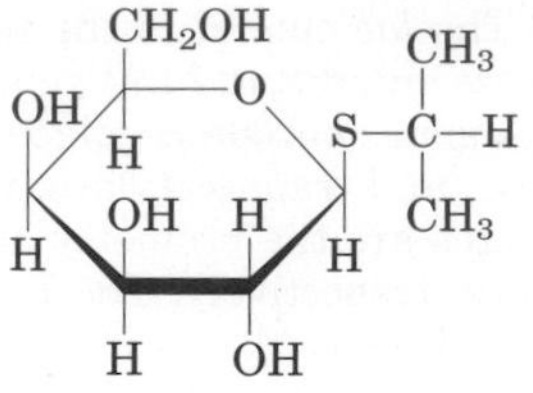

CH₂OH

Isopropylthiogalactoside
(IPTG)

Several β-galactosides structurally related to allolactose are inducers of, but not substrates for, β-galactosidase, and some are substrates but not inducers. One particularly effective and nonmetabolizable inducer of the *lac* operon often used experimentally is isopropyl-thiogalactoside (IPTG). Such nonmetabolized inducers permit the separation of the physiological function of lactose as a carbon source for growth from its function in the regulation of gene expression.

Many operons are now known in bacteria and a few have been found in lower eukaryotes. The mechanisms by which they are regulated can vary significantly from the simple model presented in Figure 27–7. Research has shown that even the *lac* operon is more complex than indicated here, with an activator protein also contributing to the overall scheme. The regulation of several well-studied bacterial operons, including *lac,* is described in more detail later in this chapter. We now consider the critical molecular interactions between DNA-binding proteins (e.g., repressors and activators) and the specific DNA sequences to which they bind.

Regulatory Proteins Have Discrete DNA-Binding Domains

Regulatory proteins generally bind to specific DNA sequences. They also bind to nonspecific DNA, but their affinity for their target sequences is generally 10^5 to 10^7 times higher. The molecular basis for this discrimination has been the subject of intensive investigation. A general conclusion is that regulatory proteins usually have discrete DNA-binding domains. In addition, the substructures within these domains that actually come in contact with the DNA fall into one of a rather small group of recognizable and characteristic structural motifs.

Before examining these protein structures, it is useful to consider the recognition surfaces on the DNA with which regulatory proteins must interact. Most of the groups that differ from one base to another and can therefore permit discrimination between base pairs are hydrogen-bond donor and acceptor groups exposed in the major DNA groove (Fig. 27–9). Most of the protein–DNA contacts that impart specificity are therefore hydrogen bonds. One notable exception is a nonpolar surface near C-5 of pyrimidines, where thymine is readily distinguished from cytosine by virtue of thymine's protruding methyl group (Fig. 27–9). Protein–DNA contacts are also possible in the minor groove of the DNA, but the hydrogen-bonding patterns here generally do not allow ready discrimination between different base pairs.

Figure 27–9 Functional groups on DNA base pairs in the major groove of DNA. The groups that can be used for base-pair recognition are shown in red for all four base pairs.

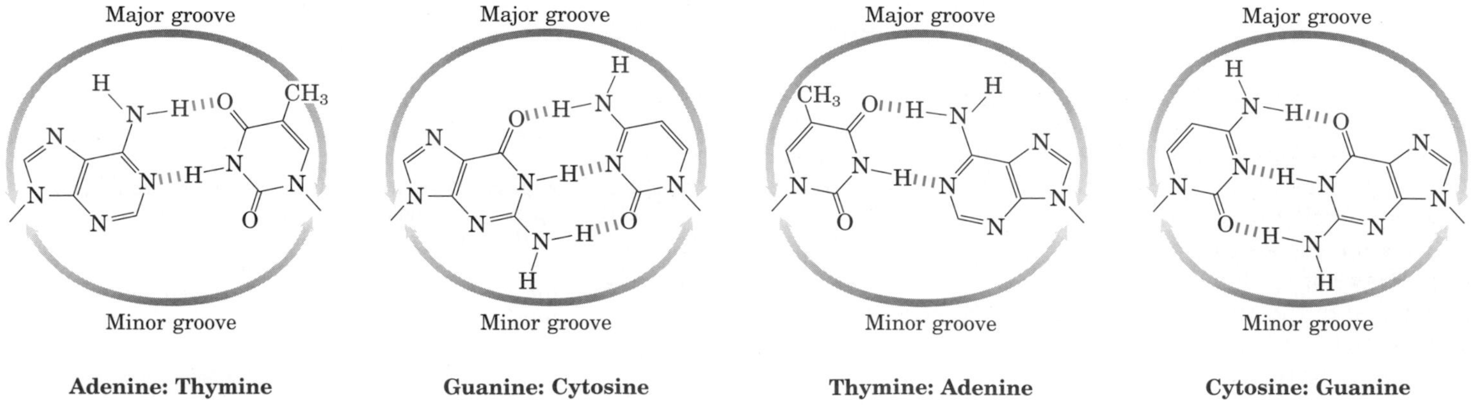

Adenine: Thymine   Guanine: Cytosine   Thymine: Adenine   Cytosine: Guanine

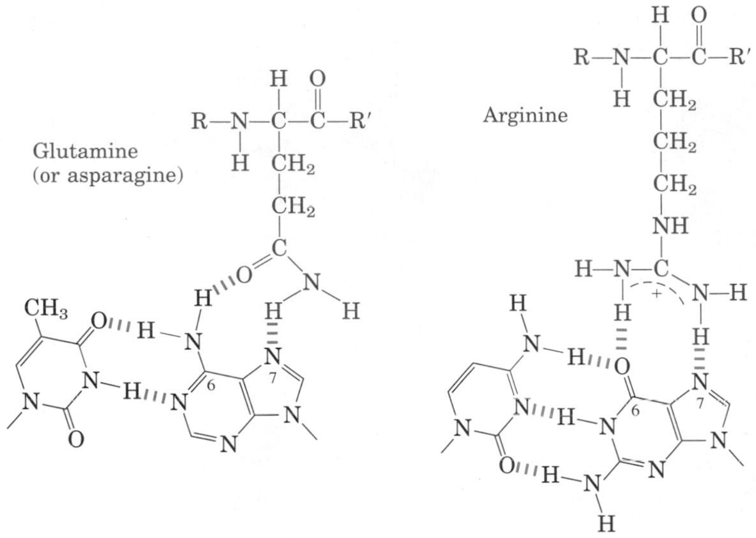

Figure 27–10 Two examples of specific amino acid–base pair interactions that have been observed in the structures of DNA-bound regulatory proteins.

Thymine: Adenine   **Cytosine: Guanine**

As for the regulatory proteins themselves, the amino acid residues whose side chains are most often found hydrogen-bonded to bases in the DNA include Asn, Gln, Glu, Lys, and Arg. Is there a simple "recognition code" in which an amino acid is always paired with a certain base? The two hydrogen bonds that can form between Gln or Asn and the N^6 and N-7 positions of adenine (Fig. 27–10) constitute a pattern that cannot form with any other base. An Arg residue can similarly form two hydrogen bonds to both N-7 and O^6 of guanine (Fig. 27–10). However, examination of the structures of many DNA-binding proteins has shown that there are multiple ways for a protein to recognize each base pair, and no simple code exists. The Gln–adenine interaction specifies A=T base pairs in some cases, whereas a van der Waals pocket for the methyl group of thymine is the mechanism used to recognize A=T base pairs in other proteins. It is not yet possible to examine the structure of a DNA-binding protein and infer the sequence of the DNA to which it binds.

The DNA-binding domains of regulatory proteins tend to be small (60 to 90 amino acid residues). Only a small subset of the amino acids within these domains actually contact the DNA, and the structure of the protein in the region where these amino acids occur is not random. Two structural motifs that play a major role in DNA binding have been found in numerous regulatory proteins: the **helix-turn-helix** motif and the **zinc finger.** Other DNA-binding motifs exist in some proteins, but the discussion here focuses on these well-studied examples.

The Helix-Turn-Helix This DNA-binding motif was the first to be studied in detail. It is the physical basis for protein–DNA interactions for many prokaryotic regulatory proteins. Closely related DNA-binding motifs also occur in some eukaryotic regulatory proteins. The helix-turn-helix motif consists of two short α-helical segments 7 to 9 amino acid residues long, separated by a β turn (about 20 amino acids total). This structure generally is not stable by itself, but it represents the reactive portion of the larger DNA-binding domain. One of the two α helices is referred to as the recognition helix, because it usually contains many of the amino acids that interact with the DNA; this helix is positioned in the major groove. The Cro repressor protein from bacteriophage 434 (a close relative of bacteriophage λ) provides a good example (Fig. 27–11).

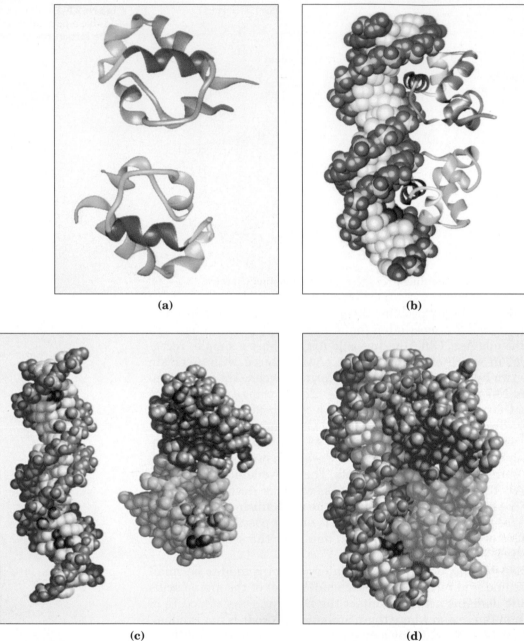

(a)   (b)

(c)   (d)

Figure 27–11 The Cro repressor of bacteriophage 434 and its interaction with DNA. Each subunit of this dimeric protein contains 71 amino acids. It is presented as a ribbon in **(a)** and **(b),** alone and complexed with its specific DNA binding site. The two subunits are shown in gray and light blue, except for the helix-turn-helix motif in each which is shown in red and yellow. The red helices are the recognition helices, which are positioned in adjacent major grooves of the DNA as seen in **(b).** The interactions between protein and DNA that allow this repressor to discriminate between its specific DNA binding site (shown here) and other DNA sequences are illustrated in **(c)** and **(d).** The protein subunits are again shown in gray and light blue; chemical groups on both the DNA and protein that interact through hydrogen bonds or van der Waals (hydrophobic) interactions are highlighted in red and orange, respectively. Discrimination is mediated by interactions between each protein subunit and four bases (the DNA binding site is a palindrome, and the interactions are the same for both subunits). One set of hydrogen bonds is formed between a Gln residue and the N^6 and N-7 of an adenine (see Fig. 27–10); another hydrogen bond is formed between the O^6 of a guanine and another Gln. In addition, van der Waals pockets on each subunit bind to the C-5 methyl groups of two adjacent thymines. The complementary interacting groups are evident in **(c),** and the complex is shown in **(d).** Many nonspecific contacts (not shown) also exist between protein and DNA in this complex. These do not contribute to discrimination between DNA sequences, but do contribute to the overall DNA-binding affinity. An interesting feature of this structure is that the DNA is bent slightly when it is bound. This occurs in the binding of many proteins to DNA (see Fig. 27–16.)

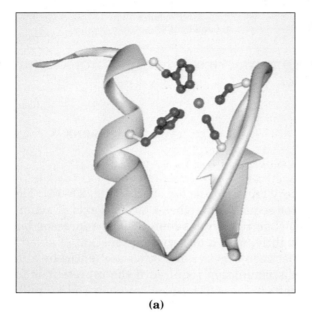

(a)

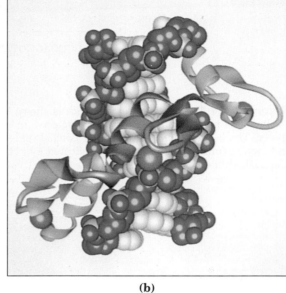

(b)

Figure 27–12 Zinc fingers. **(a)** A ribbon representation of a single zinc finger derived from the regulatory protein Zif 268. The zinc atom is in orange and the amino acid residues that coordinate it (two His and two Cys) are shown in red. **(b)** Three zinc fingers (light blue and gray) from Zif 268 are shown complexed with DNA. The zinc atoms are again shown in orange.

The Zinc Finger Zinc fingers consist of about 30 amino acid residues; four of the residues, either four Cys or two Cys and two His, coordinate a single Zn^{2+} atom (Fig. 27–12). This structural motif is found in many eukaryotic DNA-binding proteins, with several often present in a single protein. There are few, if any, known examples among prokaryotic proteins. Bacteriophage T4 has a protein, the gene 32 protein, that binds single-stranded DNA. It binds a single zinc atom within a structure that may be similar to a zinc finger. An apparent record is held by a DNA-binding protein derived from the frog *Xenopus*, which has 37 zinc fingers. The precise manner in which proteins containing zinc fingers bind to DNA may vary from one protein to the next. In some cases these structures contain the amino acid residues that are involved in sequence discrimination; in other cases the zinc fingers appear to bind DNA nonspecifically, and the amino acids required for specificity are found elsewhere in the protein. The interaction of three zinc fingers (derived from a mouse regulatory protein called Zif 268) with DNA is shown in Figure 27–12b. It should be noted that some regulatory proteins contain zinc bound within structures that are distinct from the zinc finger.

Regulatory Proteins Also Interact with Other Proteins

Regulatory proteins generally contain additional domains that are involved in interactions with RNA polymerase, other regulatory proteins, or additional copies of the same regulatory protein (Fig. 27–13). The DNA binding sites for regulatory proteins are generally inverted repeats of a short DNA sequence (a palindrome) at which two or four copies of a regulatory protein bind cooperatively, as in Figures 27–11 and 27–13.

The Lac repressor is a tetramer of identical subunits (M_r 37,000). A wild-type *E. coli* cell generally contains about ten copies of Lac repressor. The *i* gene is transcribed from its own promoter independently of the *lac* operon genes (Fig. 27–7). The repressor binds to a palindromic operator sequence that spans 22 base pairs within the larger regula-

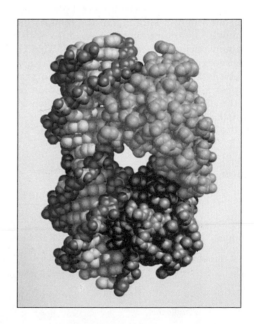

Figure 27–13 The bacteriophage λ repressor bound to DNA. The two identical subunits of the dimeric protein are shown in gray and light blue.

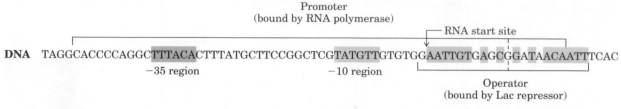

Promoter
(bound by RNA polymerase)

┌─ RNA start site

DNA TAGGCACCCCAGGCTTTACACTTTATGCTTCCGGCTCGTATGTTGTGTGGAATTGTGAGCGGATAACAATTTCAC

−35 region   −10 region

Operator
(bound by Lac repressor)

mRNA

Figure 27–14 The *lac* operator sequence is shown to illustrate its position relative to the *lac* promoter. The bases shaded beige exhibit twofold (palindromic) symmetry about the axis indicated by the dashed line.

tory region of the *lac* operon (Fig. 27–14). The symmetry of the operator sequence matches a twofold axis of symmetry in the arrangement of Lac repressor subunits. The repressor binds to the operator very tightly, with a dissociation constant of about 10^{-13} M. It discriminates between this site and other sequences by a factor of 4×10^6, a level of discrimination required if the repressor is to locate and bind specifically to these 22 base pairs among the 4.7 million or so base pairs in the *E. coli* chromosome. DNA binding is mediated by a helix-turn-helix motif within the DNA-binding domain of each repressor subunit.

Many other regulatory proteins bind to DNA as dimers, including many eukaryotic gene activators called **transcription factors.** In addition to DNA-binding domains (which often contain zinc fingers), many transcription factors have structural domains devoted to the protein–protein interactions involved in dimer formation (these dimers consist of identical or closely related proteins, as described later). Dimer formation is generally a prerequisite for DNA binding. As is the case for DNA-binding motifs, the protein structural motifs that mediate protein–protein interactions tend to fall into one of several common and recognizable patterns. Two such structural motifs that have been well characterized are the **leucine zipper** and the **basic helix-loop-helix.**

The Leucine Zipper This motif is an amphipathic α helix with hydrophobic amino acids concentrated on one side (Fig. 27–15). The hydrophobic surface is the point of contact between the two proteins making up the dimer. A striking feature of these α helices is the frequent appearance of Leu residues; they tend to occur as every seventh amino acid residue, which has the effect of arranging them in a straight line on the hydrophobic side of the α helix (Fig. 27–15). An early model for

Figure 27–15 Leucine zippers. (a) Comparison of amino acid sequences (shown here using the single-letter amino acid representations) of several leucine zipper proteins. Note the Leu (L) residues occurring every seventh residue in the dimerization region, and the number of Lys (K) and Arg (R) residues in the DNA-binding domain. (b) A leucine zipper from the yeast activator protein GCN4. Only the "zippered" α helices (white and light blue), derived from different subunits of the dimeric protein, are shown. The interacting Leu residues are shown in red and purple. All other amino acid side chains are represented by gray balls in order to expose the Leu–Leu interactions. (c) In this representation, the zipper shown in (b) has been opened up to show the alignment of Leu residues on each α helix. (d) The leucine zipper of (b) is shown with all of the amino acid side chains displayed.

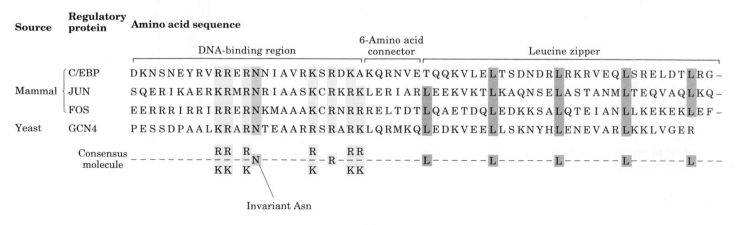

Source	Regulatory protein	Amino acid sequence
		DNA-binding region / 6-Amino acid connector / Leucine zipper
Mammal	C/EBP	DKNSNEYRVRRERNNIAVRKSRDKAKQRNVETQQKVLELTSDNDRLRKRVEQLSRELDTLRG–
	JUN	SQERIKAERKRMRNRIAASKCRKRKLERIARLEEKVKTLKAQNSELASTANMLTEQVAQLKQ–
	FOS	EERRRIRRIRRERNKMAAAKCRNRRRELTDTLQAETDQLEDKKSALQTEIANLLKEKEKLEF–
Yeast	GCN4	PESSDPAALKRARNTEAARRSRARKLQRMKQLEDKVEELLSKNYHLENEVARLKKLVGER
	Consensus molecule	– – – – – – – – RR R – – – – – R – RR – – – – L – – – – – L – – – – – L – – – – L – – – – – L – – –
		KK K N K KK

Invariant Asn

(a)

protein–protein interactions between these helices envisioned the Leu residues interdigitating, hence the name "leucine zipper." It is now known that the Leu residues on the two proteins line up side by side, and that the interacting α helices coil around each other (a coiled coil) as shown in Figure 27–15b. In regulatory proteins with leucine zippers, the DNA-binding domain is often found in an extension of the leucine-zipper α helix, which contains a high concentration of basic (Lys/Arg) residues. Leucine zippers occur in a wide range of eukaryotic regulatory proteins, and a few examples have also been found in prokaryotic proteins.

The Basic Helix-Loop-Helix A second group of eukaryotic regulatory proteins, many of which have been implicated in the control of gene expression that directs the development of multicellular organisms, share a conserved region of about 50 amino acid residues that include the determinants for both DNA binding and protein dimerization. This region can form two short amphipathic α helices linked by a variable length "loop." This helix-loop-helix (distinct from the helix-turn-helix motif associated with DNA binding) in one protein interacts with its counterpart in another to form dimers. DNA binding is again mediated by a short stretch of amino acids, rich in basic residues, that is immediately adjacent to the helix-loop-helix.

Within both classes of eukaryotic transcription factors described above, several families have been defined that are closely related in a structural sense. Within families, dimers may form between two identical proteins (a homodimer) or between two different members of the family (a heterodimer). A hypothetical family of four different leucine zipper proteins may thus form up to ten different dimeric species. In many cases the different combinations appear to have distinct regulatory and functional properties.

In addition to structural domains devoted to DNA binding and dimerization (or oligomerization), many regulatory proteins must interact with RNA polymerase and/or with other unrelated regulatory proteins. At least three different types of additional domains for protein–protein interaction have been characterized: (1) glutamine-rich, (2) proline-rich, and (3) acidic domains, the names reflecting amino acid residues that are unusually abundant. How these three structures

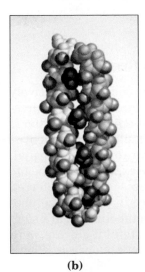

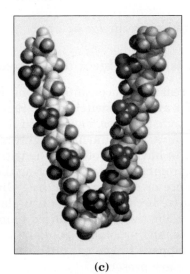

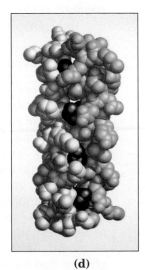

(b)   **(c)**   **(d)**

mediate protein–protein interactions is poorly understood. The various structural motifs for DNA binding and protein–protein interactions will be expanded upon with the aid of specific examples later in this chapter.

We now turn to a more detailed description of the regulatory schemes used by several prokaryotic and eukaryotic gene systems. Simple protein–DNA binding interactions are transformed into the intricate regulatory circuits so important to virtually every cellular function.

Regulation of Gene Expression in Prokaryotes

As in many other areas of biochemical investigation, the study of the regulation of gene expression was advanced earlier and faster in bacteria than in other experimental organisms. The examples of bacterial gene regulation presented here are chosen from among scores of well-studied systems, in part because of their historical significance, but primarily because they provide a good overview of the range of regulatory mechanisms employed in bacteria. Many of the lessons learned here have proved relevant to the regulation of gene expression in eukaryotes.

The lactose, arabinose, and tryptophan operons feature a representative group of regulatory proteins, but the overall regulatory mechanisms exhibited by these systems are very different. A short discussion of the SOS response in *E. coli* provides an important example of a system in which many genes scattered throughout the genome are coordinately regulated. A new level of complexity is encountered in the bacteriophage λ system. The regulatory circuit in this system oversees a molecular choice between two biological fates and provides a good model for a regulated developmental switch.

The final two systems described here represent major deviations from the focus on regulatory proteins that bind to DNA and help illustrate the diversity of mechanisms in gene regulation. The regulation of ribosomal protein synthesis focuses not on transcription but on translation; many of the proteins that coordinate the synthesis of ribosomal proteins bind to RNA rather than DNA. The regulation of phase variation in *Salmonella* provides an example of another novel mechanism for regulation: control of transcription initiation by means of genetic recombination.

To begin, we return to the *lac* operon.

The *lac* Operon Is Also Subject to Positive Regulation

The operator–repressor–inducer interactions provide an intuitively satisfying model for an on/off switch in the regulation of gene expression. Years of research, however, have shown that operon regulation is rarely so simple. Even a bacterium has a highly complex environment—too complex for it to have sets of genes sensitive to only one signal. Another major environmental factor affecting the expression of the *lac* genes is the presence or absence of glucose. Glucose is the preferred cellular energy source because of its central place in cellular metabolism. Hence, expressing the genes required to metabolize sugars such as lactose, galactose, and arabinose would be wasteful if glucose were abundant. What happens to the expression of the *lac* operon if both glucose and lactose are present?

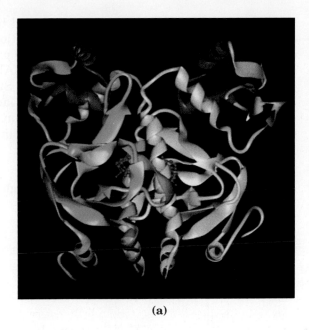

(a)

(b)

Figure 27–16 The three-dimensional structure of the CAP homodimer. **(a)** A ribbon representation with subunits shown in white and light blue. The helix-turn-helix DNA-binding motif is shown in red. Bound molecules of cAMP are shown in dark blue. **(b)** A space-filling representation of a CAP homodimer bound to DNA. Base pairs recognized by the protein are shown in green, and amino acid side chains that bind to these base pairs are shown in red. Note the bending of the DNA around the protein.

Another regulatory mechanism, called catabolic repression, has evolved to keep the genes for catabolism of lactose, arabinose, and other sugars repressed in the presence of glucose even when these secondary sugars are also present. The repressive effect of glucose is mediated by cAMP and a protein called **catabolite gene activator protein,** abbreviated **CAP.** CAP is a homodimer (subunit M_r 22,000) with binding sites for DNA and cAMP. Binding is mediated by a helix-turn-helix motif within the DNA-binding domain of the protein (Fig. 27–16). When glucose is absent, CAP binds to a specific site near the *lac* promoter (Fig. 27–17a) and stimulates RNA transcription 50-fold. CAP is therefore a positive regulatory element responsive to glucose levels, whereas the Lac repressor is a negative regulatory element responsive to lactose. The two act in concert; CAP has little effect on the system when the Lac repressor is blocking transcription, and dissociation of the repressor from the operator has little effect unless CAP is present to facilitate transcription. Stimulation by CAP is necessary because the wild-type *lac* promoter is a relatively weak promoter (Fig. 27–17b); the open complex of RNA polymerase and the promoter (Fig. 25–6) does not form readily unless CAP is present.

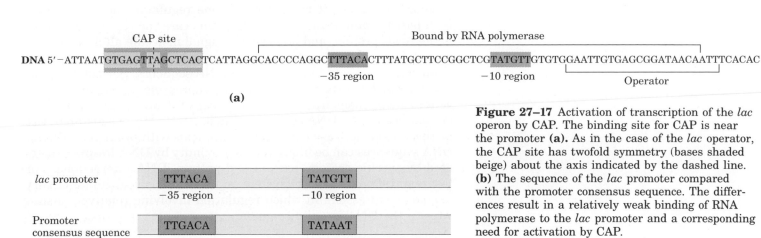

(a)

lac promoter	TTTACA	TATGTT
	−35 region	−10 region
Promoter consensus sequence	TTGACA	TATAAT

(b)

Figure 27–17 Activation of transcription of the *lac* operon by CAP. The binding site for CAP is near the promoter **(a).** As in the case of the *lac* operator, the CAP site has twofold symmetry (bases shaded beige) about the axis indicated by the dashed line. **(b)** The sequence of the *lac* promoter compared with the promoter consensus sequence. The differences result in a relatively weak binding of RNA polymerase to the *lac* promoter and a corresponding need for activation by CAP.

The effect of glucose on CAP is mediated by cAMP (Fig. 27–18). CAP binding occurs when cAMP concentrations are high and the cAMP-binding site on CAP is occupied. In the presence of glucose, the concentration of cAMP declines, preventing CAP binding and thereby decreasing the expression of the *lac* operon. Strong induction of the operon therefore requires both the presence of lactose (to inactivate the repressor) and the absence or low concentration of glucose (to increase the cAMP concentration and facilitate CAP binding).

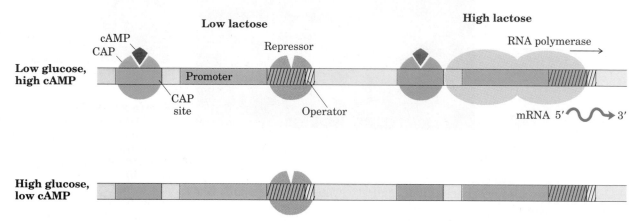

Figure 27–18 Combined effects of glucose and lactose on expression of the *lac* operon. Efficient transcription occurs only when lactose concentrations are high and glucose concentrations are low.

CAP and cAMP are involved in the coordinated regulation of many operons, primarily those that encode enzymes for the metabolism of other secondary sugars such as galactose and arabinose. Such a network of operons with a common regulator is called a **regulon.** Regulons provide a mechanism for large concerted changes in cellular functions in response to environmental changes, and they can control the action of hundreds of genes. Other examples of regulons are the heat-shock gene system that responds to changes in temperature (p. 944) and the genes that are induced in *E. coli* as part of the SOS response to DNA-damaging agents, described later in this chapter.

The *ara* Operon Undergoes Both Positive and Negative Regulation by a Single Regulatory Protein

A more complex regulatory scheme is found in the arabinose *(ara)* operon of *E. coli*. This system introduces several additional regulatory mechanisms. First, it is possible for one regulatory protein to exact both positive and negative control. In this case the regulatory protein is the AraC protein, and binding of a signal molecule alters its conformation from a repressor form that binds one DNA regulatory sequence to an activator that binds a different DNA sequence. Second, the AraC protein regulates its own synthesis by repressing transcription of its gene. This phenomenon is called **autoregulation.** Finally, the effects of some regulatory DNA sequences can be exerted from a distance; that is, these sequences are not always contiguous with promoters. Distant DNA sequences can be brought into proximity by **DNA looping,** mediated by specific protein–protein and protein–DNA interactions. This last feature makes the *ara* system an important paradigm for eukaryotic gene expression, in which regulation involving relatively distant sites in the DNA is quite common.

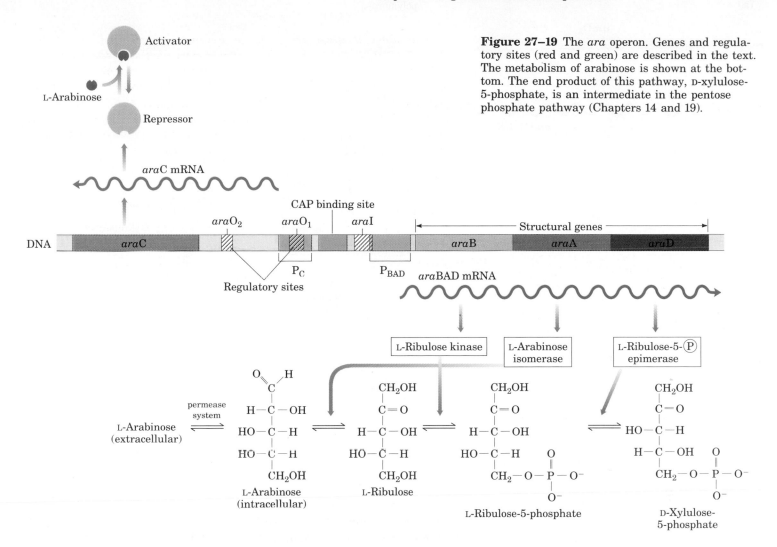

Figure 27–19 The *ara* operon. Genes and regulatory sites (red and green) are described in the text. The metabolism of arabinose is shown at the bottom. The end product of this pathway, D-xylulose-5-phosphate, is an intermediate in the pentose phosphate pathway (Chapters 14 and 19).

E. coli can use arabinose as a carbon source by converting it into xylulose-5-phosphate, an intermediate in the pentose phosphate pathway (Chapters 14 and 19). This requires the enzymes arabinose isomerase, ribulose kinase, and ribulose-5-phosphate epimerase (Fig. 27–19) encoded by the genes *ara*A, *ara*B, and *ara*D, respectively. The *ara* operon (Fig. 27–19) includes these three genes, a regulatory site including two operators ($araO_1$ and $araO_2$), another binding site for the AraC regulatory protein called *ara*I (I for inducer), and a promoter adjacent to *ara*I. The *ara*C gene is nearby and transcribed from its own promoter (near $araO_1$) in the opposite direction from the *ara*B, A, and D genes (called collectively, *ara*BAD). A CAP binding site is adjacent to the *ara* operon promoter, and transcription is modulated by CAP–cAMP as in the *lac* system. At this point the similarities largely end.

The role of AraC protein in the regulation of this system is complex (Fig. 27–20). First, it regulates its own synthesis, binding at $araO_1$ and repressing transcription of the *ara*C gene when its concentration exceeds about 40 copies per cell. Second, it acts as both a positive and a negative regulator of the *ara*BAD genes, and in this capacity it binds to $araO_2$ and *ara*I. This regulation can be summarized in the form of four metabolic scenarios: (1) Glucose is abundant and arabinose is not.

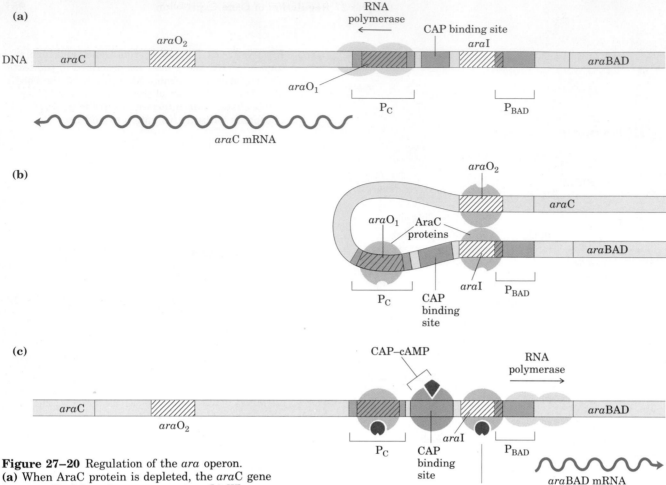

Figure 27–20 Regulation of the *ara* operon. **(a)** When AraC protein is depleted, the *ara*C gene is transcribed from its own promoter. **(b)** When arabinose levels are low and glucose levels high, AraC protein binds to both *ara*I and *ara*O₂ and brings these sites together to form a DNA loop. The operon is repressed in this state. AraC protein also binds to *ara*O₁, repressing further synthesis of AraC. **(c)** When arabinose is present and glucose concentration is low, AraC protein binds arabinose and changes conformation to become an activator. The DNA loop is opened, and the AraC protein acts in concert with CAP–cAMP to facilitate transcription.

Under these conditions, the AraC protein bound to $ara\text{O}_2$ and that bound to araI bind to each other, forming a DNA loop of about 210 base pairs. In this configuration the system represses transcription from the promoter for the araBAD genes (Fig. 27–20b). (2) Glucose is not present (or is at low levels) but arabinose is available. Under these conditions, CAP–cAMP becomes abundant and binds to its site adjacent to araI. Arabinose also binds to the AraC protein, altering its conformation. The DNA loop is opened, and the AraC protein bound at araI now becomes an activator, acting in concert with CAP–cAMP to induce transcription of the araBAD genes (Fig. 27–20c). (3) Arabinose and glucose are both abundant. (4) Arabinose and glucose are both absent. For both (3) and (4), the status of the system is not entirely clear, but it remains repressed in both cases. The *ara* operon is a complex regulatory system that provides rapid and reversible responses to changes in environmental conditions.

Genes for Amino Acid Biosynthesis Are Regulated by Transcription Attenuation

Amino acids are required in large amounts for protein synthesis, and *E. coli* has enzymes for synthesizing all of them. Not surprisingly, the genes for the enzymes needed to synthesize a given amino acid are generally clustered in an operon. These enzymes are needed, and hence the operon corresponding to an amino acid is expressed, whenever existing supplies of the amino acid are inadequate for cellular requirements. When the amino acid is in abundant supply, the biosynthetic enzymes are no longer needed and the operon is repressed.

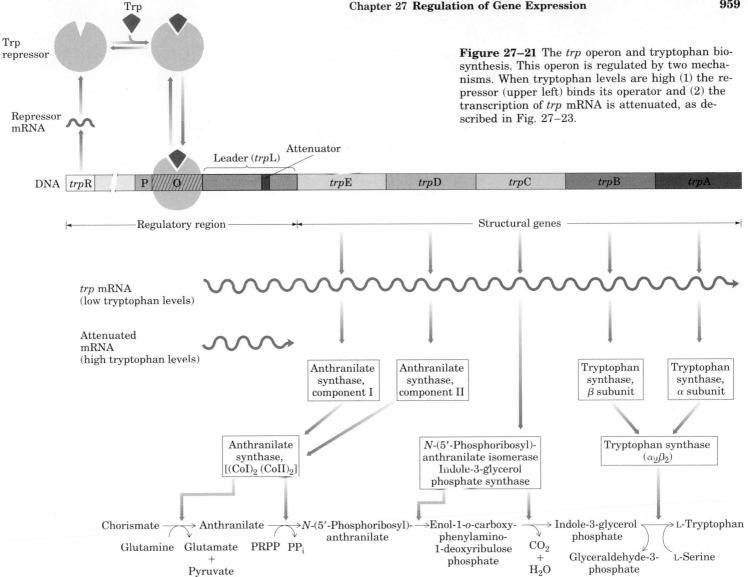

Figure 27–21 The *trp* operon and tryptophan biosynthesis. This operon is regulated by two mechanisms. When tryptophan levels are high (1) the repressor (upper left) binds its operator and (2) the transcription of *trp* mRNA is attenuated, as described in Fig. 27–23.

A well-defined example is the *E. coli* tryptophan *(trp)* operon, which includes five genes for the enzymes required to convert chorismate into tryptophan (Fig. 27–21). The mRNA from the *trp* operon has a half-life of only about 3 min, allowing the cell to respond rapidly to changing needs for this amino acid. The Trp repressor is a homodimer, with each subunit containing 107 amino acid residues (Fig. 27–22). When tryptophan is abundant, it binds to the Trp repressor, causing a conformational change that permits the repressor to bind its operator. The *trp* operator site overlaps the promoter, and binding of the repressor blocks binding of RNA polymerase.

Here, as elsewhere, this simple "on/off" circuit mediated by a repressor is not the entire regulatory story. This system responds to different tryptophan concentrations by varying the rate of synthesis of the biosynthetic enzymes over a 700-fold range. Once repression is lifted and transcription begins, the rate of transcription is fine-tuned by a second regulatory process called transcription attenuation.

Transcription attenuation describes a process in which transcription is initiated normally but is abruptly halted *before* the operon genes are transcribed. The frequency with which transcription is attenuated depends on the available concentration of tryptophan. The basis for the mechanism, as worked out by Charles Yanofsky, is the very close coupling between transcription and translation in bacteria.

Figure 27–22 Structure of the Trp repressor. The dimeric protein is shown with the helix-turn-helix DNA-binding motifs in red and bound molecules of tryptophan in blue.

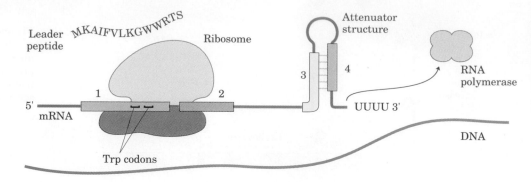

When tryptophan levels are high, the ribosome quickly translates sequence 1 (open reading frame encoding leader peptide) and blocks sequence 2 before sequence 3 is transcribed. Continued transcription leads to attenuation at the terminator-like structure formed by sequences 3 and 4.

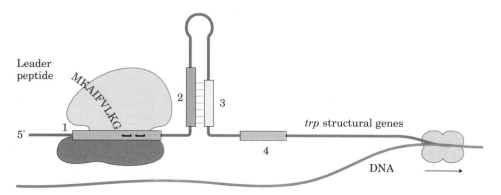

When tryptophan levels are low, the ribosome pauses at the Trp codons in sequence 1. Formation of the paired structure between sequences 2 and 3 prevents attenuation because sequence 3 is no longer available to form the attenuator structure with sequence 4.

(a)

Figure 27–23 (a) The attenuation mechanism in the *trp* operon involves four short sequences within a 162 nucleotide mRNA leader called *trp*L, preceding the *trp*E gene. Sequence 1 is a short open reading frame that encodes a small peptide called the leader peptide, translated immediately after transcription begins. Sequences 2 and 3 are complementary, as are sequences 3 and 4. The attenuator formed by sequences 3 and 4 has a structure and function similar to those of a transcription terminator (see Fig. 25–8). Note that the leader peptide itself has no other cellular function. Translation of its open reading frame has a purely regulatory function that determines which complementary sequences (2:3 or 3:4) are paired. **(b)** Sequence of the *trp* mRNA leader (*trp*L). **(c)** Pairing schemes for the interacting regions of the *trp* mRNA leader.

The *trp* operon attenuation mechanism uses signals encoded in four sequences within a 162 nucleotide **leader** region at the 5′ end of the mRNA that precedes the initiation codon of the first gene (Fig. 27–23a, b). At the end of the leader is a sequence called the **attenuator,** made up of sequences 3 and 4. Sequences 3 and 4 base-pair to form a G≡C-rich stem and loop structure followed by a series of uridylate residues, a structure that resembles a transcription terminator (Fig. 27–23c); *transcription* will halt here when this structure forms. Formation of the attenuator stem and loop depends on events occurring during the *translation* of a short open reading frame in the leader RNA that encodes a peptide of 14 amino acids (called the leader peptide because it is encoded within the leader region of the mRNA). This short gene is another of the four regulatory sequences (sequence 1 in Fig. 27–23a, b). Translation of the leader peptide begins immediately after it is transcribed, and the bound ribosome follows closely behind the RNA polymerase as transcription proceeds. Regulatory sequence 2 occurs near the end of the short open reading frame, and sequences 3 and 4, as noted above, form the attenuator stem and loop. Sequence 2 is an alternative complement for sequence 3. If sequences 2 and 3 base-pair, the attenuator structure derived from the interaction of sequences 3 and 4 cannot form and transcription continues into the *trp* biosynthetic genes (the loop formed by the pairing of sequences 2 and 3 does not obstruct transcription).

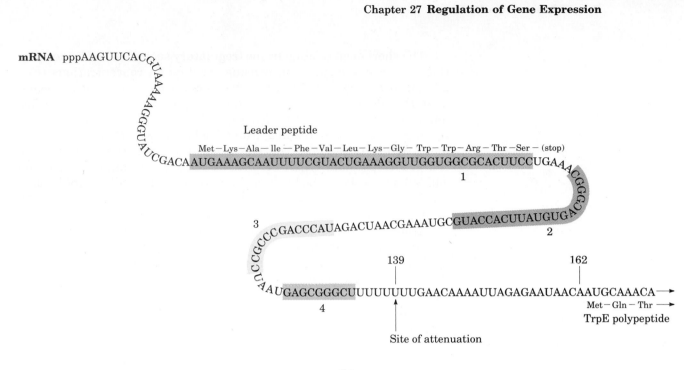

(b)

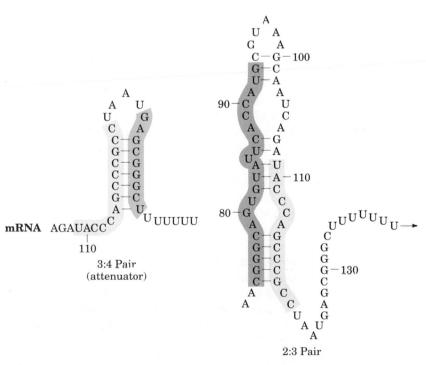

(c)

The short open reading frame (regulatory sequence 1 or leader peptide) is the key element in sensing tryptophan concentrations (Fig. 27–23). It can be thought of as a tryptophan-sensitive timing mechanism that determines whether sequence 3 pairs with sequence 4 (attenuating transcription) or with sequence 2 (allowing transcription to continue). This open reading frame includes two codons for tryptophan. When tryptophan concentrations are high, concentrations of charged tryptophan tRNA (Trp-tRNA^{Trp}) are also high. Translation will follow closely on the heels of transcription, proceeding rapidly past the Trp codons and into sequence 2 before sequence 3 is synthesized by RNA polymerase. In this case sequence 2 is covered by the ribosome and thus is rendered unavailable for pairing to sequence 3 when it is synthesized; the attenuator structure (sequences 3 and 4) is formed and transcription is halted. When tryptophan concentrations are low, however, the ribosome stalls at the two Trp codons because charged tRNA^{Trp} is unavailable. Sequence 2 remains free as sequence 3 is synthesized, these two sequences can base-pair, and transcription can proceed (Fig. 27–23). In this way, the proportion of transcripts that are attenuated increases as tryptophan concentrations increase.

Each amino acid biosynthetic operon uses a similar attenuation strategy to fine-tune biosynthetic enzymes to meet cellular requirements. The 15 amino acid leader peptide produced by the *phe* operon contains seven Phe residues. The leader peptide for the *his* operon contains seven contiguous His residues. The *leu* operon leader peptide has four contiguous Leu residues. In fact, in most amino acid biosynthetic operons (*trp* excepted), attenuation is sufficiently sensitive to be the only regulatory mechanism.

Induction of the SOS Response Requires the Destruction of Repressor Proteins

The SOS response (p. 838) is a good example of the coordinate regulation of many unlinked genes. The SOS genes are induced when the bacterial chromosome is extensively damaged; many are involved in DNA repair and mutagenesis (see Table 24–6). The key regulatory elements are a repressor, called the LexA repressor, and the RecA protein (Chapter 24).

The LexA repressor regulates the transcription of all of the SOS genes (Fig. 27–24). Induction of the SOS response involves removal of the LexA repressor, but this is not a simple dissociation from DNA in response to binding a small molecule as in the examples described above. Instead, the LexA repressor is inactivated by autocleavage into two protein fragments. Cleavage occurs at a specific Ala–Gly peptide bond that splits the protein (M_r 22,700) into two roughly equal fragments. By itself, the LexA repressor only undergoes autocleavage at elevated pH. At neutral pH, the self-digestion reaction requires RecA protein, and this RecA activity is sometimes called a coprotease activity.

It is the RecA protein that provides the link between the biological signal (DNA damage) and SOS induction. RecA protein facilitates cleavage of the LexA repressor only when RecA is bound to single-stranded DNA (see Fig. 24–31). Heavy DNA damage leads to numerous single-strand gaps in the DNA, and these gaps provide the molecular signal that activates RecA protein, leading to cleavage of the LexA repressor protein and SOS induction.

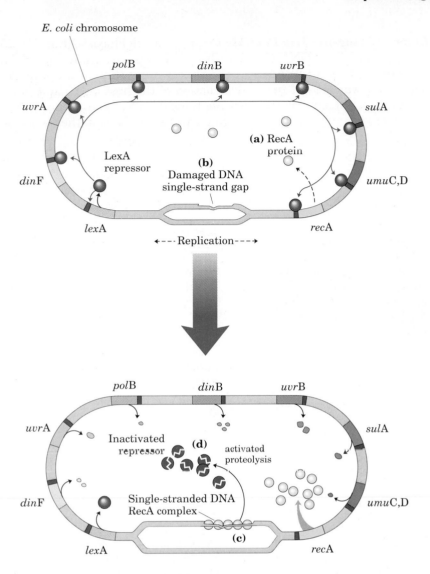

Figure 27–24 The SOS response in *E. coli*. See Table 24–6 for a description of the functions of these genes. The LexA protein is the repressor in this system, with an operator site (indicated in red) near each gene. **(a)** The *rec*A gene is not entirely repressed by the LexA repressor, and about 1,000 RecA protein monomers are normally found in the cell. **(b)** When DNA is extensively damaged (e.g., by UV light), DNA replication is halted and the number of single-strand gaps in the DNA increases. **(c)** RecA protein binds to this single-stranded DNA, activating the protein's coprotease activity. **(d)** While bound to DNA the RecA protein facilitates the cleavage and inactivation of the LexA repressor. When the repressor is inactivated, the SOS genes, including *rec*A, are induced. RecA protein levels increase 50- to 100-fold.

During induction of the SOS response, RecA protein also cleaves and thus inactivates the repressors that allow bacteriophage λ and related bacterial viruses to be propagated in a dormant (lysogenic) state within a bacterial host. These repressors have apparently evolved to mimic the LexA repressor, and they are also cleaved at a specific Ala–Gly peptide bond. This leads to replication of the virus and lysis of the cell to release the new virus particles, permitting the bacteriophage to make a hasty exit from a bacterial cell that is in distress, as described below.

Bacteriophage λ Provides an Example of a Regulated Developmental Switch

The objective of regulation of bacterial virus (bacteriophage) genes is usually the orderly assembly of new phage particles without destroying the host cell too soon. The well-studied bacteriophage λ provides an example of a complex and elegant regulatory circuit that determines the developmental fate of the virus in a given bacterial cell. This provides a paradigm for the complex problem of development in multicellular organisms.

Figure 27–25 Bacteriophage λ. **(a)** Electron micrograph of bacteriophage λ virus particles attached to an *E. coli* cell. **(b)** Two alternative fates for a bacteriophage λ infection: lysis or lysogeny. Under certain conditions (e.g., when the SOS response is induced), the lysogenic state is interrupted and the cell undergoes a lytic cycle (dashed arrow).

Lysis or Lysogeny: Two Possible Fates Bacteriophage λ (Fig. 27–25) is a medium-sized DNA phage with a chromosome containing 48,502 base pairs. It is a **temperate phage,** meaning that phage infection does not always result in the destruction of the host cell. The DNA of the invading phage has two possible fates (Fig. 27–25b): reproduction leading to generation of new phage particles and lysis of the cell **(lytic**

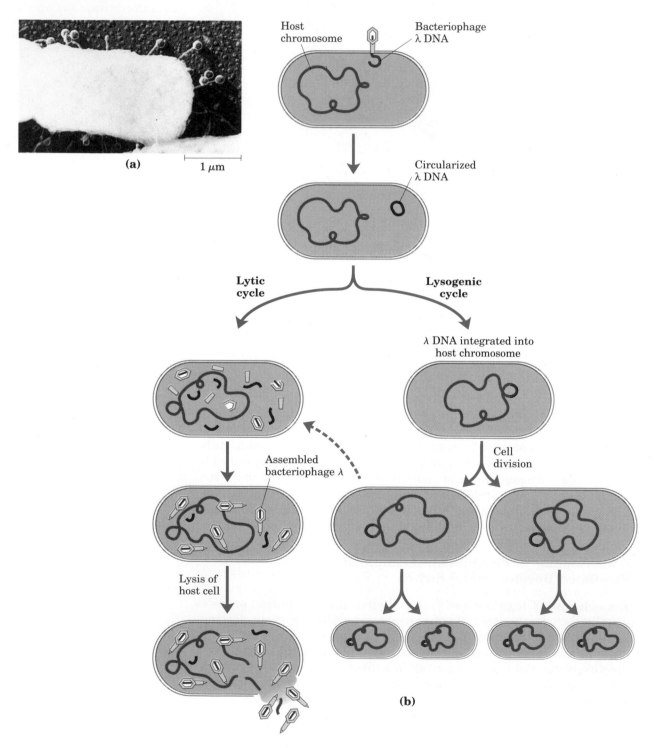

(a)   1 μm

Host chromosome   Bacteriophage λ DNA

Circularized λ DNA

Lytic cycle   Lysogenic cycle

λ DNA integrated into host chromosome

Assembled bacteriophage λ

Cell division

Lysis of host cell

(b)

cycle) or a relatively benign integration into the host chromosome where it can be replicated passively with the host DNA for many generations (lysogeny). The choice between lysis and lysogeny is governed largely by the interactions of five regulatory proteins called CI, CII, Cro, N, and Q. These proteins regulate transcription from a number of promoters in a regulatory region of the phage DNA described below. The functions of the proteins and promoters are summarized in Table 27–1.

Table 27–1 Regulatory elements of bacteriophage λ	
Regulatory element	Function
Proteins	
CI	At low concentrations a repressor of P_R and P_L and an activator of P_{RM}; at high concentrations also represses P_{RM}
CII	An activator of P_{RE} and P_{int}
Cro	At low concentrations a repressor of P_{RM}; at high concentrations also represses P_L and P_R
N	An antiterminator at t_{L1}, t_{R1}, t_{R2}, and other terminators
Q	An antiterminator for late gene transcription
Promoters	
P_R	Major rightward transcription
P_L	Major leftward transcription
P_{RM}	Transcription for *repressor maintenance*
P_{RE}	Transcription for *repressor establishment*
P_{int}	Transcription of genes for *integration* and excision

The CI, CII, and Cro proteins are approximately analogous to the types of regulatory proteins already described. The CI and Cro proteins are repressors, and the CII protein is an activator. The N and Q proteins interact directly with the *E. coli* RNA polymerase to permit readthrough (i.e., transcription) of certain transcription termination sequences built into the phage DNA genome. This activity of the N and Q proteins is referred to as **antitermination,** and it is distinct from the regulatory mechanisms described to this point.

Lysis In the lytic pathway (Fig. 27–26), phage genes are arranged in three sets according to their time of expression. Some products of genes expressed in the first stage (immediate-early) are required to permit transcription of second-stage genes (delayed-early). The products of some second-stage genes, in turn, are required to permit transcription in the last stage (late genes). In general, genes required for replication and recombination are expressed early, and genes required for assembling phage peptides and lysing the host cell are expressed late. The N and Q proteins are the key to temporal regulation of the expression of bacteriophage genes.

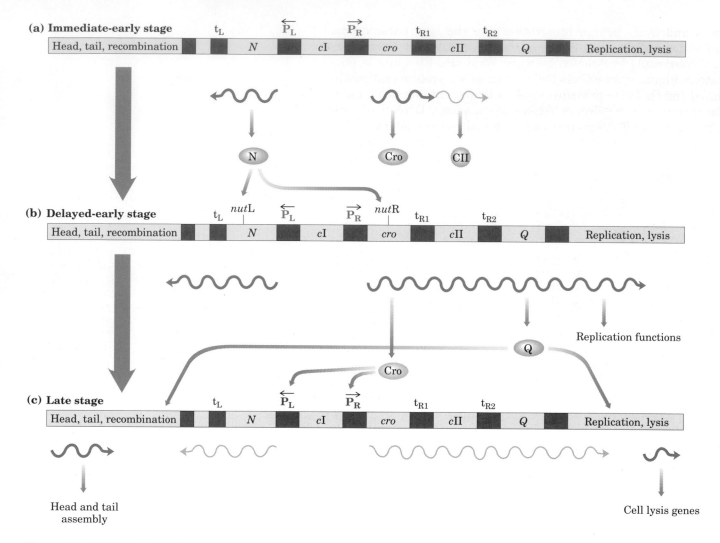

Figure 27–26 Regulation of gene expression in the lytic pathway of bacteriophage λ. A map of bacteriophage λ is shown; key genes and regulatory sites involved in the lytic pathway are highlighted. Note that this map is not drawn to scale; the regulatory region is disproportionately large. Temporal regulation of gene expression is accomplished in a cascade, in which one gene product produced at each stage is required to stimulate gene expression in the next stage. The key gene products are the N protein, produced from an immediate-early gene **(a)**, and the Q protein, produced from a delayed-early gene **(b)**. The N protein stimulates **(b)** delayed-early and the Q protein stimulates **(c)** late gene expression. The promoters P_L and P_R are critical for immediate-early and delayed-early transcription. Other regulatory sites, including the transcription terminators (t_{R1}, t_{R2}, t_L), are described in the text. **(c)** The stimulation of late gene expression by Q protein is complemented by repression of P_R and P_L transcripts by the Cro protein. The action of CII protein is described in Fig. 27–27. Components involved in activation, as well as new mRNAs synthesized at each stage, are shown in green; components involved in repression are in red.

When the phage DNA enters a host cell, the regulatory cascade begins. First, the *E. coli* RNA polymerase initiates transcription at two promoters, P_R and P_L. This immediate-early mRNA synthesis (Fig. 27–26a) is limited in both cases by transcription terminators; these are designated t_{R1} and t_{R2} for transcripts originating at P_R, and t_L for transcripts originating at P_L. The transcript from P_R includes the *cro* gene, and about half of these transcripts proceed through t_{R1} to include the *cII* gene as well (Cro and CII are considered below). The transcript from P_L includes the *N* gene.

The N protein triggers the second phase in the cascade: expression of the delayed-early genes (Fig. 27–26b). Once sufficient N protein is synthesized, it interacts with RNA polymerase, modifying it in some manner so that it overrides the three termination signals, t_L, t_{R1}, and t_{R2}. This leads to the production of longer transcripts that now include the *Q* gene and genes for proteins needed in viral replication.

The third and final phase is the expression of the late genes (Fig. 27–26c), which are transcribed from separate promoters only after the Q protein has been produced. This protein also interacts with RNA polymerase and antagonizes transcription termination at other sites (not shown in Fig. 27–26). The late genes include those for the structural proteins needed to assemble a virus particle. Once the new phages are assembled, the cell is lysed and the virus particles are freed.

The N and Q proteins have similar functions, but their mechanisms of action may differ. The N protein binds to specific DNA sequences (called *nut*L and *nut*R) located upstream of the transcription terminators (Fig. 27–26b). When RNA polymerase reaches this point it is modified (mechanism unknown) in a reaction requiring N and at least three host cell proteins. The RNA polymerase thereby acquires the ability to transcribe through many kinds of terminators. The Q protein interacts with RNA polymerase at a sequence near late promoters, where transcription pauses shortly after initiation. It modifies the RNA polymerase in a reaction requiring only Q and perhaps one other protein. The mechanism of N and Q action remains an important research area. Antitermination of RNA synthesis is also a regulatory mechanism in some eukaryotic systems.

Lysogeny Infection results in lysis only about half the time. Those cells that escape this fate are lysogenized. The lysogenic pathway is governed by another immediate-early gene product, the CII protein. The CII protein is an activator that stimulates transcription from two additional promoters, P_{RE} and P_{int} (Fig. 27–27). The transcript from P_{RE} includes the *cI* gene that encodes the CI protein, which is a repressor. If synthesized early enough, this repressor is capable of suppress-

Figure 27–27 Regulation of gene expression in the lysogenic pathway of bacteriophage λ. **(a)** RNA transcripts initiated at P_R continue through t_{R1} about 50% of the time, resulting in synthesis of CII protein as an early gene product. **(b)** The CII protein activates transcription at P_{RE} and P_{int}, resulting in production of the CI protein (a repressor) and proteins required for recombination. **(c)** The CI repressor is an antagonist of the Cro repressor (see Fig. 27–28), and it effectively shuts down bacteriophage gene expression except from P_{RM}. Early production of CII and then CI in sufficient amounts tips the balance toward lysogeny rather than lysis. Continued *cI* expression maintains the λ genome in the dormant state. (Red and green are used as in Fig. 27–26.)

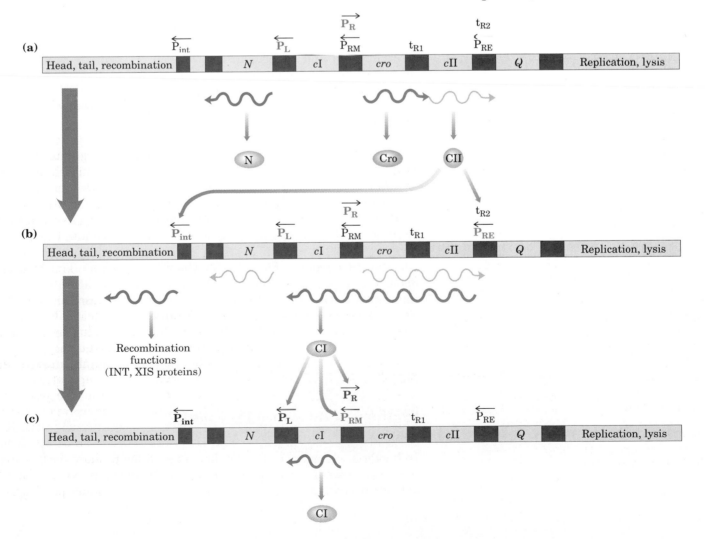

ing virtually all bacteriophage transcription except that originating at the P_{RM} promoter. The transcript from P_{int} includes genes required for the integration of viral DNA into the host chromosome through site-specific recombination (involving the INT and XIS proteins; see Fig. 24–37). Once integrated, expression of the bacteriophage genes is repressed by CI protein, and the bacteriophage genome is replicated passively with the host chromosome.

The lysogeny/lysis decision appears to be determined in part by the nutritional status of the host cell. The CII protein is subject to rapid degradation by *E. coli* proteases. The proteases are more abundant when the cell is growing rapidly in a rich medium, so that under these conditions the absence of CII (activator) limits CI (repressor) production and the scale tips in favor of lysis. When cells are starved, CII protein is elevated and the resulting production of CI protein favors the lysogenic path.

In the lysogenized state the phage is referred to as a **prophage.** To establish and maintain lysogeny, the CI protein binds to six operators, called $O_{L(1-3)}$ and $O_{R(1-3)}$ (Fig. 27–28; only the O_R operators are shown). The sequences of the six operators are different, so that CI protein binds most tightly to O_{R1} and O_{L1} and least tightly to O_{R3}. Binding to O_{L1} blocks transcription from P_L, and binding to O_{R1} and O_{R2} similarly blocks P_R. This effectively blocks all bacteriophage transcription except from the *c*I gene itself. The CI protein autoregulates its synthesis from yet another promoter, P_{RM} (P_{RE} produces the high concentration of CI protein needed to establish lysogeny; P_{RM} produces the lower concentration of CI protein needed to maintain it). O_{R3} governs transcription from P_{RM}. At low concentrations of CI protein, O_{R3} is unoccupied and CI protein is produced (CI protein bound to O_{R2} actually acts as an *activator* and stimulates RNA polymerase action at P_{RM}). The production of CI protein ceases when the concentrations are sufficient to occupy O_{R3}. The CI protein (Fig. 27–13) is also called simply the λ repressor.

The lysogenic state can continue for countless cell generations unless interrupted by an event such as treatment of the host cell by a DNA-damaging agent that induces the SOS response. The trigger that allows the prophage to emerge from the lysogenic state is a sudden reduction in the CI protein concentration as a result of self-cleavage facilitated by the RecA protein, as described above for LexA repressor. Removal of CI protein allows some transcription at P_R to produce Cro protein. The Cro protein is also a repressor, but it antagonizes the activity of CI protein (Fig. 27–28). The Cro protein binds to the same operators as CI protein, but with the reverse order of affinity. By binding tightly to O_{R3}, the Cro protein blocks further CI protein synthesis and sets in motion a series of events that leads to excision of the viral DNA from the bacterial chromosome and a lytic cycle. The system allows phage λ to make a fast exit in times of stress to the bacterial host. The interplay of proteins and DNA binding sites exhibits an extraordinary degree of sophistication and sensitivity.

Synthesis of Ribosomal Proteins Is Coordinated with rRNA Synthesis

In bacteria, changes in the cellular demand for protein synthesis are met by increasing the number of ribosomes rather than by altering the activity of individual ribosomes. In general, the number of ribosomes

Early in infection (no CI)

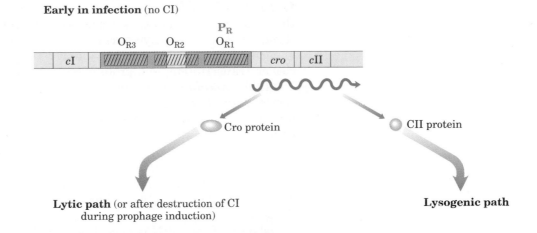

Lytic path (or after destruction of CI during prophage induction)

Lysogenic path

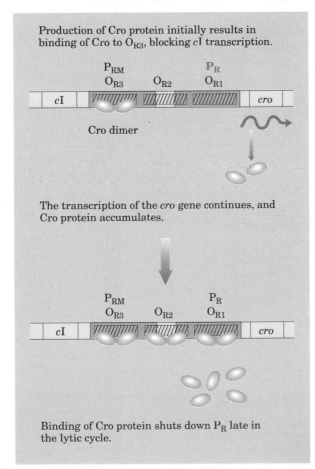

Production of Cro protein initially results in binding of Cro to O_{R3}, blocking cI transcription.

Cro dimer

The transcription of the cro gene continues, and Cro protein accumulates.

Binding of Cro protein shuts down P_R late in the lytic cycle.

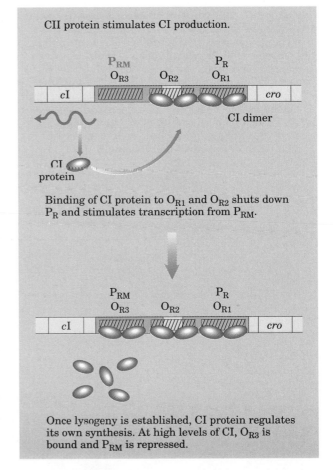

CII protein stimulates CI production.

CI dimer

CI protein

Binding of CI protein to O_{R1} and O_{R2} shuts down P_R and stimulates transcription from P_{RM}.

Once lysogeny is established, CI protein regulates its own synthesis. At high levels of CI, O_{R3} is bound and P_{RM} is repressed.

Figure 27–28 Antagonistic activities of the CI and Cro repressors at O_R in bacteriophage λ. The genes and regulatory sequences are not drawn to scale, and only the region near $O_{R(1-3)}$ is shown. The CI repressor binds the three O_R operators in the order $O_{R1} > O_{R2} > O_{R3}$. The Cro repressor binds the same three sites but has the reverse binding affinity. The CI repressor binds cooperatively so that both O_{R1} and O_{R2} are bound at low concentrations of CI.

increases as the cellular growth rate increases. At high growth rates, ribosomes constitute approximately 45% of the cell's dry weight. Because the fraction of cellular energy and matter devoted to making ribosomes is so large and the function of ribosomes is so important, it is essential for the cell to coordinate the synthesis of ribosomal proteins and rRNAs. This regulation is distinct from the mechanisms described above because it occurs largely at the level of *translation*.

The 52 genes that encode the ribosomal proteins (r-proteins) are found in at least 20 operons, each containing 1 to 11 genes. Some of these operons also contain the genes for the subunits of DNA primase (Chapter 24), RNA polymerase (Chapter 25), and protein synthesis elongation factors (Chapter 26), indicating the close coupling of replication, transcription, and protein synthesis during cell growth.

The r-protein operons are regulated primarily through a translational feedback mechanism. One r-protein encoded by each operon doubles as a **translational repressor,** which binds to the mRNA transcribed from that operon and blocks translation (Fig. 27–29). In general, the r-protein that plays the role of repressor is one that also binds directly to an rRNA. Each of the translational repressor r-proteins binds to the appropriate rRNA with higher affinity than to its mRNA, so that the mRNA will be bound and translation repressed only if the protein is present in excess over available rRNAs. In this way, translation of the mRNAs encoding r-proteins is repressed only when synthesis of the r-proteins exceeds that needed to make functional ribosomes, and the rate of r-protein synthesis is kept in balance with the available rRNAs.

The mRNA binding site for the translational repressor is near the translational start site of one of the genes in the operon, usually the first one (Fig. 27–29). In most operons this would affect only that one gene because most genes on bacterial polycistronic mRNAs have independent translation signals. In the r-protein operons, however, the translation of each gene in the operon depends upon translation of all the others. This translational coupling (mechanism unknown) permits the entire operon to be regulated by the binding of the translational repressor to a single site on the mRNA.

Figure 27–29 Structure of some ribosomal protein operons in the mRNA transcripts. The r-protein acting as translational repressor is shaded red in each case, and its site of action is indicated. Each translational repressor blocks the translation of all genes by binding to this one site on the mRNA. Genes that encode subunits of RNA polymerase are shaded yellow; genes that encode elongation factors are shaded blue. (Recall that the r-proteins of the large (50S) ribosomal subunit are designated L1 to L34; those of the small (30S) subunit, S1 to S21.)

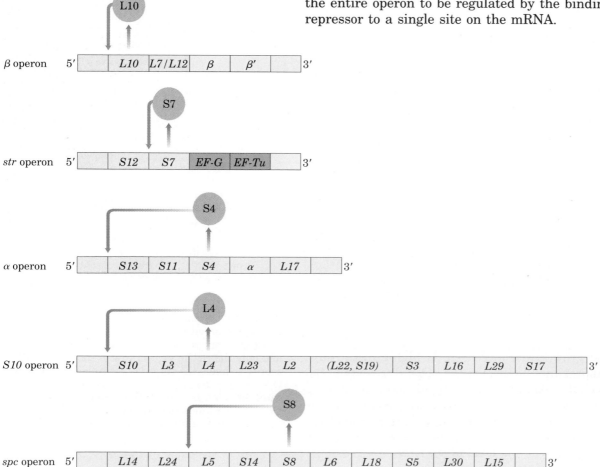

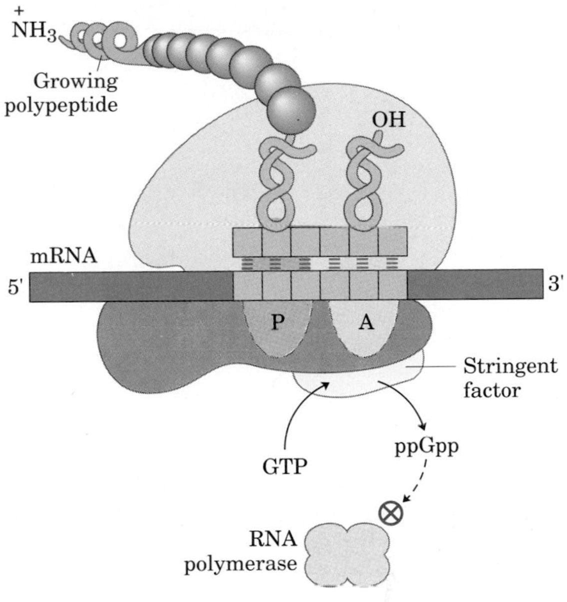

Figure 27–30 The stringent response to amino acid starvation in *E. coli* is triggered by binding of an uncharged tRNA in the ribosomal A site. A protein called stringent factor binds to the ribosome and catalyzes the synthesis of ppGpp. The signal ppGpp inhibits RNA polymerase by an unknown mechanism, reducing rRNA synthesis.

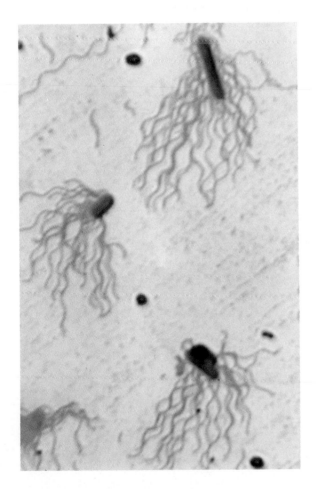

The r-protein operons are also apparently regulated at the level of transcription initiation, because transcription increases with increasing cellular growth rates. Neither the mechanism of transcriptional regulation nor the detailed relationship between transcriptional and translational regulation in this system is known.

The synthesis of r-proteins evidently is coordinated with the available rRNAs. The regulation of ribosome production therefore must ultimately reflect a regulation of rRNA synthesis. In *E. coli,* rRNA synthesis from the seven rRNA operons responds to cellular growth rate and to changes in the availability of crucial nutrients, particularly amino acids.

The regulation coordinated with amino acid concentrations is called the **stringent response** (Fig. 27–30). When amino acid concentrations are low, rRNA synthesis is halted. Amino acid starvation leads to the binding of uncharged tRNAs to the A site on ribosomes. This triggers a sequence of events that begins with the binding of a protein enzyme called **stringent factor** to the ribosome. Stringent factor catalyzes the formation of the unusual nucleotide guanosine tetraphosphate (ppGpp; see p. 354), adding pyrophosphate to the 3' position of GDP in the reaction

$$\text{GDP} + \text{ATP} \longrightarrow \text{ppGpp} + \text{AMP}$$

The abrupt rise in ppGpp in response to amino acid starvation leads to a great reduction in rRNA synthesis. It is not clear whether ppGpp binds to RNA polymerase or acts through an unidentified repressor or activator protein.

The nucleotide ppGpp, along with cAMP, belongs to a small but growing class of modified nucleotides that act as cellular second messengers (p. 354). In *E. coli,* these two nucleotides serve as starvation signals; they cause large changes in cellular metabolism by increasing or decreasing the transcription of hundreds of genes. The coordination of cellular metabolism with cell growth is very complex, and many regulatory mechanisms remain to be discovered.

Some Genes Are Regulated by Genetic Recombination

Salmonella bacteria live in the intestines of mammals and move by rotating flagella that emerge from the cell surface (Fig. 27–31). The many copies of the protein flagellin (M_r 53,000) that make up the flagella are prominent targets for the action of mammalian immune systems. To evade the immune system, *Salmonella* is able to switch between two distinct flagellin proteins once about every 1,000 cell generations in a process called **phase variation.**

Figure 27–31 The bacterium *Salmonella typhimurium,* with flagella evident.

The switch is accomplished by periodic inversion of a segment of DNA containing the promoter for a flagellin gene. The inversion is a site-specific recombination reaction (p. 845) mediated by a recombinase called Hin at specific sequences of 14 base pairs (*hix* sequences) at either end of the DNA segment (Fig. 27–32). When the DNA segment is in one orientation, the gene for H2 flagellin and another gene encoding a repressor are expressed. The repressor shuts down expression of the gene for H1 flagellin. When the DNA segment is inverted, the *H2* and repressor genes are no longer transcribed, and the *H1* gene is induced as the repressor is depleted. The Hin recombinase is encoded by the *hin* gene contained within the DNA segment that is inverted. The reaction also requires the HU protein and the FIS protein.

Figure 27–32 Regulation of flagellin genes in *Salmonella*. H1 and H2 are different flagellins. The *hin* gene encodes the recombinase that catalyzes inversion of the DNA segment that includes the H2 promoter and the *hin* gene. The recombination sites (inverted repeats) are called *hix* (shown in yellow). In one orientation *H2* is expressed **(a)**; in the opposite orientation *H1* is expressed **(b).** The interconversion between these two states is called phase variation.

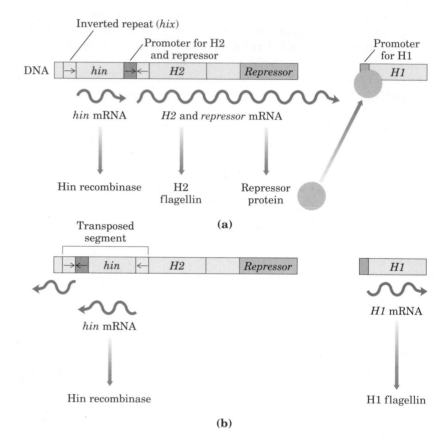

This regulatory mechanism has the advantage that it is absolute. Even low background levels of gene expression are impossible if the gene is physically separated from its promoter. An absolute on/off switch is crucial in this system, because a flagellum with even one copy of the wrong flagellar protein would be vulnerable to the action of antibodies directed against that protein. The *Salmonella* system is by no means unique. Closely related regulatory systems have been found in a number of bacterial species and bacteriophages, and recombination systems with similar functions have been found in eukaryotes (Table 27–2). Gene regulation by DNA rearrangements that move genes and/or promoters is a particularly common mechanism used by pathogens to change host range or to change surface proteins as a defense against the host immune system.

Table 27–2 Examples of gene regulation by recombination

System	Recombinase/ recombination site	Type of recombination	Function
Phase variation (*Salmonella*)	Hin/*hix*	Site-specific	Alternative expression of two flagellin genes; evades host immune system
Host range (bacteriophage Mu)	Gin/*gix*	Site-specific	Alternative expression of two sets of tail fiber genes; affects host range
Mating type switch (yeast)	HO endonuclease, RAD52 protein, other proteins/*MAT*	Nonreciprocal gene conversion*	Alternative expression of two mating types of yeast, a and α; cells of different mating types can mate and undergo meiosis
Antigenic variation (trypanosomes)	Varies	Nonreciprocal gene conversion*	Successive expression of different genes encoding the variable surface glycoproteins (VSGs). The outer surface of a trypanosome is made up of multiple copies of a single VSG, the major surface antigen. By expressing a new VSG gene at intervals, the trypanosome alters its glycoprotein coat and evades the host immune system.†

* Nonreciprocal gene conversion is a class of recombination events not discussed in Chapter 24. Genetic information is moved from one part of the genome (where it is silent) to another (where it is expressed) in a reaction similar to replicative transposition.

† Trypanosomes cause African sleeping sickness and other diseases (see Box 21-1). A cell can "change coats" more than 100 times, precluding an effective defense by the host immune system. Trypanosome infections are chronic and if untreated result in death.

Regulation of Gene Expression in Eukaryotes

In eukaryotes, as in prokaryotes, the initiation of transcription is a major regulation point for gene expression. Prokaryotes and eukaryotes also use some of the same regulatory mechanisms. However, at least three general themes encountered in eukaryotic gene regulation tend to distinguish it from the prokaryotic gene regulation already described. First, the activation of transcription is associated with multiple changes in the structure of the chromatin in the transcribed region. Second, although both positive and negative regulatory elements are found, positive regulatory mechanisms predominate in systems characterized to date. A third significant distinction is the physical separation between transcription, which occurs in the nucleus, and translation, which occurs in the cytoplasm.

The complexity of many regulatory circuits in eukaryotic cells is extraordinary. We will end with a discussion of the elaborate regulatory cascade that controls development in fruit flies.

Transcriptionally Active Chromatin Is Structurally Distinct

The effects of chromosome structure on gene regulation in eukaryotes have no clear parallel in prokaryotes. In chromosomal regions that have been activated for transcription, a variety of structural changes can be observed. The most obvious is an increased sensitivity of the DNA to nuclease-mediated degradation. When nucleases such as DNase are added to solutions of carefully isolated chromatin, they tend to cleave the DNA within the chromatin into fragments that have predictable sizes in multiples of about 200 base pairs, reflecting a regular repeating structure (the nucleosome; see Fig. 23–24). However, in actively transcribed regions, the fragments are smaller and more heterogeneous in size. Within these regions are sequences that have an especially high sensitivity to DNase I, referred to as **hypersensitive sites.** These are generally no more than 100 to 200 base pairs long and are often found within the 1,000 base pairs flanking the 5′ ends of transcribed genes (in some genes, hypersensitive sites are found farther from the 5′ end, near the 3′ end, or even within the gene itself). Many hypersensitive sites correspond to binding sites for known regulatory proteins. The relative absence of nucleosomes in these regions may facilitate binding of regulatory proteins.

The DNA in transcriptionally active chromatin also tends to be undermethylated. Methylation of cytosine residues at the 5′ position is common in eukaryotic DNA at CpG sequences (p. 347). CpG sites near many genes are generally undermethylated in tissues where the genes are expressed relative to tissues where they are not expressed.

The final significant change observed during transcription is in the histones and related proteins. Transcriptionally active chromatin tends to be deficient in histone H1, and the other core histones have a greater tendency to be modified by acetylation or by the attachment of ubiquitin (p. 936). In some cases, nucleosomes are absent in regions that are very active in transcription, such as the rRNA genes in many eukaryotic cells. The overall pattern suggests that active chromatin is prepared for transcription by the removal of potential structural barriers, but the molecular functions of all of the changes or the mechanisms by which they occur are not completely understood.

Most Eukaryotic Promoters Are Positively Regulated

In general, eukaryotic RNA polymerases have little or no intrinsic affinity for their promoters. Initiation of transcription is almost always dependent on the action of one or, more often, several activator proteins. The extensive use of positive regulatory mechanisms is probably a consequence of the larger size of the eukaryotic genome. Negative regulatory elements appear to be less common, although many eukaryotic regulatory proteins can be either activators or repressors under some circumstances.

There are at least two possible reasons for the many examples of positive regulation. First, nonspecific DNA binding of regulatory proteins becomes a more important problem in the much larger genomes of higher eukaryotes. The chance that a specific binding sequence will occur randomly at an inappropriate site also increases with genome size. One way to improve specificity is to use multiple regulatory proteins. The probability of the random occurrence of appropriate binding sites for several different proteins in proper functional juxtaposition is negligible. Using several *negative* regulatory elements will generally not improve specificity, because binding of one is sufficient to adequately block RNA polymerase action. Specificity can be improved, however, if several positive regulatory proteins must each bind specific DNA sequences and then form a complex that activates transcription. The average number of regulatory sites for a gene in a multicellular organism is probably at least five. The second reason for the use of positive regulation in a large genome is simply that it is more efficient. If the 100,000 genes in the human genome were negatively regulated, each cell would have to synthesize 100,000 different repressors in sufficient concentration to permit specific binding of each. In positive regulation, most of the genes are normally inactive (i.e., RNA polymerases

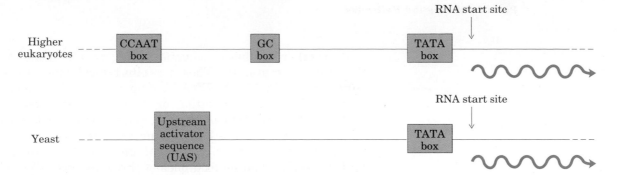

do not bind the promoters) and the cell only has to synthesize the selected group of activator proteins needed to activate transcription of the small subset of genes required in that cell.

Eukaryotic cells have three different RNA polymerases (I, II, and III; p. 863) that are specific (to a first approximation) for different classes of RNA. Messenger RNAs are generally transcribed by RNA polymerase II. Two types of generalized RNA polymerase II promoters and regulatory regions for eukaryotic genes are diagrammed in Figure 27–33, to summarize some important regulatory sequence elements. The first of these regulatory elements is a TATA box (consensus sequence TATAAAA; see Fig. 25–7). TATA boxes are located 25 to 30 base pairs from the mRNA initiation site in higher eukaryotes, although the distance is much more variable in yeast. These sequences appear to be binding sites for a transcription factor called TFIID ("TF-two-D") that is required for RNA polymerase binding. Although TATA boxes are relatively common, many genes have been found that are expressed without TATA boxes. A variety of other short sequence elements that function in the regulation of a given promoter are often found within a few hundred base pairs of the transcription start site. Two examples common to many genes are GC boxes (consensus GGGCGG) and CCAAT boxes (consensus GCCAAT). Additional regulatory sequence elements with more complex sequence structures are called **upstream activator sequences** (UASs) in yeast and **enhancers** in higher eukaryotes. For enhancer (and UAS) sequences, the location and orientation of the sequences relative to the transcription start site are relatively unimportant; they exert their regulatory effects even when moved experimentally, and they may naturally occur thousands of base pairs away from the gene being regulated.

Each of these sequence elements is recognized and bound specifically by one or more regulatory proteins (transcription factors). Regulation involves protein–protein interactions between different regulatory proteins bound at different sites and/or between regulatory proteins and RNA polymerase. Because the DNA sites bound by these proteins arc often hundreds or even thousands of base pairs away from each other or from the transcription start site, the protein–protein contact often requires DNA looping, as we have encountered in the *ara* operon in bacteria (Fig. 27–20).

A number of general transcription factors bind to RNA polymerase II and are required for recognition of the TATA box and initiation of transcription at most RNA polymerase II promoters. These are the *transcription factors* TFIIA, TFIIB, TFIID, TFIIE, and TFIIF. The first two proteins to bind in the initiation process are TFIIA and TFIID. As described above, TFIID is the protein that specifically recognizes the TATA box (Fig. 27–34). RNA polymerase II binds to the TFIIA–TFIID complex on the DNA, and the other transcription factors then bind to complete the complex. Except for TFIID, little is known about the detailed molecular functions of each of these transcription factors.

Figure 27–33 Generalized eukaryotic promoters with some typical binding sites for regulatory proteins.

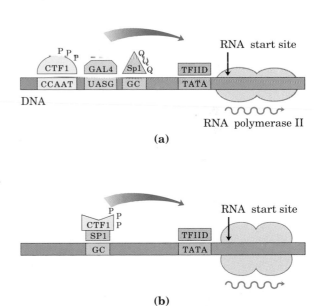

(a)

(b)

Figure 27–34 (a) Typical eukaryotic transcriptional activators that affect RNA polymerase II. TFIID is a general transcription factor. CTF1, GAL4, and Sp1 are transcriptional activators that bind to specific DNA sites. The nature of the activation domain is indicated by symbols: PPP, proline-rich; – – –, acidic; QQQ, glutamine-rich. Some or all of these proteins may activate transcription via intermediary proteins called coactivators (not shown). Note that the binding sites illustrated here are not generally found together near a single gene. **(b)** A chimeric protein with the DNA-binding domain of Sp1 and the activation domain of CTF1 activates transcription if a GC box is present.

The other regulatory sequences are generally bound by transcriptional activator proteins. These proteins typically have a distinct structural domain for specific DNA binding and one or more additional domains required for activation or interaction with other regulatory proteins. Dimerization of regulatory proteins is often mediated by domains containing the leucine zippers or helix-loop-helix structural motifs described earlier in this chapter. To illustrate activation domains, we have chosen examples of three distinct types of structural domains used for activation by this large class of proteins (Fig. 27–34).

First, the factor that binds to the GC box in vertebrates is a protein called Sp1 (M_r 80,000). The DNA-binding domain of Sp1 is found near the carboxyl terminus and contains three zinc fingers. Two other domains in Sp1 function in activation. Their mechanism of action is unclear, but these domains are notable in that 25% of the amino acid residues are Gln. A wide variety of other known activator proteins also have **glutamine-rich domains.**

Second, the CCAAT box is bound by a specific class of proteins, one of which is called CTF1 (M_r 55,000). The DNA-binding domain of CTF1 contains many basic amino acid residues, and the binding region is probably arranged as an α helix. Apparently the protein has no helix-turn-helix or zinc finger motif, and the DNA-binding mechanism remains to be clarified. The activation domain is **proline-rich,** with Pro residues accounting for more than 20% of the amino acid residues. Another protein that binds to the CCAAT box, known as C/EBP, has two activation domains; one is proline-rich and the other does not fall into the categories described here. C/EBP binds to DNA as a dimer, and it was the first protein found to dimerize by means of a leucine zipper.

A final example is the protein GAL4, which binds an upstream activator sequence called UAS_G near the genes for the enzymes of galactose metabolism in yeast. This protein contains several zinc fingers in its DNA-binding domain, near the carboxyl terminus. The activation domain is distinct in that it contains numerous acidic amino acid residues. Experiments substituting a variety of different peptide sequences for the **acidic activation domain** of GAL4 suggest that the acidic nature of this domain is critical, although its precise amino acid sequence can vary considerably. GAL4 binds to DNA as a dimer, although dimerization is mediated by a structural motif other than the leucine zipper or helix-loop-helix domains.

The distinct activation and DNA-binding domains of these regulatory proteins often act completely independently. The function of particular domains in transcriptional activator proteins has often been defined in a type of experiment called "domain-swapping." For example, the proline-rich activation domain of CTF1 can be joined (through genetic engineering; Chapter 28) to the DNA-binding domain of Sp1 to create a protein that binds to GC boxes and still activates transcription (Fig. 27–34b). The DNA-binding domain of GAL4 has similarly been replaced with the DNA-binding domain of the prokaryotic LexA repressor. This chimeric protein does not bind at UAS_G and does not activate the yeast *gal* genes, but it activates yeast transcription normally when the UAS_G sequence in the DNA is replaced by the LexA recognition site. Similar experiments have helped to map the domains required for DNA binding and activation in many activator proteins.

The mechanism by which these activators affect transcription is not yet clear. Proteins such as GAL4 with acidic activation domains may interact directly with either RNA polymerase II or TFIID. The activators Sp1 and CTF1 may act through additional bridging proteins called coactivators. The complexity of these interactions, the number of proteins involved, and the central role of these regulatory processes in the life of every eukaryote ensure that this will continue to be an area of vigorous inquiry.

Development Is Controlled by a Cascade of Regulatory Proteins

The transitions in morphology and protein composition observed in the development of a zygote into a multicellular animal or plant with many distinctly different tissues and cell types involve tightly coordinated changes in the expression of the organism's genome. More genes are expressed during early development than in any other part of the life cycle. For example, there are about 18,500 *different* mRNAs in the sea urchin oocyte, but only about 6,000 different mRNAs in the cells of typical differentiated tissues. The mRNAs present in the oocyte give rise to a cascade of events that not only regulate the expression of many genes but also determine where and when the gene products will appear in the developing organism.

Several organisms have emerged as important model systems for the study of development. These include yeasts, nematodes, fruit flies, sea urchins, frogs, chickens, and mice. Our discussion will focus on the development of fruit flies. The emerging picture of the molecular events that occur in development is particularly well advanced in fruit flies and can be used to illustrate patterns and principles of general significance.

The fruit fly, *Drosophila melanogaster,* has a complex life cycle that includes complete metamorphosis in its progression from an embryo to an adult (Fig. 27–35). Among the most important characteristics of the embryo are its **polarity** (the anterior and posterior, dorsal and ventral parts of the animal are readily distinguished) and its **me-**

Figure 27–35 The life cycle of the fruit fly *Drosophila melanogaster*. In complete metamorphosis, the adult insect is radically different in form from its immature stages; this process requires extensive "remodeling" during development. By the late embryonic stage, segments have formed from which the various structures in the adult fly will develop.

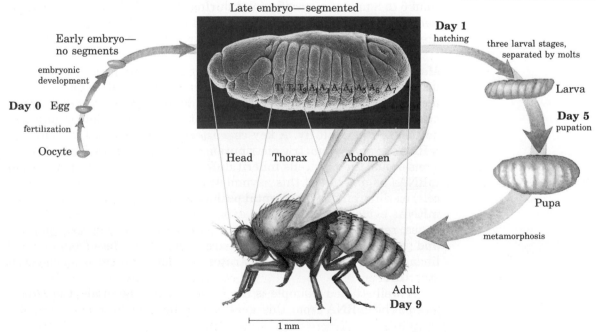

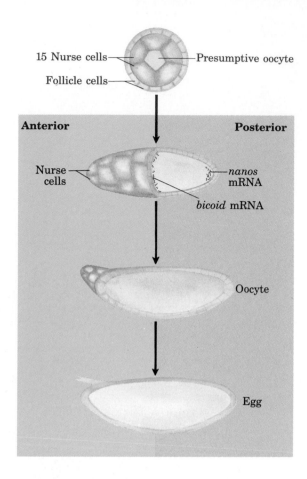

Anterior   **Posterior**

15 Nurse cells —— Presumptive oocyte

Follicle cells ——

Nurse cells —— nanos mRNA

bicoid mRNA

Oocyte

Egg

Figure 27–36 Development of a *Drosophila* egg. Maternal mRNAs (including the *bicoid* and *nanos* gene transcripts, discussed in the text) and proteins are deposited in the developing oocyte by nurse cells and follicle cells.

tamerism (the embryo body is made up of serially repeating segments, each with characteristic structural patterns). The adult fly is also segmented, and groups of these segments are organized into a head, thorax, and abdomen (Fig. 27–35). Each segment of the adult thorax has a different set of appendages. These patterns are all under genetic control. A variety of pattern-regulating genes have been discovered that dramatically affect the organization of the body, and analysis of these has provided important clues about how development is regulated.

The egg is formed from an oocyte surrounded by 15 nurse cells and a layer of follicle cells (Fig. 27–36). As the egg cell is formed prior to fertilization, mRNAs and proteins originating in the nurse and follicle cells are deposited in it; some of these play a critical role in development.

Three major classes of pattern-regulating genes specify the basic features of the *Drosophila* embryo's body and function in successive stages of development. (1) **Maternal genes** are expressed in the unfertilized egg (oocyte), and the resulting **maternal mRNAs** remain dormant until fertilization. These provide most of the proteins needed in very early development. Some of the proteins encoded by these mRNAs direct the spatial organization of the developing embryo at early stages, establishing its polarity. (2) **Segmentation genes** are transcribed from the cellular genome after fertilization and direct the formation of the proper number of body segments. There are at least three subclasses of segmentation genes that act at successive stages: **gap genes** divide the developing embryo into four broad regions, **pair-rule genes** define seven stripes, and **segment polarity genes** define 14 stripes that become the 14 segments present in a normal embryo. (3) **Homeotic genes** are expressed still later and affect the unique characteristics of individual body segments.

More than 40 regulatory genes in these three classes are known; the number is growing rapidly. They direct the development of an adult fly with a head, thorax, abdomen, the proper number of segments, and the correct appendages on each segment. Although embryogenesis takes about a day to complete, all of these genes are activated in the first 4 hours. Some of the mRNAs and proteins are present for no more than 6 to 8 min at a specific point during this period. Not surprisingly, many of these genes code for transcription factors that affect the expression of successive genes in a kind of developmental cascade.

Maternal Genes Within the *Drosophila* egg, the maternal gene products establish two axes: anterior–posterior and dorsal–ventral. Before fertilization, these genes have therefore defined the regions in the radially symmetric egg that will develop into the head/abdomen and top/bottom of the adult fly. A key characteristic of some of the maternal mRNAs is an asymmetric distribution in the egg. As the first cell divisions occur, the new cells inherit different amounts of these maternal mRNAs as a result of this asymmetry, and this in turn sets the new cells on different developmental paths. The products of these maternal mRNAs evidently regulate the expression of other pattern-regulating genes, resulting in a cascade of expressed genes. The specific pattern and sequence in which the genes are expressed differs from one cell lineage to the next and orchestrates the development of each adult structure.

A well-studied example is the *bicoid (bcd)* gene product of *Drosophila*. The mRNA from this gene is synthesized by nurse cells and deposited in the unfertilized egg near its anterior pole. During early development, this mRNA is translated and the bicoid protein diffuses through the cell, creating a concentration gradient radiating out from the anterior pole (Fig. 27–37). The bicoid protein is a transcription factor that affects the expression of a number of segmentation genes. The amounts of bicoid protein present in various parts of the developing embryo determine the cells in which a number of other genes are subsequently expressed. Changes in the shape of the bicoid protein gradient have dramatic effects on the body pattern in the resulting fly. Lack of bicoid protein results in the development of an embryo with no head or thorax and two abdomens. Another maternal mRNA, transcribed from a gene called *nanos,* has a similar role but is localized in the posterior pole of the egg. The dorsal–ventral axis in the egg is established by the combined action of at least 12 genes.

Segmentation Genes Expression of the gap genes is generally regulated by the products of one or more maternal genes. The bicoid protein, for example, appears to activate expression of a gap gene called *hunchback.* At least some of the gap genes are themselves transcription factors that affect the expression of other segmentation and

Figure 27–37 (a) A micrograph of an immunologically stained *Drosophila* egg, showing distribution of the *bicoid (bcd)* gene product. The graph measures stain intensity. This distribution is essential for normal development of the anterior structures of the animal. If the *bcd* gene is not expressed, the resulting embryo has two posteriors, as shown in **(b).**

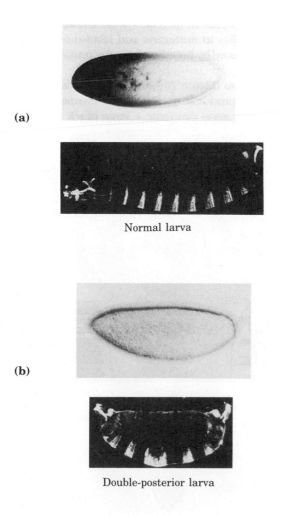

(a)

Normal larva

(b)

Double-posterior larva

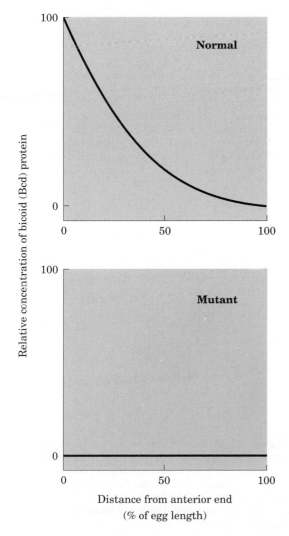

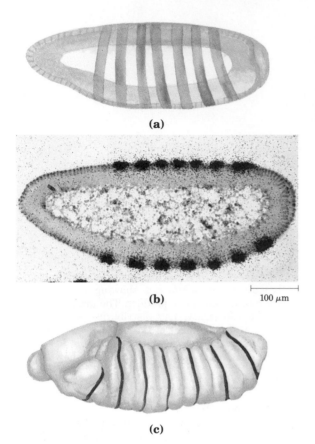

(a)

(b)

100 μm

(c)

Figure 27–38 Distribution of the *fushi tarazu (ftz)* gene product in early embryos. In the normal embryo, the gene product can be detected in seven bands around the circumference of the embryo, as shown schematically in **(a).** These bands are seen as dark spots (generated by a radioactive label) in a cross-sectional autoradiograph **(b),** and give rise to the segments shown here in red in the late embryo **(c).**

homeotic genes. One well-characterized segmentation gene is *fushi tarazu (ftz),* which belongs to the "pair-rule" subclass. When this gene is lost, the embryo develops seven double-wide segments instead of the normal 14. The mRNAs and proteins derived from the normal *ftz* gene accumulate in a striking pattern of seven stripes that encircle the posterior two-thirds of the embryo (Fig. 27–38). The stripes correspond to the positions of segments that develop later, and which are eliminated if *ftz* function is lost. The expression of pattern-regulating genes such as *ftz* (and *bcd,* expressed earlier in development) establishes a kind of chemical blueprint for the body plan that precedes the actual formation of a body structure.

Homeotic Genes Loss of homeotic genes by mutation or deletion causes the appearance of a normal appendage or body structure at an inappropriate body position. An important example is the *ultrabithorax (Ubx)* gene. When *Ubx* function is lost, the first abdominal segment develops incorrectly, having the structure of the third thoracic segment. Other known homeotic mutations cause the formation of an extra set of wings, or two legs at the position in the head where the antennae are normally found (Fig. 27–39).

The homeotic genes span long regions of DNA. The *Ubx* gene, for example, is 77,000 base pairs in length and contains introns that are as long as 50,000 base pairs. Transcription of this gene takes nearly one hour. The delay this imposes on *Ubx* gene expression is believed to be a timing mechanism involved in the temporal regulation of subsequent steps in development.

The precise nature of many of the events directed by these proteins, and in many cases the biochemical function of the proteins themselves, are unknown. A likely DNA-binding domain has been identified

(a)

(b)

(c)

(d)

Figure 27–39 The effects of mutations in homeotic genes. **(a)** Normal *Drosophila* head. **(b)** *Drosophila* homeotic mutant *(Antennapaedia)* in which antennae are replaced by legs. **(c)** Normal *Drosophila* body structure. **(d)** Homeotic mutant *(Bithorax)* in which a segment has developed incorrectly to produce an extra set of wings.

in a number of these proteins, however, which suggests that they are regulatory proteins. This domain contains 60 amino acids and is called the **homeodomain** because it was found first in homeotic genes. The DNA sequence encoding this domain is called the **homeobox.** It is highly conserved and has been identified in proteins from a wide variety of organisms. The DNA-binding segment of the domain is related to the helix-turn-helix motif.

The identification of structural determinants with identifiable molecular functions is the first step in understanding the molecular events underlying development. As more genes and their protein products are discovered, the biochemical side of this vast puzzle will slowly come together.

Summary

The expression of genes is regulated by a number of processes that affect the rates at which gene products are synthesized and degraded. Much of this regulation occurs at the level of the initiation of transcription and is mediated by regulatory proteins that either repress or activate transcription from specific promoters. Regulation by repressors and activators is called negative and positive regulation, respectively.

In prokaryotes, genes with interdependent functions are often clustered as a single transcriptional unit called an operon. The transcription of operon genes is generally blocked by the binding of a specific repressor protein at a DNA site called an operator. Dissociation of the repressor from the operator is mediated by a specific small molecule, called an inducer. These principles were first elucidated in studies of the lactose (lac) operon. The Lac repressor dissociates from the lac operator when the repressor binds to the biological inducer, allolactose.

Regulatory proteins are DNA-binding proteins that recognize specific sequences in the DNA. Most of these proteins have distinct DNA-binding domains. Within these domains, common structural motifs involved in DNA binding are the helix-turn-helix and zinc finger motifs. Regulatory proteins also contain domains for protein–protein interactions, including leucine zipper and helix-loop-helix motifs involved in dimerization and several classes of domains involved in the activation of transcription.

The lactose operon of E. coli also exhibits positive regulation by the catabolite gene activator protein (CAP). When cAMP concentrations are high (glucose concentrations are low), CAP binds to a specific site on the DNA, stimulating transcription of the lac operon and production of lactose-metabolizing enzymes. The presence of glucose depresses cAMP concentrations, restricting expression of lac (and other) genes and suppressing the use of secondary sugars. Several operons that are coordinately regulated, as with CAP and cAMP, are referred to as a regulon.

Other mechanisms of regulation are also observed in prokaryotes. In the arabinose (ara) operon, the AraC protein acts as both activator and repressor. Some repressors, as in the ara operon and the bacteriophage λ system, regulate their own synthesis (autoregulation). Some regulatory proteins in the ara system bind sites many base pairs distant from each other and interact by DNA looping mechanisms. Amino acid biosynthetic operons have a regulatory circuit called attenuation that uses a transcription termination site (the attenuator), modulating its formation in the mRNA by a mechanism that couples transcription and translation and responds to small changes in amino acid concentration. In the SOS system, multiple unlinked genes are repressed by a single type of repressor protein, and all of the genes are induced simultaneously when DNA damage triggers RecA protein–mediated proteolysis of the repressor. The bacteriophage λ has a complex regulatory circuit that oversees the choice between lysis and lysogeny. Two λ proteins, N and Q, act as antiterminators, modifying the host RNA polymerase so that it can bypass transcription termination sites. Finally, some prokaryotic genes are regulated by genetic recombination processes that physically move promoters relative to the genes being regulated. These diverse mechanisms permit very sensitive cellular responses to changes in environmental conditions.

Some regulation also occurs at the level of translation. The synthesis of ribosomal proteins in bacteria is mediated by a strategy in which one protein in each ribosomal protein operon acts as a translational repressor. The mRNA is bound by the repressor and translation is blocked only when the ribosomal protein is present in excess relative to available rRNA.

Eukaryotes employ many of the same regulatory schemes, although positive regulation appears to be more common and transcription is also accompanied by large changes in chromatin structure. Eukaryotic transcriptional activator proteins are generally required for RNA polymerase binding and activity. Some transcription factors have general functions; the TFII factors associated with RNA polymerase II, for example, are required at almost all RNA polymerase II promoters. Other transcriptional activators, unique to one gene or set of genes, have distinct domains for DNA binding and activation, and their DNA binding sites are often found hundreds of base pairs from the site where RNA synthesis begins.

Perhaps the most complex regulatory problem is the development of a multicellular animal. Here, sets of regulating genes operate in temporal and spatial succession, turning a given area of an egg cell into a predictable structure in the adult animal. Research continues into the molecular basis for this highly coordinated process.

Further Reading

General

Ingraham, J.L., Magasanik, B., Low, K.B., Schaechter, M., & Umbarger, H.E. (eds) (1987) Escherichia coli and Salmonella typhimurium, Cellular and Molecular Biology, Vol. 2, American Society for Microbiology, Washington, DC.
An excellent reference source for reviews of many bacterial operons.

Pabo, C.O. & Sauer, R.T. (1992) Transcription factors: structural factors and principles of DNA recognition. *Annu. Rev. Biochem.* **61,** 1053–1095.

Schleif, R. (1986) *Genetics and Molecular Biology,* Addison-Wesley Publishing Co., Inc., Reading, MA.
Chapters 12, 13, and 14 provide an excellent account of the experimental basis of major concepts of gene regulation in prokaryotes.

Schleif, R. (1992) DNA looping. *Annu. Rev. Biochem.* **61,** 199–223.

Struhl, K. (1989) Helix-turn-helix, zinc-finger, and leucine-zipper motifs for eukaryotic transcriptional regulatory proteins. *Trends Biochem. Sci.* **14,** 137–140.

Watson, J.D., Hopkins, N.H., Roberts, J.W., Steitz, J.A., & Weiner, A.M. (1987) *Molecular Biology of the Gene,* 4th edn, The Benjamin/Cummings Publishing Company, Menlo Park, CA.

Regulation of Gene Expression in Prokaryotes

Gottesman, S. (1984) Bacterial regulation: global regulatory networks. *Annu. Rev. Genet.* **18,** 415–441.

Jacob, F. & Monod, J. (1961) Genetic regulatory mechanisms in the synthesis of proteins. *J. Mol. Biol.* **3,** 318–356.
The operon model and the concept of messenger RNA were proposed in this historic paper.

Nomura, M., Gourse, R., & Baughman, G. (1984) Regulation of the synthesis of ribosomes and ribosomal components. *Annu. Rev. Biochem.* **53,** 75–117.

Ptashne, M., Johnson, A.D., & Pabo, C.O. (1982) A genetic switch in a bacterial virus. *Sci. Am.* **247** (November), 128–140.

Stephens, J.C., Artz, S.W., & Ames, B.N. (1975) Guanosine 5'-diphosphate 3'-diphosphate (ppGpp): positive effector for histidine operon transcription and general signal for amino acid deficiency. *Proc. Natl. Acad. Sci. USA* **72,** 4389–4393.

Yanofsky, C. (1981) Attenuation in the control of expression of bacterial operons. *Nature* **289,** 751–758.

Zieg, J., Silverman, M., Hilmen, M., & Simon, M. (1977) Recombinational switch for gene expression. *Science* **196,** 170–172.

Regulation of Gene Expression in Eukaryotes

Beardsley, T. (1991) Smart genes. *Sci. Am.* **265** (August), 86–95.
A good overview of gene regulation during development.

DeRobertis, E.M., Oliver, G., & Wright, C.V.E. (1990) Homeobox genes and the vertebrate body plan. *Sci. Am.* **263** (July), 46–52.

Guarente, L. (1988) UASs and enhancers: common mechanism of transcriptional activation in yeast and mammals. *Cell* **52,** 303–305.

Kornberg, R.D. & Lorch, Y. (1991) Irresistible force meets immovable object: transcription and the nucleosome. *Cell* **67,** 833–836.

McKnight, S.L. (1991) Molecular zippers in gene regulation. *Sci. Am.* **264** (April), 54–64.
A good description of leucine zippers.

Melton, D.A. (1991) Pattern formation during animal development. *Science* **252,** 234–241.

Ptashne, M. (1989) How gene activators work. *Sci. Am.* **260** (January), 40–47.

Pugh, B.F. & Tjian, R. (1992) Diverse transcriptional functions of the multisubunit eukaryotic TFIID complex. *J. Biol. Chem.* **267,** 679–682.

Struhl, K. (1987) Promoters, activator proteins, and the mechanism of transcriptional initiation in yeast. *Cell* **49,** 295–297.

Thummel, C.S. (1992) Mechanisms of transcriptional timing in *Drosophila. Science* **255,** 39–40.

Zlatanova, J. (1990) Histone H1 and the regulation of transcription of eukaryotic genes. *Trends Biochem. Sci.* **15,** 273–276.

Problems

1. *Negative Regulation* In the *lac* operon, describe the probable effect on gene expression of:
  (a) Mutations in the *lac* operator
  (b) Mutations in the *lac*I gene
  (c) Mutations in the promoter

2. *Effect of mRNA and Protein Stability on Regulation* An *E. coli* cell is growing in a solution with glucose as the sole carbon source. Tryptophan is suddenly added. The cells continue to grow, and divide every 30 min. Describe (qualitatively) how the amount of tryptophan synthase activity in the cell changes if:
  (a) The *trp* mRNA is stable (degraded slowly over many hours).
  (b) The *trp* mRNA is degraded rapidly, but tryptophan synthase is stable.
  (c) The *trp* mRNA and tryptophan synthase are both degraded rapidly.

3. *Functional Domains in Regulatory Proteins* A biochemist replaces the DNA-binding domain of the yeast GAL4 protein with the DNA-binding domain from the λ repressor (CI) and finds that the engineered protein no longer functions as a transcriptional activator (it no longer regulates transcription of the *gal* operon in yeast). What might be done to the GAL4 DNA-binding site to make the engineered protein functional in activating *gal* operon transcription?

4. *Bacteriophage* λ Bacteria that become lysogenic for bacteriophage λ are immune to subsequent λ lytic infections. Why?

5. *Regulation by Means of Recombination* In the phase variation system of *Salmonella,* what would happen to the cell if the Hin recombinase became more active and promoted recombination (the switch) several times in each cell generation?

6. *Transcription Attenuation* In the leader region of the *trp* mRNA, what would be the effect of:
  (a) Increasing the distance (number of bases) between the leader peptide gene and sequence 2?
  (b) Increasing the distance between sequences 2 and 3?
  (c) Removing sequence 4?

7. *Specific DNA Binding by Regulatory Proteins* A typical prokaryotic repressor protein discriminates between its specific DNA-binding site (operator) and nonspecific DNA by a factor of 10^5 to 10^6. About ten molecules of the repressor per cell are sufficient to ensure a high level of repression. Assume that a very similar repressor existed in a human cell and had a similar specificity for its binding site. How many copies of the repressor would be required per cell to elicit a level of repression similar to that seen in the prokaryotic cell? (Hint: The *E. coli* genome contains about 4.7 million base pairs and the human genome contains about 2.4 billion base pairs.)

8. *Positive Regulation* A new RNA polymerase activity is discovered in crude extracts of cells derived from an exotic fungus. The RNA polymerase initiates transcription only from a single, highly specialized promoter. As the polymerase is purified, its activity is observed to decline. The purified enzyme is completely inactive unless crude extract is added to the reaction mixture. Suggest an explanation for these observations.

Recombinant DNA Technology

Paul Berg

Stanley Cohen

Herbert Boyer

In our final chapter we describe a technology that in less than two decades has become fundamental to the advance of biochemistry. It helps to define present and future biochemical frontiers and illustrates many important principles of biochemistry. As the laws governing enzymatic catalysis, macromolecular structure, cellular metabolism, and information pathways continue to be elucidated, new research is directed at ever more complex biochemical processes. Cell division, the immune response, developmental processes in eukaryotes, vision, taste, oncogenesis, the cognitive processes in your brain as you read these words—all are orchestrated in an elaborate symphony of molecular and macromolecular interactions. As increasingly greater efforts are focused on understanding the biochemistry that underlies these processes, the real promise and implications of the biochemical journey begun in the nineteenth century become clear. Human beings not only can understand life, they can alter it.

The biochemical approach to understanding a complex biological process is to isolate and study the individual components in vitro with the goal of understanding the overall process in the whole organism. Perhaps the most fertile source for molecular insights into these processes lies in the cell's own information storehouse, its DNA. The sheer size of cellular chromosomes, however, presents us with an enormous barrier. How does one find and study a particular gene encoding a protein or RNA molecule with a molecular function we can only guess at, when that gene is only one of perhaps 100,000 genes scattered among the billions of base pairs that make up a mammalian genome? The answers began to appear in the mid-1970s.

Decades of advances in genetics, biochemistry, cell biology, and physical chemistry came together in the laboratories of Paul Berg, Herbert Boyer, and Stanley Cohen to yield techniques for locating, isolating, preparing, and studying small segments of DNA derived from much larger chromosomes. Taken together, these techniques are known as **DNA cloning.** DNA cloning has opened opportunities unimaginable just a few decades ago, including the identification and study of genes involved in almost every known biological process. These new methods are transforming basic research, agriculture, forensics, medicine, ecology, and many other fields, while at the same time presenting society with bewildering choices and serious ethical dilemmas.

Revolutionary as it is, this technology is grounded in the most fundamental biological and biochemical principles. The first two parts of this chapter outline these fundamentals, drawing on our understanding of the chemistry and enzymes of nucleic acid metabolism described in the previous five chapters. We then turn to topics that help illustrate the range of applications and the potential of this technology.

DNA Cloning: The Basics

To clone means to make identical copies; it is a term that was once restricted to the procedure of isolating one cell from a larger population of cells, then allowing it to reproduce itself to generate many identical cells. In such a way, sufficient quantities of a single cell type were made available for study. By analogy, DNA cloning involves separating a specific gene or segment of DNA from its larger chromosome and attaching it to a small molecule of carrier DNA, then replicating this modified DNA thousands or even millions of times. The result is a selective amplification of that particular gene or DNA segment. Cloning a segment of DNA, either prokaryotic or eukaryotic, entails five general procedures:

1. A method for cutting DNA at precise locations. The discovery of sequence-specific endonucleases (restriction endonucleases) provided the necessary molecular scissors.

2. A method for joining two DNA fragments covalently. DNA ligase can do this.

3. Selection of a small molecule of DNA capable of self-replication. Segments of DNA to be cloned can be joined to plasmids or viral DNAs (cloning vectors). These composite DNA molecules containing covalently linked segments derived from two or more sources are called **recombinant DNAs.**

4. A method for moving recombinant DNA from the test tube into a host cell that can provide the enzymatic machinery for DNA replication.

5. Methods to select or identify those host cells that contain recombinant DNA.

The methods used to accomplish these and related tasks are collectively referred to as **recombinant DNA technology,** or more informally as **genetic engineering.** We now turn to these methods, with emphasis on their biochemical origins.

In this initial discussion we will focus on DNA cloning in the bacterium *E. coli,* which was the first organism used for recombinant DNA work and is still the most common host cell. *E. coli* has many advantages: its DNA metabolism (and many other biochemical processes) are well understood; many naturally occurring cloning vectors such as bacteriophages and plasmids associated with *E. coli* are well characterized; and effective techniques are available for moving DNA from one bacterial cell to another. DNA cloning in other organisms will be addressed later in the chapter.

Restriction Endonucleases and DNA Ligase Yield Recombinant DNA

Particularly important to recombinant DNA technology is a set of enzymes made available by decades of research on nucleic acid metabo-

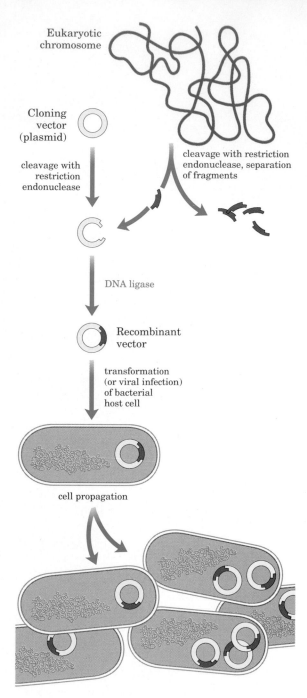

Eukaryotic
chromosome

Cloning
vector
(plasmid)

cleavage with
restriction
endonuclease

cleavage with restriction
endonuclease, separation
of fragments

DNA ligase

Recombinant
vector

transformation
(or viral infection)
of bacterial
host cell

cell propagation

Figure 28–1 Schematic illustration of DNA cloning. A fragment of DNA of interest to the researcher is obtained by cleaving a eukaryotic chromosome with a restriction endonuclease. After isolating the fragment and ligating it to a cloning vector that has also been cleaved with a restriction endonuclease, the resulting recombinant DNA is introduced into a host cell where it can be propagated (cloned). Note that the size of the *E. coli* chromosome relative to that of a typical cloning vector such as a plasmid is much greater than depicted here.

lism (Table 28–1). Two enzymes in particular lie at the heart of the general approach to generating and propagating a recombinant DNA molecule as outlined in Figure 28–1. First, **restriction endonucleases** cleave DNA at specific sequences to generate a set of smaller fragments. Second, the DNA fragment to be cloned can be isolated and joined to a suitable cloning vector using **DNA ligase** to seal the DNA molecules together. The recombinant vector is then introduced into a host cell, which "clones" it as the cell undergoes many generations of cell divisions.

Table 28–1 Some of the enzymes used in recombinant DNA technology

Enzyme(s)	Function
Type II restriction endonucleases	Cleaving DNAs at specific base sequences
DNA ligase	Joining two DNA molecules or fragments
DNA polymerase I (*E. coli*)	Filling in gaps in duplexes by stepwise addition of nucleotides to 3′ ends
Reverse transcriptase	Making a DNA copy of an RNA molecule
Polynucleotide kinase	Adding a phosphate to the 5′-OH end of a polynucleotide to label it or permit ligation
Terminal transferase	Adding homopolymer tails to the 3′-OH ends of a linear duplex
Exonuclease III	Removing nucleotide residues from the 3′ ends of a DNA strand
Bacteriophage λ exonuclease	Removing nucleotides from the 5′ ends of a duplex to expose single-stranded 3′ ends
Alkaline phosphatase	Removing terminal phosphates from either the 5′ or 3′ end or both

Restriction endonucleases are found in a wide range of bacterial species. Werner Arber discovered that their biological function is to recognize and cleave foreign DNA (e.g., the DNA of an infecting virus); such DNA is said to be *restricted*. The cell's own DNA is not cleaved because the sequence recognized by the restriction endonuclease is methylated (and thereby protected) by a specific DNA methylase. The restriction endonuclease and the corresponding methylase in a bacterium are sometimes referred to as a **restriction-modification system.** There are three types of restriction endonucleases, designated I, II, and III. Types I and III are generally large, multisubunit complexes containing both the endonuclease and methylase activities. Type I restriction endonucleases cleave DNA at random sites that can be 1,000 base pairs or more from the recognition sequence. Type III enzymes cleave the DNA about 25 base pairs from the recognition sequence. Both types of enzyme move along the DNA in a reaction that requires the energy of ATP. The type II restriction enzymes, first isolated by Hamilton Smith, are simpler, require no ATP, and cleave the DNA within the recognition sequence itself. The extraordinary utility of the type II enzymes was first demonstrated by Daniel Nathans, and these are the enzymes used most widely for recombinant DNA work.

More than 800 restriction endonucleases have been discovered in different bacterial species. Over 100 different specific sequences are recognized by one or more of these enzymes. These sequences are almost always short (four to six base pairs, occasionally more) and palindromic (see Fig. 12–20). A sampling of sequences recognized by some type II restriction endonucleases is presented in Table 28–2. Note that the name of each enzyme consists of a three-letter abbreviation of the bacterial species from which it is derived (e.g., *Bam* for B*acillus* am*yloliquefaciens*, *Eco* for E*scherichia coli*).

In a few cases, the interaction between a restriction endonuclease and its target sequence has been elucidated in exquisite molecular detail. The complex comprising the type II restriction endonuclease *Eco*RI and its target sequence is illustrated in Figure 28–2. DNA sequence recognition by *Eco*RI is mediated by 12 hydrogen bonds formed between the purines in the recognition site and six amino acid residues in the dimeric endonuclease (one Glu and two Arg residues in each subunit). Some restriction endonucleases cleave both strands of DNA so as to leave no unpaired bases on either end; these ends are often called **blunt ends** (Fig. 28–3a). Others make staggered cuts on the two DNA strands, leaving two to four nucleotides of one strand unpaired at each resulting end. These are referred to as **cohesive ends** or **sticky ends** (Fig. 28–3a) because they can base-pair with each other or with complementary sticky ends of other DNA fragments.

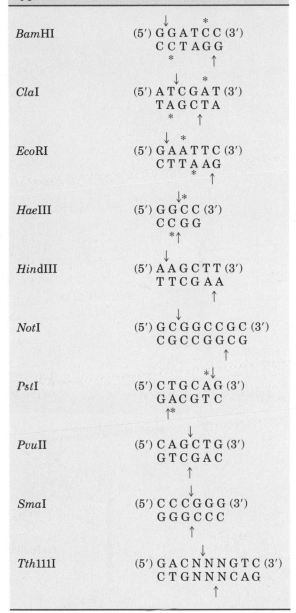

Table 28–2 Recognition sequences for some type II restriction endonucleases

Enzyme	Recognition sequence
*Bam*HI	(5') G G A T C C (3') C C T A G G
*Cla*I	(5') A T C G A T (3') T A G C T A
*Eco*RI	(5') G A A T T C (3') C T T A A G
*Hae*III	(5') G G C C (3') C C G G
*Hin*dIII	(5') A A G C T T (3') T T C G A A
*Not*I	(5') G C G G C C G C (3') C G C C G G C G
*Pst*I	(5') C T G C A G (3') G A C G T C
*Pvu*II	(5') C A G C T G (3') G T C G A C
*Sma*I	(5') C C C G G G (3') G G G C C C
*Tth*111I	(5') G A C N N N G T C (3') C T G N N N C A G

Arrows indicate the phosphodiester bonds cleaved by each restriction endonuclease. Asterisks indicate bases that are methylated by the corresponding methylase (where known). N denotes any base. The Roman numerals included in the enzyme names (e.g., *Bam*HI) distinguish different restriction endonucleases isolated from the same bacterial species rather than the type of restriction enzyme.

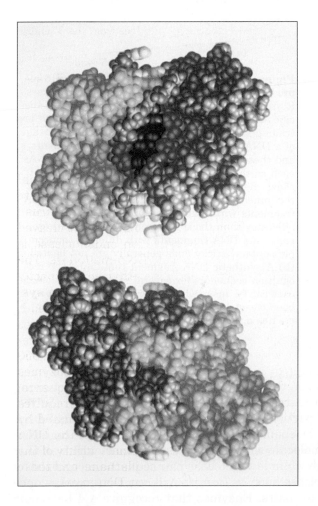

Figure 28–2 The interaction of *Eco*RI endonuclease with its target sequence. The dimeric enzyme (with its two subunits in gray and light blue) is shown bound to DNA. In the top view the DNA binding site is facing the viewer. In the bottom view the bound DNA is facing away from the viewer and is not visible. In the bound DNA, the bases that make up the recognition site for *Eco*RI are shown in red.

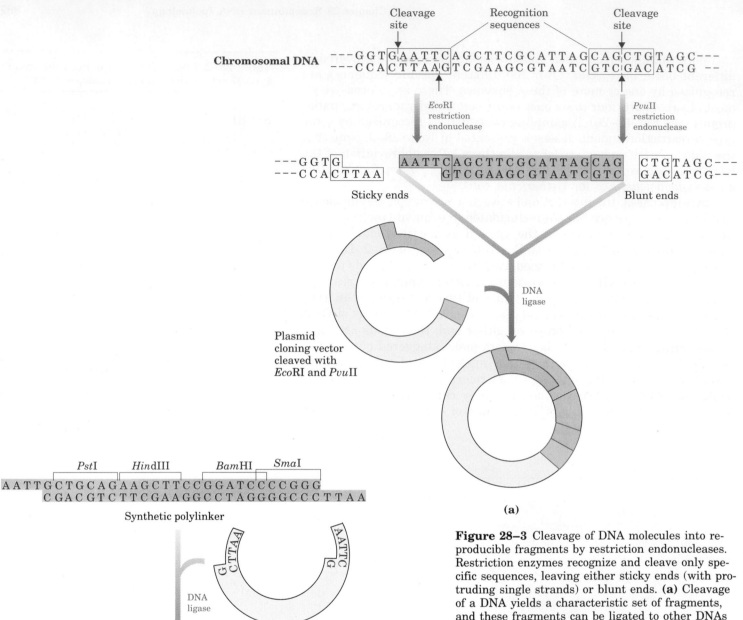

Figure 28–3 Cleavage of DNA molecules into reproducible fragments by restriction endonucleases. Restriction enzymes recognize and cleave only specific sequences, leaving either sticky ends (with protruding single strands) or blunt ends. **(a)** Cleavage of a DNA yields a characteristic set of fragments, and these fragments can be ligated to other DNAs such as the cleaved cloning vector (a plasmid) shown here. The ligation reaction is facilitated by the annealing of complementary sticky ends. DNA fragments with blunt ends are ligated at a lower efficiency than those with complementary sticky ends, and DNA fragments with different (noncomplementary) sticky ends generally are not ligated. **(b)** A synthetic DNA fragment containing the recognition sequences for several restriction endonucleases can be inserted into a plasmid after it has been cleaved by a restriction endonuclease; this creates a polylinker.

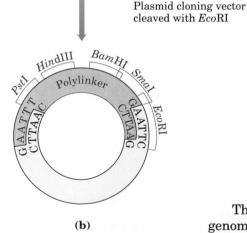

The average size of the DNA fragments produced by cleaving genomic DNA with a restriction endonuclease depends upon the frequency with which a particular restriction site occurs in a large DNA molecule; this in turn depends largely on the size of the recognition sequence. In a DNA molecule with a random sequence in which all four nucleotides are equally abundant, a 6 base pair sequence recognized by a restriction endonuclease such as *Bam*HI will occur on average once every 4^6, or 4,096, base pairs. Enzymes that recognize a 4 base pair sequence will produce smaller DNA fragments; a recognition sequence of this size would be expected to occur on average every 256 base pairs. These sequences tend to occur less frequently than this because nucleotide sequences in DNA are not random and the four nucleotides are not equally abundant. The average size of the fragments produced by restriction endonuclease cleavage of a large DNA can be increased by simply not allowing the reaction to go to completion. Such an incomplete reaction is often called a partial digest.

Once a DNA molecule has been cleaved into fragments, a particular fragment that a researcher is interested in can be separated from the others by agarose gel electrophoresis (p. 347) or HPLC (p. 122). Because cleavage of a typical mammalian genome by a restriction endonuclease may yield several hundred thousand different fragments, isolation of a particular DNA fragment by electrophoresis or HPLC is often impractical. In these cases an intermediate step in the cloning of a specific gene or DNA segment of interest is the construction of a DNA library, described later in the chapter.

When the target DNA fragment is isolated, it is joined to a cloning vector using DNA ligase. The base-pairing of complementary sticky ends greatly facilitates the ligation reaction (Fig. 28–3). Because different restriction endonucleases usually generate different sticky ends, the efficiency of the ligation step is greatly affected by the endonucleases used to generate the DNA fragments. A fragment generated by *Eco*RI generally will not be linked to a fragment generated by *Bam*HI. Blunt ends can also be ligated, albeit less efficiently.

Before ligating two DNA fragments, it is often useful to add recognition sequences for a restriction endonuclease (other than that used to create the fragments) at the junction to permit cleavage of the ligated DNA at that location later on. This is often done by inserting a synthetic DNA fragment containing the required recognition sequence between the two DNA fragments. Such a synthetic DNA fragment is generally called a **linker.** A synthetic fragment containing recognition sequences for several restriction endonucleases is called a **polylinker** (Fig. 28–3b).

The importance of sticky ends in efficiently joining two DNA fragments in a desired manner was apparent in the earliest recombinant DNA experiments. Before restriction endonucleases were widely available, some workers found that sticky ends could be generated by the combined action of the bacteriophage λ exonuclease and terminal transferase (Table 28–1). The fragments to be joined were given complementary homopolymeric tails (Fig. 28–4, p. 990). This method was used by Peter Lobban and Dale Kaiser in 1971 in the first experiments to join naturally occurring DNA fragments. Similar methods were used soon after in the laboratory of Paul Berg to join DNA segments from simian virus 40 (SV40) to DNA derived from bacteriophage λ, thereby creating the first recombinant DNA molecule involving DNA segments from different species.

Cloning Vectors Amplify Inserted DNA Segments

Three types of cloning vectors—plasmids, bacteriophages, and cosmids—are commonly used in *E. coli*. **Plasmids** (see Fig. 23–3) are circular DNA molecules that replicate separately from the host chromosome. Naturally occurring bacterial plasmids range in size from 5,000 to 400,000 base pairs.

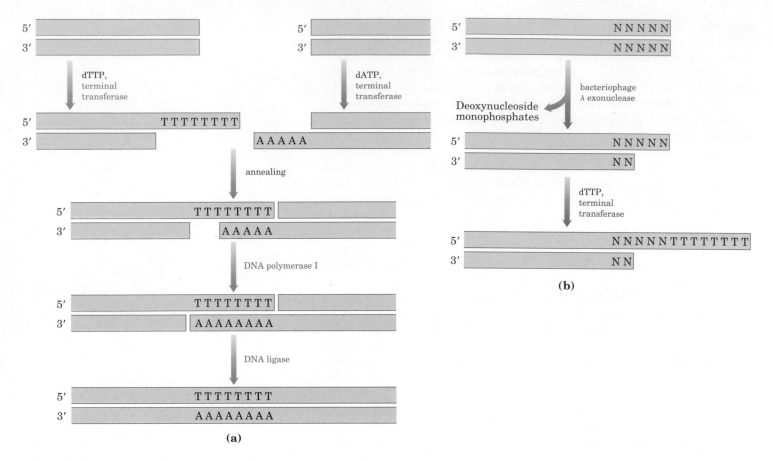

Figure 28–4 Sticky ends generated by terminal transferase can be used to join two DNA fragments. **(a)** Complementary homopolymeric tails are added to the ends of the two fragments to be joined, forming sticky ends. After annealing, the gaps are filled and the nicks sealed by the action of DNA polymerase I and DNA ligase (Chapter 24). **(b)** The optimal substrate for terminal transferase is the 3′ OH at the end of a single strand of DNA at least three nucleotides long. If the ends of the duplex DNA have a 5′ protruding single strand or are blunt ends, the λ exonuclease (which degrades DNA strands in the 5′→3′ direction) can be used to create a good substrate for terminal transferase. N denotes any base.

Plasmids can be introduced into bacterial cells by a process called **transformation.** To get the cells to take up the DNA, the cells and DNA are incubated together at 0 °C in a calcium chloride solution, then subjected to heat shock by rapidly shifting the cells to temperatures of 37 to 43 °C. For reasons not entirely understood, cells so treated become "competent" to take up the DNA. Because only a few cells take up the plasmid DNA, a method is needed for selecting those that do. The usual strategy is to build into the plasmid a gene that the host cell requires for growth under specific conditions. This makes a cell that contains the plasmid "selectable" if the cell is grown under those conditions. The gene, sometimes called a selectable marker, is often one that confers resistance to an antibiotic. Only those few cells that have been transformed by the recombinant plasmid will be antibiotic resistant and thus able to grow in the presence of the antibiotic.

Many different plasmid vectors suitable for cloning have been developed by modifying naturally occurring plasmids. Some of the important features of a cloning vector are illustrated by the *E. coli* plasmid pBR322 (Fig. 28–5): (1) the origin of replication is required to propagate the plasmid and helps maintain it at a level of 10 to 20 copies per cell; (2) two genes that confer resistance to different antibiotics allow the selection of cells that contain the plasmid or a recombinant version of it (Fig. 28–6); (3) several unique recognition sequences for different restriction endonucleases provide sites where the plasmid can be cut and foreign DNA inserted; and (4) an overall small size facilitates the plasmid's entry into cells. The efficiency of bacterial transformation decreases as plasmid size increases, and it is difficult to clone DNA segments longer than about 15,000 base pairs when plasmids are used as the vector.

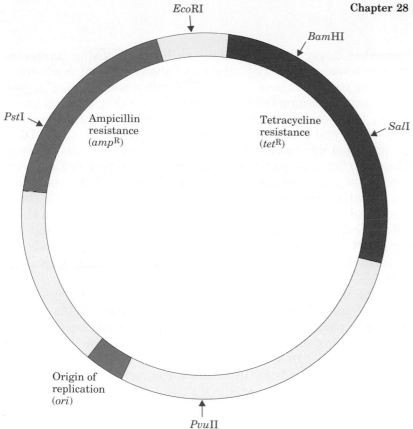

Figure 28–5 The constructed plasmid pBR322, showing the location of some important restriction sites, antibiotic-resistance genes, and the replication origin *(ori)*. This plasmid, constructed by Herbert Boyer and coworkers in 1977, was one of the early plasmids designed expressly for cloning in *E. coli.*

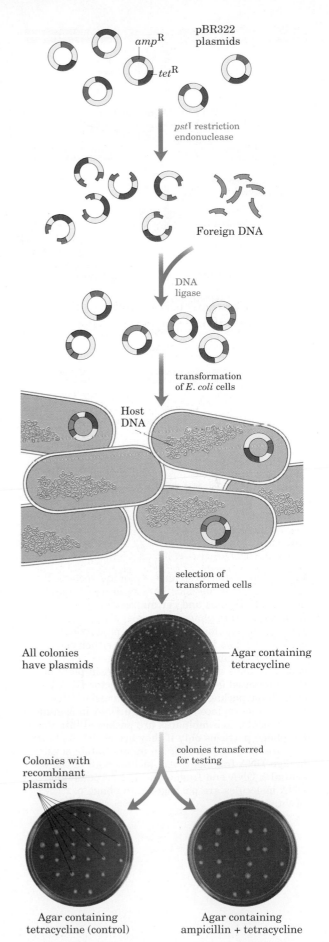

Figure 28–6 Cloning foreign DNA in *E. coli* with pBR322. If foreign DNA is inserted at the *Pst*I restriction site, the ampicillin-resistance element is disrupted and inactivated. After ligation of the DNA and transformation of *E. coli* cells, the cells are grown on agar plates containing tetracycline to select for those that have taken up a plasmid. By means of sterile toothpicks, individual colonies from these agar plates are transferred to the same position within a grid on two additional plates; one plate contains tetracycline (a control) and the other contains both tetracycline and ampicillin. Those cells that grow in the presence of tetracycline, but do not form colonies on the plate containing tetracycline plus ampicillin, contain recombinant plasmids (the ampicillin-resistance element is nonfunctional). Cells that contain pBR322 that was ligated without the insertion of a foreign DNA fragment retain ampicillin resistance and grow on both plates. Note that in this and other experiments that involve the use of two or more plate replicas an orienting mark is put on the back of each plate so that the colonies on different plates can be readily aligned.

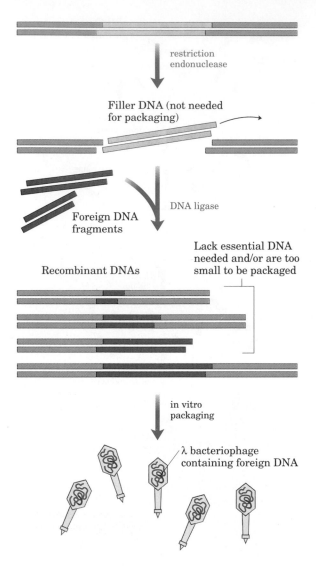

Figure 28–7 Bacteriophage λ cloning vectors. Recombinant DNA methods have been used to remove nonessential genes and certain restriction sites from the bacteriophage λ genome. The remaining genes (essential for bacteriophage production) are clustered in two large fragments at either end of the linear chromosome. Bacteriophage λ vectors generally have a piece of "filler" DNA in place of the eliminated genes to make the vector DNA large enough for packaging into phage particles. This filler can be replaced with foreign DNA in cloning experiments. Recombinants are packaged into viable phage particles only if they are of the appropriate size (i.e., they include an appropriately sized foreign DNA fragment) and include both of the essential λ DNA end fragments. The recombinant DNA molecules are packaged into phage particles in vitro.

Somewhat larger DNA segments can be cloned using **bacteriophage λ** as a vector. Bacteriophage λ has a very efficient mechanism for delivering its 48,502 base pairs of DNA into a bacterium. The general procedure for cloning DNA in bacteriophage λ (Fig. 28–7) is based on two key features of the λ genome: (1) about one-third of the genome is nonessential and can be replaced with foreign DNA, and (2) DNA will be packaged into infectious phage particles only if it is between 40,000 and 50,000 base pairs long. Bacteriophage λ vectors have been developed that can be readily cleaved into three pieces, two of which contain essential genes but which together are only about 30,000 base pairs long. Additional DNA must therefore be inserted between them to produce viable phage particles. Bacteriophage λ vectors permit the cloning of DNA fragments up to 23,000 base pairs long, and their design ensures that all viable phage particles will contain a foreign DNA fragment.

Once the bacteriophage λ fragments are ligated to foreign DNA fragments of suitable size, the resulting recombinant DNAs can be packaged into phage particles by adding them to crude bacterial cell extracts containing all the proteins needed to assemble a complete phage. This is called **in vitro packaging** (Fig. 28–7). The bacteriophage vector is now ready for insertion of the recombinant DNA into *E. coli* cells.

Cosmids are recombinant plasmids that combine useful features of both plasmids and bacteriophage λ. They are designed to permit the cloning of even larger DNA fragments (up to 45,000 base pairs). Cosmids (Fig. 28–8) are small (typically 5,000 to 7,000 base pairs), circular DNA molecules that contain (1) a plasmid origin of replication, (2) one or more selectable markers, (3) a number of unique restriction sites

Table 28–3 Types of cloning vectors used in *E. coli*

Type of vector	Method of introduction into *E. coli*	Method of propagation	Size of DNA fragment that can be cloned
Plasmids; modified by recombinant DNA techniques	Transformation; cells made competent to take up recombinant vector, then transformed cells selected using selectable marker	Plasmid replication	Up to 15,000 bp
Bacteriophage λ	Phage infection, following in vitro packaging of recombinant vector into phage particles	Phage replication	Up to 23,000 bp
Cosmids, constructed from plasmid and λ DNA genes	Either of above methods, depending on size of DNA fragment inserted; larger fragments require in vitro λ packaging	Plasmid-type replication	Up to 45,000 bp

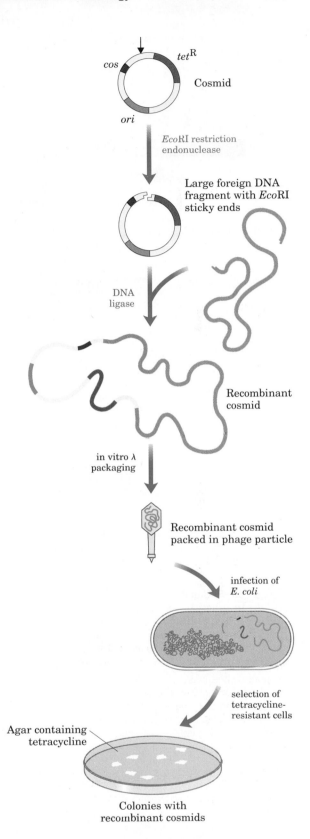

Figure 28–8 Cloning with cosmids. A cosmid contains a replication origin (*ori*) for propagation as a plasmid, and a *cos* site required for the packaging of DNA into λ phage particles. Unique restriction sites and antibiotic-resistance elements aid cloning and selection. Inserting a large foreign DNA fragment into the cosmid precludes bacterial transformation but, if the recombinant DNA molecule is a suitable size, it can be packaged into phage particles in vitro. The phage λ particles are then used to introduce the cosmid into bacterial cells, where the cosmid is propagated as a plasmid.

where foreign DNA can be inserted, and (4) a *cos* site (a DNA sequence in bacteriophage λ that is required for packaging).

Cosmids contain no other bacteriophage λ genes and can be propagated in *E. coli* like plasmids. When a large foreign DNA fragment is cloned into them, transformation of *E. coli* with these recombinant DNA constructs becomes difficult. If the cosmid contains a large enough insert of foreign DNA to be packaged into a phage particle, in vitro λ packaging systems permit the bacteriophage λ particle to be the vehicle for introducing the cosmid DNA efficiently into the bacterial cell. Once in the cell, the cosmid is again propagated as a plasmid, as it lacks the λ genes needed to make phage particles in the cell.

The three types of cloning vectors are summarized in Table 28–3.

Isolating a Gene from a Cellular Chromosome

Because a single gene is only a very small part of a chromosome, isolating a DNA fragment containing a particular gene often requires two procedures. First, a DNA library is constructed that contains many thousands of DNA fragments derived from a cellular chromosome. Second, the DNA fragment containing the gene of interest is identified by taking advantage of the one property that distinguishes it from the other DNA fragments—its sequence.

Cloning a Gene Often Requires a DNA Library

A **DNA library** is a collection of DNA fragments derived from the genome of a particular organism, with each fragment attached to a cloning vector. In short, genomic DNA is cleaved into thousands of fragments, and *all* of them are cloned. Thus, the total information content of an organism is represented by all the fragments in the library in much the same way as human knowledge is represented by all the volumes in a book library. A vector is prepared for cloning by cleaving purified vector DNA at an appropriate position with a restriction endonuclease. Genomic DNA to be cloned is reduced to appropriately sized fragments by partial digestion with a restriction endonuclease (usually the same one used to prepare the vector), followed by a procedure such as sucrose density gradient (isopycnic) centrifugation to remove fragments that are too large or too small to be cloned into the chosen vector. Fragments are then mixed with the cleaved vector DNA and ligated. The mixture is used to transform bacterial cells or is packaged

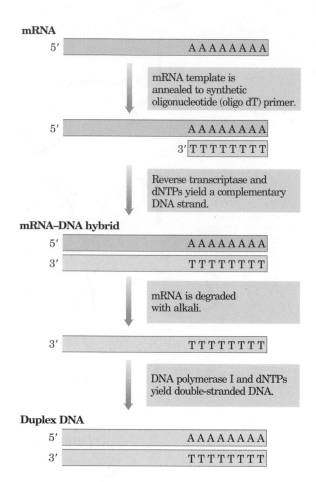

mRNA

5′   A A A A A A A A

> mRNA template is annealed to synthetic oligonucleotide (oligo dT) primer.

5′   A A A A A A A A

3′ T T T T T T T T

> Reverse transcriptase and dNTPs yield a complementary DNA strand.

mRNA–DNA hybrid

5′   A A A A A A A A

3′   T T T T T T T T

> mRNA is degraded with alkali.

3′   T T T T T T T T

> DNA polymerase I and dNTPs yield double-stranded DNA.

Duplex DNA

5′   A A A A A A A A

3′   T T T T T T T T

Figure 28–9 Constructing a cDNA library from mRNA. In practice, the mRNA from a cell will include transcripts from thousands of genes, and the cDNAs generated will be correspondingly heterogeneous. The duplex DNA produced by this method is inserted into an appropriate cloning vector.

into bacteriophage particles as described above. The final result is a large population of bacteria or bacteriophages each harboring a different recombinant DNA molecule. Ideally, nearly all of the DNA in the genome will be represented in the library, and libraries constructed in this way are referred to as **genomic libraries.** Each transformed bacterium grows into a colony or "clone" of cells, all of which have the same recombinant plasmid. In the case of bacteriophages, each type of recombinant phage creates a plaque—a clear region of lysed cells within a lawn of bacteria distributed evenly on an agar plate; all of the recombinant bacteriophages within a plaque are identical. The clone containing the particular gene a researcher is interested in must be identified within the thousands of clones in the library as described below. If the desired DNA is from a mammal with a genome of 3×10^9 base pairs of DNA and cosmids are used as cloning vectors, then the library must contain about 350,000 recombinant cosmids, each with a different insert of 35,000 to 45,000 base pairs, for there to be a 99% chance that any desired gene of unique sequence is represented in the library. The fragments in a genomic library derived from a higher eukaryote include not only genes but the noncoding DNA that makes up a large portion of many eukaryotic genomes.

A more specialized and exclusive DNA library can be constructed so as to include only those genes that are *expressed* in a given organism or even in certain cells or tissues. The critical difference between genes that are expressed and genes that are not is that the former are transcribed into RNA. The mRNA from an organism or certain cells derived from the organism is extracted, and **complementary DNAs (cDNAs)** are produced from the RNA in a multistep reaction catalyzed by reverse transcriptase (Fig. 28–9; Chapter 25). The resulting double-stranded DNA fragments are then inserted into a suitable vector and cloned, creating a population of clones called a **cDNA library.** Among the first eukaryotic genes characterized were those encoding globins, primarily because cloning of these genes was facilitated by making cDNA libraries from erythrocyte precursor cells, in which about half of the mRNA codes for globins.

Hybridization Identifies Specific Sequences in a DNA Library

After a DNA library has been generated, the challenge is to isolate one gene of interest from among the millions of other DNA segments represented in a DNA library. Because each gene has a unique nucleotide sequence, it often can be detected with the aid of a labeled (e.g., radioactive) DNA fragment that is complementary to it, called a **probe.** Typically, nitrocellulose paper is pressed onto an agar plate containing many individual bacterial colonies each containing a different recombinant DNA. Some cells from each colony adhere to the paper, forming a replica of the plate. The paper is treated with alkali to disrupt the cells and denature the DNA within, which remains localized to the region around the colony from which it came. The radioactive DNA probe is then added to the paper, annealing only to DNA containing the complementary gene. The labeled colony is visualized by autoradiography (Fig. 28–10). Detecting any nucleic acid with a labeled nucleic acid probe that is complementary to it is an application of nucleic acid hybridization (see Fig. 12–31), and many variations on this procedure have been developed to detect both DNA (Box 28–1, p. 996) and RNA.

Figure 28–10 Identifying a clone with the desired DNA segment; a schematic diagram of a hybridization method. The radioactive DNA probe hybridizes only with homologous DNA. The annealed radioactive probe is revealed by autoradiography. When the labeled colonies have been identified, the corresponding colonies on the original agar plate can be used as a source of cloned DNA for further study.

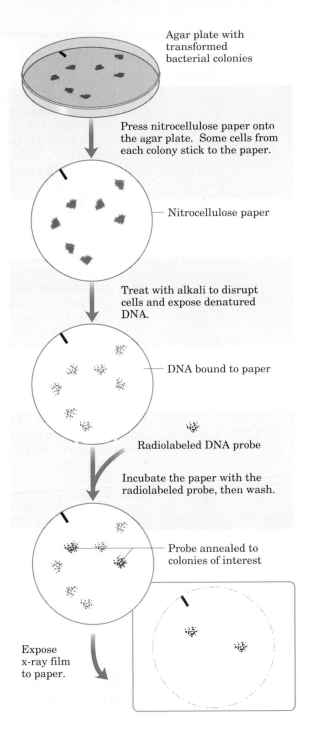

Agar plate with transformed bacterial colonies

Press nitrocellulose paper onto the agar plate. Some cells from each colony stick to the paper.

Nitrocellulose paper

Treat with alkali to disrupt cells and expose denatured DNA.

DNA bound to paper

Radiolabeled DNA probe

Incubate the paper with the radiolabeled probe, then wash.

Probe annealed to colonies of interest

Expose x-ray film to paper.

Frequently, the limiting step in cloning a gene is finding or generating a complementary strand of nucleic acid to use as a probe. The origin of a probe depends on what is known about the gene under investigation. Sometimes a homologous gene cloned from another species can be used as a probe. Alternatively, if the protein product of a gene has been purified, probes can be designed and synthesized on the basis of its amino acid sequence and a knowledge of the genetic code (Fig. 28–11).

In this work, the synthesis and sequencing methods for proteins and DNA described in previous chapters come together and complement one another. Just as the sequence of an isolated protein can be used to isolate the corresponding gene, a cloned gene can be used to isolate its protein product. The sequence of an isolated gene can be used to design and chemically synthesize a short peptide that represents part of the protein product of the gene. Antibodies to this synthetic peptide can then be used to identify and purify the protein itself (Chapter 6) for further study.

Figure 28–11 Designing a probe to detect the gene for a protein of known amino acid sequence. Because the genetic code is degenerate, there is more than one possible DNA sequence that codes for any given amino acid sequence. As the correct DNA sequence cannot be known in advance, the probe is designed to be complementary to a region of the gene with minimal degeneracy. The oligonucleotides are synthesized with the sequence selectively randomized so that some contain either of the two possible nucleotides at each position of potential degeneracy (shaded in red). In the example shown, the synthesized oligonucleotide is actually a mixture of eight different sequences; one of the eight will complement the gene perfectly. All eight will match in at least 17 of 20 positions.

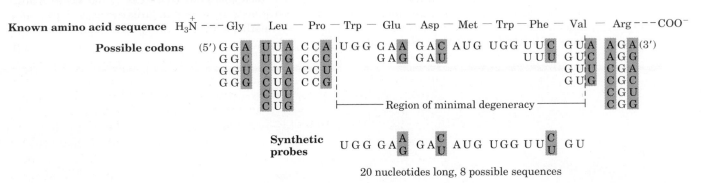

Known amino acid sequence   H₃N⁺ – – – Gly — Leu — Pro — Trp — Glu — Asp — Met — Trp — Phe — Val — Arg – – – COO⁻

Possible codons

Region of minimal degeneracy

Synthetic probes

20 nucleotides long, 8 possible sequences

BOX 28–1   A New Weapon in Forensic Medicine

Traditionally, one of the most accurate methods for placing an individual at the scene of a crime has been a fingerprint. A new identification technique has become available based on methods developed for recombinant DNA technology. This is sometimes called DNA fingerprinting (also DNA typing or DNA profiling), and in some ways it is more powerful than any other identification method.

DNA fingerprinting is based on **sequence polymorphisms** that occur in the human genome (and the genome of every other organism). Sequence polymorphisms are slight sequence differences (usually single base-pair changes) that occur from individual to individual once every few hundred base pairs, on average. Each difference from the consensus human genome sequence is generally present in only a fraction of the human population, but every individual has some of them. Some of the sequence changes affect recognition sites for restriction enzymes, resulting in variation from individual to individual in the size of certain DNA fragments produced by digestion with a particular restriction enzyme. These size differences are referred to as **restriction fragment length polymorphisms,** or RFLPs. A probe for a sequence that is repeated several times in the human genome generally identifies a few of the thousands of DNA fragments generated when the human genome is digested with a restriction endonuclease.

The detection of RFLPs relies on a specialized hybridization procedure called **Southern blotting** (Fig. 1). DNA fragments from digestion of genomic DNA by restriction endonucleases are first separated according to size by electrophoresis in an agarose gel. The DNA fragments are denatured by soaking the gel in alkali, then transferred to nitrocellulose paper in such a way as to reproduce on the paper the distribution of fragments in the gel. The paper is then immersed in a solution containing a radioactively labeled DNA probe. Fragments to which the probe hybridizes are revealed by autoradiography, using procedures similar to those described in Figure 28–10.

The genomic DNA sequences used in these tests are generally regions containing repetitive DNA (short sequences repeated thousands of times in tandem; see p. 797), which are common in the genomes of higher eukaryotes. The number of repeated units in such DNA varies from individual to individual (except in the case of identical twins). If a suitable probe is chosen, the pattern of bands in such an experiment can be distinctive for each individual tested. If several probes are used, the test can be made so selective that it can positively identify a single individual in the human population. However, the Southern blot procedure requires relatively fresh DNA samples and larger amounts of DNA than are generally present at a crime scene. To increase sensitivity, RFLP analysis is being augmented by polymerase chain reaction (PCR) methods (see Fig. 28–12), which permit vanishingly small amounts of DNA to be amplified. The improved tests allow DNA fingerprints to be obtained from a single hair, a small semen sample from a rape victim, or from samples that might be months or even many years old.

Since their introduction in 1985, these methods have been developed to the point where they are proving decisive in court cases worldwide. In the example in Figure 1, the DNA from a semen sample obtained from a rape and murder victim was analyzed along with DNA samples from the victim and two suspects. Each of the DNA samples was cleaved into fragments and separated by gel electrophoresis. Radioactive DNA probes were used to identify a small subset of these fragments that contained sequences complementary to the probe. The sizes of the fragments identified varied from one individual to the next, as seen here in the different patterns for the three individuals (victim and two suspects) tested. One rape suspect's DNA exhibits a banding pattern identical to that of a semen sample taken from the victim. One probe was used here, but three or four different probes would be used to make a positive identification. Results have been used to help both convict and acquit suspects. DNA fingerprints can also be used to establish paternity with an extraordinary degree of certainty. The results of DNA fingerprinting have been successfully challenged in some court cases because of irregularities and a lack of controls in many early examples. The far-reaching impact of this technology on court cases will nevertheless continue to grow as standards are agreed upon and the methods become widely established in forensic laboratories.

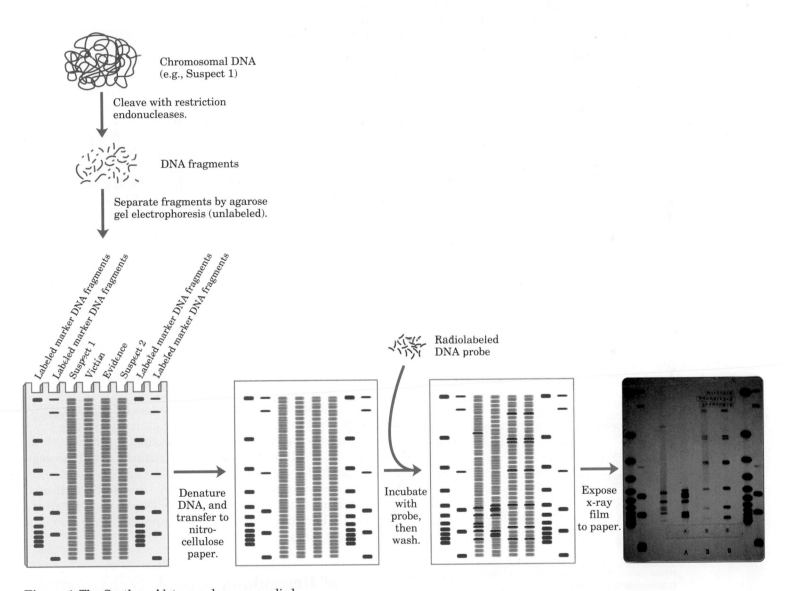

Figure 1 The Southern blot procedure, as applied to DNA fingerprinting.

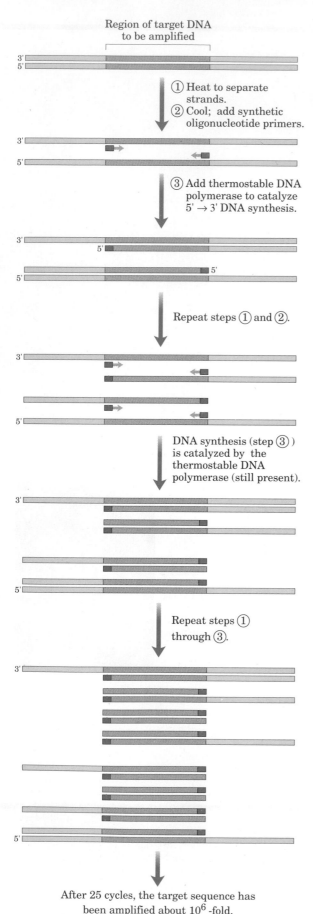

Region of target DNA
to be amplified

① Heat to separate
   strands.
② Cool; add synthetic
   oligonucleotide primers.

③ Add thermostable DNA
   polymerase to catalyze
   5' → 3' DNA synthesis.

Repeat steps ① and ②.

DNA synthesis (step ③)
is catalyzed by the
thermostable DNA
polymerase (still present).

Repeat steps ①
through ③.

After 25 cycles, the target sequence has
been amplified about 10^6-fold.

Figure 28–12 Amplifying a specific DNA segment with a polymerase chain reaction. DNA strands are separated by heating, then annealed to an excess of short synthetic DNA primers (blue) that flank the region to be amplified. After polymerization, the process is repeated for 25 or 30 cycles. The thermostable DNA polymerase *Taq*I (from *Thermus thermophilus,* bacteria that grow in hot springs) is not denatured by the heating steps.

Specific DNA Sequences Can Be Amplified

If one knows the sequence of at least part of a DNA segment to be cloned, cloning can be facilitated by amplifying the DNA segment in a process called a **polymerase chain reaction (PCR),** invented by Kary Mullis in 1984. Two oligonucleotides are synthesized, each complementary to a short sequence in one strand of the desired DNA segment and positioned just beyond the end of the sequence to be amplified; these synthetic oligonucleotides can be used as primers for replication of the DNA segment in vitro (Fig. 28–12). Isolated DNA containing the segment to be cloned is heated briefly to denature it, then cooled in the presence of a large excess of the synthetic oligonucleotide primers. A heat-stable DNA polymerase called *Taq*I and the four deoxynucleoside triphosphates are then added, and the primed DNA segment is selectively replicated. This process is repeated through 25 or 30 cycles, which can take only a few hours when automated, amplifying the DNA segment to the point where it can be readily isolated and cloned.

The PCR method is sensitive enough to detect as little as one DNA molecule in almost any type of sample. It has been used to clone DNA fragments from mummies and the remains of extinct animals such as the woolly mammoth, creating the new fields of molecular archaeology and molecular paleontology. In addition to its usefulness for cloning DNA, it is a potent new tool in forensic medicine (Box 28–1). It is also being used for detection of viral infections before they cause symptoms or elicit a detectable immune response and in prenatal diagnosis of a wide array of genetic diseases.

The Products of Recombinant DNA Technology

Normally, cloning a gene is only the first step in a much grander design. A cloned gene can be used to generate large amounts of its protein product. The amino acid sequence of the protein can be altered by introducing base-pair changes in the gene, a strategy that can be very powerful in addressing questions concerning protein folding, structure, and function. Increasingly sophisticated methods for moving DNA into and out of cells of all types are providing another avenue for studying gene function and regulation, and are allowing the introduction of new traits into plants and animals.

Our focus now turns to applications of DNA cloning, beginning with the proteins produced by cloned genes. We then describe cloning procedures used for a variety of eukaryotic cells, before finishing with an overview of the potential and implications of this technology.

Cloned Genes Can Be Expressed

Frequently it is the product of the cloned gene rather than the gene itself that is of primary interest. For example, certain proteins are immensely important for a variety of commercial, therapeutic, or research purposes. Understanding the fundamentals of DNA, RNA, and protein metabolism and their regulation in *E. coli* has made it possible to express cloned genes in order to study their protein products. Because most eukaryotic genes do not have the DNA sequence elements (promoters, etc.) required for their expression in *E. coli* cells, bacterial regulatory sequences for transcription and translation must be inserted at appropriate positions in the vector DNA relative to the eukaryotic gene itself. In some cases cloned genes are expressed so well that the protein product is overproduced, sometimes representing 10% or more of the cellular protein. Such a high concentration of a foreign protein can kill an *E. coli* cell; in these cases gene expression must be limited until a few hours before the planned harvest of the cells. Cloning vectors that have the transcription and translation signals necessary for the regulated expression of a cloned gene are often called **expression vectors.**

A variety of expression vectors have been constructed with unique restriction sites for cloning placed near a well-characterized promoter (such as the *lac* promoter from *E. coli*) and its regulatory elements (Fig. 28–13). Some of these vectors incorporate other features such as bacterial ribosome binding sites and transcription termination signals. Expression of cloned genes in bacteria and other cells has provided large quantities of many proteins important for industry and research. Some of the proteins now available for medicine and agriculture as a result of this technology are described later in this chapter.

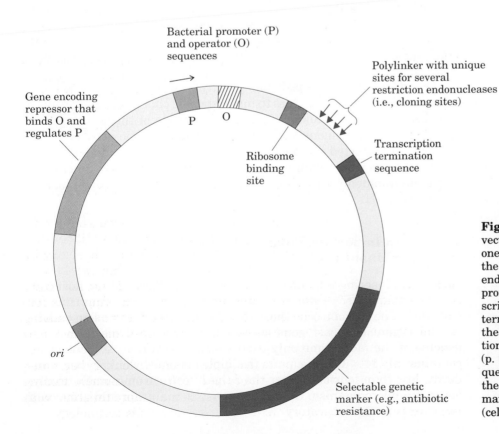

Bacterial promoter (P) and operator (O) sequences

Gene encoding repressor that binds O and regulates P

P   O

Polylinker with unique sites for several restriction endonucleases (i.e., cloning sites)

Ribosome binding site

Transcription termination sequence

ori

Selectable genetic marker (e.g., antibiotic resistance)

Figure 28–13 An example of an *E. coli* expression vector. The gene to be expressed is inserted into one of the restriction sites in the polylinker, near the promoter. The gene must be inserted with the end encoding the amino terminus proximal to the promoter. The promoter provides for efficient transcription of the inserted gene, and the transcription terminator can improve the amount and stability of the mRNA produced. The operator permits regulation by means of a repressor that binds to it (p. 944). The ribosome binding site provides sequence signals needed for efficient translation of the mRNA derived from the gene. The selectable marker allows the selection of transformed cells (cells containing the recombinant DNA).

Cloned Genes Can Be Altered

The structure of a protein can be changed by altering the DNA sequence of the cloned gene for that protein. One or a few specific amino acids may be replaced by means of **site-directed mutagenesis** (also referred to as "protein engineering"), one of the most powerful methods for studying protein structure and function. Two of the common methods for achieving this are illustrated in Figure 28–14.

Sometimes, if appropriate restriction sites closely flank the sequence to be altered, a change can be made by simply removing a DNA segment and replacing it with a synthetic one that is identical to the original except for the desired change (Fig. 28–14a). If suitably located restriction sites are not present, an approach called **oligonucleotide-directed mutagenesis** can be used to create a specific DNA sequence change (Fig. 28–14b). A short synthetic DNA strand with a specific base change is annealed to a single-stranded cloned copy of the gene. The mismatch of one base pair out of 15 to 20 does not adversely affect annealing if it is done at an appropriate temperature. This annealed strand is then used as a primer for synthesis of a complete duplex plasmid containing the mismatch. The duplex recombinant plasmid is introduced into bacteria, where the mismatch is repaired by cellular DNA repair enzymes. About half of the repair events will remove and replace the altered base, but the other half will remove and replace the normal base, retaining the desired mutation. Transformants are screened (often simply by sequencing their plasmid DNA) until a bacterial colony with the altered sequence is found. Changes can also be introduced that involve more than one base pair. Large parts of a gene can be deleted by cutting out a segment with restriction endonucleases and ligating the remaining portions to form a smaller gene. Parts of two different genes can be ligated to create new combinations. The product of such a fused gene is called a **fusion protein.** (Note that these "fusion proteins" are unrelated to the fusion proteins that participate in the process of membrane fusion, discussed in Chapter 10.) In fact, ingenious methods exist to bring about virtually any gene alteration.

Site-directed mutagenesis has greatly facilitated research on proteins by allowing investigators to make specific changes in the primary structure of a protein and to examine the effects of these changes on the folding, three-dimensional structure, and catalytic activity of the protein. Site-directed mutagenesis is also being used commercially to create proteins with enhanced activity or with the ability to function in harsh environments such as high temperatures, organic solvents, and extremes of pH.

Yeast Is an Important Eukaryotic Host for Recombinant DNA

Genetic engineering is by no means restricted to *E. coli* as the host cell. Among eukaryotes, yeasts are particularly convenient organisms for this work. The reasons echo those that recommend *E. coli* as an experimental organism: yeast genetics is a well-developed discipline; the genome of the most commonly used yeast, *Saccharomyces cerevisiae*, contains only 14×10^6 base pairs (a simple genome by eukaryotic standards, less than fourfold larger than the *E. coli* chromosome); finally, yeast is a microorganism that is very easy to maintain and grow on a large scale in the laboratory.

Figure 28–14 Two approaches to site-directed mutagenesis. **(a)** A DNA segment is synthesized and is used to replace a DNA fragment that has been removed by cleavage with a restriction endonuclease. **(b)** An oligonucleotide is synthesized with a desired sequence change at one position. This is hybridized to a single-stranded copy of the gene to be altered, and acts as primer for synthesis of a duplex DNA (with one mismatch). This DNA is then used to transform cells. Cellular mismatch repair will produce some clones with the desired sequence change.

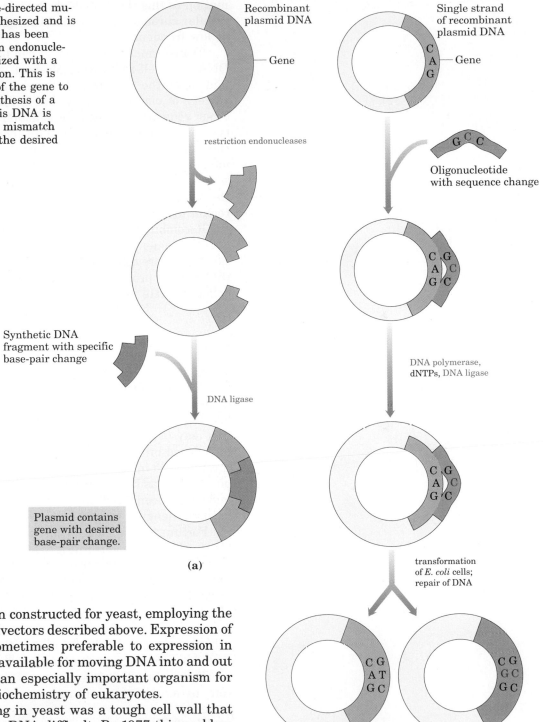

(a)

(b)

Expression vectors have been constructed for yeast, employing the same principles as for the *E. coli* vectors described above. Expression of eukaryotic genes in yeast is sometimes preferable to expression in *E. coli*. The convenient methods available for moving DNA into and out of the cells help to make yeast an especially important organism for studying many aspects of the biochemistry of eukaryotes.

The major obstacle to cloning in yeast was a tough cell wall that made the introduction of foreign DNA difficult. By 1977 this problem was overcome with methods that couple partial enzymatic degradation of the cell wall with a calcium–polyethylene glycol treatment, making the cells sufficiently porous for DNA uptake. The first transformation experiments in yeast used the yeast *leu*2 gene (which encodes β-isopropylmalate dehydrogenase; see Fig. 21–12), cloned in an *E. coli* plasmid, to transform yeast cells that had a mutant *leu*2 gene (*leu*2⁻) and thus could not grow in the absence of added leucine. Some of the cells were transformed to leu⁺ (i.e., they could synthesize leucine) following replacement of the defective *leu*2⁻ gene with the introduced *leu*2 gene by homologous recombination (Chapter 24). Transformation of this type in which the introduced DNA is integrated into a cellular chromosome is called **integrative transformation;** it occurs at low frequency.

Transformation efficiencies can be increased by introducing cloned DNA on a self-replicating plasmid. A naturally occurring yeast plasmid called the 2 micron (2μ) plasmid has been engineered to create a variety of cloning vectors that incorporate a replication origin and other sequences needed for plasmid maintenance in yeast. Another type of plasmid that provides an increase in transformation frequencies contains a yeast chromosomal origin of replication (an autonomously replicating sequence, ARS; see p. 830). Such plasmids are somewhat unstable (and are lost from the yeast population) unless they also contain a centromere that permits them to function and segregate like yeast chromosomes during cell division.

Recombinant plasmids are available that incorporate multiple genetic elements (replication origins, etc.), allowing them to be maintained in more than one species; for example, yeast or *E. coli*. Plasmids that can be propagated in cells of two or more different species are called **shuttle vectors.**

Cloning in Plants Is Aided by a Bacterial Parasite

The introduction of recombinant DNA into plants has enormous potential for agriculture, producing more nutritious and higher-yielding crops that are resistant to environmental stresses such as insect pests, disease, cold, and drought. Unlike animals, fertile plants of some species may be generated from a single transformed cell. Thus, a gene introduced into a plant cell may ultimately be transmitted to progeny through seed in successive generations. As with all systems, cloning in plant cells has its own peculiar problems. No naturally occurring plasmids have been found in plants to facilitate this process, and thus the most challenging task is getting DNA into plant cells.

Fortunately, scientists have found an important and adaptable ally in the soil bacterium *Agrobacterium tumefaciens*. The bacterium invades plants at the site of a wound, transforming plant cells near the wound and inducing them to form a tumor called a crown gall. *Agrobacterium* contains a large (~200,000 base pair) plasmid called the **Ti plasmid** (Fig. 28–15). When the bacterium contacts a plant cell, a segment of this plasmid, called the T DNA (~23,000 base pairs), is transferred from the Ti plasmid to the plant cell nucleus and is integrated at a random position in one of the plant's chromosomes during transformation. This is a rare example of DNA transfer from a prokaryote to a eukaryote; it represents a natural genetic engineering process.

The T DNA encodes enzymes that convert plant metabolites to two classes of compounds important to the bacterium (Fig. 28–16). The first class consists of the plant growth hormones, auxins and cytokinins, which stimulate growth of the transformed plant cells to form the crown gall tumor. The second is a series of unusual amino acids called opines, a food source for the bacteria. The opines are produced at high concentrations in the tumor and secreted to the surroundings. They can only be metabolized by *Agrobacterium,* using enzymes encoded elsewhere on the Ti plasmid. In this manner the bacteria monopolize available nutrients by converting them to a form that does not benefit any other organism.

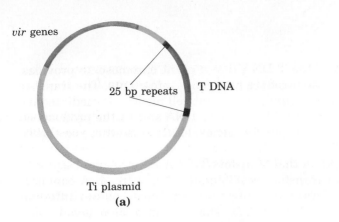

vir genes

25 bp repeats

T DNA

Ti plasmid

(a)

Figure 28–15 (a) The Ti (tumor-inducing) plasmid of *Agrobacterium tumefaciens*. **(b)** Acetosyringone, a phenolic compound produced by plants, increases in concentration in wounded plant cells and is released from the cells. *Agrobacterium* senses this compound, and the virulence *(vir)* genes on the Ti plasmid are expressed. The *vir* genes encode enzymes needed to introduce the T DNA into the genome of plant cells in the vicinity of the wound. A single-stranded copy of the T DNA is synthesized and transferred to the plant cell, where it is converted to duplex DNA and integrated into a chromosome. The T DNA encodes enzymes that synthesize growth hormones and opines (see Fig. 28–16) from common metabolites. Opines can be metabolized only by *Agrobacterium,* which uses them as a nutrient source. Expression of the T DNA genes by transformed plant cells thus leads to cell growth (tumor formation) and the diversion of plant cell nutrients to the invading bacteria.

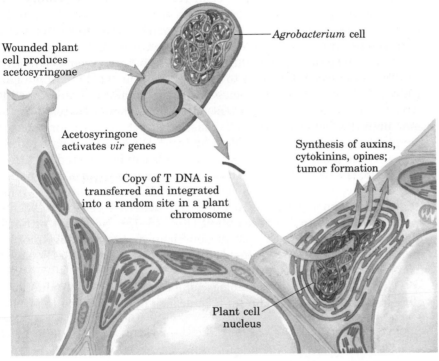

Agrobacterium cell

Wounded plant cell produces acetosyringone

Acetosyringone activates *vir* genes

Copy of T DNA is transferred and integrated into a random site in a plant chromosome

Synthesis of auxins, cytokinins, opines; tumor formation

Plant cell nucleus

(b)

Figure 28–16 Metabolites produced in *Agrobacterium*-infected plant cells. Auxins and cytokinins are growth hormones. The most common auxin is indoleacetate, derived from tryptophan. Cytokinins are adenine derivatives. Opines generally are derived from amino acid precursors; at least 14 different opines have been identified in different species of *Agrobacterium.*

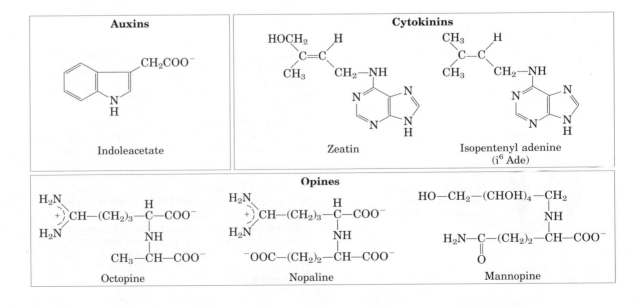

The integration of the T DNA into a plant chromosome provides the vehicle necessary to introduce new genes into plants. The transfer of T DNA from *Agrobacterium* to the plant cell nucleus is mediated by two 25 base pair repeats that flank the T DNA and by the products of several genes, called virulence *(vir)* genes, located elsewhere on the Ti plasmid (Fig. 28–15).

The bacterial system that transfers T DNA into the plant genome can be harnessed to transfer recombinant DNA instead. A common cloning strategy employs an *Agrobacterium* that contains two different recombinant plasmids (Fig. 28–17). The first is a Ti plasmid from which the T DNA segment has been removed in the laboratory. The second is an *Agrobacterium–E. coli* shuttle vector that contains the T-DNA 25 base pair repeats flanking the gene that a researcher wants to introduce and a selectable marker (often a gene that can render plant cells resistant to an antibiotic such as kanamycin). The engineered *Agrobacterium* is used to infect a leaf (Fig. 28–17). Crown galls are not formed because the genes for auxin, cytokinin, and opine biosynthetic enzymes are not present on either plasmid. Instead, the *vir* gene products from the altered Ti plasmid direct the transformation of the plant cells by the gene flanked by the T-DNA 25 base pair repeats in the second plasmid. The transformed cells can be selected by growth on kanamycin and induced with growth hormones to form new plants.

The successful transfer of recombinant DNA into plants is vividly illustrated by an experiment in which the luciferase gene from fireflies was introduced into a tobacco plant (Fig. 28–18). Note that although tobacco is not high on the list of beneficial plants needing improvement, it is often used experimentally because it is particularly easy to transform with *Agrobacterium*.

The potential of this technology is not limited to the production of glow-in-the-dark plants. The same approach has been used to produce

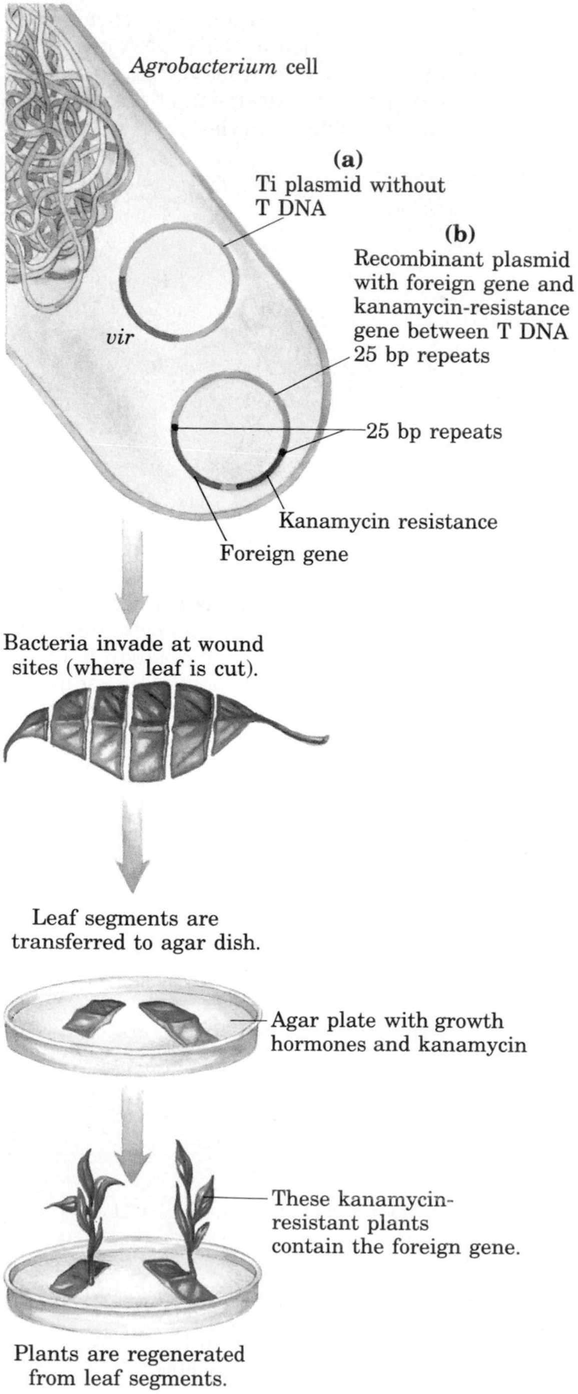

Agrobacterium cell

(a)
Ti plasmid without
T DNA

(b)
Recombinant plasmid
with foreign gene and
kanamycin-resistance
gene between T DNA
25 bp repeats

vir

25 bp repeats

Kanamycin resistance

Foreign gene

Bacteria invade at wound
sites (where leaf is cut).

Leaf segments are
transferred to agar dish.

Agar plate with growth
hormones and kanamycin

These kanamycin-
resistant plants
contain the foreign gene.

Plants are regenerated
from leaf segments.

Figure 28–17 A two-plasmid strategy to create a recombinant plant. One plasmid **(a)** is the Ti plasmid, modified so that it lacks T DNA. The other plasmid **(b)** contains the gene of interest (e.g., the gene for the insect-killing protein described in Fig. 28–19) and an antibiotic-resistance (here, kanamycin-resistance) element flanked by T-DNA 25 base pair repeats. This plasmid also contains the replication origin needed for propagation in *Agrobacterium*. The bacteria invade at the site of a wound (the edge of the cut leaf). The *vir* genes on plasmid **(a)** mediate transfer into the plant genome of the segment of plasmid **(b)** flanked by the 25 base pair repeats. New plants are generated when the leaf (with transformed cells) is placed on an agar dish with controlled levels of plant growth hormones and kanamycin. Nontransformed plant cells are killed by the kanamycin. The gene of interest and the antibiotic-resistance element are normally transferred together, and thus plants that grow in the presence of this antibiotic generally contain the gene.

Figure 28–18 A tobacco plant in which the gene for firefly luciferase is expressed. Light was produced after the plant was watered with a solution containing luciferin, the substrate for this light-producing enzyme (see Box 13–3, Fig. 2). Don't expect glow-in-the-dark ornamental plants at your local nursery anytime soon; the light is actually quite weak and this photograph represents a 24 hour exposure. The real point—that this technology allows the introduction of new traits into plants—is nevertheless elegantly made.

plants that are resistant to herbicides, plant viruses, and insect pests (Fig. 28–19). Potential benefits include increases in yields and a reduction in the need for environmentally harmful agricultural chemicals.

Cloning in Animal Cells Points the Way to Human Gene Therapy

The transformation of animal cells with foreign genetic material offers an important mechanism for advancing knowledge about the structure and function of animal genomes, as well as for the generation of animals with new traits. This potential has spawned intensive research efforts that have produced increasingly sophisticated means for cloning in animals.

Most work of this kind requires a source of individual cells. Intact tissues are often difficult to keep alive and manipulate. Fortunately, many types of animal cells can be isolated and grown in a medium in the laboratory if their growth requirements are carefully met. Many cells grown in this kind of **tissue culture** maintain the differentiated properties they had in the whole tissue for weeks or even months.

There are several common methods for introducing DNA into an animal cell. Because no suitable plasmidlike vector is available, transformation requires the integration of the DNA into a host-cell chromosome.

In **spontaneous uptake,** a calcium phosphate–DNA precipitate is taken up by the cells, perhaps by endocytosis. Generally only about one in 10^2 to 10^4 cells is transformed in this procedure. Another direct method that is sometimes more efficient is to make the cells transiently permeable to DNA by exposing them to a brief high-voltage pulse in a technique called **electroporation.**

Figure 28–19 Tomato plants engineered to be resistant to some insect larvae. The plant on the right expresses a gene for a protein toxin, derived from the bacterium *Bacillus thuringiensis*. The protein, introduced by a protocol similar to that described in Fig. 28–17, is toxic to the larvae of some moth species that destroy tomato leaves. This protein, however, is harmless to humans and other organisms. The plant on the left is not genetically altered. Both plants were exposed to equal numbers of larvae. Insect resistance has also been genetically engineered in cotton and other plants.

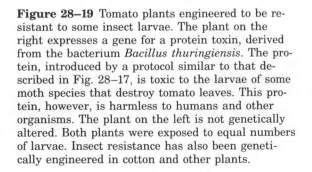

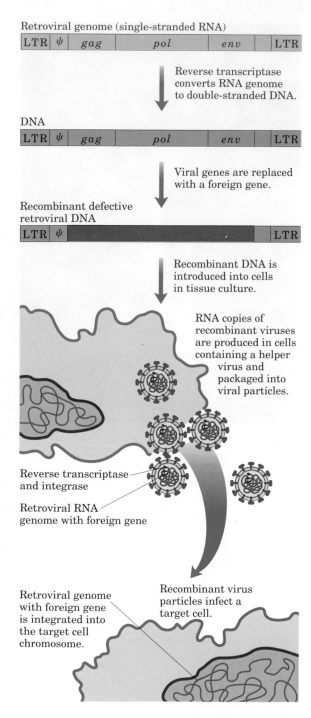

Retroviral genome (single-stranded RNA)

| LTR | ψ | gag | pol | env | LTR |

Reverse transcriptase converts RNA genome to double-stranded DNA.

DNA

| LTR | ψ | gag | pol | env | LTR |

Viral genes are replaced with a foreign gene.

Recombinant defective retroviral DNA

| LTR | ψ | | LTR |

Recombinant DNA is introduced into cells in tissue culture.

RNA copies of recombinant viruses are produced in cells containing a helper virus and packaged into viral particles.

Reverse transcriptase and integrase

Retroviral RNA genome with foreign gene

Recombinant virus particles infect a target cell.

Retroviral genome with foreign gene is integrated into the target cell chromosome.

Figure 28–20 Cloning in mammalian cells with retroviral vectors.

Microinjection entails direct injection of DNA into the nucleus of a cell with the aid of a very fine needle. For skilled practitioners this method has a high success rate, but because cells must be injected one by one, the total number that can be treated is small.

A number of eukaryotic viruses sometimes integrate their DNA into a chromosome in the host cell. Some of these, in particular certain retroviruses (p. 882), have been modified to act as **viral vectors** to introduce foreign DNA into mammalian cells. The viruses have their own mechanisms for moving nucleic acid into cells, and transformation by this route can be very efficient. A simplified map of a typical retroviral genome is shown in Figure 28–20. When the virus enters a cell, its RNA genome is converted to DNA by reverse transcriptase and is then integrated into the host genome in a reaction mediated by the viral integrase. The long terminal repeat (LTR) sequences are required for integration of retroviral DNA in the host chromosome (see Fig. 25–30), and the ψ sequence is required to package the viral RNA in viral particles.

The *gag, pol,* and *env* genes of the retroviral genome can be replaced with foreign DNA. This recombinant DNA lacks the genes required for retroviral replication and assembly of viral particles. To assemble viruses with the recombinant genetic information, the DNA must be introduced into cells in tissue culture that are infected with a helper virus, which has the genes to produce virus particles but lacks the ψ sequence required for packaging. Within the cells, the recombinant DNA is transcribed, and the RNA is packaged. The resulting viral particles therefore contain only the recombinant viral RNA and can act as vectors to introduce this RNA into target cells. Viral reverse transcriptase and integrase enzymes (produced by the helper virus) are also packaged in the viral particle and are introduced into the target cells. Once the engineered viral genome is inside a cell, these enzymes create a DNA copy of the viral RNA genome and integrate it into a host chromosome. The integrated recombinant DNA effectively becomes a permanent part of the chromosome because the virus lacks the genes necessary to produce RNA copies of its genome and package them into new virus particles. In most cases the use of recombinant retroviruses is the best method for introducing DNA into large numbers of mammalian cells.

Transformation of animal cells by any of the above techniques is problematic for several reasons. The foreign DNA is generally inserted at chromosomal locations that vary randomly from cell to cell. When the foreign DNA contains a sequence homologous to a sequence on a host chromosome, the introduced DNA is sometimes targeted to that position and integrated by homologous recombination. The nonhomologous integrants still outnumber the targeted ones, however, by factors of 10^2 to 10^5. Some of these integration events are deleterious to the cell because they occur in and disrupt essential genes. Different integration sites can also greatly affect the expression of an integrated gene, because integrated genes are not transcribed equally well everywhere in the genome. Another targeting problem involves the class of cell to be transformed. If germ-line cells are altered, the alteration will be passed on to successive generations of the organism. If somatic cells alone are affected, the alteration will affect only the treated animal.

Despite these problems, this technology has been used extensively to study chromosome structure, as well as the function, regulation, and expression of genes in eukaryotic cells. The successful introduction of recombinant DNA into an animal can again be illustrated by an experiment that altered an easily observable physical trait. The objective in this case was to alter germ-line cells in mice to create an inheritable change.

Microinjection of DNA into the nuclei of fertilized mouse eggs can produce efficient transformation (chromosomal integration). When the injected eggs are introduced into a female mouse and allowed to develop, the new gene is often expressed in some of the newborn mice. Those in which the germ line has been altered can be identified by testing their offspring. By careful breeding, a mouse line can be established in which all the mice are homozygous for the new gene or genes. Animals permanently altered in this way are called **transgenic.** This technology was used to introduce into mice the human growth hormone gene under the control of an inducible promoter. When fed a diet including the inducer, some of the mice that developed from injected embryos grew to an unusually large size (Fig. 28–21).

If mouse cells can be altered stably by recombinant DNA technology, so then can human cells. Introduction of DNA into human cells offers, for the first time, the potential for treating and even curing human genetic diseases that have been refractory to traditional therapies (Box 28–2, p. 1008). A major technological limitation in these efforts is our overall knowledge of the cellular metabolism that underlies many genetic diseases. As understanding improves, the ability to manipulate cellular metabolism by genetic engineering will improve. A contribution to this understanding may be made by the international project to sequence and map the entire human genome that will proceed through the 1990s. The technology needed to repair genetic defects brings with it the potential for altering human traits. Clearly, we are at a scientific crossroads that has far-reaching implications for the future of humankind.

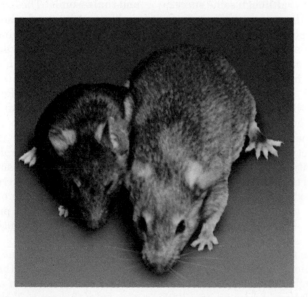

Figure 28–21 Cloning in mice. The gene for human growth hormone was introduced into the genome of the mouse on the right. Expression of the gene resulted in the greatly increased size of the mouse.

BOX 28–2   A Cure for Genetic Diseases

Human gene therapy is a reality in the 1990s. The experiments are going forward with an unprecedented level of oversight and regulation by governments and scientific review committees.

Because of the ethical issues inherent in this work, the objectives laid out by these review committees are narrowly defined; the experiments must meet strict ethical and practical criteria and are intended only to treat severe genetic disorders. First, the research is limited to somatic cells so that a treated individual cannot pass genetic alterations to offspring. Genetic engineering in human germ-line cells conjures up misguided past attempts to "improve" human beings, and evokes a wide range of objections on ethical grounds. Second, the risk to the patient must be outweighed by the potential therapeutic benefit. The inherent risk is exemplified by the possibility of random integration of DNA into a human chromosome leading to inactivation of a gene that regulates cell proliferation, effectively producing a cancer cell. For this reason the targets of the first gene therapy trials are among the most serious genetic diseases. Third, the target diseases must be limited to those that involve a known defect in a single gene, and the normal gene must be cloned and available. Fourth, the disease must involve cells that can be isolated from a patient, altered in tissue culture, and then reimplanted in the patient. This effectively limits the therapy to diseases involving cells of the skin or bone marrow, although some success has been achieved with other tissues such as liver. Fifth, the planned procedures must meet strict safety standards in animal trials before attempts are made with human beings.

The key experimental hurdle is the efficient introduction of DNA into a sufficient number of human cells in a form in which it can be expressed. Because very large numbers of cells must be transformed to have some hope of beneficial effects, research has focused on retroviral vectors (see Fig. 28–20). Expression of introduced genes has been highly variable in animal trials. In many cases, the introduced genes were expressed well in culture, then not at all when the cells were transferred to an animal. New strategies for gene expression are being developed.

Targets of human gene therapy include diseases that result from a functional lack of a single enzyme produced by a single gene (see Table 6–6). These include Lesch–Nyhan syndrome (p. 729), which occurs when hypoxanthine-guanine phosphoribosyltransferase is absent and results in mental retardation and severe behavioral problems. Two forms of severe immune deficiency, which result from a lack of adenosine deaminase (p. 729) or purine nucleoside phosphorylase, are also promising candidates. Work on correcting adenosine deaminase deficiency is already well advanced. Although these two diseases affect only a small number of people, they are very serious (people with severe immune deficiency soon die unless they are kept in a sterile environment), and in the case of adenosine deaminase deficiency the introduction of the missing gene activity into bone marrow cells does appear to have a beneficial effect.

Another effort is focused on new approaches to treating cancer. Immune-system cells known to be associated with tumors, called tumor-infiltrating lymphocytes, have been modified to produce a protein with demonstrated antitumor activity, called tumor necrosis factor (TNF). When reintroduced into a cancer patient the modified cells migrate to the tumor and the TNF they produce facilitates tumor shrinkage. Another approach is to remove and modify tumor cells themselves to produce TNF. When reintroduced into patients the modified cells stimulate the immune system to attack the cancer cells. In animal trials this approach has led to reduction or elimination of tumors and has left the animal immune to the cancer.

Additional genetic disorders that involve treatment of bone marrow cells include the genetic disorders of hemoglobin—sickle-cell anemia (p. 187) and thalassemia. These represent more formidable problems because hemoglobin is the product of more than one gene, and its expression must be limited to a small subfraction of bone marrow cells called the stem cells, which are the progenitors not only of erythrocytes but of granulocytes, macrophages, and platelets. Potential treatment of some more common genetic disorders must await development of methods to remove and replace cells from other tissues. For example, gene therapy for familial hypercholesterolemia, caused by a defect in cholesterol metabolism (p. 679) that can lead to heart attacks at an early age, requires the introduction and expression of a functional LDL receptor in hepatocytes. The prospect of curing such diseases holds great potential for alleviating human suffering.

As this technology advances, however, so does the potential to alter other physical traits. For example, the introduction and expression of a single gene from the mouse Y chromosome (the *Sry* gene) into the genome of female (XX) mouse embryos causes them to develop into male mice. The need for continued societal involvement in debating the issues generated by this technology is obvious.

Recombinant DNA Technology Yields New Products and Choices

The products of recombinant DNA technology range from proteins to engineered organisms. Large amounts of commercially useful proteins can be produced by these techniques. Microorganisms can be designed for special tasks; plants or animals can be engineered with traits that are useful in agriculture. Some products of this technology have been approved for use and many more arc in development. During the 1980s genetic engineering was transformed from a promising technology to a multibillion dollar industry. The first commercial product of recombinant DNA technology was human insulin, produced by Eli Lilly and Company and approved for human use by the U.S. Food and Drug Administration in 1982. Hundreds of companies have become involved in product development worldwide. Much of this growth has come in human pharmaceuticals, and some of the major classes of new products are listed in Table 28–4.

Table 28–4 Recombinant DNA products in medicine

Product category	Examples/Uses
Anticoagulants	Tissue plasminogen activator (TPA) activates plasmin, an enzyme involved in dissolving clots; effective in treating heart attack victims.
Blood factors	Factor VIII promotes clotting and is deficient in hemophiliacs. Use of factor VIII produced by recombinant DNA technology eliminates the risks associated with blood transfusions.
Colony stimulating factors	Immune system growth factors that stimulate leukocyte production; used to treat immune deficiencies and to fight infections.
Erythropoietin	Stimulates erythrocyte production; used to treat anemia in patients with kidney disease.
Growth factors	Stimulate differentiation and growth of various cell types; used to promote wound healing.
Human growth hormone	Used to treat dwarfism.
Human insulin	Used to treat diabetes.
Interferons	Interfere with viral reproduction; also used to treat some cancers.
Interleukins	Activate and stimulate different classes of leukocytes; possible uses in wound healing, HIV infection, cancer, immune deficiencies.
Monoclonal antibodies	Extraordinary binding specificity is used in diagnostic tests. Also used to transport drugs, toxins, or radioactive compounds to tumors as a cancer therapy; many other uses.
Superoxide dismutase	Prevents tissue damage from reactive oxygen species when tissues deprived of O_2 for short periods during surgery suddenly have blood flow restored.
Vaccines	Proteins derived from viral coats are as effective in "priming" an immune system as the killed virus more traditionally used for vaccines, but are safer. First developed was the vaccine for hepatitis B.

Erythropoietin is typical of the newer products. Erythropoietin is a protein hormone (M_r 51,000) that stimulates erythrocyte production. People with kidney disease often have a deficiency of this protein, a condition that leads to anemia. Erythropoietin produced by recombinant DNA technology can be used to treat these patients, reducing the need for repeated blood transfusions and their accompanying risks. Approved by the U.S. Food and Drug Administration in 1989, erythropoietin promises to be the most profitable pharmaceutical agent developed by recombinant DNA methods in the 1990s.

Other industrial applications of this technology are likely to continue developing. Enzymes produced by recombinant DNA technology are already used to produce detergents, sugars, and cheese. Engineered proteins are being used as food additives to supplement nutrition, flavor, and fragrance. Microorganisms are being engineered to extract oil and minerals from ground deposits, to digest oil spills, and to detoxify hazardous waste dumps and sewage. Engineered plants with improved resistance to drought, frost, pests, and disease are increasing crop yields and reducing the need for agricultural chemicals. The potential of this technology to benefit humankind and the world environment seems readily apparent yet sometimes hard to define, with the future rendered opaque by our still limited understanding of cellular metabolism and ecology.

Every major new technology comes with associated risks and a potential for unanticipated societal or environmental impact. As with the automobile and nuclear energy, economic, environmental, and ethical considerations will necessarily play an increasingly important role in determining how recombinant DNA technology is applied. One harbinger of this new relationship between biochemistry and society has been the debate in the United States and elsewhere over bovine growth hormone, which is used to increase milk production. In addition to some potential for added stress on the animals and concerns among consumers about the safety of the milk for human use, increasing milk production when a surplus already exists may have the effect of lowering prices and imposing economic hardship on dairy farmers.

Other issues raised by this technology promise to have a much broader impact. A particularly clear example can be seen in an array of new diagnostic procedures based on recombinant DNA technology. These are greatly increasing our ability to detect genetic diseases in an individual, often many years before the onset of symptoms or even before birth. The same technology that makes it possible to identify a criminal (Box 28–1) may be used to test individuals for a genetic predisposition to conditions such as Alzheimer's disease, hypercholesterolemia, asthma, and alcoholism. This information will permit better and earlier treatments, but the same information could be used to restrict individual access to health insurance (and thus health care), life insurance, and even certain jobs. The questions of who will have access to this information and how it will be used will grow in importance as more tests become widely available.

These are only some of the more straightforward examples. Release of genetically engineered organisms into the environment carries with it a level of risk that is sometimes difficult to evaluate. Human gene therapy (Box 28–2), with all of its promise, doubtless will present society with ethical dilemmas not yet anticipated. Issues of this kind must in the end foster a closer and more productive collaboration between science and the society it serves, as well as higher levels of scientific literacy in the general public, as we move toward the twenty-first century.

Summary

The study of gene structure and function has been greatly facilitated by recombinant DNA technology. The isolation of a gene from a large chromosome requires methods for cutting and joining DNA fragments, the availability of small DNA vectors that can replicate autonomously and into which the gene can be inserted, methods to introduce the vector with its foreign DNA into a cell in which it can be propagated to form clones, and methods to identify the cells containing the DNA of interest. Advances in this technology are revolutionizing many aspects of medicine, agriculture, and other industries.

The first organism used for DNA cloning was *E. coli*. Bacterial restriction endonucleases and DNA ligases provide the most important instruments for cutting DNA at specific sequences and joining DNA fragments. Bacterial cloning vectors include plasmids, bacteriophages, and cosmids. These permit the cloning of DNA fragments of different size ranges. In each case the vectors provide a replication origin for propagation in the bacterial host, and a selectable genetic element such as antibiotic resistance to facilitate the identification of cells harboring the recombinant vector. DNA is introduced into cells in viral vectors or by artificial methods that make the cell wall permeable.

The first step in cloning a gene is often the construction of a DNA library that includes fragments representing most of the genome of a given species. The library can be limited to expressed genes by cloning only the complementary DNA copies of isolated mRNAs to make a cDNA library. A specific segment of DNA can be amplified and cloned using the polymerase chain reaction. Clones containing a specific gene in a large library can be detected by hybridization with a radioactive probe containing the complementary nucleotide sequence.

Expression vectors provide the DNA sequences required for transcription, translation, and regulation of cloned genes. They allow the production of large amounts of cloned proteins for research and commercial purposes. Cloned genes also can be altered by site-directed mutagenesis, which is useful in studies of protein structure and function.

Yeast is sometimes used for cloning eukaryotic DNA and it has many of the same advantages as *E. coli*. Methods for cloning in plants and animals are producing a variety of organisms with altered traits. Plants that are resistant to disease, insects, herbicides, and drought are being produced with the aid of a natural gene transfer process promoted by the Ti plasmid of the parasitic soil bacterium *Agrobacterium tumefaciens*. Engineered DNA can be introduced into animal cells by microinjection or retroviral vectors. Such procedures have produced mice with new inheritable genetic traits. The technology extends to humans, and human gene therapy is now directed at treating human genetic diseases.

Further Reading

General

Hackett, P.B., Fuchs, J.A., & Messing, J.W. (1988) *An Introduction to Recombinant DNA Techniques*, 2nd edn, The Benjamin/Cummings Publishing Company, Menlo Park, CA.

Jackson, D.A., Symons, R.H., & Berg, P. (1972) Biochemical method for inserting new genetic information into DNA of Simian Virus 40: circular SV40 DNA molecules containing lambda phage genes and the galactose operon of *Escherichia coli*. *Proc. Natl. Acad. Sci. USA* **69**, 2904–2909.

The first recombinant DNA experiment linking DNA from two different organisms.

Lobban, P.E. & Kaiser, A.D. (1973) Enzymatic end-to-end joining of DNA molecules. *J. Mol. Biol.* **78**, 453–471.

Report of the first recombinant DNA experiment.

Sambrook, J., Fritsch, E.F., & Maniatis, T. (1989) *Molecular Cloning: A Laboratory Manual*, 2nd edn, Cold Spring Harbor Laboratory Press, Cold Spring Harbor, NY.

In addition to detailed protocols for a wide range of techniques, this three-volume set includes much useful background information on the biological, chemical, and physical principles underlying each technique.

Libraries and Gene Isolation

Arnheim, N. & Levenson, C.H. (1990) Polymerase chain reaction. *Chem. Eng. News* **68** (October 1), 36–47.

A broad overview of the technique and applications.

Arnheim, N. & Erlich, H. (1992) Polymerase chain reaction strategy. *Annu. Rev. Biochem.* **61**, 131–156.

Erlich, H.A., Gelfand, D., & Sninsky, J.J. (1991) Recent advances in the polymerase chain reaction. *Science* **252,** 1643–1651.

This issue of Science *contains several other good articles on other aspects of biotechnology.*

Neufeld, P.J. & Colman, N. (1990) When science takes the witness stand. *Sci. Am.* **262** (May), 46–53.

Potential and problems of DNA fingerprinting.

Pääbo, S., Higuchi, R.G., & Wilson, A.C. (1989) Ancient DNA and the polymerase chain reaction. The emerging field of molecular archaeology. *J. Biol. Chem.* **264,** 9709–9712.

Southern, E.M. (1975) Detection of specific sequences among DNA fragments separated by gel electrophoresis. *J. Mol. Biol.* **98,** 503–517.

The paper that introduced the Southern hybridization method.

Thornton, J.I. (1989) DNA profiling: new tool links evidence to suspects with high certainty. *Chem. Eng. News* **67** (November 20), 18–30.

Products of Recombinant DNA Technology

Bailey, J.E. (1991) Toward a science of metabolic engineering. *Science* **252,** 1668–1675.

An overview of efforts to reengineer entire metabolic pathways in microorganisms for commercial purposes.

Bains, W. (1989) Disease, DNA and diagnosis. *New Scientist* **122** (May 6), 48–51.

Botstein, D. & Shortle, D. (1985) Strategies and applications of *in vitro* mutagenesis. *Science* **229,** 1193–1201.

Buck, K. (1989) Brave new botany. *New Scientist* **122** (June 3), 50–55.

Prospects for genetic engineering in plants.

Culver, K.W., Osborne, W.R.A., Miller, A.D., Fleisher, T.A., Berger, M., Anderson, W.F., & Blaese, R.M. (1991) Correction of ADA deficiency in human T lymphocytes using retroviral-mediated gene transfer. *Transplant Proc.* **23,** 170–171.

Gene therapy of adenosine deaminase deficiency.

Hooykaas, P.J.J. & Schilperoort, R.A. (1985) The Ti-plasmid of *Agrobacterium tumefaciens:* a natural genetic engineer. *Trends Biochem. Sci.* **10,** 307–309.

Johnson, I.S. (1983) Human insulin from recombinant DNA technology. *Science* **219,** 632–637.

Koopman, P., Gubbay, J., Vivian, N., Goodfellow, P., & Lovell-Badge, R. (1991) Male development of chromosomally female mice transgenic for *Sry. Nature* **351,** 117–121.

Recombinant DNA technology is used to demonstrate that a single gene directs the development of chromosomally female mice into males.

McLachlin, J.R., Cornetta, K., Eglitis, M.A., & Anderson, W.F. (1990) Retroviral-mediated gene transfer. *Prog. Nucleic Acid Res. Mol. Biol.* **38,** 91–135.

Murray, T.H. (1991) Ethical issues in human genome research. *FASEB J.* **5,** 55–60.

This issue also contains a number of other useful papers on the human genome project.

Palmiter, R.D., Brinster, R.L., Hammer, R.E., Trumbauer, M.E., Rosenfeld, M.G., Birnberg, N.C., & Evans, R.M. (1982) Dramatic growth of mice that develop from eggs microinjected with metallothionein–growth hormone fusion genes. *Nature* **300,** 611–615.

A description of how to make giant mice.

Sulston, J., Du, Z., Thomas, K., Wilson, R., Hillier, L., Staden, R., Halloran, N., Green, P., Thierry-Mieg, J., Qiu, L., Dear, S., Coulson, A., Craxton, M., Durbin, R., Berks, M., Metzstein, M., Hawkins, T., Ainscough, R., & Waterston, R. (1992) The *C. elegans* genome sequencing project: a beginning. *Nature* **356,** 37–41.

A description of strategies being developed to sequence large genomes.

The telltale gene. *Consumer Reports* **55** (July 1990), 483–488.

A good summary of new genetic screening techniques based on recombinant DNA, with a focus on ethical dilemmas posed by the technology.

Thompson, J. & Donkersloot, J.A. (1992) N-(Carboxyalkyl)amino acids: occurrence, synthesis, and functions. *Annu. Rev. Biochem.* **61,** 517–557.

A good summary of the structure and biological functions of opines.

Verma, I.M. (1990) Gene therapy. *Sci. Am.* **263** (November), 68–84.

Watson, J.D. (1990) The human genome project: past, present, and future. *Science* **248,** 44–48.

Weatherall, D.J. (1991) Gene therapy in perspective. *Nature* **349,** 275–276.

Westphal, H. (1989) Transgenic mammals and biotechnology. *FASEB J.* **3,** 117–120.

Zambryski, P., Tempe, J., & Schell, J. (1989) Transfer and function of T-DNA genes from *Agrobacterium* Ti and Ri plasmids in plants. *Cell* **56,** 193–201.

Problems

1. *Cloning*

(a) Draw the structure of the end of a linear DNA fragment that was produced by an *Eco*RI restriction digest (include those sequences remaining from the *Eco*RI recognition sequence).

(b) Draw the structure resulting from the reaction of this end sequence with DNA polymerase I and the four deoxynucleoside triphosphates.

(c) Draw the sequence produced at the junction if two ends with the structure derived in (b) are ligated.

(d) Draw the structure produced if the structure derived in (a) is treated with a nuclease that degrades only single-stranded DNA.

(e) Draw the sequence of the junction produced if an end with structure (b) is ligated to an end with structure (d).

(f) Draw the structure of the end of a linear DNA fragment that was produced by a *Pvu*II restriction digest (as in (a)).

(g) Draw the sequence of the junction produced if an end with structure (b) is ligated to an end with structure (f).

(h) Suppose you can synthesize a short duplex DNA fragment with any sequence you desire. With such a synthetic fragment and the procedures described in (a) through (g), design a protocol that will remove an *Eco*RI restriction site from a DNA molecule and incorporate a new *Bam*HI restriction site at approximately the same location. (Hint: See Fig. 28–3.)

(i) Design four different short synthetic DNA fragments that would permit ligation of structure (a) with a DNA fragment produced by a *Pst*I restriction digest. In one of these synthetic fragments, design the sequence so that the final junction contains the recognition sequences for both *Eco*RI and *Pst*I. In the second and third synthetic fragments, design the sequence so that the junction contains only the *Eco*RI or the *Pst*I recognition sequence, respectively. Design the sequence of the fourth fragment so that neither the *Eco*RI nor the *Pst*I sequence appears in the junction.

2. *Selecting for Recombinant Plasmids*   When cloning a foreign DNA fragment into a plasmid it is often useful to insert the fragment at a site that interrupts a selectable marker (such as the tetracycline-resistance element of pBR322). The loss of function of the interrupted gene can be used to identify clones containing recombinant plasmids with foreign DNA. With a cosmid it is unnecessary to do this, yet one can easily distinguish cosmids that incorporate large foreign DNA fragments from those that do not. How are these recombinant cosmids identified?

3. *DNA Cloning*   The plasmid cloning vector pBR322 (see Fig. 28–5) is cleaved with the restriction endonuclease *Eco*RI. An isolated DNA fragment from a eukaryotic genome (also produced by *Eco*RI cleavage) is added to the prepared vector and ligated. The mixture of ligated DNAs is then used to transform bacteria, and plasmid-containing bacteria are selected by growth in the presence of tetracycline. In addition to the desired recombinant plasmid, what other types of plasmids might be found among the transformed bacteria that are tetracycline resistant?

4. *Expressing a Cloned Gene*   You have isolated a plant gene that encodes a protein in which you are interested. On the drawing below, indicate sequences or sites that you will need to get this gene transcribed, translated, and regulated in *E. coli.*

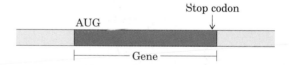

5. *Identifying the Gene for a Protein with a Known Amino Acid Sequence*   Design a DNA probe that would allow you to identify the gene for a protein with the following amino-terminal amino acid sequence. The probe should be 18 to 20 nucleotides long, a size that provides adequate specificity if there is sufficient homology between the probe and the gene.

H₃N⁺-Ala–Pro–Met–Thr–Trp–Tyr–Cys–Met–Asp–
Trp–Ile–Ala–Gly–Gly–Pro–Trp–Phe–Arg–Lys–
Asn–Thr–Lys– – –

6. *Cloning in Plants*   The strategy outlined in Figure 28–17 employs *Agrobacterium* containing two separate plasmids. Suggest a reason why the sequences on the two plasmids are not combined on one plasmid.

7. *Cloning in Mammals*   The retroviral vectors described in Figure 28–20 make it possible to integrate foreign DNA efficiently into a mammalian genome. Explain how these vectors, which lack genes for replication and viral packaging *(gag, pol, env)*, are assembled into infectious viral particles. Suggest why it is important that these vectors lack the replication and packaging genes.

Common Abbreviations in the Biochemical Research Literature

A	adenine, adenosine, or adenylate
A	absorbance
ACh	acetylcholine
ACP	acyl carrier protein
ACTH	adrenocorticotropic hormone
Acyl-CoA	acyl derivatives of coenzyme A (also, acyl-S-CoA)
ADH	alcohol dehydrogenase
adoHcy	S-adenosylhomocysteine
adoMet	S-adenosylmethionine (also, SAM)
AIDS	acquired immunodeficiency syndrome
Ala	alanine
$[\alpha]_D^{25°C}$	specific rotation
AMP, ADP, ATP	adenosine 5'-mono-, di-, triphosphate
Arg	arginine
ARS	autonomously replicating sequence
Asn	asparagine
Asp	aspartate
ATCase	aspartate transcarbamoylase
ATPase	adenosine triphosphatase
B_{12}	coenzyme B_{12}, cobalamin
BMR	basal metabolic rate
bp	base pair
1,3-BPG	1,3-bisphosphoglycerate
BPTI	bovine pancreatic trypsin inhibitor
C	cytosine, cytidine, or cytidylate
CaM	calmodulin
cAMP, cGMP	3',5'-cyclic AMP, 3',5'-cyclic GMP
CAP	catabolite activator protein
cDNA	complementary DNA
Chl	chlorophyll
CMP, CDP, CTP	cytidine 5'-mono-, di-, triphosphate
CoA	coenzyme A (also, CoASH)
CoQ	coenzyme Q (ubiquinone; also, UQ)
Cys	cysteine
D	diffusion coefficient

d	density
dADP, dGDP, etc.	deoxyadenosine 5'-diphosphate, deoxyguanosine 5'-diphosphate, etc.
dAMP, dGMP, etc.	deoxyadenosine 5'-monophosphate, deoxyguanosine 5'-monophosphate, etc.
dATP, dGTP, etc.	deoxyadenosine 5'-triphosphate, deoxyguanosine 5'-triphosphate, etc.
DEAE	diethylaminoethyl
DFP (DIFP)	diisopropylfluorophosphate
DHAP	dihydroxyacetone phosphate
DHF	dihydrofolate (also, H_2 folate)
DHU	dihydrouridine
DMS	dimethyl sulfate
DNA	deoxyribonucleic acid
DNase	deoxyribonuclease
DNP	2,4-dinitrophenol
Dol	dolichol
DOPA	dihydroxyphenylalanine
E	electrical potential
E.C.	Enzyme Commission (followed by numbers indicating the formal classification of an enzyme)
EDTA	ethylenediaminetetraacetate
EF	elongation factor
EGF	epidermal growth factor
ELISA	enzyme-linked immunosorbent assay
EM	electron microscopy
ϵ	molar absorption coefficient
ER	endoplasmic reticulum
η	viscosity
f	frictional coefficient
FA	fatty acid
FAD, $FADH_2$	flavin adenine dinucleotide, and its reduced form
FBPase-1	fructose-1,6-bisphosphatase
FBPase-2	fructose-2,6-bisphosphatase
Fd	ferredoxin

FDNB (DNFB)	1-fluoro-2,4-dinitrobenzene	G3P	glyceraldehyde-3-phosphate (also, GAP)
FFA	free fatty acid	G6P	glucose-6-phosphate
FH	familial hypercholesterolemia	GSH, GSSG	glutathione, and its oxidized form
fMet	N-formylmethionine	ΔH	enthalpy change
FMN, FMNII$_2$	flavin mononucleotide, and its reduced form	Hb, HbO$_2$, HbCO	hemoglobin, oxyhemoglobin, carbon monoxide hemoglobin
FP	flavoprotein	HDL	high-density lipoprotein
F1P	fructose-1-phosphate	H$_2$ folate	dihydrofolate (also, DHF)
F6P	fructose-6-phosphate	H$_4$ folate	tetrahydrofolate (also, THF)
Fru	D-fructose	His	histidine
ΔG	free-energy change	HIV	human immunodeficiency virus
$\Delta G^{o\prime}$	standard free-energy change	HMG-CoA	β-hydroxy-β-methylglutaryl-CoA
$\Delta G^{\ddagger}$	activation energy	hnRNA	heterogeneous nuclear RNA
ΔG_{B}	binding energy	HPLC	high-performance liquid chromatography
ΔG_p	free-energy change of ATP hydrolysis under nonstandard conditions	HRE	hormone response element
		Hyp	hydroxyproline
G	guanine, guanosine, or guanylate	I	inosine
g	acceleration due to gravity	IF	initiation factor
GABA	γ-aminobutyrate	Ig	immunoglobulin
Gal	D-galactose	IgG	immunoglobulin G
GalN	D-galactosamine	Ile	isoleucine
GalNAc	N-acetyl-D-galactosamine	IMP, IDP, ITP	inosine 5′-mono-, di-, triphosphate
GAP	glyceraldehyde-3-phosphate (also, G3P)	IR	infrared
GDH	glutamate dehydrogenase	IS	insertion sequence
GH	growth hormone	K	dissociation constant
GLC	gas-liquid chromatography	K_{a}	acid dissociation constant
Glc	D-glucose	K_{eq}	equilibrium constant
GlcA	D-gluconic acid	K'_{eq}	equilibrium constant under standard conditions
GlcN	D-glucosamine		
GlcNAc	N-acetyl-D-glucosamine (also, NAG)	K_{I}	inhibition constant
GlcUA	D-glucuronic acid	K_{m}	Michaelis–Menten constant
Gln	glutamine	K_{S}	dissociation constant
Glu	glutamate	k	rate constant
Gly	glycine	k_{cat}	turnover number
GMP, GDP, GTP	guanosine 5′-mono-, di-, triphosphate	kb	kilobase
G1P	glucose-1-phosphate	kbp	kilobase pair

α-KG	α-ketoglutarate	PBG	porphobilinogen
λ	wavelength	PC	plastocyanin; phosphatidylcholine
LDH	lactate dehydrogenase	PCR	polymerase chain reaction
LDL	low-density lipoprotein	PDGF	platelet-derived growth factor
Leu	leucine	PE	phosphatidylethanolamine
LH	luteinizing hormone	PEP	phosphoenolpyruvate
Lk	linking number	PFK	phosphofructokinase
ln	logarithm to the base e	PG	prostaglandin
log	logarithm to the base 10	2PG	2-phosphoglycerate
LTR	long terminal repeat	3PG	3-phosphoglycerate
Lys	lysine	pH	log $1/[H^+]$
M_r	relative molecular mass	Phe	phenylalanine
Man	D-mannose	PI	phosphatidylinositol
Mb, MbO$_2$	myoglobin, oxymyoglobin	PK	protein kinase; pyruvate kinase
Met	methionine	pK	log $1/K$
mRNA	messenger RNA	PLP	pyridoxal-5-phosphate
MSH	melanocyte-stimulating hormone	Pn	phosphopantetheine
mDNA	mitochondrial DNA	Pol	polymerase (DNA or RNA)
μ	electrophoretic mobility	PP$_i$	inorganic pyrophosphate
Mur	muramic acid	PQ	plastoquinone
MurNAc	N-acetylmuramic acid (also, NAM)	Pro	proline
NAD$^+$, NADH	nicotinamide adenine dinucleotide, and its reduced form	PRPP	5-phosphoribosyl-1-pyrophosphate
		$\Delta\psi$	transmembrane electrical potential
NADP$^+$, NADPH	nicotinamide adenine dinucleotide phosphate, and its reduced form	RER	rough endoplasmic reticulum
		RF	release factor; replicative form
NAG	N-acetylglucosamine (also, GlcNAc)		
NAM	N-acetylmuramic acid (also, MurNAc)	RFLP	restriction-fragment length polymorphism
NeuNAc	N-acetylneuraminic acid	Rib	D-ribose
NMN$^+$, NMNH	nicotinamide mononucleotide, and its reduced form	RNA	ribonucleic acid
		RNase	ribonuclease
NMP, NDP, NTP	nucleoside mono-, di-, and triphosphate	RQ	respiratory quotient
NMR	nuclear magnetic resonance	rRNA	ribosomal RNA
OAA	oxaloacetate	RSV	Rous sarcoma virus
P	pressure	ΔS	entropy change
P$_i$	inorganic orthophosphate	SAM	S-adenosylmethionine (also, adoMet)
pO$_2$	partial pressure of oxygen	SDS	sodium dodecyl sulfate
PAB or PABA	p-aminobenzoate	SER	smooth endoplasmic reticulum
PAGE	polyacrylamide gel electrophoresis	Ser	serine

snRNA	small nuclear RNA	Tyr	tyrosine
SRP	signal recognition particle	U	uracil, uridine, or uridylate
STP	standard temperature and pressure	UDP-Gal	uridine diphosphate galactose (also, UDP-galactose)
T	thymine, thymidine, or thymidylate		
T	absolute temperature	UDP-Glc	uridine diphosphate glucose (also, UDP-glucose)
TH	thyrotropic hormone	UMP, UDP, UTP	uridine 5'-mono-, di-, triphosphate
THF	tetrahydrofolate (also, H_4 folate)		
Thr	threonine	UQ	coenzyme Q (ubiquinone; also, CoQ)
TIM	triose phosphate isomerase	UV	ultraviolet
TLC	thin layer chromatography	V_{max}	maximum velocity
TMP, TDP, TTP	thymidine 5'-mono-, di-, triphosphate	V_0	initial velocity
TMV	tobacco mosaic virus	Val	valine
TPP	thiamine pyrophosphate	VLDL	very low-density lipoprotein
tRNA	transfer RNA	Z	net charge
Trp	tryptophan		

Chapter 2

1. (a) 625 (b) 1.1×10^5 (c) 1.9×10^{10}
2. (a) 1.1×10^4 (b) 1×10^{-4} M
3. (a) 1.1×10^{-12} g (1.1 pg) (b) 5% (c) 4.6%
4. (a) 1.3 mm; 650 times longer than the cell; the DNA must be tightly coiled. (b) 3,200
5. (a) The metabolic rate is limited by diffusion, which is limited by surface area.
   (b) 1.2×10^7 m^{-1} for the bacterium; 4×10^4 m^{-1} for the amoeba; the surface-to-volume ratio is 300 times higher in the bacterium.
6. (a) 7,900 (b) 3.1×10^{-10} m^2 (c) 2.8×10^{-9} m^2 (d) 900%
7. 2×10^6 s (about 23 days!); the rate is greater than the rate of diffusion in a large organelle, the rate is subject to regulation, and the motion is unidirectional.
8. (a) Phalloidin would interfere with cytokinesis, phagocytosis, and amoeboid movement. (b) Yes
9. Release of enzymes contained in particulate organelles would contaminate the cytoplasm with those enzymes, complicating the purification of the desired cytoplasmic enzyme, and cause proteolysis by proteases released from lysosomes.

Chapter 3

1. The vitamins from the two sources are identical; the body cannot distinguish the source.

2.

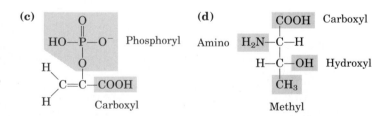

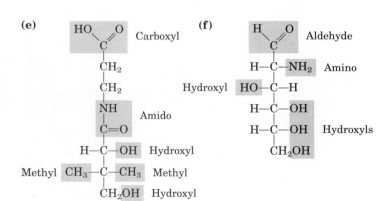

3.

The two enantiomers have different interactions with a chiral biological "receptor" (a protein).

4. Only one enantiomer of the drug was physiologically active. Dexedrine consisted of the single enantiomer, whereas Benzedrine consisted of a racemic mixture.

5. (a) 3 Phosphoric acid molecules; α-D-ribose; adenine
   (b) Choline; phosphoric acid; glycerol; oleic acid; palmitic acid (c) Tyrosine; 2 glycine molecules; phenylalanine; methionine

6. (a) CH_2O; $C_3H_6O_3$
   (b)

1   2   3

4   5   6

7   8   9

10   11   12

(c) X contains a chiral center; eliminates all but **6** and **8**.
(d) X contains an acid functional group; eliminates **8**; structure **6** is consistent with all data. (e) Structure **6**; we cannot distinguish between the two possible enantiomers.

Chapter 4

1. 3.35 mL
2. pH 1.1
3. 1.7×10^{-9} mol
4. For the equilibrium HA $\rightleftharpoons$ H$^+$ + A$^-$, the corresponding Henderson–Hasselbalch equation is: $pK_a = pH + \log [A^-]/[HA]$. When the acid is half-ionized, [HA] = [A$^-$]. Hence [A$^-$]/[HA] = 1; log 1 = 0; and $pK_a = pH$.

5. **(a)** In a zone centered about pH 9.3 **(b)** 2/3 **(c)** 10^{-2} L
(d) pH − pK_a = −2

6. **(a)** 0.1 M HCl **(b)** 0.1 M NaOH **(c)** 0.1 M NaOH

7. **(d)**; the weak base bicarbonate will titrate —OH to —O⁻, making the compound more polar and more water-soluble.

8. Stomach; the neutral form of aspirin present at low pH is less polar, and therefore more membrane permeant.

9. $NaH_2PO_4 \cdot H_2O$, 5.80 g; Na_2HPO_4, 8.23 g

10. **(a)** Blood pH is controlled by the carbon dioxide–bicarbonate buffer system, as shown in the net equation:

$$CO_2 + H_2O \rightleftharpoons H^+ + HCO_3^-$$

During *hypoventilation* the concentration of CO_2 increases in the lungs and arterial blood, driving the equilibrium to the right and raising $[H^+]$; i.e., the pH is lowered.
(b) During *hyperventilation* the concentration of CO_2 is lowered in the lungs and arterial blood. This drives the equilibrium to the left, which requires the consumption of hydrogen ions. Thus $[H^+]$ is reduced; i.e., the pH is raised from the normal 7.4 value.
(c) Lactate is a moderately strong acid (pK_a = 3.86) that completely dissociates under physiological conditions:

$$CH_3CHOHCOOH \rightleftharpoons CH_3CHOHCOO^- + H^+$$

This lowers the pH of the blood and muscle tissue. Hyperventilation is useful because it removes hydrogen ions, raising the pH of the blood and tissues in anticipation of the acid build up.

Chapter 5

1. L; determine the absolute configuration of the α carbon and compare with D- and L-glyceraldehyde.

2. (1) Glycine, (b); (2) alanine, (f); (3) valine, (f); (4) proline, (h); (5) serine, (a); (6) phenylalanine, (e); (7) tryptophan, (e); (8) tyrosine, (j); (9) aspartate, (i); (10) glutamate, (i); (11) methionine, (d); (12) cysteine, (k); (13) histidine, (g); (14) asparagine, (m); (15) arginine, (l); (16) lysine, (c)

3. **(a)** I **(b)** II **(c)** IV **(d)** II **(e)** IV **(f)** II and IV **(g)** III **(h)** III **(i)** II **(j)** V **(k)** III **(l)** IV **(m)** V **(n)** II **(o)** III **(p)** IV **(q)** V **(r)** I, III, and V **(s)** V

4. **(a)** The pI of alanine occurs at a pH above the pK for the α-carboxyl group and below the pK for the α-amino group. Hence, both groups will be present predominantly in their charged (ionized) forms.
(b) 1 in 2.2×10^7; the pI of alanine is 6.01. From Table 5–1 and the Henderson–Hasselbalch equation (p. 98), we can estimate that 1 in 4,678 carboxyl groups and 1 in 4,678 amino groups are in their uncharged form. The fraction of alanine molecules in which both groups are uncharged is the product of these numbers.

5. **(a)–(c)**

1

2

3

4

pH	Structure	Net charge	Migrates toward:
1	1	+2	Cathode (−)
4	2	+1	Cathode (−)
8	3	0	Does not migrate
12	4	−1	Anode (+)

6. 0.879 L of 0.1 M glycine and 0.121 L of 0.1 M glycine hydrochloride

7. **(a)** Asp **(b)** Met **(c)** Glu **(d)** Gly **(e)** Ser

8. **(a)** 2 **(b)** 4
(c)

9. **(a)** The structure at pH 7 is:

$pK_2 = 9.69$ $pK_1 = 2.84$

(b) The ionization of the carboxyl group in Ala and in an Ala oligopeptide forms the zwitterion. The presence of the charged α-amino group significantly and favorably affects the ionization of the carboxyl group, because of the favorable charge–charge interaction between the carboxylate anion and the protonated amino group. The protonated amino group is closer to the carboxylate anion in Ala than in an Ala oligopeptide, and the charge–charge interaction decreases as the length of the polymer increases, resulting in an increase in pK_1.
(c) The ionization of the protonated amino group in Ala and in an Ala oligopeptide destroys the favorable charge–charge interaction. Because the charged groups are closer together in Ala than in an Ala oligopeptide, it is more difficult to remove the second proton in Ala than in the oligopeptide. The longer the Ala oligopeptide, the easier it is to remove the second proton, and thus the lower is the pK_2.

10. In the absence of a protecting group on its α-amino group, an incoming activated amino acid would react not only with the growing polypeptide but also with other activated amino acid molecules. The efficiency of the addition of the amino acid to the growing polypeptide would be greatly reduced.

Chapter 6

1. 3,500 molecules

2. **(a)** 32,100 **(b)** 2

3. 1,220; 12,200

4. 75,020

5. $+2$; $+1$; 0; -2; pI = 7.84

6. Carboxyl groups; Asp and Glu

7. Lys, His, Arg; electrostatic attraction between negatively charged phosphate groups in the DNA and positively charged basic residues in the histones

8. **(a)** $(Glu)_{20}$ **(b)** $(Lys–Ala)_3$ **(c)** $(Asn–Ser–His)_5$ **(d)** $(Asn–Ser–His)_5$

9. **(a)** Specific activity after step 1 = 200 units/mg; step 2 = 600 units/mg; step 3 = 250 units/mg; step 4 = 4,000 units/mg; step 5 = 15,000 units/mg; step 6 = 15,000 units/mg
   (b) Step 4 **(c)** Step 3
   (d) Yes—the specific activity did not increase in step 6; polyacrylamide gel electrophoresis in the presence of SDS

10. **(a)** and **(b)** Phe–Val–Asn–Gln–His–Leu–CysSO$_3^-$–Gly–Ser–His–Leu–Val–Glu–Ala–Leu–Tyr–**(C)**–Leu–Val–CysSO$_3^-$–Gly–Glu–Arg–**(T)**–Gly–Phe–**(C)**–Phe–**(C)**–Tyr–**(C)**–Thr–Pro–Lys–Ala. **(C)** and **(T)** denote cleavage sites for chymotrypsin and trypsin, respectively.

11. Tyr–Gly–Gly–Phe–Leu

12.

The arrows correspond to the orientation of the peptide bonds:

Chapter 7

1. **(a)** Shorter bonds are stronger and have a higher bond order (are multiple rather than single). The C—N bond is stronger than a single bond and is midway between a single and a double bond in character. **(b)** The peptide bond is represented by two resonance structures, one single-bonded and one double-bonded. **(c)** Rotation about the peptide bond is difficult at physiological temperatures, because of its partial double-bond character.

2. The principal structural units in the wool fiber polypeptides are successive turns of the α helix, which are spaced at 0.54 nm intervals. Steaming and stretching the fiber yields an extended polypeptide chain with the β conformation, in which the distances between the R groups are about 0.70 nm. On shrinking, the fiber reassumes an α-helical structure.

3. About 40 per second

4. Repulsion among the negatively charged carboxylate groups of poly(Glu) at pHs above 6 leads to unfolding. Similarly, repulsion among the positively charged amino groups of poly(Lys) at pHs below 9 also leads to unfolding.

5. The disulfide bonds are covalent bonds, which are much stronger than the noncovalent bonds that stabilize most proteins. They serve to cross-link protein chains, increasing their stiffness, mechanical strength, and hardness.

6. Wool shrinks when polypeptide chains are converted from the extended conformation (β-pleated sheet) to the α-helical conformation in the presence of heat. The β-pleated sheets of silk, with their small, closely packed amino acid side chains, are more stable than those of wool.

7. Cystine residues (disulfide bonds) prevent the complete unfolding of the protein.

8. 30 amino acids; 89%

9. The observation that [^{14}C]hydroxyproline is not incorporated into collagen argues against route (1) but is consistent with route (2).

10. The bacterial enzyme (a collagenase) can destroy the connective-tissue barrier of the host, allowing the bacterium to invade the host tissues. Bacteria do not contain collagen.

11. At residues 7 and 19; prolines are often but not always found at bends. Between residues 13 and 24

12. External surface: Asp, Gln, Lys; interior: Leu, Val; either: Ser (see Table 5–1)

13. **(a)** Calculating the number of moles of DNP-valine formed per mole of protein gives the number of amino termini and thus the number of polypeptide chains. **(b)** 4

14. **(a)** 16,400 **(b)** There are 4 Fe atoms.

15. **(a)** Hemoglobin F **(b)** Oxygen diffuses from maternal blood to fetal blood.

Chapter 8

1. The activity of the enzyme that converts sugar to starch is destroyed by heat denaturation.

2. 2.4×10^{-6} M

3. 9.5×10^8 yr

4. **(a)** 18.8 nm **(b)** Three-dimensional folding of the enzyme brings these amino acid residues into close proximity. **(c)** The protein serves as "scaffolding" to keep the catalytic groups in a precise orientation. Also, many other interactions occur between the enzyme and its substrate, and some of the binding energy derived from these interactions contributes to catalysis.

5. The reaction rate can be measured by following the decrease in the absorption of NADH (at 340 nm) as the reaction proceeds. Determine the K_m value; using substrate concentrations well above the K_m, measure the initial rate (rate of NADH disappearance with time, measured spectrophotometrically) at several known enzyme concentrations, and make a plot of the initial rates at increasing concentrations of enzyme. The plot should be linear with a slope that provides a measure of LDH concentration.

6. $V_{max} \approx 140$ μM/min; $K_m \approx 1 \times 10^{-5}$ M

7. **(a)** 1.7×10^{-3} M **(b)** 0.33; 0.67; 0.91

8. $K_m = 2.2$ mM; $V_{max} = 0.51$ μmol/min

9. **(a)** $k_{cat} = 2.0 \times 10^7$ min^{-1} **(b)** $\Delta G^{\ddagger} = 43.3$ kJ/mol (at 37 °C or 310 K) **(c)** $\Delta G^{\ddagger} = 83.2$ kJ/mol (Hint: See p. 207)

10. 29,000; it is assumed that each enzyme molecule contains only one titratable sulfhydryl group.

11. The enzyme–substrate complex is more stable than the enzyme alone.

12. Measure the total acid phosphatase activity in the absence and presence of tartrate ion. The difference is the acid phosphatase activity from the prostate gland.

13. Acetazolamide decreases the V_{max} of the enzyme but leaves K_m unchanged, indicating that this drug is a noncompetitive inhibitor.

14. Glu^{35}: protonated; Asp^{52}: deprotonated. The pH-activity profile suggests that maximum catalytic activity occurs when Glu^{35} is protonated and Asp^{52} is deprotonated.

Chapter 9

1. The number of cis double bonds. Each cis double bond causes a bend in the hydrocarbon chain; it is more difficult to pack these bent chains in the crystal lattice.

2. Unsaturated fats (e.g., olive oil) are susceptible to oxidation by molecular oxygen.

3. Phosphatidylcholine emulsifies fat droplets.

4. (a) The sodium salts of palmitic and stearic acids, plus glycerol (b) The sodium salts of palmitic and oleic acids, plus glycerol-3-phosphorylcholine

5. 63

6. *Hydrophobic units:* (a) 2 fatty acids; (b), (c), and (d) 1 fatty acid and the hydrocarbon chain of sphingosine; (e) the hydrocarbon backbone. *Hydrophilic units:* (a) phosphoethanolamine; (b) phosphocholine; (c) D-galactose; (d) several sugar molecules; (e) alcohol group (—OH)

7. (a) Lipids that form bilayers are amphipathic molecules: they contain a hydrophilic and a hydrophobic unit. In order to minimize the hydrophobic area that is exposed to the water surface, these lipids form two-dimensional sheets with the hydrophilic units exposed to water and the hydrophobic units buried in the interior of the sheet. Furthermore, to avoid exposing the hydrophobic edges of the sheet to water, lipid bilayers close upon themselves. Similarly, if the sheet is perforated, the hole will seal because the membrane is semifluid.
(b) These sheets form the closed membrane surfaces that envelop cells and compartments within cells (organelles).

8. In order of first eluted to last eluted: cholesteryl palmitate and triacylglycerol; cholesterol and *n*-tetradecanol; the "neutral" phospholipids—phosphatidylcholine, sphingomyelin, and phosphatidylethanolamine; the charged lipids—phophatidylserine and palmitic acid.

9. Water-soluble vitamins are more rapidly excreted in the urine and are not stored effectively. Fat-soluble vitamins have very low solubility in water and are excreted only very slowly by the kidney.

10. The triacylglycerols of animal fats (grease) are hydrolyzed by NaOH, in a process known as saponification, to form soaps, which are much more soluble in water than are the triacylglycerols.

11. Eight different triacylglycerols can be constructed. All saturated (palmitic) fatty acids, all unsaturated (oleic) fatty acids, or any combination of oleic and palmitic acids can be used. Furthermore, positional isomers are possible, as the three carbons of glycerol are not equivalent. In order of increasing melting point: OOO, OOP and OPO; PPO and POP; PPP, where O = oleic, P = palmitic.

12. The definition of substances such as lipids is based, not on a common structure, but on their solubilities in water and in nonpolar solvents.

13. In order of increasing solubility: triacylglycerol; diacylglycerol; monoacylglycerol

14. Inositol-1,4,5-trisphosphate (IP_3) is highly polar and very soluble in water. Diacylglycerol is hydrophobic and is more soluble in the lipid of the membrane than in the cytosol. Thus IP_3 is more readily diffusible.

15. Both compounds should yield sphingosine and a fatty acid upon strong alkaline hydrolysis. Phosphocholine is produced only from spingomyelin; cerebroside produces one or more sugars, but no phosphate.

16. Strong alkaline hydrolysis of phosphatidylcholine released 2 fatty acids per phosphate in PC; the same treatment of sphingomyelin produces 1 fatty acid per phosphate, as well as sphingosine.

Chapter 10

1. From the known amount of lipid used, its molecular weight, and the area occupied by a monolayer (determined as shown), the area per molecule can be calculated.

2. Yes. For example, $(8 \times 109 \text{ cells/ml})(40 \text{ ml}) = 3.2 \times 1,011$ cells; their lipids cover 62 m^2, so the lipid in one cell covers 190 μm^2. Because the cell surface is 98 μm^2, the lipid can cover the cell twice—a lipid bilayer.

3. 2.1 nm. In the straight-chain form, all bond angles are about 109°. The distance between the first and third carbons is about 0.26 nm, so each —CH_2— group contributes about 0.13 nm to the chain length. For palmitate (16:0), the length is 16 × .013 nm = 2.1 nm. Two palmitates placed end to end span about 4.2 nm, approximately the width of a typical bilayer.

4. A decrease. Because cell–cell fusion requires movement of individual lipids in the bilayers, the process will occur much faster at 37 °C, when the lipids are in the "fluid" phase, than at 10 °C, when they are in the "solid" phase. The same holds for membrane protein mixing.

5. 33 kJ/mol, neglecting the effects of transmembrane electrical potential; 0.60 mol

6. 13 kJ/mol

7. Most of the oxygen consumed by a tissue is for oxidative phosphorylation, the source of most of the ATP. Therefore, about two-thirds of the ATP synthesized by the kidney is used for pumping K^+ and Na^+.

8. The rise-per-residue for an α helix (Chapter 7) is about 0.15 nm. To span a 4 nm bilayer, an α helix must contain about 27 residues; thus for seven spans, about 190 residues are required. A protein of M_r 64,000 has about 580 residues. To locate transmembrane regions, a hydropathy plot is used (see Box 10–2).

9. No; the symporter may carry more than one equivalent of Na^+ for each mole of glucose transported.

10. Salt extraction indicates a peripheral location, and inaccessibility to protease in intact cells indicates an internal location; similar to ankyrin.

11. The interactions among membrane lipids are noncovalent and reversible, and they form spontaneously.

12. The temperature of body tissues at the extremities, such as near the hooves, is generally lower than that of tissues closer to the center of the body. If lipid is to remain fluid at this lower temperature, as required by the fluid mosaic model, it must contain a higher proportion of unsaturated

fatty acids; unsaturated fatty acids lower the melting point of lipid mixtures.

13. The energetic cost of moving the highly polar, sometimes charged, head group through the hydrophobic interior of the bilayer is prohibitive.

14. At pH 7, tryptophan bears a positive and a negative charge, but indole is uncharged. The movement of the less polar indole through the hydrophobic core of the bilayer is energetically more favorable.

Chapter 11

1. **(a)**

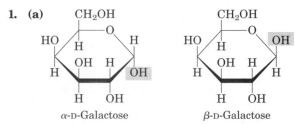

α-D-Galactose   β-D-Galactose

(b) A freshly prepared solution of α-D-galactose undergoes mutarotation to yield an equilibrium mixture of α- and β-D-galactose. Mutarotation of either the pure α or the pure β form will yield the same equilibrium mixture.
(c) 72% β form; 28% α form

2. **(a)** Measure the change in optical rotation with time.
(b) The optical rotation of the mixture is negative (inverted) relative to that of the sucrose solution.
(c) 0.63

3. Prepare a slurry of sucrose and water for the core; add a small amount of invertase; immediately coat with chocolate.

4. Sucrose is not a reducing sugar (see also the answer to Problem 12).

5. 7,840 residues/s

6. The human intestinal enzyme that splits (1→4) linkages between glucose residues is absolutely specific for the (α1→4) linkage; cellulose, with its (β1→4) linkages, cannot be digested.

7. Native cellulose consists of glucose units linked by (β1→4) glycosidic bonds. The β linkage forces the polymer chain into an extended conformation (see Fig. 11–17). A parallel series of these extended chains can form intermolecular hydrogen bonds, aggregating into long, tough, insoluble fibers. Glycogen consists of glucose units linked together by (α1→4) glycosidic bonds. The α linkage causes a bend in the chain and prevents the formation of long fibers. In addition, glycogen is highly branched. Because many of its hydroxyl groups are exposed to water, glycogen is highly hydrated, and can be extracted as a dispersion in hot water.

    The physical properties of these two polymers are well suited for their biological roles. Cellulose serves as a structural material in plants, consistent with its side-by-side aggregation into insoluble fibers. Glycogen serves as a storage fuel in animals. The highly hydrated glycogen granules with their abundance of nonreducing ends can be rapidly hydrolyzed by glycogen phosphorylase to release glucose-1-phosphate.

8. 10.8 s

9. **(a)** The branch points yield 2,3-dimethylglucose, whereas the unbranched residues yield 2,3,6-trimethylglucose.
(b) 3.74%

10. Chains of (1→6)-linked D-glucose residues with occasional (1→3)-linked branches, with about one branch for every 20 residues in the polymer

11. Yes; the empirical formula is CH_2O, typical of a carbohydrate.

12. Lactose; in sucrose, the anomeric carbons of both glucose and fructose are involved in the glycosidic bond and are not available to reduce Fehling's reagent. Lactose is a reducing sugar that converts Fe^{3+} to Fe^{2+}, which precipitates as the red oxide.

13. The rate of mutarotation (interconversion of α and β isomers) is high enough that, as the enzyme consumes β-D-glucose, α-D-glucose is further converted to the β form; eventually, all glucose is oxidized by the enzyme. Glucose oxidase is specific for glucose; it does not detect other reducing sugars (such as galactose), which do react with Fehling's reagent.

14. The negative charges on chondroitin sulfate repel each other and force the molecule into an extended conformation. Furthermore, the polar structure attracts many water molecules (water of hydration), which increases the molecular volume of the chondroitin sulfate.

15. Oligosaccharides; their subunits can be combined in more ways than the amino acid subunits of oligopeptides. Each of the several hydroxyl groups can participate in glycosidic bonds, and the configuration of each glycosidic bond may be either α or β. Furthermore, the polymer may be linear or branched.

Chapter 12

1. 0.35 mg/ml protein; 0.035 mg/ml nucleic acid

2. N-3, N-7, and N-9

3. (3′)TACGGGCATACGTAAG(5′)

4. 0.94×10^{-3} g

5. **(a)** 40° **(b)** 0°

6. The RNA helix will be in the A conformation; the DNA helix will generally be in the B conformation.

7. In eukaryotic DNA, about 5% of cytosine residues are methylated. 5-Methylcytosine can spontaneously deaminate to form thymine, and the resulting G–T pair is one of the most common mismatches in eukaryotic cells.

8. Base stacking in nucleic acids tends to reduce the absorption of UV light. Denaturation of DNA involves the loss of base stacking, and UV absorption increases.

9. One DNA contains 32% A, 32% T, 18% G, and 18% C; the other 17% A, 17% T, 33% G, and 33% C. This assumes that both are double-stranded. The DNA containing 33% G and 33% C almost certainly came from the thermophilic bacterium; its higher G≡C content makes it much more stable to heat.

Chapter 13

1. Consider the developing chick as the system; the nutrients, egg shell, and outside world are the surroundings. Transformation of the single cell into a chick drastically reduces the entropy of the system. Initially, the parts of the egg outside the embryo (the surroundings) contain complex fuel molecules (a low-entropy condition). During incubation, some of these complex molecules are converted into large numbers of CO_2 and H_2O molecules (high entropy). This increase in the entropy of the surroundings is larger than the decrease in entropy of the chick (the system).

2. **(a)** 4.75 J/mol **(b)** −7.6 J/mol **(c)** −13.7 kJ/mol

3. **(a)** 261 M **(b)** 609 M **(c)** 0.29

4. $K'_{eq} = 21$; $\Delta G^{\circ\prime} = -7.6$ kJ/mol

5. -30.7 kJ/mol

6. **(a)** -1.7 kJ/mol **(b)** -4.4 kJ/mol
   (c) At a given temperature, the value of $\Delta G^{\circ\prime}$ for any reaction is fixed and is defined for standard conditions (both fructose-6-phosphate and glucose-6-phosphate at 1 M). In contrast, ΔG is a variable that can be calculated for any set of reactant and product concentrations.

7. Less. The overall equation for ATP hydrolysis can be approximated as:

$$ATP^{4-} + H_2O \longrightarrow ADP^{3-} + HPO_4^{2-} + H^+$$

   (This is only an approximation because the ionized species shown here are the major, but not the only, forms present.) Under standard conditions (i.e., [ATP] = [ADP] = [P_i] = 1 M), the concentration of water is 55 M and does not change during the reaction. Because H^+ ions are produced in the reaction, at a higher [H^+] (pH 5.0) the equilibrium would be shifted to the left and less free energy would be released.

8. 9.6

9. **(a)** 3.8×10^{-3} M; [glucose-6-phosphate] = 8.7×10^{-8} M; no
   (b) 14 M; because the maximum solubility of glucose is less than 1 M, this is not a reasonable step.
   (c) 870 ($\Delta G^{\circ\prime} = -17$ kJ/mol); [glucose] = 1.1×10^{-7} M; yes
   (d) No; this would require such high [P_i] that the phosphate salts of divalent cations would precipitate.
   (e) By directly transferring the phosphate group from ATP to glucose, the phosphate group transfer potential ("tendency" or "pressure") of ATP is utilized without generating high concentrations of intermediates. The essential part of this transfer is, of course, the enzymatic catalysis.

10. **(a)** -12.5 kJ/mol **(b)** -14.6 kJ/mol

11. **(a)** 2.1×10^{-4} **(b)** 49.3 **(c)** 5.3×10^4

12. 10.0 kJ/mol

13. 46.0 kJ/mol

14. **(a)** 46.0 kJ/mol **(b)** 46 kg; 68% **(c)** ATP is synthesized as it is needed then broken down to ADP and P_i; its concentration is maintained at a steady state.

15. **(a)** 1.1 s
   (b) Phosphocreatine + ADP → creatine + ATP
   (c) ATP synthesis coupled to the catabolism of glucose, amino acids, and fatty acids

16. The ATP system is in a dynamic steady state; [ATP] remains constant because the rate of ATP consumption equals its rate of synthesis. ATP consumption involves release of the terminal (γ) phosphate; synthesis of ATP from ADP involves replacement of this phosphate. Hence, the terminal phosphate undergoes rapid turnover. In contrast, the central (β) phosphate undergoes only relatively slow turnover.

17. **(a)** 0.8 kJ/mol
   (b) Inorganic pyrophosphatase catalyzes the hydrolysis of pyrophosphate and drives the net reaction toward the synthesis of acetyl-CoA.

18. **(a)** 1.4×10^{-9} M **(b)** The physiological concentration (0.023 mM) is 16,000 times higher than the equilibrium concentration; this reaction does not reach equilibrium in the cell. Many reactions in the cell are not at equilibrium.

19. **(a)** NAD^+/NADH **(b)** Pyruvate/lactate **(c)** Lactate formation **(d)** -25 kJ/mol **(e)** 2.58×10^4

20. **(a)** 1.14 V **(b)** 220 kJ/mol **(c)** About 7

21. **(a)** -0.35 V **(b)** -0.32 V **(c)** -0.29 V

22. In order of increasing tendency: (a); (d); (b); (c)

23. (c) and (d)

Chapter 14

1. Net equation: Glucose + 2ATP →
   2 glyceraldehyde-3-phosphate + 2ADP + 2H$^+$
   $\Delta G^{\circ\prime} = 2.34$ kJ/mol

2. Net equation:
   Glyceraldehyde-3-phosphate + 2ADP + P_i + H$^+$ →
   lactate + 2ATP + H$_2$O
   $\Delta G^{\circ\prime} = -63$ kJ/mol

3. **(a)** $^{14}CH_3CH_2OH$ **(b)** [3-^{14}C] glucose or [4-^{14}C] glucose

4. At C-1. This experiment demonstrates the reversibility of the aldolase reaction. The C-1 of glyceraldehyde-3-phosphate is equivalent to C-4 of fructose-1,6-bisphosphate (see Fig. 14–4). The starting glyceraldehyde-3-phosphate must have been labeled at C-1. The C-3 of dihydroxyacetone phosphate becomes labeled through the triose phosphate isomerase reaction, thus giving rise to the labeling of C-3 in fructose-1,6-bisphosphate.

5. There would be no anaerobic production of ATP; aerobic ATP production would be diminished only slightly.

6. No; lactate dehydrogenase is required to recycle NAD^+ from the NADH formed during the oxidation of glyceraldehyde-3-phosphate.

7. By rapid removal of the 1,3-bisphosphoglycerate in a favorable subsequent step, catalyzed by phosphoglycerate kinase

8. **(a)** 3-Phosphoglycerate would be the product. **(b)** In the presence of arsenate there is no net ATP synthesis under anaerobic conditions.

9. **(a)** The stoichiometry of alcohol fermentation requires 2 mol of P_i per mole of glucose.
   (b) Ethanol is the reduced product formed during reoxidation of NADH to NAD^+, and CO_2 is the byproduct of the conversion of pyruvate into ethanol. Yes; pyruvate must be converted to ethanol, to produce a continuous supply of NAD^+ for the oxidation of glyceraldehyde-3-phosphate. Fructose-1,6-bisphosphate accumulates; it is formed as an intermediate in glycolysis.
   (c) Arsenate replace P_i in the glyceraldehyde-3-phosphate dehydrogenase reaction to yield an acyl arsenate, which spontaneously hydrolyzes. This prevents the formation of ATP, but 3-phosphoglycerate continues through the pathway.

10. The glucose in cells is phosphorylated to glucose-6-phosphate. Because the equilibrium of this reaction strongly favors the product, glucose that enters the cell is rapidly and irreversibly converted to glucose-6-phosphate.

11. Net equation: Glycerol + 2NAD$^+$ + ADP + P_i →
    pyruvate + 2NADH + ATP + 2H$^+$

12. **(a)** 0.029 **(b)** 316 **(c)** No. The value of the mass-action ratio is much lower than K'_{eq}, indicating that the PFK-1 reaction is far from equilibrium in cells; this reaction is slower than the subsequent reactions in glycolysis. Flux through the glycolytic pathway is largely determined by the activity of PFK-1.

13. In the absence of O_2, the ATP needs are met by anaerobic glucose metabolism (fermentation to lactate). Because aerobic oxidation of glucose produces far more ATP than does fermentation, less glucose is needed to produce the same amount of ATP.

14. **(a)** There are two binding sites for ATP: a catalytic site and a regulatory site. Binding of ATP to an allosteric site

inhibits PFK-1, either by reducing V_{max} or increasing K_m for ATP at the catalytic site.

(b) Glycolytic flux is reduced when ATP is plentiful.

(c) The graph indicates that the addition of ADP suppresses the inhibition of ATP. Because the adenine nucleotide pool is fairly constant, consumption of ATP leads to an increase in ADP levels. The data indicate that the activity of PFK-1 may be regulated by the ATP/ADP ratio.

15. **(a)** *In muscle*, glycogen is broken down to supply energy (ATP) via glycolysis. Glycogen phosphorylase catalyzes the conversion of stored glycogen to glucose-1-phosphate, which is converted to glucose-6-phosphate, an intermediate in glycolysis. During strenuous activity, skeletal muscle requires large quantities of glucose-6-phosphate. In *the liver*, the breakdown of glycogen is used to maintain a steady level of blood glucose between meals (glucose-6-phosphate is converted to free glucose).

(b) In actively working muscle, ATP flux requirements are very high and glucose-1-phosphate needs to be produced rapidly, requiring a high V_{max}.

16. 3 : 4. The value of this ratio in the cell (>100 : 1) indicates that [glucose-1-phosphate] is far below the equilibrium value. The rate at which glucose-1-phosphate is removed (through entry into glycolysis) is greater than its rate of production (by the glucogen phosphorylase reaction). This indicates that metabolite flow is from glycogen to glucose-1-phosphate, and that the glycogen phosphorylase reaction is probably the regulatory step in glycogen breakdown.

17. **(a)** Increases **(b)** Decreases **(c)** Increases

18. *Resting:* [ATP] high; [AMP] low; [acetyl-CoA] and [citrate] intermediate. *Running:* [ATP] intermediate; [AMP] high; [acetyl-CoA] and [citrate] low. Glucose flux through glycolysis increases during the anaerobic sprint, because: (1) the ATP inhibition of glycogen phosphorylase and PFK-1 is partially relieved, (2) AMP stimulates both enzymes, and (3) lower [citrate] and [acetyl-CoA] relieves their inhibitory effects on PFK-1 and pyruvate kinase, respectively.

19. The migrating bird relies on the highly efficient aerobic oxidation of fats, rather than the anaerobic metabolism of glucose used by a sprinting rabbit. The bird reserves its muscle glycogen for short bursts of energy during emergencies.

20. *Case A:* (f), (3); *Case B:* (c), (3); *Case C:* (a), (4); *Case D:* (d), (6)

21. In galactokinase deficiency, galactose accumulates; in galactose-1-phosphate uridylyltransferase deficiency, galactose-1-phosphate accumulates. The latter is more toxic.

22. The two enzymes catalyze a portion of the glycolytic pathway, converting glyceraldehyde-3-phosphate to 1,3-bisphosphoglycerate, then 1,3-bisphosphoglycerate to 3-phosphoglycerate with the formation of ATP. The ^{32}P incorporated as $^{32}P_i$ in the first step is transferred to ADP in the second step, forming [γ-^{32}P]ATP.

Chapter 15

1. **(a)**
① *Citrate synthase:*
Acetyl-CoA + oxaloacetate + $H_2O \rightarrow$ citrate + CoA + H^+
② *Aconitase:* Citrate $\rightarrow$ isocitrate
③ *Isocitrate dehydrogenase:*
Isocitrate + $NAD^+ \rightarrow \alpha$-ketoglutarate + CO_2 + NADH
④ *α-Ketoglutarate dehydrogenase:*
α-Ketoglutarate + NAD^+ + CoA $\rightarrow$
succinyl-CoA + CO_2 + NADH

⑤ *Succinyl-CoA synthetase:*
Succinyl-CoA + P_i + GDP $\rightarrow$ succinate + GTP + CoA
⑥ *Succinate dehydrogenase:*
Succinate + FAD $\rightarrow$ fumarate + $FADH_2$
⑦ *Fumarase:* Fumarate + $H_2O \rightarrow$ malate
⑧ *Malate dehydrogenase:*
Malate + $NAD^+ \rightarrow$ oxaloacetate + NADH + H^+

(b), (c) Step ① CoA, condensation; ② none, isomerization; ③ NAD^+, oxidative decarboxylation; ④ NAD^+, CoA, and thiamine pyrophosphate, oxidative decarboxylation; ⑤ CoA, phosphorylation; ⑥ FAD, oxidation; ⑦ none, hydration; ⑧ NAD^+, oxidation

(d) Acetyl-CoA + $3NAD^+$ + FAD + GDP + P_i + $2H_2O \rightarrow$
CO_2 + 3NADH + $FADH_2$ + GTP + $2H^+$ + CoA

2. **(a)** Oxidation; methanol $\rightarrow$ formaldehyde + [H—H]
(b) Oxidation; formaldehyde $\rightarrow$ formate + [H—H]
(c) Reduction; CO_2 + [H—H] $\rightarrow$ formate + H^+
(d) Reduction; glycerate + [H—H] $\rightarrow$ glyceraldehyde + OH^-
(e) Oxidation; glycerol $\rightarrow$ dihydroxyacetone + [H—H]
(f) Oxidation; $2H_2O$ + toluene $\rightarrow$ benzoate + 3 [H—H] + H^+
(g) Oxidation; succinate $\rightarrow$ fumarate + [H—H]
(h) Oxidation; pyruvate + $H_2O \rightarrow$ acetate + [H—H] + CO_2

3. **(a)** Oxidized; ethanol + $NAD^+ \rightarrow$
acetaldehyde + NADH + H^+
(b) Reduced; 1,3-bisphosphoglycerate + NADH + $H^+ \rightarrow$
glyceraldehyde-3-phosphate + NAD^+ + HPO_4^{2-}
(c) Unchanged; pyruvate + $H^+ \rightarrow$ acetaldehyde + CO_2
(d) Oxidized; pyruvate + $NAD^+ \rightarrow$
acetate + NADH + H^+ + CO_2
(e) Reduced; oxaloacetate + NADH + $H^+ \rightarrow$ malate + NAD^+
(f) Unchanged; acetoacetate + $H^+ \rightarrow$ acetone + CO_2

4. **(a)** Oxygen consumption is a measure of the activity of the first two stages of cellular respiration: glycolysis and the citric acid cycle. The addition of oxaloacetate or malate stimulates the citric acid cycle and thus stimulates respiration. **(b)** The added oxaloacetate or malate serves a catalytic role, because it is regenerated in the latter part of the citric acid cycle.

5. **(a)** 6.0×10^{-6} **(b)** 1.2×10^{-8} M **(c)** 30 molecules

6. **(a)** $^-OOCCH_2CH_2COO^-$ (succinate) **(b)** Malonate is a competitive inhibitor of succinate dehydrogenase. **(c)** A block in the citric acid cycle stops NADH formation, which stops electron transfer, which stops respiration. **(d)** A large excess of succinate (substrate) overcomes the competitive inhibition.

7. **(a)** Add uniformly labeled [^{14}C] glucose and check for the release of $^{14}CO_2$. **(b)** Equally distributed in C-2 and C-3 of oxaloacetate **(c)** An infinite number

8. **(a)** C-1 **(b)** C-3 **(c)** C-3 **(d)** Methyl group **(e)** Equally distributed in the —CH_2— groups **(f)** C-4 **(g)** Equally distributed in the —CH_2— groups

9. No. For every two carbons that enter as acetate, two leave the cycle as CO_2; thus there is no net synthesis of oxaloacetate. Net synthesis of oxaloacetate occurs by the carboxylation of pyruvate, an anaplerotic reaction.

10. **(a)** Inhibition of aconitase **(b)** Fluorocitrate; competes with citrate; by a large excess of citrate **(c)** Citrate and fluorocitrate are inhibitors of PFK-1. **(d)** All catabolic processes necessary for ATP production are shut down.

11. Net reaction:
2 Pyruvate + ATP + $2NAD^+$ + $H_2O \rightarrow$
α-ketoglutarate + CO_2 + ADP + P_i + 2NADH + $3H^+$

12. Succinyl-CoA is an intermediate of the citric acid cycle; its accumulation signals reduced flux through the cycle, calling for reduced entry of acetyl-CoA into the cycle. Citrate synthase, by regulating the primary oxidative pathway of the cell, regulates the supply of NADH and, thereby, the flow of electrons from NADH to oxygen.

13. Fatty acid catabolism increases [acetyl-CoA], which stimulates pyruvate carboxylase. The resulting increase in [oxaloacetate] stimulates acetyl-CoA consumption by the citric acid cycle, and [citrate] rises, inhibiting glycolysis at the level of PFK-1. In addition, increased [acetyl-CoA] inhibits the pyruvate dehydrogenase complex, slowing the utilization of pyruvate from glycolysis.

14. Oxygen is needed to recycle NAD^+ from the NADH produced by the oxidative reactions of the citric acid cycle. Reoxidation of NADH occurs during mitochondrial oxidative phosphorylation.

15. Toward citrate; ΔG for the citrate synthase reaction under these conditions is about -8.6 kJ/mol.

16. Steps ④ and ⑤ are essential in the reoxidation of the reduced lipoamide cofactor.

Chapter 16

1. The fatty acid portion; the carbons in fatty acids are more reduced than those in glycerol.

2. (a) 4.0×10^5 kJ (9.5×10^4 kcal) (b) 48 days (c) 0.5 lb/day

3. The first step in fatty acid oxidation is analogous to the conversion of succinate to fumarate; the second step, to the conversion of fumarate to malate; the third step, to the conversion of malate to oxaloacetate.

4. (a) $R—COO^- + ATP \longrightarrow$ acyl-AMP $+ PP_i$

   Acyl-AMP + CoA $\longrightarrow$ $R—\overset{\overset{\displaystyle O}{\|}}{C}—CoA + AMP$

   (b) Irreversible hydrolysis of PP_i to $2P_i$ by cellular inorganic pyrophosphatase

5. Yes; some of the tritium is removed from palmitate during the dehydrogenation reactions of β oxidation. The removed tritium appears as tritiated water.

6. Fatty acyl groups condensed with CoA in the cytosol are first transferred to carnitine, releasing CoA, then transported into the mitochondrion, where they are again condensed with CoA. The cytosolic and mitochondrial pools of CoA are thus kept separate, and no radioactive CoA from the cytosolic pool enters the mitochondrion.

7. (a) The carnitine-mediated entry of fatty acids into mitochondria is the rate-limiting step in fatty acid oxidation. Carnitine deficiency slows fatty acid oxidation; added carnitine increase the rate.
   (b) All of these increase the metabolic need for fatty acid oxidation.
   (c) Carnitine deficiency might result from a deficiency of Lys, its precursor, or from a defect in one of the enzymes in the biosynthetic path to carnitine.

8. Oxidation of fats releases metabolic water; 0.49 L of water per pound of tripalmitin

9. Complete oxidation of the hydrocarbon to CO_2 and H_2O

10. (a) M_r 136; phenylacetic acid (b) Even

11. Because the mitochondrial pool of CoA is small, CoA must be recycled from acetyl-CoA via the formation of ketone bodies. This allows the operation of the β-oxidation pathway, necessary for energy production.

12. (a) Glucose yields pyruvate via glycolysis, and pyruvate is the main source of oxaloacetate. Without glucose in the diet, [oxaloacetate] drops and the citric acid cycle slows.
    (b) Odd-numbered; propionate conversion to succinyl-CoA provides intermediates for the citric acid cycle.

13. (a) $CH_3(CH_2)_{12}—\overset{\overset{\displaystyle O}{\|}}{C}—CoA + 6CoA + 6FAD + 6NAD^+ + 6H_2O \rightarrow$
    7 acetyl-CoA $+ 6FADH_2 + 6NADH + 6H^+$
    (b) $CH_3(CH_2)_{16}COO^- + ATP + 9CoA + 8FAD + 8NAD^+ + 9H_2O \rightarrow$
    9 acetyl-CoA $+ AMP + 2P_i + 8FADH_2 + 8NADH + 10H^+$
    (c) $CH_3—\overset{\overset{\displaystyle OH}{|}}{CH}—CH_2COO^- + NAD^+ + ATP + 2CoA + H_2O \rightarrow$
    2 acetyl-CoA $+ AMP + 2P_i + NADH + 3H^+$

14. (a) carbonyl group (b) terminal carboxyl group (c) no label

15. $CH_3—\overset{\overset{\displaystyle OH}{|}}{CH}—CH_2COO^- + 4\tfrac{1}{2}O_2 + 25ADP + 25P_i + 25H^+ \rightarrow$
    $4CO_2 + 29H_2O + 25ATP$

16. Enz-FAD, having a more positive standard reduction potential, is a better electron acceptor than NAD^+, and the reaction is driven in the direction of fatty acyl–CoA oxidation. This more favorable equilibrium is obtained at the cost of 1 ATP; only 2 ATP are produced per $FADH_2$ oxidized in the respiratory chain (versus 3 per NADH).

17. 9 turns; arachidic acid, a 20-carbon saturated fatty acid, yields 10 molecules of acetyl-CoA, the last two both formed in the ninth turn.

18. (1) The reduction of O_2 in the respiratory chain:

$$CH_3(CH_2)_{14}COO^- + 23O_2 \longrightarrow 16CO_2 + 16H_2O$$

(2) The formation of the anhydride bond in ATP:

$$129P_i + 129ADP + 129H^+ \longrightarrow 129ATP + 129H_2O$$

19. See Fig. 16–12. [3-^{14}C]Succinyl-CoA is formed, which gives rise to oxaloacetate labeled at C-2 and C-3.

20. Methylmalonyl-CoA mutase requires the cobalt-containing cofactor formed from vitamin B_{12} (see Box 16–2).

Chapter 17

1.

(a) $^-OOC—CH_2—\overset{\overset{\displaystyle O}{\|}}{C}—COO^-$ Oxaloacetate

(b) $^-OOC—CH_2—CH_2—\overset{\overset{\displaystyle O}{\|}}{C}—COO^-$ α-Ketoglutarate

(c) $CH_3—\overset{\overset{\displaystyle O}{\|}}{C}—COO^-$ Pyruvate

(d) ⬡$—CH_2—\overset{\overset{\displaystyle O}{\|}}{C}—COO^-$ Phenylpyruvate

2. This is a coupled-reaction assay. The product of the slow transamination (pyruvate) is rapidly consumed in the subsequent "indicator reaction" catalyzed by lactate dehydrogenase, which consumes NADH. Thus the rate of disappearance of NADH is a measure of the rate of the aminotransferase reaction. The indicator reaction is monitored by observing the decrease in absorption of NADH at 340 nm with a spectrophotometer.

3. No; the nitrogen in Ala can be transferred to oxaloacetate via transamination, to form Asp.

4. (a) Phenylalanine-4-monooxygenase; a low-phenylalanine

diet **(b)** The normal route of Phe metabolism via hydroxylation to Tyr is blocked, and Phe accumulates. **(c)** Phe is transformed to phenylpyruvate by transamination, and then to phenyllactate by reduction. The transamination reaction has an equilibrium constant of 1.0, and phenylpyruvate is formed in significant amounts when phenylalanine accumulates. **(d)** Because of the deficiency in production of Tyr, a precursor of melanin, the pigment normally present in hair

5. Catabolism of the carbon skeletons of Val, Met, and Ile is impaired because of the absence of functional methylmalonyl-CoA mutase (a coenzyme B_{12} enzyme). The physiological effects of loss of this enzyme are described in Box 17–2.

6. 17 moles of ATP per mole of lactate; 15 ATP per Ala, when nitrogen removal is included

7. **(a)** $^{15}NH_2—CO—^{15}NH_2$

   (b) $^-OO^{14}C—CH_2—CH_2—^{14}COO^-$

   (c)
$$R—NH—\overset{\overset{\displaystyle ^{15}NH}{\|}}{C}—^{15}NH_2$$

   (d)
$$R—NH—\overset{\overset{\displaystyle O}{\|}}{C}—^{15}NH_2$$

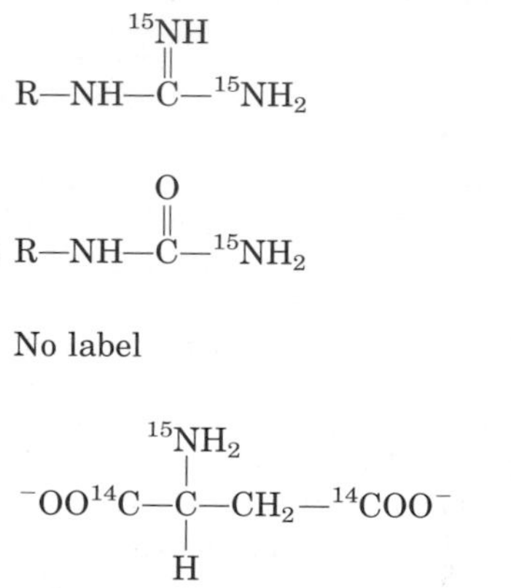

   (e) No label

   (f)
$$^-OO^{14}C—\overset{\overset{\displaystyle ^{15}NH_2}{|}}{\underset{\underset{\displaystyle H}{|}}{C}}—CH_2—^{14}COO^-$$

8. **(a)** Ile $\xrightarrow{①}$ II $\xrightarrow{②}$ IV $\xrightarrow{③}$ I $\xrightarrow{④}$ V $\xrightarrow{⑤}$ III $\xrightarrow{⑥}$ acetyl-CoA + propionyl-CoA **(b)** Step ① transamination, no analogous reaction; ② oxidative decarboxylation, analogous to the oxidative decarboxylation of pyruvate to acetyl-CoA; ③ oxidation, analogous to dehydrogenation of succinate; ④ hydration, analogous to hydration of fumarate to malate; ⑤ oxidation, analogous to dehydrogenation of malate to oxaloacetate; ⑥ thiolysis (reverse aldol condensation), analogous to thiolase reaction.

9. **(a)** Fasting resulted in low blood glucose; subsequent administration of the experimental diet led to rapid catabolism of glucogenic amino acids. **(b)** Oxidative deamination caused the rise in ammonia levels; the absence of Arg (an intermediate in the urea cycle) prevented the conversion of ammonia to urea; Arg is not synthesized in sufficient quantities in the cat to meet the needs imposed by the stress of the experiment. This suggests that Arg is an essential amino acid in the cat's diet. **(c)** Ornithine is converted to Arg by the urea cycle.

10. H_2O + glutamate$^-$ + NAD$^+$ $\longrightarrow$
α-ketoglutarate^{2-} + NH$_4^+$ + NADH + H$^+$
NH_4^+ + 2ATP^{4-} + H$_2$O + CO$_2$ $\longrightarrow$
carbamoyl phosphate^{2-} + 2ADP^{3-} + P$_i^{2-}$ + 3H$^+$
Carbamoyl phosphate^{2-} + ornithine$^+$ $\longrightarrow$
citrulline + P$_i^{2-}$ + H$^+$
Citrulline + aspartate$^-$ + ATP^{4-} $\longrightarrow$
argininosuccinate$^-$ + AMP^{2-} + PP$_i^{3-}$ + H$^+$
Argininosuccinate$^-$ $\longrightarrow$ arginine$^+$ + fumarate^{2-}
Fumarate^{2-} + H$_2$O $\longrightarrow$ malate^{2-}
Malate^{2-} + NAD$^+$ $\longrightarrow$ oxaloacetate^{2-} + NADH + H$^+$

Oxaloacetate^{2-} + glutamate$^-$ $\longrightarrow$
aspartate$^-$ + α-ketoglutarate^{2-}
Arginine$^+$ + H$_2$O $\longrightarrow$ urea + ornithine$^+$

2 Glutamate$^-$ + CO$_2$ + 4H$_2$O + 2NAD$^+$ + 3ATP^{4-} $\longrightarrow$
2 α-ketoglutarate^{2-} + 2NADH + 7H$^+$ + urea +
2ADP^{3-} + AMP^{2-} + PP$_i^{3-}$ + 2P$_i^{2-}$ (1)

Additional reactions that need to be considered are:
AMP^{2-} + ATP^{4-} $\longrightarrow$ 2ADP^{3-} (2)
O$_2$ + 8H$^+$ + 2NADH + 6ADP^{3-} + 6P$_i^{2-}$ $\longrightarrow$
2NAD$^+$ + 6ATP^{4-} + 8H$_2$O   (3)
H$_2$O + PP$_i^{3-}$ $\longrightarrow$ 2P$_i^{2-}$ + H$^+$ (4)

Summing equations (1) through (4),
2 Glutamate$^-$ + CO$_2$ + O$_2$ + 2ADP^{3-} + 2P$_i^{2-}$ $\longrightarrow$
2 α-ketoglutarate^{2-} + urea + 3H$_2$O + 2ATP^{4-}

11. A likely mechanism is:

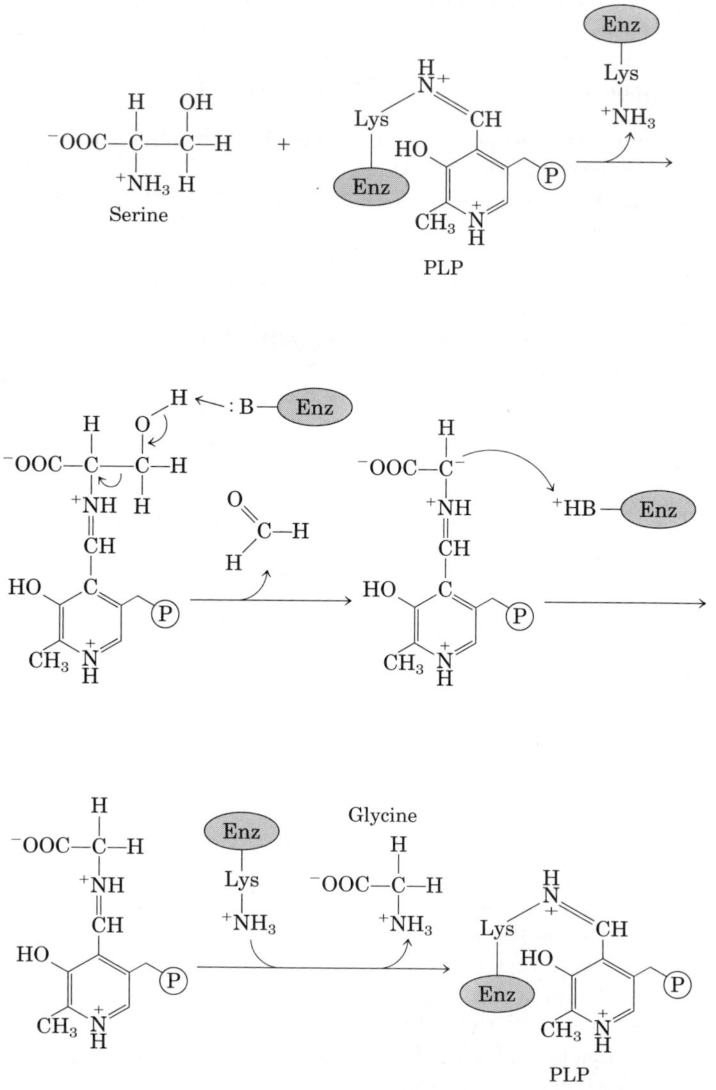

The formaldehyde (HCHO) produced in the second step reacts rapidly with tetrahydrofolate at the enzyme active site to produce N^5, N^{10}-methylenetetrahydrofolate (see Fig. 17–19).

12. **(a)** Transamination; no analogies in either pathway; cofactor: pyridoxal phosphate
(b) Oxidative decarboxylation; analogous to oxidative

decarboxylation of pyruvate to acetyl-CoA prior to entry into citric acid cycle and of α-ketoglutarate to succinyl-CoA in citric acid cycle; cofactors: NAD^+, FAD, lipoate, and thiamine pyrophosphate

(c) Dehydrogenation (oxidation); analogous to dehydrogenation of succinate to fumarate in citric acid cycle and of fatty acyl–CoA to enoyl-CoA in β oxidation; cofactor: FAD

(d) Carboxylation; analogous to carboxylation of pyruvate to oxaloacetate in citric acid cycle; cofactors: ATP and biotin

(e) Hydration; analogous to hydration of fumarate to malate in citric acid cycle and of enoyl-CoA to 3-hydroxyacyl-CoA in β oxidation; cofactors: none

(f) Reverse aldol reaction; analogous to reverse of citrate synthase reaction in citric acid cycle and identical to cleavage of β-hydroxy-β-methylglutaryl-CoA in formation of ketone bodies; cofactors: none

13. The second amino group introduced into urea is transferred from Asp, which is generated during the transamination of Glu to oxaloacetate, a reaction catalyzed by aspartate aminotransferase. Approximately one half of all the amino groups excreted as urea must pass through the aspartate aminotransferase reaction, making this the most highly active aminotransferase.

14. (a) A person on a diet consisting only of protein must use amino acids as the principal source of metabolic fuel. Because the catabolism of amino acids requires the removal of nitrogen as urea, the process consumes abnormally large quantities of water to dilute and excrete the urea in the urine. Furthermore, electrolytes in the "liquid protein" must be diluted with water and excreted. If the daily water loss through the kidney is not balanced by a sufficient water intake, a net loss of body water results.

(b) When considering the nutritional benefits of protein, one must keep in mind the total amount of amino acids needed for protein synthesis and the distribution of amino acids in the dietary protein. Gelatin contains a nutritionally unbalanced distribution of amino acids. As large amounts of gelatin are ingested and the excess amino acids are catabolized, the capacity of the urea cycle may be exceeded, leading to ammonia toxicity. This is further complicated by the dehydration that may result from excretion of large quantities of urea. A combination of these two factors could produce coma and death.

15. Ala and Gln play special roles in the transport of amino groups from muscle and from other nonhepatic tissues, respectively, to the liver.

Chapter 18

1. *Reaction (1):* (a), (d) NADH; (b), (e) E–FMN; (c) $NAD^+/$ NADH and E–FMN/$FMNH_2$
   Reaction (2): (a), (d) E–$FMNH_2$; (b), (e) Fe^{3+}; (c) E–FMN/ $FMNH_2$ and Fe^{3+}/Fe^{2+}
   Reaction (3): (a), (d) Fe^{2+}; (b), (e) UQ; (c) Fe^{3+}/Fe^{2+} and UQ/UQH_2

2. (a) $NAD^+/$NADH (b) Pyruvate/lactate (c) Lactate formation (d) −25 kJ/mol (e) 2.58×10^4

3. (a) 1.14 V (b) −220 kJ/mol (c) About 7 (d) About 4

4. From the difference in standard reduction potential ($\Delta E_0'$) for each pair of half reactions, one can calculate $\Delta G^{\circ\prime}$. The oxidation of succinate by FAD is favored by the negative standard free-energy change ($\Delta G^{\circ\prime} = -3.7$ kJ/mol). Oxidation by NAD^+ would require a large, positive, standard free-energy change ($\Delta G^{\circ\prime} = 67$ kJ/mol).

5. (a) All carriers reduced; CN^- blocks the reduction of O_2 catalyzed by cytochrome oxidase. (b) All carriers reduced; in the absence of O_2, the reduced carriers are not reoxidized. (c) All carriers oxidized (d) Early carriers more reduced; later carriers more oxidized

6. (a) The inhibition of NADH dehydrogenase by rotenone decreases the rate of electron flow through the respiratory chain, which in turn decreases the rate of ATP production. If this reduced rate is unable to meet the organism's ATP requirements, the organism dies.
   (b) Antimycin A strongly inhibits the oxidation of UQ in the respiratory chain, reducing the rate of electron transfer and leading to the consequences described in (a).
   (c) Because antimycin A blocks *all* electron flow to oxygen, it is a more potent poison than rotenone, which blocks electron flow from NADH but not from $FADH_2$.

7. (a) The rate of electron transfer necessary to meet the ATP demand increases, and thus the P/O ratio decreases.
   (b) High concentrations of uncoupler produce P/O ratios near zero. The P/O ratio decreases, and more fuel must therefore be oxidized to generate the same amount of ATP. The extra heat released by this oxidation raises the body temperature.
   (c) Increased activity of the respiratory chain in the presence of an uncoupler requires the degradation of additional fuel. By oxidizing more fuel (including fat reserves) to produce the same amount of ATP, the body loses weight. When the P/O ratio approaches zero, the lack of ATP results in death.

8. (a) The formation of ATP is inhibited. (b) The formation of ATP is tightly coupled to electron transfer: 2,4-dinitrophenol is an uncoupler of oxidative phosphorylation. (c) Oligomycin

9. (a) Glycolysis will become anaerobic. (b) Oxygen consumption will cease. (c) Lactate formation will increase. (d) ATP synthesis will shut down.

10. (a) NADH is reoxidized via electron transfer, recycling NAD^+. (b) Oxidative phosphorylation is more efficient. (c) The high mass-action ratio of the ATP system inhibits phosphofructokinase-1.

11. (a) External medium: 4.0×10^{-8} M; matrix: 2.0×10^{-8} M (b) 2:1 (c) 21 (d) No (e) From the transmembrane potential

12. (a) 1.1 μmol/s$\cdot$g (b) About 5 s; to provide a constant level of ATP, regulation of ATP production must be tight and rapid.

13. About 70 μmol/s$\cdot$g. With a steady state [ATP] of 7 μmol/g, this is equivalent to 10 turnovers of the ATP pool per second; the reservoir would last about 0.1 s.

14. The inner mitochondrial membrane is impermeable to NADH, but the reducing equivalents of NADH are transferred (shuttled) through the membrane indirectly: they are transferred to oxaloacetate in the cytosol, the resulting malate is transported into the matrix, and mitochondrial NAD^+ is reduced to NADH.

15. For the maximum photosynthetic rate, photosystem I (which absorbs light of 700 nm) and photosystem II (which absorbs light of 680 nm) must be operating simultaneously.

16. Purple sulfur bacteria use H_2S as the hydrogen donor in photosynthesis. No; O_2 is evolved because photosystem II is absent.

17. 0.57

18. DCMU blocks electron transfer between photosystem II and the first site of ATP production.

19. (a) 56 kJ/mol (b) 0.29 V

20. 5.4×10^{-78}; the reaction is highly unfavorable! In chloroplasts, the input of light energy overcomes this barrier.

21. -920 kJ/mol

22. No; the electrons from H_2O flow to the artificial electron acceptor Fe^{3+}, not to $NADP^+$.

23. About once every 0.1 s; 1 in 10^8 is excited.

24. Light of 700 nm excites photosystem I but not photosystem II; electrons flow from P700 to $NADP^+$, but no electrons flow from P680 to replace them. When light of 680 nm excites photosystem II, electrons tend to flow to photosystem I, but the electron carriers between the two photosystems quickly become completely reduced.

25. No; the excited electron from P700 returns to refill the electron "hole" created by illumination. Photosystem II is not needed to supply electrons, and no O_2 is evolved from H_2O. NADPH is not formed because the excited electron returns to P700.

Chapter 19

1. No; the transformation of two molecules of pyruvate to one molecule of glucose requires the input of energy (4ATP + 2GTP) and reducing power (2NADH), obtainable only through the citric acid cycle/oxidative phosphorylation pathway by the catabolism of amino acids, fatty acids, or other carbohydrate.

2. **(a)** In the pyruvate carboxylase reaction, $^{14}CO_2$ is added to pyruvate, but phosphoenolpyruvate carboxykinase removes the *same* CO_2 in the next step. Thus, ^{14}C is not (initially) incorporated into glucose.

(b)

3. Pyruvate carboxylase is a mitochondrial enzyme. The [^{14}C]oxaloacetate formed mixes with the oxaloacetate pool of the citric acid cycle and is equilibrated with the citric acid cycle intermediates to form a mixture of [1-^{14}C] and [4-^{14}C]oxaloacetate. Oxaloacetate labeled at C-1 leads to the formation of [3,4-^{14}C]glucose (see Problem 2).

4. PFK-1 is activated by AMP and inhibited by ATP; it regulates glycolysis. FBPase-1 is activated by ATP and inhibited by AMP; it regulates gluconeogenesis.

5. (a), (b), and (d) are glucogenic; (c) and (e) are not.

6. **(a)** The rapid increase in glycolysis; the rise in pyruvate and NADH results in a rise in lactate. **(b)** Lactate is transformed to glucose via pyruvate; this is a slower process because formation of pyruvate is limited by NAD^+ availability, the LDH equilibrium is in favor of lactate, and conversion of pyruvate to glucose is energy-requiring. **(c)** The equilibrium for the LDH reaction is in favor of lactate formation.

7. If the catabolic and anabolic pathways of glucose metabolism are operating simultaneously, futile cycling of ATP occurs, with extra O_2 consumption.

8. The observation that glycogen synthase has the lowest measured activity of the enzymes in glycogen synthesis suggests that this enzyme step is the bottleneck in the flow of metabolites, and is thus a regulatory point. This is confirmed by the observation that the stimulation of glycogen synthesis by the activation of the regulatory enzyme leads to a decrease in the concentration of the substrate of the glycogen synthase reaction (UDP-glucose) and an increase in the concentration of the product of the reaction (UDP).

9. Storage consumes 1 mol of ATP per mole of glucose-6-phosphate; this represents 0.026 (2.6%) of the total ATP available from glucose-6-phosphate metabolism (i.e., the efficiency of storage is 97.4%).

10. (c) UDP-glucose pyrophosphorylase

11. The diversion of glucose and its precursor oxaloacetate to milk production under conditions of extensive fatty acid catabolism results in ketosis. Ruminants can readily transform propionate to succinyl-CoA (via the intermediates propionyl-CoA, D-methylmalonyl-CoA, and L-methylmalonyl-CoA) and thus into oxaloacetate to avert the ketosis.

12. UDP-galactose pyrophosphorylase allows the catabolism of galactose by the following route:

Galactose + ATP $\xrightarrow{①}$ galactose-1-phosphate + ADP

Galactose-1-phosphate + UTP $\xrightarrow{②}$ UDP-galactose + PP_i

UDP-galactose $\xrightarrow{③}$ UDP-glucose

Net: Galactose + ATP + UTP $\longrightarrow$ UDP-glucose + ADP + PP_i

Step ① is catalyzed by galactokinase; ② by UDP-galactose pyrophosphorylase; ③ by galactose-4-epimerase. UDP-glucose is subsequently used to synthesize glycogen, or is hydrolyzed to UMP and glucose-1-phosphate.

13. This observation suggests that ATP and NADPH are generated in the light and are essential for CO_2 fixation; conversion stops as the supply of ATP and NADPH becomes exhausted.

14. X is 3-phosphoglycerate; Y is ribulose-1,5-bisphosphate.

15. In maize, CO_2 is fixed by the C_4 pathway worked out by Hatch and Slack, in which phosphoenolypyruvate is carboxylated rapidly to oxaloacetate (some of which undergoes transamination to aspartate) and reduced to malate. Only after subsequent decarboxylation does the CO_2 enter the Calvin cycle.

16. The isocitrate dehydrogenase reaction

17. **(a)** The equation for the lengthening of dextran by one glucose residue is:

$$\text{Sucrose} + (\text{glucose})_n \longrightarrow (\text{glucose})_{n+1} + \text{fructose}$$

(b) The fructose generated in the synthesis of dextran is readily imported and metabolized by the bacteria.

18. **(a)** Low levels of P_i in the cytosol and high levels of triose phosphate in the chloroplast **(b)** High levels of triose phosphate in the cytosol **(c)** The P_i–triose phosphate antiport system

Chapter 20

1. **(a)** The 16 carbons of palmitate are derived from 8 acetyl groups of 8 acetyl-CoA molecules. The ^{14}C-labeled acetyl-CoA gives rise to labeled malonyl-CoA.
   (b) The metabolic pool of malonyl-CoA, the source of all palmitate carbons except the first two (C-16 and C-15), does not become labeled with ^{14}C. Hence, only [15,16-^{14}C]palmitate is formed.

2. Both glucose and fructose are degraded to pyruvate in glycolysis. The pyruvate is converted to acetyl-CoA by the pyruvate dehydrogenase complex. Some of this acetyl-CoA enters the citric acid cycle, which produces reducing equivalents (NADH and NADPH). Mitochondrial electron transfer to O_2 yields ATP.

3. 8 Acetyl-CoA + 15 ATP + 14NADPH + 9H_2O $\longrightarrow$
   palmitate + 8 CoA + 15ADP + 15P_i + 14NADP$^+$ + 2H$^+$

4. **(a)** 3 Deuteriums per palmitate; all located on C-16; all other two-carbon units are derived from unlabeled malonyl-CoA. **(b)** 7 Deuteriums per palmitate; all *even*-numbered carbons except C-16

5. By using the three-carbon unit malonyl-CoA, the activated form of acetyl-CoA (recall that malonyl-CoA synthesis requires ATP), metabolism is driven in the direction of fatty acid synthesis by the exergonic release of CO_2.

6. The rate-limiting step in the biosynthesis of fatty acids is the carboxylation of acetyl-CoA catalyzed by acetyl-CoA carboxylase. High [citrate] and [isocitrate] indicate that conditions are favorable for fatty acid synthesis: an active citric acid cycle is providing a plentiful supply of ATP, reduced pyridine nucleotides, and acetyl-CoA. Citrate stimulates (increases the V_{max} of) acetyl-CoA carboxylase. Furthermore, because citrate binds more tightly to the filamentous form of the enzyme (the active form), high [citrate] drives the protomer–filament equilibrium in the direction of the active form. In contrast, palmitoyl-CoA (the end product of fatty acid synthesis) drives the equilibrium in the direction of the inactive (protomer) form. Hence, when the end product of fatty acid synthesis accumulates, the biosynthetic path is slowed down.

7. **(a)** Acetyl-CoA$_{(mit)}$ + ATP + CoA$_{(cyt)}$ $\longrightarrow$
   acetyl-CoA$_{(cyt)}$ + ADP + P_i + CoA$_{(mit)}$
   (b) 1 ATP per acetyl group **(c)** Yes

8. The double bond in palmitoleate is introduced by an oxidation catalyzed by fatty acyl–CoA desaturase, a mixed-function oxidase that requires O_2 as a cosubstrate.

9. 3 Palmitate + glycerol + 7ATP + 4H_2O $\longrightarrow$
   tripalmitin + 7ADP + 7P_i + 7H$^+$

10. In adult rats, stored triacylglycerols are maintained at a steady-state level through a balance of the rates of degradation and biosynthesis. Hence, the triacylglycerols of adipose (fat) tissue are constantly turned over, explaining the incorporation of ^{14}C label from dietary glucose.

11. Net reaction:
    Dihydroxyacetone phosphate + NADH + palmitate + oleate + 3ATP + CTP + choline + 4H_2O $\longrightarrow$
    phosphatidylcholine + NAD$^+$ + 2AMP + ADP + H$^+$ + CMP + 5P_i;
    7 ATP per molecule of PC

12. Methionine deficiency reduces the level of *S*-adenosylmethionine, which is required for the de novo synthesis of PC. The salvage pathway does not employ *S*-adenosylmethionine, but uses available choline. Thus PC can be synthesized even when the diet is deficient in Met, as long as choline is available.

13. ^{14}C label appears in three places in the activated isoprene:

$$\begin{array}{c} ^{14}CH_2 \\ \diagdown \\ C-^{14}CH_2-CH_2- \\ \diagup \\ ^{14}CH_3 \end{array}$$

14. The pathways to mevalonate and ketone bodies have steps in common. A decrease in HMG-CoA reductase results in the accumulation of HMG-CoA. This is a substrate for HMG-CoA lyase, which produces acetoacetate. Thus a decreased HGM-CoA reductase activity leads to increased production of acetoacetate.

15. **(a)** ATP **(b)** UDP-D-glucose **(c)** CDP-ethanolamine **(d)** UDP-D-galactose **(e)** Fatty acyl–CoA **(f)** *S*-Adenosylmethionine **(g)** Malonyl-CoA **(h)** Isopentenyl pyrophosphate

Chapter 21

1. The transfer potential of the methyl group of N^5-methyltetrahydrofolate is quite sufficient for the synthesis of methionine, which has an even lower methyl group transfer potential. The methyl group is activated by addition of the adenosyl group from ATP when methionine is converted to *S*-adenosylmethionine (see Fig. 17–20). Recall that *S*-adenosylmethionine synthesis is one of only two known biochemical reactions in which triphosphate is released from ATP. Hydrolysis of the triphosphate renders the reaction thermodynamically more favorable.

2. If phenylalanine hydroxylase is defective, the biosynthetic route to Tyr is blocked and Tyr must be obtained from the diet.

3. Glucose + 2CO_2 + 2NH$_3$ $\longrightarrow$ 2 aspartate + 2H$^+$ + 2H_2O

4. 5-Phosphoribosyl-1-pyrophosphate; this is the first ammonia acceptor in the purine biosynthetic pathway.

5. For Asp and Glu auxotrophs: adenine, guanine, uridine, and cytosine; for Gly auxotrophs: adenine and guanine.

6. **(a)** See Figure 17–7 for the reaction mechanism of amino acid racemization. The F atom of fluoroalanine is an excellent leaving group. Fluoroalanine causes irreversible (covalent) inhibition of alanine racemase. One plausible mechanism is:

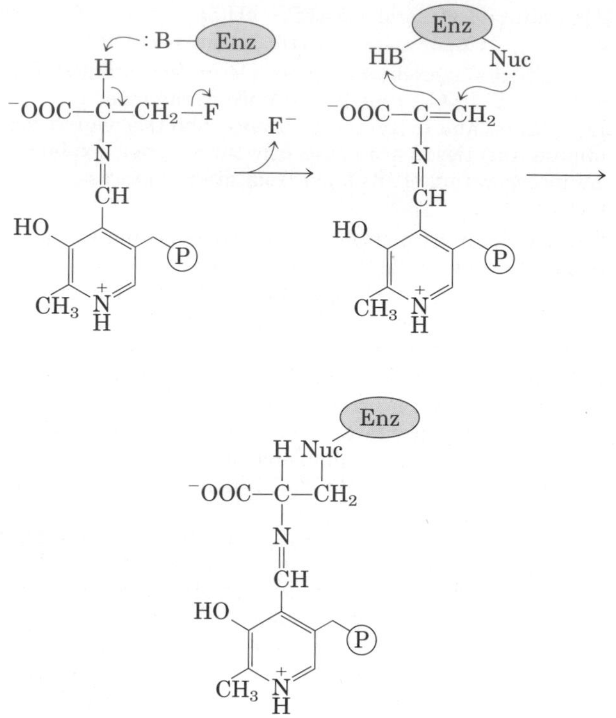

Nuc denotes any nucleophilic amino acid side chain in the enzyme active site.

(b) Azaserine (see Fig. 21–41) is an analog of Gln. The diazoacetyl group is highly reactive and forms covalent bonds with nucleophiles at the active site of a glutamine amidotransferase.

7. Organisms do not store nucleotides to be used as fuel and do not completely degrade them, but rather hydrolyze them to release the bases, which can be recovered in salvage pathways. The low C:N ratio of nucleotides makes them poor sources of energy.

8. (a) As shown in Figure 17–18, *p*-aminobenzoate is a component of N^5,N^{10}-methylenetetrahydrofolate, the cofactor involved in the transfer of one-carbon units.
(b) Sulfanilamide is a structural analog of *p*-aminobenzoate. In the presence of sulfanilamide, bacteria are unable to synthesize tetrahydrofolate, a cofactor necessary for the transformation of 5-aminoimidazole-4-carboxamide ribonucleotide (AICAR) to *N*-formylaminoimidazole-4-carboxamide ribonucleotide (FAICAR) by the addition of —CHO, and AICAR accumulates.
(c) Excess *p*-aminobenzoate reverses the growth inhibition and the ribonucleotide accumulation by competing with sulfanilamide for the active site of the enzyme involved in tetrahydrofolate biosynthesis. The competitive inhibition (Chapter 8) by sulfanilamide is overcome by the addition of excess substrate (*p*-aminobenzoate).

9. Treatment with allopurinol has two biochemical consequences. (1) Conversion of hypoxanthine to uric acid is inhibited, causing accumulation of hypoxanthine, which is more soluble and more readily excreted. This alleviates the clinical problems associated with AMP degradation. (2) Conversion of guanine to uric acid is also inhibited, causing accumulation of xanthine, which is, unfortunately, even less soluble than uric acid. This is the source of xanthine stones. Because the amount of GMP degradation is low relative to AMP degradation, the kidney damage caused by xanthine stones is less than that caused by untreated gout.

10. The bacteria in the root nodules maintain a symbiotic relationship with the plant: the plant supplies ATP and reducing power, and the bacteria supply ammonium ion by reducing atmospheric nitrogen. This reduction requires large quantities of ATP.

11. The ^{14}C-labeled orotate arises from the following pathway:

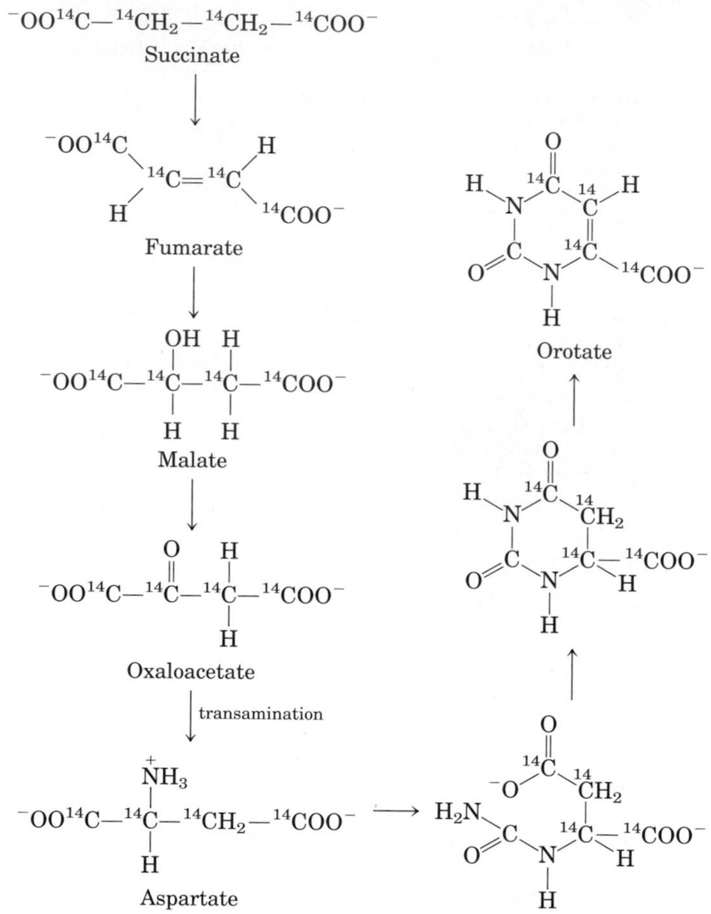

The first three steps are part of the citric acid cycle.

Chapter 22

1. Steady-state levels of ATP are maintained by phosphate transfer to ADP from phosphocreatine. 1-Fluoro-2,4-dinitrobenzene inhibits creatine kinase.

2. Ammonia is very toxic to nervous tissue, especially the brain. Excess NH_3 is removed by the transformation of glutamate to glutamine, which travels to the liver and is subsequently transformed to urea. The additional glutamate arises from the transformation of glucose to α-ketoglutarate. Additional NH_3 is removed by transamination of α-ketoglutarate to glutamate, and conversion of glutamate to glutamine.

3. From glucose, by the following route:

Glucose $\xrightarrow{\text{glycolysis}}$ dihydroxyacetone phosphate
Dihydroxyacetone phosphate $+ NADH + H^+ \rightarrow$
    glycerol-3-phosphate $+ NAD^+$

4. Pancreatic secretion is governed by the type and amount of food ingested, especially protein. Withholding dietary

protein reduces the secretion of pancreatic enzymes that can damage cellular proteins. The reduction of caloric and electrolyte intake is compensated by the intravenous administration of glucose–saline. Irritated pancreatic tissue has reduced ability to release insulin; hence, hyperglycemia results.

5. **(a)** Increased muscular activity increases the demand for ATP, which is met by increased O_2 consumption. **(b)** During the sprint, muscle transforms some glycogen to lactate (anaerobic glycolysis). After the sprint, lactate is transported to the liver where it is converted back to glucose and glycogen. This process requires ATP and thus requires O_2 consumption above the resting rate.

6. Glucose is the primary fuel of the brain, and the brain is particularly sensitive to any change in the availability of glucose for energy production. A key reaction in glucose catabolism is the thiamine pyrophosphate–dependent oxidative decarboxylation of pyruvate to acetyl-CoA, and thus a thiamin deficiency reduces the rate of glucose catabolism.

7. 1.87×10^2 m

8. **(a)** Inactivation provides a rapid means to change hormone concentrations. **(b)** A constant insulin level is maintained by equal rates of synthesis and degradation. **(c)** Other means of varying hormone concentration include changes in the rates of release from storage, of transport, and of conversion from prohormone to active hormone.

9. Because of their low solubility in lipid, water-soluble hormones cannot penetrate the cell membrane; they bind to receptors on the outer surface of the cell. In the case of epinephrine, this receptor is an enzyme that catalyzes the formation of a second messenger (cAMP) *inside* the cell. In contrast, lipid-soluble hormones can readily penetrate the hydrophobic core of the cell membrane. Once inside the cell they can act on their target molecules or receptors directly.

10. **(a)** Because adenylate cyclase is a membrane-bound protein, centrifugation sediments it into the particulate fraction. **(b)** Epinephrine stimulates the production of cAMP, a soluble substance that stimulates glycogen phosphorylase. **(c)** Cyclic AMP is the heat-stable substance; it can be prepared by treating ATP with barium hydroxide.

11. Unlike cAMP, dibutyryl-cAMP passes readily through the cell membrane.

12. **(a)** It increases the level of cAMP. **(b)** The observations suggest that cAMP regulates Na^+ permeability. **(c)** Replace lost body fluids and electrolytes.

13. **(a)** Heart and skeletal muscle lack the enzyme glucose-6-phosphate phosphatase. Any glucose-6-phosphate that is produced enters the glycolytic pathway, and under O_2-deficient conditions is converted into lactate via pyruvate. **(b)** Phosphorylated intermediates cannot escape from the cell, because the membrane is not permeable to charged species. In a "fight or flight" situation, the concentration of glycolytic precursors needs to be high in preparation for muscular activity. The liver, on the other hand, must release the glucose necessary to maintain the blood glucose level. Glucose is formed from glucose-6-phosphate and passes from the liver cells to the bloodstream.

14. **(a)** Excessive utilization of blood glucose by the liver, leading to hypoglycemia; shutdown of amino acid and fatty acid catabolism **(b)** Little circulating fuel is available for ATP requirements. Brain damage results because glucose is the main source of fuel for the brain.

15. Thyroxine acts as an uncoupler of oxidative phosphorylation. Uncouplers lower the P/O ratio, and the tissue must increase respiration to meet the normal ATP demands. Thermogenesis could also be due to the increased rate of ATP utilization by the thyroid-stimulated tissue, because the increased ATP demands are met by increased oxidative phosphorylation and thus respiration.

16. Because prohormones and preprohormones are inactive, they can be stored in quantity in secretory granules. Rapid activation is achieved by enzymatic cleavage in response to an appropriate signal.

17. Epinephrine relaxes the smooth muscle surrounding the bronchioles. It acts by stimulating the formation of cAMP in target cells. Cyclic AMP is destroyed by phosphodiesterase-catalyzed hydrolysis, but this enzyme can be inhibited by aminophylline. The administration of this drug thus prolongs and intensifies the activity of epinephrine by decreasing the rate of breakdown of cAMP.

Chapter 23

1. 372 base pairs; the actual length is probably much greater because most eukaryotic genes contain introns, which may be longer than the exons, and many also code for a leader or signal sequence in their protein products.

2. 66,461 nm. Dividing the approximate molecular weight of T2 DNA by 650 gives a length of 184,615 base pairs (recall that the approximation in Table 23–1 is 182,000); multiplying this by 0.36 nm per base pair (see Chapter 12) gives 66,461 nm. Thus the DNA is about 300 times longer than the T2 phage head.

3. Because neither the numbers of A and T residues nor the numbers of G and C residues are equal, the DNA cannot be a base-paired double helix; the M13 DNA is single-stranded.

4. No; the two strands of double-helical DNA differ in base composition (see Fig. 12–16).

5. The exons of this gene contain $3 \times 192 = 576$ base pairs. The remaining 864 base pairs are present in introns and possibly a leader or signal sequence.

6. 5,250 base pairs. In relaxed DNA, Lk is equivalent to the number of turns in the DNA helix. Multiplying 500 by 10.5 base pairs per turn gives a length of 5,250 base pairs. **(a)** No change; Lk cannot change without breaking and re-forming the covalent backbone of the DNA. **(b)** Becomes undefined; a circular DNA with a break in one strand has, by definition, no Lk. **(c)** Decrease; in the presence of ATP, gyrase underwinds DNA. **(d)** No change; this assumes that neither of the DNA strands was broken during the heating process.

7. A right-handed helix has a positive Lk; a left-handed helix (such as Z-DNA) has a negative Lk. Lowering the Lk of a closed circular B-DNA molecule by underwinding it facilitates the formation of regions of Z-DNA within certain sequences. (See Chapter 12 for a description of sequences that permit the formation of Z-DNA.)

8. The results demonstrate a fundamental structural unit in chromatin that repeats about every 200 base pairs; the DNA is accessible to the nuclease only at 200 base pair intervals. The brief treatment was insufficient to cleave the DNA at every accessible point, so that a ladder of DNA bands was created in which the sizes of the DNA fragments were in multiples of 200 base pairs (further treatment would have caused most of the DNA to be in the lowest band). The thickness of the DNA bands suggests that the distance between cleavage sites varies somewhat. For instance, not all of the fragments in the lowest band are exactly 200 base pairs long.

Chapter 24

1. If random, dispersive replication had taken place, the density of the first-generation DNAs would have been the same as actually observed, a single band midway between heavy and light DNA. In the second generation all the DNAs would again have had the same density and would have appeared as a single band, midway between that observed in the first generation and that of light DNA. However, two bands were observed in the Meselson–Stahl experiment.

2. 448,000 turns; the length of the chromosome (in base pairs) is divided by 10.5.

3. 39 min; the calculation assumes a replication rate by DNA polymerase III of 1,000 nucleotides per second (Table 24–1), and that two replication forks proceed bidirectionally from a single origin. One possibility is that each *E. coli* chromosome is replicated by four forks starting from two origins, but in fact there is only one origin in the *E. coli* chromosome. When the cells are dividing every 20 min a replicative cycle is inititiated every 20 min, each cycle beginning before the prior one is complete.

4. 28.7%; the DNA strand made from the given template strand would have A, 32.7%; G, 18.5%; C, 24.1%; T, 24.7%. The DNA strand made from the complementary template strand would have A, 24.7%; G, 24.1%; C, 18.5%; T, 32.7%. It is assumed that the two template strands are replicated completely.

5. About 2,350 to 4,700 Okazaki fragments are formed by DNA polymerase III from an RNA primer and a DNA template. Because the Okazaki fragments in *E. coli* are 1,000 to 2,000 bases long, they are firmly bound to the template strand by base pairing. Each fragment is quickly joined to the lagging strand by the successive action of DNA polymerase I and DNA ligase, thus preserving the correct order of the fragments. A mixed pool of different Okazaki fragments, detached from their template, does not form during normal replication.

6. *Leading strand:* precursors—dATP, dGTP, dCTP, dTTP; enzymes—DNA gyrase, helicase, single-strand DNA-binding protein, DNA polymerase III, topoisomerases.
   Lagging strand: precursors—ATP, GTP, CTP, UTP, dATP, DGTP, dCTP, dTTP; enzymes—DNA gyrase, helicase, single-strand DNA-binding protein, primase, DNA polymerase III, DNA polymerase I, DNA ligase, topoisomerases, and all of the primosome components listed in Table 24–4 (seven additional proteins).

7. Watson–Crick base pairing between the template and leading strand; proofreading and removal of wrongly inserted nucleotides by the 3'-exonuclease activity of DNA polymerase III. Yes—maybe; because the factors ensuring fidelity of replication are operative in both the leading and the lagging strands, the lagging strand would probably be made with the same fidelity. However, the greater number of distinct chemical operations involved in making the lagging strand compared with the leading strand might provide a greater opportunity for errors to arise.

8. Spontaneous deamination of 5-methylcytosine (see p. 344) produces thymine, and thus a G–T mismatched pair. Such G–T mismatches are among the most common mismatches in the DNA of eukaryotes. The specialized repair system restores the G≡C pair.

9. During homologous genetic recombination, a Holliday intermediate may be formed almost anywhere within the two paired, homologous chromosomes. Once formed, the branch point of the intermediate may move extensively by branch migration. In site-specific recombination, the Holliday intermediate is formed between two specific sites, and branch migration is generally restricted by heterologous sequences on either side of the recombination sites.

10. Homologous or site-specific recombination. The two can generally be distinguished by determining the nature of the sequences that were recombined and the location of the crossover event. If the recombination occurred between two long (>50 base pair) homologous sequences, and the crossover could occur anywhere within the sequences, it was homologous recombination. If the crossover occurred between two short homologous sequences and always involved the same phosphodiester bond within those sequences, it was site-specific recombination.

Chapter 25

1. 130 sec; the RNA has 6,500 bases and is transcribed at a rate of 50 nucleotides per second (see p. 859).

2. A single base error in DNA replication, if not corrected, would cause one of the two daughter cells, and all its progeny, to have a mutated chromosome. A single base error in RNA transcription would not affect the chromosome; it would lead to formation of some defective copies of one protein, but because mRNAs turn over rapidly, most copies of the protein would not be defective. The progeny of this cell would be normal.

3. 500 base pairs

4. Normal posttranscriptional processing at the 3' end (cleavage and polyadenylation) would be inhibited or blocked.

5. Because the (−) strand RNA does not encode the enzymes needed to initiate viral infection, it would probably be inert or simply degraded by cellular ribonucleases. Replication of the (−) RNA and propagation of the virus could occur only if intact RNA replicase (RNA-directed RNA polymerase) were introduced into the cell along with the (−) strand.

6. (1) The 5'→3' synthesis of a nucleic acid; (2) use of a template strand of nucleic acid, which is copied in the 3'→5' direction; (3) nucleophilic attack of a 3' hydroxyl (at the growing end of the nucleic acid) on the α phosphate of a nucleoside triphosphate, with displacement of PP_i.

7. Generally two: one to cleave the phosphodiester bond at one intron–exon junction, and the other to link the resulting free exon end to the exon at the other end of the intron. If the nucleophile in the first step were water, this step would be a hydrolytic event and only one transesterification step would be required to complete the splicing process.

8. With a built-in template, the telomerase would not move along the template in the manner of the other polymerases; instead, it might have an active site that could catalyze multiple nucleotide additions without moving (or it could be thought of as an enzyme with multiple active sites in a linear array).

9. These enzymes lack a 3'→5' proofreading exonuclease and have a high error rate; the likelihood of a replication error that would inactivate the virus is much less in a small genome than in a large one.

Chapter 26

1. (a) Gly–Gln–Ser–Leu–Leu–Ile (b) Leu–Asp–Ala–Pro (c) His–Asp–Ala–Cys–Cys–Tyr (d) Met–Asp–Glu in eukaryotes; fMet–Asp–Glu in prokaryotes

2. UUAAUGUAU, UUGAUGUAU, CUUAUGUAU, CUCAUGUAU, CUAAUGUAU, CUGAUGUAU, UUAAUGUAC, UUGAUGUAC, CUUAUGUAC, CUCAUGUAC, CUAAUGUAC, CUGAUGUAC

3. No; because nearly all the amino acids have more than one codon (e.g., Leu has six), any given polypeptide can be coded for by a number of different base sequences (see Problem 2). However, because some amino acids are encoded by only one codon and those with multiple codons often share the same nucleotide at two of the three positions, *certain parts* of the mRNA sequence encoding a protein of known amino acid sequence can be predicted with high certainty.

4. **(a)** (5′)CGACGGCGCGAAGUCAGGGGUGUUAAG(3′)
(b) Arg–Arg–Arg–Glu–Val–Arg–Gly–Val–Lys
(c) No; the complementary antiparallel strands in double-helical DNA do not have the same base sequence in the 5′→3′ direction. RNA is transcribed from only one specific strand of duplex DNA. The RNA polymerase must therefore recognize and bind to the correct strand.

5. There are two tRNAs for methionine: tRNA^{fMet}, the initiating tRNA, and tRNA^{Met}, which can insert Met in interior positions in a polypeptide. tRNA^{fMet} reacts with Met to yield Met-tRNA^{fMet}, promoted by methionine aminoacyl-tRNA synthetase. The amino group of its Met residue is then formylated by N^{10}-formyltetrahydrofolate to yield fMet-tRNA^{fMet}. Free Met or Met-tRNA^{Met} cannot be formylated. Only fMet-tRNA^{fMet} is recognized by the initiation factor IF-2 and is aligned with the initiating AUG positioned at the ribosomal P site in the initiation complex. AUG codons in the interior of the mRNA are eventually positioned at the ribosomal P site and can bind and incorporate only Met-tRNA^{Met}.

6. Allow polynucleotide phosphorylase to act on a mixture of UDP and CDP in which UDP has, say, five times the concentration of CDP. The result would be a synthetic RNA polymer with many UUU triplets (coding for Phe), a smaller number of UUC (also Phe), UCU (Ser), and CUU (Leu), and a much smaller number of UCC (also Ser), CUC (also Leu), and CCU (Pro).

7. 583 High-energy phosphate groups; this is based on 4 per amino acid residue added, except that there are only 145 translocation steps. This minimum value does not take into account any errors that are detected and corrected by the aminoacyl-tRNA synthetases. Correction of each error requires 2 high-energy phosphate groups. For glycogen synthesis, 292 high-energy phosphate groups would be used. The extra energy cost for β-globin synthesis is 291 high-energy phosphate groups, or about twice the cost of glycogen synthesis; this reflects the cost of the information content of the protein.

8. At least 20 activating enzymes, 70 ribosomal proteins, 4 rRNAs, 20 or more tRNAs, an mRNA, and 10 or more auxiliary enzymes must be made by the eukaryotic cell in order to synthesize a protein from amino acids. Synthesis of these proteins and RNA molecules in energetically expensive (as seen in Chapters 24–26). In contrast, the synthesis of an (α1→4) chain of glycogen from glucose requires only 4 or 5 enzymes (see Chapter 19).

9.

Glycine codons	Anticodons
(5′)GGU	(5′)ACC, GCC, ICC
GGC	GCC, I CC
GGA	UCC, I CC
GGG	CCC, UCC

(a) The 5′ end and the middle position **(b)** The pairings with anticodons (5′)GCC, ICC, and UCC **(c)** The pairings with anticodons (5′)ACC and CCC

10. (a), (c), (e), and (g); mutations (b), (d), and (f) cannot be the result of single-base mutations; (b) and (f) would require substitutions of two bases, and (d) would require substitutions of all three bases.

11. The two DNA codons for Glu are GAA and GAG, and the four DNA codons for Val are GTT, GTC, GTA, and GTG. A single-base change in GAA to form GTA or in GAG to form GTG could account for the Glu → Val replacement in sickle-cell hemoglobin. Much less likely are two-base changes from GAA to GTG, GTT, or GTC; and from GAG to GTA, GTT, or GTC.

12. The mutant tRNA^{Ala} would insert Ala residues at codons encoding Pro. Another type of mutation that might have similar effects is an alteration in tRNA^{Pro} that allowed it to be recognized and aminoacylated by Ala-tRNA synthetase. Most of the proteins in the cell would be inactivated, making these lethal mutations and hence never observed. This represents a powerful selective pressure for maintaining the genetic code.

13. The amino acid most recently added to a growing polypeptide chain is the only one covalently attached to a tRNA and hence is the only link between the polypeptide and the mRNA that is encoding it. A proofreading activity would sever this link, halting synthesis of the polypeptide and releasing it from the mRNA.

Chapter 27

1. **(a)** Constitutive expression of the operon; most mutations in the operator would make the repressor less likely to bind. **(b)** Either constitutive expression, as in (a), or constant repression, if the mutation destroyed the capability to bind to lactose and related compounds and hence the response to inducers **(c)** Either increased or decreased expression of the operon (under conditions in which it is induced), depending on whether the mutation made the promoter more or less similar to the consensus *E. coli* promoter (see Fig. 27–2).

2. **(a)** Tryptophan synthase levels remain high in spite of the presence of Trp. **(b)** Levels again remain high. **(c)** Levels rapidly decrease, preventing wasteful synthesis of Trp.

3. Modify the GAL4 DNA-binding site to give it the nucleotide sequence to which the CI protein normally binds (using methods described in Chapter 28).

4. The CI protein produced by the resident prophage prevents the gene expression of any bacteriophage λ that subsequently invades the lysogenic cell. The invader can neither replicate itself nor produce the recombination proteins needed to insert itself into the chromosome as a second prophage, and it is eventually degraded.

5. Each cell would have flagella made up of both types of flagellar protein, and the cell would be vulnerable to antibodies generated in response to either protein.

6. **(a)** The ribosome completing the translation of sequence 1 would no longer overlap and block sequence 2. Attenuation would become much less effective because sequence 2 would always be available to pair with sequence 3, preventing formation of the attenuator structure. **(b)** Sequence 2 would pair less efficiently with sequence 3. Attenuation would increase because the attenuator structure would be formed more often, even when sequence 2 was not blocked by a ribosome. **(c)** Attenuation would not occur, and the only regulation would be that afforded by the Trp repressor.

7. About 5,000 copies

8. A dissociable specificity factor (similar to the σ subunit of the *E. coli* enzyme) may have been purified away from the polymerase.

Chapter 28

1. **(a)** (5′)---G
        ---CTTAA
   (b) (5′)---GAATT
        ---CTTAA
   (c) (5′)---GAATTAATTC---
        ---CTTAATTAAG---
   (d) (5′)---G
        ---C
   (e) (5′)---GAATTC---
        ---CTTAAG---
   (f) (5′)---CAG
        ---GTC
   (g) (5′)---CAGAATTC---
        ---GTCTTAAG---
   (h) First, cut the DNA with *Eco*RI as in (a). At this point, one could treat the DNA as in (b) or (d), then ligate a synthetic DNA fragment with the *Bam*HI recognition sequence between the two resulting blunt ends. Another (more efficient) approach would be to synthesize a DNA fragment with the structure:

   (5′)AATTGGATCC
        CCTAGGTTAA

   This would ligate efficiently to the sticky ends generated by *Eco*RI cleavage, would introduce a *Bam*HI site, but would not regenerate the *Eco*RI site.
   (i) The four fragments (with N = any nucleotide), in order of discussion in the problem, are:
        (5′)AATTCNNNNCTGCA
             GNNNNG
        (5′)AATTCNNNNGTGCA
             GNNNNC
        (5′)AATTGNNNNCTGCA
             CNNNNG
        (5′)AATTGNNNNGTGCA
             CNNNNC

2. The cosmid DNA is not packaged into a λ phage particle unless it is joined to a foreign DNA fragment of suitable length. *All* of the viable infectious phage produced are recombinants.

3. Plasmids in which the original pBR322 was regenerated without insertion of a foreign DNA fragment; these would retain resistance to ampicillin. Also, two or more molecules of pBR322 might be ligated together with or without insertion of foreign DNA.

4. You will need a suitable bacterial promoter, a ribosome binding site for translation, and regulatory site(s) such as operators placed on the 5′ side of the gene (on the coding strand). The ribosome binding site should be immediately left (5′) of the gene, and the promoter further to the left. The regulatory elements must be located where they could affect the promoter. Many possibilities for expression and regulation are described in Chapters 25–27.

5. Focus on the amino acids with the fewest codons: Met and Trp. The best possibility is the span of DNA from the codon for the first Trp residue to the first two nucleotides of the codon for Ile. The sequence of the probe would be:

   (5′)UGGUA(U/C)UG(U/C)AUGGA(U/C)UGGAU

   The synthesis would be designed to incorporate either U or C where indicated, producing a mixture of eight 20-nucleotide probes that differ only at one or more of these positions.

6. Simply for convenience; the 200,000 base pair Ti plasmid, even when the T DNA is removed, is too large to isolate in quantity and manipulate in vitro. It is also too large to reintroduce into a cell by standard transformation techniques. Single-plasmid systems in which the T DNA of a Ti plasmid has been replaced by foreign DNA (by means of low efficiency recombination in vivo) have been used successfully, but this approach is very laborious. The *vir* genes will facilitate the transfer of any DNA between the T DNA repeats, even if they are on a separate plasmid. The second plasmid in the two-plasmid system, because it requires only the T DNA repeats and a few sequences necessary for plasmid selection and propagation, is relatively small, easily isolated, and easily manipulated (foreign DNA easily added and/or altered). It can be propagated in either *E. coli* or *Agrobacterium* and readily reintroduced into either bacterium.

7. The vectors must be introduced into a cell infected with a helper virus that can provide the necessary replication and packaging functions but cannot itself be packaged. The vectors packaged into infectious viral particles are used to introduce the recombinant DNA into a mammalian cell. Once this DNA is integrated into the chromosome of the target cell, the lack of recombination and packaging functions makes the integration very stable by preventing the deletion or replication of the integrated DNA.

Glossary

absolute configuration: The configuration of four different substituent groups around an asymmetric carbon atom, in relation to D- and L-glyceraldehyde.

absorption: Transport of the products of digestion from the intestinal tract into the blood.

acceptor control: The regulation of the rate of respiration by the availability of ADP as phosphate group acceptor.

accessory pigments: Visible light–absorbing pigments (carotenoids, xanthophyll, and phycobilins) in plants and photosynthetic bacteria that complement chlorophylls in trapping energy from sunlight.

acidosis: A metabolic condition in which the capacity of the body to buffer H^+ is diminished; usually accompanied by decreased blood pH.

actin: A protein making up the thin filaments of muscle; also an important component of the cytoskeleton of many eukaryotic cells.

activation energy ($\Delta G^{\ddagger}$): The amount of energy (in joules) required to convert all the molecules in 1 mole of a reacting substance from the ground state to the transition state.

activator: (1) A DNA-binding protein that positively regulates the expression of one or more genes; that is, transcription rates increase when an activator is bound to the DNA. (2) A positive modulator of an allosteric enzyme.

active site: The region of an enzyme surface that binds the substrate molecule and catalytically transforms it; also known as the catalytic site.

active transport: Energy-requiring transport of a solute across a membrane in the direction of increasing concentration.

activity: The true thermodynamic activity or potential of a substance, as distinct from its molar concentration.

activity coefficient: The factor by which the numerical value of the concentration of a solute must be multiplied to give its true thermodynamic activity.

acyl phosphate: Any molecule with the general chemical form R—$\underset{\underset{O}{\|}}{C}$—$OPO_3^{2-}$.

adenosine 3′,5′-cyclic monophosphate: See cyclic AMP.

adenosine diphosphate: See ADP.

adenosine triphosphate: See ATP.

adipocyte: An animal cell specialized for the storage of fats (triacylglycerols).

adipose tissue: Connective tissue specialized for the storage of large amounts of triacylglycerols.

ADP (adenosine diphosphate): A ribonucleoside 5′-diphosphate serving as phosphate group acceptor in the cell energy cycle.

aerobe: An organism that lives in air and uses oxygen as the terminal electron acceptor in respiration.

aerobic: Requiring or occurring in the presence of oxygen.

alcohol fermentation: The anaerobic conversion of glucose to ethanol via glycolysis. See also fermentation.

aldose: A simple sugar in which the carbonyl carbon atom is an aldehyde; that is, the carbonyl carbon is at one end of the carbon chain.

alkaloids: Nitrogen-containing organic compounds of plant origin; often basic, and having intense biological activity.

alkalosis: A metabolic condition in which the capacity of the body to buffer OH^- is diminished; usually accompanied by an increase in blood pH.

allosteric enzyme: A regulatory enzyme, with catalytic activity modulated by the noncovalent binding of a specific metabolite at a site other than the active site.

allosteric site: The specific site on the surface of an allosteric enzyme molecule to which the modulator or effector molecule is bound.

α helix: A helical conformation of a polypeptide chain, usually right-handed, with maximal intrachain hydrogen bonding; one of the most common secondary structures in proteins.

Ames test: A simple bacterial test for carcinogens, based on the assumption that carcinogens are mutagens.

amino acid activation: ATP-dependent enzymatic esterification of the carboxyl group of an amino acid to the 3′-hydroxyl group of its corresponding tRNA.

amino acids: α-Amino-substituted carboxylic acids, the building blocks of proteins.

amino-terminal residue: The only amino acid residue in a polypeptide chain with a free α-amino group; defines the amino terminus of the polypeptide.

aminoacyl-tRNA: An aminoacyl ester of a tRNA.

aminoacyl-tRNA synthetases: Enzymes that catalyze synthesis of an aminoacyl-tRNA at the expense of ATP energy.

aminotransferases: Enzymes that catalyze the transfer of amino groups from α-amino to α-keto acids; also called transaminases.

ammonotelic: Excreting excess nitrogen in the form of ammonia.

amphibolic pathway: A metabolic pathway used in both catabolism and anabolism.

amphipathic: Containing both polar and nonpolar domains.

amphoteric: Capable of donating and accepting protons, thus able to serve as an acid or a base.

anabolism: The phase of intermediary metabolism concerned with the energy-requiring biosynthesis of cell components from smaller precursors.

anaerobe: An organism that lives without oxygen. Obligate anaerobes die when exposed to oxygen.

anaerobic: Occurring in the absence of air or oxygen.

anaplerotic reaction: An enzyme-catalyzed reaction that can replenish the supply of intermediates in the citric acid cycle.

angstrom (Å): A unit of length (10^{-8} cm) used to indicate molecular dimensions.

anhydride: The product of the condensation of two carboxyl or phosphate groups in which the elements of water are eliminated to form a compound with the general structure R—X—O—X—R, where

$$\underset{O}{\overset{\parallel}{X}} \quad \underset{O}{\overset{\parallel}{X}}$$

X is either carbon or phosphorus.

anion-exchange resin: A polymeric resin with fixed cationic groups; used in the chromatographic separation of anions.

anomers: Two stereoisomers of a given sugar that differ only in the configuration about the carbonyl (anomeric) carbon atom.

antibiotic: One of many different organic compounds that are formed and secreted by various species of microorganisms and plants, are toxic to other species, and presumably have a defensive function.

antibody: A defense protein synthesized by the immune system of vertebrates. See also immunoglobulin.

anticodon: A specific sequence of three nucleotides in a tRNA, complementary to a codon for an amino acid in an mRNA.

antigen: A molecule capable of eliciting the synthesis of a specific antibody in vertebrates.

antiparallel: Describing two linear polymers that are opposite in polarity or orientation.

antiport: Cotransport of two solutes across a membrane in opposite directions.

apoenzyme: The protein portion of an enzyme, exclusive of any organic or inorganic cofactors or prosthetic groups that might be required for catalytic activity.

apolipoprotein: The protein component of a lipoprotein.

asymmetric carbon atom: A carbon atom that is covalently bonded to four different groups and thus may exist in two different tetrahedral configurations.

ATP (adenosine triphosphate): A ribonucleoside 5′-triphosphate functioning as a phosphate group donor in the cell energy cycle; carries chemical energy between metabolic pathways by serving as a shared intermediate coupling endergonic and exergonic reactions.

ATP synthase: An enzyme complex that forms ATP from ADP and phosphate during oxidative phosphorylation in the inner mitochondrial membrane or the bacterial plasma membrane, and during photophosphorylation in chloroplasts.

ATPase: An enzyme that hydrolyzes ATP to yield ADP and phosphate; usually coupled to some process requiring energy.

attenuator: An RNA sequence involved in regulating the expression of certain genes; functions as a transcription terminator.

autotroph: An organism that can synthesize its own complex molecules from very simple carbon and nitrogen sources, such as carbon dioxide and ammonia.

auxin: A plant growth hormone.

auxotrophic mutant (auxotroph): A mutant organism defective in the synthesis of a given biomolecule, which must therefore be supplied for the organism's growth.

Avogadro's number (N): The number of molecules in a gram molecular weight (a mole) of any compound (6.02×10^{23}).

back-mutation: A mutation that causes a mutant gene to regain its wild-type base sequence.

bacteriophage (phage): A virus capable of replicating in a bacterial cell.

basal metabolic rate: The rate of oxygen consumption by an animal's body at complete rest, long after a meal.

base pair: Two nucleotides in nucleic acid chains that are paired by hydrogen bonding of their bases; for example, A with T or U, and G with C.

β conformation: An extended, zigzag arrangement of a polypeptide chain; a common secondary structure in proteins.

β oxidation: Oxidative degradation of fatty acids into acetyl-CoA by successive oxidations at the β-carbon atom.

bilayer: A double layer of oriented am-

phipathic lipid molecules, forming the basic structure of biological membranes. The hydrocarbon tails face inward to form a continuous nonpolar phase.

bile salts: Amphipathic steroid derivatives with detergent properties, participating in digestion and absorption of lipids.

binding energy: The energy derived from noncovalent interactions between enzyme and substrate or receptor and ligand.

biocytin: The conjugate amino acid residue arising from covalent attachment of biotin, through an amide linkage, to a Lys residue.

biomolecule: An organic compound normally present as an essential component of living organisms.

biopterin: An enzymatic cofactor derived from pterin and involved in certain oxidation–reduction reactions.

biosphere: All the living matter on or in the earth, the seas, and the atmosphere.

biotin: A vitamin; an enzymatic cofactor involved in carboxylation reactions.

bond energy: The energy required to break a bond.

branch migration: Movement of the branch point in branched DNA formed from two DNA molecules with identical sequences. See also Holliday intermediate.

buffer: A system capable of resisting changes in pH, consisting of a conjugate acid–base pair in which the ratio of proton acceptor to proton donor is near unity.

calorie: The amount of heat required to raise the temperature of 1.0 g of water from 14.5 to 15.5 °C. One calorie (cal) equals 4.18 joules (J).

Calvin cycle: The cyclic pathway used by plants to fix carbon dioxide and produce triose phosphates.

cAMP: See cyclic AMP.

CAP: See catabolite gene activator protein.

capsid: The protein coat of a virion or virus particle.

carbanion: A negatively charged carbon atom.

carbocation: A positively charged carbon atom; also called a carbonium ion.

carbon fixation reactions: In photosynthetic cells, the light-independent enzymatic reactions involved in the synthesis of glucose from CO_2, ATP, and NADPH; also known as the dark reactions.

carboxyl-terminal residue: The only amino acid residue in a polypeptide chain with a free α-carboxyl group; defines the carboxyl terminus of the polypeptide.

carotenoids: Lipid-soluble photosynthetic pigments made up of isoprene units.

catabolism: The phase of intermediary metabolism concerned with the energy-yielding degradation of nutrient molecules.

catabolite gene activator protein (CAP): A specific regulatory protein that controls initiation of transcription of the genes producing the enzymes required for a bacterial cell to use some other nutrient when glucose is lacking.

catalytic site: See active site.

catecholamines: Hormones, such as epinephrine, that are amino derivatives of catechol.

cation-exchange resin: An insoluble polymer with fixed negative charges; used in the chromatographic separation of cationic substances.

cDNA: See complementary DNA.

central dogma: The organizing principle of molecular biology: genetic information flows from DNA to RNA to protein.

centromere: A specialized site within a chromosome, serving as the attachment point for the mitotic or meiotic spindle.

cerebroside: Sphingolipid containing one sugar residue as a head group.

channeling: The direct transfer of a reaction product (common intermediate) from the active site of one enzyme to the active site of a different enzyme catalyzing the next step in a sequential pathway.

chemiosmotic coupling: Coupling of ATP synthesis to electron transfer via an electrochemical H^+ gradient across a membrane.

chemotaxis: A cell's sensing of and movement toward, or away from, a specific chemical agent.

chemotroph: An organism that obtains energy by metabolizing organic compounds derived from other organisms.

chiral compound: A compound that contains an asymmetric center (chiral atom or chiral center) and thus can occur in two nonsuperimposable mirror-image forms (enantiomers).

chlorophylls: A family of green pigments functioning as receptors of light energy in photosynthesis; magnesium-porphyrin complexes.

chloroplasts: Chlorophyll-containing photosynthetic organelles in some eukaryotic cells.

chromatin: A filamentous complex of DNA, histones, and other proteins, constituting the eukaryotic chromosome.

chromatography: A process in which complex mixtures of molecules are separated by many repeated partitionings between a flowing (mobile) phase and a stationary phase.

chromosome: A single large DNA molecule and its associated proteins, containing many genes; stores and transmits genetic information.

chylomicron: A plasma lipoprotein consisting of a large droplet of triacylglycerols stabilized by a coat of protein and phospholipid; carries lipids from the intestine to the tissues.

cis and trans isomers: See geometric isomers.

cistron: A unit of DNA or RNA corresponding to one gene.

citric acid cycle: A cyclic system of enzymatic reactions for the oxidation of acetyl residues to carbon dioxide, in which formation of citrate is the first step; also known as the Krebs cycle or tricarboxylic acid cycle.

clones: The descendants of a single cell.

cloning: The production of large numbers of identical DNA molecules or cells from a single ancestral DNA molecule or cell.

closed system: A system that exchanges neither matter nor energy with the surroundings. See also system.

cobalamin: See coenzyme B_{12}.

codon: A sequence of three adjacent nucleotides in a nucleic acid that codes for a specific amino acid.

coenzyme: An organic cofactor required for the action of certain enzymes; often contains a vitamin as a component.

coenzyme A: A pantothenic acid–containing coenzyme serving as an acyl group carrier in certain enzymatic reactions.

coenzyme B_{12}: An enzymatic cofactor derived from the vitamin cobalamin, involved in certain types of carbon skeletal rearrangements.

cofactor: An inorganic ion or a coenzyme required for enzyme activity.

cognate: Describing two biomolecules that normally interact; for example, an enzyme and its normal substrate, or a receptor and its normal ligand.

cohesive ends: See sticky ends.

cointegrate: An intermediate in the migration of certain DNA transposons in which the donor DNA and target DNA are covalently attached.

colligative properties: Properties of solutions that depend on the number of solute particles per unit volume; for example, freezing-point depression.

common intermediate: A chemical compound common to two chemical reactions, as a product of one and a reactant in the other.

competitive inhibition: A type of enzyme inhibition reversed by increasing the substrate concentration; a competitive inhibitor generally competes with the normal substrate or ligand for a protein's binding site.

complementary: Having a molecular surface with chemical groups arranged to interact specifically with chemical groups on another molecule.

complementary DNA (cDNA): A DNA used in DNA cloning, usually made by reverse transcriptase; complementary to a given mRNA.

configuration: The spatial arrangement of an organic molecule that is conferred by the presence of either (1) double bonds, about which there is no freedom of rotation, or (2) chiral centers, around which substituent groups are arranged in a specific sequence. Configurational isomers cannot be interconverted without breaking one or more covalent bonds.

conformation: The spatial arrangement of substituent groups that are free to assume different positions in space, without breaking any bonds, because of the freedom of bond rotation.

conformation, β: See β conformation.

conjugate acid–base pair: A proton donor and its corresponding deprotonated species; for example, acetic acid (donor) and acetate (acceptor).

conjugate redox pair: An electron donor and its corresponding electron acceptor form; for example, Cu^+ (donor) and Cu^{2+} (acceptor), or NADH (donor) and NAD^+ (acceptor).

conjugated protein: A protein containing one or more prosthetic groups.

consensus sequence: A DNA or amino acid sequence consisting of the residues that occur most commonly at each position within a set of similar sequences.

conservative substitution: Replacement of an amino acid residue in a polypeptide by another residue with similar properties; for example, substitution of Glu by Asp.

constitutive enzymes: Enzymes required at all times by a cell and present at some constant level; for example, many enzymes of the central metabolic pathways. Sometimes called "housekeeping enzymes."

corticosteroids: Steroid hormones formed by the adrenal cortex.

cosmid: A cloning vector, used for cloning large DNA fragments; generally contains segments derived from bacteriophages and various plasmids.

cotransport: The simultaneous transport, by a single transporter, of two solutes across a membrane. See antiport, symport.

coupled reactions: Two chemical reactions that have a common intermediate and thus a means of energy transfer from one to the other.

covalent bond: A chemical bond that involves sharing of electron pairs.

cristae: Infoldings of the inner mitochondrial membrane.

cyclic AMP (cAMP): A second messenger within cells; its formation by adenylate cyclase is stimulated by certain hormones or other molecular signals.

cyclic electron flow: In chloroplasts, the light-induced flow of electrons originating from and returning to photosystem I.

cyclic photophosphorylation: ATP synthesis driven by cyclic electron flow through photosystem I.

cytochromes: Heme proteins serving as electron carriers in respiration, photosynthesis, and other oxidation–reduction reactions.

cytokinesis: The final separation of daughter cells following mitosis.

cytoplasm: The portion of a cell's contents outside the nucleus but within the plasma membrane; includes organelles such as mitochondria.

cytoskeleton: The filamentous network providing structure and organization to the cytoplasm; includes actin filaments, microtubules, and intermediate filaments.

cytosol: The continuous aqueous phase of the cytoplasm, with its dissolved solutes; excludes the organelles such as mitochondria.

dalton: The weight of a single hydrogen atom (1.66×10^{-24} g).

dark reactions: See carbon fixation reactions.

de novo pathway: Pathway for synthesis of a biomolecule, such as a nucleotide, from simple precursors; as distinct from a salvage pathway.

deamination: The enzymatic removal of amino groups from biomolecules such as amino acids or nucleotides.

degenerate code: A code in which a single element in one language is specified by more than one element in a second language.

dehydrogenases: Enzymes catalyzing the removal of pairs of hydrogen atoms from their substrates.

deletion mutation: A mutation resulting from the deletion of one or more nucleotides from a gene or chromosome.

denaturation: Partial or complete unfolding of the specific native conformation of a polypeptide chain, protein, or nucleic acid.

denatured protein: A protein that has lost its native conformation by exposure to a destabilizing agent such as heat or detergent.

deoxyribonucleic acid: See DNA.

deoxyribonucleotides: Nucleotides containing 2-deoxy-D-ribose as the pentose component.

desaturases: Enzymes that catalyze the introduction of double bonds into the hydrocarbon portion of fatty acids.

desolvation: In aqueous solution, the release of bound water surrounding a solute.

dextrorotatory isomer: A stereoisomer that rotates the plane of plane-polarized light clockwise.

diabetes mellitus: A metabolic disease resulting from insulin deficiency; characterized by a failure in glucose transport from the blood into cells at normal glucose concentrations.

dialysis: Removal of small molecules from a solution of a macromolecule, by allowing them to diffuse through a semipermeable membrane into water.

differential centrifugation: Separation of cell organelles or other particles of different size by their different rates of sedimentation in a centrifugal field.

differentiation: Specialization of cell structure and function during embryonic growth and development.

diffusion: The net movement of molecules in the direction of lower concentration.

digestion: Enzymatic hydrolysis of major nutrients in the gastrointestinal system to yield their simpler components.

diploid: Having two sets of genetic information; describing a cell with two chromosomes of each type.

dipole: A molecule having both positive and negative charges.

diprotic acid: An acid having two dissociable protons.

disaccharide: A carbohydrate consisting of two covalently joined monosaccharide units.

dissociation constant: (1) An equilibrium constant (K_d) for the dissociation of a complex of two or more biomolecules into its components; for example, dissociation of a substrate from an enzyme. (2) The dissociation constant (K_a) of an acid, describing its dissociation into its conjugate base and a proton.

disulfide bridge: A covalent cross link between two polypeptide chains formed by a cystine residue (two Cys residues).

DNA (deoxyribonucleic acid): A polynucleotide having a specific sequence of deoxyribonucleotide units covalently joined through 3',5'-phosphodiester bonds; serves as the carrier of genetic information.

DNA chimera: A DNA containing genetic information derived from two different species.

DNA cloning: See cloning.

DNA library: A random collection of cloned DNA fragments that includes all or most of the genome of a given organism; also called a genomic library.

DNA ligase: An enzyme that creates a phosphodiester bond between the 3' end of one DNA segment and the 5' end of another.

DNA looping: The interaction of proteins bound at distant sites on a DNA molecule so that the intervening DNA forms a loop.

DNA polymerase: An enzyme that catalyzes template-dependent synthesis of DNA from its deoxyribonucleoside 5'-triphosphate precursors.

DNA replicase system: The entire complex of enzymes and specialized proteins required in biological DNA replication.

DNA supercoiling: The coiling of DNA upon itself, generally as a result of bending, underwinding, or overwinding of the DNA helix.

DNA transposition: See transposition.

domain: A distinct structural unit of a polypeptide; domains may have separate functions and may fold as independent, compact units.

double helix: The natural coiled conformation of two complementary, antiparallel DNA chains.

double-reciprocal plot: A plot of $1/V_0$ versus $1/[S]$, which allows a more accurate determination of V_{max} and K_m than a plot of V_0 versus $[S]$; also called the Lineweaver–Burk plot.

E_0': See standard reduction potential.

E. coli (Escherichia coli): A common bacterium found in the small intestine of vertebrates; the most well-studied organism.

electrochemical gradient: The sum of the gradients of concentration and of electric charge of an ion across a membrane; the driving force for oxidative phosphorylation and photophosphorylation.

electrochemical potential: The energy required to maintain a separation of charge and of concentration across a membrane.

electrogenic: Contributing to an electrical potential across a membrane.

electron acceptor: A substance that receives electrons in an oxidation–reduction reaction.

electron carrier: A protein, such as a flavoprotein or a cytochrome, that can reversibly gain and lose electrons; functions in the transfer of electrons from organic nutrients to oxygen or some other terminal acceptor.

electron donor: A substance that donates electrons in an oxidation–reduction reaction.

electron transfer: Movement of electrons from substrates to oxygen via the carriers of the respiratory (electron transfer) chain.

electrophile: An electron-deficient group with a strong tendency to accept electrons from an electron-rich group (nucleophile).

electrophoresis: Movement of charged solutes in response to an electrical field; often used to separate mixtures of ions, proteins, or nucleic acids.

elongation factors: Specific proteins required in the elongation of polypeptide chains by ribosomes.

eluate: The effluent from a chromatographic column.

enantiomers: Stereoisomers that are nonsuperimposable mirror images of each other.

end-product inhibition: See feedback inhibition.

endergonic reaction: A chemical reaction that consumes energy (that is, for which ΔG is positive).

endocrine glands: Groups of cells specialized to synthesize hormones and secrete them into the blood to regulate other types of cells.

endocytosis: The uptake of extracellular material by its inclusion within a vesicle (endosome) formed by an invagination of the plasma membrane.

endonuclease: An enzyme that hydrolyzes the interior phosphodiester bonds of a nucleic acid; that is, it acts at points other than the terminal bonds.

endoplasmic reticulum: An extensive system of double membranes in the cytoplasm of eukaryotic cells; it encloses secretory channels and is often studded with ribosomes (rough endoplasmic reticulum).

endothermic reaction: A chemical reaction that takes up heat (that is, for which ΔH is positive).

energy charge: The fractional degree to which the ATP/ADP/AMP system is filled with high-energy phosphate groups.

energy coupling: The transfer of energy from one process to another.

enhancers: DNA sequences that facilitate the expression of a given gene; may be located a few hundred, or even thousand, base pairs away from the gene.

enthalpy (H): The heat content of a system.

enthalpy change (ΔH): For a reaction, is approximately equal to the difference between the energy used to break bonds and the energy gained by the formation of new ones.

entropy (S): The extent of randomness or disorder in a system.

enzyme: A biomolecule, either protein or RNA, that catalyzes a specific chemical reaction. It does not affect the equilibrium of the catalyzed reaction; it enhances the rate of a reaction by providing a reaction path with a lower activation energy.

epimerases: Enzymes that catalyze the reversible interconversion of two epimers.

epimers: Two stereoisomers differing in configuration at one asymmetric center, in a compound having two or more asymmetric centers.

epithelial cell: Any cell that forms part of the outer covering of an organism or organ.

epitope: An antigenic determinant; the particular chemical group or groups within a macromolecule (antigen) to which a given antibody binds.

equilibrium: The state of a system in which no further net change is occurring; the free energy is at a minimum.

equilibrium constant (K_{eq}): A constant, characteristic for each chemical reaction; relates the specific concentrations of all reactants and products at equilibrium at a given temperature and pressure.

erythrocyte: A cell containing large amounts of hemoglobin and specialized for oxygen transport; a red blood cell.

Escherichia coli: See *E. coli.*

essential amino acids: Amino acids that cannot be synthesized by humans (and other vertebrates) and must be obtained from the diet.

essential fatty acids: The group of polyunsaturated fatty acids produced by plants, but not by humans; required in the human diet.

ethanol fermentation: See alcohol fermentation.

eukaryote: A unicellular or multicellular organism with cells having a membrane-bounded nucleus, multiple chromosomes, and internal organelles.

excited state: An energy-rich state of an atom or molecule; produced by the absorption of light energy.

exergonic reaction: A chemical reaction that proceeds with the release of free energy (that is, for which ΔG is negative).

exocytosis: The fusion of an intracellular vesicle with the plasma membrane, releasing the vesicle contents to the extracellular space.

exon: The segment of a eukaryotic gene that encodes a portion of the final product of the gene; a portion that remains after posttranscriptional processing and is transcribed into a protein or incorporated into the structure of an RNA. See intron.

exonuclease: An enzyme that hydrolyzes only those phosphodiester bonds that are in the terminal positions of a nucleic acid.

exothermic reaction: A chemical reaction that releases heat (that is, for which ΔH is negative).

expression vector: See vector.

facilitated diffusion: Diffusion of a polar substance across a biological membrane through a protein transporter; also called passive diffusion or passive transport.

facultative cells: Cells that can live in the presence or absence of oxygen.

FAD (flavin adenine dinucleotide): The coenzyme of some oxidation–reduction enzymes; it contains riboflavin.

fatty acid: A long-chain aliphatic carboxylic acid found in natural fats and oils; also a component of membrane phospholipids and glycolipids.

feedback inhibition: Inhibition of an allosteric enzyme at the beginning of a metabolic sequence by the end product of the sequence; also known as end-product inhibition.

fermentation: Energy-yielding anaerobic breakdown of a nutrient molecule, such as glucose, without net oxidation; yields lactate, ethanol, or some other simple product.

fibroblast: A cell of the connective tissue that secretes connective tissue proteins such as collagen.

fibrous proteins: Insoluble proteins that serve in a protective or structural role; contain polypeptide chains that generally share a common secondary structure.

fingerprinting: See peptide mapping.

first law of thermodynamics: The law stating that in all processes, the total energy of the universe remains constant.

Fischer projection formulas: See projection formulas.

5′ end: The end of a nucleic acid that lacks a nucleotide bound at the 5′ position of the terminal residue.

flagellum: A cell appendage used in propulsion. Bacterial flagella have a much simpler structure than eukaryotic flagella, which are similar to cilia.

flavin-linked dehydrogenases: Dehydrogenases requiring one of the riboflavin coenzymes, FMN or FAD.

flavin nucleotides: Nucleotide coenzymes (FMN and FAD) containing riboflavin.

flavoprotein: An enzyme containing a flavin nucleotide as a tightly bound prosthetic group.

fluid mosaic model: A model describing biological membranes as a fluid lipid bilayer with embedded proteins; the bilayer exhibits both structural and functional asymmetry.

fluorescence: Emission of light by excited molecules as they revert to the ground state.

FMN (flavin mononucleotide): Riboflavin phosphate, a coenzyme of certain oxidation–reduction enzymes.

footprinting: A technique for identifying the nucleic acid sequence bound by a DNA- or RNA-binding protein.

frame shift: A mutation caused by insertion or deletion of one or more paired nucleotides, changing the reading frame of codons during protein synthesis; the polypeptide product has a garbled amino acid sequence beginning at the mutated codon.

free energy (G): The component of the total energy of a system that can do work at constant temperature and pressure.

free energy of activation ($\Delta G^{\ddagger}$): See activation energy.

free-energy change (ΔG): The amount of free energy released (negative ΔG) or absorbed (positive ΔG) in a reaction at constant temperature and pressure.

free radical: See radical.

functional group: The specific atom or group of atoms that confers a particular chemical property on a biomolecule.

furanose: A simple sugar containing the five-membered furan ring.

fusion protein: (1) A family of proteins that facilitate membrane fusion. (2) The protein product of a gene created by the fusion of two distinct genes.

futile cycle: A set of enzyme-catalyzed cyclic reactions that results in release of thermal energy by the hydrolysis of ATP.

$\Delta G^{\circ\prime}$: See standard free-energy change.

gametes: Reproductive cells with a haploid gene content; sperm or egg cells.

gangliosides: Sphingolipids, containing complex oligosaccharides as head groups; especially common in nervous tissue.

gel filtration: A chromatographic procedure for the separation of a mixture of molecules on the basis of size; based on the capacity of porous polymers to exclude solutes above a certain size.

gene: A chromosomal segment that codes for a single functional polypeptide chain or RNA molecule.

gene expression: Transcription and, in the case of proteins, translation to yield the product of a gene; a gene is expressed when its biological product is present and active.

gene splicing: The enzymatic attachment of one gene, or part of a gene, to another.

general acid–base catalysis: Catalysis involving proton transfer(s) to or from a molecule other than water.

genetic code: The set of triplet code words in DNA (or mRNA) coding for the amino acids of proteins.

genetic information: The hereditary information contained in a sequence of nucleotide bases in chromosomal DNA or RNA.

genetic map: A diagram showing the relative sequence and position of specific genes along a chromosome.

genome: All the genetic information encoded in a cell or virus.

genotype: The genetic constitution of an organism, as distinct from its physical characteristics, or phenotype.

geometric isomers: Isomers related by rotation about a double bond; also called cis and trans isomers.

germ-line cell: A type of animal cell that is formed early in embryogenesis and may multiply by mitosis or may produce, by meiosis, cells that develop into gametes (egg or sperm cells).

globular proteins: Soluble proteins with a globular (somewhat rounded) shape.

glucogenic amino acids: Amino acids with carbon chains that can be metabolically converted into glucose or glycogen via gluconeogenesis.

gluconeogenesis: The biosynthesis of a carbohydrate from simpler, noncarbohydrate precursors such as oxaloacetate or pyruvate.

glycan: Another term for polysaccharide; a polymer of monosaccharide units joined by glycosidic bonds.

glycerophospholipid: An amphipathic lipid with a glycerol backbone; fatty acids are ester-linked to C-1 and C-2 of glycerol, and a polar alcohol is attached through a phosphodiester linkage to C-3.

glycolipid: A lipid containing a carbohydrate group.

glycolysis: The catabolic pathway by which a molecule of glucose is broken down into two molecules of pyruvate.

glycoprotein: A protein containing a carbohydrate group.

glycosaminoglycan: A heteropolysaccharide of two alternating units: one is either *N*-acetylglucosamine or *N*-acetylgalactosamine; the other is a uronic acid (usually glucuronic acid). Formerly called mucopolysaccharide.

glycosidic bonds: Bonds between a sugar and another molecule (typically an alcohol, purine, pyrimidine, or sugar) through an intervening oxygen or nitrogen atom; the bonds are classified as *O*-glycosidic or *N*-glycosidic, respectively.

glyoxylate cycle: A variant of the citric acid cycle, for the net conversion of acetate into succinate and, eventually, new carbohydrate; present in bacteria and some plant cells.

glyoxysome: A specialized peroxisome containing the enzymes of the glyoxylate cycle; found in cells of germinating seeds.

Golgi complex: A complex membranous organelle of eukaryotic cells; functions in the posttranslational modification of proteins and their secretion from the cell or incorporation into the plasma membrane or organellar membranes.

gram molecular weight: The weight in grams of a compound that is numerically equal to its molecular weight; the weight of 1 mole.

grana: Stacks of thylakoids, flattened membranous sacs or discs, in chloroplasts.

ground state: The normal, stable form of an atom or molecule; as distinct from the excited state.

group transfer potential: A measure of the ability of a compound to donate an activated group (such as a phosphate or acyl group); generally expressed as the standard free energy of hydrolysis.

half-life: The time required for the disappearance or decay of one-half of a given component in a system.

haploid: Having a single set of genetic information; describing a cell with one chromosome of each type.

Haworth perspective formulas: A method for representing cyclic chemical structures so as to define the configuration of each substituent group; the method commonly used for representing sugars.

helicase: An enzyme that catalyzes the separation of strands in a DNA molecule before replication.

helix, α: See α helix.

heme: The iron-porphyrin prosthetic group of heme proteins.

heme protein: A protein containing a heme as a prosthetic group.

hemoglobin: A heme protein in erythrocytes; functions in oxygen transport.

Henderson–Hasselbalch equation: An equation relating the pH, the pK_a, and the ratio of the concentrations of the proton-acceptor (A^-) and proton-donor (HA) species in a solution.

hepatocyte: The major cell type of liver tissue.

heteroduplex DNA: Duplex DNA containing complementary strands derived

from two different DNA molecules with similar sequences, often as a product of genetic recombination.

heteropolysaccharide: A polysaccharide containing more than one type of sugar.

heterotroph: An organism that requires complex nutrient molecules, such as glucose, as a source of energy and carbon.

heterotropic enzyme: An allosteric enzyme requiring a modulator other than its substrate.

hexose: A simple sugar with a backbone containing six carbon atoms.

high-energy compound: A compound that on hydrolysis undergoes a large decrease in free energy under standard conditions.

high-performance liquid chromatography (HPLC): Chromatographic procedures, often conducted at relatively high pressures, using automated equipment that permits refined and highly reproducible profiles.

Hill reaction: The evolution of oxygen and the photoreduction of an artificial electron acceptor by a chloroplast preparation in the absence of carbon dioxide.

histones: The family of five basic proteins that associate tightly with DNA in the chromosomes of all eukaryotic cells.

Holliday intermediate: An intermediate in genetic recombination in which two double-stranded DNA molecules are joined by virtue of a reciprocal crossover involving one strand of each molecule.

holoenzyme: A catalytically active enzyme including all necessary subunits, prosthetic groups, and cofactors.

homeobox: A conserved DNA sequence of 180 base pairs encoding a protein domain found in many proteins that play a regulatory role in development.

homeodomain: The protein domain encoded by the homeobox.

homeostasis: The maintenance of a dynamic steady state by regulatory mechanisms that compensate for changes in external circumstances.

homeotic genes: Genes that regulate the development of the pattern of segments in the *Drosophila* body plan; similar genes are found in most vertebrates.

homologous genetic recombination: Recombination between two DNA molecules of similar sequence, occurring in all cells; occurs during meiosis and mitosis in eukaryotes.

homologous proteins: Proteins having sequences and functions similar in different species; for example, the hemoglobins.

homopolysaccharide: A polysaccharide made up of only one type of monosaccharide unit.

homotropic enzyme: An allosteric enzyme that uses its substrate as a modulator.

hormone: A chemical substance synthesized in small amounts by an endocrine tissue and carried in the blood to another tissue, where it acts as a messenger to regulate the function of the target tissue or organ.

hormone receptor: A protein in, or on the surface of, target cells that binds a specific hormone and initiates the cellular response.

hydrogen bond: A weak electrostatic attraction between one electronegative atom (such as oxygen or nitrogen) and a hydrogen atom covalently linked to a second electronegative atom.

hydrolases: Enzymes (proteases, lipases, phosphatases, nucleases, for example) that catalyze hydrolysis reactions.

hydrolysis: Cleavage of a bond, such as an anhydride or peptide bond, by the addition of the elements of water, yielding two or more products.

hydronium ion: The hydrated hydrogen ion (H_3O^+).

hydropathy index: A scale that expresses the relative hydrophobic and hydrophilic tendencies of a chemical group.

hydrophilic: Polar or charged; describing molecules or groups that associate with (dissolve easily in) water.

hydrophobic: Nonpolar; describing molecules or groups that are insoluble in water.

hydrophobic interactions: The association of nonpolar groups, or compounds, with each other in aqueous systems, driven by the tendency of the surrounding water molecules to seek their most stable (disordered) state.

hyperchromic effect: The large increase in light absorption at 260 nm occurring as a double-helical DNA is melted (unwound).

immune response: The capacity of a vertebrate to generate antibodies to an antigen, a macromolecule foreign to the organism.

immunoglobulin: An antibody protein generated against, and capable of binding specifically to, an antigen.

in vitro: "In glass"; that is, in the test tube.

in vivo: "In life"; that is, in the living cell or organism.

induced fit: A change in the conformation of an enzyme in response to substrate binding that renders the enzyme catalytically active; also used to denote changes in the conformation of any macromolecule in response to ligand binding such that the binding site of the macromolecule better conforms to the shape of the ligand.

inducer: A signal molecule that, when bound to a regulatory protein, produces an increase in the expression of a given gene.

induction: An increase in the expression of a gene in response to a change in the activity of a regulatory protein.

informational macromolecules: Biomolecules containing information in the form of specific sequences of different monomers; for example, many proteins, lipids, polysaccharides, and nucleic acids.

initiation codon: AUG (sometimes GUG in prokaryotes); codes for the first amino acid in a polypeptide sequence: *N*-formylmethionine in prokaryotes, and methionine in eukaryotes.

initiation complex: A complex of a ribosome with an mRNA and the initiating Met-tRNA^{Met} or fMet-tRNA^{fMet}, ready for the elongation steps.

inorganic pyrophosphatase: An enzyme that hydrolyzes a molecule of inorganic pyrophosphate to yield two molecules of (ortho) phosphate; also known as pyrophosphatase.

insertion mutation: A mutation caused by insertion of one or more extra bases, or a mutagen, between two successive bases in DNA.

insertion sequence: Specific base sequences at either end of a transposable segment of DNA.

integral membrane proteins: Proteins firmly bound to a membrane by hydrophobic interactions; as distinct from peripheral proteins.

intercalating mutagen: A mutagen that inserts itself between two successive bases in a nucleic acid, causing a frame-shift mutation.

intercalation: Insertion between two stacked aromatic or planar rings; for example, the insertion of a planar molecule between two successive bases in a nucleic acid.

interferons: A class of glycoproteins with antiviral activities.

intermediary metabolism: In cells, the enzyme-catalyzed reactions that extract chemical energy from nutrient molecules and utilize it to synthesize and assemble cell components.

intron (intervening sequence): A sequence of nucleotides in a gene that is transcribed but excised before the gene is translated.

ion channel: An integral membrane protein that provides for the regulated transport of a specific ion, or ions, across a membrane.

ion-exchange resin: A polymeric resin that contains fixed charged groups; used in chromatographic columns to separate ionic compounds.

ion product of water (K_W): The product of the concentrations of H^+ and OH^- in pure water: $K_W = [H^+][OH^-] = 1 \times 10^{-14}$ at 25 °C.

ionizing radiation: A type of radiation, such as x rays, that causes loss of electrons from some organic molecules, thus making them more reactive.

ionophore: A compound that binds one or more metal ions and is capable of diffusing across a membrane, carrying the bound ion.

iron–sulfur center: A prosthetic group of certain redox proteins involved in electron transfers; Fe^{2+} or Fe^{3+} is bound to inorganic sulfur and to Cys groups in the protein.

isoelectric focusing: An electrophoretic method for separating macromolecules on the basis of their isoelectric pH.

isoelectric pH (isoelectric point): The pH at which a solute has no net electric charge and thus does not move in an electric field.

isoenzymes: See isozymes.

isomerases: Enzymes that catalyze the transformation of compounds into their positional isomers.

isomers: Any two molecules with the same molecular formula but a different arrangement of molecular groups.

isoprene: The hydrocarbon 2-methyl-1,3-butadiene, a recurring structural unit of the terpenoid biomolecules.

isothermal: Occurring at constant temperature.

isotopes: Stable or radioactive forms of an element that differ in atomic weight but are otherwise chemically identical to the naturally abundant form of the element; used as tracers.

isozymes: Multiple forms of an enzyme that catalyze the same reaction but differ from each other in their amino acid sequence, substrate affinity, V_{max}, and/or regulatory properties; also called isoenzymes.

keratins: Insoluble protective or structural proteins consisting of parallel polypeptide chains in α-helical or β conformations.

ketogenic amino acids: Amino acids with carbon skeletons that can serve as precursors of the ketone bodies.

ketone bodies: Acetoacetate, D-β-hydroxybutyrate, and acetone; water-soluble fuels normally exported by the liver but overproduced during fasting or in untreated diabetes mellitus.

ketose: A simple monosaccharide in which the carbonyl group is a ketone.

ketosis: A condition in which the concentration of ketone bodies in the blood, tissues, and urine is abnormally high.

kinases: Enzymes that catalyze the phosphorylation of certain molecules by ATP.

kinetics: The study of reaction rates.

Krebs cycle: See citric acid cycle.

lagging strand: The DNA strand that, during replication, must be synthesized in the direction opposite to that in which the replication fork moves.

law of mass action: The law stating that the rate of any given chemical reaction is proportional to the product of the activities (or concentrations) of the reactants.

leader: A short sequence near the amino terminus of a protein or the 5' end of an RNA that has a specialized targeting or regulatory function.

leading strand: The DNA strand that, during replication, is synthesized in the same direction in which the replication fork moves.

leaky mutant: A mutant gene that gives rise to a product with a detectable level of biological activity.

leaving group: The departing or displaced molecular group in a unimolecular elimination or a bimolecular substitution reaction.

lethal mutation: A mutation that inactivates a biological function essential to the life of the cell or organism.

leucine zipper: A protein structural

motif involved in protein–protein interactions in many eukaryotic regulatory proteins; consists of two interacting α helices in which Leu residues in every seventh position are a prominent feature of the interacting surfaces.

leukotrienes: A family of molecules derived from arachidonate; muscle contractants that constrict air passages in the lungs and are involved in asthma.

levorotatory isomer: A stereoisomer that rotates the plane of plane-polarized light counterclockwise.

ligand: A small molecule that binds specifically to a larger one; for example, a hormone is the ligand for its specific protein receptor.

light reactions: The reactions of photosynthesis that require light and cannot occur in the dark; also known as the light-dependent reactions.

Lineweaver–Burk equation: An algebraic transform of the Michaelis–Menten equation, allowing determination of V_{max} and K_m by extrapolation of [S] to infinity.

linking number: The number of times one closed circular DNA strand is wound about another; the number of topological links holding the circles together.

lipases: Enzymes that catalyze the hydrolysis of triacylglycerols.

lipid: A small water-insoluble biomolecule generally containing fatty acids, sterols, or isoprenoid compounds.

lipoate (lipoic acid): A vitamin for some microorganisms; an intermediate carrier of hydrogen atoms and acyl groups in α-keto acid dehydrogenases.

lipoprotein: A lipid–protein aggregate that serves to carry water-insoluble lipids in the blood. The protein component alone is an apolipoprotein.

low-energy phosphate compound: A phosphorylated compound with a relatively small standard free energy of hydrolysis.

lyases: Enzymes that catalyze the removal of a group from a molecule to form a double bond, or the addition of a group to a double bond.

lymphocytes: A subclass of leukocytes involved in the immune response. B lymphocytes synthesize and secrete antibodies; T lymphocytes either play a regulatory role in immunity or kill foreign and virus-infected cells.

lysis: Destruction of a cell's plasma membrane or of a bacterial cell wall, releasing the cellular contents and killing the cell.

lysogeny: One of two outcomes of the infection of a host cell by a temperate phage. It occurs when the phage genome becomes repressed and is replicated as part of the host DNA; infrequently it may be induced, and the phage particles so produced cause the host cell to lyse.

lysosome: A membrane-bounded organelle in the cytoplasm of eukaryotic cells; it contains many hydrolytic enzymes and serves as a degrading and recycling center for unneeded components.

macromolecule: A molecule having a molecular weight in the range of a few thousand to many millions.

mass-action ratio: For the reaction aA + bB $\rightleftharpoons$ cC + dD, the ratio: $\dfrac{[C]^c\,[D]^d}{[A]^a\,[B]^b}$.

matrix: The aqueous contents of a cell or organelle (the mitochondrion, for example) with dissolved solutes.

meiosis: A type of cell division in which diploid cells give rise to haploid cells destined to become gametes.

membrane transport: Movement of a polar solute across a membrane via a specific membrane protein (a transporter).

messenger RNA (mRNA): A class of RNA molecules, each of which is complementary to one strand of DNA; carries the genetic message from the chromosome to the ribosomes.

metabolism: The entire set of enzyme-catalyzed transformations of organic molecules in living cells; the sum of anabolism and catabolism.

metabolite: A chemical intermediate in the enzyme-catalyzed reactions of metabolism.

metalloprotein: A protein having a metal ion as its prosthetic group.

metamerism: Division of the body into segments; in insects, for example.

micelle: An aggregate of amphipathic molecules in water, with the nonpolar portions in the interior and the polar portions at the exterior surface, exposed to water.

Michaelis–Menten constant (K_m): The substrate concentration at which an enzyme-catalyzed reaction proceeds at one-half its maximum velocity.

Michaelis–Menten equation: The equation describing the hyperbolic dependence of the initial reaction velocity, V_0, on substrate concentration, [S], in many enzyme-catalyzed reactions: $V_0 = \dfrac{V_{max}[S]}{K_m + [S]}$.

Michaelis–Menten kinetics: A kinetic pattern in which the initial rate of an enzyme-catalyzed reaction exhibits a hyperbolic dependence on substrate concentration.

microbodies: Cytoplasmic, membrane-bounded vesicles containing peroxide-forming and peroxide-destroying enzymes; include lysosomes, peroxisomes, and glyoxysomes.

microfilaments: Thin filaments composed of actin, found in the cytoplasm of eukaryotic cells; serve in structure and movement.

microsomes: Membranous vesicles formed by fragmentation of the endoplasmic reticulum of eukaryotic cells; recovered by differential centrifugation.

microtubules: Thin tubules assembled from two types of globular tubulin subunits; present in cilia, flagella, centrosomes, and other contractile or motile structures.

mitochondrion: Membrane-bounded organelle in the cytoplasm of eukaryotes; contains the enzyme systems required for the citric acid cycle, fatty acid oxidation, electron transfer, and oxidative phosphorylation.

mitosis: The multistep process in eukaryotic cells that results in the replication of chromosomes and cell division.

mixed-function oxidases (oxygenases): Enzymes, often flavoproteins, that use molecular oxygen (O_2) to simultaneously oxidize a substrate and a cosubstrate (commonly NADH or NADPH).

modulator: A metabolite that, when bound to the allosteric site of an enzyme, alters its kinetic characteristics.

molar solution: One mole of solute dissolved in water to give a total volume of 1,000 mL.

mole: One gram molecular weight of a compound. See Avogadro's number.

monoclonal antibodies: Antibodies produced by a cloned hybridoma cell, which therefore are identical and directed against the same epitope of the antigen.

monolayer: A single layer of oriented lipid molecules.

monoprotic acid: An acid having only one dissociable proton.

monosaccharide: A carbohydrate consisting of a single sugar unit.

mRNA: See messenger RNA.

mucopolysaccharide: An older name for a glycosaminoglycan.

multienzyme system: A group of related enzymes participating in a given metabolic pathway.

mutarotation: The change in specific rotation of a pyranose or furanose sugar or glycoside accompanying the equilibration of its α- and β-anomeric forms.

mutases: Enzymes that catalyze the transposition of functional groups.

mutation: An inheritable change in the nucleotide sequence of a chromosome.

myofibril: A unit of thick and thin filaments of muscle fibers.

myosin: A contractile protein; the major component of the thick filaments of muscle and other actin–myosin systems.

NAD, NADP (nicotinamide adenine dinucleotide, nicotinamide adenine dinucleotide phosphate): Nicotinamide-containing coenzymes functioning as carriers of hydrogen atoms and electrons in some oxidation–reduction reactions.

native conformation: The biologically active conformation of a macromolecule.

negative cooperativity: A phenomenon of some multisubunit enzymes or proteins in which binding of a ligand or substrate to one subunit impairs binding to another subunit.

negative feedback: Regulation of a biochemical pathway achieved when a reaction product inhibits an earlier step in the pathway.

neuron: A cell of nervous tissue specialized for transmission of a nerve impulse.

neurotransmitter: A low molecular weight compound (usually containing nitrogen) secreted from the terminal of a neuron and bound by a specific receptor in the next neuron; serves to transmit a nerve impulse.

nicotinamide adenine dinucleotide, nicotinamide adenine dinucleotide phosphate: See NAD, NADP.

ninhydrin reaction: A color reaction given by amino acids and peptides on heating with ninhydrin; widely used for their detection and estimation.

nitrogen cycle: The cycling of various forms of biologically available nitrogen through the plant, animal, and microbial worlds, and through the atmosphere and geosphere.

nitrogen fixation: Conversion of atmospheric nitrogen (N_2) into a reduced, biologically available form by nitrogen-fixing organisms.

nitrogenase complex: A system of enzymes capable of reducing atmospheric nitrogen to ammonia in the presence of ATP.

noncompetitive inhibition: A type of enzyme inhibition not reversed by increasing the substrate concentration.

noncyclic electron flow: The light-induced flow of electrons from water to $NADP^+$ in oxygen-evolving photosynthesis; it involves both photosystems I and II.

nonessential amino acids: Amino acids that can be made by humans and other vertebrates from simpler precursors, and are thus not required in the diet.

nonheme iron proteins: Proteins, usually acting in oxidation–reduction reactions, containing iron but no porphyrin groups.

nonpolar: Hydrophobic; describing molecules or groups that are poorly soluble in water.

nonsense codon: A codon that does not specify an amino acid, but signals the termination of a polypeptide chain.

nonsense mutation: A mutation that results in the premature termination of a polypeptide chain.

nonsense suppressor: A mutation, usually in the gene for a tRNA, that causes an amino acid to be inserted into a polypeptide in response to a termination codon.

nucleases: Enzymes that hydrolyze the internucleotide (phosphodiester) linkages of nucleic acids.

nucleic acids: Biologically occurring polynucleotides in which the nucleotide residues are linked in a specific sequence by phosphodiester bonds; DNA and RNA.

nucleoid: In bacteria, the nuclear zone that contains the chromosome but has no surrounding membrane.

nucleolus: A densely staining structure in the nucleus of eukaryotic cells; involved in rRNA synthesis and ribosome formation.

nucleophile: An electron-rich group with a strong tendency to donate electrons to an electron-deficient nucleus (electrophile); the entering reactant in a bimolecular substitution reaction.

nucleoplasm: The portion of a cell's contents enclosed by the nuclear membrane; also called the nuclear matrix.

nucleoside: A compound consisting of a purine or pyrimidine base covalently linked to a pentose.

nucleoside diphosphate kinase: An enzyme that catalyzes the transfer of the terminal phosphate of a nucleoside 5'-triphosphate to a nucleoside 5'-diphosphate.

nucleoside diphosphate sugar: A coenzymelike carrier of a sugar molecule, functioning in the enzymatic synthesis of polysaccharides and sugar derivatives.

nucleoside monophosphate kinase: An enzyme that catalyzes the transfer of the terminal phosphate of ATP to a nucleoside 5'-monophosphate.

nucleosome: Structural unit for packaging chromatin; consists of a DNA strand wound around a histone core.

nucleotide: A nucleoside phosphorylated at one of its pentose hydroxyl groups.

nucleus: In eukaryotes, a membrane-bounded organelle that contains chromosomes.

oligomer: A short polymer, usually of amino acids, sugars, or nucleotides; the definition of "short" is somewhat arbitrary, but usually less than 50 subunits.

oligomeric protein: A multisubunit protein having two or more identical polypeptide chains.

oligonucleotide: A short polymer of nucleotides (usually less than 50).

oligopeptide: A few amino acids joined by peptide bonds.

oligosaccharide: Several monosaccharide groups joined by glycosidic bonds.

oncogene: A cancer-causing gene; any of several mutant genes that cause cells to exhibit rapid, uncontrolled proliferation. See also proto-oncogene.

open reading frame: A group of contiguous nonoverlapping nucleotide codons in a DNA or RNA molecule that do not include a termination codon.

open system: A system that exchanges matter and energy with its surroundings. See also system.

operator: A region of DNA that interacts with a repressor protein to control the expression of a gene or group of genes.

operon: A unit of genetic expression consisting of one or more related genes and the operator and promoter sequences that regulate their transcription.

optical activity: The capacity of a substance to rotate the plane of plane-polarized light.

optimum pH: The characteristic pH at which an enzyme has maximal catalytic activity.

organelles: Membrane-bounded structures found in eukaryotic cells; contain enzymes and other components required for specialized cell functions.

origin: The nucleotide sequence or site in DNA where DNA replication is initiated.

osmosis: Bulk flow of water through a semipermeable membrane into another aqueous compartment containing solute at a higher concentration.

osmotic pressure: Pressure generated by the osmotic flow of water through a semipermeable membrane into an aqueous compartment containing solute at a higher concentration.

oxidation: The loss of electrons from a compound.

oxidation, β: See β oxidation.

oxidation–reduction reaction: A reaction in which electrons are transferred from a donor to an acceptor molecule; also called a redox reaction.

oxidative phosphorylation: The enzymatic phosphorylation of ADP to ATP coupled to electron transfer from a substrate to molecular oxygen.

oxidizing agent (oxidant): The acceptor of electrons in an oxidation–reduction reaction.

oxygen debt: The extra oxygen (above the normal resting level) consumed in the recovery period after strenuous physical exertion.

oxygenases: Enzymes that catalyze reactions in which oxygen is introduced into an acceptor molecule.

P

palindrome: A segment of duplex DNA in which the base sequences of the two strands exhibit twofold rotational symmetry about an axis.

paradigm: In biochemistry, an experimental model or example.

partition coefficient: A constant that expresses the ratio in which a given solute will be partitioned or distributed between two given immiscible liquids at equilibrium.

pathogenic: Disease-causing.

pentose: A simple sugar with a backbone containing five carbon atoms.

pentose phosphate pathway: A pathway that serves to interconvert hexoses and pentoses and is a source of reducing equivalents and pentoses for biosynthetic processes; present in most organisms. Also called the phosphogluconate pathway.

peptidase: An enzyme that hydrolyzes a peptide bond.

peptide: Two or more amino acids covalently joined by peptide bonds.

peptide bond: A substituted amide linkage between the α-amino group of one amino acid and the α-carboxyl group of another, with the elimination of the elements of water.

peptide mapping: The characteristic two-dimensional pattern (on paper or gel) formed by the separation of a mixture of peptides resulting from partial hydrolysis of a protein; also known as peptide fingerprinting.

peptidoglycan: A major component of bacterial cell walls; generally consists of parallel heteropolysaccharides cross-linked by short peptides.

peripheral proteins: Proteins that are loosely or reversibly bound to a membrane by hydrogen bonds or electrostatic forces; generally water-soluble once released from the membrane.

permeases: See transporters.

peroxisome: Membrane-bounded organelle in the cytoplasm of eukaryotic cells; contains peroxide-forming and peroxide-destroying enzymes.

pH: The negative logarithm of the hydrogen ion concentration of an aqueous solution.

phage: See bacteriophage.

phenotype: The observable characteristics of an organism.

phosphodiester linkage: A chemical grouping that contains two alcohols esterified to one molecule of phosphoric acid, which thus serves as a bridge between them.

phosphogluconate pathway: An oxidative pathway beginning with glucose-6-phosphate and leading, via 6-phosphogluconate, to pentose phosphates and yielding NADPH. Also called the pentose phosphate pathway.

phospholipid: A lipid containing one or more phosphate groups.

phosphorolysis: Cleavage of a compound with phosphate as the attacking group; analogous to hydrolysis.

phosphorylation: Formation of a phosphate derivative of a biomolecule, usually by enzymatic transfer of a phosphate group from ATP.

phosphorylation potential (ΔG_p): The actual free-energy change of ATP hydrolysis under the nonstandard conditions prevailing within a cell.

photochemical reaction center: The part of a photosynthetic complex where the energy of an absorbed photon causes charge separation, initiating electron transfer.

photon: The ultimate unit (a quantum) of light energy.

photophosphorylation: The enzymatic formation of ATP from ADP coupled to the light-dependent transfer of electrons in photosynthetic cells.

photoreduction: The light-induced reduction of an electron acceptor in photosynthetic cells.

photorespiration: Oxygen consumption occurring in illuminated temperate-zone plants, largely due to oxidation of phosphoglycolate.

photosynthesis: The use of light energy to produce carbohydrates from carbon dioxide and a reducing agent such as water.

photosynthetic phosphorylation: See photophosphorylation.

photosystem: In photosynthetic cells, a functional set of light-absorbing pigments and its reaction center.

phototroph: An organism that can use the energy of light to synthesize its own fuels from simple molecules such as carbon dioxide, oxygen, and water; as distinct from a chemotroph.

pK: The negative logarithm of an equilibrium constant.

plasma membrane: The exterior membrane surrounding the cytoplasm of a cell.

plasma proteins: The proteins present in blood plasma.

plasmalogen: A phospholipid with an alkenyl ether substituent on the C-1 of glycerol.

plasmid: An extrachromosomal, independently replicating, small circular DNA molecule; commonly employed in genetic engineering.

plastid: In plants, a self-replicating organelle; may differentiate into a chloroplast.

platelets: Small, enucleated cells that initiate blood clotting; they arise from cells called megakaryocytes in the bone marrow. Also known as thrombocytes.

pleated sheet: The side-by-side, hydrogen-bonded arrangement of polypeptide chains in the extended β conformation.

polar: Hydrophilic, or "water-loving"; describing molecules or groups that are soluble in water.

polarity: (1) In chemistry, the nonuniform distribution of electrons in a molecule; polar molecules are usually soluble in water. (2) In molecular biology, the distinction between the 5′ and 3′ ends of nucleic acids.

polyclonal antibodies: A heterogeneous pool of antibodies produced in an animal by a number of different B lymphocytes in response to an antigen. Different antibodies in the pool recognize different parts of the antigen.

polylinker: A short, often synthetic, fragment of DNA containing recognition sequences for several restriction endonucleases.

polymerase chain reaction (PCR): A repetitive procedure that results in a geometric amplification of a specific DNA sequence.

polymorphic: Describing a protein for which amino acid sequence variants exist in a population of organisms, but the variations do not destroy the protein's function.

polynucleotide: A covalently linked sequence of nucleotides in which the 3′ hydroxyl of the pentose of one nucleotide residue is joined by a phosphodiester bond to the 5′ hydroxyl of the pentose of the next residue.

polypeptide: A long chain of amino acids linked by peptide bonds; the molecular weight is generally less than 10,000.

polyribosome: See polysome.

polysaccharide: A linear or branched polymer of monosaccharide units linked by glycosidic bonds.

polysome (polyribosome): A complex of an mRNA molecule and two or more ribosomes.

porphyrin: Complex nitrogenous compound containing four substituted pyrroles covalently joined into a ring; often complexed with a central metal atom.

positive cooperativity: A phenomenon of some multisubunit enzymes or proteins in which binding of a ligand or substrate to one subunit facilitates binding to another subunit.

posttranscriptional processing: The enzymatic processing of the primary RNA transcript, producing functional mRNA, tRNA, and/or rRNA molecules.

posttranslational modification: Enzymatic processing of a polypeptide chain after translation from its mRNA.

primary structure: A description of the covalent backbone of a polymer (macromolecule), including the sequence of monomeric subunits and any interchain and intrachain covalent bonds.

primary transcript: The immediate RNA product of transcription before any posttranscriptional processing reactions.

primase: An enzyme that catalyzes the formation of RNA oligonucleotides used as primers by DNA polymerases.

primer: A short oligomer (of sugars or nucleotides, for example) to which an enzyme adds additional monomeric subunits.

probe: A labeled fragment of nucleic acid containing a nucleotide sequence complementary to a gene or genomic sequence that one wishes to detect in a hybridization experiment.

processivity: For any enzyme that catalyzes the synthesis of a biological polymer, the property of adding multiple subunits to the polymer without dissociating from the substrate.

prochiral molecule: A symmetric molecule that can react asymmetrically with an enzyme having an asymmetric active site, generating a chiral product.

projection formulas: A method for representing molecules to show the configuration of groups around chiral centers; also known as Fischer projection formulas.

prokaryote: A bacterium; a unicellular organism with a single chromosome, no nuclear envelope, and no membrane-bounded organelles.

promoter: A DNA sequence at which RNA polymerase may bind, leading to initiation of transcription.

prophage: A bacteriophage in an inactive state in which the genome is either integrated into the chromosome of the host cell or (sometimes) replicated autonomously.

prostaglandins: A class of lipid-soluble, hormonelike regulatory molecules derived from arachidonate and other polyunsaturated fatty acids.

prosthetic group: A metal ion or an organic compound (other than an amino acid) that is covalently bound to a protein and is essential to its activity.

protein: A macromolecule composed of one or more polypeptide chains, each with a characteristic sequence of amino acids linked by peptide bonds.

protein kinases: Enzymes that phosphorylate certain amino acid residues in specific proteins.

protein targeting: The process by which newly synthesized proteins are sorted and transported to their proper locations in the cell.

proteoglycan: A hybrid macromolecule consisting of a heteropolysaccharide joined to a polypeptide; the polysaccharide is the major component.

proto-oncogene: A cellular gene, usually encoding a regulatory protein, that can be converted into an oncogene by mutation.

proton acceptor: An anionic compound capable of accepting a proton from a proton donor; that is, a base.

proton donor: The donor of a proton in an acid–base reaction; that is, an acid.

proton-motive force: The electrochemical potential inherent in a transmembrane gradient of H^+ concentration; used in oxidative phosphorylation and photophosphorylation to drive ATP synthesis.

protoplasm: A general term referring to the entire contents of a living cell.

purine: A nitrogenous heterocyclic base found in nucleotides and nucleic acids; containing fused pyrimidine and imidazole rings.

puromycin: An antibiotic that inhibits polypeptide synthesis by being incorporated into a growing polypeptide chain, causing its premature termination.

pyranose: A simple sugar containing the six-membered pyran ring.

pyridine nucleotide: A nucleotide coenzyme containing the pyridine derivative nicotinamide; NAD or NADP.

pyridoxal phosphate: A coenzyme containing the vitamin pyridoxine (vitamin B_6); functions in reactions involving amino group transfer.

pyrimidine: A nitrogenous heterocyclic base found in nucleotides and nucleic acids.

pyrimidine dimer: A covalently joined dimer of two adjacent pyrimidine residues in DNA, induced by absorption of UV light; most commonly derived from two adjacent thymines (a thymine dimer).

pyrophosphatase: See inorganic pyrophosphatase.

quantum: The ultimate unit of energy.

quaternary structure: The three-dimensional structure of a multisubunit protein; particularly the manner in which the subunits fit together.

R group: (1) Formally, an abbreviation denoting any alkyl group. (2) Occasionally, used in a more general sense to denote virtually any organic substituent (the R groups of amino acids, for example).

racemic mixture (racemate): An equimolar mixture of the D and L stereoisomers of an optically active compound.

radical: An atom or group of atoms possessing an unpaired electron; also called a free radical.

radioactive isotope: An isotopic form of an element with an unstable nucleus that stabilizes itself by emitting ionizing radiation.

radioimmunoassay: A sensitive and quantitative method for detecting trace amounts of a biomolecule, based on its capacity to displace a radioactive form of the molecule from combination with its specific antibody.

rate-limiting step: (1) Generally, the step in an enzymatic reaction with the greatest activation energy or the transition state of highest free energy. (2) The slowest step in a metabolic pathway.

reaction intermediate: Any chemical species in a reaction pathway that has a finite chemical lifetime.

reading frame: A contiguous and nonoverlapping set of three-nucleotide codons in DNA or RNA.

recombinant DNA: DNA formed by the joining of genes into new combinations.

redox pair: An electron donor and its corresponding oxidized form; for example, NADH and NAD$^+$.

redox reaction: See oxidation–reduction reaction.

reducing agent (reductant): The electron donor in an oxidation–reduction reaction.

reducing end: The end of a polysaccharide having a terminal sugar with a free anomeric carbon; the terminal residue can act as a reducing sugar.

reducing equivalent: A general or neutral term for an electron or an electron equivalent in the form of a hydrogen atom or a hydride ion.

reducing sugar: A sugar in which the carbonyl (anomeric) carbon is not involved in a glycosidic bond and can therefore undergo oxidation.

reduction: The gain of electrons by a compound or ion.

regulatory enzyme: An enzyme having a regulatory function through its capacity to undergo a change in catalytic activity by allosteric mechanisms or by covalent modification.

regulatory gene: A gene that gives rise to a product involved in the regulation of the expression of another gene; for example, a gene coding for a repressor protein.

regulatory sequence: A DNA sequence involved in regulating the expression of a gene; for example, a promoter or operator.

regulon: A group of genes or operons that are coordinately regulated even though some, or all, may be spatially distant within the chromosome or genome.

release factors: See termination factors.

releasing factors: Hypothalamic hormones that stimulate release of other hormones by the pituitary gland.

renaturation: Refolding of an unfolded (denatured) globular protein so as to restore native structure and protein function.

replication: Synthesis of a daughter duplex DNA molecule identical to the parental duplex DNA.

replisome: The multiprotein complex that promotes DNA synthesis at the replication fork.

repressible enzyme: In bacteria, an enzyme whose synthesis is inhibited when its reaction product is readily available to the cell.

repression: A decrease in the expression of a gene in response to a change in the activity of a regulatory protein.

repressor: The protein that binds to the regulatory sequence or operator for a gene, blocking its transcription.

residue: A single unit within a polymer; for example, an amino acid within a polypeptide chain. The term reflects the fact that sugars, nucleotides, and amino acids lose a few atoms (generally the elements of water) when incorporated in their respective polymers.

respiration: The catabolic process in which electrons are removed from nutrient molecules and passed through a chain of carriers to oxygen.

respiratory chain: The electron transfer chain; a sequence of electron-carrying proteins that transfer electrons from substrates to molecular oxygen in aerobic cells.

restriction endonucleases: Site-specific endodeoxyribonucleases causing cleavage of both strands of DNA at points within or near the specific site recognized by the enzyme; important tools in genetic engineering.

restriction fragment: A segment of double-stranded DNA produced by the action of a restriction endonuclease on a larger DNA.

restriction fragment length polymorphisms (RFLPs): Variations, among individuals in a population, in the length of certain restriction fragments within which certain genomic sequences occur. These variations result from rare sequence changes that create or destroy restriction sites in the genome.

retrovirus: An RNA virus containing a reverse transcriptase.

reverse transcriptase: An RNA-directed DNA polymerase in retroviruses; capable of making DNA complementary to an RNA.

ribonuclease: A nuclease that catalyzes the hydrolysis of certain internucleotide linkages of RNA.

ribonucleic acid: See RNA.

ribonucleotide: A nucleotide containing D-ribose as its pentose component.

ribosomal RNA (rRNA): A class of RNA molecules serving as components of ribosomes.

ribosome: A supramolecular complex of rRNAs and proteins, approximately 18 to 22 nm in diameter; the site of protein synthesis.

ribozymes: Ribonucleic acid molecules with catalytic activities; RNA enzymes.

RNA (ribonucleic acid): A polyribonucleotide of a specific sequence linked by successive 3′,5′-phosphodiester bonds.

RNA polymerase: An enzyme that catalyzes the formation of RNA from ribonucleoside 5′-triphosphates, using a strand of DNA or RNA as a template.

RNA splicing: Removal of introns and joining of exons in a primary transcript.

rRNA: See ribosomal RNA.

S-adenosylmethionine (adoMet): An enzymatic cofactor involved in methyl group transfers.

salvage pathway: Synthesis of a biomolecule, such as a nucleotide, from intermediates in the degradative pathway for the biomolecule; a recycling pathway, as distinct from a de novo pathway.

saponification: Alkaline hydrolysis of triacylglycerols to yield fatty acids as soaps.

sarcomere: A functional and structural unit of the muscle contractile system.

satellite DNA: Highly repeated, nontranslated segments of DNA in eukaryotic chromosomes; most often associated with the centromeric region. Its function is not clear.

saturated fatty acid: A fatty acid containing a fully saturated alkyl chain.

second law of thermodynamics: The law stating that in any chemical or physical process, the entropy of the universe tends to increase.

second messenger: An effector molecule synthesized within a cell in response to an external signal (first messenger) such as a hormone.

secondary metabolism: Pathways that lead to specialized products not found in every living cell.

secondary structure: The residue-by-residue conformation of the backbone of a polymer.

sedimentation coefficient: A physical constant specifying the rate of sedimentation of a particle in a centrifugal field under specified conditions.

Shine–Dalgarno sequence: A sequence in an mRNA required for binding prokaryotic ribosomes.

shuttle vector: A recombinant DNA vector that can be replicated in two or more different host species. See also vector.

sickle-cell anemia: A human disease characterized by defective hemoglobin molecules; caused by a homozygous allele coding for the β chain of hemoglobin.

sickle-cell trait: A human condition recognized by the sickling of erythrocytes when exposed to low oxygen tension; occurs in individuals heterozygous for the allele responsible for sickle-cell anemia.

signal sequence: An amino-terminal sequence that signals the cellular fate or destination of a newly synthesized protein.

signal transduction: The process by which an extracellular signal (chemical, mechanical, or electrical) is amplified and converted to a cellular response.

silent mutation: A mutation in a gene that causes no detectable change in the biological characteristics of the gene product.

simple diffusion: The movement of solute molecules across a membrane to a region of lower concentration, unassisted by a protein transporter.

simple protein: A protein yielding only amino acids on hydrolysis.

site-directed mutagenesis: A set of methods used to create specific alterations in the sequence of a gene.

site-specific recombination: A type of genetic recombination that occurs only at specific sequences.

small nuclear RNA (snRNA): Any of several small RNA molecules in the nucleus; most have a role in the splicing reactions that remove introns from mRNA, tRNA, and rRNA molecules.

somatic cells: All body cells except the germ-line cells.

SOS response: In bacteria, a coordinated induction of a variety of genes as a response to high levels of DNA damage.

Southern blot: A DNA hybridization procedure in which one or more specific DNA fragments are detected in a larger population by means of hybridization to a complementary, labeled nucleic acid probe.

specific activity: The number of micromoles (μmol) of a substrate transformed by an enzyme preparation per minute per milligram of protein at 25 °C; a measure of enzyme purity.

specific heat: The amount of energy (in joules or calories) needed to raise the temperature of 1 g of a pure substance by 1 °C.

specific rotation: The rotation, in degrees, of the plane of plane-polarized light (D-line of sodium) by an optically active compound at 25 °C, with a specified concentration and light path.

specificity: The ability of an enzyme or receptor to discriminate among competing substrates or ligands.

sphingolipid: An amphipathic lipid with a sphingosine backbone to which are attached a long-chain fatty acid and a polar alcohol.

splicing: See gene splicing; RNA splicing.

standard free-energy change ($\Delta G°$): The free-energy change for a reaction occurring under a set of standard conditions: temperature, 298 K; pressure, 1 atm or 101.3 kPa; and all solutes at 1 M concentration. $\Delta G°'$ denotes the standard free-energy change at pH 7.0.

standard reduction potential (E_0'): The electromotive force exhibited at an electrode by 1 M concentrations of a reducing agent and its oxidized form at 25 °C and pH 7.0; a measure of the relative tendency of the reducing agent to lose electrons.

steady state: A nonequilibrium state of a system through which matter is flowing and in which all components remain at a constant concentration.

stem cells: The common, self-regenerating cells in bone marrow that give rise to differentiated blood cells such as erythrocytes and lymphocytes.

stereoisomers: Compounds that have the same composition and the same order of atomic connections, but different molecular arrangements.

sterols: A class of lipids containing the steroid nucleus.

sticky ends: Two DNA ends in the same DNA molecule, or in different molecules, with short overhanging single-stranded segments that are complementary to one another, facilitating ligation of the ends; also known as cohesive ends.

stop codons: See termination codons.

stroma: The space and aqueous solution enclosed within the inner membrane of a chloroplast, not including the contents within the thylakoid membranes.

structural gene: A gene coding for a protein or RNA molecule; as distinct from a regulatory gene.

substitution mutation: A mutation caused by the replacement of one base by another.

substrate: The specific compound acted upon by an enzyme.

substrate-level phosphorylation: Phosphorylation of ADP or some other nucleoside 5'-diphosphate coupled to the dehydrogenation of an organic substrate; independent of the electron transfer chain.

suicide inhibitor: A relatively inert molecule that is transformed by an enzyme, at its active site, into a reactive substance that irreversibly inactivates the enzyme.

suppressor mutation: A mutation that totally or partially restores a function lost by a primary mutation; located at a site different from the site of the primary mutation.

Svedberg (S): A unit of measure of the rate at which a particle sediments in a centrifugal field.

symbionts: Two or more organisms that are mutually interdependent; usually living in physical association.

symport: Cotransport of solutes across a membrane in the same direction.

synthases: Enzymes that catalyze condensation reactions in which no nucleoside triphosphate is required as an energy source.

synthetases: Enzymes that catalyze condensation reactions using ATP or another nucleoside triphosphate as an energy source.

system: An isolated collection of matter; all other matter in the universe apart from the system is called the surroundings.

telomere: Specialized nucleic acid structure found at the ends of linear eukaryotic chromosomes.

temperate phage: A phage whose DNA may be incorporated into the host-cell genome without being expressed; as distinct from a virulent phage, which destroys the host cell.

template: A macromolecular mold or pattern for the synthesis of an informational macromolecule.

terminal transferase: An enzyme that catalyzes the addition of nucleotide residues of a single kind to the 3′ end of DNA chains.

termination codons: UAA, UAG, and UGA; in protein synthesis, signal the termination of a polypeptide chain. Also known as stop codons.

termination factors: Protein factors of the cytosol required in releasing a completed polypeptide chain from a ribosome; also known as release factors.

termination sequence: A DNA sequence that appears at the end of a transcriptional unit and signals the end of transcription.

terpenes: Organic hydrocarbons or hydrocarbon derivatives constructed from recurring isoprene units. They produce some of the scents and tastes of plant products; for example, the scents of geranium leaves and pine needles.

tertiary structure: The three-dimensional conformation of a polymer in its native folded state.

tetrahydrobiopterin: The reduced coenzyme form of biopterin.

tetrahydrofolate: The reduced, active coenzyme form of the vitamin folate.

thiamine pyrophosphate: The active coenzyme form of vitamin B_1; involved in aldehyde transfer reactions.

thioester: An ester of a carboxylic acid with a thiol or mercaptan.

3′ end: The end of a nucleic acid that lacks a nucleotide bound at the 3′ position of the terminal residue.

thrombocytes: See platelets.

thromboxanes: A class of molecules derived from arachidonate and involved in platelet aggregation during blood clotting.

thylakoid: Closed cisterna, or disc, formed by the pigment-bearing internal membranes of chloroplasts.

thymine dimer: See pyrimidine dimer.

tissue culture: Method by which cells derived from multicellular organisms are grown in liquid media.

titration curve: A plot of the pH versus the equivalents of base added during titration of an acid.

tocopherols: Forms of vitamin E.

topoisomerases: Enzymes that introduce positive or negative supercoils in closed, circular duplex DNA.

topoisomers: Different forms of a covalently closed, circular DNA molecule that differ only in their linking number.

toxins: Proteins produced by some organisms and toxic to certain other species.

trace element: A chemical element required by an organism in only trace amounts.

transaminases: See aminotransferases.

transamination: Enzymatic transfer of an amino group from an α-amino acid to an α-keto acid.

transcription: The enzymatic process whereby the genetic information contained in one strand of DNA is used to specify a complementary sequence of bases in an mRNA chain.

transcriptional control: The regulation of a protein's synthesis by regulation of the formation of its mRNA.

transduction: (1) Generally, the conversion of energy or information from one form to another. (2) The transfer of genetic information from one cell to another by means of a viral vector.

transfer RNA (tRNA): A class of RNA molecules (M_r 25,000 to 30,000), each of which combines covalently with a specific amino acid as the first step in protein synthesis.

transformation: Introduction of an exogenous DNA into a cell, causing the cell to acquire a new phenotype.

transgenic: Describing an organism that has genes from another organism incorporated within its genome as a result of recombinant DNA procedures.

transition state: An activated form of a molecule in which the molecule has undergone a partial chemical reaction; the highest point on the reaction coordinate.

translation: The process in which the genetic information present in an mRNA molecule specifies the sequence of amino acids during protein synthesis.

translational control: The regulation of a protein's synthesis by regulation of the rate of its translation on the ribosome.

translational repressor: A repressor that binds to an mRNA, blocking translation.

translocase: (1) An enzyme that catalyzes membrane transport. (2) An enzyme that causes a movement, such as the movement of a ribosome along an mRNA.

transpiration: Passage of water from the roots of a plant to the atmosphere via the vascular system and the stomata of the leaves.

transporters: Proteins that span a membrane and transport specific nutrients, metabolites, ions, or proteins across the membrane; sometimes called permeases.

transposition: The movement of a gene or set of genes from one site in the genome to another.

transposon (transposable element): A segment of DNA that can move from one position in the genome to another.

triacylglycerol: An ester of glycerol with three molecules of fatty acid; also called a triglyceride or neutral fat.

tricarboxylic acid cycle: See citric acid cycle.

triose: A simple sugar with a backbone containing three carbon atoms.

tRNA: See transfer RNA.

tropic hormone (tropin): A peptide hormone that stimulates a specific target gland to secrete its hormone; for example, thyrotropin produced by the pituitary stimulates secretion of thyroxine by the thyroid.

turnover number: The number of times an enzyme molecule transforms a substrate molecule per unit time, under conditions giving maximal activity at substrate concentrations that are saturating.

ultraviolet (UV) radiation: Electromagnetic radiation in the region of 200 to 400 nm.

uncoupling agent: A substance that uncouples phosphorylation of ADP from electron transfer; for example, 2,4-dinitrophenol.

uniport: A transport system that carries only one solute, as distinct from cotransport.

unsaturated fatty acid: A fatty acid containing one or more double bonds.

urea cycle: A metabolic pathway in vertebrates, for the synthesis of urea from amino groups and carbon dioxide; occurs in the liver.

ureotelic: Excreting excess nitrogen in the form of urea.

uricotelic: Excreting excess nitrogen in the form of urate (uric acid).

V_{max}: The maximum velocity of an enzymatic reaction when the binding site is saturated with substrate.

vector: A DNA molecule known to replicate autonomously in a host cell, to which a segment of DNA may be spliced to allow its replication; for example, a plasmid or a temperate-phage DNA.

viral vector: A viral DNA altered so that it can act as a vector for recombinant DNA.

virion: A virus particle.

virus: A self-replicating, infectious, nucleic acid–protein complex that requires an intact host cell for its replication; its genome is either DNA or RNA.

vitamin: An organic substance required in small quantities in the diet of some species; generally functions as a component of a coenzyme.

wild type: The normal (unmutated) phenotype.

wobble: The relatively loose base pairing between the base at the 3′ end of a codon and the complementary base at the 5′ end of the anticodon.

x-ray crystallography: The analysis of x-ray diffraction patterns of a crystalline compound, used to determine the molecule's three-dimensional structure.

zinc finger: A specialized protein motif involved in DNA recognition by some DNA-binding proteins; characterized by a single atom of zinc coordinated to four Lys residues or to two His and two Lys residues.

zwitterion: A dipolar ion, with spatially separated positive and negative charges.

zymogen: An inactive precursor of an enzyme; for example, pepsinogen, the precursor of pepsin.

Illustration Credits

Photos and Line Art

Part I opening image NASA

Chapter 1
Figure 1–1 (a) Keith Porter/Photo Researchers, Inc.; **(b)** Jen and Des Bartlett/Photo Researchers, Inc.; **(c)** Mark Boulton/Photo Researchers, Inc.; **p. 4 (top)** UPI/Bettmann; **Figure 1–15 (a)** Courtesy of the Oriental Institute, University of Chicago; **(b)** Huntington Potter and David Dressler, Harvard Medical School, Department of Neurobiology; **p. 17 (left)** The Granger Collection; **(right)** Keystone/ The Image Works.

Chapter 2
Figure 2–3 (a) John Hansen/Photo Researchers, Inc.; **(b)** Christine Gall, University of California, Irvine; **p. 24 (top)** CNRI/Science Photo Library/Photo Researchers, Inc.; **(bottom)** Dr. Jeremy Burgess/Science Photo Library/Photo Researchers, Inc.; **p. 25** Sinclair Stammers/Science Photo Library/Photo Researchers, Inc.; **Figure 2–6 (center left)** T.J. Beveridge/Biological Photo Service; **(center right, bottom right)** Biological Photo Service; **(bottom left)** Norma J. Lange/Biological Photo Service; **Figure 2–7** Adapted from Alberts, B., Bray, D., Lewis, J., Raff, M., Roberts, K., & Watson, J.D. (1989) *Molecular Biology of the Cell,* 2nd edn, p. 19, Garland Publishing, Inc., New York; **Figure 2–10 (top right, center right)** Don W. Fawcett/Photo Researchers, Inc.; **(bottom right)** Science Source/Photo Researchers, Inc.; **Figure 2–11 (left)** Biological Photo Service; **Figure 2–12 (a)** Don W. Fawcett/Photo Researchers, Inc.; **(b)** Ursula Goodenough, Washington University, Department of Biology; **Figure 2–13 (top left)** G.F. Bahr/Biological Photo Service; **Figure 2–15 (top right)** Don W. Fawcett & Keith Porter/Photo Researchers, Inc.; **Figure 2–16 (right)** Biological Photo Service; **Figure 2–18 (a) (top)** Robert Goldman, Northwestern University Medical School, Department of Cell Biology & Anatomy; **(bottom)** K.G. Murti/Visuals Unlimited; **(b) (top)** Gary Borisy, University of Wisconsin–Madison, Department of Molecular Biology; **(bottom)** M. Schliwa/Visuals Unlimited; **(c) (top)** Robert Goldman, Northwestern University Medical School, Department of Cell Biology & Anatomy; **(bottom)** Ueli Aebi, Maurice E. Müller-Institut für Hochauflösende Elektronenmikroskopie am Biozentrumder Universität Basel; **Figure 2–23 (bottom)** Joan Peterson, University of Wisconsin–Madison, Department of Biochemistry; **Figure 2–24** Adapted from Alberts, B., Bray, D., Lewis, J., Raff, M., Roberts, K., & Watson, J.D. (1989) *Molecular Biology of the Cell,* 2nd edn, pp. 165–166, Garland Publishing, Inc., New York; **Figure 2–25 (a)** Don W. Fawcett/Photo Researchers, Inc.; **(b)** Keith Porter/Photo Researchers, Inc.; **(c)** Biophoto Associates/Photo Researchers, Inc.; **(d)** Don W. Fawcett & D. Phillips/ Photo Researchers, Inc.; **(e, f)** Science Source/Photo Researchers, Inc.; **Figure 2–26 (a)** Carol Sattler, University of Wisconsin–Madison, Department of Oncology; **(b)** Don W. Fawcett/Photo Researchers, Inc.; **(c)** From Peracchia, C. & Dulhunty, A. (1976) Low resistance junctions in crayfish. *J. Cell. Biol.,* **70,** 426. Photo courtesy Carol Sattler, University of Wisconsin–Madison, Department of Oncology; **Figure 2–28** John Finch, MRC Laboratory of Molecular Biology; **Figure 2–29** C. Dauquet, Institut Pasteur/Photo Researchers, Inc.

Chapter 3
p. 56 Bill Curtsinger/Photo Researchers, Inc.; **p. 62 (bottom)** Alfred Wolf/Explorer/Photo Researchers, Inc.; **p. 72** Historical Pictures Service; **Figure 3–19** Sigurgeir Jónasson; **Figure 3–23 (a)** François Gohier/Photo Researchers, Inc.; **(b)** Biological Photo Service.

Chapter 4
p. 81 NASA; **p. 103** Gregory G. Dimijian/Photo Researchers, Inc.

Chapter 5
Figure 5–1 Animals Animals.

Chapter 6
Figure 6–1 (a) Runk/Schoenburger/Grant Heilman Photography; **(b)** Bill Longcore/Photo Researchers, Inc.; **(c)** Mark McKenna; **(d)** Science Source/Photo Researchers, Inc.; **(e)** J.C. Stevenson/Animals Animals; **(f)** Earth Scenes/Animals Animals; **(g)** Princess Margaret Rose Orthopaedic Hospital/Science Source/Photo Researchers, Inc.; **Figure 6–4 (b)** Richard R. Burgess & Jerome J. Jendrias, University of Wisconsin–Madison, Biotechnology Center; **Figure 6–7 (b)** Patrick H. O'Farrell, University of California Medical Center, San Francisco, Department of Biochemistry and Biophysics; **p. 144 (bottom)** UPI/Bettmann; **Figure 6–9 (c)** Daniel R. Webster, Columbia University College of Physicians & Surgeons, Department of Anatomy and Cell Biology, and Gary Borisy, University of Wisconsin–Madison, Department of Molecular Biology; **p. 148 (bottom)** UPI/Bettmann; **Figure 6–16** Data provided by Lai-Su L. Yeh, Protein Identification Resource, National Biomedical Research Foundation, Georgetown University Medical Center.

Chapter 7
Figure 7–2 Adapted from Dickerson, R.E. & Geis I. (1969) *The Structure and Action of Proteins,* Benjamin/Cummings Publishing Company, Menlo Park, CA. Copyright © 1969 by Irving Geis; **p. 163 (left)** News Service, Stanford University/Photo Researchers, Inc.; **(right)** AP/Wide World Photos, Inc.; **Figures 7–5, 7–11** Adapted from Creighton, T.E. (1984) *Proteins,* p. 166. Copyright © 1984 by W.H. Freeman and Company. Reprinted by permission; **Figure 7–**

Chapter 20

Figure 20–12 (b) Daniel Lane, Johns Hopkins University, School of Medicine; **p. 661** Harvard Medical School; **p. 674 (far left, left, far right)** AP/Wide World Photos, Inc.; **(right)** UPI/Bettmann; **Figure 20–37 (b)** Courtesy of Robert L. Hamilton and the Arteriosclerosis Specialized Center of Research, University of California, San Francisco; **p. 679** UPI/Bettmann.

Chapter 21

Figure 21–3 (a) C.P. Vance/Visuals Unlimited; **(b)** Jeremy Burgess/Photo Researchers, Inc.; **p. 716 Box 21–1 Figure 1** John Mansfield, University of Wisconsin–Madison, Department of Veterinary Science; **p. 718** Massachusetts Institute of Technology Museum Collection; **Figure 21–33** Thelander, L. & Reichard, P. (1979) Reduction of ribonucleotides. *Annu. Rev. Biochem.* **48,** 133–158; **p. 730** AP/Wide World Photos, Inc.

Chapter 22

Figure 22–4 Fred Hossler/Visuals Unlimited; **Figure 22–7** D.W. Fawcett/Photo Researchers, Inc.; **p. 747** UPI/Bettmann; **p. 750 (left)** UPI/Bettmann; **(right)** AP/Wide World Photos, Inc.; **Figure 22–19 (c)** Secchi-Legaque/Photo Researchers, Inc.; **pp. 760, 763** UPI/Bettmann; **Figure 22–39** O'Malley, B.W. & Tsai, M.-J. (1992) Molecular pathways of steroid receptor action. *Biol. Reprod.* **46,** 163–167; **Figure 22–40** Schwabe, J.W.R. & Rhodes, D. (1991) Beyond zinc fingers: steroid hormone receptors have a novel structural motif for DNA recognition. *Trends Biochem. Sci.* **16,** 291–296.

Chapter 23

Figure 23–1 From Kleinschmidt, A.K., Land, D., Jackerts, D., & Zahn, R.K. (1962) *Biochem. Biophys. Acta* **61,** pp. 857–864; **Figure 23–3** Huntington Potter and David Dressler, Harvard Medical School, Department of Neurobiology; **Figure 23–4** G.F. Bahr/Biological Photo Service; **Figure 23–5** D.W. Fawcett/Photo Researchers, Inc.; **Figure 23–10** Cozzarelli, N.R., Boles, T.C., & White, J.H. (1990) Primer on the topology and geometry of DNA supercoiling. In *DNA Topology and Its Biological Effects* (Cozzarelli, N.R. & Wang, J.C., eds), pp. 139–184, figure 8A, Cold Spring Harbor Laboratory Press, Cold Spring Harbor, New York; **Figure 23–11** Saenger, W. (1984) *Principles of Nucleic Acid Structure,* p. 452, Springer-Verlag, New York; **Figure 23–12** Laurien Polder from Kornberg, A. (1980) *DNA Replication,* p. 29, W.H. Freeman, New York; **Figure 23–14** Cozzarelli, N.R., Boles, T.C., & White, J.H. (1990) Primer on the topology and geometry of DNA supercoiling. In *DNA Topology and Its Biological Effects* (Cozzarelli, N.R. & Wang, J.C., eds), pp. 139–184, figure 3B, Cold Spring Harbor Laboratory Press, Cold Spring Harbor, New York; **Figure 23–17** Cozzarelli, N.R., Boles, T.C., & White, J.H. (1990) Primer on the topology and geometry of DNA supercoiling. In *DNA Topology and Its Biological Effects* (Cozzarelli, N.R. & Wang, J.C., eds), pp. 139–184, figure 7, Cold Spring Harbor Laboratory Press, Cold Spring Harbor, New York; **Figure 23–19** Keller, W. (1975) Characterization of purified DNA-relaxing enzyme from human tissue culture cells. *Proc. Nat. Acad. Sci.* **72,** p. 2553; **Figure 23–21 (a)** James H. White, T. Christian Boles, & N.R. Cozzarelli, University of California, Berkeley, Department of Molecular and Cell Biology; **Figure 23–23 (b)** Ada L. Olins and Donald E. Olins, Oak Ridge National Laboratory; **Figure 23–24** Singer, M. & Berg, P. (1991) *Genes and Genomes, A Changing Perspective,* p. 52, University Science Books, Mill Valley, CA; **Figure 23–27 (b)** Barbara Hamkalo, University of California, Irvine, Department of Molecular Biology and Biochemistry; **Figure 23–28** D.W. Fawcett/Visuals Unlimited; **Problem 23–9** Roger Kornberg, MRC Laboratory of Molecular Biology.

Chapter 24

Figure 24–3 (a) Bernard Hirt, Institut Suisse de Recherches Experimentales sur le Cancer; **(c)** Cairns, J. (1963) *Cold Spring Har-*
bor *Symp. Quant. Biol.* **28,** 44; **p. 820** AP/Wide World Photos, Inc.; **Figure 24–11** Bramhill, D. & Kornberg, A. (1988) A model for initiation at origins of DNA replication. *Cell* **54,** 915–918; **Figure 24–17** Bruce N. Ames, University of California, Berkeley, Department of Biochemistry & Molecular Biology; **Figure 24–20** Watson, J.D., Hopkins, N.H., Roberts, J.W., Steitz, J.A., & Weiner, A.M. (1987) *Molecular Biology of the Gene,* 4th edn, p. 350, Benjamin/Cummings Publishing Company, Menlo Park, CA; **p. 839** AP/Wide World Photos, Inc.; **Figure 24–26 (a)** Bernard John, The Australian National University; **Figure 24–28 (b)** Huntington Potter and David Dressler, Harvard Medical School, Department of Neurobiology; **Figure 24–31 (a)** John Heuser, Washington University Medical School, Department of Biochemistry; **(b)** Edward Egelman and A. Stasiak, University of Minnesota; **Figure 24–33** Roca, A.I. & Cox, M.M. (1990) The RecA protein: structure and function. *Crit. Rev. Biochem. Mol. Biol.* **25,** 415–456; **Figure 24–34** West, S.C., Cassuto, E., & Howard-Flanders, P. (1981) Mechanism of *E. coli* RecA protein directed strand exchanges in post-replication repair of DNA. *Nature* **294,** 659–662.

Chapter 25

p. 862 Box 25–1 Figure 2 Carol Gross, University of Wisconsin–Madison, Department of Biochemistry; **Figure 25–8** Platt, T. (1981) Termination of transcription and its regulation in the tryptophan operon of *E. coli. Cell* **24,** 10–23; **Figure 25–11 (a)** Pierre Chambon, Laboratorie de Génétique Moléculaire des Eucaryotes, Faculté de Médecine (CNRS); **(b, c)** Chambon, P. (1981) Split genes. *Sci. Am.* **244** (May), 60–71; **Figure 25–14** Cech, T.R. (1986) RNA as an enzyme. *Sci. Am.* **255** (November), 64–75; **Figure 25–16** Konarska, M.M. & Sharp, P.A. (1987) Interaction between small nuclear ribonucleoprotein particles in formation of spliceosomes. *Cell* **49,** 763–774; **Figure 25–26 (a)** Cech, T.R. (1986) RNA as an enzyme. *Sci. Am.* **255** (November), 64–75; **(b)** Cech, T.R. & Bass, B.L. (1986) Biological catalysis by RNA. *Annu. Rev. Biochem.* **55,** 599–630; **(c)** Kim, S.-H. & Cech, T.R. (1987) Three-dimensional model of the active site of the self-splicing rRNA precursor of *Tetrahymena. Proc. Nat. Acad. Sci., USA* **84,** 8788–8792; **Figure 25–27** Cech, T.R. (1986) RNA as an enzyme. *Sci. Am.* **255** (November), 64–75; **p. 880 (left)** Courtesy of Marianne Grunberg-Manago; **(right)** AP/Wide World Photos, Inc.; **p. 882 (left)** UPI/Bettmann; **(right)** AP/Wide World Photos, Inc.; **Figure 25–32** Haseltine, W.A. & Wong-Staal, F. (1988) The molecular biology of the AIDS virus. *Sci. Am.* **259** (October), 52–62; **Figure 25–33** Kingsman, A.J. & Kingsman, S.M. (1988) Ty: a retroelement moving forward. *Cell* **53,** 333–335; **Figure 25–34** Boeke, J.D. (1990) Reverse transcriptase, the end of the chromosome, and the end of life. *Cell* **61,** 193–195; **pp. 887, 889** AP/Wide World Photos, Inc.

Chapter 26

p. 893 (top) News Office, Massachusetts General Hospital; **Figure 26–1** D.W. Fawcett/Visuals Unlimited; **p. 895 (bottom)** AP/Wide World Photos, Inc.; **p. 896** Norman Lenburg, University News & Information Service, University of Wisconsin; **Figure 26–13** Gutell, R.R., Weiser, B., Woese, C.R., & Noller, H.F. (1985) Comparative anatomy of 16-S-like ribosomal DNA. *Prog. Nucleic Acid Res. Mol. Biol.* **32,** 155–216; **pp. 909 (center), 910 (top)** UPI/Bettmann; **Figure 26–17** Kim, S.-H., Suddath, F.L., Quigley, G.J., McPherson, A., Sussman, J.L., Wang, A.H.J., Seeman, N.C., & Rich, A. (1974) Three-dimensional tertiary structure of yeast phenylalanine transfer RNA. *Science* **185,** 435–440; **Figure 26–22** Courtesy of Mark Rould, John Perona, & Thomas A. Steitz, Yale University, New Haven, CT. Photo by John Arnez; **Figure 26–30 (b)** Steven L. McKnight and Oscar L. Miller, University of Virginia, Department of Biology; **p. 929** Courtesy of Günter Blobel, Rockefeller University; **p. 930** AP/Wide World Photos, Inc.; **Figure 26–42 (c)** John Heuser, Washington University Medical School, Department of Biochemistry.

Chapter 27

p. 947 UPI/Bettmann; **Figure 27–15 (a)** McKnight, S.L. (1991) Molecular zippers in gene regulation. *Sci. Am.* **264** (April), 54–64; **Figure 27–16 (b)** Personal communication from Thomas A. Steitz. See Schultz, S.C., Shields, G.C., & Steitz, T.A. (1991) *Science* **253,** 10001. Photo by Jonathan Passner; **Figure 27–23 (b)** Watson, J.D., Hopkins, N.H., Roberts, J.W., Steitz, J.A., & Weiner, A.M. (1987) *Molecular Biology of the Gene,* 4th edn, p. 487, Benjamin/Cummings Publishing Company, Menlo Park, CA; **Figure 27–25 (a)** T.F. Anderson, E.L. Wollman, and F. Jacob/Photo Researchers, Inc.; **Figure 27–28** Ptashne, M. (1989) How gene activators work. *Sci. Am.* **260** (January), 40–47; **Figure 27–29** Nomura, M., Gourse, R., & Baughman, G. (1984) Regulation of the synthesis of ribosomes and ribosomal components. *Annu. Rev. Biochem.* **53,** 75–117; **Figure 27–31** John D. Cunningham/Visuals Unlimited; **Figure 27–34** Mitchell, P.J. & Tjian R. (1989) Transcriptional regulation in mammalian cells by sequence-specific DNA binding proteins. *Science* **245,** 371–378; **Figure 27–35 (photo)** F.R. Turner, University of Indiana, Bloomington, Department of Biology; **Figure 27–36** Melton, D.A. (1991) Pattern formation during animal development. *Science* **252,** 234–241; **Figure 27–37** Wolfgang Driever and Christiane Nüsslein-Volhard, Max-Planck-Institut; **Figure 27–38** Courtesy of Phillip Ingham, Imperial Cancer Research Fund, Oxford University; **Figure 27–39 (a, b)** F.R. Turner, University of Indiana, Bloomington, Department of Biology; **(c, d)** E.B. Lewis, California Institute of Technology, Division of Biology.

Chapter 28

p. 984 (top) UPI/Bettmann; **(center & bottom)** AP/Wide World Photos, Inc.; **Figure 28–6** Elizabeth A. Wood, University of Wisconsin–Madison, Department of Biochemistry; **p. 997 Box 28–1 Figure 1** Phil Borden/PhotoEdit; **Figure 28–18** Keith Wood, University of California, San Diego, Department of Biology; **Figure 28–19** From Fischoff, D.A. & Bowdish, K.S. (1987) Insect tolerant transgenic tomato plants. *BioTechnology* **5,** 807–812. Photo courtesy David Fischoff, Monsanto Company; **Figure 28–21** R.L. Brinster and R.E. Hammer, University of Pennsylvania, School of Veterinary Medicine.

Molecular Modeling

MOLECULAR GRAPHICS Unless indicated, all molecular graphics were produced by Alisa Zapp at the Department of Biochemistry, University of Wisconsin–Madison using MidasPlus software from the Computer Graphics Laboratory, University of California, San Francisco. See Ferrin, T.E., et al. (1988) The MIDAS display system. *J. Mol. Graphics* **6,** 13.

ATOMIC COORDINATES Unless indicated, all atomic coordinates were obtained from the Protein Data Bank at Brookhaven National Laboratory, Upton, NY. See Bernstein, F.C., Koetzle, T.F., Williams, G.J.B., Meyer, E.F., Jr., Brice, M.D., Rodgers, J.R., Kennard, O., Shimanouchi, T., & Tasumi, M. (1977) Protein data bank. A computer-based archival file for macromolecular structures. *J. Mol. Biol.* **112,** 535; and see Abola, E.E., Bernstein, F.C., Bryant, S.H., Koetzle, T.F., & Weng, J. (1987) Protein data bank. In *Crystallographic Databases—Information Content, Software Systems, Scientific Applications* (Allen, F.H., Bergerhoff, G., & Sievers, R. eds), p. 107, Data Commission of the International Union of Crystallography, Bonn.

Other atomic coordinates (indicated below) were obtained from Cambridge Crystallographic Data Centre. See Allen, F.H., Davies, J.E., Galloy, J.J., Johnson, O., Kennard, O., MacRae, C.F., Mitchell, E.M., Mitchell, G.F., Smith, J.M., & Watson, G.D. (1991) The development of versions 3 and 4 of the Cambridge structural database system. *J. Chem. Inf. Comput. Sci.* **31,** 187–204.

Some structures (noted below) were generated using Insight II, as licensed for Biosym Technologies, San Diego, CA.

DEFINE_STRUCTURE program (Version 1), as modified by Scott R. Presnell at the Computer Graphics Laboratory, University of California, San Francisco, was used to produce records for protein secondary structures. See Richards, F.M., & Kundrot, C.E. (1988) Identification of structural motifs from protein coordinate data: secondary structure and first-level supersecondary structure. *Proteins* **3,** 71.

Cover Entry 6GCH of Oct. 1990. Brady, K., et al. (1990) *Biochemistry* **29,** 7600; **Frontispiece** Personal communication from Kam Zhang & David Eisenberg, Department of Chemistry and Biochemistry, University of California, Los Angeles. See Curmi, P.M.G., et al. (1992) *J. Biol. Chem.* **267,** 16980; **Part II opening image** Personal communication from Ivan Rayment, Department of Biochemistry and Enzyme Institute, University of Wisconsin–Madison. See Fraser, R.D.B., MacRae, T.P., & Suzuki, E. (1979) *J. Mol. Biol.* **129,** 463; **Part III opening image** Entry 1GD1 version of June 1987 in the Protein Data Bank at Brookhaven National Laboratory. See Skarzynski, T., Moody, P.C.E., & Wonacott, A.J. (1987) *J. Mol. Biol.* **193,** 171; **Part IV opening image** Personal communication from Xiang-Peng Kong & John Kuriyan, Laboratory of Molecular Biophysics, Rockefeller University, New York. See Kong, X.-P., et al. (1992) *Cell* **69,** 425.

Figure 1–12 (b) ADENTP from Cambridge Crystallographic Data Centre. Kennard, O., et al. (1971) *Proc. R. Soc. Lond., Ser. A* **325,** 401; **Figure 3–7 (b, c)** Produced using Insight II; **Figure 3–8** Entry 6GCH version of Oct. 1990. Brady, K., et al. (1990) *Biochemistry* **29,** 7600; **Figure 6–8** Personal communication from David Davies, Molecular Structure Section, National Institutes of Health, Bethesda, MD. See Silverton, E.W., Mavia, M.A., & Davies, D.R. (1977) *Proc. Natl. Acad. Sci. USA* **74,** 5140; **Figure 7–1** Chymotrypsin: entry 6GCH version of Oct. 1990. Brady, K., et al. (1990) *Biochemistry* **29,** 7600; Glycine: produced using Insight II; **Figure 7–3** Entry 4TNC version of Apr. 1989. Satyshur, K.A., et al. (1988) *J. Biol. Chem.* **263,** 1628; **Figure 7–6 (c, d)** Entry 4TNC version of Apr. 1989. Satyshur, K.A., et al. (1988) *J. Biol. Chem.* **263,** 1628; **Figure 7–7** Entry 4TNC version of Apr. 1989. Satyshur, K.A., et al. (1988) *J. Biol. Chem.* **263,** 1628; **Figure 7–14 (a–d)** Personal communication from Ivan Rayment, Department of Biochemistry and Enzyme Institute, University of Wisconsin–Madison. See Fraser, R.D.B., MacRae, T.P., & Suzuki, E. (1979) *J. Mol. Biol.* **129,** 463; **Figure 7–18** Entry 1CCR version of Apr. 1987. Ochi, H., et al. (1983) *J. Mol. Biol.* **166,** 407; **Figure 7–19 (a–d)** Entry 1MBO version of Sept. 1983. Phillips, S.E.V. (1980) *J. Mol. Biol.* **142,** 531; **Figure 7–20** Cytochrome *c*: entry 1CCR version of Apr. 1987. Ochi, H., et al. (1983) *J. Mol. Biol.* **166,** 407; Lysozyme: entry 3LYM version of Oct. 1987. Kundrot, C.E. & Richards, F.M. (1987) *J. Mol. Biol.* **193,** 157; Ribonuclease: entry 1RN3 version of Dec. 1983. Borkakoti, N., Moss, D.S., & Palmer, R.A. (1982) *Acta Crystallogr., Sect. B* **38,** 2210; **Figure 7–25 (a–d)** Pyruvate kinase: entry 1PYK version of Sept. 1983. Stuart, D.I., et al. (1979) *J. Mol. Biol.* **134,** 109; Triose phosphate isomerase: entry 1TIM version of May 1984. Banner, D.W., et al. (1976) *Biochem. Biophys. Res. Commun.* **72,** 146; Cytochrome b_{562}: entry 156B version of Sept. 30, 1983. Lederer, F., et al. (1981) *J. Mol. Biol.* **148,** 427; Myohemerythrin: entry 2MHR version of Apr. 1988. Sheriff, S., Hendrickson, W.A., & Smith, J.L. (1987) *J. Mol. Biol.* **197,** 273; Carboxypeptidase: entry 5CPA version of Jan. 1987. Crees, D.C., Lewis, M., & Lipscomb, W.N. (1983) *J. Mol. Biol.* **168,** 367; Lactate dehydrogenase: entry 3LDH version of July 1989. White, J.L., et al. (1976) *J. Mol. Biol.* **102,** 759; Insecticyanin: personal communication from Hazel Holden, Department of Chemistry and Enzyme Institute, University of Wisconsin–Madison. See Holden, H.M., et al. (1987) *EMBO J.* **6**, 1565; α_1-Antitrypsin: entry 7API version of Oct. 1990. Engh, R., et al. (1989) *Protein Eng.* **2**, 407; **Figure 7–26** Entry 2HHB version of Oct. 1989. Fermi, G., et al. (1984) *J. Mol. Biol.* **175**, 159; **Figure 7–29** Oxyhemoglobin: entry 1HHO version of Jan. 1984. Shaanan, B. (1983) *J. Mol. Biol.* **171**, 31; Deoxyhemoglobin: entry 2HHB version of Oct. 1989. Fermi, G., et al. (1984) *J. Mol. Biol.* **175**, 159; **Figure 8–2** Entry 6GCH of Oct. 1990. Brady, K., et al. (1990) *Biochemistry* **29**, 7600; **Figure 8–5** Entry 7DFR version of Oct. 1990. Bystroff, C., Oatley, S.J., & Kraut, J. (1990) *Biochemistry* **29**, 3263; **Figure 8–18 (b–d)** Entry 6GCH of Oct. 1990. Brady, K., et al. (1990) *Biochemistry* **29**, 7600; **Figure 8–21** Unbound hexokinase: entry 2YKX version of Jan. 1986. Anderson, C.M., Stenkamp, R.E., & Steitz, T.A. (1978) *J. Mol. Biol.* **123**, 15; Bound hexokinase: entry 1HKG version of Oct. 1984. Steitz, T.A., et al. (1981) *Phil. Trans. R. Soc. Lond.* **293**, 43; **Figure 8–26** Personal communication from Ray Stevens, Department of Chemistry, Harvard University, Cambridge, MA; **Figure 9–1 (a)** STARAC02 from Cambridge Crystallographic Data Centre. Goto, M. & Asada, E. (1978) *Bull. Chem. Soc. Japan* **51**, 2456; **(b)** OLECAC from Cambridge Crystallographic Data Centre. Abrahamsson, S. & Ryderstedt-Nahringbauer, I. (1962) *Acta Crystallogr.* **15**, 1261; **Figure 9–2** BTRILA05 from Cambridge Crystallographic Data Centre. Gibon, V., et al. (1984) *Bull. Soc. Chim. Belg.* **93**, 27; **Figure 9–13** CHOEST20 from Cambridge Crystallographic Data Centre. Shieh, H.-S., Hoard, L.G., & Nordman, C.E. (1981) *Acta Crystallogr., Sect. B* **37**, 1538; **Figure 10–11** Entry 1PRC version of Oct. 1989. Deisenhofer, J. & Michel, H. (1989) *EMBO J.* **8**, 2149; **Figure 10–26** VALINK from Cambridge Crystallographic Data Centre. Neupert-Laves, K. & Dobler, M. (1975) *Helv. Chim. Acta* **58**, 432; **Figure 12–15 (b, c)** Produced using Insight II; **Figure 12–18** Produced using Insight II; **Figure 12–19** Personal communication from Jim Nadeau, Department of Molecular Biology, Becton Dickinson Research Center, Research Triangle Park, NC. See Crothers, D.M,. Haran, T.E., & Nadeau, J.G. (1990) *J. Biol. Chem.* **265**, 7093; **Figure 12–22 (c, d)** Produced using Insight II; **Figure 12–24** Produced using Insight II; **Figure 12–27** Entry 4TNA version of Sept. 1983. Hingerty, B., Brown, R.S., & Jack, A. (1978) *J. Mol. Biol.* **124**, 523; **Figure 14–17 (a, b)** Personal communication from Robert Fletterick, Department of Biochemistry, University of California, San Francisco; **Figure 14–20** Entry 1PFK version of July 1990. Shirakihara, Y. & Evans, P.R. (1988) *J. Mol. Biol.* **204**, 973; **Figure 18–5 (d)** Entry 4FD1 version of Oct. 1990. Stout,

C.D. (1989) *J. Mol. Biol.* **205**, 545; **Figure 19–21 (a)** Personal communication from Drs. Kam Zhang & David Eisenberg, Department of Chemistry and Biochemistry, University of California, Los Angeles. See Curmi, P.M.G., et al. (1992) *J. Biol. Chem.* **267**, 16980; **Figure 19–30** Personal communication from Kam Zhang & David Eisenberg, Department of Chemistry and Biochemistry, University of California, Los Angeles. See Curmi, P.M.G., et al. (1992) *J. Biol. Chem.* **267**, 16980; **p. 691** Personal communication from Douglas C. Rees, Department of Chemistry and Chemical Engineering, California Institute of Technology, Pasadena, CA. See Jongsun, Kim & Rees, D.C. (1992) *Science* **257**, 1677–1682; **Figure 21–4 (a, b)** Entry 2GLS version of Oct. 1989. Yamashita, M.M., et al. (1989) *J. Biol. Chem.* **264**, 17681; **Figure 22–20 (a, b)** Entry 4INS version of April 1990. Baker, E.N., et al. (1988) *Phil. Trans. R. Soc. Lond.* **319**, 369; **Figure 22–28 (b)** Entry 1CPK version of Jan. 1992. Knighton, D.R., et al. (1991) *Science* **253**, 407; **Figure 22–33** Entry 3CLN version of Jan. 1989. Babu, Y.S., Bugg, C.E., & Cook, W.J. (1988) *J. Mol. Biol.* **204**, 191; **Figure 24–1 (a, b)** Entry 1DPI Oct. 1987. Ollis, D.L., et al. (1985) *Nature* **313**, 762; **Figure 24–8 (a)** Personal communication from Xiang-Peng Kong & John Kuriyan, Laboratory of Molecular Biophysics, Rockefeller University, New York. See Kong, X.-P., et al. (1992) *Cell* **69**, 425; **p. 914,** Produced using Insight II; **Figure 26–17** Entry 4TNA version of Sept. 1983. Hingerty, B., Brown, R.S., & Jack, A. (1978) *J. Mol. Biol.* **124**, 523; **p. 907 Box 26–2 Figure 1** Produced using Insight II; **Figure 27–8 (b)** Personal communication from Xiang-Peng Kong & John Kuriyan, Laboratory of Molecular Biophysics, Rockefeller University, New York. See Kong, X.-P., et al. (1992) *Cell* **69**, 425; **Figure 27–11 (a–d)** Entry 3CRO version of Oct. 1991. Mondragon, A. & Harrison, S.C. (1991) *J. Mol. Biol.* **219**, 321; **Figure 27–12 (a, b)** Personal communication from Nikola P. Pavletich, Department of Biology, Massachusetts Institute of Technology, Cambridge, MA. See Pavletich, N.P. & Pabo, C.O. (1991) *Science* **252**, 809; **Figure 27–13** Entry 1LRD version of Jan. 1990. Jordon, S.R. & Pabo, C.O. (1988) *Science* **242**, 893; **Figure 27–15 (a–c)** Pre-release entry P2ZTA version of July 1991. O'Shea, E.K., et al. (1991) *Science* **254**, 539; **Figure 27–16 (a)** Entry 3GAP version of Jan. 1991. Weber, I.T. & Steitz, T.A. (1987) *J. Mol. Biol.* **198**, 311; **Figure 27–22** Entry 2WRP version of Jan. 1991. Lawson, C.L., et al. (1988) *Proteins: Structure, Function, and Genetics* **3**, 18; **Figure 28–2** Personal communication from John Rosenberg, Department of Crystallography, University of Pittsburgh, Pittsburgh, PA.

Index

Configurational and positional designations in chemical names (for example, 1,2-, α-, D-, L-, p-, *cis-*) are disregarded in alphabetizing; the same or similar forms used as adjectives (α carbon atom, cis form) are not disregarded. Greek letters used as adjectives are alphabetized as if they were spelled out. Word division by space or hyphen is ignored; for example, aldol condensation *follows* aldolase.

Page numbers in boldface type indicate where a structural formula is given.